Contents

Metals Handbook®
Ninth Edition

Volume 15
Casting

Prepared under the direction of the
ASM INTERNATIONAL Handbook Committee

D.M. Stefanescu, Chairman

Joseph R. Davis, Senior Editor
James D. Destefani, Technical Editor
Theodore B. Zorc, Technical Editor
Heather J. Frissell, Editorial Supervisor
George M. Crankovic, Assistant Editor
Alice W. Ronke, Assistant Editor
Diane M. Jenkins, Word Processing Specialist
Karen Lynn O'Keefe, Word Processing Specialist

Robert L. Stedfeld, Director of Reference Publications
Kathleen M. Mills, Manager of Editorial Operations

Editorial Assistance
Lois A. Abel
Robert T. Kiepura
Penelope Thomas
Nikki D. Wheaton

ASM INTERNATIONAL™

METALS PARK, OHIO 44073

First printing, September 1988

Library of Congress Cataloging-in-Publication Data

ASM INTERNATIONAL

Metals handbook.

Includes bibliographies and indexes.
Contents: v. 1. Properties and selection—[etc.]—
v. 13 Corrosion—[etc.]—
v. 15. Casting.

1. Metals—Handbooks, manuals, etc. I. ASM
INTERNATIONAL. Handbook Committee.
TA459.M43 1978 669 78–14934
ISBN 0-87170-007-7 (v. 1)
SAN 204–7586

Printed in the United States of America

Foreword

The subject of metal casting was covered—along with forging—in Volume 5 of the 8th Edition of *Metals Handbook*. Volume 15 of the 9th Edition, a stand-alone volume on the subject, is evidence of the strong commitment of ASM INTERNATIONAL to the advancement of casting technology.

The decision to devote an entire Handbook to the subject of casting was based on the veritable explosion of improved or entirely new molding, melting, metal treatment, and casting processes that has occurred in the 18 years since the publication of Volume 5. New casting materials, such as cast metal-matrix composites, also have been developed in that time, and computers are being used increasingly by the foundry industry. An entire section of this Handbook is devoted to the application of computers to metal casting, in particular to the study of phenomena associated with the solidification of molten metals.

Coverage of the depth and scope provided in Volume 15 is made possible only by the collective efforts of many individuals. In this case, the effort was an international one, with participants in 12 nations. The driving force behind the entire project was volume chairman Doru M. Stefanescu of the University of Alabama, who along with his section chairmen recruited more than 200 of the leading experts in the world to author articles for this Handbook. We are indebted to all of them, as well as to the members of the ASM Handbook Committee and the Handbook editorial staff. Their hard work and dedication has culminated in the publication of this, the most comprehensive single-volume reference on casting technology yet published.

William G. Wood
President,
ASM INTERNATIONAL

Edward L. Langer
Managing Director,
ASM INTERNATIONAL

**The Ninth Edition of Metals Handbook
is dedicated to the memory of
TAYLOR LYMAN, A.B. (Eng.), S.M., Ph.D.
(1917–1973)
Editor, Metals Handbook, 1945–1973**

Preface

The story of metal casting is as glamorous as it is ancient, beginning with the dawn of human civilization and interwoven with legends of fantastic weapons and exquisite artworks made of precious metals. It was and is involved in the two main activities of humans since they began walking the earth: producing and defending wealth. Civilization as we know it would not have been possible without metal casting. Metal casting must have emerged from the darkness of antiquity first as magic, later to evolve as an art, then as a technology, and finally as a complex, interdisciplinary science.

As with most other industries, the body of knowledge in metal casting has doubled over the last ten years. A modern text on the subject should discuss not only the new developments in the field but also the applications of some fundamental sciences such as physical chemistry, heat transfer, and fluid flow in metal casting. The task of reviewing such an extensive amount of information and of documenting the knowledge currently involved in the various branches of this manufacturing industry is almost impossible. Nevertheless, this is the goal of this Volume. For such an endeavor to succeed, only one avenue was possible—to involve in the preparation of the manuscripts as well as in the review process the top metal casting engineers and scientists in the international community. Indeed, nearly 350 dedicated experts from industry and academe worldwide contributed to this Handbook. This magnificent pool of talent was instrumental in putting together what I believe to be the most complete text on metal casting available in the English language today.

The Handbook is structured in ten Sections, along with a Glossary of Terms. The reader is first introduced to the historical development of metal casting, as well as to the advantages of castings over parts produced by other manufacturing processes, their applications, and the current market size of the industry. Then, the thermodynamic relationships and properties of liquid metals and the physical chemistry of gases and impurities in liquid metals are discussed. A rather extensive Section reviews the fundamentals of the science of solidification as applied to cast alloys, including nucleation kinetics, fundamentals of growth, and the more practical subject of interpretation of cooling curves. Traditional subjects such as patterns, molding and casting processes, foundry equipment, and processing and design considerations are extensively covered in the following Sections. Considerable attention has been paid to new and emerging processes, such as the Hitchiner process, directional solidification, squeeze casting, and semisolid metal forming. The metallurgy of ferrous and nonferrous alloys is extensively covered in two separate Sections. Finally, there is detailed information on the most modern approach to metal casting, namely, computer applications. The basic principles of modeling of heat transfer, fluid flow, and microstructural evolution are discussed, and typical examples are given.

It is hoped that the reader can find in this Handbook not only the technical information that he or she may seek, but also the prevailing message that the metal casting industry is mature but not aging. It is part of human civilization and will remain so for centuries to come. Make no mistake. A country cannot hold its own in the international marketplace without a modern, competitive metal casting industry.

It is a great pleasure to acknowledge the collective effort of the many contributors to this Handbook. The chairmen of the ten Sections and the authors of the articles are easily acknowledged, since their names are duly listed throughout the Volume. Less obvious but of tremendous importance in maintaining a uniform, high-quality text is the contribution of the reviewers. The Handbook staff of ASM INTERNATIONAL must also be commended for their dauntless and painstaking efforts in making this Volume not only accurate but also beautiful. Last but not least, I would like to acknowledge the precious assistance of my secretary, Mrs. Donna Snow, who had the patience to cope gracefully with the many tasks involved in such a complex project.

Prof. D.M. Stefanescu
Volume Chairman

Policy on Units of Measure

By a resolution of its Board of Trustees, ASM INTERNATIONAL has adopted the practice of publishing data in both metric and customary U.S. units of measure. In preparing this Handbook, the editors have attempted to present data in metric units based primarily on Système International d'Unités (SI), with secondary mention of the corresponding values in customary U.S. units. The decision to use SI as the primary system of units was based on the aforementioned resolution of the Board of Trustees and the widespread use of metric units throughout the world.

For the most part, numerical engineering data in the text and in tables are presented in SI-based units with the customary U.S. equivalents in parentheses (text) or adjoining columns (tables). For example, pressure, stress, and strength are shown both in SI units, which are pascals (Pa) with a suitable prefix, and in customary U.S. units, which are pounds per square inch (psi). To save space, large values of psi have been converted to kips per square inch (ksi), where 1 ksi = 1000 psi. The metric ton (kg × 10^3) has been shown in megagrams (Mg). Some strictly scientific data are presented in SI units only. For example, the solubility of a gas in a metal is given only in milliliters per 100 grams of metal (mL/100 g).

To clarify some illustrations, only one set of units is presented on artwork. References in the accompanying text to data in the illustrations are presented in both SI-based and customary U.S. units. On graphs and charts, grids corresponding to SI-based units appear along the left and bottom edges. Where appropriate, corresponding customary U.S. units appear along the top and right edges.

Data pertaining to a specification published by a specification-writing group may be given in only the units used in that specification or in dual units, depending on the nature of the data. For example, the typical yield strength of aluminum sheet made to a specification written in customary U.S. units would be presented in dual units, but the sheet thickness specified in that specification might be presented only in inches.

Data obtained according to standardized test methods for which the standard recommends a particular system of units are presented in the units of that system. Wherever feasible, equivalent units are also presented.

Conversions and rounding have been done in accordance with ASTM Standard E 380, with attention given to the number of significant digits in the original data. For example, an annealing temperature of 1570 °F contains three significant digits. In this case, the equivalent temperature would be given as 855 °C; the exact conversion to 854.44 °C would not be appropriate. For an invariant physical phenomenon that occurs at a precise temperature (such as the melting of pure silver), it would be appropriate to report the temperature as 961.93 °C or 1763.5 °F. In some instances (especially in tables and data compilations), temperature values in °C and °F are alternatives rather than conversions.

The policy on units of measure in this Handbook contains several exceptions to strict conformance to ASTM E 380; in each instance, the exception has been made in an effort to improve the clarity of the Handbook. The most notable exception is the use of g/cm^3 rather than kg/m^3 as the unit of measure for density (mass per unit volume).

SI practice requires that only one virgule (diagonal) appear in units formed by combination of several basic units. Therefore, all of the units preceding the virgule are in the numerator and all units following the virgule are in the denominator of the expression; no parentheses are required to prevent ambiguity.

Authors and Reviewers

G. J. Abbaschian
University of Florida
Harvey Abramowitz
Purdue University
R. Agarwal
General Motors Technical Center
Mark J. Alcini
Williams International
Robert L. Allen
Deere & Company
Richard L. Anderson
Arnold Engineering Company
John Andrews
Camden Castings Center
James J. Archibald
Ashland Chemical Company
Shigeo Asai
Nagoya University (Japan)
William H. Bailey
Cleveland Pneumatic Company
Leo J. Baran
American Foundrymen's Society, Inc.
W.J. Barice
Precision Castparts Corporation
Charles E. Bates
Southern Research Institute
Robert J. Bayuzick
Vanderbilt University
J. Beech
University of Sheffield
(Great Britain)
V.G. Behal
Dofasco Inc. (Canada)
P. Belding
Columbia Steel Casting Company
John T. Berry
University of Alabama
U. Betz
Leybold AG (West Germany)
Gopal K. Bhat
Bhat Technology International, Inc.
Yves Bienvenu
Ecole des Mines de Paris (France)
H.E. Bills
Reynolds Metals Company
Reynolds Aluminum
Charles R. Bird
Stainless Steel Foundry & Engineering Inc.
K.E. Blazek
Inland Steel Company
William J. Boettinger
National Bureau of Standards
M.A. Bohlmann
I.G. Technologies, Inc.
Charles B. Boyer
Battelle Columbus Division
Jose R. Branco
Colorado School of Mines
R. Brink
Leybold AG (West Germany)
William Brouse
Carpenter Technology Corporation
Roger B. Brown
Disamatic, Inc.
Francis Brozo
Hitchcock Industries, Inc.
Robert S. Buck
International Magnesium Consultants, Inc.
J. Bukowski
General Motors Technical Center
Wilhelm Burgmann
Leybold AG (West Germany)
H.I. Burrier
The Timken Company
Michael Byrne
Homer Research Laboratories
S.L. Camacho
Plasma Energy Corporation
Paul G. Campbell
ALUMAX of South Carolina
James A. Capadona
Signicast Corporation
C. Carlsson
Asea Brown Boveri, Inc.
James H. Carpenter
Pangborn Corporation
Sam F. Carter
Carter Consultants, Inc.
Dixon Chandley
Metal Casting Technology, Inc.
K.K. Chawla
New Mexico Institute of Technology
Dianne Chong
McDonnell Douglas Astronautics Company
A. Choudhury
Leybold AG (West Germany)
Richard J. Choulet
Steelmaking Consultant
Yeou-Li Chu
The Ohio State University
Dwight Clark
Baltimore Specialty Steels
Steve Clark
R.H. Sheppard Company, Inc.
Byron B. Clow
International Magnesium Consultants, Inc.
Arthur Cohen
Copper Development Association, Inc.
B. Cole
Fort Wayne Foundry Corporation
H.H. Cornell
Niobium Products Company, Inc.
James A. Courtois
ALUMAX Engineered Metal Processes, Inc.
Jim Cox
Hatch Associates Ltd.
D.B. Craig
Elkem Metals Company
Alan W. Cramb
Carnegie Mellon University
R. Creese
West Virginia University
T.J. Crowley
Microwave Processing Systems
Milford Cunningham
Stahl Specialty Company
Peter A. Curreri
NASA Marshall Space Flight Center
Michael J. Cusick
Colorado School of Mines
Johnathan A. Dantzig
University of Illinois at Urbana—Champaign
C.V. Darragh
The Timken Company
A.S. Davis
ESCO Corporation
Jackson A. Dean
Cardinal Service Company
Prateen V. Desai
Georgia Institute of Technology
B.K. Dhindaw
IIT Kharagpur (India)
W. Dietrich
Leybold AG (West Germany)
George Di Sylvestro
American Colloid Company
R.L. Dobson
The Centrifugal Casting Machine Company
George J. Dooley, III
United States Department of the Interior
J.L. Dorcic
IIT Research Institute
R. Doremus
Rensselaer Polytechnic Institute
G. Doughman
Casting Design and Services
B. Duca
Duca Remanufacturing Inc.
J. DuPlessis
Crucible Magnetics Division
F. Durand
Centre National de la Recherche Scientifique
Polytechnique de Grenoble (France)

William B. Eisen
Crucible Compaction Metals
Nagy El-Kaddah
University of Alabama
R. Elliott
University of Manchester
(Great Britain)
John M. Eridon
Howmet Corporation
R.C. Eschenbach
Retech, Inc.
N. Eustathopoulos
Institut National Polytechnique de Grenoble (France)
M. Evans
Cytemp Specialty Steels
Robert D. Evans
ALUMAX Engineered Metal Processes, Inc.
Daniel Eylon
University of Dayton
H.E. Exner
Max-Planck-Institut für Metallforschung
(West Germany)
Gilbert M. Farrior
ALUMAX Engineered Metal Processes, Inc.
J. Feroe
G.H. Hensley Industries Inc.
J. Feinman
Technical Consultant
Merton C. Flemings
Massachusetts Institute of Technology
S.C. Flood
Alcan International Ltd.
(Great Britain)
Victor K. Forsberg
Quanex
Robert C. Foyle
Herman-Sinto V-Process Corporation
H. Frederiksson
The Royal Institute of Technology
(Sweden)
Richard J. Fruehan
Carnegie Mellon University
B. Gabrielsson
Elwood Uddeholm Steel Company
D.R. Gaskell
Purdue University
William Gavin
Hitchcock Industries, Inc.
H. Gaye
Technical Consultant
M. Geiger
Asea Brown Boveri, Inc.
L. Gonano
National Forge Company
George Good
Ford Motor Company
George M. Goodrich
Taussig Associates, Inc.
Martha Goodway
Smithsonian Institution
P. Gouwens
CMI Novacast Inc.
J. Grach
Cominco Metals
L.D. Graham
PCC Airfoils
E.J. Grandy
H. Kramer & Company
Douglas A. Granger
Alcoa Technical Center
C.V. Grosse
Howmet Corporation
R.E. Grote
Missouri Precision Castings
Daniel B. Groteke
Metcast Associates, Inc.
Thomas E. Grubach
Aluminum Company of America
J.E. Gruzleski
McGill University (Canada)
Richard B. Gundlach
Climax Research Services
T.B. Gurganus
Alcoa Technical Center
Alex M. Gymarty
SKW Metals & Alloys, Inc.
David Hale
Ervin Industries, Inc.
T.C. Hansen
Trane Company
Michael J. Hanslits
Precision Castparts Corporation
Howard R. Harker
A. Johnson Metals Corporation
Ron Harrison
Cameron Forge Company
Richard Helbling
Northern Castings
H. Henein
Carnegie Mellon University
D.G. Hennessy
The Timken Company
John J. Henrich
United States Pipe and Foundry Company
W. Herman
Quanex
Edwin Hodge
Degussa Electronics Inc.
D. Hoffman
National Forge Company
George B. Hood
United Technologies
Pratt & Whitney
M.J. Hornung
Elkem Metals Company
Robert A. Horton
PCC Airfoils, Inc.
Daryl F. Hoyt
Wedron Silica Company
I.C.H. Hughes
BCIRA International Centre for Cast Metals Technology (Great Britain)
R. Hummer
Austrian Foundry Research Institute
(Austria)
James Hunt
Southern Aluminum Company
J.D. Hunt
University of Oxford
(Great Britain)
W.-S. Hwang
National Cheng Kung University
(Taiwan)
J.E. Indacochea
University of Illinois
K. Ito
Carnegie Mellon University
K.A. Jackson
AT&T Bell Laboratories
J.D. Jackson
Pratt & Whitney
N. Janco
Technical Consultant
H. Jones
University of Sheffield
(Great Britain)
M. Jones
Duriron Company, Inc.
J.L. Jorstad
Reynolds Aluminum
David P. Kanicki
American Foundrymen's Society
Seymour Katz
General Motors Research Laboratories
T.L. Kaveney
Technical Consultant
Avery Kearney
Avery Kearney & Company
H. Kemmer
Leybold AG (West Germany)
Malachi P. Kenney
ALUMAX Engineered Metal Processes, Inc.
Gerhard Kienel
Leybold AG (West Germany)
Dan Kihlstadius
Oregon Metallurgical Corporation
Franklin L. Kiiskila
Williams International
Ken Kirgin
Technical Consultant
David H. Kirkwood
University of Sheffield
(Great Britain)
F. Knell
Leybold AG (West Germany)
Allan A. Koch
ALUMAX Engineered Metal Processes, Inc.
G.J.W. Kor
The Timken Company
D.J. Kotecki
Teledyne McKay
Ronald M. Kotschi
Kotschi's Software & Services, Inc.
Ezra L. Kotzin
American Foundrymen's Society
R.W. Kraft
Lehigh University

W. Kurz
Swiss Federal Institute of Technology (Switzerland)
Curtis P. Kyonka
ALUMAX Engineered Metal Processes, Inc.
John B. Lambert
Fansteel
Craig F. Landefeld
General Motors Research Laboratories
Eugene Langner
American Cast Iron Pipe Company
A. Laporte
National Forge Company
David J. Larson, Jr.
Grumman Corporation
John P. Laughlin
Oregon Metallurgical Corporation
Franklin D. Lemkey
United Technologies Research Center
G. Lesoult
Ecole des Mines de Nancy (France)
Colin Lewis
Hitchcock Industries, Inc.
Don Lewis
Aluminum Smelt & Refining
Ronald L. Lewis
The Ohio State University
R. Lindsay, III
Newport News Shipbuilding
R.D. Lindsay
Plasma Energy Corporation
Stephen Liu
Colorado School of Mines
Roy Lobenhofer
American Foundrymen's Society
C.A. Loong
Noranda Research Centre (Canada)
Carl Lundin
University of Tennessee
Norris Luther
Luther & Associates
Alvin F. Maloit
Consulting Metallurgist
P. Magnin
Swiss Federal Institute of Technology (Switzerland)
William L. Mankins
Inco Alloys International, Inc.
P.W. Marshall
Technical Consultant
Ian F. Masterson
Union Carbide Corporation
Linde Division
Gene J. Maurer, Jr.
United States Industries
D. Mayton
Urick Foundry
T.K. McCluhan
Elken Metals Company
J. McDonough
Technical Consultant
J.P. McKenna
Lindberg Division
Unit of General Signal Corporation
W. McNeish
Teledyne All-Vac
Ravi Menon
Teledyne McKay
Thomas N. Meyer
Aluminum Company of America
William Mihaichuk
Eastern Alloys, Inc.
David P. Miller
The Timken Company
A. Mitchell
The University of British Columbia (Canada)
S. Mizoguchi
Nippon Steel Corporation (Japan)
G. Monzo
Elwood Uddeholm Steel Company
P. Moroz
Armco Inc.
F. Müller
Leybold AG (West Germany)
Frederick A. Morrow
TFI Corporation
C. Nagy
Union Carbide Corporation
N.E. Nannina
Cast Masters Division of Latrobe Steel
R.L. Naro
Ashland Chemical Company
E. Nechtelberger
Austrian Foundry Research Institute (Austria)
David V. Neff
Metaullics Systems
Charles D. Nelson
Morris Bean and Company
Dale C.H. Nevison
Zinc Information Center, Ltd.
Jeremy R. Newman
Titech International Inc.
Roger A. Nichting
Colorado School of Mines
I. Ohnaka
Osaka University (Japan)
Patrick O'Meara
Intermet Foundries Inc.
B. Ozturk
Carnegie Mellon University
K.V. Pagalthivarthi
GIW Industries, Inc.
H. Pannen
Leybold AG (West Germany)
J. Parks
ME International
Murray Patz
Lost Foam Technologies, Inc.
Walter J. Peck
Central Foundry Division
General Motors Corporation
Robert D. Pehlke
University of Michigan
J.H. Perepezko
University of Wisconsin—Madison
Ralph Y. Perkul
Asea Brown Boveri, Inc.
Art Piechowski
Grede Foundries, Inc.
Larry J. Pionke
McDonnell Douglas Astronautics Company
Thomas S. Piwonka
University of Alabama
Lee A. Plutshack
Foseco, Inc.
D.R. Poirier
University of Arizona
J.R. Ponteri
Lester B. Knight & Associates, Inc.
Richard L. Poole
Aluminum Company of America
William Powell
Waupaca Foundry
Henry Proffitt
Haley Industries Ltd. (Canada)
William Provis
Modern Equipment Company
Timothy J. Pruitt
Zimmer, Inc.
John D. Puckett
Nelson Metal Products Corporation
Christopher W. Ramsey
Colorado School of Mines
V. Rangarajan
Colorado School of Mines
M. Rappaz
Swiss Federal Institute of Technology (Switzerland)
Garland W. Reese
Leybold-Heraeus Technologies Inc.
J.E. Rehder
University of Toronto (Canada)
H. Rice
Atlas Specialty Steel Division (Canada)
J.E. Roberts
Huntington Alloys
C.E. Rodaitis
The Timken Company
Lynn Rogers
Ervin Industries, Inc.
Pradeep Rohatgi
University of Wisconsin—Milwaukee
Elwin L. Rooy
Aluminum Company of America
Mervin T. Rowley
Technical Consultant
Alain Royer
Pont-A-Mousson S.A. (France)
Ronald W. Ruddle
Ronald W. Ruddle & Associates
Gary F. Ruff
CMI-International
Peter R. Sahm
Giesserei-Institut der RWTH (West Germany)
Mahi Sahoo
Canadian Centre for Minerals and Energy Technology (Canada)
Robert F. Schmidt
Colonial Metals Company

Richard Schaefer
FWS, Inc.
Donald G. Schmidt
R. Lavin & Sons, Inc.
T.E. Schmidt
Mercury Marine
Division of Brunswick Corporation
Robert A. Schmucker, Jr.
Thomas & Skinner, Inc.
Rainer Schumann
Leybold Technologies Inc.
D.M. Schuster
Dural Aluminum Composites Corporation
William Seaton
Seaton-SSK Engineering, Inc.
R. Shebuski
Outboard Marine Corporation
W. Shulof
General Motors Corporation
G. Sick
Leybold AG (West Germany)
Geoffrey K. Sigworth
Reading Foundry Products
H. Sims
Vulcan Engineering Company
J. Slaughter
Southern Alloy Corporation
Lawrence E. Smiley
Reliable Castings Corporation
Cyril Stanley Smith
Technical Consultant
Richard L. Smith
Ashland Chemical Company
John D. Sommerville
University of Toronto (Canada)
Warren Spear
Technical Consultant
T. Spence
Duriron Company, Inc.
A. Spengler
Technical Consultant
D.M. Stefanescu
The University of Alabama
S. Stefanidis
I. Schumann & Company
H. Stephan
Leybold AG (West Germany)
T. Stevens
Wollaston Alloys, Inc.
D. Stickle
Duriron Company, Inc.
Stephen C. Stocks
Oregon Metallurgical Corporation
R.A. Stoehr
University of Pittsburgh
C.W. Storey
High Tech Castings
George R. St. Pierre
The Ohio State University
R. Russell Stratton
Investment Casting Institute
Ken Strausbaugh
Ashland Chemical Company
Lionel J.D. Sully
Edison Industrial Systems Center
Anthony L. Suschil
Foseco, Inc.
Koreaki Suzuki
Hiroshima Junior College (Japan)
John M. Svoboda
Steel Founders' Society of America
Julian Szekely
Massachusetts Institute of Technology
Jack Thielke
Asea Brown Boveri, Inc.
Gary L. Thoe
Waupaca Foundry, Inc.
John K. Thorne
Precision Castparts Corporation
Basant L. Tiwari
General Motors Research Laboratories
Judith A. Todd
University of Southern California
R. Trivedi
Iowa State University
Paul K. Trojan
University of Michigan—Dearborn
D. Trudell
Aluminum Company of America
D.H. Turner
Timet Inc.
B.L. Tuttle
GMI Engineering & Management Institute
Daniel Twarog
American Foundrymen's Society
Derek Tyler
Olin Corporation
A.E. Umble
Bethlehem Steel Corporation
G. Uren
Electrical Metallurgy Company
Stella Vasseur
Pont-A-Mausson (France)
John D. Verhoeven
Ames Laboratory
S.K. Verma
IIT Research Institute
Robert Voigt
University of Kansas
Vernon F. Voigt
Giddings & Lewis Machine Tool Company
Vaughan Voller
University of Minnesota
P. Voorhees
Northwestern University
Terry Waitt
Maynard Steel Casting Company
J. Wallace
Case Western Reserve University
Charles F. Walton
Technical Consultant
A. Wayne Ward
Ward & Associates
Claude Watts
Technical Consultant
Daniel F. Weaver
Pontiac Foundry, Inc.
E. Weingärtner
Leybold AG (West Germany)
D. Wells
Huntington Alloys
Charles E. West
Aluminum Company of America
J.H. Westbrook
Sci-Tech Knowledge Systems, Inc.
Kenneth Whaler
Stahl Specialties Company
Charles V. White
GMI Engineering & Management Institute
Eldon Whiteside
U.S. Gypsum
P. Wieser
Technical Consultant
W.R. Wilcox
Clarkson University
Larson E. Wile
Consultant
J.L. Wilkoff
S. Wilkoff & Sons Company
R. Williams
Air Force Wright Aeronautical Laboratories
Frank T. Worzala
University of Wisconsin—Madison
Nick Wukovich
Foseco, Inc.
R.A. Wright
Technical Consultant
Michael Wrysch
Detroit Diesel Allison Division
General Motors Corporation
R. Youmans
Modern Equipment Company, Inc.
Kenneth P. Young
AMAX Research and Development Center
William B. Young
Dana Corporation
Engine Products Division
Michael Zatkoff
Sandtechnik, Inc.

Contents

Glossary of Terms

A

acid. A term applied to slags, refractories, and minerals containing a high percentage of silica.
acidity. The degree to which a material is acid. Furnace refractories are ranked by their acidity.
acid process. A steelmaking method using an acid refractory-lined furnace. Neither sulfur nor phosphorus is removed.
acid refractory. Siliceous ceramic materials of a high melting temperature, such as silica brick, used for metallurgical furnace linings. Compare with *basic refractory.*
addition agent. (1) Any material added to a charge of molten metal in a bath or ladle to bring the alloy to specifications. (2) Reagent added to plating bath.
additive. Any material added to molding sand for reasons other than bonding, for example, seacoal, pitch, graphite, cereals.
aerate. To fluff up molding sand to reduce its density.
airblasting. See *blasting or blast cleaning.*
air channel. A groove or hole that carries the vent from a core to the outside of a mold.
air dried. Refers to the air drying of a core or mold without the application of heat.
air-dried strength. Strength (compressive, shear, or tensile) of a refractory (sand) mixture after being air dried at room temperature.
air furnace. Reverberatory-type furnace in which metal is melted by heat from fuel burning at one end of the hearth, passing over the bath toward the stack at the other end. Heat is also reflected from the roof and sidewalls. See also *reverberatory furnace.*
air hole. A hole in a casting caused by air or gas trapped in the metal during solidification.
air setting. The characteristic of some materials, such as refractory cements, core pastes, binders, and plastics, to take permanent set at normal air temperatures.
allowance. In a foundry, the specified clearance. The difference in limiting sizes, such as minimum clearance or maximum interference between mating parts, as computed arithmetically. See also *tolerance.*
alpha process. A *shell molding* and coremaking method in which a thin resin-bonded shell is baked with a less expensive, highly permeable material.
alumina. The mineral aluminum oxide (Al_2O_3) with a high melting point (refractory) that is sometimes used as a molding sand.
angularity. The angular relationship of one surface to another. Specifically, the dimensional tolerance associated with such features on a casting.
arbitration bar. A test bar, cast with a heat of material, used to determine chemical composition, hardness, tensile strength, and deflection and strength under transverse loading in order to establish the state of acceptability of the casting.
arbor. A metal shape embedded in and used to support green or dry sand cores in the mold.
arc furnace. A furnace in which metal is melted either directly by an electric arc between an electrode and the work or indirectly by an arc between two electrodes adjacent to the metal.
arc melting. Melting metal in an electric arc furnace.
as-cast condition. Castings as removed from the mold without subsequent heat treatment.
atmospheric riser. A riser that uses atmospheric pressure to aid feeding. Essentially, a *blind riser* into which a small core or rod protrudes; the function of the core or rod is to provide an open passage so that the molten interior of the riser will not be under a partial vacuum when metal is withdrawn to feed the casting but will always be under atmospheric pressure.
austenite. A solid solution of one or more elements in face-centered cubic iron (gamma iron). Unless otherwise designated (such as nickel austenite), the solute is generally assumed to be carbon.

B

back draft. A reverse taper that prevents removal of a pattern from a mold or a core from a core box.
backing board (backing plate). A second *bottom board* on which molds are opened.
backup coat. The ceramic slurry of dip coat that is applied in multiple layers to provide a ceramic shell of the desired thickness and strength for use as a mold.
bake. Heating in an oven to a low controlled temperature to remove gases or to harden a binder.
baked core. A core that has been heated through sufficient time and temperature to produce the desired physical properties attainable from its oxidizing or thermal-setting binders.
bank sand. Sedimentary deposits, usually containing less than 5% clay, occurring in banks or pits, used in coremaking and in synthetic molding sands. See *sand.*
basic refractory. A lime- or magnesia-base ceramic material of high melting temperature used for furnace linings. Compare with *acid refractory.*
batch. An amount of core or mold sand or other material prepared at one time.
bath. Molten metal on the hearth of a furnace, in a crucible, or in a ladle.
bead. (1) Half-round cavity in a mold, or half-round projection or molding on a casting. (2) A single deposit of weld metal produced by fusion.
bedding. Sinking a pattern down into the sand to the desired position and ramming the sand around it.
bedding a core. Placing an irregularly shaped core on a bed of sand for drying.
bench molding. Making sand molds by hand tamping loose or production patterns at a bench without the assistance of air or hydraulic action.
bentonite. A colloidal claylike substance derived from the decomposition of volcanic ash composed chiefly of the minerals of the montmorillonite family. It is used for bonding molding sand.
bimetal. A casting made of two different metals, usually produced by *centrifugal casting.*
binder. A material used to hold the grains of sand together in molds or cores. It may be cereal, oil, clay, or natural or organic resins.
blacking. Carbonaceous materials, such as graphite or powdered carbon, usually mixed with a binder and frequently carried in suspension in water or other liquid used as a thin facing applied to surfaces of molds or cores to improve casting finish.
blasting or blast cleaning. A process for cleaning or finishing metal objects with an air blast or centrifugal wheel that throws abrasive particles against the surface of the workpiece. Small, irregular particles of metal are used as the abrasive in gritblasting; sand, in sandblasting; and steel balls, in shotblasting.
bleed. Refers to molten metal oozing out of a casting. It is stripped or removed from the mold before complete solidification.

blended sand. A mixture of sands of different grain size and clay content that provides suitable characteristics for foundry use.
blind riser. A *riser* that does not extend through the top of the mold.
blister. A defect in metal, on or near the surface, resulting from the expansion of gas in a subsurface zone. It is characterized by a smooth bump on the surface of the casting and a hole inside the casting directly below the bump.
blow. A term that describes the trapping of gas in castings, causing voids in the metal.
blowhole. A void or large pore that may occur because of entrapped air, gas, or shrinkage; usually evident in heavy sections.
blow holes. Holes in the head plate or blow plate of a core blowing machine through which sand is blown from the reservoir into the *core box*.
bond clay. Any clay suitable for use as a *bonding agent* in molding sand.
bond strength. The degree of cohesiveness that the *bonding agent* exhibits in holding sand grains together.
bonding agent. Any material other than water that, when added to foundry sands, imparts strength either in the green, dry, or fired state.
boss. A relatively short protrusion or projection from the surface of a forging or casting, often cylindrical in shape. Usually intended for drilling and tapping for attaching parts. See also *locating boss*.
bottom board. A flat base for holding the *flask* in making sand molds.
bottom-pour ladle. A *ladle* from which metal, usually steel, flows through a *nozzle* located at the bottom.
bottom running or pouring. Filling of the mold cavity from the bottom by means of gates from the runner.
bridging. (1) Premature solidification of metal across a mold section before the metal below or beyond solidifies. (2) Solidification of slag within a cupola at or just above the tuyeres.
buckle. (1) Bulging of a large, flat face of a casting; in investment casting, caused by *dip coat* peeling from the pattern. (2) An indentation in a casting, resulting from expansion of the sand, can be termed the start of an expansion defect.
bumper. A machine used for packing molding sand in a flask by repeated jarring or jolting. See also *jolt ramming*.
burned-in sand. A defect consisting of a mixture of sand and metal cohering to the surface of a casting.
burned-on sand. A misnomer usually indicating metal penetration into sand resulting in a mixture of sand and metal adhering to the surface of a casting. See also *metal penetration*.
burnout. Firing a mold at a high temperature to remove pattern material residue.
burned sand. Sand in which the binder or bond has been removed or impaired by contact with molten metal.

C

calcium silicon. An alloy of calcium, silicon, and iron containing 28 to 35% Ca, 60 to 65% Si, and 6% Fe (max), used as a deoxidizer and degasser for steel and cast iron; sometimes called calcium silicide.
carbonaceous. A material that contains carbon in any or all of its several allotropic forms.
carbon dioxide process (sodium silicate/CO_2). A process for hardening molds or cores in which carbon dioxide gas is blown through dry clay-free silica sand to precipitate silica in the form of a gel from the sodium silicate binder.
carbon refractory. A manufactured refractory comprised substantially or entirely of carbon (including graphite).
castability. (1) A complex combination of liquid-metal properties and solidification characteristics that promotes accurate and sound final castings. (2) The relative ease with which a molten metal flows through a mold or casting die.
castable. A combination of refractory grain and suitable bonding agent that, after the addition of a proper liquid, is generally poured into place to form a refractory shape or structure that becomes rigid because of chemical action.
casting. (1) Metal object cast to the required shape by pouring or injecting liquid metal into a mold, as distinct from one shaped by a mechanical process. (2) Pouring molten metal into a mold to produce an object of desired shape.
casting defect. Any imperfection in a casting that does not satisfy one or more of the required design or quality specifications. This term is often used in a limited sense for those flaws formed by improper casting solidification.
casting section thickness. The wall thickness of the casting. Because the casting may not have a uniform thickness, the section thickness may be specified at a specific place on the casting. Also, it is sometimes useful to use the average, minimum, or typical wall thickness to describe a casting.
casting shrinkage. The amount of dimensional change per unit length of the casting as it solidifies in the mold or die and cools to room temperature after removal from the mold or die. There are three distinct types of casting shrinkage. Liquid shrinkage refers to the reduction in volume of liquid metal as it cools to the liquidus. Solidification shrinkage is the reduction in volume of metal from the beginning to the end of solidification. Solid shrinkage involves the reduction in volume of metal from the solidus to room temperature.
casting stresses. Stresses set up in a casting because of geometry and casting shrinkage.
casting thickness. See *casting section thickness*.
casting volume. The total cubic units (mm^3 or $in.^3$) of cast metal in the casting.
casting yield. The weight of a casting(s) divided by the total weight of metal poured into the mold, expressed as a percentage.
cast iron. A generic term for a large family of cast ferrous alloys in which the carbon content exceeds the solubility of carbon in austenite at the eutectic temperature. Most cast irons contain at least 2% C, plus silicon and sulfur, and may or may not contain other alloying elements. For the various forms, the word cast is often left out, resulting in *compacted graphite iron, gray iron, white iron, malleable iron,* and *ductile iron*.
cast structure. The internal physical structure of a casting evidenced by the shape and orientation of crystals and the segregation of impurities.
cavity. The mold or die impression that gives a casting its external shape.
cementite. A very hard and brittle compound of iron and carbon corresponding to the empirical formula Fe_3C, commonly known as iron carbide.
centerline shrinkage. Shrinkage or porosity occurring along the central plane or axis of a cast part.
centrifugal casting. The process of filling molds by (1) pouring metal into a sand or permanent mold that is revolving about either its horizontal or its vertical axis or (2) pouring metal into a mold that is subsequently revolved before solidification of the metal is complete. See also *centrifuge casting*.
centrifuge casting. A casting technique in which mold cavities are spaced symmetrically about a vertical axial common downgate. The entire assembly is rotated about that axis during pouring and solidification.
ceramic. Material of a nonmetallic nature, usually refractory, made from fused, sintered, or cemented metallic oxides.
ceramic molding. A precision casting process that employs permanent patterns and fine-grain slurry for making molds. Unlike monolithic investment molds, which are similar in composition, ceramic molds consist of a *cope* and a *drag* or, if the casting shape permits, a drag only.
CG iron. Same as *compacted graphite iron*.
chaplet. Metal support that holds a core in place within a mold; molten metal solidifies around a chaplet and fuses it into the finished casting.
charge. (1) The materials placed in a melting furnace. (2) Castings placed in a heat-treating furnace.
check. A minute crack in the surface of a casting caused by unequal expansion or contraction during cooling.

chill. (1) A metal or graphite insert embedded in the surface of a sand mold or core or placed in a mold cavity to increase the cooling rate at that point. (2) White iron occurring on a gray or ductile iron casting, such as the chill in the wedge test. See also *chilled iron*. Compare with *inverse chill*.

chill coating. Applying a coating to a chill that forms part of the mold cavity so that the metal does not adhere to it, or applying a special coating to the sand surface of the mold that causes the iron to undercool.

chilled iron. Cast iron that is poured into a metal mold or against a mold insert so as to cause the rapid solidification that often tends to produce a white iron structure in the casting.

clay. A natural, earthy, fine-grain material that develops plasticity when mixed with a limited amount of water. Foundry clays, which consist essentially of hydrous silicates of alumina, are used in molds and cores.

CO_2 process. See *carbon dioxide process*.

coining. (1) The process of straightening and sizing castings by die pressing. (2) A press metalworking operation that establishes accurate dimensions of flat surfaces or depressions under predominantly compressive loading.

coke. A porous, gray, infusible product resulting from the dry distillation of bituminous coal, petroleum, or coal tar pitch that drives off most of the volatile matter. Used as a fuel in cupola melting.

coke bed. The first layer of coke placed in the cupola. Also the coke used as the foundation in constructing a large mold in a *flask* or pit.

coke breeze. Fines from coke screenings, used in blacking mixes after grinding; also briquetted for cupola use.

coke furnace. Type of pot or crucible furnace that uses coke as the fuel.

cold box process. A two-part organic resin binder system mixed in conventional mixers and blown into shell or solid core shapes at room temperature. A vapor mixed with air is blown into the core, permitting instant setting and immediate pouring of metal around it.

cold chamber machine. A die casting machine with an injection system that is charged with liquid metal from a separate furnace. Compare with *hot chamber machine*.

cold cracking. Cracks in cold or nearly cold metal due to excessive internal stress caused by contraction. Often brought about when the mold is too hard or the casting is of unsuitable design.

cold lap. Wrinkled markings on the surface of an ingot or casting from incipient freezing of the surface and too low a casting temperature.

cold-setting process. Any of several systems for bonding mold or core aggregates by means of organic binders, relying on the use of catalysts rather than heat for polymerization (setting).

cold shot. (1) A portion of the surface of an ingot or casting showing premature solidification; caused by splashing of molten metal onto a cold mold wall during pouring. (2) Small globule of metal embedded in, but not entirely fused with, the casting.

cold shut. (1) A discontinuity that appears on the surface of cast metal as a result of two streams of liquid meeting and failing to unite. (2) A lap on the surface of a forging or billet that was closed without fusion during deformation. (3) Freezing of the top surface of an ingot before the mold is full.

collapsibility. The tendency of a sand mixture to break down under the pressures and temperatures developed during casting.

columnar structure. A coarse structure of parallel columns of grains, that is caused by highly directional solidification resulting from sharp thermal gradients.

combination die (multiple-cavity die). In die casting, a die with two or more different cavities for different castings.

combined carbon. Carbon in iron that is combined chemically with other elements; not in the free state as graphite or temper carbon. The difference between the total carbon and the graphite carbon analyses. Contrast with *free carbon*.

compacted graphite iron. Cast iron having a graphite shape intermediate between the flake form typical of gray iron and the spherical form of fully spherulitic ductile iron. Also known as CG iron or vermicular iron, compacted graphite iron is produced in a manner similar to that for ductile iron but with a technique that inhibits the formation of fully spherulitic graphite nodules.

constraint. Any restriction that limits the transverse contraction normally associated with a longitudinal tension, and therefore causes a secondary tension in the transverse direction.

consumable-electrode remelting. A process for refining metals in which an electric current passes between an electrode made of the metal to be refined and an ingot of the refined metal, which is contained in a water-cooled mold. As a result of the passage of electric current, droplets of molten metal form on the electrode and fall to the ingot. The refining action occurs from contact with the atmosphere, vacuum, or slag through which the drop falls. See *electroslag remelting* and *vacuum arc remelting*.

continuous casting. A process for forming a bar of constant cross section directly from molten metal by gradually withdrawing the bar from a die as the metal flowing into the die solidifies.

contraction. The volume change that occurs in metals and alloys upon solidification and cooling to room temperature.

convection. The motion resulting in a fluid from the differences in density and the action of gravity. In heat transmission, this meaning has been extended to include both forced and natural motion or circulation.

cooling stresses. Stresses developed during cooling by the uneven contraction of metal, generally due to nonuniform cooling.

cope. The upper or topmost section of a *flask*, *mold*, or *pattern*.

core. (1) A specially formed material inserted in a mold to shape the interior or other part of a casting that cannot be shaped as easily by the pattern. (2) In a ferrous alloy prepared for case hardening, that portion of the alloy that is not part of the case. Typically considered to be the portion that (a) appears light on an etched cross section, (b) has an essentially unaltered chemical composition, or (c) has a hardness, after hardening, less than a specified value.

core assembly. A complex core consisting of a number of sections.

core binder. Any material used to hold the grains of core sand together.

core blow. A gas pocket in a casting adjacent to a cored cavity and caused by entrapped gases from the core.

core blower. A machine for making foundry cores using compressed air to blow and pack the sand into the core box.

core box. A wood, metal, or plastic structure containing a shaped cavity into which sand is packed to make a core.

core dryers. Supports used to hold cores in shape during baking; constructed from metal or sand for conventional baking or from plastic material for use with dielectric core-baking equipment.

core filler. Material, such as coke, cinder, and sawdust, used in place of sand in the interiors of large cores; usually added to aid collapsibility.

coring. A variable composition between the center and the surface of a unit of structure (such as a dendrite, grain, or carbide particle) resulting from the nonequilibrium growth that occurs over a range of temperature.

core knockout machine. A mechanical device for removing cores from castings.

coreless induction furnace. An electric induction furnace for melting or holding molten die casting metals that does not utilize a steel core to direct the magnetic field.

core oil. A binder for core sand that sets when baked and is destroyed by the heat from the cooling casting.

core plates. Heat-resistant plates used to support cores during baking; may be metallic or nonmetallic, the latter being a requisite for dielectric core baking.

core print. Projections attached to a pattern in order to form recesses in the mold at points where cores are to be supported.

core sand. Sand for making cores to which a binding material has been added to obtain good cohesion and permeability after drying; usually low in clays.

core shift. A variation from the specified dimensions of a cored casting section due to a change in position of the core or misalignment of cores in assembly.

core vents. (1) A wax product, round or oval in form, used to form the vent passage in a core. Also, a metal screen or slotted piece used to form the vent passage in the core box used in a core blowing machine. (2) Holes made in the core for the escape of gas.

core wash. A suspension of a fine refractory applied to cores by brushing, dipping, or spraying to improve the surface of the cored portion of the casting.

core wires or rods. Reinforcing wires or rods for fragile cores, often preformed into special shapes.

corundum. Native alumina, or aluminum oxide, Al_2O_3, occurring as rhombohedral crystals and also in masses and variously colored grains. It is the hardest mineral except for the diamond. Corundum and its artificial counterparts are abrasives especially suited to the grinding of metals.

coupon. A piece of metal from which a test specimen is to be prepared; often an extra piece (as on a casting or forging) or a separate piece made for test purposes (such as a test weldment).

cover core. (1) A core set in place during the ramming of a mold to cover and complete a cavity partly formed by the withdrawal of a loose part of the pattern. Also used to form part or all of the cope surface of the mold cavity. (2) A core placed over another core to create a flat *parting line*.

critical dimension. A dimension on a part that must be held within the specified tolerance for the part to function in its application. A noncritical tolerance may be for cost or weight savings or for manufacturing convenience, but is not essential for the products.

Croning process. A *shell molding process* that uses a phenolic resin binder. Sometimes referred to as C process or Chronizing.

cross-sectional area. The area measured at right angles to the molten metal flow stream at any specified portion of the gating system.

crucible. A vessel or pot, made of a refractory substance or of a metal with a high melting point, used for melting metals or other substances.

crucible furnace. A melting or holding furnace in which the molten metal is contained in a pot-shaped (hemispherical) shell. Electric heaters or fuel-fired burners outside the shell generate the heat that passes through the shell (crucible) to the molten metal.

crush. (1) Buckling or breaking of a section of a casting mold due to incorrect register when the mold is closed. (2) An indentation in the surface of a casting due to displacement of sand when the mold was closed.

crush strip or bead. An indentation in the *parting line* of a pattern plate that ensures that *cope* and *drag* will have good contact by producing a ridge of sand that crushes against the other surface of the mold or core.

cupola. A cylindrical vertical furnace for melting metal, especially cast iron, by having the charge come in contact with the hot fuel, usually metallurgical coke.

curing time (no bake). The period of time needed before a sand mass reaches maximum hardness.

cut. (1) To recondition molding sand by mixing on the floor with a shovel or blade-type machine. (2) To form the sprue cavity in a mold. (3) Defect in a casting resulting from erosion of the sand by metal flowing over the mold or cored surface.

cut off. Removing a casting from the sprue by refractory wheel or saw, arc-air torch, or gas torch.

D

daubing. Filling of cracks in molds or cores by specially prepared pastes or coatings to prevent penetration of metal into these cracks during pouring.

dead-burned. Term applied to materials that have been fired to a temperature sufficiently high to render them relatively resistant to moisture and contraction.

defect. A discontinuity whose size, shape, orientation, or location makes it detrimental to the useful service of the part in which it occurs.

defective. A quality control term describing a unit of product or service containing at least one defect or having several lesser imperfections that, in combination, cause the unit not to fulfill its anticipated function.

degasification. See *degassing*.

degasifier. A substance that can be added to molten metal to remove soluble gases that might otherwise be occluded or entrapped in the metal during solidification.

degassing. (1) A chemical reaction resulting from a compound added to molten metal to remove gases from the metal. Inert gases are often used in this operation. (2) A fluxing procedure used for aluminum alloys in which nitrogen, chlorine, chlorine and nitrogen, and chlorine and argon are bubbled up through the metal to remove dissolved hydrogen gases and oxides from the alloy. See also *flux*.

dendrite. A crystal that has a treelike branching pattern, being most evident in cast metals slowly cooled through the solidification range.

deoxidation. Removal of excess oxygen from the molten metal; usually accomplished by adding materials with a high affinity for oxygen.

deoxidizer. A substance that can be added to molten metal to remove either free or combined oxygen.

deoxidizing. (1) The removal of oxygen from molten metals through the use of a suitable *deoxidizer*. (2) Sometimes refers to the removal of undesirable elements other than oxygen through the introduction of elements or compounds that readily react with them. (3) In metal finishing, the removal of oxide films from metal surfaces by chemical or electrochemical reaction.

dephosphorization. The elimination of phosphorus from molten steel.

descaling. A chemical or mechanical process for removing scale or investment material from castings.

desulfurizing. The removal of sulfur from molten metal by reaction with a suitable slag or by the addition of suitable compounds.

dewaxing. The process of removing the expendable wax pattern from an investment mold or shell mold; usually accomplished by melting out the application of heat or dissolving the wax with an appropriate solvent.

die casting. (1) A casting made in a die. (2) A casting process in which molten metal is forced under high pressure into the cavity of a metal mold.

die pull. The direction in which the solidified casting must move when it is removed from the die. The die pull direction must be selected such that all points on the surface of the casting move away from the die cavity surfaces.

die separation. The space between the two halves of a die casting die at the parting surface when the dies are closed. The separation may be the result of the internal cavity pressure exceeding the locking force of the machine or warpage of the die due to thermal gradients in the die steel.

dip coat. (1) In the solid mold technique of investment casting, an extremely fine ceramic precoat applied as a slurry directly to the surface of the pattern to reproduce maximum surface smoothness. This coating is surrounded by coarser, less expensive, and more permeable investment to form the mold. (2) In the shell mold technique of investment casting, an extremely fine ceramic coating called the first coat, applied as a slurry directly to the surface of the pattern to reproduce maximum surface smoothness. The first coat is followed by other dip coats of different viscosity and usually containing different grading of ceramic particles. After each dip, coarser stucco material is applied to the still-wet coating. A buildup of several coats forms an investment shell mold.

directional solidification. Solidification of molten metal in such a manner that feed metal is always available for that portion that is just solidifying.

discontinuity. Any interruption in the normal physical structure or configuration of a part, such as cracks, laps, seams, inclusions, or porosity. A discontinuity may or may not affect the utility of the part.

distortion. Any deviation from the desired shape or contour.

dolomite brick. A calcium magnesium carbonate ($Ca \cdot Mg(CO_3)_2$) used as a refractory brick that is manufactured substantially or entirely of *dead-burned* dolomite.

dowel. (1) A wooden or metal pin of various types used in the parting surface of parted patterns and core boxes. (2) In die casting dies, metal pins to ensure correct registry of cover and ejector halves.

downgate. Same as *sprue*.

draft. (1) An angle or taper on the surface of a pattern, core box, punch, or die (or of the parts made with them) that facilitates removal of the parts from a mold or die cavity, or a core from a casting. (2) The change in cross section that occurs during rolling or cold drawing.

drag. The bottom section of a *flask*, *mold*, or *pattern*.

draw. A term used to denote the shrinkage that appears on the surface of a casting; formerly used to describe tempering.

drawing (pattern). Removing a pattern from a mold or a mold from a pattern in production work.

draw plate. A plate attached to a pattern to facilitate drawing of a pattern from the mold.

drop. A casting imperfection due to a portion of the sand dropping from the cope or other overhanging section of the mold.

dross. The scum that forms on the surface of molten metal largely because of oxidation but sometimes because of the rising of impurities to the surface.

dry and baked compression test. An American Foundrymen's Society test for determining the maximum compressive stress that a baked sand mixture is capable of developing.

dry permeability. The property of a molded mass of sand, bonded or unbonded, dried at ~100 to 110 °C (~220 to 230 °F), and cooled to room temperature, that allows the transfer of gases resulting during the pouring of molten metal into a mold.

dry sand casting. The process in which the sand molds are dried at above 100 °C (212 °F) before use.

dry sand mold. A casting mold made of sand and then dried at ~100 °C (~220 °F) or above before being used. Contrast with *green sand mold*.

dry strength. The maximum strength of a molded sand specimen that has been thoroughly dried at ~100 to 110 °C (~220 to 230 °F) and cooled to room temperature. Also known as dry bond strength.

dual-metal centrifugal casting. Centrifugal castings produced by pouring a different metal into the rotating mold after the first metal poured has solidified. Also referred to as *bimetal casting*.

ductile iron. A *cast iron* that has been treated while molten with an element such as magnesium or cerium to induce the formation of free graphite as nodules or spherulites, which imparts a measurable degree of ductility to the cast metal. Also known as nodular cast iron, spherulitic graphite cast iron, and SG iron.

E

ejector. A pin (rod) or mechanism that pushes the solidified die casting out of the die.

ejector pin. See *ejector*.

electric arc furnace. See *arc furnace*.

electric furnace. A metal melting or holding furnace that produces heat from electricity. It may operate on the resistance or induction principle.

electrode. Compressed graphite or carbon cylinder or rod used to conduct electric current in electric arc furnaces, arc lamps, carbon arc welding, and so forth.

electroslag remelting. A *consumable-electrode remelting* process in which heat is generated by the passage of electric current through a conductive slag. The droplets of metal are refined by contact with the slag. Sometimes abbreviated ESR.

endothermic reaction. Designating or pertaining to a reaction that involves the absorption of heat. See also *exothermic reaction*.

equiaxed grain structure. A structure in which the grains have approximately the same dimensions in all directions.

ethyl silicate. A strong bonding agent for sand and refractories used in preparing molds in the investment casting process.

eutectic. (1) An isothermal reversible reaction in which a liquid solution is converted into two or more intimately mixed solids upon cooling, the number of solids formed being the same as the number of components in the system. (2) An alloy having the composition indicated by the eutectic point on an equilibrium diagram. (3) An alloy structure of intermixed solid constituents formed by a eutectic reaction.

exothermic reaction. Chemical reactions involving the liberation of heat, such as the burning of fuel or the deoxidizing of iron with aluminum. See also *endothermic reaction*.

expendable pattern. A pattern that is destroyed in making a casting. It is usually made of wax (investment casting) or expanded polystyrene (lost foam casting).

F

facing. Any material applied in a wet or dry condition to the face of a mold or core to improve the surface of the casting. See also *mold wash*.

feeder (feeder head, feedhead). A *riser*.

feeding. (1) In casting, providing molten metal to a region undergoing solidification, usually at a rate sufficient to fill the mold cavity ahead of the solidification front and to compensate for any shrinkage accompanying solidification. (2) Conveying metal stock or workpieces to a location for use or processing, such as wire to a consumable electrode, strip to a die, or workpieces to an assembler.

ferrite. An essentially carbon-free solid solution in which alpha iron is the solvent, and which is characterized by a body-centered cubic crystal structure.

ferroalloy. An alloy of iron that contains a sufficient amount of one or more other chemical elements to be useful as an agent for introducing these elements into molten metal, especially into steel or cast iron.

ferrous. Metallic materials in which the principal component is iron.

fillet. Concave corner piece usually used at the intersection of casting sections. Also the radius of metal at such junctions as opposed to an abrupt angular junction.

fillet radius. Blend radius between two abutting walls.

fin. Metal on a casting caused by an imperfect joint in the mold or die.

finish allowance. Amount of stock left on the surface of a casting for machining.

firebrick. A refractory brick, often made from *fireclay*, that is able to withstand high temperature (1500 to 1600 °C, or 2700 to 2900 °F) and is used to line furnaces, ladles, or other molten metal containment components.

fireclay. A mineral aggregate that has as its essential constituent the hydrous silicates of aluminum with or without free silica. It is used in commercial refractory products.

fired mold. A shell mold or solid mold that has been heated to a high temperature and is ready for casting.

flake graphite. Graphitic carbon, in the form of platelets, occurring in the microstructure of *gray iron*.

flash. A thin section or *fin* of metal formed at the mold, core, or die joint or parting in a casting due to the cope and drag not matching completely or where core and core print do not match.

flask. A metal or wood frame used for making and holding a sand mold. The upper part is called the *cope*; the lower, the *drag*.

flaw. A nonspecific term often used to imply a cracklike discontinuity. See preferred terms *discontinuity* and *defect*.

floor molding. Making sand molds from loose or production patterns of such size that they cannot be satisfactorily handled on a bench or molding machine, the equipment being located on the floor during the entire operation of making the mold.

flowability. A characteristic of a foundry sand mixture that enables it to move under pressure or vibration so that it makes intimate contact with all surfaces of the pattern or core box.

fluidity. The ability of liquid metal to run into and fill a mold or die cavity.

flux. (1) In metal refining, a material used to remove undesirable substances, such as sand, ash, or dirt, as a molten mixture. It is also used as a protective covering for certain molten metal baths. Lime or limestone is generally used to remove sand, as in iron smelting; sand, to remove iron oxide in copper refining. (2) In brazing, cutting, soldering, or welding, material used to prevent the formation of or to dissolve and facilitate the removal of oxides and other undesirable substances.

foundry returns. Metal in the form of gates, sprues, runners, risers, and scrapped castings of known composition returned to the furnace for remelting.

free carbon. The part of the total carbon in steel or cast iron that is present in elemental form as graphite or temper carbon. Contrast with *combined carbon*.

free ferrite. Ferrite formed into separate grains and not intimately associated with carbides as in pearlite.

freezing range. That temperature range between *liquidus* and *solidus* temperatures in which molten and solid constituents coexist.

full mold. A trade name for an expendable pattern casting process in which the polystyrene pattern is vaporized by the molten metal as the mold is poured.

G

gassing. (1) Absorption of gas by a metal. (2) Evolution of gas from a metal during melting operations or upon solidification. (3) Evolution of gas from an electrode during electrolysis.

gas holes. Holes in castings or welds that are formed by gas escaping from molten metal as it solidifies. Gas holes may occur individually, in clusters, or throughout the solidified metal.

gas pocket. A cavity caused by entrapped gas.

gas porosity. Fine holes or pores within a metal that are caused by entrapped gas or by the evolution of dissolved gas during solidification.

gate. The portion of the runner in a mold through which molten metal enters the mold cavity. The generic term is sometimes applied to the entire network of connecting channels that conduct metal into the mold cavity.

gated pattern. A *pattern* that includes not only the contours of the part to be cast but also the *gates*.

gating system. The complete assembly of sprues, runners, and gates in a mold through which metal flows to enter the casting cavity. The term is also applied to equivalent portions of the pattern.

gooseneck. In die casting, a spout connecting a molten metal holding pot, or chamber, with a nozzle or sprue hole in the die and containing a passage through which molten metal is forced on its way to the die. It is the metal injection mechanism in a *hot chamber machine*.

grain. An individual crystal in a polycrystalline metal or alloy; it may or may not contain twinned regions and subgrains.

grain fineness number. A system developed by the American Foundrymen's Society for rapidly expressing the average grain size of a given sand. It approximates the number of meshes per inch of that sieve that would just pass the sample.

grain refinement. The manipulation of the solidification process to cause more (and therefore smaller) grains to be formed and/or to cause the grains to form in specific shapes. The term refinement is usually used to denote a chemical addition to the metal, but can refer to control of the cooling rate.

grain refiner. Any material added to a liquid metal for producing a finer grain size in the subsequent casting.

grain size. For metals, a measure of the areas or volumes of grains in a polycrystalline material, usually expressed as an average when the individual sizes are fairly uniform. In metals containing two or more phases, grain size refers to that of the matrix unless otherwise specified. Grain size is reported in terms of number of grains per unit area or volume, in terms of average diameter, or as a grain size number derived from area measurements.

graphite. One of the crystal forms of carbon; also the uncombined carbon in cast irons.

graphitic carbon. Free carbon in steel or cast iron.

graphitization. The formation of graphite in iron or steel. Where graphite is formed during solidification, the phenomenon is termed primary graphitization; where formed later by heat treatment, secondary graphitization.

gravity die casting. See *permanent mold*.

gray iron. Cast iron that contains a relatively large percentage of the carbon present in the form of flake graphite.

green sand. A molding sand that has been tempered with water and is used for casting when still in the damp condition.

green sand core. (1) A *core* made of *green sand* and used as-rammed. (2) A sand core that is used in the unbaked condition.

green sand mold. A casting mold composed of moist prepared molding sand. Contrast with *dry sand mold*.

green strength. The strength of a tempered sand mixture at room temperature.

grit. Crushed ferrous or synthetic abrasive material in various mesh sizes that is used in abrasive blasting equipment to clean castings. See also *blasting or blast cleaning*.

gross porosity. In weld metal or in a casting, pores, gas holes or globular voids that are larger and in much greater numbers than those obtained in good practice.

growth (cast iron). A permanent increase in the dimensions of cast iron resulting from repeated or prolonged heating at temperatures above 480 °C (900 °F) due either to graphitizing of carbides or oxidation.

H

hardener. An alloy rich in one or more alloying elements that is added to a melt to permit closer control of composition than is possible by the addition of pure metals, or to introduce refractory elements not readily alloyed with the base metal. Sometimes called *master alloy* or rich alloy.

hearth. The bottom portions of certain furnaces, such as blast furnaces, air furnaces, and other reverberatory furnaces, that support the charge and sometimes collect and hold molten metal.

heat. A stated tonnage of metal obtained from a period of continuous melting in a cupola or furnace, or the melting period required to handle this tonnage.

heat-disposable pattern. A pattern formed from a wax- or plastic-base material that is melted from the mold cavity by the application of heat.

holding furnace. A furnace into which molten metal can be transferred to be held at the proper temperature until it can be used to make castings.

hot box process. A furan resin-base process similar to shell coremaking; cores produced with it are solid unless mandrelled out.

hot chamber machine. A *die casting* machine in which the metal chamber under pressure is immersed in the molten metal in a furnace. Sometimes called a *gooseneck* machine.

hot crack. A crack formed in a cast metal because of internal stress developed upon cooling following solidification. A hot crack is less open than a *hot tear* and usually exhibits less oxidation and decarburization along the fracture surface.

hot shortness. A tendency for some alloys to separate along grain boundaries when stressed or deformed at temperatures near the melting point. Hot shortness is caused by a low-melting constituent, often present only in minute amounts, that is segregated at grain boundaries.

hot tear. A fracture formed in a metal during solidification because of hindered *contraction*. Compare with *hot crack*.

hot top. (1) A reservoir, thermally insulated or heated, that holds molten metal on top of a mold for feeding of the ingot or casting as it contracts on solidifying, thus preventing the formation of *pipe* or *voids*. (2) A refractory-lined steel or iron casting that is inserted into the tip of the mold and is supported at various heights to feed the ingot as it solidifies.

I

impregnation. (1) Treatment of porous castings with a sealing medium to stop pressure leaks. (2) The process of filling the pores of a sintered compact, usually with a liquid such as a lubricant. (3) The process of mixing particles of a nonmetallic substance in a matrix of metal powder, as in diamond-impregnated tools.

inclusions. Particles of foreign material in a metallic matrix. The particles are usually compounds (such as oxides, sulfides, or silicates), but may be of any substance that is foreign to (and essentially insoluble in) the matrix.

induction furnace. An alternating current electric furnace in which the primary conductor is coiled and generates, by electromagnetic induction, a secondary current that develops heat within the metal charge. See also *coreless induction furnace*.

induction heating or melting. Heating or melting in an *induction furnace*.

inert gas. A gas that will not support combustion or sustain any chemical reaction, for example, argon or helium.

ingate. Same as *gate*.

ingot. A casting of simple shape, suitable for hot working or remelting.

injection. The process of forcing molten metal into the *die casting* die.

injection molding. The injection of molten metal or other material under pressure into molds.

inoculant. Materials that, when added to molten metal, modify the structure and thus change the physical and mechanical properties to a degree not explained on the basis of the change in composition resulting from their use.

inoculation. The addition of a material to molten metal to form nuclei for crystallization. See also *inoculant*.

insert. (1) A part formed from a second material, usually a metal, that is placed in the molds and appears as an integral structural part of the final casting. (2) A removable portion of a die or mold.

insulating pads and sleeves. Insulating material, such as gypsum, diatomaceous earth, and so forth, used to lower the rate of solidification. As sleeves on open risers, they are used to keep the metal liquid, thus increasing the feeding efficiency. Contrast with *chill*.

internal shrinkage. A void or network of voids within a casting caused by inadequate feeding of that section during solidification.

internal stress. See *residual stress*.

inverse chill. The condition in a casting section in which the interior is mottled or white, while the other sections are gray iron. Also known as reverse chill, internal chill, and inverted chill.

inverse segregation. Segregation in cast metal in which an excess of lower-melting constituents occurs in the earlier-freezing portions, apparently the result of liquid metal entering cavities developed in the earlier-solidified metal.

investing. The process of pouring the investment slurry into a flask surrounding the pattern to form the mold.

investment. A flowable mixture, or slurry, of a graded refractory filler, a binder, and a liquid vehicle that, when poured around the patterns, conforms to their shape and subsequently sets hard to form the investment mold.

investment casting. (1) Casting metal into a mold produced by surrounding, or *investing*, an expendable pattern with a refractory slurry that sets at room temperature, after which the wax or plastic pattern is removed through the use of heat prior to filling the mold with liquid metal. Also called *precision casting* or *lost wax process*. (2) A part made by the investment casting process.

investment precoat. See *dip coat*.

investment precoat. An extremely fine investment coating applied as a thin slurry directly to the surface of the pattern to reproduce maximum surface smoothness. The coating is surrounded by a coarser, cheaper, and more permeable investment to form the mold. See also *dip coat*.

investment shell. Ceramic mold obtained by alternately dipping a pattern set up in *dip coat* slurry and stuccoing with coarse ceramic particles until the shell of desired thickness is obtained.

J

jolt ramming. Packing sand in a mold by raising and dropping the sand, pattern, and flask on a table. Jolt squeezers, jarring machines, and jolt rammers are machines using this principle. Also called jar ramming.

jolt-squeezer machine. A combination machine that employs a jolt action followed by a squeezing action to compact the sand around the pattern.

K

keel block. A standard test casting, for steel and other high-shrinkage alloys, consisting of a rectangular bar that resembles the keel of a boat, attached to the bottom of a large riser, or shrinkhead. Keel blocks that have only one bar are often called Y-blocks; keel blocks having two bars, double keel blocks. Test specimens are machined from the rectangular bar, and the shrinkhead is discarded.

kiln. An oven or furnace for burning, calcining, or drying a substance.

knockout. (1) Removal of sand cores from a casting. (2) Jarring of an investment casting mold to remove the casting and investment from the flask. (3) A mechanism for freeing formed parts from a die used for stamping, blanking, drawing, forging or heading operations. (4) A partially pierced hole in a sheet metal part, where the slug remains in the hole and can be forced out by hand if a hole is needed.

L

ladle. Metal receptacle frequently lined with refractories used for transporting and pouring molten metal. Types include hand, bull, crane, bottom-pour, holding, teapot, shank, and lip-pour.

ladle brick. Refractory brick suitable for lining ladles used to hold molten metal.

ladle coating. The material used to coat metal ladles to prevent iron pickup in aluminum alloys. The material can only consist of sodium silicate, iron oxide, and water, applied to the ladle when it is heated.

ladle preheating. The process of heating a ladle prior to the addition of molten metal. This procedure reduces metal heat loss and eliminates moisture-steam safety hazards.

launder. A channel for transporting molten metal.

lining. Internal refractory layer of firebrick, clay, sand, or other material in a furnace or ladle.

lip-pour ladle. Ladle in which the molten metal is poured over a lip, much as water is poured out of a bucket.

liquation. Partial melting of an alloy, usually as a result of *coring* or other compositional heterogeneities.

liquation temperature. The lowest temperature at which partial melting can occur in an alloy that exhibits the greatest possible degree of segregation.

liquidus. In a phase diagram, the locus of points representing the temperatures at which the various compositions in the system begin to freeze on cooling or finish melting on heating. See also *solidus*.

loam. A molding material consisting of sand, silt, and clay, used over brickwork or other structural backup material for making massive castings, usually of iron or steel.

locating boss. A *boss*-shaped feature on a casting to help locate the casting in an assembly or to locate the casting during secondary tooling operations.

lost foam casting (process). An *expendable pattern* process in which an expandable polystyrene pattern surrounded by the unbonded sand, is vaporized during pouring of the molten metal.

lost wax process. An *investment casting* process in which a wax pattern is used.

M

macroshrinkage. Isolated, clustered, or interconnected voids in a casting that are detectable macroscopically. Such voids are usually associated with abrupt changes in section size and are caused by feeding that is insufficient to compensate for solidification shrinkage.

malleable iron. A cast iron made by prolonged annealing of *white iron* in which decarburization, graphitization, or both take place to eliminate some or all of the cementite. The graphite is in the form of temper carbon. If decarburization is the predominant reaction, the product will exhibit a light fracture surface; hence whiteheart malleable. Otherwise, the fracture surface will be dark; hence blackheart malleable. Ferritic malleable has a predominantly ferritic matrix; pearlitic malleable may contain pearlite, spheroidite, or tempered martensite, depending on heat treatment and desired hardness.

malleablizing. Annealing *white iron* in such a way that some or all of the combined carbon is transformed into graphite or, in some cases, so that part of the carbon is removed completely.

master alloy. An alloy, rich in one or more desired addition elements, that is added to a melt to raise the percentage of a desired constituent.

master pattern. A pattern embodying a double contraction allowance in its construction, used for making castings to be employed as patterns in production work.

match plate. A plate of metal or other material on which patterns for metal casting are mounted (or formed as an integral part) to facilitate molding. The pattern is divided along its *parting plane* by the plate.

melting point. The temperature at which a pure metal, compound, or eutectic changes from solid to liquid; the temperature at which the liquid and the solid are in equilibrium. See also *melting range*.

melting range. The range of temperatures over which an alloy other than a compound or eutectic changes from solid to liquid; the range of temperatures from *solidus* to *liquidus* at any given composition on a *phase diagram*.

metal penetration. A surface condition in castings in which metal or metal oxides have filled voids between sand grains without displacing them.

microsegregation. *Segregation* within a grain, crystal, or small particle. See also *coring*.

microshrinkage. A casting imperfection, not detectable microscopically, consisting of interdendritic voids. Microshrinkage results from contraction during solidification where the opportunity to supply filler material is inadequate to compensate for shrinkage. Alloys with wide ranges in solidification temperature are particularly susceptible.

misrun. Denotes an irregularity of the casting surface caused by incomplete filling of the mold due to low pouring temperatures, gas back pressure from inadequate venting of the mold, and inadequate gating.

mold. The form, made of sand, metal, or refractory material, that contains the cavity into which molten metal is poured to produce a casting of desired shape.

mold cavity. The space in a mold that is filled with liquid metal to form the casting upon solidification. The channels through which liquid metal enters the mold cavity (sprue, runner, gates) and reservoirs for liquid metal (risers) are not considered part of the mold cavity proper.

mold coating. (1) Coating to prevent surface defects on permanent mold castings and die castings. (2) Coating on sand molds to prevent metal penetration and to improve metal finish. Also called mold facing or mold dressing.

molding machine. A machine for making sand molds by mechanically compacting sand around a pattern.

molding sands. Sands containing over 5% natural clay, usually between 8 and 20%. See also *naturally bonded molding sand*.

mold jacket. Wood or metal form that is slipped over a sand mold for support during pouring.

mold shift. A casting defect that results when the parts of the mold do not match at the parting line.

mold wash. An aqueous or alcoholic emulsion or suspension of various materials used to coat the surface of a mold cavity.

mottled cast iron. Iron that consists of a mixture of variable proportions of gray cast iron and white cast iron; such a material has a mottled fracture appearance.

mulling. The mixing and kneading of molding sand with moisture and clay to develop suitable properties for molding.

N

naturally bonded molding sand. A sand containing sufficient bonding material as mined to be suitable for molding purposes.

no-bake binder. A synthetic liquid resin sand binder that hardens completely at room temperature, generally not requiring baking; used in a *cold-setting process*.

nodular graphite. Graphite in the nodular form as opposed to flake form (see *flake graphite*). Nodular graphite is characteristic of malleable iron. The graphite of nodular or *ductile iron* is spherulitic in form, but called nodular.

nodular iron. See preferred term *ductile iron*.

nominal dimension. The size of the dimension to which the tolerance is applied. For example, if a dimension is 50 mm ± 0.5 mm (2.00 in. ± 0.02 in.), the 50 mm (2.00 in.) is the nominal dimension, and the ±0.5 mm (±0.02 in.) is the tolerance.

normal segregation. A concentration of alloying constituents that have low melting points in those portions of a casting that solidify last. Compare with *inverse segregation*.

nozzle. (1) Pouring spout of a bottom-pour ladle. (2) On a hot chamber die casting machine, the thick-wall tube that carries the pressurized molten metal from the gooseneck to the die.

nucleation. The initiation of a phase transformation at discrete sites, with the new phase growing on the nuclei. See *nucleus (1)*.

nucleus. (1) The first structurally stable particle capable of initiating recrystallization of a phase or the growth of a new phase and possessing an interface with the parent matrix. The term is also applied to a foreign particle that initiates such action. (2) The heavy central core of an atom, in which most of the mass and the total positive electric charge are concentrated.

O

olivine. A naturally occurring mineral of the composition $(Mg,Fe)_2SiO_4$ that is crushed and used as a molding sand.

open hearth furnace. A reverberatory melting furnace with a shallow hearth and a low roof. The flame passes over the charge on the hearth, causing the charge to be heated both by direct flame and by radiation from the roof and sidewalls of the furnace.

open-sand casting. Any casting made in a mold that has no cope or other covering.

oxidation. A chemical reaction in which one substance is changed to another by oxygen combining with the substance. Much of the dross from holding and melting furnaces is the result of oxidation of the alloy held in the furnace.

oxidation losses. Reduction in the amount of metal or alloy through oxidation. Such losses are usually the largest factor in melting loss.

oxygen lance. A length of pipe used to convey oxygen either beneath or on top of the melt in a steelmaking furnace, or to the point of cutting in oxygen lance cutting.

P

padding. The process of adding metal to the cross section of a casting wall, usually extending from a riser, to ensure adequate feed metal to a localized area during solidification where a shrink would occur if the added metal were not present.

particle size. The controlling lineal dimension of an individual particle, such as of sand, as determined by analysis with screens or other suitable instruments.

particle size distribution. The percentage, by weight or by number, of each fraction into which a powder or sand sample has been classified with respect to sieve number or *particle size*.

parting. (1) The zone of separation between *cope* and *drag* portions of the mold or flask in sand casting. (2) In the recovery of precious metals, the separation of silver from gold. (3) Cutting simultaneously along two parallel lines or along two lines that balance each other in side thrust. (4) A shearing operation used to produce two or more parts from a stamping.

parting compound. A material dusted or sprayed on patterns to prevent adherence of sand and to promote easy separation of *cope* and *drag* parting surfaces when the cope is lifted from the drag.

parting line. (1) The intersection of the parting plane of a casting mold or the parting plane between forging dies with the mold or die cavity. (2) A raised line or projection on the surface of a casting or forging that corresponds to said intersection.

parting plane. (1) In casting, the dividing plane between mold halves. (2) In forging, the dividing plane between dies.

pattern. (1) A form of wood, metal, or other material around which molding material is placed to make a mold for casting metals. (2) A form of wax- or plastic-base material around which refractory material is placed to make a mold for casting metals. (3) A full-scale reproduction of a part used as a guide in cutting.

pattern draft. Taper allowed on the vertical faces of a pattern to permit easy withdrawal of the pattern from the mold or die.

pattern layout. A full-size drawing of a pattern showing its arrangement and structural features.

patternmaker's shrinkage. Contraction allowance made on patterns to compensate for the decrease in dimensions as the solidified casting cools in the mold from the freezing temperature of the metal to room temperature. The pattern is made larger by the amount of contraction that is characteristic of the particular metal to be used.

penetration. See *metal penetration*.

permanent mold. A metal, graphite or ceramic mold (other than an ingot mold) that is repeatedly used for the production of many castings of the same form. Liquid metal is poured in by gravity (gravity die casting).

permeability. (1) In founding, the characteristics of molding materials that permit gases to pass through them. (2) In powder metallurgy, a property measured as the rate of passage under specified conditions of a liquid or gas through a compact. (3) A general term used to express various relationships between magnetic induction and magnetizing force. These relationships are either absolute permeability, which is a change in magnetic induction divided by the corresponding change in magnetizing force, or specific (relative) permeability, which is the ratio of the absolute permeability to the permeability of free space.

phase diagram. A graphical representation of the temperature and composition limits of phase fields in an alloy system as they actually exist under the specific conditions of heating or cooling.

pinhole porosity. Porosity consisting of numerous small gas holes distributed throughout the metal; found in weld metal, castings, and electrodeposited metal.

pipe. (1) The central cavity formed by contraction in metal, especially ingots, during solidification. (2) An imperfection in wrought or cast products resulting from such a cavity. (3) A tubular metal product, cast or wrought.

pit molding. Molding method in which the drag is made in a pit or hole in the floor.

plaster molding. Molding in which a gypsum-bonded aggregate flour in the form of a water slurry is poured over a pattern, permitted to harden, and, after removal of the pattern, thoroughly dried. This technique is used to make smooth nonferrous castings of accurate size.

plunger. Ram or piston that forces molten metal into a die in a *die casting* machine. Plunger machines are those having a plunger in continuous contact with molten metal.

porosity. A characteristic of being porous, with voids or pores resulting from trapped air or shrinkage in a casting. See also *gas porosity* and *pinhole porosity*.

port. The opening through which molten metal enters the injection cylinder of a *die casting* plunger machine, or is ladled into the injection cylinder of a cold chamber machine. See also *cold chamber machine* and *plunger*.

pot. (1) A vessel for holding molten metal. (2) The electrolytic reduction cell used to make such metals as aluminum from a fused electrolyte.

pouring. The transfer of molten metal from furnace to ladle, ladle to ladle, or ladle into molds.

pouring basin. A basin on top of a mold that receives the molten metal before it enters the sprue or downgate.

precision casting. A metal casting of reproducible, accurate dimensions, regardless of how it is made. Often used interchangeably with *investment casting*.

preformed ceramic core. A preformed refractory aggregate inserted in a wax or plastic pattern to shape the interior of that part of a casting which cannot be shaped by the pattern. The wax is sometimes injected around the preformed core.

pressure casting. (1) Making castings with pressure on the molten or plastic metal, as in *injection molding, die casting, centrifugal casting*, cold chamber pressure casting, and *squeeze casting*. (2) A casting made with pressure applied to the molten or plastic metal.

primary alloy. Any alloy whose major constituent has been refined directly from ore, not recycled scrap metal. Compare with *secondary alloy*.

projected area. The area of a cavity, or portion of a cavity, in a mold or die casting die measured from the projection on a plane that is normal to the direction of the mold or die opening.

R

ramming. (1) Packing sand, refractory, or other material into a compact mass. (2) The compacting of molding sand in forming a mold.

rattail. A surface imperfection on a casting, occurring as one or more irregular lines, caused by the expansion of sand in the mold. Compare with *buckle (2)*.

recrystallization. A process in which the distorted grain structure of cold-worked metals is replaced by a new, strain-free grain structure during heating above a specific minimum temperature.

recrystallization temperature. The lowest temperature at which the distorted grain structure of a cold-worked metal is replaced by a new, strain-free grain structure during prolonged heating. Time, purity of the metal, and prior deformation are important factors.

refractory. (1) A material of very high melting point with properties that make it suitable for such uses as furnace linings and kiln construction. (2) The quality of resisting heat.

residual stress. Stress present in a body that is free of external forces or thermal gradients.

reverberatory furnace. A furnace in which the flame used for melting the metal does not impinge on the metal surface itself, but is reflected off the walls of the roof of the furnace. The metal is actually melted by the generation of heat from the walls and the roof of the furnace.

rheocasting. Casting of a continuously stirred semisolid metal slurry.

rigging. The engineering design, layout, and fabrication of pattern equipment for producing castings; including a study of the casting solidification program, feeding and gating, risering, skimmers, and fitting flasks.

riser. A reservoir of molten metal connected to a casting to provide additional metal to the casting, required as the result of shrinkage before and during solidification.

runner. (1) A channel through which molten metal flows from one receptacle to another. (2) The portion of the gate assembly of a casting that connects the sprue with the gate(s). (3) Parts of patterns and finished castings corresponding to the portion of the gate assembly described in (2).

runner box. A distribution box that divides molten metal into several streams before it enters the mold cavity.

runout. (1) The unintentional escape of molten metal from a mold, crucible, or furnace. (2) The defect in a casting caused by the escape of metal from the mold.

S

sag. An increase or decrease in the section thickness of a casting caused by insufficient strength of the mold sand of the cope or of the core.

sand. A granular material naturally or artificially produced by the disintegration or crushing of rocks or mineral deposits. In casting, the term denotes an aggregate, with an individual particle (grain) size of 0.06 to 2 mm (0.002 to 0.08 in.) in diameter, that is largely free of finer constituents such as silt and clay, which are often present in natural sand deposits. The most commonly used foundry sand is *silica*; however, *zircon*, *olivine*, *alumina*, and other crushed ceramics are used for special applications.

sandblasting. Abrasive blasting with sand. See *blasting or blast cleaning* and compare with *shotblasting*.

sand casting. Metal castings produced in sand molds.

sand grain distribution. Variation or uniformity in particle size of a sand aggregate when properly screened by standard screen sizes.

sand reclamation. Processing of used foundry sand by thermal, air, or hydraulic methods so that it can be used in place of new sand without substantially changing the foundry sand practice.

sand tempering. Adding sufficient moisture to molding sand to make it workable.

scab. A defect on the surface of a casting that appears as a rough, slightly raised surface blemish, crusted over by a thin porous layer of metal, under which is a honeycomb or cavity that usually contains a layer of sand; defect common to thin-wall portions of the casting or around hot areas of the mold.

scaling (scale). Surface oxidation, consisting of partially adherent layers of corrosion products, left on metals by heating or casting in air or in other oxidizing atmospheres.

screen. One of a set of sieves designated by the size of the openings, used to classify granular aggregates such as sand, ore, or coke by particle size.

screen analysis. See *sieve analysis*.

seam. (1) A surface defect on a casting related to but of lesser degree than a *cold shut*. (2) A ridge on the surface of a casting caused by a crack in the mold face.

secondary alloy. Any alloy whose major constituent is obtained from recycled scrap metal. Compare with *primary alloy*.

segregation. A casting defect involving a concentration of alloying elements at specific regions, usually as a result of the primary crystallization of one phase with the subsequent concentration of other elements in the remaining liquid. Microsegregation refers to normal segregation on a microscopic scale in which material richer in an alloying element freezes in successive layers on the dendrites (*coring*) and in constituent network. Macrosegregation refers to gross differences in concentration (for example, from one area of a casting to another). See also *inverse segregation* and *normal segregation*.

semipermanent mold. A permanent mold in which sand cores are used.

shakeout. Removal of castings from a sand mold. See also *knockout*.

Shaw (Osborn-Shaw) Process. See *ceramic molding*.

shell molding. Forming a mold from thermosetting resin-bonded sand mixtures brought in contact with preheated (150 to 260 °C, or 300 to 500 °F) metal patterns, resulting in a firm shell with a cavity corresponding to the outline of the pattern. Also called *Croning process*.

shift. A casting imperfection caused by the mismatch of cope and drag or of cores and mold.

shot. (1) Small, spherical particles of metal. (2) The injection of molten metal into a die casting die. The metal is injected so quickly that it can be compared to the shooting of a gun.

shotblasting. Blasting with metal *shot*; usually used to remove deposits or mill scale more rapidly or more effectively than can be done by *sandblasting*.

shrinkage. See *casting shrinkage*.

shrinkage cavity. A void left in cast metal as a result of solidification shrinkage. See also *casting shrinkage*.

shrinkage cracks. Cracks that form in metal as a result of the pulling apart of grains by contraction before complete solidification.

sieve analysis. *Particle size distribution*; usually expressed as the weight percentage retained on each of a series of standard sieves of decreasing size and the percentage passed by the sieve of finest size. Synonymous with sieve classification.

silica. Silicon dioxide (SiO_2); the primary ingredient of sand and acid refractories.

silica flour. A sand additive, containing about 99.5% silica, commonly produced by pulverizing quartz sand in large ball mills to a mesh size of 80 to 325.

skim gate. A gating arrangement designed to prevent the passage of slag and other undesirable materials into a casting.

skimming. Removing or holding back dirt or slag from the surface of the molten metal before or during pouring.

skin drying. Drying the surface of the mold by direct application of heat.

slag. A nonmetallic product resulting from the mutual dissolution of flux and nonmetallic impurities in smelting, refining, and certain welding operations. In steelmaking operations, the slag serves to protect the molten metal from the air and to extract certain impurities.

slag inclusion. Slag or dross entrapped in a metal.

slip flask. A tapered flask that depends on a movable strip of metal to hold the sand in position. After closing the mold, the strip is retracted and the flask can be removed and reused. Molds made in this manner are usually supported by a *mold jacket* during pouring.

slush casting. A hollow casting usually made of an alloy with a low but wide melting temperature range. After the desired thickness of metal has solidified in the mold, the remaining liquid is poured out.

snap flask. A foundry flask hinged on one corner so that it can be opened and removed from the mold for reuse before the metal is poured.

solid shrinkage. See *casting shrinkage*.

solidification. The change in state from liquid to solid upon cooling through the melting temperature or melting range.

solidification shrinkage. See *casting shrinkage*.

solidus. In a phase diagram, the locus of points representing the temperatures at which various compositions stop freezing upon cooling or begin to melt upon heating. See also *liquidus*.

solute. A metal or substance dissolved in a major constituent; the component that is dissolved in the solvent.

solvent. The base metal or major constituent in a solution; the component that dissolves the solvent.

sprue. (1) The mold channel that connects the *pouring basin* with the runner or, in the absence of a pouring basin, directly into which molten metal is poured. Sometimes referred to as downsprue or downgate. (2) Sometimes used to mean all gates, risers, runners, and similar scrap that are removed from castings after shakeout.

squeeze casting. A hybrid liquid-metal forging process in which liquid metal is forced into a permanent mold by a hydraulic press.

stack molding. A molding method that makes use of both faces of a mold section, with one face acting as the drag and the other as the cope. Sections, when assembled to other similar sections, form several tiers of mold cavities, and all castings are poured together through a common sprue.

stopper rod. A device in a bottom-pour ladle for controlling the flow of metal through the nozzle into a mold. The stopper rod consists of a steel rod, protective refractory sleeves, and a graphite stopper head.

stopping off. Filling in a portion of a mold cavity to keep out molten metal.

strainer core. A perforated core in the gating system for preventing slag and other extraneous material from entering the casting cavity.

stripping. Removing the pattern from the mold or the core box from the core.

styrofoam pattern. An expendable pattern of foamed plastic, especially expanded polystyrene, used in manufacturing castings by the lost foam process.

supercooling. Lowering the temperature of a molten metal below its liquidus during cooling.

superheat. Any increment of temperature above the melting point of a metal; sometimes construed to be any increment of temperature above normal casting temperatures introduced for the purpose of refining, alloying, or improving fluidity.

superheating. Raising the temperature of molten metal above the normal melting temperature for more complete refining and greater fluidity.

supersaturated. A metastable solution in which the dissolved material exceeds the amount the solvent can hold in normal equilibrium at the temperature and other conditions that prevail.

surface area. The actual area of the surface of a casting or cavity. The surface area is always greater than the *projected area*.

sweep. A type of pattern that is a template cut to the profile of the desired mold shape that, when revolved around a stake or spindle, produces that shape in the mold.

T

teapot ladle. A ladle in which, by means of an external spout, metal is removed from the bottom rather than the top of the ladle.

temper. (1) To moisten *green sand* for casting molds with water. (2) In heat treatment, to reheat hardened steel or hardened cast iron to some temperature below the eutectoid temperature for the purpose of decreasing hardness and increasing toughness. The process is also sometimes applied to normalized steel. (3) In nonferrous alloys and in some ferrous alloys (steels that cannot be hardened by heat treatment), the hardness and strength produced by mechanical or thermal treatment, or both, and characterized by a certain structure, mechanical properties or reduction in area during cold working.

thermal expansion. The increase in linear dimensions of a material accompanying an increase in temperature.

thin-wall casting. A term used to define a casting that has the minimum wall thickness to satisfy its service function.

tie bar. A bar-shaped connection added to a casting to prevent distortion caused by uneven contraction between two separated members of the casting.

tolerance. The specified permissible deviation from a specified nominal dimension, or the permissible variation in size or other quality characteristic of a part.

tramp element. Contaminant in the components of a furnace charge, or in the molten metal or castings, whose presence is thought to be either unimportant or undesirable to the quality of the casting. Also called trace element.

transfer ladle. A ladle that can be supported on a monorail or carried in a shank and used to transfer metal from the melting furnace to the holding furnace or from the furnace to the pouring ladles.

tumbling. Rotating workpieces, usually castings or forgings, in a barrel partially filled with metal slugs or abrasives, to remove sand, scale, or fins. It may be done dry or with an aqueous solution added to the contents of the barrel. Sometimes called rumbling or rattling.

tuyere. An opening in a cupola, blast furnace, or converter for the introduction of air or inert gas.

U

undercooling. Same as *supercooling*.

undercut. A recess having an opening smaller than the internal configuration, thus preventing the mechanical removal of a one-piece core.

V

vacuum arc remelting. A *consumable-electrode remelting* process in which heat is generated by an electric arc between the electrode and the ingot. The process is performed inside a vacuum chamber. Exposure of the droplets of molten metal to the reduced pressure reduces the amount of dissolved gas in the metal. Sometimes abbreviated VAR.

vacuum casting. A casting process in which metal is melted and poured under very low atmospheric pressure; a form of permanent mold casting in which the mold is inserted into liquid metal, vacuum is applied, and metal is drawn up into the cavity.

vacuum degassing. The use of vacuum techniques to remove dissolved gases from molten alloys.

vacuum induction melting. A process for remelting and refining metals in which the metal is melted inside a vacuum chamber by induction heating. The metal can be melted in a crucible and then poured into a mold. Sometimes abbreviated VIM.

vacuum melting. Melting in a vacuum to prevent contamination from air and to remove gases already dissolved in the metal; the solidification can also be carried out in a vacuum or at low pressure.

vacuum molding. See *V process*.

vacuum refining. Melting in a vacuum to remove gaseous contaminants from the metal.

vent. A small opening or passage in a mold or core to facilitate the escape of gases when the mold is poured.

vermicular iron. Same as *compacted graphite iron*.

void. A shrinkage cavity produced in castings during solidification.

V process. A molding process in which the sand is held in place in the mold by vacuum. The mold halves are covered with a thin sheet of plastic to retain the vacuum.

W

warpage. Deformation other than contraction that develops in a casting between solidification and room temperature; also the distortion that occurs during annealing, stress relieving, and high-temperature service.

wash. (1) A coating applied to the face of a mold prior to casting. (2) An imperfection at a cast surface similar to a *cut (3)*.

wax pattern. A precise duplicate, allowing for shrinkage, of the casting and required gates, usually formed by pouring or injecting molten wax into a die or mold.

white iron. *Cast iron* that shows a white fracture because the carbon is in combined form.

Y

yield. Comparison of casting weight to the total weight of metal poured into the mold.

Z

zircon. The mineral zircon silicate ($ZrSiO_4$), a very high melting point acid refractory material used as a molding sand.

zone melting. Highly localized melting, usually by induction heating, of a small volume of an otherwise solid piece, usually a rod. By moving the induction coil along the rod, the melted zone can be transferred from one end to the other. In a binary mixture where there is a large difference in composition on the liquidus and solidus lines, high purity can be attained by concentrating one of the constituents in the liquid as it moves along the rod.

Introduction and Historical Development

Section Chairman:
Joseph R. Davis, ASM INTERNATIONAL

History of Casting

Martha Goodway, Smithsonian Institution

THE CASTING OF METAL is a prehistoric technology, but one that appears relatively late in the archaeological record. There were many earlier fire-using technologies, collectively called by Wertime pyrotechnology, which provided a basis for the development of metal casting. Among these were the heat treatment of stone to make it more workable, the burning of lime to make plaster, and the firing of clay to produce ceramics. At first, it did not include smelting, for the metal of the earliest castings appears to have been native.

The earliest objects now known to have been made of metal are more than 10 000 years old (see Table 1) and were wrought, not cast. They are small, decorative pendants and beads, which were hammered to shape from nuggets of native copper and required no joining. The copper was beaten flat into the shape of leaves or was rolled to form small tubular beads. The archaeological period in which this metalworking took place was the Neolithic, beginning some time during the Aceramic Neolithic, before the appearance of pottery in the archaeological record.

Native metals were then perhaps considered simply another kind of stone, and the methods that had been found useful in shaping stone were attempted with metal nuggets. It seems likely that the copper being worked was also being annealed, because this was a treatment that was already being given stone. Proof of annealing could be obtained from the microstructures of these early copper artifacts were it not for their generally corroded condition (some are totally mineralized) and the natural reluctance to use destructive methods in studying very rare objects.

The appearance of plasters and ceramics in the Neolithic period is evidence that the use of fire was being extended to materials other than stone. Exactly when the casting of metals began is not known. Archaeologists give the name Chalcolithic to the period in which metals were first being mastered and date this period, which immediately preceded the Bronze Age, very approximately to between 5000 and 3000 B.C. Analyses of early cast axes and other objects give chemical compositions consistent with their having been cast from native copper and are the basis for the conclusion that the melting of metals had been mastered before smelting was developed. The furnaces were rudimentary. It has been shown by experiment that it was possible to smelt copper, for example, in a crucible. Nevertheless, the evidence for casting demonstrates an increasing ability to manage and direct fire in order to achieve the required melting temperatures. The fuel employed was charcoal, which tended to supply a reducing atmosphere where the fire was enclosed in an effort to reduce the loss of heat. Smelting followed.

The molds were of stone (Fig. 1). The tradition of stone carving was longer than any of the pyrotechnologies, and the level of skill allowed very finely detailed work. The stone carved was usually of a smooth texture such as steatite or andesite, and the molds produced are themselves often very fine objects, which can be viewed in museums and archaeological exhibitions. Many are open molds, although they were not necessarily intended for flat objects. Elaborate filigree for jewelry was cast in open molds and then shaped by bending into bracelets and headpieces, or cast in parts and then assembled. Certain molds, described by the archaeologist as multifaceted, have cavities carved in each side of a rectangular block of stone. Such multifaceted molds would have been more portable than separate ones and suggest itinerant founding, but they may simply represent economy in the use of a suitable piece of stone.

Table 1 Chronological list of developments in the use of materials

Date	Development	Location
9000 B.C.	Earliest metal objects of wrought native copper	Near East
6500 B.C.	Earliest life-size statues, of plaster	Jordan
5000–3000 B.C.	Chalcolithic period: melting of copper; experimentation with smelting	Near East
3000–1500 B.C.	Bronze Age: arsenical copper and tin bronze alloys	Near East
3000–2500 B.C.	Lost wax casting of small objects	Near East
2500 B.C.	Granulation of gold and silver and their alloys	Near East
2400–2200 B.C.	Copper statue of Pharoah Pepi I	Egypt
2000 B.C.	Bronze Age	Far East
1500 B.C.	Iron Age (wrought iron)	Near East
700–600 B.C.	Etruscan dust granulation	Italy
600 B.C.	Cast iron	China
224 B.C.	Colossus of Rhodes destroyed	Greece
200–300 A.D.	Use of mercury in gilding (amalgam gilding)	Roman world
1200–1450 A.D.	Introduction of cast iron (exact date and place unknown)	Europe
Circa 1122 A.D.	Theophilus's *On Divers Arts*, the first monograph on metalworking written by a craftsman	Germany
1252 A.D.	*Diabutsu* (Great Buddha) cast at Kamakura	Japan
Circa 1400 A.D.	Great Bell of Beijing cast	China
16th century	Sand introduced as mold material	France
1709	Cast iron produced with coke as fuel, Coalbrookdale	England
1715	Boring mill for cannon developed	Switzerland
1735	Great Bell of the Kremlin cast	Russia
1740	Cast steel developed by Benjamin Huntsman	England
1779	Cast iron used as architectural material, Ironbridge Gorge	England
1826	Zinc statuary	France
1838	Electrodeposition of copper	Russia, England
1884	Electrolytic refining of aluminum	United States, France

The Bronze Age

The Bronze Age began in the Near East before 3000 B.C. The first bronze that could be called a standard alloy was arsenical copper, usually containing up to 4% As, although a few objects contain 12% or more. This alloy was in widespread use and occurs in objects from Europe and the British Isles (Fig. 2) as well as the Near East.

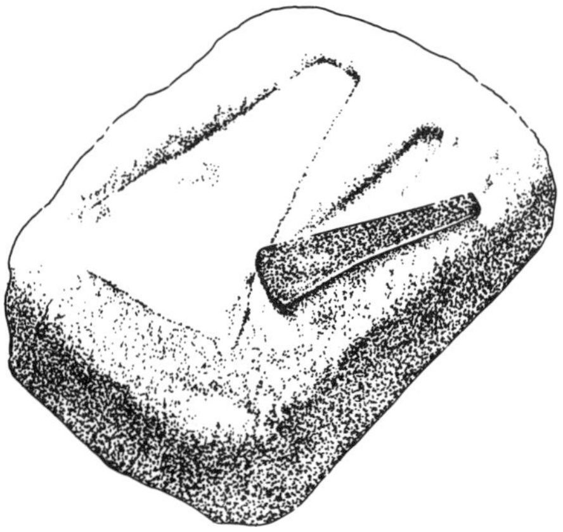

Fig. 1 Bronze Age stone mold with axe

The metal can sometimes be recognized as arsenical copper by the silvery appearance of the surface, which occurred as a result of inverse segregation of the arsenic-rich low-melting phase to the surface. This is the same phenomenon that produces tin sweat on tin bronzes, and it led earlier excavators to describe these artifacts as silver plated. A few examples of arsenic plating on tin bronze can be seen on objects from Anatolia and Egypt, but the plating method is not known.

The use of 5 to 10% Sn as an alloying element for copper has the obvious advantages of lowering the melting point, deoxidizing the melt, improving strength, and producing a beautiful, easily polished cast surface that reproduces the features of the mold with exceptional fidelity—vitally important properties for art castings (Fig. 3). There are several hypotheses to explain the development of tin bronze. One is that of the so-called natural alloy, that is, metal smelted from a mixed ore of copper and tin. Another suggests that stream tin (tin ore in the form of cassiterite) may have been added directly to molten copper. The more vexing question has been the sources of the tin, copper, and silver that have been excavated from sites in such areas as Mesopotamia, which lack local metal resources. Cornwall or Afghanistan was long thought to have been the source of this early tin, but more recent investigations have located stream tin in the Eastern Desert of Egypt and sources of copper and silver as well as tin in the Taurus mountains of south central Anatolia in modern Turkey.

Recent experiments have shown that metal cast into an open mold is sounder if the open face is covered after the mold has been filled. This observation may have led to the use of bivalve (permanent two-part) molds. They were in common use for objects having bilateral symmetry, such as axes of various designs and swords. The molds were made such that the flash occurred at the edge, which required finishing to sharpen (Fig. 4). These edges are often harder than the body of the object, evidence of deliberate work hardening. There is also evidence in the third millenium B.C. for the lost wax casting of small objects of bronze and silver, such as the stag from Alaça Hoyük, now in Ankara. This small object is also of interest because the casting sprues were left in place attached to the feet, clearly showing how the object was cast.

Although there is abundant evidence from such objects that lost wax casting was employed early in the Bronze Age, the remnants of the process, such as broken investments and master molds, have eluded researchers. Wax may well have been the material of the model; other materials may have been used, but no surviving evidence of any of these materials has been recognized. Similarly, the mold dressings used then and later remain unknown. Nevertheless, discoveries are occasionally made that greatly enlarge the geographical area in which lost wax casting is thought to have taken place. One of these discoveries occurred in 1972 at a site in England called Gussage All Saints.

At Gussage, an Iron Age (first century B.C.) factory was excavated. The lost wax process was used in this factory for the mass production of bronze bridle bits and other metal fittings for harnesses and chariots. More than 7000 fragments of clay investment molds were recovered (Fig. 5), along with crucible fragments, charcoal, slag, and other debris thought to represent the output of a single season. The bronze was leaded and in one case had been used to bronze plate a ring of carbon steel by dipping. This is the first site in Great Britain where direct evidence of lost wax casting has been found, yet the maturity of the industry suggests that earlier sites remain to be located.

The Far East

The Bronze Age in the Far East began in about 2000 B.C. more than a millenium after its origin in the Near East. It is not yet clear whether this occurred in China or elsewhere in southeast Asia, and there are vigorous efforts underway to discover and interpret early metallurgical sites in Thailand. The later date for the development of metallurgy in the Far East led to an obvious assumption that the knowledge of metal smelting and working had entered the area by diffusion from the West. This assumption was countered by mapping the geographical distribution of dated metallurgical sites in China, which indicated development in a generally east-to-west direction. The question of independent origins for the metallurgy of southeast Asia remains open.

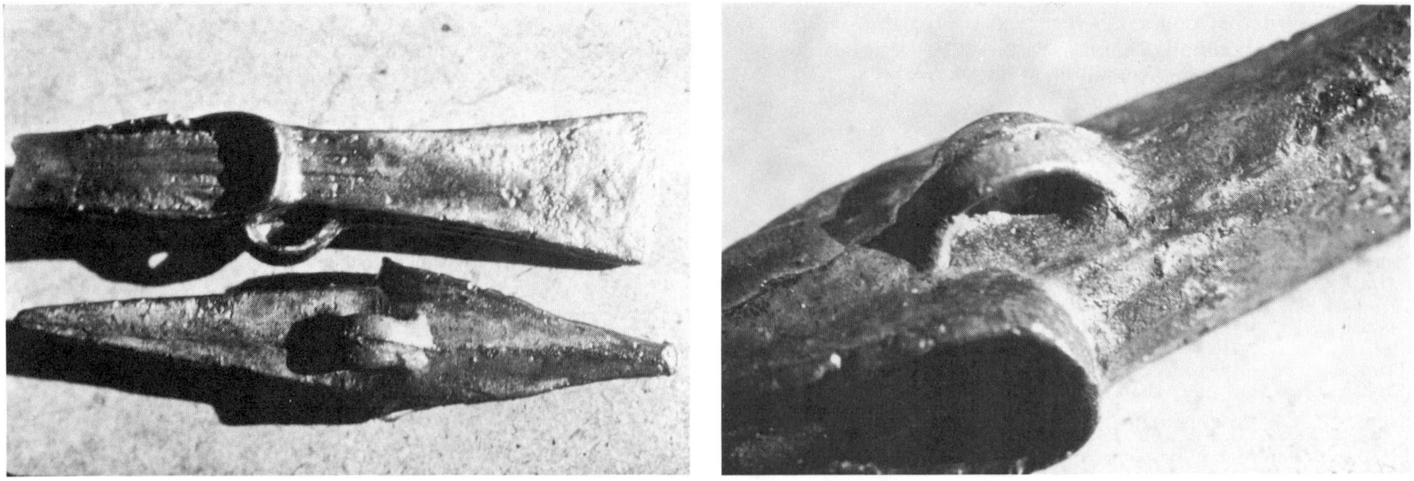

(a) (b)

Fig. 2 Top and side view (a) of arsenical copper axes from Oxfordshire, England, that appear silver plated due to inverse segregation. (b) Detail of one of the arsenical copper axes showing the joint of the bivalve (permanent two-part) mold, placed so that no core was necessary

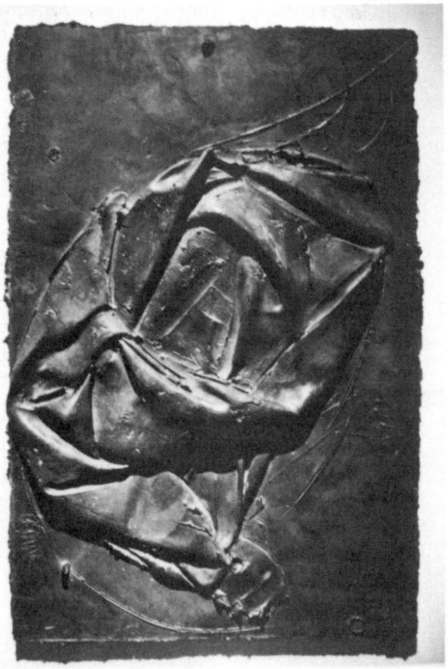

Fig. 3 Bronze panel by Giacomo Manzu for the Doors of Death to St. Peter's Basilica, the Vatican. The bronze alloy faithfully renders the texture of the surface as well as the form of the sculptor's model.

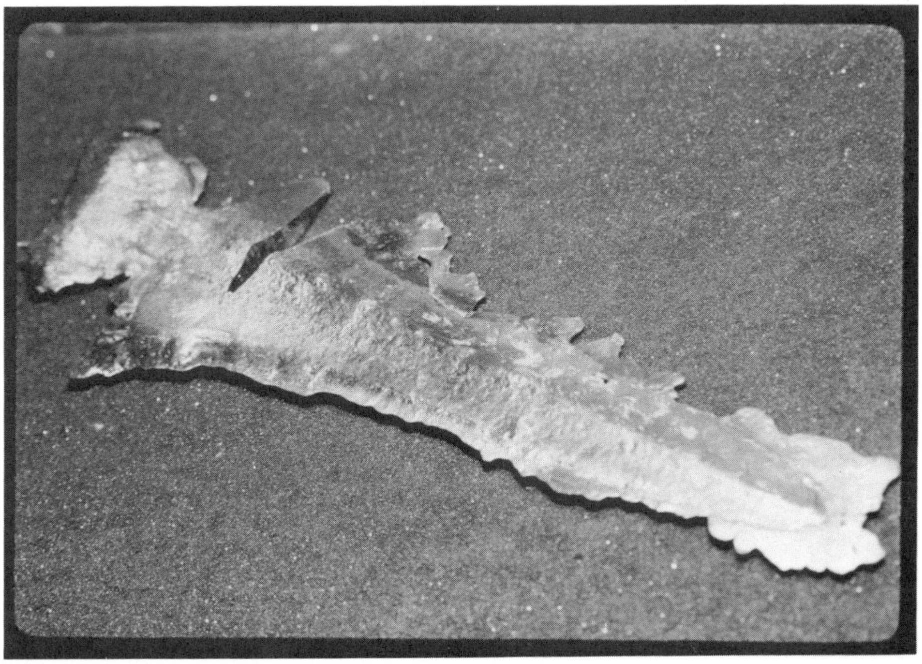

Fig. 4 A sword of typical Bronze Age design replicated by Dr. Peter Northover, Oxford, in arsenical copper using a bivalve mold. It has a silvery surface due to inverse segregation. The flash at the mold joint demonstrates the excellent fluidity of the alloy.

Casting was the predominant forming method in the Far East. There is little evidence of other methods of metalworking in China before about 500 B.C. Antique Chinese cast bronze ritual vessels were of such complexity that it was the opinion until recently that these must have been cast by the lost wax method. This had also been the opinion of Chinese scholars in recent centuries. In the 1920s, however, a number of mold fragments were unearthed at Anyang, prompting reevaluation of the lost wax hypothesis. The molds were ceramic, and they were piece molds.

Very early Bronze Age sites, approximately 2000 B.C., in Thailand present similar evidence. At one of these sites a burial was unearthed that contained the broken pieces of an apparently unused ceramic bivalve mold. The bronze founder had been buried with a piece of the mold in each hand.

The Chinese mold was a ceramic piece mold, typically of many separate parts. The wall sections of the vessels cast in these molds are quite thin and testify to very fine control over the design of the molds and pouring of the metal. The metal, usually a leaded tin bronze, was used to great effect but also in an economical manner. Parts, such as legs, which could have been cast solid, were instead cast around a ceramic core held in position in the mold by chaplets. These chaplets took several forms; some were cross shaped, others square. They were of the same alloy as the vessel but can clearly be seen in radiographs. They have occasionally become visible on the surface because their patina appears slightly different from that of the rest of the vessel.

Metal parts that in the Western tradition would have been made separately and then joined by soldering or welding were incorporated into Chinese vessels by a sequence of casting on. Handles and legs might be cast first, the finished parts set in the mold, and the body of the vessel then poured (Fig. 6). Elaborate designs demanded several such steps. An unusual feature of this way of thinking about mold making and casting metal is the deliberate incorporation of flash into the design elements.

The surface decoration of the vessels sometimes employed inlay or gilding, but even in these examples much of the decoration is cast in. Various decorative elements may have been molded from a master model, impressed into the mold with loose pieces, or incorporated by casting on metal elements. By using a leaded tin bronze, the founder increased the fluidity of the melt and consequently the soundness of the casting even in the usual thin sections. However, such a fluid melt also has a greater tendency to penetrate the joints between the pieces of the mold so as to produce flash. If the surface of the bronze is meant to be smooth, the flash must be trimmed away. The Chinese founders eventually took this casting flaw and made it a deliberate element of their design. The joints of the mold were placed in relation to the rest of the surface decoration such that the flash needed only to be trimmed to an even height to be accepted as part of the cast-in decoration.

Cast Iron

Cast iron appeared in China in about 600 B.C. Its use was not limited to strictly practical applications, and there are many examples of Chinese cast iron statuary. Most Chinese cast irons were unusually high in phosphorus, and, because coal was

Fig. 5 Fragments of a crucible (top) and a lost wax investment excavated at Gussage All Saints

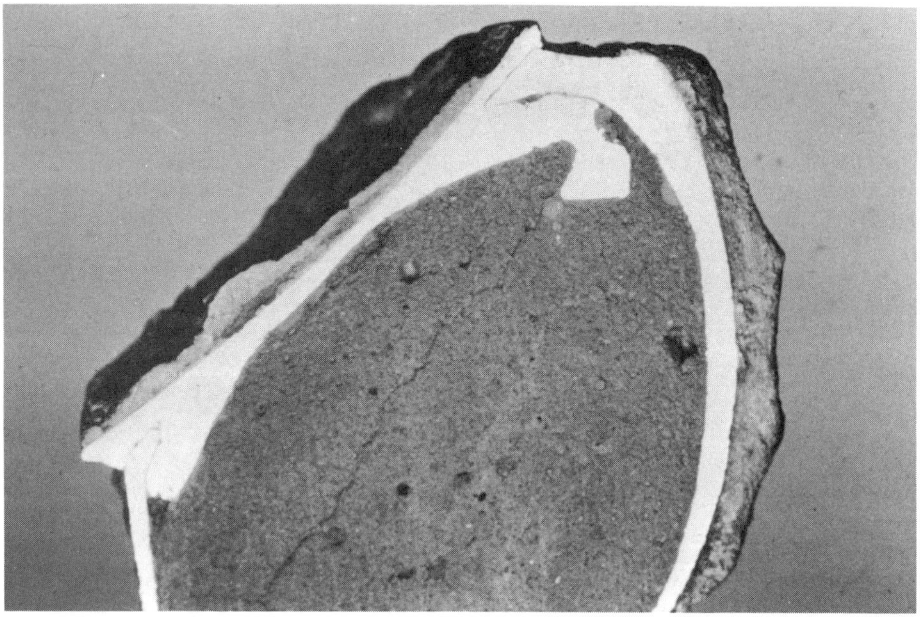

Fig. 6 Cross section of a leg and part of the attached bowl of a Chinese *ting*, a footed cauldron of the type used for cooking in China for at least 3000 years. The leg was cast around a core, which is still in place. Part of this core was excavated to allow a mechanical as well as a metallurgical joint when the leg was placed in the mold and the bowl of the vessel cast on. Source: Ref 1

often used in smelting, high in sulfur as well. These irons, therefore, have melting points that are similar to those of bronze and when molten are unusually fluid. The iron castings, like the Chinese cast bronzes, are often remarked upon for the thinness of their wall sections.

There is some dispute concerning the date of the introduction of cast iron into Europe and the route by which it came. There is less disagreement about the assumption that it was brought from the East. The generally agreed upon date for the introduction of cast iron smelting into Europe is the 15th century A.D.; it may have been earlier. At this time, cast iron was less appreciated as a casting alloy than as the raw material needed for "fining" to wrought iron, the form in which iron could be used by the local blacksmith.

The mass production of cast iron in the West, as well as its subsequent use as an important structural material, began in the 18th century at Coalbrookdale in England. Here Abraham Darby devised a method of smelting iron with coal by first coking the coal. He was successful because the local ores fortuitously contained enough manganese to scavenge the sulfur that the coke contributed to the iron. The vastly greater amounts of cast iron that could be produced by using coke rather than charcoal from dwindling supplies of timber were eventually put to use nearby in erecting the famous Iron Bridge (Fig. 7) and led to many other architectural uses of cast iron.

The dome of the United States Capitol Building is an example, as is the staircase designed by Louis Sullivan for the Chicago Stock Exchange now at the Metropolitan Museum in New York City. Cast iron architectural elements were usually painted; the Capitol dome is painted to resemble the masonry of the rest of the building. Finishes other than paint were also used. The Sullivan staircase was copper plated and then patinated to give it the appearance of having been cast in bronze. Another method suitable for interior iron work was the treatment of the surface by deliberate light rusting, followed by hydrogen reduction of the rust. This produced a velvety black adherent layer of magnetite (Fe_3O_4) that was both attractive and durable.

Granulation

Not all casting requires a shaped mold. The exploitation of surface tension led to granulation. The tiny spheres produced when small amounts of molten metal solidified without restraint were being used as decoration in gold jewelry by 2500 B.C. Granulation was primarily done in gold, silver, or the native alloy of gold and silver called electrum. Some granules were attached to copper or gilt-silver substrates. The finest work in granulation was done by the Etruscans in about the seventh century B.C. Its fineness has given it the name "dust granulation," the granules being less than 0.2 mm (0.008 in.) in diameter. Many thousands of granules were used to create the design on a single object. The Etruscan alloy was gold with about 30% Ag and a few percent of copper. The method of joining

(a) (b)

Fig. 7 The Iron Bridge (a) across the Severn River at Ironbridge Gorge. The structure was cast from iron smelted by Abraham Darby at Coalbrookdale. (b) Detail of the Iron Bridge showing the date, 1779. This was the first important use of cast iron as a structural material.

the granules varied. Sweating or soldering have both been observed, but the exact method used is often still a matter of dispute.

Tumbaga

New World metallurgy is a metallurgy almost without iron. The exception was the use of meteoric iron, which was most important among the Eskimos, who traded it all across the North. Copper-using cultures flourished further south until the sources of native copper were exhausted. There is no evidence of smelting among the native population of what is now the United States until the arrival of the Europeans.

In South America, however, the story is quite different. Early European explorers were overwhelmed by the amount of gold and silver objects they found. Many of these objects were of sheet gold or its alloys, and it has been suggested that sheet metal was viewed then as a kind of textile, as textiles in these cultures were not limited to clothing and were used for weapons and armor. The most interesting castings are of an alloy called tumbaga, which contained gold, silver, and copper in various proportions. Molds have been found (some never used) that were made by the lost wax process. After an object had been cast in tumbaga, it was pickled in a corrosive solution that attacked the silver and especially the copper and, when rinsed off, left a surface layer enriched in gold. This method of gilding is called *mise-en-couleur*, or "depletion gilding."

Africa

Africa, where sculpture is often the province of the blacksmith, presents several interesting traditions of casting. Among them are the famous Benin bronzes of Nigeria and the gold weights of Ghana, formerly the Gold Coast. Both of these traditions produced castings in brass, with the brass having a high enough zinc content to appear golden. The source of the brass, or at least that of the zinc, may well be indicated by the portrait of a Portuguese trader in a Benin bronze (Fig. 8). Recent discoveries of zinc furnaces and distillation retorts at Zawar, near Udaipur in India, as well as the very long trade routes that were opened in the 17th century, suggest the possibility that the metal may have been traded from India. The Benin bronzes were cast by the lost wax process, and the traditional method has been recorded on film.

Lost wax was also used in Ghana to make gold weights and many types of small decorative objects. Once the mold and the crucible had been made, the crucible was charged with the brass, and both mold and crucible were invested (Fig. 9). While one end of the investment was heated to the casting temperature, the mold at the other extremity was being preheated, ready to receive the metal when the investment was inverted.

Fig. 8 A Benin bronze plaque depicting a Portuguese trader of the time. The alloy is actually brass.

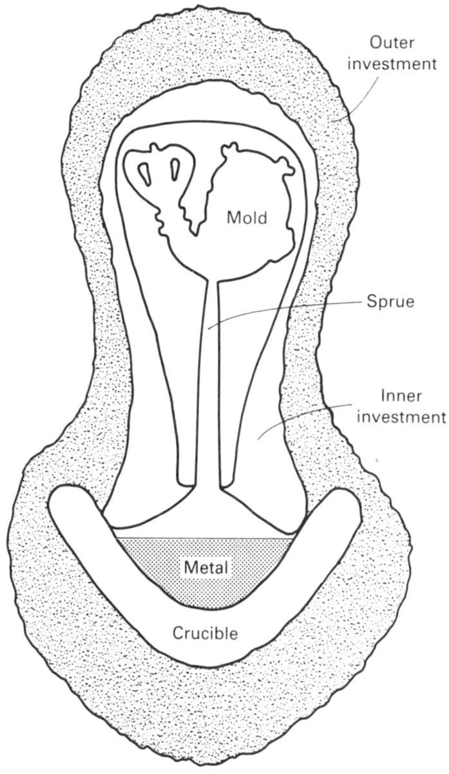

Fig. 9 Crucible and mold assembly for the lost wax casting of a small brass figure in Ghana. The metal is brought to the casting temperature, and the assembly is inverted to fill the mold. Source: Ref 2

Bells and Guns

In general, large castings were made in sections that were then bolted or welded together or were cast on sequentially. However, neither bells nor guns function well if joined and so are cast in a single pour. Large bells have traditionally represented the limits of foundry capacity. The Great Bell in the Kremlin was cast in 1735 and weighs 175 Mg (193 tons). It is now cracked. The largest bell that still sounds is the Great Bell in Beijing. It was cast early in the Ming dynasty, about 1400, and weighs 42 Mg (46.5 tons). The alloy contains 15% Sn and 1% Pb. The loudness of this bell can reach 120 dB, and on a quiet evening it can be heard 20 km (12 miles) away.

According to Theophilus, in the 12th century bells were cast into clay molds made by the lost wax process using tallow instead of wax. The clay core had to be broken out before the metal cooled, or the bell would shrink tightly around the core and crack. An iron staple to hold the clapper was placed in the mold and cast into the bell, a practice that caused many bells to crack when the iron rusted and expanded. Bell metal in Europe was a bronze usually containing 20 to 25% Sn, although bells in the Far East, which have a very different shape and sound, were cast with lower levels of tin.

Recent research has indicated that the shape of the casting as well as its integrity has a much greater effect on the tone of a bell than its alloy, and in fact close attention was given the "bell scale," the correct proportions of a bell by both Theophilus and Biringuccio.

A reproduction of the Liberty Bell was recently cast by the same foundry that cast the original. The mold was made by the same cope and core method described by Biringuccio, who also described the welding of cracked bells. A clayey loam was shaped over bricks by a strickle rotating about the axis of the bell to shape the core, and another molding board was used to shape the cavity in the cope. The alloy, containing 23% Sn, was poured at 1100 °C (2010 °F). The mold took 16 min to fill. Traditionally, the pouring rate was controlled by the sound of the liquid metal in the mold. The bell, which weighs more than 4.5 Mg (5 tons), took a week to cool.

Although large bells are usually cast of bronze, other metals have been used. Bells were cast of white iron in China, Russia, and elsewhere. After Benjamin Huntsman's development of cast steel in 1740, bells of cast steel became a specialty of Sheffield, England.

Gun barrels have been made of many materials. Cannons, albeit small ones, exist that were made of laminated leather. Laminations of welded iron strip were used to make damascene gun barrels before these were routinely cast. Gunmetal was a bronze alloy containing 10% Sn, although additional tin was added late in the pour to make up for the effects of tin sweat. Biringuccio describes gun founding in 1540. Cannons were cast around a core to form the bore. Because of its size and weight, the core required elaborate reinforcement, and it was supported in the mold on iron chaplets. Later, in 1715, Johann Maritz in Burgdorf, Switzerland, developed the boring mill that made it possible to cast cannons solid. Because cannons were cast vertically, boring removed the shrinkage along the centerline of the casting, which led to increased reliability in service. The entire sequence from mold making to milling was recorded in detail in a set of 50 watercolor paintings made by an 18th century gun founder and are known as the Royal Brass Foundry Drawings.

Art Founding

Sculpture has been made of many different materials. The earliest known are the Paleolithic Venus figurines of bone and other materials. The earliest life-size statues are of plaster. They were excavated at Ain Ghazal in Jordan in 1985 and are dated to about 6500 B.C. Early metal sculpture, like the earliest metal objects, is of worked copper sheet. The oldest (and largest) metal statue from ancient Egypt is a life-size statue of Pharoah Pepi I of the Sixth Dynasty (about 2400 to 2200 B.C.). This statue was found at Hierakopolis and is now in Cairo. It was made in several sections and is part of a group that includes a smaller figure of the pharaoh's son. The metal is copper, but because of its highly mineralized condition, there remains some doubt as to whether it was wrought or cast. The copper relief from al'Ubaid, dating from about 2600 B.C., now in the British Museum, has figures of two stags whose tines were separately cast and attached. Statuary in the round from this site was made of wrought copper sheet over a bitumen core. Cast statuary of the late third millenium B.C. includes portrait busts such as the one of Sargon of Akkad now in Baghdad.

Classical Sculpture

Most surviving large classical sculpture is in stone, but a few of the life-size bronzes known to have been cast in antiquity have survived. Some have come to light as a result of excavation or underwater finds. Greek bronzes include the Charioteer of Delphi, the Poseidon of Artemision in Athens, and the youth attributed to Lysippos now in Los Angeles. Large sculptures were piece cast by lost wax and assembled by welding. The joints were skillfully hidden by, for example, placing them along folds in drapery (Fig. 10).

(a)

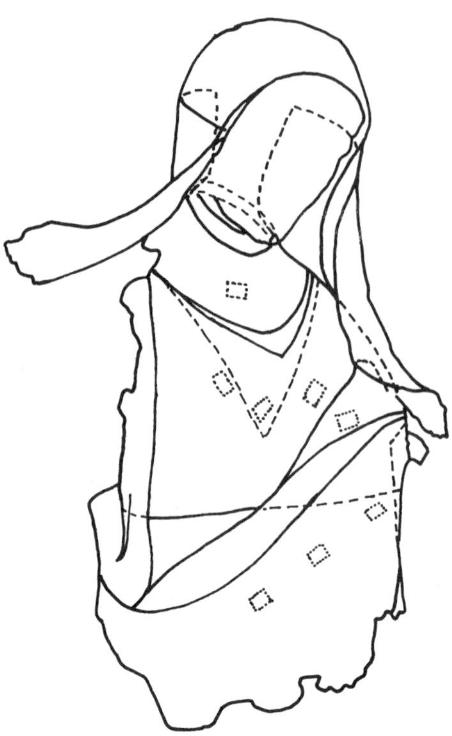

(b)

Fig. 10 Bronze statue (a), dated to the fourth century B.C., found off the coast of Turkey. Now in the museum at Izmir and known as the Lady From the Sea. (b) Assembly diagram for the precast pieces of the Lady From the Sea. Source: Ref 3

Smaller classical statuettes, thought to have been intended as votive offerings, exist in large numbers. These figures were cast head down over a core, which is usually still intact. Occasionally the design called for an additional piece, such as an extended arm, to be joined. None of the molds has been found. Vessels that earlier had been wrought, such as ewers, were also cast, as were articles of household furniture such as tripods.

One use of cast metal in the art of classical antiquity that often goes unnoticed is the use of lead in building. Lead was a relative-

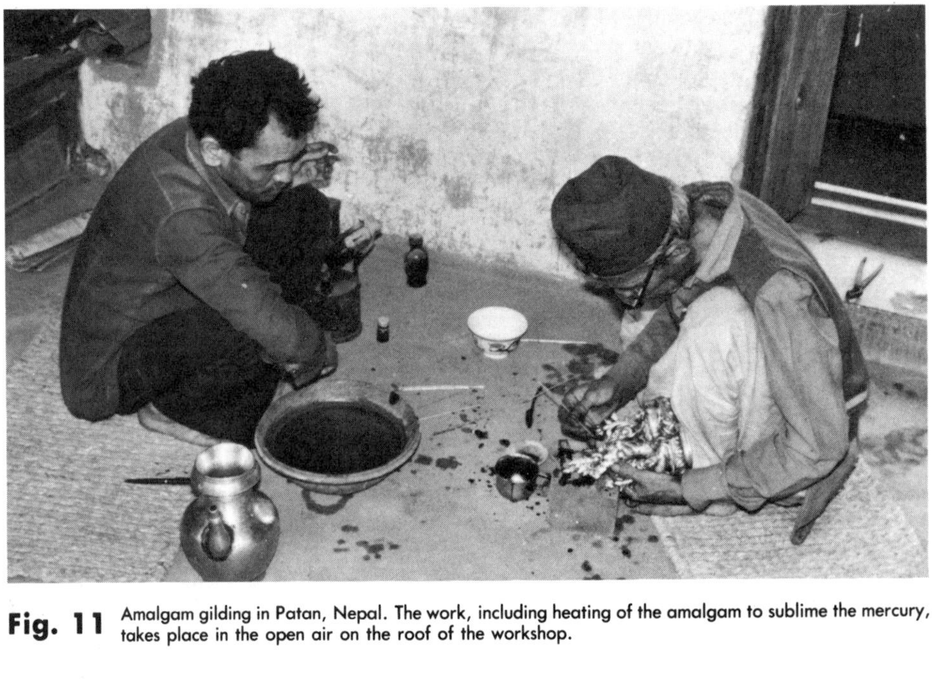

Fig. 11 Amalgam gilding in Patan, Nepal. The work, including heating of the amalgam to sublime the mercury, takes place in the open air on the roof of the workshop.

(a) (b)

Fig. 12 Overall view (a) of the Great Buddha at Kamakura, Japan, cast in high-lead tin bronze in 1252. (b) View of the face of the Kamakura Buddha showing metal losses at the joints between separate casts

ly plentiful by-product of silver smelting and refining. It was used to set the iron clamps and dowels holding the stone blocks in place, and it served to protect the stone from cracking under the pressure caused by the expansion of rusting iron. Lead was also cast as statuettes. Some classical statues were said to be cast in gold. These were more likely gilded.

Gilding

In addition to gilding by depletion, as in the case of tumbaga, gold was applied to the metal surface either as a foil or as an amalgam. Foil could be attached to the substrate in several ways. Various adhesives were used, including mercury, or the surface was given a texture so that the foil, when burnished, made a good mechanical bond with the substrate. Amalgam gilding is still being practiced (Fig. 11), although the risks of mercury poisoning have long been recognized. Experiments made while the famous Roman bronze horses from San Marco in Venice were being studied indicated that, to be successfully amalgam gilded, a bronze must be cast from an alloy low enough in tin so that the color of the metal is still coppery. Thus, a low tin analysis can be evidence that an ancient object now without gilding may have been gilded originally.

Colossal Statues

Cellini, in 1568, defined a colossal statue as one at least three times life size. The Colossus of Rhodes was a bronze statue that stood more than 30 m (100 ft) tall. Although filled with stone as ballast, it was destroyed in an earthquake in 224 B.C. The fragments remained where they fell until they were sold as scrap in 656 A.D. According to Pliny, other colossal statues were erected at Tarentum, Rome, and one at Appollonia that was later taken to Rome.

Japan boasts several *Diabutsu*, or Great Buddhas, in bronze. The Great Buddha at Nara, begun in the eighth century, is gilded and was therefore cast in a low-tin alloy. The Great Buddha at Kamakura (Fig. 12) was cast in the 13th century and contains 109 Mg (120 tons) of bronze, 18 Mg (20 tons) in the head alone. The alloy contains 9% Sn and 20% Pb. The statue was cast in place, with each section cast onto sections already in place (Fig. 13) using mechanically interlocking joints, a necessary precaution in an earthquake zone.

Modern colossal statues of bronze are cast in sections and bolted together. An example is the statue of William Penn atop the city hall in Philadelphia. Monuments very much larger than those of antiquity are wrought, not cast. Examples are the Statue of Liberty, which, like the earliest known metal objects, is of repoussé copper, and the Gateway Arch of St. Louis, which is of stainless steel.

Modern Statuary

With the Renaissance came a revival in bronze casting. Large single castings were attempted in lost wax. Cellini recommended the assistance of ordnance founders in casting them. Cellini also claimed a "secret" mold material of rotted rags in clay, although it is known that in the previous century pieces of cloth were added to the clay used for gun cores. Clearly, there was considerable exchange of techniques among

Fig. 13 View of the interior of the Kamakura Buddha showing the interlocking joints between the casts

Fig. 14 Contemporary casting of bronze into lost wax investments in Thailand

the founding specialists despite the tradition of craft secrets. Sand for molding was newly introduced from a source near Paris, and "French sand" continued to be highly recommended into the 20th century.

The 19th century saw many technical innovations, including the electroplating of copper statues and architectural elements as large as domes. The electroplating of copper on less noble metals such as cast zinc or cast iron gave these metals the surface appearance of bronze. Cast zinc was referred to as white bronze, and zinc statuary for Civil War monuments, business emblems, and the like could be ordered relatively inexpensively from catalogs of standard designs. Aluminum was more expensive, costing about as much as silver until the Hall-Heroult refining process was invented. An aluminum casting, rather than stone, was used to cap the tip of the Washington Monument in 1884, and aluminum has been occasionally used since as a statuary material.

Traditional methods of art casting continue in the 20th century (Fig. 14), but the standard "three fives" statuary bronze alloy containing 5% each of Sn, Pb, and Zn, has been replaced for occupational health reasons by silicon bronzes. An interesting variation on lost wax casting uses standard foundry sand in place of the investment, and plastic foam in place of the wax. The method is called foam vaporization and has the advantage that the model remains in place when the metal is poured, vaporizing the foam. Post World War II alloys for art casting included stainless steel, although this was more often used as welded sheet, as were the weathering steels.

REFERENCES

1. R.J. Gettens, *The Freer Chinese Bronzes*, Vol II, *Technical Studies*, Washington, DC, 1969, p 79
2. B. Menzel, *Goldgewischte aus Ghana*, Museum für Volkerkunde Berlin, Neue Folge 12, Abteilung Africa III, Berlin, 1968
3. A. Steinberg, Joining Methods on Large Bronze Statues: Some Experiments in Ancient Technology, in *Application of Science in Examination of Works of Art*, W.J. Young, Ed., Boston, 1973, p 103-138

SELECTED REFERENCES

History of Casting

- C.S. Smith, The Early History of Casting, Molds, and the Science of Solidification, in *A Search for Structure: Selected Essays on Science, Art, and History*, MIT Press, Cambridge, MA and London, 1981, p 127-173
- B.L. Simpson, *Development of the Metal Castings Industry*, American Foundrymen's Association, Chicago, 1948
- N.N. Bubtsov, *History of Foundry Practice in USSR*, Moscow, 1962; trans. Washington DC, 1975
- J. Foster, "The Iron Age Moulds From Gussage All Saints," Occasional Paper No. 12, British Museum, London, 1980

History of Metallurgy

- R.F. Tylecote, *The Early History of Metallurgy in Europe*, Longman, London, 1987
- R.F. Tylecote, *History of Metallurgy*, The Metals Society, London, 1976
- T.A. Wertime and J.D. Muhly, *The Coming of the Age of Iron*, Yale University Press, New Haven, 1980
- T.A. Wertime and S.F. Wertime, *Early Pyrotechnology: the Evolution of the First Fire-Using Industries*, Washington DC, 1982
- P. Knauth, *The Metalsmiths*, New York, 1974
- L. Aitchison, *A History of Metals*, New York, 1960

Early Treatises

- Theophilus, *On Divers Arts*, twelfth-century manuscript translated from the Latin by J.G. Hawthorne and C.S. Smith, Chicago 1963; reprinted Dover, New York, 1979
- C.S. Smith and M.T. Gnudi, trans., *The Pirotechnia of Vannocio Biringuccio*, New York, 1942; reprinted New York, 1959; and MIT Press, Cambridge, MA, and London, 1966
- Georgius Agricola, *De Re Metallica*, H.C. Hoover and L.H. Hoover, trans., London, 1912; reprinted Dover, New York, 1950
- C.R. Ashbee, trans., *The Treatise of Benvenuto Cellini on Goldsmithing and Sculpture*, London, 1888; reprinted Dover, New York, 1967
- E-t. Zen Sun and S.-C. Sun, trans., *T'ien Kung Kai Wu, Chinese Technology in the Seventeenth Century*, Pennsylvania State University Press, College Park, PA, 1966

The Far East

- R.J. Gettens, *The Freer Chinese Bronzes, Vol II, Technical Studies*, Smithsonian Institution, Washington, DC, 1969
- R.J. Gettens, Joining Methods in the Fabrication of Ancient Chinese Bronze Ceremonial Vessels, in *Application of Science in Examination of Works of Art*, W.J. Young, Ed., Boston, 1967, p 205-217
- N. Barnard, *Bronze Casting and Bronze Alloys in Ancient China*, Tokyo, 1975
- N. Barnard, The Special Character of Metallurgy in Ancient China, in *Application of Science in Examination of Works of Art*, W.J. Young, Ed., Boston, 1967, p 184-204
- W. Fong, Ed., *The Great Bronze Age of China: An Exhibition From the People's Republic of China*, Metropolitan Museum and Knopf, New York, 1980
- R. Bagley, *Shang Ritual Bronzes in the Sackler Collection*, 1987
- W. Chia-pao, A Comparative Study of the Casting of Bronze Ting-Cauldrons From Anyang and Hui-hsien, in *Ancient Chinese Bronzes and Southeast Asian Metal and Other Archaeological Artifacts*, N. Barnard, Ed., Victoria, 1976, p 17-46

- B.W. Keyser, Decor Replication in Two Late Chou Bronze Chien, *Ars Orientalis*, Vol 11, 1979, p 127-162
- R.P. Hommel, *China at Work*, New York, 1937; reprinted MIT Press, Cambridge, MA, and London, 1969
- D.B. Wagner, *Dabieshan: Traditional Iron-Production Techniques Practised in Southern Henan in the Twentieth Century*, Monograph Series No. 52, Scandinavian Institute of Asian Studies, London and Malmö, 1985
- H. Jue-ming, The Mass Production of Iron Castings in Ancient China, *Sci. Am.*, Vol 248, Jan 1983, p 120-128
- W. Rostoker, B. Bronson, and J. Dvorak, The Cast-Iron Bells of China, *Technol. Culture*, Vol 25, 1984, p 750-767

Granulation

- J. Wolters, *Granulation-Verfahren und Geschichte einer 5000 Jahrigen Schmucktechnik*, 1982
- J. Wolters, The Ancient Craft of Granulation, *Gold Bull.*, Vol 14, Munich, 1981, p 119-129

The New World

- D.T. Easby, Jr., Early Metallurgy in the New World, *Sci. Am.*, April 1966, p 72-81
- P. Bergsøe, *The Gilding Process and the Metallurgy of Copper and Lead Among the Pre-Columbian Indians*, C.F. Reynolds, trans., Ingeniorvidenskabelige Skrifter Nr. A 46, Copenhagen, 1938
- A.D. Tushingham, U.M. Franklin, and C. Toogood, *Studies in Ancient Peruvian Metalworking*, History Technology and Art Monograph No. 3, Royal Ontario Museum, Toronto, 1979
- H.N. Lechtman, The Gilding of Metals in Pre-Columbian Peru, in *Application of Science in Examination of Works of Art*, W.J. Young, Ed., Boston, 1973, p 38-52

Africa

- T. Shaw, The Making of the Igbo Vase, *Ibadan*, No. 25, Feb 1968, p 15-20
- B. Menzel, *Goldgewischte aus Ghana*, Museum für Volkerkunde Berlin, Neue Folge 12, Abteilung Afrika III, Berlin, 1968
- W. Cline, *Mining and Metallurgy in Negro Africa*, General Series in Anthropology Series No. 5, Menasha, WI, 1937

Gunfounding

- M.H. Jackson and C. de Beer, Eighteenth Century Gunfounding: The Verbruggens at the Royal Brass Foundry, A Chapter in the History of Technology, Washington DC, 1974

Cast Steels

- K.C. Barraclough, *Steelmaking Before Bessemer*, Vol 2, *Crucible Steel: The Growth of a Technology*, The Metals Society, London, 1984
- A.D. Graeff, Ed., *A History of Steel Casting*, Philadelphia, 1949
- K.C. Barraclough, *Sheffield Steel*, Historic Industrial Scenes, Hartington, UK, 1976
- P.S. Bardell, The Origins of Alloy Steels, in *History of Technology*, Vol 9, N. Smith, Ed., London, 1984, p 1-29

Sculpture

- C.S. Smith, On Art, Invention and Technology, *Technol. Rev.*, Vol 78, June 1966, p 36-41; reprinted in *A Search for Structure*, MIT Press, Cambridge, MA, and London, 1981, p 325-331
- W.A. Oddy, Materials for Sculpture in the Art of Antiquity, in *Encyclopedia of Materials Sciences and Engineering*, M. Bever, Ed., Pergamon Press/The MIT Press, Oxford, 1986, p 2794-2800

Classical Bronzes

- Pliny, *Natural History*, Book 34, sections 142 and 143
- W.A. Oddy and J. Swaddling, Illustrations of metal working furnaces on Greek vases, in *Furnaces and Smelting Technology in Antiquity*, Occasional Paper No. 48, P.T. Craddock and M.J. Hughes, Ed., British Museum, London, 1985, p 43-57
- S. Deringer, D.G. Mitten, and A. Steinberg, Ed., *Art and Technology: A Symposium on Classical Bronzes*, 1970
- A. Steinberg, Joining Methods on Large Bronze Statues: Some Experiments in Ancient Technology, in *Application of Science in Examination of Works of Art*, W.J. Young, Ed., Boston, 1973, p 103-138
- D.K. Hill, Bronze Working: Sculpture and Other Objects, in *The Muses at Work*, C. Roebuck, Ed., Cambridge, MA, and London, 1969, p 60-95
- R.T. Davis, *Master Bronzes*, Buffalo, NY, 1937

Gilding

- W.A. Oddy, L.B. Vlad, and N.D. Meeks, The Gilding of Bronze Statues in the Greek and Roman World, in *The Horses of San Marco, Venice*, London, 1979, p 182-187
- H.N. Lechtman, Ancient Methods of Gilding Silver: Examples From the Old and New Worlds, in *Science and Archaeology*, R.H. Brill, Ed., Cambridge, MA, 1971, p 2-30 and plate 1

Colossal Statuary

- M. Sekino, Restoration of the Great Buddha Statue at Kamakura, *Studies Conserv.*, Vol 10, 1965, p 30-46
- T. Maruyasu and T. Oshima, Photogrammetry in the Precision Measurements of the Great Buddha at Kamakura, *Studies Conserv.*, Vol 10, 1965, p 53-63
- K. Toishi, Radiography of the Great Buddha at Kamakura, *Studies Conserv.*, Vol 10, 1965, p 47-52

Modern Sculpture

- J.C. Rich, *The Materials and Methods of Sculpture*, New York, 1947
- J.W. Mills and M. Gillespie, *Studio Bronze Casting: Lost Wax Method*, New York and Washington DC, 1969
- A. Beale, A Technical View of Nineteenth-Century Sculpture, in *Metamorphoses in Nineteenth-Century Sculpture*, J.L. Wasserman, Ed., Cambridge, MA, 1978, p 29-55
- M.E. Shapiro, *Bronze Casting and American Sculpture, 1850-1900*, Newark, DE, 1985
- M.E. Shapiro, *Cast and Recast: the Sculpture of Frederic Remington*, Washington, DC, 1981
- L. van Zelst, Outdoor Bronze Sculpture: Problems and Procedures of Protective Treatment, *Technol. Conserv.*, Spring 1983, p 19-24
- B.F. Brown *et al.*, *Corrosion and Metal Artifacts*, Special Publication 479, National Bureau of Standards, Washington, DC, 1977
- C. Grissom, *The Conservation of Zinc Sculpture*, to be published

Development of Foundry Technology in the United States*

Ezra L. Kotzin, American Foundrymen's Society

THE ART OF IRON CASTING, as discussed in the preceding article ("History of Casting"), was introduced to Europe from China, where iron castings were being produced as early as 600 B.C. Considerable progress took place in Europe from the time of this introduction in the 15th century to the 19th century. Virtually no foundry developments, however, can be found in the Americas before 1800, and the focus of this article will be the development of foundry technology in the United States.

Before the colonization of the Atlantic Seaboard of the United States and Canada by the English, Dutch, and French in the 16th and 17th centuries, castings were not produced in North America. The North American Indians had no knowledge of metallurgy. There is some evidence that the prehistoric mound builders who preceded the Indians may have worked with melted metals. The mounds located in what is now Ohio and the Mississippi Valley have yielded certain castings. Undoubtedly, the same methods were employed as those used by the prehistoric peoples of the Middle East. For all practical purposes, however, the New World was only beginning to emerge from the Stone Age when the first Europeans landed on these shores, bringing with them their knowledge of cast metals.

Early American Foundries

Records indicate that, as a general rule, this hemisphere was explored for gold but colonized with iron. Iron first made its appearance as a result of a deliberate search by Sir Walter Raleigh, who advised of the presence of iron ore deposits on the Roanoake River in South Carolina. Samples were sent back to England, but no action was taken on the development of iron ore for nearly 40 years. In 1607, the first colony at Jamestown, Virginia, was established, and again iron ore samples were sent back to England for analysis. It was not until 1622 that there was an attempt to make use of this mineral. In that year, an iron blast furnace was established at Falling Creek, Virginia (near Richmond), with skilled melters and foundrymen from England. Unfortunately, this enterprise was completely wiped out by an Indian massacre before the furnace went into operation. Because there were no survivors, details of the project are missing.

The Saugus Iron Works

It fell to Massachusetts to have the honor, in 1642, of becoming the birthplace of the first American casting. This original American foundry was established near Lynn, Massachusetts, on the Saugus River and has been referred to in history as the Saugus Iron Works. Some details of this operation provide a picture of the typical iron foundry of that period.

The founders of the enterprise were Thomas Dexter, the mechanic and builder, and Robert Bridges, the promoter of the project. It was Bridges who took the samples of Saugus area bog ore to England and obtained the necessary financial help for starting operations. Thus was founded The Company of Undertakers for the Iron Works. The company in turn founded the small village of Hammersmith, so called because of the imported furnace and foundrymen from Hammersmith, England.

On October 14, 1642, the Saugus Iron Works was granted the exclusive right to make iron for 21 years, during which time it could freely mine or cut wood, dam streams, and set up furnaces. The Iron Works was also given public lands on which to operate tax free. The firm was allowed to sell and transport freely, and all employees were completely exempt from military duty. Members of the company could even refrain from attending church without losing their voting privileges.

With such a start, the Iron Works built a four-sided hollow stack 6 m (20 ft) high, 7.3 m (24 ft) square at its base, 2 m (6 ft) in diameter at its top, and 3 m (10 ft) in interior diameter. The blast was operated by a waterwheel. Because the wheel naturally could not function in freezing weather, no winter operations were possible. Bog ore dug from neighboring swamps was charged into the furnace, together with oyster shells for flux and charcoal for fuel. Capacity was estimated at 7.3 Mg (8 tons) of iron per week. It is interesting to note that the first metal of the new plant was made into a shaped casting. The company had retained Joseph Jenks as a master molder; he molded a cooking pot in a small mold buried in a hole in the ground. The resulting casting weighs about 1.4 kg (3 lb) and has an internal diameter of about 114 mm (4½ in.). This Saugus pot casting has been preserved and is now the property of the city of Lynn, Massachusetts.

The Saugus Iron Works should be remembered on several historic scores. Jenks, the first molder, obtained the first patent granted in the colonies for his invention of the two-handed scythe, a tool that is still made in the same original shape. Jenks was also responsible for coining the first American money, the Pine Tree coinage of the Colony of Massachusetts. In addition, because of the interest of John Winthrop, Jr., son of the governor of Massachusetts, the first firefighting equipment for Boston was made in this plant in 1654. Although the Iron Works formed the start of an industry that would eventually number over 5000 plants, the company itself never achieved greatness. It failed in 1688 as a result of litigation, nuisance suits, and the reduction of timber resources, yet the Iron Works furnished an important beginning for the new colonies.

The Spread of Foundries

New England. From this early beginning, the direct-iron blast furnace foundry spread quickly. A second works was started at Braintree near Boston in 1645. In quick succession, plants were established at Taunton in 1653 and at Concord and Raleigh

*This article was adapted with permission from B.L. Simpson, *Development of the Metal Casting Industry*, American Foundrymen's Association, 1948.

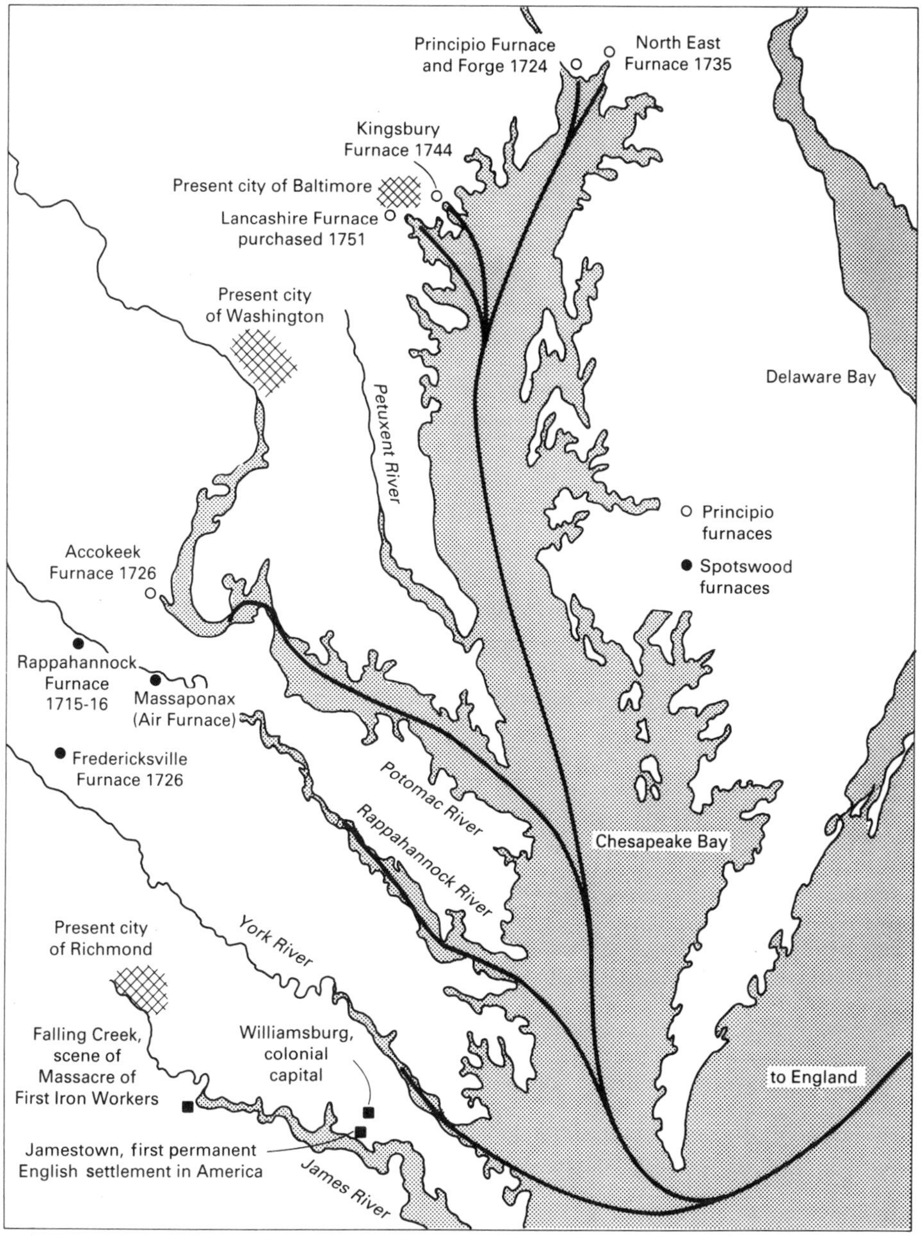

Fig. 1 Growth of iron furnaces in the Delaware, Maryland, Virginia area. Note the numerous Principio furnaces; George Washington's father, Augustine, was one of the founders of this company.

in 1657. By 1700, there were a dozen plants in eastern Massachusetts, including that of the Leonards at Raynham, and Massachusetts became the center of metalworking. Meanwhile, activity had spread to other states. John Winthrop, Jr., built Connecticut's first blast furnace foundry at New Haven. In 1658, Joseph Jenks, Jr., son of the first molder at the Saugus Iron Works, erected a plant in Pawtucket, Rhode Island. The plant burned down in 1675. Henry Leonard, who was from a noted foundry family, moved to New Jersey and established the industry in that state. Colonel Lewis Morris was also an early operator there and is said to have used the first cast iron cylinder compressed air blast in America.

Maryland and Virginia. In the southern part of Maryland and Virginia, no iron was produced for many years following the abortive attempt at Falling Creek, Virginia. However, in 1715, Virginia's governor, Colonel Alexander Spotswood, promoted iron foundry progress and established a furnace on the Rappahannock River at the junction of the Rapidan. Pig iron was hauled 24 km (15 miles) to Massaponax, where, by means of an air furnace, the metal was cast into firebacks, cooking utensils, andirons, and similar items. The quality of the iron produced by this remelt process surprised the British and is remarkable considering that the remelt process had only recently begun in France. Spotswood established another furnace at Fredericksville, Virginia, in 1726, but his efforts never reached the limits of his planning. Of the iron furnaces that were constructed in the Delaware/Maryland/Virginia area (Fig. 1), one of the most interesting is the Principio Furnace and Forge because its founders included Augustine Washington, father of George Washington.

The Principio Company was to exist for 200 years. Augustine Washington built a furnace in 1724 at the head of the Chesapeake Bay in Maryland, and his company built a chain of plants in rapid succession in and around Baltimore and in northern Virginia. By 1750, Maryland had eight operating furnaces. This furnace expansion continued until 1754, and the company was successful for years as a commercial foundry and furnace enterprise. By this time, New Jersey and Pennsylvania had entered the race to produce the iron and other castings and these states were prominent metal producers by the middle of the 18th century.

The Union Furnace. One early New Jersey plant is particularly interesting in that it still exists as a foundry and traces its origin back to 1742. The Taylor-Wharton Iron and Steel Company of High Bridge, New Jersey, is the oldest foundry and probably the oldest industrial corporation of continuous existence in the United States. This plant had its beginnings in December 1742 when a foundry furnace was erected by William Allen and Joseph Turner under the name of the Union Furnace. Shortly afterward (in 1754), another furnace—the Amesbury Furnace—was built in the same area. In 1760, Robert Taylor became works manager of Union Furnace and subsequently took over the plant in 1780. During the revolutionary war, this furnace supplied the colonial troops with guns and shot. In 1860, the name of the plant was changed to Taylor and Lange. In 1868, it became the Taylor Iron Works and later the Taylor-Wharton Iron and Steel Company.

Pennsylvania was also moving rapidly to exploit its mineral resources. Again, iron became the predominant material because of the needs of the expanding frontier and because the inhabitants relied on iron castings to carve their homes out of the wilderness. Several ironworks in eastern Pennsylvania merit discussion. In 1742, Benjamin Franklin invented the Franklin stove; he obtained his castings from a foundry known as the Warwick Furnace, located near Warwick, Pennsylvania. This stove, an invention that was soon widely adopted, was made possible because of castings, and it increased cast metal tonnage. For example, in 1742, a furnace foundry known as the Mount Joy Forge was erected in Chester County, Pennsylvania, on Earl Valley Creek. The name was later changed to Valley Forge. This foundry, which was burned by the British in 1777, served as the

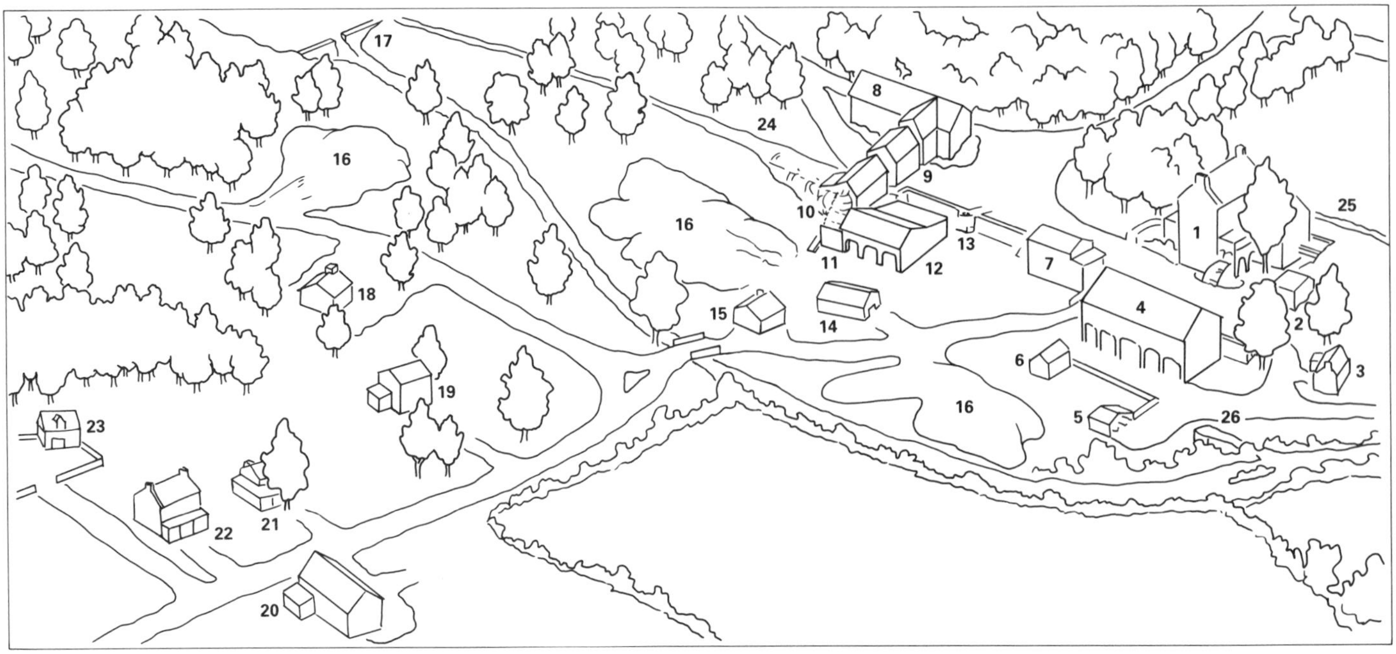

Fig. 2 Schematic of an 18th century iron plantation. 1, mansion house; 2, bakery; 3, spring house; 4, barn; 5, carriage house; 6, corn crib; 7, office; 8, charcoal storehouse; 9, furnace bridge; 10, mill wheel; 11, furnace; 12, casting house; 13, ore roaster; 14, wheelwright shop; 15, blacksmith shop; 16, slag; 17, dam; 18, schoolhouse; 19-22, tenant houses; 23, tenant barn; 24, west head race; 25, east head race; 26, tail race outlet

encampment of the American army during the winter of 1777-1778. This furnace was later rebuilt, and important early experiments on steel castings were conducted there.

Iron Plantations. No history of the American foundry industry would be complete without a description of the iron plantations, great estates that existed principally in eastern Pennsylvania in the 18th century. Dozens of these semi-industrial and partly feudal facilities had been established by 1750. A typical enterprise was Hopewell Village near Birdsboro, Pennsylvania (Fig. 2). William Bird bought this tract of approximately 10 000 acres in 1743. In 1761, his son built the furnace and developed the plantation.

At one time, nearly 1000 persons (including furnacemen, molders, miners, charcoal burners, wagonmen, and their families) lived there and derived their existence almost entirely from the production of pig iron and castings. The iron mine was located approximately 1½ km (1 mile) from the furnace, and the ore was carted in. The furnace used charcoal in vast quantities, averaging around 1500 cords of wood a year. Most of the inhabitants were woodcutters and charcoal burners, who prepared the charcoal in mounds in the forest. Iron and castings were sent to Philadelphia by boat or cart, but the inhabitants lived and worked on the plantation.

Westward Expansion. Toward the end of the 18th century, the furnaces and foundries of America began to move westward. The first ferrous foundry established west of the Alleghenies was built in Fayette County, Pennsylvania, in 1792 by William Turnbull. This foundry supplied guns and shot for General Wayne's expedition against the Indians. In the same year, the first plant in Pittsburgh was erected by George Anschultz, who made stove and grate castings. A furnace was also established at the same time on the Licking River in Bath County, Kentucky.

Far to the north, the first Canadian foundry had been installed years earlier, in 1730, at a location south of Three Rivers, Quebec. This foundry operated for 150 years.

Foundries and the Revolution. A final commentary on 18th century foundry operations in the United States concerns the connection between the foundry and the American Revolution. It is generally accepted that the stamp tax on tea and "taxation without representation" were the primary causes of the American Revolution. However, history reveals even more fundamental reasons that involve the casting of metals. In 1750, the English Parliament, envious of the growth of ironworking in the colonies, passed an act prohibiting the refining of pig iron or the casting of iron. This act also restricted the construction of any additional furnaces or forges. Pig iron could be made only if it was shipped to England, where a shortage of charcoal had seriously curtailed iron production. The act was openly resisted by early American foundrymen.

For the most part, the colonial founders joined the revolutionary cause and supported it with money, guns, and shot; other foundrymen supported it politically. Among the many foundrymen who fought in the Continental army as officers were Nathaniel Green (Rhode Island Furnace), who commanded at the Battle of Long Island; Ethan Allen (Connecticut Furnace), who commanded the Green Mountain Boys and forced the surrender of Fort Ticonderoga; and Lord Sterling (Sterling Iron Works), who served on General Washington's staff.

The production of war material was the principal task of the American foundries during the War for Independence. Many foundries made shot, shells, and cannons in great quantities, and it was through these efforts that supplies kept coming to Washington's troops. Frequently, these same founders remained unpaid. As always in time of war, foundries were military objectives, and the British directed their raids toward the destruction of the foundries. Undelivered cannons and shot were sometimes buried to keep them from falling into enemy hands in case the furnaces were captured.

During the Revolution, another figure appeared who is not usually associated with the casting of metals—Paul Revere. Before the Revolution, Revere had acquired experience in casting bronze and silver for bells and tableware and many iron articles. However, his primary metallurgical experience was obtained when he was assigned by the Continental government to work with Louis de Maresquelle (Louis Ansort), a French founder of exceptional ability. Ansort was able to soften iron by mixing metals, and he also introduced the completely bored, solid cast gun. Under this master founder, Paul Revere learned metallurgical techniques that later served him well in his further work on the malleability of copper.

After the war, Revere returned to his bell-and-fittings foundry in Boston. In an effort to improve the tonal quality of cast bells, Revere began to experiment with various coppers and copper alloys. His metallurgical success is well known today, and a company bearing his name is the direct descendant of Revere's original enterprise.

The Liberty Bell. No account of the relationship between bells and the American colonies would be complete without mention of the Liberty Bell, whose ringing in Philadelphia on July 4, 1776, announced the signing of the American Declaration of Independence. The Liberty Bell was cast by Thomas Lister of Whitechapel, London, to mark the 50th anniversary of the Commonwealth of Pennsylvania. The bell cracked twice during testing and was recast twice; its original tone has been considerably altered by the amount of copper added in the recasting. This bell casting weighs 943 kg (2080 lb) and is now preserved in Philadelphia, where it was originally hung.

The War of 1812

The War of 1812 also contributed to foundry history. During that conflict, Henry Foxall, a minister and foundryman, was making castings for the United States at Georgetown, Maryland. The British, after burning the White House in Washington, marched toward Georgetown to destroy the foundry, and Foxall vowed that if his foundry were spared he would establish a new Methodist Church. A sudden electrical storm delayed the British and then prevented them from reaching the foundry. In 1815, Foxall built the Foundry Methodist Church in Washington, D.C.

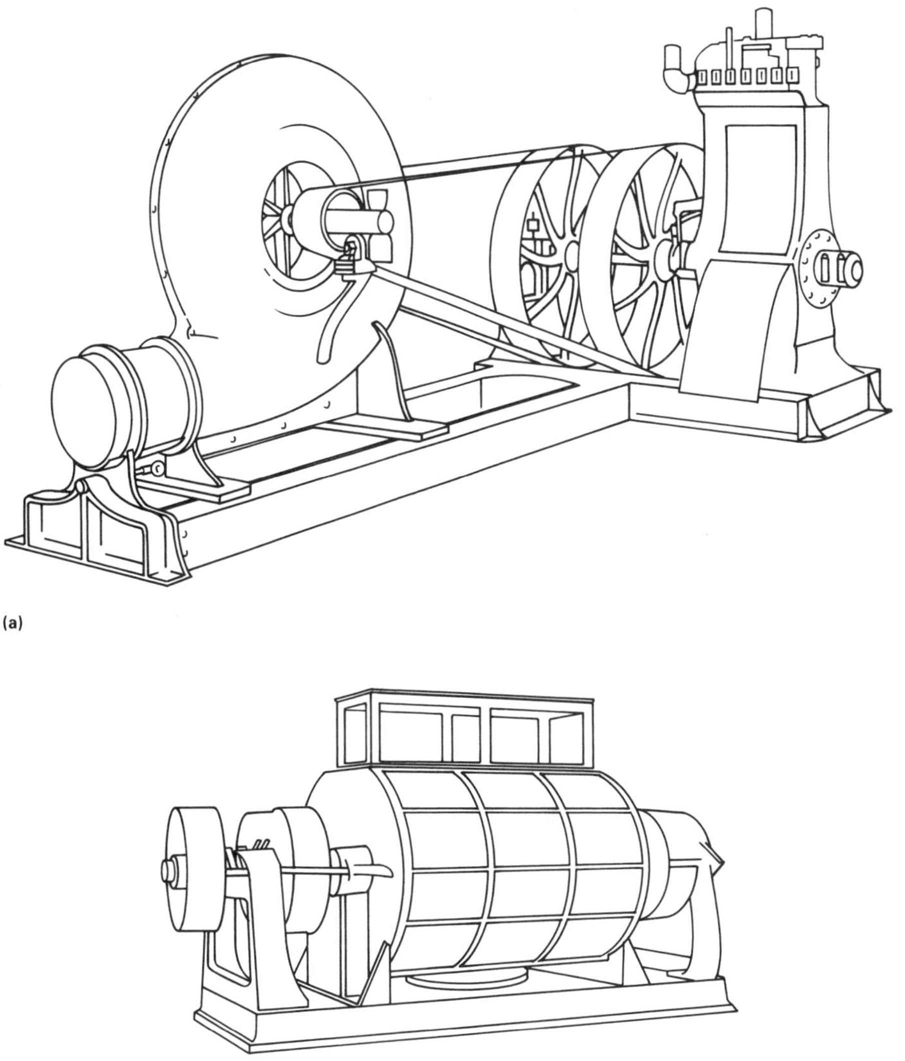

(a)

(b)

Fig. 3 Two types of blowers developed for use in the foundry industry. (a) Cyclone type. (b) Box type

Equipment Advances

Continuous melting and furnace improvements over a period of approximately 80 years during the 19th century brought to foundrymen melting tools that were superior to any previously known. With efficient and economical melting equipment, the foundry industry was able to develop a metallurgical chemistry that, coupled with the art of casting, made it possible to produce high quality parts economically. The new furnaces introduced during the 19th century did not satisfy all of the needs of the industry, but fortunately other divisions of the foundry were also progressing rapidly. The new furnaces were soon complemented by better blowers, pouring devices, microscopic analysis of metals, molding equipment, mechanical chargers, and many other tools that are commonly accepted in modern foundry practice.

Metallography was developed by Henry Clifton Sorby of Sheffield, England, in 1863. Chemical analysis had been available before this time, but Sorby was the first to polish, etch, and microscopically examine metal surfaces for analysis. Sorby first became interested in and developed microscopy as an aid to the study of meteorites. His work on the surfaces of metals soon became much more important from an industrial point of view because it enabled practicing foundrymen to supply the missing element of knowledge and to supplement their rather sketchy experience with chemical analysis.

Blowers. Metals and melting were further aided by the development of blowers designed specifically to meet foundry needs. The steam engine and the water bellows had already proved to be of great benefit to the foundries because they permitted higher temperatures and shorter melting times. Mechanical blowers entered the foundry market as commercial devices sometime after the middle of the 19th century, although homemade equipment had preceded the standardized articles for many years. These blowers were of two types: the cyclone (for example, the sturtevant blower) and the Roots (a box-type blower). These are shown in Fig. 3. Both types are well known today.

Pouring Devices. Early in the development of pouring devices, numerous mechanical aids were invented that today are highly specialized and built with standard and interchangeable parts. The shanked ladle, adapted for use by two men, appeared later in the century. Still later came the one-man ladle equipped with wheels (Fig. 4). However, when it was realized that foundry flooring was not ideal for the smooth transportation of molten metal, foundrymen began to move their ladle overhead by crane or winch. As the demand for larger castings increased during the machine age, ladles of greater capacity were required, and this increased the hazards of metal pouring. Many accidents in foundries during the early 1800s were due to improper pouring devices.

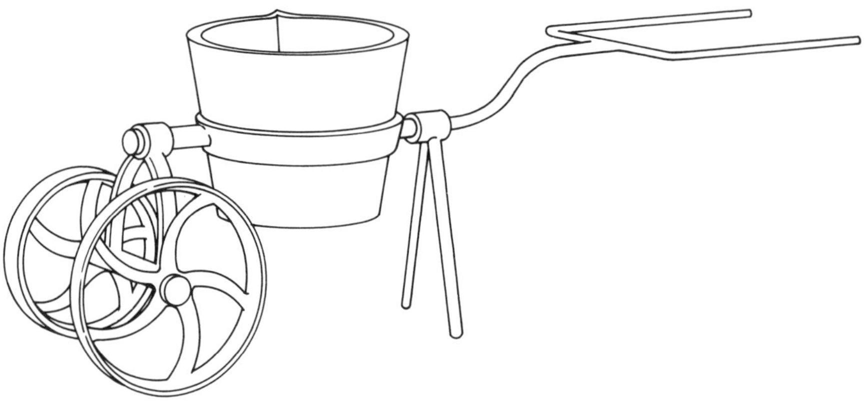

Fig. 4 Wheeled ladle for one-man operation used in the latter part of the 19th century

In 1867, James Nasmythe, the inventor of the steam hammer, came forward with a ladle that undoubtedly prevented countless metal pouring accidents. He constructed a safety foundry ladle whose tilt was controlled by gearing (Fig. 5). On the basis of greater safety and economy, the foundry industry quickly adopted this device for pouring all sizes of castings.

Molding Machines. The most important development in foundry technology was the molding machine, without which the modern foundry would be incapable of its current large-scale production. Molding machines had been the foundryman's dream for centuries, but it was not until the 19th century that such equipment actually appeared. There is a record that an unknown Englishman developed a machine in 1800 for molding screws. In this device, the pattern was backed out of the sand by lead screws of the same pitch. However, in 1837, a dependable molding machine was finally placed on the market. This was a jarring type of machine that was first made and used by the S. Jarvis Adams Company, the forerunner of the Pittsburgh Iron and Steel Foundries Company and later known as the Mackintosh-Hemphill Company. Although this machine was of rather crude design and was built for special work, it was successfully used in making a number of castings on one riser.

Molding machine designers subsequently created all manner of devices, many of which would be highly impractical today. But it was not until the 1880s that commercially viable equipment became available. These improved designs lightened the foundry task and enabled foundrymen to increase production, to produce more accurately and uniformly on a production basis, and to lower costs.

In 1896, nearly all molds were made from loose patterns and were molded by hand. The production of even one mold per hour was a laborious task. Molders used single loose patterns or gated patterns with a hard sand match. Molds were rammed by hand with sand-to-sand partings in flasks. The first high-production molding possibility came with the introduction of the drop machine, which was made for farm machinery. Sand was rammed by hand, but the half patterns (cope or drag) were drawn down through a contoured stripping plate. This drew the pattern without the aid of vibrators and made it easier to remove the mold with all the sand intact. The first machines functioned mechanically with levers and cams, but compressed air soon became the source of the jolt power.

The early squeezer machines were simple devices. The foundryman planted a vertical steel rail about 2 m (7 ft) long in the foundry floor. A cast iron table of convenient height was bolted to the rail, along with a squeeze head. The squeeze head was operated manually with a hand lever. The molds were not very hard, because of the tight (Albany) naturally bonded sands that were used. Millions of molds of stove plate were made in this manner. With the molder carefully pouring his own work, satisfactory castings were produced.

Another notable improvement in small casting work was the development of the match plate. First appearing in the literature in about 1910, match plates eliminated the hard sand match and the problems of sand-to-sand parting. Match plates, together with the air vibrator, made the jolt squeeze principle for small molds feasible. Snap flasks and steel bands were also introduced about this time.

Sandslingers. In 1914, Elmer Beardsley and Walter Piper operated a foundry in Klamath Falls, Oregon. They took a job that proved to be far beyond their capability to produce in the required time limit. They noticed, in hand molding, that when the molder had to lift out a pocket he always followed a set routine (riddle a small amount of sand into the pocket, place a nail or a gagger into the pocket, and then throw sand by the handful to fill the pocket). This was generally all the compacting required. Using this idea, they put boards into the headstock of a lathe, placed the mold beneath the lathe, and fed sand to the rotating headstock. The centrifugal force of the rotating headstock threw sand into the flask with enough force to be solidly compacted. The headstock had to be covered to direct the sand stream downward. This was the beginning of the well-known sandslinger.

The sandslinger found widespread success and is still being used in many foundries, especially large jobbing casting plants.

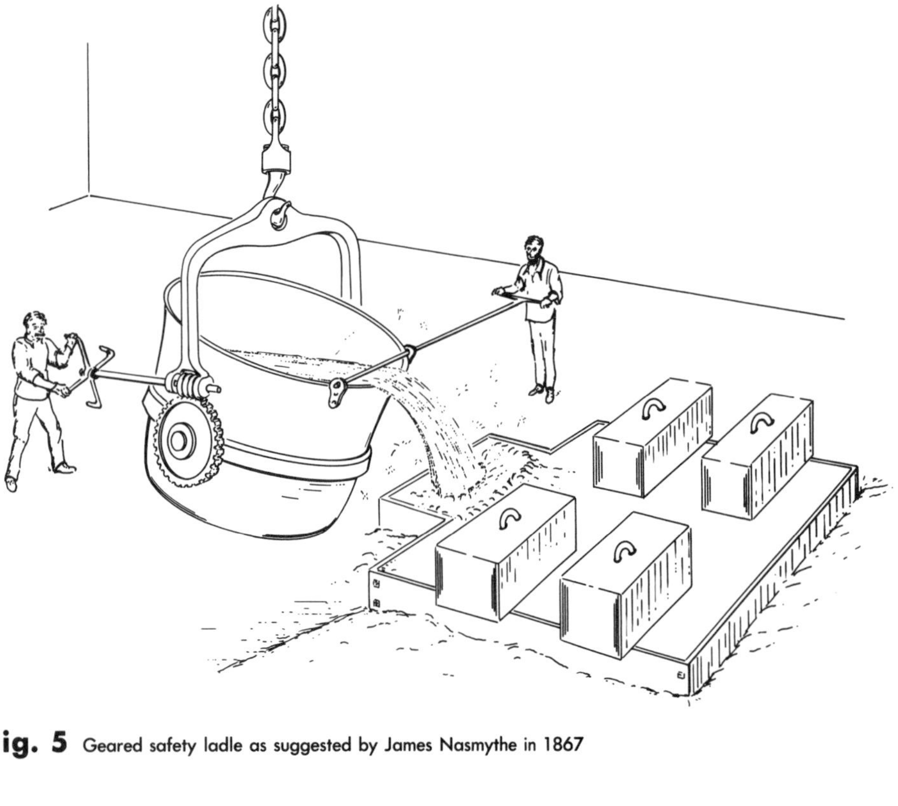

Fig. 5 Geared safety ladle as suggested by James Nasmythe in 1867

The sandslinger is the first high-pressure molding device. Mold hardness can be readily varied, depending on the amount of sand fed into the head and the speed with which the slinger moves over the work. Sandslingers have been automated and hydraulically controlled to remove much of the manual effort required in their operation.

From 1896 to approximately 1955, air-operated molding machines improved continually. Larger squeezers, higher-capacity rollovers, and bigger jolt cylinders resulted in the molding of larger flasks and in higher production. A distinct improvement occurred when a squeeze was added to the jolt rollovers. Previously, jolt rollovers were topped off with an air rammer compacting the remaining sand, thus slowing production. More information on modern molding machines is available in the section "Green Sand Molding Equipment and Processing" in the article "Sand Processing" in this Volume.

Synthetic Sands. Since the beginnings of sand molding, the art of making metal using green sand depended on the skill of the molders. Molds were made strong and hard on the outside (near the flask edge) but soft in the middle, especially the drag surface. The art of venting a mold aided the production of good castings. Naturally bonded sands contained too many clay fines and water; thus, ramming had to be held to a minimum to produce reasonably accurate castings. Synthetic sands (washed and dried silica sand to which binders such as fireclay and/or bentonite are added) appeared in the late 1920s. The use of these synthetic sands permitted molds to be produced in the green state (without being baked and dried). In the early days of using synthetic sands, the molds were still made to be hard on the outsides and quite soft in the middle. This difference in density imposed limitations on molding machines and a challenge to molding machine manufacturers.

The demand for more accurate castings led to the need for molds of uniform and higher density. Mold wall movement became a known defect of sand molds (and castings) in the 1940s and 1950s. Sand formulations continued to improve, and higher densities and mold hardnesses were available. Uniform mold hardness became a necessity for close, accurate two-pattern size castings. Because of the success of core blowers in the high production of cores, several attempts were made in 1940 to blow green sand molds, but the results were unsatisfactory. Later developments, which combined sand blowing with a hydraulic squeeze, achieved good results. More information on sand molding and foundry sands is available in the articles "Sand Molding" and "Aggregate Molding Materials" in this Volume.

Advances in Casting Alloys

At the turn of the 19th century, both in Europe and the Americas, foundry practice and foundry methods had vastly improved, yet the 19th century brought about dramatic improvements in metals, equipment, and processes. The full possibilities of iron became far better understood during this century, and gray iron was found to be the most versatile and diverse of all cast metals. As a result, the use of iron for castings was vastly increased, even though malleable iron, chilled iron, and finely cast steel were also tremendously advanced during this period.

The world was approaching the highly mechanized state in which we now find it, and the iron family and its older nonferrous relatives were destined to play vital roles in that mechanization. No modern luxuries and few modern essentials would be available today were it not for the foundry industry and the cast metals that the artisans, engineers, and inventors of the 19th century made usable and practicable.

Cupola Iron

The true birth of gray iron (a product of the cupola) and close chemical and metallurgical control were both 19th century developments. Gray iron is a true illustration of the chemistry of metal as applied to the science of casting and is described in detail in the article "Gray Iron" in this Volume. The developments that took place between 1810 and 1815 in the field of cast iron reveal this relationship. In 1810, French chemist Louis J. Broust, after extensive investigations, described cast iron as a solution of carbide of iron in iron. To modern metallurgists, this must seem extremely elemental, but when it is realized that few scientific data were available at this time, the work of the early chemists assumes its proper proportions. Also in 1810, Johns Jakob Berzelius, a Swedish chemist, produced ferrosilicon by melting silica, carbon, and iron fillings in a sealed crucible. In that same year, German physicist Wilhelm Stromeyer produced several grades of ferrosilicon in more exact experiments and proved that it was silicon and not silica in the metal.

In 1814, Karl Karsten, a German scientist and metallurgist, published the results of experiments proving that oxygen does not exist as an essential ingredient of cast iron, but that the different types of cast iron are due to the different forms of the carbon content. He then described two compounds of iron and carbon—one rich in carbon and poor in iron (graphite) and the other poor in carbon and rich in iron (white iron). Karsten was also one of the first to observe the effects of sulfur. His experiments showed that 0.05 to 0.25% S in iron made the metal hot short, while as little as 0.05% P made iron cold short. Although these researchers opened new vistas for foundrymen and metallurgists, in the final analysis it was the practical foundrymen who used this newly found information, which existed solely as pure research without practical application. Perhaps the greatest single step in this direction was the development of the cupola.

The Wilkinson cupola, which originated in England in 1794, was a great step forward, but mechanically it still fell far short of the efficient and economical melting units available to iron founders today. Records show that the early cupola had a stationary bottom with a front draw built on a stone foundation. The charge was carried up a flight of steps to the top or was thrown and shoveled from one landing or platform to another until it could be charged from the top. The shell of the cupola was generally made from castings, with an opening in front of sufficient size to permit the slag to be raked out with a hook. This cupola melted very slowly, and iron dripped out continuously into a reservoir in front, from which ladles were filled. To an experienced foundryman, this indicates the troubles the early melters encountered, yet they made quality castings.

In the United States, the cupola was introduced around 1815. In Baltimore, two of these early cupolas were still in operation in 1902. This melting unit, existing on scrap and pig iron, so widened the gap between smelting and melting that the foundry and reduction blast furnaces soon became completely separate. Merchant pig iron producers, relieved of the duties of casting metals, went on to achieve the highly specialized skill that today is theirs. Foundrymen, on the other hand, provided with a ready and reliable supply of scrap or pig iron, were able to control with greater certainty many of the variables that had proved uncontrollable.

In 1850, another important improvement was made in the cupola—the drop bottom—without which no efficient cupola could operate today. This innovation, so familiar to modern iron foundrymen, seems to have marked the beginning of modern cupola design; further developments occurred in rapid succession (see the section on "Cupolas" in the article "Melting Furnaces" in this Volume). The stack was made smaller than the crucible and was built higher to aid draft. The blast was introduced from tuyeres on opposite sides of the cupola, and a melt could now be poured in about 10 h. Some 10 years after the introduction of the drop bottom, the one-piece cupola appeared, constructed of boiler plate casing in both crucible and stack. Then came the introduction of air by means of air chambers and blast tubes. Next the cupola designers eliminated the taper to provide the same diameter from top to bottom (Fig. 6).

The first commercial cupola in the United States was the Colliau cupola, which was introduced in 1874. This cupola was prefabricated insofar as possible and was built primarily as a commercial product. It was highly successful, being a fast and econom-

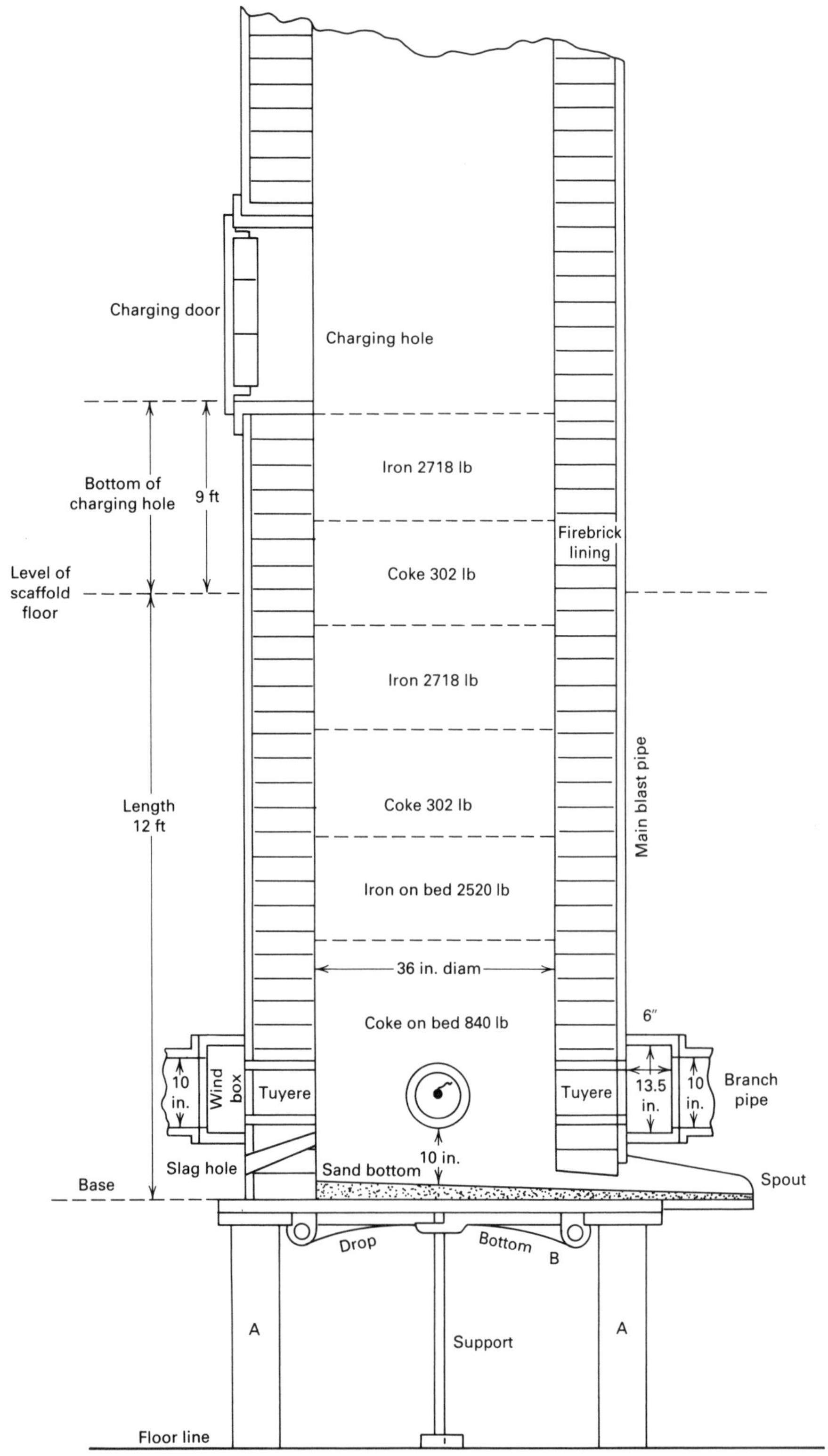

Fig. 6 Straight-sided lined cupola introduced after the middle of the 19th century

ical melting unit. When first introduced, it presented a number of improvements, such as a hot blast, and double rows of tuyeres. After the Colliau cupola, the Whiting cupola was developed. It is still sold by the company bearing the name of its inventor, John H. Whiting. In the 1880s, Whiting developed a cupola that employed two rows of tuyeres, with the lower row arranged to form an annular air inlet that distributed the blast around the entire circumference of the furnace. The tuyeres could also be adjusted vertically for changes in classes of work, type of fuel, and cupola diameter. This cupola permitted the use of either coal or coke as a fuel and was equipped with a safety alarm and blast meter. The Whiting cupola, with its standardized construction, soon proved a boon to foundrymen and was widely accepted.

Connellsville Coke

The story of the cupola and cast iron in this period would not be complete without reference to the development of Connellsville coke. Coke was in general use in Europe in 1750, but because of the heavy timber resources of the United States, it was not produced here until 1817. From that time until 1860, American foundrymen generally made coke for their own use; after 1860, coke became a commercial product. Connellsville coke was first produced in 1841 at Connellsville, Pennsylvania, in beehive ovens. The product proved so popular that the demand for Connellsville coke remained high until 1914, when by-product coke came into greater demand.

The use of coke and bituminous coal was made possible by the introduction of the hot-blast furnace in 1828 by James B. Neilson of Scotland. At first, the blast pipes ran through the furnace. The pipes were later run through a charcoal oven, and finally they were placed on top of the furnace itself. This device increased melting output from 10 to 40% so that by 1869 the production of coke and bituminous pig iron exceeded that of charcoal pig iron, and by 1875 the use of coke pig iron alone surpassed that of anthracite pig iron.

Chilled Iron

Another American development of the mid-19th century involved the introduction of chilled-iron railroad car wheels. Asa Whitney of Philadelphia obtained a basic patent in 1847 on a process for annealing chilled-iron car wheels cast with chilled tread and flange. They were satisfactory as long as cold-blast charcoal iron was used. However, trouble developed when melting practice changed from the use of anthracite coal to the hot blast.

In 1880, an attempt was made to introduce manganese for the production of chilled iron, but the resulting product did not possess adequate wear properties. Finally, a small amount of ferromanganese was introduced directly into the ladle with good results. Thus originated one of the basic products of the foundry industry—chilled car wheel iron. The excellent performance of this foundry product, as well as its wear resistance and economical properties, made possible the long-haul, heavier railroad freight loads of today.

Malleable Iron

No discussion of iron would be complete without mention of American blackheart malleable iron, as contrasted with European whiteheart malleable iron. Although credit for the progress and growth of the American malleable iron industry belongs to many, its origin lies with Seth Boyden of Newark, New Jersey.

Boyden's experiments, beginning in 1820, were primarily based on an attempt to duplicate the whiteheart European product developed by French scientist R.A.F. Réaumur. Boyden was endeavoring to lower the cost of harness hardware and wanted a strong iron that could be easily machined. Because of a larger percentage of silicon than was available in the Réaumur process, Boyden was able to produce the strength, but not the white color, of the European product. However, Boyden (either by design or accident) had shortened the annealing time to 6 or 10 days. This left free temper carbon graphitization as opposed to Réaumur's decarburization. Boyden first used a crucible and then an air furnace with a capacity of about 450 kg (1000 lb). His annealing furnace was a beehive unit into which pots loaded with castings were lowered through the top. The beehive was eventually replaced with a continuous annealing furnace having a sloping floor. This unit was known as a shoving furnace because the pots and castings were shoved in from one end and out the other.

Boyden's work was important, but the most significant advances in the malleable iron field were made by the men who developed the malleable industry to its level of prominence in the foundry picture. More information on the manufacture and properties of malleable iron is available in the article "Malleable Iron" in this Volume.

Cast Steel

Next in the chronological sequence of casting developments is steel. For centuries, the goal of manufacturing steel castings in large quantities was pursued, but inadequate equipment was the limiting factor. In addition, steel was well known centuries earlier as the Damascus and Toledo sword blades of legendary fame, but this steel was forged from the pasty masses of iron produced in the Catalan forge. The first reliable record of steelmaking was the work of Huntsman in England in 1750. Huntsman's developments in crucible refractories (the crucible process) first produced steel that could be poured as a liquid.

Steel castings are also said to have been discovered by Jacob Mayer, Technical Director of the Bochumer Verein, Bochum, Germany, sometime before 1851. Records of the Bochumer Verein Company, which is still engaged in producing steel castings, indicate that cast steel church bells were produced in 1851. Some of these bells weighed as much as 15 Mg (17 tons). Cast steel guns were also made at the Krupp Works in Germany in 1847. In all probability, much smaller steel castings were made before the bells were cast, because it is difficult to believe that such large castings would be attempted without considerable previous experience. The steel church bell castings were displayed at various expositions throughout Europe and created quite a sensation because of their fine, clear tones and the fact that their selling price was about half that of the bronze bells formerly in general use. In a park outside of Bochum, one of the early steel church bells is enshrined as a marker of steel casting history. This bell even escaped the desperate need for scrap steel in Germany during World War II.

Steel produced by the crucible process remained expensive, and it was not until the converter, the open hearth furnace, and finally the electric arc furnace were introduced that steel was produced commercially in quantities that were economically feasible. The steel for these early castings, in addition to being made in crucible furnaces, was poured in loam molds. It was not until 1845 that steel castings (steel cast to final shape) appeared on the scene. On July 14 of that year, Swiss metallurgist Johann Conrad Fischer exhibited various small castings produced from crucible steel. On July 23, 1845, Fischer applied to the British Patent Office for priority rights to a new method of making horseshoes that consisted of casting steel in sand molds.

In the United States, cast steel was produced by the crucible process in 1818 at the Valley Forge Foundry, but difficulties resulting from the lack of adequate materials—principally refractories for crucibles—caused the experiment to be abandoned. In 1831-1832, high-quality clay from Cumberland, West Virginia, enabled William Garrard of Cincinnati, Ohio, to establish the first commercial crucible steel operation in this country. It is interesting to note that the first commercial steel castings from the Garrard plant sold for 18 to 25 cents per pound and were first used as blades and guards for the McCormick reaper.

The history of steel castings in the United States begins with the Buffalo Steel Company of Buffalo, New York (later known as the Pratt and Letchworth Company). The foundry was built in 1860, and in 1861 it produced the first crucible steel made in the district. Records indicate that these first crucible steel castings were for railroad applications. Some of the first commercial steel castings produced in the United States are believed to have been made by the William Butcher Steel Works (later the Midvale Company) near Philadelphia in July of 1867. These castings are said to have been crucible steel crossing frogs and car wheels, probably for the Philadelphia and Reading Railway.

In 1870, William Hainsworth of Pittsburgh began the manufacture of cast steel cutting parts for agricultural implements using a two-pot coke-fired crucible furnace. In 1871, Hainsworth founded the Pittsburgh Steel Casting Company, which is reputed to have been the first company in the country devoted exclusively to the manufacture of steel castings. Some of the early steel foundries established in this country before 1890 are listed in Table 1.

William Kelly, whose invention of the converter clearly anticipated the work of Sir Henry Bessemer (although the latter gave his name to the unit), first operated his converter in 1851, a full 5 years before Bessemer's patent was obtained. Neither man realized that his efforts were being paralleled by the other, but it has been well established that Kelly was the first to use the converter.

Kelly developed his idea out of experiments on the refining of pig iron as a result of his inability to obtain high-quality ore at his works in Eddyville, Kentucky. His theory was that the iron in ore is not metallic but a chemical compound of oxygen and iron. He came to the conclusion that, after the metal was melted, additional fuel would not be required but that the heat generated by the union of the oxygen in air with the

Table 1 Partial listing of early steel foundries in the United States

Original	Location	Date of incorporation	Now known as
Buffalo Steel Co.	Buffalo, NY	1861	Discontinued operations
Wm. Butcher Steel Works	Nicetown, PA	July 1866	Discontinued operations
Pittsburgh Steel Casting Co.	Pittsburgh, PA	March 1871	Discontinued operations
Chester Steel Casting Co.	Chester, PA	1872	Discontinued operations
Otis Iron and Steel Co.	Cleveland, OH	*Circa* 1874	Discontinued operations
Isaac G. Johnson and Co.	Spuyten Duyvil, NY	*Circa* 1850(a)	Discontinued operations
Eureka Steel Castings Co.	Chester, PA	1877	Discontinued operations
Hainsworth Steel Co.	Pittsburgh, PA	*Circa* 1880	Discontinued operations
Old Fort Pitt Foundry	Pittsburgh, PA	*Circa* 1881	MacIntish-Hemphill Div., E.W. Bliss Co.
Solid Steel Casting Co.	Alliance, OH	August 1882	American Steel Foundries Alliance Works
Standard Steel Castings Co.	Thurlow, PA	*Circa* 1882	Discontinued operations
Johnson Steel Street Ry. Co.	Johnstown, PA	March 1883	Johnstown Corp.
Pacific Rolling Mills Co.	San Francisco, CA	1884	Discontinued operations
S.G. Flagg and Co.	Philadelphia, PA	1882–1885	Discontinued operations
Cowing Steel Castings Co.	Cleveland, OH	1882–1885	Discontinued operations
Sharon Steel Casting Co.	Sharon, PA	February 1887	Discontinued operations

(a) Began making steel castings in 1880

carbon in the metal would be sufficient to decarburize the iron. Kelly's associates, fearing for his sanity, sent him to a doctor for treatment, with the result that the doctor agreed to back Kelly in his venture. In 1851, he was able to produce a rather soft steel, but had difficulty with high carbon. In the meantime, in England, Sir Henry Bessemer was working along almost identical lines. However, he had metallurgical assistance. Robert Mushet of England developed an alloy of iron, carbon, and manganese that purified the metal and ensured the presence of enough carbon to make steel. Although Bessemer obtained American patents, Kelly proved his patent priority in 1857, and in 1866 Kelly and Bessemer joined forces.

A Kelly converter was first used in 1857-1858 at the Cambria Works at Johnstown, Pennsylvania. Later, Kelly obtained the rights to use Mushet's patent. Bessemer's first converter in the United States was installed at Troy, New York, in 1865. This converter was introduced into foundries and was further improved in 1891 by the Tropenas converter (Fig. 7), which blew over the surface of the metal rather than through it. The lower-pressure blast and resultant deeper metal bath produced better results.

The converter had scarcely gained acceptance when another furnace came into use and gave the steel industry the capacity it required. This was the Siemens-Martin open hearth, a development dating back to 1845, plus the succeeding experiments of J.M. Heath. However, this unit did not become successful until the great heat of the open hearth regenerative furnace could be supplied to it. This became possible in 1857 as a result of the invention of the gas producer, which was patented that year.

Because of its high initial cost, the first open hearth furnace was not installed in the United States until 1870. However, with its tremendous capacity, it soon surpassed the converter in tonnage until it was in turn challenged by the electric arc furnace.

The electric arc furnace (Fig. 8) was invented by Sir William Siemens, who developed an electric arc in 1878. However, it was little used until improved by Girod, Heroult, Keller, and others in 1895. An electric arc furnace was first installed in the United States in 1906 at the Holcomb Steel Company (later the Crucible Steel Company) at Syracuse, New York.

Induction electric furnaces were not introduced from Sweden until 1930, and many of the American developments of this furnace were made by Dr. E.F. Northrup and Dr. G.H. Clamer of the Ajax Metal Company, Philadelphia, Pennsylvania. Information on electric arc and induction furnaces is available in the article "Melting Furnaces" in this Volume.

Cast Alloy Steels. In 1888, the first manganese cast steel was made in the United States at the Taylor-Wharton Iron & Steel Company in High Bridge, New Jersey, under license from Robert Hadfield of England. This was also the first cast alloy steel to be produced in America, and it was used for railway crossings and switch frogs. In 1903, A.L. Marsh, an American, made an alloy of 80% Ni and 20% Cr. He was studying alloys for use as thermocouples, and he observed that this alloy had high electrical resistance and could sustain operation for extended periods at high temperature without excessive oxidation—an ideal combination for electric heating elements. This and similar high nickel-chromium alloys later became the standard for electrically heated equipment and appliances.

Before World War I, E. Maurer and B. Strauss in Germany and H. Brearley in England were considering the alloys of iron with chromium and nickel for use as pyrometer tubes and gun barrels. They noted resistance to etchants by certain compositions and realized the potential utility of such steels as stainless or rust-free in corrosive environments. By 1912, the German firm of Krupp had obtained patents on a martensitic, hardenable 14% Cr alloy and on an austenitic, nonhardenable 20Cr-7Ni alloy, which they called VM and VA, respectively. By 1916, Brearley had received patents in the United States and Great Britain on cutlery made from hardenable steels containing 9 to 16% Cr with 0.70% max C. Thus, the major classes of heat-resistant and corrosion-resistant alloys were all discovered and patented from 1905 to 1915.

The use of these materials in castings for industrial applications awaited the next decade. During World War I, urgent demands for expanded production from the infant automobile and aviation industries created the need for improved heat-treating procedures. In 1916, a patent was issued on the use of high nickel-chromium-iron electrical resistance alloys in cast carburizing boxes. These were supplied at first by foundries set up by the producers of the electrical resistance wire, and later by independent foundries formed to specialize in heat-resistant castings. At the same time, the increased output of the munitions and synthetic dye industries was making the corrosion resistance of the iron-chromium alloys attractive for handling strong oxidizers, such as nitric acid. As a result, the foundries that were making pumps and valves in carbon steel were asked to make these parts in rustless iron and stainless steel. With the end of the war and the industrial expansion that followed, the demand for both heat- and corrosion-resistant castings increased substantially.

The growth of high-alloy casting consumption has been stimulated by the continuing research objective of providing users with the materials and data needed to solve process design problems. Since 1930, there has been considerable development of new alloys and refinement of old ones as detailed in the articles found in the Sections on Ferrous Casting Alloys and Nonferrous Casting Alloys in this Volume.

Foundry Mechanization

With gray iron, malleable iron, and steel added to the foundrymen's metals for casting to shape, it follows that equipment and methods for the rapidly growing castings industry were also being given increasing attention. Molding, coremaking, sand preparation and conditioning, and metals and materials handling methods also progressed during the 1800s.

To complete the story of the development of the art and science of casting, it is only logical to trace the early progress of the tremendous mechanization of the industry. Actually, sand casting is a relatively recent development (in terms of the antiquity of casting technology), and it occupies an indispensible place in the industry. As with the development of the molding machine, foundrymen soon realized that a mixed molding material was essential. With the use of loam, the early machines for the treatment and preparation of sands aimed at compounding or grinding rather than mixing and mulling. Early in the 1870s, machines began to appear that were essentially paddle mixers. In the 1890s and early 1900s, manufacturers began to adapt equipment used by the ceramic industry. In 1912, the first muller, with individually mounted revolving mullers of varying weights, was placed on the market. Since that time, mullers that effectively coat the sand grains have been successfully used in the preparation of sand for both cores and molds.

This same period saw the first steps taken toward the development of sand screening machinery, which eventually resulted in the riddle, the magnetic separator, and the complete sand preparation plant. Mold-conveying methods, introduced about 1890, originally involved a continuous series of moving cars that looped at a steady speed from molder to pouring station and then to cool and shakeout.

Core manufacture was also standardized and equipment made available for mass production in American foundries. Long rooms filled with workers producing cores could be seen as early as 1888. All cores were racked and dried in kilns and then placed on racks to be carried to the foundry. The importance of core production soon sparked the development of baking ovens specially adapted for foundry use so that cores of different types could be baked for longer or shorter times as required. An early developer of the foundry core oven was Eli Millett in 1887. Today, coremaking, like sand molding and conveying, is a complex and highly specialized division of every

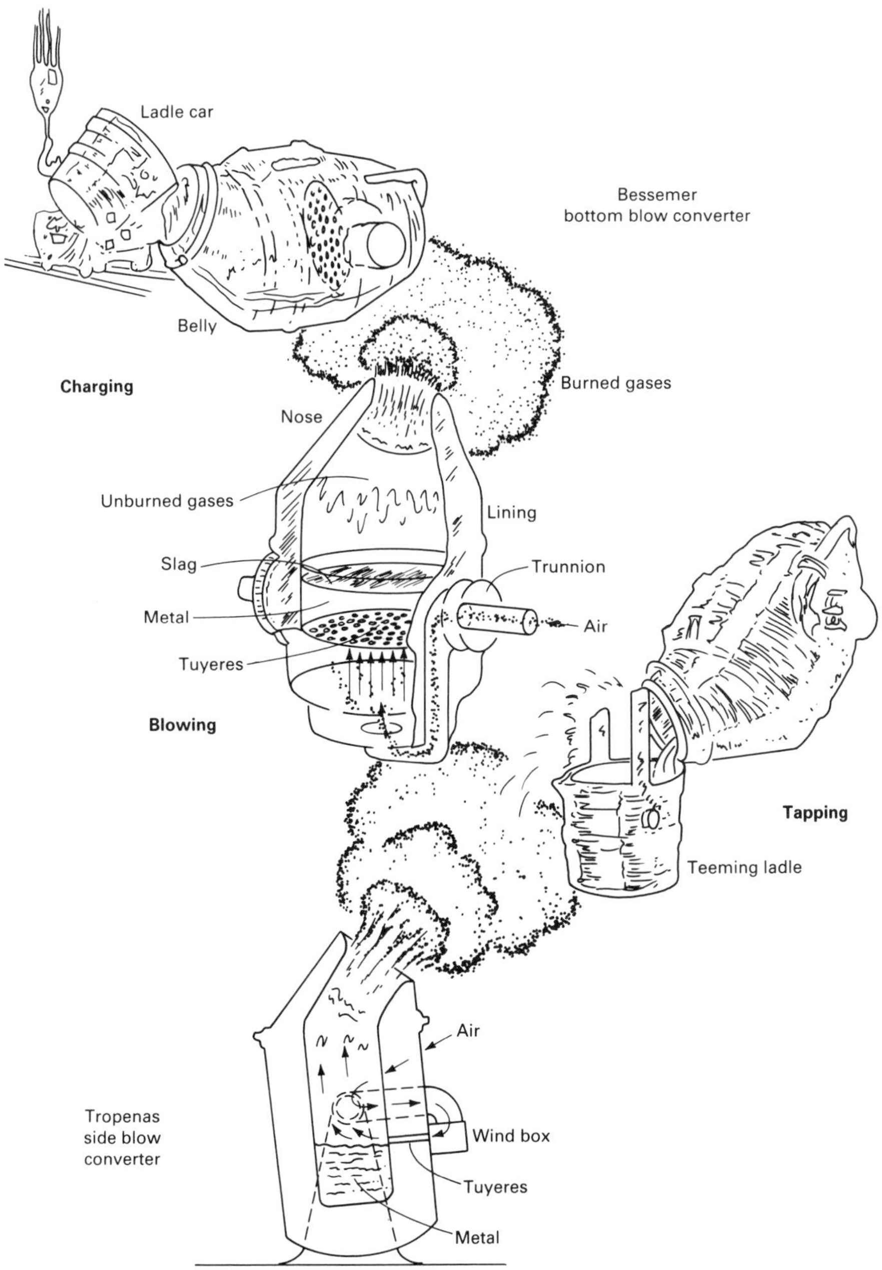

Fig. 7 The Bessemer and Tropenas converters

foundry (see the article "Coremaking" in this Volume).

Early credit for the tumbling mill must be given to the W.W. Sly Company of Cleveland, Ohio. The Sly cleaning machine was a boon to foundrymen because it enabled them to offer their customers a finished product. In addition to tumbling mills for small castings, the sand blast was developed for larger work by R.E. Tilghman of Philadelphia, Pennsylvania, in 1870 (see the article "Processing of Castings" in this Volume). The use of such equipment did not begin to broaden, however, until about 1900, when it was installed at the Logan Manufacturing Company at Phoenixville, Pennsylvania. Of the rapid improvements made since then by equipment firms, mention should be made of the American Foundry Equipment Company (now the Wheelabrator-Frye Company of Mishawaka, Indiana, and the Pangborn Corporation of Hagerstown, Maryland), whose constant innovations and engineering technology have added much to the ease and efficiency with which foundries handle many cleaning room operations.

In the 19th century, it was common for the pouring ladle to be hoisted by a jib crane located beside the furnace. The molds, arranged in a semicircle, were poured by swinging the crane from mold to mold. Modern overhead cranes have revolutionized the handling of the molten steel. Geared safety ladles (Fig. 5) were designed and built by James Nasmythe in 1867. Prior to this, bull ladles were tipped by a number of men applying leverage on large horizontal arms. Hand-shanked ladles made their appearance about the same time as the geared tilt ladles.

In the final quarter of the 19th century, industrial growth in the United States exceeded all previous experience. The mass production of machines, the new consumerism, the proliferation of steel-framed buildings, and the spread of electric power and telegraph networks all created an appetite for metals and in turn placed increasing demands on the casting industry.

The 20th Century

The 20th century began without any indication of the dramatic changes that computers and automation would bring about by the 1960s. The changes in equipment and methods would be quite obvious. As the 20th century began, the average U.S. foundry poured more tonnage than was cast throughout the world when the Nation was born. Despite so striking a transformation in the industry, man was called upon to expend far less energy. With minimal physical effort, workers produced increasingly sophisticated shapes in less time for increasingly intricate machines. As automation took over, production rates climbed until one automated foundry in the automotive field in 1967 was able to establish a consistent production figure of 6 man-hours per ton of castings. Machines replaced the labor of man and horse, and with the sudden impetus given metal manufacturing in World War I, machinery became a necessity. Without cast metal parts, the machine age could never have existed.

The metal casting industry adopted automation and did so rapidly. Characteristically, the first fully automated plant in the United States (one of the first in the world) was a Rockford, Illinois, foundry that cast hand grenades for the U.S. Army in 1918.

The history of metal casting shows that the foundryman is as eager as any manufacturer to take full advantage of inventions and even to inspire them. America's first commercial metal caster, Joseph Jenks, was awarded the first patent in America. Thomas Edison called Seth Boyden one of America's greatest inventors, for Boyden established two basic industries in America—patent leather and malleable iron. In 1851, James Bogardus's factory in Chicago, which was constructed with cast iron supports, opened the way for what many art historians considered to be America's only original contribution to the arts of the world—the skyscraper.

By 1960, less than 1% of the foundries in operation were a century old. The trend continued as huge conglomerates entered

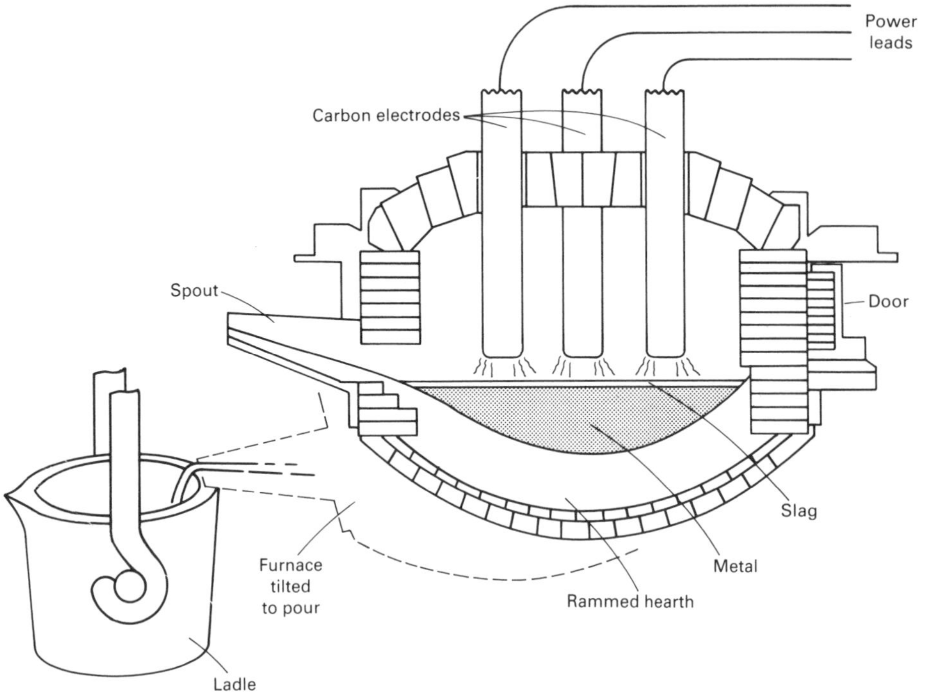

Fig. 8 Cross section of an electric arc furnace

the picture. American metal casting was big business. After a walking tour, Walt Whitman described the Nation: "Colossal foundry, flaming fires, melted metal, pounding trip hammers, surging crowds of workmen shifting from point to point, waste and extravagance of material, mighty castings; such is a symbol of America."

In the first year of the new century, foundries in the United States poured more open hearth steel than those in the United Kingdom—almost as much as the rest of the world combined. As far back as 1864, the military foundry at Old Fort Pitt (Pittsburgh) had cast a 510 mm (20 in.) smoothbore Rodman cannon weighing 52 Mg (115 000 lb). This was a hundred times bigger than the famed Urban Gun of Muhammad II that was used to fell the walls of Constantinople in 1453. Three years later, the Krupp plant in Essen, Germany, poured a 45 Mg (50 ton) cast steel cannon, and the fate of the French army in the War of 1870 was sealed.

Casting Markets. The largest consumer of metal castings, however, was not the military but the automobile industry, which in 10 years provided a greater incentive to metal casting than cannons, bells, and the steam engine had in a century and a half. Approximately 25% of all castings produced in this century have been component parts for automobiles, trucks, and tractors.

In 1924, Henry Ford made 1 million automobiles in 132 working days. Casting knowledge and the world's first mass production concept were vital to this phenomenal production increase. Automobile output in the first 10 years of the 20th century increased 3500%, with a corresponding increase in demand for castings. The mass production of trucks, tractors, and other mechanized farm and industrial equipment also heightened the demand for castings. This was followed in rapid succession by parts for such mushrooming industries as refrigeration (1930s); aviation (1940s); air conditioning (1950s); and data processing, electronics, and aerospace technology (1960s). Cast metals played a vital role in each. Major markets for castings are reviewed in the following article "Casting Advantages, Applications, and Market Size" in this Volume.

Foundry Organizations. Metallurgy began to achieve prominence in 1889 when nickel was alloyed to make a stronger steel. Although the science of metallurgy is now recognized as the basis of sound metal casting technology, in the beginning it was welcomed only by the more advanced foundry owners interested in the continuing benefits to be achieved by accepting a new technology. A group of these enterprising foundry owners arranged to form a number of industry-sponsored organizations dedicated to metal science and research and development, which, its leading members realized, could be turned to commercial advantage.

The American Foundrymen's Association (since 1948, the American Foundrymen's Society) was formed in 1896 out of the Foundrymen's Association of Philadelphia, itself only 3 years old. The New England Foundrymen's Association was formed that same year, and 1 year later the American Malleable Casting Association (changed 30 years later to the Malleable Iron Research Institute and in 1934 to the Malleable Founder's Society) was formed. In 1900, the Carnegie Research Scholarships of the Iron and Steel Institute were founded, followed by the Steel Founders' Society of America in 1902. The Foundry Equipment Manufacturers' Association (now the Casting Industry Supplier's Association) was founded in 1918, and the Gray Iron Institute was founded in 1928. Other casting associations included the American Die Casting Institute (1929), the Alloy Casting Institute (1940), the Nonferrous Founder's Society (1943), and the Foundry Educational Foundation and National Castings Council (1947). The Investment Casting Institute was founded in 1953, the Society of Die Casting Engineers in 1954, and the Ductile Iron Society in 1959. The goal of worldwide cooperation in metal casting prompted the formation in 1923 of the International Committee of Foundry Technical Associations (ICFTA, Zurich, Switzerland), which strives through 24 nations and an annual International Foundry Congress to exchange technical data.

Permanent Mold Processes. Developments in molding logically included the use of permanent molds, although the permanent mold preceded the loam mold and the sand mold by centuries. Subsequent types of permanent molds gradually appeared, but for many years they were limited in application by the metal available. Permanent molding can be defined simply as the pouring of liquid metal into a preheated metallic mold. As described in the article "Classification of Processes and Flow Charts of Foundry Operations" in this Volume, currently used permanent mold casting methods include die casting (high-pressure, low-pressure, and gravity), centrifugal casting (vertical and horizontal), and hybrid processes such as squeeze casting and semisolid metal casting.

The centrifugal casting process, which involves the pouring of molten metal into a rapidly rotating metallic mold, was developed by A.G. Eckhardt of Soho, England, in 1809. The method was soon adopted by the pipe foundries and was first used in Baltimore, Maryland, in 1848. Sir Henry Bessemer, famed for his converter, used centrifugal casting to remove gases and was the first to pour two or more metals into a single rotating mold. The centrifugal casting of steel was first attempted in 1898 at the plant of the American Steel Foundries in St. Louis, Missouri. Railroad car wheels were spun cast in 1901 at a rotation speed of 620 rpm.

Slush Casting. Following the early development of the centrifugal method, a permanent mold method known as slush casting was introduced. Slush casting is a process in which molten metal is poured into a split metal mold (generally made of bronze) until the mold is filled; then, immediately, the mold is inverted and the metal that is still liquid is allowed to run out. The time required for this casting operation is sufficient to freeze a metal shell in the mold, corresponding to the shape of the cavity wall. The thickness of the wall of the casting depends on the time interval between the filling and the inverting of the mold, as well as on the chemical and physical properties of the alloy and the temperature and composition of the mold. Usually lead and zinc alloy castings are produced by slush casting. The process is limited to the production of hollow castings (lamp bases are the principal product). More detailed information on slush casting can be found in Volume 5 of the 8th Edition of *Metals Handbook*.

Aluminum, the most abundant metal in the earth's crust, was a development of this century. Isolated in 1825, it derives its name from the Latin *alumen*, meaning bitterness. Aluminum was first exhibited in 1855, but for many years was so difficult to obtain that it was more costly than gold. In 1888, the Pittsburgh Reduction Company offered the metal in half-ton lots for $2 a pound and had difficulty attracting buyers and users until one manufacturer discovered it made good, inexpensive tea kettles. Within 5 years, the price decreased to 62 cents a pound, and by 1900 it was down to 32 cents per pound. In 1890, only 28 000 kg (62 000 lb) of aluminum was produced in the United States. Production was low until World War II, but by 1963, $635 934 000 worth of aluminum castings were shipped in the United States. In 1963, this industry, undreamed of in 1900, employed 35 970 people in 951 plants with a payroll of $221 567 000. In the first 7 months of 1968 alone, more than 412 000 Mg (450 tons) of aluminum were cast in the United States. The article "Aluminum and Aluminum Alloys" in this Volume contains more information on the processing and applications of aluminum alloys.

Magnesium. The development of magnesium as a casting metal parallels the history of aluminum (see the article "Magnesium and Magnesium Alloys" in this Volume). During World War I, magnesium sold in the United States for $5 a pound, and by 1935 only 170 Mg (375 000 lb) had been cast. By 1944, however, the industry was producing more than 39 000 Mg (43 tons) a year, a good portion of which was cast.

Die Casting. Manually operated casting machines were patented as early as 1849 (Sturgiss) and 1852 (Barr) in an effort to satisfy the insatiable demands of a growing reading public by way of rapidly cast lead type. These early inventions led to Ottmar Mergenthaler's Linotype, an automatic casting machine in which molten lead is forced by piston stroke into a metal mold. The first die casting machine bearing the Linotype name was patented in 1905 by H.H. Doehler. Two years later came E.B. Wagner's casting machine, a prototype of the now familiar hot chamber die casting machine. It was first used on a large scale during World War I for binocular and gas mask parts. Zinc alloys were used for die casting as early as 1907, but were not competitive until Price & Anderson developed the Zamak die casting alloy in 1929. Additional information on die casting machines can be found in the article "Die Casting" in this Volume.

Investment (lost wax) casting, one of the oldest casting techniques, was rediscovered in 1897 by B.F. Philbrook of Iowa, who used it to cast dental inlays. Industry paid little attention to this sophisticated process until the urgent military demands of World War I overtaxed the machine tool industry. Shortcuts were then needed to provide finished tools and precision parts, avoiding time-consuming machining, welding, and assembly.

Molding Sands and Equipment. The 20th century saw the refinement of processes and materials used in the foundry for over 400 years. Until the 1920s, sand testing consisted of squeezing a handful of sand to judge its ability to compact and stick together. Early in that period, a sand research committee of the American Foundrymen's Society began to develop sand test methods. By 1924, standards were established that covered the various properties of molding sands. A better understanding of molding sand technology has resulted in sands of a higher degree of uniformity being prepared for the repetitive green sand (clay-bonded) molding sand. This high degree of achievement could only be possible with the great advances in sand testing produced by the foremost researchers and developers of sand testing instrumentation. The current understanding of the fundamentals of clay mineralogy, sand preparation, sand compaction, and the physical properties of molding and core sands all contribute to the success of the modern foundry industry.

Ductile Irons and Austempered Ductile Irons. Continuing technological advances that seek to fulfill the need for materials capable of providing greater thermal, chemical, and mechanical properties have brought forth the development of new alloys and properties never believed possible. During World War II, the inoculation of gray iron became common practice, because high-quality cast irons replaced the scarcer steel in many castings. Shortly after the war, a new type of iron, variously known as spheroidal graphite cast iron, nodular iron, and (more universally) ductile iron, was patented and announced by the International Nickel Company. It was a major breakthrough in metallurgy because its high strength and ductility allow it to compete with malleable iron and, in certain applications, with steel.

If ductile iron is austenitized and quenched into a salt bath or a hot oil transformation bath at a temperature in the range of 320 to 550 °C (610 to 1020 °F) and held at that temperature, transformation to a structure containing mainly bainite with a minor proportion of austenite takes place. Irons so transformed are referred to as austempered ductile irons (ADI). Austempering generates a range of structures depending on the time of transformation and the temperature of the transformation bath. The properties are characterized by very high strength, with some ductility and toughness, and often an ability to work harden, giving appreciably higher wear resistance than that of other ductile irons. See the article "Ductile Irons" in this Volume for an extensive review of the properties of ductile irons and ADI.

Organic Binders. Since World War II, experimentation has been accelerated in organic and chemical sand binders for the thermosetting of molds and cores (see the article "Resin Binder Processes" in this Volume). Beginning with the Croning process (shell process), phenolics led the way to urea and the dielectric process and then to furans and urea-free resins. The continued development of binders for the production of chemically bonded cores and molds is being directed toward increasing productivity as well as achieving the dimensional repeatability necessary to meet the new challenges of net shape and near-net shape casting requirements (Table 2). Many patterns were made of epoxy resins and polyurethane and other expendables such as polystyrene.

Automation. By the late 1950s, it was obvious that a second industrial revolution had begun, consisting of machines manufacturing, repairing, and operating other machines through the control of elaborate electronic brains. The most dangerous and tedious jobs were relegated to robots programmed to lift, carry, and pour. Cupolas could now be charged and discharged not

Table 2 Development of core and mold processes

Process	Approximate time of introduction
Core oil	1950
Shell: liquid and flake	1950
Silicate/carbon dioxide	1952
Airset oils	1953
Phenolic acid-catalyzed no-bake	1958
Furan acid-catalyzed no-bake	1958
Furan hot box	1960
Phenolic hot box	1962
Oil urethane no-bake	1965
Phenolic/urethane/amine cold box	1968
Silicate ester-catalyzed no-bake	1970
Phenolic urethane no-bake	1974
Alumina phosphate no-bake	1977
Furan sulfur dioxide	1978
Polyol urethane no-bake	1978
Warm box	1982
Free radical cure sulfur dioxide	1983
Phenolic ester no-bake	1984
Phenolic ester cold box	1985

only mechanically but also automatically. These and other developments are outlined in the article "Foundry Automation" in this Volume.

Advancements During the Past Decade. The last decade has seen technological developments unfold at a rate never before experienced by this industry. In many cases, new technologies have been thrust upon the industry by a changing marketplace, a marketplace that is now demanding higher-quality and more cost-effective castings. Added to these demands is foreign competition, a force that is driving U.S. foundries toward new technology as a means of survival. All of these elements have changed the metal casting marketplace so drastically and at such a rapid rate that U.S. producers of cast components are diligently sifting through the many new technologies available in an attempt to find the ones that will provide the quality and productivity levels needed to compete in the world market.

Casting processes such as evaporative (lost) foam casting and semisolid casting, a scientific approach to the gating and risering of castings using computer simulation of solidification, the computer-aided design and manufacture of castings, integrated foundry systems (Fig. 9), the melting of metals using the plasma arc cupola, cast metal-matrix composites, and argon-oxygen decarburization for steel refining will all contribute to the continued advancement of metal casting. All of the aforementioned subjects are described in this Volume.

Looking ahead to the year 2000, the metal casting industry will continue to explore new technologies in the interests of achieving higher-quality castings that can meet the critical performance standards being imposed. This is demonstrated by the history of metal casting and by the striking fact that the industry has advanced further in the last 50 years than it has in the preceding 3000.

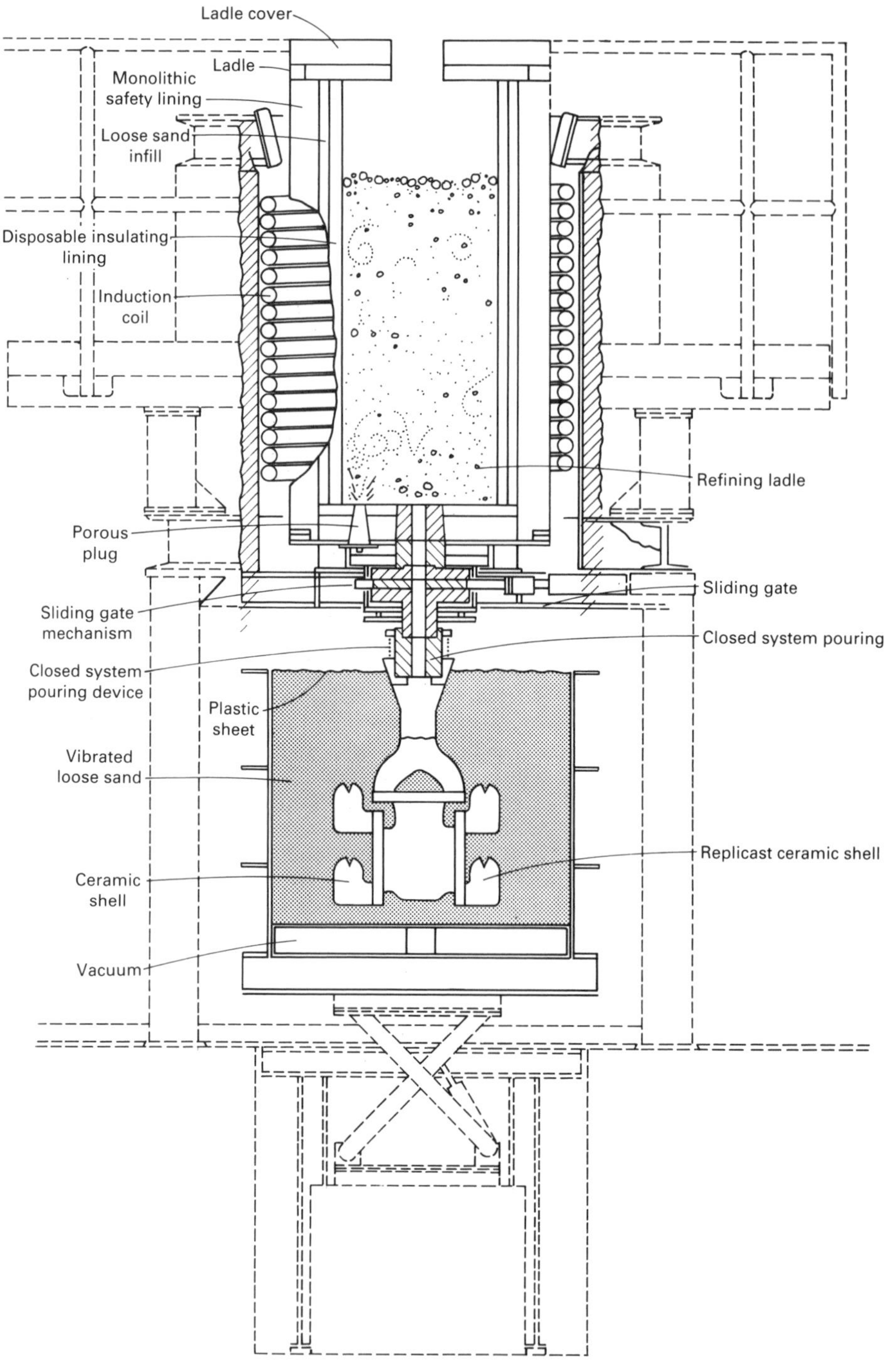

Fig. 9 Schematic showing the essential elements of an integrated Replicast casting system. See the articles "Foundry Automation" and "Replicast Process" in this Volume for additional information on integrated foundries and the ceramic shell Replicast process, respectively.

SELECTED REFERENCES

- W.H. Dennis, *100 Years of Metallurgy*, Aldine Publishing, 1964
- E. Forbes, *Paul Revere and the World He Lived in*, Houghton Mifflin, 1942
- E.N. Hartley, *A History of America's Oldest Iron and Steel Producer*, Taylor-Wharton Iron & Steel Company, 1942
- E.N. Hartley, *Iron Works on the Saugus*, University of Oklahoma Press, 1957
- E.L. Kotzin, *Metalcaster's Reference and Guide*, American Foundrymen's Society, 1972
- M. Manchester, *The Arms of Krupp*, Little, Brown, 1964
- T.A. Richard, *Man and Metals*, McGraw-Hill, 1932
- C.A. Sanders and D. Gould, *History Cast in Metal*, American Foundrymen's Society, 1976
- E.A. Schoefer, *Seventy-Five Years of Cast High Alloys*, American Society for Testing and Materials, 1982
- B. Simpson, *History of the Metalcasting Industry*, American Foundrymen's Society, 1968
- E.J. Speare, *Life in Colonial America*, Random House, 1963
- M. Whiteman, *Copper for America*, Rutgers University Press, 1971
- F.P. Wirth, *Development of America*, American Book, 1939

Casting Advantages, Applications, and Market Size

David P. Kanicki, American Foundrymen's Society

METAL CASTING is unique among metal forming processes for a variety of reasons. Perhaps the most obvious is the array of molding and casting processes available that are capable of producing complex components in any metal, ranging in weight from less than an ounce to single parts weighing several hundred tons (Fig. 1). Foundry processes are available and in use that are economically viable for producing a single prototype part, while others achieve their economies in creating millions of the same part (Fig. 2). Virtually any metal that can be melted can and is being cast.

In terms of value and volume, metal casting ranks second only to steel rolling in the metal producing industry. According to U.S. Department of Commerce statistics, metal casting remains one of the ten largest industries when rated on a value-added basis. Annually, more than 3000 U.S. foundries produce 12 to 14 million tons of castings in a variety of ferrous and nonferrous metals. The annual value of foundry products is estimated to be approaching $20 billion. This article will examine the advantages of the metal casting process, the major applications of cast components, and the technical and market trends that are shaping the foundry industry and the products it produces.

The Versatility of Metal Casting

It is estimated that castings are used in 90% or more of all manufactured goods and in all capital goods machinery used in manufacturing (Ref 1). The diversity in the end use of metal castings is a direct result of the many functional advantages and economic benefits that castings offer compared to other metal forming methods. The beneficial characteristics of a cast component are directly attributable to the inherent versatility of the casting process.

A review of Ref 2 illustrates the multifaceted nature of casting technology. Reference 2 describes in detail 38 methods for manufacturing a metal casting. These techniques are grouped into five categories. These categories, together with some of the individual processes within each group, include:

- Conventional molding processes (green sand, shell, flaskless molding)
- Precision molding and casting processes (investment casting, permanent mold, die casting)
- Special molding and casting processes (vacuum molding, evaporative pattern casting, centrifugal casting)
- Chemically bonded self-setting sand molding (no-bake, sodium silicate)
- Innovative molding and casting processes (rheocasting, squeeze casting, electroslag casting)

Most of these processes are described in detail in the Section "Molding and Casting Processes" in this Volume.

This brief sampling of casting processes illustrates the versatility currently available in the foundry industry. This diversity, in most cases, represents the continual refinements that have characterized the basic sand, ceramic, and metal mold casting methods, but others represent new approaches to producing cast metal components. Probably the most prominent example of innovative casting technology, which is receiving much attention from both producers and users, is evaporative pattern casting, often referred to as lost foam casting (Fig. 3). Lost foam processing is discussed in the article "Sand Molding" in this Volume (see the section "Unbonded Sand Molds").

It is interesting to note the number of significant developments in both molding and casting that have occurred during the past 10 years, as documented in this Handbook and in Ref 2. Some of these developments are reviewed below.

Molding Developments. Two recent developments that have proved practical and are in current use in foundries are impact molding and the Replicast process.

Impact molding can be described as a high-density green sand process. It operates on the principle of compaction by acceleration (Ref 3). By using a mixture of natural gas and/or compressed air, a controlled explosion takes place that hurls the sand grains against the pattern. The wave of energy created produces a uniformly hard yet permeable mold. Because there is little variation in mold hardness, the process is reportedly capable of producing near-net shape castings with the economies of green sand molding.

The Replicast process was developed by the Steel Castings Research and Trade Association of Sheffield, England. This process can be best characterized as a hybrid of the investment casting process (lost wax process) and evaporative pattern casting. In investment casting, a wax or plastic pattern is used to shape a ceramic shell mold, but the Replicast process utilizes an expanded polystyrene (EPS) pattern that is coated in a refractory slurry and then invested in a ceramic slurry to produce the ceramic shell (see the article "Replicast Process" in this Volume).

Although investment casting has long been recognized for its ability to produce castings with very smooth and detailed surface finishes with excellent dimensional tolerances, use of the EPS pattern is said to reduce costs by replacing the wax normally used in investment casting. One U.S. foundry that has adopted Replicast reports two major benefits (Ref 4):

- Test-design and prototype patterns can be fabricated from solid EPS to avoid die tooling costs
- For high-production die injection, expanded polystyrene has far lower material cost than wax; this is particularly important when producing larger parts

The same foundry also reports three major applications for Replicast in their operation:

- Testing casting designs and making prototypes
- Producing existing parts in an alternative alloy where the available tooling would not be suitable for the new alloy
- Making short-run replacement parts, such as replacing outdated equipment when new patterns cannot be justified

(a)

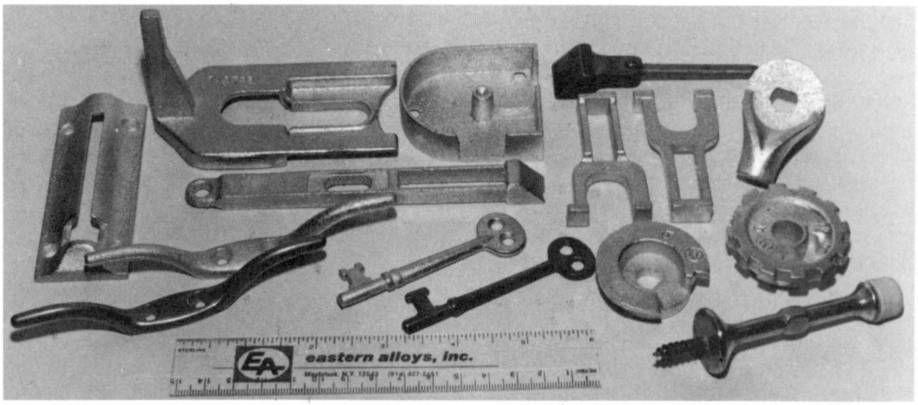

(b)

Fig. 1 The versatility of the metal casting process. (a) A 61 500 kg (135 600 lb) hot-forming die used for producing nuclear reactor pressure heads. (b) A variety of small hardware parts weighing only ounces each

Impact molding and the Replicast process serve as good examples of the continually evolving technology of metal casting. Other processes are under development. Each is aimed at meeting customer needs, offering economical alternatives to other metal forming techniques, and expanding the already wide variety of metal casting technologies.

Process Developments. Significant developments are also occurring in the area of casting operations. Two relatively new processes demonstrate the advances being made in producing clean, thin-wall ferrous and nonferrous castings. One such technique for producing aluminum castings is called the Cosworth process. The other, developed for iron and steel casting, is the FM process.

The Cosworth process was developed to meet the need for highly specialized components for the Formula One racing car engines manufactured by Cosworth Engineering, Ltd., in England (Ref 5). Zircon sand molds with a furan binder system are filled from the bottom of the mold by using an electromagnetic pump. A vertical launder is located in the middle of a holding furnace, and it moves the metal at a controlled rate into the rigid sand mold. Locating the filling tube in the middle of the furnace helps ensure that only the cleanest metal enters the mold, thus reducing the possibility of slag or dross entering the mold cavity.

With a blanket of inert gas covering the molten metal in the furnace, the molten aluminum is protected from oxygen and other gases that may lead to porosity in the casting. In addition, because the mold fill rate is closely controlled, turbulence is minimized, and this also prevents the pickup of oxygen and other gases that may lead to porosity.

Major advantages claimed for the process include yields of 85% or better, castings that are typically 10 to 12% lighter than those produced by other methods, excellent mechanical and physical properties, and the ability to specify machining allowances in the range of 1.5 to 2 mm (0.06 to 0.08 in.). Several American foundries are in the process of adopting or seriously investigating the Cosworth process for their aluminum casting operations.

The FM process has been specifically developed to produce thin-wall iron and steel castings (Ref 6). The name FM comes from *fonte mince*, meaning thin iron. Developed and used by the French firm Pont-a-Mousson, the FM process is a mold filling technique (versus mold pouring) that utilizes a controlled differential pressure to fill molds with high-melting metals, including superalloys.

Like the Cosworth process, the FM process utilizes a bottom filling technique that is controlled and yet allows for rapid filling of the mold. Any type of mold (green sand, metal, shell, and so on) can be used with the FM process. The rapid fill rates are achieved by exerting a low pressure on the liquid metal and a negative pressure on the mold and, in certain cases, on the furnace itself. Evacuation of gases from the mold is also achieved during mold filling.

Casting wall thicknesses of 2.5 to 3.0 mm (0.10 to 0.12 in.) have been obtained in gray, ductile, and alloyed cast irons and in low-carbon steels. The properties of nickel-base superalloys and high-chromium steels are also improved with this process.

Materials Developments. Research into casting materials in recent years has also produced some significant results. Among these are austempered ductile iron and aluminum-lithium investment castings.

Fig. 2 Cast iron automobile engine blocks being produced by the millions

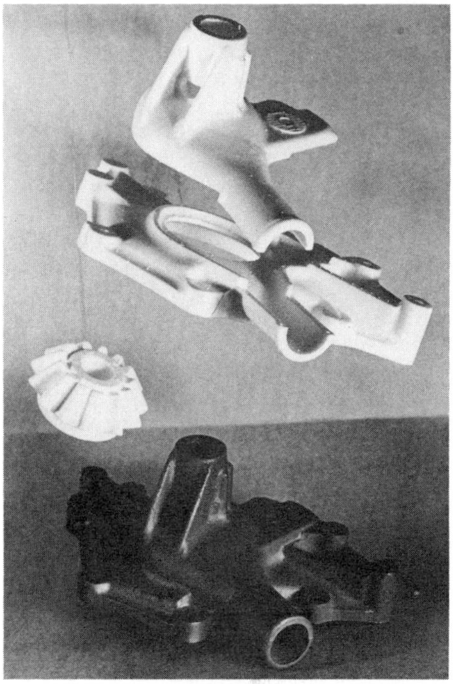

Fig. 3 Expandable polystyrene pattern and finished casting produced by lost foam casting

Austempered ductile iron, with properties that fall between those of through-hardened steels and case-carburized steels, has already begun to create new markets for cast irons because it is increasingly becoming an economical alternative to steel weldments, forgings, stampings, and other metal products (Ref 7). Some of the current applications for austempered ductile iron include gears, crankshafts, chain sprockets, stamping dies, railroad wheels, and other structural and load-bearing components (see the article "Ductile Iron" in this Volume for additional information).

Aluminum-Lithium Alloys. Research into aluminum-lithium as an investment casting material also holds promise in opening new markets to metal casting, particularly in the aircraft industry. The primary benefit of aluminum-lithium alloys is a reduction in material density, which is the major property in reducing structural weight in aircraft (Ref 8). In some applications, aluminum-lithium alloys, due to the effect of lithium on the elastic modulus of a part, exhibit an increase in stiffness of 10 to 15% and a decrease in the crack growth rate during fatigue. Research on these alloys for casting is in its early stages, but if initial findings can be confirmed and demonstrated, aluminum-lithium could lead to new aircraft markets for castings by replacing wrought aluminum-lithium products, such as forgings, extrusions, sheet, and plate.

These process and materials developments and other developments, such as the filtration of molten metals to remove oxides and other nonmetallics that can be deleterious to the final casting, represent only some of the advances currently taking place in metal casting and illustrate the versatility of the casting process. They also demonstrate the trend in the industry toward higher quality, lower costs, and new markets.

Functional Advantages

Beyond the rapidly emerging technologies that are keeping metal casting in the forefront in the metal forming industry, castings possess many inherent advantages that have long been accepted by the design engineer and metal parts user. In terms of component design, casting offers the greatest amount of flexibility of any metal forming process. The casting process is ideal because it permits the formation of streamlined, intricate, integral parts of strength and rigidity obtainable by no other method of fabrication. The shape and size of the part are primary considerations in design, and in this category, the possibilities of metal castings are unsurpassed. The flexibility of cast metal design gives the engineer wide scope in converting his ideas into an engineered part (Ref 9).

The freedom of design offered through the metal casting process allows the designer to accomplish several tasks simultaneously. These include the following (Ref 10):

- Design both internal and external contours independently to almost any requirement
- Place metal in the exact locations where it is needed for rigidity, wear, corrosion, or maximum endurance under dynamic stress
- Produce a complex part as a single, dependable unit
- Readily achieve an attractive appearance

The following list of functional advantages of castings and the metal casting process was compiled from Ref 9 and 10. These advantages illustrate why castings have been and continue to be the choice of design engineers and materials specifiers.

Rapid Transition to Finished Product. The casting process involves pouring molten metal into a cavity that is close to the final dimensions of the finished component; therefore, it is the most direct and simplest metal forming method available.

Suiting Shape and Size to Function. Metal castings weighing from less than an ounce to hundreds of tons, in almost any shape or degree of complexity, can be produced. If a pattern can be made for the part, it can be cast. The flexibility of metal casting, particularly sand molding, is so wide that it permits the use of difficult design techniques, such as undercuts and curved, reflex contours, that are not possible with other high-production processes. Tapered sections with thickened areas for bosses and generous fillets are routine.

Placement of Metal for Maximum Effectiveness. With the casting process, the optimum amount of metal can be placed in the best location for maximum strength, wear resistance, or the enhancement of other properties of the finished part. This, together with the ability to core out unstressed sections, can result in appreciable weight savings.

Optimal Appearance. Because shape is not restricted to the assembly of preformed pieces, as in welding processes, or governed by the limitations of forging or stamping, the casting process encourages the development of attractive, more readily marketable designs. The smooth, graduated contours and streamlining that are essential to good design appearance usually coincide with the conditions for easiest molten metal flow during casting. They also prevent stress concentrations upon solidification and minimized residual stress in the final casting.

Because of the variety of casting processes available, any number of surface finishes on a part are possible. The normal cast surface of sand-molded casting often provides a desired rugged appearance, while smoother surfaces, when required, can be obtained through shell molding, investment casting, or other casting methods.

Complex Parts as an Integral Unit. The inherent design freedom of metal casting allows the designer to combine what would otherwise be several parts of a fabrication into a single, intricate casting. This is significant when exact alignment must be held, as in high-speed machinery, machine tool parts, or engine end plates and housings that carry shafts. Combined construction reduces the number of joints and the possibility of oil or water leakage. Figure 4 shows a part that was converted from a multiple-component weldment into a two-part cast component.

Improved Dependability. The use of good casting design principles, together with periodic determination of mechanical properties of test bars cast from the molten metal, ensures a high degree of reproducibility and dependability in metal castings that is not as practical with other production methods. The functional advantages that metal castings offer and that are required by the designer must be balanced with the economic benefits that the customer demands. The growth of metal casting and its current stability are largely the result of the ability of the foundry industry to maintain this balancing act. The design and production advantages described above bring with them a variety of cost savings that other metal working processes cannot offer. These savings stem from four areas (Ref 10):

Fig. 4 Compressor case for a jet engine that was converted from a multiple-component weldment into a two-part cast component by a major jet engine manufacturer. The company converted all of its formerly welded cases to castings; this reduced by 27 000 the number of parts in one of its newest engines.

Table 1 Breakdown of foundries by employment and primary metal cast in 1986

Includes U.S. and Canadian plants

Employment	Gray and ductile iron	Malleable iron	Steel	Nonferrous	Total
>1000	13	1	3	6	23
500–999	23	2	17	19	61
250–499	55	2	29	60	146
100–249	200	13	95	205	513
50–99	234	8	95	282	619
20–49	304	3	115	560	982
<20	234	1	101	1326	1662
Total	**1063**	**30**	**455**	**2458**	**4006**

Source: Ref 12

- The capability to combine a number of individual parts into a single integral casting, reducing overall fabrication costs
- The design freedom of casting minimizes machining costs and excess metal
- Patterns used in casting are lower in cost compared to other types of tooling
- Castings require a comparatively short lead time for production

For these reasons and because it remains the most direct way to produce a required metal shape, metal casting will continue to be a vitally important metal forming technology. The diversity in end use in castings is also evidence of the flexibility and versatility of the metal casting process. Major casting end uses and market trends are discussed below.

Casting Market Trends and End Uses

The use of metal castings is pervasive throughout the economies of all developed countries, both as components in finished manufactured goods and as finished durable goods (Ref 1). As indicated earlier, castings are used in 90% of all manufactured goods and in all capital goods machinery used in manufacturing. They are also extensively used in transportation, building construction, municipal water and sewer systems, oil and gas pipelines, and a wide variety of other applications (Ref 11).

Industry Structure. In the broadest sense, foundries are categorized into two general groups: ferrous foundries (those that produce the various alloys of cast iron and cast steel) and nonferrous foundries (those that produce aluminum-base, copper-base, zinc-base, magnesium, and other nonferrous castings). The wide variety of ferrous and nonferrous casting alloys is extensively reviewed in the Sections "Ferrous Casting Alloys" and "Nonferrous Casting Alloys" in this Volume.

For marketing purposes, foundries are also categorized according to the nature of their operations as being either captive (producing castings for their own use) or jobbing (producing castings for sale). The market is sometimes further broken down by major casting processes when they can be readily identified or are particularly significant. This is done more often in the case of nonferrous castings; in this case, the product is usually categorized as being produced in sand, permanent mold, or die cast. Because 90% or more of all iron and steel castings are produced in some form of sand medium, distinguishing the process used is not quite as significant. The U.S. Department of Commerce produces statistics for steel investment castings.

Ferrous castings shipments are usually classified by market category. For example, iron castings are generally categorized as engineered (designed for specific, differentiated customers) and nonengineered (produced in large volumes of interchangeable units, usually consisting of ingot molds, pressure, and soil pipe).

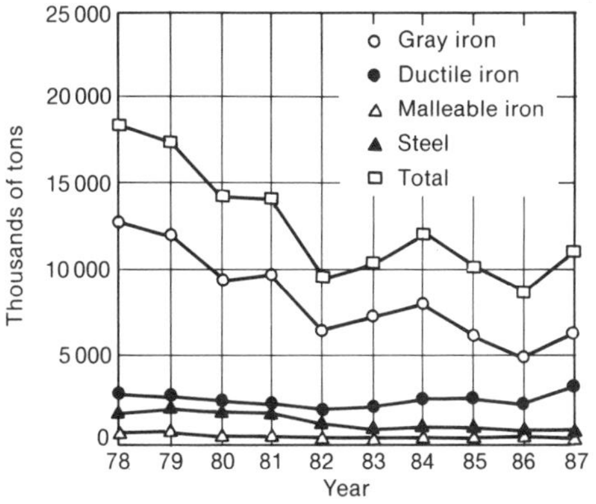

Fig. 5 U.S. ferrous casting shipments from 1978 to 1987

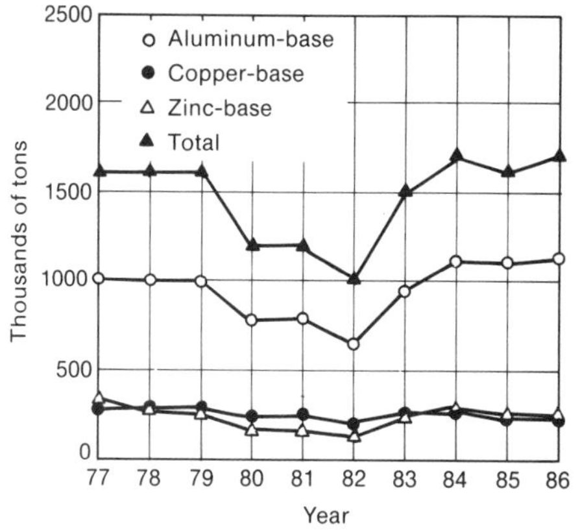

Fig. 6 U.S. nonferrous casting shipments from 1977 to 1986

The diversity among the various foundries makes it difficult to determine the exact structure of the industry. For example, it is not unusual for a single operating foundry to produce a variety of metals and alloys, both ferrous and nonferrous, in the same plant. Some also use a variety of processes in their operations. Many aluminum foundries, for instance, use both sand and permanent mold processes, and some even produce die castings in the same facility.

In addition, the foundry industry consists of a variety of large and small facilities. In terms of numbers of workers, it is estimated that nearly 80% of casting plants in the United States and Canada employ fewer than 100 persons (Ref 12). Table 1 gives a breakdown of the industry by major metal cast and employment range. Total employment in the industry is approximately 180 000.

Casting Shipments. The U.S. Department of Commerce Bureau of Census maintains the only generally available statistics on casting shipments. Statistics on most metals are available from 1943 to the present. In recent years, the accuracy of the information on gathering and reporting methodology of the Department of Commerce for casting shipments has been called into question. Part of the problem lies in the use of a standard industry code (SIC) system for gathering and reporting data.

In many cases, shipments of castings by captive foundries are reported under SIC codes other than those for foundries, because the parent company of a foundry may be categorized under a different SIC code (for example, motor vehicles—SIC 3711; agricultural equipment—SIC 3522). The SIC classifications used to report casting shipments generally include SIC 3321 (gray and ductile iron), 3322 (malleable iron), 3325 (steel foundries), and 3361-9 (nonferrous foundries and die casting plants).

For this reason, many industry observers believe that casting shipments are understated. Nonetheless, the Bureau of Census statistics are generally recognized as the best available information on the subject, and despite their shortcomings, they offer insight into the general production and shipment trends of the industry.

Ferrous castings easily make up the largest production sector of the foundry industry in terms of volume, with gray iron outdistancing the other major metals constituting this group. During the last four decades (1947 to 1986), iron and steel foundries shipped an average of 15.5 million tons of castings annually. The periods from 1947 to 1956 and 1957 to 1966 were remarkably

Table 2 Ferrous casting production from 1982 to 1986

Figures given in thousands of metric tons

Country	Year 1982	1983	1984	1985	1986
United States	8 651 455	9 313 866	10 856 200	9 991 393	8 564 725
Japan	5 735 001	5 515 999	5 128 896	5 240 261	4 860 591
West Germany	3 501 299	3 311 600	3 387 300	3 499 900	3 452 000
France	2 230 264	2 014 387	1 855 634	1 902 608	1 797 960
United Kingdom	1 621 700	1 549 100	1 490 500	...	1 222 000
Brazil	1 207 700	986 142	...	1 407 324	1 543 147
Canada(a)	671 913	781 722	981 730	856 077	865 777
Korea	639 000	762 000	805 000	848 000	934 000
Taiwan	401 990	492 150	599 122	637 660	778 930

(a) Canada's tonnage is estimated to be 75% of actual. Source: Ref 13

Table 3 Major markets for metal castings

Ranked in order of tonnage shipped. In some cases, the total of "Other major markets" is larger as a whole than the individual markets listed.

Ferrous castings		Nonferrous castings	
Gray iron Ingot molds Construction castings Motor vehicles Farm equipment Engines Refrigeration and heating Construction machinery Valves Soil pipe Pumps and compressors Pressure pipe	**Ductile iron** Pressure pipe Motor vehicles Farm machinery Engines Pumps and compressors Valves and fittings Metalworking machinery Construction machinery	**Aluminum** Auto and light truck Aircraft and aerospace Other transportation Engines Household appliances Office machinery Power tools Refrigeration, heating, and air conditioning	**Copper-base** Valves and fittings Plumbing brass goods Electrical equipment Pumps and compressors Power transmission equipment General machinery Transportation equipment
Other major markets include machine tools, mechanical power transmission equipment, hardware, home appliances, and mining machinery, oil and natural gas pumping and processing equipment	Other major markets include textile machinery, wood working and paper machinery, mechanical power and transmission equipment, motors and generators, refrigeration and heating equipment, air conditioning	Other major markets include machine tools, construction equipment, mining equipment, farm machinery, electronic and communication equipment, power systems, motors and generators	Other major markets include chemical processing, utilities, desalination, petroleum refining
Malleable iron Motor vehicles Valves and fittings Construction machinery Railroad equipment Engines Mining equipment Hardware	**Steel** Railroad equipment Construction equipment Mining machinery Valves and fittings General and special industrial machinery Motor vehicles Metalworking machinery	**Magnesium** Power tools Sporting goods Anodes Automotive	**Zinc** Automotive Building hardware Electrical components Machinery Household appliances
Other major markets include heating and refrigeration, motors and generators, fasteners, ordnance, chains, machine tools, general industrial machinery	Other major markets include steel manufacturing, spring goods, heating and air conditioning, recreation equipment, industrial material handling equipment, ships and boats, aircraft and aerospace	Other major markets include office machinery, health care, aircraft and aerospace	Other major markets include scientific instruments, medical equipment, radio and television equipment, audio components, toys, sporting goods

stable; annual shipments averaged 15.0 and 15.3 tons per year for the 20-year period. The following decade (1967 to 1976) saw a significant upturn as the average annual shipments of iron and steel castings rose to slightly more than 17.5 million tons. Iron casting shipments have averaged about 14 million tons per year during the last four decades.

The last 10 years (1977 to 1986) were trying times for American foundries. Ferrous casting shipments rose to more than 17 million tons in 1978 before falling about 9.5 million tons by 1982. Between 1983 and 1985, shipments hovered between 10 and 12 million tons. Shipments dropped again in 1986, slipping to about 8.6 million tons that year. During the decade in question, ferrous casting shipments averaged about 13 million tons per year, with iron castings accounting for nearly 12 million. Figure 5 shows ferrous casting shipments from 1978 to 1987.

Nonferrous casting shipments have followed a similar path since 1947. Copper-base casting shipments averaged nearly 500 000 tons per year between 1947 and 1956, but between 1977 and 1986, they slipped to about an average of 300 000 tons annually. Aluminum casting shipments, on the other hand, have showed slow, steady growth since the early 1950s. Between 1947 and 1956, an average of 290 000 tons of aluminum castings were shipped per year. During the last decade, this has risen dramatically to an annual average of more than 950 000 tons per year.

In general, total nonferrous casting shipments (aluminum, copper-base and zinc) averaged 1.05 million tons annually between 1947 and 1956; 1.32 million tons per year from 1957 to 1966; 1.83 million tons annually from 1967 to 1976; and 1.44 million tons during the last decade. Figure 6 illustrates nonferrous casting shipments between 1977 and 1986.

Most recently, ferrous castings have shown good improvement, increasing from 8.6 million tons shipped in 1986 to 10.9 million in 1987. Nonferrous shipments remained relatively stable, with 1.3 million tons shipped through ten months of 1987, about par for the same period in 1986.

Metal Casting Worldwide. In the case of production of both ferrous and nonferrous castings, U.S. producers are considered the world leader (excluding the Soviet Union) (Ref 1). World production of metal castings exceeded 51 million short tons in 1982, with the United States accounting for 10.5 million tons, or about 20%. This was down from the 27% share the United States enjoyed in 1979. At the same time, Japan increased its share of the world casting market from 11 to 15%. West Germany and Italy each increased their market shares between 1979 and 1982 by 1%, to 8.4 and 4.5%, respectively.

Since the end of the 1970s, casting production around the world has shown a significant decline, generally reflecting the state of manufacturing. World production of ferrous and nonferrous castings declined from 69.8 million short tons in 1979 to 51.6 million tons in 1982, a drop of about 26%. Ferrous castings fell some 27%, from 64.6 million tons in 1979 to 47.3 million tons in 1982. Nonferrous castings fared slightly better during this period, slipping about 18%, from 5.2 to 4.3 million tons. Table 2 lists the casting production trends of nine major casting producing countries between 1982 and 1986, the 5-year period after the U.S. Trade Commission conducted its study (Ref 1).

Casting End Uses. Metal castings have a great variety of end uses and are therefore largely taken for granted by the consuming public. Castings are often the hidden components of the machines and other equipment used on a daily basis, such as automobiles, lawn mowers, refrigerators, stoves, typewriters, and computers. Only in rare cases does a consumer make a conscious decision to buy a cast product unless it is readily identifiable, as in the case of cast iron or aluminum cookware, cast iron bathtubs, or ornamental products such as cast bronze sculptures.

Designers of industrial equipment and machinery, on the other hand, have long

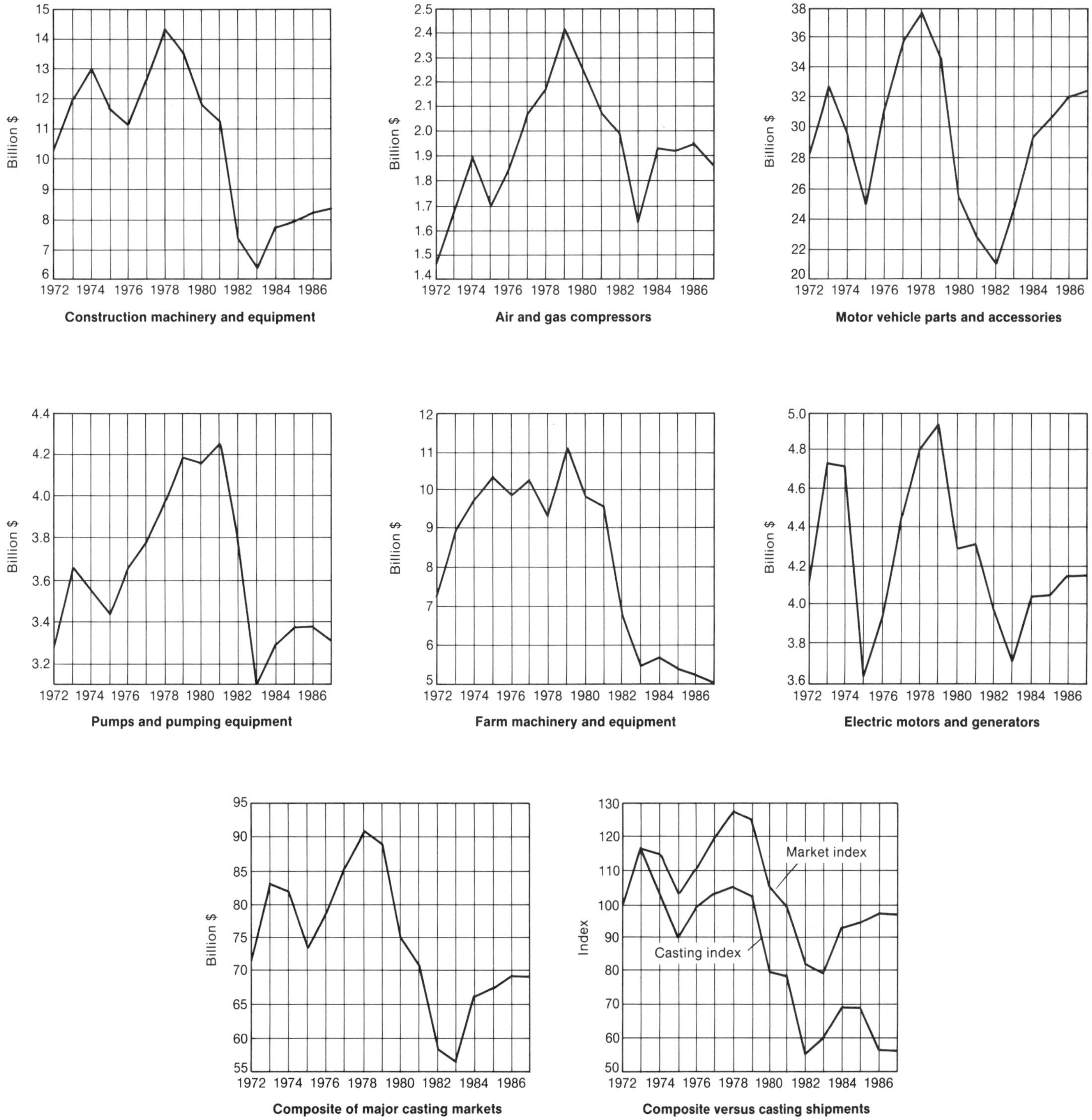

Fig. 7 Market trends of seven major casting end users from 1972 to 1987. Dollar amounts given in constant dollars. The composite market represents 15 major segments and includes the markets shown in this figure as well as such markets as valves, internal combustion engines, and heating equipment. Source: American Cast Metals Association

recognized the performance qualities of castings and regularly specify their use. These functional and economic advantages were described earlier. The major markets for cast products are listed in Table 3.

Casting Market Trends. The production of castings, as well as the markets for metal castings, has been affected by a number of factors during the last decade. Because castings are usually produced as components for machinery and equipment, shipments and the relative health of the foundry industry are closely aligned with the level of business of its major customer markets. The condition of U.S. foundries is indicated in Fig. 7, which shows the trends of seven important casting markets between 1972 and 1987.

Among the major factors that have affected the production of castings is the new and rapidly evolving technology in both processes and materials. Cast iron ingot molds, for example, which traditionally represented one of the largest tonnage markets for gray iron, are being steadily reduced to low tonnages because of the use of the efficient continuous casting process in steelmaking.

Table 4 General trends in major metal casting markets

Declining markets		Growth markets	
Gray iron motor vehicle		**Ductile iron**	
Trend	Decline from 3.4 million tons in 1979 to 2 million by 2000	Trend	Growth from 0 in 1948 to 3.1 million tons in 1979; forecast growth to over 4 million tons by 2000
Factors	Demand for less fuel consumption has meant smaller cars, smaller engines, lighter parts, near-net shape parts. Gray iron per car to decline from 600 to 300 lb	Factors	Replacing malleable iron in most market segments and some major steel casting products. Increased usage of austempered ductile iron as a replacement for carburized steel. Slight loss in passenger car utilization from 200 to 190 lb per car due to loss of differential carrier and case to front-end drive and light engines. An increase in export of pressure pipe to third world countries is expected.
Other possible trends	Intake manifold to aluminum and possibly a polymer or magnesium. Transmission case to aluminum. Head and block partially to aluminum. Camshaft partially to powderd metals and/or steel. Exhaust manifold to ductile iron and possibly fabricated stainless steel	**Aluminum castings**	
Trend	Offshore producers now supply 30% of U.S. demand. We have become offshore dependent. A domestic supply shortage could be a reality by 1989, a 9 million domestic car year.	Trend	Growth from 1.1 million tons in 1979 to 1.85 million tons in 2000
Trend	Major captive foundries will probably concentrate on blocks and heads.	Factors	Replacing iron castings in passenger cars and light trucks—now 90 lb per car, forecast to be 140 lb per car. Die castings, permanent mold castings, and sand cast aluminum are all expected to show growth. Evaporative pattern casting will show greatest growth, primarily in intake manifold and engine head.
Ingot molds		**Investment castings**	
Trend	Decline from 3.5 million tons in 1979 to 500 000 tons in 1995	Trend	Growth from 110 000 tons in 1985 to 250 000 tons annually in 2000
Factors	Continuous casting of steel eliminates the ingot. Offshore competition for steel has reduced the raw steel production in U.S. from 136 million tons to 90 million tons. Ingot mold consumption of 52 lb/ton of raw steel production is reduced to 14 lb.	Factors	Automotive applications will spur growth because of near-net shape capabilities. Commercial growth will come in valve and pump applications, but automation will be required.
Iron municipal castings		**Magnesium die castings**	
Trend	Decline from 1 million tons to 500 000 tons per year	Trend	Growth from 10 000 tons to 110 000 tons per year
Factors	Offshore competition, now 30% of demand, has been major factor. Price reductions have caused foundry closings.	Factors	New automotive usage is to provide gain if price of magnesium is competitive. Its strength-to-weight ratio is a major advantage.
Malleable iron		**Corrosion-resistant steels**	
Trend	Decline from 700 000 tons in 1979 to 50 000 tons annually	Trend	Growth from 40 000 tons in 1985 to 90 000 tons in the year 2000
Factors	Lost markets to ductile iron, aluminum and plastics. Only major usage will probably be for small electrical parts and pipe fittings.	Factors	Key factor is a recovery in the processing industries—food, desalination, petrochemical.
Steel railroad		**Machined castings**	
Trend	Decline from 1 million tons of steel casting shipments annually to 500 000 tons; had dropped to near 200 000 tons in 1983	Trend	Shipments of fully machined cast parts from noncaptive foundries are forecast to grow to 500 000 tons per year in aluminum, iron, and steel.
Factors	Freight car production declined from 90 000 in 1979 to 6000 in 1983; return to 50 000 cars is forecast	Factors	Drive by casting consumers to reduce fixed costs and break even

Source: International Metalcasters' Council

On the positive side, the increasing demand for near-net shape components is stimulating the use of investment casting, permanent molding, and other precision casting methods because of their inherent capability to reduce or minimize secondary operations, such as machining. Continued developments in directional solidification and single-crystal technology, squeeze casting, and semisolid metal casting could also contribute to increased casting production (see the article "New and Emerging Processes" in this Volume).

New materials developments are also affecting casting markets. One example is the use of plastic pipe and plumbing parts where gray iron and copper-base castings were once dominant. Other, more recent advances in engineered materials, such as polymers, resin- and metal-matrix composites, and ceramics, also hold long-term prospects for challenging the traditional metal casting and other metal forming markets. It should be noted, however, that some metal-matrix components are produced by casting processes (see the article "Cast Metal-Matrix Composites" in this Volume).

Economic and political trends have also played a role in the production of castings, as well as in all of manufacturing. The decision by the U.S. Congress in the late 1970s to set gas mileage requirements for domestically produced automobiles involved ramifications that are still being dealt with. This legislation led to a decade of the downsizing of U.S. cars. To attain the 27.5 miles/gal. (~12 km/L) goal set by Congress, U.S. automakers worked to reduce the weight of their cars. This led to lower total poundage of iron and steel, including castings. Engine blocks, heads, and other power train components—all important casting applications—were reduced in size and weight; in some cases, parts were designed out of newer automobiles. Aluminum castings were sometimes used in place of ferrous alloys.

The influx of imported castings and products that use castings has affected domestic foundries. This trend is considered to be long-term because many overseas casting operations have become firmly entrenched in U.S. markets.

Although U.S. producers of metal castings are confronted with declines in several important markets, prospects for growth are also visible. Both declining and growing markets for metal castings, as well as some of the factors affecting growth and decline, are given in Table 4.

The many factors that have influenced casting production in the United States have also affected the number of operating foundries in this country. In 1955, some 6000 foundries were operating. Today, that number stands at about 3400. A continued shakeout of marginal foundry operations is anticipated through the end of the century. However, although the base of operating foundries has decreased, American foundries have become bigger and more productive. In 1955, U.S. foundries shipped an average of 3000 tons per year; average shipments have since risen to nearly 5600 tons per foundry annually.

The last decade has been particularly tumultuous for the metal casting industry. Many of the changes that have taken place have created permanent structural shifts in foundry markets and the metal casting industry itself. These changes have caused foundries to respond with higher-quality products and innovative materials and technologies, as revealed in this article and throughout this Volume.

REFERENCES

1. "Competitive Assessment of the U.S. Foundry Industry," USITC Publication 1582, U.S. Department of Commerce, Sept 1984, p xiii
2. *Metalcasting and Molding Processes*, American Foundrymen's Society, 1981
3. G. Leslie and V. Whicker, Impact Molding Gives Deere Foundry a World-Class Edge, *Mod. Cast.*, Oct 1987, p 19-23
4. "Replicast Capability Adds Value to Foundry Offerings," News Release, Stainless Foundry & Engineering, Inc., 1987
5. D. Randall, Cosworth Process: Low-Turbulence Way to Cast High Integrity Aluminum, *Mod. Cast.*, March 1987, p 121-123
6. R. Bellocci, FM: A New Iron and Steel Casting Process, *Mod. Cast.*, Dec 1987, p 26-28
7. K. Miska, ADI Development Registers Steady Progress, *Mod. Cast.*, June 1986, p 35-39
8. T.G. Haynes, A.M. Tesar, and D. Webster, Developing Aluminum-Lithium Alloys for Investment Casting, *Mod. Cast.*, Oct 1986, p 26-28
9. *Steel Castings Handbook*, 5th ed., 1980, p 3-2
10. *Iron Castings Handbook*, 3rd ed., 1981, p 38
11. *Iron and Steel Castings: Industry Trends and End-Use Patterns*, U.S. Department of Commerce, 1986
12. Metalcasting Industry Census Guide, *Foundry Mgmt. Technol.*, April 1987
13. Census of World Casting Production, *Mod. Cast.*, American Foundrymen's Society
14. K.H. Kirgen, "The Metalcasting Industry: In a Hostile Arena," Paper presented at a meeting of the Casting Industry Suppliers Association, Rancho Mirage, CA, March 1986

Principles of Liquid Metal Processing

Section Chairman:
Seymour Katz, General Motors Research Laboratories

Introduction

Seymour Katz, General Motors Research Laboratories

THIS SECTION is intended to provide fundamental information on presolidification phenomena. Because much of the information is in the realm of physical chemistry, the first article provides a theoretical introduction to the subject. The next two articles provide pertinent thermochemical data for ferrous and nonferrous (aluminum and copper) casting alloys. The last three articles deal with important presolidification phenomena, composition control (alloy addition and impurity removal), and casting defects with origins in the liquid state (gas defects and inclusions).

In order to assist the reader, Table 1 summarizes the nomenclature used in the various articles in this Section. Additional nomenclature used throughout this Volume can be found in the tabulated Abbreviations and Symbols, which precedes the Index.

Table 1 Nomenclature used in this Section

a_i	Activity of component i relative to pure material standard state
a_i°	Activity of component i in the pure material standard state
A	Interfacial surface area
A_0	Surface area of an alloy particle
C_i	Concentration of component i
C_{P}	Heat capacity
C_{S}	Sulfide capacity
CE	Carbon equivalent
d_{p}	Particle diameter
D	Diffusion coefficient
DR	Desulfurization ratio
e_i^j	Interaction coefficient for 1 wt% standard state
E	Cell potential
f_i	Activity coefficient for 1 wt% standard state
F^*	Integral free energy of solution
F	Faraday's constant
G_i	Gibbs free energy of component i
ΔG	Integral molar free energy
$\Delta \overline{G}_i$	Partial molar free energy of component i
ΔG_i°	Molar free energy of component i
$\Delta \overline{G}^{xs}$	Excess partial molar free energy of component i
$\Delta \hat{G}_i^\circ$	Free energy of component i in solution at the 1 wt% standard state
ΔG_i° (% i)	Free energy of solution from pure material to 1 wt% standard state
G_i° (X_i)	Free energy of solution from the pure material to the hypothetical pure material standard state
h	Convective heat transfer coefficient
h_i	Activity of component i relative to the 1 wt% standard state
H_i	Molar enthalpy of component i
ΔH_i°	Standard enthalpy of formation
$\Delta \overline{H}_i$	Partial molar enthalpy of component i
$\Delta \overline{H}_i^{xs}$	Excess partial molar enthalpy of component i
J	Mass flux
k, k', k_{B}	Rate constants
k_{A}	Constant related to the absorption of an element on a surface
k_{H}	Constant related to hydrogen solubility in iron
k_{m}	Mass transfer coefficient
$K_{\mathrm{m_N}}$	Liquid-phase mass transfer coefficient
K	Equilibrium constant
m, m'	Solubility factors giving change in carbon solubility in Fe-C-X from unit mass addition of a third element
M_i	Molecular weight of component i
n_i	Moles of component i
N_i	Mole fraction of component i
p_i	Pressure of component i in the gas phase
p_i°	Pressure of component i in its standard state
P	Constant pressure conditions
$P^{\mathrm{S}}, P^{\mathrm{M}}$	Partition coefficient. Ratio alloy element concentration between liquid and solid phases in equilibrium. The superscripts S and M refer to stable and metastable products of Fe-C solidification, respectively.
Q	Heat flux
$\dot{Q}$	Gas flow rate
r, r_0	Radius of a rod
R	Universal gas constant
S	Ratio of surface area to volume
S_{c}	Ratio carbon concentration: iron alloy melt to eutectic
S_{r}	Rectified saturation degree, that is, weight fraction Fe-C eutectic formed on solidification of hypoeutectic iron
$\Delta \overline{S}_i$	Partial molar entropy of component i
$\Delta \overline{S}_i^{xs}$	Excess partial molar entropy of component i
ΔS_i°	Molar entropy of component i in the pure material standard state
t	Time
t	Celsius temperature
t_{m}	Melting time
T	Kelvin temperature
V	Gas flow rate
V'	Gas volume per tonne of metal
V_0	Volume of an alloy particle
w	Mass fraction: desulfurization to iron
W	Weight of melt
X_i	Mole fraction of component i

Greek symbols

α	Phase designation
α_{ij}	Alpha activity coefficient function for Darken's quadratic formalism
γ_i	Activity coefficient of component i relative to pure material standard state
γ_i°	Henry's law constant
δ	Boundary layer thickness
ϵ_i^j	Interaction coefficient; pure material standard state
θ	Fraction of surface sites covered by surface-active atoms
κ	Empirical constant indicating iron compositions that have the same microstructure at a given cooling rate
λ	Particle shape factor
ν_i	Stoichiometric coefficient for component i in a reaction
ρ	Density

Superscripts

B	Bulk property
e, equil	Equilibrium condition
f	Final condition
ℓ	Liquidus concentration
s	Solidus concentration
S	Surface condition
*	Interface condition

Principles of Physical Chemistry

Seymour Katz, General Motors Research Laboratories

TWO SUBJECTS that underlie all foundry processes, from melting to heat treatment, are chemical thermodynamics and chemical kinetics. Relationships derived from the disciplines are extremely useful in providing a quantitative framework for explaining and predicting chemical behavior. This includes all the chemical processes involved in preparing liquid metal for casting (going back to the winning of ores, if desired), the chemical processes involved in making molds and cores, the casting and solidification processes, the microstructural development of alloys, and the phase changes that occur after casting (for example, by heat treatment).

Thermodynamics defines the most stable chemical system (equilibrium system) that can be produced, given a set of starting conditions (compositions, temperature, pressure, and so on). An example is the phase diagram. It gives the stable phases that coexist at equilibrium, given the composition, temperature, and pressure.

While chemical thermodynamics is only concerned with the initial and final states of a system, chemical kinetics is concerned with how and at what rate the system is transformed from the initial to the final state. The practical importance of chemical kinetics implies that systems often fail to reach the most stable condition and that the path taken plays a major role in determining the speed of the reaction. In general, kinetics assumes its greatest importance when atomic movement is restricted, such as at low temperatures or in solids. Thermodynamics most often applies when atoms have high mobility, as in gases and liquids and at high temperatures. Another factor that determines the relative applicability of thermodynamics and kinetics is time. The less time available for a process, the more likely that kinetic processes will govern the chemical state of the system.

Chemical Thermodynamics

There are three thermodynamic laws that provide the basis for the relationships that describe chemical systems. These laws hold that perpetual motion and absolute zero temperature can never be achieved. The relationship between the thermodynamic laws and, for example, phase diagrams is not immediately apparent. However, the connection exists and is testimony to a great deal of brilliant theoretical work carried out over the last 100 years.

For metallurgical systems, the most important thermodynamic variables are enthalpy and Gibbs free energy. The former governs the disposition of heat, and the latter governs chemical equilibrium.

Enthalpy and Heat Capacity

The heat content of a material is termed its enthalpy H. Because the value of the enthalpy is proportional to the amount of material present, H is usually tabulated as enthalpy per unit mass. Common units of H are joules/mole, calories/gram, or Btu/pound. Thus, the total enthalpy of n_i moles of material i is:

$$(H_i)_{\text{total}} = n_i H_i \qquad \text{(Eq 1)}$$

For a mixture of materials, the total enthalpy is the sum of the individual enthalpies:

$$H_{\text{total}} = \sum_i n_i H_i \qquad \text{(Eq 2)}$$

The heat capacity C_P, given in joules/mole · K, and so on, is a measure of how H varies with temperature and is defined as:

$$C_P = \left(\frac{\partial H}{\partial T}\right)_P \qquad \text{(Eq 3)}$$

where the subscript P indicates that the values refer to constant pressure conditions.

Enthalpy and heat capacity govern the thermal state of a system and the heat generated or absorbed by chemical reactions. These variables are particularly important for establishing the energy requirements for processes, for determining rates of solidification, and for heat treatment processes. Enthalpy and heat capacity data are tabulated in Ref 1.

Gibbs Free Energy

Conditions for Equilibrium. Chemical equilibrium is a condition in which the chemical components of a system have no tendency to change. In mechanical systems, equilibrium is understood as the condition in which the potential energy is at a minimum. For chemical systems at constant temperature and pressure, an analogous statement for chemical equilibrium is that the Gibbs free energy is at a minimum, or equivalently from calculus:

$$dG(P,T,n_i\ldots) = 0 \qquad \text{(Eq 4)}$$

For systems in which chemical change can occur, dG can be expressed in partial differential form:

$$dG = \left(\frac{\partial G}{\partial T}\right)_{P,n_i\ldots} dT + \left(\frac{\partial G}{\partial P}\right)_{T,n_i\ldots} dP + \left(\frac{\partial G}{\partial n_i}\right)_{T,P,n_j\ldots} dn_i + \ldots \qquad \text{(Eq 5)}$$

where n_i stands for moles of component i. The term $(\partial G/\partial n_i)_{T,P,n_j}, \ldots$ is called the partial molar free energy $\overline{G}_i$, or the chemical potential.

It is clear from Eq 4 and 5 that, at equilibrium and at constant temperature and pressure, the following relationship exists:

$$0 = \overline{G}_i\, dn_i + \overline{G}_j\, dn_j + \ldots \qquad \text{(Eq 6)}$$

When the change in composition is dictated by a chemical reaction, stoichiometry provides a relationship between dn_i, dn_j, and so on, and Eq 6 can suitably be transformed to:

$$0 = \nu_i \overline{G}_i + \nu_j \overline{G}_j + \ldots \qquad \text{(Eq 7)}$$

where ν_i, ν_j, and so on, are the stoichiometric coefficients in the chemical reaction equation. For products and reactants, the stoichiometric coefficients are positive and negative numbers, respectively. As an example, for the chemical reaction:

$$4Al + 3O_2 = 2Al_2O_3 \qquad \text{(Eq 8)}$$

Equation 7 becomes:

$$0 = 2\overline{G}_{Al_2O_3} - 4\overline{G}_{Al} - 3\overline{G}_{O_2} \qquad \text{(Eq 9)}$$

Tabulation of Thermodynamic Data. To make thermodynamic calculations convenient, a system was developed for tabulating thermochemical data so that relatively small amounts of data could describe a wide range of chemical conditions. This was done by defining special conditions called standard states. Thus, for a chemical compound i that is an ideal gas at pressure p_i, $\overline{G}_i$ is now written:

$$\overline{G}_i = G_i^\circ + RT \cdot \ln\left(\frac{p_i}{p_i^\circ}\right) \qquad \text{(Eq 10)}$$

where G_i° is the free energy per mole of i in its standard state and p_i° is the pressure of i in its standard state. The second term on the right-hand side of Eq 10 is the free energy change to take a mole of i from its standard state p_i° to p_i. For gases, the usual standard state is $p_i^\circ = 1$ atm. Thus, p_i° is usually omitted in writing Eq 10.

Both $\overline{G}_i$ and G_i° are absolute values that cannot be measured. Changes in these variables can be measured from an arbitrary baseline. By convention, the baseline is taken as the elements in their standard states for which $(G_i^\circ)_{\text{elements}} = 0$ is defined at all temperatures. To indicate that the values are relative to a baseline, Eq 6, 7, and 10 are rewritten by replacing $\overline{G}_i$ and G_i° with $\Delta\overline{G}_i$ and ΔG_i°, respectively. Values of ΔG_i° for many compounds (called the standard free energy of formation) and phase changes have been tabulated (Ref 2, 3). Because ΔG_i° is a function of temperature, the most common tabulations are based on another thermodynamic definition of ΔG_i°, that is:

$$\Delta G_i^\circ = \Delta H_i^\circ - T\,\Delta S_i^\circ \qquad \text{(Eq 11)}$$

where ΔH_i° and ΔS_i° are the standard enthalpy and the entropy of formation, respectively. The tabulations provide ΔH_i° and ΔS_i° data. The value of ΔG_i° is calculated for any temperature using Eq 11.

Equilibrium Constant. Substituting Eq 10 into Eq 7 gives the most important relationship governing reaction equilibrium:

$$\sum_i \nu_i \cdot \Delta G_i^\circ = -RT \sum_i \ln p_i^{\nu_i} \qquad \text{(Eq 12)}$$

Here, the thermodynamic data on the left are related to actual pressures (concentrations) on the right. Equation 12 is ordinarily written in abbreviated form as:

$$\Delta G^\circ = -RT \cdot \ln K \qquad \text{(Eq 13)}$$

where K is the equilibrium constant.

The form of K is illustrated in the following. Given the gas phase chemical reaction:

$$2H_2 + O_2 = 2H_2O \qquad \text{(Eq 14)}$$

the equilibrium constant is:

$$K_{14} = \frac{p^2_{H_2O}}{p^2_{H_2} p_{O_2}} \qquad \text{(Eq 15)}$$

Activity and Condensed Phase Equilibrium. Metallurgical applications are primarily concerned with condensed phases (solids and liquids), which do not behave in the manner of an ideal gas. That is, for a condensed material, ΔG_i° does not vary with the logarithm of the concentration as might be inferred from Eq 12, where, for an ideal gas, ΔG_i° varies directly with $\ln p_i$. To preserve the useful equilibrium constant expression for nonideal, condensed phases, Eq 10 and 12 are retained in form but rewritten to enable the use of data expressing nonideality. Thus:

$$\Delta\overline{G}_i = \Delta G_i^\circ + RT \cdot \ln\left(\frac{a_i}{a_i^\circ}\right) \qquad \text{(Eq 16)}$$

$$\sum_i \nu_i\, \Delta G_i^\circ = -RT \sum_i \ln a_i^{\nu_i} \qquad \text{(Eq 17)}$$

The difference between Eq 16 and 10 and Eq 17 and 12 is that a term a_i, called activity, replaces the pressure term. The activity is a function of concentration but usually not a simple function of concentration. Analogous to the case with gases, the activity in the standard state a_i° is taken as unity. The equilibrium constant is now defined in terms of activity.

$$\ln K = \sum_i \ln a_i^{\nu_i} \qquad \text{(Eq 18)}$$

Activity Coefficients—Departure From Ideality. Common expressions for activity are:

$$a_i = \gamma_i X_i \qquad \text{(Eq 19)}$$

$$h_i = f_i\,(\%i) \qquad \text{(Eq 20)}$$

where the concentration terms X_i and $(\%i)$ are the mole fraction and weight percent, respectively, of the solute i, and γ_i and f_i are the activity coefficients that measure departure from ideality, that is, generally γ_i and $f_i \neq 1$. The terms a_i and h_i are often called the Raoultian activity and the Henrian activity, respectively, to distinguish the standard state being used. The Raoultian activity is usually used in conjunction with the pure material standard state, while the Henrian activity is used with the 1 wt% standard state (to be explained). The latter standard state and Eq 20 are used to describe dilute solutions and have been extensively applied to ferrous systems, especially steel. The pure material standard state and Eq 20 are used in most other cases.

The two standard states, their activities, and their activity coefficients are illustrated with Fig. 1, which gives the activity composition relationships for the liquid iron-nickel system. The point A represents the pure material standard state for nickel (pure nickel in its most stable form (liquid) at the temperature of concern, 1873 K). Activity, given on the ordinate, is based on Eq 19. The ideal solution line (Raoult's law line), where $a_{Ni} = X_{Ni}$, is labeled. The change in the partial molar free energy of nickel in transferring nickel to solution is given by Eq 21, which was obtained by substituting Eq 19 into Eq 16:

$$\Delta\overline{G}_{Ni} = \Delta G^\circ_{Ni} + RT \cdot \ln X_{Ni} + RT \cdot \ln \gamma_{Ni} \qquad \text{(Eq 21)}$$

In this case, $\Delta G_i^\circ = 0$ because the standard state and the baseline conditions are identical. The term $RT \cdot \ln X_{Ni}$ is the change in $\Delta\overline{G}_{Ni}$ for an ideal solution $\Delta\overline{G}^{\text{ideal}}_{Ni}$. The term $RT \cdot \ln \gamma_{Ni}$ provides the correction for the real system. As seen, $\gamma_{Ni} < 1$. Thus, $\Delta\overline{G}_{Ni}$ is more negative than $\Delta\overline{G}^{\text{ideal}}_{Ni}$. Larger negative free energy values indicate greater stability. Thus, real solutions of nickel in iron are more stable than ideal solutions of nickel in iron.

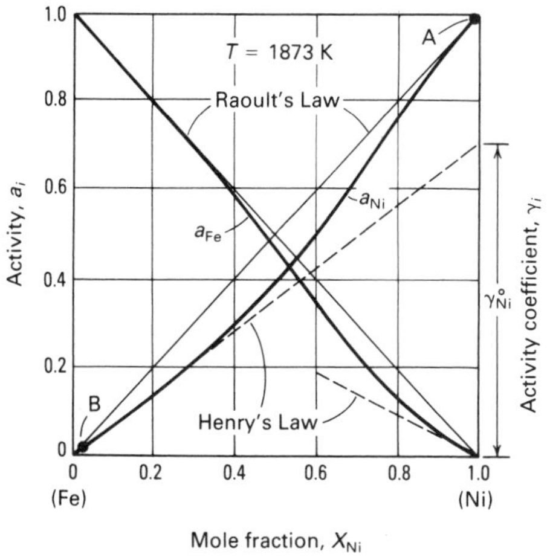

Fig. 1 Activities of iron and nickel in their binary solutions at 1873 K. Source: Ref 4, 5

A common characteristic of the activity of a component in solution is the relatively linear portion of the curve as $X_i \rightarrow 0$. This limiting slope, designated γ_i° (Fig. 1), is known as the Henry's law constant.

The 1 wt% solution standard state is marked point B in Fig. 1. It falls within or near the range in which the activity coefficient is a constant (γ_i°). By defining the standard state at point B, $f_i = 1$ in the linear region and conveniently $h_i = (\%i)$. The expression that corresponds to Eq 21 for the 1 wt% standard state is:

$$\Delta\overline{G}_{Ni} = \Delta\hat{G}^\circ_{Ni} + RT \cdot \ln(\%Ni) + RT \cdot \ln(f_{Ni}) \qquad \text{(Eq 22)}$$

In Eq 22, ΔG°_{Ni} is designated $\Delta\hat{G}^\circ_{Ni}$ to indicate that it refers to the 1 wt% standard state. Clearly, for any iron-nickel composition, $\Delta\overline{G}_{Ni}$ is the same whether it is calculated by Eq 21 or 22. Therefore, the choice is one of convenience. In this case, $\Delta\hat{G}_i^\circ \neq 0$. It is the change in free energy in taking pure liquid nickel to point B on the Henry's law line.

The difference in ΔG_i° between standard states A and B is

$$\Delta\hat{G}_i^\circ = \Delta G_i^\circ + RT \cdot \ln\left(\frac{a_B}{a_A}\right)$$
$$= \Delta G_i^\circ + RT \cdot \ln\left(\frac{M_s\,\gamma_i^\circ}{M_s + 99M_i}\right) \qquad \text{(Eq 23)}$$

where the activities for compositions A and B are Raoultian and M_s and M_i are the molecular weights, respectively, of solvent and solute. The equilibrium constant expressions (Eq 17) for the pure material and 1 wt% standard states are:

$$\sum_i \nu_i\,\Delta G_i^\circ = -RT \sum \ln(\gamma_i X_i)^{\nu_i} \qquad \text{(Eq 24)}$$

$$\sum_i \nu_i\,\Delta\hat{G}_i^\circ = -RT \sum \ln[f_i\,(\%i)]^{\nu_i} \qquad \text{(Eq 25)}$$

Very often $\Delta\overline{G}_i^\circ$ for components in the equilibrium constant are expressed in different standard states. There is no problem with this as long as appropriate values ΔG_i° or $\Delta \hat{G}_i^\circ$ are used.

Tabulations of Activity Coefficient Data. Tabulations of ΔG_i° data were already discussed. From Eq 23 to 25, it is seen that data for γ_i°, γ_i, and f_i are also needed. Tabulations of γ_i° and $\Delta \hat{G}_i^\circ$ for elements in solution in liquid iron, cobalt, and copper are given in Ref 3. Similar data for aluminum are included in the article "Thermodynamic Properties of Aluminum-Base and Copper-Base Alloys" in this Section. Data for γ_i and f_i for iron, copper, and aluminum multicomponent alloys are available in Ref 6 to 8. These data are based on a formalism (Ref 9) that is discussed in the article "Thermodynamic Properties of Aluminum-Base and Copper-Base Alloys" in this Section. Another formalism (Ref 10, 11) is used to describe the activities of carbon and silicon in multicomponent ferrous alloys in the article "Thermodynamic Properties of Iron-Base Alloys" in this Section. Most of the tabulated values were obtained from studies in dilute solution (Henry's law region). From Fig. 1 it can be seen that the shapes of the activity curves are complex. Therefore, data for the Henry's law region has only limited applicability outside the region.

To describe $\Delta \overline{G}_i$ over the entire composition range, tables giving $\Delta \overline{G}_i$ as a function of composition are used. These data are usually available only for binary alloy systems. Tables for important aluminum and copper binary systems are given in the article "Thermodynamic Properties of Aluminum-Base and Copper-Base Alloys" in this Section. Comparable data for iron are available in Ref 12. Because $\Delta \overline{G}_i$ is temperature dependent, the data are commonly given as $\Delta \overline{H}_i$ and $\Delta \overline{S}_i$, where $\Delta \overline{G}_i$ is obtained by:

$$\Delta \overline{G}_i = \Delta \overline{H}_i - T \cdot \Delta \overline{S}_i \qquad \text{(Eq 26)}$$

Examples of the use of Eq 1 to 26 will be found in succeeding articles in this Section.

Phase Diagrams

Phase diagrams describe the phases that coexist at equilibrium. Because chemical equilibrium is involved, phase diagrams have a thermodynamic basis. The complexity and variety of relationships demonstrated by phase diagrams place a detailed description beyond the scope of this article. A qualitative demonstration of the interrelationship between phase diagrams and thermodynamics will be given.

For a multiphase system, a condition of equilibrium is that the partial molar free energy or chemical potential of each component must be the same in all phases (α, β, $\cdots$):

$$\Delta \overline{G}_i^\alpha = \Delta \overline{G}_i^\beta \qquad \text{(Eq 27)}$$

Thus, for example, the chemical potential of carbon in carbon-saturated iron is the same as graphite, despite less than 5 wt% C in saturated iron. A similar situation exists with respect to silicon/aluminum-silicon alloy equilibrium, where the chemical potential of pure silicon is matched by less than 15 wt% Si in solution.

Phase diagrams describe stability relationships for compounds and solutions (solid and liquid). To understand the relationships based on thermodynamics, it is necessary to compare the phase diagram (temperature-composition plot) with the equivalent free energy-composition plot. To construct the free energy-composition plot, it must be recognized that for a binary solution the integral molar free energy is:

$$\Delta G = X_1 \cdot \Delta \overline{G}_1 + X_2 \cdot \Delta \overline{G}_2 \qquad \text{(Eq 28)}$$

Substituting Eq 16 into Eq 28 and assuming ideal solutions ($a_i = X_i$) yields F^*, the free energy change in forming 1 mol of an ideal, binary solution from components in the pure standard state (taken as the pure liquid in this case):

$$F^* = \Delta G - (X_1 \cdot \Delta G_1^\circ + X_2 \cdot \Delta G_2^\circ) = RT\,(X_1 \cdot \ln X_1 + X_2 \cdot \ln X_2) \qquad \text{(Eq 29)}$$

A plot of F^* as a function of X_2 is given in Fig. 2. Clearly, $F^* < 0$ with a minimum at $X_2 = 0.5$. For nonideal solutions ($\gamma_i \neq 1$), the curve can be higher or lower than the ideal line, and it can be asymmetrical. It can be shown that the intercepts of a tangent line drawn at any point on the curve will give $(\Delta \overline{G}_1 - \Delta G_1^\circ)$ and $(\Delta \overline{G}_2 - \Delta G_2^\circ)$ (Ref 13). These are the partial molar free energies of solution from the liquid standard state. For metallic solutions, $\Delta G_i^\circ = 0$; therefore, $\Delta \overline{G}_i$ is obtained directly from this plot.

The free energy of a phase of definite composition is represented as a point on the free energy-composition diagram. Clearly, if such a phase were in equilibrium with a solution, the equilibrium solution and the compound would have to lie along the same tangent line in order for Eq 27 to hold. This condition is illustrated in Fig. 3(a). Figure 3(b) shows the phase diagram that results from the thermodynamic constraints. In Fig. 3(a), $F^* < 0$ for pure solid phase 1 because the liquid was chosen as the standard state, and at the temperature chosen, the liquid is less stable than the solid.

The equilibrium between a liquid and a primary solid solution is illustrated in Fig. 4. From Fig. 4(a), the respective equilibrium concentrations lie along a common tangent, thus satisfying the equilibrium conditions expressed by Eq 27. The phase diagram that results from these thermodynamic conditions is given in Fig. 4(b).

For small values of X_2, either solution can theoretically exist. The solid phase is expected to be present, however, because it is more stable (lower F^*). If, by virtue of such factors as rapid cooling, the liquid phase is retained in a region where the solid is more stable, a metastable condition would exist. This is not uncommon in systems of metallurgical interest. Common examples are cementite (Fe_3C), which is metastable with respect to iron and graphite, and the glass (supercooled liquid) condition of slags, which is metastable with respect to crystalline compounds.

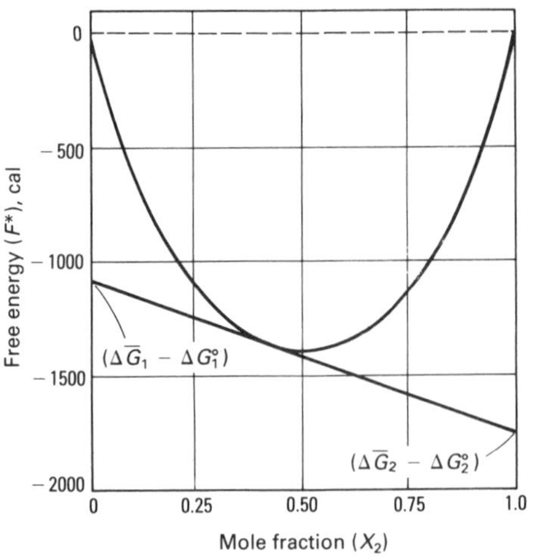

Fig. 2 Free energy of formation of an ideal solution from liquid components in the pure material standard state. Also illustrated is the graphic method of tangents to determine the partial molar free energies of solution.

Chemical Kinetics

The rates of chemical processes depend so much on local conditions that it is difficult to describe the processes in the universal terms used in thermodynamics. The rates of chemical change in solid cast metals are usually governed by solid diffusion. This subject will be examined in the Section "Principles of Solidification" in this Volume. For processes involving fluids at high temperatures, the rates of chemical reactions are inherently rapid and usually not rate limiting. The processes that generally limit the rate are nucleation of the product phase and interphase mass transport (moving material between immiscible phases).

Nucleation

The problem of nucleation is universal to the kinetics of solidification, and the subject will be treated more extensively in the article "Nucleation Kinetics" in this Volume. However, a few general comments are in order.

Nucleation of a phase in an inclusion-free metal requires supersaturation with respect to the material being precipitated. For this condition, the chemical potential of the material in the melt exceeds that in the precipitate, which is the requirement for spontaneous reaction. Initially, the diameter of the precipitated phase is in the atomic range. Thermodynamics dictates that the activity of particles increases with decreasing radius of curvature. Therefore, the activity of atomic-size particles is very high. This in turn requires very high melt supersaturation

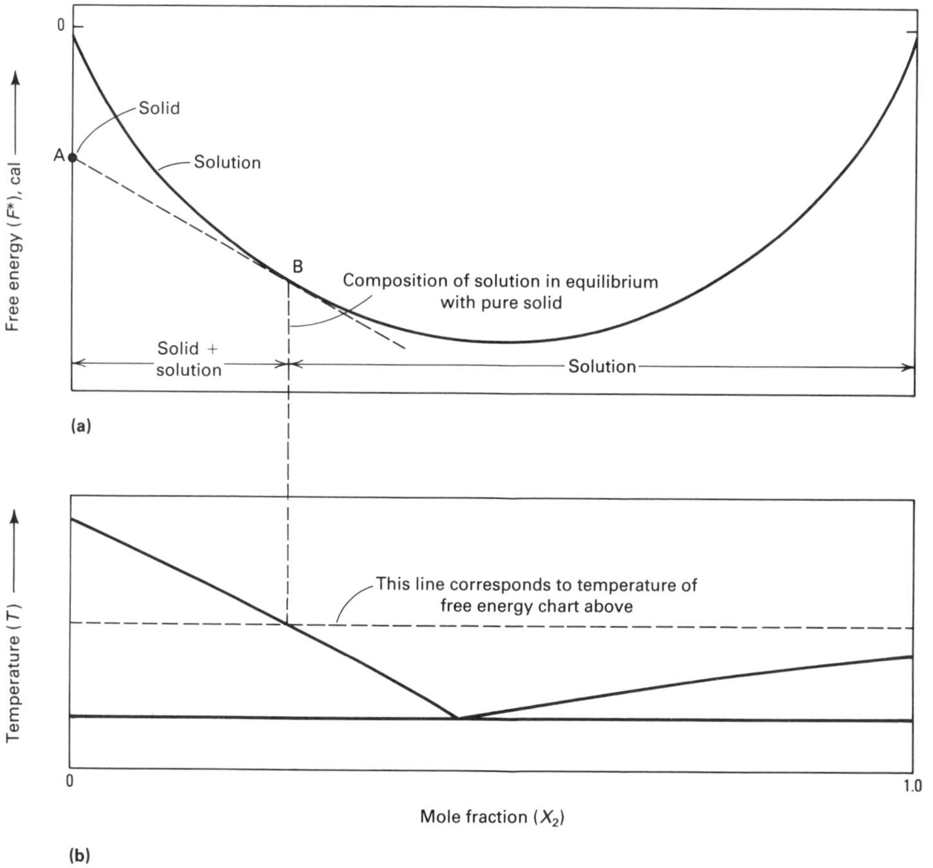

Fig. 3 Relationship between free energy-composition (a) and temperature-composition (b) diagrams for the equilibrium between a solution and a pure solid component. Source: Ref 14

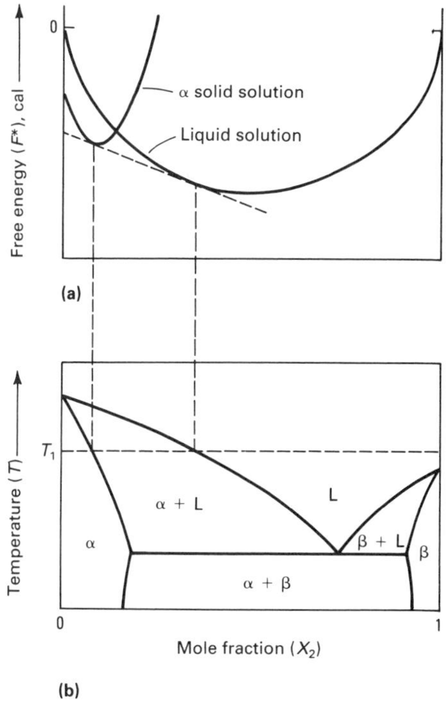

Fig. 4 Relationship between free energy-composition (a) and temperature-composition (b) diagrams for the equilibrium between a liquid and a primary solid solution. Source: Ref 15

to meet the requirement for spontaneous nucleation. This barrier to the generation of nuclei from a homogeneous melt, imposed by thermodynamics, makes it unlikely that appreciable nucleation occurs by this mechanism. Instead, nucleation is initiated on particles purposely introduced to the melt before solidification. For example, TiB_2 particles are in use to nucleate aluminum melts (Ref 16), and complex oxysulfide particles nucleate graphite in gray and ductile iron (Ref 17, 18).

Interphase Mass Transport

Because high-temperature reactions are rapid, the rate-limiting step in many cases of heterogeneous reaction is the diffusion of reactants to and from the reaction interface. This is particularly the case when the reactants are dilute. Because the reaction rate in these cases is rapid relative to rates of diffusion, the concentration of a reacting species at the reaction interface is either above or below the bulk concentration. The condition that exists depends on whether the activity of the reactive species in the second phase is higher or lower than the activity in the bulk.

The model, frequently used in quantifying mass transfer limited kinetics, assumes that there is a thin stagnant film at the interface across which a concentration gradient exists. The model is schematically represented in Fig. 5. Transport across the boundary layer is assumed to occur only by diffusion. Applying Fick's first law of diffusion, the flux of species i across the boundary layer is:

$$J = \frac{D}{\delta}(C_i^* - C_i^B) = k_m (C_i^* - C_i^B) \qquad \text{(Eq 30)}$$

where D is the diffusion coefficient, δ is the boundary layer thickness, and C_i^* and C_i^B are concentrations of the diffusion species, respectively, at the interface and in the bulk. The term D/δ is seldom used; instead, the mass transfer coefficient k_m is used to represent D/δ. When phase 2 is a fluid, δ decreases with increasing fluid velocity along the interface (that is, increased convection). Thus, k_m is a function of the fluid dynamic condition.

To determine the rate of reaction, that is, J, it is necessary to know C_i^*. This is difficult to determine experimentally. To circumvent the problem, one of two assumptions is commonly made. For extremely fast reactions in which the equilibrium constant is very large, C_i^* can be assumed to be 0. Thus, Eq 30 reduces to:

$$J = -k_m C_i^B \qquad \text{(Eq 31)}$$

This assumption has been used to describe the high-temperature gasification of coke

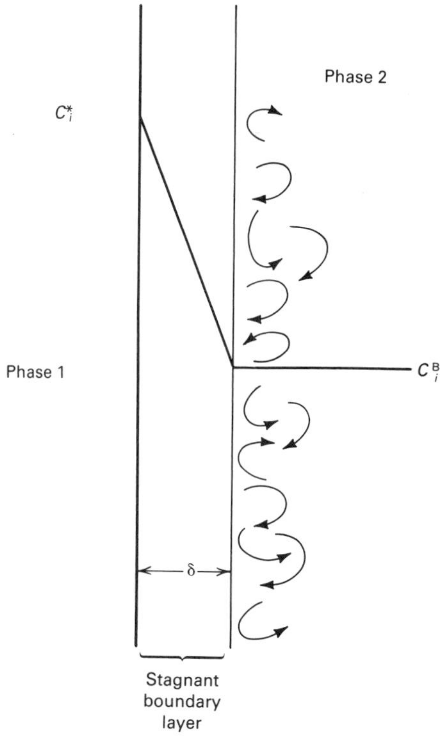

Fig. 5 Boundary layer model used for mass transfer limited kinetics

(Ref 19). The second assumption is that equilibrium between the two phases is established at the reaction interface. In this case, Eq 30 is written as:

$$J = k_m (C_i^{*\text{equil}} - C_i^B) \qquad \text{(Eq 32)}$$

The value of $C_i^{*\text{equil}}$ is determined from the equilibrium constant expression, knowing the activities of all the reactants and products except $C_i^{*\text{equil}}$. A number of examples of this treatment will be found in succeeding articles in this Section.

There are cases in which the reaction rate (rate constant k) at the interface, unimpeded by diffusion, is of the same order as the mass transfer limited rate. In this case, the observed rate and the rate constant k' are affected by both processes. This relationship is illustrated below.

Because diffusion of i through the boundary layer and the reaction of i at the interface occur sequentially, the rates of both processes must be the same at steady state. Thus:

$$J = k_m(C_i^* - C_i^B) = -k(C_i^* - C_i^{*\text{equil}}) \quad \text{(Eq 33)}$$

where the last term gives the first-order rate of reaction occurring at the interface. Eliminating C_i^* in the last two terms in Eq 33 yields:

$$J = k'\,(C_i^{*\text{equil}} - C_i^B) \quad \text{(Eq 34)}$$

where

$$\frac{1}{k'} = \left(\frac{1}{k_m} + \frac{1}{k}\right) \quad \text{(Eq 35)}$$

It is clear from Eq 34 and 35 that for the condition $k \gg k_m$, Eq 34 reduces to Eq 32. Applications in which mixed control is important are nitrogen absorption in iron alloy melts (Ref 20) and coke gasification at intermediate temperature (Ref 19).

REFERENCES

1. Y.K. Rao, *Stoichiometry and Thermodynamics of Metallurgical Processes*, Cambridge University Press, 1985, p 880-891
2. Y.K. Rao, *Stoichiometry and Thermodynamics of Metallurgical Processes*, Cambridge University Press, 1985, p 383-394
3. E.T. Turkdogan, *Physical Chemistry of High Temperature Technology*, Academic Press, 1980, p 5-26, 81, 82
4. Y.K. Rao, *Stoichiometry and Thermodynamics of Metallurgical Processes*, Cambridge University Press, 1985, p 298
5. G.R. Belton and R.J. Fruehan, The Determination of Activities by Mass Spectrometry. I. The Liquid Metallic Systems Iron-Nickel and Iron-Cobalt, *J. Phys. Chem.*, Vol 71, 1967, p 1403-1409
6. G.K. Sigworth and J.F. Elliott, The Thermodynamics of Liquid Dilute Iron Alloys, *Met. Sci.*, Vol 8, 1974, p 298-310
7. G.K. Sigworth and J.F. Elliott, The Thermodynamics of Dilute Liquid Copper Alloys, *Can. Metall. Q.*, Vol 13, 1974, p 455-461
8. G.K. Sigworth and T.A. Engh, Refining of Liquid Aluminum—A Review of Important Chemical Factors, *Scand. J. Metall.*, Vol 11, 1982, p 143-149
9. C. Wagner, *Thermodynamics of Alloys*, Addison-Wesley, 1962, p 51
10. L.S. Darken, Thermodynamics of Binary Metallic Solutions, *Trans. Metall. Soc. AIME*, Vol 239, 1967, p 80-89
11. L.S. Darken, Thermodynamics of Ternary Metallic Solutions, *Trans. Metall. Soc. AIME*, Vol 239, 1967, p 90
12. R. Hultgren *et al.*, *Selected Values of the Thermodynamic Properties of Binary Alloys*, American Society for Metals, 1973
13. Y.K. Rao, *Stoichiometry and Thermodynamics of Metallurgical Processes*, Cambridge University Press, 1985, p 285-287
14. L.S. Darken, *Thermodynamics and Physical Metallurgy*, American Society for Metals, 1950
15. L.S. Darken and R.W. Gurry, *Physical Chemistry of Metals*, McGraw-Hill, 1953, p 235
16. *Aluminum Casting Technology*, American Foundrymen's Society, 1986, p 21
17. M.J. Lalich and J.R. Hitchings, Characterization of Inclusions as Nuclei for Spheroidal Graphite in Ductile Cast Irons, *Trans. AFS*, Vol 84, 1976, p 653-664
18. B. Francis, Heterogeneous Nuclei and Graphite Chemistry in Flake and Nodular Cast Irons, *Metall. Trans. A*, Vol 10A, 1979, p 21-31
19. S. Katz, The Properties of Coke Affecting Cupola Performance, *Trans. AFS*, Vol 90, 1982, p 825-833
20. R.J. Fruehan, B. Lally, and P.C. Glaws, A Model for Nitrogen Absorption in Iron Alloy Melts, in *Fifth International Iron and Steel Congress Process Technology Proceedings*, Vol 6, The Iron and Steel Society of AIME, 1986, p 339-346

Thermodynamic Properties of Aluminum-Base and Copper-Base Alloys

Basant L. Tiwari, General Motors Research Laboratories

THIS ARTICLE will provide accessible information on the thermodynamic properties of liquid aluminum-base and copper-base alloys. Three alternative means have been used to report the thermodynamic data:

- Activities and activity coefficients
- Partial and integral molar thermal properties
- Interaction coefficients

In addition, the utility of phase diagrams will be briefly discussed. Knowledge of activities and activity coefficients is necessary in describing solution behavior and in solving problems that involve chemical equilibria. The thermal properties are useful in understanding the liquid state and in correlating data on solution behavior. The interaction coefficients provide a simple means of calculating activity coefficients in dilute solutions and are also used to correlate experimental data on dilute solutions.

Activities and Thermal Properties

The activity a of constituent i in a solution is given by the relationship:

$$\Delta\overline{G}_i = \Delta\overline{H}_i - T\Delta\overline{S}_i = RT \ln a_i \qquad \text{(Eq 1)}$$

where $\Delta\overline{G}_i$, $\Delta\overline{H}_i$, and $\Delta\overline{S}_i$ are the partial molar free energy, enthalpy, and entropy, respectively, of constituent i in solution relative to the pure constituent i at the solution temperature as the standard state; T is the temperature (in degrees kelvin), and R is the universal gas constant. Thus, from the knowledge of $\Delta\overline{H}_i$ and $\Delta\overline{S}_i$ as functions of temperature and solution composition, one may calculate $\Delta\overline{G}_i$, a_i, and γ_i (activity coefficient of component i relative to pure material standard state) for all conditions of solution. It is important to realize, however, that thermal property values vary with solution composition, unlike the properties of pure substances.

Replacing a_i in Eq 1 with $X_i\gamma_i$ yields:

$$\Delta\overline{G}_i = RT \ln X_i + RT \ln \gamma_i \qquad \text{(Eq 2)}$$

where X_i is the mole fraction of component i. In Eq 2, the term $RT \ln \gamma_i$ is called partial molar excess free energy and is expressed by the symbol $\Delta\overline{G}_i^{xs}$. It represents the contribution to $\Delta\overline{G}_i$ resulting from the departure of solution from ideal behavior. Thus, one may write:

$$\Delta\overline{G}_i^{xs} = RT \ln \gamma_i \qquad \text{(Eq 3)}$$

Corresponding to Eq 1, $\Delta\overline{G}_i^{xs}$ can also be expressed as:

$$\Delta\overline{G}_i^{xs} = \Delta\overline{H}_i^{xs} - T\,\Delta\overline{S}_i^{xs} = RT \ln \gamma_i \qquad \text{(Eq 4)}$$

However, because $\Delta\overline{H}_i$ for the ideal solution is 0 by definition, $\Delta\overline{H}_i^{xs}$ and $\Delta\overline{H}_i$ are identical. Therefore, Eq 4 can be rewritten as:

$$\Delta\overline{G}_i^{xs} = \Delta\overline{H}_i - T\,\Delta\overline{S}_i^{xs} = RT \ln \gamma_i \qquad \text{(Eq 5)}$$

Equations 1 and 5 are commonly used to calculate activities and activity coefficients from the thermal properties of the solution.

Although the partial molar thermal properties of individual constituents in a solution are most useful in solving chemical process problems, the integral properties are valuable in the discussion of the nature of solutions and also serve in some cases as the source of data on the partial properties. The integral molar free energy is related to the partial molar free energy by the following relationship:

$$\Delta G_i = X_i\,\Delta\overline{G}_i + \overline{X}_j\,\Delta\overline{G}_j + \ldots.\, X_n\,\Delta\overline{G}_n \qquad \text{(Eq 6)}$$

Similar equations can also be written to express other integral thermal properties.

In Tables 1 to 14, thermodynamic data in the form of activities, activity coefficients, partial molar thermal properties, and integral molar properties have been compiled for selected aluminum-base and copper-base alloys. The alloy systems have been chosen based on their commercial importance and on the concentration of the major alloying elements. In general, the concentrations of the alloying elements in the commercial alloys are below 10 at.%, and under this condition the interaction coefficient method described below provides a convenient means of calculating the activities of the constituents in solution. When the concentrations are close to or greater than 10 at.%, the interaction coefficient method does not yield accurate values. With this factor in mind, thermodynamic data have been selected for the following alloy systems: Al-Mg, Al-Si, Cu-Al, Cu-Ni, Cu-Sn, and Cu-Zn.

Interaction Coefficients

The use of interaction coefficients, first suggested by Wagner (Ref 3), provides a convenient means of organizing thermodynamic data on dilute solutions. Wagner proposed a Taylor series expansion for the excess partial molar free energy in order to express the logarithm of the activity coefficient of a dilute constituent in a multicomponent solution. The values of the activity coefficients calculated from this method are as accurate as the original data from which the interaction coefficients are determined, provided it is applied to solutions that are quite dilute ($X_i < 0.1$).

Consider a dilute solution of i, j, and k dissolved in a common solvent, s. If all but the first-order terms of the Taylor series expansion are neglected, the activity coefficient of solute i is expressed by the relationship:

$$\ln \gamma_i = \ln \gamma_i^\circ = X_i \left(\frac{\partial \ln \gamma_i}{\partial X_i}\right) + X_j \left(\frac{\partial \ln \gamma_i}{\partial X_j}\right) + X_k \left(\frac{\partial \ln \gamma_i}{\partial X_k}\right) \qquad \text{(Eq 7)}$$

In Eq 7, the partial derivatives are called interaction coefficients and are expressed by the symbol ϵ_i^j, where the superscript denotes the constituent that affects the activity coefficient of the subscript constituent. Thus:

$$\epsilon_i^i = \left(\frac{\partial \ln \gamma_i}{\partial X_i}\right)_{X_j} X_k = 0; X_i \to 0$$

$$\epsilon_i^j = \left(\frac{\partial \ln \gamma_i}{\partial X_j}\right)_{X_k} = 0; X_i, X_j \to 0$$

$$\ln \gamma_i = \ln \gamma_i^\circ + X_i \epsilon_i^i + X_j \epsilon_i^j + X_k \epsilon_i^k \quad \text{(Eq 8)}$$

In Eq 8, γ_i° is the limiting value of γ_i at infinite dilution:

$$\gamma_i^\circ = \lim_{X_i \to 0} \gamma_i \quad \text{(Eq 9)}$$

If the solution of i in s obeys Henry's law over a small range of concentration, then γ_i is a constant over this range, and $\epsilon_i^i = 0$.

It can be shown by using the Gibbs-Duhem equation that:

$$\epsilon_i^j = \epsilon_j^i \quad \text{(Eq 10)}$$

and this reciprocal relation is most useful in determining values of ϵ from activity data.

Values of ϵ_i^i are determined from experimental data on the binary system i-s, and values of ϵ_i^j are determined from experimental data on the ternary system i-j-s. Values of γ° and ϵ for aluminum-base and copper-base alloys are given in Tables 15 to 19.

The standard state for ϵ and γ° described above is the pure material at the temperature of the solution, and the concentration is expressed in mole fraction. In dealing with dilute solutions, however, it is common to use a hypothetical 1 wt% solution as the standard state. Under this condition, the Taylor series expansion corresponding to Eq 8 is:

$$\log f_i = e_i^i (\%i) + e_i^j (\%j) + e_i^k (\%k) \quad \text{(Eq 11)}$$

where f_i, which is the activity coefficient for 1 wt% standard state, equals $a_i/\%i$ and:

$$e_i^j = \left(\frac{\partial \log f_i}{\partial \%j}\right)_{\%k=0;\ \%i,\ \%j \to 0} \quad \text{(Eq 12)}$$

In Eq 11, f_i is the activity coefficient, and the zeroeth-order term, $\log f_i^\circ$, disappears because the activity coefficient at infinite dilution f_i° is equal to 1.

The reciprocal relationship, corresponding to Eq 10, is:

$$e_i^j = \frac{M_i}{M_j} e_j^i \quad \text{(Eq 13)}$$

where M_i is the atomic weight of i. The relationship between the two types of the interaction coefficients is:

$$e_i^j = \frac{M_s}{(2.303)\,(100)M_j} \epsilon_i^j \quad \text{(Eq 14)}$$

where M_s is the atomic weight of the solvent (Al, Cu).

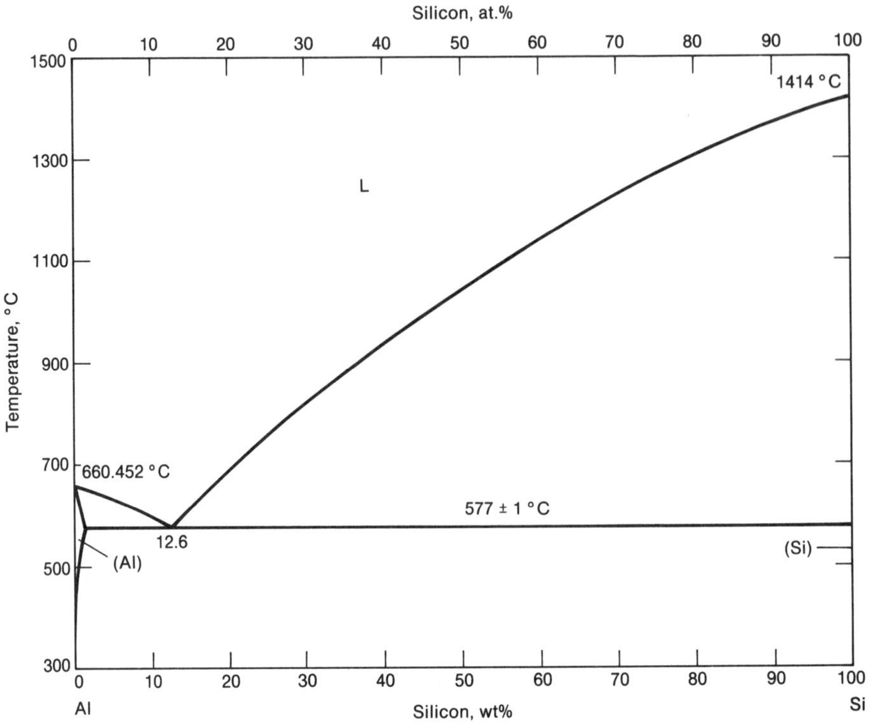

Fig. 1 The aluminum-silicon phase diagram. Source: Ref 9

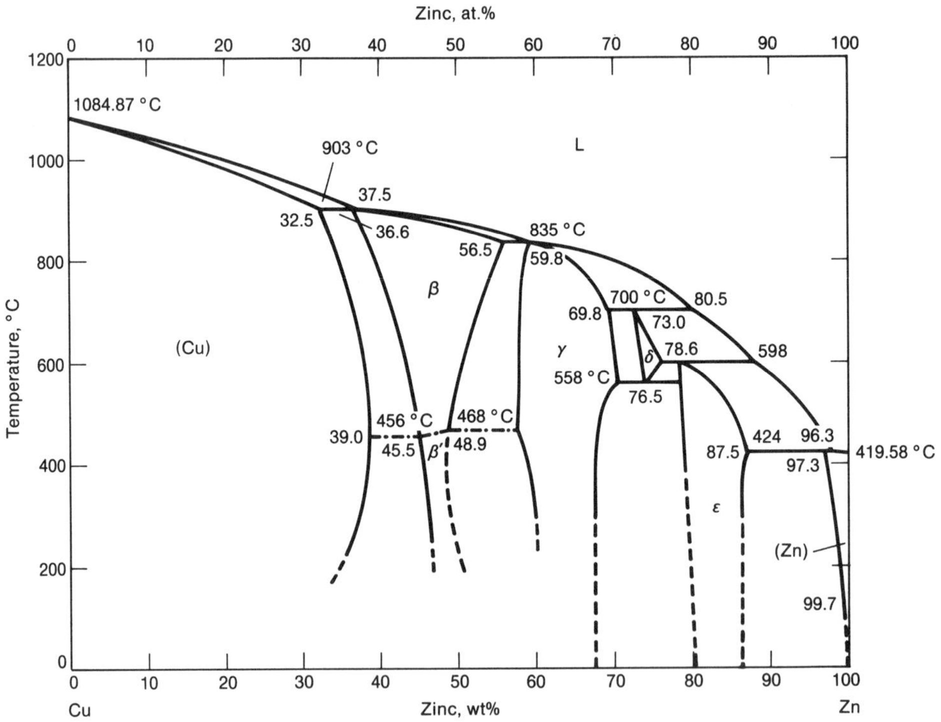

Fig. 2 The copper-zinc phase diagram. Source: Ref 9

Thermal Properties for Hypothetical Standard State

It is a common practice in solving problems on chemical equilibria to use the hypothetical pure component (X_i) or hypothetical weight percent ($\%i$) as the standard state. For this purpose, thermodynamic properties for the hypothetical standard states are needed. The values of γ° can be used to determine Gibbs free energies of mixing for elements in liquid base metal, using Eq 15 and 16:

$$\Delta G_i^\circ (X_i) = RT \ln \gamma_i^\circ \quad \text{(Eq 15)}$$

and

$$\Delta G_i^\circ (\%i) = RT \ln \frac{\gamma_i^\circ M_s}{100\, M_i} \quad \text{(Eq 16)}$$

The values of $\Delta G_i^\circ(X_i)$ and $\Delta G_i^\circ(\%i)$ thus calculated for aluminum-base and copper-base alloys are also given in Tables 15 and 17. The principal advantage of using the hypothetical standard states is that at low concentrations the activity of i can be set equal to its atom fraction or weight percent, depending on the composition scale used to express the standard state.

Phase Diagrams

A phase diagram is essentially a map that shows the relative stabilities of various phases present at equilibrium under the varying conditions of temperature and composition. Therefore, for a given composition of an alloy, a phase diagram can be used to determine the phases, which will be present at equilibrium as the melt solidifies. In addition, a phase diagram can also be used to estimate the activity of the components in liquid solution and to understand the behavior of the liquid solution (Ref 8). For example, if the diagram is for a simple eutectic system, it is likely that no strong intermetallic compound exists. If, however, a strong compound in the solid state appears in the diagram, then association of dissimilar atoms is probably occurring in the liquid. Therefore, one can predict whether a liquid solution will exhibit strong negative or positive deviations from the ideal behavior from the shapes of the liquids and azeotropes.

The phase diagrams of the Al-Si and Cu-Zn systems, which are chosen based on their commercial importance, are shown in Fig. 1 and 2. The aluminum-silicon diagram (Fig.1) represents a simple eutectic system, and the alloys containing 8.5 to 13% Si are widely used to produce automotive parts. Generally, the higher the silicon content, up to the eutectic composition (12.6% Si), the greater the fluidity and, consequently, the easier an alloy is to cast. The copper-zinc diagram (Fig. 2), however, shows a high solubility of zinc in solid copper-zinc solution and the formation of several other phases over a wide range of zinc concentrations. Copper alloys containing up to 35% Zn and significant amounts of other alloying elements are quite common. The copper-zinc phase diagram is used to adjust the compositions in order to produce alloys with the desired properties.

Table 1 Activities, activity coefficients, and partial molar thermal properties for liquid aluminum-magnesium alloys at 1073 K

Al component $Al_{(\ell)}$ = Al (in alloy)$_{(\ell)}$						
X_{Al}	a_{Al}	γ_{Al}	$\Delta\bar{G}_{Al}$, cal/g-atom	$\Delta\bar{S}_{Al}$, cal/K-g-atom	$\Delta\bar{H}_{Al}$, cal/g-atom	$\Delta\bar{G}^{xs}_{Al}$, cal/g-atom
0.9997	0.9996	0.9999	−0.9	0.0091	9	−0.2
0.9992	0.9992	1.0000	−1.7	0.0016	0	0.0
0.9989	0.9989	1.0000	−2.3	0.0105	9	0.0
0.9945	0.9940	0.9996	−12.8	0.0203	9	−0.9
0.9886	0.9876	0.9991	−26.6	0.0417	18	−1.9
0.9785	0.9791	1.0006	−45.0	0.0420	0	1.3
0.9553	0.9571	1.0020	−93.5	0.0958	9	4.3
0.8870	0.8887	1.0020	−251.6	0.2157	−20	4.3
0.8095	0.8008	0.9894	−473.7	0.3790	−67	−22.7
0.6600	0.6451	0.9775	−934.7	0.7806	−97	−48.5
0.4175	0.4115	0.9859	−1893.4	1.1548	−654	−30.3
0.2510	0.2597	1.0353	−2874.9	1.2240	−1561	74.0
0.1792	0.1652	0.9219	−3839.5	−0.2920	−4153	−173.4
0.0450	0.236	0.5251	−7988.8	−3.8840	−12157	−1373.6
Mg component $Mg_{(\ell)}$ = Mg (in alloy)$_{(\ell)}$						
X_{Mg}	a_{Mg}	γ_{Mg}	$\Delta\bar{G}_{Mg}$, cal/g-atom	$\Delta\bar{S}_{Mg}$, cal/K-g-atom	$\Delta\bar{H}_{Mg}$, cal/g-atom	$\Delta\bar{G}^{xs}_{Mg}$, cal/g-atom
0.00027	0.00022	0.8060	−18017	15.4180	−1470.8	−459.73
0.00079	0.00063	0.7974	−15696	13.9463	−729.8	−482.71
0.00104	0.00081	0.7814	−15171	12.6176	−1630.2	−525.87
0.0055	0.0048	0.8780	−11358	9.6420	−1010.9	−227.42
0.0114	0.0107	0.9353	−9683	7.4045	−1736.7	−142.53
0.0215	0.0183	0.8535	−8530	6.9708	−1049.0	−337.68
0.0447	0.0366	0.8188	−7051	5.4069	−1248.7	−424.12
0.1130	0.0929	0.8221	−5069	3.6123	−1192.1	−418.20
0.1905	0.1679	0.8813	−3806	2.7957	−805.9	−269.73
0.3400	0.3097	0.9109	−2500	1.6331	−747.6	−198.80
0.5825	0.5176	0.8885	−1405	1.2733	−38.2	−251.88
0.7490	0.6367	0.8500	−963	1.3102	443.2	−346.49
0.8208	0.7184	0.8752	−705	1.7577	1180.8	−284.20
0.9550	0.9198	0.9631	−178	1.1672	1074.3	−79.98

Source: Ref 1

Table 2 Integral thermal properties for liquid aluminum-magnesium alloys at 1073 K

	$(1-x)Al_{(\ell)} + xMg_{(\ell)} = Al_{(1-x)}Mg_{x(\ell)}$			
X_{Mg}	ΔG, cal/g-atom	ΔS, cal/K-g-atom	ΔH, cal/g-atom	ΔG^{xs}, cal/g-atom
0.00027	−5.7	0.01320	8.61	−0.34
0.00079	−14.2	0.01274	−0.58	−0.39
0.00104	−18.1	0.02365	7.29	−0.55
0.0055	−75.7	0.07366	3.34	−2.39
0.0114	−136.9	0.12577	−2.04	−3.53
0.0215	−227.3	0.19086	−22.54	−6.00
0.0447	−404.8	0.33346	−47.28	−14.91
0.1130	−796.1	0.59962	−152.48	−43.49
0.1905	−1108.7	0.83951	−207.80	−69.79
0.3400	−1466.9	1.07043	−318.19	−99.61
0.5825	−1608.9	1.22383	−295.28	−159.37
0.7490	−1442.6	1.28856	−59.77	−240.99
0.8208	−1266.9	1.39046	225.25	−264.35
0.9550	−529.7	0.93986	478.87	−138.20

Source: Ref 1

Table 3 Activities, activity coefficients, and partial molar thermal properties for liquid aluminum-silicon alloys at 1100 K

Al component $Al_{(\ell)}$ = Al (in alloy)$_{(\ell)}$				
X_{Al}	a_{Al}	γ_{Al}	$\Delta\bar{G}_{Al}$, cal/g-atom	$\Delta\bar{G}^{xs}_{Al}$, cal/g-atom
1.00	1.000	1.000	0	0
0.90	0.859	0.954	−333	−102
0.80	0.698	0.872	−786	−298
0.70	0.503	0.719	−1500	−720
0.69(a)	0.473	0.685	−1638	−826
	(±0.01)	(±0.02)	(±50)	(±50)

Si component $Si_{(\ell)}$ = Si (in alloy)$_{(\ell)}$							
X_{Si}	a_{Si}	γ_{Si}	$\Delta\bar{G}_{Si}$, cal/g-atom	$\Delta\bar{G}^{xs}_{Si}$, cal/g-atom	$\Delta\bar{H}_{Si}$, cal/g-atom	$\Delta\bar{S}_{Si}$, cal/K-g-atom	$\Delta\bar{S}^{xs}_{Si}$, cal/K-g-atom
0.00	0.000	0.040	−∞	−7060	−2502	∞	4.144
0.10	0.016	0.155	−9106	−4073	...	...	...
0.20	0.051	0.255	−6510	−2991	...	...	...
0.30	0.127	0.424	−4509	−1877	...	...	...
0.31(a)	0.147	0.473	−4194	−1634	...	...	...
	(±0.02)	(±0.06)	(±300)	(±300)	(±50)		

(a) Phase boundary. Source: Ref 2

Table 4 Integral thermal properties for liquid aluminum-silicon alloys at 1100 K

	$(1-x)Al_{(\ell)} + xSi_{(\ell)} = Al_{(1-x)}Si_{x(\ell)}$	
X_{Si}	ΔG, cal/g-atom	ΔG^{xs}, cal/g-atom
0.1	−1210	−499
0.2	−1931	−837
0.3	−2403	−1067
0.31(a)	−2430	−1076
	(±100)	(±100)

(a) Phase boundary. Source: Ref 2

Table 5 Activities, activity coefficients, and partial molar thermal properties for liquid copper-aluminum alloys at 1373 K

Al component $Al_{(\ell)}$ = Al (in alloy)$_{(\ell)}$							
X_{Al}	a_{Al}	γ_{Al}	$\Delta\bar{G}_{Al}$, cal/g-atom	$\Delta\bar{G}^{xs}_{Al}$, cal/g-atom	$\Delta\bar{H}_{Al}$, cal/g-atom	$\Delta\bar{S}_{Al}$, cal/K-g-atom	$\Delta\bar{S}^{xs}_{Al}$, cal/K-g-atom
1.0	1.000	1.000	0	0	0	0.000	0.000
0.9	0.889	0.988	−320	−32	29	0.254	0.044
0.8	0.759	0.949	−753	−144	44	0.580	0.137
0.7	0.609	0.870	−1354	−381	−60	0.942	0.234
0.6	0.441	0.735	−2235	−842	−391	1.343	0.328
0.5	0.266	0.532	−3611	−1720	−1055	1.862	0.484
	(±0.02)	(±0.04)	(±150)	(±150)	(±300)	(±0.25)	(±0.25)
0.4	0.116	0.290	−5873	−3373	−2878	2.181	0.361
0.3	0.028	0.095	−9709	−6424	−5012	3.421	1.028
0.2	0.006	0.029	−14062	−9670	−5864	5.971	2.772
0.1	0.001	0.008	−19307	−13025	−7415	8.661	4.086
0.0	0.000	0.002	−∞	−16640	−8625	∞	5.838
Cu component $Cu_{(\ell)}$ = Cu (in alloy)$_{(\ell)}$							
X_{Cu}	a_{Cu}	γ_{Cu}	$\Delta\bar{G}_{Cu}$, cal/g-atom	$\Delta\bar{G}^{xs}_{Cu}$ cal/g-atom	$\Delta\bar{H}_{Cu}$, cal/g-atom	$\Delta\bar{S}_{Cu}$, cal/K-g-atom	$\Delta\bar{S}^{xs}_{Cu}$, cal/K-g-atom
0.0	0.000	0.042	−∞	−8671	−4225	∞	3.238
0.1	0.005	0.052	−14349	−8067	−4849	6.919	2.344
0.2	0.013	0.065	−11833	−7442	−4975	4.995	1.797
0.3	0.025	0.084	−10025	−6741	−4675	3.897	1.505
0.4	0.046	0.115	−8396	−5896	−4071	3.150	1.329
0.5	0.085	0.170	−6728	−4837	−3271	2.518	1.141
	(±0.005)	(±0.01)	(±150)	(±150)	(±300)	(±0.25)	(±0.25)
0.6	0.166	0.277	−4900	−3506	−1838	2.230	1.215
0.7	0.352	0.503	−2848	−1875	−679	1.580	0.871
0.8	0.600	0.750	−1394	−785	−259	0.827	0.383
0.9	0.839	0.932	−478	−191	−65	0.301	0.092
1.0	1.000	1.000	0	0	0	0.000	0.000

Source: Ref 2

Table 6 Integral thermal properties for liquid copper-aluminum alloys at 1373 K

$(1-x)Al_{(\ell)} + xCu_{(\ell)} = Al_{(1-x)}Cu_{x(\ell)}$					
X_{Cu}	ΔG, cal/g-atom	ΔH, cal/g-atom	ΔS, cal/K-g-atom	ΔG^{xs}, cal/g-atom	ΔS^{xs}, cal/K-g-atom
0.1	−1723	−459	0.921	−836	0.275
0.2	−2969	−960	1.463	−1604	0.469
0.3	−3955	−1445	1.828	−2289	0.615
0.4	−4700	−1863	2.066	−2863	0.728
0.5	−5170	−2163	2.190	−3278	0.812
	(±150)	(±150)	(±0.1)	(±150)	(±0.1)
0.6	−5289	−2254	2.210	−3453	0.873
0.7	−4906	−1979	2.132	−3240	0.918
0.8	−3927	−1380	1.855	−2562	0.861
0.9	−2361	−800	1.137	−1474	0.491

Source: Ref 2

Table 7 Activities, activity coefficients, and partial molar thermal properties for liquid copper-nickel alloys at 1823 K

Cu component $Cu_{(\ell)}$ = Cu (in alloy)$_{(\ell)}$				
X_{Ni}	a_{Cu}	γ_{Cu}	$\Delta\bar{G}_{Cu}$, cal/g-atom	$\Delta\bar{G}^{xs}_{Cu}$, cal/g-atom
0.0	1.000	1.000	0	0
0.1	0.902	1.002	−374	8
0.2	0.814	1.017	−747	61
0.3	0.740	1.058	−1088	204
0.4	0.677	1.128	−1414	436
0.5	0.611	1.222	−1786	725
	(±0.02)	(±0.03)	(±100)	(±100)
0.6	0.534	1.334	−2275	1044
0.7	0.444	1.480	−2941	1421
0.8	0.334	1.669	−3974	1856
0.9	0.191	1.912	−5992	2349
1.0	0.000	2.227	−∞	2900
Ni component $Ni_{(\ell)}$ = Ni (in alloy)$_{(\ell)}$				
a_{Ni}		γ_{Ni}	$\Delta\bar{G}_{Ni}$, cal/g-atom	$\Delta\bar{G}^{xs}_{Ni}$, cal/g-atom
0.000		1.906	−∞	2336
0.185		1.864	−6120	2222
0.341		1.704	−3899	1931
0.455		1.517	−2851	1510
0.539		1.347	−2241	1079
0.611		1.222	−1786	725
(±0.02)		(±0.03)	(±100)	(±100)
0.682		1.136	−1337	464
0.752		1.075	−1031	261
0.826		1.032	−692	116
0.907		1.008	−353	29
1.000		1.000	0	0

Source: Ref 2

Table 8 Integral thermal properties for liquid copper-nickel alloys at 1823 K

$(1-x)Cu_{(\ell)} + xNi_{(\ell)} = Cu_{(1-x)}Ni_{x(\ell)}$					
X_{Ni}	ΔG, cal/g-atom	ΔH, cal/g-atom	ΔS, cal/K-g-atom	ΔG^{xs}, cal/g-atom	ΔS^{xs}, cal/K-g-atom
0.1	−948	240	0.652	229	−0.006
0.15	−1195	341	0.842	336	−0.003
0.2	−1377			435	. . .
0.3	−1617	. . .	. . .	596	. . .
0.4	−1745	. . .	. . .	693	. . .
0.5	−1786	. . .	. . .	725	. . .
	(±100)			(±100)	
0.6	−1742	. . .	. . .	696	. . .
0.7	−1604	. . .	. . .	609	. . .
0.8	−1349	. . .	. . .	464	. . .
0.9	−917	. . .	. . .	261	. . .

Source: Ref 2

Table 9 Activities, activity coefficients, and partial molar thermal properties for liquid copper-tin alloys at 1400 K

Cu component $Cu_{(\ell)}$ = Cu (in alloy)$_{(\ell)}$							
X_{Cu}	a_{Cu}	γ_{Cu}	$\Delta\bar{G}_{Cu}$, cal/g-atom	$\Delta\bar{G}^{xs}_{Cu}$, cal/g-atom	$\Delta\bar{H}_{Cu}$, cal/g-atom	$\Delta\bar{S}_{Cu}$, cal/K-g-atom	$\Delta\bar{S}^{xs}_{Cu}$, cal/K-g-atom
1.0	1.000	1.000	0	0	0	0.000	0.000
0.9	0.802	0.891	−613	−320	−159	0.325	0.115
0.8	0.539	0.674	−1717	−1096	−749	0.691	0.248
0.7	0.389	0.556	−2626	−1633	−1443	0.845	0.136
0.6	0.284	0.474	−3497	−2076	−1662	1.311	0.295
0.5	0.220	0.440	−4215	−2287	−1631	1.845	0.468
	(±0.025)	(±0.05)	(±300)	(±300)	(±250)	(±0.3)	(±0.3)
0.4	0.169	0.422	−4949	−2400	−1418	2.522	0.701
0.3	0.125	0.417	−5781	−2432	−1049	3.380	0.988
0.2	0.082	0.408	−6971	−2493	−640	4.522	1.324
0.1	0.038	0.379	−9104	−2698	81	6.561	1.985
0.0	0.000	0.317	−∞	−3197	1050	∞	3.034
Sn component $Sn_{(\ell)}$ = Sn (in alloy)$_{(\ell)}$							
X_{Sn}	a_{Sn}	γ_{Sn}	$\Delta\bar{G}_{Sn}$, cal/g-atom	$\Delta\bar{G}^{xs}_{Sn}$, cal/g-atom	$\Delta\bar{H}_{Sn}$, cal/g-atom	$\Delta\bar{S}_{Sn}$, cal/K-g-atom	$\Delta\bar{S}^{xs}_{Sn}$, cal/K-g-atom
0.0	0.000	0.007	−∞	−13609	−8000	∞	4.006
0.1	0.007	0.072	−13706	−7301	−5233	6.053	1.477
0.2	0.072	0.362	−7304	−2827	−1901	3.860	0.661
0.3	0.197	0.656	−4523	−1173	252	3.411	1.018
0.4	0.340	0.849	−3003	−454	706	2.649	0.828
0.5	0.467	0.934	−2119	−190	681	2.000	0.622
	(±0.05)	(±0.1)	(±300)	(±300)	(±250)	(±0.3)	(±0.3)
0.6	0.580	0.966	−1516	−95	506	1.444	0.429
0.7	0.681	0.973	−1069	−77	311	0.986	0.277
0.8	0.784	0.979	−678	−57	176	0.610	0.167
0.9	0.892	0.991	−317	−24	49	0.261	0.052
1.0	1.000	1.000	0	0	0	0.000	0.000

Source: Ref 2

Table 10 Partial molar thermal properties for liquid copper-tin alloys at 633 K

	$Sn_{(\ell)} = Sn$ (in alloy)$_{(\ell)}$			
X_{Sn}	a_{Sn}	γ_{Sn}	$\Delta\bar{G}_{Sn}$, cal/g-atom	$\Delta\bar{G}^{xs}_{Sn}$, cal/g-atom
0.98	0.965	0.985	−47	−20
0.99	0.985	0.995	−13	−7
	(±0.005)	(±0.005)	(±5)	(±5)

Source: Ref 2

Table 11 Heats of formation of solid and liquid copper-tin alloys at 723 K

			$(1-x)Cu_{(s)} + xSn_{(s)} = Cu_{(1-x)}Sn_{x(s)}$; $(1-x)Cu_{(\ell)} + xSn_{(\ell)} = Cu_{(1-x)}Sn_{x(\ell)}$; $Cu_{(\ell)} = Cu$(in alloy)$_{(\ell)}$					
X_{Sn}	Phase	ΔH, cal/g-atom	X_{Sn}	Phase	ΔH, cal/g-atom	$\Delta C_{p(a)}$, cal/K-g-atom	$\Delta\bar{H}_{Cu}$, cal/g-atom	$\Delta\bar{C}_{pCu(\ell)}$, cal/K-g-atom
0.091(b)	(Cu)	−280	0.825(b)	ℓ	−1105	1.7	. . .	. . .
			0.850	. . .	−990	. . .	. . .	. . .
0.204(b)	δ	−1300	0.900	. . .	−760	1.2	. . .	. . .
0.209	δ	−1260	0.950	. . .	−530	. . .	. . .	. . .
					(±50)			
0.244(b)	ϵ	−1686	1.000	ℓ	0		−260	1.9 (±2)
0.250		−1800	. . .	. . .	. . .	. . .	. . .	. . .
0.255(b)	ϵ	−1920	. . .	. . .	. . .	. . .	. . .	. . .
		(±100)						

(a) C_p = heat capacity. (b) Phase boundary. Source: Ref 2

Table 12 Integral thermal properties for liquid copper-tin alloys at 1400 K

	$(1-x)Cu_{(\ell)} + xSn_{(\ell)} = Cu_{(1-x)}Sn_{x(\ell)}$				
X_{Sn}	ΔG, cal/g-atom	ΔH, cal/g-atom	ΔS, cal/K-g-atom	ΔG^{xs}, cal/g-atom	ΔS^{xs}, cal/K-g-atom
0.1	−1922	−666	0.897	−1018	0.251
0.2	−2834	−979	1.325	−1442	0.331
0.3	−3195	−934	1.614	−1495	0.400
0.4	−3300	−715	1.846	−1427	0.509
0.5	−3167	−475	1.923	−1238	0.545
	(±300)	(±150)	(±0.24)	(±300)	(±0.24)
0.6	−2889	−264	1.875	−1017	0.538
0.7	−2483	−97	1.705	−784	0.491
0.8	−1937	13	1.392	−545	0.398
0.9	−1196	52	0.891	−291	0.245

Source: Ref 2

Table 13 Activities, activity coefficients, and partial molar thermal properties for liquid copper-zinc alloys at 1200 K

	Cu component $Cu_{(\ell)} = Cu$ (in alloy)$_{(\ell)}$			
X_{Zn}	a_{Cu}	γ_{Cu}	$\Delta\bar{G}_{Cu}$, cal/g-atom	$\Delta\bar{G}^{xs}_{Cu}$, cal/g-atom
0.334(a)	0.432	0.648	−2003	−1034
0.4	0.334	0.557	−2614	−1396
0.5	0.219	0.438	−3621	−1968
	(±0.05)	(±0.1)	(±400)	(±400)
0.6	0.139	0.348	−4705	−2520
0.7	0.085	0.284	−5869	−2998
0.8	0.049	0.245	−7187	−3349
0.9	0.023	0.229	−9010	−3519
1.0	0.000	0.235	−∞	−3454

Zn component $Zn_{(\ell)} = Zn$ (in alloy)$_{(\ell)}$			
a_{Zn}	γ_{Zn}	$\Delta\bar{G}_{Zn}$, cal/g-atom	$\Delta\bar{G}^{xs}_{Zn}$, cal/g-atom
0.132	0.398	−4814	−2199
0.207	0.517	−3757	−1572
0.347	0.695	−2521	−868
(±0.05)	(±0.1)	(±400)	(±400)
0.505	0.841	−1630	−412
0.657	0.939	−1002	−151
0.789	0.986	−564	−32
0.900	1.000	−250	1
1.000	1.000	0	0

(a) Phase boundary. Source: Ref 2

Table 14 Integral thermal properties for liquid copper-zinc alloys at 1200 K

	$(1-x)Cu_{(\ell)} + xZn_{(\ell)} = Cu_{(1-x)}Zn_{x(\ell)}$				
X_{Zn}	ΔG	ΔG^{xs}, cal/g-atom	x_{Zn}	ΔG, cal/g-atom	ΔG^{xs}, cal/g-atom
0.334(a)	−2942	−1423	0.6	−2860	−1255
0.4	−3071	−1466	0.7	−2462	−1005
0.5	−3071	−1418	0.8	−1889	−695
	(±400)	(±400)	0.9	−1126	−351

(a) Phase boundary. Source: Ref 2

Table 15 Standard Gibbs free energies for solution of elements in liquid aluminum

Solution reaction	$\gamma°$, 1100 K	$\Delta G_i°$ for i (X), cal/g-atom	$\Delta G_i°$ for i (%), cal/g-atom
Be(s)=Be	19.6	9404 − 2.639T	9404 − 9.607T
Bi(1)=Bi	24.5	5309 + 1.524T	5309 − 11.688T
Ca(1)=Ca	0.0086	−10 400	−10 400 − 9.93T
Cd(1)=Cd	19.9	7100 − 0.518T	7100 − 12.498T
Cu(s)=Cu	0.037	−1135 − 5.538T	−1135 − 16.385T
Fe(s)=Fe	1.6 × 10^{-4}	−27 000 + 7.18T	−27 000 − 3.411T
Ga(1)=Ga	1.1	832 − 0.52T	832 − 11.551T
Ge(1)=Ge	0.16	−2761 − 1.176T	−2761 − 12.288T
½H_2=H	. . .	. . .	11 664 + 6.523T
In(1)=In	12.3	6800 − 1.201T	6800 − 13.223T
Li(1)=Li	0.40	−5800 + 3.435T	−5800 − 3.014T
Mg(1)=Mg	0.18	−3478 − 0.30T	−3478 − 9.24T
Na(1)=Na	293	8230 + 3.798T	8230 − 5.03T
Ni(s)=Ni	0.7 × 10^{-6}	−28 280 − 2.442T	−28 280 − 13.132T
Pb(1)=Pb	115	9970 + 0.363T	9970 − 12.832T
Sb(1)=Sb	3.4	13 100 − 9.45T	13 100 − 21.589T
Si(s)=Si	0.27	9598 − 11.291T	9598 − 20.517T
Sn(1)=Sn	4.68	5845 − 2.245T	5845 − 14.333T
Zn(1)=Zn	1.92	2538 − 1.007T	2538 − 11.911T

Source: Ref 4

Table 16 Interaction coefficients for elements in liquid aluminum

i	j	ϵ_i^j	Temperature, K	i	j	ϵ_i^j	Temperature, K
Ag	Ag	−3.1	1273	Sn	Sn	6.0	1100
Cu	H	see ϵ_H^{Cu}	973–1273	Zn	Zn	−0.9	1000
H	Cu	39.0	973	Zn	Si	2.2	963–1053
H	Cu	16.6	1073	Li	Sn	−16.0	949
H	Cu	20.1	1173	Na	Si	−12	973
H	Cu	4.3	1273	Mg	Si	−9	973–1073
H	H	0	973–1273	Sn	Pb	−1.5	973–1073
H	Si	11.5	973	H	Ce	−100	800
H	Si	6.2	1073	H	Cu	15	700–800
H	Si	4.2	1173	H	Cr	−1	800–900
H	Si	1.8	1273	H	Fe	−1	800–900
Si	H	see ϵ_H^{Si}	973–1273	H	Mg	−2	700–800
Cd	Cd	−5.0	1373	H	Mn	27	800
Cu	Cu	2.2	1373	H	Ni	19	700–800
Ga	Ga	−0.3	1023	H	Th	−20	800–900
Ge	Ge	+3.0	1200	H	Ti	−42	800–900
In	In	−4.5	1173	H	Si	7.1	700–800
Mg	Mg	3.0	1073	H	Sn	0.6	800–900
Si	Si	16.0	1100				

Source: Ref 4, 6

Table 17 Standard Gibbs free energies for solution of elements in liquid copper

Element(a), i	γ_i°, 1200 °C	$\Delta G_i^\circ(X)$, cal/g-atom	ΔG_i°(wt%), cal/g-atom	Temperature, °C
Ag(l)	3.23	3900 − 0.32T	3900 − 10.52T	1100–1200
Al(l)	0.0028	−8630 − 5.84T	−8630 − 13.84T	1100
As(v)	4.8 × 10^{-4}	−22 350	−22 350 − 9.44T	1000
Au(l)	0.14	−4630 − 0.73T	−4630 − 12.09T	1175–1325
Bi(l)	1.25	5960 − 3.6T	5960 − 15.1T	1100–1300
C(graphite)	1.4 × 10^5	8550 + 17.8T	8550 + 12.0T	1100–1300
Ca(l)	5.1 × 10^{-4}	−22 200	−22 200	800–925
Cd(v)	15.6	−25 700 + 22.9T	−25 700 + 12.7T	. . .
Cd(l)	0.53	−1860	−1860 − 9.0T	. . .
Co(s)	15.4	8000	8000 − 9.0T	. . .
Cr(s)	43	11 000	11 000 − 8.72T	. . .
Fe(s)	24.1	12 970 − 2.48T	12 970 − 11.34T	1460–1580
Fe(l)	19.5	9300 − 0.41T	9300 − 9.27T	1460–1580
Ga(1)	0.034	−10 800 + 0.61T	−10 800 − 8.68T	1100–1280
Ge(l)	0.009	−16 000 + 1.52T	−16 000 − 7.5T	1255–1545
½H_2(g)	. . .	10 400 + 8.4T	10 400 − 7.5T	1100–1300
In(l)	0.41	−9550 + 4.71T	−9550 − 5.58T	700–1000
Mg(v)	0.08	−40 200 + 22.3T	−40 200 − 15.1T	650–927
Mg(l)	0.044	−8670 − 0.31T	−8670 − 7.53T	650–927
Mn(l)	0.51	−1950	−1950 − 8.83T	1244
Mn(s)	0.53	1550 − 2.31T	1550 − 11.14T	1244
Ni(l)	2.22	2340	2340 − 9.0T	. . .
Ni(s)	2.66	6500 − 2.5T	6550 − 11.5T	. . .
½O_2(g)	. . .	−20 400 + 10.8T	−20 400 − 4.43T	1100–1300
Pb(l)	5.27	8620 − 2.55T	8620 − 14.01T	1000–1300
Pd(s)	1.3	800	800 − 10.1T	1500–1600
Pt(s)	0.05	−10 200 + 0.87T	−10 200 − 10.47T	. . .
½S_2(g)	. . .	−28 600 + 13.79T	−28 600 − 6.03T	1050–1250
Sb(l)	0.014	−12 500	−12 500 − 10.4T	1000–1200
Se(v)	0.002	−18 200	−18 200 − 9.5T	1200
Si(l)	0.006	−15 000	−15 000 − 7.5T	1550
Si(s)	0.01	−2900 − 7.18T	−2900 − 14.68T	1550
Sn(l)	0.048	−8900	−8900 − 10.4T	1100–1300
Te(v)	0.0328	−10 000	−10 000 − 10.53T	1200
Ti(l)	8.5	6730 − 0.31T	6730 − 11.74T	1000–1300
V(s)	130	28 100 − 9.4T	28 100 − 18.1T	. . .
Zn(l)	0.146	−5640	−5640 − 9.1T	1150

(a) l, liquid; v, vapor; s, solid; g, gas. Source: Ref 5

Table 18 Interaction coefficients for elements in liquid copper alloys

i	j	ϵ_j^i	Temperature, °C
H	Ag	−0.5	1225
H	Al	6.2	1225
H	Au	−1.9	1225
H	Co	−3.1	1150
H	Cr	−1.6	1550
H	Fe	−2.9	1150–1550
H	Mn	−1.1	1150
H	Ni	−5.5	1150–1240
H	P	10.0	1150
H	Pb	21.0	1100
H	Pt	−8.0	1225
H	S	9.0	1150
H	Sb	13.0	1150
H	Si	4.8	1150
H	Sn	6.0	1100–1300
H	Te	−6.6	1150
H	Zn	6.8	1150
O	Ag	−0.7	1100–1200
O	Au	8.6	1200–1550
O	Co	−68	1200
O	Fe	$-4.04 \times 10^6/T$ 2183	1200–1350
O	Ni	$-36000/T + 17$	1200–1300
O	P	$-700000/T + 385$	1150–1300
O	Pb	−7.4	1100
O	Pt	38	1200
O	S	−19	1206
O	Si	−6300	1250
O	Sn	−4.6	1100
S	Au	6.7	1115–1200
S	Co	−4.8	1300–1500
S	Fe	$-25400/T + 8.7$	1300–1500
S	Ni	$-29800/T + 13$	1300–1500
S	Pt	11.5	1200–1500
S	Si	6.9	1200
Ag	Ag	−2.5	1150
Al	Al	14	1100
Au	Au	3.7	1277
Bi	Bi	$-6800/T + 1.65$	1000–1200
Ca	Ca	20	877
Fe	Fe	−5.7	1550
Ga	Ga	7	1280
Ge	Ge	13.4	1255
H	H	1.0	1123
Mg	Mg	9.8	927
Mn	Mn	6	1244
O	O	$-24000/T + 7.8$	1100–1300
Pb	Pb	−2.7	1200
S	S	$-20800/T$	1050–1250
Sb	Sb	15	1000–1200
Sn	Sn	10	1300–1320
Tl	Tl	−4.8	1300
Zn	An	4	1150
Zn	Zn	0.38	902
Zn	Zn	0.72	727
Zn	Zn	1.185	653
Zn	Zn	1.40	604

Source: Ref 4, 6

Table 19 Activity coefficients at infinite dilution in liquid metals

Solvent	Solute	γ°	Temperature, °C
Aluminum	Silver	0.38	700
		0.47	900
		0.53	1000
	Magnesium	0.88	800
Copper	Zinc	0.14	604
		0.17	653
		0.21	727
		0.24	802

Source: Ref 2, 7

REFERENCES

1. B.L. Tiwari, *Metall. Trans. A*, Vol 18A, 1987, p 1645-1651
2. R. Hultgren *et al.*, *Selected Values of the Thermodynamic Properties of Binary Alloys*, American Society for Metals, 1973
3. C. Wagner, *Thermodynamics of Alloys*, Addison-Wesley, 1952
4. G.K. Sigworth and T.A. Engh, *Scand. J. Metall.*, Vol 11, 1982, p 143-149
5. G.K. Sigworth and J.F. Elliott, *Can. Met. Quart.*, Vol 13, 1974, p 455-461
6. J.M. Dealy and R.D. Pehlke, *Trans. Met. Soc. AIME*, Vol 227, Feb 1963, p 88-94
7. J.M. Dealy and R.D. Pehlke, *Trans. Met. Soc. AIME*, Vol 227, Aug 1963, p 1030-1032
8. L.S. Darken and R.W. Gurry, *Physical Chemistry of Metals*, McGraw-Hill, 1953
9. T.B. Massalski *et al.*, *Binary Alloy Phase Diagrams*, Vol 1 and 2, American Society for Metals, 1986

Thermodynamic Properties of Iron-Base Alloys

D.M. Stefanescu, University of Alabama

BECAUSE the structures of steel and cast iron directly influence their mechanical properties, one of the most important tasks of a metallurgist is to control the solidification structure in castings. To accomplish this, it is necessary to understand and control the thermodynamics of the liquid and solid phases and the kinetics of solidification (nucleation and growth of various phases). The information linked to the practical interests that thermodynamics can provide when considering iron-base alloys encompasses a rather wide range.

Only two issues will be addressed in this article. The first is the calculation of solubility lines, relevant to the construction of phase diagrams, and the second is the calculation of the activity of various components, which allows for the determination of probability of formation and relative stability of various phases. Alloying elements will then be discussed in terms of their influence on the activity of carbon, which provides information on the stability of the main carbon-rich phases of iron-carbon alloys, that is, graphite and cementite.

Thermodynamics of Binary Fe-X Systems

Only the Fe-C and Fe-Si systems will be discussed in this article. Information on the Fe-O system is available in the article "Inclusion-Forming Reactions" in this Volume.

The Fe-C System. The phase diagram of the binary Fe-C system includes the stable (Fe-graphite) and metastable (Fe-Fe_3C) equilibria. The most recent diagram is shown in Fig. 1. The solubility of carbon in iron is described in Eq 1 to 7 (Ref 2, 3).

For the stable system, that is, with graphite as the equilibrium high-carbon phase, Eq 1 to 3 are used. For the liquid:

$$\text{wt\%C}_{max} = 1.3 + 2.57 \times 10^{-3}\, t \qquad \text{(Eq 1)}$$

in the interval 1152 to 2000 °C (2106 to 3632 °F), with t being the temperature in degrees Celsius. Using thermodynamic quantities, Eq 1 can be written as:

$$\log X_{C_{max}} = -\frac{12.728}{T} + 0.727 \log T - 3.049 \qquad \text{(Eq 2)}$$

where $X_{C_{max}}$ is the maximum solubility of graphite in liquid iron in mole fraction and T is the temperature in degrees Kelvin. Equation 2 is valid for the interval 1425 to 2300 K.

For the austenite:

$$\text{wt\%C}_{max} = -0.435 + 0.355 \times 10^{-3}\, t + 1.61 \times 10^{-6}\, t^2 \qquad \text{(Eq 3)}$$

For the ferrite:

$$\text{wt\%C}_{max} = 2.46 \times 10^3 \exp\left(-\frac{11\,460}{T}\right) \qquad \text{(Eq 4)}$$

For the metastable system, that is, with Fe_3C as the equilibrium high-carbon phase, Eq 5 to 7 apply. For the liquid:

$$\text{wt\%C}_{max} = 4.34 + 0.1874\,(t - 1150) - 200 \ln \frac{t}{1150} \qquad \text{(Eq 5)}$$

For the austenite:

$$\text{wt\%C}_{max} = -0.628 + 1.222 \times 10^{-3}\, t + 1.045 \times 10^{-6}\, t^2 \qquad \text{(Eq 6)}$$

For the ferrite:

$$\text{wt\%C}_{max} = 1.8 \times 10^3 \exp\left(-\frac{10\,908}{T}\right) \qquad \text{(Eq 7)}$$

Equations 1 to 7 can be used to calculate the maximum solubility of carbon in various phases (liquid, austenite, ferrite) as a function of temperature. For example, to calculate the carbon content of the stable eutectic, Eq 1 is used with t = 1152 °C (2106 °F) to obtain %C_{max} = 4.26%.

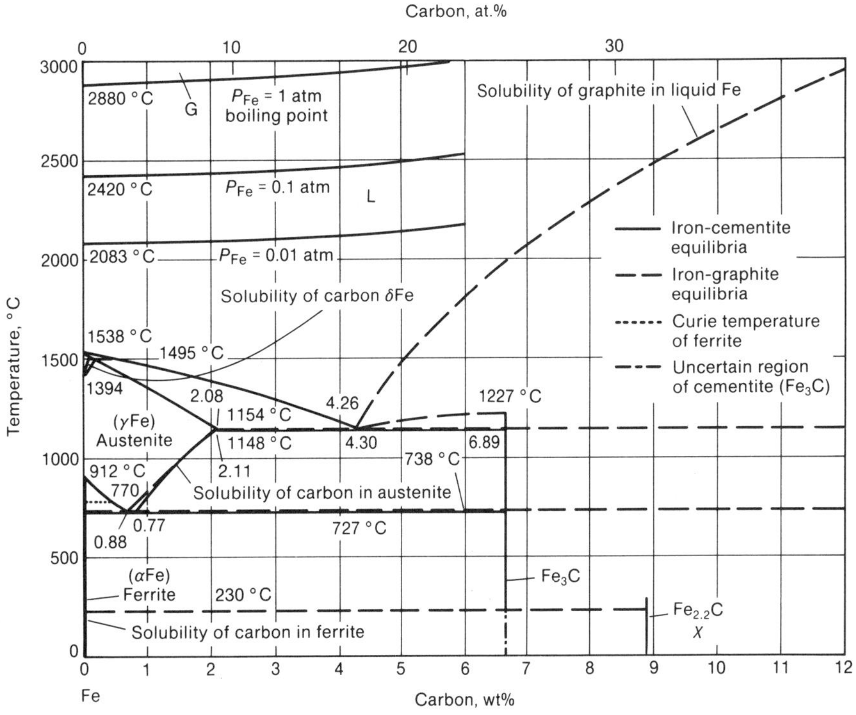

Fig. 1 Phase diagram for the binary Fe-C system. Source: Ref 1

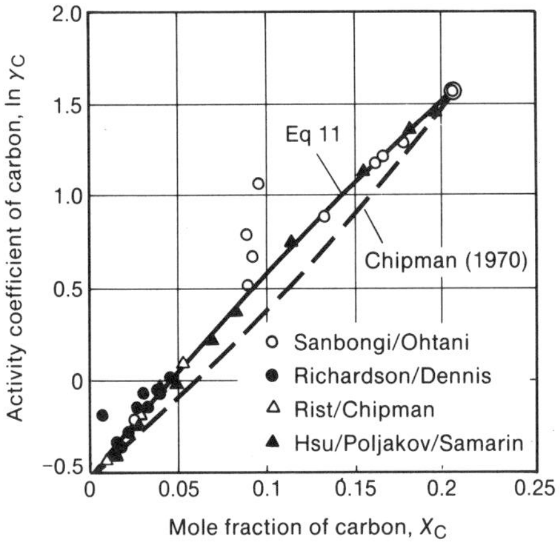

Fig. 2 Dependence of the activity coefficient of carbon on the mole fraction of carbon in liquid iron at 1550 °C (2820 °F). Source: Ref 5

Prediction of phase stability relies heavily on knowledge of the activity of various elements. In the specific case of iron-carbon alloys, an increase in the activity of carbon in the liquid parallels a greater ability of carbon to separate as graphite; that is, the activity of carbon is a measure of an increased tendency of the alloy to solidify as the stable system. On the other hand, a decreased activity of carbon reflects the carbide-promoting behavior of the system, that is, the metastable solidification tendency.

The general equation used to describe the activity of a component 2 in a component 1 according to the Darken formalism (Ref 4) is:

$$\log \frac{\gamma_2}{\gamma_2^\circ} = \alpha_{12}\,(X_2^2 - 2X_2) \qquad \text{(Eq 8)}$$

where γ_2 is the activity coefficient of component 2 = a_2/X_2 (a_2 being the activity of component 2 and X_2 the mole fraction of component 2); γ_2° is the activity coefficient of component 2 in the range of infinitely dilute solutions, that is, when %component 2 ≅ 0; it is only a function of temperature ($\gamma_2 \to \gamma_2^\circ$ as $X_2 \to 0$); and α_{12} is the function of temperature, usually of the shape A/T + B, where A, B are constants.

For the activity of C in an iron melt, Eq 8 is valid for component 1 (Fe) and component 2 (C) with (Ref 5):

$$\alpha_{Fe-C} = -\left(\frac{1270}{T} + 1.74\right) \qquad \text{(Eq 9)}$$

$$\log \gamma_C^\circ = \frac{1180}{T} - 0.87 \qquad \text{(Eq 10)}$$

Using the natural logarithm (base e) instead of the common logarithm (base 10), Eq 8 can be written as follows for the Fe-C systems:

$$\ln \gamma_C = \frac{2714}{T} - 2 + \left(\frac{2920}{T} + 4.01\right)\left(2X_C - X_C^2\right) \qquad \text{(Eq 11)}$$

Equation 11 is in good agreement with experimental data, as shown in Fig. 2.

The Fe-Si System. Equation 8 can be used to calculate the activity of silicon in the Fe-Si system, with the following remarks (Ref 5): $X_{Si} \leq 0.2$; component 1 is iron; component 2 is silicon:

$$\alpha_{Fe-Si} = -\left(\frac{153}{T} + 3.02\right) \qquad \text{(Eq 12)}$$

$$\log \gamma_{Si}^\circ = 0.914 - \frac{6863}{T} \qquad \text{(Eq 13)}$$

As shown in Fig. 3, calculations with Eq 12 and 13 are in good agreement with experimental data (note that the equation in Fig. 3 results from introducing Eq 12 and 13 into Eq 8 and using a natural logarithm rather than a common logarithm).

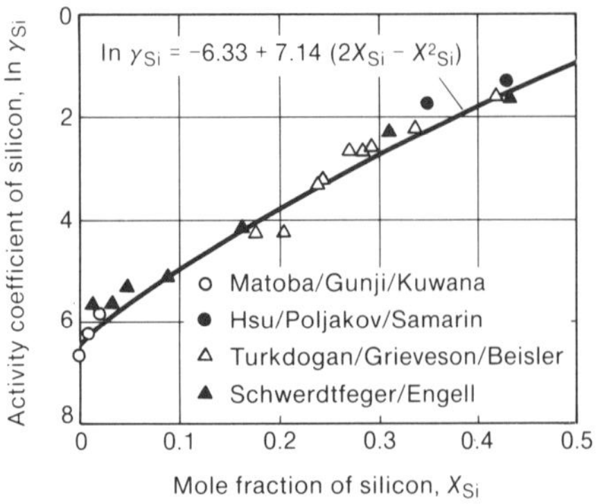

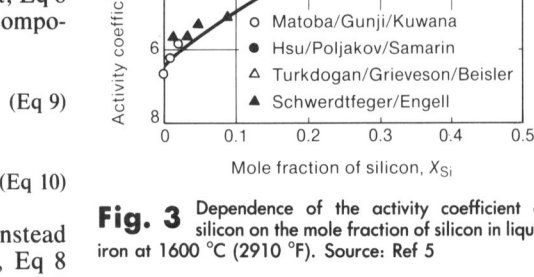

Fig. 3 Dependence of the activity coefficient of silicon on the mole fraction of silicon in liquid iron at 1600 °C (2910 °F). Source: Ref 5

Thermodynamics of Ternary Iron-Base Alloys

Using a Taylor series expansion for the excess partial molar free energy, one researcher derived an expression for the activity coefficient of component 2 in a multicomponent solution, which for dilute solutions can be written as (Ref 6):

$$\ln f_2 = \ln \frac{\gamma_2}{\gamma_2^\circ} X_2\,\epsilon_{22} + X_3\,\epsilon_{23} + X_4\,\epsilon_{24}, \ldots \qquad \text{(Eq 14)}$$

where γ_2 is the activity coefficient of component 2 dissolved in the three-component system, γ_2° is the activity coefficient of component 2 in the range of infinitely dilute solution, and ϵ_{22}, ϵ_{23}, ϵ_{24}, . . . are coefficients defined as:

$$\epsilon_{22} = X_1 \xrightarrow{\text{limit}} 1 \frac{\partial \ln f_2}{\partial X_2}$$

$$\epsilon_{23} = X_1 \xrightarrow{\text{limit}} 1 \frac{\partial \ln f_2}{\partial X_3} \qquad \text{(Eq 15)}$$

For infinitely dilute solutions, that is, $X_2 \to 0$, $X_3 \to 0$:

$$\epsilon_{23} = \epsilon_{32}$$

Table 1 Interaction coefficients in ternary iron-base alloys for carbon, hydrogen, nitrogen, oxygen, and sulfur at 1600 °C (2910 °F)

Element j	e_{Cj}	e_{Hj}	e_{Nj}	e_{Oj}	e_{Sj}
Aluminum	0.043	0.013	−0.028	−3.9	0.035
Boron	0.24	0.05	0.094	−2.6	0.13
Carbon	0.14	0.06	0.13	−0.13	0.11
Cobalt	0.008	0.002	0.011	0.008	0.003
Chromium	−0.024	−0.002	−0.047	−0.04	−0.011
Copper	0.016	<0.001	0.009	−0.013	−0.008
Manganese	−0.012	−0.001	−0.02	−0.021	−0.026
Molybdenum	−0.008	0.002	−0.011	0.004	0.003
Nitrogen	0.11	. . .	0	0.057	0.01
Niobium	−0.06	−0.002	−0.06	−0.14	−0.013
Nickel	0.012	0	0.01	0.006	0
Oxygen	−0.34	−0.19	0.05	−0.20	−0.27
Phosphorus	0.051	0.011	0.045	0.07	0.29
Sulfur	0.046	0.008	0.007	−0.133	−0.028
Silicon	0.08	0.027	0.047	−0.131	0.063
Tin	0.041	0.005	0.007	−0.011	−0.004
Titanium	. . .	−0.019	−0.53	−0.31	−0.072
Vanadium	−0.077	−0.007	−0.093	−0.14	−0.016
Tungsten	−0.006	0.005	−0.002	−0.009	0.01
Zirconium	. . .	. . .	−0.63	(−3.0)	−0.052

Source: Ref 8, 9

Table 2 Free energies of solution in liquid iron at 1 wt%

M_s, liquid iron at 1600 °C (2910 °F)

Element i	γ_i°	ΔG_s (cal · g · atom^{-1})
Aluminum (l)	0.029	−15 100 − 6.67T
Carbon (g)	0.57	5400 − 10.10T
Cobalt (l)	1.07	240 − 9.26T
Chromium (s)	1.14	4600 − 11.20T
Copper (l)	8.6	8000 − 9.41T
½H_2(g)	. . .	8720 + 7.28T
Manganese (l)	1.3	976 − 9.12T
½N_2(g)	. . .	860 + 5.71T
Nickel (l)	0.66	−5000 − 7.42T
½O_2(g)	. . .	−28 000 − 0.69T
½P_2(g)	. . .	−29 200 − 4.60T
½S_2(g)	. . .	−32 280 + 5.60T
Silicon (l)	0.0013	−31 430 − 4.12T
Titanium (s)	0.038	−7440 − 10.75T
Vanadium (s)	0.1	−4950 − 10.90T
Tungsten (s)	1.2	7500 − 15.20T
Zirconium (s)	0.043	−8300 − 11.95T

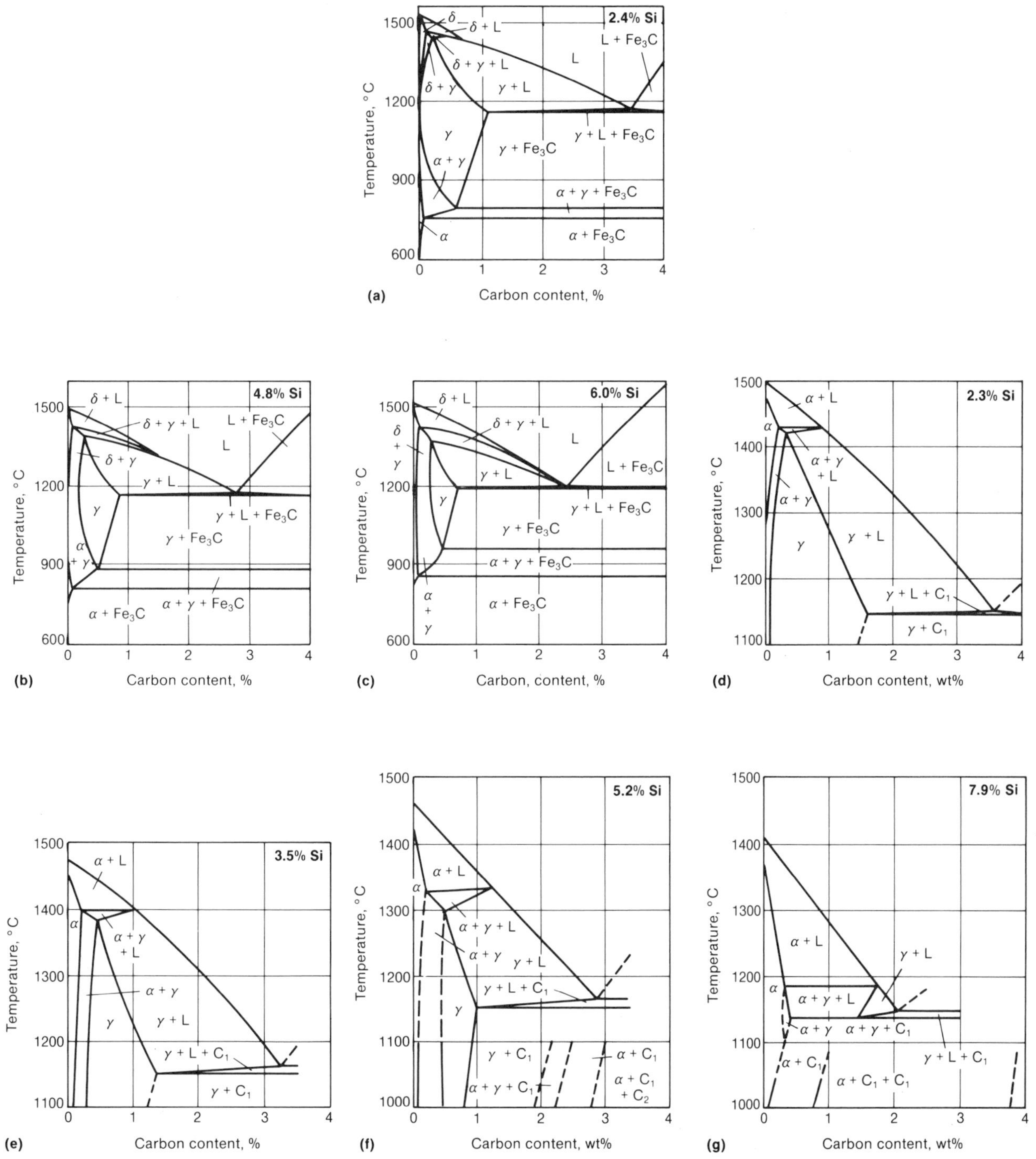

Fig. 4 Influence of silicon content on the solubility lines and equilibrium temperatures of the Fe-C system. (a) to (c) Source: Ref 11. (d) to (g) Source: Ref 10

Simplification of the Darken formalism for infinite dilutions gives the interrelationship between ε and α, as follows (Ref 7):

$$\frac{\epsilon_{22}}{2.303} = -2\alpha_{12}$$

$$\frac{\epsilon_{33}}{2.303} = -2\alpha_{13}$$

$$\frac{\epsilon_{23}}{2.303} = \frac{\epsilon_{32}}{2.303} = \alpha_{23} - \alpha_{12} - \alpha_{13} \qquad \text{(Eq 16)}$$

The solute concentration is often given in weight percent. For a ternary system, Eq 14 is written:

$$\log f_i = e_{ii}\,[\%i] + e_{ij}\,[\%j] \qquad \text{(Eq 17)}$$

where the interaction coefficient e is related to ε by:

$$\epsilon \;= 230\,(M_i/M_1)\,e_{ii} + [(M_1 - M_i)/M_1]$$
$$\epsilon^{ii}_{ij} = 230\,(M_j/M_1)\,e_{ij} + [(M_1 - M_j)/M_1] \qquad \text{(Eq 18)}$$

Data on interaction coefficients in liquid iron for carbon, hydrogen, nitrogen, oxygen, and sulfur are given in Table 1. For metallurgical reactions involving dilute solutions, it is often necessary to change the standard state from pure component to that of 1 wt% in solution. The free energy change is given by:

$$\Delta G_s = RT \ln \frac{0.5585\,\gamma_i^\circ}{M_i} \qquad \text{(Eq 19)}$$

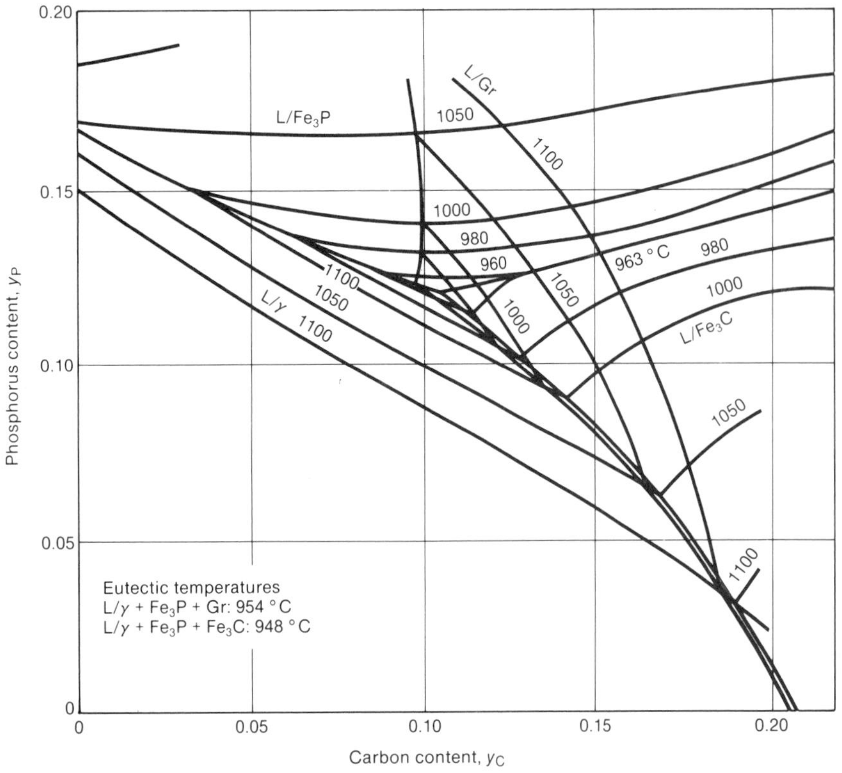

Fig. 5 Calculated liquidus surfaces in the Fe-C-P phase diagram. Source: Ref 12

Free energies of solution ΔG_s of selected elements in liquid iron are given in Table 2.

Thermodynamics of Ternary Fe-C-*X* Systems

Detailed equilibrium diagrams of the Fe-C-B, Fe-C-Cr, Fe-C-Cu, Fe-C-Mn, Fe-C-Mo, Fe-C-N, Fe-C-Ni, Fe-C-Si, and Fe-C-W systems are available in Ref 10. Only the Fe-C-Si, Fe-C-P, and Fe-C-Mn systems will be discussed in this article.

The Fe-C-Si System. The addition of silicon to a binary iron-carbon alloy decreases the stability of Fe_3C, which is already metastable, and increases the stability of ferrite (the α field is enlarged, and the γ field is constricted). The equilibrium diagrams in Fig. 4 show that as the silicon content in the Fe-C-Si system increases, the carbon contents of the eutectic and eutectoid decrease ($\%C_{max}$ decreases), while the eutectic and eutectoid temperatures increase (Ref 11). The activity coefficient, γ_2, of component 2, dissolved in a three-component system as a function of the concentration X_2 and X_3 of dissolved components 2 and 3, is described by Eq 20 (Ref 7):

$$\log \frac{\gamma_2}{\gamma_2^\circ} = -2\,\alpha_{12}X_2 + (\alpha_{23} - \alpha_{12} - \alpha_{13})\,X_3 + \alpha_{12}X_2^2 + \alpha_{13}X_3^2 + (\alpha_{12} + \alpha_{13} - \alpha_{23})\,X_2X_3 \quad \text{(Eq 20)}$$

where α_{12}, α_{13} are the modified interaction parameters of the two-component systems, α_{23} is the modified interaction parameter that describes the activity of both dissolved components, and γ_2° is the activity coefficient of component 2 in the range of infinitely dilute solution.

Starting from Eq 20, the activity of carbon and silicon in an Fe-C-Si melt can be calculated as a function of temperature and concentration with the following relationships (Ref 5):

$$\log a_C = \frac{1180}{T} - 0.87 + \left(\frac{2540}{T} + 3.48\right) X_C - \left(\frac{1270}{T} + 1.74\right) X_C^2 + \log X_C + \left(\frac{1423}{T} + 4.75\right) X_{Si} - \left(\frac{153}{T} + 3.02\right) X_{Si}^2 - \left(\frac{1423}{T} + 4.76\right) X_C X_{Si} \quad \text{(Eq 21)}$$

$$\log a_{Si} = -\frac{6863}{T} + 0.914 + \left(\frac{306}{T} + 6.04\right) X_{Si} - \left(\frac{153}{T} + 3.02\right) X_{Si}^2 + \log X_{Si} + \left(\frac{1423}{T} + 4.76\right) X_C - \left(\frac{1270}{T} + 1.74\right) X_C^2$$

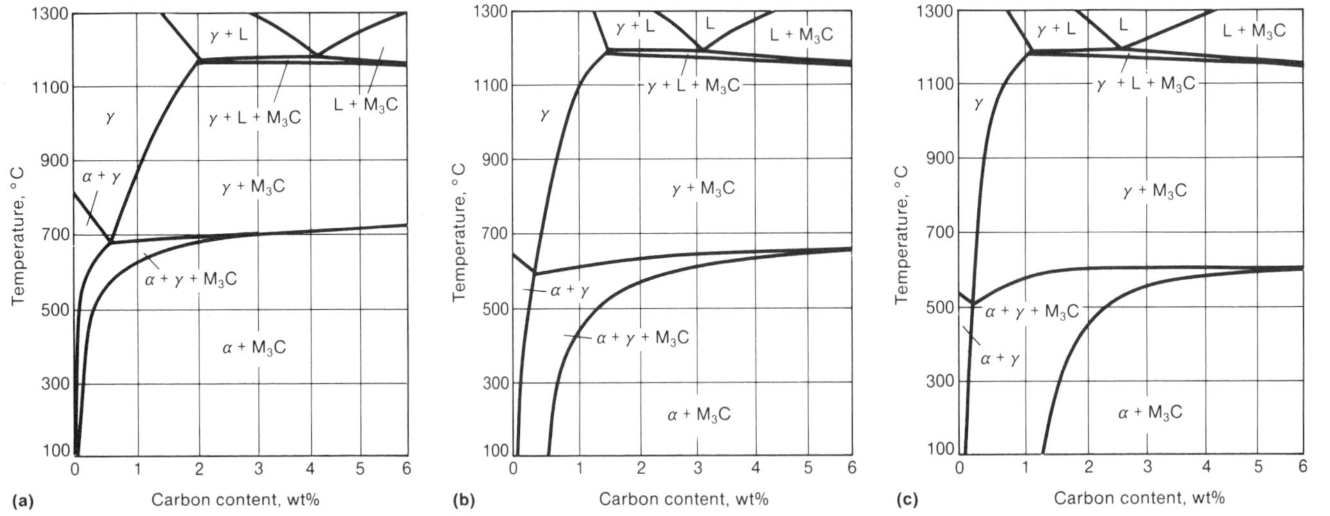

Fig. 6 Vertical sections through the Fe-C-Mn ternary phase diagram at 4.92 (a), 12.8 (b), and 19.7 (c) wt% Mn

Table 3 Influence of a third element on the temperature change of critical points on the Fe-C diagram

+, increase; −, decrease

Element	Maximum solubility of carbon in austenite, °C/wt%X Metastable (E)	Stable (E′)	Eutectoid, °C/wt%X Metastable (S)	Stable (S′)	Eutectic, °C/wt%X Metastable C	Stable C′	Eutectic, °C/at.%X(a) Metastable C	Stable C′
Silicon	−10 to 15	+2.5	+8	0–30	−10 to 20	+4	−3.25	10.91
Copper	−2	+5.2	...	−10	−2.3	+5	−3.32	8.41
Aluminum	−14	+8	+10	+10	−15	+8	−7.74	0.91
Nickel	−4.8	+4	−20	−30	−6	+4	−1.33	6.50
Cobalt	...	...	...	...	...	...	−2.23	1.48
Chromium	+7.3	−	+15	+8	+7	−	4.97	−10.23
Manganese	+3.2	−2	−9.5	−3.5	+3	−2	−2.33	−7.15
Molybdenum	−	−	+	+	−	−	−8.90	−12.36
Tungsten	−	−	+	+	−	−	−12.13	−15.55
Boron	...	...	...	...	...	...	−15.47	−18.53
Nickel	...	...	...	...	...	...	19.67	16.87
Titanium	...	...	...	...	...	...	−16.91	−18.91
Vanadium	+6–8	...	+15	+	6–8	−	...	...
Phosphorus	−180	−180	+	+6	−37	−30	16.05	−16.98
Sulfur	...	...	...	...	...	...	...	−18.53

(a) Calculated results. Source: Ref 3, 13

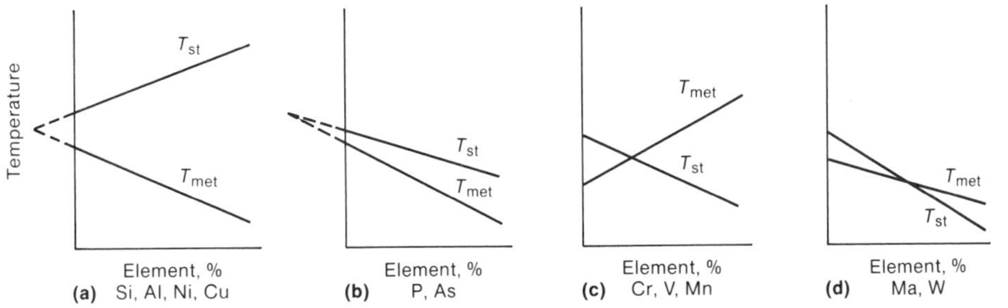

Fig. 7 Classification of the influence of a third element in the Fe-C-*X* system on the graphite- or carbide-promoting tendency in cast iron, based on their influence on the eutectic stable and metastable temperatures. (a) Strong graphitizers. (b) Weak graphitizers. (c) Strong carbide stabilizers. (d) Weak carbide stabilizers

$$-\left(\frac{1423}{T} + 4.76\right) X_C X_{Si} \quad \text{(Eq 22)}$$

The Fe-C-P System. Because the partition coefficient of phosphorus between solid and liquid iron is low, phosphorus segregates into the liquid during solidification, and the melt becomes supersaturated in phosphorus, which results in the formation of Fe_3P. Therefore, the four possible solid phases of the stable and metastable systems are austenite (γ), Fe_3P, Fe_3C, and graphite. The calculated liquidus surfaces for the four phases and the eutectic lines formed by their interactions are shown in Fig. 5, which is a superimposition of the γ-Fe_3P-Fe_3C and the γ-Fe_3P-graphite (Gr) diagrams. Two ternary eutectic points are shown in Fig. 5. To the left is $L \rightarrow \gamma + Fe_3P + G_r$ at 954 °C (1749 °F) and $y_C = 0.099$, $y_P = 0.123$. To the right is $L \rightarrow \gamma + Fe_3P + Fe_3C$ at 948 °C (1738 °F) and $y_C = 0.106$, $y_P = 0.123$. Here, $y_P = X_P/(1 - X_C)$, and $y_C = X_C/(1 - X_C)$, where X_P is the weight fraction of phosphorus and *X* is the weight fraction of carbon. In general, it is agreed that phosphorus does not have a strong graphite- or carbide-stabilizing influence.

The Fe-C-Mn System. The influence of increasing manganese contents in the Fe-C-Mn system over the equilibrium phase diagram is shown in Fig. 6. It is evident that manganese:

- Decreases the eutectoid temperature
- Increases the eutectoid interval, that is, the range of temperature and carbon contents over which α, γ, and carbide can coexist
- Decreases the carbon content in the eutectoid and in the eutectic
- Increases the eutectic temperature (by about 3 °C, or 5 °F, for each 1% Mn)

The carbide M_3C, which is $(FeMn)_3C$, is stable over a wide range of manganese and iron contents, extending at 1000 °C (1832 °F) from Mn_3C to nearly pure Fe_3C. At high manganese contents (for example, above 40% Mn), other manganese carbides such as M_5C_2, M_7C_3, $M_{23}C_6$, and $M_4C_{1+x}(\epsilon)$ can form (Ref 10).

Influence of a Third Element on the Equilibrium Temperatures. Various elements, dissolving in liquid and solid iron phases, change the equilibrium temperatures on the Fe-C diagram, as indicated in Table 3. This results in increased or decreased α and γ fields.

Of particular interest for the case of cast iron is the influence of third elements on the stable-metastable eutectic interval ($T_{st} - T_{met}$). In general, elements that increase the $T_{st} - T_{met}$ interval promote graphite formation, while those that decrease the interval promote carbide formation. Based on their specific influence on the $T_{st} - T_{met}$ interval, third elements can be classified into the following four groups (Fig. 7):

- Strong graphitizers that increase T_{st} and decrease T_{met}, such as silicon, aluminum, nickel, and copper (Fig. 7a)
- Weak graphitizers that decrease both T_{st} and T_{met} but increase the $T_{st} - T_{met}$ interval overall, such as phosphorus and arsenic (Fig. 7b)
- Strong carbide stabilizers that decrease T_{st} but increase T_{met}, such as chromium, vanadium, and manganese (Fig. 7c)
- Weak carbide stabilizers that decrease both T_{st} and T_{met}, such as molybdenum and tungsten (Fig. 7d)

Real values for the $\Delta T^{st}_{met} = T_{st} - T_{met}$ interval for a number of elements can be

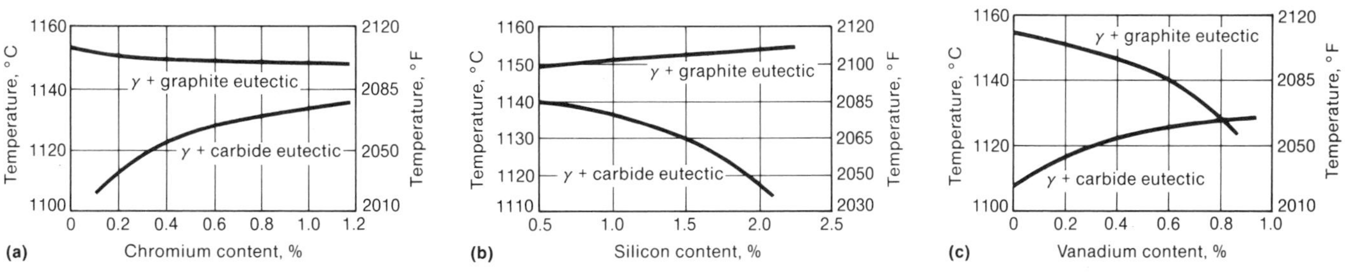

Fig. 8 Influence of chromium (a), silicon (b), and vanadium (c) on the equilibrium eutectic temperatures of cast irons

Table 4 Equilibrium partition coefficients of a third element *X* in the Fe-C-*X* system

Element *X*	$P_X^{\gamma/L}$ calc	$P_X^{\gamma/L}$ exp	$P_X^{Fe_3C/L}$ calc	$P_X^{Fe_3C/L}$ exp	P^S	P^M	$\Delta P = P^M - P^S$
Silicon	1.71	1.72	0	0.05	1.55	0.78	−0.77
Copper	1.57	1.62	0.12	0.08	1.43	0.78	−0.65
Aluminum	. . .	1.15	. . .	0.03	1.05	0.55	−0.50
Nickel	1.46	1.61	0.43	0.32	1.33	0.90	−0.43
Cobalt	1.18	1.13	0.59	0.60	1.07	0.85	−0.21
Chromium	0.53	0.55	1.96	1.95	0.48	1.32	0.84
Manganese	0.70	0.75	1.03	1.21	0.64	0.90	0.26
Molybdenum	0.41	0.38	0.60	0.84	0.37	0.52	0.15
Tungsten	0.23	0.42	0.42	0.88	0.21	0.33	0.12
Boron	0.06	. . .	0.22	. . .	0.06	0.15	0.09
Nitrogen	2.04	2.04	2.12	. . .	1.86	2.09	0.23
Titanium	0.04	. . .	0.09	0.27	0.04	0.07	0.03
Phosphorus	0.15	. . .	0.08	0.09	0.14	0.11	−0.03
Sulfur	0.06	. . .	. . .	. . .	0.06	. . .	. . .

(a) calc, calculated; exp, experimental. Source: Ref 13

calculated with the data given in Table 3, which also includes data on the influence of some elements on the eutectoid temperature and on the maximum solubility of carbon in austenite. Two examples of the use of data in Table 3 are provided below.

Example 1: Calculation of the value of the $T_{st} - T_{met}$ eutectic interval for a cast iron having 2% Si, 0.5% Mn, and 1% Cu by weight:

$$T_{st} = 1154\ °C + 4\%\ Si - 2\%\ Mn + 5\%\ Cu - 1166\ °C$$

$$T_{met} = 1148\ °C - 15\%\ Si + 3\%\ Mn - 2.3\%\ Cu - 117.2\ °C$$

$$\Delta T_{Fe-C} - 1154 - 1148 = 6\ °C$$

$$\Delta T_{Fe-C-Mn-Cu} = 1166 - 1117.2 = 48.8\ °C$$

Example 2: Calculation of the temperatures of the point of maximum carbon solubility in austenite, T_E and $T_{E'}$, for the above iron:

$$T_E = 1154\ °C - 10\%\ Si + 3.2\%\ Mn - 2\%\ Cu - 1133.5\ °C$$

$$T_{E'} = 1148\ °C + 2.5\%\ Si - 2\%\ Mn + 5.2\%\ Cu - 1157.2\ °C$$

Some typical experimental results for silicon, chromium, and vanadium are shown

Table 5 Solubility factors of various third elements for carbon-saturated Fe-C-*X* melts

Third element *X*: Atomic No.	Third element *X*: Symbol	$\Delta X_C^X = m \cdot X_X$(a): *m* exp	*m* calc	Validity	$\Delta\%C^X = m' \cdot \%X$: *m'* exp	*m* calc	Validity	$\Delta\%C_C^X = m'_\gamma\%X$: m'_γ	Validity
5	B	−0.575	−0.51	<0.04	−0.539	−0.462	<1.0	. . .	. . .
6	C	−0.67	−0.68	. . .	−0.610	−0.622	. . .	. . .	. . .
13	Al	−0.52	−0.51	<0.05	−0.220	−0.215	<2.0	+0.04	<2
14	Si	−0.71	−0.68	<0.007	−0.310	−0.294	<5.5	−0.11	<6
15	P	−0.81	−0.85	<0.048	−0.331	−0.349	<3.0	−0.35	<0.4
16	S	−1.00	−1.02	<0.004	−0.405	−0.414	<0.4	−0.08	<5
22	Ti	+0.508	+0.44	<0.013	+0.159	+0.138	<1.6	−2.08	<0.8
23	V	+0.359	+0.33	<0.1	+0.105	+0.097	<11.0	+0.18	<1
24	Cr	+0.215	+0.22	<0.13	+0.064	+0.062	<10.0	−0.07	<20
25	Mn	+0.105	+0.11	<0.30	+0.028	+0.029	<25.0	+0.006	<60
26	Fe	0.0	0.0	. . .	0.0	0.0	. . .	. . .	. . .
27	Co	−0.10	−0.11	<0.20	−0.027	−0.029	<40.0	−0.017	<20
28	Ni	−0.20	−0.22	<0.10	−0.051	−0.055	<8.0	−0.07	<6
29	Cu	−0.313	−0.33	<0.02	−0.076	−0.080	<3.8	+0.014	<5
32	Ge	−0.677	−0.66	<0.1	−0.144	−0.141	<14.0	. . .	. . .
33	As	−0.734	−0.77	<0.07	−0.151	−0.159	<10.0	. . .	. . .
41	Nb	+0.51	+0.33	<0.045	+0.058	+0.030	<9.0	. . .	. . .
42	Mo	+0.24	+0.22	<0.14	+0.014	+0.012	<2.0	−0.3	<4
44	Ru	0.0	0.0	<0.008	−0.020	−0.023	<1.5	. . .	. . .
50	Sn	−0.70	−0.66	<0.02	−0.110	−0.104	<4.5	. . .	. . .
51	Sb	−0.79	−0.77	<0.05	−0.119	−0.117	<15.0	. . .	. . .
52	Te	−0.85	−0.88	<0.005	−0.122	−0.126	<1.5	. . .	. . .
73	Ta	+0.49	+0.33	<0.03	+0.004	−0.009	<11.0	. . .	. . .
74	W	+0.256	+0.22	<0.05	−0.015	−0.01	<18.0	−0.12	<12
76	Os	0.0	0.0	<0.013	−0.035	−0.035	<4.0	. . .	. . .
78	Pt	−0.224	−0.22	<0.015	−0.052	−0.052	<6.0	. . .	. . .
79	Au	−0.333	−0.33	<0.024	−0.060	−0.060	<10.0	. . .	. . .
92	U	+0.53	+0.55	<0.01	−0.006	−0.005	<5.0	. . .	. . .

(a) exp, experimental; calc, calculated. Source: Ref 2

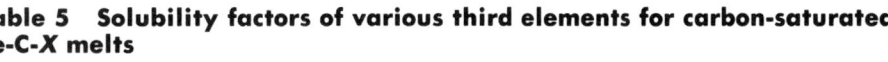

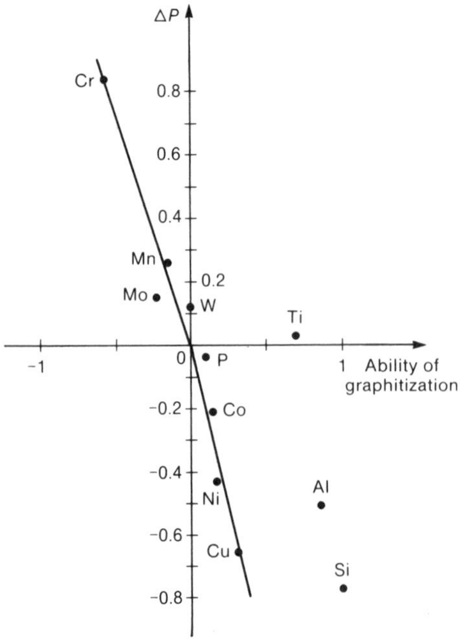

Fig. 9 Correlation between the partition coefficient, ΔP, and the graphitizing influence of a third element in the Fe-C-*X* system. Source: Ref 13

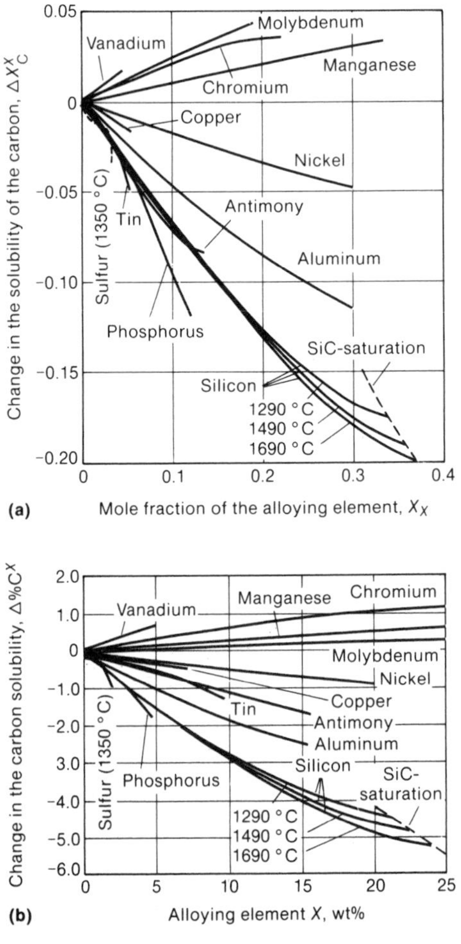

Fig. 10 Influence of third elements, *X*, on the solubility of carbon in molten iron. (a) In mole fraction. (b) In weight percent. Source: Ref 2

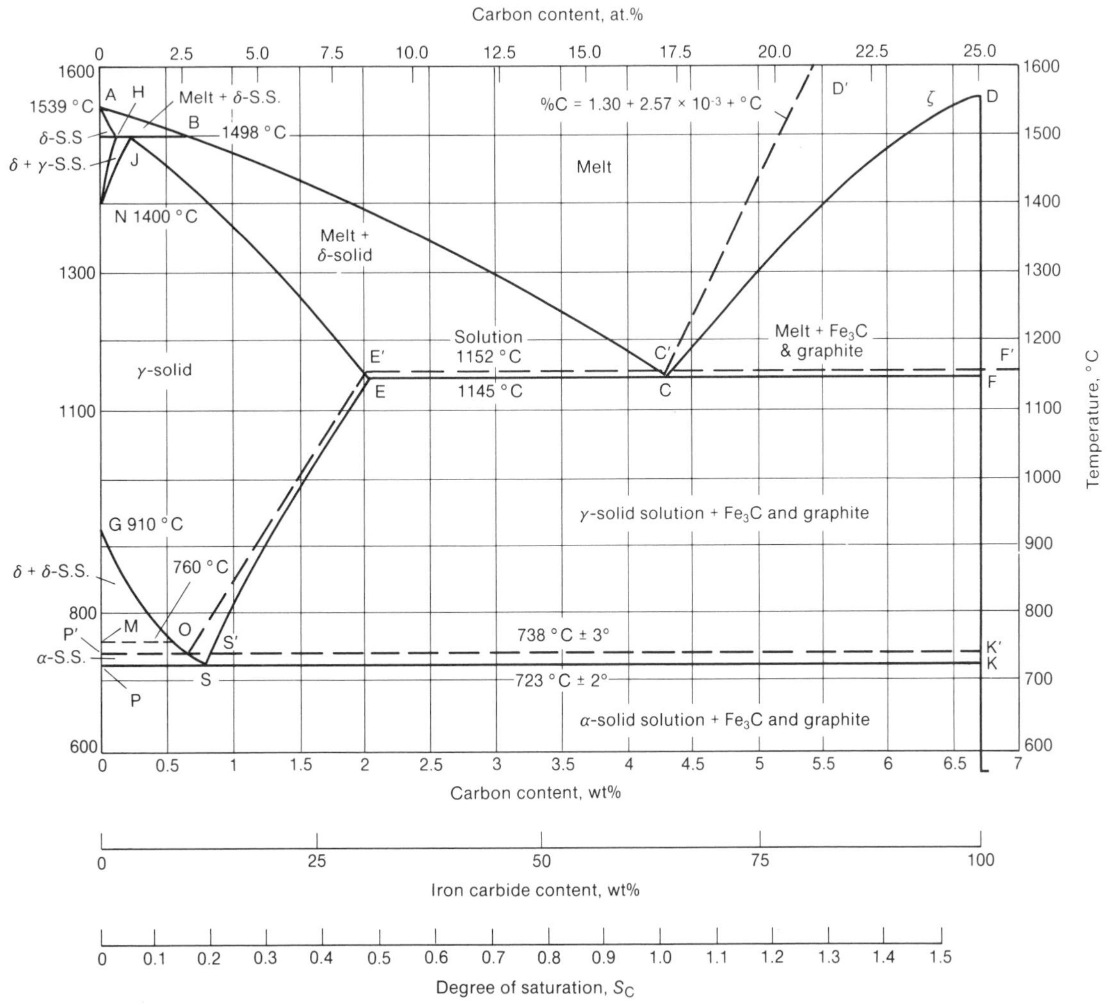

Fig. 11 The Fe-C diagram with letter notations. S.S., stainless steel

in Fig. 8. These results are in agreement with theoretical predictions.

Partition of a Third Element Between Various Phases in the Fe-C-*X* System. It has been established from the thermodynamic stability of phases involved in the eutectic solidification of Fe-C-X alloy that the differences in the equilibrium partition coefficients ΔP ($=P^M - P^S$, where P^M and P^S are the partition coefficients of element X between liquid and eutectic metastable or stable solid), the partition of element X between austenite and liquid, $P_X^{\gamma/L}$, and the partition of element X between cementite and austenite, $P_X^{Fe_3C/\gamma}$, are related to one another and have similar effects on graphitization during eutectic solidification (Ref 13). Calculated and experimental data for partition coefficients of various elements in phases of iron-carbon alloys are given in Table 4.

Figure 9 illustrates the correlation between the graphitizing influence of a third element (expressed by the normalized addition of each element required to increase or decrease the chill depth by a given amount) and the ΔP value. It is quite obvious that a definite correlation exists, although silicon, aluminum, and titanium deviate from the line. These three elements have a strong affinity for nitrogen; this may explain their higher-than-expected graphitizing influence. Indeed, it is accepted that the nitrides may act as nuclei for the stable eutectic, promoting gray solidification. This is why experiments show that titanium, for example, can act either as a carbide or graphite promoter, depending on the nitrogen content.

Influence of a Third Element on Carbon Solubility in the Fe-C-*X* System. The influence of a third element on the solubility of carbon can be expressed by (Ref 2):

$$\Delta X_C^X = X^X_{C_{max}} - X_{C_{max}} \text{ (mole fraction)} \quad \text{(Eq 23a)}$$

$$\Delta \%C^X = \%C^X_{max} - \%C_{max} - \%C_{max} \text{ (wt\%)} \quad \text{(Eq 23b)}$$

where ΔX_C^X and $\Delta\%C^X$ represent the increase or decrease of carbon solubility in mole fraction or weight percent, respectively; $X_{C_{max}}$ and $\%C_{max}$ represent the saturation concentration in the Fe-C system calculated from Eq 1 or 2; and $X^X_{C_{max}}$ and $\%C^X_{max}$ are the saturation concentration in the Fe-C-X system.

The changes in carbon solubility resulting from additions of third elements, X, are shown in Fig. 10. These data are valid in the range from 1200 to 1700 °C (2190 to 3090 °F), with the exception of silicon and sulfur at high concentrations. In the region of low concentration of the third element ($\%X = 0$ to 5%), the change in the solubility of carbon can be represented by the temperature-independent linear equation:

$$\Delta X_C^X = m \cdot X_X \quad \text{(Eq 24a)}$$

or

$$\Delta \%C^X = m' \cdot \%X \quad \text{(Eq 24b)}$$

where m and m' are solubility factors. In Eq 24, m and m' as well as ΔX_C^X and $\Delta\%C^X$ are positive for an increase in the solubility of carbon and negative in the other case. It is also generally accepted that carbide-promoting elements increase the solubility of carbon (activity coefficient of carbon in solution is decreased), while graphite-promoting elements decrease it (activity coefficient of carbon in solution is increased).

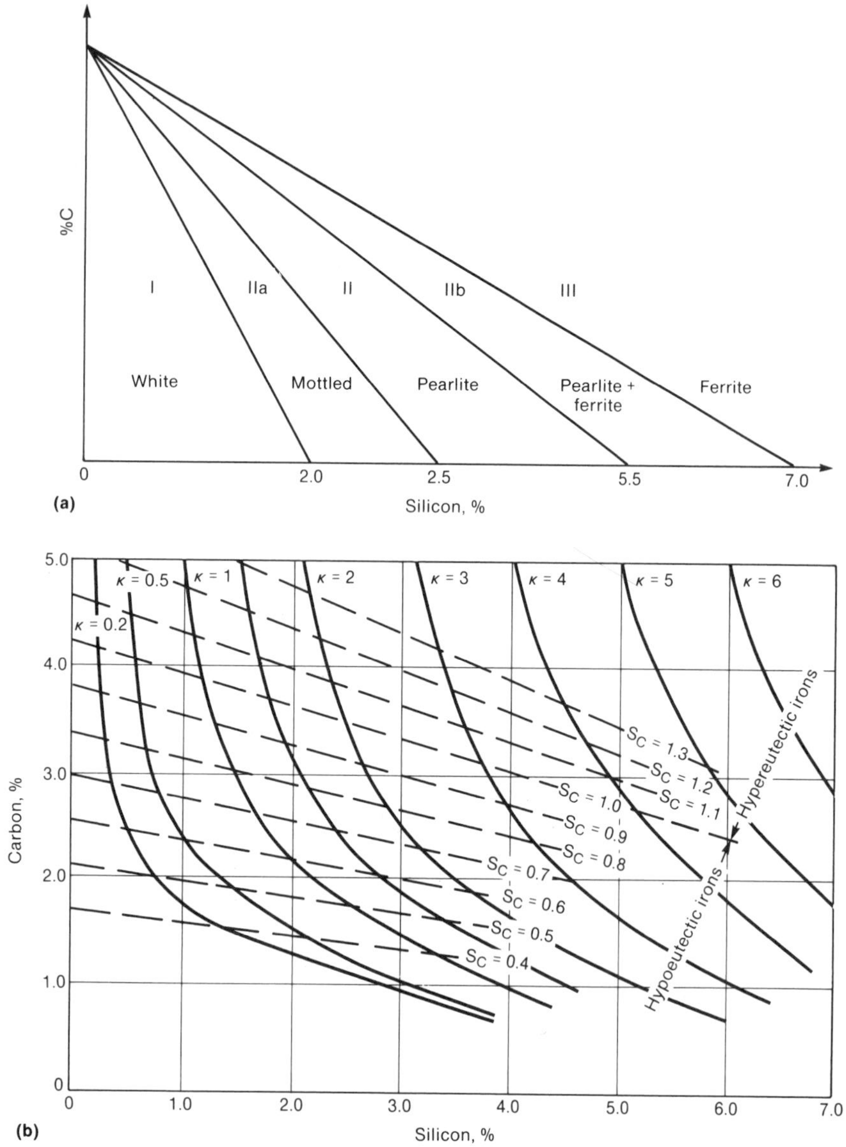

Fig. 12 Structural diagrams for cast iron. (a) Maurer diagram. (b) Laplanche diagram. See also Fig. 13. Source: Ref 14

Therefore, third elements that have a negative solubility factor promote graphitization, while elements that have a positive solubility factor promote carbide formation. The value of the solubility factor is proportional to their effect. It must be noted, however, that the accuracy of the prediction of the behavior of elements based on solubility factor is questioned in Ref 13 for elements that have a strong influence on the austenite liquidus.

As indicated in Table 5, experimental values are close to the theoretical values for many elements. Equations similar to Eq 24 can be used to calculate the change in the maximum solubility of carbon in austenite upon the addition of a third element:

$$\Delta\%C_E^X = m'\gamma \cdot \%X \quad \text{(Eq 25)}$$

Data for m'_γ are also given in Table 5.

Thermodynamics of Multicomponent Fe-C Systems

Carbon Solubility in Multicomponent Systems. Assuming that the values in Table 5 for m and m', which were determined for the ternary Fe-C-X system, can be extended to multicomponent systems (in this case, the effect of alloying elements on the solubility of carbon can be considered to be additive, as shown in Eq 27, at least in the region of low concentrations), the saturation concentration of carbon in multicomponent systems can be calculated as (Ref 2):

$$\%C_{max}^{Si,Mn,P,S,...} = \%C_{max} + \Delta\%C^{Si,Mn,P,S,...} \quad \text{(Eq 26)}$$

which is Eq 23b rewritten for multicomponent systems. In turn:

Table 6 κ lines for various structural ranges

Bar diameter mm	in.	κ for mottled structure	κ for pearlitic structure	κ for pearlitic-ferritic structure
30	1.2	0.65–0.85	0.85–2.05	2.05–3.10
20	0.8	0.75–1.10	1.10–2.25	2.25–3.40
10	0.4	1.05–1.50	1.50–2.35	2.235–3.50

$$\%C_{max}^{Si,Mn,P,S,...} = \%C^{Si} + \Delta\%C^{Mn} + \Delta\%C^{P} + \Delta\%C^{S} + \ldots \quad \text{(Eq 27)}$$

Then, using Eq 1 and data from Table 5, Eq 26 becomes:

$$\%C_{max}^{Si,Mn,P,S,...} = 1.3 + 2.57 \times 10^{-3}\,t - 0.31\%\,Si + 0.027\%\,Mn - 0.33\%\,P - 0.4\%\,S \pm \ldots \quad \text{(Eq 28)}$$

Saturation Degree and Carbon Equivalent. The Fe-C diagram with letter notations for critical points (Fig. 11) will be used in this section. It is well known that mechanical properties of cast iron strongly depend upon the amount and shape of graphite. In turn, for hypoeutectic gray irons, the amount of graphite depends on the amount of eutectic.

The amount of eutectic can be calculated with the lever rule. For hypoeutectic irons one can write:

$$S_r = \frac{\text{Amount of eutectic}}{\text{Amount of eutectic + amount of austenite}} = \frac{\%C_{anal} - \%C'_E}{\%C_{anal} - \%C'_E + \%C'_C - \%C_{anal}} = \frac{\%C_{anal} - \%C'_E}{\%C'_C - \%C'_E} \quad \text{(Eq 29)}$$

where S_r is the rectified saturation degree, $\%C_{anal}$ is the analyzed carbon content of cast iron, and $\%C'_C$ and $\%C'_E$ represent the carbon content of the eutectic and of the austenite, respectively, at the eutectic temperature in the multicomponent system (Fig. 11). Assuming the solubility factors can be extrapolated to the hypoeutectic region, $\%C'_C$ can be calculated with Eq 28. A similar approach can be used to calculate $\%C'_E$ using data in Table 5 for m'_γ:

$$\%C'_E = 2.11 - 0.11\%\,Si - 0.35\%\,P + 0.006\%\,Mn - 0.08\%\,S \quad \text{(Eq 30)}$$

In foundry practice, the following simplified form of Eq 29 is used:

$$S_C = \frac{\%C_{anal}}{\%C'_C} = \frac{\%C_{anal}}{4.26 - 0.31\%\,Si - 0.33\%\,P + \ldots} \quad \text{(Eq 31)}$$

For more accuracy, $\%C_{max}$ as a function of temperature rather than $\%C_{max} = 4.26$ can be used, where the temperature is the eutectic temperature of the multicomponent system. Further simplification gives:

$$S_C = \frac{\%C_{anal}}{4.25 - 0.3\,(Si + P)} \quad \text{(Eq 32)}$$

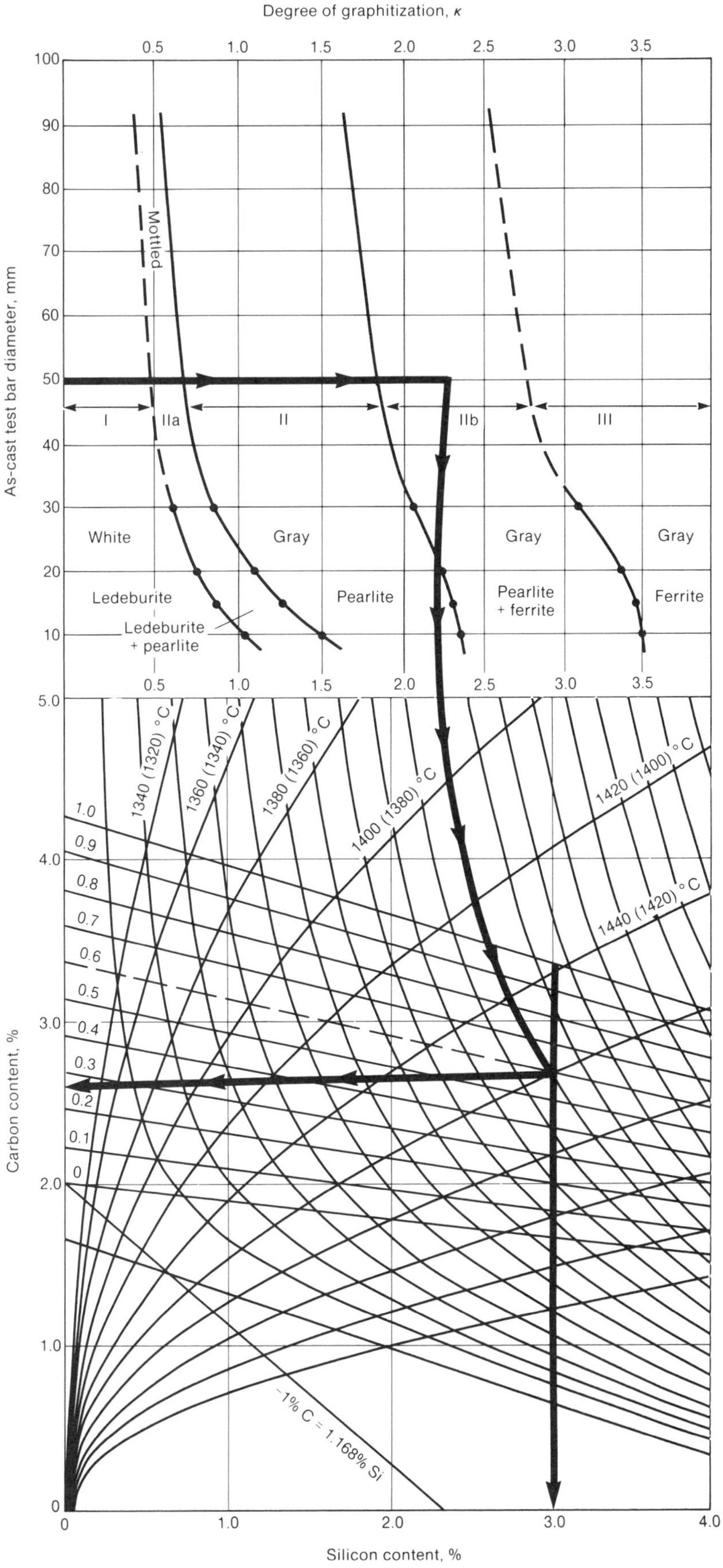

Fig. 13 Patterson and Doepp structural diagram for cast iron. See the corresponding text for details. Source: Ref 15

where $S_C < 1$ for hypoeutectic irons, $S_C = 1$ for eutectic irons, and $S_C > 1$ for hypereutectic irons.

Saturation degree is used in most of the European literature. In the Anglo-Saxon literature and foundry practice, carbon equivalent rather than saturation degree is used. Carbon equivalent (CE) can be calculated as:

$$\begin{aligned} CE &= \%C_{anal} - \Delta\%C^{Si} - \Delta\%C^{Mn} - \Delta\%C^{P} \\ &\quad - \Delta\%C^{S} - \ldots \\ &= \%C_{anal} + 0.31\%\ Si + 0.33\%\ P \\ &\quad - 0.027\%\ Mn + 0.4\%\ S \end{aligned} \qquad \text{(Eq 33)}$$

A eutectic iron has CE = 4.26%. It must be noted that although S_r gives directly the amount of eutectic in the structure (for example, $S_r = 0.9$ means 90% eutectic), S_C and CE, although easier to calculate, do not allow for exact estimation of the amount of eutectic.

Structural Diagrams

For the practicing metallurgist involved in cast iron production, one of the main applications of the thermodynamics of the Fe-C system is the calculation of structure-composition correlations. The so-called structural diagrams are used for this purpose.

The first and simplest structural diagram is the Maurer diagram (Fig. 12a), in which the as-cast structure is considered to be solely the product of the carbon and silicon contents of cast iron. Because it was recognized long ago that solidification rate, as well as chemical composition, plays a significant role in the formation of the as-cast structure of castings, especially for cast iron, Laplanche (Ref 14) further developed the Maurer diagram to include the influence of cooling rate through the section size (or bar diameter) of the casting (Fig. 12b). The κ lines on the diagram are lines of same structure (same degree of graphitization) at a given cooling rate. The value κ is an empirical function of composition, as follows:

$$\kappa = \tfrac{4}{3}\,Si\left(1 - \frac{5}{3C + Si}\right) \qquad \text{(Eq 34)}$$

The correlation between κ and cooling rate is given in Table 6. Thus, if one wants to produce a pearlitic iron in castings with equivalent section sizes of 10 to 30 mm (0.4 to 1.2 in.) bar diameters, Table 6 would suggest a κ of 0.85 to 2.35. For example, κ = 1.8 is then chosen. A saturation degree, S_C, is also selected depending on the required mechanical properties (for example, $S_C = 0.8$), and the intersection of the κ = 1.8 with $S_C = 0.8$ will then give the carbon and silicon contents required for the iron.

A further contribution to structural diagrams was made by Patterson and Doepp

(Ref 15). This diagram (Fig. 13) encompasses not only structure, composition, and cooling rate but also a broad range of mechanical properties. Part of this diagram is based on Laplanche's κ lines.

As an example of the way this diagram is used, assume it is desired to pour a 50 mm (2 in.) diam cast iron bar (or equivalent section size) with a pearlitic-ferritic structure. A heavy horizontal line is drawn from the *y*-axis at the 50 mm (2 in.) diam bar location to the middle of zone IIb. This corresponds to a determination of $\kappa = 2.2$. This κ line can be followed to the lower diagram until it intersects a preselected saturation degree—in this case $S_C = 0.6$. The composition is now read as C = 2.6% and Si = 3%, together with the equilibrium temperature for the $(SiO_2) + 2[C] \leftrightarrows [Si] + \{CO\}$ reaction, which is 1460 °C (2660 °F). An additional explanation on the significance of this reaction can be found in the article "Solidification of Eutectic Alloys" in this Volume (see the section "Cast Irons"). It must be pointed out that foundry variables such as superheating and holding during melting, charge composition (raw materials), and inoculation can cause significant deviation from these calculations.

REFERENCES

1. T.B. Massalski *et al.*, Ed., *Binary Alloy Phase Diagrams*, Vol 1, ASM INTERNATIONAL, 1986
2. F. Neumann, The Influence of Additional Elements on the Physico-Chemical Behavior of Carbon in Carbon Saturated Molten Iron, in *Recent Research on Cast Iron*, H.D. Merchant, Ed., Gordon and Breach, 1968, p 659
3. N.G. Girsovitch, Ed., *Spravotchnik po tchugunomu litja* (Cast Iron Handbook), Mashinostrojenie, 1978
4. L.S. Darken, The Thermodynamics of Ternary Metallic Solutions, *Trans. TMS*, Vol 239, 1967, p 80
5. F. Neumann and E. Dötsch, Thermodynamics of Fe-C-Si Melts With Particular Emphasis on the Oxidation Behavior of Carbon and Silicon, in *The Metallurgy of Cast Iron*, B. Lux *et al.*, Ed., Georgi Publishing, 1975, p 31
6. C. Wagner, *Thermodynamics of Alloys*, Addison-Wesley, 1952
7. L.S. Darken, Thermodynamics of Ternary Metallic Solutions, *Trans. TMS*, Vol 239, 1967, p 90
8. E.T. Turkdogan, *Physical Chemistry of High Temperature Technology*, Academic Press, 1980
9. G.K. Sigworth and J.F. Elliott, The Thermodynamics of Liquid Dilute Iron Alloys, *Met. Sci.*, Vol 8, 1974, p 298
10. *Metallography, Structures and Phase Diagrams*, Vol 8, 8th ed., *Metals Handbook*, American Society for Metals, 1973, p 400-416
11. E. Piwowarsky, *Hochwertiger Gusseisen*, 2nd ed., Springer-Verlag, 1958
12. M. Hillert and P.O. Söderholm, White and Gray Solidification of the Fe-C-P Eutectic, in *The Metallurgy of Cast Iron*, B. Lux *et al.*, Ed., Georgi Publishing, 1975, p 197
13. A. Kagawa and T. Okamoto, Partition of Alloying Elements on Eutectic Solidification of Cast Iron, in *The Physical Metallurgy of Cast Iron*, H. Fredriksson and M. Hillert, Ed., North-Holland, 1985, p 201
14. H. Laplanche, The Maurer Diagram and Its Evolution and a New Structural Diagram for Cast Iron, *Foundry Trade J.*, No. 1669-1671, 1948, p 191, 225, 249
15. W. Patterson and R. Doepp, *Giessereiforschung*, Vol 21 (No. 2), 1969, p 91

Composition Control

Seymour Katz, General Motors Research Laboratories

COMPOSITION CONTROL is vital for the production of quality castings because the most common raw material used for iron and aluminum castings is scrap. By its nature, scrap is of uncertain composition, both with respect to concentrations of desirable alloy elements and the presence of deleterious elements. Thus, despite all precautions in melting, adjustments in melt composition are required. Composition control can be divided into two areas: alloy addition and melt purification. Both will be discussed in this article. The chemical and physical processes involved in alloy addition do not vary much with base metal; therefore, the subject can be treated generically. On the other hand, because melt purification processes tend to be material-specific, each base metal is treated separately.

Kinetics of Alloy Additions

There are two fundamental mechanisms governing the kinetics of alloy additions, depending on whether the alloy is liquid or solid at the temperature of the metal bath. These processes are schematically illustrated in Fig. 1.

Dissolution of Alloy With Melting Point Lower Than Bath Temperature

For the case where the alloy addition is liquid at the melt temperature (Fig. 1, top row sequence), addition of the cold alloy at time t_0 is quickly followed by formation of a layer of frozen base metal on the alloy surface (t_1) due to heat extraction by the alloy. At longer times (t_2), the shell thickness increases, while melting occurs at the alloy surface. As the interior of the alloy heats up, the thermal demand decreases. As a result, the frozen base metal shell begins to redissolve while the thickness of the liquid annulus increases (t_3). In the final stage, beginning at t_4, the redissolution of the base metal shell is complete, and as a result, the liquid alloy held in the annular space is released to the melt. Any alloy that remains solid is exposed directly to the liquid metal bath and is progressively incorporated into the melt by the surface melting and convection. Dissolution is completed at t_m (melting time).

The length of the incubation period ($t_0 - t_4$) and the length of the alloy dissolution period ($t_4 - t_m$) are governed by heat transfer. Conditions favoring rapid alloy dissolution are high superheat temperatures, small alloy particle size (large surface for heat transfer), and low alloy thermal conductivity. The heat flux to the alloy particle is given by:

$$Q = h\,(T_L^B - T_L^*) \qquad \text{(Eq 1)}$$

where Q is the heat flux (J · m^{-2} · s^{-1}), h is the convective heat transfer coefficient (J · m^{-2} · s^{-1} · K^{-1}), T_L^B is the temperature of the bulk liquid (in degrees Kelvin), T_L^* is the melting point temperature of the alloy at the solid/liquid interface (in degrees Kelvin), and $(T_L^B - T_L^*)$ is the superheat temperature (in degrees Kelvin).

During the incubation period, the superheat temperature is less than that in the alloy dissolution period because T_L^* corresponds to the melting point of the base metal and alloy in the former and latter cases, respectively. For steel and aluminum melts, superheat temperatures in the incubation period are generally smaller (20 to 100 K) than for cast iron (200 to 300 K). In the alloy dissolution period, the superheat temperature for a given alloy addition increases with increasing melting point of the base metal. Thus, superheat temperatures, and therefore dissolution rates, increase in the following order: aluminum < copper < cast iron < steel.

Melting times t_m can be estimated by assuming that t_m is equal to the total heat requirement for melting, divided by the rate at which heat is supplied from the bath (Ref 1), that is:

$$t_m = \frac{\Delta H \rho_0 V_0}{Q A_0} \qquad \text{(Eq 2)}$$

where ΔH is the heat supplied to 1 g of alloy at the time it dissolves and ρ_0, V_0, and A_0 are the density, volume, and surface area of the alloy particle. The variable Q is defined by Eq 1.

Equation 2 is most accurate when applied at the extremes of possible superheat temperatures. For small superheat temperatures, it can be assumed that by the time the frozen shell redissolves the alloy is a liquid at the melting point of the base metal. At high superheat temperatures, it can be assumed that no frozen layer forms and that the alloy liquefies at its melting point.

These two cases have been examined for the dissolution of a 100 mm (4 in.) particle diameter (d_p) sphere of silicomanganese in a

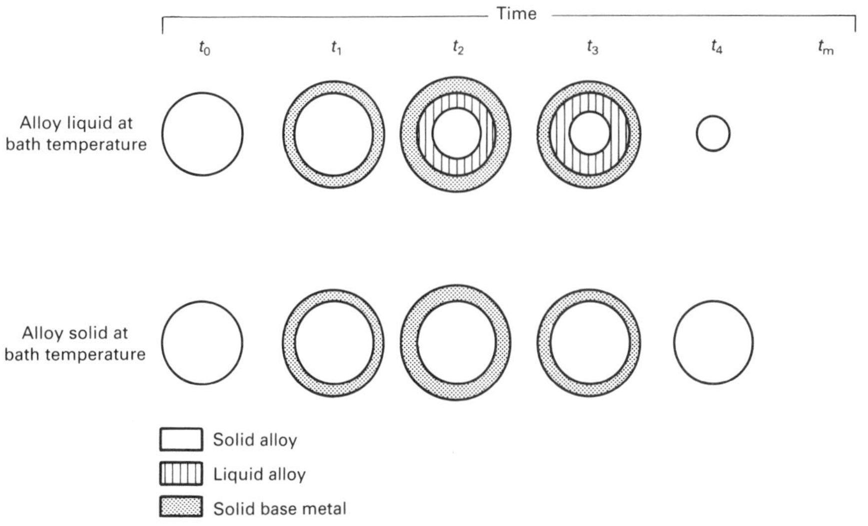

Fig. 1 Two kinetic paths for melting and/or dissolving alloy additions in liquid metal baths. Source: Ref 1

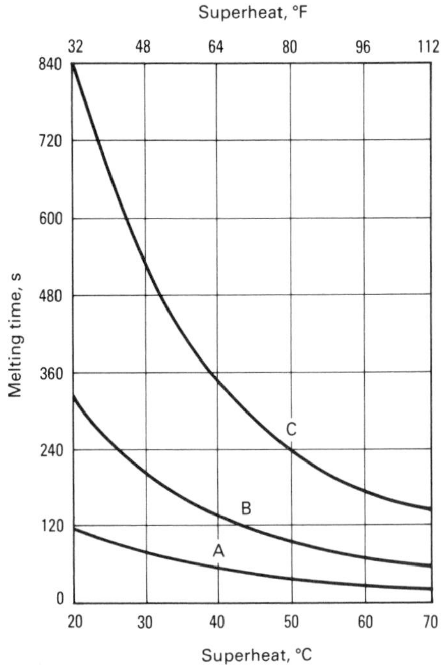

Fig. 2 Predicted melting/dissolution times for silicomanganese spheres in quiescent steel baths versus steel superheat temperature. Particle diameters of the spheres are as follows: A, 25 mm (1 in.); B, 50 mm (2 in.); C, 100 mm (4 in.). Source: Ref 1

steel melt (T_L^B = 1823 K). Data for the calculations are given in Table 1. The very long melting time of 1200 s obtained by restricting the calculations to a small superheat temperature (20 K) is in good agreement with the 900 s obtained with a more sophisticated model (Fig. 2). The difference in t_m for the two extreme cases considered in Table 2 underscores the importance of superheat temperature in alloy dissolution processes. This is clearly illustrated in Fig. 2, in which the role of particle diameter is also demonstrated. It is apparent from Fig. 2 that the large value of t_m in Table 1 for the case ($T_L^B - T_L^*$) = 20 and d_p = 100 mm can be substantially reduced with moderate changes in ($T_L^B - T_L^*$) and d_p.

The range of melting times obtained with the addition of an alloy to different base metals is examined below. Consider the addition of pure aluminum to melts of steel, cast iron, copper, and aluminum under conditions where no frozen shell is formed. In this case, T_L^* is the same (933 K), but T_L^B varies. Pertinent data for the calculation and the calculated melting times are given in Table 2. It is clear from Table 2 that the duration of alloy melting becomes a more serious problem as the melting point of the base metal decreases.

The short melting times predicted for steel and cast iron melts in Table 2 highlight another important consideration in alloy addition, namely, immersion time due to alloy buoyancy. Because iron is denser than common alloy additions such as carbon, silicon, and aluminum, buoyant forces prevent these alloys from being easily submerged for times approaching t_m. This is illustrated in Fig. 3, which plots immersion times versus the density ratio, $\rho_{alloy}/\rho_{base\ metal}$. For 10 mm (0.4 in.) aluminum particles (density ratio: 0.39) that are dropped a distance of 3 m (10 ft) to the surface of an iron bath, the immersion time is less than 1 s, while the melting time (from Table 2) is 1.9 s. Immersion times can be increased by entraining particles in strong recirculating flows or by wire and rod feeding (Ref 2, 3). These conditions have higher associated heat transfer coefficients, which further facilitate dissolution.

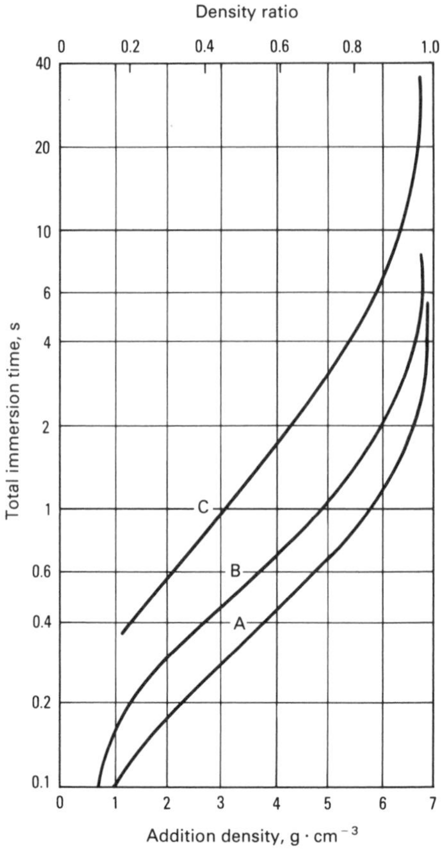

Fig. 3 Total immersion times versus density ratio for alloys entering a stagnant steel bath at 7.67 m · s^{-1} (25.2 ft · s^{-1}) (drop height: 3.0 m, or 10 ft). Particle diameters are as follows: A, 10 mm (0.4 in.); B, 50 mm (2 in.); C, 250 mm (10 in.). Source: Ref 1

Dissolution of Alloy That Is Solid at Bath Temperatures

The processes involved in the dissolution of this type of addition are schematically illustrated in the bottom row sequence in Fig. 1. As in the previous case, heat transfer controls the duration of the incubation period ($t_0 - t_4$). However, in this case, mass transfer is the controlling factor in the alloy dissolution period ($t_4 - t_m$). In general, mass transfer controlled dissolution is slower than heat transfer controlled dissolution, and the number of variables controlling the rate is greater. For mass transfer control, melt temperature and fluid dynamic considerations are important, as in the heat transfer controlled case. However, melt composition and diffusion variables are also important. The flux of alloy element to the melt is given by:

$$J = k_m (C_L^* - C_L^B) \qquad \text{(Eq 3)}$$

where J is the mass flux of alloy element (g · cm^{-2} · s^{-1}), k_m is the mass transfer coefficient (cm · s^{-1}), C_L^B is the concentration of the alloy element in the bulk melt (g · cm^{-3}), and C_L^* is the concentration of the alloy element in the melt at the solid/liquid interface (g · cm^{-3}). To illustrate mass transfer limited dissolution processes, two industrially important examples will be examined below.

Carbon Dissolution in Cast Iron. Carbon is the most important alloy element in cast iron. The addition of carbon to iron-carbon melts is a common foundry practice that is necessary for control of cast iron composition. The rate at which carbon dissolves in cast iron is controlled by the rate of diffusion of carbon from carbon-saturated iron at the carbon/iron interface into the bulk liquid (Ref 4-6). Thus, Eq 3 can be written:

$$J = k_m (C_{sat}^* - C_L^B) \qquad \text{(Eq 4)}$$

where C_{sat}^* indicates carbon-saturated conditions at the carbon/melt interface. The

Table 1 Data for melting time calculations for a 100 mm (4 in.) diameter sphere of silicomanganese in a steel melt

Condition	ΔH, J/g	ρ_0, g/cm^3	V_0/A_0, cm	h, J/g K	T_L^*, K	($T_L^B - T_L^*$), K	t_m, s
Frozen shell	387	5.6	1.67	0.15	1803	20	1200
No frozen shell	307	5.6	1.67	0.15	1395	418	45

Source: Ref 1

Table 2 Data for melting time calculations for a 10 mm (0.4 in.) diameter aluminum sphere in different liquid metal melts

Base metal	ΔH, J/g	ρ_0, g/cm^3	V_0/A_0, cm	h, J/g K	T_L^*, K	($T_L^B - T_L^*$), K	t_m, s
Steel	576	2.7	0.167	0.15	1823	890	1.9
Cast iron	576	2.7	0.167	0.15	1723	790	2.2
Copper	576	2.7	0.167	0.15	1323	290	6.0
Aluminum	576	2.7	0.167	0.15	1023	90	19.2

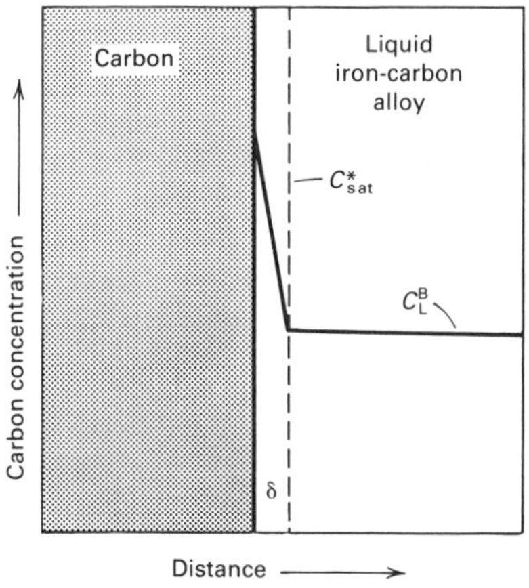

Fig. 4 Carbon concentration profiles for a carbon rod dissolving in an iron-carbon melt

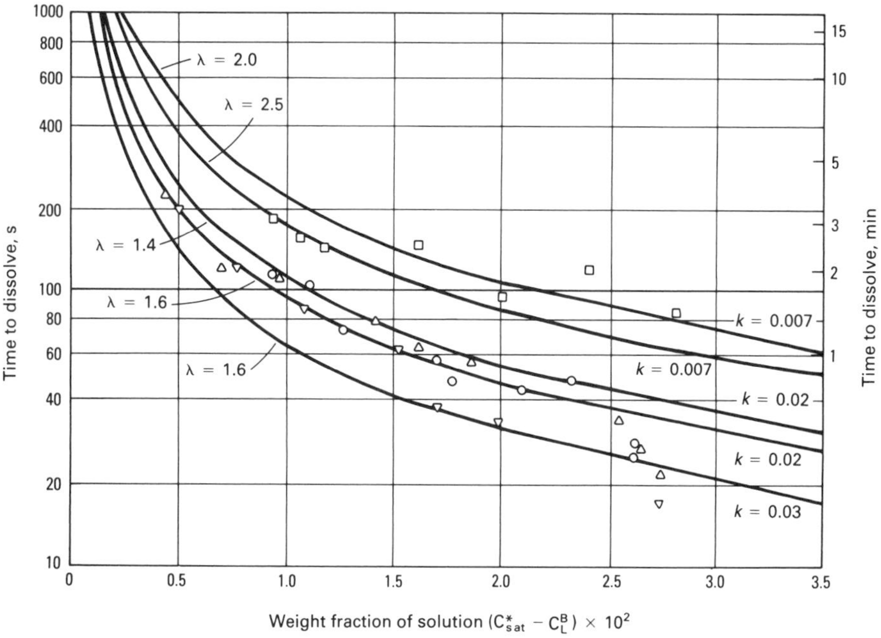

Fig. 5 Time of dissolution of carbon particles (−10 +14 mesh) versus (C^*_{sat} − C^B_L). Source: Ref 6

conditions in the melt are qualitatively described in Fig. 4. Referring to Fig. 4 and Eq 4, the flux of carbon, and therefore the dissolution rate, increases with decreasing bulk carbon concentration C^B_L, increasing carbon concentration at saturation C^*_{sat} (increasing bath temperature), and decreasing boundary layer thickness δ. The boundary layer thickness is a function of the fluid velocity, and it affects the value of k_m.

The effect of the difference ($C^*_{sat} - C^B_L$) on the dissolution rate of carbon is illustrated in Fig. 5 with data for the dissolution of graphite granules in induction-stirred cast iron melts. Theoretical curves are drawn for different rate constants k and particle shape factors λ, where λ is defined as:

$$\lambda = \frac{S_{particle}}{S_{sphere}} \qquad \text{(Eq 5)}$$

where $S_{particle}$ is the surface-to-volume ratio of the graphite particles and S_{sphere} is the equivalent surface-to-volume ratio if the particles are spherical.

The role of boundary layer thickness (fluid dynamics) on dissolution kinetics is illustrated in Fig. 6, which indicates the variation of the rate constant (dissolution rate) with the peripheral velocity (rotational speed) of a cylindrical graphite rod. The line drawn through the data has a slope of 0.7, which is expected from earlier correlations (Ref 7).

Dissolution of Steel in Iron-Carbon Melts. Another well-established case of

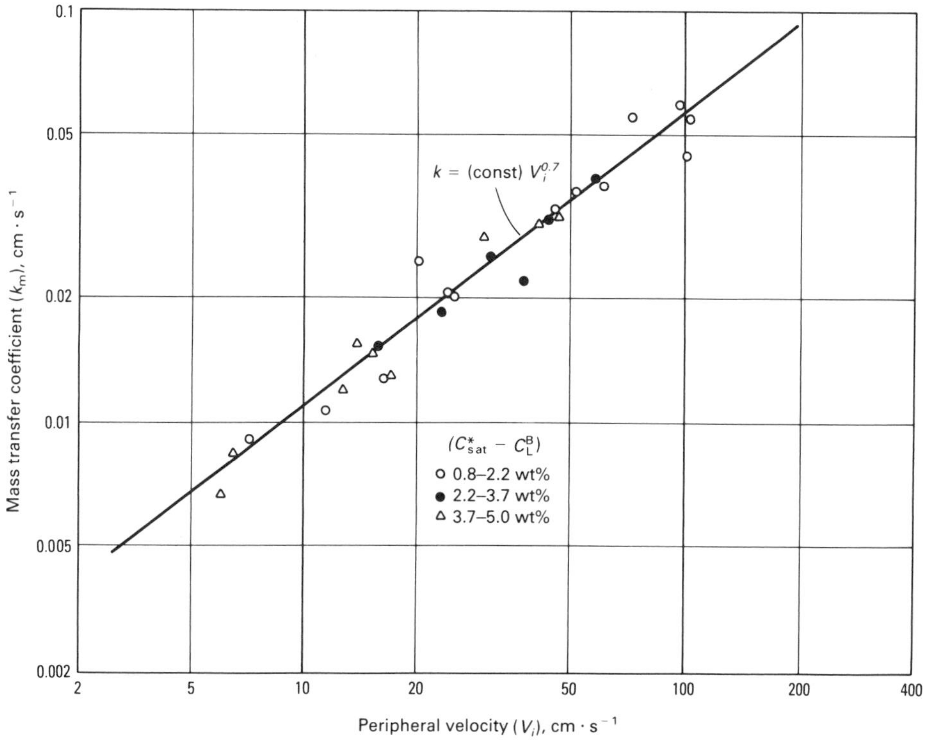

Fig. 6 Experimental mass transfer coefficient versus peripheral velocity for the dissolution of a rotating carbon rod in an iron-carbon melt. Source: Ref 5

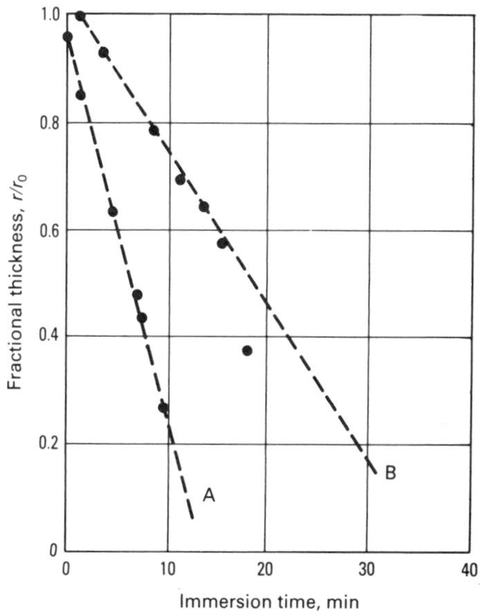

Fig. 7 Melting curves for 75 mm (3.0 in.) diam steel rods immersed in iron-carbon baths. A, high (4.6% C) iron-carbon bath; B, low (2.1% C) iron-carbon bath. Plotted is the fractional thickness (the fraction of the rod diameter remaining undissolved) versus immersion time. Source: Ref 10

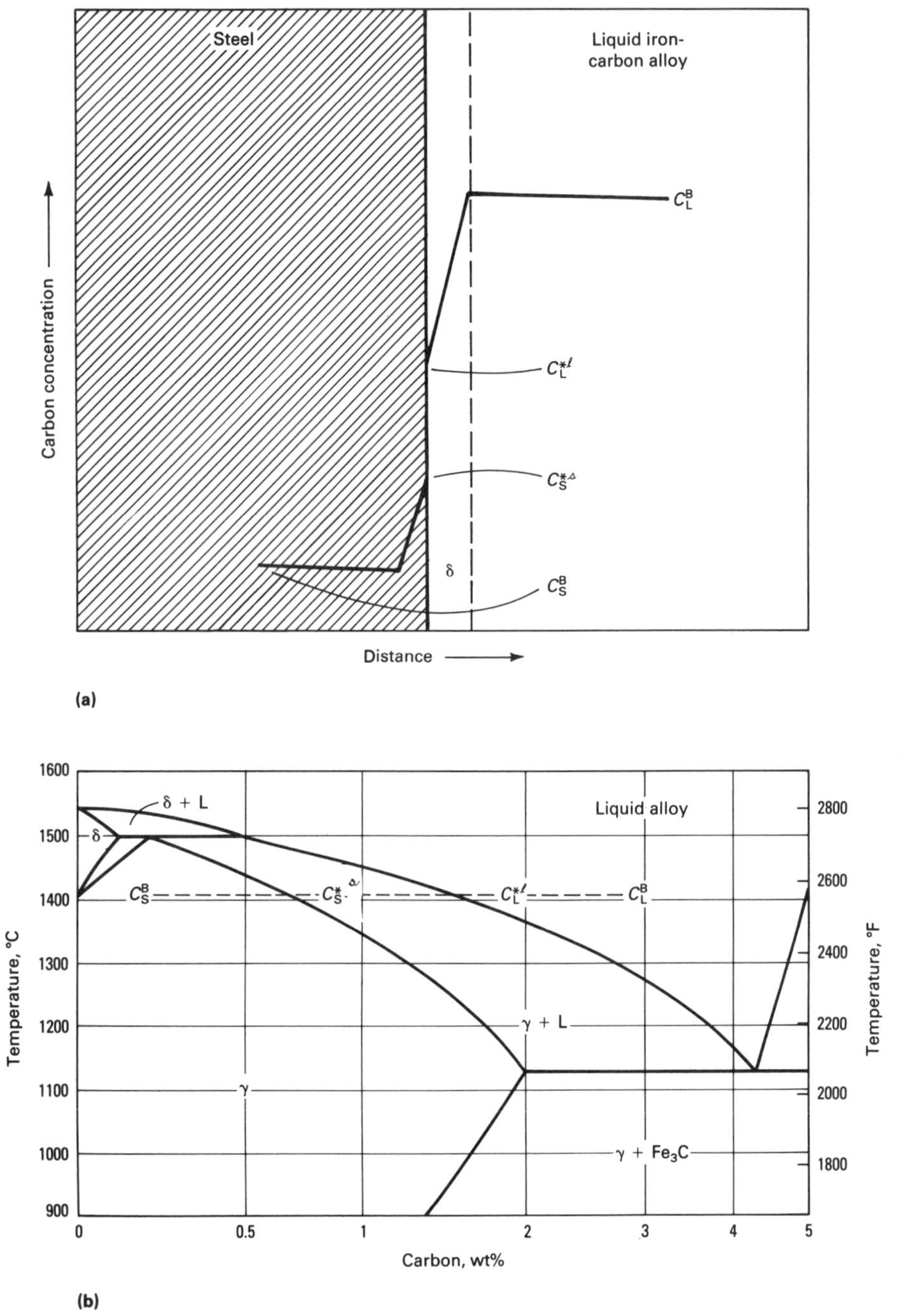

Fig. 8 Carbon concentration profiles (a) for a steel rod dissolving in an iron-carbon melt. (b) Iron-carbon phase diagram defining the carbon concentrations noted in (a). Source: Ref 10

mass transfer controlled dissolution is the dissolution of steel in iron-carbon melts (Ref 8-11). This process is critical in the electric melting of steel and cast iron. For this case, the rate of steel dissolution increases with melt carbon concentration (Fig. 7), melt temperature (Ref 10), and bath velocity (Ref 8, 9, 11) (affecting the boundary layer thickness). The data can be accounted for on the basis that carbon diffuses from the bulk of the melt to the steel/liquid iron interface, where steel is isothermally converted to a liquid with a liquidus composition, $C_L^{*\ell}$.

The conditions existing in the bath are illustrated in Fig. 8. Figure 8(a) shows the carbon concentration profiles in the liquid bath in the vicinity of a dissolving steel bar. The meaning of the carbon concentration notation is indicated in Fig. 8(b). As indicated in Fig. 8(a), carbon diffuses across the liquid boundary layer, while equilibrium conditions are maintained at the solid/liquid interface. Thus, the carbon concentrations in the solid and the liquid at the interface are the isothermal solidus C_S^{*s} and the liquidus $C_L^{*\ell}$ concentrations, respectively. Dissolution takes place by isothermal conversion of the steel into a liquid, that is, by a chemical melting process. It is clear from Fig. 8(b) that as temperature increases, $C_L^{*\ell}$ decreases significantly. Thus, the driving force for diffusion increases with temperature.

Based on Eq 3 and Fig. 4, the flux of carbon entering the steel is:

$$J = k_m (C_L^B - C_L^{*\ell}) \quad \text{(Eq 6)}$$

The carbon flux can also be defined in terms of a carbon balance:

$$J = -\left(\frac{\rho_S^B C_L^{*\ell}}{\rho_L^{*\ell}} - C_S^B\right)\frac{dr}{dt} \quad \text{(Eq 7)}$$

where ρ_S^B and $\rho_L^{*\ell}$ are the respective densities of solid steel and liquid iron at the liquidus composition, C_S^B is the bulk carbon concentration of solid steel, and dr/dt is the rate of change of the radius or thickness of the steel with time. By combining Eq 6 and 7 and integrating between the limits $r = r_0$ at $t = 0$ and $r = 0$ at $t = t_m$, the isothermal melting time t_m is given as:

$$t_m = \frac{(\rho_S^B C_L^{*\ell}/\rho_L^{*\ell}) - C_S^B}{k_m (C_L^B - C_L^{*\ell})} r_0 \quad \text{(Eq 8)}$$

In laboratory studies, Eq 8 accurately predicted steel dissolution rates in a stagnant bath (Ref 10). Research has suggested that melting times are reduced by about a factor of two in induction-stirred melts (Ref 9). Other work has indicated that t_m can be reduced by an order of magnitude with strong convection (Ref 11).

On the basis of Eq 8 and the mass transfer correlation for rapidly dissolving rods (Ref 12), correlations were developed relating t_m to bath temperature and to carbon concentration (Ref 10). This is given in Fig. 9 for a steel thickness of 2.5 mm (0.098 in.). Because t_m is directly proportional to the thickness of the steel, Fig. 9 can be used to estimate t_m for a wide range of steel thickness by making the appropriate thickness correction. Checks of the predictions from Fig. 9 in production operations have indicated that predicted t_m values are realistic (Ref 10).

Purification of Metals

The structure and properties of cast metals are sensitive to numerous impurities. Because purification of melts generally adds considerable cost to castings, the lowest cost and surest defense against contamination is careful selection of scrap. Purification is generally reserved for elements that are so pervasive that avoidance is impossible. This is exemplified by sulfur and oxygen removal from cast iron and steel, respectively, and removal of alkali and alkaline earth elements from aluminum. Because of their immediate importance, aspects of the physical chemistry of these processes are reviewed.

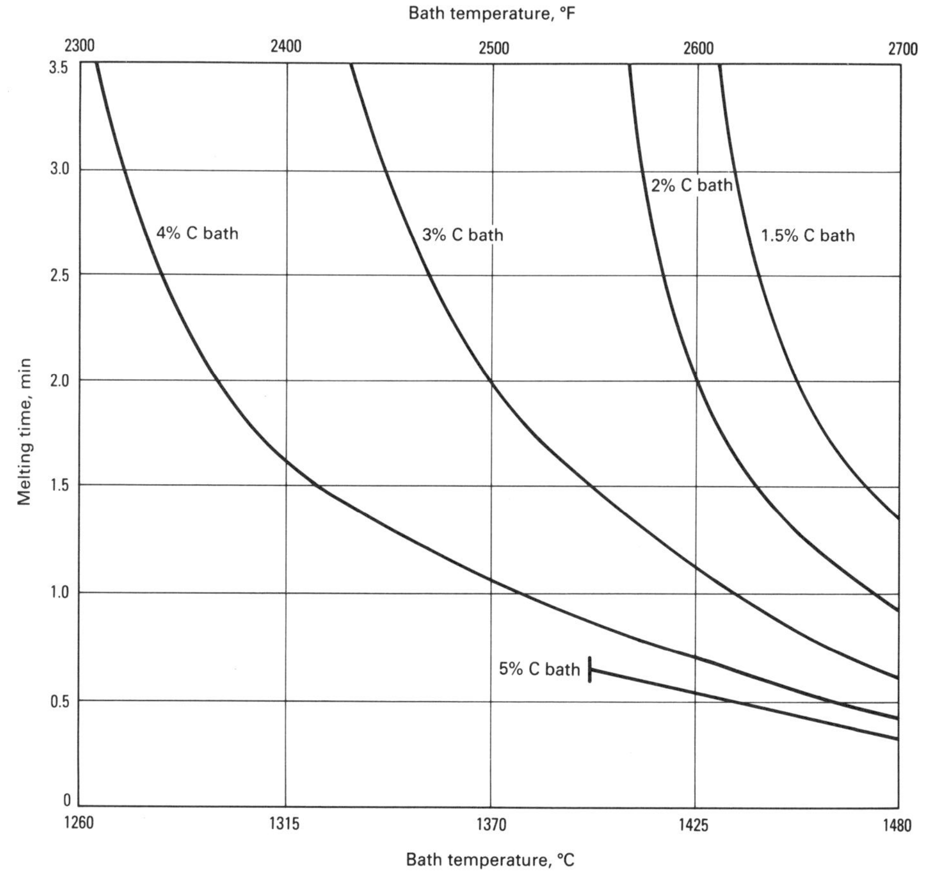

Fig. 9 Predicted melting times for vertical steel plates and cylinders, 2.5 mm (0.098 in.) scrap thickness and 0.1% C composition, $[(\%S)_i = 0.03]$ immersed in stagnant iron-carbon baths at different bath temperatures. Source: Ref 10

Ferrous Melts

One of the most important processes involved in cast iron and steel production is desulfurization. For steels, desulfurization is necessary to reduce the level of inclusions, leading to stronger and more fatigue-resistant steels. For cast iron, desulfurization is practiced in the manufacture of ductile iron castings in order to develop spherical graphite morphology. Ductile iron is used in applications where high fracture toughness is needed.

Sulfur is removed from iron and steel when the metals are liquid. Although a variety of reagents are employed to remove sulfur, namely, calcium, magnesium, sodium, and rare earths, the most important is calcium. Common forms of calcium include the metal; alloys such as calcium silicon (CaSi); the oxide, calcium oxide (CaO); and the carbide, calcium carbide (CaC_2). Despite the use of various forms of calcium, the governing chemical reaction in all cases appears to be the same (Ref 13, 14):

$$CaO + S = CaS + O \quad \text{(Eq 9)}$$

The equilibrium constant for the reaction (Eq 9) is:

$$K_9 = \frac{a_{CaS}\, h_O}{a_{CaO}\, h_S} \quad \text{(Eq 10)}$$

For both cast iron and steel, target sulfur concentrations after desulfurization are in the range of 0.006 to 0.010% S. In electric-melted cast iron and steel, the sulfur levels before desulfurization are 0.02 to 0.03% S,

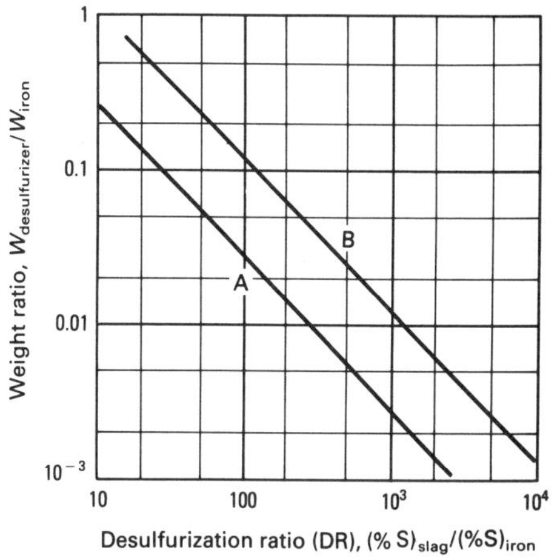

Fig. 10 Weight fraction desulfurizer required to achieve given desulfurization ratios. Plots for electric-melted iron and steel (curve A) assume initial sulfur concentrations $[\%(S)_i = 0.03]$ different from those of cupola-melted iron $[(\%S)_i = 0.10]$ (curve B).

while input sulfur levels to the cupola are generally much higher—0.1 to 0.2% S.

Requirements for Desulfurization. For reasons related to reaction kinetics and thermodynamics, the final sulfur $(\%S)_f$ concentration achieved in desulfurization processes depends on three factors:

- Initial sulfur concentrations $(\%S)_i$
- Amount of desulfurizer used, usually expressed as the weight fraction of desulfurizer to metal, W
- Extraction efficiency of the desulfurizer, which is measured by the desulfurization ratio (DR), that is, the ratio of sulfur concentrations: desulfurizer to metal

The three variables can be related as follows through a mass balance on sulfur:

$$(\%S)_f = \frac{(\%S)_i}{1 + W(DR)} \quad \text{(Eq 11)}$$

For liquid-desulfurizing slags that are not saturated with respect to calcium sulfide (CaS), the maximum value of DR, that is, the equilibrium value, can be predicted from thermodynamic considerations:

$$(DR)_{max} = \left(\frac{(\%S)_{slag}}{(\%S)_f}\right)_{max} = \frac{C_S f_S K_{14}}{K_{15} h_O} \quad \text{(Eq 12)}$$

where C_S is the slag sulfide capacity, defined as (Ref 15):

$$C_S = (\%S)_{slag}\left(\frac{p_{O_2}}{p_{S_2}}\right)^{1/2} \quad \text{(Eq 13)}$$

Table 3 Theoretical weight ratios: desulfurizer to iron to achieve 0.008% S in various systems

Type of melt	Slag composition	T, K	C_S (× 10^4)(a)	f_S	h_O (× 10^4)(b)	$(DR)_{max}$(c)	W
A. cast iron—cupola	44 CaO-15MgO-5Al_2O_3-36SiO_2	1773	2.7	5	1.3	90	0.13
B. cast iron—cupola (ladle desulfurized)	CaO_{sat}-Al_2O_3	1773	59	5	5.0	417	0.027
C. cast iron—electric	CaO_{sat}-Al_2O_3	1773	59	5	5.0	417	0.0065
D. steel—electric	CaO_{sat}-Al_2O_3	1873	316	1	0.45	5280	0.00052

(a) C_S is obtained from optical basicity correlations (Ref 15). f_S is based on iron composition (Ref 16). (b) Case A: h_O is governed by Si/SiO_2 equilibrium based on respective concentration in iron and slag (Ref 17). Cases B and C: h_O is governed by Si/SiO_2 equilibrium with $a_{SiO_2} = 1$ due to ladle exposure to air (Ref 18). Case D: h_O is governed by Al/Al_2O_3 equilibrium with % Al = 0.05 (Ref 19), assumed no contact with air. (c) K_{14} and K_{15} data from Ref 20

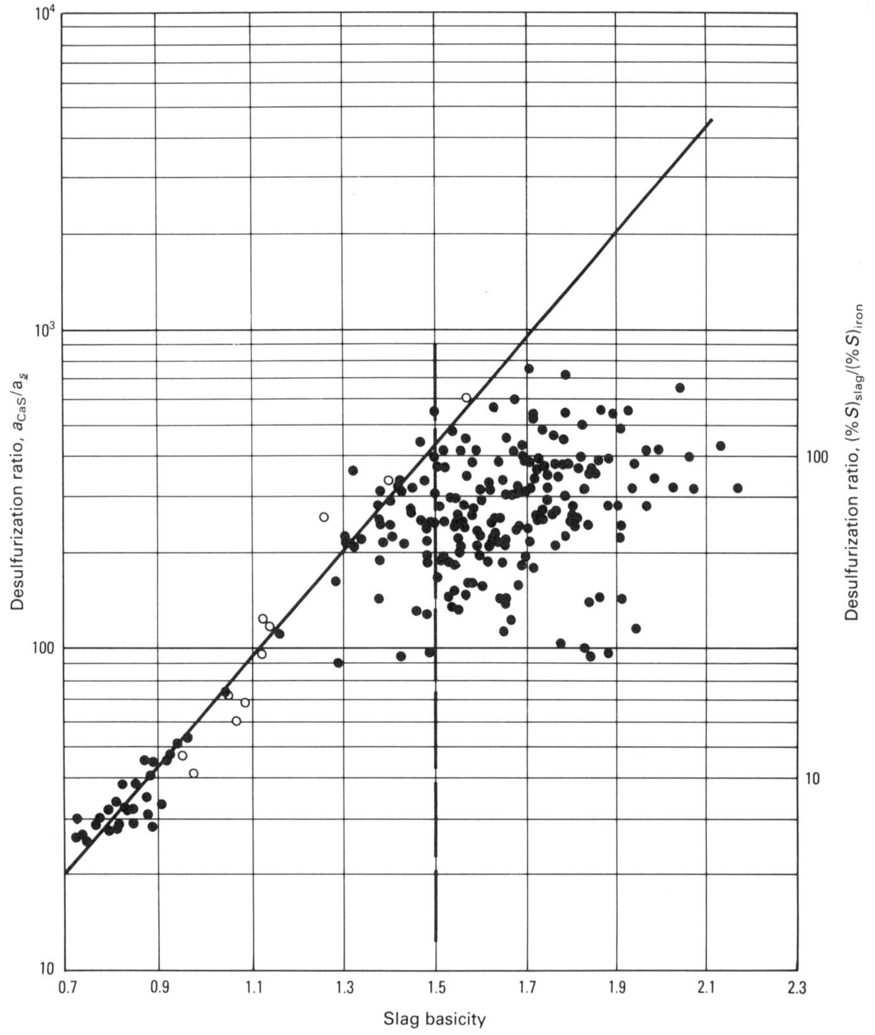

Fig. 11 Desulfurization ratio-basicity comparisons for cupola (closed circles) and laboratory data (open circles). Equilibrium values are indicated by the angled line. The vertical line indicates the slag basicity above which slags are saturated at 1500 °C (2730 °F), with respect to dicalcium silicate. This is the point at which the observed DR should equal $(DR)_{max}$. Source: Ref 14

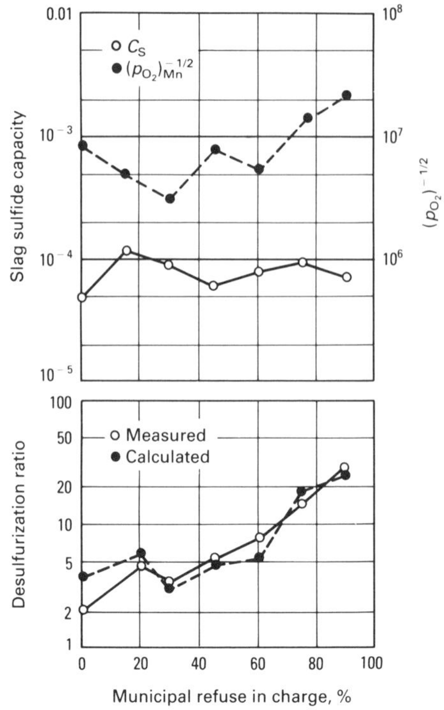

Fig. 12 Slag sulfide capacities, oxygen activities, and desulfurization ratios, measured and calculated, for a cupola operated with municipal ferrous refuse as a charge material. Source: Ref 21

and K_{14} and K_{15} are the respective equilibrium constants for the reactions:

$$\frac{1}{2}O_2 = \underline{O} \quad \text{(Eq 14)}$$

$$\frac{1}{2}S_2 = \underline{S} \quad \text{(Eq 15)}$$

and f_S is the activity coefficient for 1 wt% S in the standard state.

Equation 12 can be derived from Eq 10. Using known input and desired output sulfur values, Eq 11 gives the required desulfurization ratios as a function of weight fraction desulfurizer. These data are plotted in Fig. 10 for two cases. In the electric-melting case, initial and final sulfur concentrations were assumed to be 0.03 and 0.008% S, respectively, In the cupola-melting case, the equivalent concentrations were 0.10% S and 0.008% S. The cupola line applies for both cupola iron and ladle-desulfurized cupola iron.

The relationships illustrated in Fig. 10 are useful in defining systems that will provide the necessary desulfurization conditions. Four systems are compared in Table 3. These cover cupola- and electric-melted cast iron and electric-melted steel. Also examined are two liquid slag systems: a basic cupola slag (dicalcium silicate saturated) and a CaO-saturated $CaO\text{-}Al_2O_3$ slag.

Table 3 shows that:

- The best desulfurizing cupola slags have lower sulfide capacity and desulfurization ratio than slags used in ladle desulfurization; the consequence is the need for much larger quantities of slag*
- Compared to steel, cast iron desulfurization benefits from higher f_S because of the presence of relatively high concentrations of carbon and silicon in cast iron
- Ladle desulfurization systems that are exposed to air suffer higher h_O and, as a result, poorer desulfurization than might otherwise be anticipated

*The actual differences are less than those indicated. Calcium sulfide has only limited solubility in $CaO\text{-}Al_2O_3$ slags. As a result, a minimum $W = 0.01$ is needed to maintain the CaS-unsaturated condition.

For the cupola slag case in Table 3, the thermodynamically predicted value of $(DR)_{max} = 90$ is in good agreement with measured data (Fig. 11). This indicates that the cupola desulfurization process operates close to equilibrium levels. Further evidence for this is given in Fig. 12, which plots desulfurization data for a cupola operated with varying amounts of municipal ferrous refuse in the charge. The upper portion of Fig. 12 plots C_S and oxygen activity. The latter is expressed as the partial pressure of oxygen. The lower portion of Fig. 12 is a comparison of (DR), measured and calculated. The good agreement found is evidence that near-equilibrium conditions existed.

For the cases discussed above, the oxygen activity in cupola iron has been found to be governed by the reaction (Ref 17, 21):

$$\underline{Mn} + \underline{O} = MnO \quad \text{(Eq 16)}$$

Therefore, the overall cupola desulfurization reaction is:

$$CaO + \underline{S} + \underline{Mn} = CaS + MnO \quad \text{(Eq 17)}$$

Considerably lower sulfur could be achieved (Ref 14) if h_O were governed by:

$$\underline{C} + \underline{O} = CO \quad \text{(Eq 18)}$$

However, equilibrium for this reaction has not been observed.

This discussion has concerned CaS-unsaturated slags. However, a desulfurizing slag, saturated with CaS, can in many cases continue to desulfurize as long as CaO is

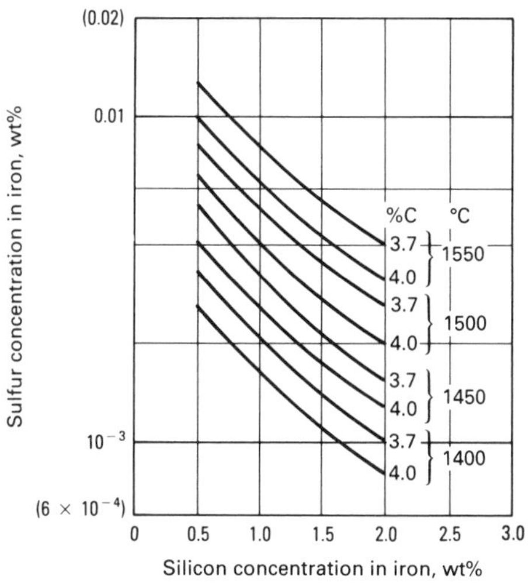

Fig. 13 Equilibrium sulfur concentrations for CaO desulfurization calculated with Eq 19 and 20, assuming $a_{CaO} = a_{CaS} = a_{SiO_2} = 1$. Source: Ref 14

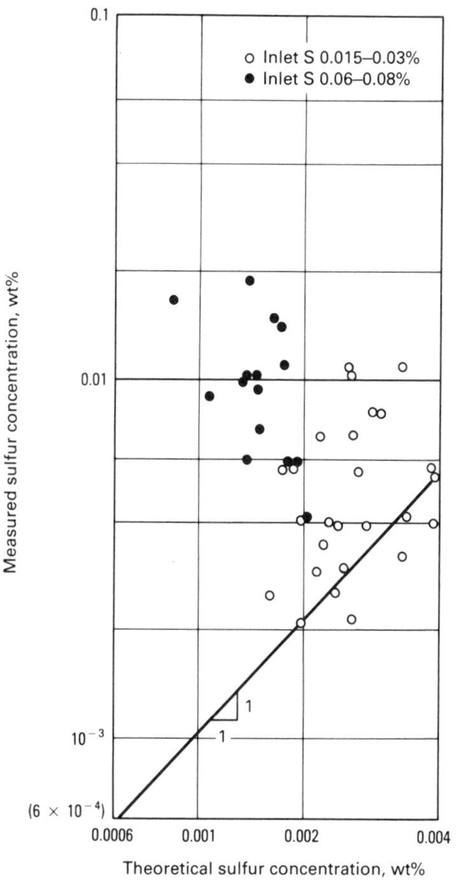

Fig. 14 Comparisons of the sulfur concentrations in production-continuous desulfurization using CaC_2 with the sulfur concentrations from Eq 19 and 20. Open circles indicate electric-melted iron. Closed circles indicate cupola iron. Source: Ref 14

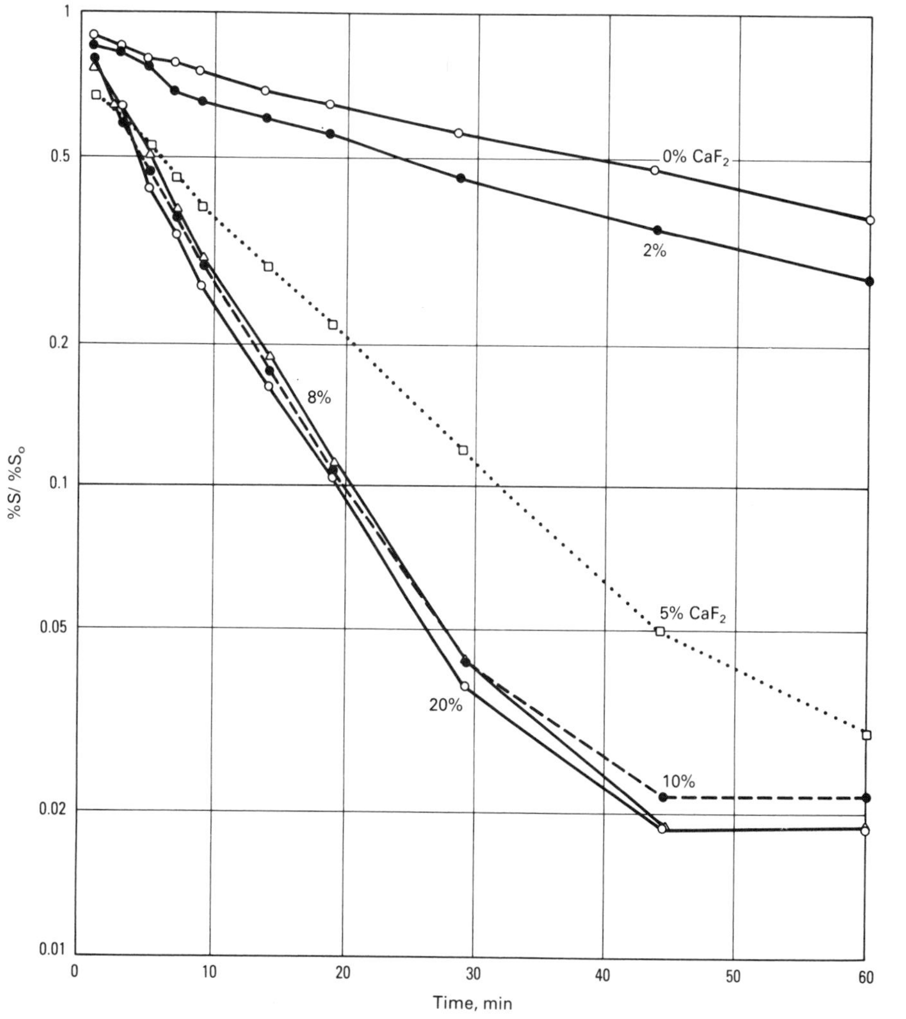

Fig. 15 Desulfurization rates of carbon-saturated iron, containing 0.4% Si, with CaO and varying amounts of CaF_2 at 1450 °C (2640 °F). Source: Ref 22

present. In this case, the ultimate sulfur levels are not as low as those for CaS-unsaturated slags. Nevertheless, CaS-saturated slags can produce sulfur levels that are more than sufficient for cast iron and steel applications. In fact, CaS-saturated conditions are often employed in industrial applications because much less desulfurizer is required.

In addition to the CaS-saturated liquid slags, totally solid slags such as CaO and CaC_2 are in the same category because CaS does not form solid solutions with these materials. Also in this category are desulfurizers containing small amounts of liquid phase. These materials possess the desirable properties of liquid and solid slags. That is, they combine the fast reaction rates of liquid slags with the large desulfurizing capacities (high CaO concentration) of solid slags.

The final sulfur concentrations attainable under CaS-saturated conditions can be obtained from Eq 10 by setting $a_{CaS} = 1$. For ladle desulfurization, the most important industrial desulfurizers are also CaO-saturated, that is, $a_{CaO} = 1$. Applying the condition of double saturation to Eq 10 yields:

$$(\%S)_f = \frac{h_O}{K_9 f_S} \quad \text{(Eq 19)}$$

In the ladle desulfurization of cast iron, h_O is governed by the reaction (Ref 14):

$$\tfrac{1}{2}\underline{Si} + \underline{O} = \tfrac{1}{2}SiO_2 \quad \text{(Eq 20)}$$

where $a_{SiO_2} = 1$. This is attributed to the exposure of the iron to air (Ref 14). Sulfur concentrations obtained with Eq 19, assuming silicon-silicon dioxide (Si-SiO_2) equilibrium and $a_{SiO_2} = 1$, are given in Fig. 13. In continuous ladle-desulfurization processes, equilibrium sulfur levels are achieved when input sulfur levels are low, but are only approached when input levels are high. This is illustrated in Fig. 14 with desulfurization data for CaC_2 (Ref 14). Other desulfurizers such as CaO-CaF_2 behave similarly.

High desulfurization rates are needed for effective desulfurization in the short times required. For CaO-base desulfurizers, the presence of relatively small amounts of liquid phase (<25 vol%) significantly increases the rate of desulfurization. This is illustrated in Fig. 15, in which the rates of desulfurization by CaO with varying amounts of calcium fluoride (CaF_2) are compared. The

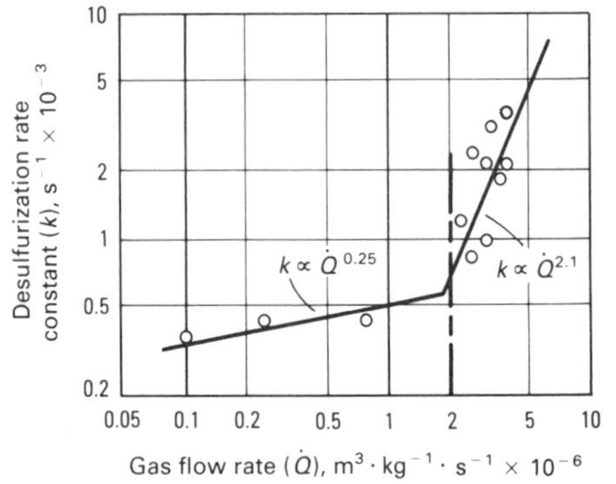

Fig. 16 Dependence of CaO desulfurization rate constant on the rate of gas flow through a reactor. Source: Ref 29

faster rates obtained with CaF_2 additions were due to the formation of a $CaO\text{-}CaF_2$ liquid phase that provided a path for CaS diffusion from the reaction interface into the porous CaO particle. This prevented the normal rapid development of an impervious CaS coating on the CaO surface (Ref 23, 24). A similar explanation, involving liquid-phase formation, was used to account for the higher CaO desulfurization rates of steel when aluminum was concurrently added (Ref 25, 26).

Another rate-controlling variable in ladle desulfurization is bath agitation. For gas-stirred melts, the desulfurization rate constant k is a function of gas flow rate $\dot{Q}$, that is, $k \propto \dot{Q}^n$. For well-dispersed solid-liquid mixtures, the rate of diffusion-controlled interphase reaction is a relatively weak function of $\dot{Q}$, that is, $n \sim 0.2$ to 0.4 (Ref 27-29). For a poorly dispersed system, such as a ladle of iron with a cover slag of desulfurizer, agitation has a much greater influence on the rate constant, with n typically in the range of 1.0 to 2.5 (Ref 27-29). This is due to the entrainment of increasing amounts of top slag into the liquid metal with increasing $\dot{Q}$. This effect is illustrated in Fig. 16, which plots the apparent mass transfer coefficient for desulfurization versus the gas flow rate.

Deoxidation. In the manufacture of steel, FeO-saturated conditions (~0.2% O) are approached as a result of oxygen blowing to remove carbon and silicon as the respective oxides. To make a steel product, the oxygen levels are subsequently lowered to 0.005 to 0.02% O by reactive alloy additions of electropositive elements. The low oxygen concentrations are necessary to maintain the number of oxide inclusions at a suitable low level, to prevent the formation of CO blowholes, and to ensure effective desulfurization.

For single-element deoxidation, expressed by:

$$x\underline{M} + y\underline{O} = M_xO_y \qquad \text{(Eq 21)}$$

the equilibrium oxygen concentration in the liquid steel, obtained by rearranging the equilibrium constant expression, is:

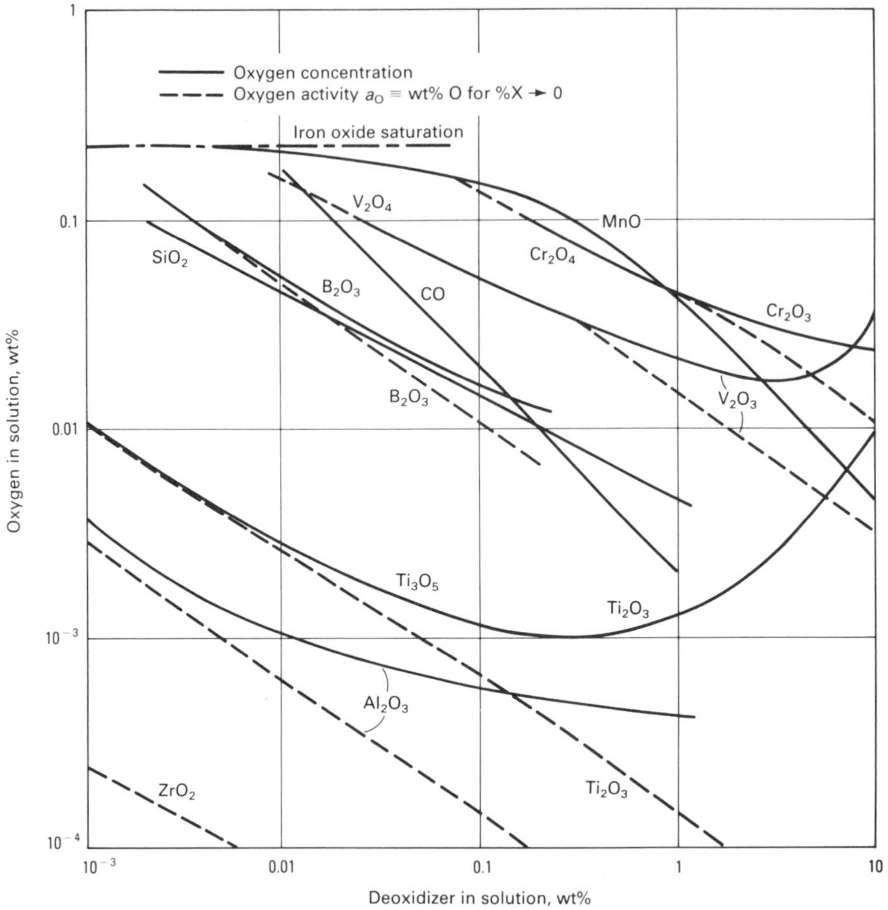

Fig. 17 Deoxidation equilibria in liquid iron alloys at 1600 °C (2910 °F). Source: Ref 30

$$(\%O) = \left(\frac{a_{M_xO_y}}{K_{22}(\%M)^x f_M^x f_O^y}\right)^{y^{-1}} \qquad \text{(Eq 22)}$$

The calculated equilibrium values of oxygen solubility in liquid iron at 1600 °C (2910 °F), based on Eq 22, are given in Fig. 17 for several elements as a function of the concentration of the element, assuming $a_{M_xO_y} = 1$. As seen among the common deoxidizers, aluminum produces the lowest oxygen. Although not shown, rare-earth metals produce even lower oxygen (Ref 30). Figure 17 plots data for oxygen concentration and oxygen activity. In all cases, oxygen activity decreases with increasing levels of deoxidizer, but the oxygen concentration can go through a minimum. This occurs in cases where the interaction coefficient e_O^M is a large negative number, and as a result, f_O decreases significantly even at relatively low concentrations of M. To obtain exact values for [%O] or h_O, equations for K_{22} can be found in Ref 13 and 20.

As indicated by Eq 22, the oxygen concentration can be reduced if a complex deoxidation product is formed so that $a_{M_xO_y} < 1$. Common deoxidation systems of this type are Si-Mn, Al-Si-Mn, or Al-CaO. An advantage, in addition to lower oxygen, is that less deoxidizer is required in solution to achieve a given level of oxygen. This is illustrated in Fig. 18, which plots the oxygen activity of iron as a function of the concentration of CaO in the calcium aluminate inclusion. Separate lines are given for aluminum concentrations ranging from 0.001 to 0.05% Al. Also indicated is the oxygen activity when pure aluminum oxide (Al_2O_3) is the reaction product. The data in Fig. 18 show that an oxygen activity of 4 ppm can be produced at three aluminum levels—0.002%, 0.005%, or 0.01%—depending on whether the respective deoxidation product was CaO-saturated calcium aluminate, $CaAl_2O_4$-saturated calcium aluminate, or Al_2O_3.

The levels of total oxygen measured in steel melts considerably exceed equilibrium levels (Ref 20). Two possible explanations are the kinetic limitations in the deoxidation reaction and the ineffective separation of deoxidation products from the melt. Consideration of the first possibility suggests that the most important kinetic limitation would be oxygen and/or deoxidant solute diffusion to inclusions in the melt (Ref 20). Considering the presence of 10^5 to 10^7 inclusion particles in a cubic centimeter of melt, deoxidation to equilibrium levels of 10 to 20 ppm would require a few minutes. Appreciably longer times would be needed

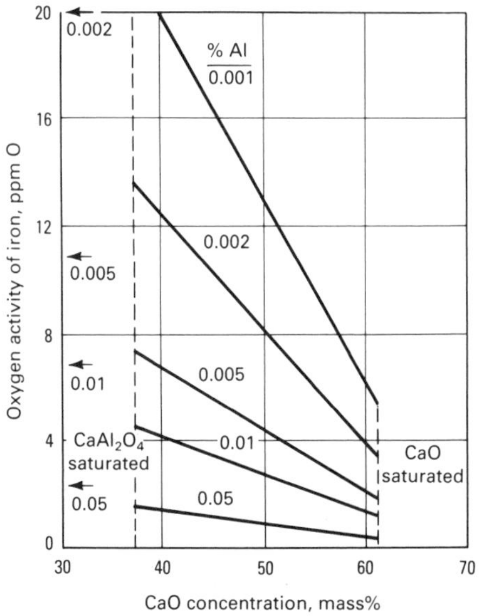

Fig. 18 Concentration of oxygen and aluminum in liquid steel in equilibrium with calcium aluminate at 1600 °C (2910 °F); arrows indicate oxygen values when the reaction product is pure Al_2O_3. Source: Ref 31

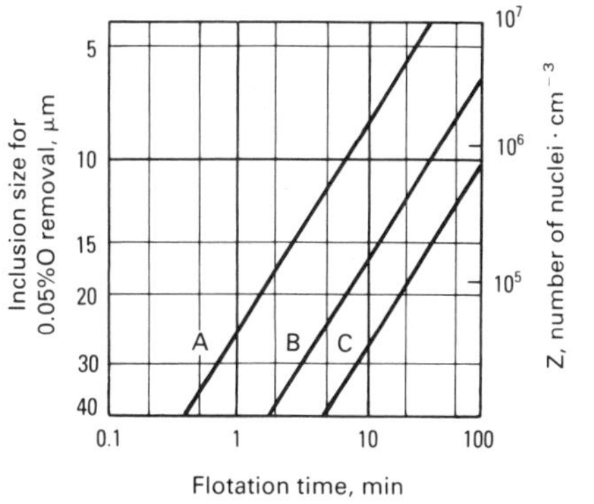

Fig. 19 Calculated time of flotation of inclusions in stagnant melts as a function of inclusion size. Melt depth: A, 50 mm (2 in.); B, 500 mm (20 in.); C, 2000 mm (80 in.). Source: Ref 20

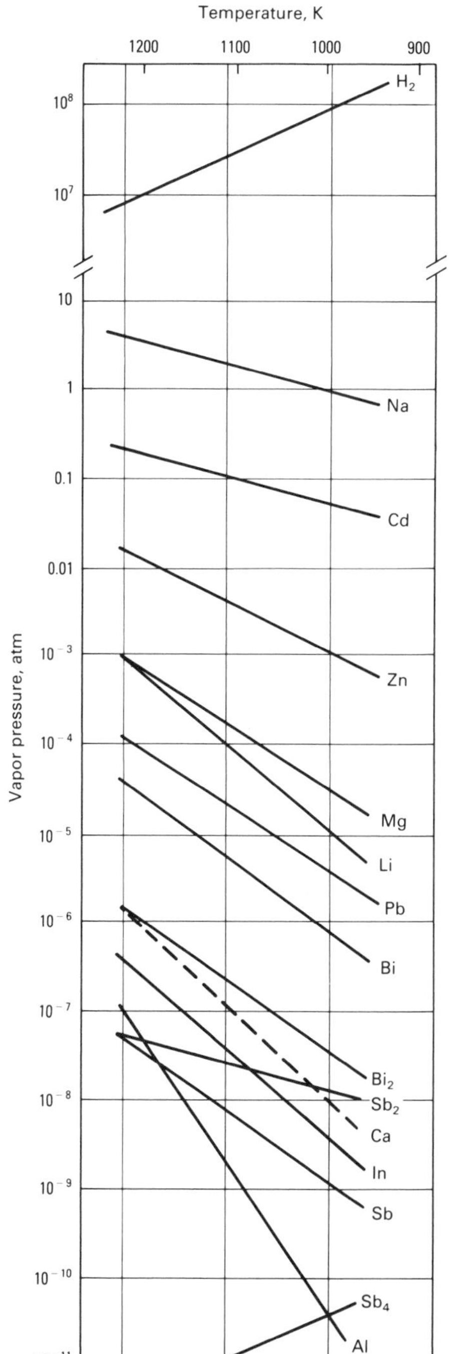

Fig. 20 Calculated vapor pressures of selected elements dissolved in aluminum at the hypothetical 1 wt% standard state. Source: Ref 35

to reduce oxygen to less than 1 ppm, as with rare-earth additions. For most cases, oxide formation appears faster than oxide particle separation from the bath. Thus, for applications that depend on dissolved oxygen, such as desulfurization or CO blowhole formation, equilibrium oxygen values can be assumed to exist after relatively short reaction periods. For considerations of the inclusion content of steel, the kinetics of particle separation need to be considered.

The limiting case of particle flotation at velocities obtained from Stokes' law calculations is shown in Fig. 19. Inclusion size versus time of flotation is plotted. The particles produced during deoxidation are generally small (<5 μm, or 200 μin.); accordingly, separation times are long. The times indicated in Fig. 19 are representative of conditions in a quiescent bath. Times are considerably shorter for stirred melts because articles grow by collision and coalescence to sizes of 30 to 100 μm (Ref 20). The efficiency of coalescence of particles is material dependent. The energy of adhesion of particles is determined by the wetting characteristics of the particle, measured by wetting angle. For example, the force of adhesion of Al_2O_3 particles with a wetting angle of 140° is twice that for SiO_2 with a wetting angle of 115°. On this basis, Al_2O_3 inclusions are expected to cluster more easily than SiO_2 inclusions, a phenomenon that has been observed in practice (Ref 20).

Aluminum Melts

The principal alloying elements in all cast irons, namely, carbon and silicon, are always the same, but the situation for aluminum is different because there are a number of different important aluminum alloy systems, such as Al-Si, Al-Cu, Al-Mg, and Al-Zn (Ref 32). As a result, the preparation of casting alloys from aluminum scrap holds greater risk of contamination than cast iron alloy preparation. In addition to the different elements present in aluminum alloy classes, minor elements are commonly present that are beneficial to some alloy systems but harmful to others. These elements must also be controlled in alloy preparation.

An example of minor element control is the case of silicon modifiers: sodium, strontium, antimony, and phosphorus. These elements favorably alter the nucleation and growth kinetics of silicon in aluminum-silicon alloys to produce a more compact form of the precipitate. Contamination of sodium- or strontium-modified alloys with alloys containing phosphorus or antimony negates the modification. Therefore, control must be exercised over these elements.

Another contaminant that can have positive and negative impact on aluminum castings is hydrogen. Because aluminum reacts readily with atmospheric moisture, hydrogen contamination is common in aluminum alloys (Ref 32). Although the solubility of hydrogen in liquid aluminum is only of the order of 1.0 ppm, most of the hydrogen precipitates during solidification, producing 0.03 volume fraction of gas porosity. This is a serious problem for castings having high strength requirements. On the other hand, for castings that are not subjected to high stress, hydrogen-generated porosity is often a desirable condition because it is used to counter the large solidification shrinkage (0.065 volume fraction) associated with aluminum castings.

The control of cast aluminum alloy composition is difficult. Few but the largest foundries attempt to process purchased scrap; instead they rely on secondary alloy producers with special equipment to perform this function (Ref 33). Because of the wide range of elements involved in aluminum alloy casting, the first line in composition control is the separation of scrap into alloy type and the accurate characterization of the composition of the batch by chemical analysis. Next, compatible batches are combined, and necessary alloy additions are made. This process avoids the difficult task of removing unwanted elements. Chemical purification is commonly practiced for hydrogen and for elements that are more electropositive than aluminum, namely, the alkali and alkaline earth elements.

Table 4 Approximate theoretical minimum metal compositions to be realized by the vacuum treatment of aluminum

T = 1000 K.

Element	Composition, ppm
Sodium	3.7×10^{-7}
Cadmium	7.0×10^{-6}
Zinc	3.3×10^{-4}
Magnesium	1.3×10^{-2}
Lithium	3.4×10^{-2}
Lead	0.11
Bismuth	0.41
Calcium	39
Indium	110
Antimony	310

Source: Ref 35

These separations are usually performed in a single process, with small bubbles of inert or reactive gas removing the undesired elements by evaporation and/or oxidation (Ref 34).

Refining Aluminum by Evaporation Treatment. Some elements dissolved in aluminum have higher vapor pressures than aluminum and can therefore be separated from aluminum melts by inert gas flushing or vacuum treatment. For the evaporation reaction:

$$m\underline{M} = M_m(g) \qquad \text{(Eq 23)}$$

the equilibrium constant is:

$$K_{24} = \frac{p_{M_m}}{(h_M)^m} \qquad \text{(Eq 24)}$$

and for the condition $h_M = 1$, that is, at the hypothetical 1 wt% standard state, the vapor pressure is only a function of K. Using the thermodynamic data given in Table 15 in the section "Aluminum-Base and Copper-Base Alloys" of the article "Thermodynamic Properties of Liquid Metals" in this Volume, the vapor pressures of selected elements were calculated. The results are plotted in Fig. 20. The vapor pressure of aluminum is also shown because it represents a practical lower limit to which impurity vapor pressures can be reduced without incurring significant losses of aluminum. Making this assumption with regard to limiting the impurity vapor pressure, it can be seen in Table 4 that elements with higher vapor pressures than calcium can be effectively removed.

Volatile impurities approach their equilibrium vapor pressures in purge gas bubbles (Ref 35); therefore, the rate of impurity removal depends on the vapor pressure of the element and the volumetric gas flow rate (see the article "Gases in Metals" in this Volume). For elements other than hydrogen, the volumetric ratio of purge gas to impurity vapor is so large that the economics of the process are jeopardized (Ref 35). For alkali and alkaline earth metals, this situation is rectified by adding a reactive gas to the purge gas.

Refining Aluminum With Reactive Gases. Alkali and alkaline earth metals form more stable halides and oxides than

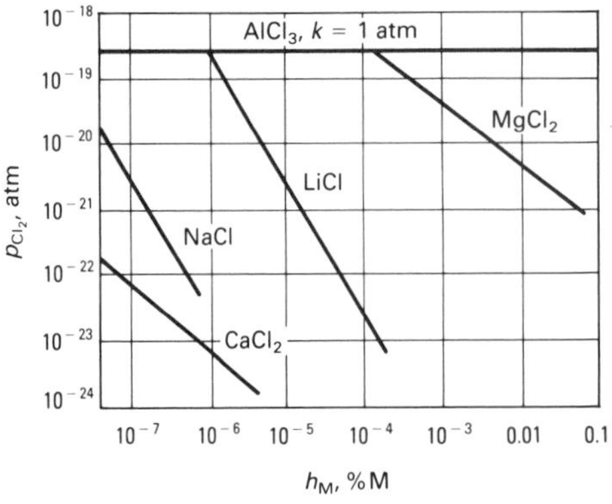

Fig. 21 Calculated equilibria for the ternary Al-M-Cl system at 1000 K. Source: Ref 35

aluminum; therefore, by adding F_2, Cl_2, or O_2 to the purge gas, these elements can be separated from aluminum. Because the reaction products in this case are condensed phases, the rate of impurity removal no longer depends on $\dot{Q}$ but on the gas/liquid interfacial area, where the reactions take place. For the generalized reaction between a metal M and halogen or oxygen X_2 gas:

$$M + xX_2 = \frac{1}{m} M_mX_{2mx} \qquad \text{(Eq 25)}$$

the equilibrium constant is:

$$K_{26} = \frac{(a_{M_mX_{2mx}})^{1/m}}{h_M(p_{X_2})^x} \qquad \text{(Eq 26)}$$

and the equilibrium X_2 pressure is:

$$p_{X_2} = \left[\frac{(a_{M_mX_{2mx}})^{1/m}}{K_{26}h_M}\right]^{1/y} \qquad \text{(Eq 27)}$$

If p_{X_2} for an impurity element is less than p_{X_2} for the corresponding reaction with aluminum, then impurity removal is theoretically possible. Conversely, if p_{X_2} for the impurity element is greater than p_{X_2} for aluminum, the aluminum compound will separate in preference to the impurity. Figures 21 and 22 plot p_{X_2} versus h_M, respectively, for reactions with chlorine and fluorine. The oxide data were not included, because oxygen is much less effective than the halides (Ref 35). The data in Fig. 21 and 22 assume $a_{M_mX_{2mx}} = 1$.

Table 5 compares the minimum impurity concentrations that are possible by treatment with Cl_2, F_2, and O_2. For all the cases

Table 5 Approximate theoretical minimum contents to be realized by refining with reactive gases

T = 1000 K

Element	Metal content, ppm, remaining after oxidation with: Oxygen	Chlorine	Fluorine
Calcium	<4000	3×10^{-5}	5.7×10^{-4}
Lithium	<270	0.01	0.046
Magnesium	1160	1.4	4.9
Sodium	<950	9×10^{-5}	0.19

Source: Ref 35

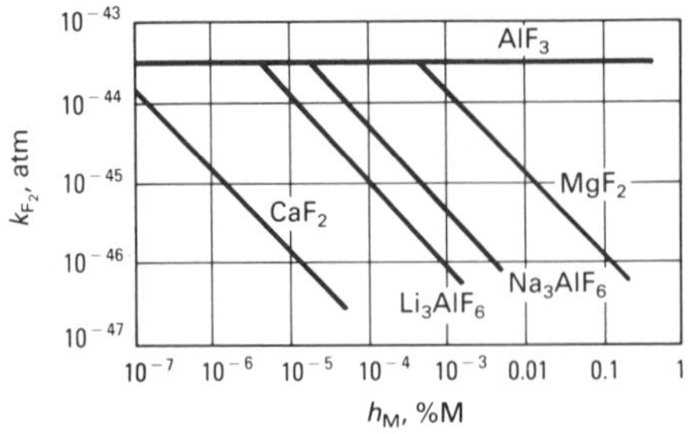

Fig. 22 Calculated equilibria for the ternary Al-M-F system at 1000 K. Source: Ref 35

examined, chlorine appears to be the best reagent. Chlorine is the most commonly used reagent for removing these elements from aluminum melts. In practice, the thermodynamically predicted levels are not achieved (Ref 35).

The use of halogens has a disadvantage in terms of the environmental precautions that need to be taken. This has prompted the development of numerous electrochemical separation processes.

Electrochemical Refining of Aluminum. An electrochemical process for separating magnesium from molten aluminum is illustrated in Fig. 23. Through the application of a potential, magnesium in the aluminum melt (anode) dissolves in the electrolyte, while pure magnesium deposits at the cathode. As seen at left in Fig. 23, densities of the three phases form a convenient system for cell construction, with the aluminum phase having the highest density and the magnesium phase having the lowest. Therefore, liquid magnesium is easily removed from the top of the cell. An advantage of the electrochemical method is that the recovered metallic magnesium has high commercial value compared to the $MgCl_2$ produced by reactive gas separations.

The open-circuit cell potential $\mathscr{E}$ for this system can be estimated from the Nernst equation:

$$\mathscr{E} = -\frac{RT}{2\mathscr{F}} \ln \frac{(a_{Mg})_{anode}}{(a_{Mg})_{cathode}} \qquad \text{(Eq 28)}$$

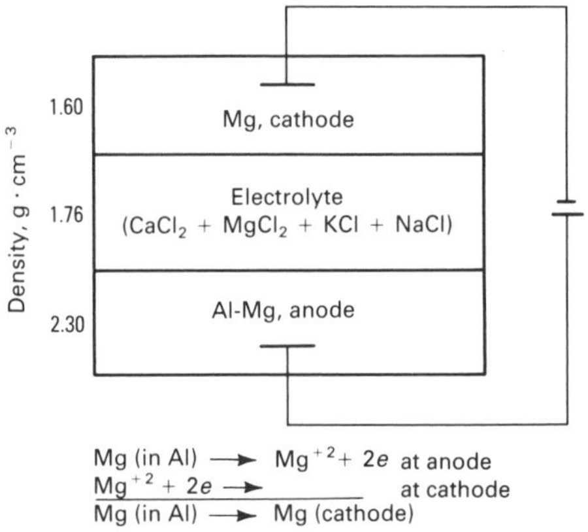

Fig. 23 Schematic of an electrochemical magnesium separation apparatus. Source: Ref 36

where $\mathcal{F}$ is Faraday's constant and $(a_{Mg})_{cathode} = 1$. For producing alloys with less than 0.1% Mg, as required for a number of important casting alloys, such as 319 and 380, Eq 28 predicts that a cell potential of 0.32 V is required (Ref 36). This voltage must not be greatly exceeded, because aluminum will be transferred from anode to cathode at a potential of 0.5 V.

REFERENCES

1. R.I.L. Guthrie, Addition Kinetics in Steelmaking, in *Proceedings of the 35th Electric Furnace Conference*, Iron and Steel Society of AIME, 1978, p 30-41
2. F.A. Mucciardi and R.I.L. Guthrie, Aluminum Wire Feeding in Steelmaking, *Trans. ISS*, Vol 3, 1983, p 53-59
3. J.W. Robison, Jr., "Ladle Treatment With Steel-Clad Metallic Calcium Wire," Paper 35, presented at *Scaninject III*, Part II, MEFOS, Lulea, Sweden, 1983
4. L. Kalvelage, J. Markert and J. Pötschke, Measurement of the Dissolution of Graphite in Liquid Iron by Following the Buoyancy, *Arch. Eisenhüttenwes.*, Vol 50, 1979, p 107-110
5. R.G. Olsson, V. Koump, and T.F. Perzak, Rate of Dissolution of Carbon in Molten Fe-C Alloys, *Trans. Met. Soc. AIME*, Vol 236, 1966, p 426-429
6. O. Angeles, G.H. Geiger, and C.R. Loper, Jr., Factors Influencing Carbon Pickup in Cast Iron, *Trans. AFS*, Vol 74, 1968, p 3-11
7. M. Eisenberg, C.W. Tobias, and C.R. Wilke, Mass Transfer at Rotating Cylinders, *Chem. Eng. Prog. Symp. Series*, Vol 51, 1955, p 1-16
8. R.G. Olsson, V. Koump, and T.F. Perzak, Rate of Dissolution of Carbon Steel in Molten Iron-Carbon Alloys, *Trans. Met. Soc. AIME*, Vol 233, 1965, p 1654-1657
9. R.D. Pehlke, P.D. Goodell, and R.W. Dunlap, Kinetics of Steel Dissolution in Molten Pig Iron, *Trans. Met. Soc. AIME*, Vol 233, 1965, p 1420-1427
10. R.I.L. Guthrie and P. Stubbs, Kinetics of Scrap Melting in Baths of Molten Pig Iron, *Can. Metall. Q.*, Vol 12, 1973, p 465-473
11. K. Mori and T. Sakuraya, Rate of Dissolution of Solid Iron in Carbon-Saturated Liquid Iron Alloys With Evolution of CO, *J. Iron Steel Inst. Japan*, Vol 22, 1982, p 964-990
12. P.T.L. Brian and H.B. Hales, Effects of Transpiration and Changing Diameter on Heat and Mass Transfer to Spheres, *AIChE J.*, Vol 15, 1969, p 419-425
13. R.J. Fruehan, *Ladle Metallurgy: Principles and Practices*, Iron and Steel Society of AIME, 1985, p 7
14. S. Katz and C.F. Landefeld, Cupola Desulfurization, in *Cupola Handbook*, American Foundrymen's Society, 1984, p 351-363
15. I.D. Sommerville, The Capacities and Refining Capabilities of Metallurgical Slags, in *Foundry Processes: Their Chemistry and Physics*, S. Katz and C.F. Landefeld, Ed., Plenum Press, 1988, p 101-133
16. G.K. Sigworth and J.F. Elliott, The Thermodynamics of Liquid Dilute Iron Alloys, *Met. Sci.*, Vol 8, 1974, p 298-310
17. S. Katz and H.C. Rezeau, The Cupola Desulfurization Process, *Trans. AFS*, Vol 87, 1979, p 367-376
18. S. Katz, D.E. McInnis, D.L. Brink, and G.A. Wilkinson, The Determination of Aluminum in Malleable Iron From Measured Oxygen, *Trans. AFS*, Vol 88, 1980, p 835-844
19. R.J. Fruehan, *Ladle Metallurgy: Principles and Practices*, Iron and Steel Society of AIME, 1985, p 8
20. E.T. Turkdogan, *Physical Chemistry of High Temperature Technology*, Academic Press, 1980
21. S. Katz and V.R. Spironello, Effect of Charged Aluminum on Iron Temperature, Silicon Recovery and Desulfurization in an Iron-Producing Cupola, *Trans. AFS*, Vol 92, 1984, p 161-172
22. C.F. Landefeld and S. Katz, Kinetics of Iron Desulfurization by CaO-CaF_2, in *Proceedings of the Fifth International Iron and Steel Congress*, Vol 6, Iron and Steel Society of AIME, 1986, p 429-439
23. S. Katz and C.F. Landefeld, Plant Studies of Continuous Desulfurization with CaO-CaF_2-C, *Trans. AFS*, Vol 93, 1985, p 215-228
24. S. Katz and B.L. Tiwari, A Critical Overview of Liquid Metal Processing in the Foundry, in *Foundry Processes: Their Chemistry and Physics*, S. Katz and C.F. Landefeld, Ed., Plenum Press, 1988, p 1-52
25. J. Niederinghaus and R.J. Fruehan, Desulfurization Mechanisms for CaO-Al and CaO-CaS in Carbon Saturated Iron, *Metall. Trans. B*, to be published
26. E.T. Turkdogan, Physiochemical Phenomena of Mechanisms and Rates of Reaction in Melting, Refining and Casting of Foundry Irons, in *Foundry Processes: Their Chemistry and Physics*, S. Katz and C.F. Landefeld, Ed., Plenum Press, 1988, p 53-100
27. S. Asai and I. Muchi, Fluid Flow and Mass Transfer in Gas Stirred Ladles, in *Foundry Processes: Their Chemistry and Physics*, S. Katz and C.F. Landefeld, Ed., Plenum Press, 1988, p 261-292
28. S.-H. Kim and R.J. Fruehan, Physical Modelling of Liquid/Liquid Mass Transfer in Gas Stirred Ladles, *Metall. Trans. B*, Vol 18B, 1987, p 381-390
29. J. Ishida, K. Yamaguchi, S. Sugiura, N. Demukai, and A. Notoh, *Denki Seiko*, Vol 52, 1981, p 2-8
30. E.T. Turkdogan, Ladle Deoxidation, Desulfurization and Inclusions in Steel—Part I: Fundamentals, *Arch. Eisenhüttenwes.*, Vol 54, 1983, p 1-10
31. E.T. Turkdogan, Slags and Fluxes for Ferrous Ladle Metallurgy, *Ironmaking Steelmaking*, Vol 12, 1985, p 64-78
32. *Aluminum Casting Technology*, American Foundrymen's Society, 1986
33. *Recycled Metals in the 1980's*, National Association of Recycling Industries, 1982
34. J.H.L. Van Linden, R.E. Miller, and R. Bachowski, Chemical Impurities in Aluminum, in *Foundry Processes: Their Chemistry and Physics*, S. Katz and C.F. Landefeld, Ed., Plenum Press, 1988, p 393-409
35. G.K. Sigworth and T.A. Engh, Refining of Liquid Aluminum—A Review of Important Chemical Factors, *Scand. J. Metall.*, Vol 11, 1982, p 143-149
36. B.L. Tiwari and R.A. Sharma, Electrolytic Removal of Magnesium From Scrap Aluminum, *J. Met.*, Vol 36 (No. 7), 1984, p 41-43

Gases in Metals

Richard J. Fruehan, Carnegie Mellon University

GAS POROSITY is one of the most serious problems in the casting of cast iron, aluminum, and copper. It is generally caused by the evolution of gases during the casting and solidification process. The gases may be the result of a reaction between the casting sand or mold and the metal, or they may result from the evolution of gases dissolved in the liquid metal during solidification.

An example of a chemical reaction evolving gas and causing porosity is the reaction of moisture (H_2O) in the sand with elements in the cast iron, such as carbon, silicon, aluminum, or iron itself. For example, aluminum in cast iron often causes porosity problems, and the reaction responsible for the gas evolution is:

$$3H_2O + 2\underline{Al} = Al_2O_3 + 3H_2 \qquad \text{(Eq 1)}$$

where the underlining of aluminum ($\underline{Al}$) indicates that it is dissolved in the metal. The free energy for reaction (Eq 1) is highly negative even for very low concentrations of aluminum, indicating that the thermodynamics for the reaction are highly favorable. Similar reactions can be written for silicon, carbon, and iron. For carbon, carbon monoxide is also evolved, according to the reaction:

$$H_2O + \underline{C} = H_2 + CO \qquad \text{(Eq 2)}$$

For the casting of aluminum, the reaction is even more favorable. Whether or not the reaction will actually occur depends on many complex factors, including liquid-phase mass transfer and surface tension; therefore, its occurrence is difficult to predict. However, if there is excessive moisture in the sand, the reaction is highly likely to occur.

The other major cause of gas porosity is the evolution of dissolved gases during casting. For example, liquid cast iron may have dissolved hydrogen and nitrogen. The solubility of these gases in the solid may be less than that in the liquid, and the gases may therefore be evolved during solidification. Whether or not this will occur depends on the amount of hydrogen and nitrogen present, the alloy being cast, chemical kinetics, and the surface tension of the alloy. Similarly, hydrogen can dissolve in liquid aluminum, but its solubility is much lower in solid aluminum and can cause porosity. For copper, the evolution of hydrogen, water vapor, or carbon monoxide could cause gas porosity.

In this article, the solubilities of the common gases present in cast iron, aluminum, and copper will be reviewed. The kinetics of the relevant reactions and the reactions during solidification will be discussed. Finally, possible methods of control or removal of the dissolved gases will be analyzed. The discussion will primarily focus on cast iron because more is known about the thermodynamics and kinetics of iron than other elements. However, the same basic principles also apply to aluminum and copper.

Cast Iron (Ref 1)

Solubility of Hydrogen and Nitrogen. The gases in cast iron that can cause porosity are hydrogen and nitrogen. Castings from iron produced in a cupola are nearly saturated with nitrogen; therefore, the presence of nitrogen is a major concern (Ref 2). Hydrogen and nitrogen dissolve in liquid iron alloys as atomic species according to:

$$\tfrac{1}{2}H_2 = \underline{H} \qquad \text{(Eq 3)}$$

$$\tfrac{1}{2}N_2 = \underline{N} \qquad \text{(Eq 4)}$$

The solubilities of both elements in liquid and solid iron alloys have been extensively measured, and the results have been adequately summarized (Ref 3). The solubility of nitrogen and hydrogen in liquid iron is decreased by carbon. The nitrogen and hydrogen contents in iron-carbon alloys in equilibrium with 1 atm pressure of the respective gases are shown in Fig. 1 as a function of carbon content. The solubility of nitrogen at 1823 K decreases from about 450 ppm for pure iron to about 80 ppm for an alloy containing 4.5% C. The solubilities of hydrogen for the same metals are 24 and 10 ppm, respectively. The hydrogen solubility decreases with temperature and at eutectic temperature is 6.5 ppm; temperature has only a small effect on the solubility of nitrogen. Silicon decreases the solubility of nitrogen even further. For a 4C-1Si melt, the solubility of nitrogen is about 75 ppm.

When the cast iron solidifies, it forms austenite and graphite or cementite. The

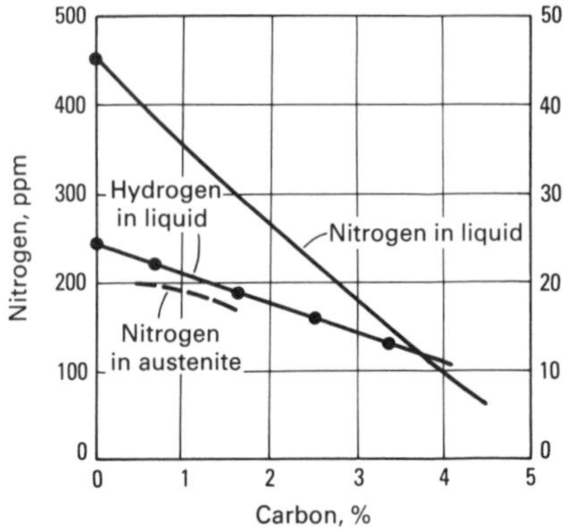

Fig. 1 Solubility of hydrogen and nitrogen (1 atm) in iron-carbon alloys at 1823 K in the liquid and 1500 K for austenite

solubility of nitrogen in austenite (Ref 3) is also shown in Fig. 1. For austenite containing about 1.8% C, the solubility is about 150 ppm. Therefore, when the liquid cast iron freezes, the solubility in the solid is higher than that in the liquid; this interesting phenomenon will be discussed later in detail. There is some solubility of nitrogen in cementite (Ref 4). The exact amount is not known, but is about the same as that in austenite. The solubility of hydrogen in austenite is about 7 ppm; therefore, the solubility of hydrogen in the liquid and solid are about equal, and there is little segregation of hydrogen during solidification (Ref 3).

Kinetics of Gas-Liquid Reactions. The absorption of hydrogen and nitrogen is controlled by one of the following steps:

- Diffusion of the gas to the surface
- Chemical reaction on the surface
- Diffusion of the element in the liquid away from the surface

For the desorption of a gas, the same three steps are important but occur in reverse order. The process need not be controlled by only one of the above steps; it can be controlled by two or more steps in series. For liquid cast iron, nitrogen is the more important of the gases, and it will be considered in detail.

Gas diffusion is usually not rate controlling for nitrogen absorption from the atmo-

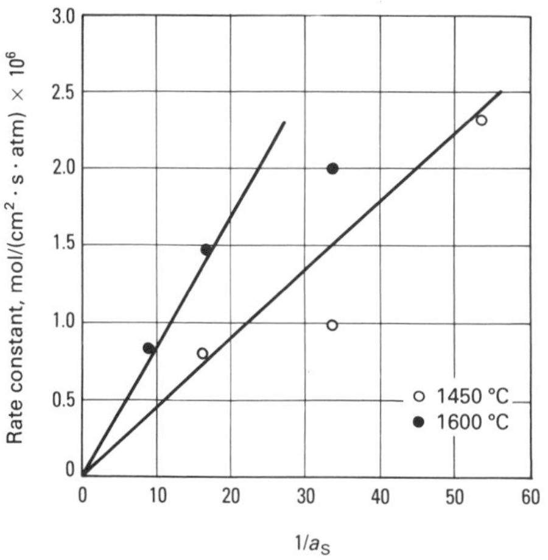

Fig. 2 Effect of sulfur in the rate of the nitrogen reaction on Fe-C_{SAT}-S alloys

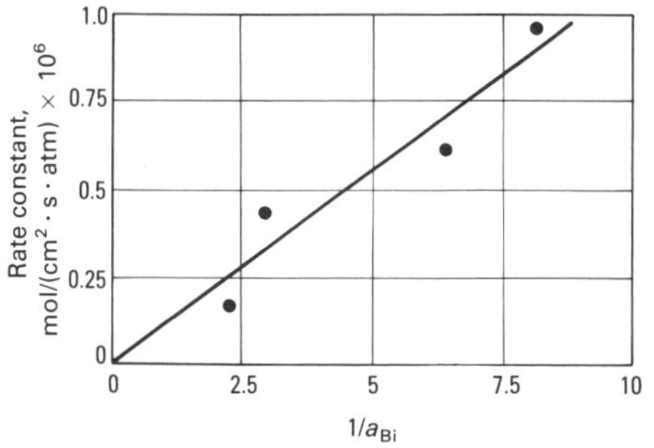

Fig. 3 Effect of bismuth on the nitrogen reaction with Fe-C_{SAT}-Bi alloys at 1450 °C (2640 °F)

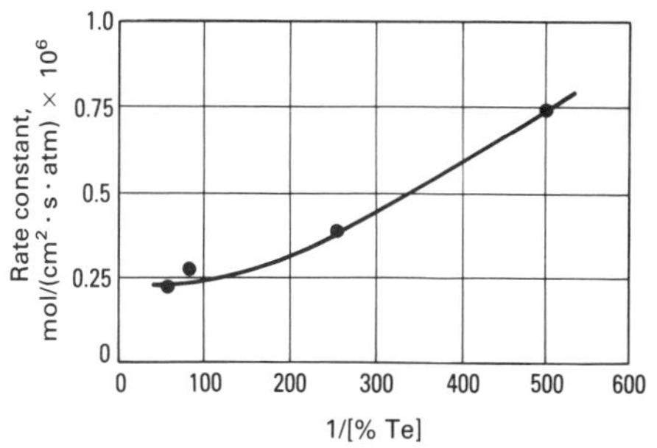

Fig. 4 Effect of tellurium on the nitrogen reaction with Fe-C_{SAT}-Te alloys at 1450 °C (2640 °F)

sphere because of the pressure of N_2; consequently, the driving force for diffusion is high. As a result, one of the other steps is slower and rate controlling. However, for nitrogen or hydrogen removal by an inert gas, diffusion of N_2 or H_2 must be considered. The flux of nitrogen J_{N_2} away from the surface is given by:

$$J_{N_2} = \frac{k_m}{RT}(p^S_{N_2} - p^B_{N_2}) \quad \text{(Eq 5)}$$

where k_m is the mass transfer coefficient, $p^B_{N_2}$ is the pressure of nitrogen in the bulk gas, and $p^S_{N_2}$ is the surface pressure of nitrogen in equilibrium with the melt where:

$$p^S_{N_2} = K^2_N f^2_N (\%N)^2 \quad \text{(Eq 6)}$$

where K_N is the equilibrium constant for Eq 4 and f_N is the activity coefficient of nitrogen with respect to 1 wt%. However, in general, gas diffusion for these reactions is faster than the chemical reaction or liquid-phase diffusion and can be neglected in most cases.

The rate of the chemical reaction is controlled by the dissociation of the nitrogen molecule on the surface, and the rate is given by:

$$\frac{dn_{N_2}}{dt} = k_B A(1 - \theta)(p_{N_2} - p^e_{N_2}) \quad \text{(Eq 7)}$$

where k_B is the rate constant for pure iron, A is the surface area, $p^e_{N_2}$ is the equilibrium nitrogen pressure given by an expression similar to Eq 6, and $(1 - \theta)$ is the fraction of vacant sites not occupied by surface-active elements. Certain elements are surface active on liquid iron in that they lower the surface tension of iron by covering most of the surface. For example, oxygen and sulfur are surface active on iron, and for bulk concentrations of 0.03% O or 0.03% S, over 90% of the surface sites will be covered by oxygen or sulfur. These elements therefore retard the rate of chemical reaction. At high coverage, the surface-active element $(1 - \theta)$ is inversely proportional to the activity a of the element. Therefore, the rate is given by:

$$\frac{dn_{N_2}}{dt} = \frac{k_B k_A A}{a}(p_{N_2} - p^e_{N_2}) \quad \text{(Eq 8)}$$

where k_A is related to the adsorption coefficient of the element on the surface and the quantity $k_B k_A/a$ is the overall rate constant k. For dilute solutions, the activity of the surface-active element is proportional to its concentration. Therefore, the overall rate constant is inversely proportional to the weight percent of the surface-active element in iron.

Liquid-Phase Mass Transfer. If the rate is controlled by liquid-phase mass transfer, the flux of nitrogen atoms is given by:

$$J_N = k_{m_N}(C^B_N - C^S_N) \quad \text{(Eq 9)}$$

where k_{m_N} is the liquid-phase mass transfer coefficient of nitrogen and C^B_N and C^S_N are the bulk and surface concentrations of nitrogen. The integrated form of Eq 7 is:

$$\ln\left[\frac{\%N - \%N_e}{\%N_o - \%N_e}\right] = -\frac{A\rho k_{m_N} t}{W} \quad \text{(Eq 10)}$$

where ρ is the density of iron, W is the weight of the metal, and $\%N_e$ and $\%N_o$ are the equilibrium and initial nitrogen contents, respectively.

The rate is often controlled by two processes in series—usually the chemical reaction and liquid-phase mass transfer. The rate in this case can be obtained by equating the fluxes given by Eq 8 and 9 and solving for the surface concentration as indicated in Ref 5.

The rate of the nitrogen reaction with iron alloys containing oxygen, sulfur, chromium, and other elements has been measured by many investigators, and there is good agreement in most cases (Ref 6-11). However, the rate for carbon-saturated iron containing other elements was not investigated until recently (Ref 12).

In this work, an isotope exchange technique was used to measure the rate of dissociation of the nitrogen molecule (N_2). In particular, the effects of carbon, sulfur, phosphorus, lead, tin, bismuth, and tellurium on the rate were investigated.

Sulfur, phosphorus, lead, bismuth, and tellurium decreased the rate, while carbon and tin had no significant effect. For example, the effect of sulfur is shown in Fig. 2 as the rate constant versus the reciprocal of the activity of sulfur (Ref 11). The activity of sulfur is in weight percent, and the activity coefficient of sulfur in carbon-saturated iron is 6.3. The rate for carbon-saturated iron with no sulfur is about 10^{-5} mol/cm² · s · atm at 1450 °C (2640 °F). For example, for 0.009% S ($1/a_s = 18$), the rate is decreased to 8×10^{-7} mol/cm² · s · atm. Bismuth and tellurium had an even larger effect (Ref 12), as shown in Fig. 3 and 4. The activities are relative to pure bismuth and tellurium, respectively, which have larger deviations from ideal behavior. As little as 50 ppm Te reduced the rate by 90%, and 50 ppm Bi decreased it by 80%.

On the other hand, carbon is not surface active and does not affect the rate at moderate sulfur contents (0.017% S) up to 4.5% C (Ref 12), as shown in Fig. 5. Therefore, the initial rate of nitrogen formation is primarily controlled by chemical kinetics at the surface, and surface-active elements, such as sulfur and tellurium, can reduce the rate. This information can be helpful in controlling the rate of the reaction. For example, if it is desired to remove nitrogen by argon gas flushing, the concentrations of these elements should be as low as possible. On the other hand, it may be possible to retard nitrogen evolution during solidification by the deliberate addition of these elements.

Reactions During Solidification. During the solidification of most simple iron alloys, there is enrichment of the alloying element because there is greater solubility in the liquid than in the solid. The difference in the solubilities is expressed by the partition ratio $k^{S/L}$, which is determined from the slopes of the solidus to liquidus lines on the phase diagram. For most simple alloys, the partition ratio is less than 1, resulting in liquid enrichment during solidification. For example, for an Fe-0.02N alloy, the last portion of liquid to solidify will be enriched and will have a concentration of 0.045% N, which is the solubility of nitrogen in liquid iron at 1 atm (Ref 13). Therefore, even

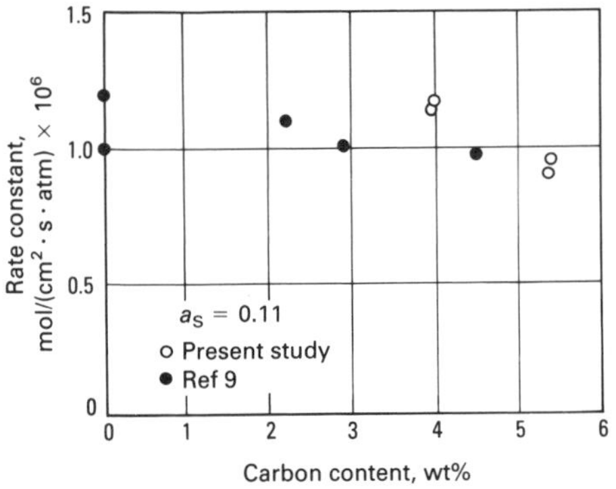

Fig. 5 Effect of carbon on the nitrogen reaction on Fe-C-S alloys at constant sulfur activity

though the liquid alloy is far from being saturated with nitrogen, during solidification, the concentration will increase to the point where the solubility is exceeded and nitrogen gas will be evolved.

Researchers have measured the partition ratio for nitrogen in high-carbon iron alloys (Ref 4). The partition ratio was found to be 1.9 and 2.2 for stable and metastable eutectic solidification, indicating that there is greater solubility in the solid than in the liquid. Therefore, in this case, as the alloy solidifies, nitrogen is actually enriched in the solid, and the nitrogen content decreases in the liquid. It may appear that, because there is no enrichment during solidification, the thermodynamic pressure of nitrogen cannot increase. The thermodynamic pressure is defined by an expression similar to Eq 6. Carbon greatly increases the activity coefficient of nitrogen in the liquid and decreases its solubility; therefore, the thermodynamic pressure may increase because the carbon content of the liquid is increasing during solidification.

For example, for an alloy containing 3.8% C that is saturated at 1500 °C (2730 °F) with 110 ppm N just prior to eutectic solidification, the concentration in the liquid will decrease to 97.5 ppm because of enrichment in the austenite. However, the solubility of nitrogen in the liquid has decreased to 90 ppm primarily because of the increase in the carbon content. Consequently, the thermodynamic pressure of nitrogen exceeds 1 atm, and the evolution of nitrogen may cause a pinhole.

When determining the possibility of pinhole formation, one must consider the total thermodynamic pressure of all the gases, and if the total pressure p_T exceeds 1 atm, a gas pinhole may result. For example, for hydrogen and nitrogen, the total pressure is given by:

$$p_T = K_H^2 f_H^2 (\%H)^2 + K_N^2 f_N^2 (\%N)^2 \qquad \text{(Eq 11)}$$

where K_i is the equilibrium constant for the gas reaction and f_i is the activity coefficient. At the end of solidification, the concentrations will be as indicated in Eq 11. Even a small amount of hydrogen may be important. For example, in the case of an Fe-3.8C alloy containing only 4 ppm H, the hydrogen content in the liquid will increase slightly, because of the enrichment, to about 4.5 ppm just prior to the eutectic reaction. However, even for this small amount of hydrogen, its thermodynamic pressure is 0.4 atm. Therefore, if the thermodynamic pressure of nitrogen exceeds 0.6 atm, a gas pinhole may result. Calculations indicate that iron with as little as 80 ppm N and 4 ppm H may have a gas pressure exceeding 1 atm.

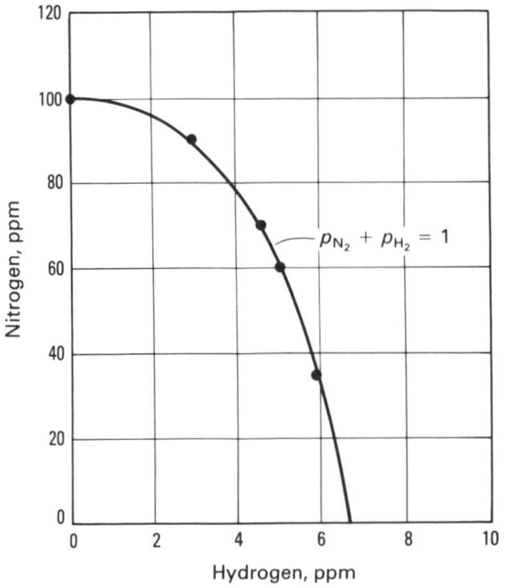

Fig. 6 The thermodynamic pressure of nitrogen and hydrogen at the end of solidification equal to 1 atm for an Fe-3.8C alloy

Figure 6 shows a plot of the total pressure of hydrogen and nitrogen just prior to final solidification equal to 1 atm as a function of bulk nitrogen and hydrogen contents in an Fe-3.8C alloy. If the hydrogen and nitrogen contents are such that they are below the line, the pressure will not exceed 1 atm, and a pinhole due to gas evolution will not occur; if above the line and the pressure exceeds 1 atm, a pinhole from this source is possible.

For pinholes to develop, it is also necessary to nucleate a bubble. Bubble nucleation is a complex phenomenon and is influenced by surface tension. For melts with low surface tension, bubble nucleation is favored. On the other hand, elements that reduce the surface tension also reduce the rate of N_2 formation. Therefore, it is difficult to predict the net effect on the probability of bubble formation of an alloying element that reduces the surface tension of iron.

Nitrogen and Hydrogen Removal by Inert Gas Flushing. The preceding discussion indicates that nitrogen and hydrogen evolution during solidification may be a cause of pinholes in castings. Therefore, it would be desirable to remove a portion of these gases if possible. One method is by argon bubbling in the desulfurization reactor or ladle. The desulfurization reactor is a continuous reactor with metal flowing in and out almost continuously. However, for the present calculation it will be assumed to be a batch reactor. Because this is only an order of magnitude calculation, this assumption is not unreasonable.

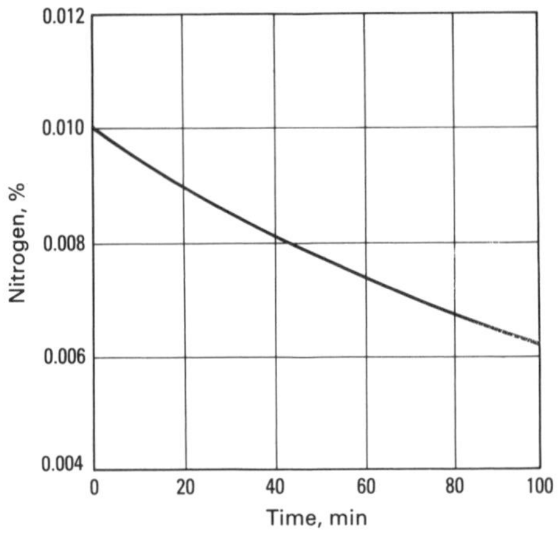

Fig. 7 Rate of removal of nitrogen from an Fe-3.8C alloy with argon bubbling at 0.005 m³/s (10 scfm) in a 10 Mg (11 ton) reactor

When argon is bubbled through the melt, nitrogen atoms combine on the bubble surface to form N_2. The rate is controlled by the chemical reaction on the surface and the liquid-phase mass transfer of nitrogen to the surface in series. For the purpose of the present calculations, the following were assumed:

- 10 Mg (11 ton) reactor 1 m (3.3 ft) deep
- Argon flow rate of 0.005 m^3/s (10 scfm) through the melt
- Initial nitrogen content of 100 ppm
- 60 mm (2.4 in.) diam bubbles

The rate constant for iron containing a relatively small amount of sulfur is about 1.5×10^{-6} $mol/cm^2 \cdot s \cdot atm$, and the mass transfer coefficient k_m is about 0.1 cm/s. The gas velocity, retention time, and surface area were estimated as done previously (Ref 5). Equations 8 and 9 were solved simultaneously, and the results are given in Fig. 7. The calculations demonstrate that the rate is truly mixed control, as indicated by the surface concentration being between zero and the bulk concentration. The results indicate that it would take about 40 min to remove 20 ppm; if the argon flow were doubled, it would still take over 20 min to remove 20 ppm. Although these calculations are rather crude, they indicate that it would be difficult to remove nitrogen by argon bubbling.

For hydrogen, the chemical reaction and mass transfer are considerably faster than for nitrogen. If equilibrium between the metal and the gas bubbles leaving the system is assumed, it is possible to calculate the amount of hydrogen that can be removed by gas bubbling from thermodynamic considerations alone. The equations are developed similarly to those for steel (Ref 13), taking the form:

$$\frac{1}{[H]} - \frac{1}{[H_o]} = k_H V \quad \text{(Eq 12)}$$

where [H] and $[H_o]$ are the hydrogen content after bubbling and the initial hydrogen content, respectively, and V is the total volume of gas used per ton of metal. The constant k_H is related to the solubility of hydrogen and therefore depends on the alloy composition. For example, 10 Mg (11 tons) of an Fe-4.5%C melt at 1773 K containing 4 ppm H bubbling argon for 20 min at 0.005 m^3/s (10 scfm) will reduce the hydrogen content to 1.6 ppm. The above calculation assumes equilibrium and therefore the fastest rate possible and may be an overestimation of the amount of hydrogen removed. Generally, the efficiency of the flushing gas would be roughly 50% for the conditions encountered in practice. However, the calculation does indicate that it is possible to remove significant amounts of hydrogen by gas bubbling.

Aluminum

The gas primarily responsible for porosity in aluminum casting is hydrogen. Hydrogen can enter the liquid alloy by the reaction of the aluminum with water vapor:

$$H_2O(v) + \tfrac{2}{3}\underline{Al} = \tfrac{1}{3}Al_2O_3 + 2\underline{H} \text{ (in Al)} \quad \text{(Eq 13)}$$

Details of the reactions of hydrogen with aluminum alloys and the factors influencing hydrogen removal can be found in Ref 14.

Hydrogen Solubility and Reactions During Solidification. Researchers have measured the solubility of hydrogen in cubic centimeters per 100 g of alloy or in weight percent in equilibrium with 1 atm of hydrogen (Ref 15, 16). These results are given in Fig. 8. Alloying elements affect the solubility of hydrogen. The solubilities for selected alloys at 750 °C (1382 °F) are given in Table 1. Generally, the common alloying elements decrease the solubility of hydrogen.

Because the solubility of hydrogen is significantly higher in the liquid aluminum as compared with the solid, there will be enrichment of the liquid during solidification. Assuming no solid diffusion and complete

Table 1 The solubility of hydrogen in aluminum and its alloys

Alloy	Hydrogen solubility, ppm
Pure aluminum	1.20
Al-7Si-0.3Mg	0.81
Al-4.5Cu	0.88
Al-6Si-3.5Cu	0.67
Al-4Mg-2Si	1.15

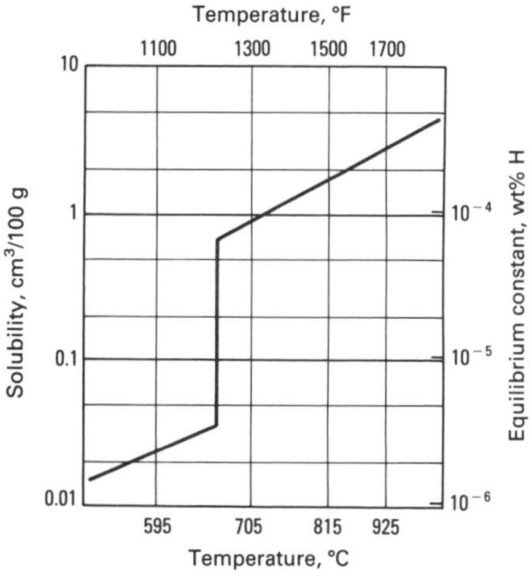

Fig. 8 Solubility of hydrogen (p_{H_2} = 1 atm) in aluminum

liquid diffusion, at the end of solidification there will be a large increase in the hydrogen concentration. Therefore, even when the hydrogen content is far below the solubility for the bulk liquid during solidification, the concentration in the enriched liquid will increase, and the solubility limit may be exceeded. For example, consider pure aluminum containing 0.4 ppm H at 750 °C (1382 °F). By the time the alloy is 90% solidified, the concentration in the remaining liquid will be about 3.6 ppm H, which will exceed the hydrogen solubility for 1 atm, resulting in the formation of hydrogen bubbles.

Inert Gas Flushing. It is possible to remove hydrogen from liquid aluminum by inert gas (argon or nitrogen) flushing. As discussed previously for cast iron, it is possible to calculate the fastest possible rate, or the minimum amount of purge gas needed to remove hydrogen. The general form of the equation is:

$$\frac{1}{[H]} - \frac{1}{[H]_o} = \frac{k_H V t}{W} \quad \text{(Eq 14)}$$

where [H] and $[H_o]$ are the hydrogen content in ppm at time t and the initial hydrogen content, respectively; W is the weight (in kilograms); t is the time (in minutes); k_H is a constant related to the solubility; and V is the flow rate (in $m^3_{(STP)}/s^{-1}$). In practical units, at 750 °C (1380 °F):

$$k_H = 6.2 \times 10^4 \frac{kg \cdot m^3_{(STP)}}{(ppm) \cdot s} \quad \text{(Eq 15)}$$

For example, for 10 Mg (2.2 × 10^4 lb) of aluminum, a gas purge rate of 0.01 m^3/s (0.35 ft^3/s) for 100 s could reduce the hydrogen from 0.2 to 0.09 ppm. As the hydrogen content decreases, it becomes increasingly difficult to remove hydrogen (Ref 14), as indicated in Fig. 9. As the hydrogen content decreases, the ratio R of the purge gas to the hydrogen gas removed increases from about 15 at 0.4 ppm to over 500 at 0.1 ppm (1 cm^3/100 g = 0.9 ppm). It should be emphasized

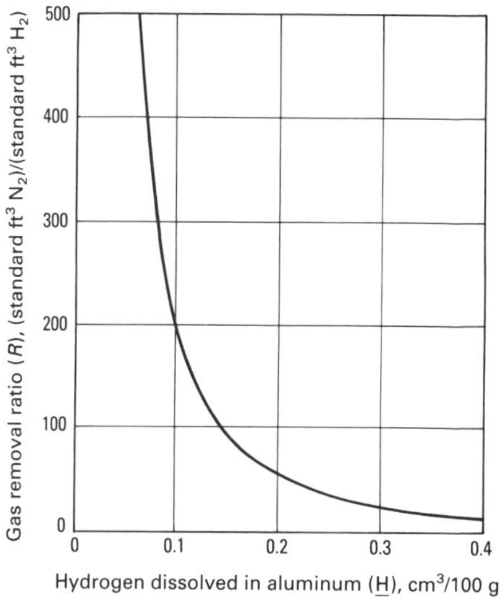

Fig. 9 Gas removal ratio for equilibrium in aluminum at 760 °C (1400 °F)

that this calculation indicates the theoretical minimum amount of purge gas required. Due to kinetic factors, the amount of purge gas can be significantly higher.

There are two limiting cases with respect to hydrogen removal: the thermodynamic limit discussed above and the diffusion in the liquid film boundary layer surrounding the gas bubble. For the limiting case of diffusion control, the rate equations are similar to those discussed previously for nitrogen in iron, and the rate can be expressed by:

$$\ln \frac{\%H}{\%H_i} = -\frac{k_{m_H} \rho A}{W} t \quad \text{(Eq 16)}$$

where $\%H_i$ is the initial hydrogen content, ρ is the liquid density, k_{m_H} is the mass transfer coefficient for hydrogen, A is the bubble surface area, and W is the weight of the melt. Both A and k_{m_H} increase with decreasing bubble size; consequently, the rate is sensitive to bubble size, with the hydrogen removal favored by small bubbles.

In general, the rate will be controlled by both diffusion and the thermodynamic limit in series. In an analysis of the case of mixed control, it was found that the purging efficiency—the ratio of the actual amount of gas required to the theoretical minimums given by Eq 14—depends on the bubble size and hydrogen content (Ref 17), as shown in Fig. 10. Both the interfacial area and the mass transfer coefficient increase with decreasing bubble size; consequently, the rate of hydrogen removal is enhanced by having small bubbles. The use of porous plugs helps improve the rate by providing small bubbles, but the bubbles generally coalesce in the bath, reducing the beneficial effect. The use of an impeller with a porous plug improves the rate further by dispersing the bubbles.

It has also been found that the addition of chlorine (Ref 18) or freon (Ref 19) improves

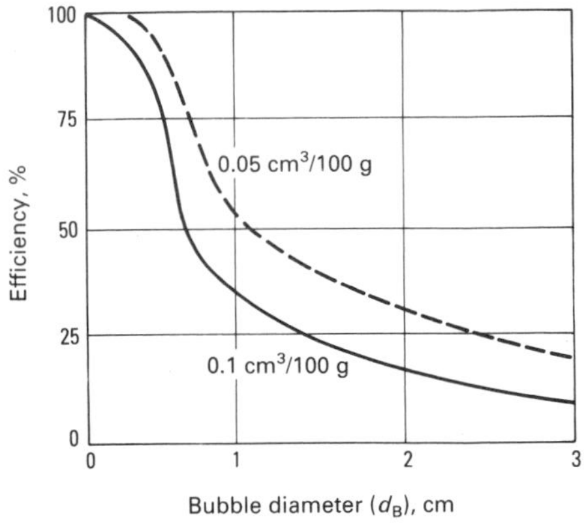

Fig. 10 Purging gas efficiency as a function of bubble size for 0.5 cm^3/100 g and 0.1 cm^3/100 g of hydrogen

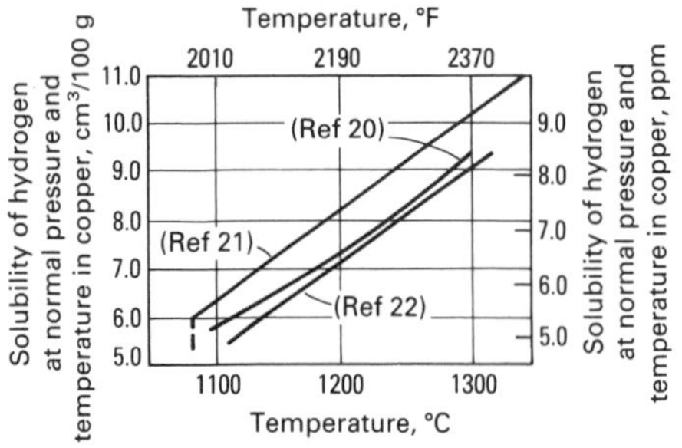

Fig. 11 Solubility of hydrogen in liquid copper as a function of temperature (p_{H_2} = 1 atm)

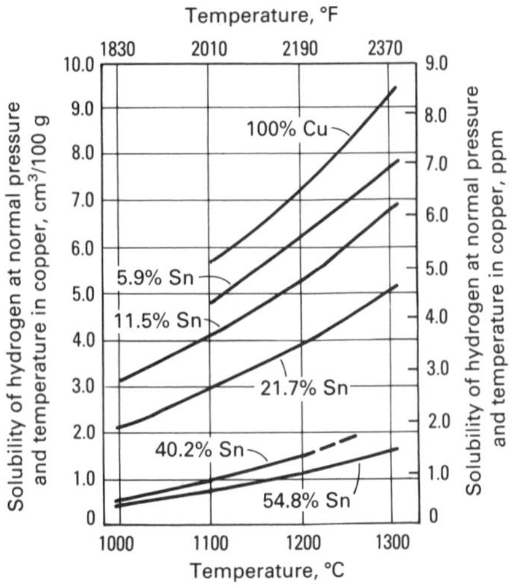

Fig. 12 Solubility of hydrogen in copper and copper-tin alloys (p_{H_2} = 1 atm) as a function of temperature

the rate in some cases. These halogen gases increase the mass transfer coefficient k_m and the rate. However, if the rate is being primarily controlled by the thermodynamic limit, the addition of these gases will have little effect. It is also possible that the halogen is actually taking part in a hydrogen removal reaction:

$$2\underline{H} + Cl_2 = HCl \qquad \text{(Eq 17)}$$

Equation 17, under some circumstances, is thermodynamically favorable. This will increase the amount of hydrogen that can be present in the inert gas bubble; that is, hydrogen can be present as H_2 or hydrogen chloride gas, thus increasing the rate of removal.

Copper and Copper Alloys

Gas porosity is a problem in the casting of copper and copper alloys. The dissolved gas that is most important is hydrogen. However, gaseous compounds such as water vapor, carbon monoxide, and sulfur dioxide can also evolve during solidification.

Hydrogen Solubility and Reactions. Hydrogen dissolves in copper and copper alloys as hydrogen atoms. The solubility in terms of cubic centimeters of H_2 per 100 g of metal is shown in Fig. 11 (1 cm^3/100 g = 0.9 ppm) (Ref 20-22). Recent work (Ref 23) has confirmed the results of earlier researchers (Ref 20), and these are the current accepted values. The most common alloying element for copper is tin, and as shown in Fig. 12, tin decreases the solubility of hydrogen (Ref 20). For example, in a 50Sn-50Cu alloy at 1200 °C (2190 °F), the solubility of hydrogen is only about 1 ppm as compared to 6.5 ppm for pure copper. On the other hand, nickel increases the solubility of hydrogen. For example, for a 10Ni-90Cu alloy, the solubility is about 10 ppm.

Hydrogen can enter the copper directly from the hydrogen in the atmosphere, but most likely it enters according to:

$$2Cu + H_2O = Cu_2O + 2\underline{H} \qquad \text{(Eq 18)}$$

$$H_2O = 2\underline{H} + \underline{O} \qquad \text{(Eq 19)}$$

Equations 18 and 19 are favored by copper with low oxygen contents. If there are high concentrations of alloying elements with more stable oxides than cuprous oxide, such as tin, the alloying element may react to put hydrogen into solution, for example:

$$\underline{Sn} + H_2O = SnO + 2\underline{H} \qquad \text{(Eq 20)}$$

The main cause of porosity in the casting results from the evolution of hydrogen or water vapor during solidification. The hydrogen and oxygen contents increase because of enrichment during solidification, as described previously for cast iron. As the concentrations increase toward the end of the solidification, hydrogen or water vapor is evolved by the reverse reactions of Eq 18 and 19, which become thermodynamically favorable because the hydrogen and oxygen contents in the last liquid to solidify are high. This mechanism is supported by the observation that the voids in the casting are normally distributed along the grain boundaries, which are the last parts to solidify.

Other gases that can cause porosity are sulfur dioxide and carbon monoxide, which also form during solidification according to Eq 21 and 22, respectively:

$$S + 2\underline{O} = SO_2 \qquad \text{(Eq 21)}$$

$$\underline{C} + \underline{O} = CO \qquad \text{(Eq 22)}$$

The thermodynamic pressure of carbon monoxide gas can be very high even at moderate oxygen and carbon contents because the solubility of the carbon in copper is limited and therefore its chemical activity is high.

Porosity in Copper Alloys. In copper ingots, the gas porosity usually results from H_2 and water evolution. The evolution of as little as 1 ppm H results in a gas evolution in 44% of the volume of the metal. If the sulfur content exceeds 0.05%, the evolution of sulfur dioxide may also contribute to gas porosity. Only if the oxygen content exceeds 0.01% is carbon monoxide considered to be the cause of porosity (Ref 24). Because phosphorus deoxidizes copper significantly, the oxygen content of phosphorized copper is very low and water, carbon monoxide, and sulfur dioxide do not contribute to porosity; only H_2 need be of concern.

Copper-zinc alloys rarely have problems associated with gas porosity, primarily because zinc deoxidizes the metal. Zinc may also remove dissolved gases because of its high vapor pressure, which results in the zinc vapor flushing hydrogen out of the melt. For copper-tin alloys, hydrogen is the major concern because tin significantly increases the solubility of hydrogen.

Degassing of Copper Alloys. The two primary methods of removing dissolved gases from copper alloys are oxidation-reduction and inert gas flushing. Although vacuum degassing is theoretically possible, it is rarely used because it is not cost effective.

In oxidation-reduction, the first step is to remove the hydrogen by oxidation using an oxidizing slag or oxygen-rich copper. Then, just prior to pouring, the melt is deoxidized by adding phosphorus or other deoxidizers. Calcium boride, boron carbide, and lithium have also been used for deoxidation (Ref 25). A charcoal cover is often used for deoxidation and protection of the melt from reoxidation.

For copper alloys that contain strong deoxidizers, such as phosphorus, zinc, and tin, it is not possible to remove hydrogen by oxidation, because these elements will form stable oxides. For these alloys, inert gas flushing with nitrogen is commonly used. The theoretical minimum amount of inert gas is determined by using relationships similar to those derived for iron and aluminum. In practice, 150 to 250 L (5.3 to 8.8 ft^3) of nitrogen per 1000 kg (2200 lb) of copper alloy is generally recommended (Ref 26).

Overcoming Gas Porosity

Gas in cast iron, aluminum, and copper can be a major problem. The porosity results from reactions with the environment or the evolution of gases during solidification. In determining if gas evolution can cause porosity, the thermodynamics for the reactions must be favorable. In this article, the solubilities of the common gases and the reactions during solidification were discussed. Because of enrichment during solidification, the concentration of the elements increases to the extent that the thermodynamic pressure of the dissolved gases may exceed 1 atm. In addition to the thermodynamics, kinetic factors and interfacial energies determine if the gases will evolve.

The most common method for removing hydrogen from aluminum and copper, which can also be used for removing hydrogen and nitrogen from cast iron, is inert gas flushing. The theoretical minimum amount of flushing gas required for removing hydrogen and nitrogen can be computed. In practice, the amount of inert gas required is somewhat higher because of kinetic factors. The kinetics of inert gas flushing can be improved by the dispersion of small bubbles throughout the melt.

REFERENCES

1. R.J. Fruehan, in *Proceedings of the Physical Chemistry of Foundry Processes Symposium* (Warren, MI), General Motors Corporation, 1986
2. S. Katz and C. Landefeld, General Motors Research, private communication, 1986
3. *Making, Shaping and Treating of Steel*, 10th ed., United States Steel Corporation, 1985
4. A. Kagawa and T. Okamoto, *Trans. Jpn. Inst. Met.*, Vol 22 (No. 2), 1981, p 137
5. R.J. Fruehan, B. Lally, and P.C. Glaws, in *Proceedings of the Fifth International Iron and Steel Congress* (Washington, DC), Iron and Steel Society of AIME, 1986
6. R.D. Pelke and J. Elliott, *Trans. TMS-AIME*, Vol 227, 1963, p 894
7. M. Inouye and T. Choh, *Trans. JISI*, Vol 8, 1968, p 134
8. R.J. Fruehan and L.J. Martonik, *Metall. Trans. B*, Vol 11B, 1980, p 615
9. P.C. Glaws and R.J. Fruehan, *Metall. Trans. B*, Vol 16B, 1985, p 551
10. P.C. Glaws and R.J. Fruehan, *Metall. Trans. B*, Vol 17B, 1986, p 317
11. M. Byrne and G.R. Belton, *Metall. Trans. B*, Vol 14B, 1983, p 441
12. F. Tsukihashi and R.J. Fruehan, submitted to *Trans. JISI*, 1987
13. R.J. Fruehan, *Ladle Metallurgy: Principles and Practices*, Iron and Steel Society of AIME, 1985
14. G.K. Sigworth, *Trans. AFS*, 1987, p 73
15. W.R. Opie and W.J. Grant, *Trans. AIME*, Vol 188, 1950, p 1234
16. C.E. Ramsley and H. Neufeld, *J. Inst. Met.*, Vol 74, 1947-1948, p 559
17. G.K. Sigworth and T.A. Engh, *Metall. Trans. B*, Vol 13B, 1982, p 447
18. J. Botor, *Metal. Odlev*, Vol 6, 1980, p 21
19. J. Botor, *Aluminum*, Vol 56, 1980, p 519
20. M.B. Bever and C.F. Floe, *Trans. AIME*, Vol 156, 1944, p 149
21. A. Sieverts and W. Krumbharr, *Ztch. Phys. Chem.*, Vol 74, 1910, p 277
22. P. Roentgren and F. Moeller, *Metallwertschaft*, Vol 13, 1934, p 81
23. E. Kato, H. Ueno, and T. Orimo, *Trans. Jpn. Inst. Met.*, Vol 11, 1970, p 351
24. R.H. Waddington, *J. Inst. Met.*, 1948-1949, p 311
25. E.R. Thews, *Metall.*, June 1956, p 431
26. W.A. Baker and F.C. Child, *J. Inst. Met.*, Vol 70, 1944, p 349

Inclusion-Forming Reactions

Paul K. Trojan, University of Michigan—Dearborn

INCLUSIONS can be defined as nonmetallic and sometimes intermetallic phases embedded in a metallic matrix. They are usually simple oxides, sulfides, nitrides, or their complexes in ferrous alloys and can include intermetallic phases in nonferrous alloys. In almost all instances of metal casting, they are considered to be detrimental to the performance of the cast component.

For example, mechanical properties can be adversely influenced by inclusions, which act as stress raisers. There is no set pattern for the effect, but some properties are more sensitive to the presence of inclusions than others. Elongation or reduction in area is usually modified more significantly than ultimate tensile strength. As a result, ductility specifications are common quality control indices in cast products.

Similar observations are found with porosity due to gas or occluded shrinkage. The loss in casting properties measured by a tensile test may then reflect a combination of defects; of these defects, inclusions are only a single but very important source of poor performance.

Because imperfections become areas of higher stress concentration, the percentage of property loss becomes greater when the strength requirement is higher. An important example is the decrease in fracture toughness when inclusions are present in higher-strength lower-ductility alloys. Similar pronounced property degradation is observed in tests that reflect slow, rapid, or cyclic strain rates, such as creep, impact, and fatigue testing.

In all cases, the degree of property loss will depend on the parameters that control the characteristics of all microstructures. These metallurgical phase controls are the nature, amount, size, shape, distribution, and orientation of the phases. If inclusions are to be controlled, these characteristics must be suitably modified to minimize their impact on casting performance.

Although inclusions are generally considered to be detrimental, their intentional introduction in larger quantities can lead to unique dispersion-strengthened materials. An example would be the dispersion of insoluble silicon carbide whiskers or alumina (Al_2O_3) fibers in a molten aluminum alloy that is subsequently cast into a final shape by squeeze casting methods (Ref 1) (see the article "Cast Metal-Matrix Composites" in this Volume).

Inclusion Types

There are essentially two classifications for all inclusions:

- Exogenous: those derived from external causes
- Indigenous: those that are native, innate, or inherent in the molten metal treatment process

Exogenous Inclusions. Slag, dross, entrapped mold materials, and refractories are examples of inclusions that would be classified as exogenous. In most cases, these inclusions would be macroscopic or visible to the naked eye at the casting surface (Fig. 1). When the casting is sectioned, they may also appear beneath the external casting surface if they have had insufficient time to float out or settle due to density differences with respect to the molten metal.

Indigenous inclusions include sulfides, nitrides, and oxides derived from the chemical reaction of the molten metal with the local environment. Such inclusions are usually small and require microscopic magnification for identification. They are often uniformly distributed within the microstructure. However, a grain-boundary distribution of indigenous inclusions (Fig. 2) can be particularly damaging to mechanical properties.

There are cases in which indigenous inclusions such as oxides may react with mold materials to promote an exogenous inclusion variety that otherwise might not appear in the absence of the indigenous oxide. To circumvent the strict definition as indigenous or exogenous, a number of investigators prefer to catalog all inclusions as either macroinclusions or microinclusions. Although such definitions describe the antici-

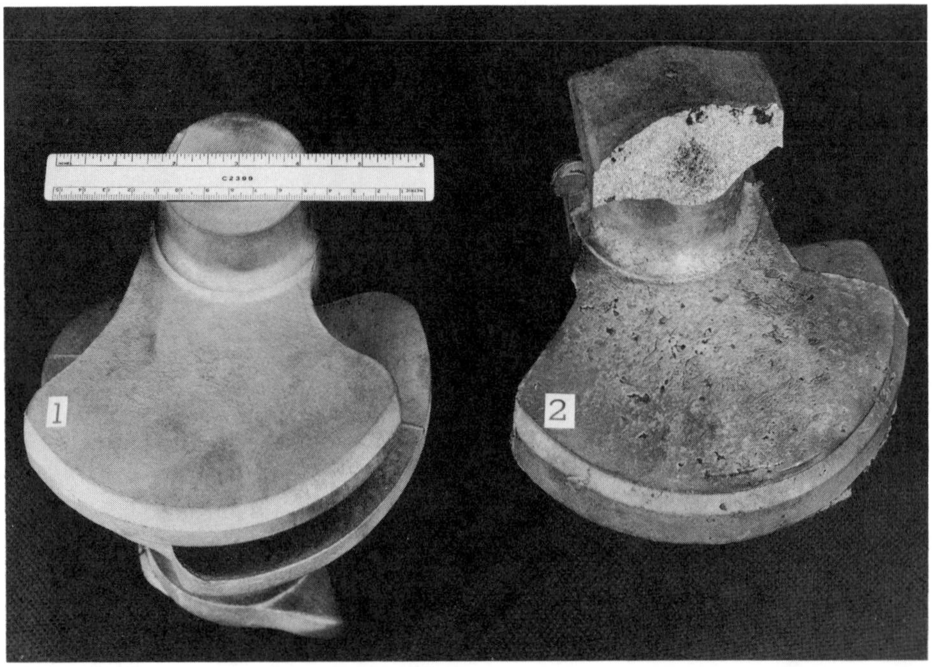

Fig. 1 Ductile iron crankshaft segment essentially free of exogenous inclusions (1, left) and with numerous exogenous inclusions (2, right). Low pouring temperature and poor mold filling practice were the cause of the inclusions in part 2.

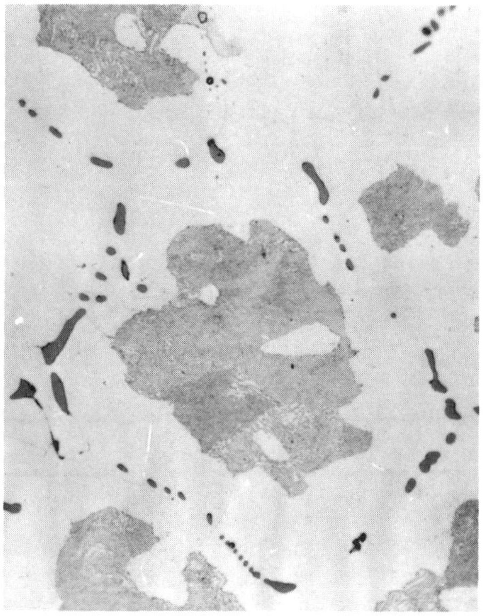

Fig. 2 Type II indigenous sulfide inclusions in a 0.25% C steel. Etched using 2% nital; 500×

pated effect on mechanical properties, there can be a loss in source identification of the inclusion. Whatever definition is used, control of inclusions requires an understanding of their origin and the associated physical chemistry.

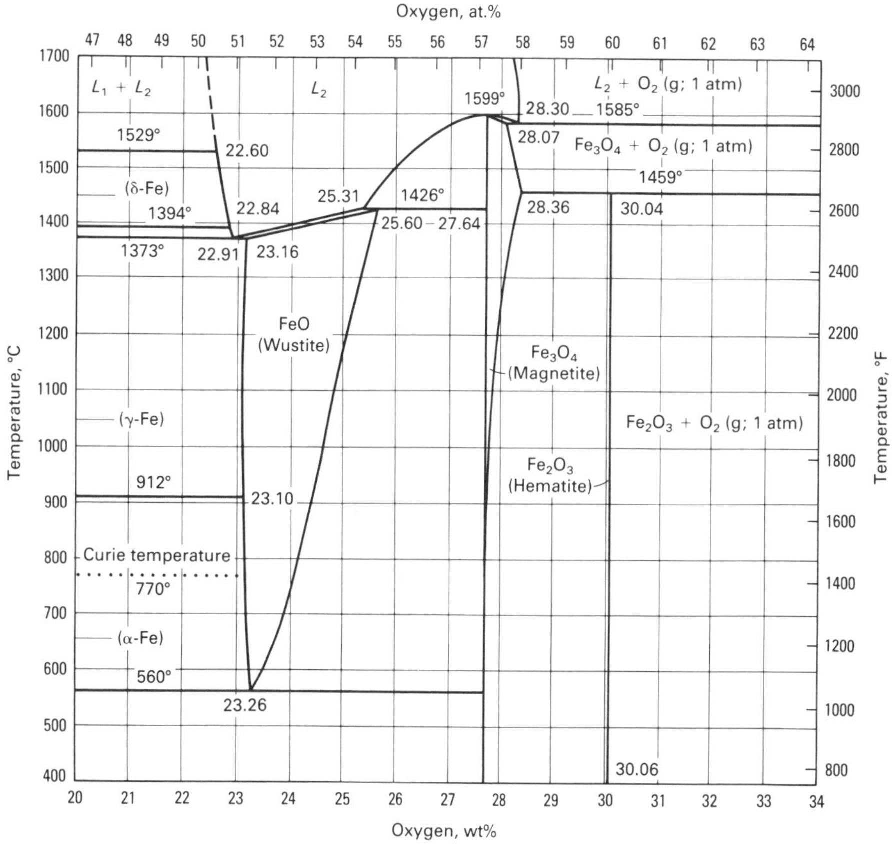

Fig. 3 Iron-oxygen phase diagram

Physical Chemistry

Phase diagrams, thermochemical relationships, and reaction rates must all be considered for proper understanding of inclusion formation. The general concepts of each, along with their practical significance, are established in the following sections in this article.

Phase Equilibria. Phase diagrams map the occurrence of phases to be expected under equilibrium conditions. Equilibrium also requires long periods of time to achieve. This is certainly not the case in castings, in which even moderate cooling rates may inhibit equilibrium phases.

For example, in the Fe-O system shown in Fig. 3, wustite (FeO) produced at high temperatures should break down to α-iron and magnetite (Fe_3O_4) at 560 °C (1040 °F). However, iron scales may show considerable amounts of retained FeO at room temperature.

A further limitation is that pure elements seldom exist in the casting process, and more than two elemental species are usually present. In Fig. 3, pure oxygen is one component while air (20% O) is more likely to be the oxidizing source during casting. Furthermore, a cast steel will contain carbon and other elements such as silicon or aluminum that will have a greater affinity for oxygen than iron. Therefore, phase diagrams become a tool for understanding, but the predicted equilibrium for inclusion occurrence may not always be observed for the casting cooled under nonequilibrium conditions.

Thermochemistry. Any chemical reaction will occur if the products are at a lower energy level than the reactants. The energy measurement is the chemical free energy, which can be treated simply as the energy of chemical position. Free energy values can be obtained directly from numerous sources or can be calculated from tabulated enthalpy, entropy, and heat capacity data. An equilibrium is then established between the products and reactants such that very little remaining reactant indicates a more or less complete reaction. A more thorough discussion of chemical free energy is provided in the article "Principles of Physical Chemistry" in this Volume and in standard textbooks on chemical thermodynamics.

By way of example, assume pure molten iron contains dissolved oxygen at 1600 °C (2910 °F). A small amount of aluminum is added to the melt, and the chemical reaction is:

$$2Al + 3O \rightarrow Al_2O_3(s)$$

where the (s) indicates that Al_2O_3 is solid at 1600 °C (2910 °F). (The melting point of Al_2O_3 is 2072 °C, or 3762 °F.)

For the reaction to occur, the standard free energy exchange at 1600 °C (2910 °F) must be negative. However, the quantity of each component in the reaction is also important, and the chemical activity of each species must be known. An overabundance of aluminum will react stoichiometrically with oxygen, and the remainder may go into solid solution or react with the surroundings.

In the above example, the Al_2O_3 product is insoluble in the molten iron and can result as inclusions trapped in the solidified casting. It is then possible to calculate the free energy exchange for the occurrence of various inclusion chemistries. Their relative stability is dictated by the most negative value of free energy. In the above example, iron oxide would not occur, because Al_2O_3 has a lower free energy of formation. In fact, iron oxide would be predictably reduced by aluminum to form Al_2O_3. As with phase equilibria, thermochemical calculations are again based on equilibrium data, and castings may not be at equilibrium throughout the solidification process.

Kinetics. There are several reasons why inclusion-forming reactions are usually incomplete and why reaction rates therefore become the most important practical consideration. Much of the understanding of reaction rates is associated with the following observations:

- Not all species diffuse at the same rate; when coupled with the dependence of diffusion rates on temperature, nonuniform inclusion formation and distribution might be anticipated

- Chemical free energy is also temperature dependent and may not vary in the same fashion for all of the species under consideration
- Available phase equilibrium data may predict the relative solubility of inclusion-forming species in the liquid state, but may not treat decreasing solubility in the cooling liquid or the effects of the solidification process on the equilibrium
- The surface energy of species will change with temperature and will influence the ability of an inclusion to float, sink, agglomerate, or react secondarily with its surroundings
- Reaction products at a surface can become a barrier to further reaction and can block completion of the chemical reaction

Experimental kinetic data can be presented in several ways, including mathematical relationships associated with activation energy and first-order kinetic reaction rate theory. Perhaps more familiar are isothermal and continuous cooling transformation graphical interpretations consistent with nucleation and growth-controlled phenomena.

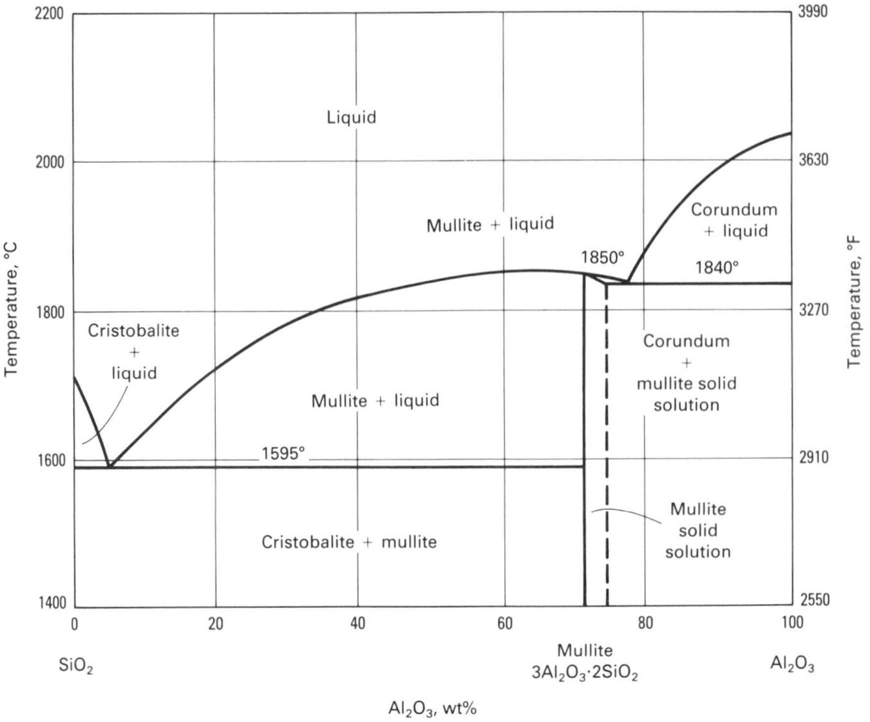

Fig. 4 SiO_2-Al_2O_3 phase diagram. Source: Ref 2

Control of Inclusions

The previous discussion suggests several practical means of controlling the occurrence of inclusions in any cast metal. All of the control methods can be cataloged as chemical, mechanical, or a combination of these. Several of the more common techniques are discussed separately below.

Chemistry Control. Although control of alloy chemistry may be obvious, composition cannot always be modified to minimize inclusions. The metal composition may be inherently reactive with the local environment. For example, the presence of magnesium is required in nodular cast iron; the propensity for inclusions is greater than for the same nominal chemistry of a gray cast iron.

Similarly, the existence of inclusions is accepted because the alternative is worse. Aluminum added to steels may result in Al_2O_3 inclusions, but the lack of deoxidation if aluminum were not used would result in gas holes with a greater negative effect on casting performance.

However, this does not imply that if a small addition is good then more is better. Phosphorus added to a copper-base alloy can be an effective deoxidizer, but excess phosphorus can be very aggressive and can react with mold materials commonly used with this alloy family. Therefore, the importance of chemistry control cannot be overemphasized if inclusions are to be controlled.

Mold/Metal Interface and Refractory Control. Another chemical variable that can lead to inclusions is that of the mold materials and refractories. The SiO_2-Al_2O_3 phase diagram is shown in Fig. 4. It is obvious that the solidus temperature increases with Al_2O_3 content. Even if a refractory is to be used well below the solidus temperature, a high-Al_2O_3 refractory may still be preferred because Al_2O_3 plus mullite would be present rather than the silica (SiO_2) plus mullite that would be present at compositions less than 72% Al_2O_3. The SiO_2 does not possess the same thermochemical stability as Al_2O_3; therefore, aggressive liquid metals or oxides may react more readily with SiO_2.

Higher temperatures and longer holding times at temperature increase the possibility for reaction and the production of inclusions associated with refractory wear and erosion. Therefore, melt overheating, especially for long periods of time, is to be avoided.

The scale on scrap surfaces can also lead to refractory erosion and exogenous inclusions. Rust on ferrous scrap can be considered a hydrated iron oxide that, when heated and dehydrated, can form low-melting complex oxides with the refractories.

Inclusion difficulties have also been known to increase when adhering mold materials remain on charged revert scrap. The severity depends on the nature of the mold bonding agent; bentonite and organically bonded mold materials would display different tendencies toward refractory attack.

The above observations can be applied to reaction between the mold materials and the molten metal. This is not to be confused with metal penetration, in which molten metal merely fills the void spaces between grains of the mold material. However, the solution to the problem—for example, a mold wash—may be the same. In this case, the wash fills the voids between sand grains and isolates the mold material from contact with a potentially aggressive molten metal or oxide. Longer solidification times merely increase the possibility for mold-metal reaction and the accompanying inclusion-forming tendency.

Separation Techniques. The choice for inclusion separation is between removal before the molten metal enters the mold or use of a removal system as a component part of the mold. Gating systems will be discussed in the next section in this article. In both cases, inclusion control is basically mechanical.

External methods include simple skimming of ladle slag or dross to minimize exogenous inclusion sources before pouring of the castings. A teapot ladle can be very effective for decanting clean metal from beneath a layer of floating slag or dross.

Screens or sieves have been used in gating systems to filter inclusions. In general, metallic filters suffer from the tendency to alloy with the poured metal, and they are difficult to separate from the charge returns. Ceramic foam filters with consistent pore size and structural integrity are being increasingly used for those alloys from which definite cost benefits can be derived (Ref 3-5) (see also the article "Nonferrous Molten Metal Processes" in this Volume).

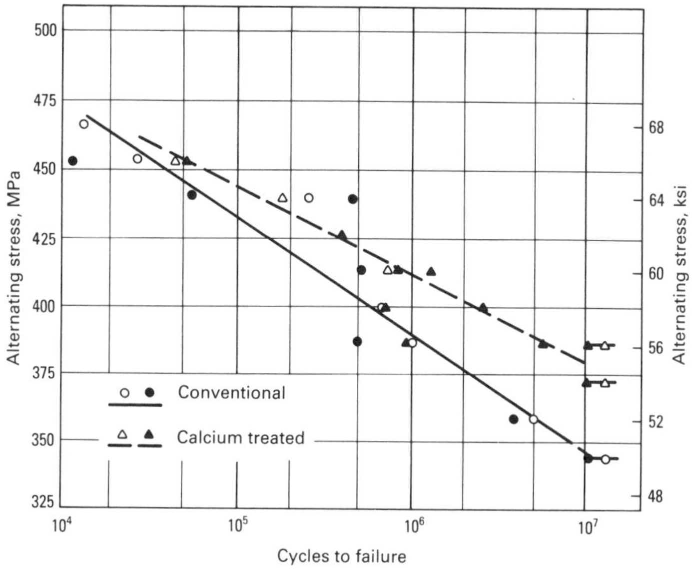

Fig. 5 Comparison of axial fatigue data for untreated and calcium-treated rolled ASTM A516 steel. 51 mm (2 in.) thick plates tested with alternating stress ratio of 0.1. Source: Ref 8

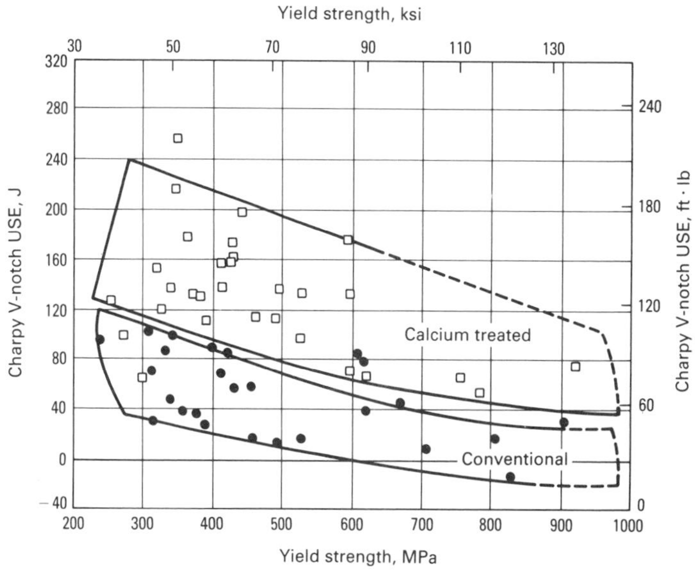

Fig. 6 Comparison of Charpy V-notch upper shelf energies (USEs) for several grades and thicknesses of untreated and calcium treated steel. Source: Ref 9

Gating Methods. Gating and pouring practice has been known to create and to solve inclusion problems in castings. If exogenous inclusions leave the ladle because of inadequate upstream control, the gating system should be designed to remove such inclusions in the pouring basin or in the runner system.

Too often, a pouring basin is not designed to fill easily or to remain filled, and inclusions that might float are forced into the casting cavity. Whirl gates can be used to collect the inclusions and to ensure a better opportunity to have clean metal enter the casting proper (Ref 6). In addition, runners can have dead-end extensions that collect the first metal to enter the mold, which is the most likely to contain slag or dross (Ref 7).

Pressurized gating systems are often recommended for preventing the aspiration of air that can cause the formation of inclusions during the pouring process. This technique is more important when the metal or alloy has a great tendency toward inclusion formation because of so-called secondary oxidation. More detailed information on correct gating is available in the article "Gating Design" in this Volume; many of the gating practices discussed in that article are used to minimize inclusions.

Inclusion Shape Control. As pointed out previously, mechanical properties suffer when inclusions are present, with a more pronounced degradation in strain rate sensitive applications. For example, an inclusion at the surface of a casting subjected to cyclic strain can become a source for fracture initiation in fatigue.

The most conservative approach in design is to determine the smallest inclusion at the surface that might be present without casting rejection. With no variation in inclusion continuity or size, a spherical shape (lower microstress concentration) would be less damaging than more angular shapes. Inclusions beneath the surface would exhibit the same dependence on shape for fracture propagation rather than initiation.

As shown in Fig. 5, calcium treatment of a steel provides longer life under cyclic loading due to inclusion shape modification. Although the data in Fig. 5 are for a rolled constructional steel, similar results would be anticipated for a cast steel, where anisotropy also exists due to directional solidification.

The advantage of a spherical inclusion shape is possibly more evident in Charpy V-notch impact data for constructional steels (Fig. 6). SEM fractographs show a pronounced difference in deformation mode with an inclusion shape modification (Fig. 7). Local ductility increases adjacent to the inclusions when the shape is spherical.

This would then suggest that if inclusions are to occur and cannot be completely removed, a spherical shape is the least detrimental. Inclusion shape control, therefore, is a useful technique that is often applied to more demanding casting applications in which added processing costs can be justified. The principle is a control of surface energy between the inclusion and the metal—normally through suitable chemistry modification of the metal. By analogy, the technique may not be unlike the addition of magnesium to cast iron to change the graphite shape from flakes to nodules.

Inclusions in Ferrous Alloys

Some applications for inclusion minimization or shape control are immediately obvious. For example, proprietary addition agents are available for modifying the viscosity of slag or dross, thus facilitating their ease of removal. On the other hand, treatment of steels and cast irons is likely to be different, and a few examples, along with the application of appropriate physical chemistry, are given below.

Steels

Oxides. The amount of carbon in iron controls the amount of dissolved oxygen (Fig. 8). The curve has been calculated from thermochemical data with variable CO/CO_2 ratios. Experimental points would generally lie above the theoretical curve, but the relationship between carbon and oxygen would be the same. Low carbon in solution, as in many cast steels, results in higher dissolved oxygen.

The solid solubility of oxygen in iron is low, and during solidification, oxygen is rejected to the remaining liquid, where it can react with carbon to form CO and result in gas porosity. Although this is allowed to occur in steel ingots (induced porosity that is "healed" during subsequent processing), the amount of dissolved oxygen in steel castings is reduced by additions of aluminum or silicon. The resultant oxides are insoluble and form inclusions.

Because Al_2O_3 has a lower free energy of formation than SiO_2, aluminum is a more effective deoxidizer than silicon (Fig. 9). The FeO content is an index of the amount of dissolved oxygen in the steel where the oxygen can also be reduced by carbon, as shown in Fig. 8. At a given carbon content, the oxygen can then be reduced by aluminum or silicon to still lower levels, because the free energies for SiO_2 and Al_2O_3 are

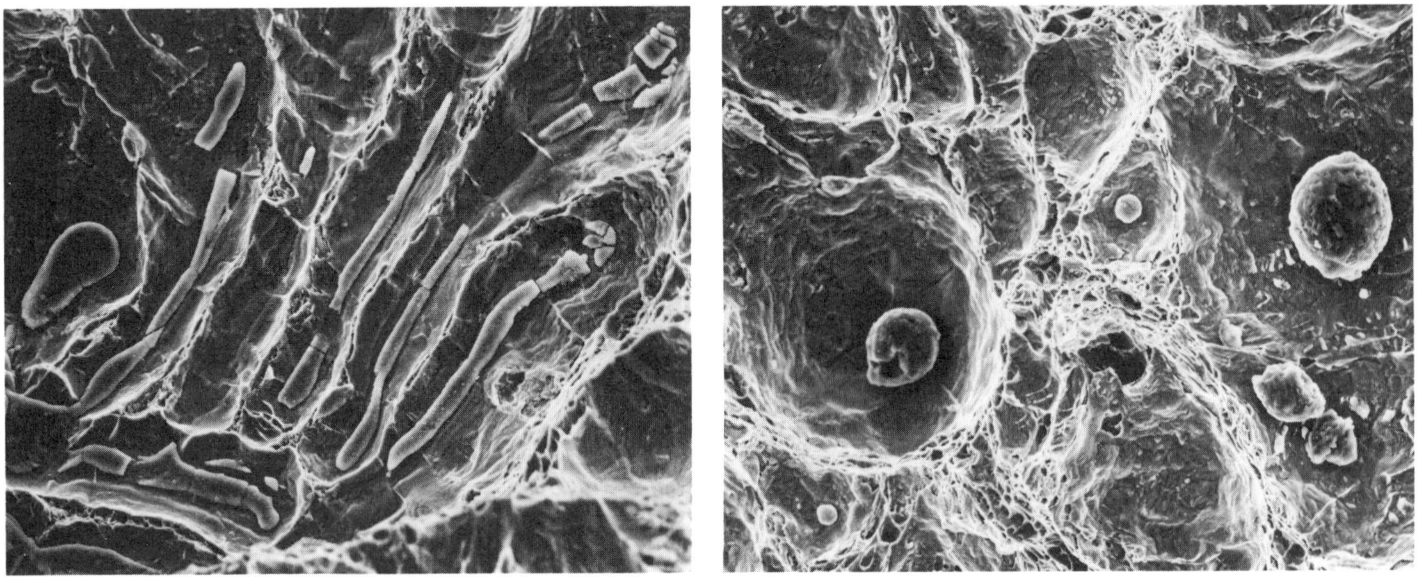

(a) (b)

Fig. 7 SEM fractographs showing the effect of calcium treatment on the fracture morphology of ASTM A633C steel impact specimens. (a) Untreated steel with type II manganese sulfide inclusions showing evidence of brittle fracture. (b) Calcium treated steel with spherical inclusions; the fracture is ductile. Courtesy of A.D. Wilson, Lukens Steel Company

lower than those for CO or CO_2 at the reaction temperature.

Once the oxide forms, the inclusion removal rate depends on the nucleation of particles, their growth, ultimate agglomeration, and rise to the surface. Although there is a significant difference in density between the solid particles and the liquid metal, a simple application of Stokes' law is difficult. The problem is dynamic and is influenced by the presence of other indigenous or exogenous inclusion species. Furthermore, gravitational effects are different while holding the treated metal in the ladle, pouring through air with an opportunity for the metal to reoxidize, and during solidification with the associated thermal convective currents. Therefore, it should not be surprising that oxide inclusions can become trapped in the final casting.

Sulfides. Sulfur is also often present in steels and is controlled by the nature of the slag-refining processes in the furnace. In general, as the process becomes more reducing (that is, the oxygen content of the metal is lower) and the slag more basic, lower sulfur will be present. Because the sulfur content is not reduced to zero, manganese is always present in steels to produce the more desirable manganese sulfide (MnS) inclusions rather than FeS inclusions with lower melting point.

The resulting inclusions, although usually identified as manganese sulfides, may have several distributions and shapes, depending on the presence of other elements and the deoxidation procedure.

Complex Inclusions. The various elements compete to form compounds with oxygen and sulfur. Furthermore, alloying elements in excess of those necessary to form compounds will modify the surface energy of the resultant inclusions. The shape and distribution of the inclusions will then also vary, depending on the variables discussed at the beginning of this article.

Historically, inclusions have been classified into three general types, as shown in Fig. 10. Of these, Type II has been demonstrated as particularly damaging to mechanical properties. The argument has been that

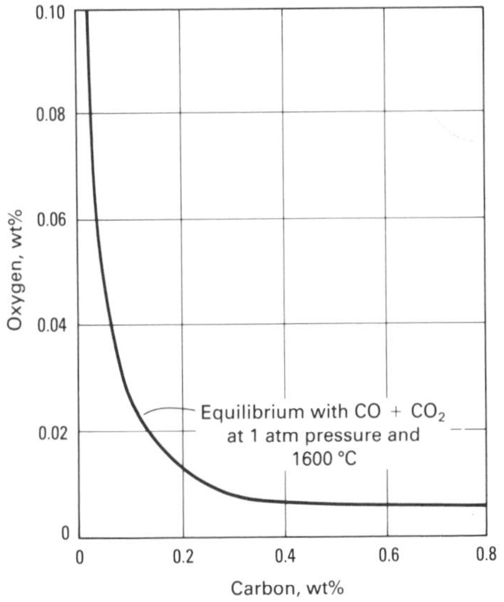

Fig. 8 Carbon-oxygen equilibrium at 1600 °C (2910 °F). Source: Ref 10

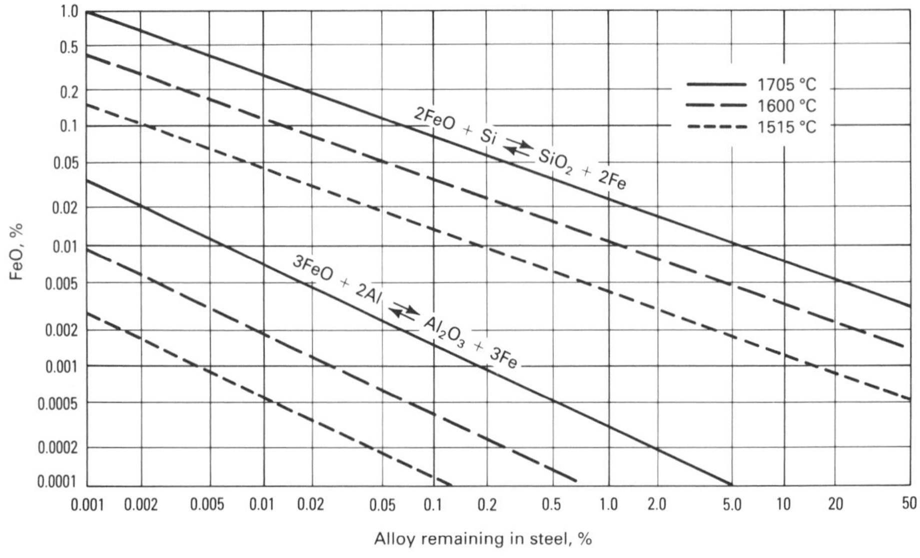

Fig. 9 Equilibrium between FeO and dissolved aluminum and silicon. Source: Ref 11

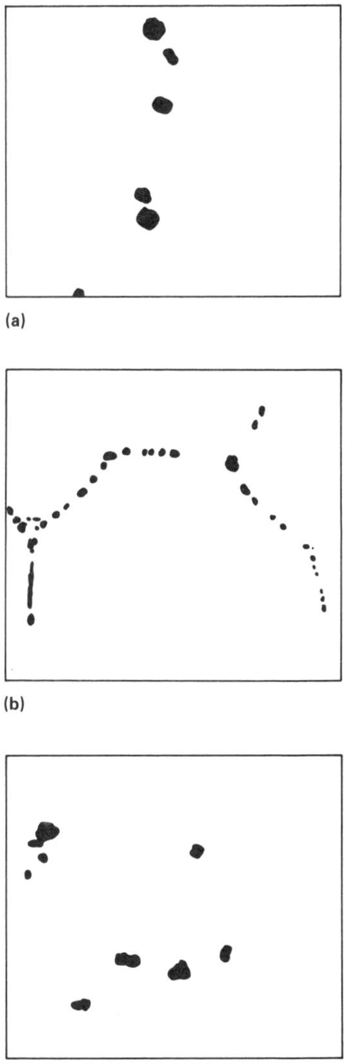

Fig. 10 Illustrations of common inclusion shapes and distributions found in steel. (a) Type I: globular silicates and oxides (no aluminum deoxidizer). (b) Type II: continuous eutectic sulfides and Al_2O_3 at grain boundaries (low aluminum addition). (c) Type III: duplex irregularly shaped sulfides and Al_2O_3 (large aluminum additions). Source: Ref 12, 13

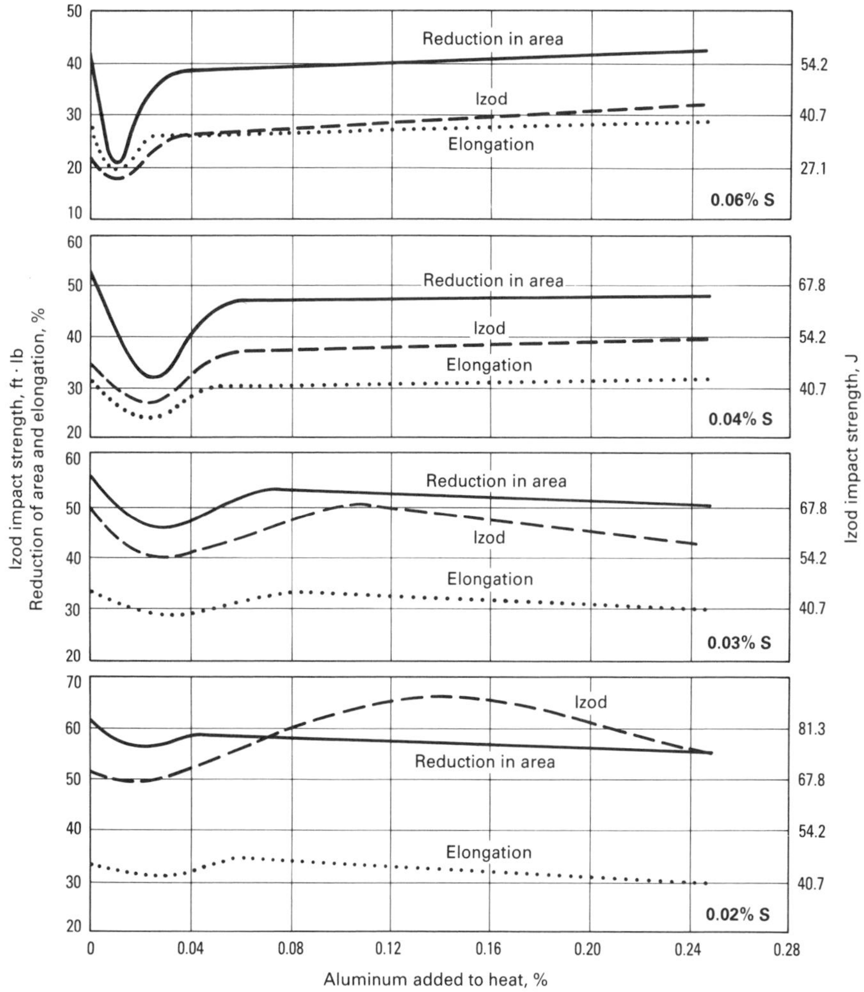

Fig. 11 Combined effect of aluminum and sulfur on the ductility and impact strength of cast and normalized medium-carbon steel. Source: Ref 11-13

sulfur, oxygen, and deoxidizer play a combined role in the inclusion type. Figure 11 shows this effect and the resulting decrease in structure-sensitive properties such as ductility and impact strength. The minimum in the curves is attributed to the formation of Type II inclusions.

The role of aluminum in the modification of inclusion shape and distribution also depends on the oxygen-sulfur equilibrium shift when the aluminum is added. Therefore, complex inclusions such as oxy-sulfides may be present in a cast steel, but can be controlled by the quantity of oxide-sulfide forming elements. A more complete discussion of oxide inclusion formation mechanisms and the relationship to mechanical properties is given in Ref 14.

Inclusion shape control becomes a consideration especially if fracture toughness, impact strength, or fatigue resistance must be optimized. Magnesium and calcium additions can be successfully used for shape control, that is, to produce more spherical shapes (Ref 15). It is noteworthy that both magnesium and calcium form stable sulfides and oxides as indexed by the free energy of formation of the compounds.

Several production methods have been employed for calcium or magnesium additions, either as the metallic elements or as compounds. Because of the high vapor pressure of the pure metals and to a lesser extent the compounds at the addition temperatures, special procedures have evolved, including injection at the bottom of a filled ladle. In this case, the ferrostatic head provides a pressure that suppresses the boiling action and gives a longer reaction dwell time.

Rare earth elements added in conjunction with calcium, magnesium, and aluminum have been reported to enhance the spheroidization and fineness of the inclusions by breaking up the Al_2O_3 galaxies (Ref 16, 17).

Nitrides. Just as there are elements that preferentially form oxides or sulfides, there are differences in the free energy of formation of the nitrides. These elements control the amount of nitrogen that can be dissolved in both the solid and liquid states. For example, vanadium and manganese increase nitrogen solubility, while carbon and phosphorus decrease nitrogen solubility (Ref 12).

The effect of aluminum on nitrogen solubility has been extensively studied, and criteria that result in the precipitation of aluminum nitrides have been developed (Fig. 12).

Other nitride-forming elements, such as titanium, zirconium, and vanadium, would show similar but numerically different results.

It is not uncommon to consider these nitride inclusions to be beneficial; they peg austenite grain boundaries, and this results in grain refinement. On the other hand, brittleness results when the nitrides become more or less continuous.

Exogenous Inclusions. Numerous examples of inclusions occur because of the

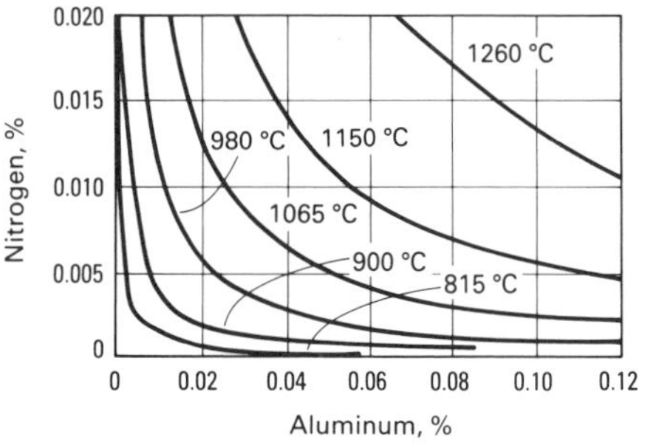

Fig. 12 Isothermal precipitation of aluminum nitride. Compositions to the right or above each isotherm cause precipitation. Source: Ref 12

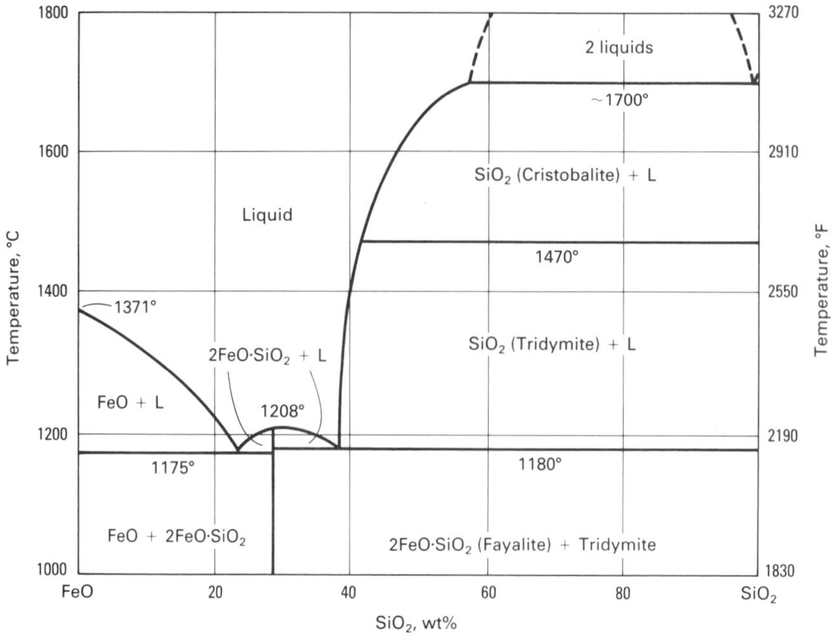

Fig. 13 FeO-SiO_2 phase diagram. Source: Ref 10

reaction between the molten steel and the adjacent environment. A few of these are considered below.

When a steel is killed with aluminum, residual aluminum remains in solution. If the metal contacts a refractory that contains free SiO_2, the aluminum can reduce the SiO_2 because of the lower free energy of formation of the Al_2O_3. The eroded or partially dissolved refractory becomes a source of inclusions, as does the Al_2O_3 that is produced. Careful control of refractory chemistry and good pouring practice minimize this problem.

Similar reactions can occur at the mold/metal interface, where the first equilibrium of importance is iron-oxygen-silicon. When conditions are oxidizing, FeO is present, and the corresponding form for silicon is SiO_2. The FeO-SiO_2 phase diagram is shown in Fig. 13. The most significant observation here is the formation of lower-melting liquid phases as the amount of FeO is increased.

With minimal oxidation, burned-on sand and metal penetration will occur. However, when the oxidation level increases, local macroinclusions may form, especially if poor pouring practice, segregated bentonite, nonuniform mold ramming, and excessive mold moisture exist. A number of the problems and solutions associated with exogenous inclusions in steel have been summarized in Ref 18.

Cast Irons

Gray Cast Iron. The earlier discussion of the carbon-oxygen equilibrium (Fig. 8) suggests that dissolved oxygen should not be a problem in cast irons because of their high carbon contents. Therefore, oxide inclusions should not occur in gray cast iron. However, because the silicon content of this material is also significant, the carbon-silicon-oxygen equilibrium is of interest.

The temperature dependencies of the free energies of formation for SiO_2 and CO are different for the two species. At temperatures above approximately 1535 °C (2795 °F), CO is the more stable compound; below this temperature, SiO_2 is more stable. This can be observed on the surface of a cast iron melt that appears clean at high temperatures and yet develops a slag covering at lower temperatures.

The significance of the equilibrium shift is that skimming the slag from a melt surface may be unnecessary at high temperatures, because little may exist. Inclusions are then found in the casting because of pouring at lower temperatures and exposure of the metal to reoxidation during pouring. These inclusions are often silicates derived from the normal equilibrium change with temperature. Once the molten iron is in the mold cavity, there is less opportunity to form the silicates because conditions are usually less oxidizing.

Nitride and sulfide inclusions also exist in gray cast iron, but are not as severe a problem as in steel because of the inherently lower dissolved oxygen content of the cast iron. Special gray cast iron chemistries containing alloying elements with an affinity for nitrogen, oxygen, or sulfur may result in inclusions by further reaction with the refractories, mold, or atmosphere. The remedies are usually similar to those given for steel.

Ductile Cast Iron. A number of foundries erroneously assumed that ductile (nodular) cast iron would be as easy to process as gray cast iron because the major difference was merely a lower sulfur plus a modest magnesium addition. Although many characteristics were found to be different, one of the more perplexing was the formation of dross usually not found in gray cast iron. The result was the so-called cope-side defect that was very prevalent in early ductile iron castings.

Magnesium forms a very stable oxide and can therefore reduce oxides of lower stability as determined by their free energy of formation. Furthermore, the alloys used to produce the nodular graphite found in ductile iron contain considerable amounts of silicon and are low in density, which allows them to float on ladle surfaces rather than reacting completely. These observations help to explain how dross particles form and can become entrapped in castings.

In general, there are three types of dross inclusions in ductile cast iron (Ref 19). Type I consists of large dross particles generally derived from oxidation of the silicon and magnesium used in the nodularizing alloys. The dross particles are present on the treated liquid-metal surface and can be removed by skimming. Silica does not exist in the pure form; rather, magnesia (MgO) and forsterite (Mg_2SiO_4) are the predominant species. This suggests a driving force from the high-MgO end of the phase diagram shown in Fig. 14. Under some circumstances, especially after a late silicon inoculation, the dross also contains fayalite (2FeO·SiO_2) (see Fig. 13). Fayalite and forsterite form a complete series of solid solutions known as the olivines and are used as molding materials.

Type II inclusions are in a more or less stringlike configuration in conjunction with decayed graphite shapes. Chemically, they are predominantly MgO with some iron and silicon atom replacement for the magnesium in the MgO lattice. Reaction with the oxidizing components in the mold material is one source for this type of inclusion, as is an excess of nodularizing agent. Their position is usually just beneath the casting surface. Control of the mold atmosphere and the amount of nodularizer usually minimizes the defect.

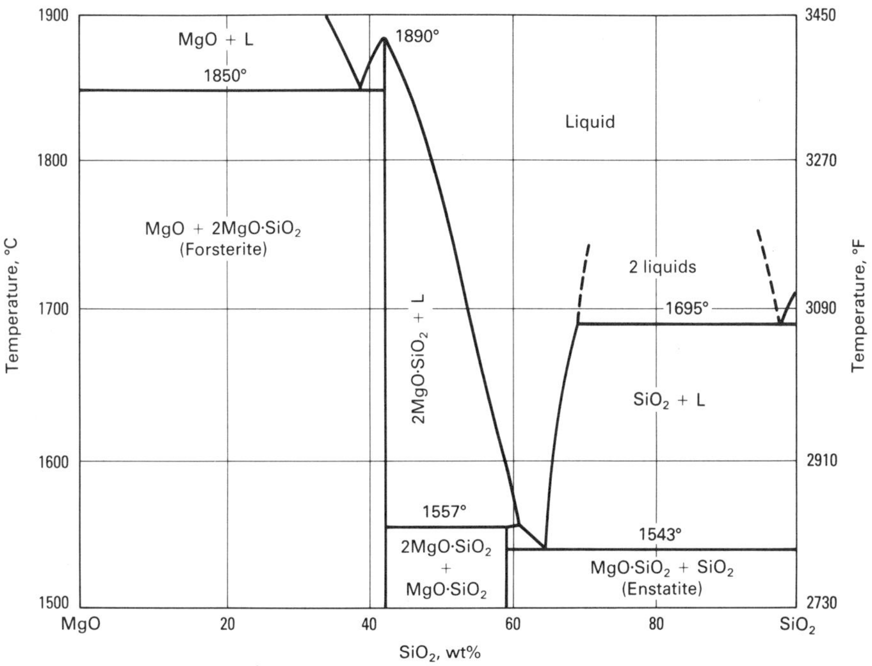

Fig. 14 MgO-SiO_2 phase diagram. Source: Ref 10

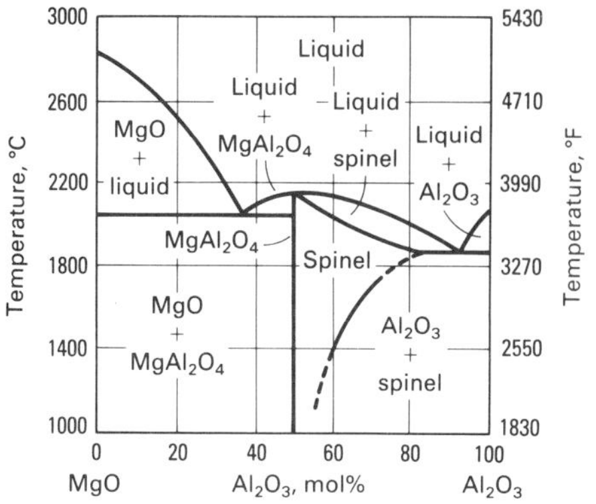

Fig. 15 MgO-Al_2O_3 phase diagram. Source: Ref 2

Type III inclusions are fine dispersions of MgO and MgS, occasionally mixed with Mg_2Si. These components would normally be expected only if the oxygen content in the immediate vicinity is very low. Their normal uniform distribution does not usually detract from the overall performance of the casting. However, undissolved nodularizer may be a source of Type III inclusions because Mg_2Si is the active magnesium source in the nodularizing agent.

Finally, the examples provided in this section are not intended to represent all of the inclusions to be anticipated in every ferrous alloy. On the other hand, they are sufficiently representative so that analogies for occurrence and remedy can be derived for the alloys not discussed.

Inclusions in Nonferrous Alloys

Several important cases are given below for aluminum, copper, and magnesium alloys. The principles discussed in this section can be extended to include other alloy systems. In the alloys discussed, both nonmetallic and intermetallic compounds will be considered deleterious.

Aluminum Alloys

Oxides. It is nearly impossible to prevent dross formation in the melting of aluminum alloys. The most common form of dross contains large quantities of Al_2O_3 and, when agglomerated, can be removed from the melt surface by skimming. Pouring operations should employ skimming ladles to prevent larger dross particles from entering the mold. Furthermore, a gating practice must be used that minimizes Al_2O_3 formation during mold filling.

The fact that Al_2O_3 forms is not surprising, because the free energy of formation of Al_2O_3 is very low. Furthermore, the specific gravity of Al_2O_3 is not greatly different from that of the molten aluminum alloys. Although more dense, Al_2O_3 forms at an outside surface first if the melt is exposed to the atmosphere. The tenacious skin will float unless it is broken up into small particles, and it can become quite thick when melts are held for extended periods in the furnace.

When small oxide particles are formed, they may not sink, but may become suspended in the melt. In high-quality castings, these particles must be removed; ceramic filters have been used to remove the particles, as mentioned earlier in this article (Ref 3).

The degradation of mechanical properties also depends on the size and distribution of oxide inclusions. Therefore, solidification rate is very important for minimizing property loss. When the casting solidifies rapidly, as in thin sections or in permanent molds, the oxide inclusions are smaller and more uniformly dispersed. This observation is consistent with the general constraints of nucleation and growth where less time during solidification not only creates smaller particles but also prevents their agglomeration and growth.

Complex Oxides. Other elements in the aluminum alloys may react with oxygen sources and produce complex oxides with Al_2O_3 as one of the components. This is the case in magnesium-containing alloys, in which the spinel $MgAl_2O_4$ may form (Fig. 15). Here MgO has a lower free energy of formation than Al_2O_3; the spinel results when they form together.

Again, precautions in pouring, gating, and minimizing the reoxidation of the metal are important in inclusion control. The use of filtering systems is also advantageous, especially in high-strength castings in which fracture toughness characteristics are important.

Other Inclusion Types. Inclusions in aluminum alloys can occur from grain-refining operations in which the grain-refining additions may not be adequately controlled. Titanium, vanadium, and boron are used in various combinations to produce grain refinement, and the diborides can become undesirable inclusions. Again, the fact that a small addition is beneficial does not mean that larger additions will provide greater rewards.

Many of the commercial aluminum alloy compositions have been developed to exploit their age-hardening characteristics in conjunction with an alloyed matrix. They then become susceptible to the presence of seemingly minor elements such that undesirable intermetallic particles can be formed during cooling. In those cases of precipitation during solidification, the particles are placed in the category of inclusions.

The intermetallic inclusions are particularly troublesome in aluminum alloys containing large quantities of zinc or magnesium. Complexes can be formed with the transition elements.

These observations suggest that the remedy is close control of the metal chemistry. There are many examples in which control is inadequate and the problem is not identified until mechanical properties are marginally out of specification. Even aluminum carbide and iron-aluminum compounds have degraded properties by the formation of inclusions. Although the solubilities are low, the problem is additive in that scrap/revert procedures can cause a gradual increase in the impurity levels, and inclusion problems become more severe. The hazards associated with recycling then become ap-

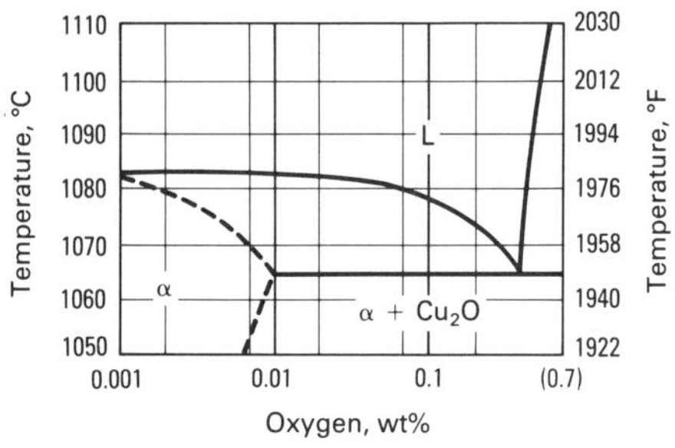

Fig. 16 A portion of the copper-oxygen equilibrium diagram. Source: Ref 20

parent, especially when the final product is sensitive to the elements present and when strict mechanical property specifications are imposed.

Copper Alloys

A portion of the copper-oxygen system is shown in Fig. 16. The eutectic shows that the last liquid to solidify may contain 0.39% O even though the initial melt may contain 0.01% O. Therefore Cu_2O inclusions may be more predominant internal to the casting (liquid enrichment in oxygen due to slow diffusion) and may also be present at grain boundaries. The key to control is to add elements that combine with dissolved oxygen (that is, to form oxides with free energies lower than that of Cu_2O).

Examples are lithium, boron, magnesium, or phosphorus, where control of the excess addition becomes important to modification of the electrical conductivity in the high-copper alloys. The excess deoxidizer goes into solid solution. A more thorough discussion of gas solubility is available in Ref 20 (see also the article "Gases in Metals" in this Volume).

A number of the copper alloys are susceptible to dross formation, and the normal precautions in pouring and gating are absolutely necessary to minimize exogenous inclusions. The drosses are usually complex oxides of copper, zinc, tin, lead, or aluminum in the aluminum bronze family.

Iron is added as both a dispersion strengthener and as a grain refiner in aluminum and manganese bronzes. However, iron can become an undesirable inclusion in the other copper alloys. When a foundry pours several different alloy chemistries, scrap segregation becomes important to inclusion control. Consideration can also be given to alternative grain-refining agents such as zirconium.

In some cases, the intermetallic inclusions are rich only in iron, and complexes are formed. This is especially true when alloys are high in zinc, such as the high-strength yellow brasses. The problem becomes more severe when a melt is held close to its liquidus for extended periods. In this case, the solubility of the complex intermetallics is lower, and their formation is enhanced.

High-zinc alloys offer fewer problems with dissolved oxygen because of the vaporization of zinc and the purging action of the melt. Oxides as inclusions are then a lesser problem. However, zinc oxide (ZnO) can react with refractories and mold materials to produce inclusions during pouring.

In addition to careful pouring and gating practices, exogenous inclusions can be removed by filtering processes. However, the practice has not been as well received for copper alloys as for aluminum alloys.

Although not strictly an inclusion from the previous definition, lead in many of the copper alloys can form oxides that will react with mold bonding agents such as bentonite. These lead complexes can become leachable in standard tests, and waste sand streams that contain lead compounds may be classified as toxic with the attendant difficulty in disposal (Ref 21).

Magnesium Alloys

Of the nonferrous alloys considered in this section, magnesium has the lowest free energy of formation for its oxide. Therefore, the complete prevention of oxide inclusions is impossible, and a number of methods have been developed to minimize oxide inclusion formation. Agents that evolve gas to partially insulate the melt from the atmosphere and fluxes that agglomerate the oxide (plus other impurities that are to be removed) are two examples of control of the melting practice.

Again, great care must be exercised in gating to minimize reoxidation. The magnesium alloys are light in weight, and many of the potential exogenous inclusions are collected as sludge in the bottom of the furnace and ladle. Desludging operations are then necessary. The use of filtering systems, whether metallic or ceramic, is common to all of the higher-quality magnesium castings. As with aluminum alloys, excess grain-refining agents can result in undesirable inclusions that can have complex chemistries because of secondary reactions.

The fluxes themselves must be reactive in order to perform their intended function. Excess flux can then react with the atmosphere, refractories, or the mold materials to produce inclusions. On the other hand, insufficient flux causes metal reaction with the atmosphere to produce oxides and nitrides that may be difficult to remove if not in an agglomerated form. Temperature, time at temperature, alloy composition, pouring practice, and gating procedures dictate the optimum fluxing process.

So-called fluxless melting has been suggested for magnesium alloys in order to overcome the environmental and metallurgical problems associated with fluxes (Ref 22). Fluxless melting uses nonreactive gaseous atmospheres above the melt. The most effective protective atmosphere is CO_2 plus the addition of trace inhibitors such as SO_2 and SF_6.

REFERENCES

1. L. Ackermann, J. Charbonnier, G. Desplanches, and H. Kaslowksi, "Properties of Reinforced Foundry Alloys," *Trans. AFS*, Vol 94, 1986, p 285-290
2. R.A. Flinn and P.K. Trojan, *Engineering Materials and Their Applications*, 3rd ed., Houghton-Mifflin, 1986
3. F.R. Mollard and N. Davidson, Experience With Ceramic Foam Filtration of Aluminum Castings, *Trans. AFS*, Vol 88, 1980, p 595-600
4. L.A. Aubrey, J.W. Brockmeyer, and P.F. Wieser, Dross Removal From Ductile Iron With Ceramic Foam Filters, *Trans. AFS*, Vol 93, 1985, p 171-176
5. L.A. Aubrey, J.W. Brockmeyer, P.F. Wieser, I. Dutta, and A. Ilhan, Cast Steel Improvement by Filtration With Ceramic Foam Filters, *Trans. AFS*, Vol 93, 1985, p 177-182
6. P.K. Trojan, P.J. Guichelaar, and R.A. Flinn, An Investigation of Entrapment of Dross and Inclusions Using Transparent Whirl Gate Models, *Trans. AFS*, Vol 74, 1966, p 462-469
7. D.G. Schmidt, Gating of Copper Base Alloys, *Trans. AFS*, Vol 88, 1980, p 805-816
8. A.D. Wilson, Calcium Treatment of Plate Steels and Its Effect on Fatigue and Toughness Properties, *Proc. 11th Annual Offshore Technology Conference*, OTC 3465, May 1979, p 939-948
9. A.D. Wilson, Effect of Calcium Treatment on Inclusions in Constructional Steels, *Met. Prog.*, Vol 121 (No. 5), April 1982, p 41-46
10. *The Making, Shaping and Treating of Steel*, 9th ed., United States Steel Corporation, 1971
11. C.W. Briggs, *The Metallurgy of Steel Castings*, McGraw-Hill, 1946
12. R.A. Flinn, *Fundamentals of Metal Casting*, Addison-Wesley, 1963
13. C.E. Sims and F.B. Dahle, The Effect of Aluminum on the Properties of Medium Carbon Cast Steel, *Trans. AFS*, Vol 46, 1938, p 65-132
14. W.O. Philbrook, M.C. Flemings, L.H. Van Vlack, A. Gittins, and J. Lankford, Oxide Inclusions in Steel, A series of five papers, *Int. Met. Rev.*, Sept 1977, p 187-228
15. *Low Sulfur Steel*, Symposium proceedings, AMAX Inc., 1986
16. S.K. Paul, A.K. Chakrabarty, S. Basu, Effect of Rare Earth Additions on the Inclusions and Properties of Ca-Al Deoxidized Steel, *Metall. Trans. B*, Vol 13B (No. 2), June 1982, p 185-192
17. M. Olette, G. Gatellier, Effect of Calcium, Magnesium or Rare Earth Addi-

tions on Cleanness of Steel, *Rev. Metall. Cah. Inf. Tech.*, Vol 78 (No. 12), Dec 1981, p 961-973
18. R.A. Flinn, L.H. Van Vlack, and G.A. Colligan, Mold-Metal Reactions in Ferrous and Nonferrous Alloys, *Trans. AFS*, Vol 94, 1986, p 29-46
19. D.R. Askeland and P.K. Trojan, The Approach to Equilibrium and Dross Formation in Nodular Cast Iron, *Trans. AFS*, Vol 77, 1969, p 344-352
20. *Casting Copper Base Alloys*, American Foundrymen's Society, 1984
21. T.R. Ostrom, B.P. Winter, and P.K. Trojan, Lead Transfer From Copper Base Alloys Into Molding Sand, *Trans. AFS*, Vol 93, 1985, p 757-762
22. J.W. Fruehling and J.D. Hanawalt, Protective Atmospheres for Melting Magnesium Alloys, *Trans. AFS*, Vol 77, 1969, p 159-164

Principles of Solidification

Section Chairmen:
W. Kurz, Swiss Federal Institute of Technology
D.M. Stefanescu, The University of Alabama

Abbreviations and symbols used in this Section

Supplementary nomenclature, such as acronyms, units of measure, and derived units, can be found in the Abbreviations and Symbols listing in the back of this Volume.

a atomic distance between crystallographic planes parallel to interface; jump distance

a_0 length scale related to interatomic distance; molecular diameter

A_m area occupied by one mole at the interface

A_{nL} nucleant-liquid interfacial area

A_{nS} nucleant-solid interfacial area

A_{SL} solid-liquid interfacial area

C_l number of atoms per cubic meter in the liquid

C_P^s specific heat of the sample

C_a number of surface atoms of the nucleation site per unit volume of liquid

C_{i_∞} uniform level of solute that exists at sufficiently large distance from interface

C_B composition of ideal solution

C_E eutectic composition

C_L solute composition in the liquid

C_L^* liquid concentration in mutual equilibrium across a plane solid-liquid interface

C_{max} maximum composition

C_{min} minimum composition

$C(n)$ metastable equilibrium concentration of clusters of a given size

$C(n_{cr})$ concentration of critical clusters

C_o initial alloy composition

C_P heat capacity

C_S solid composition of an alloy; solute composition in the solid

$\bar{C}_S$ mean solid composition

C_S^* solid concentration in mutual equilibrium across a plane solid-liquid interface

ΔC_L change in liquid solubility

ΔC_S change in solid solubility

D ingot diameter

d particle-solid distance

d_S contact distance between particles

d_0 minimum separation distance between particle and solid ($\approx 10^{-5}$ cm)

d_1 minimum separation ($\approx 10^{-7}$ cm)

D diffusion coefficient

D_C^γ diffusivity of carbon in austenite

D_{eff} effective diffusion coefficient

D_L liquid diffusivity

$D_{L,i}$ liquid diffusivity of solute i

D_S diffusion coefficient in the solid

e_P particle charge

e_S solid charge

F(f) function of volume fraction f_α and f_β

f_E volume fraction of the eutectic phase

f_S volume fraction of the solid

F faceting factor

F_r repulsive force

f_α volume fraction of α phase

f_β volume fraction of β phase

$f(\theta)$ shape factor

ΔF change in free energy

g acceleration due to gravity

G temperature gradient

G_L temperature gradient in the liquid; molar free energy in the liquid

G_M average temperature gradient in the two-phase region

G_S temperature gradient in the liquid; molar free energy of the solid

ΔG_{cr} activation barrier for nucleation

ΔG_m molar free energy change of mixing

$\Delta G(r)$ free energy change to form a cluster of size r

h height above substrate

h contact distance between particles

H_L enthalpy of the liquid

H_S enthalpy of the solid

ΔH heat of solidification

ΔH_f^A latent heat of pure solvent A

ΔH_m heat of mixing

ΔH_s enthalpy of sublimation per mole

ΔH_v enthalpy of vaporization per mole

i solute

I(P) Ivantsov function

K interface curvature; permeability

k equilibrium partition coefficient; Boltzman constant; solute distribution coefficient

k_{ef} effective partition coefficient

k_N nonequilibrium partition coefficient

k_v dependence of partition coefficient on velocity

K thermal conductivity; modulus constant

K_L thermal conductivity of the liquid

K_P thermal conductivity of the particle at the interface

K_S thermal conductivity of the solid

L liquid

l_i characteristic diffusion length

l_t characteristic conduction length

L length; latent heat per unit volume

ℓ cell length

ℓ^β thickness of the beta phase

m_L liquidus slope

m_S solidus slope

m_α slope of liquidus line of α phase

m_β slope of liquidus line of β phase

n planar nucleant substrate

n_{cr} number of atoms in a cluster

P pressure; particle

P Péclet number

P_c solute Péclet number

Pr Prandtl number

Q activation energy for liquid diffusion; quality index

r particle radius; radius of any particle with no irregularities

r_b radius of any particle irregularity or bump

r_{cr} critical cluster size

r_{Gr} radius of graphite

r_P paraboloid tip radius

r_γ radius of austenite

r^* spherical radius

R gas constant; growth rate

R_{cr} critical interface growth rate

R_{Gr} rate of growth of graphite

R_{screw} growth by screw dislocation

R_{step} growth on the step of a defect boundary

R_{2D} growth by two-dimensional nucleation

S solid

S_{cr} number of atoms surrounding a cluster

S_L entropy of the liquid

SR segregation ratio

S_S entropy of the solid

ΔS entropy of fusion per unit volume

ΔS_m entropy change upon mixing

t time

t_f local solidification time

T temperature

$\dot{T}$ cooling rate

T_b dendrite base temperature

T_c transition temperature

T_E eutectic temperature

T_f equilibrium temperature; furnace temperature

T_{f_K} equilibrium temperature between liquid and solid across interface with curvature K

T_G growth temperature

T_L liquidus temperature

T_{L_∞} liquidus temperature related to bulk liquid concentration

T_n temperature of the standard sample

T_p peritectic temperature

T_S solidus temperature

T_{sol} solidus temperature

T_∞ actual temperature of the bulk liquid

T^* actual temperature of the moving interface

ΔT undercooling

ΔT_c average chemical undercooling of the interface; undercooling at columnar front

ΔT_K kinetic undercooling; curvature undercooling

ΔT_n critical undercooling for nucleation on a substrate

ΔT_o freezing range of alloy ($T_L - T_S$)

T_s temperature of the sample

$\langle \Delta T \rangle$ mean undercooling

u velocity of interdendritic liquid

u bulk liquid velocity

u_n flow velocity normal to the isotherms

U velocity of isotherms

U_{TL} liquidus isotherm velocity

v velocity

v_a absolute velocity for planar interface stability

v_{cr} critical velocity

V_a atomic volume

V_m^γ molar volume of austenite

V_m^{Gr} molar volume of graphite

V_s volume of the sample

V_{SC} spherical cap volume

V_0 atomic volume

X^{Gr} molar fraction of carbon in graphite

$X^{\gamma/Gr}$ molar fraction of austenite at the austenite/graphite boundary

$X^{\gamma/L}$ molar fraction of austenite at the austenite/liquid boundary

z^* ratio between number of near-neighbor atoms in plane of interface and total number of near-neighbor atoms in the bulk

α thermal diffusivity; shape factor of the interface

α_L thermal diffusivity of the liquid

α_S thermal diffusivity of the solid

γ austenite; interfacial energy

Γ Gibbs-Thomson coefficient; capillary constant

δ_c thickness of liquid diffusion boundary layer; solute boundary layer ahead of the interface

δ_i thickness of the diffusion boundary layer

δ_{PL} interface energy between particle and liquid

δ_{PS} interface energy between particle and solid

δ_{SL} interface energy between solid and liquid

δ_t thickness of the thermal boundary layer

$\Delta\rho$ density difference between liquid and the particle

η viscosity of the liquid; heat transfer coefficient

η^0 viscosity of the suspending fluid

η^* effective viscosity

θ contact angle

λ cell spacing, dendrite arm spacing, lamellar spacing or eutectic spacing

$\bar{\lambda}$ mean value of the primary and secondary dendrite arm spacing

$\langle \lambda \rangle$ mean spacing

λ_{br} critical spacing (diverging lamellae)

λ_{ex} extremum spacing

λ_{min} critical spacing (converging lamellae)

λ_2 secondary dendrite arm spacing

λ_1 primary dendrite arm spacing

μ chemical potential; viscosity

ν_{SL} jump frequency associated with atom jumps from the liquid to join the cluster

ρ_L density of the liquid

ρ_P particle density

σ solid-liquid interfacial tension; interfacial energy

σ_{LS} liquid-solid interfacial energy

ϕ regularity constant

Ω net interatomic interaction

Ω_c solutal supersaturation at the interface

∇ vector differential operator

Nucleation Kinetics

J.H. Perepezko, University of Wisconsin—Madison

NUCLEATION PROCESSES play a key role in the solidification of castings by controlling to a large extent the initial structure type, size scale, and spatial distribution of the product phases. During many solidification processes, the size scale of critical nucleation events is too small and the rate of their occurrence too rapid for accurate observation by direct methods. Nonetheless, nucleation effects in the solidification microstructure exert a strong influence on the grain size and morphology as well as the compositional homogeneity. The final microstructure is also modified by the crystal growth, fluid flow, and structural coarsening processes that are important in the later stages of ingot freezing.

In large bulk castings, the solidification temperature corresponding to the onset of freezing is often close to but slightly less than the melting point or equilibrium liquidus temperature. The offset of the solidification temperature with respect to the equilibrium temperature is called the undercooling or supercooling, ΔT, and it plays a vital role in the overall description of the initial stage of the solidification that is controlled by nucleation. The level of melt undercooling at the onset of solidification is important to consider in developing an understanding of the variety of structural modifications and grain-refining practices in common casting alloys and is the basis of more recent solidification processing technologies using rapid solidification methods. In this article, selected highlights of thermodynamic relationships during solidification and nucleation kinetics behavior will be discussed in connection with the basis of nucleation treatments, such as grain refinement and inoculation, to provide a summary of nucleation phenomena during casting.

Thermodynamics of Solidification

Macroscopic Solids. Throughout the analysis of solidification, thermodynamics is used to judge the alloy phase constitution, to describe the solidification path and composition changes in terms of partition coefficients and the slopes of the liquidus and solidus phase boundaries, and to account for the free energy changes involved in various crystallization processes. In most castings, a full global equilibrium associated with a uniform phase composition does not occur throughout the ingot. However, it is often valid to take the liquid and solid compositions at the solidification front to be represented by the liquidus and solidus compositions determined by the tie line end points at a given temperature on the phase diagram. This concept is called local interfacial equilibrium, and it has been successfully applied to the description of casting processes in which the rate of diffusion, particularly in the solid, is too sluggish to achieve uniform phase compositions (Ref 1).

The consideration of nucleation and the level of melt undercooling introduce another type of departure from full equilibrium that is known as metastable equilibrium. At full equilibrium, thermodynamics predicts that solidification is not possible. Only when there is a departure from full liquid-solid equilibrium will solidification be possible. For solidification to occur, this departure brings the liquid into an undercooled state in which it is metastable because of the absence of one or more stable solid phases. The change from a stable to a metastable state occurs in a continuous manner without an abrupt change in the physical properties, such as molar volume or heat capacity. Therefore, metastable states can exhibit a true reversible equilibrium.

The thermodynamic description of solidification can be quantified by recalling that for a pure metal at the equilibrium temperature, T_f, the free energy change is:

$$G_S - G_L = (H_S - H_L) - T_f(S_S - S_L) = 0$$

or

$$\Delta G_f = \Delta H_f - T_f\,\Delta S_f = 0 \quad \text{(Eq 1)}$$

so that:

$$\Delta S_f = (\Delta H_f)/T_f \quad \text{(Eq 2)}$$

where G_S and G_L, H_S and H_L, and S_S and S_L are the molar free energy, enthalpy, and entropy for the solid and liquid, respectively.

At T, a temperature other than T_f, $\Delta G_f \neq 0$, and Eq 1 and 2 can be combined to yield:

$$\Delta G_f = \frac{\Delta H_f\,(T_f - T)}{T_f} = \frac{\Delta H_f\,\Delta T}{T_f} \quad \text{(Eq 3)}$$

when the small correction due to heat capacity terms is neglected. Based on Eq 3, there is a direct relationship between ΔG_f and the undercooling ΔT. The relationship in Eq 3 is represented in Fig. 1, in which the relative stability of the two phases (stable α and metastable β) that form from the liquid without composition change is shown as a function of temperature. The behavior is also approximately followed for stable and metastable eutectics such as Al/Al_3Fe and Al/Al_6Fe and Fe/C and Fe/Fe_3C (Ref 2). An important feature illustrated in Fig. 1 is that the melting of a metastable phase always occurs at a lower temperature than the stable-phase melting.

For alloys, the free energy of a phase is a function of both temperature and composition. When components are mixed together to form an alloy, the molar free energy change, ΔG_m, can be estimated from thermodynamic solution models (Ref 3). The simplest approach is to assume that the heat of mixing, ΔH_m, is zero and that the entropy change upon mixing, ΔS_m, is given by the ideal entropy of mixing. This is the description of an ideal solution, and it is often a useful approximation for dilute alloys. For an ideal solution of composition C_B:

$$\Delta G_m = \Delta H_m - T\,\Delta S_m = RT\,[C_B \ln C_B + (1 - C_B)\ln(1 - C_B)] \quad \text{(Eq 4)}$$

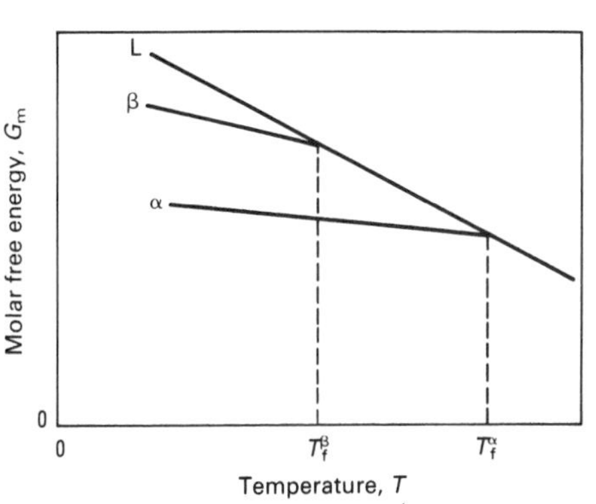

Fig. 1 Molar free energy versus temperature curves for a single component with a stable α phase and a metastable β phase, both at equilibrium temperature T_f. $T_f^\beta < T_f^\alpha$ in every case.

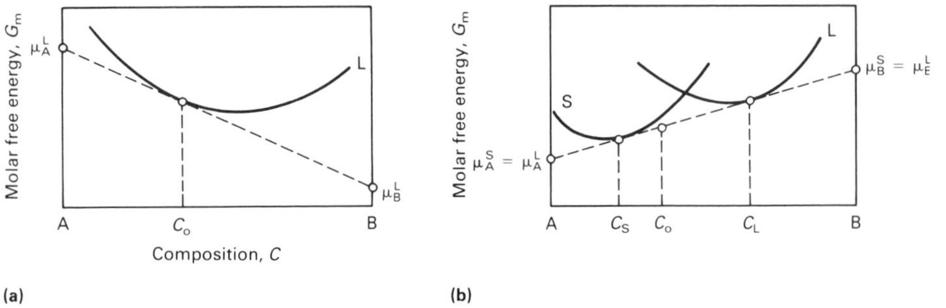

Fig. 2 Alloy phase equilibrium diagrams showing the chemical potentials of the liquid and solid components. (a) Variation of the molar free energy of mixing with composition for a liquid alloy solution. The chemical potentials, μ_A^L and μ_B^L, for the solution at composition C_o are given by the intercepts on the pure component axes. (b) Molar free energy versus composition diagram for two-phase liquid-solid equilibrium in an alloy of composition C_o. Solid of composition C_S is in equilibrium with liquid of composition C_L, and the chemical potentials of A and B are equal in the solid and liquid.

With more concentrated alloys, a regular solution model is often applied in which the heat of mixing is given by the symmetrical term $\Omega C_B(1 - C_B)$, where Ω reflects the net interatomic interaction. For regular solutions, ΔS_m is taken as the ideal solution term in Eq 4.

The description of equilibrium between solid and liquid phases over a range of alloy composition and temperature is represented in terms of the component chemical potentials μ_A and μ_B for each phase. The chemical potential is related to the molar free energy at a given composition by the standard thermodynamic definition (Ref 3):

$$\mu_A = G_m - C_B \left(\frac{\partial G_m}{\partial C_B} \right)_{T,P} \quad \text{(Eq 5a)}$$

$$\mu_B = G_m + (1 - C_B) \left(\frac{\partial G_m}{\partial C_B} \right)_{T,P} \quad \text{(Eq 5b)}$$

As illustrated in Fig. 2(a), the definitions in Eq 5a and b indicate that a tangent to a molar free energy curve at composition C_o intersects the pure component axes at μ_A and μ_B. With this description, alloy phase equilibrium is represented by an equality of chemical potentials for each of the components in the equilibrated phases so that $\mu_A^L = \mu_A^S$ and $\mu_B^L = \mu_B^S$, as shown in Fig. 2(b). The geometric interpretation of alloy phase equilibrium is expressed in terms of the common tangent construction, which is also shown in Fig. 2(b). The molar free energy of the liquid or the solid phase is also related to the chemical potential by the expression:

$$G_m = (1 - C_B)\, \mu_A + C_B\, \mu_B \quad \text{(Eq 6)}$$

and provides a useful form in representing the free energy changes involved in the formation of solid phases from an alloy liquid solution.

Free energy versus composition diagrams provide a useful method of analyzing the free energy relationships for alloy solidification reactions. Some of the relationships are shown in Fig. 3 for a temperature below the eutectic. For temperatures below the eutectic, the (α + L) and (β + L) equilibria are metastable with respect to the stable (α + β) phase mixture. In the phase diagram, the metastable two-phase conditions can be examined in terms of an extension of the respective solidus and liquidus phase boundaries below the eutectic, as noted in Fig. 3. For large excursions below the eutectic, the phase boundary extrapolations may not follow a simple linear extension. The relationships illustrated in Fig. 3 also show the common characteristic for a metastable phase to exhibit a higher solubility for solute than the equilibrium stable-phase solution. Similar behavior is found for the cases of peritectic and monotectic solidification reactions (Ref 2).

When one or both of the stable solid phases in a eutectic or peritectic reaction are suppressed, metastable solid phases may develop and yield metastable eutectics and peritectics (Ref 2). Figure 4 shows the modified phase diagram for a metastable peritectic. The stable Al/Al_3Ti and metastable Al/Al_xTi peritectic reactions may be relevant to the operation of grain refining in aluminum alloys (Ref 4, 5).

To provide a description of compositional segregation plus the influence of constitutional supercooling and solute redistribution during two-phase eutectic growth, phase diagram information such as the partition coefficient, k, or the liquidus slope, m_L, is often used in alloy solidification analysis (Ref 6, 7). For dilute solutions, the Vant'Hoff relation for liquid-solid equilibrium holds and relates k and m_L as:

$$k = 1 - \frac{m_L\, \Delta H_f^A}{R(T_f^A)^2} \quad \text{(Eq 7)}$$

where ΔH_f^A and T_f^A are the latent heat and melting point, respectively, of pure solvent, A. The partition coefficient can be evaluated separately for a dilute solution by:

$$k = \exp \left[\frac{\Delta H_f^B}{R} \left(\frac{1}{T_f^B} - \frac{1}{T_f^A} \right) \right] \quad \text{(Eq 8)}$$

where ΔH_f^B and T_f^B are the latent heat and melting point, respectively, of component B.

Microscopic Solids. The preceding discussion of solidification thermodynamics applies to cases in which the solid phases are of macroscopic size. There are several important situations during nucleation, dendritic solidification, or eutectic growth at high velocity in which the solid is of microscopic size or exhibits a sharply curved boundary with the liquid. The free energy of a small particle increases inversely with its size or radius of curvature. The increased free energy for the microscopic solid compared to the macroscopic solid results in a decrease in melting temperature, as expected for metastable equilibrium behavior. For

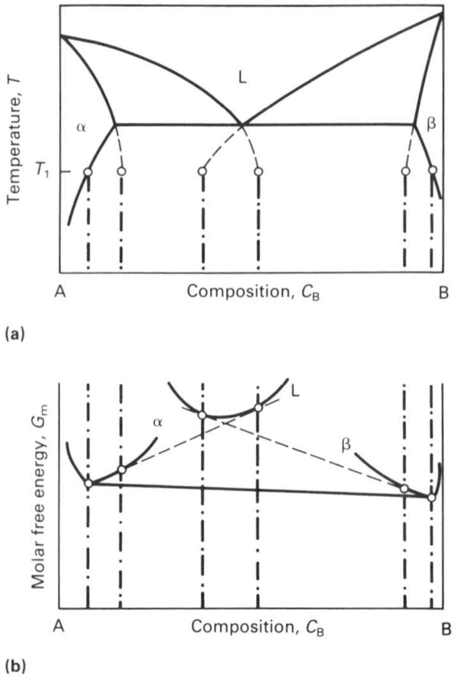

Fig. 3 Eutectic phase diagram (a) illustrating the metastable extensions of the liquidus and solidus phase boundaries to a temperature T_1 below the eutectic. (b) The molar free energy versus composition diagram at T_1 shows the metastable (α + L) and (β + L) equilibria by dashed common tangents and the stable (α + β) equilibrium the solid common tangent.

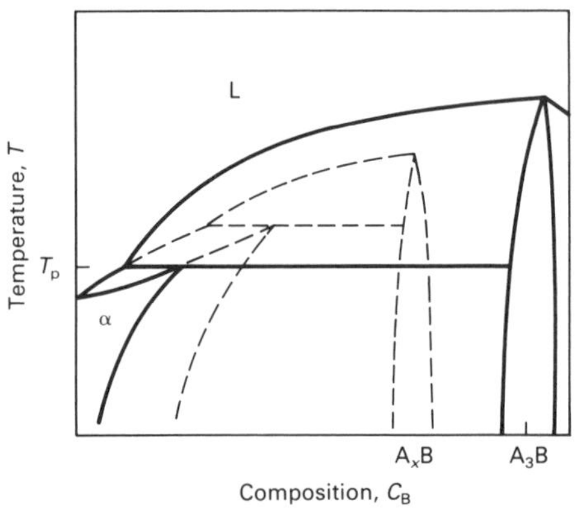

Fig. 4 Modified phase diagram for a metastable peritectic. When a stable A_3B phase is suppressed, the stable peritectic reaction involving A and A_3B may be replaced by a metastable peritectic reaction between A and A_xB. The peritectic temperature T_p and α phase solubility for the metastable peritectic are higher than for the stable peritectic case.

a pure component, the free energy increment due to curvature K is given by:

$$\Delta G_k = \sigma K\, V_m \qquad \text{(Eq 9)}$$

where V_m is the molar volume and σ is the interfacial energy. For a sphere of radius, r, $K = 2/r$. In terms of Eq 9, the melting of a microscopic particle of pure component with radius r, $T_f(r)$, is depressed from the melting point of a macroscopic particle by:

$$T_f(r) = T_f - \frac{2\sigma\, V_m}{(r\, \Delta S_f)} \qquad \text{(Eq 10)}$$

For alloys, the influence of curvature can be treated by including a term based on Eq 9 to the chemical potential for the solid (Eq 6). This will shift the liquidus and solidus phase boundaries, but in a manner that results in a negligible change in k and m_L (Ref 6).

Nucleation Phenomena

Nucleation during solidification is a thermally activated process involving a fluctuational growth in the sizes of clusters of solids (Ref 8). Changes in cluster size are considered to occur by a single atom addition or by removal exchange between the cluster and the surrounding undercooled liquid. At small cluster sizes, the energetics of cluster formation reveal that the interfacial energy is dominant, as can be observed by noting that the ratio of surface area to volume of a sphere is $3/r$. For the smallest sizes, clusters are called embryos; these are more likely to dissolve than grow to macroscopic crystals. In fact, the excess interfacial energy due to the curvature of small clusters is the main contribution to the activation barrier for solid nucleation. This accounts for the kinetic resistance of liquids to crystallization and is manifested in the frequent observation of undercooling effects during solidification (Ref 9, 10).

Homogeneous Nucleation

The principal features of nucleation phenomena and the kinetics of the rate process during solidification can be illustrated in the simplest terms by using the capillarity model to evaluate the kinetic factors (Ref 8). With this approach, it is useful to examine first the case of homogeneous nucleation in which solid formation occurs without the involvement of any extraneous impurity atoms or other surface sites in contact with the melt. As a further simplification, only the case of isotropic interfacial energy is treated, but it should be recognized that anisotropic behavior can yield faceted cluster shapes. The energetics of cluster formation for a spherical geometry can be expressed in terms of a surface and a volume contribution as:

$$\Delta G(r) = 4\,\pi r^2 \sigma + \tfrac{4}{3}\,\pi r^3\, \Delta G_V \qquad \text{(Eq 11)}$$

where $\Delta G(r)$ is the free energy change to form a cluster of size r and $\Delta G_V = \Delta H_f\, \Delta T / T_f V_m$.

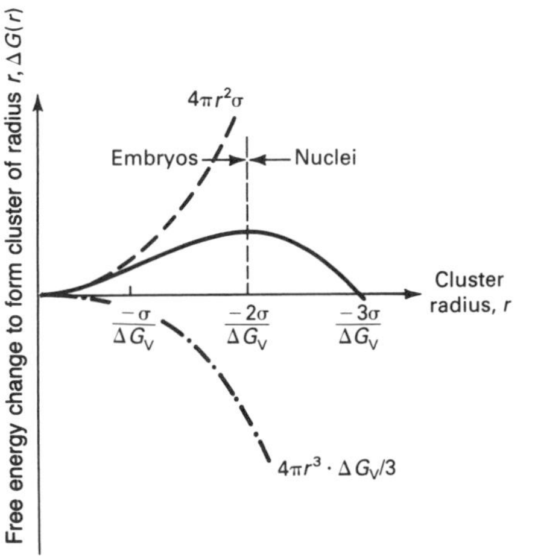

Fig. 5 Free energy change for cluster formation as a function of cluster size. The surface contribution to $\Delta G(r)$ is $4\pi r^2\sigma$, and the volume contribution is $4\pi r^3 \cdot \Delta G_V/3$; r_{cr} occurs at $-2\sigma/\Delta G_V$.

The relationship in Eq 11 is characterized in Fig. 5, in which the activation barrier for nucleation, ΔG_{cr}, is reached at a critical size r_{cr} (that is, $d\Delta G(r)/dr = 0$), as given by:

$$r_{cr} = -\frac{2\sigma}{\Delta G_V} = -\frac{2\sigma\, T_f V_m}{\Delta H_f \cdot \Delta T} \qquad \text{(Eq 12)}$$

and

$$\Delta G_{cr} = \frac{16\,\pi\,\sigma^3}{3\,\Delta G_V^2} = \frac{16\,\pi\,\sigma^3\, T_f^2\, V_m^2}{3\,\Delta H_f^2\, \Delta T^2} \qquad \text{(Eq 13)}$$

At increasing values of undercooling, r_{cr} is reduced ($r_{cr} \propto \Delta T^{-1}$) and G_{cr} is reduced more rapidly ($\Delta G_{cr} \propto \Delta T^{-2}$). A cluster is often considered to reach the stage of a nucleus capable of continued growth with a decreasing free energy when the size r_{cr} is achieved, but in fact stable nucleus growth ensues when the cluster size exceeds r_{cr} by an amount corresponding to $(\Delta G_{cr} - kT)$ in Fig. 5 (Ref 8). The relationship between cluster size and the number of atoms in a cluster, n_{cr}, is expressed by $(n_{cr} V_a) = 4/3\pi r_{cr}^3/3$, where V_a is the atomic volume.

To relate cluster energetics and fluctuational growth to the rate of nucleation, it is necessary to describe the cluster population distribution. Because the mixture of clusters in an undercooled melt is a dilute solution, the entropy can be described in terms of an ideal solution. The metastable equilibrium concentration of clusters of a given size, $C(n)$, is then given by (Ref 8):

$$C(n) = C_\ell \exp\left(\frac{-\Delta G(n)}{kT}\right) \qquad \text{(Eq 14)}$$

where C_ℓ is the number of atoms per cubic meter in the liquid and $\Delta G(n)$ is given by Eq 11 when r is converted to n, as noted above.

If solid nucleation is regarded as the growth of clusters past the critical size, then the resulting cluster flux or the nucleation rate I (in $m^3 \cdot s^{-1}$) can be represented kinetically by the product:

$$I = \nu_{SL}\, S_{cr}\, C(n_{cr}) \qquad \text{(Eq 15)}$$

where ν_{SL} is the jump frequency associated with atom jumps from the liquid to join the cluster and can be estimated from the liquid diffusivity, D_L, and jump distance, a, as in (D_L/a^2); S_{cr} is the number of atoms surrounding a cluster that is roughly $(4\pi r_{cr}^2/a^2)$; and $C(n_{cr})$ is the concentration of critical clusters. The full expression for the steady-state nucleation rate is then:

$$I = \left(\frac{D_L}{a^2}\right)\left(\frac{4\pi\, r_{cr}^2}{a^2}\right) C_\ell \exp\left(\frac{-\Delta G_{cr}}{kT}\right) \qquad \text{(Eq 16)}$$

For typical metals, $C_\ell \sim 10^{28}$ m^{-3}, $D_L \sim 10^{-9}$ m^2/s, and $a_o \sim 0.3 \times 10^{-9}$ m so that:

$$I \simeq 10^{40} \exp\left(-\frac{16\,\pi\,\sigma^3\, T_f^2\, V_m^2}{3k \cdot \Delta H_f^2 T \cdot \Delta T^2}\right) \qquad \text{(Eq 17)}$$

and shows a rather steep temperature dependence, as illustrated in Fig. 6. At high temperature, the temperature dependence of I is dominated by the driving free energy term, which is contained within the exponential dependence on the activation barrier, and I can vary by about a factor of five per degree Celsius. Equation 16 or 17 indicates that the maximum nucleation rate occurs at $(1/3)T_f$. Over a wide temperature range, the liquid diffusivity cannot be taken as constant, but can instead be represented as an Arrhenius function, $D_L = D_o \exp(-Q/kT)$, where D_o is a constant and Q is the activation energy for liquid diffusion. At large undercooling or low temperatures, and especially for glass-forming alloys, the diffusivity is often obtained from the viscosity in the form $D_L = (kT/3\pi a\, n_o) \exp -[A/(T - T_o)]$, where n_o, A, and T_o are constants. In evaluating nucleation rates, accurate values for the activation barrier are important because of the exponential dependence of I on ΔG_{cr}. The evaluation of ΔG_V is based on thermodynamic data or reasonably accurate models (Ref 11). However, separate, independent measurements of liquid-solid interfacial energy, σ_{LS}, at the nucleation temperature are not available, and calculations based on model-dependent estimates of σ_{LS} are of uncertain accuracy.

Heterogeneous Nucleation

Homogeneous nucleation is the most difficult kinetic path to crystal formation because of the relatively large activation barrier for nucleus development (ΔG_{cr}). To overcome this barrier, classical theory predicts that large undercooling values are required, but in practice an undercooling of only a few degrees or less is the common observation with most castings. This behavior is accounted for by the operation of heterogeneous nucleation in which foreign bodies such as impurity inclusions, oxide films, or crucible walls act to promote crystallization by lowering ΔG_{cr}. Because only a single nucleation event is required for the freezing of a liquid volume, the likelihood of

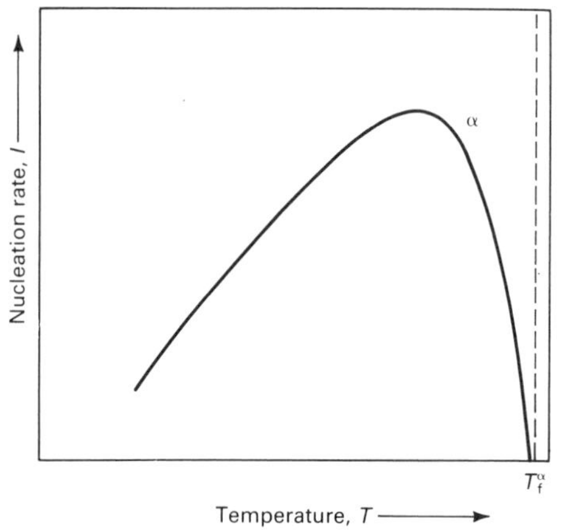

Fig. 6 Steady-state nucleation rate as a function of undercooling below the melting point. At low undercooling, the nucleation rate is primarily controlled by driving free energy; at high undercooling, nucleation is limited by diffusional mobility.

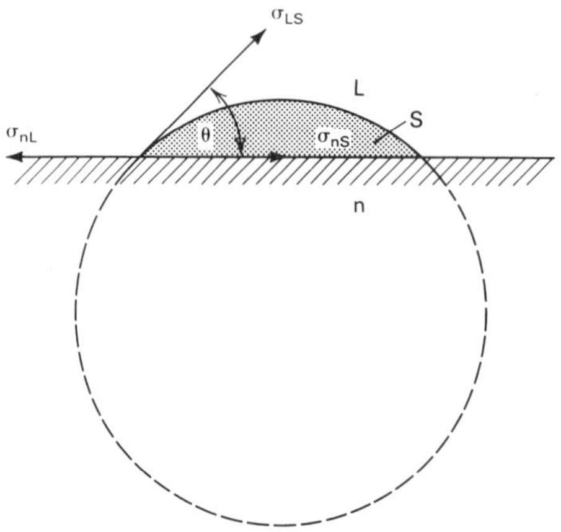

Fig. 7 The interfacial energy, σ, relationships among a planar nucleant substrate (n), a spherical sector solid (S), and the liquid (L). The interfacial regions are designated by the subscripts LS (liquid-solid), nL (nucleant-liquid), and nS (nucleant-solid).

finding a heterogeneous nucleation site in contact with a bulk liquid is great. Indeed, it has been estimated that even in a sample of high-purity liquid metal there is a nucleant particle concentration of the order of about 10^{12} m^{-3} (Ref 9). Only by using special sample preparation methods to isolate the melt from internal and external nucleation sites by subdivision into a fine droplet dispersion has it been possible to achieve undercoolings in the range of 0.3 to 0.4 T_f (Ref 10).

The action of heterogeneous nucleation in promoting crystallization can be visualized in terms of the nucleus volume that is substituted by the existing nucleant, as illustrated schematically in Fig. 7. For a nucleus that wets a heterogeneous nucleation site with a contact angle θ, the degree of wetting can be assessed in terms of $\cos\theta = (\sigma_{nL} - \sigma_{nS})/\sigma_{LS}$, where the interfacial energies are defined in Fig. 7. As θ approaches 0, complete wetting develops; as θ approaches 180°, there is no wetting between the nucleus and the nucleant (which is inert), and the conditions approach homogeneous nucleation.

The energetics of heterogeneous nucleation can be described by a modification of Eq 11 to account for the different interfaces and the modified cluster volume involved in nucleus formation. In terms of the cluster formation shown in Fig. 7, the free energy change during heterogeneous nucleation is expressed by:

$$\Delta G(r)_{het} = V_{SC}\,\Delta G_V + A_{SL}\,\sigma_{SL} + A_{nS}\,\sigma_{nS} - A_{nL}\,\sigma_{nL} \quad \text{(Eq 18)}$$

where V_{SC} is the spherical cap volume and A_{SL}, A_{nS}, and A_{nL} are the solid-liquid, nucleant-solid, and nucleant-liquid interfacial areas, respectively. When the volume and relevant interfacial areas are expressed in terms of the geometry of Fig. 7, the evaluation of ΔG_{cr} for heterogeneous nucleation yields:

$$\Delta G_{cr}(\text{het}) = \frac{16\,\pi\,\sigma_{LS}^3}{3\,\Delta G_V^2}\left[\frac{2\text{–}3\cos\theta + \cos^3\theta}{4}\right] = \Delta G_{cr}(\text{hom})[f(\theta)] \quad \text{(Eq 19)}$$

so that the barrier for homogeneous nucleation is modified by $f(\theta)$, the shape factor, during heterogeneous nucleation. Over the variation in θ ranging from complete wetting (θ = 0°) to nonwetting (θ = 180°), $f(\theta)$ or $[\Delta G_{cr}(\text{het})/\Delta G_{cr}(\text{hom})]$ varies from 0 to 1.0, as illustrated in Fig. 8. It is important to note that the value of r_{cr} for the curvature of the critical nucleus does not change in the classical analysis of heterogeneous nucleation, but the reduction in the activation barrier shown in Fig. 8 has a significant influence on the nucleation rate. This feature is also described in Fig. 8, in which the spherical cap size as measured by the ratio (h/r) is plotted as a function of θ. The dimension h is the height above the substrate.

Following an analysis similar to that developed for homogeneous nucleation based on Eq 15, an expression can be obtained for the rate of heterogeneous nucleation. In this case, the value of the surface area, S_{cr}, for a spherical cap geometry is estimated by $[2\pi r_{cr}^2\,(1 - \cos\theta)/a^2]$. In addition, the concentration of critical clusters is represented in terms of the number of surface atoms of the nucleation site per unit volume of liquid, C_a, which is of the order of 10^{20} m^{-3}. The heterogeneous nucleation rate expression I_{het} ($m^{-3} \cdot s^{-1}$) is given by:

$$I_{het} = \left(\frac{D_L}{a^2}\right)\left(\frac{2\,\pi\,r_{cr}^2\,(1 - \cos\theta)}{a^2}\right) C_a \exp\left[\left(-\frac{\Delta G_{cr}(\text{hom})}{kT}\right) f(\theta)\right] \quad \text{(Eq 20)}$$

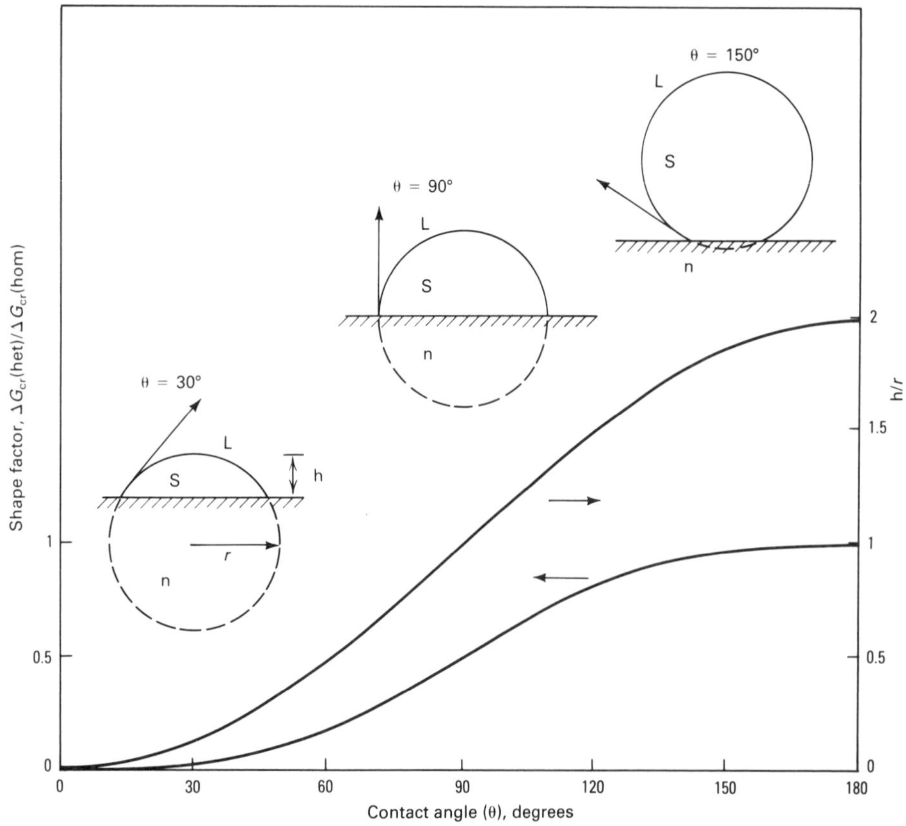

Fig. 8 The variation in shape factor $f(\theta)$ or $[\Delta G_{cr}(\text{het})/\Delta G_{cr}(\text{hom})]$ and spherical cap size, h/r, as a function of the contact angle, θ

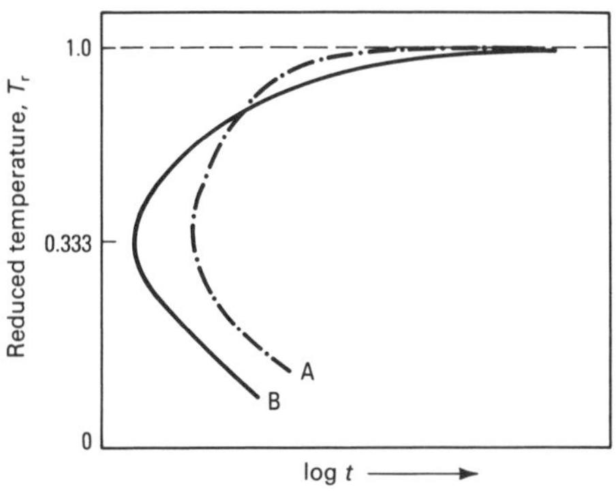

Fig. 9 Comparison between heterogeneous nucleation (A) and homogeneous nucleation (B) in terms of the relative transformation kinetics below the melting point. The reduced temperature $T_r = T/T_f$ and time $t \propto I^{-1}$.

or for typical metals:

$$I_{het} \simeq 10^{30} \exp\left[-\frac{16\,\pi\,\sigma_{LS}^3\,T_f^2\,V_m^2 f(\theta)}{3k \cdot \Delta H_f^2 \cdot T \cdot \Delta T^2}\right] \quad \text{(Eq 21)}$$

The comparison between heterogeneous and homogeneous nucleation kinetics is illustrated in Fig. 9 by a schematic time-temperature transformation diagram. Although only a single transformation curve (that is, C-curve) is shown in Fig. 9 for heterogeneous nucleation, in reality there will be as many curves as the number of heterogeneous nucleation sites. Each curve for heterogeneous nucleation will be distinguished by a catalytic potency, that is, $f(\theta)$, and a site density (Ref 10). To attain homogeneous nucleation conditions, it is clear that all heterogeneous nucleation sites must be removed or bypassed kinetically.

Inoculation Practice

There is a wide commercial application of cast alloy treatments that modify the initial solidification characteristics to provide a means for effective control of grain size and morphology. Inoculation is the common approach to grain refinement, and it involves the introduction of nucleating agents to a melt either externally in the form of a fine dispersion or through internal means by phase or chemical reactions that result in the formation of a solid reaction product. Although the success of this approach is established, an understanding of the actual mechanism by which inoculating agents operate still remains somewhat uncertain. There is widespread experience with inoculation, and several models have been developed for the process, primarily (but not exclusively) in connection with aluminum alloy melts (Ref 12-24). It is not possible to discuss all of this work in this article; instead, the key features of inoculation will be treated in relation to specific features of nucleation during solidification.

In commercial practice, inoculants are used as additions to numerous molten alloys to yield fine grain sizes in castings. In cast iron, ferrosilicon is added to nucleate graphite; zirconium or carbon can be added to magnesium alloys; iron, cobalt, or zirconium is added to copper-base alloys; phosphorus is introduced to aluminum-silicon alloys to refine the silicon phase size; arsenic or tellurium can be added to lead alloys; and titanium is added to zinc-base systems. Other additives, such as sodium in aluminum-silicon alloys, are used to modify the growth morphology. In addition, there is a large body of experience on the grain refinement of aluminum-base alloys by aluminum-titanium and aluminum-titanium-boron melt additions. To illustrate the principal features of inoculation and the physical mechanisms that have been developed to account for the effectiveness of inoculating agents, some of the experience with aluminum alloys will be used as a basis for discussion. The basic features of inoculation in other alloy systems follow the pattern that appears to apply to aluminum alloys.

The basic requirements for an efficient nucleant can be assessed from consideration of nucleation theory (Ref 25-27). To promote the formation of crystals on a nucleant, the interface between the nucleant and the liquid should be of higher energy than that between the nucleant and the crystal solid. A means of maximizing this condition is to provide a nucleant crystal relationship that is associated with a good crystallographic fit between the respective crystal lattices. In fact, the potency of a given catalyst is believed to be increased for decreasing values of relative lattice disregistry, with the degree of undercooling decreasing with a lowering of lattice disregistry. In addition to the basic requirement of lattice compatibility, the melt should tend to wet the surface of the nucleant, and for a uniform refining, the solid nucleant should remain widely dispersed in the liquid.

As an ancillary condition, the onset of nucleation should not be followed by a rapid and an isotropic crystal growth, because this will permit the full effect of the potential nucleants to be realized. In alloys, the growth restriction can be achieved as a consequence of solute segregation during freezing, which allows for nucleants throughout the melt to become effective and favors an equiaxed fine grain structure (Ref 27). At the onset, it should be noted that the theoretical description of an efficient nucleant should serve as a basis for predicting useful inoculants. However, in practice, this approach has not been successful; instead, nucleation theory has been primarily used to rationalize the identification of a useful grain refiner.

Within the general requirements for effective nucleants, there appear to be two general classes of compounds that are effective in aluminum-base alloys. Similar compound particle types apply to other common casting alloys. The first group includes Al_3Ti, Al_3Zr, and Al_7Cr and can be considered to be associated with a peritectic reaction. The second group comprises compounds such as TiC, TiB_2, and AlB_2, which are added either intentionally or which result from a chemical reaction in the melt with residual impurities. It is useful to note that the undercooling required to activate solidification is usually less than about 5 °C (9 °F). Despite the rather extensive discussion concerning the mechanism of action and the activity of these proposed nucleants, the strongest evidence for the operation of compound nucleant is derived from the observed orientation relationships.

Table 1 lists disregistry calculations along two directions, *a* and *b*, that are normal to each other and are located in the plane in which nucleation is considered to initiate, as illustrated in Fig. 10. For the most part, the crystallographic relationships that have been observed involve planes of relatively close packing between the nucleant and aluminum with disregistries of usually less than about 10%. The issue of specific nucleant identity is not insignificant considering that in commercial melts the rather broad spectrum of background catalysts may often result in solidification at undercoolings of about 5 °C (9 °F). Indeed, only about 1 to 2% of all potential nucleant particles in a master alloy addition result in the formation of grains. Furthermore, it has not been clearly established that these proposed nucleants operate singly or in association with each other or with background nucleants in the melt.

Among the nucleant-forming elements listed in Table 1, titanium and, in particular, titanium associated with boron in the master alloy form with a typical composition (Al-5Ti-1B); this is the most effective grain refinement addition in use for aluminum. In the commercial application of grain-refining agents involving titanium and boron, it is well known that an optimum contact time (that is, the holding time necessary after the addition of refining agents to obtain maximum grain refinement) can vary considerably among different master alloy compositions. As shown in Fig. 11, this feature is generally observed to occur where the optimum contact time interval (period A-B) for minimum grain size is followed by contact time intervals that yield large grain sizes during the interval B-C, which is called a fading process.

Grain Refinement Models

It is generally viewed that Al_3Ti crystals are mainly responsible for grain refinement at alloy compositions above 0.15% Ti, that is, within the peritectic range. For compositions outside the peritectic range, where

Table 1 Observed characteristics of potential nucleant compounds

Compound	Crystallographic relationships	Lattice disregistry, % *a* direction	Lattice disregistry, % *b* direction	Ref	Undercooling (ΔT) (a) °C	Undercooling (ΔT) (a) °F	Ref
Al_3Ti	(221)Al‖(001)Al_3Ti; [010]Al‖[113]Al_3Ti	0	−0.25	22	. . .	. . .	22
	(001)Al‖(001)Al_3Ti; [100]Al‖[100]Al_3Ti	5.2	5.2	22, 19	3–5	5–9	22
$Al_3(Ti,B)$	(011)Al‖(001)Al_3Ti; [$0\bar{1}\bar{1}$]Al‖[110]Al_3Ti	−0.6	5.2	22	0	0	22
	($\bar{1}\bar{1}1$)Al‖(110)Al_3Ti	−5.1	−14.3	14	. . .	. . .	. . .
AlB_2	(111)Al‖(001)AlB_2	−5.9	−5.0	14	0.5–1.0	0.9–1.8	22
TiB_2	None observed	. . .	. . .	14	0–0.3	0–0.5	14
TiC	(001)Al‖(001)TiC	6.5	6.5	21	0	0	12

(a) For peritectic compounds, ΔT is referenced to T_p; for other compounds, ΔT is referenced to T_f.

the Al_3Ti phase is not stable under equilibrium conditions, there are basically two interpretations of the grain refinement effect: the carbide-boride view and the peritectic reaction theory.

The carbide-boride model suggests that a compound such as AlB_2, TiB_2, or TiC is responsible for the nucleation of aluminum crystals, because the Al_3Ti particles that are introduced into the master alloy are expected to dissolve rapidly. The borides are hexagonal structures, and the carbide is a cubic structure with a relatively close correspondence (Table 1) with the aluminum lattice. Although it is considered that a boride or a carbide is the site of nucleation, the Al_3Ti particles do play a role in the overall grain refinement process. The dissolution of Al_3Ti is required, for example, to provide a constitutional deterrent (that is, excess titanium) to rapid crystal growth. Within this viewpoint, the establishment of an optimum contact time can be related to the initial dissolution of Al_3Ti and to the possible conditioning of the nucleant surface. The subsequent fading reaction is related to the agglomeration and settling of nucleant particles (Ref 15).

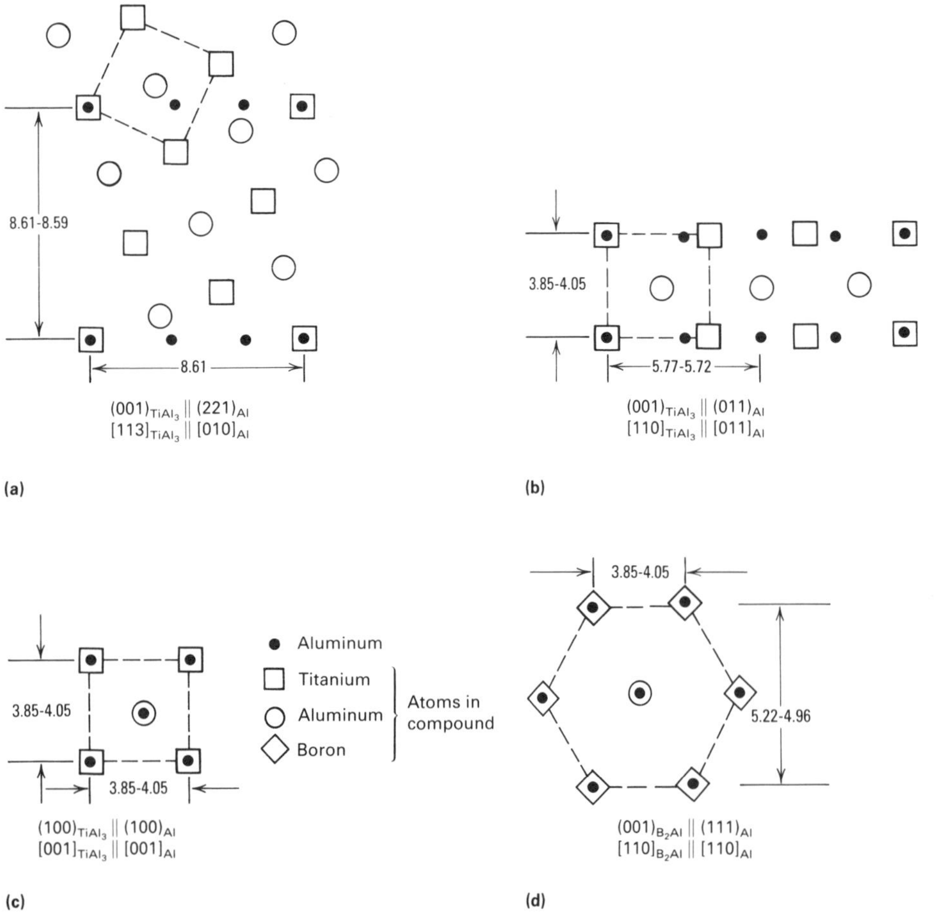

Fig. 10 Orientation relationships found among some potential nucleants and aluminum. The dashed lines delineate the compound nucleant lattice. Lattice dimensions given in angstroms

Although there is evidence for the stability of boride or carbide particles in aluminum melts and reports of observations of these particles in ingots, other observations have indicated a number of problems in attributing nucleation solely to the action of the boride or carbide particles. For example, a consistent orientation relationship between Al_3Ti and aluminum can be found; however, no consistent relationship between TiB_2 and aluminum has been detected even though the orientation indicated in Table 1 for AlB_2 would be expected to develop for TiB_2 with a lattice disregistry of −5.9 and −5.8%, respectively, in the *a* and *b* directions. In addition, boride particles have often been observed to be located at grain boundaries rather than within grain centers, indicating that the borides were present as insoluble material and inactive particles for nucleation during freezing (Ref 14).

Peritectic Reaction Theory. There is an alternate viewpoint of the nucleation behavior during grain refinement that was originally developed on the basis of a peritectic reaction, but this viewpoint has subsequently evolved into a model with more general scope (Ref 13). Although many types of particles are believed to be active in the nucleation of aluminum crystals at undercoolings less than about 5 °C (9 °F), the presence of Al_3Ti is believed to offer a more active catalyst requiring little or no undercooling. Thus, Al_3Ti crystals, when present, will predominate in the grain refinement nucleation. For peritectic alloys such as the one illustrated by the solid lines in Fig. 4, it is clear that Al_3Ti can form upon cooling below the liquidus and can promote the formation of aluminum crystals at the peritectic temperature T_p. At alloy compositions below the peritectic range, a grain refinement effect is expected to be reduced by the loss of Al_3Ti particles, which are added externally in master alloy form. Indeed, independent Al_3Ti particles in liquid aluminum that is not saturated with titanium dissolve within several minutes.

On the other hand, boride particles, that is, TiB_2 or $(Al,Ti)B_2$, are observed to be

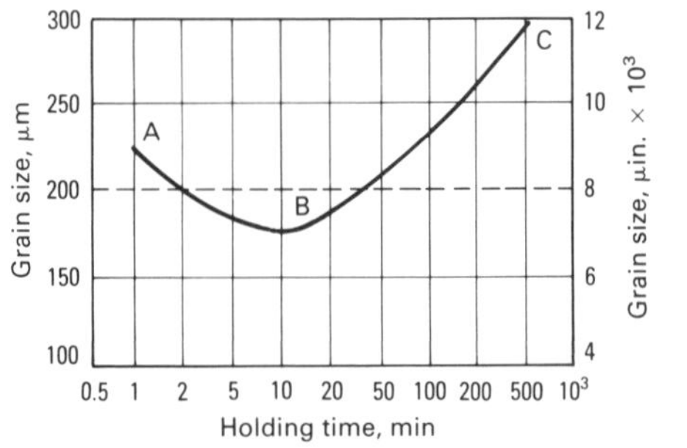

Fig. 11 The influence of holding time after the introduction of a grain-refiner master alloy on the degree of grain refinement. The period A-B represents the contact time, and the period B-C refers to a fading process.

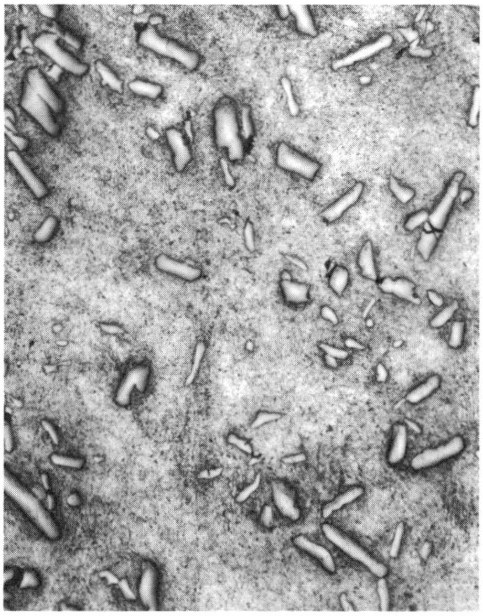

Fig. 12 Photomicrograph of an Al-6Ti master alloy illustrating the range of size and shape of Al_3Ti particles

reasonably stable in melts and offer a fine particulate dispersion. To account for the beneficial effect of incorporating boron into the master alloy compositions and also the observation of the interaction between TiB_2 and Al_3Ti, it appears likely that TiB_2 particles may provide an effective substrate for the nucleation of Al_3Ti for compositions spanning the peritectic reaction. During cooling of the melt or holding (that is, the contact time period), the TiB_2 particles will react slowly with liquid aluminum to form an $(Al,Ti)B_2$ solution and to liberate excess titanium locally to provide for the formation of a sheath of Al_3Ti as a coating on the $(Al,Ti)B_2$ particles (Ref 19). In this way, the boride particle is believed to act as a substrate for the formation of Al_3Ti particles at low titanium levels, that is, even in alloys with compositions below the peritectic range. Of course, this process is a transitory one, and with time a full equilibrium will develop and accompany the loss of the Al_3Ti coating as required to account for a fading effect. With this viewpoint, the presence of boride enhancement of grain refinement is attributed to the effect of the boride and TiB_2 in retarding the dissolution of Al_3Ti in the melt, but the actual mechanism for this process may be complex.

Master Alloy Processing

Another important aspect of the grain refinement reaction that has not received much consideration relates to the influence of preparation conditions during master alloy processing and structural differences on the effectiveness of nucleant particles (Ref 19, 20).

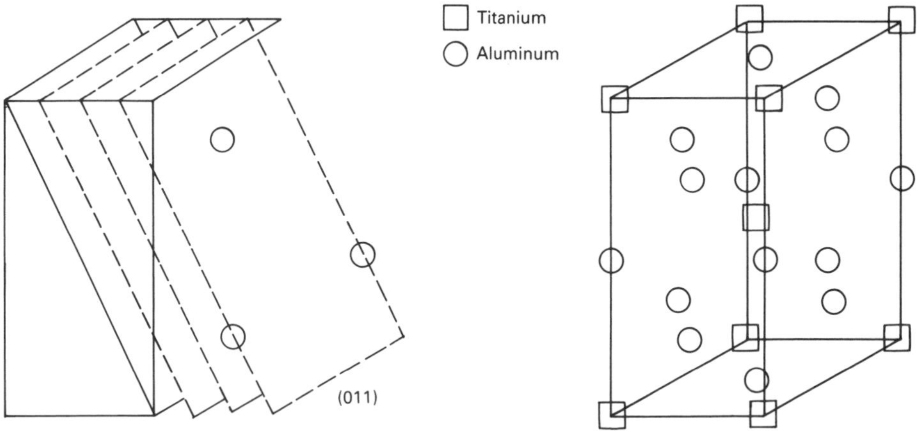

Fig. 13 The Al_3Ti structure and the arrangement of the (011) planes. Three successive (011) planes identified by dashed lines contain only aluminum atoms and offer a low disregistry to the atom arrangement on (012) planes of aluminum.

This feature has taken on new significance with the observation that Al_3Ti particles may appear with different morphologies (Fig. 12) and may exhibit several possible crystallographic orientation relationships with aluminum crystals. The degree of refinement is observed to be dependent on the Al_3Ti crystal morphology. Essentially, two types of Al_3Ti crystals are observed to be present: a flake type exposing (001) planes to the melt that forms at high preparation temperatures (>760 °C, or 1400 °F) and a blocky form that tends to form at low temperatures. Each particle morphology is found to differ in nucleating potency.

The blocklike crystals expose (011) planes to the melt, as indicated in Fig. 13, for a [(011)Al_3Ti ∥ (012) Al] relationship with aluminum. For flake-type crystals, the crystallographic axes of Al_3Ti and aluminum are closely parallel, that is, (001)Al_3Ti ∥ $(00\bar{1})$ Al. Although the latter relationship also gives a relatively small disregistry, the (011) Al_3Ti surface as indicated in Fig. 13 is occupied by three successive layers of aluminum atoms, which can be viewed as somewhat distorted (012) aluminum lattice planes. The formation of aluminum crystals on such a surface merely requires a growth (that is, epitaxy) of this existing distorted lattice at little or no undercooling below the peritectic temperature T_p. The (001) Al_3Ti surface, on the other hand, contains titanium atoms so that a simple growth of the existing

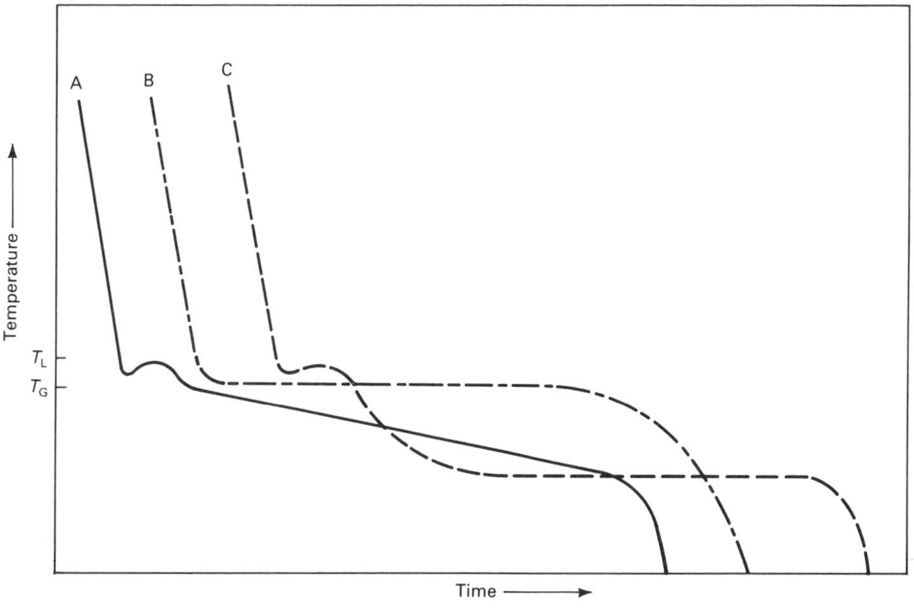

Fig. 14 Representative types of cooling curves for an aluminum melt treated with different master alloys to yield varying levels of grain refinement. A, high degree of grain refinement coincides with Al_3Ti crystals present in melt; B, poor grain refinement due to few Al_3Ti crystals in melt; C, intermediate case in which there are a sufficient number of Al_3Ti crystals but an insufficient local titanium concentration to promote grain refinement. The characteristic peritectic hump for all three curves lies between the growth temperature, T_G, and the equilibrium liquidus temperature, T_L. Source: Ref 18

surface is not sufficient to generate an aluminum crystal, and a lower nucleating potency may be expected and is observed. In addition, the observation of different morphologies and twinning for Al_3Ti crystals may account for the variety of orientation relationships that have been observed between Al_3Ti and aluminum. Other particles, such as TiB_2, TiC, TiN, ZrB_2, and TaB_2, have been examined, but the undercooling required to initiate solidification was minimized when Al_3Ti particles were present.

Thermal Analysis Techniques. Grain-refining characteristics can also be conveniently assessed through thermal analysis techniques. In examining cooling curves after master alloy dilution, three main groups of characteristic curves are often observed, as shown in Fig. 14. The characteristic curve A in Fig. 14 is detected for a high degree of grain refinement and is related to a large number of aluminum crystals that form on surviving Al_3Ti particles at T_p. With continued cooling upon reaching the growth temperature, T_G, the aluminum crystals can grow, consuming the liquid and raising the temperature to the equilibrium liquidus temperature, T_L. This reaction creates a peritectic hump, which is an indication that Al_3Ti crystals are present in the melt. The type A cooling curve in Fig. 14 was detected with blocky Al_3Ti crystals following a short holding time but also with flake-type Al_3Ti after a long contact time of 1 h or more. With a cooling curve of the type B configuration in Fig. 14, relatively few Al_3Ti crystals remain; therefore, poor grain refinement occurs. A cooling curve of the type C configuration in Fig. 14 represents an intermediate case in which there are a sufficient number of Al_3Ti crystals, but the local titanium concentration is not sufficient to allow for growth at high temperatures, a condition that yields poor grain refinement.

In addition to these factors, new observations concerning the possible effect of metastable phases produced in Al-Ti and Al-Zr systems on grain refinement have been reported (Ref 4, 5, 28). In aluminum-titanium and aluminum-zirconium alloys cooled at relatively high rates (1 °C/s, or 1.8 °F/s), a reaction has been reported above the peritectic isotherm in which a metastable phase forms and can be associated with existing intermetallic particles, that is, Al_3Ti or Al_3Zr. During holding of the liquid, the metastable phase decomposes, but the time periods involved are of the same order of magnitude as those obtained in grain refinement.

Although the results of recent work represent a significant improvement in the description of grain refinement and in the improved characterization of grain refinement agents and their effectiveness, the overall understanding of the grain refinement of aluminum or other metallic melts by compound particles remains incomplete. Most examinations have focused on the study of one particular particle, but it is clear that in commercial purity alloys several different particles or alloying elements can be present and can interact to obscure the actual nucleant identity. This is sometimes observed in terms of the poisoning of a grain refiner (Ref 23). In addition, the interaction among processing parameters is an unresolved point, but is likely to be significant.

REFERENCES

1. J.C. Baker and J.W. Cahn, in *Solidification*, American Society for Metals, 1971, p 23
2. J.H. Perepezko and W.J. Boettinger, *Proc. Mater. Res. Soc.*, Vol 19, 1983, p 223
3. D.R. Gaskell, *Introduction to Metallurgical Thermodynamics*, 1973
4. J. Cisse, H.W. Kerr, and G.F. Bolling, *Metall. Trans.*, Vol 5, 1974, p 633
5. L. Arnberg, L. Backerud, and H. Klang, *Met. Technol.*, Vol 9, 1982, p 14
6. M.C. Flemings, *Solidification Processing*, McGraw-Hill, 1974
7. W. Kurz and D.J. Fisher, *Fundamentals of Solidification*, Trans Tech, 1974
8. J.K. Lee and H.I. Aaronson, in *Lectures on the Theory of Phase Transformations*, H.I. Aaronson, Ed., The Metallurgical Society, 1975, p 83
9. J.H. Hollomon and D. Turnbull, *Prog. Met. Phys.*, Vol 4, 1953, p 333
10. J.H. Perepezko, *Mater. Sci. Eng.*, Vol 65, 1984, p 125
11. W.J. Boettinger and J.H. Perepezko, in *Rapidly Solidified Crystalline Alloys*, S.K. Das, B.H. Kear, and C.M. Adam, Ed., The Metallurgical Society, 1986, p 21
12. A. Cibula, *J. Inst. Met.*, Vol 76, 1949-1950, p 321; Vol 80, 1951-1952, p 11
13. F.A. Crossley and L.F. Mondolfo, *Trans. AIME*, Vol 141, 1951, p 1143
14. I.G. Davies, J.M. Dennis, and A. Hellawell, *Metall. Trans.*, Vol 1, 1970, p 275
15. G.P. Jones and J. Pearson, *Metall. Trans. B*, Vol 7B, 1976, p 223
16. J.A. Marcantonio and L.F. Mondolfo, *J. Inst. Met.*, Vol 98, 1970, p 23
17. I. Maxwell and A. Hellawell, *Acta Metall.*, Vol 23, 1975, p 895
18. J.H. Perepezko and S.E. LeBeau, in *Aluminum Transformation—Technology and Applications*, American Society for Metals, 1982, p 309
19. L. Arnberg, L. Backerud, and H. Klang, *Met. Technol.*, Vol 9, 1982, p 1, 7
20. M.M. Guzowski, G.K. Sigworth, and D.A. Senter, *Metall. Trans. A*, Vol 18A, 1987, p 603
21. J. Cisse, G.F. Bolling, and H.W. Kerr, *J. Cryst. Growth*, Vol B-14, 1972, p 777
22. L.F. Mondolfo, in *Grain Refinement in Castings and Welds*, G.J. Abbaschian and S.A. David, Ed., The Metallurgical Society, 1983, p 3
23. D.A. Granger, "Practical Aspects of Grain Refining Aluminum Alloy Melts," Laboratory Report 11-1985-01, Aluminum Company of America, 1985
24. J.L. Kirby, in *Aluminum Alloys—Physical and Mechanical Properties*, E.A. Starke and T.H. Sanders, Ed., Vol 1, Engineering Materials Advisory Service Ltd., 1986, p 61
25. D. Turnbull and B. Vonnegut, *Ind. Eng. Chem.*, Vol 44, 1952, p 1292
26. P.B. Crosley, A.W. Douglas, and L.F. Mondolfo, *The Solidification of Metals*, The Iron and Steel Institute, 1968, p 10
27. A. Hellawell, *Solidification and Casting of Metals*, The Metals Society, 1979, p 161
28. E. Nes and H. Billdal, *Acta Metall.*, Vol 25, 1977, p 1031

Fundamentals of Growth

Basic Concepts in Crystal Growth and Solidification

G. Lesoult, Ecole des Mines de Nancy, France

CRYSTAL GROWTH AND SOLIDIFICATION in metal castings is largely a function of atomic mobility. Thermal and kinetic factors must be considered when determining whether crystal growth will be inhibited or accelerated. Whether spherical or needlelike in configuration, the metal particles behave differently depending on their location within the composition: in the liquid, at the liquid/solid interface, or in the solid. In addition, metals such as aluminum and copper have only one structure (face-centered cubic, fcc); on the other hand, metals such as iron and cobalt can have different crystal structures at different temperatures (for example, iron can be fcc and body-centered cubic, bcc).

Liquid and Solid State

Atomic Mobility. The solidification of metals results in an enormous and abrupt decrease in atomic mobility. The dynamic viscosity η of pure liquid metals near their melting temperature T_f is comparable to that of water at room temperature, that is, of the order of 10^{-3} Pa · s (10^{-2} P), as shown in both Table 1 and Fig. 1. On the other hand, the following observations can be made:

- In the solid state, metals and alloys have a high tensile strength
- Pure metals resist stresses of the order of 10^4 Pa (1.5 psi) near the melting point
- The decrease in ductility of commercial alloys several hundred degrees below the solidus temperature is due to the presence of liquid films in the segregated zones

The self-diffusion and chemical diffusion of alloying elements and impurities are much slower in the solid than in the liquid, as shown in Table 2 for both iron- and aluminum-base alloys. In liquid iron for example, diffusion coefficients range from 10^{-9} m²/s (1.1×10^{-8} ft²/s) for the slowest solutes to 10^{-7} m²/s (1.1×10^{-6} ft²/s) for hydrogen. The diffusion coefficient range between the slowest solutes and hydrogen is larger in ferrite and is maximum in austenite.

Heat release during solidification is large—approximately 270 MJ/tonne (116 Btu/lb) for steel. The higher the melting point of the metal, the larger the latent heat of fusion (Table 1). Therefore, solidification processing is initially a matter of extracting large quantities of heat quickly.

Table 1 Physical properties of pure metals relevant to solidification

Property	Iron (δ)	Copper	Aluminum
Dynamic viscosity η of liquid at T_f, 10^{-3} Pa · s	5.03	3.05	1.235
Melting point, T_f, K	1809	1356	933
Enthalpy of fusion per mole, J/mol	13 807	13 263	10 711
Enthalpy of fusion per volume, J/m³ · 10^9	1.93	1.62	0.95
Heat capacity C_P of liquid at T_f, J/K · m³ · 10^6	5.74	3.96	2.58
Heat capacity C_P of solid at T_f, J/K · m³ · 10^6	5.73	3.63	3.0
Thermal conductivity of liquid K_L at T_f, W/m · K	35	166	95
Thermal conductivity of solid K_S at T_f, W/m · K	33	244	210
Thermal diffusivity of liquid α_L at T_f, m²/s · 10^{-6}	6.1	42	37
Thermal diffusivity of solid α_S at T_f, m²/s · 10^{-6}	5.8	67	70
Density of liquid ρ_L at T_f, g/cm³	7.024	7.937	2.368
Density of solid ρ_S at T_f, g/cm³	7.265	8.350	2.548
Solidification shrinkage $(V^\ell - V^s)/V^s$, %	3.55	5.20	7.59
Solid/liquid interfacial tension σ, MJ/m²	251	237	158
Faceting factor F for the most dense planes	1.78	1.96	2.19
Molar volume of the solid V_m, cm³/mol	7.69	7.61	10.59
Entropy of fusion per mole ΔS_f, J/K · mol	7.63	9.78	11.48

Table 1 also lists heat capacities C_P, thermal conductivities K, and thermal diffusivities α in solid (α_S) and liquid (α_L) at the melting temperature for iron, aluminum, and copper. Even in the liquid, thermal diffusivities are more than 100 times larger than coefficients of chemical diffusion. Moreover, relative differences in thermal diffusivity between liquid and solid are small compared to the related differences in chemical diffusion coefficients.

Solidification Shrinkage. Most metals shrink when they solidify. Solidification shrinkage ranges from 3 to 8% for pure metals (Table 1). It may result in the formation of voids (microporosity and shrinkage) during solidification. Thermal contraction of the solid during subsequent cooling may increase the risk of shrinkage if care is not exercised in casting of the metal.

Several commercial foundry alloys, based on simple eutectic alloys, form nonmetallic phases during solidification that are atomically less dense than the melt—for example, graphite in iron-carbon alloys or silicon in aluminum-silicon alloys. In the case of gray cast iron, the precipitation of austenite from the melt is associated with shrinkage, while the growth of graphite from the same melt is associated with volume expansion (Table 2). The sign of the resultant volume change is uncertain; the alloy may shrink or expand upon solidification, depending on the composition of the melt.

Solubility. When the heterogeneous thermodynamic equilibrium between liquid and solid is achieved, most crystalline solid solutions contain a smaller amount of each solute than the related liquid solutions. This difference in composition, combined with slow solid-state diffusion, results in various segregation patterns in cast alloys. Table 2 lists the values of the equilibrium partition coefficient (or ratio) k, which is defined as follows:

$$k = \frac{C_S^*}{C_L^*} \qquad \text{(Eq 1)}$$

where C_S^* and C_L^* are the solid and liquid concentrations, respectively, in mutual equilibrium across a plane solid/liquid interface. Tabulated values are valid only for very dilute binary alloys.

Adding solutes in a metallic melt usually decreases the liquidus temperature T_L, that is, the temperature at which the alloy begins to solidify. Table 2 lists values of the liq-

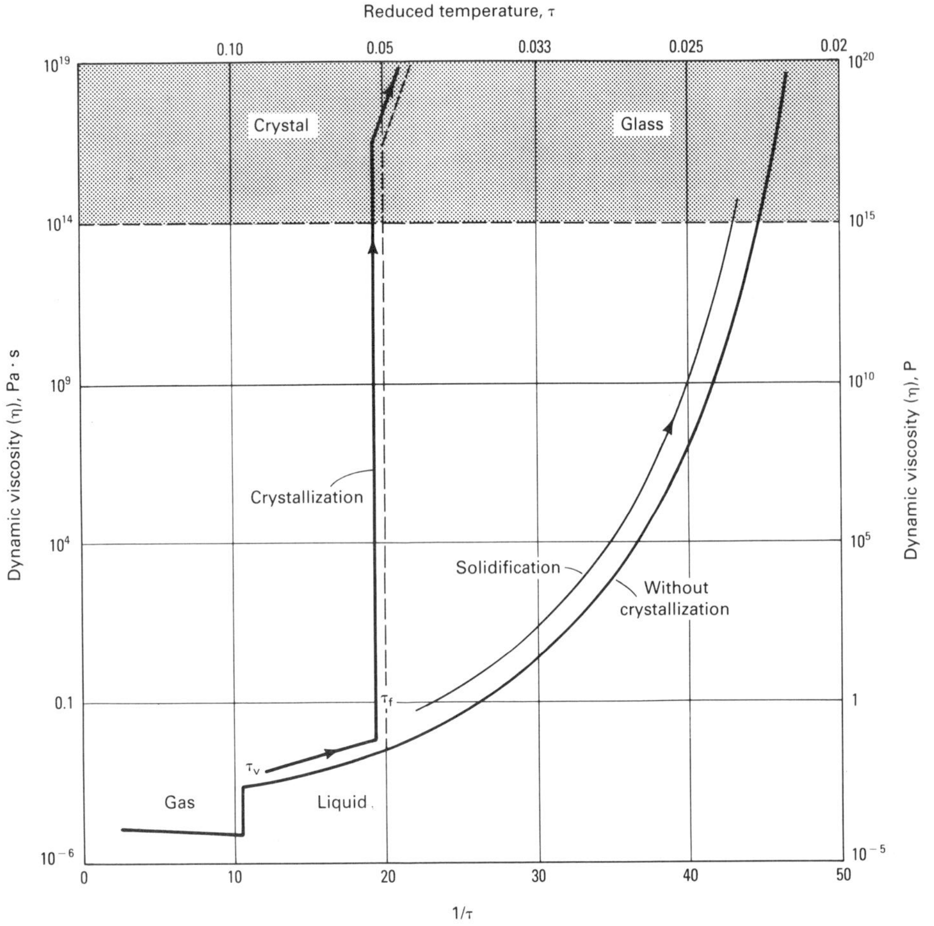

Fig. 1 Dynamic viscosity of metals as a function of reduced temperature $\tau = R \cdot T/\Delta H_v$, where ΔH_v is the enthalpy of vaporization per mole. The melting point corresponds to τ_f. Source: Ref 1

Table 2 Physical properties of alloys relevant to solidification

Base metal	Diffusion coefficient at T_f, m^2/s: Iron, Liquid	Iron, δ	Iron, γ(a)	Aluminum, Liquid	Aluminum, Solid
Self-diffusion	1.4×10^{-8}	2.3×10^{-11}	6.5×10^{-13}	...	1.2×10^{-12}
Diffusion of solutes					
Carbon	2.32×10^{-8}	5.72×10^{-9}	5.53×10^{-10}	...	...
Manganese	3.83×10^{-9}	2.57×10^{-11}	5.02×10^{-13}	...	...
Nickel	5.47×10^{-9}	2.65×10^{-11}	2.47×10^{-13}	...	...
Copper	...	...	...	4.9×10^{-9}	1.95×10^{-12}
Silicon	...	...	...	4×10^{-9}	5×10^{-12}

Dilute alloys	Partition ratio k and liquidus slope m_L: k in δ iron	$-m_L$, K for δ iron	k	$-m_L$, K
Hydrogen	0.3	70 700(a)	...	...
Carbon	0.11	8000	...	...
Manganese	0.73	500	0.7–0.9	100
Silicon	0.77	1300	0.13	600
Copper	0.70	...	0.14	260

Interface	Interfacial tension, σ, of graphite/iron interface(b), MJ/m^2: Without O and S	With O and S
Basal plane	1460	1270
Prismatic plane	1720	845

(a) Extrapolated, not measured, values to compare magnitude. (b) Change of partial molar volume of carbon during solution of graphite: 70%

uidus slope m_L for some very dilute iron- and aluminum-base alloys. The liquidus slope is defined as follows:

$$m_L = \frac{dT_L}{dC_L} \qquad \text{(Eq 2)}$$

Solid/Liquid Interface

The first models to describe the solid/liquid interface and its motion at the atomic scale were strongly influenced by the physical models that successfully describe the behavior of the crystal/vapor surfaces. Crystal growth from the vapor can proceed either by two-dimensional nucleation or by lateral motion in steps of atomic height around screw dislocations emerging at the surface. This is because these surfaces are atomically sharp. In this case, the average thermal energy available for the surface atoms to escape from the crystal (that is, $R \cdot T$, where R is the gas constant) is too small compared to the difference in cohesive energy between crystal and vapor (that is, the enthalpy of sublimation per mole ΔH_s). Physicists relate the sharpness of the crystal/vapor surfaces to the high value of the ratio $\Delta H_s/R \cdot T$ (Ref 2).

The solid/liquid interface is a region between two condensed phases in which interatomic cohesive energies are comparable. Interactions between solid and liquid atoms across the interface must be taken into account to describe this interface properly at the atomic scale; enthalpy of fusion is not a proper quantity for this condition. Physicists successfully correlate the sharpness of the solid/liquid interfaces of pure substances to a faceting factor F, which is defined as follows (Ref 3):

$$F = \frac{z^*}{1 - z^*} \cdot \frac{A_m \cdot \sigma}{R \cdot T_f} \qquad \text{(Eq 3)}$$

where z^* is the ratio between the number of near-neighbor atoms in the plane of the interface and the total number of near-neighbor atoms in the bulk ($z^* = 0.5$ for dense planes of fcc crystals), A_m is the area occupied by 1 mol at the interface, σ is the solid/liquid interfacial tension, and T_f is the melting temperature. Pure metals are expected to grow from their own melts with a diffuse interface as opposed to a sharp interface, because their faceting factors are approximately equal to the critical value of 2 (Table 1).

Interface Equilibrium. The solid/liquid interfacial tension can be considered excess energy associated with any area of the solid/liquid interface. It causes the change in equilibrium melting point, often called Gibbs-Thomson undercooling, when the interface is curved. For pure substances, this curvature undercooling ΔT_K is given by:

$$\Delta T_K = T_f - T_{f_K} = \Gamma \cdot K \qquad \text{(Eq 4)}$$

where T_{f_K} is the equilibrium temperature between liquid and solid across an interface whose curvature is K and where Γ is the Gibbs-Thomson coefficient. The curvature can be expressed as:

$$K = \frac{1}{r_1} + \frac{1}{r_2} \qquad \text{(Eq 5)}$$

where r_1 and r_2 are the principal radii of curvature; K is defined such that a positive undercooling is associated with a portion of the solid/liquid interface that is convex toward the liquid.

The Gibbs-Thomson coefficient is given by:

$$\Gamma = \frac{\sigma \cdot V_m}{\Delta S_f} \quad \text{(Eq 6)}$$

where V_m is the molar volume of the solid and ΔS_f is the entropy of fusion per mole.

For most metals, Γ is of the order of 10^{-7} K · m. Therefore, the curvature undercooling might often appear negligible; it is of the order of 0.05 K for a spherical portion of the interface where the radius of curvature is equal to 10 μm. However, this undercooling is very important for describing the nucleation stage and the morphological stability of growth shapes (dendritic and eutectic structures). Table 1 lists σ for pure iron, aluminum, and copper; it can be assumed to be isotropic for estimating the undercooling. Table 1 also lists T_f, V_m, and ΔS_f.

Equation 4 can also be used to estimate the effect of the curvature on the liquidus temperature of an alloy with a given composition. Conversely, it can be used to calculate the change in equilibrium composition of both liquid and solid solutions in mutual contact at a fixed temperature across a curved interface. The changes in liquid and solid solubilities are then given by:

$$\Delta C_{L,K} = C_{L(T,K)} - C_{L(T,K=\infty)} = \frac{K \cdot \Gamma}{m_L} \quad \text{(Eq 7a)}$$

$$\Delta C_{S,K} = C_{S(T,K)} - C_{S(T,K=\infty)} = \frac{K \cdot \Gamma}{m_S} \quad \text{(Eq 7b)}$$

where m_S is the solidus slope.

Some nonmetallic eutectic phases, such as graphite or silicon in foundry alloys, exhibit strong anisotropies of interfacial tension in contact with the melted alloy. In this case, surface-active solutes or impurities can significantly modify the interfacial tensions and their degree of anisotropy (Table 2) (Ref 4).

Interface Kinetics. Solidification can be described as a succession of atomic events that occur in series; it includes heterogeneous chemical reactions at the interface. At a diffuse interface, as for pure metals, liquid atoms can become solid atoms at almost every lattice site. The interface then moves more or less continuously, and the growth is said to be continuous. The kinetics of continuous growth have been described by using the rate theory of classical chemistry (Ref 5). This theory leads to the following estimate for the kinetic undercooling ΔT_k in the case of pure substances:

$$\Delta T_k = T_{f_k} - T^* = \left(\frac{a}{D_L}\right)\left(\frac{R \cdot T_f}{\Delta S_f}\right) v \quad \text{(Eq 8)}$$

where T_{f_k} is the equilibrium temperature defined by Eq 4, T^* is the actual temperature of the moving interface, D_L is the liquid diffusivity, a is the atomic distance between the crystallographic planes parallel to the interface, and v is the velocity of the interface.

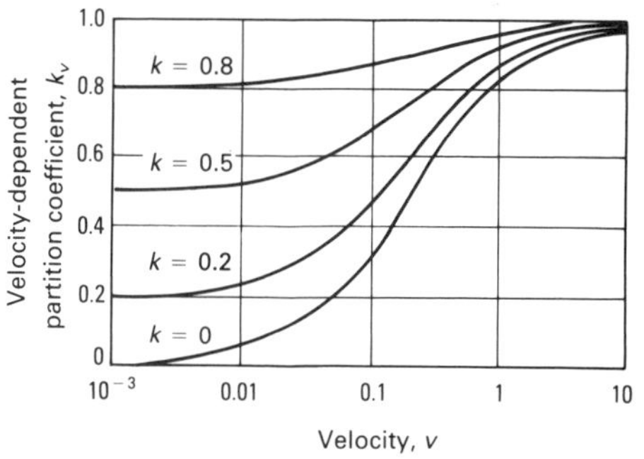

Fig. 2 Dependence of k_v on v according to Eq 9 when a_o = 5 nm and $D_L = 10^{-9}$ m²/s

For most metals, the proportionality coefficient between the kinetic undercooling and the velocity is of the order of 1 K · s/cm, according to Eq 8. For most usual solidification conditions, velocities do not exceed the order of 10^{-4} m/s (3.3×10^{-4} ft/s) or 0.36 m/h (1.18 ft/h), and the kinetic undercooling does not exceed 10^{-2} K. Therefore, unlike other major contributions, kinetic undercooling is often neglected. This simplification is probably invalid in the case of rapid solidification, even for pure metals.

Kinetic undercooling might be much higher than that predicted by Eq 8 in the case of crystals growing from a dilute solution such that the interface is atomically sharp. This might occur for graphite growing from iron melts or for silicon growing from aluminum-silicon melts.

Although the mean value of the kinetic undercooling is often negligible, the anisotropy of the faceting factor predicted by Eq 3 and the related anisotropy of the kinetics of the growth of metallic crystals from the melt are observable. For a given driving force, the growth velocity of the interfaces parallel to atomically dense planes is smaller than any other growth velocity. For an fcc crystal, for example, the dense {111} planes will tend to spread laterally more or less rapidly and form pyramids that have fourfold symmetry axes, that is, ⟨100⟩ axes. The vertex of each pyramid will then move along a ⟨100⟩ axis during growth. Therefore, the well-known fact that dendrites of cubic metals exhibit ⟨100⟩ axes and branches is probably related to their slight growth anisotropies.

Departure from interface equilibrium during growth results not only in a kinetic undercooling but also a dependence of the partition coefficient on velocity. The following functional form has been proposed to describe the effect of velocity on k_v where $k < 1$ (Ref 6):

$$k_v = \frac{C_S^*}{C_L^*} = \frac{k + (a_o/D_L) \cdot v}{1 + (a_o/D_L) \cdot v} \quad \text{(Eq 9)}$$

where a_o is a length scale related to the interatomic distance; its value is estimated to be 0.5 to 5.0 nm (5 to 50 Å). Figure 2 shows how the partition coefficient changes monotonically from its equilibrium value to unity as growth velocity increases according to Eq 9.

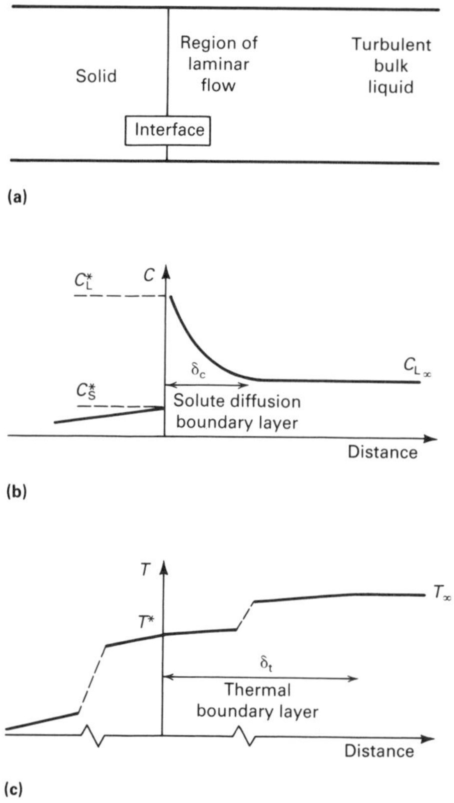

Fig. 3 Constrained growth during plane front solidification (a) and schematic profiles of solute concentration (b) and temperature (c)

Mass and Heat Transport

Basic Phenomena. Solidification typically involves heat and mass transport phenomena on the micro- and macroscopic levels. Because this Section is devoted to the formation of microstructures in cast metals, this discussion will focus on the problems of microscopic heat and mass transfer near the interface.

From the viewpoint of mass transfer, it is useful to distinguish among the zones of plane front solidification illustrated in Fig. 3(a) and (b):

- The solid, where only diffusion is effective, although it is usually very slow
- The interface, where heterogeneous chemical reactions occur
- A boundary layer of thickness δ_i in the liquid, where diffusion is the only mechanism effective for solute i to move perpendicular to the interface
- The bulk liquid, where convection is also effective

For metallic alloys, solid and liquid can usually be assumed to be in mutual equilibrium locally across the interface, except for rapid solidification (Fig. 1 and 2). Therefore, for a given interface temperature, the

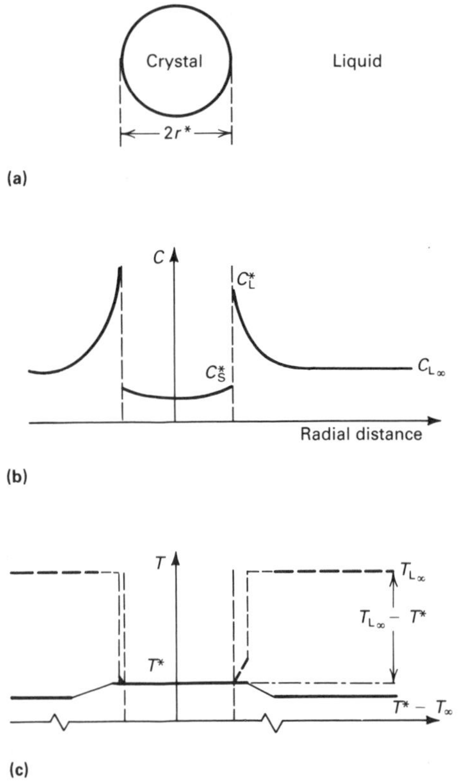

Fig. 4 Free growth of a spherical crystal (a) and schematic profiles of solute concentration (b) and both liquidus temperature T_L (dashed line) and actual temperature T (full line) (c)

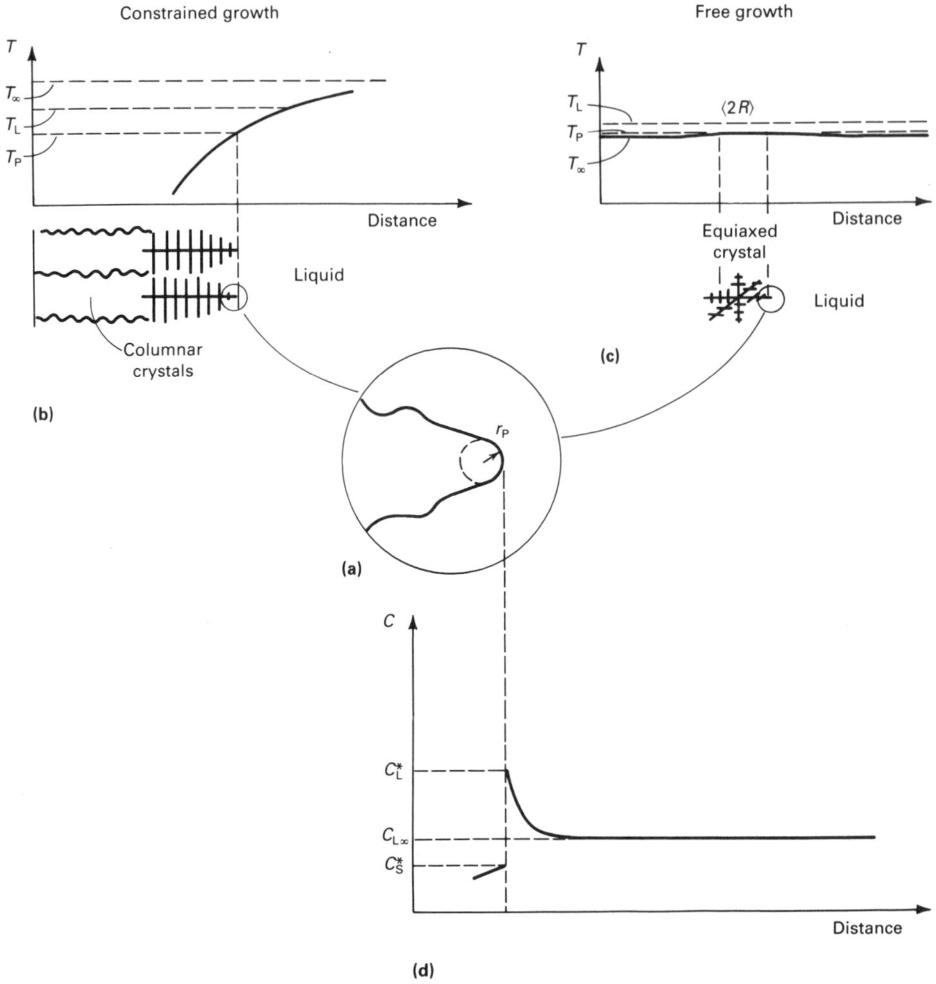

Fig. 5 Growth of a needle (a) and temperature profiles of constrained growth (b) and free growth (c). A profile of solute concentration in both cases is shown in (d).

partition of solute between liquid and solid will be calculated by taking into account the equilibrium phase diagram and the curvature effect (Eq 7).

In the case of plane front solidification, the thickness of the diffusion boundary layer, when controlled by turbulent convection, can be estimated as follows (in SI units):

$$\delta_i = 5.6 \times 10^{-3} \cdot L^{0.1} \cdot \left(\frac{\eta}{\rho_L}\right)^{0.57} \cdot D_{L,i}^{0.33} \cdot U^{-0.9} \quad \text{(Eq 10)}$$

where L is a characteristic length for the convective flow parallel to the interface, η is the dynamic viscosity of liquid, ρ_L is the density of liquid, and U is the mean liquid velocity parallel to the interface.

When the thickness δ_i is large enough, the solute builds up ahead of the interface (if $k < 1$), while a uniform level C_{i_∞} of the solute exists at a sufficiently large distance from the interface. The thickness of the region where solute i is built up is scaled by the characteristic diffusion length $l_i = D_{L,i}/v$ when smaller than δ_i.

From the viewpoint of heat transfer, one can discern three separate zones: the solid, the thermal boundary layer in the liquid, and the bulk liquid. The thickness of the thermal boundary layer δ_t can be calculated as for δ_i by substituting α_L for D_L in Eq 10. Because thermal diffusivities α are much larger than chemical diffusion coefficients for metals, the thermal boundary layer is always thicker (by a factor of 10) than the diffusion boundary layer.

When the thickness δ_t is large enough, the thickness of the region close to the interface where the release of latent heat is detectable can be measured by the characteristic conduction length $l_t = \alpha/v$. Then, because of the large difference in orders of magnitude between D_L and α_L, the temperature profile approaches linear characteristics (Fig. 3c), while the composition profile approaches an exponential curve (Fig. 3b) when both are plotted against the distance from the interface.

Regardless of the shape of the interface, the rejection of the solute always triggers changes in the liquid composition and the interface temperature. Therefore, it is useful to define the dimensionless solutal supersaturation at the interface Ω_c as the ratio of the change in liquid concentration at the interface $C_L^* - C_{L_\infty}$ to the equilibrium concentration difference $C_L^* - C_S^*$ at the temperature of the interface (Ref 7):

$$\Omega_c = \frac{C_L^* - C_{L_\infty}}{C_L^* - C_S^*} \quad \text{(Eq 11)}$$

It is instructive to compare supersaturations in various growth conditions.

Plane Front Solidification. In terms of heat transfer, plane front solidification requires a high positive temperature gradient. The case in which heat flow is opposite to the direction of growth is often referred to as constrained growth.

When the bulk liquid is at rest, the thickness of the diffusion boundary layer is infinite. Moreover, when a stationary state is achieved, the solid formed approaches the composition of the liquid far ahead of the interface. Then, supersaturation simply equals unity:

$$\Omega_c = 1 \quad \text{(Eq 12)}$$

When the bulk liquid is stirred, the thickness of the diffusion boundary layer δ_c can be of the order of the diffusion length, l_c (Fig. 3). In this case, the solute content of the solid formed is lower than that of bulk liquid. It is then useful to define an effective partition coefficient k_{ef}:

$$k_{ef} = \frac{C_S^*}{C_{L_\infty}} \quad \text{(Eq 13)}$$

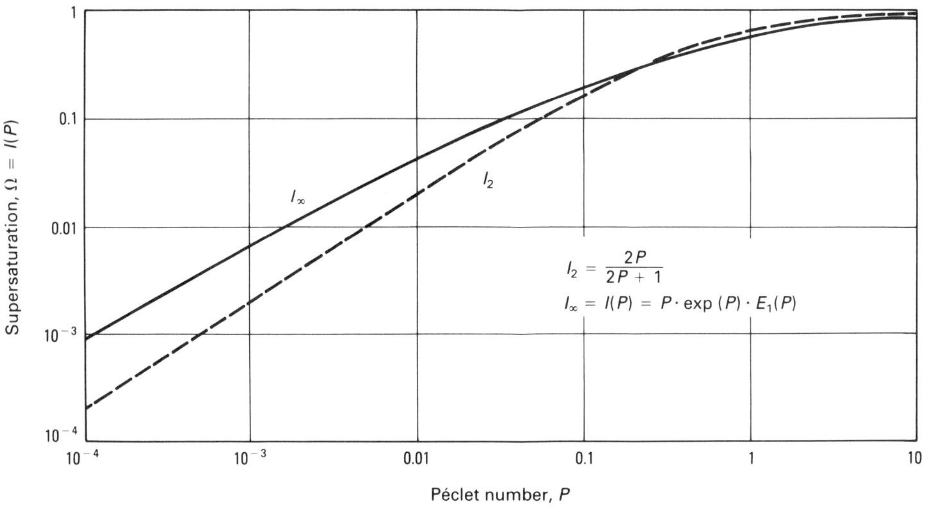

Fig. 6 Analytical approximation I_2 for $\Omega = I(P)$ according to Eq 19b. I_∞ is the exact solution. Source: Ref 7

The supersaturation becomes:

$$\Omega_c = \frac{1 - (k/k_{ef})}{1 - k} \qquad \text{(Eq 14)}$$

When a quasi-stationary state is achieved, the effective partition coefficient is related to the solidification and stirring conditions (Ref 8):

$$k_{ef} = \frac{k}{k + (1 - k) \cdot \exp(-\delta_c/l_c)} \qquad \text{(Eq 15)}$$

It changes monotonically from the equilibrium value k to unity as the ratio $\sigma_c l_c$ (or the product $v \cdot \delta_c$) increases from 0 to infinity. At the same time, supersaturation increases from 0 to unity. Therefore, stirring can greatly decrease supersaturation by reducing δ_c, and it increases segregation in the solid.

Growth of a Sphere. Figure 4 shows a schematic illustration of a crystal growing from a melt. This situation, often referred to as free growth, requires the crystal to be hotter than the surrounding liquid and the radial heat flux to have the same direction as that of the growth.

If a quasi-stationary growth could be achieved under such conditions, solute mass transfer by liquid diffusion would only impose the following supersaturation:

$$\Omega_c = 2 \cdot P_c \qquad \text{(Eq 16)}$$

where P_c is the solute Péclet number associated with the spherical crystal of radius r^*:

$$P_c = \frac{r^* \cdot dr^*/dt}{2 \cdot D_L} \qquad \text{(Eq 17)}$$

Moreover, heat balance around a crystal that grows quasi-steadily makes the Péclet number P proportional to the actual undercooling of the bulk liquid $T_{L_\infty} - T_\infty$, provided $(1 - k) \cdot (T_f - T_\infty)$ is not too small:

$$P_c \simeq \frac{1}{2 \cdot (1 - k)} \cdot \frac{T_{L_\infty} - T_\infty}{T_f - T_\infty} \qquad \text{(Eq 18)}$$

where T_{L_∞} is the liquidus temperature related to the bulk liquid concentration C_{L_∞}, and T_∞ is the actual temperature of the bulk liquid. Thus, quasi-stationary growth of a spherical crystal, if achievable, should proceed with a constant solute supersaturation that should be proportional to the actual undercooling of the bulk liquid.

Growth of a Needle. The situations described previously are either difficult to achieve or transient. Plane front solidification is possible only for low imposed velocities, and the nondendritic growth of spherical crystals is possible only for low undercoolings and small radii. Otherwise, the interface will tend to form needles that are often described as paraboloids of revolution (Fig. 5).

The mathematical solution of the diffusion problem for a steady-state growing paraboloid was derived by Ivantsov, who deduced the following relationship among supersaturation, the paraboloid tip radius r_P, and the growth velocity (Ref 9):

$$\Omega_c = I(P_c) \qquad \text{(Eq 19a)}$$

where $I(P) = P \cdot \exp(P) \cdot E_1(P)$ is the Ivantsov function, $E_1(P) = \int (e^{-u}/u)du$ is the exponential integral function, and $P_c = v \cdot r_P/2D_L$ is the Péclet number associated with the moving tip of the paraboloid.

When simple analytical expressions are needed, the following estimate is useful in the range $P_c > 0.1$ (Fig. 6) (Ref 10):

$$\Omega_c \simeq \frac{2 \cdot P_c}{2 \cdot P_c + 1} = I_2 \qquad \text{(Eq 19b)}$$

For a given alloy composition and for a given growth velocity, the smaller the tip radius, the smaller the solute supersaturation at the tip of the needle. The needle can be either hotter or colder than its liquid surroundings depending on whether the growth is free or constrained (Fig. 5).

REFERENCES

1. D. Turnbull, The Liquid State and the Liquid-Solid Transition, *Trans. AIME*, Vol 221, 1961, p 422
2. W.K. Burton and N. Cabrera, Crystal Growth and Surface Structure, *Dis. Faraday Soc.*, Vol 5, 1949, p 33
3. A. Passerone, N. Eustathopoulos, J.C. Joud, and P. Desre, Equilibrium Atomic Roughness at Solid-Liquid Interfaces of Pure Metals, *Mater. Chem.*, Vol 1, 1976, p 45
4. R.H. McSwain, C.E. Bates, and W.D. Scott, Iron-Graphite Surface Phenomena and Their Effects on Iron Solidification, *Trans. AFS*, Vol 82, 1974, p 85
5. W.B. Hillig and D. Turnbull, The Theory of Crystal Growth in Undercooled Pure Liquids, *J. Chem. Phys.*, Vol 24, 1956, p 914
6. M.J. Aziz, Model for Solute Redistribution During Rapid Solidification, *J. Appl. Phys.*, Vol 53 (No. 2), 1982, p 1158
7. W. Kurz and D.J. Fisher, *Fundamentals of Solidification*, Trans. Tech., 1986
8. J.A. Burton, R.C. Prim, and W.P. Slichter, The Distribution of Solute in Crystals Grown From the Melt, *J. Chem. Phys.*, Vol 21 (No. 11), 1953, p 1987
9. G.P. Ivantsov, Thermal and Diffusion Processes in Crystal Growth, *Dokl. Akad. Nauk SSSR*, Vol 58, 1947, p 567
10. M. Hillert, The Role of Interfacial Energy During Solid State Phase Transformations, *Jernkontorets Ann.*, Vol 141, 1957, p 757

Solidification of Single-Phase Alloys

R. Trivedi, Iowa State University
W. Kurz, Professor, Swiss Federal Institute of Technology, Switzerland

The solidification process by which a liquid metal freezes in a mold plays a critical role in determining the properties of the as-cast alloy. Even when the final object is obtained by the mechanical forming of ingots, the solidification structures of ingots often influence the properties of the object. The influence of the solidification process on properties arises primarily because of the following effects:

- The initial uniform composition in liquid becomes nonuniform as the liquid transforms to solid
- Different solidification conditions give rise to different microstructures of the solid
- Many casting defects, such as porosity and shrinkage, depend on the manner in which the alloy is solidified in a mold

Two important factors that control solidification microstructures are the composition of the alloy and the heat flow conditions in the mold. These two factors will be described first, and their influence on the microstructure and the accompanying solute segregation profiles will then be discussed in this section. The solidification structures of ferrous, nonferrous, and superalloy single-crystal single-phase alloys are discussed in the other articles in this Section.

Alloy Composition

An alloy consists of a base metal to which other elements are added to give the desired properties. In this discussion, only binary alloys that solidify into a single-phase structure will be considered. When an element is added to the base metal, it significantly alters the solidification process. A pure metal has a specific melting point T_m, while an alloy freezes over a range of temperatures. This freezing range is generally represented by a phase diagram, as shown in Fig. 1. The liquidus line represents the temperature at which the liquid alloy begins to freeze, and the freezing process is complete when the solidus temperature is reached, if the solidification occurs close to equilibrium conditions or below the solidus under nonequilibrium conditions.

When an alloy of uniform liquid composition C_o is cooled, it begins to solidify when the temperature of the liquid reaches the liquidus temperature T_L, if nucleation occurs readily. The composition of the solid that forms at T_L will be different from the composition of the liquid, and it is given by the composition on the solidus line at temperature T_L, as shown in Fig. 1. The ratio of the solid to liquid composition at a given temperature is called the solute distribution coefficient k. For dilute solutions, the solidus and the liquidus lines are generally assumed to be straight lines in which case k is a constant that does not depend on temperature. Figure 1 shows a phase diagram in which $k < 1$.

The first solid that forms at temperature T_L will have a composition kC_o, which is lower than the liquid composition C_o. Thus, the excess solute rejected by the solid will give rise to a solute-rich liquid layer at the interface. As the alloy is cooled further, the liquid composition increases. This increase in liquid composition, along with the lowering of temperature, gives rise to solute segregation patterns in the solid if the diffusion of solute in the solid is not very rapid (Ref 1).

The buildup of solute in liquid requires diffusion of solute in liquid for further growth. For efficient distribution of the solute in liquid, the interface may change its shape. Thus, the actual solute segregation pattern is dictated by the shape of the interface. In addition to the solute transfer, the interface shape is governed by the effective removal of the latent heat of fusion. This heat flow problem will be described below.

Heat Flow Conditions

The thermal field in a casting is very important in determining the microstructure of the cast alloy. Two distinctly different heat flow conditions may exist in a mold.

In the first case, the temperature gradients in the liquid and the solid are positive such that the latent heat generated at the interface is dissipated through the solid.

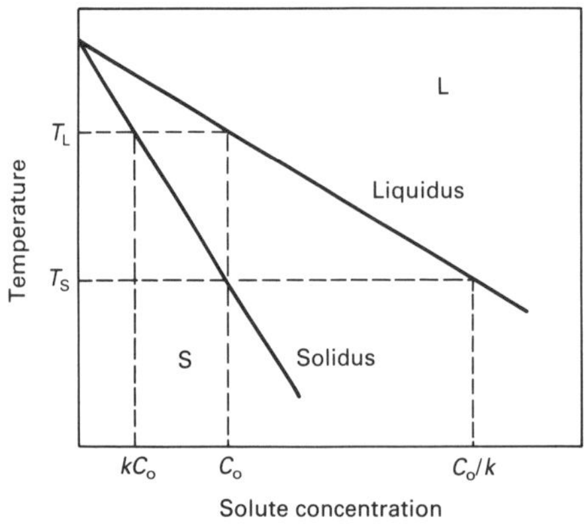

Fig. 1 A single-phase region of a phase diagram showing the liquidus and the solidus lines

Such a temperature field gives rise to directional solidification and results in the columnar zone in a casting.

In the second case, an equiaxed zone exists if the liquid surrounding the solid is undercooled so that a negative temperature gradient is present in the liquid at the solid/liquid interface. In this case, the latent heat of fusion is dissipated through the liquid. Such a thermal condition is generally present at the center of the mold.

A positive temperature gradient in the liquid at the interface gives rise to a planar solid/liquid interface for pure metals. However, for alloys, the shape of the interface is dictated by the relative magnitudes of the interface velocity and the temperature gradients in the solid and liquid at the interface. For given temperature gradients and composition, four different interface morphologies can exist, depending on the velocity (Fig. 2). Below some critical velocity v_{cr}, a planar solid/liquid interface will be present. However, above v_{cr}, the planar interface becomes unstable and forms a cellular, a cellular dendritic, or a dendritic interface. For the heat flow condition in which the solid grows in an undercooled melt, a dendritic structure is present, as shown in Fig. 3. Such equiaxed dendrites form for pure metals as well as for alloys.

Interface Velocity Below Critical Velocity

Planar interface growth occurs only under directional solidification conditions and, for alloys, only under low growth rate or high-temperature gradient conditions. To describe quantitatively the condition under which a planar interface growth can occur, consider an interface that is moving at a constant velocity v, with heat flowing from the liquid to the solid under temperature gradients G_L and G_S in liquid and solid, respectively, at the interface. Under steady-state growth conditions, the interface temperature corresponds to the solidus temperature T_S in Fig. 1. At this temperature, the interface composition in liquid is C_o/k, which is larger (for $k < 1$) than the liquid composition C_o far from the interface so that a solute-rich layer exists in the liquid ahead of the interface (Fig. 4). This liquid composition profile gives rise to a variation in the liquidus temperature with distance, as indicated by T_f in Fig. 4. If the actual liquid temperature lies below the liquidus profile, a region of supercooled liquid is present ahead of the interface. This supercooled region is indicated in Fig. 4.

The stability of a planar interface can be determined by examining whether any small bump on the interface will grow or decay. If the supercooling ahead of the interface increases with distance, then any small bump will see a larger supercooling and may grow faster, which will make the planar interface unstable. To avoid supercooling ahead of

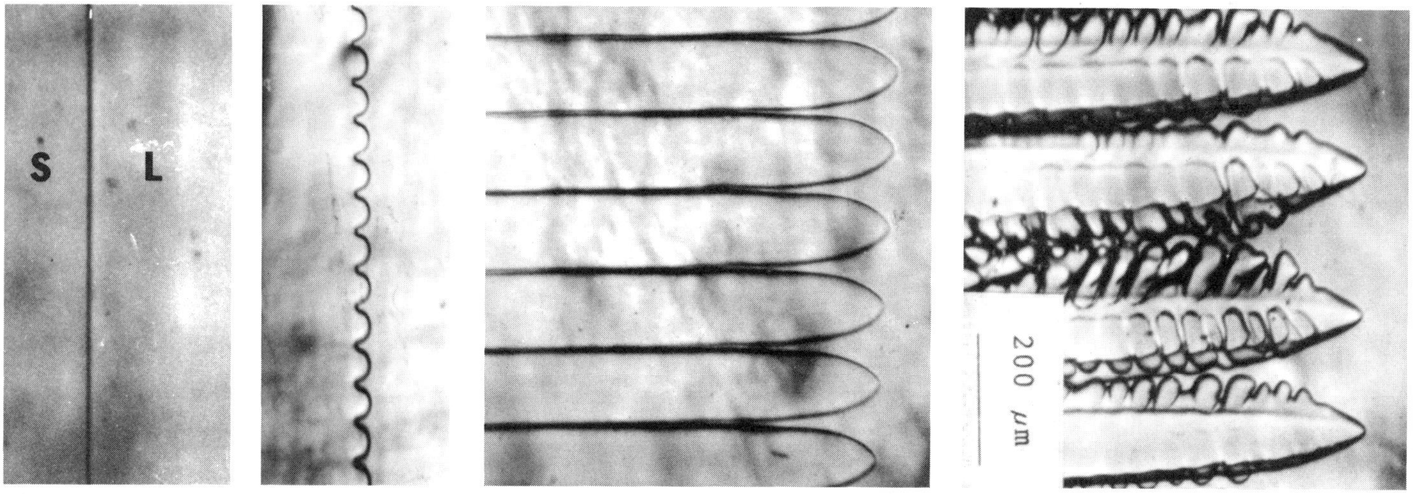

(a) (b) (c) (d)

Fig. 2 Effect of increasing growth rate on the shape of the solid/liquid interface in a transparent organic system, pivalic acid-0.076 wt% ethanol, solidified directionally at G = 2.98 K/mm (75.7 K/in.). (a) v = 0.2 μm/s (8 μin./s). (b) v = 1.0 μm/s (40 μin./s). (c) v = 3.0 μm/s (120 μin./s). (d) v = 7.0 μm/s (280 μin./s)

the interface, the actual temperature gradient in the liquid (line 2, Fig. 4) must be equal to or larger than the gradient of the liquidus profile at the interface (line 1, Fig. 4). This condition for the planar interface stability, known as the constitutional supercooling criterion (Ref 2), is given by:

$$G_L \geq \frac{v \cdot \Delta T_o}{D} \quad \text{(Eq 1)}$$

Fig. 3 Formation of equiaxed crystals at the center of the mold during the solidification of transparent ammonium chloride-water mixture

where D is the diffusion coefficient of solute in liquid and ΔT_o is the freezing range of the alloy, that is, $\Delta T_o = T_L - T_S$ in Fig. 1.

The above constitutional supercooling criterion does not take into account the effect of the temperature gradient in the solid. It also neglects the effect of interfacial energy, which may be significant because the formation of a bump is accompanied by an increase in the interfacial area. A more detailed model of the planar interface stability is given by a linear stability analysis (Ref 3). In such a model, a planar interface is perturbed infinitesimally, as shown in Fig. 5, and the change in amplitude of the perturbation with time is examined. If the amplitude decreases with time, the planar interface is stable. Thus, for the planar interface stability, the velocity at point A must be smaller than the velocity at point B. Such an analysis shows that a planar interface will be stable below a critical velocity v_{cr} and above a certain velocity v_a, where v_a is known as the absolute velocity for the planar interface stability.

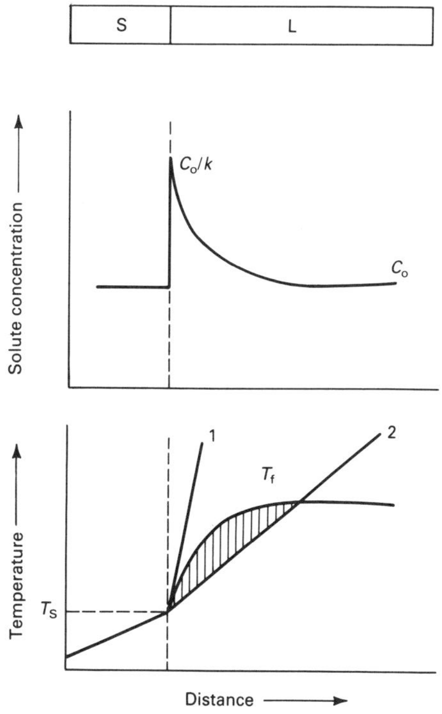

Fig. 4 Constitutional supercooling diagram. The solute concentration profile in the liquid gives rise to the variation in the equilibrium freezing temperature T_f of liquid near the interface. The actual temperature in liquid is given by line 1, and the slope of T_f at the interface is given by line 2. A supercooled liquid exists in the shaded region.

The velocity v_a is given by:

$$v_a = \frac{D \cdot \Delta T_o}{\Gamma \cdot k} \quad \text{(Eq 2)}$$

where $\Gamma = \gamma/\Delta S$ is the capillarity constant in which γ is the interfacial energy and ΔS is the entropy of fusion per unit volume. For typical metallic systems with $D \simeq 10^{-9}$ m^2/s, ΔT_o = 5 K, $\Gamma = 10^{-7}$ K · m, and k = 0.2, one obtains v_a = 0.25 m/s. This velocity is large, and it can be obtained by the laser or electron beam scanning technique (Ref 4). In casting, the velocity is significantly smaller than v_a, so that the important planar interface stability criterion of interest is $v < v_{cr}$. The linear stability analysis gives v_{cr} as:

$$v_{cr} = \frac{G \cdot D}{\Delta T_o} + f(\Gamma, G, \Delta T_o, k) \quad \text{(Eq 3)}$$

The second term on the right-hand side of Eq 3 is generally less than 10% of the first term, so that the planar interface growth condition can be approximated as:

$$v < \frac{G \cdot D}{\Delta T_o} \quad \text{(Eq 4)}$$

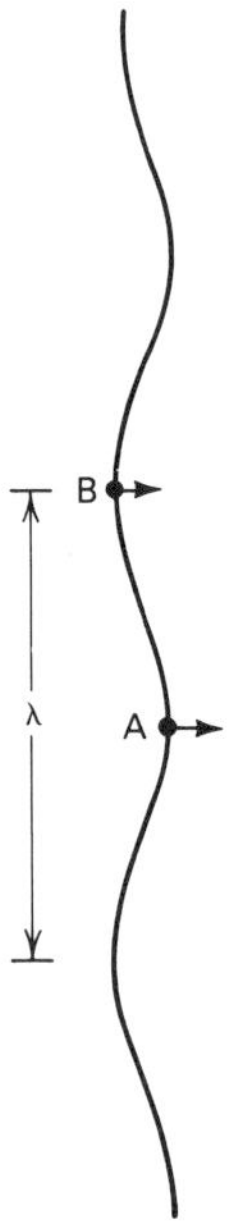

Fig. 5 Perturbed shape of the interface with wavelength λ. The linear stability analysis determines the condition when the velocity of point A is less than that of point B.

where G is the conductivity-weighted average temperature gradient at the interface:

$$G = \frac{K_L \cdot G_L + K_S \cdot G_S}{K_L + K_S} \quad \text{(Eq 5)}$$

where K_S and K_L are the thermal conductivities of solid and liquid, respectively.

This stability criterion is similar to the constitutional supercooling criterion except that the proper temperature gradient is G rather than G_L. Under most casting conditions, G_L is very small compared to G_S. In addition, for metallic systems, $K_S > K_L$. Thus, the stability analysis, not the constitutional supercooling criterion, should be used to obtain the appropriate planar stability criterion for cast microstructures. For most metallic systems, the diffusion coefficient is about 10^{-9} m²/s. Thus for $\Delta T_o = 5$ K and $G = 10^4$ K/m, the planar interface growth will occur only when the velocity is less than $2 \cdot 10^{-6}$ m/s. Because typical velocities in castings are larger than this value, most practical metallurgical alloys rarely solidify with a planar interface in a mold.

Interface Velocity Exceeding Critical Velocity

Cellular and Cellular Dendritic Structures. Under directional solidification conditions, a cellular or a cellular dendritic interface is observed when the interface velocity exceeds the critical velocity for the planar interface growth. For velocities just above v_{cr}, the cellular structures that form have two important characteristics. First, the length of the cell is small, and it is of the same order of magnitude as the cell spacing (Fig. 2b). Second, the tip region of the cell is broader, and the cell has a larger tip radius. At higher velocities, a cellular dendritic structure forms (Fig. 2c) in which the length of the cell is much larger than the cell spacing. Also, the cell tip assumes a sharper, nearly parabolic shape, which is similar to the dendrite tip shape so that the term cellular dendritic is used to characterize this structure (Ref 5-7).

The formation of cellular structures gives rise to solute segregation in the solid. The tip of the cell is at a higher temperature than the base of the cell. Thus, for $k < 1$, the solid that forms at the cell tip will have a lower composition than the solid that forms at the cell base. This microsegregation profile is approximately characterized by the normal freeze, or Scheil, equation:

$$C_S = k \cdot C_o (1 - f_S)^{k-1} \quad \text{(Eq 6)}$$

where f_S is the volume fraction of solid, which is 0 at the cell tip and 1 at the cell base. Equation 6 is derived under the assumptions that k is constant and that the composition of liquid is uniform in a small-volume element in the direction perpendicular to the growth direction. Equation 6 also assumes that the diffusion in the solid is negligible, so that it predicts C_S to be infinity at the base of the cell.

Equation 6 is useful for nonequilibrium solidification when the phase diagram shows the presence of a higher-composition second phase that can nucleate in the intercellular region. For example, for systems with eutectic phase diagrams, the maximum composition in the single phase corresponds to kC_E, where C_E is the eutectic composition. Thus, once this composition is achieved, the intercellular liquid will freeze with a eutectic structure, as shown in Fig. 6. For this case, Eq 6 can be used to predict the volume fraction of the eutectic phase f_E:

$$f_E = \left(\frac{C_E}{C_o}\right)^{1/(k-1)} \quad \text{(Eq 7)}$$

For single-phase solidification in which the diffusion in the solid is important, it is preferable to describe microsegregation by the segregation ratio (SR). The segregation ratio is defined as the ratio of the maximum solid composition (at the cell base) to the minimum solid composition (at the cell tip). If ℓ is the length of the cell, G_M the average temperature gradient in the two-phase region, and ΔT the cell tip undercooling, then:

$$SR = 1 + \frac{\ell \cdot G_M}{\Delta T} \quad \text{(Eq 8)}$$

As the cellular structure becomes cellular dendritic or dendritic, ℓ increases sharply and ΔT decreases under normal solidification conditions. Thus, the segregation ratio will increase significantly upon the formation of cellular dendrites or dendrites.

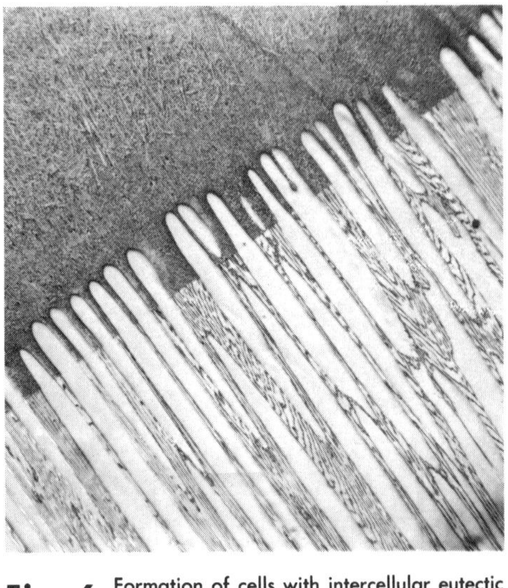

Fig. 6 Formation of cells with intercellular eutectic in the directionally solidified Sn-20Pb alloy. $G = 31$ K/mm (79 K/in.) and $v = 1.2$ μm/s (48 μin./s). The nearly flat eutectic interface is at the eutectic temperature.

Figure 2(b) shows that all cells have the same orientation; therefore, once the base of each individual cell solidifies, a single grain is obtained. This single grain, however, will have a microsegregation pattern that reflects the periodicity of the cells. This cell spacing λ is important because the time required to homogenize the solid depends on λ and is given by (Ref 8):

$$t = \frac{0.47 \cdot \lambda^2}{D_S} \quad \text{(Eq 9)}$$

where t is the time required to homogenize to 1% of the original composition difference and D_S is the diffusion coefficient in the solid. The spacing λ decreases with velocity for both the cellular (Ref 9) and the cellular dendritic structures (Ref 6). However, λ increases sharply at the cellular-to-cellular-dendritic transition, as shown in Fig. 7.

Cellular structures are observed in castings only for heat flow conditions that produce directional solidification and for alloys with very small freezing ranges. Thus, cellular structures are important for very dilute alloys or for alloys that are close to the eutectic composition.

Dendritic Structures. A dendritic structure is formed when the interface velocity is increased beyond the cellular dendritic regime. Dendritic structures are characterized by the formation of sidebranches (Fig. 2d). These sidebranches, as well as the primary dendrite, grow in a preferred crystallographic direction, for example, ⟨100⟩ for cubic metals, so that cubic metals exhibit fourfold sidebranches. A three-dimensional view of dendrites in metals is difficult to observe because only parts of dendrites that intersect the plane of polish are visible. Figure 8 shows a three-dimensional view of cobalt dendrites in a cobalt-samarium-

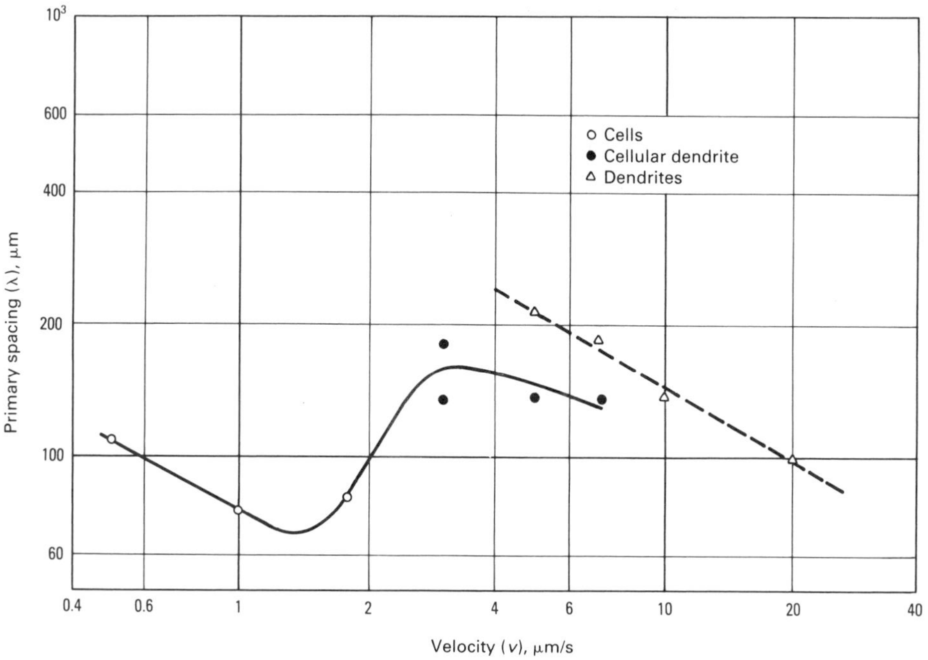

Fig. 7 The effect of velocity on cellular, cellular dendrite, and primary dendrite spacings in the pivalic acid-ethanol system. Source: Ref 6

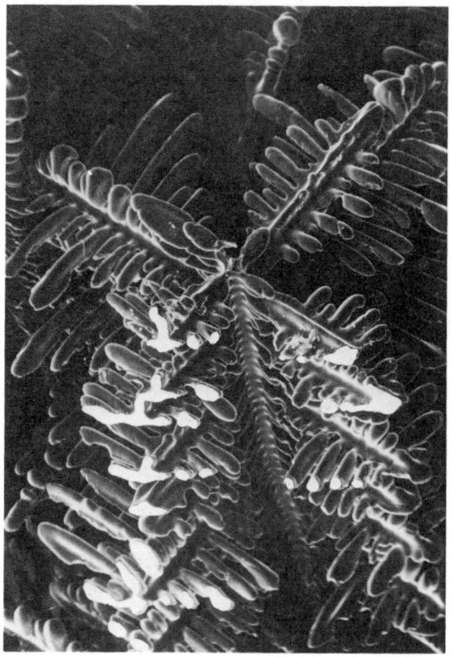

Fig. 8 Directionally solidified peritectic cobalt-samarium-copper alloy showing primary cobalt dendrites when the $Co_{17}Sm_2$ matrix is etched away. The cut surfaces in the foreground indicate the structure that would be observed on the plane of polish if the matrix were not etched away.

copper alloy in which the matrix is etched away. The cut surfaces in the foreground are those that are typically observed in a polished section of a solid alloy.

The formation of secondary dendrite arms is clearly seen for a dendritic structure in a transparent alloy (Fig. 9). The secondary arms form very close to the dendrite tip, and the first few sidebranches are uniformly spaced. However, the secondary arm spacing increases as the base of the dendrite is approached (Ref 10). Initial coarsening occurs by the competition in the growth process among secondary arms. However, once the diffusion fields of their tips interact with those of the neighboring dendrite, the growth of the secondary arms is reduced, and a coarsening process to reduce interfacial energy begins (Ref 10). The final secondary arm spacing near the dendrite base is significantly larger than that near the dendrite tip. This final secondary arm spacing controls the microsegregation profile in the solidified alloy. This microsegregation pattern is analogous to that discussed for the cellular structure, except that the periodicity of segregation is controlled by the final secondary arm spacing and not by the primary spacing.

Because the secondary arm coarsening requires solute diffusion, the coarsening process is negligible once the interdendritic liquid has solidified. Thus, the final value of secondary spacing, λ_2, is determined by the total time that a given secondary branch spends in contact with the liquid because the diffusion coefficient of the solute is significantly larger in the liquid than in the solid. This total time spent by a secondary branch in the two-phase region is known as the local solidification time, t_f. For directional solidification, $t_f = L/v$, where L is the length of the primary dendrite and v is the imposed velocity. The following relationship between λ_2 and t_f is predicted:

$$\lambda_2 = a t_f^{1/3} \qquad \text{(Eq 10)}$$

where a is a constant that depends on system parameters such as the diffusion coefficient in liquid, solid-liquid interfacial energy, and the equilibrium freezing range of the alloy (Ref 1, 7).

Another important microstructural parameter for dendritic structures is the length of the dendrite, which is given by the relationship:

$$L = \frac{T_L - T_b}{G_M} \qquad \text{(Eq 11)}$$

where T_L is the liquidus temperature of the alloy and T_b is the base temperature of the dendrite. Under most casting conditions, the dendrite tip undercooling is quite small, so that the tip temperature has been approximated as T_L.

For alloy systems that exhibit a eutectic phase transformation, a eutectic structure will form at the base of the dendrite, as shown in Fig. 10. Thus, in eutectic systems, with negligible solid diffusion, $T_b = T_E$. For dilute alloys, in which appreciable solid diffusion occurs, a single phase is observed in which T_b is between the solidus and the eutectic temperatures. For this case, a lower limit on dendrite length can be estimated by taking $T_b = T_S$, as has been verified in the iron-carbon system for low carbon steels. Thus, the dendrite length will be directly proportional to the equilibrium freezing range and inversely proportional to the average temperature gradient in the two-phase region. For a large freezing range alloy, with ΔT_o = 50 K, a temperature gradient of 0.5 K/mm (13 K/in.) will give a dendrite that is 100 mm (3.9 in.) long. Such long dendrites are more prone to breakup

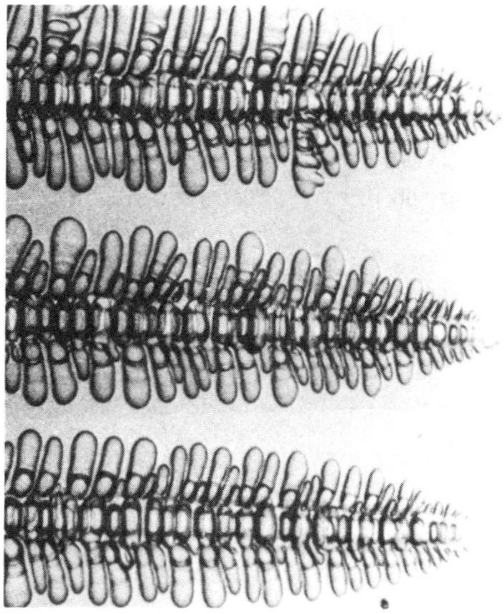

Fig. 9 Dendritic structure in a directionally solidified transparent organic system, succinonitrile-4.0 wt% acetone. G = 6.7 K/mm (170 K/in.) and v = 6.4 μm/s (260 μin./s). The secondary dendrite arm spacing increases with the distance behind the tip.

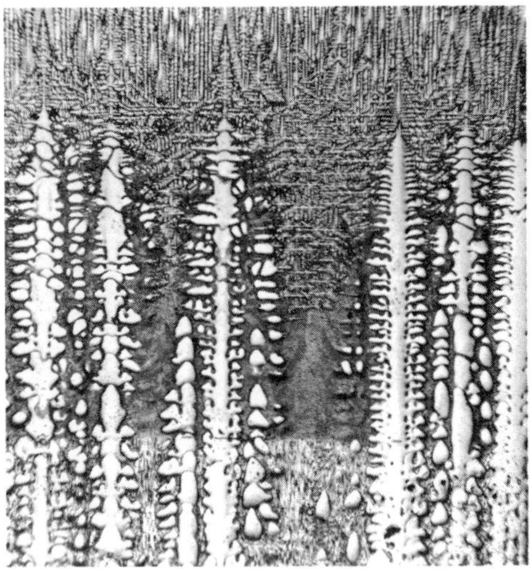

Fig. 10 Longitudinal section of Sn-20Pb alloy, directionally solidified at v = 11.8 μm/s (472 μin./s) under G = 31 K/mm (790 K/in.). A eutectic interface is observed between the dendrites.

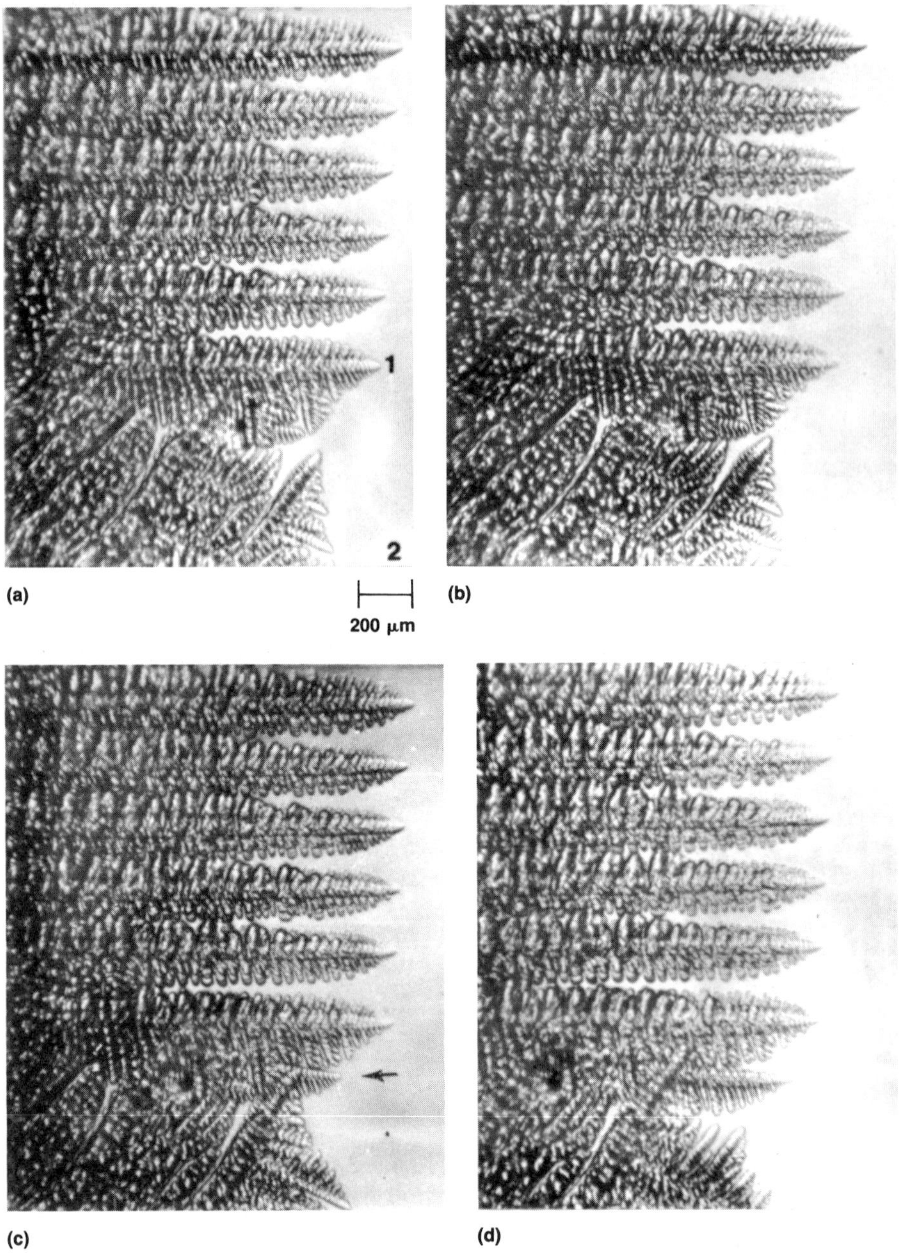

Fig. 12 Mechanism of chill to columnar transition. Region 1, in which dendrites are favorably oriented, will expand by converting a tertiary branch into a new primary dendrite. Such expansion continues until region 2 is eliminated. All dendrites in region 1, after solidification, become a single grain.

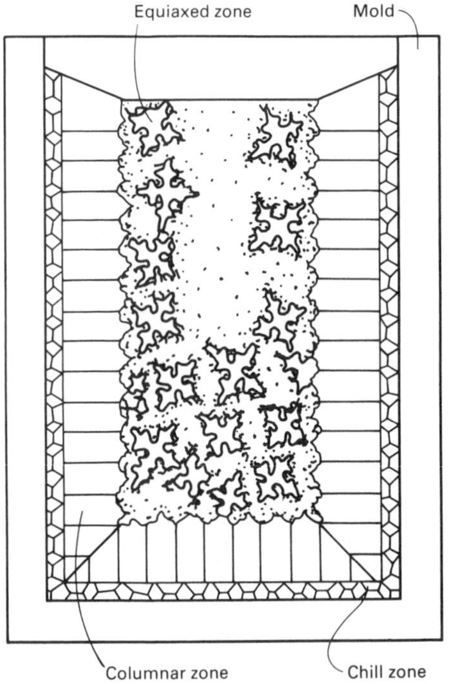

Fig. 11 Schematic of microstructure zone formation in castings. Directional solidification conditions give rise to a columnar zone, while an equiaxed zone is formed at the center where the liquid is undercooled.

during growth and can influence the final microstructure of the casting.

Most practical metallurgical alloys solidify with a dendritic structure in a mold. Both columnar and equiaxed dendrites may be present in a casting (Fig. 11). As the liquid metal is poured into the mold, solid nuclei appear at the mold wall that give rise to an equiaxed chilled zone. Some of these crystals are then favorably oriented for growth under directional heat flow conditions. The actual mechanism by which this preferred growth occurs is illustrated in Fig. 12. The dendrites in region 1 are favorably oriented with respect to heat flow compared to those in region 2. As the growth proceeds, region 1 expands by creating new primary dendrites from the tertiary branches of the dendrite at the junction of regions 1 and 2. Region 2 will be eliminated if the dendrites below it are also favorably oriented. Because all of the dendrites in a given region have formed from the same initial crystal, they will give rise to one grain when the solidification is complete. These favorably oriented grains produce a columnar zone in a casting, and they exhibit [100] texture in cubic metals.

For alloy solidification, equiaxed dendritic crystals are observed at the center of the mold (Fig. 11). This phenomenon can be explained in the following way: Because of the convection effects present in the melt, the temperature ahead of the columnar dendrite in the later stages of solidification is nearly constant, and it approaches the dendrite tip temperature. Because the columnar dendritic tip is slightly undercooled, a small undercooling also exists at the central melt region of the casting. Thus, if any crystals or efficient nuclei are present, they can grow in this undercooled liquid, creating equiaxed crystals. Observations in transparent alloys show that the nuclei for the

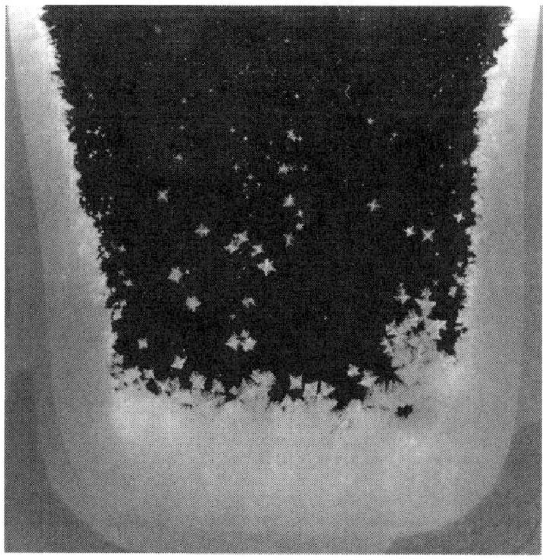

Fig. 13 A model experiment showing the microstructure formation during the freezing of an ammonium chloride-water system in a mold. The broken dendrite branches are transported to the center by convection in the liquid, where they form an equiaxed structure.

equiaxed zone come from the detached dendrite arms that are carried to the center of the mold by convection currents (Fig. 13). The dendrite breakup occurs easily if the dendrites are thin and very long, as in the case of large freezing range alloys. In addition, if long dendrites are present in the columnar zone, the feeding problem becomes critical because the liquid must be transported from the tip region to the base region through a complex secondary branch structure. Consequently, it may not be possible to avoid small shrinkage voids or microporosity in alloys with very large freezing ranges.

REFERENCES

1. W. Kurz and D.J. Fisher, *Fundamentals of Solidification*, Trans. Tech. Publications, 1984
2. W.A. Tiller, K.A. Jackson, J.W. Rutter, and B. Chalmers, The Redistribution of Solute Atoms During the Solidification of Metals, *Acta Metall.*, Vol 1, 1953, p 428
3. W.W. Mullins and R.F. Sekerka, Stability of a Planar Interface During Solidification of a Dilute Binary Alloy, *J. Appl. Phys.*, Vol 35, 1964, p 444
4. W. Kurz and R. Trivedi, Recent Advances in the Modelling of Solidification Microstructures, in *Solidification Processing*, The Metals Society, 1988
5. K. Somboonsuk, J.T. Mason, and R. Trivedi, Interdendritic Spacing: Part I. Experimental Studies, *Metall. Trans. A*, Vol 15A, 1984, p 967
6. M.A. Eshelman, V. Seetharaman, and R. Trivedi, Cellular Spacings—I. Steady State Growth, *Acta Metall.*, Vol 36, 1988
7. M.C. Flemings, *Solidification Processing*, McGraw-Hill, 1974
8. J.D. Verhoeven, *Fundamentals of Physical Metallurgy*, John Wiley & Sons, 1975
9. J.D. Hunt, *Primary Dendrite Spacing in Solidification and Casting of Metals*, Book 192, The Metals Society, 1979
10. S.C. Huang and M.E. Glicksman, Fundamentals of Dendritic Solidification, *Acta Metall.*, Vol 29, 1981, p 701

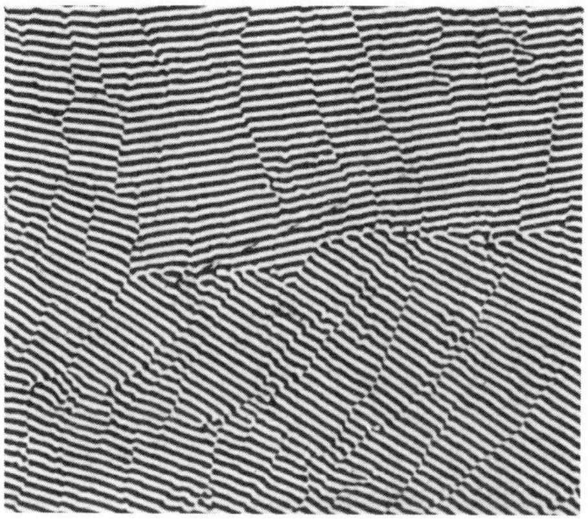

Fig. 1 Example of a lamellar eutectic microstructure ($Al-Al_2Cu$) with approximately equal volume fractions of the phases. Transverse section of a directionally solidified sample. As-polished

Solidification of Eutectics

P. Magnin and W. Kurz, Swiss Federal Institute of Technology, Switzerland

Alloys of eutectic composition make up the bulk of cast metals. The reason for their widespread use can be found in the unique combination of good castability (comparable to that of pure metals), relatively low melting point (minimizing the energy required for production), and interesting behavior as "composite" materials.

Eutectic Morphologies

Eutectic structures are characterized by the simultaneous growth of two or more phases from the liquid. Three or even four phases are sometimes observed growing simultaneously from the melt. However, because most technologically useful eutectic alloys are composed of two phases, only this type will be discussed in this section.

Eutectic alloys exhibit a wide variety of microstructures, which can be classified according to two criteria:

- Lamellar or fibrous morphology of the phases
- Regular or irregular growth

Lamellar and Fibrous Eutectics. When there are approximately equal volume fractions of the phases (nearly symmetrical phase diagram), eutectic alloys generally have a lamellar structure, for example, $Al-Al_2Cu$ (Fig. 1). On the other hand, if one phase is present in a small volume fraction, this phase will in most cases tend to form fibers, for example molybdenum in NiAl-Mo (Fig. 2).

In general, the microstructure obtained will usually be fibrous when the volume fraction of the minor phase is lower than 0.25, and it will be lamellar otherwise. This is because of the small separation of the eutectic phases (typically several microns) and the resulting large interfacial area (of the order of 1 m^2/cm^3) that exists between the two solid phases. The system will therefore tend to minimize its interfacial energy by choosing the morphology that leads to the lowest total interface area. For a given spacing (imposed by growth conditions), the interface area is smaller for fibers than for lamellae at volume fractions below 0.25.

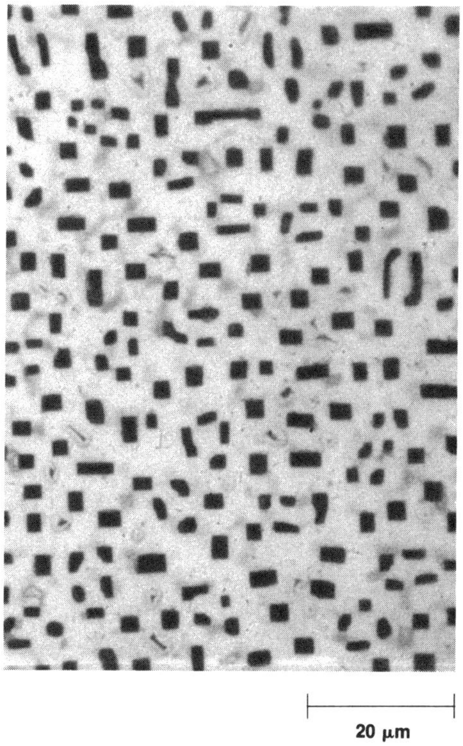

Fig. 2 Example of a fibrous eutectic microstructure with a small volume fraction of one phase (molybdenum fibers in NiAl matrix). Transverse section of a directionally solidified sample. As-polished. Courtesy of E. Blank

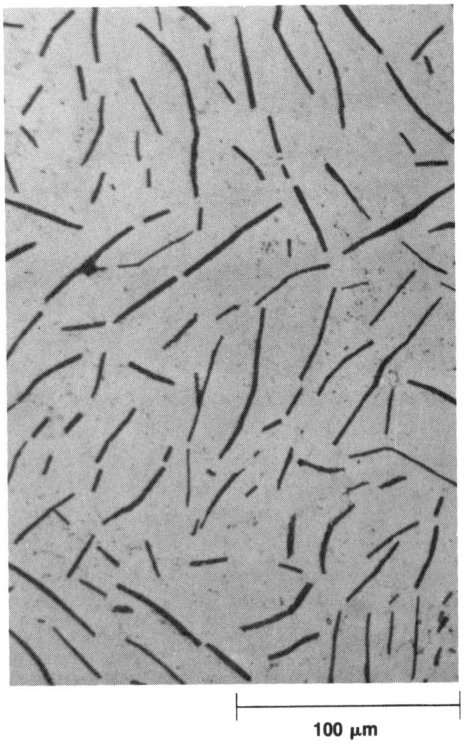

Fig. 3 Microstructure of a gray cast iron showing flake graphite. Transverse section etched with nital

However, when the minor phase is faceted*, a lamellar structure may be formed even at a very low volume fraction, because the interfacial energy is then considerably lower along specific planes, along which the lamellae can be aligned. This is the case in gray cast iron, where the volume fraction of the graphite lamellae is 7.4% (Fig. 3).

Many eutectic microstructures can be classified as lamellar or fibrous, but there is an important exception, namely, spheroidal graphite cast iron (Fig. 4). In this case, there is no cooperative eutecticlike growth of both phases; instead, there is separate growth of spheroidal graphite particles as a primary phase (at least during the initial stages), together with austenite dendrites. This special case of eutectic growth (divorced growth) is discussed further in the article "Ductile Iron" in this Volume. Cast iron often exhibits intermediate microstructural forms, such as vermicular or chunky graphite.

Regular and Irregular Eutectics. If both phases are nonfaceted (usually when both are metallic), the eutectic will exhibit a regular morphology. The microstructure is then made up of lamellae or fibers having a

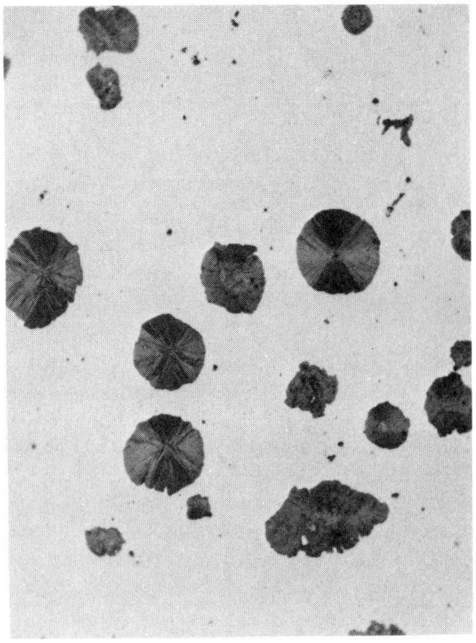

Fig. 4 Graphite in spheroidal cast iron, which results from the divorced growth of the phases. Etched with nital. 130×

high degree of regularity and periodicity, particularly in unidirectionally solidified specimens (Fig. 1).

On the other hand, if one phase is faceted, the eutectic morphology often becomes irregular (Fig. 3 and 5). This is because the faceted phase grows preferentially in a direction determined by specific atomic planes. Because the various faceted lamellae have no common crystal orientation, their growth directions are not parallel, and the formation of a regular microstructure becomes impossible. The two eutectic alloys of greatest practical importance—iron-carbon (cast iron) and aluminum-silicon—belong to this category.

Although the examination of metallographic sections of irregular eutectics seems to reveal many dispersed lamellae of the minor phase, these lamellae are generally interconnected in a complex three-dimensional arrangement. In the foundry literature, such eutectic grains are often referred to as eutectic cells. In the solidification literature, the term cell defines a certain interface morphology (see Fig. 13b in related discussion below); therefore, the term eutectic grain will be used throughout this section.

The regularity of some eutectic microstructures can be used to make *in situ* composites. By using a controlled heat flux to achieve slow directional solidification, it is possible to obtain an aligned microstructure throughout the entire casting. When one of the phases is particularly strong, as in the case of TaC fibers in the Ni-TaC eutectic, the mechanical properties of the alloy can be enhanced in the growth direction. In contrast, an equiaxed microstructure can be formed by inoculation, and there is no long-range orientation.

Fig. 5 Irregular "Chinese script" eutectic consisting of faceted Mg_2Sn phase (dark) in a magnesium matrix. Etched with glycol. 250×

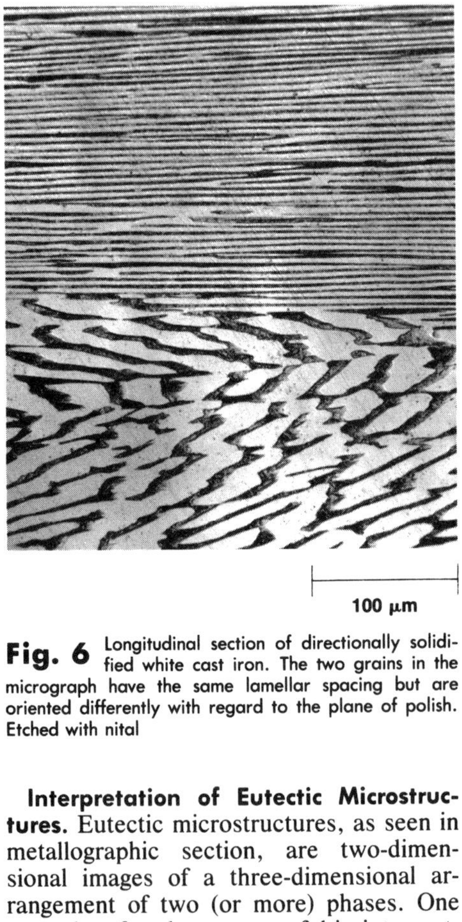

Fig. 6 Longitudinal section of directionally solidified white cast iron. The two grains in the micrograph have the same lamellar spacing but are oriented differently with regard to the plane of polish. Etched with nital

Interpretation of Eutectic Microstructures. Eutectic microstructures, as seen in metallographic section, are two-dimensional images of a three-dimensional arrangement of two (or more) phases. One must therefore be very careful in interpreting these metallographic sections. For example, Fig. 6 shows a longitudinal section of a directionally solidified lamellar eutectic (white cast iron) covering two different grains. Despite their different appearances, the two grains have the same lamellar spacing. However, the sectioning plane is perpendicular to the lamellae of one grain but is at a small angle with respect to the lamellae of the other grain. Therefore, in directionally solidified samples, the lamellar spacing of eutectic microstructures must always be

*Growth of faceted phases occurs on well-defined atomic planes, thus creating planar, angular surfaces (facets). Faceted substances are generally characterized by an entropy of fusion (ratio of the molar entropy of fusion to the gas constant R) higher than 2.0. Typical examples of faceted phases are graphite, silicon, and intermetallic compounds. Additional information is available in Ref 1.

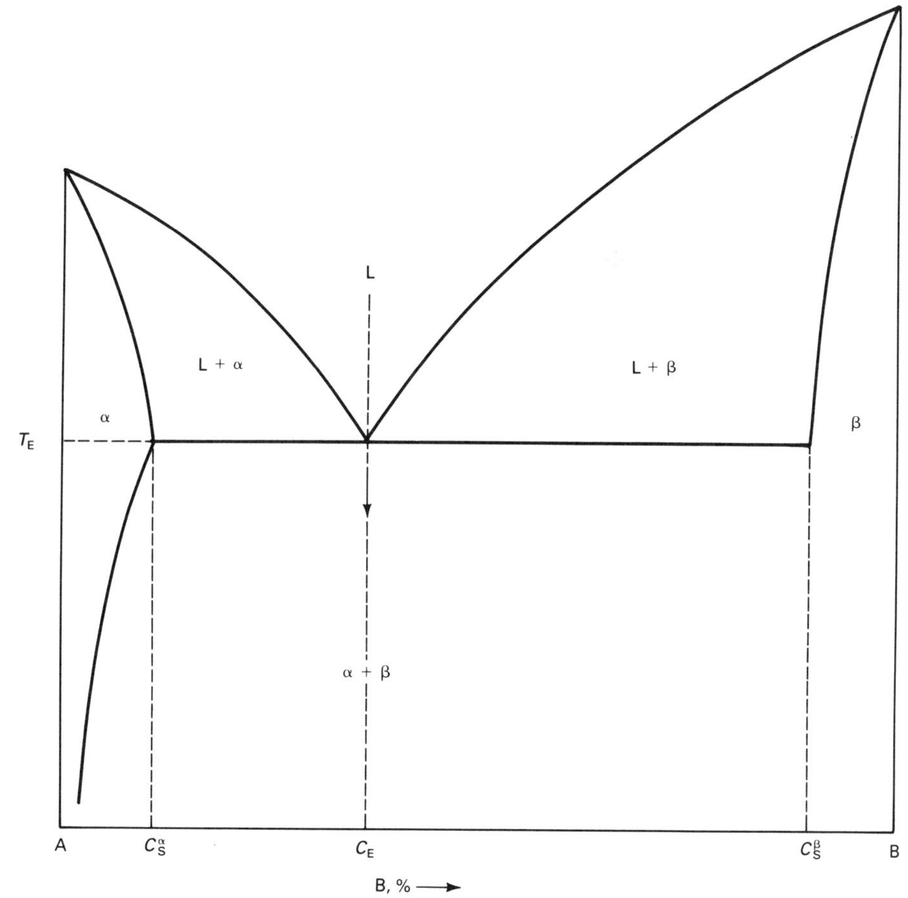

Fig. 7 Schematic eutectic phase diagram. See text for explanation.

measured perpendicular to the growth direction. In a casting containing equiaxed grains, only a mean spacing can be measured. Eutectic grains are often difficult to identify, as can be seen in Fig. 1.

Solidification and Scale of Eutectic Structures

Figure 7 shows a schematic eutectic phase diagram. When a liquid L of eutectic composition C_E is frozen, the α and β solid phases solidify simultaneously when the temperature of the melt is below the eutectic temperature T_E. A variety of geometrical arrangements can be produced. For simplicity, the case of a lamellar microstructure is considered in this discussion; the solidification of fibers can be described in terms of similar mechanisms. Because eutectic growth is essentially solute diffusion controlled, there is no fundamental difference between equiaxed and directional solidification. Therefore, the mechanisms described are valid for both cases.

Regular Eutectic Growth. During eutectic solidification, the growing α phase rejects B atoms into the liquid because of their lower solubility with respect to the liquid concentration. Conversely, the β phase rejects A atoms. If the α and β phases grow separately, solute rejection would occur only in the growth direction. This involves long-range diffusion. Therefore, a very large boundary layer would be created in the liquid ahead of the solid/liquid interface, as shown in Fig. 8(a).

However, during eutectic solidification, the α and β phases grow side by side in a cooperative manner; the B atoms rejected by the α phase are needed for the growth of the β phase, and conversely. The solute then needs only to diffuse along the solid/liquid interface from one phase to the other (Fig. 8b). The solute buildup in the liquid ahead of the growing solid/liquid interface is considerably lowered by this sidewise diffusion (diffusion coupling), thus being thermodynamically favorable (see Appendix 1). This is the fundamental reason for the occurrence of eutectic growth. As can be seen in Fig. 8(b), the smaller the lamellar spacing λ, the smaller the solute buildup, if the driving force for diffusion provided by the concentration gradient remains constant.

On the other hand, at the three-phase junction α/β/L, the surface tensions must be balanced to ensure mechanical equilibrium (Fig. 9). This imposes fixed contact angles, leading to a curvature of the solid/liquid interface. This curvature is thermodynamically disadvantageous. Because the contact angles are material constants, this curvature is higher when the lamellar spacing is small.

The scale of the eutectic structure is therefore determined by a compromise between two opposing factors:

- Solute diffusion, which tends to reduce the spacing
- Surface energy (interface curvature), which tends to increase the spacing

The lamellar spacing λ and the growth undercooling ΔT (defined as the difference

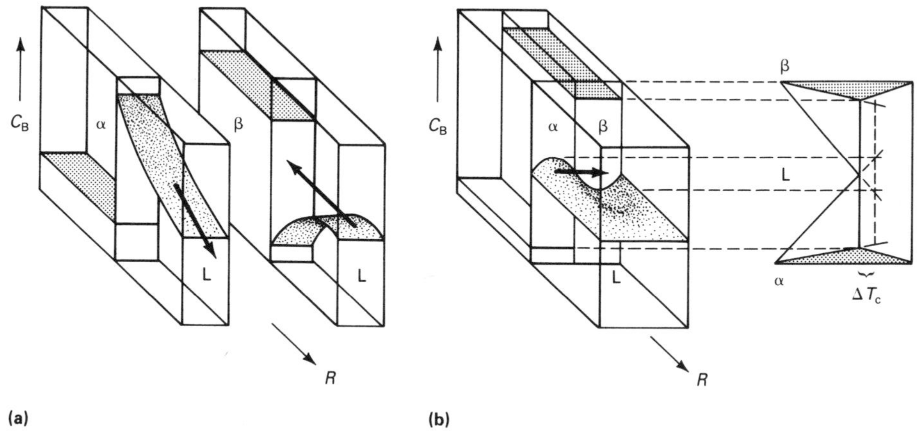

Fig. 8 Diffusion fields ahead of the growing α and β phases in isolated (a) and coupled (b) eutectic growth. The dark arrow represents the flux of B atoms. Source: Ref 1

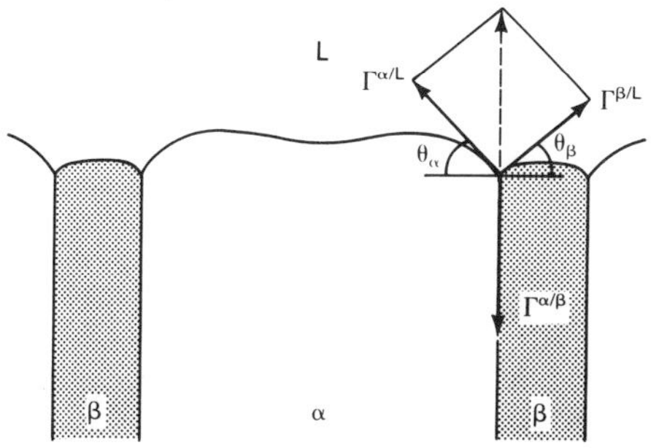

Fig. 9 Surface tension balance at the three-phase (α/β/L) junction, and the resulting curvature of the solid/liquid interface

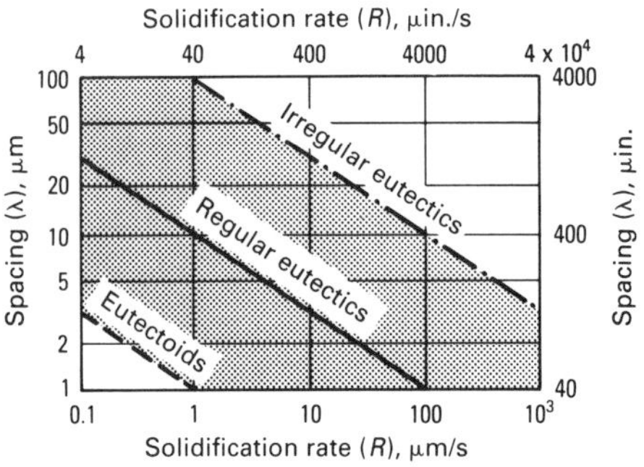

Fig. 10 Typical spacings of eutectics and eutectoids as a function of growth rate. Source: Ref 1

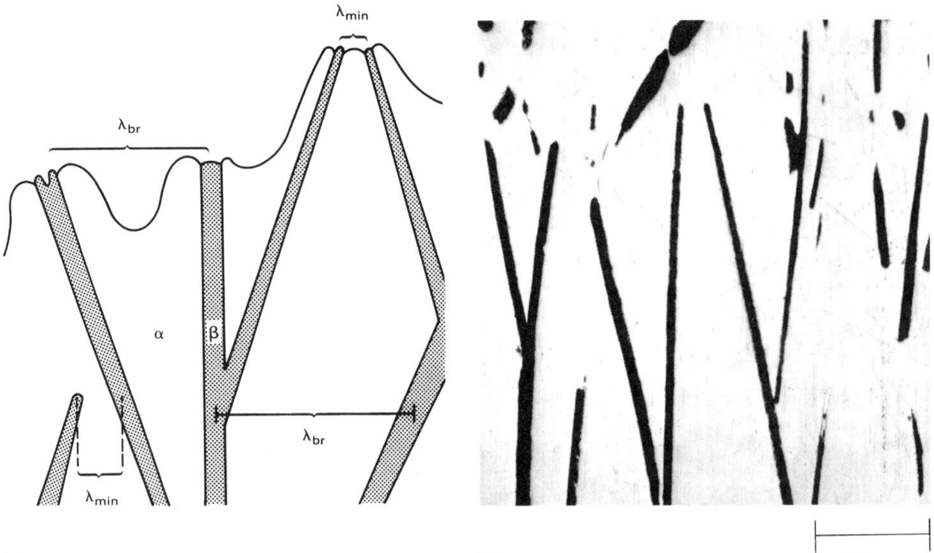

(a) (b) 100 µm

Fig. 11 Growth of irregular eutectics. (a) Schematic of branching of the faceted phase at λ_{br}, termination at λ_{min}, and the corresponding shape of the solid/liquid interface. (b) Iron-carbon eutectic alloy directionally solidified at R = 0.017 µm/s. Branching was induced by a rapid tenfold increase in R. Longitudinal section. As-polished. Source: Ref 4

between the eutectic temperature T_E and the actual interface temperature during growth) are given by (Ref 2, 3):

$$\lambda = \frac{\phi K_1}{\sqrt{R}} \qquad \text{(Eq 1)}$$

$$\Delta T = \frac{(\phi + 1/\phi)}{2} K_2 \sqrt{R} \qquad \text{(Eq 2)}$$

where R is the solidification rate (velocity at which the solid/liquid interface advances), K_1 and K_2 are constants related to the material properties (see Appendix 1), and ϕ is a regularity constant whose value is close to unity for regular eutectics. Figure 10 shows typical values for the $\lambda(R)$ relationship (Eq 1). It can be seen that regular eutectics have spacings between the coarse ones of irregular eutectics and the fine ones of eutectoids. In the latter, the effect of diffusion on spacing is more marked because it occurs in a solid phase.

The scale of the eutectic microstructure depends on the solidification rate, not directly on the cooling rate. The reason is that the thermal gradient has a negligible effect on the size of the eutectic microstructure (Ref 4). Because the cooling rate is the product of the solidification rate and the thermal gradient, two growth conditions characterized by the same cooling rate but with different thermal gradients lead to different solidification rates and therefore to different spacings. An important characteristic of regular eutectic growth is that the lamellae (or fibers) are parallel to the heat flow direction during solidification and perpendicular to the solid/liquid interface.

Irregular Eutectic Growth. Irregular eutectics grown under given growth conditions exhibit an entire range of spacings because the growth direction of the faceted phase (for example, graphite in cast iron or silicon in aluminum-silicon) is determined by specific atomic orientations and is not necessarily parallel to the heat flux. In this case, growth involves the following mechanism: When two lamellae converge, the growth of one simply ceases when λ becomes smaller than a critical spacing λ_{min} because the local interface energy becomes too large (Ref 5). Thus, the spacing is increased. This mechanism is illustrated in Fig. 11. Conversely, diverging lamellae can grow until another critical spacing, λ_{br}, is reached. When this occurs, one of the lamellae branches into two diverging lamellae, thus reducing the spacing. Growth of an irregular eutectic thus occurs within the range of interlamellar spacings between λ_{min} and λ_{br}.

It can be shown that the growth temperature of the region of small λ is higher than that in the large λ zones. The solid/liquid interface is therefore nonisothermal; that is, its shape is irregular (Fig. 11a) and is the opposite of the isothermal planar solid/liquid interface that characterizes regular eutectic growth (Fig. 9 and 12).

A mean spacing $\langle\lambda\rangle$ and a mean undercooling $\langle\Delta T\rangle$ can be defined and are still given by Eq 1 and 2. In this case, ϕ (the ratio of the mean spacing $\langle\lambda\rangle$ to the minimum undercooling spacing, which is close to λ_{min}; see Appendix 1) is greater than unity. Therefore, the spacings and undercoolings obtained are higher than those observed in regular eutectics (Fig. 10).

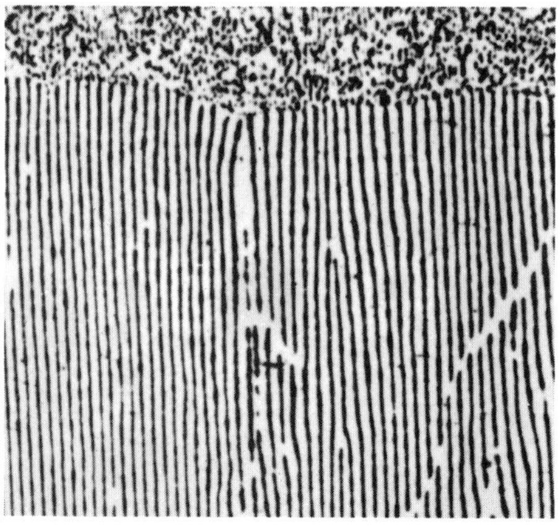

Fig. 12 Nearly planar solid/liquid interface of a regular cadmium-tin eutectic as revealed by quenching. Etched with ferric chloride. 210×

Competitive Growth of Dendrites and Eutectics

The solidification of a binary alloy of exactly eutectic composition was examined earlier in this section. In this case, provided the growth is regular, the solid/liquid interface is planar. However, when alloy composition departs from eutectic or when a third alloying element is present, the interface can become unstable for the same reason as in the case of a simple solid/liquid interface. As shown in Fig. 13, two types of morphological instability can develop: single-phase and two-phase.

A single-phase instability (Fig. 13a) leads to the solidification of one of the phases in the form of primary dendrites plus interdendritic eutectic. This situation is primarily observed in off-eutectic alloys because one phase becomes much more constitutionally undercooled than the other. For example, during the solidification of a hypoeutectic alloy, the α phase is heavily undercooled because the liquidus temperature at that composition is much higher than T_E (Fig. 7). The α phase can therefore grow faster (or at higher temperature) than the eutectic.

A two-phase instability (Fig. 13b) is characterized by cellularlike growth and leads to the appearance of eutectic colonies. This situation is observed when a third

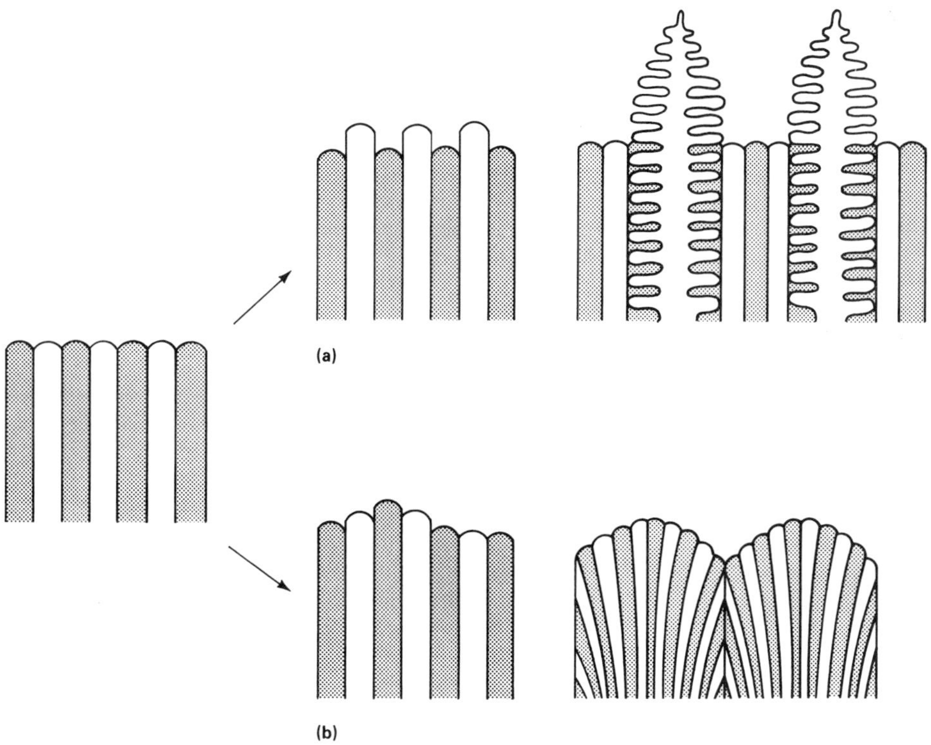

Fig. 13 Types of instability of a planar solid/liquid eutectic interface. (a) Single-phase instability leading to the appearance of dendrites of one phase. (b) Two-phase instability leading to the appearance of eutectic cells or colonies in the presence of a third alloying element

alloying element that partitions similarly at both the α/L and β/L interfaces produces a long-range diffusion boundary layer ahead of the solid/liquid interface, thus making the growing eutectic interface constitutionally undercooled with respect to this element.

Coupled Zone of Eutectics. The eutectic-type phase diagram appears to indicate that microstructures consisting entirely of eutectic can be obtained only at the exact eutectic composition. In fact, experimental observations show that purely eutectic microstructures can be obtained from off-eutectic alloys over a range of growth conditions (Ref 6). On the other hand, dendrites can sometimes be found in alloys with the exact eutectic composition if the growth rate is high. This is of considerable practical importance because the properties of a casting can be significantly changed when single-phase dendrites appear.

To explain these observations, one must consider the growth mechanisms of the competing phases (Ref 7). Because of the differing growth characteristics of eutectics and dendrites, the solidification of eutectic (high-efficiency diffusion coupling) can be faster than the isolated growth of one phase (primary dendrites), even for off-eutectic alloys. In this case, the dendrites are overgrown, and a purely eutectic microstructure is obtained over a range of off-eutectic compositions (the volume fraction of both phases in this case is determined by alloy composition and is therefore different from that obtained in the eutectic alloy). Conversely, if one of the phases (for example, β) is faceted, the growth of this phase (and consequently of the eutectic) is slower at a given undercooling. Dendrites of α phase may then grow more rapidly than the eutectic at the eutectic composition; purely eutectic microstructures are obtained only in hypereutectic alloys.

The temperature of a growing eutectic solid/liquid interface is a function of the growth rate. This relationship is used, together with the dendrite tip temperatures of α and β primary crystals, to establish the coupled zone. In the diagrams shown in Fig. 14, each point below the eutectic temperature is associated with a solidification rate

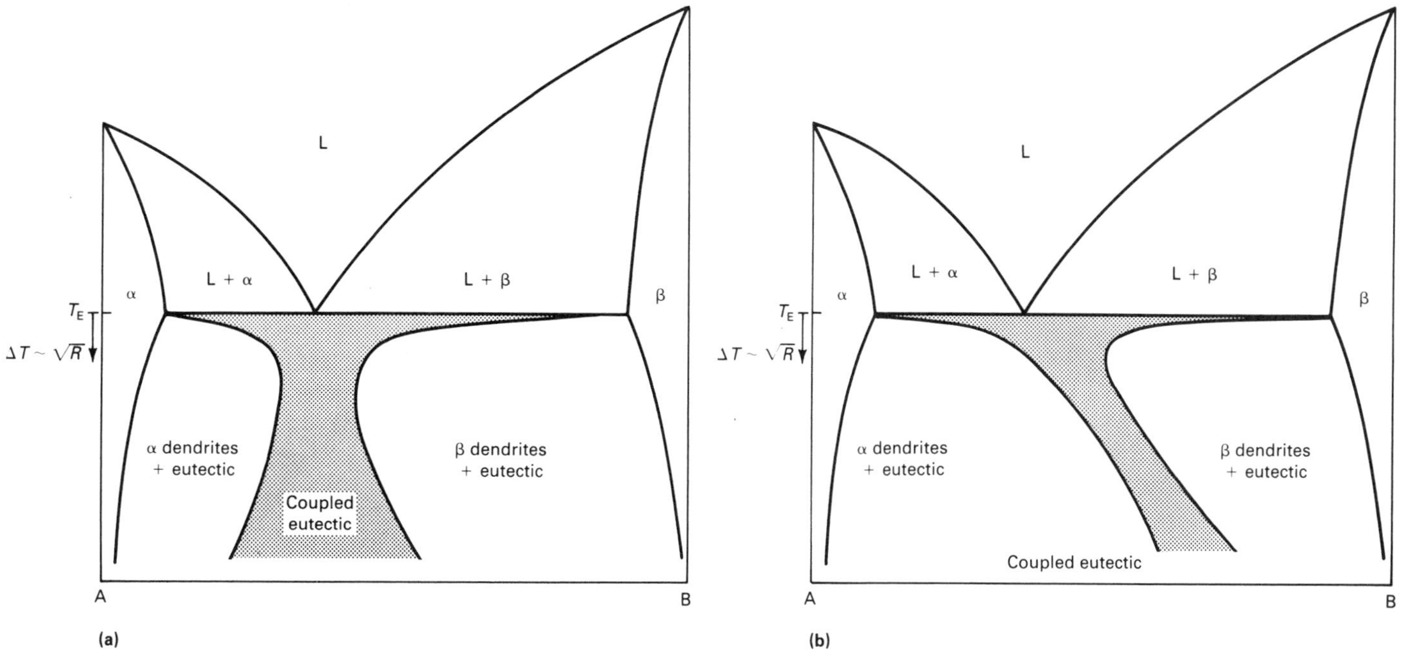

Fig. 14 Coupled zones (shaded regions) on eutectic phase diagrams. The coupled zones represent the interface temperature (solidification rate) dependent composition region in which a completely eutectic structure is obtained. (a) Nearly symmetrical coupled zone in regular eutectic. (b) Skewed coupled zone in an irregular eutectic. In both cases, the widening of the coupled zone near the eutectic temperature is observed only in directional solidification (positive thermal gradient).

through Eq 2 (that is, the lower the temperature, the higher the solidification rate). The coupled zone (a shaded region) then represents the solidification rate dependent composition region in which the eutectic grows more rapidly (or at a lower undercooling) than α- or β-phase dendrites. This zone corresponds to an entirely eutectic microstructure. Outside the coupled zone, the microstructure consists of primary dendrites and interdendritic eutectic.

Figure 14(a) shows the coupled zone of a regular eutectic system. In this case, a purely eutectic microstructure is obtained at the eutectic composition for all growth conditions. However, in the case of the skewed coupled zone of an irregular eutectic (Fig. 14b, where β is the faceted phase), the alloy composition must be carefully chosen as a function of the growth rate imposed by the casting process if a completely eutectic microstructure is required. For example, the composition of cast iron or aluminum-silicon alloys must often be hypereutectic if one wants to eliminate metal dendrites, especially when using high solidification rate casting techniques.

Appendix 1: Simplified Theory of Eutectic Growth

Solute Diffusion. As shown in Fig. 8, the growing eutectic phases reject solute into the liquid. Because of this solute buildup, the composition at the interface departs from the eutectic concentration. Figure 8 shows that in this case the equilibrium temperature between the liquid and the α or β phase is lower than the eutectic temperature T_E. The average chemical undercooling of the interface ΔT_c is proportional to the amplitude of the composition variation at the interface. The latter is proportional to the rate of rejection of solute (that is, to the solidification rate R) and to the distance (the lamellar spacing λ) over which diffusion must occur. Therefore:

$$\Delta T_c = K_c \lambda R \qquad \text{(Eq 3)}$$

where K_c is a constant related to the material properties whose value is given by (Ref 2):

$$K_c = \frac{m\, C_o\, F(f)}{D} \qquad \text{(Eq 4)}$$

where $m = m_\alpha m_\beta/(m_\alpha + m_\beta)$ and m_α and m_β are the slopes of the liquidus lines of the α and β phases, respectively (defined so that both are positive); C_o is the length of the eutectic tieline ($C_S\beta - C_S\alpha$, see Fig. 7); and D is the diffusion coefficient of the solute in the liquid. Here $F(f)$ is a function of the volume fractions f_α and f_β of the phases and can be approximated for a lamellar eutectic by:

$$F(f) \approx 0.335\,(f_\alpha f_\beta)^{0.65} \qquad \text{(Eq 5)}$$

For α-phase fibers, $F(f)$ can be approximated by:

$$F(f_\alpha) \approx 4.908 \cdot 10^{-3} + 0.3122 f_\alpha + 0.6918 f_\alpha^2 - 2.604 f_\alpha^3 + 3.238 f_\alpha^4 - 1.619 f_\alpha^5 \qquad \text{(Eq 6)}$$

Capillarity Effects. The equilibrium transformation temperature between a solid and a liquid phase is a function of the curvature of the solid/liquid interface. The change ΔT_r in this equilibrium temperature (with respect to that for a planar interface) is termed the curvature undercooling and is proportional to the interface curvature; it is positive when the solid phase is convex.

The surface tension balance among α, β, and L (Fig. 9) governs the shape of the solid/liquid interface at the three-phase junction. The interface is therefore characterized by an average (positive) curvature of the solid/liquid interface that is inversely proportional to the lamellar spacing λ. The curvature undercooling can then be expressed as:

$$\Delta T_r = \frac{K_r}{\lambda} \qquad \text{(Eq 7)}$$

where K_r is a material constant given by (Ref 2):

$$K_r = 2m\delta\left(\frac{\Gamma_\alpha \sin\theta_\alpha}{f_\alpha m_\alpha} + \frac{\Gamma_\beta \sin\theta_\beta}{f_\beta m_\beta}\right) \qquad \text{(Eq 8)}$$

where Γ_α and Γ_β are the Gibbs-Thomson coefficients (ratio of surface energy to entropy of fusion) and θ_α and θ_β are the contact angles at the three-phase junction for the α and β phases, respectively, as defined in Fig. 9. The parameter δ is equal to unity for lamellar eutectics and equal to $2\sqrt{f_\alpha}$ for fibrous eutectics (fibers of α phase).

Operating Range of Eutectics. The total undercooling ΔT is given by:

$$\Delta T = \Delta T_c + \Delta T_r = \frac{K_c \lambda R + K_r}{\lambda} \qquad \text{(Eq 9)}$$

Equation 9 is shown in Fig. 15. To determine the spacing, one still requires another criterion. It has been shown that an entire range of spacings is stable under given growth conditions (Ref 2, 8). However, the system will try to grow not too far from equilibrium, that is, close to the minimum undercooling. This is equivalent to the maximum growth rate for a given undercooling (Fig. 15b). Whenever possible, that is, in regular eutectics, the eutectic microstructure will therefore adopt a lamellar spacing close to that corresponding to the minimum undercooling. However, this is not possible in the case of irregular eutectics, because a range of lamellar spacings is observed (Fig. 16). The ratio φ of the mean spacing of an irregular eutectic to the extremum spacing λ_{ex} (which is close to λ_{min}) is a material constant and is independent of the growth conditions for normal growth rates (Ref 4). Setting $d\Delta T/d\lambda$ equal to 0 in Eq 9 and setting $\langle\lambda\rangle$ equal to φ λ_{ex} leads to the relationships:

$$\langle\lambda\rangle^2 R = \phi^2 \frac{K_r}{K_c} \qquad \text{(Eq 10)}$$

$$\frac{\langle\Delta T\rangle}{\sqrt{R}} = \left(\phi + \frac{1}{\phi}\right)\sqrt{K_r K_c} \qquad \text{(Eq 11)}$$

$$\langle\Delta T\rangle \langle\lambda\rangle = (\phi^2 + 1) K_r \qquad \text{(Eq 12)}$$

Equations 10 to 12 describe the growth of both regular ($\phi \approx 1$) and irregular ($\phi > 1$) eutectics. In the latter case, the value of φ can be approximated (supposing that β is the faceted phase) by (Ref 4):

$$\phi = 0.5 + \left[F'(f_\beta)\left(1 + \frac{f_\beta\, m_\beta}{f_\alpha\, m_\alpha} \cdot \frac{\Gamma_\alpha \sin\theta_\alpha}{\Gamma_\beta \sin\theta_\beta}\right)\right]^{-1/2} \qquad \text{(Eq 13)}$$

where $F'(f_\beta) \approx 0.03917 + 0.6047 f_\beta - 1.413 f_\beta^2 + 2.171 f_\beta^3 - 1.236 f_\beta^4$.

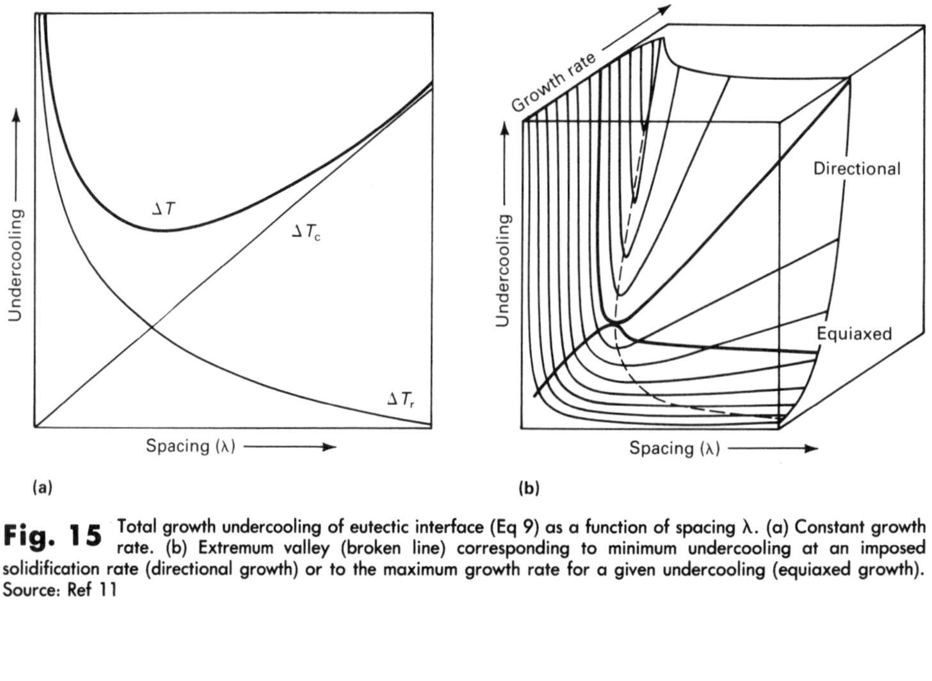

Fig. 15 Total growth undercooling of eutectic interface (Eq 9) as a function of spacing λ. (a) Constant growth rate. (b) Extremum valley (broken line) corresponding to minimum undercooling at an imposed solidification rate (directional growth) or to the maximum growth rate for a given undercooling (equiaxed growth). Source: Ref 11

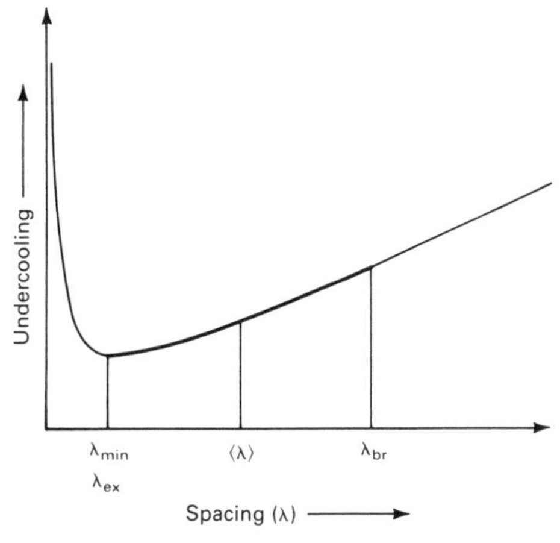

Fig. 16 Operating range of spacings (between λ_{min} and λ_{br}) and undercoolings for irregular eutectics

Appendix 2: Eutectic Growth at Very High Solidification Rates

Rapid solidification processing (for example, the laser surface treatment of eutectic alloys) has led to very interesting properties. The microstructures obtained are extremely fine, and spacings as small as 10 nm have been observed. It can be shown that Eq 10 to 12 are no longer valid in this case (Ref 9). There are three main reasons for this, as follows.

First, at very high solidification rates, the sidewise diffusion shown in Fig. 8(b) becomes less effective because, as a result of the very high surface energy involved, the lamellar spacing decreases only as $1/\sqrt{R}$ (Eq 1), while the diffusion distance in the growth direction decreases as $1/R$. Therefore, the thermodynamic advantage of coupled eutectic growth as compared to separate growth (Fig. 8) is diminished and can even disappear completely.

Second, the high growth undercooling at high R values lowers the interface temperature where the diffusion occurs. The diffusion coefficient of the solute in the liquid decreases strongly when the temperature is lowered, thus slowing the growth rate and reducing the spacing.

Third, the parameter C_o (Eq 4) is a function of the metastable phase diagram below the eutectic temperature and can differ significantly from its equilibrium value at high undercoolings. In addition, at very high solidification rates, solute segregation has no time to occur. The solute atoms are then trapped in the growing phases (Ref 10), and a metastable supersaturated phase is formed. Because of these phenomena, there is a critical solidification rate (for many systems, of the order of 0.1 to 1 m/s) beyond which the eutectic microstructure can no longer form (Ref 9).

REFERENCES

1. W. Kurz and D.J. Fisher, *Fundamentals of Solidification*, Trans. Tech. Publications, 1984
2. K.A. Jackson and J.D. Hunt, *Trans. AIME*, Vol 236, 1966, p 1129
3. H. Jones and W. Kurz, *Z. Metallkd.*, Vol 72, 1981, p 792
4. P. Magnin and W. Kurz, *Acta Metall.*, Vol 35, 1987, p 1119
5. D.J. Fisher and W. Kurz, *Acta Metall.*, Vol 28, 1980, p 777
6. F. Mollard and M.C. Flemings, *Trans. AIME*, Vol 239, 1967, p 1534
7. W. Kurz and D.J. Fisher, *Int. Met. Rev.*, Vol 24, 1979, p 177
8. S. Strässler and W.R. Schneider, *Phys. Cond. Matter.*, Vol 17, 1974, p 153
9. R. Trivedi, P. Magnin, and W. Kurz, *Acta Metall.*, Vol 35, 1987, p 971
10. M.J. Aziz, *J. Appl. Phys.*, Vol 53, 1982, p 1158
11. P.H. Shingu, *J. Appl. Phys.*, Vol 50, 1979, p 5743

Solidification of Peritectics

H. Fredriksson, The Royal Institute of Technology, Sweden

Peritectic reactions or transformations are very common in the solidification of metals. Many interesting alloys undergo these types of reactions—for example, iron-carbon and iron-nickel-base alloys as well as copper-tin and copper-zinc alloys.

Phase diagrams are very instructive when describing peritectic phase transformations. Figure 1 shows a phase diagram with a peritectic reaction. This diagram shows that, under equilibrium conditions, all alloys to the left of I will solidify to α crystals. Similarly, all alloys to the right of III will solidify to β crystals. Alloys between II and III first solidify to α crystals and then transform to stable β crystals. Alloys between I and II also solidify to α crystals, but they are partially transformed to β crystals later.

The volume fraction of each phase will be given by the lever rule if the alloy solidifies under equilibrium conditions (the lever rule is defined in the article "Interpretation and Use of Cooling Curves (Thermal Analysis)" in this Volume). In most cases, the lever rule will not give the volume fraction of the different phases. This is because the kinetics as well as the diffusion rate in the solid phases are determining the time for reaching equilibrium.

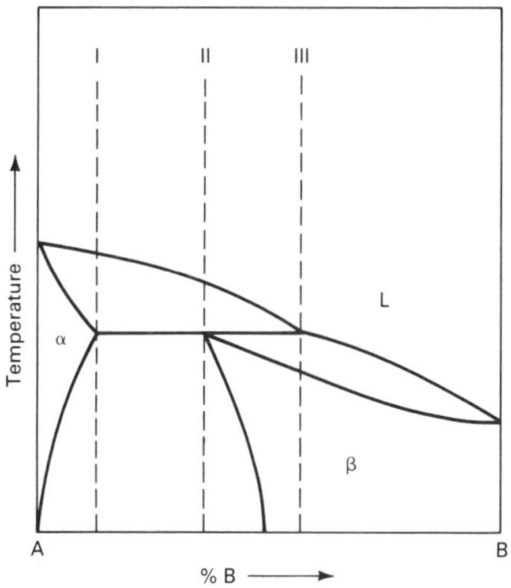

Fig. 1 Phase diagram with a peritectic reaction

In this section, the kinetics involved in different peritectic systems will be discussed, and the definitions introduced by Kerr to distinguish between a peritectic reaction and a peritectic transformation will be used (Ref 1). During a peritectic reaction, all three phases (α, β, and liquid) are in contact with each other. In the peritectic transformation, the liquid and the α phase are isolated by the β phase. The transformation takes place by long-range diffusion through the secondary β phase. Later in this section, the reaction in multicomponent systems, particularly in iron-base alloys, will be discussed. The influence of different nucleation conditions for the secondary phase will also be addressed. Finally, the possibility of precipitating metastable β crystals instead of α crystals in an alloy with a composition to the left of point III in Fig. 1 will be reviewed. Additional information is available in the article "Peritectic Structures" in Volume 9 of the 9th Edition of *Metals Handbook*.

Peritectic Reaction

Depending on surface tension conditions, two different types of the peritectic reactions can occur:

- Nucleation and growth of the β crystals in the liquid without contact with the α crystals
- Nucleation and growth of the β crystals in contact with the primary α phase

In the first case, the secondary phase is nucleated in the liquid and does not contact the primary phase. This is because of the surface tension conditions. Following nu-

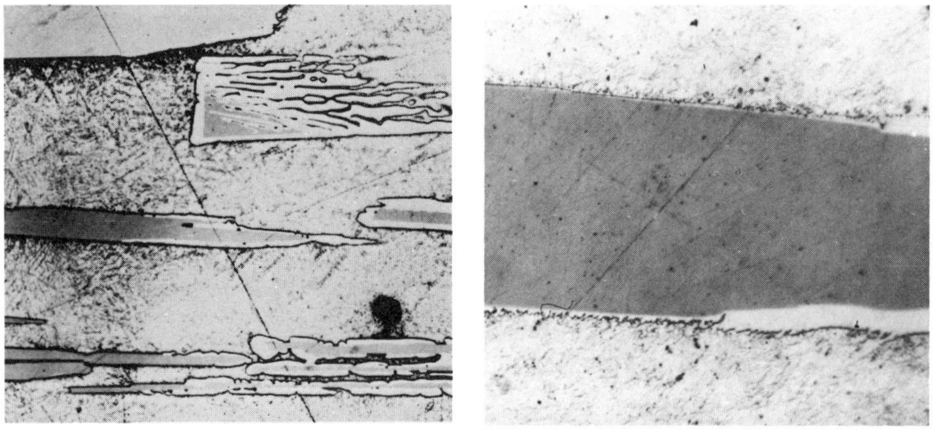

Fig. 2 Peritectic reaction (a) in a unidirectionally solidified Cu-70Sn sample. (b) Larger magnification of the middle section of (a)

cleation, the secondary phase grows freely in the liquid. At the same time, the primary phase will dissolve. The secondary phase will not develop a morphology similar to a precipitated primary phase. This type of peritectic reaction has been observed for the reaction $\gamma + L \rightarrow \beta$ in the Al-Mn system (Ref 2). There has also been a tendency for the secondary phase to grow around the primary phase at increasing cooling rates. Similar reactions have been observed in Ni-Zn (Ref 3) and Al-U (Ref 4) systems.

In the second type of reaction, which is the most common, nucleation of the secondary β phase occurs at the interface between the primary α phase and the liquid. A lateral growth of the β phase around the α phase then takes place. This type of reaction is illustrated in Fig. 2 and 3. Figure 2 shows the growth process in a unidirectionally solidified copper-tin alloy containing 70% Sn, in which the ϵ phase and the liquid react to produce the η phase. Figure 3 shows how the precipitated primary phase will partially dissolve by diffusion of solute through the liquid from the secondary phase boundary to the primary phase boundary.

A remelting process is normally much faster than a solidification process and there are no forces hindering dissolution. The dissolution will therefore occur at the same rate as the precipitation of the secondary phase. Using the maximum growth rate theory described in Ref 2, it can be shown that the thickness of the secondary phase is influenced by the growth rate and by the surface tension σ, where the expression $\sigma^{L/\beta} + \sigma^{\alpha/\beta} - \sigma^{L/\alpha}$ is the dominating factor. The larger this expression, the thicker the β layer will be.

Nucleation of the secondary phase on the primary phase is sometimes favored. When this occurs, a number of crystals will be formed around the primary phase. This type of reaction has been reported in Ref 3 for the Cd-Cu system and is illustrated in Fig. 4.

The Peritectic Transformation

The thickness of the β layer will normally increase during subsequent cooling. There are three reasons for this:

- Diffusion through the β layer
- Precipitation of β directly from the liquid
- Precipitation of β directly from the α phase

The precipitation of β directly from the liquid and the solid depends on the shape of the phase diagram and the cooling rate. The diffusion process through the β layer depends on the diffusion rate, the shape of the phase diagram, and the cooling rate.

To explain the transformation process, an isothermal case will first be analyzed. This has been discussed in Ref 5 and 6. From

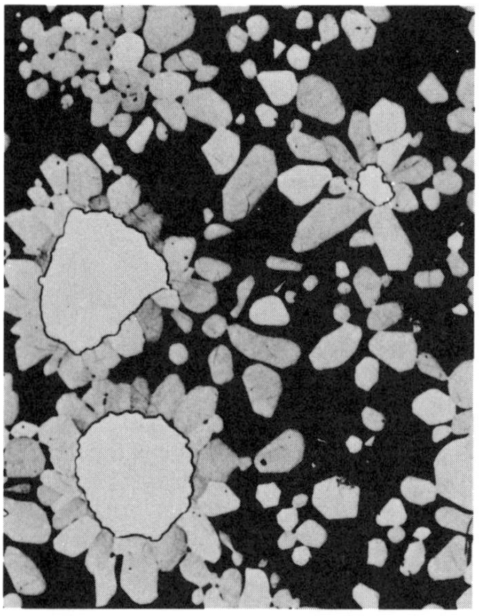

Fig. 4 Microstructure in a Cd-10Cu sample that has passed a peritectic reaction. The primary Cu_5Cd_8 crystals are white, the dark matrix is cadmium, and the peritectically formed $CuCd_3$ is gray.

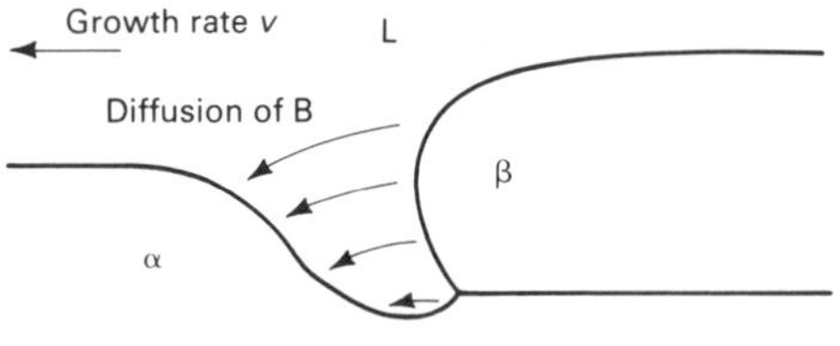

Fig. 3 Peritectic reaction by which the secondary solid β phase grows along the surface of the primary solid α phase

those calculations, one can assume that the growth of the β layer is controlled by the diffusion through it at a temperature just below the peritectic temperature. The diffusion process and the concentration profile and its relation to the phase diagram are illustrated in Fig. 5.

A mass balance gives the following:

$$\alpha \rightarrow \beta\colon D^{\beta}\,\frac{(x^{\beta/L} - x^{\beta/\alpha})}{\ell^{\beta}} = \frac{d\ell^{\beta/\alpha}}{dt}\cdot(x^{\beta/\alpha} - x^{\alpha/\beta}) \qquad \text{(Eq 1)}$$

and

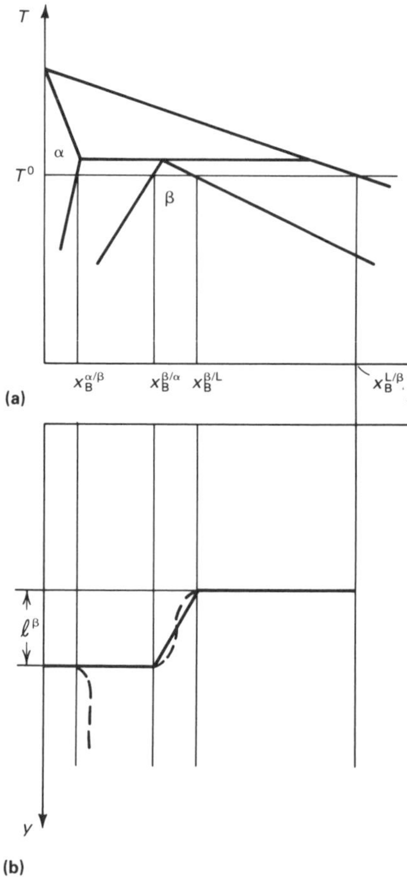

Fig. 5 The peritectic transformation in a system with a high diffusion rate in the β phase. (a) The phase diagram. (b) Concentration distribution. The dashed lines in the concentration profile are for a system with a low diffusion rate in the β phase where the volume fraction of β increases with decreasing temperature. See Eq 6 and corresponding text. See also Fig. 7(b).

$$\mathrm{L} \rightarrow \beta\text{: } D^{\beta}\,\frac{(x^{\beta/L} - x^{\beta/\alpha})}{\ell^{\beta}} = \frac{d\ell^{\beta/L}}{dt}\cdot(x^{L/\beta} - x^{\beta/L}) \qquad \text{(Eq 2)}$$

where ℓ^{β} is the thickness of the β phase, t is the time, and D^{β} is the diffusion coefficient in the β phase. All other terms are concentrations that are defined in Fig. 5:

$$\frac{d\ell^{\beta}}{dt} = \frac{d\ell^{\beta/\alpha}}{dt} + \frac{d\ell^{\beta/L}}{dt} = \frac{D^{\beta}}{\ell^{\beta}}\cdot(x^{\beta/L} - x^{\beta/\alpha})\cdot\left(\frac{1}{(x^{\beta/\alpha} - x^{\alpha/\beta})} + \frac{1}{(x^{L/\beta} - x^{\beta/\alpha})}\right) \qquad \text{(Eq 3)}$$

Integrating Eq 3 results in:

$$(\ell^{\beta})^2 = D^{\beta}\cdot(x^{\beta/L} - x^{\beta/\alpha})\cdot\left(\frac{1}{(x^{\beta/\alpha} - x^{\alpha/\beta})} + \frac{1}{(x^{L/\beta} - x^{\beta/L})}\right) \qquad \text{(Eq 4)}$$

Equations 3 and 4 show that the growth rate increases with increasing undercooling. For example, at the peritectic temperature, the expression $(x^{\beta/L} - x^{\beta/\alpha})$ is 0 and increases with increasing undercooling. Equations 3 and 4 also show that the growth rate is dependent on the diffusion coefficient. For substitutionally dissolved alloying elements in face-centered cubic metals, the diffusion coefficient near the melting point is of the order of $\leq 10^{-13}$ m²/s. In such a case, the growth rate will be very low and the time for the peritectic transformation will be unrealistically large. In a normal casting process, the reaction rate will be so low that the amount of β phase formed by the peritectic reaction will be negligible in comparison with the precipitation of β from the liquid. For body-centered cubic metals and for interstitially dissolved alloying elements, the diffusion rates are much higher than for substitutionally dissolved elements in face-centered cubic metals. The diffusion process has in this case a much larger influence on the precipitation process.

The peritectic transformation in iron-carbon alloys will be analyzed in the following paragraphs. The iron-carbon phase diagram is very similar to the diagram illustrated in Fig. 5(a). A detailed numerical calculation of the transformation for one alloy has been performed (Ref 7). However, a simpler analytical model will be used in this discussion (Ref 8). The model is the same as the isothermal model described earlier, but is used with the assumption that the boundary conditions change during cooling. Equation 3 can therefore be used. In Eq 3, $d\ell^{\beta}/dt$ can be expanded to $(d\ell^{\beta}/dT)(dT/dt)$, where dT/dt is the cooling rate of the sample. Using the phase diagram, the concentrations can be transferred to a temperature. If the slope of the lines in the phase diagrams and the cooling rate are both assumed to be constant, Eq 3 can be simplified in the following way:

$$\ell^{\beta} d\ell^{\beta} = \mathrm{A}\cdot\frac{dt}{dT}\cdot dT \qquad \text{(Eq 5)}$$

where A is a constant. By integrating Eq 5 under the assumption of a constant cooling, one can obtain a relation that can be used to calculate the temperature interval under which the peritectic transformation takes part. This has been done in Ref 8, and the result of those calculations are shown in Fig. 6. Figure 6 shows that the peritectic reaction in iron-carbon alloys is very rapid and is finished at a maximum of 6 to 10 K below the peritectic temperature. The shape of the curve is a result of the change in the volume fraction of ferrite with the carbon content. The curve is in agreement with the results of other experiments (Ref 7, 9).

In addition to the effect of the growth of the β layer due to diffusion, the effect of the phase diagram must also be considered. This was not taken into consideration when deriving Eq 3 and in the above discussion for iron-carbon alloys. The effect of the phase diagram is illustrated in Fig. 5(a) and 7(a). Figures 5(a) and 7(a) show two different types of phase diagrams, in which the slopes of the α/β regions are different. In the first case (Fig. 5a), the β layer will increase in thickness both at the interface α/β and at the interface β/L. In the second case (Fig. 7a), the β layer will increase only at the side against the liquid and will decrease at the side against the α phase. These two cases are somewhat more difficult to treat theoretically than the isothermal example or the case involving a high diffusion rate.

The concentration profiles for the transformation are given in Fig. 5(b), as shown by the dashed lines, and 7(b). Equation 3 will now be changed in the following way:

$$\frac{d\ell^{\beta}}{dt} = \frac{D^{\beta}}{(x^{\alpha/\beta} - x^{\beta/\alpha})}\cdot\left(\frac{dx^{\beta}}{dy}\right)_{y=\ell^{\alpha/\beta}} + \frac{D^{\beta}}{x^{L/\beta} - x^{\beta/L}}\left(\frac{dx^{\beta}}{dy}\right)_{y=\ell^{\beta/L}} + \frac{D^{\alpha}}{x^{\alpha/\beta} - x^{\beta/\alpha}}\left(\frac{dx^{\alpha}}{dy}\right)_{y=\ell^{\alpha/\beta}} + \frac{\ell^{L}\,dx^{L}}{x^{L/\beta} - x^{\beta/L}} \qquad \text{(Eq 6)}$$

where the first and second terms on the right-hand side describe the increase in thickness due to diffusion into the β phase and α phase, respectively, from the boundary α/β. The third term describes the increase in thickness due to diffusion into the β phase from the boundary β/L. The last term is the increase in the β phase due to the changing of the composition in the liquid dx^{L} with a decrease in the temperature dT according to the phase diagram.

Solving Eq 6 is a difficult task that can be accomplished only by using numerical methods. One simplification of this equation has been performed (Ref 2). It was assumed that the diffusion distance corresponded to the thickness of the β phase and that there are no concentrating gradients in the α phase. The results of these calculations for a Cu-20Sn alloy are shown in Fig. 8. A close agreement was achieved between the experiments and the theory.

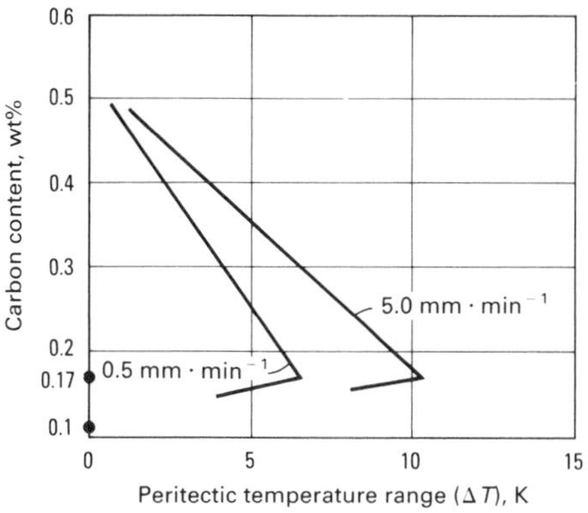

Fig. 6 Temperature range of peritectic reaction in iron-carbon alloys as a function of carbon content and the solidification rate. The temperature gradient G is 6000 K/m.

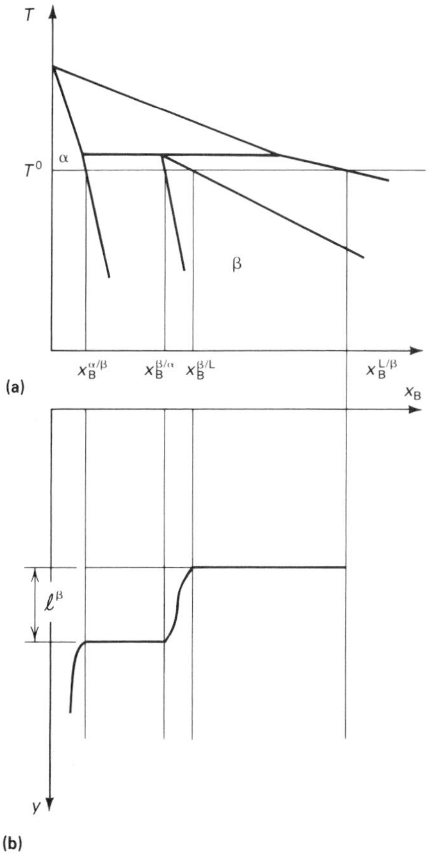

Fig. 7 The peritectic transformation during continuous cooling in a system with a low diffusion rate and where the volume fraction of β increases with decreasing temperature. (a) The phase diagram. (b) Concentration distribution

Equation 6 has been solved in Ref 8 by assuming that diffusion into the β phase could be described in the same manner as in Ref 10 for back diffusion during segregation at a primary precipitation. The model was used to calculate the concentration distribu-

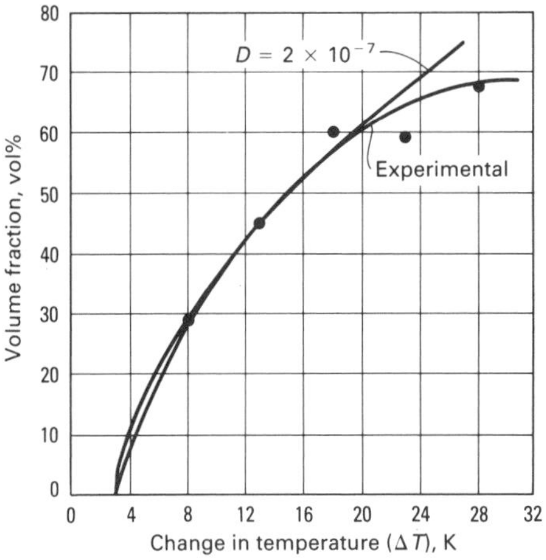

Fig. 8 Thickness of the secondary phase layer as function of temperature below the peritectic temperature in the Cu-Sn system. The solidification rate was 100 mm/min. The diffusion units are given in cm^2/s. The volume fraction is defined as ℓ^{β}/λ where λ corresponds to the dendrite arm space.

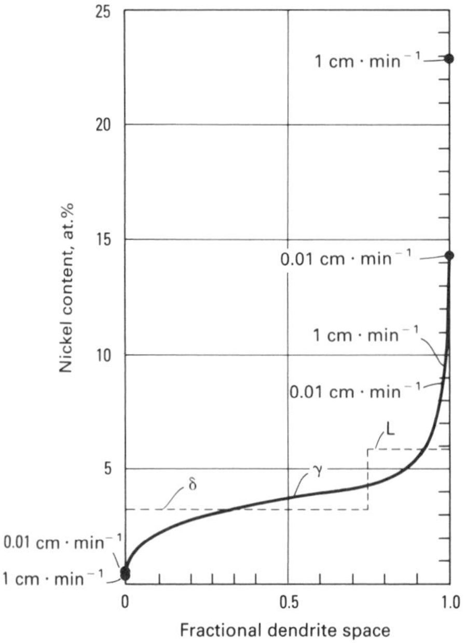

Fig. 9 Nickel distribution after peritectic reaction in a steel containing 4 wt% Ni. The temperature gradient was 60 K/cm. Calculations were made at different solidification rates. The dotted line shows the nickel distribution at the start of the peritectic reaction. δ is primary ferrite, γ is austenite. Source: Ref 11

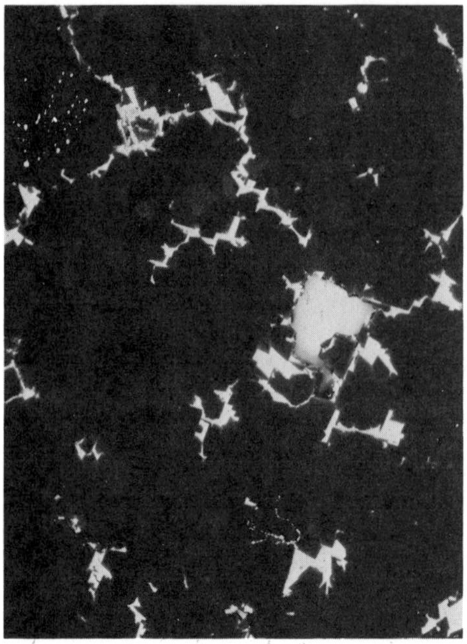

Fig. 10 Microstructure with two peritectic envelopes in a Cd-25Ni alloy. Shown are nickel crystals (dark gray) with a β layer (black) and γ layer (light gray). The matrix (white) is cadmium. Source: Ref 3

tion of austenite for iron-nickel alloys, as shown in Fig. 9. These calculations are in agreement with experiments reported in Ref 11 and 12.

Cascades of Peritectic Reaction

The theoretical analysis shows that the rate of the peritectic transformation is influenced by the diffusion rate and the extension of the β phase region in the phase diagram. If the diffusion rate is small, the peritectic transformation will be negligible compared to the peritectic reaction. The thickness of the β phase envelope surrounding the α phase is determined by the peritectic reaction followed by an increase in thickness due to a precipitation directly from the liquid.

In many systems, one peritectic reaction at a high temperature is followed by one or more peritectic reactions at lower temperatures. If the diffusion rate is low in the initially formed peritectic layer, a second peritectic layer can be formed when the second peritectic temperature is reached. This type of series of peritectic reactions, referred to as a cascade, has been studied in Ref 3. The resulting microstructure formed in a Cd-25Ni alloy is shown in Fig 10.

Primary Metastable Precipitation of Beta

In many systems, the secondary β phase has been observed to form as a primary phase for alloys with a composition on the left-hand side of point III in Fig. 1. This is especially true for iron-base alloys (Ref 11, 12).

The possibility of forming a metastable β phase directly from the liquid will now be examined. Different cooling rates will be chosen, and binary iron-nickel alloys will be considered. In Fig. 11, the metastable extensions of the equilibrium between liquid and ferrite and between liquid and austenite are represented by broken lines. The melting point of pure iron as austenite is 11 °C (20 °F) below the melting point of pure iron as ferrite. It can be seen that there is a difference in the partition coefficient between ferrite and liquid on one hand and austenite and liquid on the other.

The growth rates were calculated for needles of ferrite and austenite. The alloy composition is chosen to the left of the peritectic point, and the calculations of the growth rates were carried out for different undercoolings.

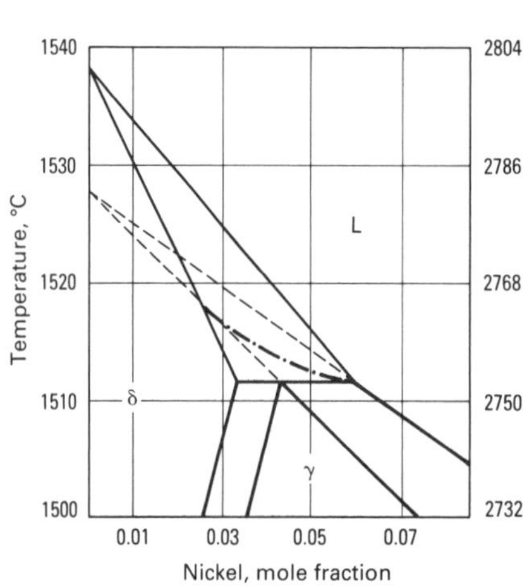

Fig. 11 The binary Fe-Ni system. The broken line shows the undercooling at which ferrite, δ, and austenite, γ, grow at the same rate. Source: Ref 13

Only ferrite can form at low undercoolings. Austenite can also form if the temperature is chosen just below the extension of its liquidus line, but ferrite grows faster and should therefore dominate in the solidification process. However, below a critical line, austenite will have the highest growth rate and may dominate, although the temperature is still above the peritectic temperature. This kinetic advantage of austenite is due to its partition coefficient being smaller than that for ferrite. The critical line is indicated by a broken line in Fig. 11. The partition coefficient of an element can be closer to unity for ferrite than for austenite

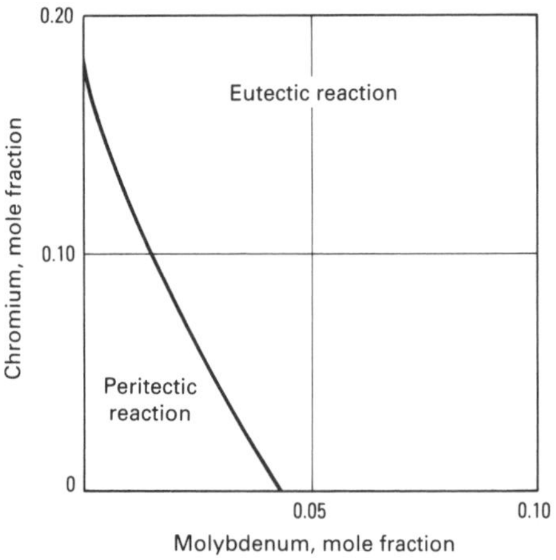

Fig. 12 The transition from a peritectic to a eutectic reaction as a function of chromium and molybdenum content in a stainless steel containing 11.9% Ni

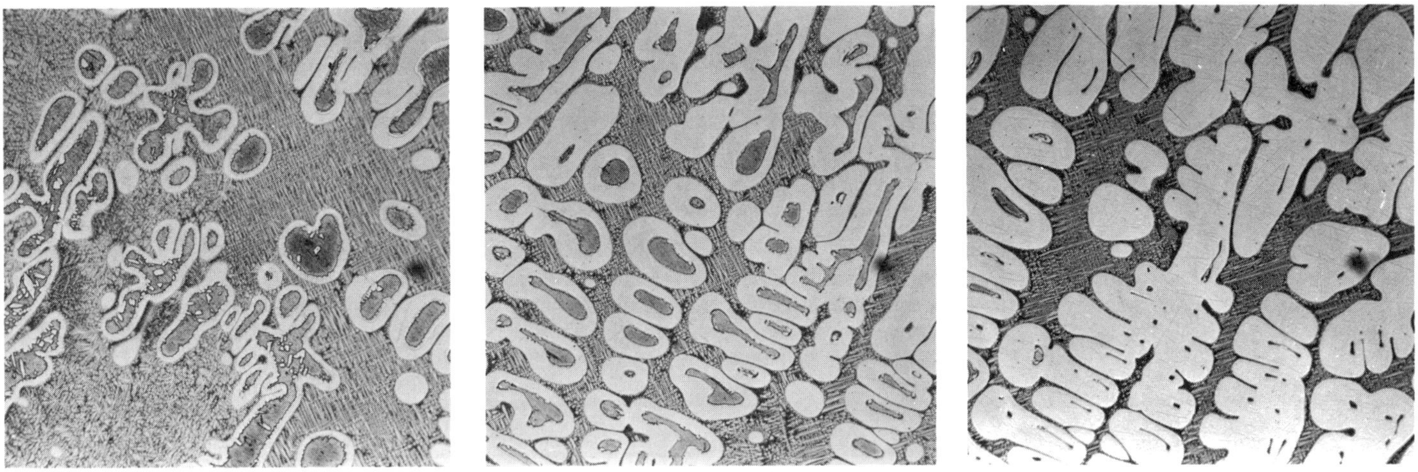

(a) (b) (c)

Fig. 13 Three stages of a peritectic reaction in a unidirectionally solidified high-speed steel. (a) First stage structure. Dark gray is austenite, white is ferrite. The mottled structure is quenched liquid. (b) Subsequent peritectic transformation of (a). (c) Further peritectic transformation of (a) and (b). Dark gray in the middle of the white ferrite is newly formed liquid. Source: Ref 15

in ternary alloys and ferrite will then have the kinetic advantage and will be favored by a high cooling rate.

It has also been reported that an aligned structure can be formed in peritectic systems (Ref 14). However, it has been argued that this type of structure is formed when α and β primary phases are growing at the same rate (Ref 2). In addition, the composition of the liquid must be chosen so the sum of the volume fraction of the two solid phases will be unit.

Peritectic Transformations in Multicomponent Systems

Alloys often consist of more than two alloying elements. However, very little information is given in the literature about the peritectic reaction in multicomponent alloys. Recent investigations of iron-base alloys have shown that peritectic reactions are very common in stainless steels (Ref 11, 15). The peritectic reaction in these alloys gives the same type of distribution as that shown in Fig. 9. In stainless steels, the peritectic reaction will transfer to a eutectic reaction if the chromium content is increased to 20% or more. This transition is also influenced by the molybdenum content, as shown in Fig. 12.

Both chromium and nickel are substitutionally dissolved elements. Iron-base alloys often consist of carbon with some other elements. Carbon is interstitially dissolved and has a very high diffusion rate. The other alloying elements are primarily substitutionally dissolved with very low diffusion rates. This gives rise to transformations that are determined by the movement of the substitutional elements, and carbon is distributed according to equilibrium conditions. As a result, a normal peritectic transformation does not occur. To fulfill the criterion that carbon should follow the equilibrium conditions, liquid must be formed at the border between ferrite and austenite. This reaction is illustrated in Fig. 13. This type of reaction has been both experimentally and theoretically analyzed in Ref 15.

REFERENCES

1. H.W. Kerr, J. Cisse, and G.F. Bolling, On Equilibrium and Non-Equilibrium Peritectic Transformation, *Acta Metall.*, Vol 22, 1974, p 677
2. H. Fredriksson and T. Nylén, Mechanism of Peritectic Reactions and Transformations, *Met. Sci.*, Vol 16, 1982, p 283-294
3. G. Petzow and H.E. Exner, Zur Kenntnis Peritektischer Umwandlungen, *Radex Rundsch.*, 3/4, 1967, p 534-539
4. S. Uchida, "Systematik and Kinetik Peritektischer Umwandlund," Ph.D. thesis, Technical University, 1980
5. M. Hillert, Eutectic and Peritectic Solidification, in *Solidification and Casting of Metals*, The Metals Society, 1979, p 81-87
6. D.H. St. John and L.M. Hogan, The Peritectic Transformation, *Acta Metall.*, Vol 25, 1977, p 77-81
7. Y.K. Chuang, D. Reinisch, and K. Schwerdtfeger, Kinetics of the Diffusion Controlled Peritectic Reaction During Solidification of Iron-Carbon-Alloys, *Metall. Trans. A*, Vol 6A, 1975, p 235-238
8. H. Fredriksson and J. Stjerndahl, Solidification of Iron-Base Alloys, *Met. Sci.*, Vol 16, 1982, p 575-585
9. J. Stjerndahl, "The Solidification Process of Iron Base Alloys," thesis, Department of Casting of Metals, Royal Institute of Technology, 1978
10. H.D. Body and M.C. Flemings, Solute Redistribution in Dendritic Solidification, *Trans. Met. Soc. AJIME*, Vol 236, 1966, p 615-624
11. H. Fredriksson, The Solidification Sequence in an 18-8 Stainless Steel, *Metall. Trans.*, Vol 3, 1972, p 2989-2997
12. H. Fredriksson and J. Stjerndahl, On the Formation of a Liquid Phase During Cooling of Steel, *Metall. Trans. B*, Vol 6, 1975, p 661
13. H. Fredriksson, Segregation Phenomena in Iron-base Alloys, *Scand. J. Metall.*, Vol 5, 1976, p 27-32
14. W.J. Boettinger, The Structure of Directionally Solidified Two Phase Sn-Cd Peritectic Alloys, *Metall. Trans.*, Vol 5, 1974, p 2023-2031
15. H. Fredriksson, The Mechanism of the Peritectic Reaction in Iron-Base Alloys, *Met. Sci.*, Vol 11, 1976, p 77-86
16. H. Fredriksson, Transition From Peritectic to Eutectic Reaction in Iron-Base Alloys, in *Solidification and Casting of Metals*, Publication 192, The Metals Society, 1977, p 131-136

Columnar to Equiaxed Transition

S.C. Flood, Alcan International Ltd., Great Britain
J.D. Hunt, University of Oxford, Great Britain

In general, as-cast metal exhibits three distinct zones of grain structures:

- A chill zone of very small crystals produced by rapid cooling at the extreme edge
- A zone of long, thin columnar crystals lying along the direction of heat flow and stretching in from the chill zone
- A region of roughly spherical equiaxed crystals at the center of the casting

All three zones may not be present in a particular case; however, when a casting contains columnar and equiaxed grains, the transition between the two morphologies (the columnar to equiaxed transition) is usually narrow, and the columnar and equiaxed zones are quite distinct. A great deal of effort has been devoted to understanding the mechanisms behind the development of macrostructure during solidification because the grain structure influences the properties of a casting and the worked metal inherits characteristics from the as-cast state. This section will focus on the formation of the equiaxed zone as the crucial process that determines macrostructure.

In the absence of an equiaxed zone, the casting will be wholly columnar. Equiaxed grains grow ahead of the columnar dendrites, and the columnar to equiaxed transition occurs when these equiaxed grains are sufficient in size and number to impede the advance of the columnar front. The extent of the equiaxed zone is the result of competition between the columnar and the equiaxed grains. The formation of an equiaxed zone requires both:

- The presence of nuclei in the bulk
- Conditions that promote their growth relative to the columnar dendrites

In the 1960s and 1970s, the tendency was to explain trends in the columnar to equiaxed transition only in terms of different mechanisms for the supply of equiaxed nuclei. It was assumed that the nuclei will develop into an equiaxed zone if they are present. The ability and tendency of the grains to grow were not considered. However, if the equiaxed growth is slow relative to the columnar or if it is restricted to a narrow undercooled region ahead of the columnar front, then, although equiaxed nuclei are present, the columnar growth could still dominate the macrostructure; it would simply absorb the small equiaxed grains as it advances. Grain growth was neglected until recently, but over the last few years, many papers have considered this aspect of equiaxed zone formation. The important processes of sedimentation and convective motion of the grains have yet to be covered satisfactorily.

Influence of Casting Parameters

Macrostructure is found to be affected by such factors as superheat, alloy system, composition, fluid flow, mechanical disturbance, inoculation, and the addition of grain refiner, and casting size. The important trends are summarized in Table 1. The various mechanisms and models for the columnar to equiaxed transition need to be discussed and evaluated with reference to these experimental observations.

Origin of the Equiaxed Nuclei

The production of an equiaxed zone requires the existence of small crystallites, or nuclei, in the bulk during freezing. Three mechanisms for the provision of these nuclei have been proposed, and they are used to form the basis for the discussion of the effect of the casting parameters on columnar to equiaxed transition:

- Constitutional supercooling driven heterogeneous nucleation
- Big bang mechanism
- Dendrite detachment mechanism

Experimental observations were interpreted as proving one mechanism to be valid at the expense of the other two. In retrospect, however, the view that the nucleation of grains in the bulk occurs by a single mechanism is unrealistically limited, because both the big bang and dendrite detachment mechanisms are strongly supported by the experimental evidence.

Constitutional Supercooling Driven Mechanism. It has been proposed that sufficient constitutional supercooling (CS) might be produced ahead of the growth front to cause the heterogeneous nucleation of equiaxed grains (Ref 23). Researchers have predicted that constitutional supercooling exists ahead of a planar front when (Ref 24):

$$\frac{G_L D}{v} < \frac{-m_L C_o (1 - k)}{k} \qquad \text{(Eq 1)}$$

where G_L is the temperature gradient in liquid, v is the velocity, m_L is the liquidus slope, C_o is the initial alloy composition, k is the distribution (or equilibrium partition) coefficient, and D is the solutal liquid diffusion coefficient.

The right-hand side of Eq 1 is the constitutional supercooling parameter. Constitutional supercooling also occurs ahead of cells and dendrites, but is less than that ahead of a planar front because, in cells and dendrites, solute is rejected laterally, as well as forward.

The fact that constitutional supercooling exists in the bulk melt during the freezing of an alloy is not disputed. It can be easily measured and is usually necessary for the growth of grains ahead of the columnar front (pure materials do not freeze in an equiaxed fashion under normal casting conditions, although they can be made to do so by vigorous stirring both to fragment the growth front and to remove the bulk superheat).

Furthermore, CS-driven heterogeneous nucleation is obviously important when an

Table 1 Casting parameters affecting macrostructure

Casting variables	Effects	Ref
Superheat	Increased superheat increases the extent of columnar growth; the trend is less noticeable in large castings.	1–8
Alloy system	Low values of $-m_L(1 - k)C_o/k$ favor columnar structure; high values, equiaxed structure.	7, 9–11
Composition	Increased alloying content (C_o) tends to decrease the extent of columnar region; some investigators report that the columnar region is not a simple function of alloy concentration.	5–12
Fluid flow (natural or forced)	Increased fluid flow decreases the extent of the columnar region.	1–3, 5, 13–16
Inoculation grain refining	Producing nuclei is not sufficient to give an equiaxed zone. Grain size is cooling rate dependent. Grain-refining additions can reduce the extent of columnar growth.	5, 6, 11, 17–19
Mechanical vibration	Mechanical vibration promotes grain refinement and can extend the equiaxed zone.	20, 21
Size	Superheat has less effect on the grain structure of large castings. Increasing cross section yields increasing proportion of equiaxed grains. Most sensitive to height of casting	3, 7, 22

efficient substrate, such as a grain refiner or inoculant, is present. However, the evidence that it is generally the crucial mechanism determining the columnar to equiaxed transition is not convincing. This evidence falls into two groups:

First, many workers have reported correlations at the columnar to equiaxed transition between C_o and a combination of G_L and v, namely, G_L/v or $G_L/v^{0.5}$ (Ref 12, 25-30). Because these quantities affect the degree of constitutional supercooling ahead of the columnar front (Eq 1), these correlations were thought to provide evidence for the CS-driven nucleation mechanism.

Second, a correlation between the constitutional supercooling parameter and macrostructure is observed experimentally. Low values of the constitutional supercooling parameter were found to give large-grain columnar structures, while high values produced equiaxed structures (Ref 9). It has been shown that the size of the equiaxed zone was related to the constitutional supercooling parameter (Ref 10).

However, the relationship among C_o, G_L, and v at the columnar to equiaxed transition and the dependence of the equiaxed zone on the constitutional supercooling parameter can be rationalized in terms of the growth of the grains, and it does not necessarily indicate the influence of constitutional supercooling on heterogeneous nucleation. The variables G_L, v, and C_o affect the growth of the equiaxed grains by controlling the degree and extent of the undercooled liquid in the bulk. A simple steady-state growth analysis suggests that some relationship of the form $G_L/v \propto C_o$ at the columnar to equiaxed transition is not unexpected, owing to the influence of these variables on the growth of the grains (see the section "Predicted Versus Observed Behavior" in this article). Furthermore, the same combination of variables that is the constitutional supercooling parameter also appears in the growth analysis, via the kinetic expression for the dendrite tips (Ref 11, 31-34).

The CS-driven nucleation mechanism has been criticized because constitutional supercooling in an alloy must exist ahead of the columnar front at an early stage of freezing (because the liquid temperature gradient drops rapidly and a solute layer is soon established ahead of the columnar front), but the columnar to equiaxed transition does not occur until some time later (Ref 35, 36). Chalmers has shown, through the mechanical isolation of the center of a casting with a metal cylinder, that the CS nucleation mechanism is not solely responsible for equiaxed growth (Ref 35). The center solidified as fine, equiaxed grains without the cylinder, but when the center was mechanically isolated, these grains were replaced by fewer and coarser grains, even though the center was still constitutionally supercooled. In addition, another researcher has shown that equiaxed grains can be produced in a melt in the absence of heterogeneous nuclei prior to solidification (Ref 37, 38).

Big Bang Mechanism. In this mechanism, equiaxed grains result from the predendritic nuclei formed during pouring by the initial chilling action of the mold. The grains are then carried into the bulk by fluid flow and survive until the superheat has been removed (Ref 35). The survival of chill nuclei until the superheat is dissipated is quite likely at moderate superheats, because of the large latent heat of solidification of metals.

Big bang nucleation has been observed in cooled, saturated NH_4Cl solution (Ref 39, 40). A sharp change was noted, with increasing superheat, from a very large number to zero crystals remaining in the liquid after pouring. When nuclei produced by the initial chill survived, they grew into equiaxed crystals and settled to occupy only the bottom part of the casting.

Predendritic grains have been observed trapped in the columnar and equiaxed zones (Ref 41). They are rounded and smooth, as would be expected if they had been in contact with liquid for a long time. However, the origin of these nuclei is uncertain; although they could have originated during pouring, they might have been produced by dendrite remelting. The existence of the chill zone has been viewed as evidence in support of big bang nucleation.

Unlike the CS theory, the big bang theory can explain the effects of superheat and convection in the early stages of casting. Variations in superheat and convection do not significantly alter the onset and extent of constitutional supercooling in the melt, yet they do exert considerable influence on the cast structure (Ref 1, 2). Increasing the pouring temperature reduces the size of the equiaxed zone and coarsens the grain sizes, and reducing the convection by the introduction of a static magnetic field can eliminate the equiaxed zone altogether. The big bang mechanism provides two viable explanations:

- Increasing the superheat diminishes the chilling of the melt upon pouring and increases the time taken for the bulk superheat to dissipate; consequently, fewer nuclei are produced upon pouring; and fewer still survive to grow into equiaxed grains (Ref 40)
- Decreasing convection reduces the number of nuclei formed at the edges of a casting that reach the center

Chalmers suggested that mechanically isolating the center of the casting alters the macrostructure by obstructing the flow of chill nuclei (Ref 35). However, another researcher modified Chalmers experiment by leaving a gap at the bottom of the central cylinder (Ref 42). Thus, the grain structure at the center was shown to be the same whether the casting was filled by pouring down the central tube or into the outer section. This cast doubt on the Chalmers big bang interpretation because the two pouring arrangements (down the central tube or into the outer section) should have washed different numbers of chill nuclei into the central section if the mechanism were operative. Another weakness of the big bang theory is its inability to account for equiaxed zone formation in the absence of a chilled mold (Ref 37).

Dendrite Detachment Mechanism. Other researchers noticed that convective mixing or stirring during solidification of organic analogs produced a large number of nuclei in the liquid (Ref 39). It was therefore postulated that fluctuations in the growth rate caused dendrite arms to melt off and then float into the center (Ref 39). Remelting due to surface energy occurred under isothermal conditions, but recalescence, either locally or throughout the entire casting, was thought to be the main mechanism for dendrite arm detachment. Remelting was promoted by the presence of sufficient solute to alter appreciably the melting point of the solvent. It was proposed that convection could cause dendrite detachment mechanically because the yield point of the metal is negligible near its melting point (Ref 43). It was also demonstrated that side-arm remelting can occur at very low interdendritic fluid flow velocities (of the order of 40 $\mu m \cdot s^{-1}$, or 1600 $\mu in. \cdot s^{-1}$) (Ref 44).

The surface dendrite layer at the top of a casting is an important source of equiaxed grains (Ref 39, 45). Other researchers thought that the coarse equiaxed structure produced in their large castings, with high superheats and an applied constant magnetic field, was caused by grains showering from the top; the superheat and lack of convection would have been detrimental to the survival and creation of big bang nuclei and detached dendrite arms (Ref 3). Dendrite fragments rejected ahead of the columnar front as a result of density-driven interdendritic flow and subsequent channeling are a further source of equiaxed grains (Ref 46).

Dendrite detachment is consistent with the influence of convection. Reducing the convection in the bulk by applying a static magnetic field (Ref 1, 2) or by utilizing the Coriolis force (Ref 13) reduces the extent of the equiaxed zone and coarsens the grains, or even eliminates the zone completely. Conversely, increasing the convection by imposing an alternating magnetic field promotes the equiaxed zone and refines the grain structure (Ref 14). Convection should favor dendrite detachment by mechanical means or local remelting and then transport the dendrite fragments ahead of the front. In delayed field experiments, researchers have found that the big bang and dendrite detachment mechanisms both operate (Ref 1). However, in the absence of pouring turbulence, the removal of the magnetic field when the central temperature reached the freezing plateau did not produce an equiaxed zone (Ref 2). This implied that the dendrite detachment mechanism did not operate. However, of the three nucleation mechanisms, only dendrite detachment can explain the formation of an equiaxed zone when heterogeneous nuclei were absent and the mold exerted no chilling effect (Ref 37).

Factors Supporting All Three Mechanisms. Constitutional supercooling would seem not to drive the nucleation of equiaxed grains except in the presence of a grain refiner or other efficient substrate. In the following section, it is seen that the effect of composition is probably due to its influence on the growth of the grains, not to its influence on their nucleation. There is strong evidence to support the big bang and dendrite detachment theories. Neither of the two mechanisms is compatible with all of the observations, but a combination of the two is consistent with most of them. In particular, the big bang theory can effectively explain the superheat effect (in terms of nucleation, but growth arguments can also rationalize the effect of superheat to some extent), and dendrite detachment is frequently observed in analogs and is consistent with the influence of convection on macrostructure. The big bang and dendrite detachment mechanisms would appear to be very efficient. Many big bang nuclei are probably produced upon pouring if the mold is cold, and it is likely that the dendrite detachment mechanism will provide a large number of nuclei during solidification because of the ease of fragmentation of the dendrites. Additional information on nucleation, solidification, and structure is available in the article "Nucleation Kinetics" and other sections of the article "Fundamentals of Growth" in this Volume.

Growth of Equiaxed Grains

Two modes of equiaxed growth have been observed (Ref 4, 40, 47-49):

- Grains in the bulk that sediment out to form a pile at the base of the casting which then impede the advancing columnar front
- Equiaxed grains attach themselves to the columnar front and then start to develop some columnar characteristics

Sedimentation produces the fully equiaxed zone, and adhesion gives the branched columnar structure (Ref 4, 40, 47-49). The combination of sedimentation and adhesion leads to a macrostructure of the form shown schematically in Fig. 1. The branched columnar zone increases with height because the front is obstructed at a later time by the sedimenting grains at higher positions. Sedimentation is due to the change in the density of the grain as a result of solidification shrinkage (~6%), not because of solute rejection during freezing, because most of the rejected solute is trapped interdendritically within the envelope of the grain.

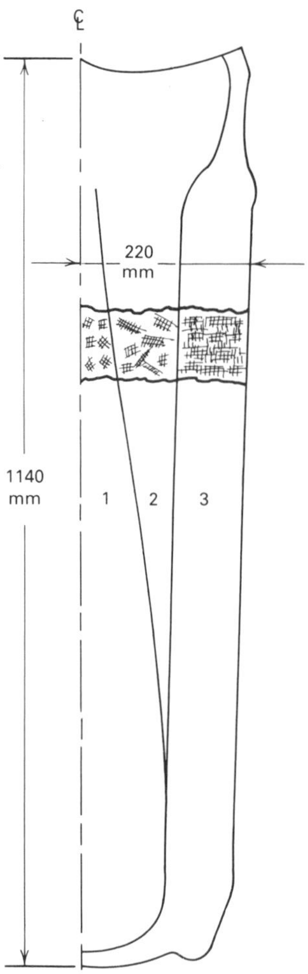

Fig. 1 Structure diagram of a 2.7 kg (6.0 lb) steel ingot cross section from a 220 × 220 mm (8.7 × 8.7 in.) square ingot that is 1140 mm (44.9 in.) long. Diagram illustrates equiaxed (1), branched columnar (2), and columnar (3) crystals. Source: Ref 6

Analog investigations have shown that equiaxed growth starts at an early stage in an undercooled layer just ahead of the columnar dendrite tips, before the superheat has been removed from the center of the casting (Ref 40, 50, 51). Calculations support this and indicate that two regions can arise (Ref 5, 52):

- Equiaxed growth continues to be confined to a narrow undercooled layer; the dimensions of the layer then determine the extent of the equiaxed growth
- Superheat is removed from the center at an early stage; equiaxed growth then occurs right to the center

In the first case, the equiaxed growth might be sufficient to obstruct the columnar dendrites. Growth then continues by the inward movement of an equiaxed front. The continuation of the equiaxed nature of the front requires a sufficiently frequent rate of adhesion of equiaxed crystals from the bulk; otherwise, the attached equiaxed grains at the front will develop a columnar character (Ref 6).

An equiaxed zone forms when the equiaxed grains in the bulk are sufficient in number and grow rapidly enough to obstruct the columnar front. There is a growth competition between the columnar and equiaxed grains. The crucial factors that determine the outcome are the degree and extent of the constitutional supercooling in the liquid, and the velocity of the columnar front. It is possible for equiaxed nuclei to exist ahead of a columnar front and yet not develop into an equiaxed zone because of the conditions being unfavorable for their growth.

An equiaxed zone is encouraged by a shallow temperature gradient in the bulk. Several investigations have shown that the equiaxed zone can be reduced or suppressed by maintaining higher temperatures in the melt and by reducing natural convection (Ref 13, 15). For example, the thermal gradient and bulk temperature were both found to be greater in space experiments than on earth; and nuclei introduced ahead of a columnar front in space did not grow, while on earth the same casting arrangement yielded a large equiaxed zone (Ref 16). It was also suggested that electromagnetic stirring promotes equiaxed growth not by increasing the dendrite detachment but by flattening the thermal profile in the bulk and increasing the rate of heat transport, because, even without stirring, natural convection is sufficient to create a large number of nuclei (Ref 53).

Growth Models

Early work on the columnar to equiaxed transition concentrated on the production of new crystals, but recently more attention has been paid to their growth. Lipton *et al.* (Ref 7), Fredriksson and Olsson (Ref 6), and Flood and Hunt (Ref 5, 52) have produced models of the competition between the columnar and equiaxed grains. In all three models, a columnar front advances into a melt, and equiaxed grains grow ahead of it. The important result of these models is that they reproduce the experimentally observed trends in the columnar to equiaxed transition by considering the growth of the equiaxed grains and without recourse to nucleation arguments. A fixed number of potential grains per unit volume was assumed to exist in the bulk at all times throughout a simulation. The discussion that follows points out the salient features of these models.

Description of Columnar Front. Lipton *et al.* and Fredriksson and Olsson specify the velocity with a $t^{-0.5}$ relation and then deduce the front undercooling from an expression linking velocity and undercooling. Lipton *et al.* use the Kurz-Fisher expression (Ref 54), and Fredriksson and Olsson use a parabolic relation.

On the other hand, Flood and Hunt calculate the velocity and undercooling dynamically throughout the calculation. They do not fix these parameters *a priori*, because this might prejudice the outcome of the competition between the columnar and equiaxed growth. Unlike the other two models, the Flood-Hunt model accounts for the thermal interaction between the front and the equiaxed grains. The front velocity and undercooling are obtained by invoking continuity of heat flow at the columnar dendrite tips in conjunction with Burden-Hunt velocity relation (which reduces to a $v \propto (\Delta T)^2$ relation in the limit of low gradient) (Ref 31-34). The columnar dendrites are modeled as Scheil shapes truncated at a varying undercooling, that is, the solid fraction in the columnar region is described by the Scheil equation up to the tip temperature. The Scheil assumption is good because the mixing of the solute is nearly complete within a very short distance behind the dendrite tips.

Description of Equiaxed Grains. Lipton *et al.* assume the grains to be spherical, with a growth rate controlled by the rate of diffusion of latent heat into the liquid. They relate the final grain size to the square root of the mean temperature difference between the grain and the surrounding liquid, and the duration of this undercooling.

Flood and Hunt and Fredriksson and Olsson adopt a different and arguably more realistic treatment. They describe the grains as bundles of dendrites growing at a rate dependent on the square of the local bulk undercooling. This law introduces the solute diffusion away from the equiaxed dendrite tips. The dendrite tips are assumed to be at the local bulk undercooling, which is a good approximation owing to the high thermal diffusivity of metals; the thermal diffusion fields of neighboring grains will soon overlap. Flood and Hunt and Fredriksson and Olsson include the latent heat liberated by the equiaxed grains. The former assume that the internal solid fraction of the grains is the Scheil fraction for the local temperature; that is, they assume complete interdendritic mixing. On the other hand, Fredriksson and Olsson assume it to be a constant value (of 0.3). The grains are assumed to be stationary in the bulk, but Flood and Hunt argue that, when they consider convective mixing in the bulk, they are in effect allowing for complete mixing of liquid and grains.

Flood and Hunt accounted for the impingement of neighboring equiaxed grains by an Avrami-type treatment (Ref 55). Impingement was separated from the kinetics of growth through the concept of an extended volume fraction of grains.

Treatment of Thermal Transport. Lipton *et al.* and Fredriksson and Olsson consider convective heat transport in the bulk, describing this with the laminar boundary layer theory. Flood and Hunt consider both diffusive and convective heat flow in the liquid.

Criterion for the Columnar to Equiaxed Transition. Lipton *et al.* and Fredriksson and Olsson apply thermal criteria. Lipton *et al.* found that, in an organic analog system, the latent heat evolved by growing equiaxed grains suppressed the thermal boundary layer ahead of the columnar front and that a rise in the temperature of the thermal boundary layer accompanied the onset of the columnar to equiaxed transition (Ref 56). They believe that this rise in temperature corresponded to the overlap of the thermal fields of neighboring grains, which they considered would occur when a grain had grown to a critical size of one-tenth of the intergranular spacing. Consequently, Lipton *et al.* allow the columnar front to advance until it reaches a grain that has grown to this critical size, at which point the columnar to equiaxed transition is said to have occurred. However, as indicated earlier, in metallic systems, the thermal diffusivity is so high that the neighboring grains would probably interact thermally long before reaching the critical size.

On the other hand, Fredriksson and Olsson assume that the columnar to equiaxed transition occurs when the bulk temperature falls to a minimum prior to recalescence. This was based on an experimental correlation and was rationalized by associating recalescence with equiaxed grains that had developed to a size sufficient to obstruct the columnar front. Fredriksson and Olsson thought that this recalescence would also increase the temperature difference between the front and the bulk, causing greater convection, which in turn would enhance crystal multiplication and accelerate the obstruction of the front. However, the increasing temperature difference across the thermal boundary layer would appear not to fit the findings of Lipton *et al.*

Flood and Hunt adopt a criterion based on a probability argument. The columnar to equiaxed transition is assumed to occur when a certain volume fraction of grains exists ahead of the columnar front.

Predicted Versus Observed Behavior. Fredriksson and Olsson and Flood and Hunt calculated cooling curves that show recalescence accompanying equiaxed growth. The nature of the latent heat evolution produced a more curved cooling curve with equiaxed growth than with columnar growth. This difference in character has been noted experimentally (Fig. 2).

All three growth models predict a decreasing columnar range with increasing alloy content and decreasing superheat, which suggests that these trends do not necessitate a nucleation based mechanism but can be explained in terms of growth. The compositional dependence enters through the velocity-undercooling expression and, in the case of Flood and Hunt, also through the shape of the columnar dendrites (Ref 57). The superheat effect is due to the initial influence of the pouring temperature on growth conditions. Lipton *et al.* and Fredriksson and Olsson calculated a larger and more realistic superheat trend than Flood and Hunt, but this could have resulted from modeling a different size of casting, employing a different dendrite growth expression, or imposing different cooling rates. However, Flood and Hunt speculated that the lack of a dramatic superheat effect in their results supports the big bang mechanism; consideration of only grain growth is not sufficient to reproduce the influence of superheat. Flood and Hunt also predict a columnar to equiaxed transition sooner after the dissipation of the superheat than Fredriksson and Olsson. However, this would appear to be the case because Flood and Hunt modeled a larger number of grains per unit volume and used a higher kinetic constant (causing higher growth rate).

Flood and Hunt showed that flattening the thermal gradient in the bulk by convection reduces the columnar range by increasing the bulk undercooling and promoting equiaxed growth. Introducing a temperature-dependent nucleation rate into a growth model postponed equiaxed growth until the heterogeneous nucleation undercooling was achieved. If the columnar front undercooling did not fall below this value, then the casting was fully columnar.

Fredriksson and Olsson forecast an increase in the columnar range with increased height or width of the casting, with the columnar to equiaxed transition being more sensitive to the width than to the height. These results are consistent with the findings of other researchers (Ref 22). Lipton *et al.* point out that the columnar to equiaxed transition will occur only if the total solidification time is significantly longer than the time taken for the dissipation of the superheat. This means that small billets and thin slabs will solidify as predominantly columnar, while large sizes will reveal a large portion of equiaxed crystals.

Growth Parameters

The size of an equiaxed grain in the growth models is determined by:

- The time interval between the grains starting to grow and the arrival of the columnar front
- The undercoolings that are experienced during this period

Recognizing this, it is possible to perform a simple steady-state analysis of equiaxed growth that provides some insight into the columnar to equiaxed transition (Ref 11,

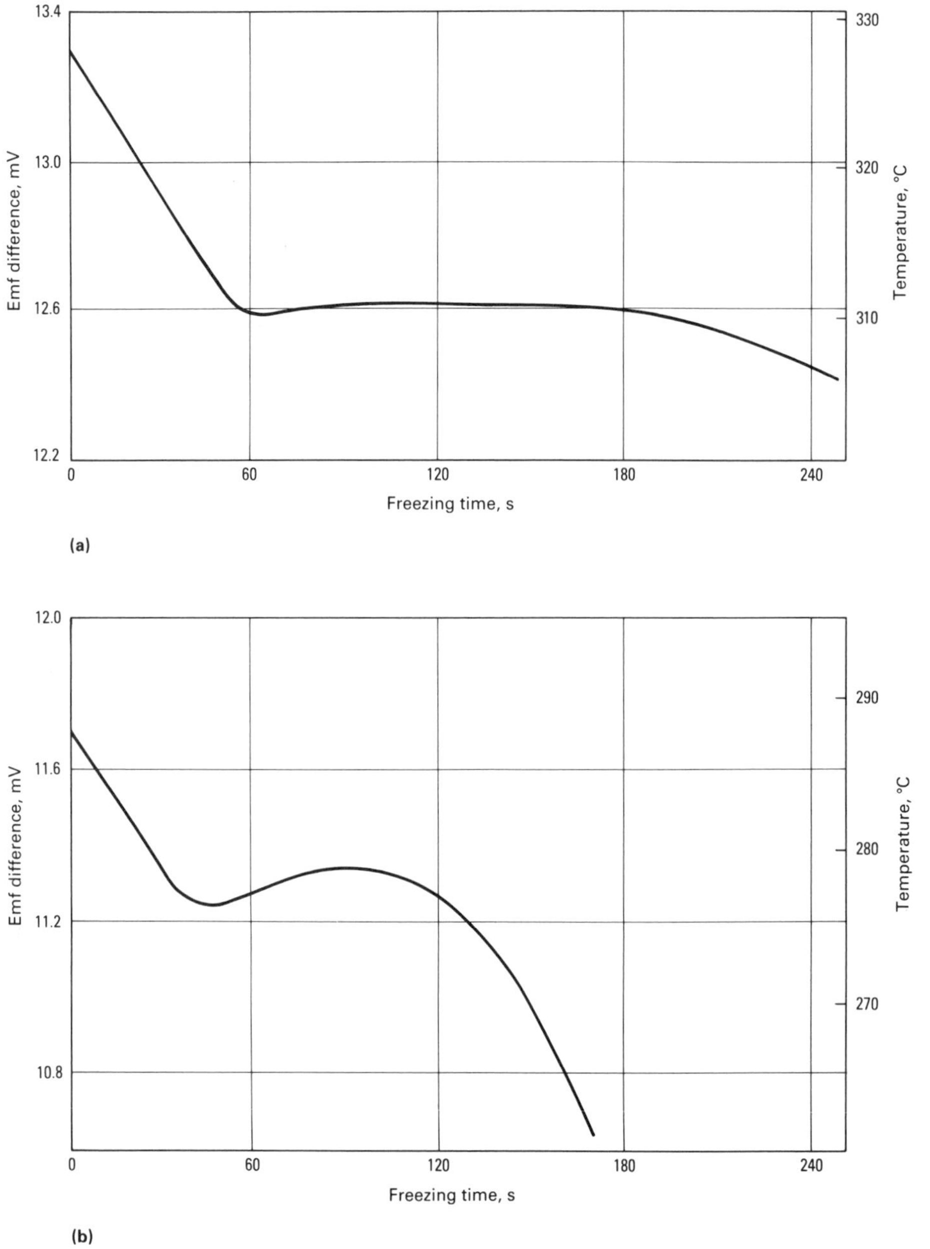

Fig. 2 Comparison of the configuration of cooling curves for a columnar (a) and fully equiaxed (b) growth. (a) The first casting, having a Pb-2.25 Sb concentration and a 0.360 °C · s^{-1} cooling rate. (b) The fourth casting, having a Pb-6.50 Sb concentration and a 0.320 °C · s^{-1} cooling rate. Source: Ref 50

57). It suggests that the structure will be fully columnar when:

$$G_L > 0.617\,(100\,N_o)^{1/3}\left[1-\left(\frac{\Delta T_n}{\Delta T_c}\right)^3\right]\Delta T_c \quad \text{(Eq 2)}$$

and fully equiaxed when:

$$G_L < 0.617\,(N_o)^{1/3}\left[1-\left(\frac{\Delta T_n}{\Delta T_c}\right)^3\right]\Delta T_c \quad \text{(Eq 3)}$$

and

$$\Delta T_c = \left[-8\,\Gamma\,m_L\,(1-k)\,\frac{C_o v}{D}\right]^{1/2} \quad \text{(Eq 4)}$$

where N_o is the number of nuclei per unit volume, ΔT_c is the undercooling at the columnar front, ΔT_n is the critical undercooling for nucleation on a substrate, and Γ is the Gibbs-Thomson coefficient.

These inequalities enable us to rationalize the columnar to equiaxed transition under different casting conditions. For example, in a roll caster in which the heat transfer coefficient is very high, both the velocity and the temperature gradient will be high; therefore, the extent of the equiaxed zone will critically depend on the number of heterogeneous nucleation sites. In contrast, in a sand casting in which there are low gradients and velocities, equiaxed growth will depend less on the quality of grain refiner and more on its efficiency.

The compositional dependence of the structure in the steady-state analysis stems from the quantity $-m_L\,(1-k)C_o$ in the expression for the columnar front undercooling. This quantity, which is introduced through the velocity-undercooling relation for the dendrite tips, is in fact the temperature difference between the liquidus and solidus at a composition of $k\,C_o$ (and very nearly the constitutional supercooling parameter). At steady state, therefore, Hunt points out that in a peritectic system such as iron-carbon, the tendency to form equiaxed crystals will increase with increasing carbon concentration in the δ region until austenite begins to form, when it will decrease, but will then eventually increase at higher concentrations (see the section "Solidification of Peritectics" in this article) (Ref 11). Lipton *et al.* also discuss the ferrite-austenite phase change and note that the undercooling of the columnar front will adapt to the change in properties (mainly m_L and k) and alter the columnar to equiaxed transition (Ref 7).

Fredriksson and Olsson discuss the influence of convection and the sedimentation of grains on the columnar to equiaxed transition, neither of which have been treated adequately to date (Ref 6). They pointed out that, prior to the removal of the superheat in the bulk, convection would cause remelting of many of the free crystals growing in an undercooled layer near the front by transporting them into the center and that, after the removal of superheat, convection would cause crystal multiplication. Sedimentation of crystals would tend to remove free crystals from the melt, and this was thought to have caused a decrease in a measured cooling curve following the recalescence at the columnar to equiaxed transition.

Rappaz and Thévoz have considered the growth of equiaxed grains in detail (Ref 58). They modeled the solute rejection from an individual grain during growth and accordingly corrected the solid fraction within the grain and the supersaturation in the bulk. Flood and Hunt neglected this rejection and assumed, first, that the solid fraction within a grain is given by the Scheil equation and, second, that the bulk composition is constant throughout solidification. Rappaz and Thévoz described the solute field in the liquid, within and outside a grain, by assuming a uniform liquid composition within the grain, by applying a symmetry condition at half the intergranular distance, and by invoking solute conservation. The temperature and composition within a grain were coupled by assuming equilibrium, and a grain was considered to be isothermal. Cooling curves were calculated for a single grain and could be fitted, by adjusting parameters, to provide a good match with experimental results.

However, the computational effort required by their full numerical treatment caused Rappaz and Thévoz to devise a simpler, more approximate analysis so that they could scale up their simulations from a single grain to an entire casting (Ref 59). This analytical model agrees well with the full calculations; it accounts for the solute in a diffusion profile at the edge of a grain and provides an expression for solid fraction that is dependent on grain size and dendrite tip velocity. Thévoz *et al.* (Ref 17) used this solid fraction expression in place of the Scheil assumption in a model of equiaxed growth similar to that of Flood and Hunt (Ref 5, 57). They calculated cooling curves showing a recalescence and predicted a variation in grain size across the casting owing to different local cooling rates. They used a rather complex nucleation model. Early work by other researchers considered the growth of nondendritic spheres, and a similar dependence of grain size on cooling rate was obtained by using a classical heterogeneous law (Ref 18).

The Rappaz-Thévoz model probably overestimates the rejection of solute from an equiaxed grain. Solute is not rejected over the entire surface of the envelope of a grain. Much of the solute will be trapped interdendritically, and the solute profile at the leading dendrite tips will probably extend over only a few tip radius lengths. The Rappaz-Thévoz model, however, is the first to introduce the accumulation of solute in the bulk. The corrections that it introduces into the Flood-Hunt analysis will be significant toward the end of solidification, because then the solute loss from a grain will have the greatest effect on the solid fraction calculation and the solute fields of neighboring grains will overlap and decelerate growth.

Rappaz and Thévoz introduced a slowing factor into their dendrite velocity relation to fit their calculated cooling curves to the measured ones. Indeed, the growth of equiaxed tips has been measured in analogs as being considerably slower than that predicted by the usual expressions for velocity. This might suggest that a more complex and better analysis of the growth of equiaxed dendrite tips should be developed. However, regardless of the validity of such an analysis, the uncertainty in the material parameters (that is, surface energy and diffusion coefficient) is so great that an error of one or two orders of magnitude in velocity could be expected. It might be suggested that fitting the velocity-undercooling relation is pragmatically more useful than applying a model of dendrite growth when simulating real castings.

Lipton *et al.* have proposed a more sophisticated treatment of the equiaxed dendrite tip (Ref 60). They consider the coupled Ivantsov solutions for the temperature and composition profiles at an isolated unconstrained dendrite tip and then obtain an operating point by invoking marginal stability. The model should probably be used with caution, however, because it considers an isolated dendrite rather than an array.

Witzke *et al.* treated the redistribution of solute ahead of a columnar front and attempted to calculate the degree of constitutional supercooling in the bulk (Ref 61). The forward gradient term in the solutal transport equation was neglected because, it was argued, it is small compared to another term, yet it is this neglected term that rejects solute into the bulk. In addition, solute would accumulate in the fluid as it passes the front, but this is not included in the treatment. This accumulation would cause the front to melt back and curve. The boundary layer theory predicts that the thermal boundary layer, and therefore the undercooled region, will increase with depth in a casting, and this could explain the observation that crystals formed first near the base of an analog casting (Ref 50, 51).

Columnar Versus Equiaxed Grain Growth

The importance of the relative growth rates of columnar and equiaxed grains has been recognized in recent years. Growth models can predict all the main trends in the columnar to equiaxed transition. The development of an equiaxed zone is dependent on the presence of equiaxed nuclei and is affected by the thermal growth conditions in the bulk and the speed of the columnar front. The effect of composition on the columnar to equiaxed transition would seem to be due to the dependence of the dendrite growth rate on composition. Convection can promote equiaxed growth by lowering the thermal gradient. Experimental evidence suggests that both the big bang and dendrite detachment mechanisms apply and are efficient processes for the creation of equiaxed grains. Although the superheat effect can be predicted by a consideration of growth alone, the big bang mechanism is probably partly responsible. The CS-driven heterogeneous nucleation mechanism is important only when a grain refiner or efficient substrate is present.

The main areas of macrostructural development which have yet to be treated in sufficient detail are the effect of fluid flow during pouring and growth plus the effect of the sedimentation of the equiaxed crystals.

REFERENCES

1. D.R. Uhlmann, T.P. Seward III, and B. Chalmers, *Trans. AIME*, Vol 236, 1966, p 527
2. J.A. Spittle, G.W. Delamore, and R.W. Smith, in *The Solidification of Metals*, Publication 110, Iron and Steel Institute, 1968, p 318
3. R. Morando, H. Biloni, G.S. Cole, and G.F. Bolling, *Metall. Trans.*, Vol 1, 1970, p 1407
4. R.D. Doherty and D. Melford, *J. Iron Steel Inst.*, Vol 204, 1964, p 1131
5. S.C. Flood and J.D. Hunt, *J. Cryst. Growth*, Vol 82, 1987, p 552; see also *Modeling of Casting and Welding Processes II*, American Institute of Mining, Metallurgical and Petroleum Engineers, 1983, p 207; *Modeling and Control of Casting and Welding Processes III*, American Institute of Mining, Metallurgical and Petroleum Engineers, 1986, p 607
6. H. Fredriksson and A. Olsson, *Mater. Sci. Technol.*, Vol 2, 1986, p 508
7. J. Lipton, W. Kurz, and W. Heinemann, *Concast Technol. News*, Vol 22, 1983, p 4
8. B. Chalmers, *Principles of Solidification*, John Wiley and Sons, 1964
9. L.A. Tarshis, J.L. Walker, and J.W. Rutter, *Metall. Trans.*, Vol 2, 1971, p 2589
10. R.D. Doherty, P.D. Cooper, M.H. Bradbury, and F.J. Honey, *Metall. Trans. A*, Vol 8A, 1977, p 397
11. J.D. Hunt, *Mater. Sci. Eng.*, Vol 65, 1984, p 75
12. G.S. Cole and G.F. Bolling, *Trans. AIME*, Vol 242, 1968, p 153
13. G.S. Cole and G.F. Bolling, *Trans. AIME*, Vol 239, 1967, p 1824
14. G.S. Cole and G.F. Bolling, *Trans. AIME*, Vol 236, 1966, p 1366
15. G.S. Cole and G.F. Bolling, *Trans. AIME*, Vol 233, 1968, p 1568
16. J.M. Papazian and T.Z. Kattamis, *Metall. Trans. A*, Vol 11A, 1980, p 483
17. Ph. Thévoz, Zon Jie, and M. Rappaz, in *Proceedings of the Third International Conference on Solidification Processing* (Sheffield, U.K.), Institute of Metals, 1988
18. I. Maxwell and A. Hellawell, *Acta Metall.*, Vol 23, 1975, p 229
19. J. Leszezynski and N.J. Petch, *Met. Sci.*, Vol 8, 1974, p 5
20. J. Campbell, in *Solidification Technology in the Foundry and Casthouse*, The Metals Society, 1980, p 61
21. J. Campbell, *Int. Met. Rev.*, Vol 26, 1981, p 71
22. I. Laren and H. Fredriksson, *Scand. J. Metall.*, Vol 1, 1972, p 59-69
23. W. Winegard and B. Chalmers, *Trans. ASM*, Vol 46, 1954, p 1214
24. W.A. Tiller, K.A. Jackson, J. Rutter, and B. Chalmers, *Acta Metall.*, Vol 1, 1953, p 428
25. T.S. Plaskett and W.C. Winegard, *Trans. ASM*, Vol 51, 1959, p 222
26. H. Biloni and B. Chalmers, *J. Mater. Sci.*, Vol 3, 1968, p 139
27. D. Walton, *Trans. ASM*, Vol 51, 1959, p 222
28. G.S. Cole, *Can. Metall. Q.*, Vol 8, 1969, p 189
29. R. Elliot, *Br. Foundryman*, Vol 9, 1964, p 389

30. W.A. Tiller, *Trans. AIME*, Vol 224, 1962, p 448
31. M.H. Burden and J.D. Hunt, *J. Cryst. Growth*, Vol 22, 1974, p 99
32. M.H. Burden and J.D. Hunt, *J. Cryst. Growth*, Vol 22, 1974, p 109
33. M. Tassa and J.D. Hunt, *J. Cryst. Growth*, Vol 34, 1976, p 38
34. J.D. Hunt, *Solidification and Casting of Metals*, The Metals Society, 1977, p 3
35. B. Chalmers, *J. Aust. Inst. Met.*, Vol 8, 1962, p 225
36. F.R. Hensel, Ph.D thesis, University of Berlin, 1929
37. J.L. Walker, private communication, 1966
38. J.L. Walker, in *Transactions of the Sixth Vacuum Metallurgy Conference*, New York University Press, 1963, p 33
39. K.A. Jackson, J.D. Hunt, D. Uhlmann, and T.P. Seward III, *Trans. AIME*, Vol 236, 1966, p 149
40. M.H. Burden, D. Phil. thesis, University of Oxford, 1973
41. H. Biloni and B. Chalmers, *Trans. AIME*, Vol 233, 1965, p 373
42. R.T. Southin, in *The Solidification of Metals*, Publication 110, Iron and Steel Institute, 1968, p 305
43. S. O'Hara and W.A. Tiller, *Trans. AIME*, Vol 239, 1967, p 497
44. M.R. Bridge, Ph.D thesis, University of Sheffield, 1981
45. R.T. Southin, *Trans. AIME*, Vol 239, 1967, p 220
46. R.J. McDonald and J.D. Hunt, *Trans. AIME*, Vol 245, 1969, p 1993
47. H. Fredriksson and M. Hillert, *Metall. Trans.*, Vol 3, 1972, p 565
48. P. Salmon-Cox and J. Charles, *J. Iron Steel Inst.*, Vol 201, 1963, p 863
49. A. Kohn, *Mem. Sci. Rev. Met.*, Vol 60, 1963, p 711
50. S.E. Kisakurek, *J. Mater. Sci.*, Vol 19, 1984, p 2289
51. S. Witzke, J.-P. Riquet, and F. Durand, *Mem. Sci. Rev. Met.*, 1979, p 701
52. S.C. Flood and J.D. Hunt, *J. Cryst. Growth*, Vol 82, 1987, p 543
53. M.R. Bridge and G.D. Rogers, *Metall. Trans. B*, Vol 15B, 1984, p 581
54. W. Kurz and D.J. Fisher, *Acta Metall.*, Vol 29, 1980, p 11
55. J.W. Christian, in *The Theory of Transformations in Metals and Alloys, Part I*, Pergamon Press, 1975, p 17
56. J. Lipton, W. Heinemann, and W. Kurz, *Arch. Eisenhüttenwes.*, Vol 55, 1984, p 195
57. S.C. Flood, D. Phil. thesis, University of Oxford, 1985
58. M. Rappaz and Ph. Thévoz, *Acta Metall.*, Vol 35, 1987, p 1487
59. M. Rappaz and Ph. Thévoz, in *Proceedings of the Third International Conference on Solidification Processing* (Sheffield, U.K.), Institute of Metals, 1988
60. J. Lipton, M.E. Glicksman, and W. Kurz, *Mater. Sci. Eng.*, Vol 65, 1984, p 57
61. S. Witzke, J.P. Riquet, and F. Durand, *Acta Metall.*, Vol 29, 1981, p 365

Microsegregation and Macrosegregation

I. Ohnaka, Department of Materials Science and Processing, Osaka University, Japan

All metallic materials contain solute elements or impurities that are randomly distributed during solidification. The variable distribution of chemical composition on the microscopic level in a microstructure, such as dendrites and grains, is referred to as microsegregation. Variation on the macroscopic level is called macrosegregation. Because these segregations generally deteriorate the physical and chemical properties of materials, they should be kept to a minimum.

Equilibrium Phase Diagram and Equilibrium Partition Coefficient

The solute elements in alloys are redistributed during the solidification process so that the chemical potential in the liquid and solid phases is equalized (Ref 1). As the solidification proceeds under the equilibrium condition, the solute compositions in the solid, C_S, and the liquid, C_L, vary along the solidus and liquidus lines, respectively (Fig. 1). The ratio C_S/C_L is referred to as the equilibrium partition or distribution coefficient, k.

In the equilibrium condition, the liquid and solid composition can be calculated by:

$$C_L = \frac{C_o}{[1 - (1 - k) f_S]} \qquad \text{(Eq 1)}$$

where C_o is the initial alloy composition and f_S is the solid fraction, and by:

$$C_S = k\, C_L \qquad \text{(Eq 2)}$$

After solidification ($f_S = 1$), the solute composition is designated C_o; theoretically, no microsegregation occurs at this point. In actuality, however, the equilibrium solidification rarely takes place, because the solute diffusion is not so rapid.

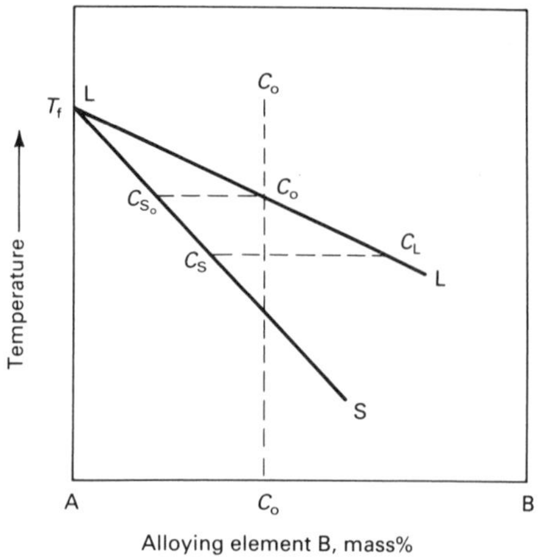

Fig. 1 Schematic of an equilibrium phase diagram of a binary alloy. The liquidus is represented by line LL; the solidus by line LS.

Solute Redistribution in Nonequilibrium Solidification

If the solute redistribution in a volume element between platelike dendrites as shown in Fig. 2 is considered and if negligible undercooling at the solid/liquid interface (local equilibrium condition) with no net flow of solute through the volume element is assumed, the liquid composition can be calculated as follows.

First, in the case of $D_S = 0$ and $D_L = \infty$ (indicating no diffusion in the solid and complete mixing in the liquid), where D_S is the solid diffusivity and D_L is the liquid diffusivity, Eq 3, which is often called Scheil's equation, holds for any solid morphology (Ref 2, 3):

$$\frac{C_L}{C_o} = (1 - f_S)^{(k-1)} \qquad \text{(Eq 3)}$$

Second, in the case of $D_S \neq 0$ (finite diffusion in the solid) and $D_L = \infty$, Eq 4 to 6 have been proposed (Ref 4-6):

$$\frac{C_L}{C_o} = (1 - \psi f_S)^{(k-1)/\psi} \qquad \text{(Eq 4)}$$

where

$$\psi \equiv 1 - \frac{2Bk}{1+2B}$$

where B is the back diffusion coefficient in the solid phase

$$B = \frac{4 D_S t_f}{\lambda^2}$$

where D_S is the diffusion coefficient in the solid phase, t_f is the local solidification time, and λ is the dendrite arm spacing.

A more accurate or exact solution for this model has been obtained (Ref 5). Equation 4 and a similar equation given in Ref 6 approximate the exact solution below $f_S <$ 0.9. Equation 4 is applicable not only to

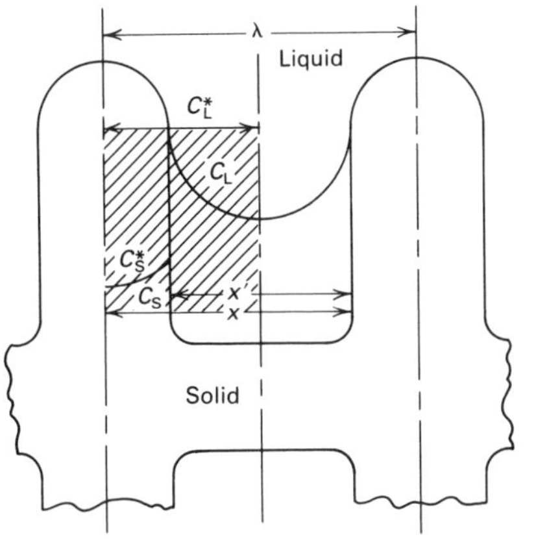

Fig. 2 Solute distribution in a volume element (crosshatched area) between dendrites

platelike dendrites but also to columnar dendrites if 2B in Eq A is doubled. It also agrees with Eq 1 for D_S or B $\gg$ 1 and with Eq 3 for D_S or B $\ll$ 1, respectively. The Brody-Flemings equation (Ref 7) is not applicable for B > 0.5.

Third, there is a solid-state diffusion and solute buildup ahead of the solid-liquid interface. An analytical solution for this actual case has not been obtained.

However, in the case where $D_S = 0$ and a solute boundary layer controls the solute transfer in the liquid, the effective partition coefficient k_{ef} has been derived for semi-infinite volume-element and steady-state conditions (Ref 8):

$$k_{ef} \equiv \frac{C_S^*}{C_o} = \frac{k}{k + (1 - k)\exp(-R\delta_c/D_L)} \quad \text{(Eq 5)}$$

where C_S^* is the solid composition at the interface, R is the growth rate, and δ_c is the solute boundary layer ahead of the interface.

The coefficient k_{ef} can be used for k in Eq 3:

$$\frac{C_L}{C_o} = (1 - f_S)^{(k_{ef} - 1)} \quad \text{(Eq 6)}$$

and

$$C_S = k_{ef} \cdot C_L \quad \text{(Eq 7)}$$

Although Eq 5 cannot be directly applied to dendritic solidification, it is useful for an understanding of the formation of microsegregation. Further, it is applied to evaluate macrosegregation in single-crystal growth.

Microsegregation

In practice, microsegregation is usually evaluated by the Microsegregation Ratio, which is the ratio of the maximum solute composition to the minimum solute composition after solidification, and by the amount of nonequilibrium second phase in the case of alloys that form eutectic compounds. Some data and an isoconcentration profile for an Fe-25Cr-19Ni columnar dendrite (Ref 9) are given in Table 1 and Fig. 3.

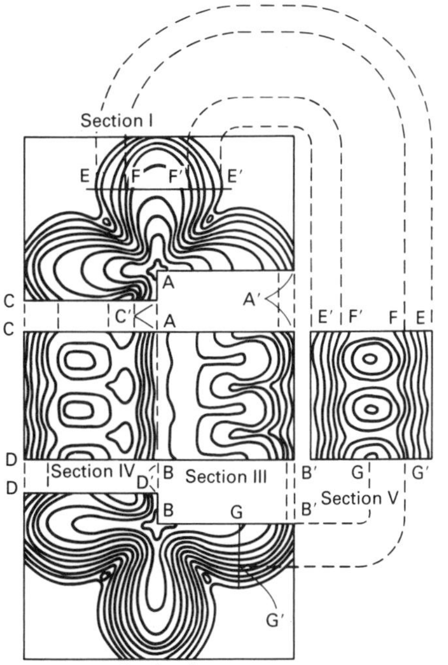

Fig. 3 Isoconcentration profile in an Fe-25Cr-20Ni columnar dendrite

Equation 3 is often used in the case of the lower back diffusion parameter B (for example, for aluminum alloy castings), and Eq 1 and 4 are used in the case of higher B (for example, for steel castings). However, it is not easy to estimate the real microsegregation as listed in Table 1, because the real phenomena are very complicated and the solid composition after solidification cannot be calculated by Eq 4 if the finite solid diffusion is not negligible. The following points should be considered.

Solidification Mode and Structure. Microsegregation varies considerably with the history of the growth of the solid. For example, microsegregation often increases with cooling rate in the case of equiaxed dendritic solidification, but it decreases in the case of unidirectional dendritic solidification. This is because, in the former case, the liquid composition is rather uniform in the interdendritic liquid, and Eq 4 is applicable. In the latter case, the solute buildup on the dendrite tip cannot be neglected, and equations such as Eq 5 or the Solari-Biloni equation (Ref 10), which considers the solute buildup ahead of the dendrite tip and dendrite curvature, should be used. Alternatively, a numerical calculation is necessary. Estimating the microsegregation in an equiaxed globular grain structure requires information on its formation mechanism and the history of the grain (that is, dendrite melt-off and settling in the liquid).

Morphology of the Dendrite and Diffusion Path. In the case where solid-state diffusion is not negligible, the diffusion path or the morphology of the solid is very important. Although Eq 4 can be applied to the volume element in a primary or secondary dendrite array, the real diffusion occurs three dimensionally in both dendrites. Therefore, careful attention is required to determine the dendrite spacing λ. One method is to employ the mean value of the primary and secondary dendrite arm spacing, $\bar{\lambda} = (\lambda_1 + \lambda_2)/2$ (Ref 6). Equation 8 is also recommended (Ref 4):

$$\frac{C_L}{C_o} = (1 - \psi_1 f_{S_1})^{[(k_1 - 1)/\psi_1]} \left(\frac{1 - \psi_2 f_S}{1 - \psi_1 f_{S_1}}\right)^{[(k_2 - 1)/\psi_2]} \quad \text{(Eq 8)}$$

where the subscripts 1 and 2 are used for the state $f_S \le f_{S_1}$ and $f_S > f_{S_1}$, respectively. Thus, if the diffusion path changes from the primary dendrite to the secondary dendrite at the fraction solid f_{S_1}, then λ_1 and λ_2 are used for Ψ_1 and Ψ_2, respectively.

Phase Transformation. If a phase transformation occurs during solidification, the microsegregation can change considerably

Table 1 Microsegregation ratio (numbers without dimension) and amount of nonequilibrium second phase (mass%)

Microsegregation ratio	Alloys, mass%
Mo (1.4–2.0)	Carbon steel (0.3–0.4C)
Cr (1–5, increases with C up to 1.4%)	Fe-(1–3)Cr-C
Mo (2.7–3.8), Cr (1.4–1.5)	1.2Cr–0.25Mo steel
Ni (1.2–1.4), Cr (1.3–1.5), Mo (2.6–3.8)	2.8Ni–0.8Cr–0.5Mo steel
Mn (1.3–1.8)	1.5Mn steel
Ni (1.06–1.07), Cr (1.3)	18Cr–8.6Ni stainless steel
Ni (1.1), Cr (1.1–1.3)	25Cr–19Ni stainless steel
Si (1.8–3.1), Mn (1.3–1.8)	19Cr–15Ni stainless steel
P (36 for cooling rate $\dot{T}$ = 0.083 K/s)	22Cr–20Ni stainless steel
P (30 for cooling rate $\dot{T}$ = 0.167 K/s)	
P (15 for cooling rate $\dot{T}$ = 0.833 K/s)	
Al (1.9–2.0), Ti (2.1–2.2)	Ni-5Al-13Ti
V (1.3)	Ti-(2–10)V
Sn (1.6–3.7, decreases with growth rate	Cu-8Sn
Cu (1–2 vol%)	Al-2Cu
Cu (3.9 area% for equiaxed structure; 1.5–3 area% for columnar structure)	Al-4.5Cu
Cu (2.8 vol%)	Al-6.5Cu
Cu (4.1 vol%)	Al-6.5Cu-0.26V
Cu (4 vol%)	Al-6.5Cu-0.1Ti
Mg (4–7 area% for equiaxed structure; 1–4 area% for columnar structure)	Al-10.4Mg
P (25–26.5 mass%)	Fe-4P

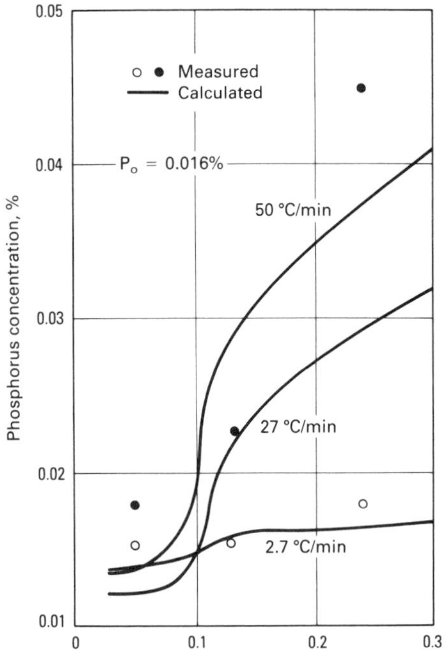

Fig. 4 Effect of carbon concentration and cooling rate on phosphorus concentration in an Fe-C-0.016P dendrite upon cooling to 1537 K

because the equilibrium partition coefficient varies with phase. For example, the k of phosphorus in steel castings is 0.13 in ferrite and 0.06 in austenite. Therefore, as shown in Fig. 4, in steel castings the peritectic reaction, which is affected by carbon composition, greatly affects microsegregation (Ref 11). If it is assumed that the phase change occurs at $f_S = f_{S_1}$, then Eq 8 is also applicable, but it may result in a large error. A more accurate estimation of microsegregation requires numerical calculations that take into consideration the amount of change in each phase (Ref 11, 12).

Effect of Third Solute Element. Figure 5 shows that the partition coefficient is affected by the third solute element (Ref 13). In aluminum alloys, chromium decreases the partition coefficient of magnesium. Further, it should be noted that the dendrite morphology varies with solute elements resulting in a different diffusion effect.

Dendrite Coarsening. Because coarsening or remelting of the dendrites occurs during solidification, the dendrite spacing is not constant, and the resolved solid dilutes the liquid composition. Although a numerical analysis has been performed, such effects have not been made clear (Ref 14).

Movement of the Liquid Phase. In many cases, the interdendritic liquid does not remain stationary but moves by solidification contraction or by thermal and solutal convections, resulting in varying degrees of microsegregation.

Temperature and Concentration Dependency of Diffusion Coefficient. Physical properties such as D_S and D_L are temperature and concentration dependent. Because the diffusion coefficient may vary by an order of magnitude in the case of a large solidification interval, both values must be closely monitored. This can be determined by numerical calculation.

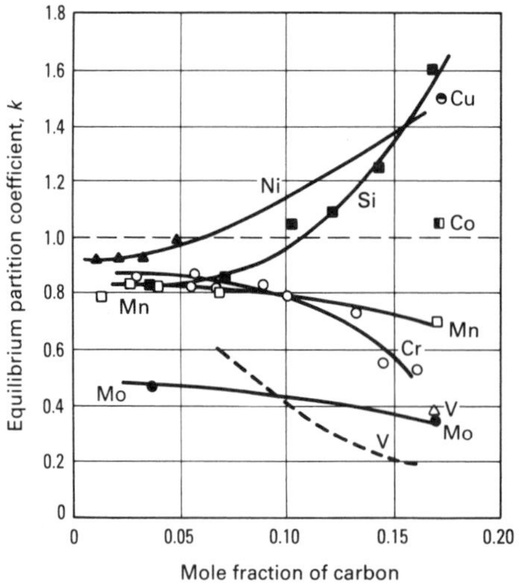

Fig. 5 Variation of equilibrium partition coefficient with third solute elements in an iron-carbon alloy

Undercooling. In actual use, undercooling at the dendrite tip does exist. However, it is not a factor that affects microsegregation in typical solidification processes, with the exception of welding and unidirectional solidification (Ref 13).

Other Effects. When a very high temperature gradient exists (for example, over 40 K/mm), the Soret effect, which considers solute transport to be a function of a temperature gradient, becomes a factor (Ref 15).

Microsegregation in Rapid Solidification Processing

In rapid solidification processing, the solid growth rate can be very high, resulting in a completely different solute distribution. If the atomic motions responsible for interface advancement are much more rapid than those necessary for the solute element to escape at the interface, microsegregation-free or diffusion-free solidification can occur (Ref 16). The nonequilibrium partition coefficient k_N is considered to increase monotonically with velocity (Ref 17):

$$k_N = \frac{k + [(R \cdot \Lambda)/D_i^*]}{1 + [(R \cdot \Lambda)/D_i^*]} \qquad \text{(Eq 9)}$$

where D_i^* is the interface interdiffusivity (Ref 17) and Λ is the interatomic spacing of the solid. However, the solute trapping is also dependent on solute concentration, and the complete equation has yet to be formulated.

Macrosegregation

Macrosegregation is caused by the movement of liquid or solid, the chemical composition of which is different from the mean composition. The driving forces of the movement are:

- Solidification contraction
- Effect of gravity on density differences caused by phase or compositional variations
- External centrifugal or electromagnetic forces
- Formation of gas bubbles
- Deformation of solid phase due to thermal stress and static pressure
- Capillary force

Macrosegregation is evaluated by:

- Amount of segregation (ΔC): $\Delta C = \overline{C}_S - C_o$
- Segregation ratio or index: C_{max}/C_{min} or $(C_{max} - C_{min})/C_o$
- Segregation degree (in percent): $100\ \overline{C}_S/C_o$

where C_o is the initial alloy composition, $\overline{C}_S$ is the mean solid composition at the location measured, and C_{max} and C_{min} are the maximum and minimum compositions, respectively. For example, the following carbon segregation index has been empirically obtained for steel ingots (except hot top) (Ref 18):

$$(C_{max} - C_{min})/(C_o D)\ (\%) = 2.81 + 4.31\ \text{H/D} + 28.9\ (\%\text{Si}) + 805.8\ (\%\text{S}) + 235.2\ (\%\text{P}) - 9.2\ (\%\text{Mo}) - 38.2\ (\%\text{V}) \qquad \text{(Eq 10)}$$

where D and H are ingot diameter and height in meters, respectively.

Macrosegregation is especially important in large castings and ingots, and it is also a factor in some aluminum or copper alloy castings of small and medium size. Various types of macrosegregation and their formation mechanisms are described below, mainly for the case of $k < 1$. In the case of $k > 1$, similar but converse results are obtained.

Plane Front Solidification. When plane front solidification occurs, as in single-crystal growth, the formation mechanism of macrosegregation is rather simple, and Eq 6 can be applied. A schematic of the typical solute distribution is shown in Fig. 6. The solid composition of the initially solidified portion is low and has an approximate value of kC_o, which gradually increases with time because of diffusion as the solute is pushed ahead, resulting in a higher concentration at the finally solidified portion or the ingot center. This segregation is termed normal segregation. As seen from Eq 5, the degree of normal segregation increases with decreasing growth rate (R) or the solute boundary layer thickness (δ_c), which decreases with increasing intensity of the liquid flow.

Further, changes in the growth rate during solidification results in a segregation as shown in Fig. 6. If the growth rate increases

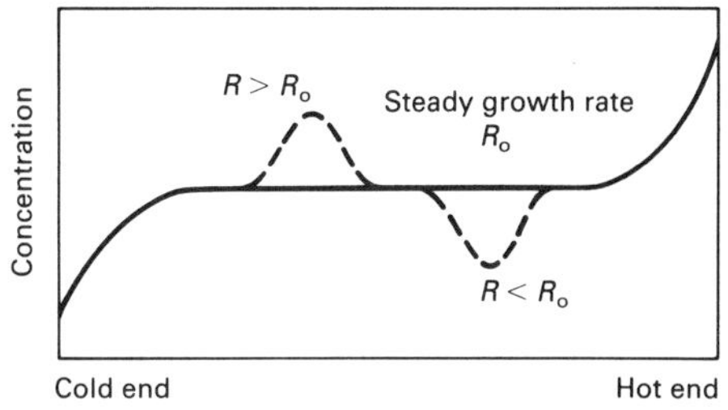

Fig. 6 Typical solute distribution in plane front solidification where R is the growth rate

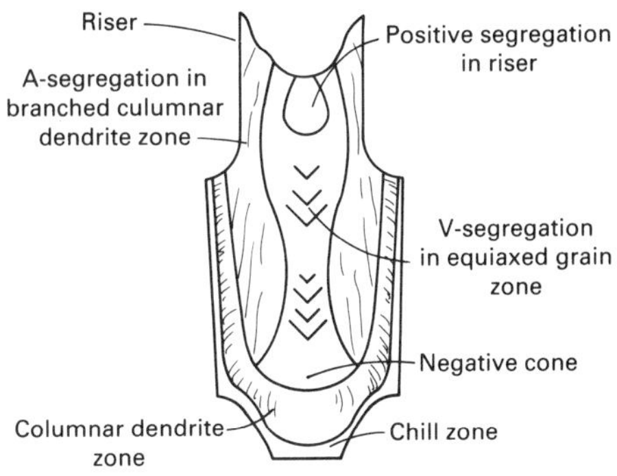

Fig. 7 Typical macrosegregation observed in steel ingots. A-segregation and V-segregation are discussed later in this article.

suddenly from the steady state to a higher rate, then a larger effective partition coefficient is realized, resulting in a concentration higher than the mean composition, which is termed positive segregation. (The normal segregation is a type of positive segregation.) Conversely, a sudden decrease in the growth rate R results in a solute-poor band or negative segregation. If R or δ_c varies periodically, then periodical composition change, which is termed banding or solidification contour, occurs. In either case, it is essential to consider the fluid flow and the solute boundary layer δ_c, which typically ranges from 0.1 to 1 mm (0.004 to 0.04 in.) for the single-crystal growth of metals.

Gravity segregation is caused by the settling or floating up of solid and liquid phases having a chemical composition different from the mean value. For example, the initially solidified phase or melted-off dendrites settle in the bottom of the casting because they are of higher density than the liquid. This phenomenon can be the source of negative cone, which often occurs in steel ingots, as shown in Fig. 7. If lighter solids such as nonmetallic inclusions and kish or spheroidal graphites are formed, they can float up to the upper part of the casting.

In steel castings, the interdendritic liquid is often lighter than the bulk liquid and floats up, resulting in positive segregation in the upper part of the casting. Although various types of macrosegregation are caused by the gravity effect, the compositional change between the upper and lower parts of a casting due to the simple gravity effect is called gravity segregation. In centrifugal casting, centrifugal force simulates gravity and can cause compositional changes between the internal and external parts of the casting.

Liquid Flow Induced Segregations in the Mushy Region. If only the liquid flows* in the mushy region, where a concentration gradient exists, macrosegregation occurs. Equation 12 can be derived by assuming the local equilibrium condition and the constant liquidus slope and by neglecting the dendrite curvature effect (Ref 20):

$$\frac{\partial f_S}{\partial t} = \left(\frac{1-\beta}{1-k}\right)(1-f_S)\left(1 + A - \frac{u_n}{U}\right)\cdot\frac{1}{C_L}\cdot\frac{\partial C_L}{\partial t} \qquad \text{(Eq 11)}$$

where

$$\beta = \frac{\rho_S - \rho_L}{\rho_S}$$

and ρ_S and ρ_L represent the density of the solid and the liquid, respectively; u_n is the flow velocity normal to the isotherms; and U is the velocity of isotherms. In this equation, $A = 0$ corresponds to zero diffusion in a solid and:

$$A = \frac{kf_S}{(1-\beta)(1-f_S)} \qquad \text{(Eq 12)}$$

corresponds to complete diffusion in a solid.*

If it is assumed that there is no shrinkage ($\beta = 0$) and no fluid flow ($u_n = 0$), Eq 12 is integrated to give either the equilibrium or Scheil equations (Eq 1 and 3) depending on the choice of A. For example, in the case of no diffusion in the solid ($A = 0$), Eq 12 is integrated to:

$$\frac{C_L}{C_o} = (1 - f_S)^{[(k-1)/\xi]} \qquad \text{(Eq 13)}$$

where $\xi \equiv (1 - \beta)(1 - u_n/U)$ and k are assumed to be constant.

The following results can be seen from Eq 13:

- In the case of $\xi = 1$ or $u_n/U = -\beta/(1 - \beta)$, Eq 13 is identical to Eq 3 and the average composition of the solid is C_o, which means that no macrosegregation occurs
- In the case of $\xi > 1$ or $u_n/U < -\beta/(1 - \beta)$, C_L becomes lower than the value calculated by Eq 3, which indicates that negative segregation may occur
- In the case of $0 < \xi < 1$ or $1 > u_n/U > -\beta/(1 - \beta)$, positive macrosegregation occurs. For example, because at the mold wall $u_n = 0$, it results in positive segregation described below

The segregation shown in Fig. 8, which is called inverse segregation, is where solute concentration is higher in the earlier freezing portion (Ref 21). This is caused by the solute-enriched interdendritic flow due to solidification contraction, which is the main driving force, plus the liquid density increase during cooling.** If a gap is formed between the mold and the solidifying casting surface, then the interdendritic liquid is often pushed into the gap by static pressure or by the expansion due to the formation of gas bubbles or graphite in the liquid. This results in severe surface segregation or in exudation, a condition in which a solute-rich liquid covers the casting surface and forms solute-rich beads.

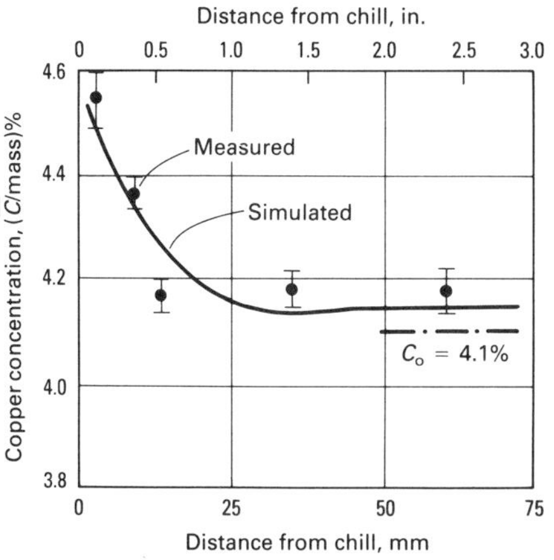

Fig. 8 Inverse segregation in an Al-4.1Cu ingot with unidirectional solidification

Changes in liquid velocity may cause segregation. For example, the change in the shape of the casting shown in Fig. 9 can change the velocity, resulting in a segregation, as shown in Fig. 10 (Ref 21).

The interdendritic fluid flow is also caused by the change in liquid density due to solute redistribution and cooling (solutal and thermal convection). For example, Fig. 11 illustrates the fluid flow in a horizontally solidifying aluminum-copper ingot, resulting in a segregation, as shown in Fig. 12 (Ref 21).

The bulk liquid flow can penetrate the mushy region or dendrite array and sweep out the solute-rich interdendritic liquid, resulting in a negative segregation. This is called the washing effect and is thought to be the primary mechanism for the white band, a type of negative segregation often observed in electromagnetically stirred continuous castings. Some researchers claim that the main mechanism is the change in growth rate due to stirring (Ref 23). The washing effect can also be the cause of positive or normal segregation, and the following effective partition coefficient is proposed (Ref 24):

$$k_{ef} = 1 - (1.33 \times 10^{-4})(1 - k)(1 - f_{sh})\frac{u}{U_{TL}} \qquad \text{(Eq 14)}$$

*It is usually assumed that the flow follows the D'Arcy law, that is, $\mathbf{u} \propto K\nabla P/(\mu f_L)$ where $\mathbf{u}$ is the velocity of the interdendritic liquid, K is the permeability (Ref 19), P is the pressure, and μ is the liquid viscosity.

**Even in the case of equiaxed grain structure, a similar inverse segregation can occur if the grains do not move as much as is the case in vertically solidified aluminum-copper ingots (Ref 22). However, quite different segregation occurs if the solid moves.

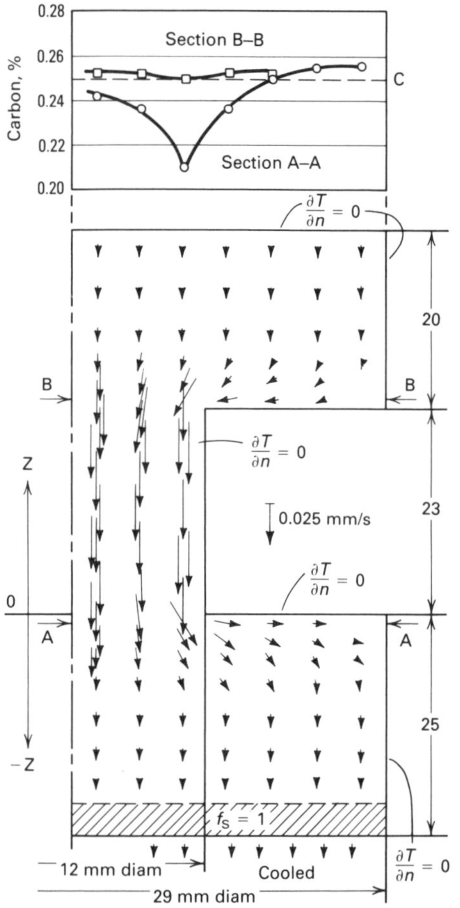

Fig. 9 Simulated fluid flow at 50 s after cooling and macrosegregation in an Fe-0.25C specimen

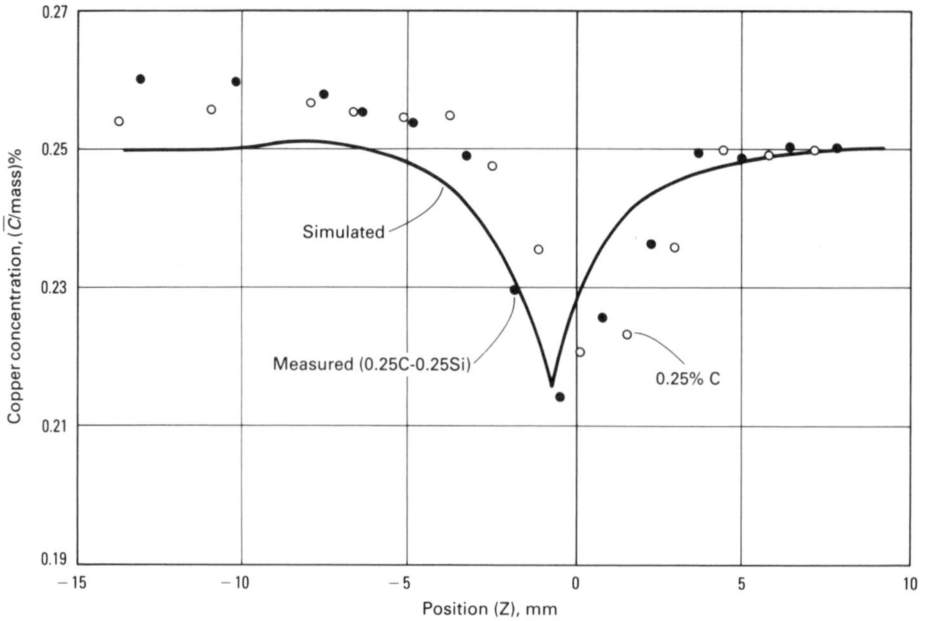

Fig. 10 Macrosegregation observed along Z-direction (cross-sectional mean value) for an Fe-0.25C specimen

where f_{sh} is the maximum fraction solid below which the washing effect acts, u is the bulk liquid velocity, and U_{TL} is the liquidus isotherm velocity. Although Eq 14 was derived experimentally assuming complete substitution of the bulk liquid for the interdendritic liquid, it has practical applications, especially in the continuous casting industry. However, the washing effect is usually coupled with convection in the mushy region and is strongly dependent on the liquid density change and the permeability (Ref 21, 24).

In continuous slab casting, bulging causes interdendritic flow and results in rather sharp and thin positive segregation at the ingot center. This is called centerline segregation.

The formation of an equiaxed grain structure by electromagnetic stirring or other methods can considerably decrease the segregation because both the solid and liquid are free to move. In this case, Eq 11 does not apply.

Inhomogeneous Solid Distribution and Channel Segregation. Actually, the solid is not uniformly distributed, and liquid pockets often form in the mushy region because of the preferential growth of some dendrites and/or because of the agglomeration of equiaxed dendrites at the advancing interface (Ref 26), as shown in Fig. 13. These liquid pockets may become a semi-macrosegregation or a spot segregation, which is a positive segregation several hundred microns in diameter often observed in the equiaxed region of steel castings continuously cast using an electromagnetic stirring device.

This inhomogeneous distribution of solid phase may be the origin of a preferred flow channel because the flow resistance of the channel connecting such liquid pockets is small. Further, once the liquid flows and if $u_n/U > (1 + A)$, then $\partial f_S/\partial t$ becomes negative in Eq 11. This means that remelting occurs and that the channel becomes larger. If bubbles are formed in the channel, it accelerates the flow velocity, resulting in

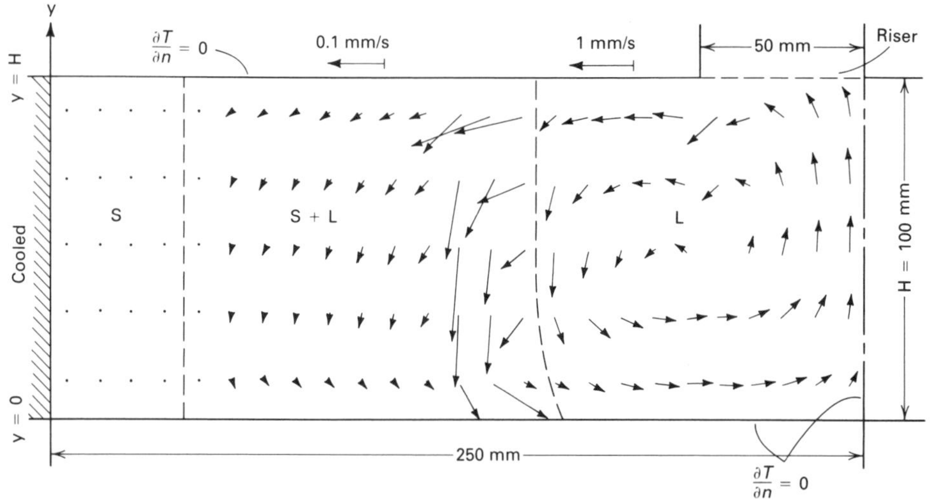

Fig. 11 Simulated fluid flow at 400 s after cooling in a horizontally solidified Al-4.4Cu ingot

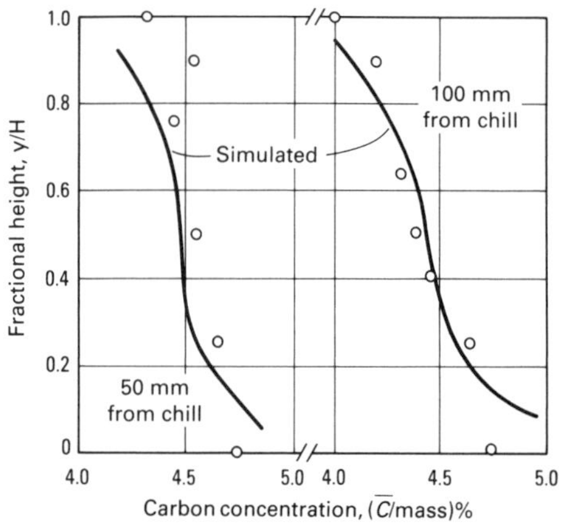

Fig. 12 Solute distribution in the Al-4.4Cu ingot shown in Fig. 11

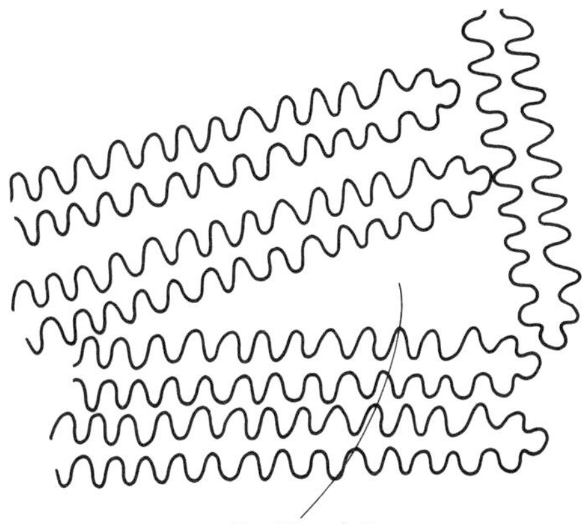

Fig. 13 Liquid pocket in the mushy region

enlargement of the channel (pores observed after solidification are often caused by solidification contraction). This is the mechanism of channel segregation. In large steel ingots, rodlike solute-enriched streaks such as the A-shape illustrated in Fig. 7 are often observed; this is termed A-segregation (inverse V-segregation or a ghost). In unidirectionally solidified ingots, the channel segregation is called a freckle.

A practical criterion for A-segregation formation in steel ingots is (Ref 27):

$$\dot{T}U_{0.35}^{1.1} < A_c \quad \text{(Eq 15)}$$

where $\dot{T}$ is the cooling rate and $U_{0.35}$ is the velocity of isotherm of fraction solid 0.35 and A_c is a constant dependent on the alloy (Ref 28). If the preferred channel is formed, the solute-rich liquid easily floats up or down, resulting in a severe positive segregation in the upper or lower part of the casting.

The following steps can be effective in preventing channel segregation:

- Increasing the cooling rate or reconditioning the solute element to form a dense packing of dendrites. For example, in steels, lowering the silicon content by carbon deoxidation results in smaller dendrite arm spacing and lower permeability (that is, higher flow resistance)
- Adjusting the alloying element, which minimizes the change in the solute-rich liquid density (see Eq 10)
- Electromagnetic stirring to form equiaxed grains and to obtain a uniform and dense solid distribution

Solid Phase Movement and Segregation. If equiaxed grains are formed or if dendrite melt-off occurs, the solid may migrate because of flow or solidification contraction. In tall steel ingots with equiaxed grain structure, for example, V-shaped solute-rich regions, consisting of blurred rodlike streaks, appear periodically; this is termed V-segregation (Fig. 7). In this case, the equiaxed grains move because of the solidification contraction, and the V-shaped slip faces are periodically formed by the viscoelastic motion of the equiaxed grains (Ref 29-31). Because the slip plane contains a loose grain structure, the permeability is greater and preferential flow channels may be formed as described in the case of A-segregation. The solutal convection may also result in a similar segregation, but it is not as severe.

If an external force acts on the mushy region where the solid fraction is high (for example, $f_S = 0.7$), then the grains or the dendrites often open and attract the interdendritic liquid, resulting in a positive segregation. This is termed healing in shape castings and is termed internal cracking in continuous castings. If the interdendritic liquid is insufficient, a hot tear occurs. In centrifugal casting, a periodic external force, such as vibration, may affect the mushy region, resulting in a periodic segregation banding.

REFERENCES

1. M.C. Flemings, *Solidification Processing*, McGraw-Hill, 1974, p 272
2. G.H. Gulliver, *J. Inst. Met.*, Vol 9, 1913, p 120
3. E. Scheil, *Z. Metallkd.*, Vol 34, 1942, p 70
4. I. Ohnaka, *Trans. ISIJ*, Vol 26, 1986, p 1045
5. S. Kobayashi, *Tetsu-to-Hagané (J. Iron Steel Inst. Jpn.)*, Vol 71, 1985, p S199, S1066
6. T.W. Clyne and W. Kurz, *Trans. AIME*, Vol 12A, 1981, p 965
7. H.D. Brody and M.C. Flemings, *Trans. TMS-AIME*, Vol 236, 1966, p 615
8. G.F. Bolling and W.A. Tiller, *J. Appl. Phys.*, Vol 32, 1961, p 2587
9. M. Sugiyama, T. Umeda, and J. Matsuyama, *Tetsu-to-Hagané (J. Iron Steel Inst. Jpn.)*, Vol 63, 1977, p 441
10. M. Solari and M. Biloni, *J. Cryst. Growth*, Vol 49, 1980, p 451
11. Y. Ueshima, S. Mizoguchi, T. Matsumiya, and H. Kajioka, *Metall. Trans. B*, Vol 17B, 1986, p 845
12. H. Fredriksson, *Solidification and Casting of Metals*, The Metals Society, 1979, p 131
13. Z. Morita and T. Tanaka, *Trans. ISIJ*, Vol 23, 1983, p 824; Vol 24, 1984, p 206; and private communication
14. D.H. Kirkwood, *Mater. Sci. Eng.*, Vol 65, 1984, p 101
15. J.D. Verhoeven, J.C. Warner, and E.D. Gibson, *Metall. Trans.*, Vol 3, 1972, p 1437
16. J.C. Baker and J.W. Chan, *Solidification*, American Society for Metals, 1970, p 23
17. M.J. Aziz, *J. Appl. Phys.*, Vol 53, 1982, p 1158; *Appl. Phys. Lett.*, Vol 43, 1983, p 552
18. J. Comon, Paper presented at the Sixth International Forgemaster's Meeting, (NJ), Oct 1972
19. D.R. Poirier, *Met. Trans. B*, Vol 18B, 1987, p 245
20. M.C. Flemings, *Solidification Processing*, McGraw-Hill, 1974, p 244
21. I. Ohnaka and M. Matsumoto, *Tetsu-to-Hagané (J. Iron Steel Inst. Jpn.)*, Vol 73, 1987, p 1698
22. H. Kato and J.R. Cahoon, *Metall. Trans. A*, Vol 16A, 1985, p 579
23. M.R. Bridge and G.D. Rogers, *Metall. Trans. B*, Vol 15B, 1984, p 581
24. T. Takahashi, K. Ichikawa, M. Kudo, and K. Shimabara, *Trans. ISIJ*, Vol 16, 1976, p 263
25. F. Weinberg, *Metall. Trans. B*, Vol 15B, 1984, p 681
26. M.R. Bridge, M.P. Stephenson, and J. Beech, *Met. Technol.*, Vol 9, 1982, p 429
27. K. Suzuki and T. Miyamoto, *Tetsu-to-Hagané (J. Iron Steel Inst. Jpn.)*, Vol 63, 1977, p 53
28. H. Yamada, T. Sakurai, T. Takenouchi, and K. Suzuki, in *Proceedings of the 11th Annual Meeting* (Dallas), American Institute of Mining, Metallurgical, and Petroleum Engineers, Feb 1982
29. M.C. Flemings, *Scand. J. Metall.*, Vol 5, 1976, p 1
30. H. Sugita, H. Ohno, Y. Hitomi, T. Ura, A. Terada, K. Iwata, and K. Yasumoto, *Tetsu-to-Hagané (J. Iron Steel Inst. Jpn.)*, Vol 69, 1983, p A193
31. H. Inoue, S. Asai, and I. Muchi, *Tetsu-to-Hagané (J. Iron Steel Inst. Jpn.)*, Vol 71, 1985, p 1132

Behavior of Insoluble Particles at the Solid/Liquid Interface

D.M. Stefanescu, The University of Alabama
B.K. Dhindaw, IIT Kharagpur, India

The problem of the behavior of insoluble particles at the solid/liquid interface has received the attention of theoreticians since the publication of a paper on the subject in 1964 (Ref 1). Only recently, however, has it been recognized that the problem is relevant to systems of practical significance. For example, porosity results from the incorporation of gaseous bubbles evolved during solidification or generated at the mold/metal interface in castings. If liquid droplets are considered, typical examples are phosphides in cast iron or inclusions in steel, which are incorporated into intergranular regions. In addition, structure formation in monotectic alloys would be explained based on liquid particle behavior at the interface. Finally, spheroidal graphite in cast iron, inclusions in steel, particulate *in situ* composites such as iron-vanadium carbide alloys, and particulate metal-matrix composites are examples in which solid particles interact with the solid/liquid interface during solidification (Ref 2, 3).

Basically, when a moving solidification front intercepts an insoluble particle, it can either push it or engulf it. Engulfment occurs through the growth of the solid over the particle, followed by enclosure of the particle in the solid. If, for various reasons, the solidification front breaks down into cells, dendrites, or equiaxed grains, two or more solidification fronts can converge on the particle. In this case, if the particle is not engulfed by one of the fronts, it will be pushed in between two or more solidification fronts and will finally be entrapped in the solid at the end of local solidification.

It is considerably easier to understand particle behavior at the solid/liquid interface in directional solidification processes, in which particles can only be pushed or engulfed, when a planar interface is maintained. In multidirectional solidification (castings), particles can be pushed, engulfed, or entrapped.

This article will discuss the variables of the process. The available theoretical and experimental work for both directional and multidirectional solidification will also be reviewed.

Particle Behavior in Directional Solidification

The advantage of using directional solidification while studying this problem lies in the possibility of achieving a variety of

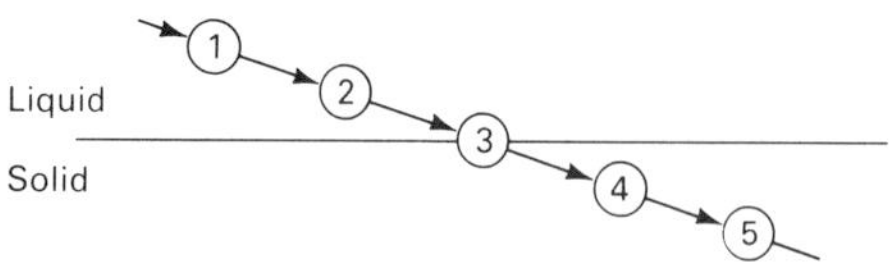

Fig. 1 Schematic for thermodynamic calculations of particle entrapment. Source: Ref 4

interface morphologies, such as planar, cellular, or dendritic. Because the nature of the solid/liquid interface plays a major role in particle behavior, the analysis of the process will be structured based on the type of interface.

Planar Interface

There are two basic theoretical approaches to the study of particle behavior at a solid/liquid interface: thermodynamic and kinetic.

The Thermodynamic Approach. Researchers have considered the case of a single particle being engulfed at the liquid/solid interface, assuming that the solid interface remained planar and neglecting buoyancy forces (Fig. 1). As the particle moves from position 2 to 3, the change in free energy per unit area is:

$$\Delta F_{23} = \tfrac{1}{2}(\sigma_{PS} - \sigma_{PL}) - \tfrac{1}{4}\sigma_{SL} \quad \text{(Eq 1)}$$

Similarly, when moving from 3 to 4, the change in free energy is:

$$\Delta F_{34} = \tfrac{1}{2}(\sigma_{PS} - \sigma_{PL}) + \tfrac{1}{4}\sigma_{SL} \quad \text{(Eq 2)}$$

where σ is the interface energy between particle (P), liquid (L), and solid (S), in various combinations.

The net change in free energy during engulfment is:

$$\Delta F_{net} = \Delta F_{23} + \Delta F_{34} = \sigma_{PS} - \sigma_{PL} \quad \text{(Eq 3)}$$

If $\Delta F_{net} < 0$, engulfment is to be expected; for $\Delta F_{net} > 0$, pushing should result.

The kinetic approach is based on the simple idea that as long as a finite layer of liquid exists between the particle and the solid, the particle will not be engulfed. In other words, for a particle to be pushed, mass transport in the liquid layer is required between the particle and the solid. The concept of a critical interface rate R_{cr}, below which particles are pushed and above which particles are engulfed, was postulated.

For a particle to be pushed, a repulsive force must exist between the particle and the solid. The nature of this repulsive force is not known, although several possibilities are supported by various investigators.

It has been suggested that the repulsive force may result from the variation in surface free energy $\Delta\sigma$ when the particle approaches the interface (Ref 1):

$$\Delta\sigma = \sigma_{PS} - (\sigma_{PL} + \sigma_{LS}) \quad \text{(Eq 4)}$$

which varies with the particle-solid distance d according to Eq 5:

$$\Delta\sigma = \Delta\sigma_0 \left(\frac{d_0}{d}\right)^n \quad \text{(Eq 5)}$$

where d_0 is the minimum separation distance between particle and solid, n is an exponent equal to 4 or 5, and $\Delta\sigma_0 = \Delta\sigma$ at $d = d_0$. Coulomb forces may also be responsible for a repulsive force F_r:

$$F_r = \frac{e_S e_P}{d^2} \quad \text{(Eq 6)}$$

where e_S and e_P are the charges of the solid and the particle, respectively.

Indeed, it has been demonstrated that particles are electrically charged (Ref 1). However, the researchers dismissed the influence of these charges on the repulsive force on the grounds that no correlation was found between the average charge on the particles in a system and the critical velocity.

Van der Waals type forces were considered to be responsible for the repulsive force in another theoretical treatment (Ref 5):

$$F_r = \frac{\pi B_3 r}{d_0^2} \quad \text{(Eq 7)}$$

where B_3 is a constant $\cong 10^{-7}$ J, r is particle radius, and $d_0 \cong 10^{-5}$ cm. In general, it seems that the repulsive force is described by an equation of the type:

$$F_r = \frac{B}{d_0^n} \quad \text{(Eq 8)}$$

where B is a constant. Repulsive forces of the type found in ionic crystals would obey Eq 8 with $n = 2$ (Ref 6), while dispersion forces between molecules, as in liquids or gases, will follow the same law with $n = 7$ (Ref 7).

Assuming that mass transport in the particle-solid gap occurs only by diffusion, that the repulsive force results from differences in surface tension (Eq 5), that there is no viscous drag for small particles, and that viscous drag exists for large particles, researchers have derived Eq 9 and 10 for the critical interface growth rate R_{cr} (Ref 1). For small particles:

$$R_{cr} = \tfrac{1}{2}(n + 1)\frac{L a_0 V_0 D}{k T r_b^2} \quad \text{(Eq 9)}$$

For large particles:

$$R_{cr} = \frac{d_S h L a_0 d_1}{6\eta r r_b^2 n} \left[\left(1 + \frac{6\eta r n(n + 1)V_0 D}{d_S h d_1 k T}\right)^{1/2} - 1\right] \quad \text{(Eq 10)}$$

where L is the latent heat per unit volume, a_0 is the molecular diameter, V_0 is the

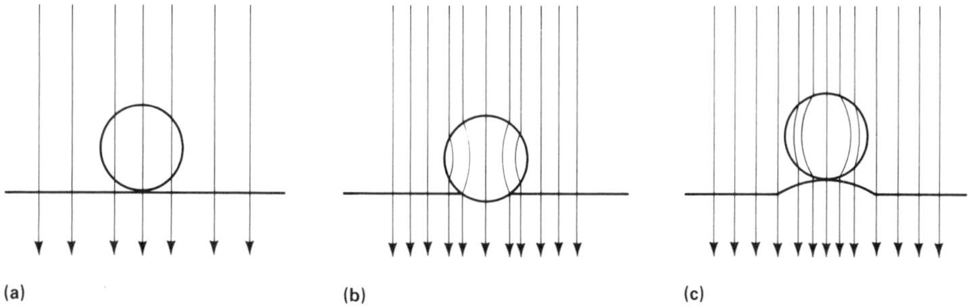

Fig. 2 Influence of thermal conductivity K of particle (K_P), liquid (K_L), and solid (K_S) on the shape of the solid/liquid interface. (a) Flat interface; $K_P = K_L$ and K_S. (b) Engulfment; $K_P > K_L$ and K_S. (c) Bump formation (pushing); $K_P < K_L$ and K_S

atomic volume, D is the liquid diffusivity, k is the Boltzman constant, T is temperature, r_b is the radius of any particle irregularity or bump (for particles without irregularities, $r_b = r$), η is the viscosity of the liquid, d_1 is minimum separation (10^{-7} cm), and h and d_S are contact distances. The contact distances d_1, d_S, and h cannot be calculated, but must be estimated for different systems.

Other researchers have assumed that mass transport occurs by diffusion and fluid flow, that the particle does not wet the solid and has the same thermal conductivity as the liquid, and that the repulsive force is again described by Eq 5 (Ref 8). They derived Eq 11 to 14. For small, smooth particles ($r < r_b$):

$$\eta^2 R^2 r^3 = N \frac{4\psi(\alpha)}{9\pi} kT\,\sigma_{SL} a_0 \qquad \text{(Eq 11)}$$

For bigger particles with bumps ($r \geq r_b$):

$$\eta^2 R^2 r_b^3 + \frac{2\psi(\alpha)}{9\alpha} g\Delta\rho a_0 \eta R r^3 r_b = \frac{4\psi(\alpha)}{9\pi} kT\sigma_{SL} a_0 \qquad \text{(Eq 12)}$$

For very large particles ($r \gg r_b$):

$$\eta R r^3 = \frac{2\alpha NkT\sigma_{SL}}{\pi r_b g \Delta\rho} \qquad \text{(Eq 13)}$$

In the absence of bumps, Eq 14 can be written:

$$\eta R r^4 = \frac{2\alpha kT\sigma_{SL}}{\pi g\Delta\rho} \qquad \text{(Eq 14)}$$

where N is the number of points at which the particle is in contact with the interface ($N = 1$ for a grain surface or flat interface, $N = 2$ for a grain boundary, and $N = 3$ for a triple point), α is a shape factor of the interface ($\alpha = 0$ for a flat interface and $\alpha = 1$ for an interface having the same curvature as a particle), $\psi(\alpha)$ is a function of α requiring assumption for calculation, g is acceleration due to gravity, and $\Delta\rho$ is the density difference between the liquid and the particle.

Other researchers assumed mass transport by fluid flow only, repulsion due to molecular forces (Eq 7), and attraction due to the drag on the particle by the viscous melt (Ref 5). They defined two characteristic lengths:

$$\lambda = \left(\frac{V_0\sigma_{SL}}{\Delta SG}\right)^{1/2}$$

$$l = \left(\frac{B_3 V_0}{\Delta SG}\right)^{1/4} \qquad \text{(Eq 15)}$$

where ΔS is the entropy of melting and G is temperature gradient.

Small particles are then defined as having $r < \lambda^2/l$, while large particles have $r > \lambda^2/l$. Equations 16 and 17 were derived for small and large particles, respectively:

$$R_{cr} = \frac{0.14\,B_3}{\eta r}\left(\frac{\sigma_{SL}}{B_3 r}\right)^{1/3} \qquad \text{(Eq 16)}$$

$$R_{cr} = \frac{0.15\,B_3}{\eta r}\left(\frac{\Delta SG}{B_3 V_0}\right)^{1/4} \qquad \text{(Eq 17)}$$

All the above approaches assume a planar liquid/solid interface, only one particle at the interface, and thermal conductivity of the particle K_P equal to that of the liquid, K_L.

Equation 18 takes into account the difference in thermal conductivity between particle and liquid (Ref 9):

$$R_{cr} = \frac{\Delta\sigma_0 d_0}{6\,(n-1)\,\eta r}\left[2 - \frac{K_P}{K_L}\right] \qquad \text{(Eq 18)}$$

Analysis of Eq 18 shows that the governing variables are $\Delta\sigma$, which can be positive or negative, n (defined in Eq 5), and K_L/K_P, which can be greater than or less than 1. Depending on their relative values, the particles can be either pushed or engulfed. Therefore, as shown in Fig. 2, Eq 18 combines the thermodynamic criterion (Ref 4) with the thermal conductivity criterion (Ref 8).

The thermal conductivity criterion implies that when $K_L \gg K_P$ a bump is formed, and the particle can roll over. The bump will then remelt, but another bump will be formed adjacent to the new position of the particle. Therefore, the interface will cease to be flat, and the particle will be continuously pushed, making engulfment impossible. Indeed, calculations of R_{cr} using Eq 18 for the Al-SiC system for a particle radius of 50 μm (2000 μin.) result in $R_{cr} > 1280$ μm/s (51 200 μin./s), a rate at which the interface obviously cannot be flat. In fact, interface destabilization is so extensive that cellular or dendritic solidification will occur. Particles can then be incorporated into the solid by entrapment rather than by engulfment, as will be shown later in this section.

However, if $K_P \gg K_L$, a trough will form (Fig. 2). However, $\Delta\sigma$ and η will also play a major influence in determining the value of R_{cr}.

The role of thermal conductivity on particle behavior has also been emphasized through the empirical heat diffusivity criterion, $(K_P C_P \rho_P / K_L C_L \rho_L)^{1/2}$ (Ref 10). When this ratio is greater than 1, particles are supposed to be engulfed. Good agreement has been found between experimental data and predictions (Ref 2).

Experimental Results. The existence of a critical interface rate R_{cr} has been documented experimentally for both liquid particles (xylene and orthoterphenyl in water) and solid particles (silver iodide, graphite, magnesia, silt, silicon, tin, diamond, nickel, iron oxide, and zinc in orthoterphenyl, salol, and thymol) of various shapes (spherical and irregular) and sizes (1 to 300 μm) (Ref 1). Other researchers have investigated tungsten and copper particles in water, as well as aluminum, silver, copper, silica, tungsten, and tungsten carbide particles in salol (Ref 8, 11). A more accurate analysis of experimental results on glass, Teflon, polystyrene, nylon, and acetal particles (10 to 200 μm in diameter) in biphenyl and naphthalene has concluded that the transition between pushing and engulfment is not sharp, but rather that, for a given system, there are three modes of particle behavior (Ref 4):

- At high rates, particles are engulfed instantly
- At intermediate rates, particles are pushed some distance before being engulfed
- At low rates, particles are pushed along continuously

Computer curve fitting analysis of experimental results has shown that the critical rate depends on particle radius r according to:

$$R_{cr} \cdot r^n = \text{constant} \qquad \text{(Eq 19)}$$

where the exponent n ranges from 0.28 to 0.90.

The validity of the thermal conductivity criterion has been confirmed experimentally for a number of metallic particles (tungsten, tantalum, molybdenum, iron, nickel, and chromium) in tin and bismuth (Ref 12). Equations 1 to 19 have been derived for a single particle ahead of a planar interface. When several particles are considered, the interface is expected to exhibit a series of bumps and troughs, eventually resulting in interface breakdown. Therefore, the concept of critical interface rate becomes increasingly difficult to use.

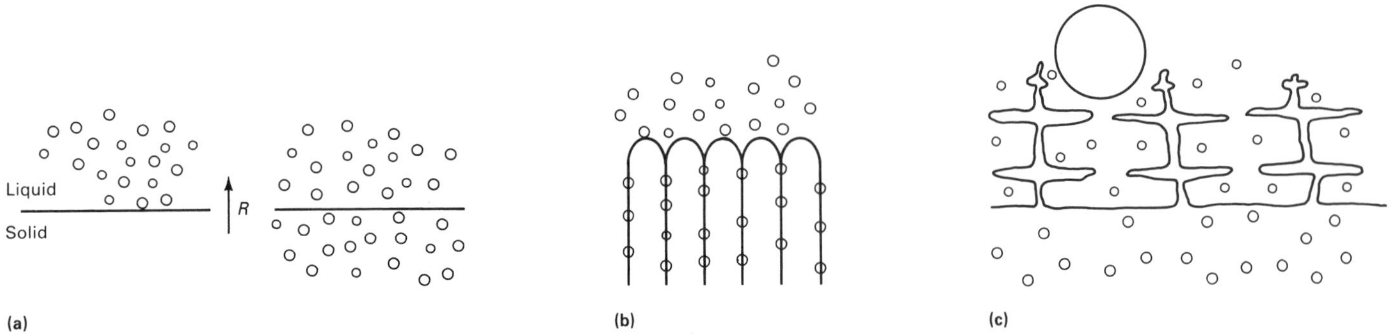

Fig. 3 Influence of interface shape on pushing or entrapment of particles. (a) Planar interface can result in pushing (left) or engulfment (right). (b) Cellular interface showing pushing at interface and entrapment between cells. (c) Dendritic interface; small particles are entrapped in interdendritic spaces while large particles are pushed. Source: Ref 9

To summarize, a number of process variables can be listed. A first group comprises those included in Eq 18, as follows:

- Particle radius r
- Viscosity of the liquid η
- Surface energy among particle, liquid, and solid (σ_{PL}, σ_{PS}, σ_{LS})

It must be noted that $\Delta\sigma$ can be altered by the surface modification of particles (for example, coating or heat treatment) or by changing the chemistry of the melt through the addition of surface-active elements (Ref 13).

A second group of variables, although effective in single particle-planar interface systems, was not considered in the theoretical work summarized in this discussion, because of obvious complications in calculations. The second group consists of:

- Particle shape
- Particle aggregation, that is, gas, liquid, or solid
- Convection level in the liquid
- Density of liquid (ρ_L) and particle (ρ_P)

Equations 1 to 19 have been derived on the assumption of spherical particles. As expected, particles of nonspherical shape will tend to position themselves in front of the moving interface to expose the smallest cross sectional area to the interface, thus minimizing the repulsive force.

Particle aggregation may be important in the process to the extent that particle deformation at the interface may influence R_{cr}. A complicating and difficult-to-quantify variable is the convection level in the liquid. Particle size becomes significant when discussing the importance of convection because for small particles (<15 μm) even microconvection resulting from Brownian forces must be taken into account. For large particles, the primary influences are buoyancy-driven convection (that is, convection deriving from solutal and thermal differences within the liquid) and convection deriving from the Stokes forces imposed by the liquid on the particles (flotation or sedimentation of particles). Buoyancy-driven convection depends on the gravity level. Of course, the difference in density between particles and the liquid will play a significant role in Stokes convection, as shown for the Fe-graphite and Fe-VC systems (Ref 14).

Cellular and Dendritic Interfaces

In metal-ceramic systems of practical importance, such as metal-matrix composites, alloys rather than pure metals, and several particles rather than one particle, must be considered. Because of the impurity of the liquid and the influence of particles, the interface will break down to become cellular, and more likely dendritic. All process variables that influence R_{cr} in the case of one particle-planar interface system will obviously play a role in multiparticle-rough interface (metal-ceramic) systems.

A third group of process variables, particularly important in multiparticle-rough interface systems, consists of:

- Liquid/solid interface shape
- Volume fraction of particles at the interface
- Temperature gradient ahead of the interface

The influence of liquid/solid interface shape on particle behavior during directional solidification is shown in Fig. 3. If the interface is planar, pushing results in clean metal, while engulfment results in metal-matrix composites or oxide dispersion strengthened materials (Fig. 3a). If the interface is cellular, particles can be pushed at the tips of the cells while they are entrapped at cell boundaries, as observed for particles in the ice-water (Ref 1), Pb-1Sn-iron (Ref 15), and water-nylon systems (Ref 15), resulting in an aligned particulate composite (Fig. 3b). When the interface breaks down into dendrites, a more complex situation occurs, with small particles being entrapped in the interdendritic spaces, while large particles are possibly being pushed by the dendrite tips (Fig. 3c). This is actually contrary to the theoretical laws discussed previously.

The primary cause of interface breakdown, assuming constant growth rate, is the purity level of the metal. Obviously, in metal-matrix composites, the high impurity level favors interface breakdown. Additional sources of perturbation are the particles themselves. Figure 4 shows a comparison between the quench interfaces of two Al-2Mg alloys, directionally solidified under identical conditions (Ref 9). The main difference is that the one shown in Fig. 4(a) did not have silicon carbide particles, while the one shown in Fig. 4(b) did have carbides. The effect of silicon carbide particles in further disturbing the cellular interface is evident.

This behavior can be explained by the perturbation of the thermal field ahead of the interface, as discussed earlier in the model that includes the contribution of thermal conductivity. According to Eq 18, R_{cr} becomes too high for engulfment because $K_L \gg K_P$ for this particular case; however, entrapment is possible because of the breakdown of the interface (Fig. 4b). Therefore, it is rather difficult to extrapolate equations such as those discussed earlier, which are derived for planar interfaces, to the interpretation of alloy systems where the planarity of the interface is susceptible to disruption by either the particles or the solidification conditions.

The volume fraction of particles must also be considered. At high volume fraction of particles in the melt, viscosity becomes a function of the amount of particles. A first approximation is given by Einstein's equation:

$$\eta^* = \eta^0 \left(1 + \frac{5}{2}\phi\right) \qquad \text{(Eq 20)}$$

where η^* is effective viscosity, η^0 is the viscosity of the suspending fluid, and ϕ is the volume fraction of particles.

Because $R_{cr} \sim 1/\eta$ (see Eq 18), it is expected that engulfment becomes easier as the volume fraction builds up. This trend has been qualitatively confirmed for silicon carbide particles in Al-6.1Ni alloys (Ref 16).

Although Eq 17 hints at some possible influence of temperature gradient G, it is difficult to rationalize such an influence in single particle-planar interface systems. On the contrary, experimental evidence for metal-ceramic systems documents the role

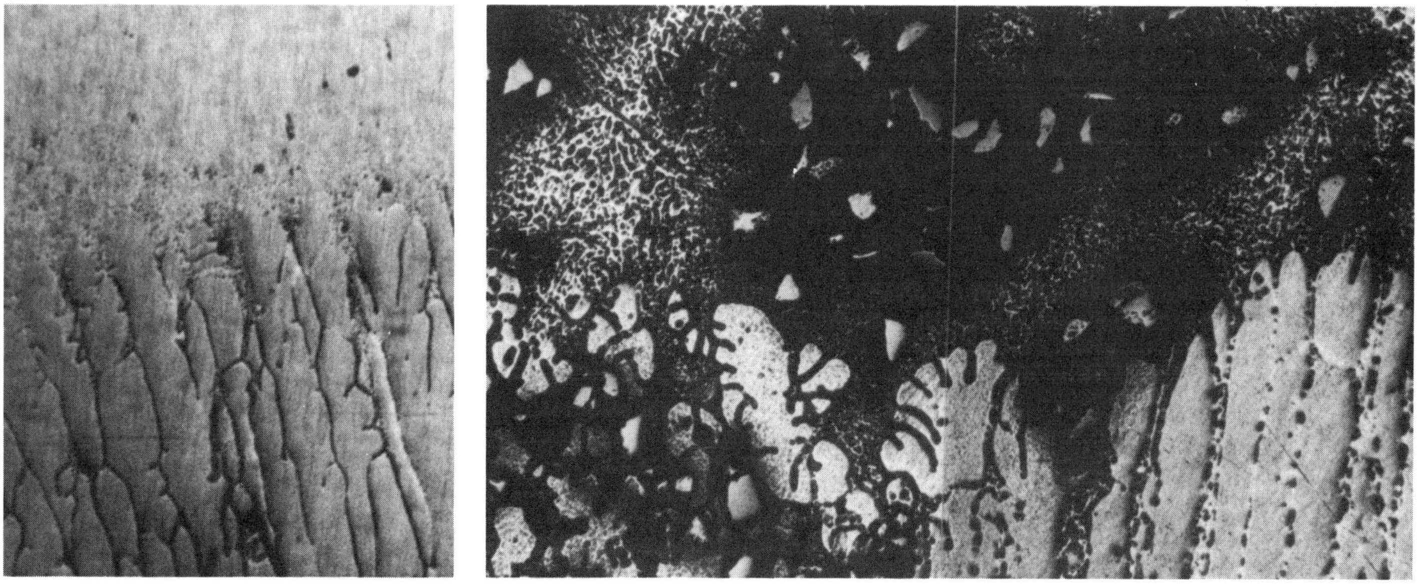

(a) (b)

Fig. 4 Effect of silicon carbide particles on the morphology of the solid/liquid interface in Al-2Mg alloys. Solidification conditions were identical for (a) and (b); the solidification direction was upward in each case. (a) Alloy without carbides showing unperturbed cellular-dendritic interface. (b) Alloy with carbides and highly perturbed cellular-dendritic interface

of G rather convincingly. Figure 5 shows that increasing G favors pushing over entrapment for aluminum-silicon carbide composites (Ref 9). This could again be rationalized in terms of interface stability. As G increases, the planar interface becomes more stable and particles are pushed because engulfment is not possible.

In general, the influence of most of the above variables on the critical interface rate of entrapment in metal-ceramic systems can be summarized as follows:

$$R_{cr} = f\left(\frac{\Delta\sigma}{\eta\phi}, \frac{1}{r}, \frac{K_L}{K_P}, G\right) \quad \text{(Eq 21)}$$

Although this correlation cannot be used for calculations and does not take interface shape into account, it can be used as a guide for manipulating parameters to achieve particle entrapment in metal-ceramic systems, because all parameters are measurable and can be used as experimental variables.

Particle Behavior in Multidirectional Solidification

Castings solidify under multidirectional heat flow conditions at much faster rates than those required for planar interface and are mostly made from alloys rather than pure metals. Therefore, solidification in castings normally proceeds by multidirectional growth of dendrites, followed by the solidification of the interdendritic liquid, or by equiaxed growth, with the intergranular liquid solidifying last. Under these conditions, it is still possible to generate some qualitative information on the behavior of particles in front of the melt interface, but any quantification is very difficult. A number of issues must be addressed:

- Solidification structure (dendritic or equiaxed)
- Structure/fineness versus particle size
- Agglomeration of particles

Dendritic Solidification. The first situation to be considered is that when particles are of the order of magnitude of secondary dendritic arm spacing or smaller. If the

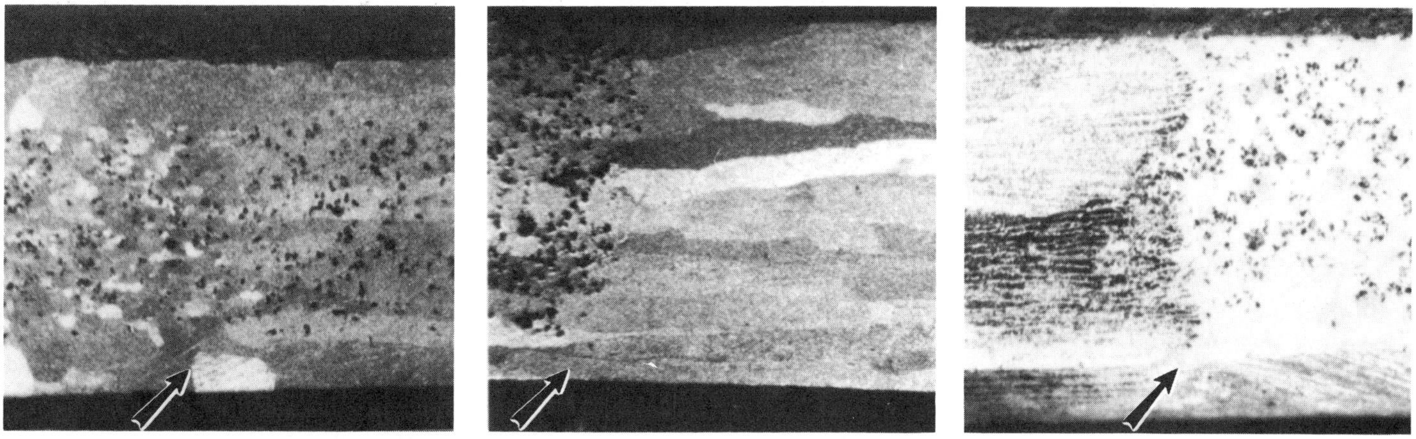

(a) (b) (c)

Fig. 5 Effect of temperature gradient G at the interface on the behavior of 10 to 150 μm silicon carbide particles in an Al-2Mg alloy. Solidification direction in all three cases was upward; start of directional solidification in (a) and (b) is indicated by an arrow. (a) G = 74 °C/cm (340 °F/in.); particles are entrapped. (b) G = 95 °C/cm (435 °F/in.); particles are pushed then entrapped due to volume buildup. (c) G = 117 °C/cm (535 °F/in.); particles are pushed and accumulate in the quench zone. In (c), the transition from directional solidification to the quench zone is indicated by an arrow.

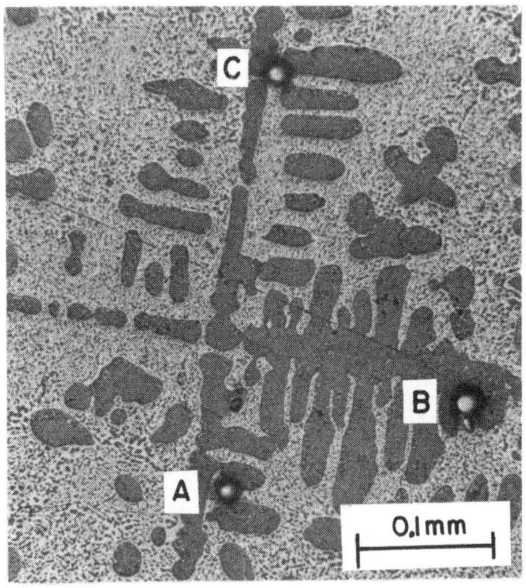

Fig. 6 Iron particles entrapped in interdendritic regions in a cast Pb-50Sn alloy

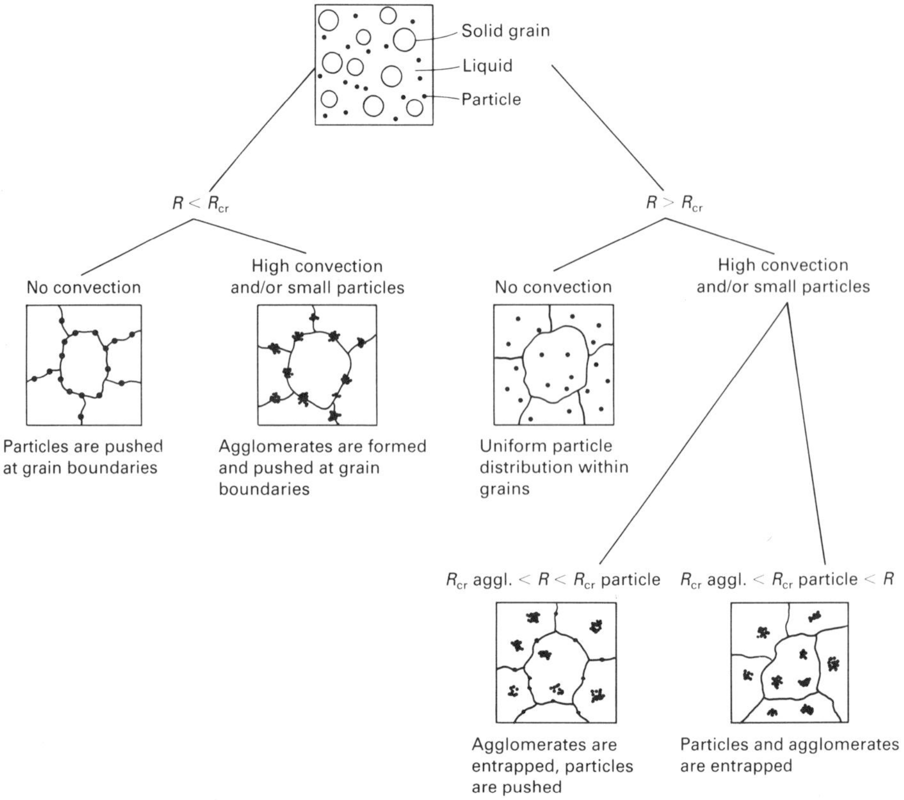

Fig. 7 Possible structures in multidirectional solidification as a function of solidification rate, convection level, and particle size

liquid does not wet the particles, their engulfment is not possible at normal solidification rates in castings. The particles are then pushed by the growing dendrites and are eventually entrapped in the interdendritic regions solidifying at the end (Fig. 6).

When the particles are larger than the interdendritic spacing, some pushing may occur by the tip of the dendrites, again with eventual entrapment in the interdendritic space. Because this is a totally random behavior, it may be possible to achieve good particle distribution throughout the matrix, if proper volume fraction of the particles and processing parameters for the composite are used (Ref 17).

Equiaxed Solidification. Figure 7 shows a schematic of possible resulting structures for the case in which particle size is much smaller than grain size. The concept of critical interface rate R_{cr} must now be applied at the level of microvolumes, where grain/liquid interface may be considered as planar.

It is expected that nonwetting particles, which in general will fit into the $R < R_{cr}$ situation, will be pushed toward the grain boundaries. A higher level of convection may yield agglomeration of particles. The agglomerates will also be pushed in the intergranular liquid.

However, if $R > R_{cr}$, which is anticipated when good wetting exists between particles and liquid, particle engulfment is expected, resulting in a uniform distribution of particles within the grains. High levels of convection can cause further complication (Fig. 7). Similar behavior, as for dendritic solidification, is expected for the case of particle size of the same order of magnitude or larger than grain size.

ACKNOWLEDGMENT

This work has been supported by grant No. NAGW-10 from the Center for the Space Processing of Engineering Materials at Vanderbilt University and by grant No. NAG8-070 from NASA-Marshall Space Flight Center.

REFERENCES

1. D.R. Uhlmann, B. Chalmers, and K.A. Jackson, Interaction Between Particles and a Solid-Liquid Interface, *J. Appl. Phys.*, Vol 35 (No. 10), 1964, p 2986
2. P.K. Rohatgi, R. Asthana, and S. Das, Solidification, Structures and Properties of Cast Metal-Ceramic Particle Composites, *Int. Met. Rev.*, Vol 31 (No. 3), 1986, p 115
3. K.C. Russell, J.A. Cornie, and S.Y. Oh, Particulate Wetting and Particle: Solid Interface Phenomena in Casting Metal Matrix Composites, in *Interfaces in Metal-Matrix Composites*, A.K. Dhingra and S.G. Fishman, Ed., The Metallurgical Society, 1986, p 61
4. S.N. Omenyi and A.W. Neumann, Thermodynamic Aspects of Particle Engulfment by Solidifying Melts, *J. Appl. Phys.*, Vol 47 (No. 9), 1976, p 3956
5. A.A. Chernov, D.E. Temkin, and A.M. Melnikova, Theory of the Capture of Solid Inclusions During the Growth of Crystals From the Melt, *Sov. Phys. Crystallogr.*, Vol 21 (No. 4), 1976, p 369
6. R. Sprul, *Modern Physics*, John Wiley & Sons, 1956, p 193
7. J.O. Hirschfelder, C.F. Curtiss, and R.B. Bird, *Molecular Theory of Gases and Liquids*, John Wiley & Sons, 1954
8. G.F. Bolling and J. Cissé, A Theory for the Interaction of Particles With a Solidifying Front, *J. Cryst. Growth*, Vol 10, 1971, p 56
9. D.M. Stefanescu, B.K. Dhindaw, S.A. Kacar, and A. Moitra, Behavior of Ceramic Particles at the Solid-Liquid Metal Interface in Metal Matrix Composites, *Metall. Trans.*, in print
10. M.K. Surappa and P.K. Rohatgi, *J. Mater. Sci.*, Vol 16 (No. 2), 1981, p 562
11. J. Cissé and G.F. Bolling, The Steady-State Rejection of Insoluble Particles by Salol Grown From the Melt, *J. Cryst. Growth*, Vol 11, 1971, p 25
12. A.M. Zubko, V.G. Lobanov, and V.V. Nikonova, *Sov. Phys. Crystallogr.*, Vol 18, 1973, p 239
13. B.K. Dhindaw, A. Kacar, and D.M. Stefanescu, Entrapment/Pushing of Particles During Directional Solidification of Aluminum-Silicon Carbide Metal Matrix Composites, in *Advanced Materials and Processing Techniques for Structural Applications*, Proceedings of the ASM Europe Conference, Paris, 1987
14. D.M. Stefanescu, M. Fiske, and P.A. Curreri, Behavior of Insoluble Particles During Parabolic Flight Solidification Processing of Fe-C-Si and Fe-C-V Alloys, in *Proceedings of the 18th Inter-*

national SAMPE Technical Conference, Vol 18, J.T. Hoggatt et al., Ed., Society for the Advancement of Material and Process Engineering, 1986, p 309
15. C.E. Schvezov and F. Weinberg, Interaction of Iron Particles With a Solid-Liquid Interface in Lead and Lead Alloys, *Metall. Trans. B*, Vol 16B, 1985, p 367
16. B.K. Dhindaw, A. Moitra, and D.M. Stefanescu, Directional Solidification of Al-Ni/ SiC Composites During Parabolic Trajectories, *Metall. Trans.*, in print
17. D.M. Schuster, M. Skibo, and F. Yep, SiC Particles Reinforced Aluminum by Casting, *J. Met.*, Vol 39 (No. 11), 1987, p 60

Low-Gravity Effects During Solidification

Peter A. Curreri, NASA, Marshall Space Flight Center
D.M. Stefanescu, University of Alabama

Solidification processes are strongly influenced by gravitational acceleration through Stokes flow, hydrostatic pressure, and buoyancy-driven thermal and solutal convection. Stokes flow of second-phase particles in off-eutectic and off-monotectic alloys and in particulate metal-matrix compositions severely limits casting composition. Porosity in an equiaxed casting is dependent on the hydrostatic pressure.

Buoyancy-independent solidification within the gravitational field at the earth's surface is accomplished only within strict limits. In one dimension, strong magnetic fields can dampen convection, and density gradients can be oriented with gravity for stability. However, magnetic flow dampening in one direction increases flow velocity (segregation and so on) in the transverse direction. Opposition of thermal and solutal convection for many alloy compositions makes stabilization of convection by orientation infeasible even in one dimension.

Space flight provides solidification research with the first long-duration access to microgravity. Supporting commercial and academic interest in solidification in space are several short-duration free-fall facilities. These include drop towers (4 s, 0.0001 *g*, where $g = 9.80\ m/s^2$, or 32.2 ft/s^2), parabolic aircraft flight (30 s at 0.01 *g*; 1 min at 1.8 *g* repetitive cycles), and suborbital sounding rockets (5 min, 0.0001 *g*) (Fig. 1).

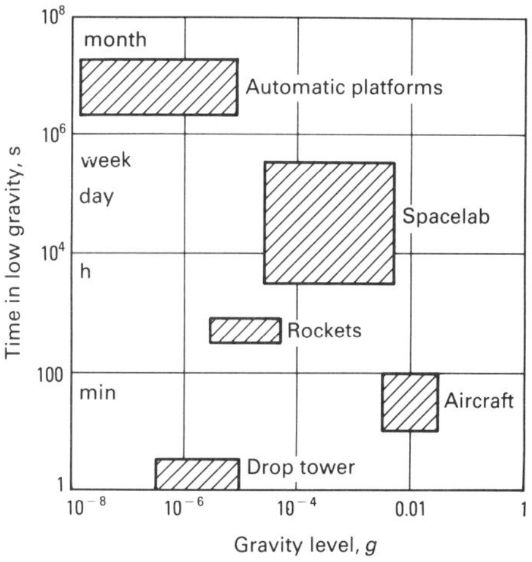

Fig. 1 Available low-*g* experimental systems

Convection and the Melt Temperature Field in the Liquid

Thermal Convection and Temperature Gradient. Solidification progresses through the melt as a result of the extraction of heat from the liquid at the solid/melt interface. Thermal gradients during solidification are therefore present in the melt, causing density gradients that, under a gravitational field, result in buoyancy-driven convective flows. Convective flow, for constant furnace conditions, can modify the liquid thermal gradient G_L at the solid/liquid interface.

The sensitivity of the melt thermal profile to convection is a function of the physical properties of the liquid. A dimensionless number from the Navier-Stokes equations that is useful in determining the sensitivity of a melt thermal profile in response to convective flow is the Prandtl number *Pr*, which is defined by:

$$Pr = \frac{\mu C_P}{K} \qquad \text{(Eq 1)}$$

where μ is viscosity (in $N \cdot s/m^2$), C_P is specific heat (in $J/kg \cdot K$), and *K* is thermal conductivity (in $W/m \cdot K$).

At low *Pr*, the convective transport of momentum dominates the convective transfer of heat; thus, thermal convection has less influence on the temperature field. The Prandtl number for liquid metals is generally much less than 1 (of the order of $Pr = 0.01$). Ammonium chloride/water melts (often used as transparent analogs for metal solidification) have $Pr = 6$.

The response of the temperature fields to fluid flow in semi-infinite plates is illustrated schematically in Fig. 2 for high and low *Pr*. The heat flux is along the *z*-axis, that is, upward, so that if conduction alone is considered the temperature profile can be represented by the dashed line, showing the hot temperature T_{hot} at the bottom and the cold temperature T_{cold} at the top. Convective velocity, represented by the dotted curve, results in mass transport to the right in the upper half of the plate compensated by mass transport to the left in the lower half, which produces a convective roll. The effect is higher for low *Pr* (Fig. 2b) than for high *Pr* (Fig. 2a). Nevertheless, materials with low *Pr*, such as liquid metals, have normally high *K*, which reduces the effect of mass transport on the heat transport (temperature profile). The overall result on the temperature profile is indicated by the solid curves. It predicts a rather limited effect of convective flow for metals (Fig. 2b) as compared to some organic or semiconducting materials (Fig. 2a).

For a given composition and growth velocity, alloy microstructure and solid/liquid interface morphology are dependent on G_L; therefore, the influence of convection on the liquid thermal profile has been extensively modeled. One configuration examined is unidirectional solidification in the

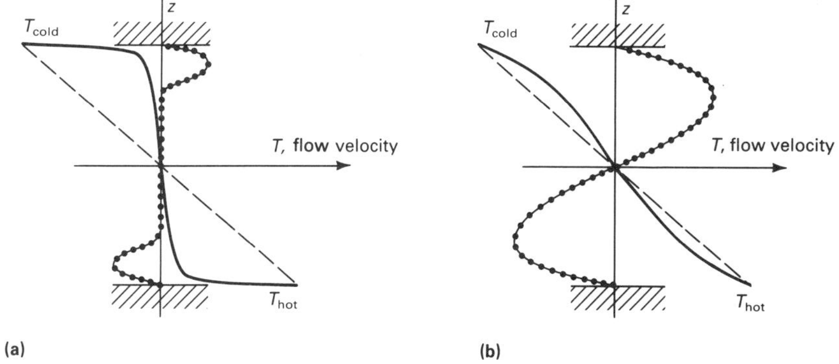

Fig. 2 Schematic of the sensitivity of the temperature field to fluid flow for melts with high and low *Pr*. (a) $Pr \gg 1$. (b) $Pr \ll 1$. Dashed line represents temperature profile from thermal conduction only. Dotted curve is convective velocity, and solid curve is the corresponding temperature profile. Source: Ref 1

vertically stabilized configuration, that is, solidification antiparallel to gravity (less dense hotter fluid upward).

Temperature field numerical results for gallium-doped germanium (Pr = 0.01) show that the temperature isotherms in the melt at the crucible center line compress (higher G_L) at high gravity because of convective flow (Ref 2). The thermal gradient at the crucible wall decreases with higher convection. This results from the convective flow of hot liquid toward the crystal/melt interface center that is compensated for by the flow of cold liquid away from the solidification interface at the crucible walls. Thus, under strong convection, the minimum G_L across the solid/liquid interface increases.

Solutal Convection and Temperature Gradient. At the growing interface during alloy solidification, solute of differing density from the bulk liquid is often rejected. Under a gravitational field (even if the system is stable for thermal convection), this results in solutal convection when the solute is less dense than the liquid. Thermal and solutal density gradients combine to cause thermosolutal convection. For metal alloys, the solute convection contribution to the fluid flow is normally large compared to the thermal convection contribution, causing solutal convective flow to dominate.

The laser interferograms of solidifying ammonium chloride shown in Fig. 3 illustrate the development of solutal convection (Ref 3). When the ampule of liquid ammonium chloride is placed on a cold block, solidification begins upward. Growth enriches solute in the liquid ahead of the solid/liquid interface, thereby decreasing its density relative to the bulk melt. The solute-rich layer builds up until a plume of less dense solute breaks away from the interface and travels upward in the melt. The solutally driven convective flow decreases the liquid thermal gradient at the solid/liquid interface. The effect on the temperature field is amplified by the high Pr number (Pr = 6) of the system. The experiment was repeated in low gravity during aircraft parabolic flight (10^{-2} g) and the G_L in low gravity was 15% greater (relative to G_L in 1 g) (Ref 4). As expected from Fig. 2, low-gravity experiments with metallic alloys (low Pr) find less dependence of G_L on convective flow.

Convection and Solute Redistribution

The solute concentration in the solid during alloy solidification differs (unless the partition coefficient is 1) from that in the liquid ahead of the solidification interface. When the liquid ahead of the solidification interface is continually well mixed by convective flow, the resulting solute concentration in the solid (if diffusion in the solid is negligible) continually increases. However, if the solute rejected into the liquid is transported only by diffusion (and Soret diffusion is negligible), a solute-enriched boundary layer forms in the liquid ahead of the solid/liquid interface and increases in solute concentration as solidification progresses until a steady state is reached.

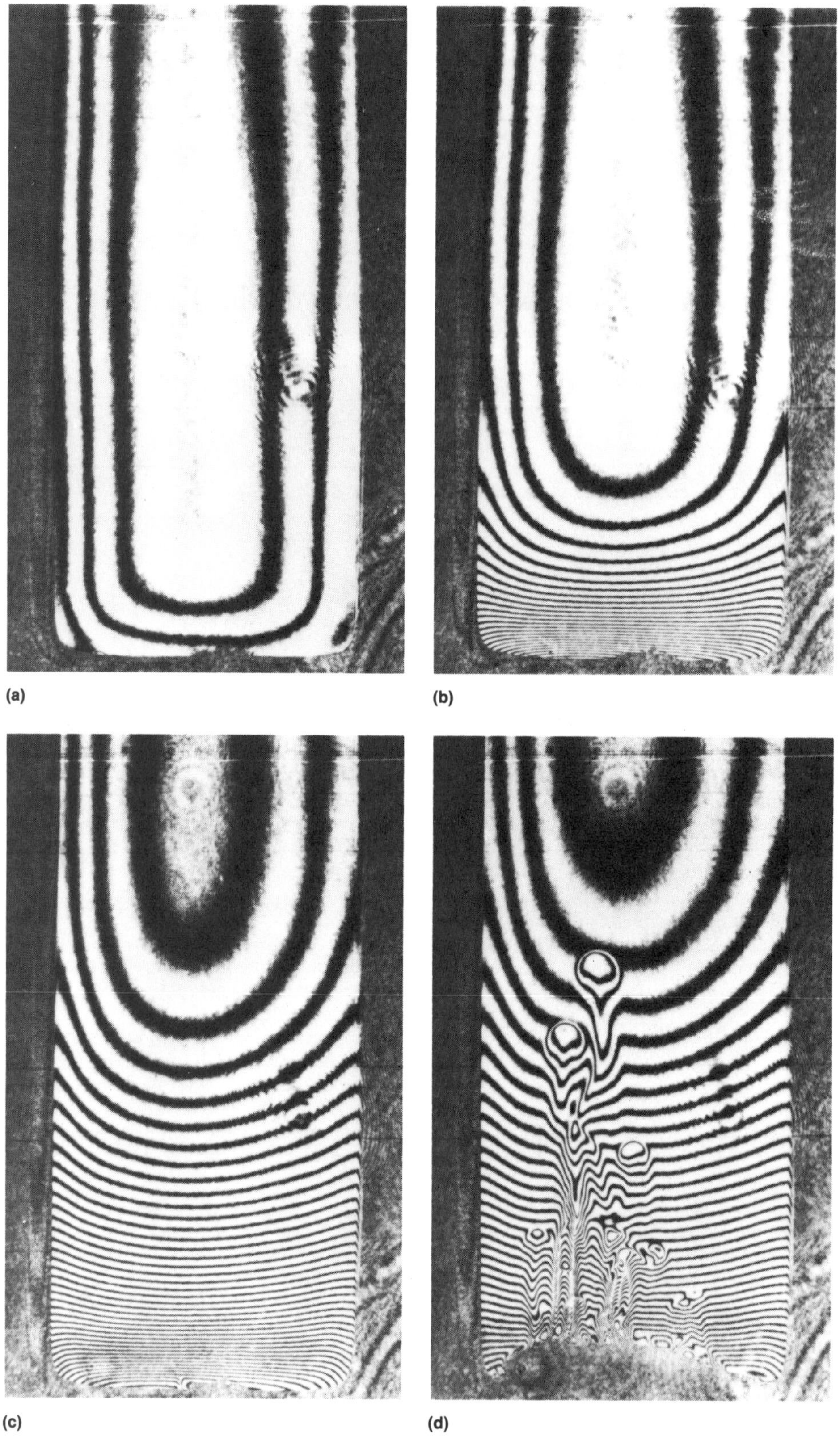

Fig. 3 Laser interferograms of an ammonium chloride/water transparent model during initial solidification upward in normal gravity showing the development of solutal convective plumes at elapsed times of 0.0 s (a), 30 s (b), 90 s (c), and 120 s (d)

Table 1 Effect of low gravity for on-eutectic interphase spacing

Alloy composition	Low-gravity solidification effect on interphase spacing, %
Lamellar eutectics	
Al_2Cu/Al	No change
Fe_3C/Fe	−25
Fibrous eutectics	
MnBi/Bi	−50
InSb/NiSb	−20
Al_3Ni/Al	17

Source: Ref 7

Alloy solidification in the presence of a steady-state boundary layer can, in contrast to the example with strong convection, solidify with a nearly constant composition of the solid. Low-gravity planar growth experiments with tellurium-doped indium-antimony (Skylab, 1972) first demonstrated improved crystal solute homogeneity (over that for growth in 1 *g*) by steady-state diffusion boundary layer controlled growth.

Stagnant Film Models. Variable-thickness diffusion boundary layer (stagnant film) models are often used to assess the influence of convective flow on the solute distribution during solidification. Convective flow is assumed to mix the solute completely in the melt outside the diffusion boundary layer. The boundary layer thickness decreases with increasing convective flow. Inside the diffusion boundary layer, the model assumes that solute transport is by diffusion only. For intermediate convective flow velocity, the model assumes a dynamic diffusion boundary layer that can be calculated via an effective (convective-dependent) partition ratio. At high convective flow, the boundary layer approaches 0, yielding a solid solute concentration that varies insignificantly with solidified fraction until the end of solidification is approached.

Stagnant film models are often used in the literature for simplicity of calculation. The assumption that convective flow does not penetrate the diffusion boundary layer, however, leads to incorrect predictions. The transverse diffusion boundary layers for on-eutectic growth or secondary dendrite arms are much smaller than the momentum boundary layer that the model considers to be affected by convective flow. Therefore, stagnant film models falsely predict that even vigorous convective flow will have no effect for on-eutectic or secondary dendritic spacings.

Numerical Solution of Solute Redistribution. Finite-element methods more accurately model convective flow at the melt/solid interface and the resulting solute segregation. Numerical predictions are qualitatively similar to those of the stagnant film models (Ref 2). However, finite-element calculations reveal that a stagnant boundary layer does not exist in the presence of convection. Flow within the diffusion boundary layer has a strong convective flow contribution. This contribution must be carefully considered in analyzing the effect of convection on microstructure.

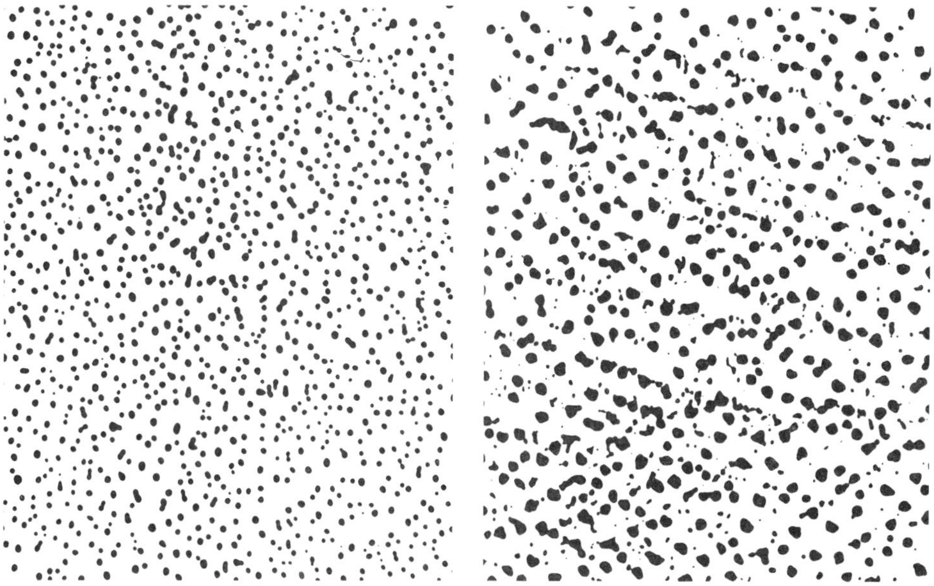

(a) (b)

Fig. 4 Transverse sections of bismuth/manganese bismuth rod eutectic solidified in low-gravity (SPAR IX sounding rocket) (a) and a 1 *g* control sample (b). Source: Ref 5

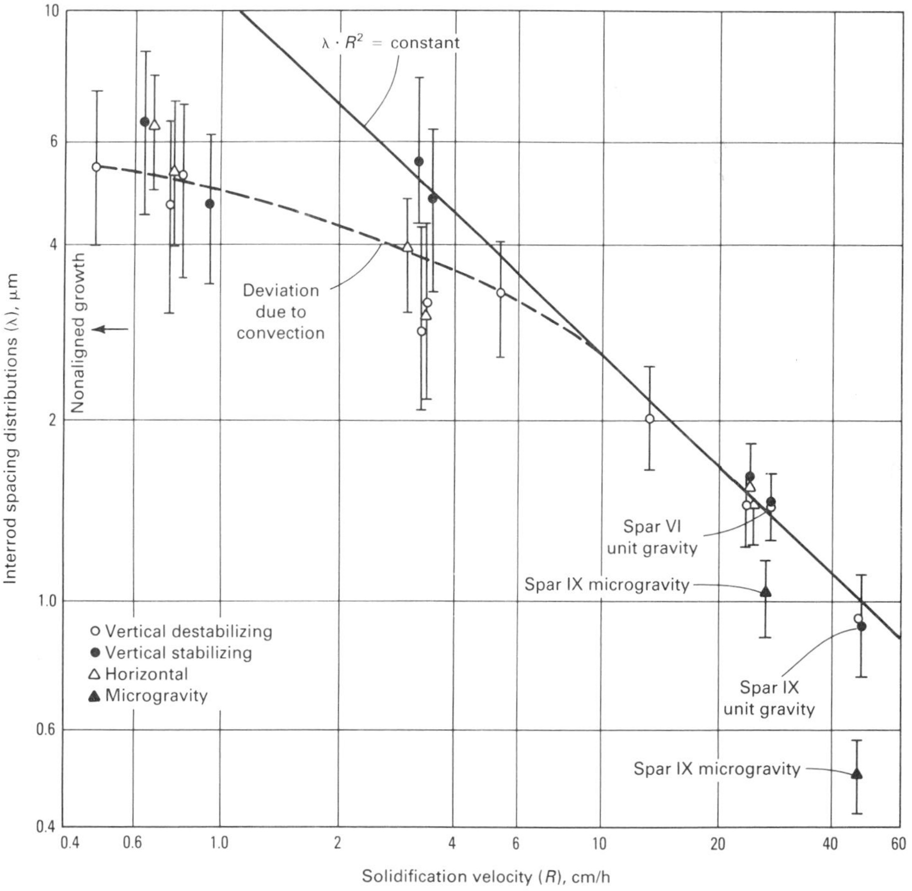

Fig. 5 Interrod spacing for bismuth/manganese bismuth versus velocity for samples solidified parallel and antiparallel to gravity on earth and for low-gravity solidified samples. Source: Ref 6

Solidification of Composites in Low-Gravity Eutectic Alloys

Buoyancy-driven convection and Stokes flow strongly affect eutectic solidification. The casting of off-eutectic alloys with independently nucleated primary phase (under normal gravity) results in severe macrosegregation due to Stokes flow. An example is the kishing of graphite in castings of hypereutectic gray iron. Solidification in zero gravity eliminates Stokes flow. Low-gravity experiments with off-eutectic iron-carbon alloys have shown that primary graphite flakes or nodules that float away from the interface in normal gravity are incorporated into the solidifying interface under low gravity (Ref 4). Thus, the solidification of off-eutectic castings with independently nucleated primary particles in zero gravity eliminates macrosegregation due to Stokes flow.

Cooperative growth of on-eutectic alloys has been shown to be strongly influenced by convection. On-eutectic MnBi/Bi was solidified in low gravity on Space Processing Applications (sounding) Rocket flights SPAR VI (R = 6 mm/min, or 0.24 in./min) and SPAR IX (R = 8.3 mm/min, or 0.33 in./min) (Ref 5). Sample microstructures, compositions, and properties were compared to 1 g controls (solidified under identical conditions except for gravity). The eutectic interphase spacing (relative to 1 g controls) for a low-gravity solidified sample decreases by over 50%. This is evident in the transverse sections shown in Fig. 4. A decrease in rod diameter is also apparent for low-gravity solidified samples. The volume fraction of MnBi rods in low-gravity samples is also smaller (about 7%). Thermal data reveal increased interfacial undercooling (3 to 5 °C, or 5 to 9 °F) during low-gravity solidification. Low-gravity solidification produces samples with increased intrinsic coercivity (resistance to demagnetization) (Ref 6).

Phase spacing can also be refined in 1 g by increasing the solidification rate R. Figure 5 shows the interrod spacings λ as a function of R under various processing conditions. It is evident that the orientation of the sample in 1 g has no effect on interrod spacing. At higher R, the spacing appears to obey $\lambda \cdot R^2$ = constant, with the exception of finer spacing for low-gravity samples.

Table 1 includes these first results and the data reported from subsequent studies. Interphase spacing refinement during low-gravity solidification is observed for both fibrous (MnBi/Bi and InSb/NiSb) and lamellar (Fe_3C/Fe) on-eutectic compositions. Gravity independence of lamellar spacing is reported for aluminum-copper and a decrease in spacing for low gravity is reported for Al_3Ni/Al. Although results for each alloy were reproducible, there is no obvious trend.

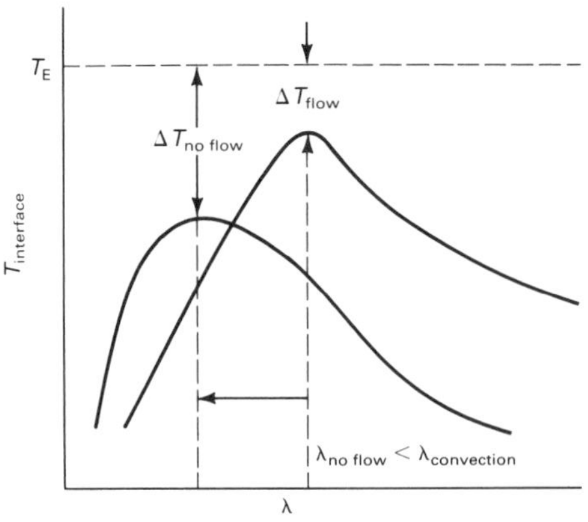

Fig. 6 Predicted evolution of undercooling-interphase spacing relationship with forced convective flow parallel to the solid/liquid interface from the Quenisset-Naslain model. Source: Ref 8

The mechanism for the influence of low gravity during on-eutectic solidification on phase spacing has yet to be established. The importance of convective flow is demonstrated for MnBi/Bi by the duplication of low-gravity spacing and interfacial undercooling through solidification in strong (0.3 T, or 3 kG) magnetic field. The effect of gravity on convective flow is relatively well understood. The challenge is to develop a theory that adequately relates convective flow at low gravity to the mechanisms controlling eutectic phase spacing. Several approaches being pursued to develop the needed theory are discussed below in relation to experimental findings.

Analysis of Convective Flow for On-Eutectic Growth. A simple stagnant film model predicts that the presence (or absence) of convection will not affect cooperative on-eutectic growth. Solute redistribution for on-eutectic growth occurs on the scale of the lamellar spacing, which is much smaller than the convective flow momentum boundary layer (of the order of D_L/R, where D_L is the liquid diffusivity). Thus, for steady-state on-eutectic growth at fixed volume fraction, convective flow is not expected to affect lamellar spacing.

Table 2 Effect of low gravity on dendrite spacing

Composition	g–g_{low}	Low-gravity solidification relative dendrite spacings
NH_4Cl-4%H_2O	1–0.0001	+30% secondary +10% tertiary
Al-Cu	1–0.00001	+150% primary
Pb-Sn	1–0.0001	+50% secondary
MAR-M246	1.8–0.01	+50% secondary
Fe-C-Si-P	1.8–0.01	+20% secondary
PWA-1480	1.8–0.01	+20% primary
Al-Cu	1–0.001	+500% primary

More rigorous analysis has demonstrated the invalidity in the above conclusions (Ref 8). The actual flow fields (diffusive and convective) in the liquid ahead of the eutectic solid/liquid interface are numerically calculated. A series of curves for lamellar spacing versus interfacial undercooling for a given R can be determined for different magnitudes of forced convective flow (Fig. 6). Growth is assumed to be preferred at the extremum (minimum interfacial undercooling), which then defines the eutectic spacing for a given R.

The theory predicts that forced convection decreases interfacial undercooling and increases interphase spacing. The theory semiquantitatively predicts phase spacing at high convective flows for Ti/Ti_5Si_3, MnBi/

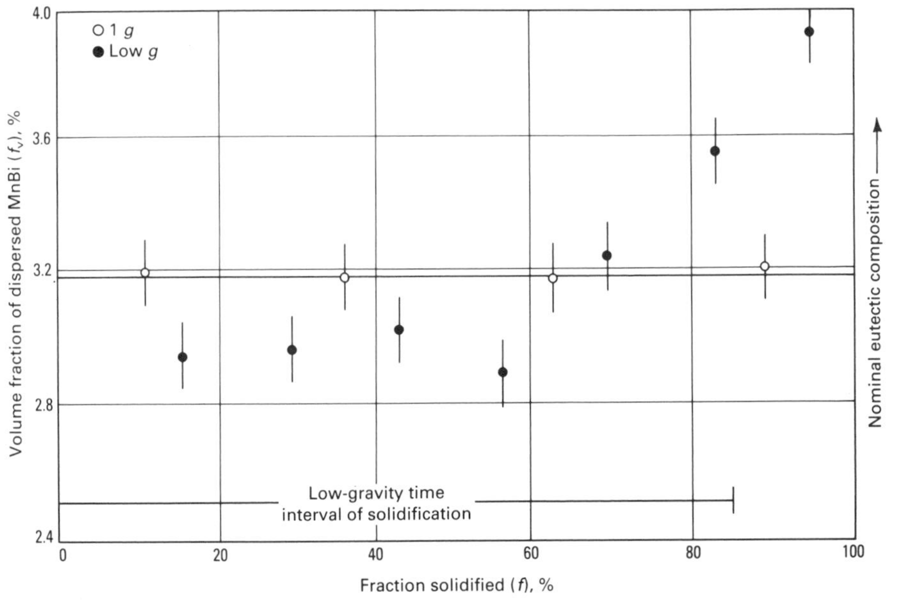

Fig. 7 Solidification temperature (undercooling) data for MnBi/Bi solidified in low gravity relative to 1 g controls. Source: Ref 6

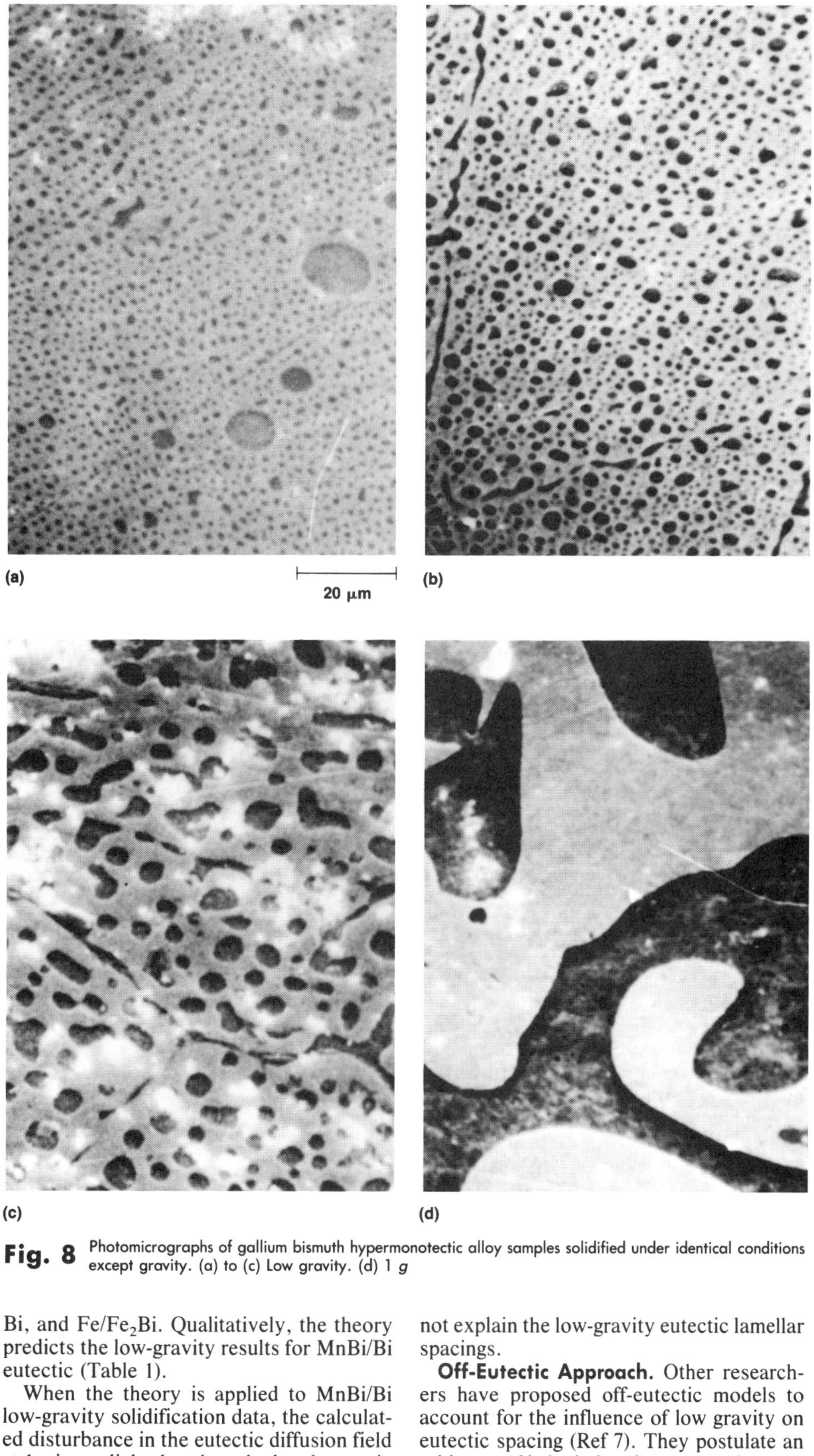

Fig. 8 Photomicrographs of gallium bismuth hypermonotectic alloy samples solidified under identical conditions except gravity. (a) to (c) Low gravity. (d) 1 *g*

Bi, and Fe/Fe_2Bi. Qualitatively, the theory predicts the low-gravity results for MnBi/Bi eutectic (Table 1).

When the theory is applied to MnBi/Bi low-gravity solidification data, the calculated disturbance in the eutectic diffusion field at 1 *g* is so slight that the calculated eutectic phase spacings for solidification in 1 *g* and in space are essentially equivalent (Ref 9). Thus, the calculated effect of convective flow on the liquid concentration field does not explain the low-gravity eutectic lamellar spacings.

Off-Eutectic Approach. Other researchers have proposed off-eutectic models to account for the influence of low gravity on eutectic spacing (Ref 7). They postulate an arbitrary 1% deviation from eutectic composition. Boundary layer theory is used to predict the effect of convection on the volume fraction term in equations for eutectic spacing.

Off-eutectic models offer several advantages. The solute redistribution boundary layer is of the order of D_L/R, which is much larger than that (eutectic spacing λ) for on-eutectic cooperative growth. Convection has a much more pronounced effect on the concentration field ahead of the solid/liquid interface. The sign of the composition deviation from eutectic determines the sign of the low-gravity induced phase spacing change (low-gravity spacing that is smaller for hypereutectic and larger for hypoeutectic). No change, that is, insensitivity to convection, is also predicted for some materials.

Manganese bismuth/bismuth eutectic samples experience considerable solid/liquid interfacial undercooling during directional solidification in low gravity. Under equivalent solidification conditions (except in 1 *g*), MnBi/Mn essentially solidifies at the eutectic temperature. Volume fractions f_V for MnBi/Bi data show on-eutectic solidification in normal gravity and bismuth-rich off-eutectic solidification in low-gravity (Fig. 7).

Due to the lack of published data, it is not known if the other alloys in Table 1 also experience increased eutectic solid/liquid interface undercooling in low gravity. Alloy-dependent undercooling in low gravity could explain the data in Table 2.

The off-eutectic model (Favier and de-Goer) tested (Ref 7) with the f_V flight data for MnBi/Bi (Fig. 7) predicts $\lambda_{1g} = 1.05\ \lambda_{0g}$. This is only 10% of the observed change. Therefore, the gravity-dependent change in f_V, using the Jackson-Hunt expression (and the assumptions that the volume fraction, liquid diffusivity, and temperature change at constant velocity are constant), cannot explain low-gravity eutectic spacings.

Diffusion/Atomic Transport. Neither on-eutectic nor off-eutectic convection models predict the data in Table 1. Convective effects on thermal gradient and growth rate also fail to explain the low-gravity eutectic spacing.

Another approach examines the influence of gravity on solidification through microconvections driven by microscopic concentration and temperature gradients (Ref 7). These microconvections are independent of the previously discussed macroconvection but are indistinguishable from collective atomic motion, that is, liquid diffusion. The effective liquid diffusion coefficient D_{eff} in normal gravity consists of the intrinsic diffusion coefficient plus an atomic transport component due to microconvection. In low-gravity liquid-metal diffusion experiments for zinc and tin, it was found that D_{eff}(0 *g*) is less than D_{eff}(1 *g*) by 10 to 60% (Ref 7). A decrease in D_{eff} of this order could explain the magnitude of lamellar spacing decrease found for low-gravity solidified Fe/Fe_3C and Bi/MnBi (Ref 7). The low-gravity data for on-eutec-

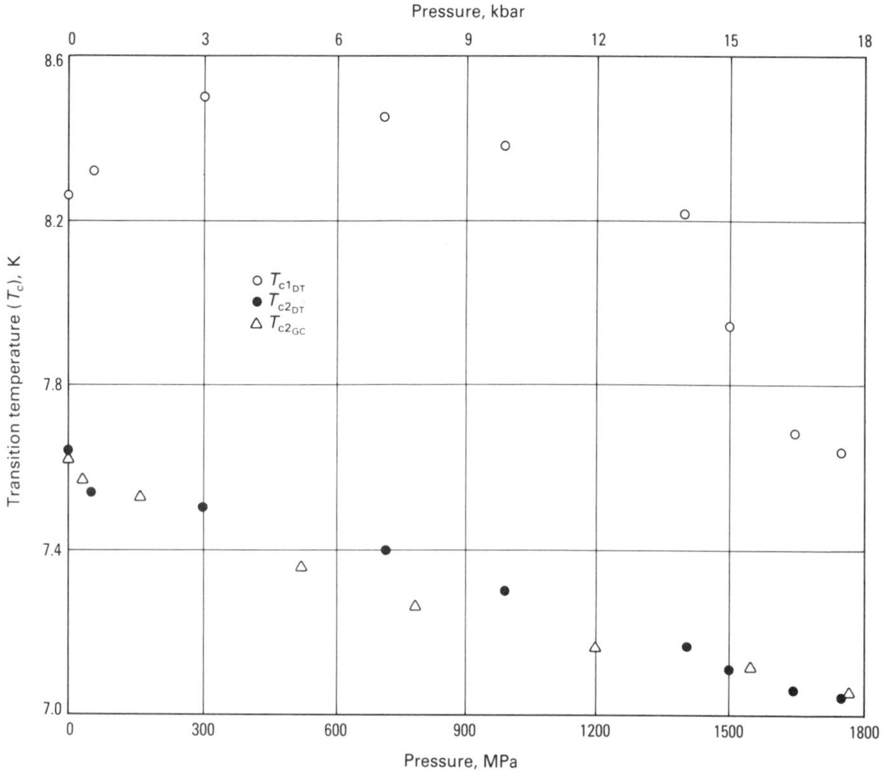

Fig. 9 Superconducting transition temperature T_c for gallium bismuth hypermonotectic samples showing higher T_c phase in low-gravity sample. Ground control (GC) sample: solidified in 1 g; drop tower (DT) sample: solidified in the drop tower (10^{-4} g). Source: Ref 11

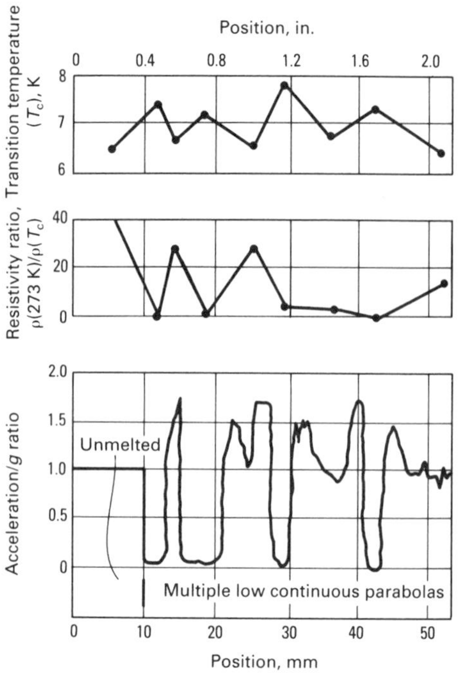

Fig. 10 Superconducting transition temperature, resistivity ratio, and gravitational acceleration ratio (g = 9.8 m/s^2) during solidification for an aluminum-indium-tin monotectic sample solidified during KC-135 aircraft low-gravity maneuvers. Source: Ref 12

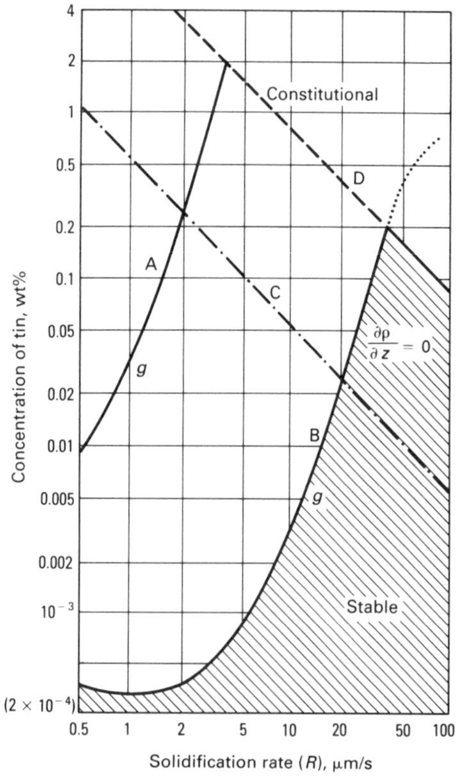

Fig. 11 Calculated stability diagram for critical concentration of tin versus solidification rate for lead-tin (grown in the vertical-stabilized Bridgman configuration). Curves (A, at 10^{-4} g; B, at 1 g) are the composition above which thermosolutal convection instability occurs. Curve C is the neutral density criterion. Curve D represents the morphological instability criterion. G_L = 200 K/cm. Source: Ref 14

tic spacing for aluminum-copper and Al_3Ni/Al, however, have yet to be convincingly explained by this approach.

Monotectic alloys contain a miscibility gap or dome in their phase diagrams. During off-monotectic alloy solidification in the temperature region through the miscibility gap, two immiscible liquids exist in equilibrium. Melt processing in normal gravity results in density-driven Stokes coalescence of the liquid droplets and massive segregation similar to that experienced for oil and water dispersions. Powder metallurgical (P/M) methods are therefore used to prepare hypermonotectic compositions. However, the interfacial purity necessary to exploit enhanced electronic property applications of hypermonotectic alloys is difficult to maintain with P/M processing. Thus, off-monotectic alloys have had only limited technical importance.

Low-gravity drop-tower experiments with gallium-bismuth have demonstrated the feasibility of producing high volume fraction immiscible alloys with finely dispersed microstructures (Fig. 8) by low-gravity solidification (Ref 10). Subsequent low-gravity experiments have identified a number of nonbuoyancy-driven coalescence mechanisms. Surface wetting of the crucible by the minority liquid, thermal migration of droplets, interfacial energy differences between the liquid phases and the solid, surface tension driven convection, and nucleation kinetics must all be carefully controlled to obtain fine dispersions of hypermonotectic alloys in low gravity.

Hypermonotectic solidification in low gravity has often resulted (relative to a sample solidified in 1 g) in samples with enhanced electronic properties. Low-gravity solidified gallium-bismuth samples exhibit unusual resistivity versus temperature characteristics and possess a superconducting phase with a higher transition temperature T_c (Fig. 9). Skylab experiments for lead-antimony-zinc and gold-germanium hypermonotectic alloys also report a phase with higher T_c (than that of 1 g control samples) for a sample solidified in low gravity. An aluminum-indium-tin hypermonotectic alloy that was directionally solidified during low-gravity parabolic maneuvers also yielded a higher T_c minority phase present only in low-gravity sections (Fig. 10). The low-gravity sections exhibit semimetal temperature versus resistance characteristics, while the 1 g and high-gravity sections are metallic.

Ceramic Metal-Matrix Composites. The melt processing of ceramic metal composites, although less commonly employed than powder metallurgy, has some important advantages. These include better matrix particle bonding, control of the matrix solidification, and processing simplicity. Melt processing is, however, severely limited by gravity-driven segregation.

Low-gravity processing eliminates Stokes forces that dominate the segregation process for large (>1 μm, or 40 μin.) particles. This allows the study of normally masked surface energy driven processes. Small-particle suspensions (<1 μm, or 40 μin., where Brownian collisions dominate aggregation) can be solidified without buoyancy-driven convective flows. Low-gravity experimental results can be used to modify ground processing for improved properties, to develop space construction processes (for example, welding of oxide-strengthened composites in space), or to produce unique space-processed materials.

The first low-gravity ceramic metal-matrix composite experiments were performed on Skylab. (Low-gravity results for composite processing are reviewed in Ref 13.) The gravity-driven segregation of silicon carbide whiskers in silver decreased, yielding a composite with improved (relative to normal gravity processing) mechanical properties. These results were confirmed by low-gravity sounding rocket experiments. The solidification of metal-matrix composites with large (1 to 160 μm, or 40 to 6400 μin.) ceramic particles has also yielded more homogeneous dispersions in low gravity. Surface tension driven segregation and the effects of grain boundaries can dominate segregation low gravity. Composites with small (0.1 to 0.5 μm, or 4 to 20 μin.) oxide particles were found to solidify with a more homogenous dispersion and with fewer particle-free areas in low gravity. Because Stokes forces (sedimentation and flotation) are not strong for this size of particle, the improved homogeneity was hypothesized to result from the lack of convective flows.

Morphological Stability of the Solid/Liquid Interface in Low Gravity

The solute composition at the solid/melt interface is dependent on convective flow. Therefore, convection by stagnant film models can be related to the constitutional supercooling criterion for solid/liquid interfacial morphological stability. More detailed analysis shows that it is critical to consider both thermal and solutal convective flow. Numerical analysis has been reported for coupled thermosolutal convection and morphological stability for a lead-tin alloy solidified in the vertical-stabilized Bridgman configuration (Ref 14). The calculated stability diagram for thermosolutal convective flow is shown in Fig. 11. Although the constitutional supercooling criterion predicts stability in the entire field under the constitutional curve in Fig. 11, when thermosolutal convection is taken into account, the stability field is decreased to the shaded region. In low gravity (10^{-4} *g*), calculation predicts higher stability for constant boundary conditions.

Morphological Stability

The influence of gravity on the planar-to-cellular transition for an iron-carbon-silicon-phosphorus alloy has been determined in experiments on directional solidification during multiple-aircraft parabolic low-gravity maneuvers (Ref 4). Sample composition, gradient and growth rate were selected such that the interface was only marginally morphologically stable during solidification in 1 *g*. Samples were then solidified under the same conditions except during low-gravity maneuvers (continuous cycling of about 25 s at 0.01 *g* and 1 min at 1.8 *g*). The solid/liquid interface became more unstable in low-gravity and could be made to shift from planar (or cellular) to equiaxed solidification morphology during the shift in aircraft gravity from 1.8 to 0.01 *g* (Fig. 12). Solidification experiments in space also found greater destabilization for the Al-1.0Cu interface in low gravity (Ref 15).

Cellular and Equiaxed Growth

Eutectic cell (grain) size for austenite-flake graphite eutectic iron-carbon-silicon-phosphorus has been shown to increase in low gravity during directional solidification in aircraft low-gravity maneuvers (Ref 4). The cell diameter for nickel-base superalloy was two times greater for low-gravity solidification on Spacelab than that for 1 *g* control samples (Ref 16).

Dendritic Growth and Segregation

Dendritic growth processes are critical to the most common alloy casting techniques. Dendrite arm spacings correlate with mechanical strength. Interdendritic macrosegregation causes nonhomogeneous casting properties, yet controlled segregation is used for precipitation-strengthening pro-

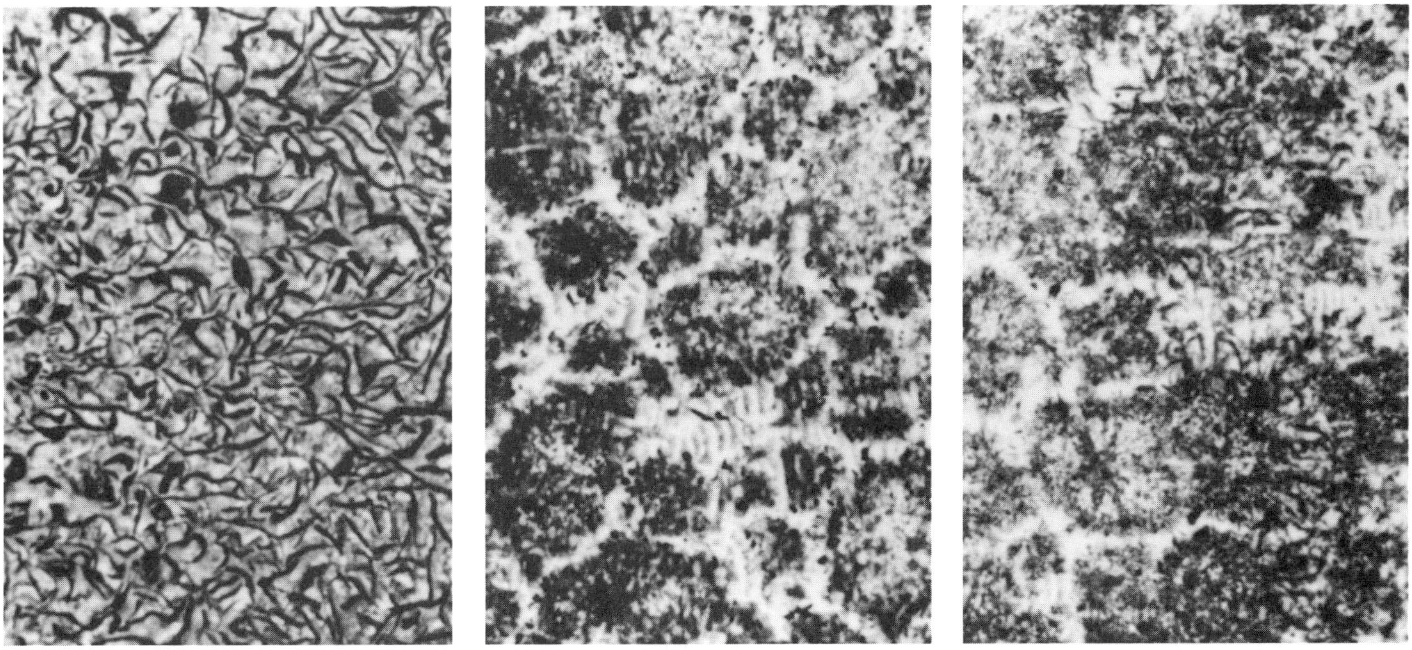

(a) (b) (c)

Fig. 12 Shift from planar morphology (a) to equiaxed solidification morphology (b and c) at the high-gravity (1.8 *g*) to low-gravity (0.01 *g*) transition. Microstructures shown are type A graphite (a), independent nucleation type D graphite (b), and independent nucleation eutectic grains (c).

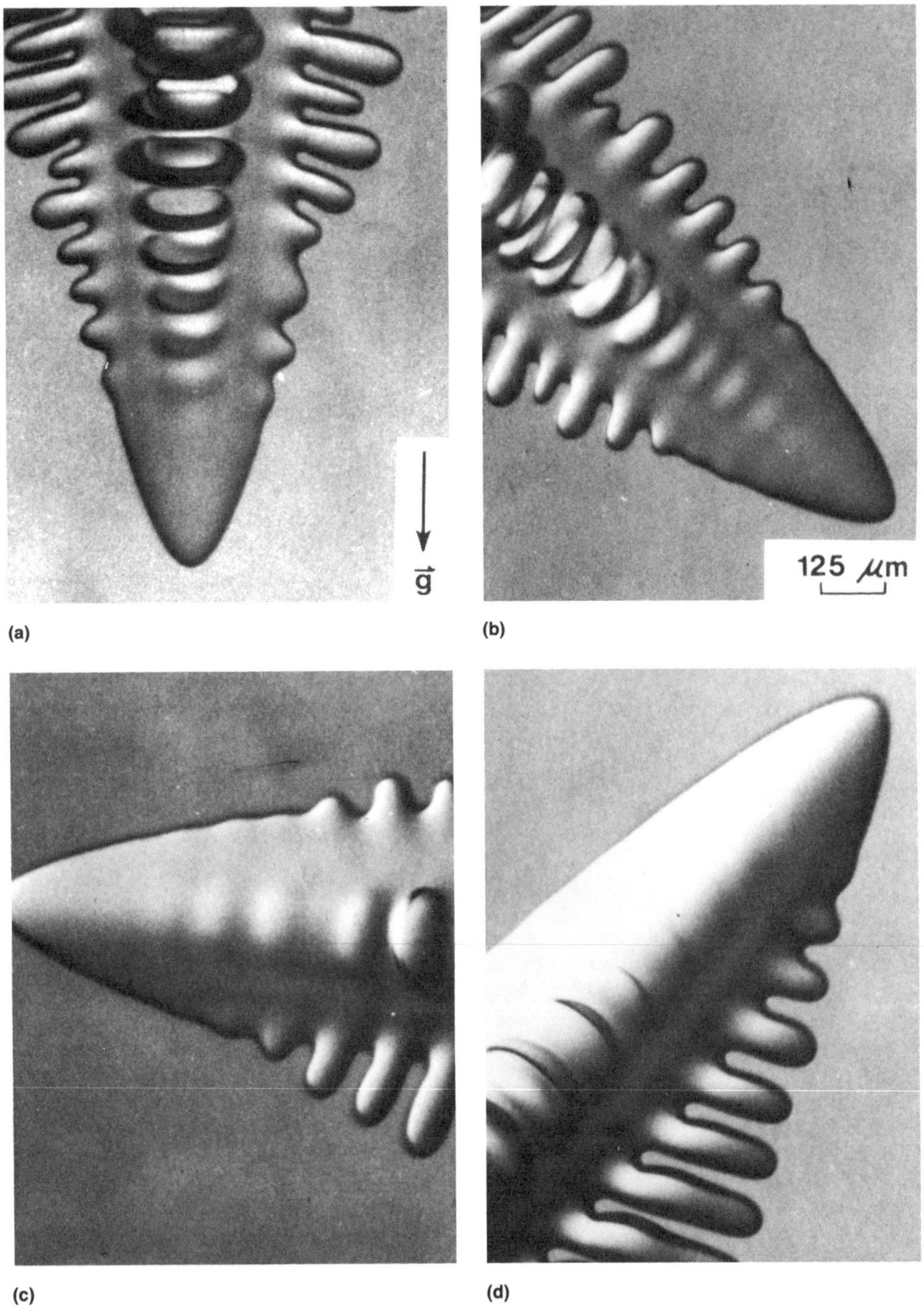

Fig. 13 Influence of orientation of growth direction relative to gravity on the dendritic morphology of succinonitrile (transparent analog model) showing suppression of sidebranches for inclinations above 90° to gravity. Orientation relative to direction of g is 0° (a), 45° (b), 90° (c), and 135° (d).

cesses. Buoyancy-driven convective flow in the liquid perturbs the metal alloy solute concentration profiles (and to a lesser degree, temperature profiles) that control microstructure and properties. Therefore, gravity has a strong influence on the processes of dendritic growth.

Dendrite Tip Growth Morphology and Growth Velocity. Single dendrites of pure succinonitrile have been studied as a function of undercooling, tip growth rate, and orientation with respect to gravity (Ref 17). Gravity-independent diffusion-limited growth occurs at undercooling of 1 to 9 °C (2 to 16 °F). However, at undercoolings less than 1 °C (2 °F), morphology and growth rate are dependent on dendritic orientation to gravity (Fig. 13). The growth rate perturbations observed at small undercooling are consistent with the enhanced transfer of latent heat by natural convection. Low-gravity experiments with Al-0.4Cu have shown that convection can influence dendritic morphology and crystallographic orientation in metallic alloys (Ref 15).

Gravity and Dendritic Spacing. Table 2 lists the results of low-gravity solidification studies on dendrite spacings (Ref 18). The effect of low gravity on secondary dendrite spacings was studied on sounding rockets (5 min at 0.0001 g) for ammonium chloride/water metal-model material and for lead-tin binary alloy, and on parabolic aircraft maneuvers for iron-carbon and MAR-M246 superalloy. Primary dendrite spacings have been studied (Ref 19) by solidification on orbital laboratories for aluminum-copper (Fig. 14) and during parabolic aircraft maneuvers for iron-carbon and PWA-1480 superalloy (Fig. 15). In all studies to date in which differences have been noted, regardless of alloy complexity, low-gravity solidification has resulted in coarser dendritic spacing compared to solidification in normal gravity.

Local solidification times t_f for aluminum-copper and Nimonic-90 (cobalt-nickel) alloys have been found to decrease with increasing forced convective flow velocity (Ref 18). Forced convection also decreases primary dendrite spacing. Aluminum-copper secondary dendrite spacings have also been observed to decrease with decreasing t_f (Ref 18).

Convection influences t_f by improving mold-metal heat extraction and by increasing solute transfer. The behavior of local solidification time in low-gravity dendritic solidification remains to be established. Continued increase in t_f with the elimination of natural convection in low gravity should increase dendrite spacings.

Buoyancy-Driven Interdendritic Solute Transport. A primary dendrite spacing difference of a factor of ten has been noted between ammonium chloride/water metal-model grown parallel (down) and that grown antiparallel (up) to gravity (Ref 18). Ammonium chloride, as the metal analog in the ammonium chloride/water model, freezes dendritically. Water (analogous to the alloying element) is the rejected solute. It is believed that the spacing difference is due to buoyancy-driven fluid flow. For upward growth, less dense rejected solute (water-rich) is driven from the interdendritic region by buoyancy-driven flow. For downward growth, the solute density gradient is convectively stable (similar to conditions in low gravity); solute transport, and therefore dendrite spacing, is limited by diffusion.

The influence of gravity on solute transport, and therefore interdendritic constitutional supercooling, can explain the adjustment of primary spacing to gravity level for a superalloy during solidification through multiple aircraft parabolas (Fig. 15 and 16). Buoyancy-driven flow is expected to increase interdendritic solute transport. This is consistent with available low-gravity data (Table 2), in which all tested alloy types result in unchanged or increased primary spacing.

Dendrite Coarsening. The expected effect of low gravity on the coarsening rate is dependent on the coarsening mechanism. Zero gravity eliminates convective trans-

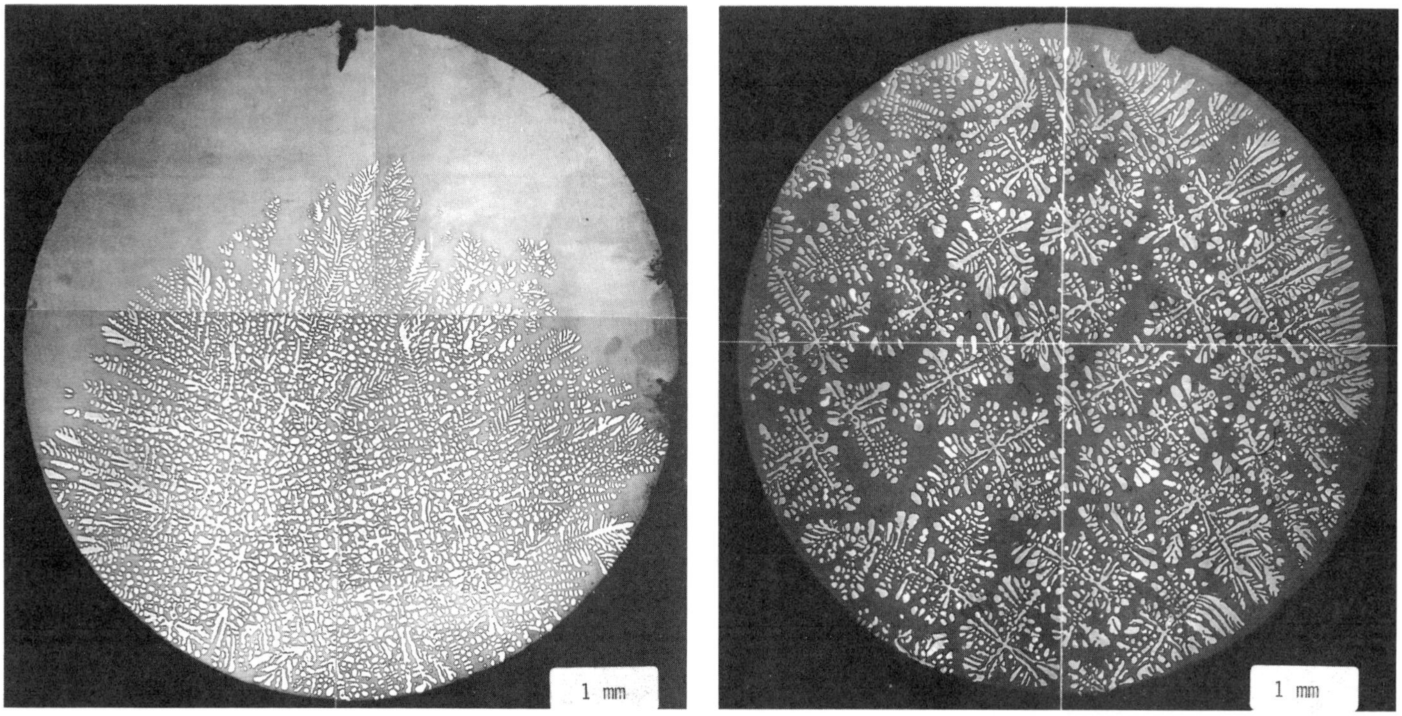

(a) (b)

Fig. 14 Transverse sections of Al-26Cu alloy grown vertically in 1 *g* (a) and in orbit on Spacelab (b). The dendrites in the low-gravity solidified sample have coarser, more regularly spaced dendrites and negligible radial segregation. Courtesy of D. Camel and J.J. Favier, Commissariat A L'Energie Atomique (Grenoble, France)

port of interdendritic rejected solute. Increased average solute concentration reduces dendrite curvature and therefore reduces dendrite coarsening by the arm coalescence model. However, thinning of dendrite arms will cause increased coarsening with increased solute concentration for the arm-melting model.

Dendrite arm coarsening for aluminum-copper alloy has been measured by interface quenching experiments in an orbital low-gravity laboratory (Ref 21). Initial results show a lower dendrite coarsening rate in low gravity. This confirms earlier suggestions that dendrite arm coarsening for aluminum-copper occurs primarily by coalescence (Ref 18).

Interdendritic Fluid Flow—Macrosegregation. Thermal solutal convection and solidification shrinkage produce interdendritic fluid flow that is directly related to casting macrosegregation and porosity. Interdendritic flow perturbs concentration and temperature fields that determine local solid composition in the casting. Aluminum-copper dendritic solidification perpendicular to the gravity vector has been analyzed as a function of the magnitude of gravity. Macrosegregation under normal gravity causes copper fraction across the casting to vary by 45%. Because flow due to solidification shrinkage is perpendicular to the thermal isotherms, theory predicts negligible macrosegregation in zero gravity (Ref 22). The low-gravity dendritic directional solidification of off-eutectic Pb-Sb, Bi/Mn-Bi, Al-Cu (Fig. 17), and Sn-Pb has resulted in negligible macrosegregation, indicating diffusive solute transport (Ref 18).

Macroscopic Dendritic Solid/Liquid Interface Shape. The dendritic solidification of off-eutectic alloy in the convectively stable configuration (both the solid and denser liquid down) minimizes buoyancy-driven macrosegregation. This configuration, however (except for compositions where thermal and solutal buoyancy forces balance), produces interfacial steepling and dendritic clustering casting defects (Ref 23). The denser solute-rich liquid settles into solid/liquid interfacial depressions. The increase in solute concentration causes the depressions to deepen. The interfacial curvature increasingly exceeds the curvature of the isotherm, producing interfacial steepling. Thus, for dendritic directional solidifica-

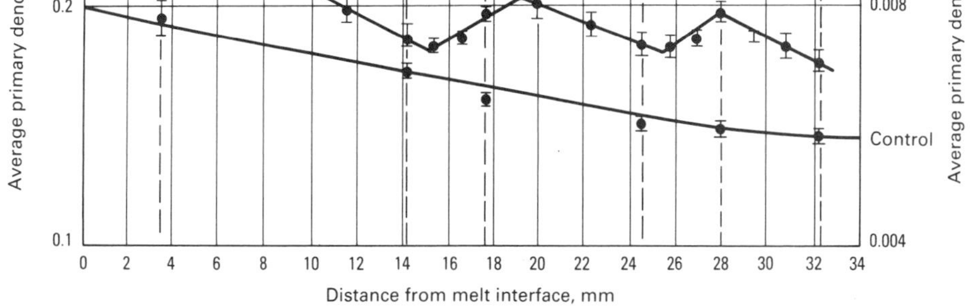

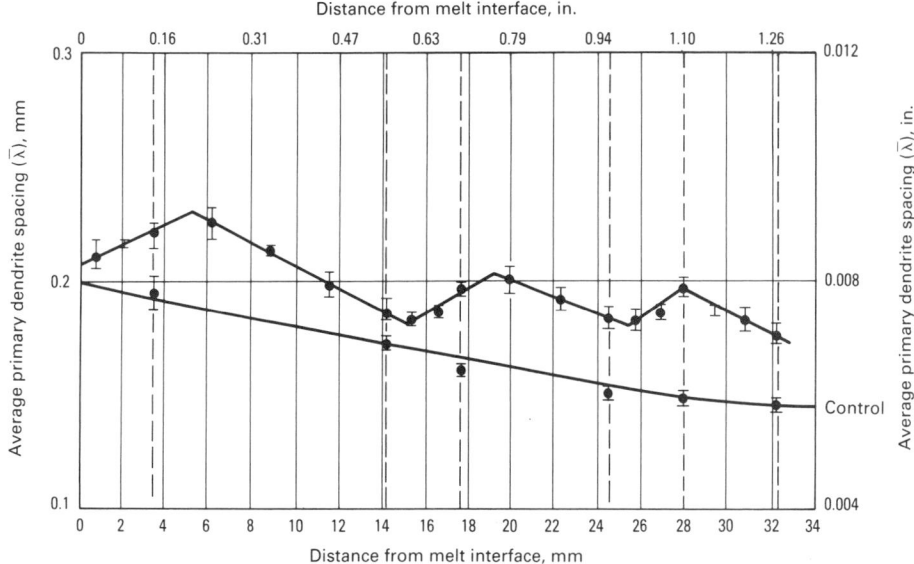

Fig. 15 Average primary dendrite spacing versus solidification distance for PWA-1480 nickel-base superalloy directionally solidified through multiple low-gravity maneuvers. Growth rate is 8 mm/min (0.31 in./min). Source: Ref 20

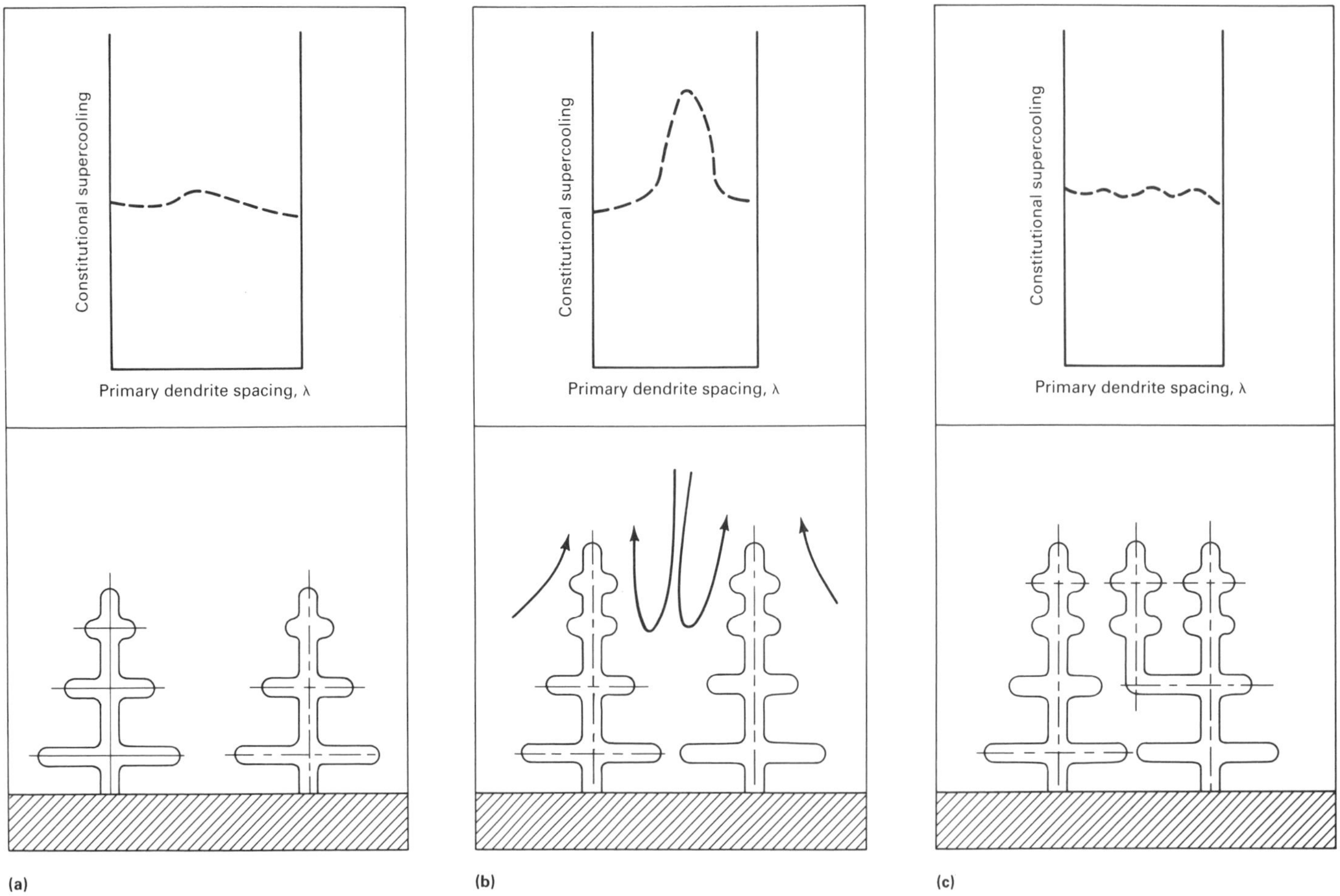

Fig. 16 Mechanism for decreasing primary dendrite spacing in directional solidification during the low-gravity to high-gravity portion of aircraft parabolas. (a) Dendritic growth in low gravity. Solute transport is by diffusion. (b) Dendritic growth continues in high gravity. Solute transport is increased by buoyancy-driven convection. Interdendritic constitutional supercooling increases. (c) Ternary arm driven by increased constitutional supercooling grows to the dendrite tip growth front, decreasing interdendritic spacing. Source: Ref 18

tion, the convectively unstable configuration results in macrosegregation, and the convectively stable configuration results in solid/liquid interface steepling, producing dendritic clustering in the solidified casting.

Channel-Type Segregates. Ingots of commercial alloys often contain defects consisting of vertical solute-rich lines of refined grain structure termed channel-type segregates or freckles. Superalloy and transparent model studies have demonstrated that this casting defect results from interdendritic thermal solutal convective flow (Ref 18). Recent process models agree quantitatively with experimental data. When interdendritic flow solutally enriches a particular mushy zone area, local dendritic remelting occurs. The increased interdendritic channel size results in decreased resistance to flow and further remelting. This process results in local solute enrichment of the order of 30%. The debris of remelted dendrite arms act as nuclei refining the channel grain structure. In low gravity, the elimination of thermosolutal convective flow removes the driving force for channel-type segregate formation.

Grain Multiplication and Casting Grain Structure

At the chill surface of a casting, convection is a major cause of grain multiplication. Conversely, chill zone formation in castings can be prevented by damping convective flow (Ref 18). Transparent model studies have shown that convective pulses at the dendrite roots cause dendrite remelting. Dendrite fragments are swept from the solid/liquid interface by the convective flow, which simultaneously dissipates superheat. The fragments nucleate new grains that grow in the equiaxed region of the casting. Experiments with metal alloys and with model materials (Fig. 18) have shown that the grain refinement by this mechanism increases under forced convection; conversely, it is absent in low gravity (Ref 24).

REFERENCES

1. F. Rosenberger and G. Muller, *J. Cryst. Growth*, Vol 65, 1983, p 102
2. C.J. Chang and R.A. Brown, *J. Cryst. Growth*, Vol 63, 1983, p 350
3. M.H. Johnston and R.B. Owen, *Metall. Trans. A*, Vol 14A, 1983, p 2164
4. D.M. Stefanescu, P.A. Curreri, and M.R. Fiske, *Metall. Trans. A*, Vol 17A, 1986, p 1121-1130
5. D.J. Larson and R.G. Pirich, Influence of Gravity Driven Convection on the Directional Solidification of Bi/MnBi Eutectic Composites, in *Materials Processing in the Reduced Gravity Environment of Space*, G.E. Rindone, Ed., Materials Research Society, 1982, p 523-532
6. R.G. Pirich, "Space Processing Applications Rocket (SPAR) Project, SPAR IX, Final Report," Technical Memorandum 82549, National Aeronautics and Space Administration, 1984, p 1-46
7. P.A. Curreri, D.J. Larson, and D.M. Stefanescu, Influence of Convection on Eutectic Morphology, in *Solidification Processing of Eutectic Alloys*, Proceedings of the 1987 Fall Meeting, The Metallurgical Society, 1988
8. J.M. Quenisset and R. Naslain, *J. Cryst. Growth*, Vol 54, 1981, p 465-474
9. V. Baskaran and W.R. Wilcox, *J.*

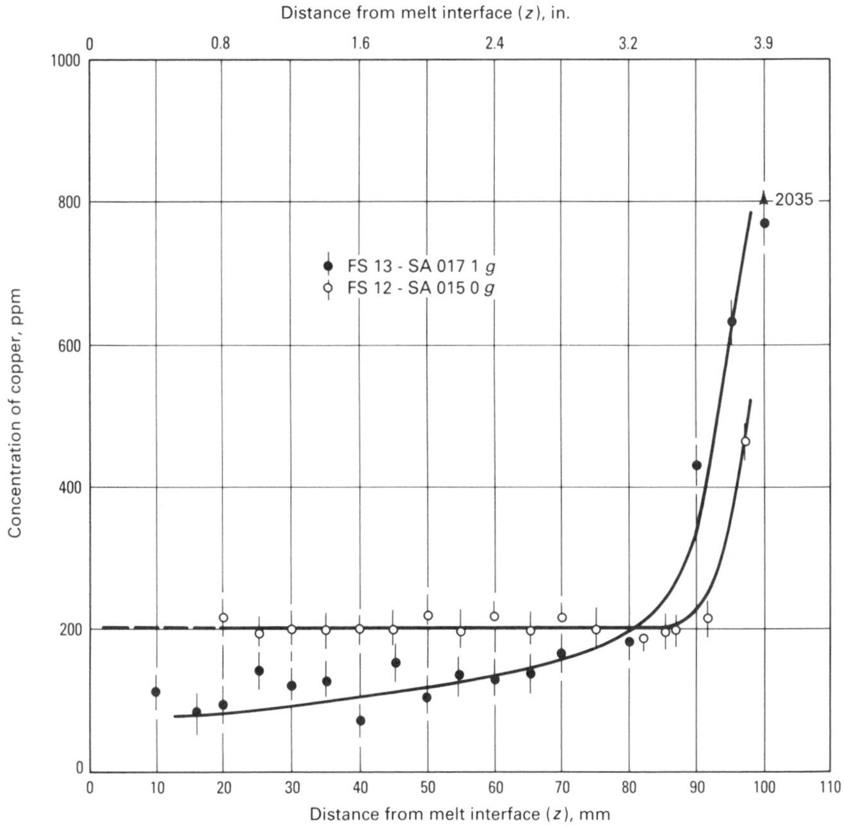

Fig. 17 Copper concentration versus distance for directionally solidified aluminum-copper samples showing decreased macrosegregation for the low-gravity solidified sample. Source: Ref 15

Cryst. Growth, Vol 67, 1984, p 343-352
10. L.L. Lacy and G.H. Otto, *AIAA J.*, Vol 13, 1975, p 219
11. M.K. Wu, J.R. Ashburn, C.J. Torng, P.A. Curreri, and C.W. Chu, Pressure Dependence of the Electrical Properties of GaBi Solidified in Low Gravity, in *Materials Processing in the Reduced Gravity Environment of Space*, Vol 87, R.H. Doremus and P.C. Nordine, Ed., North-Holland, 1987, p 77-84
12. M.K. Wu, J.R. Ashburn, P.A. Curreri, and W.F. Kaukler, *Metall. Trans. A*, Vol 18A, 1987, p 1515
13. H.U. Walter, *Fluid Sciences and Material Sciences in Space*, Springer-Verlag, 1987
14. S.R. Coriell, M.R. Cordes, W.J. Boettinger, and R.F. Sekerka, *J. Cryst. Growth*, Vol 49, 1980, p 22
15. J.J. Favier, J. Berthier, Ph. Arragon, Y. Malmejac, V.T. Khryapov, and I.V. Barmin, *Acta Astro.*, Vol 9, 1982, p 255-259
16. H.J. Sprenger, Skin Technology—Directional Solidification of Multiphase Alloys, in *Scientific Results of the German Spacelab Mission D1*, P.R. Sham, R. Jansen, and M.H. Keller, Ed., DFVLR, 1987, p 46-59
17. M.E. Glicksman, N.B. Singh, and M. Chopra, Influence of Diffusion and Convective Transport on Dendritic Growth in Dilute Alloys, in *Materials Processing in the Reduced Gravity Environment of Space*, G.E. Rindone, Ed., Materials Research Society, 1982, p 523-532
18. P.A. Curreri, J.E. Lee, and D.M. Stefanescu, Dendritic Solidification of Alloys in Low Gravity, in *Proceedings of the Second International Symposium on Experimental Methods for Microgravity Materials Research*, 117th TMS-AIME Annual Meeting, The Metallurgical Society, to be published
19. D. Camel, J.J. Favier, M.D. Dupouy, and R. le Maguet, Microgravity and Low-Rate Dendritic Solidification, in *Scientific Results of the German Spacelab Mission D1*, P.R. Sham, R. Jansen, and M.H. Keller, Ed., DFVLR, 1987, p 25-35
20. M.H. McCay, J.E. Lee, and P.A. Curreri, *Metall. Trans. A*, Vol 17A, 1986, p 2301-2303
21. H.M. Tensi and J.J. Schmidt, Influence of Thermal Gravitational Convectional on Solidification Processes, in *Scientific Results of the German Spacelab Mission D1*, P.R. Sham, R. Jansen, and M.H. Keller, Ed., DFVLR, 1987, p 5
22. A.L. Maples and D.R. Poirier, *Metall.*

(a) (b) (c) (d) (e)

Fig. 18 Sn-15Pb sample showing grain refinement at higher gravity. Sample solidified at lower gravity: (a) 0 *g*. (b) 1 *g*. Sample solidified at high gravity in a centrifuge: (c) 1¾ *g*. (d) 3½ *g*. (e) 5 *g*

Trans. B, Vol 15B, 1984, p 163-172

23. J.D. Verhoeven, J.T. Mason, and R. Trivedi, *Metall. Trans. A*, Vol 17A, 1986, p 991-1000
24. M.H. Johnston and R.A. Parr, Low-Gravity Solidification Structures in the Tin-15 Wt% Lead and Tin-3 Wt% Bismuth Alloys, in *Materials Processing in the Reduced Gravity Environment of Space*, G.E. Rindone, Ed., Materials Research Society, 1982, p 651-656

SELECTED REFERENCES

- R.J. Naumann and H.W. Herring, *Materials Processing in Space: Early Experiments*, National Aeronautics and Space Administration, 1980
- *Proceedings of the Third Space Processing Symposium Skylab Results*, Vol I, Report M-74-5, National Aeronautics and Space Administration, 1974
- H.U. Walter, *Fluid Sciences and Material Sciences in Space*, Springer-Verlag, 1987

Solidification of Eutectic Alloys

Aluminum-Silicon Alloys

Douglas A. Granger, Alcoa Technical Center
R. Elliott, University of Manchester, Great Britain

SOME 238 COMPOSITIONS for foundry aluminum alloys have been registered with the Aluminum Association. Although only 46% of this total consists of aluminum-silicon alloys, this class provides nearly 90% of all the shaped castings manufactured. The reason for the wide acceptance of the 3*xx.x* alloys can be found in the attractive combination of physical properties and generally excellent castability. Mechanical properties, corrosion resistance, machinability, hot tearing resistance, fluidity, and weldability are considered the most important.

From the standpoint of applications, the 3*xx.x* series can be subdivided into the binary nonheat-treatable types and the heat-treatable age-hardening alloys, which may contain magnesium, copper, and nickel alone or in combination. Compositions of the most commonly used aluminum-silicon alloys are presented in Table 1, and their physical characteristics are summarized in Table 2.

The attributes of the 3*xx.x*, alloys are best exemplified by the most widely used alloy of the class—A356. Table 3 lists the properties that can be obtained by different casting processes, heat treatments, and small adjustments in composition. Detailed information on the properties and applications of aluminum casting alloys is given in the article "Aluminum and Aluminum Alloys" in this Volume.

The properties of aluminum-silicon alloys are strongly dependent on the casting process used, the chemical additions made to control eutectic structure, primary silicon and grain structure, and molten metal treatment to reduce hydrogen gas content and to remove inclusions. Each of these topics will be discussed separately in this section.

Grain Structure

In general, a small equiaxed grain structure is preferred in aluminum alloy castings because it improves such casting attributes as resistance to hot tearing and mass feeding and because it enhances most mechanical properties and surface finishing characteristics. Property improvements largely result from an increase in soundness, although an important contributor is the overall increase in the homogeneity of an equiaxed casting compared to a nonuniform columnar-grain structure. A more homogeneous structure results in less segregation so that the casting responds better to heat treatments, and defects such as porosity and intermetallic constituents are more uniformly distributed and therefore less harmful.

The principal manner in which reducing grain size increases resistance to hot tearing is by lowering the coherency temperature (Ref 1); however, in aluminum-silicon alloys the combination of a large volume fraction of eutectic liquid and relatively low coherency temperature means that, in practice, foundry alloys with greater than about 5% Si already exhibit a low cracking tendency compared to an Al-4.5Cu alloy (Ref 2). In Fig. 1, susceptibility to hot cracking is indicated by a hot tearing index of less than

Table 1 Compositions of common aluminum-silicon alloys

Alloy	Product(a)	Nominal composition, %(b) Cu	Mg	Mn	Si	Others
355.0	S, P	1.2	0.50	0.50 max	5.0	0.15Ti
A356.0	S, P	. . .	0.35	0.35 max	7.0	. . .
A357.0	S, P	. . .	0.60	0.03 max	7.0	0.15Ti, 0.04Be
360.0	D	. . .	0.50	0.35 max	9.5	. . .
380.0	D	3.5	. . .	0.50 max	8.5	. . .
390.0	D	4.5	0.60	0.10 max	17.0	. . .
413.0	D	. . .	. . .	0.35 max	12.0	. . .
B443.0	S, P	. . .	. . .	0.50 max	5.2	. . .

(a) S, sand casting; P, permanent mold casting; D, die casting. (b) All compositions contain balance of aluminum.

Table 2 Characteristics of some permanent mold and sand casting aluminum-silicon alloys

Rankings are relative to other alloys in the same casting category. 1, best; 5, worst

	Foundry characteristics				Product characteristics						
Alloy	Resistance to hot cracking	Fluidity	Solidification shrinkage	Pressure tightness	Corrosion resistance	Machining	Polishing	Electroplating	Anodizing appearance	Hot strength	Weldability
Permanent mold castings											
355.0	1	2	2	1	3	3	3	2	4	2	2
A356.0	1	2	1	1	2	3	3	1	4	3	2
A357.0	1	2	1	1	2	3	3	1	4	3	2
B443.0	1	1	2	1	2	5	4	2	4	4	1
Sand castings											
C355.0	1	1	1	1	3	3	3	1	4	2	2
A356.0	1	1	1	1	2	4	5	2	4	3	2
A357.0	1	1	1	1	2	3	4	2	4	2	1
B443.0	1	1	1	1	2	5	5	2	5	4	1

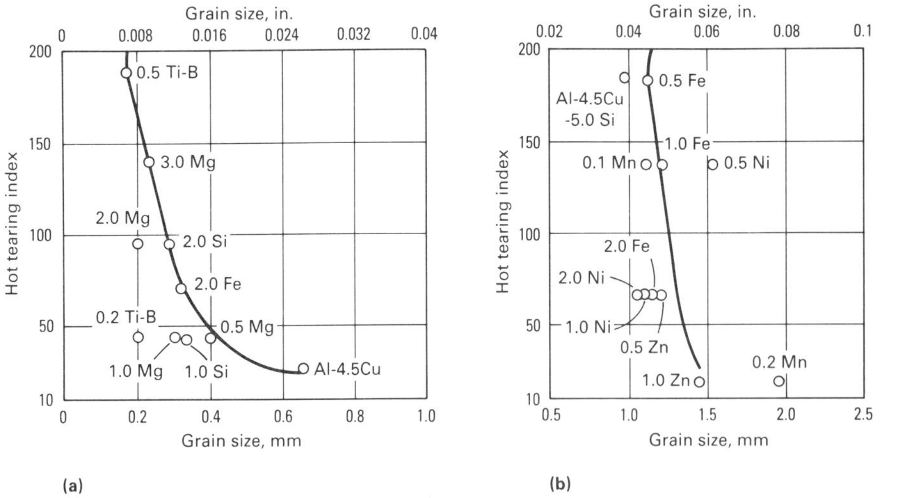

Fig. 1 Effect of grain size on the hot tearing tendency of Al-4.5Cu (a) and Al-4.5Cu-5Si (b) alloys with various additives. Source: Ref 2

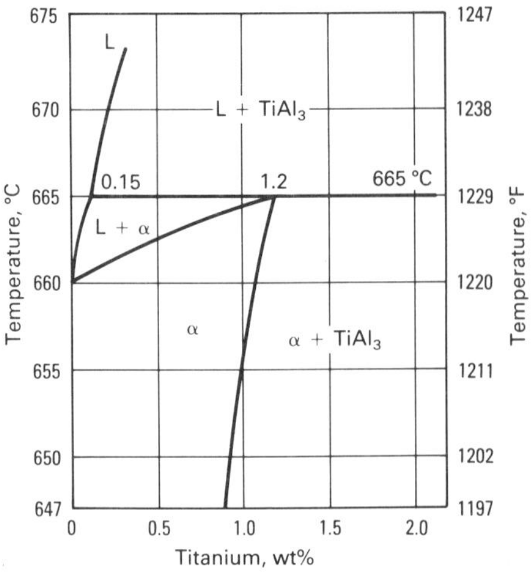

Fig. 2 Aluminum-titanium binary phase diagram

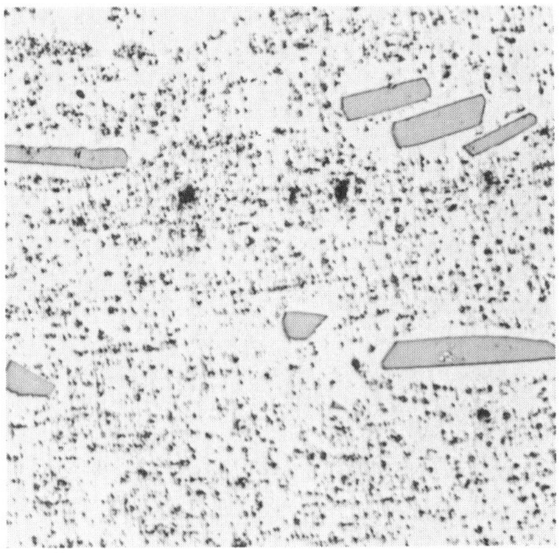

Fig. 3 Micrograph showing large, soluble $TiAl_3$ particles and small, insoluble $(Ti,Al)B_2$ particles in Al-5Ti-1B master alloy. As polished. 250×

50. This occurs in an Al-4.5Cu alloy with a grain size of approximately 350 μm (0.014 in.) to a grain size of approximately 1300 μm (0.052 in.) in an Al-4.5Cu-5Si alloy (Ref 3).

Mass feeding and reduced shrinkage porosity are claimed for grain refinement, although there is a lack of substantive evidence in the literature (Ref 4). In aluminum-silicon alloys, the most direct and quantifiable impact of a reduction in grain size (and the accompanying increase in grain-boundary area) is a more uniform distribution of gas porosity and eutectic colonies. For the same hydrogen content, there is a measurable reduction in the volume of gas porosity in a grain-refined casting compared to a nonrefined casting (Ref 5). The reduced volume fraction of gas porosity is also reported to improve resistance to fatigue crack growth (Ref 6).

Grain Structure Control. Reduction in grain size is brought about by the efficient heterogeneous nucleation of the α-aluminum phase. This can be achieved through crystal multiplication using mechanical or fluid flow forces to detach dendrite arms, but in practice chemical additives are used to provide the necessary nuclei.

Other factors that affect the ultimate cast grain size are alloy composition, freezing rate, temperature gradient in the melt, and casting method. The commercially available chemical additives, usually provided in the form of aluminum-base master alloys, perform well over the range of alloy compositions and casting conditions used in practice.

Mechanism of Grain Refinement. An efficient heterogeneous nucleus for α-aluminum is one that will provide a surface for growth at, or just above, the liquidus temperature of the alloy. It is generally acknowledged that the $TiAl_3$ constituent meets all the necessary criteria (Ref 7). The aluminum-titanium binary alloy contains $TiAl_3$, but because $TiAl_3$ is soluble in molten aluminum, it is necessary to add titanium to levels greater than 0.15% in order to retain grain-refining effectiveness. At this concentration, the peritectic point is exceeded (Fig. 2), and the first phase formed in the alloy upon solidification is the $TiAl_3$ constituent. However, this high level of titanium can give rise to coarse intermetallic particles, which are detrimental to mechanical properties. Moreover, the reduction in grain size is not as great as can be achieved with the aluminum-titanium-boron master alloys. These master alloys, which contain 3 to 10% Ti and a Ti:B ratio ranging from about 3 to 50, comprise soluble $TiAl_3$ particles and insoluble boride particles in an aluminum matrix (Fig. 3). The ternary master alloys are effective at titanium levels significantly lower than the peritectic level, typically 0.01 to 0.03% Ti; this suggests that the boride particles play a dominant role in nucleation (Ref 8).

Despite much study over several decades, the mechanism of grain refinement is not completely clear. It is argued that even at low titanium levels the peritectic mechanism is operative because boron, even at very low levels, shifts the peritectic point to the left (Ref 9). Recent studies have shown that commercial aluminum-titanium-boron master alloys contain $(Al,Ti)B_2$ particles that are effective centers for the surface concentration of titanium atoms (Ref 10). Titanium boride (TiB_2) particles are not effective nuclei, and the successful master alloys contain an excess of titanium over that required to form TiB_2 stoichiometrically. Thus, an aluminum-titanium-boron master alloy operates by releasing soluble and insoluble particles in the melt. As the soluble aluminide phase releases titanium into the melt, the $(Al,Ti)B_2$ particles become increasingly active nuclei for α-aluminum.

Grain-Refining Agents. A number of commercial master alloys are available for

Table 3 Minimum tensile properties of Alloy A356.0 castings produced using various processes

Casting process	Condition	Ultimate tensile strength, MPa	Ultimate tensile strength, ksi	Yield strength, MPa	Yield strength, ksi	Elongation, %
Sand cast	As-cast	130	19	50	7.3	5
Sand cast	T6	234	34	166	24	3.5
Permanent mold	As-cast	160	23	80	11.6	3
Permanent mold	T61	255	37	179	26	5
Squeeze cast	As-cast	195	28	124	18	15
Squeeze cast	T6	300	44	250	36.3	10

Table 4 Characteristics of elements used as silicon modifiers in aluminum-silicon alloys

Element	Atomic radius ratio(a)	Melting temperature, K	Vapor pressure at 1000 K Pa	Vapor pressure at 1000 K atm	Free energy of oxide formation (ΔG_{oxide}) at 1000 K, kJ/mol	Comments on modifying capability
Barium	1.85	998	5.07	5×10^{-5}	−482	Moderate modifying power
Calcium	1.68	1112	26.3	2.6×10^{-4}	−509	Weak modifying action
Strontium	1.84	1042	101.3	0.001	−480	Moderate modifying power; optimal addition: 0.01–0.02%. Good resistance to fade; semipermanent effect; much easier to store and handle than sodium; overmodification not as serious as with sodium, but can increase porosity
Sodium	1.58	371	2×10^{4}	0.2	−367	Very potent modifier with 0.1% addition. Dissolves rapidly; fades rapidly due to evaporation; difficult to store and handle; difficulty in controlling additions can lead to overmodification; can increase porosity
Cerium	1.56	1071	10^{-11}	10^{-16}	−497	Can be conveniently added as mischmetal; requires up to 60 min for dissolution; very resistant to fade; modification effect increases up to 2% additions; reduces dendrite arm spacing in hypoeutectic alloys
Aluminum	1.22	933	5.4×10^{-6}	5.3×10^{-11}	−457	. . .
Silicon	1.00	1683	9×10^{-12}	8.9×10^{-17}	−354	. . .

(a) Ratio of atomic radius of element to atomic radius of silicon

use as grain-refining agents. Where it is the practice to add titanium above the peritectic level, that is, in the range of 0.1 to 0.2%, aluminum-titanium binary alloys containing up to 10% Ti can be obtained in the form of cast waffles for furnace additions. Also in use are briquettes consisting of titanium powder compacted with salts such as KCl and KBF_4.

For controlling grain structure at lower titanium addition levels, the aluminum-titanium-boron master alloys are available as both cast waffle and chopped rod. The rod form, which is used in the primary industry for continuous addition to the molten metal stream during ingot casting, has a superior metallurgical structure in terms of the constituent particle size and distribution.

Silicon Modification

It is standard practice to refine the eutectic structure as well as the grain structure of aluminum-silicon casting alloys. A moderate improvement in mechanical properties is guaranteed with structural integrity when the silicon eutectic phase is refined with arsenic, antimony, or sulfur (Ref 11). The more usual and more effective treatment is structure modification of the silicon phase, although on occasion there may be an increased susceptibility to porosity. Modification occurs naturally at rapid solidification rates (quench modification), but requires a modifying agent at the slower solidification rates typical of sand casting (impurity modification). Elements in groups I and IIa and the rare earths europium, lanthanum, cerium, praseodymium, and neodymium modify, but only sodium and strontium produce a strong modifying action at the low concentration required for commercial application (Table 4). Both modifying actions transform the flake eutectic silicon into a fibrous form, producing a compositelike structure with increased ultimate tensile strength, ductility, hardness, and machinability.

Mechanism of Modification

The finer silicon-phase distribution in the modified structure is evident from structures D and E in Fig. 4. Region C in Fig. 4 is the coupled zone. This defines the compositional and growth temperature (velocity) limits within which it is necessary to solidify in order to obtain a completely eutectic structure in aluminum-silicon alloys. The zone retains its shape, but narrows as the temperature gradient in the liquid is reduced (see the section "Solidification of Eutectics" in the article "Fundamentals of Growth" in this Volume). Modification was originally attributed to the repeated nucleation of the eutectic silicon phase at a reduced temperature (Ref 12, 13). It is now established that, although the nucleation temperature is depressed, the silicon phase grows continuously without repeated nucleation at an increased undercooling (Ref 14, 15). The aluminum phase is not affected structurally by modification, and there is evidence that both sodium and strontium are concentrated in the silicon phase. Consequently, modification is now considered to be associated primarily with a change in the silicon-phase growth mechanism. This change is induced either at high growth velocities or by a modifier at all but extremely low velocities, and it results in a change from a faceted to a more isotropic growth morphology.

Growth Characteristics

Faceted morphologies occur in high-entropy phases and are a consequence of anisotropic growth. The solid/liquid interface of a high-entropy phase is smooth on an atomic scale. Silicon is a borderline case with (111) and possibly (100) interfaces in this category. Growth normal to a (111) silicon interface occurs by the lateral propagation of intrinsic steps across the interface at a rate determined by the rate of step generation. Primary silicon phase often displays a faceted octahedral morphology that is bound by slow-growing (111) faces, as shown in structure S′ in Fig. 4. Growth rate anisotropy is increased when growth normal to certain planes is enhanced by such defects as twins and screw dislocations, which create self-perpetuating steps at the growth front (Ref 16). This occurs in silicon when the twin-plane reentrant edge mechanism (TPRE) operates and enhances growth in the ⟨211⟩ directions, resulting in hexagonal plate and starlike primary silicon morphologies (structure S, Fig. 4).

These silicon growth characteristics are in evidence at the duplex solid-liquid eutectic growth front, and the silicon eutectic phase in an unmodified alloy develops several distinct growth habits, depending on the growth velocity and temperature gradient in the liquid at the interface, as shown in Fig. 5. The ratio of temperature gradient G to growth velocity v (G/v ratio) in region A is high enough for the eutectic phases to grow independently to produce large, faceted silicon crystals in an aluminum solid-solution matrix. Growth in region B occurs by a more coupled, short-range diffusion process. The silicon phase leads at the interface and develops a [100] preferred growth direction. Its morphology varies from closely packed rods at high-temperature gradients to a variety of faceted forms at lower gradients (structure B, Fig. 5). The angular silicon structure is prominent in the

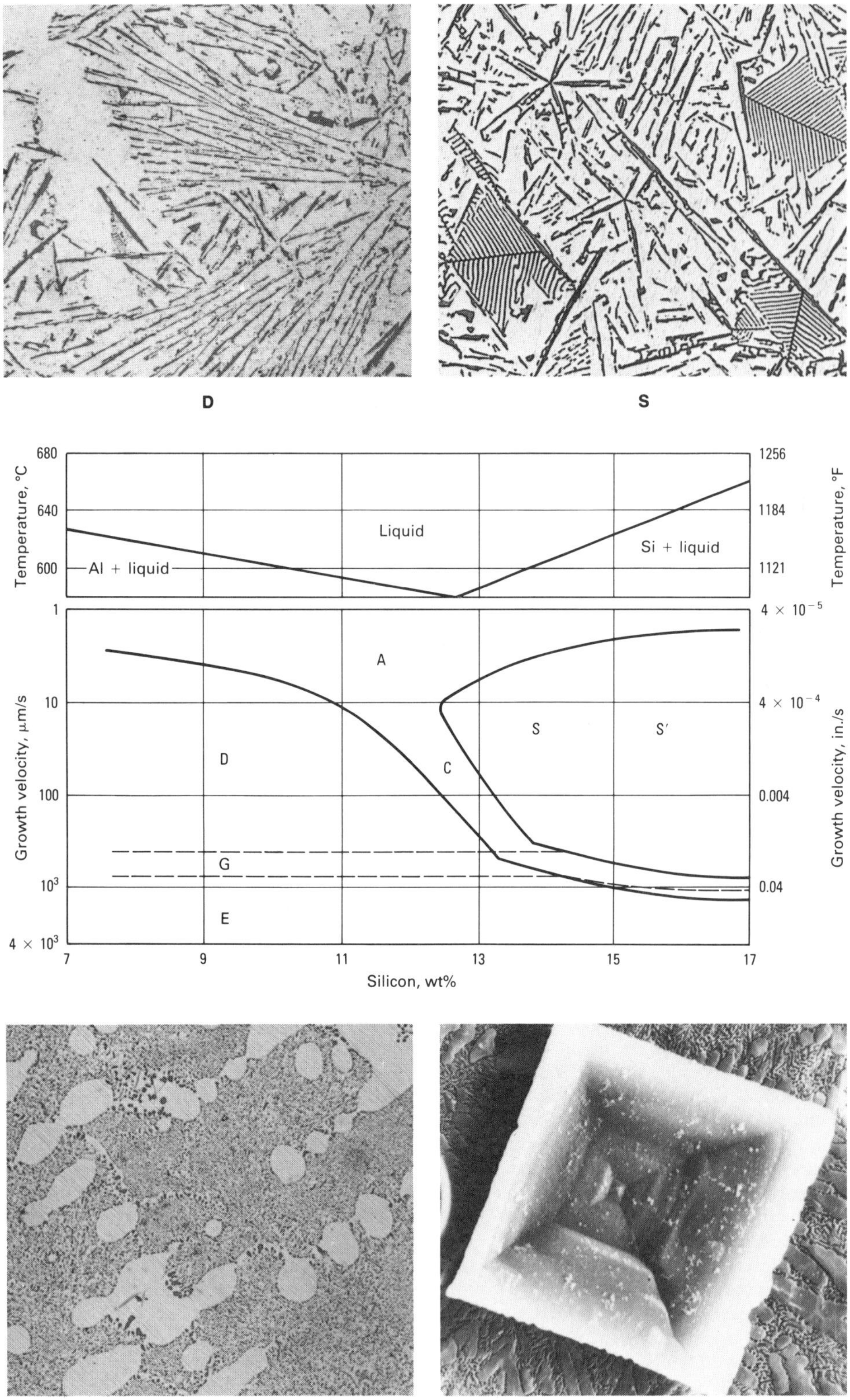

Fig. 4 Coupled zone diagram for aluminum-silicon alloys determined from directional solidification studies with a temperature gradient in the liquid of 125 °C/cm (570 °F/in.). Region A, massive, rod, and angular eutectic silicon and complex regular silicon. Region C, coupled zone (see Fig. 5). Region D, angular and flake eutectic silicon and aluminum dendrites. Region G, flake-to-fibrous eutectic silicon transformation. Region E, fibrous eutectic silicon and aluminum dendrites. Region S, eutectic silicon and complex regular and starlike primary silicon. Region S′, eutectic silicon and complex regular, starlike, and polyhedral primary silicon. Micrographs: S, complex regular and starlike primary silicon with flake eutectic silicon; 100×. D, typical unmodified structure of a 413 alloy showing flake eutectic silicon and aluminum dendrites; 100×. E, typical chill-modified structure of a 413 alloy showing fibrous eutectic silicon and a higher volume fraction of finer aluminum dendrites than structure D; 100×. S′, scanning electron micrograph showing a (100) section through an octahedral primary silicon particle revealing four {111} planes; 1500×

lower areas of region B. Interconnected silicon flakes form between the angular silicon particles in region B + C, and the flake structure predominates in region C + B, which covers the thermal conditions encountered in sand casting. The flakes have {111} habit, and their flat surfaces are parallel to internal {111} twins. They lead the aluminum phase at the nonisothermal growth front and branch by splitting or displacement twinning to achieve the spacing characteristics of the growth velocity. Successive {111} twinning on the surface of a flake permits a direction change through any angle while maintaining growth in a [112] direction. All the active {111} planes are cozonal (Ref 18).

Quench Modification

Quench-modified eutectic forms in region G of Fig. 4 and in region G + B′ of Fig. 5. Transmission electron microscopy has shown that the silicon fibers have smooth external surfaces and that many are twin free (Ref 19, 20). These features are consistent with isotropic fibrous growth at an atomically rough interface, as shown in Fig. 6. The silicon phase can branch readily and no longer leads at the interface. The fibers are much finer than slowly grown flakes and are finer than impurity-modified fibers. The diffusion process is short range and approximates that of a normal rod structure.

Chemical Modification. The efficiency of impurity modification is evident from Fig. 7, which shows how the structures of directionally solidified aluminum-silicon alloys change with increasing strontium content. Transmission electron microscopy has shown that fibers in impurity-modified alloys are microfaceted to varying degrees and are heavily twinned on up to four {111} systems. Multiple twinning creates many reentrant steps on the silicon interface, facilitating growth by the TPRE mechanism at an isothermal interface, with branching achieved much more easily than with the flake morphology. The change from flake to fibrous growth is a consequence of impurity incorporation into the silicon phase by adsorption at the silicon growth front. A hard sphere model has been used to show how an impurity atom of the appropriate size can force a step propagating across an interface (as occurs in flake growth) to miss one regular close-packed position and, by falling into the next alternative stacking sequence, create a twin. This size factor requirement (atomic radius of modifier/atomic radius of silicon > 1.64) is thought to be the first and principal requirement for impurity modification (Ref 21).

Modifiers and Their Side Effects

Modifying Agents. The observations recorded in Table 4 show that factors other than atomic size control the efficiency of a

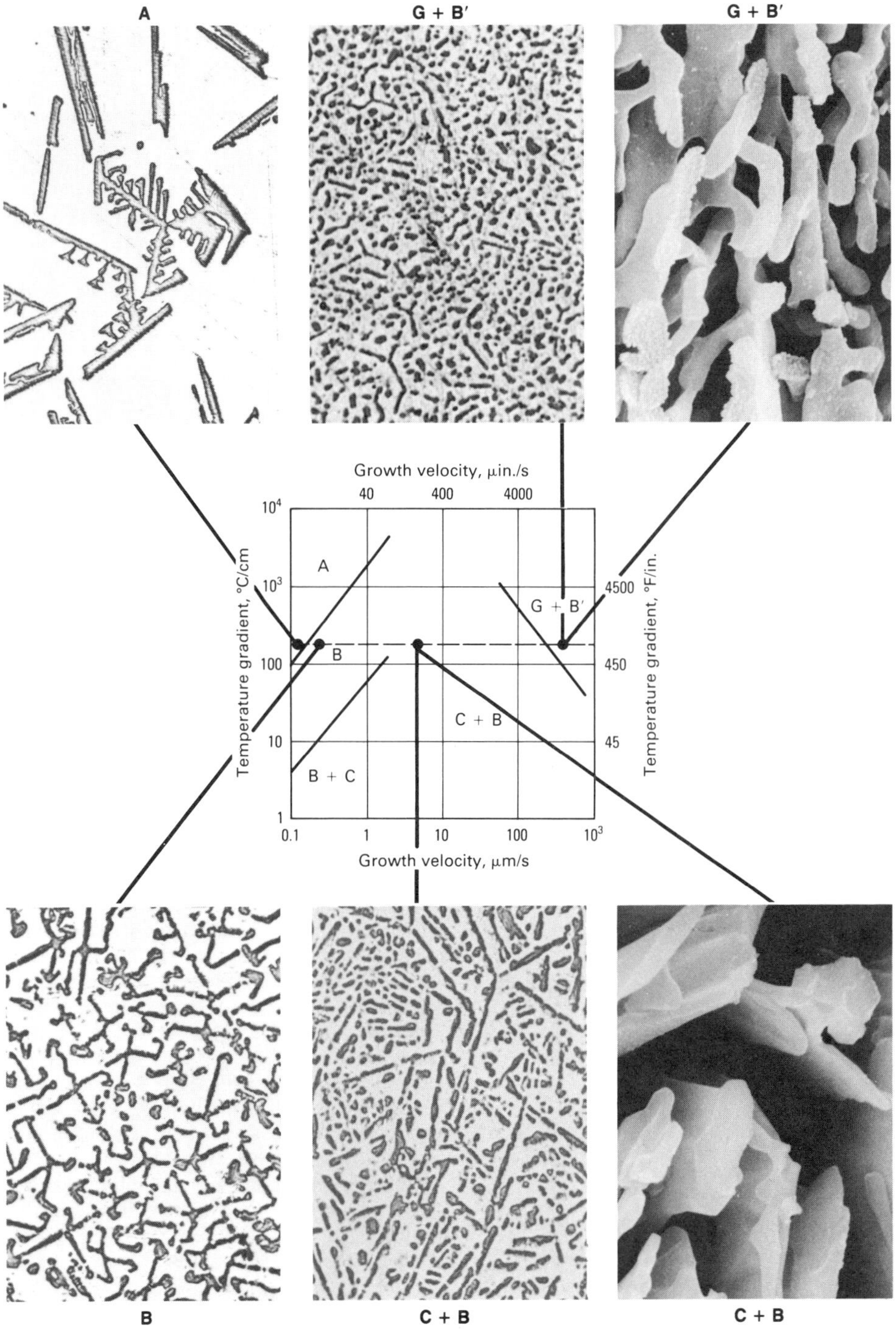

Fig. 5 Eutectic silicon morphologies found in the coupled zone as a function of growth velocity and temperature gradient in the liquid at the growth interface. Region A, massive, faceted eutectic silicon in an aluminum matrix. Region B, rod and rod with faceted sideplate eutectic silicon in an aluminum matrix. Region B + C, angular silicon with some flake eutectic silicon in an aluminum matrix. Region C + B, mainly flake eutectic silicon with some angular silicon in an aluminum matrix. Region G + B′, quench-modified fibrous silicon with some modified angular silicon eutectic in an aluminum matrix. Micrographs: A, massive faceted eutectic silicon; 100×. B, rod with faceted sideplate eutectic silicon; 100×. C + B, mainly eutectic flake silicon with some angular silicon, 100×, and scanning electron micrograph (lower right) showing angular silicon and flake eutectic silicon; 1500×. G + B′, quench-modified fibrous silicon and modified angular silicon, 100×, and scanning electron micrograph (upper right) showing quench-modified fibrous silicon; 1500×. Source: Ref 17

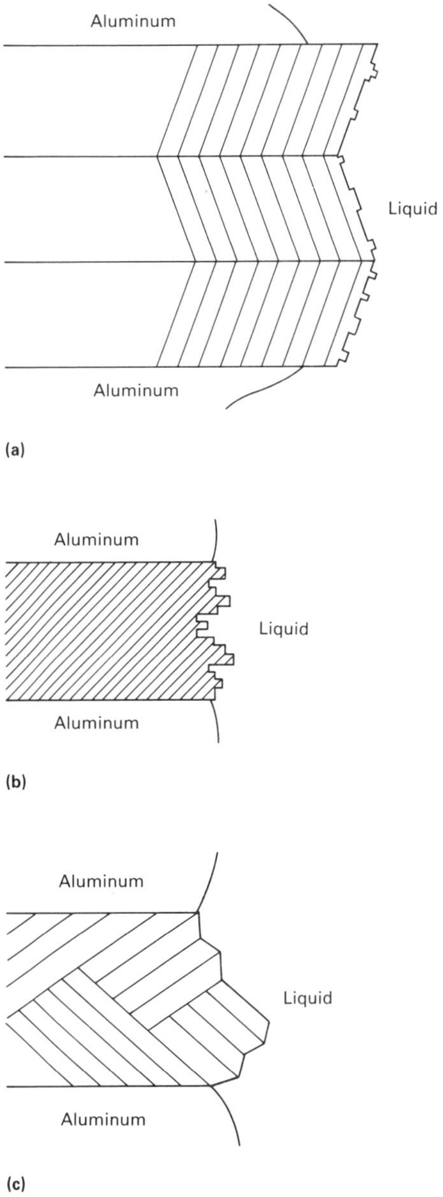

Fig. 6 Schematic of the eutectic silicon growth interface. (a) Unmodified flake silicon. Widely spaced {111} twins probably make the propagation of intrinsic steps across (111) interfaces the significant growth mechanism. (b) Quench-modified fibrous silicon. An atomically rough, twin-free eutectic silicon interface at which atomic additions are made randomly. Branching occurs readily by overgrowth. (c) Impurity-modified fibrous silicon. Finely spaced {111} twins mean that TPRE is the major step source during silicon growth. If twinning occurs equally on four {111} systems, a [100] fiber growth direction results.

modifier. A low melting point and a high vapor pressure promote rapid dispersion of the modifier in the melt, but a high vapor pressure will encourage fade by evaporation. Oxidation loss will be a problem with modifiers having a free energy of oxide formation higher than that of aluminum. A low solid solubility and a wide miscibility gap having a monotectic point at a very low concentration of modifier, as in the aluminum-sodium system, produce a large increase in modifier concentration at the growth front and a powerful modifying effect. Therefore, sodium dissolves and disperses rapidly in the melt without oxidation, but fades quickly (<20 min) and within this time provides a powerful modifying action. Strontium dissolves quickly, and although it oxidizes slightly, it has a greater resistance to fade than sodium, producing a more permanent but weaker modifying action. A

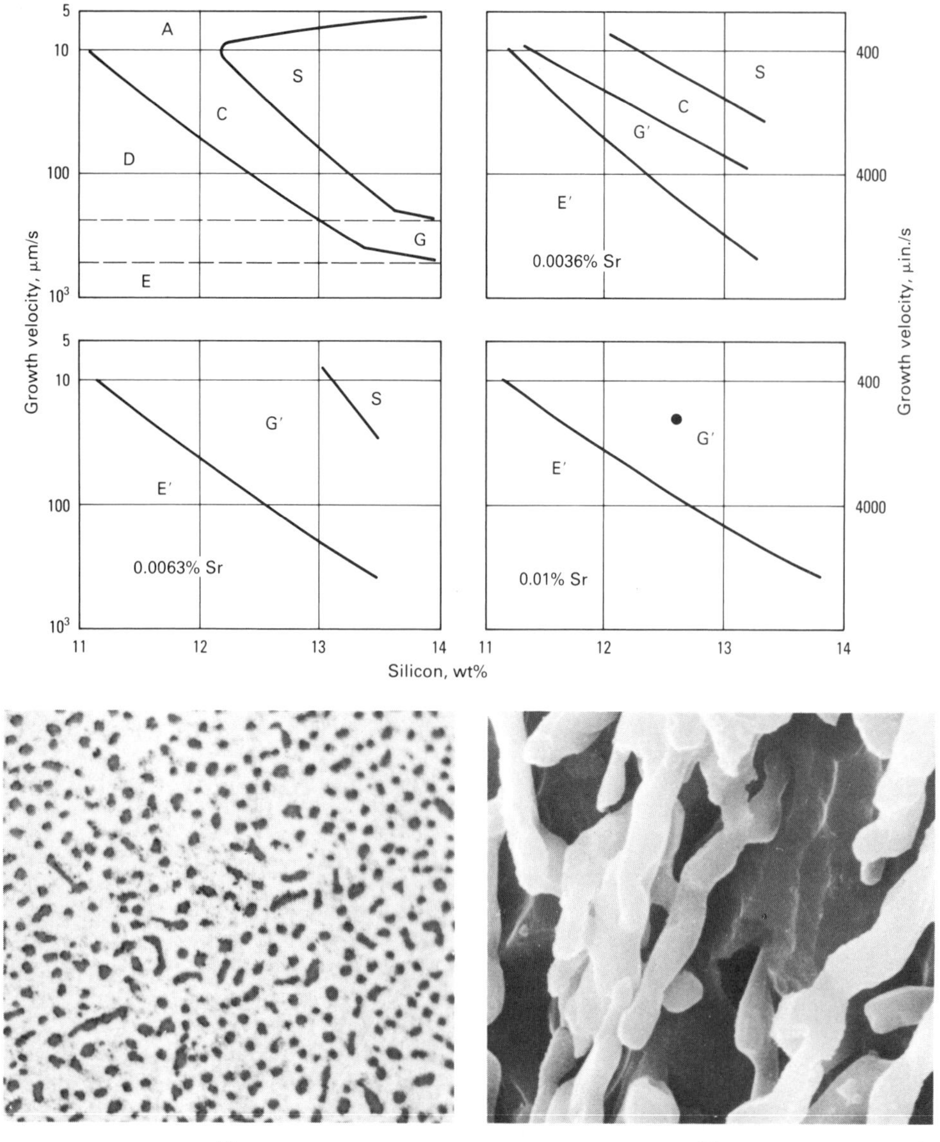

Fig. 7 Change in the coupled zone diagram of aluminum-silicon alloys with increasing strontium additions as determined from directional solidification studies with a temperature gradient in the liquid of 125 °C/cm (570 °F/in.). Regions A, C, D, E, G, and S are shown in Fig. 4. Region G′, impurity-modified fibrous silicon; Region E′, impurity-modified fibrous silicon and aluminum dendrites. Micrographs: G′ strontium-modified fibrous silicon, 300×, and scanning electron micrograph (right) showing strontium-modified fibrous silicon; 2500×

sodium addition of more than 0.02 wt% causes the overmodification associated with AlSiNa compound formation and a reduction in mechanical properties. Strontium does not exhibit the same complex overmodification behavior, but undesirable compounds of the Al_2SrSi_2 type in A356 alloys containing more than 0.05 wt% Sr contribute to a decrease in mechanical properties (Ref 22).

Sodium can often be added as a metal in preweighed, sealed aluminum cans or as sodium compounds in cover fluxes. Strontium does not present the same storage and handling problems, and it is usually added as a low-strontium master alloy such as Al-10Sr or Al-14Si-10Sr, a high-strontium master alloy (for example, Al-90Sr), or a pure metal. Master alloys are usually supplied in waffle form, but several aluminum-base master alloys containing 3 to 10% Sr are supplied in rod form for rapid dissolution. Waffles dissolve slowly and require 30 to 45 min for optimum modification. Metallic strontium additions are smaller, dissolve quicker, and introduce less iron into the alloy than master alloy additions. Strontium should be added in the temperature range of 670 to 720 °C (1240 to 1330 °F) in A356 alloys. Low-strontium master alloys should be added at high melt temperatures for rapid dissolution with high recovery. High-strontium master alloys should be added to a melt with minimal superheat (Ref 22).

Hydrogen Pickup and Loss. The subject of hydrogen pickup in aluminum-silicon melts is highly controversial, especially when one considers the effect of modifiers on the rate of hydrogen pickup from moisture in the atmosphere. The literature sometimes presents conflicting reports—for example, sodium and strontium both result in "gassy" melts and that they have no effect. This difficulty can arise because the gas content is often inferred from the amount of porosity formed upon casting rather than from a direct measurement of the hydrogen content of the melt. However, because both sodium and strontium form oxides of strong chemical stability and because numerous foundrymen report commercial problems with gas pickup, there is probably a tendency toward increased hydrogen pickup in modified alloys. The problem apparently becomes more severe at higher melt temperatures and at higher modifier concentrations.

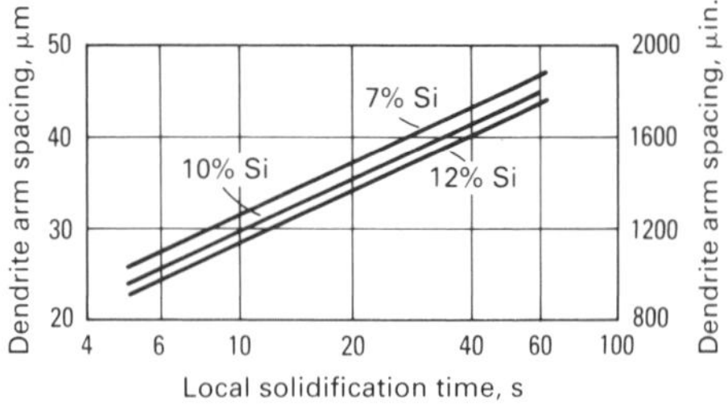

Fig. 8 Variation of dendrite arm spacing with local solidification time for Al-7Si, Al-10Si, and Al-12Si alloys. Local solidification time is defined as the time elapsed for the temperature to decrease from the liquidus to the solidus temperature at a particular location on the casting.

Therefore, the benefits of eutectic modification are offset to some extent by hydrogen pickup, accompanying oxide film formation, and porosity generation. This is particularly true of the most effective chemical modifiers—sodium and strontium. An advantage of the eutectic refiner antimony is that porosity formation can be alleviated by gas fluxing with an inert gas or an inert gas-chlorine mixture to obtain low hydrogen levels without loss of antimony (Ref 23). In contrast, any active gas, such as chlorine or freon, rapidly removes sodium or strontium from a melt.

Whenever a modifying agent is used, there is an increase in hydrogen in the melt. Hydrogen pickup occurs when the oxide film on an aluminum melt is broken and a reaction takes place with moisture in the ambient atmosphere. Therefore, to minimize pickup, modifiers (and refiners) must be added with as little disturbance to the melt surface as possible. The addition of sodium, even when carefully added using the vacuum-packed form, introduces gas to the melt. Similarly, strontium-containing master alloys are a source of hydrogen, but this is reduced by the use of high-purity aluminum-strontium binary alloys and particularly the strontium-base master alloys, which are melted and cast under conditions that minimize hydrogen contamination.

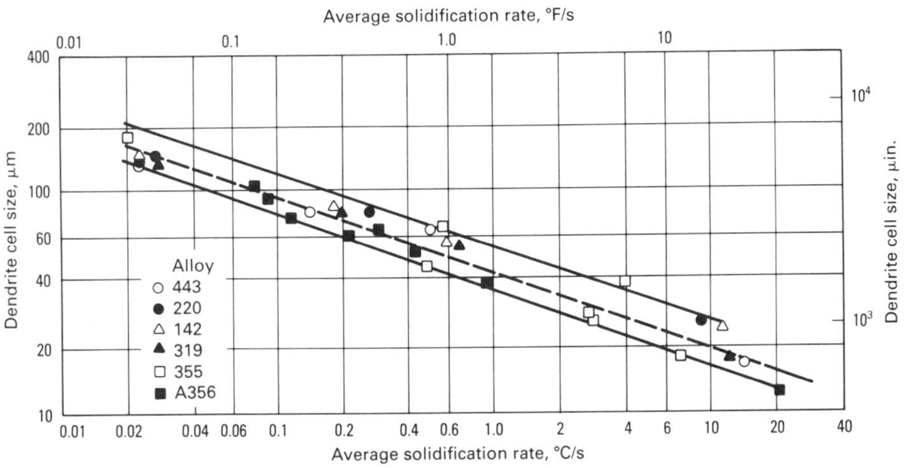

Fig. 9 Dendrite cell spacing as a function of cooling rate for some aluminum casting alloys

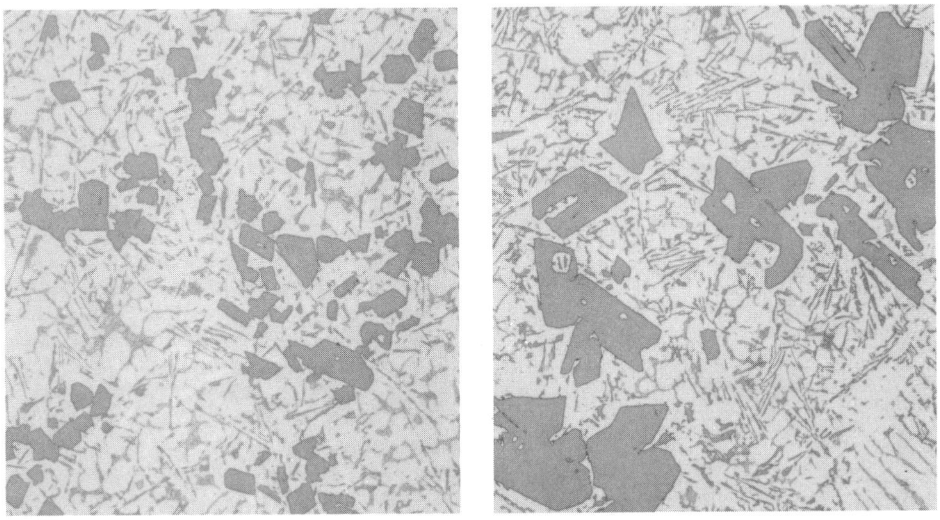

(a) (b)

Fig. 10 Micrographs showing structures of phosphorus-refined 390 alloy (a) and unrefined 390 alloy (b). Both 250×

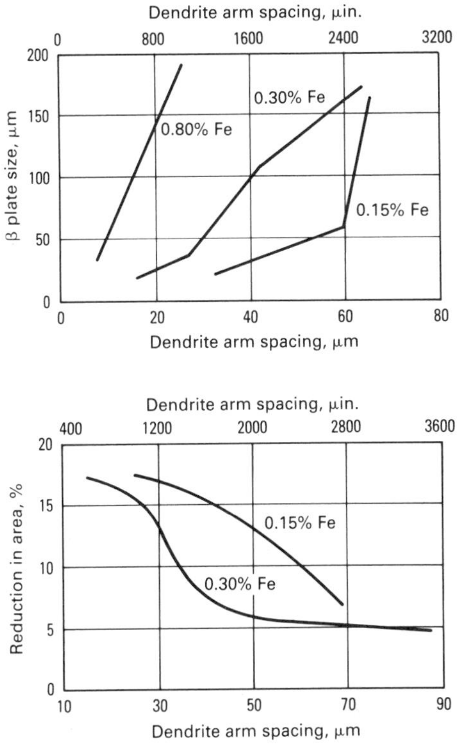

Fig. 11 Maximum β plate length and reduction in area of Al-7Si-0.3Mg alloys containing various amounts of iron as function of dendrite arm spacing

Excess hydrogen can be removed from modified melts by the use of inert gas fluxing techniques, although it appears more difficult to achieve low levels (<0.10 cm^3/100 g) in strontium-containing melts than in sodium-containing melts. However, in both cases, the very act of molten metal fluxing accelerates the loss of the modifier from the melt.

It is widely reported that hydrogen introduced into a melt by strontium is more difficult to remove than that introduced by sodium. These reports are often based on observations of porosity instead of the hydrogen content of the melt, and this can be misleading. For example, in A356 melts with identical hydrogen contents—one modified with strontium and the other nonmodified—the strontium-modified melt will have a higher volume fraction porosity for the same solidification conditions (Ref 5).

Fluidity. One of the principal attributes of the aluminum-silicon alloys is their excellent fluidity, or the ability to fill a mold cavity. Fluidity is a complex characteristic that is influenced by surface tension, viscosity, alloy freezing range, melt cleanliness, superheat, and solidification conditions. Chemical modifiers are generally considered to be detrimental to fluidity despite the fact that all the chemical modifiers commonly in use decrease surface tension. This anomaly is due to the overwhelming influence of the surface oxide film, which can increase the surface tension of pure aluminum threefold. Therefore, the most surface active modifier—sodium—has been found to reduce fluidity by about 10% (Ref 24), while the available evidence on strontium-modified melts indicates that the fluidity is the same as that of the nonmodified melts.

Other Structural Features

Other important structural features are secondary dendrite arm spacing or dendrite cell spacing in hypoeutectic alloys (Ref 25), the size and distribution of primary silicon phase in hypereutectic alloys, and the type and distribution of intermetallic phases. For all practical purposes, the dendrite cell spacing, as defined in Ref 25, is equivalent to the dendrite arm spacing and is controlled by the local solidification time (the time elapsed for the temperature to fall from the liquidus to the solidus at a particular location in the casting).

Dendrite Structure. Aluminum dendrite cells grow as the casting cools through the solidification temperature range after heterogeneous nucleation of the α-aluminum phase. Spacings in aluminum-silicon alloys are shown in Fig. 8. For comparison, Fig. 9 shows dendrite cell sizes as a function of cooling rate for some aluminum-base foundry alloys.

Primary silicon particles impart wear resistance to hypereutectic alloys such as 390. Optimum wear resistance and good machinability with good castability require a fine and uniform distribution of primary silicon particles. This can be achieved with the high rate of solidification in pressure die casting, but it is not good practice. All alloys should be treated to refine the primary silicon phase with phosphorus, usually in the form of copper-phosphorus additions. The phosphorus reacts with aluminum to produce a fine dispersion of AlP, which is a very effective heterogeneous nucleate for silicon. The primary

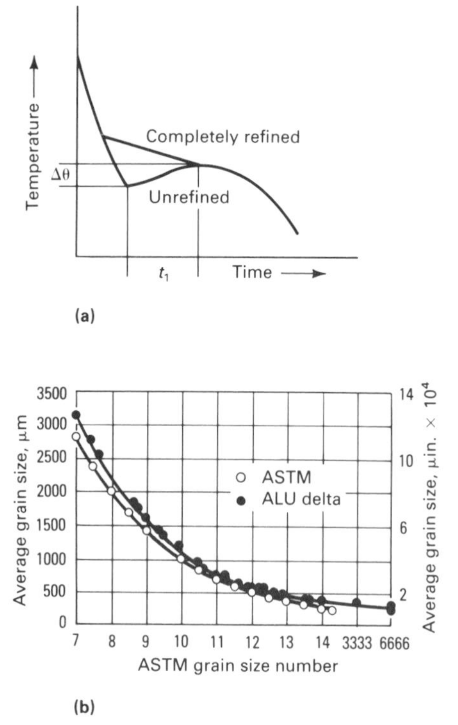

Fig. 12 Cooling curve (a) close to the liquidus for completely grain-refined and unrefined hypoeutectic aluminum-silicon casting alloys. (b) Correlation between average grain size determined metallographically and the grain size number displayed by one instrument for an Al-7Si alloy. Source: Ref 29

silicon-phase morphology in refined and unrefined 390 alloy is shown in Fig 10.

Intermetallic phases can seriously impair the mechanical properties of aluminum-silicon casting alloys. They form as a result of overmodification, excessive titanium from grain-refining additions, or impurities. They can be controlled by good founding practice and by limiting or neutralizing impurities. Iron in the form of plates of $FeSiAl_5$ (or β phase) is important in this respect (Ref 26). The sizes of the β plates and their effect on mechanical properties depend on iron content and cooling rate (dendrite arm spacing), as shown in Fig. 11.

Structural Assessment

Rapid on-line melt structural assessment for grain refinement and degree of modification is becoming an essential part of quality control. The off-line optical examination of separately cast, polished and etched samples is the traditional method of structural assessment, but the development of microprocessor technology has made possible the introduction of rapid on-line techniques based on electrical resistivity (Ref 27), and in particular, thermal analysis (Ref 28-33). In the latter, grain refinement is assessed by examining the thermal response at the liquidus temperature and the modification that occurs at the eutectic temperature. Direct, differential, or derived cooling

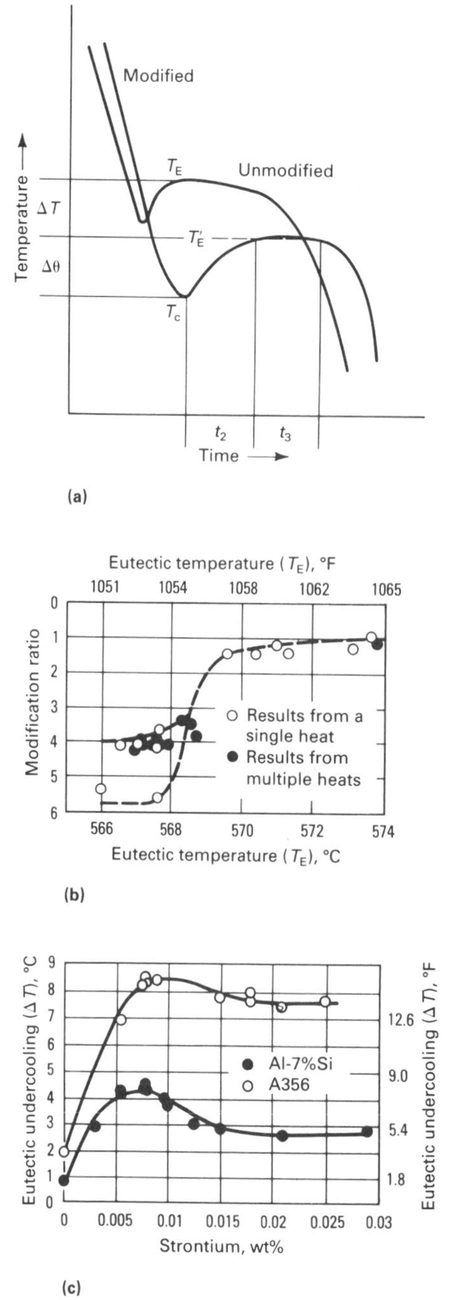

Fig. 13 Thermal parameters (a) on the cooling curve of unmodified and modified aluminum-silicon casting alloys close to the eutectic temperature. (b) Correlation between level of modification and eutectic arrest temperature in A356 alloys. Modification levels: 1, unmodified; 6, very fine structure. Source: Ref 16. (c) Effect of increasing strontium content on the value of ΔT in Al-7Si and A356 alloys. Source: Ref 29

curves and heat-of-solidification measurements can be used.

The instruments currently in use analyze a cooling curve obtained from a small melt sample solidified under standardized conditions in a sand cup. One instrument assesses grain refinement in hypoeutectic alloys in terms of the barrier to nucleation of the aluminum primary phase. A grain size number is computed using an algorithm based

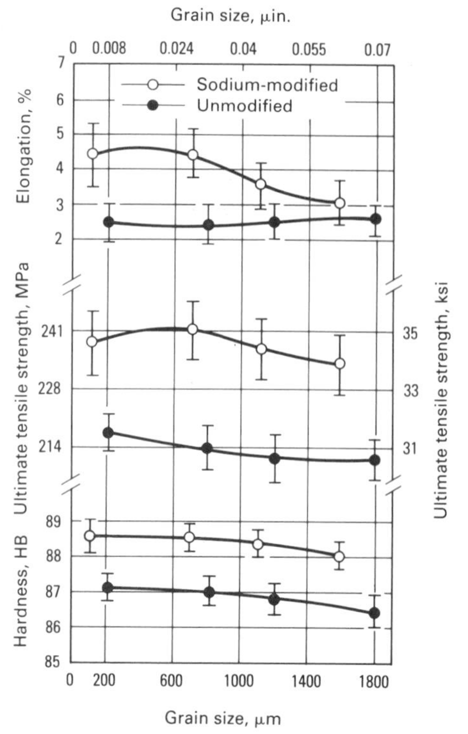

Fig. 14 Mechanical properties of A356 alloy as-cast test bars. Source: Ref 3

on the undercooling for nucleation Δθ and the duration of the nucleation event t_1 (Fig. 12a). The relationship among the computed grain size number, the measured grain size, and the ASTM grain size number for an Al-7Si alloy is shown in Fig. 12(b). The shape of the cooling curve at the liquidus in hypereutectic alloys is not as sensitive. Consequently, silicon primary particle refinement is usually assessed by measuring the solidification start temperature and a scale calibrated with similar measurements for refined and unrefined alloys.

The cooling curve parameters sensitive to modification and recorded by an instrument are shown in Fig. 13. The rating system (Ref 28) uses the ΔT value as an index for modification (Fig. 13). Figure 13(c) shows that ΔT does not increase continuously with strontium content and that it differs in magnitude from alloy to alloy. It is now realized that ΔT is not a comprehensive modification parameter, because it reflects not only modification but also the influence of alloying elements and impurities. Recently, revised algorithms based on ΔT, t_2, and t_3 (Ref 29) and on T_E, T_c, Δθ, and ΔT (Ref 30) have been shown to be capable of assessing modification in terms of unmodified, undermodified, modified, and overmodified states. The total heat of solidification is sensitive to the degree of modification. For A356 alloys, it is 95 cal/g for an unmodified alloy; it increases to 98 cal/g for an undermodified alloy, reaches a maximum value of 100 to 103 cal/g for a fully modified alloy, and decreases to 98 cal/g when overmodified.

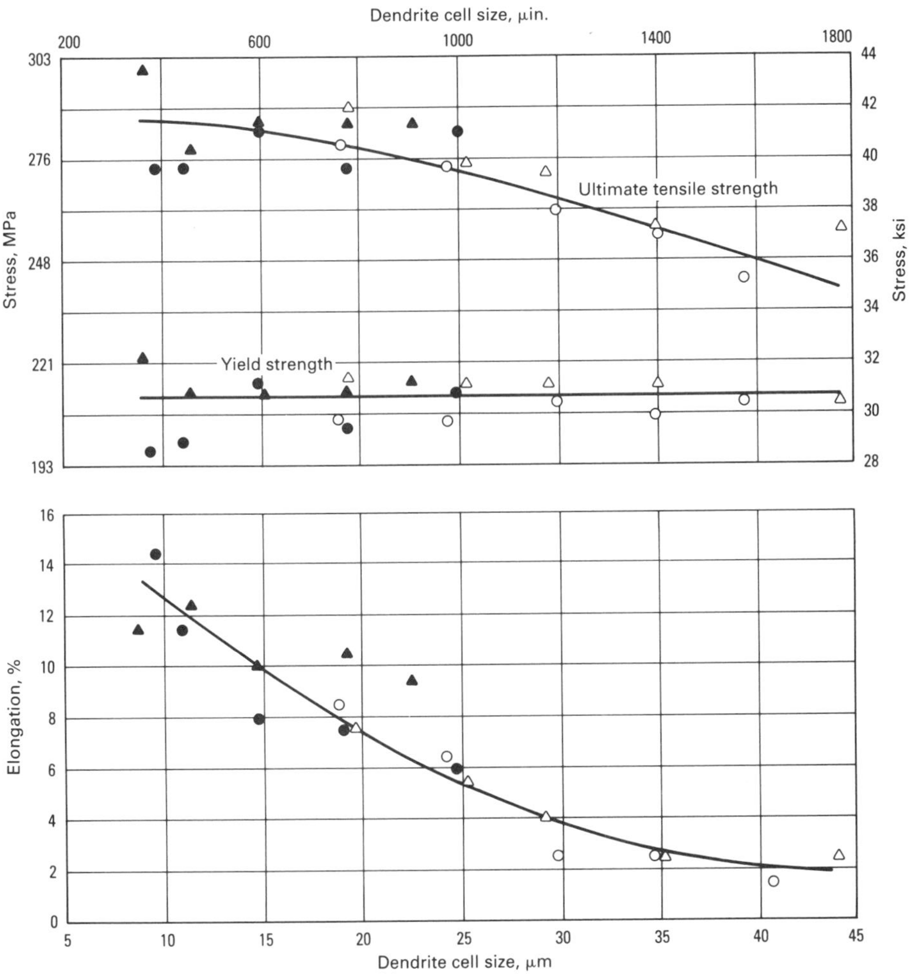

Fig. 15 Relationship between dendrite cell size and tensile properties for alloy A356-T62. The different symbols indicate results from different heats. Source: Ref 25

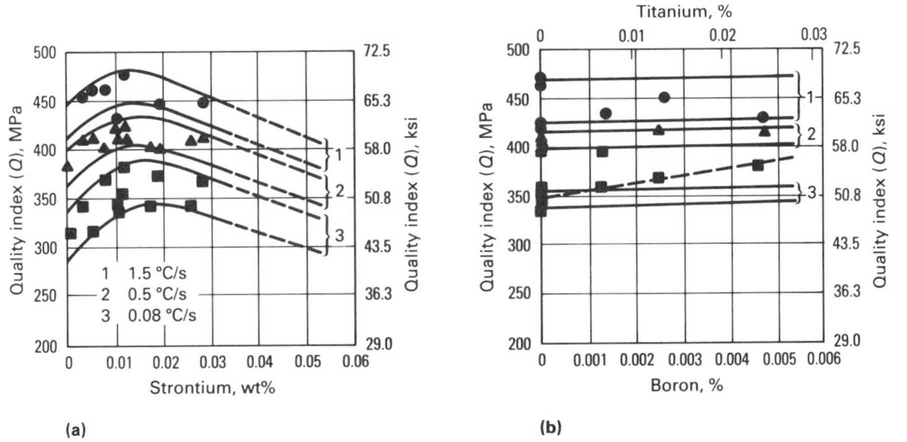

Fig. 16 Variation of *Q* with strontium (a) for three cooling rates in A356 alloy. Property measurements made on cast material after heat treatment to develop T6 temper. (b) Influence of titanium-boron grain refiner on strontium-modified A356-T6 as a function of titanium content. Source: Ref 34

Effect of Structure on Properties

Aluminum-silicon alloys exhibit a highly desirable combination of characteristics, such as castability, weldability, good corrosion resistance, wear resistance, and machinability. Mechanical properties are principally controlled by the cast structure. Therefore, for a given composition (and heat treatment), it is the ability to optimize grain size, eutectic structure, cell size, casting soundness, and the size and distribution of intermetallics that determines the properties.

As discussed earlier, achieving an optimum combination of mechanical properties requires recognition and control of the undesirable side effects of modification and grain refining. The effects of the principal structural modifiers taken singly and in combination are illustrated in the examples outlined below.

The modification of the eutectic phase in hypoeutectic and eutectic aluminum-silicon alloys allows an optimum combination of strength and elongation (Ref 14). Both under- and overmodification result in a degradation of these properties, as shown below for a sodium-modified, sand cast eutectic alloy:

	Ultimate tensile strength		Elongation, %
	MPa	ksi	
Specification	160 min	23 min	5
Unmodified alloy	77	11	2
Partially modified alloy	147	21	4
Fully modified alloy	188	27	8
Overmodified alloy	167	24	6

The key attributes of strontium, compared to those of sodium, are its greater resistance to loss from a melt and greater tolerance (in terms of amount) to overmodification.

Grain Refinement. There appears to be no strong effect of grain size reduction on mechanical properties in hypoeutectic alloys (Ref 3). Figure 14 illustrates a small loss in tensile properties in permanent mold cast A356 test bars with increasing grain size. The effect is greatest in the modified alloy, in which elongation decreases from 4½ to 2½%, with an increase in grain size from 200 to 1600 μm (0.008 to 0.064 in.).

Dendrite cell size, which is controlled by solidification rate, is an important indicator of the mechanical properties of a casting. As shown in Fig. 15, ultimate tensile strength and elongation are enhanced by a reduction in cell spacing. It must be realized that increasing solidification rate (decreasing cell size) also has other beneficial attributes, such as reduction in gas porosity, smaller intermetallic particle size, more refined eutectic structure, and greater solute saturation. Therefore, all improvements in properties cannot be attributed solely to a smaller cell spacing.

Eutectic Structure and Cell Size. Hypoeutectic alloys such as A356 can be considered composites that consist of dendrite fibers, which can be strengthened by heat treatment, and a eutectic matrix strengthened by modification. The mechanical properties of the composite depend on volume fraction and fiber spacing (dendrite arm spacing). The mechanical properties of A356 are often expressed by a quality index Q, where:

$$Q = \text{UTS} + 150 \log \text{elongation}$$

where UTS is ultimate tensile strength and elongation is measured in percent.

Figure 16(a) shows the variation of Q with strontium for three cooling rates in A356, and Fig. 16(b) shows the effect of Al-5Ti-1B grain refiner on strontium-modified A356 as a function of titanium and boron content. Figures 16(a) and (b) show the influence of cooling rate (dendrite arm spacing) and strontium modification on Q values and the fact that grain refinement has an effect only at the lowest cooling rates.

REFERENCES

1. M.C. Flemings and S. Metz, *Trans. AFS*, Vol 78, 1970, p 453-460
2. S. Oya, T. Fujii, M. Ortaki, and S. Baba, *J. Jpn. Inst. Light Met.*, Vol 34, 1984, p 511-516
3. H.T. Lu, L.C. Wang, and S.K. Kung, *J. Chin. Foundrymen's Assoc.*, Vol 29, 1981, p 10-18
4. G.K. Sigworth, in *International Molten Aluminum Processing Conference*, American Foundrymen's Society, 1986, p 75-89
5. Q.T. Fang, P.N. Anyalebechi, R.J. O'Malley, and D.A. Granger, in *Proceedings of the Third International Solidification Conference* (Sheffield, U.K.), Institute of Metals, 1987
6. U. Honma and S. Oya, *Aluminium*, Vol 59, 1983, p E169-172
7. L. Bäckerud, *Light Met. Age*, Oct 1983, p 3-7
8. G.K. Sigworth and M.M. Guzowski, *Trans. AFS*, Vol 93, 1985, p 907-912
9. L. Mondolfo, in *Proceedings of the Third International Solidification Conference* (Sheffield, U.K.), Institute of Metals, 1987
10. R. Kiusalaas, "Relation Between Phases Present in Master Alloys of the Al-Ti-B Type," Thesis, University of Stockholm, 1986
11. G. Nagel and R. Portalier, *AFS Int. Cast Met. J.*, Vol 5, 1980, p 2
12. R.W. Smith, *Solidification of Metals*, Publication 110, Iron and Steel Institute, 1968, p 224
13. P.B. Crossley and L.F. Mondolfo, *Mod. Cast.*, Vol 49, 1966, p 53
14. R. Elliott, *Eutectic Solidification Processing*, Butterworths, 1983
15. M.D. Hanna, Shu-Zu Lu, and A. Hellawell, *Metall. Trans. A*, Vol 15A, 1984, p 459
16. D.R. Hamilton and R.G. Seidensticker, *J. Appl. Phys.*, Vol 31, 1960, p 1165
17. A. Hellawell, *Prog. Mater. Sci.*, Vol 15, 1973, p 1
18. L.M. Hogan and M. Shamsuzzoha, in *Proceedings of the Third International Solidification Conference* (Sheffield, U.K.), Institute of Metals, 1987
19. Shu-Zu Lu and A. Hellawell, *J. Cryst. Growth*, Vol 73, 1985, p 316
20. A.O. Atasoy, Ph.D. thesis, University of Manchester, 1979
21. Shu-Zu Lu and A. Hellawell, *Metall. Trans. A*, Vol 18A, 1987, p 1721
22. J.E. Gruzleski, M. Pekguleryuz, and B. Closset, in *Proceedings of the Third International Solidification Conference* (Sheffield, U.K.), Institute of Metals, 1987
23. G. Nagel and R. Scalliet, *Trans. AFS*, Vol 86, 1979
24. F.R. Mollard, M.C. Flemings, and E.F. Niyama, *J. Met.*, Nov 1987, p 34-37
25. R.E. Spear and G.R. Gardner, *Trans. AFS*, Vol 71, 1963, p 209-215
26. O. Vorren, J.E. Evensen, and T.B. Pedersen, *Trans. AFS*, Vol 92, 1984, p 459
27. H. Oger, B. Closset, and J.E. Gruzleski, *Trans. AFS*, Vol 91, 1983, p 17
28. D. Apelian, G.K. Sigworth, and K.R. Whaler, *Trans. AFS*, Vol 92, 1984, p 297
29. D. Apelian and J.A. Cheng, *Trans. AFS*, Vol 94, 1986, p 797
30. B. Closset, K. Pirie, and J.E. Gruzleski, *Trans. AFS*, Vol 92, 1984, p 123
31. J. Charbonnier, *Trans. AFS*, Vol 92, 1984, p 907
32. S. Argyropoulos, B. Closset, J.E. Gruzleski, and H. Oger, *Trans. AFS*, Vol 91, 1983, p 351
33. C.M. Yen, W.J. Evans, R.M. Nowicki, and G.S. Cole, *Trans. AFS*, Vol 93, 1985, p 199
34. B. Closset and J.E. Gruzleski, *Trans. AFS*, Vol 90, 1982, p 453

Cast Iron

D.M. Stefanescu, University of Alabama

Cast iron is a binary Fe-C or a multicomponent Fe-C-X alloy that is rich in carbon and exhibits a considerable amount of eutectic in the solid state. Two such possible eutectics may result, as follows:

- If solidification occurs according to the metastable diagram, Fe-Fe_3C, the white eutectic or austenitic (γ), iron carbide (Fe_3C) forms
- If solidification follows the stable diagram iron-graphite (a significant amount of silicon is required for this to occur), the gray eutectic, austenite-graphite (Gr), results

Depending on composition, cooling rate, and liquid treatment, it is also possible to produce a mixed white-gray eutectic called mottled structure. The two basic types of eutectic are very different, with mechanical properties such as strength, ductility, and hardness varying over very large intervals as a function of the type and the amount of eutectic formed. To understand the mechanism of the solidification of cast iron, it is necessary first to discuss the structure of liquid iron-carbon alloys.

Structure of Liquid Iron-Carbon Alloys

Observations from x-ray, neutron diffraction, and sound velocity measurements on liquid binary iron-carbon alloys at temperatures approximately 20 °C (35 °F) above the liquidus indicate that for up to 1.8% C the distance between nearest iron neighbors r_I, as well as the number of nearest neighbors N_I in the first coordination sphere, increases (Fig. 1). Above 1.8% C, the distance remains constant, while the number of nearest neighbors continues to grow.

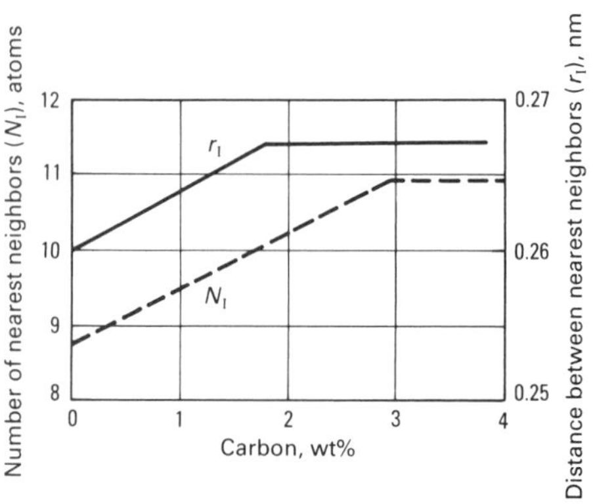

Fig. 1 Variation of distance between nearest neighbors (r_I) and the number of nearest neighbors (N_I) as a function of the percentage of carbon in the iron-carbon alloy. Source: Ref 1

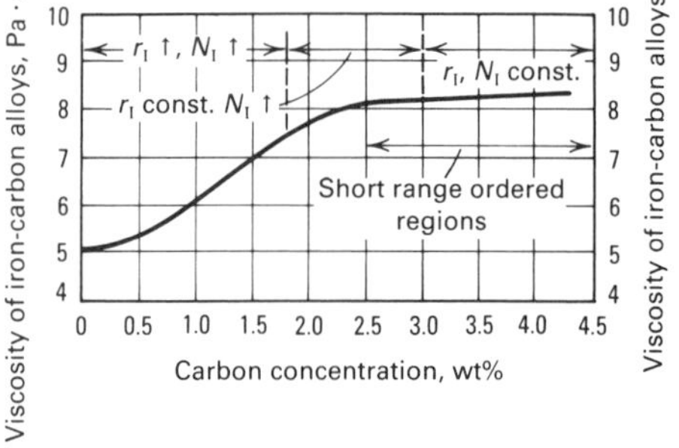

Fig. 2 Viscosity of iron-carbon alloys as a function of carbon concentration at a temperature ≈20° above liquidus. Source: Ref 2

Above 3.5% and up to 5.5% C, both the distance and the number of neighbors remain constant. Above 3.5% C, short-range

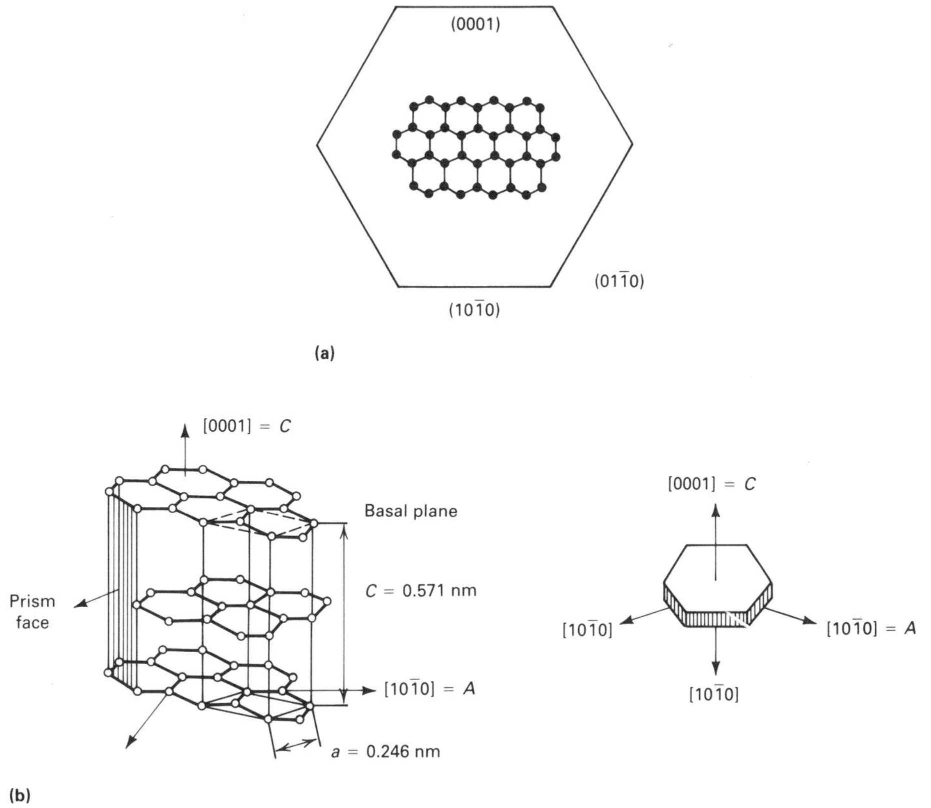

Fig. 3 Crystalline structure of graphite. (a) Crystal of graphite bounded by (0001) and (10$\bar{1}$0) type planes; the hexagonal arrangement of the atoms within the (0001) plane is shown relative to the bounding (10$\bar{1}$0) faces. (b) Hexagonal structure of graphite showing the unit cell (heavy lines). Source: Ref 5

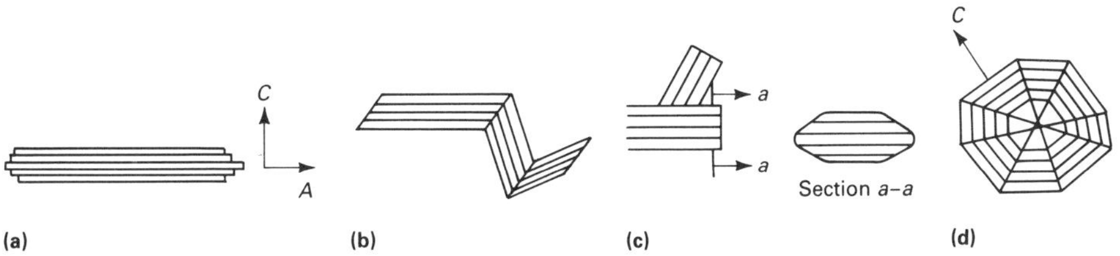

Fig. 4 Schematic of graphite types occurring in the austenite-graphite eutectic. (a) Flake graphite. (b) Compacted/vermicular graphite. (c) Coral graphite. (d) Spheroidal graphite.

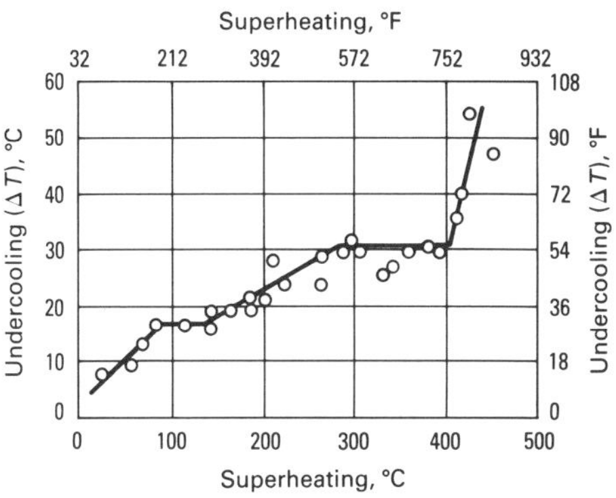

Fig. 5 Relationship between superheating and maximum undercooling in flake graphite cast iron. Source: Ref 6

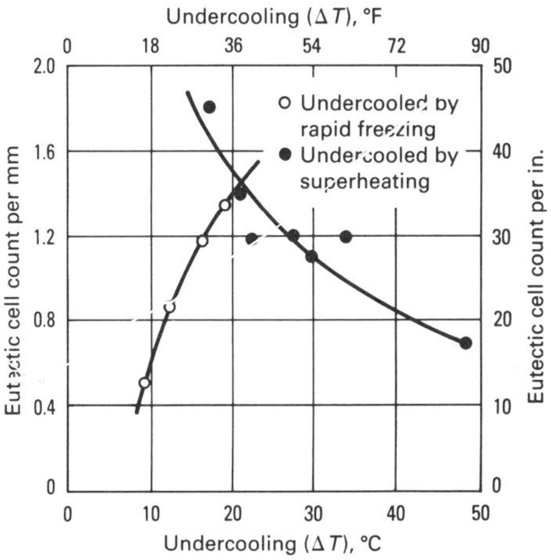

Fig. 6 Influence of undercooling on eutectic cell count in flake graphite cast iron. Source: Ref 7

order regions rich in carbon exist in the melt. This means that the melt becomes more dense with the addition of carbon. A maximum packing density is reached at 3% C, and it remains constant at higher carbon concentrations. The excess carbon forms carbon-rich regions (nonhomogeneities) in the melt.

Viscosity measurements (Fig. 2) show a correlation between viscosity and percentage of carbon. This correlation can be further explained in terms of increased viscosity as the interatomic distance becomes smaller.

Liquid iron-carbon alloys with low carbon content (< 3.5% C, that is, steels and cast irons poor in carbon) are microscopically homogeneous. Liquid iron-carbon alloys with high carbon (>3.5% C, that is, cast irons rich in carbon) are colloidally dispersed systems with microgroups of carbon in liquid solution. The nature of these microgroups is not clear. It is hypothesized that they are either Fe_3C clusters (Ref 3) or C_n clusters (Ref 1, 4).

Because the nucleation energy for Fe_3C is smaller than that for graphite, it is thermodynamically possible for the carbon-rich regions to exist as Fe_3C clusters. Other investigators consider the C_n clusters to be stable (Ref 3). Their size is supposed to be in the range of 1 to 20 μm (40 to 800 μin.), and it increases with the carbon equivalent, lower silicon content, and lower holding time and temperature. It is to be expected that the carbon-rich configurations existing in molten iron-carbon alloys are in dynamic equilibrium and that they diffuse within the melt.

Nucleation of Eutectic in Cast Iron

Structure of Graphite in the Austenite-Graphite Eutectic. The graphite phase is a faceted crystal bounded by low index planes. For graphite crystallizing from an iron-carbon melt, the normally observed bounding planes are (0001) and {10$\bar{1}$0}, as shown in Fig. 3(a). The crystallographic structure of graphite and the possible growth directions, *A* and *C*, are shown in Fig. 3(b). Because unstable growth occurs on {10$\bar{1}$0} planes, the edges of the platelike graphite crystals are not well defined. Graphite growing out of liquid iron-carbon alloys has a layer-type structure, with strong covalent bonds (4.19×10^5 to 5×10^5 J/mol, or 3.09×10^5 to 3.70×10^5 ft · lbf/mol) between atoms in the same layer. There is a trielectronic bond of each atom with its neighbors, while the fourth electron is common for the layer giving the metallic properties of graphite. Weak molecular forces exist between layers (4.19×10^3 to 8.37×10^3 J/mol, or 3.09×10^3 to 6.17×10^3 ft · lbf/mol). The prism plane is a high-energy plane at which impurities absorb preferentially. Strength and hardness are higher in the *C* direction of the graphite crystal.

Complete destruction of the graphite structure occurs only at about 4000 °C (7230 °F). This explains the presence of some graphite aggregates in molten iron even at

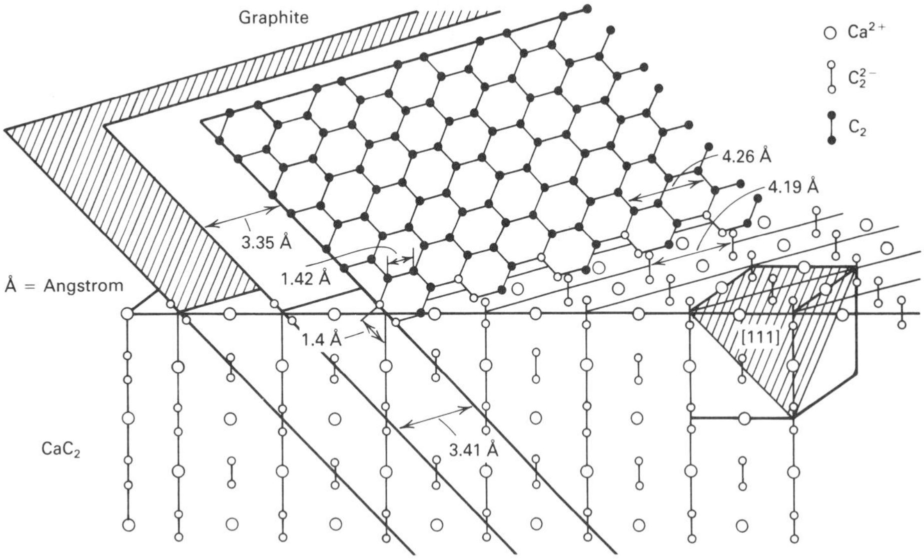

Fig. 7 Epitaxial growth of graphite on a CaC_2 crystal. Source: Ref 10

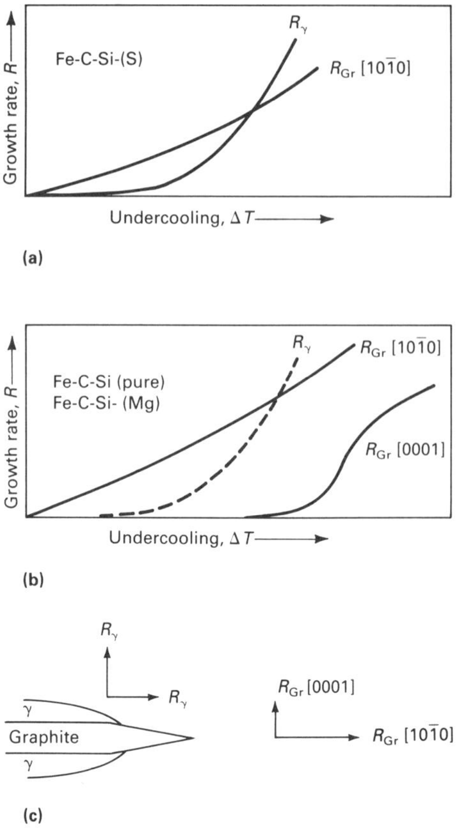

Fig. 9 Suggested growth rate curves for austenite and graphite in the iron-carbon-silicon alloys. (a) Austenite-FG eutectic. (b) Austenite-SG eutectic. (c) Different graphite growth rates in lamellar growth. R_{Gr} [10$\bar{1}$0] determines the lamellar habits of graphite flakes, while R_{Gr} [0001] gives the rate of thickening for the flake. Source: Ref 5

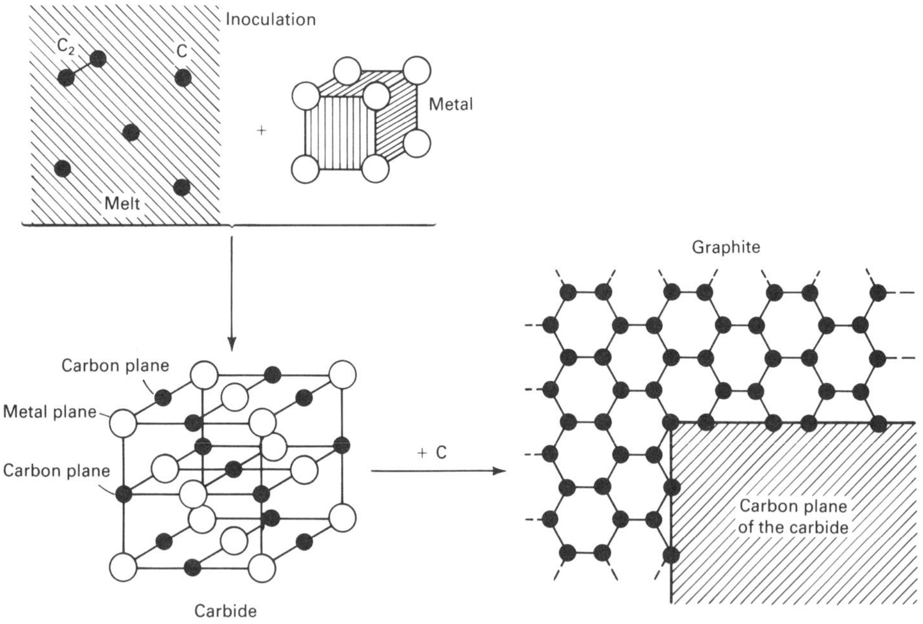

Fig. 8 Schematic of the inoculation of cast iron by metals to form saltlike carbides. Source: Ref 10

temperatures considerably higher than the liquidus temperature.

Depending on the chemical composition and the temperature gradient/growth rate ratio, G/R, and/or cooling rate, $G \cdot R$, a variety of graphite shapes can solidify as part of the austenite-graphite eutectic. Basically, they are as follows:

- Flake (actually plate) graphite (FG)
- Compacted/vermicular graphite (CG)
- Coral graphite
- Spheroidal (nodular) graphite (SG)

A schematic of these graphite types, showing the traces of the {10$\bar{1}$0} planes, is given in Fig. 4.

Nucleation of the Austenite-Flake Graphite Eutectic. Experimental evidence indicates that various types of nuclei become effective at various temperatures. Indeed, if only one type of nuclei would be active in a cast iron melt, an undercooling-superheating curve will show a single arrest for the temperature region over which the nuclei become effective. However, as shown in Fig. 5, a number of steps are observed in the relationship, suggesting that various foreign nuclei become effective as superheating is increased. Increasing the superheating apparently destroys the effective nuclei. Consequently, the number of eutectic grains (eutectic cells) decreases as the superheating increases, and as a result undercooling increases. However, when undercooling, ΔT, is increased at constant superheat by increasing the cooling rate or the growth rate, R, the eutectic cell count will increase (Fig. 6). Mathematically, this is shown by $r^* \propto 1/\Delta T$, where r^* is the critical size radius. As ΔT increases, r^* decreases, yielding an increasing number of critical size nuclei as well as more eutectic cells.

Homogeneous nucleation is improbable in cast iron because typical undercoolings are much smaller than required by the theory of homogeneous nucleation (1 to 10 °C, or 2 to 20 °F), as opposed to 230 °C, or 415 °F). Nevertheless, some C_n clusters (as discussed earlier in this section) or undissolved graphite may act as heterogeneous nuclei for graphite formation. When two separate melt charges of similar composition containing white iron and gray iron are compared, the resulting structure provides support for the idea that some type of residual graphite serves as graphite nuclei during solidification (Ref 8). Further, additions of graphite to the melt enhance nucleation because the eutectic cell count is increased while the chilling tendency decreases (Ref 9).

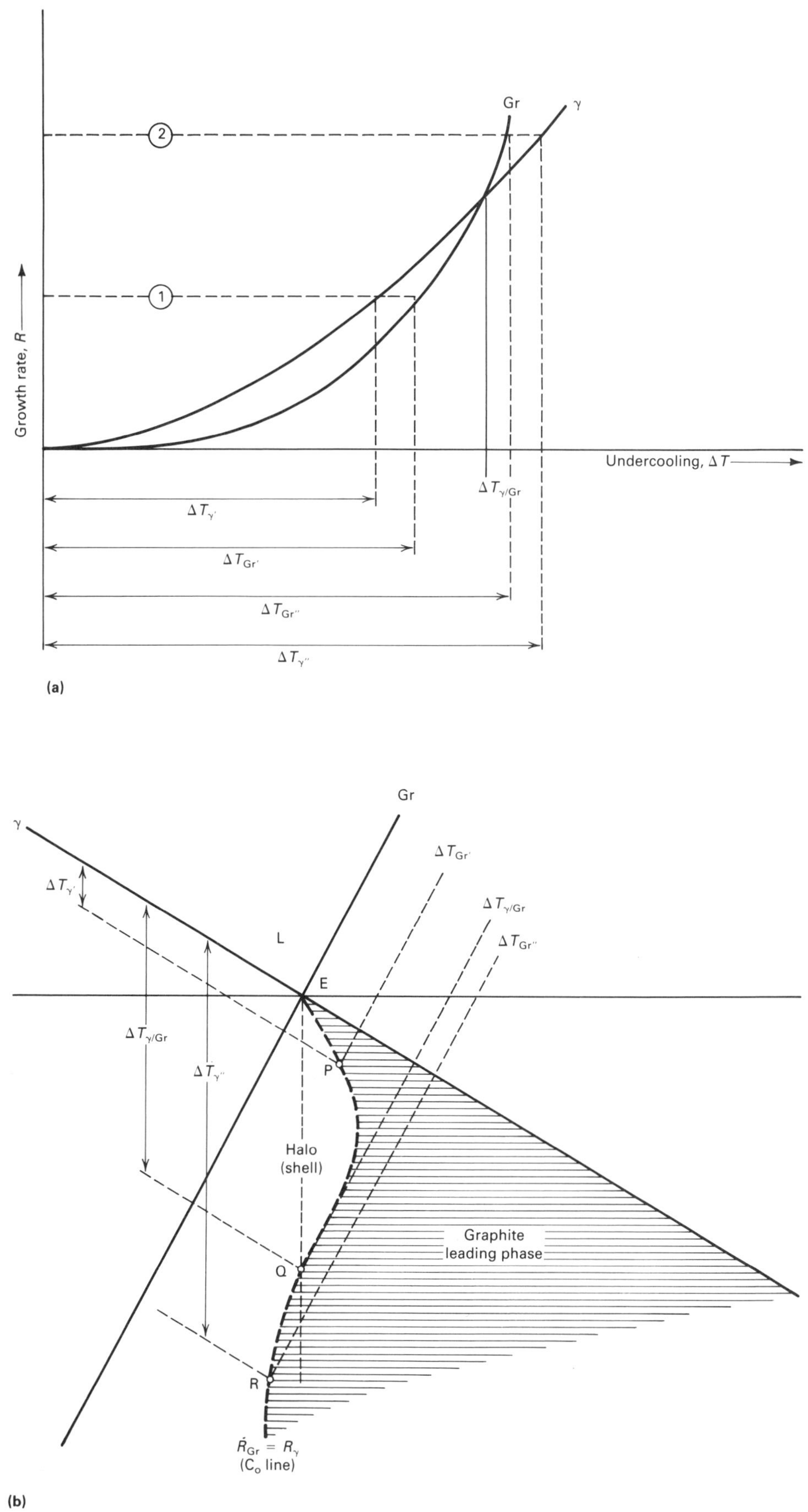

Fig. 10 Constructions of coupled growth line in iron-carbon-silicon alloys. (a) γ and Gr growth curves. (b) Construction of the equal growth rate curve $R_\gamma = R_{Gr}$. Source: Ref 22

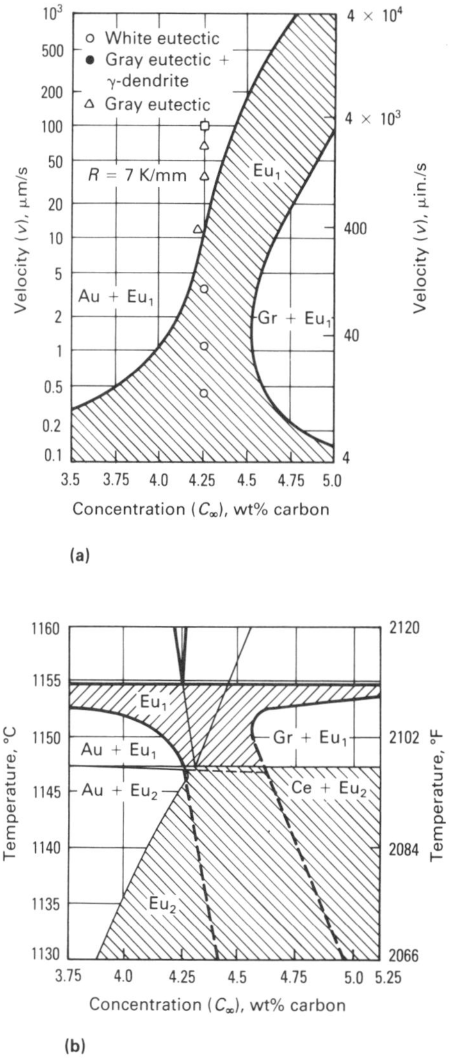

Fig. 11 Coupled zone of directionally grown iron-carbon eutectic. (a) Growth rate-composition diagram showing calculated curves and experimental points. (b) Temperature-composition diagram with superimposed phase diagrams, stable (Eu_1) and metastable (Eu_2). Au, austenite dendrites; Gr, primary graphite plates; Ce, primary cementite plates. Source: Ref 23

A rather wide variety of compounds have been claimed to serve as nuclei for FG cast iron, including oxides (for example, silicon dioxide) or silicates, sulfides, nitrides (boron nitride), carbides (for example, Al_4C_3), and intermetallic compounds (Ref 8). In addition, a number of metals, such as sodium, potassium, calcium, strontium, barium, and yttrium and the lanthanides, act as inoculants in FG iron and can therefore be considered to play a significant role in the heterogeneous nucleation of the austenite-FG eutectic (Ref 8, 10-12). High silicon concentrations in the melt can also contribute to the heterogeneous nucleation of graphite.

The heterogeneous nucleation of FG iron occurs mainly on silicon dioxide particles (Ref 13). This assertion is based on experimental results as well as on the following theoretical considerations. If the oxygen dissolved in the iron is in equilibrium with

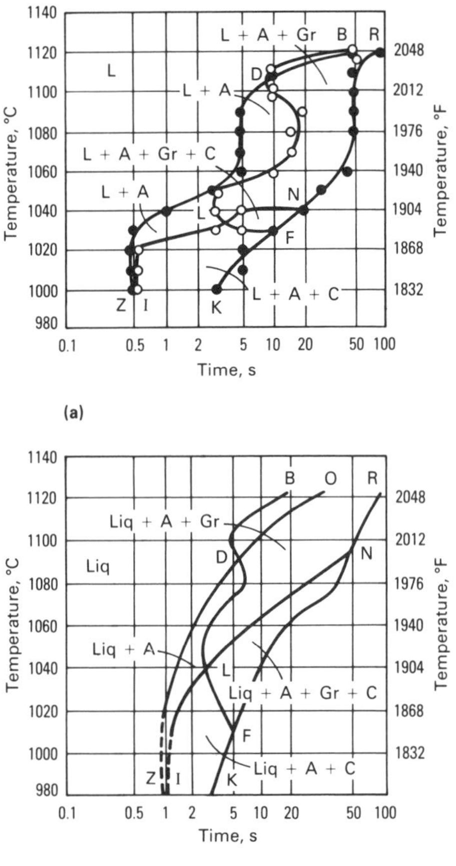

Fig. 12 Time-temperature-transformation diagrams for isothermal solidification of eutectic cast iron. (a) FG cast iron. (b) SG cast iron. L, liquid; A, austenite; C, cementite; Gr, graphite. Source: Ref 7

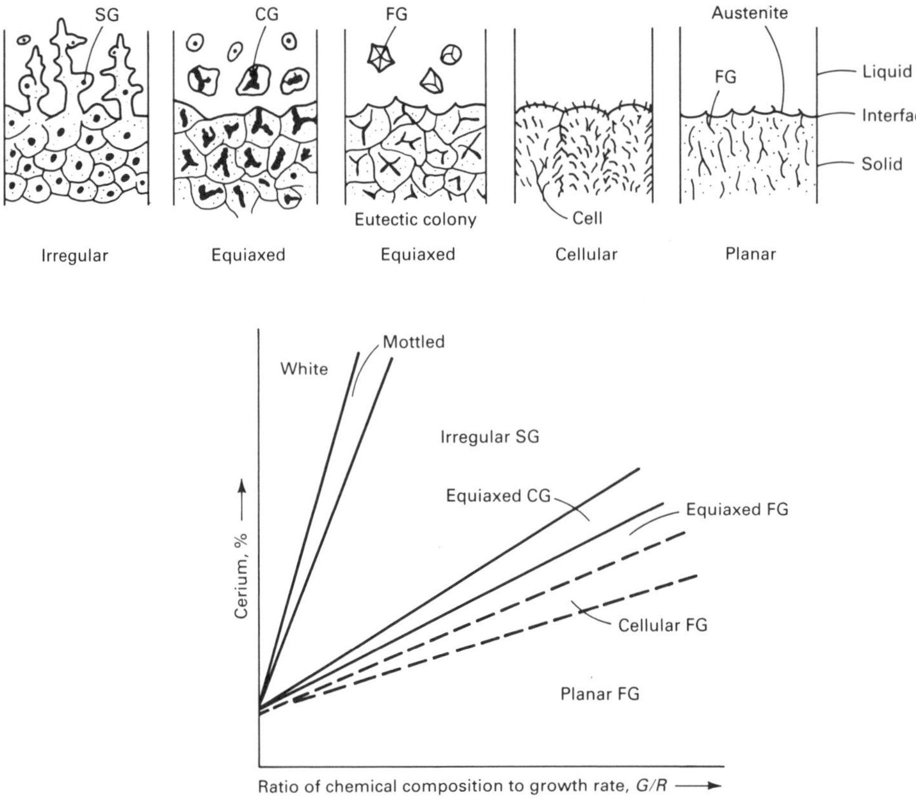

Fig. 13 Schematic showing the influence of composition (% cerium), *G*, and *R* over the eutectic morphology of iron-carbon-silicon alloys. Source: Ref 24

carbon in an iron-acid lining-slag-atmosphere system, Eq 1 can be written:

$$(SiO_2) + 2[C] \rightleftharpoons [Si] + 2\{CO\} \qquad \text{(Eq 1)}$$

The equilibrium temperature for Eq 1 can be calculated as:

$$\log \frac{[Si]}{[C]^2} = -\frac{27\,486}{T} + 15.47 \qquad \text{(Eq 2)}$$

where T is the temperature in degrees K.

Below the equilibrium temperature, solid silica is formed by precipitation deoxidation; above the equilibrium temperature, silica is decomposed. Silica serves as nuclei for graphite growth. The formation of silica particles in turn requires nucleation, which can be promoted by inoculation. Therefore, the inoculation of FG iron can be described as a deoxidation process with the following requirements:

- Enough oxygen must be available in the melt
- The addition of elements with high affinity to oxygen, such as calcium, aluminum, and barium, to form nuclei for silicon dioxide precipitation (heterogeneous catalysis)
- Inoculation temperature must be a maximum of 50 °C (90 °F) above the equilibrium temperature to avoid dissolution of the silicon dioxide particles
- The cooling rate after inoculation must be low enough to allow sufficient time for silicon dioxide formation, but high enough to prevent coalescence and segregation of silicon dioxide

A different theory of heterogeneous nucleation in FG iron has also been proposed (Ref 10). Based on the study of the influence of the inoculation of pure iron-carbon and iron-carbon-silicon alloys with pure metals, it was concluded that the addition of elements capable of forming saltlike carbides increases the number of eutectic cells and therefore the number of nuclei in the cast iron. Saltlike carbides are carbides containing the ion C_2^{2-}. Typical saltlike carbides of metals that have been proved to act as nuclei are as follows:

Group I	Group II	Group III
$NaHC_2$	CaC_2	YC_2
KHC_2	SrC_2	LaC_2
	BaC_2	

These carbides have a number of properties that support the hypothesis of their role as nuclei:

- Due to the strong ionic bond between carbon and the metal, they will not dissolve in the melt but will precipitate as insoluble particles
- They are regions of concentration in the melt as compared with the bulk composition of the liquid and can serve as heterogeneous nuclei
- They have carbon planes in their structures that can serve as a basis for the epitaxial growth of graphite from the melt (Fig. 7)

A schematic of the mechanism of inoculation by metals forming saltlike carbides is shown in Fig. 8. A metal from the I, II, or III Group in the periodic table is added to the melt and reacts with carbon to form a saltlike carbide, M_zC. Graphite then grows heterogeneously from the melt on the carbon planes of the carbide.

Nucleation of the Austenite-Spheroidal Graphite Eutectic. Because most of the inoculants used in FG iron treatment are also effective for SG irons, it is reasonable to assume that similar mechanisms and substrates are active in both types of irons. Extensive transmission electron microscopy and scanning electron microscopy studies have been done to identify the composition of graphite nuclei in SG irons, with a rather wide variety of results. Numerous types of compounds have been found in the middle of graphite spheroids; therefore, it

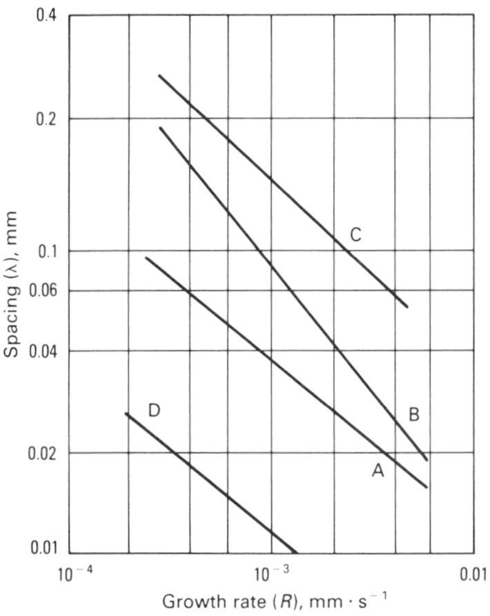

Fig. 14 λ-*R* relationships in FG cast iron. A, Lakeland $\lambda = 3.8 \times 10^{-5} \cdot R^{-0.5}$ cm; B, Nieswaag and Zuithoff $\lambda = 0.56 \times 10^{-5} \cdot R^{-0.78}$ cm (0.004% S); C, Nieswaag and Zuithoff $\lambda = 7.1 \times 10^{-5} \cdot R^{-0.57}$ cm (>0.02% S); D, Jackson and Hunt $\lambda = 1.15 \times 10^{-5} \cdot R^{-0.5}$ cm (theoretical). Source: Ref 5

was hypothesized that they could act as nuclei. Some of these compounds are as follows: $3MgO \cdot 2SiO_2 \cdot 2H_2O$ (cristobalite) (Ref 14), $xMgO \cdot yAl_2O_3 \cdot zSiO_2$, $xMgO \cdot ySiO_2$, $xMgO \cdot ySiO_2 \cdot zMgS$ (Ref 15), MgS (Ref 16), Te + Mn + S (Ref 17), and lanthanide sulfides (Ref 18).

A rather extensive work on what appeared to be substrates for spheroidal graphite in chilled ductile iron concluded that the substrates are duplex sulfide-oxide inclusions with a diameter of 1 μm (40 μin.). The core is probably made of Ca-Mg or Ca-Mg-Sr sulfides, while the outer shell is made of complex Mg-Al-Si-Ti oxides with a spinel structure (Ref 19).

The orientation relationships were established as follows. For the nucleus core/nucleus shell:

$$(110)\text{ sulfide} \parallel (111)\text{ oxide}$$
$$[10\bar{1}0]\text{ sulfide} \parallel (2\bar{1}\bar{1})\text{ oxide}$$

For the nucleus shell/graphite:

$$(111)\text{ oxide} \parallel (0001)\text{ graphite}$$
$$[1\bar{1}0]\text{ oxide} \parallel [10\bar{1}0]\text{ graphite}$$

The x-ray diffraction data showed that the first few graphite layers, adjacent to the (111) oxide, had a dilated lattice (0.264 nm, or 2.64 Å, instead of 0.246 nm, or 2.46 Å). It was suggested that the spacing within the graphite layers decreases away from the oxide until unconstrained spacing is reached. Dislocations were frequently observed in the matrix, and it was suggested that these were generated to relieve some of the elastic strain in the graphite layers adjacent to the oxide.

These observations allow for a theory of nucleation of spheroidal graphite similar to the catalytic theory of nucleation of flake graphite on silicon dioxide, as discussed previously. It can be assumed that the nucleation process begins with the formation of complex sulfides that serve as nuclei for complex oxides, which in turn serve as nuclei for spheroidal graphite.

Structure of Iron Carbide in the Austenite-Iron Carbide Eutectic. The iron carbide (cementite) consists of an orthorhombic unit cell with 12 iron atoms and 4 carbon atoms per cell and therefore has a carbon content of 6.7 wt%. Its density is 7.6 g/cm^3.

Nucleation of the Austenite-Iron Carbide Eutectic. Very little information is available on the nature of the nuclei of the white eutectic. Nevertheless, it is accepted that the nucleation of Fe_3C occurs at lower undercooling than that of graphite (Ref 7). This is supported by the slopes of the solubility lines in the phase diagram, which shows that the supersaturation of cementite increases faster than that of graphite, resulting in an increased probability for the nucleation of cementite as the temperature is lowered. Cooling curve data also show that the solidification of white iron begins at a lower temperature than that of gray iron. There is some evidence that silicon dioxide and aluminum oxide can serve as substrates for the growth of the $Fe-Fe_3C$ eutectic and that the nature of the substrate can influence the morphology of the eutectic (Ref 20).

Nucleation of Primary Phases in Cast Iron

Depending on the carbon equivalent (or saturation degree), the primary phase can be either austenite for hypoeutectic cast iron or graphite for hypereutectic cast iron.

Primary Austenite. The formation and growth characteristics of austenite dendrites have received considerably less attention from investigators than eutectic cell count and morphology simply because dendrites are not readily discernible in the structure. Nevertheless, it has been demonstrated that inoculants that are effective in increasing the cell count (such as calcium-titanium, strontium, and 75% FeSi) have little effect on the solidification pattern of primary austenite dendrites (Ref 21).

A number of elements, such as titanium, vanadium, and aluminum, were proved to increase the number of dendrites and their length when compared with the base iron.

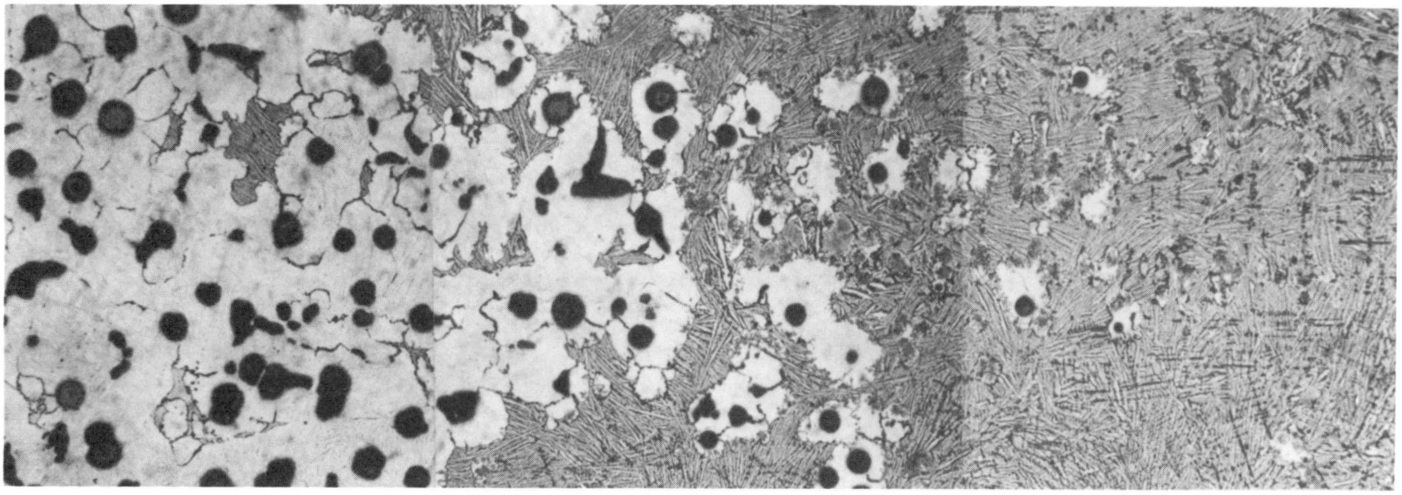

Fig. 15 Microstructure of the solid/liquid interface for a directionally solidified hypereutectic cerium-treated spheroidal graphite cast iron. Source: Ref 25

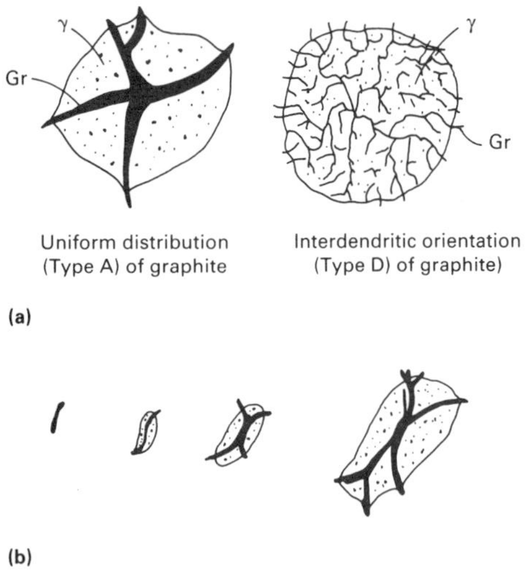

Fig. 16 Schematic of solidification of flake graphite. (a) Typical eutectic colonies (cells). (b) Growth sequence for a eutectic colony

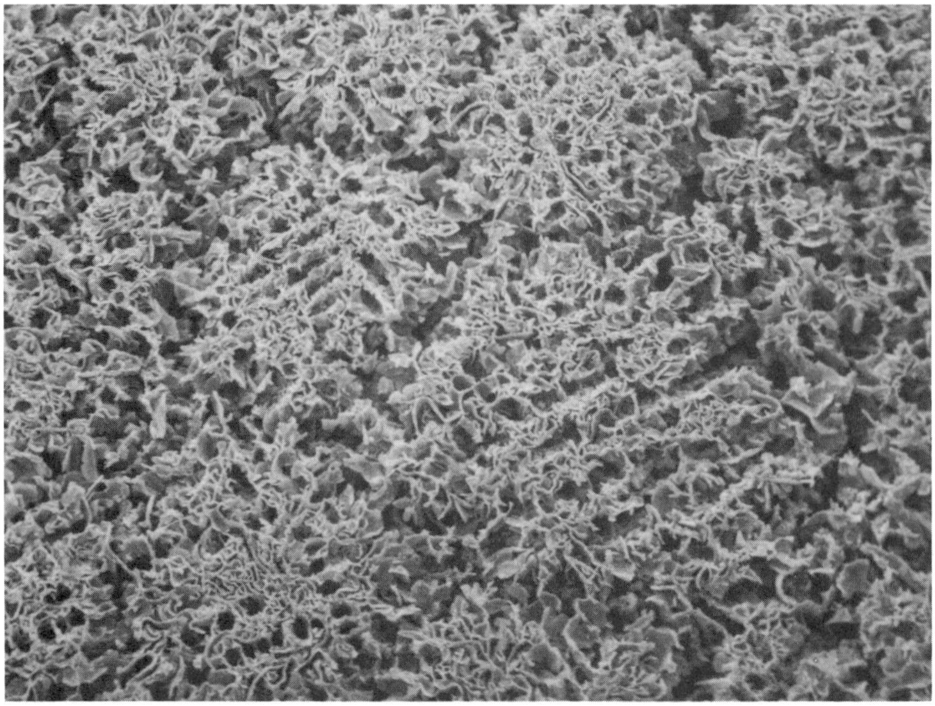

Fig. 17 SEM photomicrograph showing graphite, eutectic cell, and prior dendrite structure in gray cast iron. 200×. Courtesy of Gary F. Ruff, CMI International

For titanium and vanadium, this was attributed to the formation of carbides, nitrides, and carbonitrides, which then act as substrates for austenite solidification. The influence of aluminum was not explained. Other elements, such as cerium and boron, although being instrumental in increasing the number and length of dendrites, were not considered to be nucleants for austenite. It was suggested that these elements restrict the growth of the eutectic cells, which subsequently results in larger undercooling for eutectic solidification. This means that more time (a larger temperature interval) is available for the nucleation and growth of austenite (Ref 21).

Primary Graphite. It is reasonable to assume that nuclei that are active in the solidification of eutectic graphite also serve as nuclei for hypereutectic primary graphite. Nevertheless, it must be noted that inoculation does not seem to increase significantly the number of eutectic cells in hypereutectic FG iron. On the other hand, inoculation is quite effective in hypereutectic SG iron.

Growth of Eutectic in Cast Iron

Coupled Zone in Cast Iron

The degree and type of eutectic growth that occurs in cast iron can be determined by using tools such as growth rate curves to locate coupled zone regions, isothermal time-temperature diagrams to gage susceptibility to carbide formation, and growth rate-composition plots to ascertain parameters that affect both directional and multidirectional solidification. As shown in Fig. 14 in the section "Solidification of Eutectics" in the article "Fundamentals of Growth" in this Volume, the coupled growth region of the eutectic in cast iron is asymmetric. It is possible to construct a theoretical coupled zone for gray iron from the condition of equal growth rate of the austenite (γ) and graphite (Gr) phases (Ref 22). First one must consider the growth rate curves for γ and Gr in the Fe-C system (Fig. 9). For FG iron, the growth rate of austenite, R_γ, and that of graphite along the $[10\bar{1}0]$ direction, $R_{Gr}[10\bar{1}0]$, intersect; therefore, a coupled zone can be constructed. For SG iron, where the predominant growth direction is along [0001], the two rates, R_γ and $R_{Gr}[0001]$, do not intersect, which means that coupled growth is impossible.

Figure 10(a) shows the point at which the graphite and γ growth curves for gray iron cross and are equal ($\Delta T_{\gamma/Gr}$); at points 1 and 2 on the curves, the growth rates are the same, but the undercooling differs. Figure 10(b) shows the phase diagram in the vicinity of the eutectic point, with $\Delta T_{\gamma'}$ and $\Delta T_{\gamma''}$ drawn for growth conditions represented by the points of Fig. 10(a). Point Q in Fig. 10(b) corresponds to $\Delta T_{\gamma/Gr}$ where $R_\gamma = R_{Gr}$. At the left of the $R_{Gr} = R_\gamma$ line, primary γ dendrites will form, while at the right of this line, graphite will be the leading phase. Coupled eutectic growth can only occur for combinations of undercoolings and compositions on the dotted line $R_{Gr} = R_\gamma$.

The transition from a fully eutectic to a eutectic plus dendrite structure in pure iron-graphite alloys of eutectic composition has been determined, and the γ-iron and graphite-iron eutectic boundaries have been calculated (Fig. 11).

Isothermal Solidification

Solidification studies are usually performed athermally because of the high rate of the liquid-solid transformation. Nevertheless, useful information can be extracted from the isothermal time-temperature-transformation diagrams shown in Fig. 12. It is apparent that SG iron is more susceptible to carbide formation than FG iron. Graphite precipitates earlier in SG iron than in FG iron at all undercoolings, although the time interval for complete gray solidification is smaller in FG irons.

Growth in Directional Solidification

It is quite obvious from the previous discussion that undercooling and composition must be carefully selected to achieve coupled growth of the eutectic in cast iron. The γ-FG eutectic is a coupled irregular eutectic of the faceted (Gr)/nonfaceted (γ) type. The γ-SG eutectic is a divorced eutectic.

The basic parameters affecting the morphology of the eutectic are the G/R ratio and composition. As shown in Fig. 13, it is possible to achieve a variety of structures in cast iron when varying G/R and/or the level of impurities (for example, cerium). The gray eutectic can solidify with a planar interface. The theoretical relationship $\lambda^2 R$ = constant (Jackson and Hunt), where λ is the spacing, is not obeyed in the growth of the γ-FG eutectic. Different experimental

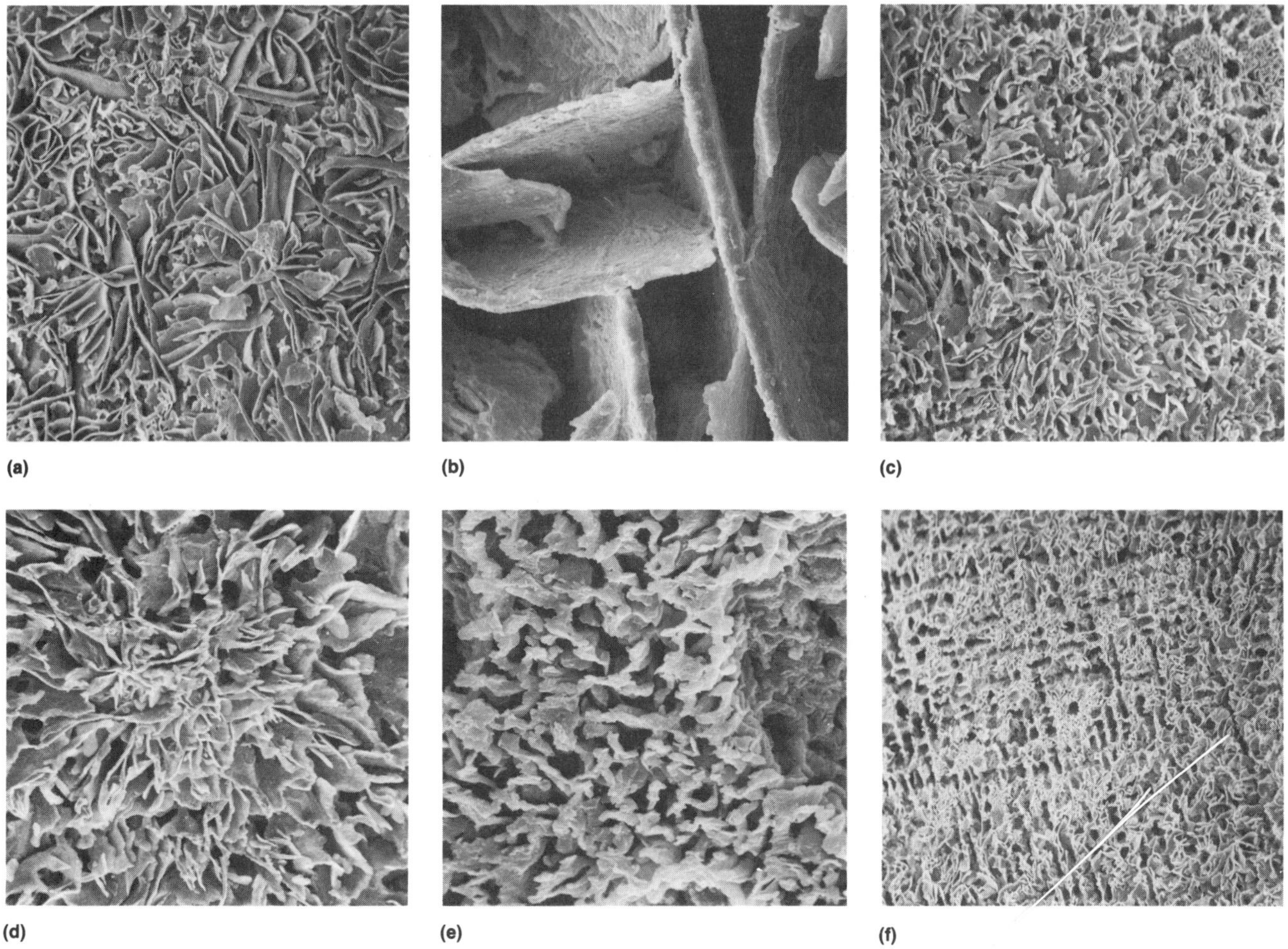

Fig. 18 SEM photomicrographs illustrating variety of flake graphite structures present in gray cast iron: Type A structure at (a) 100× and (b) 430×; Type B structure at (c) 100× and (d) 430×; (e) Type D at 2100×; and (f) Types D (fine) and E (coarse) at 100×. Courtesy of Gary F. Ruff, CMI International

results are compared with the theoretical behavior in Fig. 14.

The departure of experimental values from the theoretical line is a clear indication of irregular rather than regular cooperative growth in FG cast iron. For G/R ratios selected to produce a planar interface during the directional solidification of FG iron, it has been measured that the growth constant in the $R = \mu \cdot \Delta T^2$ equation is $\mu = 8.7 \times 10^{-8}$ m · s^{-1} · K^{-2} (Ref 23).

In the case of SG iron, even for hypereutectic irons, the graphite spheroids do not grow in independent austenite envelopes, but rather are associated with austenite dendrites (Fig. 15). The eutectic austenite is dendritic and cannot be distinguished from the primary austenite. Spheroidal graphite precipitates directly from the melt, becomes enveloped in an austenite shell that is very soon incorporated into the dendrites, and then grows together with the dendrites (Ref 25, 26).

Growth in Multidirectional Solidification

Flake (Lamellar) Graphite Eutectic. The austenite-FG eutectic solidifies with the formation of eutectic colonies (cells) that are more or less spherical in shape. It is generally thought that each eutectic cell is the product of a nucleation event. The eutectic cell is made of interconnected graphite plates surrounded by austenite. The degree of ramification of graphite within the cell depends on undercooling, with higher undercooling resulting in more graphite branching (Fig. 16). The leading phase during the eutectic growth is the graphite. Graphite spacing is determined by the same parameters as for regular eutectics (see the section "Solidification of Eutectics" in the article "Fundamentals of Growth" in this Volume), with branching occurring as a response to interface instability. In turn, interface instability is determined by localized changes in composition, convection currents, crystallographic orientation different from the heat extraction direction, and a change in temperature gradient.

Figure 17 shows a scanning electron microscopy (SEM) photomicrograph of graphite eutectic cell and prior austenitic dendrite structures in gray cast iron.

The variations in graphite structures have been classified, together with the length of the flakes, by standards that have been utilized for many years. Flake graphite in gray cast iron can be designated as:

- Type A, uniform distribution, random orientation
- Type B, rosette grouping, random orientation
- Type C, superimposed flake sizes, random orientation
- Type D, interdendritic segregation, random orientation
- Type E, interdendritic segregation, preferred orientation

The formation of the eutectic flake graphite (Types A, B, C, and D) is greatly influenced by the amount by which the iron melt cools below the equilibrium temperature for the austenite-graphite eutectic before appreciable solidification occurs. Gray iron with Type A graphite undergoes only small amounts of undercooling, whereas that with Type D graphite undercools significantly below this equilibrium temperature. The undercooling that occurs with Type B graphite is intermediate between the two, producing fine graphite flakes, like Type D, in the center of the eutectic cells or rosette

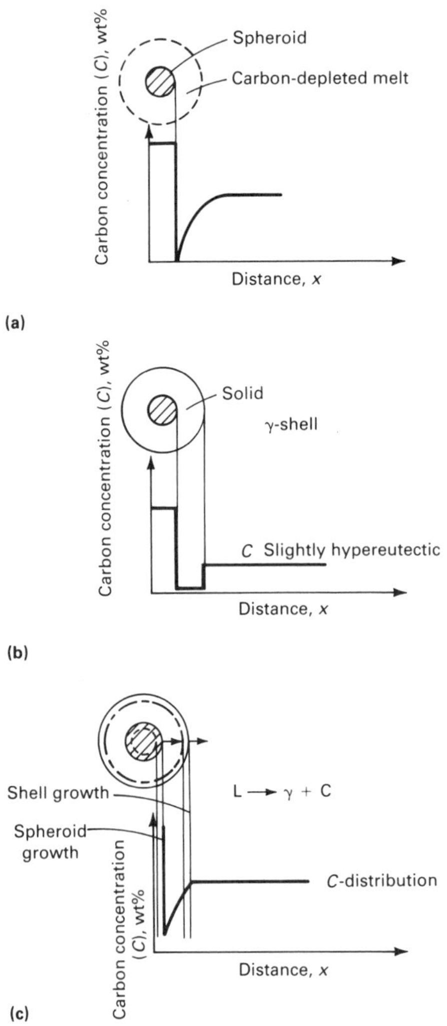

Fig. 19 Isothermal growth of a graphite spheroid within an austenite shell and growth of the shell with a smooth interface. (a) Growth of spheroidal graphite in contact with melt. (b) Envelopment by austenite. (c) Growth of spheroidal graphite within the austenite shell. Source: Ref 22

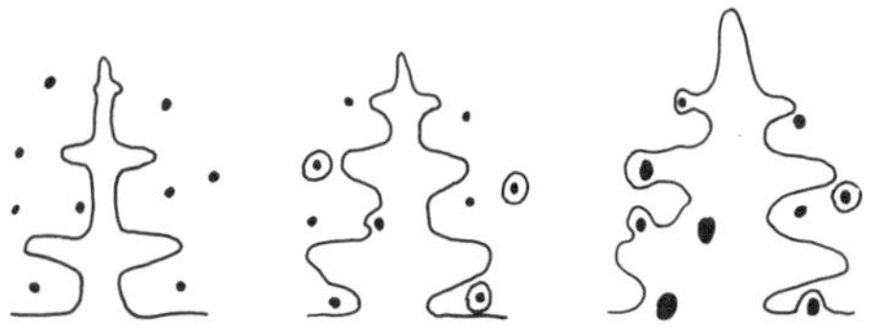

Fig. 20 Schematic illustrating the progression of growth in austenite-SG eutectic

and a coarser type like Type A at the outer cell boundaries. Type E graphite occurs in strongly hypoeutectic gray irons with carbon equivalents well below 4.3%. Figure 18 shows examples of Type A, B, D, and E flake graphite structures in gray cast irons at various magnifications.

Calculations using data from continuous cooling experiments given in Ref 27 allowed estimation of the growth constant μ in the range of 7.25×10^{-8} to 9.5×10^{-8} m · s^{-1} · K^{-2}, which is of the same magnitude as the values given previously for the case of directional solidification.

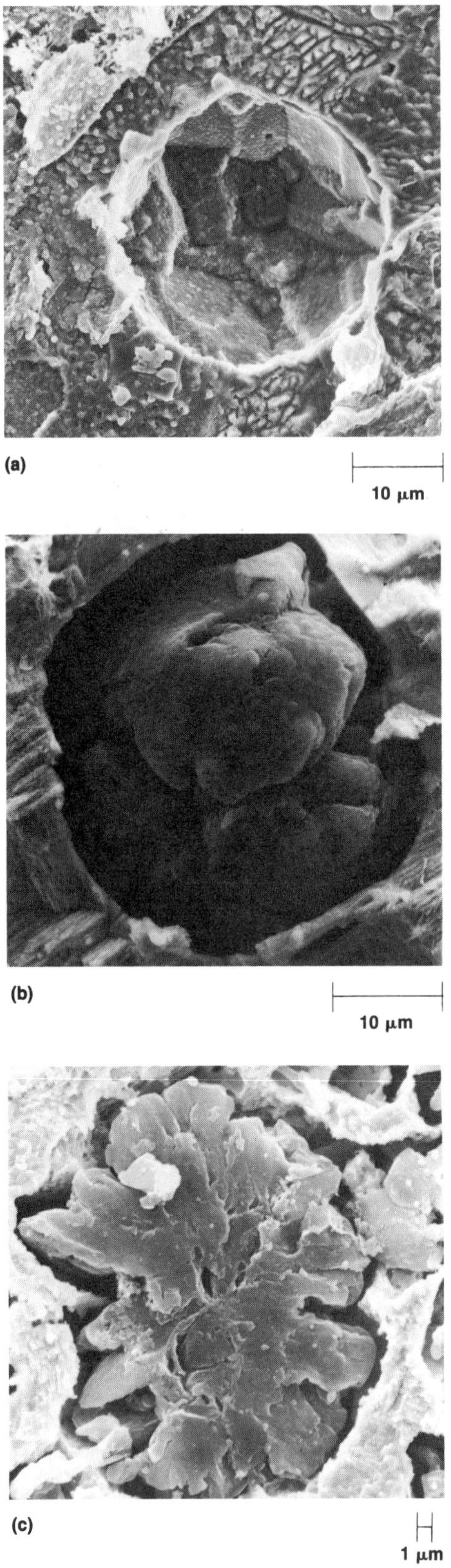

Fig. 21 SEM micrographs of deep-etched spheroidal graphite samples showing a fractured graphite spheroid (a). Nodularity decreases from (a) through (c).

The solidification of off-eutectic FG iron will start with the nucleation and growth of a primary phase, which is austenite dendrites for hypoeutectic iron or primary graphite plates for hypereutectic iron. It is not unreasonable to assume that the primary phases may play a role in the nucleation of the eutectic, with further growth of the eutectic occurring, as experimental evidence shows, on the primary phases.

Spheroidal Graphite Eutectic. Growth of the austenite-SG eutectic is more complicated and less understood than that of the γ-FG eutectic, although a good number of theories have been proposed. As discussed previously in this section, the γ-SG eutectic is a divorced eutectic. It has been rather widely accepted that the growth of this eutectic begins with nucleation and the growth of graphite in the liquid, followed by early encapsulation of these graphite spheroids in austenite shells (envelopes). A schematic of the process is shown in Fig. 19. Graphite nucleation and growth deplete the melt of carbon in the vicinity of the graphite; this creates conditions for austenite nucleation and growth around the graphite spheroid. Once the austenite shell is formed, further growth of graphite can occur only by solid diffusion of carbon from the liquid through the austenite.

Calculations of the diffusion-controlled growth of graphite through the austenite shell were originally made based on Zener's growth equation for an isolated spherical particle in a matrix of low supersaturation (Ref 28). Equation 3 was derived:

$$R_{Gr} = \frac{V_m^{Gr}}{V_m^{\gamma}} D_c^{\gamma} \frac{r_\gamma}{r_{Gr}(r_\gamma - r_{Gr})} \frac{X^{\gamma/L} - X^{\gamma/Gr}}{X^{Gr} - X^{\gamma/Gr}} \quad \text{(Eq 3)}$$

where R_{Gr} is rate of growth of graphite; V_m^{Gr} and V_m^{γ} are the molar volumes of graphite and austenite, respectively; D_c^{γ} is the diffusivity of carbon in austenite; r_{Gr} and r_γ are the radii of graphite and austenite, respectively; $X^{\gamma/L}$ and $X^{\gamma/Gr}$ are the molar fractions of austenite at the austenite/liquid and austenite/graphite boundaries, respectively; and X^{Gr} is the molar fraction of carbon in graphite. It has also been shown that $r_\gamma = 2.4\, r_{Gr}$. A slightly different approach, based on a steady-state diffusion model of carbon through the austenite shell (Ref 29), also allowed derivations for R_{Gr} and r_γ.

However, recent research has shown that the solidification mechanism of SG iron is more complicated and that austenite dendrites play a significant role in eutectic solidification (Ref 25, 26). The eutectic austenite is dendritic and can scarcely be distinguished from primary austenite dendrites. The sequence of solidification is as follows:

- At the eutectic temperature, austenite dendrites and graphite spheroids nucleate independently in the liquid

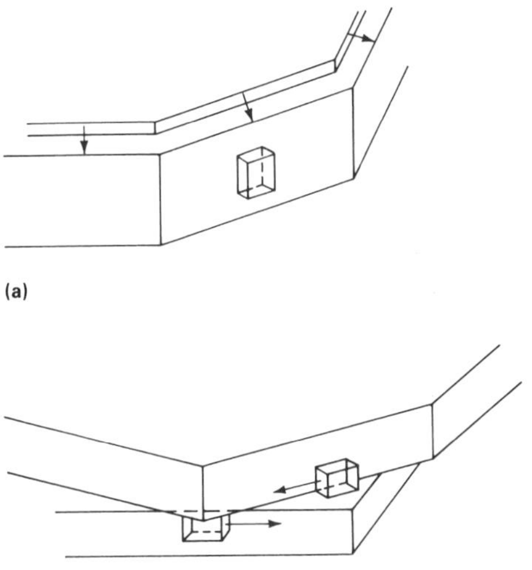

(a)

(b)

Fig. 22 Growth of graphite in the ⟨10$\bar{1}$0⟩ direction. (a) Growth by two-dimensional nucleation on (10$\bar{1}$0) faces illustrates that steps on (0001) surfaces will advance only as far as bounding crystal edges. (b) Growth from step to twist boundary illustrates that the (1010) faces grow by nucleation of planes at the step. Source: Ref 32

- Limited growth of spheroidal graphite occurs in contact with the liquid
- Flotation or convection then determines the collision of spheroidal graphite with the austenite dendrites
- Graphite encapsulation in austenite can occur before or immediately after the contact between graphite and austenite dendrites
- Further growth of graphite occurs by carbon diffusion through the austenite shell

This sequence is illustrated schematically in Fig. 20.

Graphite shape in SG iron is not perfectly spheroidal. Depending on cooling rate and chemical composition, significant deviations from the true spheroidal shape can be obtained (various nodularities) (Fig. 21).

It remains now to explain the reasons for the change of the habits of graphite from lamellar to spheroidal. Many theories have been proposed (see, for example, review works in Ref 5 and 8), but because of space limitations, only a few will be discussed in this section.

Many theories capitalize on the observation that the graphite/liquid surface energy is higher in SG iron than in FG iron. These theories explain spheroidal graphite formation by either simply implying that a sphere will have less free surface energy than a lamella with the same volume above a certain critical interface energy (Ref 30) or by suggesting that the high interface energy will curve the growing crystal in order to decrease the energy/volume ratio, resulting in spheroidal rather than lamellar graphite (Ref 31).

The defect growth of graphite theory explains spheroidal graphite formation based on the possible growth mechanisms of faceted crystals such as graphite (Ref 32). Three growth mechanisms are considered: two-dimensional nucleation (Fig. 22a), step of a defect (twisted) boundary (Fig. 22b), and screw dislocation. The first two mechanisms are governed by exponential laws and apply to the (10$\bar{1}$0) surface, but the third is governed by a parabolic law and applies to the (0001) surface of the graphite crystal.

When weak, reactive impurities such as sulfur are present in the melt, a contaminated environment occurs. These elements change the edge energy of steps, resulting in a relative position change of the growth rates involved, as shown in Fig. 23(a). The curve for growth on the step of a defect boundary, R_{step}, is at a lower undercooling than those for growth by two-dimensional nucleation, R_{2D}, or by screw dislocation, R_{screw}.

In a pure environment such as an iron-carbon-silicon alloy with no sulfur contamination, the growth rate curves are displaced to higher undercooling (Fig. 23b). In a melt of sufficient purity, or when increasing cooling rate, the higher degree of undercooling may allow growth with R_{screw} so that graphite spheroids can form. This has been achieved experimentally for pure nickel-carbon alloys by increasing the cooling rate of the melt, or for ultrapure iron-carbon alloys by cooling slowly in a vacuum.

In an environment with reactive impurities (for example, magnesium), the impurity will react with the surface, and the growth at a step of a twist boundary will be neutralized. Only the curves for R_{2D} and R_{screw} are left, and they are displaced to greater undercoolings (Fig. 23c).

Another theory relates graphite shape in cast iron with undercooling (kinetic plus constitutional) during solidification (Ref 5). As shown in Fig. 24, each graphite form has its own temperature for growth, which is achieved by a specific cooling rate and composition. Steps on surfaces can change graphite morphology from plate to rod. With an increase in undercooling, pyramidal instabilities will occur on the faces of the graphite crystal. At undercoolings of 29 to 35 °C (50 to 65 °F), instabilities occur on the (10$\bar{1}$1) faces of the pyramid, and it is suggested that graphite spheroids form at these undercoolings. Finally, at large undercoolings of 40 °C (70 °F), the growth form noted is a pyramidal one, bounded by (10$\bar{1}$1) faces. These pyramidal crystals are part of the series of imperfect forms observed particularly in thick-wall SG iron castings.

Imperfect spheroidal graphite shapes, such as those in Fig. 21(b) and (c), are due to incorrect kinetic undercooling and an imbalance of impurities influencing constitutional undercooling. An insufficient amount of impurities influencing kinetic undercooling (for example, magnesium) will favor the formation of intermediate graphite shapes, such as compacted/vermicular, while an insufficient amount of impurities influencing constitutional undercooling (for example, lead) will perturb the surface of the spheroid and promote the formation of protuberances.

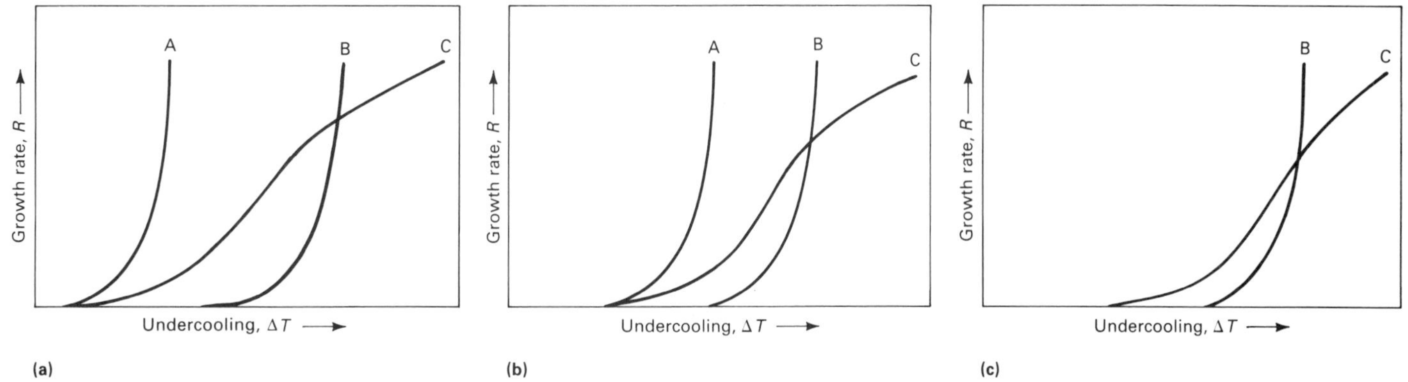

Fig. 23 Suggested ΔT-R correlation for (10$\bar{1}$0) and (0001) crystal faces of graphite growing in various environments. (a) Contaminated environment (for example, sulfur-containing iron-carbon-silicon alloy). (b) Pure environment (for example, iron-carbon-silicon alloy). (c) Environment with reactive impurities (for example, magnesium-containing iron-carbon-silicon alloy). Three growth mechanisms are discussed: A, on the step of the defect boundary (R_{step}); B, two-dimensional nucleation (R_{2D}); C, screw dislocation (R_{screw}). Source: Ref 32

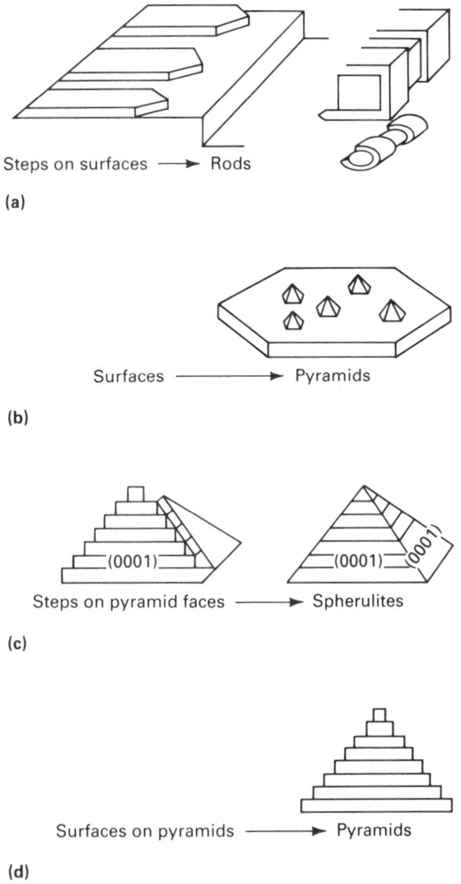

Fig. 24 Correlation among the different types of instability observed in graphite growth and growth morphologies with increasing undercooling, ΔT. (a) ΔT = 4 °C (7 °F). (b) ΔT = 9 °C (16 °F). (c) ΔT = 30 °C (54 °F). (d) ΔT = 40 °C (72 °F). Source: Ref 5

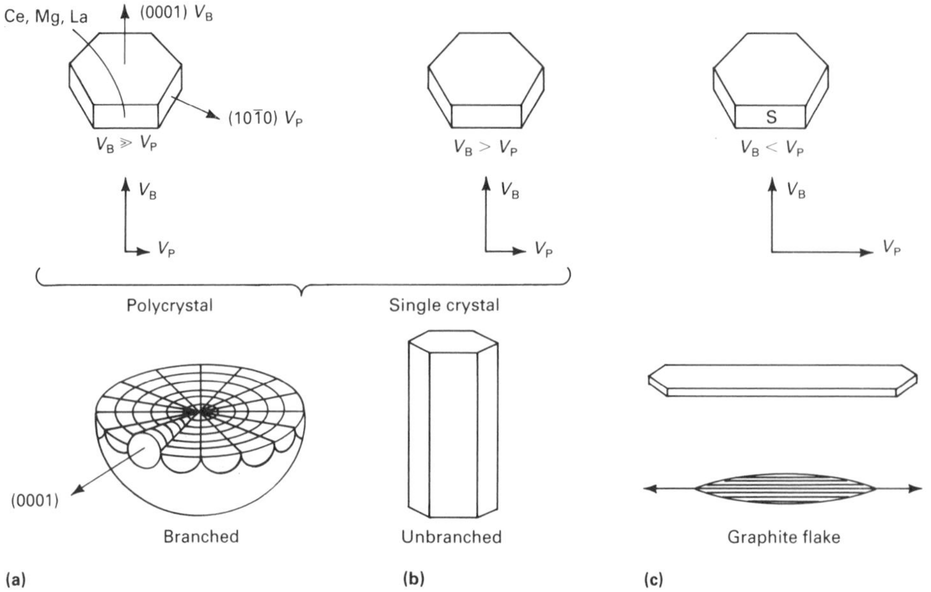

Fig. 25 Schematic of the change in the growth rate of graphite due to the absorption of foreign atoms in spheroidal graphite eutectic. Three variations of an iron-carbon-silicon cast iron are as follows. (a) With nodularizer added as reactive impurity environment. (b) Pure environment. (c) Contaminated environment in which surface-active elements such as oxygen and sulfur are absorbed into system. For (a) and (b), density in the basal plane, V_B is greater than the density in the prism face, V_P, and either branch polycrystalline or unbranched single crystals result. For (c), $V_B < V_P$. Sulfur adsorption makes prism faces the most densely packed, and graphite flakes are subsequently formed.

Surface Adsorption Theory. This somewhat older theory of graphite growth postulates that the change from lamellar to spheroidal graphite occurs because of the change in the ratio between growth on the (10$\bar{1}$0) face and growth on the (0001) face of graphite (Ref 33). For equilibrium conditions, the Gibbs-Curie-Wulf law states that the crystalline phase with the higher interface energy has a slow rate of growth in the normal direction. Bravais's rule stipulates that the growth rate in the direction normal to a plane is inversely proportional to the density of atoms located on the plane.

Accordingly, it follows that under equilibrium conditions the crystallographic plane with the highest density of atoms has the lowest interface energy and the minimum growth rate in a direction perpendicular to the plane. Nevertheless, under the nonequilibrium conditions prevailing during the solidification of cast iron, kinetic considerations become important. Assuming growth by two-dimensional nucleation, the highest rate of growth will be experienced by the face with the higher density of atoms, where the probability for nucleation is higher.

Therefore, in a pure environment, the highest growth rate will be in the (0001) direction of the graphite crystal (Fig. 25), resulting in the formation of unbranched single crystals (coral graphite). In a contaminated environment, surface-active elements such as sulfur or oxygen are absorbed on the high-energy plane (10$\bar{1}$0), which has fewer satisfied bonds. Subsequently, the (10$\bar{1}$0) plane face achieves a lower surface energy than the (0001) face, and growth becomes predominant in the (10$\bar{1}$0) direction, resulting in lamellar (plate) graphite. Finally, the reactive impurities (such as magnesium, cerium, and lanthanum) in an environment scavenge the melt of surface-active elements (sulfur, diatomic oxygen, lead, antimony, titanium, and so on), after which they also block growth on the (10$\bar{1}$0) prism face. A polycrystalline spheroidal graphite results.

Compacted/Vermicular Graphite Eutectic. The sequence of growth of compacted/vermicular graphite during the eutectic transformation is shown schematically in Fig. 26, based on experimental data from Ref 34 on rapidly quenched samples from successive stages during the solidification process. It can be seen that at the beginning graphite precipitates as spheroids, which then degenerate during growth and subsequently develop into compacted graphite. Compacted graphite develops as interconnected segments within an austenitic matrix. Typical compacted graphite is shown in Fig. 27.

It was proposed that the growth of compacted graphite occurs by the twin/tilt of

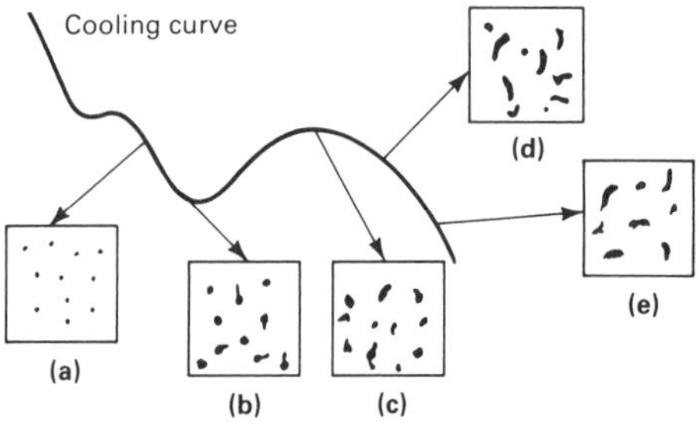

Fig. 26 Schematic of the sequence of development of compacted/vermicular graphite: (a) small spheroids; (b) and (c), some spheroids have tails; (d) compacted graphite plus spheroidal graphite; and (e) compacted graphite

boundaries (Ref 35). The initiation of a twin/tilt is related to the unstable growth of the (10$\bar{1}$0) interface, which can be induced by the presence of some reactive elements. The formation of a (10$\bar{1}$2) twin/tilt plane is shown in Fig. 28. It was further hypothesized that when insufficient reactive element is present, the tilt orientation of twin boundaries can alternate, resulting in typical compacted graphite (Fig. 29a). On the other hand, when there is enough impurity in the melt, the tilt orientation is singular, and spheroidal graphite may occur (Fig. 29b).

Based on this growth mechanism, compacted graphite can be considered to grow either by transition from flake graphite (Fig. 30a) or by degeneration of spheroidal graphite (Fig. 30b). The degree of interconnection of the compacted graphite within the eutectic cell is also apparent from Fig. 30. Based

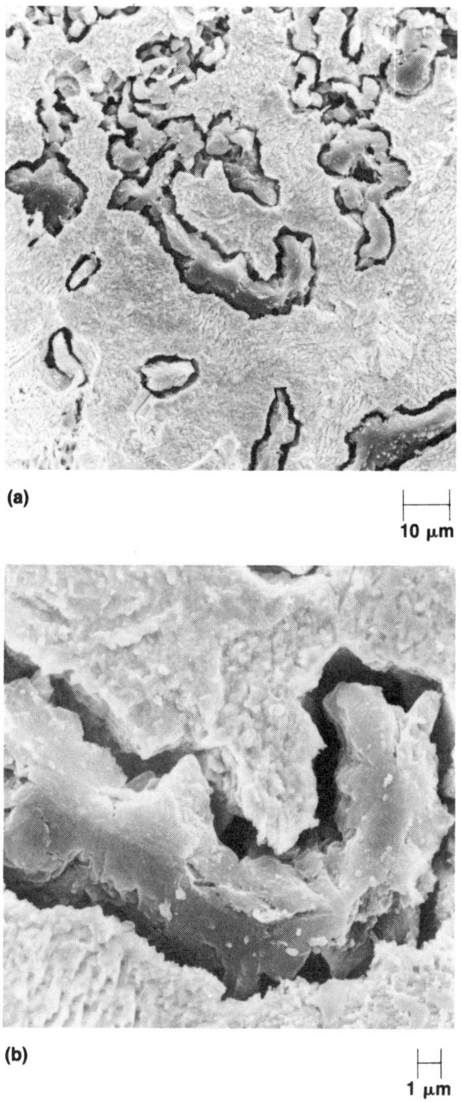

Fig. 27 SEM micrograph of deep-etched CG iron samples. (a) 560×. (b) 2800×

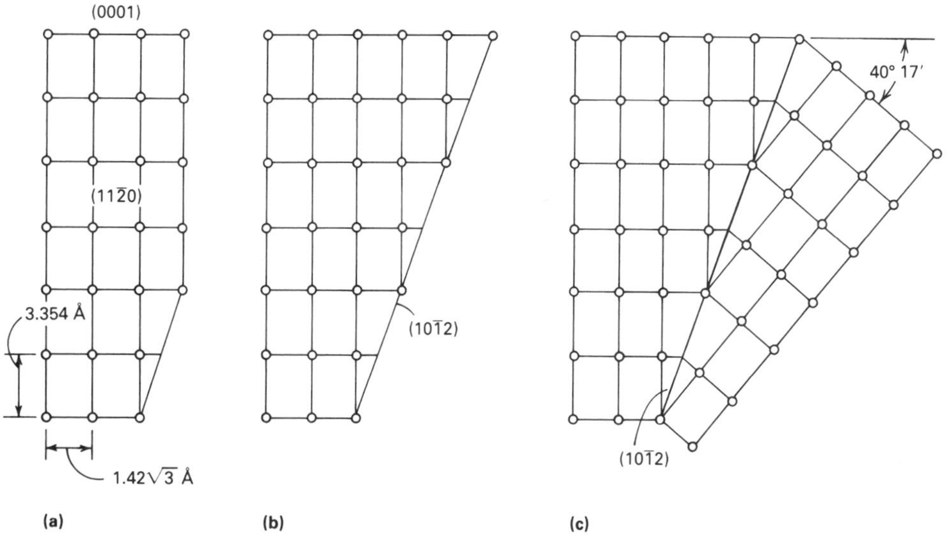

Fig. 28 Diagram of sequences (a) through (c) involved in the formation of a twin/tilt plane for graphite growing in the [1010] direction. Source: Ref 35

on cooling curve configuration, it is reasonable to assume that the growth rate of compacted graphite is similar to that of spheroidal graphite, which is not surprising because the growth direction is supposedly in the [0001] direction of the graphite crystal (see the article "Compacted Graphite Irons" in this Volume).

Austenite-Iron Carbide Eutectic. Growth of the austenite-iron carbide (ledeburite) eutectic begins with the development of a cementite plate on which an austenite dendrite nucleates and grows (Fig. 31). This destabilizes the Fe_3C, which then grows through the austenite. As a result, two types of eutectic structure develop: a lamellar eutectic with Fe_3C as the leading phase in the edgewise direction, *a*, and a rodlike eutectic in the sidewise direction, *c*.

Cooling rate has a significant influence on the morphology of the γ-Fe_3C eutectic. At moderate undercoolings, ledeburite structure is expected. High cooling rates, as obtained in quenching experiments, produce a degenerated eutectic structure dominated by Fe_3C plates. A coarse mixture of Fe_3C and γ fills the spaces between the Fe_3C plates. This structure obviously does not result from cooperative growth (Ref 7). A platelike carbide structure associated with equiaxed eutectic grains can be obtained by increasing the undercooling by superheating, by decreasing the silicon content, or by adding chromium or magnesium.

The gray or white solidification mode of cast iron apparently depends on the relative nucleation probability and the growth rates of the graphite and Fe_3C phases. In turn, this will be a function of the cooling rate and chemistry of the alloy. As shown in Fig. 32, only the graphite eutectic can nucleate and grow between the eutectic temperature for the gray eutectic (1153 °C, or 2107 °F) and that for the white eutectic (1148 °C, or 2098 °F). Below 1148 °C (2098 °F), both eutectics can occur. The growth rate of the γ-Fe_3C eutectic rapidly exceeds that of the γ-Gr eutectic, and at a temperature of approxi-

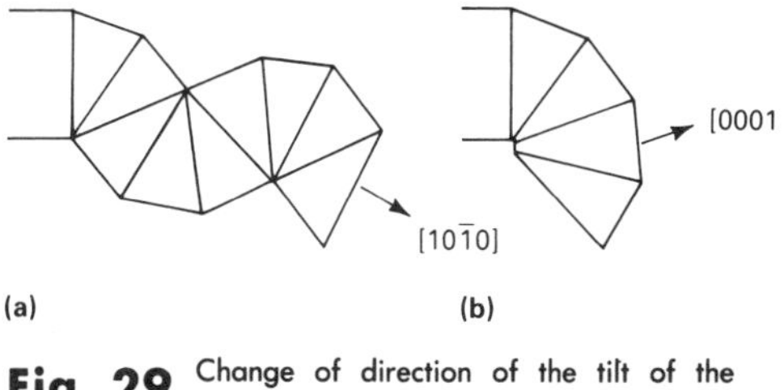

Fig. 29 Change of direction of the tilt of the boundary due to the amount of reactive impurity. (a) Insufficient spheroidization. (b) Sufficient spheroidization. Source: Ref 35

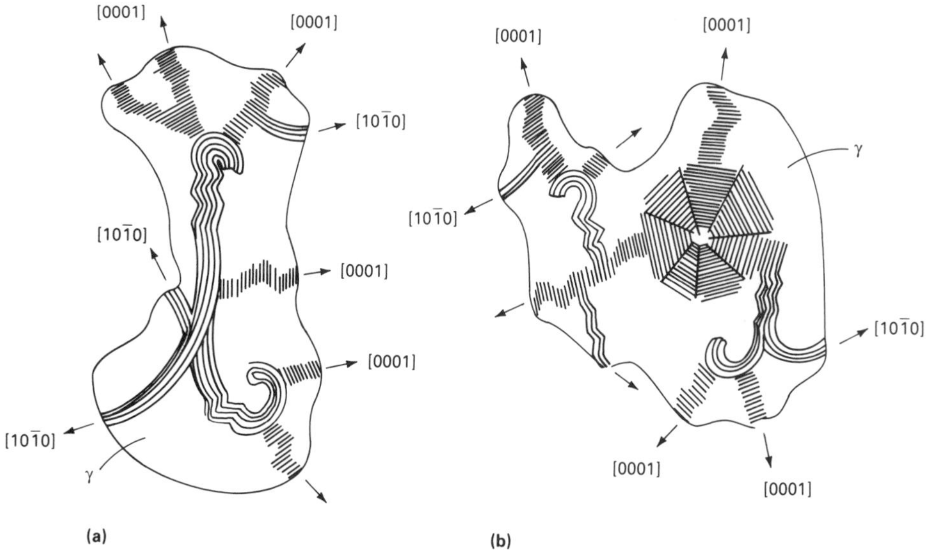

Fig. 30 Transition from flake to compacted graphite (a) and from spheroidal to compacted graphite (b) based on the twin/tilt of boundaries growth mechanism. Source: Ref 36

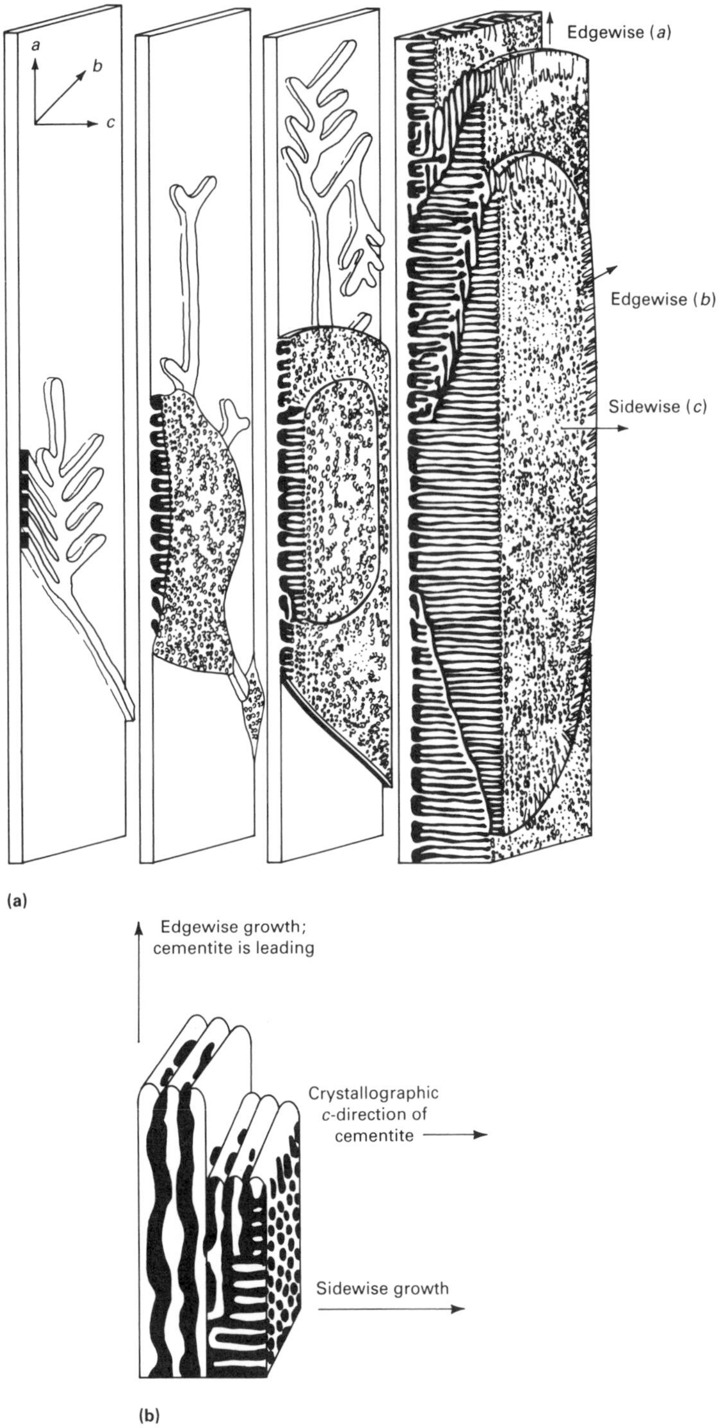

Fig. 31 Schematic illustrating growth of ledeburite (austenite iron carbide) eutectic. (a) Lamellar eutectic with cementite as the leading phase in the edgewise, *a*, direction. (b) Rodlike eutectic in the sidewise, *c*, direction. Source: Ref 37

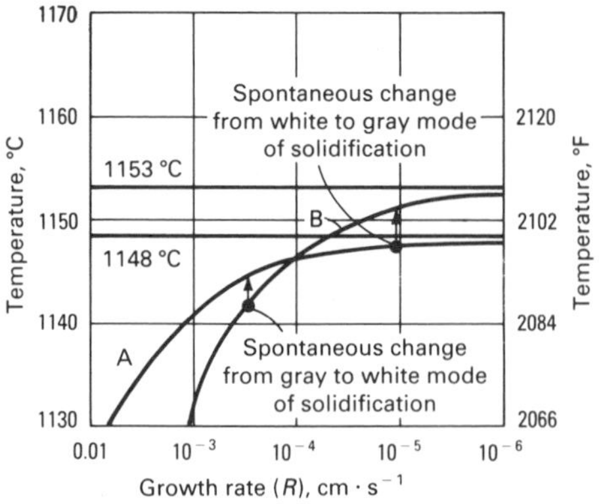

Fig. 32 Gray-to-white eutectic transition as a function of undercooling and growth rate. A, white structure; B, gray structure. Source: Ref 38

mately 1140 °C (2085 °F), the gray-to-white transition occurs (Ref 5, 38).

Cooling Curve Analysis

Cooling curves have been extensively used in the experimental study and production control of cast irons. The history of solidification can be traced on the cooling curve. A detailed description of the principles involved in this analysis and its application in cast iron technology is given in the articles ''Interpretation and Use of Cooling Curves (Thermal Analysis)'' and ''Compacted Graphite Irons'' in this Volume.

REFERENCES

1. S. Steeb and U. Maier, Structure of Molten Fe-C Alloys by Means of X-ray and Neutron Wide Angle Diffraction as well as Sound Velocity Measurements, in *The Metallurgy of Cast Iron*, B. Lux *et al.*, Ed., Georgi Publishing, 1975, p 1
2. W. Krieger and H. Trenkler, *Arch. Eisenhüttenwes.*, Vol 42 (No. 3), 1971, p 175
3. A.A. Vertman *et al.*, *Liteinoe Proizvod.*, No. 10, 1964
4. L.S. Darken, ''Equilibria in Liquid Iron With Carbon and Silicon,'' Technical Publication 1163, American Institute of Mining, Metallurgical, and Petroleum Engineers, 1940, p 1
5. I. Minkoff, *The Physical Metallurgy of Cast Iron*, John Wiley & Sons, 1983
6. W. Patterson and D. Ammann, Solidification of Flake Iron-Graphite Eutectic in Gray Iron, *Giesserei*, Jan 1959, p 1247
7. H.D. Merchant, Solidification of Cast Iron—A Review of Literature, in *Recent Research on Cast Iron*, H.D. Merchant, Ed., Gordon and Breach, 1968, p 1
8. J.F. Wallace, Effects of Minor Elements on the Structure of Cast Iron, *Trans. AFS*, Vol 83, 1975, p 363
9. S.L. Liu, C.R. Loper, Jr., and S. Shirvani, Inoculation of Gray Cast Irons With Carbonaceous Materials, *Trans. AFS*, Vol 93, 1985, p 501
10. B. Lux, Nucleation of Graphite in Fe-C-Si Alloys, in *Recent Research on Cast Iron*, H.D. Merchant, Ed., Gordon and Breach, 1968, p 241
11. J.V. Dawson and S. Maitra, Recent Research on the Inoculation of Cast Iron, *Br. Foundryman*, April 1967, p 117
12. D.M. Stefanescu, Inoculation of Gray Iron With Sodium and Barium, *Giesserei-Prax.*, No. 24, 1972, p 429
13. W. Weiss, The Importance of Deoxidation in the Crystallization of Cast Iron, in *The Metallurgy of Cast Iron*, B. Lux *et al.*, Ed., Georgi Publishing, 1975, p 69
14. M.B. Zeedijk, Identification of the Nu-

clei in Graphite Spheroids by Electron Microscopy, *J. Iron Steel Inst.*, July 1965, p 737
15. P. Poyet and J. Ponchon, Contributions to the Study of Graphite Nucleation in Cast Iron for Ingot Molds, *Fonderia*, No. 277, 1969, p 183
16. J.C. Mercier *et al.*, Inclusions in Graphite Spheroids, *Fonderia*, No. 277, 1969, p 191
17. H. Nieswaag and A.J. Zuithoff, "The Occurrence of Nodules in Cast Iron Containing Small Amounts of Tellurium," Paper presented at the 37th International Foundry Congress, Brighton, England, Sept 1970
18. R.J. Warrick, Spheroidal Graphite Nuclei in Rare Earth and Magnesium Inoculated Irons, *Trans. AFS*, Vol 76, 1968, p 722
19. M.M. Jacobs, T.J. Law, D.A. Melford, and M.J. Stowell, Basic Processes Controlling the Nucleation of Graphite Nodules in Chill Cast Iron, *Metals Technology*, Vol 1, Part II, No. 1974, p 490
20. J. Rickard and I.C.H. Hughes, Eutectic Structure in White Cast Iron, *BCIRA J.*, Vol 9 (No. 1), 1961, p 11
21. G.F. Ruff and J.F. Wallace, Control of Graphite Structure and Its Effect on Mechanical Properties of Gray Iron, *Trans. AFS*, Vol 84, 1976, p 705
22. B. Lux, F. Mollard, and I. Minkoff, On the Formation of Envelopes Around Graphite in Cast Iron, in *The Metallurgy of Cast Iron*, B. Lux *et al.*, Ed., Georgi Publishing, 1975, p 371
23. H. Jones and W. Kurz, Growth Temperatures and the Limits of Coupled Growth in Unidirectional Solidification of Fe-C Eutectic Alloys, *Metall. Trans. A*, Vol 11A, 1980, p 1265
24. D.M. Stefanescu, P.A. Curreri, and M.R. Fiske, Microstructural Variations Induced by Gravity Level During Directional Solidification of Near-Eutectic Iron-Carbon Type Alloys, *Metall. Trans. A*, Vol 17A, 1986, p 1121
25. S.K. Biswal, D.K. Bandyopadhyay, D.M. Stefanescu, unpublished research
26. A. Rickert and S. Engler, Solidification Morphology of Cast Irons, in *The Physical Metallurgy of Cast Iron*, Vol 34, H. Fredriksson and M. Hillert, Ed., Proceedings of the Materials Research Society, North Holland, 1985, p 165
27. H. Fredriksson and S.E. Wetterfall, A Study of Transition From Undercooled to Flake Graphite in Cast Iron, in *The Metallurgy of Cast Iron*, B. Lux *et al.*, Ed., Georgi Publishing, 1975, p 227
28. S.E. Wetterfall, H. Fredriksson, and M. Hillert, Solidification Process of Nodular Cast Iron, *J. Iron Steel Inst.*, May 1972, p 323
29. K.C. Su, I. Ohnaka, I. Yamauchi, and T. Fukusako, Computer Simulation of Solidification of Nodular Cast Iron, in *The Physical Metallurgy of Cast Iron*, Vol 34, H. Fredriksson and M. Hillert, Ed., Proceedings of the Materials Research Society, North Holland, 1985, p 181
30. H. Geilenberg, A Critical Review of the Crystallization of Graphite From Metallic Solutions After the "Surface Tension Theory", in *Recent Research on Cast Iron*, H.D. Merchant, Ed., Gordon and Breach, 1968, p 195
31. J.P. Sadocha and J.E. Gruzleski, The Mechanism of Graphite Spheroid Formation in Pure Fe-C-Si Alloys, in *The Metallurgy of Cast Iron*, B. Lux *et al.*, Ed., Georgi Publishing, 1975, p 443
32. I. Minkoff and B. Lux, Graphite Growth From Metallic Solution, in *The Metallurgy of Cast Iron*, B. Lux *et al.*, Ed., Georgi Publishing, 1975, p 473
33. K. Herfurth, Investigations Into the Influence of Various Additions on the Surface Tension of Liquid Cast Iron With the Aim of Finding Relationships Between the Surface Tension and the Occurrence of Various Forms of Graphite, *Freiberg Forschungs*, Vol 105, 1965, p 267
34. E.N. Pan, K. Ogi, and C.R. Loper, Jr., Analysis of the Solidification Process of Compacted/Vermicular Graphite Cast Iron, *Trans. AFS*, Vol 90, 1982, p 509
35. P. Zhu, R. Sha, and Y. Li, Effect of Twin/Tilt on the Growth of Graphite, in *The Physical Metallurgy of Cast Iron*, Vol 34, H. Fredriksson and M. Hillert, Ed., Proceedings of the Materials Research Society, North Holland, 1985, p 3
36. X. Den, P. Zhu, and Q. Liu, Structure and Formation of Vermicular Graphite, in *The Physical Metallurgy of Cast Iron*, Vol 34, H. Fredriksson and M. Hillert, Ed., Proceedings of the Materials Research Society, North Holland, 1985, p 141
37. M. Hillert and H. Steinhauser, The Structure of White Iron Eutectic, *Jernkontorets Ann.*, Vol 144, 1960, p 520
38. M. Hillert and V.V. Subba Rao, "The Solidification of Metals," Publication 110, The Iron and Steel Institute, 1968

Interpretation and Use of Cooling Curves (Thermal Analysis)

H. Fredriksson, The Royal Institute of Technology, Sweden

THERMAL ANALYSIS is a classical method of determining phase diagrams. By melting and cooling an alloy of known composition and registering the temperature-time curves, the liquidus temperature for the respective alloy can be determined. Figure 1 shows an idealized view of the relationship between the cooling curves and the phase diagram for an alloy A-B. The starting point for solidification is given by a change in the slope of the cooling curve due to the evolution of the heat of solidification. This starting point is normally used to define the liquidus temperature. The method is described more fully in Ref 1.

The idealized behavior illustrated in Fig. 1 is not seen in experimentally determined temperature-time curves. For example, Fig. 2, an experimentally determined cooling curve for an iron-carbon alloy with 3.2% C, shows that the formation of austenite and the eutectic reaction occur at a temperature well below the liquidus and the eutectic temperatures, respectively. The reason for this depends on the nucleation and growth of the crystals in the liquid. In this article, this effect will be analyzed, and the use of thermal analysis in industrial processes and in research will be discussed.

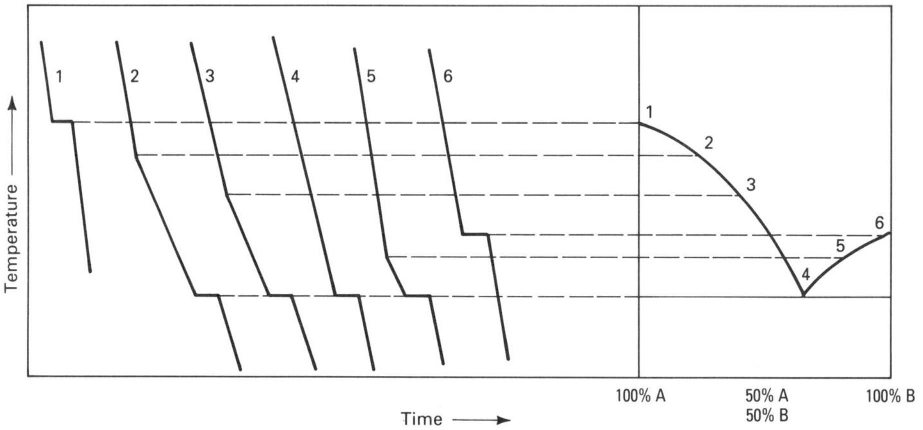

Fig. 1 Relationship between a cooling curve and phase diagram for alloy A-B

Quantitative Thermal Analysis

Cooling curves describe a balance between the evolution of heat in the sample and the heat transport away from the sample. It is often easiest to describe this in a mathematical form. However, heat transport away from the liquid is often very complex and can be described in many different ways because of the experimental conditions. Currently, heat transport is often described by numerical methods. In the following discussion, only the expression dQ/dt will be used to define the heat transport away from the experimental setup. This heat transport must be equal to the evolution of heat in the sample. Before and after the solidification process, this is balanced by a loss of heat capacity. Before the start of the solidification process, this can be written in analytical form as:

$$\frac{dQ}{dt} = V \cdot \rho \cdot C_p \frac{dT}{dt} \qquad \text{(Eq 1)}$$

where V is the volume of the sample, ρ is the density of the liquid, C_p is the heat capacity, and dT/dt is cooling rate of the liquid.

From a cooling curve similar to that shown in Fig. 2, dT/dt is the slope of the curve before solidification begins. By evaluating the slope as a function of time in an experiment, dQ/dt can be determined if ρ and C_p are known. This can later be used in the analysis of the temperature-time curve.

When the liquid cools to the liquidus temperature, crystals are nucleated and begin to grow. The heat balance from Eq 1 will now be:

$$\frac{dQ}{dt} = \left(V \cdot \rho \cdot C_p + \rho \Delta H \frac{df}{dT}\right) \frac{dT}{dt} \qquad \text{(Eq 2)}$$

where ΔH is the heat of solidification and df/dT is the volume fraction of solid formed at a changing temperature. The change in shape of the temperature-time curve now depends on the volume fraction of solid formed. The larger the rate of increase of volume fraction of solid, the smaller the slope (dT/dt) of the temperature-time curve. The volume fraction of solid formed is related to the number of crystals and their growth rate.

Crystals are normally nucleated when the temperature of the liquid reaches the liquidus temperature as well as the eutectic temperature. In spite of this, the temperature of the liquid will decrease further and initially the curve can be extrapolated to show where no crystals are growing. After a

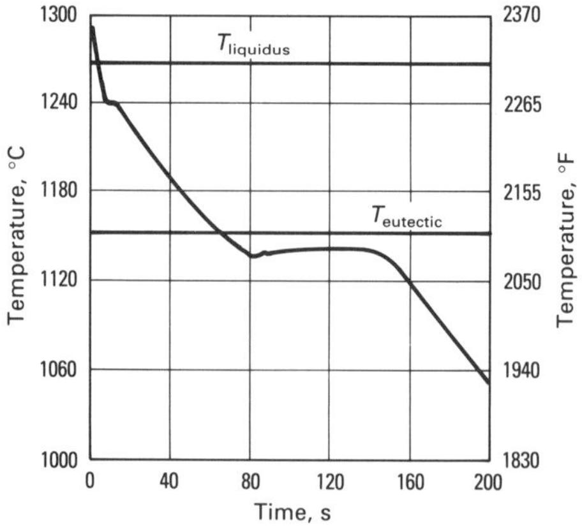

Fig. 2 Cooling curve for an iron-carbon alloy with 3.2% C. The solidification starts with a primary precipitation of austenite, followed by a eutectic reaction.

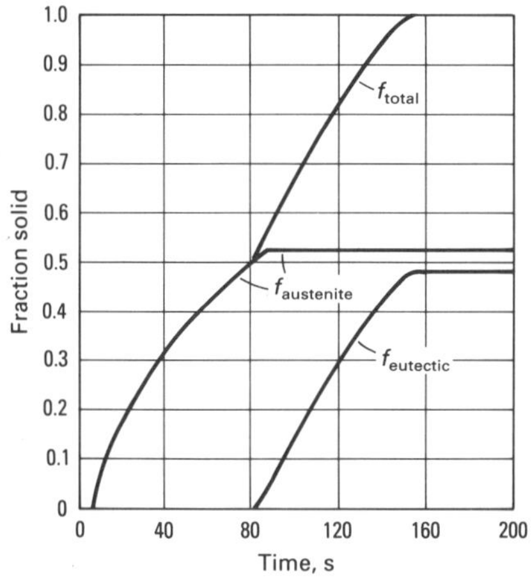

Fig. 3(a) The fraction of solid phases as a function of time for the alloy with a cooling curve shown in Fig. 2

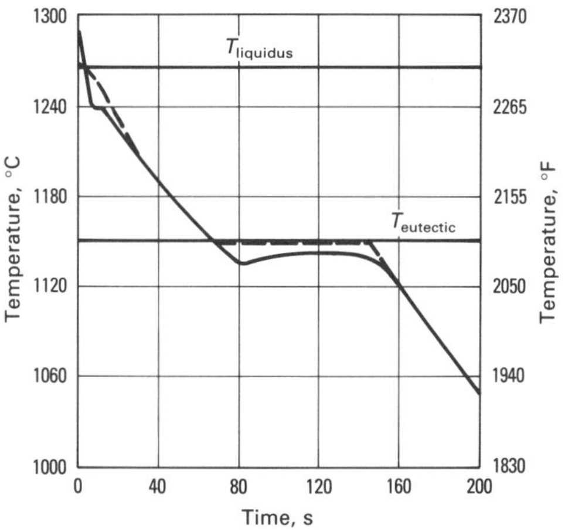

Fig. 3(b) Deviation of the cooling shown in Fig. 2 resulting from equilibrium solidification condition (see dashed line)

certain time, the slope of the curve begins to decrease, slowly at first and then more rapidly. In the early stages of crystal formation, the growing crystals are small (as is heat development) and therefore have no effect on the bulk liquid temperature.

The growth rate is also low because of the low supercooling close to the liquidus temperature. The supercooling increases when the temperature decreases; therefore, the growth rate will increase as the area of the crystal increases. The result is that more heat is given off and the slope of the temperature-time curve begins to decrease. The change in the slope will begin earlier and will proceed faster. The more crystals there are in the liquid, the earlier the change in slope starts, and the faster it will proceed. When the temperature increases, the growth rate of the free crystals decreases again, but they are so large by this time that the volume solidifying per unit of time will also be large. As such, the temperature will continue to increase until the growth rate and the growth area of the crystals offset the heat extraction. The larger the area of the growing crystals, the lower the growth rate and the closer the temperature of the bulk liquid to the liquidus or eutectic temperature.

Deviation From Equilibrium

The solidification process illustrated by the cooling curve in Fig. 2 begins with a primary precipitation of austenite, followed by a eutectic reaction. In this case, the liquid was cast into a sand mold. The heat extraction from the liquid can be described by Chvorinov's law and follows the expression:

$$\frac{dQ}{dt} = \frac{K}{\sqrt{t}} \qquad \text{(Eq 3)}$$

where t is the time and K the so-called modulus constant. The value K can be determined by using Eq 1 and the slope of the cooling curve before solidification begins. The fraction of solid can be determined by using Eq 2 and 3 and the data given in Fig. 2. The results of these calculations are shown in Fig. 3(a). This type of analysis has been performed and is discussed in Ref 2 to 4; it can be used to analyze the deviation from solidification under equilibrium conditions.

The Fe-3.2C alloy discussed earlier in this article can be used as an example to illustrate this analysis. According to the iron-carbon phase diagram, the liquidus temperature for this alloy is 1268 °C (2314 °F). At this temperature, dendritic crystals of austenite are formed. Because carbon has a very high diffusivity in austenite, the lever rule* can be used to determine the fraction of solid as a function of temperature. The rule is derived with respect to temperature, and the derivate df/dT is obtained. This derivate is inserted into Eq 2. The temperature-time curve is then easily calculated. The result is shown in Fig. 3(b) as a dashed line. The deviation from the real curve and that calculated under equilibrium conditions is greatest at the beginning of the precipitation of austenite. After the initial stage during the cooling down to the eutectic temperature, the lever rule will describe the fraction of solid formed. The eutectic reaction for the equilibrium case occurs without any undercooling. This derivation from equilibrium will be interpreted below.

The austenite dendrites can be assumed to nucleate randomly in the liquid. They grow in leading tips in a starlike fashion. The secondary dendrite arms form behind the tips. The growth of the dendrite tips is dependent upon undercooling. This undercooling is determined by the growth rate, and it increases with increasing growth rate. The growth rate of the tips is determined by the heat extraction and by the number of growing crystals. When the dendrite crystals have collided, a network of dendrite arms is formed.

The solidification now proceeds by a thickening of the dendrite arms. Similarly, eutectic colonies are randomly nucleated in the liquid; they grow spherically and collide with each other. When the collision occurs, the growth area decreases, and the temperature of the liquid decreases.

Theoretical models that combine growth equations, statistical laws for the collision of crystals, and heat extraction laws have been developed in recent years to describe the above discussion (Ref 2-6). In the future, these models will probably be used to predict the structure formed during the solidification process and the resultant properties of the material.

Differential Thermal Analysis

Thermal analysis has a long history of use in research. To increase accuracy, the experiments are normally carried out by so-called differential thermal analysis.

Simplified Thermal Analysis. Before discussing the theoretical basis of differential thermal analysis, a simplified experimental setup will be considered. This setup consists of a crucible with a melt placed inside a furnace. A thermoelement is placed in the center of the melt to measure the temperature of the sample T_s. Another thermocouple is placed at the wall of the furnace; the purpose of this thermocouple is to regulate the temperature of the furnace T_f as a function of time. The furnace is assumed to cool at a constant rate in degrees Kelvin per minute.

Figure 4 shows the temperature-time cycle for an aluminum-copper alloy that contains 5% Cu. The furnace has cooled during the experiment at a rate of 10 K/min (~285 °C/min, or 510 °F/min). Figure 4 shows that the temperature of the sample is somewhat higher than that of the furnace. This divergence increases slightly with time. A stationary condition, that is, the same cooling rate for the sample and the furnace, never occurs; the cooling rate of the sample is always somewhat lower than that of the furnace. The melt begins to solidify at the point designated t_{start}. Heat is emitted by the sample and the temperature difference between the sample and the furnace increases. A hump sometimes appears in the temperature-time curve of the sample. The maximum point, or the deviation point, on the cooling curve is usually used to establish the liquidus temperature.

The solidification rate, or the reaction rate, decreases after the initial period, and the cooling rate of the sample increases.

*The lever rule can be applied to any two-phase field of a phase diagram to determine the amounts of the different phases present at a given temperature in a given alloy. A horizontal line, referred to as a tie line, represents the lever, and the alloy composition represents its fulcrum. The intersection of the tie line with the boundaries of the two-phase field fixes the compositions of the coexisting phases, and the amounts of the phases are proportional to the segments of the tie line between the alloy and the phase compositions.

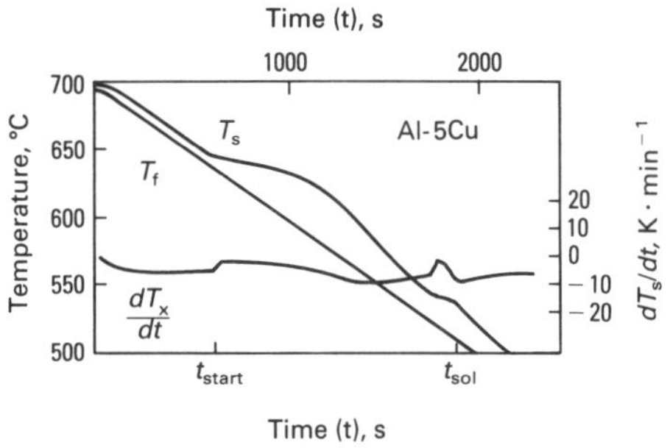

Fig. 4 Thermal analysis of an Al-5Cu alloy cooled at a rate of 10 K/min (~285 °C/min, or 510 °F/min). Source: Ref 6

Toward the end of solidification, or when everything has solidified, the cooling rate of the sample becomes greater than that of the furnace. This is a result of the large temperature difference that exists between the sample and the furnace when a considerable amount of solidification heat is given off. Afterward, the cooling rate of the sample approaches that of the furnace.

It is often very difficult to determine the end of the solidification process and to determine the solidus temperature t_{sol} during the experiment. To facilitate evaluation of the liquidus and solidus temperatures, the temperature-time curve of the sample is derived in most cases.

In general, the derivation of the temperature-time curve for the sample shows a discontinuity when the heat of solidification ceases to develop. The solidus temperature is generally defined by this discontinuity. This discontinuity becomes less pronounced in cases in which the solidification interval is large and the reaction rate is low toward the end of solidification.

The reaction rate is normally evaluated by a simple mathematical analysis. This is described by:

$$\left(\frac{dT_s}{dt}\right) \rho \cdot C_p^s \cdot V_s + \Delta H \cdot \rho \cdot V_s \cdot \frac{df}{dt} = \eta(T_s - T_f) \qquad \text{(Eq 4)}$$

where dT_s/dt is the cooling rate of the sample, C_p^s is the specific heat of the sample, V_s is the volume of the sample, ΔH is the heat of solidification, df/dt is the volume fraction of solid formed per unit time, T_s is the sample temperature, T_f is the furnace temperature, and η is the heat transfer coefficient.

The second term on the right-hand side in Eq 4 is 0 before solidification begins. The heat transfer coefficient η can be evaluated at this time by measuring dT_s/dt, T_s, and T_f. The reaction rate df/dt can be determined by using that value and by remeasuring the temperature and its first-time derivate during solidification.

Differential Thermal Analysis. Figure 5 shows a schematic setup for differential thermal analysis. The sample to be analyzed and the standard sample, with a thermocouple in each, are placed in a furnace. The temperature of the furnace and the cooling rate are controlled and measured with a separate thermocouple. The standard sample and the analysis sample must have the same temperature at the beginning of the experiment. During the cooling cycle, the samples will undergo one or more phase transformations that give off heat. This emission of heat implies that, during the time of reaction, the temperature of the analysis sample differs from that of the standard sample. The temperature difference between the analysis sample and the standard sample is recorded at the same time the furnace temperature is recorded.

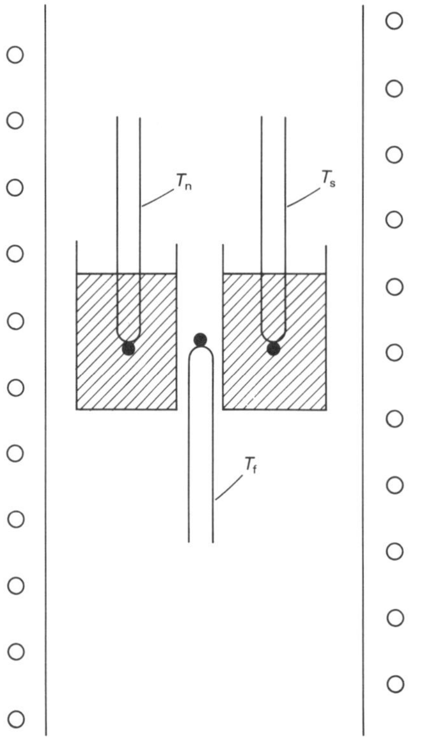

Fig. 5 Schematic of principal setup for differential thermal analysis equipment. T_n is the standard temperature, T_s is the sample temperature, and T_f is the furnace temperature.

Using the experimental setup shown in Fig. 5, one normally registers the difference in temperature between the standard sample and the analysis sample, not the difference between the sample and the furnace. Therefore, only $(T_s - T_n)$ and T_s are known, where T_n is the temperature of the standard sample. In order to be able to calculate the reaction velocity, the term $(T_s - T_f)$ in Eq 4 is replaced with the term $(T_s - T_n) + (T_n - T_f)$. For the standard sample, it is true that $\eta(T_n - T_f)$ is equal to:

$$\rho \cdot C_p^s \cdot V_n \frac{dT_n}{dt} \qquad \text{(Eq 5)}$$

where dT_n/dt is the cooling rate of the standard sample. Replacing this in Eq 4 gives:

$$\rho \cdot C_p^s \cdot V_s \frac{dT_s}{dt} + (\Delta H) \cdot \rho \cdot V_s \frac{df}{dt} = \eta(T_s - T_n) + \rho \cdot C_p^n V_n \cdot \frac{dT_n}{dt} \qquad \text{(Eq 6)}$$

If the heat of solidification ΔH is known, the reaction rates can be calculated by measuring the temperatures and the cooling rates. If the heat transfer coefficient is known, the heat of solidification can also be determined by integrating Eq 6 over the entire solidification interval (from $f = 0$ to $f = 1$).

Simplifications of Eq 6 are often found in the literature when the reaction rate and the heat of solidification are determined. It is assumed that the first term on the left-hand side is equal to the last term on the right-hand side, which leads to the following approximate relationship:

$$\frac{df}{dt} = \frac{\eta}{(\Delta H)\, V_s}(T_s - T_n) \qquad \text{(Eq 7)}$$

Equation 7 shows a direct proportionality between the measured temperature difference and the reaction rate.

In Eq 7, $\eta/\Delta H V_s$ can be considered constant. The difference in temperature between the sample and the normal $(T_s - T_n)$ is measured and is thus known as function of time. The constant can be determined by integrating Eq 6 from $f = 0$ to $f = 1$. This integration is often done graphically.

Equation 7 is also often used in interpreting differential thermal analysis data. However, one should keep in mind that doing so introduces many faults. This can be demonstrated by a comparison between Eq 6 and 7. For example, it is assumed that C_p^s, V_s, V_n, and C_p^n are equal. This is often not the case. To avoid this problem, the sample itself can be used as an imaginary standard sample. The temperature difference can then be determined by a simple computer analysis (Ref 6). To determine the heat of solidification, the heat transfer coefficient η must be determined by a series of reference experiments such as that described for the simple type of experiment in Eq 4.

Determining Liquidus and Solidus Temperature

Cooling curves can be used in many different ways and are often used to determine the liquidus and solidus temperatures. In such applications, the plateau temperature shown in Fig. 2 is often used. However, as indicated earlier, this temperature does not correspond to the liquidus temperature, because of the kinetics during crystal growth. To increase the accuracy, the cooling curve is often derived. In this case, the temperature at which the derivate changes is used as the liquidus temperature. As indicated earlier, this will not provide the correct liquidus temperature. The only way to obtain the correct liquidus temperature is to perform a series of experiments at different cooling rates. The plateau temperature can

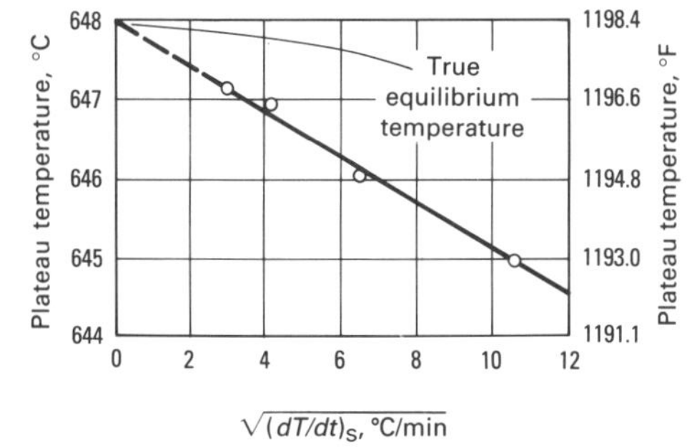

Fig. 6 Liquidus temperature determined for an aluminum-copper alloy. Source: Ref 7

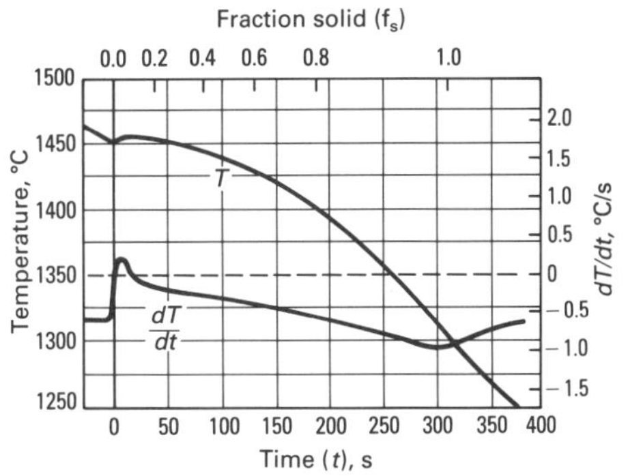

Fig. 7 Temperature-time curve for an iron-base alloy with 1.01% C, 0.25% Si, and 0.46% Mn. The dashed line represents the cooling rate $dT/dt = 0$. Source: Ref 8

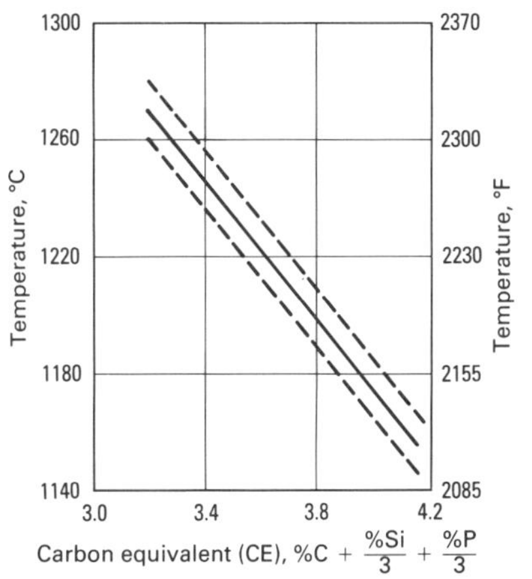

Fig. 8 Thermal analysis used for determining the chemical composition of cast irons

then be plotted as a function of the cooling rate. The liquidus temperature is determined by extrapolating the line to zero cooling rates. An example of this method is shown in Fig. 6.

It is more difficult to determine the solidus temperature than the liquidus temperature. Toward the end of the solidification process, the derivative of the cooling curve often shows a discontinuity (Fig. 7). The solidus temperature is defined in general by this discontinuity. However, the temperature evaluated in this way varies with the cooling rate much more than the liquidus temperature does. This is due to the microsegregation that occurs during solidification. This microsegregation is very much influenced by the cooling rate (Ref 9). Therefore, thermal analysis by cooling is nearly useless for determining the equilibrium solidus temperature. This temperature can be determined only by analyzing the melting process on heating curves.

Thermal analysis is often used in industrial processes. One of the most common applications is determination of the chemical composition of cast iron. In this case, a linear relationship has been found between the liquidus temperature and the carbon equivalent, CE, which is defined as:

$$CE = \%C + \frac{\%Si}{3} + \frac{\%P}{3} \quad \text{(Eq 8)}$$

An example of the use of thermal analysis for determining the chemical composition of cast irons is shown in Fig. 8.

Given the present knowledge of thermal analysis, it should be noted that these diagrams can be used with accuracy only if the same conditions are prevalent every time. This is due to the influence of the cooling rate and the nucleation frequency on the plateau temperature. Casting temperature, for example, can influence the results. It is well known that the superheat of the liquid influences the number of crystals. The higher the temperature, the lower the number of growing crystals. The result will be a lower liquidus temperature. However, in most cases, a foundry will use the same melting procedure consistently and can therefore determine chemical composition easily and accurately by thermal analysis, provided a curve similar to the one shown in Fig. 8 is developed for their specific conditions.

REFERENCES

1. K.W. Andrews, *Physical Metallurgy*, Vol 1, William Clowes & Sons, 1973
2. H. Fredriksson and S.E. Wetterfall, *The Metallurgy of Cast Iron*, Proceedings of the Second International Symposium on the Metallurgy of Cast Iron, Geneve, May 1974, Georgi Publishing, 1974
3. H. Fredriksson and I.L. Svensson, *Computer Simulation of the Structure Formed During Solidification of Cast Iron*, Proceedings of the Third International Symposium on the Metallurgy of Cast Iron, North Holland, 1984
4. D.M. Stefanescu and C. Kanetkar, *Computer Modelling of the Solidification of Eutectic Alloys: The Case of Cast Iron*, Proceedings of the Society for Metals and Materials Science, Computer Simulation Division, The Metallurgical Society (Toronto, Canada), 1985
5. H. Fredriksson and A. Olsson, Mechanism of Transition from Columnar to Equiaxed Zones in Ingots, *Mater Sci. Tech.*, Vol 2, 1986, p 508-516
6. H. Fredriksson and B. Rogberg, *Met. Sci.*, Vol 13, 1979, p 685-690
7. L. Bäckerud and B. Chalmers, *Trans. Met. Soc. AIME*, Vol 245, 1969, p 309-318
8. *A Guide to the Solidification of Steels*, Jernkontorets, Stockholm, 1977
9. M.C. Flemings, *Solidification Processing*, McGraw-Hill, 1974

Patterns

Section Chairman:
Robert Voigt, The University of Kansas

Patterns and Patternmaking

Robert C. Voigt, University of Kansas

A PATTERN is a form made of wood, metal, plastic, or composite materials around which a molding material (usually prepared sand) is formed to shape the casting cavity of a mold. Most patterns are removed from the completed mold halves and used repeatedly to make many duplicate molds. Expendable patterns of such materials as wax or expanded polystyrene are made in quantity and are used only once to produce an individual mold (see the articles "Investment Casting" and "Replicast Process" and the discussion on the evaporative foam process in the article "Sand Molding" in this Volume). The pattern equipment (tooling) needed to make a casting includes the pattern and may also include one or more core boxes. Core boxes are used to make refractory inserts or cores that are placed within a mold cavity to form internal cavities or passageways within the casting. The selection of the type of pattern equipment used to make a casting depends on many factors, including the number of castings to be produced, the size and shape of the casting, the molding or casting process to be used, and other special requirements such as the dimensional accuracy required.

Patternmaking begins with the dimensions of the casting required; however, the design and dimensions of the resultant pattern equipment must incorporate other features and take into account various pattern allowances (Ref 1, 2). Patterns must be oversized to correct for the contraction of the metal upon cooling and any extra metal to be removed from the machined surfaces of a casting. A mold parting line must be selected, and taper or draft must be included on the vertical faces of the pattern to permit the removal of reusable patterns from the mold. Patterns also incorporate provisions for gating and risering (rigging), which facilitate the flow (feeding) of metal into the mold cavity. For castings to be made with cores, the pattern and core boxes must include projections called core prints, which are used to support and locate the cores in the mold cavity. Figure 1 illustrates a typical pattern and mold configuration.

The cost of pattern equipment for a given casting can vary greatly, depending on pattern material, type of pattern, and sometimes the dimensional accuracy required. However, because the pattern equipment is only a part of the moldmaking process, the least expensive pattern is not necessarily the most economical. Additional pattern cost and quality often lead to lower end costs by a reduction in molding costs and pattern repair costs and by an improvement in overall casting quality.

The patterns and patternmaking techniques described in this article are for patterns used to make expendable molds (primarily sand molds). However, the principles described also apply to the dies used for other metal casting processes such as die casting and permanent mold casting (see the articles "Die Casting" and "Permanent Mold Casting" in this Volume).

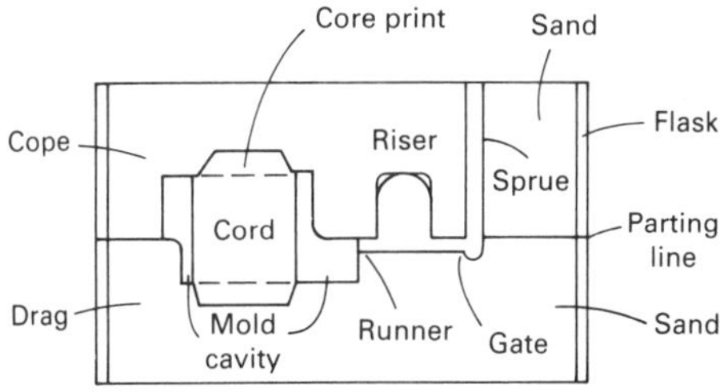

Fig. 1 Schematic of a sand mold. The pattern is used to form the mold cavity, the core print for locating the core, the gate, the runner, the riser, and the sprue. A separate core box is used to make the sand core that is inserted into the parted mold before pouring.

Types of Patterns

The type of pattern used for a specific application depends primarily on the number of castings required, the molding or casting process to be used, the size of the pattern, and the casting tolerances that are required. The stage of development of a casting design is also a factor. If the casting (or its rigging) is likely to be redesigned, an inexpensive prototype pattern is often used first.

The choice of pattern material is also closely tied to the molding or casting process and the type of pattern to be used. Figure 2 illustrates the most commonly used basic reusable pattern types for a simple pattern shape—a loose pattern, a match plate pattern, and cope and drag patterns. Also indicated are the typical materials and service lives for each pattern type. It should be noted that many factors affect the number of molds that can be made from an actual production pattern before it must be reworked. The pattern lives indicated in Fig. 2 are intended only as general guidelines. Actual pattern life can vary considerably.

Loose patterns, also called one-piece patterns or solid patterns, are the simplest, least expensive type of reusable pattern and are suited only to very low-quantity production. This type of pattern (Fig. 2a) is most appropriate for experimental or prototype castings and is only rarely used for short-run production castings. Molding with a loose pattern requires more manual operations and a much higher degree of molder skill than molding with other pattern types. Many foundry molding departments are not set up to use loose patterns. This increases the cost per mold and results in variations in casting quality from mold to mold compared to moldmaking with other pattern types. A loose pattern can be made in one piece or in two pieces (split patterns). The split in the two-piece pattern corresponds to the parting line in the pattern. One-piece loose patterns are often used when one side of the pattern is a flat plane that can serve as the parting line, while two-piece loose patterns are more commonly used when the shape to be cast is more complex. Gates and risers for loose patterns can be molded by hand or can be constructed with loose pieces and molded with the pattern.

Molding with one-piece loose patterns often requires considerable time and skill on the part of the molder. The pattern is placed or mounted on a flat mold board or a shaped follow board to support the pattern during molding and to define the parting line. The drag, or bottom, half of the mold is molded first, the mold is turned over, and the mold board is removed. Next the cope, or top, half of the mold is made on top of the drag, the cope is removed, and the pattern is removed from the drag. The gates and risers are then manually cut into the sand mold, and the completed mold is closed. The riser contact can be attached to the pattern so that the mold cavity surface is not damaged

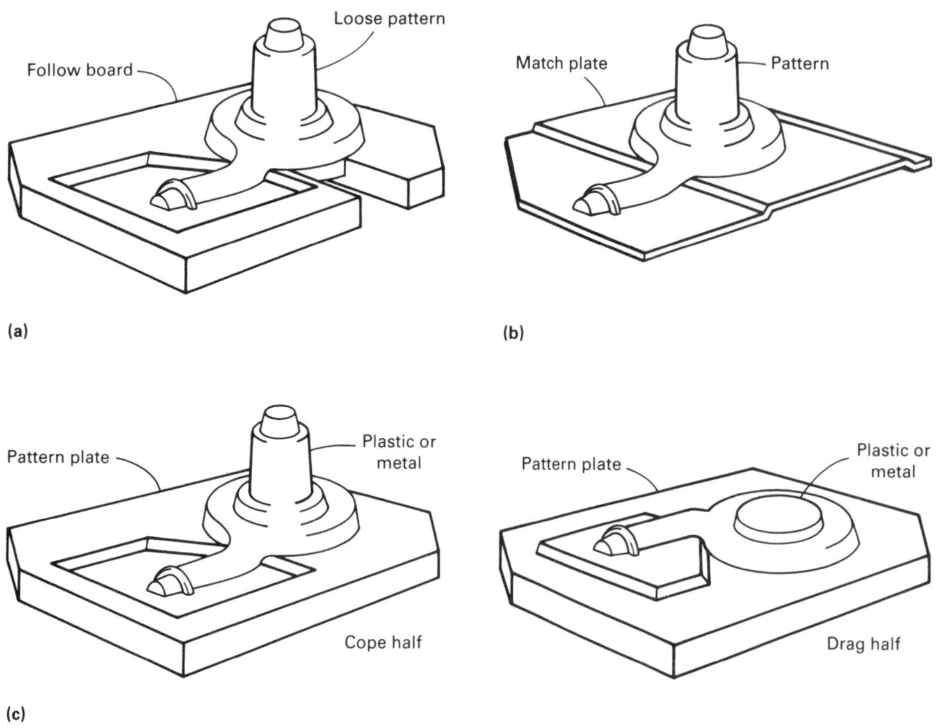

Fig. 2 Three types of patterns used to produce a water pump casting in various quantities. (a) Wood pattern on a follow board, good for 20 to 30 castings. (b) Match plate pattern for up to 50 000 castings depending on material. (c) Cope and drag plastic (up to 20 000 castings) or metal (100 000 or more castings) pattern

when cutting the gates and risers. A flask is usually placed around the completed mold to ensure alignment of the mold halves and to prevent the escape of molten metal during pouring. Two-piece loose patterns are molded in a similar manner except that the split pattern usually incorporates a flat parting plane that simplifies the molding of the cope and drag. Cope and drag mold halves can be made separately, and the split pattern can be removed from the respective mold halves. In this case, care must be taken to ensure alignment of the mold cavities at the parting line.

Match plate patterns (Fig. 2b) are split patterns in which the cope and drag portions are mounted on opposite sides of a plate, called the match plate, conforming to the parting line. The pattern, as well as the associated gating and risering system, is usually made separately and then mounted on the match plate, but can be cast integrally with the plate. The size of the match plate corresponds to the size of the flask used to make the final mold. Flask pin guides are used to ensure accurate alignment of the match plate pattern in the flask and to assist in pattern removal from the mold. Multiple patterns of small parts can be mounted on a single match plate. Common gates and risers on the match plate can often be shared by the multiple patterns on the match plate. Interlocking features can be added to the cope and drag sides of the match plate to ensure accurate alignment of the cope and drag mold halves after removal of the pattern.

Match plate patterns are used for moderate to high-volume production of small- and medium-size castings with considerable dimensional accuracy. The molding operation is simplified considerably by the use of match plate patterns. The decrease in molding costs and increased mold quality offset higher pattern costs for higher-volume castings. Large patterns are usually not match plate patterns, because of the limitations on flask sizes and the difficulties of molders or molding machines in handling large match plate patterns. Because significant pattern wear may occur over the life of a high-production match plate pattern, the selection of pattern material with adequate wear resistance is important. The match plate pattern design often must be rigid enough to withstand high squeeze pressures during automated molding (Ref 3, 4).

If the pattern has a flat parting surface, cope and drag halves can be made separately and fastened to the plate. However, most metal match plate patterns, whether they have a flat parting surface or not, are cast integrally with the plate. In integral casting, the cope and drag mold halves are shaped from a master pattern and then aligned so that the mold cavities face each other but remain separated by a distance equal to the thickness specified for the metal plate. When molten metal is poured, it fills the pattern halves of the mold (usually a plaster mold) and the space that separates them. The resulting pattern is a metal plate with the cope half of the pattern on one side and the drag half on the other. Even when precision molding methods are used to cast a match plate pattern, considerable straightening, machining, and polishing are usually necessary before a metal match plate pattern is ready for use.

Cope and drag patterns (Fig. 2c) are similar to match plate patterns except that the cope and drag portions of a split pattern are mounted or integrally cast on separate plates. Cope and drag patterns are preferred over match plate patterns for high-volume production or for the production of large castings. Pattern cost is higher than for match plate patterns, but the total molding cost per casting may be lower because the cope and drag mold sections can be made simultaneously. Automated vertical parting line molding machines require cope and drag pattern plates because a cope and drag impression is simultaneously made on opposite faces of the same block of sand. Successive blocks of sand are then pushed together side by side. The mold cavity is formed from the cope impression from one block of sand and the drag impression from the next block of sand. Thus, the equivalent of one mold cavity can be made for each block of sand (see the article "Sand Molding" in this Volume). Because the cope and drag are mounted on separate plates, accurate alignment of the two patterns is essential. The design and manufacture of cope and drag patterns are identical to those of match plate patterns.

Special Patterns. A wide variety of special pattern types and special patternmaking techniques can have specific uses in the foundry. Master patterns are used for casting metal patterns and for making plastic patterns. Multiple patterns are cast from the master to make match plate or cope and drag patterns. Special dimensional allowances must be made when making master patterns (see the section "Pattern Allowances" in this article). A wide variety of techniques and pattern materials can be used to create the master pattern and to form the production patterns from the master pattern. Reverse master patterns (negative patterns) are also used with certain pattern materials to create production patterns. The basic steps for producing production patterns from master patterns are illustrated in Fig. 3.

A pattern need not always be a complete replica of the final casting. Molds for very large castings are often made from an assembly of chemically bonded sand mold sections (cores) made in numerous core boxes. This is referred to as an all-core mold. In this case, a single pattern is not required; rather, the pattern is made up of a core box for each section of the mold. These large mold assemblies are made on the foundry floor or in large pits built into the foundry floor (Ref 5). For example, the mold for a large gear can be made from a sectional pattern or core box that is a

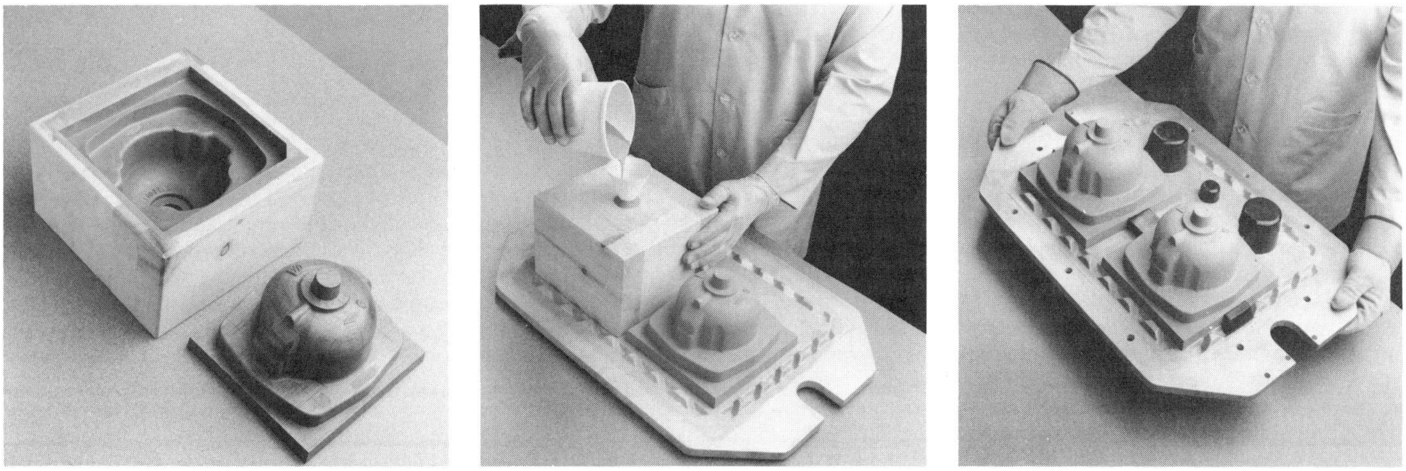

(a) (b) (c)

Fig. 3 Steps in producing a match plate pattern from a master pattern. (a) Wood master pattern is constructed with appropriate shrinkage factors included. A reverse from the master pattern. (b) Reverse mold is used to cast the polyurethane pattern on the aluminum match plate. (c) Reverse mold is removed, and a wood gating system is added. Gating can also be cast on with the pattern or made from aluminum stock. Courtesy of W. Allen, Allen Pattern of Michigan

wedge-shaped segment of the hub, spokes, rim, and teeth. The complete gear mold is assembled from successively made wedge segments. In rare cases, molds can be made by template-shaped patterns that are rotated around a spindle to form a mold surface (called a sweep) or translated along an edge (called screeding) to form an irregular shape. Both sweeping and screeding are manual moldmaking techniques that have been largely replaced by molds that are assembled from core sections produced in core boxes.

Core Boxes. Cores are primarily used to form internal passageways or cavities in castings. Cores can also be used to form sections of the mold for a cast shape that has external projections or negative draft, which would prevent the pattern from being withdrawn from the mold. Core boxes are the reverse patterns or reverse molds used to make the cores that become part of the final mold assembly. The term pattern equipment refers to the pattern itself and to all of the core boxes needed to produce the cores that will be part of the final mold assembly. Because a given casting may have a complex internal shape or many internal passageways, the cost of producing core boxes may exceed that of the associated mold pattern. An example of an intricate full split aluminum core box used in a core blowing machine is shown in Fig. 4. Because many coremaking processes require the use of a gas catalyst to cure the core and binder, special venting considerations must be included in the core box design. Ejection pins can also be included to aid in removing the core from the core box halves.

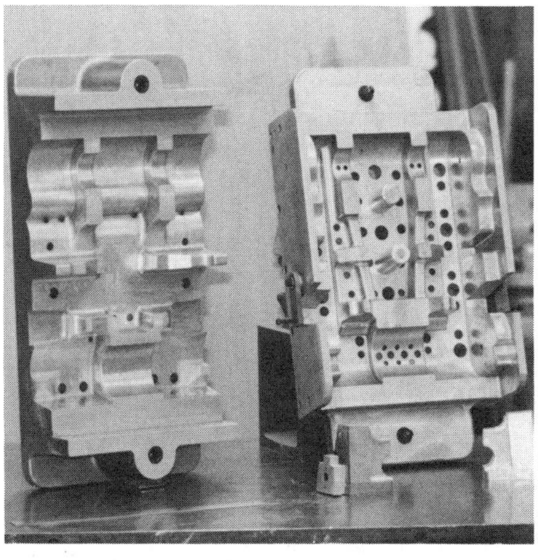

Fig. 4 Full split aluminum core box designed to eliminate pasting of core halves. From *Steel Castings Handbook*, 5th ed., Steel Founders' Society of America, 1980

The type of core box used and the material used to make a core box can vary greatly, just as for patterns. Core box construction depends on the type of coremaking process used and the number of cores required. Sand cores that develop strength during the high-temperature curing of thermal-setting binders (shell hot box and warm box cores) require heat-conducting metal core boxes. (Core sand binder systems and coremaking machines are discussed in the article ''Coremaking'' in this Volume.) Core boxes can also be subjected to considerable abrasive wear and pressure when core blowing machines deliver sand into the boxes at high air pressures. If cores are symmetrical, they can be pasted together from two core halves produced in a less expensive single half-core box. However, if higher tolerances and higher production rates are required, full split core boxes are used. As for patterns, the lowest cost core box is not necessarily the most economical. Proper casting design to minimize the use of cores is important not only for minimizing core box cost but also for decreasing molding costs and the overall cost of a casting. Casting design to minimize or eliminate cores is discussed in the Section ''Design Considerations'' in this Volume.

Expendable Patterns. One-piece patterns made of low-melting-temperature materials such as wax or materials that readily vaporize such as expanded polystyrene (EPS) can also be used as inexpensive expendable patterns. One such pattern is needed for each casting to be made.

Two common casting techniques that use expendable patterns are investment casting and evaporative process casting. In investment casting, a wax pattern is made in metal dies and coated with multiple layers of a ceramic or plaster slurry that hardens to form a thin rigid mold. The mold is heated to fire the ceramic and to melt out the wax, leaving an empty mold cavity of the desired shape.

Close-tolerance precision castings can be made using the investment casting process (see the article ''Investment Casting'' in this Volume).

In evaporative process casting, an EPS pattern is coated with a thin refractory slurry and placed in an unbonded sand mold. When molten metal is poured into the mold, the advancing metal vaporizes the polystyrene and fills the resulting cavity (see the discussion on evaporative foam casting in the article ''Sand Molding'' in this Volume). Evaporative process casting can be used for prototype or one-of-a-kind castings as well as for high-production castings. Prototype patterns are fabricated from stock foam pieces that are shaped and glued together. Production patterns are made by expanding polystyrene beads to final shape in multiple-piece metal dies. Expanded polystyrene has also been used to anchor multiple cores together in complex core assemblies prior to placement in a mold cavity (Ref 6). During pouring of the mold, the EPS pattern vaporizes, leaving the multiple sand cores in their proper position. This helps to ensure accurate placement of the cores.

Even though a new pattern is necessary for each casting made using expendable pattern processes, there are unique advantages to these processes. Casting design flexibility is greatly increased because the pattern does not have to be designed with parting lines, draft allowances, cores, or other features that are necessary when reusable patterns have to be removed from the mold to create the mold cavity. Complex multiple-piece patterns and sometimes the attachment of gates, risers, and runners are usually accomplished with the aid of adhesives.

Specialized molding presses and auxiliary molding equipment are used to make both wax and EPS patterns. The molds, or negative patterns, are usually machined from cast aluminum. The parting lines for wax and EPS pattern molds must be machined to closer tolerances than those generally required for green sand patterns in order to minimize parting line flash on the pattern. It is difficult to compare overall costs between EPS patterns and reusable patterns. Expanded polystyrene pattern costs are often offset by reduced molding and coremaking costs. A further discussion of wax and expendable polystyrene pattern materials can be found in the section "Expendable Pattern Materials" in this article.

Dies and permanent molds are reusable reverse patterns into which molten metal is poured to make a casting. They have much in common with core boxes and with the reverse patterns or dies used to make wax or EPS expendable patterns. Many of the pattern characteristics described in this article for sand casting with reusable patterns also apply to the design and manufacture of permanent and semipermanent mold and dies used in the permanent mold casting process and the die casting process. These casting processes are discussed in the article "Permanent Mold Castings" in this Volume.

Additional Features of Patterns

The primary purpose of the pattern is to produce a mold cavity of the desired shape. However, to produce successful castings, additional pattern features and pattern design considerations are also important. These include parting line considerations, patterns for gates and risers, core prints, and locating points. Pattern allowances for ensuring a dimensionally correct final pattern are discussed in the next section of this article.

Parting Line. It is necessary to separate (part) the completed mold halves and to withdraw the pattern from the mold for most sand casting processes. Selection of the most convenient and economical pattern parting line (mold parting surface) and design of the pattern so that it can be readily removed from the mold are important pattern design considerations. On a flat pattern plate, the parting surface is a simple plane. Many complexly shaped castings, however, require irregular parting surfaces to facilitate mold parting. The pattern match plate or cope and drag plates are used to establish the parting surface. Irregular parting surfaces can be readily incorporated with these types of patterns. (The term parting line is also used to refer to the very small fin (flash) or line left on the final casting at the parting surface where the two mold halves join.) Some complex pattern shapes cannot be readily removed from the parted mold halves unless the casting is redesigned or cores are used. By selecting the best parting line for a given casting to facilitate pattern removal from the mold, the use of cores to modify the external shape of the casting can be minimized (see the Section "Design Considerations" in this Volume).

These important considerations of mold parting surface and pattern withdrawal from the mold are of concern only when using reusable patterns. Expendable pattern molding processes use one-piece molds for which parting lines and pattern withdrawal are not of concern. This can result in additional casting design flexibility when using the expendable pattern casting process or investment casting techniques.

Gates and risers (rigging) are important factors in the production of quality castings and should be fully considered in the design and construction of patterns. Gating provides for the efficient flow of the molten metal into the mold cavity. Risers are large reservoirs of molten metal attached to the casting. They provide sufficient liquid metal to the casting as it solidifies to prevent internal shrinkage cavities or shrinkage defects in the casting. Gates and risers are commonly attached directly to the pattern and pattern plate and are molded with the pattern. For prototype castings, gates and risers can be made from loose pieces or cut into the molding sand by hand. On production match plate patterns, the gates and risers can be integrally cast on the pattern plate. The proper design of pattern gates and risers depends on the casting shape, the type of metal to be poured, and the molding and casting processes to be used. The location, size, and shape of gates and risers should be determined by an experienced foundry engineer, possibly with computer software design assistance. This is often a trial-and-error process, especially for complexly shaped castings. Pattern gating and risering is often modified after initial castings are made and evaluated, not only to produce sound, dimensionally accurate castings but also to increase casting yield efficiency.

Core Prints. When the production of a cast shape requires cores, provision must be made on the pattern for core prints. Core prints are portions of the pattern that locate and anchor the core in the proper position in the mold. The core print is added to the pattern (and removed from the core box), but does not appear on the final casting, because it is filled with the core itself. Core prints are illustrated in Fig. 1.

Locating Points. The foundry, the pattern shop, and the machine shop all use locating points or locating surfaces on the casting to check casting dimensions. Subsequent machining operations can also use the locating points in establishing the position of the machined surfaces relative to the rest of the casting.

Pattern Allowances

Although a pattern is used to produce a casting of desired dimensions, it is not dimensionally identical to the casting. A number of allowances must be made on the pattern to ensure that the finished casting is dimensionally correct, to ensure that the pattern can be effectively removed from the mold, and to allow for cores to be firmly anchored.

Shrinkage allowance is the correction factor built into the pattern to compensate for the contraction of the metal casting as it solidifies and cools to room temperature. The pattern is intentionally made larger than the final desired casting dimensions to allow for solidification and cooling contraction of the casting. This allowance for contraction is sometimes called patternmaker's shrinkage. The total contraction is volumetric, but is usually expressed linearly. Table 1 lists commonly used pattern shrinkage allowances for various casting alloys. Because different shrinkage allowances must be used for the individual types of metals cast, it is not possible to use the same pattern equipment for different cast metals without expecting dimensional changes. It must be emphasized that these shrinkage allowances are only guidelines and can only be applied with considerable knowledge of the actual casting design and the foundry molding techniques to be used.

The patternmaker's shrink rule is a special scale that eliminates the need to compute the amount of the shrinkage allowance that must be provided on a given dimension. For example, on a 10.5 mm/m (⅛ in./ft) patternmaker's shrink rule, each meter (foot) is 10.5 mm (⅛ in.) longer, and each graduation on the shrink rule is proportionately longer than its conventional length. Double shrinkage allowances must sometimes be made if a master pattern is first

Table 1 Typical pattern shrinkage allowances for various casting materials

Patternmaker and foundry should be consulted before shrinkage is specified.

Alloy being cast	Allowance	Approximate shrinkage, %	Shrinkage allowance mm/m	Shrinkage allowance in./ft
Steel	1 in 64	1.6	15/7	3/16
Gray cast iron	1 in 100	1.0	8/4	1/10
Ductile cast iron	1 in 120	0.8	7/8	3/32
Aluminum	1 in 77	1.3	13/1	5/32
Brass	1 in 70	1.4	14/4	11/64
Magnesium	1 in 77	1.3	13/1	5/32

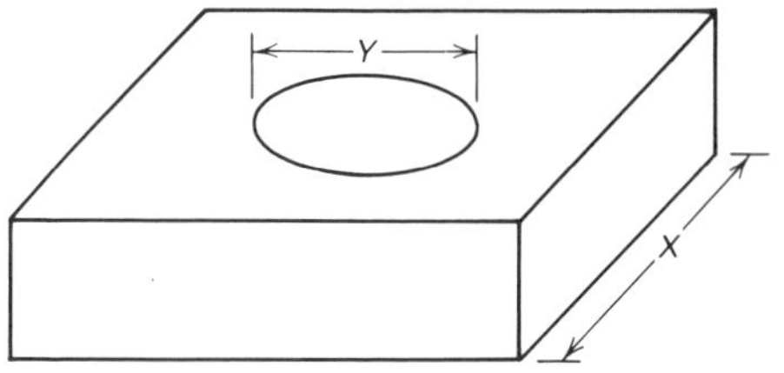

Fig. 5 Casting design that would require different shrinkage allowances than those normally used. Dimension *X* can contract freely, while dimension *Y* is restrained by the core used to make the hole.

made in wood and then used to make a metal match plate or cope and drag production pattern. For example, an aluminum pattern made from a wood master pattern would require a double shrinkage allowance on the wood pattern if a steel casting is to be made. The total shrinkage allowance on the wood pattern would then provide for the shrinkage of the aluminum pattern casting and of the steel casting made from the aluminum production pattern.

However, it is important to remember that pattern shrinkage allowances such as those listed in Table 1 cannot be used blindly to predict actual casting shrinkage. Other factors, such as casting shape, section thickness, and mold rigidity, significantly influence the actual change in dimensions as the casting solidifies and cools to room temperature. The resistance of the mold to the normal contraction of the casting and the use of different molding materials and methods may require that different shrinkage allowances be used for different patterns (Ref 7, 8). The same pattern used to make castings in a hand-rammed green sand mold, by high-pressure green sand molding, and in a high-strength chemically bonded sand mold will produce castings of varying dimensions because of the great differences in mold wall dilation during casting and solidification.

It is even possible that several different shrinkage allowances will be needed in one pattern, depending on constraint conditions. For example, in Fig. 5, two different contraction situations exist. Along dimension *X*, the casting has virtually no constraint to contraction, and the pattern should be made correspondingly larger along the *X* dimension surfaces. Dimension *Y*, however, is restrained from contraction by the core used to make the center hole and will require little or no shrinkage allowance on the pattern dimensions.

Because casting design and foundry methods have a great influence on the required shrinkage allowance and on the shrinkage characteristics of the metal itself, the direct involvement of the foundry producing the castings in the determination of the required shrinkage allowances is essential. Where tight dimensional tolerances are required, an excessive shrinkage allowance can be used when making the initial pattern. The pattern dimensions are then corrected when the precise shrinkage that actually occurred in prototype castings can be measured.

Draft is taper allowed on the vertical faces of a pattern to permit its removal from the sand or other molding medium without tearing of the mold walls. (On vertically parted molds, draft is required on horizontal pattern surfaces.) The amount of draft required depends on the shape and size of the casting, the molding process used, the method of mold production, and the condition of the pattern. A draft angle of approximately 1.5° (16 mm/m, or 0.2 in./ft) is often added to design dimensions. The draft angle may be higher when manual molding techniques are used. Interior surfaces usually require somewhat more draft than exterior surfaces, and deep pockets or cavities may require considerably more draft.

In special cases, draft can even be eliminated. Vertical walls can sometimes be drawn if the pattern is smooth and clean and if the drawing equipment is properly aligned. Both liquid and dry proprietary parting compounds can assist in drawing the mold from the pattern. Patterns for V-process molding (vacuum molding; see the discussion in the article "Sand Molding" in this Volume) can be made without draft because the smooth plastic film that covers the sand mold face offers almost no resistance to drawing the pattern from the mold (Ref 9). Expendable patterns, such as EPS patterns, can also be designed with straight walls (no draft) or even with negative draft because they do not have to be withdrawn from the mold. This can offer the casting designer greater shape flexibility without the use of expensive cores.

The machine finish allowance provides for sufficient excess metal on all cast surfaces that require finish machining. The required machine finish allowance depends on many factors, including the metal cast, the size and shape of the casting, casting surface roughness and surface defects that can be expected, and the distortion and dimensional tolerances of the casting that are expected. Accurate patterns combined with automated molding can often produce close-tolerance castings with a minimum machine finish allowance that can reduce final machining costs considerably. Table 2 lists suggested machine finish allowances. However, the machining allowance for a given casting may vary greatly, depending on customer requirements and casting process capabilities. More demanding dimensional requirements should be discussed with the producing foundry prior to specification.

Distortion Allowances. Certain cast shapes, such as large flat plates and dome- or U-shaped castings, sometimes distort when reproduced from straight or perfect patterns. This distortion is caused by the nonuniform contraction stresses during the solidification of irregularly shaped designs. Minor distortions are normally corrected by mechanically pressing or straightening the casting, but if distortions are consistent and prominent, the pattern shape can be intentionally changed to counteract the casting distortions. The "distorted" pattern will then produce a distortion-free casting. Prior consultation with the foundry is necessary to review their experience with the warpage distortion of similarly shaped castings.

Pattern Materials

A wide variety of both traditional materials and advanced composite materials are used in the manufacture of reusable patterns and core boxes (Ref 10-13). These pattern materials vary greatly in their physical characteristics, the techniques used for manufacturing and repairing the patterns, and the applications for which they are best suited. Pattern material selection is based on many factors, including the number of castings to be made, the dimensional accuracy required, the molding or coremaking process to be used, and the size and shape of the casting. General pattern standards concerning the use and specification of patterns and pattern materials have been published by the National Association of Pattern Manufacturers (Ref 14), but they are not very widely used. Table 3 lists some general characteristics and properties for four commonly used pattern materials: wood, aluminum, cast iron, and polyurethane. Improvements in pattern performance and durability often require that patterns be made of a combination of materials (such as wood with metal or plastic inserts

Table 2 Suggested pattern machine finish allowances

Pattern size, mm (in.)	Allowances, mm (in.) Bore	Surface	Cope side
For cast irons			
Up to 152 (6)	3.2 (1/8)	2.4 (3/32)	4.8 (3/16)
152–305 (6–12)	3.2 (1/8)	3.2 (1/8)	6.4 (1/4)
305–510 (12–20)	4.8 (3/16)	4.0 (5/32)	6.4 (1/4)
510–915 (20–36)	6.4 (1/4)	4.8 (3/16)	6.4 (1/4)
915–1524 (36–60)	7.9 (5/16)	4.8 (3/16)	7.9 (5/16)
For cast steels			
Up to 152 (6)	3.2 (1/8)	3.2 (1/8)	6.4 (1/4)
152–305 (6–12)	6.4 (1/4)	4.8 (3/16)	6.4 (1/4)
305–510 (12–20)	6.4 (1/4)	6.4 (1/4)	7.9 (5/16)
510–915 (20–36)	7.1 (9/32)	6.4 (1/4)	9.6 (3/8)
915–1524 (36–60)	7.9 (5/16)	6.4 (1/4)	12.7 (1/2)
For nonferrous alloys			
Up to 76 (3)	1.6 (1/16)	1.6 (1/16)	1.6 (1/16)
76–152 (3–6)	2.4 (3/32)	1.6 (1/16)	2.4 (3/32)
152–305 (6–12)	2.4 (3/32)	1.6 (1/16)	3.2 (1/8)
305–510 (12–20)	3.2 (1/8)	2.4 (3/32)	3.2 (1/8)
510–915 (20–36)	3.2 (1/8)	3.2 (1/8)	4.0 (5/32)
915–1524 (36–60)	4.0 (5/32)	3.2 (1/8)	4.8 (3/16)

Source: Ref 1

Table 3 Characteristics of pattern materials

Characteristic	Pattern material Wood	Aluminum	Cast iron	Polyurethane
Machinability	E	G	F	G
Wear resistance	P	G	E	E
Strength	P	G	E	F
Repairability	E	F	G	E
Corrosion resistance	E	E	P	E

E, Excellent; G, Good; F, Fair; P, Poor

on wear surfaces), coated materials, or composite materials.

Wood is widely used for both master patterns and production patterns because it is inexpensive and easy to shape (Ref 1). However, wood patterns, especially those constructed of pine, are susceptible to shrinkage or swelling and warpage due to changes in atmospheric moisture. Figure 6 shows how atmospheric humidity can affect the dimensions of woods used in patternmaking. Mahogany is a popular pattern material because it has good strength and wear resistance compared to pine and slightly more dimensional stability. As Fig. 6 shows, the ideal moisture content of lumber as-received is 5%. When the initial moisture content of unprotected mahogany or northern pine is increased by an additional 5% (due to an atmospheric humidity of 50%), a volumetric expansion of 1.5% results. At an atmospheric humidity of 75%, the volumetric expansion may be 3% or more. From these data, it is apparent that appreciable changes in humidity can result in loss of dimensional accuracy, which translates to a corresponding loss in the dimensional accuracy of castings. Proper construction of wood patterns can inhibit warpage somewhat.

Pattern and core box sections made of white pine can be joined with glue and nails; pattern sections made of hardwoods, which are more durable than white pine, are usually joined by bolts, screws, and glue (Ref 1). This construction resists the cyclic stresses set up by the expansion and contraction of the wood. Pattern sections are sawed from suitable lumber, machined, and sanded to the required dimensions (considerations for draft and shrinkage allowances will be discussed later in this article). Good construction also calls for three or more coats of pattern lacquer, which minimizes moisture absorption and the resultant volumetric changes in the pattern. Automotive primer paints can also be used to finish wood patterns.

Patterns can also be made from compressed wood laminates (primarily plywood). These hardwood laminates weigh about twice as much as untreated wood, but have a high strength-to-weight ratio. In addition to their strength, they are tough and have much more resistance to abrasion and moisture than untreated wood. They are readily machined with conventional wood patternmaking machinery, with some adjustment in speeds, feeds, and cutting tools. The life of wood laminate patterns can be considerably longer than that of conventional wood patterns. Wood laminate pattern sections are joined by screws, bolts, or epoxy resins or can be fully machined from a block of material. When subjected to high humidity for long periods of time, these patterns swell so slowly that dimensional changes and warping are negligible for most applications. Protective coatings such as lacquers make wood laminate patterns nearly impervious to the humidity conditions typical of foundry operations.

Metal or plastic inserts or pieces can also be a part of wood production patterns. These materials are often used in high-wear regions of the pattern to improve pattern life and durability. Repair of wood patterns is commonly done with wood inserts or plastic resins (epoxies or polyurethanes). Worn areas can be rebuilt and then finished to the desired final dimensions. Polyurethane pattern block materials (machinable plastic boards) are also available for producing master patterns with even higher dimensional accuracy, stability, and pattern life than wood. Patterns can be formed and constructed with standard woodworking equipment and techniques.

Metal pattern equipment is particularly well suited to long production runs. Compared to wood patterns, metal patterns and core boxes are more costly, stronger, more abrasion resistant, and are dimensionally stable under changing humidity. They can be fabricated for use with automated high-pressure molding equipment, can be made to high dimensional tolerances, and can be used for a large number of castings before pattern repair or rework is necessary.

Metal patterns and core boxes are normally made from aluminum alloys, gray iron, ductile iron, or steel. Metal patterns are required for shell molds and cores and molds from sands bonded with thermal-setting resins. Patterns used for this purpose are typically made of cast iron because aluminum tends to gall and wear quickly at elevated temperatures. The properties on which metal pattern material selection is based include wear resistance, dimensional stability, machinability, and the ability to provide a smooth surface finish after machining.

Metal patterns are either cast to shape in precision plaster or sand molds made from wood master patterns (cast to shape tooling) or machined directly from metal stock. Multiple-production pattern shapes can be made from one master pattern.

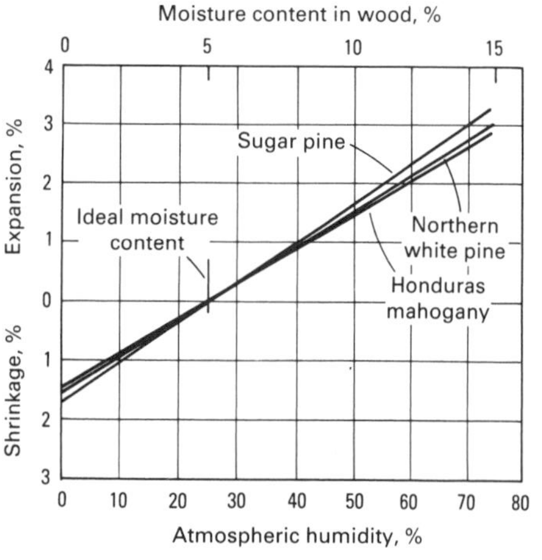

Fig. 6 Effects of moisture content of wood and atmospheric humidity on the expansion and shrinkage of three pattern woods

Metal patterns can be cast to shape in plaster (ceramic) or sand molds made from wood master patterns. The master pattern must be oversized to account for both the shrinkage of the metal pattern as it cools and the final shrinkage of the metal casting as it cools in the mold made from the metal pattern (double shrinkage allowance). After metal patterns are cast, gating and flash must be removed, and considerable machining, filling of surface imperfections, and polishing are required so that the final pattern surfaces are smooth and free of imperfections. These finishing operations may require 50% or more of the time needed for pattern construction (Ref 6).

Precision ceramic molding processes with dimensional accuracies of ±0.001 mm/mm (0.001 in./in.) can be effectively used to produce cast-to-size patterns with good as-cast surface finishes and minimum final machining (Ref 15). Even when precision molding methods are used to cast the original pattern plates or core boxes to close tolerances, some final machining and polishing may be needed. Aluminum pattern equipment is usually made by pressure casting processes to reduce the amount of finishing required.

Metal patterns and core boxes can also be machined directly from metal stock without the intermediate steps involved in producing a master pattern and the initial casting of semifinished pattern plates. The expanding use of computer numerical control (CNC) machining and computer-aided design data bases for parts dictates that direct machining of pattern equipment will continue to grow (see the section "Automation in Pattern Design and Manufacture" in this article).

The abrasion resistance of metal patterns can be improved even further by the use of abrasion-resistant coatings. Both electrolytically deposited hard chromium platings and electroless nickel platings can increase the life of metal pattern equipment. Polyurethane plastic inserts adhesively bonded to high-wear locations on metal patterns can also increase metal pattern life.

Plastic Pattern Equipment. The versatility, desirable properties, and cost effectiveness of plastic pattern equipment have resulted in the continued growth of plastics for patterns and core boxes. Plastics are primarily used for small- and intermediate-size patterns. Plastic patterns are more durable but are also usually more expensive than wood patterns. The use of plastic patterns and core boxes is increasing because plastic is replacing wood patterns in medium-volume applications and is replacing metal high-production pattern equipment. Multiple plastic patterns can be inexpensively made from one master pattern (usually wood). Plastic patterns have excellent dimensional stability and can be produced with less skill and expense than comparable metal patterns. They have been successfully used in high-pressure molding machines (Ref 11). In addition, the polyurethane and epoxy resins typically used for patterns have the following desirable characteristics:

- Better compressive strength, bending strength, abrasion resistance, and impact resistance than wood
- Resistance to environmental and chemical attack
- Good bonding to other pattern materials for repairs and inserts
- Easy release from molding sands

When properly made, plastic patterns shrink very little; therefore, the molds do not require double shrinkage allowance, as in the production of metal patterns from master patterns. This simplifies construction and results in a considerable cost savings for designs with intricate contours.

Epoxy patterns are suitable for small patterns and core boxes for limited production. Larger or higher-production patterns often require glass fiber reinforcement of the epoxy resin for increased strength and durability. Epoxy pattern life can also be extended with metal coatings. The polyurethane elastomers are used much more extensively than the epoxies and are becoming one of the most widely used pattern materials. The urethanes are more wear resistant and less brittle than the epoxies. They are used for complete patterns and core boxes, for loose pieces on wood or metal pattern plates, as a thin pattern face skin material, and as wear-resistant inserts on wood or metal patterns.

Techniques for producing plastic patterns depend on the type of pattern required, the type of resin system used, and the choice of reinforcement or backing, if any. Plastic patterns and gates and risers are usually mounted on metal pattern plates, but one-piece pattern and pattern plates can be manufactured. Because of the high cost of the resins and the distortion caused by the heat generated during the curing process, less expensive rigid backing materials are used whenever possible for medium and large patterns and core boxes. Metal reinforcing frames can also be used to stiffen and support large pattern plates for high-pressure molding equipment. Four common techniques for producing plastic pattern equipment are solid casting, casting with a core, various layup techniques, and direct machining of plastic board stock.

Solid patterns are manufactured using either a reverse master pattern or a reverse plaster mold made from a master pattern (usually wood). A sealer and a release agent are applied to the negative pattern or negative mold surface to prevent the resin from bonding to the mold. The resin system (usually two-part or three-part) is poured into the mold and allowed to harden. Curing times and temperatures vary greatly, depending on the resin and specific resin formulation used. Cured patterns can be readily stripped and mounted on pattern boards.

Fillers are often added to resins to reduce the amount of resin needed to make a pattern, to provide strengthening reinforcement, and/or to control shrinkage. A wide variety of mineral or metal powders are used as filler materials. Other low-density filler materials can be used when strength is not critical for large patterns. By reducing the amount of casting resin for a given volume with fillers, the temperature rise during the early stages of curing can be lessened to minimize overall pattern shrinkage and distortion.

A variation of the solid casting method (face casting) uses a 1 to 3 mm (0.04 to 0.12 in.) face or skin layer of resin that is allowed to cure partially. The remainder of the mold is then filled with a less expensive backing mix that is heavily reinforced with fillers such as silica sand. Thicker solid molds can also be built up in this way rather than with resins alone. Thin, brush-on surface (gel) coats can also be applied to wood patterns to produce inexpensive patterns with considerably longer pattern life than wood alone.

Patterns With a Core. In this method, an undersize core (plug) of wood, aluminum, plastic foam, or some other inert material is supported in the reverse pattern mold. The resin is poured or injected under pressure to fill the clearance space between the mold and the core and then cured to create the pattern surface. The use of the core minimizes the amount of resin needed for a given pattern volume, which minimizes shrinkage during curing. This method is most applicable for small- and medium-size castings of simple shape (Ref 16, 17).

Composite Patterns. A wide variety of techniques have been developed for building composite pattern equipment using plastic resins. Resin skins can be backed or supported by layers of glass fiber reinforcement. The rest of the mold is then filled with a suitable backing material using methods described previously.

Machined Patterns. Patterns, particularly master patterns, can also be fabricated from machinable polyurethane boards using the same patternmaking techniques and patternmaking tools used for wood patterns. The additional cost of using urethane stock rather than wood is justified when patterns must be made to close dimensional tolerances. Patterns of this type are usually machined using CNC machine tools. Large patterns of this type can be fabricated by using the machinable plastic board only as a facing material, along with a foam core or other backing material to support the urethane face to be machined.

Pattern Coatings. The performance and service lives of wood, metal, and plastic patterns can be improved with an appropriate wear-resistant coating. Standard elec-

Table 4 Pattern coating materials and methods used to enhance wear resistance

Coating/method	Applicability to pattern substrate material: Wood	Epoxy	Aluminum	Cast iron
Metal spraying				
Tin-zinc alloys	X	X	...	...
Tin-bismuth alloys	X	X	...	...
Aluminum-zinc alloys	...	X	...	...
Electrolytic plating				
Nickel	...	...	X	X
Chromium	...	X	X	X
Electroless plating				
Nickel	...	X	X	X

trolytic plating methods, electroless nickel plating, and metal spraying (metallizing) can all be used (Ref 18, 19). Table 4 illustrates the coating methods that are applicable for the various pattern substrates.

Figure 7 illustrates pattern wear rates for coated and uncoated pattern materials. In the investigation that yielded these data, laboratory sand erosion tests were conducted to evaluate the relative wear resistance of various coated and uncoated pattern materials. The test method was thought to simulate the impact abrasion that occurs on patterns from core blowers or other sand-blowing equipment. These results may not simulate other pattern abrasive wear conditions, because wear rates for different materials depend strongly on the angle of attack of the abrading particle. The relative wear resistance of the various pattern materials is indicated by the percentage of weight loss during abrasion testing. Materials with less weight loss are more abrasion resistant and would likely have longer pattern service lives. The formulations of the epoxy and urethane materials have also been improved since these data were obtained; this may affect the results indicated for these materials. The methods for applying metal coatings to the various pattern substrates will be briefly described. More detailed descriptions of these coating techniques are available in Volume 5 of the 9th Edition of *Metals Handbook*.

In electrolytic plating, the coating is electrolytically deposited on patterns negatively charged (anode) in aqueous solutions containing metal salts. Pattern coating thicknesses usually range from 0.05 to 0.13 mm (0.002 to 0.005 in.). Coating thickness may vary somewhat for a pattern, depending on the distance from the electrode.

In electroless nickel plating, a nickel alloy containing 8 to 10% P is chemically plated from a hot aqueous solution of nickel salts and activators. A typical plating cycle consists of 6 to 8 h in an 80 °C (180 °F) bath. The coatings produced have a high hardness and low coefficient of friction. Average

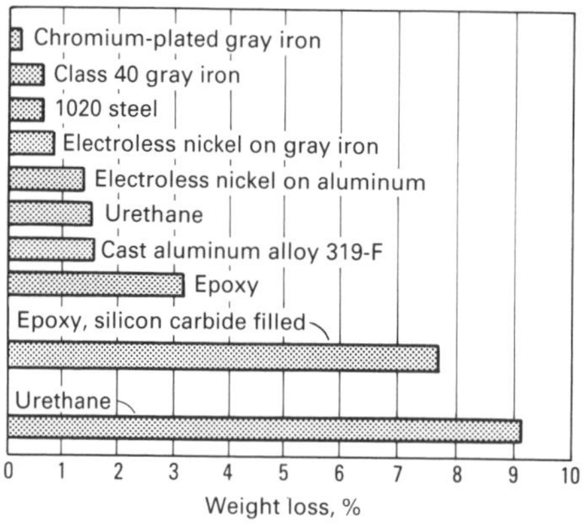

Fig. 7 Average weight losses (percent) of pattern materials tested for 12 h in silica erosion. Source: Ref 20

coating thicknesses range from 0.08 to 0.13 mm (0.003 to 0.005 in.) or more. The as-deposited hardness of electroless nickel can be substantially increased by subsequent heat treatment at temperatures between 290 and 400 °C (550 and 750 °F). Epoxy pattern equipment can also be electroless nickel coated.

Metal Spraying (Metallizing). Metal spraying is the most rapid and least expensive method of applying a thin metal coating on a wood, epoxy, or metal substrate. Because a sprayed metal coating can be applied uniformly or nonuniformly by directing the spray pattern, metal spraying can also be used to rebuild areas of patterns where excessive wear has taken place. Arc spray or plasma spray guns are used to spray atomized liquid metal onto a pattern. Low-melting alloys can even be sprayed on wood patterns. Coating thicknesses of 1.5 to 2 mm (0.060 to 0.080 in.) are common except when pattern repair necessitates thicker coatings. Sprayed metal coatings are often more nonuniform than other plated coatings and therefore may require more finishing and polishing. Metal spraying can also be used to coat a negative mold (usually epoxy coated with a release agent), which is then backed up with cast epoxy. The metallized epoxy pattern is then removed from the negative mold.

Pattern Repair and Rebuilding. Pattern rebuilding must be carried out on a regular basis to repair patterns that have been damaged by use or abuse and to build up worn areas of the pattern that are no longer producing castings within tolerance. The rebuilding procedures used depend on the pattern material and the service life required for the rebuilt pattern. Epoxies and urethanes are widely used to rebuild all types of pattern materials. Metal spraying or weld overlaying can also be used to build up worn areas of production metal patterns.

Pattern Handling and Storage. Because significant investments are made in pattern equipment, pattern handling and storage must be carefully considered by the foundry. Improper storage and handling can cause pattern deterioration and damage, particularly for wood patterns because humidity and temperature changes can cause severe distortion. The foundry typically has many patterns in inventory. Obsolete or discontinued patterns should be routinely destroyed to make room for new patterns.

Expendable Pattern Materials. Wax patterns used for investment casting and EPS patterns used for the evaporative casting process are the two most common expendable pattern materials. Aspects of these pattern materials and pattern processing are discussed in the following sections of this article. Details on the investment casting and evaporative foam casting processes are available in the articles "Investment Casting" and "Sand Molding" in this Volume.

Wax Patterns. Various types of specially formulated waxes are used to make wax patterns. Patterns are made by injecting liquid or semisolid wax into a multiple-piece pattern die. The wax is allowed to solidify, the die is parted, and the wax pattern is removed. Injection waxes are usually blended from selected waxes, resins, and modifiers (Table 5). A typical pattern wax formulation might be:

Ingredient	Composition, %
Hard wax(a)	40
Microcrystalline wax	25
Soft resinous plasticizers(b)	15
Hard resins(b)	20
Antioxidant	0.05

(a) Microcrystalline, amorphous. (b) Amorphous

Many important properties and characteristics of pattern waxes are measured and controlled, including the following:

- Solidification temperature range (freezing range)
- Ash content
- Viscosity
- Strength and hardness
- Softening temperature

Depending on the number of patterns to be cast at one time and the complexity of the wax pattern dies, some wax pattern assembly may be required in order to attach the gating system and sprue. Patterns are typically joined to the feeding system by dip sealing to create pattern clusters surrounding a central sprue or tree.

Expanded Polystyrene Patterns. Expanded polystyrene is polystyrene containing a hydrocarbon blowing agent (usually pentane). When heat is applied, the dense polystyrene beads soften, and the blowing agent vaporizes to form a low-density foam. Patterns for prototype or short-run castings are machined and assembled from EPS shapes and adhesively bonded together.

Table 5 Materials used in formulating waxes for expendable patterns

Material	Melting point M or softening point S, °C (°F)
Hard waxes	
Vegetable wax (candelilla)	M: 67 (152)
Vegetable wax (carnauba)	M: 89 (192)
Mineral wax (montan)	M: 86 (187)
Synthetic wax (nonchlorinated)	M: 85.5 (186)
Synthetic wax (chlorinated)	M: 93 (199)
Microcrystalline waxes	
Petroleum derivative	M: 79 (175)
Petroleum derivative	M: 63 (145)
Beeswax, USP	M: 64 (147)
Soft resinous plasticizers	
Rosin derivatives	S: 43–71 (110–160)
Terpene resins	S: 40 (104)
Coal-tar resins	S: 25–75 (77–167)
Petroleum hydrocarbon resins	S: 70 (158)
Chlorinated resins	S: 60–73 (140–163)
Elastomers	S: Wide range
Hard resins	
Rosin derivatives	S: 138–193 (280–380)
Terpene resins	S: 85 (185)
Coal-tar resins	S: 70–107 (158–224)
Petroleum hydrocarbon resins	S: 95–100 (203–212)
Chlorinated resins	S: 98–110 (208–230)
Modifiers	
Synthetic wax (nonchlorinated)	S: 70–105 (158–221)
Elastomers	S: 149 (300)
Polyethylene resins	S: 99–108 (210–226)

Production quantities of EPS patterns are molded to shape in equipment similar to that used to make wax patterns. The use of the evaporative foam casting process is rapidly increasing (Ref 21-25).

The production of EPS patterns for the foundry involves two major processing steps: preexpansion and molding. During preexpansion, the EPS beads are heated to soften them, and the blowing agent vaporizes, causing the bead to expand to a predetermined bulk density close to that of the final pattern bulk density. The preexpanded beads are blown loosely into a pattern cavity, where additional heat (steam) is again used to soften the beads and cause them to expand further to fill the void spaces between the beads and to bind them together. The molded pattern is then cooled to harden the pattern for removal from the die cavity. Patterns are often made from pattern pieces that are adhesively bonded together.

The final bulk density of the EPS pattern is a very important consideration. Increasing the bulk density increases the strength, durability, and dimensional stability of the pattern during storage. However, low pattern bulk densities are usually required to minimize the amount of carbon picked up by the molten metal when it is poured into the mold over the EPS pattern to make a casting (particularly for cast steel). Common EPS pattern bulk densities used for the various cast metals are:

Metal cast	Recommended EPS pattern density, kg/m³	lb/ft³
Aluminum	24.0–25.6	1.5–1.6
Brass, bronze	20.0–21.6	1.25–1.35
Gray iron	20.0	1.25
Steel	17.6	1.10

The pattern dimensions must take into account the fact that the EPS patterns themselves will shrink with age. Patterns will shrink a maximum of 8.3 mm/m (0.100 in./ft) over a period of 30 days. The dimensional variations that take place during the first 2 days after pattern manufacture are shown in Fig. 8. Metal shrinkage in evaporative foam casting is generally the same as for other conventionally bonded sand molds and must be considered along with the pattern shrinkage when designing the EPS mold. Shrinkage of EPS patterns depends on many factors, including the time period between pattern fabrication and pattern use, the size and bulk density of the foam, the section size, the pentane (expander) content, and the storage environment. Expanded polystyrene patterns are coated with a thin, ceramic slurry coating that is allowed to dry before molding.

In a variation of the evaporative foam process known as the ceramic shell or Replicast process, a heavier ceramic shell is built up around a high-density EPS pattern (Ref 24). As with wax patterns, the EPS pattern is then removed by firing before the molten metal is poured into the ceramic shell. This prevents carbon pickup in the molten metal before solidification that cannot be avoided with conventional evaporative foam techniques. More information on the ceramic shell process is available in the article "Replicast Process" in this Volume.

Selection of Pattern Type

The type of pattern or core box used and the pattern or core box material used for a given set of castings depend on the following fundamental factors:

- The number of castings to be produced
- The molding or coremaking process to be employed
- The casting design
- The dimensional tolerances required

The life and cost of a pattern can both vary dramatically, depending on the pattern material and the type of pattern equipment.

In the developmental stage of a pattern design, only a few prototype castings need to be produced before modifications are made to the pattern dimensions or to the gating and risering. If such revisions are likely, an inexpensive wood pattern is often used first. This will enable engineering changes to be made quickly and inexpensively. After the design and the tolerances of the casting have been approved, a permanent pattern is selected based on production quantity and the molding or coremaking processes to be used.

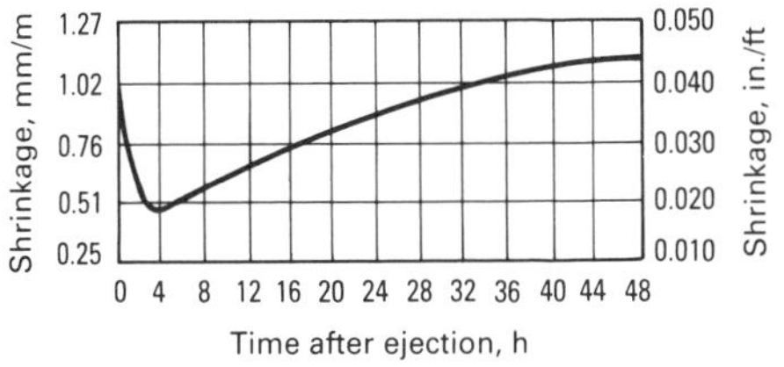

Fig. 8 Shrinkage of EPS patterns after ejection from the pattern mold. Patterns are approximately 1.27 mm (0.050 in.) undersize when ejected from the mold. Source: Ref 21

Costs that are influenced by pattern equipment selection depend largely on the pattern material and pattern type and are dictated by production quantity. Expensive pattern equipment can often be justified if production quantities are high. The complexity of the pattern and the quality of the material used to make the pattern generally increase with the number of castings to be produced from one set of patterns. For example, an unmounted or loose softwood pattern could be used only for very limited production before it would require repair or replacement. A similar pattern made from a more durable material and mounted on a pattern board would increase the useful life of the pattern dramatically. Table 6 lists the basic types of patterns, indicating typical pattern materials and pattern lives for each during automatic green sand molding. Figure 9 indicates the general relationship between casting cost and production rate for various reusable pattern materials.

Foundries also use various molding processes, casting processes, and coremaking processes. Therefore, during molding, patterns and core boxes can be subjected to differing degrees of abrasion, temperature, and stress that affect the pattern type and choice of pattern material.

For example, the stresses that the pattern encounters during automated high-pressure green sand molding necessitate the use of high-strength pattern materials and rigid pattern construction. When large green sand molds are made with sand slingers, the rapid abrasive wear of softer pattern materials is a major concern. Shell molds or cores can be made only from metal patterns because they are heated to temperatures of approximately 260 °C (500 °F). In V-process molding, a thin polyethylene sheet prevents the sand mold from contacting the pattern during molding, and this allows inexpensive wood patterns to have an almost indefinite pattern life.

Tolerances. The type of pattern used also has a significant effect on the casting tolerances that can be obtained and maintained. In general, final casting tolerances can be held within tighter limits as the rigidity and durability of the pattern equipment in-

Table 6 Approximate life before repair of various pattern materials

Approximate number of castings produced before pattern repairs: Pattern	Core box	Materials
Small castings (largest dimension: ≤610 mm, or 24 in.)		
200	300	Hardwood patterns and core boxes
2000	2000	Hardwood patterns and core boxes, wearing surfaces faced with metal
6000	6000	Aluminum patterns and core boxes; plastic match plate patterns; urethane patterns or urethane-lined magnesium-framed core boxes
100 000	100 000	Cast iron patterns and core boxes
Medium castings (largest dimension: ≤1.8 m, or 72 in.)		
100	100	Hardwood patterns and core boxes
1000	750	Hardwood patterns and core boxes, wearing surfaces faced with metal
3000	3000	Aluminum patterns and core boxes; urethane inserts in wear areas
Large castings (largest dimension: >1.8 m, or 72 in.)		
50	50	Softwood patterns and core boxes
200	150	Softwood patterns with exposed projections metal faced; softwood core boxes, metal faced
500	500	Hardwood patterns reinforced with metal; hardwood, metal-faced core boxes

Source: Ref 8

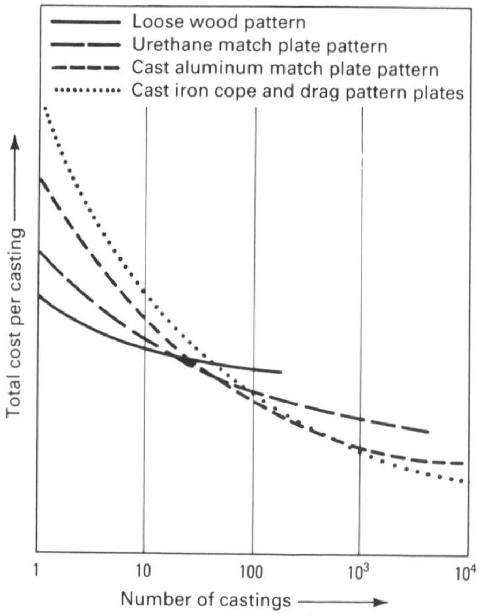

Fig. 9 Generalized casting costs versus production quantity for four pattern materials

crease. When close tolerances are required, it may be desirable to construct the pattern equipment so that it can be dimensionally adjusted based on the results of prototype castings. These adjustments may involve refitting and remachining.

It is very difficult to compare the cost of castings made from EPS patterns to similar castings made from permanent patterns. The additional casting design flexibility and reduced molding costs associated with the evaporative foam casting process may offset high pattern and pattern die costs, resulting in lower total casting cost.

Automation in Pattern Design and Manufacture

The continued development of computer-aided design and manufacturing technology has had a direct effect on both pattern design and patternmaking methods (Ref 26-29). The influence of CAD/CAM technology will continue to grow as it becomes more affordable for the small foundry and the pattern shop. Maximum benefit can be achieved with CAD/CAM technologies if pattern design, inspection, and manufacture are fully integrated.

This integrated patternmaking approach begins with a geometric data base of the part to be cast. The data base may already exist or may have to be constructed from drawings. The pattern designer adds a parting line and draft and machining allowances at his CAD workstation. The part dimensions can be automatically scaled up to the designed pattern dimensions that included the shrinkage allowance by inputting the appropriate scale factor. At the current stage of development of CAD software systems, necessary pattern features such as the parting line and draft must be added interactively by the pattern designer. To complete the pattern design, appropriate gates and risers can be added by using one of a number of software packages (see the articles "Riser Design" and "Gating Design" in this Volume). These procedures result in a software pattern geometry data base that can be easily used and modified.

Solidification modeling software can be used to model the solidification of the final resultant casting from geometry data alone before even prototype patterns are made (see the articles in the Section "Computer Applications in Metal Casting" in this Volume). Pattern dimensions, gating, and riser design can all be evaluated and readily modified by adjusting the pattern geometry data base. The use of solidification modeling programs can dramatically reduce or eliminate pattern prototyping stages. A prototype pattern and final pattern design changes and/or modifications may still be necessary, but these can also be easily performed by adjusting the pattern data base. For example, a different pattern shrinkage could be readily accommodated by rescaling the data base model.

Numerical control (NC) machining of patterns to very close tolerances can also be readily performed using the pattern geometry data base. Numerical control programs for automatic machining of the desired pattern shape can be automatically or interactively generated from the pattern data base. The NC part programmer can retrieve the pattern geometry model and use that model to construct the appropriate cutter paths for pattern machining. Graphical simulation of the NC machining operation on the CAD terminal can be used to verify the correctness of the NC part program before actual pattern machining takes place. Any future modifications to the pattern geometry can be readily incorporated. Changes are made to the pattern data base, and the modified NC output is used to modify the appropriate pattern dimensions. Multiple impression tooling can be made with the assurance that each impression is dimensionally accurate. An example of the benefits that can be obtained using CAD/CAM patternmaking techniques compared to conventional patternmaking techniques is shown in Table 7. A 17% reduction in the time necessary to produce a final pattern is shown.

Complete NC machining of patterns will continue to expand as the benefits of CAD/CAM technology continue to be exploited. The close dimensional tolerances and complex geometries required on many patterns are more suited to NC pattern machining than to manual construction by patternmakers and machinists. Numerical control machining can be used to machine master pattern equipment from which production patterns and core boxes are made. However, it is more common for production patterns and core boxes to be machined directly from NC data, with the NC pattern data

Table 7 Comparison of traditional and CAD/CAM patternmaking times for an aluminum investment die

Technique	Time
Traditional methods (Accuracy of pattern: ±0.01 in.)	
1) Design	50 h
2) Wooden model	50 h
3) Duplicating	40 h
4) Other machining	85 h
5) Polishing and finishing	40 h
Total	**265 h**
CAD/CAM methods (Accuracy of pattern: ±0.002 in.)	
1) CAD/CAM design	40 h
2) NC cutter path development	40 h
3) CNC machining	25 h
4) Other machining	85 h
5) Polishing and finishing	30 h
Total	**220 h**

base itself serving as the master pattern. Although CAD/CAM has made inroads in the tooling and die industries, the technology is just beginning to be used for the design and manufacture of metal casting patterns.

Verification of pattern and casting dimensions is also a tedious process when traditional manual layout techniques are used. An automated coordinate-measuring machine is an extremely accurate inspection tool that can dramatically improve the speed and reliability of pattern inspection (Ref 27). Such machines provide absolute measurement capabilities in three dimensions simultaneously; this eliminates the need to make tedious comparisons between the pattern and gage blocks or other length standards. With data processing capabilities, the pattern need not even be aligned with the axes of the coordinate measuring machine, but is digitally aligned. Data storage and printout capabilities allow easy measurement of many important pattern features. For example, cope and drag pattern plate alignment can be readily determined from comparisons of the locating pin and bushing positions for each pattern plate. Pattern dimensions or pattern inserts can also be inspected at specific intervals during pattern use with little downtime for monitoring pattern wear.

REFERENCES

1. *Patternmaker's Manual*, American Foundrymen's Society, 1986
2. E. Hamilton, *AFS Patternmaker's Guide*, American Foundrymen's Society, 1976
3. W.A. Blower, Pattern Design for High Pressure Molding, *Trans. AFS*, Vol 78, 1980, p 313-316
4. Patterns for High Pressure Molding, *Foundry*, Vol 98 (No. 10), Oct 1971, p 83-84
5. E.L. Kotzin, *Metalcasting and Molding Processes*, American Foundrymen's Society, 1981
6. R.L. Allen, Cold Box Design Specializing in Urethane Lined Tooling, *Trans. AFS*, Vol 85, 1977, p 323-326
7. D.R. Dreger, Smooth No-Draft Castings, *Mach. Des.*, May 25, 1978, p 63-65
8. *Steel Castings Handbook*, 5th ed., P.F. Weiser, Ed., Steel Founders' Society of America, 1980
9. J.L. Gaindhar, C.K. Jain, and K. Subarathnamatah, Prediction of Pattern Dimensions for V-Process Precision Castings Through Response Surface Methodology, *Trans. AFS*, Vol 94, 1986, p 343-349
10. R. Brown, Plastic Patterns for High Pressure Molding, *Mod. Cast.*, Vol 73, Nov 1983, p 41-43
11. J.W. Francis, Practical Patternmaking Techniques for the Foundry Industry, *Br. Foundryman*, Vol 73 (No. 9), 1980, p 258-264
12. J.W. Francis, Practical Patternmaking Techniques for the Foundry Industry, *FWP J.*, Vol 24 (No. 6), 1984, p 29-44
13. J.D. Pollard, Materials for Today's Patternshop—A Personal Choice, *Foundry Trade J.*, Vol 158 (No. 3306), May 23, 1985, p 415-419
14. "Pattern Standards," National Association of Pattern Manufacturers
15. M.J. Sneden, Who's Afraid of Cast-To-Size Tooling, *Trans. AFS*, Vol 85, 1977, p 9-14
16. J. Sheeham and D. Richardson, The HMP Process—A New Method for Producing Plastic Patterns, *Trans. AFS*, Vol 92, 1984, p 203-208
17. G. Anderson and T.J. Crowley, Precision Foundry Tooling Utilizing CAD/CAM and the HMP Process, *Trans. AFS*, Vol 93, 1985, p 895-900
18. M.K. Young, Sprayed Metal Foundry Patterns for Short Run and Production Equipment, *Trans. AFS*, Vol 88, 1980, p 217-223
19. J.R. Henry, Metallic Coatings for Patterns and Coreboxes, *Mod. Cast.*, April 1983, p 22-24
20. R.D. Maier and J.F. Wallace, Pattern Material Wear or Erosion Studies, *Trans. AFS*, Vol 84, 1977, p 161-166
21. R.H. Immel, Expandable Polystyrene and Its Processing Into Patterns for the Evaporative Casting Process, *Trans. AFS*, Vol 87, 1979, p 545-550
22. M.K. Siebel and E.L. Kotzin, Evaporative Pattern Castings: The Process and Its Potential, *Mod. Cast.*, Vol 76, Jan 1986, p 31-34
23. A.J. Clegg, Expanded-Polystyrene Molding—A Status Report, *Foundry Trade J. Int.*, Vol 9 (No. 30), June 1986, p 51-69
24. M.C. Ashton, S.G. Sharman, and A.J. Brookes, The Replicast CS (Ceramic Shell) Process, *Foundry Trade J. Int.*, Vol 7 (No. 21), March 1984, p 33-42
25. A.M. Arzt and P.M. Bralower, Questions About EPC Vaporize With Proper Practice, *Mod. Cast.*, Vol 77, Jan 1987, p 21-24
26. D.R. Westlund, G.R. Anderson, and A.G. Anderson, Applying CAD/CAM to Foundry Tooling, *Mod. Cast.*, Jan 1984, p 20-24
27. J.D. Taylor, Coordinate Measuring Machine Application in the Pattern Shop, *Trans. AFS*, Vol 88, 1980, p 195-198
28. J.R. Woods, NC-CNC and CAM in the Pattern Shop, *Trans. AFS*, Vol 91, 1983, p 743-746
29. D.B. Welbourn, CAD/CAM Plays a Major Role in Foundry Economics, *Mod. Cast.*, Vol 77 (No. 9), Sept 1987, p 40-42

Molding and Casting Processes

Section Chairman:
Thomas S. Piwonka, The University of Alabama

Classification of Processes and Flow Charts of Foundry Operations

Thomas S. Piwonka, University of Alabama

CASTING PROCESSES have existed since prehistoric times (see the article "History of Casting" in this Volume). Over the years a wide variety of molding and casting methods have been developed, because the only limitation is human ingenuity. These methods will be introduced and classified in this article. More detailed information on each process can be found in the subsequent articles in this Section.

Casting Processes (Ref 1)

Figure 1 shows a simplified flow diagram of the basic operations for producing a sand casting. There are variations from this flow sheet depending on the type of material cast, the complexity of the component shape, and the quality requirements established by the customer. There are also many alternative methods of accomplishing each of these tasks.

The right side of Fig. 1 begins with the task of patternmaking. The article "Patterns and Patternmaking" in this Volume describes in detail the various pattern materials and considerations necessary in producing a quality pattern. A pattern is a specially made model of the component to be produced, used for producing molds.

Generally, sand is placed around the pattern and, in the case of clay-bonded sand, rammed to the desired hardness. In the case of chemical binders, the mold is chemically hardened after light manual or machine compaction. Molds are usually produced in two halves so that the pattern can be easily removed. When these two halves are reassembled, a cavity remains inside the mold in the shape of the pattern. Mold-making equipment and processing are described in the article "Sand Processing" in this Volume.

Internal passageways within a casting are formed by the use of cores. Cores are parts made of sand and binder that are sufficiently hard and strong to be inserted in a mold. Thus, the cores shape the interior of a casting, which cannot be shaped by the pattern itself. The patternmaker supplies core boxes for the production of precisely dimensioned cores. These core boxes are filled with specially bonded core sand and compacted much like the mold itself. Cores are placed in the drag, or bottom section, of the mold, and the mold is then closed by placing the cope, or top section, over the drag. Mold closing completes the production of the mold, into which the molten metal is then poured. Procedures for making cores are described in detail in the articles "Resin Binder Processes" and "Coremaking" in this Section.

Casting production begins with melting of the metal (left side, Fig. 1). Molten metal is then tapped from the melting furnace into a ladle for pouring into the mold cavity, where it is allowed to solidify within the space defined by the sand mold and cores. Melting, refining, and pouring of castings are described in the following articles in the Section "Foundry Equipment and Processing" in this Volume:

- "Melting Furnaces"
- "Vacuum Melting and Remelting Processes"
- "Degassing Processes (Converter Metallurgy)"
- "Degassing Processes (Ladle Metallurgy)"
- "Nonferrous Molten Metal Processes"
- "Automatic Pouring Systems"

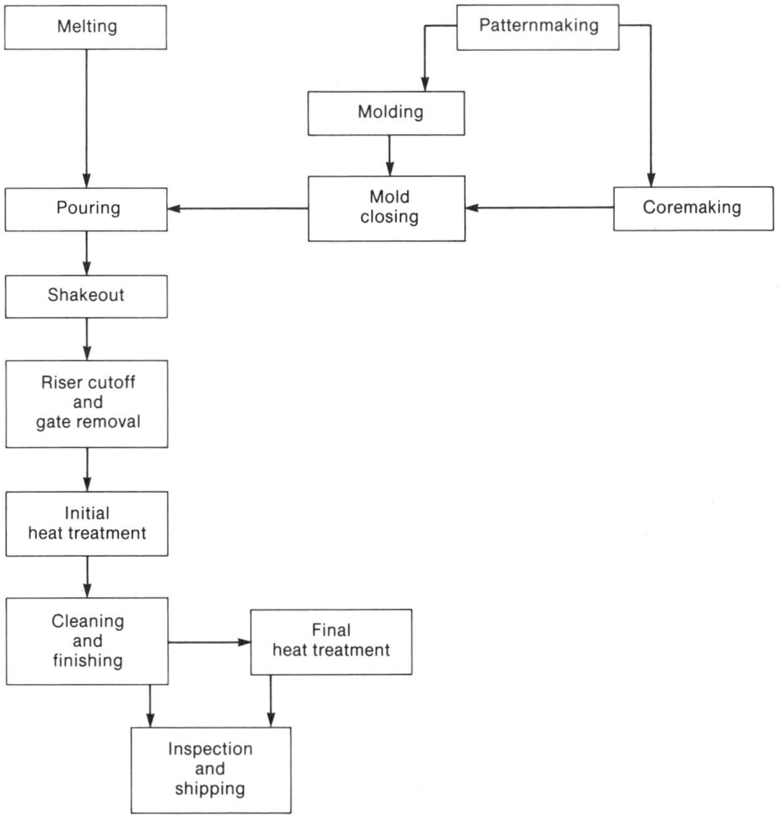

Fig. 1 Simplified flow diagram of the basic operations for producing a steel casting. Similar diagrams can be applied to other ferrous and nonferrous alloys produced by sand molding. Source: Ref 1

Table 1 Classification system for foundry processes based on mold type

The processes listed are all shape casting methods. Continuous casting, which is also covered in this Volume, is not listed.

Expendable Mold Processes

- Permanent patterns
  - Clay/water bonds (green sand molding)
    - Silica sand
    - Olivine sand
    - Chromite sand
    - Zircon sand
  - Heat-cured resin binder processes
    - Shell process (Croning process)
    - Furan hot box
    - Phenolic hot box
    - Warm box (furfuryl/catalyst)
    - Oven bake (core oil)
  - Cold box resin binder processes
    - Phenolic urethane
    - Furan/SO_2
    - Free radical cure (acrylic/epoxy)
    - Phenolic ester
  - No-bake resin binder processes
    - Furan (acid catalyzed)
    - Phenolic (acid catalyzed)
    - Oil urethane
    - Phenolic urethane
    - Polyol urethane
  - Silicate and phosphate bonds
    - Sodium silicate/CO_2
    - Shaw process (ceramic molding)
    - Unicast process (ceramic molding)
    - Alumina phosphate
  - Plaster bonds
    - Gypsum bond
  - No bond
    - Magnetic molding
    - Vacuum molding
- Expendable patterns
  - Foamed patterns
    - Lost foam casting
    - Replicast process
  - Wax patterns (investment casting)
    - Ethyl silicate bonded block molds
    - Ethyl silicate bonded ceramic shell molds
    - Colloidal silica bond
    - Plaster bond
    - Counter-gravity low-pressure casting

Permanent Mold Processes

- Die casting
  - High-pressure die casting
  - Low-pressure die casting
  - Gravity die casting (permanent mold)(a)
- Centrifugal casting
  - Vertical centrifugal casting
  - Horizontal centrifugal casting
- Hybrid processes
  - Squeeze casting
  - Semisolid metal casting (rheocasting)
  - Osprey process(b)

(a) When sand or plastic complex cores are used instead of metal cores, the term semipermanent mold casting should be used. (b) The Osprey process, which involves the production of preforms by a buildup of sprayed (atomized) metal powder, is discussed in the article "Spray Deposition of Metal Powders" in Volume 7 of the 9th Edition of *Metals Handbook*.

After it has solidified, the casting is shaken out of the mold, and the risers and gates are removed. Risers (also called "feeders") are shapes that are attached to the casting to provide a liquid-metal reservoir and control solidification. Metal in the risers is needed to compensate for the shrinkage that occurs during cooling and solidification. Gates are the channels through which liquid metal flows into the mold cavity proper. Heat treatment, cleaning and finishing, and inspection follow. These steps are outlined in the article "Processing of Castings" and in the articles on specific metals and alloys in the Sections "Ferrous Casting Alloys" and "Nonferrous Casting Alloys" in this Volume.

Classification of Molding and Casting Processes

Foundry processes can be classified based on whether the molds are permanent or expendable. Similarly, subclassifications can be developed from patterns, that is, whether or not the patterns are expendable. A second subclassification can be based on the type of bond used to make the mold. For permanent molding, processes can be classified by the type of mechanism used to fill the mold. Table 1 provides one possible classification system for the molding and casting processes described in this Section. Permanent pattern, expendable pattern, and permanent mold processes, as categorized in Table 1, are summarized below.

Permanent Pattern Processes. As indicated in Table 1, a number of processes use permanent patterns. Of these processes, however, green sand molding is the most prevalent. The typical steps involved in making a casting from a green sand mold are shown in Fig. 2 and described below (Ref 1).

The sequence begins with a mechanical drawing of the desired part. Patterns are then produced and mounted on pattern plates. Both the cope and drag patterns include core prints, which will produce cavities in the mold to accommodate extensions on either end of the core. These extensions fit solidly into the core prints to hold the core in place during pouring. The gate or passageway in the sand mold through which the molten metal will enter the mold cavity is usually mounted on the drag pattern plate. Locating pins on either end of the pattern plates allow for accurately setting the flask over the plate.

Cores are produced separately by a variety of methods. Figure 2 shows the core boxes, which are rammed with a mixture of sand and core binder (see the article "Coremaking" in this Section). If the cores must be assembled from separately made components, they are pasted together after curing. They are then ready to be inserted into the sand mold.

The mold is made by placing a flask (an open metal box) over the cope pattern plate. Before molding can begin, risers are added to the pattern at predetermined points to control solidification and supply liquid metal to the casting to compensate for the shrinkage that takes place during cooling and solidification. Thus, any shrinkage voids form in the risers, and the casting will be sound. A hole or holes (called sprues) must also be formed in the cope section of the mold to provide a channel through which the molten metal can enter the gating system and the mold cavity.

The cope half of the mold is produced by ramming sand into the flask, which is located on the pattern plate with pins. The flask full of sand is then drawn away from the pattern board, and the riser and sprue pieces removed.

A flask is subsequently placed over the drag pattern plate using the locating pins on the plate. Sand is rammed around the pattern, and a bottom board is placed on top of the flask full of sand. The pattern, flask, and bottom board are then rolled over 180°, and the pattern is withdrawn.

The completed core is set into the core prints in the drag half of the mold and the cope half of the mold is set on top of the drag. Proper alignment of the mold cavity in the cope and drag portions of the mold is ensured by the use of closing pins, which align the two flasks. The flasks can be clamped together, or weights can be placed on top of the cope, to counteract the buoyant force of the liquid metal, which would otherwise tend to float the cope off the drag during pouring.

Metal is then poured into the mold cavity through the sprue and allowed to solidify. The casting is shaken from the sand and appears as shown in Fig. 2, with the sprue, gating system, and risers attached. Following shakeout, the flasks, bottom boards, and clamps are cycled back to the molding station while the casting is moved through the production process. When the gates and risers are removed from the casting, they are returned to the furnace to be remelted. After cleaning, finishing, and heat treating, the castings are ready for shipment.

Expendable pattern processes use polystyrene patterns (lost foam casting and Replicast process) or wax patterns (see discussion below on investment casting). Both of these foundry processes are increasing in use.

The investment casting process has been known for at least 6000 years, but its use for the production of commercial castings has grown considerably during the second half of the 20th century. The process is also referred to as the lost wax process and as precision casting. The term precision implies high accuracy of dimensions and tight tolerances. Investment casting also yields smoother, high-integrity surfaces that require little or no machining, depending on the application.

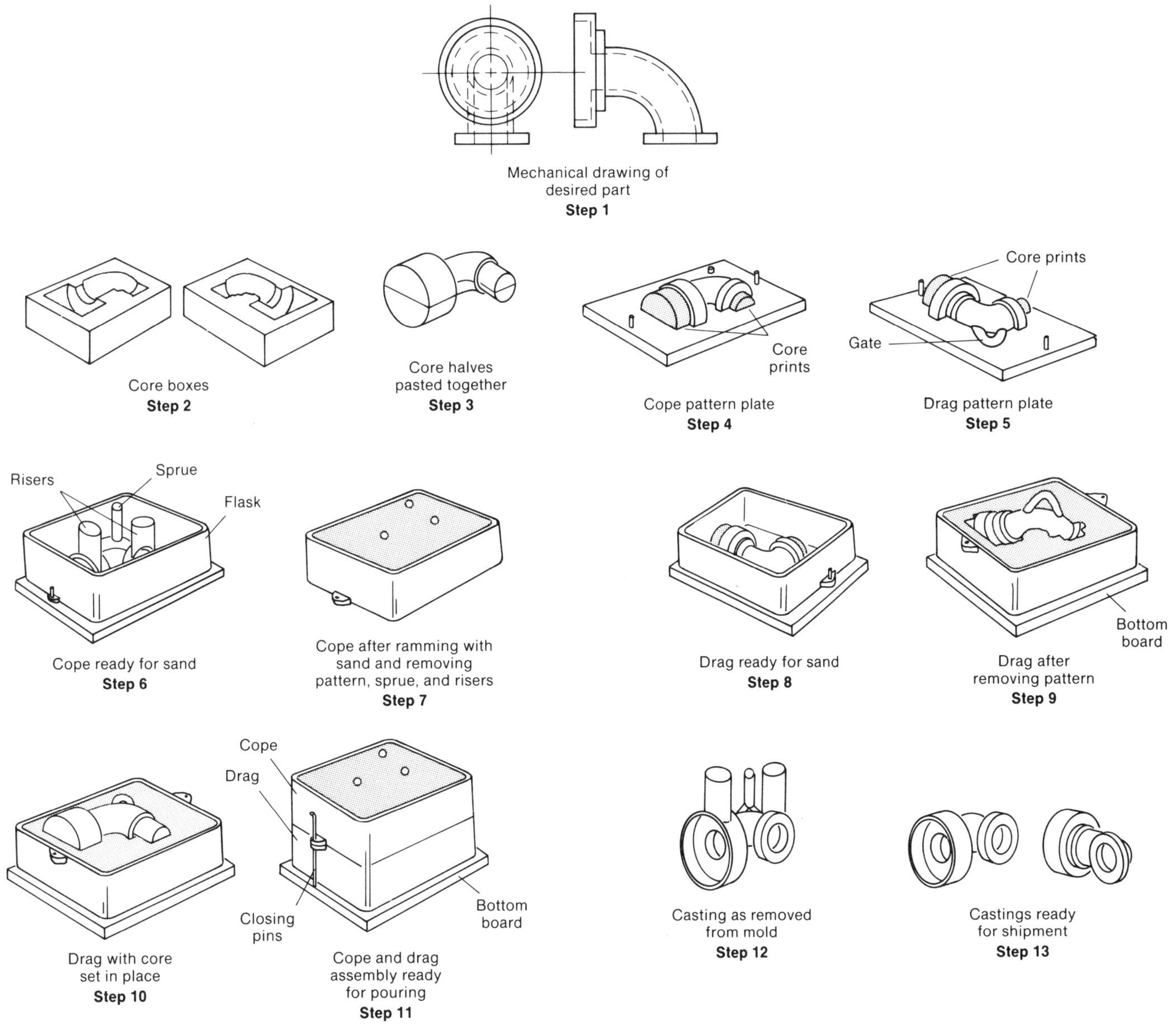

Fig. 2 Basic steps involved in making a casting from a green sand mold

The basic steps of the investment casting process are as follows:

- Production of heat-disposable patterns, usually made of wax or wax/resin mixtures
- Assembly of these patterns onto a gating system
- Investing, or covering, the pattern assembly with ceramic to produce a monolithic mold
- Melting out the pattern assembly to leave a precise mold cavity
- Firing the ceramic mold to remove the last traces of the pattern material, to fire the ceramic and develop the high-temperature bond, and to preheat the mold ready for casting
- Casting (pouring)
- Shakeout, cutoff, and finishing

These basic process steps are outlined schematically in Fig. 3. Detailed information on each of these processing steps can be found in the article "Investment Casting" in this Section.

Although it has a wide variety of applications, investment casting is particularly favored for the production of parts for gas turbine blades and vanes (nickel and cobalt alloys) and aircraft structural components (titanium, superalloys, and 17-4 PH stainless steel). The application of directional solidification (DS) and single-crystal (SC) technology to investment casting has also increased interest and use. Detailed information on developments in DS/SC technology can be found in the following articles in this Volume:

- "Solidification of Single-Phase Alloys"
- "New and Emerging Processes" (see the section "Directional and Monocrystal Solidification")
- "Vacuum Melting and Remelting Processes"
- "Nickel and Nickel Alloys"

Examples of investment castings for critical applications are shown in Fig. 4(a) to 4(c).

Permanent mold processes involve the use of metallic (ferrous) or solid graphite molds. On a volume basis, die casting, centrifugal casting, and permanent mold (gravity die) casting are the most important. Each of these is covered in detail in this Section. As indicated in Table 1, however, a number of hybrid processes, such as squeeze casting and semisolid metal processing have been developed that use permanent molds. Figure 5 shows a flowchart

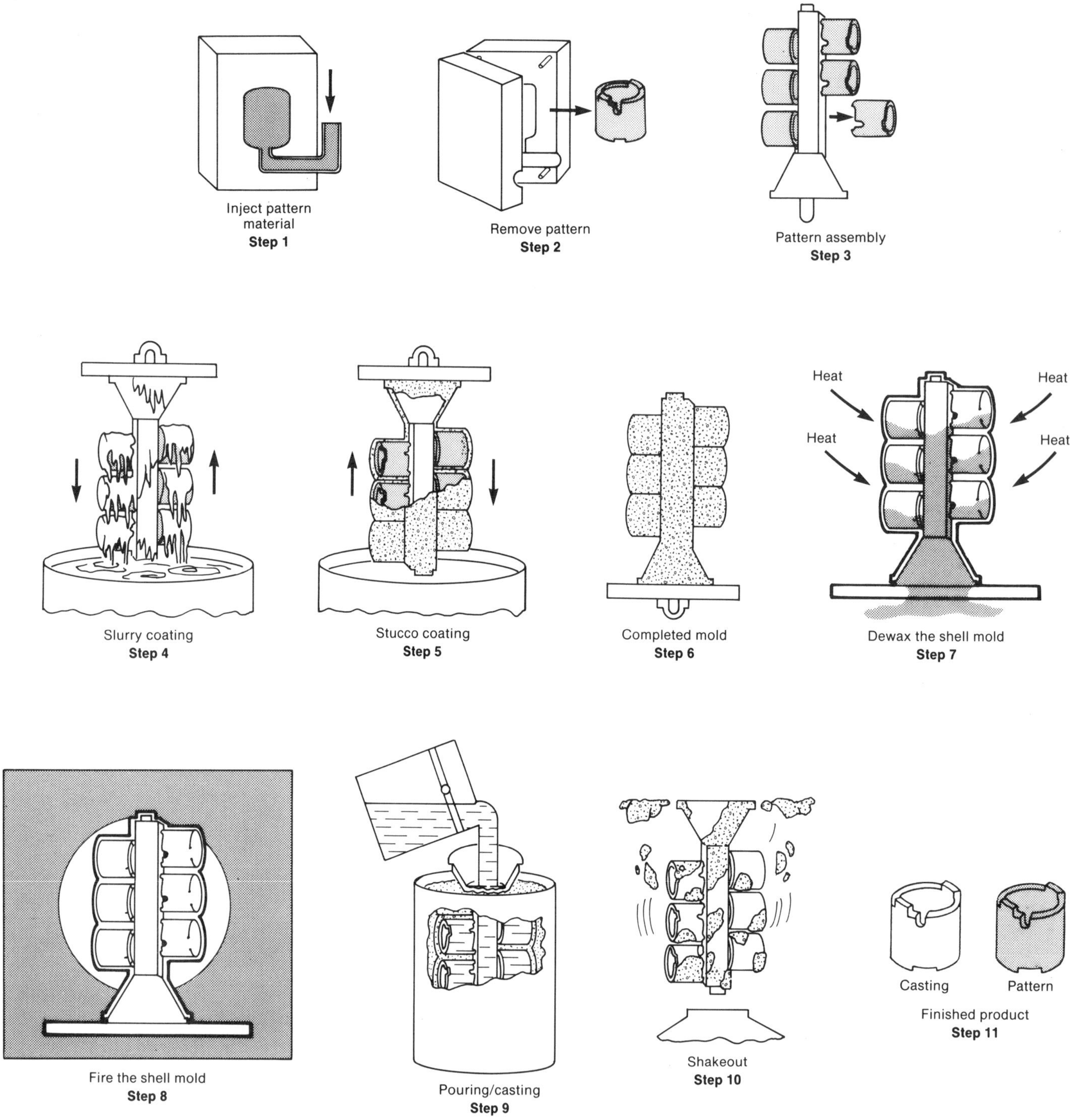

Fig. 3 Basic steps involved in investment casting

of operations for the rheocast method of semisolid casting. This process involves vigorous agitation of the melt during the early stages of solidification to break up the solid dendrites into small spherulites. The benefits provided by semisolid forming processes, as well as the microstructures produced by these methods, are discussed in the articles "New and Emerging Processes" (see the section on "Semisolid Metal Casting and Forging") and "Zinc and Zinc Alloys" in this Volume.

REFERENCE

1. P.F. Wieser, Ed., *Steel Castings Handbook*, 5th ed., Steel Founders' Society of America, 1980, p 1-2 to 1-5, 10-11

SELECTED REFERENCES

- E.L. Kotzin, *Metalcasting & Molding Processes*, American Foundrymen's Society, 1981
- *Molding Methods and Materials*, American Foundrymen's Society, 1962
- C.F. Walton and T.J. Opar, Ed., *Iron Casting Handbook*, Iron Castings Society, 1981

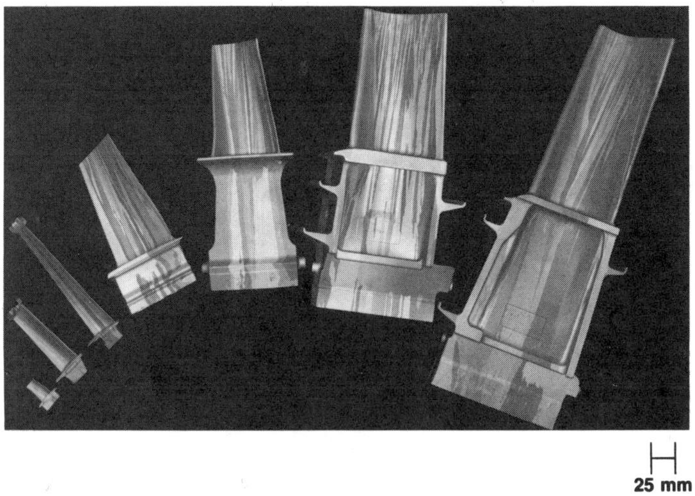

Fig. 4(a) Directionally solidified land-based turbine blades made from investment cast nickel-base superalloys. Courtesy of Howmet Corporation, Whitehall Casting Division

Fig. 4(b) Radial and axial turbine wheels made from investment cast Mar-M-247 nickel-base superalloy. Courtesy of Howmet Corporation, Whitehall Casting Division

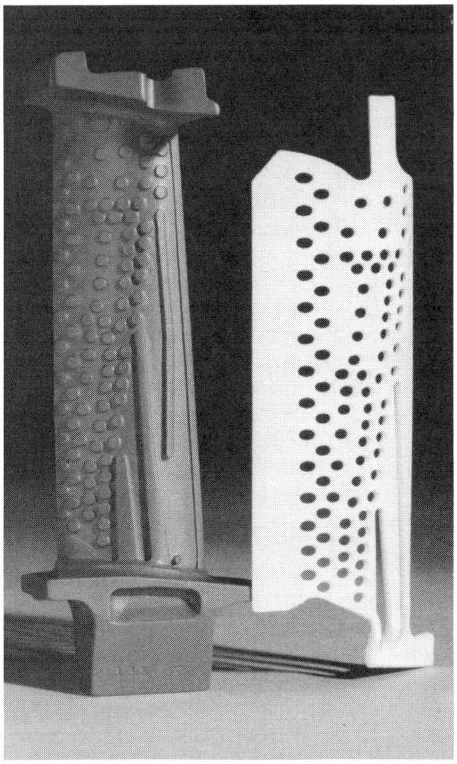

Fig. 4(c) Investment cast turbine blade with convex wall removed showing complex core

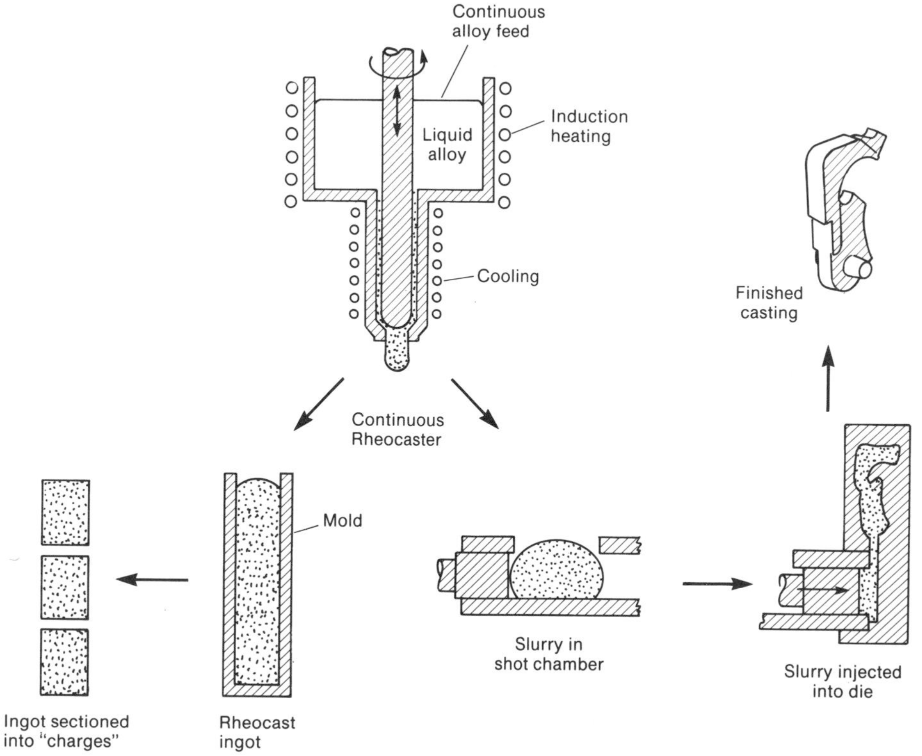

Fig. 5 Rheocast process. Source: M.C. Flemings, Massachusetts Institute of Technology

Aggregate Molding Materials

Thomas S. Piwonka, University of Alabama

THE CASTING PROCESS involves the pouring of molten metal into a mold; therefore, the mold material and molding method must be selected with care. Most castings are made in sand molds because metallic molds wear out too quickly to be economical for ferrous metals production. For low and medium production runs, the lower tooling costs of sand molding give it an overwhelming cost advantage over permanent molds or die casting.

Selection of the molding material and its bonding system depends on the type of metal being poured, the type of casting being made, the availability of molding aggregates, the mold and core making equipment owned by the foundry, and the quality requirements of the customer. A thorough understanding of all of these factors is necessary to optimize the molding system used in the foundry.

This article will discuss the various materials used to produce molds and cores for sand casting. These materials include sands, clays, additions to sand mixes, and plastics. The principles that explain how these materials are bonded together are discussed in the articles "Bonds Formed in Molding Aggregates" and "Resin Binder Processes" in this Section. Additional information on the preparation, mulling, handling, and reclamation of sands is available in the article "Sand Processing" in this Volume.

Sands

The refractory molds used in casting consist of a particulate refractory material (sand) that is bonded together to hold its shape during pouring. Although various sands can be used, the following basic requirements apply to each (Ref 1):

- Dimensional and thermal stability at elevated temperatures
- Suitable particle size and shape
- Chemically unreactive with molten metals
- Not readily wetted by molten metals
- Freedom from volatiles that produce gas upon heating
- Economical availability
- Consistent cleanliness, composition, and pH
- Compatibility with binder systems

Many minerals possess some of these features, but few have them all.

Silica Sands

Most green sand molds consist of silica sands bonded with a bentonite-water mixture. (The term green means that the mold, which is tempered with water, is not dried or baked.) The composition, size, size distribution, purity, and shape of the sand are important to the success of the moldmaking operation.

Sands are sometimes referred to as natural or synthetic. Natural sands contain enough naturally occurring clays that they can be mixed with water and used for sand molding. Synthetic sands have been washed to remove clay and other impurities, carefully screened and classified to give a desired size distribution, and then reblended with clays and other materials to produce an optimized sand for the casting being produced. Because of the demands of modern high-pressure molding machines and the necessity to exercise close control over every aspect of casting production, most foundries use only synthetic sands.

Composition. Foundry sands are composed almost entirely of silica (SiO_2) in the form of quartz. Some impurities may be present, such as ilmenite ($FeO\text{-}TiO_2$), magnetite (Fe_3O_4), or olivine, which is composed of magnesium and ferrous orthosilicate [$(Mg,Fe)_2SiO_4$]. Silica sand is used primarily because it is readily available and inexpensive. However, its various shortcomings as a foundry sand necessitate the addition of other materials to the sand mix to produce satisfactory castings, as described later in this article.

Quartz undergoes a series of crystallographic transitions as it is heated. The first, at 573 °C (1064 °F), is accompanied by expansion, which can cause mold spalling. Above 870 °C (1598 °F), quartz transforms to tridymite, and the sand may actually contract upon heating. At still higher temperatures (>1470 °C, or 2678 °F), tridymite transforms to cristobalite.

In addition, silica reacts with molten iron to form a slag-type compound, which can cause burn-in, or the formation of a rough layer of sand and metal that adheres to the casting surface. However, because these problems with silica can be alleviated by proper additions to the sand mix, silica remains the most widely used molding aggregate.

Shape and Distribution of Sand Grains. The size, size distribution, and shape of the sand grains are important in controlling the quality of the mold. Most mold aggregates are mixtures of new sand and reclaimed sand, which contain not only reclaimed molding sand but also core sands. In determining the size, shape, and distribution of the sand grains, it is important to realize that the grain shape contributes to the amount of sand surface area and that the grain size distribution controls the permeability of the mold.

As the sand surface area increases, the amount of bonding material (normally clay and water) must increase if the sand is to be properly bonded. Thus, a change in surface area, perhaps due to a change in sand shape or the percentage of core sand being reclaimed, will result in a corresponding change in the amount of bond required.

Rounded grains have a low surface-area-to-volume ratio and are therefore preferred for making cores because they require the least amount of binder. However, when they are recycled into the molding sand system, their shape can be a disadvantage if the molding system normally uses a high percentage of clay and water to facilitate rapid, automatic molding. This is because rounded grains require less binder than the rest of the system sand.

Angular sands have the greatest surface area (except for sands that fracture easily and produce a large percentage of small grains and fines) and therefore require more mulling, bond, and moisture. The angularity of a sand increases with use because the sand is broken down by thermal and mechanical shock.

The subangular-to-round classification is most commonly used, and it affords a compromise if shape becomes a factor in the sand system. However, control of grain size distribution is more important than control of grain shape. The grain size distribution, which includes the base sand size distribution plus the distribution of broken grains and fines from both molding sand and core sands, controls both the surface area and the packing density or porosity of the mold.

Table 1 Screen scale sieves equivalent

USA series No.	Tyler screen scale sieves, openings per lineal inch	Sieve opening, mm	Sieve opening, μm	Sieve opening, in., ratio $\sqrt{2}$, or 1.414	Permissible variation in average opening, ±mm	Wire diameter, mm
6	6	3.35	3350	. . .	0.11	1.23
8(a)	8(a)	2.36	2360	0.0937	0.08	1.00
12	10	1.70	1700	0.0661	0.06	0.810
16(a)	14(a)	1.18	1180	0.0469	0.045	0.650
20	20	0.850	850	0.0331	0.035	0.510
30	28	0.600	600	0.0234	0.025	0.390
40	35	0.425	425	0.0165	0.019	0.290
50	48	0.300	300	0.0117	0.014	0.215
70	65	0.212	212	0.0083	0.010	0.152
100	100	0.150	150	0.0059	0.008	0.110
140	150	0.106	106	0.0041	0.006	0.076
200	200	0.075	75	0.0029	0.005	0.053
270	270	0.053	53	0.0021	0.004	0.037

Note: A fixed ratio exists between the different sizes of the screen scale. This fixed ratio between the different sizes of the screen scale has been taken as 1.414, or the square root of 2 ($\sqrt{2}$). For example, using the USA series equivalent No. 200 as the starting sieve, the width of each successive opening is exactly 1.414 times the opening in the previous sieve. The area or surface of each successive opening in the scale is double that of the next finer sieve or one-half that of the next coarser sieve. That is, the widths of the successive openings have a constant ratio of 1.414, and the areas of the successive openings have a constant ratio of $\sqrt{2}$. This fixed ratio is very convenient; by skipping every other screen, a fixed ratio of width of 2 to 1 exists.
(a) These sieves are not normally used for testing foundry sands. Source: Ref 2

Table 2 Typical calculation of AFS grain fineness number

Size of sample: 50 g; AFS clay content: 5.9 g, or 11.8%; sand grains: 44.1 g, or 88.2%

USA sieve series No.	Amount of 50 g sample retained on sieve, g	Amount of 50 g sample retained on sieve, %	Multiplier	Product
6	none	0.0	3	0
12	none	0.0	5	0
20	none	0.0	10	0
30	none	0.0	20	0
40	0.20	0.4	30	12
50	0.65	1.3	40	52
70	1.20	2.4	50	120
100	2.25	4.5	70	315
140	8.55	17.1	100	1710
200	11.05	22.1	140	3094
270	10.90	21.8	200	4360
Pan	9.30	18.6	300	5580
Total	**44.10**	**88.2**		**15 243**

Source: Ref 2

The porosity of the mold controls its permeability, which is the ability of the mold to allow gases generated during pouring to escape through the mold. The highest porosity will result from grains that are all approximately the same size. As the size distribution broadens, there are more grains that are small enough to fill the spaces between large grains. As grains break down through repeated recycling, there are more and more of the smaller grains, and the porosity of the mold decreases.

However, if the porosity of the mold is too great, metal may penetrate the sand grains and cause a burn-in defect. Therefore, it is necessary to balance the base sand distribution and continue to screen the sand and use dust collectors during recycling to remove fines and to determine the proper bond addition. Most foundries in the United States use the American Foundrymens' Society (AFS) grain fineness number as a general indication of sand fineness. The AFS grain fineness number of sand is approximately the number of openings per inch of a given sieve that would just pass the sample if its grains were of uniform size, that is, the weighted average of the sizes of grains in the sample. It is approximately proportional to the surface area per unit weight of sand exclusive of clay.

The AFS grain fineness number is determined by taking the percentage of sand retained on each of a series of standard screens, multiplying each by a multiplier, adding the total, and then dividing by the total percentage of sand retained on the sieves (Ref 2). Table 1 lists the series of sieves used to run the standard AFS standard sieve analysis. A typical calculation of the AFS fineness number, which includes the multiplier factor, is given in Table 2.

It is important to understand that various grain distributions and grain shape classifications can result in similar grain fineness numbers. Table 3 provides a sample sieve analysis demonstrating that two sands assigned the same AFS grain fineness number can have very different grain size distributions.

Preparation of Sands. The production of sand for the foundry industry requires a series of mining and refining steps to yield pure, consistent sands (Ref 3). The actual production flow sheets vary with the source of the sand, but in general they include mining, one or more scalping operations to remove roots and pebbles, and then repeated washing and desliming operations to remove naturally occurring clays. The sand is screened and/or classified and then prepared for shipment to the foundry.

Zircon

Zircon is zirconium silicate ($ZrSiO_4$). It is highly refractory and possesses excellent foundry characteristics (Ref 2). Its primary advantages are a very low thermal expansion, high thermal conductivity and bulk density (which gives it a chilling rate about four times that of quartz), and very low reactivity with molten metal. Zircon requires less binder than other sands because its grains are rounded. The very high dimensional and thermal stabilities exhibited by zircon are the reasons it is widely used in steel foundries and investment foundries making high-temperature alloy components.

Olivine

Olivine minerals (so called because of their characteristic green color) are a solid solution of forsterite (Mg_2SiO_4) and fayalite (Fe_2SiO_4). Their physical properties vary with their chemical compositions; therefore, the composition of the olivine used must be specified to control the reproducibility of the sand mixture. Care must be taken to calcine the olivine sand before use to decompose the serpentine content, which contains water (Ref 4).

The specific heat of olivine is similar to that of silica (Ref 5), but its thermal expansion is far less. Therefore, olivine is used for steel casting to control mold dimensions. Olivine is somewhat less durable than silica (Ref 1), and it is an angular sand.

Table 3 Similarity in AFS grain fineness number of two sand samples with different grain size distributions

USA sieve No.	Percentage retained: Sand A	Percentage retained: Sand B
6	0.0	0.0
12	0.0	0.0
20	0.0	0.0
30	1.0	0.0
40	24.0	1.0
50	22.0	24.0
70	16.0	41.0
100	17.0	24.0
140	14.0	7.0
200	4.0	2.0
270	1.7	0.0
Pan	0.3	1.0
Total	**100.0**	**100.0**
AFS grain fineness No.	60.0	60.0

Source: Ref 2

Chromite

Chromite ($FeCr_2O_4$), a black, angular sand, is highly refractory and chemically unreactive, and it has good thermal stability and excellent chilling properties (Ref 1). However, it has twice the thermal expansion of zircon sand, and it often contains hydrous impurities that cause pinholing and gas defects in castings. It is necessary to specify the calcium oxide (CaO) and silicon dioxide (SiO_2) limits in chromite sand to avoid sintering reactions and reactions with molten metal that cause burn-in (Ref 4).

Aluminum Silicates

Aluminum silicate (Al_2SiO_5) occurs in three common forms: kyanite, sillimanite, and andalusite. All break down at high temperatures to form mullite and silica (Ref

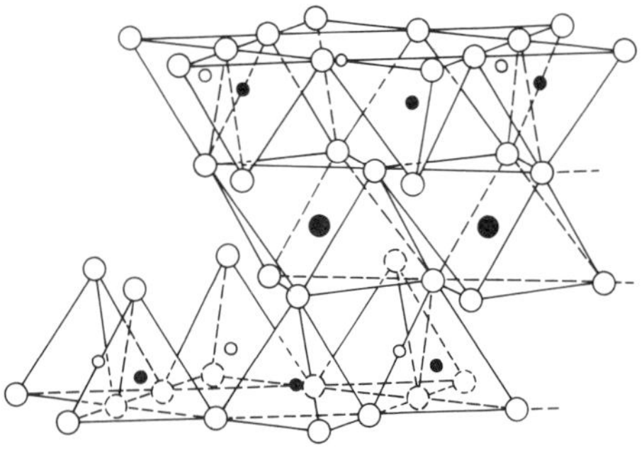

Fig. 1 Structure of montmorillonite. Large closed circles are aluminum, magnesium, sodium, or calcium. Small closed circles are silicon. Large open circles are hydroxyls. Small open circles are oxygen.

1). Therefore, aluminum silicates for foundry use are produced by calcining these minerals. Depending on the sintering cycle, the silica may be present as cristobalite or as amorphous silica. The grains are highly angular. These materials have high refractoriness, low thermal expansion, and high resistance to thermal shock. They are widely used in precision investment foundries, often in combination with zircon.

Clays

Bonds in green sand molds are produced by the interaction of clay and water. Each of the various clays has different properties, as described below.

Bentonites

The most common clays used in bonding green sand molds are bentonites, which are forms of montmorillonite or hydrated aluminum silicate. Montmorillonite is built up of alternating tetrahedra of silicon atoms surrounded by oxygen atoms, and aluminum atoms surrounded by oxygen atoms, as shown in Fig. 1. This is a layered structure, and it produces clay particles that are flat plates. Water is adsorbed on the surfaces of these plates, and this causes bentonite to expand in the presence of water and to contract when dried.

There are two forms of bentonite: Western (sodium) and Southern (calcium). Both are used in foundry sands, but they have somewhat different properties.

Western Bentonite. In Western bentonite, some of the aluminum atoms are replaced by sodium atoms. This gives the clay a net negative charge, which increases its activity and its ability to adsorb water. Western bentonite imparts high green and dry strengths to molding sand, and it has advantages for use with ferrous alloys, as follows.

First, Western bentonite develops a high degree of plasticity, toughness, and deformation, along with providing good lubricity when mulled thoroughly with water. Molding sand bonded with plasticized Western bentonite squeezed uniformly around a pattern produces excellent mold strengths.

Second, because of its ability to swell with water additions to as much as 13 times its original volume, Western bentonite is an excellent agent between the sand grains after compaction in the mold. It therefore plays an important role in reducing sand expansion defects.

Finally, Western bentonite has a high degree of durability. This characteristic allows it to be reused many times in a system sand with the least amount of rebonding additions.

In using Western bentonite, it is important to control the clay/water ratio. Failure to do so can result in stiff, tough, difficult-to-mold sand with poor shakeout characteristics. Although these conditions can be alleviated by adding other materials to the molding sand, control of the mixture is preferable.

Southern Bentonite. In Southern bentonites, some of the aluminum atoms are replaced by calcium atoms. Again, this increases the ion exchange capability of the clay. Southern bentonite is a lower-swelling clay, and it differs from Western bentonite in the following ways:

- It develops a higher green compressive strength with less mulling time
- Its dry compressive strength is about 30 to 40% lower
- Its hot compressive strength is lower, which improves shakeout characteristics
- A Southern bentonite bonded sand flows more easily than Western bentonite and can be squeezed to higher densities with less pressure; it is therefore better for use with complex patterns containing crevices and pockets

Use of Southern bentonite also requires good control of the clay-water mixture. Southern bentonite requires less water than Western bentonite and is less durable.

In practice, it is common to blend Western and Southern bentonites together to optimize the sand properties for the type of casting, the molding equipment, and the metal being poured. Examples of the effect of mixing bentonites on various sand properties are shown in Fig. 2. At high temperatures, bentonites lose their adsorbed water and therefore their capacity for bonding. The superior high-temperature properties of Western bentonite are due to the fact that it retains water to higher temperatures than Southern bentonite (Ref 6). However, if the sand mix is heated to more than 600 °C (1110 °F), water is driven out of the clay crystal structure. This loss is irreversible, and the clay must be discarded.

Fireclay

Fireclay consists essentially of kaolinite, a hydrous aluminum silicate that is usually combined with bentonites in molding sand. It is highly refractory, but has low plasticity. It improves the hot strength of the mold and allows the water content to be varied over greater ranges. Because of its high hot strength potential, it is used for large castings. It is also used to improve sieve analysis by creating fines whenever the system

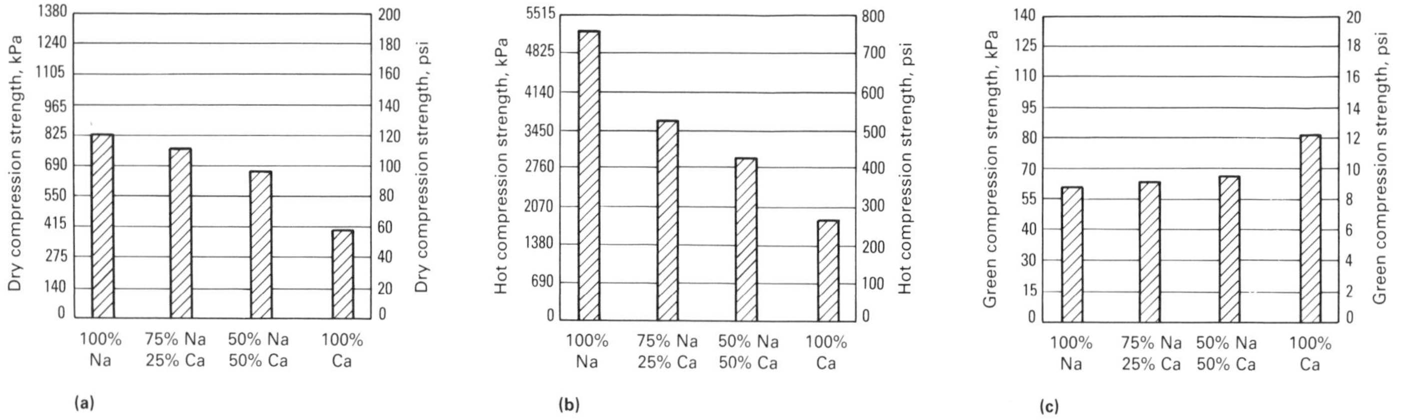

Fig. 2 Effect of blending sodium and calcium bentonites on molding sand properties. (a) Dry compression strength. (b) Hot compression strength at 900 °C (1650 °F). (c) Green compression strength

does not have an optimum wide sieve distribution of the base sand. However, because of its low durability, its use is generally limited. In addition, the need for fireclay can usually be eliminated through close control of sand mixes and materials.

Other Additions to Sand Mixes

As noted above, silica sand, although inexpensive, has some shortcomings as a molding sand. If done properly, the addition of other materials can alleviate these deficiencies.

Carbonaceous Additions. Carbon is added to the mold to provide a reducing atmosphere and a gas film during pouring that protects against oxidation of the metal and reduces burn-in. Carbon can be added in the form of seacoal (finely ground bituminous coal), asphalt, gilsonite (a naturally occurring asphaltite), or proprietary petroleum products. Seacoal changes to coke at high temperatures expanding three times as it does so; this action fills voids at the mold/metal interface. Too much carbon in the mold gives smoke, fumes, and gas defects, and the use of asphalt products must be controlled closely because their overuse waterproofs the sand.

The addition of carbonaceous materials will give improved surface finish to castings. Best results are achieved with such materials as seacoal and pitch, which volatilize and deposit a pyrolytic (lustrous) carbon layer on sand at the casting surface (Ref 7).

Cellulose is added to control sand expansion and to broaden the allowable water content range. It is usually added in the form of wood flour, or ground cereal husks or nut shells. Cellulose reduces hot compressive strength and provides good collapsibility, thus improving shakeout. At high temperatures, it forms soot (an amorphous form of carbon), which deposits at the mold/metal interface and resists wetting by metal or slags. It also improves the flowability of the sand during molding. Excessive amounts generate smoke and fumes and can cause gas defects. In addition, if present when the clay content drops too low, defects such as cuts, washes, and mold inclusions will occur in the castings.

Cereals, which include corn flour, dextrine, and other starches, are adhesive when wetted and therefore act as a binder. They stiffen the sand and improve its ability to draw deep pockets. However, use of cereals makes shakeout more difficult, and excessive quantities make the sand tough and can cause the sand to form balls in the muller. Because cereals are volatile, they can cause gas defects in castings if used improperly.

Plastics

Plastic materials, or resins, are widely used in metal casting as binders for sand, particularly for cores of all sizes and production volumes, and for low-volume high-accuracy molding. Generally, these materials fall into three categories:

- Those composed of liquid polymeric binders that cross link and set up in the presence of a catalyst (thus transforming from a liquid to a solid)
- Those composed of two reactants that form a solid polymeric structure in the presence of a catalyst
- Those that are heat activated

Fluid-to-solid transition plastics are primarily furfuryl alcohol-base binders that are cured with acid catalysts. The polymers coat the sand when in the liquid form and are mixed with the liquid catalyst just before being placed in the core box. Alternatively, the catalyst can be delivered to the mix as a gas once the sand mix is in the core box.

Reaction-based plastics include phenolics (phenol/aldehyde), oil/urethanes, phenolic/polymeric isocyanates, and polyol/isocyanate systems. Curing catalysts include esters, amines, and acids, which can be delivered to the core mix either as liquids or gases.

Heat-activated plastics are primarily thermoplastics or thermosetting resins such as novolacs, furans (furfuryl alcohols), phenols, and linseed oils. They are applied as dry powders to the sand, and the mix is heated, at which time the powders melt, flow over the sand, and then undergo a thermosetting reaction. Alternatively, they may consist of two liquids that react to form a solid in the presence of heat.

Most binder systems are proprietary. The major ingredients are often mixed with nonreactive materials to control the reaction rate. The reactants are often dissolved in solvents to facilitate handling. Although various materials and schemes are used to form organic bonds in mold and core making, the technology rests on only a few compounds.

The presence of so many different systems allows casting producers to tailor the bonding system to the particular application. However, selection of the bonding system requires care. Care must also be taken in controlling process parameters because the systems are sensitive to variations in temperature and humidity. Consideration must also be given to environmental issues in the selection of the system because some organic systems emit noxious odors and fumes. More detailed information on organic binders can be found in the article "Resin Binder Processes" in this Volume.

REFERENCES

1. T.E. Garnar, Jr., *AFS Cast Met. Res. J.*, Vol 2, June 1978, p 45
2. Particle Size Distribution of Foundry Sand Mixtures, in *Mold and Core Test Handbook*, American Foundrymens' Society, 1978, p 4-1 to 4-14
3. F.P. Goettman, *Trans. AFS*, Vol 83, 1975, p 15
4. E.L. Kotzin, *Trans. AFS*, Vol 90, 1982, p 103
5. K. Kubo and R.D. Pehlke, *Trans. AFS*, Vol 90, 1982, p 405
6. F. Hofmann, *Trans. AFS*, Vol 93, 1985, p 377
7. I. Bindernagel, A. Kolorz, and K. Orths, *Trans. AFS*, Vol 83, 1975, p 557

Bonds Formed in Molding Aggregates

Thomas S. Piwonka, University of Alabama

MOLDING AGGREGATES must be held together, or bonded, to form a mold. By far the most common types of bonds are those formed from sand, clay, and additives. These materials are described in the previous article "Aggregate Molding Materials" in this Section. Organic bonds, described briefly here and in detail in the following article "Resin Binder Processes," also have a substantial part of the market for core making.

Silica-Base Bonds

Because of the abundance and low cost of clays, green sand molds are normally clay bonded, but various forms of silica can also be used in bonding molding aggregates.

Clay-Water Bonds. As noted in the article "Aggregate Molding Materials," bentonites are not electrically neutral and can therefore attract water molecules between the clay plates. Water is also adsorbed on the quartz surfaces. Thus, there is a network of water adsorbed on sand and clay particles that is set up throughout the molding sand. If the clay covers each sand grain entirely, then clay-water bridges form between grains (Ref 1). In the case in which the clay coverage is nonuniform, similar bridges are formed.

The clay-water bond can also be explained in terms of the specific surface area of the clay, the type and strength of the water bond at the clay surface, and the hydration envelopes of the adsorbed cations (Ref 2, 3). Clay particles hold adsorbed cations on their surfaces. The bonding of cations on clay particles is weak, and ion exchange is possible in the presence of appropriate electrolytes. Therefore, clay particles and ions are surrounded by electric force fields that direct the water dipoles (the water is polarized at the clay surface) and bind the water network. The field strength decreases with increased distance from the surface of the clay, so that the dipoles closest to the clay surface are bonded most strongly. Beyond the distance at which the force field is effective, the water behaves as a liquid and has no bonding action.

There is an ideal water content at which all of the water is polarized and active in the bonding process (because the water added to activate the clay bond is called temper water, this is known as the temper point). Above this water content, some of the water will exist as liquid water, which is not involved in bonding. Below this value, there is insufficient water to develop the bond fully. At the temper point, the green strength of the sand is at its maximum, and additions of water beyond this point decrease the strength of the sand/clay/water mixture. The effect of this can be seen quite clearly in Fig. 1.

Colloidal silica bonds are used in investment casting. Colloidal silica particles are about 4 to 40 nm in diameter and form a sol in water. Their stability is determined by surface charge, pH, particle size, concentration, and electrolyte content (Ref 4). The silica is spherical and amorphous, and it contains a small amount of a radical, such as a hydroxide, to impart a negative charge to each particle so that they repel each other and do not settle out. When water evaporates from these sols (as happens when the mold layers are dried), the silica particles are forced close enough together for hydroxyl groups to condense, splitting out the water and forming siloxane bonds between the aggregates (Ref 5).

The molds made from colloidal silica are dried in air and have enough strength to retain their shapes. However, they must be fired at an elevated temperature (>815 °C, or 1500 °F) to develop a strong silica ceramic bond. Each mold system has an optimum silica content for maximum mold strength. More detailed information on colloidal silica bonds can be found in the article "Investment Casting" in this Volume.

Ethyl Silicate. An alternate silica bond can be produced from hydrolyzed ethyl silicates. These are precipitation bonds, such as (Ref 4):

$$[n\ Si(OH)_4] \rightleftharpoons [SiO(OH)_{2n} + nH_2O]$$

The precipitated silicate bond is a gel that comes out of suspension by a change in binder ion concentration.

Hydrolyzed ethyl silicate is manufactured by the reaction of silicon tetrachloride with ethyl alcohol. Two types of ethyl silicate are commonly available. Ethyl silicate 30, the first type, is a mixture of tetraethyl orthosilicate and polysilicates containing about 28% silica. Ethyl silicate 40, the second type, is a mixture of branched silicate polymers containing about 40% silica.

Slurries formed from these ethyl silicates and mold aggregates, such as fused silica or zircon, are very sensitive to changes in pH. The slurries are normally kept at a pH of around 3. To gel them around a pattern, they are exposed to ammonia vapor, and their pH increases to 5, where they gel. The shells are then fired, and the ethyl alcohol evaporates or burns off, causing the silica binders to condense and form the silica bond. Ethyl silicate molds have an advantage over colloidal silica in that they do not require long drying times between dips and can be used for monolithic block molds. However, the mold strength of these molds is much less than that of colloidal silicate bonded molds because of the fine craze cracking that occurs during firing. On the other hand, this fine network of cracks is also responsible for the high dimensional reproducibility of castings made in block molds. Additional information on ethyl silicate molds can be found in the article "Investment Casting" in this Volume.

Sodium Silicate Bonds. The sodium silicate process is another method of forming a bond made up of a silicate polymer. In this case, carbon dioxide is used to precipitate sodium from what is essentially silicic acid containing large quantities of colloidal sodium. The reaction is:

$$Na_2O\ 2SiO_2 + CO_2 \rightleftharpoons Na_2CO_3 + 2SiO_2$$

Continued gassing gives:

$$Na_2O\ 2SiO_2 + 2CO_2 + H_2O \rightleftharpoons 2Na_2HCO_3 + 2SiO_2$$

This shows that continued gassing dehydrates the amorphous silica gel and increases the strength of the mold (Ref 6).

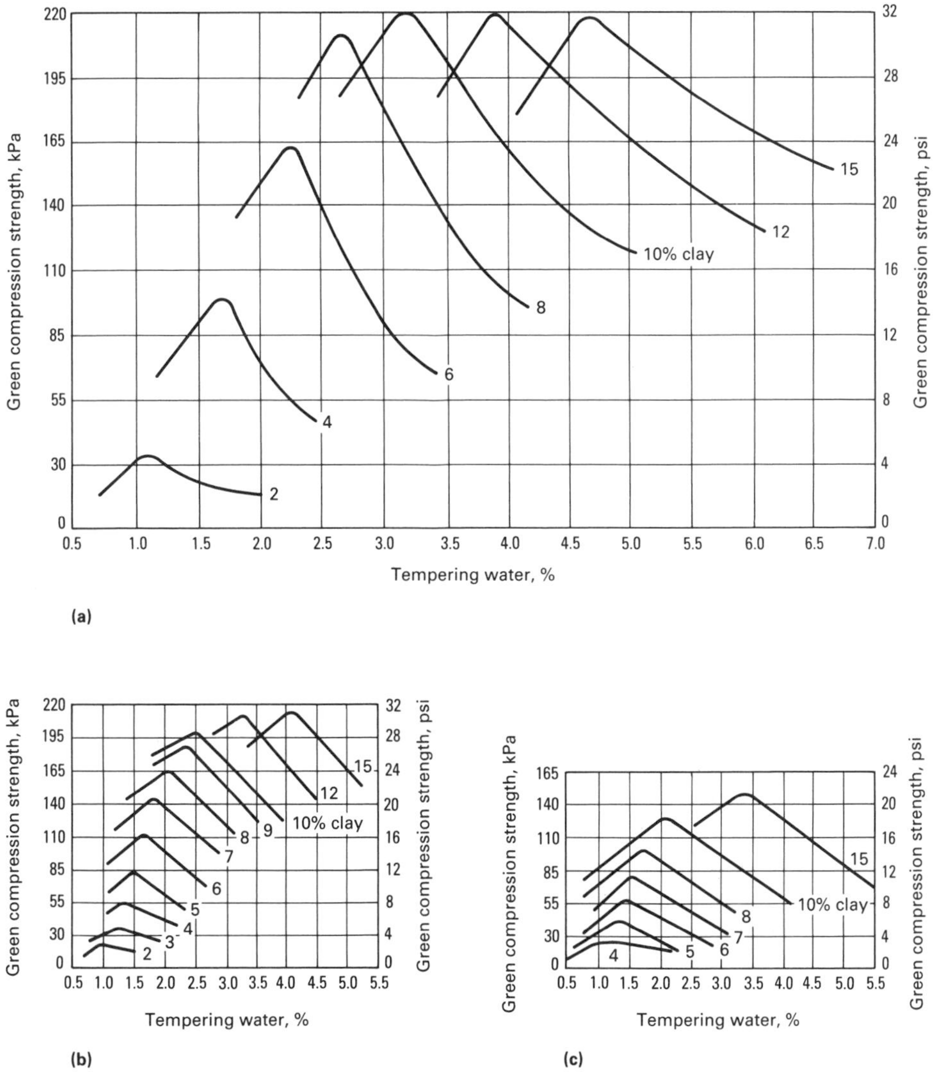

Fig. 1 Variation of mold properties with water content. (a) Southern bentonite. (b) Western bentonite. (c) Kaolinite. Source: Ref 1

Sodium silicate molds are widely used for large cores and castings where there is a premium on mold hardness and dimensional control. The bond breaks down easily at high temperatures and therefore facilitates shakeout. The silicate-bonded sand, after pouring and shakeout, can be reclaimed by mechanical means, and up to 60% of the reclaimed sand can be reused. Wet reclamation of silicate sand systems is also possible. Additional information on sodium silicate molds can be found in the article "Sand Molding" (see the Section "Bonded Sand Molds") in this Volume.

Phosphoric Acid Bonds

Phosphoric acid bonds are used in both ferrous and nonferrous precision casting to produce monolithic molds. They are a reaction-type bond with the general form:

$$[MO + H_3PO_4 \rightleftharpoons M(HPO_4) + H_2O]$$

where M is an oxide frit or mixture of frits. The pH must be controlled carefully and kept acidic (Ref 4). The powdered metal oxide hardener is dry blended with the sand, and the liquified phosphoric acid is then incorporated. The coated sand is compacted into core or pattern boxes and allowed to harden chemically before removal. A similar procedure for producing phosphate bonds is described in the article "Sand Molding" (see the Section "Bonded Sand Molds") in this Volume.

Organic Bonds

Organic bonds are used in resin-bonded sand systems. These systems vary widely. The sand is coated with two reactants that form a resin upon the application of heat or a chemical catalyst. The resin is a solid plastic that coats the sand so that it holds its shape during pouring. A thorough review of organically bonded systems can be found in the following article "Resin Binder Processes" in this Section.

REFERENCES

1. R.F. Grim and F.L. Cuthbert, Engineering Experiment Station Bulletin 357, University of Illinois, 1945
2. G.A. Smiernow, E.L. Doheny, and J.G. Kay, *Trans. AFS*, Vol 88, 1980, p 659
3. D. Boenisch, *Tonindustrie Zeitung*, Vol 86, 1962, p 237
4. W.F. Wales, *Trans. AFS*, Vol 81, 1973, p 249
5. R.L. Rusher, *AFS Cast Met. Res. J.*, Dec 1974, p 149
6. J. Gotheridge, *Trans. AFS*, Vol 87, 1979, p 669

Resin Binder Processes*

James J. Archibald and Richard L. Smith, Ashland Chemical Company

THE FOUNDRY INDUSTRY uses a variety of procedures for casting metal parts. These include such processes as permanent mold casting, centrifugal casting, evaporative pattern casting, and sand casting, all of which are described in the Section "Molding and Casting Processes" in this Volume. In sand casting, molds and cores are used. Cores are required for hollow castings and must be removed after the metal has solidified.

Binders were developed to strengthen the cores, which are the most fragile part of a mold assembly. Although the use of binders in mold production is increasing, most sand casting employs green sand molds, which are made of sand, clay, and additives (green sand molding is described in the section "Bonded Sand Molds" in the article "Sand Molding" in this Volume).

Inorganic binders, such as clay or cement, are materials that have long been used in the production of foundry molds and cores. This article is limited to organic resin-base binders for sand molding. In practice, these binders are mixed with sand, the mixes are compressed into the desired shape of the mold or core, and the binders are hardened, that is, cured, by chemical or thermal reactions to fixate the shapes. Typically, 0.7 to 4.0 parts (usually 1 to 2 parts) of binder are added to 100 parts of sand.

Classification of Resin Binder Processes

Although a wide variety of resin binder processes are currently used, they can be classified into the following categories:

- No-bake binder systems
- Heat-cured binder systems
- Cold box binder systems

In the no-bake and cold box processes, the binder is cured at room temperature; in the shell molding, hot box, and oven-bake processes, heat cures are applied. Selection of the process and type of binder depends on the size and number of cores or molds required, production rates, and equipment. Properties of the various binder systems are described below and compared in Tables 1 to 3. Figure 1 summarizes the trends in resin binder usage in the foundry industry.

No-Bake Processes

A no-bake process is based on the ambient-temperature cure of two or more binder components after they are combined on sand. Curing of the binder system begins immediately after all components are combined. For a period of time after initial mixing, the sand mix is workable and flowable to allow the filling of the core/mold pattern. After an additional time period, the sand mix cures to the point where it can be removed from the box. The time difference between filling and stripping of the box can range from a few minutes to several hours, depending on the binder system used, curing agent and amount, sand type, and sand temperature.

Furan Acid Catalyzed No-Bake. Furfuryl alcohol is the basic raw material used in furan no-bake binders. Furan binders can be modified with urea, formaldehyde, phenol, and a variety of other reactive or nonreactive additives.

The great variety of furan binders available provides widely differing performance characteristics for use in various foundry applications. Water content may be as high as 15% and nitrogen content as high as 10% in resins modified with urea. In addition, zero-nitrogen and zero-water binders are available. The choice of a specific binder depends on the type of metal to be cast and the sand performance properties required. The amount of furan no-bake binders used ranges between 0.9 and 2.0% based on sand weight. Acid catalyst levels vary between 20 and 50% based on the weight of the binder. The speed of the curing reaction can be adjusted by changing the catalyst type or percentage, given that the sand type and temperature are constant. The furan no-bake curing mechanism is shown in Fig. 2.

Furan no-bake binders provide high dimensional accuracy and a high degree of resistance to sand/metal interface casting defects, yet they decompose readily after the metal has solidified, providing excellent shakeout. Furan no-bake binders also exhibit high tensile strength, along with the excellent hot strength needed for flaskless no-bake molding. They often run sand-to-metal ratios of as low as 2:1.

Phenolic Acid Catalyzed No-Bake. Phenolic resins are condensation reaction products of phenol(s) and aldehyde(s). Phenolic no-bake resins are those formed from phenol/formaldehyde where the phenol/formaldehyde molar ratio is less than 1. Again, as with furan no-bakes, these resins can be modified with reactive or nonreactive additives.

Table 1 A comparison of properties of no-bake binder systems

	Process(a)							
	Acid catalyzed		Ester cured					
Parameter	Furan	Phenolic	Alkaline/ phenolic	Silicate	Oil urethane	Phenolic urethane	Polyoliso-cyanate	Alumina phosphate
Relative tensile strength	H	M	L	M	H	M	M	M
Rate of gas evolution	L	M	L	L	M	H	H	L
Thermal plasticity	L	M	M	H	L	L	L	L
Ease of shakeout	G	F	G	P	P	G	E	G
Humidity resistance	F	F	E	P	G	G	G	P
Strip time, min(b)	3–45	2–45	3–60	5–60	2–180	1–40	2–20	30–60
Optimum (sand) temperature, °C (°F)	27 (80)	27 (80)	27 (80)	24 (75)	32 (90)	27 (80)	27 (80)	32 (90)
Clay and fines resistance	P	P	P	F	F	P	P	F
Flowability	G	F	F	F	F	G	G	F
Pouring smoke	M	M	L	N	H	M	M	N
Erosion resistance	E	E	E	G	F	G	P(e)	G
Metals not recommended	(c)	...	...	...	...	(d)	(e)	...

(a) H, high; M, medium; L, low; N, none; E, excellent; G, good; F, fair; P, poor. (b) Rapid strip times required special mixing equipment. (c) Use minimum N_2 levels for steel. (d) Iron oxide required for steel. (e) Use with nonferrous metals

*Adapted with permission from P.R. Carey *et al.*, Updating Resin Binder Processes—Part I through IX, *Foundry Mgmt. Technol.*, Feb 1986

Table 2 Comparison of properties of the heat-cured binder systems

	Process(a)				
		Hot box			
Parameter	Shell process	Furan	Phenolic	Warm box	Oven bake (core oil)
Relative tensile strength	H	H	H	H	M
Rate of gas evolution	M	H	H	M	M
Thermal plasticity	M	L	M	L	M
Ease of shakeout	F	G	F	G	G
Humidity resistance	E	F	G	G	G
Cure speed	H	H	M	H	L
Resistance to overcure	G	F	F	F	P
Optimum core pattern temperature, °C (°F)	260 (500)	260 (500)	260 (500)	175 (350)	205 (400)
Clay and fines resistance	F	P	P	P	F
Flowability	E	G	F	G	F
Bench life of mixed sand	E	F	F	F	G
Pouring smoke	M	M	M	M	M
Metals not recommended	N	(b)	Steel	(b)	(c)

(a) H, high; M, medium; L, low; N, none; E, excellent; G, good; F, fair; P, poor. (b) Use minimum N_2 levels for steel. (c) Iron oxide required for steel

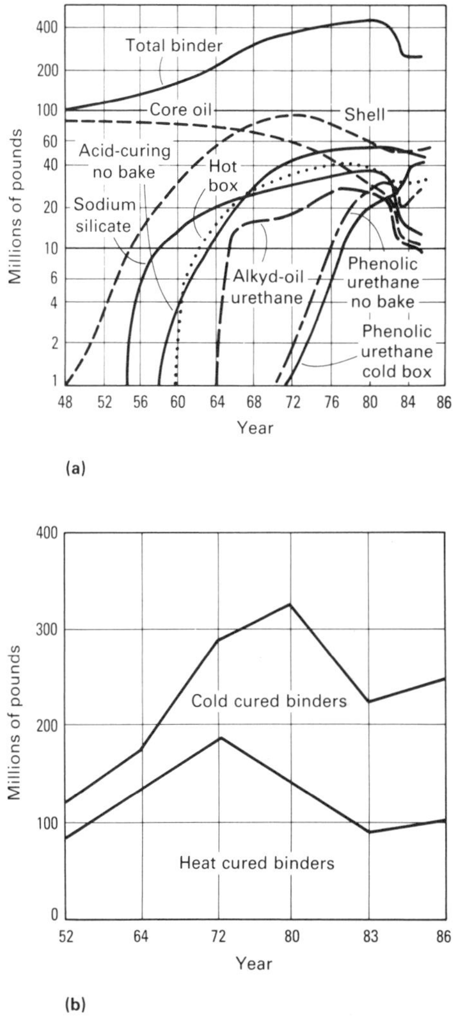

Fig. 1 Market status of resin binder processes. (a) Trends in foundry sand binder consumption showing great variations in volume. These variations have been accompanied by machinery changes. The 1985 figures are extrapolated. (b) Heat cured versus cold cured binders (U.S. foundry consumption). Source: Ashland Chemical Company

These resins are clear to dark brown in appearance, and their viscosities range from medium to high. Sand mixes made with these resins have adequate flowability for the filling of mold patterns or core boxes. Resins of this type contain free phenol and free formaldehyde. Phenol and formaldehyde odors can be expected during sand-mixing operations.

One disadvantage of acid-cured phenolic no-bake resins is their relatively poor storage stability. Phenolic binders are usually not stored for more than 6 months. Phenolic resole resins contain numerous reactive methylol groups, and these are generally involved in auto-polymerization reactions at ambient or slightly elevated temperatures. The storage period can be considerably longer during the winter months if the temperature of storage remains at 20 °C (70 °F) or lower. The viscosity advances as the binder ages.

The catalyst needed for the phenolic no-bake resin is a strong sulfonic acid type. Phosphoric acids will not cure phenolic resins at the rate required for most no-bake foundry applications.

The phenolic no-bake reaction mechanism is:

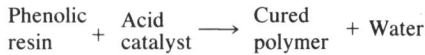

Phenolic resin + Acid catalyst → Cured polymer + Water

The catalyst initiates further condensation of the resin and advances the cross-linking reaction. The condensation reactions produce water which results in a dilution effect on the acid catalyst that tends to slow the rate of cure. Because of this effect, it is necessary to use strong acid catalysts to ensure an acceptable rate of cure and good deep set properties.

Ester-Cured Alkaline Phenolic No-Bake. The ester-cured phenolic binder system is a two-part binder system consisting of a water-soluble alkaline phenolic resin and liquid ester co-reactants. The reaction mechanism is:

Alkaline phenolic resin + Ester co-reactant
→ Suspected unstable intermediate
→ Splits to form:
  Polymerized phenolic resin
  Alkaline salts and alcohol

A secondary reaction is thought to occur when the partially polymerized resin contacts heat during the pouring operation, yielding an extremely rigid structure.

The viscosity of the ester-cured phenolic is similar to that of the acid-catalyzed phenolic no-bakes. It has a shelf life of 4 to 6 months at 20 °C (70 °F). Typically, 1.5 to 2.0% binder based on sand and 20 to 25% co-reactant based on the resin are used to coat washed and dried silica sand in most core and molding operations.

Both the resin and co-reactant are water soluble, permitting easy cleanup. Physical strength of the cured sand is not as high as that of the acid-catalyzed and urethane no-bakes at comparable resin contents. However, with care in handling and transporting, good casting results can be obtained. The distinct advantages of the ester-cured phenolic no-bake systems are the reduction of veining in iron castings and excellent erosion resistance.

Silicate/Ester-Catalyzed No-Bake. This no-bake system consists of the sodium silicate binder and a liquid organic ester that

Table 3 Comparison of properties for cold box binder systems

	Process(a)				
Parameter	Phenolic urethane	SO_2 process (furan/SO_2)	FRC process acrylic/epoxy	Phenolic ester	Sodium silicate CO_2
Relative tensile strength	H	M	H	L	L
Rate of gas evolution	H	L	H	L	M
Thermal plasticity	L	N	L	L	H
Ease of shakeout	G	E	G	G	P
Moisture resistance	M	H	M	M	L
Curing speed	H	H	H	M	M
Resistance to overcure	G	G	G	G	P
Optimum temperature, °C (°F)	24 (75)	32 (90)	24 (75)	24 (75)	24 (75)
Clay and fines resistance	P	P	P	P	F
Flowability	G	G	E	F	P
Bench life of mixed sand	F	G	E	F	F
Pouring smoke	H	L	M	L	N
Erosion resistance	G	E	F	E	G
Veining resistance	F	F	G	G	F
Metals not recommended	(b)	...	(c)	...	...

(a) H, high; M, medium; L, low; N, none; E, excellent; G, good; F, fair; P, poor. (b) Iron oxide required for steel. (c) Binder selection available for type of alloy

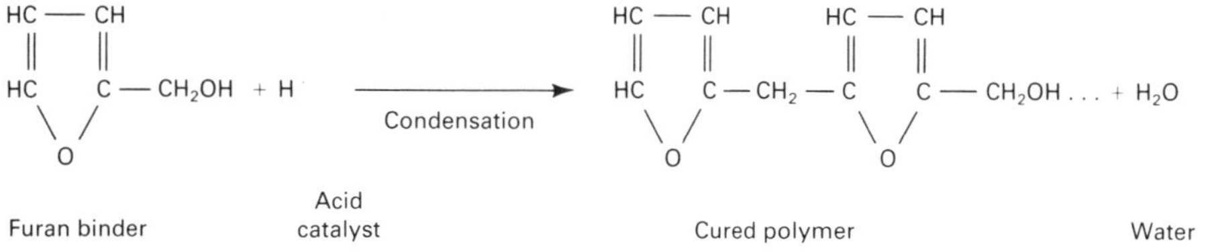

Fig. 2 The furan acid-catalyzed no-bake curing mechanism

functions as the hardening agent. High-ratio binders with SiO_2:Na_2O contents of 2.5 to 3.2:1 are employed for this process, and mixtures usually contain 3 to 4% binder. The ester hardeners are materials such as glycerol diacetate and triacetate, or ethylene glycol diacetate; they are low-viscosity liquids with either a sweet or acetic acid-like smell. The normal addition rate for the ester hardener is 10 to 15% based on the weight of sodium silicate and should be added to the sand prior to the addition of the silicate binder.

The curing rate depends on the SiO_2:Na_2O ratio of the silicate binder and the composition of the ester hardener. Suppliers produce blends of ester hardeners giving work times that are controllable from several minutes to 1 h or longer. The hardening reaction, involving the formation of silica gel from the sodium silicate, is a cold process, and no heat or gas is produced. When added to a sand mixture containing the alkaline sodium silicate, the organic esters hydrolyze at a controlled rate, reacting with sodium silicate to form a silica gel that bonds the aggregate. A simplified version of this curing mechanism is:

Sodium silicate (Na_2SiO_3) + Liquid ester hardener ⟶ Cured polymer

Mixed sand must be used before hardening begins. Material that has exceeded the useful work time and feels dry or powdery should be discarded. The use of sand past the useful bench life will result in the production of weak, friable molds and cores that can result in penetration defects.

Curing takes several hours to complete after stripping. Large molds may need 16 to 24 h. Strengths can be higher than those of CO_2-cured molds, and shelf life is better. Although shakeout is easier than with CO_2-silicate systems, it is not as good as the other no-bake processes outlined in this article.

Odor and gaseous emission levels are low during mixing, pouring, cooling, and shakeout, but depend on the extent of organic additives in the mix. Casting defects such as veining and expansion are minimal. Burn-on and penetration are generally more severe than for the other no-bake systems and can be controlled by sand additives and a wash practice.

Oil urethane no-bake resins (also known as oil-urethane, alkyd-urethane, alkyd-oil-urethane, or polyester-urethane) are three-component systems that consist of Part A, an alkyd oil type resin; Part B, a liquid amine/metallic catalyst; and Part C, a polymeric methyl di-isocyanate (MDI) (the urethane component).

The three-part system uses the Part B catalyst to achieve a predictable work/strip time. It can be made into a two-part system by preblending Parts A and B when the amount of the Part B catalyst added to the resin controls the work/strip time. Part A can also be modified for better coating action, improved performance in temperature extremes, or better strippability.

Part A is normally used at 1 to 2% of sand weight. The Part B catalyst, whether added as a separate component or preblended with Part A, is 2 to 10% by weight of Part A. The Part C isocyanate is always 18 to 20% by weight of Part A.

Although the oil urethane no-bake system is easy to use, the curing mechanisms are difficult to understand because there are two separate curing stages and two curing mechanisms. When the three components are mixed together on the sand, the polyisocyanate (Part C) quickly begins to cross-link with the alkyd oil resin (Part A) at a rate controlled by the urethane catalyst component of Part B, as shown in Fig. 3. This action produces a urethane coating on the sand with enough bonded sand strength to strip the pattern and handle the core or mold.

The other stage of the curing reaction is similar to a paint-drying mechanism in which oxygen combines with the alkyd-oil resin component and nearly polymerizes it fully at room temperature to form a tough urethane bond. The metallic driers present in the Part B catalyst accelerate the oxygenation or drying (slowly at room temperature or quickly at 150 to 205 °C, or 300 to 400 °F), but because the full cure is oxygen dependent, section size and shape, along with temperature, determine how long it takes to attain a complete cure.

The alkyd-oil urethane mechanism is a two-stage process involving:

Alkyd + NCO (polymeric isocyanate) (partial cross-link) + Urethane catalyst → Alkyd urethane

Alkyd + O_2 + Metallic driers → Rigid cross-linked urethane

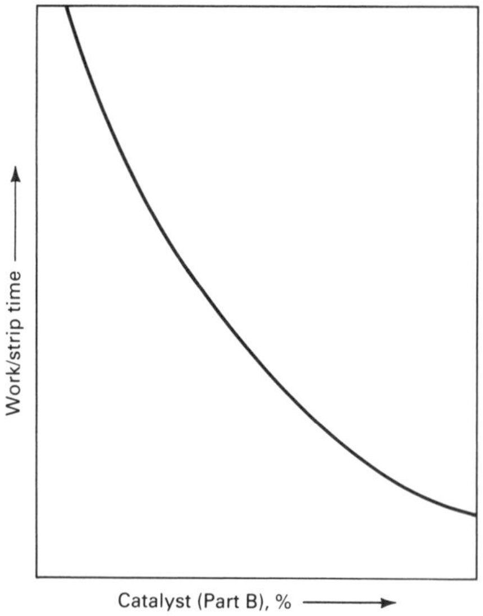

Fig. 3 Effect of increasing oil urethane system (Part B) (catalyst) on work time and strip time

For maximum cure and ultimate casting properties, the mold or core should be heated to about 150 °C (300 °F) in a forced air oven for 1 h.

The oil urethane no-bake system, with its unique two-stage cure, results in unmatched stripping characteristics and provides foundrymen with a good method of producing large cores and molds that require long work and strip times.

The phenolic urethane no-bake (PUN) binder system has three parts. Part I is a phenol formaldehyde resin dissolved in a special blend of solvents. Part II is a polymeric MDI-type isocyanate, again dissolved in solvents. Part III is an amine catalyst that, depending on strength and amount, regulates the speed of the reaction between Parts I and II. The chemical reaction between Part I and Part II forms a urethane bond and no other by-products. For this reason and because air is not required for setting, the PUN system does not present the problems with through-cure or deep-set found in other no-bake systems. A simplified version of the curing mechanism for phenolic urethane no-bake systems is:

Liquid phenolic resin Part I + Liquid polyisocyanate Part II + Liquid amine catalyst Part III = Solid resin + heat

Phenolic urethane no-bake binders are widely used for the production of both ferrous and nonferrous castings and can be successfully used for high-production operations or jobbing shops because of their chemical reaction time and ease of operation.

Although many types of mixers can be used with PUN binders, zero-retention high-speed continuous mixers are the most widely used. Because the mixing takes place rapidly, the fast strip times (as fast as 30 s) of the PUN system can be utilized in practice. No mixed sand is retained in the mixer to harden after it is shut off. Further, the mixers can be coordinated with pattern movement, sand compaction, stripping operations, and mold or core finishing and storage to create a simple manual or fully automated no-bake loop.

Total binder level for the PUN system is 0.7 to 2% based on the weight of sand. It is common to offset the ratio of Part I to Part II at 55:45 or 60:40. The third-part catalyst level is based on the weight of Part I. Depending on the catalyst type and strip time required 0.4 to 8% catalyst (based on Part I) is normally added.

Compaction of the mixed sand can be accomplished by vibration, ramming, and tucking. The good flowability of PUN sand mixes allows good density with minimum effort. Because the PUN system cures very rapidly, the time required for the compacted pattern to reach rollover and strip must coincide with the setup or cure time of the sand mix.

For certain ferrous applications (most commonly steels), the addition of 2 to 3% iron oxide to the sand mix can improve casting surface finish. This addition is also beneficial in reducing lustrous carbon defects by promoting a less reducing mold atmosphere. The PUN resin system contains about 3.0 to 3.8% N (which is about 0.04% based on sand). To reduce the chance of nitrogen-related casting defects, the Part I to Part II ratio can be offset 60:40 in favor of the Part I because substantially all the nitrogen is in Part II. It has also been shown that as little as 0.25% red iron oxide is effective in suppressing the ferrous casting subsurface porosity associated with nitrogen in the melt and/or evolved from the PUN binder.

The polyol-isocyanate system was developed in the late 1970s for aluminum, magnesium, and other light-alloy foundries. Previously, the binder systems used in light-alloy foundries were the same as those used for the ferrous casting industry. The lower pouring temperatures associated with light alloys are not sufficient to decompose most no-bake binders, and removal of cores from castings is difficult. The polyol-isocyanate system was developed to provide improved shakeout.

The nonferrous binders are similar to the PUN system described previously. Part I is a special polyol designed for good thermal breakdown dissolved in solvents. Part II is a polymeric MDI-type isocyanate, again dissolved in solvents. Part II is an amine catalyst that can be used to regulate cure speed.

The chemical curing reaction of the polyol-isocyanate system is as follows:

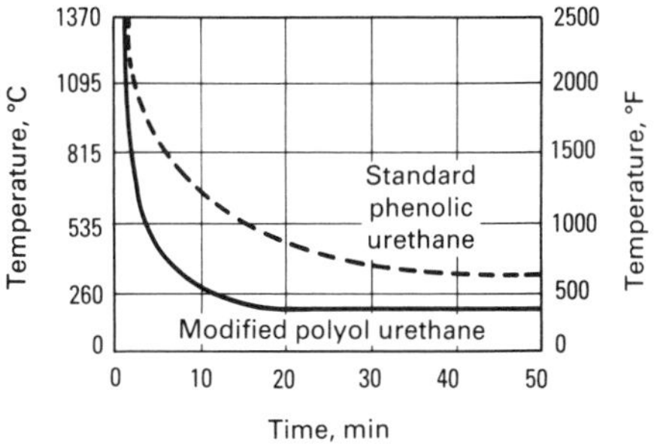

Fig. 4 Collapsibility of polyol urethane compared to that of phenolic urethane

$$\text{Liquid polyol resin} + \text{Liquid polyisocyanate} = \text{Solid resin} + \text{heat}$$

In practice, polyol-isocyanate binders are used in much the same way as the PUN binders they evolved from. One difference is that the system does not require a catalyst. Several phenol formaldehyde (Part I) resins are available that provide strip times from 8 min to over 1 h. For maximum control, however, an amine (Part III) catalyst can be used to regulate strip times to as fast as 3 min.

For light-alloy applications, binder levels range from 0.7 to 1.5% based on sand. Part I and Part II should be used at a 50:50 ratio for best results. Reactivity, strengths, and work-time-to-strip-time ratio are affected by the same variables as the PUN binders. Because of the fast thermal breakdown of the binder (Fig. 4), the polyol-urethane system is not recommended for ferrous castings.

Alumina-Phosphate No-Bake. Alumina-phosphate binders consist of an acidic, water-soluble alumina-phosphate liquid binder and a free-flowing powdered metal oxide hardener. Although this system is classified as a no-bake process (Table 1), both of its parts are inorganic; all other no-bake systems are organic or, in the case of silicate/ester systems, inorganic and organic. More detailed information on phosphate-bonded systems can be found in the article "Sand Molding" in this Volume (see the section "Bonded Sand Molds").

Shell Process

In the shell process, also referred to as the Croning process, the sand grains are coated with phenolic novolac resins and hexamethylenetetramine. In warm coating, dissolved or liquid resins are used, but in hot coating, solid novolac resins are used. The coated, dry, free-flowing sand is compressed and cured in a heated mold at 150 to 280 °C (300 to 535 °F) for 10 to 30 s. Sands prepared by warm coating cure fast and exhibit excellent properties. Hot-coated sands are generally more free flowing with less tendency toward caking or blocking in storage.

Novolac Shell-Molding Binders. Novolacs are thermoplastic, brittle, solid phenolic resins that do not cross-link without the help of a cross-linking agent. Novolac compositions can, however, be cured to insoluble cross-linked products by using hexamethylenetetramine or a resole phenolic resin as a hardener. A simplified version of the Novolac curing mechanism is:

$$\text{Novolac} + \text{Hexamethylenetetramine} \xrightarrow{\text{Heat}} \text{Cured polymer} + \text{ammonia}$$

Phenol-formaldehyde novolac resins are the primary resins used for precoating shell process sand. These resins are available as powder, flakes, or granules or as solvent- or waterborne solutions. A lubricant, usually calcium stearate (4 to 6% of resin weight) is added during the resin production or the coating process to improve flowability and release properties. Hexamethylenetetramine, 10 to 17% based on resin weight, is used as a cross-linking agent.

Producing Cores and Molds. The shell-resin curing mechanism involves the transition from one type of solid plastic to another—thermoplastic to thermosetting. This physical conversion must be completed during a brief period of the shell cycle before the heat (necessary to cure the resin) begins to decompose the binder. Pattern temperatures are typically 205 to 315 °C (400 to 600 °F). Operating within the ideal temperature range provides a good shell thickness, optimum resin flow, and minimal surface decomposition. Higher pattern temperatures of 275 to 315 °C (525 to 600 °F) are often successfully used to make small cores, because the shell cycle is short enough that little surface definition is lost by decomposition of the resin at the pattern interface during the relatively brief cure cycle generally needed.

Various additives are used during the coating operation for specific purposes. They include iron oxide to prevent thermal cracking, to provide chill, and to minimize gas-related defects.

The shell process has some advantages over other processes. The better blowability and superior flowability of the lubricant-containing shell sand permits intricate cores to be blown and offers excellent surface reproduction in shell molding. Because the bench life of the coated shell sand is indefinite, machines do not require the removal of process sand at the end of a production period.

Storage life of cured cores or molds is excellent. A variety of sands are usable with the process, and nearly all metals and alloys have been successfully cast using shell sand for cores and molds.

Hot Box and Warm Box Processes

In the hot box and warm box processes, the binder-sand mixture is wet. A liquid thermosetting binder and a latent acid catalyst are mixed with dry sand and blown into a heated core box. The curing temperature depends on the process. Upon heating, the catalyst releases acid, which induces rapid cure; therefore, the core can be removed within 10 to 30 s. After the cores are removed from the pattern, the cure is complete as a result of the exothermic chemical reaction and the heat absorbed by the core. Although many hot box cores require postcuring in an oven to complete the cure, warm box cures require no postbake oven curing.

Hot Box Binders. Conventional hot box resins are classified simply as furan or phenolic types. The furan types contain furfuryl alcohol, the phenolic types are based on phenol, and the furan-modified phenolic has both. All conventional hot box binders contain urea and formaldehyde. The furan hot box resin has a fast cure compared to that of the phenolic-type system and can therefore be ejected faster from the core box. Furan resin also provides superior shakeout and presents fewer disposal problems because of the lack of phenol. Typical resin content is 1.5 to 2.0%.

A simplified hot box reaction mechanism is:

$$\text{Liquid resin} + \text{Catalyst} + \text{Heat} = \begin{matrix}\text{Solid resin}\\ + \text{ water}\\ + \text{ heat}\end{matrix}$$

Catalyst selection is based on the acid demand value and other chemical properties of the sand. Sand temperature changes of 11 °C (20 °F) and/or variations of ±5 units in the acid demand value of the sand require a catalyst adjustment to maintain optimum performance. When a liquid catalyst is used, many operations have winter and summer grades that can be mixed together during seasonal transitions. Both chloride and nitrate catalysts are used. The chloride catalyst is the more reactive. Therefore, the chloride is the winter grade, and the nitrate the summer grade.

Hot box resins have a limited shelf life and increase in viscosity with storage. If possible, containers should be stored out of the sun in a cool place and used on a first-in-first-out basis. Hot box catalysts have indefinite storage lives.

Pattern temperature should not vary more than 28 °C (50 °F). Measurements should be made at the highest and lowest points across the pattern. Most production shops run hot box pattern temperatures of 230 to 290 °C (450 to 550 °F), but the ideal temperature is between 220 and 245 °C (425 and 475 °F). The most common mistake made with the hot box process is to run too high a pattern temperature, which causes poor core surfaces. This condition results in a friable core finish that is especially detrimental to thin-section cores.

The color of the core surface shows how thoroughly the core is cured and is a good curing guideline. The surface should be slightly yellow or very light-brown—not dark brown or black. Overall, the phenolic and furan hot box resins are extensively used in the automotive industry for producing intricate cores and molds that require good tensile strengths for low cost gray iron castings.

Warm Box Binders. The warm box resin is a minimum-water (<5%) furfuryl alcohol-type resin (furfuryl alcohol content: ~70%) that is formulated for a nitrogen content of less than 2.5%. Because the resin/sand mix exhibits a high degree of rigid thermoset properties when fully cured, little or no post strip distortion or sagging occurs. High hot and cold tensile properties are characteristic of warm box sands and generally permit a binder level between 0.8 to 1.8%, or 20% less than the conventional hot box resin content.

Warm box catalysts are copper salts based primarily on aromatic sulfonic acids in an aqueous methanol solution. The catalyst amount used is 20 to 35%, based on resin weight. These catalysts are unusual in that they impart an excellent latent property (unreactive at room temperature) to the binder system, but still form strong acids when heated. They promote a thorough curing action at temperatures at approximately 65 °C (150 °F) or higher.

A simplified warm box curing reaction mechanism is:

$$\begin{matrix}\text{Furan}\\ \text{resin}\end{matrix} + \begin{matrix}\text{Latent}\\ \text{acid } (H^+)\end{matrix} \xrightarrow{\text{Heat}} \text{Cured furan binder}$$

The binder components remain stable when mixed together in the proper ratios in sand until activated by heat, which decomposes the catalyst and releases the acid that causes the resin to polymerize.

The pattern temperatures used range from 150 to 230 °C (300 to 450 °F). The optimum temperature of 190 °C (375 °F) is about 55 °C (100 °F) below the operating temperature for hot box binders. Low resin and catalyst viscosity combine to produce a flowable sand mix. Castings produced with warm box binders exhibit casting features that are very similar to those of a furan no-bake system. Good dimensional accuracy and excellent erosion resistance are observed with warm box binders.

Oven-Bake Processes/ Core-Oil Binders

Core-oil binder is used in combination with a water-activated cereal to produce a coated sand mix that has green strength. Green strength permits the wet sand mix to be blown or hand rammed into a simple vented, relatively low-cost core box at room temperature and to retain its shape when removed from the pattern.

The uncured plasticlike cores are generally placed on a flat board or a dryer plate (a supporting structure to maintain the shape of the core) for oven drying. This process translates into a fast method of producing cores or molds. Except for the subsequent drying operations, the cores are in effect made almost as fast as they are blown.

Types of Core Oil. A binder system that uses water and cereal to develop green strength and then cures or dries in a hot, forced air oven is normally referred to as a core-oil process. Several types of binders fall into this category. Linseed or vegetable oil binders account for most of the volume, but urea formaldehyde and resole phenolic resins are also used. The sand-coating procedure, sand formulation, coremaking techniques, and general foundry procedures are chemically different, but are similar for all types of core-oil processes.

Urea formaldehyde is noted for its excellent shakeout and is used in aluminum foundries and shops that operate dielectric curing ovens instead of the hot, forced air ovens. Because of its high nitrogen content and low hot strength, urea formaldehyde has found rather limited application outside of nonferrous shops.

Additives. Core mixes generally contain 1% or less cereal, based on sand weight. The cereal is kept to a minimum because it generates gas. Normally, when more green strength is required for core stripping and/or handling, the cereal is mulled along with the water for a longer time.

Small additions of Southern and/or Western bentonite (up to ½% based on sand weight) to cereal have also proved useful for developing green strength. An additional benefit is that bentonite evolves far less gas than cereal does.

Water is added to the mix to activate the cereal and to create green strength. The amount must be controlled to develop optimum properties, as indicated in Table 4. Baked strength, green strength, and baking rate are influenced by moisture content. A 2% water addition gives optimum results.

Operational Considerations. A number of points are essential for effective use of core-oil systems:

- *Basic Mix.* A standard core-oil mix contains about 1% cereal, 1 to 3.5% water, 1% binder, and 0.1% of a flowability/release agent
- *Mixing Order.* As indicated in Table 5, the sequence of additions to the muller has a significant effect on core properties. The best order of addition to the sand clearly is (1) cereal and dry additives, (2) water, (3) oil, and (4) flowability/release agent

Table 4 Effects of moisture on core-oil sand mixes

1% cereal, 1% oil, 90 min bake at 200 °C (400 °F), AFS 62 GPN silica sand

Property	Percentage of moisture							
	0	0.5	1.0	1.5	2.0	3.0	4.0	5.0
Green strength, kPa (psi)	1.4 (0.2)	3.4 (0.5)	6.2 (0.9)	8.9 (1.3)	9.6 (1.4)	7.6 (1.1)	6.9 (1.0)	6.2 (0.9)
Tensile strength, kPa (psi)	1205 (175)	1450 (210)	1825 (265)	2275 (330)	2480 (360)	2345 (340)	2135 (310)	1725 (250)
Percentage baked	100	100	100	100	100	97	88	78
Scratch hardness index	60	71	82	92	96	95	90	88

- *Oven Drying*. Forced hot air is the principal means of curing core-oil binders. Heat causes the binder in the core-oil sand mix to cross-link and provide strength. The proper combination of oven temperature, drying humidity, and time determines the final strength, dimensional stability, and surface finish of the core

It is important to note that mixing cores produced by other binder processes in the same drying oven with uncured core-oil cores usually weakens the non-core-oil cores because of the steam evolved from the water used in the core-oil process

Cold Box Processes

The term cold box process implies the room-temperature cure of a binder-sand mixture accelerated by a vapor or gas catalyst that is passed through the sand. Several different cold box systems are currently used, employing different binders and gas or vapor catalysts—for example, triethylamine or dimethylethylamine for phenolic urethane binders, sulfur dioxide for furan and acrylic epoxy binders, methyl formate for ester-cured alkaline phenolic binders, and carbon dioxide for silicate binders.

Phenolic Urethane Cold Box (PUCB). The process uses a three-part binder system consisting of a phenolic resin (Part I), a polymeric isocyanate (Part II), and a tertiary amine vapor catalyst (Part III). Sand is coated with the Part I and Part II components and compacted into a pattern at room temperature. The tertiary amine catalyst is introduced through vents in the pattern to harden the contained sand mix. The catalyst gas cycle is followed by an air purge cycle that forces the amine gas through the sand mass and removes residual amine from the hardened core. It is recommended that the exhaust from the core box be scrubbed chemically to remove the amine.

The reactive component of the PUCB Part I is phenolic resin. It is dissolved in solvents to yield a low-viscosity resin solution to facilitate coating the sand and blending it with the second component. Part II is a polymeric MDI-type isocyanate that again is blended with solvents to form a low-viscosity resin solution. The hydroxyl groups provided by the phenolic resin in Part I react with the isocyanate groups in Part II in the presence of the amine catalyst to form the solid urethane resin. It is this urethane polymer that bonds the sand grains together and gives the PUCB process its unique properties.

A simplified curing reaction mechanism for the PUCB process is:

$$\text{Phenolic resin} + \text{Polyisocyanate} \xrightarrow{\text{Vapor amine catalyst}} \text{Urethane}$$

The urethane reaction does not produce water or any other by-product. The system contains 3 to 4% N, which is introduced from the Part II polymeric isocyanate component. The organic resins and solvents in the PUCB system make it high in carbon content, which contributes to the formation of lustrous carbon and a reducing mold atmosphere during casting.

The PUCB process can be used with all of the sands commonly used for coremaking in the foundry industry. Some consideration must be given, however, to the effects of sand temperature, chemistry, and moisture content on the resin performance of PUCB. The ideal sand temperature is 20 to 25 °C (70 to 80 °F). Lower temperature can reduce mixing efficiency and increase cure times. Higher sand temperatures reduce gassing cycles and the amount of catalyst required, but shorten the usable life of the coated sand mix.

A maximum sand moisture content of 0.2% is acceptable for the PUCB process at room temperature (~20 °C, or 70 °F), but when the sand temperature rises to 30 °C (90 °F), the moisture content of the sand must be kept at less than 0.1% for the process to function properly.

All types of popular sand-mixing equipment can be used with the PUCB process. A sand delivery system that causes the least amount of aeration is best. Typically, 1.5% total binder, consisting of equal parts of Part I and Part II components, is used on a washed and dried sand for ferrous castings. Many foundries prefer to offset the ratio and use slightly more Part I for various technical reasons. For the casting of aluminum, magnesium, and other low pouring temperature alloys, binder levels of 1% and less are used to facilitate shakeout.

The volatile liquid tertiary amines commonly used to cure the PUCB binders are triethylamine or dimethylethylamine. Various designs of generators vaporize and blend these amines with carrier gas and deliver them to the core machine. The best generators provide a consistently high concentration of amine to facilitate fast, predictable cure cycles.

The exhaust from the core box is delivered to a chemical scrubber, and the amine is removed by reacting it with dilute sulfuric acid to form an amine sulfate salt. In larger foundry operations, concentrated liquor from the scrubber has been recycled through chemical processing to convert it back into usable amine, thus providing economic and ecological advantages.

Certain sand additives can be used with the PUCB system to eliminate specific casting defects. Veining in ferrous and brass castings can be substantially reduced by the addition of 1 to 2% proprietary clay-sugar blends or 1 to 3% iron oxide.

Black and red iron oxide additions of 2 to 3% are recommended for steel castings. Red iron oxide at levels as low as 0.25% can be effective in eliminating binder-induced subsurface pinhole porosity in alloys prone to those defects.

SO_2 Process (Furan/SO_2). The sulfur dioxide (SO_2) process can be described as a rapid-curing, gas-activated, furan no-bake. Various furfuryl alcohol-base resins, blended with an adhesion promoter, are used to coat the sand in the range of 0.9 to 1.5%. Organic hydroperoxides at 30 to 50% by

Table 5 Sequence of muller additions to core-oil sands

1% cereal, 1% oil, 1.5% water, 90 min bake at 200 °C (400 °F), silica sand

Property	Percentage of moisture						
	All at once	1. Cereal 2. Oil 3. Water	1. Cereal 2. Water 3. Oil	1. Oil 2. Cereal 3. Water	1. Oil 2. Water 3. Cereal	1. Water 2. Oil 3. Cereal	1. Water 2. Cereal 3. Water
Green strength, kPa (psi)	6.9 (1.0)	2.0 (0.3)	10.3 (1.5)	3.4 (0.5)	4.1 (0.6)	3.4 (0.5)	7.6 (1.1)
Tensile strength, kPa (psi)	1930 (280)	1345 (195)	2495 (362)	1725 (250)	1450 (210)	1380 (200)	2275 (330)
Scratch hardness index	89	80	97	84	84	86	91

weight of the resin are added to the sand mixture and/or blended with the resin. Methanol-diluted silane (5 to 10%, based on resin weight) is used to increase strength, to improve shelf life, and to increase humidity resistance.

Once the sand is in place, SO_2 gas is introduced. It reacts with the peroxide and water in the furan resin, causing an *in situ* formation of a complex group of acids that cures the furan binder.

The simplified curing reaction mechanism for the SO_2 process is:

$$\text{Furan binder} + H^+ \xrightarrow{\text{Heat}} \underset{\text{(polycondensation)}}{\text{Cross-linking}} + \text{dehydration}$$

The curing reaction begins when the SO_2 first contacts the peroxide and continues even after removal from the pattern. As the catalyzed furan/SO_2 resin bonded sand ages, the water of condensation resulting from the furan polymerization continues to dissipate from the still-curing sand until the reaction is complete. Upon curing, the sand changes from a light color to dark green or black.

The recommended sand temperature for the SO_2 process is 25 to 40 °C (80 to 100 °F). Lower temperatures reduce cure speed and may produce partially cured cores. Higher temperatures promote evaporation of solvents and reduce mixed-sand bench life. Hot purging with air at 95 °C (200 °F) is required to achieve optimum cure.

The SO_2 process develops approximately 20 to 50% of its overall tensile strength upon ejection from the core box. It then builds strength rapidly to 85 to 95% of overall tensile strength after about 1 h. Bench life of the mixed sand is 12 to 24 h. In typical core blowing operations, relatively low blow pressures of 275 to 415 kPa (40 to 60 psi) are possible because excellent flowability is characteristic of the system.

The free radical cure (FRC) process includes all acrylic and acrylic-epoxy functional binders. The binders are cured using an organic hydroperoxide and SO_2. A variety of acrylic-epoxy binders have been developed for both ferrous and nonferrous applications, ranging from 100% acrylic binders to approximately 30:70 ratios of acrylic-epoxy blends. Sand performance and casting properties are influenced by the ratio of acrylic to epoxy functional components present in the binder system.

Acrylic binders are based on acrylic functional components. When combined with small amounts of organic hydroperoxides, acrylic binders can be cured through the application of a wide range of concentrations of sulfur dioxide with inert gas carriers such as nitrogen (1 to 100% SO_2).

Acrylic binders are primarily used in light-metal applications because of their good shake-out properties; however, they have specific applications in ferrous metals where veining defects are troublesome with other binder systems. When acrylic binders are used in ferrous applications, a refractory coating and nonturbulent gating design are recommended to reduce the threat of erosion.

Acrylic-epoxy binders are blends of acrylic and epoxy functional components. The acrylic-epoxy binder systems offer alternatives to existing cold box systems. The process utilizes a two-part liquid binder system consisting of (1) an unsaturated resin and/or monomer and (2) an organic hydroperoxide with an epoxy resin. The mixed sand, once exposed to SO_2 gas, yields a cured polymer. A simplified version of this reaction mechanism is:

$$\text{Epoxy resin} + \begin{array}{c}\text{Unsaturated polymer}\\ \text{and monomer}\end{array} \xrightarrow[\begin{array}{c}SO_2\text{ organic}\\ \text{hydroperoxide}\end{array}]{} \text{Cured polymer}$$

Varying the acrylic-epoxy composition influences such core- and moldmaking properties as rigidity, cure speed, SO_2 consumption, humidity resistance, tensile and transverse strengths, and core release. In addition, casting properties such as veining resistance, shakeout, erosion resistance, and surface finish can be influenced by changing the acrylic-epoxy composition.

The FRC process employs a variety of binder systems, including specific systems for ferrous and nonferrous casting applications. Binder levels range from 0.6 to 1.8%, depending on the type of sand used and the type of metal poured.

Because the FRC process components do not react until SO_2 is introduced to the pattern, the prepared sand mix has an extremely long bench life when compared to other cold box and hot box binder systems. This feature minimizes waste sand, provides for consistent flowability, and, most important, decreases the machine downtime because sand magazines, hoppers, and mixers do not have to be cleaned on a daily basis.

Tooling and equipment requirements are similar to those of the other cold box processes. Gassing units used for the phenolic-urethane or the SO_2 process are replaced or simply converted for use with the FRC system. Substitutions of caustic solutions for acid solutions are made in the wet scrubbers when the FRC process replaces the phenolic urethane system.

Disposal of SO_2 Gas. Scrubbing of the gas is accomplished with a wet scrubbing unit that utilizes a shower of water and a sodium hydroxide. A 5 to 10% solution of sodium hydroxide at a pH of 8 to 14 provides efficient neutralization of the SO_2 and prevents the by-product (sodium sulfite) from precipitating out of solution. Higher sodium hydroxide concentrations will cause precipitation of the neutralized product. Stoichrometrically, 0.58 kg (1.27 lb) of sodium hydroxide is required to neutralize 0.45 kg (1.0 lb) of SO_2.

Casting dimensional accuracy and resistance to veining of ferrous and nonferrous metals are influenced by the thermal expansion of sand and by the hot distortion characteristics of the binder system. The antiveining feature of specific FRC binders has eliminated the need for specialty sands, sand additives, and special slurry applications in iron and steel castings. Further, the lack of nitrogen and the absence of water minimize the nitrogen and/or hydrogen pinholing porosity formation often associated with cold box systems containing water or nitrogen.

The phenolic ester cold box (PECB) process was introduced to the foundry industry in 1984. A two-part system, it consists of a water-soluble alkaline phenolic resole resin and a volatile ester vapor co-reactant. Sand is coated with the phenolic resin and blown into the core box. The liquid ester co-reactant is vaporized and injected as gas through the sand mix. The theoretical reaction is as follows:

$$\begin{array}{c}\text{Alkaline}\\ \text{phenolic resin}\end{array} + \begin{array}{c}\text{Ester}\\ \text{co-reactant}\end{array} \xrightarrow[\begin{array}{c}\text{Reactive}\\ \text{intermediate}\end{array}]{} \begin{array}{c}\text{Polymerized}\\ \text{phenolic resin}\end{array}$$

Because the ester is consumed in the curing reaction, purging of excess ester vapor can be accomplished with the minimum volume of purge air. However, purge air helps to distribute the ester vapor throughout the sand mix.

Methyl formate is the preferred ester for curing the phenolic resin because it is volatile and vaporized more easily than other esters. Methyl formate is readily available and relatively inexpensive.

The alkaline-phenolic binder is a low-viscosity (0.1 to 0.2 Pa · s) liquid at typically 50 to 60% solids in aqueous solution. The system contains less than 0.1% N and produces a less reducing mold atmosphere than the cold box binder systems.

The PECB system is affected by the physical characteristics of the sand, such as grain fineness, grain shape, and screen distribution. The best strengths are achieved with high-purity, washed and dried, round-grain silica sands. Because of the alkaline nature of the resin, however, it is not very sensitive to sand acid demand value. The fact that it is a water-soluble aqueous resin makes the system less sensitive to moisture, and it can tolerate up to 0.3% water in the sand.

The PECB system can be mixed using conventional mullers and continuous mixers. Binder levels vary, depending on sand type, but 1.75 to 2.5% resin is typically used with washed and dried silica sands. For sufficient handling strength, somewhat higher binder levels are required than for other gas-cured organic binder systems.

Methyl formate is volatilized in generating equipment designed especially for the PECB process. Because the methyl formate is not a catalyst, but a co-reactant of the system, the generating equipment must be capable of delivering a large volume of highly concentrated vapor to promote cure.

Stoichiometrically, about 15% methyl formate based on phenolic resin is required to harden the binder/sand mixture. In practice, the methyl formate requirement ranges from 30 to 80% and is largely dependent on venting and negative exhaust on the tooling. Lower gassing pressures and longer curing times promote the most efficient use of the ester co-reactant.

Despite the low handling strengths characteristic of the PECB system, castings made from all alloys show good surface finish. The erosion resistance and veining resistance of the PECB system are better than those of the phenolic urethane and phenolic hot box systems. A coating is recommended to control penetration defects. The PECB system does not usually require sand mix additives such as iron oxides or sugars to reduce veining and to control nitrogen defects. Because of the alkaline nature of the PECB sand, care must be taken so that it does not contaminate sand systems that are sensitive to pH change, especially if the reclaimed sand is used for other binder processes.

Sodium Silicate/CO_2 System. This system consists of liquid sodium silicate and CO_2 gas, and it is an inorganic system. Silicate binders are odorless, nonflammable, suitable for all types of work (high production to large molds), applicable to all types of aggregates, produce no noxious gases upon mixing/molding/coring, and produce a minimum of volatile emissions at pouring/cooling/shakeout. More detailed information on sodium silicate/CO_2 systems can be found in the article "Sand Molding" in this Volume (see the section "Bonded Sand Molds").

ACKNOWLEDGMENT

The authors would like to acknowledge the Technical and Research Departments of Ashland Chemical's Foundry Products Division for their help in preparing this article.

Sand Molding

SAND MOLDING (CASTING) is one of the most versatile of metal-forming processes, providing tremendous freedom of design in terms of size, shape, and product quality. Sand molding processes are classified according to the way in which the sand is held (bonded). For the purposes of this Handbook, sand molding processes have been categorized as:

- *Resin Binder Processes*. These organically bonded systems include no-bake binders, heat-cured binders (the Shell process and warm box, hot box, and oven-bake processes), and cold box binders. Each of these systems is described in the articles "Resin Binder Processes" and "Coremaking" in this Volume
- *Bonded sand molds* are based on inorganic bonds and include such processes as green sand molding, dry sand molding, skin dried molds, and loam molding, sodium silicate-carbon dioxide systems, and phosphate bonded molds
- *Unbonded Sand Molds*. With unbonded sand molding processes, dry, unbonded, free-flowing sand surrounds the pattern. Lost foam processing, which uses expandable polystyrene patterns, and vacuum molding, are examples of unbonded sand molds. Lost foam molds for large castings are sometimes backed up with a no-bake binder system

This article will describe the latter two categories. More information on sand molding equipment and processing can be found in the article "Sand Processing" in this Volume.

Bonded Sand Molds

Patrick O'Meara, Intermet Foundries Inc.
Larson E. Wile, Consultant
James J. Archibald and Richard L. Smith, Ashland Chemical Company
Thomas S. Piwonka, University of Alabama

According to the American Foundrymen's Society (AFS), approximately 90% of all castings produced annually in the United States are processed by sand molding (Ref 1). This section will review a number of sand molding methods that use bonded sand (see classification system described above) with emphasis on green sand molds, the most widely used molding method for small-to-medium castings in all metals.

Green Sand Molds

The phrase green sand refers to the fact that the medium has been tempered with water for use in the production of molds (temper water, which is the water added to activate clay/water bonds in green sand molds, is described in the article "Bonds Formed in Molding Aggregates" in this Volume). As will be described below, the control of a green sand process requires an understanding of the interaction of the various parameters normally measured in a system sand.

Process Control Requirements

A realistic approach to sand control is to target those system variables with which actual control can be implemented and realized. Put in more simple terms, one must control those system parameters that are directly affected by actions taken on the foundry floor. Clay and water are the primary additives of a system sand. The functions they perform are measured by determining the clay content and the percent of compactability of the prepared sand. Seacoal, cellulose, and starches may also be added to the sand. These organic components of the system sand are normally measured by the percent volatile and/or the total combustible test.

The percent volatile test measures the volatile content of the system sand at a specified temperature, usually 650 °C (1200 °F). The total combustible test is conducted by burning the samples of system sand at an elevated temperature, normally 1010 °C (1850 °F). Detailed procedures for these tests can be found in the AFS *Mold and Core Test Handbook*.

The remaining parameters measured on a system sand, such as green strength and permeability are secondary controls. They should be tracked using trend line analysis techniques. This type of analysis allows the monitoring of the variable in question over an extended time so that subtle changes in the magnitude of the variable can be detected. Significant changes in these secondary parameters indicate equipment problems, changes in raw material quality or consistency, and/or changes in product mix being made in the system sand.

Sand Systems

Types of Sand. Sand for green sand molding is composed of various ingredients, each with a specific purpose. The most basic of these ingredients is the base sand itself. The most predominant type of base sand is silica sand. It is classified in two categories: naturally bonded and synthetic sand.

The naturally bonded sand (or bank sand, as it is sometimes called) contains clay-base contaminants. These naturally occurring clays are the result of sedimentation deposits produced during the formation of the sand deposit. The use of this type of sand as a green sand molding medium is determined by the type of metal being cast, economics, casting quality, and the degree of consistency demanded by the final product.

Synthetic sand is composed of base sand grains of various grain distributions. Bonding agents are added to these base sands to produce the desired molding characteristics. The major base sand in this category is silica, although zircon, olivine, and chromite are used for special applications.

Controlling Sand Properties. Sand grain structure is a very important characteristic in the selection of a base sand. The selection dictates the ultimate mold permeability and density, and both of these parameters are critical to the production of quality castings.

When molten metal is introduced into a green sand mold, gases and steam are generated as a result of the thermal decomposition of the binder and other additives or contaminants that are present. If the permeability of the mold is not sufficient to allow the escape of the generated gases, mold pressures will increase, impeding the flow

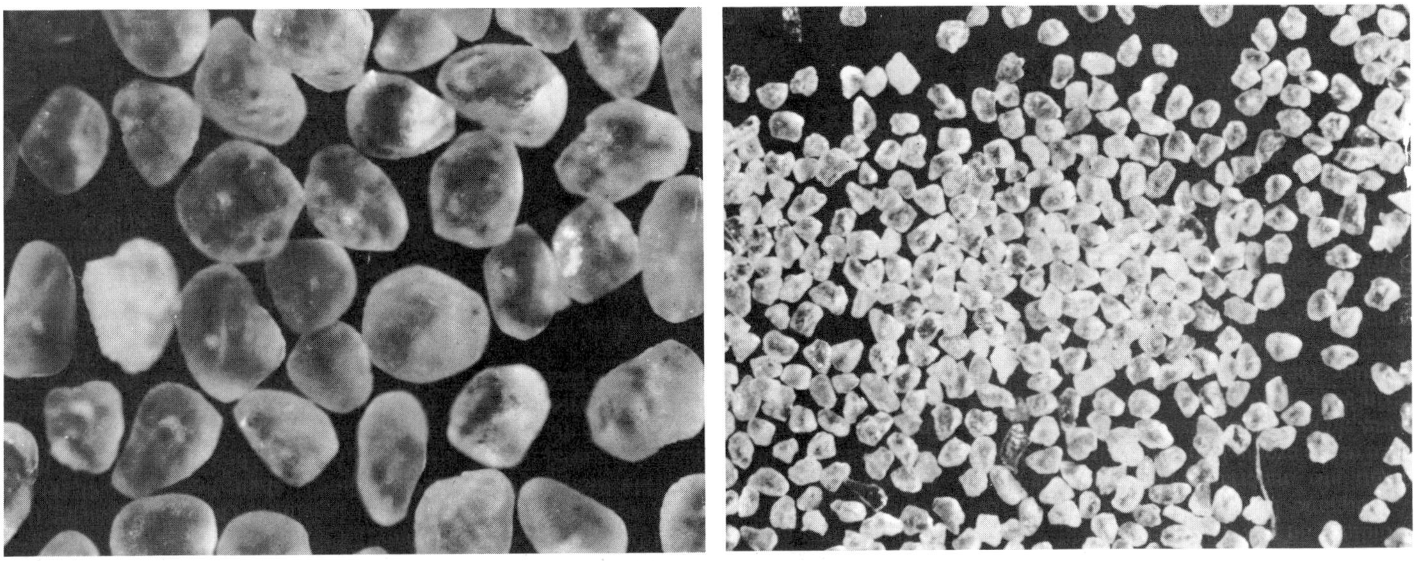

Fig. 1 Two sizes of rounded sand grains. 35×

of molten metal, or even causing the metal to be blown from the mold. Thus, the selection of a base sand that provides adequate mold porosity is very important.

Because resistance to gas flow increases as the size of the pores (voids) between the sand grains decreases, the minimum porosity required is determined by the volume of the gas generated within the mold cavity. In like turn, the selection of the base sand is determined by the total amount of gas produced within the mold cavity, as well as by surface finish requirements.

The fact that gas is generated within the mold cavity is not always a disadvantage. Pressures within the mold from the generation of gases help prevent metal penetration into the sand. This minimizes burned-on sand grains and resulting problems associated with cleaning and machining the casting. Thus, a balance between mold permeability and gas generation must be maintained. For example, if mold permeability is low because of the fineness of the base sand, the sand additives should be those conducive to the production of a low volume gas. On the other hand, if permeability is high, it is advantageous to select materials that yield higher levels of gas.

Permeability is controlled by the amount and size of the voids between densely packed sand grains. The size of the voids is determined by the size, size distribution, shape, and packing pattern of the grains. Figure 1 illustrates two sizes of rounded sand grains. Figure 2 shows that the voids in a mold face are large for a coarse sand and small for a fine sand, although the total void area per cubic unit of volume is almost the same for both sands. However, these distribution criteria also govern the dimensional stability of the base sand.

A green sand mold must withstand the erosion caused by the metal impinging on and flowing over the sand surface. If the individual sand grains are not held firmly in place during metal flow, the result will be loose sand grains that will wash into the casting cavity and cause a defective casting. Sand grains are held in place by a combination of two mechanisms: a wedging action in which the sand grains are mechanically locked to adjacent grains, and the clay-water bond established between the grains. The combined action of these two mechanisms forms the basis of the sand strength developed in the mold cavity. The best sand condition for optimum mold strength and density development is produced by sand grains that show

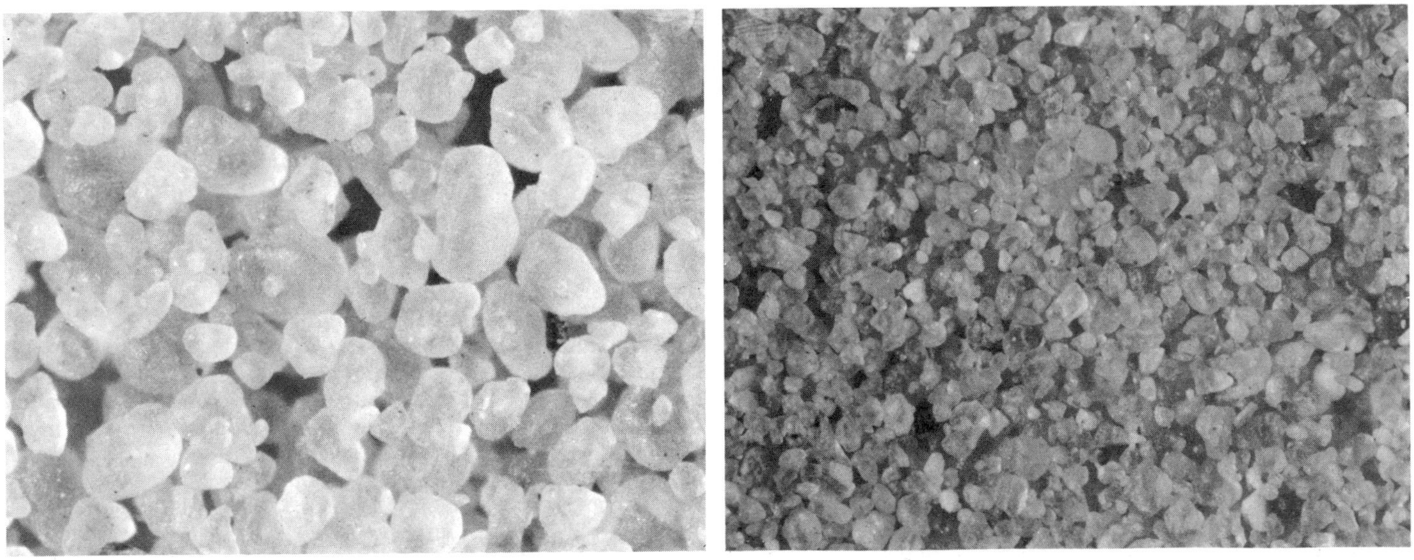

Fig. 2 Sizes of pores in faces of molds made from coarse sand and from fine sand. 35×

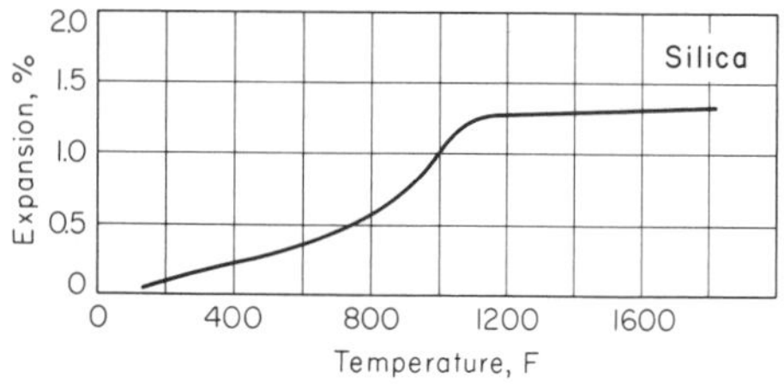

Fig. 3 The effect of temperature on the expansion of silica

a normal distribution over four or more adjacent screen sizes.

As molten metal is introduced into the mold cavity, heat is transferred from the molten metal to the adjacent sand grains, causing the sand grains to expand. Between 425 and 600 °C (800 and 1110 °F), silica undergoes a phase change from alpha to beta, which is accompanied by a rapid increase in volumetric size (Fig. 3). Each sand grain must be allowed to expand, or the mold surface will be altered or destroyed, with resultant loss in casting quality. Therefore, the silica sand grains must not be compacted so densely or rammed so tightly that they are unable to expand without disrupting the mold surface.

Four methods for optimizing the dimensional stability of the base sand are:

- Selection of a base sand aggregate suitable for a dimensionally stable mold surface. Generally, this will be a four-screen sand, although three-screen sands may be used for certain castings
- Addition of carbonaceous additives such as seacoals and cellulose to the green sand system. The thermal decomposition of these additives creates voids, which allow for the expansion of the silica sand grains
- Increasing clay content to develop higher green strengths, which tend to produce more stable molds
- Controlling the mold density produced by the molding equipment

If it is necessary to use a mold material with less thermal expansion than silica, alternate materials such as zircon, olivine, chromite, or calcined clay may be chosen. Zircon and chromite have the additional advantage of possessing higher heat transfer capabilities. Calcined clay is sometimes used in the production of very large castings in dry sand molds because of its extremely low thermal expansion.

The sand-to-metal ratio for a given mold influences the required pore or void size. The amount of heat transferred to the sand is a function of pouring temperature, volume of metal poured, and amount of time the sand is exposed to the elevated temperatures. These same conditions dictate the volume of gas generated for a given sand formulation. Therefore, for large, heavy castings with high pouring temperatures, a sand with large pores is preferred. For small castings, a sand with smaller pores is the sand of choice.

Finer sands with smaller pores may have reduced ability to allow the decomposition gases to escape. However, they do improve the surface finish and enhance the reproduction of pattern detail. Sand of a single mesh size distribution provides the best venting action, but affords the least protection against erosion or expansion defects. It should be noted that fine sand may require higher amounts of bonding agents (clay, water) because of the higher surface area that must be coated. This further aggravates the gas generation problems because the increased level of bonding agents generates increased amounts of gas that have to permeate the less permeable mold.

The selection of a suitable base sand is a compromise at best. The optimum selection is a multiscreen sand with adequate permeability for the metal and geometry being poured. Factored into the decision also are the economics of the raw materials and the surface finish and casting quality required.

Clays for Green Sand Molding

Green sand additives can be divided into two categories, clays and carbonaceous materials. The major purpose of the clays is to function as a bonding agent to hold together the sand grains during the casting process. The carbonaceous materials aid dimensional stability of the mold, surface finish, and cleanability of the finished casting.

Types of Clay. Clays normally used in green sand molding are of three general types:

- *Montmorillonite,* or bentonite clays. These are subdivided into two general types: Western, or sodium, bentonite; and Southern, or calcium, bentonite. The two clays differ in their chemical composition as well as in their physical behavior within a system sand
- *Kaolinite,* or fireclay as it is normally called
- *Illite,* a clay not widely used. The material is derived from the decomposition of certain shale deposits

The most significant clays used in green sand operations are the bentonites. Western and Southern bentonites differ in chemical makeup and, thus, their physical characteristics also. In general, Western bentonite develops lower green strength and higher hot strength than the same amount of Southern bentonite. Southern bentonite, at the same concentration, produces higher green strength and lower hot strength. This phenomenon is sometimes confused with what is referred to as durability.

Controlling Clay Properties. All clays can be made plastic and will develop adhesive qualities when mixed with the proper amounts of water. All clays can be dried and then made plastic again by the addition of water, provided the drying temperature is not too high. However, if the temperature does become too high, they cannot be replasticized with water. It is this third condition that dictates the durability of the clay in a system sand.

All clays, regardless of type, develop both adhesive and cohesive properties when mixed with water. The amount of adhesive or cohesive property depends on the amount of water added. When the water content is low, the cohesive properties are enhanced and the clays tend to cohere, or stick to themselves, rather than adhere, or stick to the sand grains to be bonded. With high water additions, the converse is true.

In addition to having different bonding and durability characteristics, the various clays have very distinctive behavior patterns as a result of their differing physical characteristics. System sands formulated with high levels of Western bentonite have high levels of hot strength. A system sand formulated with an equivalent level of Southern bentonite will have a significantly lower hot strength. In addition, the flowability of the two sands is different because of the greater swelling tendency of the Western bentonite clays compared to that of the Southern bentonite materials. Therefore, the proper formulation of clay materials for a green sand system must take into consideration the flowability requirements as well as the shakeout requirements of the sand.

The ratio of clay to water is of critical importance in optimizing the properties of clays. The shear strength of a clay-water mixture is representative of the green strength of the compacted sand, because it is the shear strength of the films of clay coating the sand grains that bonds the sand together. This parameter is controlled by the amounts of water and clay added to the mixture and is measured by monitoring the pressure required to extrude various clay-water mixtures through a fixed orifice (Fig. 4).

In a sand and clay mixture, water is absorbed by the clay up to its maximum capacity. Any additional water is carried as free water in the system sand and does not contribute to bonding. Therefore, high water content clay yields low shear strength. As water content is decreased, a sharp rise in shear strength occurs. The free water content in bentonite clays is normally in the range of 28 to 40%, and for fireclays it is from 15 to 20%.

While the ratio of clay to water in a sand mixture controls the ultimate strength of the sand mixture, the origin of the clay has a significant contribution on the strength potential. Clays from different geographic regions, even though they may be classified as being the same, have different strength curves. However, many of these differences are minimized by modern techniques used in the mining of the clays.

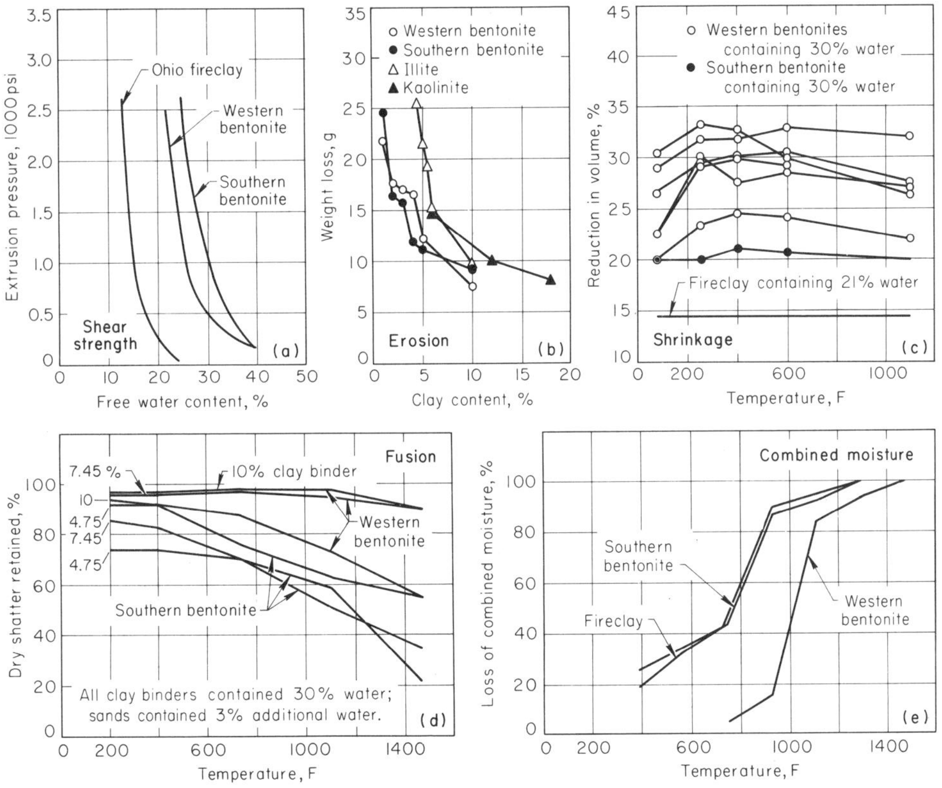

Fig. 4 The effect of several variables on the efficiency of clay used as a bonding agent in sand molds. (a) Relationship of shear strength, as measured by pressure required for extruding a continuous worm of clay through an orifice, to water content, for three clays. (b) Effect of type and quantity of clay on erosion of sand-clay mixtures. (c) Effect of temperature on shrinkage of various types of clays. (d) Effect of mold temperature during casting on fusion of clay binder. (e) Effect of temperature on loss of combined moisture in clays

Clay quality is generally measured against the amount required to develop a specified green strength in a sand mixture. Care must be taken when evaluating a clay in this fashion because of the effects of water on strength. The term "the sticky point" defines the point of transition from predominantly cohering properties to those of adhesion. Clays selected for foundry use should have compositions near their "sticky point," or the state at which their cohesive and adhesive properties are balanced.

Once the type of clay is determined for the system sand, economic considerations must be evaluated, because the geographic location of the foundry will, in part, dictate the type, or the combination, of clays used in the operation.

Western bentonite requires a higher energy output to develop its properties than do the Southern bentonites. The fireclays contribute little to green property development, but contribute dramatically to dry and hot property development. The proper combinations of clays allow the formulation of a system sand conducive to the production of quality castings.

System Formulation

Of utmost importance in controlling a green sand system is the selection and consistency of the raw materials introduced into the system sand. Acquisition of the basic raw materials should be from reputable sources only, that is, those that have ongoing quality improvement programs, including the understanding and application of statistical process control techniques. This is critical for a successful control program: Inconsistency in the raw materials used in the system results in sand variations that no amount of attention or corrective action can overcome.

Next in importance is the condition of the sand processing equipment. This includes the sand muller or mixer, sand cooling equipment, and dust collection equipment. It is important to coat the individual sand grains with a uniform thickness of the bonding agent; this governs the physical property development of the sand. The coating action, in turn, is controlled by the condition of the mixing and/or the muller equipment. Failure to monitor the equipment and to maintain it adds appreciably to variations in the sand and a loss of casting quality. More detailed information on equipment for green sand processing can be found in the article "Sand Processing" in this Volume.

Third is the identification of the critical primary and secondary control parameters. The primary control parameters for a system sand are:

- Determination of the organic components measured by the total combustible and/or percent volatile tests
- Determination of clay content measured by the methylene blue titration method
- Percent compactability of the sand controlled by the molding machine

For the majority of foundries, primary controls are limited to the system clay and the content of water and carbonaceous material (secondary control parameters are discussed below).

Actual sampling of a system sand should be accomplished as close to the point in time of use as is practical without compromising worker safety. By so doing, corrective actions can be carried out prior to the molding operation. Tests should be conducted when applicable according to standard procedures outlined in the AFS *Mold and Core Test Handbook*.

The clay content of the system sand is normally measured by the methylene blue titration method. This method of determining clay content is based on the ability of a test sample of the system sand to absorb the methylene blue dye. In this test, the dye is added to the test sample by a buret. The end point of the titration is read by the technician as a "bursting halo" when a drop of the test material is placed onto a hardened piece of filter paper with a stirring rod. The "halo" is an indication that the dye-absorbing ability of the clay has been reached. The amount of dye required to reach the end point (measured in milliliters) is compared against a known standard mixture.

Care should be exercised in the use of the methylene blue titration test because of its vulnerability to operator error. Provision should be made to obtain a test analysis of each revolution of the system sand. However, it is more important to react properly to the available test results than to be concerned with the quantity of the available test data. Clay control can be enhanced by close monitoring of clay additions. Simply knowing what goes into the sand can result in a significant reduction in system sand variations.

Raw Material Additions

The sand in a green sand molding system is primarily made up of recycled, reclaimed, and reused sand. The rejuvenation of this sand is the principal function of the sand preparation system. Sand for the green sand molding system is recovered from the shakeout, cooled, cleaned, and screened. New sand is added to compensate for that which has been lost from spillage or carried away in deep pockets of the casting. Clay, water, and other additives are introduced to bring the sand mix to specification.

Maintaining Sand System Quality. Because the system sand is basically made up of recycled sand, it consists not only of mold sand, but also of core sand. The size and shape of core sand, and the binders used for core sands, are frequently quite different from those used for molding sands, and this must be considered for maintaining the sand system. Also, over time, the sand in the system breaks down as a result of mechanical attrition and thermal cycling, and therefore changes in size, size distribution, and shape. Daily variations in the product mix affect the ratio of core sand to molding sand being recycled. Raw or virgin sand additions to the system should be made to dilute the contaminating effects of residual core binders.

The introduction of additives to a system sand based on a programmed approach greatly reduces system variations. Additives are consumed at identifiable rates. Core sand dilution can be determined as well as any raw virgin sand additions. Losses due to sand carry-out, dust collection loss and material handling can be calculated or approximated. By monitoring sand system variations, a predictable quantity of bonding material can be determined and added to the system. This approach helps minimize system variations and also reduces the necessity for clay analysis as a control tool.

New or reclaimed sand additions to a system sand are of critical importance. Their most important function is to reduce the concentration of contaminants in the system sand. The origin of these contaminants may be alloy by-products or contamination from a core process. A second function is to maintain the total system volume.

All ingredients added to a system sand should be added gravimetrically. Volumetric or timed additions of dry additives are generally not consistent enough to ensure adequate control and only contribute to overall consistency problems. This applies to the addition of a new or reclaimed sand as well as to the return system sand for a batch system.

Most system sands contain additives other than clays; monitoring those additions is critical to the production of quality castings. Carbon additives are either single- or multiple-component. The relative concentration of the carbon additive must remain constant for a laboratory-derived measurement to be of any value. For example, casting finish produced in a system sand with a total combustible value of 3.50% derived from a carbon additive based totally on seacoal can differ drastically from a similar system sand with an equivalent total combustible value derived from a multiple-component carbon additive. In cases in which the benefits can be justified, the use of preblended sand additives is recommended.

Compactibility is the only physical characteristic of a system sand that a molding machine realizes, and it is of paramount importance that it be controlled within a tight band. The desired range will be specific to each foundry. The range chosen should allow optimum strength of the clay as well as adequate moisture levels to minimize sand friability while satisfying the requirements of the molding equipment. The secondary control parameters are as important as the primary, but must be looked at over a period of time. It is necessary to realize that problems develop gradually in a system sand. With this in mind, the use of trend line analysis technique as a tool to a better understanding of the effects of subtle changes in a system sand can be of significant benefit. The specific secondary tests that should be run on a system sand include green compressive strength, permeability, moisture, AFS clay, screen distribution, and AFS washed fineness tests (use the AFS *Mold and Core Test Handbook*). The mulling efficiency should be monitored and charted on an ongoing basis.

While testing a system sand is very important, of equal importance is knowledgeable and responsible review of the data and corrective actions when necessary. Routine review of all test results should include representation from all the various disciplines within the foundry.

Included in the control program for a green sand system should be routine monitoring of the dust collection system. Particular attention should be given to the screen distribution, methylene clay content test, and the total combustible level test. Deviations from normal operating levels indicate equipment malfunctions that require corrective action.

Screen distribution is important. For system sands comprised of only base sand (plus new-sand additions and core sand dilution), if the grain shape remains constant, changes in screen distribution will occur slowly, a major contributor being the condition of the dust collector system. On the other hand, in those systems that have an influx of various types of base sand from varying core processes and sand sources, changes in the base sand distribution will be more dramatic and will cause dimensional problems unless preventive measures are implemented. For these systems, screen distribution should be tested more frequently.

Influence of Molding Equipment. The type of molding equipment used is also critical in the selection of a green sand system. Green sand molding can be divided into three basic types:

- Low-density and low-pressure molding includes manually operated jolt-squeeze units. The green compressive strengths of the sand for these units is generally in the low to mid teens (given in pounds per square inch)
- Medium-density units include automatic or semiautomatic units with rigid flasks that combine jolting action and hydraulic squeeze pressures
- High-pressure or high-density units produce molds with hardness values in the high 90s*. These units incorporate flaskless molding technology and use hydraulic pressure and other energy sources for sand compaction

The third group, that is, the high-pressure or high-density molding methods, dominate the high-production and highly automated foundry today. This type of green sand molding lends itself to both flask and flaskless system designs. Some of the major criteria that dictate whether a system is designed as a flask or a flaskless unit is the amount of metal to be poured in each mold and the total mold area that is required. It is not within the scope of this article to deal with all the factors that influence the selection of a molding system. However, all of the systems in this category can be used for high-production repetitive work. Dimensional control of castings with these processes is quite good, and the economics of their operation have yet to be matched by any other. Unfortunately, these systems represent a significant capital investment.

As automation and production rates have increased, the capability for the precise placement of the sand on the pattern has diminished. Thus, a critical characteristic for today's high-production sands is the capability of uniformly covering the pattern surface before application of the compaction energy. Therefore, sand flowability and consistency are of utmost importance in the formulation of a system sand.

Sand Reclamation

The economics of a foundry operation require sand reclamation to reduce the costs associated with new sand and the costs of landfill use, and to reduce the problems associated with the control of environmentally undesirable contaminants in the discarded sand.

In addition, tangible operational advantages result from sand reclamation. These begin with the ability to select the best sand for the casting process, knowing that most of it will be reclaimed during operation. In addition, the use of reclaimed sand reduces the number of variables that must be controlled, and provides operational consistency over a period of time. Sand grain shape and distribution and binder system bonding are more uniform, thus reducing sand defects. A properly designed sand reclamation system begins with green sand and converts it to a product very similar to new sand. Figure 5 shows the appearance of sand before casting, after molding, and after reclamation.

*Hardness in this situation refers to green surface hardness as measured by the C scale hardness tester described in the AFS *Mold and Core Test Handbook*. It should be noted that this scale is not related to the Rockwell C scale hardness indication.

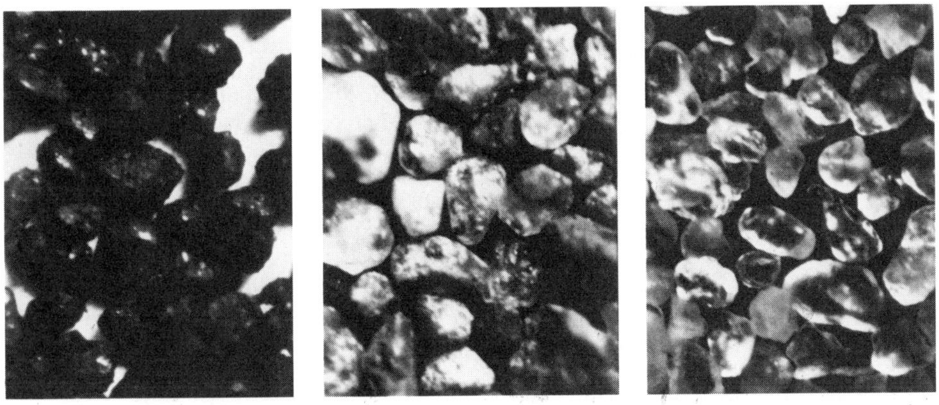

(a) (b) (c)

Fig. 5 Influence of sand reclamation on the appearance of green sand. (a) After molding (no reclamation). (b) After thermal reclamation. (c) New sand

Sand reclamation begins with the removal of tramp and foreign materials, such as core rods, metal spills, slag, and paper, and the disintegration of lumps of sand. Then organic and inorganic binders are removed by attrition (scrubbing) and/or thermal methods. Dead clay is removed as fines. The sand is then brought up to specification by the addition of new sand, clay, and other sand additives.

Sand reclamation systems must be selected with regard to their cost, the specifications for the system sand, system capacity, compatibility with the sand system, metal being poured, and core mixes being used. It is important for the foundry to have a clear understanding of its needs in sand reclamation before calling in vendors. A variety of reclamation systems are described below. More detailed information can be found in the article "Sand Processing" in this Volume (see the section on "Sand Reclamation").

Wet Washing/Scrubbing. The cores of large castings can be removed by high-velocity jets of water. In the process, the cores are broken down into grains, and some binder is removed. Excess molding sand can be added and washed simultaneously. If the shakeout system is dry, the sand is charged into an agitator system where the solid content is held between 25 and 35%. Excess molding sand may be blended with the core knockout material. A similar system uses intensive scrubbing with a solids content of 75 to 80% and units in series. This latter method is superior because of closer and more frequent grain-to-grain contact. After washing, the sand is classified and may be used either wet (naturally drained to 4 to 5% moisture) and added to a system sand, dried for cores, or used for facing sand. Facing sand mixes derive their name from the fact that they are used in limited quantities against the face of the pattern or core box. They have properties usually different from those of the backup or system sand and contain additives not otherwise present. Facing sands are designed to perform special functions, such as providing higher green strength for lifting deep pockets, higher deformation for limited draft patterns, and special carbons that enhance skin drying.

The wet system has limitations in that only a portion of the binder, clay, and carbon is removed. The product, however, is excellent for use as a makeup sand in systems.

Dry Scrubbing/Attrition. This method is widely used, and there is a large variety of equipment available in price ranges and capacities adaptable to most binder systems and foundry capacities. Dry scrubbing may be divided into pneumatic, mechanical, and combined thermal-calcining/thermal-dry scrubbing systems.

In pneumatic scrubbing, grains of sand are agitated in streams of air normally confined in vertical steel tubes called cells. The grains of sand are propelled upward and rub and impact each other, thus removing the binder. In some systems, grains are impacted against a steel target. Banks of tubes may be used depending on capacity and degree of cleanliness desired. Retention time can be regulated, and fines are removed through dust collectors.

In mechanical scrubbing, the equipment available offers foundrymen a number of techniques for consideration. An impeller may be used to accelerate the sand grains at a controlled velocity in a horizontal or vertical plane against a metal plate. The sand grains impact each other and metal targets, thereby removing the binder. The speed of rotation has some control on impact energy. The binder and fines are removed by exhaust systems, and screen analysis is controlled by air gates and/or air wash separators. Additional equipment options include:

- A variety of drum types with internal baffles, impactors, and disintegrators to reduce lumps to grains and to remove binder
- Vibrating screens with a series of decks for reducing lumps to grains, with recirculating features and removal of dust and fines
- Shot-blast cleaning equipment that may be incorporated with other specially designed units to form a complete casting cleaning/sand reclamation unit
- Vibro-energy systems that use synchronous and diametric vibration. Separation of the binder from the sand grains is caused by frictional and compressive forces. One special, advantageous feature of these systems is the small number of moving parts

Thermal-Calcining/Thermal-Dry Scrubbing Combinations. These systems offer the best reclamation for the organic and clay-bonded systems. Grain surfaces are not smooth; they have numerous crevices and indentations. The application of heat with sufficient oxygen oxidizes the binders or burns them off. In attrition, only because there is no contact in the crevices, the binder remains. Heat offers the simplest method of reducing the encrusted grains of molding sand to pure grains. Both horizontal and vertical rotary kiln and fluidized bed systems are available.

In the horizontal rotary kiln, material is fed into one end (usually the cold one) and moved progressively through the heat zone by rotation assisted by baffles, flights, or other mechanical means. Some mechanical scrubbing also occurs. Some systems incorporate heat exchanger technology to considerably reduce the energy required. The latest technology also includes provision for recovery of metal entrained in the sand and collection and detoxification of the process wastes for suitable nonhazardous waste disposal.

Several fluidized bed system designs are available. Some use preheating chambers and hot air recuperation. A drying compartment may also be added. Sand is introduced into the top (preheating) chamber of the reactor and is lifted by the hot air stream from below until it assumes some of the characteristics of a fluid. The hot air coming in contact with the sand grains burns the organics and calcines the clay. At the same time, some attrition takes place. A correct pressure differential must be maintained between the compartments if more than one is used in order to ensure downward flow of sand; otherwise gravity flow must be provided. Fluidizing is a very good method for cooling sand when using cool air.

Multiple-Hearth Furnace/Vertical Shaft Furnace. The multiple-hearth furnace consists of circular refractory hearths placed one above the other and enclosed in a refractory-lined steel shell. A vertical rotating shaft through the center of the furnace is equipped with air-cooled alloy arms containing rabble blades (plows) that stir the sand and move it in a spiral path across each hearth. Alternate hearths are "in" or "out." That is, sand is repeatedly moved outward from the center of a given hearth to the periphery, where it drops through holes to the next hearth. This action gives excellent contact between sand grains and the heated gases. Material is fed into the top of the furnace. It makes its way to the bottom in a zigzag fashion, while the hot gases rise countercurrently, burning the organic material and calcining clay, if one or both are present. Discharge can be directly from the bottom hearth into a tube cooler, or other cooling methods may be used The units are best suited to large tonnages, that is, five tons or more. They are extremely rugged and relatively maintenance free.

Combinations of systems may also be used, for example, thermal methods followed by dry attrition scrub to remove calcined clay from molding sand or undesirable chemicals and oxides from core processes. Also, commercial centers for sand reclamation are in operation and may be used by smaller foundries.

Dry Sand Molding

The essential difference between dry sand and green sand molding is that the moisture in the mold sand is removed prior to pouring the metal. Dry sand molding is more applicable to medium and large castings than to small castings. The molds are stronger and more rigid than green sand molds. They can therefore withstand more handling and resist the static pressure of molten metal, which may cause green sand molds to deform or swell. In addition, they may be exposed to the atmosphere for long periods of time without detrimental effect. Such exposure may be necessary for placing and fitting a large number of cores.

Seacoal is the most common carbon material used in green sand; pitch is the most common in dry sand. Other materials in dry sand are gilsonite, cereal (corn flour), molasses, dextrine, glutrin, and resin. These additives thermoset at the baking/drying temperature (150 to 315 °C, or 300 to 600 °F) to produce high dry strength and rigid mold walls. The base sand is normally coarser than in green sand to facilitate natural venting and mold drying.

Dry Sand Molding Methods. As described below, dry sand molds are made by a variety of methods.

Large sand compaction machines of the jolt, roll-over, and draw-type or jolt-only-type compact the sand in conjunction with tucking, hand peening, and air ramming.

Sand Slingers. These machines throw and compact the sand by means of centrifugal force. A variety of sizes are available. Some supplemental hand ramming may be required.

Floor Molding. This molding method uses larger flasks normally requiring the services of an overhead crane. The molds

Fig. 6 Dry sand pit mold for stationary diesel engine

are made by a combination of mechanical equipment (slinger), hand peening, and pneumatic hand-operated rammers. Sand must be placed in the flasks in layers, and care must be taken by the molder to make certain that each layer knits and adheres to the other and is of uniform hardness.

Pit Molding. This method is used for very large castings when flasks are impractical. Pits are normally constructed of concrete walls and sometimes floors to withstand great pressures during pouring. Because the drag part of a pit cannot be rolled over, the sand under the pattern must be rammed or bedded in, or the bottom must be constructed with dry sand cores. A bed of coke, cinders, or other means of venting the pit bottom must be provided. Once the pattern is in place, mold-making procedures are the same as those for floor molding.

Curing Sand Mixtures. The sand mixture for dry sand molds can be cured (hardened) by baking in an oven or using forced hot air or stoves placed in the mold.

Mold Coating. Dry sand molds are coated with refractory washes. Water and/or solvent carriers using graphite, silica, or zircon are the most frequently used. It is common to apply several coats. An example of a dry sand mold being coated is shown in Fig. 6.

Equipment used for dry sand molding must be strong and rigid. Flasks are equipped with crossbars that normally extend to within several inches of the pattern. The molding sand is reinforced with hooks and gaggers (L-shaped steel bars) that are usually coated with a clay slurry to enhance bonding to the sand. Cores may be suspended in the cope by means of threaded rods and bolts. Venting of the molds and cores is important. Drilling holes in the side walls of flasks facilitates venting. Large flasks and pit molds must be tightly secured with heavy clamps and weights to prevent runout, which could be extremely dangerous and cause casting loss as well.

Cooling of Castings. Large castings should be cooled slowly in order to prevent internal stresses and/or cracking of the casting. It is sometimes possible to have the molds or cores so rigid that castings will hot tear. The use of inert filler materials placed a safe distance from the casting surface and/or the hollowing out of heavy mold sections will help to prevent such problems.

Skin-Dried Molds

Almost all dry sand molding has been replaced by the no-bake molding (see the article "Resin Binder Processes" in this Volume). Sometimes the pattern may be faced with a no-bake sand mixture and then backed up with a green/dry sand mixture. An intermediate (between green sand and dry sand) type of molding referred to as skin drying is sometimes used. The process is similar to dry sand molding in that the same type of sand mixtures and equipment are used. After coating the surface with a refractory wash, the molds are dried to a depth of 6 to 12 mm (0.25 to 0.5 in.). Skin-dried molds have some characteristics of green sand molds and some of dry sand molds such as ease of shakeout and firm mold face, respectively.

Loam Molding

Loam molding is one of the oldest methods known. It requires the skill of an experienced craftsman, and is seldom used today because of the lack of these craftsmen and the advent of the no-bake systems that have a number of advantages over loam molding, especially greatly reduced production time. Loam molding is particularly adapted to medium or large castings, usually those of circular shape.

The equipment required is relatively simple. No pattern is used; however, a flask or pit may be necessary to support the finished mold. Replacing the pattern is one or more sweeps conforming to the shape of the mold/core. The sweep is attached to an upright, rotating spindle. A bed plate, a cover plate, and arbors for cores are also necessary items.

Materials for making the molds consist of:

- Soft bricks capable of absorbing water
- Molding sand that is quite coarse (sometimes referred to as gravel) and containing a high percentage of clay
- Sawdust, straw, hay, hair, or cloth clippings may be used. These materials contribute to mold strength and burn during casting, thus enhancing collapsibility (disintegration of mold/core)

The molding sand is mixed with water to which molasses may be added until a thick

Table 1 Typical screen analysis of a loam molding sand

Screen No.	Percent retained
6	4.5
12	12.5
20	25.3
30	15.0
40	10.4
50	7.7
70	3.0
100	2.4
140	1.4
200	0.7
270	0.6
Pan (−270)	1.2
AFS clay	15.3

loam slurry resembling concrete is obtained. This mixture is used along with the bricks to build up the general shape, using the same technique that a mason would use. Once the general shape is attained, the bricks are covered with a layer of loam, and the sweep is used to form the outside surface. A layer of loam from 6 to 25 mm (0.25 to 1 in.) thick remains. The mold/core is then baked until thoroughly dry, after which a refractory coating is applied. During the drying process, numerous small cracks form on the surface of the low moisture content clay. These small cracks produce a surface that is very receptive to the refractory coating, which may be brushed, swabbed, sprayed, or actually troweled on. After application, the coating is brushed with molasses and water. A camel hair brush may be used to impart an exceptionally smooth surface. The coating must be dried prior to assembly. Loam molds impart a superior peel to castings.

Before the advent of spun cast pipe in metal molds, large-diameter pipe was made in pits (pit cast pipe). The cores for these molds were made from loam. Hay rope was wrapped around a metal mandrel and the loam slurry was applied, smoothed, and dried, after which a refractory coating was applied.

A typical screen analysis of a loam molding sand is given in Table 1. A casting made from a loam molded cope is shown in Fig. 7.

Fig. 7 Casting produced by loam molding

Silicate and Phosphate Bonded Molds

Sodium Silicate/CO_2 System. Typical silicate binders are odorless, nonflammable, suitable for all types of work (high production to large molds), and applicable to all types of aggregates. They produce no noxious gases on mixing/molding/coring and only a minimum of volatile emissions at pouring/cooling/shakeout.

The amount of silicate binder used for cores and molds varies from 3 to 6%, depending on the type of sand, grain fineness, and degree of sand contaminants. The type of metal poured and its temperature, and the amount of erosion resistance the core or mold will have to withstand are additional factors. A clean, rounded sand grain of 55 fineness requires approximately 2.5 to 3% of binder. As the sand fineness increases, the amount of binder that must be used to coat each grain increases. Thus, a sand of 120 to 140 fineness requires from 1.5 to 3.0% more binder than a sand of 55 fineness.

Either continuous-type or batch-type mixers can be used with sodium silicates. Overmixing should be avoided. Mixtures normally have a good bench life, which becomes shorter when higher-ratio silicates are used. Hoppers of mixed sand should be covered with plastic sheets or damp sacking to prevent premature hardening (crusting).

The initial tensile strength of cores gassed for 5 s with CO_2 varies from 255 to 310 kPa (37 to 45 psi), depending on the binder used. When the core stands, strength increases to a maximum of about 670 to 1380 kPa (100 to 200 psi) in 24 h. This increase is due to some dehydration of unreacted silicates and continued gelling of the silicate. Under normal conditions, cores and molds do not deteriorate and can be stored. An exception may occur during periods of high humidity when sand mixtures are unable to achieve maximum strength by dehydration during storage after gassing. Strength deterioration in high humidity conditions is even more prevalent when the binder formulation or the sand mixture contains organic additives such as sugars, starches, or carbohydrates. Under these conditions, a short heating cycle is necessary to obtain the desired strength. When a core is hardened by carefully controlled gassing, and further hardened by dehydration and polymerization during the subsequent 24 h, good strengths are maintained over a long storage period.

Washes may be necessary on cores and molds made by the silicate/CO_2 process to prevent burned-on sand and metal penetration. Alcohol-base and solvent-type washes are normally used; isopropyl alcohol is preferred to methyl alcohol because it is less toxic and has a higher boiling point. Specially prepared graphite and zircon refractory pastes, which may be diluted with alcohol or solvent in the foundry, are commercially available. Washes on cores and molds promote peel, aid collapsibility, improve casting surface finish, and aid in resisting moisture absorption during periods of high humidity. Water-base washes can be used if care is taken to dry the core thoroughly immediately after coating. This procedure must be carried out with care due to the softening effect of the water on the silicate bond.

Additives may be used in sodium silicate bonded sand mixtures to improve shakeout or collapsibility. Sugars are commonly used for this purpose; they can be compounded with the silicate binder or added separately to a sand mixture.

Small cores and molds are successfully made with core blowers or shooters combined with gassing stations operating on predetermined cycles and automatically controlled. For these applications, CO_2 gas is injected through a hollow pattern or double-wall corebox, or by means of a mandrel in a core, or through a hood covering the box.

Larger cores can be cured by means of lance pipes of about 5 mm (3/16 in.) in diameter that are open at the lower end. Using a rod, holes are made about 150 mm (6 in.) apart; the lance is inserted into each hole successively, and the gas is applied for 10 to 15 s at about 170 kPa (25 psig). The gas permeates and cures an area of sand having about a 75 mm (3 in.) radius around the hole. Large cores and molds may be gassed using specially designed, gasketed covers or hoods that fit over the flask or box. It is important to place vents properly to ensure the flow of gas through all parts of the mass.

Coreboxes and patterns may be made of wood, metal, or plastic and should be washed regularly to prevent sticking problems caused by a buildup of sodium silicate.

The silicate-bonded sand, after pouring and shakeout, may be reclaimed by mechanical means; up to 60% of the reclaimed sand can be reused. Wet reclamation of the silicate sand system is possible, but requires a significant amount of water to scrub the sand. Recently, new methods of attrition reclamation combined with low-temperature thermal methods have shown some promise (see the discussion on sand reclamation in this article).

Phosphate-Bonded Molds. This inorganic binder system, which consists of an acidic, water-soluble, liquid phosphate binder and a powdered metal oxide hardener, was designed to comply with air quality control regulations. Because its components are inorganic, fumes, smoke, and odor are reduced at pouring and shakeout. The phosphate no-bake binder system has shakeout properties superior to those of the silicate/ester catalyzed no-bake systems described in the article "Resin Binder Processes" in this Volume. The bonded sand can be reclaimed easily by either shot blast or dry-attrition reclamation units.

The hardener component is an odorless, free-flowing powder. It must be kept dry; in contact with water, it slowly undergoes a mildly alkaline hydration reaction that alters its chemical reactivity and physical state. Under normal ambient conditions, the material is not hydroscopic. Flow properties of the powdered hardener are good. Standard powder feeding equipment can be used to disperse the hardener into sand mixers.

Curing characteristics of the phosphate no-bake binder system depend on the ratio of hardener to binder. Varying the level of hardener can typically control strip times from 25 min to more than 1 h. Recommended phosphoric acid-base binder levels are from 2.5 to 3.0% for molds and 3.5 to 4.0% for core production. The hardener level should be kept within 18 to 33% of the binder weight for best results.

Sand type also affects cure speed. Strongly alkaline sands such as olivine tend to accelerate the cure rate. Zircon forms an extremely strong and stable bond with phosphate binders, and as a result shakeout is more difficult.

High-quality defect-free castings can be produced using the phosphate binder system for molds and cores with a variety of metals, including gray and ductile irons and various steels. Erosion resistance of both washed and unwashed molds is excellent. Veining resistance is good on unwashed surfaces and can be controlled with the proper coating selection.

Unbonded Sand Molds

Murray Patz, Lost Foam Technologies, Inc.
Thomas S. Piwonka, University of Alabama

Casting processes that use unbonded sand molds are viable alternatives to conventional green sand molding processes. This section will review lost foam casting and vacuum molding both of which offer considerable advantages. A brief description is also provided for the magnetic molding process, which borrows from lost foam technology. Additional information on these processes can be found in Ref 1 to 12.

Lost Foam Casting*

The lost foam casting process originated in 1958 when H.F. Shroyer was granted a patent for the cavity-less casting method using a polystyrene foam pattern imbedded in traditional green sand, which was not removed before pouring of the metal (patent 2,830,343). The polystyrene foam pattern left in the sand mold is decomposed by the molten metal. The metal replaces the foam pattern, exactly duplicating all of the features of the pattern. Early use of the process was limited to one of a kind rough castings because the foam material used was coarse and hand fabricated and because the packed green sand mold would not allow the gases from the decomposing foam pattern to escape rapidly from the mold (the trapped gases usually resulted in porous castings).

The most significant breakthrough in the lost foam process came in 1964 with the issuance of a patent to T.R. Smith for the utilization of loose, unbonded sand as a casting media (patent 3,157,924). With this development, it was becoming clear to the foundry industry that the lost foam casting process was an emerging technology deserving of attention.

In the 1960s and 1970s, most of the lost foam casting activity took place in automotive company research facilities. Very few production castings were produced during this time. However, use of the lost foam casting process has been increasing rapidly since the expiration of the Smith patent in 1981. Currently, many casting facilities are dedicated strictly to the lost foam process.

Process Technique

Because the casting is an exact representation of the polystyrene foam pattern, the first critical step in the lost foam process is to produce a high-quality foam pattern. As will be described below, surface quality, fusion, dimensional stability, and foam pattern density are key control areas.

The foam pattern is prepared for casting by attaching it to a gating system (sometimes molded as part of the pattern) of

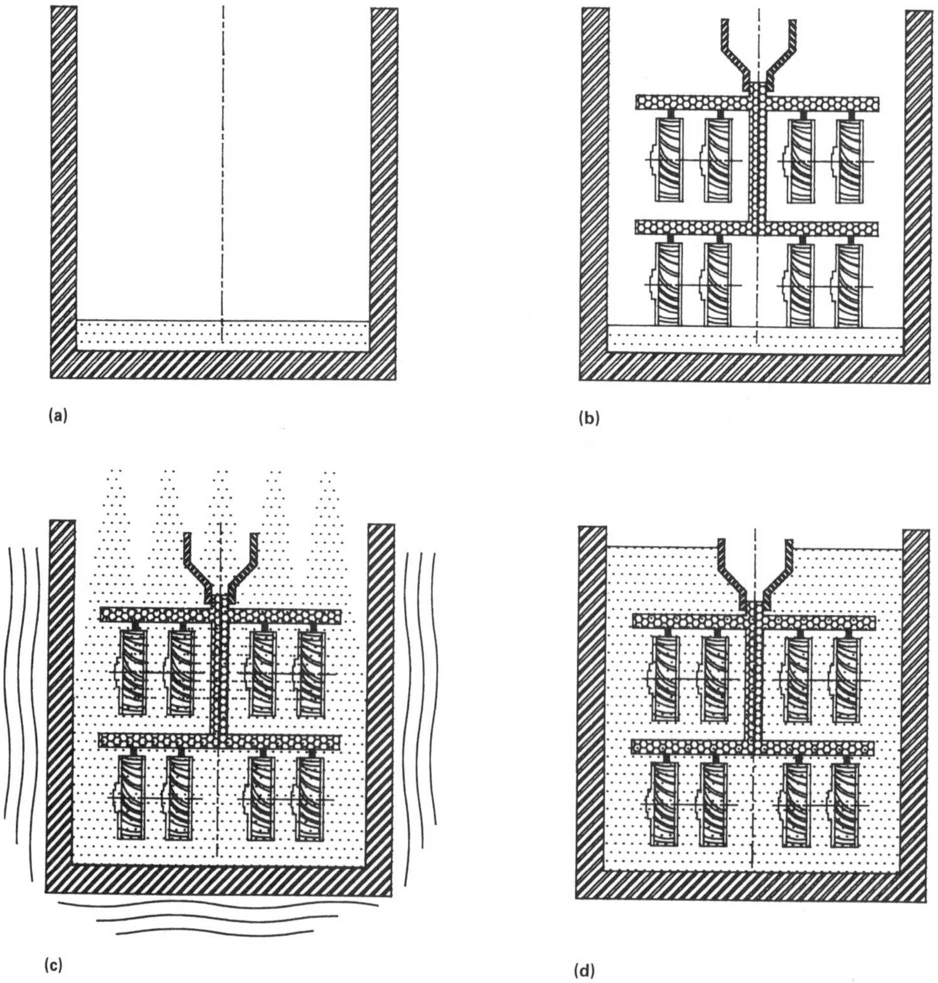

Fig. 1 Lost foam pattern system. (a) Flask that contains a 25 to 75 mm (1 to 3 in.) sand base. (b) Positioning the pattern. (c) Flask being filled with sand, which is subsequently vibratory compacted. (d) Final compact ready for pouring

*Lost foam casting is also referred to in the literature as evaporative pattern casting, evaporative foam casting, the lost pattern process, the cavity-less expanded polystyrene casting process, expanded polystyrene molding, or the full mold process.

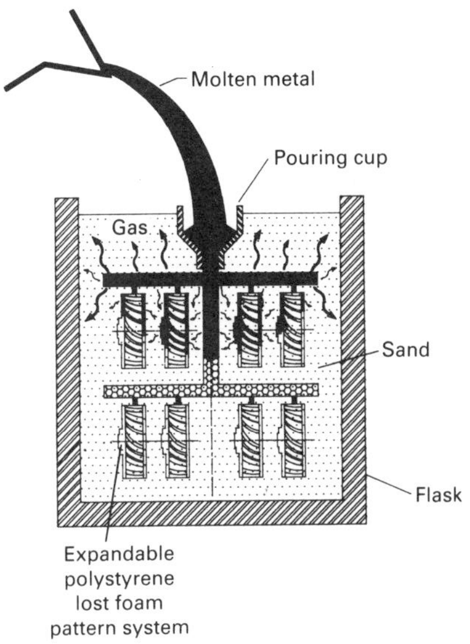

Fig. 2 Pouring of a lost foam casting

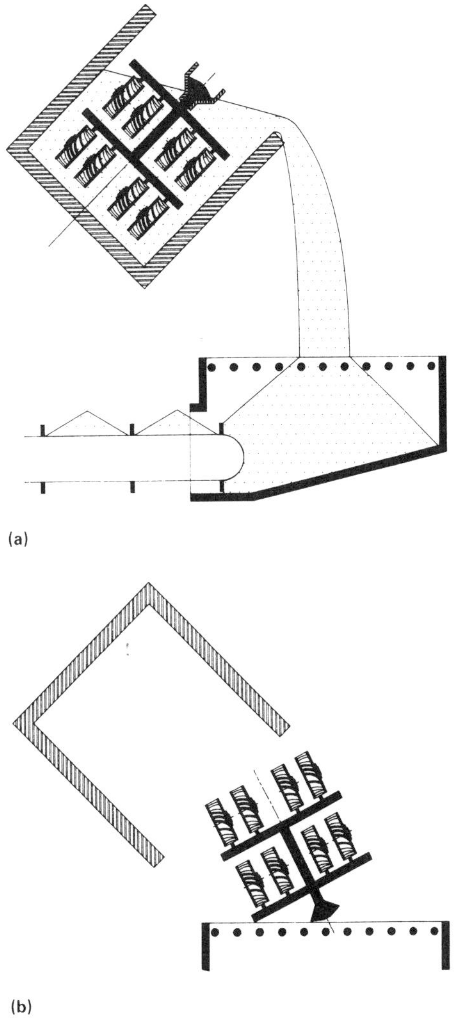

Fig. 3 Processing of completed (cooled) casting. (a) Flask is tipped, and the sand is recycled. (b) Casting is ready for degating and cleaning.

material of the same type and density. The pattern system with gating is then coated inside and out with a permeable refractory coating. Once the coating is dry, the pattern system is ready for investment into a one-piece sand flask. Investment of the pattern is achieved by positioning the pattern system in the flask, which has a 25 to 75 mm (1 to 3 in.) bed of sand in the bottom of the flask.

Once the pattern system is properly positioned, loose, unbonded sand is introduced in and around it (Fig. 1). The flask is vibrated to allow the loose sand to flow and compact in and around all areas of the pattern. A pouring basin or sprue cup is usually positioned around the exposed foam downsprue. When compaction (sand densification) is complete, the flask is moved into the pouring area, and molten metal is poured into the pouring basin (sprue cup). The metal vaporizes the foam pattern, precisely duplicating all the intricacies of the pattern (Fig. 2). The casting is then allowed to cool for approximately the same amount of time as with green sand. The flask is usually tipped over, allowing the loose sand to fall away from the casting (Fig. 3a). This sand is collected for reuse and the casting is ready for degating and cleaning (Fig. 3b).

Processing Parameters

The Foam Pattern. Expandable polystyrene (EPS) has been, and will probably continue to be, the preferred material for manufacturing lost foam patterns. There are other foam materials under development that show some promise, but to date their use is quite limited. There are several grades of EPS, as indicated in Table 1. Grades T and X are preferred because they give the foam pattern molder the ability to produce the smoothest surfaces and the thinnest sections possible for lost foam patterns.

Expandable polystyrene weighs approximately 640 kg/m^3 (40 lbf/ft^3) in its raw state. For the EPS to be useful in the manufacture of lost foam patterns, its bulk density must be reduced to a level between 16 and 27 kg/m^3 (1.0 and 1.7 lbf/ft^3), as indicated in Table 2. This is achieved through a process known as preexpansion. The raw EPS is introduced into a heated chamber and kept in constant motion for even distribution of heat throughout the batch of material. The plastic beads soften, and the gas within the beads expands; this increases the diameter of each individual bead and thus reduces the bulk density (a good analogy is popping corn).

The volume ratio at a finished density of 16 kg/m^3 (1.0 lbf/ft^3) is approximately 40:1. After a period of time in the preexpander, the material is discharged and weighed to check density. The density of the foam pattern for a given cast product is critical. For consistent casting results, EPS density must be controlled within ±2% of target density. This is achieved by monitoring and adjusting time and temperature in the preexpander.

Once the preexpanded EPS is discharged from the preexpander, it is placed in intermediate storage to cool and stabilize. After the stabilizing process is complete (usually 6 to 12 h, depending on the type of bead used), the preexpanded EPS is conveyed to a hopper attached to a pattern-molding press. This sequence is illustrated schematically in Fig. 4.

Pattern molding for the lost foam process can be grouped into four major functions: filling, fusion, cooling, (stabilization), and ejection.

Filling. The preexpanded stabilized material at the desired density is fed from the hopper on the press to the mold (or pattern die). A vented fill gun is used to feed the material, as illustrated in Fig. 5.

Fusion. After the mold cavity is full, heat is added by passing steam through the material in the cavity; this reinitiates the expansion process and softens the material as in the preexpansion process (Fig. 6a). The material expands again, filling the air voids between the individual EPS beads and fusing the beads together into a solid mass to form the desired foam pattern.

Cooling. During the fusion portion of a molding cycle, the molded part exerts pressure against the cavity walls. If the part has not been cooled to reduce the internal pressure of the molded part, the part will continue to expand after ejection. This condition is known as postexpansion. To eliminate postexpansion, it is necessary to cool the mold cavity, thus reducing the internal pressure of the molded part to a point at which it can be ejected and still maintain its dimensional integrity. Cooling is usually accomplished by spraying water on the back of the mold wall cavity (Fig. 6b).

Ejection. Once the part has been cooled, the press can be opened and the part ejected (Fig. 6c). This can be done pneumatically or mechanically.

Table 1 Grades of expandable polystyrene

Bead grades	Raw bead diameter, mm	Raw bead diameter, in.	Diameter at 24 kg/m^3 (1.5 lbf/ft^3), mm	Diameter at 24 kg/m^3 (1.5 lbf/ft^3), in.	Typical use
A	0.83–2.0	0.033–0.078	2.5–5.9	0.097–0.231	Insulation
B	0.58–1.2	0.023–0.047	1.7–3.5	0.068–0.138	Packaging
C	0.33–0.71	0.013–0.028	1.0–2.1	0.040–0.082	Cups (coffee)
T	0.25–0.51	0.010–0.020	0.74–1.5	0.029–0.058	Cups/lost foam
X	0.20–0.33	0.008–0.013	0.61–1.0	0.024–0.040	Lost foam

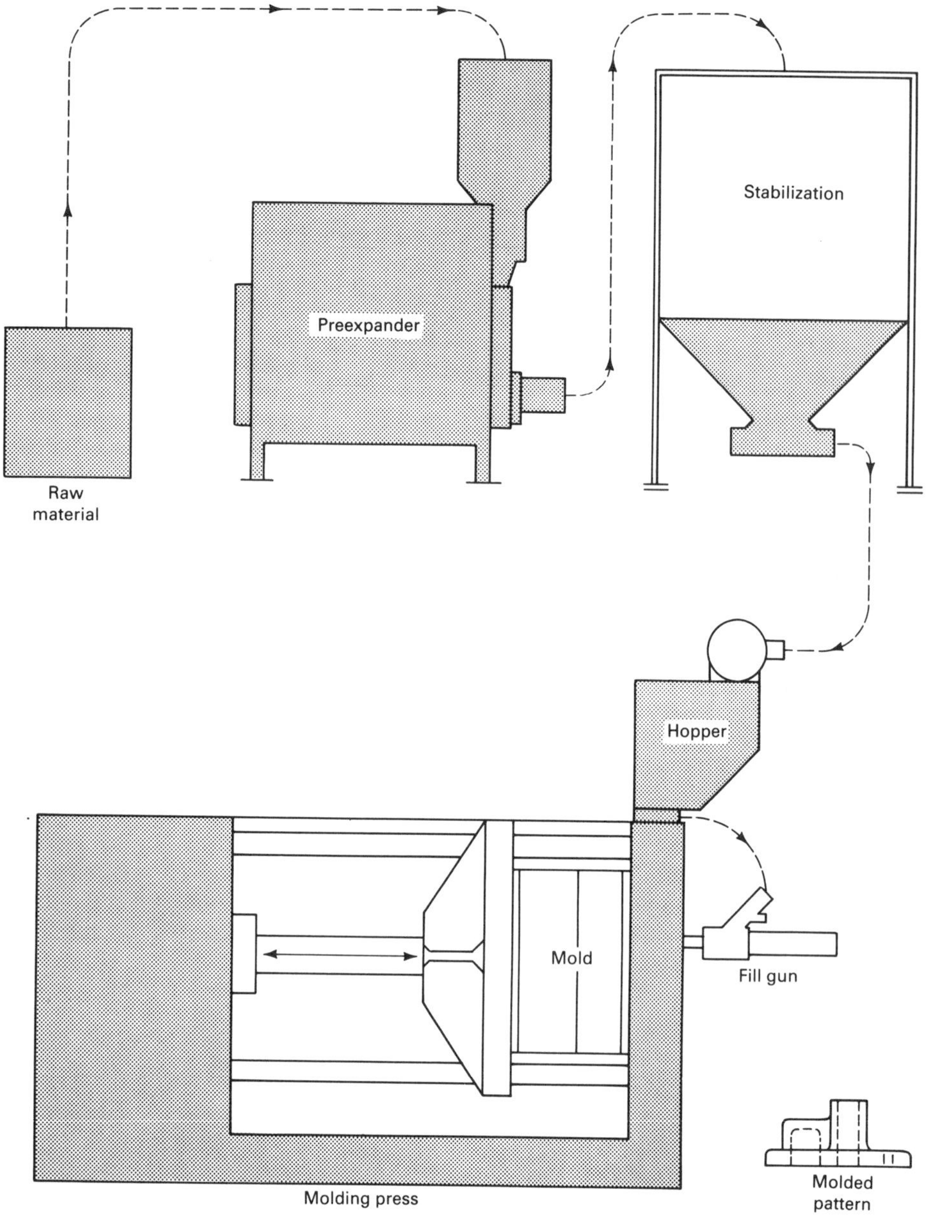

Fig. 4 Sequence of operations for producing a foam pattern

Pattern Assembly. Lost foam patterns typically consist of multiple pieces that must be assembled to form a completed pattern and gating system. The most widely accepted method of assembling patterns is to glue them together with a hot melt adhesive. The adhesive used is specifically formulated for the lost foam process. The temperature at which it is applied to the foam patterns is much lower than the application temperatures of other commercial hot melt adhesives. Lost foam hot melts are generally applied between 120 and 130 °C (245 and 270 °F). To ensure consistent dimensional control and adhesive joint quality, an automatic or semiautomatic gluing procedure is preferred. Figure 7 illustrates a typical pattern assembly sequence.

For prototypes and short run quantities, hand-operated glue printing and assembly jigs can be used. Other methods of pattern assembly can be used in specific applications. These may include hand-applied rubber cement, liquid and contact adhesives, robotically applied airset adhesives, and foamed hot melt procedures. Whichever adhesive material is chosen, it must be suitable for use with lost foam patterns and must be compatible with the molten metal during the casting operation.

The vaporization rate and ash content of the adhesive must be approximately the same as those of the EPS pattern. The adhesive must not adversely affect the dimensional integrity of the pattern. Care must be taken to control all aspects of pattern assembly to maintain consistency. Poor assembly techniques will result in increased casting scrap and dimensional control problems. Thorough inspection of the completed foam pattern system is essential before proceeding to the next step in the process.

Coating Types. It is possible to produce a lost foam casting without using a refractory coating on the foam pattern, but for maximum process latitude, it is advantageous to coat the pattern system. This specialized coating serves two purposes. First, it provides a barrier between the smooth surface of the pattern and the coarse surface of the sand. Second, it provides controlled permeability, which allows the gaseous products created by the vaporizing foam pattern to escape through the coating and into the sand away from the cast metal (Fig. 8).

Lost foam coatings are available in various permeabilities. Lower-permeability coatings are preferred in aluminum castings with high surface-area-to-volume ratios—for example, intake manifolds. Heavy-section aluminum castings and other nonferrous castings may require medium- or high-permeability coatings. Ferrous castings generally require higher permeability than nonferrous castings. Silica is the refractory of choice. Its combination of low cost and excellent heat transfer characteristics makes it attractive for use with most metals.

Zircon and olivine coatings are also available, but are not widely used. The choice of coating carrier is limited because the carrier must be compatible with the foam pattern. Hydrocarbons and chlorinated solvents will attack the EPS. Water is the most widely used carrier.

Application of the Coating. Proper preparation and control of the coating is absolutely essential for the consistent production of high-quality lost foam castings. The coating, when mixed properly, can be applied to the pattern system by dipping, brushing, spraying, or flow coating.

Brushing, spraying, and flow coating are used on very large patterns. Dipping is the preferred method for small- and medium-size patterns. Because of the extreme difference in density between the foam pattern system and the coating slurry, large patterns are sometimes difficult to immerse.

Table 2 Typical pattern density requirements for lost foam casting

Metal	Metal pouring temperatures		EPS pattern density	
	°C	°F	kg/m³	lb/ft³
Aluminum	705–790	1300–1450	24–27	1.5–1.7
Brass/bronze	1040–1260	1900–2300	20–21.6	1.25–1.35
Gray iron	1370–1455	2500–2650	≤20	≤1.25
Steel	1595–1650	2900–3000	≤17.6	≤1.10

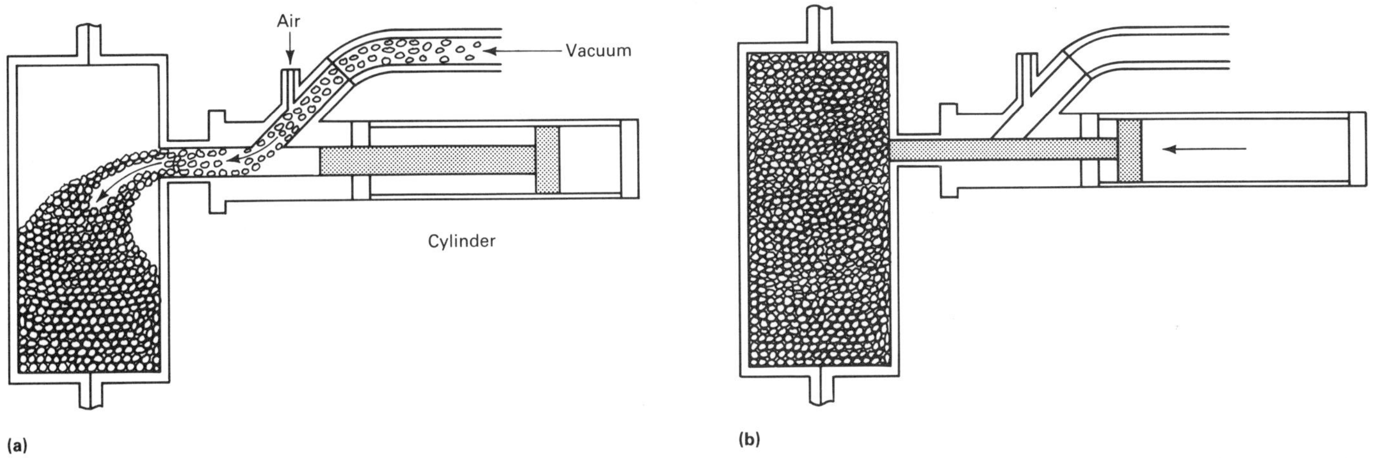

Fig. 5 Fill gun used to feed preexpanded stabilized material to molding press. (a) Open cylinder with pattern material transported by air/vacuum. (b) Cylinder closed prior to fusion

Pattern orientation and coating fluidity must facilitate pattern coverage in blind areas. The entire pattern system must be covered inside and out with the exception of the exposed downsprue area, on which molten metal will be poured.

Foam pattern cluster systems must be handled carefully during coating and drying. The weight of the wet coating may cause the patterns to distort or break. Medium- and high-production lost foam facilities find automated coating systems very helpful in controlling the consistency of the coating.

The next step is to dry the coated pattern. This can be done at ambient conditions within 24 h, or the pattern can be force dried in a drying oven or heated room. Forced drying is done at oven temperatures of 50 to 65 °C (120 to 150 °F), along with numerous changes of air per hour and possibly dehumidification. This method of drying coated foam patterns usually takes 2 to 6 h. Microwave drying in the final stage can be used in very high production applications.

Investing the Foam Pattern in Sand. As indicated in Table 1, the coated pattern system is placed in a one-piece flask. Loose, dry, unbonded sand is introduced into the flask generally through a sand raining system. This system gently fills the flask with sand, minimizing the possibility of pattern distortion due to side movement of the sand against the pattern. During the filling of the flask, the flask is isolated from the flask conveying system and vibrated with a high-frequency compaction system either from the bottom or the side of the flask. Variable frequency and amplitude may be required to compact the sand fully in both the internal and external areas of the foam pattern system.

Sand System. The sand most commonly used is silica sand—subangular to round. Different sizes are used for different applications. Generally, AFS grain fineness number 35-3 screen distribution is used in ferrous applications, while AFS 45-3 screen distribution is used in most nonferrous systems (methods for determining sand grain fineness number are discussed in the article "Aggregate Molding Materials" in this Volume). Most sand systems, in addition to the compaction and sand fill stations, incorporate a pouring area, cooling area, dump-out area, sand return system, and sand cooler classifier system. To maintain consistent sand permeability, it is necessary to screen or separate the fine particles that accumulate. The sand must be cooled before being returned to the flask fill area. Foam pattern distortion and adhesive softening may occur if the temperature of the sand introduced in the flask is higher than 50 °C (120 °F).

Pouring a lost foam casting is much like pouring metal in other sand casting methods. The pouring basin or pour cup must remain full throughout the pour. Metal feed is controlled by the vaporization rate of the foam pattern system. A positive head of metal must be maintained (Fig. 2). Failure to do so may result in a mold collapse and scrapped castings or a partial mold collapse, thus providing the opportunity for sand and coating inclusions in the casting. Metal temperature at the time of pouring is critical and must be controlled tighter than with most other sand casting methods.

After pouring, the castings are allowed to cool at approximately the same rate as green sand castings. The flasks are moved into the dump-out area, sand and castings

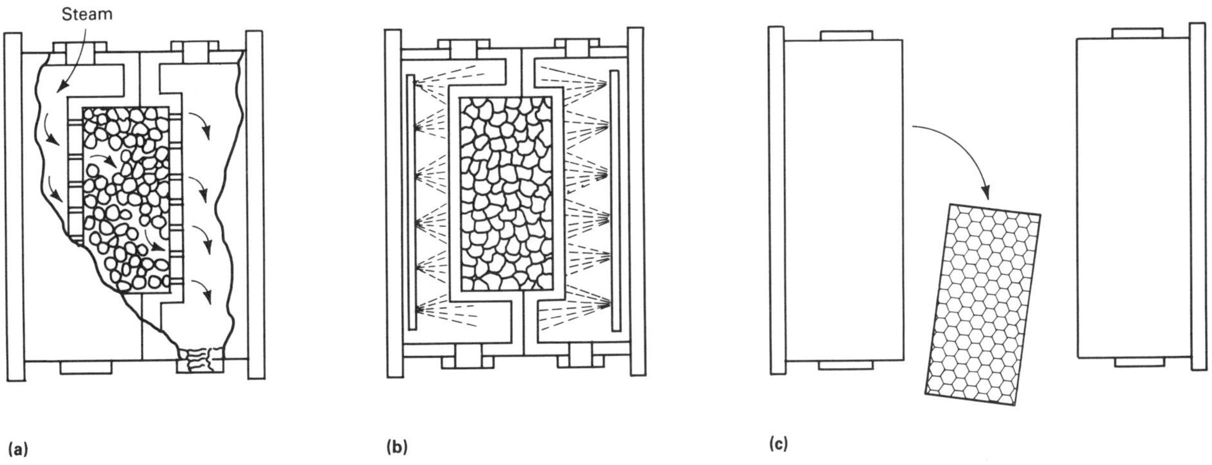

Fig. 6 Schematic showing fusion (a), cooling (b), and ejection (c) of lost foam pattern material

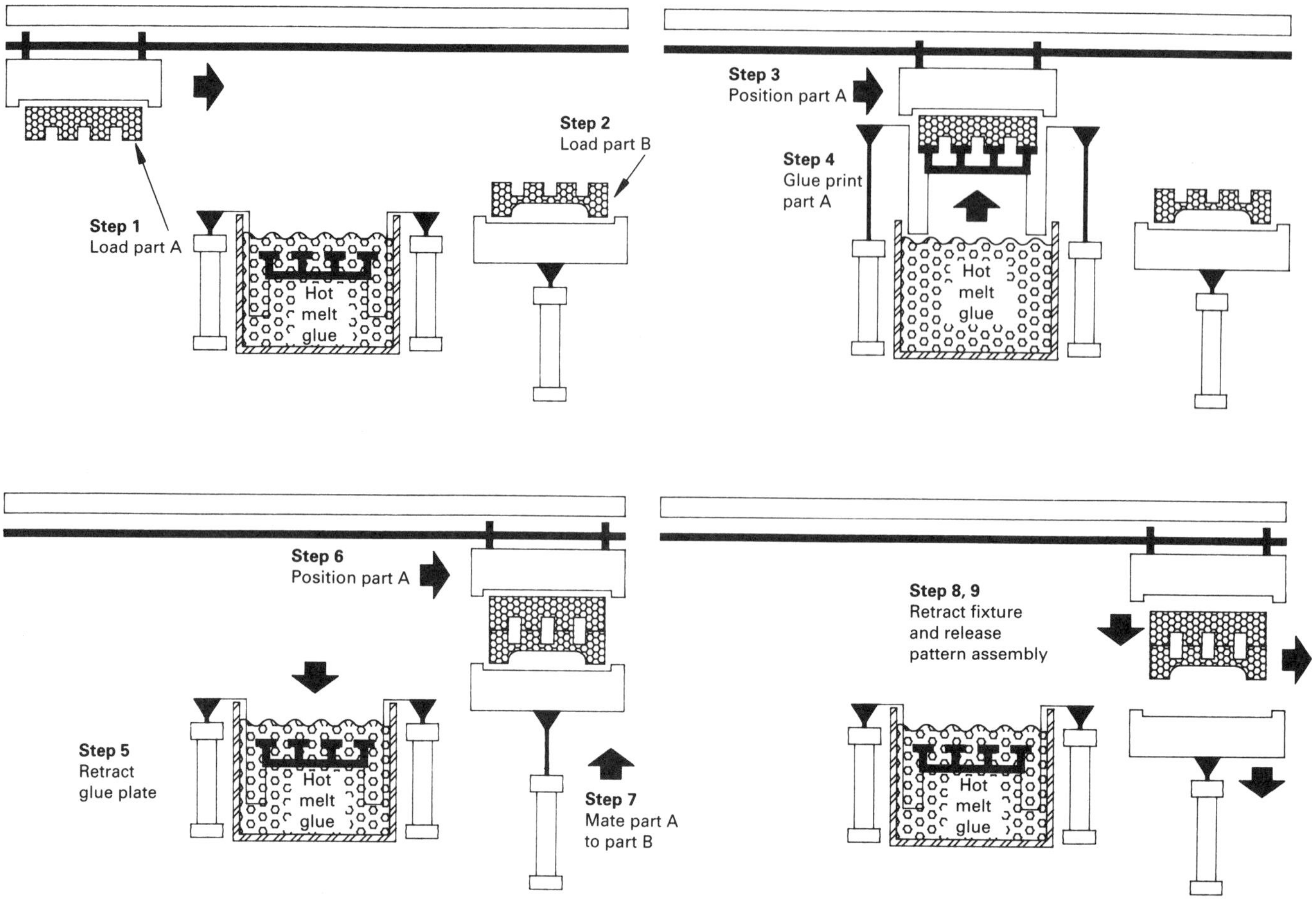

Fig. 7 Pattern assembly sequence showing nine steps performed in the pattern assembly machine

are separated (Fig. 3), and the castings are ready for degating and cleaning. Conventional cleaning methods, such as sandblasting or shotblasting, can be used. Vibratory media cleaning systems may be necessary to remove the coating, which adheres to internal cast surfaces. Water quenching will greatly assist the cleaning of aluminum castings. Methods for cleaning castings are reviewed in detail in the article "Processing of Castings" in this Volume.

Advantages of Lost Foam Casting

The benefits of lost foam processing are numerous, and most are obvious to those who have witnessed the simplicity of the process. In lost foam casting, there is no mold parting line and there are no cores. One-piece flasks are readily moved.

The use of untreated, unbonded sand makes the sand system economical and easy to manage. There is less maintenance on sand handling equipment. Less sand is needed in the system, and the recycling of the sand is an ecological advantage.

These advantages lead to others that are equally desirable. Casting cleaning is greatly reduced and is sometimes eliminated except for removal of the wash coating that transfers from the pattern to the casting. Casting yield can be considerably increased by pouring into a three-dimensional flask with the castings gated to a center sprue.

Another advantage of lost foam casting is its ability to reduce the labor and the skill required in the casting process. In addition to reduced cleaning and the elimination of the need for core setting or parting line matches, handling of the flasks and castings is more readily automated with the lost foam process because locations are predictable.

A lost foam casting facility has the ability to produce a variety of castings in a continuous and timely manner. Lost foam foundries can pour diverse metals with very few changeover problems, and this adds to the versatility of the foundry. Response to customer demands can also be very rapid with this type of manufacturing system.

Further benefits of lost foam casting result from the freedom in part design offered by the process. Assembled patterns can be used to make castings that cannot be produced by any other high-production process. Part development costs can be reduced because of the ability to prototype with the foam. Product and process development can be kept in-house.

Finally, cast-in features and reduced finishing stock are usually benefits of using the lost foam process. Inserts can be cast into the metal, and bimetallic castings can be made.

Magnetic Molding (Ref 12)

Magnetic molding is similar to the lost foam process in that an EPS pattern is used. However, the unbonded sand used as the mold backing aggregate is replaced with magnetic iron or steel shot that is bonded by a magnetic field. The process consists of positioning the EPS pattern in the flask and encasing it with iron particles between 0.5

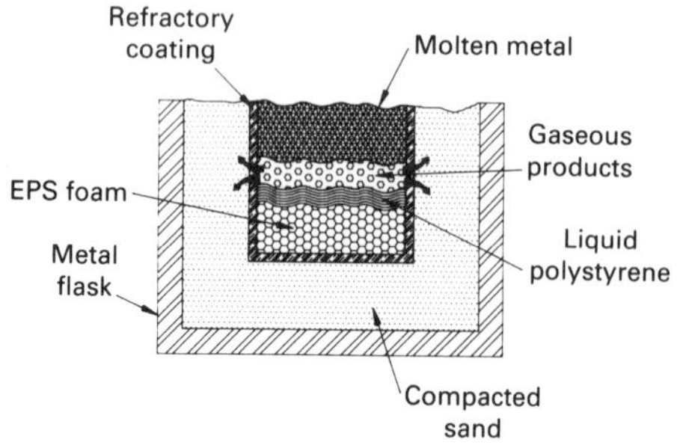

Fig. 8 Reactions taking place during a lost foam pouring operation

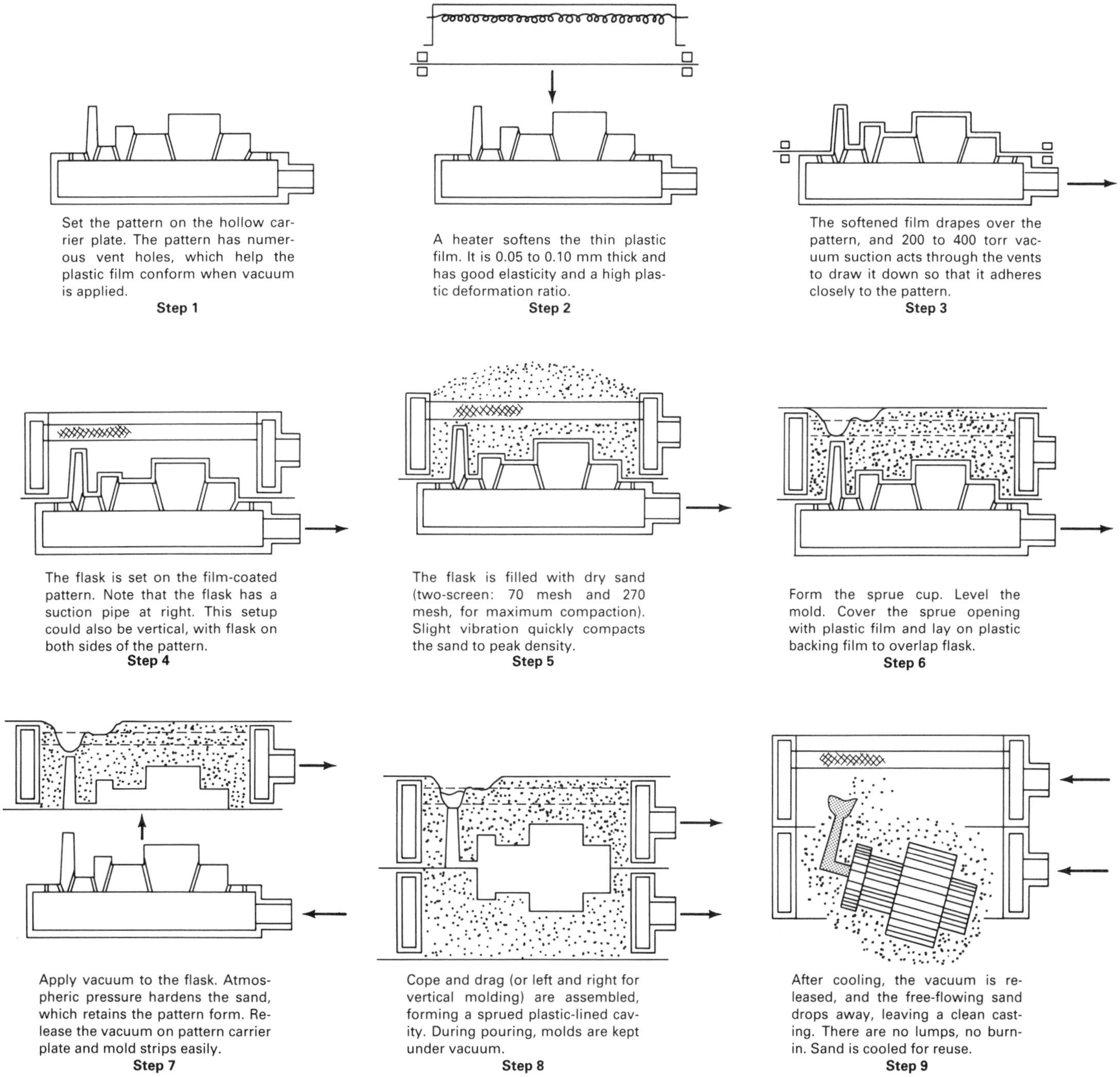

Set the pattern on the hollow carrier plate. The pattern has numerous vent holes, which help the plastic film conform when vacuum is applied.
Step 1

A heater softens the thin plastic film. It is 0.05 to 0.10 mm thick and has good elasticity and a high plastic deformation ratio.
Step 2

The softened film drapes over the pattern, and 200 to 400 torr vacuum suction acts through the vents to draw it down so that it adheres closely to the pattern.
Step 3

The flask is set on the film-coated pattern. Note that the flask has a suction pipe at right. This setup could also be vertical, with flask on both sides of the pattern.
Step 4

The flask is filled with dry sand (two-screen: 70 mesh and 270 mesh, for maximum compaction). Slight vibration quickly compacts the sand to peak density.
Step 5

Form the sprue cup. Level the mold. Cover the sprue opening with plastic film and lay on plastic backing film to overlap flask.
Step 6

Apply vacuum to the flask. Atmospheric pressure hardens the sand, which retains the pattern form. Release the vacuum on pattern carrier plate and mold strips easily.
Step 7

Cope and drag (or left and right for vertical molding) are assembled, forming a sprued plastic-lined cavity. During pouring, molds are kept under vacuum.
Step 8

After cooling, the vacuum is released, and the free-flowing sand drops away, leaving a clean casting. There are no lumps, no burn-in. Sand is cooled for reuse.
Step 9

Fig. 9 Elementary sequences in producing V-process molds

and 0.1 mm (0.02 and 0.004 in.) in diameter (supported by periodic vibrating and/or tilting), then magnetizing the molding material and pouring the molten metal while the evolving gases are drawn off through the base of the flask. The magnetic field is turned off after solidification and cooling, resulting in immediate shakeout. The free-flowing magnetic shot molding material is returned to its point of origin after cooling, dedusting, and metal splash removal.

The increased heat conductivity of the molding material results in a finer grain structure in the cast metal. Other advantages include the absence of a chemical binder, reductions in dust and noise levels, full mechanization or automation of the process, and the elimination of molding activities normally used (such as ramming and jolting).

To date, most of the research on magnetic molding has been carried out in the Soviet Union and Germany on cast irons, carbon and low-alloy steels, high-chromium steels, and copper-base alloys (Ref 5-8). Investigations are still being conducted to determine the influence of process variables such as vibration time, magnetic field strength, and time of application on the dimensional accuracy of the castings produced. In addition, work is required to establish more clearly the magnetic properties of the iron powder to provide accurate information on the values of magnetic field strength required for the process (Ref 4).

Vacuum Molding (Ref 10-12)

The vacuum molding process, or V-process, is a sand molding process in which no binders are used. Instead, the sand is positioned between two sheets of thin plastic

Table 3 Plastic films used for the V-process

Type of film	Density, g/cm³	Melting point °C	°F
Low-density polyethylene	0.920	88–90	190–194
High-density polyethylene	0.960	94–97	201–207
Nylon	1.13	215–222	419–432
Polypropylene	0.90–0.91	160–170	320–338
Ionomer	0.93–0.94	72–75	162–167
EVA(a)	0.940	58	136
Polyvinylchloride	1.450	56–90	133–194

(a) Ethylene-vinylacetate co-polymer. Source: Ref 10

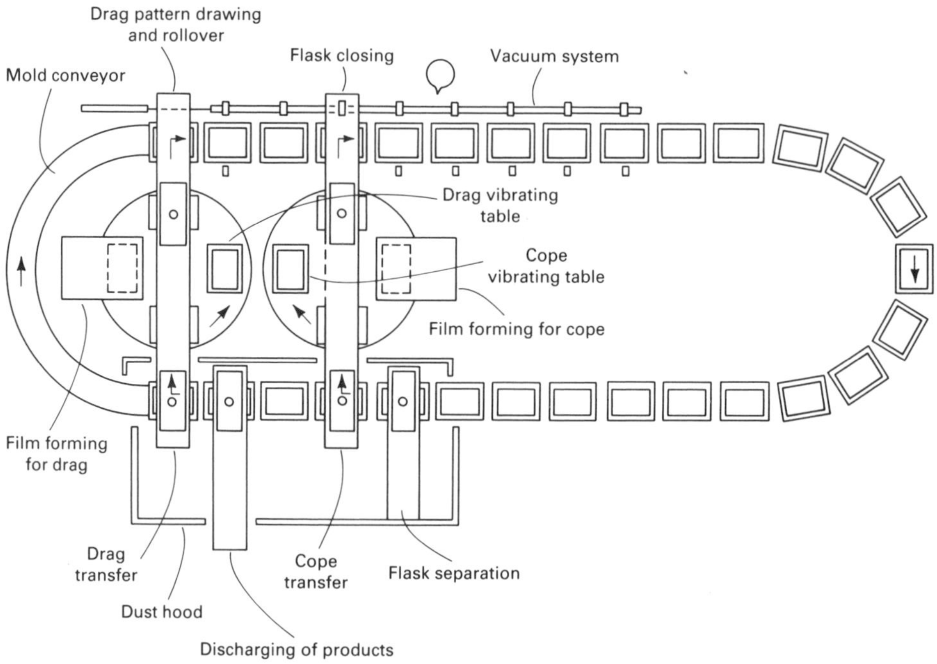

Fig. 10 High-speed automatic V-process production line. Production: 60 complete molds per hour. Flask size: 1300 × 1100 × 400 mm (50 × 40 × 15 in.). Amount of sand: 120 Mg (130 tons/h). Number of operators: 5

and is held in place by the application of a vacuum. Originally developed in Japan for the production of castings with high surface-area-to-volume ratios, the process is now licensed for use worldwide, and it has been successfully used to cast all metals that are normally cast in conventional green sand mold mixes.

Sequence of Operations. The steps in the process are shown in Fig. 9. A specially constructed pattern is built on a hollow carrier plate. The pattern contains a number of small vent holes. A thin (0.05 to 0.10 mm, or 0.002 to 0.004 in.) plastic film is then heated and stretched over the pattern and the sprue pin. A moderate vacuum (27 to 53 kPa, or 200 to 400 torr) is applied to the pattern plate, and this draws the film tightly to the pattern by suction through the vent holes.

A specially constructed hollow-wall flask containing a vacuum connection is placed around the pattern, filled with sand, and vibrated at a low power level. A two-screen sand (70 and 270 mesh) is recommended for producing high-density molds. The excess sand is then struck off, and the top of the sprue pin is wiped clear of sand. A sheet of plastic film is placed over the top of the sand, and vacuum is applied to the flask. The vacuum holds the sand tightly in place, producing an extremely hard (AFS mold hardness 90+) dense mold. When the vacuum is released on the pattern carrier plate, the pattern is easily withdrawn.

Cope and drag are assembled while maintaining the vacuum, and the casting is poured. After the casting cools, the vacuum is released in the flasks, and the casting and loose sand are removed. The castings separate easily from the sand, which requires only cooling and screening for reuse.

Before pouring, the plastic that covers portions of the mold cavity that would normally be open to the atmosphere (such as the tops of risers, vents, and sprues) is removed to ensure that the mold cavity remains at atmospheric pressure during pouring in order to maintain the pressure differential between the mold and the mold cavity (otherwise the mold will collapse). Care must be taken to ensure that the gases generated during pouring exit to the atmosphere, not to the sand and vacuum system.

Plastic Film Characteristics. The plastic films used for vacuum molding are thermoplastic compositions, while thermosetting resins are used as foundry sand binders (see the article "Resin Binder Processes" in this Volume). Plastics that have been evaluated for their applicability to vacuum sealing are listed in Table 3, along with some representative properties. The most commonly used material is an ethylene-vinylacetate co-polymer (EVA) with vinylacetate contents ranging from 14 to 17%.

The characteristics of the plastic film must be suitable for vacuum forming so that pattern contours will be closely replicated without film shape defects (folds and creases) or rupturing, and the original sheet should be free of pinholes, tears, air bubbles, or blemishes. The important characteristics for easy shaping to pattern contour are both elastic and plastic behavior in proper balance. These properties are a function of film composition, temperature, thickness, stress, rate of loading, direction of loading, and the presence of notch inducers. More detailed information on the film characteristics and test methods used to evaluate these properties is available in Ref 10.

Advantages. The V-process offers a number of benefits. The lack of a binder means that fumes and dust are reduced and that no sand mixing system is necessary. No moldmaking machines are needed, and shakeout and sand reclamation are simplified. Patterns are identical to those used for green sand molding, except that vent holes must be judiciously placed around the patterns for plastic film adhesion. Pattern wear is reduced because the sand never makes contact with the pattern surface. Casting surfaces and dimensions are excellent.

The V-process is not normally used for high-production runs, because of the number of functions that must be performed; however, production rates of 60 to 100 molds per hour have been achieved. Figure 10 shows the layout for a high-speed automated vacuum molding line for the production of medium-size castings at a rate of 60 molds per hour. Vacuum molding is often used in the floor molding of relatively large patterns (flask sizes ≥750 × 750 mm, or 30 × 30 in., are typical) in jobbing foundries, where production rate is five to ten molds per hour.

As the plastic film vaporizes with a reducing gas, metal fluidity is often improved in V-process molds. However, because the molds contain no moisture, solidification rates are slower. The high mold rigidity increases riser efficiency because there is virtually no mold wall movement.

REFERENCES

1. E.J. Sikora, Evaporative Casting Using Expendable Polystyrene Patterns and Unbonded Sand Casting Techniques, *Trans. AFS*, Vol 86, 1978, p 65
2. M.C. Ashton, S.G. Sharman, and A.J. Brookes, The Replicast (Full Mold) and CS (Ceramic Shell Process), *Trans. AFS*, Vol 92, p 271-280
3. N. Moll and D. Johnson, A New Moldable Foam for Casting Ferrous Materials Without Lustrous Carbon Defects, in *Advanced Casting Technology*, ASM

INTERNATIONAL, 1987, p 175-185
4. A.J. Clegg, Development of a Magnetic-Molding Process Test-Rig, *Foundry Trade J.*, Vol 138, 1975, p 833-841
5. *Full Mold Casting*, Conference Proceedings of the Institute of Foundry Problems, Ukranian S.S.R. Academy of Sciences, 1979 (in Russian)
6. E.I. Belov and B.V. Rabinovich, The Structure of Magnetic-Mold Iron Castings, *Liteinoe Proizvod.*, Vol 7, 1982, p 11-12 (in Russian)
7. P.D. Afanasov, Effect of Casting Quality on Properties of Magnets, *Liteinoe Proizvod.*, Vol 2, 1981, p 11 (in Russian)
8. U.E. Zalcman and A.E. Mikelson, Magnetic Molds for Casting Production, *Giessereitechnik*, Vol 21 (No. 5), 1975, p 170-179 (in German)
9. K. Muckhoff, Present Position of the Magnetic Molding Process, *Giesserei*, Vol 61 (No. 5), 1974, p 104-109 (in German)
10. Y. Kubo, K. Nakata, and P.R. Gouwens, Molding Unbonded Sand With Vacuum—The V-Process, *Trans. AFS*, Vol 81, 1973, p 529-544
11. T. Miura, Casting Production by V-Process Equipment, *Trans. AFS*, Vol 84, 1976, p 233-236
12. E. Kotzin, *Metalcasting and Molding Processes*, American Foundrymen's Society, 1981, p 143-147, 162

Coremaking

Larson E. Wile, Consultant
Ken Strausbaugh, James J. Archibald, and Richard L. Smith, Ashland Chemical Company
Thomas S. Piwonka, University of Alabama

COREMAKING involves coating a refractory aggregate with binder, compacting the coated sand into the desired shape, and then curing (hardening) the compacted mass so that it can be handled. Refractory coatings are also often applied to the core surface to prevent sand-adhering defects on the casting and to yield a smoother casting finish.

Selection of a Core System

Although a wide variety of methods and materials are used for making cores, coremaking processes are generally categorized according to curing method and binder type. Detailed information on these processes can be found in the article "Resin Binder Processes" in this Volume. The three major systems used for coremaking are listed below:

Cold Box Processes

- Phenolic urethane/amine
- Furan/SO_2
- (Acrylic/epoxy)/SO_2 (free radical cure process)
- Sodium silicate/CO_2
- Phenolic ester

Heat-Cured Processes

- Shell process
- Core oil (oven bake)
- Phenolic hot box
- Furan hot box
- Furan (furfuryl alcohol) warm box

No-Bake Processes

- Furan/acid
- Phenolic/acid
- Alkaline phenolic/ester
- Silicate/ester
- Phenolic urethane/amine
- Polyol isocyanate
- Alumina phosphate

The cold- and heat-cured (heat-activated) processes lend themselves exceptionally well to medium- and high-production applications. Examples of high-production operations are foundries that cast automotive, pipe fitting, air conditioning, and refrigeration components (Fig. 1). The no-bake processes are generally used for short-run production quantities.

Both cold- and heat-activated cores are cured in the core box, thus maintaining excellent dimensional accuracy. Cold processes utilize gases that are forced through the compacted sand mixture to cure the core (Fig. 2). Heat-cured processes require the core box to be heated to 175 to 290 °C (350 to 550 °F) prior to introduction of the prepared sand. The no-bake process uses binder systems that consist of chemicals that, when mixed together in sand, cure without the introduction of an external agent, such as heat. The process is well suited to the fabrication of larger cores.

In the core-oil (oven-bake) process, sand is mixed with corn flour and/or clay and water along with the binder to give sufficient strength to the mass of sand so that it can be immediately removed from the core box onto core plates. The cores are then hardened by baking at 200 to 260 °C (400 to 500 °F). The oven-bake process does not require complicated curing equipment, and materials are inexpensive; but dimensional accuracy is more difficult to maintain.

The type of metal to be poured must be considered in selecting a core process. For example, steel and certain high-alloy cast irons are sensitive to urea because of the tendency to form gas defects due to the amount of nitrogen present. In making aluminum castings, the pouring temperature may not be high enough to break down the binder system adequately for easy removal from the castings. Although silicate cores usually collapse sufficiently when used with aluminum and steel, their use in cast iron may be a problem because of a secondary bond that forms at iron-pouring temperatures. Cores for magnesium castings require inhibitors that must be compatible with the binder system.

Wood, plastic, and metal core boxes can be used with the cold processes, while only metal can be used with the heat-activated systems. However, once a system is established, metal core boxes and machines designed for one type of heat-activated process can usually be used for alternate heat-activated processes. The same is true for cold processes.

Although equipment is available for blowing or shooting core weights of up to 540 kg (1200 lb), most cores of this size and above use one of the no-bake processes. Smaller cores are usually made using cold, warm, or hot box techniques unless they have high surface-area-to-volume ratios, in which case the required equipment would be cumbersome.

Elevated temperatures accelerate chemical reactions. Summer temperatures may adversely affect some of the hot systems to the point where they actually harden prematurely. In cold systems, the silicon dioxide (SO_2) types possess the longest working times.

The core sand will eventually become system sand. Therefore, suitability for reclamation and compatibility with the sand reclamation and sand system used must be considered.

Mixing

Sand mixing is easily accomplished for all of the processes except the oven-bake and shell processes. Special mullers are required to mix the sand and dry ingredients adequately with water and binders to develop the necessary strength to permit baking process cores to be removed from the core boxes in the uncured state.

Most shell process sands are hot coated in paddle/plow-type mixers using solid or liquid novolac resins. Cold and warm coating, both of which utilize powdered or liquid novolac resins, have declined.

With the hot coating process, a solid flake or liquid novolac resin is dispersed into a sand that has been preheated to 130 to 175 °C (265 to 350 °F). Hardener (contents ranging from 10 to 17% based on the weight of the resin) is first dissolved in water, and this solution is added after the resin is dispersed, and a lubricant is added near the end of the cycle.

The moisture content of the sand mix will range from 1.5 to 2.5%. Immediately after the addition of the water/hardener solution, cold air is blown into the mixer to cool and

(a) (b)

Fig. 1 Cores and corresponding castings produced by the cold box process. (a) Large transmission case and resulting gray iron casting for an agricultural equipment manufacturer. (b) Disk brake rotor and turbine cores and castings. Courtesy of Ashland Chemical Company

dry the sand. Whatever process is selected, the coating process is carried out in a paddle-type mixer that is usually equipped with a system (fluid bed, screen, or cyclone) to cool and complete the drying of the coated sand.

Various types of blending equipment are used to disperse the binders uniformly over the sand used in the other processes. Batch or continuous mixers can be employed to coat the sand used in the cold or heat-activated processes. If a batch mixer is used, the sand and additives are weighed or volumetrically measured before being introduced into a mix. There the mixture is agitated by a revolving mixer or by vibratory motion for 20 s to 5 min. Prepared sand is then discharged and transferred to the coremaking station.

A continuous mixer is almost a necessity for a no-bake coremaking operation, but is also adequate for mixing sand for the cold or heat-activated processes, with the exception of the shell process. Sand and additives are continuously metered into a mixing trough or chamber that contains revolving mixers of various designs, and prepared sand is discharged. In the no-bake process, the sand is discharged directly into the core box, where it is immediately compacted. Prepared sand for the cold and heat-activated processes is generally discharged into a storage hopper above a coremaking machine.

Compaction

Cores must be sufficiently dense to obtain good grain-to-grain bonding of the sand mass and to prevent metal penetration between the sand grains. Prepared sand can be hand rammed into the core box, and this is the usual procedure for no-bake cores; but mechanical methods are required for high-production operations.

Core blowing machines are commonly used to compact sand into the core boxes. Prepared sand is added to a chamber, which is then sealed except for exit ports (blow ports) that are placed directly over inlet ports (investment areas) of the core boxes. The core boxes are equipped with air relief

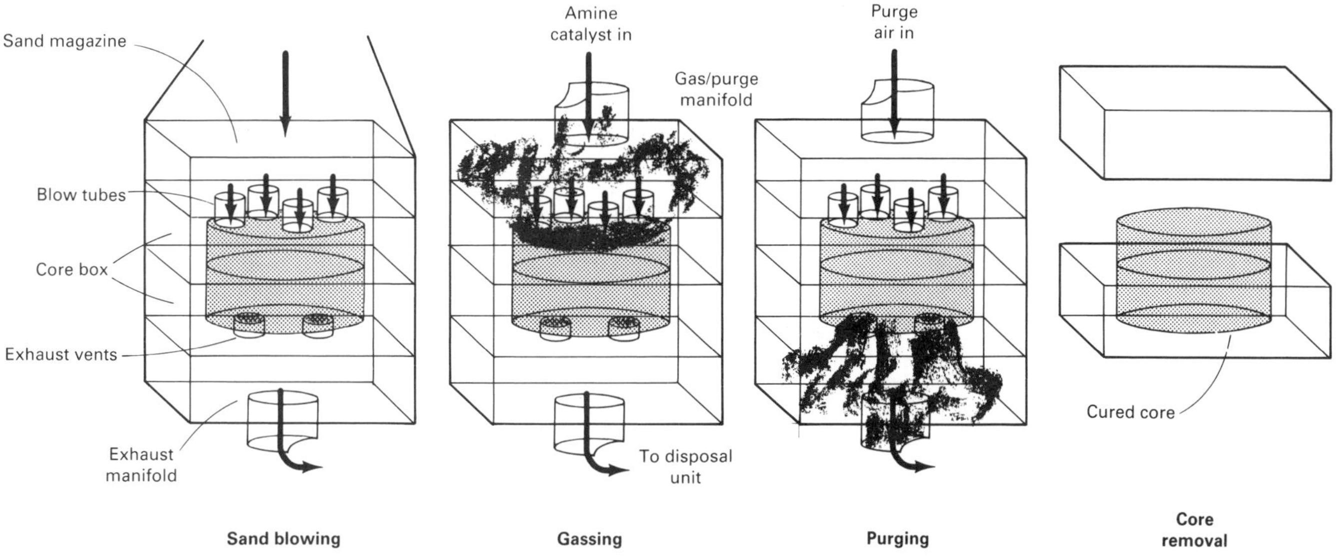

Fig. 2 The cold box coremaking process. The wet sand mix, prepared by mixing sand with the two-component liquid resin binder, is blown into the core box. The core box is then situated between an upper gas input manifold and a lower gas exhaust manifold. The catalyst gas enters the core box through the blow ports or vents and passes through the core, causing almost instantaneous hardening of the resin-coated sand. The core is ready for ejection from the core box after purging with clean air for a few seconds. After the catalyst gas passes through the core, it leaves the core box through vents into the exhaust manifold. From the gas exhaust manifold, the catalyst gas is piped to an appropriate disposal unit.

vents. A large volume of compressed air at 275 to 700 kPa (40 to 100 psi) is suddenly introduced above the sand in the chamber, forcing the sand into the core box through the investment area. The carrier air is exhausted through the vents in the core box. The blow air is then turned off, the excess pressure in the blow chamber is exhausted, and the core boxes are disengaged from the blow ports. All these steps are generally electrically or mechanically controlled and require only seconds to complete.

Release Agents

All core binders are sticky, but some are less so than others. It is this characteristic that requires the use of release agents/parting compounds on core boxes, regardless of the type of core box material. The desirability of having a dense core compounds the problem of core release, because binder-coated sand is packed or blown against the surface with force, which enhances sticking. In the heat-activated processes, the binder melts or softens and migrates to the heated surface.

It is necessary to start with a good core box; no amount of release agent will overcome roughness and scoring. The core box should be chromium plated if it is to be used for high production, and it should be broken in gradually. If a release agent is functioning properly, some buildup will occur; it should be removed frequently because it adversely affects core dimensions and quality. One method of removing buildup is blasting with ground walnut hulls, corn cobs, or other soft media. Solvents can also be used. Minimum amounts of release agents should be used, and when adding them during mixing, they should be added near the end of the cycle. The release agent should be chemically compatible with the binder system.

Oleic acid diluted with kerosene (usually 10:1) is used as a release agent for core-oil cores. It is added to the sand mixture and used to keep the core boxes clean. The amounts used are 125 to 250 mL/450 kg (1 pt to 1 qt/1000 lb) of sand.

Shell and other heat-cured processes use emulsified silicone release agents diluted in water. A low surface tension silicone should be used for spraying core boxes. Stearates, usually calcium, are added to shell sand during the coating process. Amounts recommended are 2 to 5%, based on resin solid weight. Small amounts of core oil or silicone emulsion can be added to hot box sand during mixing.

The no-bake and cold box processes use a wide variety of release agents. Small-to-medium size high-production core systems will normally use liquid types, which can be sprayed through jet nozzles and integrated into an automatic cycle or sprayed with a hand gun. No-bake systems normally use liquids, suspensions in liquids, or dry materials for larger cores. The suspension systems usually use a solvent carrier that evaporates readily. Materials can be sprayed, brushed, dusted, or hand rubbed onto the core boxes. Ingredients include (in many combinations) water, alcohol, chlorinated solvents, mica, talc, silicones, aluminum powder, graphite, and soybean oil (lecithin).

Curing

Curing of the compacted mass of sand and binders varies according to the process chosen. In the cold box processes, the appropriate gas for the process (carbon dioxide, sulfur dioxide, tertiary amines, or methyl formate) is forced through the compacted sand (Fig. 2). Chemical reactions then occur that create rigid bonds between sand grains. Excess curing gas is purged from the cured core with compressed air, and the cores are removed from the core box as hardened cores.

In the heat-activated process, prepared sand is blown into a heated core box. The heat releases curing agents from latent hardeners in the sand mix, and these curing agents then cure the binder.

Time alone is required for the binders in the no-bake cores to react, and after sufficient time, generally from 30 s to 1 h, the cores are rigid enough to be removed from the boxes. Additional curing may continue for several hours.

Cores from the core-oil process are placed in ovens after stripping from the core box and are baked for 1 h or longer, depending on section thickness. They are removed from the oven when thoroughly cured and allowed to cool to ambient temperature.

Coating

Coatings are frequently applied to cores to enhance the casting surface finish and to reduce casting defects at the mold/metal interface. Coatings accomplish this by having a higher refractory value than the sand and/or by forming an impermeable barrier between the metal and the core. Most coatings are formulated with five major components:

- Refractory materials
- Carrier system
- Suspension system
- Binder system
- Chemical modifiers

Refractory materials for coatings can be made of any of the following materials:

Zirconium oxide	Carbon (many forms)
Zirconium silicate	Silicon dioxide
Magnesium oxide	Magnesium/calcium oxide
Olivine	Mullite (aluminum silicate)
Chromate	Mica
Pyrophyllite	Iron oxide
Talc	Magnesite

These materials are blended in many combinations to make up a broad selection of proprietary coatings. Silica, alumina, and carbon are the most commonly used, and water, alcohol (isopropyl, methanol), naphtha (petroleum distillate), chlorinated solvents (inhibited 1,1,1-trichloroethane, or chlorothene NU, and methylene chloride), and aliphatic hydrocarbon are used as liquid carriers.

The carrier allows the refractory to be applied evenly to the sand mold or core. After the coating has been applied, the carrier must be removed. When the carrier is water, it can be torched or oven dried. Alcohol carriers are commonly ignited and allowed to burn off. The volatile chlorinated hydrocarbons readily air dry and require neither torching nor oven drying.

Suspension System. The function of the suspension system is to maintain the refractory material uniformly dispersed in solution. With water carriers, sodium bentonite is commonly used, while organic bentonite or bentones are generally used with alcohol and chlorinated hydrocarbon carriers.

The binder system in a coating is usually an organic resin, and it behaves similarly to the resins used to bond sands. The amount of binder used depends on the density and fineness of the refractory. It is used sparingly to avoid casting surface defects.

Chemical modifiers to the coating include surfactants to enhance wettability, antifoaming agents, and bactericides.

Coating Benefits. Most of the reasons for using coatings center around reducing casting costs by improving sand peel and reducing mold/core reaction, reducing or eliminating metal penetration (burn-in and/or burn-on), and reducing or eliminating veining. These reasons are directly associated with casting cleaning costs. Coatings are also used to reduce machining time and tool wear by having a clean smooth surface, to improve casting appearance, to facilitate handling (especially where low-strength cores may be required because of shakeout conditions), or to promote chill or increase hardness in metal (for which selenium or tellurium paste is used). Coatings should not be used indiscriminately. No coating will compensate for a poor-quality core.

Blending and Applying Coatings. All the advantages of selecting the proper coating, or wash, can be lost if it is not properly blended. Coating ingredients are available in powder, paste, slurry forms, and ready-to-use forms.

In preparing a core wash, the ingredients are mixed to a slurry with an electric or air-driven mixer. The mixing tank should be twice as high as its diameter, should be equipped with flow-through baffles spaced at approximately 120°, and should have an impeller shaft with two blades. The slurry must be mixed until creamy and lump free, with ingredients in suspension/solution. The Baumé must be kept 25 to 30° higher than application requires and further diluted prior to use. The carrier should be added in steps to avoid overdilution, which will adversely affect suspension. Coatings should be mixed a specified time before use (usually 12 to 24 h) to permit tempering of the suspension and bonding agents, and overmixing should be avoided.

The methods of application will depend on a number of factors, such as the size and shape of the core, plant layout (if a continuous process is used), and production requirements. For example, small cores would not lend themselves to spraying but to dipping in groups using fixtures and/or baskets. Conversely, large cores in a production line could move continuously on a conveyor through a dip tank to a drain station and then to a drying station. Regardless of the method of application, obtaining a smooth, even coat is of paramount importance.

Water- or solvent-base washes can be used for furan, furan/phenolic, urethane cold box and no-bakes, phenolic esters, and baked core-oil cores. However, water-base washes should not be used on silicate or phosphate-bonded cores, because such washes will degrade the binder.

The best practice for drying coatings is to apply coatings on cured cores and then dry them by appropriate means, such as ovens, burn-off, or warm air circulation. If cores are cold, they should be heated before or immediately after application. If cores are coated green (incomplete cure), heat should be applied immediately. Evaporation of the solvent is important for avoiding deterioration of the core binder.

Core Venting

Venting is necessary, and the general rule is to provide the maximum allowable venting. To understand the fact that gases must be dispersed, one need only consider the volume change that occurs in the transition of a liquid or solid to a gas.

The principal gases resulting from the breakdown of the chemical binder systems have been found to be hydrogen, carbon monoxide, carbon dioxide, methane, and nitrogen (Ref 1). In venting, however, the volume of gas is more important than its composition.

Care must be taken to lead all vents to core prints and to provide exits from the mold with minimum pressure buildup. In addition, vents must be kept far enough away from the surface to prevent metal breakthrough. As coarse a sand mix as good practice permits should be used. The permeability of sand is related to its coarseness, assuming a normal distribution. The coarser material will have the highest permeability and the most natural venting.

Cut, punched, filed, or scratched vent passages can be made by removing sand during coremaking, after curing, and before assembly. Mandrels can be used to hollow out cores, thus saving sand, or small holes can be drilled into cured cores.

Strands of wax and/or plastic tubing can also be used. These materials are available in a variety of sizes. In the heat-activated processes, some of these materials melt and are absorbed into the sand, leaving a passage. Permeable plastic tubes can be used in cold box processes. In very large cores, the center can be filled with dry unbonded sand, coke, slag, or gravel, thus avoiding gas generation. A secondary advantage of using these materials is that they replace the same volume of mixed sand and thus reduce cost.

REFERENCE

1. C.E. Bates and W.D. Scott, AFS Research Progress Report: Decomposition of Resin Binders and the Relationship Between Gases Formed and the Casting Surface Quality, *Trans. AFS*, Vol 84, 1976, p 793-804

Plaster Molding

Revised by Charles D. Nelson, Morris Bean and Company

PLASTER MOLD CASTING is a specialized casting process used to produce nonferrous castings that have greater dimensional accuracy, smoother surfaces, and more finely reproduced detail than can be obtained in sand molds or coated permanent molds. The four generally recognized plaster mold processes are:

- Conventional plaster mold casting
- Match plate pattern plaster mold casting
- The Antioch process
- The foamed plaster process

Applications

Near-net shape castings such as compressor wheels, impellers, components of electronics gear, tire molds, aerospace fuel pump systems, and heat exchanger systems are examples of castings produced by plaster mold casting. Machining costs would be exceedingly high if these parts were cast in most other types of molds.

For many parts, only a small portion of the total area requires the capabilities of plaster molding. Composite molds are more suitable for such parts.

Although about 90% of the plaster mold castings weigh less than 9 kg (20 lb) each, castings weighing as much as 34 kg (75 lb) are being produced in substantial quantities. To illustrate the capabilities of the technique, an aluminum alloy casting weighing 1815 kg (4000 lb) has been successfully cast in a plaster mold.

Advantages. Most of the advantages to be gained from the use of plaster molds have to do with casting quality:

- As-cast surface finishes of 50 to 125 RMS are readily obtained. By using flexible rubber patterns and incorporating the advantages of lost wax techniques, slurry molding successfully produces intricate designs
- The dimensional accuracy of the castings is good (about equal to the accuracy of castings made in investment molds)
- Thin sections can be cast because the castings normally cool quite slowly due to the insulating qualities of the plaster; wall thickness limits are a function of casting design, but selected areas 0.64 to 1.02 mm (0.025 to 0.040 in.) thick are typical
- Slow cooling minimizes warpage and promotes uniformity of structure and mechanical properties in the castings
- Plaster molds are especially well suited to the use of chills, which permit close control of thermal gradients in the molds; this optimizes mechanical properties desired in isolated areas of a complex casting

Disadvantages in using plaster molds are largely related to molding and casting procedures:

- Cost is high, partly because of the lengthy processing procedures and partly because the mold materials are not reusable
- Mold processing requires more equipment than is required for most other molding processes
- Because of the lengthy processing procedures needed to make plaster molds, duplicate (or multiple) pattern equipment, as well as duplicates of other processing equipment, may be required in order to meet production schedules
- With the exception of the foamed plaster process, the permeability of plaster molds is inherently low when compared to the sand casting process; therefore, it is usually necessary to apply pressure or vacuum during pouring or to provide greater permeability by a special procedure

Plaster Mold Compositions

In all of the processes for making plaster molds and cores, the principal mold ingredient is calcium sulfate. Other materials used to enhance such mold properties as green and dry strength, permeability, and castability include cement, ceramic talc, fiberglass, sand, clay, Wollastanite, pearlite, and fly ash.

Cores are typically made of the same material and by the same process as molds, but cores are sometimes made of other materials, such as shell molding sand. Cores for the match plate plaster mold process are made of sand. Specific mold and core formulations, as well as techniques for making the slurries, are discussed under each process in this article.

Characteristics of Calcium Sulfate. Calcium sulfate exists as two different hydrates and in an anhydrous form. In its dihydrate form ($CaSO_4 \cdot 2H_2O$), it is known as gypsum. (Commercially, all three forms of calcium sulfate are sometimes referred to as gypsum.) When heated above an equilibrium temperature of 128 °C (262 °F), gypsum loses three-fourths of its water to form the hemihydrate ($CaSO_4 \cdot \frac{1}{2}H_2O$), which is plaster of paris. Similarly, when the plaster of paris is heated above an equilibrium temperature of 163 °C (325 °F), all of the water of crystallization is removed, and anhydrous calcium sulfate ($CaSO_4$) is formed. The above temperatures are based on laboratory studies of pure calcium sulfate under equilibrium conditions. Temperatures at which the hydrates are formed depend greatly on vapor pressure. For practical purposes, temperatures well above equilibrium temperatures are generally used.

If it is exposed to moisture at a temperature lower than 100 °C (212 °F), the anhydride formed near 163 °C (325 °F) will absorb water and will revert to the hemihydrate; the anhydride produced by heating to about 540 °C (1000 °F) will absorb water less rapidly. Similarly, when plaster of paris (the hemihydrate) is mixed with water, gypsum (the dihydrate) is reestablished, yielding a coherent solid in a few minutes. This reaction is termed setting. Prolonged exposure of the hemihydrate to moist air at temperatures below 100 °C (212 °F) results in a slow conversion to the dihydrate.

In the production of a plaster mold, plaster of paris is mixed with water to form a slurry. This slurry is immediately poured over a pattern or into a core box, where it sets, forming a solid mold or core composed of gypsum with free water distributed throughout the plaster mold.

The next stage is to dry the plaster in an oven to remove the excess water. Various drying temperatures are used, depending on the plaster molding process. If partings and waxes are used, the temperatures must be high enough to remove any waxes.

Plaster molds have low heat capacity. Therefore, the cooling rates for castings made in such molds are low. Figure 1 shows that the freezing time for a test casting made in a conventional plaster mold is more than four times as long as that for one made in a hard-rammed green sand mold.

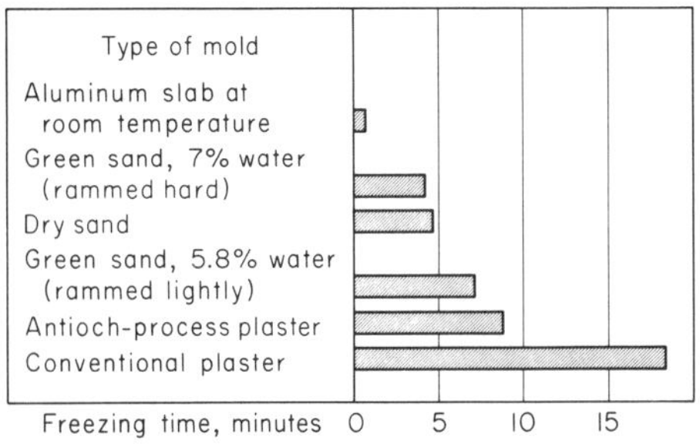

Fig. 1 Comparison of freezing times for identical test castings poured in various types of molds

Slow cooling has advantages and disadvantages. It permits better feeding, particularly in thin sections, as well as the replication of intricate detail that would be difficult to obtain with fast cooling. However, it slows production by lengthening the time between pouring and shake out, and for most alloys it results in castings of lower strength. This latter disadvantage can usually be eliminated by the use of chills in specific areas.

Patterns and Core Boxes

Metal and Rubber Patterns. Patterns and core boxes for plaster mold casting are commonly made of metal (aluminum alloy, brass, or zinc alloy). Flexible rubber patterns are widely used for producing intricate molds, such as the molds used for casting wheels having angular vanes. Flexible rubber patterns having as much as 30° negative draft can be withdrawn without damaging molds that have high green strength.

Cast epoxy resin patterns are useful for some purposes. For example, when several patterns are required, they can be made quickly by molding them from epoxy resin in metal master dies. Cast epoxy patterns also have greater dimensional stability and longer production life.

Wood Patterns. Because patterns for production casting by plaster mold processes must withstand repeated exposure to liquid slurries, wood patterns usually are not adequate. Wood patterns, although not extensively used in the production of castings, are specified as patterns for match plate pattern molds. Wood patterns must be sealed, or they will swell and distort.

Mold Drying Equipment

Plaster molds made by any process must be dried. The most common drying temperatures range from 120 to 260 °C (250 to 500 °F), but temperatures to 870 °C (1600 °F) have been used under carefully controlled conditions. Once the optimum temperature for a specific application has been determined, it must be closely maintained—usually within ±6 °C (±10 °F).

After this initial requirement has been met, the choice of equipment depends largely on the temperature to be used and the number of molds to be processed. For drying at 120 to 260 °C (250 to 500 °F), either batch-type or conveyor core ovens are satisfactory.

Higher-temperature drying requires furnaces similar to those used for the heat treating of steel. Either batch-type or conveyor furnaces can be used, but conveyor furnaces are generally preferred, especially for drying large quantities of molds. The molds must be cooled to at least 205 °C (400 °F) before they are removed from the furnace.

Flasks

Flasks are usually made of low-carbon steel and aluminum. They vary in size in accordance with the size of the pattern, the number of identical molds to be produced, and the number of patterns in a flask.

When the flask is to hold only one pattern half, especially when only a few molds are required, a simple, bottomless flask is placed on a mold board, and the pattern half is positioned within it on the board, ready to receive the slurry. When many identical molds are required, especially when two or more patterns are placed in a single flask, a flask with a flat bottom that serves as a mold board is used; the pattern halves are arranged on the flask bottom. In the Antioch process, for small single-pattern molds, pattern and flask are often an integral unit (see the section "The Antioch Process" in this article). The flask for a given application should be large enough to allow space for the same thickness of plaster surrounding the pattern as would be needed if the mold material were dry sand.

Standardization of flask sizes is important in production operations because this simplifies tooling, especially when vacuum pouring is used. Standardization of flask size is also desirable for pouring the slurry and drying the molds. In some foundries, it is common practice to select as standard a flask size that is suitable for the largest casting that will normally be poured. When smaller castings are to be made, a family mold practice is often used. In this practice, two or more patterns (sometimes as many as 16) are positioned in the flask.

Metals Cast in Plaster Molds

Only nonferrous metals are cast in plaster molds. Ferrous alloys are not suitable, because most of them are poured at temperatures that would melt the calcium sulfate the molds are made of. Calcium sulfate undergoes a phase transformation at 1195 °C (2180 °F) and melts at 1450 °C (2640 °F).

Aluminum. All of the aluminum alloys that can be successfully cast in sand molds are suitable for casting in plaster molds. The more readily castable alloys—43, A344, 355, 356, and 357—are preferred.

Copper. Most of the coppers and copper alloys that can be successfully cast in sand molds can be cast in plaster molds; again, the more castable alloys are preferred (see the article "Copper and Copper Alloys" in this Volume). Copper alloys containing more than approximately 5.0% Pb are generally not recommended for casting in plaster molds, because the higher-lead alloys react with some mold compositions, resulting in poor surfaces on the castings and defeating one of the objectives of plaster mold casting.

Magnesium alloys are not recommended for plaster mold casting. A reaction is likely between the magnesium alloys and the mold material. In particular, magnesium alloys will react with any free water that remains in the mold and will cause an explosion.

Zinc alloys are frequently cast in plaster molds, most often for prototype castings. The die casting alloys AG40A and AC41A are often used, but a proprietary alloy whose coefficient of thermal expansion is very close to that of aluminum alloys is frequently cast. Master patterns appropriate for this zinc alloy or for aluminum can be made according to a single shrinkage rule.

Nominal compositions of the three zinc alloys mentioned are:

Alloy	Composition, %			
	Al	Cu	Mg	Zn
AG40A	4.10	0.10 max	0.035	rem
AC41A	4.10	1.00	0.055	rem
Proprietary	12.00	0.87	0.02	rem

Except for allowances for shrinkage, the technique for making the molds is the same as that for aluminum and zinc alloys.

Conventional Plaster Mold Casting

Melting practice equipment and methods for metal to be poured into plaster molds are the same as those used for preparing the metal for pouring into other types of molds. A brief outline of conventional plaster molding is given in the following sequence of operations for producing conventional plaster molds.

- Mix dry ingredients
- Add dry ingredients to water
- Soak (2 to 4 min)
- Mix (2 to 5 min)
- Coat patterns (or core boxes)
- Pour slurry
- Set at room temperature (15 min)
- Remove pattern
- Dry molds (or cores)
- Assemble cores and mold halves

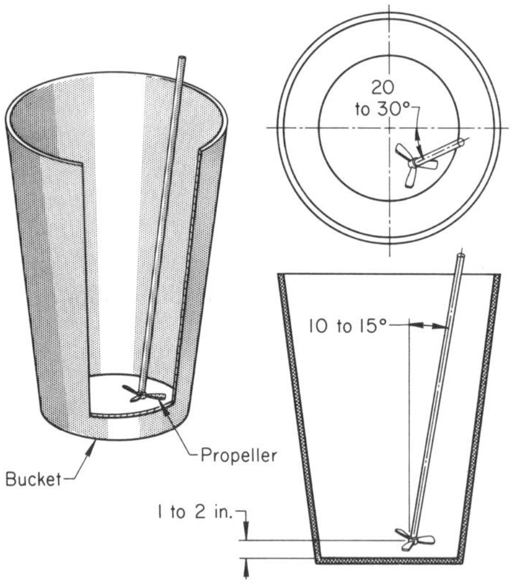

Fig. 2 Preferred shape of bucket and position of propeller used in the batch mixing of plaster slurries for conventional plaster molds

Composition and Preparation of Dry Ingredients. Dry ingredients ready to mix with water are available as proprietary compositions. For small operations, it is usually cost effective to purchase these ready-to-mix dry ingredients. Compositions vary considerably, but the gypsum content (typically, a blend of gray and white gypsum) is commonly 50 to 60%, by weight, of the dry mixture. The rest of the mixture consists of two or more filler materials, such as wollastanite, ceramic talc, fly ash, pearlite, and sand. Portland cement, hydrocals, fiberglass, and terra alba are also sometimes used to enhance the strength of the mold and to improve setting behavior.

Mixing the Slurry. Equipment requirements for mixing the slurry are very precise. The principal components of a batch-type mixer are a bucket and a propeller (Fig. 2). The bucket height should be equal to or slightly greater than the top diameter, and the bottom diameter should be approximately two-thirds of the top diameter.

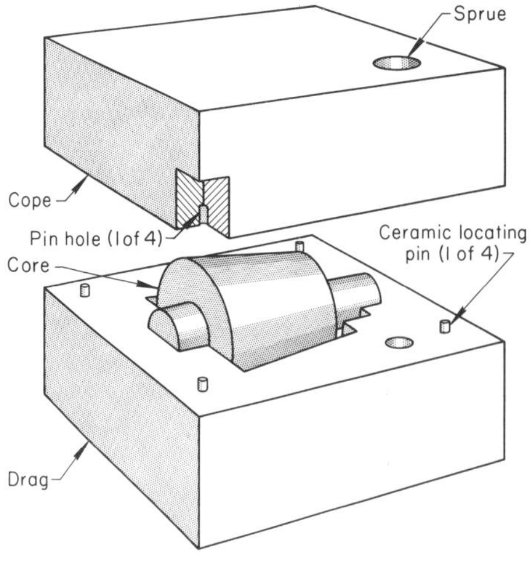

Fig. 3 Plaster mold and core showing location pins for matching the cope and drag sections

Preparing the Pattern. Because the slurry begins to set as soon as it is mixed, it must be poured over the pattern (or into the core box) almost immediately, and the pattern must be ready to receive the slurry.

Pouring the Slurry. When the slurry is poured into the flask, the lip of the bucket should be kept close to the pattern. The slurry should be poured at a constant rate and made to flow over the pattern rather than splash on it.

Set time varies somewhat, depending on the slurry composition, but 30 min is usually the maximum set time. After the slurry has set, the pattern and flask are removed, and the drying cycle is started as soon as is practical.

Drying the Molds. All conventional plaster molds must be dried enough to expel the free water and most of the water of crystallization to a depth of at least 13 mm (½ in.) below the surface. Some foundries prefer to have the molds completely calcined.

Oven drying should begin as soon as is practical after the mold has been removed from the flask. Molds that have partially dried by standing at room temperature are more susceptible to cracking than those that are oven dried immediately after setting. If the molds must stand at room temperature for some time (such as over a weekend or even overnight), they should be covered with a damp cloth or stored in a humid atmosphere.

During drying, the mold should be uniformly supported on its edge or face. Common practice is to place the mold on a perforated flat metal plate, a rigid metal grid, or some other type of level support. If smooth plates are used for support, they should be covered with a thin layer of talc or similar material to allow movement of the mold during drying without the danger of its cracking or warping.

Time and temperature cycles used for drying conventional plaster molds vary widely among foundries. Temperature may vary from about 175 to 870 °C (350 to 1600 °F), and time from 45 min to 72 h. The fact that furnaces are operated at high temperatures (760 to 870 °C, or 1400 to 1600 °F) does not mean that all areas of the mold must reach this temperature range, but the interior of the mold must be at a temperature of at least 105 °C (220 °F), to ensure removal of the free water. (Mold temperatures are measured by thermocouples in the center of mold sections.) The specific time and furnace temperature required for drying a specific mold usually must be determined by experimentation. Once established, the time-temperature cycle must be rigidly controlled for best reproducibility.

Mold Assembly. After the mold has been removed from the oven, cores are placed in the drag half. Depending on mold complexity, this can be done while the molds are still hot or after they have cooled to room temperature. Ceramic or plaster pins are placed in the holes provided in the drag half. The cope half is then lowered so that the pins protruding from the drag enter the matching pin holes in the cope, as shown in Fig. 3.

Preheating. Following assembly, the mold is ready for preheating. Some foundries preheat all conventional plaster molds to a preestablished temperature (commonly 120 °C, or 250 °F) before pouring the metal. Other foundries preheat the molds only in specific applications for which preheating has proved advantageous. Preheating of the mold can help to minimize defects or to obtain better replication of fine detail in the casting.

Pouring Practice. A dried plaster mold made by the conventional process has extremely low permeability—about 1 to 2 AFS, compared with 80 and upward for sand molds. Because of this low permeability, either vacuum assist or pressure is usually required for the pouring of molds. Simple gravity pouring is rarely done.

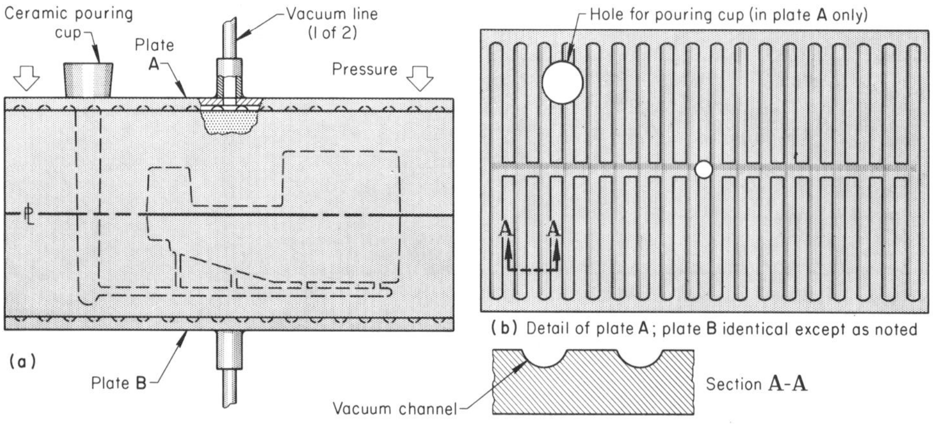

Fig. 4 Typical setup for vacuum-assist pouring of a conventional plaster mold casting. (a) Side view of conventional plaster mold positioned between upper and lower plates for pouring with vacuum assist. (b) Details of a top plate showing vacuum channels

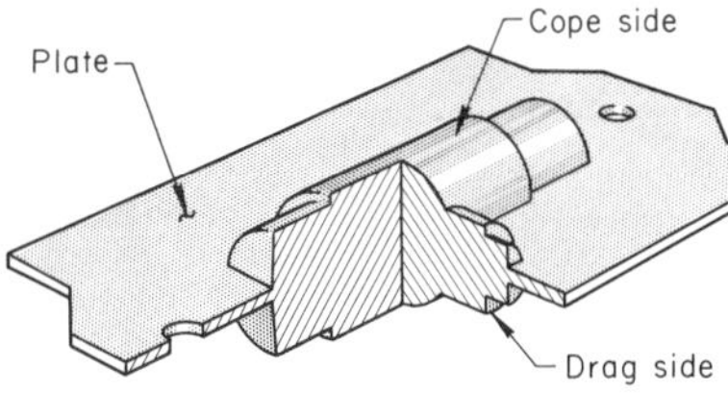

Fig. 5 A typical metal match plate pattern

Most foundries use vacuum assist for pouring conventional plaster molds. A typical setup for vacuum-assist pouring is shown schematically in Fig. 4(a). In this setup, the mold is supported by a fixed bottom plate (plate B, Fig. 4a).

Casting of Match Plate Patterns

Match plate patterns are cast by a particular adaptation of the conventional method of plaster molding. Changes in details of the conventional method ensure high accuracy and smooth surface finish, which are required in metal match plate patterns.

Common dimensional requirements for match plate patterns are:

- Match between the cope and drag, within 0.25 mm (0.010 in.)
- Parallelism between the two halves, within 0.51 mm (0.020 in.)

A surface finish of 2.5 μm (100 μin.) or better is obtained.

Sizes and Weights of Match Plate Patterns. A match plate pattern consists of a cope section and a drag section separated by a plate. These three components are cast as an integral unit. A typical match plate pattern is shown in Fig. 5.

Metal match plate patterns vary considerably in size and weight. Patterns weighing as much as 180 kg (400 lb) each have been cast, although these are unusually large. Match plate patterns seldom weigh more than 45 kg (100 lb) each, and most weigh from 8.2 to 16 kg (18 to 35 lb).

A single plate may hold one pattern or many patterns. The thickness of the plate for small patterns having one pattern per plate is usually 9.5 mm (3/8 in.). The plate thickness increases as the size and weight of the pattern increase; a thickness of 19 mm (3/4 in.) would be used for a plate measuring 1.1 × 1.1 m (44 × 44 in.). Plate thickness is also increased to provide adequate stiffness and bend resistance in designs involving a stepped parting line separating cope and drag mold sections.

Mold Composition. Materials for molds for match plate patterns are available as proprietary dry mixtures ready for mixing with water. However, some foundries prefer to make their own mixtures.

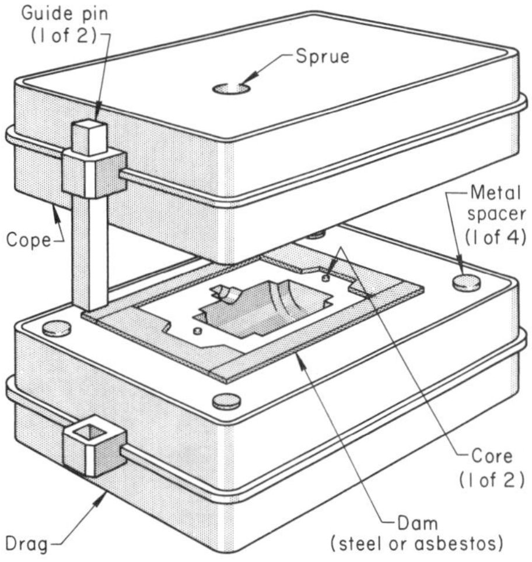

Fig. 6 Assembly details of a match plate mold

Master patterns for match plate patterns are usually made of wood (for economy), but they can be made of metal. Wood patterns must be lacquered or otherwise coated to prevent the absorption of water from the slurry.

Separate master patterns are made for the cope and the drag. The plate portion is developed by the technique described in the section "Mold Assembly" in this article, rather than by means of a pattern. Before the slurry is poured, the patterns are coated with a release agent.

Flasks for the casting of match plate pattern molds are bottomless boxes. They differ from flasks for conventional plaster mold casting in two respects. First, standardization of flask size is seldom feasible in match plate pattern mold work because pattern sizes vary widely. However, standardization is less important because molds are not required in production quantities. Second, because the mold remains in the flask through the drying and metal-pouring operations, provision for matching the cope and drag sections must be incorporated into the flask rather than the pattern as in conventional operations. The use of guide pins for the matching of mold sections is illustrated in Fig. 6.

Mixing and Pouring the Slurry. Equipment and procedures for mixing the slurry and pouring it over the patterns are essentially the same as those for the conventional process (see Fig. 2 and accompanying text). Common practice is to make several vents in the mold with a nail or wire immediately after the slurry has been poured. The vents serve two purposes. Vent holes traversing the entire mold sections of dry molds provide openings for the escape of steam and other gases when the metal is poured. Through-vents in wet molds facilitate the separation of mold halves and the removal of patterns by acting as channels for the injection of compressed air.

Set Time. A slurry will set in 20 to 35 min. After the slurry sets, the patterns are removed, and the molds are dried.

Drying. Molds for match plate patterns, like molds for conventional plaster castings, should be dried as soon as possible after the plaster has set. High-temperature drying cannot be used for match plate molds, because it results in unacceptable distortion and size change. Match plate molds are usually dried at 120 to 205 °C (250 to 400 °F) for 12 to 72 h. Size and section thickness of the molds determine the length of time in the drying ovens. The center of the thickest section of the mold should reach at least 105 °C (220 °F) before drying is stopped. The permeability of a match plate mold at this stage is approximately the same as that of a conventional mold (1 to 2 AFS).

Mold Assembly. After the mold halves have reached room temperature, the cope and drag sections are matched as shown in Fig. 6. The assembly of a match plate mold is significantly different from that of a conventional plaster mold. Cores, if used, are positioned in the drag mold section. Cores are used if holes are required in the plate. Internal cores are used only in large match plate molds, so that the patterns will be lighter and can be handled more easily. Cores are made of sand.

In heavy sections, it is common practice to insert aluminum chills with nail-like protrusions into the plaster to make an internally cast chill. It is common to have 50 to 100 chills in a large mold with varying sections. The cope section, aligned by guide pins, is lowered onto the drag (Fig. 6). It is then raised for a distance equal to the desired thickness of the plate. Metal spacers of this thickness are inserted at each corner, and dams of steel or Fiberfrax are placed so as to form the desired outer contour of the plate (Fig. 6). When the metal is poured, it flows outward in the space between the cope and drag sections and forms the plate portion of the match plate pattern.

Metals Used. Match plate patterns are cast from aluminum alloys, most frequently Alloys 355 and 356. One foundry using a blend of equal quantities of 356 and 319 reported greater ductility and better machinability than could be obtained if either alloy were used alone. Ductility is important in making match plate patterns because the patterns often require straightening.

Pouring Practice. Because of the low permeability of a match plate mold, some assistance is required to fill the mold quickly and completely. Pressure assist is used rather than vacuum, partially because pressure equipment is more adaptable to a wider variety of flask sizes than vacuum equipment is. However, some foundries use a combination of pressure and vacuum assist.

The equipment for pressure casting is illustrated in Fig. 7. The first step in its use

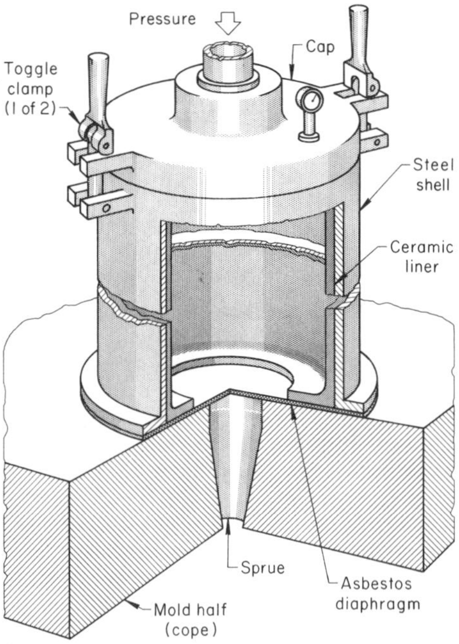

Fig. 7 Cylinder used for pouring a casting with pressure assist

is to place a diaphragm of sheet Fiberfrax (about 3.2 mm, or ⅛ in., thick) over the sprue in the mold. A ceramic-lined cylinder is then placed on the diaphragm, as shown in Fig. 7. A predetermined amount of metal is poured into the cylinder, and a cap is placed on the cylinder and clamped tight. The cap is attached to a source of compressed air. After the cap has been secured in position, the air valve is opened. Air pressure against the molten metal ruptures the Fiberfrax diaphragm and forces the metal into the mold cavity. A pressure of 10 to 17 kPa (1½ to 2½ psi) is kept on the sprue for about 30 min.

The Antioch Process

The Antioch process (U.S. Patent 2,220,703) was developed to overcome the principal limitations of conventional plaster molds and cores without sacrificing the advantages of the plaster mold process. If undried molds are partially dehydrated and then allowed to rehydrate without being disturbed, gypsum crystals slowly recrystallize into granules about the size of sand grains, and the mold acquires a porous structure of relatively high permeability. Permeability is held within a range of 15 to 30 AFS; the permeability of conventional plaster molds is 1 to 2 AFS. Recrystallization does not take place at the surface of the mold, because not enough water is present. Therefore, the surface remains smooth.

In addition to the greater permeability developed by the dehydration-rehydration process, the molds produced have greater heat capacity than conventional plaster molds because they are composed of approximately 50% sand. Figure 1 shows that the freezing time for a casting in an Antioch process mold is only about 20% longer than the freezing time for an identical casting in a lightly rammed green sand mold and is less than half the time required for a casting in a conventional plaster mold.

Unlike conventional molds, Antioch process molds do not shrink. In fact, they expand slightly—from 0.001 to 0.0025 mm/mm (0.001 to 0.0025 in./in.)—during processing.

Because of their porous structure, the molds have low dry strength. This characteristic, in promoting early collapse of cores as the casting cools, minimizes hot tears in the castings. For very large molds, the low dry strength sometimes necessitates the use of internal reinforcement, which is achieved with hardware cloth or core rods such as those used in making sand cores. When possible, reinforcement is avoided because of the difference in expansion between the reinforcing metal and the molding material.

After setting, but before the dehydration-rehydration treatment, Antioch process molds have relatively high green strength. When flexible rubber patterns are used, this high green strength permits the withdrawal of patterns having a severe back draft without damage to the mold. This makes the Antioch process particularly well suited to the production of molds for parts having angular, bladelike sections—rotor sand nozzles, for example.

In addition to the cost, which is high for all plaster molds, the major disadvantage of Antioch process molds is the long processing time required to make the mold. This occupies expensive equipment for long periods.

Mold and Core Materials. The dry mixture for Antioch process molds and cores consists of silica sand, white molding plaster, talc No. 2, and a small amount of material (such as portland cement) for expansion control. The typical mixture given in Table 1 varies somewhat among different foundries. However, once a formulation has been established in a specific foundry, it is retained for all castings, regardless of their size and shape. Optimum results are obtained by weighing all ingredients accurately. Only by consistent use of a specific formulation is it possible to obtain maximum reproducibility.

Processing. The sequence of operations for producing Antioch process molds is given below:

- Mix dry ingredients
- Add dry ingredients to water
- Soak (1 to 3 min)
- Mix (2 to 4 min)
- Coat patterns (or core boxes)
- Pour slurry
- Set at room temperature (15 to 20 min)
- Remove pattern
- Dehydrate in autoclave (6 to 12 h)
- Rehydrate in air (14 h)
- Dry molds (or cores)
- Assemble cores and mold halves

Table 1 Typical composition of dry material for Antioch process molds

Ingredient	Weight, kg	Weight, lb
Washed silica sand (AFS 50 is typical)	20.4	45
White molding plaster	18.6	41
Talc No. 2	3.6	8.0
Portland cement	0.2	0.4
Sodium silicate	0.4	0.8
Western bentonite	1.1	2.5
Terra alba	1.5	3.4

Slurry is made by mixing 45 kg (100 lb) of the above dry mixture with 24 kg (54 lb) of water.

The set time for a slurry formulated from a composition such as that shown in Table 1 will be approximately 15 to 20 min. Set time can be decreased by adding up to 3% terra alba and heating the water. For example, the minimum set time of 6 to 7 min is achieved by adding 3% terra alba and using water at 32 °C (90 °F). The temperature and humidity of the surrounding atmosphere have very little influence on set time, although an atmospheric temperature of 21 to 27 °C (70 to 80 °F) is preferred.

Dehydration. The time between setting of the slurry and the beginning of the dehydration cycle is not extremely critical if steps are taken to prevent the mold from drying out. If the set molds are covered with damp cloths, they can be held overnight, or sometimes even over a weekend, without a significant impact on subsequent dehydration. If the molds are placed in humidity cabinets, they can be stored for longer periods before dehydrating. However, the dehydration cycle should begin soon after the pattern is removed if the mold cannot be kept moist.

For dehydration, the molds are placed on suitable racks in a standard autoclave. The autoclave is sealed, and steam is admitted. The autoclave is operated with a steam pressure of 105 kPa (15 psi) for 6 to 8 h. For extremely large molds, it is operated for 12 h. The autoclave is then opened, and the molds are removed.

Rehydration. The mold is permitted to remain at room temperature for 14 h. After rehydration, the mold is ready for drying.

Drying temperature ranges from 175 to 230 °C (350 to 450 °F), and drying time from 1 to 70 h. Drying time depends mainly on the size of the mold and the temperature used. The center of the mold must reach a temperature of at least 120 °C (250 °F). This can be accomplished considerably more quickly at an oven temperature of 230 °C (450 °F) than at 175 °C (350 °F).

Regardless of the cycle used, it is important that the same cycle be used for all molds of the same size. Maximum reproducibility (of dimensions, in particular) can be achieved only through close control of the cycle.

Mold assembly is essentially the same as that described for conventional plaster molds. After the molds have cooled to room temperature, cores (if used) are placed in the drag, and the cope is placed over the drag and core assembly. Matching is done by means of locating pins. The pins used for matching Antioch process molds are usually 13 to 19 mm (½ to ¾ in.) in diameter. Even when molds are permitted to remain in their flasks, guide pins on the sides of the flasks are seldom used for matching.

Metals Cast. All of the aluminum alloys that can be cast in other types of plaster molds can be cast in Antioch process molds. Most copper-base alloys can be cast in Antioch process molds. Yellow brass is the copper alloy most often cast. The Antioch process is seldom used for alloys that must be poured at temperatures above about 1040 °C (1900 °F).

Pouring Practice. It is generally possible to pour castings in Antioch process molds by gravity, using gating systems that are similar to those used for sand molding. Molds are usually at room temperature when pouring begins. Vacuum assist can be applied when difficulty is encountered in replicating fine detail or when thin sections have not filled properly. The technique is generally the same as that for pouring conventional molds with vacuum assist.

Foamed Plaster Molding Process

The foamed plaster process offers a means of achieving greater mold permeability than can be obtained in conventional plaster molds and at much improved dry strength over Antioch plaster molds. This gain is achieved by adding a foaming agent, such as alkyl aryl sulfonate, either to the dry ingredients before mixing or to the liquid slurry, as a separately generated foam mix. A special method of mixing foams the slurry with many fine air bubbles, thus decreasing the density and increasing the volume of the slurry. In general, with regard to the composition of the metal poured, casting size, and casting shape, the applicability of foamed plaster molds is the same as that of plaster molds made by other procedures.

Characteristics. Foamed plaster molds have smooth surfaces with air cells just below the surface. During setting and subsequent drying of the molds, these air cells become interconnected, thus permitting the escape of gases as the metal is poured. The permeability of a foamed plaster mold depends mainly on the volume increase from the addition of air when the slurry is mixed. For most molds, a volume increase of 50 to 100% is recommended. This increase usually results in a mold permeability of approximately 15 to 30 AFS for dried molds.

Equipment for mixing will vary somewhat with the slurry used. The type described below has proved suitable for mixing a proprietary dry mixture with water.

The mixer must be capable of beating air into the slurry and producing air cells no larger than about 0.25 mm (0.01 in.) in diameter. Large air cells are not permitted, because they break under pressure from molten metal, resulting in casting defects. Proper mixing can be accomplished with several types of mixers. Regardless of the type of mixer used, the greater the power input, the finer the mold structure (the smaller the air cells) and the smoother the surface of the mold.

The bucket should be similar to that shown in Fig. 2, but the mixing device is a round, two-ply, 3 mm thick (⅛ in. thick) rubber disk (which can be made from 3 mm, or ⅛ in., two-ply rubber belting) attached to a shaft, as shown in Fig. 8. The diameter of the shaft is not critical. Other mixer designs that are successfully used for mixing air into foamed plaster incorporate a 127 to 152 mm (5 to 6 in.) diam perforated disk mounted on a shaft that is tilted slightly from normal.

Processing. The following sequence of operations is typical of the routing required to produce a component by the foamed plaster molding process:

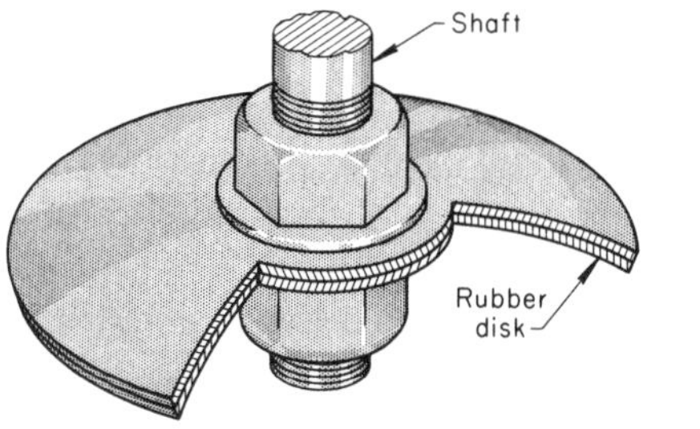

Fig. 8 Two-ply rubber disk attached to shaft for the high-speed mixing of foamed plaster slurries

- Mix foamed plaster dry mix with 32 to 43 °C (90 to 110 °F) soft water to a consistency of 80 to 100 AFS. The mixer used should have two speeds: a high speed (~1750 rpm) to mix the dry material and water thoroughly and a slower speed (~800 to 1000 rpm) to beat air into the slurry and to obtain a smooth, creamy slurry. Mixing time is varied to obtain 100% volume increase
- Pour slurry
- Set time varies from 20 to 40 min. Set time is a function of water temperature, mixing time (high speed), dry material setting characteristics, and consistency. Changes in variables to decrease set time also serve to decrease working time
- Strip patterns
- Dry in standard type dryers at temperatures from 175 to 260 °C (350 to 500 °F) for 8 to 16 h depending on mold size. Foam plaster is very insulating, and thermal gradients can set up dimensional change, thus causing mold cracking. Avoid conditions that will cause thermal shock at mold temperatures above 120 °C (250 °F)
- Assemble and cast; molds can be poured from room temperature up to the dryer temperature

Metals Cast. For the most part, all aluminum alloys can be cast, but the aluminum-magnesium alloys are the most compatible.

Pouring Practice. All conventional foundry pouring techniques are applicable to foamed plaster molds. The insulating qualities of foamed plaster make it very suitable as a riser insulation material.

Ceramic Molding

CERAMIC MOLDING techniques are based on proprietary processes (two of which are discussed in this article) that employ permanent patterns and fine-grain zircon and calcined, high-alumina mullite slurries for molding. Except for distribution of grain size, the zircon slurries are comparable in composition to those used in ceramic shell investment molding (see the article "Investment Casting" in this Volume). Like investment molds, ceramic molds are expendable. However, unlike the monolithic molds obtained in investment molding, ceramic molds consist of a cope and a drag or, if the casting shape permits, a drag only.

Both investment molding and ceramic molding can produce castings with fine detail, smooth surfaces, and a high degree of dimensional accuracy. The ceramic mold surface has refractory properties that enable it to withstand high metal pouring temperatures (such as those necessary for steel and heat-resistant alloys), give it excellent thermal stability, and do not permit burn-in.

Applications

Ceramic molding has two principal applications. The first is the production of precision castings that require patterns too large and unwieldy for molding with expendable wax or plastic patterns. The second principal application is the production of castings in limited quantities for which permanent wood patterns may be more economical and require less lead time to make than the metal pattern dies required for molding wax or plastic patterns.

Ceramic molding is intended for the production of castings of high quality, not only in terms of their dimensional accuracy and surface finish but also in terms of soundness and freedom from nonmetallic inclusions. In general, the capabilities of ceramic shell investment molding and ceramic molding are similar, and the selection of one process over the other is largely dependent on the size of the casting, the quantities required, and the molding costs involved. In some applications, depending on casting shape, permanent patterns used in ceramic molding may provide greater dimensional accuracy than wax patterns, primarily because wax expands during melt-out. Permanent patterns are also less susceptible to damage and distortion in handling than wax or plastic patterns.

The principal disadvantage of ceramic molding in its early stages of development was the high cost of the molding materials. These materials were used in large quantities for making the slurry for all-ceramic molds and could not be reclaimed. This disadvantage was greatly diminished by the development of less expensive, composite ceramic molds, in which chamotte (an aluminous fireclay) serves as a backup for a ceramic facing, thus greatly decreasing the amount of binder and zircon or mullite required. Furthermore, as much as 80 to 90% of the backup molding material can be reclaimed.

Suitable Work Metals. Both ferrous and nonferrous alloys are cast in ceramic molds, but ferrous applications are more numerous and account for the major tonnage of castings produced. Alloys of aluminum, copper (especially beryllium copper), nickel, and titanium are nonferrous alloys that are suitable for ceramic mold casting; suitable ferrous alloys include ductile iron, carbon and low-alloy steels, stainless steels, and tool steels.

Types of Parts Cast. The range of products cast in ceramic molds is closely related to the alloy used. For example, among the products cast from tool steels (especially H12, H13, and A2) are forging dies and punches; trimming dies; die inserts; dies for hot upsetting; extrusion bridges, spiders, and cups; and inserts for die casting dies. Products cast from stainless steel include food-machinery components; valves for the chemical, pharmaceutical, and petroleum industries; glass molds; aircraft structural components; and hardware for atomic reactors and aerospace vehicles. Typical copper alloy castings include food-machinery parts, marine hardware, and decorative trim items used in architectural applications.

Complex castings with thin sections ($\leq$1.6 mm, or 1/16 in.) can be cast in ceramic molds, using special techniques. Casting surfaces have excellent smoothness—usually 3.2 μm (125 μin.) or better.

Dimensional accuracy depends primarily on the accuracy of the pattern. Metal contraction can normally be predicted closely enough to provide castings within the following tolerances:

- ±0.08 mm (±0.003 in.) on dimensions up to 25 mm (1 in.)
- ±0.13 mm (±0.005 in.) on dimensions over 25 to 75 mm (1 to 3 in.)
- ±0.38 mm (±0.015 in.) on dimensions over 75 to 203 mm (3 to 8 in.)
- ±0.76 mm (±0.030 in.) on dimensions over 203 to 381 mm (8 to 15 in.)
- ±1.14 mm (±0.045 in.) on dimensions over 381 mm (15 in.)

For dimensions across the parting line, an additional tolerance of ±0.25 to ±0.51 (±0.010 to ±0.020 in.) must be provided.

Closer tolerances are readily obtainable by reworking the pattern after a test casting has been compared dimensionally with the pattern. Reproducibility is excellent; once the pattern has been established, subsequent castings will be extremely consistent throughout the life of the pattern.

Shaw Process

The Shaw process relates to two distinctly different types of ceramic molds: a one-piece all-ceramic mold and a composite ceramic mold consisting of an inexpensive fireclay backup material with a relatively thin facing of ceramic slurry. Selection of mold type depends almost exclusively on the size of the casting and the cost of the mold material. Many small castings can be produced economically in one-piece all-ceramic molds, because the amount of expensive ceramic slurry needed for the mold is moderate and the additional pattern and labor costs for composite molding cannot be justified.

Patterns. Two sets of patterns are commonly required for the production of composite molds: a set of oversize preform patterns for molding the coarse backup material and a second set of patterns, representative of the dimensional accuracy desired in the casting, for molding the ceramic facing. Thus, for cope and drag molding, a total of four patterns may be required.

The preform patterns are made 2.4 to 9.5 mm (3/32 to 3/8 in.) oversize to allow for the thickness of the ceramic facing. In some applications, the preform pattern can be eliminated by using the final pattern, backed with a sheet of plastic, cardboard, or felt of suitable thickness, to mold the preform impression in the coarse fireclay backup. Because solid ceramic molds are made without backup material, only the final patterns are required for molding.

The pattern materials used for ceramic molding are conventional: wood, epoxy, aluminum, brass, tool steel, and cast iron. Wood, epoxy, and aluminum are most widely used. Prior to use, wood patterns are coated with a sealant that will not dissolve in alcohol.

To ensure a high degree of accuracy in the alignment of cope and drag, by mechanically meshing certain surfaces, it is common practice in ceramic molding to incorporate parting blocks and locators in the cope and drag halves of patterns. These are molded into both the backup and the ceramic facing, together with the casting contours.

Preparing the Backup. The backup refractory commonly consists of a coarse-grain chamotte (an aluminous fireclay that has been calcined at a high temperature) and a sodium silicate binder. A typical screen analysis for the chamotte is as follows:

Mesh	Cumulative percentage
6	Trace
8	5–9
12	27–41
16	60–68
30	94
70	99

The refractory and binder are mixed in a muller or an auger to the desired consistency and are rammed or vibrated over a pattern assembly. The pattern assembly consists of a preform pattern (or a final pattern covered with a suitable spacer material) mounted on a molding board and carefully located in a conventional flask, together with the required sprues. Before use, the preform pattern and other elements of the pattern assembly are coated or sprayed with a wax-silicone mixture to facilitate parting.

Carbon dioxide gas is diffused into the mold material for hardening. One method of doing this consists of piercing the mold material in several locations with small-diameter rods, thus creating channels into which the gas can be introduced. More uniform gassing can be accomplished in a vacuum chamber. After the gassing and hardening operation, the preform pattern is removed from the mold.

Preparing the Ceramic Facing. The principal ingredient of the ceramic facing is usually finely pulverized zircon, calcined mullite, or a mixture of both, although fused silica, magnesium oxide, and other refractory flours have also been used. Typical screen analyses for two calcined mullite flours are:

Mullite flour	Cumulative % on mesh size of: 100	200	325
A	3	35–65(a)	. . .
B	2	17–32	30–45(a)

(a) Remainder passes through this mesh size.

A typical slurry contains a mixture of 75% zircon and 25% calcined mullite combined with a hydrolyzed ethyl silicate binder in proportions of approximately 0.91 kg (2 lb) of refractory blend to 100 mL (3.4 oz) of binder. To this mixture is added a small amount of gelling agent, which causes the slurry to set in a predetermined period of time, usually 3 to 4 min. Immediately after this addition, the slurry is ready for pouring.

Pouring the slurry over the pattern is always a gravity operation. However, when pattern detail is critical, the pattern and slurry are placed in a vacuum chamber, where the ungelled slurry is vacuumed to remove entrapped air bubbles.

Immediately prior to use, the final pattern is coated with a wax-silicone mixture to facilitate parting, and the backup mold is placed over it. The slurry is poured through pouring gates until the level rises to invest the final pattern. In a few minutes, the chemical gelling action is complete, leaving the green ceramic with a consistency like that of vulcanized rubber, and the pattern is stripped from the mold. The rubberlike consistency is critical because it facilitates the separation of molds from patterns that have no draft, straight sides, extremely fine detail, protruding pins, or holes. Stripping is sometimes done mechanically to ensure that withdrawal of the pattern will not damage the mold. Removal of the pattern does not cause the mold to expand or contract. After stripping, the mold is ready for burn-off.

All-Ceramic Mold Casting. An unbacked, all-ceramic mold is made entirely from slurry of the type used for facing a composite ceramic mold. The procedure for pouring an all-ceramic mold is shown in Fig. 1. The pattern, affixed to a flat plate, is surrounded by a flask, into which the ceramic slurry is poured. After the chemical gelling action is complete, the green mold is stripped from the pattern, and the flask, which is built with a small amount of draft, is removed from the mold. The mold is then ready for burn-off.

Burn-off (Mold Stabilization). After gelling, the ceramic mold or mold facing is ignited with a torch and burns until most volatiles are consumed. During burn-off, the ceramic develops a microcrazed pattern, which is a three-dimensional network of microscopic cracks induced by the rapid evaporation of the alcohol in the slurry and by solid-phase reactions. Microcrazing is characterized by jagged ceramic particles separated by minute fissures or air gaps. These fissures are small enough to prevent molten metal from penetrating the mold surface, but they are large enough to permit the venting of air and other gases and to accommodate the expansion of the ceramic particles when they are in contact with the molten metal. Thus, microcrazing is advantageous in promoting dimensional stability without detracting from surface finish (see the section "Mold Stabilization" in this article).

Baking is the final stage of processing. All remaining volatiles are removed, and the colloidal silica left by the hydrolyzed ethyl silicate binder forms a high-temperature bond of SiO_2 that is stable nearly to its melting point (1710 °C, or 3110 °F), thus providing enough mold strength to resist washout, or erosion, by molten metal. To perform these functions satisfactorily, baking must be done at not less than 480 °C (900 °F). Baking composite molds at more than 650 °C (1200 °F) may cause a differential expansion between the facing and backup layers due to the contraction of the soda-rich silica bond in the backing material. This can produce distortion in the mold cavity. Nevertheless, some foundries bake composite molds at temperatures as high as 815 to 980 °C (1500 to 1800 °F) without harmful effects.

Two methods of baking are currently in use. In one method, termed skin heating, the mold is directly heated from above with electric or gas-torch heaters. This method has the advantage of fully curing the ceramic facing without overheating the backup material. It is a rapid technique, and it permits molds to be cured in conventional steel flasks.

Heating the molds in an electric or gas-fired furnace is the other method of baking. Furnace baking has the advantage of removing all moisture and volatiles from the backup material as well as from the facing, thus minimizing the risk of bleeding moisture from the backup to the facing after mold assembly. Such bleedback can produce surface defects in the casting.

The normal baking period for molds is 4 to 6 h at the prescribed baking temperature. After baking, molds are allowed to cool to an appropriate temperature for pouring.

Pouring. Before pouring, molds are usually checked for cleanliness and temperature. Mold temperature at the start of pouring ranges from 40 to 540 °C (100 to 1000 °F), depending on casting shape and the alloy being poured. Cope and drag halves are assembled, and a layer of cement is usually troweled on to seal the mold around the parting line. For safety, heavier molds are reinforced by fastening a steel band around the outside of the mold. This assists in handling of the mold and prevents seepage of hot metal in the event that the mold wall cracks during pouring. Finally, ceramic pouring tubes and exothermic or insulating riser sleeves are set up in the cope.

The mold is clamped firmly, and metal is poured either directly from the melting furnace or from a ladle. Pouring must proceed as rapidly as possible to avoid cold shuts and gas voids. After the metal has been poured, additional exothermic material is sprinkled on the risers to keep them fluid and to prevent shrinkage tears.

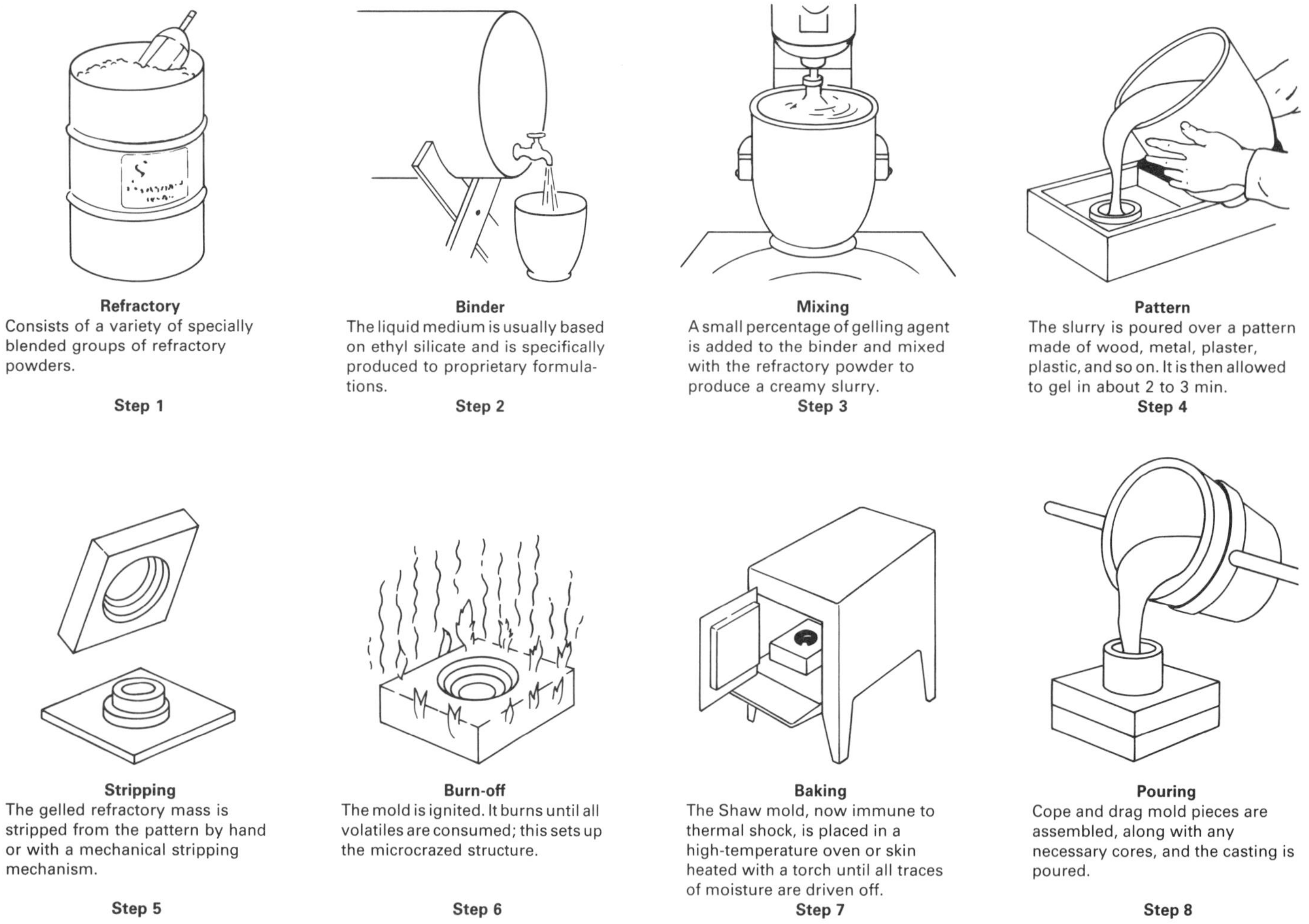
Refractory
Consists of a variety of specially blended groups of refractory powders.
Step 1
Binder
The liquid medium is usually based on ethyl silicate and is specifically produced to proprietary formulations.
Step 2
Mixing
A small percentage of gelling agent is added to the binder and mixed with the refractory powder to produce a creamy slurry.
Step 3
Pattern
The slurry is poured over a pattern made of wood, metal, plaster, plastic, and so on. It is then allowed to gel in about 2 to 3 min.
Step 4
Stripping
The gelled refractory mass is stripped from the pattern by hand or with a mechanical stripping mechanism.
Step 5
Burn-off
The mold is ignited. It burns until all volatiles are consumed; this sets up the microcrazed structure.
Step 6
Baking
The Shaw mold, now immune to thermal shock, is placed in a high-temperature oven or skin heated with a torch until all traces of moisture are driven off.
Step 7
Pouring
Cope and drag mold pieces are assembled, along with any necessary cores, and the casting is poured.
Step 8

Fig. 1 Sequence of operations used in the Shaw all-ceramic mold process

Composite Ceramic Mold. This type of mold is a sodium silicate/carbon dioxide cured and fired chamotte backing that significantly decreases the cost of refractory materials. A preform pattern and a finish pattern are needed to make composite molds (Fig. 2). Cope and drag preforms are made 2.4 mm ($3/32$ in.) oversize on all mold cavity surfaces to allow for the thickness of the ceramic facing material.

The cope and drag flasks must be accurately aligned, and at each stage of mold production, alignment holes are drilled and reamed to a jig. As a further safeguard, a sand strip is cast into the insides of the flasks to keep the hardened preforms from shifting during handling. Preform patterns are then mounted on a pin lift molding table. The flasks are aligned by dowel pins and the alignment holes on the table before they are fastened to the patterns by quick-acting clamps.

The sodium silicate refractory mixture contains a percentage of invert sugar, which aids shakeout of the mold after pouring. The refractory mixture is rammed around a preformed pattern (step 1, Fig. 2), and the rammed half mold is placed in a bell jar for 10 to 20 s, during which it is gassed with a commercial grade carbon dioxide (step 2). While the first preform is being gassed, the operator rams the second flask with the composite mixture. By this time, the first preform is hardened and is removed from the bell jar. The operator repeats the gassing procedure for the second preform. The gassed preform mold halves are then stripped from the preform patterns, and each hardened preform mold is aligned with the final pattern bolted to the molding table. The flasks are fastened to the patterns by the quick-acting clamps.

The Shaw ceramic slurry needed for the facing of the composite mold is virtually the same as that used for the solid body mold—a specially blended formula of refractory aggregates, ethyl silicate-base binder, and a gelling agent. The thick, creamy slurry is then poured into the gap between the final pattern and the hardened preform material (steps 3 to 6). The slurry not only fills all of the details of the mold cavity but also penetrates the aggregate voids of the surface of the porous backing, creating a strong mechanical bond.

Gelling of the slurry requires 2 to 3 min, and the fine ceramic facing of the composite mold is formed. The composite cope and drag are unclamped and lifted from the patterns by a foot-operated stripping plate. The molds are then placed on a conveyor for movement through a torching zone, where the volatiles are ignited and consumed (step 7). The microcrazed structure of the Shaw mold occurs during this burn-off period. The cope and drag molds are moved directly into a high-pressure torching zone for 5 to 10 min; in this zone, the molds are brought to incandescence. The copes and drags are assembled using the alignment pins to ensure accuracy. The molds can be assembled either hot or cold.

Unicast Process

The Unicast process differs from the Shaw process principally in the method of mold stabilization employed. Mold stabilization refers to the treatment given the fine ceramic facing of a composite mold, or the total mass of an all-ceramic mold, shortly after it has set and while it is still green.

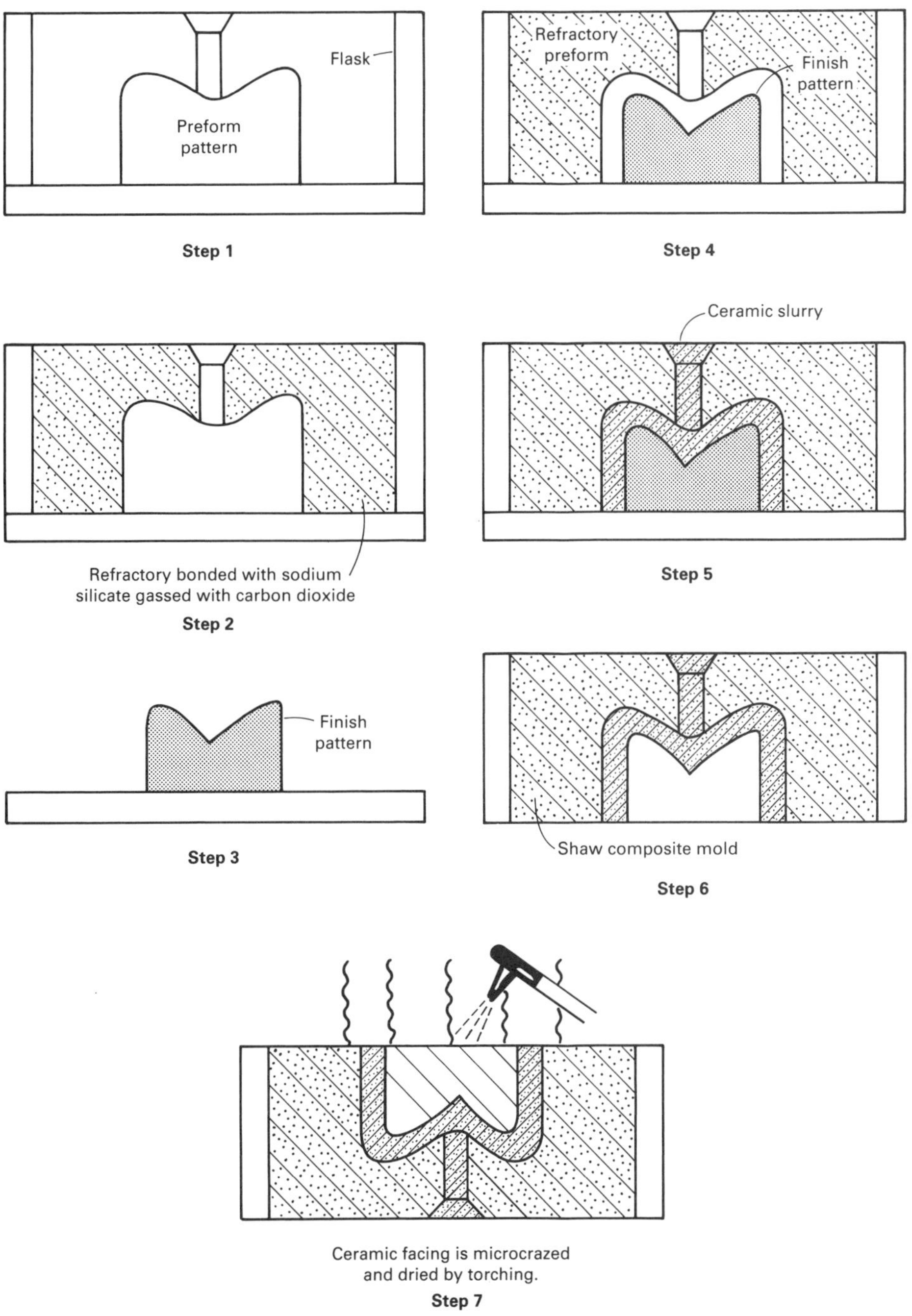

Fig. 2 Diagram of Shaw composite mold process. See text for details.

Advantages. The Unicast process provides accurate dimensional control, and warpage or distortion is almost entirely eliminated. As a result of the stabilized molecular growth, the mold develops a cellular or spongelike structure caused by the interstitial separation of particles. This structure provides for gas venting and particle expansion during casting. The advantages of this structure are metal soundness, ease of casting long, thin sections, and reductions in traditional mold venting. Molds can be poured hot or at room temperature without any change in thermal characteristics. Directional solidification of the cast alloy is fully controlled, resulting in high physical properties and outstanding metal soundness.

The major advantages of solid ceramic mold casting include the following:

- Only a short lead time is necessary due to simple tools and a fast mold build cycle
- A modest installation cost can be maintained using simple equipment
- The manufacturing cost is considerably lower than that of machining or other alternate methods
- Patterns are usually inexpensive and quickly prepared
- The die life of cast tooling is longer than that of dies made from wrought materials
- Cast tooling can frequently be used as-cast without further machining
- The process is suitable for impossible-to-machine grain and texture details for large injection mold bodies
- The process is feasible for casting sizes ranging from small to very large

Composite Mold. Another difference between the Shaw and Unicast processes relates to the sequence followed in preparing composite molds. In the Shaw process, preparation of the coarse mold backup precedes pouring of the fine ceramic facing. In the Unicast process, this sequence is reversed (see the section "Typical Process Sequence" in this article). The castings produced by the two processes are comparable in quality and surface finish, and several foundries are licensed to use both processes.

Although the Unicast process commonly uses a solid ceramic mold for smaller and short-run castings, it generally employs a composite-ceramic mold for large parts and for those produced in high volume. In this case, the Unicast ceramic is used as a 6.4 to 13 mm (¼ to ½ in.) facing backed by a specially blended, coarse-sand aggregate for support. This technique sharply reduces mold cost by conserving the ceramic material, and it permits the mechanized production of molds.

Patterns and Flasks. The Unicast process employs conventional pattern equipment throughout. The patterns must be of good quality because the slightest flaw or defect in the pattern will be reproduced on the final casting. The reproduction of detail is such that even fingermarks on a pattern coating are likely to be transferred to the casting. Most of the common pattern materials are used, including the denser hardwoods, plaster, aluminum, plastics, and hard rubber.

Alloys Cast. Alloys normally poured in Unicast molds are about evenly divided between ferrous and nonferrous. Of the ferrous alloys, about 60% are irons, and the rest are principally stainless steels and tool steels. Of the nonferrous alloys, about 30% are alloys of cobalt, titanium, or uranium, and about 20 to 30% are alloys of aluminum, magnesium, or zinc. The rest are principally copper alloys, with beryllium copper predominating.

Slurries. The fine ceramic slurries used in the Unicast process are comparable to those used in the Shaw process and consist of mixtures of a finely pulverized ceramic, such as zircon, and an organic gelling binder, such as hydrolyzed ethyl silicate. Accordingly, the chemical reactions that take place during gelation are the same as those

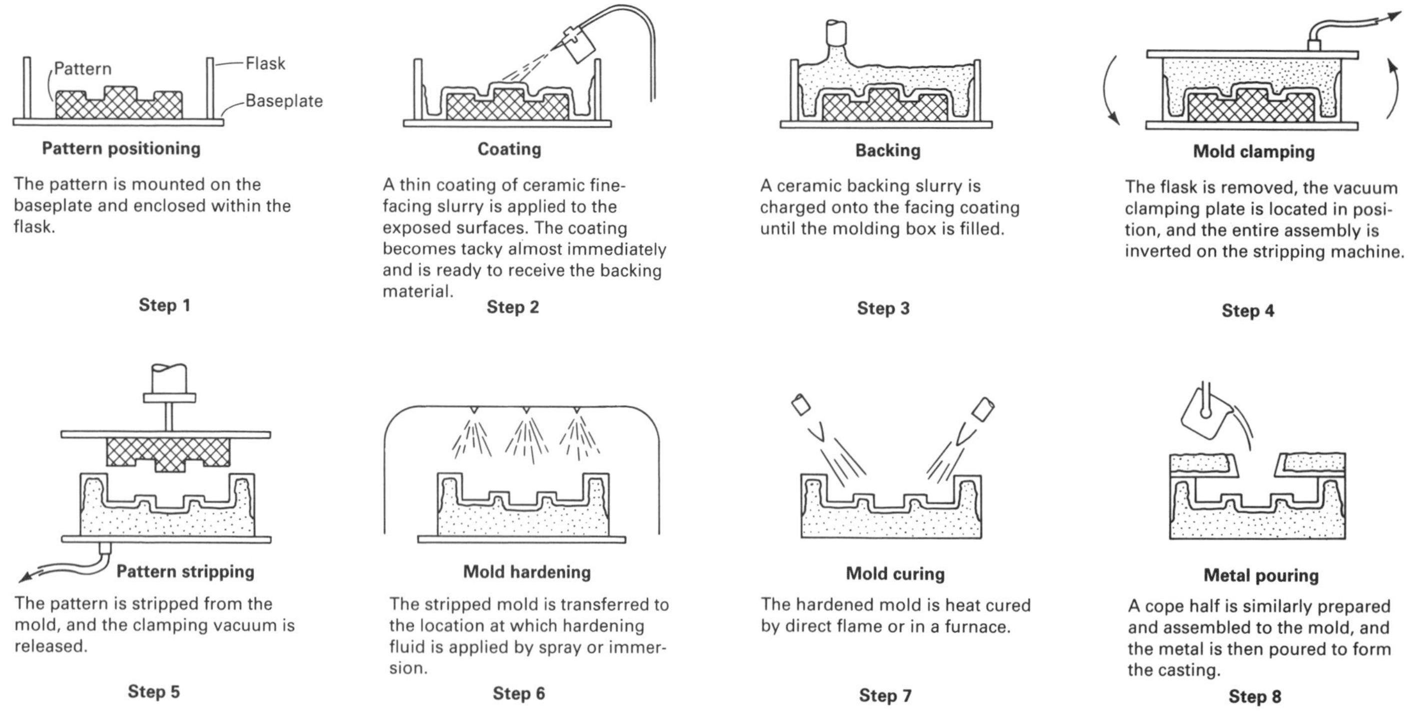

Pattern
Flask
Baseplate
Pattern positioning
The pattern is mounted on the baseplate and enclosed within the flask.
Step 1
Coating
A thin coating of ceramic fine-facing slurry is applied to the exposed surfaces. The coating becomes tacky almost immediately and is ready to receive the backing material.
Step 2
Backing
A ceramic backing slurry is charged onto the facing coating until the molding box is filled.
Step 3
Mold clamping
The flask is removed, the vacuum clamping plate is located in position, and the entire assembly is inverted on the stripping machine.
Step 4
Pattern stripping
The pattern is stripped from the mold, and the clamping vacuum is released.
Step 5
Mold hardening
The stripped mold is transferred to the location at which hardening fluid is applied by spray or immersion.
Step 6
Mold curing
The hardened mold is heat cured by direct flame or in a furnace.
Step 7
Metal pouring
A cope half is similarly prepared and assembled to the mold, and the metal is then poured to form the casting.
Step 8

Fig. 3 Steps involved in producing a Unicast process casting

in the Shaw process. Backup slurries, used in composite mold preparation, are coarse and less expensive ceramic grains mixed with a refractory binder that will harden in the presence of carbon dioxide.

Mold stabilization refers to the treatment given the fine ceramic slurry after it has gelled but while it is still green. The methods used for mold stabilization in the Shaw and Unicast processes are distinctly different, although the chemical reactions involving the binder are the same.

When the binder becomes a hard, rubbery gel, it consists of a liquid and a solid phase. The liquid phase is largely free alcohol, and the solid phase is a form of polysilicic acid. By itself, polysilicic acid is unstable and rapidly forms new chemical chains that are of larger mass than the original molecule. Molecular growth in the green state causes the suspended ceramic particles to move or separate, thus producing a pattern of microscopic cracks. Unless molecular growth is arrested in a short period of time, the cracks can become excessive and form macrocracks.

In the Shaw process, rapid evaporation of the alcohol resulting from hydrolysis or gelling reaction is accomplished immediately after (or very soon after) gelling by ignition or firing of the mold. This operation, referred to as burn-off, is usually performed by igniting the mold facing with a torch, as shown in Fig. 1. Burn-off follows removal of the pattern from the mold. It arrests excessive molecular growth, which may be catalyzed by moisture in the air or by other contaminants. This reaction must be avoided to ensure optimum binding properties and to avoid excessive cracking of the mold.

In the Unicast process, the gelled mold is immersed in a liquid hardening bath or vapor atmosphere that permits aging and internal stabilization of the mold. Various hardening media can be used, depending on the composition of the mold binder and the final curing step. Thus, commercial ethyl alcohol can be used with an ethyl silicate binder because the binder evolves ethyl alcohol as a result of the hydrolysis action.

Alternatively, the bath can be a liquid such as acetone, kerosene, benzene, or other hydrocarbon with which the evolved alcohol is readily miscible. The hardening bath is more reactive when hot; the heat provides good results even when the bath liquid is not chemically identical to the evolved alcohol. Even boiling water can be used; the temperature of the water drives the volatiles out of the mold to be dissolved in the water.

A typical aging period for the mold in the bath is 10 to 15 min. The aging reaction is enhanced when the pH of the hardening bath or atmosphere is maintained at a level somewhat higher than that of the mold and, preferably, on the alkaline side of neutral pH. Decreasing acidity serves to accelerate setting and gelation. Organic amines or other generally alkaline ammonium compounds can be added to the bath for this purpose.

The precise function of the hardening medium is not completely understood. Aging is enhanced when the gelled mold material is entirely immersed in a liquid or vapor that is substantially identical to the evolved alcohol, as compared to burn-off or to isolation of the mold surface from air by application of a wax or other suitable impermeable coating.

Final curing, which follows the hardening bath, is usually done in an oven maintained at about 980 °C (1800 °F), but direct flame heating can be used. The cured molds are then transferred to the melting station for pouring.

Typical Process Sequence. In a typical molding and casting sequence (Fig. 3), a pattern, mounted on a baseplate, is enclosed within a flask. A thin (4.8 to 9.5 mm, or 3/16 to 3/8 in.) coating of fine-grain ceramic slurry is applied to exposed pattern surfaces by spraying or similar means. The coating becomes tacky almost upon contact and is ready to receive backing material. An inexpensive, coarse backup slurry is poured rapidly over the facing coat until the flask is filled. The slurry is chemically controlled to set to a semirigid state within 2 to 3 min, at which time the upper surface of the mold is leveled with the flask edges with a striking tool. The flask is removed, and a vacuum clamping plate is placed in position on top of the mold. The entire assembly is inverted on a stripping machine, the pattern is stripped from the mold, and the clamping vacuum is released.

The stripped mold is transferred to the chemical hardening tank or chamber, where it is immersed in, or sprayed with, hardening fluid. It is then cured by heating with direct flame impingement or in a furnace.

Investment Casting

Robert A. Horton, PCC Airfoils, Inc.

IN INVESTMENT CASTING, a ceramic slurry is applied around a disposable pattern, usually wax, and allowed to harden to form a disposable casting mold. The term disposable means that the pattern is destroyed during its removal from the mold and that the mold is destroyed to recover the casting. Related processes, in which ceramic slurries are poured against permanent patterns to make cope and drag molds, are covered in the articles "Plaster Molding" and "Ceramic Molding" in this Volume.

There are two distinct processes for making investment casting molds: the solid investment (solid mold) process and the ceramic shell process. The ceramic shell process has become the predominant technique for engineering applications, displacing the solid investment process. By 1985, fewer than 20% of non-airfoil investment castings and practically no airfoil castings (the largest single application of investment casting) were being made by the solid investment process (Ref 1). Today the solid investment process is primarily used to produce dental and jewelry castings and has only a small role in engineering applications, mostly for nonferrous alloys. Because of its declining role and because it was discussed in the article "Investment Casting" in Volume 5 of the 8th Edition of *Metals Handbook*, the solid investment process will not be covered in this article. Only the ceramic shell process will be described. The basic steps in this process are illustrated in Fig. 1.

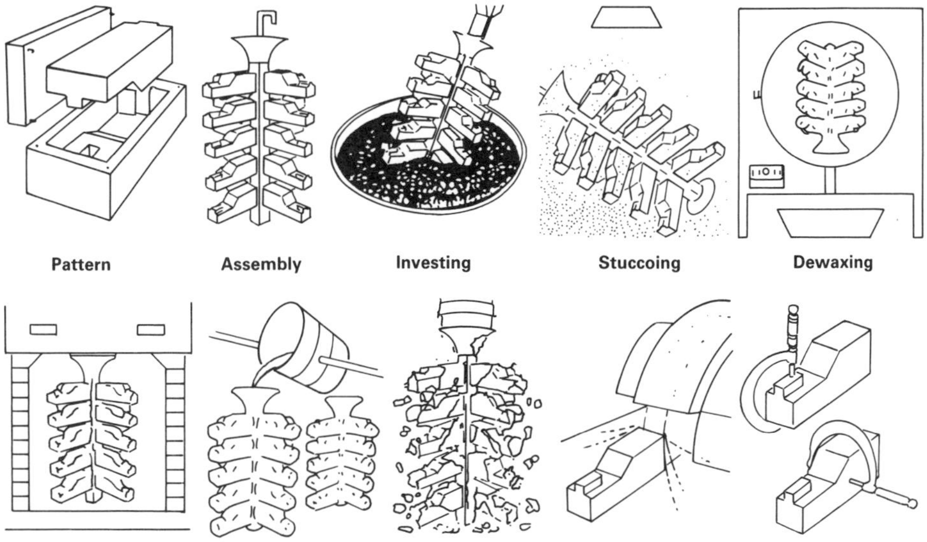

Fig. 1 Steps in the investment casting process

Pattern Materials

Pattern materials for investment casting can be loosely grouped into waxes and plastics. Waxes are more commonly used. Plastic patterns (usually polystyrene) are frequently used in conjunction with relatively thin ceramic shell molds in a process known as Replicast CS (see the article "Replicast Process" in this Volume).

Waxes

Wax is the preferred base material for most investment casting patterns, but blends containing only waxes are seldom used. Waxes are usually modified to improve their properties through the addition of such materials as resins, plastics, fillers, plasticizers, antioxidants, and dyes (Ref 2).

The most widely used waxes for patterns are paraffins and microcrystalline waxes. These two are often used in combination because their properties tend to be complementary. Paraffin waxes are available in closely controlled grades with melting points varying by 2.8 °C (5 °F) increments; melting points ranging from 52 to 68 °C (126 to 156 °F) are the most common. The low cost of these waxes, combined with their ready availability, convenient choice of grades, high lubricity, and low melt viscosity, accounts for their wide usage. However, applications are limited somewhat by their brittleness and high shrinkage. Grades designated fully refined should be selected for pattern waxes. Microcrystalline waxes tend to be highly plastic and lend toughness to wax blends. Hard, nontacky grades and soft, adhesive grades are available. Microcrystalline waxes are available with higher melting points than the paraffins and are often used in combination with paraffin.

Ozocerite is an imported mineral wax that can serve a similar function. Fisher-Tropsch waxes are synthetic hydrocarbon waxes resembling paraffins, but are available in grades that are much harder and higher melting. They improve hardness and rigidity and promote faster setup in pattern-making operations. Polyethylene waxes (of much lower molecular weight than the usual polyethylene plastics) function in much the same manner as the Fisher-Tropsch waxes, but are available in even higher melting grades (to over 132 °C, or 270 °F).

Candelilla is an imported vegetable wax that is moderately hard and slightly tacky. It has relatively low thermal expansion and somewhat less solidification shrinkage than the hydrocarbon waxes. Candelilla is useful for hardening paraffin and raising its softening point. Unfortunately, it has been subject to shortages and wide price fluctuations.

Carnauba is another imported vegetable wax. It has a fairly high melting point (83 °C, or 181 °F) and a low coefficient of thermal expansion. Carnauba is very hard, nontacky, and brittle. It is rather costly, which tends to limit its use.

Beeswax is a natural insect wax that melts at approximately 64 °C (147 °F). It is widely used for modeling and is also useful in pattern blends, contributing some of the same characteristics as comparable microcrystalline waxes. It is believed to be the original wax used for lost wax casting, but today its use is limited by its cost relative to other waxes.

Palmitic and stearic acids are fatty acids that have good wax properties. They have low viscosity when melted, relatively low thermal expansion in the solid state, relatively low solidification shrinkage, and good compatibility with a wide range of materials. Still other waxes that have been used are montan wax (both acid and ester types), fatty acid amides, ceresin, stearone, and hydrogenated castor oil.

The waxes listed represent those most commonly used for pattern blends. Of these, the paraffins and microcrystalline waxes are the most popular. Many other waxes are available and are at least potentially useful. In many cases, a particular wax is not used simply because a less expensive wax offers similar properties.

Waxes are reasonably priced, yet they provide a good balance of properties not readily obtainable in other materials. They are easily blended to suit different requirements and are compatible with other materials that can improve their properties still further. Their low melting points and low melt viscosities make them easy to compound, inject, assemble into clusters, and melt out without cracking the thin ceramic shell molds. These properties allow waxes to be injected at low temperatures and pressures, and this, combined with their lack of abrasiveness, leads to lower tooling costs.

Additives. Plain waxes possess many useful properties, but there are two important areas in which they are deficient:

- Strength and rigidity, especially where very fragile patterns need to be made
- Dimensional control, especially with regard to surface cavitation resulting from solidification shrinkage during and after pattern injection

Improvements can be made in these areas with nonwaxy additives.

The strength and toughness of waxes are improved by the addition of well-known high molecular weight plastics such as polyethylene, ethyl cellulose, nylon, ethylene vinyl acetate, and ethylene vinyl acrylate. These materials are highly viscous at wax-working temperatures, which tends to limit the amounts that can be used.

Nevertheless, these amounts are sufficient to make significant improvements in the handling characteristics of patterns and clusters. Polyethylene is widely used because it is economical and compatible with a wide spectrum of waxes. Ethyl cellulose has had some use, but it is much more limited in terms of compatibility and is more expensive. Polystyrene is rarely used because of its incompatibility with the commonly used waxes. Ethylene vinyl acetate and ethylene vinyl acrylate are newer materials that are finding increased application. Nylon has had only small usage.

Solidification shrinkage, which causes surface cavitation, is reduced somewhat by the plastics mentioned above. The effect, however, is limited by the low amounts that can be used before viscosity becomes excessive. Greater effects can be obtained by adding resins and fillers. Resins are used in most present-day pattern waxes. Fillers are used more selectively, and this leads to the description of pattern waxes as being either filled or unfilled.

The resins used for shrinkage reduction are of lower molecular weight than the plastics used for toughening*. The most useful resins soften gradually and continuously with increasing temperature, and they do not exhibit the large solid-to-liquid expansions during heating, or the reverse contractions during cooling, that characterize the solid-to-liquid transformation in waxes. Thus, resins reduce this effect in direct proportion to the amount used. The amount used can be quite large because resins yield relatively low viscosity melts at wax-processing temperatures.

Suitable resins include the numerous rosin derivatives (esterified, polymerized, hydrogenated, and dehydrogenated) as well as such tree-derived resins as Burgundy Pitch, dammar, and the terpene resins. Additional resins include coumarone-indene, various hydrocarbon resins from petroleum, and a number of coal tar resins. The resins listed encompass a wide range of softening points and viscosity-versus-temperature relationships. They vary at room temperature from soft and tacky to hard and brittle, and they may have widely different compatibilities with particular waxes and plastics. All these factors must be considered and properly balanced in using resins in pattern waxes.

Fillers. The solidification shrinkage of waxes can also be reduced by mixing in powdered solid materials called fillers. These are insoluble in, and higher melting than, the base wax, and they produce an injectable suspension when the mixture is molten.

Because they do not melt, fillers do not contribute to the solidification shrinkage of the mixture, which is reduced in proportion to the amount used. Figure 2 shows the manner in which fillers reduce expansion during heating (as well as corresponding shrinkage upon cooling). Fillers that have been used in pattern waxes include polystyrene, various dicarboxamides and related compounds, isophthalic acid, pentaerythritol, and hexamethylenetetramine. The filler should preferably be in the form of small, equally sized spheres (Ref 5). Several such fillers have been developed, including spherical polystyrene, hollow carbon microspheres, and spherical particles of thermosetting plastic. In a variation of the filler concept, water has been emulsified into molten wax to serve the same function as the solid fillers (Ref 6). This type of wax is commercially available.

*The terms resins and plastics are used in accordance with the definitions proposed for pattern wax technology in Ref 2.

Several other additives can be used in pattern waxes. Colors in the form of oil-soluble dyes are used to enhance appearance, to provide identification, and to facilitate inspection of molded patterns. Antioxidants can be used to protect waxes and resins subject to thermal deterioration. Oils and plasticizers have been used to alter injection properties and, in certain cases, to reduce shrinkage.

As a result of the above considerations, the compositions of present-day pattern waxes (unfilled) typically fall into the ranges given below:

Ingredients	Composition, %
Waxes (usually more than one)	30–70
Resins (one or two)	20–60
Plastic (one)	0–20
Other	0–5

Filled waxes usually have similar base materials, with 15 to 45% filler added.

Wax Selection. In formulating wax pattern materials, a number of processing areas and other concerns must be addressed. These are listed below, along with the properties or considerations appropriate to each:

- *Injection*: Softening point, freezing range, rheological properties, ability to duplicate detail, surface, and setup time
- *Removal, handling, and assembly*: Lubricity, strength, hardness, rigidity, impact resistance, stability, and weldability
- *Dimensional control*: Thermal expansion/shrinkage, solidification shrinkage, cavitation tendency, distortion, and stability
- *Mold making*: Strength, wettability, and resistance to binders and solvents
- *Mold dewaxing and burnout*: Softening point, viscosity, thermal expansion, thermal diffusivity, and ash content
- *Miscellaneous*: Cost, availability, ease of recycling, toxicity, and environmental factors

Assistance in formulating pattern waxes can be found in Ref 2, 7, and 8. Reference 8 provides an extensive listing of pattern ingredients and their properties. A number of standard tests are available to facilitate the effort (Ref 9-11). The application of thermal analysis methods to wax testing is being investigated (Ref 7, 12), and this approach warrants extension for both control and development work. The Investment Casting Institute has provided a test die for evaluating waxes and wax injection presses (Ref 13, 14). No simple laboratory tests are available that can evaluate the suitability of a

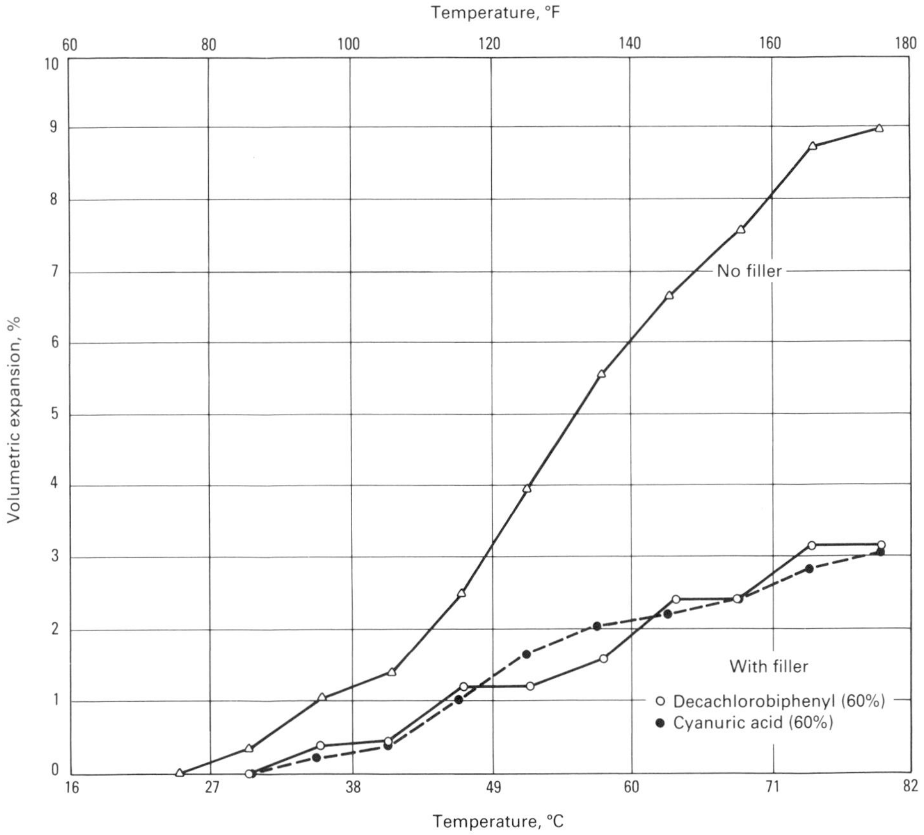

Fig. 2 Effects of filler materials on the volumetric thermal expansion of a pattern wax. Source: Ref 3, 4

pattern wax. The best approach usually consists of laboratory tests for screening purposes combined with practical evaluation under such productionlike conditions as injection behavior, handling characteristics, dimensional consistency, and dewaxability.

Plastics

Next to wax, plastic is the most widely used pattern material. A general-purpose grade of polystyrene is usually used.

The principal advantages of polystyrene and other plastics are their ability to be molded at high production rates on automatic equipment and their resistance to handling damage even in extremely thin sections. In addition, because polystyrene is very economical and very stable, patterns can be stored indefinitely without deterioration. Most wax patterns deteriorate with age and eventually must be discarded.

A disadvantage that limits the use of polystyrene is its tendency to cause mold cracking during pattern removal, a condition that is worse with ceramic shell molds than with solid investment molds. Other common plastics, such as polyethylene, nylon, ethyl cellulose, and cellulose acetate, are similar in this regard. In addition, tooling and injection equipment for polystyrene is more expensive than for wax. As a result, polystyrene finds only limited use. A recent survey of 72 investment casting plants in the United States revealed that fewer than 20% had machines for injecting polystyrene and that wax machines outnumbered plastics machines by almost 13 to 1 (Ref 15).

Perhaps the most important application for polystyrene is for small, delicate airfoils. Patterns for such castings are incorporated into composite wax/plastic assemblies for integral rotor and nozzle patterns. These airfoils are extremely thin and delicate; they would be too fragile if molded in wax because even a single chip or crack will cause an expensive casting to be rejected. Polystyrene solves this problem, and the use of wax for the rest of the assembly makes it feasible to process such large patterns without excessive mold cracking. The other important use for polystyrene is for small patterns running in large quantities. However, this application has declined as ceramic shell molds have replaced solid investment molds.

Other Pattern Materials

Urea-base patterns were developed in Europe and have found some application there (Ref 16). They resemble plastics in that they are very hard and strong and require high-pressure injection machines. An advantage of urea patterns is that they can be easily removed without stressing ceramic shell molds; the pattern is simply dissolved out in water or an aqueous solution. Another water-soluble pattern material is based on mixtures of low-melting salts (Ref 17). Patterns based on paradichlorobenzene and naphthalene have found some application in Japan (Ref 18). Foamed polystyrene has long been used for gating system components, and it is also being used for patterns in the Replicast Process (see the article "Replicast Process" in this Volume).

Patternmaking

Patterns for investment casting are made by injecting the pattern material into metal molds of the desired shape. Small quantities of patterns can also be produced by machining (see the section "Machining of Patterns" in this article).

Injection of Wax Patterns

Wax patterns are generally injected at relatively low temperatures and pressures in split dies using equipment specifically designed for this purpose. Patterns can be injected in the liquid, slushy, pastelike, or solid condition. Injection in the solid condition is often referred to as wax extrusion. Temperatures usually range from about 43 to 77 °C (110 to 170 °F), and pressures from about 275 kPa to 10.3 MPa (40 to 1500 psi). Liquid waxes are injected at higher temperatures and lower pressures; solid waxes, at lower temperatures and higher pressures. In some cases, the same wax can be injected under some, or all, of these conditions, but it is often beneficial to tailor a wax for one particular condition.

Equipment for wax injection ranges from simple and inexpensive to sophisticated and costly. It can be as simple as a pneumatic unit with a closed, heated reservoir tank that is equipped with a thermostat, pressure regulator, heated valve, and nozzle and is connected to the shop air line for pressurization. A small die is held against the nozzle with one hand while the valve is operated with the other. Such machines are limited by available air pressure (usually $<$690 kPa, or 100 psi) and are therefore generally used to inject liquid wax. Such equipment is satisfactory for a large variety of small hardware parts that are commonly made as investment castings.

Parts that are more demanding in terms of size, complexity, or dimensional requirements are made on hydraulic machines. These machines provide higher pressures for improved injections as well as high clamping pressures to accommodate large dies and high injection pressures. They can also be operated with very low pressures, as is often necessary when injecting around thin ceramic cores. Hydraulic wax injection machines are sized according to their clamping force capability, with models ranging in capacity from 9.1 to 363 Mgf (10 to 400 tonf) or more.

Machines for manual or semiautomatic operation generally have horizontal platens that clamp the die closed (one stationary, one movable). Injection is horizontal along the parting line; a second vertical nozzle is sometimes available. Models are available with C-frame construction and with two-, three-, and four-post H-frame construction.

Machines for automatic injection have the platens mounted vertically, with injection through the stationary platen and one-half of the die or at the parting line. A reservoir tank is usually provided for liquid and slushy wax, having either slow agitation or recirculation to keep the wax uniform and to retain solid material (wax or filler) in suspension. Reservoir tanks are sometimes omitted, and the machines are connected directly to a central wax supply.

Machines for paste wax injection require preconditioned wax in a metal cylinder. The cylinder is inserted into the machine, and the wax is pushed out by a piston. Solid injection (extrusion) machines require pre-tempered wax billets. Paste and solid injection both require a separate unit (oven or tempering bath) for tempering the cylinders or billets. Liquid injection usually requires separate units for melting and tempering the wax before it is introduced into the reservoir tank.

The parameters customarily controlled include wax and nozzle temperature, pressure, flow rate, and dwell time. Other available features, which may or may not be considered standard, include control of flow rate acceleration and deceleration, purging of the nozzle with fresh wax before each shot, nozzle drool control, unlimited shot capacity, die handling means, nozzle position adjustment, water-cooled platens, and multistation operation.

Reference publications that are available include a guide for troubleshooting wax injection operations (Ref 19) and an atlas of pattern defects (Ref 20). Other recommended publications are given in Ref 21 to 26.

Injection of Plastic Patterns

Polystyrene patterns are generally injected at temperatures of 177 to 260 °C (350 to 500 °F) and pressures of 27.6 to 138 MPa (4 to 20 ksi) on standard plastic injection machines. These are hydraulic machines with vertical water-cooled platens that carry the die halves, and horizontal injection takes place through the stationary platen. Polystyrene granules are loaded into a hopper, from which they are fed into a plasticizing chamber (barrel). Modern machines have a rotating screw that reciprocates within the heated barrel to prepare a shot of material to the proper consistency and then inject it into the die. Some older machines having plungers instead of screws are still in use.

The use of multicavity dies, as well as the running of a number of dies at once, is commonplace. Coupled with automatic or semiautomatic operation, this results in extremely high productivity.

The same parameters are controlled as for wax injection. The development of wax-base compositions that perform well in plastic injection machines has permitted the advantages of wax patterns to be combined with the advantages of plastic injection for certain applications (Ref 27).

Machining of Patterns

When only one or a few patterns are required, as for prototype and experimental work, it is expedient to machine them directly from wax or polystyrene. This avoids the time and expense involved in making pattern tooling. Special waxes have been specifically developed for this application (Ref 28).

Pattern Tooling For Investment Casting

Investment casting permits various potential tooling options that are made possible by the low melting point, good fluidity, and lack of abrasiveness of waxes. This often represents an important competitive advantage. For a given part configuration, anticipated production requirements, choice of pattern material, and available patternmaking equipment, the selection is based on a consideration of cost, tool life, delivery time, pattern quality, and production efficacy. The methods in use can be grouped into three basic categories: machining, forming against a positive model (using a variety of methods), and casting into a suitable foundry mold.

Many materials can be used, including soft metals (lead-bismuth-tin alloys), zinc alloys, brass, bronze, beryllium copper, nickel-plate, steel, rubber, plastic, metal-filled plastic, plaster, and combinations of these. All of these are suitable for wax patterns, but plastic patterns usually require steel or beryllium copper tooling. Soft metals and nonmetallics are often used for temporary tooling.

Machined Tooling. Most production tooling is made by machining. The early investment casting industry favored tooling made from master models, but as parts became larger and more complex and production methods more demanding, machined tooling became dominant. Computer-aided design, electric discharge machining, and computer numerical controlled machine tools are commonly used.

Aluminum is preferred for most wax tooling, steel, for plastic tooling. Aluminum is economical to machine, has good thermal conductivity, and is conveniently light. Brass or steel inserts can be used in areas subject to wear. The higher strength of steel is not usually needed for wax injection; instead, steel is primarily used for plastic injection for small parts that run in sufficiently large quantities to justify its higher cost. For polystyrene, this quantity is usually around 10 000 pieces or more; for very small parts, perhaps as low as 5000.

Tool steels are most frequently used for plastic dies. AISI type P20 is extensively used because it can be purchased prehardened, then machined, and finally put into use without heat treatment. AISI A2 and O1 are also used. Both are machined in the annealed state, then hardened and tempered. Carbon steels such as 1020 are sometimes used for intermediate quantities. Holder blocks, shoes, and mold bases used to hold the plastic injection molds can be made of prehardened alloy steel (for example, 4140).

Tooling made against a positive model includes a variety of methods for forming metal or metal-faced tooling, such as spraying (Ref 29), pressure casting (Ref 30), cold and warm hobbing (Ref 31, 32), electroforming (Ref 33, 34), and (potentially) gas plating (Ref 35). It also includes the casting of such nonmetallic materials as plaster (usually used to back up a spray metal facing), plastics (Ref 36), and rubber (Ref 37), as well as the vulcanizing of solid rubber under pressure (Ref 38).

All these methods begin with a positive model in the shape of the final investment casting; the model is machined oversize to include the appropriate shrink factors. The methods are economical because it is generally (although not always) less expensive to machine a positive model than to machine a negative cavity of identical shape. Furthermore, with the exception of hobbing, the model can be made of a material that is easier to machine. The various methods of making dies from the master model are relatively inexpensive, so that the combined cost of machining the model and making the die is usually less than that for full machining. When multiple cavities are needed or where molds must be replaced, the same master can be used, resulting in further economies.

Cast Tooling. Steel and beryllium copper are frequently used for cast tooling. Aluminum and zinc alloys have also been used. Wax can be cast against a master model to produce a pattern, which is then used to make an investment cast cavity. Better as-cast accuracy can be achieved if the model is coated with only a thin layer of wax. This is used to investment cast a steel shell, which is then backed up with cast aluminum (Ref 39). The ceramic mold process can also be used to cast injection cavities (Ref 40).

Both casting and hobbing can produce excellent tooling that is suitable for the wax or plastic injection of relatively simple parts, such as many small hardware castings. Machined tooling is also used for many of the same applications, as well as for complex parts, such as airfoils, and for large structural parts. Soft metal tooling and nonmetallic tooling are primarily used for short runs and experimental work or for temporary tooling. However, under favorable conditions, runs of up to 40 000 parts have been reported for cast bismuth-tin tooling (Ref 41). Cast, hobbed, and machined dies used for wax injection often last indefinitely. Rubber molds are regularly used for making patterns for jewelry casting.

Pattern and Cluster Assembly

Large patterns for investment casting are set up and processed individually, but small-to-medium ones are assembled into multipattern clusters for economy in processing. Clusters of aircraft turbine blades, for example, may range from 6 to 30 parts. For small hardware parts, the number may run into the dozens or even the hundreds. Most patterns are injected complete, including their casting gates. However, very large or complex parts can be injected in segments, which are assembled into final form. The capacity of injection machines and the cost of tooling are important considerations. Gating components, including pour cups, are produced separately, and patterns and gating are assembled to produce the pattern cluster. Standard extruded wax shapes are often used for gating, especially for mock-up work. Preformed ceramic pour cups are often used in place of wax ones. Most assembly is done manually.

Pattern Assembly

Wax components are readily assembled by wax welding using a hot iron or spatula or a small gas flame. The wax at the interface between two components is quickly melted, and the components are pressed together until the wax resolidifies. The joint is then smoothed over. A hot melt adhesive can be used instead of, or in addition to, wax welding. A laser welding system was recently described for this purpose (Ref 42).

Manual wax welding requires a fair degree of skill and considerable attention to detail. Fixtures are essential to ensure accurate alignment in assembling patterns, and they are often useful for cluster assembly. Patterns must be properly spaced and aligned. Joints must be strong and completely sealed with no undercuts. Care must be taken to avoid damaging patterns or splattering drops of molten wax on them.

Polystyrene pattern segments can be assembled by solvent welding. The plastic at the interface is softened with the solvent, and the parts are pressed together until bonded. The polystyrene becomes very tacky when wet with solvent (such as methyl ethyl ketone) and readily adheres to itself. Frequently, only one of the two halves needs to be wet. The assembly of polystyrene to wax is done by welding, with only the wax being melted.

Automation. Most assembly and setup operations are still performed manually, but some automation is being introduced. In one application, a robot is used to apply sealing compound in the assembly of patterns for integrally cast nozzles having 52 to 120 airfoils apiece. In the area of cluster setup, a few units for automatic assembly are being offered commercially, one of which has been described in great detail (Ref 43). Clusters are assembled automatically by welding pairs of wax patterns on opposite sides along the length of a wax runner bar, using heated copper blades and spring pressure. Standard gate designs in a range of size increments eliminate the need for special fixturing. The unit is suitable for most small-to-medium size castings.

Cluster Design

The design of the cluster is a critical factor because it affects almost every aspect of the process. Factors to be considered include ease of assembly, number of pieces processed at a time, ratio of metal poured to castings shipped, handling strength, ease of mold forming and drying, wax removal, liquid metal flow, filling of thin sections, feeding of shrinkage, control of grain size and shape (when specified), shell removal, ease of cutoff and finishing, and available equipment and processes.

Three requirements are essential:

- Providing a cluster that is properly sized and strong enough to be handled throughout the process
- Meeting metallurgical requirements
- Providing separate specimens for chemical or mechanical testing (when required)

Once these essentials are satisfied, other factors are adjusted to maximize profitability. The process is very flexible, and foundries approach this goal in various ways. Some tailor the cluster design to each individual part to maximize parts per cluster and metal usage. Others prefer standardized clusters to facilitate handling and processing. Where close control of grain is required, whether for equiaxed, columnar (directionally solidified), or single-crystal casting, circular clusters are often used to provide thermal uniformity. The specific casting process used, such as counter-gravity casting, directional solidification (see the article "New and Emerging Processes" in this Volume), or centrifugal casting (see the article "Centrifugal Casting"), may dictate the basic features of cluster design.

The critical aspects of cluster design are gating and risering, which are discussed in the articles "Riser Design" and "Gating Design" in this Volume. Basic concepts of risering borrowed from sand casting, such as progressive solidification toward the riser, Chvorinov's rule and its extensions, solidification mode, and feeding distance as a function of alloy and section size, also apply to investment casting.

However, feeding distances tend to be longer in hot investment molds than in sand molds (Ref 44, 45). Separate risers are sometimes used, but more often the gating system also serves the risering function, especially for the myriad of small parts that are commonly investment cast. The use of wax clusters permits great flexibility in the design of feeding systems. Process development clusters are readily mocked up for trial. Extruded wax shapes are easily bent into feeders that can be attached to any isolated sections of the part that are prone to shrinkage. Once proved, they can be incorporated into tooling if this is cost effective. If not, they can be applied manually during cluster assembly. This capability makes it practical to cast very complex parts with high quality and makes it feasible to convert fabrications assembled from large numbers of individual components into single-piece investment castings at great cost savings.

Considerable effort is being directed toward applying heat transfer models based on finite-element and finite-difference methods to the analysis of solidification and the design of feeding systems for castings, including investment castings (Ref 46). Although this has not yet had a significant impact on production engineering activities, eventual implementation and widespread use is anticipated. A specific application to a complex investment casting has recently been described (Ref 47).

Manufacture of Ceramic Shell Molds

Investment shell molds are made by applying a series of ceramic coatings to the pattern clusters. Each coating consists of a fine ceramic layer with coarse ceramic particles embedded in its outer surface. A cluster is first dipped into a ceramic slurry bath. The cluster is then withdrawn from the slurry and manipulated to drain off excess slurry and to produce a uniform layer. The wet layer is immediately stuccoed with relatively coarse ceramic particles either by immersing it into a fluidized bed of the particles or by sprinkling the particles on it from above.

The fine ceramic layer forms the inner face of the mold and reproduces every detail of the pattern, including its smooth surface. It also contains the bonding agent, which provides strength to the structure. The coarse stucco particles serve to arrest further runoff of the slurry, help to prevent it from cracking or pulling away, provide keying (bonding) between individual coating layers, and build up thickness faster.

Table 1 Nominal compositions and typical properties of common refractories for investment casting

Data are for comparison only, are not specifications, and may not describe specific commercial products.

Material	Nominal composition, %	Crystalline form	Approximate theoretical density g/cm^3	lb/in.^{3}	Relative leachability(a)	Approximate melting point °C	°F	PCE temperature(b) °C	°F	pH	Color
Aluminosilicates											
42%	Al_2O_3-53SiO_2	Mixture	2.4–2.5	0.086–0.090	Poor	...	...	1750	3180	6.5–7.8	Gray to tan
47%	Al_2O_3-49SiO_2	Mixture	2.5–2.6	0.090–0.094	Poor	...	...	1760	3200	6.5–7.8	Gray to tan
60%	Al_2O_3-36SiO_2	Mixture	2.7–2.8	0.097–0.10	Poor	...	...	1820	3310	6.5–7.8	Gray to tan
70%	Al_2O_3-25SiO_2	Mixture	2.8–2.9	0.10–0.104	Poor	...	...	1865	3390	6.5–7.8	Gray to tan
73%	Al_2O_3-22SiO_2	Mixture	2.8–2.9	0.10–0.104	Poor	...	...	1820	3310	6.5–7.8	Gray to tan
Alumina	99% + Al_2O_3	Trigonal	4.0	0.144	Poor	2040	3700	...	...	8.5–8.9	White
Fused silica	99.5% SiO_2	Typically 97% + amorphous	2.2	0.079	Good	1710	3110	...	...	6.0–7.5	White
Silica-quartz	99.5% SiO_2	Hexagonal	2.6	0.094	Good	1710	3110	...	...	6.4–7.5	White to tan
Zircon	97% + $ZrSiO_4$	Tetragonal	4.5	0.162	Moderate	2550	4620	...	...	4.7–7.0	White to tan

(a) Poor: slight reaction in hot concentrated alkali; Good: soluble to very soluble in hot concentrated alkali or hydrofluoric acid; Moderate: reacts with hot concentrated alkali solutions. (b) PCE, pyrometric cone equivalent. Source: Ref 48

Each coating is allowed to harden or set before the next one is applied. This is accomplished by drying, chemical gelling, or a combination of these. The operations of coating, stuccoing, and hardening are repeated a number of times until the required mold thickness is achieved. The final coat is usually left unstuccoed in order to avoid the occurrence of loose particles on the mold surface. This final, unstuccoed layer is sometimes referred to as a seal coat.

Mold Refractories

The most common refractories for ceramic shell molds are siliceous, for example, silica itself, zircon, and various aluminum silicates composed of mullite and (usually) free silica. These three types in various combinations are used for most applications. Alumina has had some use for superalloy casting, and this application has increased with the growth of directional solidification processes. Alumina is generally considered too expensive and unnecessary for commercial hardware casting. Silica, zircon, aluminum silicates, and alumina find use for both slurry refractories and stuccos. Table 1 lists the typical properties of these materials, and Fig. 3 shows thermal expansion curves.

Other refractories, such as graphite, zirconia (ZrO_2), and yttria (Y_2O_3), have been suggested for use with reactive alloys. Still other materials have been proposed for specific purposes. These specialized applications are summarized in Ref 49.

Silica is generally used in the form of silica glass (fused silica), which is made by melting natural quartz sand and then solidifying it to form a glass. It is crushed and screened to produce stucco particles, and it is ground to a powder for use in slurries. Its extremely low coefficient of thermal expansion imparts thermal shock resistance to molds, and its ready solubility in molten caustic and caustic solutions provides a means of removing shell material chemically from areas of castings that are difficult to clean by other methods.

Silica is also used as naturally occurring quartz. This is the least expensive material used to any extent. However, its utility is limited by its high coefficient of thermal expansion and by the high, abrupt expansion at 573 °C (1063 °F) accompanying its α-to-β-phase transition. As a result, shells containing quartz must be fired slowly, a practice most foundries find inconvenient.

Zircon occurs naturally as a sand, and it is used in this form as a stucco. It is generally limited to use with prime coats because it does not occur in sizes coarse enough for stuccoing backup coats. It is also ground to powder (and sometimes calcined) for use in slurries, often in conjunction with fused silica and/or aluminum silicates. Its principal advantages are high refractoriness, resistance to wetting by molten metals, round particle shape, and availability.

Aluminum silicates for investment casting are made by calcining fireclays or other suitable materials to produce a series of products ranging in alumina content from about 42 to 72%, with the remainder being silica plus impurities. Refractoriness and cost increase with alumina content. The only stable compound between alumina and silica at elevated temperatures is mullite ($3Al_2O_3{\cdot}2SiO_2$), which contains 72% alumina. Mixtures containing less than 72% alumina produce mullite plus free silica. The latter is usually in the form of silica glass, although some crystalline silica in the cristobalite form may be present. As the alumina content increases, the amount of mullite increases and free silica decreases until, at about 72% alumina, the material contains only mullite. Fired pellets of these materials are crushed or ground and carefully sized to produce a range of powder sizes for use in slurries, and granular materials for use as stuccos.

Alumina is produced from bauxite ore by the Bayer process. It is more refractory than silica or mullite and is less reactive toward many alloys than the siliceous refractories. Its use is primarily confined to superalloy casting, in which these properties can be used to advantage. The usual grades are tabular, which has been calcined just below the melting point, and fused, which has been electrically melted. The latter is slightly denser and slightly purer.

Binders

The commonly used binders are also siliceous and include colloidal silica, hydrolyzed ethyl silicate, and sodium silicate. Hybrid binders have also been developed, and alumina or zirconia binders are used for some processes.

Colloidal silica is the most widely used. It is manufactured by removing sodium ions from sodium silicate by ion exchange. The product consists of a colloidal dispersion of virtually spherical silica particles in water. The dispersion is stabilized by an ionic charge, which causes the particles to repel one another, thus preventing agglomeration. The stabilizing ion is usually sodium (up to 0.6%), although ammonia can also be used. In either case, the product is alkaline.

Colloidal silica can also be stabilized at an acid pH, but such products are not widely used. The most popular grades are sodium-stabilized with a silica content of 30% and an average particle size of either 7 or 12 nm. They are either used at this 30% level or diluted with water to reduce the silica content to 18 to 30%. Coherent gels having excellent bonding properties are formed by adding ionic salts that neutralize the ionic charge or by concentrating the sol (as by drying a coating).

Colloidal silica is an excellent general-purpose binder. Its main disadvantage is that its water base makes it slow drying, especially in inaccessible pockets or cores.

Ethyl silicate is produced by the reaction of silicon tetrachloride with ethyl alcohol. The basic compound formed is tetraethylorthosilicate ($Si(OC_2H_5)_4$). This corresponds to a theoretical silica content of 28.8%. The grade used for investment applications is designated ethyl silicate 40, and it consists of a mixture of ethyl polysilicates averaging

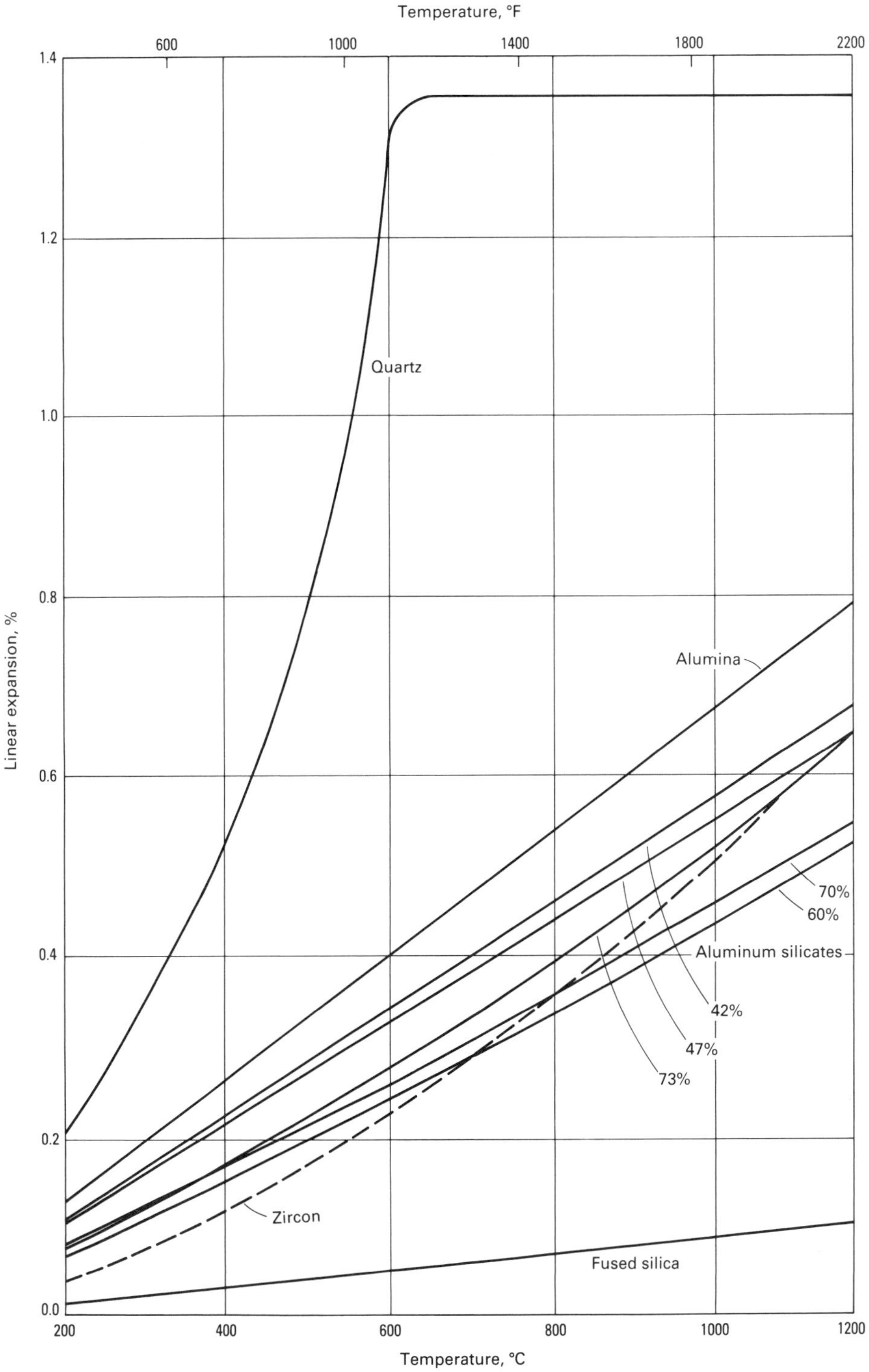

Fig. 3 Linear thermal expansion of some refractories common to investment casting. Source: Ref 48

five silicon atoms per molecule and having a silica content of around 40%.

By itself, ethyl silicate has no bonding properties. It is converted to ethyl silicate binder by reaction with water (hydrolysis). The reaction is usually carried out in ethyl alcohol, which serves as a mutual solvent, using an acid catalyst such as hydrochloric acid. The reaction is fairly rapid and highly exothermic. Its progress can be monitored with temperature measurements. The reaction forms complex silicic acids that are capable of condensing to form coherent gels having good bonding properties. This process is promoted by drying (concentrating) or by adding an alkali such as ammonia. The result is similar in many ways to that obtained using colloidal silica. Prehydrolyzed grades of ethyl silicate having reasonable shelf life are also available, making it unnecessary for foundries to perform this chemical operation themselves.

Ethyl silicate, with its alcohol base, dries much faster than colloidal silica. It is, however, more expensive and poses fire and environmental hazards. It is best used in applications involving rapid dipping cycles. Ethyl silicate slurries are readily gelled by exposure to an ammonia atmosphere; this permits dips to be applied very quickly, yet still provides proper drying because of the high volatility of the alcohol.

Hybrid binders have been developed in an effort to combine the advantages of colloidal silica and ethyl silicate. The water required for hydrolysis of the ethyl silicate is supplied by using colloidal silica, providing an additional source of silica for improved strength. Less flammable solvents can be substituted for alcohol. The resulting product presents a desirable combination of properties (Ref 50).

Liquid sodium silicate solutions are used where a very inexpensive binder is desired. Upon evaporation, these binders form a strong, glassy bond. Sodium silicate binders have poor refractoriness, which greatly limits their sphere of application. In addition, they are not resistant to the steam atmosphere of dewaxing autoclaves. Nevertheless, they have found some use, sometimes in conjunction with colloidal silica or ethyl silicate.

Other Binder Materials. The operation of directional solidification processes, which subject the mold to high temperatures for relatively long times, along with the introduction of even more reactive superalloys, has led to interest in more refractory binders. Colloidal alumina and colloidal zirconia binders have been made available for this purpose (Ref 51, 52). Both, however, are inferior to colloidal silica in room-temperature bonding properties.

Other Ingredients

Wetting Agents. In addition to the binder and refractory, slurries generally contain anionic or nonionic wetting agents, such as sodium alkyl aryl sulfonates, sodium alkyl sulfates, or octylphenoxy polyethoxyethanol, to promote wetting of the pattern or prior slurry coats. These are generally used in amounts of 0.03 to 0.3% by weight of the liquid. Wetting agents are sometimes omitted from ethyl silicate/alcohol slurries and from water-base backup slurries.

Antifoam Compounds. Where wetting agents are used, especially in prime coats, an antifoam compound may be included to suppress foam formation and to permit air bubbles to escape. Commonly used defoamers are aqueous silicone emulsions and liquid fatty alcohols such as *n*-octyl alcohol and 2-ethyl hexyl alcohol. Depending on type, these formulations are effective in very low concentrations of 0.002 to 0.10%, based on the liquid weight.

Miscellaneous Constituents. Organic film formers are sometimes used to improve the green strength and resilience of the dried coating or to enhance the coating ability of the slurry. Aqueous polyvinyl acetate emulsions, polyvinyl alcohol, and ammonium alginate have been used for this purpose. Small additions of clay have also been used to improve coating characteristics. Where close control of grain size is required, as for equiaxed superalloy casting, a nucleating agent (grain refiner) is added to the prime slurry in amounts ranging from about 0.5 to 10% by weight of the slurry refractory (Ref 53-55). Preferred nucleating agents are refractory cobalt compounds such as aluminates, silicates, titanates, and oxides.

Slurry Formulation

The actual percentage composition of ceramic shell slurries depends on the particular refractory powder, type and concentration of binder, liquid vehicle, and desired slurry viscosity. Composition generally falls in the following broad range by weight:

Ingredient	Composition, %
Binder solids	5–10
Liquid (from binder or added)	15–30
Refractory powder	60–80

Other optional ingredients, when present, are used in the amounts already given. Slurry compositions are usually proprietary, but three published formulas for zircon slurries are given in Table 2.

Two extremely important properties of ceramic shell molds are green strength and fired strength. A fundamental study of the effects of composition and processing parameters on these properties is described in Ref 56 and 57, which also provide additional examples of slurry compositions. This work was further amplified by another researcher (Ref 58). The effects of the usual refractory powders and stuccos on hot strength were investigated, and it was found that strength after firing is not a good indication of hot strength (Ref 59). However, it is still an important property because it affects the shell removal operation.

Slurry Preparation and Control

Preparation. Slurries are prepared by adding the refractory powder to the binder liquid, using sufficient agitation to break up agglomerates and thoroughly wet and disperse the powder. Viscosity initially tests excessively high because of air entrainment and lack of particle wetting; therefore, mixing is continued until the viscosity falls to its final level before the slurry is put into use. Continued stirring is required in production to keep the powder from settling out of suspension. Either rotating tanks with baffles or propeller mixers are used for this purpose—but generally at a lower agitation level than that used for the initial dispersion.

Control procedures for slurries vary considerably among foundries, reflecting in part the wide range of specifications that different shops work to, depending on their product line. The most prevalent controls are the measurement of the initial ingredients and the viscosity of the slurry. The latter is measured by a flow cup, such as a No. 4 or 5 Zahn cup, or a rotating viscometer of the Brookfield type.

Other parameters that are often controlled include slurry temperature, density, and pH, all of which are easily determined. More involved and time consuming is the determination of the actual composition of the slurry in terms of water content, silica binder, and refractory powder. This type of analysis, therefore, is performed less often and is sometimes reserved for troubleshooting rather than control. In the ceramic retention test, which is used by relatively few foundries, the actual weight of slurry adhering to a standard test plate under controlled conditions is determined.

Some properties of the finished ceramic shells that can be monitored include weight, modulus of rupture (green and/or fired), and permeability. Equipment for determining the hot strength of shells is relatively rare in the industry and is largely used for research. Test procedures for raw materials, slurries, molds, and cores are described in Ref 48, and the problems involved in slurry control are discussed in Ref 60 to 62.

Cluster Preparation

Before dipping, pattern clusters are usually cleaned to remove injection lubricant, loose pieces of wax, or dirt. Cleaning is accomplished by rinsing the pattern clusters in an appropriate solution, such as a water solution of a wetting agent, a solvent that does not attack the wax, or a solvent mixture capable of attacking the wax in a controlled manner to produce a fine uniform etching action that promotes slurry adhesion without affecting the cast surface. An additional rinse can be used to remove the cleaning agent. Another procedure involves dipping the cluster into a liquid that deposits an ultrathin refractory oxide coating that renders the surface hydrophilic; thus, the cluster is readily wetted by the ceramic slurry (Ref 63). These different solutions can sometimes be used in combination.

The clusters are usually allowed to dry before dipping. Drying produces a chilling effect, which causes unwanted contraction in pattern dimensions; therefore, the clusters are generally allowed to stand until they return to room temperature.

Coating and Drying

Dipping, draining, and stuccoing of clusters are carried out manually, robotically, or mechanically. Companies are increasingly using robots in order to heighten productivity, to process larger parts and clusters, and to produce more uniform coatings. When robots are introduced, they are often programmed to reproduce the actions of skilled operators. Dedicated mechanical equipment can sometimes operate faster, especially with standardized clusters, and finds some application. Most dipping is done in air. Dipping under vacuum is very effective for coating narrow passageways and for eliminating air bubbles, but it is not widely practiced.

The cleaned and conditioned cluster is dipped into the prime slurry and rotated; it is then withdrawn and drained over the slurry tank with suitable manipulation to produce a uniform coating. Next the stucco particles are applied by placing the cluster in a stream of particles falling from an overhead screen in a rainfall sander or by plunging the cluster into a fluidized bed of the particles. In the fluidized bed, the particles behave as a boiling liquid because of the action of pressurized air passing through a porous plate in the bottom of the bed.

Generally, prime slurries contain finer refractory powder, are used at a higher viscosity, and are stuccoed with finer particles than backup coats. These characteristics provide a smooth-surfaced mold capable of resisting metal penetration.

Backup coats are formulated to coat readily over the prime coats (which may be somewhat porous and absorbent), to provide high strength, and to build up the

Table 2 Formulations and properties of three types of zircon slurries

	Constituents										Properties		
	Colloidal silica (30%)		Water		Zircon powder		Wetting agent		Silica		Density		
Slurry	L	gal.	L	gal.	kg	lb	cm^3	Fl oz	kg	lb	g/cm^3	$lb/in.^3$	Viscosity, s(a)
1(b)	9.5	2.5	3.8	1.0	45	100	10	0.34	...	...	2.7–2.75	0.097–0.099	8–10
2(c)	11.4	3	...	...	45	100	10	0.34	...	...	2.9–2.95	0.104–0.106	9–11
3(d)	9.5	2.5	3.8	1.0	41	90	10	0.34	4.5	10	2.65–2.7	0.095–0.097	9–11

(a) No. 4 Zahn cup. (b) A general-purpose slurry. (c) A high-strength slurry used for ceramic shell molds for castings with heavy sections and for applications in which high wax pressure during meltout and high metal pressure are used. (d) Modified with silica; used with castings with small cored holes from which cores are removed by leaching with molten alkali salts required thickness with a minimum number of coats. One or two additional prime coats are sometimes used before starting the backup coats. The number of coats required is related to the size of the clusters and the metal weight to be poured and may range from as few as 5 for small clusters to as many as 15 or more for large ones. For most applications, the number ranges from 6 to 9.

Between coats, the slurries are hardened by drying or gelling. Air drying at room temperature with circulating air of controlled temperature and humidity is the most common method. Drying is usually carried out on open racks or conveyors, but cabinets or tunnels are sometimes used.

Drying is complicated by the high thermal expansion/contraction characteristics of waxes. If drying is too rapid, the chilling effect causes the pattern to contract while the coating is wet and unbonded. Then, as the coating is developing strength and even shrinking somewhat, the wax begins to expand as the drying rate declines and it regains temperature. This can actually crack the coating. Therefore, relative humidity is generally kept above 40%. For normal conditions, a relative humidity of 50% has been recommended as ideal (Ref 64). The essential point is that the maximum temperature differential between the wax and the drying air should not be excessive. Shop experience indicates that 4 to 6 °C (7.2 to 10.8 °F) is a practical maximum (Ref 65).

The slurry does not have to be thoroughly dried between coats, but it must be dry enough so that the next coat can be applied without washing off the previous one. An alternative technique is to harden the slurry by chemical gelation, which can be accomplished without drying. This is most successful with ethyl silicate binders, which can be surface gelled by exposure to an ammonia atmosphere. Hardening by chemical gelation permits the minimum time between coats. Gelling can also be accomplished by adding gelling agents to the stucco or by alternating alkaline and acidic slurries or negatively charged and positively charged slurries in the dipping sequence. Shells that are gelled without drying are generally weaker even when subsequently dried. In addition, the weight of the unevaporated solvent can be excessive and can cause the wax cluster to yield or even break in some cases.

Manufacture of Ceramic Cores

Ceramic coring is widely used in investment casting to produce internal passageways in castings. Investment casting cores are either self-formed (produced in place during the mold building operation) or preformed (made separately by an appropriate ceramic forming process).

Self-Formed Cores

Self-forming requires that the wax patterns already have the openings corresponding to the passageways desired in the castings. For simple shapes, this is accomplished by the use of metal pull cores in the pattern tooling. For shapes in which a simple pull core cannot be extracted, soluble wax cores are made, placed in the pattern tooling, and the pattern injected around them. The soluble core is then dissolved out in a solution that does not affect the wax pattern, such as an aqueous acid. Soluble cores are generally made from a solid polyethylene glycol, with a powdered filler such as sodium bicarbonate or calcium carbonate. The fillers dissolve in the acid with vigorous gas evolution, which provides agitation to speed the dissolution process.

Where openings are large enough, self-formed cores are made in the normal course of shelling, but where openings are deep or narrow, the shelling process must be modified to accommodate the special requirements of the cores. This may require special attention to slurry viscosity, stucco particle size, and drying. Vacuum dipping is an excellent tool in this case. Ethyl silicate/alcohol slurries reduce drying problems. The self-forming of cores is extensively practiced for small hardware castings, for which the cost of a preformed core would be prohibitive.

Preformed Cores

Self-formed cores have severe limitations, which are overcome by using preformed cores. Preformed cores require separate tooling and can be produced by a number of ceramic forming processes. Simple tubes and rods are commonly extruded from silica glass. Many cores, especially larger ones, are produced from ethyl silicate slurries that are cast by gravity or injected under pressure into aluminum dies and then gelled. The technology is similar to ceramic molding processes (see the article "Ceramic Molding" in this Volume).

Many cores are made by injection molding; a smaller number, by transfer molding. Ceramic injection molding is similar to the plastics injection molding described earlier, except that a mixture of fine ceramic powder in an organic vehicle is injected, usually into hardened tool steel dies. The organic vehicle, which also functions as a green binder, employs such thermoplastic materials as polyethylene, ethyl cellulose, shellac, resins, waxes, and subliming organic compounds such as naphthalene and paradichlorobenzene. After forming, the cores are subjected to a two-stage heat treatment. In the first stage, the organics are removed without disrupting the core (which is sometimes a lengthy process), and in the second stage, the core is sintered to its final strength and dimensions.

Transfer molding is similar to injection molding. The primary difference is that a preform of thermosetting plastic and ceramic powder is softened and injected into a die, where it is cured (thermally set) under heat and pressure.

Although a considerable number of materials have been proposed for ceramic cores, they are usually made of fused silica, sometimes with additions of zircon. Fused silica cores provide satisfactory refractoriness and are readily leached from the castings in molten caustic baths, aqueous solutions of caustic in open pots or autoclaves, and hydrofluoric acid solutions.

Preformed cores are usually used by placing them in the pattern die and injecting wax around them. For some simple shapes, the cores are inserted into the patterns immediately after the patterns are injected. Cores expand differently from the shell molds in which they are used because of differences in composition and heating rates; therefore, cores of any size must be provided with slip joints in the mold (Ref 66). The factors to be considered in using ceramic cores are discussed in Ref 66.

Pattern Removal

Pattern removal is often the operation that subjects the shell mold to the most stresses, and it is a frequent source of problems. These arise from the fact that the thermal expansions of pattern waxes are many times those of the refractories used for molds (compare Fig. 2 and 3). When the mold is heated to liquify the pattern, this expansion differential leads to enormous pressure that is capable of cracking and even destroying the mold. In practice, this problem can be effectively circumvented by heating the mold extremely rapidly from the outside in. This causes the surface layer of wax to melt very quickly before the rest of the pattern can heat up appreciably. This molten layer either runs out of the mold or soaks into it, thus providing space to accommodate expansion as the remainder of the wax is heated. Melt-out tips open to the outside are sometimes provided or holes are drilled in the shell to relieve wax pressure.

Even with these techniques, the shell is still subjected to high stress; therefore, it should be as strong as possible. It should also be thoroughly dried before dewaxing. To achieve thorough drying, shells are subjected to 16 to 48 h of extended drying after the last coat; this drying is sometimes enhanced by the application of vacuum or extremely low humidity.

A number of methods have been developed to implement the surface melting concept (Ref 67), but only two have achieved widespread use: autoclave dewaxing and high temperature flash dewaxing. In addition, hot liquid dewaxing has found some use among smaller companies seeking to minimize capital investment.

Autoclave dewaxing is the most widely used method. Saturated steam is used in a jacketed vessel, which is generally equipped with a steam accumulator to ensure rapid pressurization, because rapid heating is the key to success. Autoclaves are equipped with a sliding tray to accommodate a number of molds, a fast-acting door with a safety lock, and an automatic wax drain valve. Operating pressures of approximately 550 to 620 kPa (80 to 90 psig) are reached in 4 to 7 s. Molds are dewaxed in about 15 min or less. Wax recovery is good. Polystyrene patterns cannot be melted out in the autoclave.

Flash dewaxing is carried out by inserting the shell into a hot furnace at 870 to 1095 °C (1600 to 2000 °F). The furnace is usually equipped with an open bottom so that wax can fall out of the furnace as soon as it melts. Some of the wax begins to burn as it falls, and even though it is quickly extinguished, there is greater potential for deterioration than with an autoclave. Nevertheless, wax from this operation can be reclaimed satisfactorily. Flash dewaxing furnaces must be equipped with an afterburner in the flue or some other means to prevent atmospheric pollution. Polystyrene patterns are readily burned out in flash dewaxing, but polystyrene often can cause extensive mold cracking unless it is embedded in wax in the pattern (as in integral nozzle patterns) or unless the polystyrene patterns are very small.

Liquid dewaxing requires the minimum investment in equipment. Hot wax at 177 °C (350 °F) is often used as the medium. Other liquids can also be used. Cycles are somewhat longer than for autoclave and flash dewaxing, and there is a potential fire hazard.

Mold Firing and Burnout

Ceramic shell molds are fired to remove moisture (free and chemically combined), to burn off residual pattern material and any organics used in the shell slurry, to sinter the ceramic, and to preheat the mold to the temperature required for casting. In some cases, these are accomplished in a single firing. Other times, preheating is performed in a second firing. This permits the mold to be cooled down, inspected, and repaired if necessary. Cracked molds can be repaired with ceramic slurry or special cements. This will not heal cracks, but will seal them and reinforce the shell so that it can be successfully cast. Many molds are wrapped with ceramic-fiber blanket at this time to minimize the temperature drop that occurs between the preheat furnace and the casting operation or to provide better feeding by insulating selected areas of the mold.

Gas-fired furnaces are used for mold firing and preheating, except for molds for directional solidification processes, which are preheated in the casting furnace with induction or resistance heating. Batch and continuous pusher-type furnaces are most common, but some rotary furnaces are also in use. Most furnaces have conventional firebrick insulation, but furnaces with ceramic-fiber insulation and luminous wall furnaces (Ref 68) are also in use. The latter two provide very fast heat up and cool down as well as good fuel economy. Conventional and ceramic-fiber furnaces can be equipped with ceramic recuperators for greater fuel economy and reduced smoke emissions (Ref 69). Good circulation is essential, especially in preheat furnaces, to provide uniform temperatures. Approximately 10% excess air is provided in burnout furnaces to ensure the complete combustion of organic materials.

Burnout is commonly conducted between 870 and 1095 °C (1600 and 2000 °F). Many molds can be placed directly into the hot furnace, but others may have to be loaded at a low temperature and heated gradually. This includes molds made with crystalline silica and molds in which some parts are shielded from radiation. Preheat temperatures vary over a large range above and below burnout temperatures, depending on part configuration and the alloy to be cast. Common ranges are 150 to 540 °C (300 to 1000 °F) for aluminum alloys, 425 to 650 °C (800 to 1200 °F) for many copper-base alloys, and 870 to 1095 °C (1600 to 2000 °F) for steels and superalloys. Molds for the directional solidification process are preheated above the liquidus temperature of the alloy being cast.

Melting and Casting

Melting Equipment. In the past, the small carbon arc rollover furnace was the workhorse of the early investment casting industry (Ref 70), but this furnace has been gradually displaced by the induction furnace, which is more flexible and economical to operate. Induction furnaces are used in conjunction with various casting methods developed for the specific requirements of investment casting. Today most investment casting uses induction melting.

The furnaces used are of the coreless type and generally employ preformed crucibles for melting (or monolithic linings for large sizes). They consist of a water-cooled copper coil that surrounds the melting crucible, a power supply to energize the coil with alternating current, and the appropriate electrical controls. The alternating current flowing through the coil sets up an alternating magnetic field that induces eddy currents in the metal charge. The heating effect is self-generated within the charge as a result of its resistance to the current flow. Earlier models used spark gap converters or motor-generator sets as power supplies, but today solid-state power supplies are standard.

Most induction furnaces for investment casting have capacities ranging from 7.7 to 340 kg (17 to 750 lb). They are usually tilting models, although some small rollover models are used where very rapid mold filling is required. Melting rates of 1.36 kg/min (3 lb/min) are common, but this depends on the relative size of the power supply and the melt. Induction furnaces can be employed for melting in air, inert atmosphere, or vacuum. They are extensively used for melting steel, iron, cobalt and nickel alloys, and sometimes copper and aluminum alloys. The crucibles typically used are magnesia, alumina, and zirconia, which are made by slip casting, thixotropic casting, dry pressing, or isostatic pressing. For vacuum casting, there is some use of disposable liners, which are made by conventional ceramic shell techniques or by the slip casting of fused silica.

Gas-fired crucible furnaces and, more recently, electrical resistance furnaces are used for aluminum casting. Gas-fired furnaces using glazed clay-bonded graphite crucibles are satisfactory for aluminum and copper alloy castings, and these furnaces are inexpensive. Silicon carbide crucibles are sometimes used for copper-base alloys. Resistance furnaces eliminate combustion products and help reduce hydrogen porosity. Magnesium can be melted in gas-fired furnaces using low-carbon steel crucibles.

Consumable-electrode vacuum arc skull furnaces are used for melting and casting titanium in a process originally developed at the U.S. Bureau of Mines (Ref 71). A billet of the alloy to be cast serves as the consumable electrode and is arc melted under vacuum into a water-cooled copper crucible. The crucible has a permanent, thin skull of solidified alloy on its inner face, and this skull prevents the crucible from contaminating the melt. Molten alloy accumulates in the crucible until there is a sufficient quantity to fill the mold, at which time it is poured. Furnaces of various sizes are available and are capable of melting up to 522 kg (1150 lb). The number of such installations is small because only a few companies make titanium investment castings, but this reflects the technical difficulties involved rather than any lack of a market.

Electron beam melting has been used in Europe as an alternative to vacuum arc melting for the casting of titanium (Ref 72). It is being used by an aircraft engine manufacturer in the United States for melting superalloys for directionally solidified and single-crystal casting (Ref 73). The arrangement is basically similar to that used for vacuum arc skull melting except that the electron beam is used in place of the vacuum arc.

Casting Methods. Both air and vacuum casting are important in investment casting. Most vacuum casting is investment casting, although there is some use of rammed graphite molds in vacuum arc furnaces for casting titanium. Most castings are gravity poured.

Air casting is used for such commonly investment cast alloys as aluminum, magnesium, copper, gold, silver, platinum, all types of steel, ductile iron, most cobalt alloys, and nickel-base alloys that do not contain reactive elements. Zinc alloys, gray iron, and malleable iron are usually not investment cast for economic reasons, but if they were, they would be air cast.

Vacuum casting provides cleaner metal and often superior properties, and sometimes this is an incentive to cast some of the normally air-melted alloys in vacuum. However, its major use is for alloys that cannot be satisfactorily cast in air, such as the γ'-strengthened nickel-base alloys, some cobalt alloys, titanium, and the refractory metals. Batch and semicontinuous interlock furnaces are satisfactory for vacuum casting, but the latter are clearly preferred.

A major advantage of investment casting is its ability to cast very thin walls. This results from the use of a hot mold, but is further enhanced by specific casting methods, such as vacuum-assist casting, pressurized casting, centrifugal casting, and counter-gravity casting.

In vacuum-assist casting, the mold is placed inside an open chamber, which is then sealed with a plate and gaskets, leaving only the mold opening exposed to the atmosphere. A partial vacuum is drawn within the chamber and around the mold. The metal is poured into the exposed mold opening, and the vacuum serves to evacuate air through the porous mold wall and to create a pressure differential on the molten metal, both of which help to fill delicate detail and thin sections.

Rollover furnaces are pressurized for the same purpose. The hot mold is clamped to the top with its opening in register with the furnace opening, and the furnace is quickly inverted to dump the metal into the mold while pressure is applied using compressed air or inert gas.

Centrifugal casting uses the centrifugal forces generated by rotating the mold to propel the metal and to facilitate filling. Vacuum arc skull furnaces discharge titanium alloy at a temperature just above its melting point, and centrifugal casting is usually needed to ensure good filling. Dental and jewelry casting use centrifugal casting to fill thin sections and fine detail. These three applications represent the major uses for centrifugal casting in investment casting.

Counter-gravity casting is discussed in the article "New and Emerging Processes" in this Volume. Counter-gravity casting is also an excellent method of filling thin sections, and it is not limited to use with investment molds.

Postcasting Operations

Postcasting operations represent a significant portion (often 40 to 55%) of the cost of producing investment castings. A standard shop routing is provided for each part, and important savings can be realized by specifying the most cost-efficient routing. For example, it is often cost effective to detect and eliminate scrap early to avoid wasting finishing time, even if this means including an extra inspection operation. Alternative methods may be available for performing the same operation, and the most efficient one should be selected. The actual sequence in which operations are performed can be important. Some specifications require verification of alloy type, and this is done before parts are removed from the cluster.

Knockout. Some of the shell material may spall off during cooling, but a good portion usually remains on the casting and is knocked off with a vibrating pneumatic hammer or by hand. Brittle alloys require special attention. Part of the prime coat sometimes remains adhered to the casting surface, and bulk shell material may remain lodged in pockets or between parts. This is removed in a separate operation, usually shotblasting. Clusters are hung on a spinner hanger inside a blasting cabinet or are placed on a blasting table. If cores are to be removed in a molten caustic bath, the entire cluster can be hung in the bath at this point, and the remaining refractory can be removed along with the cores. High-pressure water blasting (69 MPa, or 10 ksi) is sometimes used instead of mechanical knockout.

Cut-Off. Aluminum, magnesium, and some copper alloys are cut off with band saws. Other copper alloys, steel, ductile iron, and superalloys are cut off with abrasive wheels operating at approximately 3500 rpm. Torch cutting is sometimes used for gates that are inaccessible to the cutting wheel. Some brittle alloys can be readily tapped off with a mallet if the gates are properly notched. Shear dies have also been used to remove castings from standardized clusters. Following cutoff, gate stubs are ground flush and smooth using abrasive wheels or belts. These are also used for other finishing operations, along with small hand grinders equipped with mounted stones.

Core Removal. Where the opening size permits, cores can be removed by abrasive or water blasting. If abrasive or water blasting cannot be used, the cores can be dissolved out. This can be accomplished in a molten caustic bath (sodium hydroxide) at 480 to 540 °C (900 to 1000 °F), in a boiling solution of 20 to 30% sodium hydroxide or potassium hydroxide in an open pot, or in the same solutions in a high-pressure autoclave. Hydrofluoric acid can also be used with alloys that are inert to it, such as platinum and many cobalt superalloys.

Heat Treatment. Both air and vacuum heat treatments are extensively performed as needed to meet property requirements. The vacuum heat treatment of stainless steels (instead of in air) is sometimes practiced to achieve a clean, bright surface and to avoid the need for any further cleaning. Before they are heat treated, single-crystal castings must be handled very carefully to avoid recrystallization during subsequent heat treatment. They must not be dropped, allowed to hit one another, or subjected to a blasting operation until heat treated.

Abrasive Cleaning. Blast cleaning is also used to remove scale resulting from core removal or heat treatment. Both pneumatic and airless (centrifugal) blasting machines are employed. Metallic abrasives (steel or iron grit or shot) and ceramic abrasives (silica sand, aluminum oxide, garnet, and staurolite) are all commonly used, depending on the application.

The hardness and angular shape of alumina produce rapid cleaning, which is especially cost effective when work is being blasted by hand. Alumina blasting is very effective at opening up and exposing surface defects, thus facilitating subsequent finishing or inspection operations. However, it is rather expensive, and less costly materials such as garnet, staurolite, or silica sand are often used instead. Metal shot and glass beads provide good peening action and produce shiny surfaces, both of which are sometimes desired. Equipment used for blasting includes hand cabinets as well as automatic units with rotating tables, spinner hangers, or rotating tumblers to hold the work.

Another type of abrasive finishing process is mass finishing. In this process, castings are treated with chemical finishing compounds and synthetic abrasives of specific geometric shapes, such as cones, triangles, and stars, to produce deburring, fine finishing, and radii blending in vibratory, tumbling, or centrifugal machines.

Miscellaneous Operations. Hot isostatic pressing is becoming increasingly important for the densification of castings to eliminate porosity, to improve fatigue, ductility, and other properties, and to reduce property scatter (see the section "Hot Isostatic Pressing of Castings" of the article "Processing of Castings" in this Volume). Its most important application in investment casting is for titanium. Applications for steel, superalloys, and aluminum are more selective but still very important.

Machining is often performed on investment castings, although the amount may be minimal, and there are many applications in which no machining is required. Machining is generally confined to selected areas requiring closer dimensions than can be achieved by casting, and it is occasionally used to incorporate some detail that is less expensive to machine in than to cast. The machining and welding of investment castings are discussed in Ref 74. Welding can be used to repair and join castings—primarily large, structural castings. Other methods used to improve dimensional accuracy are broaching and coining, as well as abrasive grinding. Straightening operations are performed when required, either manually or using hydraulic presses, with suitable fixtures. Castings can be heated to facilitate straightening.

Chemical finishing treatments are also used. The various treatments include acid pickling to remove scale, passivation treatments for stainless steel, chemical milling to remove the α case on titanium, and chemical treatment to apply an attractive satin finish to aluminum or to polish stainless steel.

Inspection and Testing

Alloy Type Test. The first inspection operation performed is often an alloy verification test. This test is conducted on the cluster before cutoff. A spectrometer or x-ray analyzer is used to verify that the correct alloy has in fact been poured.

Visual Inspection. An early visual inspection is essential so that obvious scrap does not get passed on to expensive finishing or inspection operations. Some commercial parts require only visual inspection.

Liquid fluorescent penetrant inspection is extensively used for nonmagnetic alloys. It can also be used for magnetic alloys, but a magnetic penetrant is generally specified instead. The liquid fluorescent penetrant detects defects on, or open to, the surface, such as porosity, shrinkage, cold shuts, some inclusions, dross, and cracks of any origin (hot tears, knockout, grinding, heat treat, straightening).

In this technique, the surface is properly cleaned, and a liquid penetrant with low surface tension and low viscosity is applied and is drawn into the defects by capillary action. Excess liquid is wiped away, a developer is applied that functions as a blotter to draw the liquid out, and the area is examined visually in a dark enclosure under black (near ultraviolet) light, which reveals defects that cannot be detected visually.

Applicable specifications include ASTM E 165 and MIL-1-6866. Turbine blades, which represent one of the most demanding applications for investment casting, are often subjected to a thermal cycle before the final fluorescent penetrant inspection; this is done to open very small defects that might otherwise go undetected.

Because the defects detected are often only surface defects, they can sometimes be removed by various finishing operations, and such rework operations are commonly attempted. They are generally successful if the defect is not so deep that dimensional or surface finish requirements are violated in removing it.

Magnetic particle inspection is used to detect the same types of defects as fluorescent penetrant inspection, but is preferred for use on ferromagnetic alloys. The method involves surface preparation, magnetization of the casting, and application of either a liquid suspension of magnetic particles (wet method) or fine magnetic iron particles (dry method). The presence of a defect causes a leakage field that attracts the magnetic particles and causes them to cling to the defect and define its outline. Colored particles and fluorescent particles (for viewing under black light) are available. Applicable specifications include ASTM E 125 (Reference Photographs), ASTM E 109 (Dry Particle), ASTM E 138 (Wet Particle), and MIL-STD-1949.

X-ray radiography is used to detect differences in material density or thickness to reveal such internal defects as dross, shrinkage, gas holes, inclusions, broken cores, and core shift. Many parts receive 100% inspection; others are inspected according to an appropriate sampling plan. Even when x-ray inspection is not required, it can be used as a foundry control tool to aid in establishing a satisfactory gating system, to determine the effect of process changes, to monitor process reliability, and to troubleshoot foundry problems. X-ray inspection is sometimes used to examine wax patterns containing delicate ceramic cores to ensure that the cores were not broken during the pattern injection operation. Applicable specifications for castings are ASTM E 192, MIL-C-2175, and MIL-STD-453. Real time x-ray inspection is beginning to be used in the larger investment foundries serving the aircraft industry.

Formerly, defects within the casting detected by x-ray radiography could be repaired only by grinding and welding the affected area, but this procedure was not always cost effective and could not be applied to all alloys or applications. Currently, many castings, including those that cannot be welded, are being repaired by hot isostatic pressing.

Miscellaneous Inspection Methods. Hardness testing is widely used to verify the response of castings to heat treatment. Chemical analysis is generally controlled through the use of certified master heats. Mechanical properties are determined on separately cast test bars or test specimens mounted on production clusters. Specimens machined from castings are used for process development and periodic audits.

Grain size is regularly checked on many equiaxed castings, following chemical or electrolytic etching. Even where grain inspection is not specified, it is often used as part of the process development effort. Electrolytic etching is also used for examining and detecting grain defects in directionally solidified and single-crystal castings. The orientation of single-crystal castings is determined by Laue back-reflection x-ray diffraction.

Pressure tightness tests are conducted for a variety of applications. Dimensional inspection ranges from manual checks with a micrometer or simple go/no-go gages to the use of coordinate-measuring machines and automatic three-dimensional inspection stations capable of checking a sculptured surface in a continuous sweep. Wall thickness on many cored turbine blade castings is determined ultrasonically. Nodularity is checked metallographically on the first and last cluster poured from heats of ductile iron. Metallography is an essential part of process development for high-performance castings.

Design Advantages of Investment Castings

The challenge in designing for investment casting is to make full use of the enormous capability and flexibility inherent in the process to produce parts that are truly functional and cost effective. Often such parts will also be more aesthetically pleasing. The principal advantages of investment casting that permit this challenge to be met will be discussed in this section.

Table 3 Some applications of investment casting

Aircraft engines, air frames, fuel systems	Machine tools
Aerospace, missiles, ground support systems	Materials handling equipment
Agricultural equipment	Metalworking equipment
Automotive	Oil well drilling and auxiliary equipment
Baling and strapping equipment	Optical equipment
Bicycles and motorcycles	Packaging equipment
Cameras	Pneumatic and hydraulic systems
Computers and data processing	Prosthetic appliances
Communications	Pumps
Construction equipment	Sports gear and recreational equipment
Dentistry and dental tools	Stationary turbines
Electrical equipment	Textile equipment
Electronics, radar	Transportation, diesel engines
Guns and small armaments	Valves
Hand tools	Wire processing equipment
Jewelry	

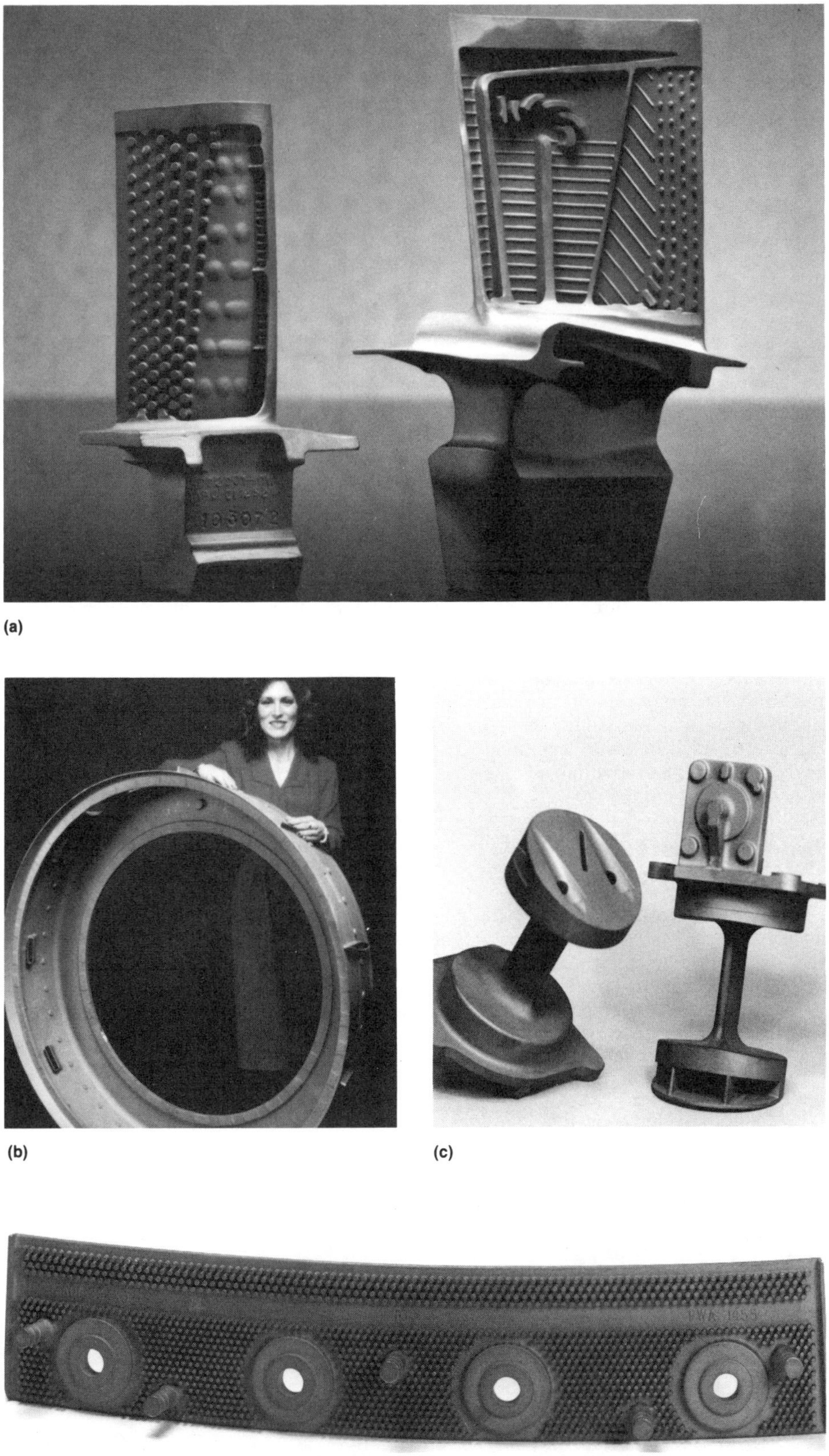

(a) (b) (c) (d)

Fig. 4 Some aircraft and aerospace applications for investment castings. (a) Single-crystal turbine blades investment cast using complex ceramic cores. Courtesy of Pratt and Whitney Aircraft. (b) 17-4PH stainless steel fan exit case; weight: 96 kg (212 lb). Courtesy of Precision Castparts Corporation. (c) Aircraft fuel sensor strut cast in 17-4PH stainless steel. Both ceramic and soluble cores were used to produce the complex internal passages. Courtesy of Northern Precision Casting Company. (d) Aircraft combustion chamber floatwall. The large number of small posts required high mold and pouring temperatures. Courtesy of Pratt and Whitney Aircraft

Complexity. Almost any degree of external complexity, as well as a wide range of internal complexity, can be achieved, and in certain cases, the only limitation is the state of the art in ceramic core manufacturing. As a result, parts previously manufactured by assembling many individual components are currently being made as integral castings at much lower costs and often with improved functionality. This ability to produce complexity with ease can even benefit simple parts, which can be redesigned to save weight without loss of strength by providing I- or H-sections, or thin walls with ribs. Thus, many parts from competitive processes can be converted to investment casting.

Freedom of Alloy Selection. Any castable alloy can be used, including ones that are impossible to forge or are too difficult to machine. Further, the cost of the alloy is less important in the final price of an investment casting than it is in many other metalforming processes; therefore, an upgraded alloy can often be specified (especially if the part is redesigned to save weight) at little or no increase in price.

Close Dimensional Tolerances. The absence of parting lines and the elimination of substantial amounts of machining by producing parts very close to final size give investment casting an enormous advantage over sand casting and conventional forging.

The availability of prototype and temporary tooling is a major advantage in the design and evaluation of parts. The direct machining of wax patterns or the use of any of the quick, inexpensive tooling methods described earlier facilitates timely collaboration between the designer and the foundry to produce parts that are functional and manufacturable. This capability is simply not found in such competitive processes as die casting, powder metallurgy, or forging. Further, temporary tooling used in the design phase can often serve for production while the market is tested or permanent tooling is constructed.

Reliability. The long-standing use of investment castings in aircraft engines for the most demanding applications has fully demonstrated their ability to be manufactured to the highest standards.

Wide Range of Applications. In addition to complex parts and parts that meet the most severe requirements, investment casting also produces many very simple parts competitively. This capability is often made possible by the low tooling costs associated with investment casting. Investment castings are competitively produced in sizes ranging from a few grams to more than 300 kg (660 lb), and the upper limits continue to increase.

Design Recommendations

The following recommendations provide a useful guide to the design of investment castings:

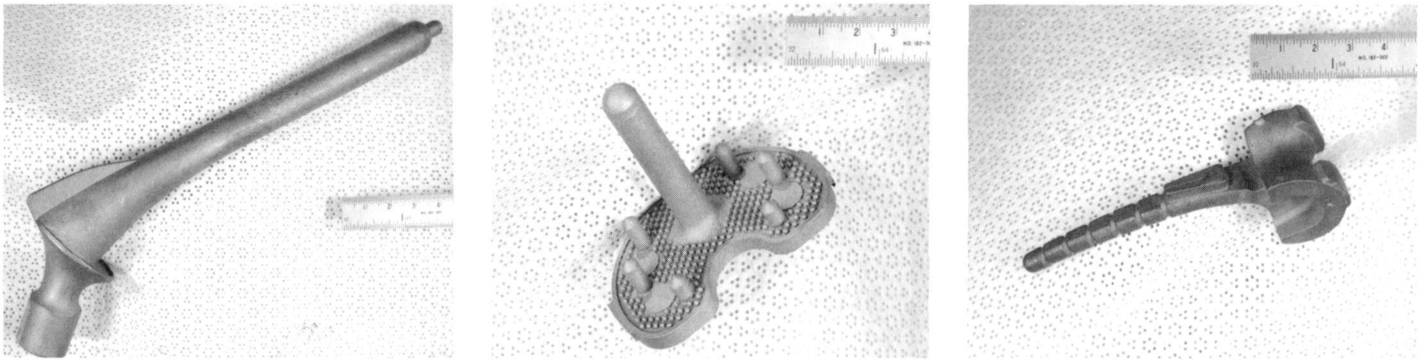

(a) (b) (c)

Fig. 5 Biomedical applications for investment castings. (a) Whiteside hip-femoral prosthesis. (b) Whiteside II-C knee-tibial base. (c) London elbow-humeral prosthesis. All cast in ASTM F75 cobalt-chromium-molybdenum alloy; all courtesy of Dow Corning Wright

- Look for ways to implement the advantages mentioned above
- Focus on final component cost rather than casting cost
- Design parts to eliminate unnecessary hot spots through changes in section sizes, use of uniform sections, location of intersections, and judicious use of fillets, radii, and ribs
- Promote good communication between foundry and customer to resolve questions of functionality versus producibility
- Use prototype castings to resolve questions of functionality, producibility, and cost
- Do not overspecify; permit broader-than-usual tolerances wherever possible
- Use the data in Ref 75 as a guide to the dimensional tolerances attainable
- Indicate datum planes and tooling points on drawings; follow ANSI Y14.5M for dimensioning and tolerancing

More detailed information is available in Ref 75 and 76.

Applications

Applications for investment castings exist in most manufacturing industries. A partial list is given in Table 3. The largest applications are in the aircraft and aerospace industries, especially turbine blades and vanes cast in cobalt- and nickel-base superalloys as well as structural components cast in superalloys, titanium, and 17-4-PH stainless steel. Examples of current applications for investment castings are shown in Fig. 4 to 6. All were cast in ceramic shell molds using wax patterns unless otherwise noted.

Special Investment Casting Processes

The investment casting process is highly flexible and can handle a great variety of parts with the same basic method. However, specialized versions of the process can be highly effective in reducing costs on particular types of parts. Two variations that are being increasingly applied are the Shellvest system and the Replicast CS process.

The Shellvest system provides a unique method of manufacturing small parts in high quantities at low cost. To provide large numbers of patterns at high production rates and low cost, automatic plastic injection machines are used, along with a special plastic/wax pattern material. Wax patterns are used when production requirements do not justify plastic tooling.

Pattern assemblies are made by attaching the patterns to large-diameter horizontal cardboard drums instead of conventional wax sprues. As shown in Fig. 7, the drum has a through handle for handling and for mounting in the moldmaking equipment. Each end of the drum is closed with an end plate, one of which is removable. The drum is wrapped with a thin layer of cardboard having corrugations on its inner surface and a smooth outer surface. The corrugated cardboard is attached with masking tape. Over the cardboard is a layer of thin, per-

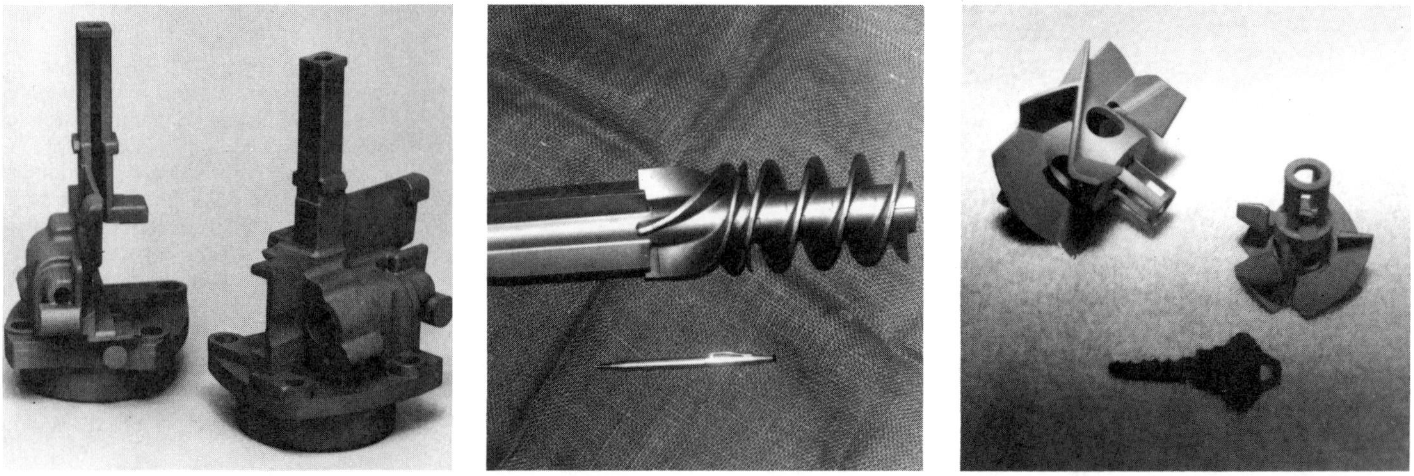

(a) (b) (c)

Fig. 6 Miscellaneous applications for investment castings. (a) Nosepiece for nailgun cast in 8620 alloy steel. The part is used as-cast. Courtesy of Northern Precision Casting Company. (b) Ni-Resist Type II cast iron inducer for deep well oil drilling. Courtesy of Bimac Corporation. (c) Small 17-4PH turbine vanes; the smaller weighs 71 g (2.5 oz); the larger, 185 g (6.5 oz). Courtesy of K.W. Thompson Tool Company, Inc.

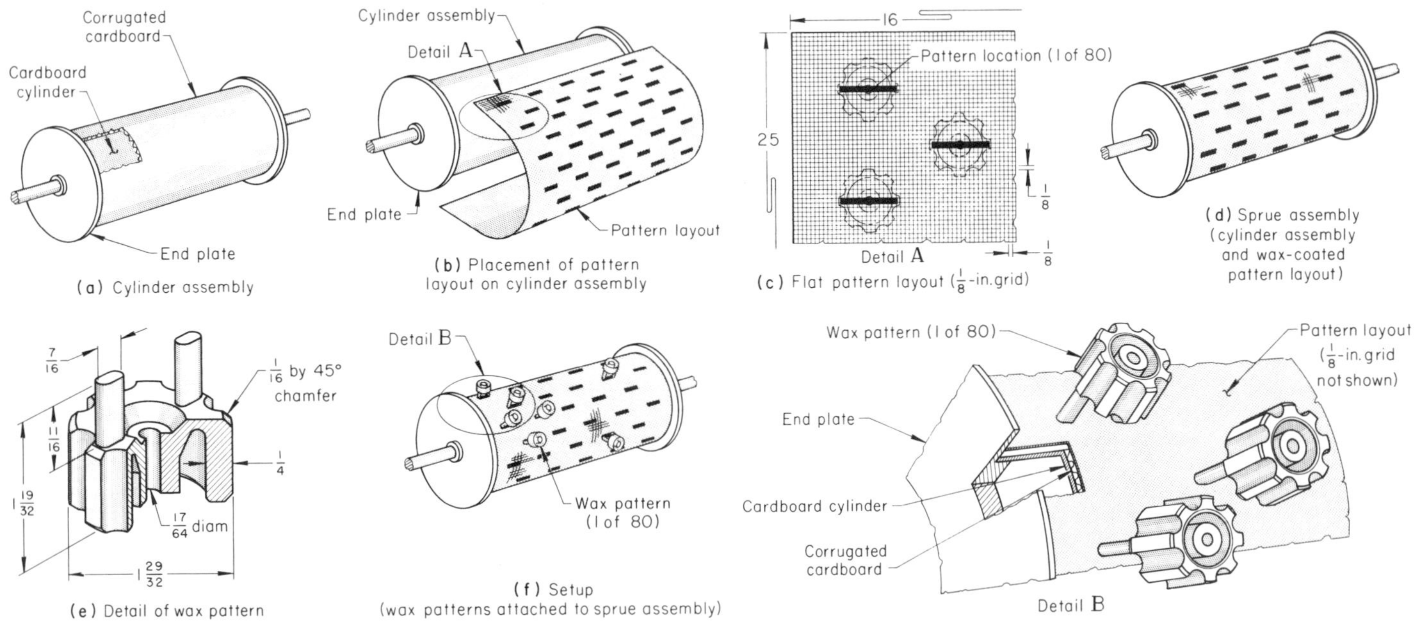

Fig. 7 Schematic showing the Shellvest system for investment casting of small parts. Dimensions given in inches

forated paper called a gate master. As shown in Fig. 7, the gate master has markings printed on it that indicate the cross section of each pattern gate and exactly where it is to be mounted on the drum. This greatly facilitates the operation of mounting patterns on the drum and ensures proper spacing and orientation.

The completed drum assembly is rotated through a bath of molten wax to apply a thin wax coating to which the patterns are attached. Individual patterns are attached by wax welding. Splitters can be used that divide the drum into several segments so that up to four molds can be formed simultaneously (Ref 77).

The completed pattern assembly is mounted horizontally in a vacuum slurry tank. The assembly is lowered into the tank so that its lower surface is immersed in the ceramic slurry. Here it rotates through the slurry while the chamber is under sufficiently high vacuum to boil the water in the slurry. The combination of rotation and vacuuming flushes air out of the system and enables the slurry to enter and coat the finest detail on the patterns, including core openings.

After the coating operation is complete, the assembly is raised out of the slurry and rotated in air to drain excess slurry back into the bath. It is then spun to provide a very uniform coating, after which it is immediately transferred to a fluidized bed for stuccoing. Here it is again mounted horizontally and rotated so that its bottom surface passes through the fluidized stucco particles. After stuccoing, the assembly is dried, and the sequence is then repeated four to six times to build up the required shell thickness.

When shelling is complete, one end plate is removed, and the cardboard tube is pulled out. Next the corrugated cardboard liner and the gate master paper are removed, and the setup readily separates into individual molds according to the number of splitters used. Each mold includes a cylinder with a large number of individual pattern cavities gated into its surface. The molds are dewaxed in a conventional steam autoclave. This operation is greatly facilitated by the fact that the gate of each pattern is immediately accessible to the steam. This provides very rapid melting and release of pressure, so that the plastic/wax material is readily removed without cracking the shell.

The ceramic shell mold is prepared for casting by firing it at the required preheat temperature, thus combining burnout and preheating. The mold is then placed over a ceramic-coated resin-bonded sand core in a vacuum-assist casting furnace (Ref 78). The diameter of the core is selected so that the annular space between the mold and the core will contain sufficient metal to feed all of the castings that are gated into this area. This is called a hollow-sprue gating system. It permits the use of large-diameter sprues carrying large numbers of parts, while still maintaining a very favorable ratio of gating metal to castings. Casting is carried out by the vacuum-assist method described previously.

After casting, the resin-bonded sand core disintegrates from the heat of the metal and falls out. The casting then proceeds through the conventional finishing and inspection operations.

The Replicast CS process uses conventional investment molding techniques to form a thin ceramic shell mold around a foamed polystyrene pattern of the type used in the lost foam casting process (see the discussion on the evaporative foam process in the article "Sand Molding" in this Volume). Information on the Replicast CS process is available in the article "Replicast Process" in this Volume.

REFERENCES

1. C.H. Schwartz, Coated Stucco Solves Ceramic Shell Drying Problems, *Mod. Cast.*, June 1987, p 31
2. R.A. Horton, "Formulating Pattern Waxes," Paper presented at the 31st Annual Meeting, Investment Casting Institute, Oct 1983
3. P. Solomon, Disposable Pattern, Composition for Investment Casting, U.S. Patent 3,887,382, 1975
4. P. Solomon, Disposable Pattern, Composition for Making Same and Method of Investment Casting, U.S. Patent 3,754,943, 1973
5. J. Booth, Pattern Waxes for Investment Casting, *Foundry Trade J.*, 6 Dec 1962, p 707
6. E. Faulkner, F. Hocking, and J. Aherne-Heron, Wax Emulsion for Use in Precision Casting, U.S. Patent 3,266,915, Aug 1966
7. D. Mills, "Second Thoughts on Wax Testing and Its Relationship to the Injection of Investment Patterns," Paper presented at the 11th Annual Conference, Bournemouth, England, British Investment Casters' Technical Association, May 1971
8. S. Rau, "Formulation, Preparation and Properties of Waxes for Wax Patterns," Paper presented at the Investment Casting Institute Advanced Technical Sem-

inar, Cleveland, Case Institute of Technology, 1963
9. *Standard Test Procedures—Pattern Materials*, Investment Casting Institute, 1979
10. *Acceptance Tests for Wax Pattern Materials*, British Investment Casters' Technical Association, 1979
11. H. Bennet, Test Methods Applicable to Various Waxes (ASTM, Amerwax, USP), in *Industrial Waxes*, Vol II, Chemical Publishing, 1975, p 297
12. M. Koenig, "Quality Control of Investment Casting Waxes by Thermal Analysis," Paper presented at the 25th Annual Meeting, Investment Casting Institute, 1977
13. J. Roberts, "Standardized Wax Testing Die," Paper presented at the 28th Annual Meeting, Investment Casting Institute, 1980
14. R. Bird, R. Brown, H. Eck, and W. Wurster, "Reports on the Use of the Institute's Test Die," Paper presented at the 30th Annual Meeting, Investment Casting Institute, 1982
15. State of the Investment Casting Industry, *Precis. Met.*, Nov 1982, p 39
16. H.S. Strache, "In-line Production of Precision Castings up to 100 kg. in Weight," Paper presented at the 12th European Investment Casters' Conference, Eindhoven, Holland, Sept 1967
17. K. Suzuki and O. Hiraiwa, U.S. Patent 3,131,999, 1964
18. K. Ugatta, Y. Morita, and Y. Mine, Investment Casting Method, U.S. Patent 3,996,991, 1976
19. *Wax Injection Trouble Shooting Guide*, Investment Casting Institute, 1979
20. *Atlas of Pattern Defects*, Investment Casting Institute, 1983
21. C. S. Treacy, *Some Notes on Waxes for Investment Castings*, Investment Casting Institute, 1964
22. V. Stanciu, *Quality of Wax Patterns as a Function of Wax Preparation and Injection*, European Investment Casters' Federation, 1970
23. I. Malkin, *The Effect of Die Temperature and Wax Pressure on Cavitation of Wax Patterns*, Investment Casting Institute, 1969
24. P.L. Wilkerson, Improved Patternmaking Techniques in Investment Casting, *Precis. Met. Mold.*, Oct 1974, p 40
25. J. Hockin, "Wax Pattern Dimensional Variation in Automatic Injection Machines," Paper presented at the 27th Annual Meeting, Investment Casting Institute, 1979
26. *Wax Injection Technology: A Guide to the Influence Injection Variables Have on Pattern Contraction*, Blayson-Olefines, 1967
27. C.H. Watts and M. Daskivich, Process and Material for Precision Investment Casting, U.S. Patent 3,263,286, 1966
28. R.A. Horton and E.M. Yaichner, Pattern Material Composition, U.S. Patent 4,064,083, 1977
29. Moulds For Prototype Production, *Foundry Trade J.*, 6 Aug 1964, p 167
30. W.F. Davenport and A. Strott, How to Set Up a Precision Casting Foundry, *Iron Age*, April 1951, p 90
31. I. Thomas and E. Spitzig, When and How to Hob, *Mod. Plast.*, Feb 1955, p 117
32. S. Richards, "Die Design for Productivity: Multi-Cavity Dies," Paper presented at the 26th Annual Meeting, Investment Casting Institute, Oct 1978
33. F.L. Siegrist, Have You Considered Electroforming?, *Met. Prog.*, Oct 1964
34. F.L. Siegrist, Electroforming With Nickel—A Versatile Production Technique, *Met. Prog.*, Nov 1964
35. W.C. Jenkin, Gas Plating for Molds and Dies, *Plast. World*, March 1967, p 82
36. N. Wood, Epoxies Key to Complex, Accurate Patterns, *Foundry*, Nov 1968, p 149
37. H.T. Bidwell, Patternmaking, in *Investment Casting*, Machinery Publishing, 1969
38. Rubber Molds Reduce Wax Pattern Tooling Costs, *Precis. Met.*, March 1979, p 66
39. R.A. Horton, J.H. Simmons, and T.R. Bauer, Method of Making Tooling, U.S. Patent 4,220,190, 1980
40. R. Schoeller, Building Injection Molds by Casting, *Plast. Des. Process.*, Aug 1975, p 13
41. Cast Dies for Investment Patterns, *Foundry*, April 1974, p 28
42. D. Draper, T.L. Cloniger, J.M. Hunt, J.F. Holmes, T. Mersereau, and M. Hosler, Laser Welding of Wax Patterns for Investment Casting, *Mod. Cast.*, Oct 1987, p 24
43. F. Ellin, "Automatic Pattern Assembly Machine," Paper presented at the 33rd Annual Meeting, Investment Casting Institute, 1985
44. H. Present and H. Rosenthal, Feeding Distance of Bars in Investment Molds, *Trans. AFS*, Vol 69, 1961, p 138
45. E.M. Broard *et al.*, "Feeding Distance of 410 Stainless Steel Cast in Phosphate Bonded Solid Investment Molds," Subcommittee Report on Casting and Solidification, Investment Casting Institute, May 1963
46. H.D. Brody, "Numerical Analysis of Heat and Fluid Flow in Sand and Investment Castings," Paper presented at the Solidification Processing Seminar, Dedham, MA, U.S.-Japan Cooperative Program, June 1983
47. A.A. Badawy, P.S. Raghupathi, and A.J. Goldman, Improving Foundry Productivity With CAD/CAM, *Comput. Mech. Eng.*, July/Aug, 1987, p 38
48. *Ceramic Test Procedures*, Investment Casting Institute, 1979
49. R.C. Feagin, "Casting of Reactive Metals Into Ceramic Molds," Paper presented at the Sixth World Conference on Investment Casting, Washington, 1984
50. D.J. Ulaskas, "Hybrid Binder Systems," Paper presented at the 29th Annual Meeting, Scottsdale, AZ, Investment Casting Institute, 1981
51. K.A. Buntrock, "Development Work in Aluminosilica, Colloidal Alumina and Polysilicate Binders," Paper presented at the 29th Annual Meeting, Scottsdale, AZ, Investment Casting Institute, 1981
52. R.C. Feagin, "Alumina and Zirconia Binders," Paper presented at the 29th Annual Meeting, Scottsdale, AZ, Investment Casting Institute, 1981
53. R.A. Horton, R.L. Ashbrook, and R.C. Feagin, Making Fine Grained Castings, U.S. Patent 3,019,497, 1962
54. R.A. Horton, R.L. Ashbrook, and R.C. Feagin, Making Fine Grained Castings, U.S. Patent 3,157,926, 1964
55. "Mason Nucleating Compounds," Data Sheet, Engineered Materials, Inc.
56. R.L. Rusher, Strength Factors of Ceramic Shell Molds, Part I, *AFS Cast Met. Res. J.*, Dec 1974, p 149
57. R.L. Rusher, Strength Factors of Ceramic Shell Molds, Part II, *AFS Cast Met. Res. J.*, March 1975
58. M.O. Roberts, "Factors Affecting Shell Strength," Paper presented at the 25th Annual Meeting, Investment Casting Institute, Oct 1977
59. J.D. Jackson, "Evaluation of Various Mold Systems at High Temperature," Paper presented at the 26th Annual Meeting, Investment Casting Institute, Oct 1978
60. P.R. Taylor, "A Rationalized Approach to Slurry Composition Control for Ceramic Shell Moulding," Paper presented at the Fourth World Conference on Investment Casting, Investment Casting Institute, British Investment Casters' Technical Association, European Investment Casters' Federation, Amsterdam, 1976
61. P.R. Taylor, "Quality Control of Slurry Materials for Ceramic Shell Moulding," Paper presented at the 26th Annual Meeting, Phoenix, AZ, Investment Casting Institute, 1978
62. P.R. Taylor, "Problems of Primary Coat Quality for Control of Ceramic Shell Molds," Paper presented at the 27th Annual Meeting, Investment Casting Institute, 1979
63. R.A. Horton, "Methods and Materials for Treating Investment Casting Patterns," U. S. Patent 3,836,372, 1974
64. L.M. Dahlin, Colloidal Silica Systems for Investment Shells, *Precis. Met.*, Jan 1969, p 81
65. M.G. Perry and A.J. Shipstone, "Research Into the Drying of Ceramic Shell

Moulds,'' Paper presented at the Second World Conference of Investment Casting, Investment Casting Institute, British Investment Casters' Technical Association, European Investment Casters' Federation, June 1969
66. S. Uram, ''Commercial Applications of Ceramic Cores,'' Paper presented at the 26th Annual Meeting, Investment Casting Institute, Oct 1978
67. P. Breen, ''The Dewaxing and Firing of Ceramic Shell Moulds,'' Paper presented at the 12th Annual Conference, Eastbourne, England, British Investment Casters' Technical Association, May 1973
68. J.R. Keough, ''An Innovative Approach to Mold Firing,'' Paper presented at the 33rd Annual Meeting, Investment Casting Institute, 1985
69. L.R. Ceriotti and J.J. Przyborowski, ''Use of Ceramic Recuperators on Investment Casting Mold Burnout Furnaces,'' Paper presented at the 33rd Annual Meeting, Investment Casting Institute, 1985
70. A.W. Merrick, Founding Apparatus and Method, U.S. Patent 2,125,080, 1938
71. R.E. Beal, F.W. Wood, J.O. Borg, and H.L. Gilbert, ''Production of Titanium Castings,'' Report RI 5625, U.S. Bureau of Mines, Aug 1956
72. P. Magnier, ''Titanium Investment Castings for Aircraft Applications,'' Paper presented at the 12th Annual Conference, Eastbourne, England, British Investment Casters' Technical Association, May 1973
73. J.D. Jackson and M.H. Fassler, ''Developments In Investment Casting,'' Paper presented at the 33rd Annual Meeting, Investment Casting Institute, 1985
74. H.T. Bidwell, Machining Investment Castings, *Machinery*, Jan 1970, p 60
75. H.T. Bidwell, Design and Application of Investment Castings, in *Investment Casting*, Machinery Publishing, 1969
76. R.H. Herrmann, Ed., *How To Design And Buy Investment Castings*, Investment Casting Institute, 1959
77. C.H. Watts and R.A. Horton, Method of Investment Casting, U.S. Patent 3,424,227, 1966
78. C.H. Watts and R.A. Horton, Methods of Casting, U.S. Patent 3,336,970, 1967

Replicast* Process

R.E. Grote, Missouri Precision Castings
Thomas S. Piwonka, University of Alabama

THE REPLICAST PROCESS was developed to overcome several shortcomings of the early evaporative pattern (lost foam) casting processes—primarily the formation of lustrous carbon defects in the castings and carbon pickup in steel castings. The process has two variations:

- Replicast FM (full mold), which is nearly identical to the lost foam process
- Replicast CS (ceramic shell), which is similar to investment casting in that it employs a ceramic shell to surround the pattern

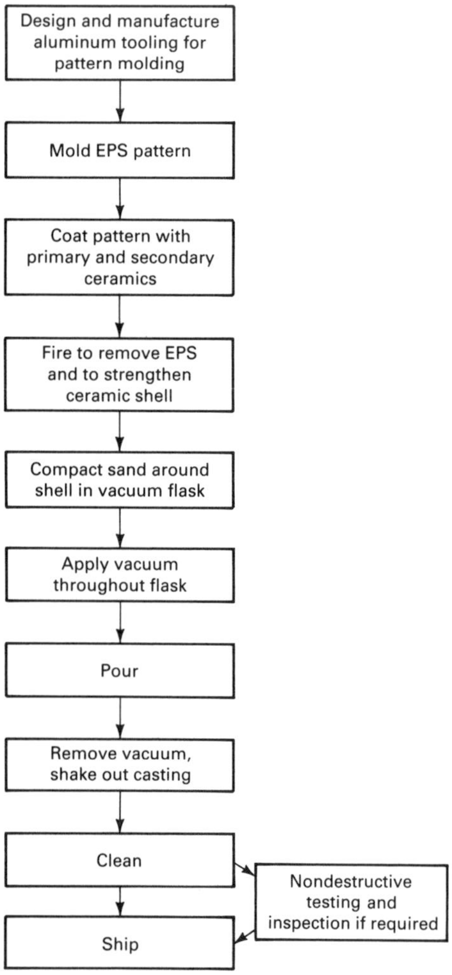

Fig. 1 Flow chart of the Replicast CS process. Source: Ref 1

The difference between the Replicast CS process and investment casting is that in the Replicast process the pattern is made from expanded polystyrene (EPS) and is surrounded by a much thinner shell than in investment casting. Figure 1 shows a flow chart of the Replicast CS process, and Table 1 compares the essential features of the Replicast CS and investment casting processes.

The Replicast CS process is the one that is commonly referred to as Replicast. More information on lost foam casting is available in the article "Sand Molding" in this Volume (see the discussion of the evaporative foam process in the section "Unbonded Sand Molds").

Process Details

Pattern Production. Aluminum tooling must be developed in order to produce the EPS pattern. The tooling is made from cast aluminum and incorporates risers, cored passages, and so on, wherever possible. The necessity of incorporating styrofoam fill openings, vents, steam, air, and water lines makes the tooling complex and more expensive than metal green sand equipment. However, the use of insert-type tooling can significantly reduce tooling costs.

Patterns are produced from size "T" high-density polystyrene beads. The beads are preexpanded using steam or vacuum, passed into the aluminum tooling, and expanded again to fill the tool cavity and to bond with each other. The high-density EPS used in the Replicast process yields a better casting surface finish than the low-density material used for lost foam patterns.

The following steps are required for the production of EPS patterns:

- Close and hydraulically clamp molding box halves
- Blow in preexpanded styrofoam beads
- Close off styrofoam fill guns, then inject steam to expand and bond the beads
- Stop the steam and run water through cooling lines

Table 1 Essential features of the Replicast CS and investment casting processes

Feature	Replicast CS	Investment casting
Pattern manufacture	Partially expanded EPS beads are blown into aluminum tooling and completely expanded. Finished patterns are lightweight, have high density, and provide good surface finish and excellent dimensional accuracy.	Softened wax is injected at high pressure into a metal tool. The wax is subject to shrinkage and deformation, and it is expensive and heavy. It is reclaimable to some degree.
Shell manufacture	Successive coats of refractory slurry and stucco are applied. Three or four coats are required, resulting in a relatively light and easy-to-handle shell. Firing at 925–1000 °C (1700–1830 °F) for 5 min removes the EPS pattern and hardens the shell.	Successive coats of refractory slurry and stucco are applied. Five to ten coats are required; completed shells are often heavy and difficult to handle. Firing at 1000 °C (1830 °F) for 20 min removes the residual wax and hardens the shell.
Pouring	Thin ceramic shell is surrounded by loose sand vibrated to maximum bulk density, and the vacuum is applied during pouring to prevent shell breakage.	Metal is frequently poured into hot, unsupported shells; breakage is possible.

Source: Ref 2

*Replicast is a registered trademark of the Steel Castings Research and Trade Association (SCRATA), Sheffield, England.

Fig. 2 Completed EPS pattern assemblies ready for ceramic coating. Courtesy of R.E. Grote, Missouri Precision Castings

- Open the mold box
- Eject pattern from cavity

The finished EPS pattern has a density of 10.75 to 12.90 g/cm^3 (2.5 to 3.0 lb/ft^3) and a smooth, hard surface. Minimal shrinkage of the EPS pattern occurs with age. Coating the EPS pattern with ceramic immediately after fabrication produces no detrimental effects.

Pattern Assembly. In some cases, the EPS pattern is complete as-molded and is sent to the ceramic coating area. More frequently, additional work is needed to produce a complete pattern assembly. Risers can be glued to the pattern, and ceramic pouring cones can be added. Gluing is done by spray, brush, or glue gun. A low ash content hot-melt glue is used for brush or gun application. A low-melting wax is used to fill and smooth the glue joints. Figure 2 shows completed pattern assemblies ready for the ceramic coating area.

Ceramic Coating and Firing. The next step in the Replicast process is the formation of a thin ceramic shell over the EPS pattern assembly. This is done by using techniques similar to those found in investment casting. The pattern assembly is first dipped into a ceramic slurry and then stuccoed using a granular refractory in a fluidized bed or rain sander. The shell is allowed to air dry for a minimum of 1 h. This process is repeated until the desired shell thickness is achieved. Thickness depends on product size, shape, and section thickness; shell thicknesses of 3.2 to 4.8 mm (1/8 to 3/16 in.) are common.

Completed shells are then fired at 925 to 1000 °C (1700 to 1830 °F) to remove the EPS pattern material completely and to harden the ceramic. The shell is then embedded in an unbonded sand mold to support the thin ceramic shell and to prevent breakage. This allows the use of shells thinner than those employed for investment casting.

Pouring. The completed ceramic shells are transported to the foundry, where several shells are placed into a molding flask. The flask incorporates a plenum chamber through which air can be extracted to create a vacuum. The flasks are filled with unbonded loose sand, which is vibrated to achieve maximum bulk density. A vacuum is applied to the flask immediately before pouring. Pouring is then accomplished using conventional techniques. The vacuum can be switched off a few minutes after pouring.

Cleaning. After cooling, the castings are shaken out and moved to a cleaning area. The brittle ceramic shell fractures easily and tends to break off the casting surface. After initial abrasive blast cleaning, conventional cleaning techniques are implemented (see the section "Blast Cleaning of Castings" in the article "Processing of Castings" in this Volume).

Process Capabilities

Outstanding features of the Replicast process include:

- Surface finish comparable to that obtainable in conventional investment casting
- Elimination of cores through the use of core inserts in patternmaking tooling; patterns can often be produced in one piece, including risers
- Improved casting yields because of the elimination of downsprues, runners, and so on
- High as-cast quality levels with regard to casting integrity and dimensional accuracy

Dimensional Accuracy

The dimensional accuracy of Replicast parts is comparable to that of investment castings. Based on measurements of Replicast parts of various sizes, the following tolerances on linear dimensions are recommended:

Dimension, mm (in.)(a)	Tolerance, mm (in.)
2.5–102 (0.1–4)	±0.25 (±0.010) or 0.75% of dimension, whichever is greater
104–305 (4.1–12)	±0.76 (±0.030)
307–610 (12.1–24)	±1.52 (±0.060)

(a) Nonrisered surfaces

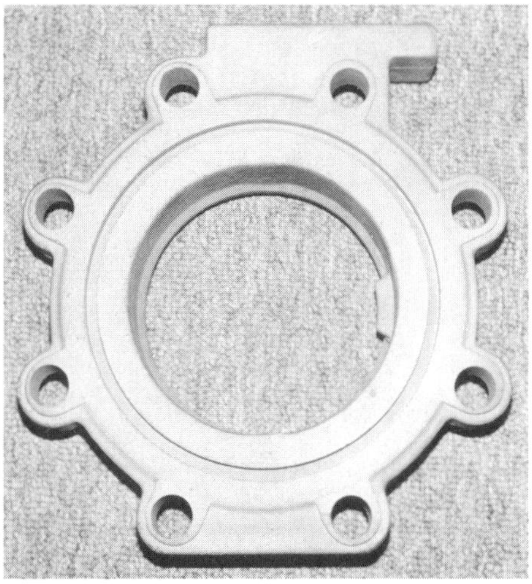

(a)

(b)

Fig. 3 Two applications of Replicast steel parts. (a) CF-8M stainless steel 150 mm (6 in.) butterfly valve body. The part is approximately 305 mm (12 in.) in outside diameter, 64 mm (2 1/2 in.) thick, and weighs 11 kg (25 lb). Note as-cast bolt holes and O-ring groove. (b) 8640 steel drive sprocket for armored personnel carrier. The part is 510 mm (20 in.) in outside diameter, 114 mm (4 1/2 in.) thick, and weighs 28 kg (62 lb). Sprocket teeth are used as-cast, resulting in a significant savings in machining costs. Courtesy of R.E. Grote, Missouri Precision Castings

With no wear occurring on tooling, long-term dimensional reproducibility is readily obtainable.

The smooth surface finish of Replicast parts, combined with near zero draft molding capabilities, can result in the elimination of considerable machining. Cast bolt holes, virtually flat flange faces, and blind and through holes as small as 9.6 mm (3/8 in.) in diameter can be attained. For those surfaces requiring finish machining, an allowance of 1.6 to 3.2 mm (1/16 to 1/8 in.) is usually sufficient.

Applications

Because the EPS pattern material is completely removed from the ceramic shell during firing, the Replicast process is suitable for pouring any metal or alloy without fear of carbon pickup. The process is commonly used to produce carbon, low-alloy, and stainless steel castings. Figure 3 illustrates two typical applications.

REFERENCES

1. R.E. Grote, Replicast Ceramic Shell Molding: The Process and Its Capabilities, *Trans. AFS*, Vol 94, 1986, p 181-186
2. M.C. Ashton, S.G. Sharman, and A.J. Brookes, The Replicast FM (Full Mold) and CS (Ceramic Shell) Process, *Trans. AFS*, Vol 92, 1984, p 271-280

Rammed Graphite Molds

Dan Kihlstadius, Oregon Metallurgical Corporation

RAMMED GRAPHITE is used to produce molds for the casting of reactive metals and alloys such as titanium and zirconium (see the articles "Titanium and Titanium Alloys" and "Zirconium and Zirconium Alloys" in this Volume). Because such metals and alloys react vigorously with silica, conventional molding sands cannot be used.

Production of Rammed Graphite Molds

Mold Mixture. Rammed graphite molds are manufactured from a mixture consisting primarily of finely divided graphite having a closely controlled particle size and size distribution. To this mixture, nominally 10% water, 10% pitch, 7% baumé syrup, and 3% starch are added. These ingredients are required for good coating of the graphite grains and for the development of optimal mold properties. The mold mixture has the appearance of green sand (Fig. 1). The mold mixture is pneumatically tamped (rammed) around a pattern, stripped from the pattern, air dried, baked, and fired.

Air drying influences both the green and baked strengths of the mold (Fig. 2). Molds are air dried on racks; after 24 h of air drying, the mold reaches about 13% of its maximum green strength. Maximum green strength is achieved after 94 h of air drying.

Baking. After air drying, molds are baked at 175 °C (350 °F). Baking time depends on mold section thickness and geometry. All moisture must be removed during baking to prevent steam generation and potential mold cracking during firing.

Firing removes all of the remaining volatile carbonaceous materials from the mold and develops final bonds. The strength and hardness of the mold increase with firing temperature (Fig. 3); best results are obtained at minimum firing temperatures of 870 °C (1600

Fig. 1 Graphite mold mixture, which resembles green sand in appearance

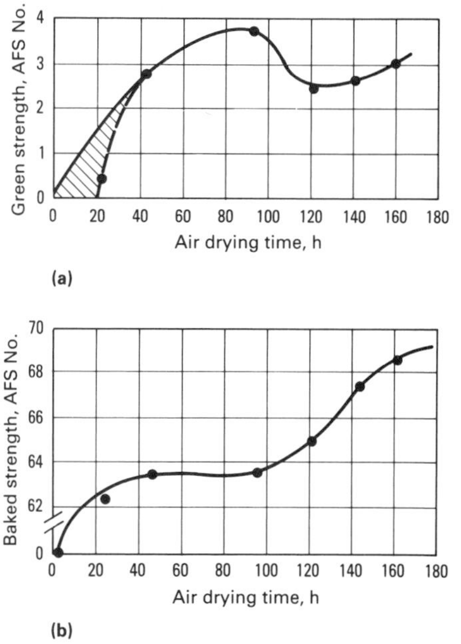

Fig. 2 Effect of air drying time on green strength (a) and baked strength (b) of graphite molds

°F). Firing is performed in a reducing atmosphere to prevent oxidation of the graphite. The fired mold is rigid and is similar to ceramic in texture and hardness (see the article "Ceramic Molding" in this Volume).

Storage. After firing, molds should be stored in a controlled environment (minimum temperature of 25 °C, or 75 °F, and relative humidity below 40%) to prevent moisture absorption. Moisture absorption is particularly rapid at high relative humidities and low temperatures.

Molding Problems

Although a correct binder system can be developed for any graphite particle size distribution, mold permeability will vary with the type and amount of binder used. Mold shrinkage can be predicted and controlled if ramming pressure and binder composition are constant.

Graphite mixing problems may also be related to the pitch and syrup ingredients. Good mixing should result in uniform coating of the graphite grains with pitch and the water-syrup mixture. Pitch or syrup in excess of that required for coating of the graphite particles may cause balling of the mold mixture. Pitch balls are aggregates of excess graphite and graphite fines containing a greater-than-average amount of syrup. Mold mixtures become more susceptible to balling as the percentage of reclaimed graphite fines in the mixture increases. Therefore, control of graphite particle size distribution is crucial.

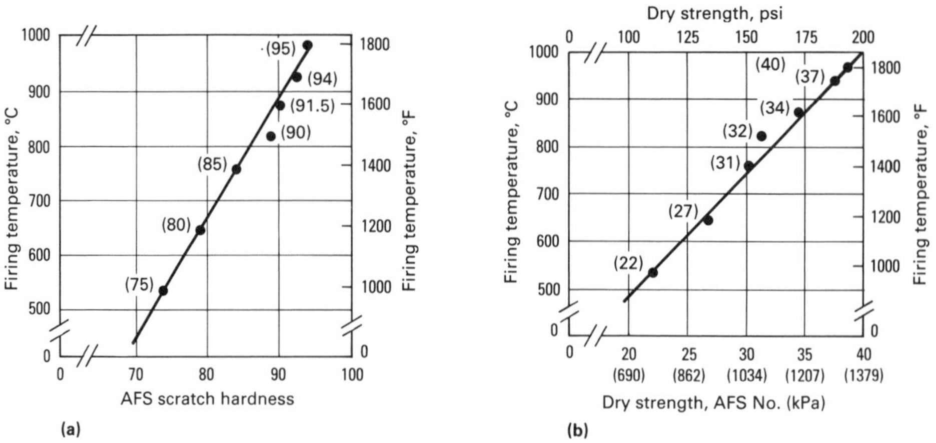

Fig. 3 Effect of firing temperature on the hardness (a) and strength (b) of graphite molds

Permanent Mold Casting

Revised by Charles E. West and Thomas E. Grubach, Aluminum Company of America

IN PERMANENT MOLD CASTING, sometimes referred to as gravity die casting, a metal mold consisting of two or more parts is repeatedly used for the production of many castings of the same form. The liquid metal enters the mold by gravity. Simple removable cores are usually made of metal, but more complex cores are made of sand or plaster. When sand or plaster cores are used, the process is called semipermanent mold casting.

Permanent mold casting is particularly suitable for the high-volume production of castings with fairly uniform wall thickness and limited undercuts or intricate internal coring. The process can also be used to produce complex castings, but production quantities should be high enough to justify the cost of the molds. Compared to sand casting, permanent mold casting permits the production of more uniform castings, with closer dimensional tolerances, superior surface finish, and improved mechanical properties. Figure 1 shows castings made by the permanent mold process.

Permanent mold casting has the following limitations:

- Not all alloys are suitable for permanent mold casting
- Because of relatively high tooling costs, the process can be prohibitively expensive for low production quantities
- Some shapes cannot be made using permanent mold casting, because of parting line location, undercuts, or difficulties in removing the casting from the mold
- Coatings are required to protect the mold from attack by the molten metal

Metals that can be cast in permanent molds include the aluminum, magnesium, zinc, and copper alloys and hypereutectic gray iron. Practical sizes of permanent mold castings differ according to materials cast, part configuration, and number of parts needed.

Aluminum Alloys. In high production, permanent mold castings weighing up to 70 kg (150 lb) have been made from aluminum alloys in casting devices. However, much larger castings can be produced. For example, aluminum alloy engine blocks with a trimmed weight of 354 kg (780 lb) have been produced in a four-section permanent mold having a vertical parting line.

Magnesium alloys, despite their comparatively low castability, have been cast in permanent or semipermanent molds to produce relatively large and complex castings. For example, an 8 kg (17.7 lb) housing for an emergency power unit was poured from Alloy AZ91C in a semipermanent mold. In another application, 24 kg (53 lb) spoolhead castings 760 mm (30 in.) in diameter were produced from Alloy AZ92A in a two-segment permanent mold with vertical parting.

Copper alloy permanent mold castings weighing more than 9 kg (20 lb) can rarely be justified.

Gray Iron. The production of gray iron castings in permanent molds is seldom practical when the castings weigh more than 13.6 kg (30 lb).

Casting Methods

Manually operated permanent molds may consist of a simple book-type mold arrangement (Fig. 2a). For castings with high ribs or walls that require mold retraction without rotation, the manually operated device shown in Fig. 2(b) can be used. With either type of device, the mold halves are separated manually after releasing the eccentric mold clamps.

Semiautomatic Devices. For high-volume production, manual drives are replaced by two-way air or hydraulic mechanisms. These units can be programmed to open and close in a preset cycle. Therefore, the operation is automatic except for pouring of the metal and removal of the castings.

Figure 3 shows an automatic casting device equipped with automatic mold, core, and insert-setting components; the only manual operations are pouring of the metal and placing the inserts in the setter. For simple castings, this type of device can be fully automatic.

Devices for Horizontal Parting and Tilt Casting. The mold parting for the casting devices shown in Fig. 2 and 3 is in a vertical plane, which is often the preferred position for mold opening and casting removal. Many castings, however, are best poured with the parting in a horizontal plane. Some castings with horizontal parting are poured using devices with a tilting mechanism; thus pouring is done with the parting in the horizontal plane, and the mold position is then changed to permit removal of the casting with parting in the vertical plane.

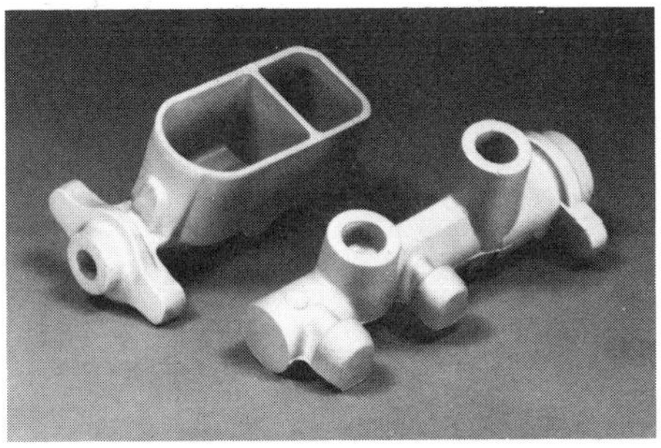

(a)

(b)

Fig. 1 Two examples of aluminum alloy castings made in permanent molds. (a) Alloy 356 brake master cylinder. (b) Alloy 356 basketball backboard. Courtesy of Stahl Specialty Company

Some castings are partially poured with the parting in the horizontal plane and then slowly rotated while the pouring is completed. For example, in casting a frame for a duplicating machine, pouring with the parting in a vertical plane required a long drop of molten metal, which resulted in splashing and severe turbulence and produced unacceptable castings. To eliminate the long drop of molten metal, the mold was placed in a tilt device that allowed the metal to be introduced into the mold with the parting oriented horizontally. The main body of the casting was poured with the mold in this position. The lower offset segment of the casting was then gently filled as the mold was rotated 90° to place the parting in a

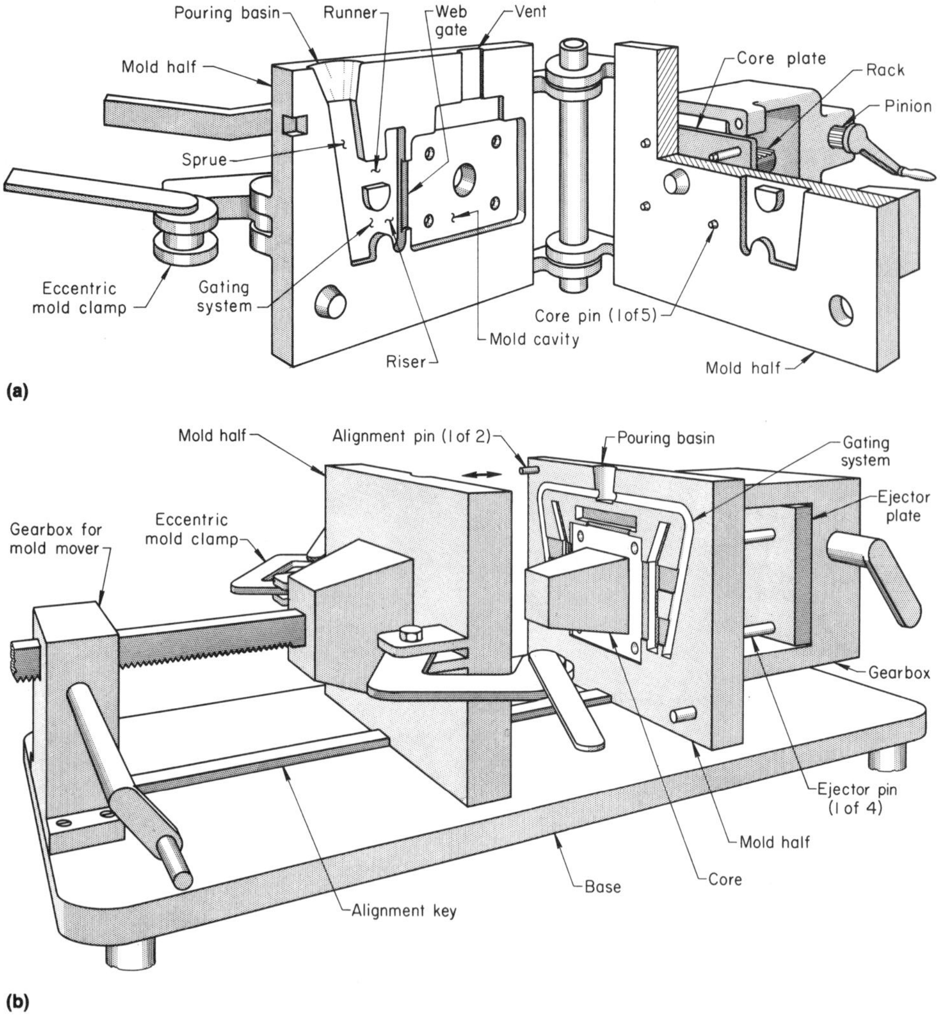

Fig. 2 Two types of manually operated permanent mold casting machines. (a) Simple book-type mold for shallow-cavity castings. (b) Device with straight-line retraction for deep-cavity molds

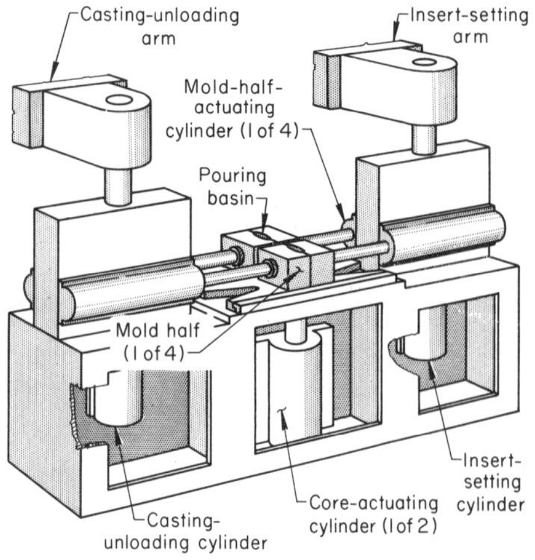

Fig. 3 Automatic permanent mold casting machine with a setter for cast-in inserts

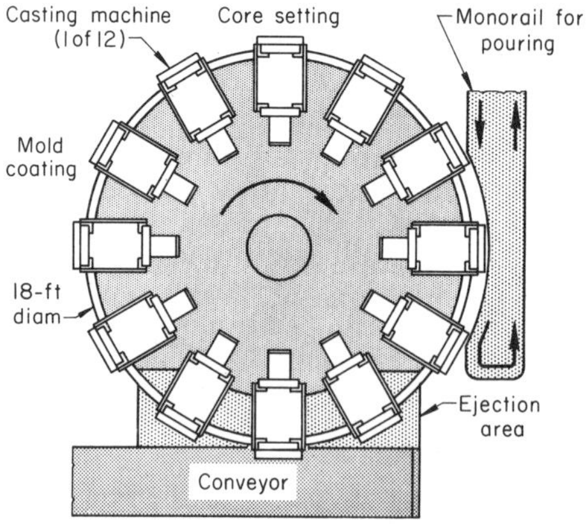

Fig. 4 Schematic of a 12-machine turntable for automatic permanent mold casting

vertical plane. At this position, the casting was allowed to solidify and was then removed from the mold.

Turntables. Small, lightweight castings can be poured and removed manually, but manual handling becomes increasingly difficult as pouring temperatures rise and casting weight increases. The casting process must then be automated. This is often done by using casting machines mounted on a turntable.

A commonly used turntable accommodates 12 casting devices, as shown in Fig. 4, and completes one revolution in 2 to 7 min. Steps in the casting process, which include pouring, mold coating, core setting, solidification, and ejection, are completed progressively as the casting devices pass through the various work stations. With this type of equipment, maximum production rates are achieved when all 12 devices have identical molds. However, different molds can be used in all 12 devices.

In low-pressure die casting, the mold is placed in a casting device above a sealed airtight chamber that contains a crucible holding molten metal (Fig. 5a). A fill tube extends from the mold down into the molten metal bath. The casting is made by pressurizing the chamber and forcing the metal up into the mold. The metal in the fill tube acts as the riser; this gives low-pressure die casting very favorable casting yields. The low-pressure method lends itself to automation, and it usually runs at lower mold temperatures and with shorter cycle times than conventional gravity-poured permanent mold methods. The rapid solidification rates associated with low-pressure casting result in castings with finer grain size, smaller dendrite arm spacings, and enhanced mechanical properties.

Vacuum casting is similar to low-pressure casting, except a vacuum is created within the mold cavity and the metal is pulled rather than pushed into the mold (Fig. 5b). Excellent mechanical properties and high production rates are often realized in vacuum casting because of the low mold temperatures associated with the method. As with low-pressure die casting, the metal in the fill tube acts as a riser, and excellent metal yields are obtainable. The process lends itself to automation, and the result is the ability to produce large quantities of high-quality castings at a competitive price. The process is usually associated with smaller castings and requires specialized, complex mold designs to induce the vacuum properly.

In constant-level pouring, the mold is placed in a device, and as the metal is poured into the mold, the mold is lowered at a rate consistent with the rate of fill, effectively pouring the metal at one constant level. If the metal source, rather than the mold, is in the device, the metal source is raised and the mold remains stationary. This method virtually ensures lamellar flow and greatly reduces oxide formation.

In centrifugal casting, cylindrical or symmetrically shaped castings are poured using the centrifugal force of a spinning mold to force the metal into the mold. The sprue is located at the center of rotation. The force generated by the spinning of the mold helps the metal fill thin casting sections and maintains good contact between the metal and the mold. This provides a

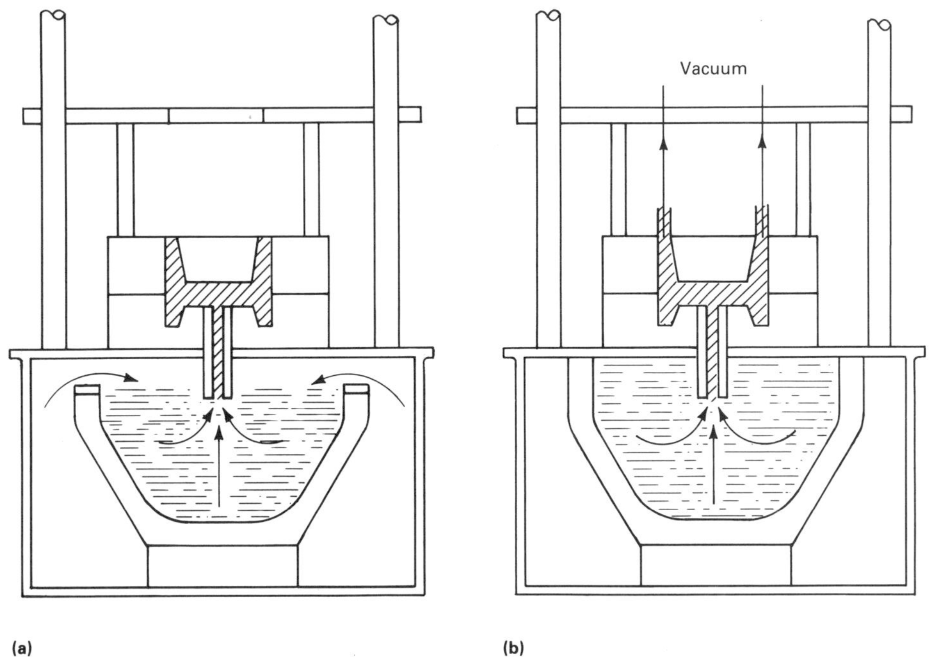

Fig. 5 Schematics of low-pressure (a) and vacuum casting (b) units used with permanent molds

higher rate of heat flow and a more rapid solidification rate, resulting in increased mechanical properties. Multiple molds can be used in centrifugal casting, and care should be taken to design gating systems that produce little or no turbulence. Segregation can be a problem with alloys containing immiscible or heavy elements, such as tin in the 850 aluminum alloys. These alloys require special casting practices. More information on centrifugal casting is available in the article so titled in this Volume.

Squeeze casting is another variation of the permanent mold process. It consists of pouring a specific amount of metal into the lower half of a mold, closing the mold, and then allowing the metal to solidify under pressure. One of the advantages of squeeze casting is the casting yield, because no gating system is required. More information on this process is available in the section "Squeeze Casting" of the article "New and Emerging Processes" in this Volume.

Continuous casting is usually considered an ingot-making process (see the article "Continuous Casting" in this Volume). However, it can also be used to make small, solid cylindrical castings. It is primarily used with the 850 series aluminum-tin alloys. Continuous casting can be done either horizontally or vertically, but vertical casting is the most common.

The process consists of a highly chilled mold with a movable base. The metal is poured into the mold as the base is lowered. As the solidified portion of the casting is lowered out of the mold, the water cooling the mold flows onto the casting, increasing the solidification rate of the casting. The molten metal flowing into the top of the mold acts as the casting riser. The lack of a gating and risering system provides for a good casting yield, and the rapid solidification rate gives continuous cast material good mechanical properties.

Mold Design

A simple permanent mold is shown in the book-type casting device illustrated in Fig. 2. Here the two mold halves are hinged on a pin and aligned. The mold cavity with the mold halves closed determines the shape of the casting. The casting is poured by means of the sprue and runners to the riser, which is provided with a web gate to the mold cavity. The cavity is vented to allow air to escape. The plate-shape cavity shown in Fig. 2 required five core pins, which were moved by means of the manually driven gearbox mounted on the back of the right-hand mold half.

In operation, the mold halves are closed and locked. Metal is then poured to fill the gating system and the mold cavity. After the metal has solidified, the mold is opened, leaving the casting on the core pins. The core pins are withdrawn, and the casting is removed manually.

The mold shown in Fig. 2 is designed with the parting vertical and in a single plane. This mold could also be designed for horizontal parting or with parting in two or more planes, and instead of side gating, it could have bottom gating (Fig. 6a).

The mold shown in Fig. 2(b) is also designed for vertical parting and side gating. However, because of the deep cavity and correspondingly long core required, a hinged-type mold cannot be used. The mold shown is opened and closed by straight-line movement of one mold half to and away from the other mold half, which remains fixed.

Undercuts on the outside of a casting complicate mold design and increase casting cost because additional mold parts or expendable cores are needed. Complicated and undercut internal sections are usually made more easily with expendable cores than with metal cores, although collapsible steel cores or loose metal pieces can sometimes be used instead of expendable cores.

Isolated heavy sections completely surrounded by thin areas should be avoided. Thin deep ribs should also be avoided because they are likely to cause cold shuts and misruns. Adequate draft must be allowed in order to prevent ribs from sticking and breaking off in the mold.

Casting Ejection. Only the most simple permanent mold castings can be ejected from the mold with no mechanical help. Most castings are ejected by well-distributed ejector pins, or are confined in one mold half during opening of the mold and then ejected by the withdrawal of the retaining core or cores. It is important that the casting remain in the correct mold half until ready for ejection.

The number of castings per mold is a major consideration in designing the mold; the objective is to have the optimum number of cavities per mold that will yield acceptable castings at the lowest cost. Except for very small and thin castings, the machine cycle time increases as the weight of the metal being cast per mold increases. However, these increases are not directly proportional. A mold with the maximum number of cavities will often produce more castings per unit of time than a mold with a smaller number of cavities that was designed to operate on a shorter cycle. This is because there is a minimum solidification time for every casting, regardless of the number of cavities in the mold. The number of rejects sometimes increases as the number of cavities is increased, but this is usually offset by the greater productivity. For relatively simple castings, the cavities can be placed one above the other. This permits maximum use of the face area of the mold. However, for more complex castings, especially those for which there are significant projections in the cavities, it is usually necessary to gate each cavity individually.

For low-pressure casting, the mold parting plane will be horizontal, and the gating system shown in Fig. 6(a) will be replaced by the system shown in Fig. 6(b). In Fig. 6(b), the metal enters the mold through a centrally located fill tube and is distributed to the casting cavities by the gating system shown. The gating system also acts as a riser. An alternate system (Fig. 6c) is often

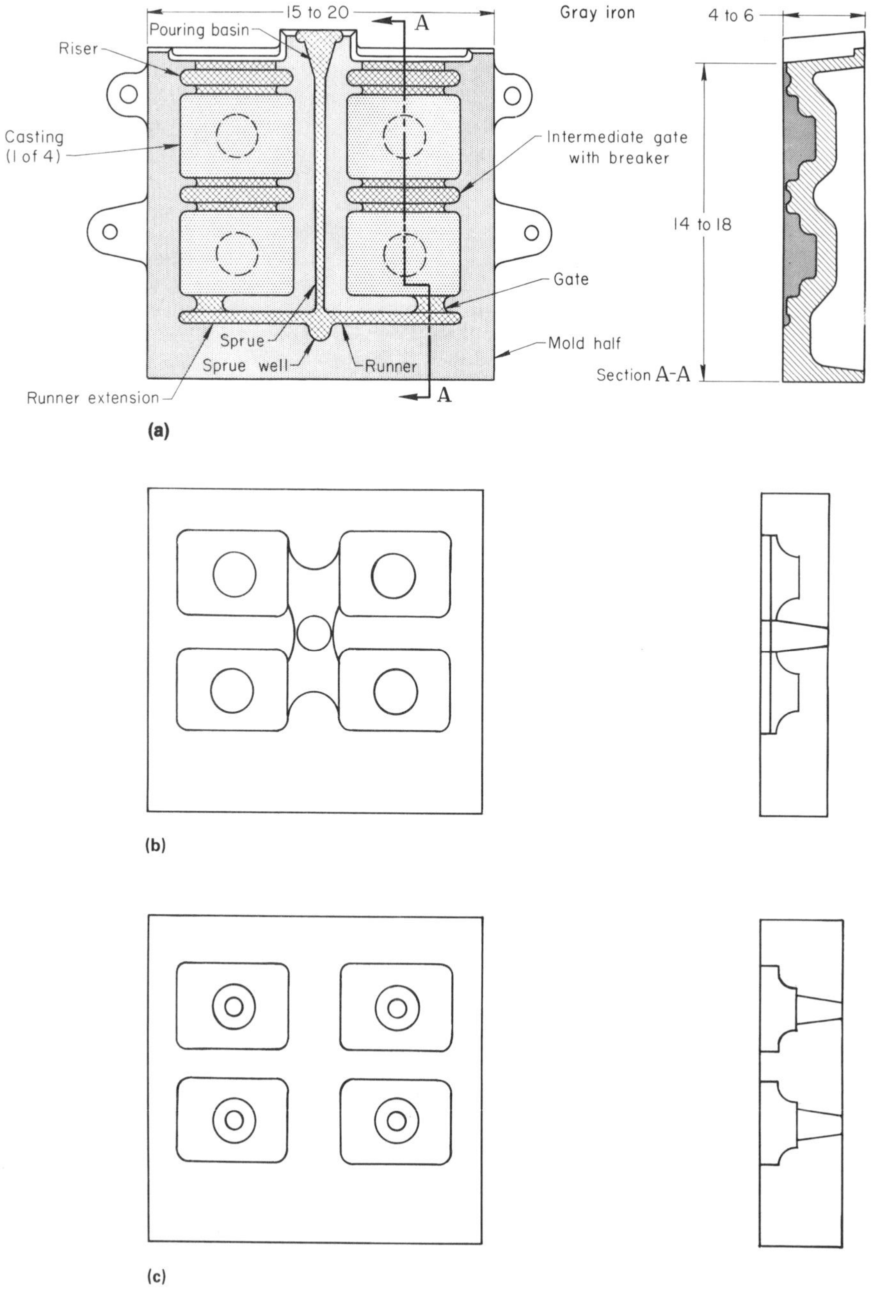

Fig. 6 Bottom-gated permanent mold (a) with stacked cavities for four castings. (b) Multicavity mold with low-pressure gating system. (c) Alternate gating system for low-pressure or vacuum casting

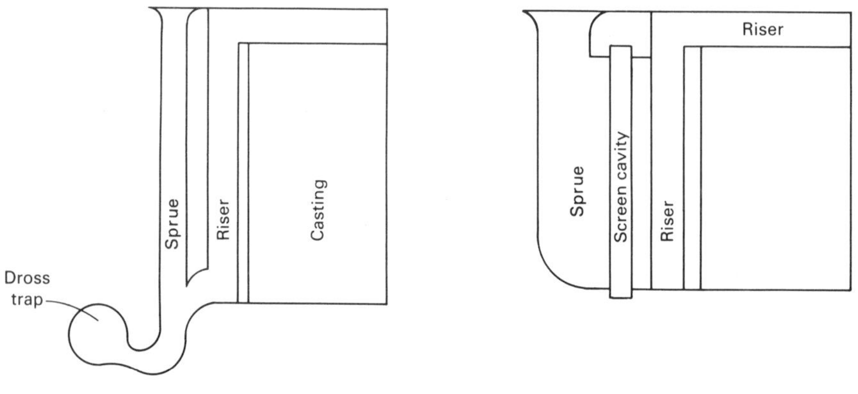

Fig. 7 Gating systems with dross traps or screens

used in low-pressure and vacuum casting. Here the metal enters the cavity directly, reducing the need for an additional gating system, decreasing feeding distances, and increasing casting yield.

Progressive Solidification. Alloys should be cast so that solidification takes place progressively toward the risers, which are generally to one side or on top of the casting. To achieve this solidification pattern, thinner sections of the casting should be away from the gating system, and heavy sections should be adjacent to it. Rib sections and thin walls vent and fill more easily when they are vertical, but filling a vertical mold cavity may promote turbulence in the molten metal, resulting in excessive dross; consequently, the mold should be tilted when being poured. To avoid oxide contamination, dross traps or filters are placed in the gating system as shown in Fig. 7. Steel or fiberglass can be used to filter the metal and reduce metal velocity. Tilt molding can be used for large, vertically gated castings.

In designing a permanent mold, the part is laid out in the desired orientation, and the mold is designed around it, allowing sufficient space for gating, for the seal needed to prevent metal leakage, and for coring and mold inserts. It is common practice to contour the back of the mold so that its exterior conforms roughly with the cavity. This permits a more even temperature distribution and heat dissipation. For castings with heavy sections, the adjacent mold sections are generally heavier. For aluminum castings, a ratio of three or four mold wall thicknesses to one casting wall thickness is often used, but a mold wall this thin cannot always be used in making thin-wall castings without jeopardizing mold stability. Ribbing is often used to stiffen the mold structure, but excessive ribbing can cause distortion by increasing the temperature differential between inner and outer mold surfaces.

Temperature stability must also be considered in the design of molds for thin-wall castings. Thin-wall castings are sensitive to temperature changes and will misrun readily. Consistent, relatively high mold temperatures are required; this necessitates the use of thicker, more massive molds.

Vents. The gap that exists between the mold halves after closing is sometimes large enough to permit air to escape and thus prevent misruns and cold shuts. Frequently, however, vents must be added to allow the air to escape as the mold is filled. Mold coatings can also be used for this purpose. Some form of venting also must be provided in die casting (see the article "Die Casting" in this Volume).

Gating Systems

Permanent mold castings can be gated from the top, side, or bottom. Single or multiple gates can be used.

Top Gating. In this arrangement, the sprue and the riser are usually the same. Thin sections are placed farthest from the gate so that directional solidification is toward the gate. After pouring, the gate functions as a riser. The metal in the riser solidifies last, thus ensuring sound metal throughout the casting.

Side gating is frequently used, particularly for aluminum castings. In this gating system, the riser is at the top of the casting. The gate extends up the side of the casting to nearly 90% of its height, which ensures that the metal at the top of the casting and in the riser is hotter than the first metal to enter the mold. Thin sections should be placed remote from the gate and riser. The direction of solidification is from the mold cavity toward the gate and riser so that shrinkage porosity is minimized.

Gating systems for permanent molds are less flexible than those for sand molds and are nearly always located in parting planes. Gating must supply metal fast enough to fill all sections of the casting with minimal turbulence. Gating systems that permit minimal turbulence are especially important in the casting of aluminum and magnesium alloys because turbulence creates excessive amounts of oxide, which may cause defective castings. Molds for aluminum and magnesium alloys are poured in the vertical position or tilted from vertical. With this method, air is readily displaced and vented off at the mold parting.

Bottom Gating. If bottom-gating systems are not properly designed, the last metal into the casting (the hottest metal) will be at the bottom of the casting. This will interfere with gravity feeding and progressive solidification and will produce castings with shrinkage porosity. The same is true, even with proper gating, if the casting is too thick. Bottom gating in conventional permanent molds is usually used on castings with thinner sections.

With low-pressure or vacuum casting, bottom gating is the most widely used system. The difference is that in these methods the metal is moved into the mold by a pressure differential. This pressure differential allows the metal to feed into the casting during solidification.

In addition to supplying liquid metal to compensate for casting shrinkage, risers reduce the velocity of the metal before it enters the cavity and help sustain the mold temperature. The number of points at which metal is admitted to the mold cavity depends on the section thickness and the distance the metal must flow. Excessive flow through too few inlets may result in hot spots and consequent shrinkage. Sprues are usually restricted in area to choke off and prevent dross and air from entering the cavity. Because gating systems are often easier to enlarge than to reduce and because oversize gating can slow the cycle, it is common practice to start with small gates and then to enlarge them if necessary.

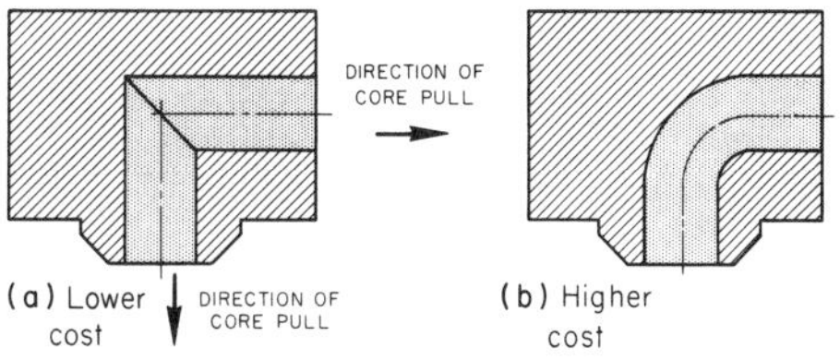

Fig. 8 Two types of cavities requiring different coring methods. (a) Corner radii must be sacrificed if metal cores are to be used. (b) Corner radii can be obtained with sand or plaster cores, but casting cost will be higher than in (a).

Misruns have several possible causes, including entrapped air, low mold temperature, or low pouring temperature. Misruns are often caused by a combination of two or more such conditions. Adequate venting is the simplest way to eliminate entrapped air. Increasing the mold temperature or the pouring temperature, or both, may eliminate misruns, but is likely to cause other defects, such as porosity. One common approach is to increase the mold temperature at critical locations by applying heat from an external source (antichill) or by using an insulating type of mold coating in the specific location of the mold to prevent the liquid metal from chilling in a specific area.

Shrinkage porosity associated with gating can be extremely difficult to eliminate because casting design sometimes precludes ideal metal feeding. This may necessitate complete redesign of the gating system and a change in the casting operation.

Dross. Gating systems often must be revised to eliminate excessive formation of dross. Fiberglass, metals, and ceramics are used in gating systems to remove dross, slow metal velocity, and help prevent further oxide generation.

Cores

The cores used in the permanent mold process can be of gray iron, steel, sand, or plaster. Metal cores can be movable or stationary. Stationary cores must be perpendicular to the parting line to permit removal of the casting from the mold, and they must be shaped such that the casting is readily freed. Metal cores not perpendicular to the parting line must be movable so that they can be withdrawn from the casting before it is removed from the mold.

When sand or plaster cores are used in a permanent mold, the process is referred to as semipermanent molding. With sand or plaster cores, which are expendable, more complex shapes can be produced than with metal cores.

The two cored passageways shown in Fig. 8 illustrate the limitations of cores in permanent mold castings. The cored passageway shown in Fig. 8(a) could be produced with metal cores. Sand or plaster cores would be needed to provide the radius of the passageway shown in Fig. 8(b), which would also be more expensive to produce.

Cored holes in permanent mold castings can usually be held to closer tolerances, on both size and location, than in sand castings. Movable cores and stationary cores can both be machined to close dimensions and can be accurately located. Draft requirements for holes formed by steel cores are:

Hole dimensions, mm (in.) Depth	Diameter	Draft allowance per side, degrees
6.4 (¼)	6.4 (¼)	10
12.7 (½)	25.4 (1)	5
25.4 (1)	51 (2)	3
51 (2)	102 (4)	3
102 (4)	203 (8)	3

These drafts apply to all nonferrous metals cast in permanent molds.

Steel cores require the same coating as that applied to the mold; the dimensional accuracy of cavities made from coated steel cores is affected by the same factors that affect the accuracy of dimensions formed by coated metal molds. Unless a core is stationary, the clearance that must be provided to permit its withdrawal from the mold permits core movement, which may affect dimensions.

In terms of dimensional accuracy, sand cores have approximately the same limitations in permanent molds as in sand molds. However, when a sand core is located in a sand mold, some mold material may be disturbed or displaced. Because a permanent mold is more rigid, it provides a more accurate seat for locating a core.

Sand cores permit the casting of tortuous passages or of chambers or passages that are larger in section than the opening through the wall of the casting. Because sand cores are expendable, their removal presents no major problem.

Plaster cores are used in semipermanent molds to provide a surface finish better than that obtainable with sand or coated metal cores. For example, in a wave-guide casting requiring a smooth inside surface, the use of permeable plaster cores provides a surface finish better than that which could be obtained with coated metal cores. Plaster also has excellent insulating qualities and does not prematurely chill the metal in thin sections. With regard to dimensional accuracy, plaster cores offer no superiority over sand cores.

Collapsible Cores. It is preferable to avoid designs that require the use of collapsible metal cores (multiple-piece cores) in permanent molds. These cores add to the cost of the castings, and their assembly and removal increase production time. Furthermore, dimensional variations can result from the use of collapsible cores because

Table 1 Recommended permanent mold materials

Casting alloy	Number of pours: 1000	10 000	100 000
For small castings (25 mm, or 1 in., maximum dimension)			
Zinc	Gray iron; 1020 steel	Gray iron; 1020 steel	Gray iron; 1020 steel
Aluminum, magnesium	Gray iron; 1020 steel	Gray iron; 1020 steel	Gray iron with AISI H14 inserts; 1020 steel
Copper	Gray iron	Gray iron	Alloy cast iron
Gray iron	Gray iron(a)	Gray iron(a)	Quantity not poured
For medium and large-size castings (up to 915 mm, or 36 in., maximum dimension)			
Zinc	Gray iron; AISI H11(b)	Gray iron; AISI H11(b)	Gray iron; AISI H11(b)
Aluminum, magnesium	Gray iron	Gray iron	Gray iron; AISI H11 or H14(c)
Copper	Alloy cast iron	Alloy cast iron	Alloy cast iron(d)
Gray iron	Gray iron(a)	Gray iron(a)	Quantity not poured

(a) Same composition as being poured. (b) AISI H11 is used when polish is required on medium-size castings. (c) For medium-size parts; recommended materials for large parts are gray iron with H11 inserts or solid H11 die steel. (d) For medium-size parts; large parts are not poured in this quantity.

Table 2 Recommended materials for small cores (<76 mm, or 3 in., in diameter and 255 mm, or 10 in., long) for permanent molds

Casting alloy	Recommended core materials(a)
Zinc	Sand, plaster, gray iron, 1020 steel
Aluminum, magnesium	1010 or 1020 steel, sand, plaster, H11 die steel or equivalent(b), carbon(c)
Copper	Sand, 1020 steel, gray iron, plaster(d), graphite(c)
Gray iron	Sand, graphite, carbon, gray iron

(a) Materials are listed in descending order of preference. (b) Hardened to 42–45 HRC. (c) For use with relatively few pours. (d) For casting of aluminum bronzes

they cannot be positioned as securely in the mold as single-piece cores or because of movement of core segments when the casting metal is poured. When castings must be designed to be made with collapsible cores, the designer should allow the loosest tolerances possible.

Despite the disadvantages of collapsible cores, they are extensively used in making certain castings. For example, nearly all of the aluminum pistons made for the automobile industry in the United States are cast in permanent molds using five-piece collapsible metal cores.

Selection of Mold and Core Materials

Four principal factors affect the selection of materials for permanent molds and cores:

- The pouring temperature of the metal to be cast
- The size of the casting
- The number of castings per mold
- Cost of the mold material

These variables form the basis for the recommendations given in Tables 1 to 3.

As indicated in Table 1, gray iron is the most commonly used mold material. Aluminum or graphite molds are sometimes used for the small-quantity production of aluminum and magnesium castings, and graphite or carbon liners on steel are sometimes used for molds for casting copper alloys (see the section "Solid Graphite Molds" in this article).

With aluminum or magnesium casting alloys, it is not unusual to obtain 100 000 castings, or more, per mold; however, molds for copper or gray iron casting alloys have a shorter life because of the higher pouring temperatures required. Gray iron molds without tool steel inserts are satisfactory for long production runs of aluminum and magnesium castings that will be machined extensively and for which surface finish is not a major consideration. In the casting of zinc, well over 100 000 pours are possible in a gray iron mold (die casting is usually selected to produce zinc castings in such large quantities).

Mold Inserts. Full or partial mold cavity inserts of the same material as the mold, or of a different material, are sometimes used to obtain longer mold life, or to simplify machining, handling, or replacement. Inserts can also be used for venting, cooling thin walls, and heating portions of the mold or the full cavity area. Inserts made of cast-to-shape gray iron are used for casting complex aluminum and magnesium parts that range in surface area from 320 to 2900 cm^2 (50 to 450 $in.^2$). Tolerances on these parts range from ±0.76 to ±1.5 mm (±0.030 to ±0.060 in.). Inserts last for 5000 to 20 000 pours, depending on casting complexity.

Core materials are recommended in Tables 2 and 3 on the basis of performance over a wide range of coring requirements for small and large cores. An expendable core is used when the location or shape of the core does not permit its removal from the casting or when an intricate design can be obtained at less cost with materials for such cores. These materials are listed below in order of increasing preference:

- Sand (oil-bonded or resin-bonded, shell, carbon dioxide-silicate)
- Plaster
- Graphite and carbon

Mold parts can be cast to shape when the castings to be made are not complicated. Cast molds are suitable for applications for which gray iron or AISI H11 die steel is recommended in Table 3. Such cast molds are most often used when production requirements are large and when mold replacement or shipping schedules require several of the same molds.

Machining of the mold cavity is often a more significant factor in the cost of a mold than the cost of the mold material. The relative machining costs of identical molds fabricated from different materials increase as shown below. In this listing, machining costs for a gray-iron mold are taken as 1.0, and the costs include heat treatment when required:

Gray iron	1.0
Cast low-alloy steel	1.44
Gray iron with tool steel inserts	1.73
Cast tool steel	2.16

Table 3 Recommended materials for large cores (>76 mm, or 3 in., in diameter and 255 mm, or 10 in., long) for permanent molds

Casting alloy	Material for indicated number of pours(a): 1000	10 000	100 000
Zinc	Gray iron, 1020 steel	Gray iron, 1020 steel	Gray iron, 1020 steel
Aluminum, magnesium	Gray iron, gray iron with 1020 steel inserts(b), sand, plaster(b)	Gray iron(b), gray iron with 1020 or H11 inserts(b), sand, plaster(b)	Gray iron(b), gray iron with H11 inserts(b), H11 die steel
Copper	Sand	Sand	Quantity not poured
Gray iron	Sand, graphite, carbon, gray iron	Sand, graphite, carbon, gray iron	Quantity not poured

(a) Materials are listed in descending order of preference. (b) Except for openings with complex shapes, which require expendable sand cores

Mold Life

Mold life can vary from as few as 100 pours to as many as 250 000 pours (or even more), depending on the variables discussed later in this section. A mold for an aluminum piston, for example, can be expected to produce 250 000 castings before requiring repair. After the production of 250 000 more castings, the repaired mold will require a major overhaul. With repeated repairs and overhauls, the mold can produce as many as 3.5 million castings before being discarded. However, a piston mold, with its relatively simple design, will have a much longer life than a mold that requires elaborate internal coring and external inserts, such as a cylinder-head mold.

Mold life is likely to be longer in the casting of magnesium alloys than in the casting of aluminum alloys of similar size and shape; this is because molten magnesium does not attack ferrous metal molds. However, the difference in mold life for magnesium alloys depends to a great extent on the effectiveness of the mold coating used. In the casting of gray iron, mold life is expected to be short compared to the casting of similar shapes from aluminum alloys.

Molds are often fabricated from cast iron because casting the mold close to the finished shape can decrease machining costs. In addition, cast iron is much more resistant to attack by molten aluminum than steel. Steel, however, is weldable and easier to repair than cast iron. Therefore, steel molds are often used for high-production castings.

Major variables that affect the life of permanent molds are:

- *Pouring temperature*: The hotter the casting metal is poured, the hotter the mold is operated, which leads to rapid weakening of the mold metal
- *Weight of casting*: Mold life decreases as casting weight increases
- *Casting shape*: Mold walls are required to dissipate more heat from castings having thick sections than from those having thin sections. When there is a significant variation in the section thickness of a casting, a temperature differential is set up among different portions of the mold. As the temperature differential increases, mold life decreases
- *Cooling methods*: Water cooling is more effective than air cooling, but it substantially decreases mold life
- *Heating cycles*: Generally, a continuous run, in which the mold is maintained at a uniform temperature, provides maximum mold life. Repeated heating and cooling over a wide temperature range will shorten mold life
- *Preheating the mold*: This is done to operating temperature with a gas flame or electric heaters, and it greatly increases mold life. Thermal shock is one of the principal causes of mold failure
- *Mold coating*: This protects the mold from erosion and soldering by preventing the metal from contacting mold surfaces, thus increasing mold life
- *Mold materials*: See Table 1
- *Storage*: Improper storage can lead to excessive rusting and pitting of mold surfaces, which will reduce mold life
- *Cleaning*: The common practices for cleaning molds are abrasive blasting, dipping in caustic solution, and wire brushing. Dipping in caustic can be hazardous to the operator. Wire brushing and abrasive blasting can cause excessive mold wear if not carefully controlled. Glass beads are the safest abrasive blast material; their use minimizes dimensional changes due to erosion from the abrasive blast
- *Gating*: A poor gating system can greatly reduce mold life by causing excessive turbulence and washout at the gate areas
- *Method of mold operation*: Although the same materials are used to make molds and cores for both automatically operated equipment and hand-operated equipment, the life of the tool materials on hand-operated equipment is shorter because of the abuse the tooling must withstand. Tools for automatic equipment may last up to twice as long as for hand-operated equipment
- *End use of casting*: If the structural function of a casting is more important than its appearance, a mold can be used for more pourings before being discarded

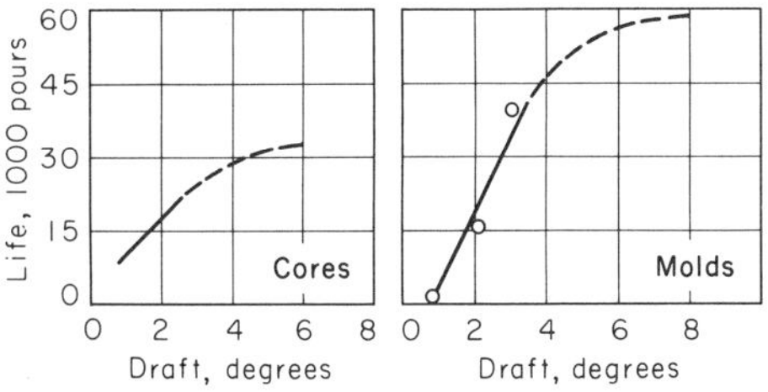

Fig. 9 Effect of draft angle on core and mold life in the permanent mold casting of aluminum alloys

Influence of Mold Design. In addition to the above factors, mold design has a marked effect on mold life. Variation in mold wall thickness causes excessive stress to develop during heating and cooling, which in turn causes premature mold failure from cracking. Abrupt changes in thickness without generous fillets also cause premature mold failure. Small fillets and radii lead to reduced mold life because checking and cracking, as well as ultimate failure, often start at these points.

Usually, less draft is required on external mold surfaces than on internal mold surfaces because of the shrinkage in the casting. A 5° draft is desirable, but 2° on external and 3° on internal mold surfaces can be used. Lower draft angles, however, decrease the number of castings that can be made between mold repairs. The effect of draft angle on the life of cores and molds used for producing aluminum alloy castings is shown in Fig. 9.

Projections in the mold cavities contribute greatly to reduced mold life. These projections become extremely hot, which increases the possibility of extrusion, deformation, and mutilation when the casting is removed. It is sometimes possible to extend mold life by using inserts to replace worn or broken projections.

Mold Coatings

A mold coating is applied to mold and core surfaces to serve as a barrier between the molten metal and the surfaces of the mold while a skin of solidified metal is formed. Mold coatings are used for five purposes:

- To prevent premature freezing of the molten metal
- To control the rate and direction of solidification of the casting and therefore its soundness and structure
- To minimize thermal shock to the mold material
- To prevent soldering of molten metal to the mold
- To vent air trapped in the mold cavity

Types. Mold coatings are of two general types: insulating and lubricating. Some coatings perform both functions. A good insulating coating can be made from (by weight) one part sodium silicate to two parts colloidal kaolin in sufficient water to permit spraying. The lubricating coatings usually include graphite in a suitable carrier. Typical compositions of 15 mold coatings are listed in Table 4. Coatings are available as proprietary materials.

The various requirements of a mold coating are not always obtained with one coating formulation. These requirements are often met by applying different coatings to various locations in the mold cavity.

Coating Requirements. To prolong mold life, a coating must be noncorrosive. It must adhere well to the mold and yet be easy to remove. It must also keep the molten metal from direct contact with the mold surfaces.

A mold coating must be inert to the cast metal and free of reactive or gas-producing materials. If insulation is needed to prevent thin sections, gates, and risers from solidifying too quickly, fireclay, metal oxides, diatomaceous earth, whiting (chalk), soapstone, mica, vermiculite, or talc can be added to the mold coating. Graphite is added if accelerated cooling is needed. Lubricants, which facilitate removal of castings from molds, include soapstone, talc, mica, and graphite (Table 4).

Coating Procedure. The mold surface must be clean and free of oil and grease. The portions to be coated should be lightly sand blasted. If the coating is being applied with a spray, the mold should be sufficiently hot (205 °C, or 400 °F) to evaporate the water immediately.

For optimum coating retention, a primer coat of water wash should be applied before spraying the mold coating. Water wash is a very dilute solution of a mold coating. Dilute kaolin makes an excellent primer. An acceptable alternative is a 20 to 1 dilution of the coating to be sprayed. The high water content of the water wash very lightly oxidizes the mold surface and provides a sub-

Table 4 Typical compositions of coatings for permanent molds

	Composition, % by weight (remainder, water)									
		Insulators				Lubricants				
Coating No.	Sodium silicate	Whiting	Fire-clay	Metal oxide	Diatomaceous earth	Soap-stone(a)	Talc(a)	Mica(a)	Graphite	Boric acid
1	2	...	4	...	...	...	...	...	1	...
2	8	...	4	...	...	...	...	...	...	...
3	...	7	...	...	...	...	...	...	...	7
4(b)	12	9	...	...	...	...	...	...	...	...
5	5	11	...	2	...	...	5	...	...	...
6	9	...	4	...	...	14	...	...	...	...
7	11	...	...	17	...	...	...	...	...	...
8	...	...	4	...	...	23	...	5	...	...
9	7	...	1	...	...	23	...	2	...	...
10	23	...	...	...	...	...	20	...	...	...
11	30	...	...	...	5	...	10	...	...	...
12	18	...	...	...	41	...	...	...	...	...
13	8	...	...	60	...	...	...	...	...	...
14	7	...	...	...	...	...	62	...	...	...
15	20	53	...	...	...	...	...	...	...	...

(a) Serves also as an insulator. (b) Plus silicon carbide, 2% by weight, for wear resistance

strate for subsequent layers to stick to. The water wash should be sprayed until the dark color of the mold starts to disappear. Lubricating materials or coatings are not acceptable as primers. Lubricants can be sprayed over insulating coatings, but insulating coatings will not adhere to lubricants.

The coating can be applied by spraying or brushing. It must be thick enough to fill minor surface imperfections, such as scratches. It should also be able to dry with a smooth texture on mold areas of light draft that form ribs and walls in the casting, and it must dry with a rough texture on large, flat areas of the mold to permit entrapped air to escape. The most pleasing cast surfaces are obtained when the coating has a matte or textured finish, which is most often obtained by spraying. Extremely smooth coatings should be avoided because they increase the formation of oxide skins. Thin successive layers are applied until the coating reaches the desired thickness, up to a maximum of 0.8 mm (1/32 in.).

Thick coatings are especially useful on the surfaces of sprues, runners, and risers because they provide more insulation than thinner coatings and result in slower metal freezing. However, they are more likely to flake off and should not be used on the surfaces of casting cavities. Thick coatings are applied by dabbing with a paint brush and adhere better if applied over an initial thin spray coat. It is mandatory that the coating be thoroughly dry before a casting is made, or an explosion will result.

Coating life varies considerably with the temperature of the metal being cast, the size and complexity of the mold cavity, and the rate of pouring. Some molds require recoating at the beginning of each shift; others may run for several shifts with only spot repairs or touchups before recoating is needed. Light abrasive blasting is used to prepare the coating for touchup or to remove old coats. To maintain maximum feeding with the mold, risers, runners, and gates should be recoated about every second time the casting cavity is recoated.

Mold Coatings for Specific Casting Alloys. The metal being cast has a major influence on the type of coating selected. Lubricating coatings are usually used for the casting of aluminum and magnesium. Relatively complex mixtures are sometimes used. For the casting of copper alloys, because of their high pouring temperatures and their solidification characteristics, an insulating type of mold coating is generally required.

The mold coatings used in the production of gray iron castings are divided into two categories: an initial coating, which is applied before the mold is placed in production, and a subsequent coating of soot (carbon), which is applied prior to each pouring. The initial coating consists of sodium silicate (water glass) and finely divided pipe clay, mixed in a ratio of about 1 to 4 by volume with enough water (usually about 15 parts by volume) to allow spraying or brushing. This mixture is applied to molds heated to 245 to 260 °C (475 to 500 °F).

The secondary coating is a layer of soot (carbon) deposited on the mold face and cavities each time the mold is to be poured. The soot is formed by burning acetylene gas delivered at low pressure (3.5 to 5.2 kPa, or 0.5 to 0.75 psig) so that a maximum amount of soot is produced and a minimum of heat is generated. It can be applied either manually or by automatic burners. This soot layer provides insulation between the mold and the casting, permitting easy removal of the castings from the mold, and it prevents chilling of the castings. It also provides a seal between the mold faces to minimize leakage. The thickness of the soot deposit is 0.10 to 0.25 mm (0.004 to 0.010 in.).

Mold Temperature

If the mold temperature is too high, excess flash develops, castings are too weak to be extracted undamaged, and mechanical properties and casting finish are impaired. When mold temperature is too low, cold shuts and misruns are likely to occur, and feeding is inhibited, which generally results in shrinkage, hot tears, and sticking of the casting to molds and cores.

The variables that determine mold temperature include:

- *Pouring temperature*: The higher the pouring temperature, the higher the temperature of the mold
- *Cycle frequency*: The faster the operating cycle, the hotter the mold
- *Casting weight*: Mold temperature increases as the weight of molten metal increases
- *Casting shape*: Isolated heavy sections, cored pockets, and sharp corners not only increase overall mold temperature but also set up undesirable thermal gradients
- *Casting wall thickness*: Mold temperature increases as the wall thickness of the casting increases
- *Mold wall thickness*: Mold temperature decreases as the thickness of the mold wall increases
- *Thickness of mold coating*: Mold temperature decreases as the thickness of the mold coating increases

After the processing procedure has been established for a given casting operation, mold coating, cycle frequency, chills, and antichills have significant effects on mold temperature. Mold coating is difficult to maintain at an optimum thickness, primarily because the coating wears during each casting cycle and because it is difficult to measure coating thickness during production. The most widely used method for controlling coating thickness is periodic inspection of the castings. Improper coating thickness is reflected by objectionable surface finish and loss of dimensional accuracy.

Preheating of Molds. In many casting operations, molds are preheated to their approximate operating temperature before the operation begins. This practice minimizes the number of unacceptable castings produced during establishment of the operating temperature.

Molds can be preheated by exposure to direct flame, although this method can be detrimental to the molds because of the severity and nonuniformity of heat distribution. Customized heaters are often built for molds. Preheating of the mold face in an oven is the best method because the thermal gradients are of smaller magnitude. Unfortunately, this is usually impractical for larger molds. Final mold operating temperatures are achieved after the first few production cycles.

Control of Mold Temperature

Optimum mold temperature is the temperature that will produce a sound casting in the shortest time. For an established process cycle, temperature control is largely achieved through the use of auxiliary cooling or heating and through control of coating thickness.

Auxiliary cooling is often achieved by forcing air or water through passages in mold sections adjacent to the heavy sections of the casting. Water is more effective, but over a period of time scale can coat the passages, thus necessitating frequent adjustments in water flow rates. Without cleaning, the flow of water eventually stops. Water passages should be checked and cleaned each time a mold is put into use.

The problem of scale formation has been solved in some plants by the use of recirculating systems containing either demineralized water or another fluid such as ethylene glycol. However, such systems are rarely used.

Water flow is regulated manually to each mold section with the aid of a flowmeter. A main shutoff valve is used to stop the water flow when the casting process is interrupted. Adjusting the rate of water flow to control the solidification rate of a heavy section permits some leeway in the variation of wall thickness that can be designed into a single casting. In addition to the control of water flow, the temperature of the inlet water (or any other coolant that might be used) affects the performance of the mold cooling system.

If water or another liquid coolant is used, it must never be allowed to contact the metal being poured, or a steam explosion will result. The intensity of a steam explosion increases as metal temperature increases. In addition, water will react chemically with molten magnesium.

A mold coating of controlled thickness can equalize solidification rates between thin and heavy sections. Chills and antichills can be used to adjust solidification

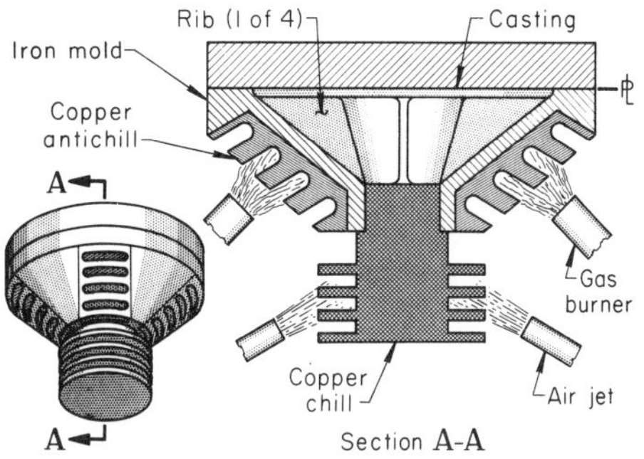

Fig. 10 Use of air-cooled chills and flame-heated antichills to equalize cooling rates in casting sections of varying thickness

rates further, so that freezing proceeds rapidly from thin to intermediate sections and then into heavy sections, and finally into the feeding system.

Chills are used to accelerate solidification in a segment of a mold. This can be done by directing cooling air jets against a chill inserted in the mold (Fig. 10) or, more simply, by using a metal insert without auxiliary cooling. Chilling can also be achieved by removing some or all of the mold coating in a specific area to increase thermal conductivity. Chills can be used to increase production rate, to improve metal soundness, and to increase mechanical properties.

Antichills. An antichill serves to slow the cooling in a specific area. Heat loss in a segment of a permanent mold can be reduced by directing an external heating device, such as a gas burner, against an antichill inserted in the mold (Fig. 10). The same effect can be produced by the use of insulating mold coatings.

Pouring Temperature

Permanent mold castings are generally poured with metal that is maintained within a relatively narrow temperature range. This range is established by the composition of the metal being poured, casting wall thickness, casting size and weight, mold cooling practice, mold coating, and gating systems used.

Low Pouring Temperature. If pouring temperature is lower than optimum, the mold cavity will not fill, inserts (if used) will not be bonded, the gate or riser will solidify before the last part of the casting, and thin sections will solidify too rapidly and interrupt directional solidification. Low pouring temperature consequently results in misruns, porosity, poor casting detail, and cold shuts. Sometimes only a small increase in pouring temperature is needed to prevent cold shuts.

High pouring temperature causes casting shrinkage and mold warpage. Warpage leads to loss of dimensional accuracy. In addition, variations in metal composition may develop if the casting metal has components that become volatile at a high pouring temperature. High pouring temperature also decreases solidification time (thus decreasing production rate) and almost always shortens mold life.

Pouring Temperatures for Specific Metals. The pouring temperature for aluminum alloys usually ranges from 675 to 790 °C (1250 to 1450 °F), although thin-wall castings can be poured at temperatures as high as 845 °C (1550 °F). Once established for a given casting, pouring temperature should be maintained within ±8 °C (±15 °F). If this control of pouring temperature cannot be maintained, the cooling cycle must be adjusted for the maximum temperature used. Internal mold cooling can be controlled by means of solenoid valves actuated by thermocouples inserted in the mold walls.

For magnesium alloys, the normal temperature range for pouring is 705 to 790 °C (1300 to 1450 °F). Thin-wall castings are poured near the high side of the range; thick-wall castings, near the low side. However, as for any permanent mold casting, pouring temperature is governed by the process variables listed in the section "Mold Temperature" in this article, and some experimentation is often required to establish the optimum pouring temperature for a specific casting. Once established, the pouring temperature should be controlled within ±8 °C (±15 °F).

Copper alloys are poured at 980 to 1230 °C (1800 to 2250 °F), depending on the alloy as well as the process variables discussed in the section "Mold Temperature" in this article. Once the temperature is established for a specific set of conditions, it should be controlled within ±15 °C (±25 °F).

The fluidity of gray iron is excellent, and little difficulty is experienced at pouring temperatures of 1275 to 1355 °C (2325 to 2475 °F). Excessive pouring temperatures can cause flashing and leaking due to mold distortion. As the pouring temperature increases, there is a rapid increase in defects caused by local hot spots on the cavity surface and insufficient soot coverage.

Because the temperature of the molten iron decreases considerably between the time that the first and last machines are serviced, it is usually necessary to deliver the metal to the casting area in a transfer ladle. The metal in this transfer ladle is delivered at a higher temperature than that suitable for pouring. To obtain the desired pouring temperature, small amounts of chill (foundry scrap of the same metal) are added to the pouring ladle as needed. If several machines are being serviced, the metal may have cooled sufficiently so that no chilling is required by the time the last machine is serviced.

Removal of Castings From Molds

After a casting has solidified, the mold is opened and the casting is removed. To facilitate release of the casting from the mold, a lubricant is often added to or sprayed over the mold coating. The use of as much draft as permissible on all portions of the casting facilitates ejection. For many castings, ejector pins or pry bars must be used. Core pins and cores should be designed so that they do not interfere with the removal of castings from the mold.

Aluminum alloy castings require at least a 1° draft for mechanical ejection from the mold prior to manual removal (the more draft, the easier the ejection). For castings with low draft angles, the mold coating usually contains a lubricating agent (usually graphite) to prevent sticking.

Magnesium alloy castings are subject to cracking when removed from the mold because the metal is hot short. Therefore, the use of adequate draft is mandatory. On ribs, a draft of 5° is an absolute minimum. However, 10° is recommended and will result in fewer ejection difficulties. In addition, because of the danger of cracking, extreme care should be taken to avoid side thrust when removing cores that must be retracted before the mold is opened.

Copper alloy castings will stick in the molds for any of several reasons, but insufficient draft is usually the primary reason. Draft requirements vary from less than ½ to as much as 5°, depending on alloy, depth of cavity, dimensional and tolerance requirements, and general mold layout (location and number of parting planes). Normally, if draft angles of 4 to 5° are acceptable, castings do not stick in the mold. If tighter dimensional control is required (necessitating smaller draft angles), castings may stick. Sticking can be prevented by providing for mechanical ejection or by increasing draft on noncritical areas.

Casting Design

The design of permanent mold castings for production to acceptable quality at the lowest cost involves many considerations that apply to any method of casting (see the article "Casting Design" in this Volume). For example, casting sections should be as uniform as possible, without abrupt changes in thickness. Heavy sections should not be isolated and should be fed by risers. Tolerances should be no closer than necessary. In addition to these general considerations, the following aspects of design are particularly applicable to the low-cost production of sound permanent mold castings:

- Insofar as possible, all locating points should be in the same half of the mold cavity; in addition, locating points should be kept away from gates, risers, parting lines, and ejector pins
- The use of cored holes less than 6.4 mm (¼ in.) in diameter should be avoided, even though cored holes 3.2 mm (⅛ in.) in diameter or smaller are sometimes possible
- Draft angles in the direction of metal flow on outside surfaces may vary from 1 to more than 10°, and internal draft from slightly less than 2 to 20°. However, using minimum draft increases casting difficulty and cost. Internal walls can be cast without draft if collapsible metal cores are used, but this practice increases cost
- Nuts, bushings, studs, and other types of inserts can often be cast in place. The bond between inserts and casting can be essentially mechanical, metallurgical, or both
- Under conditions of best control, in small molds, allowance for machining stock can be less than 0.8 mm (1/32 in.). However, maintaining machining allowance this low usually increases cost. Generally, it is more practical to allow 0.8 to 1.6 mm (1/32 to 1/16 in.) of machining stock for castings up to 250 mm (10 in.) in major dimension and to allow up to 3.2 mm (⅛ in.) for larger castings
- The designer should not expect castings to have a surface finish of better than 2.5 μm (100 μin.) under optimum conditions. Ordinarily, casting finish ranges from 3 to 7.5 μm (125 to 300 μin.), depending on the metal being cast

The producibility of a casting can often be improved by avoiding abrupt changes in section thickness. Heavy flanges adjacent to a thin wall are especially likely to cause nonuniform freezing and hot tears; in such cases, redesign of the casting may be necessary. The minimum section thickness producible at reasonable cost varies considerably with the size of the casting and the uniformity of wall thicknesses in the casting.

Dimensional Accuracy

The dimensional accuracy of permanent mold castings is affected by short-term and long-term variables. Short-term variables are those that prevail regardless of the length of run:

- Cycle-to-cycle variation in mold closure or in the position of other moving elements of the mold
- Variations in mold closure caused by foreign material on mold faces or by distortion of the mold elements
- Variations in thickness of the mold coating
- Variations in temperature distribution in the mold
- Variations in casting removal temperature

Long-term variables that occur over the life of the mold are caused by:

- Gradual and progressive mold distortion resulting from stress relief, growth, and creep
- Progressive wear of mold surfaces primarily due to cleaning

Dimensional variations can be minimized by keeping heating and cooling rates constant, by operating on a fixed cycle, and by maintaining clean parting faces. It is particularly important to select mold cleaning procedures that remove a minimum of mold material.

Mold Design. The mold thickness and the design of the supporting ribs both affect the degree of mold warpage at operating temperatures. Supporting ribs on the back of a thin mold will warp the mold face into a concave form. This mold design error can alter casting dimensions across the parting line by as much as 1.6 mm (1/16 in.). Adequate mold lockup will contribute to the control of otherwise severe warpage problems.

Mold erosion resulting from metal impingement and cavitation due to improper gating design both contribute to rapid weakening of the mold metal and to heat checking. These mold design errors contribute to rapid dimensional variation during a long run. Mechanical abrasion due to insufficient draft or to improperly designed ejection systems also contributes to the rapid variation of casting dimensions.

Sliding mold segments require clearance of up to 0.38 mm (0.015 in.) to function under varying mold temperatures. This clearance and other mechanical problems associated with sliding mold segments contribute to variations in casting dimensions. Sand cores further aggravate the problem.

Mold Operation. Metal buildup from flash can prevent the mold halves from coming together and can cause wide variations in dimensions across the parting line, even in a short run. Mold coatings on the cavity face are normally applied in thicknesses from 0.076 to 0.15 mm (0.003 to 0.006 in.). Poor mold maintenance can allow these coatings to build to more than 1.5 mm (0.060 in.) thick, causing extreme variation in casting dimensions. Inadequate lubrication of sliding mold segments and ejector mechanisms will contribute to improper mold lockup and consequent variation in casting dimensions. Variation in the casting cycle and in metal temperature will contribute to dimensional variations.

Wear Rates. The dimensions of many mold and core components change at a relatively uniform rate; therefore, it is possible to estimate when rework or replacement will be required. To maintain castings within tolerances, it is sometimes necessary to select mold component materials on the basis of their wear resistance.

Surface Finish

The surface finish on permanent mold castings depends mainly on:

- *Surface of the mold cavities*: The surface finish of the casting will be no better than that of the mold cavity. Heat checks and other imperfections will be reproduced on the casting surface

- *Mold coating*: Excessively thick coatings, uneven coatings, or flaked coatings will degrade casting finish
- *Mold design*: Enough draft must be provided to prevent the galling or cracking of casting surfaces. The location of the parting line can also affect the surface finish of the casting
- *Gating design and size*: These factors have a marked effect on casting finish because of the influence on the rate and smoothness of molten metal flow
- *Venting*: The removal of air trapped in mold cavities is important to ensure smooth and complete filling
- *Mold temperature*: For optimum casting surface finish, mold temperatures must be correct for the job and must be reasonably uniform
- *Casting design*: Surface finish is adversely affected by severe changes of section, complexity, requirements for change in direction of metal flow, and large flat areas

Casting Defects

The defects that can occur in permanent mold castings are porosity, dross, nonmetallic inclusions, misruns, cold shuts, distortion, and cracking. Aluminum alloy castings are subject to all of these defects.

Magnesium alloy castings can have the same defects as aluminum alloy castings. In addition, magnesium alloys are more likely to be hot short.

Copper alloy castings are also susceptible to most of the defects common to aluminum and magnesium. Because of the high pouring temperatures, heat checking of the mold cavities is an added problem. Copper alloy castings often stick in the molds; this can sometimes be prevented by redesigning the mold cavity.

Cost

The total cost of a permanent mold casting includes the cost of metal, labor, fuel, supplies, maintenance of molds and other equipment, and inspection.

Manual Versus Automated Methods. Manually operated equipment is generally more economical for low production quantities, but automated molding invariably costs less for medium-to-high production quantities.

Cost Versus Quantity. Permanent mold casting is primarily used for medium and high production, although the process is sometimes used advantageously for low production. Cost per casting or per pound invariably decreases as quantity increases.

Permanent Mold Versus Sand Casting. The permanent mold process is often selected in preference to sand casting or another alternative process primarily because of the lower cost per casting, but there are often added benefits. For some castings, a minor design change can permit a change from sand casting to permanent mold casting that results in a considerable cost savings.

When castings must be machined, the significant cost is often not that of the casting itself but of the final machined product. Permanent mold casting is often economical because it permits a reduction in the number of machining operations required or in the amount of metal removed.

Solid Graphite Molds*

Permanent molds can be machined from solid blocks of graphite instead of steel. The low coefficient of thermal expansion and superior resistance to distortion of graphite make it attractive for the reproducible production of successive castings made in the same mold.

Because graphite oxidizes at temperatures above 400 °C (750 °F), molds would wear out quickly even if used for nonferrous casting. To protect the molds and to extend their service lives, they are usually coated with a wash, which is normally made of ethyl silicate or colloidal silica. Molds typically show wear by checking or by forming minute cracks in their surface.

Graphite permanent molds are used for a variety of products (notably bronze bushings and sleeves), and graphite chills are often inserted in molds to promote progressive or directional solidification. The use of graphite as a permanent mold material is perhaps best demonstrated in the casting of chilled iron railroad car wheels (the Griffin wheel casting process), as shown in Fig. 11. Graphite is a particularly suitable mold material for this process. It produces castings with closer tolerances than can be achieved with sand molding, and the high thermal conductivity of graphite chills the metal next to the mold face very efficiently, giving it a wear-resistant white iron structure.

*This section was prepared by Thomas S. Piwonka, University of Alabama.

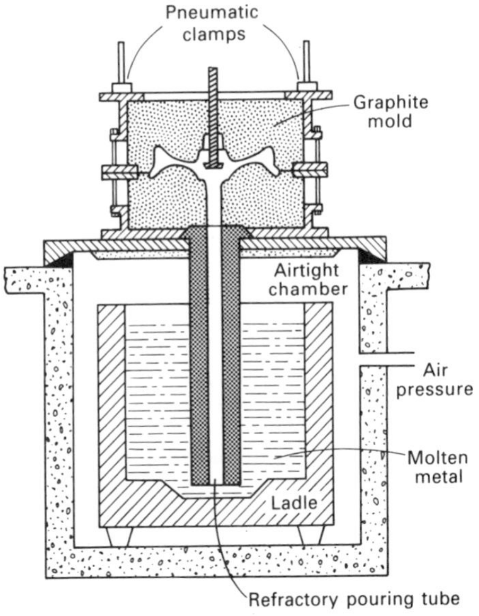

Fig. 11 Schematic of the Griffin wheel casting process. See text for details.

However, because graphite erodes easily, pouring the metal into molds from the top under the influence of gravity causes unacceptable mold wear. As a result, the process was developed so that the mold is positioned over a ladle of molten metal placed in an airtight chamber. A refractory pouring tube extends from deep in the ladle up to the bottom of the graphite mold. Pouring is carried out by pressurizing the metal in the ladle by means of increasing the pressure in the chamber. This forces metal up the pouring tube into the mold. When the casting is filled, a plunger blocks the pouring tube, the air pressure is released, and the metal that remains in the pouring tube drains back into the ladle. The mold and casting are then transferred to a cooling conveyor, and another mold is placed in position over the ladle.

The process is highly automated, and it has a high casting yield with little gate removal required. Because metal is drawn from the bottom of the ladle, there is little chance for slag to be entrained in the castings, which are clean and have excellent surfaces. Casting quality is further controlled by the ease of regulating the flow of metal into the mold. The technique has been used to make ferrous castings weighing up to 410 kg (900 lb).

Die Casting

Lionel J.D. Sully, Edison Industrial Systems Center

DIE CASTING is characterized by a source of hydraulic energy that imparts high velocity to molten metal to provide rapid filling of a metal die. The die absorbs the stresses of injection, dissipates the heat contained in the metal, and facilitates the removal of the shaped part in preparation for the next cycle. The hydraulic energy is provided by a system that permits control of actuator position, velocity, and acceleration to optimize flow and force functions on the metal as it fills the cavity and solidifies.

Die Casting Processes

The variety in die casting systems results from trade-offs in metal fluid flow, elimination of gas from the cavity, reactivity between the molten metal and the hydraulic system, and heat loss during injection. The process varieties have many features in common with regard to die mechanical design, thermal control, and actuation. Four principal alloy families are commonly die cast: aluminum, zinc, magnesium, and copper-base alloys (Table 1). Lead, tin, and, to a lesser extent, ferrous alloys can also be die cast. The three primary variations of the die casting process are the hot chamber process, the cold chamber process, and direct injection.

The hot chamber process is the original process invented by H.H. Doehler. It continues to be used for lower-melting materials (zinc, lead, tin, and, more recently, magnesium alloys). Hot chamber die casting places the hydraulic actuator in intimate contact with the molten metal (Fig. 1). The hot chamber process minimizes exposure of the molten alloy to turbulence, oxidizing air, and heat loss during the transfer of the hydraulic energy. The prolonged intimate contact between molten metal and system components presents severe materials problems in the production process.

The cold chamber process solves the materials problem by separating the molten metal reservoir from the actuator for most of the process cycle. Cold chamber die casting requires independent metering of the metal (Fig. 2) and immediate injection into the die, exposing the hydraulic actuator for only a few seconds. This minimal exposure allows the casting of higher-temperature alloys such as aluminum, copper, and even some ferrous alloys.

Direct injection extends the technology used for lower-melting polymers to metals by taking the hot chamber intimacy to the die cavity with small nozzles connected to a manifold, thus eliminating the gating and runner system. This process, however, is still under development.

Process control in die casting to achieve consistent high quality relates to timing, fluid flow, heat flow, and dimensional stability. Some features are chosen in die and part geometry decisions and are therefore fixed; others are defined by the process at the machine and can be adjusted in real time. All are related and therefore must be dealt with in parallel; the best die castings result from an intimate interrelationship between product design and process design.

Product Design for the Process

Product design and die design are intimately related. The principal features of a die casting die are illustrated in Fig. 3. The

Table 1 Compositions of selected die casting alloys

Alloy	Principal alloying elements, %(a)								
	Al	Cu	Fe	Mg	Mn	Pb	Si	Sn	Zn
Aluminum alloys									
A360	rem	0.60	1.0	0.40–0.60	. . .	. . .	9.0–10.0	. . .	0.40
A380	rem	3.0–4.0	1.0	0.10	. . .	. . .	7.5–9.5	. . .	2.9
A383	rem	2.0–3.0	1.0	0.10	. . .	. . .	9.5–11.0	. . .	2.9
A384	rem	3.0–4.5	1.0	0.10	. . .	. . .	10.5–12.0	. . .	2.9
B390	rem	4.0–5.0	1.0	0.5–0.65	. . .	. . .	16.0–18.0	. . .	1.4
A413	rem	1.0	1.0	0.10	. . .	. . .	11.0–13.0	. . .	0.40
518	rem	0.25	1.1	7.6–8.5	. . .	. . .	0.35	. . .	0.15
Copper alloys									
C85800	0.25	57 min	0.50	. . .	0.25	1.5	0.25	1.50	31 min
C87900	0.15	63 min	0.40	. . .	0.15	0.25	0.75–1.25	0.25	rem
C87800	0.15	80 min	0.15	0.01	0.15	0.15	3.75–4.25	0.25	rem
Magnesium alloys									
AZ91B	8.3–9.7	0.35	. . .	rem	0.13	. . .	0.50	. . .	0.13
AM60A	5.5–6.5	0.35	. . .	rem	0.13	. . .	0.50	. . .	0.22
AS41A	3.5–5.0	0.06	. . .	rem	0.20	. . .	0.50	. . .	. . .
Zinc alloys									
AC40A	3.9–4.3	0.10	0.075	0.025–0.05	. . .	0.004	. . .	0.002	rem
AG41A	3.5–4.3	0.25	0.10	0.02–0.05	. . .	0.005	. . .	0.003	rem
Alloy 7	3.9–4.3	0.75–1.25	0.075	0.03–0.06	. . .	0.004	. . .	0.002	rem
ILZRO 16	3.5–4.3	0.75–1.25	0.10	0.03–0.08	. . .	0.005	. . .	0.003	rem

(a) Maximum, unless range is given or otherwise indicated. More detail on composition ranges and minor constituents is available in Volume 2 of the 9th Edition of *Metals Handbook*. rem, remainder

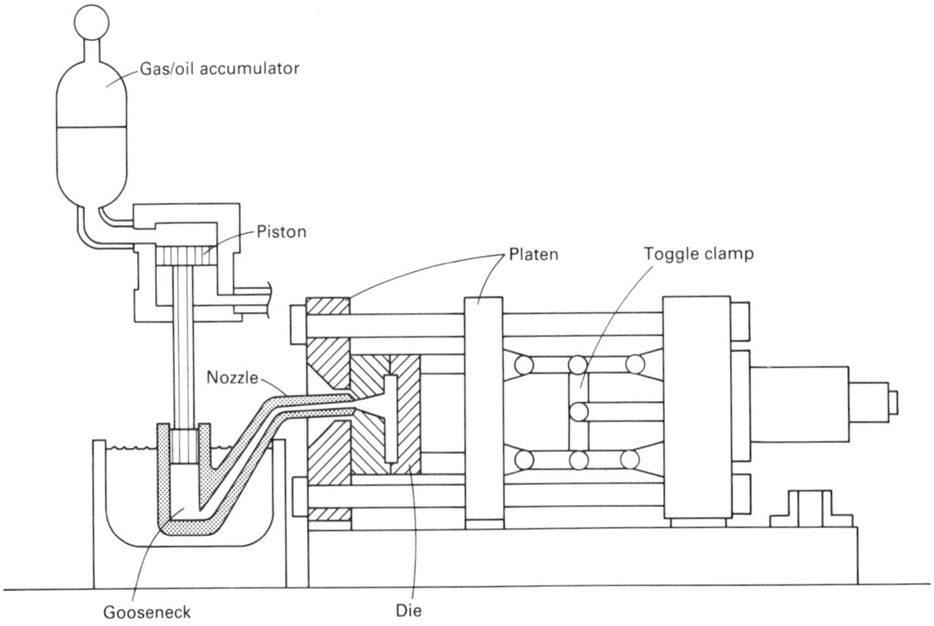

Fig. 1 Schematic showing the principal components of a hot chamber die casting machine

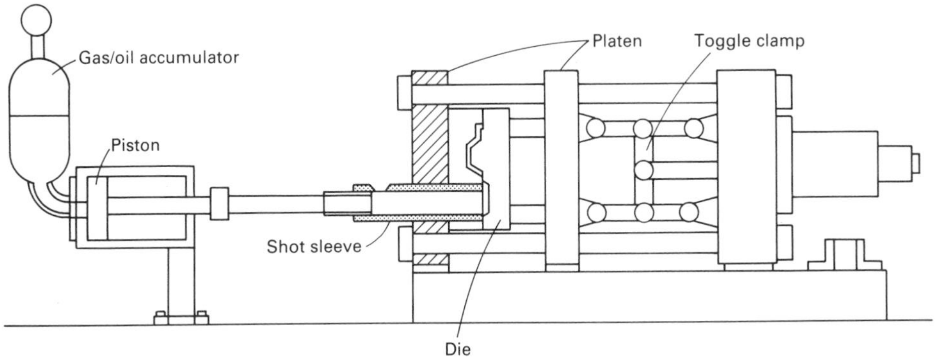

Fig. 2 Schematic showing the principal components of a cold chamber die casting machine

high-speed nature of the process allows the filling of thin-wall complex shapes at high rates (of the order of 100 parts per hour per cavity). This capability places additional demands on the casting designer because traditional feeding of solidification shrinkage is almost impossible. The inability to feed in the traditional sense demands that machining stock be kept to a minimum; high-integrity surfaces should be preserved.

A factor in cost is the parting line topology. The parting line is the line on the casting generated by the separation between one die member and another. The simplest and lowest-cost die has a parting line in one plane. Casting design should be adjusted if possible to provide flat parting lines. Draft is required on the die casting walls perpendicular to the parting line or in the direction of die motion (Fig. 4). An important characteristic of good design is uniform wall thickness, which is necessary for obtaining equal solidification times throughout the casting. Die castings have wall thicknesses of about 0.64 to 3.81 mm (0.025 to 0.150 in.), depending on casting shape and size (Table 2). Bosses, ribs, and filleted corners always cause local increases in section size. In particular, bosses that must be machined require consideration of the entire product-manufacturing cycle. The machinist will find it easier to drill into a solid boss; cored bosses may require floating drill heads in order to align the drill with the cast tapered hole that preserves the high-integrity skin of the casting.

Cores and slides provide side motions for undercuts. A core body is generally round and buried within the cover or ejector die. A slide body has a rectangular or trapezoidal shape and crosses the parting line of the die. As with the cover and ejector dies, the impression steel is often separate from the holder steel. Cores and slides are actuated by various methods, including hydraulic cylinders, rack and pinion, and angle pins. Innovative die design permits radial die motion at a price of die expense. There are die casting processes that use complex-shaped disposable cores similar to those in other gravity casting processes. Cores and slides provide the casting designer with tremendous flexibility at the expense of an increase in die complexity. A standard set of cores—fixed core pins for small holes that are screwed in, or bolted-in inserts—can be used to reduce die construction cost and to permit rapid replacement.

Loose Pieces and Inserts. In certain cases, a reentrant shape needs to be cast into the part where there is no space for core/slide mechanisms. In such a case, the die designer can use a loose piece. A loose piece is placed in the die before each shot is made. It is then ejected from the die with the casting and separated manually or by fixture. Although it provides design flexibility, the load/unload sequence required for loose pieces slows the process, thus increasing cost.

Similarly, the die casting process can allow the part designer great flexibility in local material properties by the use of cast-in inserts of other materials, such as steel, iron, brass, and ceramics. The bond between insert and casting is physical, not chemical, in nature. Therefore, the insert should be clean and preheated. The insert should be designed to prevent pullout or rotation under working loads; knurling, grooves, hexagons, or flats are commonly used for this purpose. Proper support of hollow inserts will prevent crushing of the insert under the high metal injection pressure. The wall thickness of the casting surrounding an insert should be no less than 2.0 mm (0.080 in.) to prevent cracking by shrinkage, hot tearing, and excessive residual stresses.

Trimming. The die cast part is ejected from the die with a variety of appendages (gates, overflows, vents, flash, and robot grasping lugs) that must then be removed. This secondary process is called trimming. Although trimming can be done manually, the high production rates characteristic of die casting demand automation. Trim presses are used to remove the excess material. Castings are often trimmed immediately after the casting process because their higher temperature reduces the strength of the metal.

Trimming conditions directly influence the design of the part and the die casting process, especially gating and parting line definition. Trimming is facilitated by flat parting lines. The relatively rough edge that results from trimming may be acceptable and is often left as is. In some cases, this rough edge is not acceptable and must be removed by machining or grinding. The direction of flash must be such that the edge is machinable.

Dimensional variation is determined by die design, the accuracy of die construction, and process variation. The most accurate

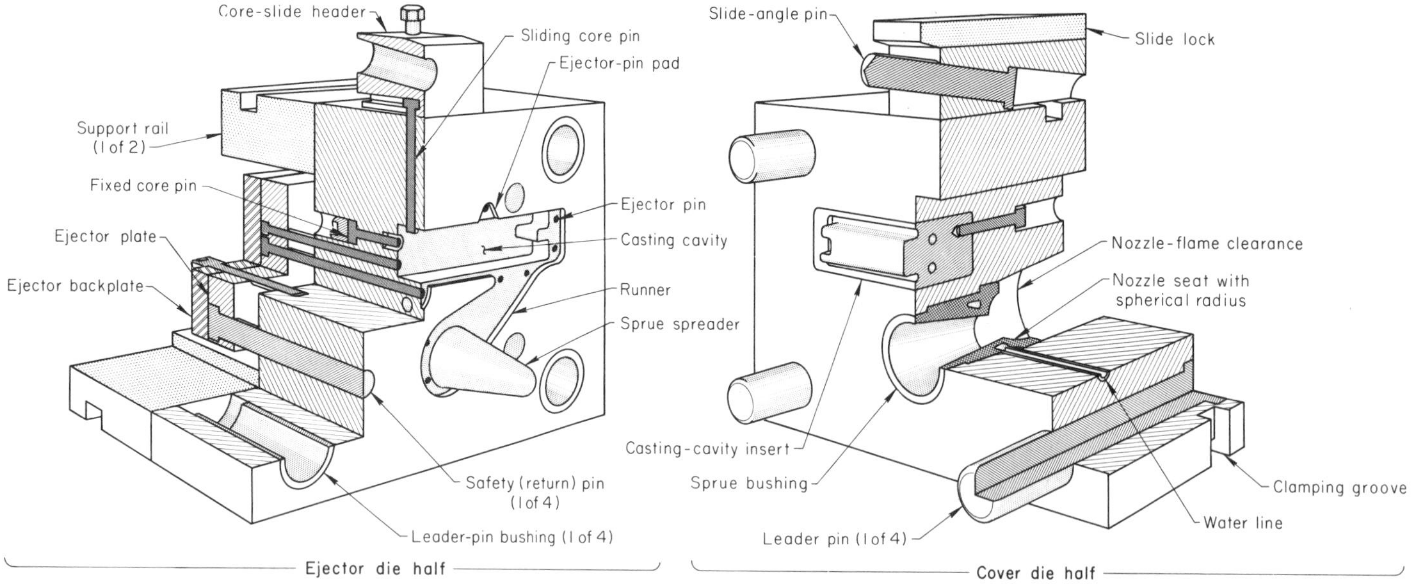

Fig. 3 Components of a single-cavity die casting die for use in a hot chamber machine

dies are those machined using computer numerical control methods. Close control of alloy composition, temperature casting, time, and injection pressure will lead to more consistent casting dimensions. The minimum variation in dimensions is required for those features contained entirely within one die half. Table 3 lists the tolerances on linear dimensions recommended by the American Die Casting Institute (ADCI); Tables 4 and 5 list additional tolerances recommended by ADCI. Therefore, machining locators should ideally be placed in the same die half. Tolerances are a function of casting size and projected area. Features across parting lines have added variation because of the accuracy of repeated die closing. Die temperature, machine hydraulic pressures, and die cleanliness are the principal factors to be controlled. Finally, further dimensional variation occurs if the feature is in a moving die member such as a slide or core.

In summary, a cost-effective die casting demands proper attention to the dimensional variation of the process. Inattention to dimensional factors will lead to an inability to provide consistent products within economic process conditions. The product designer and the die caster must therefore initiate a dialog early in the product cycle.

Gating

The first step in the process sequence is the supply of the molten alloy to the casting machine and its injection into the die. The fluid flow is divided into three considerations: metal injection, air venting, and feeding of shrinkage.

Metal Injection

The distinguishing characteristic of the die casting process is the use of high-velocity injection. The short fill time (of the order of milliseconds) allows the liquid metal to move a great distance despite a high rate of heat loss. The elements of a typical metal gating system are illustrated in Fig. 5.

Proper process performance depends on the delivery of molten metal with high quality as defined by temperature, composition, and cleanliness (gas content and suspended solids). The molten alloy is prepared from either primary ingot or secondary alloys. A melting furnace is used to provide the proper temperature and to allow time for chemistry adjustment and degassing. The alloy is often filtered during transfer to a holding furnace at the casting machine.

The Injection Chamber. Three components make up the injection chambers used for the three types of die casting: the shot sleeve, the gooseneck, and the nozzle (Fig. 1, 2). The cold chamber shot sleeve (Fig. 2) is unique. Initially, it is only partially filled to prevent splashing and to allow for metering error, and it must be filled by slow piston movement to avoid wave formation and air entrainment. Then, for all three chambers, the hydraulic piston rapidly accelerates the molten metal to the desired velocity for injection (Fig. 6). Most die casting machines provide the ability to control the piston acceleration in a linear fashion. Parabolic velocity curves are also available on some controls. This phase of injection can be accomplished in several steps. The third phase of injection is activated as the cavity is close to being filled. This intensification phase draws on an accumulator of high-pressure hydraulic fluid or multiplies pressure using conventional

Table 2 Minimum section thicknesses for die castings

Surface area of casting(a)		Minimum section thickness for: Tin, lead, and zinc alloys		Aluminum and magnesium alloys		Copper alloys	
cm²	in.²	mm	in.	mm	in.	mm	in.
Up to 25	Up to 3.875	0.635	0.025	0.81	0.032	1.52	0.060
25–100	3.875–15.5	1.02	0.040	1.27	0.050	2.03	0.080
100–500	15.5–77.5	1.52	0.060	1.78	0.070	2.54	0.100
Above 500	Above 77.5	2.03	0.080	2.54	0.100	3.05	0.120

(a) Area of a single main plane

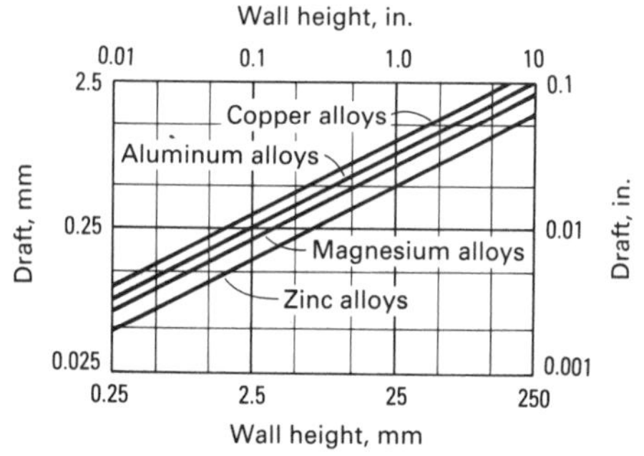

Fig. 4 Minimum drafts required for inside walls of die castings made from four different types of casting alloys

Table 3 Recommended tolerances on as-cast linear dimensions of die castings

Additional tolerances are listed in Tables 4 and 5.

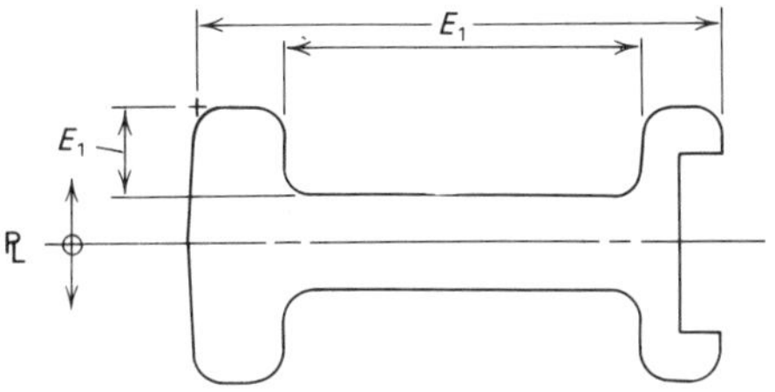

The tolerance on a dimension E_1 will be the value shown in the tables for dimensions between features formed in the same die part. The tolerance must be increased for dimensions of features formed between moving die parts (see Tables 4 and 5).

Length of dimension E_1, in.	Basic tolerance (in.) for: Zinc alloy castings	Aluminum and magnesium alloy castings	Copper alloy castings	Additional tolerance(a) (in.) for each additional inch of dimension E_1 for: Zinc alloy castings	Aluminum and magnesium alloy castings	Copper alloy castings
Noncritical dimensions						
Up to 1	±0.010	±0.010	±0.014	...	...	...
1–12	...	...	...	±0.0015	±0.002	±0.003
Above 12	...	...	...	±0.001	10.001	...
Critical dimensions						
Up to 1	±0.003	±0.004	±0.007	...	...	...
1–12	...	...	...	±0.001	±0.0015	±0.002
Above 12	...	...	...	±0.001	±0.001	...

(a) Example: an aluminum alloy casting would have a tolerance of ±0.010 in. on a critical 5.000 in. dimension E_1 (that is, the basic tolerance of ±0.004 in. + 4 (0.0015) = ±0.010 in.). Source: Ref 1

piston intensifiers. This increases the pressure on the metal to force the rapidly freezing alloy into incipient shrinkage cavities.

Sprues and Runners. The sprue provides a smooth transition from the shot sleeve or nozzle and promotes high cooling heat flow after injection is complete. The runner carries the flowing metal from the injection chamber to the desired location(s) on the casting periphery. Runners are not used in direct injection. Heat loss, unnecessary turbulence, and die erosion can be minimized by proper attention to basic hydraulic principles when designing runners. Typical runners are therefore round or nearly square trapezoidal in section to minimize surface area and heat loss. There is a distinct change in section from the thick runner to the thin gate. A change in flow direction also often occurs. The approach section is the means for achieving these two needs. The shape of this section of the flow channel often provides the name of the gate, for example, chisel gate or fan gate. The use of tapered tangential runners eliminates this approach feature.

The gate is the controlling entry point into the casting. The gate serves a fluid flow need, but it must later be removed from the casting by trimming. Therefore, the gate cross section should be the smallest in the gating system. The cross section is determined by the desired fill time and flow rate that the casting machine can provide. A number of methods are available for calculating gate area; these are discussed below.

The shape of the part is primarily governed by the end use, not by fluid flow considerations. Indeed, the die casting process excels in very complex near-net shape configurations. The high-velocity inertia-driven flow, combined with rapid heat loss and partial freezing during fill, eliminates the possibility of a rigorous fluid mechanics solution. However, the die caster must always attempt to understand the flow in the part cavity.

The overflow is the final component in the fluid flow system. Although they add to the weight of remelt, overflows do serve a variety of purposes. They can act as a reservoir for metal to be removed from the cavity, and they can provide an off-casting location for ejection pins, robot holds, or instrumentation points.

Gating System Design. Several methods are available for designing gating systems. Design of the gating system is always a compromise. Unlike the flow of polymers or metal flow in forging, the high-velocity metal flow of die casting, combined with heat loss and simultaneous solidification, cannot be rigorously solved with computational methods. Therefore, various methods have been developed to provide the die caster with tools to address the problem on a sound, consistent basis. All of these methods attempt to take into account the influence of the following key variables:

- Part shape
- Internal quality
- Surface quality
- Mechanical properties
- Die temperature
- Die erosion
- Die material
- Die venting
- Metal temperature
- Metal fluidity
- Metal heat content
- Metal microstructure

Since the invention of the die casting process, many die castings have been successfully made with gating systems designed by experience only. Each company has a reservoir of this closely guarded experience. Trial-and-error adjustments at the casting machine are frequently part of the learning. However, the decline of the presence of the artisan in the foundry is forcing a move toward analytically based gating design, but the analysis base is still tempered with the fine tuning of experience. This is especially true in gate location and local angle

Table 4 Recommended parting line tolerances for die castings

Tolerances given in this table are to be added to the basic tolerances given in Table 3. See also Table 5.

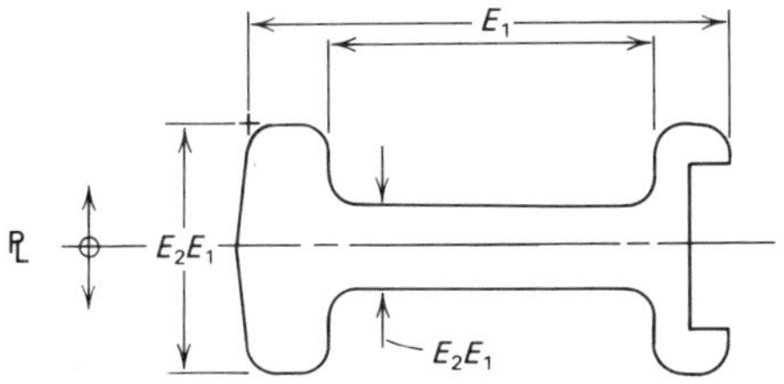

The tolerance on dimensions such as E_2 E_1, which are perpendicular to the parting plane, will be the value shown in the table plus the linear tolerance from Table 3. The value chosen from the table depends on the projected area of the part. Additional tolerances in the case of other moving die parts are shown in Table 5.

Projected area of casting, in.^{2}(a)	Additional tolerance(b) (in.) for: Zinc alloy castings	Aluminum and magnesium alloy castings	Copper alloy castings
Up to 50	±0.004	±0.005	±0.005
50–100	±0.006	±0.008	...
100–200	±0.008	±0.012	...
200–300	±0.012	±0.015	...
300–500	10.016	±0.020	
500–800	±0.020	±0.025	...
800–1200	±0.025	±0.030	...

(a) Projected area is the area of the part in the parting plane. (b) Example: an aluminum die casting with a projected area of 75 in.^{2} would have a tolerance of ±0.018 in. on a critical 5.000 in. dimension E_2E_1 (that is, ±0.008 in. for 75 in.^{2} plus the basic linear tolerance of 0.010 in.). See Table 3. Source: Ref 2

Table 5 Recommended additional tolerances for die castings produced in dies with moving parts

Tolerances in this table should be used in conjunction with those listed in Table 3. See also Table 4.

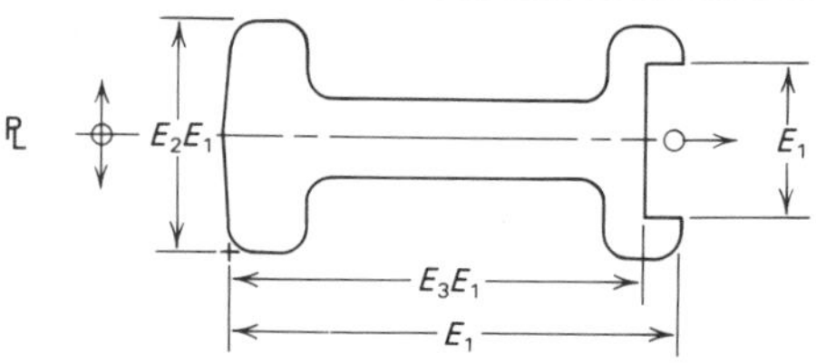

The tolerance on dimensions such as E_3 E_1 will be the value shown in the table plus the linear tolerance from Table 3. The value chosen from the table depends on the projected area of the portion of the die casting formed by the moving die part perpendicular to the direction of movement.

Projected area of die casting, in.^{2}	Additional tolerance(a) (in.) for: Zinc alloy castings	Aluminum and magnesium alloy castings	Copper alloy castings
Up to 10	±0.004	±0.005	±0.010
10–20	±0.006	±0.008	. . .
20–50	±0.008	±0.012	. . .
50–100	±0.012	±0.015	. . .
100–200	±0.016	±0.020	. . .
200–350	±0.020	±0.025	. . .
350–600	±0.025	±0.030	. . .
600–1000	±0.030	±0.035	. . .

(a) Example: An aluminum alloy casting formed using a moving die part and having a projected area of 75 in.^{2} would have a tolerance of ±0.025 in. on a critical 5.000 in. dimension E_3E_1 (that is, ±0.015 in. for 75 in.^{2} plus ±0.010 in. on linear dimensions). See Table 3. Source: Ref 3

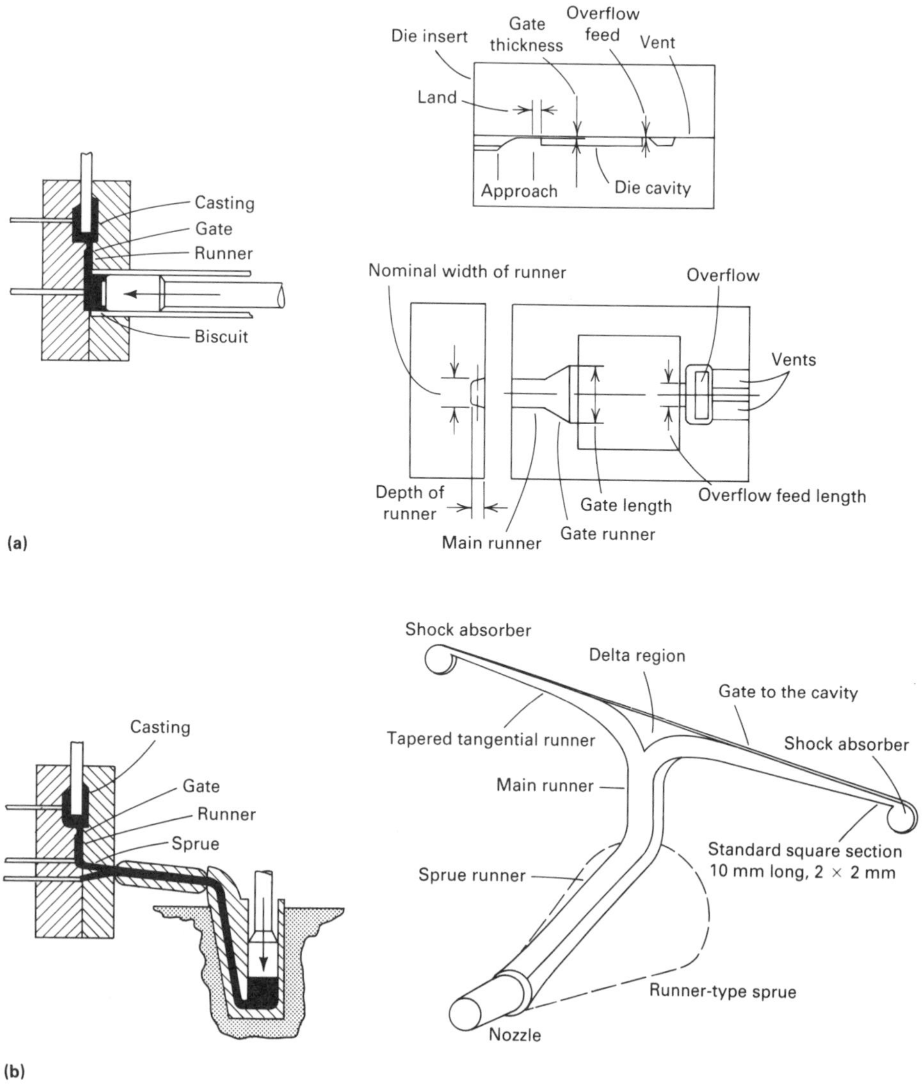

Fig. 5 Schematics showing gating systems for cold chamber (a) and hot chamber (b) die casting machines

of entry, which are directly affected by part shape and secondary operations.

One of the first analysis methods was the ADCI/DCRF Nomograph (Fig. 7), which solves geometric relationships for the bulk flow design. The selection of a fill time for the casting is based on experience and experiment. The limited selection of plunger diameters for a given machine restricts the design. The cold chamber process links the volume of metal to plunger diameter by filling the shot sleeve about two-thirds full. The nomograph is used to develop a required volume fill rate Q.

It has recently been recognized that the ability of the casting machine to provide this metal volume flow, while keeping the dies closed during injection, must be considered. The tool that has been developed for this purpose is called the P-Q^2 diagram (Fig. 8). It can be shown that the pressure P on the metal and hydraulic system is proportional to the square of the injection velocity and therefore the volume flow rate Q. The line with the negative slope is the machine characteristic line. The characteristic line moves as shown with changes in hydraulic pressure, shot valve throttling, and plunger diameter. The line that starts at the origin of the graph is a measured relationship of pressure to flow rate for the particular casting and gate being cast. The effect of adjusting the gate area is shown.

Optimization of these various parameters for the casting and machine provides the process engineer with a powerful tool for process definition and debugging. A number of microcomputer programs are available to

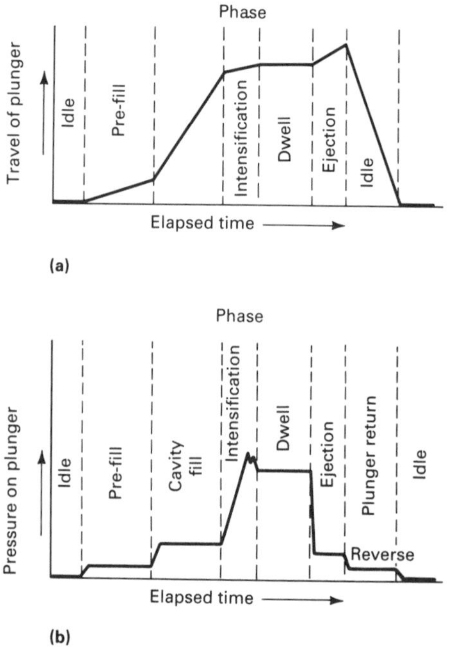

Fig. 6 Curves for plunger travel versus time (a) and plunger pressure versus time (b) indicating the various phases of a shot

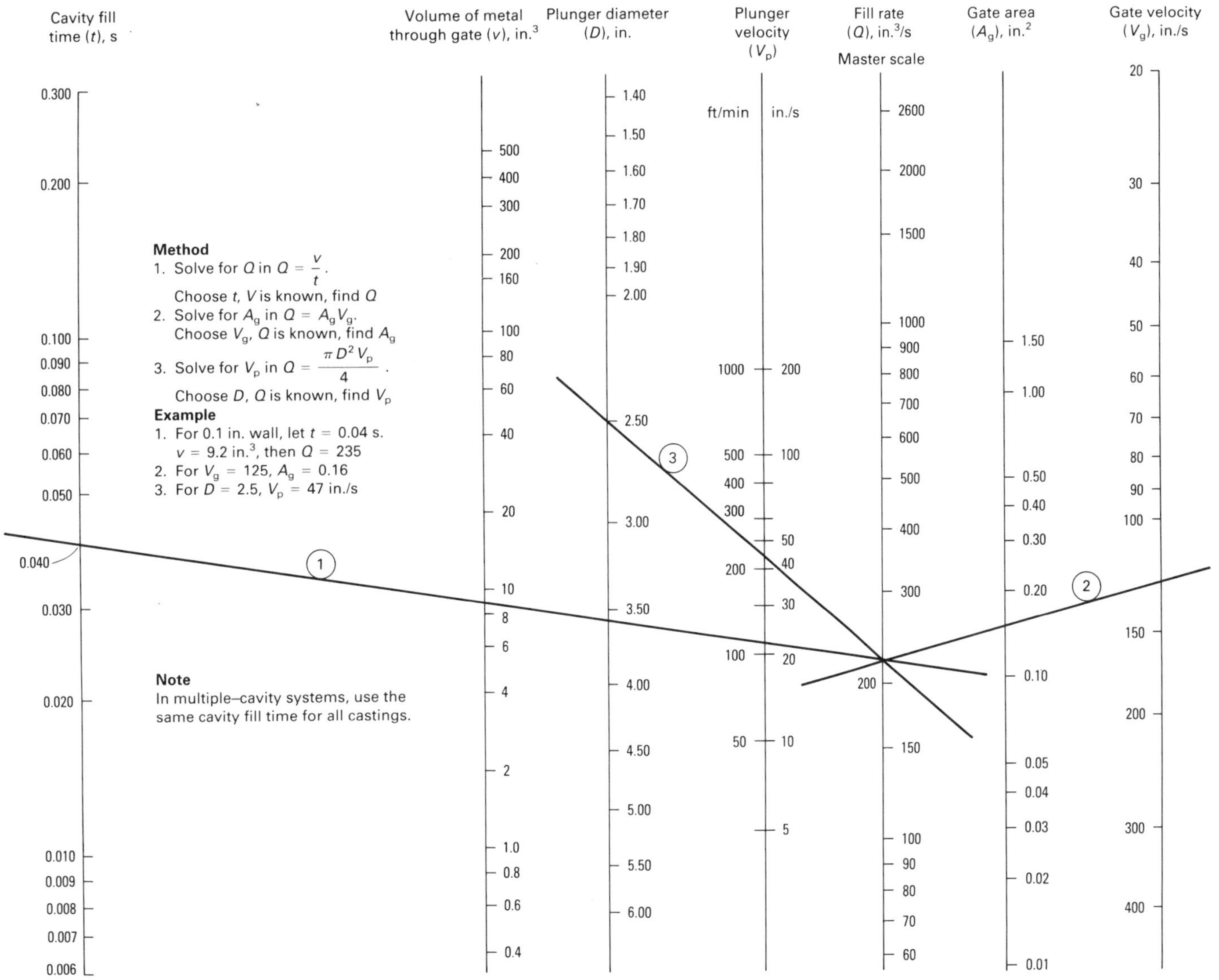

Fig. 7 Nomograph used to determine the volume fill rate Q required for different casting process parameters

base gate design on this hydraulic approach (see the Selected References at the end of this article).

Air Venting

The subject of air venting is often neglected in die casting texts, but it can have a profound effect on product quality. Back pressure directly affects fluid flow. The molten metal injected into the die must displace the gas initially within the die cavity or absorb it as compressed bubbles. The common belief that die castings cannot be heat treated is an indication that they often contain entrained high-pressure gas pockets. Effective air removal eliminates entrained gas, resulting in optimum metal properties and even heat-treatable castings.

Gas Displacement. The most common method of gas displacement is the use of thin channels called vents that are open to the outside of the die. The total area of vents should be approximately 20% of the gate area. The machined channels are no more than 0.18 to 0.38 mm (0.007 to 0.015 in.) thick to ensure that the metal will freeze off before the edge of the impression block is reached. The turbulence of flow makes complete removal of cavity gases difficult by displacement venting. Vent placement is a matter of experience, with final design often done by trial and error on the casting machine.

An alternative to displacement venting is the extraction of the air ahead of the metal by a vacuum system using an accumulator. Once the die is closed and the shot hole sealed, the vacuum valve is opened and the air in the cavity is drawn into the accumulator. The vacuum valve is either closed after a preset time or by metal impingement, which shuts the valve to prevent metal from filling the vacuum system. The accumulator is then pumped down for the next cycle.

Another method of eliminating entrained gas bubbles is to rely on the reactivity of the molten alloy with oxygen. The normal air in the cavity can be displaced by a pure oxygen purge. When the molten metal is injected, the gaseous oxygen combines with the metal to form a finely distributed solid oxide. This is known as the pore-free process.

The overflow can also act as a reservoir for molten alloy that has flowed through the cavity. One method of removing entrained gas from the cavity relies on the transfer of some of the gas into the overflow. However, for this method to be effective, the proportion of overflow volume to casting volume must be relatively large.

Feeding of Shrinkage

Practically all casting processes except die casting and squeeze casting rely on a massive reservoir or riser of molten metal to feed incipient shrinkage. There is no such riser in a die casting die. Metal pressure intensification at the end of the injection cycle is the only force used to feed shrink-

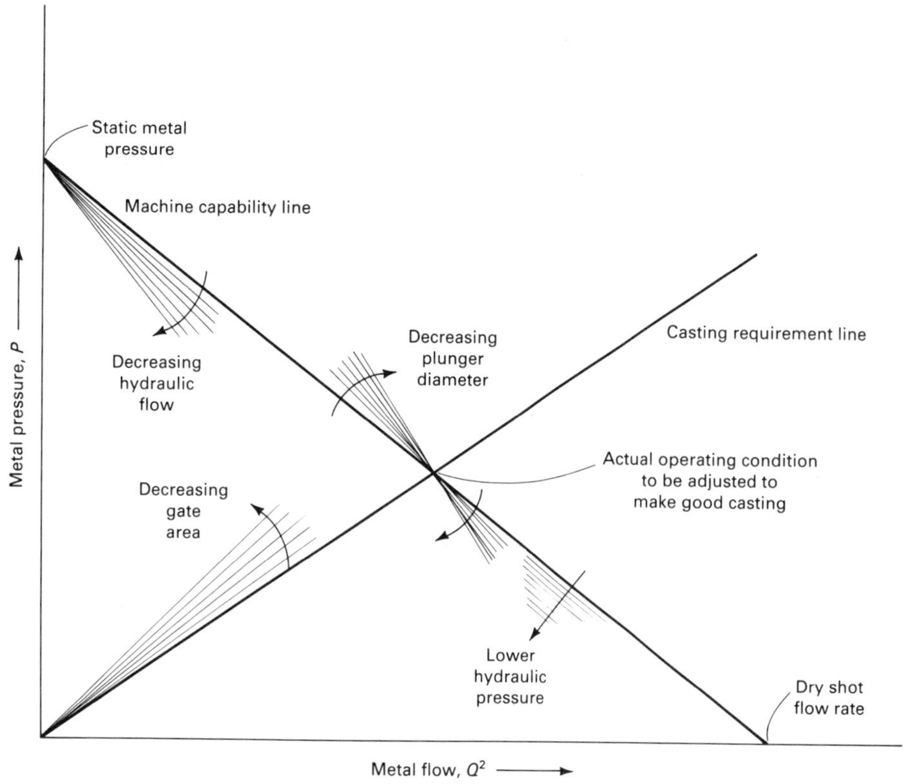

Fig. 8 Metal pressure versus metal flow (P-Q^2) diagram used to determine or optimize process parameters in die casting

age. The largest thermal section in cold chamber die casting is the biscuit and the runner adjacent to the shot sleeve and up to the gate (see Fig. 5a). Given the very thin walls of many die castings, the effective feeding distance from the gate is very short. Therefore, the die caster must rely on other techniques to assist in minimizing shrinkage.

For example, near-eutectic alloys with short freezing ranges are preferred. Inoculation (grain refinement) ensures a small eutectic cell size for well-dispersed microshrinkage that is not interconnected. The remnant shrinkage might also be moved to an uncritical area. This is accomplished by manipulating metal flow and die temperatures. A mechanical solution is the use of pressure pins, which are small, hydraulically actuated cores that are driven into the freezing casting. Timing and pressure are critical to the proper use of pressure pins. This approach adds to the complexity of the die.

Interaction of Fluid Flow With Dies and Machines

The forces generated by injection of the molten metal are a key factor in die design. Although the forces are highly dynamic in nature, most dies are designed for peak static loads. The geometric factor that usually determines the size of the casting machine that must be used is the combined projected area of the casting, runner, and gating system. The force that must be contained by the machine is this area multiplied by the peak metal pressure during the injection cycle. This is the peak of the intensification pressure (Fig. 6b). Because metal pressures under intensification can exceed 69 MPa (10 ksi), the force pushing the die open can be more than 450 000 kg (over 1 million lbf) for a large casting, based on projected area. This force determines the locking capacity needed to keep the die closed. Die casting machines are available with locking forces ranging from 450 kN to 35 MN (50 to 4000 tonf). The tooling cost and process cost increase with projected area and machine size. The distribution of forces within the tie bars is also important; the objective is to load each tie bar equally.

If the die uses slides and cores, these must also resist the force of the metal. This is usually accomplished by a mechanical lock. The closing actuator should not be expected to perform this function.

The thermal effects of molten metal flow in the die are a major factor in determining die life, casting surface quality, and many internal quality parameters. Impingement of metal upon the die, excessive turbulence, and cavitation in the flow increase the local heat load on the die. The result can be premature die failure from thermal fatigue.

In summary, the fluid flow aspects of die casting are a major determinant in product quality and die life. Therefore, fluid flow, or gating, can control the economic success of a particular die casting product. The flow is complex, and success depends on experience to aid an evolving analysis technology.

Heat Removal

Once the molten alloy is in the die cavity, the heat in the alloy must be removed to allow solidification and subsequent cooling to occur; the die acts as a heat exchanger. A number of heat flow paths are available for this transfer, as shown in Fig. 9. As with any heat flow problem, the geometry of the heat exchanger is critical; in this case, heat exchanger geometry is dictated by the part geometry. Longer cooling times are achieved by utilizing heavy sections, distance from cooling lines, and shapes with large volume-to-surface-area ratios (or large section modulus). Another critical result of shape is the possibility of large differences in cooling times creating stresses within the die casting. If cooling stresses become too high under the rigid constraint of the die, hot tears and cracks can result.

The rate of heat removal is such that cooling is significant even during the filling process. Therefore, superheat and latent heat are transferred to the die before the alloy is stationary. The metal remote from the gate will be partially solid before the end of die filling. This transient heat flow results in more heat being transferred to the die near the gate.

Control of Heat Flow. The most important factor in controlling heat flow is the die/casting interface. The interface heat transfer coefficient is affected by a number of factors, such as shape, surface finish, lubrication, and steel oxide state. However, die material can also be used to influence heat flow. Conductive materials such as tungsten-base alloys and beryllium copper are used as inserts to promote high heat flow rates in local areas.

The location of waterlines is a key control of heat flow. The complex shapes of many die castings restrict the die designer in the placement of cooling lines. The basic factors controlling waterline effectiveness are location, coolant flow rate, temperature and pressure, and scale buildup. Placement of waterlines must be done in design. Coolant flow is a process control tool; scale buildup in waterlines is routinely controlled as part of preventive maintenance.

Cooling water can be replaced by hot oil to provide a source of heat in a die. Heat is often needed in configurations that suffer from excess temperature loss in filling. Hot oil and hot water systems can be effectively used to preheat a die before casting begins. This will promote longer die life.

In contrast, external die sprays between shots are one method of obtaining rapid heat flow where internal cooling is insufficient. As in the case of preheating by making shots, external sprays are very detri-

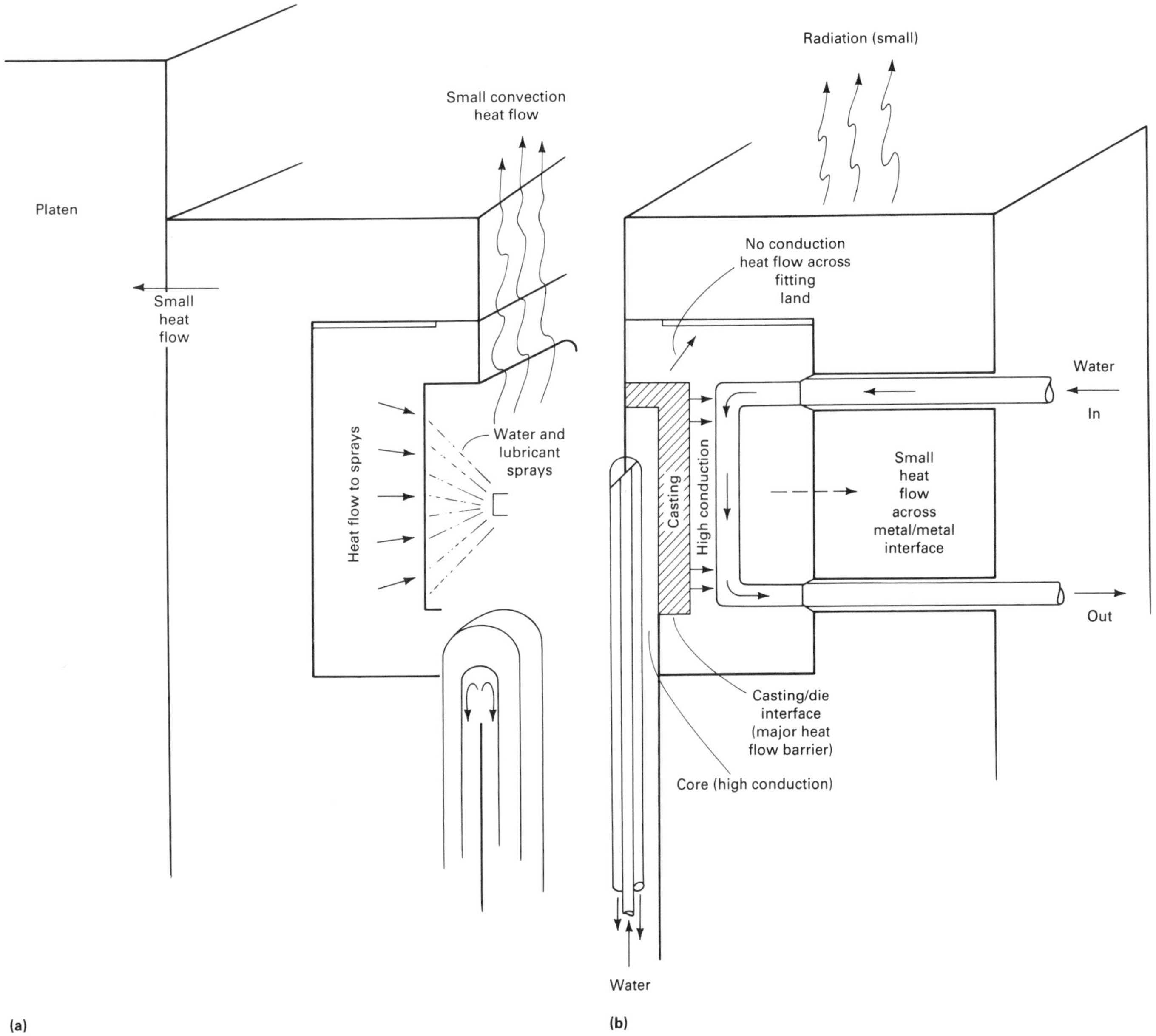

Fig. 9 Various heat flow paths available in die casting. (a) Die open for service. (b) Die closed after shot

mental to die life. The trade-off in this case is between the cost required to maintain the die steel (including downtime) and the reduced cycle time for each part.

Analysis of Cooling Requirements. Several analytical methods are available for analyzing the cooling requirements of a die for a particular part. These tools vary widely in the detail and the time required for analysis. There are very simple methods that can be done quickly, and these are used for preliminary analysis. There are also methods that require more time and effort but provide the die designer with the detail necessary to obtain effective process definition.

A simple overall heat balance is the most basic tool that is used. In this technique, the heat content is calculated for one molten alloy shot between the injection temperature and the ejection temperature. This is the heat that must be removed within one cycle from the casting and the die. The proportion of the heat removed by cooling lines can then be determined. The size of preheating units required can also be found by using this simple method and taking into account the die properties.

There are also methods that use purely geometric considerations. In this case, the geometric shapes of part and die segments are considered in the conduction equation for the die. Computer-aided design (CAD) software is available to make this method simple to use once a surface model is created.

Analysis of cooling channel placement, along with hydraulic considerations for circuiting, is in general use for plastic injection molds. This method can also be adapted for die casting cooling design, and again, software is available in conjunction with a number of CAD packages.

Diffusion solutions, such as finite-element and finite-difference methods, represent the most detailed solution to the heat transfer design problem in die casting die design. The use of these tools will become more common as the cost of computation decreases. The sophistication of automated mesh generation must also improve if models are to be created and run within the short lead times often required for die design. The increased use of CAD for product design, where three-dimensional geometry is available to the die designer from the beginning, will also promote the use of these tools for initial design and for process improvement and optimization.

Ejecting the Casting

After the casting has solidified and cooled, it must be removed from the die. Slides and cores generally must be withdrawn first, followed by opening of the die by ejector die motion. As the die opens, the casting is ejected from the die by the ejector pin. This is done by stopping the ejector plate while the die continues to retract or by using an actuator driven by hydraulic cylinder or rack and pinion. The casting is then taken from the pins manually, by dropping through the base of the machine under gravity, or by a robot or extractor.

Ejector pin size and location are governed by part geometry, especially depth of draw, draft, and surface contact area. The function of the casting can have considerable bearing on the allowable ejector pin location, and this is a subject for negotiation between the caster and the product designer. Bosses are often provided to ensure that there is a flat surface upon which the ejector pin can push. The mechanical design aspects of the ejection system must address stability and guidance during travel. Similarly, the die base in which the ejection system is contained must be strong enough to withstand the locking forces of the casting machine and metal flexure without undue deflection. Support pillars under the cavity may be required.

Preparation for the Next Cycle

With the casting removed from the die, external water sprays can be used to cool the die. Compressed air blow off is used to blow residual water and any other loose flash from the internal die surfaces. This is followed by a lubricant spray to provide an insulating release agent between metal and die. A lubricant is also applied to the shot cylinder. The die can then be closed in preparation for the next cycle. The combined ejection and preparation phases can be up to 40 to 50% of the process cycle, as compared to 20% for plastic injection molding. The time required for ejection and preparation depends on casting size and die complexity.

Die Casting Defects

Effective die design (fluid flow, heat flow, and mechanical design) and a well-defined process capability will combine to produce quality die castings. Proper attention to die casting process control will result in consistently high quality parts. The establishment of a stable process is the combined result of a partnership in casting design, die design, and process engineering.

Defects will occur if process variation is too broad. Defects are caused by three basic sources:

- Mechanical problems in the die
- Metallurgical problems in the molten alloy
- The interaction of heat flow and fluid flow

Care is necessary in order to identify the cause of a specific defect because several outward appearances can have different sources.

Mechanically induced defects such as galling or drag marks on the casting surface occur during ejection of the casting and are usually caused by insufficient draft in the die. Galling will be aligned with the direction of relative motion of the casting and its adjacent die segment. Lack of draft can be an error in die building, but it is readily corrected. Die design can cause lack of draft by inadequate specification, poor ejector system alignment, and inadequate slide or core alignment. Improper machine setup with uneven tie bar loading can cause the die to shift upon closing and opening and therefore create galling. Distorted or cracked castings are the result of extreme cases of poor mechanical design. Although a new die may be free of such mechanical defects, the normal wear and tear of the process may eventually lead to these defects. Proper attention to preventive maintenance will minimize such behavior.

Another class of mechanically induced defects can result from improper injection system performance. Machine-monitoring systems are available to measure and control critical timing, pressures, and velocities. These can be monitored on a shot-by-shot basis or on a periodic audit plan. In the event of loss of process control, these injection parameters should be the first to be checked. Changes to the die should be the last to be checked because they are the most difficult to diagnose and the most costly to implement.

Metallurgical Defects. Proper control of the quality of the molten metal is of primary importance with regard to metallurgical defects. The four principal factors are alloy composition, dissolved gas content, entrained solids (such as oxides and intermetallic compounds), and improper temperatures. The results can be poor fluidity, die soldering, shrinkage porosity, hot cracking, and gas porosity. Metallurgical factors interact directly with the primary causes of casting defects: heat flow and fluid flow. A die casting process that experiences unexplained variations over time in the quantity of defects may be receiving poor-quality metal. As with a deterioration in the mechanical behavior of the die, metal quality is one of the first process parameters to be checked.

Interaction of Heat Flow and Fluid Flow. The die casting die is primarily a means of shaping a molten volume of metal and removing enough heat to permit ejection. Improper interaction of fluid and heat flow can lead to poor casting quality. The rapid fill time, the complex part geometries, and the high heat transfer rates of die casting combine to form a complex set of potential causes of defects. Flow/heat defects can be built into the die by poor design of the gating, venting, and thermal die layout. However, once the process is proven and the die is properly maintained, these parameters should not produce variations in the process.

Cold shuts are seams in the casting where two streams of metal have come together but have not fused. Cold shuts occur when solidification progresses too far in the flow streams, resulting in insufficient fluidity at the seam for mixing. The fixed causes of cold shuts are poor flow patterns due to inappropriate gating, excessive back pressure due to inadequate venting, and excessively thin walls in the part design. Process variables that cause the appearance of cold shuts are low die or metal temperatures, inadequate injection pressures, and excessive fill times.

Accurate volume measurement is important for precise process control. Insufficient volume can result in a very short biscuit (the cylinder left in the shot sleeve), premature freezing, and loss of effectiveness in injection pressure intensification. Excess volume can prolong cold chamber casting cycles because the biscuit is the last section of the casting to solidify.

Gas porosity consists of discrete, separate holes that have two sources: entrained air and, less frequently, dissolved gas. The latter source is entirely a metallurgical control problem, while the former has a variety of process causes. Built-in causes of excessive entrained gas are:

- Too empty a shot sleeve (excessive diameter or length)
- Inadequate venting
- Excessive use of lubricant
- Residual moisture from sprays
- Poor metal flow patterns that prevent venting

If the gas porosity is adjacent to the surface, a blister may form immediately following ejection as the high-pressure gas deforms the weak metal. The secondary factor is excessive die temperature on the blister side.

Shrinkage porosity is a series of interconnected holes created by a lack of feed metal at the end of solidification. Shrinkage is confined to the thermal center of a section, which can extend to the casting surface if local die temperatures are excessive.

Shrinkage porosity can be built in by poor casting design that contains large sections, excessive die heating from poor metal flow distribution, and inadequate internal die cooling. Process variations in metal temperature, die temperature, inadequate injection pressures, and poor cooling from clogged cooling lines can result in shrinkage porosity.

Soldering is the adherence of the molten metal to the die surface; this results in tearing of the casting surface upon ejection. The condition appears where impingement of the flowing metal causes local overheating of the die. If the overheating is extreme, the molten metal stream will erode the die surface. This is most often seen at or near the gate. Poor gate design is the primary cause of soldering, combined with inadequate die cooling; poor die surface polishing can aggravate the problem. Process variations in lubricant application, die surface maintenance, or temperature control can lead to soldering and die erosion.

Molten aluminum is very aggressive toward unprotected die steels. If the iron content of the aluminum alloy is too low, the molten material tends to dissolve the die steel during injection. This tendency is especially pronounced under conditions that promote soldering and die erosion. This special soldering problem can be avoided by maintaining the iron content of aluminum die casting alloys between 0.8 and 1.1%.

Heat check fins are replica die cracks created by thermal fatigue. Thermal fatigue cracking (heat checking) is the result of the temperature cycles experienced at the die surface. High stresses may be built into the die by the design of the casting. Sharp corners and other stress raisers should be removed if at all possible. Excessive temperatures, combined with local plastic deformation, creep, and phase changes in the steel, eventually lead to crack formation. Improper heat treatment will aggravate phase-induced failure. Dies for the low-melting metals rarely fail by thermal fatigue. Copper-base and ferrous castings rapidly cause checking of the die surface. Two process abuses are the most frequent irritants: making shots in a cold die, and excessive use of external water sprays.

Postcasting Operations

Die Trimming. There are two postcasting operations that are unique to die casting. Die castings are taken from the die with flash, gates, and overflows attached to them. Trimming of these extraneous elements can be done by hand for low production volumes, but is usually done with a hydraulic trim press for high-production parts. The press is set up with two hand-switch interlocks to ensure that the operator's hands are out of the press after loading/unloading. Robot operation is also possible. Two types of trim dies are used: push-through and reverse-trim. Commercially available die sets are often used. The trim knives are hardened to greater than 50 HRC, and provision is made for self-cleaning or blow-off of flash to keep the trim dies clean.

Impregnation. Porosity in a die casting can lead to a lack of pressure tightness. Although porosity can be minimized by proper process design or control, it is sometimes necessary or cost effective to fill voids by using an impregnation process. Sodium silicate and anaerobic organic compounds are among the impregnants available. The typical procedure is as follows:

- Clean the casting
- Place in an autoclave and draw a vacuum of 710 mm (28 in.) of mercury for 15 min
- Introduce the sealant and apply hydrostatic pressure for 15 min
- Pump out and remove the castings
- Wash and dry

REFERENCES

1. "Linear Dimension Tolerances for Die Castings," ADCI-E1-83, American Die Casting Institute
2. "Parting Die Tolerances," ADCI-E2-83, American Die Casting Institute
3. "Moving Die Part Tolerances," ADCI-E3-83, American Die Casting Institute

SELECTED REFERENCES

- H.H. Doehler, *Die Casting*, McGraw-Hill, 1951
- E.A. Herman, *Die Casting Dies, Designing*, Society of Die Casting Engineers, 1985
- E.A. Herman, *Heat Flow in the Die Casting Process*, Society of Die Casting Engineers, 1985
- A. Kaye and A. Street, *Die Casting Metallurgy*, Monograph in Materials, Butterworths, 1982
- "Metal Flow Predictor," Computer Program, American Die Casting Institute
- "METLFLOW," Computer Program, Moldflow Australia Pty. Ltd.
- H.H. Pokorny and P. Thukkaram, *Gating Die Casting Dies*, Society of Die Casting Engineers, 1984
- "Product Standard for Die Casting," American Die Casting Institute
- "Runner Design," Computer Program, American Die Casting Institute

Centrifugal Casting

Horizontal Centrifugal Casting

Alain Royer and Stella Vasseur, Pont-á-Mousson S.A., France

HORIZONTAL CENTRIFUGAL CASTING is used to cast pieces having an axis of revolution. The technique uses the centrifugal force generated by a rotating cylindrical mold to throw the molten metal against the mold wall and form the desired shape.

The first patent on a centrifugal casting process was obtained in England in 1809. The first industrial use of the process was in 1848 in Baltimore, when centrifugal casting was used to produce cast iron pipes. In the 1890s, the principles already known and proved for liquids in rotation about an axis were extended to liquid metals, and the mathematical theory of centrifugal casting was developed in the early 1920s.

Horizontal centrifugal casting was first used mainly to manufacture thin-wall gray iron, ductile iron, and brass tubes. Improvements in equipment and casting alloys made possible the development of a flexible and reliable process that is both economical and capable of meeting stringent metallurgical and dimensional requirements. Cylindrical pieces produced by horizontal centrifugal casting are now used in many industries. Of particular importance are large-diameter thick-wall bimetallic and specialty steel tubes used in the chemical processing, pulp and paper, steel, and offshore petroleum production industries.

Equipment

A horizontal centrifugal casting machine must be able to perform four operations accurately and with repeatability:

- The mold must rotate at a predetermined speed
- There must be a means to pour the molten metal into the rotating mold
- Once the metal is poured, the proper solidification rate must be established in the mold
- There must be a means of extracting the solidified casting from the mold

Figure 1 shows a common design for a horizontal centrifugal casting machine. Many variations of this basic design are in use. Details may vary; for example, there are different types of drive systems, carrying rollers, and so on.

Molds

Molds consist of four parts: the shell, the casting spout, roller tracks, and end heads. The mold assembly is placed on interchangeable carrying rollers that enable the use of different mold diameters and fine adjustments. Molds are cooled by a water spray, which can be divided into several streams for selective cooling.

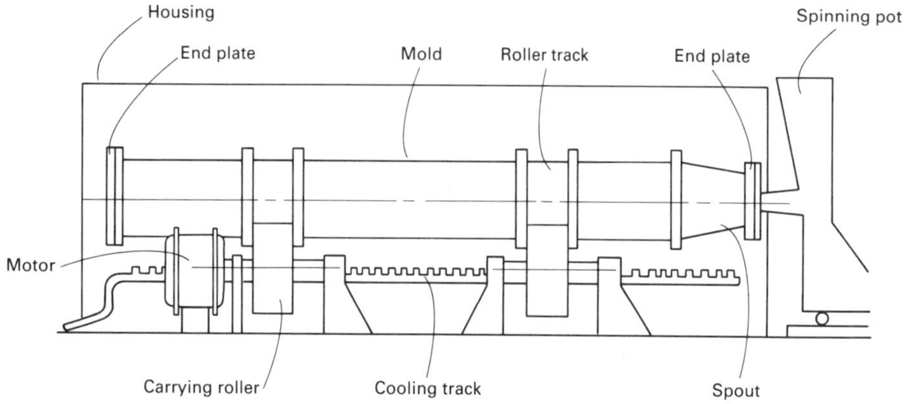

Fig. 1 Schematic of a common design for a horizontal centrifugal casting machine

Different types of molds are generally used according to the geometry and quantity of castings needed and the characteristics of the metal or alloy being cast. Molds can be either expendable (a relatively thin case lined with sand) or permanent.

Expendable molds lined with sand are widely used in centrifugal casting, especially for producing relatively few castings. A single mold case can be used with different thicknesses of sand linings to produce tubes of various diameters within a limited range.

Green sand is commonly used as the liner in expendable molds. Various mixtures and binders are used—for example, a mixture of 60% silica sand and 40% calcined and crushed asbestos or sand bound with resin. Phenolic binders are also used with silica sand. One proprietary process uses a mixture of sand, silica flour, bentonite, and water.

Dry sand molds can also be employed; in this case, the sand is pressed down around a pattern having the same dimensions as the casting. Hardening is sometimes accomplished with carbon dioxide.

Mold washes of various compositions are used with sand molds. The wash hardens the mold surface and minimizes erosion of the mold by molten metal.

Permanent Molds. The most common materials for permanent molds are steel, copper, and graphite.

Steel molds are used to cast large quantities and for some casting alloys that require specific solidification conditions. Steel molds are sensitive to thermal shock; alumina- or zirconia-base mold sprays are used to lessen thermal shocks to the mold and to improve the mold surface. Mold coatings are also important in regulating the solidification rates of some casting materials. Other ceramic coating materials are beginning to be employed.

Copper molds are sometimes used for their high thermal conductivity. Their relatively high cost and the difficulty of calculating the correct dimensions of these molds limit their field of application.

Graphite Molds. Because of their relatively low cost, graphite molds can be an economical alternative to sand in the production of small quantities of parts. Graphite is the mold material of choice in the casting of 80% Cu bronzes, high-phosphorus brasses, and other copper alloys. Graphite has excellent thermal conductivity

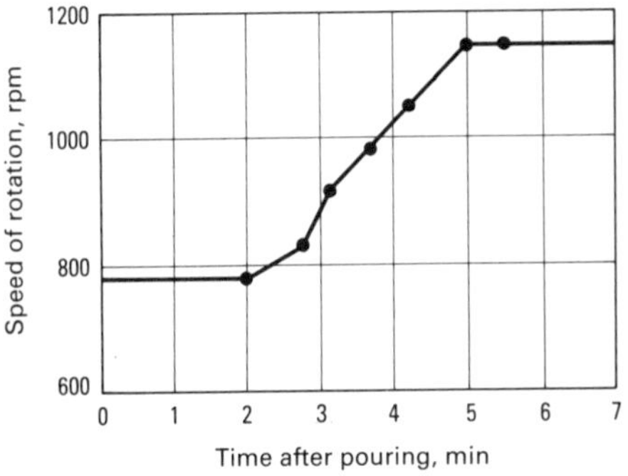

Fig. 2 Typical cycle of rotation in horizontal centrifugal casting

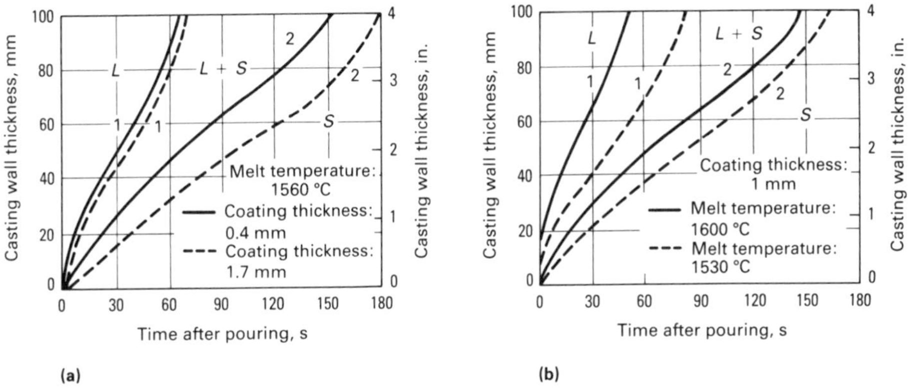

Fig. 3 Effect of mold coating thickness (a) and molten metal temperature (b) on solidification in horizontal centrifugal casting. Numbers 1 and 2 indicate liquidus and solidus curves, respectively.

and resistance to thermal shock, and it is easily machined. Care must be taken, however, to maintain the mold well below the oxidation temperature of graphite.

Table 1 Common material combinations used in bimetallic centrifugally cast tubes

Outer material	Inner material	Typical applications
Chilled iron	Gray iron	Bearing rollers
27% Cr cast iron	Ni-Resist cast iron	Mill rollers
Low-alloy steel	Manganese-molybdenum steel	Continuous casting rollers
Low-alloy steel	Ni-Hard cast iron, martensitic or 27% Cr white iron	Wear-resistant applications
Chromium-nickel white iron	Gray ductile iron	Mill rollers
Chromium white iron	Gray iron	Mill rollers
Stainless steel	Gray iron	Rollers for pulp and paper industry
Low-alloy steel	Pearlitic gray iron, stainless steel, or superalloy	Liners, pipelines for corrosives

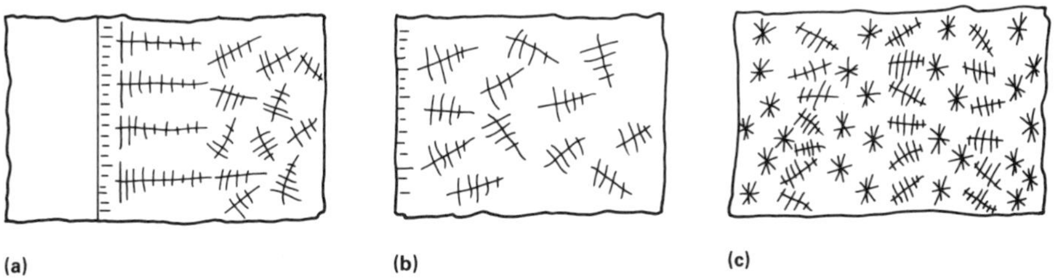

Fig. 4 Three types of as-cast structures seen in centrifugally cast ferrous alloys. (a) Fine columnar skin, large well-oriented columnar grains, and equiaxed area. (b) Completely equiaxed structure sometimes observed in ferritic steels. (c) Equiaxed bands of varying grain size. This type of structure is thought to be caused by machine vibrations.

Casting Techniques

Pouring. Molten metal can be introduced into the mold at one end, at both ends, or through a channel of variable length. Pouring rates vary widely according to the size of the casting being produced and the metal being poured. For example, a pouring rate of 1 to 2 Mg/min (1.1 to 2.2 tons/min) has been used to cast low-alloy steel tubes 5 m (200 in.) long and 500 mm (20 in.) in outside diameter with 50 mm (2 in.) thick walls. Pouring rates that are too slow can result in the formation of laps and gas porosity, while excessively high rates slow solidification and are one of the main causes of longitudinal cracking.

Casting Temperatures. The degree of superheat required to produce a casting is a function of the metal or alloy being poured, mold size, and physical properties of the mold material. The following empirical formula has been suggested as a general guideline to determine the degree of superheat needed:

$$L = 2.4\Delta T + 110 \qquad \text{(Eq 1)}$$

where L is the length of spiral fluidity (in millimeters) and ΔT is the degree of superheat (in degrees centigrade). The use of Eq 1 for ferrous alloys results in casting temperatures that are 50 to 100 °C (120 to 212 °F) above the liquidus temperature. In practice, casting temperatures are kept as low as possible without the formation of defects resulting from too low a temperature.

A high casting temperature requires higher speeds of rotation to avoid sliding; low casting temperatures can cause laps and gas porosity. Casting temperature also influences solidification rates and therefore affects the amount of segregation that takes place.

Mold Temperature. Numerous investigators have studied the relationship between initial mold temperature and the structure of the resultant casting. Initial mold temperatures vary over a wide range according to the metal being cast, the mold thickness, and the wall thickness of the tube being cast. Initial mold temperature does not affect the structure of the resultant casting as greatly as the process parameters discussed above do.

Speed of Rotation. Generally, the mold is rotated at a speed that creates a centrifugal force ranging from 75 to 120 g (75 to 120 times the force of gravity). Speed of rotation is varied during the casting process; Fig. 2 illustrates a typical cycle of rotation. The cycle can be divided into three parts:

- At the time of pouring, the mold is rotating at a speed sufficient to throw the molten metal against the mold wall
- As the metal reaches the opposite end of the mold, the speed of rotation is increased
- Speed of rotation is held constant for a time after pouring; the time at constant speed varies with mold type, metal being cast, and required wall thickness

The ideal speed of rotation causes rapid adhesion of the molten metal to the mold wall with minimal vibration. Such conditions tend to result in a casting with a uniform structure.

As the molten metal enters the mold, a pressure gradient is established across the tube thickness by centrifugal acceleration. This causes alloy constituents of various densities to separate, with lighter particles such as slags and nonmetallic impurities gathering at the inner diameter. The thickness of these impurity bands is usually limited to a few millimeters, and they are easily removable by machining.

Too low a speed of rotation can cause sliding and result in poor surface finish. Too high a speed of rotation can generate vibrations, which can result in circumferential segregation. Very high speeds of rotation may give rise to circumferential stresses high enough to cause radial cleavage or circular cracks when the metal shrinks during solidification.

Solidification

In horizontal centrifugal casting, heat is removed from the solidifying casting only through the water-cooled mold wall. Solidification begins at the outside diameter of the casting, which is in contact with the mold, and continues inward toward the casting inside diameter. Several parameters influence solidification:

- The mold, including the mold material, its thickness, and initial mold temperature
- The thickness and thermal conductivity of the mold wash used
- Casting conditions, including degree of superheat, pouring rate, and speed of rotation
- Any vibrations present in the casting system

Thermal Aspects of Solidification. It appears that the mold-related parameters listed above have relatively little influence on solidification. Large variations in mold thickness, however, could become significant.

The parameters with the greatest effect are the degree of superheat in the molten metal and the thickness of the mold wash employed. Both of these process variables affect local solidification conditions and therefore modify the structure of the casting. Figure 3 illustrates the general effects of mold wash thickness and degree of superheat on solidification rates. Charts such as Fig. 3 can be used to predict total solidification time. They are especially useful in determining the optimal casting conditions for bimetallic tubes based on the type of bond required.

Metallurgical Aspects of Solidification. The as-cast structures obtained in the horizontal centrifugal casting of steels vary according to composition. Regardless of the phase or phases that solidify first, certain features are common to the structures of centrifugally cast ferrous alloys (Fig. 4a):

- Very thin, fine columnar skin
- Well-oriented columnar structure adjacent to the skin
- More or less fine equiaxed structure

In the case of steels that solidify as ferrite, the columnar areas may be nonexistent if superheat and mold wash thickness are low (Fig. 4b). In steels that solidify as austenite, it is relatively easy to obtain well-oriented 100% columnar structures.

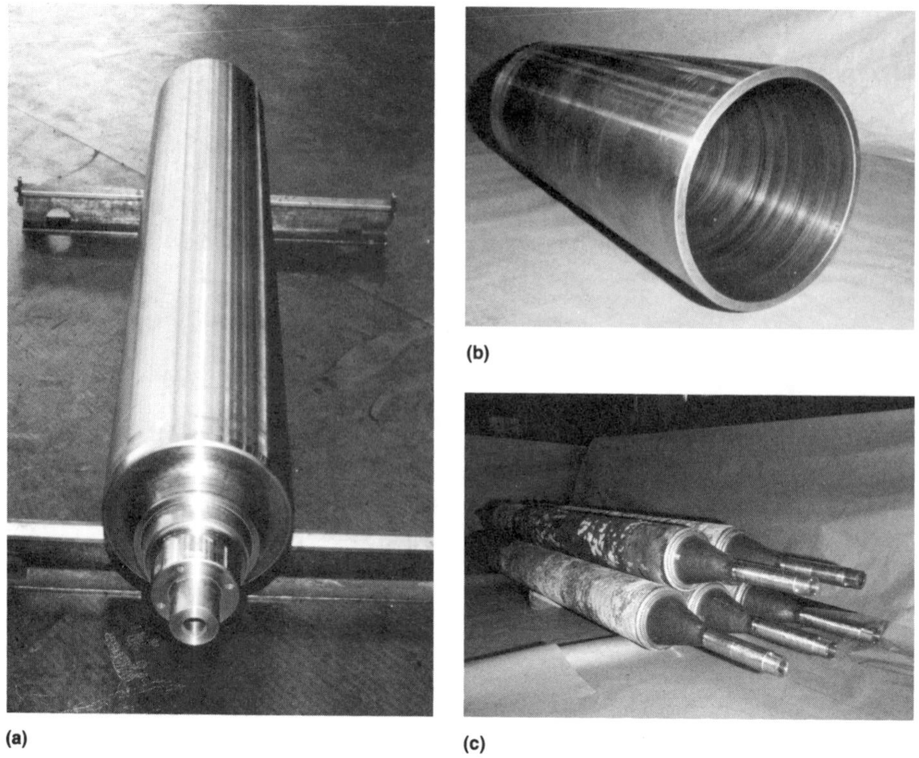

Fig. 5 Three applications for centrifugally cast parts in the iron and steel industry. (a) Continuous casting roller. (b) Winding spool. (c) Annealing furnace rollers

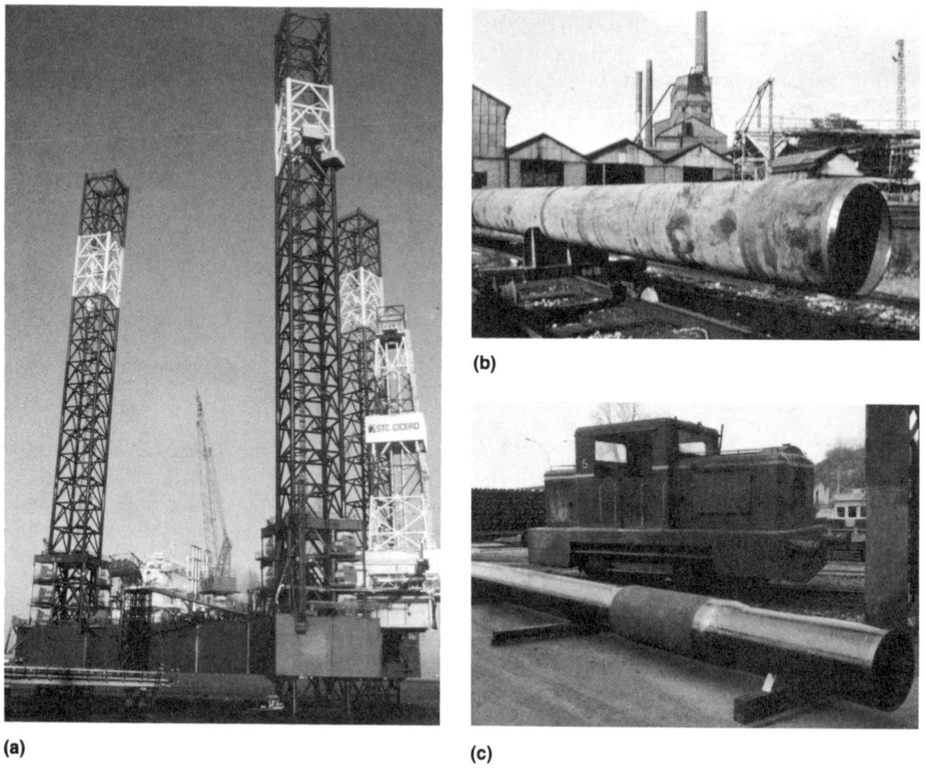

Fig. 6 Offshore petroleum production applications for centrifugally cast parts. (a) Jackup leg. (b) Risers. (c) Buckle-crack arrestor

A phenomenon specific to horizontal centrifugal casting is the formation of equiaxed bands through the entire thickness of the casting (Fig. 4c). A plausible explanation for this phenomenon is linked to machine vibrations, which may cause recirculation of the molten metal during solidification.

As in static casting, the rejection of solute ahead of the solidification front leads to microsegregation and to a progressive enrichment of the remaining liquid. Carbon steels are particularly sensitive to this effect; carbon, sulfur, and phosphorus contents must be limited to avoid local precipitation of carbides and sulfides.

Process Advantages

Flexibility in Casting Composition. Horizontal centrifugal casting is applicable to nearly all compositions with the exception of high-carbon steels (0.40 to 0.85% C). Carbon segregation can be a problem in this composition range.

Wide Range of Available Product Characteristics. The metallurgical characteristics of a tubular product are mainly characterized by its soundness, texture, structure, and mechanical properties. Centrifugal castings can be manufactured with a wide range of microstructures tailored to meet the demands of specific applications.

The high degree of microstructural control possible with horizontal centrifugal casting results in great flexibility in selecting properties for specific applications. Tubes can be manufactured with resistance to elevated temperatures, corrosion resistance, thermal fatigue resistance, low-temperature ductility, and so on. Centrifugally cast parts have a high degree of metallurgical cleanliness and homogeneous microstructures, and they do not exhibit the anisotropy of mechanical properties evident in rolled/welded or forged tubes.

Dimensional Flexibility. Horizontal centrifugal casting allows the manufacture of pipes with maximum outside diameters close to 1.6 m (63 in.) and wall thicknesses to 200 mm (8 in.). Tolerances depend on part size and on the type of mold used.

(a)

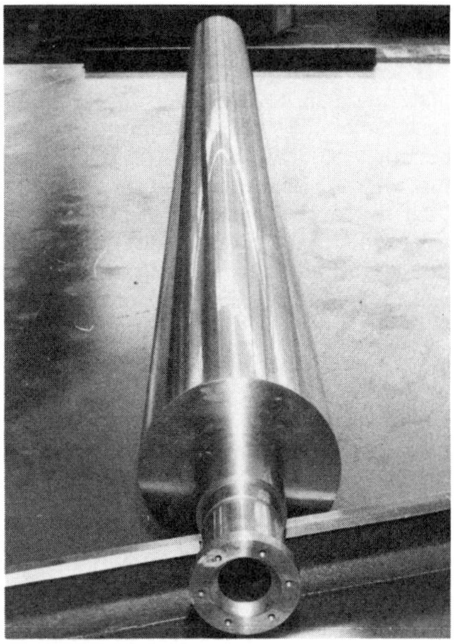

(b)

(c)

Fig. 7 Miscellaneous applications for centrifugal castings. (a) Hydraulic cylinders. (b) Float glass roller. (c) Exterior columns of Beaubour Museum, Paris

Materials

As mentioned at the beginning of this article, any material that can be statically cast can also be centrifugally cast. Materials that are currently being processed by horizontal centrifugal casting include high-strength low-alloy steels, duplex stainless steels, and chromium-molybdenum alloy steels. Also of importance are bimetallic tubes. These materials will be discussed briefly in this section. Specific applications for centrifugally cast tubes are discussed in the section "Applications" in this article.

High-strength low-alloy steels are important in offshore structures that must withstand extreme climatic conditions, for example, offshore drilling equipment in the North Sea. These materials must be weldable, with ductile-to-brittle transition temperatures below −40 °C (−40 °F). Likely candidate materials for use under such conditions are manganese-molybdenum steels with microalloying additions of vanadium, nickel, or niobium. Heavy-wall weldable pipes with good mechanical properties can be produced from such materials by horizontal centrifugal casting. Several proprietary alloys have been approved for use in offshore oil applications in the North Sea.

Duplex stainless steel tubes can be readily produced by horizontal centrifugal casting in any section size. Castings retain strength at temperatures to 600 °C (1110 °F), have good ductility, and are weldable without special precautions.

Chromium-molybdenum alloy steel tubes produced by horizontal centrifugal casting have homogeneous structures and are metallurgically sound. They are resistant to thermal fatigue and wear and have good toughness.

Bimetallic tubes with metallurgical (rather than mechanical) bonds can be readily produced by horizontal centrifugal casting. They are most commonly produced by successively casting one alloy inside the other. Bimetallic tubes are used for two primary reasons: to reduce cost by using an exotic material bonded to a less expensive backing material, and to obtain combinations of properties that could not be achieved by other methods.

There are no general rules regarding what materials can be combined in centrifugally cast bimetallic tubes, although it may be beneficial to cast the inner layer of such tubes from a material that is more fusible than the outer material. In addition, relatively thin inner layers should be manufactured from alloys with coefficients of thermal expansion smaller than that of the outer alloy. In this way, the thin inner layer is put into compression, making it more resistant to cracking. Table 1 lists materials that are commonly combined in bimetallic tubes, as well as their applications.

Applications

The flexibility of the horizontal centrifugal casting process, in terms of both materials and the wide range of part sizes that can be produced, has led to the application of centrifugally cast parts in many industries. Some of the most common applications are outlined briefly in this section.

Iron and Steel Industry. Centrifugally cast parts are used in the production of iron and steel for continuous casting rollers, rolling mill rolls, furnace rollers, special pipelines, winding spools, and other applications. Some of these uses are shown in Fig. 5.

Petroleum Production. Offshore production platforms in the oil and gas industry use centrifugally cast tubes in various applications (Fig. 6). Hot extruded bimetallic tubes are used in pipelines and gathering systems.

Other applications for centrifugally cast tubes include hydraulic cylinders, rollers for glass production, pipelines for the transport of abrasive materials, rollers in the pulp and paper industry, tubes for the chemical processing industry, foundation piles, and building columns. Some of these applications are shown in Fig. 7.

Vertical Centrifugal Casting

R.L. Dobson, The Centrifugal Casting Machine Company

There are essentially two basic types of centrifugal casting machines: the horizontal type, which rotates about a horizontal axis, and the vertical type, which rotates about a vertical axis. Horizontal centrifugal casting machines are generally used to make pipe, tubes, bushings, cylinder sleeves (liners), and cylindrical or tubular castings that are simple in shape (see the previous section "Horizontal Centrifugal Casting" in this article). The range of application of vertical centrifugal casting machines is considerably wider. Castings that are not cylindrical, or even symmetrical, can be made using vertical centrifugal casting. The centrifugal casting process uses rotating molds to feed molten metal uniformly into the mold cavity. Directional solidification provides for clean, dense castings with physical properties that are often superior to those of the static casting processes.

Centrifugal castings are produced by pouring molten metal into a rotating or spinning mold. The centrifugal force of the rotating mold forces the molten metal against the interior cavity (or cavities) of the mold under constant pressure until the molten metal has solidified. Cylindrical castings are generally preferred for the centrifugal casting process. Tubular castings produced in permanent molds by centrifugal casting usually have higher yields and higher mechanical properties than castings produced by the static casting process. Centrifugal casting is the most economical method of producing a superior-quality tubular or cylindrical casting with regard to casting yield, cleaning room cost, and mold cost.

Centrifugal castings can be best described as isotropic, that is, having equal properties in all directions. This is not true of a forging. By utilizing the outstanding advantage created by the centrifugal force of rotating molds, castings of high quality and integrity can be produced because of their high density and freedom from oxides, gases, and other nonmetallic inclusions. An economic advantage of centrifugal castings is the elimination or minimization of gates and risers.

All metals that can be cast by static casting can be cast by the centrifugal casting process, including carbon and alloy steels, high-alloy corrosion- and heat-resistant steels, gray iron, ductile and nodular iron, high-alloy irons, stainless steels, nickel steels, aluminum alloys, copper alloys, magnesium alloys, nickel- and cobalt-base alloys, and titanium alloys. Nonmetals can also be cast by centrifugal casting, including ceramics, glasses, plastics, and virtually any material that can be made liquid or pourable.

Sand molds, semipermanent molds, and permanent molds can be used for the centrifugal casting process. Selection of the type of mold is determined by the shape of the casting, the degree of quality needed, and the production (number of castings) required.

Centrifugal Casting Processes

There are three types of centrifugal casting:

- True centrifugal casting
- Semicentrifugal casting
- Centrifuge centrifugal casting

True centrifugal casting is used to produce cylindrical or tubular castings by spinning the mold about its own axis. The process can be either vertical or horizontal, and the need for a center core is completely eliminated. Castings produced by this method will always have a true cylindrical bore or inside diameter regardless of shape or configuration. The bore of the casting will be straight or tapered, depending on the horizontal or the vertical spinning axis used. Castings produced in metal molds by this method have true directional cooling or solidification from the outside of the casting toward the axis of rotation. This directional solidification results in the production of high-quality defect-free castings without shrinkage, which is the largest single cause of defective sand castings.

Semicentrifugal casting is used to produce castings with configurations determined entirely by the shape of the mold on all sides, inside and out, by spinning the casting and mold about its own axis. A vertical spinning axis is normally used for this method. Cores may be necessary if the casting is to have hollow sections. Directional solidification is obtained by proper gating, as in static casting. Castings that are difficult to produce statically can often be economically produced by this method, because centrifugal force feeds the molten metal under pressure many times higher than that in static casting. This improves casting yield significantly (85 to 95%), completely fills mold cavities, and results in a high-quality casting free of voids and porosity. Thinner casting sections can be produced with this method than with static casting. Typical castings of this type include gear blanks, pulley sheaves, wheels, impellers, and electric motor rotors.

Centrifuge centrifugal casting has the widest field of application. In this method, the casting cavities are arranged about the center axis of rotation like the spokes of a wheel, thus permitting the production of multiple castings. Centrifugal force provides the necessary pressure on the molten metal in the same manner as in semicentrifugal casting. This casting method is typically used to produce valve bodies and bonnets, plugs, yokes, brackets, and a wide variety of various industrial castings.

Mold Design

Gating Practice. The practicality of casting a part using the semicentrifugal or cen-

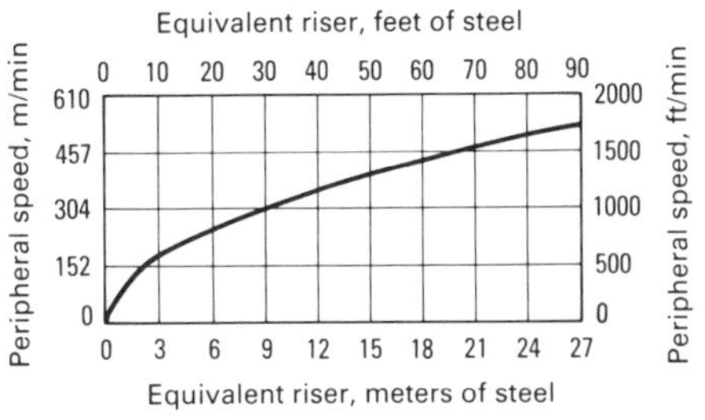

Fig. 1 Relationship between peripheral speed in centrifugal casting and equivalent riser

Table 1 Comparison of casting yields for miscellaneous static and centrifugal castings

Casting	Casting weight kg	lb	Static yield, %	Centrifugal yield, %
A	27.7	61	39.0	79.7
B	18.1	40	53.4	74.1
C	7.7	17	49.2	75.5
D	14.1	31	38.0	67.4
E	27.2	60	36.3	77.0
F	14.3	31.5	41.5	64.3
G	28.6	63	40.0	79.3
H	20.0	44	36.4	73.3
I	5.9	13	22.2	73.8

trifuge centrifugal process is determined by casting configuration. The gating system of centrifugal castings usually employs a single gate, which combines the function of gate and riser. Centrifugal force greatly magnifies the feeding action of the riser and produces a greater metal density in the casting than would otherwise result. Figure 1 shows that a centrifugal casting mold spinning at a peripheral velocity of 305 m/min (1000 ft/min) at the outer diameter of the casting(s) will have an equivalent head of 9 m (30 ft). This is equivalent to a pressure of 703 kPa (102 psi). A typical static casting has a head less than 0.3 m (1 ft) high. The centrifugal force acting on the molten metal will provide better feeding action than a static cast head; therefore, it is possible to feed molten metal into and through lighter and thinner mold sections into heavier mold sections much more easily than in static casting.

Certain casting configurations that do not inherently produce directional solidification can be made to directionally solidify by certain molding practices. Pads, auxiliary gates, chills, or blind heads can be used. However, a casting of nonuniform section thickness might require such elaborate mold methods that molding and cleaning costs become too high.

Sand Molds. When sand molds are used, particularly green sand, it is usually necessary to begin pouring the molten metal at a slow mold-spinning speed. When the mold is partially or wholly poured, the spinning speed is increased to that required to prevent or reduce erosion of the mold cavities from the molten metal. Molds should almost always be poured while rotating, even if the speed of rotation is only 5 rpm. This ensures the proper distribution of hot and cold metal in the mold for the optimum feeding action.

Green Sand Molds. Centrifugal castings can be made in green sand or dry sand molds. When green sand molds are used, flasks (preferably round) are required. Three methods can be used to fasten the green sand mold to the table of the centrifugal casting machine.

In the first method, two pins are fastened to the table over which the flask is lowered. The cope and drag are held together by clamps in the usual manner of static casting.

In the second method, a device similar to a lathe chuck is fastened to the table. The cope and drag are clamped together as in static casting. The green sand mold is lowered onto the table, and the clamps are tightened to secure the mold.

In the third method, the green sand mold is transported on a roller conveyor into a flask body, which is fastened to the table. Rollers are also used in the casting machine to facilitate movement of the green sand mold. The flask body opens in half on a hinge. A cover, which holds a runner cup, serves to distribute equally the pressure from the clamping arrangement to the mold. The clamping arrangement, which is part of the flask body, keeps the cope and drag in firm contact.

Dry Sand Molds. Flasks are not required when dry sand molds are used. Two methods are available for handling dry sand molds on the casting machine.

In the first method, the dry sand molds are placed in a jacket somewhat similar to that used in conjunction with snap flasks. The jacketed mold is then carried to the casting machine and lowered to the table. Clamps are used to hold the cover onto the mold and to fasten the mold firmly to the table.

In the second method, the dry sand molds are transported on a roller conveyor into the casting machine, in which rollers are incorporated. Using a flask body of this type, both dry sand and green sand molds can be used interchangeably in the same casting machine.

Molding Costs. A green sand mold prepared for static casting generally costs the same as a green sand mold prepared for centrifugal casting. If dry sand molds are used for centrifugal casting rather than green sand, the molding cost may be higher, resulting in a reduced cost savings. There is an appreciable cost savings in the cleaning room because risers need not be removed from castings. This reduces the amount of cutting, chipping, and grinding required to clean the castings. The additional cleaning room savings might be offset by operating costs, depreciation of the centrifugal casting equipment, and possible increased molding costs due to the use of dry sand molds. An increase in yield primarily due to the elimination of riser heads is indicated in Table 1, which compares casting yields for static and centrifugal processes.

Other Sand Molding Considerations. The requirements for a centrifugal sand mold or flask are more stringent than those for static cast molds because concentricity with the spinning axis must be more exact to prevent an imbalance that could cause vibration. However, most casting speeds required for centrifugal sand molding are relatively slow compared to true centrifugal casting in permanent molds. Molds made up of sections using dry sand cores are also used for centrifugal casting. The sand used can be sodium silicate, dry sand, chromite sand, or any molding material with sufficient strength to withstand the forces imposed by the spinning speed. It is desirable to use a suitable mold wash with most sand molds to reduce or minimize mold erosion from the molten metal. Excessive mold erosion generally does not occur, because of the relatively slow spinning speeds used.

Permanent Molds. Two basic types of permanent molds are in use: metal molds and graphite or carbon molds. Because of the faster heat extraction from permanent molds, there is usually an increase in the quality (especially properties) of the castings produced in this type of mold.

Metal Molds. A large number of cast iron molds are still used for centrifugal casting. However, steel molds are more common and much safer. Cast iron molds can be dangerous because of defects that may occur during the mold casting process and the fact that cast iron has relatively low tensile strength. In addition, when cast iron molds are water cooled, there is always the danger of cracking, rupture, or breakage. Most metal molds used in centrifugal casting are water cooled on the outside diameter using high-velocity water jets to increase solidification rates.

Steel molds are recommended for centrifugal casting. It is very important that the molds be perfect and free of any defects. The molds are machined all over to obtain a smooth surface and to ensure dynamic balance. A machined surface finish of 3 μm (125 μin.) RMS is recommended on the mold bore and outside diameter. If there are any defects on the portion of the mold that

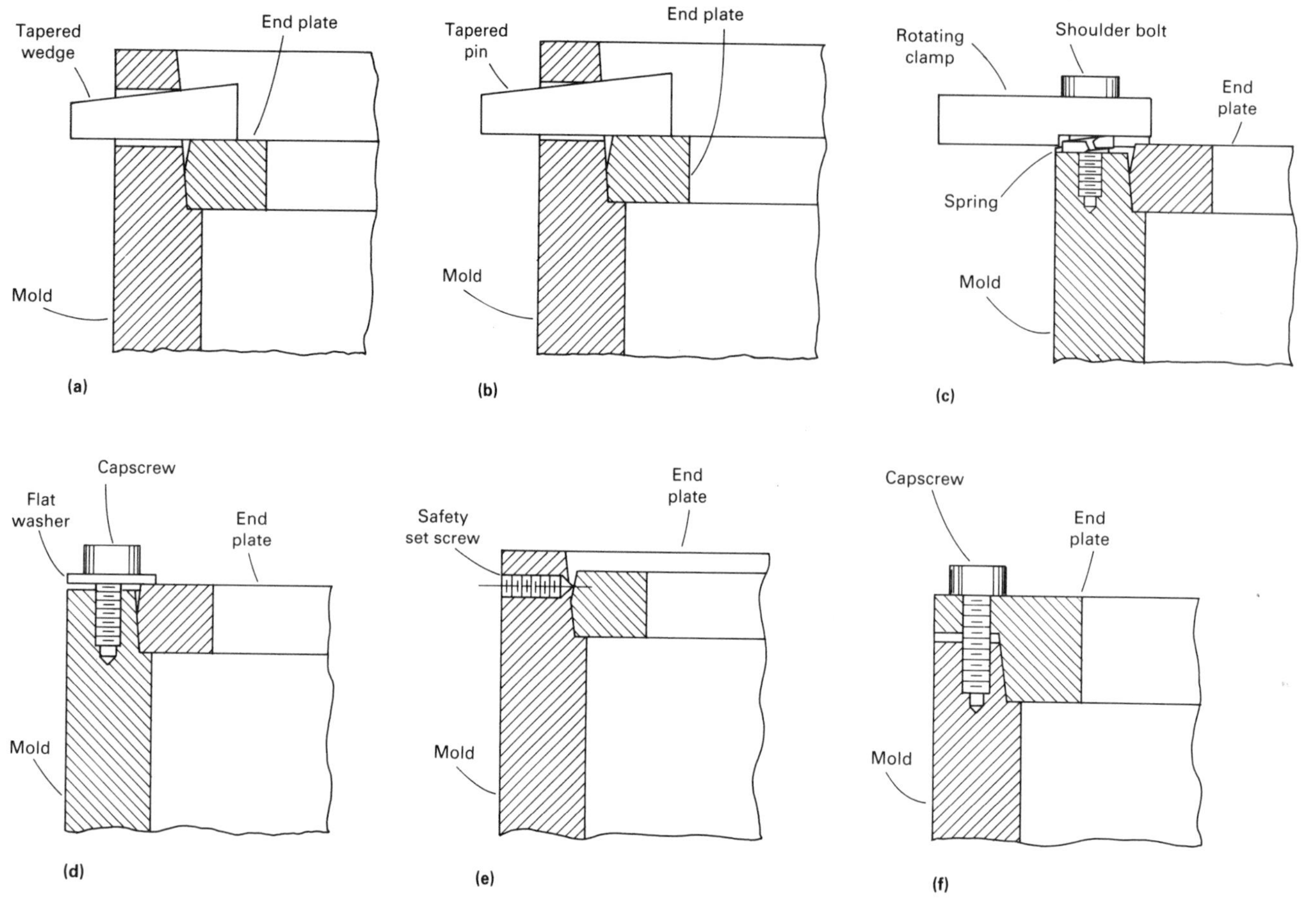

Fig. 2 Six end plate designs used in centrifugal casting. (a) through (c) are the most common.

contacts the casting, it will be difficult, if not impossible, to remove the casting from the mold. Steel molds should always be thoroughly stress relieved (minimum temperature of 620 °C, or 1150 °F) before the finish-machining operations. In the case of small, simple molds machined from heavy-wall seamless steel tubing, it has been standard practice not to stress relieve, because there is no apparent difference in the operational life of the mold. Stress relieving effectively prevents warpage in most molds and therefore prolongs the service life of the mold.

A molybdenum disulfide lubricant is recommended for all threaded fasteners used on a mold, such as mold lock clamps, end plate bolts, mold attaching bolts, and mold centering bosses.

Various mold end plate designs (Fig. 2) can be used, depending on individual preference and the size of the mold being used.

It is recommended that low-carbon steels be used for centrifugal molds. Alloy steels and steels with more than 0.30% should not be used, because they are subject to cracking due to thermal shock and heat stresses from the casting of the molten metal. Satisfactory mold materials are 1018, 1020, or ASTM A106, grade A, steel. It is usually most convenient to make small molds from either hot-rolled solid bar stock or from heavy-wall seamless steel tubing. Larger-diameter molds can be made from forgings or hot-rolled bar stock with the center trepanned or from centrifugally cast heavy-wall steel tubing.

Metal Mold Design. Machining allowances for castings depend on melting practices, the condition and thickness of the mold insulation coating, the volumetric shrinkage of the metal being cast, and the susceptibility of the metal to casting defects such as dross formation. The suggested machining allowances given in Table 2 can be used as a basis for mold design. With these minimum allowances, the mold can be remachined to allow for more machining allowance as determined by experience and the specific melting practices used.

Patternmaker's shrinkage is the shrinkage allowance built into the pattern to compensate for the change in dimensions caused by the contraction of the cast metal as it cools. This shrinkage allowance is the factor with which the moldmaker is concerned. Different casting alloys have different shrinkage rates. Although average values for unrestricted cooling conditions are available (Table 3), it must be remembered that these can vary considerably according to the resistance offered by molds and cores.

Mold Wall Thickness. Figure 3 can be used as a general guideline for determining the proper mold wall thickness for water-cooled steel molds. Wall thicknesses may vary depending on the specific equipment and application involved. Mold wall

Table 2 Minimum machining allowances for design of centrifugal casting molds

Casting size		Allowance on casting OD (each side)		Allowance on casting ID (each side)					
				Irons		Bronzes		Steels	
mm	in.	mm	in.	mm	in.	mm	in.	mm	in.
50–200	2–8	1.6	1/16	1.6	1/16	3.2	1/8	4.8	3/16
200–300	8–12	2.5	3/32	2.5	3/32	5	3/16	7.5	9/32
300–500	12–20	4	5/32	4	5/32	8	5/16	12	15/32
500–700	20–28	6	7/32	6	7/32	12	7/16	18	21/32
700–900	28–353	8	5/16	8	5/16	16	5/8	24	15/16
900–1100	35–43	10	3/8	10	3/8	20	3/4	30	1 1/8

Table 3 Guidelines for shrinkage allowances for various metals and alloys

Metal or alloy	Shrinkage allowance in./ft	mm/m
Aluminum alloys	5/32	13
Aluminum bronze	1/4	21
Yellow brass (thick sections)	5/32	13
Yellow brass (thin sections)	3/16	16
Gray cast iron(a)	1/10–5/32	8–13
White cast iron	1/4	21
Tin bronze	3/16	16
Gun metal	1/8–3/16	11–16
Lead	5/16	26
Magnesium	1/4	21
Magnesium alloys (25%)	3/16	16
Manganese bronze	1/4	21
Copper-nickel	1/4	21
Nickel	1/4	21
Phosphor bronze	1/8–3/16	11–16
Carbon steel	3/16–1/4	16–21
Chromium steel	1/4	21
Manganese steel	5/16	26
Tin	1/4	21
Zinc	5/16	26

(a) Shrinkage in cast iron depends mainly on the speed with which the casting cools. Greater shrinkage results in more rapid cooling, less shrinkage from slower cooling.

thickness is independent of casting wall thickness for water-cooled steel molds. Casting wall thicknesses exceeding 305 mm (12 in.) have been made using these recommended mold wall thicknesses.

End plate thicknesses depend on casting wall thickness and end plate material. Commonly used end plate materials are ASTM A36 steel, 1015 steel, or 1018 steel plate. In cases of extremely large diameter plates (>450 mm, or 18 in.), the end plate material can be pressure vessel quality ASTM A285 grade C or A515 grade 70.

For maximum adhesion of the mold wash to flat surfaces of the mold top and bottom end plates, it is advisable to have a machined surface finish of about 9 to 13 μm (350 to 500 μin.) RMS, with the lay approximately circular relative to the center of the plate. The opening in the top plate should be at least 13 mm (½ in.) smaller in diameter than the casting inside diameter but large enough to permit the molten metal to enter the mold.

The end plate recess in the mold provides an effective seal to prevent leakage of the molten metal from the mold. The amount of end plate recess used depends on mold inside diameter and can be as high as the mold inside diameter plus 25 mm (1 in.) for mold inside diameters of 150 mm (6 in.) or more.

End Plate Mold Lock Design. Various removable end plates are shown in Fig. 2. These end plate types are recommended for all vertical molds. Preferred end plate types are those shown in Fig. 2(a), (b), and (c). Some end plates can be turned over to prolong their service lives, and it is recommended that they be turned over between each cast.

Mold end plate diametral clearance fit into the mold recess diameter should be a minimum of 1.2 mm (3/64 in.). This clearance is effective for end plates up to about 380 mm (15 in.) in diameter. On larger end plates, clearances as large as 2.5 mm (3/32 in.) may be necessary for easy removal and assembly. The standard draft angle of 3° will permit easy removal and assembly.

The number of fasteners or mold clamps or wedges used depends on mold diameter. Recommended numbers of fasteners are:

- Three for molds up to 250 mm (10 in.) bolt circle or outside diameter
- Four for molds from 250 to 380 mm (10 to 15 in.) bolt circle or outside diameter
- Five for molds from 380 to 480 mm (15 to 19 in.) bolt circle or outside diameter

Large diameter molds or molds for extremely heavy wall castings (>50 mm, or 2 in., in wall section) require additional fasteners for safe operation.

End Plate Wedge and Pin Design. Wedges for holding the end plate secure in the mold are principally of two different types: the tapered wedge and the tapered pin. The maximum allowable stress for tapered wedges and pins is 69 MPa (10 ksi). Wrought steel bar stock, such as cold-rolled carbon steel or stainless steel, can be used; cast pins or wedges should not be used.

The following allowable safe loads can be used regardless of the wedge or pin material:

Holder	Maximum safe load kgf	lbf
Tapered wedge	1840	4060
Tapered pin	2350	5180

Mold Adapter Tables. A mold adapter table (Fig. 4) is usually furnished or required with a vertical centrifugal casting machine to facilitate attachment of the molds to the machine. Also furnished or required with the mold adapter table is a mold centering boss or index boss/plug for aligning and centering the mold so that it is concentric with the spinner shaft of the machine.

Two methods can be used to secure the mold to the adapter table. One method is to allow for a flanged extension of the bottom mold plate with holes for bolting directly into the table. The other method is to use a flanged extension of the mold bottom plate with dog clamps fastened to the table in the same manner in which a part is held on a mill table or vertical lathe.

The vertical centrifugal casting machine is available with a water-cooled bottom mold plate. Water cooling requires the machining of radial grooves or slots in the bottom of the bottom mold plate for water passage. Water cooling is sometimes advantageous for extremely heavy wall castings, which can transfer excessive heat into the bottom of the mold.

Pretreatment for Metal Molds. All molds, end plates, and surfaces to which mold wash is to be applied should be treated before use in the following manner:

- Preheat mold to approximately 150 °C (300 °F)

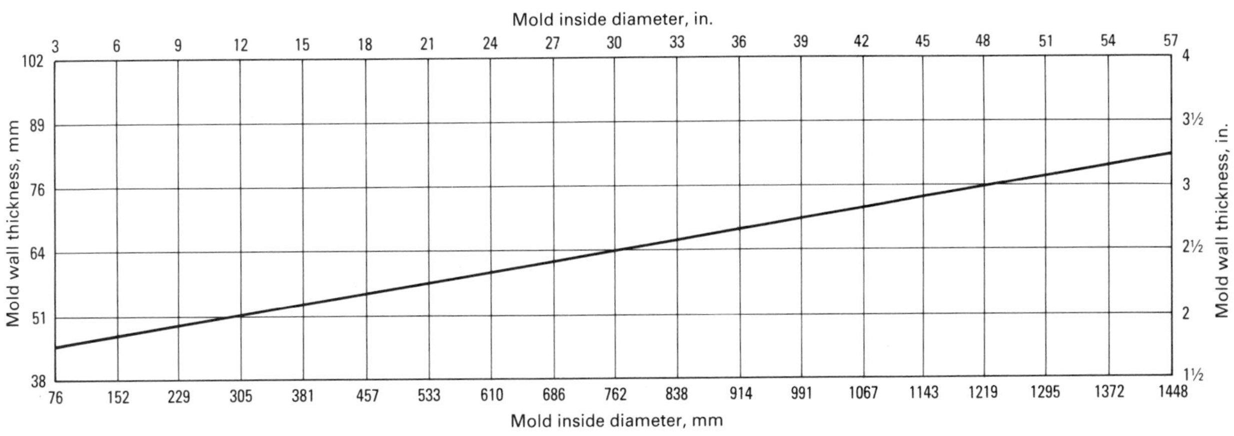

Fig. 3 Mold wall thickness as a function of mold inside diameter. Values are recommendations from one manufacturer for steel molds.

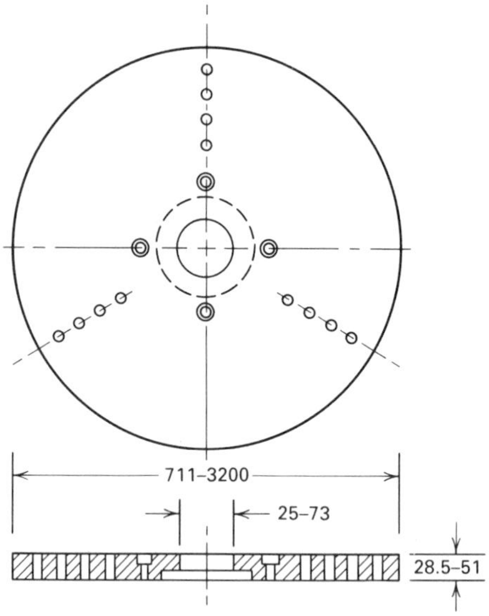

Fig. 4 Adapter table for mounting of molds on the centrifugal casting machine. Dimensional ranges are given in millimeters.

- Swab inside surface of mold with concentrated aqueous ammonium persulfate
- Flush mold with water to remove all contaminants. The mold can be put into service after this treatment

For new molds, the following treatment is recommended before use:

- Preheat mold to 205 to 260 °C (400 to 500 °F)
- Spray with a coating of mold wash approximately 0.8 mm (1/32 in.) thick
- Brush out mold wash, and repeat spray procedure. The mold should then be brushed out and resprayed a third time

The mold is ready for use after this treatment. If the wash still does not adhere, the mold should be cleaned again with ammonium persulfate as described above.

Mold Wash for Permanent Molds. The mold wash, when used on a permanent mold, serves as a refractory insulating material. When applied in sufficient thickness, the mold wash insulates the mold, thus reducing the surface temperature of the mold and increasing its useful life. A wide variety of centrifugal casting mold washes are commercially available, including silica, zirconia, and alumina washes. The centrifugal mold wash must be inert to the molten metal being cast.

The insulating characteristic of the mold wash is necessary to retard or slow the initial solidification rate in order to eliminate the formation of cold shuts, laps, droplets, and so on, and to produce a high-quality homogeneous outside surface on the casting. In most cases, a mold wash coating thickness of 0.8 mm (1/32 in.) is desired to obtain satisfactory castings. Spraying equipment is available for applying centrifugal casting mold washes.

Water Cooling of Permanent Steel Molds. The temperature of the mold increases with each casting poured. The mold can be operated at temperatures to 370 °C (700 °F). However, the usual operating temperatures of the mold should range from 150 to 260 °C (300 to 500 °F). To maintain this mold temperature for successive casting production, water cooling of the outside surface of the mold is generally employed. The high velocity of the water impinging upon the mold surface will prevent the formation of an insulating steam barrier, which would actually inhibit the extraction of heat from the mold. Quick-opening valves must be used to prevent warpage of the mold, and the spinning of the mold should be interlocked with the water valves to prevent spraying water on a mold that is not spinning and ultimately warping an expensive mold. In practice, the water cooling is usually activated immediately upon completion of the pouring and allowed to remain on long enough to permit the mold to be sprayed with the mold wash at the proper temperature range for the next casting cycle, after extracting the solidified casting.

Graphite and Carbon Molds. The choice between using a graphite or a carbon mold depends primarily on the availability to the user. Graphite has a higher rate of heat conductivity than carbon and is therefore sometimes used because of the desired metallurgical properties of the finished casting.

Graphite can be easily machined into a variety of intricate mold forms with a very good surface finish. Graphite molds have excellent chill characteristics, with thermal conductivity three times that of iron and a specific heat about double that of iron. The chilling ability of a material is roughly equal to the product of its mass multiplied by its specific heat. Graphite is nonreactive with most molten metals. In casting phosphorus bronze, graphite molds are not burned into as iron molds are. Carbon pickup is negligible in casting low-carbon stainless steel because of the quick chilling ability of the graphite, which almost instantaneously solidifies a skin layer of steel and therefore makes carbon pickup impossible. In certain cases where rapid chilling is not desired, an insulating mold wash is used. Graphite molds are extremely resistant to thermal shock, with a thermal conductivity about three times that of steel and a low Young's modulus. The strength of graphite molds increases with temperature.

Graphite molds are generally designed with two considerations: minimum wall thickness and weight ratio of mold to casting. A steel sleeve or master die holder is usually employed with a heat shrink-fit over the graphite mold. The steel is preheated to 425 °C (800 °F) for this shrink-fit; therefore, there is considerable compression upon the graphite mold. The graphite mold wall thickness must be sufficient to withstand the stress of the shrink-fit without breaking. For heat shrink-fit applications, the graphite mold wall should not be thinner than 19 mm (3/4 in.). The second factor is the minimum weight ratio of mold to casting to reduce the effects of graphite oxidation and to obtain the proper chilling effect. There are many factors involved in arriving at this ratio, but in general the ratio should be a minimum of 0.75. The factors that would influence a larger ratio include the heavy load imposed because of rapid succession of casts, greater chill depth desired, availability of graphite mold stock, and general flexibility in machining.

The graphite mold can be machined before or after the metal jacket or steel sleeve has been encased around the graphite. The expected life of a graphite mold depends on three major factors: stripping time, total cycle time, and permissible rebores.

Stripping time should be as long as possible to permit the casting to shrink away from the mold, particularly if the casting diameter is less than 150 to 200 mm (6 to 8 in.). With this precaution, the casting, with burrs, will not score the graphite mold wall nearly as much as when extracting the casting immediately after solidification when there is only negligible casting shrinkage.

Total cycle time between pours should be as long as possible to allow the graphite mold to cool to 95 to 150 °C (200 to 300 °F), thus reducing the time that the graphite will be above its oxidizing temperature of 425 to 480 °C (800 to 900 °F).

Permissible rebores are very influential in determining total mold life. If a casting size is of such tolerance that rebored molds cannot be considered and if there is no larger size of bushing or casting for which the mold can be used, it is obvious that total mold life will be considerably shortened.

Process Details

Casting Inside Diameters. When making castings on a vertical centrifugal casting machine, the inside diameter (bore) of the casting will be tapered in accordance with the following formula:

$$n = 264\sqrt{\frac{h}{r_1^2 - r_2^2}} \qquad \text{(Eq 1)}$$

where n is speed of rotation (in revolutions per minute), r_1 is the inside radius at the top of the casting (in inches), r_2 is the inside radius at the bottom of the casting (in inches), and h is casting height (in inches).

Actually, if the length of the casting does not exceed approximately twice its inside diameter, the amount of taper will be negligible. The optimal speed of rotation results in a centrifugal force of 75 g (75 times the force of gravity) on the inside diameter. It can be seen from Eq 1 that too slow a speed

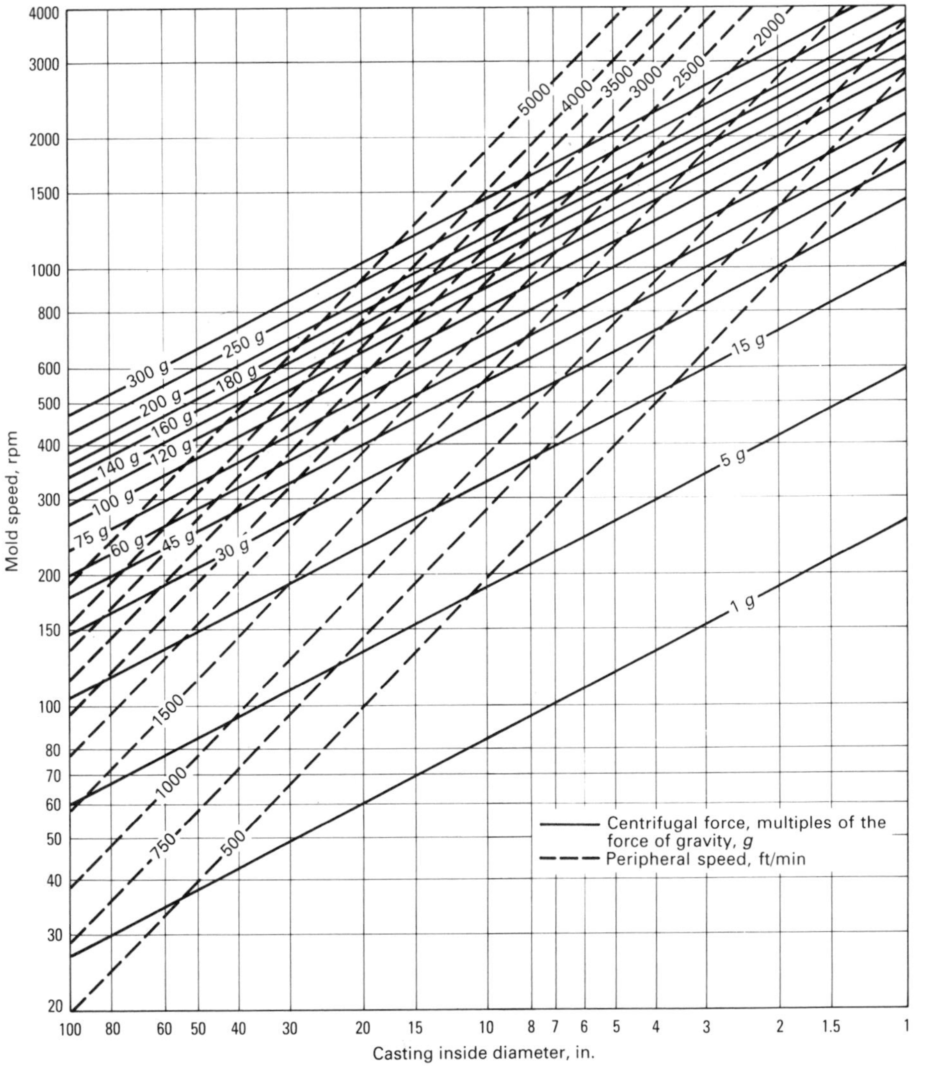

Fig. 5 Nomograph for determining mold speed based on the inside diameter of the casting and the required centrifugal force. See text for example of use.

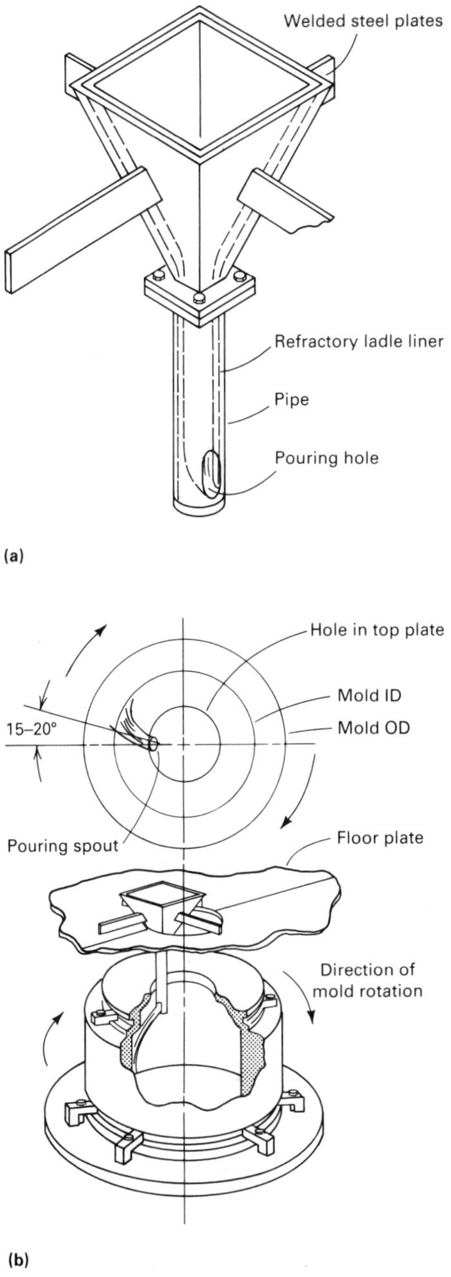

Fig. 6 Typical pouring spout design (a) and pouring spout positioning during casting (b)

of rotation will result in excessive taper on the inside diameter of the casting.

There are castings for which it is desirable to cast the inside diameter with a predictable taper. Using Eq 1, the exact speed can be calculated to produce an approximate given taper on the inside diameter of the casting.

Speed of Rotation. To establish a temperature gradient of the molten metal from the outside diameter toward the center of rotation (that is, directional solidification), it is usually necessary for the mold to be spinning when the metal is poured. In some cases, in order to eliminate defects such as erosion and dirt in sand molds, it is desirable to pour at a slow speed of rotation. However, true centrifugal castings having a wall section of 12.7 mm (½ in.) or less must be poured at spinning speed because the metal in this thin section solidifies quickly.

Nomographs are available for determining the proper speed of rotation for centrifugal casting. However, Eq 2 can be used to calculate spinning speed:

$$g = 0.0000142Dn \quad \text{(Eq 2)}$$

where g is the centrifugal force (in pounds per pound or number of times gravity), D is the inside diameter of the casting (in inches), and n is the speed of rotation (in revolutions per minute). Equation 2 can be easily manipulated to solve for speed.

Mold Speed Curves. Mold speeds are determined by the inside diameter of the castings to be made. The mold speed curve shown in Fig. 5 is based on the inside diameter of the casting. The length of the casting is not considered in determining mold speed.

For example, the mold speed for producing a casting 100 mm (4 in.) in outside diameter by 75 mm (3 in.) in inside diameter at a centrifugal force of 60 g is calculated as follows. Find the 3 in. diameter at the bottom of the curve. Move vertically from this point until the 3 in. line intersects the diagonal line marked 60 g. From this intersection, move directly to the right-hand edge of the curve; the speed of rotation of the mold in this case should be 1150 rpm.

Caution. From the standpoint of safety for vertical centrifugal casting, it is highly recommended that the g force acting on the outside diameter of the mold be considered. It is safe practice to limit this force to approximately 200 g on the outside diameter of the mold. After the proper mold speed is determined from the mold speed curve, this speed should be checked with the mold outside diameter to limit the g force on the outside diameter to less than 200 g. If it is found that the force is more than 200 g, the

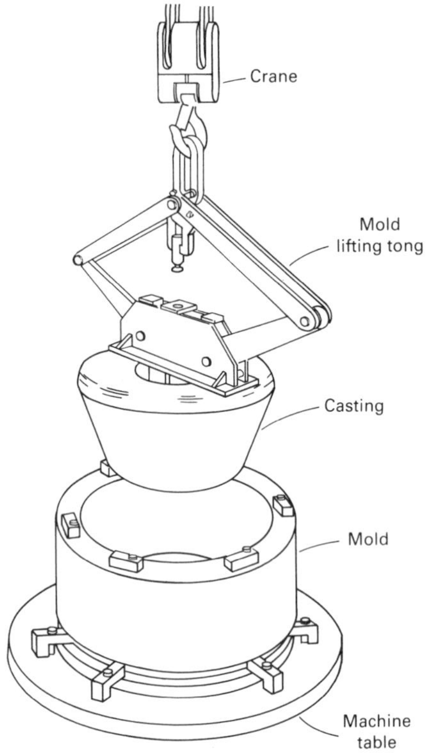

Fig. 7 Lifting tong assembly used to extract centrifugal castings from the mold

speed of rotation should be slowed so that 200 *g* is not exceeded on the mold outside diameter.

Pouring Techniques. For permanent molds, the metal is generally poured about 40 °C (100 °F) higher than the temperature used for the same casting if poured statically in a sand mold. This is because of the more rapid chilling effect of permanent molds.

The pouring rates required for successful permanent mold centrifugal casting are quite high compared to those for static casting in sand molds. It is particularly important that the initial rate of pour at the beginning be very high to prevent cold laps and cold shuts. For most types of centrifugal castings weighing less than 45 kg (100 lb), a pour rate of about 9 kg/s (20 lb/s) is recommended. For castings weighing up to 450 kg (1000 lb), an initial pour rate of 9 to 23 kg/s (20 to 50 lb/s) is recommended. For castings weighing more than 450 kg (1000 lb), pour rates of 45 to 90 kg/s (100 to 200 lb/s) are recommended. When pouring into a vertically spinning mold, it is important to introduce the molten metal into the mold in such a way as to prevent or minimize turbulence of the molten metal, which can cause splashing, spraying, or droplets and can result in undesirable casting defects.

Although many vertical centrifugal castings can be poured directly into the mold from the ladle to produce a quality centrifugal casting, it is more often desirable to use a pouring funnel. With a pouring funnel, the nozzle can be lined to the required diameter so that, with a certain riser height of molten metal in the funnel, a controlled pour rate can be obtained for a particular casting weight. In addition, with a pouring funnel, the entry of molten metal into the mold can be made to impinge upon the body of the mold with initial metal flow in the direction of mold rotation. This type of pouring will provide superior casting quality by minimizing or eliminating any upsetting turbulence in the flow of molten metal that might cause defects. Figure 6 shows a pouring funnel and funnel position.

Table 4 Load capacities of the vertical centrifugal casting machines of one manufacturer

All loads are based on balanced loads for thrust loads only and are calculated based on an L_{10} bearing life of 100 000 h.

Machine speed, rpm	Capacity of indicated proprietary model number: Model AS, Model C, kg	lb	Model VS, Model VSC, Model E, kg	lb
100	6930	15 270	21 260	46 875
200	6930	15 270	17 275	38 085
300	6130	13 520	15 355	33 850
400	5620	12 395	14 025	30 925
500	5260	11 595	13 140	28 970
600	4980	10 980	12 400	27 345
700	4755	10 485	11 810	26 040
800	4570	10 075	11 370	25 065
900	4410	9 720	11 075	24 415
1000	4270	9 415	10 690	23 565
1100	4150	9 150	...	...
1200	4045	8 915	...	...
1300	3950	8 710	...	...
1400	3860	8 510	...	...
1450	3825	8 430	...	...

Extraction of Castings. Commercially available casting pulling tongs (Fig. 7) are recommended for extracting vertical centrifugally cast castings. These pullers engage onto the inside diameter of the casting and are used to lift the casting from the mold.

Defects in Centrifugal Casting

Segregation banding occurs only in true centrifugal casting, generally where the casting wall thickness exceeds 50 to 75 mm (2 to 3 in.). It rarely occurs in thinner-wall castings. Banding can occur in both horizontal and vertical centrifugal castings.

Bands are annular segregated zones of low-melting constituents, such as eutectic phases and oxide or sulfide inclusions. They are characterized by a hard demarcation line at the outside edge of the band that usually merges into the base metal of the casting.

Most alloys are susceptible to banding, but the wider the solidification range and the greater the solidification shrinkage the more pronounced the effects may be. Banding has been found when some critical level of rotational speed is attained, and it has been associated with very low speeds, which can produce sporadic surging of molten metal. Therefore, both mechanisms may be involved. Minor adjustments to casting operation variables, such as rotational speed, pouring rate, and metal and mold temperatures, will usually reduce or eliminate banding.

Various theories have been presented to account for banding. One holds that vibration is the principal cause and that during solidification a zone of low-melting liquid exists immediately adjacent to the main crystal growth. Nucleation can occur, and if disturbed by vibration, banding results. This theory further states that growth takes place from these new nuclei in such a manner as to form a sandwich of liquid metal surrounded by solid metal, which is isolated from the liquid bath at the bore.

Another theory proposes that banding can be caused by variations in gravitational force between the top and bottom of the mold and that centrifugal separation of the constituents of the metal occurs once per revolution during the period of solidification. This theory does not bear close examination because it assumes that solidification will occur in a matter of seconds even in very thick sections. In reality, solidification takes place over a period of many minutes in thick sections.

Another explanation for banding supposes that the process is far less complicated and that there are irregularities in the flow of the liquid metal (incipient laps) as it enters the rotating mold. As the metal enters the mold, it forms a tubular casting that tends to solidify in the normal manner; however, if the additional liquid metal arrives at a particular position too late, the initial liquid metal has already partially solidified. This would result in a distinct lap, cold shut, or lamination and/or banding.

Raining. In a horizontal machine, raining can occur if the mold is rotated at too low a speed or if the metal is poured into the mold too fast. In this phenomenon, the metal actually rains or falls from the top of

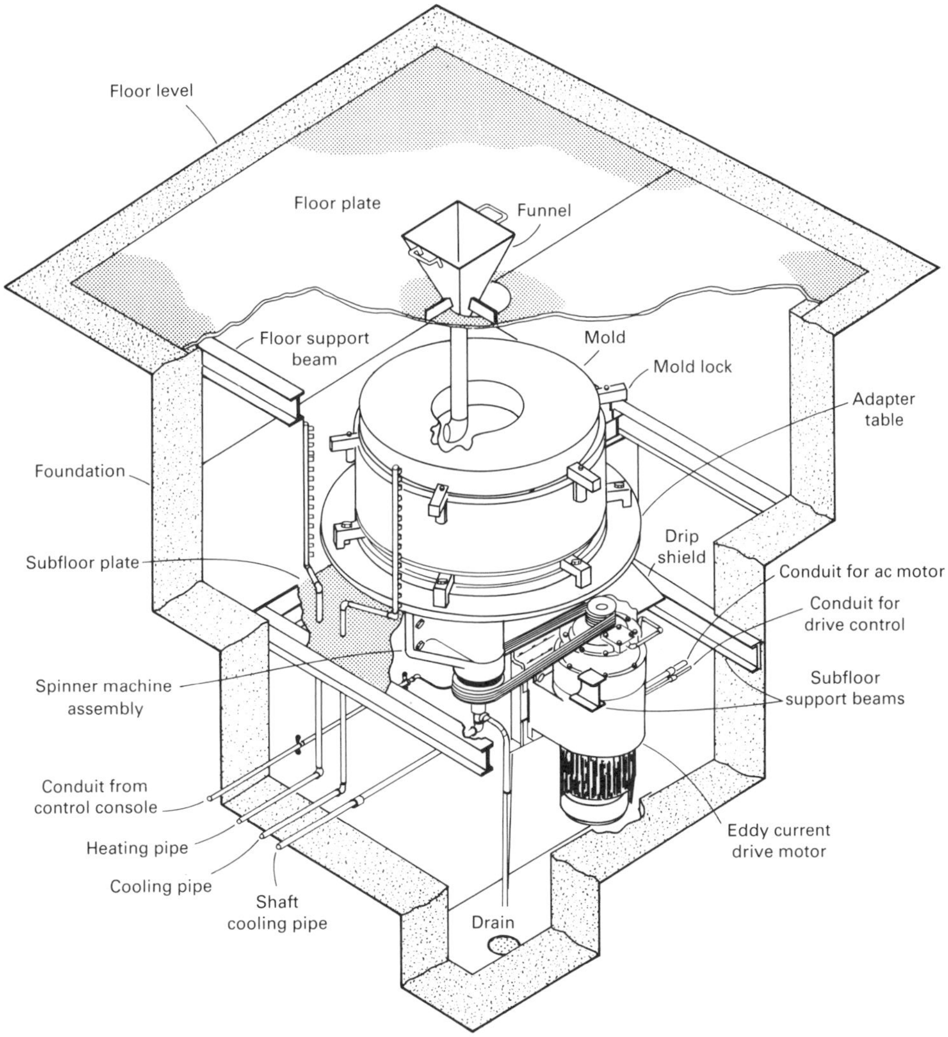

Fig. 8 Typical installation of a vertical centrifugal casting machine. The equipment is controlled from a remote console (not shown).

the mold to the bottom. More information on raining is available in the section "Horizontal Centrifugal Casting" in this article.

Vibration Defects. Vibration can cause a laminated casting. It can be held to a minimum by proper mounting, careful balancing of the molds, and frequent inspection of rollers, bearings, and other vital parts.

Equipment

Vertical centrifugal casting machines are used for producing bushings and castings that are relatively large in diameter and short in length. The usual maximum length of the casting is two times the outside diameter of the casting. Vertical axis machines are also used for producing castings of odd or asymmetrical configurations. Table 4 lists the load capacities of vertical centrifugal casting machines from one manufacturer.

Safety. The importance of safety in machine design, mold design, installation, and operation cannot be overemphasized. A centrifugal casting machine and installation should incorporate all of the safety factors possible. Centrifugal force increases directly as the square of the speed of rotation and directly as the length of the radius from the axis of rotation. Centrifugal force can be tremendous and very destructive. Speeds of rotation should never exceed those required to produce the casting. Molds must be centered on the spinning axis as accurately as possible and must be statically or dynamically balanced if necessary.

The method of attachment of both bottom and top cover plates to the mold is of great importance for withstanding and containing the force of the molten fluid metal while spinning. All clamping arrangements should be designed to tighten with, or to be unaffected by, centrifugal force. Molds should be firmly clamped to the table because unbalanced molds may fly off the table during operation. Adequate safety guards with interlocks should be used around all machines to protect workers from molten metal, which can spray from the mold if too much metal is poured.

Heat expansion can occur suddenly and rapidly in the mold body, top and bottom mold plates, and even into the centrifugal casting machine. This sudden expansion can put bending and shearing stresses into fasteners and other retention and clamping devices. A thorough understanding of the forces involved in centrifugal casting is necessary to ensure the utmost in safety for all concerned.

Moisture in a sand mold or moisture in the mold wash can turn into steam when the molten metal contacts it, and the resulting forces would be incalculable. Most vertical centrifugal casting machines should be installed completely below floor level for maximum operator safety. Figure 8 illustrates such an installation.

Continuous Casting

Robert D. Pehlke, University of Michigan

THE ADVANTAGES OF CONTINUOUS CASTING in primary metals production have been recognized for more than a century. In recent decades, a dramatic growth of this processing technology has been realized in both ferrous and nonferrous metal production. The principal advantages of continuous casting are a substantial increase in yield, a more uniform product, energy savings, and higher manpower productivity. These advantages and the ease of integration into metals production systems have led to the wide application of continuous casting processes.

Historical Aspects of Continuous Casting

One of the earliest references to continuous casting is a patent granted in 1840 to George Sellers, who had developed a machine for continuously casting lead pipe (Ref 1). There is some indication that this process had been underway before Sellers' patent, which was directed toward improvement of this continuous casting process. The first work on continuous casting of steel was by Sir Henry Bessemer, who patented a process for "manufacture of continuous sheets of iron and steel" in 1846 and made plant trials on continuous casting of steel in the 1890s (Ref 2).

Although continuous casting had its start before the beginning of the 20th century, it was not until the mid-1930s in Germany that commercial production of continuously cast brass billets was introduced. Sigfried Junghans, an active inventor of casting technology, provided many improvements in the process, in particular the introduction of the oscillating-mold system to prevent the casting from sticking to the mold. Further development of the process for the casting of nonferrous metals continued, including the installation of processing units in North America. Mold lubrication in the form of oil, or, more recently, low-melting slag powders, was introduced. Taper of the mold to compensate for metal shrinkage on solidification provided improved heat transfer and, more importantly, fewer cracks.

In 1935, a plant with casting rolls for continuous production of brass plates was operated at Scovill Manufacturing in the United States and the Vereinigte Leichtmetallwerke in 1936 started a semicontinuous casting machine for aluminum alloys. Immediately after World War II, commercial development of continuous casting of steel began in earnest, with pilot plants at Babcock and Wilcox Company (United States), Low Moor (Great Britain), Amagasaki (Japan), Eisenwerk Breitenfeld (Austria), BISRA (Great Britain), and Allegheny Ludlum Corporation (United States). These were followed by production plants for casting billets in the West and stainless slabs in the Soviet Union and Canada (the latter at Atlas Steels) (Ref 3).

The Schneckenburger and Kung patent on the curved strand was filed in Switzerland in 1956 and production was commercialized with a billet machine at Von Moosche Eisenwerke (Switzerland) in 1963. In 1961, at Dillingen Steelworks (West Germany), the first vertical-type large slab machine with bending of the strand to horizontal discharge was started up. In 1964, Shelton Iron and Steel (Great Britain) was the first new steelworks to turn out its entire production by continuous casting, consisting of four machines with 11 strands for medium to very large bloom sizes, and operating in connection with Kaldo converters. That same year, the first Concast S-type curved-mold machine for large slabs was started up at Dillingen Steelworks (West Germany). The height of this type of machine was less than 50% of the corresponding height of a vertical type of machine. In the same year, a bow-type slab caster was presented by Mannesmann (West Germany). In 1968, McLouth Steel (United States) started up four curved-strand slab casting machines immediately after the first four-strand low head caster for large slabs in the West was started up at the Weirton Steel Division of National Steel (United States). Subsequently, National pioneered the casting of slabs for tinplate applications (Ref 3).

Continuous Casting of Steel

Continuous casting of steel is entering a new era of development, not only with respect to its increasing application in the production process, but also in its own evolution as a process and its interaction with other processes in steel manufacture. Continuous casting output has shown an accelerating growth curve. More than 50% of current world steel production is continuously cast, and continuous casting in Japan exceeds 80%. The advantages of the process, along with its developments and current challenges for improvement, are outlined in the following sections.

General Description of the Process

The purpose of continuous casting is to bypass conventional ingot casting and to cast to a form that is directly rollable on finishing mills. The use of this process should result in improvement in yield, surface condition, and internal quality of product when compared to ingot-made material.

Continuous casting involves the following sequence of operations:

- Delivery of liquid metal to the casting strand
- Flow of metal through a distributor (tundish) into the casting mold
- Formation of the cast section in a water-cooled copper mold
- Continuous withdrawal of the casting from the mold
- Further heat removal to solidify the liquid core from the casting by water spraying beyond the mold
- Cutting to length and removing the cast sections

A diagram showing the main components of a continuous casting machine is presented in Fig. 1. Molten steel in a ladle is delivered to a reservoir above the continuous casting machine called a tundish. The flow of steel from the tundish into one or more open-ended, water-cooled copper molds is controlled by a stopper rod-nozzle or a slide gate valve arrangement. To initiate a cast, a starter, or dummy bar, is inserted into the mold and sealed so that the initial flow of steel is contained in the mold and a solid skin is formed. After the mold has been filled to the desired height, the dummy bar is gradually withdrawn at the same rate that molten steel is added to the mold. The initial liquid steel freezes onto a suitable attachment of the dummy bar so

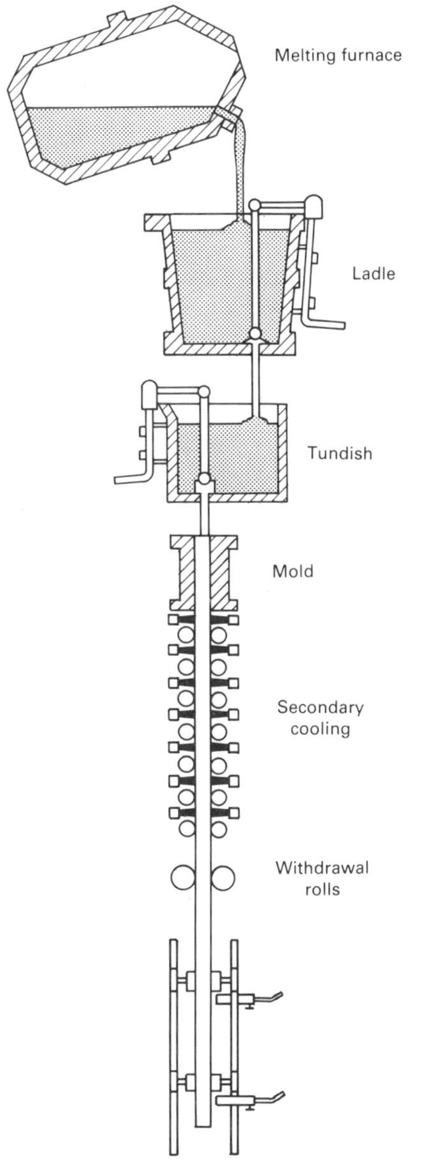

Fig. 1 Main components of a continuous casting strand. Source: Ref 4

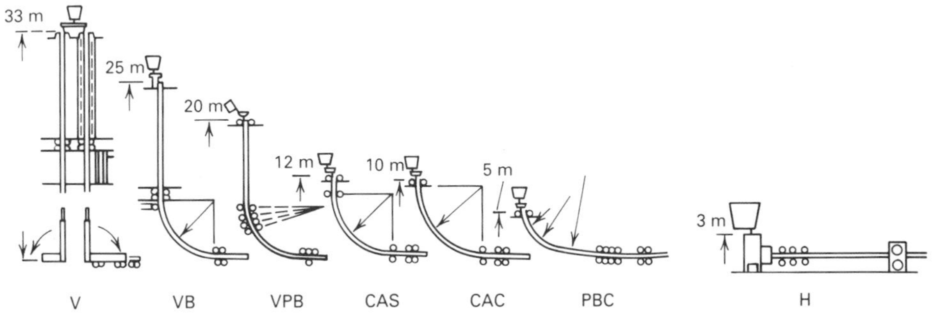

Fig. 2 Principal types of continuous casting. V, vertical; VB, vertical with bending; VPB, vertical with progressive bending; CAS, circular arc with straight mold; CAS, circular arc with curved mold; PBC, progressive bending with curved mold; H, horizontal. Source: Ref 5

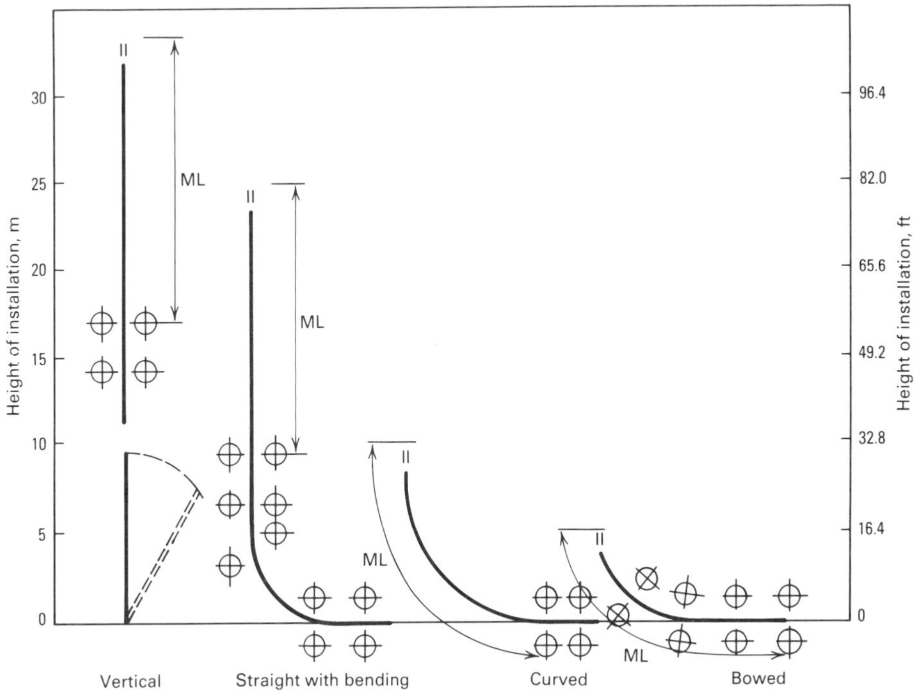

Fig. 3 Four basic designs for continuous casting machines. ML, metallurgical length. Source: Ref 6

that the cast strand can be withdrawn down through the machine. Solidification of a shell begins immediately at the surface of the copper mold. The length of the mold and the casting speed are such that the shell thickness is capable of withstanding the pressures of the molten metal core upon exiting from the copper mold. To prevent sticking of the frozen shell to the copper mold, the mold is normally oscillated during the casting operation and a lubricant is added to the mold. The steel strand is mechanically supported by rolls below the mold where secondary cooling is achieved by spraying cooling water onto the strand surface to complete the solidification process. After the strand has fully solidified, it is sectioned into desired lengths by a cutoff torch or shear. This final portion of the continuous casting machine also has provision for disengagement and storage of the dummy bar.

Several arrangements are now in commercial use for the continuous casting of steel. The types of continuous casting machines in use include vertical, vertical with bending, curved or S-strand with either straight or curved mold, curved strand with continuous bending, and horizontal. Examples of the principal types of machines currently producing slabs are shown in Fig. 2.

Most of the original continuous casting machines for steel were vertical machines. Vertical machines with bending and curved strand machines, although more complicated in their construction, were developed to minimize the height of the machine and allow installation in existing plants without modification of crane height. Four basic caster designs for slabs are shown in Fig. 3, with an indication of the required installation height and the corresponding solidification distance or metallurgical length (ML).

Plant Layout

The design and layout of a steelmaking facility often focus initially on continuous casting. The optimum plant layout varies markedly from one installation to another. One major factor in the configuration is whether or not the steelmaking complex is being constructed on a greenfield site or being added ("shoehorned") into an existing works. Many of the major integrated steel works in Japan were constructed as greenfield installations during the period from 1960 to 1975. Most of the minimills constructed throughout the world, and in particular in the United States, were also built on greenfield sites.

In building a greenfield site, the plant layout should incorporate two major features: a smooth and well-organized arrange-

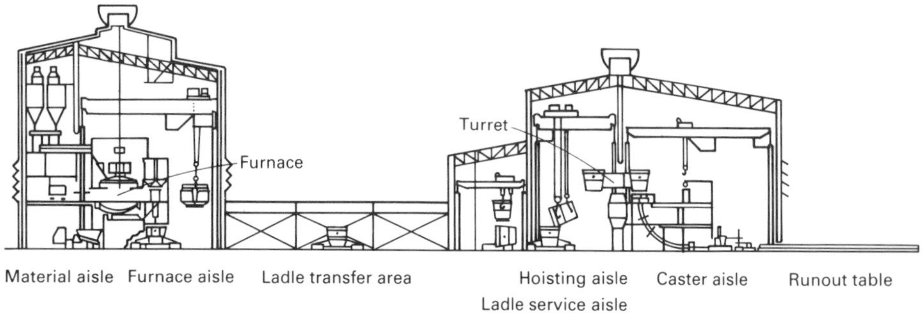

Fig. 4 Proposed melt shop capable of producing 1.2 million Mg (1.32 million tons) of billets annually. Source: Ref 9

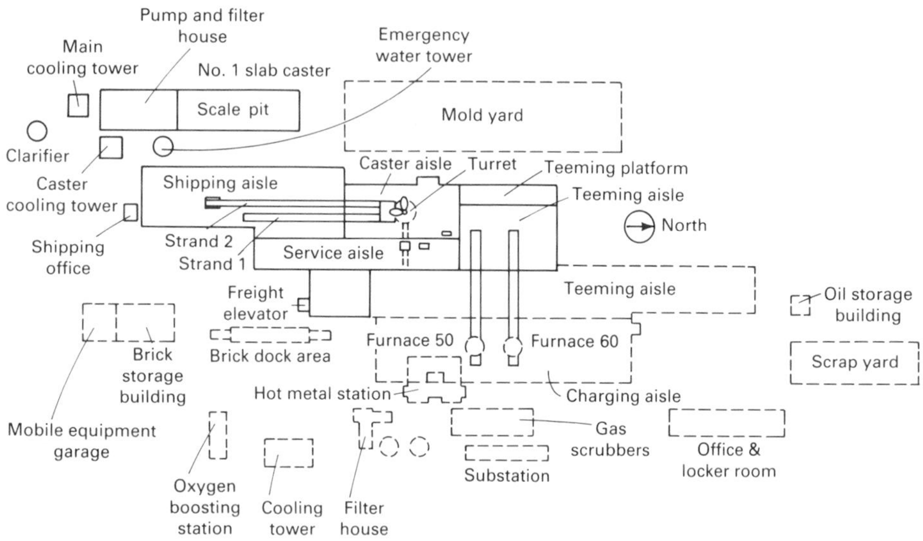

Fig. 5 Plant layout at the Inland Steel Company. Source: Ref 10

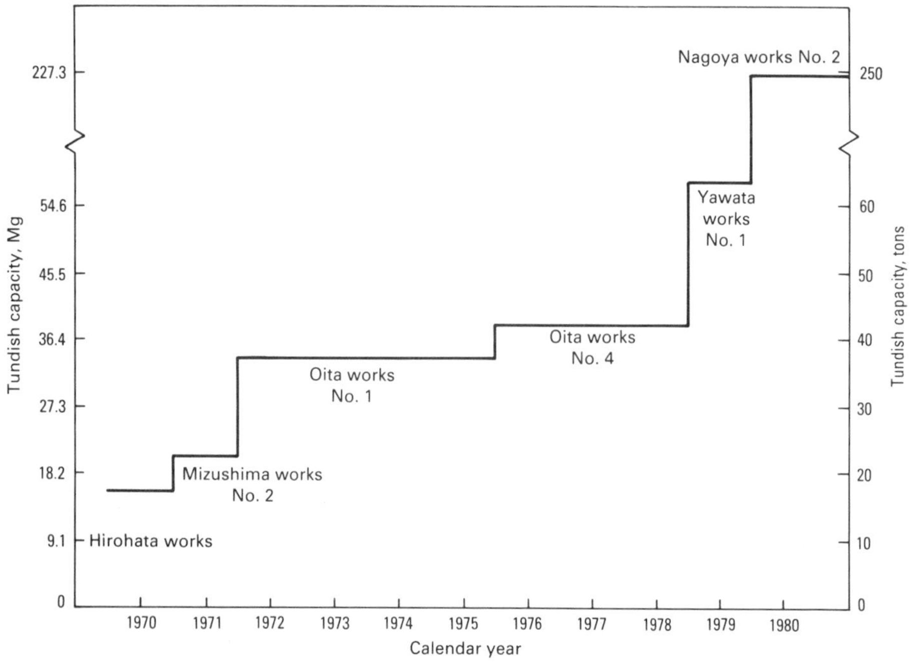

Fig. 6 Increases in capacity of tundishes for a two-strand slab caster. Source: Ref 11

ment for material handling and flow, and the capacity for future expansion. Generally, these plants are designed for 100% continuous casting, and no ingot facilities are included (Ref 7). Nearly all of the recently built minimills incorporate one or more electric furnaces and provide for 100% continuous casting of billets. Chaparral Steel Company at Midlothian, TX is an excellent example (Ref 8). A profile of a proposed large-scale electric furnace billet casting plant is shown in Fig. 4.

A twin-strand slab caster was shoehorned into the Number 4 basic oxygen furnace shop at Inland Steel Company. The addition of this caster substantially increased the output of this facility, where ingot casting was the rate-limiting step in production. The arrangement of this installation showing the caster and ingot facilities is presented in Fig. 5.

An important characteristic of the plant layout, and in particular of the material-handling facilities, is the concept that the continuous casting machine cannot wait. This design and operating concept has had a dramatic impact on steelmaking operations, which have now become a synergism to the continuous casting facilities. Because of this shift in priorities, marked improvements in productivity have been developed for continuous casting, as outlined below.

Dramatic increases in energy costs, as well as the desire for higher productivity, led to the development of the "hot connection". Substantial energy savings can be achieved by directly charging the hot continuously cast slab or billet to the reheating furnaces of the rolling mill. The latest installations have included direct in-line hot rolling of the cast product.

Process Development and Machine Design

A number of methods for distributing liquid steel to the mold or to several molds of the casting machine have been investigated. Use of a tundish with appropriate flow control has been found to be a superior method for the production of quality steel. Considerable effort has been directed toward improvement of refractories and development of methods for preventing nozzle blockage. Geometric arrangements in tundishes, including the use of dams and weirs, have provided suitable fluid flow characteristics to maximize the separation of nonmetallic inclusions, resulting in improved quality. Another factor in separation of nonmetallic inclusions in the tundish is tundish size because of its effect on residence time. Figure 6 indicates the trend in tundish size for major slab casting installations over the past 20 years.

Reoxidation of the molten steel is to be avoided. The use of refractory shrouds between the ladle and the tundish and the

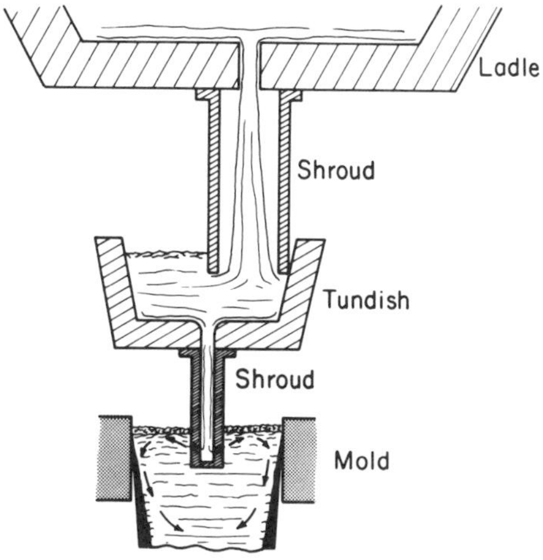

Fig. 7 Pouring shrouds from ladle to tundish to mold. Source: Ref 12

tundish and the mold have been adopted for slab casting (Fig. 7).

One of the difficulties encountered in continuous casting of small sections has been the protection of the pouring stream from tundish to mold because of the inapplicability of a pouring tube and the mechanical difficulty caused by the oscillation of the mold relative to the fixed tundish. In one instance, this has been overcome by the use of a flexible bellows (Fig. 8) and successfully applied to the continuous casting of special product quality steel bars. Recently, further development has been made in the use of ceramic shrouds to protect pouring streams on billet machines.

Protection of the surface of the molten steel pool in the tundish has been achieved through the use of suitable synthetic oxide slags, inert-gas blanketing, and a refractory cover to seal the tundish. The metal level in the tundish, as well as the fluid flow pattern, is important in avoiding the ingestion of the slag layer into the metal stream flowing downward into the mold.

The tundish is supported with either two tundish cars or a two-position turret. Frequently, the double turret provides for changing of the tundish while continuing the casting process, which allows extended sequence casting. The changing of the tundish not only occurs at an appropriate time, but also allows a change in steel composition with a minimal length of transition in the cast product.

The flow-through water-cooled copper mold is the key element of the casting machine. Special attention has been given to problems associated with the design and material requirements for molds. A number of different designs have been used, including thin-wall tube-type molds, solid molds, and molds made from plate. Plate molds were found to provide excellent mold life and to avoid the necessity for manufacture of molds from solid copper blocks. Steel and brass, as well as copper, have been used for molds, but the most outstanding material is nearly pure copper with small additions of elements that promote precipitation hardening or raise the recrystallization temperature, because both effects apparently provide longer mold life. Mold coatings are applied to extend service life. A chromium coating is generally used, often with an intermediate layer of nickel for improved coherence.

The most suitable length for a continuous casting mold has been found to be 510 to 915 mm (20 to 36 in.), a range that seems to remain constant regardless of section size. This surprising result can be explained by the higher rates of heat removal achieved with smaller sections and higher casting rates. Also, a thinner skin can be permitted for smaller sections exiting from the mold than for larger sections because bulging of the solidifying shell will be less severe. At higher casting rates, the use of an increased taper in the mold is necessary to maintain high heat removal rates, particularly for the narrow faces of slab molds.

Oscillating molds, as used by Junghans, have been adopted almost universally, although fixed molds can be successfully used with efficient lubricating systems. The oscillation is usually sinusoidal, a motion that can be achieved easily with simple mechanical arrangements. A fairly short stroke and a high frequency are used to provide a short period of "negative strip" during each oscillation, in which the mean downward velocity of the mold movement is greater than the speed of withdrawal of the casting strand in the casting direction. Oscillating frequencies are being increased from 50 to 60 cycles per minute (cpm) up to 250 to 300 cpm, with the benefits of shallower oscillation marks, less cracking, and reduced conditioning requirements.

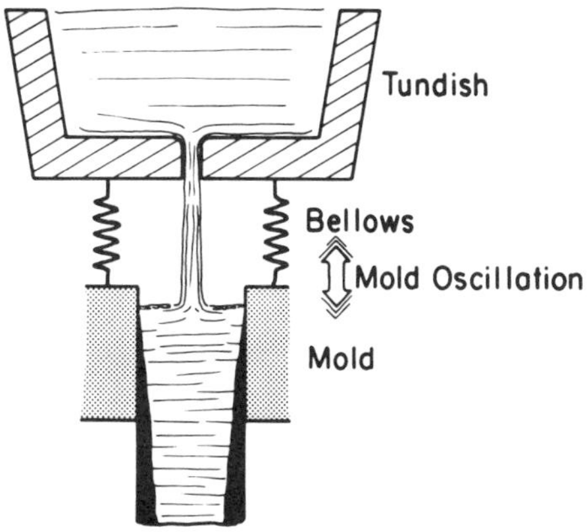

Fig. 8 Bellows between tundish and mold for casting billets. Source: Ref 12

Molten metal in the slab mold is normally covered with a layer of mold powder to protect the metal from reoxidation and absorb inclusions. The powder has a low melting point and flows over the liquid steel to provide mold lubrication and to help control heat transfer. Rapeseed oil, which has since been replaced by synthetic oils, has typically been used to prevent sticking to the mold in billet casting. Metal flow rates are matched with slab casting speeds using a stopper rod in the tundish, a slide gate, or a metering nozzle just above the shroud to control the delivery rate. Billets are normally cast with fixed metering nozzles, and the strand speed is adjusted to any changes in steel flow rate.

Support of the thin steel shell exiting the mold is required, particularly for slab cast-

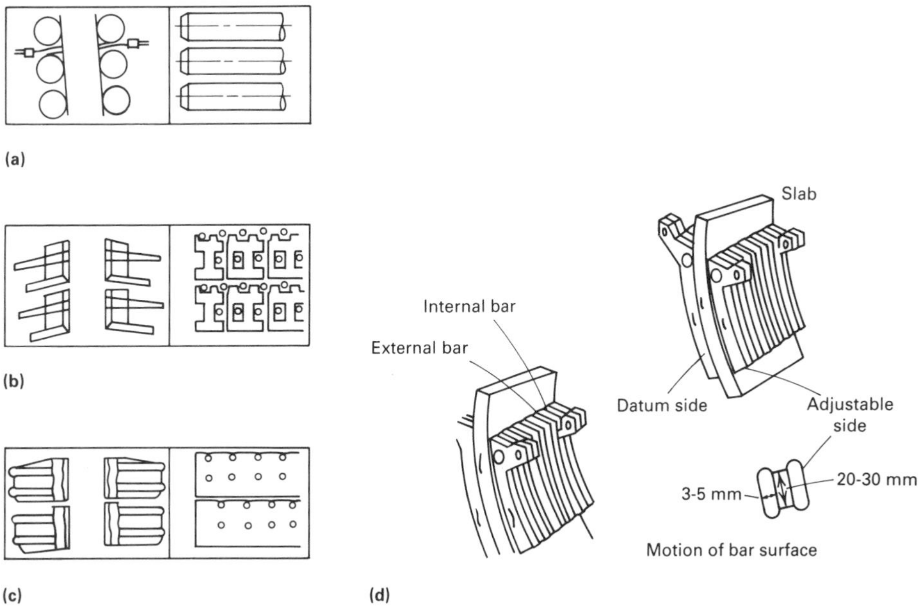

Fig. 9 Four types of shell supporting systems used in continuous casting. (a) Roll. (b) Cooling grid. (c) Cooling plate. (d) Walking bar. Source: Ref 13

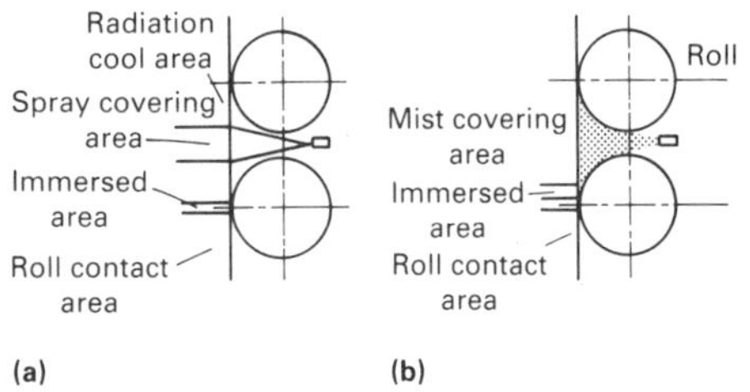

Fig. 10 Comparison of spray systems used in continuous casting. (a) Conventional spray. (b) Air-water mist spray. Source: Ref 14

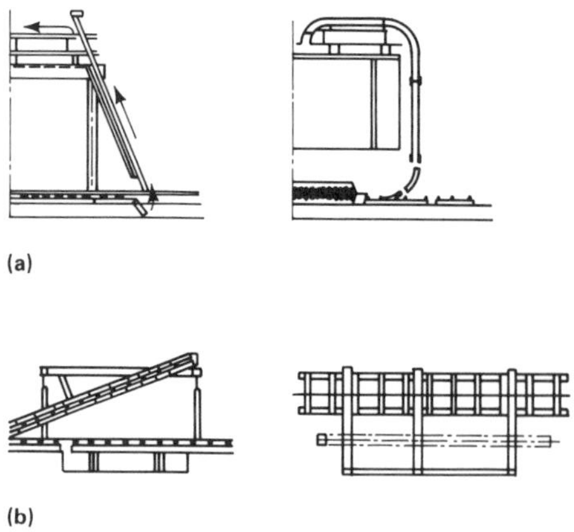

Fig. 11 Dummy bar arrangement. (a) Top feeding. (b) Bottom feeding. Source: Ref 17

ers. Several systems have been used (Fig. 9), all of which provide intensified direct water cooling.

Water spraying (secondary cooling) is critical to the process in that maximum cooling should be accomplished, but overcooling and large temperature increases must be avoided. The amount of heat removed by water sprays depends on the volume of the water, its temperature, and in particular the method of delivery, including spray pressure. Pressure is important in that the spray should be sufficiently intense to penetrate the blanket of steam on the surface of the solidifying strand. The thermal conductivity of steel is relatively low; consequently, as the surface temperature decreases and the shell thickness increases, cooling water has less influence on the solidification characteristics. The interrelationship between support rolls and the spray water and its delivery characteristics is quite important, particularly for the casting of wide slabs.

Mixed spraying of air and water develops a unique secondary cooling system with a uniform water droplet size. Air-mist cooling is illustrated in Fig. 10, along with the conventional water spray, as applied to slab casting. Although more costly, the air-mist spray system has a more uniform spray pattern and intensity that offers excellent spray cooling control. Decreases in transverse and longitudinal cracking with air-mist cooling have been reported (Ref 15, 16).

The straightening operation on curved strand casters has required special design and operating control. In general, temperatures at or above 900 to 1050 °C (1650 to 1920 °F) have been recommended to avoid conditions under which certain grades of steel have limited ductility and are susceptible to cracking. Multipoint and four-point straighteners have reduced imposed strains, and compression casting systems have reduced tensile stresses. Uniform temperature of the strand, including corners, which tend to cool more quickly, is required.

The dummy bar, which is used to stopper the mold for the initiation of a cast, can be inserted from above or below, depending on the individual installation. Some arrangements are shown in Fig. 11. The top feeding arrangement, which "chases" the last metal cast through the machine, offers a productivity advantage.

Productivity Improvements

While continual increases in casting speeds over the years have led to improvement in casting machine productivity, the most dramatic factor has been sequence casting, that is, continuous-continuous casting. Perfection of this development has required extraordinary achievement in the design of ladle and tundish handling systems, and in design and maintenance considerations for long-term operation with processing times that extend for several days. A summary of sequence casting records in Japan has been presented by T. Harabuchi (Ref 18) and is reported in Table 1. In comparison, the Jan 1983 casting record at Great Lakes Steel Corporation in the Detroit district involved a large slab caster (240 mm × 2.5 m, or 9½ × 99 in.), which cast 402 ladles and 83 160 Mg (91 480 tons) of steel continuously. In addition to the bulk steel handling requirements of sequence casting, the ability to change nozzles, shrouds, and tundishes at frequent intervals is also required. The string of casts at Great Lakes Steel Corporation involved 35 tundish changes and 177 shroud changes over a period of 13⅔ days. This record has been exceeded more recently at the Gary Works of United States Steel International, Inc.

While it has been shown that sequence casting extending beyond five or six heats does not dramatically increase productivity, provided a reasonable turnaround time and interactive scheduling with the steelmaking facilities exist, the capability for long-term sequence casting represents the opportunity for increased productivity with high quality at an essentially steady state operation of the caster.

An important characteristic of a casting operation with regard to productivity is the percentage of total clock time that steel is being processed in the machine. These percentages in high productivity casters can exceed 90% for frequently used casting operations. The turnaround time required for the dummy bar to restart a cast is one of the factors in producing steel-in-mold results, but under ideal conditions scheduled maintenance will be a major factor, and other items, such as problems in steelmaking or mechanical difficulty on the strand, will be minimized.

In the past and in many present installations, interruption of continuous casting production is required in order to make a width change. Several developments that avoid this requirement are now in operation. In one such arrangement, used at Great Lakes Steel Corporation, a very wide slab was cast and then slit into two or three optimum widths. Another approach is taken at Oita Works of Nippon Steel Corporation, where a sizing mill adjusts the slabs to various desired widths. Another, more versatile, system involves the use of a variable width mold in which the taper and width are adjusted continuously during the casting process. These techniques have permitted the adoption of sequence casting with a minimum mold inventory based on width.

Another barrier to increasing productivity of continuous casting has been the accommodation of ladle-to-ladle composition changes with adoption of sequence casting. Under ideal circumstances, as when a plant produces a narrow range of compositions,

Table 1 Record of sequence casting in Japan

Product	Company	Continuous casting plant	Heats per cast	Weight of steel per cast, Mg	Weight of steel per cast, tons	Time of achievement
Slab	Nippon Kokan	Keihin	270	22 523	24 775	Sept 1974
Slab	Nippon Kokan	Fukuyama	244	54 360	59 796	Aug 1981
Slab	Kawasaki Steel	Mizushima No. 5	204	51 000	56 100	Nov 1980
Slab	Nippon Steel	Yamata No. 2	186	30 199	33 219	July 1974
Bloom	Sumitomo Metals	Wakayama No. 3	1129	160 110	176 124	July 1982
Bloom	Daido Steel	Chita	320	19 091	21 000	Feb 1984
Bloom	Sumitomo Metals	Kokura	281	17 648	19 413	April 1982
Bloom	Nippon Steel	Kamishi	202	17 857	19 643	Jan 1979
Billet	Osaka Steel	Okajima	214	17 455	19 200	July 1978
Billet	Kokko Steel	Main Plant	174	5 063	5 569	Aug 1978
Billet	Funabashi Steel	Main Plant	99	5 535	6 089	Oct 1975

Source: Ref 18

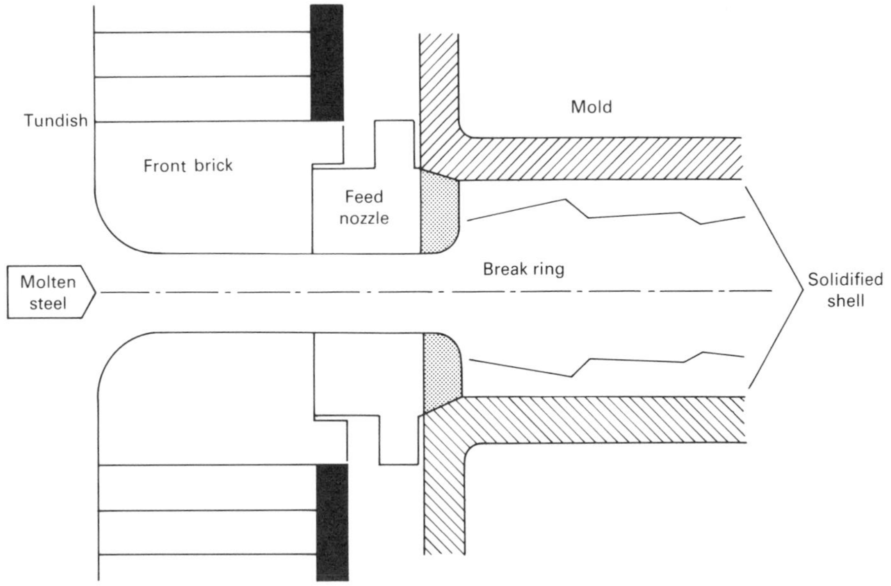

Fig. 12 Tundish and mold arrangement for horizontal caster. Source: Ref 22

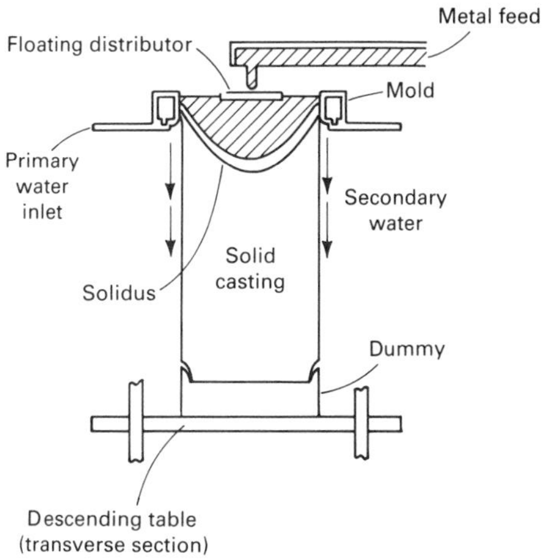

Fig. 13 Conventional vertical direct-chill casting arrangement. Source: Ref 22

often with overlapping compositional requirements, compositional changes can be slowly stepped through each grade to accommodate the desired sequence of heats. However, when substantial variations in composition must be accommodated, physical barriers in the form of steel plates have been used to provide isolation of the grade changes in the strand. In this way, the transition zone can be minimized, and compositional changes can be accommodated without substantial losses in yield or quality.

Quality Improvements

The operation of a continuous casting strand to produce good-quality steel uniformly and reliably on a routine basis can be the most valuable asset of the process. Development work continues in an effort to improve control and production reliability, particularly with regard to avoidance of inclusions and cracking for internal and surface quality.

Internal cracking is less important for those products (for example, sheet) that have large reduction ratios. Radial cracks, center looseness or centerline cracking, minor amounts of gas evolution, and other internal defects are not deleterious in heavily rolled products. In the manufacture of heavy plates, however, these conditions can represent serious product defects.

For aluminum-killed steel, subsurface inclusions are usually in the form of aluminum oxide. In some cases, surface scarfing can be effective in removing these inclusions, which could provide a surface defect in a final rolled product. Unfortunately, this surface conditioning results in substantial yield losses. It has been reported (Ref 4) that use of multiport shrouded nozzles can produce a fluid flow action in slab molds, which brings inclusions in contact with the molten mold powder covering to flux and dissolve oxide particles. This provides a clean subsurface zone that requires no scarfing. However, this method leads to the entrapment of complex nonmetallic inclusions containing mold powder. Clean steel practices in combination with nozzle configurations optimized for specific casting conditions are necessary to minimize nonmetallic inclusions. As noted above, the control of fluid flow and promotion of separation in ladle, tundish, and mold can provide a substantial reduction in large inclusions.

Depending on steel grade, cracking can result from intensive cooling or deformation in the casting strand. Surface cracking can be minimized by proper control of lubrication and cooling within the mold, and cooling and alignment in the upper spray zone. Other critical factors in cracking are control of spray cooling to avoid surface temperature rebound (resulting in midway cracks) or nonuniformity of cooling across or along the strand. Control of temperature distribution at the straightener and proper roll gap settings for slabs and blooms have also been noted as being very important.

Horizontal Continuous Casting

Horizontal casting systems have been explored for many years, but only in the recent past have they been successfully applied to steels. Oldsmobile Division of General Motors Corporation started casting 95 mm (3.75 in.) steel bars in a horizontal mold in 1969 (Ref 19, 20), with uninterrupted casts of up to 24 h. This is believed to be the longest sustained continuous casting operation for steel up to that time.

Voest-Alpine is developing a horizontal continuous casting machine for steels, and the Nippon Kokan (NKK) Steel Company of Japan has been producing horizontally cast steel sections up to 210 mm (8 in.) square (Ref 21). These systems depend on a refractory nozzle or "break ring" at the entrance end of a stationary water-cooled metallic mold, and an intermittent motion for translating the bar, each forward stroke being followed by a discrete rest or dwell. A cross section of the tundish and mold arrangement of the NKK machine is shown in Fig. 12. Casting of larger sections is being explored throughout the world, as are continuous casting processes for high-speed casting of thin sections. As the technologies for cleaner steels evolve, these processes will come closer to commercial adoption.

Nonferrous Continuous Casting

The early development of continuous casting processes occurred to a large extent at production installations for nonferrous alloys.

Direct-Chill Casting

The principal casting process for light metals is the direct-chill process (Ref 22-25). The vertical direct-chill casting process was patented by Alcoa in 1942 (Ref 26), and is shown schematically in its present form in Fig. 13. The process can directly prepare billets for extrusion, blocks for rolling, and sheet for fabrication, thus eliminating intermediate mechanical working processes by casting near-net shapes.

Most direct-chill casting capacity is of the vertical type for semicontinuous casting, but more importance is being assumed by the continuous horizontal direct-chill casting process (Fig. 14). The section sizes in which aluminum alloys are cast range from 1.5 × 0.5 m (5 × 1.6 ft) blocks for rolling to 5 to 30 mm (0.2 to 1.2 in.) thick by 2 m (6.6 ft) wide for plate and strip. There is considerable economic advantage in wide strip casting. This processing, which is far ahead

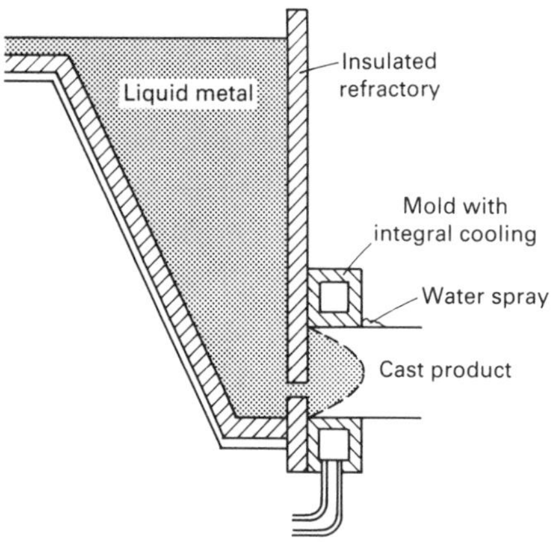

Fig. 14 Horizontal direct-chill casting system. Source: Ref 22

of steel continuous casting of wide strip, could portend the future for steel.

The key operating requirement in direct-chill casting is that a sufficiently strong shell be developed in the limited time of contact with the mold to retain the interior molten pool. Withdrawal rates of up to 0.2 m/min (0.66 ft/min) can be achieved in conventional casters. Withdrawal speeds of 2.5 m/min (8.2 ft/min) for a 10 mm (0.4 in.) thick section have been reported for horizontal casters producing pure aluminum strip (Ref 27). Pure aluminum or dilute alloys are easier to cast than higher alloys with wide freezing ranges. Higher casting speeds have led to problems in maintaining casting shape and have also caused higher internal stresses in the solidified ingot. Control of heat extraction rates is required to limit the extent of these difficulties.

Vertical direct-chill casting is used extensively to produce rectangular slabs and cylindrical billets of copper alloys, and, to a lesser extent, of pure copper. A diagram of an entire vertical direct-chill unit for casting

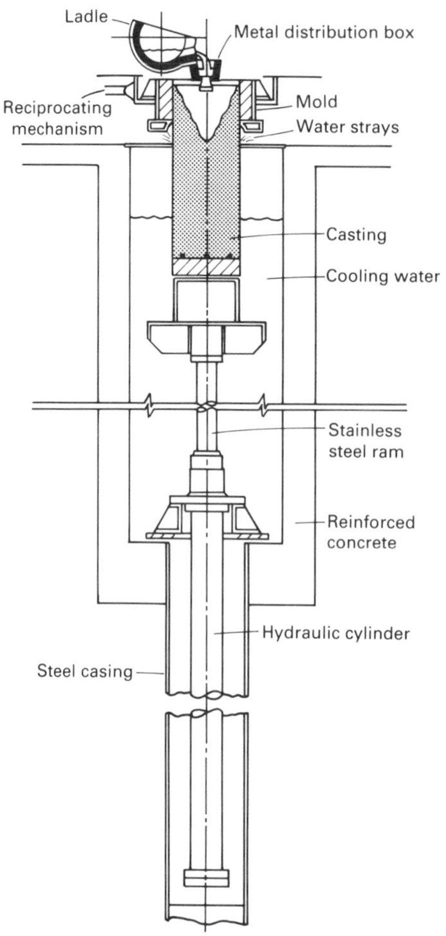

Fig. 15 A direct-chill unit for casting copper alloy slabs. The slabs produced by this unit are approximately 9 m (29.5 ft) long. Source: Ref 28

copper slabs is shown in Fig. 15. For copper and copper-base alloys, the liquid metal is poured through a water-cooled, oscillating, graphite-lined collar or mold. The graphite

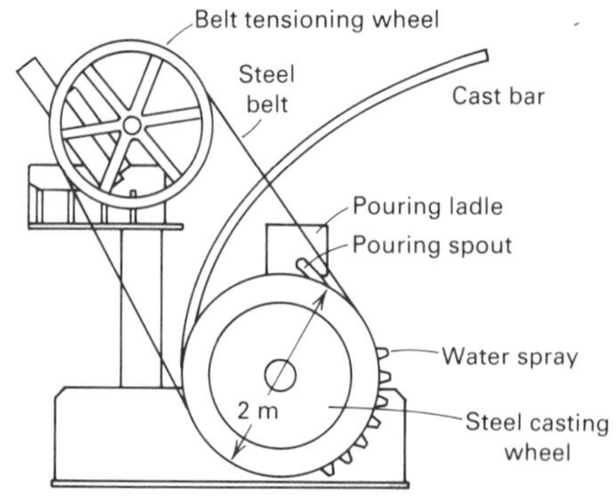

Fig. 16 The Properzi casting machine. Source: Ref 30

produces a smooth surface on the casting and minimizes oxidation.

Other Processes

There are numerous types of continuous casting processes that commercially produce nonferrous metals, principally aluminum and copper alloys (Ref 29). These processes can be characterized by their products. The more common processes are discussed below.

Wheel-and-Band Machines. Rod and bar are continuously cast on wheel-and-band machines, such as the Properzi process (Fig. 16), or the more recent Southwire casting system, which is illustrated in Fig. 17. These processes involve casting between the circumference of a large copper-rimmed wheel containing the mold configuration and a steel band (Fig. 18). The metal solidifies in the gap as the wheel and band rotate through a portion of a circular path, which includes water sprays. Casting rates of 32 Mg/h (35 tons/h) for a 450 × 450 mm

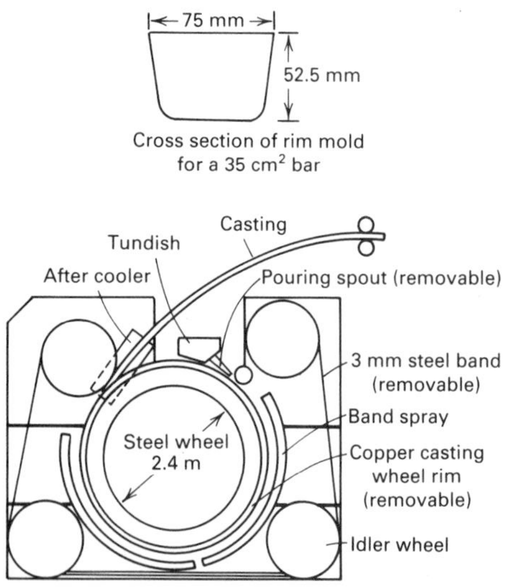

Fig. 17 The Southwire wheel-and-band continuous casting machine. Source: Ref 31

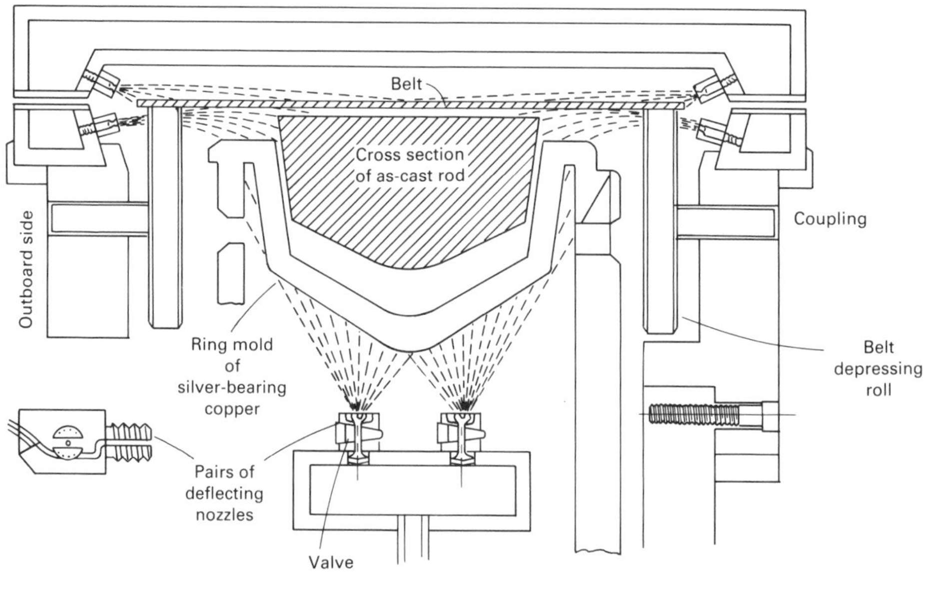

Fig. 18 The Properzi ring mold, belt, and as-cast rod. Source: Ref 32

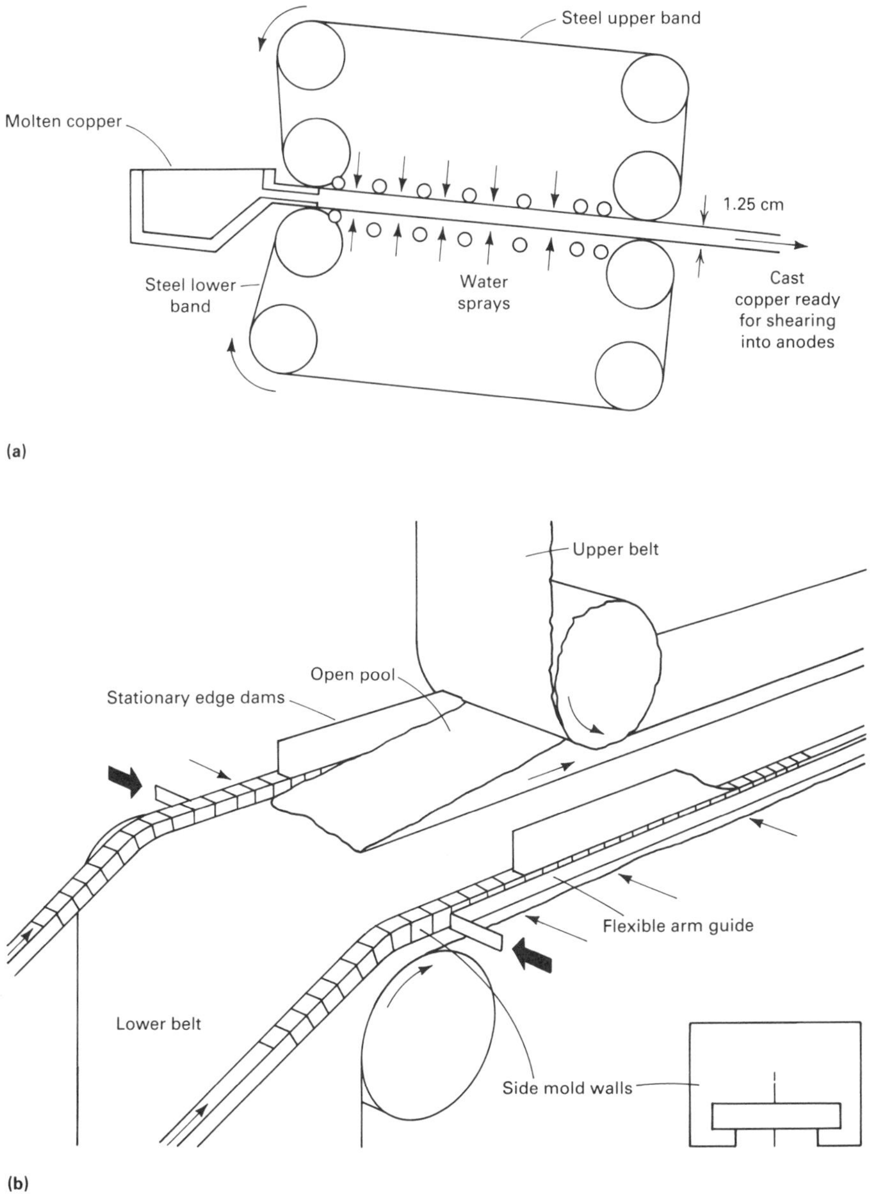

Fig. 19 The Hazelett machine for continuous casting of copper anode strip. Source: Ref 33

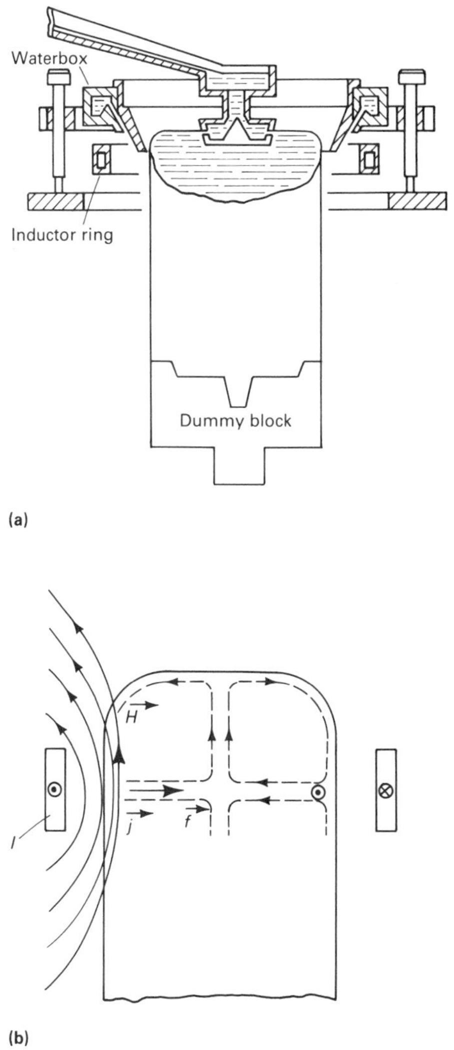

Fig. 20 The electromagnetic casting process. (a) Equipment. (b) Physical principles involved. *H*, magnetic field strength; *I*, inductor current; *j*, eddy and current density; *f*, volume force

(18 × 18 in.) copper billet have been reported (Ref 31). The Southwire process has recently been applied to the production of steel billets.

Wide, thick strip is normally produced on a twin band machine, such as the Hazelett caster (Fig. 19).

The bands are separated by edge dams on each side of the castings; these dams can be moved to set the width of the strip.

Twin roll casting has been used for the production of wide, thin aluminum strips. Several twin roll processes are used to cast 6 to 12 mm (0.24 to 0.48 in.) thick by 1500 to 2000 mm (59 to 79 in.) wide strip.

Many product quality problems relate to mold and air gap formation, as well as interaction of the ingot shell and liquid core. These could be eliminated by moldless casting, wherein an electromagnetic field supports the liquid metal until it enters the direct-quench zone. Such a process was developed in the Soviet Union (Ref 34). Electromagnetic casting is currently being used in the aluminum industry (Ref 35) and has been developed for copper-base alloys (Ref 36). The process is shown in Fig. 20.

Future Developments

The emphasis on the further development of continuous casting will focus on control systems and automation, with the objective of maintaining high quality and high productivity. Accomplishing this will include monitoring metal quality and ensuring that all aspects of the process are under proper control. Various operating parameters, such as mold, slag, or flux levels in ladle, and tundish, also will be monitored directly. Sensors will be developed to control automatically the process for proper cooling in the mold, first zone, and secondary cooling systems. Hot inspection techniques will be developed that will provide a direct measure of product quality as it leaves the casting machine and moves to hot-rolling processing.

Also, current worldwide efforts on steel processing are in the area of the belt casting of thin slabs, the roll casting of thin strip, and the electromagnetic casting of steel.

ACKNOWLEDGMENTS

The author gratefully acknowledges the contributions to this article made by C.R. Jackson and N.T. Mills of Inland Steel Company, R. Lincoln of Chaparral Steel Company, K. Schwaha of Voest-Alpine

A.G., and T. Harabuchi of Nippon Steel Corporation.

REFERENCES

1. G. Sellers, U.S. Patent 1908, 1840
2. H. Bessemer, "On the Manufacture of Continuous Sheets of Malleable Iron and Steel Direct from Fluid Metal," Paper presented at the Iron and Steel Institute Meeting, Oct 1891; also, *J. Met.*, Vol 17, (No. 11), 1965, p 1189-1191
3. *Continental Iron and Steel Trade Report*, The Hague, Aug 1970
4. R. Clark, *Continuous Casting of Steel*, Institute for Iron and Steel Studies, 1970, p 7
5. T. Ohnishi, in *Proceedings of the Nishiyama Memorial Lecture*, 80-9, p 45
6. R.W. Joseph and N.T. Mills, "A Look Inside Strand Cast Slabs," Association of Iron and Steel Engineers, May 1975
7. T. Haribuchi *et al.*, Technical Report 294, Nippon Steel, Jan 1978
8. R. Lincoln, "Melt Shop Reprofit at Chaparral/Goal 450 000 Tons of Prime Billets, in *Proceedings of the Electric Furnace Conference*, Vol 36, American Institute of Mining, Metallurgical, and Petroleum Engineers, 1978, p 168-171
9. F.M. Wheeler and A.G.W. Lamont, Current Trends in Electric Meltshop Design, *Proceedings of the Electric Furnace Conference*, Vol 36, American Institute of Mining, Metallurgical, and Petroleum Engineers, 1978, p 139-147
10. C.R. Jackson and L.R. Schell, Fifteen Years of Looking—One Year of Operating, The Start-Up and First Year's Operation of Inland's No. 1 Slab Caster, in *Proceedings of the Open Hearth Conference*, Vol 57, American Institute of Mining, Metallurgical, and Petroleum Engineers, 1974, p 55-66
11. T. Haribuchi, Nippon Steel Corporation, private communication, May 1984
12. R.D. Pehlke, Reoxidation of Liquid Steel, *Radex Rundsch. Heft 1/2*, Jan 1981, p 349-367
13. C.R. Jackson, Inland Steel Company, private communication, May 1983
14. Report 77, Steelmaking Committee of ISIJ, Nippon Steel
15. M. Tokuda *et al.*, *Iron Steel Inst. Jpn.*, Vol 83, p 919
16. Y. Kitano *et al.*, *Iron Steel Inst. Jpn.*, Vol 84, p 179
17. K. Schwaha, Voest-Alpine, private communication, May 1984
18. T. Harabuchi, Nippon Steel Corporation, private communication, May 1984
19. F.J. Webberre, R.G. Williams, and R. McNitt, "Steel Scrap Reclamation Using Horizontal Strand Casting," Research Publication GMR-11, General Motors Corporation, Oct 1971
20. W.G. Patton, GM Casts In-Plant Scrap Into In-Plant Steel, *Iron Age*, Dec 1971, p 53-55
21. F.G. Rammerstorfer *et al.*, "Model Investigations on Horizontal Continuous Casting," Paper presented at the Voest-Alpine Continuous Casting Conference, 1984
22. C.M. Adam, Overview of D.C. Casting, in *Proceedings of the 1980 Conference on Aluminum-Lithium Alloys*, The Metallurgical Society, 1981, p 39-48
23. E.F. Emley, *Int. Met. Rev.*, Vol 21, 1976, p 75
24. D.M. Lewis, *Metall. Rev.*, Vol 6, 1961, p 143
25. C. Baker and V. Subramanian, DC and Continuous Casting, in *Proceedings of the 1978 Symposium on Aluminum Transformation Technology and Applications*, American Society for Metals, 1980, p 335-388
26. U.S. Patent 301,027, 1942
27. G. Moritz and F.O. Ostermann, *J. Inst. Met.*, Vol 100, 1972, p 301
28. J. Newton, *Extractive Metallurgy*, John Wiley & Sons, 1959, p 505
29. A.K. Biswas and W.G. Davenport, chapter 17 in *Extractive Metallurgy of Copper*, Pergamon Press, 1976
30. M.S. Stanford, Wire Rod Production Alternatives, *Copper*, Vol 1 (No. 4), 1967, p 11-15
31. D. Barnes, H. Nomura, Y. Arakida, and M. Watanabe, Development of Southwire's SCR System and Its Automation at Hitachi Wire Rod Company, Ltd., *Continuous Casting*, K.R. Olen, Ed., Iron and Steel Society and American Institute of Mining, Metallurgical, and Petroleum Engineers, ISS-AIME, 1973, p 93-121
32. W.L. Finlay, *Silver-Bearing Copper*, Corinthian Editions, 1968, p 222
33. R.W. Hazelett and C.E. Schwartz, "Continuous Casting Between Moving Flexible Belts," Paper presented at the AIME Annual Meeting, American Institute of Mining, Metallurgical, and Petroleum Engineers, 1964
34. Z.N. Getseleev, Casting in an Electromagnetic Field, *J. Met.*, Vol 23 (No. 10), 1971, p 38-39
35. C. Vives and R. Ricou, "Experimental Study of Continuous Electromagnetic Casting of Aluminum Alloys," *Metall. Trans. B*, Vol 16B, 1985, p 377-384
36. D.E. Tyler, B.G. Lewis, and P.D. Renschen, Electromagnetic Casting of Copper Alloys, *J. Met.*, Vol 37 (No. 9), 1985, p 51-53

SELECTED REFERENCE

- *Continuous Casting*, Vol 1-4, The Iron and Steel Society, 1983-1988

New and Emerging Processes

Counter-Gravity Low-Pressure Casting

Dixon Chandley, Metal Casting Technology, Inc.

ENGINEERS have made various attempts, since early in this century, at drawing metal up into molds against the flow of gravity to produce castings. Molds were typically put into a vacuum box with a fill pipe extending from the mold out of the box. The fill pipe was submerged in molten metal, and a vacuum was generated around the mold, causing the metal to rise into the mold. Metal molds were generally used, and by World War II, many premium-quality aluminum castings were made in such molds by counter-gravity filling. Metal molds were vented to permit the vacuum to exhaust the mold cavities properly and to draw in the metal. In the early 1970s, new methods were developed for counter-gravity casting into nonmetal permeable molds, first using ceramic investment molds and later using low-temperature bonded sand molds. The processes described in this article are covered by U.S. and other patents and are in high-volume production in nine countries and fourteen companies worldwide. Counter-gravity low-pressure casting processes are widely known by acronyms; therefore, these acronyms will be used here for convenience. Counter-gravity low-pressure casting processes include:

- Counter-gravity low-pressure casting of air-melted alloys (CLA)
- Counter-gravity low-pressure casting of vacuum-melted alloys (CLV)
- Check valve casting (CV)
- Counter-gravity low-pressure air-melted sand casting (CLAS)

The CLA Process

The CLA process (Fig. 1) is used to cast metals that are normally melted in air. The economies of the process are based on casting more parts per mold, high gating efficiencies (because most of the gating metal flows back into the furnace), and fewer casting defects (especially melt inclusions, which are reduced because the fill pipe is always submerged into clean metal). Thin-wall parts (wall thickness as small as 0.75 mm, or 0.03 in.) are easily made at high volume and low cost.

Applications. The CLA process is used to make parts from all types of alloys for many industries. For the automotive industry, steering system components, transmission parts, diesel precombustion chambers, rocker arms, mounts, and hinges are made.

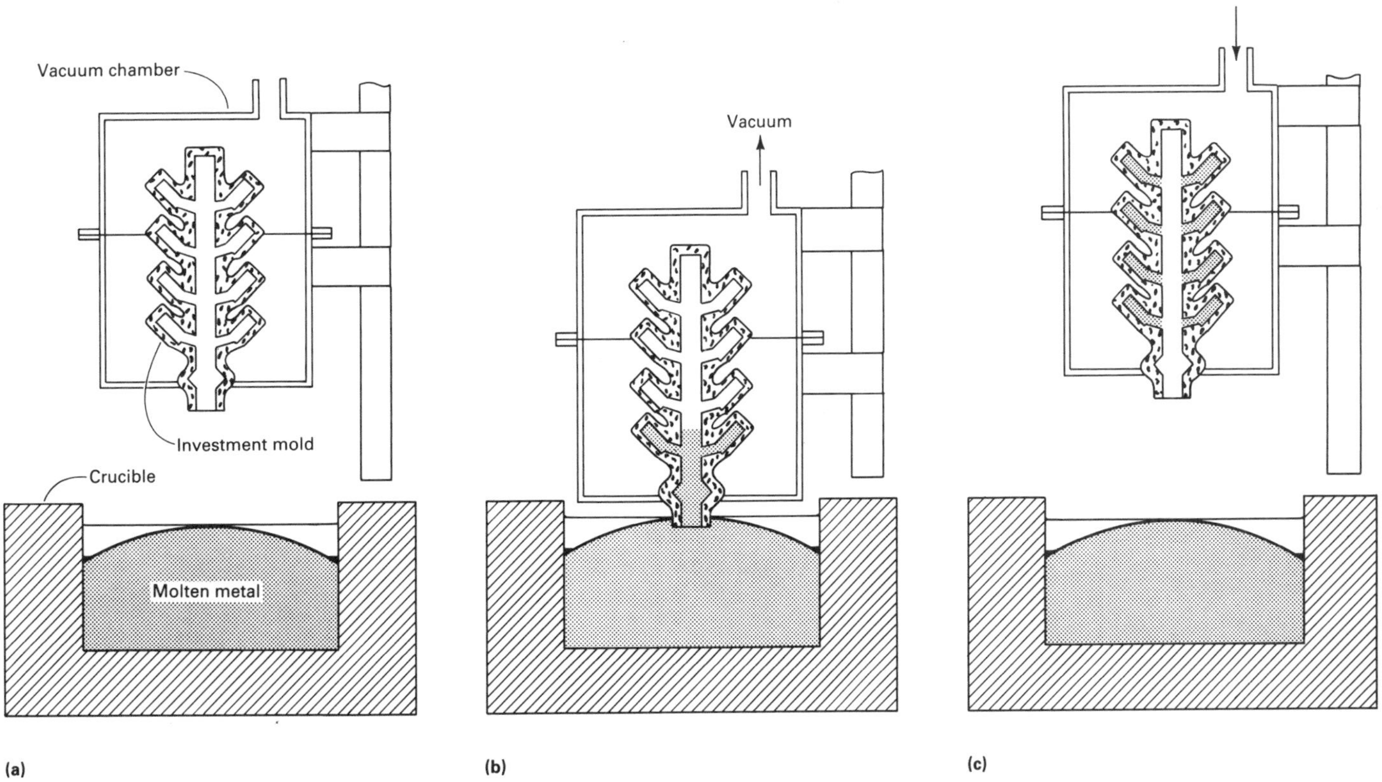

Fig. 1 Schematic of the operations in the CLA process. (a) Investment shell mold in the casting chamber. (b) Mold lowered to filling position. (c) Mold containing solidified castings; most of the gating has flowed back into the melt.

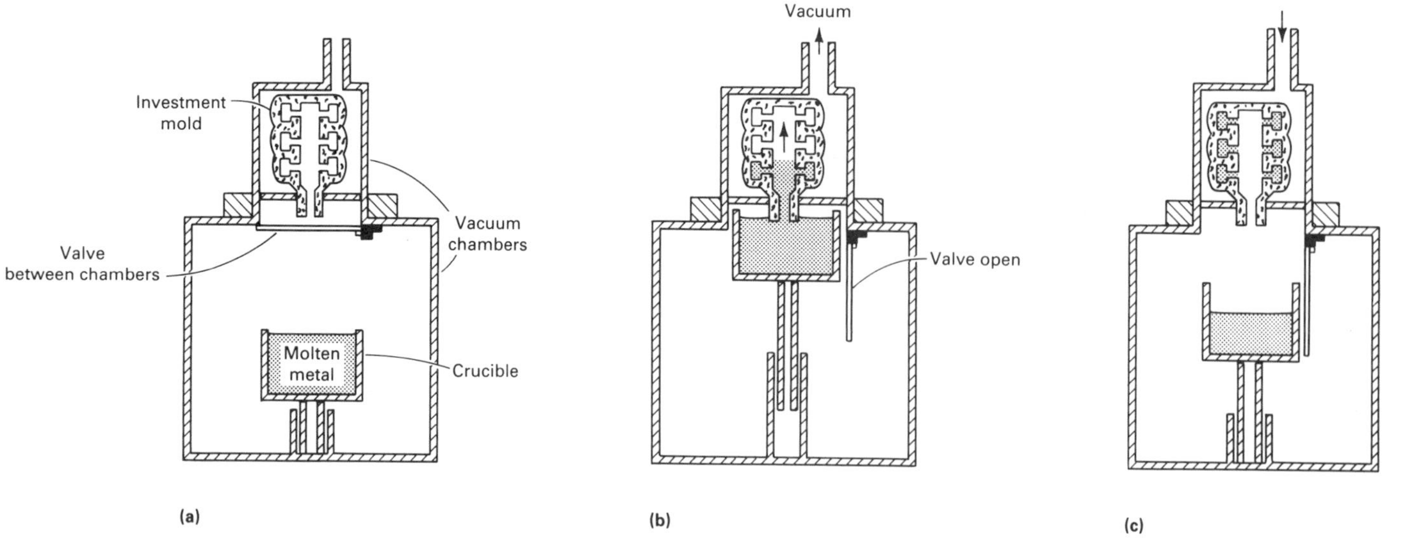

Fig. 2 Schematic showing steps in the CLV process. (a) Metal is melted in vacuum, and the hot mold is introduced into a separate upper chamber. A vacuum is then created in the second chamber. (b) Both chambers are partially flooded with argon, the valve between the chambers is opened, and the fill pipe enters the molten metal. Additional vacuum is then applied to the upper chamber to draw metal upward. (c) The vacuum is released after the parts are solidified, and the remaining molten metal in the gating system returns to the crucible.

Among the components produced for the aircraft and aerospace industries are temperature probes, fuel pump impellers, missile wings, brake parts, pump housings, and structural parts. Other applications include golf club heads, innumerable machine parts, wood router tool bits, tin snip blades, small wrenches, lock parts, gun parts, valves and fittings, and power tools.

The CLV Process

The CLV process (Fig. 2) is used for alloys containing reactive metals, especially the superalloys, which may contain aluminum, titanium, zirconium, and hafnium. It can be seen that the advantages of the CLA process also apply to the CLV process. The outstanding features of this process include the ability to fill thin sections and to make castings that are free of the small oxides that plague gravity pouring methods. It is possible to make castings of large area in wall thicknesses down to 0.5 mm (0.02 in.) and without the small oxides that would render such parts defective.

Applications. For gas turbine engines, the CLV process provides parts with the lowest level of melt oxide inclusions for alloys such as MAR-M 509 and René 125, which are noted for hafnium and zirconium inclusions when gravity poured. The process enables a new approach to the design of jet engine burner cans, in which the rolled and welded sheet metal design has been replaced with an assembly of thin-wall (0.5 mm, or 0.02 in.) castings shaped for maximum heat transfer and mechanically assembled to reduce thermal fatigue. Such burner cans provide a much higher temperature capability. Important cost and quality improvements have been achieved in cast airfoils, turbine seals, conduits, clamps, and turbocharger wheels.

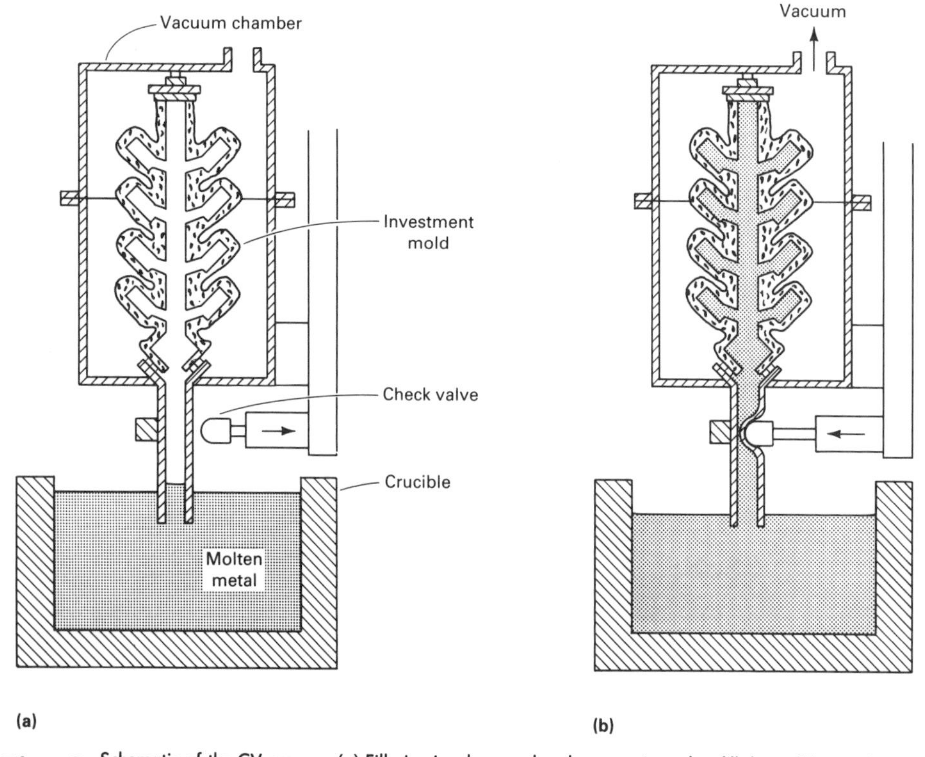

Fig. 3 Schematic of the CV process. (a) Fill pipe is submerged and vacuum is used to fill the mold as in the CLA process. (b) When the mold is filled, a check valve crimps the fill pipe shut. Metal is trapped in the mold, which is then moved away and allowed to solidify as in gravity-filled molds.

The CV Process

The CV process (Fig. 3) is used for castings that are too thick to solidify in the manner required by the CLA or CLV process. As shown in Fig. 3, this process uses a flexible fill pipe, which is crimped shut by a check valve when the mold is filled. The CV process provides a good fill of thin castings, along with the improved metal cleanliness of the other processes, and it applies to all molding methods.

Applications of the process include large missile wings, valve bodies, hinges, and other large parts with varying section thickness.

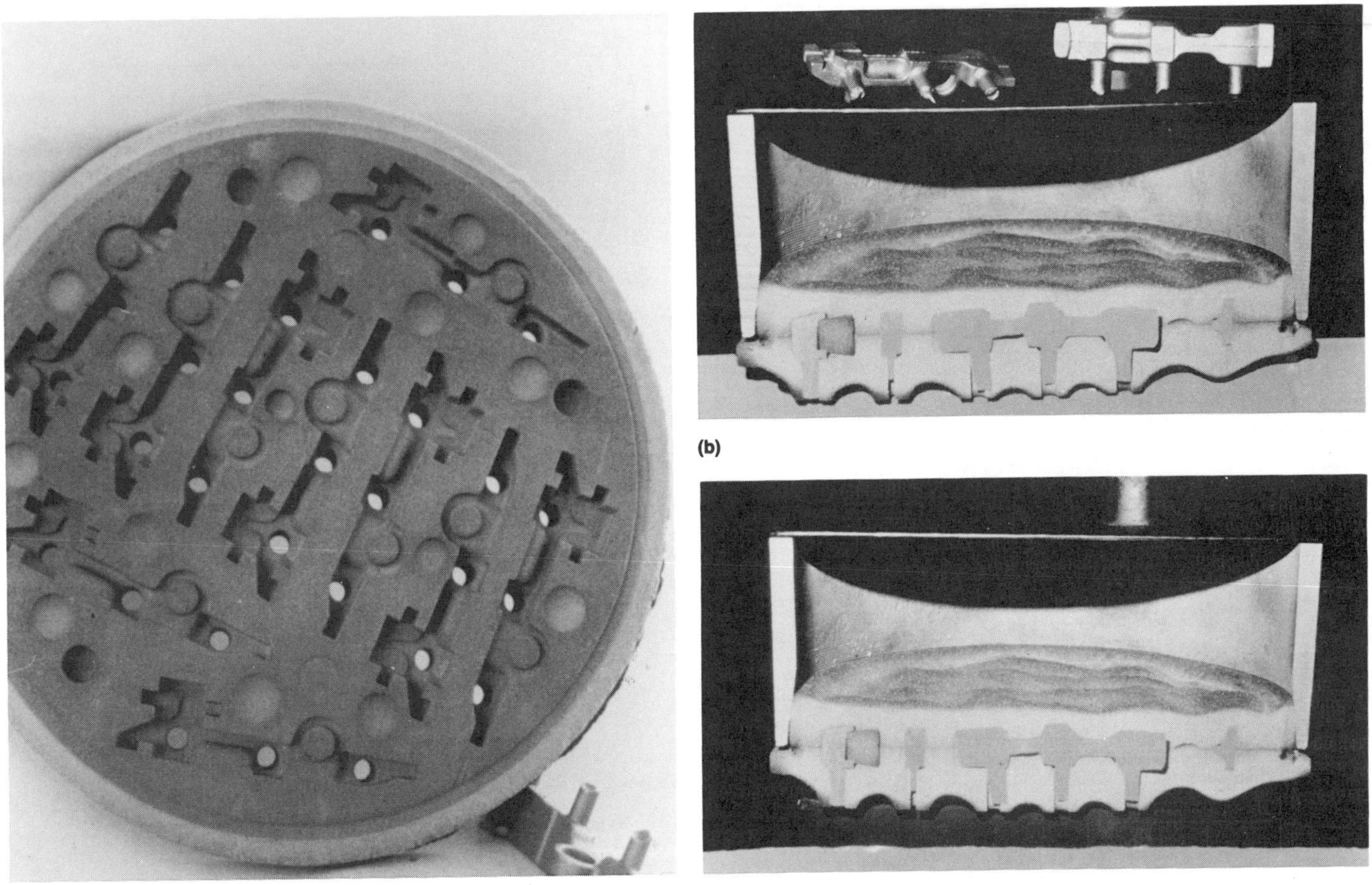

(a) (b) (c)

Fig. 4 Essential features of the CLAS process. (a) View of CLAS mold showing high pattern density and open bottom gates. The mold shown uses direct gating into parts. (b) Cross section of mold and casting head during mold filling. (c) Cross section of mold and casting head after solidification. Blind risers can be used for thick-wall castings.

The CLAS Process

The CLAS process (Fig. 4) is used for sand casting and is quite different from the other processes discussed above. It has a unique ability to make thin castings at low vacuum levels. Because the metal is taken from the clean portion of the melt and because castings are made at low metal temperatures, melt inclusions are very low in volume as compared to those found in gravity casting.

Applications. Connecting rods and thin, hollow cam shafts for automotive use can be made with wall thicknesses of only 1.5 mm (0.06 in.) in steels and cast irons. Stainless steel truck wheel centers that are lighter than aluminum wheels can be made, owing to the thin walls that are possible with this process.

Directional and Monocrystal Solidification

Thomas S. Piwonka, Metal Casting Technology Center, University of Alabama

Most castings are employed in applications in which stress fields are isotropic, but there are a few important uses in which the stresses are primarily unidirectional along a single axis. In such cases, casting practices that enhance the properties along that axis have been developed. Two of the most widely used methods are directional solidification and monocrystal (single-crystal) casting.

Directional Solidification

The process used to manufacture directionally solidified castings with a columnar structure requires careful control to ensure that castings which are of acceptable quality are produced. Specialized furnaces are used, and mold design is quite different from that used for conventional investment castings.

Metallurgical Effects of Columnar Structures

An important application of directional solidification is in the manufacture of blades (rotating parts) for gas turbine engines (Fig. 1). These components are subjected to high stresses along their major axes, as well as high temperatures. Because grain boundaries are weaker than grains at high temperatures, it is logical to align them parallel to the axis of principal stress to minimize their effect on properties.

Materials. The alloy that was originally used in directionally solidified turbine components was MAR M-200, a nickel-base alloy containing 12.5% W (Ref 1). The solidified structure consisted of tungsten-rich dendrites with high strength and creep resistance that grew to the length of the casting. The grain-boundary material, which was parallel to the dendrites, was strong enough to withstand the transverse stresses on the components. The properties of the directionally solidified alloy were far superior to those of the equiaxed alloy, as shown in Fig. 2. Other alloys have since been designed to make use of the process.

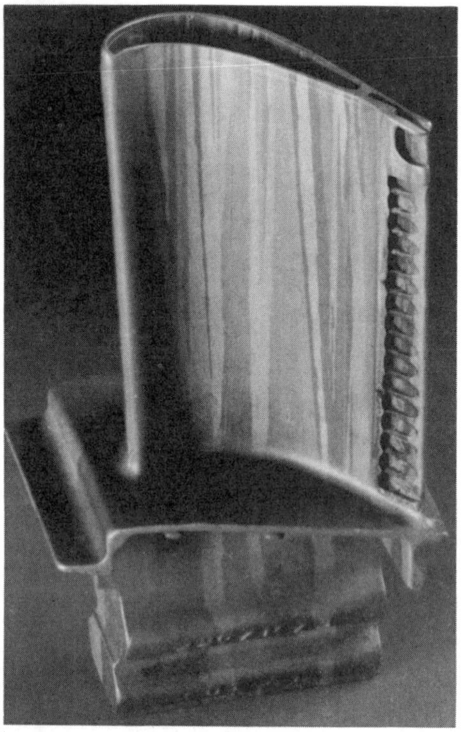

Fig. 1 Directionally solidified thin-wall turbine blade cast from Alloy CM 247 LC (modified MAR M-247 composition)

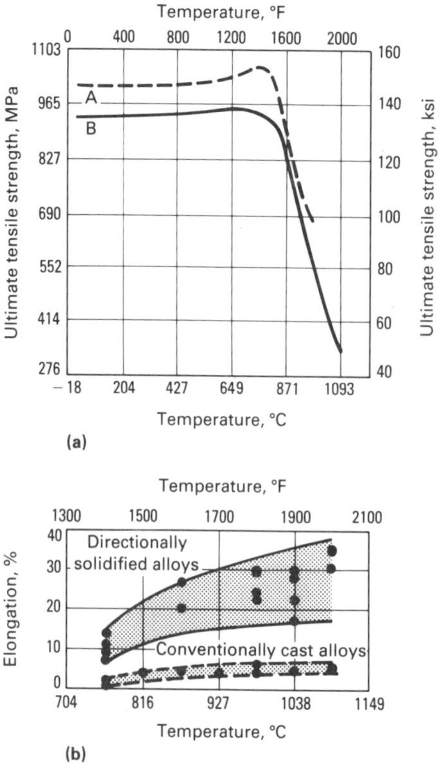

Fig. 2 Comparison of properties of directionally solidified and conventionally cast alloys. (a) Ultimate tensile strength of MAR M-200 alloy. Curve A, directionally solidified; curve B, conventionally cast. Source: Ref 1. (b) Average rupture elongation of various alloys

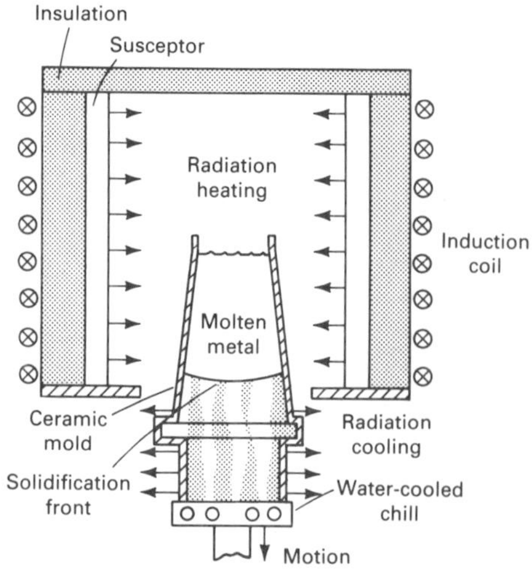

Fig. 3 Schematic showing the directional solidification process. Source: Ref 2

In columnar structures, the primary dendrites are aligned, as are the grain boundaries. The primary dendrites form around spines of the highest-melting constituent to freeze. As freezing continues, the solid rejects solute into the residual liquid (segregation occurs) until the final low-melting eutectic has frozen at the grain boundaries. Because segregation products collect in the grain boundaries, it is important to consider the composition of these grain boundaries in directional structures.

An ideal composition for directional solidification is one in which the primary dendrites form around a strong spine, while the grain boundaries also retain their strength. A poor alloy is one in which the segregation products form embrittling phases, especially adjacent to secondary dendrite arms, which are normal to the primary stress axis.

Heat Flow Control

To obtain a directionally solidified structure, it is necessary to cause the dendrites to grow from one end of the casting to the other. This is accomplished by removing the bulk of the heat from one end of the casting. To this end, a strong thermal gradient is established in the temperature zone between the liquidus and solidus temperatures of the alloy and is passed from one end of the casting to the other at a rate that maintains the steady growth of the dendrite, as shown in Fig. 3. If the thermal gradient is moved through the casting too rapidly, nucleation of grains ahead of the solid/liquid interface will result; if the gradient is passed too slowly, excessive macrosegregation will result, along with the formation of freckles (equiaxed grains of interdendritic composition) (Ref 3). Therefore, the production of directionally solidified castings requires that both the thermal gradient and its rate of travel be controlled. For the case of nickel-base alloys, thermal gradients of 36 to 72 °C/cm (165 to 330 °F/in.) have been found to be effective (Ref 4), and rates of travel of 30 cm/h (12 in./h) can be used. There is, however, no upper limit on the allowable gradient, and higher gradients usually produce better castings than lower gradients. The lower limit on the thermal gradient is a function of alloy composition and casting geometry.

The most effective way to control heat flow is to use a thin-wall mold, such as an investment casting mold, that is open at the bottom. The mold is placed on a chill (which is usually water cooled) and heated above the liquidus temperature of the alloy. Molten metal is poured into the mold, and the mold is cooled from the chilled end by withdrawing the mold from the mold-heating device.

The chill is used to ensure that there is good nucleation of grains to start the process. Because of the low thermal conductivity of nickel-base alloys, the thermal effect of the chill extends only about 50 to 60 mm (2 to 2.4 in.) (Ref 5, 6). Although the grains originally nucleate with random orientations, those with the preferred growth direction normal to the chill surface grow and crowd out the other grains. Therefore, those grains that grow through the casting are all aligned in the direction of easiest growth. For nickel-base alloys, the preferred growth direction is <001>; therefore, in castings made of these alloys, the grains are aligned in the <001> direction. Passing the thermal gradient through the casting at a uniform rate ensures that the secondary dendrite arm spacing is uniform throughout the casting (Ref 5).

Processing of Directionally Solidified Castings

In the most common directional solidification process, an investment casting mold, open at the bottom as well as the top, is placed on a water-cooled copper chill and raised into the hot zone of the furnace (Fig. 4). The mold is heated to a temperature above the liquidus temperature of the alloy to be poured. Meanwhile, the alloy is melted (usually under vacuum) in an upper chamber of the furnace. When the mold is at the proper temperature and the charge is molten, the alloy is poured into the mold. After a pause of a few minutes to allow the grains to nucleate and begin to grow on the chill, during which the most favorably oriented grains are established, the mold is withdrawn from the hot zone to the cold zone.

Furnaces. The furnace shown in Fig. 4 has a relatively small chill diameter (140 mm, or 5.5 in.) to enhance the thermal gradient, a resistance-heated hot zone, and an unconventional melting method in which the charge melts through a plate in a bottom-pour crucible instead of being poured.

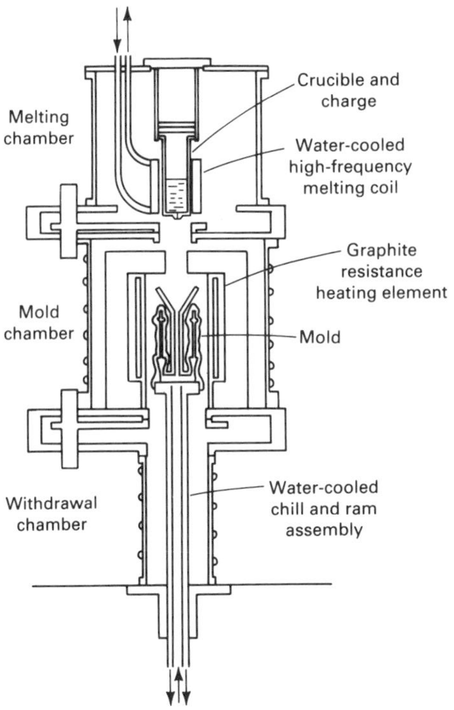

Fig. 4 Configuration of one type of directional solidification furnace. Source: Ref 7

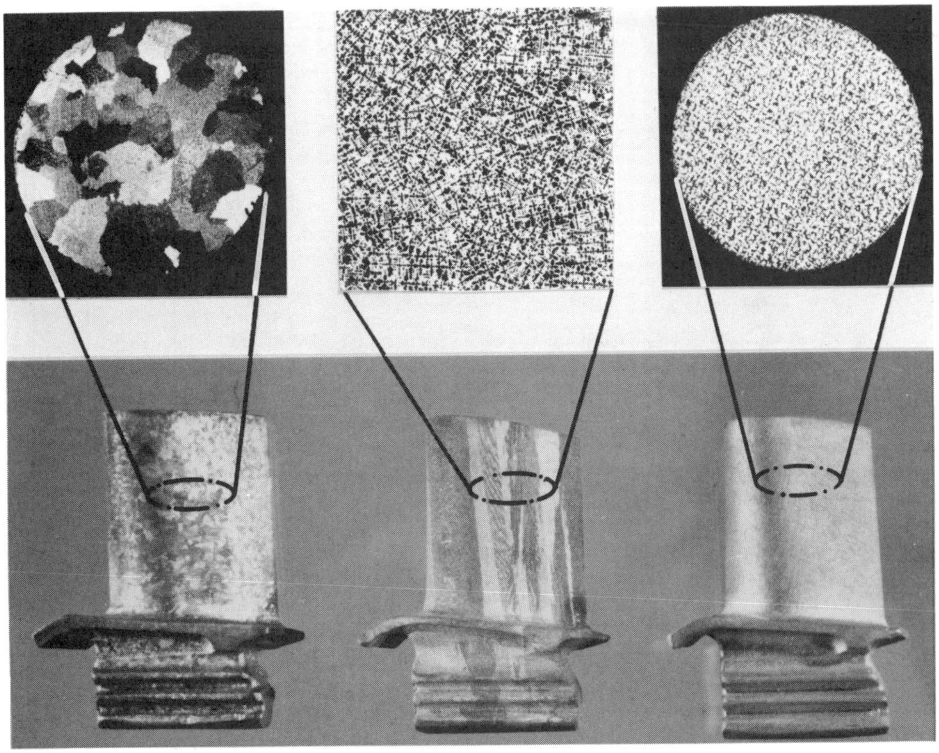

Fig. 5 Comparison of microstructures in (from left) equiaxed, directionally solidified, and single-crystal blades. Courtesy of P.R. Sahm, Giesserei-Institut der RWTH (West Germany)

However, other furnace designs use larger chill plates (up to 500 mm, or 20 in.), induction-heated graphite susceptors in their hot zones, and conventional pouring to produce these castings. Additional information on furnaces and other equipment for directional solidification is available in the section "Vacuum Induction Remelting and Shape Casting" in the article "Vacuum Melting and Remelting Processes" in this Volume.

Gating. Castings can be gated either into the top of the mold cavity or the bottom. Bottom gating heats the mold just above the chill and sets up a very high gradient that encourages well-aligned dendrites. Particular care is taken to keep the transition between the hot and cold zones as sharp as possible through the use of radiation baffles made of refractory materials; these baffles are placed at the chill level between the hot and cold zones.

Mold Design. In designing molds for the process, consideration must be given to the orientation of the part on the cluster. Because heat transfer is by radiation, parts must be placed to minimize shadowing. Internal radiation baffles are sometimes added to the mold, particularly around the center downsprue, to distribute radiation energy to those parts of the mold that would otherwise be shadowed, and some furnace designs use a heating source or a cooling baffle around the center downpole (the chill is designed with a circular cutout at its center) to increase the gradient. Because castings solidify directionally, it is possible to stack them on top of each other to increase the number of castings that can be made in each heat.

Process Control. A very high degree of control must be exercised over the process; therefore, the furnaces are highly automated. Completely automated furnaces (which charge, melt, heat the mold, pour, hold, and withdraw according to a programmed cycle) are commonly used, and even in those furnaces in which melting is done manually the solidification (withdrawal) cycle is automated. Thermocouples are placed within the mold cavity on large clusters to ensure that the molds are at the proper temperature before pouring.

Withdrawal rates during solidification are not necessarily constant. Large differences in section size in specific castings change the solidification rate, and the withdrawal rate can be changed to compensate for this. In selecting a solidification cycle for a hollow part, the effect of the core must be included. Cores lengthen the time required to preheat the mold and slow the withdrawal rate because the heat they contain must also be removed in the process.

Defects Unique to Directional Solidification

Directionally solidified castings are routinely inspected by etching their surfaces and examining the surface visually for defects. The most obvious defect is the presence of an equiaxed or misoriented grain. Equiaxed grains are most often freckles, which are caused by segregation of eutectic liquid that is less dense than the bulk liquid in many alloys. This liquid forms jets within the mushy zone, and as these jets freeze they form equiaxed grains. Freckles are usually cured by increasing the thermal gradient and solidification rate in the casting.

Misoriented grains occur when the temperature ahead of the interface falls below the liquidus temperature and new grains nucleate. These grains will have a random orientation, but because they are growing in gradient, they will be columnar. They can be eliminated by increasing the gradient.

Shrinkage is sometimes encountered on the upper surfaces of directionally solidified castings. There is no way to feed these surfaces; the addition of risers to these surfaces usually interferes with radiation heat transfer from another part of the casting. The most common solution is to invert the casting in order to minimize the surface area that is susceptible to shrinkage.

Microporosity may occur in directionally solidified castings if the length of the mushy zone (length of the casting that is between the liquidus and solidus temperatures during solidification) becomes too great for feed metal to reach into the areas where solidification is taking place. Increasing the thermal gradient (which shortens the length of the mushy zone) usually solves this problem.

Mold or Core Distortion. A frequent cause of scrap in directionally solidified

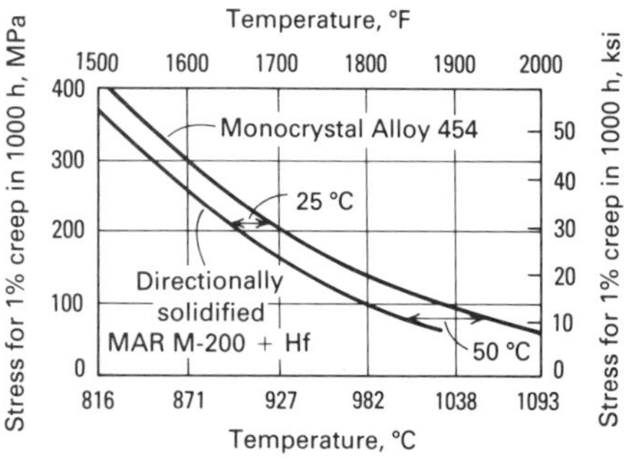

Fig. 6 Creep strength of monocrystal Alloy 454 compared with that of directionally solidified MAR M-200 + Hf. Source: Ref 2

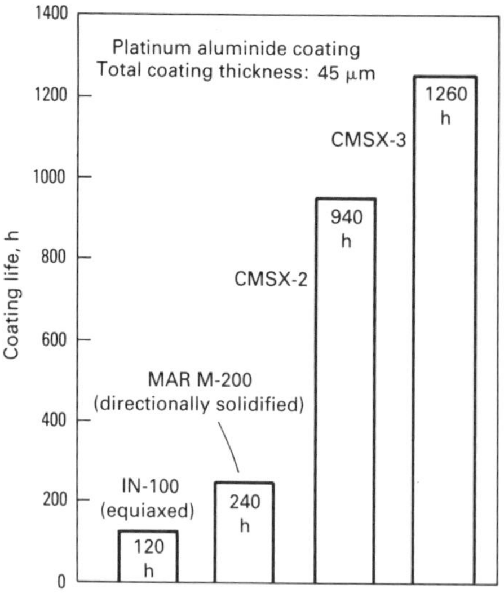

Fig. 7 Effect of grain structure on coating life in hot corrosion testing. Test consisted of exposure to hot salt at 850 °C (1562 °F) and oxidizing atmosphere at 1000 °C (1832 °F). Source: Ref 11

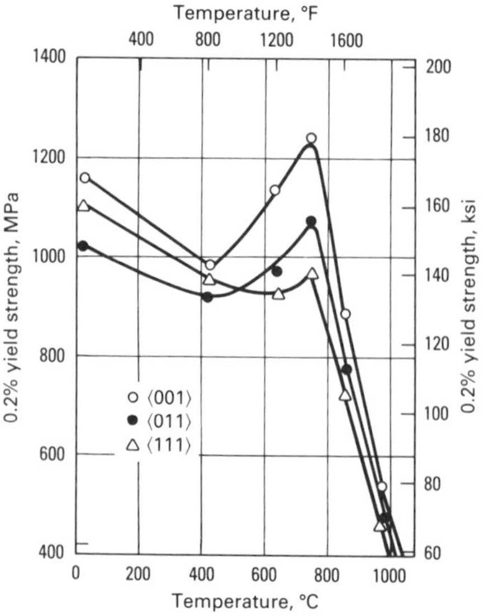

Fig. 8 Yield strength of monocrystal PWA 1480 alloy as a function of temperature and orientation. Source: Ref 13

castings results from mold or core distortion. Because the mold and core are held at high temperatures for long times while the casting solidifies, it is possible for the mold or core to sag or to undergo local allotropic transformations of the refractory materials from which they are made. The resulting changes in mold or core dimensions are reflected in the casting dimensions. Careful control of the core and mold composition, their uniformity, and the firing conditions under which they are made is required in order to avoid these dimensional problems.

Other Directional Solidification Methods

The process described above is the most widely used method of producing directionally solidified castings. However, other methods can be used as well. For example, instead of using electrical means to heat the mold before pouring, the mold can be invested with an exothermic material (Ref 8). When ignited, the material burns in the classic thermit reaction, heating the mold to a temperature above the liquidus of the alloy. The mold and hot exothermic material are placed on a water-cooled copper chill, and the alloy is poured. Both the exothermic material and the metal are cooled by the chill, thus causing solidification to proceed directionally. This process is limited by the properties of the exothermic mixture, and it is most useful for small solid parts. It can also be used in sand molds, for which the grain size and alignment specifications are not as stringent as they are for aerospace castings.

A variation on the mold withdrawal method described above, which is used when very high thermal gradients are desired (as in the production of directionally solidified eutectic alloys), is the liquid metal cooling process (Ref 9). In this method, the mold is immersed in a bath of a liquid metal, such as tin. Heat is removed from the mold by conduction; in addition, the liquid metal bath is an extremely efficient baffle for the radiation in the hot zone.

Monocrystal Casting

It was early recognized that if columnar-grain castings could be produced, the production of castings that contained only a single crystal (more accurately, a single grain or primary dendrite) could be produced by suppressing all but one of the columnar grains (Ref 10). Such a casting is compared with equiaxed and directionally solidified castings in Fig. 5. The fact that the castings consisted of a single crystal removed the limitations on transverse strength imposed by the grain boundaries, but overall properties were only slightly improved.

Metallurgy of Monocrystal Casting

Many alloys contain elements added as grain-boundary strengtheners. These elements lower the incipient melting temperature and therefore limit the temperature at which the alloys can be solutionized.

After it was recognized that alloys having no grain boundaries need no grain-boundary strengtheners, and therefore can be solutionized at higher temperatures, development of high-temperature nickel-base monocrystal alloys began in earnest (Ref 2, 11, 12). These alloys have better high-temperature properties than conventionally cast or directionally solidified alloys because they can precipitate a higher percentage of the γ' strengthening phase (Fig. 6). In addition, because they have no grain boundaries, monocrystal alloys have improved corrosion resistance (Fig. 7).

Applications of monocrystal castings must take into consideration the fact that many alloy systems are anisotropic; that is, their properties vary with crystallographic orientation, as shown in Fig. 8. This means that designers must design with this in mind and that castings must be produced with specific orientations (a tolerance of ±5° of the required orientation is often specified). Further, orientation control may be re-

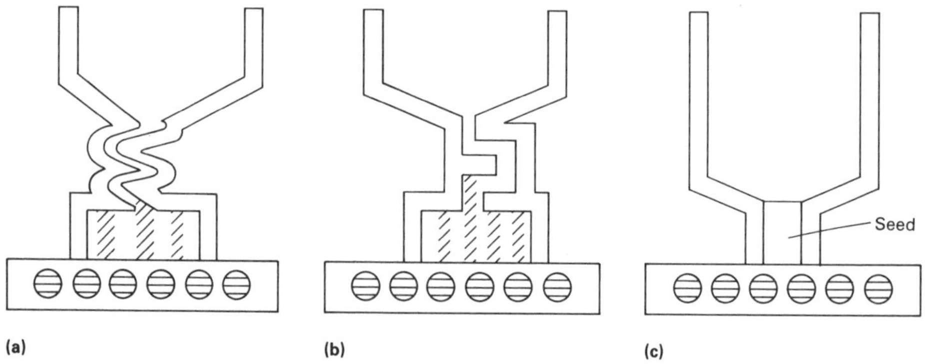

Fig. 9 Three methods of producing monocrystal castings. (a) Use of a helical mold section. (b) Use of a right-angle mold section. (c) Seeding. Source: Ref 14

quired in more than one crystallographic direction.

Monocrystal Casting Processing

Monocrystal castings are produced using techniques similar to those used for directionally solidified castings, with one important difference: A method of selecting a single, properly oriented grain is required. Three methods are most commonly used, as shown in Fig. 9.

Helical Mold Sections. In the first method, a helical section of mold is placed between the chill and the casting. Only the most favorably oriented grains are able to grow through this helix because all others are intercepted by the helix wall. Eventually, only one grain is left to emerge from the helix to form the casting. In this method, only the primary orientation of growth can be controlled, and it will be the preferred growth direction (<001>) for nickel-base alloys.

Right-Angle Mold Sections. The second method uses a series of right-angle bends in the helix. Because growth takes place along the preferred growth direction in each of the arms of the selector, the grain that emerges tends to be doubly oriented. If an orientation other than the preferred growth direction is desired, the casting can be tilted on the selector.

Seeding, the third method, is particularly useful for an orientation other than the preferred growth direction. Use of seeding requires that the seeds be prepared and placed in the mold before the casting is poured.

Molds. Because monocrystal alloys have higher incipient melting temperatures than conventional alloys, mold preheat temperatures will normally be higher for their manufacture than for columnar-grain castings. Therefore, mold composition control is of particular importance in the production of these castings.

Testing and Inspection. In addition to surface etching, monocrystal castings are inspected by using back reflection Laué techniques to determine crystallographic orientation. Defects in monocrystal castings are usually the same as those found in columnar castings, and remedial actions are also generally the same. Some monocrystal alloys are susceptible to local recrystallization from rough handling or solidification-induced strains and must be given a stress-relief heat treatment before solution heat treatment.

REFERENCES

1. F.L. VerSnyder and M.E. Shank, *Mater. Sci. Eng.*, Vol 6, 1970, p 213
2. M. Gell, D.N. Duhl, and A.F. Giamei, The Development of Single Crystal Superalloy Turbine Blades, in *Superalloys 1980*, American Society for Metals, 1980, p 205
3. S.M. Copley, A.F. Giamei, S.M. Johnson, and M.F. Hornbecker, *Metall. Trans.*, Vol 1, Aug 1970, p 2193
4. F.L. VerSnyder, *High Temperature Alloys for Gas Turbines 1982*, Reidel, 1982, p 1
5. T.S. Piwonka and P.N. Atanmo, in *Proceedings of the 1977 Vacuum Metallurgy Conference*, Science Press, 1977, p 507
6. S. Morimoto, A. Yoshinari, and E. Niyama, in *Superalloys 1984*, The Metallurgical Society, 1984, p 177
7. M.J. Goulette, P.D. Spilling, and R.P. Anthony, in *Superalloys 1984*, The Metallurgical Society, 1984, p 167
8. G.S. Hoppin, M. Fujii, and L.W. Sink, Development of Low-Cost Directionally-Solidified Turbine Blades, in *Superalloys 1980*, American Society for Metals, 1980, p 225
9. P.M. Curran, L.F. Schulmeister, J.F. Ericson, and A.F. Giamei, in *Proceedings of the Second Conference on In-Situ Composites*, Xerox, 1976, p 285
10. B.J. Piearcey, U.S. Patent 3,494,709
11. K. Harris, G.L. Erickson, and R.E. Schwer, in *Superalloys 1984*, The Metallurgical Society, 1984, p 221
12. D.A. Ford and R.P. Arthey, in *Superalloys 1984*, The Metallurgical Society, 1984, p 115
13. D.M. Shah and D.N. Duhl, in *Superalloys 1984*, The Metallurgical Society, 1984, p 105
14. G.K. Bouse and J.R. Mihalisin, *Superalloys, Composites and Ceramics*, Academic Press, to be published

Squeeze Casting

J.L. Dorcic and S.K. Verma, IIT Research Institute

Squeeze casting, also known as liquid-metal forging, is a process by which molten metal solidifies under pressure within closed dies positioned between the plates of a hydraulic press. The applied pressure and the instant contact of the molten metal with the die surface produce a rapid heat transfer condition that yields a pore-free fine-grain casting with mechanical properties approaching those of a wrought product.

The squeeze casting process is easily automated to produce near-net to net shape high-quality components. The process was introduced in the United States in 1960 and has since gained widespread acceptance within the nonferrous casting industry. Aluminum, magnesium, and copper alloy components are readily manufactured using this process. Several ferrous components with relatively simple geometry—for example, nickel hard-crusher wheel inserts—have also been manufactured by the squeeze casting process. Despite the shorter die life for complex ferrous castings requiring sharp corners within the die or punch (tooling), the process can be adopted for products where better properties and/or savings in labor or material costs are desired.

Advantages of Squeeze Casting (Ref 1-11)

With the current emphasis on reducing materials consumption through virtually net shape processing and the demand for higher-strength parts for weight savings, the emergence of squeeze casting as a production process has given materials and process engineers a new alternative to the traditional approaches of casting and forging. By pressurizing liquid metals while they solidify, near-net shapes can be achieved in sound, fully dense castings.

The near-net and net shape capabilities of this manufacturing process are key advantages. Tolerances of ±0.05 mm (±0.002 in.) are not uncommon for nonferrous castings, with yields of 100% in a number of applications. Improved mechanical properties are additional advantages of squeeze cast parts.

Squeeze casting has been successfully applied to a variety of ferrous and nonferrous alloys in traditionally cast and wrought compositions. Applications include aluminum alloy pistons for engines and disk brakes; automotive wheels, truck hubs, barrel heads, and hubbed flanges; brass and bronze bushings and gears; steel missile components and differential pinion gears; and a number of parts in cast iron, including ductile iron mortar shells.

Squeeze casting is simple and economical, efficient in its use of raw material, and has excellent potential for automated operation at high rates of production. The process generates the highest mechanical properties attainable in a cast product. The microstructural refinement and integrity of squeeze cast products are desirable for many critical applications.

Process Description (Ref 1-3)

As shown in Fig. 1, squeeze casting consists of metering liquid metal into a preheated, lubricated die and forging the metal while it solidifies. The load is applied shortly after the metal begins to freeze and is maintained until

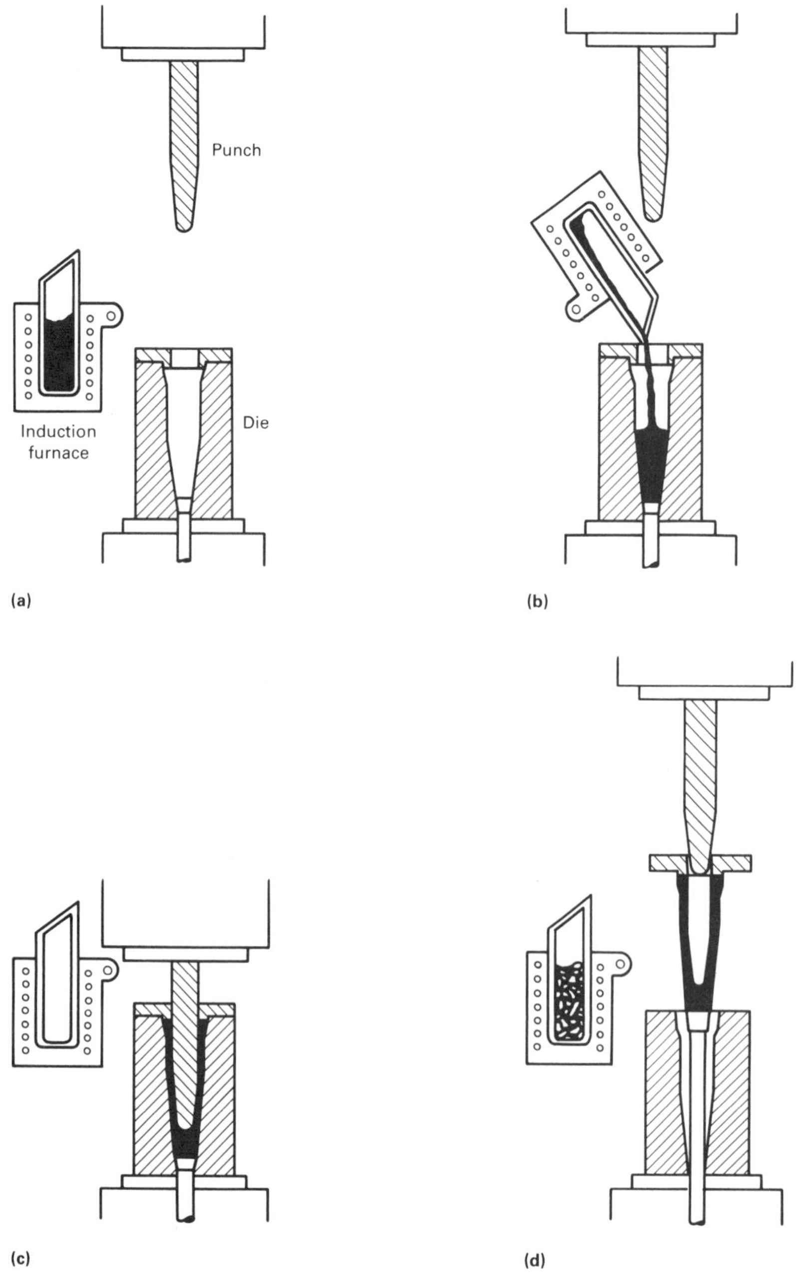

Fig. 1 Schematic illustrating squeeze casting process operations. (a) Melt charge, preheat, and lubricate tooling. (b) Transfer melt into die cavity. (c) Close tooling, solidify melt under pressure. (d) Eject casting, clean dies, charge melt stock.

the entire casting has solidified. Casting ejection and handling are done in much the same way as in closed die forging.

The high pressure applied (typically 55 to 100 MPa, or 8 to 15 ksi) is enough to suppress gas porosity except in extreme cases, for which standard degassing treatments are used. The tendency toward shrinkage porosity is limited by using a bare minimum of superheat in the melt during pouring. This is possible in squeeze casting because melt fluidity, which requires high pouring temperatures, is not necessary for die fill, the latter being readily achieved by the high pressure applied. In heavy sections of the casting, which are particularly prone to the incidence of shrinkage porosity, the applied pressure squirts liquid or semiliquid metal from hot spots into incipient shrinkage pores to prevent pores from forming. Alloys with wide freezing ranges accommodate this form of melt movement very well, resulting in sound castings with a minimum of applied pressure.

The squeeze casting cycle starts with the transfer of a metered quantity of molten metal into the bottom half of a preheated die set mounted in a hydraulic press (Fig. 1). The dies are then closed, and this fills the die cavity with molten metal and applies pressures of up to 140 MPa (20 ksi) on the solidifying casting.

Process Variables

There are a number of variables that are generally controlled for the soundness and quality of the castings. The variable ranges discussed in the following sections vary with the alloy system and part geometry being squeeze cast.

Melt Volume. Precision control of the metal volume is required when filling the die cavity. This ensures dimensional control.

Casting temperatures depend on the alloy and the part geometry. The starting point is normally 6 to 55 °C (10 to 100 °F) above the liquidus temperature.

Tooling temperatures ranging from 190 to 315 °C (375 to 600 °F) are normally used. The lower range is more suitable for thick-section casting. The punch temperature is kept 15 to 30 °C (25 to 50 °F) below the lower die temperature to maintain sufficient clearance between them for adequate venting. Excess punch-to-die clearance allows molten metal to be extruded between them, eroding the surface.

Time delay is the duration between the actual pouring of the metal and the instant the punch contacts the molten pool and starts the pressurization of thin webs that are incorporated into the die cavity. Because increased pouring temperatures may be required to fill these sections adequately upon pouring, a time delay will allow for cooling of the molten pool before closing of the dies to avoid shrink porosity.

Pressure levels of 50 to 140 MPa (7.5 to 20 ksi) are normally used; 70 MPa (10 ksi) is generally applied, depending on part geometry and the required mechanical properties. There is an optimum pressure for each of the systems after which no added advantages in mechanical properties are obtained.

Pressure duration varying from 30 to 120 s has been found to be satisfactory for castings weighing 9 kg (20 lb). However, the pressure duration is again dependent on part geometry. Applied pressure after composite solidification and temperature equalization will not contribute any property enhancements and will only increase cycle times.

Lubrication. For aluminum, magnesium, and copper alloys, a good grade of colloidal graphite spray lubricant has proved satisfactory when sprayed on the warm dies prior to casting. Care should be taken to avoid excess buildup on narrow webs and fin areas where vent holes or slots are used. Care must also be taken to prevent plugging of these vents. For ferrous castings, ceramic-type coatings are required to prevent welding between the casting and the metal die surfaces.

Quality Control

In general, the process parameters are optimized for each component geometry to

Table 1 Comparative properties of commercial wrought and cast alloys

Alloy	Process	Tensile strength MPa	Tensile strength ksi	Yield strength MPa	Yield strength ksi	Elongation, %
356-T6 aluminum	Squeeze casting	309	44.8	265	38.5	3
	Permanent mold	262	38.0	186	27.0	5
	Sand casting	172	25.0	138	20.0	2
535 aluminum (quenched)	Squeeze casting	312	45.2	152	22.1	34.2
	Permanent mold	194	28.2	128	18.6	7
6061-T6 aluminum	Squeeze casting	292	42.3	268	38.8	10
	Forging	262	38.0	241	35.0	10
A356 T4 aluminum(a)	Squeeze casting	265	38.4	179	25.9	20
A206 T4 aluminum(a)	Squeeze casting	390	56.5	236	34.2	24
CDA 377 forging brass	Squeeze casting	379	55.0	193	28.0	32.0
	Extrusion	379	55.0	145	21.0	48.0
CDA 624 aluminum bronze	Squeeze casting	783	113.5	365	53.0	13.5
	Forging	703	102.0	345	50.0	15.0
CDA 925 leaded tin bronze	Squeeze casting	382	55.4	245	35.6	19.2
	Sand casting	306	44.4	182	26.4	16.5
Type 357 (annealed)	Squeeze casting	614	89.0	303	44.0	46
	Sand casting	400	58.0	241	35.0	20
	Extrusion	621	90.0	241	35.0	50
Type 321 (heat treated)	Squeeze casting	1063	154.2	889	129.0	15
	Forging	1077	156.2	783	113.6	7

(a) Added as additional reference information only. Source: Ref 3, 7, 10

be squeeze cast. Maintaining established optimized parameters is critical to the quality and reproducibility of the squeeze cast components. Failure to do so can result in one or more of the following defects.

- Oxide inclusions
- Porosity
- Extrusion segregation
- Centerline segregation
- Blistering
- Cold laps
- Hot tearing
- Sticking
- Case debonding
- Extrusion debonding

The root causes of these defects, as well as methods for their control, are described below.

Oxide inclusions result from the failure to maintain clean melt-handling and melt-transfer systems. To minimize the likelihood of introducing metallic inclusions, filters should be included within the melt-transfer system, or molten metal turbulence should be minimized when filling the die cavity. Preventing foreign objects from entering open dies is also helpful.

Porosity and voids can occur when insufficient pressure is applied during squeeze casting operations. A general rule of thumb is to apply a pressure of 70 MPa (10 ksi), although sound castings have been produced with pressures as low as 50 MPa (7.5 ksi). Porosity and/or voids are usually eliminated by increasing the casting pressure when the other variables are optimized.

Extrusion Segregation. The relative microsegregation that occurs in squeeze cast components is much less than that in other cast components. However, the regions filled by back extrusion are rich in solute; these areas are the last to solidify within a casting. The extrusion segregation can lead to local variations in mechanical and corrosion properties. Such defects can be avoided by designing dies properly, by using a multiple gate system, by increasing die temperature, or by decreasing delay time before die closure.

Centerline segregation is a defect that is normally encountered with high-alloy wrought aluminum alloys at lower solute temperatures. As solidification begins on the die walls, the liquid phase becomes more concentrated with the lower-melting solute, which is trapped within the center areas of the extruded projections or more massive areas of the casting. Such defects are avoided by increasing die temperature, by minimizing die closure time, or by selecting an alternative alloy.

Blistering. Air or gas from the melt that is trapped below the surface during turbulent die filling forms blisters on the cast surface upon the release of pressure or during subsequent solution heat treatments. Methods of avoiding such defects include degassing the melt and preheating the handling transfer equipment, using a slower die closing speed, increasing the die and punch venting, and reducing the pouring temperature.

Cold laps are caused by molten metal overlapping previously solidified layers,

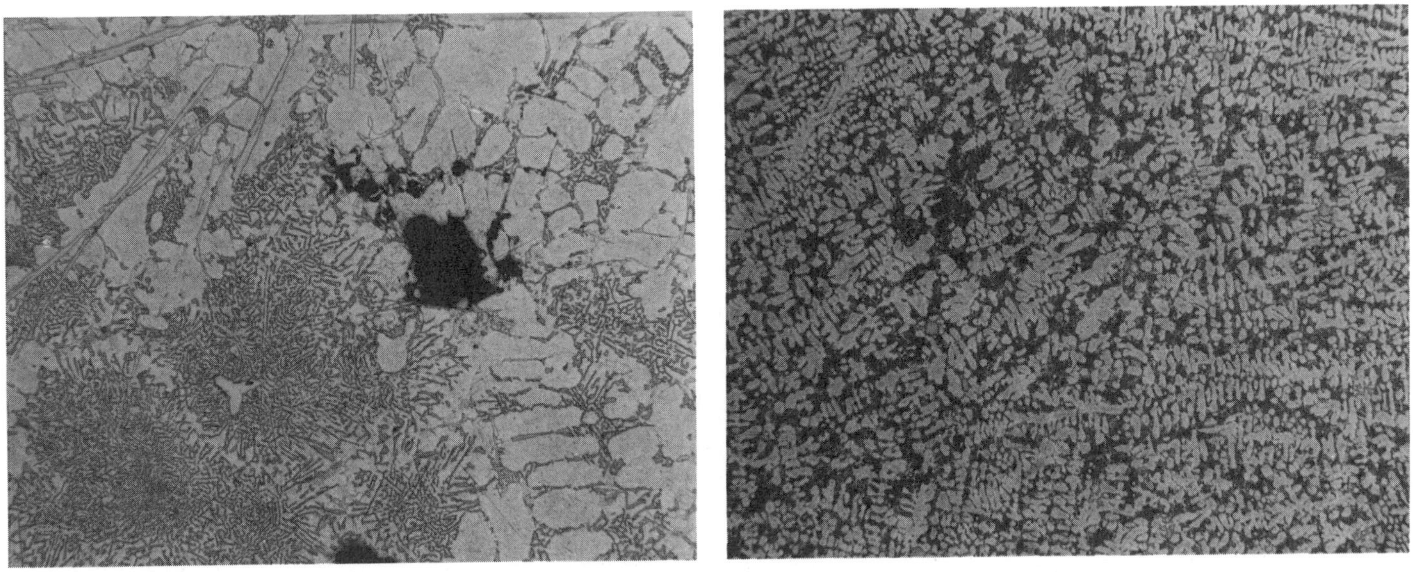

(a) (b)

Fig. 2 Comparison of permanent mold cast (a) and squeeze cast (b) E-132 aluminum near the edge in contact with the punch. The microstructure shown in (a) is coarse dendritic; that in (b) is preferred ultrafine dendritic. Both 80×

Fig. 3 Typical ferrous and nonferrous parts produced by squeeze casting. Dome in center measures 423 mm (16.5 in.) in outside diameter and weighs 29.5 kg (65 lb). Courtesy of IIT Research Institute

with incomplete bonding between the two. To alleviate cold laps, it is necessary to increase the pouring temperature or the die temperature. Reducing the die closure time has also been found to be beneficial.

Hot tearing takes place in alloys that have an extended freezing range (for example, off eutectic composition). When solid and liquid coexist over a wide range of temperatures, contraction of the solid around the rigid mold surface can initiate rupture in partially solidified regions. The methods used to avoid hot tearing in squeeze cast products include reducing the pouring temperature, reducing the die temperature, increasing the pressurization time, and increasing the draft angles on the casting.

Sticking. A thin layer of casting skin adheres to the die surface because of rapid cycling of the process without sufficient die/punch cooling and lubrication. To avoid sticking, it is recommended to decrease die temperature or pouring temperature.

Case debonding is found only in high-silicon alloys when an extremely fine-grain case 0.51 to 2.0 mm (0.020 to 0.080 in.) thick is formed on the surface and peels off during subsequent machining or cleaning operations. It is caused by extreme chilling of the outer skin of the casting against a cold punch or die. This problem can be overcome by increasing the tooling temperature or the pouring temperature. Decreasing the die closure time may also help eliminate case bonding.

Extrusion debonding takes place when the casting has deeply extruded details and the metal remains in the open die for a long period of time before it is extruded to fill the die cavity. The oxide present on the partially solidified crust in the die remains there after the melt has been extruded around it, resulting in the absence of a metal-to-metal bond at oxide stringer locations. Extrusion debonding can be prevented by increasing the tooling temperature or the pouring temperature. Decreasing the die closure time can reduce the oxide formation on the semiliquid metal present in the die.

Microstructure (Ref 1-6)

In addition to the densification achieved, there are several reasons why squeeze casting produces castings with superior properties. Even moderately applied pressure causes intimate contact between the solidifying casting and the die for a tenfold increase in heat transfer rate over permanent mold casting. This results in relatively fine grains in the casting. Fine grain size is also promoted by the large number of nuclei formed because of the low casting temperature and the elevated pressure. Furthermore, because die filling in squeeze casting does not require high melt fluidity, a number of wrought alloys can be squeeze cast. Again, pressurized solidification with rapid heat transfer tends to minimize the segregation that wrought alloys are usually prone to. As indicated in Table 1, the tensile properties of ferrous and nonferrous materials produced by squeeze casting are generally comparable to those of forgings.

In a recent investigation, a side-by-side comparison was made between squeeze casting and permanent mold casting for E-132 aluminum components. As seen in Fig. 2, the squeeze casting is sounder and has a pore-free, fine-grain, nearly equiaxed microstructure as compared to that of the permanent mold casting. In particular, a thin case, which is characterized by an unusually fine cast structure, forms to 2.0 mm (0.080 in.) below the punch. This is caused by a combination of high pressure (resulting in undercooling and a greater number of nucleation sites) and rapid heat extraction into the punch. In practice, squeeze castings made with a ±0.76 mm (±0.030 in.) tolerance on the as-cast head location can be machined to the finished tolerances and still retain more than 0.51 mm (0.020 in.) of the ultrafine-grain case.

Product Applications

The squeeze casting process has been explored for a number of applications using various metals and alloys. The parts shown in Fig. 3 include an aluminum dome, a ductile iron mortar shell, and a steel bevel gear. Other parts that have been squeeze cast include stainless steel blades, superalloy disks, aluminum automotive wheels and pistons, and gear blanks made of brass and bronze. Recently, this process has also been adopted to make composite materials at an affordable cost (see the article "Cast Metal-Matrix Composites" in this Volume). A porous ceramic preform is placed in the preheated die, which is later filled with the liquid metal; pressure is then applied. The pressure, in this case, helps the liquid metal infiltrate the porous ceramic preform, giving a sound metal ceramic composite. The technological breakthrough of manufacturing metal-ceramic composites, along with the ability to make complex parts by a near-net shape squeeze casting process, suggests that this process will find application where cost considerations and physical properties of alloys are key factors.

REFERENCES

1. S.K. Verma and J. Dorcic, "Squeeze Casting Process for Metal-Ceramic Composites," Paper presented at the International Congress and Exposition, Detroit, MI, Society of Automotive Engineers, Feb 1987
2. S. Rajagopal, Squeeze Casting: A Review and Update, *J. Appl. Metalwork.*, Vol 1 (No. 4), 1981, p 3-14
3. M.A.H. Howes, "Ceramic Reinforced Metal Matrix Composites Fabricated by Squeeze Casting," Paper presented at the Advanced Composite Conference, Dearborn, MI, American Society for Metals, Dec 1985
4. S. Rajagopal *et al.*, "Squeeze Casting of Aluminum Alloy Heavy-Duty Pistons," Final Report IITRI-M08086-1, Illinois Institute of Technology Re-

search Institute, June 1981
5. Y. Nishida and H. Matsubara, *Br. Foundryman*, Vol 69, 1976, p 274-278
6. O.G. Epanchintsev, *Russ. Cast. Prod.*, May 1972, p 188-189
7. R.E. Spear and G.R. Gardner, *Trans. AFS*, Vol 71, 1963, p 209
8. J.C. Benedyk, Paper 86, *Trans. SDCE*, Vol 8, 1970
9. J.C. Benedyk, Paper CM71-840, Society of Manufacturing Engineers, 1971
10. R.F. Lynch, R.P. Olley, and P.C.J. Gallagher, *Trans. AFS*, Vol 83, 1975, p 561-568
11. R.F. Lynch, R.P. Olley, and P.C.J. Gallagher, *Trans. AFS*, Vol 83, 1975, p 569-576

Semisolid Metal Casting and Forging

Malachi P. Kenney, James A. Courtois, Robert D. Evans, Gilbert M. Farrior, Curtis P. Kyonka, and Alan A. Koch, ALUMAX Engineered Metal Processes, Inc.
Kenneth P. Young, AMAX Research & Development Center

Semisolid metalworking, also known as semisolid forming, is a hybrid manufacturing method that incorporates elements of both casting and forging. It was based on a discovery made at the Massachusetts Institute of Technology (MIT) in the early 1970s. Processes based on the discovery were identified by MIT as rheocasting, thixocasting, or stir casting (Ref 1). Today it is a two-step process for the near-net shape forming of metal parts using a semisolid raw material that incorporates a unique nondendritic microstructure (Fig. 1).

The key to the process is shown in Fig. 2, in which the semisolid slug has been cut with a spatula while free standing (that is, without containers), thus demonstrating the thixotropic nature of the material. The thixotropic properties permit the material to be handled by robotic devices in the semisolid condition, allowing process automation and precision controls while increasing productivity.

The major commercial semisolid metalworking activity is in the semisolid forging of a variety of aluminum alloy parts for military, aerospace, and automotive applications. In addition, there is moderate copper alloy production for electrical and fluid-handling use. Further, semisolid metalworking technology has been demonstrated to be applicable to most engineering alloy families, including zinc (Ref 2), magnesium (Ref 3), copper (Ref 4), ferrous (Ref 5), titanium (Ref 6), and superalloys (Ref 7).

The unique nondendritic microstructure and the initial processes are protected by a series of patents that began with awards to M.C. Flemings and associates at MIT in the 1970s and has continued with awards to a number of individuals and organizations. A number of corporations have investigated and tested the MIT processes. Universities in Great Britain, Europe, and the United States have also conducted studies in the technology. Along with MIT, these include Delaware and Virginia (USA), Sheffield (Great Britain), and Aachen (West Germany), which have been studying the applications to high-temperature alloys for a number of years. The patent list continues to grow as the technology broadens its application in the emerging field of metal-matrix composites, in which semisolid metalworking has been a key process since the early 1970s (see the article "Cast Metal-Matrix Composites" in this Volume).

Background

Basic Discovery. Semisolid metal forming is based on a discovery made during research on hot tearing undertaken at MIT in the early 1970s. Seeking to understand the magnitude of the forces involved in deforming and fragmenting dendritic growth structures, MIT researchers constructed a high-temperature viscometer. They poured model lead-tin alloys into the annular space created by two concentric cylinders and measured the forces transmitted through the freezing alloy when the outer cylinder was rotated. During the course of these experiments, it was discovered that when the outer cylinder was continuously rotated the semisolid alloy exhibited remarkably low shear strength even at relatively high fractions solidified. This unique property was attributed to a novel nondendritic (that is, spheroidal) microstructure.

The research expanded, and the MIT engineers coined the term rheocasting to describe the process of producing this unique microstructure (a schematic of the rheocast process is shown in Fig. 5 of the article "Classification of Processes and Flow Chart of Foundry Operations" in this Volume) (Ref 5). They showed that sheared and partially solidified alloys could be assigned an apparent viscosity and that they possess many of the characteristics of thixotropy (Fig. 3). Most notably, the semisolid alloys displayed viscosities that depended on shear rate and that rose to several hundred, even thousands, of poise (approaching the consistency of table butter) when at rest and yet decreased to less than 5.0 Pa · s or 50 P (poise) (the range of machine oils) upon vigorous agitation or shearing. For the first time, therefore, these results afforded an opportunity to control the viscosity of alloy melts from that of fully liquid to any desired upper limit.

Potential Benefits. The MIT researchers were quick to identify several potential benefits that could result from forming pro-

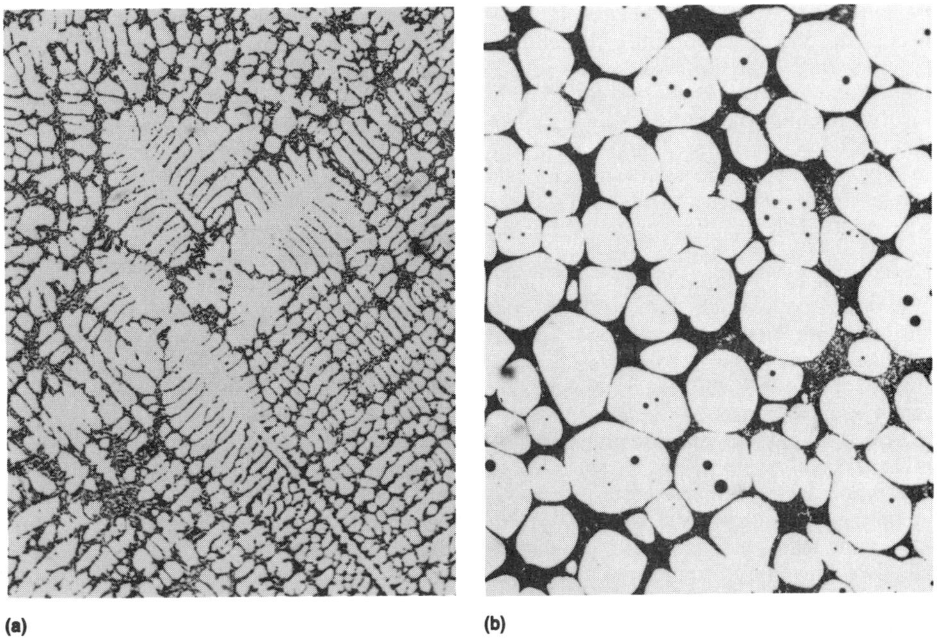

(a) (b)

Fig. 1 Comparison of dendritic conventionally cast (a) and nondendritic semisolid formed (b) microstructures of aluminum alloy 357 (Al-7Si-0.5Mg). Both 200×

Fig. 2 Semisolid aluminum billet being cut by a spatula

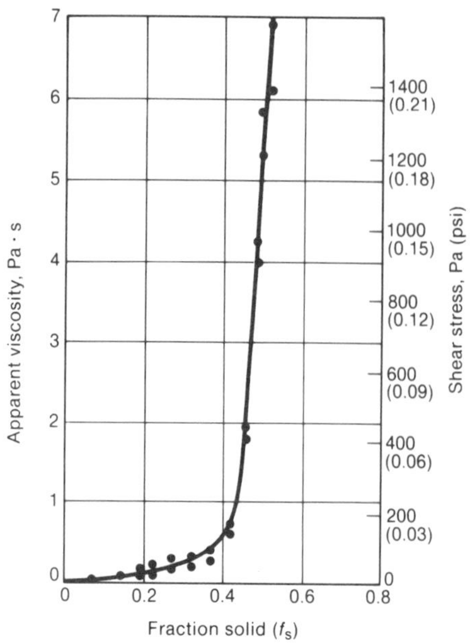

Fig. 3 Apparent viscosity and shear stress of model Sn-15Pb semisolid melts as a function of fraction solid at constant shear rate. Source: Ref 13

cesses utilizing semisolid metal and that would differentiate these processes from conventional casting (Ref 8). First, and particularly significant for higher-melting alloys, semisolid metalworking afforded lower operating temperatures and reduced metal heat content (reduced enthalpy of fusion). Second, the viscous flow behavior could provide for a more laminar cavity fill than could generally be achieved with liquid alloys. This could lead to reduced gas entrainment. Third, solidification shrinkage would be reduced in direct proportion to the fraction solidified within the semisolid metalworking alloy, which should reduce both shrinkage porosity and the tendency toward hot tearing.

In addition, the MIT team showed that the viscous nature of semisolid alloys provided a natural environment for the incorporation of third-phase particles in the preparation of particulate-reinforced metal-matrix composites (Ref 9). Here the enhanced viscosity of semisolid metalworking alloys would serve to entrap the reinforcement material physically, allowing time to develop good bonding between the reinforcement and the matrix alloy.

As these ideas unfolded, research into the nature of semisolid alloys progressed, and it became apparent that bars could be cast from semisolid fluids possessing the rheocast nondendritic microstructure. The final freezing of these bars captures this microstructure. The bars then represented a raw material that could be heated at a later time or a remote location to the semisolid temperature range to reclaim the special rheological characteristics.

This process, using semisolid alloys heated from specially cast bars, was termed thixocasting by the MIT inventors (Ref 10). This distinguished it from rheocasting, which has come to be known as the process used for producing semisolid structures and/or forming parts from slurry without an intermediate freezing step.

The efforts at MIT to continue the development of semisolid metalworking were supported by the U.S. government under the Advanced Research Projects Agency and the Defense Advanced Research Projects Agency. This work was directed to a machine casting process for ferrous alloys. Thousands of ferrous components were successfully formed using semisolid metalworking ingots as part of these programs.

Semisolid Metalworking Processes

Forming in the semisolid state requires that a metal or alloy have a roughly spherical and fine-grain microstructure when it enters the forming die. There are two general approaches to the process: rheocasting, in which slurry is produced in a mixer and delivered directly into the die, and semisolid forging, in which a billet is cast in a mold equipped with a mixer (which creates the spherical microstructure during casting) and is stored for subsequent use. If the volume of billet material required is known, the slug weight can be readily determined; later, a slug cut from the billet is heated to the semisolid state and formed in a die. Normally, the cast billet is forged when 30 to 40% is liquid; in the slurry process, 60 to 70% of the material is liquid.

There have been several attempts in the United States and abroad to commercialize rheocasting, but none of these ventures is known to have been commercially successful (Ref 11, 12). On the other hand, semisolid forging, which exploits the manufacturing advantages of thixotropic semisolid alloy bars, began commercial production in 1981. It is now a rapidly expanding commercial process. The production of raw material has been brought to full commercial realization, and the use of semisolid forged parts is broadening in the aerospace, automotive, military, and industrial sectors.

This section will primarily discuss semisolid forging. Therefore, it will describe both raw material production and part-forming techniques.

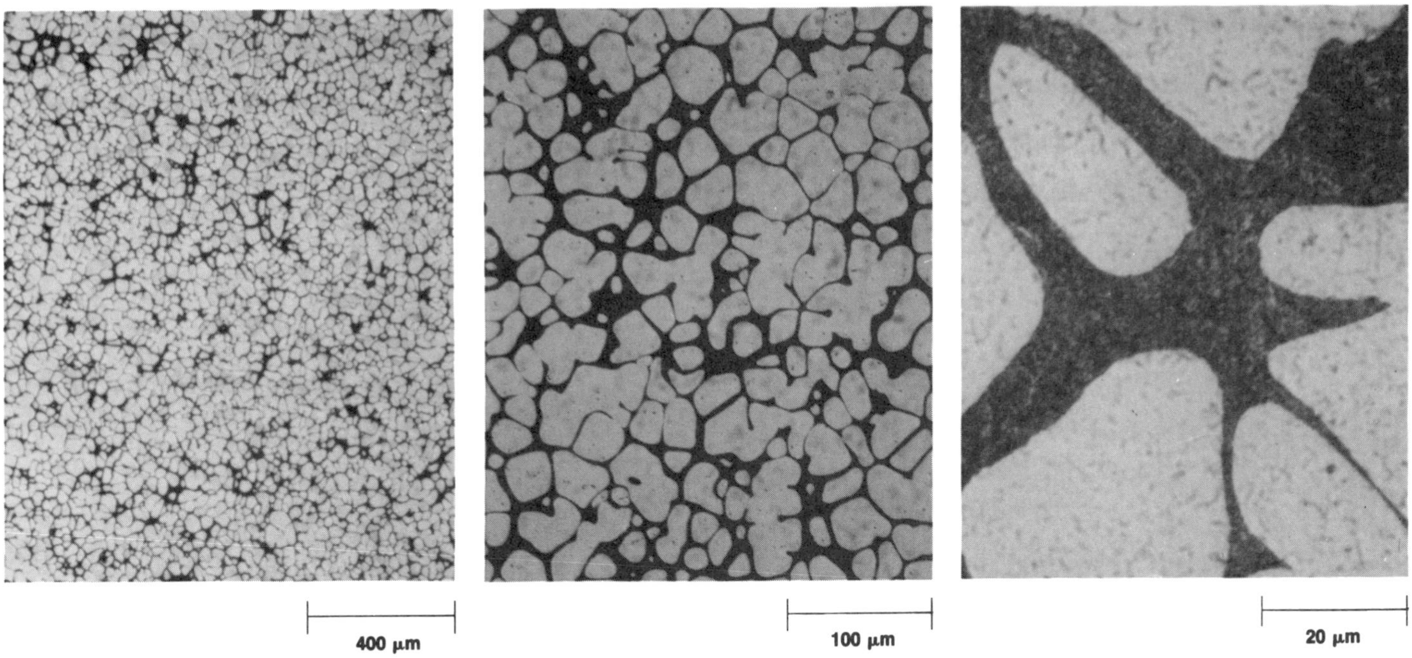

Fig. 4 Magnetohydrodynamically cast 357 aluminum alloy billet shown at increasingly higher magnifications. Note the fine eutectic which is unresolved at lower magnifications.

Raw Material Production: Casting Processes

Raw material for semisolid metal forming requires the special microstructures illustrated in Fig. 1. When semisolid, this structure comprises solid particles in the form of globules or spheroids suspended in a matrix of lower-melting alloy liquid. Recapturing this structure in materials heated from the solid state requires the retention of some residual microsegregation to provide differential melting between solid and liquid phases. This condition is almost always present in commercial alloys.

Original Rheocasting Approach. The original MIT research showed that a very effective method of producing the semisolid metalworking microstructure was to mechanically agitate an alloy vigorously during the solidification process. The evolution of this concept at MIT ranged from simple eggbeater-type mixers in slowly cooled crucibles to complex gas-shielded high-temperature continuous mixing systems for steels and superalloys (Ref 5).

Over the course of several years, others became involved and much was learned of the mechanical mixing approach. In at least two cases, large-scale systems have been operated at the pilot level, producing aluminum ingots 150 mm (6 in.) in diameter or larger. The equipment was very large and cumbersome, however, and none is known to be operating commercially (Ref 11).

Key Rheocasting Parameters. The research discussed above and the accompanying basic work showed that average solid-

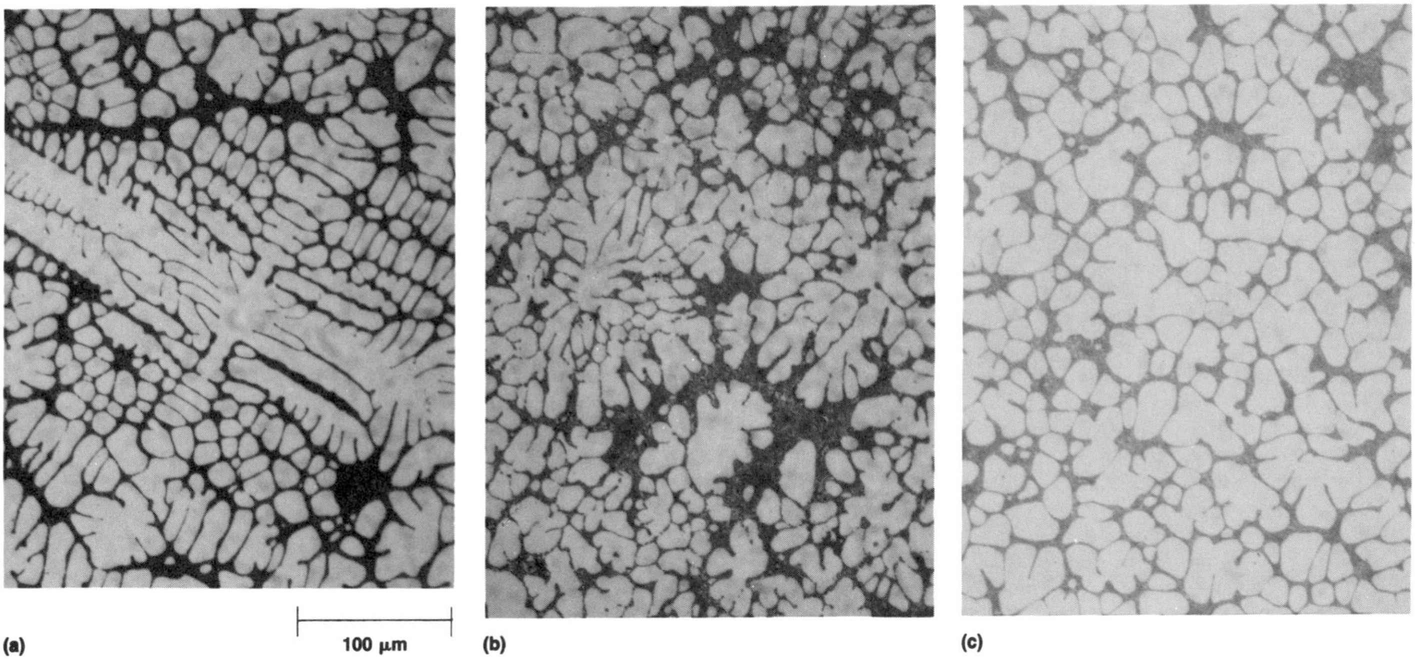

Fig. 5 Effect of stirring (shear) on microstructure. (a) No stirring. (b) and (c) Increasing stirring

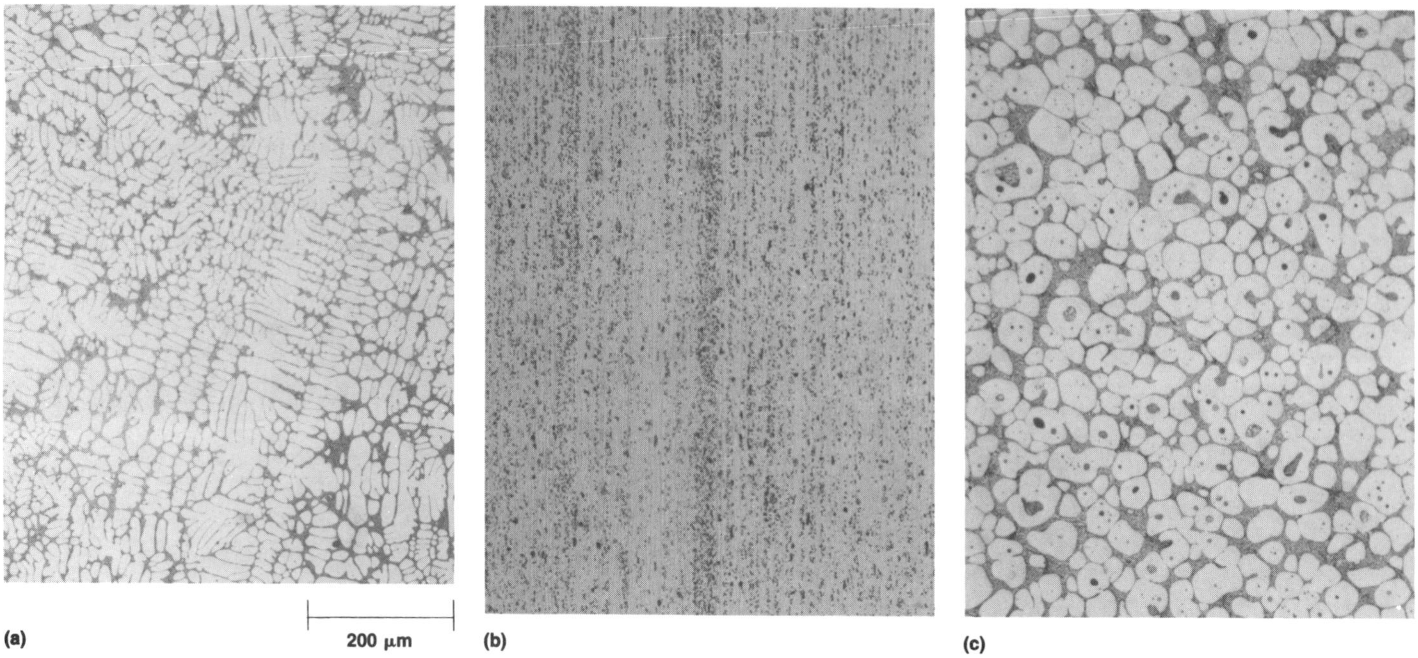

Fig. 6 SIMA processing route for semisolid raw material. (a) Cast structure. (b) Extruded (hot worked) structure. (c) Structure after straining and reheating

particle diameters were primarily controlled by solidification rate, while the optimization of the spheroidal shape was controlled by shear rate (mixing). High solidification rates produced finer particle diameters, and high shear rates produced more rounded particle shapes with less clustering of particles (Ref 13). Typically, solid-particle diameters ranged from 100 to 400 μm (0.004 to 0.016 in.) for normal solidification rates.

Alternative Rheocasting Approaches. After the original announcements and publications by the MIT researchers, a number of alternative approaches to the production of the semisolid raw material were proposed by other researchers. Although several of these techniques build upon the mechanical agitation approach (Ref 12, 14), such as the Gircast process described in the article "Zinc and Zinc Alloys" in this Volume, others utilize a passive stirring technique for stimulating turbulent flow through cooling channels (Ref 15, 16). At least one approach uses isothermal holding to induce particle coarsening. Most of these alternatives appear to be confined to the laboratory, although one or two have been demonstrated at a pilot production level. To date, none has shown economic viability.

In all of the above, difficulties appear to have been encountered with regard to metal quality, contamination, process control, and economics. Although passive stirring offers some simplicity of design relative to mechanical mixing systems, it eliminates control of the key variable of shear rate (Ref 17), thus placing crippling restrictions on flexibility and throughput and, consequently, economics.

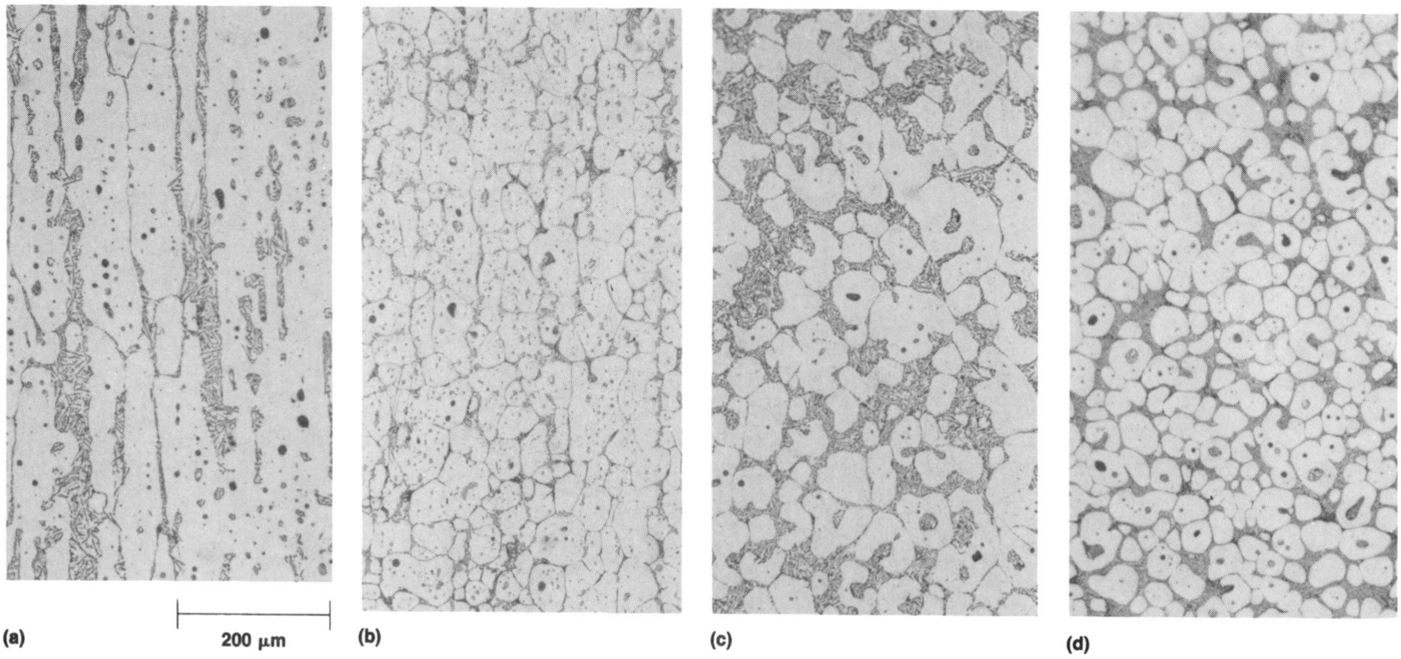

Fig. 7 Impact of strain on the microstructure of reheated SIMA aluminum alloy 357. (a) No strain. (b) to (d) Increasing strain

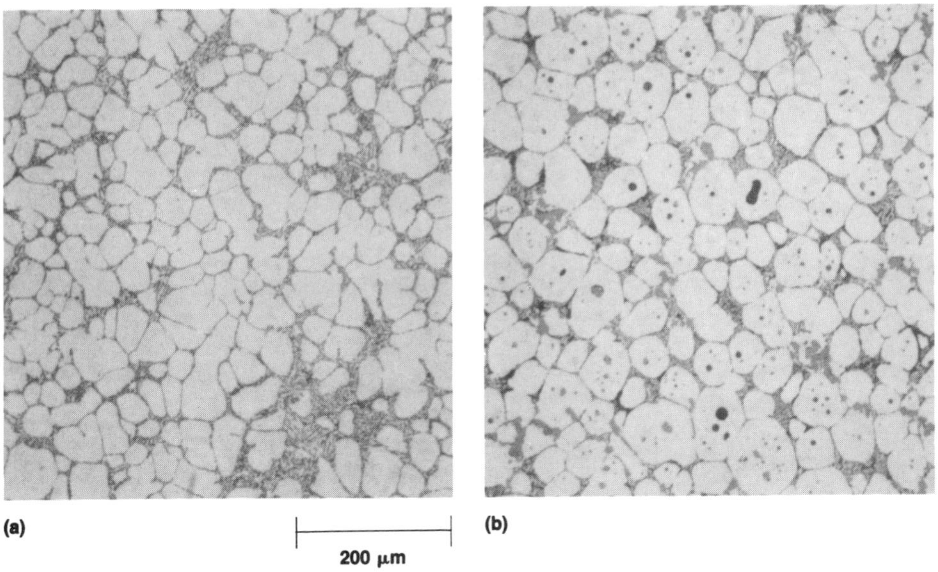

Fig. 8 Microstructures of semisolid forged aluminum alloy 357. (a) As MHD cast. (b) Forged

The early laboratory studies of semisolid formed metals focused on the structures obtained, and there was little concern for the occasional oxide inclusion or entrapped residue from the rotor or other mixer structures. The entrainment of atmospheric gases was also incidental to the achievement of the new morphology, and because the mechanical mixer produced good structure, other mixers or methods were of less importance to the early investigators.

Magnetohydrodynamic (MHD) Casting. The industrial application of semisolid metalworking to metal parts used in military, aerospace, automotive, or other high-quality or safety-critical applications demands integrity of the materials. Equivalence to the materials used in the conventionally cast or forged parts is a minimum specification. The MHD caster was developed to meet these requirements. The development of this equipment recognized the need for the exclusion of gases, oxides, and nonmetallic inclusions and the avoidance of other discontinuities. It also showed that there is an important relationship between the stirring shear rate and the solidification rate, and this relationship determines the type of semisolid metalworking microstructure that is generated (Ref 18).

The design of the MHD casting system incorporates the feed of filtered and degassed metal into the direct chill mold well below a quiescent surface in the delivery vessel, or tundish (see the article "Continuous Casting" in this Volume). The metal, near the freezing point in the mold, is vigorously stirred by a dynamic electromagnetic field, which creates the necessary shearing action. At the same time and location, controlled conductive heat transfer through the mold wall to a surrounding water jacket induces freezing. The MHD casting process therefore provides the ability to control precisely the shearing action and the rate of heat removal and thus deliver the desired solidified microstructure with a grain size that is normally about 30 μm (0.001 in.). This compares favorably with the 100 to 400 μm (0.004 to 0.016 in.) grain size produced by mechanical mixers.

Today, MHD casting systems are installed in the primary aluminum plants of one company on both vertical and horizontal continuous casters, producing a variety of fine-grain semisolid metalworking aluminum alloys from 38 to 152 mm (1.5 to 6 in.) in diameter. The MHD equipment is essentially superimposed on the mature direct chill casting technology. The MHD bar, therefore, benefits from superior metal quality, which is now routinely achieved while generating the high-quality semisolid metalworking microstructure at competitive casting rates (Fig. 4).

Casting Process Controls. The MHD casting process has been engineered to deliver aluminum billets suitable for semisolid forging at commercially competitive rates. The essential requirements of the billet are a fine-grain nondendritic microstructure and freedom from oxides, nonmetallic inclusions, and gas. The casting equipment was engineered with a special concern for providing multiple controls over the heat extraction/freezing rate and the shearing action. Figure 5 shows the effects of one of these variables—stirring power (shearing of dendrites). Other casting variables include:

- Melt chemistry
- Melt temperature
- Cast metal temperature
- Cast metal flow rate
- Coolant temperature
- Coolant flow rate
- Lubricant flow rate
- MHD degassing practice
- MHD filter technique
- Mold thermal conductivity

Proper control of the casting variables will result in billet properties that consistently meet the quality specifications for MHD billets. Before use in semisolid forging, billet material is inspected metallographically and chemically to ensure conformance to standards for chemistry, morphology, grain size, porosity, nonmetallic inclusions, and oxides. Other raw material characteristics that may be important in semisolid forging include electrical conductivity and surface emissivity.

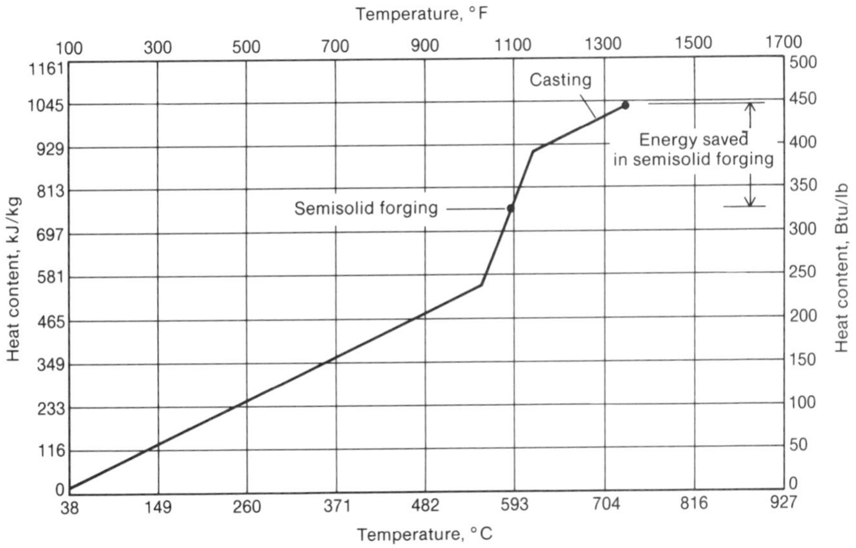

Fig. 9 Heating curves showing energy savings possible in semisolid forging versus casting of 357 aluminum alloy

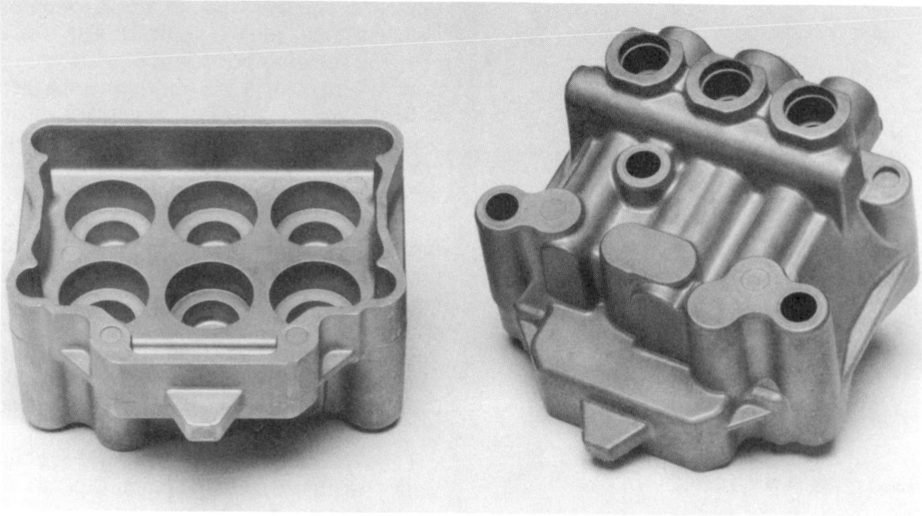

Fig. 10 Semisolid forged hydraulic brake valve

Wrought Processes

Small-diameter materials (<38 mm, or 1.5 in.) and some wrought alloys are difficult and/or expensive to cast, and an alternative method is needed for the production of raw material for forging. A number of researchers have studied the possibilities of solid-state processing.

One such process proposed the use of mixed powders (Ref 19), but a more effective approach utilizing wrought processing technology was perfected in 1981. This second technique has been named strain induced, melt activated (SIMA) by its inventors (Ref 20).

The SIMA process relies on the development of process steps in which the basic hot-worked metal alloy extrusions or rolled bars are subjected to additional cold work (Fig. 6). When sufficient strain is induced and the material is heated to the semisolid condition, the structure transforms to an extremely fine, uniform, nondendritic spherical microstructure (Fig. 7). This process has not been fully explained in fundamental terms, but it is likely allied to recrystallization processes and may be partly explained by concepts proposed in Ref 21.

The semisolid forging of SIMA-processed bar has been demonstrated for a wide variety of alloys, including aluminum, magnesium, copper, and ferrous alloys, and has been in production since 1983 for aluminum and copper alloys. Strain-induced melt-activated processing represents a cost-effective and readily available source of small-diameter raw material for semisolid forging. The process is also technically operable for larger sizes, but the costs of processing are not competitive with those of MHD casting in most metal alloys.

Semisolid Metal Forming. Semisolid forming is a three-step process (Ref 22). First, the billet is cut to the appropriate weight. Second, the cut billet (slug) is heated to the desired semisolid condition. Lastly, the semisolid slug is placed in the die and forged. The critical elements of the process are the forging press, the dies, the raw material, and the heating method. Figure 8 compares the microstructures of MHD cast and forged samples of aluminum alloy 357.

Forging presses may vary, but the ability to control the forming speed and the pressure precisely is essential if the press is to be used to semisolid forge a variety of parts. Depending on part size, geometry, alloy, and quality specified, forming speeds may range from a few inches per second to over 1270 mm/s (50 in/s) and mold pressures from a few hundred pounds per square inch to 140 MPa (10 tsi) or more.

Forging Dies. Closed dies for producing aluminum components are normally made of AISI H13 tool steel (for copper alloys, H21) hardened and drawn to a hardness of approximately 45 to 48 HRC. Die cavities are ground or electric discharge machined; tolerances and shrink rule are approximately the same as for die casting. Polishing is frequently advised to improve metal flow, to ease part ejection, and to optimize surface quality. Care in the design and fabrication of forge tooling is critical to the forging process.

Raw material consists of an MHD billet cut to a length that provides the desired mass (called a slug). Slug size is determined by the weight of the part to be formed. The MHD billet diameters that are commercially available range from 40 to 150 mm (1.5 to 6.0 in.). Ratios of slug length to diameter between 1 and 2 are usually found to be the easiest to heat and handle, although shorter and longer slugs are also used.

Heating the MHD Slug. A variety of methods can be used to heat the slug. Selection of the method depends on the alloys involved, the costs of different forms of energy, and the levels of heat control required. Furnaces can employ convection or radiant electric or gas heat, electrical induction, or resistance. The time and energy requirements may vary widely, depending on the alloy and the size of the slug. The size of the system varies accordingly. The large production systems deliver up to 1815 kg/h (4000 lb/h) of 16 kg (35 lb) slugs, and smaller heating systems deliver 16.3 kg/h (36 lb/h) of slugs that weigh about 40 g (1.4 oz) each.

Uniformity of slug condition (heat content and fraction solid) from surface to center, top to bottom, and slug to slug is essential to successful part production. This demands consistency in all elements involved in energy input, energy loss, and thermal transfer. Some of these factors are characteristic of the raw material, and others are elements of the heating system and its control. Experience shows that strict control of these elements will result in the delivery of raw material into the die ready to form with a heat content tolerance of ±1%. This type of control is well within the state of the art and permits semisolid forging production to deliver consistent products.

The energy required to heat aluminum alloys for casting (ignoring losses) is 35% greater (on a per-pound basis) than the energy required to heat the same aluminum alloy to the condition ready for semisolid forging (Fig. 9). When this is adjusted for the yield differences, the actual advantage of semisolid forging can be observed to be substantially greater.

Part Forming. The actual forging sequence is completed within a few tenths of a second. The material flow begins with the impact of the ram on the slug, and the thixotropic nature of the MHD slug allows the metal to flow into the cavity at very low pressures. Only at the end of the forming stroke does the pressure increase to the selected level to form the fully dense component.

The time of containment under pressure depends on alloy and part dimensions and is normally only a few seconds in duration because heat transfer to the die is extremely efficient at the high pressure. At the completion of the forming cycle, dies are opened and the part is ejected.

Semisolid forging systems are completely automated today, forming parts weighing from 20 g (0.7 oz) to 13.6 kg (30 lb) at production rates that range from 360 to 120 parts per hour, respectively, in single-cavity tools. Shuttle presses equipped with multicavity dies increase the rate proportionately.

Heat Treatment. All current semisolid forged aluminum parts are being heat treated to either a T5 or T6 temper, although most parts meet specifications with the less expensive T5 treatments. This results from

Table 1 Mechanical properties of typical semisolid forged aluminum parts

Aluminum alloy	Temper	Mechanical properties: Ultimate tensile strength, MPa	ksi	Tensile yield strength, MPa	ksi	Elongation, %	Hardness, HB
206	T7	386	56.0	317	46.0	6.0	103
2017	T4	386	56.0	276	40.0	8.8	89
2219	T8	352	51.0	310	45.0	5.0	89
6061	T6	330	47.8	290	42.1	8.2	104
6262	T6	365	52.9	330	47.9	10.0	82
7075	T6	496	72.0	421	61.0	7.0	135
356	T5	234	34.0	172	25.0	11.0	89
356	T6	296	43.0	193	28.0	12.0	90
357	T5	296	43.0	207	30.0	11.0	90
357	T6	358	52.0	290	42.0	10.0	100

Table 2 Comparison of semisolid forging and permanent mold casting for the production of aluminum automobile wheels

See Example 1.

Process	Characteristic: Weight direct from die or mold, kg	lb	Finished part weight, kg	lb	Production rate per die or mold, pieces per hour	Aluminum alloy	Heat treatment	Ultimate tensile strength, MPa	ksi	Yield strength, MPa	ksi	Elongation, %
Semisolid forging	7.5	16.5	6.1	13.5	90	357	T5	290	42	214	31	10
Permanent mold casting	11.1	24.5	8.6	19.0	12	356	T6	221	32	152	22	8

Fig. 11 Semisolid forged aluminum alloy 357 automobile wheels

one of the special characteristics of semisolid forging—the rapid quench the part receives from contact with the die in the forging operation. This press quench, along with the very fine spherical grains, can be used to advantage with some aluminum alloys to avoid the expensive solution treatment and quench and to achieve higher mechanical properties with only the aging treatment.

When the semisolid forged aluminum part is ejected from the die, it is normally between 205 and 425 °C (400 and 800 °F). The temperature variation from part to part within a production run can be reasonably controlled to within ±10 °C (±18 °F) and the temperature range for any part can be established by the system operator. Temperature variations depend on dwell time control, die temperatures, and the type and volume of die lubricant. Section thicknesses are always another consideration. Most installations employ a cold water quench following ejection, and partially because of the low temperature of the part, care in quenching will prevent distortion.

Advantages and Limitations of Semisolid Forging. Two key manufacturing advantages of semisolid forging are, first, the degree of automation that can be incorporated into the material handling systems, as well as the heating and forging operations, and, second, the precision that is routine in the control of the process. In summary:

- Precision controls and automation provide reproducibility
- Lower temperature and short dwell time lead to longer die life
- Short process cycle and automation result in higher productivity
- Fluid flow filling die cavity under high final pressure enables the filling of thinner sections and the forming of lighter parts
- Precision steel dies produce near-net shape parts that require less machining
- Semisolid heating requires only 65% of the energy required for casting
- Applicability to a broad range of alloys results in broader parts applications
- Semisolid charge produces less liquid/solid shrinkage and less microporosity
- Laminar flow in the die excludes gas entrapment and porosity
- Rapid quench in the forging press avoids expensive solution treatment to obtain higher properties

Conversely, semisolid forging is not for all applications. Some limitations include:

- Specially prepared raw material
- Higher cost raw materials
- Few sources of raw material
- Expensive tooling of special design
- Special and expensive capital equipment
- Highly skilled staff required

Table 3 Comparison of the characteristics of semisolid forged and machined electrical parts

See Example 2.

Process	Aluminum alloy	Raw material diameter, mm	Raw material diameter, in.	Raw material weight, g	Raw material weight, oz	Finished part weight, g	Finished part weight, oz	Material yield, %	Production rate (primary operation), pieces per hour	Heat treatment	Ultimate tensile strength, MPa	Ultimate tensile strength, ksi	Yield strength, MPa	Yield strength, ksi	Elongation, %
Semisolid forging	SIMA 6262	19	0.75	25	0.88	23	0.81	92	300	T6	345	50	276	40	10
Machining	6262	57	2.25	245	8.64	23	0.81	9.4	200	T6	345	50	276	40	10

Semisolid Forging Applications

The advantages of semisolid forging have enabled it to compete effectively with a variety of conventional processes in a number of different applications. In this section, a number of automotive, aerospace, and industrial applications of semisolid forged parts are outlined with data on alloys, yield, production rates, mechanical properties, and performance requirements. Figure 10 shows a complex high-pressure hydraulic brake valve typical of applications for semisolid forged components.

Semisolid forged parts have replaced conventional forgings, permanent mold and investment castings, impact extrusions, machined extrusion profiles, parts produced on screw machines, and, in unusual circumstances, die castings and stampings. Applications include automobile wheels, master brake cylinders, antilock brake valves, disk brake calipers, power steering pump housings, power steering pinion valve housings, engine pistons, compressor housings, steering column mechanical components, airbag containment housings, power brake proportioning valves, electrical connectors, and various covers and housings that require leak-tight integrity. Table 1 lists mechanical properties of selected aluminum alloys used in these components.

Reproducible high integrity is a key in this diversity, but this could be offset by the higher cost of the raw material—MHD billets. The near-net shape ability of the process reduces both the weight required and the machining time. When the higher production rate is added to these other advantages, the process becomes cost-effective for many applications.

Example 1: Aluminum Automobile Wheels. Aluminum automobile wheels (Fig. 11) have been produced by permanent mold casting (gravity and low pressure), squeeze casting, and fabrications of castings or stampings welded to rolled rims. Semisolid forging is a more recent process. Table 2 compares the characteristics of aluminum automobile wheels produced by semisolid forging and permanent mold casting. In addition to an economic advantage, semisolid forging offers other advantages that will be discussed below.

Lighter Weight. The ability to form thinner sections without heavy ribs to aid in filling the cavity allows a wheel to be semisolid formed nearer to net size with light ribs on the brake side. This results in a finished wheel that is up to 30% lighter than a cast wheel of the same style.

Consistent Quality. The forging process employs a high-quality, specially prepared (MHD) billet with an engineered metallurgical structure, closely controlled chemistry, and consistent casting variables, supplying an extremely consistent raw material with complete traceability. The wheel-forming process is computer controlled and automated with precise control of the heating and forging process variables, making the entire process adaptable to statistical process control.

Structure and Properties. The semisolid forged wheel is fine grained, dense structured, and formed to close tolerances in precision tooling in which the temperature is controlled to provide consistent forging conditions. This provides consistency in part dimensions and metallurgical properties. Forging in the semisolid state avoids the entrapment of air or mold gas, and the high fraction of solid material, together

Fig. 12 Aluminum alloy (SIMA 6262) electrical connectors produced by semisolid forging

Table 4 Comparison of a permanent mold cast aluminum part with a semisolid forged replacement

See Example 3.

Process	Weight (as forged or cast), kg	Weight (as forged or cast), lb	Weight (finish machined), kg	Weight (finish machined), lb	Production rate, pieces per hour	Aluminum alloy	Heat treatment	Ultimate tensile strength, MPa	Ultimate tensile strength, ksi	Yield strength, MPa	Yield strength, ksi	Elongation, %
Semisolid forging	0.45	1.0	0.39	0.85	150	357	T5	303	44	228	33	8
Permanent mold casting	0.76	1.67	0.45	1.0	24	356	T6	290	42	214	31	8

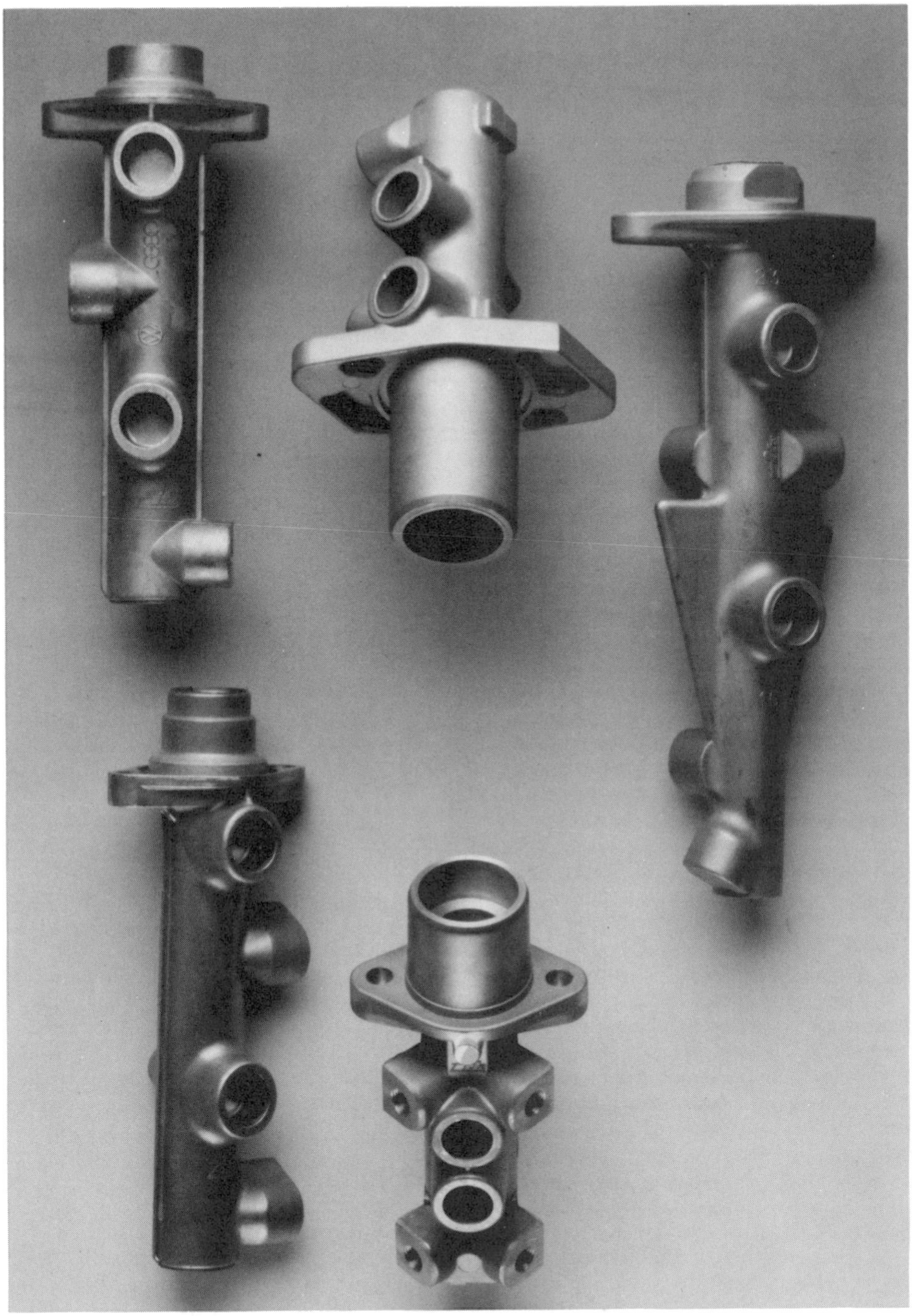

Fig. 13 Aluminum alloy 357 automotive brake master cylinders produced by semisolid forging

with the high pressure after forming, reduces the microporosity due to liquid/solid shrinkage. Unlike conventional forgings, the wheel properties are isotropic, reflecting the nondendritic structure of the high-performance aluminum alloy 357 used in the billet.

Design Versatility. The ability to form thin sections (roughly one-quarter to one-half the thickness of casting) permits not only a reduction in the weight of the wheel, but also allows the designer to style the wheel with thinner ribs/spokes and finer detail. Forming in the semisolid state under very high final pressure provides part surfaces and details that reflect the die surfaces. Therefore, the designer has a selection of surface conditions to enhance the style and can obtain exact replication of the fine detail designed in the die.

Example 2: Electrical Connector Multiconductor. Military/aerospace electrical connectors (Fig. 12) are highly stressed in service. As a result, the qualification of these parts involves extensive and severe functional testing under various environmental conditions. Production quality control procedures are equally stringent to ensure conformance to specifications and performance under load.

Before semisolid forging, these parts were machined (on screw machines) from extruded aluminum alloy 6262-T9 bar. Today, a large number of these parts are semisolid forged from SIMA 6262 bar and finish machined after the T6 heat treatment. The characteristics of semisolid forged and machined parts are compared in Table 3.

The semisolid forged component possessed all major keyways and locating devices with tolerances at least equivalent to machined parts. The savings in material, plus the higher production rate, makes the semisolid forged part a very economical selection.

After full T6 heat treatment, the parts are finish machined and anodized. The parts are tested per MIL-C-38999 and meet all specified performance criteria. These tests include bending, vibration, durability, thermal shock, and impact.

Example 3: Aluminum Brake Master Cylinder. Some permanent mold cast brake master cylinders (Fig. 13) are cast essentially solid in order to place solidification shrinkage defects in the center bore, where they are subsequently machined away. The energy crisis motivated the move to lighter automobiles to improve fuel economy. This has caused the conversion of many ferrous automobile components into aluminum. Brake master cylinders have been difficult and expensive to convert, and until the application of semisolid forging, the only acceptable production methods were low-pressure or gravity permanent mold casting. Table 4 compares a permanent mold cast aluminum part, cored as indicated above, with a semisolid forged replacement.

Semisolid forging utilizes a two-cavity indirect forming approach and MHD cast alloy 357 slugs cut from 76 mm (3 in.) diameter bar. All major holes are sufficiently cored for tapping or other finishing operations. After semisolid forging, T5 heat treatment, and final machining, parts are subjected to a 9.7 MPa (1400 psi) nitrogen leak test and extensive endurance testing, which includes 300 000 to 1 million hydraulic cycles that must be endured without a sign of wear.

Example 4: Valve Bodies. Automotive, aerospace, refrigeration, and industrial

Table 5 Comparison of weights for semisolid forged and extruded parts

See Example 4.

Process	Weight (as forged or cut off, including kerf) kg	lb	Relative raw material cost per pound (foundry ingot = 1)	Weight of finished valve kg	lb
Semisolid forging	0.32	0.7	1.2	0.27	0.6
Extrusion	0.57	1.25	1.4	0.39	0.87

Fig. 14 Aluminum alloy valve bodies machined from extruded profiles

Fig. 15 Copper alloy C36000 (free-cutting brass) plumbing fittings produced by semisolid forging

valves are frequently manufactured by machining, drilling, and tapping the hydraulic passages and valve seats in a section cut from an extruded aluminum profile. Valves typical of such production are shown in Fig. 14.

Semisolid forging production can be competitive with this efficient manufacturing method by the substitution of semisolid forging for extrusion, cutoff, milling, and drilling. The semisolid forged valve body is cored and sculpted to minimize metal weight. Finishing requires tapping and drilling the smallest holes, requiring less than one-half the time needed to machine the extruded valve. A typical comparison of weights is given in Table 5.

Example 5: Brass Plumbing Parts. Brass plumbing components (Fig. 15) were semisolid forged using a closed die to produce a net shape part complete with threads. Strain-induced melt-activated processed Alloy C36000 (free-cutting brass) is heated to the semisolid condition before direct forging. After forging, the part is unscrewed and tumbled to remove minor flash and the die lubricant. Die life of 40 000 pieces is typical for this type of application. Total weight reduction of 50% is frequent when replacing conventional forgings, and 60% weight reduction is common when replacing castings. This weight reduction is an important part of the cost improvement.

Quality Factors

As with all processes, it is possible to form parts when all variables are not controlled within specified limits and when discontinuities result. Some discontinuities that may be observed in semisolid forging are outlined below and are shown in Fig. 16.

Blisters (Fig. 16a) are occasionally observed on the surfaces of aluminum parts that are subjected to the 540 °C (1000 °F) solution heat treatment. They may result from one of several causes, the most frequent of which is an excessively high metal flow rate where metal is sprayed ahead of the main mass flow. It coats die surfaces, and the following mass entraps minute amounts of mold gas between the skin and the bulk of the mass.

Another cause may be the contamination of die surfaces by substances that are contained in various lubricants and hydraulic fluids. These become contained in the part surface. Later, when the temperature rises, the yield strength drops, and the pressure of the contained gas deforms the surface. The blister usually originates in a zone that is 0.5 to 2.0 mm (0.020 to 0.080 in.) below the surface. Blisters can be avoided by ensuring that all die surfaces are free of contamination and by slowing metal velocities, increasing gate areas, and reducing ram velocities.

Cold shuts, folds, and laps (Fig. 16b) are infrequent. These defects occur when two metal surfaces meet but fusion is incomplete. This may be caused when an oxide film is entrapped or when the flow of metal in two approaching surfaces is slowed by entrapped air, allowing solidification to proceed before the surfaces have fully fused. Such problems can be avoided by changing the metal flow pattern and/or increasing localized venting.

Non-fill (Fig. 16c) occurs infrequently when cold semisolid metal, premature

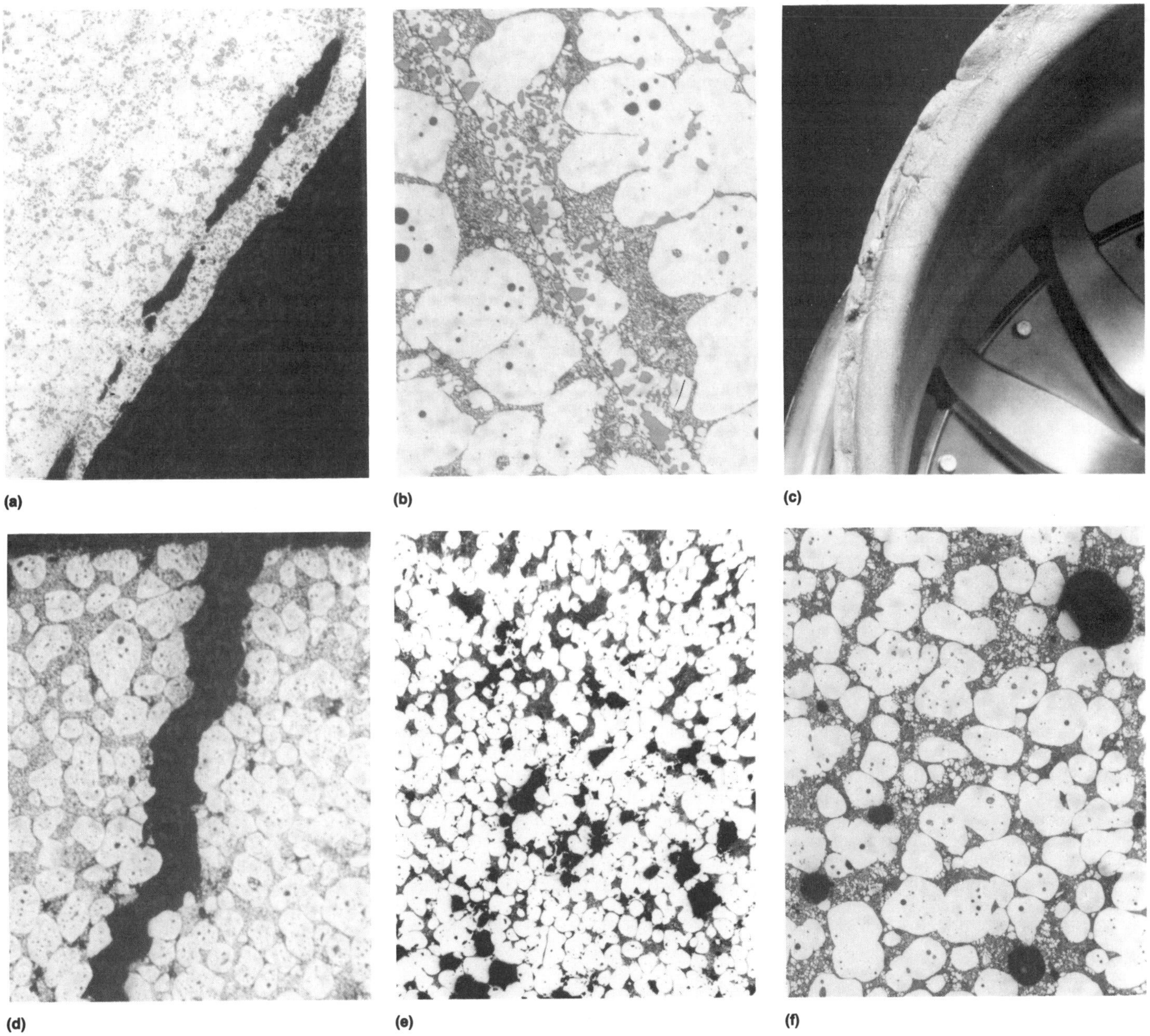

Fig. 16 Defects that may be encountered in semisolid forging. (a) Surface blisters. 75×. (b) Cold shuts. 225×. (c) Non-fill. 0.5×. (d) Hot tears. 75×. (e) Shrinkage porosity. 40×. (f) Gas porosity. 100×

freezing, low pressure or metal velocity, or entrapped mold atmosphere prevents the complete filling of the die cavity. It can be corrected by increasing the liquid fraction in slug, raising die temperatures, increasing venting, and adjusting hydraulics for higher pressure or velocity.

Surface blow occurs occasionally when moisture (lubricant, hydraulic fluid) trapped on the die surface by the semisolid metal generates gas pressure, forcing the semisolid metal surface away from the die. Surface blow is identified by a shiny surface area or an indentation in the part. It can be corrected by ensuring that all die surfaces are clean and dry.

Hot tears (Fig. 16d) are infrequent in casting alloys, but are more frequent in wrought alloys. They occur when the solidified part is constrained by the die while solid-state shrinkage induces strains that exceed the hot ductility. Hot tears may be internal or external. They can be avoided by part/die redesign, higher metal pressure in the die, reduced dwell time, or changing alloys.

Shrinkage porosity (Fig. 16e) can occur when heavy and/or thicker sections solidify after their feeding sections. It can be prevented by adjusting die temperatures and feeding sections and by higher time and pressure during dwell after forging.

Gas porosity (Fig. 16f) is very infrequent. It can be caused by excessively high gate velocities, resulting in excessive turbulence in the metal flow and entrapping mold atmosphere. Gas porosity can be avoided by reducing velocity or opening the gates.

Flow lines occur occasionally where metal flows through a heavy section or past a filled region. They can be observed visually unaided on polished sections of a part, but are difficult to resolve at higher magnifications. The zone between a stationary surface and the moving material evidences slightly higher eutectic concentration.

Metal-Matrix Composites (MMC)

As discussed earlier in this section, the viscous behavior of semisolid alloys allows for substantial benefits in particle incorporation processes. This was recognized by the early MIT inventors, who coined the term compocast to describe the application of rheocasting techniques to the inclusion of fibers and particulate in metal alloy matrices (Ref 9).

In 1981, one company began a program in which tons of MMCs were mixed and formed into housings for lighting fixtures. These composites, mixtures of about 20% by volume of SiO_2 (silica sand) and aluminum, were produced with a continuous process in an effort to substitute an inexpensive filler (that is, foundry sand) for the increasingly expensive aluminum (Ref 23). The process was successful and, the product met all expectations, including cost. However, the price of aluminum decreased and so did the need for this early MMC.

A number of organizations are currently engaged in developing methods for the creation of MMCs, but regardless of the preparation techniques, semisolid forging offers a unique advantage for forming near-net shape parts from MMCs. Therefore, semisolid forging represents a viable process for shaping these difficult-to-cast MMC materials and can achieve a high degree of accuracy. This is particularly beneficial when the MMC material presents difficulties in machining. A number of prototype applications of semisolid formed MMCs are currently under evaluation. Detailed information on other foundry production methods for MMCs is available in the article "Cast Metal-Matrix Composites" in this Volume.

Safety

The potential hazards of semisolid forging are similar to those encountered in other high-speed forging operations. These include mechanical and electrical hazards.

Mechanical Hazards. Semisolid forming equipment typically operates at high speed and is capable of exerting high forces. Therefore, pinch points and personnel access should be shielded, and the equipment is quite capable of automatic operation.

Electrical Hazards. Induction heating coils typically operate at high voltage (>1000 V) and frequency. Coils and leads should be well protected and shielded from personnel contact. Induction coils should also be shielded from inadvertent contact with the workpiece, particularly in the last stages of heating when the possibility exists for workpiece slumping or collapse in extreme circumstances. This could cause coil arcing.

REFERENCES

1. M.C. Flemings and K.P. Young, "Rheocasting," Yearbook of Science and Technology, McGraw-Hill, New York, 1978
2. H. LeHuy, J. Masounave, and J. Blain, *J. Mater. Sci.*, Vol 20 (No. 1), Jan 1985, p 105-113
3. F.C. Bennett, U.S. Patent 4,116,423, 1978
4. K.P. Young, R.G. Riek, J.F. Boylan, R.L. Bye, B.E. Bond, and M.C. Flemings, *Trans. AFS*, Vol 84, 1976, p 169-174
5. K.P. Young, R.G. Riek, and M.C. Flemings, *Met. Technol.*, Vol 6 (No. 4), April 1979, p 130-137
6. B. Toloui and J.V. Wood, *Biomedical Materials*, Vol 55, Materials Research Society, 1986
7. J. Cheng, D. Apelian, and R.D. Doherty, *Metall. Trans. A*, Vol 17A (No. 11), Nov 1986, p 2049-2062
8. R. Mehrabian and M.C. Flemings, *Trans. AFS*, Vol 80, 1972, p 173-192
9. R. Mehrabian, A. Sato, and M.C. Flemings, *Light Metals*, Vol 2, The Metallurgical Society, 1975, p 175-193
10. R.G. Riek, A. Vrachnos, K.P. Young, and R. Mehrabian, *Trans. AFS*, Vol 83, 1975, p 25
11. U. Feurer and H. Zoller, "Effect of Licensed Consection on the Structure of D.C. Cast Aluminum Ingots," Paper presented at the 105th AIME Conference, Las Vegas, NV, The Metallurgical Society, 1976
12. J. Collot, "Gircast-A New Stir-Casting Process Applied to Cu-Sn and Zn-Al Alloys, Castability, and Mechanical Properties," *Proceedings of International Symposium on Zinc-Aluminum Alloy*, Canadian Institute of Mining and Metallurgy, 1986, p 249
13. P.A. Joly and R. Mehrabian, *J. Mater. Sci.*, Vol 2 (No. 1), Jan 1976, p 393
14. A.C. Arruda and M. Prates, *Solidification Technology in the Foundry and Cast House*, The Metals Society, 1983
15. R.L. Antona and R. Moschini, *Metall. Sci. Technol.*, Vol 4 (No. 2), Aug 1986, p 49-59
16. G.B. Brook, *Mater. Des.*, Vol 3, Oct 1982, p 558-565
17. K.P. Young, U.S. Patent 4,565,241, 1986
18. K.P. Young, D.E. Tyler, H.P. Cheskis, and W.G. Watson, U.S. Patent 4,482,012, 1984
19. R.M.K. Young and T.W. Clyne, *Powder Metall.*, Vol 29 (No. 3), 1986, p 195-199
20. K.P. Young, C.P. Kyonka, and J.A. Courtois, U.S. Patent 4,415,374, 1983
21. R.D. Doherty, H.I. Lee, and E.A. Feest, *Mater. Sci. Eng.*, Vol 65, 1984, p 181-189
22. K.P. Young, U.S. Patent 4,687,042, 1987
23. M.P. Kenney, K.P. Young, and A.A. Koch, U.S. Patent 4,473,107, 1984

Foundry Equipment and Processing

Section Chairman:
Thomas Prucha, CMI International

Sand Processing

Green Sand Molding Equipment and Processing

Roger B. Brown, Disamatic, Inc.

GREEN SAND MOLDING is one of many methods available to the foundryman for making a mold into which molten metal can be poured. Green sand molding and chemically bonded sand molding are considered to be the most basic and widely used moldmaking processes. The molding media for the two methods are prepared quite differently. Chemically bonded media are prepared by coating grains of sand with a binder that is later cured by some type of chemical reaction. Green sand media are prepared by coating the grains of sand with binder that is later shaped into a rigid mass by the application of force.

Green sand molding is the least expensive, fastest, and most common of all the currently available molding methods. The mixture of sand and binder can be used immediately after the mixing process that coats the sand grains. Although the time taken to shape the mold is of importance in some cases, for the purposes of this article, the forming process can be considered to be almost instantaneous. This section will cover the preparation, mulling, delivery, fabrication, and handling of green sand molds.

Materials

Green sand as a molding medium consists of a number of different materials that must be present in varying amounts and grades in order to produce the desired results for a specific type of casting. The increased demands for casting accuracy and integrity have caused an increase in the use of high-pressure, high-density molding machines. Except for relatively few applications (such as thin-section castings), fireclay (kaolin) is unsuitable for this type of molding (Ref 1). The montmorillonite, or bentonite, clays are used primarily because of their increased durability when heated, higher bonding strength, and plasticity.

Bentonite Clays. The bentonites can be classified as two distinct types: sodium (western) bentonite and calcium (southern) bentonite. The properties derived from the use of each vary widely. The use of one in preference to the other depends on the castings to be made, the system being run, and the economics of the total situation. Fortunately, these bentonites are compatible and can be blended in any ratio to tailor sand properties to the specific requirements of a system (casting condition).

Molding Methods

Green sand molds can be made in a number of ways. The optimum method depends on the type of casting, its size, and the required production. When only a few castings are required, it may be more economical to have a loose pattern made and to have the mold made by hand. Hand ramming is the oldest and slowest method of making a mold. Unfortunately, it is becoming increasingly difficult to locate a foundry with hand molding skills. In most cases, the pattern will at least be mounted on some kind of board to facilitate fabrication of the molds. There are two basic types of green sand molds: flask and flaskless.

Flask Molds. A flask can be defined as the container that is positioned on the pattern (or platen in some cases) and into which the prepared sand is placed before the molding operation. Although there are flasks termed slip flasks, which are slid up the mold as the depth of compacted sand becomes deeper, the most common types of flasks are snap flasks and tight flasks.

Snap flasks are usually square or rectangular. Diagonal corners are held together with cam-action clamps. The clamps are moved to the open position after the mold is made, the cores have been set, and the cope (the upper half of a mold) has been placed on the drag (the lower half of a mold); this allows for easy removal of the finished mold.

Tight flasks are designed as one-piece units that have no clamps. This type of flask remains with the mold during the pouring operation and, normally, until the shakeout operation.

Regardless of the type used, the flask must become more rigid as molding pressure increases. Flexing or movement in the sidewall of the flask will adversely affect the accuracy of mold dimensions, flask-to-pattern alignment, and flask-to-flask alignment during closing of the mold halves.

Flaskless Molds. During the last few years, flaskless molding equipment has become increasingly popular, especially when molds of less than 160 kg (350 lb) are being considered. As the name implies, the flaskless molding machine has no flask. Rather, the flask is replaced with a box or molding chamber that is an integral part of the molding machine.

Flaskless molding gives the foundryman additional versatility in the molding operation. With flaskless molding, a number of things are simplified. Until the advent of flaskless molding, most molds were filled with sand by gravity. With the tighter and more repeatable tolerances that result from the molding chamber being an integral part of the molding machine, other filling methods become more practical. In addition, the parting line of the mold need not be horizontal, but can then be easily placed in the vertical orientation.

First-Generation Machines

This section will discuss jolt-type, jolt squeeze, and sand slinger molding machines.

Jolt-type molding machines (Fig. 1) operate with the pattern mounted on a pattern plate (or plates), which in turn is fastened to the machine table. The table is fastened to the top of an operating air piston. A flask is placed on the pattern and is positively located by pins relative to the pattern. The flask is filled with sand, and the machine starts the jolt operation. This is usually accomplished by alternately applying and releasing air pressure to the jolt piston, which causes the flask, sand, and pattern to lift a few inches and then fall to a stop, producing a sharp jolt. This process is repeated a predetermined number of times, depending on sand conditions and pattern configuration. Because the sand is compacted by its own weight, mold density will be substantially less at the top of a tall pattern. The packing that results from the jolting action will normally be augmented by some type of supplemental compaction, usually hand or pneumatic ramming. When ramming is complete, push-off pins, bearing against the bottom edges of the flask, lift the

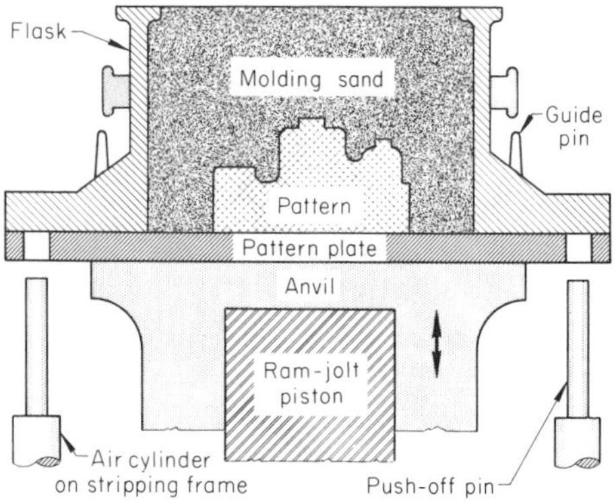

Fig. 1 Primary components of a jolt-type molding machine

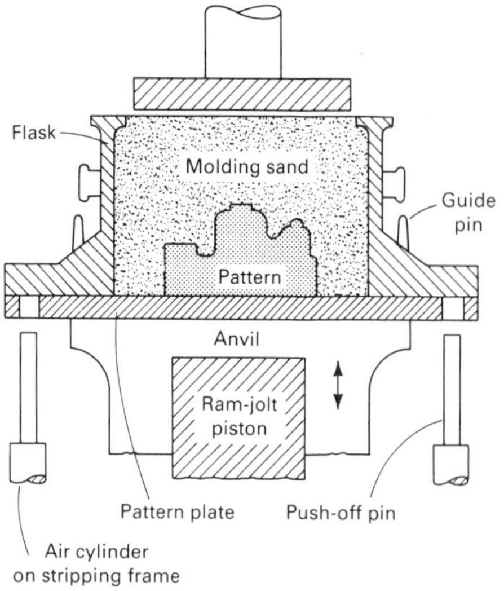

Fig. 2 Jolt squeeze molding machine with solid squeeze heads

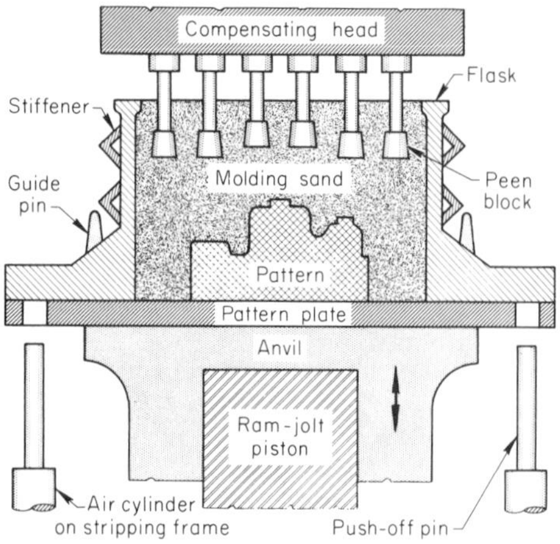

Fig. 3 Jolt squeeze molding machine with compensating heads

flask and completed mold half off the pattern. Various mechanisms are used to lift the mold from the pattern and turn it over (in the case of the drag mold) or turn it for finishing operations (in the case of the cope mold).

Jolt squeeze molding machines operate in much the same manner as jolt-type molding machines. The main difference is that the supplemental compaction takes place as the result of a squeeze head being forced into the molding flask, thus compacting the loose sand at the top. The required pressure can be applied pneumatically or hydraulically. In many cases, the squeeze head will be one piece (Fig. 2) and may even have built-up areas to provide more compaction in deep areas that are hard to ram. In other cases, the squeeze head may be of the compensating type, which consists of a number of individual cylinders, each exerting a specified force on the rear mold face (Fig. 3). Some machines exert the same force on all areas of the mold, while other machines allow the operator to adjust squeezing pressure in zones. Jolt squeeze machines are available in many sizes and are suitable for many different purposes and production levels. They can be operated manually or automatically. The operator has the option of independently adjusting the number of jolts from zero to any number and adjusting the squeeze pressure from zero up to pressure that is considered excessive. Hand or pneumatic ramming is often combined with this process; supplemental ramming normally takes place after jolting but before squeezing.

Sand slinger molding machines deliver the sand into the mold at high velocity from a rotating impeller. Molds made by this method can have very high strengths because a very dense mold can be made. Density is a function of sand velocity and the thickness through which the high-velocity sand must compact previously placed sand. Sand slingers may or may not be portable. Some ride on rails to the mold, while others have the molds brought to the slinger. Generally speaking, larger molds have the slinger brought to the mold, while smaller molds are brought to the molding station.

Although slingers are useful in producing larger molds, it should be noted that the sand entry location and angle are critical to the production of good molds. Entry location is controlled by the operator, while entry angle and, to some extent, location are controlled by internal adjustment. It is extremely important that these adjustments be maintained in accordance with the appropriate maintenance manual. Error can and does lead to soft spots in the mold or to excessive pattern wear. A considerable amount of operator skill is required to achieve consistent results. Additional information on green sand molding can be found in the articles "Aggregate Molding Materials," "Sand Molding," and "Coremaking" in this Volume.

A number of variations are possible in the above methods. Smaller patterns (resulting in smaller molds) can be constructed such that both the cope and drag impressions are mounted on opposite sides of the same plate. These squeezer or matchplate patterns (Fig. 4) are often used to produce molds with any combination of hand ramming, jolting, and squeezing, just as cope and drag patterns are (Fig. 5).

Second-Generation Machines

Foundry technology has progressed rapidly since the mid-1950s, and molding methods have been a large part of this progression. It was not until the early 1960s that high-pressure molding machines were developed. Depending on design, this new generation of molding machine would accept either match plate or cope and drag patterns and can be of the tight flask or flaskless configuration.

Along with increased levels of foundry technology comes the demand for more accurate and higher-integrity castings (see the article "Casting Design" in this Volume). In any case, modern molding machines, metal technology, sand technology, and support equipment technology assist the foundryman in supplying these demands.

Rap-jolt machines were among the first of the newer high-pressure molding machines. These machines are similar in many respects to jolt squeeze machines. Rap-jolt machines have the option of jolting the mold as described above and/or rapping the mold. Rapping is accomplished by rapidly striking the bottom of the platen on which the pattern is mounted with a weight. The force imparted to the platen/flask/mold combination may not exceed 1 g, or separation between the flask and pattern will occur. Therefore, there is very little if any vertical movement of the pattern and flask. This method allows for the possibility of squeezing and rapping simultaneously. Some machines of this type allow the operator to jolt prior to the rap-jolt operation. Depending on the individual molding machine, any one or any combination of the operations can be

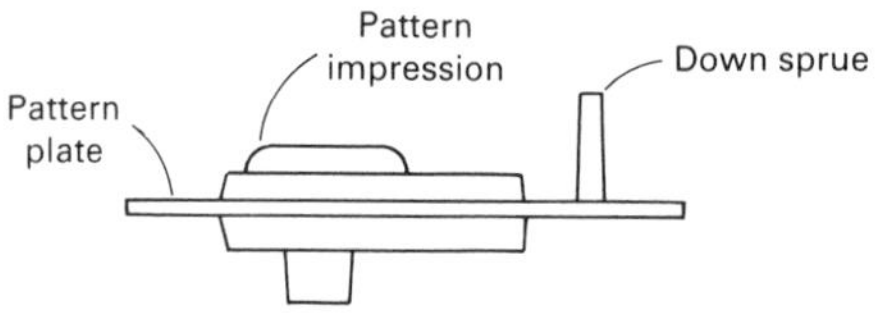

Fig. 4 Primary components of a match plate (squeezer) pattern

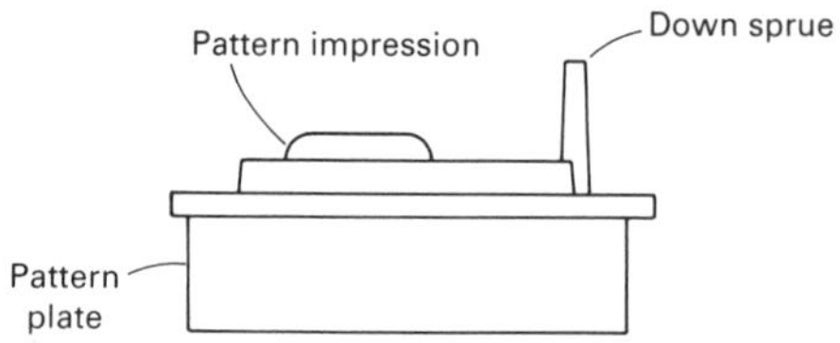

Fig. 5 Drag half of a cope and drag pattern

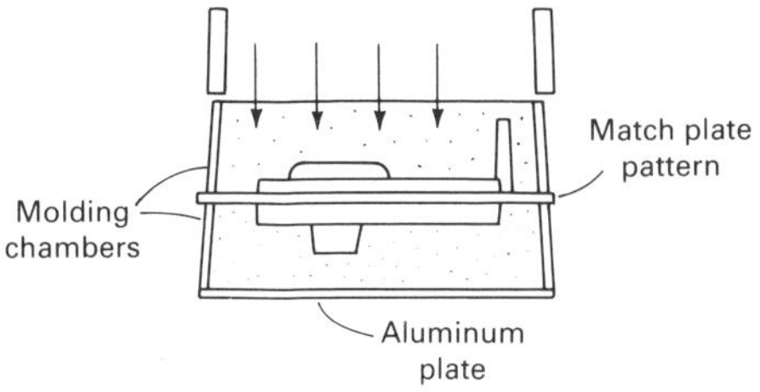

Fig. 6 Gravity-fill pressure squeeze molding machine using match plate patterns

used to make the mold. The equipment described thus far has all made use of some type of flask—either the snap or tight flask configuration.

Match Plate Pattern Machines. Automatic molding machines that use match plates have been used in both tight flask and flaskless designs. Because the patterns do not have the strength to withstand the pressure exerted during compaction without flexing, both the cope and drag must be squeezed simultaneously.

Some match plate machines (Fig. 6) fill both the cope and drag by gravity. This type of machine will close up the molding chambers to the pattern and then rotate the assembly so that the drag surface of the pattern is facing up. Sand is then dropped into the drag chamber, and a sealing plate (usually aluminum) is inserted. The molding chamber/pattern assembly is then rotated so that the cope pattern face is up, and the cope chamber is filled with sand. The mold is then compacted by squeezing, the molds are withdrawn from the pattern, and the pattern is removed. The open mold is then available for any finishing work or core setting. The mold is then closed and removed from the molding chambers.

Other match plate machines fill cope and drag molding chambers simultaneously by blowing the sand into the cavity (Fig. 7). After the blowing operation, the mold is compacted by a squeezing operation. After squeezing, the mold halves are withdrawn from the pattern and are available for any necessary finishing or core setting operations. Depending on the design of the machine, it may or may not be necessary to add to the machine cycle time to complete these operations.

Match plate pattern machines are available in tight flask and flaskless designs. These machines normally utilize gravity fill of both cope and drag molds. The cope is filled in much the same manner as for a flaskless machine. The drag is filled by sealing the bottom of the drag flask prior to the gravity-fill operation. The drag flask, still sealed, is then closed to the pattern, and the mold is compacted by squeezing. The squeeze pressure is applied by individual cylinders, each covering a small area of the mold.

Cope and Drag Machines. Automatic molding machines that use cope and drag

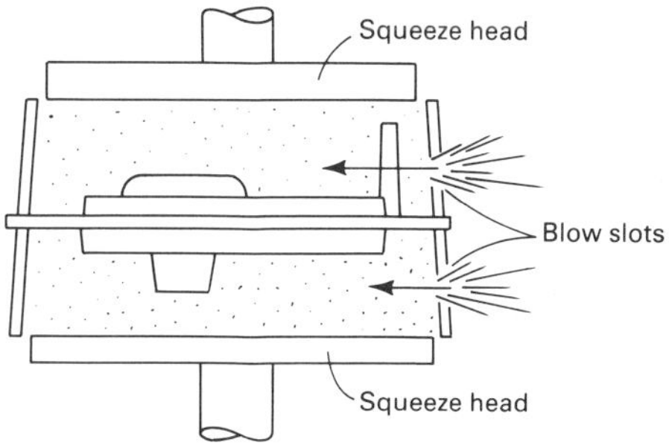

Fig. 7 Blow-fill pressure squeeze molding machine using match plate patterns

patterns can also be utilized in tight flask and flaskless designs. Because the patterns normally do not have the strength to withstand the pressure exerted during compaction without flexing, the pattern plates are usually mounted against a platen or grid. In most cases, the cope and drag mold halves are filled and compacted with the pattern facing up. Except in the case of special finishing operations to the cope half of the mold, it is not necessary to rotate either the patterns or the cope half of the mold. However, it is necessary to turn the drag half of the mold over to allow for setting of cores, close up, and pouring.

Most of the first automatic molding machines were automated versions of the rap-jolt machine mentioned earlier. The automation of rollover, transportation, and in some cases core setting has greatly increased the rate at which these machines produce molds.

A relatively common method of compaction in tight flask machines utilizes pressure from one or several compensating squeeze heads, as shown in Fig. 3. This pressure is normally adjustable in order to optimize the molding conditions. The mold halves can be filled by gravity or the sand blown in using air pressure.

Pressure Wave Method. More recent designs utilize pressure wave technology as the compaction method. These designs normally fill the flasks with sand by gravity. The top of the mold is sealed by a chamber. The chamber then emits a pressure wave, either by rapid release of air pressure or by an explosion of a combustible gas mixture (Fig. 8). As the pressure wave hits the back side of the mold, the sand grains are accelerated toward the pattern. The pattern immediately stops the downward movement of the sand grains, causing the kinetic energy of the mass to compact the sand. Molds made using this method are most dense at the pattern face and progressively less dense as distance increases from the pattern face. There is no need for additional compaction by the application of squeeze pressure.

Horizontal flaskless molding machines are a relatively recent design. The patterns are mounted in these machines on a hollow pattern carrier (Fig. 9). A grid supports the

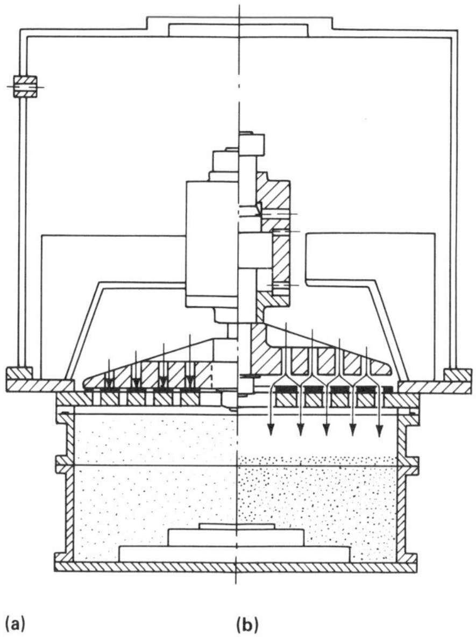

Fig. 8 Pressure wave molding machine that compacts sand by the rapid release of air pressure or an explosive combustible gas mixture. Part (a) shows the mold filled by gravity prior to being compacted by the pressure wave at (b).

underside of the pattern to avoid flexing during compaction. The molding chambers are formed by the pattern, the four sides of the molding chamber, and a plate with a sand injection slot. Vacuum is used to evacuate the chamber formed by the pattern carrier and the pattern plates. Vents in the pattern carrier and pattern plates allow the vacuum into the molding chambers, which causes sand to flow into the molding chambers. Upon completion of the filling sequence, the mold is compacted by squeeze pressure and the molds are withdrawn from the pattern. The pattern carrier retracts as the drag half of the mold swings out for blow out and/or core setting while another mold is being produced. Because the molds are produced in the same attitude as they will be used, there is no need to turn either half of the mold over.

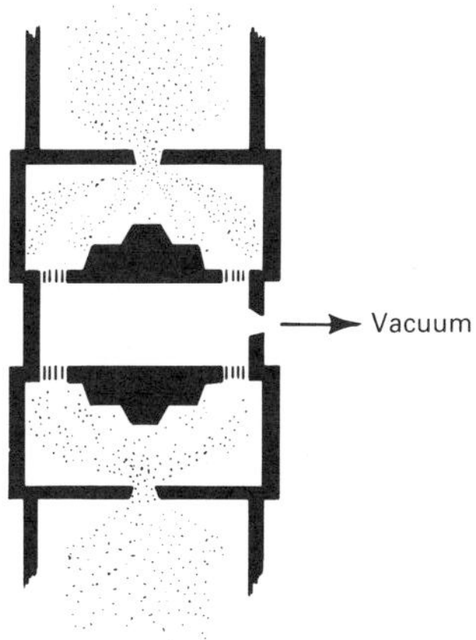

Fig. 9 Vacuum-fill pressure squeeze machine that uses cope and drag patterns

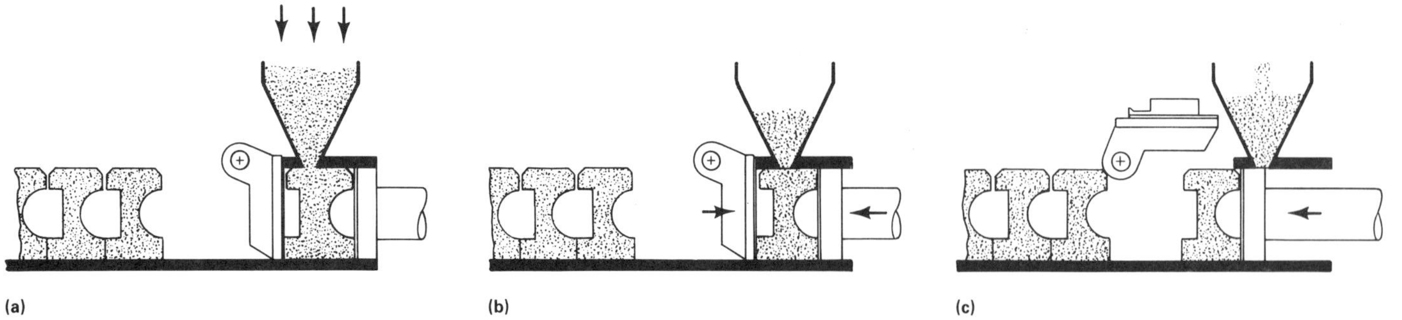

Fig. 10 Blow-fill pressure squeeze molding machine making vertically parted molds. (a) Molding chamber filled with sand. (b) Sand compacted by squeeze pressure. (c) Finished sand mold pushed out of molding chamber

Vertically parted molding machines have been commercially available since the mid 1960s. Like their horizontal counterparts, vertical machines have undergone a number of design changes as electronic technology has improved. Molds are made in these machines by closing the ends of a four-sided chamber with the patterns, which in turn are mounted on platens. The top chamber wall has a slot through which molding sand is blown. After the molding chamber is filled with sand, it is subsequently compacted by squeeze pressure (Fig. 10). Blow and squeeze pressure are both adjustable to optimize molding conditions. After compaction, one of the platens with its mounted pattern swings out of the way, allowing the other platen and pattern to push out the newly made mold to join with previously made molds. At this position, the mold is available for core setting. Blow off is accomplished automatically. Some models are capable of porting vacuum to the back side of the pattern to assist in the filling of deep pockets.

The vertically parted molding machines that are available are flaskless by nature. However, many deliver the mold to a device that will provide added physical support for the mold sides, thus increasing their flexibility.

Molding Media Preparation

The sand in the metal casting process forms a continuous loop, as shown in Fig. 11 (see also the article "Classification of Processes and Flow Chart of Foundry Operations" in this Volume). It is therefore difficult to determine exactly where it begins. For the purposes of this article, sand preparation will be discussed first. Sand, clay, water, and carbonaceous materials are charged into the mixing device. The length of time these ingredients are left in the mixing device is best determined by the type of device used and the desired sand properties. These devices are called either mullers or mixers. As with molding equipment, different types of mixing equipment can be used. Most foundries use either a continuous or a batch-type muller.

Continuous Muller. In a continuous muller (Fig. 12), the sand is fed to the muller into one bowl, and it exits through a door in the other bowl. In most cases, sand is fed into the muller in a regulated, continuous stream, and discharge is controlled based on the power draw of the muller motor. As power draw reaches a predetermined level, the discharge door opens for a short period, allowing some of the sand to leave the muller. On average, sand will pass through both bowls twice before it is ejected. It is possible, however, that a small percentage of sand will pass directly from the input to the exit in one pass. This type of muller is designed to produce large quantities of sand continuously.

Batch-Type Muller. Although not a new design, the batch-type muller (Fig. 13) can produce high-quality molding sand. It is equipped with plows to move the sand mass under the large, weighted rolling wheels, which are vertically oriented. This kneading action provides the capability of consistent control but not short cycle times.

The high-speed batch muller shown in Fig. 14 has been adapted to meet the requirements of both high-production molding lines and jobbing foundries. Sand is plowed from the floor of the muller up to the position where the rubber-tired wheels knead the sand against the rubber-lined sidewall. However, not all of the mulling action takes place as a result of the wheels; much of the mulling action takes place as a result of the amount of sand in the muller. For maximum mulling efficiency, the weight of sand charged into this type of muller should not be too small. The manufacturer should be able to provide the necessary information. This type of muller also offers the possibility of cooling the aggregate. A blower can be installed that will force air into the bottom of the unit. As the air flows through the aggregate, it picks up moisture and is exhausted through the top of the muller. By adding sufficient amounts of water, the sand can be evaporatively cooled.

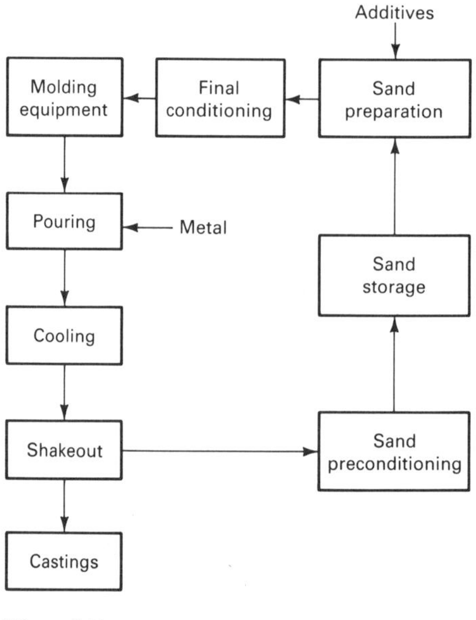

Fig. 11 Flow chart of a metal casting system

Intensive Mixer. Another type of mixer that is increasing in use is the intensive mixer (Fig. 15). The intensive mixer utilizes a rotating bowl into which the sand is charged. Inside the rotating bowl are one or two driven rotors that are available with different designs, depending on requirements. To avoid internal buildup, the unit is equipped with a scraper that directs sand away from the sidewalls and back toward the center of the mixer. This type of mixer is available in either batch or continuous configurations.

Mixing Variables. Regardless of the type of mixing/mulling equipment chosen, muller input and output must be closely controlled. Clay, combustible material, return sand, and new sand must all be added consistently and in amounts indicated by sand test results. Some sand systems are run on a volumetric basis, while others are run on a weight basis. The preferred method is to add return sand and additives by weight, thus affording closer control. Water additions are controlled in a number of different ways. Some equipment samples the sand going into the muller and bases water additions on testing those samples for heat and moisture content. Other water addition equipment samples the sand inside the mul-

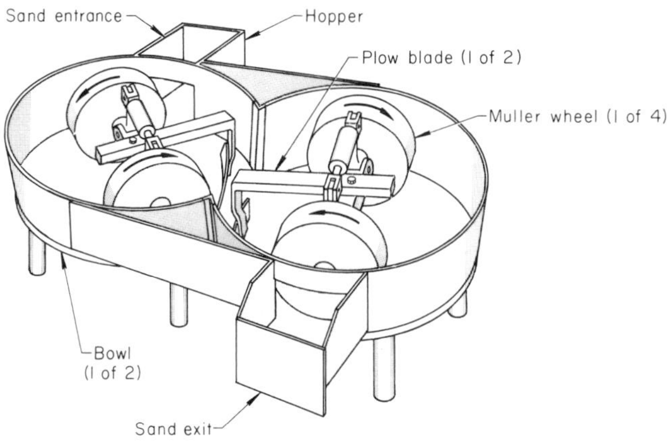

Fig. 12 Primary components of a continuous muller

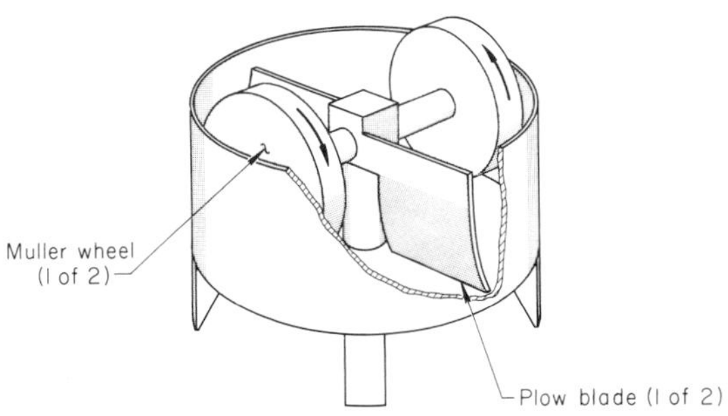

Fig. 13 Essential components of a conventional vertical wheel batch-type muller

ler and adds additional amounts of water as needed. Still other systems use a combination of these. Regardless of the type of equipment used, a minimum of 80% of the total added water should go into the muller immediately before the sand or at least with the sand to allow the maximum amount of time for cooling (if applicable) and/or clay activation and to provide the maximum control (Ref 2).

Most mulling equipment does not discharge prepared sand in its most flowable condition. Even if it did, prepared sand will normally be routed through at least one hopper (probably two or more) and will be transferred from one belt conveyor to another. Each time the sand is dropped into a hopper or onto another belt, some prepacking takes place. To provide a more flowable sand at the pattern face, operators of manual molding machines may, at the expense of time, riddle the first sand that enters the flask. In addition, again at the expense of time, an operator may even hand tuck the deeper pockets. Automatic molding equipment affords little or no opportunity for this special treatment. For these reasons, each molding station should be equipped with a good aerator to perform the final conditioning of the prepared sand. The aerator should be located on the last belt feeding the molding equipment in order to avoid any subsequent prepacking of the aggregate.

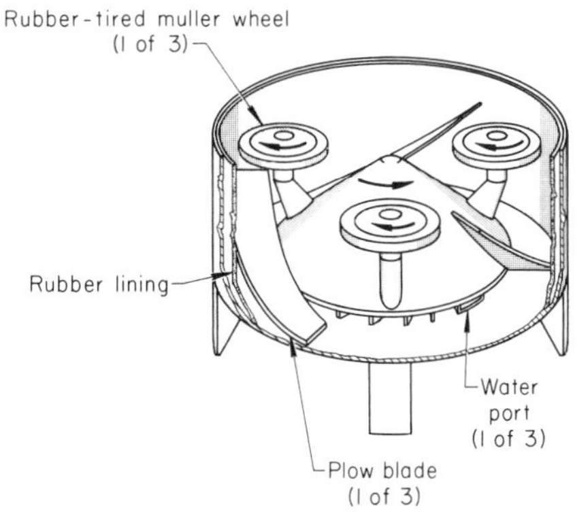

Fig. 14 High-speed batch-type muller with horizontal wheels

It is suggested that prepared sand always be conveyed to the molding equipment by belt conveyors. Other methods of conveying sand have been known to introduce unwanted moisture into the aggregate or to scrub clay from the coated sand grains. Prepared sand systems should always be designed with the receiving hoppers large enough to accept the total amount of sand from the feeding equipment. For example, the prepared sand hopper at the molding machine should always have enough capacity to receive all the sand from the muller (or another hopper). Thus, belts can be kept running, minimizing the drying time of the prepared sand on the belt. In no case should sand that will be used for molding be allowed to dry on a delivery belt. At best, this drying will be intermittent and will lead to inconsistent results at the molding station.

The requirements of dimensional stability and casting integrity do not tolerate poor or inconsistent control of molding sand. The old hand squeeze method of testing is not capable of controlling sand within the necessary specifications. Manual operation of the muller can and does lead to incorrect assumptions and corrections. For example, a sand that feels as though it does not have enough body indicates a system that is beginning to run low on clay. The operator is then likely to add additional water. When this additional water is added, the sand may seem to have the correct body, but in fact, it will have an excessive amount of moisture. Defects resulting from oxidation will increase, green sand pockets will be harder to fill, mold wall movement will increase and cause shrinkage defects, casting surfaces will become rougher and shakeout may become more difficult.

Sand test samples should be taken at the point where the sand enters the molding machine. Samples taken from other points, such as at the muller discharge, will provide a false indication of production conditions.

The properties tested often vary from foundry to foundry, as will the necessary frequency of testing. Those properties often include (but are not limited to) moisture content, active clay content, loss on ignition, grain size distribution, compression strength, permeability, and compactability. In some cases, green shear and wet tensile tests are also performed. All these tests provide the operating personnel with useful information.

Although the primary function of this section is not sand testing, it is necessary to mention another sand test that is widely used in Europe and is gaining acceptance in the United States. It is recommended that green tensile tests be conducted along with the other tests. Green tensile strength does not follow the same conditions that apply to green compression and shear strength. Although not documented, tensile strength decreases more rapidly and earlier than the other strength properties. This, combined with the fact that many molding defects (stickers, drops, and so on) are the result of poor tensile strength, makes the test well worth running. The test has not found widespread use, probably because it is more difficult to conduct and it requires expensive additions to the sand test equipment. Fortunately, another test has been developed whose results have a numerical relationship to green tensile. This test is spalling (splitting) strength. The manufacturers of sand test equipment can supply the necessary information on this test.

Molding Problems

Molding machines have been discussed in the section ''Molding Methods'' in this article. The use of these machines has been the subject of much discussion and, in some cases, disagreement. There are a number of pitfalls the operator can fall into. The first and most common involves compaction force. Most machines utilize hydraulic pressure to apply the compaction force. This pressure is adjustable, usually over a wide

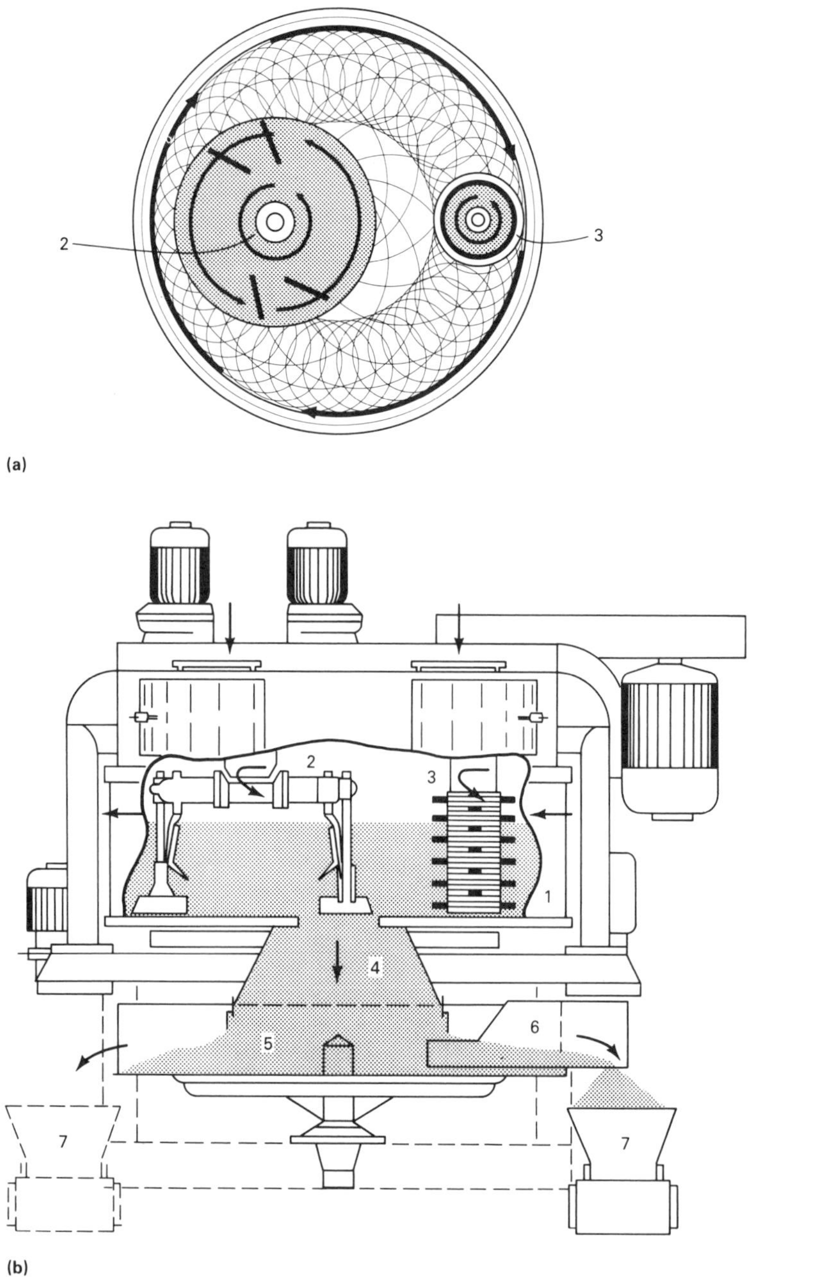

Fig. 15 Top (a) and side (b) views of an intensive mixer. The top view illustrates the loop pattern created by the mixing pan (1) rotating clockwise and the rotating mixing tools [movable mixing star (2) and stationary wall scraper] rotating either clockwise or counterclockwise while mounted eccentrically inside the mixing pan. The optional high-energy rotor (3) intensifies the mixing action. Other components include the discharge opening (4), rotary discharge table (5), discharge plough (6), and either one or two discharge chutes (7).

range. In 1970, the Green Sand Molding Committee, 80-M, of the American Foundrymen's Society accepted the definition of high-pressure molding as the pressure exerted on the mold face being equal to or greater than 700 kPa (100 psi). The committee also accepted the definition of high-density molding as a uniformly compacted mold with a mold hardness of 85 or greater as measured with a B-scale mold hardness tester (Ref 3). Unfortunately, it is often mistakenly held that greater pressure means better molds. Although a certain amount of pressure is necessary to compact the mold to arrive at the necessary strength and stability, it can easily be overdone. As pressures on the mold face exceed 1050 kPa (150 psi) and especially as they near 1400 kPa (200 psi), a number of detrimental effects can be noticed.

Springback. As mentioned earlier, the muller is used to coat the sand grains evenly with clay that has been made plastic by water absorption. Ideally, this coating of clay will completely cover each sand grain. The clay, expanded and made plastic by the water, is compacted by the molding machine with enough pressure to cause the clay that coats one grain to cohere to the clay that coats an adjacent grain, thus forming clay bridges. As compaction pressure is increased, these bridges gain more area and become stronger, at least up to a point.

As with most plasticized materials, the clay does have a certain amount of memory. As compaction pressure is released, the clay will attempt to recover some of its former shape. The mold will grow very slightly as pressure is released, even when pressures are quite low. This springback cannot be avoided (Ref 4).

As compaction pressures increase above 1050 kPa (150 psi) and approach 1400 kPa (200 psi), pattern release from the mold can become a serious problem. Not only does the mold swell slightly, causing pockets to become tight in the pattern, but the growth can also reach such a magnitude that the clay bridges are partially broken (Ref 5). When these bridges are fractured, tensile strength suffers drastically as greater compaction pressure is applied. The deeper pockets then become increasingly difficult to draw.

Expansion Defects. Tensile strength is not the only problem with high compaction pressures. As compaction pressure increases, more bentonite is squeezed into the voids that exist between the sand grains. When the grain-to-grain distance becomes too small, there is insufficient bentonite/water to contract as the sand grains expand during the heating by pouring. The result is expansion defects such as rattails, buckles, and scabs.

Similar problems can exist with machines that use some form of the pressure wave principle. The problem of springback may not be quite as great, because the mold becomes progressively less dense as distance away from the pattern face increases. However, if the compaction energy becomes too high, the sand grains may actually be stripped off of the bentonite coating, especially on the pattern face, giving rise not only to expansion defects but also to casting surface finish problems.

Parting Sprays. The tendency of the mold to stick to the pattern is often combatted by spraying the pattern with some kind of parting spray (usually proprietary). This helps draw the pattern from the completed mold and greatly reduces the likelihood of a buildup of bentonite and fines on the pattern surfaces. Unfortunately, there is a tendency to use too much of the spray. Like compaction pressure, the right amount of parting spray is advantageous, but excessive amounts cause a number of problems, including stickers, rough casting finish, penetration, blows, and sand inclusions. Generally, the pattern should be sprayed with a very light coating once every second or third mold. The operating foundryman should experiment with the optimum spray frequency and spray no more than is absolutely necessary.

Pattern Heating. Although the use of a parting spray is effective, a better first step is to heat the patterns to 6 to 11 °C (10 to 20 °F) above the temperature of the molding sand. Cold patterns will cause the moisture in the molding sand to condense on the pattern face, which makes the sand stick to the pattern. Many molding machines do not have any provision for heating the pattern during the production run. Whether the molding machine has the heating capability or not, the pattern should be preheated to the proper temperature prior to the start of the production run to assist in a rapid start-up.

Mold Finishing

After the mold has been compacted and the pattern removed, the mold is ready for the finishing operations. These operations usually consist of blowing out any loose sand, looking for any molding defects, drilling the sprue cup (if applicable), and setting any necessary cores. Once the finishing operations are complete, the cope can be accurately placed on the drag and the mold sent to the pouring station.

Mold blowoff is one of the areas that can cause surface finish problems if not properly controlled. Excessive amounts of air blown onto the mold can cause localized drying of the mold surface. On the other hand, if the air contains large amounts of moisture, the mold face can become excessively wet, giving rise to rough finish and burn-in penetration, just as excessive amounts of water in the molding sand or excessive amounts of parting spray will.

Core setting is another part of the process that has undergone tremendous change in the last few years. The increased demand for more accurate castings has affected cores and core setting just as it has molding. With mold hardness in the 85+ range, it is no longer possible to press an oversize core into undersize core prints. From the viewpoint of casting accuracy as well as cleaning room costs, it is equally unacceptable to place undersize cores in an oversize core print. Modern core processes allow the possibility of making the same size core every time, just as the same size mold can be made every time. Thus, it becomes apparent that every cavity in the corebox and every impression on the pattern must be as close as possible to the same dimensions. This obviously places greater demands on the supplier of patterns and coreboxes as well as on the process itself.

Mold closing is the next step in the operation. Some molding machines close the mold inside the machine, while others close the mold just outside the machine. Still others utilize a separate piece of equipment to perform this function totally external to the molding machine. The halves should not be allowed to remain separated any longer than absolutely necessary. Separation of the mold halves for excessive periods of time will allow the mold faces to dry, and this can lead to cuts, washes, and a general degradation of the casting surface.

Transportation of the mold to the closing station is critical. Jarring of the mold can cause green sand pockets to break away from the mold. In some cases, the pocket may not break away until the mold is poured; thus, the mold, the cores, and the metal are wasted. Rough transportation can easily cause heavy cores to shift off location, which can cause errors in casting dimensions, broken cores, and excessive metal around coreprints.

Mold closing is just as critical as mold transportation, and the same basic rules apply. The mold must be treated as smoothly and gently as possible to avoid the same type of defects. The mold guiding mechanism is as important at mold closing as it is during manufacture of the mold. Too often the accuracy and smoothness with which the mold is closed is overlooked. Again, the results can be drops and core movement as well as mold shift, crush, casting dimension problems, and so on.

After the mold is closed, mold transportation again becomes important. The finished mold must be carefully transported to the pouring area, or problems such as those already mentioned will likely occur.

After the mold has been poured, the molten metal must be given time to solidify and cool to the proper temperature before it is removed from the mold. During the solidifying process, the mold halves must be held solidly together. Any movement will introduce the possibility of casting inaccuracies or increased demands for feed metal or both. The loss of casting accuracy has obvious consequences. The requirement for additional feed metal has the consequence that shrinkage cavities may form and will probably not be evident from the casting exterior. Cooling time after solidification is critical for many casting/alloy combinations. Insufficient cooling time can lead not only to dimensional problems due to lack of casting rigidity but also to hardness and internal stress problems, even to the point of cracking the casting.

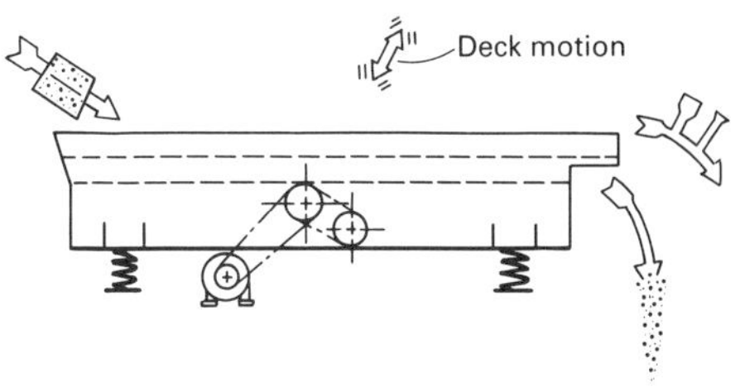

Fig. 16 Flat deck vibratory type shakeout device

Shakeout

After the castings have cooled sufficiently, they can be shaken out, that is, separated from the sand mold. Shakeout devices are available in a number of different configurations. Many of the devices available are of the flat deck, vibratory type. They range from normal intensity, frequency, and travel to high-intensity units that utilize a very short travel but high frequency. Some shakeout units are rotary in nature and, depending on design, can also provide the added function of cooling the sand. Another type is the vibratory barrel.

Deck-type shakeouts (Fig. 16) are available in a number of different configurations for various applications. The first is the stationary type. Stationary refers to the casting and sprue, not the shakeout itself. This type of shakeout is normally used by bringing the mold to the shakeout device; therefore, its primary application is for larger molds and low-to-medium production lines. The deck-type shakeout is also available as a unit that provides the function of conveying the castings from one end of the unit to the other. As mentioned earlier, either type is available in a variety of strokes, intensities, and frequencies. Selection of the shakeout is a function of casting design. Heavier castings can be quite successfully run using a longer-stroke shakeout, while thin-wall castings may require a short-stroke high-frequency unit to prevent breakage or damage to the casting.

Rotary-type shakeouts (Fig. 17) are also available in different configurations. The sand may exit at the same end that the sand and castings enter the unit, or it may exit at the opposite end. This type of shakeout also provides the function of conveying the castings from one end of the unit to the other.

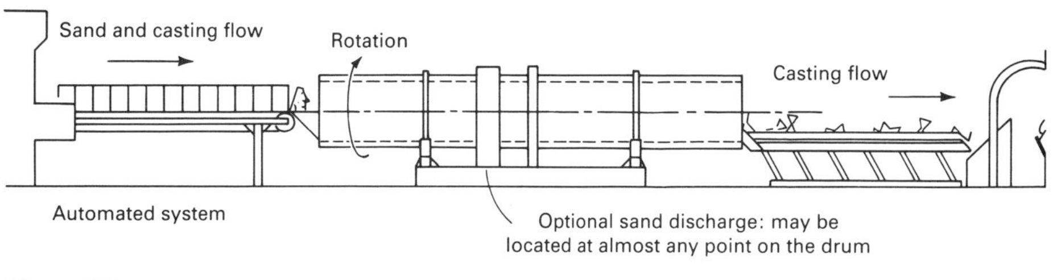

Fig. 17 Rotary-type shakeout system

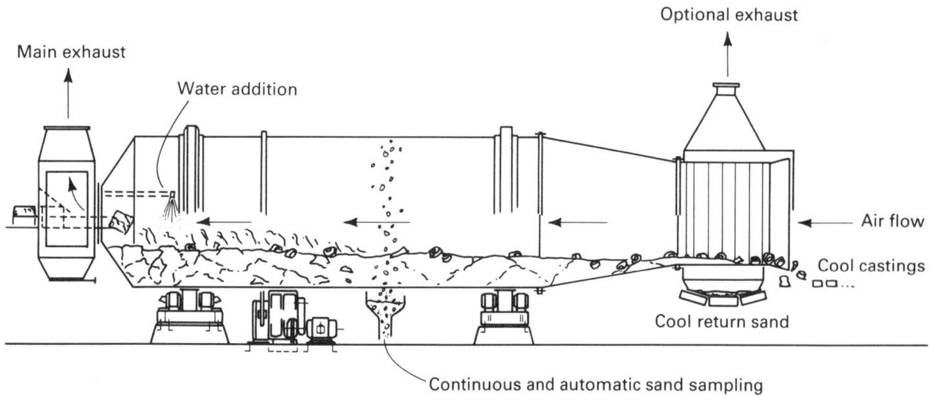

Fig. 18 Rotary plus cooling type shakeout system in which the castings and water-cooled mold sand are separated at the drum exit

Rotational speed is adjustable on most units to allow flexibility in shakeout intensity. In general, as rotational speed decreases, intensity decreases and castings are less likely to be damaged. Light thin-section castings may not be suitable for this type of shakeout. Although the castings themselves may not damage each other, the sprue is sometimes heavy enough that it can damage the castings.

Rotary plus cooling type shakeouts are also available in a configuration that not only holds the sand and castings together for an extended period but also affords the opportunity to cool the molding aggregate. This type of device is designed such that the castings and sand are held together throughout the length of the drum (Fig. 18). The castings and sprue aid in the breakdown of lumps. Sand temperature samples are normally taken somewhere along the length of the drum to determine the amount of water necessary for cooling the sand. The cool sand in turn cools the castings, often down to a temperature that can be comfortably handled at the exit of the drum. Sand and castings are separated at the exit of the drum. As with rotary shakeouts, sprue can damage certain types of castings, especially as wall sections become thinner. In-mold cooling time can become more critical when the castings and sand are kept together in the cooling device. When castings are too hot, hardness problems can result. In some cases, stresses can also be introduced into the castings because of the rapid quenching of the casting in the molding sand.

Vibrating drum type shakeouts (Fig. 19) combine the operating principles of rotating drum and vibrating deck units. The vibrating section is round in cross section, but it does not rotate. Instead, a rotating action is imparted to the sand and castings by the vibratory action. As the drum vibrates, material is constantly agitated to produce particle migration in both axial and transverse directions. The drum can be designed to provide a very rapid blending action or a gentle folding action, depending on process requirements. Because air can be forcibly exhausted from the drum and because the surface of the sand within the drum is constantly changing, a limited amount of cooling is possible. Additional information on shakeout is available in the Section "Shakeout and Core Knockout" in the article "Processing of Castings" in this Volume.

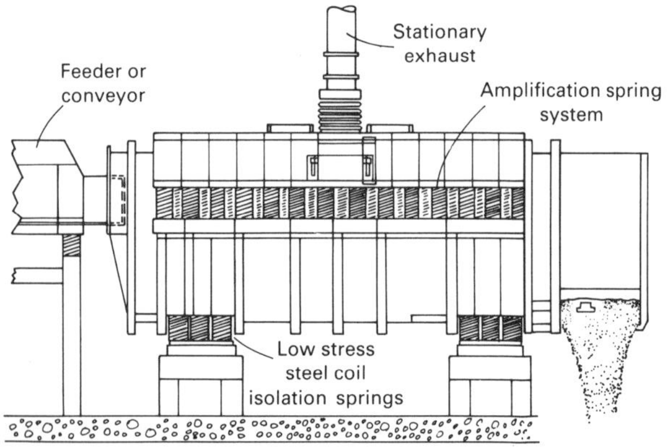

(a)

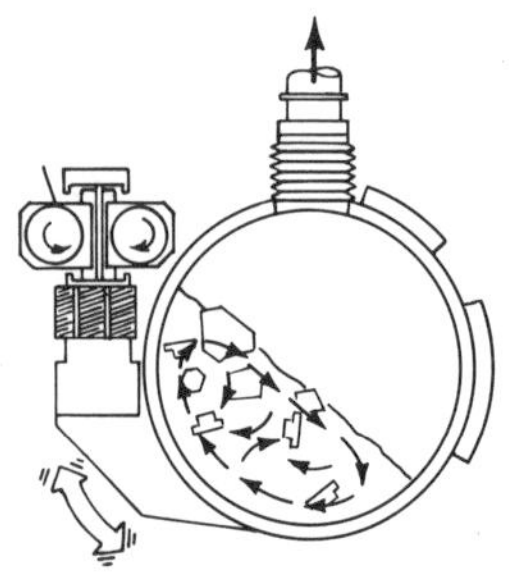

(b)

Fig. 19 Front (a) and side (b) views of a vibratory drum type shakeout system

Sand/Casting Recovery

What happens to the sand after shakeout is of great importance to the design and operation of the system (see the section "Sand Reclamation" in this article). Historically, the sand is returned from the shakeout to a storage bin, where it is kept until the next time it is mixed with additional clay, water, and carbonaceous materials. Unfortunately, sand-to-metal ratios of 3:1 to 6:1 are quite common. Sand-to-metal ratios in this range, combined with cooling times that allow the castings to become cool enough to separate from the sand, can easily create return sand temperatures of 120 °C (250 °F) and above (Ref 6).

High sand temperatures cause innumerable problems not only with regard to molding and surface finish but also for the system itself. Bentonite does not absorb water and become plastic to develop the necessary cohesive and adhesive strengths when sand is above 45 to 50 °C (115 to 120 °F). Therefore, the molding sand must be below these temperatures long enough for the muller to provide the necessary input of energy to coat the sand grains properly. Hot sand, usually above 50 °C (120 °F), is difficult to temper and bond, and when above 70 °C (160 °F), hot sands are impossible to rebond (Ref 7).

Unfortunately, sand is not easily cooled, especially in the quantity necessary to keep a molding line running. Molding sand is a relatively good insulator and therefore tends to hold heat for long periods of time. Storage quantity is therefore not the answer. Not only does sand stored in a bin hold its heat for long periods of time but it also cools from the outside toward the center. As it cools in this manner, moisture tends to migrate toward the cooler sand, which causes it to cake on the outside walls. As time goes on, the caking on the outside wall becomes thicker until only a small portion of the sand is actually being circulated through the system. Vibrators and bin poppers have been designed and can be of some help in combatting this rat holing tendency of return sand bins, but the ideal situation would be to cool the sand prior to storage.

Evaporative cooling is the only practical method of cooling the amount of sand needed in green sand systems. Hot sands must therefore have water added in amounts that exceed those required for tempering if both cooling and tempering are to take place. In addition, an ample supply of air must be present to carry away the heated water vapor.

The cooling of molding sand may be regarded as a two-stage process, although no sharp line separates the stages. At temperatures in excess of approximately 70 °C (160 °F), added water causes a flash evaporation cooling effect (Ref 6). Temperature will continue to decrease fairly rapidly to about 60 °C (140 °F), but more slowly after that. As sand temperature approaches ambient temperature, further cooling becomes more difficult and time consuming. Conditions of high ambient temperature, especially when combined with high ambient humidity, can substantially reduce the effectiveness of cooling devices. Therefore, ambient conditions should be considered carefully when sand systems are being designed or modified.

Some mullers have the capability of blowing air through the sand mixture and will cool the sand very effectively. However, there are some disadvantages to this method. It must be kept in mind that mulling (coating sand grains with bentonite) does not take place until the mixture is cool enough to be tempered and bonded (Ref 6). Cooling time must therefore be added to the mulling time. Although western bentonite provides the mold stability needed by most foundries, it does require more time and energy to absorb water and develop the necessary properties (Ref 8). Thus, the job of the muller or mixer becomes even more difficult and time consuming. The storage bin will still have the tendency to rat hole, thus returning sand more quickly and hotter to the muller and further aggravating the situation. Control of solid additives and water becomes more difficult as the molding sand becomes hotter. However, this is not an impossible situation; this method is used quite effectively in a number of foundries.

A few steps can be taken to provide some amount of cooling to the return sand in an existing system and to keep equipment costs as low as possible. For example, water can be fogged on the return sand, preferably as early as possible. Chains can then be dragged through the aggregate and/or plows can be used to turn the mixture over. Additional air can be introduced by fans or other sources to enhance cooling. Elevators can be vented to enhance air flow, but this provides little help because the sand is being conveyed in solid buckets. The only assistance realized will be at the transfer points. Although these and similar methods do help to reduce return sand temperature, they are generally of only marginal value. An effective job of cooling return sand normally requires the addition of water, along with forced air being blown or pulled through the aggregate by some type of auxiliary cooling device.

A number of auxiliary cooling devices are available that utilize forced air for evaporative cooling. These units should always be placed as close as possible to the casting shakeout. In fact, one type of unit, the shakeout—cooling drum, combines the functions of shakeout and sand cooling. Cooling the sand at or near the shakeout enables tighter control, reduces the tendency toward rat holing in the return sand bin, and reduces the demand for cooling on the muller. Because many muller designs make no provision for cooling, adequate external cooling is not only desirable but necessary.

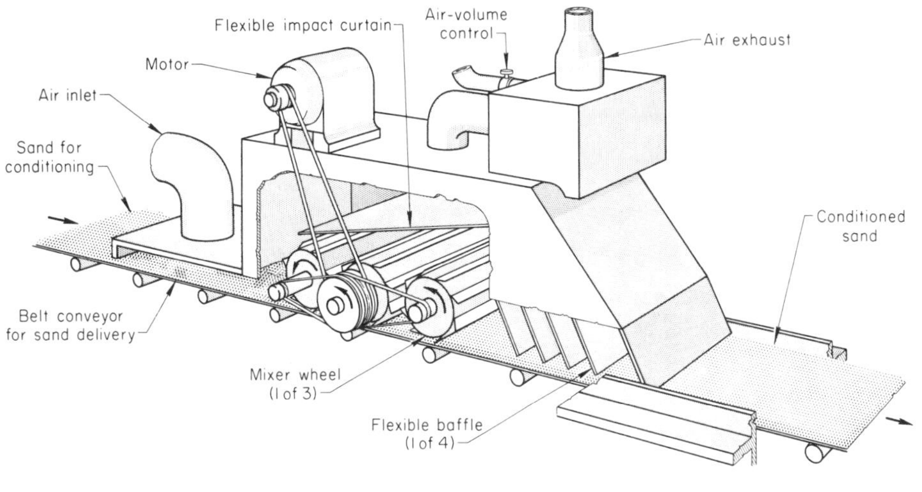

Fig. 20 Mechanized sand cooler used in high-production molding lines

Cooling the sand as early as possible reduces the total cycle time of the muller by reducing or eliminating the time necessary for cooling and provides a method for making mulling time more efficient. Southern bentonite can be mulled in very quickly if the aggregate temperature is low enough. As mentioned earlier, western bentonite is not mulled in very quickly, because it must swell by such a large amount (Ref 8). For this reason, it is advisable to keep the bentonite swelled and as active as possible. Many of the auxiliary cooling devices can be controlled to the point where the level of return sand moisture will be such that the western bentonite will remain activated. Normally, a retained moisture level of 1.8 to 2.0% will not only keep bentonites activated but will also reduce the amount of dusting at transfer points, thus reducing the load on dust collection equipment.

Cooling Devices. As mentioned earlier, it is possible to realize some cooling by adding water to a return sand belt and then using some method of turning the sand over at various places along the length of the belt. There are mechanized devices (Fig. 20) that perform similar functions and provide air flow through the sand. The effectiveness of these methods is often somewhat limited because of conveyor belt lengths; as belt lengths become shorter, the method becomes less effective. Difficult sand temperature problems will require more serious measures.

Drums used as cooling units are among the oldest of the effective devices (Fig. 21). A cooling drum does not keep the sand and

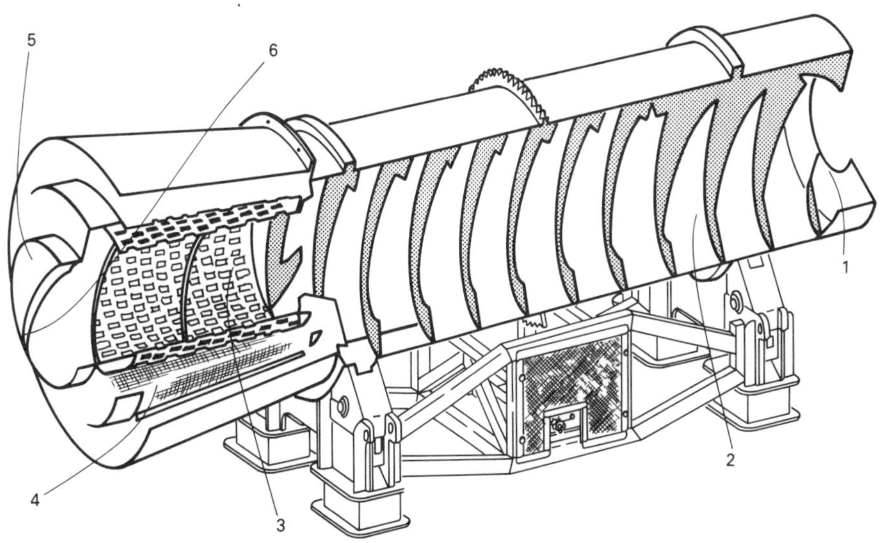

Fig. 21 Cutaway view of a sand cooling drum system. Sequence of operations proceeds from right to left: 1, hot shakeout and spill sand enter, and helical flights convey sand forward to begin blending process; 2, cascading effect provides sand cooling as well as sand homogenization; 3, blended and cooled sand is discharged onto perforated cylinder, which screens off tramp metal and core butts while passing sand; 4, replaceable screen passes sand to discharge onto conveyor; 5, lumps that do not pass final screen carry across to lifter paddles for discharge into overburden chamber

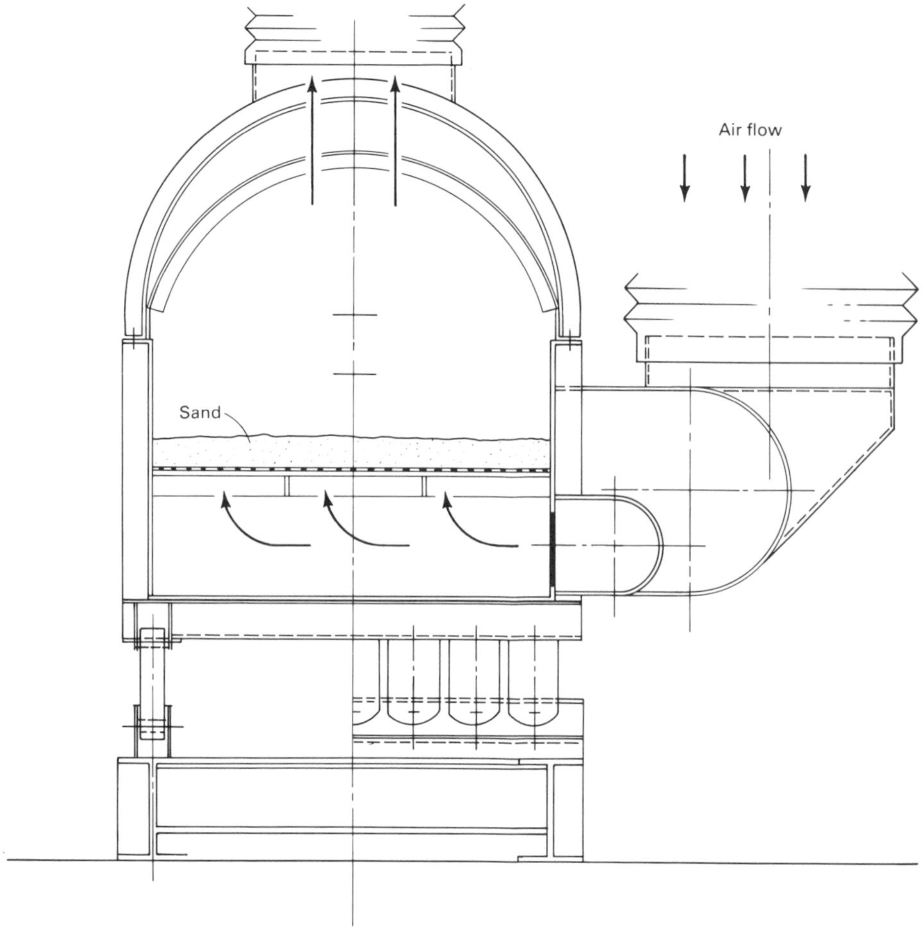

Fig. 22 Schematic of a fluid bed cooler

castings together; instead, this is a separate piece of equipment through which sand from the shakeout flows. As with other cooling devices, water must be added to the molding sand to allow the air moving through the drum to provide the necessary cooling by evaporation.

The fluid bed cooler (Fig. 22) is a vibratory type of conveyor through which the sand flows in a more or less continuous but controlled stream. Air is pumped through the sand from underneath, causing the necessary evaporation and cooling.

Figure-Eight Cooler. Similar to the continuous muller shown in Fig. 13, the figure-eight cooler is designed so that air can be pumped through it and provide the necessary cooling. This device has been used directly above the muller, but a more desirable location would be as close to the shakeout as possible for the reasons already mentioned.

Regardless of the equipment used, it is necessary to control the moisture additions so that sufficient moisture is available for cooling and bentonite activation without getting the return sand so wet that problems will be experienced with plugging up of the sand system. The movement of air through the aggregate will almost certainly remove some of the finer material. The higher the velocity of air movement, the better the cooling, but also the greater the loss of that fine material. The loss of a certain amount of that material (such as dead, burnt clay and ash) can be beneficial. Unfortunately, a number of beneficial materials can also be lost, such as the finer grains of sand, coal dust, and bentonite. Any cooling device should be planned with a solids separator on the exhaust air so that these materials can be collected and fed back into the system at a controlled rate. This will improve surface finish, and trapping and using the bentonites and coal dust will provide economic benefits.

Metal Separation and Screening. The shakeout does the primary job of separating the sand from the sprue and castings. Smaller pieces of metal can easily slip through the grating of the shakeout device and be processed along with the sand. This will cause casting defects, and it may damage the equipment. Therefore, it is advisable to remove as much of the tramp metal as possible. When magnetic metals such as most irons and steels are being cast, the job is relatively easily accomplished with magnets. The suggested practice is to install an over-belt magnet somewhere along the length of a conveyor belt and a pulley magnet at the discharge end of the same belt. Placing both magnets on the same belt allows more complete separation of the magnetic particles.

Nonmagnetic alloys present a different problem. Devices are available that separate the metallic particles based on density differences, but the most common method is to use screens. Multiple screens are often used, and the mesh size from screen to screen becomes progressively finer.

Lumps are found in all sand systems and consist of system sand or core parts that have not been sufficiently heated to break down the binder. For this reason, it is necessary to have a good screen in all systems. The opening size in the screen should be as fine as is practical for the system involved.

Two basic types of screens are in use: flat deck and rotary. The flat deck type is usually vibratory in nature and has the added function of providing further lump reduction as well as the screening function. The rotary type of screen is normally a large barrel that continually rotates. The exterior of the barrel has the desired size of holes in it to provide the screening action. Because of the tumbling action within the screen, lump reduction similar to that obtained with the vibrating flat deck can be expected. In both cases, the size of the screen should be as fine as is practical. After the sand has been cooled, the tramp metal removed, and the core butts and lumps removed, the sand is ready to be returned to the storage hopper to be used again.

Computer-Aided Manufacture

Recent years have seen a rapid advancement in the use of data processing units and data communication. These advancements have made possible almost complete and instantaneous record keeping and, equally important, trend recognition.

The technology is advancing rapidly; there are systems currently in place that record on a continual basis the amounts of return sand, new sand, bentonite (or premix), and water that go into each batch of sand. In many cases, mixing time and maximum current draw of the muller are also recorded. With some systems, compactability can also be recorded. In any case, output data, such as compactability and muller current draw, can be stored for a period of time, and a trend analysis can be done automatically.

Molding machines have also become more sophisticated. With microcomputers and programmable controllers being used to control machine movements, it is possible to read the pattern number automatically when the pattern is installed. Using information that had previously been stored in the memory of the computer or controller, the molding machine can optimize its molding parameters for the individual pattern.

A hypothetical case will illustrate the extent of the available information. During a shift, a new pattern is installed on the molding machine. The operator tells the machine that 1250 molds are needed. Optimum molding parameters, poured weight, necessary cooling time, and so on, have already been determined during earlier runs and stored in the computer. At any point during the run, the operator or someone operating a distant host computer can query the molding machine to find out which mold is going to reach shakeout next, how much cooling time it had, how much metal is required to complete the production run, how much time will be required to complete the production run based on existing molding rates, how many cores will be required to complete the run, how many molds have been made and/or poured, and so on.

These outputs can be used as control signals. More water or less water can be added to the sand cooler when sand from the new molds reaches the cooling device. Molding sand compresses more in the molding chamber/flask as sand becomes wetter (higher compactability), thus trend analysis can be done by recording mold compression during compaction, and the resulting information can be fed back to the sand preparation equipment. The exact position required for an automatic pouring device can be set by the molding machine. Daily production data reports can be printed out that will give information on each run; this information includes the number of castings, production rate, productivity, number of cored molds, and reasons for downtime (such as waiting for sand or metal).

In the event of machine difficulty, the machine can help troubleshoot itself. It is not only possible but practical to allow the molding machine to exchange data with a remote location (via telephone lines) if assistance in troubleshooting is needed.

The quantity of information that is available and transmittable depends on the mechanical and electronic design of the equipment. Some units are designed to allow one-way communication (output), while others are designed to allow two-way communication (output and input). In the latter case, it is possible for a remote location to control some or all inputs to the production equipment. These remote locations can consist of keyboard inputs from a host computer or even data output from other pieces of equipment.

The type of information available (either as inputs or outputs), the form the information is in, and the communication protocols may vary greatly among manufacturers. It is therefore necessary to research the technical information available from each manufacturer to determine the best way for the various pieces of equipment to communicate and the best way to handle the information obtained. Additional information on the role of computers in the manufacture of green sand molds is available in the Section "Computer Applications in Metal Casting" in this Volume.

REFERENCES

1. V.K. Gupta and M.W. Toaz, New Molding Techniques: A State of the Art Review, *Trans. AFS*, Vol 86, 1978, p 519-532
2. M.J. O'Brien, Cause and Effect in Sand Systems, *Trans. AFS*, Vol 82, 1974, p 593-598
3. *High Pressure Molding*, 1st ed., American Foundrymen's Society, 1973
4. D. Boenisch, "Strength Problems in High Pressure Compacted Sand Molds," Paper presented at the Disamatic Convention, Disamatic Inc., 1971, p 69-84
5. D. Boenisch and B. Koehler, Sand Compaction and Grain Rupture in High Pressure Molding Machines, *Giesserei*, Vol 63 (No. 17), Aug 1976, p 453-464
6. J.S. Schumacher and R.W. Heine, The Problem of Hot Molding Sands—1958 Revisited, *Trans. AFS*, Vol 91, 1983, p 879-888
7. C.A. Sanders, *Foundry Sand Practice*, American Colloid Company, 1973, p 441
8. J.S. Schumacher, R.A. Green, G.D. Hanson, D.A. Hentz, and H.J. Galloway, Why Does Hot Sand Cause Problems?, *Trans. AFS*, 1974, p 181-188

Sand Reclamation

Michael Zatkoff, Sandtechnik, Inc.

Reclamation is defined by the American Foundrymen's Society (AFS) Sand Reclamation and Reuse Committee 4-S as the physical, chemical, or thermal treatment of a refractory aggregate to allow its reuse without significantly lowering its original useful properties as required for the application involved. To achieve this objective, one must evaluate the type of sand entering the reclamation system, the binder system used, and the area for its reuse.

This section will provide a brief review of sand reclamation systems for both chemically bonded (resin bonded) sands and clay-bonded sands (green sands). Detailed information on sand molding principles and processes can be found elsewhere in this Volume.

Reclamation of Chemically Bonded Sand

The primary requirement of any reclamation system is to remove the resin coating around the sand grains. This involves abrasion and attrition to break the bond, as well as classification to remove the fines that are generated. The three basic reclamation systems are thermal, dry, and wet. Selection of a system depends greatly on the type of organic binder to be removed from the sand grains. More detailed information on organically bonded sand systems can be found in the article "Resin Binder Processes" in this Volume.

Wet Reclamation Systems

Wet reclamation systems were used for clay-bonded system sands in the 1950s, but are now used for silicate binder systems only. Silicate systems are very difficult to reclaim by dry processes and are impossible to reclaim in thermal systems. This is because silicate is an inorganic system that melts rather than burns in the furnace.

The complete system includes lump-breaking and crushing equipment, an attrition unit, wet scrubber, dewatering system, and dryer. The systems require about one pound of water per pound of sand reclaimed, and in some cases the water can be discharged directly into municipal sewer lines. Most installations allow 100% reuse of the reclaimed sand, with makeup sand as the only new sand addition.

Dry Reclamation Systems

Many factors determine the degree of cleanliness required in a reclaimed sand. These factors include the type of resin system used for rebonding, the sand-to-metal ratio, the type of metal poured, the condition of the reclaimed sand, the type of new sand used, and the ratio of new sand to reclaimed sand.

Attrition reclaimers break down the sand lumps to a smaller grain size. Some fines are removed, but the binder is not removed completely from the surfaces of the sand grains. In most cases, these units produce a sand that requires a higher concentration of new sand when the attritor is coupled with a sand scrubber, as described below.

Additional scrubbing is sometimes required, and there are basically two types of scrubbers: mechanical and pneumatic. Selection between the two types is primarily a question of wear, ease of maintenance, and energy consumption because the units pro-

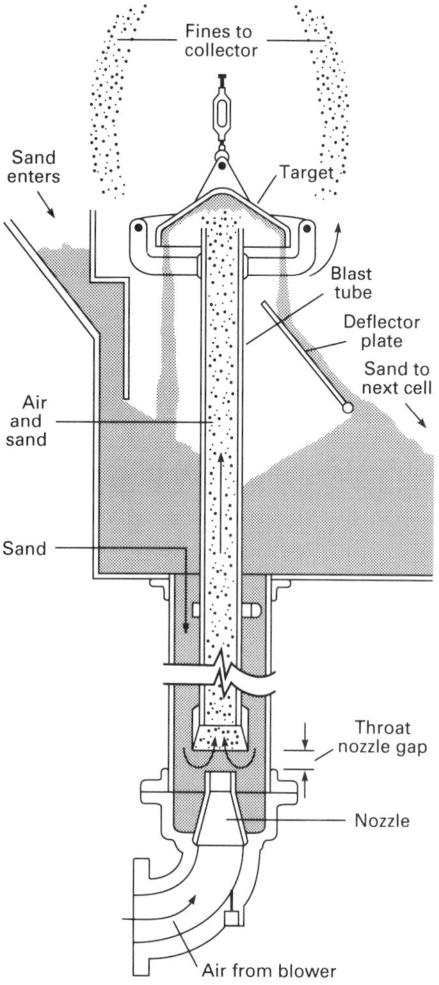

Fig. 1 One cell of a pneumatic scrubber

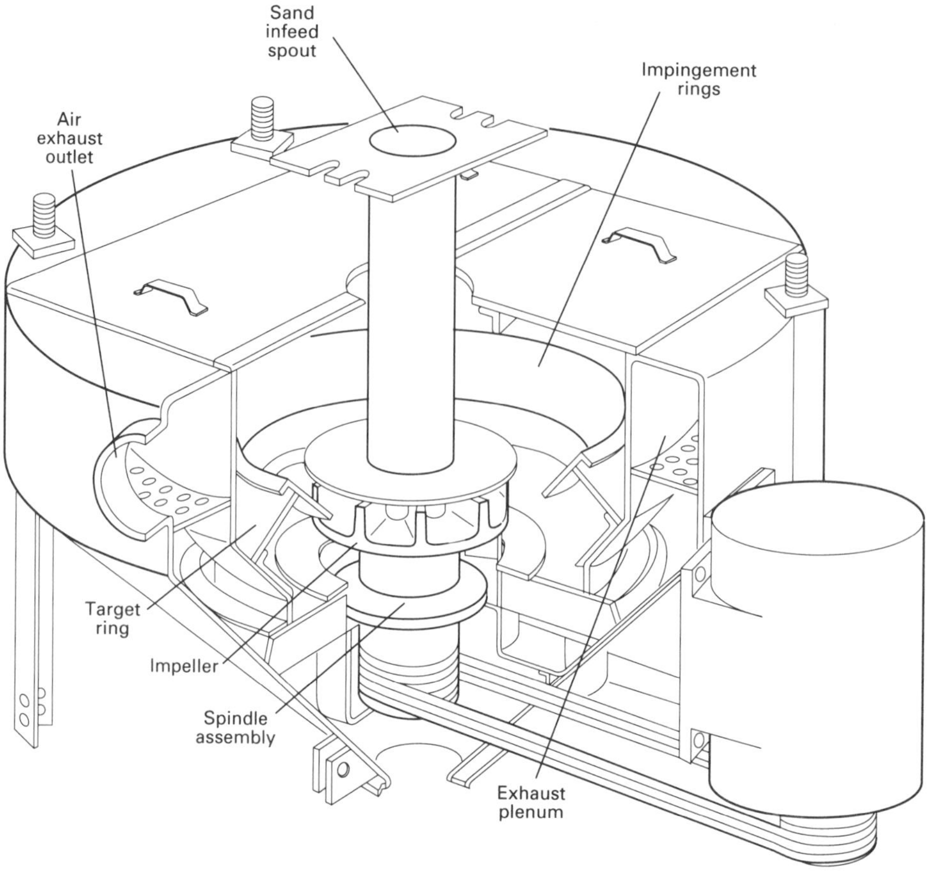

Fig. 2 Horizontal mechanical scrubber

vide comparable performance in terms of scrubbing action.

Pneumatic Scrubbing. Figure 1 shows one cell of a pneumatic scrubber. Sand is introduced by gravity at the top of the unit, and it flows down around the blast tube. High-volume low-pressure air from a turbine blower flows through the nozzle and lifts the sand up through the blast tube to the target plate. The sand grains undergo intense attrition in the tube by impacting on each other; further attrition occurs at the target as binder is removed from the sand grains. These fines and resin husks are then removed from the system by a classification dust collection system. Scrubbed sand falls from the target and is deflected to the next cell or is kept within the same cell for further scrubbing. The degree of cleanliness attained is determined by the retention time in the cells (controlled by the deflector plate) and the number of cells. Sand exiting the final cell should be screened to remove any foreign material that may be present in the refuse sand.

A conscientious maintenance program must be followed for these scrubbers. High-wear parts are the impellers, targets, and tubes. Improperly maintained units will not yield a consistent reclaimed product. The dust collection system is of equal importance and must also be properly maintained. Excess fines in the sand increase residual binder and decrease sand permeability.

Mechanical Scrubbing. There are two types of mechanical scrubbers: horizontal and vertical. Figure 2 shows a horizontal scrubber. Clean sand (crushed, with metal removed) is fed into the center of the unit and thrown against the target ring at a controlled velocity by the impeller. Some sand-on-sand attrition takes place, but the intense scrubbing occurs at the target ring. The exhaust plenum surrounds the target ring to remove dust and binder husks. These units can be arranged in sequence for additional scrubbing.

Figure 3 shows a vertical mechanical scrubber. Sand enters the center of the impeller and is thrown upward at a target plate. Attrition takes place as the sand hits the target. The sand falls away and exits into an air wash separator, where the fines are exhausted from the sand. For additional scrubbing, this unit can also be operated in series, as shown in Fig. 3. As with pneumatic scrubbers, the impellers and targets are high-wear parts. The units must be properly maintained to yield a consistent product. The dust collection system must also be properly maintained.

Process Controls. There are a number of tests that can be performed on the reclaimed sand from the scrubber. Two of the more important tests are loss on ignition* and screen distribution. These tests are good measures of the operating efficiency of the classifiers. A build-up of sand on the fine screens, such as 200, 270, and 300 mesh, is usually associated with an increase in loss on ignition, which is a problem that is most likely attributable to the classifier. An increase in loss on ignition without the accompanying shift in screen distribution will indicate that binder removal in the scrubber is low and that maintenance may be required on the unit. Most manufacturers supply a troubleshooting guide.

Temperature control is critical when working with chemically bonded sand. The heat of the reclaimed sand at the discharge point is affected by three factors. The first is the temperature of the sand at shakeout. This will vary with the sand-to-metal ratio, the type of metal poured, and the amount of lump reduction from the shakeout. Secondly, heat will be generated within the reclamation system itself. All the components will gradually heat up, thus increasing the temperature of the sand. The third factor

*The loss on ignition test involves firing a 50 g sand sample at approximately 980 °C (1800 °F) to determine the amount of carbonaceous material burned off.

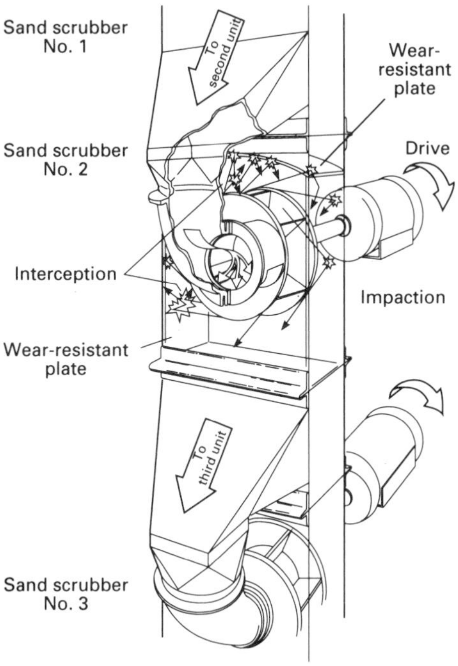

Fig. 3 Vertical mechanical scrubber

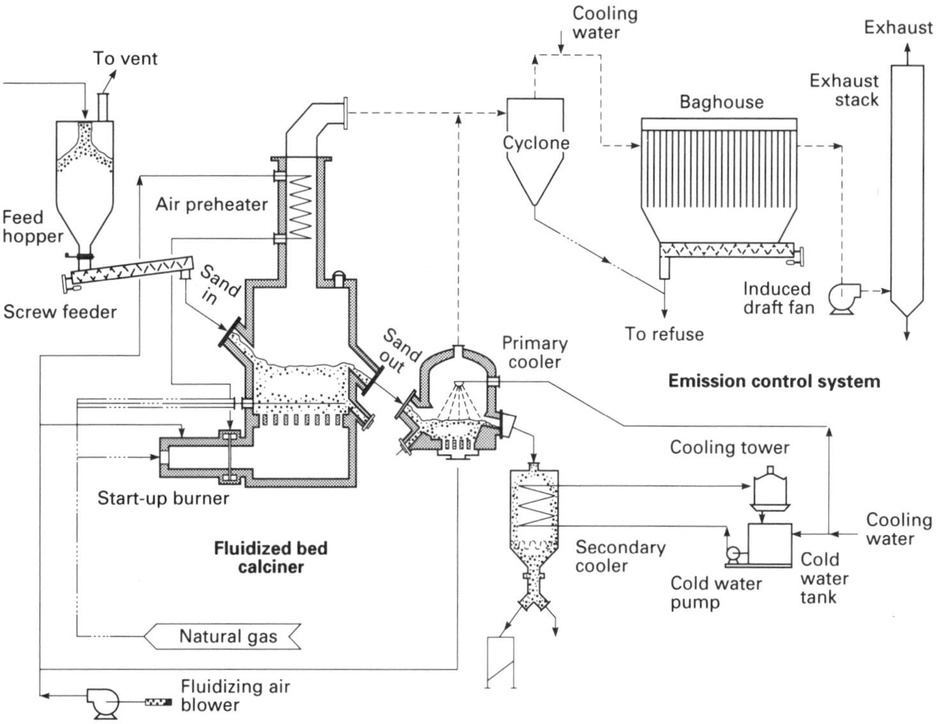

Fig. 4 Fluidized bed reclamation unit

that affects the temperature is the season of the year. In the summer, for example, there are more problems with hot sand. Therefore, based on these factors, a sand heater or cooler may be a necessary addition to the total system.

Blending of the sand from the scrubbers with new sand is very important. Blending is done to help replace the sand that is lost in the casting and reclamation processes and to limit the effects of residual binder on the sand grains. The sand should be measured and blended thoroughly to ensure that there are no concentrations that would cause undesirable effects on the castings. Most of the sand reclaimed from these units can be used in blends of 80% reclaimed sand to 20% new sand. Again, the exact ratio will be determined by the metal poured and the type of resin system used.

Thermal Reclamation

Thermal reclamation of chemically bonded sand is achieved by bringing the sand to a sufficiently high temperature over the proper time period to ensure complete combustion of the organic resin and material in the sand. If the proper temperature and atmosphere are not maintained in the unit, the organic resin will volatilize and send volatile organic carbons up the emission stack. This would be a violation of clean air standards. Currently, the two types of thermal units available for sand reclamation are rotary drums and fluidized bed furnaces. For these units to be cost effective, they must be operated 20-24 h per day and make maximum use of energy recuperation techniques.

The rotary drum has been in use since the 1950s for the reclamation of shell and chemically bonded sands. The direct-fired rotary drum is a refractory-lined steel drum that is mounted on casters. The feed end is elevated to allow the sand to flow freely through the unit. The burners can be at either end of the unit with direct flame impingement on the cascading sand; flow can be either with the flow of solids or counter to it.

Table 1 Screen analysis of a base silica sand before and after reclamation

See also Fig. 5(a).

Screen analysis	Before reclaim	Reclaimed sand	Cyclone and dust collector
20	...	...	...
30	Trace	Trace	...
40	2.3	2.4	...
50	28.9	30.7	0.1
70	35.4	36.5	3.0
100	22.2	21.6	9.2
140	7.8	6.4	41.2
200	2.9	1.8	21.2
270	0.4	0.3	9.3
Pan	0.2	0.1	15.7
Total	**100.1**	**99.8**	**99.7**
GFN(a)	58.69	56.30	144.99
Yield	...	93.9%	6.1%

(a) American Foundrymen's Society grain fineness number. See the article "Aggregate Molding Materials" in this Volume for explanation. Source: Ref 1

Table 2 Screen analysis of a base olivine sand before and after reclamation

See also Fig. 5(b).

Screen analysis	Before reclaim	Reclaimed sand	Cyclone and dust collector
20	...	...	...
30	...	...	...
40	1.0	0.4	...
50	34.9	27.1	0.4
70	40.3	37.1	4.9
100	16.8	22.6	6.7
140	3.3	7.3	17.6
200	2.1	3.6	15.9
270	0.8	1.1	15.1
Pan	0.8	0.4	39.1
Total	**100.0**	**99.6**	**99.7**
GFN(a)	56.41	63.31	195.25
Yield	...	77.5%	22.5%

(a) American Foundrymen's Society grain fineness number. Source: Ref 1

Table 3 Screen analysis of a base chromite sand before and after reclamation

See also Fig. 5(c).

Screen analysis	Before reclaim	Reclaimed sand	Cyclone and dust collector
20	0.3	0.2	...
30	2.8	1.9	...
40	17.4	14.1	...
50	28.3	26.9	...
70	25.2	25.2	0.7
100	17.1	18.5	4.2
140	6.6	8.3	13.3
200	1.7	3.4	19.7
270	0.2	0.8	15.3
Pan	0.1	0.4	46.9
Total	**99.7**	**99.7**	**100.1**
GFN(a)	51.53	56.97	215.25
Yield	...	82.7%	17.3%

(a) American Foundrymen's Society grain fineness number. Source: Ref 1

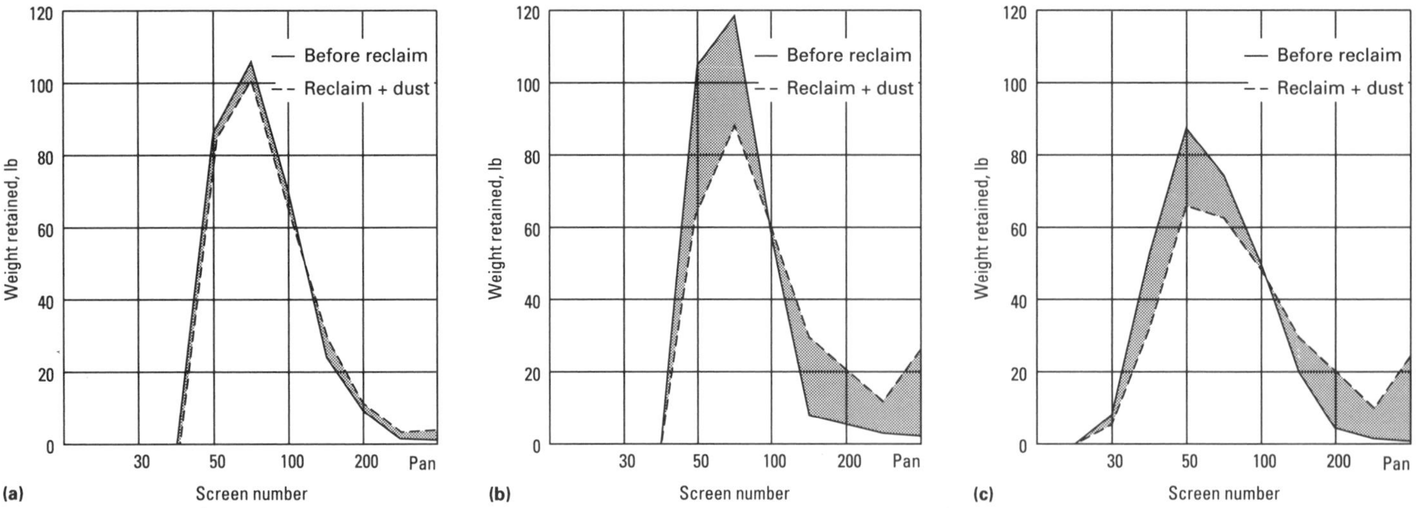

Fig. 5 Relationship between the before-reclamation sand weight distribution and the reclaimed sand/dust distribution for 135 kg (300 lb) silica (a), olivine (b), and chromite (c) sand samples. See Tables 1, 2, and 3, respectively, for screen analyses of these sands. Source: Ref 1

In indirect-fired units, the drum is mounted on casters in the horizontal position and is surrounded by refractory insulation. Burners line the side of the drum, with the flames in direct contact with the metal drum. The feed end is elevated to allow the sand to flow freely through the unit, and in some cases flights (paddles connected by chains) are welded to the inside to assist material flow. The advantage of the rotary drum is its lower capital cost. Its disadvantages are high heat losses, use of moving parts at high temperatures, short refractory life, poor control of material flow, and poor atmospheric control.

Fluidized bed units have been in use for the reclamation of clay and chemically bonded sands since the 1960s. The fluidized bed calciner illustrated in Fig. 4 consists of a cylindrical, brick-lined, vertical combustion chamber. Sand that has been crushed is taken from the surge (feed) hopper by means of an adjustable, closed screw feeder and is fed into the unit. In the bottom, a hot sand bed is kept fluidized by the use of a combustion air blower. This blower controls the output of the unit and ensures the availability of sufficient air for combustion. The fluidized bed uses two sets of burner systems. The start-up burner brings the bed of sand up to operating temperature. After bed temperature is reached, the second system is energized. This system consists of gas lances positioned around the perimeter of the unit and inserted directly into the bed. It maintains the sand bed temperature to within ±8 °C (±15 °F).

The intense mixing of the sand and the hot gases provides for combustion of the hydrocarbons and residual resin at an appropriate temperature and retention time. The fluidized bed is heated by a postcombustion zone that ensures complete combustion of the waste gases without the use of an afterburner. After the postcombustion zone, an induced-draft fan pulls the dust and hot gases from the unit through the dust collector to ensure constant throughput requirements and efficient emission control. The sand then exits by gravity feed to the cooling and classifying systems. The disadvantage of the unit is its higher capital cost. The advantages are long refractory life, low heat losses (energy consumption), no moving parts at high temperatures, and control of material flow, temperature, and atmosphere.

Reclamation of Clay-Bonded System Sand

The reclamation of clay-bonded molding sand (green sand) allows the reuse of sand in any area of the foundry, including the core room. This practice has been common in Japan for the past 20 years, but has been adopted in the United States only recently. The process combines the different pieces of equipment described above. These reclamation plants must be designed with a total system approach to ensure proper integration of the various pieces of equipment.

After proper crushing and metal removal, the sand is transported to a storage or feed hopper to be metered into the sand calciner. Temperature control in the calciner is very important. If the sand in the unit becomes too hot, the clays in the sand will fuse to the sand grains; this makes clay removal difficult. If the temperature is too low, pollution problems will result.

After calcining, the sand is cooled to an appropriate temperature. Sand exiting the cooler is transported to a scrubber/classifier, as described for chemically bonded

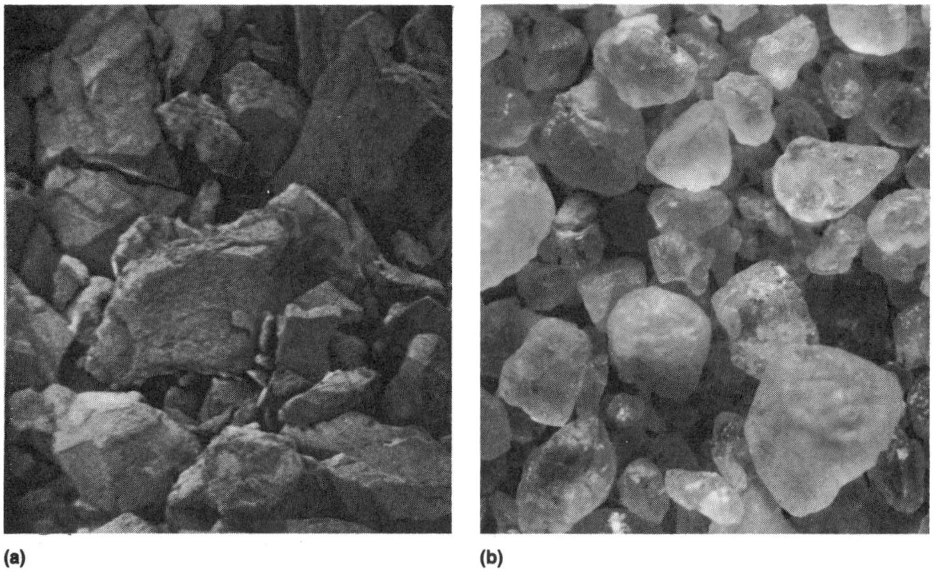

Fig. 6 Comparison of angular olivine sand grains (a) and rounded silica sand grains (b). Courtesy of M.J. Granlund, National Engineering Company (retired)

(a) (b)

Fig. 7 Chromite sand before (a) and after (b) reclamation. Note the smaller, more rounded grains due to reclamation. SEM. Both 50×. Courtesy of M.J. Granlund, National Engineering Company (retired)

sand. In most cases, additional classification is required for proper mineral separation. The final product is then transported to the storage silo. More detailed information on the reclamation of clay-bonded green sand systems can be found in the article "Sand Molding" in this Volume (see the section "Bonded Sand Molds").

Process Controls. The most important tests for clay-bonded reclaimed sand are screen distribution, AFS clay, acid demand value (ADV), and pH.

The screen distribution test indicates the operating efficiency of the scrubber/classifier units.

The AFS clay test indicates the effectiveness of the postprocessing equipment in mineral separation.

The ADV-pH Test. Some lake sands contain quantities of calcium carbonate. The carbonate registers in the ADV-pH test because the material is water soluble. After thermal treatment, the calcium carbonate converts to calcium oxide. This will result in a higher pH because the oxide is much more reactive, while the ADV number may decrease or remain the same.

Reclamation Effects on Base Sands (Ref 1)

As described in the article "Aggregate Molding Materials" in this Volume, a wide variety of sands are used in sand molding processes. Each differs in composition, particle size and distribution, purity, shape, and hardness. These properties are not only important to successful moldmaking and coremaking operations but also influence the reclamation process.

To determine the effects of reclamation on the surface area, shape, and yield of sand samples, a number of raw sands, without resin coating, were reclaimed through the use of a pneumatic scrubber (Fig. 1). The entire 135 kg (300 lb) sample was split through a large sand splitter before reclamation to obtain a representative sample. The representative sample of the before-reclamation sand was retained. After reclamation, the sand was weighed to determine the yield, and again the total reclaimed sample was split and the sample retained. The cyclone dust collector material was then weighed, split, and sampled.

The results demonstrated that the reclaimed sand was similar in screen analysis to the before-reclamation sand. Tables 1 to 3 give the screen analyses of the before-reclamation sand, the reclaimed sand, and the cyclone dust collector sample, as well as the yield for silica, olivine, and chromite sands.

Figures 5(a) to (c) show the relationships between the before-reclamation sand weight distribution and the reclaimed sand and dust distribution. The vertical scales in Fig. 5 are not in percent retained on the screen, but in the weight retained on each screen (pounds per screen). The shaded areas show the amount of change in the total distribution. The greater the shaded area, the greater the change in the total sand retained on any individual screen.

The rounding of angular sands due to reclamation results in a sand with less surface area and a sand that packs to a more dense configuration (less permeable molds). This is clearly the case for reclaimed olivine sands, which are extremely angular and have a higher hardness than silica sands. Olivine and silica sands are compared in Fig. 6. Figure 7 shows chromite sand before and after reclamation.

REFERENCE

1. M.J. Granlund, Base Sand Reclamation, *Trans. AFS*, Vol 92, 1984, p 177-198

Melting Furnaces

Electric Arc Furnaces

Nick Wukovich, Foseco, Inc.

THE ELECTRIC ARC FURNACE made its appearance as a production tool at the beginning of the 20th century. These early furnaces had capacities of 910 to 14 000 kg (1 to 15 tons). Currently, the electric arc furnace is regarded as one of the primary melting tools used by foundries and steel mills. Electric arc furnaces are used as melters and holders in duplex operations and as melting and refining units. This article will focus on the construction and operation of these furnaces and their auxiliary equipment in the steel metals industry.

Power Supply

The current applied to the electric arc furnace is supplied by a local electric utility company and passes through an electrical substation that is designed specifically for the furnace(s) to be supplied. A simplified diagram is shown in Fig. 1.

The power supplied to the furnace during melting is provided by an electrical arc established from three carbon or graphite electrodes. During the meltdown portion of the heat, the three arcs or phases act as a single phase. Two arcs can strike the furnace charge and draw current without a current going to the third arc or electrode. Because of the type of scrap used and the arc lengthening and shortening that take place, there is a great fluctuation in the current during meltdown, which causes a significant variation in the electrical supply system. If the same power source is the supply for other plant power, a flickering of lights and voltage fluctuation will be noticed on machinery and electrical equipment. During the refining cycle or after meltdown, the arcs tend to stabilize, partially because of a slag cover on the liquid metal and the flatness of the bath. In addition, the arcs have been shortened to direct their energy in a smaller area.

A great amount of energy is produced during this meltdown and refining of ferrous metals. Controls for harnessing and directing the energy of the arcs are required in order to produce molten iron and steel in the electric furnace without destroying the furnace refractories. The energy requirements for melting various carbon levels in iron or steel are shown in Fig. 2.

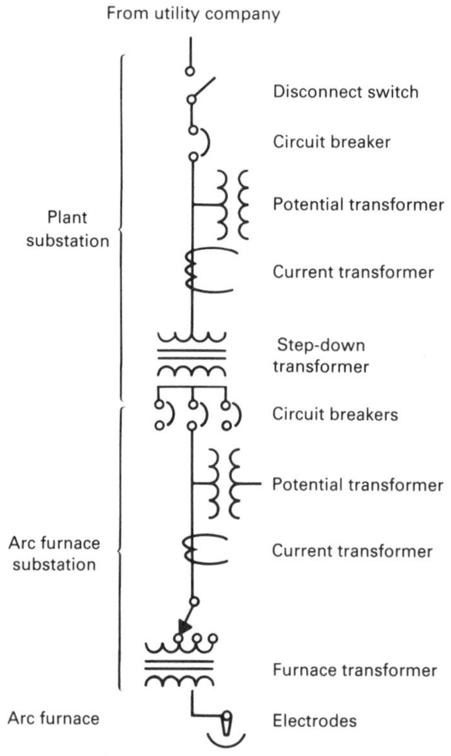

Fig. 1 Schematic of electrical network for the electric arc furnace

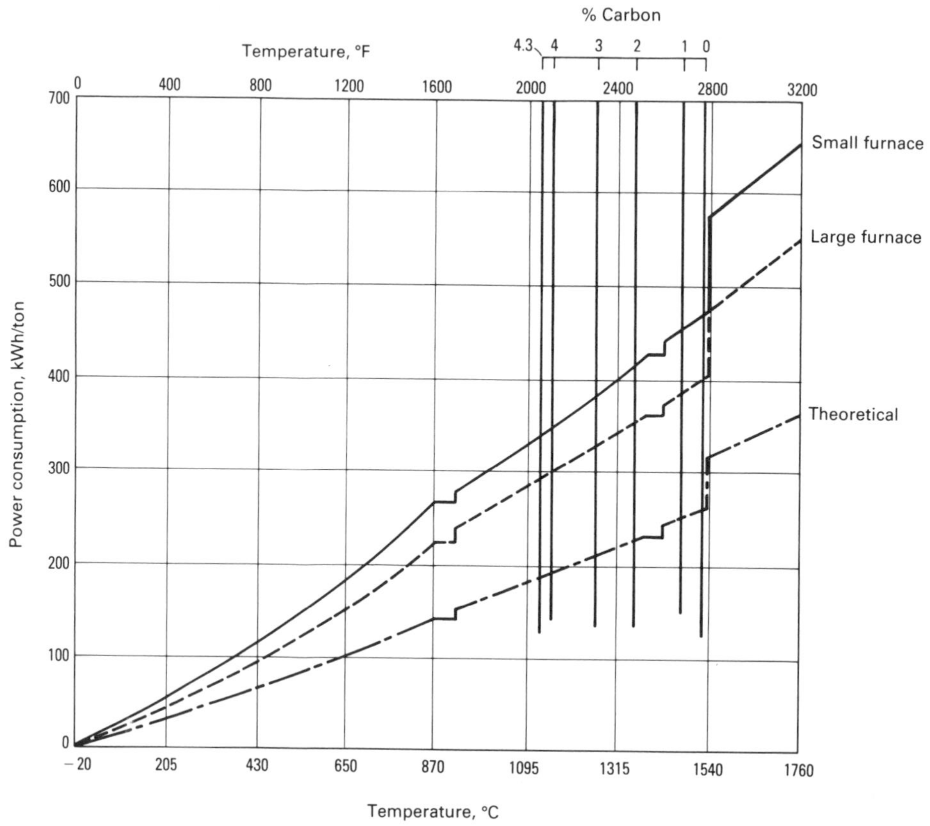

Fig. 2 Power consumed in melting iron and steel in the electric arc furnace. Values will vary depending on scrap, transformer, lining, and so on. The melting point of pure iron (0.0% C) is 1535 °C (2795 °F); of iron containing 4.3% C, 1130 °C (2066 °F).

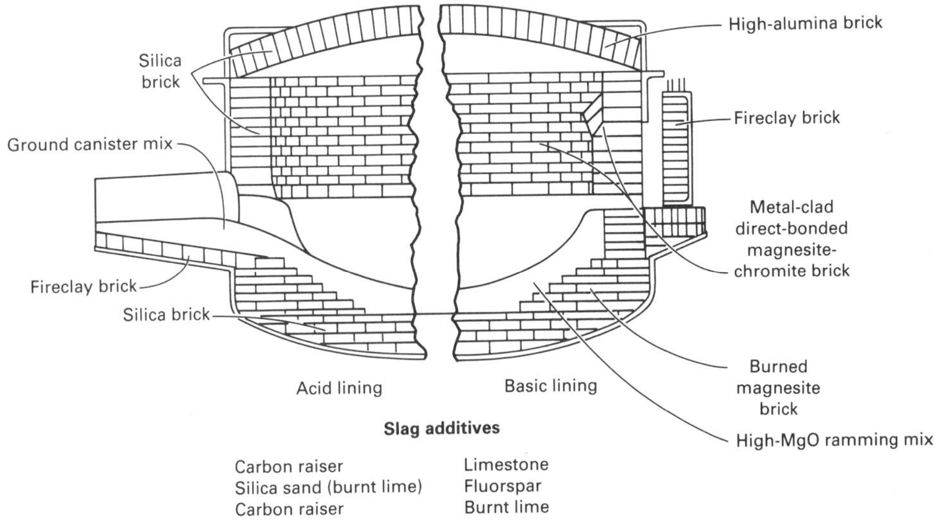

Fig. 3 Cutaway view of electric arc furnace showing the typical refractories used

Power Factor. The efficiency with which power is transferred to the melt is called the power factor. The power factor, PF, is the ratio of watts, W, divided by volt-amperes, VA:

$$PF = \left(\frac{W}{VA}\right) \cdot 100$$

One method of measuring the average power factor over a period of time requires the use of two separate meters: a watt-hour meter to accumulate the useful power and a var-hour meter to accumulate the reactive power. The var-hours (in a given time) divided by the watt-hours (in the same time) will give the tangent of the power factor angle. The cosine of this angle is the power factor. The local utility company can be contacted for this information and for recommendations on the effect on energy usage.

Arc Furnace Components

The list of equipment associated with an electric arc furnace can be extensive. This list is assembled as if a proposal for the furnace has been made.

The locations for the furnace foundation and pits or elevated platforms are selected. Factors that also must be considered include the location of the furnaces in the plant, flow and access to raw materials, storage of melt materials, ladle construction, ladle preheating, crane runways, water, air, electrical lines, transformers, laboratory, and temperature equipment.

This list is not all-inclusive. Fume and dust collection equipment can be added, as well as used-refractory and slag removal equipment. For the melting portion, a supply of oxygen, hoses, pipe for oxygen blowing, gages, valves, thermocouples, and so on, will be needed.

The furnace itself is the primary concern after the location, cement work, and electrical supply have been chosen. The electrical system for the furnace includes the transformer and the control panel. Very small furnaces (<910 kg, or 1 ton) are not used in current production schedules; heats of these sizes are normally made in induction furnaces. Furnaces are described as supplying so many tons or pounds per heat under normal conditions, but larger heats can be made by modifying the furnace. This is done by stopping off the tap hole and raising the breast on the working and charge door (if the furnace has one).

The furnace is lined with refractories that determine the type of melting practice (acid or basic) that will be used (Fig. 3). Table 1 lists various refractory furnace materials used in acid and basic melting practices. Depending on whether the process is basic or acid, slag color may be sufficient to determine the iron oxide content of the slag, as shown below:

Color	FeO, %	CaO/SiO_2
Acid slag		
Gray-black	30–40	...
Dark green	20–30	...
Streaked dark green	15–20	...
Blue green (jade green)	<15	...
Basic slag		
Black	14.85	3.33
Dark brown-gray	11.50	2.19
Light brown	1.05	1.91
Gray	1.05	1.88
White	0.56	2.01
White	0.51	2.53

The silica content of basic slag can profoundly affect the melt. Silica ties up to two to three times its mole weight of basic slag formers, thus reducing basicity and the ability of the formers to desulfurize the steel. The effect of silica content on basic slag can be summarized as follows:

- SiO_2, P, and S—acid in nature
- Silica more acid than P or S
- CaO and MgO—basic in nature
- Reaction in basic slag when silica is excessive:

$$2CaO + SiO_2 \rightarrow 2CaO{\cdot}SiO_2$$

$$2MgO + SiO_2 \rightarrow 2MgO{\cdot}SiO_2$$

$$2CaO{\cdot}SiO_2 + 2MgO{\cdot}SiO_2 \rightarrow 2(CaO{\cdot}MgO{\cdot}SiO_2)$$

The size of the furnace can vary from 450 to 91 000 kg (0.5 to 100 tons) or larger. Typical furnace data are given in Table 2.

The electrode can be graphite or carbon. The graphite electrode has greater current capacity than the carbon unit for the same size electrode. Table 3 lists some data on electrode sizes, weights, and current capacities.

The electrodes are supported on three separate arms and held to the electrode arm by a spring-clamped, pneumatically released electrode holder. On older furnaces, the electrode holder had a mechanical wedge-type clamp; on newer, larger furnaces, power-operated clamping devices are used. The electrode arms support the electrodes, holders, bus tubes, and transformer power cables. Smaller bus tubes and power cables can be used if they are air or water cooled.

The electrodes are raised and lowered by manual or automatic controls. These controls can be operated manually when the bottoms of the electrodes are raised to the roof or when electrodes are shifted. The electrodes are raised and lowered automatically during the melting and refining process (Fig. 4).

The furnace designer will make the electrode circle as small as possible to increase the speed of melting. This will intensify the heat in the delta section of the roof and can cause excessive refractory wear in this area. The use of water-cooling rings around the electrode ports can help to reduce this wear.

The electrodes should be vertical from all sightings around the furnace. When the electrodes are not vertical, the distance and arc flare on the sidewalls are not uniform, and a hot spot may cause the refractory wear to be greater in the areas closer to the electrodes. In addition, because the arc works between the electrodes and the bath, the current per phase can fluctuate (Fig. 5).

The roof structure is arched to be self-supporting and to keep the greatest distance between the arc action in the delta section. Refractory roofs start with a roof ring that is designed as a channel or an angle beam that is rolled or cast into a ring. The ring generally has a water-cooling chamber on all or a portion of its height (Fig. 6). Newer designs of water-cooled roofs are cast in gray iron except for the delta section and the exhaust and oxygen lance holes on large roofs, and they are

Table 1 Refractories as a source of slag based on typical composition

Type	Composition, %: SiO_2	Al_2O_3	Cr_2O_3	TiO_2	CaO	MgO	Fe_2O_3	Other	Softening point: °C	°F
Fireclay										
Super duty	49–56	40–44	. . .	1.5–2.5	. . .	. . .	. . .	2.5–4.0	1745–1765	3175–3210
Medium duty	57–70	25–38	. . .	1.3–2.1	. . .	. . .	. . .	4.0–7.0	1660–1685	3020–3065
Semi silica	72–80	18–26	. . .	1.0–1.5	. . .	. . .	. . .	1.0–3.0	1640–1685	2985–3065
Alumina type (high)										
50%	41–47	47½–52½	. . .	2.0–2.8	. . .	. . .	. . .	3.0–4.0	1785	3245
60%	31–37	57½–62½	. . .	2.0–3.3	. . .	. . .	. . .	3.0–4.0	1805–1820	3280–3310
70%	20–26	67½–72½	. . .	3.0–4.0	. . .	. . .	. . .	3.0–4.0	1820–1850	3310–3360
80%	11–15	77½–82½	. . .	3.0–4.0	. . .	. . .	. . .	3.0–4.0	1865	3390
90%	7½–9	89–91	. . .	0.4–0.8	. . .	. . .	. . .	1.0–2.0	1930	3505
Mullite	18–34	60–78	. . .	0.5–3.1	. . .	. . .	. . .	1.0–3.0	1850	3360
Corundum	0.2–1.0	98–99.5	. . .	Trace	. . .	. . .	. . .	0.3–1.0	2000	3630
Silica type										
Silica super duty	95–97	0.15–0.35	. . .	. . .	2.5–3.5	. . .	0.3–2.2	0.02–0.10	1680–1700	3060–3090
Conventional	94–97	0.45–1.20	. . .	. . .	1.8–3.5	. . .	0.3–0.9	0.10–0.30	1635–1665	2975–3025
Basic type										
Chrome	3.0–6.0	15–34	28–33	. . .	. . .	14–19	11–17	1.0–2.0	1290–1425	2350–2600
Magnesite	0.7–1.0	0.3–1.5	. . .	. . .	1.0–3.5	85–93	0.3–7.0	0.5–1.0	1480–1675	2700–3050
Magnesite high periclase	0.5–5.0	0.2–1.0	. . .	. . .	0.5–1.5	92–98	0.2–1.0	0.0–0.6	1595–1705	2900–3100
Chrome-magnesite	4–8	16–27	18–28	. . .	0.7–1.5	27–53	. . .	. . .	1595–1675	2900–3050

Table 2 Typical dimensions and capacities of foundry-size electric arc furnace components

Shell dimensions (inside): Diameter, m	Diameter, ft	Depth, m	Depth, ft	Melting rate, kg/h	Melting rate, lb/h	Refractory dimensions: Sidewall, mm	Sidewall, in.	Bottom, mm	Bottom, in.	Liquid-metal capacity: Normal, kg	Normal, lb	Maximum, kg	Maximum, lb	Electrode diameter, mm	Electrode diameter, in.	Transformer capacity, kVA
1.5	5.0	1.14	3.75	450–640	1 000–1 400	254	10	305	12	1 360	3 000	1 810	4 000	75	3	1 000–1 500
1.8	6.0	1.4	4.5	910–1 270	2 000–2 800	254	10	381	15	2 810	6 200	3 750	8 260	152	6	1 500–2 000
2.20	7.25	1.7	5.5	1360–1 910	3 000–4 200	254	10	432	17	4 260	9 400	5 670	12 500	178	7	2 000–3 000
2.4	8.0	1.8	6.0	1810–2 540	4 000–5 600	254	10	457	18	6 080	13 400	8 120	17 900	203	8	3 000–5 000
2.67	8.75	2.01	6.58	2270–3 180	5 000–7 000	368	14.5	483	19	9 070	20 000	12 100	26 700	229	9	4 000–6 000
2.74	9.0	2.06	6.75	2720–3 810	6 000–8 400	368	14.5	483	19	9 840	21 700	13 200	29 000	229	9	4 000–6 000
3.05	10.0	2.3	7.5	3630–5 440	8 000–12 000	368	14.5	508	20	14 500	32 000	19 100	42 000	254	10	5 000–9 000
3.35	11.0	2.51	8.25	5440–8 160	12 000–18 000	368	14.5	508	20	22 700	50 000	29 900	66 000	305	12	7 500–12 500
3.81	12.5	2.84	9.33	7260–10 900	16 000–24 000	368	14.5	533	21	29 900	66 000	39 900	88 000	356	14	10 000–15 000

Table 3 Electrode sizes, weights, and current capacities

Diameter × length, mm	Diameter × length, in.	Approximate current capacity, A	Approximate weight of electrode and nipple, kg	lb
Graphite electrodes				
102 × 1015	4 × 40	1 800–3 300	13.6	30
152 × 1220	6 × 48	3 500–5 800	37.6	83
152 × 1520	6 × 60	3 500–5 800	47.1	104
178 × 1220	7 × 48	4 400–7 500	46.3	102
178 × 1520	7 × 60	4 400–7 500	59.9	132
203 × 1220	8 × 48	5 500–9 300	62.1	137
203 × 1520	8 × 60	5 500–9 300	77.1	170
229 × 1520	9 × 60	6 700–11 300	96.6	213
254 × 1220	10 × 48	8 000–13 300	97.1	214
254 × 1520	10 × 60	8 000–13 300	118	261
305 × 1520	12 × 60	11 300–17 000	171	378
305 × 1830	12 × 72	11 300–17 000	201	444
356 × 1520	14 × 60	18 000–25 000	228	502
356 × 1830	14 × 72	18 000–25 000	273	602
Carbon electrodes				
203 × 1520	8 × 60	2 500–4 500	79.8	176
254 × 1520	10 × 60	3 100–6 300	125	275
305 × 1520	12 × 60	4 500–7 900	178	392
356 × 1520	14 × 60	5 400–10 000	240	528
356 × 1830	14 × 72	5 400–10 000	293	647

cooled by a network of steel pipes running throughout the structure. The delta section and electrode ports are usually constructed of a high-alumina refractory brick, rammed material, or precast shapes.

The roofs of all modern electric furnaces are designed for top charging. The roof and electrodes are raised several inches above the top of the furnace shell and swung to one side. The charge is dropped into the furnace, and the roof is swung back into place. The raising of the roof is raised by electrohydraulic systems. Schematics of roof-lifting devices are shown in Fig. 7.

Tilt Control. The electric furnace is tapped out by tilting the furnace into a position in which the metal and slag will run out of a hole above the slag line in the furnace wall down into a spout and into a ladle. Tapping of the furnace requires that the roof and electrodes remain attached to the furnace shell; therefore, safety locks or stops are found welded to the shell and roof ring. The tilting mechanism can be mechanical for very small furnaces or can be motor

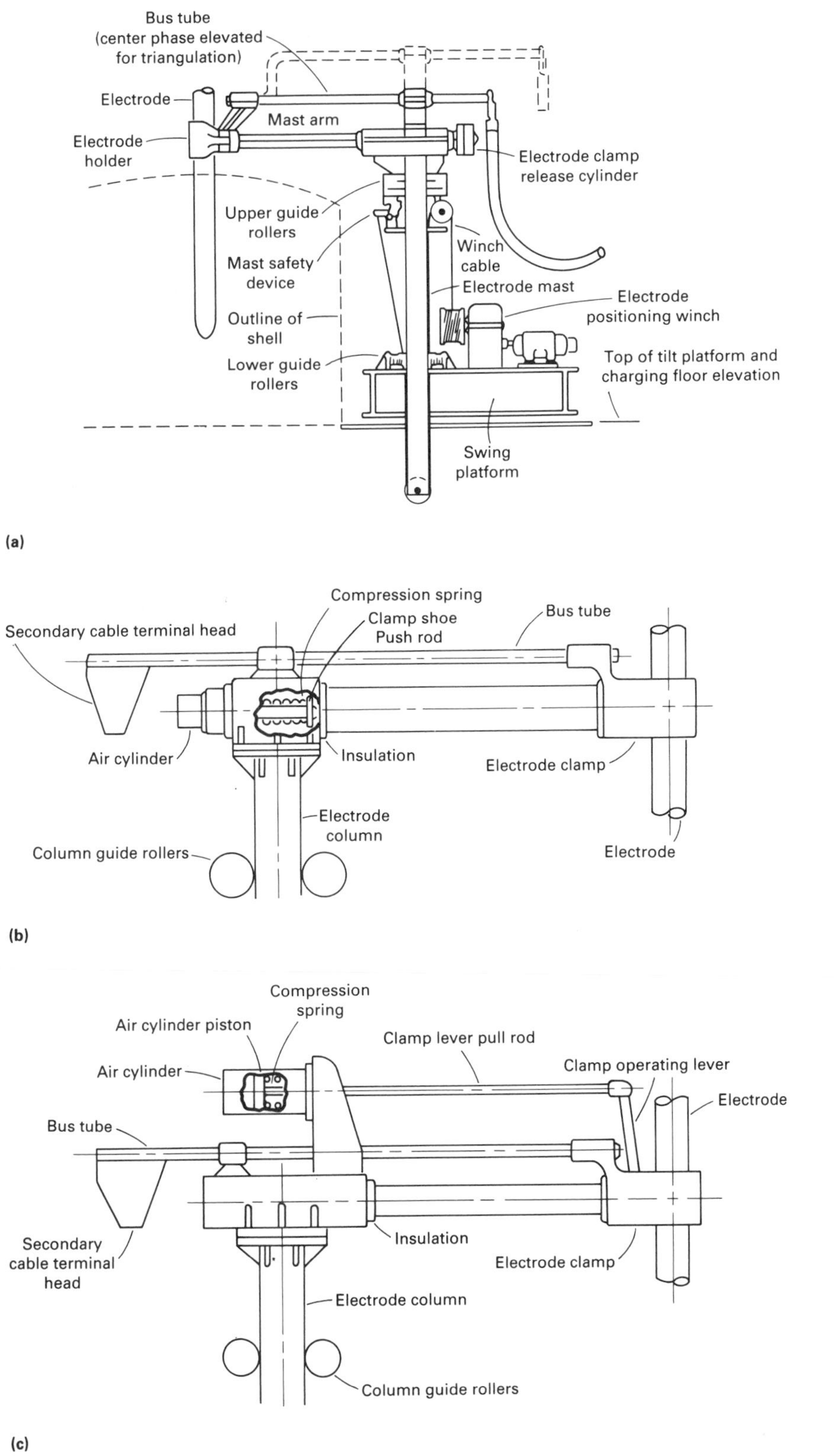

Fig. 4 Schematics of apparatuses for raising and lowering electrodes. (a) Arrangement of electrode and mast. (b) Electrode arm arrangement with power-operated electrode holder in which clamp shoe is held in place by a push rod mechanism. (c) Typical arrangement of electrode arm using power-operated electrode holders; pressure on clamp shoe is exerted by spring-operated pull rod.

driven (either by a screw pitman or a rack-and-pinion mechanism) or hydraulic, and it has rockers attached to the furnace and rails attached to the foundation. Emergency equipment is usually provided to pour the heat from the furnace in case of a power failure.

The furnace shell is typically made of carbon steel sections that are rolled and welded to the diameters listed in Table 2. The diameter and height define the furnace size. The furnace bottom is formed and welded to complete the pot. The shell will have door openings cut into it and a cooling frame welded around the door. The door-lifting supports will also be welded onto the shell. The furnace rockers and other lifting equipment attachments are welded onto the shell. With the rockers and door aligned, the spout opening is cut out of the shell and the spout frame attachment welded onto the shell. To ensure that the furnace is properly grounded, straps, rods, or mesh is attached to the bottom on the inside and ground on the outside prior to lining of the furnace. The choice of refractories will depend on the melting practice selected (Fig. 3 and Table 1). New furnaces being built in the 1980s have seen the use of water-cooled panels in the furnace sidewalls above the slag line (Fig. 8).

Spout and Tap Hole. The length and shape of the furnace spout have been a source of controversy. The melter and metallurgist want the shortest possible spout, while the design engineer is confronted with a network of electrodes, crane cables, pit dimensions, and ladle and tap hole dimensions needed for positioning the ladle in order to obtain a fast and smooth tap from the furnace, using an acceptable spout.

New designs and changing the locations of tap holes may reduce the physical and mechanical problems encountered with the spout. One of these alternatives is the side slide gate spout (Fig. 9), and another is eccentric bottom tapping with the use of a slide gate (Fig. 10).

Tap-Out and Back-Slagging Pits. The design of the furnace pit should include sufficient depth for the slag boxes, and along with this depth, provision must be made for preventing ground water from entering the pit. Sump pumps are needed to minimize the presence of water and to ensure an environment that is safe from steam explosions. The length and width of the pit should be sufficient to allow movement of the ladle; this is necessary if the furnace slag is to be kept out of the ladle so that ladle metallurgy or secondary refining can be done following tap out.

Water-Cooling System. Water cooling of the furnace is becoming common practice. Before this innovation, water cooling was used on:

- Electrode holders, arms, and clamps
- Roof rings and electrode rings in the delta section
- Around the door
- Cooling panels positioned on the outside of the furnace to cool the furnace wall refractories

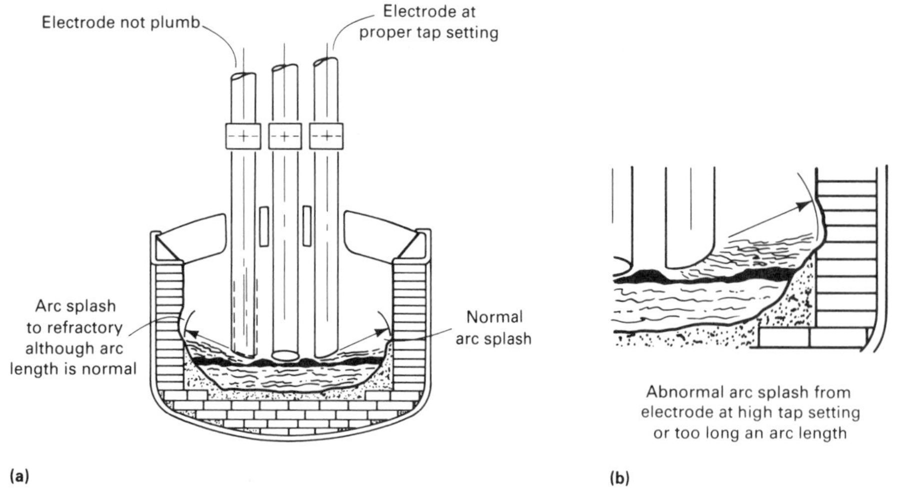

Fig. 5 Effect of arcs on furnace refractories due to electrodes not being vertical (a) on section left of center and (b) running too high a tap after meltdown

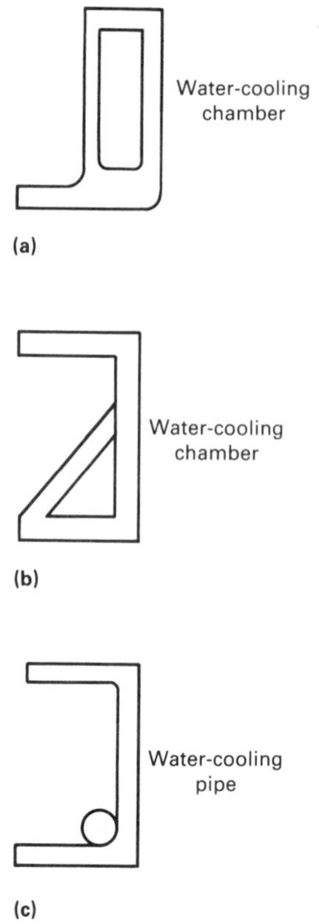

Fig. 6 Typical roof ring with cooling face. (a) Designed for use with special-shaped or cut starter brick. (b) Designed for use with skewback starter brick. (c) Designed for use with special-shaped or cut starter brick

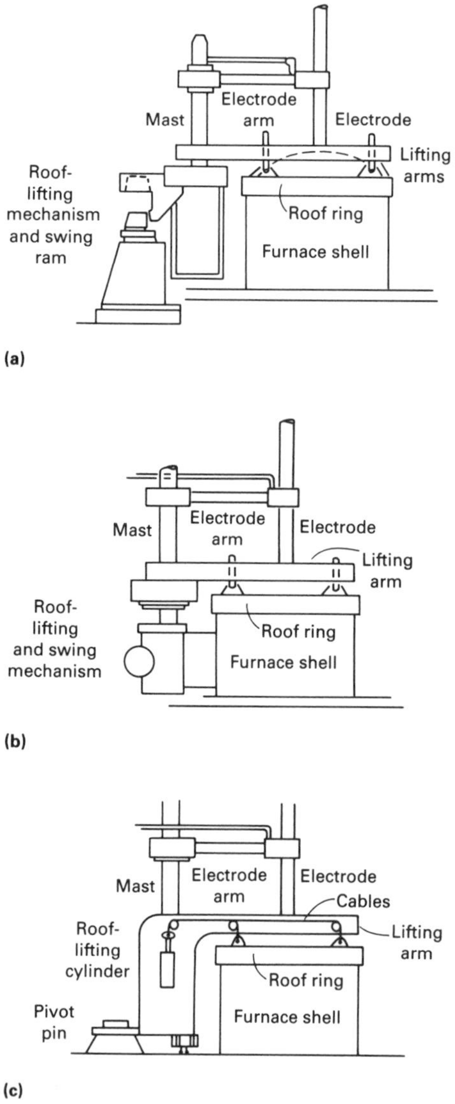

Fig. 7 Roof-lifting devices and swing mechanisms used on top-charging electric furnaces. (a) Roof-lifting device and swing arm mounted separately from furnace. (b) Roof-lifting device and swing mechanism attached to furnace. (c) Roof lifting by cylinder and cable/swing system by rotating roof with a kingpin arrangement

Preheat and Furnace Scrap Burners. Preheating scrap with natural gas or flue gases helps reduce the electrical cost of melting a charge. On the electric furnace, this preheating may also help stagger furnace operation to keep electrical demand peaks to a minimum.

Furnace oxygen/natural gas burners can help melt the charge and reduce electrical charges and electrode usage while shortening the melt time from tap to tap. Improper use of these burners can cause excessive oxidation of the heat and the electrodes and can reduce the life of the refractory lining.

Fume and dust collection at the electric furnace has been a required practice in melting during the last 25 to 30 years. The complexity of the fume and dust collection equipment varies from a simple fourth-hole connector to an entire separate room for the melting furnace in which the fumes and dust

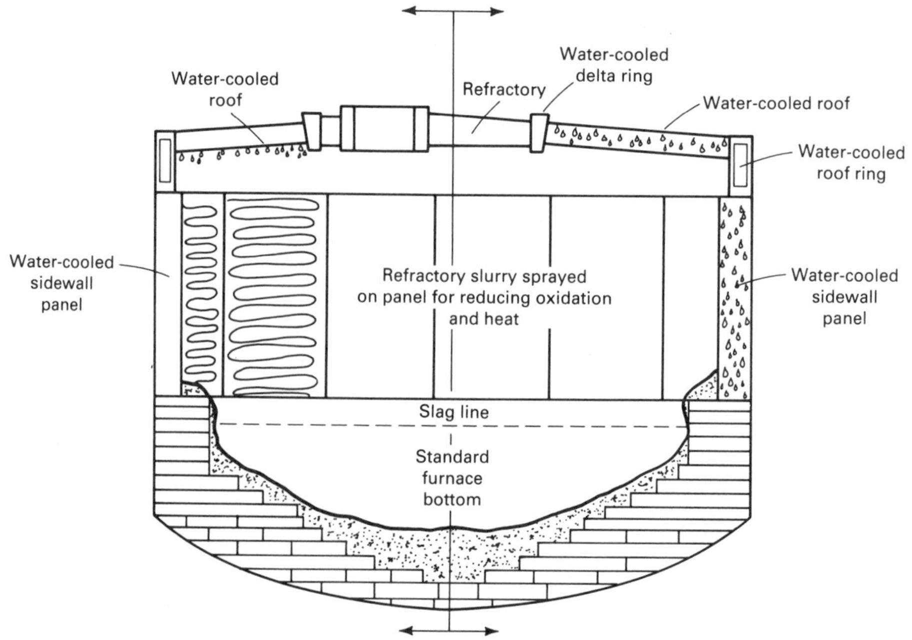

Fig. 8 Furnace design incorporating a water-cooled roof and upper sidewalls. The location of the water-cooling passages depends on manufacturer's suggestions.

Fig. 9 Furnace slide valve used to hold back slag

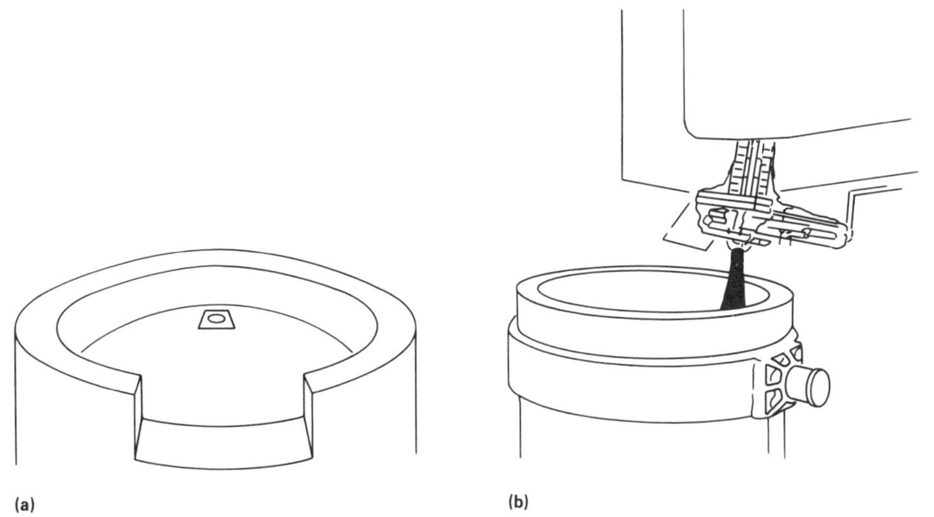

Fig. 10 Eccentric bottom tap hole furnace (a) with slide gate attached to furnace bottom (b)

are collected and where the furnace is tapped into a ladle that is brought to the furnace and then removed through a dust protection door. Figure 11 shows typical dust and fume collectors.

The charge bucket is sized to fit within the refractory walls of the furnace, and is open at the top with a lifting bale that folds back or is removable from two trunions. The bottom of the charge bucket can be designed to drop the charge in one of several ways:

- The clam shell bottom is actuated with a trip lever
- The orange peel bottom, which is set in a stand to fold the leaves back and is tied with a rope, burns off when the charge bucket is placed into a hot furnace
- The preheat charge bucket uses a trip lever and after being charged with scrap is preheated to some temperature above 260 °C (500 °F) to up to 815 °C (1500 °F)

Figure 12 shows a typical charge loaded in two different types of charge buckets.

The operations control panel has seen many improvements in the last 20 to 30 years—from electrodes that were hand operated to electrode control motors to the current solid-state computerized technology. The heart of the melting operation of the furnace comes from this panel. The panel is usually set into the wall of the transformer vault next to the furnace where the operator can see his controls and observe the furnace operation. As shown in Fig. 13, the panel will typically have the following connections and instruments:

- A, clock, for time of day and logging actions taken at the furnace during the heat
- B, arc lamps, one for each electrode. The melter observes the intensity of these lamps indicating the arc action; from the lamp intensity and flickering, he can tell if the electrodes have a proper arc or if one or all are dipping into the bath
- C, megawatt meter indicating the total power being used by the three phases during melting
- D, megavar meter indicating total reactive power used during melting
- E, kilowatt-hour meter showing electrical energy consumed during melting
- F, voltmeter that reads actual phase-to-ground and phase-to-phase voltage; a se-

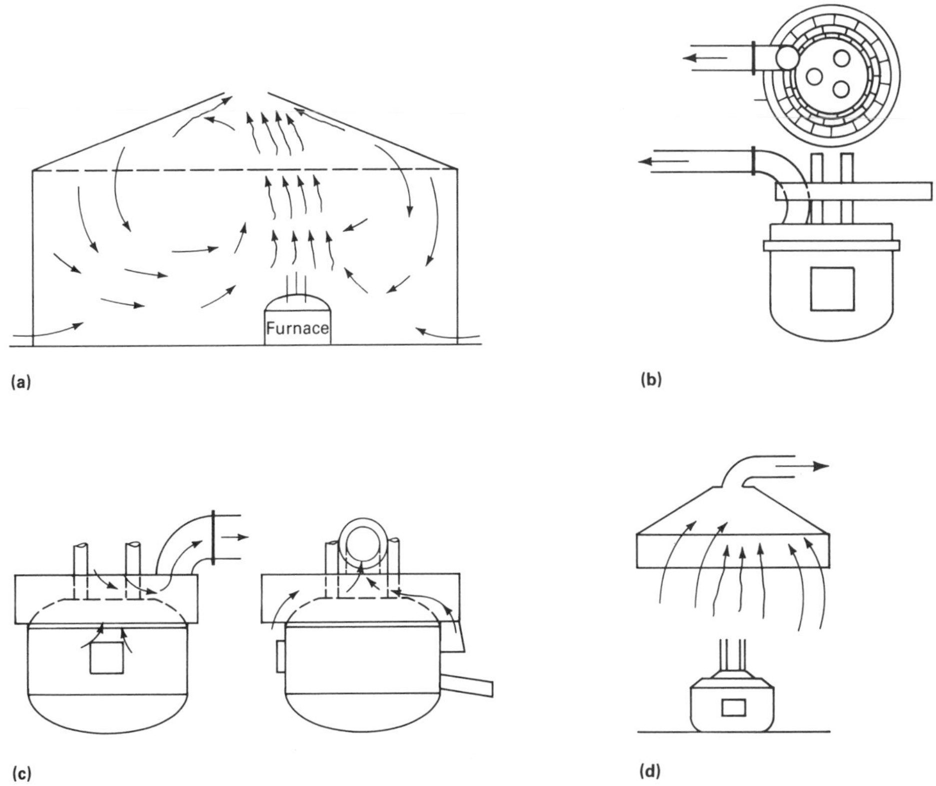

Fig. 11 Four methods of dust and fume control in electric furnaces. (a) Prepollution control ventilation for dust and fume removal. (b) Direct furnace dust and fume collection (both front view and top view are shown). (c) Total furnace hood for fume and dust collection. (d) Canopy hood for fume and dust collection

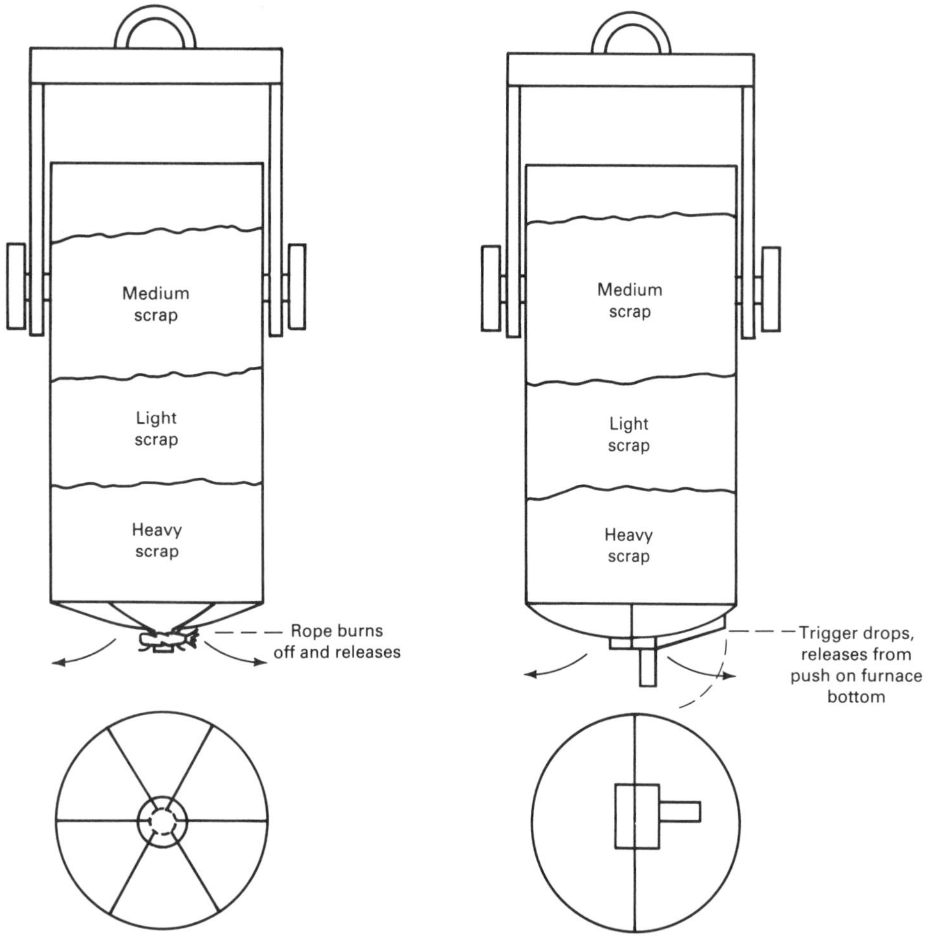

Fig. 12 Schematics of foundry charge buckets. (a) Orange peel bottom. (b) Clam shell bottom

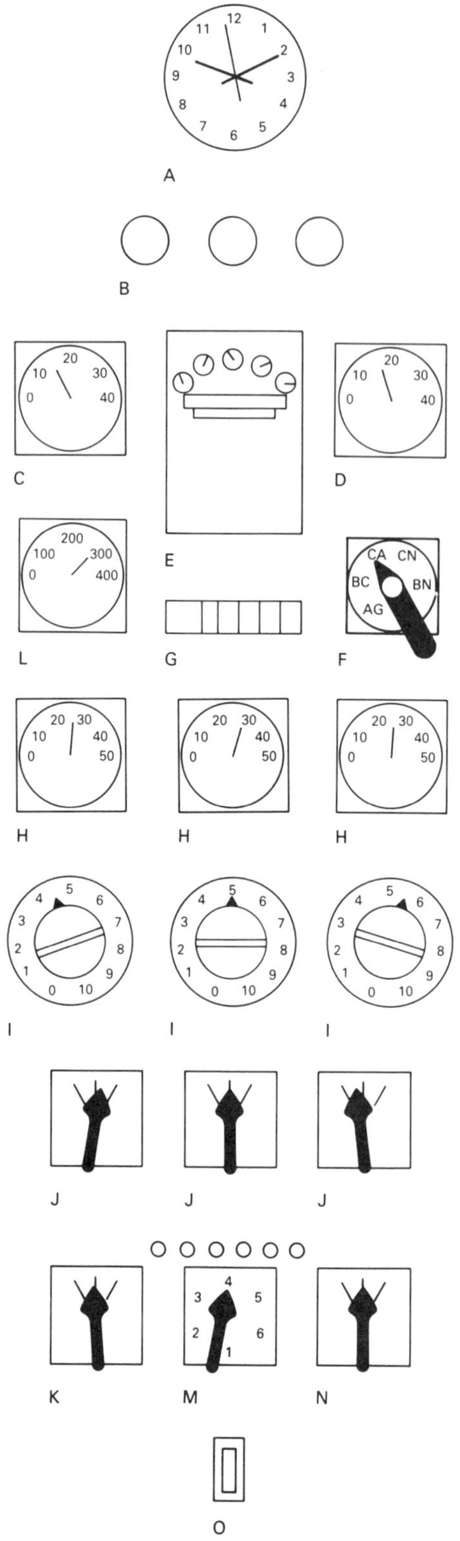

Fig. 13 Typical components of an operations control panel. See text for discussion.

lector switch allows the operator to control phase-to-ground or phase-to-phase voltage

- G, kilowatt-hour register for a readout of data on kilowatt-hour meter, which the operator should reset to zero after each heat
- H, three ammeters connected to show the secondary current in amperes for each phase
- I, three rheostats for regulating the distance between the electrode tip and the charge material or bath
- J, three switches for the individual electrode raising and lowering control
- K, gang-control switch for raising and lowering the electrodes
- L, voltmeter switch with multiple positions
- M, tap-change switch with four to six positions that connect the wattmeter across the secondary phases to ground; newer panels may have additional features
- N, power circuit breaker that disconnects power to the furnace by remotely stopping the high-voltage furnace power supply; this must be open during off-load tap changing
- O, control switch that cuts all power to the furnace controls; this switch is turned off when the furnace is not in operation or when maintenance is being performed

Ladles may be part of the equipment assigned to the electric furnace, or they may be assigned to the pouring crew or the mold department. Nevertheless, the heat is not finished until the metal is removed from the furnace.

In foundries and steel mills, the ladles are refractory lined—usually with an alumina lining. More than 72% alumina is used if a clean steel low in inclusions due to ladle refractory is desired. In one application, magnesite or olivine is used in situations where the metal and slag will attack the refractory, as in Hadfield manganese steel.

In foundry operations, the type of ladle used varies. For holding ladles, bottom-pour, teapot, and lip-pour ladles are used. For pouring ladles, bottom-pour, teapot shanks, lip-pour shanks, and drum ladles are used. The ladle linings have changed from refractory lining to the one-heat type with insulating linings made of insulating board or of one-piece designs on shank ladles to eliminate preheating. The repeated use of the refractory reduces cast steel cleanliness and quality, but the board ladles have overcome this problem (Fig. 14).

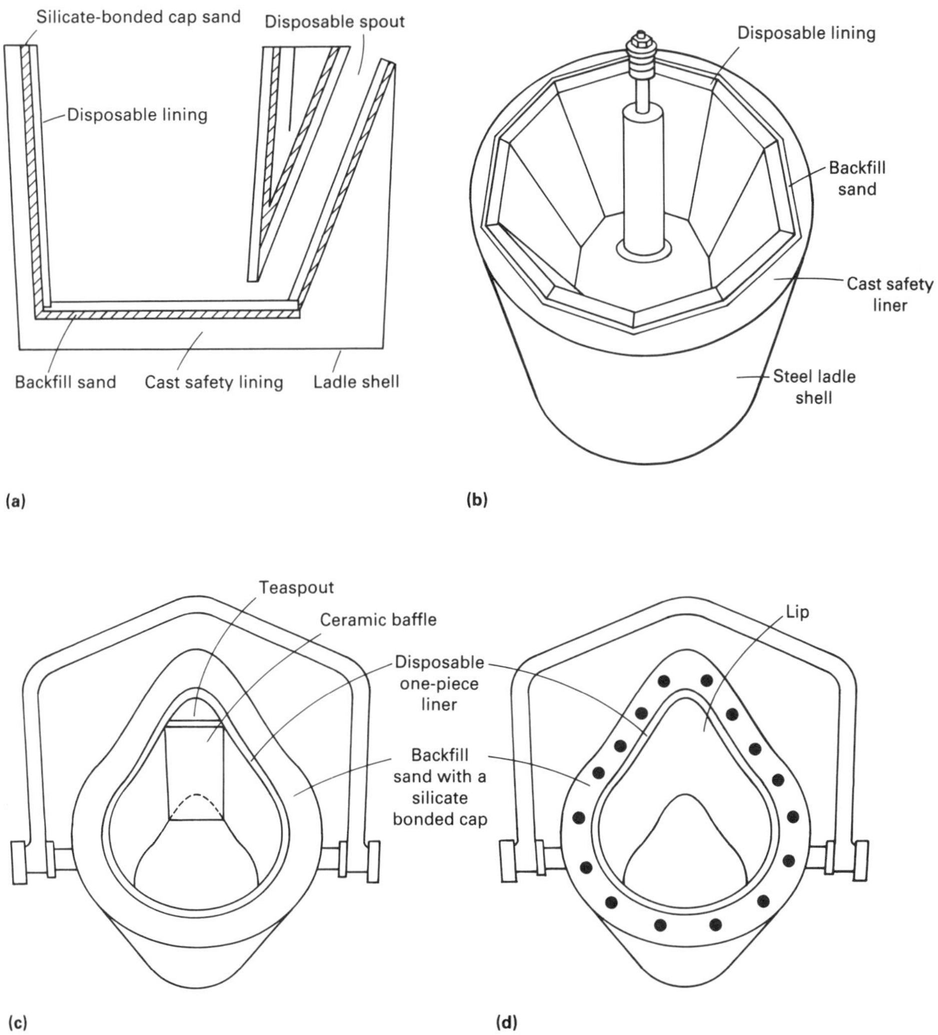

Fig. 14 Typical types of insulating ladles with cold disposable liners used in foundry operations. (a) Teapot. (b) Bottom pour. (c) Teapot shank. (d) Lip-pour shank

Precision Weighing and Laboratory Testing Equipment

Scales. Charge and alloy scales are required for weighing the charge and alloy additions. The charge scale can be the standard beam scale type or the newer load cell type. These scales should be checked daily or weekly to ensure accurate weighing of the furnace melt. The alloy scale can be a small spring or a large balance beam type capable of measuring in pounds and ounces. These should be periodically checked for accuracy.

Quality Control Testing Equipment. The demands for faster and more accurate chemical analysis methods have spawned advancements in the spectrographic, x-ray defraction, and scanning electron microscope analysis of metals. Carbon, sulfur, alloy, and residual analysis is quick and accurate. Gas analysis, oxygen probes, and nitrogen and hydrogen analysis equipment is within the reach of the average melt shop to assist in quality control. Temperature control immersion thermocouples that are easy to use and recorders that are accurate to within 3 or 6 °C (5 or 10 °F) are available.

Scrap and Alloy Storage

Purchase scrap, shop returns, and alloys should be stored in a covered area. Wet scrap will produce small errors in weight and increase the gas content of the melt. Purchase scrap should be stored by size and the density of its bulk. Scrap should be stored according to specification or alloy content. Alloys and melt additions, that is, slag formers, fluxes, and deoxidizers, should be organized by usage.

Acid Melting Practice

The melting operation from charge to tap out will vary, depending on the type of charge materials and alloy additions used to produce steel in the acid melting practice. The furnace lining will be made from silica or alumina refractories. These refractories determine the type of slag generated during melting, along with the dirt or silica sand included in the charge. The acid practice is a fast oxidizing operation that does not remove phosphorus or sulfur from the melt. Therefore, charge materials with low phosphorus and sulfur levels should be selected in order to meet the final chemical analysis of the steels produced.

The acid melting practice is selected for the following reasons:

- Flexibility in producing small quantities of steel to various chemical compositions
- Lower refractory costs due to the initial cost and the refractory life
- Lower power consumption as compared to a two-slag basic practice
- Greater tonnage per unit of time

Acid Steelmaking

Steelmaking practices can be classified as partial-oxidation or complete-oxidation operations. These practices are differentiated by the amount of carbon removal during oxidation compared to the final desired carbon content of the melt. The advantages of acid melting practices utilizing partial oxidation are:

- Rapid steel production
- Good refractory life due to lower iron oxide content in the slag
- Lower power input
- Lower electrode consumption
- No recarburization required

The disadvantages include:

- Insufficient removal of gases
- Occasional poor fluidity due to oxides
- Occasional poor mechanical properties

The advantages of acid melting practices utilizing complete oxidation are:

- Low gas content
- Good fluidity
- Good mechanical properties
- Elimination of nearly all silicon and most of the manganese to a uniform base metal to which additions can be calculated more closely

The disadvantages are:

- Slower steel production
- Shorter refractory life due to higher iron oxide content in the slag
- Higher production cost

In partial oxidation, the charge carbon content is 0.05 to 0.15% above the finishing carbon content. If the carbon is above this level, a short blow with the oxygen lance is required. No slag-forming materials are added to the bath or the charge, and a fast melting practice is used. The rust and scale in the charge, along with the oxidation of the charge during melting, are sufficient to produce a light boil. When melting is complete, the carbon content will be about 0.25 to 0.35%, and the silicon content will be 0.15 to 0.20%; the actual value depends on the amount of oxidation from the charge and the oxidation during meltdown. By the time the laboratory report on the melted down carbon and manganese is prepared, the carbon content will have dropped another two or three points, and by the time the tap temperature is reached, the carbon will have dropped so that any alloy additions will bring the carbon content to the desired finishing range.

Slag additions are seldom required. The slag will be 3 to 5% of the charge weight because of silica dirt and refractory erosion.

Foundries using partial oxidation have minimal mechanical test properties to meet. They use a good grade of scrap, and their primary concern is the high production of molten steel from their furnace capacity. This practice is losing favor because of quality standards and product liability laws affecting the finished product.

Complete Oxidation. Most steel foundries using an acid melting method employ the complete-oxidation practice. This is because nearly all steel castings must meet minimum mechanical test requirements, along with magnetic-particle, visual, and radiographic inspection.

In this process, rapid melting is necessary to start a boil during meltdown, followed by oxidation with the use of the oxygen lance. Some melters add iron ore or mill scale to the charge to obtain a partial boil during meltdown. This boiling action or oxidation is accomplished with the following reaction:

$$C + FeO \rightarrow CO$$

Slag-forming materials are not added to the charge or the bath. Slag is formed from silica and dirt on the charge and erosion from the furnace refractories. A fluid black slag should be obtained by prudent additions of mill scale to ensure consistent silicon and manganese contents at meltdown.

Meltdown Period. The meltdown carbon content should be 0.20 to 0.40% above the aimed carbon level. The meltdown carbon level will drop 20 to 40 points after oxygen lancing to about 0.20% C. The silicon and manganese contents at meltdown will be about 0.13 to 0.20% and 0.25 to 0.35%, respectively. After oxygen lancing, the silicon content will be 0.02 to 0.10%, and the manganese content will be 0.07 to 0.15%. A vigorous boil (oxygen lancing) will eliminate carbon, most of the hydrogen, and small amounts of nitrogen from the bath; the elimination of oxygen depends on the carbon-iron oxide reaction during this period. The slag should be green when the boil is completed. After the boil, silica sand can be added to increase the slag volume and to reduce its iron oxide content. Lime is sometimes added to the slag to reduce some of the iron oxide in the slag. Most slag formers are added to the bath during the refining period. Lime is sometimes added to thin the slag.

To minimize the loss of oxidizable elements—silicon and manganese—during melting, the percentages of these elements in the charge should be low. Overoxidizing of the heat should be avoided, and alloying elements should be added immediately before tap out.

The Refining Period. After decarburization, the boil will slow to a gentle roll, and recarburization and slag additions can be made if required. The recarburizers may be low-silicon, low-phosphorus pig iron, sorel pig iron, high-carbon steel, broken electrodes, petroleum coke, and so on, and occasional dipping of the carbon electrodes. The addition of carbon gives a carbon boil:

$$C + FeO \rightarrow CO + Fe$$

This helps keep the gas content down. Small additions of ferrosilicon and ferromanganese or silicomanganese are made to the bath at the end of this boil to hold the bath at the desired carbon level. This is known as a block. The heat should be tapped as soon as possible after this procedure and after the addition of any required alloying elements.

Deoxidation is required for preventing the reaction of dissolved oxygen and carbon in producing gaseous products such as CO or CO_2, into which hydrogen can find a nuclei and into which it can accumulate and form pinholes in the cast product. The first of these additions is silicon in the form of ferrosilicon or silicomanganese and then aluminum, a powerful deoxidizer. Special deoxidizers such as titanium and zirconium help tie up nitrogen, while complex deoxidizers containing silicon, aluminum, calcium, and other elements help modify inclusions.

Melting Heats and Raw Materials

The preceding discussion is a general description of the acid melting process. This section will discuss the heat and the materials used.

The charge will consist of 30 to 50% shop returns (heads, gates, and scrap castings) and 50 to 70% purchase scrap (plate, clean punchings, railroad wheels, and so on). The weight of each group of materials is recorded on the furnace log. The charge distribution into the furnace will normally be heavy scrap and casting on the bottom, followed by light scrap and purchased plate. Nonoxidizable materials such as nickel and molybdenum can also be added with the scrap charge. Carbon raiser additions to the furnace, if desired, are normally done before the charge is placed into the furnace. The roof is swung into place after an inspection of the furnace top to ensure that the scrap does not contact the roof ring and the roof is then locked down.

Melting should proceed as soon as the charge is in the furnace. The No. 1 tap will normally be used for meltdown. To protect the roof from excessive erosion, the No. 2 tap can be used to start the electrodes into the charge, followed by a change back to the No. 1 tap. This procedure will reduce roof wear in the delta section. A pool of metal will form on the bottom of the furnace as the electrodes burrow through the charge. Once the electrodes have reached this pool of metal, the remainder of the charge will be melted by radiation from the pool and the arc splash off of the electrodes. Rust and any iron ore added to the charge will help in forming a watery slag with the sand, eroded refractory, and slag remaining from the previous heat to form an oxidizing slag on the pool of metal produced from the melting portion of the heat. When the charge is completely melted, the tap is changed to No. 2 or lower to reduce arc splash on the furnace walls and roof. A laboratory sample is taken to determine the carbon level of the bath in preparation for the blow-down, or oxygen lancing.

The oxidation following meltdown should not begin until a temperature of at least 1540 °C (2800 °F) is reached. When heats containing chromium are to be made, temperatures of 1595 to 1620 °C (2900 to 2950 °F) should be reached before lancing to minimize the loss of chromium to the slag, from which it cannot be recovered. Depending on the amount of carbon, the bath temperature, and the amount of oxygen present in the bath, a decrease of 0.08 to 0.10% C will occur for each minute of oxygen injection. Generally, a steel pipe is used as the lance (ceramic-coated steel pipes are also available). The pipe must be clean and free of grease and oil. The blow time is recorded on the furnace log. After the oxygen blow, a lab sample is taken, and the heat is prepared for finishing and tap out.

When the lab test is returned to the melter, adjustments to the carbon level will be made by recarburizing or by additional lancing to the required carbon level. A final check on carbon is made, and the heat is blocked with ferrosilicon. This silicon addition will bring the bath silicon content up to about 0.10% or slightly higher to stop any remaining carbon boil. Final additions are made to the furnace to adjust the bath to the final chemistry. These additions may include ferrosilicon, ferromanganese, ferrochromium, ferromolybdenum, silicomanganese, nickel, or other alloys. Following these additions, the bath is heated for a few minutes to ensure that the alloys have melted and that the desired tap temperature has been attained.

The bath is quickly tapped out into a ladle and the refractory is preheated to minimize temperature loss and skulling from the refractories. Any alloy additions required to meet the final chemistry can be added to the ladle before tap out. The furnace should be tapped so as to hold back the furnace slag and to give a clean full metal stream. The furnace spout should be smooth, clean, and dry in order to obtain a full metal stream without a spray into the ladle.

Deoxidation of the steel in the ladle can be accomplished in several ways. One method is to add the aluminum as ingots to the bottom of the ladle just before tap out with other ladle additions. In another method, a steel bar with stars or doughnut ingots attached to it is plunged into the ladle when the ladle is one-half to three-quarters full. Small additions of less than the weight of an ingot are made with aluminum shot or wire clippings weighed into a bag and tossed into the metal stream at tap out. The amount of aluminum used for deoxidation ranges from 1.0 to 1.5 kg/metric ton (2 to 3 lb/ton); the higher carbon heats require less aluminum, while the lower carbon heats require more aluminum because of their higher oxygen levels in the steel. All alloy additions should be recorded on the furnace log with the wattmeter reading, along with any comments on the heat.

Other deoxidizers are often used to provide a complex deoxidation of the steel. These include titanium and zirconium to tie up nitrogen plus calcium and barium to modify inclusions. The type and amounts of these supplemental deoxidizers are determined according to the need and choice of each foundry.

After tap out, the furnace should be inspected for slag line and bottom erosion and should be patched with silica sand or a mixture of low-grade alumina and silica. This prepares the furnace for the next heat.

Loss of Alloys. In the acid melting process, alloying elements that are oxidizable are not completely recoverable, and care must be taken to estimate their loss when the additions are made to the bath and the ladle, as well as the loss while the heat is being melted. This loss of alloying element can be compensated for based on the data given in Table 4.

Table 4 Alloy loss in acid melting practice

Element	Loss from charge after blow, %	Loss when making addition, %	Required element weight to be added as furnace or ladle addition to make up desired weight, %
Manganese	80–85	20–30	125
Silicon	90–95	5	95
Nickel	0.0–2	0.0–2	100
Chromium	35–50	0.0–3	For less than 1% Cr, add 150% to charge; over 1% Cr, add 1/3 after blow for a total of 120% for >4% Cr alloy.
Molybdenum	0.0–2	0.0–2	100
Niobium	0.0–2	0.0–10	100
Copper	0.0–2	0.0–2	100

Alloy Additions. The typical analysis of alloy additions made to the electric furnace is shown in Table 5. Because this is an incomplete list, alloy manufacturers should be consulted for an expanded list and additional detail with regard to residual elements, material size, and packaging.

Basic Melting Practice

The melting operation from charge to tap out will vary, depending on the type of

Table 5 Analysis of alloy additions for the electric furnace

Alloy(a)	Composition, % C	Mn	Si	Cr	Ni	Mo	Cu	Nb	Al	Ca	P	S	V	Fe	Ba
50% ferrosilicon	...	...	47–51	...	...	...	...	...	1.25 max	0.20 max	0.04	...	...	...	...
75% ferrosilicon	...	...	74–79	...	...	...	...	...	1.25 max	0.60 max	...	...	...	...	...
Standard ferromanganese	7.0	78 min	1.0 max	...	...	...	...	...	...	...	0.35 max	...	...	...	...
67% silicomanganese	1.50–2.00	65–68	16–18.5	...	...	...	...	...	...	...	0.20 max	...	...	...	...
Pig iron	4.0–4.5	0.30–0.50	1.00 max	...	...	...	...	...	...	...	...	...	...	...	...
Low-carbon ferrochromium	0.025 max	...	2.00 max	67–72	...	...	...	...	...	...	...	...	...	...	...
Molybdenum oxide	...	...	4–7	...	...	55–60	0.70 max	...	Fe + Al	...	...	...	...	...	...
Low-carbon ferromolybdenum	0.12 max	0.25 max	1.5 max	...	...	65 min	0.70 max	...	...	...	...	...	...	...	...
High-carbon charge chromium	5–7	...	2–5	65–70	...	...	...	...	...	...	0.03 max	0.07 max	...	...	...
Standard ferromolybdenum	2.5 max	0.25 max	1.5 max	...	...	60 min	0.80 max	...	...	...	...	...	...	...	...
Electrolytic nickel	...	...	...	...	99.9	...	...	...	...	...	...	...	...	...	...
Electrolytic manganese	0.005 max	99.9	0.001	...	...	...	...	...	0.002 max	...	...	...	...	...	...
Ferroniobium	0.10	...	1.3	...	...	...	...	55.0 + 3.0Ta	...	...	...	...	...	...	...
Copper	...	...	...	...	...	...	99.7	...	...	...	...	...	...	...	...
Ferrovanadium	0.20 max	...	1.50 max	...	...	...	...	...	...	...	...	...	55–60	...	...
Armco iron	...	...	...	...	...	...	...	...	...	...	...	...	...	98.0	...
Manganese calcium-silicon	...	14–20	49–55	...	...	...	...	...	1.75 max	14–20	...	...	...	...	...
Calcium silicon	...	...	60–65	...	...	...	...	...	...	28–32	...	...	...	5.0 max	...
Calsibar® alloy	1.00 max	...	57–62	...	...	...	...	...	1.5 max	14–17	0.035	...	...	7.0 max	14–18
Hypercal® alloy	...	...	38–40	...	...	...	...	...	19–21	10–13	...	...	...	...	9–12

(a) Aluminum is not included in this list, but it is the primary deoxidizer used in steelmaking. The melter needs to understand that there are various grades of aluminum. For example, primary aluminum in the form of ingots and shot has a purity of 99%, while Grades I, II, III, and IV can contain 86–96% Al plus other nonferrous alloy elements. Care should be taken to purchase and add this deoxidizer on the weight contained and not to buy on lowest cost basis.

charge materials and the alloy additions used to produce steel in the basic melting practice. The lining of the furnace bottom and sidewalls is made of basic materials, such as dolomite, magnesite, chrome magnesite, bricks, and/or rammable materials. The roof is generally made of high-alumina brick, and the delta section is made of rammable material. Silica materials or materials high in silica will cause melting and refining problems and will attack the basic lining materials. The slag additions for basic melting are lime and high-lime-containing minerals.

The basic practice can remove phosphorus from the steel if a double-slag practice is employed, but the results and the percentage of removal have been disappointing. Sulfur removal with the basic practice can be very successful if a very high lime-to-silica ratio (2.5 to 3.5) is maintained in the refining slag or if a synthetic desulfurizer is used at tap out in the ladle. Because of the ability of slag to remove phosphorus (by oxidation) and sulfur (by reduction), lower-quality scrap can be used in the charge. A great portion of the oxidizable elements can also be recovered from the lime slag during refining.

For these reasons, the basic melting practice is the preferred method for the production of highly alloyed steels, stainless steels, high-manganese steels, and specialty alloys. However, the basic steelmaking practice will be slower than the acid melting practice when its abilities to reduce the phosphorus and sulfur from the steel and to reduce alloys from the slag are employed to capacity.

Basic Steelmaking

The basic steelmaking practice is divided into two parts: the oxidizing period and the reducing (refining) period. Both periods are important because of the reactions that take place during melting. Of the two, the reducing portion of the melt is of greater importance.

The charge for the foundry practice will consist of 30 to 50% shop returns, such as heads, gates, and casting scrap; the remainder will be 50 to 70% purchase scrap. Along with these materials, a carbon raiser and lime or limestone are added to the charge. The amount of carbon raiser added should be sufficient to increase the meltdown to 0.20 to 0.45% C over the finish carbon level. In addition to the charge to the furnace, a small amount of limestone or lime is normally added to the furnace bottom along with the recarburizer. (It is sometimes considered preferable to add lime to the middle or top of the melt.) Heavy scrap is first placed on the bottom, followed by light scrap and finally medium scrap on the top of the charge. Every effort is made to get the entire charge into the furnace at one time to reduce the time loss caused by delays in back charging the required weight into the furnace. When stainless steels are being produced, a greater portion of the chromium addition is held out of the charge to conserve chromium, to help keep carbon levels down, and to cool the bath after the carbon boil when temperatures may reach 1870 to 1925 °C (3400 to 3500 °F) to achieve concentrations of 0.02 to 0.08% C.

Because of the negative effect of silica in the basic practice, a conscious effort should be made to keep sand out of the charge. Oily and wet scrap should also be avoided in the charge materials to minimize gas pickup.

The oxidizing period will require the addition of lime to form a slag on the bath during melting until about 1.5 to 2.0% CaO is present in the charge. Once the charge has been melted and temperatures of 1540 to 1595 °C (2800 to 2900 °F) have been reached, the oxygen lancing can begin. If phosphorus is to be removed, another approach requiring a double-slag practice should be used.

The following conditions must be met to remove phosphorus from the melt during the oxidation period:

- Temperature below 1540 °C (2800 °F), because the process is reversible with high temperature
- High iron oxide content in the bath and the slag; the use of iron oxide in the early stages of meltdown is required as a source of oxygen to oxidize the phosphorus
- Low silicon and silica in the bath and the slag
- A high lime ratio to negate the effect of the silica in the slag; because this slag must also be fluid, fluorspar additions may be required. High concentrations of manganese oxides also hinder the satisfactory removal of phosphorus

The desired reaction for the phosphorus is:

$$2P + 5FeO + 3CaO \rightarrow 3CaO{\cdot}P_2O_5 + 5Fe$$

Tricalcium phosphate is formed in the slag and the bath; it must be removed to lower the total phosphorus content. This is accomplished by raking off the slag. Limited success is achieved in reducing the phosphorus content, as manifested by the film visible around the sidewalls of the bath following slag removal. This film may contain a major portion of the tricalcium phosphate remaining in the bath, which, if left in the furnace, allows for only minor phosphorus reductions and the reversion of phosphorus to the bath with higher temperatures.

After the oxidizing slag has been removed, a new lime slag must be added to the melt. Any oxidizable element lost to the first slag must also be added back to the bath during the reducing and final tap out.

In the single-slag process, oxygen lancing will remove silicon, carbon, manganese, and chromium from the bath, with manganese and chromium partially reduced from the slag in the reducing period. To minimize chromium loss, a high temperature is required before lancing; therefore, a starting oxygen lancing temperature of 1595 °C (2900 °F) is suggested.

The silicon and manganese contents after oxygen lancing will be between 0.01 to 0.05% and 0.03 to 0.20%, respectively, with carbons in the 0.10 to 0.20% range. An addition of ferrosilicon to bring the melt chemistry to 0.10% Si will stop the boil.

Reducing or refining of the heat follows the oxygen lancing of the bath. At this time, a reduction in sulfur can be made, or the practice can proceed quickly without adjusting the slag to reduce the bath sulfur. With the temperature of the bath increased after lancing, additions of lime or calcium oxide and fluorspar are made to the furnace to prepare for sulfur reduction and alloy recovery. Sulfur reduction depends on:

- A high lime-to-silica ratio in the slag, called the V-ratio
- High temperature
- Low oxygen content in the steel and slag

The reaction can be expressed as follows:

$$CaO + FeS \rightarrow FeO + CaS$$

This reaction is reversible. An increase in silica will reduce the availability of CaO to react and thus retard the reaction, while high FeO will limit the amount of sulfur that can be removed.

The slag in the basic melting practice is the key to the production of quality steel. The reduction of iron oxide by the use of coal, aluminum, graphite, or low-sulfur coke is often done to deoxidize the slag and to reduce the oxide content. With this deoxidation, manganese and chromium in the slag are returned to the bath.

After a check on the melt or bath chemistry, alloy additions are made to adjust the chemistry with ferrosilicon, silicomanganese, and ferromanganese, along with chromium, nickel, molybdenum, and other alloys if required. At tap out, final alloy additions are made to the ladle, and the heat is deoxidized with aluminum as the primary deoxidizer plus special deoxidizers as desired.

Melting Heats and Raw Materials

The following discussion is a brief review of the basic melting practice. A heat and the raw materials used will also be described.

The charge is placed into the furnace with the heavy scrap carbon raiser and limestone on the bottom, the light scrap in the middle, and the medium scrap on the top. Nonoxidizable materials such as nickel and molybdenum can also be charged into the furnace at this time. The roof is swung into place with an inspection of the top of the charge to ensure that the scrap does not contact the roof ring, and it is then locked down.

Melting should proceed as soon as the charge is in the furnace on the No. 2 tap to burrow a hole into the scrap, followed by a change to the No. 1 tap until the charge is melted. The initial use of the No. 2 tap will reduce wear on the furnace delta section.

A pool of metal will form on the bottom of the furnace as the electrodes burrow through the charge. Once the electrodes have reached this pool, the charge will be melted by radiation from the pool and the arc splash off the electrodes. Rust on the scrap, along with oxidizing scrap from the melting charge, will react with the lime and eroded refractories to make a watery slag. When the charge is completely melted, additional limestone or lime is added to the furnace to increase the volume of slag for an estimated 1½ to 2% of the charge weight.

At this point, the melt can be processed as a single-slag heat without phosphorus removal, or it can be processed as a double-slag heat where phosphorus removal is attempted. This procedure is described earlier in this section. If a double-slag practice is chosen, the oxygen content of the slag and the bath must be increased while holding down the bath temperature. The old slag must then be removed and a new slag built on the bath. Oxygen lancing can be done before or after this stage to lower the carbon content to the required level, or it could have been done to increase the oxygen content of the bath.

Reducing the Heat. The slag must be reduced in iron oxide, and because only a small amount of slag has been added to the furnace up to this point, another 1½ to 3% of charge is added to this slag, along with a measured amount of fluorspar. This additional slag volume will dilute the iron oxide content in the slag to reduce the oxygen content and to improve the ability of the slag to retain sulfur. A deoxidizer such as coke, coal, or aluminum can be added to the slag if desired to speed the reaction. Sulfur removal and alloy recovery are the primary concerns while the bath is being heated in preparation for finishing. If a specific sulfur level is needed, further laboratory testing and lime additions may be required for achieving a higher lime-to-silica ratio, along with fluorspar to thin the slag. If the sulfur level is acceptable, the heat, blocked with ferrosilicon and alloy additions, is finished, and a temperature check is made prior to tap out.

The melt is tapped into a ladle that has been preheated if it is refractory lined, or it is tapped into a cold ladle liner of insulating basic material or boards. Any further alloy additions are made to the ladle prior to tap out. The furnace should be tapped to hold back the slag and to provide a clean full stream. The furnace spout should be clean, smooth, and dry to obtain a full metal stream without a spray into the ladle.

An additional opportunity for sulfur removal presents itself at this point. A desulfurizing agent can be added to the ladle during tap out to reduce the sulfur content by 20 to 50%.

Deoxidation of the steel should be accomplished in the ladle; this can be done in several ways. One method is to add the aluminum as ingots to the bottom of the ladle and tap onto it. Another method is to attach stars or doughnuts of aluminum to a steel rod and plunge the rod into the ladle when the ladle is one-third to three-quarters full. Small additions of aluminum should be in the form of shot or wire clippings weighed and placed in a bag and thrown into the stream while tapping. The amount of aluminum used for deoxidation is about 1.0 to 1.5 kg/metric ton (2 to 3 lb/ton), although higher carbon heats require less and lower carbon heats require more because of the oxygen levels in the steel.

After tap out, the furnace should be inspected for slag line and bottom erosion and then patched with dolomite or other materials. Patching is done by shoveling in or gunning on the patch material. This prepares the furnace for the next heat.

Loss of Alloys. A wide range of melting techniques are used in the basic melting practice by foundries—for example, single slag, double slag, deoxidation of slags, desulfurization, and alloy production. Therefore, the individual foundry and melter will need to adjust the recovery and additions to meet particular needs and final chemical analysis based on experience.

New Technology

The direct current arc furnace has been introduced into steel mills and should be changing the arc melting practices in the future from the three-phase arc furnace to the single-arc direct arc practice (Fig. 15). The direct current arc furnace has not made its appearance in the steel foundry yet, and it has been put into operation in only a few steel mills.

Those steel mills using the direct current arc furnace have lower electrode consumption and slightly lower operating costs. The direct current arc furnace was developed for two main purposes:

- An alternative to the conventional three-phase ac electric arc
- For the final stage in practices for melting prereduced material for the production of liquid iron

Since its development, this furnace has been used in the production of ferrotungsten, ferrosilicon, carbon, low-alloy steel, and stainless steels.

An existing three-phase furnace can be modified into a direct current arc furnace by the addition of a thyristor converter to its controls and the incorporation of a water-cooling system. Modification of the existing furnace requires the removal of the two outboard electrodes and the positioning

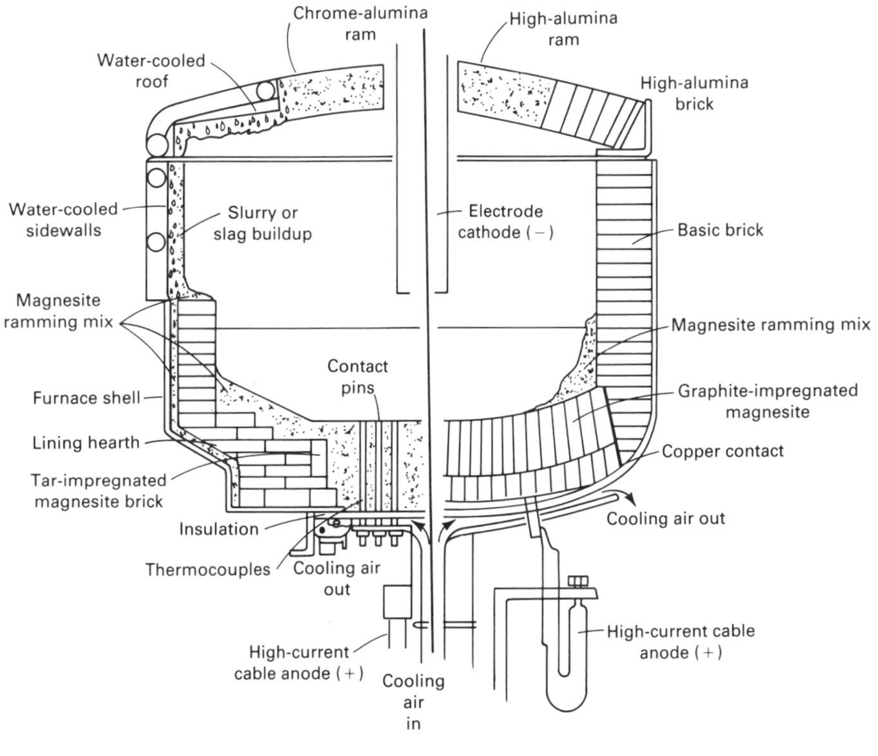

Fig. 15 Cutaway view showing the basic components of a direct current arc furnace

of the center electrode to the center of the roof. It has also been shown that a larger electrode will require additional hardware to conduct current and to provide support. The electrode is the cathode in this system. In the three-phase system, current passes from one electrode to another through the charge; in the direct current system, the current passes from the electrode through the charge or bath to the cathode located at the bottom of the furnace.

The current from the bottom of the furnace passes through graphite- or tar-impregnated refractories to a copper base plate to the outside cables, or through pins of steel connected to outside current cables (Fig. 15). The heat distribution in the direct arc is 360°, while in the three-phase furnace, the heat pattern is off the edge of the electrode, causing localized refractory heating and erosion. The electrode usage per ton of steel is reported to be 50 to 60% of a conventional three-phase furnace.

The shell bottom of the furnace needs to be modified to accept the current-carrying loads that will result and should be insulated to protect it from shorting out. Thermocouples and cooling air will be required for monitoring overheating or loss of current-carrying ability in the bottom. The furnace foundation may also require modification to accept power cables and air cooling ducts.

Patching of the refractory is limited to the slag line, sidewalls, and roof. Noncurrent refractories on the bottom will cause poor conduction. Another consideration to be made by the melter is the heat sequence, because direct arc melting requires long start-ups or a heel of metal over the bottom to make electrical contact.

Ladle metallurgy has been introduced for the removal of sulfur and for the microalloying of metals.

Argon stirring of heats of steel in ladles as a method to float out some inclusions found in the heat at tap out is becoming more popular. The need for dry gas is very important because argon containing more than 5 ppm of moisture will induce gas pickup.

Calcium wire injection has recently been introduced to modify inclusions in steel along with argon stirring to improve steel cleanliness.

Induction Furnaces

Ralph Y. Perkul, Asea Brown Boveri, Inc.

Induction furnaces have gradually become the most widely used means for melting iron and, increasingly, nonferrous alloys as well. The key to the ready acceptance of this type of furnace has been its excellent metallurgical control coupled with its relatively pollution-free operation.

Currently, induction furnaces are available in a wide variety of sizes. Coreless units range in capacity from a few pounds, favored by the precision cast metal producers, to 68 Mg (75 tons) powered at 21 000 kW. Channel-type units have been built with a capacity of over 180 Mg (200 tons) powered at 4000 kW per inductor.

In the past, large, high-powered induction furnaces operated at 60 Hz, the frequency of the incoming power. Thanks to major breakthroughs in electronic, solid-state frequency conversion techniques, it is now possible to build a highly efficient furnace operating at medium frequencies (70 to 5000 Hz). Coreless units of 7000 kW in power for ferrous metals and 4000 kW for nonferrous metals have been built in medium frequencies, and even channel-type units have recently been equipped with solid-state, medium frequency power supplies.

The advantage of medium frequency is that the power density of the furnace system can be increased substantially without increasing its size. Much more melting power can be applied, while maintaining the stirring action at desirable levels. Heat losses, which are a function of furnace size, are reduced, and overall efficiency of the system is improved. The furnace is also easier to operate because a single potentiometer will typically provide efficient and stepless control of the power.

Types of Furnaces

There are two classifications of induction furnaces: coreless and channel. Cross sections of each are shown in Fig. 1 and 2.

In a coreless furnace, the refractory-lined crucible is completely surrounded by a water-cooled copper coil, while in the channel furnace the coil surrounds only a small appendage of the unit, called an inductor. The term "channel" refers to the channel that the molten metal forms as a loop within the inductor. It is this metal loop that forms the secondary of the electrical circuit, with the surrounding copper coil being the primary. In a coreless furnace, the entire metal content of the crucible is the secondary.

Although both employ similar electrical principles, the two types of furnaces are quite different in their capabilities and operation. The coreless furnace is more widely used for melting and superheating, whereas a channel furnace is better suited for superheating, duplexing, and holding.

The channel furnace has been used by nonferrous foundries for many years. Recently, however, this type of furnace has been increasingly replaced by the coreless and resistance furnace in aluminum-melting applications. It is still commonly used for melting copper alloys.

Technological innovations in the channel furnace have been slower in developing than advances in the coreless furnace. Only a few prototype units with a medium frequency power supply have been installed thus far. Many furnaces still rely on 60 Hz power supplies and use power-regulating autotransformers for control.

Inherent in the channel concept is the fact that only the small amount of metal in the inductor channel receives energy through the surrounding coil. This necessitates superheating the metal in the loop and pumping it out of the channel into the main vessel, where it mixes with the cold metal and heats it by convection.

The ratio of heat transfer is limited, as is the stirring action in the main vessel. This is the primary reason this type of furnace has seen only limited service as a melter. The power that can be connected to a vessel is much smaller than for a similarly sized coreless vessel. The mild stirring action also makes it difficult to get alloy and carbon additions dissolved and mixed rapidly, particularly important in the production of synthetic irons.

In addition, the channel furnace must be started with a supply of molten metal, and a metal heel must be continuously maintained in the furnace. The furnace must not be shut down even for holidays or for other extended periods, as the metal will freeze and considerable refractory damage will be incurred.

The channel furnace is more economical to manufacture in large sizes and uses electric power more efficiently than a coreless furnace. If more power is desired, additional inductors can be connected to the main vessel.

Several geometric shapes of vessels have been used. These include configurations that are described as Vertical (upright cylindrical), Drum (horizontal cylindrical), and Low Profile (spherical). Typically, all of these units are hydraulically tilted to permit the emptying of the contents.

The coreless furnace is completely surrounded by a spiral of copper tubing of a special cross section that features a water-cooling channel in its center. It must also provide good electrical coupling and strength to withstand substantial electromagnetic forces, yet with provision for thermal expansion.

Because the magnetic field is transmitted in all directions, special vertical laminations

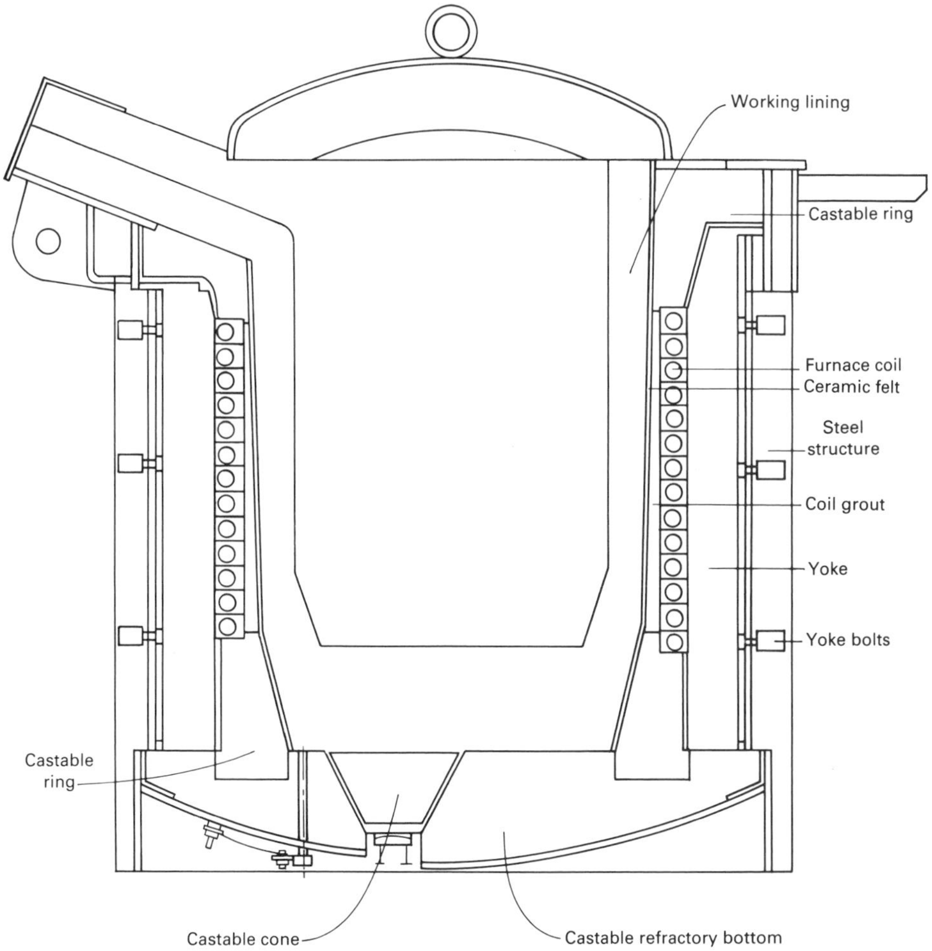

Fig. 1 A cross section of a coreless-type induction furnace showing water-cooled copper induction coil and key structural components. The entire molten metal bath (which serves as the secondary) is surrounded by the coil (the primary) that encircles the working lining.

of transformer iron, which form the magnetic yokes, are evenly spaced around the circumference of the coil to provide additional strength and to pick up the stray flux that would otherwise heat up the furnace frame. Coreless furnaces are typically tilted about the spout through a 95 to 100° angle to empty their contents. The medium-frequency units can be completely emptied for alloy changes or factory shutdowns and quickly started up again with a cold charge. It is not necessary to maintain a molten heel or to use precast starter blocks, as is the case with the 60 Hz units.

Electromagnetic Stirring

When alternating current is applied to an induction coil, it produces a magnetic field, which in turn generates a current flow through the charge material, heating and finally melting it. The amount of energy absorbed by the charge depends on the magnetic field intensity, electrical resistivity of the charge, and the operating frequency.

A second magnetic field is created by the induced current in the charge. Because these two fields are always in opposite directions, they create a mechanical force that is perpendicular to the lines of flux and cause metal movement, or stirring, when the charge is liquified. The mechanical force stays perpendicular to the field only in the center of the coil; on both ends of the coil it changes direction. The metal is pushed away from the coil, moves upward and downward, and flows back. Figure 3 shows the four-quadrant stirring. It is this stirring that allows excellent alloy and charge absorption and aids in producing a melt that is both chemically and thermally homogeneous.

The stirring is directly determined by the amount of power induced and is inversely proportional to the square root of the frequency. Therefore, the higher the power and the lower the frequency, the more intense the stirring.

The reasons for an increased interest in induction melting for aluminum applications are many, but perhaps most significant is the question of metal loss. Because of the inherent electromagnetic stirring action in the coreless furnace, charge materials immediately become immersed in the bath, minimizing oxidation losses. This effect is particularly pronounced when melting light-gage scrap. Thus, the growing emphasis on recycling favors induction melting as a cost-effective alternative to other metal melting processes.

Power Supplies

In order to meet the range of melting needs, the proper power supply must be selected. Because the vast majority of existing furnaces use line frequency, this arrangement will be discussed first (see Fig. 4).

Protection of the furnace and the power system, phase balancing, and power correction are required in order to operate an effective installation. The typical 60 Hz (line frequency) power supply basically consists of:

- Primary switchgear
- Furnace transformer
- Starting resistor and bypass contactor (for installations with off-load tap changers)
- Phase balancing system
- Power factor correction capacitor bank
- Instrumentation
- Controls and supervision

A primary transformation from the high-voltage grid of the local utility is usually required because the voltage applied to the coil is between 500 and 3000 V.

Because of its high inductance, an induction furnace coil has a low power factor. In order to bring the power factor to near unity, capacitors are installed and connected in parallel to the furnace coil. When starting a furnace, in order to reduce the inrush, a soft start is provided.

When the transformer with an on-load tap changer is installed, the tap changer travels down to a lower tap whenever the furnace is switched off. This provides a start with a low voltage, resulting in low inrush current. When an off-load tap changer is applied, a starting resistor bank with bypass contactors is provided.

Because the induction furnace is a single-phase furnace but its feed is commonly a three-phase line, a phase balancing system consisting of capacitors and reactors has to be provided. In ferrous applications, refractory lining wear decreases the coupling distance between the coil and charge, resulting in a power increase. Because power input is limited, the voltage has to be decreased in order to keep power from exceeding its maximum. This is typically done by using steps on a transformer tap changer. As the lining wears and the power increases, the voltage has to be reduced by going to a lower tap.

Additionally, in order to compensate for the amount of metal in the furnace and the condition of the refractory lining, a portion

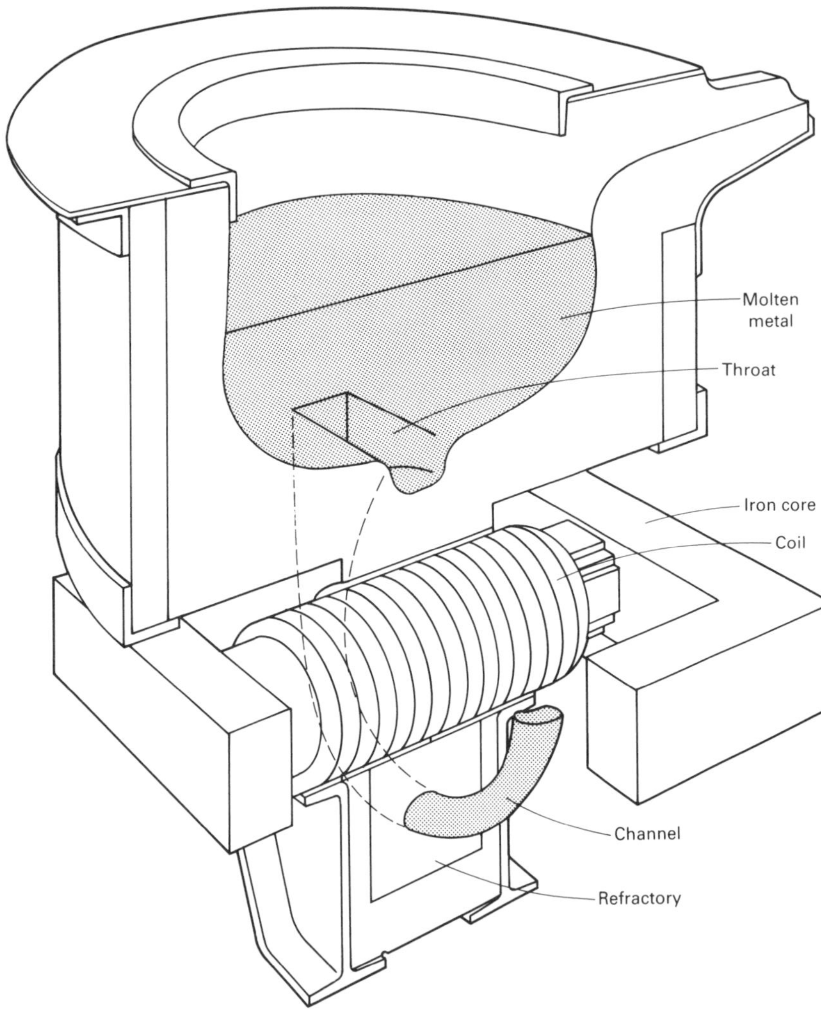

Fig. 2 A cross section of a channel-type induction furnace showing the water-cooled copper induction coil which is located inside of a 360° loop formed by the throat and channel portion of the molten metal vessel. It is the channel portion of the loop which serves as the secondary of the electrical circuit in which the copper coil is the primary.

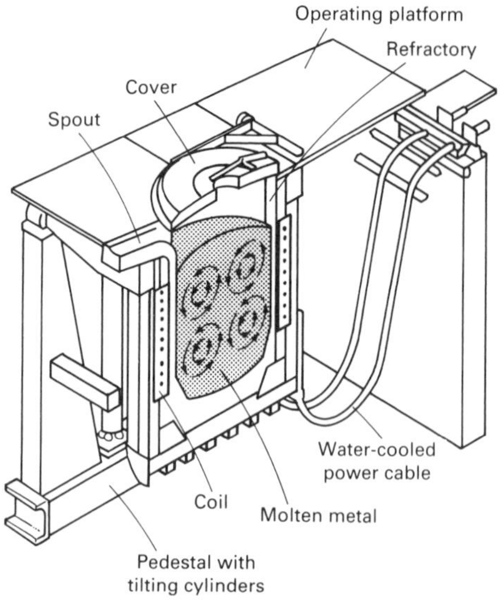

Fig. 3 A cross-sectional view of a coreless-type induction furnace illustrating four-quadrant stirring action, which aids in producing homogeneous melt

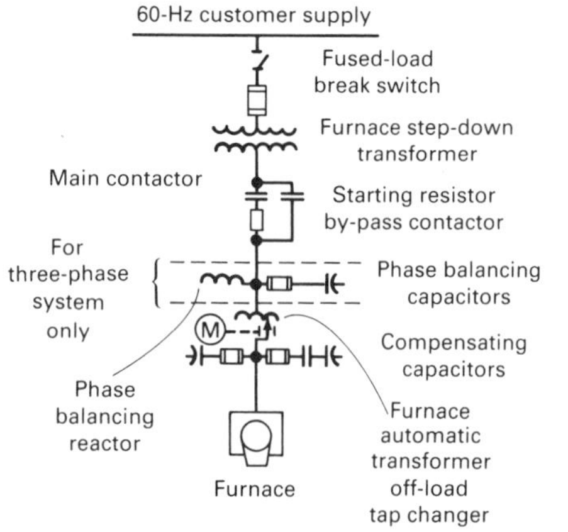

Fig. 4 Modifying 60-Hz line frequency supplied by utility company to serve as power supply for induction furnaces

of the capacitor bank that is installed in parallel with the coil needs to be switched by adjusting the coarse and fine settings on the electrical panel. In the modern medium-frequency power supplies, (see Fig. 5), many of the functions described are now performed automatically or have been eliminated altogether.

The typical conversion is in an ac-dc-ac sequence. While frequency conversion entails electrical losses, modern power supplies incorporate solid-state frequency converters, and the increase in thermal efficiency usually outweighs the conversion losses. Modern converters approach an efficiency of 97%. In the power conversion commonly used, the incoming 60 Hz ac power is rectified to dc and then chopped by an inverter to a higher frequency.

Series and Parallel Inverters. Two types of inverters are typically used, namely, series and parallel. In each case the furnace and compensating capacitor bank (fixed, in this case) form part of a resonance circuit.

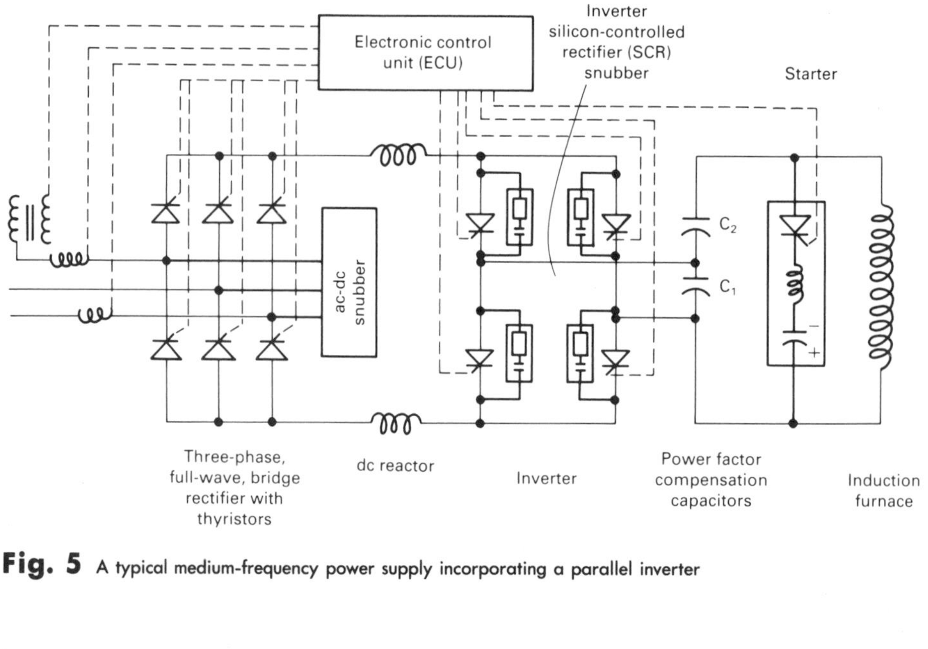

Fig. 5 A typical medium-frequency power supply incorporating a parallel inverter

In a series inverter, capacitors are connected in series with the induction coil, requiring, at low power, a matching transformer. Power to the furnace is controlled by detuning the inverter. It is unnecessary to control the rectifier, thus allowing the use of diodes.

In a parallel inverter, the capacitor bank is connected in parallel with the induction coil. Power control is accomplished by detuning the inverter in the full-power range. Thyristors in the rectifier act to lower the dc voltage, if required, and provide half-cycle protection, thereby essentially becoming an electronic circuit breaker. This characteristic, coupled with the smaller number of components in the circuit, have made the parallel system the most widely used inverter in industry.

In both series and parallel circuits, it is possible to get power control that allows full power to be drawn regardless of the level of the charge in the furnace or the condition of the refractory lining. The midfrequency converters operate on a variable frequency. The load-commutated converter increases frequency as the charge contents of a furnace increase while it simultaneously compensates for lining wear. No capacitor switching is required as in the line frequency units. In addition, a single potentiometer offers stepless and infinite control, eliminating the need for tap changer systems for power control.

Modern induction furnaces take advantage of the latest technology, and many new installations are equipped with programmable controllers in lieu of relay logic. An even more recent advance has been the use of industrial computers for controlling the furnace and auxiliaries.

With the variety of software that is currently available, the operator can choose screens such as systems data, automatic lining sintering, automatic cold starting of the furnace after a prolonged shutdown, automatic melting programs for various alloys, continuous calculated-temperature monitoring, fault detection, demand limiting, tabulation of energy consumption, melting cost data, metal inventory control, and chemistry adjustment. Also, by tying the computer to support systems such as charge make-up and hot-metal delivery, the utilization of the furnace can be increased.

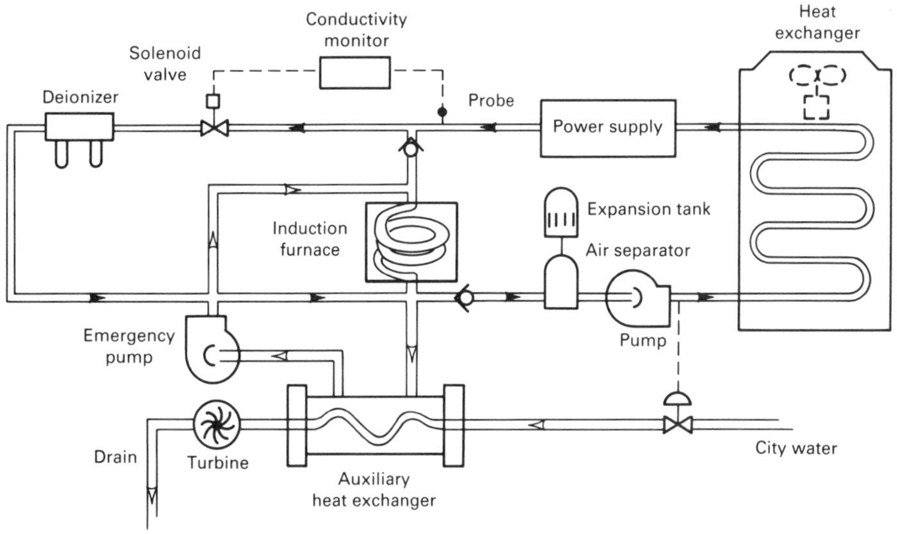

Fig. 6 A layout of a closed-loop cooling system showing routing of furnace-cooling water under both normal (black arrows) and emergency conditions (white arrows). Standby emergency pump is activated if the primary pump fails.

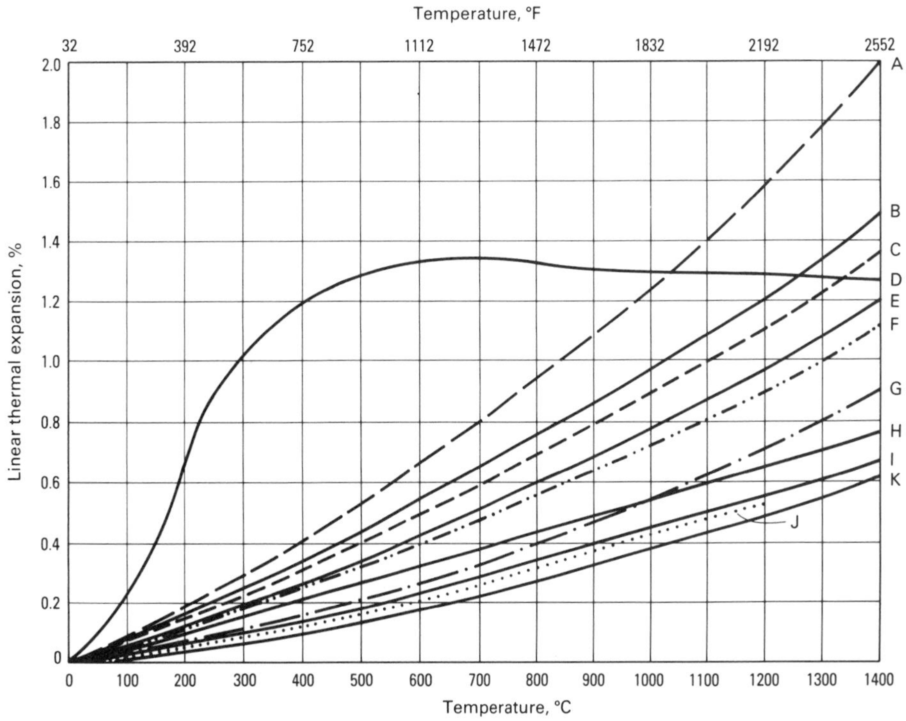

Fig. 7 Thermal expansion curves of various refractory brick oxide materials used for linings in induction furnaces: A, magnesia; B, chrome magnesia; C, chromite; D, silica; E, zirconia; F, corundum 99; G, corundum 90; H, fireclay; I, sillimanite; J, zircon; K, silicon carbide

Packaging of Systems. Until the early 1970s, the electrical components of coreless induction furnaces were typically assembled on site in a concrete vault. Field construction costs made this installation technique financially prohibitive except for the largest of furnaces (usually above 18 Mg, or 20 tons in capacity) for which a vault is still used. Even in this size range, modern component assembly techniques have allowed for the power supply to be packaged into steel modules at the equipment manufacturing plants.

For furnaces up to 3.6 Mg (4 tons) in capacity, current practice is to completely assemble, wire, and pipe all components on a steel frame, to test, and subsequently to ship the unit as a single module, thus minimizing installation and start-up costs. This packaging concept applies to furnaces in the 4.5 to 18.1 Mg (5 to 20 ton) range, as well, although several modules may be required because of their size. Still, they are equipped with quick-connect fittings for fast and easy field assembly.

For furnaces with power above 1250 kW, a step-down transformer from the main distribution voltage is normally used. Below 1250 kW, a 480 V distribution line backed by a disconnecting device can be used. The step-down transformers, when required, can be of the oil-filled outdoor type, which is the least expensive, or the indoor dry or silicon-filled type.

Water Cooling Systems

Because the furnace coils and power supplies of the induction furnaces need to be cooled, a closed-loop water system is generally called for. In a medium-frequency

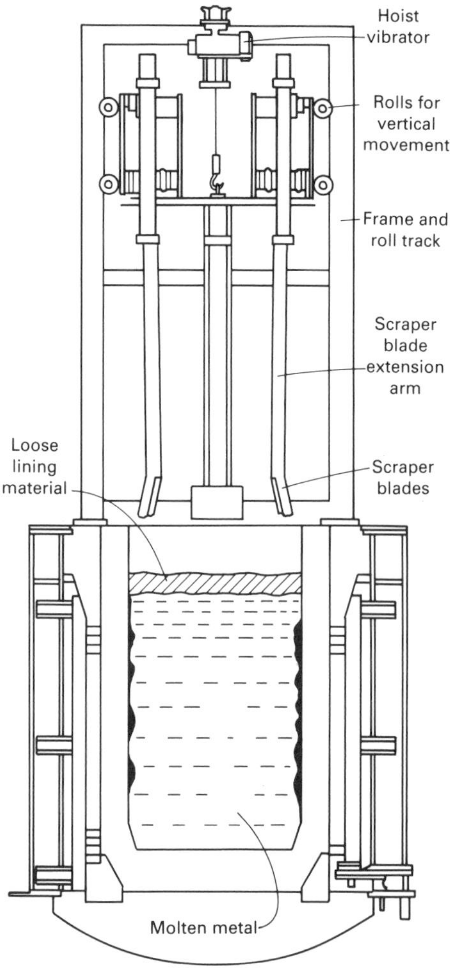

Fig. 8 A cross section of a patented mechanical scraper used to clean the refractory lining of induction furnaces that melt nonferrous metals

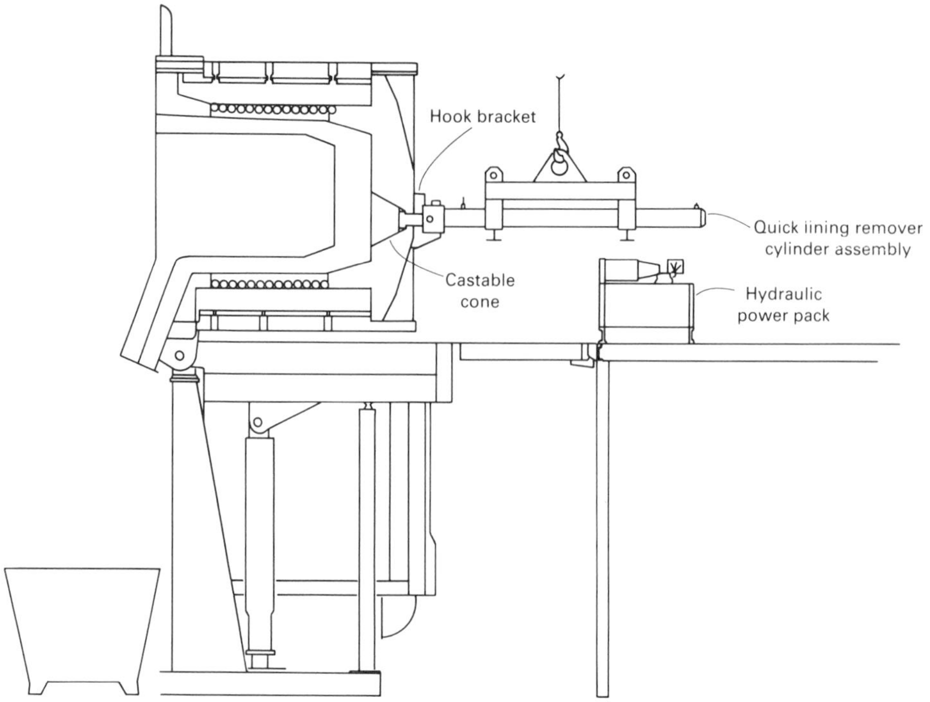

Fig. 9 The principal components of a patented push-out device used to extract deteriorated refractory lining from coreless induction furnaces

power supply, the electronic components of the system contain a dc leg; it is therefore important to have clean, demineralized water circulating, typically with conductivity set at 50 μS (50 μmho) in order to prevent electrolysis from taking place.

The heat exchanger to be used depends entirely on local circumstances such as climate, availability of raw water, and so on. The most widely used are the evaporative, the water-to-water, and the air-to-water types. The latter has been frequently adapted to heat recovery systems. In those applications the hot water from the furnace power supply and coil is piped to a location in the plant where heat is required. The water is circulated through a tube bundle, and a fan dispenses the warm air to the surroundings. Considering that up to 10% of the power from a channel furnace and 20% of the power from a coreless furnace is lost to the water system, heat recovery systems for induction furnaces are very practical.

In all coreless- and channel-type furnaces, the water is circulated through the power supply and then routed through the coil and on to the heat exchanger. Typically, two pumps are provided, one as a standby with an automatic switchover, because it is extremely important to circulate water through the coil continuously, to prevent it from being damaged. To protect induction furnace equipment during momentary power interruptions or prolonged power failures, manufacturers provide a special pressure-operated valve that, when activated, allows emergency deployment of city water to circulate through the coil. Such a system is shown in Fig. 6.

In addition, with channel furnaces, standby generators are often employed to maintain some power in the inductor. If the metal in the channel freezes, considerable damage to the refractory lining may occur.

Lining Material

The selection of lining material is determined by the metallurgical requirements, the operating temperatures, and the type of operation. Refractories used in induction furnaces are oxides of minerals. Typically used are silica (SiO_2), alumina (Al_2O_3), or magnesia (MgO) linings. From a chemical point of view, the silica is classified as an acid, alumina as a neutral, and magnesia as a basic material.

Silica is the clear choice in iron melting, because it does not readily react with the acid slag typically produced in high-silica iron, has an extremely forgiving expansion curve, and is easy and economical to use. Alumina is the usual choice for aluminum-melting furnaces.

Each of the refractory groups mentioned above has a different thermal expansion characteristic, as shown in Fig. 7. As can be seen, alumina and magnesia have a nearly linear expansion. Silica, however, completes its expansion at approximately 815 °C (1500 °F) and remains constant at higher temperatures. With the selection of silica lining, the furnace can be shut down, cooled, and restarted without running the risk of metal penetrating thermal cracks.

Magnesia is very sensitive to thermal shock and has the greatest expansion. Once cracked, this type of lining is not likely to seal itself, thus resulting in metal penetration.

Alumina has approximately one-half the expansion of magnesia and is therefore less crack sensitive. It is important, however, to heat the furnace lining slowly when undergoing a cold start. This programmed start-up allows the cracks to seal before the metal adhering to the side walls melts and penetrates. This applies particularly to aluminum-melting furnaces because the metal melting temperature is relatively low.

In the case of aluminum and some copper alloys, the oxide and metal buildup on the crucible not only poses a potential refractory penetration problem, but also effectively increases the lining thickness and reduces the power that goes into the furnace. If not removed, this layer, which contains a small amount of metal, absorbs energy and becomes superheated. The superheated metal can penetrate the lining and cause premature failure. For these reasons, the furnace crucible walls must be cleaned periodically.

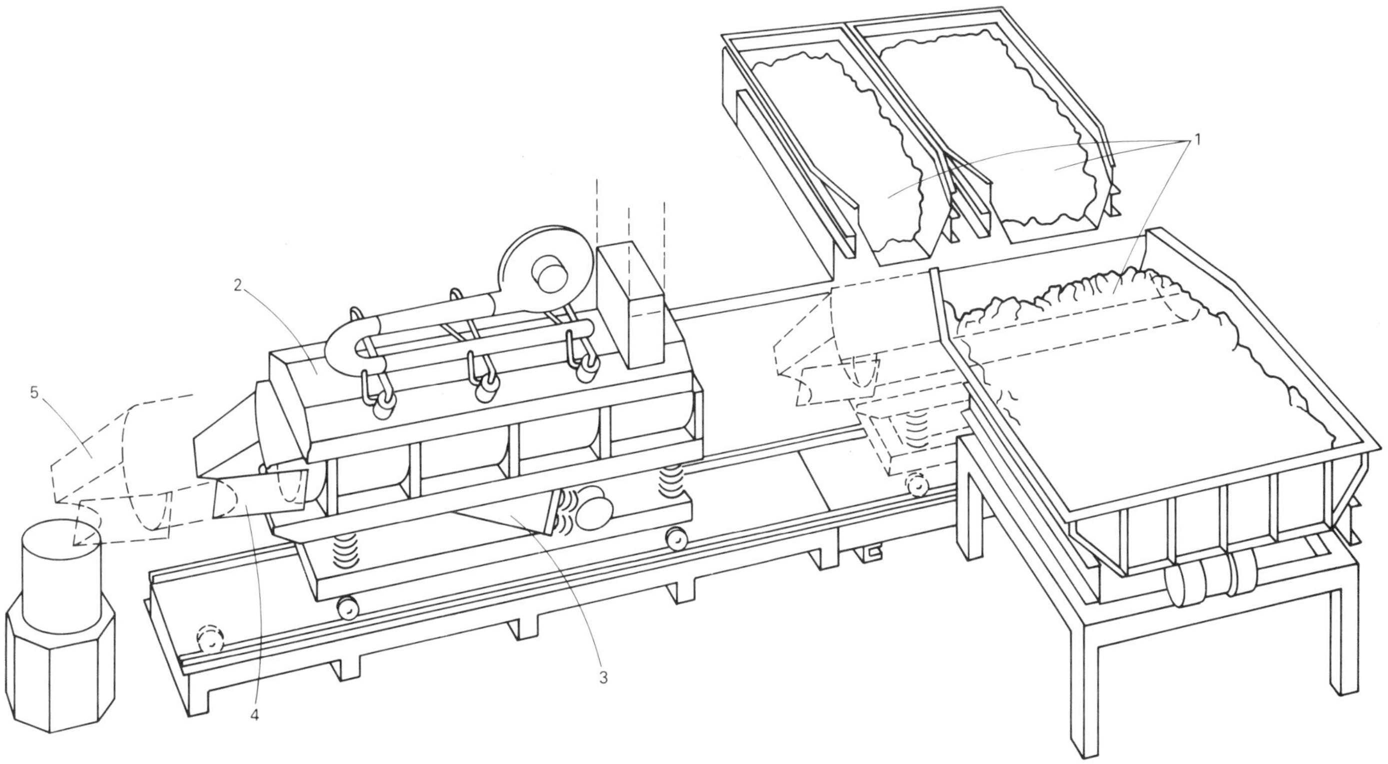

Fig. 10 The major components of a dryer used to preheat induction furnace charge material. Integrated scrap preheat process combines (1) weigh hoppers, (2) preheat hood, (3) material transfer mechanism, and (4 and 5) furnace-charging apparatus into a single automated process.

Crucible Wall Scrapers. The task of scraping the crucible wall has been made easier with the introduction of a patented mechanical scraper (Fig. 8), which quickly and easily cleans the sides in a full, hot furnace. The device essentially consists of a number of oscillating blades assembled on a steel frame. To start the operation, the scraper is placed on top of the furnace and the oscillating tools travel down the depth of the crucible. Radial pressure on the furnace is adjustable. The loose materials float to the top of the bath and are skimmed off.

Ramming Mixes. Most induction furnaces operating in iron foundries are, as stated previously, lined with a silica ramming mix. Alumina, magnesia, or zircon are typically used for melting steel and alloys. Alumina is typically used for aluminum.

The current practice is to use granular materials. For example, the silica ramming mix is granular material combined with the sintering agent boric acid (H_2BO_3). The amount of boric acid depends on the operating temperature and the wall thickness of the crucible.

The flux or sintering agent fuses the silica particles to each other, forming a hard, glazed, ceramic surface. Ideally, the lining, when sintered, consists of one-third hard, sintered ceramic; one-third fritted material, in which fine particles are sintered and larger particles can still be identified; and one-third loose particles.

This loose material has its purpose. When starting up a furnace from a cold start, the lining naturally expands. The loose material provides a cushion to handle the pressure.

All granular materials are rammed on the furnace bottom using an electric vibrator equipped with a tamping plate. For the side walls, the lining material is spread behind a lining form that is vibrated with a pneumatic vibrator attached to it.

For ferrous applications, the lining form is consumed in the first heat. For nonferrous applications, the lining form can be removed and reused. It may be tapered or collapsible.

The sintering process itself takes some time because it is necessary to heat the lining at the rate of approximately 110 °C/h (200 °F/h) until a temperature of about 30 °C (50 °F) above operating temperature is achieved. This temperature is held for 1 h before the furnace is ready for production.

When the linings are to be replaced, they must be chipped out with pneumatic hammers, and the lining procedure must be repeated. A patented device is now available that eliminates the need to chip out the spent linings manually, saving both labor and time.

The key components of this device, a schematic of which is provided in Fig. 9, are a locking plate that is bolted to the bottom of the furnace and a hydraulic ram assembly that is brought into position and latched to the plate. Once activated, the hydraulic ram moves forward, exerting pressure on a specially designed refractory cone installed in the bottom of the furnace. This distributes the force over the cross section of the lining. The push-out is aided by a slip joint between the coil grout and the working lining. The whole operation takes but a few minutes, with additional time being saved because the cool-down time is cut in half.

Melting Operations

There are two distinct ways of operating a coreless induction furnace. One method is a batch operation, in which the entire contents of the furnace are emptied and the unit is recharged with solids, typically using a conveyor or bucket charger. The other, more accepted, method involves a tap-and-charge operation, in which a portion of the furnace contents, typically one-third to one-half, is tapped and the identical weight in solids is recharged using a charge bucket or charge conveyor.

Batch operation has now become feasible because medium-frequency power supplies allow the furnace to be started with ordinary scrap, rather than with starting blocks.

In terms of electrical consumption, the batch-melting method provides a 7½% higher efficiency in iron-melting furnaces and a 4½% higher efficiency in nonferrous-melting units over the tap-and-charge method. This is because the cold charge material couples with the magnetic flux more effectively than does molten metal. In addition to having greater electrical efficiency, a medium-frequency batch operation allows the

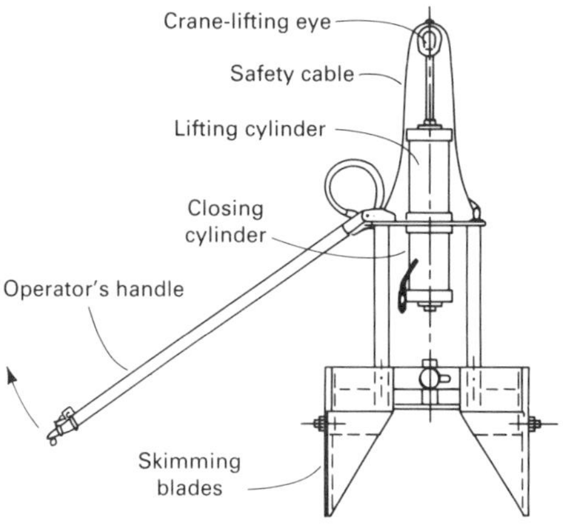

Fig. 11 A mechanical clamshell-type power skimmer used to remove slag before each tap of an induction furnace

drying of wet or oily charge material in the furnace, minimizing the need for a scrap-drying system.

The tap-and-charge operation is still the most popular approach, however, because the chemistry is much easier to control because less than one-half of the furnace contents are replenished at any one time. Additionally, the refractory lining lasts longer because of less thermal cycling. Finally, in order to use a batch operation effectively, the contents of the unit have to be quickly and completely discharged, usually into a holding furnace. This option may not be possible if small amounts of molten metal need to be taken out at defined times and the capital and operational costs of a holding furnace cannot be economically justified.

In either method of operation, it is particularly important to use the power supply efficiently and keep the cover closed to minimize potentially high radiation losses.

Utilization is calculated in terms of the time that the power is on. It is obvious that power cannot be on constantly, because certain operations, such as charging, slagging, sampling, and tapping are best and most safely done while the power is off. The objective is to keep the power on at least 75% of the time. Sometimes a second furnace added to the power supply improves utilization, because the power can be switched from one furnace to the other while the above-mentioned operations take place. Utilization can also be improved by quickly charging and tapping to minimize waiting time and heat losses as well as by increasing the charge size.

Although medium frequency has made the need for heavy charge materials obsolete, it is still good practice to choose materials that have a reasonably high bulk density, are relatively clean and dry, and are not too long. In order to avoid bridging, the length of the pieces should be no more than two-thirds the diameter of the crucible. If longer pieces such as casting gating systems and structural steel sections are available, they should be broken or cut first.

The high bulk density is important because the new charge must be contained in the space vacated by the tapped-out molten metal. If low-density charge materials are used, several charging cycles may be necessary, reducing furnace utilization.

Dry charge material is best suited for melting operations. Oils on the charge surface generate fumes from the furnace, which may need to be exhausted. Moisture on the charge also can be a problem, because a steam explosion can result. If dry materials are not readily available, a charge dryer is often employed to drive off any moisture remaining on the charge. The most common dryer consists of a refractory-lined hood equipped with gas burners, which heat the solids on a vibrating conveyor by means of forced convection of the hot gases through the material. Figure 10 shows the various major components of the system. It is important to get the preheated charge into the furnace quickly to take advantage of the thermal energy in the material and to reduce the electrical energy output required for melting. The average temperature used for drying is typically 315 °C (600 °F); for preheating, it is 540 °C (1000 °F). When applied properly, the preheating of the charge material can increase the melt rate by 10 to 15%. Direct impingement of the flames on the charge material, particularly if it is thin, should be avoided because oxidation losses can occur.

When dirty material is charged into the furnace, slag forms from the dirt and rust in the scrap as well as from the foundry return sand. Slag is undesirable in induction furnaces because it absorbs energy and erodes the lining. It is, therefore, removed each cycle with a skimmer to prevent slag accumulation.

Mechanical Skimmers. Furnace operators can increase the power supply utilization by the use of mechanical skimmers (Fig. 11). This is particularly important in large furnaces because mechanical slagging units can minimize the time that power is either turned off or reduced for skimming during each charge-to-tap cycle. A power skimmer can easily boost the output and productivity of the melting furnace.

A skimmer is typically mounted on a jib. This tool has a long handle, which keeps the operator farther away from the hot furnace. The best and most effective type of slagger is the clamshell type. The skimmer is positioned over the furnace and lowered into the slag. With an automatic raking motion, the jaws of the unit close around the slag, which is then lifted clear of the bath and deposited into a container. The blades are typically dipped into a slurry to reduce slag adhesion during the next charge-to-tap cycle.

Reverberatory Furnaces and Crucible Furnaces

Avery Kearney, Avery Kearney & Company

Reverberatory and crucible furnaces are extensively used for the batch melting of ferrous and nonferrous metals. This section will focus on nonferrous metals such as aluminum, copper, zinc, and magnesium, with emphasis on the melting of aluminum. Either type of furnace can individually serve as a melter or a holder or as a combined melter/holder prior to pouring the casting.

Crucible furnaces typically have a maximum holding capacity of 1.4 Mg (3000 lb) of aluminum; reverberatory furnace capacities range from 0.91 to 90 Mg (2000 to 200 000 lb) of aluminum. However, because crucible furnaces can isolate the gas or electric heat source from direct contact with the molten metal bath, they do provide the advantage of minimizing contaminants in the melt when purity is a primary concern.

Reverberatory Furnaces

Reverberatory or hearth furnaces for foundry use have been extensively redesigned since the fuel shortages of the mid-1970s. Rising fuel costs gave the industry good reason to look closely at melting techniques. Since then, many of the well-accepted ideas have been thoroughly tested and replaced with newer, better, more efficient, and more quality-oriented techniques. The answers and evaluations for determining the best technology in each specific situation are far from complete. Foundry use of hearth furnaces is primarily limited to aluminum alloys.

Types of Hearth Furnaces

The term reverberatory furnace is a throwback to the World War II era, when the most common fuels for hearth aluminum melting were coal or coke (Fig. 1). This necessitated locating the heat source or fire box at one end of the bath and the flue at the other end. The flue (stack) was designed to be strong enough to draw the flame over the bath. The roof was slanted to bounce or reverberate the flame (like sound echoes) off the ceiling, down across the metal to be melted in the molten bath, and then onto the flue. A primary objective was to heat the

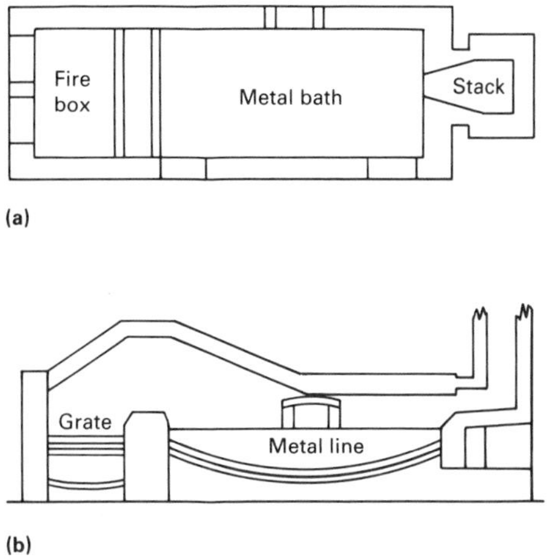

Fig. 1 An early version of a coal-fired melter hearth furnace used to produce aluminum castings. (a) Plan view. (b) Vertical cross section

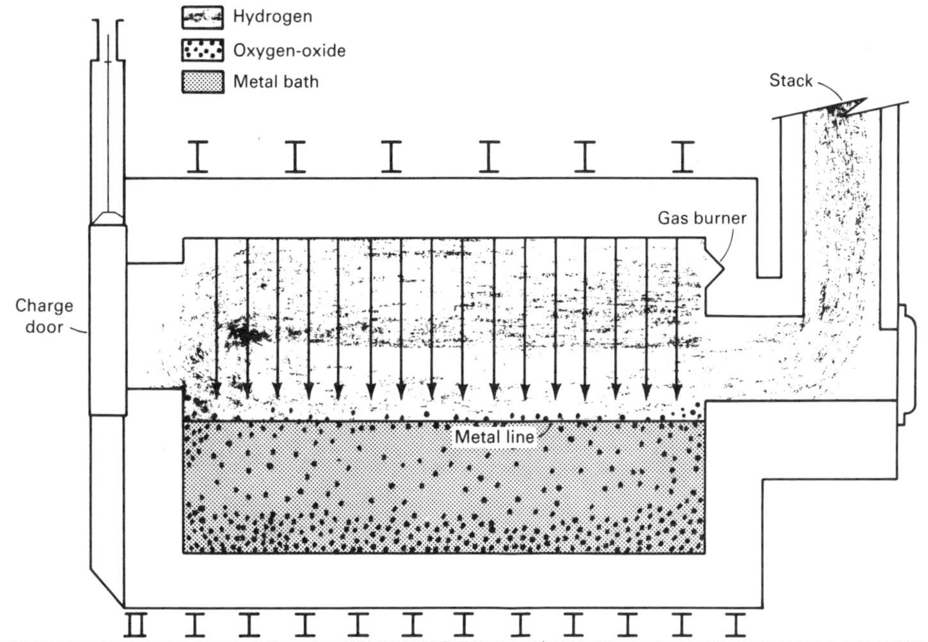

Fig. 2 Schematic of a wet hearth reverberatory furnace heated by conventional fossil fuel showing the position of the hydrogen and oxygen gases relative to the molten metal bath. Arrows indicate heat radiated from top of furnace chamber.

refractory walls and ceiling and then use these surfaces to radiate heat to melt the charge. This same flame pattern and the method of using a flame to heat the bath were implemented when the fuel used was changed from coke to oil and gas.

This older concept of aluminum metal melting is still used in many foundries and smelters. However, the current trend is to use roof burners or side burners directed toward the metal surface. In essence, the trend is to heat the metal not the furnace for efficient operation.

Reverberatory furnaces are direct fuel fired and can be designated in either the wet hearth or the dry hearth configuration. In a wet hearth furnace, the products of combustion are in direct contact with the top of the molten bath, and heat transfer is achieved by a combination of convection and radiation (Fig. 2). In a dry hearth furnace, the charge of solid metal is positioned on a sloping hearth above the level of the molten metal so that the charge is completely enveloped by the hot gases (Fig. 3). Heat is rapidly absorbed by the solid charge, which melts and subsequently drains from the sloping hearth into the wet holding basin or chamber.

Figure 4 illustrates the wide array of burner flame pattern configurations available for use in reverberatory furnaces. Electrically heated furnaces using resistance heating elements are also available.

Modern aluminum foundries use numerous furnace designs. Figure 5 shows a schematic of a wet hearth reverberatory furnace charged by means of a charging ramp. In this setup, two burners are mounted in the end wall opposite the flue. In updated furnace designs, the burners are mounted on either side of the flue to increase the time that the flame remains in the furnace chamber. In this way, fuel efficiency is improved. The capacity of this type of furnace is typically 9.1 to 45 Mg (20 000 to 100 000 lb).

Another widely used furnace for the melting of aluminum is the melter/holder furnace. The capacity of these units varies from 2.3 to 18 Mg (5000 to 40 000 lb). Burners in the furnace roof provide efficient heating and unobstructed floor space for furnace operation all around the perimeter of the furnace. Units of this type are often used to supply metal to several permanent mold molding stations.

Figure 6 shows a schematic of a small electric resistance melter/holder furnace for melting aluminum. Metal quality is improved with this type of unit because it eliminates the problem of combustion products and because it provides a low-temperature heating source. Practical sizes range from 0.91 to 45 Mg (2000 to 100 000 lb).

A tilting reverberatory furnace is illustrated in Fig. 7. The burner is located on the

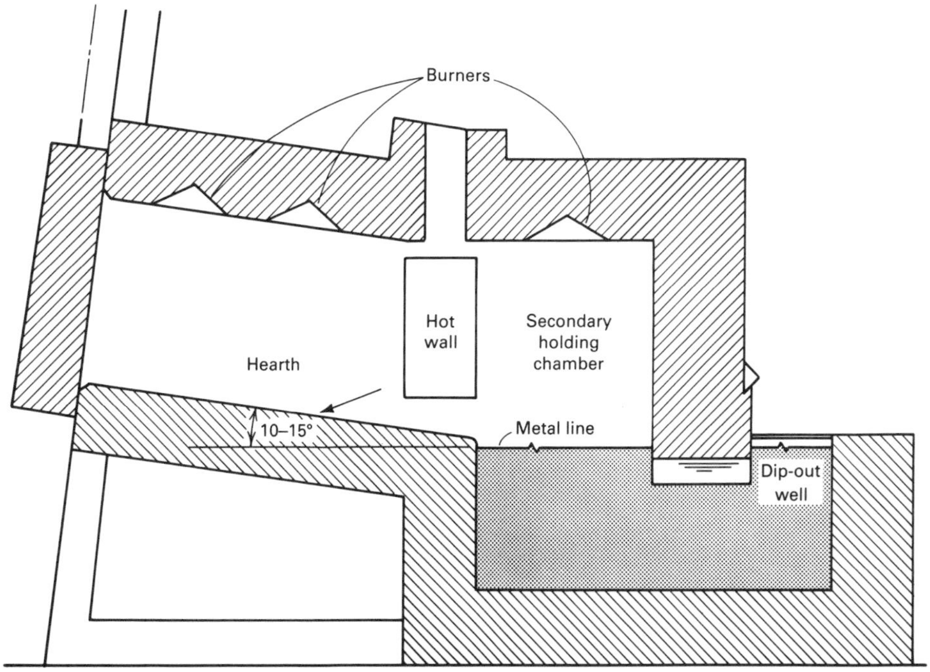

Fig. 3 Schematic of a radiant-fired dry hearth reverberatory furnace illustrating the position of the sloping hearth relative to the molten bath

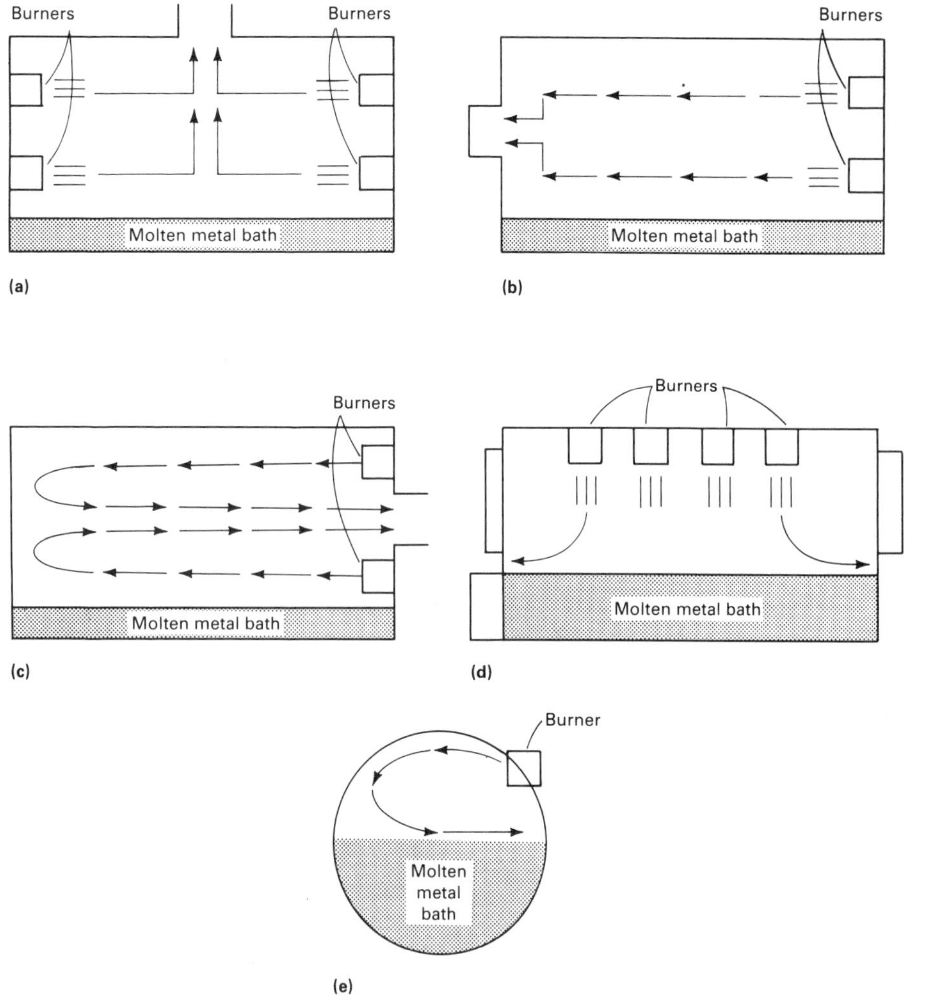

Fig. 4 Front views of the numerous types of burner flame patterns used in reverberatory furnaces. (a) Center flue arrangement. (b) Straight through firing. (c) W-fired double pass. (d) Roof mounted. (e) Barrel type

axis of rotation at the end that is opposite to the charge well. This design provides simpler and safer tapping or metal removal than furnaces that have plugs and tap blocks. When used to melt aluminum, such a furnace typically has a capacity of 2.0 to 9.1 Mg (4400 to 20 000 lb) and can melt at a rate of 1.8 Mg/h (4000 lb/h).

Figure 8 shows a cutaway view of a dry hearth melting/holding furnace that serves as a metal supply to casting units. These furnaces supply excellent-quality metal. The fact that the molten metal does not get heated above the pouring temperature avoids many problems encountered in other types of melting. These units are extensively used in automobile wheel foundries and other high-volume quality foundry applications. One problem area relates to cleaning of the dry hearth. The flame and high temperatures can cause oxide and other melt residue to accumulate into large refractory deposits on this hearth. Regular and thorough cleaning will easily control this problem. Capacities of this style of furnace range from 1.8 to 9.1 Mg (4000 to 20 000 lb).

A unique hearth design that carefully addresses many of the hearth operation and design problems that are encountered in small furnaces is shown in Fig. 9. This melter/holder furnace has a hinged roof, which facilitates access for hearth cleaning. A stack ingot and heavy scrap preheater arrangement provides high fuel efficiencies, good metal temperature control, and charging safety features. The optional dry hearth provides the metal quality advantage offered by melter/holder furnaces. The roof-mounted burner system yields maximum use of floor space around the furnace. As with many of the other newer furnaces, a high-efficiency insulation provides a cool furnace exterior. Capacities range from 0.91 to 6.8 Mg (2000 to 15 000 lb).

Table 1 Heat required for melting various alloys

Metal	Heat required kJ/kg	Heat required Btu/lb
Iron	1240	534
Aluminum	1060	457
Copper	675	290
Zinc	280	120
Tin	116	50
Lead	65	28

Aluminum Alloy Melting Practice

The equipment and processes required for melting aluminum alloys vary widely and take into consideration the unique requirements of aluminum alloy melting. The following list describes equipment, energy, and metal property requirements that must be considered to melt aluminum cost effectively.

Energy Requirements for Melting. Very large quantities of heat are required for the melting of aluminum alloys. Table 1 compares the energy per unit mass needed to heat and to melt various metals when the furnace is operated from a cold start. It is interesting to note that the amount of heat required to melt 0.45 kg (1 lb) of aluminum (at 660 °C, or 1220 °F) is over 85% as much heat as that required to accomplish the same task with 0.45 kg (1 lb) of iron at 1260 °C (2300 °F). Careful consideration of the heat transfer problem in the design and operation of equipment is required to provide efficient melting.

Table 2 lists the amounts of heat and the percentages of total heat required to heat aluminum to various temperatures. From a practical standpoint, this indicates the importance of preheating; that is, it takes 41% of the furnace energy to heat aluminum to

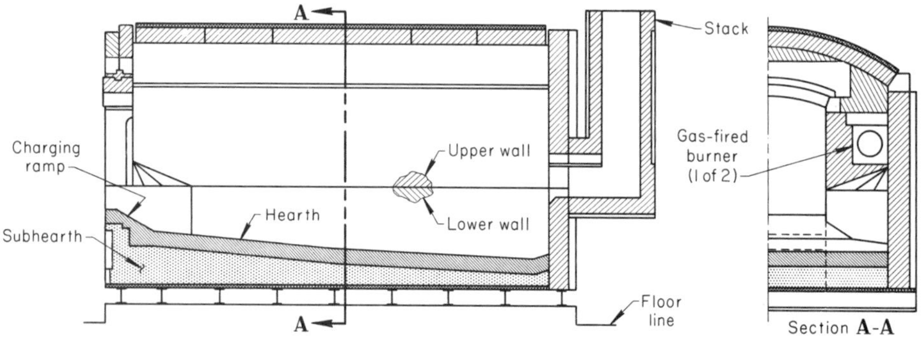

Fig. 5 Large reverberatory furnace charged through a ramp

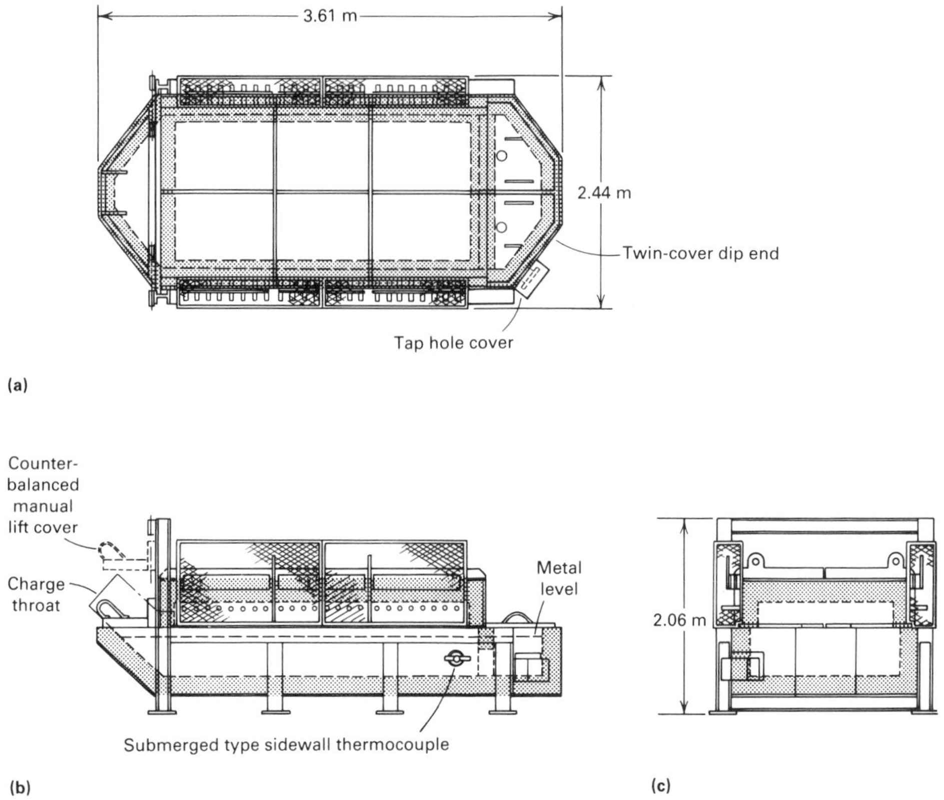

Fig. 6 Schematic of a small (4.42 Mg, or 9750 lb, capacity) electric resistance melter/holder reverberatory furnace for melting aluminum. Top (a), front (b), and side (c) views are shown.

425 °C (800 °F). Therefore, with heated charges, one can almost double the capacity of the melt furnace. At the same time, much shorter meltdown times are developed.

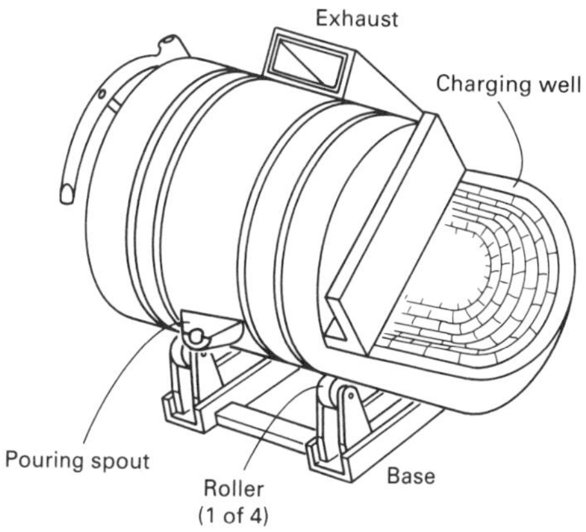

Fig. 7 Simplified view of a small tilting reverberatory furnace

High Chemical Activity. This property of aluminum requires special precautions, including:

- Careful isolation from moisture and water; under certain conditions, violent explosions can occur
- Consideration of possible thermite reactions; silicon, iron, and other metallic oxides are often reduced by molten aluminum, resulting in bath contamination and other problems that require special consideration in refractory selection
- Heavy firing and/or direct flame impingement, under some conditions, will cause thermiting or aluminum oxidation on the bath surface

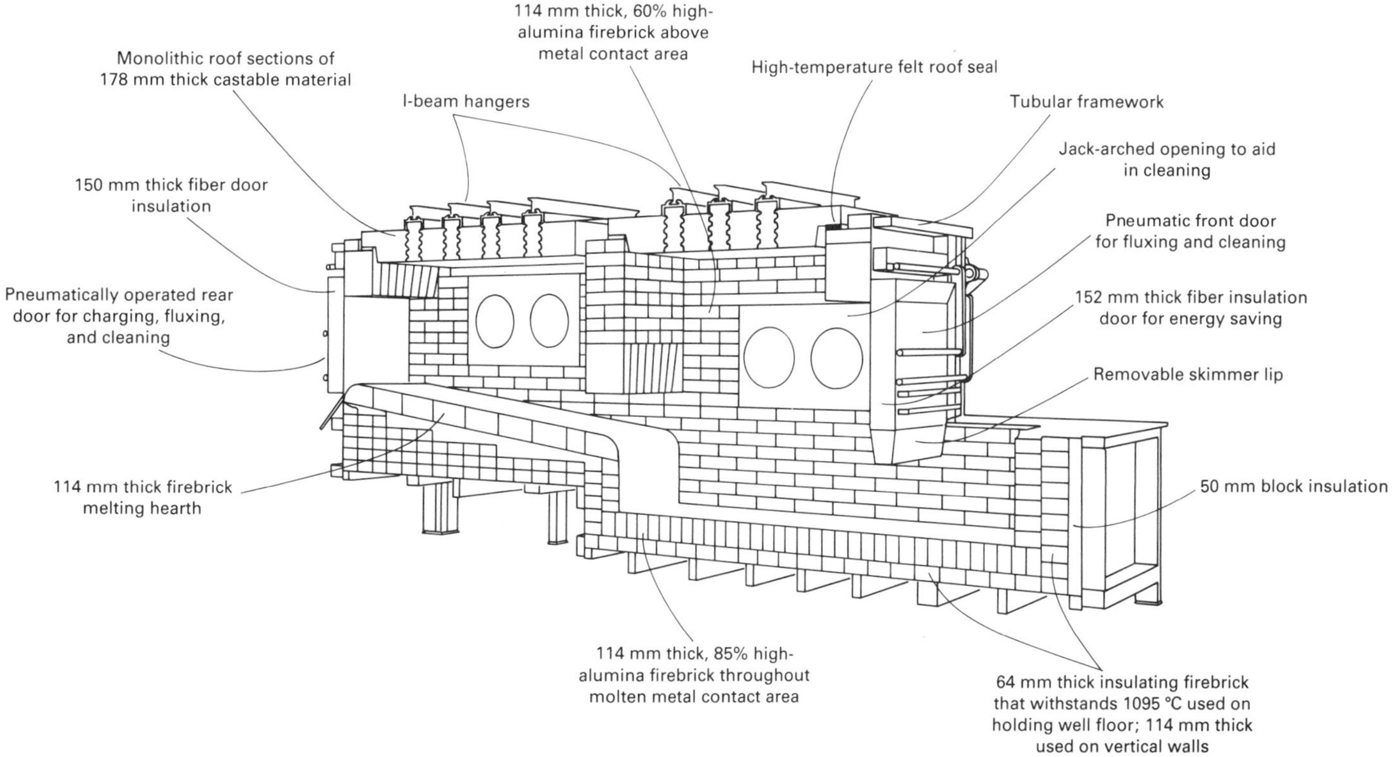

Fig. 8 Cutaway view of a combination of dry hearth/holder reverberatory furnace used as a source of molten metal for the production of aluminum castings

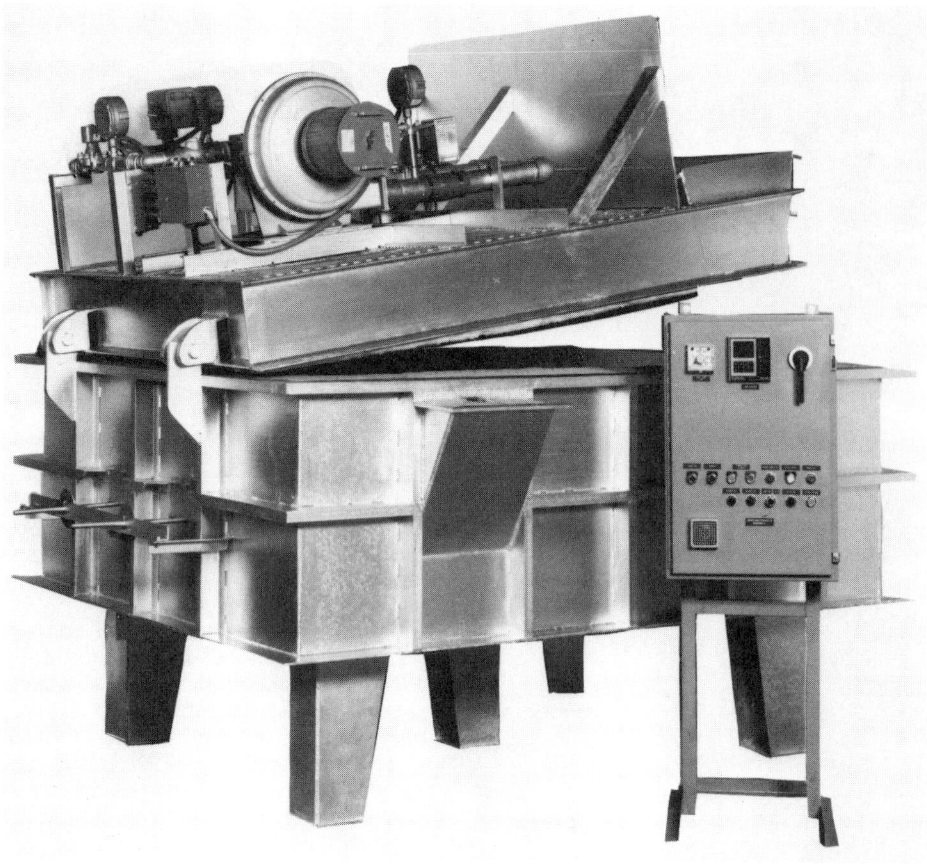

Fig. 9 State-of-the-art melter/holder reverberatory furnace. Courtesy of Courtland Furnace

Bath Stratification. The hot molten metal, being less dense, tends to float toward the top of the bath, while the cooler molten metal and the unmelted solids remain on the furnace bottom. Variations of up to 110 °C (200 °F) between the top and bottom of a 686 mm (27 in.) deep furnace have been documented.

Natural circulation or convection currents in aluminum baths are very poor. Cold molten metal is denser or heavier than hot molten metal. For this reason, there is no natural tendency to develop circulation. By design, many furnaces are heated through the surface, and the surface tends to get hotter. In addition, with the cold charge at the bottom, the molten metal there tends to stay cold. This is an inherent problem in the design and operation of hearth furnaces that must be solved to provide efficient heating of the molten metal.

Refractory deposits are classified as furnace floor deposits and melt line deposits. Both are discussed below.

Furnace Floor Deposits. Upon melting, the solid first becomes a slush or a solid liquid mixture. The more fluid component of this slush will drain off and often will leave a residue or sludge consisting of iron, manganese, chromium, silica, and copper. These high melting temperature alloy components are the last to become liquid. If substantial quantities or substantial depth of this unmelted alloy or sludge is allowed to collect on the furnace bottom, this sludge will slowly change into an insoluble, highly refractory monolithic growth on the furnace floor. With such continuous abusive furnace operation, the deposit can seriously decrease the furnace capacity. Solutions to this problem include limiting bath depth, frequent stirring, and the use of mechanical or electromagnetic circulators. Circulating pumps can reduce the temperature differential between the top and bottom of a 686 mm (27 in.) deep bath to about 4 °C (7 °F) and can increase melting rate by 30%.

Table 2 Heat input as a function of aluminum alloy temperature

Temperature rise, °C	Temperature rise, °F	Specific energy required, kJ/kg	Specific energy required, Btu/lb	% of total	Cumulative %
20–425	70–800	484	208	41	41
425–660	800–1220	186	80	16	57
Liquid to solid	Liquid to solid	395	170	33	90
660–770	1220–1420	116	50	10	100

Melt Line Deposits. Many aluminum oxides, sand charged with scrap, and other light materials will float and collect on the top of the molten aluminum bath. Often, refractory slag compositions are formed and deposited on the furnace walls at the metal line. Regular periodic cleaning (once per shift) is usually required to control this problem in furnaces in which considerable metal melting is done.

Dross By-Product. The melting of aluminum results in a residue of oxides or dross that must be controlled. This residue varies in density. Some of it floats, and other forms remain either in the bath or sink to the furnace bottom.

This dross accumulation is minimized where dry hearth melting is used. Here the newly melted liquid that separates from the oxides flows into the bath, leaving most oxides and other insoluble material on the dry hearth.

However, in melting in which solids are charged into the bath, oxides lighter than the melt will float; those heavier or the same weight as the bath will remain in the bath to be removed later. Dross causes four serious problems:

- Molten aluminum in small beads becomes entrapped in floating dross and is often lost at heavy costs; recovery processes (fluxing) can be used to reduce this loss
- With surface-fired furnaces, the floating drosses can become hot enough to light-off or start the burning of aluminum; this burning or thermiting of aluminum can result in large melt loss costs. Aluminum burning releases considerable heat and can produce foundry data that would indicate low fuel costs; in reality, all commercial fuels are less costly than aluminum as heat sources
- Large surface oxides in the surface-fired furnaces aggravate the furnace sidewall deposit or buildup problem
- Depending on metal quality requirements, excessive melt drossing often makes the subsequent fluxing, degassing, filtration, and other quality improvement processing more difficult and less reliable

Optimizing metal quality and prolonging furnace life involve consideration of five factors. These are temperature control, refractory temperature variations, furnace charging techniques, burner maintenance, and general preventive maintenance.

Temperature Control as Related to Metal Quality. Aluminum, being very active chemically, becomes even more active as the temperature of the molten metal is increased. Therefore, aluminum foundries should use the lowest practical melt temper-

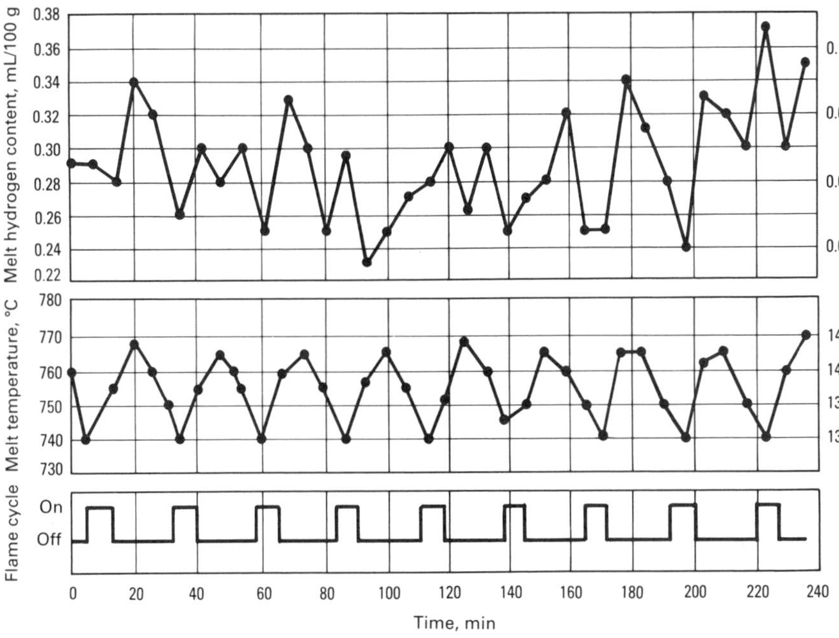

Fig. 10 Effect of metal temperature and hydrogen content on casting porosity

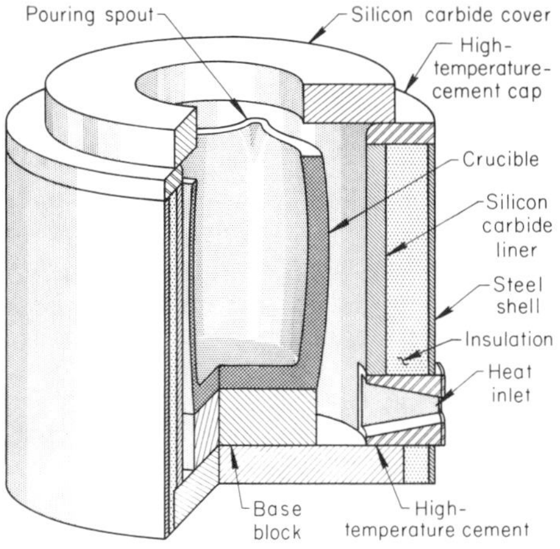

Fig. 11 Typical lift-out version of a stationary crucible furnace specifically adapted to the foundry melting of small quantities (<140 kg, or 300 lb) of copper alloys

atures at all times. The general rule is to heat the bath to no more than 60 °C (100 °F) above pouring temperature; it should be noted that the oxidation rate of aluminum doubles between 770 to 800 °C (1420 to 1470 °F). Melt quality control procedures are often substantially reduced in effectiveness when they are used at higher-than-necessary temperatures. This includes procedures for hydrogen porosity control, grain refinement, silicon constituent modification, and loss of highly active alloy additives such as magnesium, beryllium, boron, sodium, and lithium.

An example of this temperature-dependent activity is shown in Fig. 10. The melt temperature and melt hydrogen content are related to the time when the furnace flame is on and when it is off. Figure 10 clearly shows that a repeating sequence occurs. First, the heat or flame comes on, and then within a couple of minutes, the temperature begins to rise. Shortly after that, the hydrogen content of the bath increases. The reverse occurs after the flame is turned off. Hydrogen content is increased by 50% (from 0.24 to 0.36 mL/100 g, or 0.07 to 0.1 in.^{3}/lb), with the temperature cycle between 740 and 770 °C (1365 and 1420 °F). It is interesting that the gas content is also instantaneously lost as the temperature is lowered.

Similarly, the beneficial effects of grain refiners and silicon constituent modifiers fade more rapidly or are lost sooner at high temperatures than at lower temperatures. Unlike hydrogen porosity, fading grain refinement or modification is not regained simply by lowering melt temperatures.

Effect of Refractory Temperature Variations on Furnace Life. Continuous operation without weekend or other frequent cool-down periods does much to extend furnace life and to minimize maintenance problems. Cooling of the refractory from operating temperatures to room temperatures causes considerable damage, such as cracks and loosening of anchors.

The best preventive maintenance procedure is close and careful condition examinations during weekly drain and clean operations. Holes or weaknesses in the refractory and wall or hot spots in the floor are usually obvious. Contact temperature indicators are useful. Use of optical or photo infrared equipment to monitor external furnace surface temperatures is an excellent method of keeping a record of developing hot spots or weak furnace wall, floor, and roof areas.

The start-up procedure for cold furnaces should always include heating of the empty furnace to a high temperature prior to charging. This allows the refractory to heat, to expand, and to close all cracks that have opened during cooling. If the metal is melted in a cold furnace, these cracks will be filled with molten metal and will weaken the refractory. This can substantially reduce the life of the furnace.

Effect of Furnace Charging Techniques on Melt Quality. The rough or aggressive charging of sharp ingot, sow, or scrap into an empty furnace can and often does cause considerable damage to furnace floors and walls. Light scrap is often charged first to cushion and to protect the furnace floors and walls.

It is essential that water and moisture in furnace charges be avoided. Serious and even fatal accidents have resulted when metal that contains water was loaded into hearth furnaces. To prevent this and to reduce fuel costs substantially, preheated charges should be used as often as possible. Preheating of the scrap can reduce melt time by as much as 60%.

Alloy additions often require special care. Furnace charging techniques for alloy additions are covered in the four points that follow.

First, active light metal additions will float and, without care, will burn on the surface without producing the intended effect. Alloy additions such as magnesium, sodium, calcium, and lithium should be submerged with a tool to prevent excessive losses.

Second, silicon will also float and needs to be stirred into the melt before excessive surface oxidation occurs. This can oxidize to a protective surface that will not allow the silicon metal to dissolve.

Third, heavy metals such as copper and lead will sink to the bottom of the furnace (the coldest part of hearth furnaces) and will often be difficult to dissolve. For the fastest alloying cycle, these should be suspended high in the bath (using a basket, for example).

Finally, zinc and tin cause special problems because they are both heavy and have low melting and vaporization temperatures. Improper charging of these metals can drill a hole in the furnace floor (zinc drilling). Large slabs that are charged directly will go to the melt floor; the zinc then melts and vaporizes. The vapor then escapes into the porous refractory and condenses when it gets to the cooler portions of the floor. Here it condenses, solidifies, and expands to crack the refractory. This process can and will bore a hole through the furnace floor and cause furnace failure. To prevent this, metals such as zinc and tin should be charged using a steel basket held high above the melt furnace floor.

Burner Maintenance to Optimize Metal Quality. As a result of past fuel shortages and increased fuel costs, many important improvements have been made in the design and operation of melt furnace burners. In most cases, these improvements have resulted from better air-fuel mixing, flame control, and the use of preheated, constant air temperature feed to the burners. A particularly interesting burner development uses a pulsating flame on the metal surface for more efficient heat transfer. Contrary to past practice, very significant efficiency and quality gains have been realized by directing the flame on the bath surface.

Careful maintenance is often required in order to reap benefits of these improvements. Clean burner blocks are particularly important, because these units usually control air flow and flame shape. The furnace air-to-fuel mixture or furnace atmospheric condition is also very important.

Particular maintenance effort should be directed toward burners and related components in order to achieve efficient furnaces that produce quality aluminum alloys for good castings. A neutral or reducing atmosphere should be used because an oxidizing atmosphere will cause a dramatic increase in metal loss if the gases impinge directly on the metal surface.

A regular preventive maintenance program should include operation and condition checks on the following items:

- Burner block and refractory conditions
- Burner systems. Particular attention should be paid to deposits on and in the burners
- Fuel-to-air ratios. Unless constant-temperature air is used for combustion, fuel-to-air ratios should be checked carefully. With any major difference in air temperature (for example, seasonal changes), air volume must be adjusted to supply the proper amount of oxygen for efficient fuel use and melting
- Tap hole condition (inside and out)
- Flue or stack condition, particularly at the exit from the furnace chamber
- Housekeeping in the working and safety lane areas

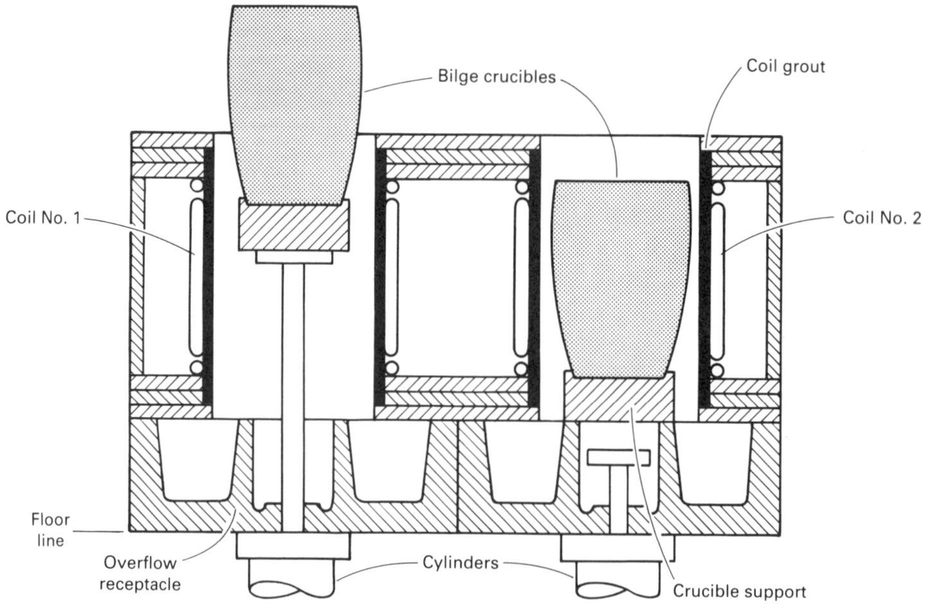

Fig. 12 Cross section of a double pushout stationary crucible furnace

Safety

All operations dealing with molten metal must involve careful consideration of operator safety. A thorough discussion of the safety precautions involved in foundry melting is highly recommended for any and all melt room activities. The three areas of metal melting that require special attention (furnace operation, furnace design, and personnel protection) are discussed below.

Furnace Operation. Most accidents occur during the charging and removal of molten metal. Worker exposure to splashes and potential explosions should be examined closely. Training programs are often required so that operators have a good understanding of all necessary precautions and safety measures. Such measures include the following:

- Whenever possible, carefully preheated charges should be used. This will avoid the possibility of loading water- or moisture-containing charges
- All areas around a melting furnace should be clear of obstructions to provide quick evacuation or escape routes in case of emergency
- Clean, dry, and carefully coated ladles and other tools should be used at all times
- Thorough preheating procedures should be followed before any tool is used in the bath
- Safety equipment and instructions for its use should be prominently displayed in case of furnace leaks or failures
- A well-conceived plan should be implemented to minimize potential hazards and to prepare personnel for any emergency
- Few of the materials used in most foundry hearth furnace operations can be considered hazardous. National Institute of Occupational Safety and Health programs require that Material Safety Data Sheets be readily available and used

Furnace Design. A principal consideration in furnace design should be to minimize worker exposure to splashing or exploding metal. Most furnace designers avoid floor-level charging wells as a concession to this safety consideration; instead, they use dip-out and charge wells that are waist high or higher. This enables the operator to avoid most splashes and spills of molten metal.

In the past, the furnace-tapping operation was a high-risk activity. Today, however, with the availability of insulated fiber cones or plugs, much of the problem has been overcome. Protective clothing should be worn, and at least two well-protected operators should be on hand when a furnace is

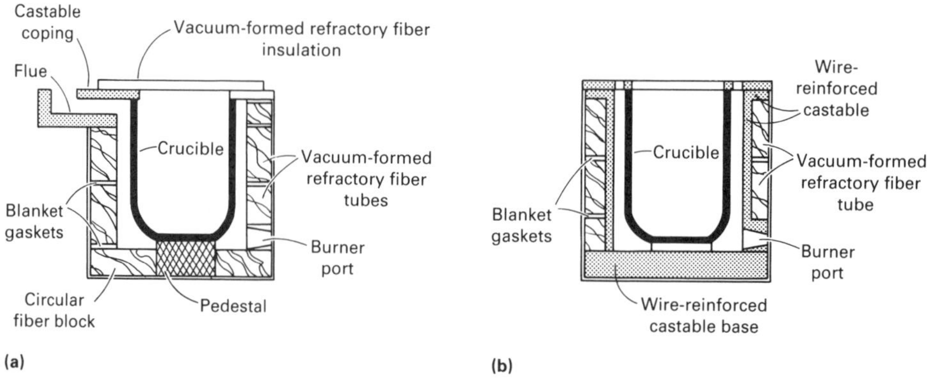

Fig. 13 Stationary crucible furnaces equipped for hand-ladle dip-out pours. (a) Side flue configuration. (b) Configuration with cover plate openings on outside diameter of crucible top

being tapped. When an oxygen lance is used to open a tap hole, extreme care and full protective clothing and head gear should be used.

Personnel Protective Equipment and Clothing. Protection from flying molten metal due to splashes and minor explosions is the key to the selection of proper clothing and safety equipment. The use of wool shirts and trousers is a long-standing solution to these hazards. A metal splash causes the apparel to emit the odor of burning wool, and quicky warns the wearer of a potential problem. Clothes made of acrylic and other synthetic fibers should never be worn in the melting area, because they easily burn or form a hot liquid that can produce very serious burns. A 0.5 kg (1 lb) quantity of aluminum at 675 °C (1250 °F) that is plunged directly into water will produce an explosion equivalent to 1.4 kg (3 lb) of TNT.

Crucible Furnaces

Few if any metallurgical or industrial processes have a more distinguished history than the crucible melting of metals. Crucibles or earthenware smelting pots that were used to refine copper from King Solomon's mines can still be found in that area of ancient Israel. Some pots had capacities of up to 0.40 m^3 (14 ft^3). The Wootz steels of India, the famous Damascus sword blades, and the equally famous steels of Toledo, Spain, were produced all using crucible melting processes.

The modern crucible process offers industry a great deal of flexibility and a wide variety of options with regard to metals, melt sizes, fuels, smelting, and processing techniques. Aluminum, brass, bronze, copper, ductile or gray iron, steel, magnesium, Monel, nickel, refractory alloys, steel, and other metals and alloys are produced using the crucible melting processes. Crucible capacity can vary from mere ounces for laboratory melts to 1.4 Mg (3000 lb) for the melting of aluminum alloys. Fuel choices include coal, coke, electricity, commercial gases (natural, propane, producer, and so on), and fuel oil. Crucible liners are often used for low- or high-frequency electric induction furnaces.

Foundry metal melting and handling operations that utilize crucible equipment include furnace melting, pouring ladle equipment (either separately heated or as a ladle liner), and liners for electric induction melting furnaces. Metal-processing operations include in-line filtration, fluxing, and hydrogen gas content control. Crucible melting can also be used as a component of quality adjustment equipment between melter and casting furnaces to alter the chemistry and temperature of the melt for control of such factors as porosity content prior to casting. Crucible melting is a simple and flexible process. Crucible furnaces can generally be started or shut down at a moment's notice. Normally, there is no stand-by charge for labor costs between uses of this equipment.

Furnace Types

Foundry crucible furnaces can be classified as stationary, tilting, or movable.

Stationary furnaces utilize a crucible set inside a refractory-lined shell that has heating equipment attached. Normally, molten metal is dipped out with hand ladles, or the crucible containing the molten metal is removed from the shell for the casting operation.

A stationary crucible furnace built to accommodate a lift-out crucible is shown in Fig. 11. By using tong and shank tools to lift and carry the bilge-shaped crucible to the mold, pouring of the casting is done directly from the crucible without the need for potentially damaging metal transfers. The maximum crucible size normally used in this manner holds 140 kg (300 lb) of aluminum or copper. The weight and the strength of the crucible limit this type of operation. A similar stationary crucible furnace designed for use with a bowl shape and for hand ladle dip out has a capacity of up to 1.4 Mg (3000 lb) of aluminum alloy. The double pushout furnace shown in Fig. 12 uses induction coils instead of fuel gases to heat the melt and hydraulic cylinders to raise and lower the crucibles within the coils.

Some stationary crucible furnaces are equipped for hand-ladle dip-out pours. For most metals, it is imperative that the flame and flue gases do not contact the charge. Two methods of controlling this critical factor are illustrated in Fig. 13. The first is the side flue configuration used in an aluminum melter installation (Fig. 13a), and the second is the cover plate configuration used for a brass foundry furnace (Fig. 13b). The cover plate setup incorporates holes on the outside diameter of the crucible top. Also shown is the extensive use of highly insulating fiber refractory material. This material has a low heat content and serves to conserve fuel and to minimize the temperature override that often occurs with the heavier refractory furnace linings. The added refractory flame face is required to accommodate the higher melt temperature.

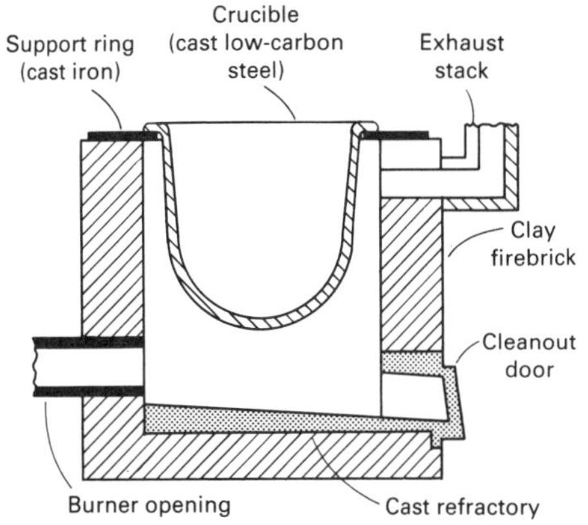

Fig. 14 Cross section of a stationary fuel-fired furnace used for the open crucible melting of magnesium alloys

Steel crucibles can also be used in stationary crucible furnaces, as shown in Fig. 14. The high heat conductivity of the steel offers fuel and meltdown time savings. Steel and ductile iron crucibles are extensively used for magnesium and zinc melting. In these molten alloys, there is little attack of the crucible walls.

On the other hand, aluminum alloys will attack the steel crucible walls by forming sludge (combination of iron, manganese, and chromium), which eventually causes

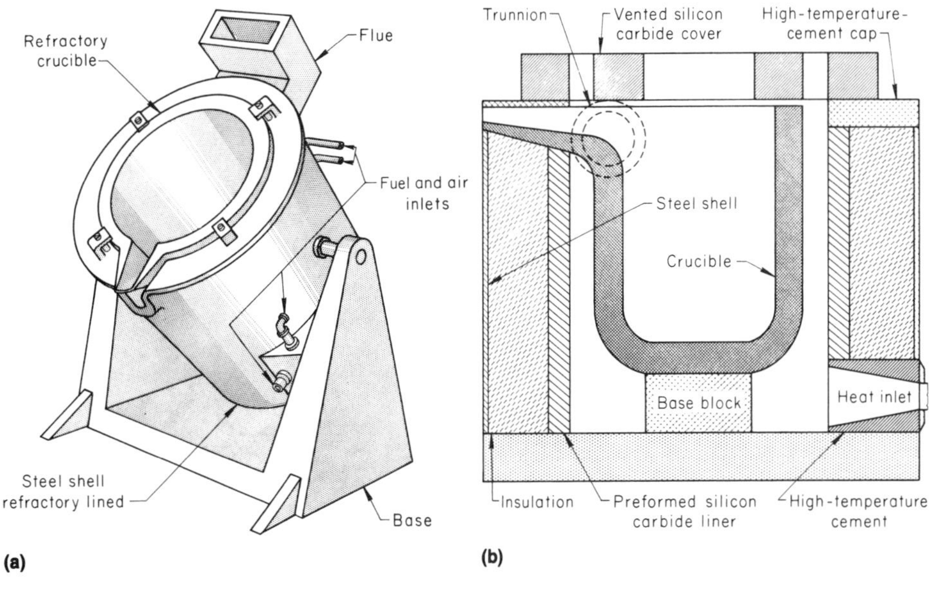

Fig. 15 Two variations of a tilting crucible furnace. (a) Center axis. (b) Lip axis

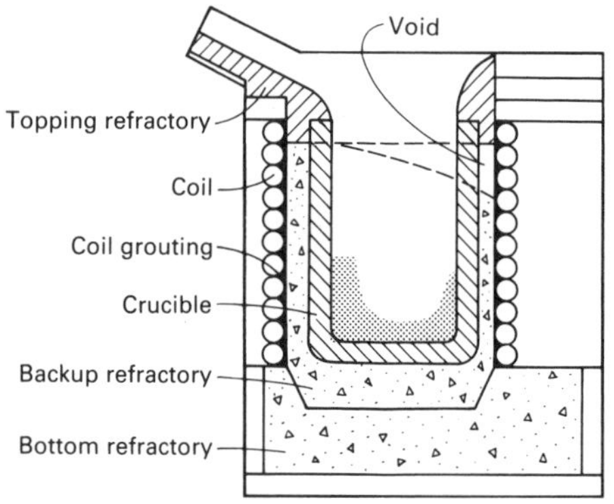

Fig. 16 Cross section of a tilt furnace for the high-frequency induction melting of brass and bronze alloys. Crucible is of clay-graphite composition. Also shown are the locations of the molten metal buildup and the voids in the backup refractory, which shorten crucible life.

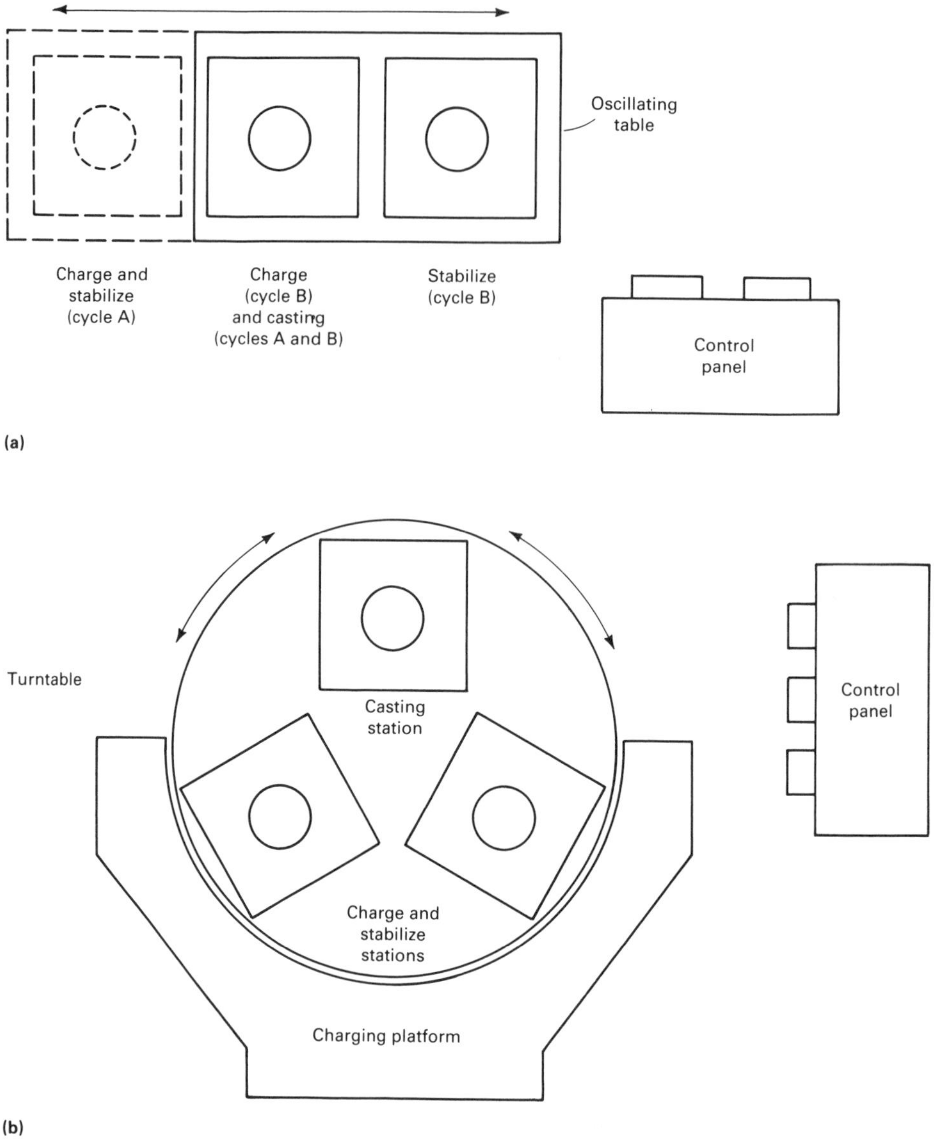

Fig. 17 Two versions of an automated modular melting operation used to produce castings. (a) Shuttle. (b) Turntable. Either method can use two to four separate crucible-type melters.

hot spots in the castings produced from this melt. However, there have been successful aluminum foundry uses of steel crucibles. This requires a thorough refractory coating on the crucible surfaces in contact with molten metal and careful cleaning of the rust that tends to accumulate in the firing chamber (for example, flame contact with the crucible causes rust and scaling). This can collect in the firing chamber. A special clean-out door is used to remove this residue.

Both aluminum and magnesium can react explosively with iron oxide. The use of electrical resistance furnaces and steel crucibles reduces the seriousness of this problem, and long-term use of steel crucibles and resistance furnaces has been successful in the casting of aluminum alloys.

Tilting furnaces incorporate a crucible that is supported in the shell such that both are free to rotate on an axis. This allows the molten metal to be poured into ladles or molds. The axis of the pour can be either at the center of gravity or at the lip of the furnace. This design simplifies the handling of molten metal in many foundries.

Tilting crucible furnaces are designed to simplify metal removal and have been extensively used in numerous foundries for many years. Figure 15(a) illustrates a center-axis furnace; Fig. 15(b) shows a cross section of a lip-axis furnace. Tilting crucible furnaces are useful where castings require more metal than can be easily handled by stationary crucible furnaces. These units are more flexible and can easily handle production requirements that are too small for the efficient operation of larger hearth furnaces.

Figure 16 illustrates the use of a crucible liner for an induction melting tilt furnace. Use of the crucible liner overcomes the problems that would occur as a result of the strong erosive currents existing at the crucible sidewalls. The low electrical conductivity of clay-graphite crucibles makes them ideal for use in such applications. A gradual buildup of molten metal in the heel and voids in the backup refractory, both caused by repeated tilting of the furnace, are detrimental to crucible life (see the discussion "Furnace Design Considerations" in this section).

Crucible liners, which provide a durable wear-resistant surface for molten metal, are often used in the brass, steel, and ductile iron foundry industries. This application for the foundry use of crucibles is similar to that for the induction furnace liners shown in Fig. 16.

Movable Furnaces and/or Pouring Ladles. In this category, either melter or holder furnaces are cycled between the in-line pouring station and the melt and/or the postmelting temperature and quality adjustment operation. The requirements of new highly automated pouring methods have encouraged new thinking in many areas. The modular melting method is particularly interesting. This method uses two-, three-, or four-crucible melters that are arranged on either a shuttle or a turntable, as diagrammed in Fig. 17. In operation, one furnace is always in a holding mode. This furnace is considered to be in the pour position. The remaining furnace positions are used alternately for recharging, stabilizing, and metal treatment. This method eliminates the need for the transfer of molten metal within the foundry. When automatic pouring is used, temperature stability of ±3 °C (±5 °F) can be maintained while casting with robotic pouring equipment. A three-crucible modular melter is capable of preparing, melting, and delivering up to 650 kg/h (1425 lb/h) of molten metal to the casting unit.

In recent years, the quality of refractory crucibles has improved dramatically. Since the 1940s, for example, the practice of glazing the crucibles and a more thorough factory curing of crucibles have eliminated the need for a 3 to 4 day curing or heat-up operation at the foundry before actual crucible use. Modern crucible fabrication, curing, and inspection methods have also done much to improve the quality and length of crucible life.

Furnace Design Considerations

Furnace design and operation are critical for efficient operation. Key points are discussed in the following sections.

Proper Furnace Atmosphere. With few exceptions, a slightly oxidizing flame should be used. With aluminum alloys, this is a pale yellow color. For copper-base alloys, this will be tinged with green (due to the oxidation of the copper). A traditional test for the correct furnace atmosphere is to place a cold piece of zinc in the flame. A yellowish color indicates an oxidizing flame, and a black color indicates a reducing flame. The objective is to obtain a flame having a tinge of yellow, which indicates the presence of the desirable, slightly oxidizing atmosphere.

Oxidizing conditions were maintained in old coke-fired crucible furnaces at the crucible walls by burning the coke in a 75 mm (3 in.) space around the crucible sidewall. The burning coke maintained a slightly oxidizing atmosphere at the walls and ensured good heat transfer.

Burner Control. For efficient melting and optimum crucible life, it is necessary to have even heating of the crucible and to avoid hot and cold areas around the sides of the crucible bowl. Some furnace designers use two or four burners to minimize this problem. This will contribute to longer crucible life, better fuel economy, faster melting, and lower noise levels. Precise periodic adjustment of the burner(s) is essential to efficient crucible furnace operation.

Special care is required to avoid overheating of the bath. Approximately 80 to 95% of the heat required to prepare a heat for the foundry is used to heat the charge to the melting temperature and then to convert the solid to a liquid. Very little heat is required to raise the temperature of the newly melted liquid to the required pouring temperature. This can cause problems.

For rapid melting of the crucible contents, high heat inputs are normally used. As soon as the charge becomes entirely liquid, little or no additional heat is required. If the equipment or the operator does not immediately turn the heat down or off, the bath temperature can become far higher than required. For most metals, this causes quality control problems. Currently, most foundries depend on their furnace operators to control this harmful variable.

Burner Design. Because of the wide diversity in crucible applications utilizing many different fuels, a summary of these data is well beyond the scope of this discussion. It should be noted that developments in the design and operation of burners have significantly improved this furnace component. Any new installations of significance probably should include recent developments in both burner designs and burner control units.

Charging Techniques. The most efficient melting (in terms of heat transfer or efficient use of fuel) is achieved with faster melt times, and the highest metal quality is obtained when a heel of metal is used to start the melting. Here the heat travels easily from the flame through the crucible wall and into the metal. A crucible that contains only a solid charge provides little contact between the crucible wall and the metal to be melted. As a result, there is a considerable time lapse between charging and the initial melting of the charge. For efficient melting, a good heel size is roughly 25% of the crucible capacity.

A second point concerning charging is the need to keep a loose charge. When a cold charge is tightly jammed into the crucible, the expansion of this charge upon heating will often break or crack the crucible.

The third point concerns the need for gentle charging of all heavy or sharp-cornered material. Crucibles are fragile and easily broken. Many crucibles have been broken by rough and careless charging with heavy and/or sharp ingot or scrap.

Handling of Crucibles During Melting and Pouring. It is important to keep in mind that hot crucibles filled with molten metal are fragile and therefore deserve careful handling. Care should be taken to set them on sand or other nonconductive surfaces. Placing a hot crucible full of molten metal on a steel plate or damp floor can easily break or seriously weaken the bottom of the crucible.

Cleaning of crucibles is important in preventing premature failure. In many, if not most, metal-melting operations, slag or dross buildups develop at the metal surface line, and heavy unmelted slush residue collects on the bottom. Both of these residues shorten crucible life by causing hot and weak wall areas. The poor and uneven heat transfer around the crucible walls causes hot spots and uneven stresses within the crucible walls. The metallic residues that will expand faster than the crucible upon heating can easily rupture the wall. At times, these residues expand and fracture the crucibles during the normal heating and cooling cycles (Fig. 16). Cleanliness and careful workmanship on the melt room floor should be encouraged. However, in many foundries, the tradition and environment of this work area tend to foster an attitude of disdain among personnel for the meticulous work required for efficient operation. This is the source of many foundry problems. Quality begins with the melting and handling of metal, and failure in this portion of the operation can doom the final casting sequence.

Using the correct metal melting technique and equipment for each specific melt will produce high quality castings. In recent years, the introduction of statistical process control (SPC) methods to foundry operations in order to monitor each step of the casting production process has been helpful in keeping scrap losses to minimum levels previously unattainable.

Cupolas

Sam F. Carter, Carter Consultants, Inc.

The cupola is basically a cylindrical shaft furnace that burns coke intensified by the blowing of air through tuyeres (nozzles). Alternate layers of metal, along with replacement coke, are charged into the top. In its descent, the metal is melted by direct contact with the countercurrent flow of hot gases from the coke combustion. The molten metal collects in the well, where it is discharged for use by intermittent tapping or by continuous flow.

Melting in a cupola dates back several centuries to old cask-type units that were blown with hand-pumped bellows. The cupola as we know it today was patented in 1794 by John Wilkinson. Prior to the 1950s, the cupola was the predominant furnace for iron melting because of its simplicity and low cost of melting.

Through the years, the cupola has been progressively refined toward more efficient combustion, better control, and more flexibility in the use of more varied raw materials. However, the cupola emits considerable amounts of smoke and particulate matter. The emission has been cleaned to meet the highest environmental standards, but this has required much engineering effort and considerable expense. During the 1950s, with environmental laws requiring elaborate pollution control equipment, the cupola lost much of its advantage of low cost and simplicity, especially on small cupolas, for which pollution equipment was relatively more expensive.

In the meantime, progressive improvements were made in electric induction furnace equipment. Because induction furnaces required little or no pollution control equipment and yielded low-sulfur iron that was satisfactory for the fast-growing ductile iron industry, electric induction furnaces replaced many cupolas, especially in small-tonnage operations and where ductile iron or several types of iron were produced. Electric induction melting was found to have a lower melting cost under many of these conditions. The number of cupolas has decreased from 4000 in the early 1950s to less than 1400 today. However, the cupola remains the predominant tonnage melter with over 60% of the iron tonnage still melted in cupolas.

In high-tonnage operations, the cupola remains the most reliable source of the continuous high volumes of iron needed to satisfy multiple molding lines of high-production foundries or multiple casting machines of centrifugal pipe producers. The continuous stream of cupola-melted iron is not interrupted by charging, which is done independently upstairs on the back side, nor by slag removal, because the slag flows off to the side continuously from a front or rear slag spout without interrupting the melting process.

In such high-tonnage continuous-demand situations, the cupola has remained the lowest cost method of melting under the prevailing costs of coke and electric power. The cupola can melt a wide range of bulk scrap. Cupola iron is more difficult to control, but is favorably characterized by a low tendency toward chilling and good fluidity, machinability, and shrinkage characteristics due to natural nucleation resulting from continuous contact with the coke.

Progressive Refinements in Cupola Equipment

Since the early days of the simple cupola, continuing technological refinements have been progressively applied to the cupola to improve efficiency, flexibility, and control.

A preheated air blast was found to intensify the combustion reaction, thus increasing melting rate, temperature, and carbon pickup with the same amount of coke. Alternatively, the amount of coke can be reduced for the same melt rate, temperature, and carbon pickup.

Some of the early hot blast systems could preheat the incoming air only 150 to 260 °C (300 to 500 °F), but exhibited the advantages of the preheated blast. The most successful hot blast systems have been externally fired with gas and are capable of 540 to 650 °C (1000 to 1200 °F) preheat. Higher blast temperatures continue to improve performance, but they complicate the construction of the hot blast equipment.

Recuperative hot blast systems, in recent years, have efficiently utilized the heat content in the effluent gas to preheat the blast air and have lowered effluent carbon monoxide content to very desirable environmental levels.

Oxygen enrichment and injection, which adds 1 to 4% O_2 to the blast air, also intensifies the combustion reaction by increasing temperature and melting rate.

Divided-blast cupolas divide the air between two rows of tuyeres in controlled proportions. In many cases, this improves combustion efficiency to accomplish either coke reduction or increased melt rate and temperature with the same coke ratio.

Humidity control of the blast air has been employed on some cupolas to eliminate the effect of varying humidity where very close control is important.

Injection of fine coke through the tuyeres has proved effective in increasing carbon content and in using fine coke as a substitute for premium coke.

Basic slags made possible the melting of very low sulfur iron directly from the cupola, and the higher carbon pickup made possible the use of more steel scrap. By regulating slag basicity, a wide range of compositions can be obtained from a wide range of materials.

Water cooling has extended the length of the melting period to 24 h and has permitted a single cupola stack to be banked and operated for weeks before requiring repairs during a weekend shutdown. Water cooling by removing the effect of refractory on the slag chemistry permits the same cupola to operate with a choice of basic, neutral, or acid slag simply by regulating the amount of basic flux added.

Pollution control equipment was progressively developed to meet the most rigid environmental standards, either with high-energy wet scrubbers or with dry baghouse collectors.

Duplex electric holders have ensured instant supply during variations in demand and have improved the consistency of composition and temperature. Large electric channel induction furnaces have become popular as receivers for large cupolas. These electric channels can boost temperature or maintain temperature over down periods.

Continuous analyses of top gases has made it possible to monitor combustion efficiency and to signal any water leaks to avoid explosive gas combinations.

Computer applications have been increasingly utilized to record, monitor, and progressively control more cupola conditions.

Cupola Construction and Operation

This section will discuss the shell, intermittent or continuous tapping, tuyere and blower systems, refractory lining, water-cooled cupolas, emission control systems, and storage and handling of the charge composition.

Shell

A vertical, cylindrical steel shell is the container of the combustion and melting operation, as shown in Fig. 1. The diameter is chosen for the required melting rate.

Standard commercial cupolas are available in 15 sizes, as indicated in Table 1. Sizes range in diameter from No. 1 to 12:

- No. 1: 813 mm (32 in.) shell lined to 584 mm (23 in.) ID with a melting capacity of 0.9 to 1.8 Mg/h (1 to 2 tons/h)
- No. 12: 2.74 m (108 in.) shell lined to 2.13 m (84 in.) ID with a melting capacity of 23 to 36 Mg/h (25 to 40 tons/h)

Larger cupolas have been custom designed to 3.81 m (150 in.) in diameter with a melting capacity of 64 to 91 Mg/h or 70 to 100 tons/h.

A charge door opening at a vertical distance usually four to five times the diameter receives the charges of metal, coke, and flux dumped from drop-bottom buckets, side-dump skips, or shaker conveyors.

Drop-bottom doors at the bottom of the stack facilitate quick emptying of the stack at the end of an operation to expedite refractory repairs and the resumption of melting.

The foundation must be adequate for the size and load.

A sand bottom is rammed over the closed bottom doors to contain the melted iron in the well. This sand bottom, or refractory, is sloped toward a taphole that is properly sized and maintained for the outward flow of metal.

Intermittent or Continuous Tapping

Iron flow can be intermittent or continuous. Early cupolas were intermittently tapped by opening and closing the taphole as in a blast furnace. Slag was accumulated and flowed out a back slag hole at a higher level.

Most modern cupolas are tapped continuously, with clean metal flowing under a refractory knife and over a dam. This special trough skims off the lighter slag on top. This slag then flows continuously down a side spout into a water-sluicing trough and containers for handling. The levels of metal dam and slag notch must be carefully calculated and maintained to the proper relationship with blast pressure inside the cupola.

The dry-bottom cupola is a European development in which iron and slag are not collected in the well of the cupola but run into a special vessel outside and beside the cupola. In the special outside vessel, the accumulated metal and slag are separated as they flow out through the usual front slagging dam and spout. The advantages claimed for this setup include more consis-

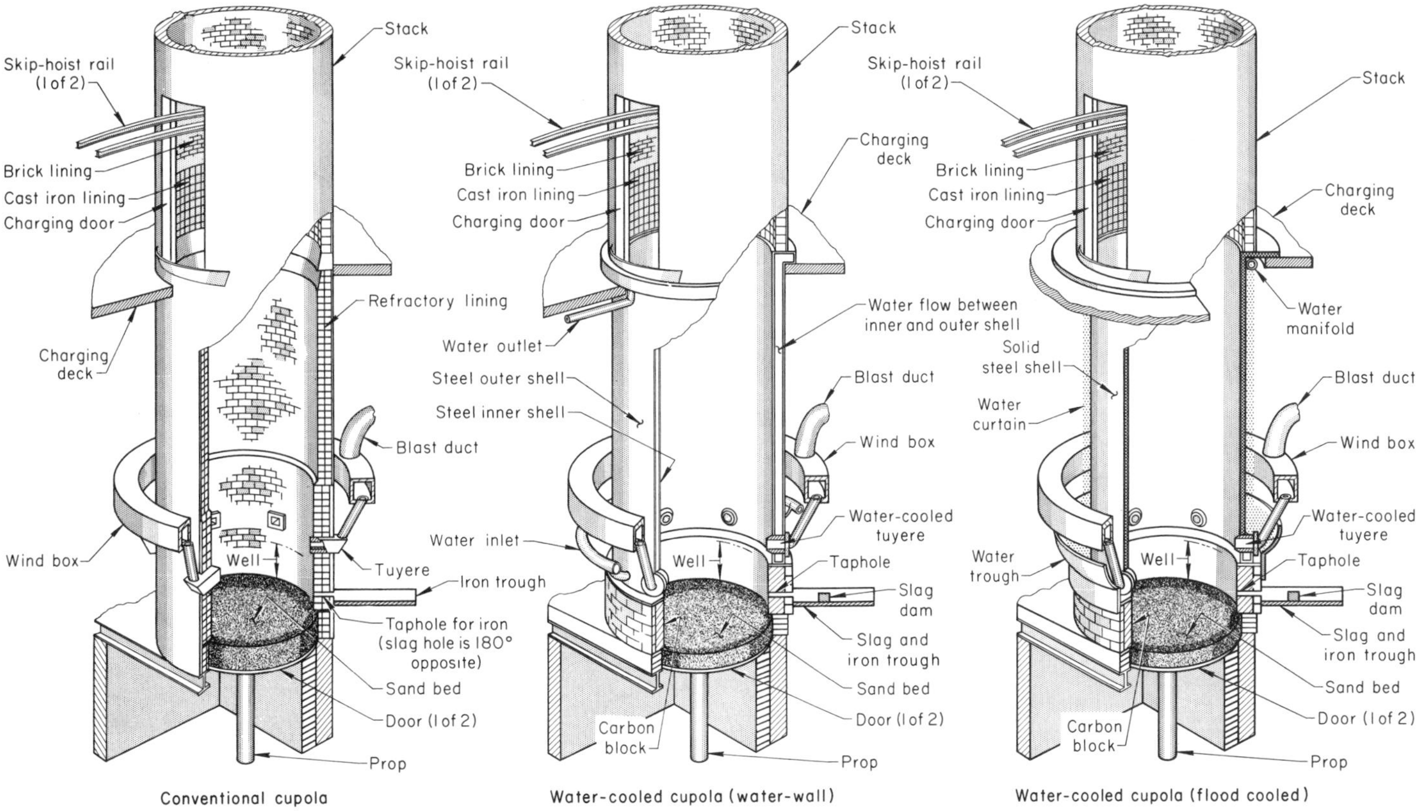

Fig. 1 Sectional views of conventional and water-cooled cupolas. The conventional type shown is refractory lined. Water-cooled types incorporate either an enclosed jacket or an open cascade flow.

tent iron analysis and easier repair of refractories.

Tuyere and Blower Systems

Tuyeres. Air for the combustion of coke is introduced through tuyeres equally spaced around the cupola, with the number and size proportioned to cupola size for proper air volume, pressure, and velocity. Total tuyere area usually ranges from 3 to 10% of cupola area for optimum air velocity for penetration and combustion efficiency. On large, water-cooled hot blast cupolas, the desirable air velocity range is 4600 to 7600 m/min (15 000 to 25 000 ft/min). On water-cooled cupolas, the tuyeres are projecting water-cooled copper nozzles that protrude into the cupola 152 to 457 mm (6 to 18 in.) and are inclined downward 10 to 15°.

Blowers for generating the necessary blast air volume and pressure have been of the positive displacement, fan, or centrifugal type. The latest preference on large cupolas has been toward two or three multistage centrifugal blowers with automatic air weight control.

The air required to melt 900 kg (1 ton) of iron is 620 to 680 m^3 (22 000 to 24 000 ft^3) STP or 770 to 835 kg (1700 to 1840 lb). Air

Table 1 Range of sizes for conventional cupolas

Cupola size	Shell diameter		Minimum thickness of lower lining(a)		Diameter of inside lining		Area inside lining		Average melt rate(b)		Air through tuyeres(c)		Suggested blower selection(d) Discharge		Suggested blower selection(d) Pressure	Approximate holding capacity of cupola well(e)	
	mm	in.	mm	in.	mm	in.	m^2	$in.^2$	kg/h	ton/h	m^3/min	ft^3/min	m^3/min	ft^3/min	oz	kg	lb
1	813	32	114	4.5	584	23	0.268	415	0.91–1.8	1–2	26	910	29	1040	20	258	570
2	914	36	114	4.5	686	27	0.369	572	0.91–1.8	1–2	37	1290	40	1430	24	372	820
2.5	1040	41	178	7	686	27	0.369	572	1.8–2.7	2–3	37	1290	40	1430	24	526	1160
3	1170	46	178	7	813	32	0.519	804	2.7–3.6	3–4	51	1810	57	2000	28	699	1540
3.5	1295	51	178	7	940	37	0.693	1075	3.6–4.5	4–5	69	2420	76	2700	28	903	1990
4	1420	56	178	7	1065	42	0.893	1385	4.5–5.4	5–6	88	3100	98	3450	32	1030	2280
5	1600	63	229	9	1145	45	1.026	1590	6.3–7.3	7–8	102	3600	113	4000	32	1180	2610
6	1675	66	229	9	1220	48	1.167	1809	8.2–10	9–11	116	4100	127	4500	36	1540	3390
7	1830	72	229	9	1370	54	1.477	2290	11–13	12–14	147	5200	163	5740	36	1840	4050
8	1980	78	229	9	1525	60	1.824	2827	14–15	15–17	181	6400	201	7100	36	2230	4910
9	2135	84	229	9	1675	66	2.207	3421	16–18	18–20	218	7700	244	8600	42	2650	5840
9.5	2290	90	229	9	1830	72	2.627	4071	19–21	21–23	261	9200	289	10 200	42	3100	6840
10	2440	96	229	9	1980	78	3.082	4778	22–24	24–26	303	10 700	337	11 900	42	3610	7960
11	2590	102	305	12	1980	78	3.082	4778	24–26	27–29	303	10 700	337	11 900	48	...	...
12	2745	108	305	12	2135	84	3.575	5542	27–29	30–32	354	12 500	394	13 900	48	...	...

(a) Proportional to length of heats. (b) Specific melt rate will depend on coke required for thermal requirement and amount of oxygen available in blast. (c) Based on 99.3 m^3/m^2 (325 ft^3/ft^2) of area inside lining and type of cooling. (d) When auxiliary equipment is added to blast system or when piping is long or intricate, more pressure capacity may be required. Density and/or height of charge material is also a factor. (e) Based on 305 mm (12 in.) average depth of metal—distance sand bottom to slag spout from the blower is transmitted into tuyeres through a duct system and windbox. Back pressure normally ranges from 3.3 to 8.5 N (0.75 to 1.9 lbf) or more, depending on charge density and stack height.

Refractory Lining

Conventional cupolas are lined with refractories to protect the shell against abrasion, heat, and oxidation. Lining thicknesses range from 114 to 305 mm (4.5 to 12 in.). The most popular lining is fireclay brick or block. As the heat progresses, the refractory lining in the melting zone is progressively fluxed away by the high temperature and oxidizing atmosphere.

Early general practice was to drop the bottom and repair the lining at the end of the day's melt, restoring it to original dimensions. Generally, a pair of cupolas were alternately used each day to facilitate the repair and preparation of the furnace lining after each day of use.

Several methods have been satisfactorily used for this refractory patching. Among them are combinations of fireclay brick and mortar as well as monolithic ramming mixtures of clay, ganister, and sand.

The newest and most widely used method of lining repair uses a pneumatic gun to blow a mixture of clay and ganister through a nozzle containing water that becomes entrained in the stream as it exits the mixing nozzle.

Basic Refractories for Basic Slag. The advent and growth of ductile iron in the 1950s made it desirable to find a way to remove sulfur inside the cupola using a sufficiently basic slag, as is accomplished in basic steelmaking furnaces. Obviously, a highly basic slag could not be maintained in contact with the acid refractories. In addition, because the lining consumed during melting is a major source of slag, the lining must be basic or neutral in order to contribute an excess of basic constituents to the slag.

After considerable experimentation, successful basic refractories were developed for use in cupolas. Originally, linings were either magnesite or dolomite brick. Basic patch materials composed of dolomite or magnesite aggregate were developed that could be rammed or blown in using the pneumatic gun. Basic refractories cost several times more than comparably sized acid clay refractories, which increase the need for water cooling on basic slag cupolas.

Water-Cooled Cupolas

When the melting operations of high-production foundries were extended to 16 to 20 h, the continuous attack on the refractory lining left little or no lining remaining in the melting zone after hard blowing for two shifts. The need for two- and three-shift melting operations encouraged the development of water-cooled cupolas. Water cooling was first used as a backup to the refractory lining.

Because of the increasing dependence on water cooling, it was found that the refractory lining could be eliminated from the melting zone and that the cupola could be operated continuously without a lining for a full week, dropping the bottom on the weekend for minor refractory repairs to the well, taphole, and trough. The liningless water-cooled cupola has its shell offset at the tuyere level to accommodate adequate lining thickness in the well area (usually lined with carbon refractories) below the tuyeres.

When iron was needed for only two shifts, the liningless water-cooled cupola was banked during a down shift and readily brought back on stream when needed. It was then learned that these water-cooled cupolas could be banked over a weekend and operated continuously for 2 to 4 weeks before dropping the bottom for repairs. Some of the largest cupolas are equipped with double tapholes and double rear siphon slag spouts to enable the cupola to be operated continuously for long periods.

Two types of water cooling have replaced the consumable refractory lining located in the melting zone (Fig. 1b and c). Figure 1(a) shows the refractory lining of the conventional refractory-lined cupola. Figure 1(b) shows a sectional view of an early water-cooled cupola design that uses an enclosed jacket or tank through which water is circulated. Water introduced into the enclosed jacket must be at sufficient pressure to maintain a velocity and flow pattern that provide adequate cooling and prevent the buildup of steam pockets and hot spots. A major problem with the enclosed jacket is a buildup of deposits which reduces cooling effectiveness. The enclosed jacket also limits accessibility when a hot spot occurs.

A more widely used cooling system incorporates open external water sprays or cascade cooling (Fig. 1c). At the top of the liningless area, a water curtain is applied to the shell, usually by spray rings encircling the cupola, with a neoprene wiper to even out the distribution or with an overflow moat. The water curtain cools the shell as it cascades down the tapered shell and is collected in a trough where the shell is offset at the well. Figure 2 shows a cutaway view of a water-cooled cascade flow cupola.

One advantage of the open external cooling is that the shell can be blasted at intervals to remove excess deposit and to maintain thermal conductivity. In addition, any areas of inadequate cooling are made apparent by visible steaming, which can be easily corrected with increased water flow to the area.

Water-cooled cupolas have greatly extended the duration of melting operations and have reduced refractory repair costs. On basic cupolas, water cooling is especially attractive for reducing the consumption of the higher cost basic refractories. In addition, with the need for expensive pollution control systems, water cooling made it possible to reduce the number of cupola stacks, which meant reduced investment in emission control equipment.

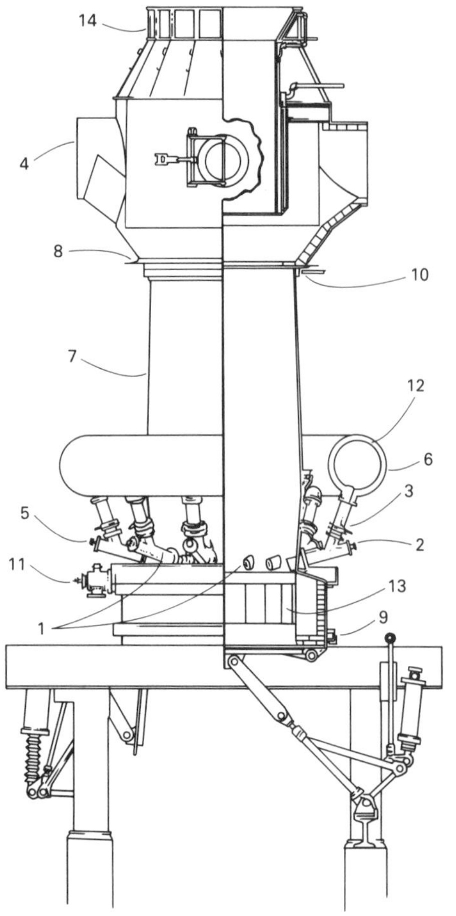

Fig. 2 Cutaway view of an open cascade flow water-cooled liningless cupola showing component parts. 1, water-cooled tuyeres; 2, tuyere elbows; 3, control dampers; 4, gas take-off; 5, sight port; 6, detached windbox; 7, welded body; 8, dry expansion joint; 9, selective well cooling; 10, upper spray ring; 11, overflow spout; 12, windbox lining; 13, carbon lining; 14, upper stack, independently supported

However, these advantages of water cooling were achieved at some loss in thermal efficiency. Higher heat losses through the water-cooled shell necessitate the use of more coke to maintain equivalent temperature and melting conditions.

In recent years, some plants have installed thin refractory linings inside the water-cooled shell. With the water-cooled backup, the lining thickness is maintained for a full week of operation. In addition, the thin lining provides sufficient insulation to reduce heat loss and coke use. This combination of water cooling behind a thin lining is gaining acceptance throughout industry.

Emission Control Systems

Equipment has been successfully engineered to enable the present-day cupolas to

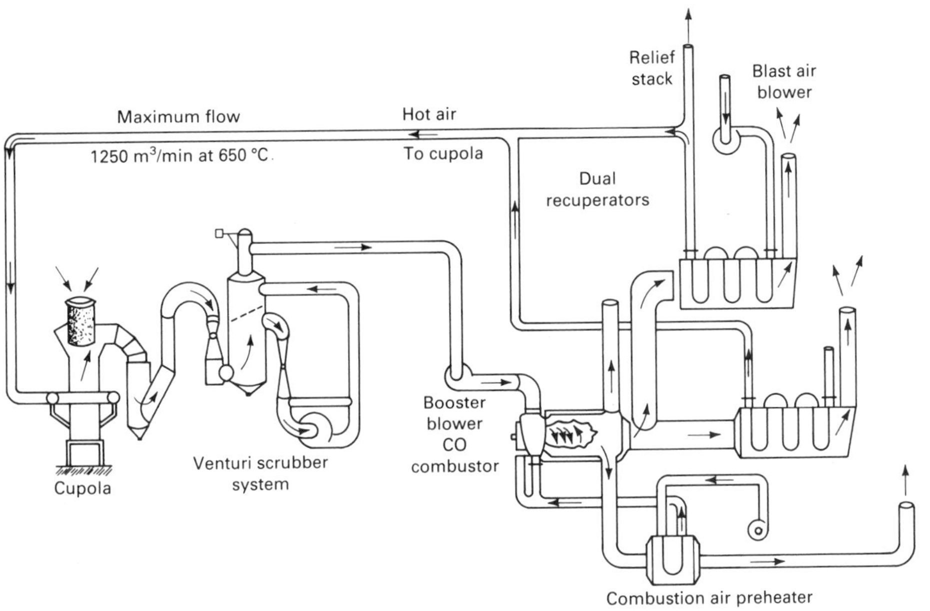

Fig. 3 Recuperative hot blast cupola with wet scrubber emission control

meet the most rigid environmental regulations. The two most common types of collectors currently used are the high-energy wet scrubber and the dry baghouse.

High-Energy Wet Scrubber. In this unit, high-velocity water jets impinge upon effluent particles passing through a venturi scrubber. The resultant sludge is piped to a remote settling pond, where the sludge is dewatered. Figure 3 illustrates a wet scrubber emission control system connected to a recuperative hot blast system.

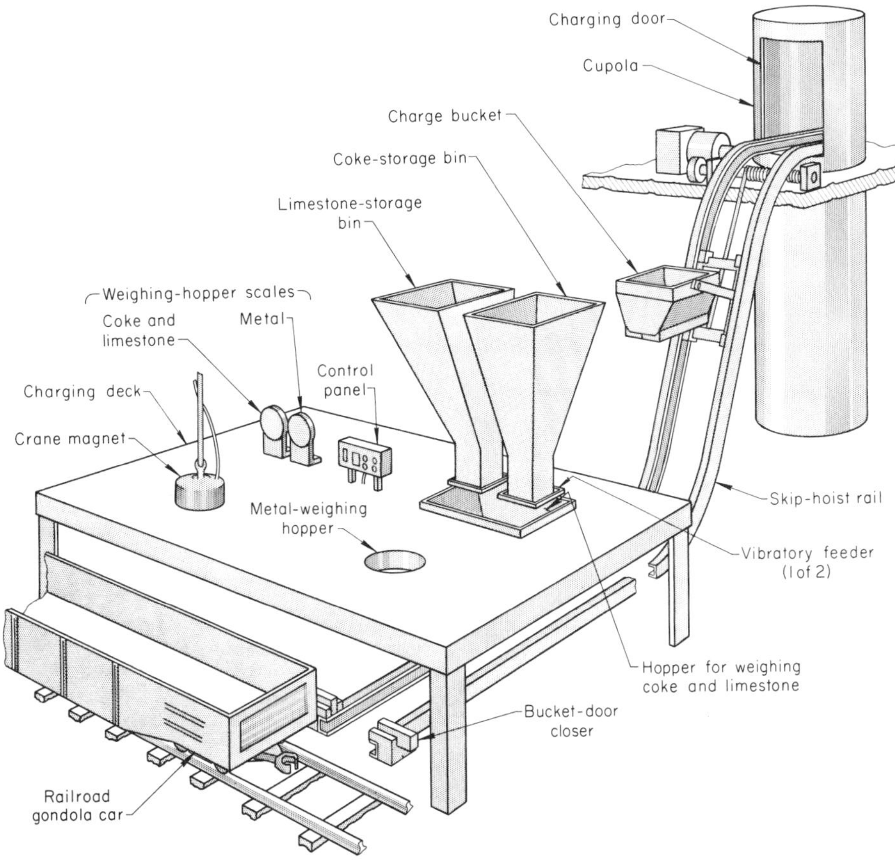

Fig. 4 Typical mechanized system for charging a cupola

Dry Baghouse. In this unit, the effluent is collected in a series of interconnected fabric bags.

Electrostatic precipitators have also been used on some cupolas primarily in Europe. Recently developed precipitators show more promise in controlling emissions.

Storage and Handling of Charge Components

Essential to efficient cupola operation is a good material storage and handling system. Figure 4 illustrates the essential components of such a system. Metal charge components are unloaded into storage bins using a crane magnet and transferred to a weigh hopper to designated weight values as charge components are progressively added.

Coke and limestone from overhead storage bins are discharged into a weigh hopper and into the charge bucket. The charge bucket with all weighed amounts of metal, ferroalloys, coke, and flux is elevated to the charge door and dumped either with an overhead crane or with a skip hoist up an inclined track.

Charge Materials

This section will discuss coke specifications, the metal charge, trace element control, scrap limitations, ferroalloys, charge calculations, and fluxes for fluidizing slag.

Coke Specifications

As the major fuel, coke is the most important charge material. High-quality, foundry grade coke is essential for optimum cupola performance. Good foundry grade coke requires longer coking time and more careful blending of coals and is therefore higher in price than other grades of coke.

Foundry coke must be strong and sufficiently impact resistant to maintain adequate lump size to provide permeability, blast penetration, and air availability to the center of a large cupola. Experiments and production experience have shown that performance is optimized when coke size is within a range proportional to cupola diameter, preferably 8 to 12% of cupola diameter.

Fine coke should be screened out because it fills interstitial space, limits air distribution, and overloads the emission system. On large cupolas, the largest available coke size is preferred, but on small cupolas, coke screened to proper size ranges provides the best performance.

One foundry, when operating four sizes of cupolas, ordered coke screened to three size ranges:

- For a 711 mm (28 in.) cupola: 50 × 127 mm (2 × 5 in.) coke
- For 914 and 1220 mm (36 and 48 in.) cupolas: 75 × 178 mm (3 × 7 in.) coke

- For two 2130 mm (84 in.) cupolas: ≥127 mm (≥5 in.) coke

Coke should have low ash content (<10%) and a chemical structure that provides for optimum carbon pickup and combustion efficiency. Low sulfur content (<0.70% S) is required for preventing excessive sulfur pickup. Volatiles should be less than 1.0% and fixed carbon above 90% for maximum cupola performance.

Efforts have been directed toward the economical use of limited amounts of less expensive substitutes for quality coke, such as briquetted fine coke, anthracite coal, injected fine coke, and various substitute fuels. However, all of these substitutes require some sacrifice in cupola efficiency, which must be weighed against fuel cost savings. The amount of coke in the charge usually ranges within 8 to 16% of the metal charge, depending on the temperature required, the percentage of steel scrap used, and many other variables.

Metal Charge

The modern cupola can melt a wide range of charge materials, mostly scrap iron and steel. Through its efficient use of scrap materials, the casting industry contributes to environmental cleanup and resource utilization.

Foundry returns, such as gates, runners, and internally generated scrap castings, usually constitute 30 to 50% of the pouring weight and should be remelted in the proportions generated.

Pig iron is considered a premium charge material because of its uniformity of composition and size and its lack of undesirable trace elements. In the early years, pig iron was the predominant charge component for ensuring quality iron with sufficient carbon content, and it was free of detrimental tramp elements. As economics encouraged the use of more scrap and less pig iron, pig iron has remained the best material for restoring iron quality whenever excessive tramp elements or poor-quality iron are produced.

Cast iron scrap has been increasingly used as faster spectrometer analyses have given assurance of chemical control and as more knowledge has been gained about the characteristics of various types of scrap. With the cooperation of scrap dealers and their trade organization, cast iron scrap has been separated and sold in 10 to 15 designated types with specific composition ranges and inclinations. Through careful blending, anticipation of trends, avoidance of excessive proportions of certain extremes, and fast spectrometer checks on trace elements, high proportions of cast iron scrap have been successfully used in some of the highest-quality irons.

Steel scrap in select grades was first melted in low percentages to lower the carbon content of gray iron in order to meet the higher strength levels of class 30 and class 40 irons. In low-carbon malleable irons, steel scrap in considerable proportions is needed to obtain a low carbon content.

The Scrap Institute has classified steel scrap into 15 to 20 categories. Based on experimentation and experience, adjustments can be made to compensate for the effects of size, thickness, and surface-area-to-volume ratio affecting the various reactions in the cupola.

With the advent of ductile iron that requires the lowest possible phosphorus content, steel scrap became a more attractive charge material for cost reduction. The basic cupola allows the adjustment of fluxes and slag chemistry to permit the melting of 50 to 80% steel scrap.

Trace Element Control

The use of considerable miscellaneous scrap increases the likelihood of traces of many undesirable elements getting into the iron melt. These extraneous elements may be alloyed in some scrap parts; may be attached to the iron as bearings, die castings, and so on; may be found in chromium plating; or may be present in paint or coatings.

Zinc and Cadmium. The more reactive zinc and cadmium will volatilize out of the iron, but if highly concentrated in the effluent, they can cause the baghouse dust or wet scrubber sludge to earn a hazardous material rating and cause expensive disposal costs.

Lead and bismuth can be sufficiently recovered to cause detrimental effects on graphite and properties, especially in ductile iron, and must be avoided. In addition, a high concentration of lead can cause the collected effluent to be designated as a hazardous waste.

Aluminum, zirconium, and titanium are essentially oxidized out into the slag, but the initial presence of aluminum within critical ranges can increase hydrogen absorption and gas porosity tendencies.

Other elements include inadvertent alloying elements such as chromium, tin, copper, molybdenum, and nickel, which increase pearlite content and contribute to strength and hardness. When pearlite is needed for increased strength and hardness, these trace alloys become part of the total alloy factor, reducing the amount of alloys to be added. However, these alloys should be avoided in ferritic ductile iron, in which pearlite is undesirable.

Scrap Limitations

A wide range of scrap metals can be melted in the cupola. However, some forms of scrap metal are not practical to use in a cupola charge. Cast iron borings charged directly into the cupola would be excessively oxidized if not blown out of the stack. With briquetting, limited proportions can be used. Steel turnings are more impractical. Thin sheet steel cannot be used without being compressed into bundles. Slugs and compact punchings tend to clog interstitial coke passages and hinder combustion efficiency. Very thick steel parts and large-diameter shafts are not completely melted within the melting zone, and this results in lower iron temperatures and cupola problems.

Ferrous Alloys

Silicon, which is added to allow for oxidation losses and to yield the desirable final silicon content, is usually added in any of the following forms:

- Lump ferrosilicon: 50 or 75% Si
- Silvery pig: 16 to 20% Si
- Silicon briquets: approximately 40% Si with bricks containing a known weight of silicon that can be counted rather than weighed
- Silicon carbide: provides deoxidizing influence in addition to silicon

Manganese can be in the following forms to obtain the net desirable final manganese content:

- Spiegel: 20% Mn
- Ferromanganese: usually 80% Mn
- Manganese briquets

Charge Calculations—Least Cost Mix

Based on the market price of various types of scrap available and the safety precautions taken for the castings produced, the first decision is to determine the percentage of various scrap types that can be safely used. A metal charge is then calculated using preferred percentages of available types of cast iron and steel scrap, pig iron, and returns. The charge input of each element from each component is totalled and allowances calculated for anticipated losses and gains. Ferrous alloys are calculated to complete the necessary silicon and manganese contents.

Determining the lowest cost charge combination previously required several time-consuming trial-and-error calculations. Today, the computer can calculate the least cost mix quickly.

Fluxes for Fluidizing Slag

The addition of limestone or dolomite is necessary to fluidize the slag. Natural sources of slag such as coke ash, oxidized silicon, and lining consumed, as well as any sand and dirt on the scrap, are predominantly acid and produce a very viscous slag without the addition of sufficient basic flux.

Limestone or dolomite additions of 2 to 5% are generally needed in an acid cupola to contribute enough basic constituents of calcium oxide and magnesium oxide to impart sufficient fluidity for the slag to flow freely through the coke spaces, into the well, and down the front slagging trough. In the basic slag cupola, either with basic refractories or neutral water cooling, more basic flux is added to net a sufficient excess of basic slag constituents.

Limestone is a natural calcium carbonate stone. A lump size of 50 × 19 mm (2 × ¾ in.) is preferable. Silica is a neutralizing impurity and should be less than 2%. Several special or supplementary fluxes are occasionally used.

Soda ash (Na_2CO_3) is a strong slag liquefier used as briquets on the bed to supplement the limestone.

Fluorspar (CaF_2) is a powerful fluidizer used to clean up a cupola clogged with excessively viscous acid slag. In highly basic slags, fluorspar is needed to maintain fluidity. However, excessive fluorspar can cause environmental problems.

Calcium carbide (CaC_2) serves the multiple functions of deoxidizing, raising carbon, raising temperature, and providing a strong calcium oxide (CaO) flux contribution.

Combustion Reactions and Zones

In the cupola, heat for melting is generated by the combustion of carbon in the coke to carbon dioxide and carbon monoxide mixtures as the blast air penetrates the cupola and reacts with the hot coke. As the air enters from the tuyeres and contacts the first hot coke, oxygen in the air reacts with carbon in the coke according to the reaction:

$$C + O_2 = CO_2$$

As the gases penetrate the coke bed, oxygen is progressively depleted, along with the formation of carbon dioxide, and a zone of maximum temperature up to 1650 to 1705 °C (3000 to 3100 °F) is created.

As the gas high in carbon dioxide moves up through the excess coke, some of the carbon dioxide reacts endothermally to form a balance of carbon monoxide and carbon dioxide. This secondary reaction is:

$$CO_2 + C = 2CO$$

The top effluent gas typically yields 10 to 14% CO_2 and 15 to 10% CO. Effluent gas with carbon dioxide content on the high side and a carbon monoxide content on the low side indicates a more oxidizing net condition from hard blowing and/or a minimum coke percentage or reactivity. This condition will give the highest melting rate, but will also produce lower iron temperature, less carbon pickup, and more oxidation of silicon.

Effluent gas higher in carbon monoxide and lower in carbon dioxide indicates a less oxidizing net condition from more coke and/or softer blowing. This less oxidizing condition will melt less iron but at higher temperature with less oxidation of silicon and greater carbon pickup.

Cupola Zones

Cupola zones are designated functionally as follows:

- Preheat zone in the upper stack, in which the charge metal is preheated progressively by the ascending hot gases as the metal descends into the melting zone
- Melting zone is generally located 508 to 1270 mm (20 to 50 in.) above the tuyeres, where maximum temperature is generated and the metal charge is melted
- Well zone collects the melted metal and slag as it drips through the coke, accumulates in the well, and flows out the taphole

Chemical Zones. There are several chemical zones in the cupola. The zone immediately out from the tuyeres is very oxidizing until the excess oxygen is burned. The hottest melting zone where high carbon dioxide content prevails is slightly oxidizing. A reducing zone develops in the upper melting and preheat zones where coke and carbon monoxide predominate.

As it descends, the metal charge melts and drips through the coke bed and passes through varying combinations of these reducing, neutral, and oxidizing zones. The net effect is slightly oxidizing, but it varies depending on coke ratio, coke reactivity, and blast air volume and velocity.

Because of a net oxidizing influence, silicon loss is generally 8 to 15% in the acid cupola and 25 to 40% in the basic slag cupola. Silicon in the charge is calculated to make allowance for the expected loss.

The oxidation loss of manganese is 10 to 20% in the acid cupola and 5 to 10% in the basic cupola. During melting, sulfur is absorbed from the sulfur present in the coke.

Carbon is generally absorbed to maintain an equilibrium carbon saturation level. Hypereutectic pig iron generally loses some carbon. Low-carbon steel scrap is carbon hungry and continuously absorbs carbon. More coke is needed for high-steel charges to provide the dissolved carbon. Return and cast iron scrap absorb less carbon because they contain considerable carbon and are less hungry for absorption.

The molten bath that accumulates in the well has further contact with coke and further opportunity to absorb carbon, depending on many factors. In addition, some slag reaction can occur because the molten slag and molten metal have further opportunity to react toward equilibrium. Although cupola control is complex, an understanding of the various chemical interactions, accurate records, and diligent correlation of experiences will provide good control, quality, and efficiency.

Coke Bed

Preparation of the coke bed is vital to the initial start of the melt and to the maintenance of the melt at the specified level in the cupola combustion zones. Before metal is charged, a bed of good-quality, sized coke is ignited and blown to a red heat by any of several methods. The height of the bed coke is adjusted to a specific height above the tuyeres (usually within 1020 to 1525 mm, or 40 to 60 in.) that is determined by experience and based on blast volume, cupola diameter, and desired melting conditions. Bed flux and metal charges are then added, the blowers are turned on, and the melting begins within a few minutes.

With a high bed, melting occurs higher in the stack, slower and hotter with less oxidation and more carbon pickup. A low bed melts lower, faster, and colder with more oxidation.

Charge coke is intended to replace coke burned in the melting of each charge in order to maintain the effective melting zone within a desirable position in the combustion zones. If the bed drops too low during the heat, booster coke additions may be needed to restore operating bed height.

Control Principles

Melt Rate. A cupola is expected to maintain a melting rate sufficient to meet the needs of mold production. The cupola size must be chosen for the midrange of production demand. A general rule is that 4.5 kg (10 lb) of iron can be melted per hour for each square inch of cross-sectional area.

By increasing or decreasing the blast, melt rate can be varied ±20% with reasonably good efficiency and control. Varying the hot blast and/or oxygen injection can extend this control range.

Beyond these limits, overblowing a cupola increasingly oxidizes the iron, causing lower carbon and silicon levels and decreasing the temperature. Underblowing will lower the temperature because the blast will not penetrate the cupola with sufficient air distribution for optimum combustion efficiency.

Iron Temperature. The iron temperature at the cupola spout must be adequate to survive handling losses and to be poured into the molds at sufficient temperature and fluidity to produce sound castings. Iron temperature is measured by optical pyrometers, thermocouples, and continuously recording radiamatic instruments.

Iron temperature is primarily influenced by the coke concentration in relation to the metal charge. If coke percentage is increased from 8 to 12% of the metal charge, the iron temperature is usually increased approximately 65 °C (120 °F), while the melting rate is reduced 20 to 25%, especially if air volume is maintained at the same level. Increasing the preheat temperature on the blast air or increasing the percentage of oxygen injected will instantaneously increase metal temperature.

Carbon and Silicon. Carbon and silicon concentrations are most important in the control of iron properties because both elements are strong graphitizers. In malleable iron, carbon and silicon concentrations must be kept low to encourage a carbidic structure as-cast and to avoid graphitization until heat treatment.

In gray iron, sufficient concentrations of carbon and silicon must be maintained, within control limits, to avoid chill and to encourage graphitization in desirable flake form and a structure most suitable for the section size of the castings and strength levels specified. In ductile iron, sufficient carbon and silicon concentrations must be maintained within controlled ranges for graphitization in nodular form and control of matrix structure. In gray iron, and more critically in ductile iron, a postinoculating ladle addition of ferrosilicon is essential to nucleate most favorable graphite precipitation and structure.

Silicon is controlled from the cupola with the addition of ferrosilicon alloy, allowing for the anticipated oxidation loss of total silicon in the charge. Carbon content is primarily controlled by the percentage of steel scrap in the charge. Because steel contains relatively little carbon, increases in the percentage of steel decrease carbon content. Some carbon is absorbed as steel scrap melts in the presence of coke but does not have time to reach carbon saturation equilibrium; therefore, the net effect is a lowering of the carbon level.

This carbon absorption rate is influenced by coke properties, coke ratio, and slag basicity. In addition, carbon pickup is increased by increases in blast temperature and oxygen injection as well as by fine coke injection. The need for immediate increase in final carbon can be quickly realized from an increase in one of these supplementary boosters of combustion.

Sulfur in gray iron is not detrimental until the level is above 0.120% or possibly 0.150% S if manganese is maintained at a sufficient ratio in relation to sulfur. Sulfur in ductile base iron should be below 0.03% and preferably below 0.015%, because base sulfur reacts with the magnesium from nodulizing alloys and complicates magnesium control and iron cleanliness.

Manganese is needed in gray iron at an amount of 1.7 times sulfur plus an additional 0.30% to ensure that the sulfur is precipitated as manganese sulfide rather than iron sulfide. In pearlitic ductile iron, manganese can contribute beneficially to the total alloys needed to produce a pearlitic matrix. However, in ferritic ductile iron, manganese should be as low as possible to minimize pearlite in the as-cast structure.

Phosphorus is controlled by a variety of charge materials that are low in phosphorus.

In gray iron, phosphorus contents should be below 0.20% and even below 0.10% because high phosphorus levels are detrimental to machinability and soundness in complex shapes. In ductile iron, phosphorus content below 0.10% and preferably 0.05% is desirable for the maximum impact resistance and ductility potential of ductile iron. For these low phosphorus levels, considerable pig iron is sometimes required. Steel scrap is generally the lowest phosphorus scrap material and is extensively used with ductile iron.

Residual Alloys and Tramp Elements. The effects of various inadvertent alloys and tramp elements should be understood, and control ranges and maximum levels should be established. Rapid analysis by modern spectrometers makes it possible to monitor trace element levels and to coordinate these levels closely with the composition of incoming raw materials. Any scrap source contributing excessive concentrations of detrimental elements should be removed from the cupola charge.

Certain tramp elements detrimental to graphite forms, such as lead and bismuth, must be controlled well below the maximum level allowed. Titanium in trace amounts is beneficial to gray iron flake graphite, but is detrimental to the nodular graphite forms needed in ductile iron. Inadvertent alloys such as chromium, copper, tin, and molybdenum, if maintained within controlled ranges and combinations, can contribute to the pearlitic matrix structure of high-strength gray irons and pearlitic ductile iron. However, for ferritic ductile iron with the highest ductility, acceptable limits for residual alloys are much lower.

Control Tests and Analyses

The chill test was one of the earliest control tests and is still a good indicator of the graphitizing balance of an iron. A chill specimen is cast either as an elongated wedge in a sand mold or as a bar cast against a metal chiller on one side.

After a few minutes, the test specimen is broken, and the depth of carbidic white iron chill indicates the carbon, silicon, and total balance of graphitizing elements against carbide-forming elements. This chill depth signals the need for more or less silicon and/or carbon or an adjustment in the postinoculation.

Wet chemical analysis provides a guide for subsequent iron control, but analysis time is too long for immediate correction.

Eutectic arrest methods were developed to provide, within a few minutes, carbon equivalent and carbon content with reasonable accuracy, as well as calculated silicon content. These rapid determinations make possible the immediate corrective adjustments of carbon and silicon and with much better composition control.

Spectrometer equipment greatly facilitates iron control, with complete analysis of important elements within a few minutes. In large plants, samples are conveyed to the laboratory in pneumatic tubes, and analyses are reported to the melting center by phone, autowriter, or computer readouts within a few minutes.

Mechanical tests of several types are conducted to ensure the uniformity of iron properties and compliance with specifications. Tensile tests of gray iron ensure that the iron meets the strength level expected from the specified class of iron. On ductile iron, tensile tests verify that the specified grade is being met. Each grade specifies ultimate tensile strength, yield strength, and percentage of elongation. Brinell hardness tests of castings monitor uniformity and general matrix structure.

On ductile iron, rapid microscopic evaluation of each treatment batch is essential to ensure graphite nodularity before castings reach the shakeout operation and lose their identity. Nodular graphite structure in a casting can be verified with properly calibrated ultrasonic test equipment. Ultrasonic testing of all castings is performed by some companies to ensure nodularity on critical parts or to evaluate a questionable lot.

Basic Slag Composition Effects

In conventional acid slag operation, no sulfur removal or refinement is expected from the slag. Enough basic flux, such as limestone, is added to fluidize the acid sources of slag and mechanically cleanse the cupola for efficient combustion. Tables 2 and 3 give the compositions of several acid cupola slags with a predominance of acid silicon dioxide and a ratio of basic constituents ($CaO + MgO)/SiO_2$ of less than 1.

Within normal ranges and adequate fluidity, the acid slag exerts little effect on iron chemistry. However, very viscous high silicon dioxide slags, because of insufficient flux, tend to cling to the coke and tend to clog interstitial air passages and the coke surface, resulting in lower iron temperature, lower carbon content, and increased oxidation. Heavy flux additions yield a very fluid neutral or mildly basic slag that is corrosive to acid refractories but cleanses the coke, resulting in higher carbon and lower sulfur than normal.

Basic cupola operation was developed to yield low-sulfur iron directly from the cupola by supplying sufficient excess of basic constituents so that the slag will absorb and remove sulfur from the iron. Sulfur removal from the iron into the slag increases with slag basicity, volume, and fluidity, and low iron oxide content improves the desulfurizing potential of the slag. Table 4 lists the compositions of various basic cupola slags.

Table 2 Acid slag compositions

Acid slag	Acid constituent SiO_2	Neutral Al_2O_3	Basic constituents: CaO+MgO	MnO	FeO
A	37.9–52.2	6.2–22.9	19.8–44.3	1.7–3.6	5.0–15.6
B	35–45	5–26	30–40	1.3–4.0	5–8
C	32.0	15.0	38.0	2.5	8.0
D	...	...	...	...	...
E	41.1–51.7	...	18.3–36.8	...	0.4–10.8
F	37.0–65.0	9.7–23.4	7.7–40.7	1.0–23.4	0.9–44.4
G	40–50	10–20	25–38	1–5	1–8

Table 3 Typical variations in acid cupola slags

Description	Cupola size, mm	Cupola size, in.	Steel, %	Coke, %	Flux(a), %	Slag composition, %: SiO_2	Al_2O_3	FeO(b)	MnO	CaO	MgO
Average fluidity	2135	84	33	7.5	1.9	47.1	12.1	6.9	4.6	22.0	1.6
Average fluidity	1220	48	22	12.8	3.0	46.2	11.0	1.1	1.4	37.2	1.0
Viscous slag	1220	48	22	12.8	2.2	52.1	12.4	1.5	1.9	29.2	0.9
Fluid and corrosive	2135	84	14	11.0	3.5	39.6	10.1	3.8	2.9	38.8	2.9
Overblown cupola 150 $m^3/min/m^2$ (490 $ft^3/min/ft^2$)	533	21	50	14.0	5.0	45.6	18.5	16.1	2.8	15.4	1.0

(a) Limestone. (b)Total iron as FeO including small portion of Fe_2O_3

Some investigators consider the effective basicity ratio to be $(CaO + MgO)/(SiO_2 + Al_2O_3)$. Others find that aluminum oxide is more nearly amphoteric and neutral with little effect as an acid constituent and therefore consider the basicity ratio $(CaO + MgO)/SiO_2$ to be a more accurate and simple guide.

When sufficiently basic and low in ferrous oxide, basic slags can reach a level of 1.0 to 2.0% S that has been removed from the metal. Metal sulfur contents as low as 0.004 to 0.02% have been obtained from a basic slag cupola.

In addition to sulfur removal, the basic slag increases carbon pickup as basicity increases. Increased carbon pickup permits the use of higher proportions of steel scrap, which is economically attractive in producing ductile iron. However, the highly basic slags that yield the desired, very low sulfur levels can yield carbon levels sometimes higher than wanted. Consequently, control is very sensitive on a highly basic cupola, and silicon loss is much higher—in the range of 25 to 40%. For these reasons, some basic slag operations in water-cooled cupolas have been shifted to neutral or acid slag operations, with desulfurization performed externally by one of several desulfurizing methods.

Desulfurization Methods

Several successful desulfurizing techniques have been developed in the search for methods of obtaining low-sulfur base iron for ductile iron. In earlier years, the removal of some excessive sulfur in gray iron was accomplished by the forehearth or ladle addition of sodium carbonate (soda ash), but refractory attack made it impractical to reach very low sulfur levels.

Calcium carbide has proved to be the most effective desulfurizing agent for low sulfurs. Lime is also an effective desulfurizing agent, but it requires more volume, time, and agitation. Small percentages of fluorspar and various salts improve the desulfurizing effectiveness of lime.

Metal stirring or agitation is essential for promoting optimum desulfurization. The spent carbide or lime slag containing sulfur is then removed from the metal. Disposal of the carbide slag requires attention to safety and environmental considerations.

The following stirring methods have been found to be effective:

- The injection of carbide with inert gas through graphite or refractory lances that supply the carbide and simultaneously stir the melt
- A porous plug in the bottom of a ladle through which inert gas is bubbled, with desulfurizing agent fed onto the surface with a vibrating feeder
- The shaking ladle system places a ladle of iron in a gyrating shaking mechanism that stirs in the desulfurizing agent
- The refractory stirrer system rotates a refractory paddle in the bath, stirring the carbide or lime into contact with the iron

Injection and porous plug desulfurizing systems can be either batch type, treating individual ladles, or they can be a continuous system, with the cupola stream flowing through a special forehearth outfitted with a porous plug and a carbide feed (or injection lance). The spent carbide or lime overflows or is raked off, and the desulfurized metal flows continuously into the holder or duplex electric furnace.

With any of these stirring methods, additions of 0.50 to 1.0% CaC_2 have desulfurized irons of 0.07 to 0.10% S, as-melted, down to a final 0.01 to 0.02% S. A temperature loss of 30 to 55 °C (50 to 100 °F) can be expected. Lime alone or with various salts can be used as the desulfurizing agent; the efficiency is lower than that with carbide, but there are fewer environmental problems.

Computer Use in Cupola Operation

The ultimate refinement in cupola operation is the use of the computer to record, monitor, and control many of the important variables in cupola operation. Computer applications are continually being expanded. Some of the earliest and most significant applications have been:

- Least cost charge calculations by quickly calculating the charge cost of various combinations of available raw materials; the lowest cost charge can thus be determined
- Metal weight compensation; in the dropping of metal components from crane magnets into weigh hoppers, any overshooting or undershooting can be balanced out by the computer on subsequent charges
- Automatic weighing of coke, flux, and ferrous alloys has been done by computer

Table 4 Compositions of various basic cupola slags

Basic slag	Composition, %: SiO_2	Al_2O_3	CaO	MgO	MnO	FeO	S	CaF	Basicity(a)
A	30–35	8–10	42–47	9–11	1.5	1.5	1.0	2.0	...
B	33.4	13.3	34.3	12.7	2.7	1.5	0.99	0.6	1.01
	28.6	6.5	41.2	19.6	1.2	1.8	0.56	...	1.73
	32.4	11.7	39.0	15.1	0.6	1.2	0.62	...	1.23
	30.6	7.8	42.3	15.2	0.8	0.7	0.70	...	1.51
	21.0	...	54.1	9.5	0.3	1.3	1.07	1.0	1.27
	21.4	22.3	47.6	8.5	0.4	0.3	2.20	1.1	1.39
C	28.1	7.1	45.1	16.0	...	0.8	0.81	...	1.91
	22.6	5.2	48.8	18.6	...	1.2	0.54	...	0.94
	37.0	12.7	28.8	17.7	...	1.1	0.31	...	1.15
	27.7	7.4	25.4	31.9	...	3.9	0.30	...	1.67
D	28.3	6.0	50.0	8.7	...	0.6	1.36	...	1.71
	27.7	7.9	43.6	15.1	...	0.4	1.45	...	1.48
	27.6	11.4	37.2	20.4	...	0.4	2.01	...	1.50
	28.0	9.6	41.8	14.6	...	1.1	...	...	...

(a) $CaO + MgO/SiO_2 + Al_2O_3$

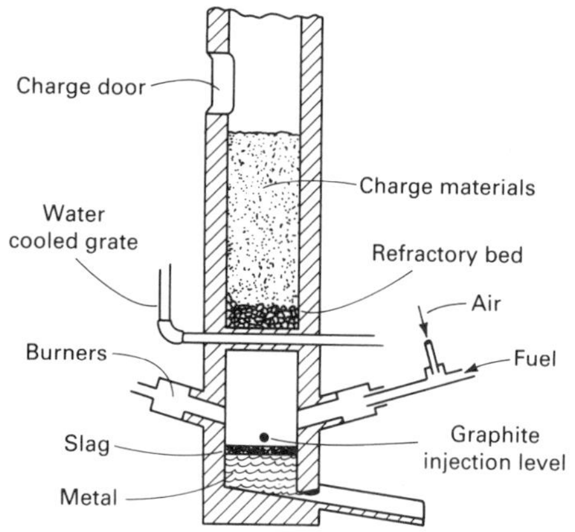

Fig. 5 Schematic of the Taft cokeless cupola

- Adjustments in coke weight to compensate for atmospheric humidity, as well as adjustments in coke weight for overshooting or undershooting on metal charge weights
- Continuous analysis and recording of top gas composition; in addition, alert signals are activated when explosive levels of hydrogen are approached from water leaks
- Display of chemical analyses with statistical process control limits and alert signals when any element is approaching limits
- The indicating, recording, or controlling of any data important to maintenance or control

Further computer applications are continually being directed toward better control of iron content and cupola performance.

Specialized Cupolas

Cokeless Cupola. In the 1970s, a cokeless cupola fired with gas or oil was developed in Europe and was put into operation in several countries. It was a refractory-lined shaft furnace with a water-cooled grate supporting a bed of refractory spheres, as illustrated in Fig. 5.

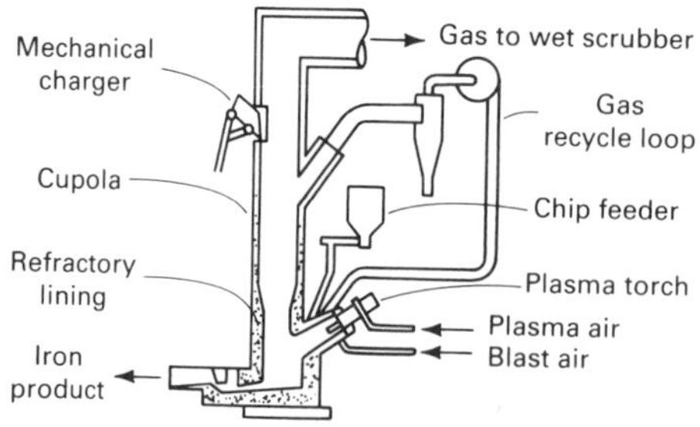

Fig. 6 Schematic of a plasma-fired pilot cupola

Heat for melting is generated by either gas or oil burners located below the grate and refractory bed. The iron melted in a cokeless cupola has the advantage of having a low sulfur content but also has a low carbon content, both of which are due to the absence of coke. In the Taft cupola, the iron is carburized to a specified carbon content by injecting a fine carburizer (either petroleum coke or graphite) into the iron in the well.

Pollution from the cokeless cupola is minimal and reportedly would meet environmental standards in some countries without expensive emission control equipment. In countries where coke must be imported at a high price and oil or gas is readily available, the cokeless cupola is economically attractive. Production experiences have been reported on small cupolas in the 4.5 to 9.1 Mg/h (5 to 10 ton/h) range, with no practical experience reported on large cupolas. Fuel cost can be attractive in some countries, but the high cost of refractory sphere replacements and the restrictions on the percentage of steel scrap used limit the attractiveness of this equipment in many raw material markets. In the United States, some pilot heats have been run on a gas fired cupola. This project was sponsored by the Gas Research Institute, but no reports are available.

Plasma-Fired Cupola. In the 1980s, a pilot plant plasma-fired cupola was built, and a number of experimental heats were run with some encouraging results. In the 762 mm (30 in.) ID cupola, a single 2 MW plasma torch was projected into the cupola at tuyere level, as shown in Fig. 6.

In the water-cooled torch, an electric arc from a high-amperage direct current is maintained across a gap between two copper sleeves, and the arc is rotated by a magnetic field. Air (or gas) is passed through the electric arc, where it is ionized and heated to very high temperatures unattainable with fossil fuels.

The approximately 30 experimental heats conducted thus far have indicated several advantages of using a plasma-fired cupola:

- Coke could be considerably reduced by supplementing electric power from the torch
- Lower-quality coke could be used; blast furnace coke, fine waste coke, and anthracite coal have melted successfully
- Direct melting of borings and turnings proved practical because of lower air velocity and lower oxidation level
- The generation of silicon from injected sand and coke breeze
- Iron reduced from as much as 20% iron oxide injected at tuyere level

Many possibilities of the plasma-fired cupola merit further exploration. Economically, the most attractive immediate application is the direct melting of 70% or more iron borings or steel turnings. A large plasma-fired cupola is being installed in an automotive foundry scheduled to start operation in 1988.

SELECTED REFERENCES

- *Conference on Cupola Operation*, American Foundrymen's Society, June 1980
- *Cupola Design, Operation, and Control*, BCIRA, 1979
- *Cupola Handbook*, American Foundrymen's Society, 1975

Vacuum Melting and Remelting Processes

Chairman: Gerhard Kienel, Leybold AG, West Germany

Vacuum Induction Melting (VIM)

A. Choudhury and H. Kemmer, Leybold AG, West Germany

THE PRODUCTION OF LIQUID METAL under vacuum in an induction-heated crucible is a tried and tested process. It has its origins in the middle of the 19th century, but the actual technical breakthrough occurred in the second half of the 20th century.

Vacuum induction melting can be used to advantage in many applications, particularly in the case of the complex alloys employed in aerospace engineering. The following advantages have a decisive influence on the rapid increase of metal production by vacuum induction melting:

- Flexibility due to small batch sizes
- Fast change of program for different types of steels and alloys
- Easy operation
- Low losses of alloying elements by oxidation
- Achievement of very close compositional tolerances
- Precise temperature control
- Low level of environmental pollution from dust output
- Removal of undesired trace elements with high vapor pressures
- Removal of dissolved gases, for example, hydrogen and nitrogen

Vacuum induction melting is indispensable in the manufacture of superalloys, which must be melted under vacuum or in an inert gas atmosphere because of their reactivity with atmospheric oxygen and nitrogen. The process is suitable for the production of high-purity metals under oxygen-free atmosphere. This limits the formation of nonmetallic oxide and nitride inclusions. However, problems can arise in the case of alloying elements with high vapor pressures, such as manganese.

Figure 1 shows the various installation and application possibilities of the vacuum induction furnace. The casting weight can vary from a few kilograms to 30 Mg (33 tons), depending on whether the furnace is being used for precision casting or for the production of ingots or electrodes for further remelting.

Metallurgy of the Vacuum Induction Furnace

In contrast to the electric arc furnace (ladle metallurgy), there are many different factors that arise with induction furnace melting and significantly affect the metallurgical process. In a VIM furnace, slag would be transported to the crucible wall by the characteristic bath movement. The result is

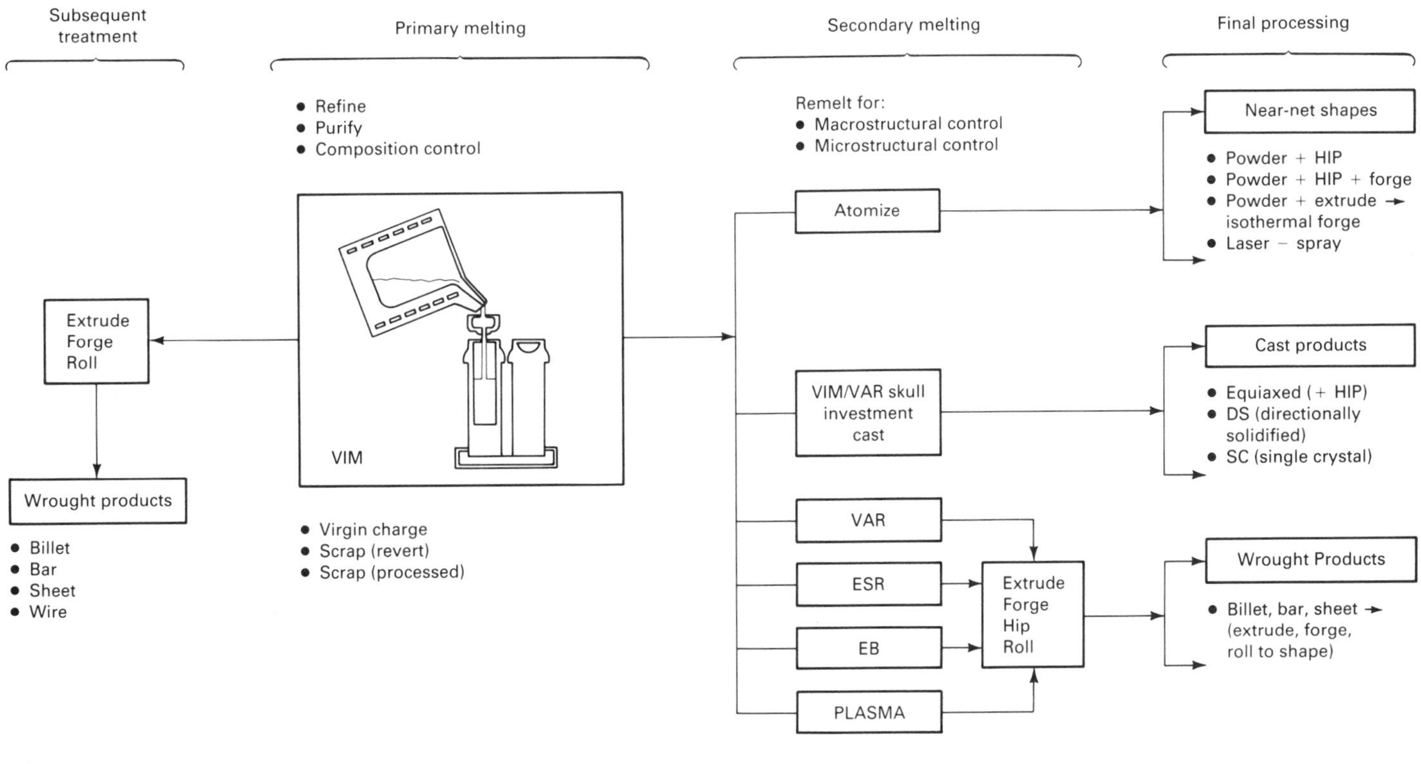

Fig. 1 Potential processing routes for products cast from VIM ingots or electrodes. Source: Ref 1

that the slag solidifies at the wall and therefore has an insufficient reaction with the melt. For this reason, metallurgical operations such as dephosphorization and desulfurization are limited.

Crucible Materials. In addition to the slag movement, the rammed crucible wall lining is susceptible to significantly higher erosion than the brick wall of a ladle. Therefore, VIM metallurgy is primarily limited to the pressure-dependent reactions, such as carbon, oxygen, nitrogen, and hydrogen, and the evaporation of undesired elements with high vapor pressures, such as copper, lead, bismuth, tellurium, and antimony. Figure 2 shows a typical process profile for the vacuum induction melting of nickel- and cobalt-base superalloys.

The crucible material has an extraordinary effect on the metal/slag reaction because the ceramic outer wall reacts with the liquid metal and with the slag. It is therefore beneficial to use a slag that is more or less saturated with the oxides of the crucible lining in order to minimize heavy slag attack of the lining.

The refractory material used for the crucible lining is based on oxides such as Al_2O_3, MgO, CaO, or ZrO_2 (Table 1). The lining is almost always rammed and sintered; prefabricated brick is used in larger furnaces. Dried silicate, combined with small oxide additions, appears to be very suitable for crucible lining because of its thermal characteristics. Because of an irreversible thermal expansion of 8% above 1000 °C (1830 °F), a high densification of the lining takes place during sintering. For this reason, this active lining is suitable for foundries. The behavior of the lining refractory with regard to stability at high temperature under vacuum must also be considered.

Table 1 Typical refractories used to line VIM crucibles

Refractory	Maximum melt temperature °C	Maximum melt temperature °F	Refractory density g/cm³	Refractory density lb/in.³	Resistance to thermal shock	Applications
MgO	1600	2910	2.8	0.101	Good	Superalloys, high-quality steels
Al_2O_3	1900	3450	3.7	0.134	Good	Superalloys, high-quality steels
MgO-spinel	1900	3450	3.8	0.138	Poor	Superalloys, high-quality steels
Al_2O_3-spinel	1900	3450	3.7	0.134	Relatively good	Superalloys, high-quality steels
ZrO_2	2300	4170	5.4	0.195	Poor	Superalloys, high-quality steels
Graphite	2300	4170	1.5	0.054	Excellent	Copper, copper alloys

The carbon monoxide partial pressures at which the lining oxides are reduced by carbon in the melt are:

Lining material	Carbon monoxide partial pressure, kPa	Carbon monoxide partial pressure, torr
CaO	0.04	0.3
ZrO_2	0.13	1.0
MgO	0.53	4.0
Al_2O_3	0.53	4.0
SiO_2	81.1	610.0

Source: Ref 1

It is evident that CaO represents the stable crucible material, while SiO_2 is reduced at relatively high pressure. Apart from that, SiO_2 can be reduced by alloying elements such as manganese, chromium, aluminum, titanium, or zirconium. This can cause a heavy chemical erosion of the lining, accompanied by an undesired silicon pickup in the melt. Because of these undesired reactions with a silica lining, a spinel-forming basic refractory material is used for the melting of high-grade steels and superalloys. The spinel formation during sintering leads to volume growth; therefore, as with the oxidic material, a densification of the rammed lining takes place.

Trace Element Removal. The removal of undesired volatile trace elements such as arsenic, antimony, tellurium, selenium, bismuth, and copper, from the vacuum induction furnace is of considerable practical importance. These elements must be held to very low concentrations to avoid the risk of premature part failure, particularly for the production of superalloys for critical appli-

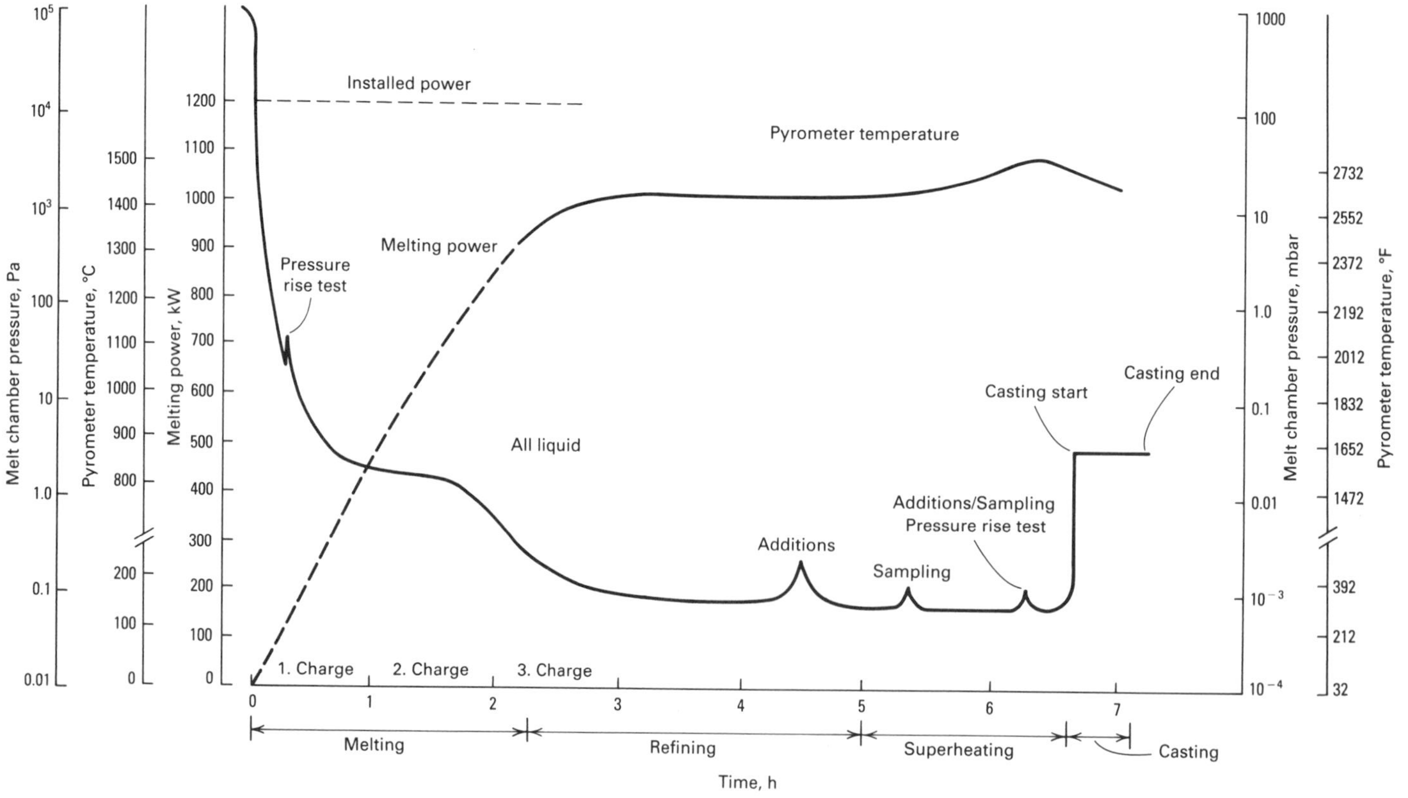

Fig. 2 Typical VIM melt protocol for nickel- and cobalt-base superalloys

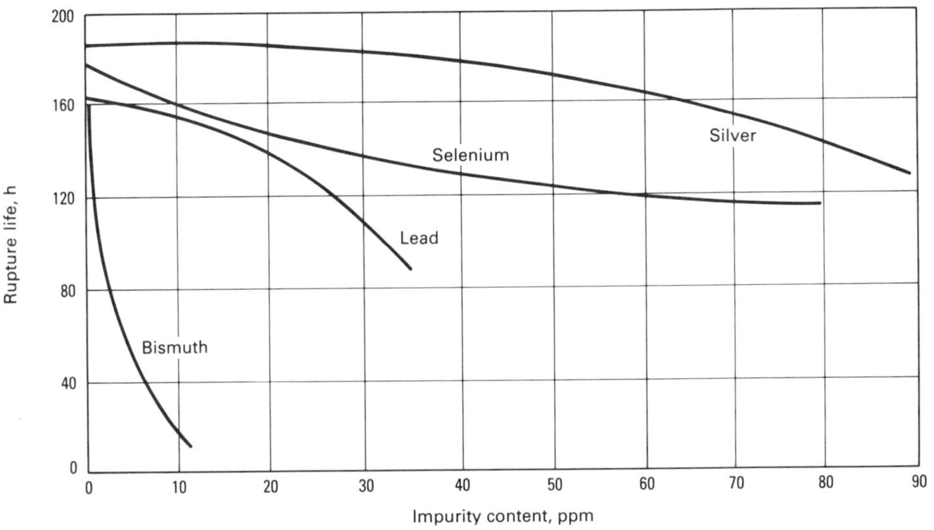

Fig. 3 Effect of trace elements on the stress rupture properties of Alloy 718. Test conditions: 650 °C (1200 °F), 690 MPa (100 ksi). Source: Ref 2

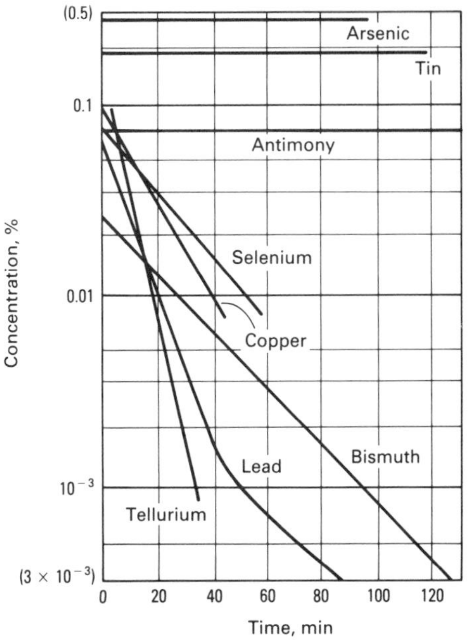

Fig. 4 Evaporation of trace elements from Ni-20Cr melts under vacuum. Source: Ref 3

cations, such as jet engine parts. Figure 3 shows the influence of some trace elements on the stress rupture properties of Alloy 718. Because of the high vapor pressures of most of the undesirable trace elements, they can be kept to very low levels by distillation during melting under vacuum.

Figure 4 shows how the different trace elements behave under vacuum. In a nickel-chromium melt, arsenic, antimony, and tin cannot be reduced over the gas phase, while copper, lead, selenium, and bismuth can be reduced to a level far below 10 ppm.

Degassing. Nitrogen and hydrogen can be decreased via the gas phase because the solubilities of these gases at constant temperature are directly proportional to their partial pressures.

Nitrogen Removal. As long as no strong nitride-forming elements are present in the melt, there is no difficulty in reducing the nitrogen content to 20 ppm (Ref 4-6). When nitride-forming elements such as chromium, vanadium, aluminum, and titanium are present, the activity of the nitrogen is very low; therefore, removal of nitrogen under high vacuum is difficult. If low nitrogen contents are desired in alloys containing nitride-forming elements, raw materials low in nitrogen must be used.

Figure 5 shows the reduction in nitrogen levels obtained in the vacuum induction melting of a die steel. From an average level of about 400 ppm at the beginning of treatment, the nitrogen has been reduced to a level of 50 ppm. Additional argon purging during melting of the steel did not improve nitrogen removal. In the case of iron-nickel melts, however, the additional argon purging reduces nitrogen to much lower values.

Hydrogen Removal. The solubility of hydrogen at atmospheric pressure and at 1600 °C (2910 °F) in iron or nickel melts is in the range of 30 ppm. During vacuum induction melting, the hydrogen can be removed to very low concentrations. Figure 6 shows the hydrogen contents of a die steel before and after vacuum treatment.

It is evident from Fig. 6 that final hydrogen contents below 1 ppm can be routinely achieved. The reduction in hydrogen during the degassing treatment amounts to approximately 80%. Argon purging had no perceptible influence.

The bath agitation in the case of this furnace, which is operating at normal frequency, is sufficient for hydrogen removal. Similar results were also obtained for the VIM processing of superalloys.

Deoxidation of the melt in the vacuum induction furnace can be done via the gas phase, as described below:

$$C + O \rightarrow CO \text{ (gas)}$$

The final attainable oxygen content for a given carbon content is directly proportional to the carbon monoxide partial pressure.

The theoretical equilibrium values for oxygen are below 1 ppm in carbon-containing iron or nickel melts at 1600 °C (2910 °F) and at a pressure of 0.1 Pa (10^{-3} mbar, or 7.5×10^{-4} torr). However, the actual values are up to one order of magnitude higher. These higher oxygen contents result from impurities in the crucible lining, crucible outgassing and leakage, and above all the fact that the reaction does not reach equilibrium with decreasing oxygen content because of the difficulty of carbon monoxide nucleation.

In addition, the hydrostatic pressure of the liquid metal in the melt must be taken into account because only a relatively small bath surface area is in direct contact with the prevailing vacuum pressure. However, the influence of the liquid-metal hydrostatic pressure can be minimized through the use of additional agitation effects.

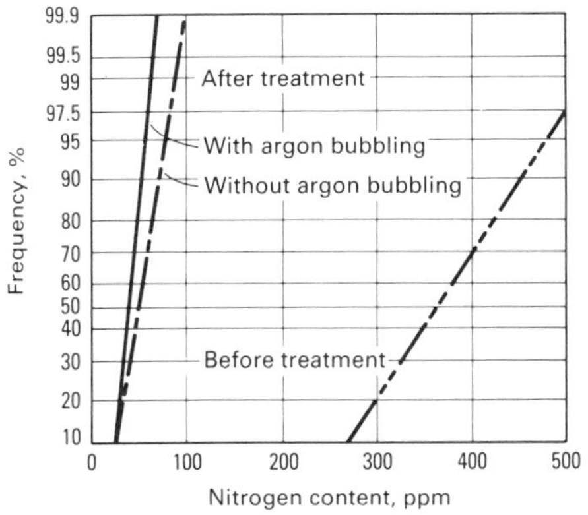

Fig. 5 Reduction in nitrogen content of X38CrMoV51 (Fe-0.38C-5.2Cr-1.3Mo-0.4V-1Si-0.4Mn) die steel after VIM processing

The carbon monoxide reaction occurs in two stages (Ref 1). The first stage is boiling, that is, the formation of carbon monoxide bubbles within the melt along with a strong bath agitation as a result of this gas formation. The second stage is desorption, in which no more carbon monoxide bubbles form inside the melt and carbon monoxide formation takes place only at the bath surface.

Deoxidation with carbon is also a decarburization reaction, which is used for the production of low-carbon high-chromium steels. By using the carbon monoxide reaction under vacuum, very low carbon contents can be achieved without a noticeable loss of chromium. The oxygen content in such alloys is higher because of the decrease in oxygen activity due to chromium. In this case, precipitation deoxidation using aluminum, silicon, or titanium must be applied.

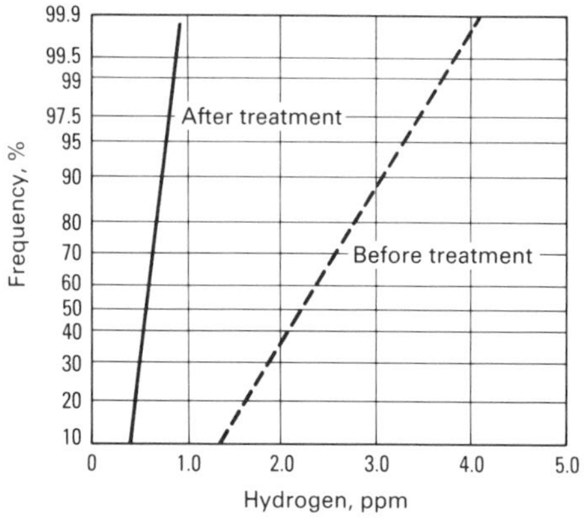

Fig. 6 Reduction in hydrogen content of X38CrMoV51 die steel after vacuum induction degassing

Cleanliness of VIM Melts. Freedom from oxide inclusions is extremely important for vacuum induction melted alloys, particularly for superalloys with extremely high strength properties at higher operating temperatures. Figure 7 shows the effect of oxygen content on the rupture lives of two superalloys.

Cleanliness can be significantly improved if a reactive liquid slag capable of absorbing oxides and sulfides is in contact with the melt. Vacuum induction furnaces are generally not operated with active slags. Therefore, reaction products can precipitate only on the crucible walls, and the melt may not be as clean as with other processing methods.

Because the various alloys produced in vacuum induction furnaces must meet the highest quality requirements and because the vacuum induction furnace is primarily a consolidation unit and only secondarily a refining unit, the following methods are used to produce clean melts:

- Selection of a more stable refractory material for the crucible lining
- Rinsing of the melt with inert gas
- Minimizing the contact time of the melt in the crucible
- Exact temperature control to minimize crucible reactions with the melt
- Suitable deslagging and filtering techniques during pouring
- Conception of a suitable tundish and launder system for good oxide removal

For particular applications, however, the quality of the material produced by vacuum induction melting will not be sufficient to satisfy the highest quality requirements with respect to cleanliness and primary structure. In this case, the VIM-produced material must undergo another remelting process.

Production of Nonferrous Materials

Apart from melting high-grade steels and superalloys, vacuum induction melting is being increasingly used for the production of nonferrous metals and alloys. Table 2 shows some examples for possible use in nonferrous metallurgy.

Aluminum alloys with additives such as zirconium, titanium, beryllium, cerium, tellurium, and cadmium must be melted under vacuum or under inert gas atmosphere because of their high reactivity with air and in some cases their toxicity. Aluminum-lithium alloys are also candidates for VIM processing.

Copper Alloys. The production of high-purity copper having less than 2 ppm O can be accomplished only in a vacuum induction furnace. Oxygen content influences the electrical conductivity of copper alloys; the lower the oxygen content, the higher the electrical conductivity (Ref 8). For the production of oxygen-free copper, melting and casting must be carried out under vacuum.

Selective Evaporation of Alloying Elements. The use of vacuum metallurgy is primarily linked with degassing and decarburization. A side effect of these treatments is the evaporation of elements with high vapor pressures. In nonferrous metallurgy, this effect is used for the distillation of metals—for example, for the separation of lead and zinc in lead refining, in zinc production, and for the reduction of magnesium and nonalkali metals. Similarly, copper can be refined from copper scrap by using the vacuum to evaporate volatile elements such as lead and zinc (Ref 9).

Furnace Design and Casting Technologies

Beginning with the original furnace design (Fig. 8), VIM equipment has continued to improve to adapt to special requirements. Equipment can be combined in various ways for maximum productivity. For example, in some designs, the mold chamber can be prepared for casting without disturbing the melting operations, and vice versa. In addition, the smaller vacuum chambers of modern equipment reduce the pumping capacity required and/or allow low pressures to be reached faster.

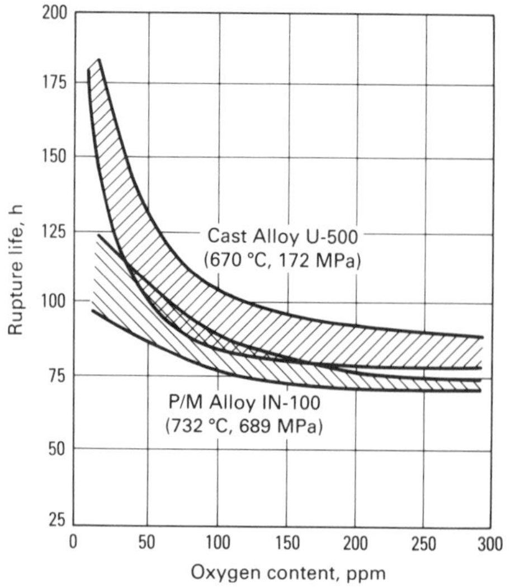

Fig. 7 Effect of oxygen on the rupture life of two superalloys. Source: Ref 7

Vacuum Induction Degassing and Pouring (VIDP). A new development has led to the VIDP furnace design (Fig. 9). This furnace is a further development of the conventional VIM furnace, which offers all the possibilities of metallurgical and process technology. The VIDP furnace design employs a melting and treatment unit that enables the use of different casting techniques. The modular concept allows for connecting it to casting units for ingot casting, horizontal and vertical continuous casting, or powder production. Because of the smaller volume of the VIDP furnace compared to the VIM furnace and significantly lower desorption and leakage rates, it is possible to obtain very low pressures with lower pumping capacity. The lower part of the furnace can be decoupled and replaced rapidly; therefore, the vacuum induction degassing and pouring furnace enables faster replacement of different furnace vessels with changes of alloy.

Table 2 Possible applications for VIM in the processing of nonferrous metals and alloys

Metal or alloy	Metallurgical results	Achieved by
Aluminum, aluminum alloys	Low oxygen content, effective removal of oxides	Vacuum tank degassing without additions of polluting gases or materials
High-purity metals	Oxygen-free copper (2 ppm O)	Melting and pouring under steady, controlled atmosphere or vacuum; gas stirring; carbon as reducing agent
"New" alloys	Al-Mg, Al-Ce, Al-Zr, Al-Hf alloys; copper alloys with beryllium, cobalt, titanium, zirconium, and lithium; avoidance of lithium losses in production of Al-Li alloys	Furnace and casting system under inert gas atmosphere; vacuum-tight, highly automated furnace
Metal production and recycling....	Production of calcium, barium, lithium, strontium, and magnesium; recycling of hard metals; purification of alloys or melts; removal of zinc and lead	Thermic/alumino-thermic vacuum distillation; leaching with zinc and subsequent vacuum distillation melting; vacuum distillation

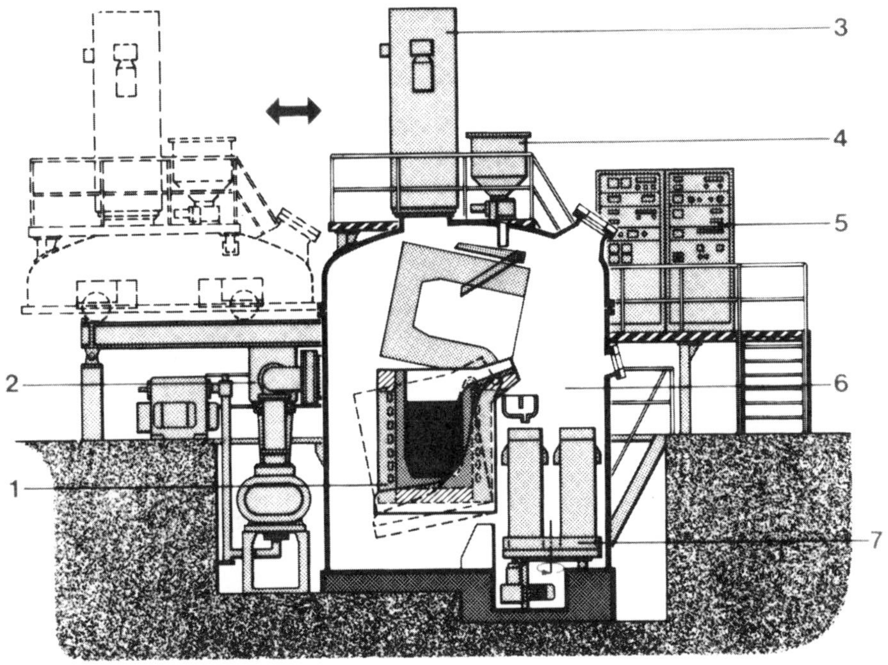

Fig. 8 Schematic of a typical VIM furnace. 1, furnace insert; 2, vacuum pumping system; 3, bulk charger; 4, fine charger; 5, control cabinets; 6, melting and mold chamber; 7, mold turntable

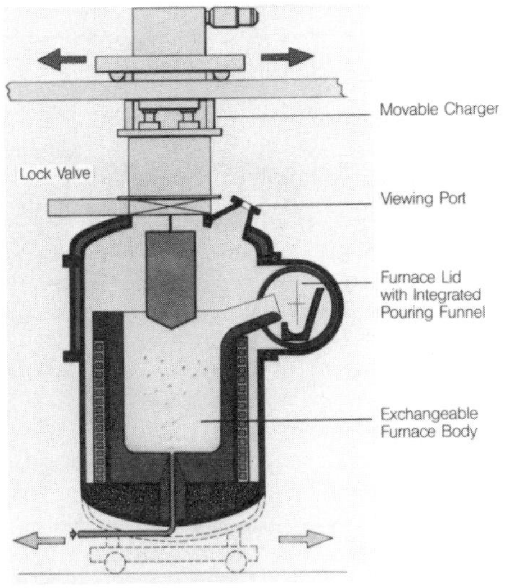

Fig. 9 Schematic of the VIDP furnace

Melting bath agitation is caused by the induction coil itself, depending on the power input and the installed frequency. However, use of the coil to assist in the bath agitation is basically limited to the meltdown period. Degassing, for example, is improved by means of good bath agitation. However, this can be done with the induction coil alone only if there is a high-power input and therefore strong bath heating. This high temperature increase is not always metallurgically desirable.

For this reason, two agitation systems are offered in state-of-the-art vacuum induction furnaces, and one or the other can be used according to requirements. They are:

- Electromagnetic agitation with an additional coil
- Agitation by argon purging through the bottom of the crucible

The difference between the electromagnetic agitation by a separate coil and the agitation caused by the electromagnetic forces of the main coil lies in the fact that the bath can be circulated at 50 to 60 Hz without temperature increase. A second transformer is of course required. The second method of bath agitation is derived from the principle of argon purging usually applied in ladle metallurgy (see the section "Argon Oxygen Decarburization" in the article "Degassing Processes (Converter Metallurgy)" in this Volume).

Argon purging has a long history of use in open induction melting furnaces. A characteristic of this purging is that the porous plug is not in direct contact with the liquid melt. Instead, the plug is covered with refractory material identical to that used for the crucible lining. A new development (Fig. 10) has the basic difference that the porous plug is in direct contact with the melt and is suitable for operating under vacuum. This ensures that the admitted purging gas flows through the melt and does not take the path of least resistance through the lining of the crucible.

Both agitation processes offer advantages with regard to higher reaction rates, correct temperature adjustment, homogenization with simultaneously lower erosion of the crucible lining, and better cleaning of the melt. Figure 11 shows the three different agitation methods operating in a typical VIM process sequence. The agitation effect at 200 to 500 Hz in the melting phase assists in shortening the melting time. In the refining and superheating periods, it is better to use electromagnetic agitation at 50 to 60 Hz or argon purging through a porous plug.

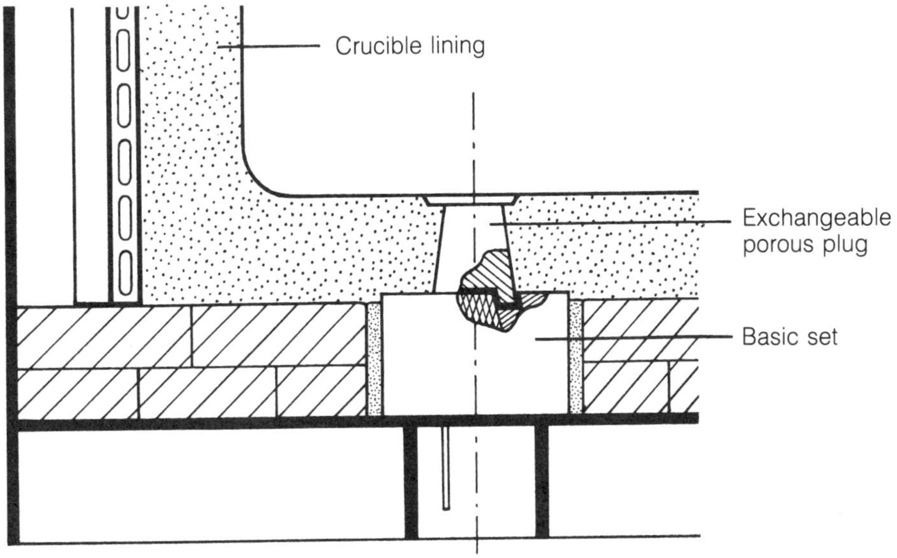

Fig. 10 Argon purging system for VIM furnaces

Process Technology and Automation

The installation of programmable control in combination with a process computer for automation provides better reproducibility of the melts. In this way, increasing metallurgical demands for cleanliness and homogeneity can be met.

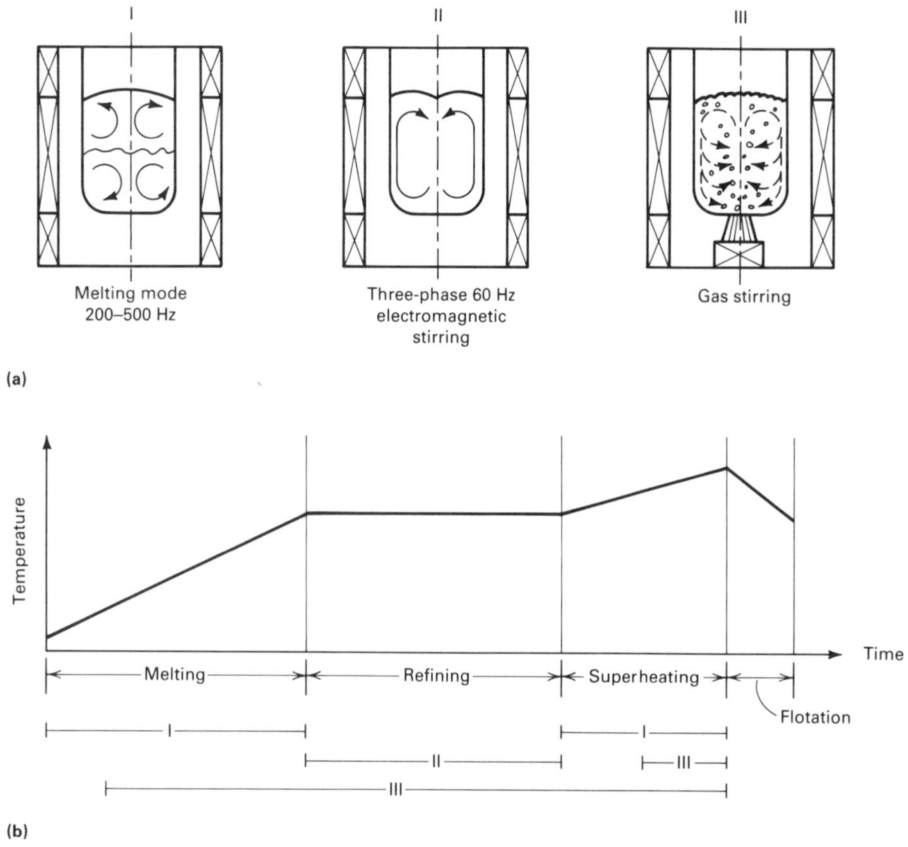

Fig. 11 Melting (a) and stirring (b) modes of the VIM process

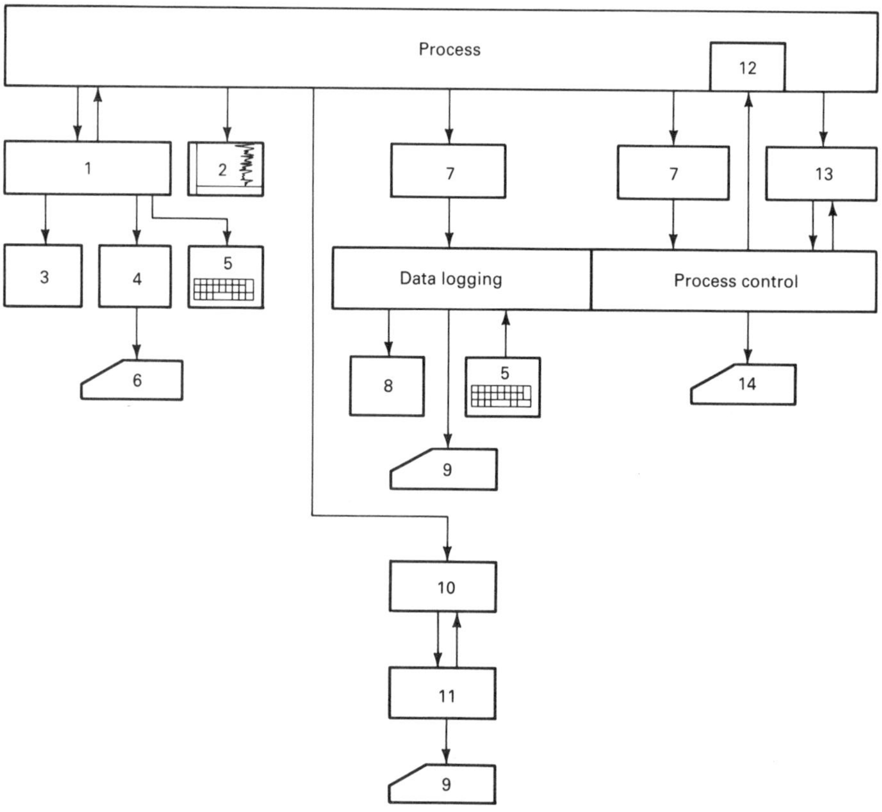

Fig. 12 Block diagram of modern VIM process control system. 1, PLC control; 2, recorder; 3, mimic diagram; 4, monitor screen; 5, keyboard; 6, fault printer; 7, analog/digital output; 8, monitor screen; 9, printer; 10, mass spectrometer; 11, personal computer; 12, melting power supply; 13, pyrometer; 14, graphic plotter

In addition, close composition tolerances can be achieved, in which all of the data necessary for the process are registered, stored, and processed. The basic concept is 100% automation of the VIM process by using a combination of individual modules, such as control and regulation of the furnace, model calculations, and calculations for alloying additions.

For a number of years, these systems have been successfully used for controlling the vacuum system, recording the melting parameters, and recognizing and diagnosing defects. Figure 12 illustrates schematically a modern VIM process control.

A relatively new method is the use of a video screen display to provide a visual of the process. This technology will replace the conventional methods of recording the melting process, which involve the use of recorders and manual melting logs. Moreover, with computerized systems, all process control data can be displayed, printed out, or stored. With the installation of discontinuous or continuous bath temperature measurement, which is linked to a computer (theoretical thermal model), melting power can be automatically adjusted to give prespecified bath temperatures.

Using charge and alloy calculations, it is easy to achieve the required chemical analysis with minimal cost. For this application, in the ideal case, the computer is linked to the analysis system computer so that the additives can be calculated and, if necessary, weighed and added directly after the analysis. If this alloy calculation is laid out as a charge calculation, the charges and the amounts of scrap necessary can be optimized. The complete calculation system enables a calculation of the alloying elements, starting with the amount of scrap needed and proceeding to the necessary alloy addition at the end of the refining period.

A computer-controlled mass spectrometer system, specially developed for the VIM process, can make a significant contribution to the optimization of the melting process and its economics. Using a computer evaluation of the gas composition in the furnace chamber, information about the state of the degassing process and of the chemical reactions in the melt can be continuously obtained.

The melting process can thus be controlled very exactly. Process parameters that can be closely controlled include:

- Determination of the correct time for the addition of chemically active elements and the sequence of addition of such elements
- Determination of the process steps, such as completion of the refining period and the suitable time for tapping
- Advance detection of leaks in the furnace chamber, cooling water system, or hydraulic system

- Monitoring of the state of degassing of the crucible refractory lining

REFERENCES

1. J.W. Pridgeon *et al.*, in *Superalloys Source Book*, American Society for Metals, 1984, p 201-217
2. W.B. Kent, *Int. Vac. Sci. Technol.*, Vol 11 (No. 6), 1974, p 1038-1046
3. P.P. Turillon, in *Transactions of the Sixth International Vacuum Metallurgy Conference* (Boston), American Vacuum Society, 1983, p 88
4. G.A. Simkovich, *Int. Met.*, Vol 253 (No. 4), 1966, p 504-512
5. H. Katayam *et al.*, in *Proceedings of the Seventh International Conference on Vacuum Metallurgy* (Tokyo), The Iron and Steel Institute of Japan, 1982, p 933-940
6. A. Choudhury *et al.*, *World Steel and Metalworking Manual*, Vol 9, 1987-1988, p 1-6
7. P. Hupfer, Fachberichte Hüttenpraxis Metallverarbeitung, 1986, p 773-781
8. O. Kamado *et al.*, Method of Producing Electrical Conductor, European Patent 0121152, 1986
9. J.G. Krüger, *Proceedings of the Fifth International Vacuum Metallurgy Conference* (Munich), 1976, p 75-80

Vacuum Induction Remelting and Shape Casting

R. Brink, Leybold AG, West Germany

Vacuum precision casting furnaces are used to remelt alloys that have already been treated in vacuum. The remelted alloy is then precision cast into preheated ceramic molds. Heating and melting in such furnaces almost always takes place by induction, although some electron beam melters and vacuum arc furnaces are used. Electron beam furnaces offer the advantage of ceramic-free melting, but they do not permit a sufficiently accurate and reproducible temperature control for charges of more than approximately 5 to 7 kg (11 to 15 lb). Vacuum arc furnaces are used for titanium precision casting.

Furnaces

The vacuum induction precision casting furnace is a two-chamber furnace with a melting chamber and a mold chamber (Fig. 1). The mold chamber is preferably placed under the melting chamber so that, in connection with another chamber for the charge or melting material, semicontinuous operation with minimal cycle times is possible.

For cobalt- or nickel-base superalloys, the melting chamber pressure is of the order of 0.01 Pa (10^{-4} mbar, or 7.5×10^{-5} torr), while the mold lock would be evacuated to approximately 5 Pa (0.05 mbar, or 0.038 torr) within 1 min. To date, equipment for melting charges weighing 15 to 150 kg (33 to 330 lb) has been manufactured.

Processes

Until the 1970s, most furnaces were configured for the so-called equiaxed casting (that is, for the production of polycrystalline castings with equiaxed grains). Directional solidification (DS) and single-crystal (SC) casting are rapidly gaining acceptance. Figure 2 compares the structures obtained by these casting methods, and both methods are discussed in the section "Directional and Monocrystal Solidification" of the article "New and Emerging Processes" in this Volume.

In the past, furnaces for casting were often modified for use with the DS process. The current trend for production furnaces, however, is toward dedicated, single-purpose melting equipment. Multipurpose furnaces will still be required in the future, but the changeover from equiaxed to DS processes involves necessary compromises that introduce too many limitations with respect to technology and economy. Therefore, multipurpose furnaces are almost exclusively used for research and development or as pilot furnaces.

Products

In addition to the casting of equiaxed, DS, and SC blades for gas turbine engines, vacuum precision casting can be used for large parts (melt weights up to 1000 kg, or 2200 lb) or for the large-scale production of small parts. Examples of large parts produced using vacuum precision casting include compressor sections of jet engines and structural parts for stationary turbines.

A typical small casting produced by this method is a turbocharger wheel for automotive engines (Fig. 3). Automotive turbocharger wheels weighing 400 to 500 g (0.9 to 1.1 lb) are produced in automatic precision casting machines. Total cycle time, including 40 s for melting, is typically 90 to 100 s. In the production of such small parts, the crucible is often integrated into the mold. Casting yields of 80 to 90% are obtained.

Process Automation

Because of the increasing demand for productivity and constant casting quality under reproducible conditions, single-purpose furnaces for equiaxed casting are being automated in all decisive process steps. The primary objective is reproducibility, especially regarding temperature control during melting and casting. As a result of these requirements, modern precision casting furnaces are usually equipped with process automation devices for melting, melting bath temperature measurement, degassing, mold positioning, and casting. Automated units are outfitted with extensive monitoring devices which collect data for statistical process control.

Experiences with the automatic melting unit, which is controlled by a two-color pyrometer, confirm good measurement and control accuracy. For example, temperature differences of less than 5 K between the two-color pyrometer and a thermocouple were found to persist after ten charges had been processed. This reproducibility also enhances the endurance of ceramic melting crucibles.

Similar positive results have also been obtained with the automatic casting device. With this unit, single steps such as melt washing, preheating of the crucible lip, and tilting can be preprogrammed. The pouring process itself is adjustable against several parameters that can be preprogrammed individually, including:

- Pouring temperature
- Mold temperature
- Mold positioning
- Pouring height

The pouring profile or tilting of the crucible in order to fill the mold as rapidly as possible can also be preprogrammed. These data, which vary depending on the casting process chosen, are fed into the computer manually or automatically and can then be called up as often as necessary.

The objective of developments such as those outlined above is fully automated computer-controlled furnaces in which the manual portion of the process sequence is limited to inserting the molds and the material to be melted. This goal has been realized, especially for DS and SC furnaces. With computer software and data recording, process parameters such as temperatures, times, and solidification rates remain freely selectable.

Equipment Developments. The development of certain furnace components was essential to the optimization and automation of precision casting furnaces for equiaxed castings. Examples of such developments include:

- Hydraulic, rotary coaxial feed cables for the induction coil

Fig. 1 Computer-controlled vacuum precision casting furnace showing the mold chamber

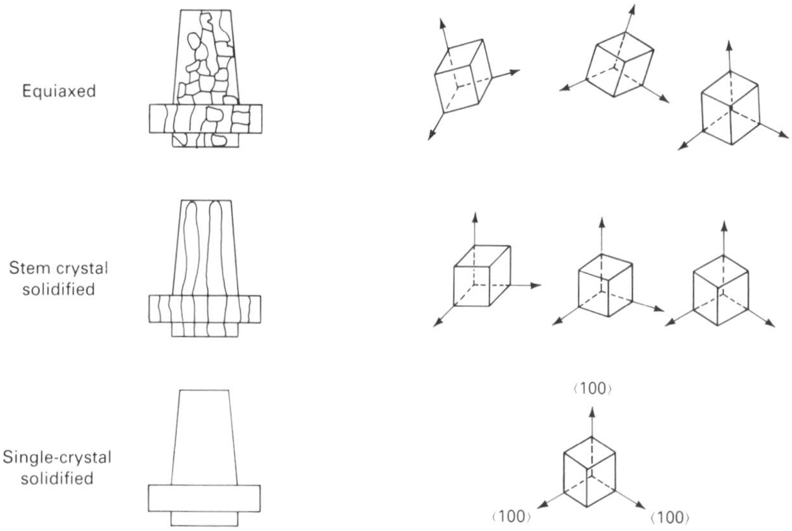

Fig. 2 Types of solidification in nickel-base alloy blades for gas turbines. (a) Equiaxed, randomly oriented grains. (b) Directionally solidified structure with all grains oriented the same. (c) Single crystal with preferred ⟨100⟩ orientation

- Self-supporting induction coils, which can be easily and quickly exchanged and adapted to the use of crucible containers
- Crucible containers for melt weights to 25 kg (55 lb)
- Swiveling lock valves between the melting chamber and the mold chamber to facilitate mold transfer between the chamber and the casting position

Although furnaces for equiaxed casting can be used for DS and SC processes with some modifications, certain compromises must be made that result in less-than-optimum operation. The furnace modifications required for DS and SC casting will be outlined in the next section of this article.

Special Features of DS and SC Furnaces

Furnaces for directional solidification and single-crystal casting incorporate several features that distinguish them from furnaces used for equiaxed casting. These features include zone heaters, special motors for mold withdrawal movement, and water-cooled copper chills.

The zone heaters used in DS and SC furnaces consist of two or more resistance heating elements or graphite induction heating elements that can be independently controlled. The use of zone heating ensures that temperature gradients of at least 60 to 80 K/cm (150 to 200 K/in.) are obtained at the solidification front; temperature gradients of 30 K/cm (75 K/in.) are common in equiaxed casting. Macroscopic geometric distortions at the solidification front are also minimized by the high temperature gradients.

The motors used to lower the mold out of the heating system are extremely low vibration units. Low vibration is particularly important for single-crystal casting. The use of such motors is based on experience with Czochralski and Bridgman furnaces used to grow single crystals of semiconducting materials.

Water-cooled copper chills are used directly under the mold. They promote directional solidification and are particularly important for securing high temperature gradients in SC casting.

Furnaces for directional solidification and single-crystal casting also use bottom pouring of the melt rather than the tilt pouring used in equiaxed casting. This considerably reduces inclusions in precision cast parts.

Furnaces for DS and SC casting, like those for the casting of equiaxed parts, also have components that are designed to optimize the process. These include:

- One-shot crucibles, which are loaded along with the furnace charge. Such cru-

Fig. 3 Precision cast turbocharger wheels for automotive engines. From left: mold with integrated crucible, bar stick, cast part, machined turbocharger wheel

cibles give almost abrasive-free melting at rates of about 1.5 kg/min (3.3 lb/min) and result in minimal inductive bath movement

- Temperature control of melting through the use of a two-color pyrometer and a thermocouple. In this way, both melt bath temperature and casting spout temperature can be monitored
- Easily accessible mold locks with movable chill plates allow reproducible mold positioning and adjustment

Electroslag Remelting (ESR)

A. Choudhury and F. Knell, Leybold AG, West Germany

The rapid development of modern technology places ever greater demands on engineering alloys. These increasing demands can scarcely be met by classical steelmaking processes. Metallurgists were required to develop new steel-refining processes that guarantee uniformly high quality. One of the refining processes used is electroslag remelting with consumable electrodes in a water-cooled copper crucible, usually under normal atmosphere.

The theory behind this process was known in the 1930s and was the subject of a U.S. patent (Ref 1), but a general breakthrough for this process took more than 30 years. Intensive studies carried out in the Soviet Union, Germany, United Kingdom, Austria, and Japan after World War II made the use of electroslag remelting possible on a production scale (Ref 2-8). In contrast to vacuum arc remelting (VAR), the remelting in the ESR process (Fig. 1) does not occur by striking an arc under vacuum. In electroslag remelting the ingot is built up in a water-cooled mold by melting a consumable electrode immersed in a superheated slag.

The heat required is generated by an electrical current (usually ac) flowing through the liquid slag, which provides the electrical resistance. As the slag temperature rises above the liquidus temperature of the metal, the tip of the electrode melts. The molten metal droplets fall through the liquid slag and are collected in the water-cooled mold. During the formation of the liquid film, the metal is refined and cleaned of contaminants, such as oxide particles. The high degree of superheat of the slag and of the metal favors the metal/slag reaction. Melting in the form of metal droplets greatly increases the metal/slag interface surface area. The intensive reactions between metal and slag result in a significant reduction in sulfur and nonmetallic inclusions. The remaining inclusions are very small and are evenly distributed in the remelted ingot.

Another special feature of the ESR process, as in vacuum arc remelting, is the directional solidification of the ingot from bottom to top. The macrostructure is marked by an extraordinarily high density and homogeneity as well as by the absence of segregations and shrinkage cavities (Ref 4, 5).

The homogeneity of the ingot also results in uniform mechanical properties in the longitudinal and transverse directions after hot working. Because of the absence of macrosegregations or a heterogeneity in the distribution of nonmetallic inclusions, the yield of good ingot material is increased; the additional cost of the remelting process is therefore justified. In addition, the very clean and smooth surface of the ESR ingot, which is specific to this process, helps to reduce production costs because surface conditioning before hot working is not necessary.

Remelting of Steels

There are a number of publications that deal primarily with improvements in the physical properties of remelted steel (Ref 2-8). However, they discuss not only the directional solidification of the metal but also the influence of liquid, superheated, and therefore reactive slags (Ref 4, 9). The extent and direction of the metallurgical reactions in the ESR process are determined by the steel composition, the slag used, and the atmosphere used (inert gas, air, etc.).

The principal feature of the ESR process is the slag bath. A continuous transport of liquid metal through the slag takes place. During this transport, the slag and the metal compositions change, according to the kinetic and thermodynamic conditions. To perform its intended function, the slag must have some well-defined properties, for example:

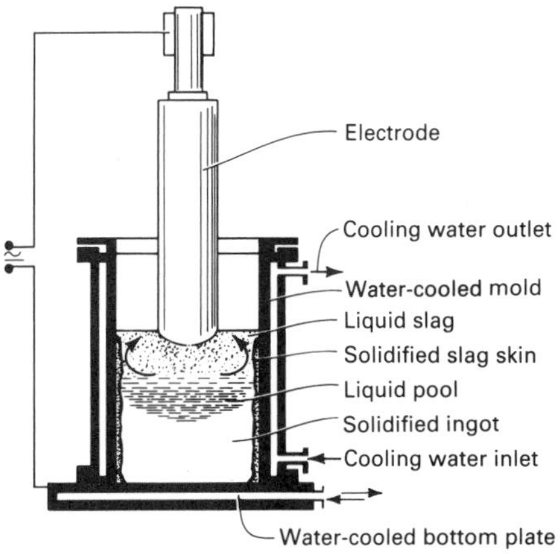

Fig. 1 Schematic of the electroslag remelting process

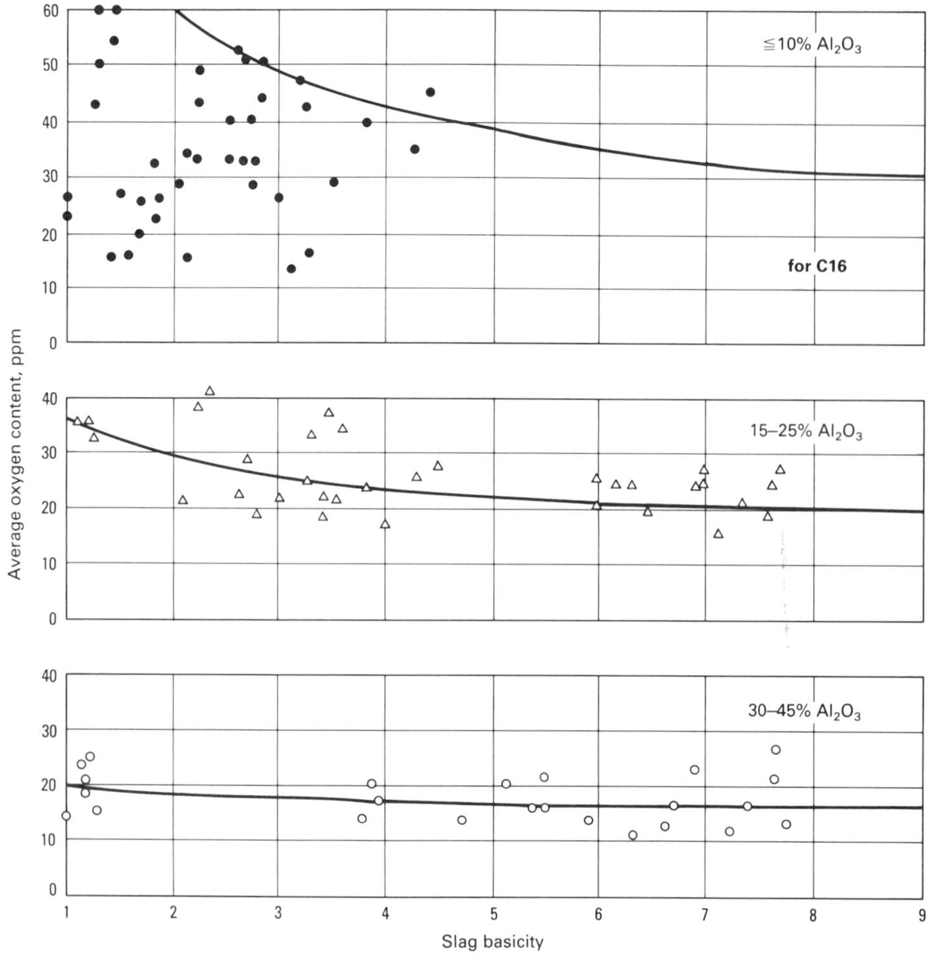

Fig. 2 Effect of slag composition on the oxygen content of the remelted ingot. Slag basicity is the ratio of CaO to SiO_2.

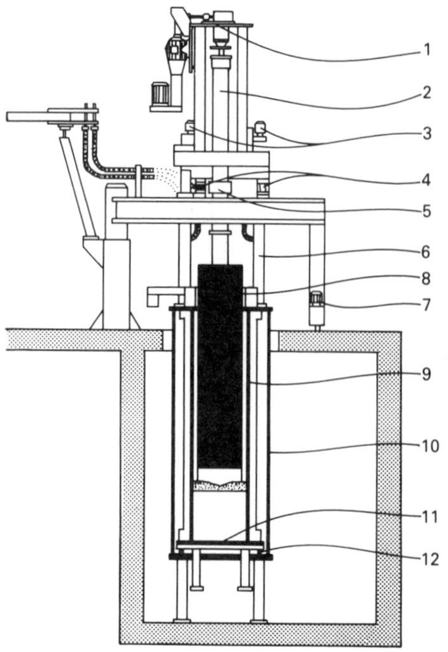

Fig. 3 Schematic of an ESR furnace with a stationary mold. 1, RAM drive system; 2, electrode RAM; 3, XY adjustment; 4, load cell system; 5, sliding contact; 6, four bus tubes; 7, pivoting drive; 8, electrode; 9, mold assembly; 10, coaxial bus tube; 11, base plate; 12, multicontacts

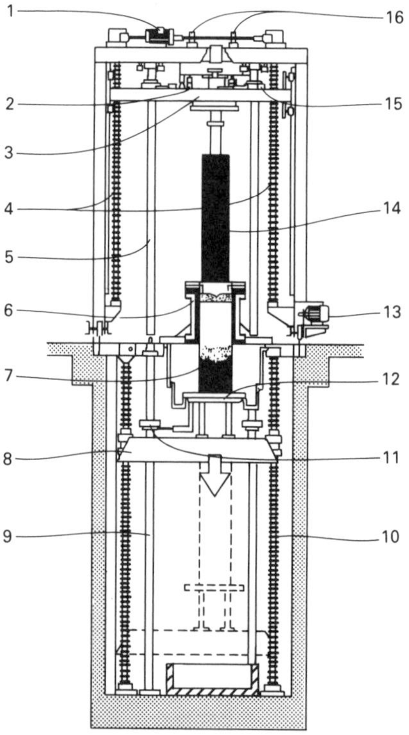

Fig. 4 Schematic of an ESR furnace with retractable bottom plate. 1, electrode drive system; 2, load cell system; 3, stinger crosshead; 4, two ball screws; 5, three bus tubes; 6, mold assembly; 7, ingot; 8, ingot withdrawal table; 9, three bus tubes; 10, three ball screws; 11, three sliding contacts; 12, base plate; 13, furnace head drive; 14, electrode; 15, three sliding contacts; 16, XY adjustment

- Its melting point must be lower than that of the metal
- It must be electrically efficient
- Its composition should be such that the desired reactions, such as removal of sulfur and oxides, are ensured
- It must have suitable viscosity at remelting temperature

Slags for electroslag remelting are usually based on calcium fluoride (CaF_2), lime (CaO), and alumina (Al_2O_3). Magnesia (MgO), titania (TiO_2), and silica (SiO_2) are also added, depending on the alloy to be remelted. Table 1 shows the compositions of some common ESR slags.

Table 1 Compositions of slags commonly used in ESR

Slag designation	Composition, % CaF_2	CaO	MgO	Al_2O_3	SiO_2	Comments
100F	100	...	...	...	...	Electrically inefficient; use when oxides are not permissible.
70F/30	70	30	...	...	...	Difficult starting; high conductivity; use when aluminum is not allowed; risk of hydrogen pickup
70F/20/0/10	70	20	...	10	...	Good general-purpose slags; medium resistivity
70F/15/0/15	70	15	...	15	...	
50F/20/0/30	50	20	...	30	...	
70F/0/0/30	70	...	...	30	...	Some risk of aluminum pickup; good for avoiding hydrogen pickup; higher resistivity
40F/30/0/30	40	30	...	30	...	Good general-purpose slags
60F/20/0/20	60	20	...	20	...	
80F/0/10/10	80	10	...	10	...	Moderate resistivity; relatively inert
60F/10/10/10/10	60	10	10	10	10	Low-melting, "long" slag
0F/50/0/50	...	50	...	50	...	Difficult starting; electrically efficient

Source: Ref 3

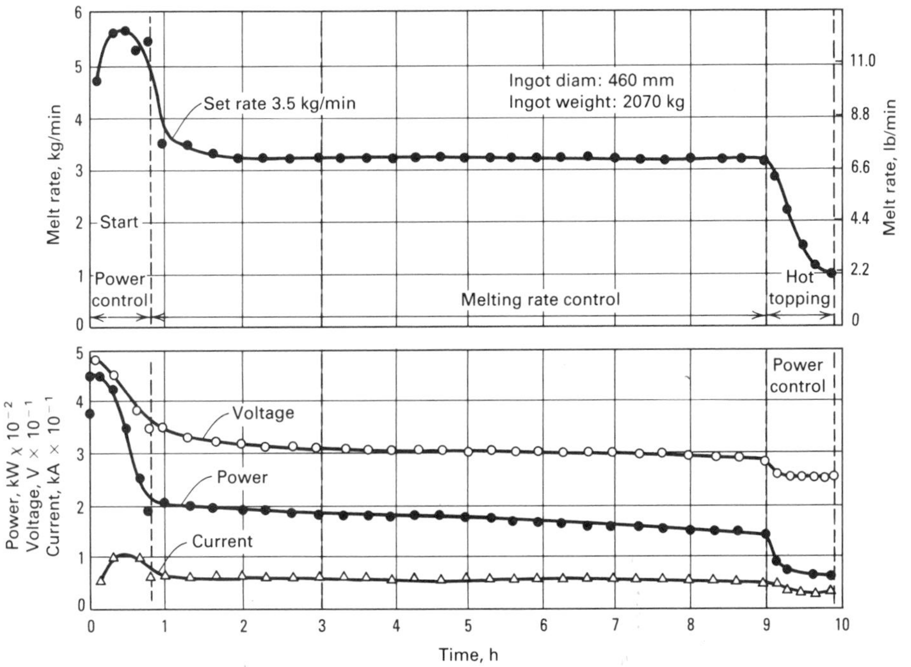

Fig. 5 ESR process parameters for the melting of a nickel-base alloy

Behavior of Sulfur. One of the primary advantages of the ESR process for steel is the good desulfurization of the metal. The final desulfurization is determined by two reactions. The first is the metal/slag reaction, in which sulfur is transferred from the metal to the slag:

$$FeS + CaO \rightarrow CaS + FeO \qquad \text{(Eq 1)}$$

The second reaction is the slag/gas phase reaction. In this case, the sulfur absorbed by the slag is removed by the oxygen of the gas phase in the form of gaseous sulfur dioxide:

$$CaS + \tfrac{3}{2}O_2 \text{ (gas)} \rightarrow CaO + SO_2 \text{ (gas)} \qquad \text{(Eq 2)}$$

It is evident that a saturation of the slag with sulfur does not take place; therefore, the desulfurization capacity of the slag remains intact throughout the entire remelting process. With a highly basic slag ($CaO/SiO_2 > 3$), more than 80% of the sulfur can be removed.

Fig. 6 Control desk of a modern ESR furnace

Behavior of Oxygen. As mentioned previously, the ESR process is usually carried out under a normal air atmosphere. Oxidation of the metal is unavoidable. Oxygen can be transferred into the metal in several ways:

- Oxidation of the electrode surface above the slag bath
- Oxidation on the slag surface of elements with variable valences, such as iron and manganese
- Oxides attached to the electrode surface

Transfer of oxygen into the slag also takes place, and oxygen is transferred into the metal according to Eq 1. This results in losses of easily oxidizable elements, such as aluminum and silicon, during remelting. To counteract this oxidation, the slag should be continuously deoxidized, preferably with aluminum. With proper slag composition and remelting techniques, oxygen contents of less than 20 ppm in unalloyed steel are possible. Figure 2 shows the influence of slag composition on the final oxygen content of remelted ingots.

Ingot Solidification

The thermodynamics of solidification of an ESR ingot are the same as in the VAR process (see the section "Vacuum Arc Remelting (VAR)" in this article). The solidification structure of an ingot of a given composition is a function of the solidification rate and temperature gradient at the solid/liquid interface. For example, when remelting with a consumable electrode in a water-cooled copper mold, the solidification rate is constant because of the constant heat transfer through the cooling water; therefore, a relatively high temperature gradient at the solidification front must be maintained during the entire remelting period to achieve a directional dendritic structure.

Even in the case of directional dendritic solidification, the microsegregations increase with increasing dendrite arm spacing. For optimum results, the objective should be a solidification structure with dendrites oriented as close to parallel to the ingot axis as possible.

However, this is not always possible. To achieve a good ingot surface, there is a minimum energy requirement, and therefore a minimum melting rate, which is a function of ingot diameter. This means that the melting rate for large-diameter ingots cannot be maintained for axis-parallel dendrite growth. Figure 2 in the section "Vacuum Arc Remelting (VAR)" in this article shows the melting rates for different

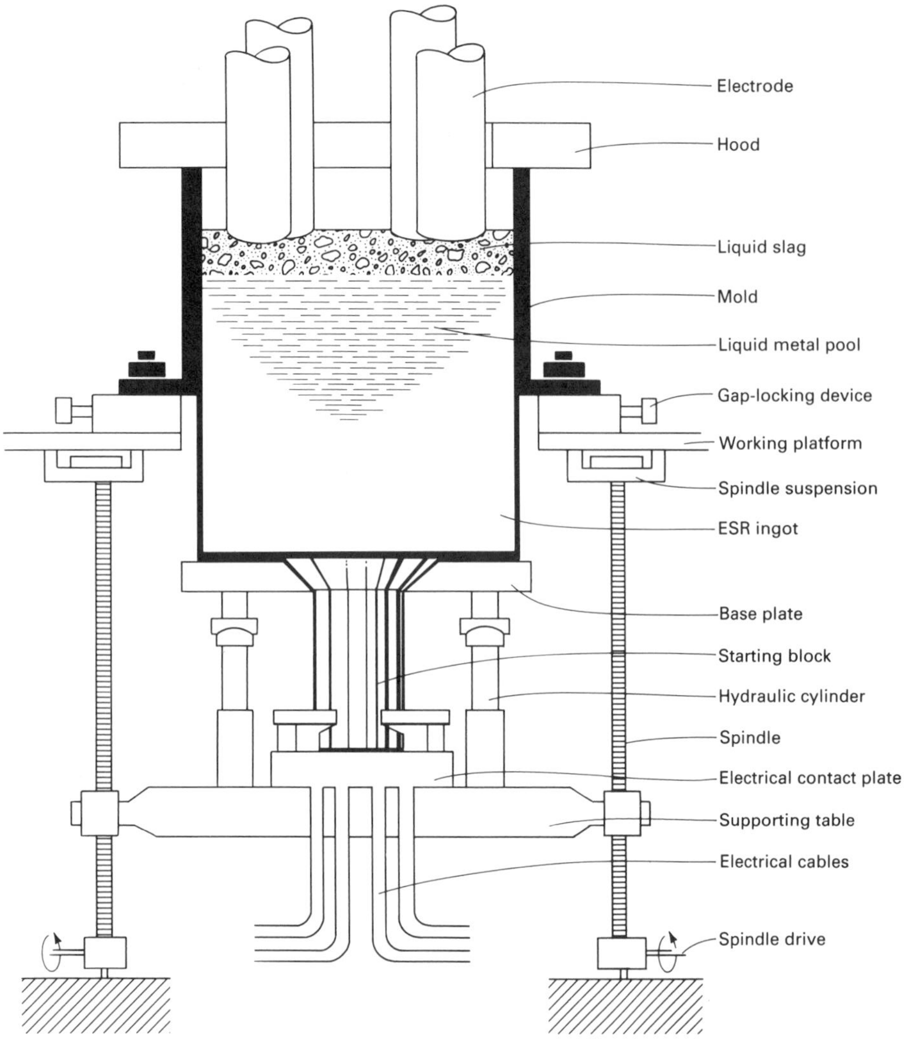

Fig. 7 Schematic of an ESR furnace for the manufacture of large (160 Mg, or 176 ton) ingots

steels and alloys as a function of ingot diameter. The data are empirical values obtained from experience in operation. These melting rates resulted in low microsegregation while achieving reasonably good surface quality.

Ingot Defects. In spite of directional dendritic solidification, various defects, such as the formation of tree ring patterns, freckles, and white spots, can occur in a remelted ingot. These defects also occur in VAR ingots and are discussed in more detail in the section "Vacuum Arc Remelting (VAR)" in this article.

Current State of Plant Design

Significant advances have been made in the past few years in plant design in the area of process control and coaxial current supply. Figure 3 shows the basic design of a modern ESR furnace with a fixed mold for an ingot weight of 20 Mg (22 tons) and an ingot diameter of 1000 mm (39.4 in.). Figure 4 shows a schematic of an ESR furnace equipped with a retractable bottom plate.

A fully coaxial furnace design is required for the remelting of segregation-sensitive alloys in order to prevent melt stirring by stray magnetic fields. In the case of the retractable bottom plate furnace, current feedback does not take place through a closed copper pipe; instead, it is accomplished with four symmetrically arranged current feedback tubes.

Automation of Process Control. The ESR process, like vacuum arc remelting, is fully automatic. Examples of process parameters for the remelting of a nickel-base alloy are shown in Fig. 5. Melting was accomplished in a fully automatic operational mode. Figure 6 shows a modern ESR control desk.

Electroslag Remelting of Heavy Ingots

At the end of the 1960s, the concept of using ESR plants to manufacture large ingots weighing up to 350 Mg (385 tons) gained acceptance (Ref 10). Increasing demands for energy and the trend toward larger electrical power generating units required cast ingots weighing 100 Mg (110 tons) and more for the manufacture of generator and turbine shafts. With ESR, it is possible to achieve higher yields and to avoid such internal defects as porosity, macrosegregation, and the accumulation of nonmetallic inclusions.

Figure 7 shows a schematic of a large ESR furnace, which was brought on line in West Germany in 1971 to manufacture ingots 2300 mm (90.5 in.) in diameter and 5000 mm (197 in.) in length (Ref 11-14).

Ingots weighing up to 160 Mg (176 tons) can be manufactured using a retractable bottom plate. Several electrode changes are necessary to produce such large ingots.

Remelting of large ingots of this type in retractable bottom plate furnaces demands the solution of specific technical and metallurgical problems, such as:

- Ensuring consistent remelting over a period of several days without interruption
- Prevention of steel and slag breakout when remelting with a retractable bottom plate
- Achieving a good ingot surface
- Ensuring directional solidification to avoid macrosegregation and shrinkage cavities
- Controlling the steel and slag compositions over the entire ingot height
- Adjusting for low hydrogen content
- Adjusting for a low aluminum content (<0.010%) for rotor steels; this has a decisive influence on the creep rupture properties of the rotors

After these problems had been solved, heavy rotors for electrical generators could be manufactured from ESR ingots. Figure 8 shows a 160 Mg (176 ton) ESR ingot before forging.

Interior defects such as macrosegregation, shrinkage cavities, and nonuniform distribution of inclusions can be avoided in large ingots only if directional solidification is ensured over the entire ingot section. By maintaining the correct melting rate and temperature of the slag, directional solidification can be achieved for ingot diameters of 2300 mm (90.5 in.) (Ref 12). Accordingly, the ESR ingot is free of macrosegregation in spite of the large diameter (Fig. 9).

The cleanliness and homogeneity of ESR ingots result in excellent mechanical properties. Figure 10 shows the superior toughness of ESR steel compared to conventionally manufactured steel.

Process Variations

Two variations of electroslag remelting have recently been developed: remelting

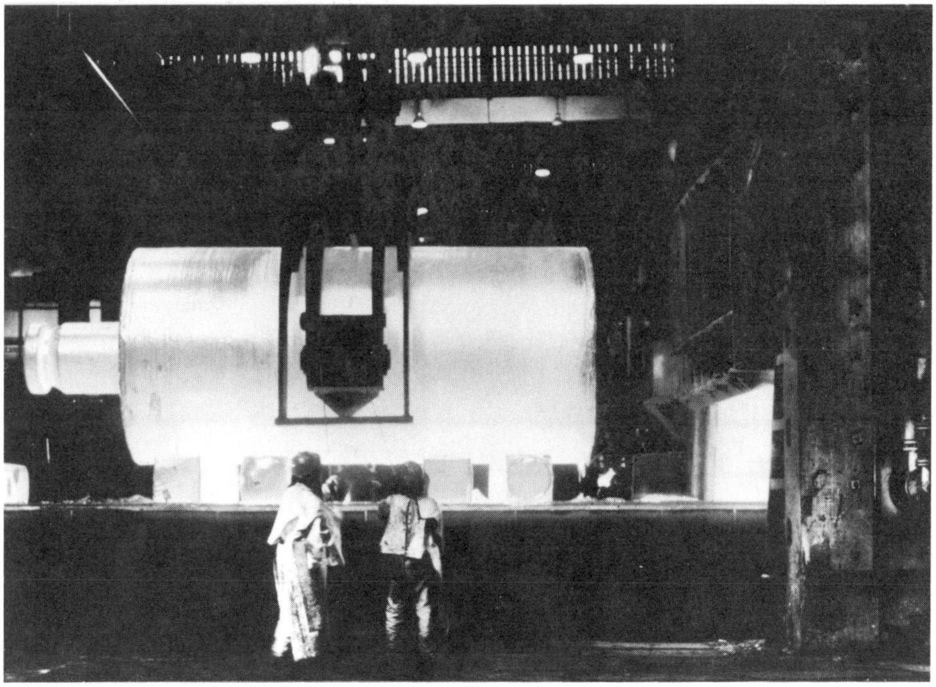

Fig. 8 160 Mg (176 ton) ESR ingot before forging into a generator rotor

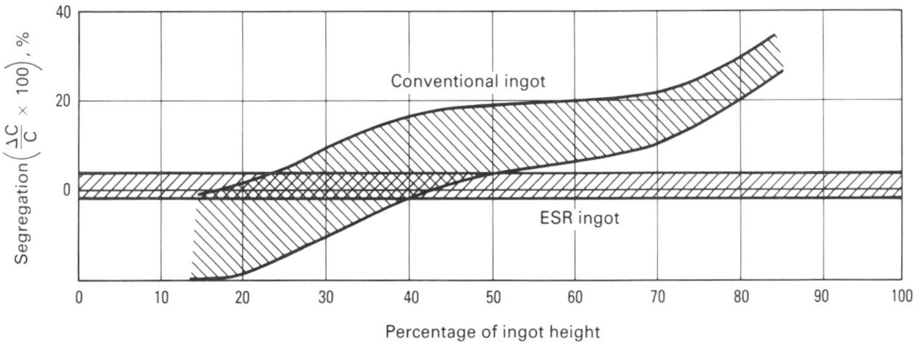

Fig. 9 Comparison of carbon segregation in conventional and ESR alloy steel ingots. Specimens were taken from the ingot axis along the entire ingot height.

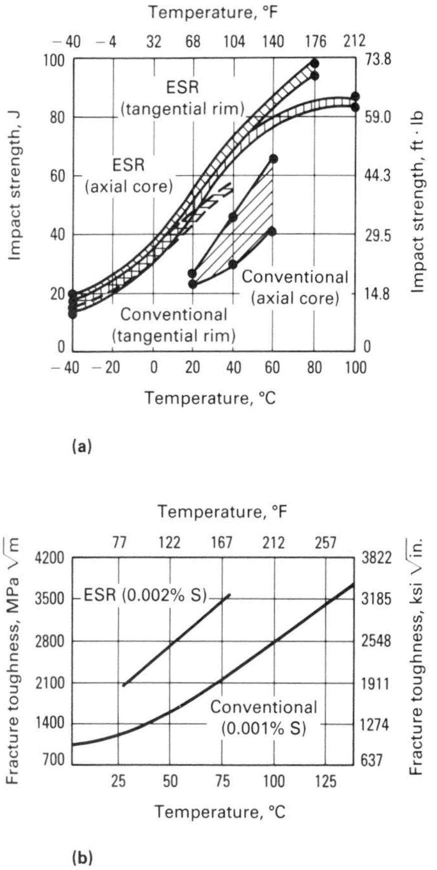

Fig. 10 Comparison of properties of steel rotor forgings made from ESR and conventionally melted ingots. (a) Impact strength of grade X22CrMoV121. (b) Fracture toughness of grade 30CrMoNiV511. Specimen orientation and location are indicated next to curves. Source: Ref 13-15

under high pressure (DESU) and remelting under reduced pressure (VAC-ESR).

Remelting Under High Pressure. The strength of austenitic steels can be increased by alloying with nitrogen as long as the nitrogen is interstitially dissolved in a stable condition. Nitrogen solubility, however, is limited in the case of melting under atmospheric pressure for a given steel composition.

Because the solubility of nitrogen in iron is proportional to the nitrogen partial pressure in the furnace atmosphere, there is the possibility of alloying nitrogen into the melt under higher pressure and allowing it to solidify. The ESR process is suitable for this technique. The first trials in this area were carried out in 1971 (Ref 16).

For some years, a high-manganese steel has been successfully remelted in a DESU plant under a pressure of 4.2 MPa (610 psi). Up to 0.5% interstitial solubility of nitrogen in steel has been achieved. The nitrogen-alloyed austenitic steels are used for the manufacture of retaining rings in generators.

Electroslag remelting under vacuum is another interesting process. Alloy 718 has been remelted in a VAC-ESR pilot plant. A fluoride-free slag based on lime and alumina was used. Because the remelting is carried out under vacuum, problems of oxidation of the melt do not arise. In addition, dissolved gases, such as hydrogen and nitrogen, can be removed. Thus, the advantages of both ESR and VAR are combined in one process (Ref 17).

REFERENCES

1. R.K. Hopkins, Method and Apparatus for Producing Cast Metal Bodies, U.S. Patent 2,380,238, 1945
2. B.E. Raton, B.I. Medower, and W.E. Raton, Ein neues Verfahren—Das Elektroabgiessen von Blöcken, *Bull. tech. Informazil NTO*, No. 1, Maschprom., Kiew, 1956
3. G. Hoyle, *Electroslag Processes*, Applied Science, 1983
4. W. Holzgruber and E. Plöckinger, Metallurgische und verfahrenstechnische Grundlagen des Elektroschlacke-Umschmelzens von Stahl, *Stahl Eisen*, Vol 88, 1968, p 638-648
5. M. Wahlster, A. Choudhury, and K. Forch, Einfluss des Umschmelzens nach Sonderverfahren auf Gefüge und einige Eigenschaften von Stählen, *Stahl Eisen*, Vol 88, 1968, p 1193-1202
6. M. Wahlster and A. Choudhury, Beitrag zum Elektroschlacke-Umschmelzen von Stählen, *Rheinstahl Technik*, Vol 5, 1967, p 31-37
7. M. Wahlster, G.H. Klingelhöfer, and A. Choudhury, Neue metallurgusche und technologische Ergebnisse einer 10 t ESU-Anlage, *Radex-Rundsch.*, Vol 2, 1970, p 99-111
8. G.H. Klingelhöfer, A. Choudhury, and E. Königer, Etude comparée des caractéristique des aciers elaborés à l'air et des aciers refondus selon le procédé sous laiter electroconducteur (ESR)

pour les cylindres de laminage à froid et les cylindres calandreurs, *Rev. Metall.*, Vol 67, 1970, p 512-522
9. H.J. Klingëlhofer, P. Mathis, and A. Choudhury, Ein Beitrag zur Metallurgie des Elektroschlackeumschmelzverfahrens, *Arch. Eisenhüttenwes.*, Vol 42, 1971, p 299-306
10. H. Löwenkamp, A. Choudhury, R. Jauch, and F. Regnitter, Umschmelzen von Schmiedeblöcken nach dem Elektro-Schlacke-Umschmelzverfahren und dem Vakuumlichtbogenofenverfahren, *Stahl Eisen*, Vol 93, 1973, p 625-635
11. R. Jauch, A. Choudhury, H. Löwenkamp, and F. Regnitter, Herstellung grosser Schmiedeblöcke nach dem Elektroschlacke-Umschmelzverfahren, *Stahl Eisen*, Vol 95, 1975, p 408-418
12. A. Choudhury, R. Jauch, and H. Löwenkamp, Primärstruktur und Innenbeschaffenheit herkömmlicher und nach dem Elektro-Schlacke-Umschmelzverfahren hergestellte Blöcke mit einem Durchmesser von 2000 and 2300 mm BA für Forschung und Technologie, Kennzeichen NTS 23
13. R. Jauch, A. Choudhury, F. Tince, and H. Steil, Herstellung und Verarbeitung von ESU-Blöcken bis zu 160 t, *Stahl Eisen*, Vol 101, 1981, p 41-44
14. A. Choudhury, R. Jauch, and F. Tince, Low Frequency Electroslag Remelting of Heavy Ingots With Low Aluminium Content, in *Proceedings of the Sixth International Conference on Vacuum Metallurgy* (San Diego), 1979, p 785-794
15. A. Choudhury, R. Jauch, H. Löwenkamp, and F. Tince, Application of the Electroslag Remelting Process for the Production of Heavy Turbine Rotors from 12 %-Cr-Steel, *Stahl Eisen*, Vol 97, 1977, p 857-868
16. Ch. Kubisch, Druckstickstoffstähle—eine neue Gruppe von Edelstählen Berg.—und Hüttenmännische Monatshefte, 1971, p 84-88
17. A. Choudhury, "Vacuum Electroslag Remelting—A New Process for Better Products," Paper presented at the Vacuum Metallurgy Conference, Pittsburgh, PA, June 1986

Vacuum Arc Remelting (VAR)

A. Choudhury and E. Weingärtner, Leybold AG, West Germany

The vacuum arc remelting process was the first commercial remelting process for superalloys. It was used in the late 1950s to manufacture materials for the aircraft industry. The primary feature of vacuum arc remelting is the continuous melting of a consumable electrode by means of a dc arc under vacuum. The molten material solidifies in a water-cooled copper mold.

The basic design of the VAR furnace has remained largely unchanged over the years; however, significant advances have been made in the fields of control and regulation of the process with the object of achieving a fully automatic melting procedure. This in turn has had a decisive positive influence on the metallurgical properties of the products. The manufacture of homogeneous ingots with minimal segregation requires careful control of remelting parameters. Of these, the melting current density has the largest influence on the melting bath geometry and conditions of solidification.

Process Advantages

The primary benefits of melting a consumable electrode under vacuum are:

- Removal of dissolved gases, such as hydrogen and nitrogen
- Minimizing the content of undesirable trace elements having high vapor pressures
- Improvement of cleanliness by removal of oxides
- Achievement of directional solidification of the ingot from bottom to top in order to avoid macrosegregation and to minimize microsegregation

Oxide inclusion removal is optimized because of the relatively short reaction paths during melting of the hot electrode end and because of a good drop dispersion in the plasma arc. Oxide removal is achieved by chemical and physical processes. Less stable oxides or nitrides are thermally dissociated or are reduced by carbon present in the alloy and are removed by conversion into the gas phase. However, in superalloys and in high-alloy steels, the nonmetallic inclusions (for example, alumina and titanium carbonitride) are very stable. The removal of these inclusions during remelting takes place by flotation.

The remaining inclusions in the solidified ingot are small and are evenly distributed in the cross section. The solidification structure of an ingot of a given composition is a function of the local solidification time and the temperature gradient at the liquid/solid interface, as shown schematically in Fig. 1.

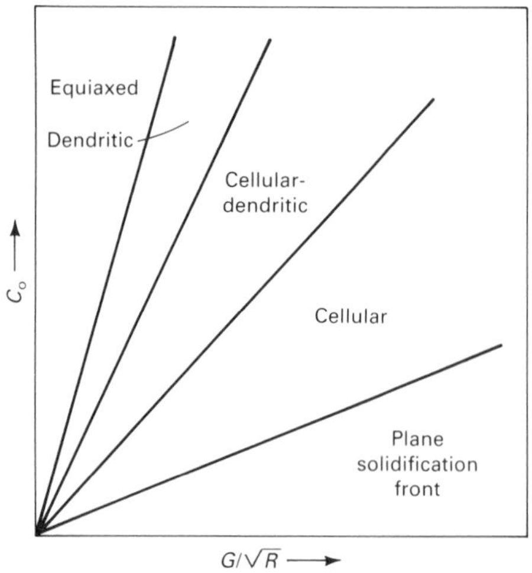

Fig. 1 Primary structure of a VAR ingot as a function of temperature gradient in the melt *G*, solidification rate *R*, and chemical analysis of the melt C_o. Source: Ref 1

To achieve a directed dendritic primary structure, a relatively high temperature gradient at the solidification front must be maintained during the entire remelting period. The growth direction of the cellular dendrites corresponds to the direction of the temperature gradient or the direction of the heat flow at the moment of solidification at the solidification front. The direction of heat flow is always perpendicular to the solidification front or, in the case of a curved interface, perpendicular to the tangent. The growth direction of the dendrites is a function of the metal pool profile during solidification. For remelting processes in water-cooled copper molds, the pool profile has a rotationally symmetric paraboloidal shape in the first approximation. The pool depth increases with melting rate.

The gradient of the dendrites, with respect to the ingot axis, increases with melting rate. In extreme cases, the growth of the directed dendrites can come to a stop. The ingot core then solidifies nondirectionally in equiaxed grains, which leads to segregation and microshrinkage. Even in the case of directional dendritic solidification, the microsegregation increases with dendrite arm spacing. A solidification structure with dendrites parallel to the ingot axis yields optimal results. However, this is not always possible. A good ingot macrostructure requires a minimum energy input and, accordingly, a minimum melting rate. Optimal melt rates and energy inputs depend on ingot diameter. This means that the necessary melting rate for large-diameter ingots cannot be maintained for axis-parallel crystallization.

Figure 2 shows melting rates for various steels and alloys as a function of ingot diameter. These are empirical values that were obtained from experience in operation. These melting rates gave low microsegregation while achieving acceptable surface quality. The importance of pool depth was also investigated by numerous researchers (Ref 2-6).

Ingot Defects

In spite of directional dendritic solidification, such defects as tree ring patterns, freckles, and white spots can occur in a remelted ingot. This can lead to rejection of the ingot, particularly in the case of superalloys.

Tree ring patterns can be identified in a macroetched transverse section as light-etching rings. They usually represent a negative crystal segregation. Tree ring patterns seem to have little effect on superalloy material properties (Ref 7).

Tree ring patterns are the result of a wide fluctuation in the remelting rate. In modern remelting plants, however, the remelting rate is maintained at the desired value by means of an exact control of the melting rate during operation, so that the melting rate exhibits no significant fluctuations.

Freckles and white spots have a much greater effect on material properties than tree ring patterns, especially in the case of superalloys. Both defects represent an important cause of the premature failure of turbine blades in aircraft engines.

Freckles are dark-etching circular or nearly circular spots that are generally rich in carbides or carbide-forming elements. The formation of freckles is usually a result of a high pool depth or movement of the rotating liquid pool. The liquid pool can be set into motion (rotation) by stray magnetic fields during remelting. Freckles can be avoided by maintaining a low pool depth and by avoiding disturbing magnetic fields through the use of a coaxial current supply.

White spots are typical defects in VAR ingots. They are recognizable as light-etched spots on a macroetched surface. They are low in alloying elements, for example, titanium and niobium in Alloy 718. There are several mechanisms that could account for the formation of white spots (Ref 8, 9):

- Relics of unmelted dendrites of the consumable electrode
- Pieces of the crown that fall into the pool and are not dissolved or melted
- Pieces of the shelf region transported into the solidifying interface

All three of the above mechanisms, individually or combined, can be considered as sources of white spots. This indicates that white spots cannot be completely avoided during vacuum arc remelting. To minimize the frequency of occurrence of these defects, the following conditions should be observed during remelting:

- Use the maximum acceptable metal rate permitted by the ingot macrostructure
- Use a short arc gap to minimize crown formation and to maximize arc stability
- Use a homogenous electrode free of cavities and cracks

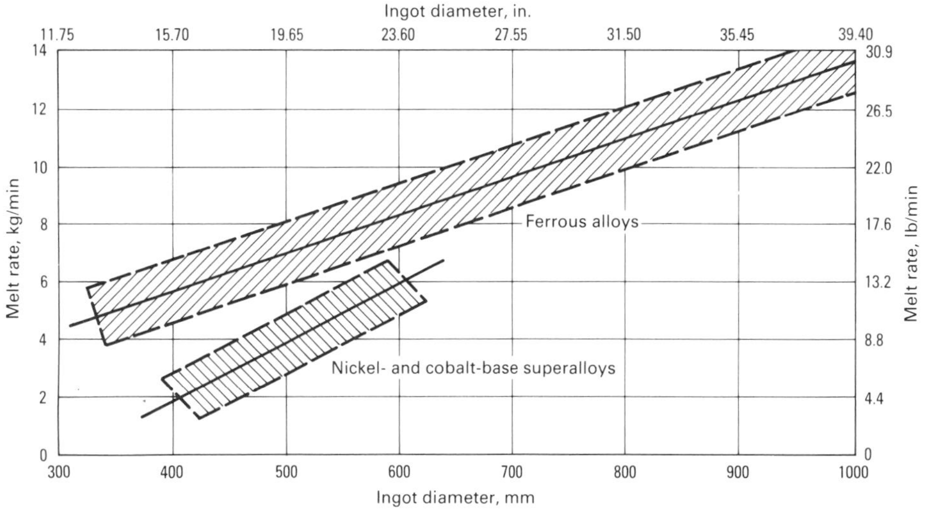

Fig. 2 Typical VAR melting rates for various steels and nickel- and cobalt-base superalloys

Process Variables

Atmosphere. The heat generated by melting of the metal in vacuum arc remelting results from the electric arc between the consumable electrode and the liquid pool on the top of the ingot. The pressure in the remelting vessel is usually of the order of 0.1 to 1 Pa (10^{-3} to 10^{-2} mbar, or 7.5×10^{-4} to 0.0075 torr). In some exceptional cases, the melting is also carried out under inert gas with a pressure up to 10 kPa (100 mbar, or 75 torr). Evaporation losses of volatile alloying elements are minimized at this pressure.

Melt Rate. As mentioned earlier, the melt rate is an important factor in the quality of the ingot macrostructure. A modern VAR furnace is therefore equipped with a load cell system to measure the weight of the electrode at a particular interval of time. The actual values of the melt rate are compared by computer with the desired set values. Any difference between the measured melt rate and the desired value is eliminated by the proper accommodation of the power input.

Figure 3 shows the melt rate and the melting current at start-up, during steady-state melting, and during hot topping. Start-up and hot topping are usually controlled based on time. The melting phase is controlled based on weight.

Hot topping begins when a preselected residual electrode weight is reached. A computer controls the melting parameters, which are stored in the form of up to 250 different recipes in the computer. Figure 4 shows a simplified version of the computer furnace control. With this, melting rate can be controlled with a precision of better than ±2%. The computer also provides documentation in the form of tables and graphs for the relevant process parameters.

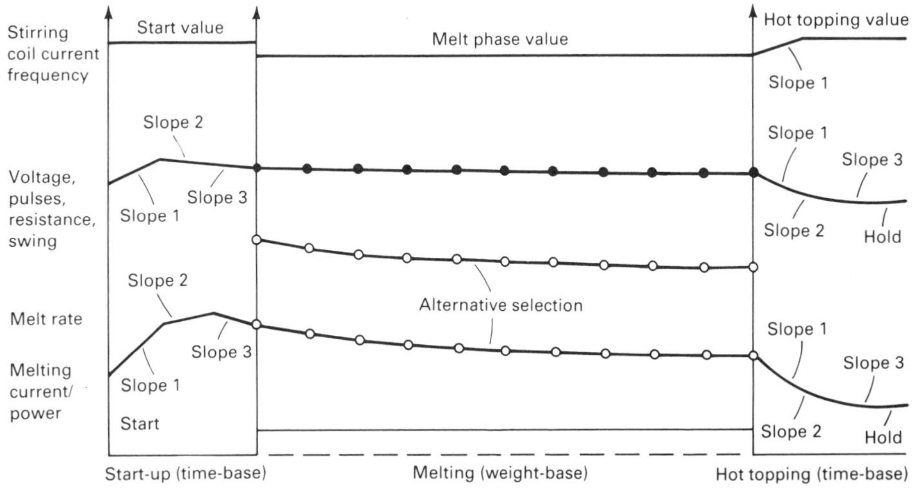

Fig. 3 Process control parameters and set point functions in vacuum arc remelting

Remelting Variations Under Vacuum

Apart from the remelting of a consumable electrode in a water-cooled copper crucible, there is a recent development of the vacuum arc remelting process, namely vacuum arc double electrode remelting (VADER). Figure 5 shows the basic design of the VADER process with a static crucible. The arc is struck between the two horizontal electrodes that are to be melted.

As in vacuum arc remelting, the metal drops fall into a water-cooled copper mold. Bath temperature, and therefore pool depth, can be very closely controlled. Remelting can be done with minimal superheating; segregation is thus minimized. The advantages of the VADER process over vacuum arc remelting are as follows (Ref 10):

- Very low or no superheating of the pool and high rate of nucleation, producing a fine grain structure
- The lowest possible influence of magnetic fields on melting bath movement
- No condensation formation due to evaporation of liquid elements on the crucible walls
- Lower specific energy consumption
- Good ultrasonic testability due to the fine macrostructure of the ingot

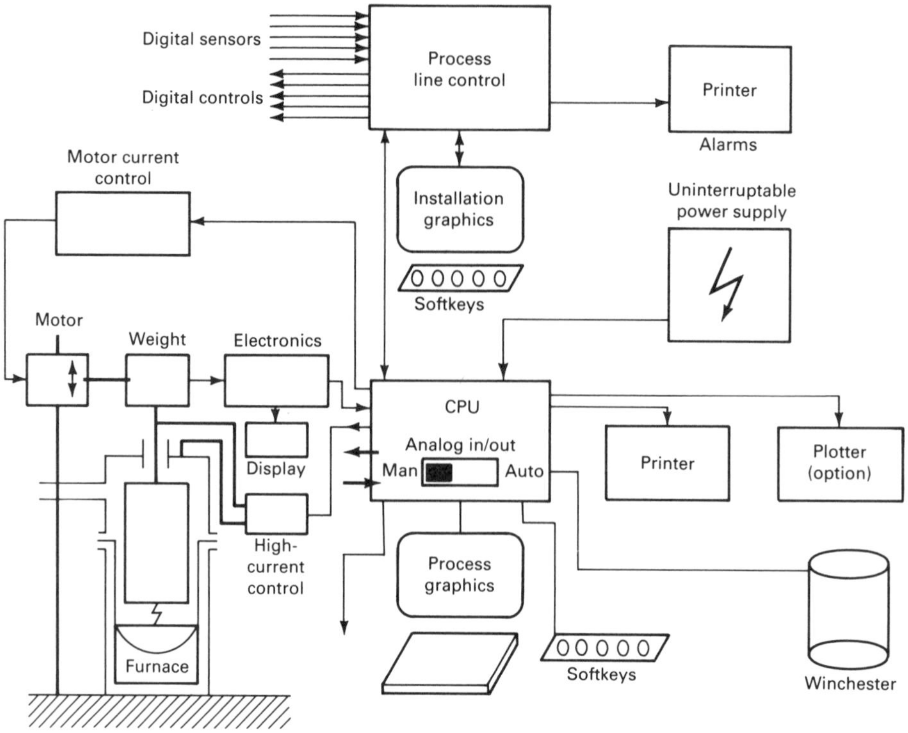

Fig. 4 Schematic of automatic melt control system

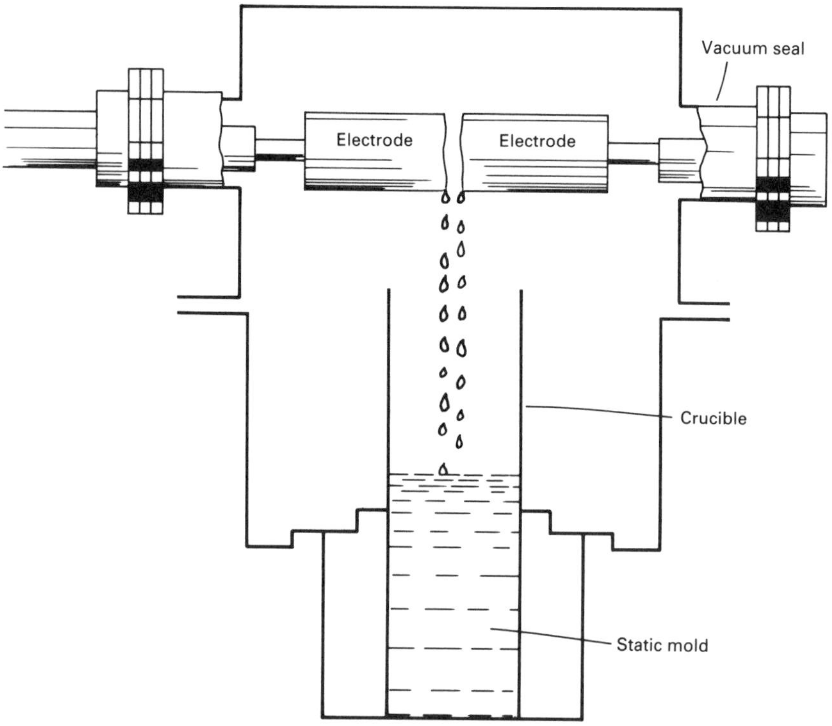

Fig. 5 Schematic of the VADER process

REFERENCES

1. W.A. Tiller and J.W. Rutter, *Can. J. Phys.*, Vol 311, 1956, p 96
2. W.H. Sutton, in *Proceedings of the Seventh International Vacuum Metallurgy Conference* (Tokyo), The Iron and Steel Institute of Japan, 1982, p 904-915
3. J. Preston, in *Transactions of the Vacuum Metallurgy Conference*, American Vacuum Society, 1965, p 366-379
4. A.S. Ballentyne and A. Mitchell, *Ironmaking Steelmaking*, Vol 4, 1977, p 222-238
5. S. Sawa *et al.*, in *Proceedings of the Fourth International Vacuum Metallurgy Conference* (Tokyo), The Iron and Steel Institute of Japan, 1974, p 129-134
6. J.W. Troutman, in *Transactions of the Vacuum Metallurgy Conference*, American Vacuum Society, 1968, p 599-613
7. R. Schlatter, *Giesserei*, Vol 61, 1970, p 75-85
8. A. Mitchell, in *Proceedings of the Vacuum Metallurgy Conference*, Pittsburgh, PA, 1986, p 55-61
9. F.J. Wadier, in *Proceedings of the Vacuum Metallurgy Conference*, Pittsburgh, PA, 1984, p 119-128
10. J.W. Pridgeon, F.M. Darmava, J.S. Huntington, and W.H. Sutton, in *Superalloys Source Book*, American Society for Metals, 1984

Vacuum Arc Skull Melting and Casting

F. Müller and E. Weingärtner, Leybold AG, West Germany

Titanium investment casting has recently gained the same importance as the precision casting of superalloys (see the article "Titanium and Titanium Alloys" in this Volume). Titanium skull melting originated at the Bureau of Mines in Albany, Oregon. The first castings were made in 1953, although possibilities had been announced as early as 1948 and 1949. In the late 1950s, this technology was applied by research institutes, which were looking for a practical means of liquefying and pouring uranium into graphite molds, for example, to produce uranium carbide.

An early industrial vacuum arc skull melter was built in 1963 for the continuous production of uranium carbide. This furnace had a crucible volume of approximately 0.01 m^3 (0.35 ft^3) and used a nonconsumable graphite electrode to liquefy the uranium pellets fed into the crucible. The crucible tilting system was hydraulically driven. The molds were stationary. It was not until 1973 that one of the first skull melters for titanium went into operation; this furnace started production in 1974 in West Germany.

State-of-the-art titanium vacuum arc skull melting furnaces are often equipped with turntable systems for centrifugal casting (up to 350 rpm). Casting weights of more than 1000 kg (2200 lb) are possible. Vacuum arc skull melting and casting is used for many titanium investment castings for aircraft, aerospace, medical, and chemical applications. Electron beam skull melting is also used for titanium alloys (see the section "Electron Beam Melting and Casting" in this article).

Furnaces

Vacuum arc skull casting furnaces basically consist of a vacuum-tight chamber in which a titanium or titanium alloy electrode is driven down into a water-cooled copper crucible. The dc power supply provides the fusing current needed to strike an electric arc between the consumable electrode and the crucible. Because the crucible is water cooled, a solidified titanium skull forms at the crucible surface, thus avoiding direct contact between melt and crucible.

Once the predetermined amount of liquid titanium is contained in the crucible, the electrode is retracted, and the crucible is tilted to pour the melt into the investment casting mold positioned below. For optimum mold filling, the mold can be preheated and/or rotated on a centrifugal turntable.

Figure 1 shows the operating principle of a modern 50 kg (110 lb) vacuum arc skull melting furnace. At an operating pressure of approximately 1 Pa (10^{-2} mbar, or 0.075 torr), the specific working current ranges from approximately 1 kA/kg for small furnaces to about 0.2 kA/kg for large pouring weights. This batch-type skull melting furnace allows for cycle times of approximately 1 h for a full 50 kg (110 lb) pumping/melting/casting cycle, and in principle three consecutive pours can be obtained from one electrode. This furnace basically consists of a vacuum chamber, an arc voltage-controlled electrode drive system, a skull crucible, a centrifugal casting system with stepless adjustable turntable speed, an automatically sequenced vacuum pump system, a power supply, and an electrical control system with control desk.

The cylindrical vacuum chamber is equipped with two large dished doors that support the crucible with the tilting mechanism, the mold platform with the centrifugal casting system, and the casting tundish with its cover. The crucible support system with an additional detachable device is also used for electrode loading. The chamber is jack-

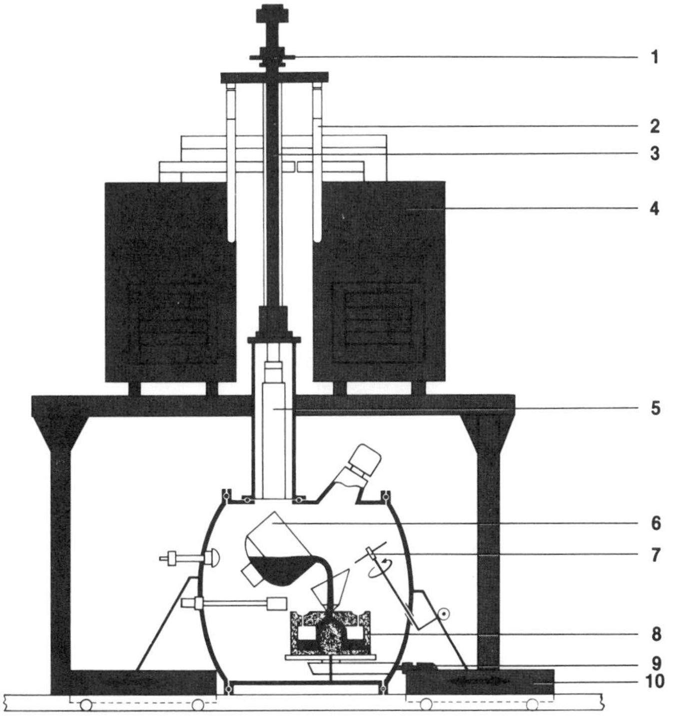

Fig. 1 Schematic of a modern 50 kg (110 lb) vacuum arc skull melting and casting furnace. 1, fast retraction system; 2, power cables; 3, electrode feeder ram; 4, power supplies; 5, consumable electrode; 6, skull crucible; 7, tundish shield; 8, mold arrangement; 9, centrifugal casting system; 10, chamber lid carriage

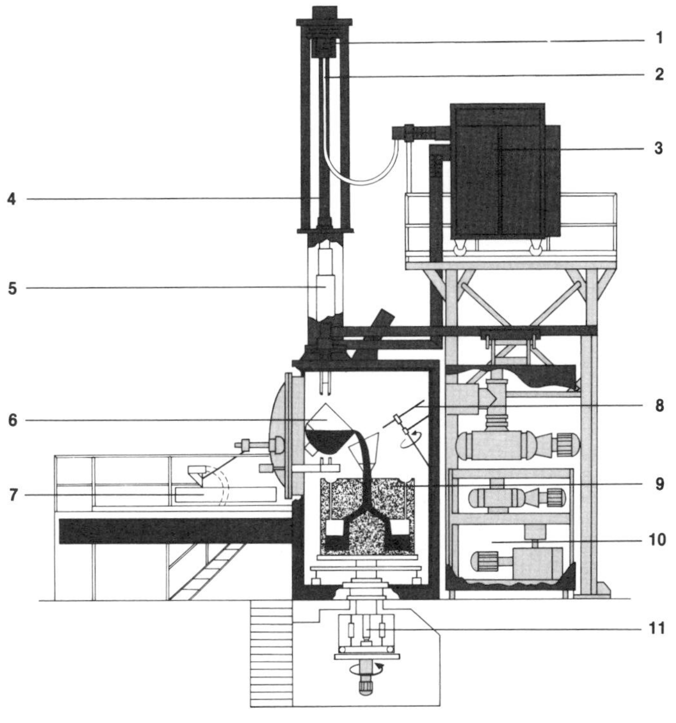

Fig. 2 Schematic of a modern semicontinuously operating vacuum arc skull melter for charge weights of up to 1000 kg (2200 lb). 1, fast retraction system; 2, power cables; 3, power supplies; 4, electrode feeder ram; 5, consumable electrode; 6, skull crucible; 7, crucible carriage; 8, tundish shield; 9, mold arrangement; 10, vacuum pumping system; 11, centrifugal casting system

eted for water cooling in regions that are subject to heat radiation.

A top flange with a throat carries the electrode chamber and the electrode feeding system. Viewing ports allow for video monitoring of the melting and pouring. A vacuum pumping port is located in the cylindrical portion of the chamber.

Figure 2 shows a semicontinuously operated vacuum arc skull melter for charge weights of up to 1000 kg (2200 lb). The principal difference between this furnace and the smaller model (Fig. 1)—apart from capacity-related layout features—is the rectangular vacuum chamber. Again, the crucible and the tilting mechanism are carried by a dished door.

In this furnace design, the centrifugal casting system is introduced from the bottom of the chamber to allow the horizontal connection of a separate cooling chamber, if desired. Molds are loaded from the back side into the chamber and can be discharged through a front door, which also allows the use of a cooling and charging chamber with lock valves for continuous mold transport flow.

Modern vacuum arc skull melting furnaces are usually equipped with:

- Coaxial power feed directly to the skull crucible to avoid electromagnetic fields that can disturb the melt bath
- Programmable control systems for crucible tilting to allow repeatable pouring profiles for consistent parameters
- Highly accurate electrode weighing system for precise determination of pouring weights
- *XY* adjustment system for coaxial positioning of the electrode in the skull crucible
- Compact air-cooled power-supply modules of high capacity for achieving high melt rates, thinner skulls, and correspondingly increased yields above 80%
- Proven vacuum pumping and measuring systems
- Forced argon cooling systems for faster mold cooling

Electron Beam Melting and Casting

W. Dietrich and H. Stephan, Leybold AG, West Germany

Electron beam melting and casting technology is accepted worldwide for the production of niobium and tantalum ingots weighing up to 2500 kg (5500 lb) in furnaces with electron beams of 200 to 1500 kW. Another application in East Germany and other Soviet bloc countries is the production of steel ingots weighing 3.3 to 18 Mg (3.6 to 20 tons) using electron beams of up to 1200 kW. Furnaces of up to 2400 kW in electron beam power have been used since 1982 for recycling titanium scrap to produce 4.8 Mg (5.3 ton) slabs 1140 mm (45 in.) wide. Furnaces of 200 to 1200 kW are used to refine nickel-base superalloys. Other metals, such as vanadium and hafnium, are melted and refined in furnaces between 60 and 260 kW. Approximately 150 furnaces with melting powers ranging from 20 to 300 kW are in operation in research facilities. These furnaces are used in the development of new grades and purities of conventional and exotic metals and alloys, for example, uranium, copper, precious metals, rare-earth alloys, intermetallic materials, and ceramics. The total power of installed electron beam melting and casting furnaces worldwide was approximately 25 000 kW at the end of 1987.

Electron beam melting and casting includes melting, refining, and conversion processes for metals and alloys. In electron beam melting, the feedstock is melted by impinging high-energy electrons. Electron beam refining takes place in vacuum in the pool of a water-cooled copper crucible, ladle, trough, or hearth. In electron beam refining, the material solidifies in a water-cooled continuous casting copper crucible or in an investment ceramic or graphite mold. This technology can be used for all materials that do not sublimate in vacuum.

Competing processes include sintering (for example, for refractory metals), vacuum arc melting and remelting (for reactive metals and superalloys), and electroslag melting and vacuum induction melting (for superalloys, specialty steels, and nonferrous metals). Some advantages and limitations of the competing vacuum processes are given in Table 1. Additional information on some of these processes is available in the sections "Vacuum Arc Remelting (VAR)," "Electroslag Remelting (ESR)," and "Vacuum Induction Melting (VIM)" in this article.

Electron Beam Melting and Casting Characteristics

The characteristics of electron beam melting and casting technology are:

- The flexibility and controllability of the process temperature, speed, and reaction
- The use of a wide variety of feedstock

Table 1 Comparison of characteristics of electron beam melting and competing processes

Metal	Sintering: Advantages	Sintering: Limitations	Vacuum arc melting: Advantages	Vacuum arc melting: Limitations	Electron beam melting: Advantages	Electron beam melting: Limitations
Tungsten, molybdenum	Small grain size; most often used	Refining limited; small batches; high energy consumption	Moderate grain size; acceptable workability; large ingots; low energy consumption	Refining limited; costly electrode preparation; melting dangerous	Highest possible purity; economical feedstock preparation; large ingots; low energy consumption	Large grain size; brittle product; very rarely applied
Tantalum, niobium	Small grain size; good workability	Same as above; rarely applied	Alloying; moderate grain size; large ingots; low energy consumption	Refining limited; expensive electrode; melting dangerous	Same as above; most frequently used	Alloying limited
Hafnium, vanadium	...	Same as above	Alloying during remelting	Almost no refining; costly electrode preparation; melting dangerous	Good refining; economical feedstock preparation and ingot production; most often used	High melting costs
Zirconium, titanium	...	Not used	Very low contamination; wide range of alloying possible; large ingots; low energy consumption; economical melting	Limited refining; expensive feedstock preparation; only round ingots	Economical feedstock preparation; refining of high-density inclusions; melting of slabs, ingots, and rods; high production rate; low energy consumption	Alloying limited; material losses from splatter; high furnace investment

materials in terms of material quality, size, and shape

- The different methods of material processing available
- Product quality, size, and quantity

Contamination of the product is avoided by melting in a controlled vacuum and in water-cooled copper crucibles (Fig. 1).

The energy efficiency of electron beam processing exceeds that of competing processes because of the control of the beam spot dwell time and distribution at the areas to be melted or maintained as liquid. In addition, unnecessary heating of the ingot pool, as occurs in vacuum arc remelting, for example, is avoided. Power losses of the electron beam inside the gun and between the gun nozzle and the target are very small, but approximately 20% of the beam power is lost because of beam reflection, radiation of the liquid metal, and heat conductivity of the water-cooled trough and crucible walls.

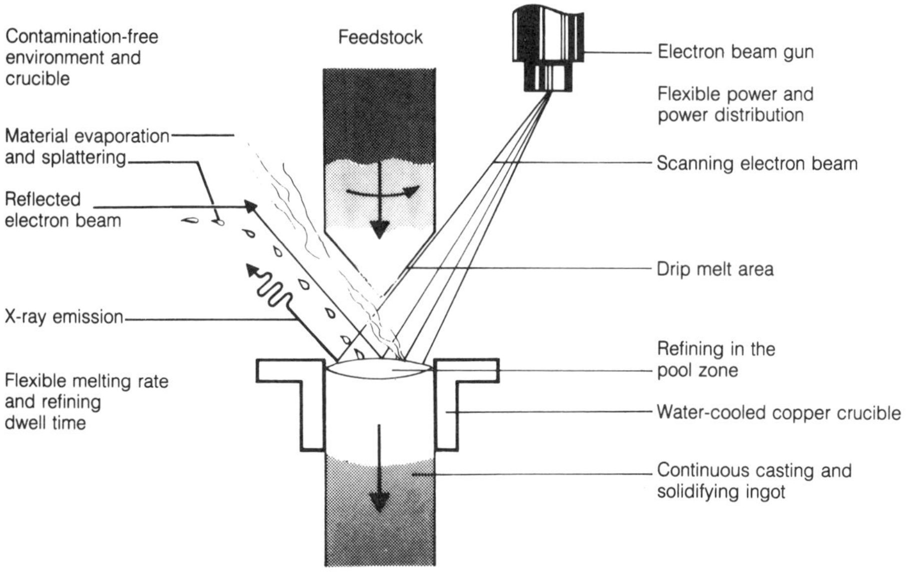

Fig. 1 Schematic of the electron beam melting process

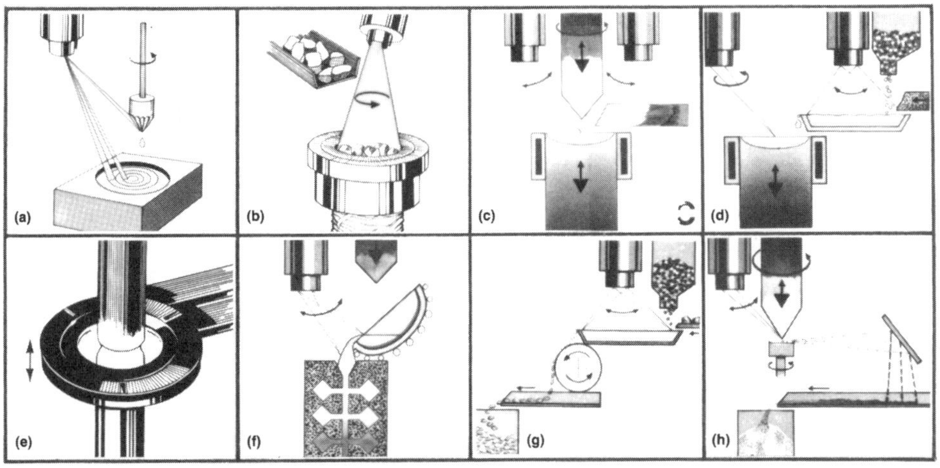

Fig. 2 Examples of electron beam melting and casting processes. (a) Button melting with controlled solidification for quantitative determination of low-density inclusions. (b) Consolidation of raw material, chips, and solid scrap to consumable electrodes for vacuum arc or electron beam remelting. (c) Drip melting of horizontally or vertically fed feedstocks. (d) Continuous flow refining/melting. (e) Floating zone melting. (f) Investment casting. (g) Pelletizing (manufacture of pellets from scrap and other materials for scrap recycling). (h) Atomization and granulation of refractory and reactive metals

Electron Beam Melting and Casting Processes

From the large variety of electron beam melting and casting processes shown in Fig. 2 only the processes illustrated in (a), (c), (d), and (f) are related to processes used in foundry technology:

- Button melting processes for the quality control of steel and superalloy cast parts to control the content of low-density inclusions
- Drip melting process for the preparation of refractory and reactive metal feedstock material for electron beam and VAR skull melting and casting
- Continuous flow melting process for the feedstock refining of superalloys for VIM and electron beam casting
- Electron beam investment casting process

Electron Beam Heat Source Specifications

For all electron beam melting and casting processes, except for the crucible-free floating zone melting process, Pierce-type electron beam guns with separately evacuated beam generating and prefocusing rooms are the key components of the furnaces used. The essential features of these guns are:

- Large power range of 0 to 1200 kW
- Long free beam path of 250 to 1500 mm (10 to 60 in.) and the adjustable beam power distribution
- Beam deflection angle of ±45° and spot frequency up to 500 Hz

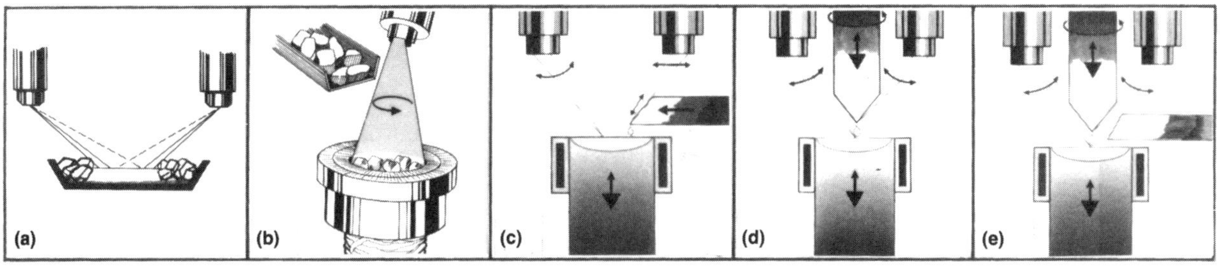

Fig. 3 Schematics of electron beam consolidation and drip melting processes. (a) Consolidation of coarse and solid scrap. (b) Continuous consolidation of raw material, chips, and solid scrap by direct feeding into a continuous casting crucible. (c) Drip melting of horizontally fed compacts, sintered bars, or consolidates for initial melting of reactive and refractory metals. (d) Drip melting of vertically fed vacuum induction melted or conventionally melted electrodes. (e) Drip melting of horizontally and vertically fed materials for the production of alloys from feedstocks with very different melting points

Table 2 Melting and refining data of refractory and reactive metals and alloy steels gained in laboratory and pilot production furnaces

Metal	Feedstock size, mm (in.)	Ingot diameter, mm (in.)	Ingot weight, kg (lb)	Integral melting rate, kg/h (lb/h)	Electron beam power of second melt, kW	Operating vacuum pressure of last melt, Pa (torr)	Total specific melting energy, kW · h/kg	Material yield, %	Interstitial elements in feedstock and final ingot, ppm C	O	N	H	Hardness, HB
Tungsten	40 (1.6) diam	...	...	...	...	...	...	...	70	4100	30	10	...
		60 (2.4)	37 (80)	10.5 (23)	119	8×10^{-3} (6×10^{-5})	10.3	93.1	10	8	5	1	200
	55 (2.2) diam,	...	...	...	...	...	...	...	200	5000	50	20	...
	100 (4) long	115 (4.5)	200 (440)	20 (44)	300	2×10^{-2} (1.5×10^{-4})	9.9	90	10	8	7	1	210
Tantalum	60 (2.4) diam	...	...	...	...	...	...	...	100	1200	140	10	...
		80 (3.2)	65 (145)	16.7 (37)	130	8×10^{-3} (6×10^{-5})	6.0	92	10	75	30	1	65
	60 (2.4) square,	...	...	...	...	...	...	...	...	650	13	10	...
	160 (6.3) long	160 (6.3)	523 (1150)	38.4 (85)	371	3×10^{-3} (2×10^{-5})	8.30	91.8	6	15	13	2	69
Molybdenum	100 (4) diam	...	...	...	...	...	...	...	170	810	50	10	...
	100 (4) diam	100 (4)	64 (140)	12.5 (27.5)	130	8×10^{-4} (6×10^{-6})	10.4	95	12	10	10	2	140
	120 (4.7) square,	...	...	...	...	...	...	...	200	750	60	10	...
	180 (7.1) long	180 (7.1)	408 (900)	50.2 (111)	290	10^{-3} (7.5×10^{-6})	5.2	96.8	10	12	11	2	140
Niobium	80 (3.2) square,	...	...	...	...	...	...	...	160	5220	554	35	...
	150 (6) long	150 (6)	227 (500)	17.6 (39)	240	10^{-2} (7.5×10^{-5})	12.3	87	12	106	60	5	77
	120 (4.7) square,	...	...	...	...	...	...	...	80	4500	330	40	...
	180 (7.1) long	180 (7.1)	326 (720)	13.2 (29)	218	5×10^{-3} (3.8×10^{-5})	15.9	96.8	6	111	52	8	66
Hafnium	60 (2.4) square	...	...	...	...	...	...	...	...	1870	95	2	...
	80 (3.2) diam	80 (3.2)	40 (90)	1.7 (3.7)	80	5×10^{-3} (3.8×10^{-5})	38	93.1	...	170	25	1	160
	100 (4) diam	...	...	...	...	...	...	...	500	900	100	2	...
	100 (4) diam	130 (5.1)	173 (380)	7.5 (16.5)	110	4×10^{-3} (3×10^{-5})	14.7	93.5	100	200	50	1	170
Zirconium	60 (2.4) square,	...	...	...	...	...	...	...	...	950	95	30	...
	100 (4) long	100 (4)	19.5 (43)	14.3 (31.5)	80	8×10^{-3} (6×10^{-5})	4.6	88.2	...	545	30	3	120
	60 (2.4) square,	...	...	...	...	...	...	...	...	950	95	30	...
	150 (6) long	150 (6)	179 (395)	73 (161)	250	2×10^{-2} (1.5×10^{-4})	2.6	...	...	735	48	9	125
Vanadium	60 (2.4) diam,	...	...	...	...	...	...	...	...	1045(a)	210(a)	16(a)	...
	80 (3.2) long	80 (3.2)	7.7 (17)	2.8 (6.2)	80	8×10^{-3} (6×10^{-5})	3.1	91	...	235(a)	95(a)	13(a)	30–100
Titanium	100 (4) diam	...	...	...	...	...	...	...	...	2180	100	30	...
Ti-6Al-4V	...	100 (4)	28.2 (62)	45.2 (100)	87	0.8 (6×10^{-3})	1.91	99	...	1850	80	16	...
Ti-8Al-1Mo-1V	100 (4) diam	...	...	...	...	...	...	...	...	890	70	...	...
		100 (4)	28.2 (62)	22.5 (50)	60	0.4 (3×10^{-3})	2.66	98	...	730	50	...	...
4340 steel	80 (3.2) diam	...	...	...	...	...	...	...	3900	63	100	0.3	...
		150 (6)	31 (70)	10 (22)	52	4×10^{-3} (3×10^{-5})	1.67	93	4300	2.4	26	0.08	...
		150 (6)	31 (70)	80 (176)	80	2.0 (0.015)	1.0	99	3622	10	78	0.10	...

(a) The reproducibility of the refining data could not be confirmed.

- Usable vacuum pressure range between 1 and 0.0001 Pa (10^{-2} and 10^{-6} mbar, or 7.5×10^{-3} and 7.5×10^{-7} torr)
- Reliability of the gun and cathode system

The power control and distribution system allows a very accurate distribution of beam power and energy for achieving the required heating for material melting, superheating, refining, and electrothermal effects.

Button Melting for Quality Control

The button melting process (Fig. 2a) serves to control the quality of feedstock materials for investment casting and to produce casting samples. In contrast to the conventional electron beam melting process, this process is not used for refining, but only for flotation and concentration of low-density inclusions. During the eight-step process, the sample is heated and drip melted. Low-density inclusions are floated to the surface and concentrated in the center of the pool of molten metal during controlled solidification by computer-controlled reduction of beam power and circular electrothermal stirring. The concentrated impurities can be identified and evaluated by conventional metallographic methods, but the size of the raft gives the first indication of the quantity of impurities in the metal.

Most button melting furnaces are completely automated and microprocessor controlled to guarantee process reproducibility. Melting is usually carried out in the pressure range of 1 to 0.001 Pa (10^{-2} to 10^{-5} mbar, or 7.6×10^{-3} to 7.6×10^{-6} torr).

Drip Melting

The drip melting processes (Fig. 3c and d) are primarily used for the production of clean, mostly ductile ingots of refractory and reactive metals or of specialty steels. The feedstock for the first melt (Fig. 3c) can be compacted sponge, granular, powder, or scrap, which might be presintered in a vacuum heating furnace. In some cases, loose raw materials can be consolidated in a water-cooled copper trough (Fig. 3a). The consolidated ingot can then be fed horizontally for drip melting. Raw material that is continuously consolidated in a water-cooled copper crucible with a retractable bottom plate (Fig. 3b) can be fed horizontally or vertically for drip melting. In both consolidation processes, only 20 to 80% of the material is melted. Refining and losses of material by splattering are negligible.

Drip melting of horizontally fed compacts is the most frequently used process for the production of ingots from refractory or reactive metals. The resulting ingot is of sufficient purity, but has an area of inhomogeneity caused by the shadow of the horizontally fed bar. Two or more electron guns are used in drip melting to make use of reflected electron beams and to reduce evaporation and splattering. The end of a compact is welded to the front of the following one to avoid dropping semisolid material into the pool. Table 2 lists processing parameters that have been successfully used to electron beam melt various reactive and refractory metals and 4340 alloy steel.

Vertical Feeding of Ingots. Refractory and reactive metal ingots of high purity, homogeneity, and smooth surface are remelted by vertical feeding (Fig. 3d). The molten metal droplets run down the conical, rotating electrode tip, are refined, and then drop into the pool center. The crucible pool is normally of the same diameter as the electrode but is sometimes smaller or larger. It is kept in the liquid state to allow final refining and to guarantee ingot homogeneity. Because two or more electron guns are used, the entire pool can be equally bombarded; thus, shadow effects of the electrode can be eliminated.

Simultaneous melting of horizontally and vertically fed electrodes (Fig. 3e) can be used for the production of critical alloys. In this case, the feedstock should be of the desired purity.

Fig. 4 Drip melting of 330 mm (13 in.) square steel billet in a 1100 kW single-gun furnace. Melt rate: 1000 kg/h (2200 lb/h). Courtesy of VEB-Edelstahlwerk, East Germany

Horizontally Fed Ingots. Drip melting of horizontally fed material with a single electron gun (Fig. 4) is used for refining some steel alloys in East Germany and other Soviet bloc countries. In this process, the feedstock size is smaller than the pool diameter to minimize the shadow effect of the horizontally fed bar. In production units, feeding can be carried out from two opposite sides.

Other Process Considerations. To ensure the production of clean, homogeneous metals and alloys in electron beam drip melting furnaces, various aspects of material processing and handling must be controlled. Key considerations include:

- Dimensions and quality of the feedstock, and the feeding system used
- Ingot cooling and unloading during melting of another ingot
- Passivation and removal of condensates from the melt chamber
- Planning of melt sequences to minimize the number of furnace cleanings required
- Routine preventive furnace maintenance to ensure reliability
- Operator skill in operation of the furnace
- Material yield and energy consumption

Equipment for Drip Melting

The essential equipment groups required for drip melting—melting furnaces, control systems, and power supply units—are all important for achieving optimum productivity.

The melting furnace (Fig. 5) includes the electron beam gun as the heat source, material feeding and ingot withdrawal systems, a crucible for material solidification, and a vacuum system to maintain the low pressure. Process observation, both visually and with video systems, is possible through viewports. The melt chamber flanges are equipped with x-ray absorbing steel boards, and interlocking systems prevent operation failures and accidents.

The control system allows the adjustment and control of such operating process parameters as electron beam power, operating vacuum level, material feed rate, and ingot withdraw speed. The control system also records and logs the process data.

Power Supply Units. One or more high-voltage power supply units are needed to supply the electron beam guns with the required continuous voltage (30 to 40 kV). The beam power of each gun can be adjusted between zero and maximum power with an accuracy of ±2%.

Other Equipment. Large production furnaces are equipped with lock-valve systems to allow simultaneous melting and unloading of ingots without breaking the vacuum in the melt chamber. Production is thus limited only when the condensate remaining in the melt chamber requires cleaning or when a different alloy is to be melted.

Characteristics of Electron Beam Drip Melted Metals

Electron beam melted and refined material is of the highest quality. The amount of interstitials present is very low, and trace elements of specific high vapor pressure can also be reduced to very low values (Ref 1, 2).

Reactive and Refractory Metals

Tantalum and niobium ingots have smooth surfaces and are of sufficient ductility that they can be cold worked, and sheets and wires can be produced.

Tungsten and molybdenum ingots are also of the highest possible purity, but the ingots are brittle because of the very large grain size and the concentration of impurities at grain boundaries.

Hafnium. Electron beam melted hafnium is of higher ductility than the vacuum arc remelted metal (Ref 3). The main application of electron beam melted hafnium is as control elements for submarine nuclear reactors.

Vanadium is refined by electron beam drip melting. The aluminothermically produced feedstock is drip melted in several steps. During this procedure, the ingot diameter is reduced at each step by approximately 30 to 40 mm (1.2 to 1.6 in.) to obtain an ingot 30 to 40 mm (1.2 to 1.6 in.) in diameter, regardless of the initial ingot diameter. The clean vanadium ingots are primarily used in nuclear reactor applications (Ref 4).

Applications for electron beam melted refractory and reactive metals are listed in Table 3.

Steels

The purity and properties of electron beam melted steels are in some respects better than those of vacuum arc and electroslag remelted steels, but the processing costs are higher. The electron beam melting of steel is primar-

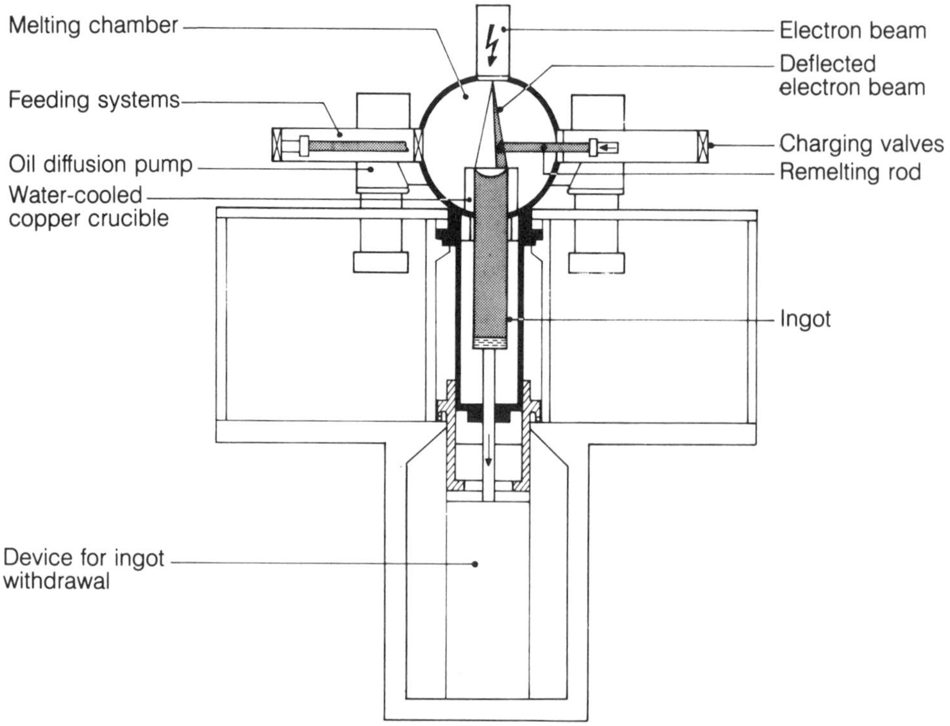

Fig. 5 Single-gun 1200 kW furnace for horizontal drip melting of steels. Melting rates of up to 1100 kg/h (2425 lb/h) are possible.

Table 3 Principal applications for vacuum arc remelted (VAR), electron beam melted (EB), and powder metallurgy (P/M) reactive and refractory metal ingots

Metal	Applications
Reactive metals, VAR and EB melting	
Hafnium	Flash bulbs and glow discharge tubes for the electronics industry; control rods and breakoff elements in submarine nuclear reactors
Vanadium	Targets for high deposition rate sputtering processes in the electronics industry; breakoff elements, fixtures, and fasteners in nuclear reactors; standards for basic research; alloying element for certain high-purity alloys
Zirconium	Getter material in tubes in the electronics industry; stripes for flash bulbs; fuel claddings, fasteners, and fixtures for nuclear reactors
Titanium	Components for bleaching equipment and desalination plants in the chemical industry; superconductive wires; turbine engine disks, blades and housings, rain erosion boards, landing legs, wing frames, missile cladding, and fuel containers in the aircraft and aerospace industries; shape memory alloys; biomedical fixtures and implants; corrosion resistant claddings
Refractory metals, EB melting and P/M	
Tungsten	Heating elements, punches and dies, and nonconsummable electrodes for arc melting and gas tungsten arc welding for metal processing equipment; targets for x-ray equipment and high sputtering rate devices such as very large-scale integrated circuits, cathodes and anodes for electronic vacuum tubes in the electronics industry; radiation shields in the nuclear industry; cladding and fasteners for missile and reentry vehicles
Tantalum	Condensers, autoclaves, heat exchangers, armatures, and fittings for the chemical industry; electrolytic capacitors for the electronics industry; surgical implants; fasteners for aerospace applications
Molybdenum	Dies for conventional and isothermal forging equipment; electrodes for glass melting; targets for x-ray equipment; cladding and fasteners for missile and reentry vehicles
Niobium	Superconductive wire for energy transmission and large magnets for the electrical and electronics industries; heavy ion accelerators and radio frequency cavities for nuclear applications; components for aircraft and aerospace applications

ily used in East Germany and other Soviet bloc countries. The resulting ingots are up to 1000 mm (40 in.) in diameter and weigh up to 18 Mg (20 tons). The furnaces used have been in operation since 1965, and have beam powers of up to 1200 kW. Larger furnaces for the production of ingots weighing up to 30 to 100 Mg (33 to 110 tons) are under construction (Ref 5).

The essential advantage of the electron beam melting of steel is the drastic reduction of metallic and nonmetallic impurities and interstitial elements (Ref 6, 7). The principal applications for electron beam melted steels are in the machinery industry for parts for which high wear resistance and long service life are required. The extended service lives of the parts and the reduced manufacturing time (for example, less surface polishing is required for electron beam melted steel) can justify the higher material costs.

The electron beam melting of steel and superalloys can become much more economical when melting and refining are done by continuous flow melting or cold hearth refining. These melting and refining methods reduce energy costs and minimize material losses.

Continuous Flow Melting

The continuous flow melting process (cold hearth refining process) (Fig. 6) was developed approximately 10 years after drip melting (Ref 8). Continuous flow melting is mainly used for refining specialty steels and superalloys and for refining and recycling reactive metal scrap, especially Ti-6Al-4V from high-density tungsten carbide tool tips (Ref 9).

Principles of Continuous Flow Melting

Continuous flow melting (Fig. 7) is the most flexible vacuum metallurgical melting process. It is a two-stage process in which the first step (material feeding, melting, and refining) takes place in a water-cooled copper trough, ladle, or hearth. In the second step, solidification occurs in one of several round, rectangular, or specially shaped water-cooled continuous copper crucibles. Both process steps are nearly independent from each other; they are linked only by the continuous flow of the liquid metal stream. The major refining actions are carried out in the hearth, but some postrefining takes place in the pool of the continuous casting crucible, similar to the drip melting of horizontally fed billets. Refinement in continuous flow melting occurs by vacuum distillation in the hearth pool, superheating, and stirring of the molten metal pool.

Removal of Impurities. Most impurities with densities lower than that of the melt (for example, metalloids in steels and superalloys) can be segregated by flotation and formed into a slag raft. The raft is then held in place by either mechanical or electrothermal

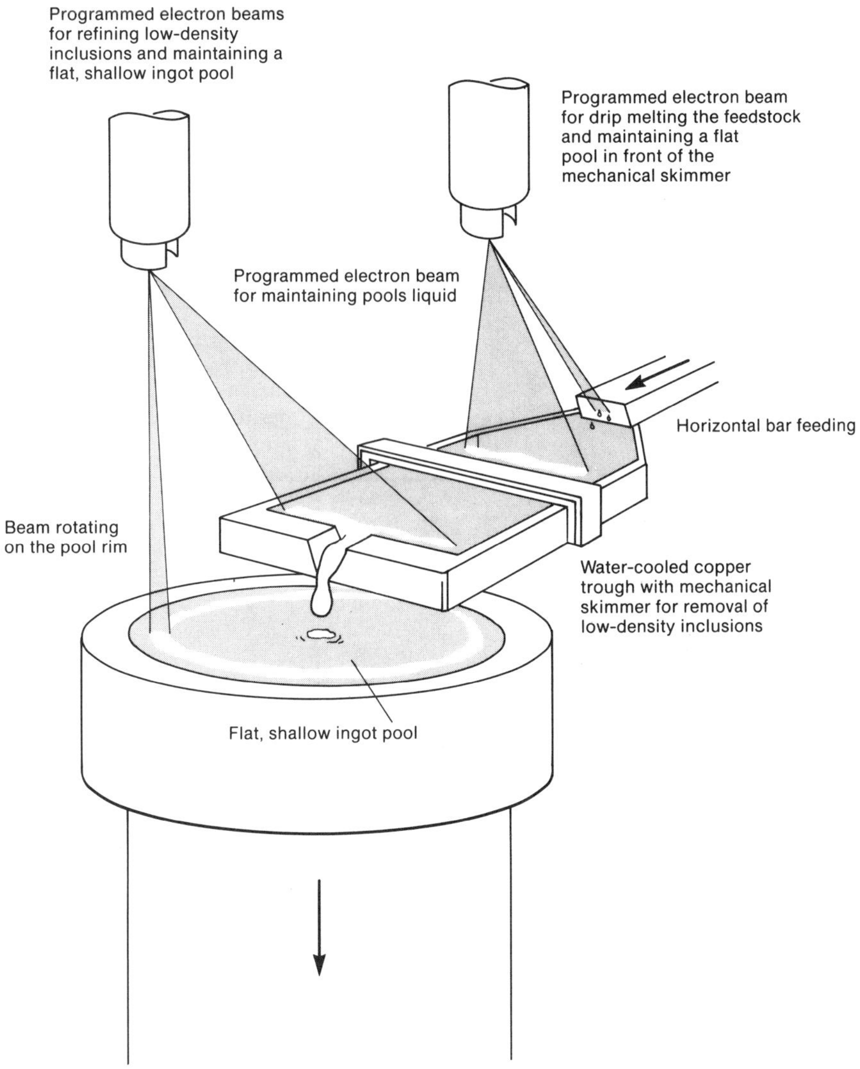

Fig. 6 Schematic of the continuous flow melting process

means. Impurities denser than the melt, such as tungsten carbide tool tips in titanium, are removed by sedimentation. Inclusions with densities such that efficient flotation or sedimentation does not occur can be partially removed by adhesion to the slag raft.

Hearth dimensions are based on the type and amount of refining required. For example, hearths for vacuum distillation should be nearly square and relatively deep to allow sufficient melt stirring. For flotation refining, the hearth should be long and narrow (for superalloys, approximately 10 mm, or 0.4 in., of hearth length for each 100 kg/h, or 220 lb/h, of melt rate is recommended). Hearths for titanium alloy scrap recycling can be relatively short if all the materials can be transported to the pool of the hearth rather than to the ingot pool.

Feeding. Material feeding criteria include 100% homogenous material transportation to avoid uncontrolled evaporation of alloying elements and correct feeding into or above the hearth pool. Horizontal feeding of compacted, premelted, or cast material is most often used. Loose scrap and raw material are used only when compaction is too expensive. Feeding of liquid metal was used in one of the first continuous flow melting furnaces to produce a ferritic steel in a vacuum induction furnace (Ref 10). Postrefining was carried out in a cascade of five hearths 1.5 m (60 in.) long and 1 m (40 in.) wide.

Casting and Solidification. The criteria for material casting and solidification include the shape of the final product and the solidification rate required to avoid ingot tears or other defects and to ensure a homogeneous ingot structure. The multiple casting of small ingots is sometimes used, especially when forging is impossible because of the brittleness of the solidified material (for example MCrAly wear-resistant coating alloys). The casting of round and rectangular ingots and slabs is common practice, and the continuous casting of hollow ingots is also being used (Ref 11). The casting of segregation-free ingots and ingots with a fine grain size is under development to improve the workability of superalloys (Ref 12, 13).

Continuous Flow Versus Drip Melting

Table 4 compares the essential features of drip melting and continuous flow melting. Generally, continuous flow melting is used for all refractory metals, superalloys, and specialty steels, especially when flotation or sedimentation of inclusions is required. Drip melting is used for refractory metals because of their high melting points and the resulting high heat losses to the water-cooled copper crucible. Depending on production quantity, double or triple drip melting may require less energy than a single continuous flow melt of some materials, such as niobium.

Refining and Production Data

Data on continuous flow electron beam melting and refining in laboratory and pilot production furnaces are given in Table 5. The data demonstrate the effectiveness of the process in reducing impurities and interstitial elements. It can also be seen that the selective evaporation of chromium from superalloys can be controlled by the distribution of beam power at the trough pool and by controlling trough pool area and melt rate. The selective evaporation of aluminum from Ti-6Al-4V alloy is much more difficult to control; additional aluminum must be used to compensate for the aluminum evaporated.

Table 4 Comparison of the characteristics of drip melting and continuous flow melting

Characteristic	Refractory metals	Reactive metals, superalloys, and specialty steels
Power density	High	Soft; smoothly distributed
Inclusions	Irrelevant	Must be removed
Ingot shape and structure	Round; coarse grain	Round or flat; fine grain, segregation-free
Mass production	Low	High
Competitive economical processes	Vacuum arc remelting	Vacuum arc remelting; electroslag remelting
Preferred method	Drip melting	Continuous flow melting

Equipment for Continuous Flow Electron Beam Melting

The equipment required for continuous flow melting is different from that used in drip melting mainly because of the trough and the somewhat larger melting chamber. In addition, because of the materials often melted in the continuous flow process (superalloys and titanium alloys), additional instrumentation is often provided. This may include an ingot pool level control system,

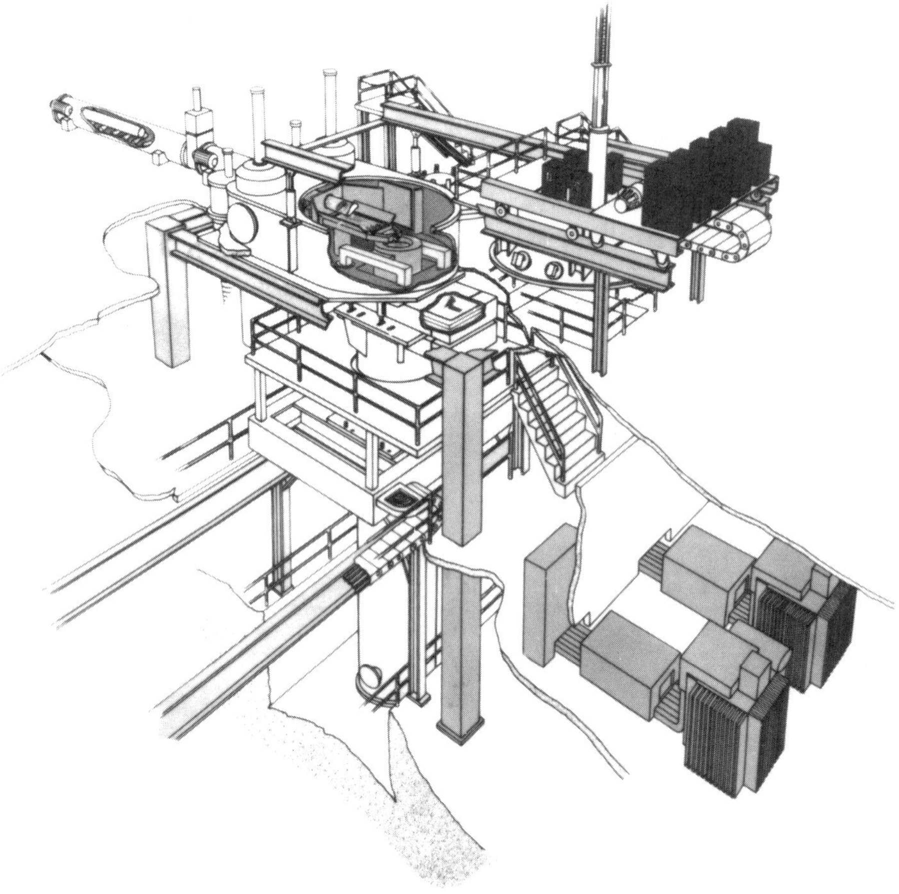

Fig. 7 Four-gun 1200 kW combined electron beam drip melting and continuous flow melting furnace

metal vapor and partial pressure analyzers, a two-color temperature control system, and a data logging system.

Accurate beam power distribution is achieved in two- or three-gun furnaces by microprocessor control, which allows the splitting of a single beam to 64 locations and the adjustment of dwell time at each location between 0.01 and 1000 s. The beam spot at each of the 64 locations can be scanned over an elliptical or rectangular area. With such systems, the required refining can be achieved without unnecessary power consumption and evaporation of alloying elements (Ref 15).

Process observation is accomplished with a video monitoring system. Samples can be obtained from both the trough pool and the ingot pool for nearly continuous control of material quality.

Feeding systems for continuous flow furnaces must maintain homogeneity along the length of the feed material. The trough and crucible should be easily accessible for convenient maintenance, especially when different alloys are to be melted in the same furnace.

Characteristics of Continuous Flow Melted Materials

Titanium ingots and slabs can be produced from titanium scrap contaminated with tungsten carbide tool tips. The electron beam melted product contains tungsten car-

Table 5 Refining and production data for the continuous flow melting of reactive and refractory metals and stainless steels in laboratory and pilot production furnaces

Metal	Feedstock size, mm (in.)	Trough size, mm (in.)	Ingot size, mm (in.)	Ingot weight, kg (lb)	Melt rate, kg/h (lb/h)	Electron beam power, kW	Operating pressure, Pa (torr)	Specific melting energy, kW · h/kg	Composition of feedstock and product: C, ppm	O, ppm	N, ppm	H, ppm	Al, %	V, %	Cr, %
Hafnium	60 (2.4) square	120 × 250	100 (4) diam	83.0 (183)	40 (88)	180	4×10^{-2}	4.5	...	900	...	...	...	...	...
		(5 × 10)					(3×10^{-4})		...	600	...	...	...	...	...
Zirconium	100 (4) square	120 × 300	150 (6)	90.5 (200)	42 (92.5)	185	3.5×10^{-2}	4.4	...	950	95	30	...	...	...
		(5 × 12)					(2.6×10^{-4})		...	540	30	3	...	...	...
Zirconium	80 (3.2) square	120 × 300	100 (4)	40.2 (89)	80 (176)	140	3.5×10^{-2}	1.75	...	4000	800	10	...	...	...
		(5 × 12)					(2.6×10^{-4})		...	1520	210	3	...	...	...
Vanadium	50 (2) square	120 × 300	100 (4)	...	20 (44)	130	1.5×10^{-2}	6.5	...	1045	210	10	...	99	...
		(5 × 12)					(1.1×10^{-4})		...	277	50	3	...	99	...
Ti-6Al-4V	Swarf	120 × 300	150 (6)	62.6 (138)	40 (88)	122	2×10^{-2}	3.0	400	2600	110	84	6.0	4.0	...
		(5 × 12)					(1.5×10^{-4})		200	2700	110	22	4.4	4.2	...
Ti-6Al-4V	Solid scrap	120 × 300	150 (6)	62.6 (138)	70 (154)	140	7×10^{-2}	2.0	1520	1520	75	15	6.0	4.0	...
		(5 × 12)					(5.3×10^{-4})		...	1320	76	8	4.8	4.1	...
Ti-6Al-4V	125 (5) diam	150 × 400	2 × 75 (3)	2 × 32 (70.5)	91 (200)	147	6×10^{-2}	1.61	...	...	...	...	6.0	4.0	...
		(6 × 16)	diam				(4.5×10^{-4})		...	...	...	...	3.6	4.3	...
Commercially pure titanium	160 (6.3)	150 × 250	100 × 400	96.4 (213)	86.3 (190)	148	6×10^{-2}	1.71	...	...	...	...	...	...	...
		(6 × 10)	(4 × 16)				(4.5×10^{-4})		...	...	...	...	...	...	...
Commercially pure titanium	Sponge	150 × 500	100 × 400	103.0 (227)	41.2 (91)	226	8×10^{-2}	5.5	...	...	...	...	...	...	...
		(6 × 20)	(4 × 16)				(6×10^{-4})								
Stainless steel	150 (6) diam	150 × 400	2 × 75 (3)	2 × 55 (121)	136 (300)	144	6×10^{-2}	1.06	701	97	155	...	...	...	18.25
		(6 × 16)	diam				(4.5×10^{-4})		536	33	68	...	...	...	18.11
Alloy 718	133 (5.2) diam	150 × 400	2 × 75 (3)	2 × 57 (126)	136 (300)	156	6×10^{-2}	1.15	417	14	52	...	...	...	19.11
		(6 × 16)	diam				(4.5×10^{-4})		363	17	34	...	0.72	...	18.73
AISI type 316 stainless steel	150 (6) diam	150 × 400	3 × 65 (2.6)	3 × 41.5	136 (300)	156	6×10^{-2}	1.15	...	...	...	...	...	...	...
		(6 × 16)	diam	(91.5)			(4.5×10^{-4})								

Source: Ref 14

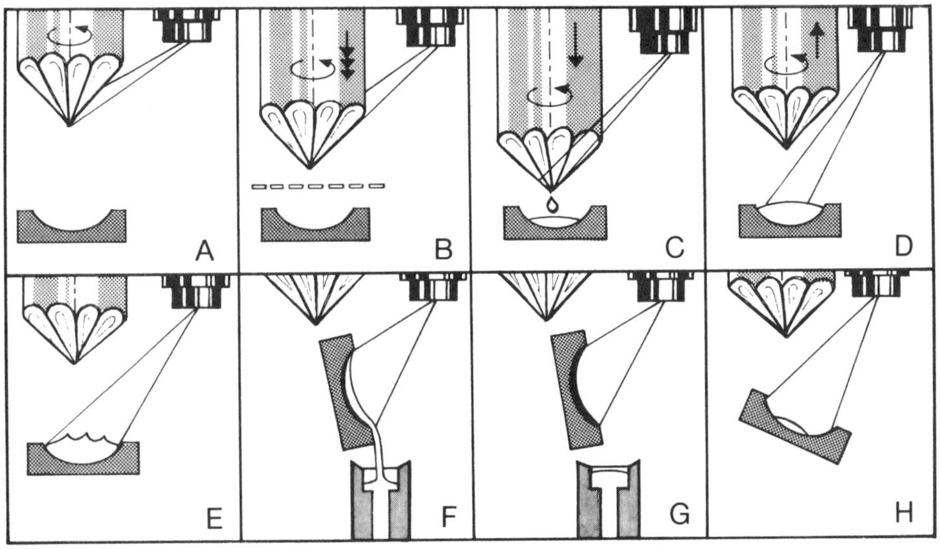

Step	Function	Time, s	Electron beam power, kW	Specific electron beam energy kW/kg	Specific electron beam energy kW/lb
A..........	Heating of electrode tip and positioning into melting area	4	80	0.20	0.090
B..........	Distribution of electron beam power to the electrode only	0.5	80	0.02	0.009
C..........	Drip melting of rotating electrode with simultaneous droplet counting and weight control	35	80	1.72	0.78
D..........	Retraction of electrode and distribution of increased power to the melt pool	0.5	100	0.74	0.33
E..........	Pool superheating	12	100	0.74	0.33
F..........	Pouring with simultaneous pool superheating	0.7	100	0.04	0.33
G..........	Washing of crucible for reduction of skull weight	6.0	60	0.22	0.10
H..........	Cleaning of the spout lip to obtain reproducible pouring streams; tilting back of crucible	1.0	60	0.04	0.02

Fig. 8 Procedure and process data for the electron beam melting and casting of 0.45 kg (1 lb) superalloy parts. Cycle time: 60 s; total electron beam energy: 3.01 kW · h/s

bide particles no larger than 0.7 mm (0.028 in.) in diameter. Oxygen content can be reduced by fitting titanium sponge compacts around a forged ingot or slab. Continuous flow melted titanium ingots can be directly remelted in a VAR furnace.

Superalloys. Continuous flow electron beam melted superalloy ingots 150 to 200 mm (6 to 8 in.) in diameter are often used in VIM investment casting furnaces. Such ingots are nearly free of nonmetallic inclusions and trace elements (Ref 16). The simultaneous continuous casting of twin, triple, or multiple ingots is under development in a 200 kW pilot furnace. Multi-ingot casting can become economical when 90 Mg (100 tons) or more of an alloy is being produced annually (Ref 17).

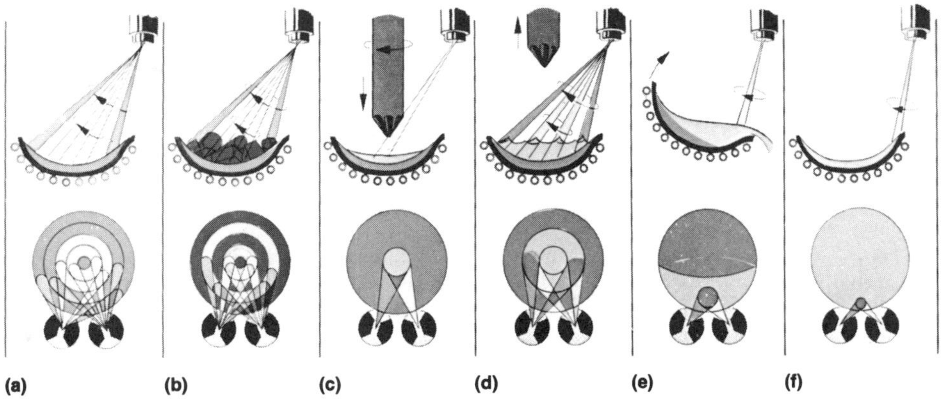

Fig. 9 Procedure and process data for the electron beam melting and casting of titanium and titanium alloy parts starting from scrap and consumable electrodes. (a) Melting of skull with two beams, each scanning and jumping on two circles (500 kW, 7 min). (b) Melting of scrap with two beams, each scanning and jumping on different circles (500 kW, 7 min). (c) Melting of consumable electrode with two concentrated beams (500 kW, 6 min). (d) Superheating of the pool with two beams, each scanning and jumping on two different circles (500 kW, 3 min). (e) Superheating of the metal stream during pouring with two concentrated beams (500 kW, 10 s). (f) Cleaning of the mold spout with two slightly scanning beams (10 kW, 2 min)

Investment Casting Using Electron Beam Melting

Investment casting with electron beam heat sources from water-cooled copper skull crucibles has been used for the production of superalloy turbine parts with directionally solidified or monocrystal structures and for titanium parts with equiaxed structures. In both applications, the feedstock is clean material that is melted with no major refining effect and without picking up any contamination. The process competes for superalloys with vacuum induction melting and investment casting and for titanium with vacuum arc skull melting and casting.

Characteristics of Electron Beam Investment Casting

With electron beam melting, it is possible to superheat the casting material just before and during pouring to increase the fluidity of the metal. This is not possible with vacuum arc skull melting. The extremely short metal flow path between the crucible spout and the mold funnel reduces erosion of the mold, and the controllable pouring speed allows the use of small mold funnels for reducing revert material. An essential advantage of electron beam titanium casting is the possibility of using premelted VAR ingots and solid, clean in-house recycling scrap separately or simultaneously to reduce material costs. The decisive advantage of electron beam casting technology is the possibility of process automation, which guarantees reproducibility of quality and a high production rate. Therefore, electron beam melting is used only when a high production rate is demanded. The obvious disadvantage is the decreasing metal temperature during pouring caused by the temperature gradient of the melt in the water-cooled copper crucible; for this reason, the process cannot be used for the production of superalloy cast parts with very specific equiaxed structures.

Process Technology and Equipment for Superalloy Castings

For the economical mass production of superalloy turbine parts, the electron beam casting process and equipment shown in Fig. 8 and described in Table 6 meet the required process and product specifications. The pool temperature before casting can be adjusted between 1650 and 1900 °C (3000 and 3450 °F), depending on part requirements. The adjusted tem-

Table 6 Processing data for the mass production of directionally solidified and monocrystal superalloy castings in 120 and 300 kW electron beam casting machines

Machine	Maximum electron beam power, kW	Ultimate vacuum pressure, Pa (torr)	Leak and degassing rate, mbar · L/s	Electrode dimensions, mm (in.)	Number of electrodes per magazine	Casting weight, kg (lb)	Average weight deviation, %	Cycle time, s	Pool surface temperature at pouring, °C (°F)	Average temperature deviation, %	Material loss by evaporation and splattering, %
120 kW	120	2×10^{-4} (1.5×10^{-6})	2×10^{-4}	125 (5) diam; 1600 (63) long	12	0.45 (1)	1.8	60	1850 (3360)	0.3	~1.5
300 kW	300	2×10^{-4} (1.5×10^{-6})	2×10^{-4}	125 (5) diam; 1600 (63) long	12	8 (18)	1.5	440	1850 (3360)	0.3	~1.5

perature can be reproduced with an accuracy of 0.3%.

The accuracy of the part weight was 1.8% for small parts (450 g, or 1 lb) and 1.5% for larger parts (8 kg, or 18 lb). The cycle time for smaller parts was 58 ± 1.5 s; for larger parts, 300 ± 10 s.

Reduction in chromium content was less than 0.1%. Feedstock load is twelve electrodes 125 mm (5 in.) in diameter and 1600 mm (63 in.) long. The electrode magazine can be reloaded without venting the melt chamber. Exchange of water-cooled copper crucibles, for changing alloys or part weight, can also be accomplished without venting the melt chamber.

More than 50% of the evaporating and splattering material can be condensed and collected at the condensate plate. This plate can be replaced through a lock-valve system.

Vacuum pressure during melting and pouring is in the range of 0.001 to 0.0001 Pa (10^{-5} to 10^{-6} mbar, or 7.5×10^{-6} to 7.5×10^{-7} torr). Leak and wall degassing rates are below 2×10^{-4} mbar · L/s.

Melting and pouring within one electrode cycle can be done automatically. Movement of an electrode from the storage position to the melting position is semiautomatic.

The distance between the copper crucible pouring spout and the mold funnel was less than 40 mm (1.6 in.) during pouring. The mold funnel cross section was 20 to 30 mm (0.8 to 1.2 in.) for casting small parts.

These results can be achieved with a computer-controlled electron beam melting furnace using the process data given in Table 6. The fundamental equipment groups are the means for accurate beam power distribution at the electrode tip and crucible pool and its programmable power level during the many steps of process procedure, and integrated droplet counter and balance to control the material to be drip melted, and the weight of the remaining skull, which will be remelted again and again at each step. The high degree of process reproducibility is achieved by controlling pool surface temperature, crucible washing, and crucible spout lip cleaning. Crucible washing and spout lip cleaning minimize the skull weight remaining in the crucible and ensure a reproducible beam pattern for pouring in a small mold funnel. A computer-controlled production casting system that uses electron beam melting is described in detail in Ref 18.

Process Technology and Equipment for Titanium Castings

The production of titanium castings in electron beam furnaces is economical when large quantities of the same size and weight must be manufactured and when the entire specification range of alloy compositions can be used. Electron beam casting is economical and superior to the competing VAR skull melting and nonconsumable arc melting processes; a blend of solid scrap and consumable electrodes can be used because the crucible spout can be cleaned by electron beam melting after each pour. Figure 9 illustrates the beam control possible with microprocessor control. Another advantage of this process is the possibility of superheating the pool just before pouring and the metal stream during pouring. During the melting and casting process, the feedstock can be loaded into the scrap feeder or into the electrode lowering device without influencing the process cycle. Mold loading, transportation, positioning below the crucible spout, spinning there with up to 500 rpm, and unloading of the cooled casting are completely automated.

REFERENCES

1. R.E. Lüders, Tantalum Melting in a 800 kW EB Furnace, in *Proceedings of the Bakish Conference on Electron Beam Melting and Refining* (Reno, NV), R. Bakish, Ed., 1983, p 230-244
2. J.A. Pierret and J.B. Lambert, Operation of Electron Beam Furnace for Melting Refractory Metals, in *Proceedings of the Bakish Conference on Electron Beam Melting and Refining* (Reno, NV), R. Bakish, Ed., 1984, p 208-218
3. H. Sperner, Hafnium, *Metallwis. Technik.*, Vol 7 (No. 16), 1962, p 679-682
4. R. Hähn and J. Krüger, Refining of Vanadium Aluminium Alloys to Vanadium 99.9% by EB Melting, in *Proceedings of the Bakish Conference on Electron Beam Melting and Refining* (Reno, NV), R. Bakish, Ed., 1986, p 53-67
5. J. Lambrecht, D. Rumberg, and K.H. Werner, Stand der Technologie der Stahlerzeugung im Elektronenstrahlofen mit Blockmassen bis 100 t, *Neue Hütte*, Vol 10, Oct 1984
6. C.E. Shamblen, S.L. Culp, and R.W. Lober, Superalloy Cleanliness Evaluation Using the EB Button Melt Test, in *Proceedings of the Bakish Conference on Electron Beam Melting and Refining* (Reno, NV), R. Bakish, Ed., 1983
7. F. Hauner, H. Stephan, and H. Stumpp, Ergebnisse bei Elektronenstrahl-Schmelzen von gerichtet erstarrten Knopfproben zur Identifizierung von nichtmetallischen Einschlüssen in hochreinen Superlegierungen, *Metall.*, Vol 2 (No. 40), 1986, p 2-7
8. C.d'A. Hunt and H.R. Smith, Electron Beam Processing of Molten Steel in Cold Hearth Furnace, *J. Met.*, Vol 18, 1966, p 570-577
9. H.R. Harker, Electron Beam Melting of Titanium Scrap, in *Proceedings of the Bakish Conference on Electron Beam Melting and Refining* (Reno, NV), R. Bakish, Ed., 1983, p 187-190
10. C.d'A. Hunt, H.R. Smith, and B.C. Coad, The Combined Induction and Electron Beam Furnace for Steel Refining and Casting, in *Proceedings of the Vacuum Metallurgy Conference* (Pittsburgh, PA), June 1969, p 1-22
11. H.R. Harker, The Present Status of Electron Beam Melting Technology, in *Proceedings of the Bakish Conference on Electron Beam Melting and Refining* (Reno, NV), R. Bakish, Ed., 1986, p 3-7
12. C.d'A. Hunt, J.C. Lowe, and T.H. Harrington, Electron Beam, Cold Hearth Refining for the Production of Nickel and Cobalt Base Superalloys, in *Proceedings of the Bakish Conference on Electron Beam Melting and Refining* (Reno, NV), R. Bakish, Ed., 1984, p 295-304
13. H. Stephan, R. Schumann, and H.J. Stumpp, Production of Superclean Fine-Grained Superalloys for Improvement of the Workability and Engine Efficiency by EB Melting and Refining Methods, in *Proceedings of the Eighth International Conference on Vacuum Metallurgy* (Linz), 1985, p 1219-1309
14. H. Stephan, Production of Ingots and Cast Parts From Reactive Metals by Electron Beam Melting and Casting, in *Proceedings of the Third Electron Beam Processing Seminar* (Stratford, UK), 1974, p lb1-lb69
15. H. Ranke, V. Bauer, W. Dietrich, J.

Heimerl, and H. Stephan, Melting and Evaporation With the Newly Developed Leybold-Heraeus 600 kW EB Gun at Different Pressure Levels, in *Proceedings of the Bakish Conference on Electron Beam Melting and Refining* (Reno, NV), R. Bakish, Ed., 1985
16. M. Krehl and J.C. Lowe, Electron Beam Cold Hearth Refining for Superalloy Revert for Use in Foundry Production, in *Proceedings of the Bakish Conference on Electron Beam Melting and Refining* (Reno, NV), R. Bakish, Ed., 1986, p 286-296
17. M. Romberg, R. Schumann, H. Stephan, and H. Stumpp, Electron Beam Melting and Refining of Superalloys for Ingot and Barstick Production, in *Proceedings of the Bakish Conference on Electron Beam Melting and Refining* (Reno, NV), R. Bakish, Ed., 1986
18. J. Mayfield, Computer-Controlled Production Gains, *Aviat. Week Space Technol.*, 3 Dec 1979, p 1-5

Plasma Melting and Casting

H. Pannen and G. Sick, Leybold AG, West Germany

Plasma melting is a material processing technique in which the heat of a thermal plasma is used to melt the feed material. A thermal plasma is considered a suitable heat source if high temperature and a defined gas atmosphere are needed to melt the material before subsequent processing, such as solidification or atomization. The products are usually ingots, slabs, castings, or powders.

Plasma Torch

Plasma Generation. An electric current passing through an ionized gas leads to a phenomenon known as gaseous discharge. To generate this kind of plasma, electrical breakdown of the gas must be accomplished. Breakdown (that is, the creation of charge carriers) establishes a conducting path between a pair of electrodes in gases that are insulators at room temperature. Other plasmas can be created by electrodeless radio frequency (RF) discharges, microwaves, shock waves, lasers, or high-energy particle beams.

In this section, only thermal (hot) plasmas will be considered. The term thermal differentiates these plasmas from cold (nonequilibrium) plasmas, which are characterized by high electron temperatures and low sensible temperatures of the heavy particles. Cold plasmas are produced in various types of glow discharges, low-pressure RF discharges, and corona discharges. Cold plasmas are not suitable for melting processes.

Most plasma generators (plasma torches) for melting processes use an electric arc to produce gaseous discharges. The characteristics of an electric arc include relatively high current densities, low cathode fall, and high luminosity of the column.

A typical potential distribution along an arc is shown in Fig. 1 (Ref 1). The column cross section at the cathode is smaller than at the anode, with contrary current and power densities. This fact influences the design principle of plasma torches.

Torch Design. Plasma torches can be used in either the transferred or the nontransferred mode. In the nontransferred mode, the cathode and the anode are inside the torch. This mode is generally suitable only for the melting of nonconductive materials and has the disadvantage of lower efficiency compared to the transferred mode. In the transferred mode, one electrode is inside the torch, and the counterelectrode is the material to be melted. Figure 2 shows two typical design principles for torches in the transferred mode: the tungsten tip design and the hollow copper electrode design.

In the tungsten tip design (Fig. 2a), the torch electrode is connected as the cathode because high current densities are required. The cathodes are usually made of tungsten, along with small additions of thoria to lower the thermionic work function of tungsten. Still, electron emission requires high electrode temperatures (3500 to 6000 K) at the attachment of the arc. Therefore, the cathode material is liquid at the arc attachment

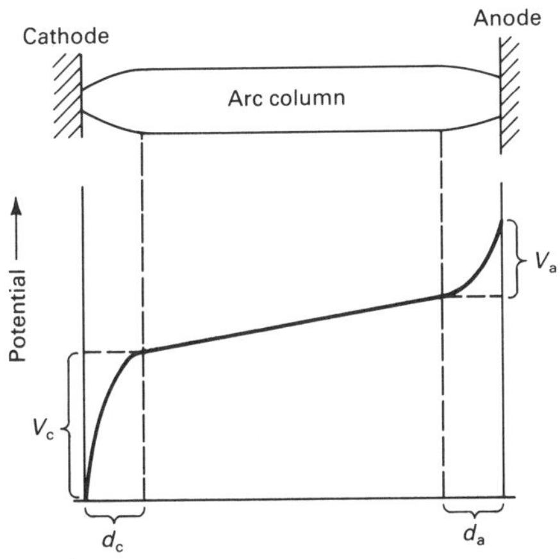

Fig. 1 Typical potential distribution along a plasma arc. V_a, anode voltage; V_c, cathode voltage; d_a, anode current density; d_c, cathode current density

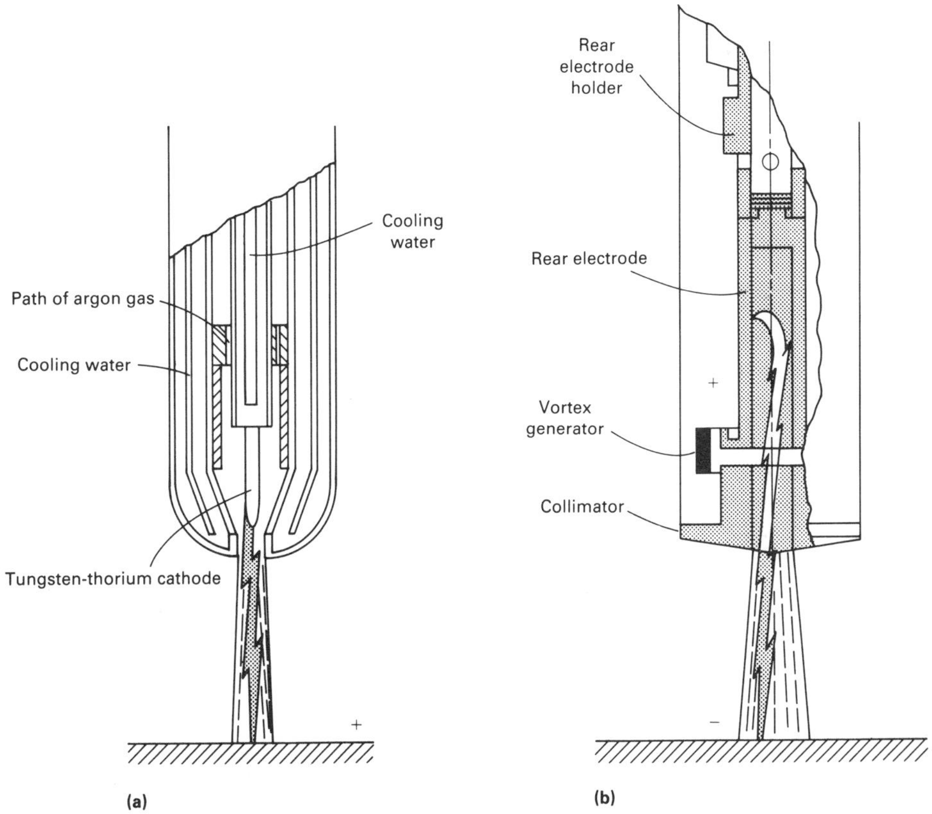

Fig. 2 Design concepts for plasma arc torches in the transferred mode. (a) Torch with tungsten tip and concentric gas flow. (b) Torch with hollow copper electrode and vortex generator

Table 1 Characteristics and operating parameters of furnaces for plasma melting processes

Typical industrial furnace	Feed material	Product form and size, mm (in.)	Typical melt rate, kg/h (lb/h)	Specific power consumption, kW · h/kg	Plasma gas	Typical furnace pressure	Torch design principle and power	Ref
Plasma consolidation furnaces								
Daido Steel PCCF, Japan	Titanium scrap and sponge	Ingot 355–430 (14–17) diam, 3000 (120) long; density >90%	250–260 (550–570)	1.4–1.6	Argon	≥Atmospheric	Tungsten tip with external ignition bar; 6 × 90 kW	2
Retech furnace at OREMET, United States	Titanium scrap and sponge	Ingot 700 (27.5) diam, 3800 (150) long; density 98%	250–670 (550–1480)	0.89–2.4	Argon	≥Atmospheric	Hollow copper electrode with graphite liner; 600 kW	3
Plasma arc remelting furnaces								
Soviet type I	Titanium, titanium alloys	Ingot 125 (5) diam, 500 (20) long	50–70 (110–155)	2.0–2.5	...	≥Atmospheric	Probably tungsten tip; 500 kW	4
Soviet type II	Specialty steels, heat-resistant alloys	Ingot 150–200 (6–8) diam, 1200 (47) long; slabs 70 × 300 × 1200 (2.8 × 12 × 47)	90–135 (200–300)	1.5–2	...	≥Atmospheric	Probably tungsten tip; 500–1000 kW	4
Soviet "large scale"	Specialty steels, heat-resistant alloys	Ingots 650 (25.5) diam, 1500–2300 (60–90) long	400–900 (880–1985)	1.2–1.8	...	≥Atmospheric	Probably tungsten tip; 300–3600 kW	4
Plasma cold hearth melting furnaces								
ULVAC furnace at Nippon Stainless Steel, Japan	Titanium scrap and sponge	Slab 245 × 1125 × 2450 (9.6 × 44.3 × 96.5)	330 (730)	9	...	1–13 Pa (0.01–0.13 mbar)	Six hollow cathode torches; 700 kW on hearth; 1170 kW on crucible	5
Retech pilot-scale furnace, United States	Titanium scrap and sponge, superalloys	Ingot 195 (7.7) diam	90–227 (200–500)	1.3–3	Helium; gas recycling	≥Atmospheric	Hollow copper electrode, vortex stabilized; 200 kW on hearth, 100 kW on crucible	6
Plasma cold crucible melting and casting furnaces								
Leybold AG	Titanium scrap and sponge, superalloys	Casting ingots	380 (840)	1.3	Argon; gas recycling	50 kPa (500 mbar)	Hollow copper electrode, vortex stabilized; 650 kW	7

point even with water cooling at the rear of the cathode. For this reason, oxidizing gases generally cannot be used in direct contact with the tungsten tip. Argon shielding is one solution. Further, it should be kept in mind that portions of the small liquid pool at the cathode tip are ejected into the material being melted. The high melting point of tungsten can result in inclusions in the product; such defects are not tolerable in high-strength metals such as titanium alloys and superalloys for aircraft applications.

With argon as the plasma gas, the service life of the cathode in the tungsten tip design is typically between 30 and 150 h. The inert gas consumption and plasma column stabilization of such torches are low.

The hollow copper electrode design is shown in Fig. 2(b). The electrode material, usually copper alloyed with chromium or zirconium, is intensively water cooled. Copper is a field emitter of electrons (as opposed to tungsten, a thermionic emitter) because its boiling point is substantially below that required for thermionic electron emission. Current densities at the attachment points on cold cathodes are double those for thermionic emission on hot cathodes. Cathode spots are of high intensity, rapidly moving on the surface of the cold cathode and causing erosion by locally vaporizing copper. For this reason, the polarity is usually reversed in plasma torches with cold electrodes. The rear electrode acts as the anode, and the workpiece or melt is the cathode. The anodic arc attachment is smoother, and erosion is reduced. The plasma-forming gas is blown in tangentially between the rear electrode and the nozzle to create a swirl or vortex to stabilize the arc and to rotate the arc attachment in the rear electrode. With the movement of the arc attachment, a substantially lower power density can be achieved in the rear electrode.

Electrode life of 1000 h has been demonstrated in torches using argon as the plasma gas. All inert and reactive gases, or mixtures of these, can generally be used in such torches, but the influence of the gases on electrode erosion varies. The gas consumption of vortex-stabilized torches is higher than that of other designs.

The efficiency of plasma torches is less affected by design principles than by process parameters and by the type of plasma gas used. Efficiency is defined as P_w/P_t, where P_t is the electrical power input of the torch and P_w is the thermal power applied to the workpiece or melt. The efficiency of a 650 kW torch measured in a cold furnace ranges from approximately 20 to 40%, depending on the plasma gas used, furnace pressure, and the distance between torch and workpiece.

The highest efficiency is obtained with helium as the plasma gas and a short distance between material and torch. The power loss in torches with hollow water-cooled electrodes is approximately 20%. Remaining losses are due to radiation and convection of the plasma column between torch and workpiece. These losses are primarily a function of the distance between torch and workpiece, but radiative losses of the plasma column also increase with pressure.

Furnace Equipment

Plasma melting furnaces usually consist of double-wall, vacuum-tight, water-cooled constructions that can withstand the radiative and convective heat transfer from the plasma column. Furnaces also usually have material feeding systems. These can consist of vibratory or rotary feeders with bins (for bulk materials) or bar feeding systems. If the furnace is to operate continuously, the feed material must be "locked in" to prevent disturbance of the furnace atmosphere.

The furnace must also have a vacuum pump so that it can be evacuated before backfilling. If the furnace is operated under reduced pressure, an offgas pump with an on-line gas control valve is required.

Atmosphere Control

Plasma melting furnaces are usually operated under slightly positive pressure to prevent the potential atmospheric contamina-

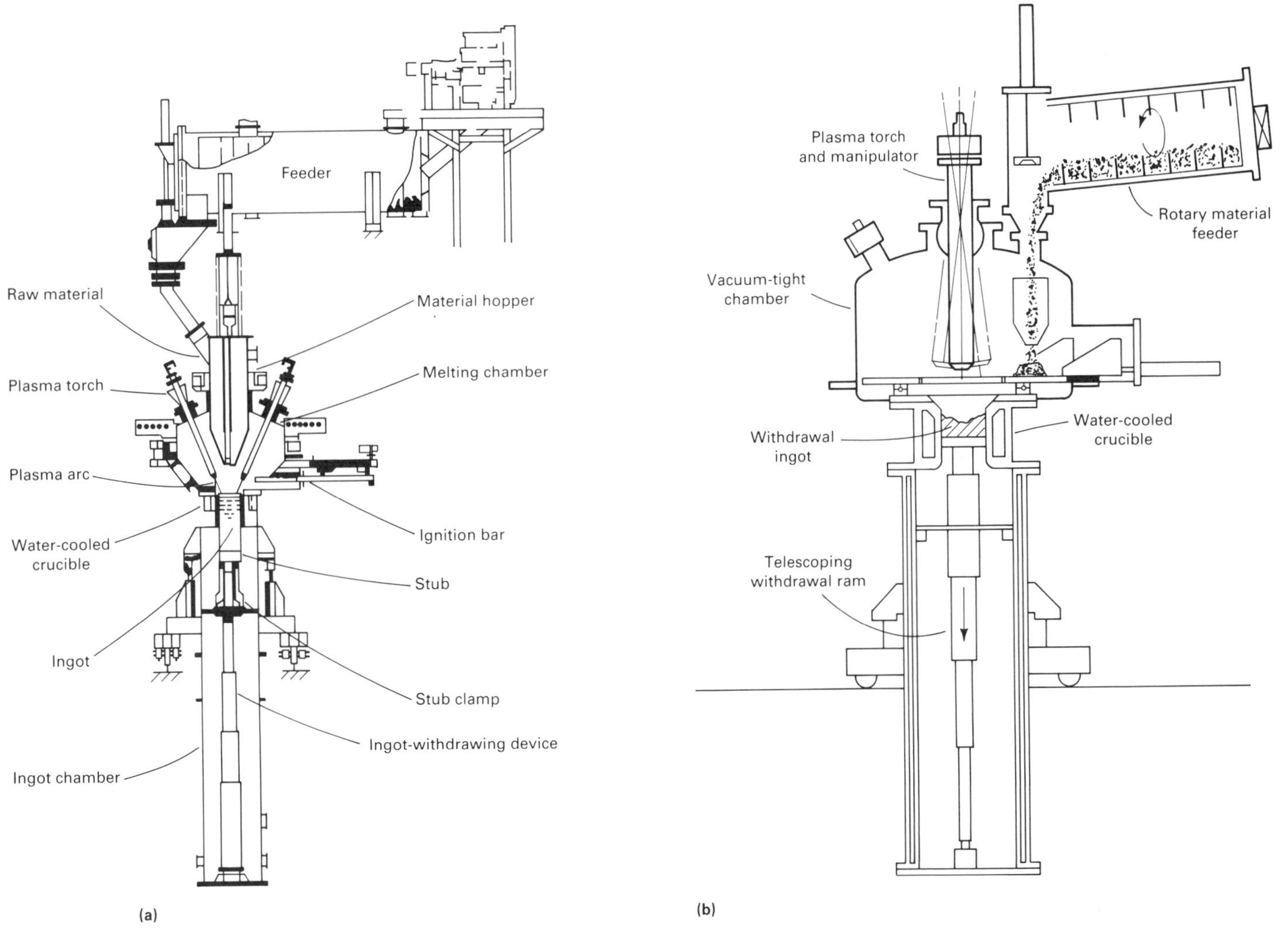

Fig. 3 Plasma consolidation furnaces. (a) Japanese furnace with six torches and total power of 540 kW. Source: Ref 6. (b) Retech furnace in the United States with one torch. Source: Ref 3

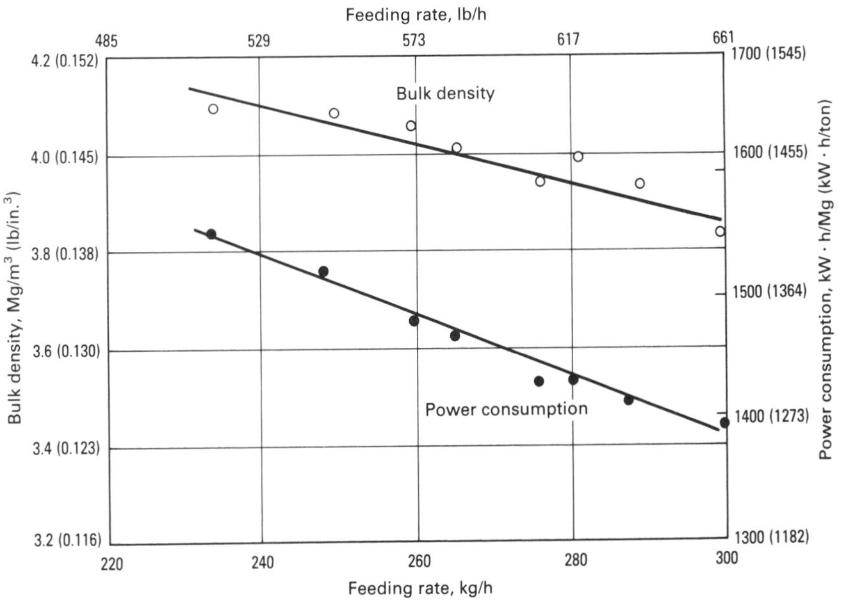

Fig. 4 Relationship among raw material feeding rate, ingot bulk density, and specific power consumption in the plasma consolidation of titanium. Source: Ref 2

tion by oxygen and nitrogen. However, state-of-the-art furnaces are vacuum tight and can be operated at pressures between 5 and 200 kPa (50 and 2000 mbar, or 38 and 1500 torr). Vacuum tightness is essential because the back diffusion of oxygen, nitrogen, and moisture through small leaks can be easily demonstrated even with positive pressure in the furnace. Plasma torches are normally operated with argon plasma gas. A typical gas purity of 99.999% indicates 10 ppm gaseous impurities and corresponds to an absolute pressure of 1 Pa (0.01 mbar, or 0.0075 torr) in a vacuum process.

Additional sources of atmospheric contamination are the moisture and gases in the feed material, desorption of the furnace wall, and leaks in the furnace construction. In the case of metals with high affinities for oxygen or nitrogen, these gases are totally absorbed by the melt and will be found as oxide or nitride inclusions in the product.

To obtain a clean initial furnace atmosphere, the vacuum-tight plasma furnace is evacuated with a mechanical pump to final

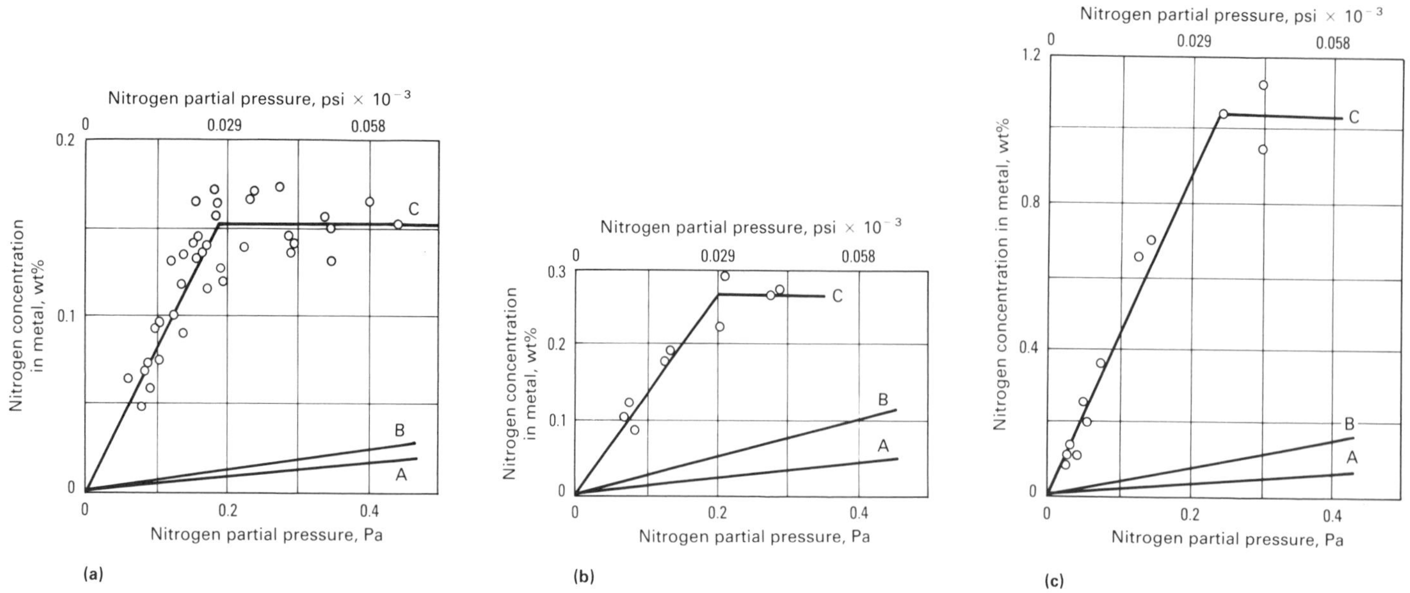

Fig. 5 Nitrogen concentrations in various metals as a function of the partial pressure of nitrogen after plasma arc remelting. (a) Iron. (b) 16-25-6 stainless steel. (c) Austenitic stainless steel. Curves A and B, induction heating; minimum and maximum nitrogen concentrations, respectively. Curve C, plasma arc remelting. Source: Ref 4

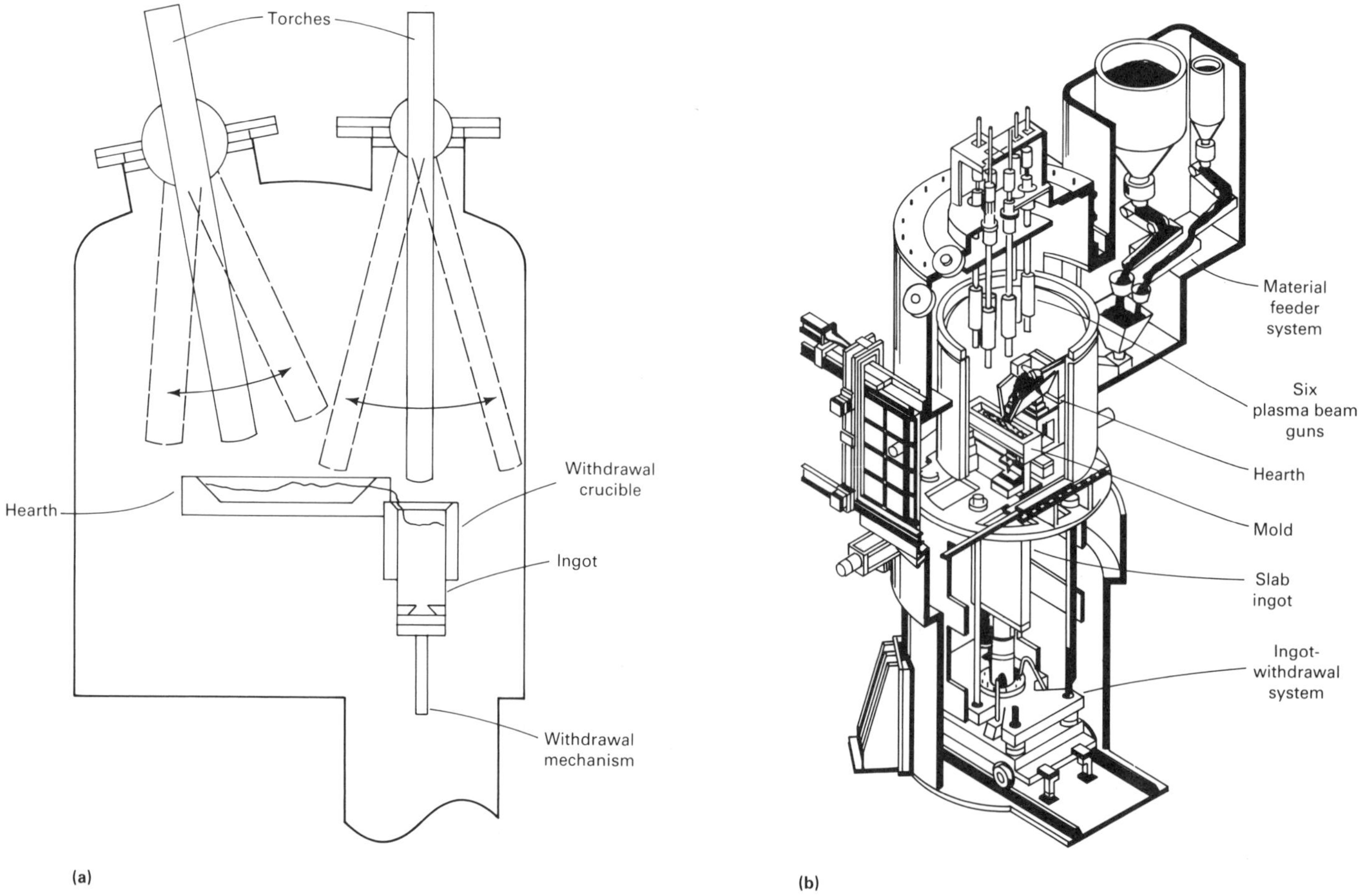

Fig. 6 Two plasma cold hearth melting furnaces. (a) Pilot-scale furnace with two tiltable plasma torches. Total power: 400 kW. (b) ULVAC plasma beam furnace with six fixed plasma beam guns. Total power: 2000 kW. Source: Ref 3

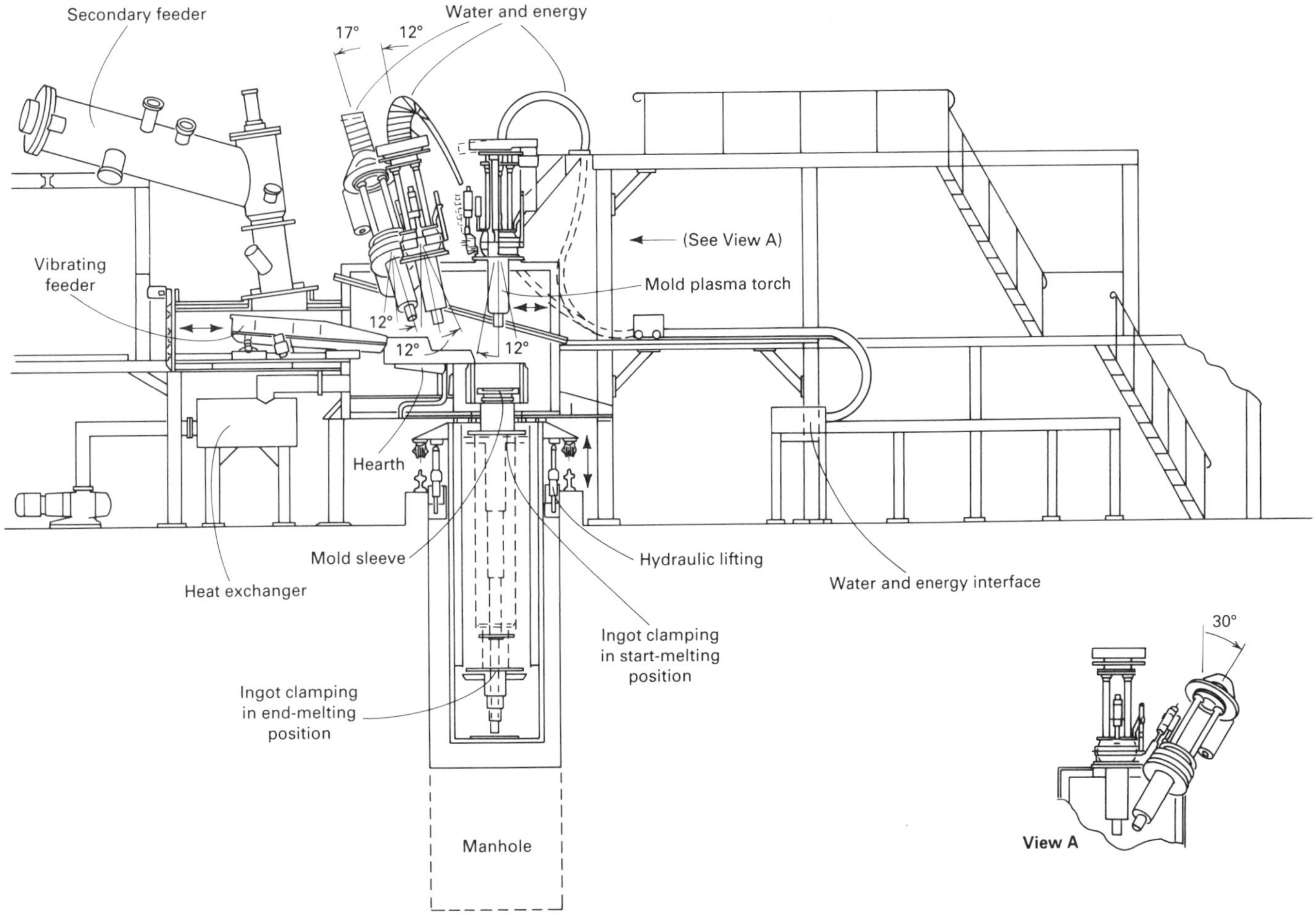

Fig. 7 Plasma cold hearth melting concept for the recycling of titanium scrap

pressure (typically about 1 Pa, or 0.01 mbar, or 0.0075 torr). The furnace walls can be degassed with the help of warm water running between the double walls of the furnace to enhance moisture desorption. The furnace is then backfilled with high-purity inert gas up to 1 kPa (10 mbar, or 7.5 torr), evacuated again to 1 Pa (0.01 mbar, or 0.0075 torr), and backfilled to operating pressure. This procedure dilutes residual gas impurities by a factor of 1000.

Processes

Application of Plasma Melting Processes. The selection of a suitable heat source for melting, remelting, casting, and atomizing reactive metals, titanium, and superalloys is of great interest to the manufacturer and user of plasma furnace equipment. Skull melting processes in water-cooled copper crucibles are often the only melting technology for these metals (see the section "Vacuum Arc Skull Melting and Casting" in this article). Plasma torches, however, are the only nonconsumable heat sources for melting under high inert gas pressures in skull crucibles. High pressure is an essential requirement for preventing the selective evaporation of alloying elements that are characterized by high activity coefficients and/or vapor pressures, such as chromium and manganese in superalloys and aluminum in titanium alloys.

Various plasma melting processes have been developed, including plasma consolidation, plasma arc remelting, plasma cold hearth melting, and plasma casting. Plasma cold hearth melting and plasma cold crucible casting are thought to have the most potential for industrial application. Table 1 lists data on feed materials, operating parameters, and products of furnaces for some of these processes.

Plasma Consolidation. Consolidation implies that low-density feed material is converted into a high-density product. Plasma consolidation is used for titanium scrap and sponge to produce an electrode for further remelting in vacuum arc remelting (VAR) furnaces. The material is usually fed directly into a water-cooled copper crucible and melted by plasma heat from one or several plasma torches. The water-cooled copper baseplate of the crucible is continuously withdrawn. The shape and depth of the liquid pool in the crucible depend on feed rate and power input. Under these conditions, 100% melting cannot be ensured; some material can be trapped at the liquid/solid interface before melting. Figure 3 shows two typical plasma consolidation furnaces.

Figure 4 shows ingot density as a function of the raw material feed rate and power consumption in the plasma consolidation of titanium. Plasma consolidation processes can reduce manufacturing costs for the preparation of titanium VAR electrodes, which are usually plasma melted from pressed sponge or chip compacts.

With plasma consolidation, metal chlorides can be removed from titanium sponge to sufficiently low levels. Tungsten carbide tool tips in titanium chips cannot be removed and eventually form high-density inclusions in the product. Therefore, the conventional x-ray inspection of scrap is still necessary. Low-density inclusions such as titanium nitride particles have not been found after plasma consolidation of contaminated titanium scrap and subsequent VAR remelting (Ref 3). Plasma consolidation has also been successfully applied to the processing of titanium aluminide (TiAl), a low-density alloy with excellent high-temperature service properties (Ref 3).

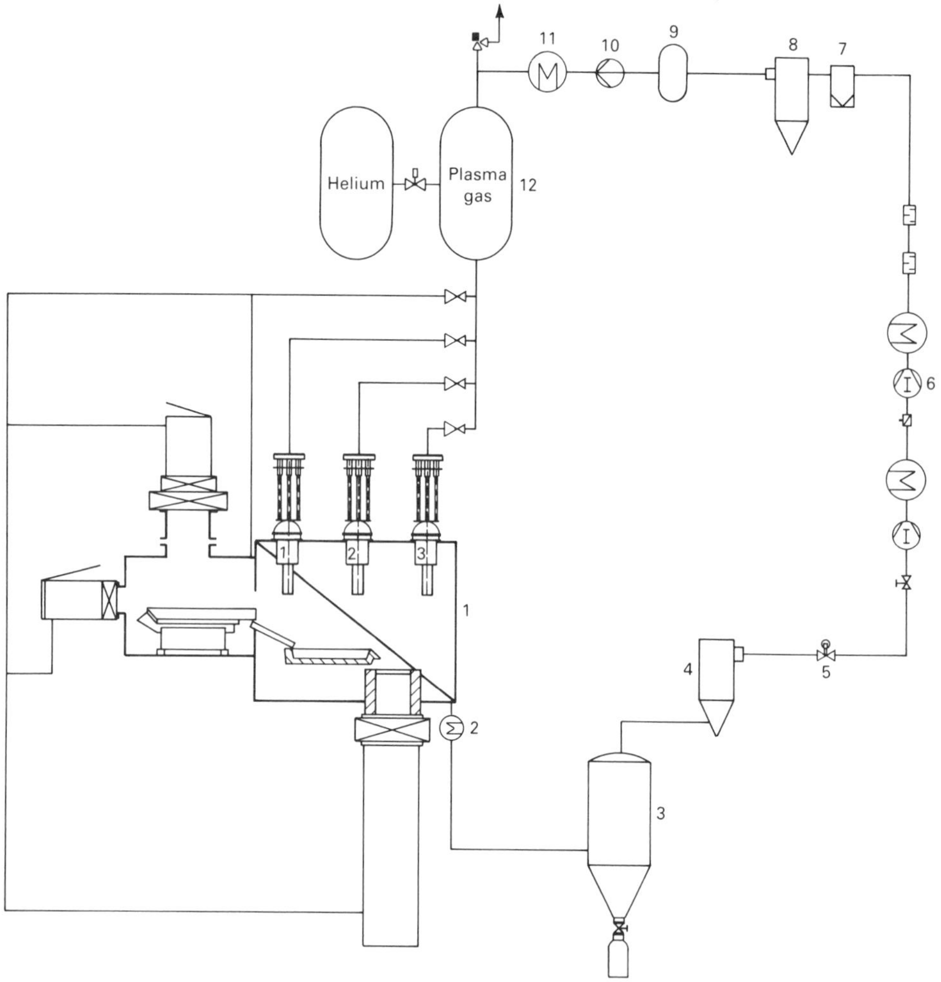

Fig. 8 Schematic of plasma gas-recycling system. 1, plasma furnace; 2, off-gas heat exchanger; 3, cyclone separator; 4, mechanical filter; 5, motor-actuated valve; 6, vacuum pump set; 7, chemical filter; 8, fine filter; 9, buffer; 10, compressor; 11, heat exchanger; 12, gas storage tank

Plasma Arc Remelting. An electrode that has already been melted by a primary melting process such as electric arc or induction melting can be plasma arc remelted into a water-cooled withdrawal crucible. The major objectives for plasma arc remelting are:

- To obtain directional solidification without changing the chemical composition of the feed material
- To improve cleanliness by removal, size reduction, shape control, and redistribution of inclusions
- To lower gas impurity content
- To alloy with nitrogen

Plasma arc remelting is used most frequently in the Soviet Union to produce high-temperature alloys, bearing steels, superalloys, and titanium alloys. Plasma arc remelting can be advantageous for the nitrogen alloying, deoxidation, or desulfurization of nonreactive alloys. An example of successful reactive melting is the nitrogen alloying of steel during plasma arc remelting with argon-nitrogen mixtures as the plasma gas (Ref 4). Figure 5 shows nitrogen concentrations in various metals as a function of nitrogen partial pressure in the gaseous phase after plasma arc remelting. The data indicate that excited molecules of nitrogen in the plasma react with the metal (Ref 4).

In plasma cold hearth melting, bars, sponge, or scrap can be continuously fed into a water-cooled copper trough, usually called a hearth, and melted by means of plasma heat. The liquid metal flows over the lip of the hearth into a withdrawal mold. The plasma cold hearth melting process is adapted from the electron beam cold hearth refining process, which is a well-developed refining and casting method for superalloys. Plasma heat, instead of an electron beam, is employed to minimize selective evaporative losses that occur during electron-beam melting in the pressure range of 0.01 to 1 Pa (10^{-4} to 0.01 mbar, or 7.5×10^{-5} to 0.0075 torr).

In the hearth, high-density inclusions from tungsten carbide tool tips will sink down to the skull. Low-density inclusions can be dissociated or dissolved in the base metal during interaction with the superheated melt and the plasma jet.

Process optimization is necessary to prevent short circuiting of the hearth by unmelted particles and to establish sufficient dwell time at high melting rates. The power distribution of one or more plasma torches required for smooth melting and fluid flow can be ensured by automated torch-moving devices.

Figure 6 shows a pilot plasma cold hearth furnace (Fig. 6a) and the only industrial-scale plasma cold hearth melting furnace currently in operation (Fig. 6b). The plasma heat source of the industrial furnace is unconventional in that there is no mechanism provided for torch movement. The torches, called plasma electron beam guns, use hollow tantalum cathodes and operate at pressures of 1 to 10 Pa (0.01 to 0.1 mbar, or 0.0075 to 0.075 torr). The unusually high specific power consumption of this furnace reflects the relatively low efficiency of the heat source used.

Figure 7 shows a modern plasma cold hearth melting furnace concept with a double-wall, water-cooled, vacuum-tight furnace chamber and three plasma torches for the production of ingots 710 mm (28 in.) in diameter. Two torches serve the hearth, and the third the ingot mold. The torches use hollow water-cooled copper electrodes. The high gas consumption of this torch design means that an effective gas recycling system must be provided, especially if helium is used as the plasma gas.

The operating principle of a gas-recycling system is shown in Fig. 8. Key technology in this system includes reactive filters, oil-free high-capacity pump sets, and compressors. If necessary, the recycling system can be supplemented with cleaning systems for oxygen, carbon dioxide, nitrogen, hydrogen, and moisture.

Plasma cold crucible casting is an alternative process to vacuum arc skull melting and casting or electron beam skull melting and casting, and it can overcome some of the disadvantages of these processes. The disadvantages of vacuum arc skull melting in the melting and casting of titanium and superalloys are:

- Control of superheating is not possible; increased power results in increased melting rate instead of superheating
- Expensive electrodes must be used instead of the high-quality unconsolidated revert material that can be used for plasma cold crucible casting

For small batches of superalloys (up to 8 kg, or 18 lb), electron beam skull casting is a well-established technique. Electron beam melting and casting is rarely used for large castings of either superalloys or titanium alloys. This is because of the shallow melt pool in the crucible due to lack of stirring

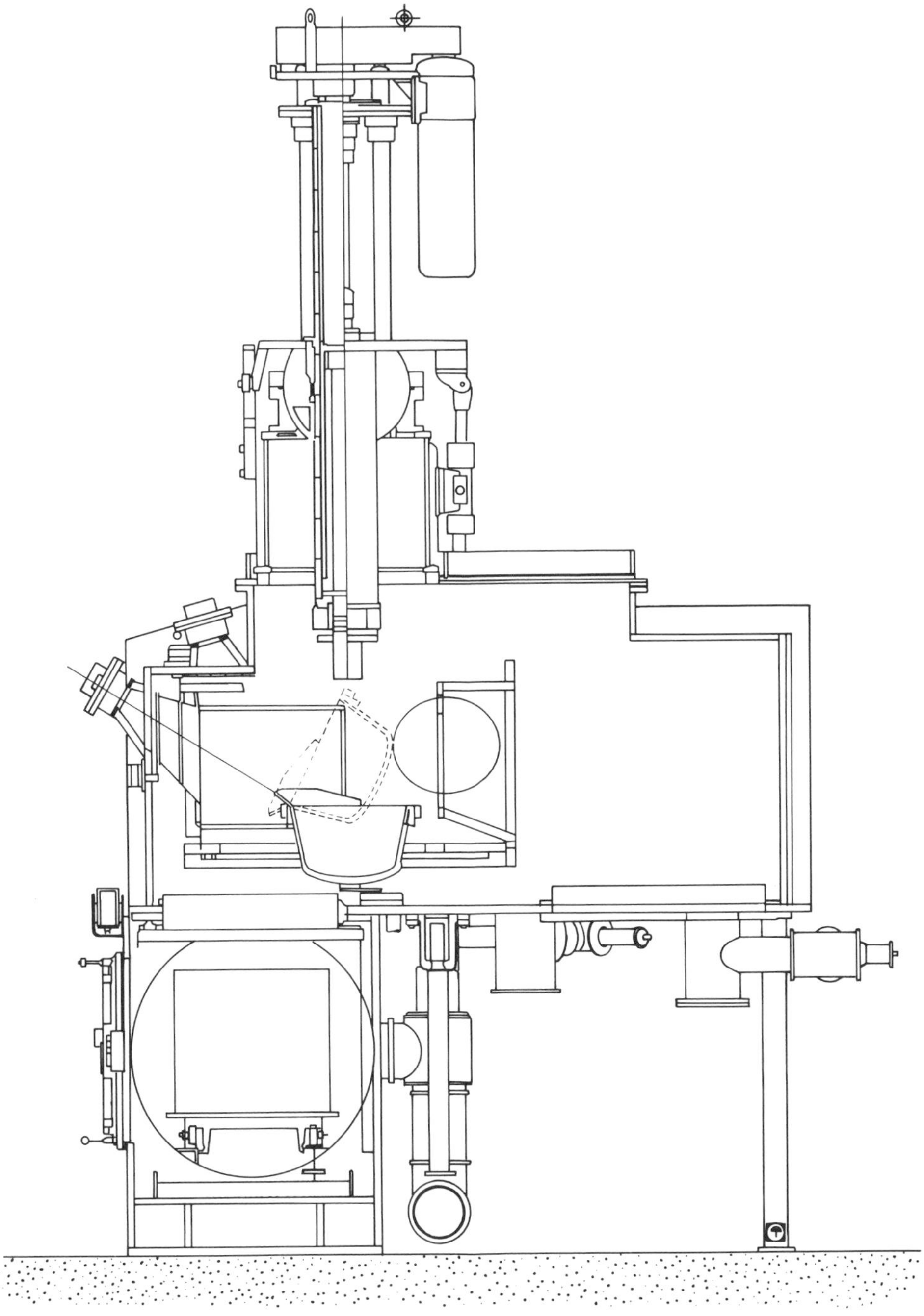

Fig. 9 Industrial-size plasma cold crucible casting furnace with 650 kW torch power

action and the potential for selective evaporation of alloying elements during melting.

Figure 9 shows a sectional view of an industrial-size plasma cold crucible casting furnace. The furnace consists of a cylindrical, horizontally installed mold chamber with lockchambers at both sides and the melt chamber on top of the mold chamber. A programmable torch-moving device is installed on top of the melt chamber. The pressure in the furnace during the melting operation can be preselected; pressures typically range from 30 to 50 kPa (300 to 500 mbar, or 225 to 375 torr). The melting operation can be started at every pressure in this range. Solid material can be charged continuously into the water-cooled copper crucible, and additional material can be fed into the crucible during melting. The melt is poured into stationary molds or centrifugal casting molds situated in the mold chamber. The pouring stream can be superheated during casting.

The typical casting weight of this furnace is 27 kg (60 lb) of titanium with a 650 kW plasma torch. To achieve a deep molten pool, a high heat flow from the plasma torch to the melt must be established in combination with high stirring velocity in the melt to enhance the effective heat transfer coefficient. The stirring action in the melt can be regulated by controlling current, gas flow, and torch-to-melt distance. Continuous feeding of the metal to be melted is required in order to achieve a deep pool.

In contrast to vacuum arc skull melting, the crucible need not be inspected after each melt. The power densities in the plasma attachment area are adjusted to avoid damaging the crucible if it is inadvertently exposed to the plasma for short periods of time.

REFERENCES

1. E. Pfender, M. Boulos, and P. Fauchais, Methods and Principles of Plasma Generation, in *Plasma Technology in Metallurgical Processing*, J. Feinman, Ed., Iron and Steel Society, 1987, p 27-47
2. T. Yagima *et al.*, Development of the Plasma Progressive Casting Process and Its Application for Titanium Melting, in *Titanium 1986: Products and Applications*, Vol II, Proceedings of the Technical Program from the 1986 International Conference, Titanium Development Association, 1987, p 985-993
3. S. Stocks and D. Hialt, Plasma Consolidation of Large Diameter Titanium Electrodes, in *Titanium 1986: Products and Applications*, Vol II, Proceedings of the Technical Program from the 1986 International Conference, Titanium Development Association, 1987, p 918-927
4. G.K. Bhat, Plasma Arc Remelting, in *Plasma Technology in Metallurgical Processing*, J. Feinman, Ed., Iron and Steel Society, 1987, p 163-174
5. K. Murase, T. Suzuki, T. Kijima, H.Takei, and Y. Yoneda, Production of Titanium Slab Ingot in Vacuum Plasma Electron Beam Furnace, in *Proceedings of the Electron Beam Melting and Refining State of the Art 1986 Conference*, Bakish Materials Corporation, 1986, p 184-194
6. R.C. Eschenbach, in *Proceedings of the 1986 Vacuum Metallurgy Conference on Specialty Metals Melting and Processing* (Pittsburgh, PA), Iron and Steel Society, 1986
7. G. Sick, in *Proceedings of the 1986 Vacuum Metallurgy Conference on Specialty Metals Melting and Processing* (Pittsburgh, PA), Iron and Steel Society, 1986, p 179-186
8. N.A. Barcza, Application of Plasma Technology to Steel Processing, in *Plasma Technology in Metallurgical Processing*, J. Feinman, Ed., Iron and Steel Society, 1987, p 131-148

Degassing Processes (Converter Metallurgy)

Chairman: Gerhard Kienel, Leybold AG, West Germany

CONVERTERS are used in secondary metallurgy to refine melts outside the primary metallurgical melting unit. Various converters are available that apply the bottom blowing of oxygen-inert gas mixtures. These bottom-blowing converters use different types, numbers, and arrangements of injection nozzles and the following gases:

- Argon as a cooling inert gas with a purity ranging from 85 to 99.99%
- Nitrogen as a cooling inert gas with a purity of 99 to 99.9%
- Argon-oxygen and nitrogen-oxygen mixtures in the case of the argon oxygen decarburization converter
- Dry air as a diluted reactive gas
- Oxygen as a reactive gas with a purity of 80 to 99.5%
- Steam as a cooling reactive gas
- Carbon dioxide as a diluted reactive gas

This article will review three converter designs. These are the argon oxygen decarburization vessel; the oxygen top and bottom blowing converter, which is an extension of argon oxygen decarburization technology; and the vacuum oxygen decarburization converter. Additional information on degassing processes can be found in the article "Degassing Processes (Ladle Metallurgy)" in this Volume.

Argon Oxygen Decarburization (AOD)

Ian F. Masterson, Union Carbide Corporation—Linde Division

Argon oxygen decarburization (AOD) is a secondary refining process that was originally developed to reduce material and operating costs and to increase the productivity of stainless steel production. In addition to its economic merits, argon oxygen decarburization offers improved metal cleanliness, which is measured by low unwanted residual element contents and gas contents; this ensures superior mechanical properties. The AOD process is duplexed, with molten metal supplied from a separate melting source to the AOD refining unit (vessel). The source of the molten metal is usually an electric arc furnace or a coreless induction furnace. Foundries and integrated steel mills utilize vessels ranging in nominal capacity from 1 to 160 Mg (1 to 175 tons).

Although the process was initially targeted for stainless steel production, argon oxygen decarburization is used in refining a wide range of alloys, including:

- Stainless steels
- Tool steels
- Silicon (electrical) steels
- Carbon steels, low-alloy steels, and high-strength low-alloy steels
- High-temperature alloys and superalloys

Fundamentals

In the AOD process, oxygen, argon, and nitrogen are injected into a molten metal bath through submerged, side-mounted tuyeres. The primary aspect of the AOD process is the shift in the decarburization thermodynamics that is afforded by blowing with mixtures of oxygen and inert gas as opposed to pure oxygen. References 1 to 3 contain detailed discussions of the thermodynamics and the dilution principle and its influence on the refining of chromium-bearing steel.

To understand the AOD process, it is necessary to examine the thermodynamics governing the reactions that occur in the refining of stainless steel, that is, the relationship among carbon, chromium, chromium oxide (Cr_3O_4), and carbon monoxide (CO). The overall reaction in the decarburization of chromium-containing steel can be written as:

$$\tfrac{1}{4}Cr_3O_4 + \underline{C} \rightleftharpoons \tfrac{3}{4}\underline{Cr} + CO(g) \qquad \text{(Eq 1)}$$

The equilibrium constant, K, is given by:

$$K = \frac{(a_{Cr})^{3/4}\,(P_{CO})}{(a_{Cr_3O_4})^{1/4}\,(a_C)} \qquad \text{(Eq 2)}$$

where a and P represent the activity and partial pressure, respectively.

At a given temperature, there is a fixed, limited amount of chromium that can exist in the molten bath that is in equilibrium with carbon. By examining Eq 2, one can see that by reducing the partial pressure of CO, the quantity of chromium that can exist in the molten bath in equilibrium with carbon increases. The partial pressure of CO can be reduced by injecting mixtures of oxygen and inert gas during the decarburization of stainless steel. Figure 1 illustrates the relationship among carbon, chromium, and temperature for a partial pressure of CO equal to 1 and 0.10 atm (1000 and 100 mbar, or 760 and 76 torr). The data shown in Fig. 1 indicate that diluting the partial pressure of CO allows lower carbon levels to be obtained at higher chromium contents with lower temperatures.

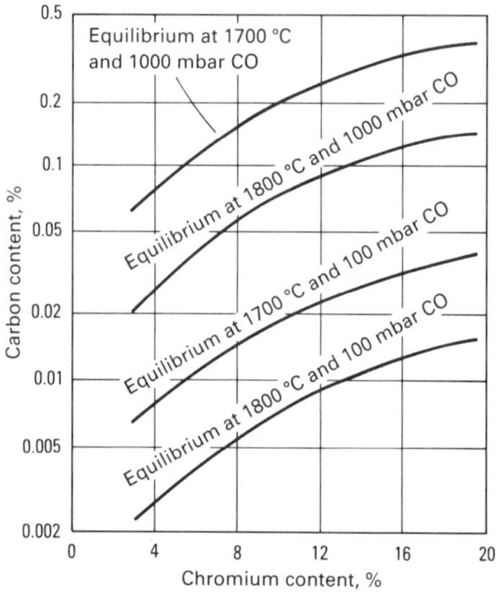

Fig. 1 Carbon-chromium equilibrium curves

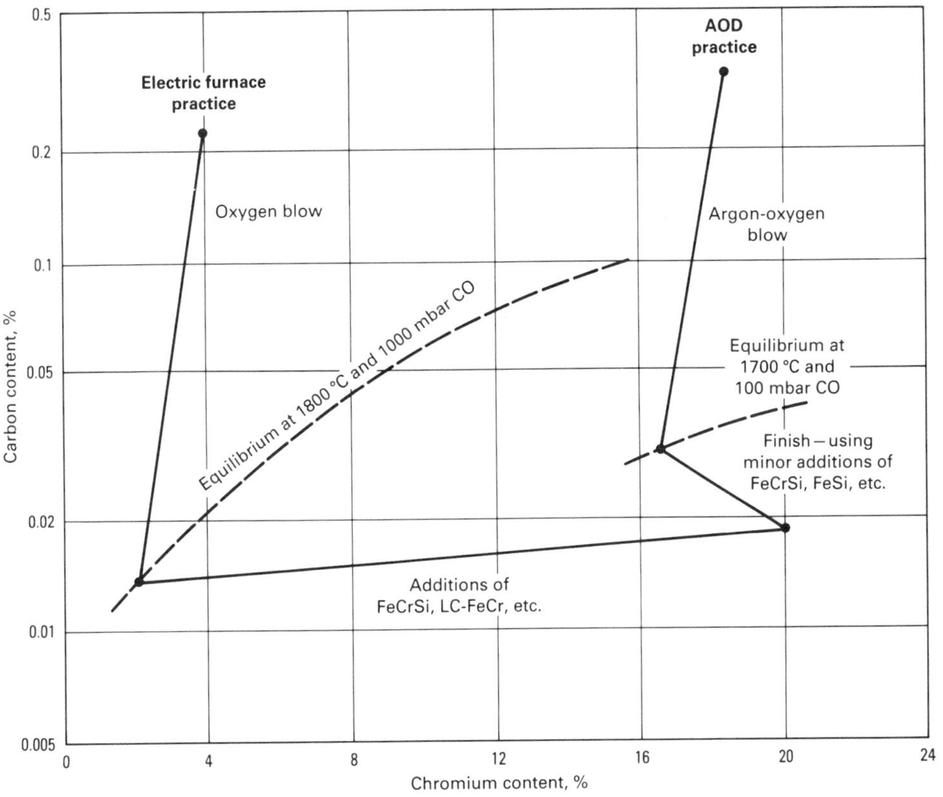

Fig. 2 Composition changes in refining type 304-L stainless steel using electric arc furnace practice and argon oxygen decarburization

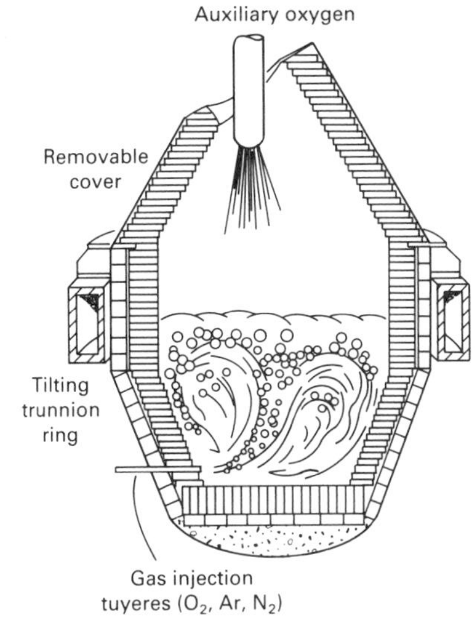

Fig. 3 Schematic of argon oxygen decarburization vessel

In refining stainless steel, it is generally necessary to decarburize the molten bath to less than 0.05% C. Chromium is quite susceptible to oxidation; therefore, prior to the introduction of the AOD process, decarburization was accomplished by withholding most of the chromium until the bath had been decarburized by oxygen lancing. After the bath was fully decarburized, low-carbon ferrochromium and other low-carbon ferroalloys were added to the melt to meet chemical specifications.

Dilution of the partial pressure of CO allows the removal of carbon to low levels without excessive chromium oxidation. This practice enables the use of high-carbon ferroalloys in the charge mix, avoiding the substantially more expensive low-carbon ferroalloys. Figure 2 compares the refining steps in the two processes.

Equally important, however, are the processing advantages offered by submerged blowing. These include excellent slag/metal and gas/metal contact, superior decarburization kinetics, and 100% utilization of the injected oxygen by reaction with the bath. In addition, submerged injection allows the accurate control of end point nitrogen and the removal of dissolved gases (nitrogen and hydrogen) and nonmetallic inclusions.

Equipment

The processing vessel consists of a refractory-lined steel shell mounted on a tiltable trunnion ring (Fig. 3). With a removable, conical cover in place, the vessel outline is sometimes described as pear shaped. Several basic refractory types and various quality levels of the refractories have gained widespread acceptance (Ref 4, 5). Dolomite refractories are used in most AOD installations; magnesite chromium refractories are predominant in small (<9 Mg, or 10 ton) installations and are used almost exclusively in Japan.

As seen in Fig. 3, process gases are injected through submerged, side-mounted tuyeres. The number and relative positioning of tuyeres are determined in part by vessel size, range of heat sizes, process gas flow rates, and the types of alloys refined. Process gases are oxygen, nitrogen, argon, and in some cases carbon dioxide (CO_2). The most recent AOD installations include the use of top-blown oxygen (one-third of all AOD installations have this capability). Oxygen top- and bottom-blowing converters are described in the discussion "Oxygen Top and Bottom Blowing (OTB)" in this section.

The specific volume of AOD vessels ranges from 0.4 to 0.8 m^3/Mg (16 to 32 ft^3/ton). The gas control system supplies the process gases at nominal rates of 0.5 to 3 m^3/min/Mg (15 000 to 6000 ft^3/h/ton). The system accurately controls the flow rates and monitors the amount of gas injected into the bath to enable the operator to control the process and keep track of the total oxygen injected.

Normally, a shop has three interchangeable vessels. At any given time, one of the vessels is in a tiltable trunnion ring refining steel, a second vessel is at a preheating station, and the third vessel is being relined. The vessel in the trunnion ring can be replaced with a preheated vessel in less than 1 h.

The control of process gases, vessel activities, and ancillary equipment can range from manual to fully automated. Most installations are equipped with a computer to assist in process control by calculating the required amount of oxygen as well as alloying additions. Some installations have computer control systems capable of sending set points and flow rates to the gas control systems.

Processing

Stainless Steels. Charge materials (scrap and ferroalloys) are melted in the melting furnace. The charge is usually melted with the chromium, nickel, and manganese concentrations at midrange specifications. The carbon content at meltdown can vary from 0.50 to 3.0%, depending on the scrap content of the charge. Once the charge is melted down, the heat is tapped, and the slag is removed and weighed prior to charging the AOD vessel.

In the refining of stainless steel grades, oxygen and inert gas are injected into the bath in a stepwise manner. The ratio of oxygen to inert gas injected decreases (3:1, 1:1, 1:3) as the carbon level decreases. Once the aim carbon level is obtained, a reduction mix (silicon, aluminum, and lime) is added. If extra-low sulfur levels are desired, a second desulfurization can be added. Both of these steps are followed by an argon stir. After reduction, a complete chemistry sample is usually taken and trim

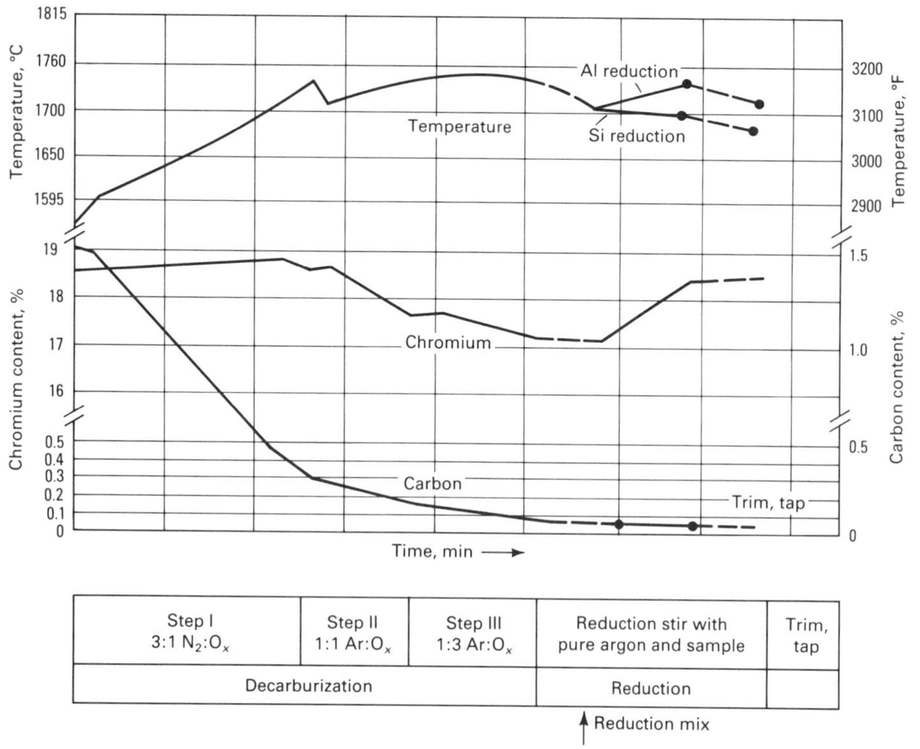

Fig. 4 Schematic of stainless steel refining cycle for small (<27 Mg, or 30 ton) vessels showing the relationship between carbon and chromium contents as a function of time and temperature

additions made following analysis. Figure 4 illustrates the relationships among carbon, chromium, temperature, and the various processing steps for refining a typical type 304 stainless steel using argon oxygen decarburization.

Another critical aspect of AOD refining is the ability to predict when to change from nitrogen to argon to obtain the aim nitrogen specification. Table 1 lists the process parameters, carbon removal efficiencies, and carbon removal rates for the AOD refining of type 304 stainless steel. The point during refining when the oxygen-to-inert-gas ratio is lowered is based on carbon content and temperature. The ratios and carbon switch points are designed to provide optimum carbon removal efficiency without exceeding a bath temperature of 1730 °C (3150 °F).

Carbon and Low-Alloy Steels. The refining of carbon and low-alloy steels involves a two-step practice: a carbon removal step, followed by a reduction/heating step. The lower alloy content of these steels eliminates the need for injecting less than a 3:1 ratio of oxygen to inert gas. Once the aim carbon level is obtained, carbon steels are processed similarly to stainless steels. Figure 5 illustrates carbon content and temperature relationship for the AOD refining of carbon and low-alloy steels. Because the alloy content of these grades of steel is substantially lower than that of stainless steel and because the final carbon levels are generally higher, there is no thermodynamic or practical reason for using an oxygen, inert-gas ratio of less than 3:1.

Oxidation measurements indicate that all of the oxygen reacts with the bath and that none leaves the vessel unreacted. By monitoring and recording the oxygen consumption during refining, very close control of end point carbon is achieved. Because the oxygen and inert gases are introduced below the bath and at sonic velocities, there is excellent bath mixing and intimate slag/metal contact. As a result, the reaction kinetics of all chemical processes that take place within the vessel are greatly improved.

Decarburization. In both stainless and low-alloy steels, the dilution of oxygen with inert gas results in increased carbon removal efficiencies without excessive metallic oxidation. In stainless grades, carbon levels of 0.01% are readily obtained.

Chemistry Control. The excellent compositional control of AOD-refined steel is indicated in Table 2 for a ten-heat series of high-strength low-alloy steel. The injection of a known quantity of oxygen with a predetermined bath weight enables the steelmaker to obtain very tight chemical specifications.

Desulfurization. Sulfur levels of 0.01% or less are routinely achieved, and levels less than 0.005% can be achieved with single slag practice. When extra-low sulfur levels are required, a separate slag treatment for 3 min is sufficient.

Slag Reduction. During oxygen injection for carbon removal, there is some metallic oxidation. Efficient slag reduction with stoichiometric amounts of silicon or aluminum permits overall recoveries of 97 to 100% for most metallic elements. Chromium recovery averages approximately 97.5%, and the recovery of nickel and molybdenum are approximately 100%.

Nitrogen Control. Degassing in argon oxygen decarburization is achieved by inert gas sparging. Each argon and CO bubble leaving the bath removes a small amount of dissolved nitrogen and hydrogen. Final nitrogen content can be accurately controlled by substituting nitrogen for argon during refining. Nitrogen levels as low as 25 to 30 ppm can be obtained in carbon and low-alloy steels, and 100 to 150 ppm N can be obtained in stainless steels. The ability to obtain aim nitrogen levels substantially reduces the need to use nitrided ferroalloys for alloy specification, and this also minimizes the use of argon. Hydrogen levels as low as 1.5 ppm can be obtained.

Property Improvements. It has been well documented that AOD-refined steels exhibit significantly improved ductility and toughness, along with impact energy increases of over 50% (Ref 6-8). These improved properties result from a decrease in the number and size of inclusions. The capability to produce low gas content steel with exceptional microcleanliness, along with alloy savings, is the primary factor for the growth of the AOD process for refining stainless, carbon, and low-alloy steels.

Table 1 Processing parameters for AOD refining of type 304 stainless steel

Total gas flow rate: ~1.5 m^3/min/ton (3500 ft^3/h/ton)

Carbon level, %	Gas ratios			Partial pressures, atm			CRE, %(a)	CRR, ppm/min(b)	Temperature	
	O_2	N_2	Ar	Ar	N_2	CO			°C	°F
3.0–0.7	4	1	...	...	0.14	0.86	80	1200	1510	2750
1–0.25	3	1	...	...	0.20	0.80	65	830	1700	3100
0.25–0.12	1	1	...	...	0.53	0.47	45	430	1700	3100
0.12–0.04	1	...	3	0.83	...	0.17	30	133	1730	3150
0.04–0.01	1	...	8	0.96	...	0.04	15	26	1730	3150

(a) CRE, carbon removal efficiency. (b) CRR, carbon removal rate

Oxygen Top and Bottom Blowing

Oxygen top and bottom blowing (OTB), also referred to as combined blowing, is an extension of AOD technology for refining steel. During OTB, oxygen is injected into the molten steel through a top lance during

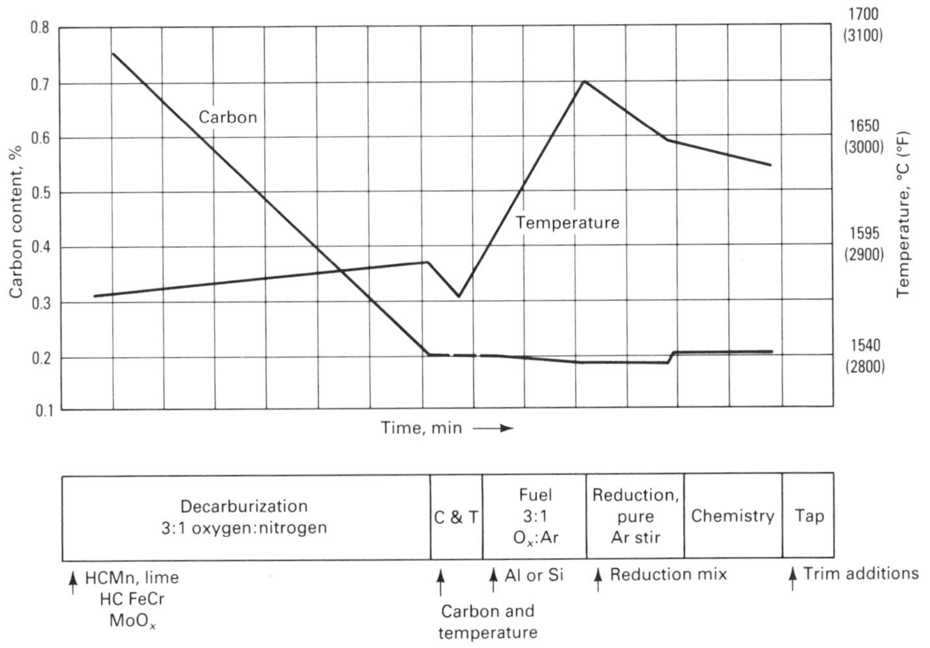

Fig. 5 Schematic of carbon and low-alloy steel argon oxygen decarburization refining

Table 2 Composition control of AOD-refined high-strength low-alloy steel

Element	Aim	Mean	Standard deviation
Carbon	0.27	0.264	0.006
Manganese	0.80	0.83	0.03
Silicon	0.50	0.49	0.04
Phosphorus	...	0.019	0.002
Sulfur	...	0.002	0.001
Molybdenum	0.28	0.28	0.01
Chromium	0.50	0.56	0.07
Nickel	...	0.13	0.03
Aluminum	0.05	0.043	0.009

carbon removal, and inert gases, such as argon, nitrogen, or carbon dioxide, are injected with oxygen through submerged tuyeres or alternate forms of gas injectors such as canned bricks, porous bricks, or thin pipes set in the refractory brick. The top-blown oxygen can react with the bath, reducing refining times, or with CO. The combustion of CO above the surface of the bath increases the thermal efficiency of the refining process and decreases the quantity of silicon and aluminum required for reduction of metallic oxides.

If the top-blown oxygen system is designed so that more than 65% of the oxygen reacts with the bath, the system is referred to as hard blown; if less than 65% of the oxygen reacts with the bath, it is referred to as soft blown. As a general guideline, a top-blown oxygen system installed in AOD vessels with a nominal capacity greater than 45 Mg (50 tons) will be hard blown, smaller vessels will be soft blown. In hard-blown top oxygen systems, the flow rate of oxygen through the top lance will be between 50 and 150% of the oxygen injected through tuyeres or injectors. In soft-blown systems, the top oxygen flow rate is between 50 and 100% of the oxygen flow through the tuyeres or injectors.

Because foundry AOD vessels have a nominal capacity less than 45 Mg (50 tons) the top-blown oxygen systems are designed to be soft blown. This design maximizes the thermal efficiency of the smaller AOD vessels, reducing the quantity of fuel and reduction material required during refining. Top-blown oxygen flow rates in foundry AOD vessels range from 50% of the bottom oxygen flow rate to 120%.

More detailed information on OTB technology can be found in Ref 9 and 10. Figures 3 and 4 in the following section "Vacuum Oxygen Decarburization" compare OTB with various refining processes.

REFERENCES

1. F.D. Richardson and W.E. Dennis, Effect of Chromium on the Thermodynamic Activity of Carbon in Liquid Iron-Chromium-Carbon Metals, *J. Iron Steel Inst.*, Nov 1953, p 257-263
2. R.J. Choulet, F.S. Death, and R.N. Dokken, Argon-Oxygen Refining of Stainless Steel, *Can. Metall. Q.*, Vol 10 (No. 2), 1971, p 129-136
3. R.B. Aucott, D.W. Gray, and C.G. Holland, The Theory and Practice of the Argon-Oxygen Decarburizing Process, *J. W. Scot. Iron Steel Inst.*, Vol 79 (No. 5), 1971-1972, p 98-127
4. D.A. Whitworth, F.D. Jackson, and F.F. Patrick, Fused Basic Refractories in the Argon-Oxygen Decarburization Process, *Ceram. Bull.*, Vol 53 (No. 11), 1974
5. D. Brosnan and R.J. Marr, "The Use of Direct Burned Dolomite Brick in the AOD Vessel," Paper presented at the Electric Furnace Conference, Detroit, MI, Iron and Steel Society, Dec 1977
6. P.A. Tichauer and L.J. Venne, AOD, a New Process for Steel Foundries, Promises Better Properties, Less Repair, *33 Magazine*, April 1977, p 35-38
7. D. Heckel, J.P. Wiencek, and N. Netoskie, "Metallurgical Aspects of the AOD," Paper presented at the Electric Furnace Conference, Detroit, MI, Iron and Steel Society, Dec 1977
8. F.J. Andreini and S.K. Mehlman, "Quality Aspects of Foundry AOD," Paper presented at the Electric Furnace Conference, Detroit, MI, Iron and Steel Society, Dec 1979
9. S.K. Mehlman, Ed., *Mixed Gas Blowing*, Proceedings of the Fourth Process Technology Conference, Iron and Steel Society of AIME, American Institute of Mechanical Engineers, 1984
10. L.G. Kuhn, Ed., Mixed Gas Blowing—A New Era of Pneumatic Steelmaking, Issue of *Iron and Steel Maker*, Aug 1983

Vacuum Oxygen Decarburization (VODC)

Wilhelm Burgmann, Leybold AG, West Germany

Vacuum oxygen decarburization can be carried out in a converter vessel (VODC) or in the ladle (VOD). The parameters that favor the choice of ladles as reaction vessels include the following:

- Tapping, treatment, and teeming are done in the same reaction vessel. Thus, there are no temperature losses due to any final transfer of the melt, and the high level of cleanliness achieved during the treatment can be preserved up to teeming
- The use of electric power as an inexpensive energy source permits the highest flexibility in the melt shop. The ladle unit can act as a time buffer between the melting unit and the casting stand

Information on VOD ladle technology can be found in the article "Degassing Processes (Ladle Metallurgy)" in this Volume.

The primary reasons for selecting the converter as the VOD reaction vessel are as follows:

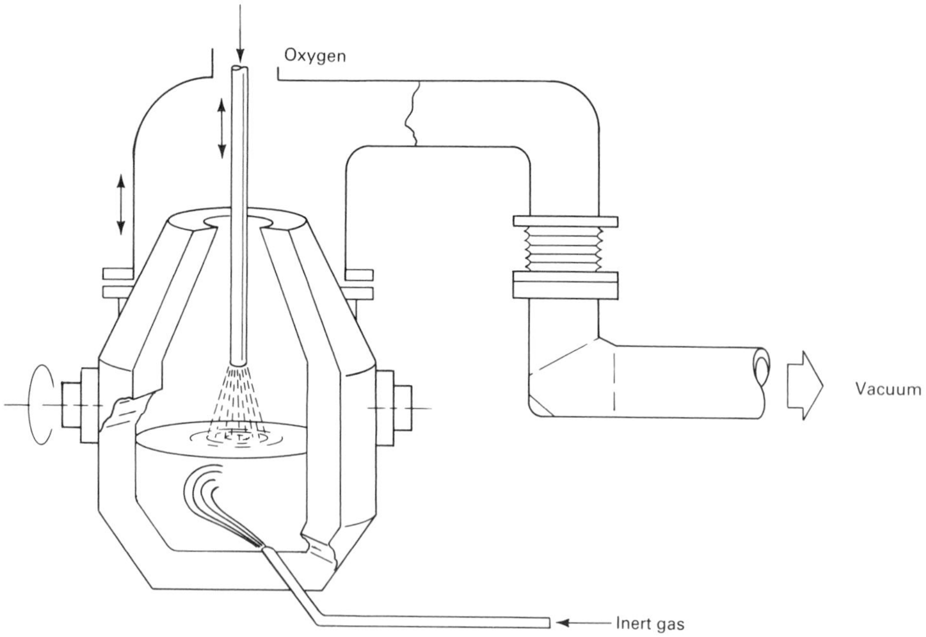

Fig. 1 Schematic layout of a vacuum oxygen decarburization converter

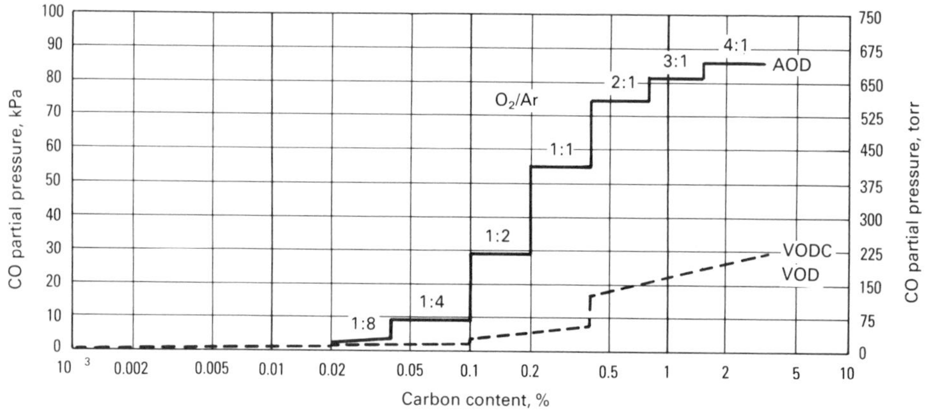

Fig. 2 Comparison of CO partial pressure for the AOD, VODC, and VOD processes

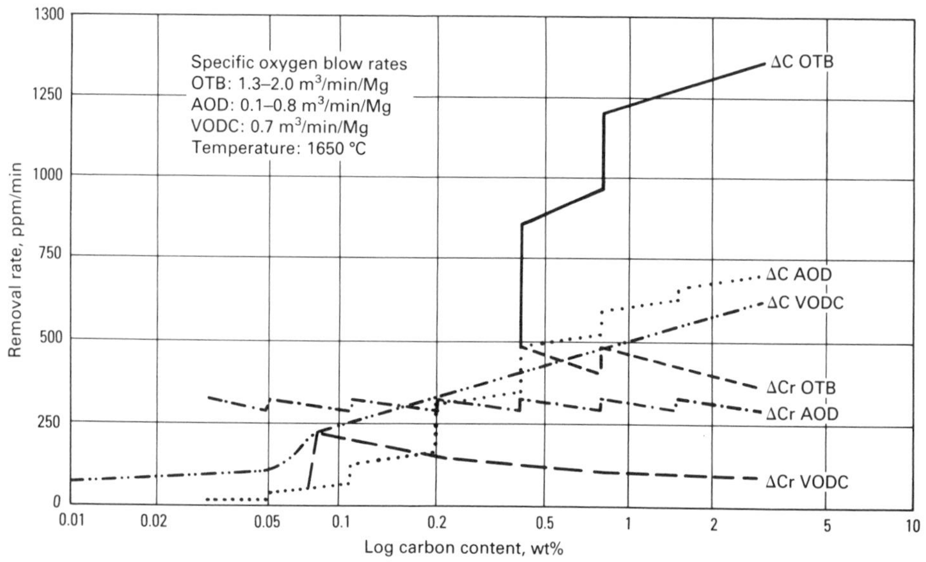

Fig. 3 Carbon removal rates and chromium oxidation rates for the VODC, OTB, and AOD processes

- Initial carbon contents are as high as possible, together with high oxygen blow rates
- Ease of deslagging and strong stirring action
- Shop restrictions such as a limited number of ladles, prohibited use of slide gates, and restricted crane capacity in the casting bay make converter technology more attractive

This section will review both the equipment and the processing parameters for vacuum oxygen decarburization, which is used for the production of low-alloy steels, stainless steels, and superalloys.

Equipment and Processing Parameters

Equipment. Vacuum oxygen decarburization converters are similar to argon oxygen decarburization (AOD) and oxygen top blown (OTB) converters in terms of design and the tilting device used. (AOD and OTB converters were discussed in the previous section "Argon Oxygen Decarburization (AOD)" in this article.) Bottom blowing is, however, restricted to the introduction of small amounts of inert gas through simple pipes, thus avoiding the special erosion-resistant refractory material used around tuyeres. Flue gas handling is easier and is incorporated into the vacuum system. In terms of vessel design, the conical converter top is closed by a vacuum hood with an oxygen lance feed through and vacuum addition lock (Fig. 1).

Because the VODC system is closed and no air enters the vessel, permanent control of the decarburization rate and the carbon level in the bath can be maintained and monitored with a flue gas analyzing device (Ref 1, 2). Pollution control for carbon monoxide (CO) and dust is also incorporated into the system.

Processing Parameters. Figures 2 and 3 show the relationship among CO partial pressure, gas blow rates, and carbon and chromium oxidation rates for the VODC, VOD, AOD, and OTB processes. As shown, low chromium losses and high carbon removal rates are possible using vacuum oxygen decarburization converters. The high carbon removal rates are due to the higher oxygen yield of the vacuum oxygen decarburization converter and the fact that oxygen dilution is not required. The amount of bottom blown inert gas is only about 2 to 5% of the oxygen volume.

With a specific oxygen blow rate of 0.7 m^3/min/Mg (~25 ft^3/min/ton), the decarburization rate exceeds 600 ppm/min at high carbon levels (Fig. 3). In smaller VODC heats, higher blow rates of the order of 10 m^3/min (~350 ft^3/min) are maintained, leading to a carbon removal of 1500 ppm/min in 5 Mg (5.5 ton) units.

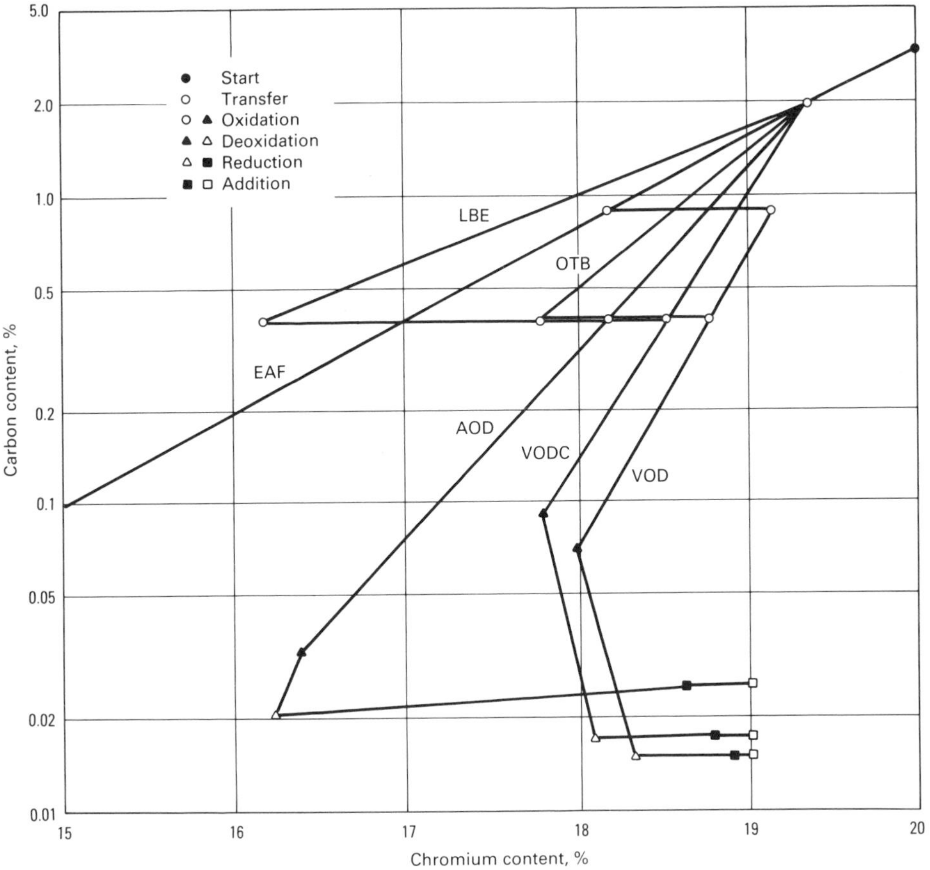

Technique	Total gas flow m³/min/Mg	Total gas flow ft³/min/ton	Inert gas, %
Electric arc furnace (EAF)	1.2	46	0
Lance bubbling equilibrium (LBE)	2.5	95	1
Oxygen top blowing (OTB)	2.0	77	10
Argon oxygen decarburization (AOD)	1.0	40	20–90
Vacuum oxygen decarburization converter (VODC)	0.7	27	2–5
Vacuum oxygen decarburization (VOD-ladle)	0.4	15	0.5–4

Fig. 4 Processing steps for achieving proper chromium and carbon levels in stainless steel using various refining techniques with varying gas blow rates and inert gas additions

Figures 2 and 3 demonstrate clearly the low chromium losses and the high ratio of carbon-to-chromium removal rate due to the very low CO partial pressure achieved in the VODC process. As in any vacuum process, the lowest carbon levels (from 50 to 500 ppm) are reached in stainless steel heats very quickly (6 to 10 times the carbon removal rate of AOD) without applying excessive heat and excessive amounts of inert gas. As shown in Fig. 3, temperatures of about 1650 °C (3000 °F) are common. Inert gas volumes are generally less than about 40 L/min/Mg (1.5 ft^3/min/ton).

Figure 4 shows the relationship between the oxygen blow rates and the inert gas addition for various refining techniques, as well as the resulting chromium and carbon levels achieved at specific process steps. Figure 4 also shows the extent to which decarburization is accompanied by chromium oxidation. Carbon and nitrogen levels below 100 ppm are achieved with corresponding chromium yields of over 95% before reduction and 98% after reduction (Ref 3, 4). Sulfur levels are at 10 to 30 ppm for any grade of steel without any sacrifice in the degassing effect. Hydrogen levels are below 2 ppm.

The operating conditions necessary for achieving low levels of sulfur and oxygen are the same as those in all strong stirring converter processes (AOD, OTB, VODC). A disadvantage of the vacuum oxygen decarburization converter, as with all converter processes, is that gas pickup of the melt is encountered during tapping into the teeming ladle. Optimum cleanliness and castability therefore require a subsequent ladle treatment, as described in the article "Degassing Processes (Ladle Metallurgy)" in this Volume.

REFERENCES

1. R. Heinke and S. Köhle, Computer Control of the VOD Process, in *Proceedings of the Seventh International Conference on Vacuum Metallurgy* (Tokyo), Section 34-2, Iron and Steel Institute of Japan, 1982, p 1356-1363
2. L. Tolnay, I. Sziklavari, and G. Karoly, Method of Regulating the Endpoint of the Vacuum Decarburization Process Developed for the Production of Acid Resistant Steel Grades With Extra Low Carbon Content in ASEA-SKF Ladle Metallurgy Plant, in *Proceedings of the Eighth International Vacuum Metallurgy Conference* (Linz), M1.3, Eisenhütte Osterreich (Vienna), 1985, p 606-615
3. P. Tennilae, M. Kaivola, and M. Walter, Systems Technology and Operating Results of VODC Converter, in *Proceedings of the First Electr. Steel Congress* (Aachen), F4, Verein Deutscher Eisenhütterleute (VDEh, Düsseldorf), 1983, p 1-12
4. W. Burgmann, O. Wiessner, and G. Reese, Vacuum Converter Technology for the Production of Superalloys, Stainless and Low Alloy Steels, in *Proceedings of the Vacuum Metallurgy Conference on Special Metals and Melting Process* (Pittsburgh, PA), Iron and Steel Society, 1985, p 3-10

Degassing Processes (Ladle Metallurgy)

Chairman: Gerhard Kienel, Leybold AG, West Germany

Vacuum Ladle Degassing (VD)

Wilhelm Burgmann, Leybold AG, West Germany

THE DEVELOPMENT OF ARGON INJECTION devices and the progress in ladle metallurgy regarding porous plugs, tuyeres, refractory material, heating technologies, slide gates, and so on, have significantly contributed to static ladle vacuum degassing technology. This section will review the equipment associated with vacuum ladle treatment, along with some processing results.

Equipment and Processing

Two basic setups can be utilized during ladle degassing in vacuum. The ladle, containing argon injection or induction coils, can be placed into an independent vacuum vessel (Ref 1), or if argon injection/induction coils are needed to stir the melt (Fig. 1), a vacuum ladle with a directly applied vacuum hood can be used. The vacuum vessel can be either stationary with a vacuum connection (Fig. 2a) or movable on a buggy (Fig. 2b).

The vessel is equipped with a ladle stool for quick and safe placement of the ladle into the vessel. The vessel also has an inert gas connecting device for the ladle. The ladle has the required freeboard (Fig. 3) and is covered by a ladle hood, which retains metal splashes and heat. There is enough space left below the ladle bottom to receive the molten metal in the event of a breakthrough of the ladle.

The vacuum vessel cover is equipped with the necessary control devices for adding slag constituents and alloying and cooling material and for conducting temperature measurement, sampling, and observation to ensure uninterrupted molten metal processing. For optimum metallurgical results, all of these activities must be executed without breaking the vacuum. Devices are also available for the addition of bulky cooling material and for the injection of aluminum, calcium, calcium-silicon, graphite, and ferroalloys.

Vacuum degassing is used to melt heats ranging in weight from 15 to 350 Mg (16.5 to 390 tons). Its success depends on treatment time, which in turn depends on the previous soaking of the ladle lining and the surface-to-volume ratio of the melt. The mean temperature decrease during vacuum treatment varies from 6 °C/min (11 °F/min) for 15 Mg (16.5 ton) heats to 0.7 °C/min (1 °F/min) for 300 Mg (330 ton) heats

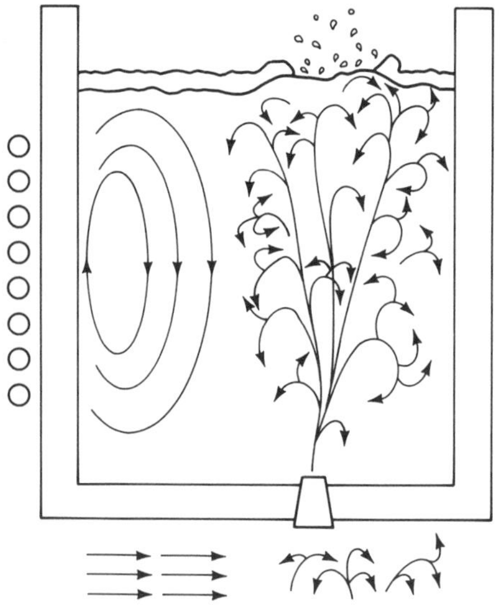

Fig. 1 Flow pattern of inductive and inert gas injection stirring in ladles. Left: regular flow (no particle agglomeration). Right: irregular flow (increased particle agglomeration)

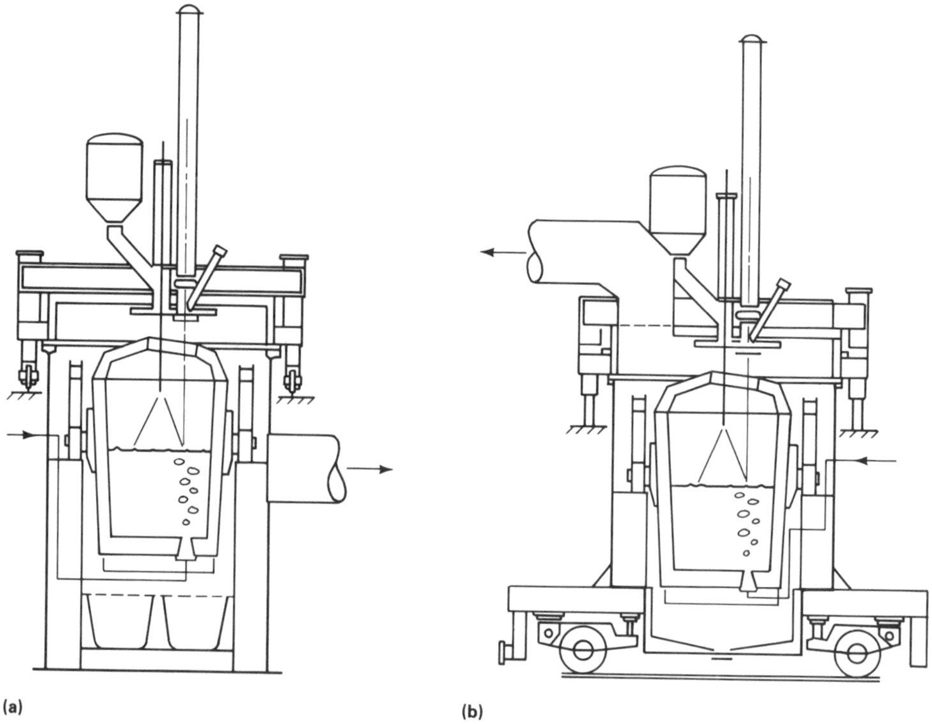

Fig. 2 Schematics of stationary (a) and movable (b) vacuum vessels

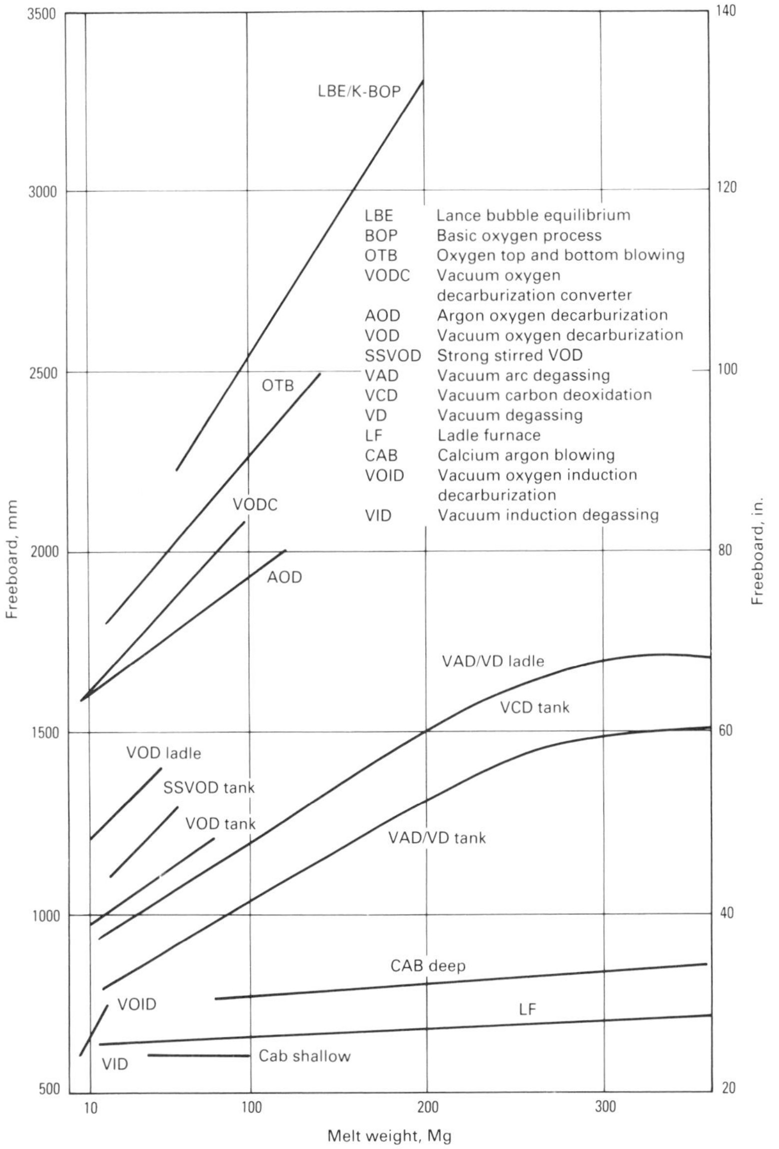

Fig. 3 Freeboard requirements in secondary metallurgy

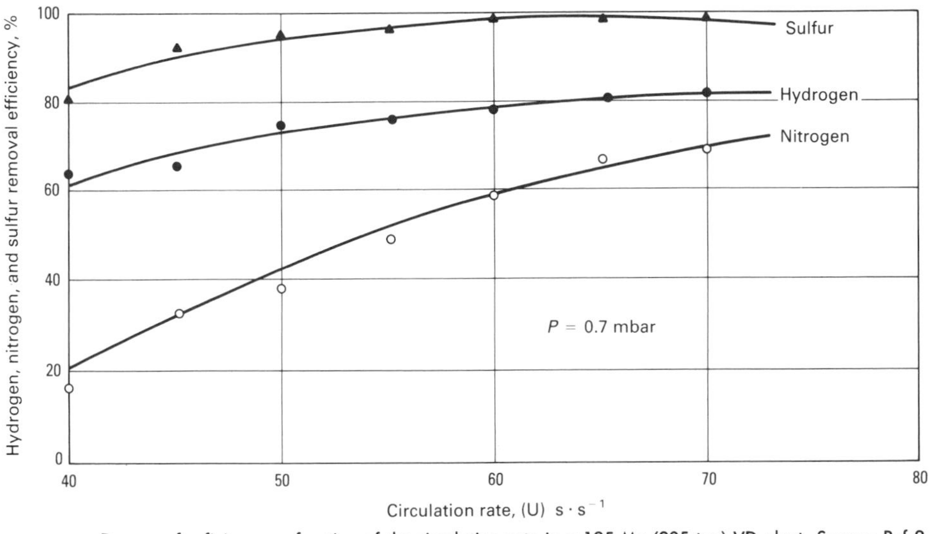

Fig. 4 Degree of refining as a function of the circulation rate in a 185 Mg (205 ton) VD plant. Source: Ref 2, 3

during the time the ladle lining is in thermal equilibrium.

It has not been possible to achieve satisfactory degassing results with the presence of slag. However, this has changed because of:

- The progress made in the fabrication of basic linings and in the corresponding lining and heating technology
- The availability of low-silica active slag mixtures
- The development of vacuum technology that prohibits foaming of the slag and permits the injection of large quantities of inert gas up to 10 L/min/Mg (0.3 ft^3/min/ton) to optimize stirring

Such improvements have led to simultaneous desulfurization rates of 95%, hydrogen removal of over 85%, nitrogen removal of 60%, and aluminum contents below 200 ppm. Figure 4 shows the removal of sulfur, hydrogen, and nitrogen as a function of gas circulation rate for a 180 Mg (200 ton) VD treatment. Vacuum treatment results for several low-alloy steel heats are given in Table 1.

Vacuum Degassing Procedures. There are two vacuum degassing stirring procedures. In the first, hydrogen, nitrogen, and sulfur are eliminated by strong stirring, that is, with 5 to 10 L/min/Mg (0.15 to 0.3 ft^3/min/ton). In this case, the oxygen level is decreased only to 30 to 50 ppm total oxygen, with the major portion of this being the suspended oxidic inclusions. In the second procedure, the total oxygen level is reduced to 10 to 25 ppm by a soft stirring phase in combination with a previously conducted aluminum deoxidation procedure. This requires a very low stirring rate, a strongly basic slag, and sufficient temperature reserve for an extended treatment. If the thermal reserve is not adequate, an additional heat supply, for example, in the form of electric arcing at low voltage, is necessary. Additional heating by electric arcing is described later in this article (see the sections "Ladle Furnace and Vacuum Arc Degassing" and "Plasma Heating and Degassing").

The distribution of stirring energy should be uniform. For strong stirring, inductive stirring combined with inert gas stirring is required. Inductive stirring alone is sufficient for gentle stirring if the flow pattern is irregular enough to create turbulences that will produce inclusion agglomeration. The differences in the flow patterns created by inductive and inert gas stirring are shown in Fig. 1.

REFERENCES

1. R. Schumann and W. Schwarz, Ladle Metallurgy Comparison of Vacuum La-

dles and Vacuum Vessels, *Metall. Plant Technol.*, Vol 4, 1985, p 10-18
2. K.H. Bauer and H. Wagner, Operational Experience With a Vacuum Ladle Degassing in a BOF Shop, *Stahl Eisen*, Vol 107, 1987, p 426-430
3. T. Hsiao, T. Lehner, and B. Kjellberg, *Scand. J. Met.*, Vol 9 (No. 3), 1980, p 10

Table 1 Parameters and results of vacuum ladle degassing of low-alloy steel

EAF, electric arc furnace; see Fig. 3 for explanations of other abbreviations.

Parameter	EAF/VD, 35–40 Mg (38–44 ton)	LF/VD, 40–42 Mg (44–46 ton)	LF/VD, 75–80 Mg (82–88 ton)	BOP/VD, 180–190 Mg (200–210 ton)
Initial hydrogen, ppm	3–14	5–10	5–6	3–7
Final total hydrogen, ppm	0.4–1.0	0.6–1.5	1.0–1.5	0.9–1.6
Initial nitrogen, ppm	100–150	120–150	80–120	50–70
Final nitrogen, ppm	30–60	50–100	45–90	25–35
Oxygen activity, ppm	5–10			1–2
Final total oxygen, ppm	20–25	15–25	20–40	10–20
Total aluminum, ppm	300–450	200–400	100–300	100–400
Dissolved calcium, ppm	. . .	. . .	. . .	20–30
Initial sulfur, ppm	100–200	300–400	130–200	150–200
Final sulfur, ppm	40–50	100–200	20–30	5–10
Temperature loss, °C (°F)	50–60 (90–110)	30–40 (55–70)	30–40 (55–70)	35–50 (65–90)
Alloying additions, kg/Mg (lb/ton)	2–6 (4–12)	0–0.5 (0–1.0)	0.5–1.0 (1.0–2.0)	1–3 (2–6)
Slag additions, kg/Mg (lb/ton)	20–25 (40–50)	0 (0)	1 (2)	7–9 (14–18)
Total amount of slag, kg/Mg (lb/ton)	20–25 (40–50)	20–25 (40–50)	18 (30)	15–22 (30–44)
Argon volume, m^3/Mg (ft^3/ton)	0.03–0.05 (0.95–1.6)	0.02–0.03 (0.64–0.95)	0.15–0.18 (4.8–5.8)	0.12–0.15 (3.9–4.8)
Duration of treatment, min	12–16	15–16	20–30	25–30

Vacuum Oxygen Decarburization (VOD)

Wilhelm Burgmann, Leybold AG, West Germany

Vacuum oxygen decarburization differs from vacuum degassing, which is conducted under reducing conditions, in that decarburization is an oxidizing process that not only increases dissolved oxygen content in the melt by solid oxygen carriers but also introduces gaseous oxygen through a top-blowing lance that permits higher decarburization rates. The VOD process is primarily used for decarburizing stainless steel melts or other high-chromium bearing alloys; however, a VOD plant can be used for all types of steel grades and alloys when decarburization, degassing, desulfurization, or chemical heating with oxygen is required (Ref 1-3). Chemical heating with oxygen is also called vacuum oxygen heating.

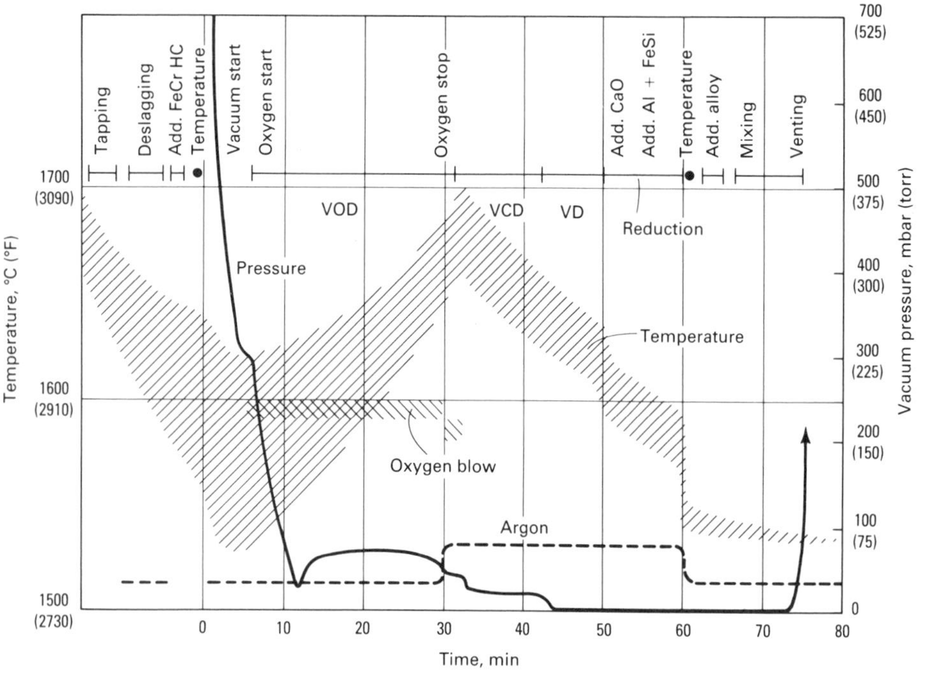

Fig. 1 Typical sequence of operations during VOD ladle treatment

The VOD process is based on the fact that decarburization is strongly favored under reduced pressure with respect to other oxidizing reactions. Thus, chromium is protected against oxidation, as discussed in the section "Argon Oxygen Decarburization" of the article "Degassing Processes (Converter Metallurgy)" in this Volume. This applies to a limited extent to manganese and silicon if the oxidation rates are kept sufficiently low.

Equipment and Processing

A modern VOD plant for heats ranging in weight from 15 to 90 Mg (16.5 to 100 tons) consists of the following equipment:

- Basic-lined ladle with a freeboard of 1 to 1.5 m (3 to 5 ft), a porous plug in the bottom for inert gas injection, a slide gate, and refractory-lined ladle hood tightly sealed on the ladle
- Vacuum vessel with ladle stool, an inner lining, a safety basin for receiving a metal breakthrough, a vacuum connection, and a vessel cover with accessories and water-cooled splash protection
- Top-blowing oxygen lance
- Probing device for in-process temperature measurement and sampling
- Movable guide pipe for feeding either aluminum wire or wire filled with such materials as calcium, calcium-silicon, titanium, graphite, and fluxes
- Observation port with television camera
- Vacuum pump set (steam ejectors, water ring pump, or mechanical pumps) with a suction capacity such that during oxygen blowing the vacuum is held below 10 kPa (100 mbar, or 75 torr), and oxidizing as well as reducing vacuum treatments can be performed below 0.1 to 1 kPa (1 to 10 mbar, or 0.75 to 7.5 torr). The pump must also be adjusted to handle carbon monoxide gas safely and to remain operational under a heavy dust load of the carried gas of up to 0.4 kg/m^3 (0.4 oz/ft^3).

Oxygen blow rates vary from 5 to 25 m^3/min (175 to 900 ft^3/min), depending on heat size and initial carbon content. The specific blow rate ranges from 0.2 to 1.4 m^3 O_2/min/Mg (6 to 45 ft^3 O_2/min/ton); the highest

Table 1 Process parameters and metallurgical results in VOD ladle plants

Parameter	EAF/VOD, 15 Mg (16.5 ton)	EAF/VOD, 27 Mg (30 ton)	EAF/VOD, 30 Mg (33 ton)	LF/VOD, 40 Mg (44 ton)	EAF-AOD-VOD 65 Mg (72 ton)	EAF-AOD-VOD 65 Mg (72 ton)
Chromium, %	18	18	18	12	18	17
Nickel, %	10	...	11	...	10	...
Molybdenum, %	...	2	2	...	...	...
Initial carbon, %	1.1–1.9	0.6–1.1	0.8–1.0	0.5–0.6	0.6–0.8	0.7–0.9
Final carbon, ppm	200–600	80–180	300–400	50–100	10–20	10–30
Initial nitrogen, ppm	...	300–430	...	150–200	400–800	300–600
Final nitrogen, ppm	...	40–80	140–240	50–85	50–70	50–80
Chromium yield, %	...	97.6	96.6	96.3	99.2	99.0
Titanium yield, %	...	40–60	74.5	...	...	...
Manganese yield, %	...	60–80	85.0	...	70.0	80.0
Oxygen blow rate, m^3/min (ft^3/min)	5–8 (175–280)	5–14 (175–495)	5–10 (175–355)	5–15 (175–530)	10–15 (355–530)	10–16 (355–565)
Specific oxygen rate, m^3/min/Mg (ft^3/min/ton)	0.53 (17)	0.52 (16.7)	0.34 (11)	0.37 (12)	0.23 (7.5)	0.25 (8.0)
Inert gas rate, m^3/min (ft^3/min)	0.03 (1.0)	0.05–0.25 (1.75–9.0)	0.05–0.10 (1.75–3.5)	0.1–0.2 (3.5–7.0)	0.3–0.6 (11–21)	0.4–0.7 (14–25)
Ar/O_2 ratio, %	0.5	0.3–5.0	0.5–2.0	0.7–4.0	2.0–6.0	2.5–7.0
Ladle freeboard, m(ft)	0.8–1.0 (2.5–3.0)	1.0–1.2 (3.0–4.0)	1.2–1.5 (4.0–5.0)	1.5 (5.0)	1.6 (5.2)	1.6 (5.2)
Slag addition, kg/Mg (lb/ton)	20–25 (40–50)	10–12 (20–24)	30–40 (60–80)	10 (20)	25 (50)	25 (50)
Alloy addition, kg/Mg (lb/ton)	12–20 (24–40)	10–15 (20–30)	15–25 (30–50)	5–8 (10–16)	10–12 (20–24)	7–10 (14–20)

specific blow rates are achieved with the smallest heat sizes.

The inert gas blow rates vary strongly with the process and the desired results; that is, absolute rates range from 0.1 to 1.2 m^3/min (3.5 to 45 ft^3/min), and specific rates are 2 to 7% of oxygen flow.

According to equilibrium data for the Fe-Cr-C-O system, the lower the chromium losses, the lower the partial pressure of the formed $CO + CO_2$ (equilibrium conditions for chromium-containing steels are described in the section "Argon Oxygen Decarburization" of the article "Degassing Processes (Converter Metallurgy)" in this Volume). However, the main oxygen-blowing period of the VOD process is performed at an intermediate vacuum pressure range of 3 to 8 kPa (30 to 80 mbar, or 20 to 60 torr), which is governed by the suction capacity of the installed vacuum pump and the freeboard available in the ladle. The chromium loss at given carbon levels and temperatures depends on the prevailing kinetic conditions at the blowing spot. With an intense inert gas stirring from the ladle bottom, the chromium loss can be reduced to nearly zero by the strong stirring VOD process, as described in Ref 4.

In most VOD plants, the chromium yield at the end of blowing is 90 to 95% and is increased to 98 to 99% during the subsequent reducing phase by the combination of low vacuum pressure and the addition of reducing agents such as aluminum and silicon. The yield of chromium also depends on the initial and the final carbon contents of the melt.

The limited freeboard of VOD ladles and the strongly exothermic effect of the process dictate that the starting carbon content and the oxygen blow rate be limited. A typical sequence of operations during a VOD treatment is shown in Fig. 1. Achievable results reported from several plants that utilize VOD capability are summarized in Table 1.

REFERENCES

1. H.G. Bauer, K. Behrens, and M. Walter, Review on VOD/VAD-Treatments, in *Proceedings of the Seventh International Conference on Vacuum Metallurgy* (Tokyo), The Iron and Steel Institute of Japan, Section 33-1, 1982, p 1314-1331
2. W. Burgmann, J.-S. Yie, and M. Kaivola, Operational VOD Results in Ladles and Converters, in *Proceedings of the Seventh International Conference on Vacuum Metallurgy* (Tokyo), The Iron and Steel Institute of Japan, Section 34-1, 1982, p 1348-1355
3. G. Pietschmann, G. Scharf, W. Burgmann, and M. Velikonja, The Freital Ladle Metallurgy Plant: Equipment and Results, in *Proceedings of the Eighth International Conference on Vacuum Metallurgy* (Linz/Austria), Eisenhütte Österreich, 1985, p 582-605
4. H. Kaito, T. Othani, and S. Iwaoka, Production of Super Ferritic Stainless Steels by S.S.-VOD Process, *Tetsu-to-Hagané,* 1977, p 139-156

Ladle Furnace and Vacuum Arc Degassing (LF/VD-VAD)

Wilhelm Burgmann, Leybold AG, West Germany

Secondary metallurgy techniques are energy consuming, and any extended ladle treatment leads to heat losses. As a result, the goal has always been to perform such operations as vacuum degassing, desulfurization, and alloying with simultaneous heating so that the ladle treatment could be carried out with no evident heat loss. Simul-

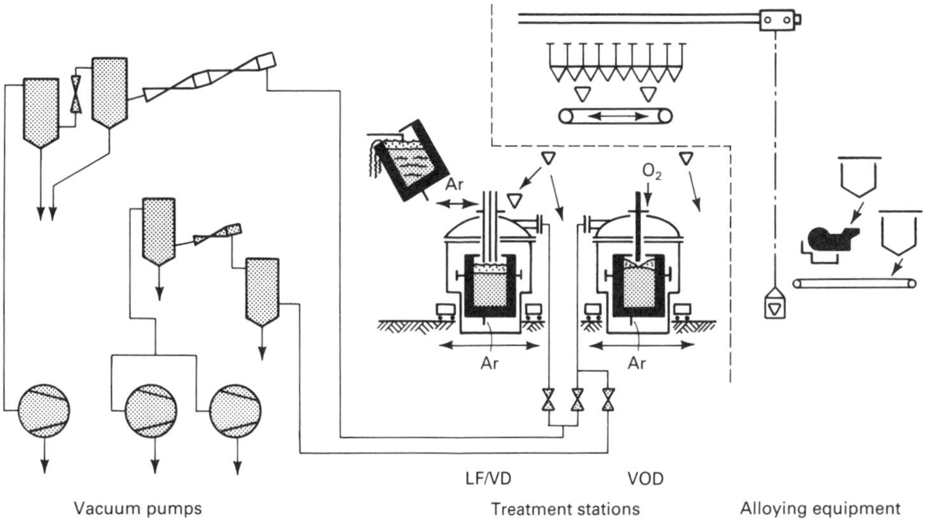

Fig. 1 Schematic of a combined ladle furnace and vacuum degassing plant

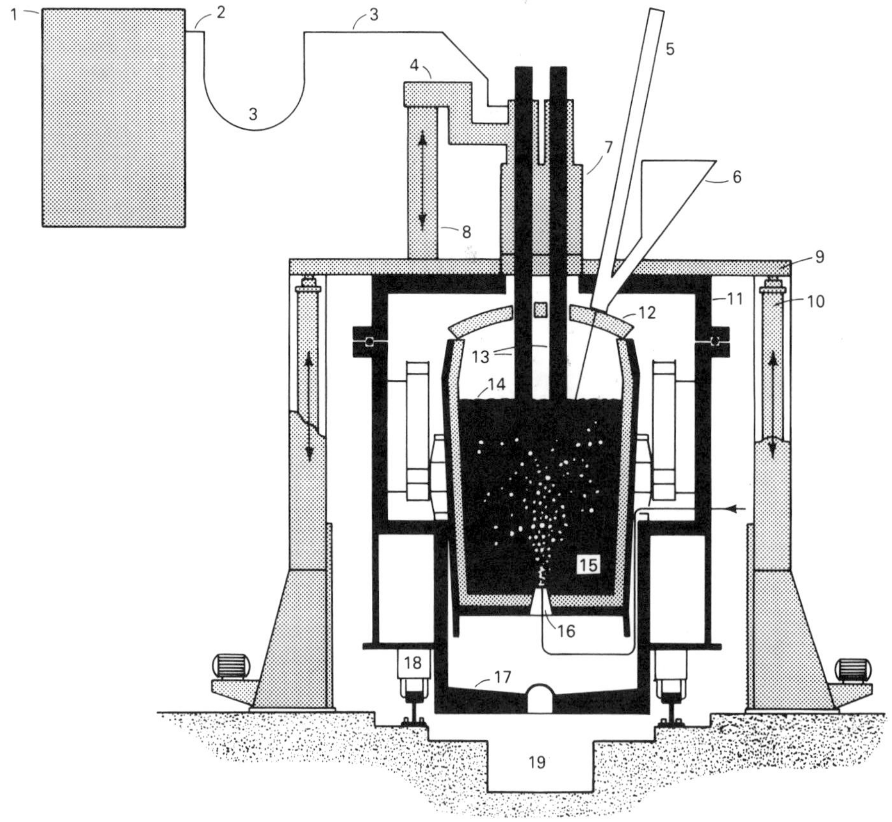

Fig. 2 Schematic of a vacuum arc degassing plant layout. 1, transformer; 2, bus bars; 3, flexible cables; 4, electrode arms; 5, temperature-measuring and sampling device; 6, alloying lock; 7, electrode housing; 8, electrode masts; 9, operating platform; 10, lifting device for platform; 11, vacuum cover; 12, ladle heat shield; 13, electrodes; 14, slag; 15, melt; 16, argon plug; 17, vacuum vessel; 18, vessel drive; 19, emergency pit

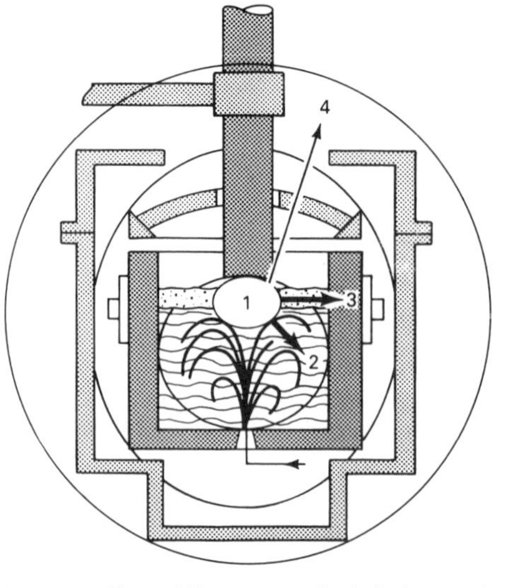

Fig. 3 Thermal flow pattern of a ladle furnace. 1, arc; 2, melt; 3, slag, electrode tip, and lining; 4, vessel, cover, busbars, and environment

taneous heating and degassing is difficult and previously was possible only by vacuum induction degassing or partially by vacuum arc degassing (VAD); however, the combination of an arc-heating Ladle Furnace (LF) with a vacuum degassing (VD) station meets operational requirements and produces excellent metallurgical results, as will be summarized in this section.

Equipment and Processing

Several types of LF/VD-VAD layout designs are possible. Although vacuum ladles do not provide the best operational reliability (Ref 1), they are essential if induction coils are needed for stirring. In this case, the ladle and coil are placed on a tiltable ladle car that moves from the open loading position underneath the vacuum hood and/or underneath the three-phase arc-heating ladle furnace. With the incorporation of an induction coil, this design is also known as the ASEA/SKF process.

For vacuum vessels, there are two principal designs. The first is the stationary vessel, and the second is the track-mounted movable vessel.

The stationary vessel is enclosed in a mobile vacuum cover or a swiveling ladle furnace cover. For tandem stationary vessels, each is enclosed in a moving vacuum cover, while both are served by a common swiveling ladle furnace, a common vacuum pump, and common alloying equipment. This design permits brief tap-to-tap times of about 30 to 45 min. A schematic of a combined LF/VD plant is shown in Fig. 1.

The movable vacuum vessel moves with the vacuum ladle car to several treatment platforms that contain the vacuum cover, arc-heating ladle furnace, plasma torches (in the case of ladle plasma heating units), injection stands, and alloying and wire feeding station. There are single-stand layouts that permit both arc heating and vacuum degassing in addition to other secondary metallurgy treatments, such as alloying, injection, and stirring, at the same treatment station. The movable vacuum vessel is transported to this treatment station, where arc heating and vacuum degassing can be performed by ladle furnace vacuum degassing or vacuum arc degassing.

For combined ladle furnace and vacuum degassing, arcing is carried out at the pressure created by a pollution control device; this underpressure is ~0.5 kPa (~5 mbar, or 3 torr). The protective atmosphere is created under a partially closed vessel lid, leaving only the openings for three electrodes, alloying, sampling, process control, and flue gas suction. The atmosphere composition is controlled by the carbon monoxide formed by the reaction of graphite electrodes with oxidic slag and the inert gas introduced through porous plugs in the ladle bottom or under the ladle furnace roof. For vacuum treatment, the electrodes are lifted above the vessel lid, and the lid opening is closed by a vacuum cover.

When vacuum arc degassing is used, complete protection of the electrodes is provided by telescopic water jackets and inflatable top seals that permit simultaneous arcing, vacuum treatment, and inert gas stirring. However, arcing under reduced pressure is restricted to about 30 kPa (300 mbar, or 225 torr). Below this pressure, the gas atmosphere would be ionized, resulting in severe damage to the equipment. It should be noted that 30 kPa (300 mbar, or 225 torr) arcing is feasible only if the arc voltage is kept below 100 V and the electrode regulation is reliable.

The arc length increases with the arc voltage and with the reduced pressure. Therefore, pressures are not applied below 80 kPa (800 mbar, or 600 torr), because the extreme arc length cannot be fully immersed in the slag. Submerged arcing is one precondition for the high thermal efficiency of this process. Therefore the VAD process is also divided into two steps: one arcing step at 60 to 90 kPa (600 to 900 mbar, or 450 to 700 torr), followed by a vacuum procedure carried out at below 100 Pa (1 mbar, or 0.7 torr) with the electrodes lifted.

Ladle Furnace Design. The basic design and equipment of any arc-heated ladle furnace are similar to those of electric arc melting furnaces, except that no tilting movement is required and ladle furnaces are frequently in a gantry design (Fig. 2). Electrode regulation is accomplished by either electromechanics or hydraulics. The hydraulic fluid is a water emulsion, water glycole, or oil. Direct current ladle furnaces

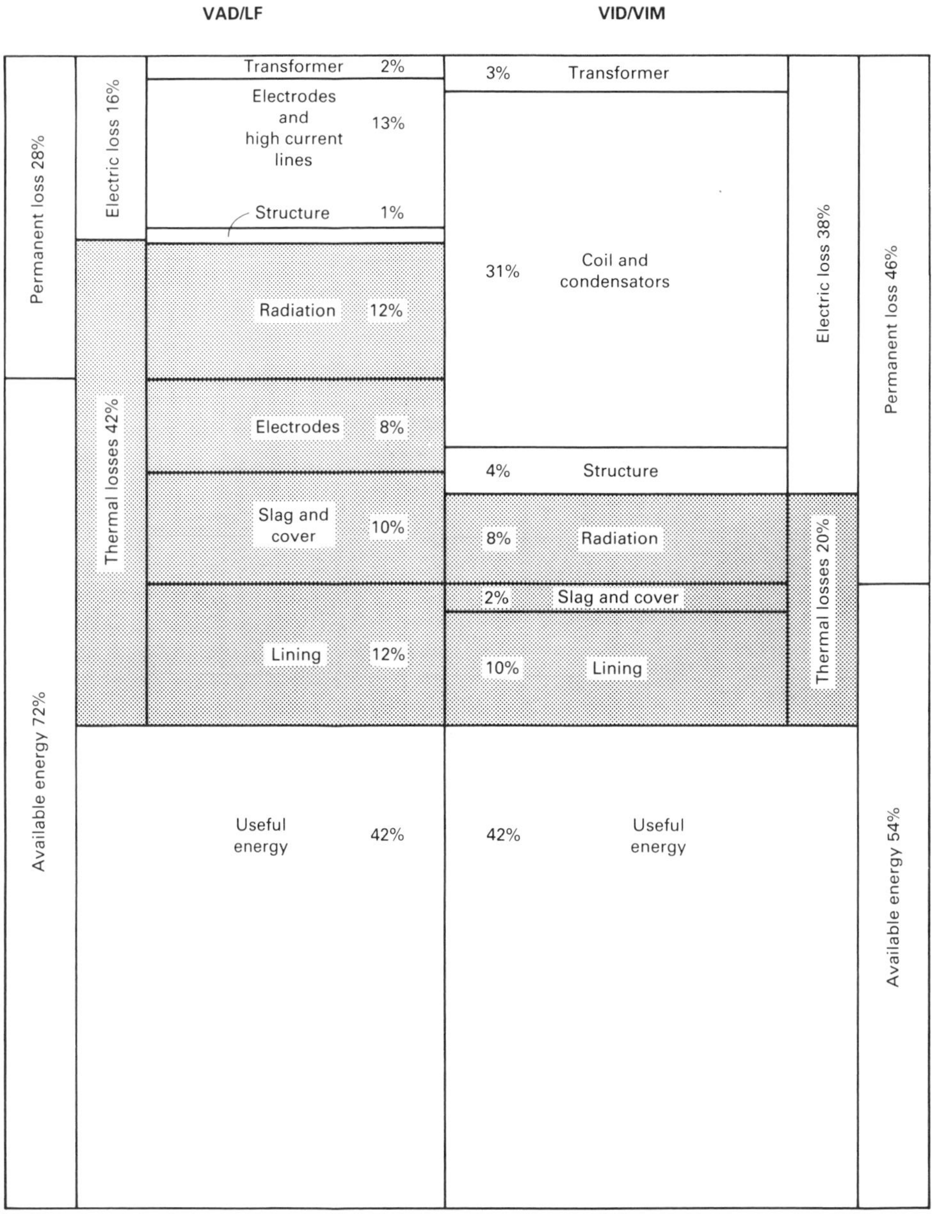

Fig. 4 Thermal energy balances for arc-heated and induction-heated processes

are not manufactured, but if very small ladle furnace units are required, a monophase alternating current design with two electrodes is available. Such units are similar to those described in the article "Vacuum Melting and Remelting Processes" (see the section "Electroslag Remelting") in this Volume.

Stirring, by inductive coils or inert gas, is necessary in any ladle furnace design. However, the ladle furnace has requirements that are very different from those of the electric arc melting furnace. Ladle furnace requirements include the following:

- Controls governing electrode movement must be more sensitive because of the inert gas bubbling through the ladle bottom
- The arc must be immersed in the foaming slag; this requires short arcs and therefore a low voltage, a low voltage-to-current ratio, and a low total inductivity

The refractory wear of immersed arcing is concentrated at the wall/slag zone. It is proportional to the amount of specific arc power brought into the slag and the intensity of slag cooling during treatment.

For ladle furnaces, secondary voltages of 180 to 280 V, combined with a 2 to 3 mΩ furnace plus transformer reactance, leads to arc voltages below 100 V. The resulting relatively high current density is 10 A/mm^2 (6500 A/in.^{2}) in rigid bars and pipes, 7 A/mm^2 (4500 A/in.^{2}) in flexible water-cooled cables, and 0.40 A/mm^2 (250 A/in.^{2}) in graphite electrodes.

The response time resulting from electrode control is defined as the time required to dissolve a short circuit by lifting the electrode about 10 mm (3/8 in.). This response time may vary between 100 and 400 ms, according to the electrode control system applied. This value is important because shorter response times result in less carbon pickup and higher thermal yields.

Table 1 Data comparison for various ladle heating operations

	VAD	LF	ASEA/SKF(a)	VID	VID, Calidus/IRSID(b)	Plasma	VOH
Method of heating	Arc	Arc	Arc	Induction	Induction	Plasma	Oxygen + Al
Method of stirring	Gas injection	Gas injection	Induction plus gas	Induction plus gas	Induction	Gas	Gas
Working pressure, mbar (torr)	500–800 (375–600)	1020 (770)	1020 (770)	0.1–1 (7.5 × 10^{-2} to 7.5 × 10^{-1})	1–10 (7.5 × 10^{-1} to 7.5)	1020 (770)	1–100 (7.5 × 10^{-1} to 75)
O_2 partial pressure, mbar (torr)	10–20 (7.5–15)	20–50 (15–38)	20–50 (15–38)	0.02–0.2 (1.5 × 10^{-2} to 1.5 × 10^{-1})	0.2–2 (1.5 × 10^{-1} to 1.5)	10–20 (7.5–15)	0.2–20 (1.5 × 10^{-1} to 15)
Heating rate, K/min	3–5	3–5	2–4	5–15	2–7	0.2–1	20–50
Melt weight, Mg (ton)	30–150 (33–165)	15–300 (16.5–330)	15–250 (16.5–275)	0.5–15 (0.55–16.5)	0.3–25 (0.33–27.5)	50–300 (55–330)	5–300 (5.5–330)
Installed power, MVA	5–20	5–40	5–25	0.5–3	0.3–5	1–6	...
Lining thickness							
bricked, mm (in.)	200–350 (8–14)	200–300 (8–12)	150–250 (6–10)	80–100 (3–4)	...	200–300 (8–12)	200–300 (8–12)
rammed, mm (in.)	...	...	...	...	100–150 (4–6)	...	...

(a) Source: Ref 1. (b) Source: Ref 2, 3

The heating rate is also affected by heat transfer, particularly during the first 10 to 30 min of treatment. Heat losses from the melt to the different furnace components are illustrated in Fig. 3, which shows that the heating rate depends on the heat flow from the arc (area 1) to the different parts surrounding the melt (area 2). Even in the case of thermal equilibrium with the lining (area 3), permanent thermal and electrical losses (area 4) are present such that the best thermal yield under a steady state of heat transfer is only 50 to 60%. The thermal balance reached at a near steady state for arc heating (VAD/LF) and induction heating (vacuum induction melting and degassing, VIM/VID) is shown in Fig. 4. Figure 5 shows a typical treatment cycle and associated parameters for a 75 Mg (80 ton) VAD-heated melt.

Independent of thermal yield, the physical boundary for heat transfer by arc, plasma, electron beams, and oxygen gas streams is that the melt evaporates at the heat transfer spots. Table 1 compares the various types of heating processes described in this section, in addition to vacuum induction degassing, plasma heating, and vacuum oxygen heating (VOH), which are described elsewhere in this article.

REFERENCES

1. R. Schumann and W. Schwarz, Ladle Metallurgy Comparison of Vacuum Ladles and Vacuum Vessels, *Metall. Plant Technol.*, Vol 4, 1985, p 10-18
2. R. Vasse, H. Gaye, J.P. Motte, and Y. Fautrelle, Thermal and Metallurgical Efficiencies in Ladle Induction Furnace Treatments, in *Proceedings of the Seventh International Vacuum Metallurgy Conference* (Tokyo), The Iron and Steel Institute of Japan, Section 25-2, 1982, p 1125-1132
3. C. Lechevalier, J.P. Motte, and G. Forestier, Procédé de Chauffage par Induction de L'acier Liquide en Poche, *Rev. Mét.*, Vol 10, 1980, p 791-797

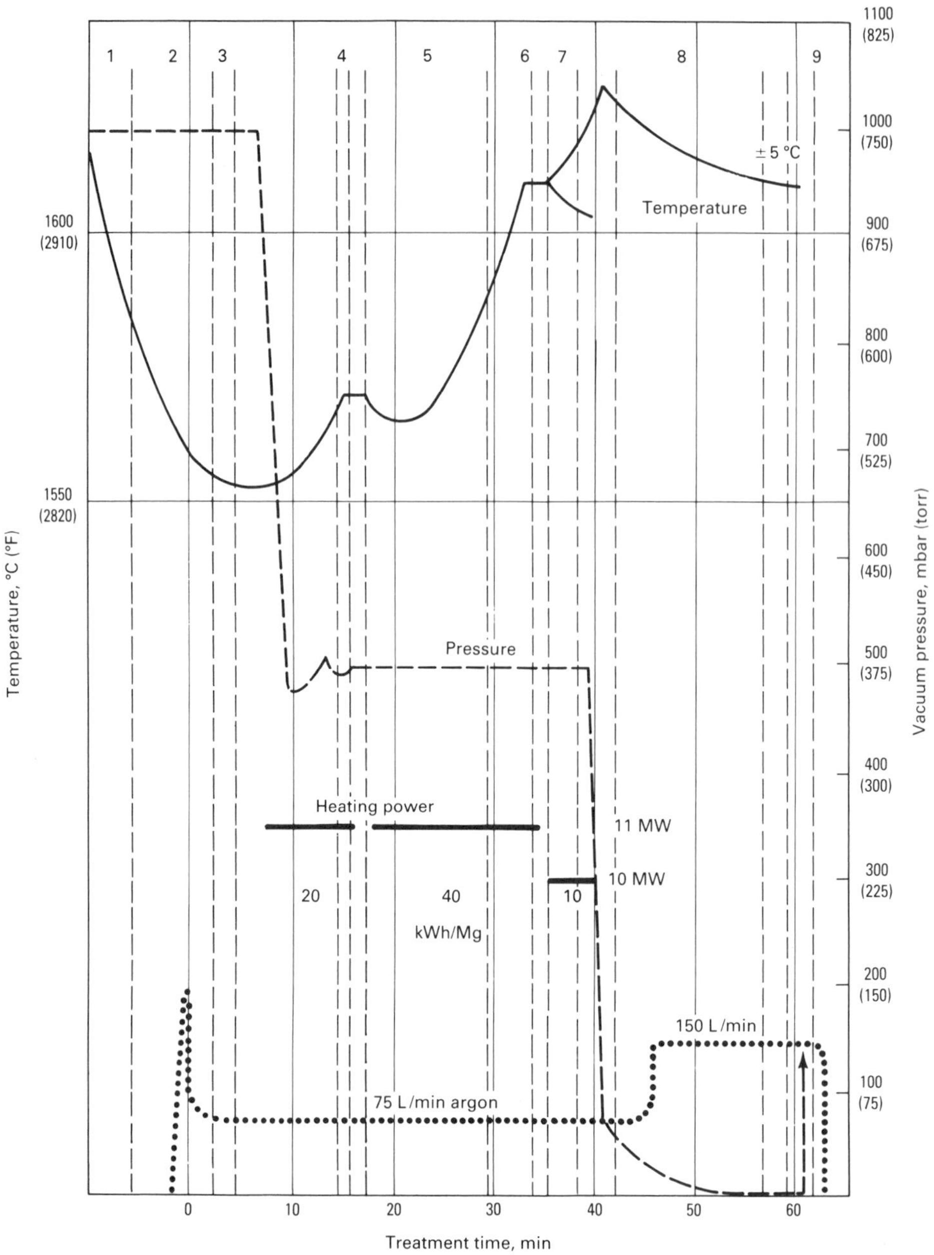

Fig. 5 Treatment cycle of a 75 Mg (80 ton) VAD-heated melt. 1, tapping; 2, additions of C + Al + 0.5% CaO; 3, temperature check; 4, sampling and temperature check; 5, addition of 1.5% alloys + 0.5% CaO; 6, temperature check; 7, additions + 0.5% CaO; 8, degassing; 9, venting

Vacuum Induction Degassing (VID)

H. Kemmer and U. Betz, Leybold AG, West Germany

Thermal balance is extremely critical for the secondary metallurgical treatment of small heats. In most cases, excessive chemical heating through aluminum additions is the only method of compensating for temperature losses during the secondary metallurgical treatment of melts below 15 Mg (16.5 tons). This technique, however, has various disadvantages, such as increased aluminum oxide (Al_2O_3) inclusion content in the final product and the high equipment costs associated with aluminum heating and oxygen (O_2) blowing. In such cases, vacuum induction degassing, a new secondary steelmaking technology, also permits economical vacuum treatment for small heat sizes ranging from 1 to 15 Mg (1.1 to 16.5 tons). Table 1 compares the metallurgical applications of vacuum induction degassing with other heating and melting operations.

Equipment and Processing

Vacuum induction degassing units consist of a simple induction furnace, designed for vacuum operation, that is equipped with a bell-type vacuum cover (Fig. 1). The furnace crucible is either rammed or lined with brick and has a porous plug in the bottom. A frequency converter power supply provides

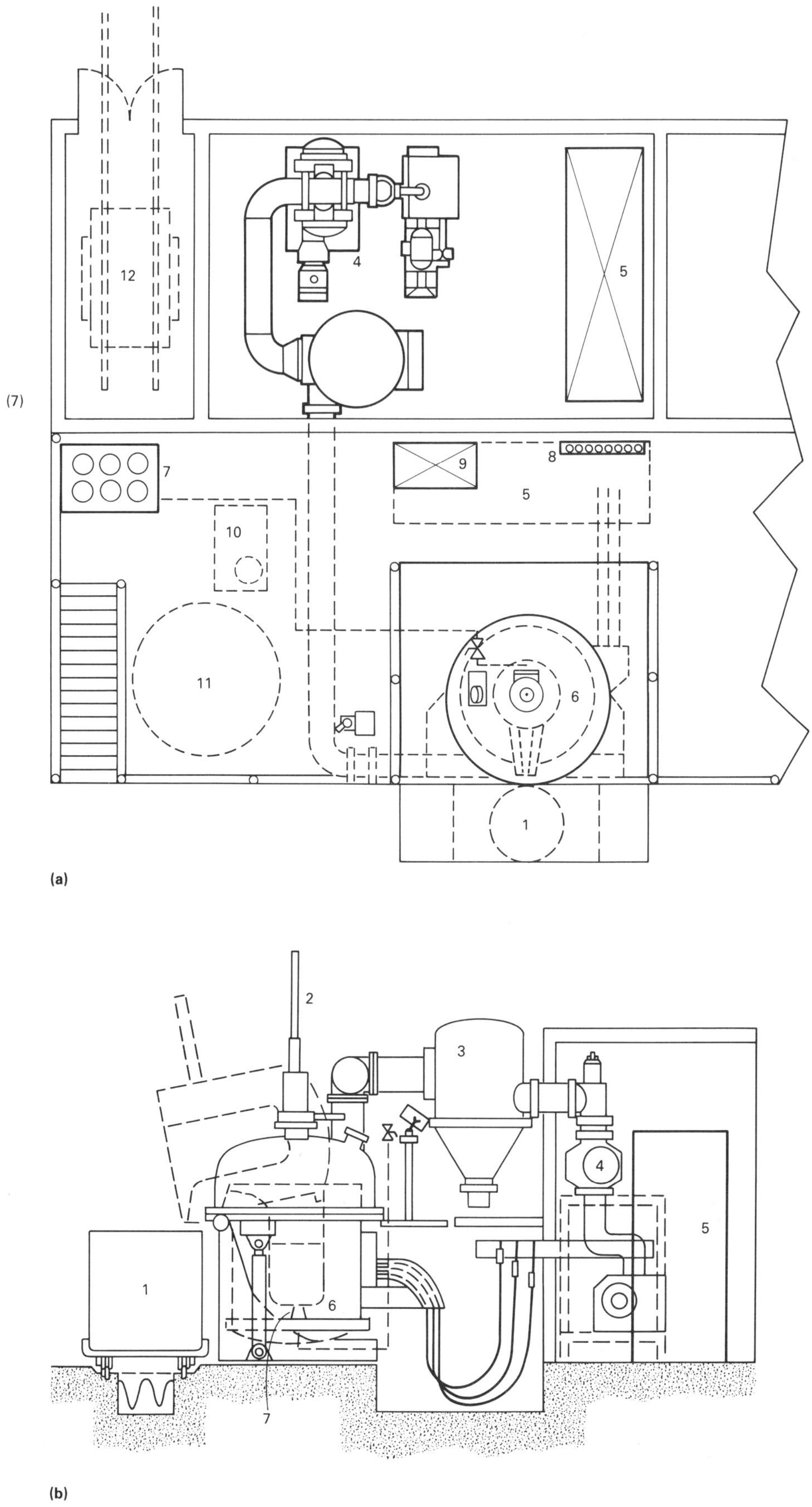

Fig. 1 Schematic of a VID unit. (a) Top view. (b) Side view. 1, mold, die, or ladle; 2, charging device; 3, filter; 4, vacuum pumping system; 5, melt current supply; 6, vessel with vacuum bell; 7, gas-purging set; 8, cooling water manifold; 9, control cabinet; 10, hydraulic unit; 11, lid station; 12, transformer

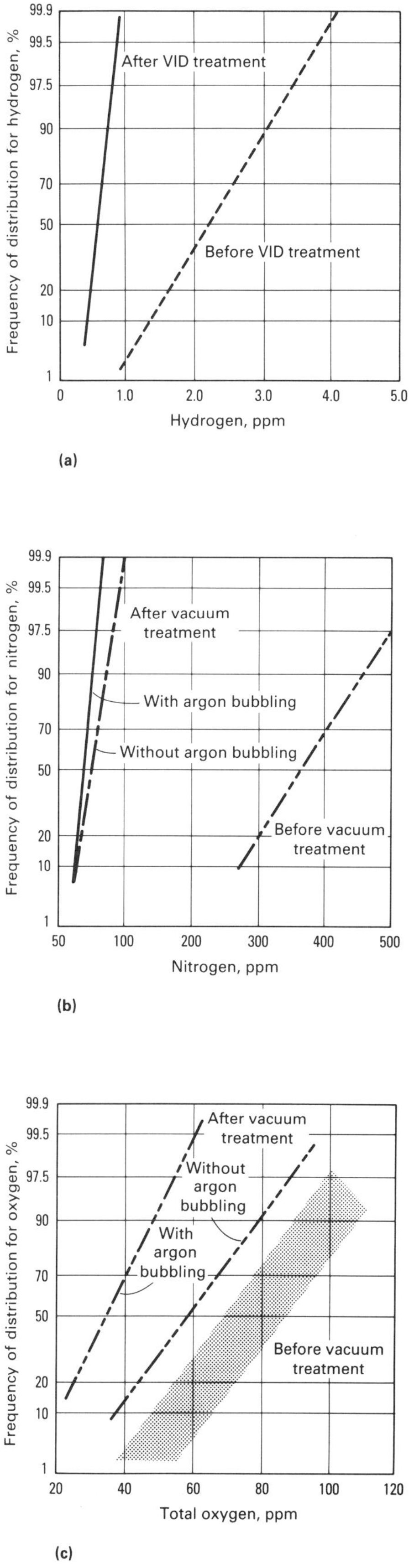

Fig. 2 Effect of VID processing on the hydrogen (a), nitrogen (b), and total oxygen (c) contents of X 38 CrMoV 51 die steel (Fe-0.38C-1.0Si-0.40Mn-5.2Cr-1.3Mo-0.40V)

Table 1 Comparison of applications and operational characteristics of various heating and melting processes

Benefit/application	Process(a)				
	VD	LF/VD	VOD	VODC	VID
Melt improvement					
Hydrogen removal	P	P	S	S	P
Nitrogen removal	P	P	P	P	P
Deoxidation	P	P	S	S	P
Vacuum carbon deoxidation	P	P	P	P	P
Desulphurization	P	P	P	P	P
Inclusion removal	P	P	S	S	P
Extralow carbon	...	...	P	P	P
Temperature adjustment	...	P	...	...	P
Savings					
Basic materials	S	S	P	P	S
Alloy material	P	P	P	P	P
Operational capabilities					
Relief of primary melting system	P	P	P	P	P
Increased production capacity	P	P	P	P	P
Buffer function	...	P	...	...	P
Collection of molten metal	...	P	...	...	P
Split melts	...	P	...	...	P

(a) P, primary benefit or application; S, secondary benefit or application. VD, vacuum degassing; LF/VD, ladle furnace vacuum degassing; VOD, vacuum oxygen decarburization (ladle metallurgy); VODC, vacuum oxygen decarburization (converter metallurgy); VID, vacuum induction degassing

the energy for the induction coil to control the temperature of the melt.

With the vacuum cover removed, the furnace is charged with either liquid or solid. For vacuum treatment, the vacuum cover is placed on the furnace body by crane. A mechanical pumping system is used to achieve an operating pressure of around 0.01 to 0.1 kPa (0.1 to 1 mbar, or 0.075 to 0.75 torr).

The process is primarily applied to tool and die steels, bearing steels, chromium steels, nickel-chromium steels, and nickel-base alloys to remove dissolved gases such as hydrogen and nitrogen and to remove oxide inclusions from the melt. Figure 2 shows the results of VID treatment on the hydrogen, nitrogen, and oxygen contents of a die steel. In Fig. 2(a), the use of vacuum induction degassing lowered the hydrogen content from more than 2.2 ppm to 0.60 ppm (hydrogen removal of 50%). With proper metallurgical treatment, the nitrogen content can also be decreased. Figure 2(b) shows that the nitrogen content was lowered from more than 370 ppm to approximately 50 ppm (nitrogen removal of 50%).

With vacuum induction degassing, oxide inclusions can be removed by applying metallurgically active slags as an absorbing buffer. In addition, VID units are equipped with argon bubbling systems, which accelerate and improve the removal of nonmetallic inclusions. Changes in oxygen content as a result of vacuum induction degassing, with and without argon bubbling, are shown in Fig. 2(c). The final total oxygen content when using VID with argon bubbling is 30% lower than VID without argon bubbling, which demonstrates the benefits of soft argon rinsing.

Plasma Heating and Degassing

H. Pannen and G. Sick, Leybold AG, West Germany

Plasma heating is a process that employs plasma arc torches to supply electric heat to a ladle of molten metal either to maintain the temperature or to raise the temperature of the melt. Until recently, heat losses during ladle metallurgy treatment were compensated for by means of graphite electrode arc heating or induction heating. However, for numerous steel grades, especially those with extremely low carbon contents, graphite electrodes cannot be used, because of the risk of carbon pickup. The demand for nonconsumable heat sources has increased interest in high-power plasma torches as heat sources, which can be used with minimum modification to the existing ladle design. Prior to their integration into ladle metallurgy stations, plasma torches had been used primarily for scrap melting and heat loss compensation before and during continuous casting.

In this section, ladle furnace concepts applicable to plasma technology will be outlined. Details on metallurgical reactions during the high-power plasma primary melting of various steel grades are available in Ref 1 to 4.

Equipment and Processing

The plasma torches currently used for steel melting are of the transferred arc type. Two very different design principles are used:

- Tungsten tip design either in alternating current (ac) or direct current (dc) operation
- Hollow copper electrode design in dc operation

Details on plasma torch design can be found in the article "Vacuum Melting and Remelting Processes" (see the section "Plasma Melting and Casting") in this Volume. Technical data on plasma torches used for steel melting and ladle heating are summarized in Table 1.

A three-phase ac plasma ladle furnace unit employed in plasma heating and degassing applications is shown in Fig. 1. These units do not need a bottom electrode but do require more sophisticated power supplies than dc plasma torches do in order to limit the torch current and to ensure a quick, uninterrupted current pass through zero.

A plasma power supply with a thyristorized three-phase ac controller has been developed; in this unit, the amperages of all three torches are equal and restricted even in asymmetrical operation. With a thyrister power supply, the current passes steeply through zero, and the arc remains stable and is not likely to extinguish.

To apply the three-phase plasma system to ladle heating, where the ladle freeboard remains constant, a more simple method for current limitation can be used because of the lack of rapid voltage changes. Figure 2 shows a schematic of a simplified power supply with reactor, an ignition power supply, and the necessary controls for ac plasma torch heating in ladle furnaces. Power rating and torch size for the three-phase plasma heating of ladle furnaces as a function of ladle capacity is shown in Fig. 3. It is obvious from Fig. 3 that compensation for heat losses in ladles is possible with available plasma torch technology. Figure 3 also shows that 20 kA torches are required to achieve heating rates of 4 K/min in a 200 Mg (220 ton) ladle.

Direct Current Plasma Torches. A dc plasma torch unit with hollow copper electrodes and vortex stabilization of 5.5 MW (1250 V, 4400 A) rated power has been applied to maintain or slightly increase the

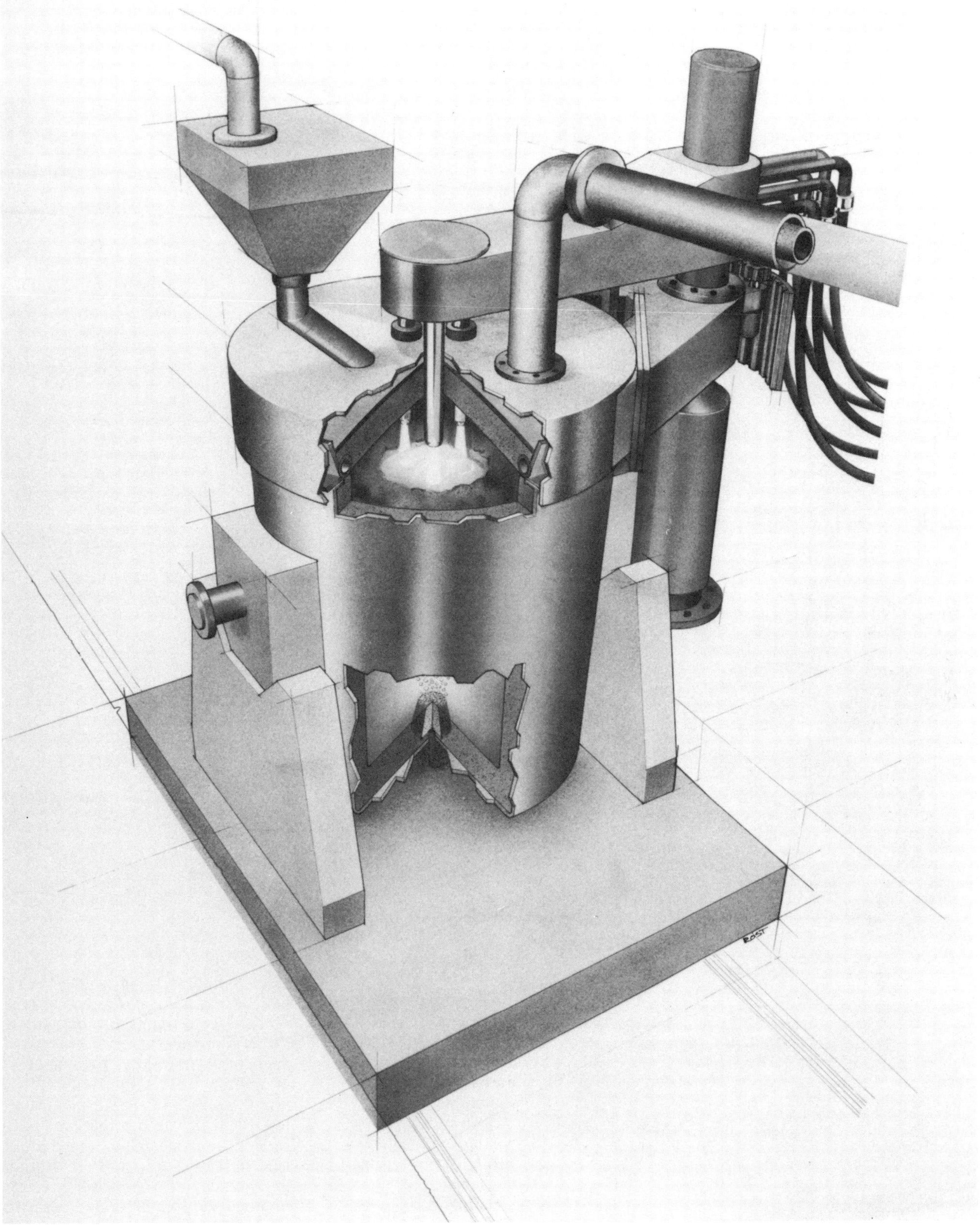

Fig. 1 Three-phase ac plasma ladle furnace

Table 1 Data on high-power plasma torches for steel melting and ladle heating

Parameter	Installation No. 1	Installation No. 2	Installation No. 3	Installation No. 4
Design principle	Tungsten tip	Tungsten tip	Hollow copper electrode	Tungsten tip
Type of torch	Transferred arc	Transferred arc	Transferred arc	Transferred arc
Current	ac	dc	dc	dc
Maximum amperage, kA	6	9	5.0	10.8
Voltage, V	400	600	1000	690
Power rating, MW	2.4	5.4	5.0	7.5
Gas consumption	9 m^3 Ar/h (320 ft^3Ar/h)	11.25 m^3/h (400 ft^3/h)	N_2: 2–3 m^3/min (80–100 ft^3/min)	15–22 m^3/h (530–775 ft^3/h)
Electrode life, h	100	...	> 1000 (Ar) > 250 (N_2)	...

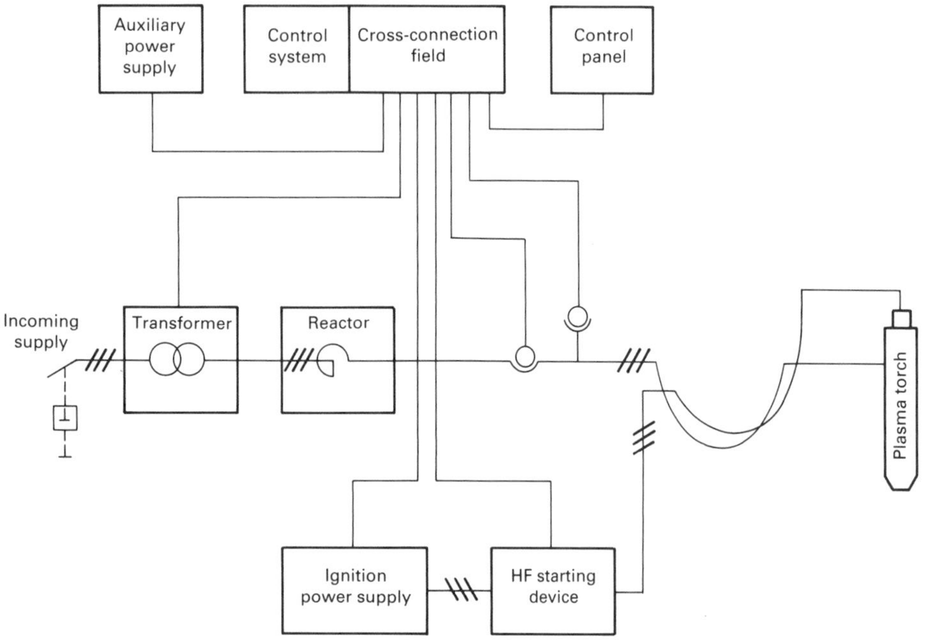

Fig. 2 Schematic of a simplified power supply with reactor, ignition power supply, and necessary controls for ac plasma torch heating in ladle furnaces. Source: Ref 5

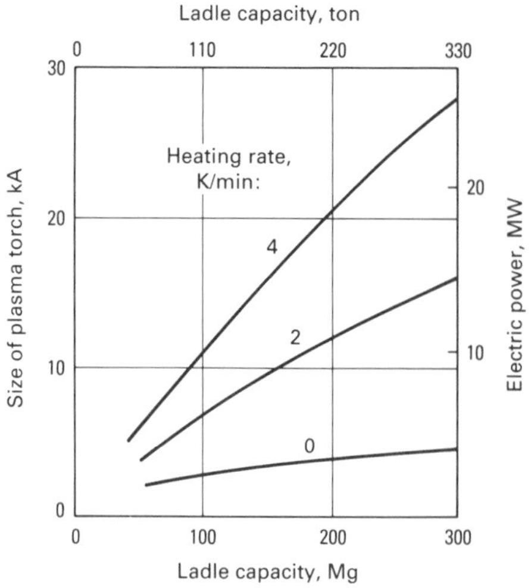

Fig. 3 Power rating and torch size for the three-phase plasma heating of ladle furnaces (arc voltage: 300 V). Source: Ref 6

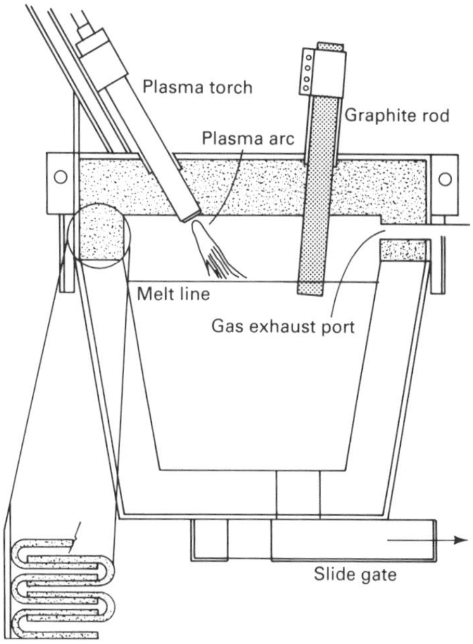

Fig. 4 Schematic of a plasma ladle reheater

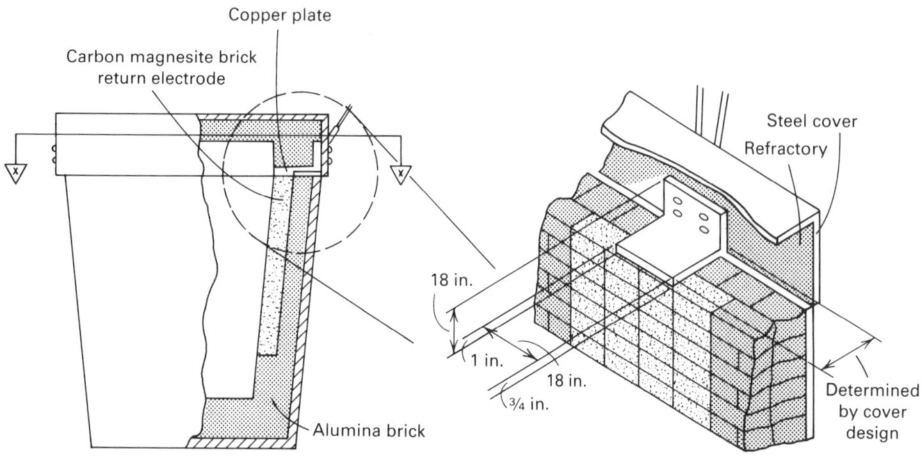

Fig. 5 Schematic of conductive ladle and cover with carbon magnesite bricks as the counter electrode

steel temperature in a 200 Mg (220 ton) ladle when aborts are encountered in the vertical continuous casting process (Ref 7). Plasma heating has proved to be an acceptable alternative to dumping the ladle of molten steel into a mixer, where the temperature and metal chemistry are altered and the heat is rendered out-of-spec for continuous casting (temperature, or the lack of heat, is often the reason for aborting or recycling a heat). Achieving the homogenization of metal temperature in the ladle resulted in increased yield (+4 to 5%) of cast metal.

The plasma ladle reheater described in Ref 7 is a refractory-lined ladle cover with one or more plasma arc torches mounted on top (Fig. 4). Also mounted on top of the unit is a return electrode mechanism. The return electrode can be a graphite bar that is immersed in the melt from the top. It can consist of graphite bricks installed on the side or at the bottom of the ladle, or it can be another plasma arc torch. In an alternate design, the return electrode consists of a conductive layer of carbon-magnesite bricks (Fig. 5). When the cover is placed on the ladle, a copper plate makes contact with the electrically conducting bricks and serves to conduct the plasma current to the grounded cover.

Effect of Nitrogen. For most steel processing, nitrogen is often selected as the plasma gas. It is known from laboratory scale investigations that nitrogen pickup in liquid steel under argon (Ar)/nitrogen (N_2)/oxygen (O_2) plasma gas mixtures is strongly affected by oxygen and nitrogen partial pressure in the plasma gas (Ref 8). The presence of oxygen leads to significantly higher nitrogen contents in steel. This is of particular importance for the nitrogen alloying of some stainless steel grades and for the application of nitrogen plasma for the ladle reheating of nitrogen-sensitive steel grades, where minimum nitrogen pickup in the steel is the goal. Figures 6 and 7 show nitrogen pickup in two steel grades as a function of oxygen and nitrogen partial pressure in the plasma gas.

The use of low-cost nitrogen as the plasma gas is recommended for vortex-stabilized hollow electrode plasma torches. These types of torches are characterized by high gas flow and long electrode lifetimes (~300 h). There are no technical restrictions

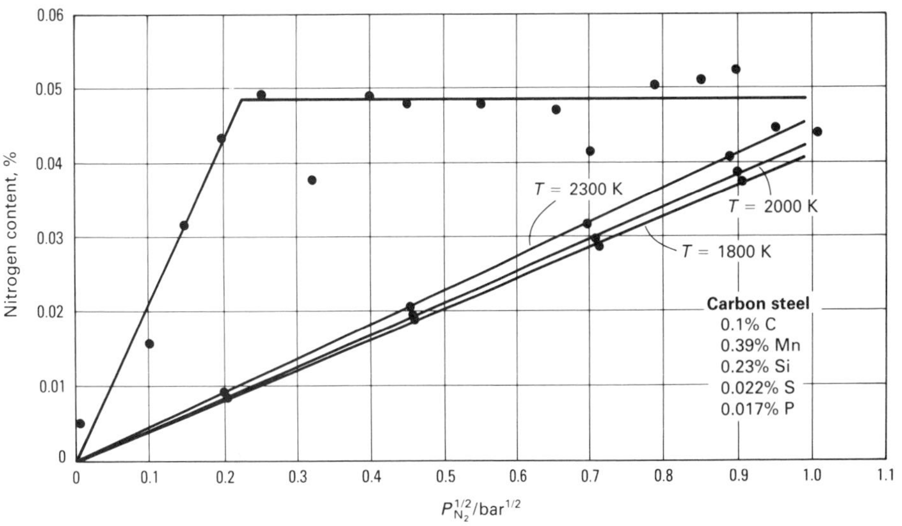

Fig. 6 The influence of nitrogen partial pressure in plasma gas (Ar/N_2) on the nitrogen pickup of a plain carbon steel (P_{total} = 1 bar). Source: Ref 8

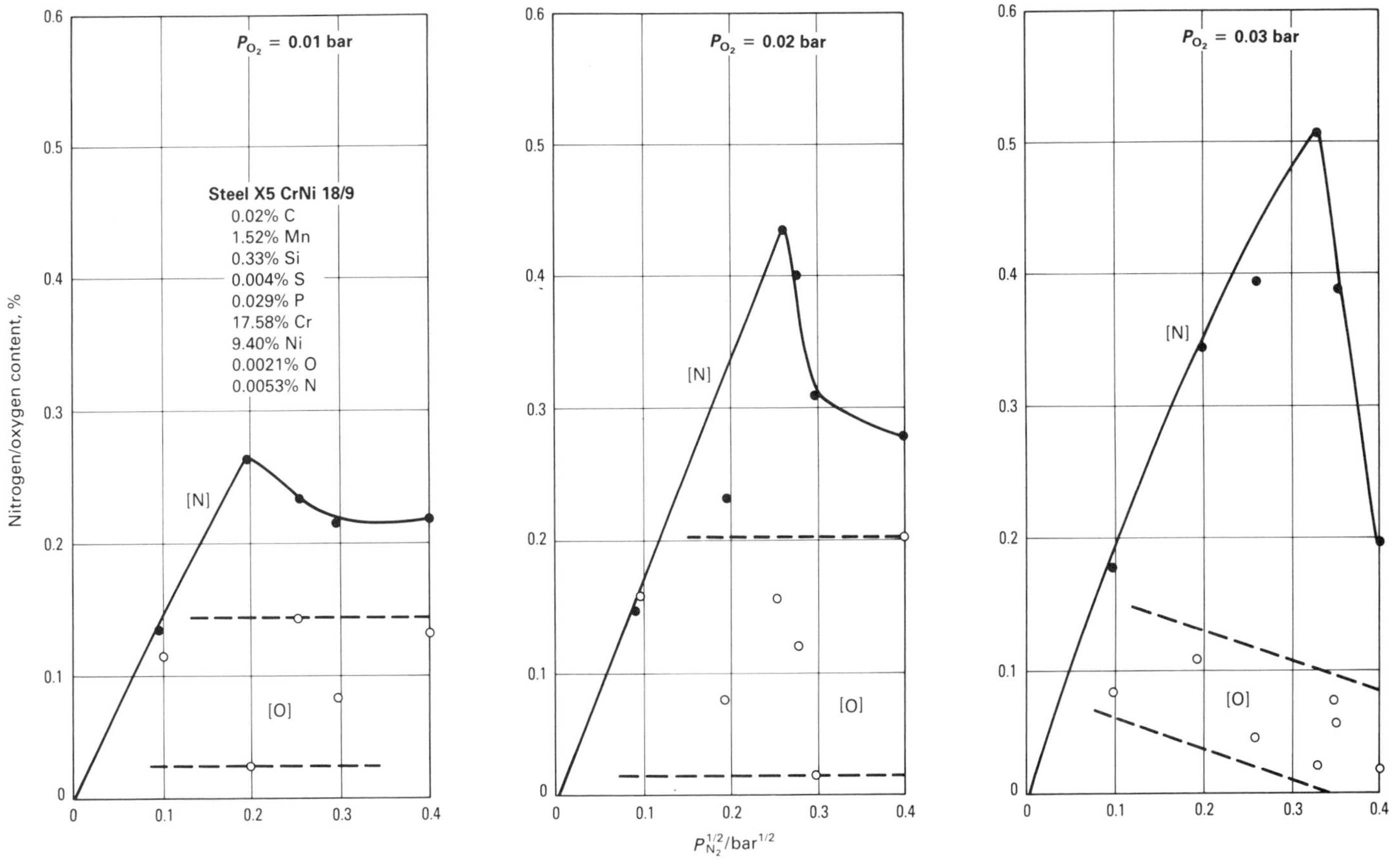

Fig. 7 The influence of increasing oxygen partial pressure in plasma gas on the nitrogen pickup in an austenitic stainless steel. Source: Ref 8

concerning the type of plasma gas used for these torches; even air can be used. Hollow electrode torches are extremely durable and are not affected by splashes of molten metal.

The state of the art in plasma technology for ladle metallurgy permits ladle sizes of approximately 60 Mg (65 ton) to be processed. Research and development is ongoing in this field. Figure 8 shows the layout for a modern plasma ladle metallurgy station.

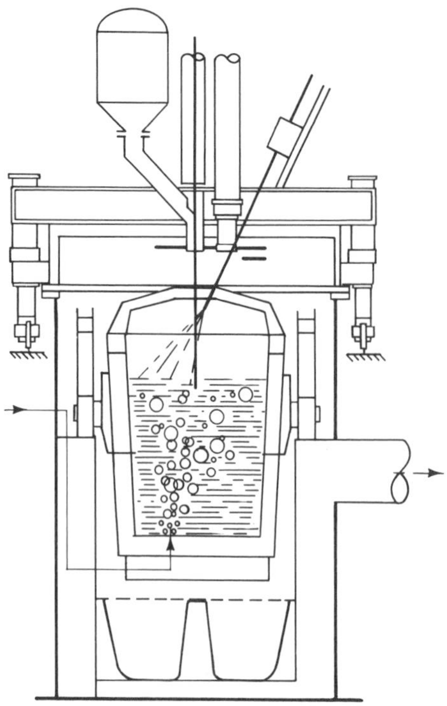

Fig. 8 Schematic of a vacuum plasma heating and degassing station concept for ladle metallurgy

REFERENCES

1. K. Skuin, W. Lachner, and L. Stephan, 10 Jahre Plasmametallurgie im VEB Edelstahlwerk Freital Freiberger, *Forschungshefte*, B 238, 1983, p 98-108
2. G. Scharf, L. Stephan, and W. Praske, Herstellung von Sonderstählen und Legierungen im Plasmaofen, *Neue Hütte*, Vol 2, 1937, p 48-51
3. D. Neuschütz and H.-O. Rossner, "Heating Molten Steel in an Inert-Gas Atmosphere Using 3-Phase a.c. Plasma Torches," Paper presented at the-Deutsch-Japanisches seminar, Düsseldorf, May 1987
4. O. Steipe, D. Peterek, and W. Lugscheider, "Results of Steel Production in a 45 t Plasma Furnace in Linz 8, Paper presented at the International Conference on Vacuum Metallurgy, Linz, 1985
5. K.-H. Heinen, R. Heinke, H.-G. Kunze, D. Neuschütz, and H.-O. Rossner, Entwicklungsstand der Drehstromplasmatechnik and Aufbau eines 20 MVA Plasmaofens bei, *Krupp Fachberichte Hüttenpraxis Metallweiterverarbeitung*, Vol 24 (No. 10), 1986, p 934-939
6. D. Neuschütz, H.-O. Rossner, H.-J. Bepper, and J. Hartwig, Development of 3-phase a.c. plasma furnace at Krupp, *Iron Steelmaker*, May 1985, p 27-33
7. S.L. Camacho, Plasma Ladle Reheater in Steelmaking, *Iron Steelmaker*, Vol 16, July 1986
8. T. ElGammal, G. Hinds, W. Häsing, and J. Vetter, Nitrogen Pick-Up in Liquid Steel Under Nitrogen Containing Plasma Gas, in *Proceedings of the Seventh International Conference on Vacuum Metallurgy* (Tokyo), 1982

Nonferrous Molten Metal Processes

David V. Neff, Metaullics Systems

IN THE MELTING of commercial nonferrous metals and alloys (aluminum, magnesium, copper, zinc, and lead), various auxiliary molten metal processing steps are necessary other than melting and alloying. For example, many nonferrous molten metal treatment technologies exist that improve metal quality through the control of impurities.

Generally, the historic practices of fluxing, metal refining, deoxidation, degassing, and grain refining have been used, and they apply to virtually all nonferrous metal systems. In addition, molten metal pumping and filtration are two somewhat newer but now commonly practiced technologies in nonferrous molten metal processing.

This article will describe the molten metal processing methodologies currently used in conventional nonferrous molten metal operations. These process methodologies pertain not only to foundry melting and casting but also to smelting, refining, and in certain cases mill product operations.

Fluxing

The term fluxing is used in this article to represent all additives to, and treatments of, molten metal in which chemical compounds or mixtures of such compounds are employed. These compounds are usually inorganic. In some cases, metallic salts are used in powder, granulated, or solid tablet form and may often melt to form a liquid when used. They can be added manually or can be automatically injected, and they can perform single or, in combination, various functions, including degassing, cleaning, alloying, oxidation, deoxidation, or refining. The term fluxing also includes the treatment of nonferrous melts by inert or reactive gases to remove solid or gaseous impurities. Fluxes are commonly used to some extent with virtually all nonferrous molten metal operations in both the foundry and in the production of mill products.

Fluxing of Aluminum Alloys

In aluminum melting, and especially in the remelting of foundry returns or other scrap, oxide formation and nonmetallic impurities are common. Impurities appear in the form of liquid and solid inclusions that persist through melt solidification into the casting. Inclusions can originate from dirty tools, sand and other molding debris, sludge (iron-chromium-nickel intermetallic compounds commonly found in die casting alloys), metalworking lubricant residues, and the oxidation of alloying elements and/or the base metal.

The term fluxing, in the broadest sense, applies to a treatment technique to the melt containing such impurities and inclusions as those mentioned above. Fluxing of the melt facilitates the agglomeration and separation of such undesirable constituents from the melt.

Fluxing is temperature dependent. The temperature must be high enough to achieve good physical separation or the desired chemical reaction. At sufficiently high temperatures, the fluidity of both the metal and the fluxing agent is likely to be very high, which provides for good contact between the two and better reactivity.

Flux Composition. The specific compounds or chemical reagents used in fluxes depend on the specific purpose of the flux. Most fluxing compounds consist of inorganic salt mixtures. The various constituents of these salts or other materials in the flux serve to:

- Form low-melting high-fluidity compounds at use temperature, as is the case with sodium chloride (NaCl)-potassium chloride (KCl) mixtures
- Decompose at use temperature to generate anions, such as nitrates, carbonates, and sulfates, capable of reacting with impurity constituents in the melt. This creates impurity metal oxides or other compounds with densities different from that of the base melt and facilitates physical separation
- Act as fillers to lower the cost per pound or to provide a matrix or carrier for active ingredients or adequately cover the melt
- Absorb or agglomerate reaction products from the fluxing action

Not all components are necessary or found in each flux. Table 1 lists the properties of numerous materials commonly used in commercial aluminum fluxes and fluxing reactions.

In general, the choice of which components to use in a flux depends on the temperature of use, whether the flux is to

Table 1 Characteristics of some materials used as fluxes for aluminum alloys

Material	Chemical formula	Density, g/m^3		Melting point		Boiling point	
		Solid	Liquid	°C	°F	°C	°F
Aluminum chloride	$AlCl_3$	2.440	1.31	190	374	182.7(a)	360.9
Aluminum fluoride	AlF_3	3.070	...	1040	1904	...	...
Borax	$Na_2B_4O_7$	2.367	...	741	1366	1575	2867
Calcium chloride	$CaCl_2$	2.512	2.06	772	1422	1600	2912
Calcium fluoride	CaF_2	3.180	...	1360	2480	...	...
Carnalite	$MgCl_2KCl$	1.60	1.50	487	909	...	...
Zinc chloride	$ZnCl_2$	2.910	...	262	504	732	1349.6
Zinc fluoride	ZnF_2	4.84	...	872	1602	...	...
Cryolite	$3NaFAlF_3$	2.97	...	1000	1832	...	...
Lithium chloride	LiCl	2.068	1.50	613	1135	1353	2467.4
Lithium fluoride	LiF	2.295	1.80	870	1598	1676	3048.8
Magnesium chloride	$MgCl_2$	2.325	...	712	1314	1412	2573.6
Magnesium fluoride	MgF_2	3.00	...	1396	2545	2239	4062.2
Potassium chloride	KCl	1.984	1.53	776	1711	1500 (a)	2732
Potassium fluoride	KF	2.480	...	880	1616	1500	2732
Potassium borate	$K_2B_2O_4$	...	...	947	1737	...	...
Potassium sulfate	K_2SO_4	2.662	...	1076	1969	...	...
Potassium carbonate	K_2CO_3	2.290	...	891	1636	...	...
Sodium chloride	NaCl	2.165	1.55	801	1474	1413	2575.4
Sodium fluoride	NaF	2.790	1.91	980	1796	1700	3092

(a) Sublines at indicated temperature. Source: Ref 1

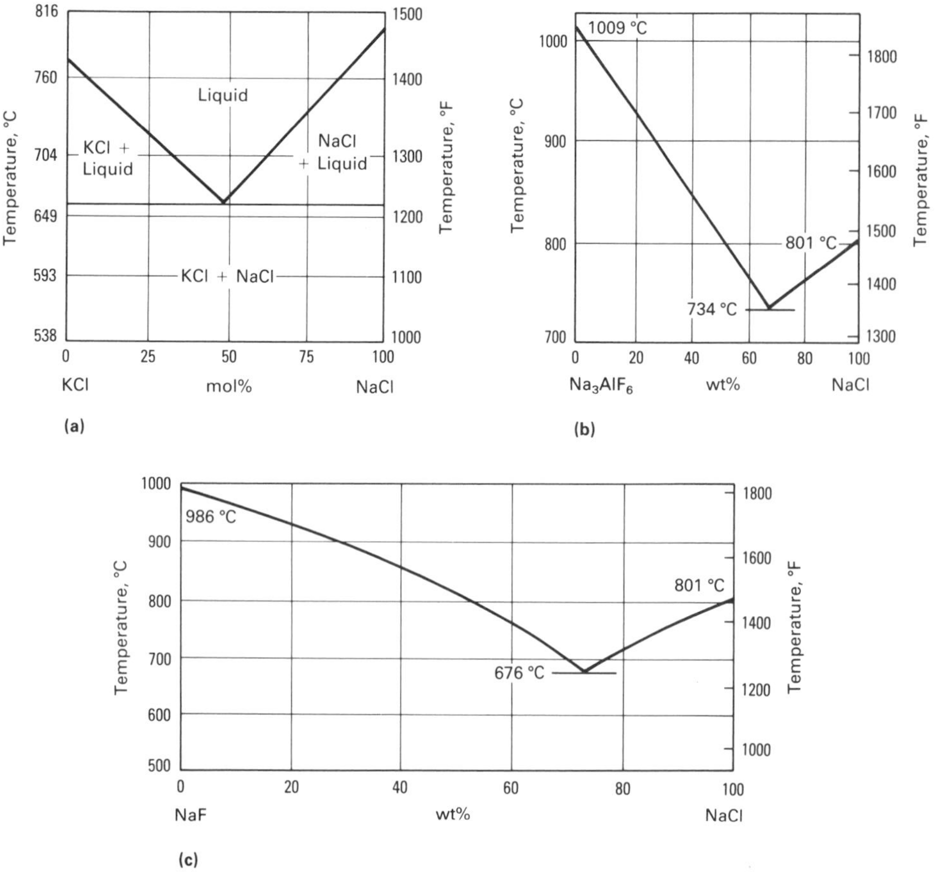

Fig. 1 Binary phase diagrams for three common aluminum fluxing compounds. Actual compositions are selected on the basis of melting point, cost, and performance. (a) KCl-NaCl. (b) Na_3AlF_6-NaCl. (c) NaF-NaCl. Source: Ref 2

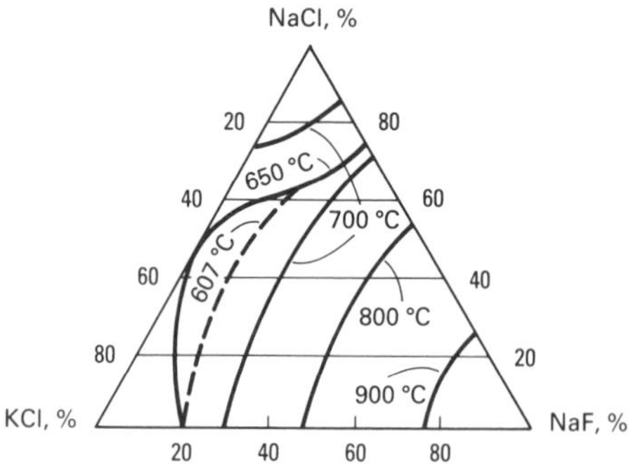

Fig. 2 Ternary phase diagram for a common commercial cleaning flux composition for aluminum alloys. Source: Ref 2

provide a molten cover or a solid cover, the desired reactivity, and the specific alloy chemistry. For example, sodium-bearing fluxes should not be used with aluminum-magnesium alloys in order to avoid sodium salt flux decomposition and the attendant sodium contamination of the melt. Sodium fluxes should also not be used with hypereutectic aluminum-silicon alloys refined with phosphorus because an easily formable sodium phosphide compound will render the phosphorus ineffective as a refiner.

Types of Fluxes. Four principal types of fluxes are used for aluminum alloys. They are cover fluxes, cleaning fluxes, drossing fluxes, and refining fluxes. Wall-cleaning fluxes are also employed, but these are usually sprayed onto furnace walls rather than added to the melt.

Cover fluxes are designed to be used primarily with smaller (pot, crucible) furnaces to provide a physical barrier to oxidation of the melt or to serve as a cleanser for alloys, scrap foundry returns, or fresh ingot being charged. Such fluxes are dry, or they may be of such composition as to be molten on the melt surface, thereby wetting any materials being charged. This helps agglomerate loose debris, and it minimizes the oxidation of new surfaces. Such fluxes are often called wet fluxes. Figure 1 shows three binary phase diagrams for components commonly used in cover fluxes.

Cleaning fluxes are usually higher in chloride salt compound content and usually contain fluorides to facilitate wetting of the oxide inclusions for easier separation from the melt. Figure 2 illustrates a ternary phase diagram for the ingredients of a commonly used flux.

Drossing fluxes are designed to promote separation of the aluminum oxide (Al_2O_3) dross layer that forms on the surface of the melt from the molten metal. Drosses and liquid or solid metal are usually intermingled in the dross layer. The drossing fluxes are designed to react with Al_2O_3 in the slag or dross layer and to recover metal. These fluxes usually contain compounds capable of reacting exothermically, such as double fluorides and salt compounds that melt (as a eutectic mixture) and increase wettability. The exothermic reaction raises the temperature, thus increasing the local fluidity. The fluorides wet and dissolve thin oxide films according to the general reaction $6Na_2SiF_6 + 2Al_2O_3 \rightarrow 4Na_3AlF_6 + 3SiO_2 + 3SiF_4$. With sufficient mechanical agitation through rabbling with a rake, these films will be broken long enough to release entrapped metal. These fluoride salts, however, cannot dissolve massive Al_2O_3 particles (only electrolysis would suffice, as is used for Al_2O_3 reduction in making primary aluminum).

Drossing fluxes are used to great advantage in the aluminum industry to reduce the rich metallic content of the dross (Ref 3). Untreated dross may contain 60 to 85% free metal, which, if allowed to burn or thermite, will convert to unrecoverable Al_2O_3. Untreated dross is often transferred to separate dross coolers to reduce this reaction. While on the melt surface, however, the dross can be reacted with an exothermic drossing flux and substantial metal recovered, reducing the overall metal content of the dross to about 30%. Considerable cost savings results because proper fluxing action will deliver perhaps 50% metal directly back into the melt. Figure 3 compares rich dross high in metal content with the dry, powdery residue low in metallics after proper fluxing.

Refining fluxes contain compounds that break down at use temperature and are thermodynamically favorable to reaction with certain metallic elements in the base melt. For example, certain chlorine-containing compounds will react in molten aluminum containing magnesium, calcium, lithium, sodium, and potassium to form insoluble chlorides of these metals, which will then partition through density differential to the dross phase, where they can be removed by skimming. Sophisticated technologies for impurity refining are addressed in detail in the section "Melt Refining of Aluminum" in this article.

Wall-cleaning fluxes contain compounds that help soften the oxide buildup that occurs on furnace walls. These fluxes can often be applied with a typical refractory gunning device.

Proper Use of Fluxes. Fluxing compounds must be used at temperatures specified by the flux manufacturer to ensure proper reactivity capability. Fluxing compounds are usually hygroscopic to a certain extent and therefore must be stored in a safe, dry place. They should never be used

(a)

(b)

Fig. 3 Comparison of metal-rich (a) and powdery, low-metallic dross (b) from aluminum or zinc alloys. (b) is obtained by proper fluxing.

if they are wet, because of the possibility of an explosive reaction. In addition, moisture-wet fluxing material will introduce hydrogen into the aluminum melt, which has great solubility for hydrogen in the liquid state. Tools (shovels, scrapers, rakes, perforated spoons) used to handle fluxes should be kept clean and dry and should always be preheated above 95 °C (200 °F) before use to avoid spitting and popping when they come into contact with the melt.

Flux should be applied in accordance with the manufacturer's recommendations, which have evolved over many years of experience. Cover fluxes can be spread over the melt by hand or with a shovel. Drossing fluxes usually need to be worked into the dross layer. The flux should be thoroughly mixed with this dross layer by rabbling with a rake. Care must be taken not to bring up too much new molten metal into this dross layer when rabbling.

When any flux is used, it should be carefully separated from the melt before pouring or casting. Thorough skimming is required to prevent flux inclusions from becoming entrapped within the cast metal.

In any situation where flux is used, a quiescent time for the bath is recommended to allow adequate settling of heavy inclusions or floating out of lighter-density fluxing salts and flux-wetted oxide inclusions. Optimal settling time may vary from 5 to 10 min for a small crucible melt to 1 to 2 h for a large (45 Mg, or 50 ton) furnace. More information on the fluxing of aluminum alloys is available in Ref 4 to 6 (see also the article "Aluminum and Aluminum Alloys" in this Volume).

Fluxing of Magnesium Alloys

Magnesium and its alloys are exceptionally susceptible to oxidation, melt loss, and fires because of the extreme reactivity between magnesium and oxygen. Protection is therefore always required when melting this alloy family. Historically, this has involved the so-called flux process, which uses salt fluxes as a cover, or more recently the fluxless process, which uses inert gas. Although the latter is now preferred for quality, environmental, and cost reasons, there is still considerable use of the flux process in many magnesium alloy melting operations (see the article "Magnesium and Magnesium Alloys" in this Volume).

Molten magnesium oxidizes readily to form a magnesium oxide (MgO) film. This film is easily disturbed, and discontinuous MgO liquid film inclusions readily wet and coat solid charge materials and can also entrain liquid metal. Fluxes are therefore used to protect the melt from oxidation, to agglomerate nonmetallic inclusions originating with the charge, and to break up and collect the oxide inclusions and skins that may form during melting. These fluxes are usually low-melting mixtures of halide salts capable of wetting both solid and liquid metal surfaces. In addition, they specifically wet MgO skins to aid in their removal. Because the density of molten magnesium alloy is so low (about 1.55 g/cm^3, or 0.056 lb/in.^{3}), it is most desirable and practical for magnesium fluxes to have a higher density than the liquid metal, facilitating better mixing of flux and metal within the melt for maximum efficiency. These fluxes then collect at the bottom of the melting vessel as a sludge that must be drawn off or removed periodically.

Types of Fluxes. A typical flux composition includes approximately 49% $MgCl_2$, 27% KCl, 20% barium chloride ($BaCl_2$), and 4% CaF_2 (Ref 7). The magnesium and potassium chloride salts provide the low-melting eutectic; the fluoride, the surface wettability and chemical reactivity with magnesium oxide; and the heavy barium chloride salt constituent, the density component to effect mixing and sludging capability for separation. Magnesium chloride in the flux also appears to minimize or extinguish burning by providing a thin-film layer on the metal surface, thus retarding magnesium/oxygen reactivity. Other useful cover fluxes include a simple mixture of sulfurous compounds with fluoborate salts or boric acid. However, these merely prevent surface burning and do not clean the melt.

Certain magnesium alloys contain rare-earth elements (mischmetal) as an alloy requirement. Fluxes containing substantial quantities of $MgCl_2$ tend to react with these rare-earth elements. For this reason, most, if not all, of the $MgCl_2$ in the flux can be replaced by more stable $CaCl_2$. However, it is desirable to retain a small amount of $MgCl_2$ to avoid burning on the surface. Therefore, suitable fluxes for rare-earth alloys contain $MgCl_2$ in slight excess only of the CaF_2 content. One flux composition for rare-earth alloys is 28% $BaCl_2$, 26% $CaCl_2$, 10% KCl, 16% NaCl, and 20% MgF_2 (Ref 7).

Use of Fluxes. All flux materials should be kept clean and dry and should be stored in their original containers. All tools used with fluxes should be clean, dry, and preheated to drive off any surface moisture and to minimize thermal shock when placed into the melt.

When melting only ingot or very clean scrap, only a dry flux powder cover to prevent burning or a liquid flux sufficient to melt and form a continuous flux blanket over the molten metal surface need be used. This can be as little as 1% by weight of the melt. However, more flux may be required when melting dirty charge materials, such as gates and runners with oxide and sand debris, scrap castings, or die casting scrap containing lubricants and other contaminants. Indeed, a fluid cover flux is desirable to agglomerate impurities as the dirty charge melts.

The chloride-containing salts used as cover fluxes are hygroscopic and will react in time with moisture in the air to form HCl and oxychloride. This is especially true with $MgCl_2$. Therefore, a flux cover will lose fluidity and begin to crust over, and it is also possible for fine oxide particles to become airborne. When holding molten magnesium alloy for any period of time, it is necessary to replenish the flux cover with fresh flux.

It is often desirable when melting ingot and scrap materials to charge directly into an already molten pool of flux. In this way, the flux precoats the solid metal as it melts to prevent the burning of clean metal, to clean obvious debris, and to prevent further oxidation (Ref 8).

Flux refining is widely used. In this technique, additional flux is mixed into the melt to come into greater contact with the oxide skins and other nonmetallic inclusions (silica sand, and so on) to render them more easily removable. Stirring is usually accomplished with graphite, cast iron, low-carbon steel, or 400-series stainless steel tools. Overmixing, agitation, or violent stirring must be avoided, because the flux globules will break up into a too-fine dispersion that will retard the desired settling of dross or sludge. It is important to allow sufficient settling time when using the flux process for melt refining so that the salt-wetted inclusions can settle into the sludge properly. Settling time is usually at least 10 to 20 min, but it varies with the size of the melt, the nature of the charge, and the amount of flux added.

In addition, the temperature at which flux refining is carried out must be carefully controlled. A normal temperature of about 705 °C (1300 °F) is preferred; this provides the optimum surface energy characteristics between liquid flux and molten metal to achieve good inclusion agglomeration without excessive flux dispersion into fine globules. Lowering the temperature of the melt from 705 °C (1300 °F) to closer to the casting temperature after stirring also helps achieve better flux/metal separation.

Flux/metal separation may be difficult in some magnesium alloys containing zirconium along with iron, zinc, and magnesium. Insoluble metallic compounds of these elements may be present in concentrations exceeding the solubility limit. These insoluble compounds generally collect at the flux/metal interface and favor continued dispersion of the flux globules rather than recoalescence and settling to the sludge (Ref 9).

As with the use of fluxes in other alloy systems, good housekeeping, sludge removal, skimming, and proper handling and pouring practices of the molten alloy must be observed to minimize contamination by flux inclusions. More detailed information on the basis of fluxing processes used for magnesium alloys is available in Ref 7.

Fluxless Melting. The problems associated with the flux melting and refining of magnesium alloys—for example, melt loss, inclusions, and flux disposal problems—have led to the development of the fluxless process. The first protective atmosphere used sulfur dioxide (SO_2). This gas, however, is toxic and malodorous (and thus requires containment and ventilation), and it was necessary to direct the gas atmosphere into direct contact with the molten metal surface to provide satisfactory protection. In addition, SO_2 presents the possibility of an explosion hazard in order to provide satisfactory protection. Therefore, an improved system was still needed.

Sulfur hexafluoride (SF_6) is now commonly used in the fluxless melting process. This gas satisfies the objective of an alternative to salt fluxes, thus eliminating the possibility of salt flux inclusions and high melt loss. In addition, the corrosive nature of salt flux fumes is avoided. The SF_6 gas is also nontoxic, odorless, and safe (the explosion hazard associated with SO_2 is not present with SF_6).

It is generally believed that SF_6 reacts with molten magnesium to form a continuous layer of MgF_2 on the surface of the melt, serving as a barrier to prevent oxygen from reaching the melt surface and to prevent burning (Ref 10). Because the density of SF_6 is greater than that of air, it readily covers the melt surface and displaces air and oxygen from just above the melt. Very low concentrations of SF_6 are required to achieve the desired result, and it is usually mixed with a less expensive inert carrier gas such as carbon dioxide (CO_2) or nitrogen. Only 0.5 to 1.0% SF_6 is needed. Above 0.8%, SF_6 may cause unwanted corrosion problems with iron and steel equipment used in the melting process and in the environment (Ref 11). It is also necessary to provide proper hooding or sealing of the furnace for optimum economy and to prevent oxygen and air ingestion from drafts.

Fluxing of Copper Alloys

Fluxing practice in copper alloy melting and casting encompasses a variety of different fluxing materials and functions. Fluxes are specifically used to remove gas or prevent its absorption into the melt, to reduce metal loss, to remove specific impurities and nonmetallic inclusions, to refine metallic constituents, or to lubricate and control surface structure in the semicontinuous casting of mill alloys. The last item is included because even these fluxes fall under the definition of inorganic chemical compounds used to treat molten metal.

Types of Fluxes. Fluxes for copper alloys fall into five basic categories: oxidizing fluxes, neutral cover fluxes, reducing fluxes (usually graphite or charcoal), refining fluxes, and semicontinuous casting mold fluxes.

Oxidizing fluxes are used in the oxidation-deoxidation process; the principal function here is control of hydrogen gas content. This technique is still practiced in melting copper alloys in fuel-fired crucible furnaces, where the products of combustion are usually incompletely reacted and thus lead to hydrogen absorption and potential steam reaction (see the section "Degassing of Copper Alloys" in this article). The oxidizing fluxes usually include cupric oxide or manganese dioxide (MnO_2), which decompose at copper alloy melting temperatures to generate the oxygen required. Figure 4 illustrates the effectiveness of oxidizing fluxes in reducing porosity due to hydrogen and in improving mechanical properties for a tin bronze alloy.

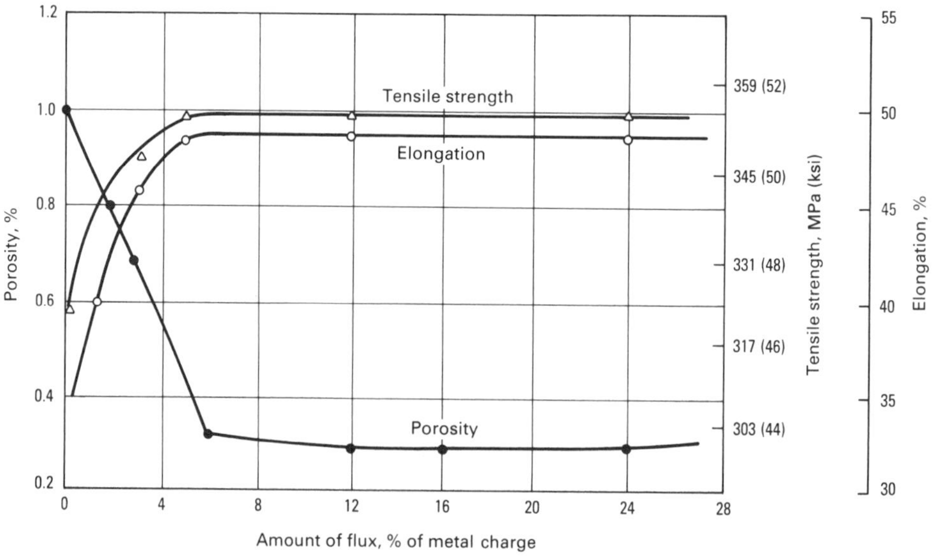

Fig. 4 Effect of amount of flux used on the porosity and mechanical properties of cast tin bronze alloy. Source: Ref 12

Neutral cover fluxes are used to reduce metal loss by providing a fluid cover. Fluxes of this type are usually based on borax, boric acid, or glass, which melts at copper alloy melting temperatures to provide a fluid slag cover. Borax melts at approximately 740 °C (1365 °F). Such glassy fluxes are especially effective when used with zinc-containing alloys, preventing zinc flaring and reducing subsequent zinc loss by 3 to 10%. The glassy fluid cover fluxes also agglomerate and absorb nonmetallic impurities from the charge (oxides, molding sand, machining lubricants, and so on). As with aluminum alloys, fluxes containing reactive fluoride salts (CaF_2 and NaF) can strip oxide films in copper-base alloys, thus permitting entrained metallic droplets to return to the melt phase. Table 2 indicates the effectiveness of this type of flux in reducing melt loss in yellow and high-tensile brass. For red brasses, however, it may not be proper to use a glassy flux cover, because such a cover will prevent or limit beneficial oxidation of the melt (see the section "Degassing of Copper Alloys" in this article). Use of a glassy cover flux can sometimes result in reduced alloy properties (Ref 12).

Oxide films in aluminum and silicon bronzes also reduce fluidity and mechanical properties. Fluxes containing fluorides, chlorides, silica, and borax provide both covering and cleaning, along with the ability to dissolve and collect these objectionable oxide skins. Chromium and beryllium-copper alloys oxidize readily when molten; therefore, glassy cover fluxes and fluoride salt components are useful here in controlling melt loss and achieving good separation of oxides from the melt.

Reducing fluxes containing carbonaceous materials such as charcoal or graphite are used on higher-copper lower-zinc alloys. Their principal advantage lies in reducing oxygen absorption of the copper and reducing melt loss. Low-sulfur, dry, carbonaceous flux materials should always be used with copper alloys to avoid gaseous reactions with sulfur or with hydrogen from contained moisture. However, carbonaceous materials will not agglomerate nonmetallic residues or provide any cleaning action when melting fine or dirty scrap. For this reason, a glassy cover must also be used in the latter case. Table 2 indicates the beneficial effects of a glassy cover when melting brass turnings.

Use of Fluxes. As with all fluxes, the flux materials for copper alloys should be stored in a clean dry place and should be kept covered to prevent contamination from unwanted materials. Manufacturer's recommendations should always be followed. The tools used must be kept clean and dry. Because many flux materials are hygroscopic, tools with flux residue from previous use will pick up moisture. Any residual flux should be cleaned from the tool, and the tool should be preheated above 105 °C (225 °F) before its reuse. Proper rabbling must be done when required for optimum flux reactivity, and fluxes should be thoroughly skimmed from the melt surface before or during pouring to prevent flux entrapment during the casting process. Unlike aluminum or magnesium, most copper fluxes have significantly lower density than the base metal; this permits easy gravity separation before skimming and facilitates skimming.

Melt Refining of Copper Alloys. It is possible to remove many metallic impurity constituents from copper alloys through the judicious use of fire refining (oxidation). According to standard free energy of reaction (Fig. 5), elements such as iron, tin, aluminum, silicon, zinc, and lead are preferentially oxidized before copper during fire refining (Ref 13), and there is an order of preference for their removal (Fig. 6). These metallic impurities are thus rendered removable if the oxide product formed can be adequately separated from the melt phase itself. A wet cover flux such as borax is useful with fire refining because it will agglomerate the impurity metal oxides formed and minimize the metal content of the dross.

The need to refine specific metallic impurities is highly dependent on and variable with the specific alloy system being refined. An alloying element in one family of copper alloys may be an impurity in another, and vice versa. In red brass (Cu-5Zn-5Pb-5Sn; UNS/CDA C83600), the el-

Table 2 Effect of various slags and covers on losses in the melting of high-tensile and yellow brass

Alloy	Melting conditions	Metal temperature °C	Metal temperature °F	Melting time, min	Gross loss, %	Net loss, %
Yellow brass	No lid or cover	1085	1985	50	2.8	1.8
Yellow brass	Charcoal	1087	1989	49	1.1	0.6
Yellow brass	Glassy cover flux	1100	2012	62	0.9	0.4
High-tensile brass	No lid or cover	1090	1994	65	2.5	1.2
High-tensile brass	Charcoal	1095	2003	60	1.9	0.7
High-tensile brass	Cleaning cover flux	1090	1994	54	0.6	0.3

Source: Ref 12

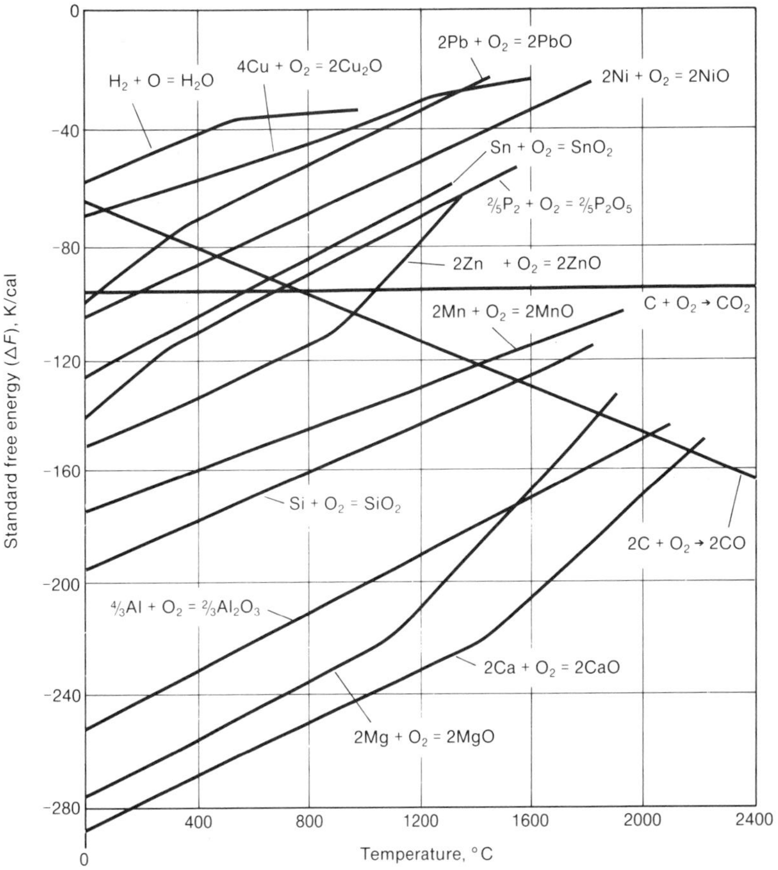

Fig. 5 Free energy changes for various metal oxidation reactions. Source: Ref 12

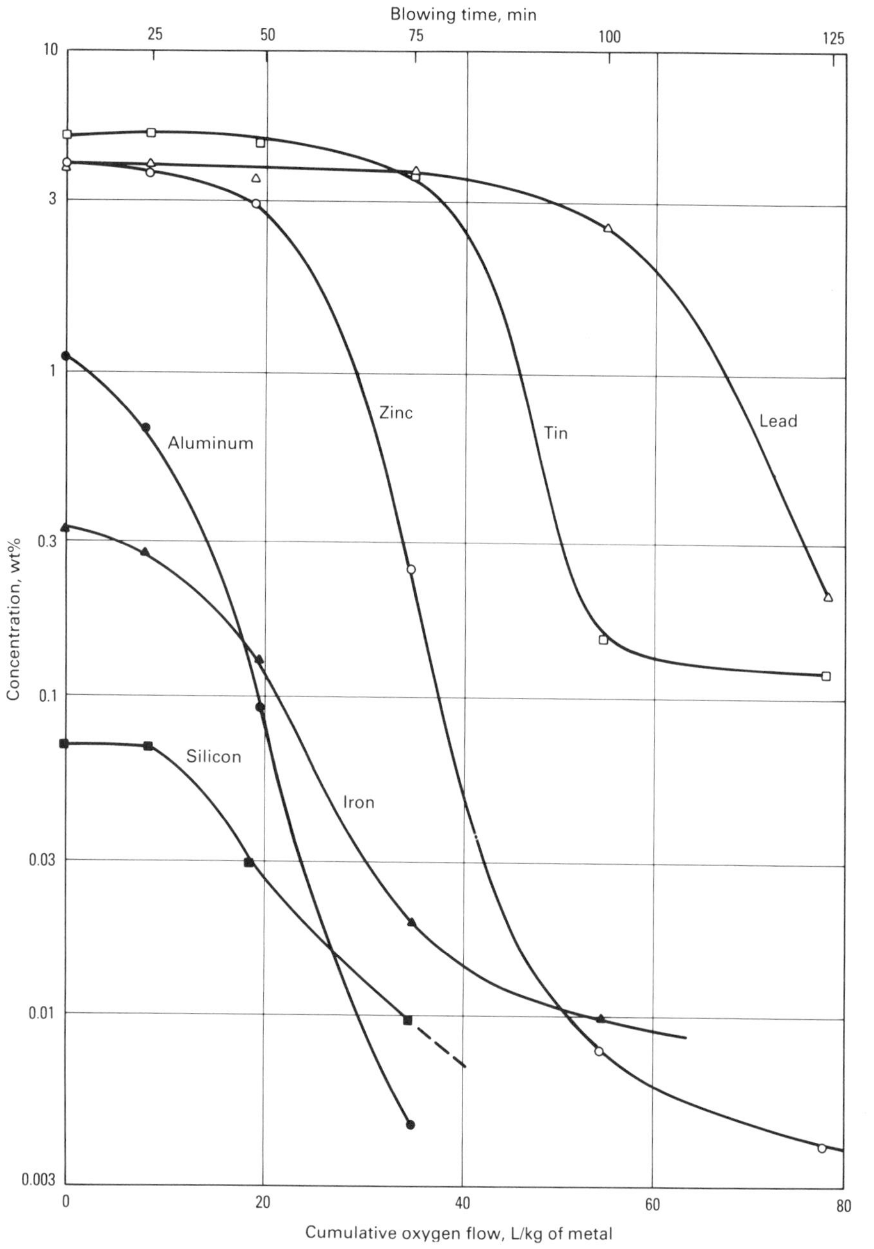

Fig. 6 Effect of fire refining (oxygen blowing) on the impurity content of molten copper. Source: Ref 14

ements lead, tin, and zinc are used for alloying, while aluminum, iron, and silicon are impurities. In aluminum bronzes, on the other hand, lead, tin, and zinc become contaminants, while aluminum and iron are alloying elements.

Foundries typically do little melt refining, leaving this assignment to the secondary smelter supplier of their foundry ingot. However, there may be certain instances when additional refining capability is necessary within the foundry or mill. Table 3 gives the results of fire refining a melt of C83600 with aluminum, silicon, and iron contaminants under a variety of flux covers.

Fire refining (oxidation) can be used to remove impurities from copper-base melts roughly in the following order: aluminum, manganese, silicon, phosphorus, iron, zinc, tin, and lead. Nickel, a deliberate alloying element in certain alloys but an impurity in others, is not readily removed by fire refining, but nickel oxide can be reduced at such operating temperatures. Mechanical mixing or agitation during fire refining improves the removal capability by increasing the reaction kinetics. Removal is limited, however, and in dilute amounts (<0.05 to 0.10%) many impurities cannot be removed economically.

Oxygen-bearing fluxes can be effective in removing certain impurities, although they are less efficient than direct air or oxygen injection. Figure 7 demonstrates the effect of increasing the copper oxide content of a flux in removing iron and zinc from phosphor bronze.

Lead has been removed from copper alloy melts by the application of silicate fluxes or slags. The addition of phosphor copper or the use of a phosphate or borate slag flux cover and thorough stirring improves the rate of lead removal, as shown in Fig. 8 (Ref 15).

Sulfur, arsenic, selenium, antimony, bismuth, and tellurium can occur as impurities in copper alloy scrap, foundry ingot, and prime metal through incomplete refining of metal from the ore, electronic scrap, other scrap materials, or cutting lubricant. These impurities can largely be controlled by application and thorough contacting with fluxes containing sodium carbonate (Na_2CO_3) or other basic flux additives such as potassium carbonate (K_2CO_3). Figure 9 demonstrates the ability of Na_2CO_3 fluxes plus fire refining in eliminating arsenic, bismuth, and antimony from copper.

Sulfur is a harmful impurity in copper-nickel or nickel silver alloys. It can be removed from these materials by an addition of manganese metal or magnesium.

Aluminum is often a contaminant in copper alloy systems, particularly the leaded tin bronzes and red brasses. Porosity and lack of pressure tightness result when the aluminum content is as little as 0.01%. Aluminum can be removed by a flux containing oxidizing agents to oxidize the aluminum, and fluoride salts to divert the Al_2O_3 from the melt and render it removable. Silicon can also be removed, but only after the aluminum has reacted. As much as 0.3% contaminant can be reduced to less than 0.01%, ensuring pressure tightness, by using a flux consisting of 30% NaF, 20% CaF_2, 20% Na_3AlF_6, 20% Na_2SO_4, and 10% Na_2CO_3 at an addition rate of 1 to 1.5% for 10 min at about 1100 °C (2010 °F) (Ref 17). As usual, the flux must be intimately mixed with the melt to ensure good reactivity. The melt should then be allowed to settle, and the flux residue or slag layer should be thoroughly skimmed.

Borax is useful as a flux constituent for refining to provide adequate fluidity and to agglomerate flux-reacted impurity oxides without excessive entrapment and loss of alloying elements. The borax mineral fluxes razorite and colemanite are commonly used in secondary smelting practice in converting copper alloy brass and bronze scrap to specified-composition foundry ingot.

Chlorine fluxing also has potential for refining impurities from copper-base melts, particularly magnesium, aluminum, manganese, zinc, iron, lead, and silicon. However, very little chlorine refining is practiced commercially; the process may be cost effective only when removing aluminum.

Mold Fluxes. Certain mold-lubricating fluxes have been used in the direct chill

Table 3 Effect of fire refining under various fluxes on impurity levels in leaded red brass

50 kg (110 lb) heats were melted under 1 kg (2.2 lb) of flux at 1150 °C (2100 °F).

Melt No.	Flux	Refining time, min	Oxygen used, liters per kilogram of metal	Amount of zinc in refined metal, %	Impurities in refined metal, %		
					Al	Si	Fe
1	Borax	0	0	5.49	0.05	0.08	0.38
		10	1.5	5.66	0.015	0.05	0.38
		20	2.9	5.00	0.011	0.02	0.29
2	Borax-25% sand	0	0	6.30	0.14	0.13	0.37
		10	1.5	6.24	0.017	0.12	0.42
		20	2.9	6.45	0.007	0.12	0.41
3	Borax-20MnO	0	0	7.40	0.21	0.26	0.47
		10	1.5	7.35	0.05	0.19	0.52
		20	3.1	7.33	0.017	0.12	0.51
4	$70CaF_2$-$20Na_2SO_4$-$10Na_2CO_3$	0	0	7.47	0.10	0.08	0.62
		10	1.5	7.07	<0.005	0.02	0.55
		20	3.1	6.69	<0.005	<0.005	0.45
5	30NaF-$20CaF_2$-20KF-$20Na_2SO_4$-$10Na_2CO_3$	0	0	7.35	0.23	0.38	0.70
		10	1.8	6.74	0.005	0.17	0.83
		20	3.5	7.20	<0.005	0.08	0.88

Source: Ref 14

semicontinuous casting of brass and copper alloys into semifinished wrought shapes. These fluxes serve to protect the metal from oxidation during casting. They also act as lubricants so that the solidifying skin separates easily from the mold wall as the solidifying billet or slab moves downward from the mold during casting. Especially in brass alloys, zinc flaring and zinc oxide (ZnO) formation on the melt surface reduce lubricity, causing tearing and other undesirable skin defects during solidification that are detrimental to subsequent forming operations. Fluxing compounds are used on the melt surface feeding the mold to alleviate these problems. Fluxes used to alleviate this problem usually contain borax, fluoride salts, soda ash, and eutectic salt mixtures to ensure that the flux is molten as the cast continues. In the direct chill casting of higher-copper alloys, graphite may also be present in such fluxes. Because solid flux particles can cause inclusions in the solidifying skin, the flux must be free of coarse particles and must melt quickly.

Fluxing of Zinc Alloys

In the melting of clean, pure zinc and zinc alloys, there is little need for cover fluxes to protect the melt because zinc does not oxidize appreciably or absorb hydrogen at normal melting temperatures. However, chloride-containing fluxes that form fluid slag covers can be used to minimize melt loss if they are carefully skimmed from the melt before pouring. When melting dirty metal, scrap returns, and the like, both a cover flux and a reactive flux are advantageously used to separate entrapped metal from oxides.

The principal alloying elements present in zinc die casting alloys include aluminum, magnesium, and copper. Manganese and silicon may be present as impurities in the remelt; chromium and nickel are often seen when plated scrap is remelted. In most cases, zinc die casting alloys can tolerate these impurity amounts up to the limit of their solubility during solidification (generally <0.02%). The quantities of these elements present in excess of solubility limits will form oxides and/or complex intermetallic compounds, especially when iron is present (from melting or casting vessels). It is possible to collect some of these impurity constituents, particularly oxides, in the dross phase; therefore, a fluid slag or flux cover is beneficial. At typical zinc melt operating temperatures, various mixtures of zinc chloride ($ZnCl_2$), KCl, and NaCl form the basis of such fluxes. These fluxes will also agglomerate nonmetallic residues and contaminants from dirty scrap and will help cleanse charge materials during remelting (Ref 18).

The zinc alloy drosses that form during melting, holding, turbulent transfer, and remelt operations can contain as much as 90% entrapped finely divided free metallic zinc, in addition to oxides, intermetallic compounds, and other debris. An exothermic drossing flux, usually containing nitrate salts and silicofluoride double salts, can be used to recover much of this entrapped zinc. The exothermic reaction, assisted by rabbling or raking the flux into intimate contact with the dross phase, serves to raise the local temperature, increase fluidity, decrease the surface tension of the oxide skin, and chemically reduce ZnO.

Proper fluxing procedures will permit more than half of the entrapped zinc to be recovered, yielding only about 35 to 40% unrecoverable and oxidized fines. The amount of flux to be added is often determined empirically and is usually 1 to 4.5 kg (2 to 10 lb) per ton of metal being treated, but this may vary considerably. Suitable melt temperatures are usually above 480 °C

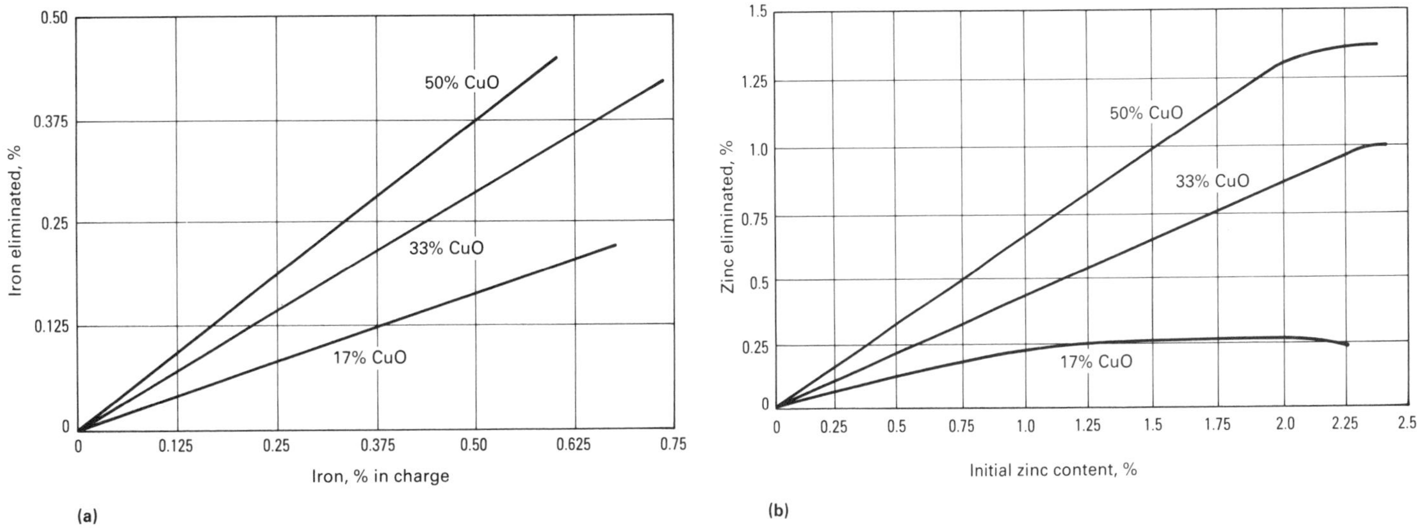

Fig. 7 Influence of oxygen content of flux in reducing iron (a) and zinc (b) impurities in phosphor bronze. Source: Ref 12

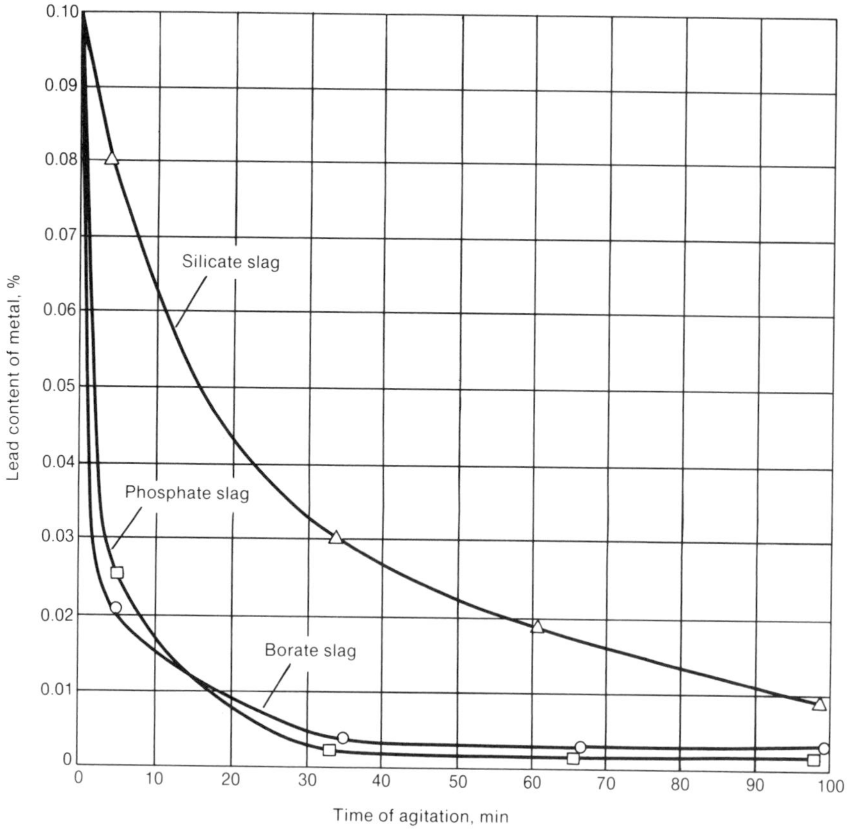

Fig. 8 Removal of lead from oxidized Cu-0.1Pb melts at 1150 °C (2100 °F) by different slags (fluxes) 2% of charge weight. Melt stirred with nitrogen at 2 L/min. Source: Ref 15

(900 °F). As the flux is worked into the dross, the rich metallic appearance will give way to a drier, duller, and more powdery residue ready for final skimming (Fig. 3).

The dross reclamation process is often carried out directly on zinc remelt pot furnaces. Alternatively, a separate dross mixer consisting of a vessel and a rotating paddle is used in which hot drosses and flux are added, mixing begins, and the fluid metal being recovered runs out the bottom and is collected in an ingot mold beneath the dross mixer.

Galvanizing Fluxes

Galvanizing is the practice of applying a protective zinc or zinc alloy coating on iron or steel surfaces. The zinc oxidizes or is attacked preferentially and thus protects the underlying metal from corrosion.

In hot dip galvanizing, the articles to be coated are immersed in molten zinc in a galvanizing tank or kettle at a temperature ranging from 450 to 460 °C (840 to 860 °F). A molten salt flux is usually maintained on the surface of the kettle, serving a number of important functions. The flux serves as a preheating and drying medium to reduce sputtering and explosions, and it keeps the galvanizing zinc metal itself free from oxides. The flux also precleans the article being galvanized.

The principal ingredients of galvanizing fluxes include ZnCl, ammonium chloride (NH_4Cl), zinc ammonium chloride ($ZnCl_2 \cdot NH_3$), and occasionally other chloride salts used as extenders. Typical reactions that take place in the flux during the galvanizing process include:

$$2NH_4Cl + Zn \rightarrow H_2 + NH_3 + ZnCl_2 \cdot NH_3 \quad \text{(Eq 1)}$$

$$2NH_4Cl + FeO \rightarrow H_2O + 2NH_3 + FeCl_2 \quad \text{(Eq 2)}$$

$$FeCl_2 + Zn \rightarrow Fe + ZnCl_2 \quad \text{(Eq 3)}$$

$$2NH_4Cl + ZnO \rightarrow H_2O + NH_3 + ZnCl_2 \cdot NH_3 \quad \text{(Eq 4)}$$

Equation 1 explains the occurrence of zinc consumption, hydrogen fires in the flux, and ammonia odors. Equations 2 and 3 show how the flux dissolves iron oxide and how elemental iron is freed both to alloy with zinc (desirable) and to form dross (undesirable). When fresh flux containing the active ingredient NH_4Cl is added to partially spent flux, the high-melting viscous basic zinc chlorides (shown as ZnO) are converted to the lower-melting zinc ammonium chlorides. Figure 10 shows the phase diagram of the $ZnCl_2/NH_4Cl$ system. At a normal operating temperature of 455 °C (850 °F), NH_4Cl would vaporize. However, the liquidus boundary indicates that as the mixture becomes richer in $ZnCl_2$, the mixture of chloride can contain 3 to 4% NH_4Cl and remain a liquid rather than become a gas. Foaming agents (glycerin) are often added to slow the loss of NH_4Cl due to vaporization.

The working galvanizing flux is therefore metastable. Because NH_4Cl is lost through vaporization and is used up in reactions, fresh flux containing NH_4Cl must be added continuously to keep the flux active. The viscosity of the flux will also increase because of the loading or buildup of high-melting zinc and iron oxides and debris carried in with the steel articles. The addition of fresh flux reduces viscosity by converting these compounds to ammoniated zinc chloride.

Aluminum is often added (usually $<0.1\%$) to galvanizing baths to serve as a brightener for the zinc coating on the workpiece. Aluminum oxidizes readily and will also form $AlCl_3$. The viscosity of the flux is increased with the presence of Al_2O_3, and $AlCl_3$ formation depletes the flux of active NH_4Cl.

Flux Injection

Flux injection is a relatively new process in which fluxing compounds are introduced into the molten metal by a mechanical device using an inert gas carrier. Typical components of the flux injection process, shown schematically in Fig. 11, consist of a flux feeder supply, a mixing chamber for the inert gas carrier and the flux, and a lance for injecting the flux. Appropriate nozzle mixing technology and the use of special lance materials and insulated construction have been developed by the manufacturers of flux injection equipment to overcome clogging and premature melting or reaction of flux materials within the lance.

Current manual application of fluxing compounds usually involves surface application and subsequent manual raking or rabbling of the flux to achieve good mixing and reactivity. Essentially, the manual process is labor intensive and is often merely a surface or near-surface treatment limited to the dross and the molten metal immediately adjacent to the dross layer. Flux injection permits better reactivity and mixing through the bubbling action of the inert gas carrier, and it also permits refining of the full molten metal bath because it submerges the fluxing materials to substantial depths within the melt. The fluxing materials can thus be used to full efficiency. The injection process can be automated for specific injection amounts and treatment times, thus minimizing waste. Additional benefits of the flux injection process include reduced treatment times over manual methods, less metal loss in the dross, and fewer emissions.

It has been reported that all aspects of molten metal treatment are possible with flux injection. Most of these techniques are being practiced in the aluminum industry, though not extensively. There has been little activity thus far with other nonferrous

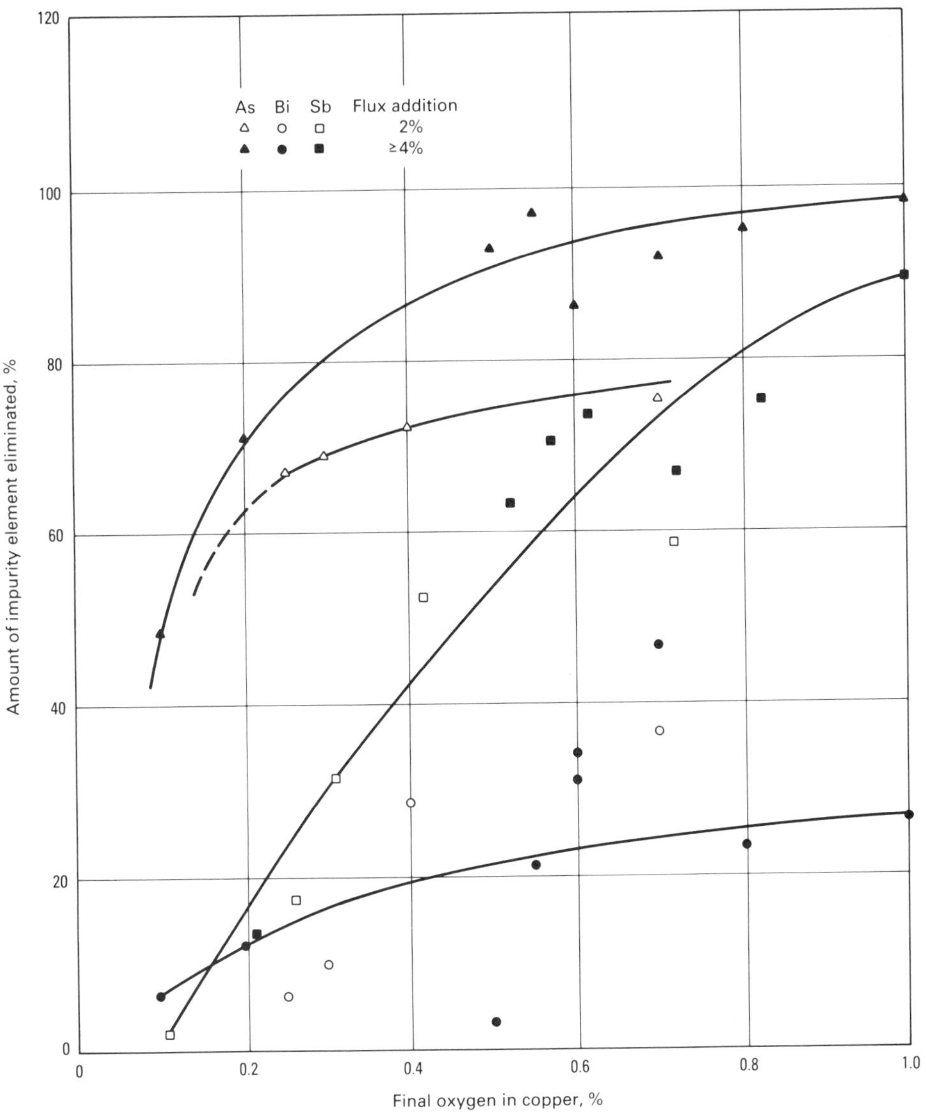

Fig. 9 Effect of fire refining and use of Na_2CO_3 flux on removal of arsenic, bismuth, and antimony impurities from copper. Source: Ref 16

metals. Capabilities of flux injection are summarized below.

Degassing can be accomplished with the carrier gas itself or with an inert gas plus compounds such as hexachloroethane. The hexachloroethane decomposes to form chlorine, which then forms an $AlCl_3$ gas bubble that collects hydrogen gas as it ascends to the melt surface. At an addition rate of 0.2%, a 10-min flux injection treatment is reported to be capable of reducing hydrogen content from 0.35 mL/100 g to as little as 0.05 mL/100 g (Fig. 12a). This level of hydrogen is more than acceptable for all aluminum foundry alloy applications.

Nonmetallic inclusions can be removed by injecting salt fluxes that dewet these particles from the molten aluminum. Varying amounts of flux can be added in the range of 0.2 to 0.5% by weight of the entire melt. A settling time or standing time of the melt (usually 5 min for small foundry melts) is required to allow flux particles and nonmetallic inclusions to agglomerate and float to the surface for removal.

Hypoeutectic Alloy Modification. In aluminum alloys with less than 12% Si, the as-cast structure is very coarse, with an acicular silicon eutectic phase. Modifiers are therefore added to cause this structure to be a more rounded, dispersed phase with better mechanical properties. Modification is usually achieved by using sodium, strontium, calcium, or antimony in the form of pure metals, master alloys, or salt fluxes that contain these modifiers. Currently, injection fluxes containing sodium salts are available and are used at an addition rate of 0.2 to 0.4% to achieve modification.

The modification process, however, is sensitive to several variables, and reversion of the structure back to the unmodified state is well recognized. Higher silicon contents, temperatures, turbulence, standing times, and melt surface areas favor reversion.

Grain Refinement. The flux injection process can be used to grain refine aluminum melts. Conventional applications would use master alloys or salts containing titanium and boron, which decompose and recombine to form a fine titanium diboride particle that serves as a nucleating agent during the solidification of the cast structure. Titanium carbides have also been used. It has been reported that a titanium addition in the range of 0.45 to 0.90% and a carbon level of 0.003 to 0.005% are sufficient to form the titanium carbide sites required for grain nucleation (Ref 21). The dispersion capabilities of the fluxing injection process enhance the effective use of whichever titanium compound and grain-refining mechanism is employed.

Alkali Metal Removal. Calcium, sodium, lithium, and magnesium impurities can be removed from aluminum alloy melts to a large extent by employing salt fluxes and/or reactive gas mixtures that contain chlorine and fluorine. Such halogen compounds react preferentially with the impurities to form chlorides and/or fluorides with different densities, which will then separate from the base melt. Figure 12(b) shows the results of alkali impurity removal with a 0.36% by weight flux injection.

Hypereutectic Silicon Refinement. Hypereutectic aluminum alloys (>12% Si) form a primary silicon phase upon solidifying. Although these alloys depend on the properties provided by silicon (strength, wear resistance), it is important to avoid coarse primary silicon structures, which cause reduced mechanical properties and poor machinability. Refinement of this structure is therefore usually accomplished through an addition of 0.007 to 0.01% P, depending on silicon content. Pure phosphorus is an explosion and fire hazard, and the commercially available phoscopper shot often employed adds copper to the melt. A nickel-phosphorus compound (Ni_3P) is reported to provide favorable phosphorus recovery and reaction.

It is expected that advances in flux injection technology will give rise to additional possibilities, including perhaps multipurpose fluxes that will perform several molten metal treatment functions in a single injection. Although currently limited to smaller foundry applications, flux injection is being explored on a larger scale to produce aluminum alloys by the metal powder or flux injection of such alloying elements as magnesium, silicon, manganese, copper, and iron (Ref 22-24).

Circulation of Aluminum Melts (Forced Convection)

In the melting of aluminum and its alloys, forced convectional heat transfer within the

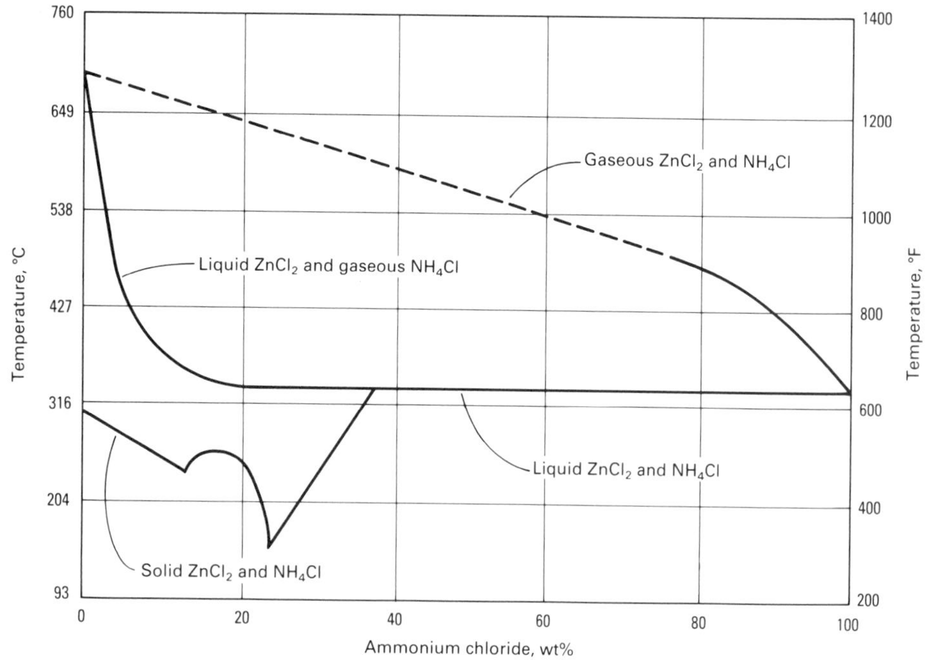

Fig. 10 Binary phase diagram for the system $ZnCl_2/NH_4Cl$. Source: Ref 19

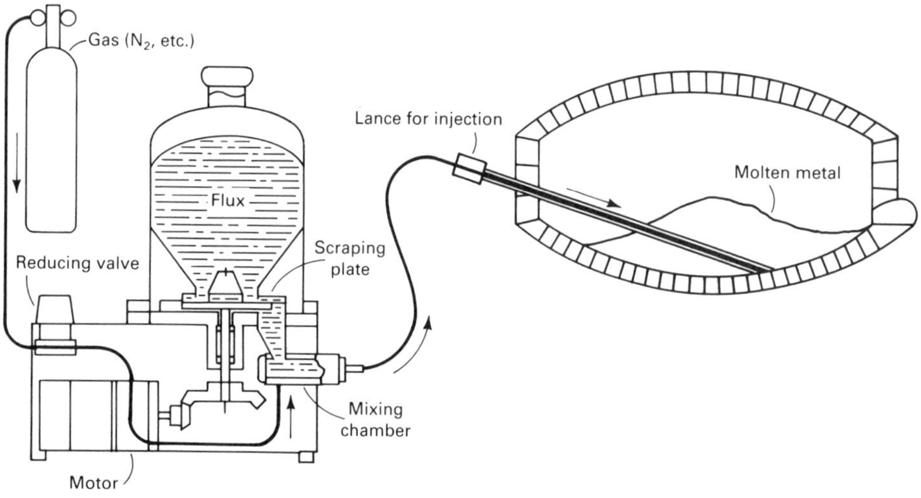

Fig. 11 Schematic of the flux injection process. Source: Ref 20

melt often plays an important role. There is a considerable latent heat of fusion (95 cal/g) associated with taking aluminum at its melting point from a solid to a liquid. Forced convection assists this transition, particularly for larger volumes of metal, by better mixing of the solid metal with already molten metal, thus reducing time and energy input into the melting process. Other nonferrous alloys, such as zinc, magnesium, and copper, could also benefit from molten metal circulation, but pumping systems for circulation are not usually employed. Mechanical stirrers are effective, particularly for zinc, owing in part to the lower heat of fusion of zinc and the smaller melt sizes usually used.

Heat Transfer in an Aluminum Furnace. In a gas- or oil-fired aluminum reverberatory furnace, heat is transferred to the bath from the burners by radiation (see the discussion of reverberatory furnaces in the article "Melting Furnaces" in this Volume). The furnace roof and walls are heated and radiate this heat energy onto the bath surface. High wall and roof temperatures create a significant thermal head to the melting furnace. Radiant heat is absorbed by the molten metal or charge to the extent of its heat capacity. Other types of heat transfer (conduction and convection) must take place to transmit thermal energy further, either in the solid charge or the molten metal. Conduction is a fundamental property of the material in question (aluminum or an aluminum alloy). Convection—heat transfer in a gas or liquid by moving masses of matter—is influenced by momentum and velocity within the molten metal. Natural convection takes place because of the density and thermal gradients within a molten metal bath, but it is extremely small in magnitude compared to forced convection.

Forced convection enables the radiant heat energy absorbed by the bath surface to move, that is, to be transmitted throughout the bath depth, so that a steady heat flux is developed and continuing heat transfer to the surface occurs by radiation. Without such forced convection, the surface layer of the molten metal becomes heat saturated, and the surface temperature rises. Ultimately, excessive surface temperature will result in oxidation and dross buildup, which reduces energy efficiency. Forced convection is produced by the circulation of the molten metal bath, either by stirring (in small pot furnaces, crucibles, and so on) or by using a molten metal circulation pump (in larger furnaces of 2000 to 110 000 kg or 5000 to 250 000 lb, capacity). Forced convection improves the heat transfer rate and energy efficiency.

The Molten Metal Pump (Ref 25, 26). Forced circulation is usually best achieved in large furnaces by using a submerged discharge molten metal pump. Although several types of pumps have appeared over the years (electromagnetic, pressure), the centrifugal pump is the most common type currently used. These pumps are usually installed in the pump wells of the open-well reverberatory furnaces used for remelting in large aluminum foundries, die casting operations, secondary smelters producing foundry ingot, and in mill products casthouses. Figure 13 shows a typical installation.

Forced circulation, which provides the highest convectional heat transfer, is related to the kinetic energy imparted to, and thus the velocity of, the moving molten metal. A centrifugal pump creates a steady flow and a discharge velocity in direct proportion to the speed of impeller rotation. Pumps operating on other principles are generally not capable of producing steady fluid flows or discharge velocities as high as those associated with centrifugal pumps. Accordingly, these systems are not as efficient or as capable of producing maximum forced circulation as a centrifugal pump. For example, one electrically driven centrifugal pump can create a discharge velocity measured in excess of 5.6 m/s (1100 ft/min), while certain electromagnetic pumps can create an equivalent velocity component of about 0.76 m/s (150 ft/min) (Ref 27).

Advantages of Forced Circulation. Centrifugal circulation pumps increase energy efficiency, productivity, and quality. Improved melting rates, reduced energy con-

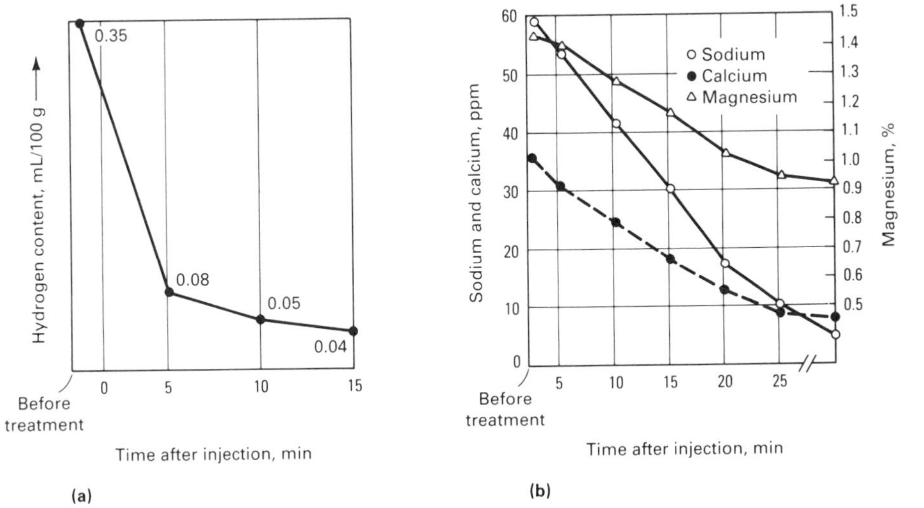

Fig. 12 Effectiveness of flux injection in removing hydrogen (a) and alkali metals (b) from an aluminum alloy melt. Source: Ref 20

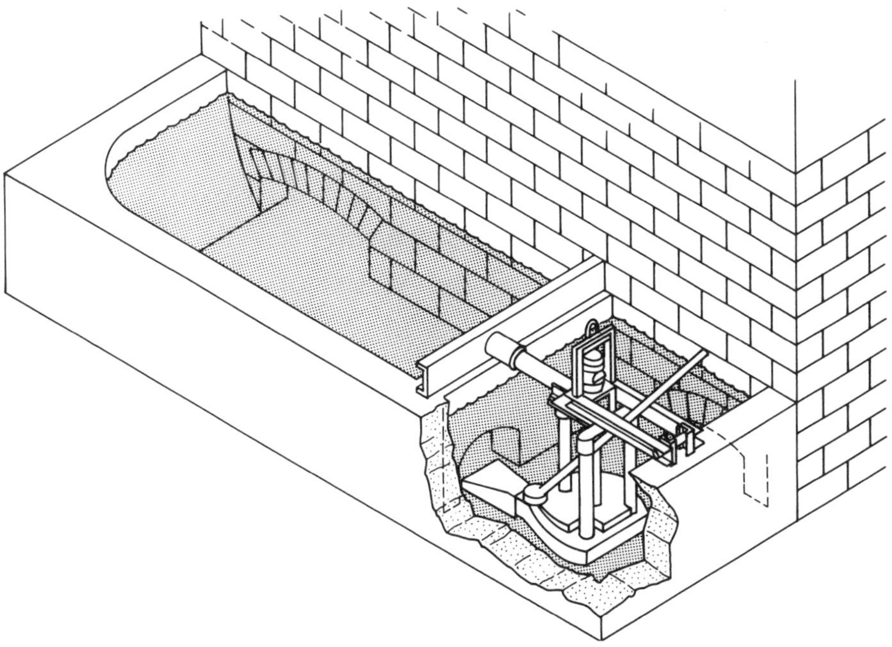

Fig. 13 Typical placement of a molten metal gas injection/circulation pump in an open-well reverberatory aluminum-melting furnace

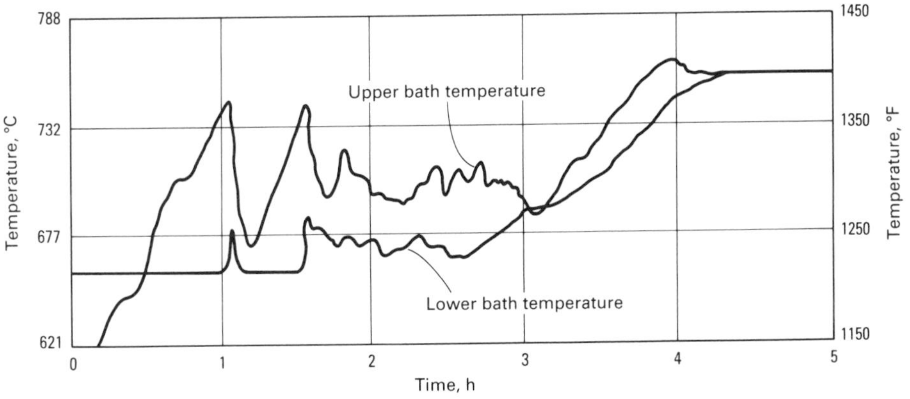

Fig. 14 Temperature homogeneity obtained in a large (86 Mg, or 95 ton) aluminum-melting furnace using a circulation pump. Source: Ref 25

sumption, elimination of stratification in the bath, and longer furnace refractory life are among the potential benefits (Ref 25, 26).

For example, 50 to 100% improvements in melt rates in remelt operations are not uncommon. In one installation, a 4.5 Mg (5 ton) charge was increased to 10 Mg (11 tons) in the same melting time. In many cases, melt cycles in large furnaces have been cut in half. Lesser improvements, although still quite cost effective, have also been realized in many installations.

Energy consumption (as measured by energy used per unit weight of aluminum produced) has also been significantly reduced. Use of a circulation pump has commonly resulted in energy savings ranging from 1160 to 4600 kJ/kg (500 to 2000 Btu/lb). Overall energy consumption in open-well scrap melting can be reduced to well below 4600 kJ/kg (2000 Btu/lb) with a pump and prudent furnace management. Smaller furnaces (<18 Mg, or 20 ton, capacity) in foundry operations can show energy usage as low as 2300 kJ/kg (1000 Btu/lb) using molten metal pumping, combined with other technologies such as heat recuperation and regenerative burner systems. Preheating the charge can also substantially reduce the energy input required for melting.

A circulation pump provides better mixing of the melt and charge, and it overcomes stratification in bath temperature and alloying inhomogeneity. In an uncirculated, large melting furnace, it is common to experience a temperature differential of 28 °C (50 °F) per foot of bath depth from bottom to top. Circulation pumps have reduced temperature variation between the top and bottom of the melt to less than 10 °C (18 °F). Figure 14 shows the temperature homogenization that has been achieved by using a centrifugal circulation pump even in a very large (86 Mg, or 95 ton) furnace. In smaller furnaces, homogenization can be achieved more quickly. Holding furnaces also benefit from the alloy concentration and temperature homogenization provided by a circulation pump.

The improved heat transfer produced by forced convection also results in lower thermal heads. Reduced roof, sidewall, and bath surface temperatures mean reduced energy costs, less dross, and longer furnace life.

Most centrifugal pumps have traditionally been powered by air motors. However, air motors require substantial lubrication and maintenance, and variable incoming air line pressure is also a potential problem. Consequently, electric drive systems (Fig. 15) are currently replacing air-powered systems in many applications. Operating costs are lower than those for air motors, and more controllable, variable-speed operation permits smoother running of the pump and less maintenance.

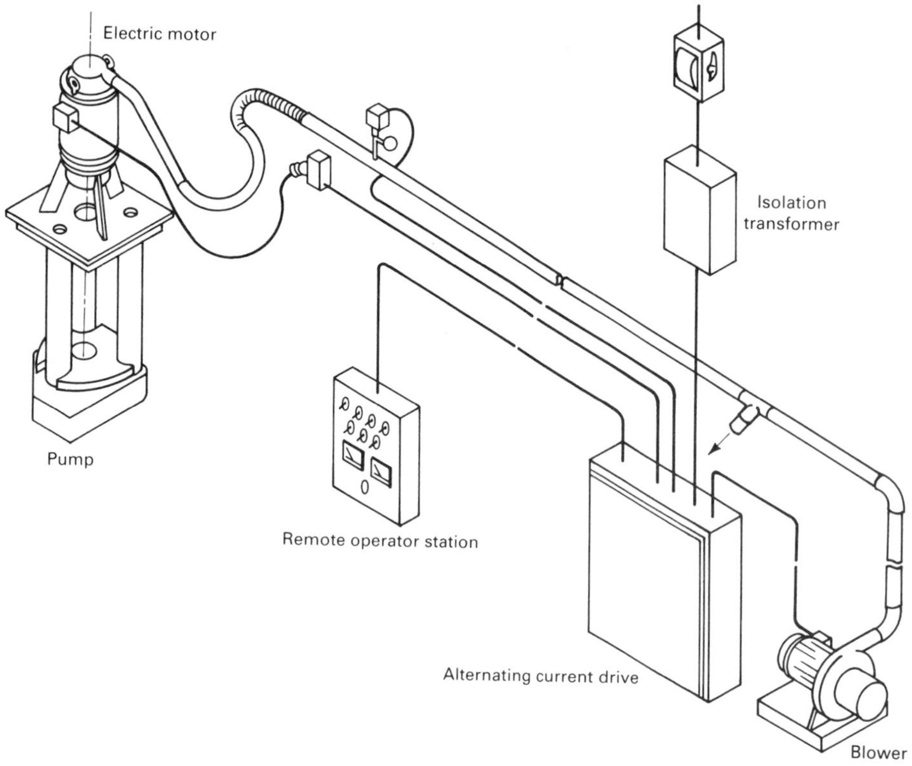

Fig. 15 Schematic of an electrically driven centrifugal circulation pump

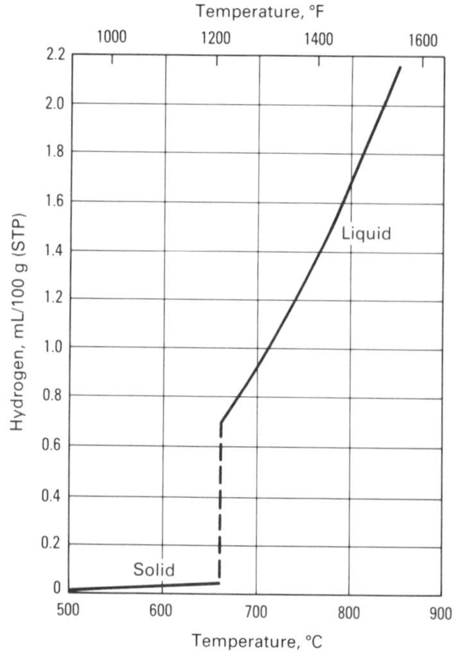

Fig. 16 Solubility of hydrogen in aluminum. Source: Ref 30

The principal advantage of an electric drive system, however, is in the higher output energy created. Typically, an air motor will produce 650 to 700 rpm in the pump, while an electric drive system can achieve 1100 to 1200 rpm. This results in substantially higher discharge velocities and therefore more forced convection.

Applications. Circulation pumps can be used to advantage in the remelting of scrap materials such as gates and risers, scrap castings, turnings and borings, used beverage containers, mill scrap, siding, and old extrusions. In addition, any large solid material (ingot, pig) for remelt can be more easily melted using forced convectional circulation. Circulation pumps are also commonly used in secondary smelting, in the production of cast extrusion billets, and in large foundries. Circulation pumps can also be used even with much smaller furnaces to facilitate the remelting of gates, runners, and scrap castings, along with foundry ingot.

Recently, pumps have been applied to direct-charged closed hearth furnaces (Ref 28, 29). In this case, a separate pump well is added to the furnace. The charge, often large pigs or ingot plus scrap, is placed directly into the hearth area of the furnace.

The advantages of using circulation pumping in this type of furnace include increased melt rate and energy efficiency improvements near 15%. Moreover, with gas injection/circulation pumps (described later in this article), it is possible to degas the melt *in situ*. This avoids an additional degassing step using flux wands, which usually requires opening of the furnace door with the attendant temperature loss. Quality improvement has been realized, along with productivity gains, in such applications.

Factors Affecting Pump Performance. Because of the hostile nature of the molten aluminum environment (heat, oxidation, corrosion), the principal materials used in the construction of molten metal pumping systems are usually graphite and ceramics. These materials provide the thermal shock resistance, corrosion resistance, and stability needed to survive the hostile conditions. However, care must be exercised during handling. Preheating and/or slow immersion into the molten metal are required. Impellers can become plugged with dross, sludge, rocks, and steel contaminants. Not only is pump performance reduced but shaft breakage can also occur.

Fluxing salts will generally attack graphite, silicon carbide, and oxide ceramics (alumina, silica), causing a buildup of corrosion products. At the melt surface, residual fluxes and moisture present in the air will combine with oxides to form a tenacious corrosion product that could alter pump performance and limit pump life. Consequently, a preventive maintenance schedule of periodic hot cleaning of the molten metal pump is usually prescribed by the manufacturer to remove buildups from the intake, discharge, and melt line on the pump.

Degassing of Aluminum Alloys

Aluminum and its alloys are very susceptible to hydrogen absorption in the molten state. Solubility is much lower in the solid state (Fig. 16). Hydrogen is the only gas that is appreciably soluble in molten aluminum. Because of the affinity of the metal for oxygen, the principal source of hydrogen absorption will be the reduction of water vapor from the atmosphere in contact with the melt:

$$3H_2O + 2Al \rightarrow 6H + Al_2O_3$$

The solubility of hydrogen gas increases exponentially as the temperature is raised. Therefore, it is imperative that excessive temperatures be avoided during melting to minimize gas absorption, especially if a standard degassing treatment is used based on a given initial expected gas content (Ref 6).

Alloy Sensitivity to Hydrogen Absorption

Various aluminum alloys have different sensitivities to hydrogen absorption and ensuing gas porosity if removal is not accomplished. The major alloying elements in aluminum foundry alloys are silicon, copper, and magnesium. Increasing silicon and copper contents decrease hydrogen solubility, although the effect is greater for copper (Ref 31). Magnesium, on the other hand, increases hydrogen solubility, and recent studies have shown that lithium behaves similarly (Ref 32).

Theory of Hydrogen Removal

Various thermodynamic and kinetic factors govern the removal of hydrogen from molten aluminum, and these factors have been extensively studied. Gas removal is easier or quicker at lower temperatures and higher concentrations, following the thermodynamic solubility changes with temperature. Kinetic factors, however, determine whether the hydrogen concentration will follow thermodynamic solubility and how close the concentration will come to equilibrium.

Hydrogen is removed by (Ref 33):

- Hydrogen transport in the melt to the vicinity of an inert gas bubble by a combination of convection and diffusion
- Diffusive transport through a thin, stagnant layer of fluid, called a boundary layer, surrounding the bubble
- The chemical adsorption onto and subsequent desorption from the bubble surface
- Diffusion of hydrogen as a gas inside the bubble of purge gas
- Escape of hydrogen from the melt surface or at refractory vessel walls

Consequently, hydrogen removal is dependent on mass transfer coefficients. Two fundamental results of theoretical investigations that have been borne out by actual practice are that:

- Smaller purge gas bubbles produce more efficient degassing results
- Chlorine usually improves degassing efficiency when medium or large purge gas bubbles (>6 mm, or 0.25 in.) are present because the mass transfer coefficient is increased

A full discussion of the scientific aspects is beyond the scope of this article, but additional information on hydrogen removal is available in Ref 33 to 36.

Sources of Hydrogen in Aluminum

There are many potential sources of hydrogen in aluminum, including the furnace atmosphere, charge materials, fluxes, external components, and reactions between the molten metal and the mold.

Furnace Atmosphere. The fuel-fired furnaces sometimes used in melting can generate free hydrogen because of the incomplete combustion of fuel oil or natural gas.

Charge Materials. Ingot, scrap, and foundry returns may contain oxides, corrosion products, sand or other molding debris, and metalworking lubricants. All these contaminants are potential sources of hydrogen through the reduction of organic compounds or through the decomposition of water vapor from contained moisture.

Fluxes. Most salt fluxes used in aluminum melt treatment are hygroscopic. Damp fluxes can therefore result in hydrogen pickup in the melt from the decomposition of water.

External Components. Furnace tools such as rakes, puddlers, skimmers, and shovels can deliver hydrogen to the melt if they are not kept clean. Oxides and flux residues on such tools are particularly insidious sources of contamination because they will absorb moisture directly from the atmosphere. Furnace refractories, troughs and launders, mortars and cements, sampling ladles, hand ladles, and pouring ladles also are potential sources of hydrogen, especially if refractories are not fully cured.

Metal/Mold Reactions. If metal flow is excessively turbulent during the pouring process, air can be aspirated into the mold. If the air cannot be expelled before the start of solidification, hydrogen pickup can result. Improper gating can also cause turbulence and suctioning. Excessive moisture in green sand molds can provide a source of hydrogen as water turns to steam.

Effects of Hydrogen

When the dissolved hydrogen in molten aluminum cannot be thoroughly expelled from the casting product as it solidifies, porosity defects will result. Porosity can be very fine, widely dispersed, or localized in those areas of the casting that are the last to solidify if the entrapped gas concentration is relatively low. Hydrogen microporosity may not be particularly harmful except when pressure tightness is required. On the other hand, some gas porosity defects can be fairly large, and they appear as blowholes and cracks.

It is important to differentiate between true gas porosity and shrinkage porosity. The former is due to the presence of trapped gas, and such defects are usually rounder and shinier upon metallographic examination. The latter is due simply to inadequate feed metal during final solidification. Shrinkage porosity is usually more irregular in shape than gas porosity (Ref 37).

Porosity defects result in a loss of mechanical properties in aluminum alloys. Figure 17 shows the effect of hydrogen porosity on the mechanical properties of alloy 356.

A certain amount of gas retained in aluminum during solidification can be beneficial. Permanent mold castings often contain some unremoved or unexpelled hydrogen or sometimes gas is even added. This retained gas offsets gross solidification and aids in complete mold filling and does not detract from surface characteristics or minimum mechanical properties required for most commercial castings. The more rapid chilling and solidification rate of permanent mold casting versus sand casting facilitates the acceptance of hydrogen; the faster solidification rate usually results in finer grain size and enhanced mechanical properties. The same is true for die casting, in which extremely rapid solidification under pressure greatly reduces the tendency toward hydrogen porosity.

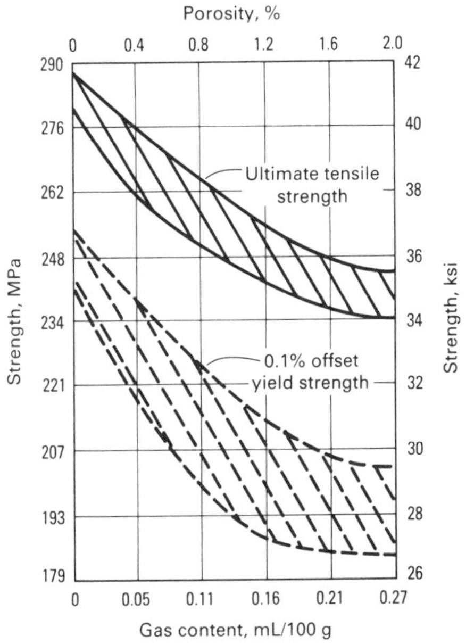

Fig. 17 Influence of gas content on the tensile and yield strengths of aluminum alloy 356. Source: Ref 38

Testing for Hydrogen

The Straube-Pfeiffer test (reduced-pressure test) for the measurement of hydrogen is a qualitative method that has been commonly used. A small sample of melt is placed in a crucible in a chamber, which is then evacuated to a specified pressure and held until the metal solidifies. Bubble evolution during solidification and the appearance of the sample surface are indicators of the amount of hydrogen present. A mushroom head indicates a gassy sample, while a recessed surface would indicate a relatively gas-free sample. The solidified sample is often sectioned and the evident porosity compared with a series of standards (Fig. 18). It is important to specify the reduced pressure used for repeated tests in order to make valid comparisons. Figure 19 shows the effect of different pressures on the observed response for samples containing the same amount of gas. These standards can be correlated with analytically determined specific hydrogen results or with individual foundry and casting acceptance criteria. If such standards are used, they must be evaluated in terms of each individual alloy, casting, and foundry experience and practice.

In the Straube-Pfeiffer test the sample behavior is influenced by general melt cleanliness as well as by hydrogen. Inclusions will nucleate hydrogen gas bubble formation during solidification; therefore, the test does not provide an absolute measurement or effect of dissolved hydrogen alone.

The Straube-Pfeiffer test is also used to determine the sample density D under specific test conditions and to correlate the

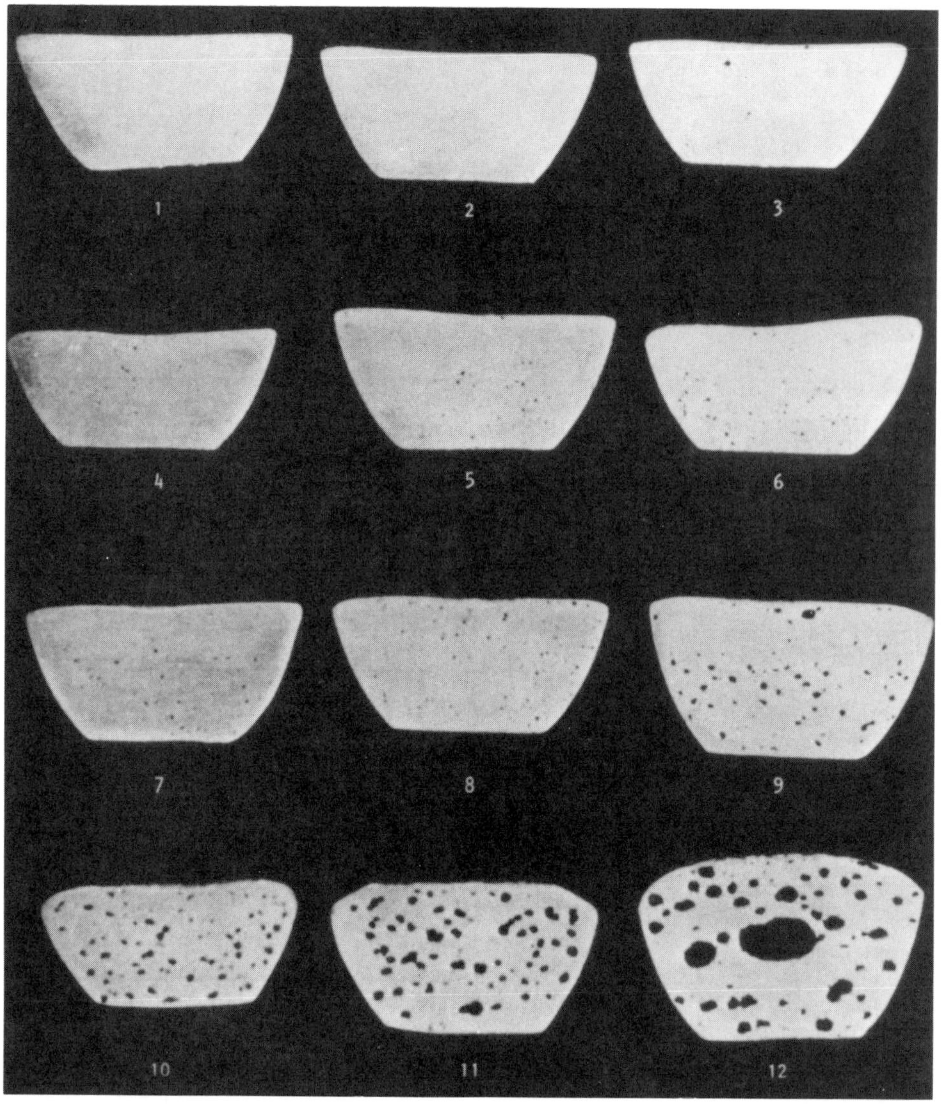

Fig. 18 Standards for comparison with reduced-pressure test specimen used to check the amount of hydrogen gas present in aluminum alloys. Such standards must be developed for a given alloy and set of test conditions. Courtesy of Wabash Alloys

result with the known full density of the alloy (Ref 39). Hydrogen gas content can then be determined from the difference in densities plus a correction for standard temperature and pressure (STP) conditions, because the gas forms in the sample during testing at a reduced pressure and at the freezing temperature of the alloy. The calculation is:

$$D = \frac{\text{Weight of metal + weight of gas}}{\text{Volume of metal + volume of gas}} \qquad \text{(Eq 5)}$$

The weight of the gas can be neglected. Therefore:

$$\frac{1}{D} = \frac{\text{Volume of metal + volume of gas}}{\text{Weight of metal}} \qquad \text{(Eq 6)}$$

The first term on the right-hand side of Eq 5 is simply the known theoretical density D_0, of the alloy in question. The measured volume of gas V_m is therefore given by:

$$V_m = \frac{1}{D} - \frac{1}{D_0} \qquad \text{(Eq 7)}$$

where V_m is in milliliters per 100 g of metal.

The measured volume of gas V_m must then be corrected to the volume of gas at standard conditions V_{STP} using Boyle's and Charles's laws:

$$V_{STP} = \left(\frac{\text{Test pressure}}{760}\right) \cdot \left(\frac{273}{273 + \text{alloy freezing temperature}}\right) \cdot V_m \qquad \text{(Eq 8)}$$

where test pressure is in millimeters of mercury (torr) and alloy freezing temperature is in degrees Celsius. Finally:

$$V_{STP} = K\left(\frac{1}{D} - \frac{1}{D_0}\right) \qquad \text{(Eq 9)}$$

where K is a constant and V_{STP} is in milliliters per 100 g of metal.

For each alloy and set of test conditions, a different K must be calculated. For a given alloy, however, it is possible to develop a relationship between density and actual hydrogen content. Densities of samples with varying gas contents can be measured by the Archimedes principle of weighing in air and water:

$$D = \frac{W_a}{W_a - W_w} \qquad \text{(Eq 10)}$$

where W_a and W_w are the weights in air and water, respectively.

Nomographs can then be developed from derived results such that the hydrogen content can be determined directly from weight determination of the reduced-pressure test specimens rather than by mathematical manipulation. Figure 20 shows a nomograph developed for aluminum alloy 1100 that relates observed first bubble pressure with hydrogen content.

Quantitative Reduced-Pressure Technique (Ref 41). The standard reduced-pressure test gives only a semiquantitative measure of hydrogen content, which is usually then visually interpreted and correlated with comparison samples exhibiting varying degrees of porosity. Inclusions nucleate porosity effects; therefore, the reduced-pressure test does result in the determination of true hydrogen content.

An instrument is available that measures hydrogen directly. A constant mass (100 g, or 0.22 lb, of metal) is placed in the chamber, and the pressure is rapidly reduced to a specific predetermined value. The sample is then allowed to solidify under vacuum in isolation from the pump.

As hydrogen gas is evolved during solidification, the gas partial pressure is measured by a calibrated Pirani gage whose output is converted continuously to a digital display of hydrogen content. The Pirani gage operates on the principle that at low pressure the thermal conductivity of any gas varies with pressure.

A test can be completed in about 5 min, and results with several foundry casting alloys have been shown to be accurate within 5%. However, high hydrogen readings have been obtained for wrought aluminum alloys, possibly because of the fundamental differences in oxide structure and hydrogen retention or occlusion capabilities of wrought alloy compositions.

The Telegas instrument (Ref 42) is a shop floor instrument for hydrogen determination that has been extensively used for many years (Fig. 21). Its primary application is in the wrought industry or for research purposes. The unit operates with a nitrogen carrier gas, which dissolves hydrogen from the melt. The nitrogen is bubbled through the melt for a sufficiently long period of time such that equilibrium is achieved with respect to hydrogen dissolved in the melt versus hydrogen dis-

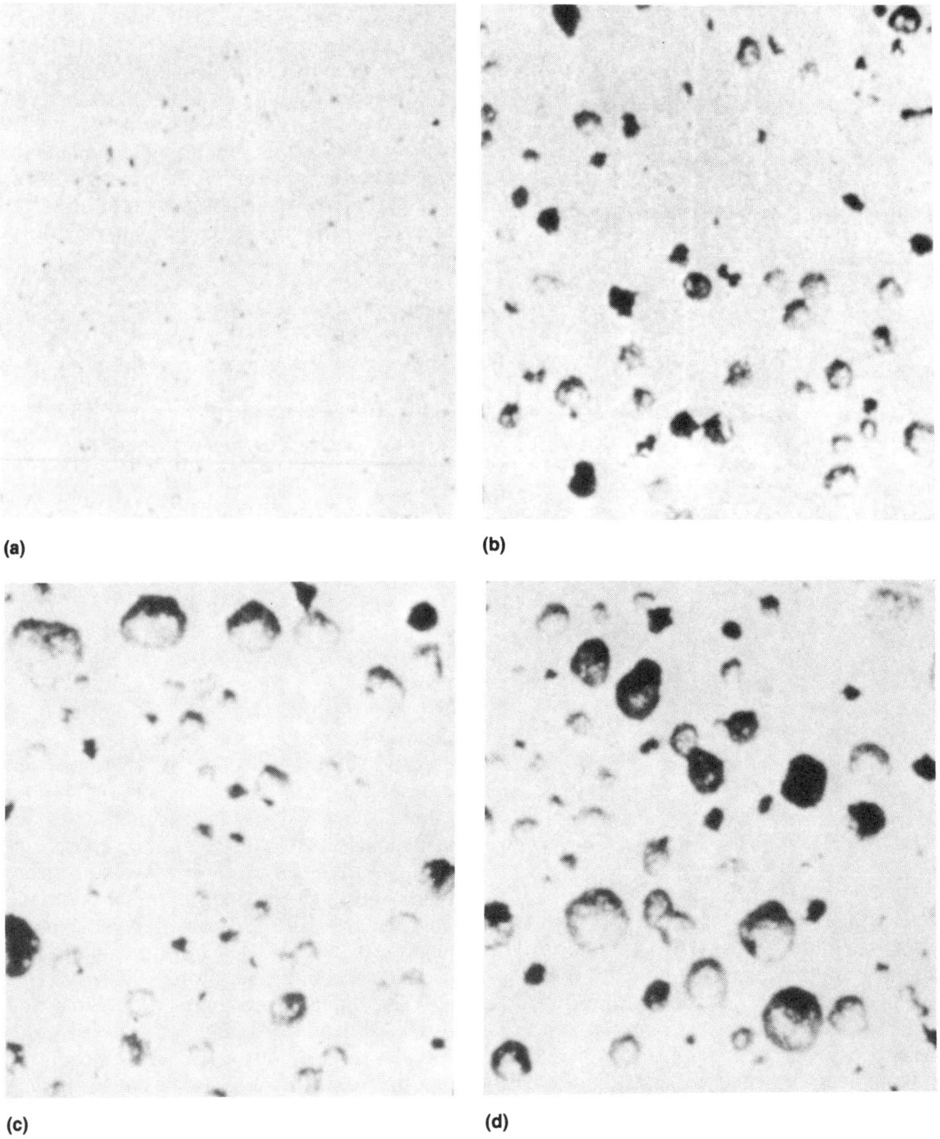

Fig. 19 Effect of pressure on the appearance of aluminum alloy reduced-pressure test samples obtained from the same melt. (a) Pressure: 100 kPa (760 torr). (b) 12 kPa (90 torr). (c) 6 kPa (45 torr). (d) 3 kPa (22.5 torr). Courtesy of Wabash Alloys

solved in the nitrogen. The hydrogen within the nitrogen carrier gas is then measured using a Wheatstone bridge circuit and a thermal conductivity cell known as a katharometer. The instrument is precalibrated, and the result is correlated with known standards of premixed gases containing a known level of hydrogen.

Instrument readings are converted into hydrogen content by using a family of curves relating to hydrogen concentration. Correlations are available that cover the normal melt temperature range of 675 to 750 °C (1245 to 1380 °F) and hydrogen concentrations of 0.05 to 0.40 mL/100 g. Correction factors also exist for different alloy compositions because of their greater or lesser sensitivity to hydrogen. The Telegas instrument performs well when it is properly operated, and it has been found to compare very favorably with the vacuum hot extraction analytical method, with a maximum standard deviation of 0.023 mL/100 g.

Although the Telegas instrument is not difficult to operate, the ceramic probes used with this instrument are very fragile. To avoid thermal shock to the probes, they must be preheated and immersed with extreme caution.

Vibrated Vacuum Gas Test (Ref 43). The standard vacuum solidification test provides only a qualitative measure of metal quality as influenced by both hydrogen gas content and inclusions. The vibrated vacuum gas test provides several features that improve upon the standard test and sort out hydrogen effects versus inclusion-nucleated porosity. A larger-capacity pump and more sensitive vacuum gage are used to produce a better and more reproducible vacuum. A preheated outgassed refractory sample cup permits slower solidification. A controlled vibrator assists the nucleation of gas bubbles, promotes coalescence, and reduces the surface tension of the molten sample.

Results of the vibrated vacuum test have shown good differentiation between the hydrogen content effects (presence of gas bubbles) of filtered (fewer inclusions) versus unfiltered metal (Ref 43). Figure 22 shows the improvement in metal quality at any test pressure for filtered metal, correlating with lower hydrogen content as verified by hot extraction analysis. Therefore, the vibrated vacuum test can provide a more sensitive determination of metal quality as affected by hydrogen alone.

Methods of Degassing

Several methods are available for degassing aluminum. The simplest method is to hold the metal at a lower temperature for some period of time during which the hydrogen solubility is lower and thus provide for natural outgassing. In addition, casting conditions that permit solidification with natural expulsion of hydrogen without the attendant porosity are also desirable. However, neither of these is always possible. Deliberate degassing of the melt can be accomplished in the following ways:

- Gas purging or gas flushing
- Tableted flux degassing
- Mechanical mixer degassing

Gas Purging. The simplest method of degassing molten aluminum is by injecting a purging gas or gas mixture under pressure through a flux tube, pipe, lance, or wand. The tubes themselves are either ceramic-coated cast iron or steel pipe, or more often graphite.

A purge gas serves to collect hydrogen because of the lower partial pressure of the hydrogen afforded by the collector bubble versus the surrounding melt. Hydrogen diffuses into the purge gas bubble, which rises to the surface of the melt and is expelled into the atmosphere.

Purge gas can be either inert (argon or nitrogen) or reactive (chlorine and Freon 12). Reactive gases are used in small concentrations of under 10% along with an inert gas. Reactive gases undergo an actual chemical reaction with the melt. Chlorine reacts with molten aluminum to form a gaseous $AlCl_3$, which then serves as the purging gas. In the case of Freon, fluorine reacts to form AlF_3, a solid phase.

Both chlorine and fluorine have very favorable effects on bubble surface tension and in wetting out inclusions from the melt. Therefore, gas purging that uses either of these two elements can be expected to improve degassing efficiency through more effective coalescence of hydrogen and hydrogen inclusion complexes. However, although chlorine is very effective as a purge gas, it is also very noxious, and consider-

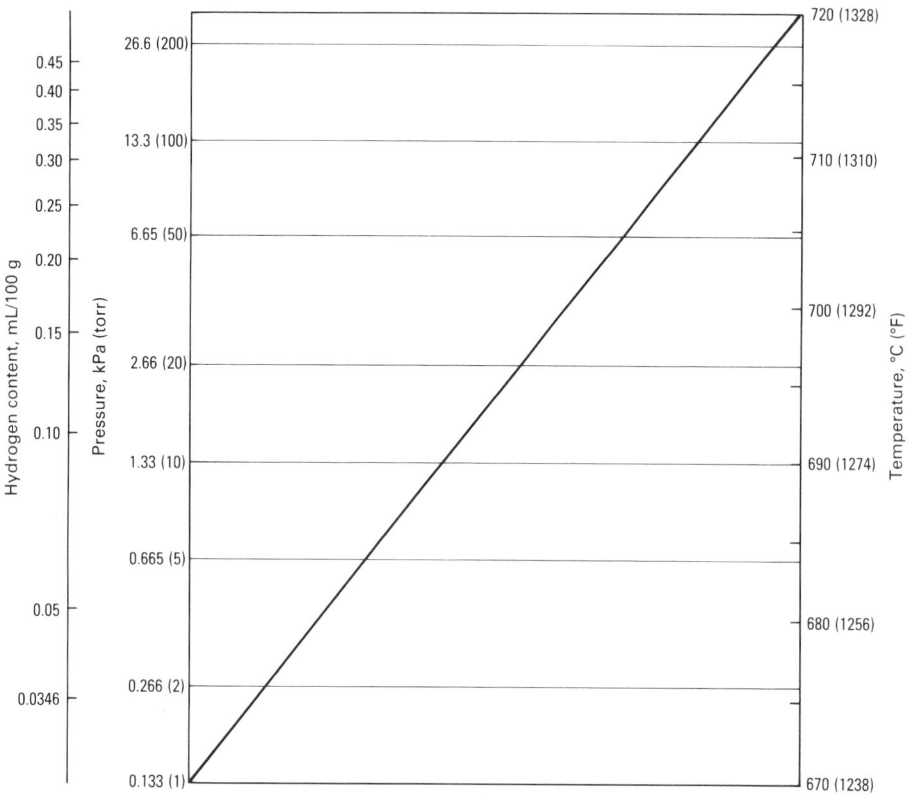

Fig. 20 Nomograph correlating reduced-pressure reading at appearance of first bubble with actual hydrogen content of aluminum alloy 1100. Melt temperature, 730 °C (1350 °F); gas flow rate, 6 L/min (0.2 scfm) Source: Ref 40

able effluent may result. Therefore, it is almost always used with an inert gas.

A mixture of nitrogen, chlorine, and carbon monoxide (CO) has been successfully used for degassing (Ref 44). This trigas mixture and Freon are both reported to have lower gaseous emissions or effluent than either chlorine alone or chlorine plus inert gas mixtures (Ref 44, 45).

Results of Hydrogen Removal With Purging Gas. In general, chlorine gas or mixtures with chlorine are more beneficial than inert gases or Freon alone. Figure 23(a) shows the relative effects of three different purging gases or mixtures, while Fig. 23(b) compares the degassing achieved with argon and nitrogen.

Vacuum fluxing is not used on any significant commercial basis in North America, but is practiced in Europe. Recent results have shown that this technique is beneficial (Ref 48). In large (45 Mg, or 50 ton) melts, a series of alloys (99.5Al, Al-0.5Mg-0.5Si, and Al-5Zn-1.1Mg-0.3Cu) was refined using argon fluxing for 30 min through porous plugs. Pressure was maintained at 7 Pa (0.001 psi). Hydrogen contents were reduced from 0.16 to 0.11 ppm, 0.24 to 0.12 ppm, and 0.17 to 0.13 ppm for the three alloys, respectively. One particular advantage of vacuum fluxing is the minimal amount of dross formed.

Hexachloroethane Degassing (Ref 49, 50). Perhaps the most common method of degassing in foundry applications is the use of hexachloroethane (C_2Cl_6) tablets. The tablet decomposes in the aluminum melt to form gaseous $AlCl_3$. The rising $AlCl_3$ gas bubbles then collect hydrogen gas and deliver the gas to the melt surface for release. The tablets may also contain salt fluxes to help wet oxide inclusions within the melt and thus enable some removal of the hydrogen gas associated with inclusions.

To be fully effective, the tablets must be kept dry and should be plunged with clean, dry tools (preferably a perforated bell-shaped plunger to permit easy gas bubble escape). The plunge should be to the full depth of the melt, and the tablets should be held in position until a gentle bubbling on the melt surface subsides. If this is not done, the lighter-density tablets will rise and float on the melt surface during decomposition, providing little or no effective degassing of the entire melt.

Tableted products are available for virtually any melt size, although their principal application is for smaller furnaces and ladles within the foundry casting industry. When properly used, C_2Cl_6 provides very effective degassing. However, the somewhat noxious odor of the raw tablet—as well as the sometimes strong effluent during use, particularly improper use—creates environmental difficulties that have forced many foundries to discontinue use of the tablets.

Rotary Degassing. The use of simple gases or gas mixtures in lance or flux tube degassing, although reasonably effective, is not necessarily fully efficient. The ability of this simple process to degas depends on the bubble size generated and the amount of contact area between the degassing bubbles and the melt. The bubbles are often large, and very little mixing occurs in large melts or vessels when flux tubes or lances are used.

A significant improvement was the development of rotary gas injection systems. These systems originated in the aluminum primary and mill products casting industry and are used as in-line treatment systems in auxiliary vessels between furnace and ladle or furnace and casting pit.

The principle of the rotary injector system (Fig. 24) is that gas is injected into the column or shaft of a rotating member and is released through fine openings at the base or rotor. When rotating at speeds of 300 to 500 rpm, the disperser shears the incipient gas bubble released, producing a wide dispersion of very fine bubbles for degassing. The high surface-area-to-volume ratio of degassing bubbles provides greatly increased contact area and therefore increased reaction kinetics, resulting in more efficient degassing. The rotating member also disperses the bubble array more uniformly within the reaction vessel to provide more thorough mixing of metal with the purge gas. Figure 25 shows a typical rotary degassing system used in primary and mill products casting operations. Such systems are sized according to metal volume and flow rate conditions, and either single gases such as argon, nitrogen, and chlorine or mixtures of these are used in the gas mixing and control system. The level of hydrogen removal depends on gas flow rate and initial hydrogen content as well as metal flow rate, but most commercial systems can achieve final hydrogen contents well below 0.15 mL/100 g (Fig. 26).

Degassing performance is determined by several interrelated factors, including initial hydrogen content, metal flow rate, vessel size (which determines residence time in conjunction with flow rate), purge gas flow rate, mixing capability, and specific alloy (thermodynamic factors and mass transfer coefficients). In general, all such systems have achieved satisfactory degassing results (Ref 51-56).

Rotary degasser systems have recently been downsized for foundry applications with good results (Ref 47, 57). The degassing system is inserted directly into a crucible or ladle of metal, and the metal is degassed in a batch process before pouring.

There are optimum combinations of purge gas volume, ratio of impeller diameter to crucible diameter, and shaft revolutions per minute for best efficiency of degassing without excessive turbulence. An impeller/

Fig. 21 The Telegas unit for hydrogen determination. Courtesy of Aluminum Company of America

crucible diameter ratio of 20 to 25%, a 350 to 450 rpm shaft speed, and a gas flow rate of 18 to 20 standard cubic feet per hour (SCFH) have been found to be appropriate for one unit. Variations in impeller design, however, may alter these parameters. Gas removal results achieved with rotary degassing compare very favorably with those achieved by lance treatment. This is illustrated in Fig. 27 for a 180 kg (400 lb) heat of Alloy A357.

Other Mechanical Degassing Systems. The circulation gas injection pump (see the section ''Demagging of Aluminum'' in this article) can also be used for degassing. The high speed (600 to 1100 rpm) of the impeller shears the gas bubbles exiting the injection tube and creates a very fine and widely dispersed bubble pattern. In large reverberatory furnaces, the melt can be degassed through the injection and pumping action without the need for additional furnace fluxing with wands. In this manner, degassing efficiency and energy efficiency are both improved because furnace doors need not be opened, as in flux tube or lance fluxing. Large (45 Mg, or 50 ton) melts have been successfully degassed to levels of 0.15 mL/100 g and below with the gas injection pump.

Another device being used for degassing in the primary industry is a swirling tank reactor. The metal flows in a swirling pattern counter to gas flow. The gas is injected through a series of nozzles, as shown schematically in Fig. 28. Typical results obtained with this process are shown in Fig. 29.

Porous Plug Degassing. Another method that can be used to degas by creating fine bubbles is the use of a porous disperser on

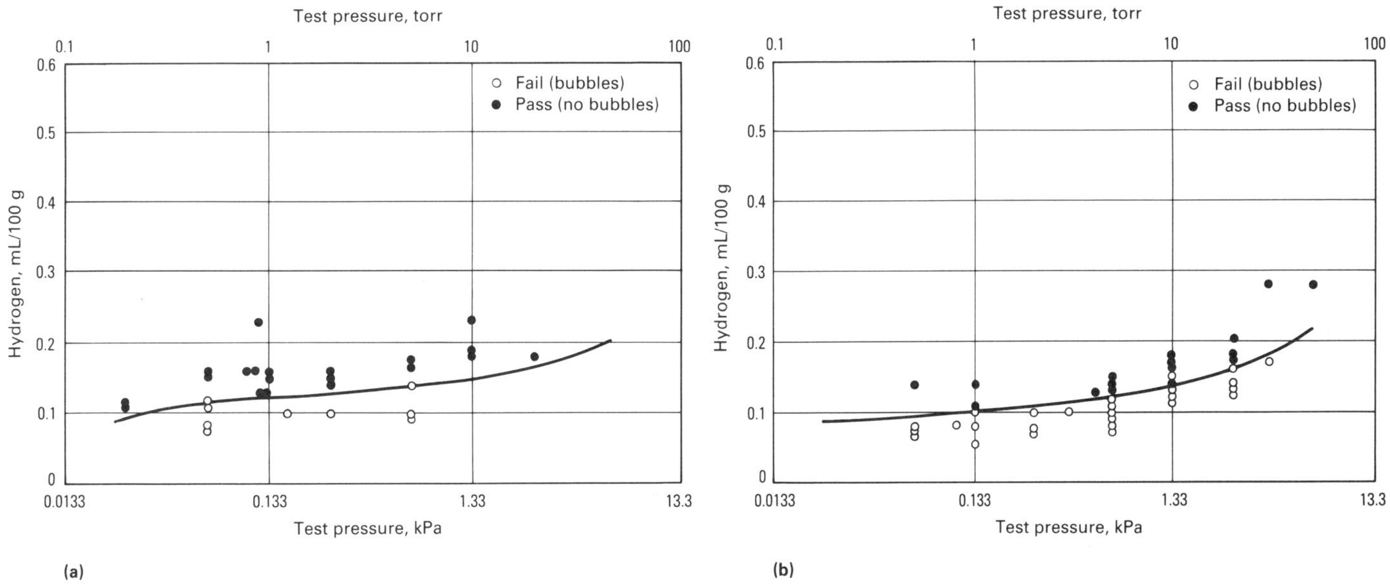

Fig. 22 Hydrogen content as a function of vibrated vacuum gas test pressure for unfiltered (a) and filtered (b) aluminum alloy. Source: Ref 43

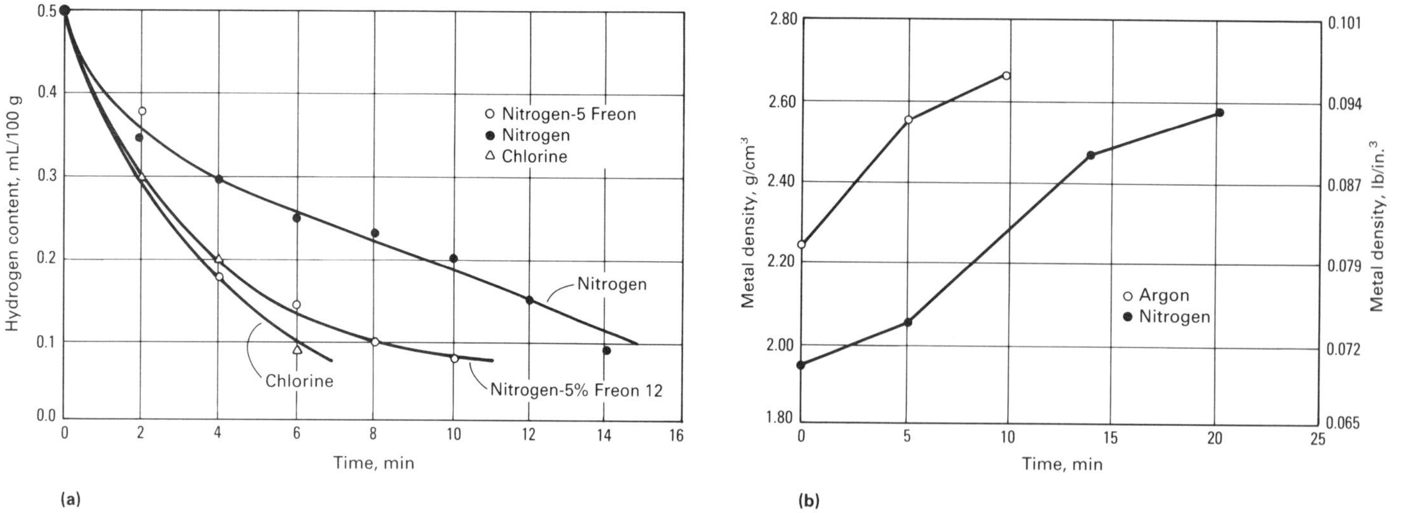

Fig. 23 Relative efficiencies of purge gases in the degassing of aluminum alloy 356 (a) and aluminum alloy 319 (b) at constant gas flow rates. Metal density in (b) increases as gas content of the metal decreases. Source: Ref 46, 47

the end of a flux tube or a fixed porous plug (Fig. 30). Porous plugs are graphite or ceramic materials with a very fine, interconnected porosity through which gas can be conveyed. These plugs can be installed into ladle bottoms, furnaces, or auxiliary treatment vessels. The fine porosity of the plug material allows a fine bubble size to be generated (comparable to that of rotary degassers). However, degassing efficiency is strongly related to bubble residence time, gas flow rate, depth of the melt, size of the vessel, and the ability to keep the pores in the plug open for full gas emission. Porous plug degassing can be more economical than rotary degassing systems, in which the rotors and shafts, usually graphite, require periodic replacement as they wear and oxidize.

Figure 31 compares the degassing obtained by using identical gas flows for porous plug, lance, and rotary degassing systems. The porous plug is more effective than a lance, and satisfactory degassing results can be obtained in many applications by controlling vessel size, gas flow rate, and residence time.

A combination in-line degassing/filtration system using porous plugs has recently been introduced (Ref 59). The degassing chamber is designed to accommodate all the above-mentioned variables. This system (Fig. 32) has produced final hydrogen contents of 0.06 mL/100 g at low metal flow rates (40 kg/min, or 90 lb/min) and 0.14 mL/100 g at flow rates near 330 kg/min (725 lb/min).

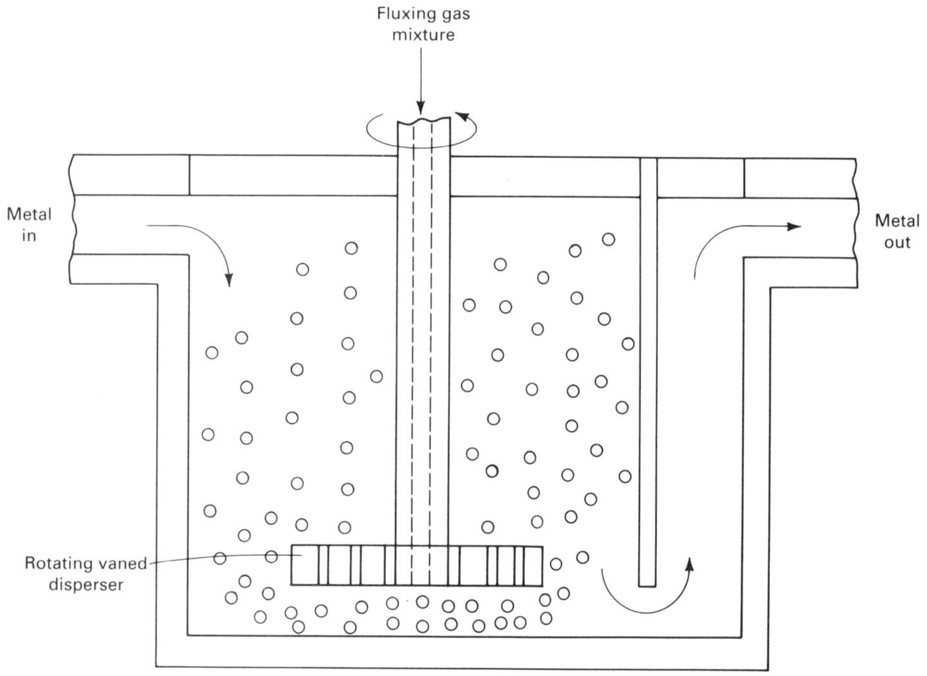

Fig. 24 Schematic of an in-line rotary degassing unit

Degassing of Magnesium

Magnesium behaves similarly to aluminum in its ability to dissolve hydrogen easily in the molten state (Fig. 33). It too will suffer from porosity when cast if the hydrogen is not removed or naturally expelled during solidification. Aluminum and zinc, which are typical alloying elements in magnesium alloys, lower this solubility somewhat. Figure 34 indicates that much more gas is evolved from magnesium alloys containing zinc and aluminum than from pure magnesium upon solidification. Gas porosity occurs in solidified castings when the residual hydrogen exceeds 15 mL/100 g.

Zirconium is also a major constituent of certain magnesium alloys and has a marked effect on hydrogen solubility. Hydrogen solubility decreases as the zirconium content of the alloy increases (Ref 7). Because of this low solubility even in the liquid state, hydrogen porosity is minimal in castings of magnesium-zirconium alloys. However, excess hydrogen in the melt readily precipitates as zirconium hydride, thus depleting the alloy of zirconium in solid solution. It is therefore advisable to minimize hydrogen absorption into the melt and/or to employ degassing methods even with this alloy.

Sources of gas in magnesium alloy melts are the same as those for aluminum—atmospheric conditions, fluxes, tools, and so on. Flux covers or antiburning agents do not necessarily preclude hydrogen absorption during melting.

Like aluminum alloys, molten magnesium metal and alloys can be degassed with hexachloroethane, chlorine, or nitrogen. Hexa-

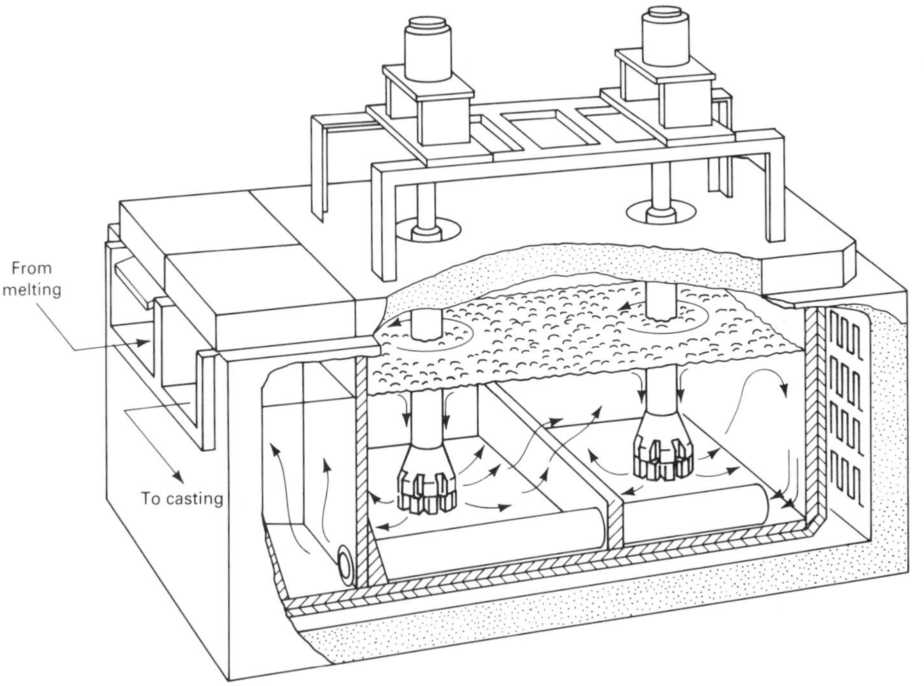

Fig. 25 Schematic of spinning nozzle inert flotation (SNIF) rotary degassing system for in-line aluminum refining. Source: Linde Division of Union Carbide Corporation

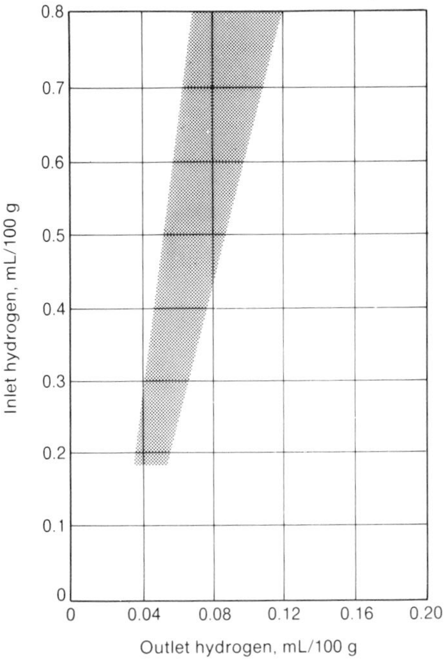

Fig. 26 Hydrogen removal results with the SNIF rotary degasser shown in Fig. 25. Similar results can be obtained with comparable commercially available equipment. Source: Ref 51

chloroethane has the added advantage of supplying carbon through the decomposition of the hexachloroethane, with the carbon serving as a grain refiner (see the section "Grain Refining of Magnesium Alloys" in this article). Chlorine or chlorine/nitrogen gas purging is optimally performed at temperatures of 725 to 750 °C (1335 to 1380 °F). In this range, liquid $MgCl_2$ will also be formed, which will retard burning on the surface. Liquid $MgCl_2$ wets the oxides and the magnesium melt surface, facilitating hydrogen escape. Pure chlorine provides for more liquid $MgCl_2$ formation than either gas mixtures or hexachloroethane and therefore creates better degassing. Fluxes higher in $MgCl_2$ that are used as covers during degassing also aid in hydrogen desorption at the flux/metal interface and in subsequent expulsion to the atmosphere (Ref 60).

The degassing of magnesium alloys in crucible or pot-type furnaces should be carried out with clean, dry lances, preferably graphite, with thorough mixing for 5 to 10 min. Pressures and volumes should not exceed those required to achieve a mild, gentle roll or small bubble (≤25 mm, or 1 in., in

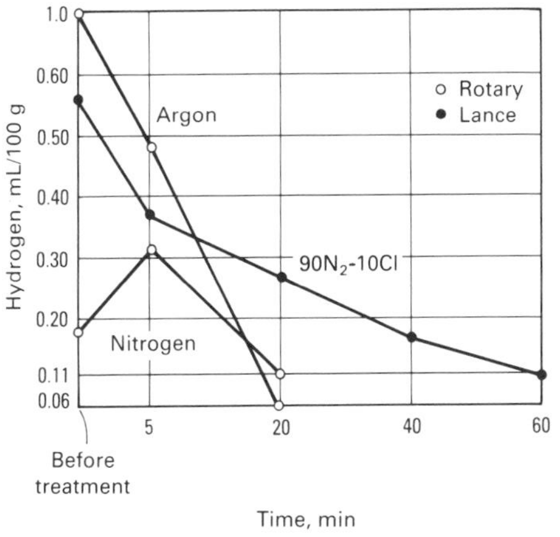

Fig. 27 Comparison of efficiency of rotary degassing and lance degassing of aluminum alloy A357. The rotary system used a 150 mm (6 in.) rotor at 375 rpm. Melt weight: 180 kg (400 lb). Source: Ref 57

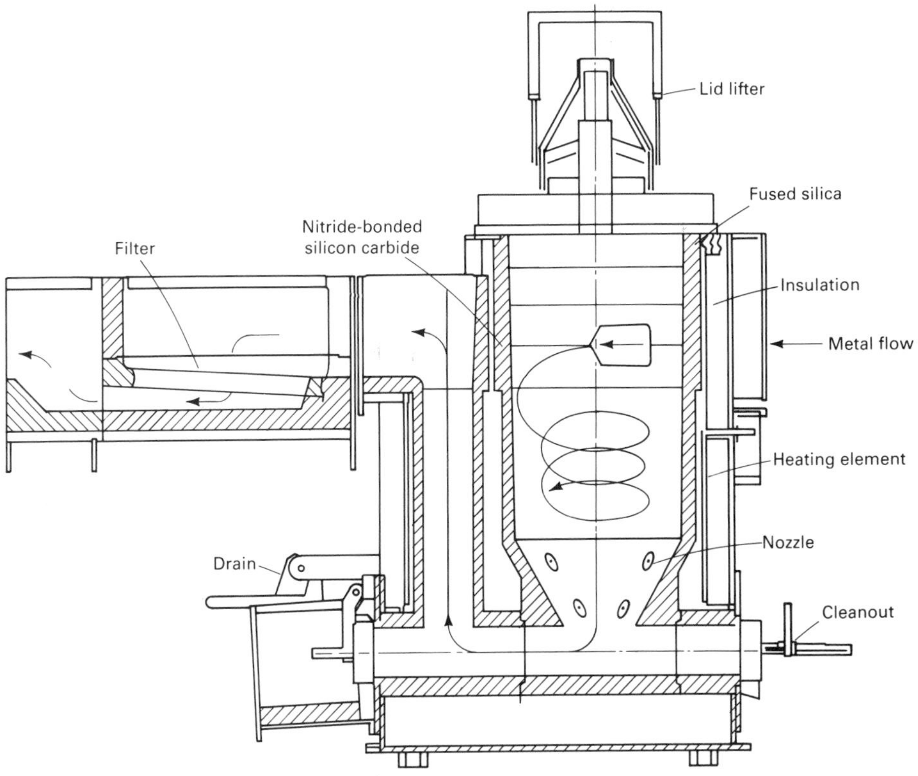

Fig. 28 Schematic of the metal in-line treatment (MINT) degassing system. Source: Selee Corporation

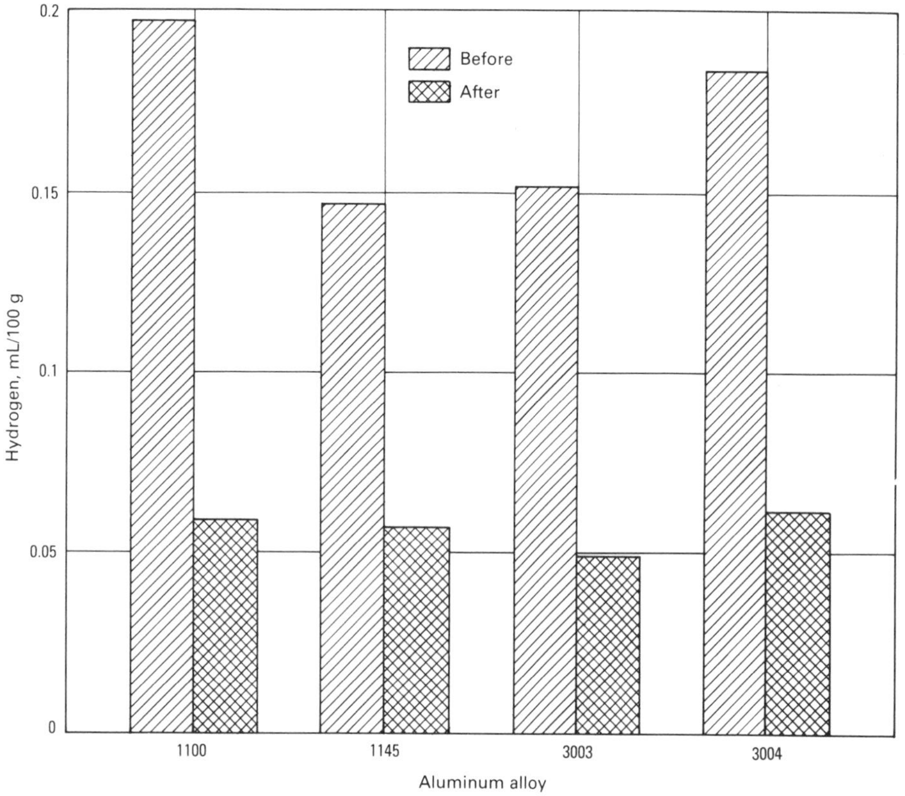

Fig. 29 Efficiency of the MINT III degassing system shown in Fig. 28 using a gas mixture of Ar-0.5/2Cl. Source: Ref 58

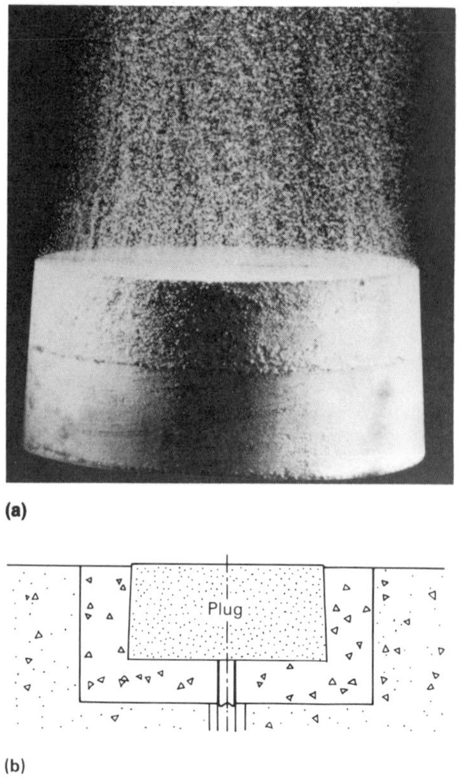

Fig. 30 Porous plug degassing unit (a) showing fine bubbles produced in water. (b) Plug installation in vessel bottom

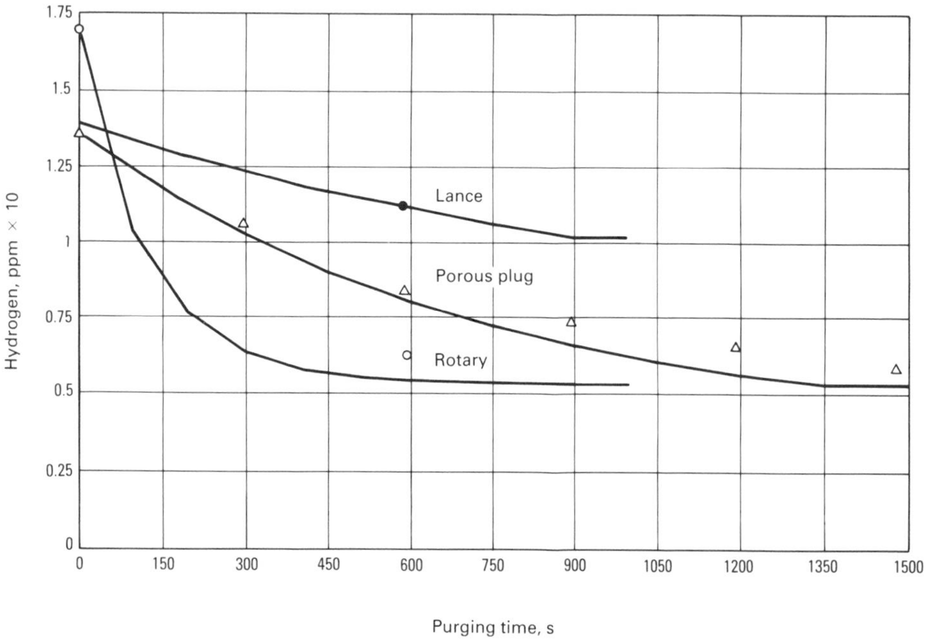

Fig. 31 Comparison of lance, porous plug, and rotary degassing efficiencies for identical gas flows. Source: Ref 36

diameter) on the melt surface. Rare earth containing magnesium alloys should only be degassed when required with nitrogen or argon, because chlorination will readily remove these expensive alloying elements.

Degassing of Copper Alloys

In the melting and casting of many copper alloys, hydrogen gas absorption can occur because of the generous solubility of hydrogen in the liquid state of these alloys. The solid solubility of hydrogen is much lower; therefore, the gas must be rejected appropriately before or during the casting and solidification process to avoid the formation of gas porosity and related defects (excessive shrink, pinholing, blowholes, and blistering). These defects are almost always detrimental to mechanical and physical properties, performance, and appearance. Different copper alloys and alloy systems have varying tendencies toward gas absorption and subsequent problems. Other gases, particularly oxygen, can cause similar problems (see the section "Deoxidation of Copper Alloys" in this article).

Sources of Hydrogen Porosity

There are many potential sources of hydrogen when melting copper alloys. For the most part, these are identical to the sources of hydrogen in aluminum melting (see the section "Sources of Hydrogen in Aluminum" in this article).

Gas Solubility

Hydrogen is the most obvious gas to be considered in copper alloys. Figure 35 shows the solubility of hydrogen in molten copper. As in aluminum and magnesium alloys, solubility is reduced in the solid state; therefore, the hydrogen must be removed prior to casting or rejected in a controlled manner during solidification. Al-

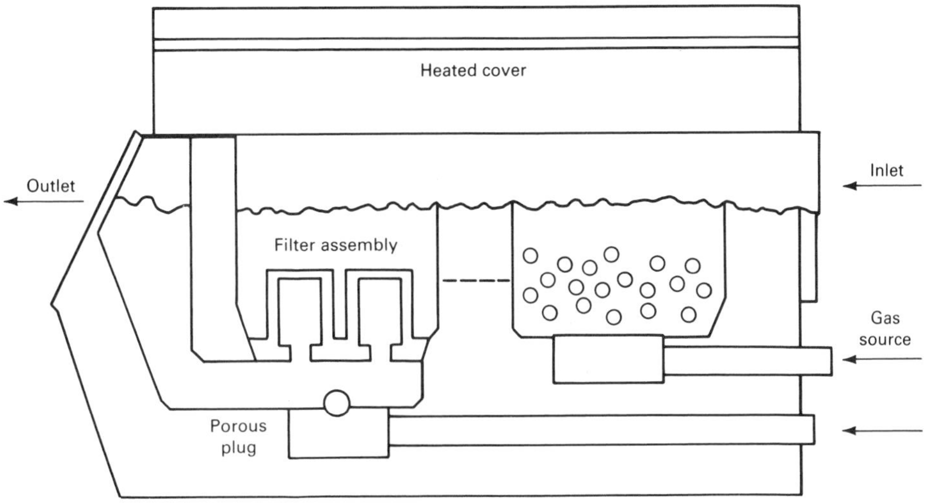

Fig. 32 Schematic of proprietary combination in-line degassing/filtration system. Source: Ref 59

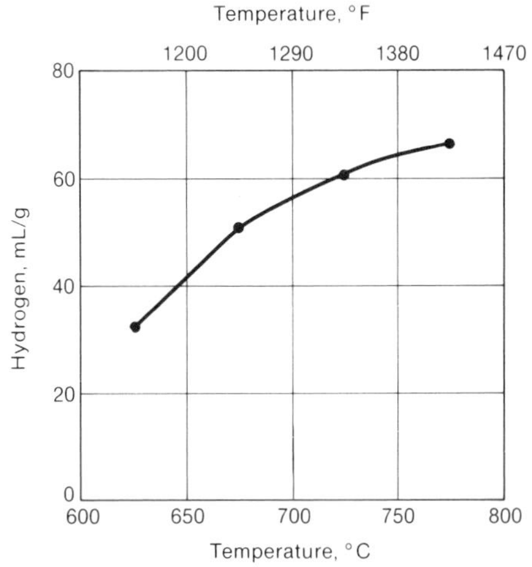

Fig. 33 Solubility of hydrogen in molten magnesium at 101 kPa (760 torr). Source: Ref 7

loying elements have varying effects on hydrogen solubility (Fig. 36).

Oxygen also presents a potential problem in most copper alloys. In the absence of hydrogen, oxygen alone may not cause problems, because it has limited solubility in the melt. However, it forms a completely miscible liquid phase with the copper in the form of cuprous oxide (Fig. 37). During solidification, the combination of cuprous oxide and hydrogen can give rise to casting porosity resulting from the steam reaction (discussed later).

Sulfur gases have significance in primary copper through the smelting of sulfide ores and in the remelting of mill product scrap containing sulfur-bearing lubricants. Sulfur dioxide is the most probable gaseous product. Foundry alloys and foundry processing usually do not experience sulfur-related problems unless high-sulfur fossil fuels are used for melting.

Carbon can be a problem, especially with the nickel-bearing alloys. The nickel alloys have extensive solubility for both carbon and hydrogen. Carbon can be deliberately added to the melt, along with an oxygen-bearing material such as nickel oxide. The two components will react to produce a carbon boil, that is, the formation of CO bubbles, which collect hydrogen as they rise through the melt (Ref 63). If not fully removed, however, the residual CO can create gas porosity during solidification.

Nitrogen does not appear to be detrimental or to have much solubility in most copper alloys. However, there is some evidence to suggest that nitrogen porosity can be a problem in cast copper-nickel alloys (Ref 64).

Water vapor can exist as a discrete gaseous entity in copper alloys (Ref 61, 65, 66). Water vapor is evolved from solidifying copper alloys, which always have some residual dissolved oxygen. When the oxygen becomes depleted, hydrogen is produced as a separate species.

The type and amount of gas absorbed by a copper alloy melt and retained in a casting depend on a number of conditions, such as melt temperature, raw materials, atmosphere, pouring conditions, and mold materials. Table 4 lists the gases that can be found in a number of copper alloy systems.

Testing for Gases

There are essentially three ways to determine the presence of gas in a copper alloy melt. The easiest and simplest is a chill test on a fracture specimen. In this method, a

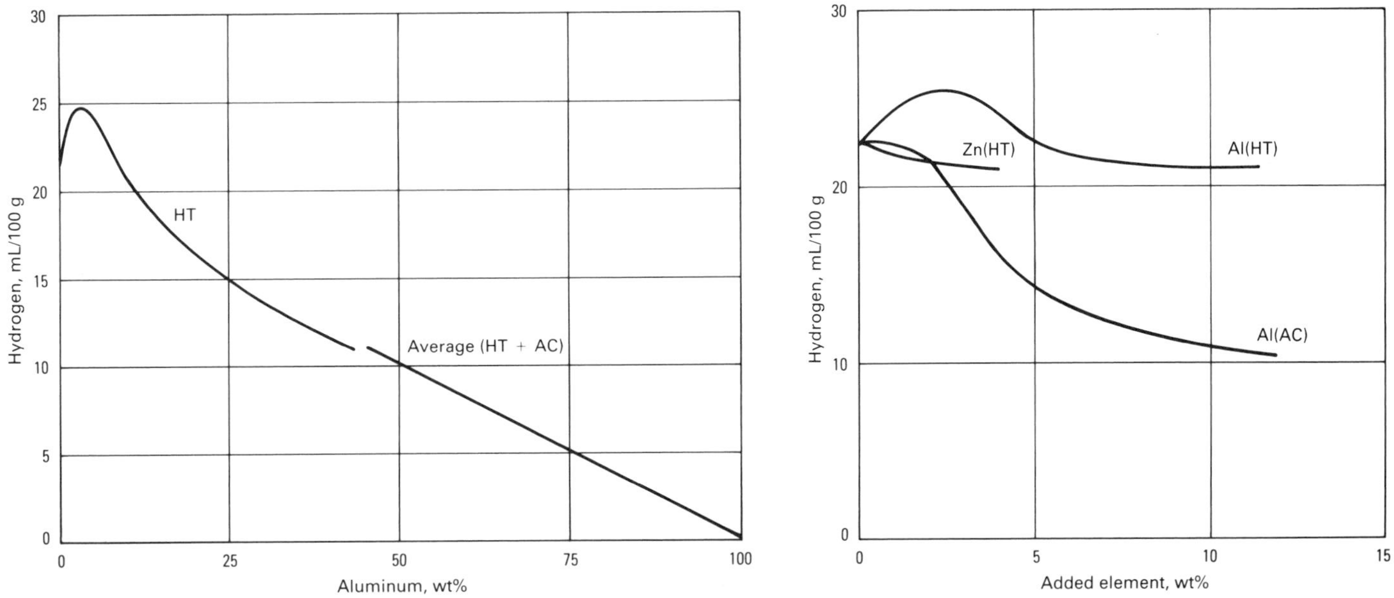

Fig. 34 Effect of alloying elements on hydrogen solubility in liquid magnesium. AC, as-cast; HT, heat treated. Source: Ref 7

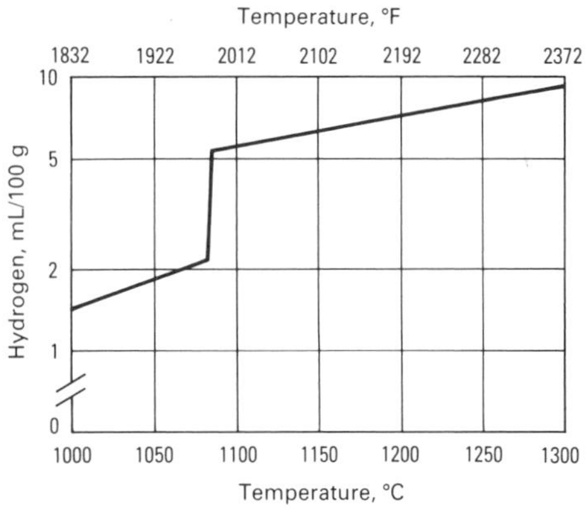

Fig. 35 Solubility of hydrogen in copper. Source: Ref 61

standard test bar is poured and allowed to solidify. The appearance of the fractured test bar is then related to metal quality standards (Ref 67).

The second method is a reduced-pressure test similar to the Straube-Pfeiffer test used for aluminum alloys. In recent years, considerable development and definition have been given to this test as adapted for copper alloys, because the different classes of copper alloys have different solidification characteristics that affect the response to the test (Ref 61, 68-70). The objective of the test is to use a reduced pressure only slightly less than the dissolved gas pressure so that the surface of the sample mushrooms but does not fracture (Fig. 38). The test apparatus is shown schematically in Fig. 39. The key to establishing controllable results, that is, unfractured but mushroomed test sample surfaces, is to provide wave front freezing. This is accomplished by proper selection of vessel materials to achieve the correct solidification characteristics. Because the freezing ranges of copper alloys vary from short to wide, different materials must be used. The various material combinations that have proved successful are described in Ref 61.

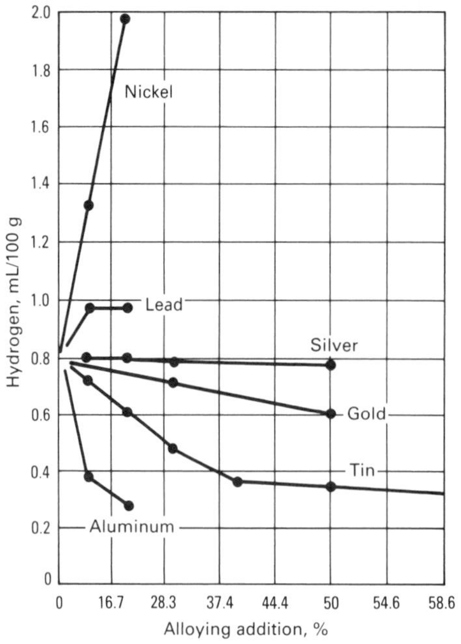

Fig. 36 Effect of alloying elements on the solubility of hydrogen in copper. Source: Ref 62

Degassing Methods

Oxidation-Deoxidation Practice. The steam reaction previously mentioned is a result of both hydrogen and cuprous oxide being present in the melt. These constituents react to form steam, resulting in blowholes during solidification as the copper cools:

$$2H + Cu_2O \rightarrow H_2O\ (\text{steam}) + 2Cu$$

The proper deoxidation of a copper alloy melt will generally prevent this steam reaction, although excessive hydrogen alone can still cause gas porosity if it is not expelled from the casting before the skin is completely solidified. Fortunately, there is a mutual relationship between hydrogen and oxygen solubility in molten copper (Fig. 40). Steam is formed above the line denoting equilibrium concentration, but not below. Consequently, as the oxygen content is raised, the capacity for hydrogen absorption decreases. Therefore, it is useful to provide excess oxygen during melting to preclude hydrogen entry and then to remove the oxygen by a deoxidation process to prevent further steam reaction during solidification.

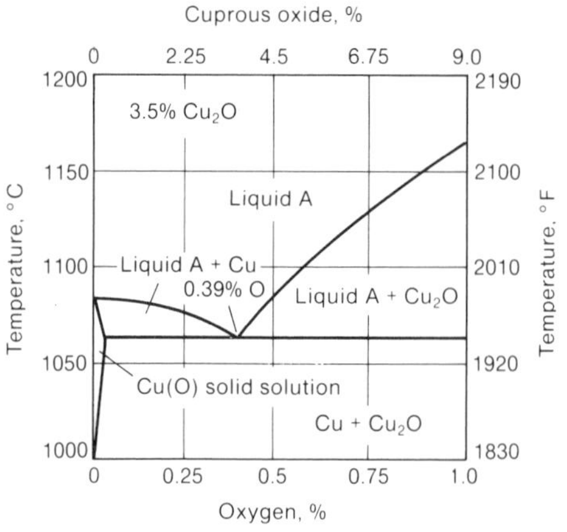

Fig. 37 Copper-oxygen phase diagram. Source: Ref 62

This relationship gave rise to the Pell-Walpole oxidation-deoxidation practice for limiting hydrogen, especially when fossil fuel fired furnaces were used for melting. The melt was deliberately oxidized using oxygen-bearing granular fluxes or briquetted tablets to preclude hydrogen absorption during melting. The melt was subsequently deoxidized to eliminate any steam reaction with additional hydrogen absorbed during pouring and casting.

Zinc Flaring. The term zinc flare applies to those alloys containing at least 20% Zn. At this level or above, the boiling or vaporization point of copper-zinc alloys is close to the usual pouring temperature, as shown in Fig. 41 (Ref 62). Zinc has a very high vapor pressure, which precludes hydrogen entry into the melt. Zinc may also act as a vapor purge, removing hydrogen already in solution. Furthermore, the oxide of zinc is less dense than that of copper, thus providing a more tenacious, cohesive, and protective oxide skin on the melt surface and increasing resistance to hydrogen diffusion. In zinc flaring, the melt temperature is deliberately raised to permit greater zinc vapor formation. Because some zinc loss may occur, zinc may have to be replenished to maintain the correct composition.

Inert Gas Fluxing. With gas fluxing, an inert collector or sparger gas such as argon or nitrogen is injected into the melt with a graphite fluxing tube. The bubbling action collects the hydrogen gas that diffuses to the bubble surface, and the hydrogen is removed as the inert gas bubbles rise to the melt surface. The reaction efficiency depends on the gas volume, the depth to

Table 4 Summary of gases found in copper alloys

Alloy family	Gases present	Remarks
Pure copper	Water vapor, hydrogen	Approximate hydrogen/water vapor ratio of 1. Higher purity increases the amount of water vapor and lowers hydrogen.
Copper-tin-lead-zinc alloys	Water vapor, hydrogen	Lead does not affect the gases present. Higher tin lowers total gas content. Increased zinc increases the amount of hydrogen, with a loss in water vapor.
Aluminum bronzes	Water vapor, hydrogen, CO	The presence of 5 wt% Ni in alloy C95800 causes CO to occur rather than water vapor. Lower aluminum leads to higher total gas contents.
Silicon brasses and bronzes	Water vapor, hydrogen	Approximate hydrogen/water vapor ratio of 0.5. Increased zinc decreases hydrogen and increases water vapor.
Copper-nickels	Water vapor, hydrogen, CO	All three gases are present up to 4 wt% Ni, after which only CO and hydrogen are present. Hydrogen increases with increasing nickel up to 10 wt% Ni, but is decreased at 30 wt% Ni.

Source: Ref 61

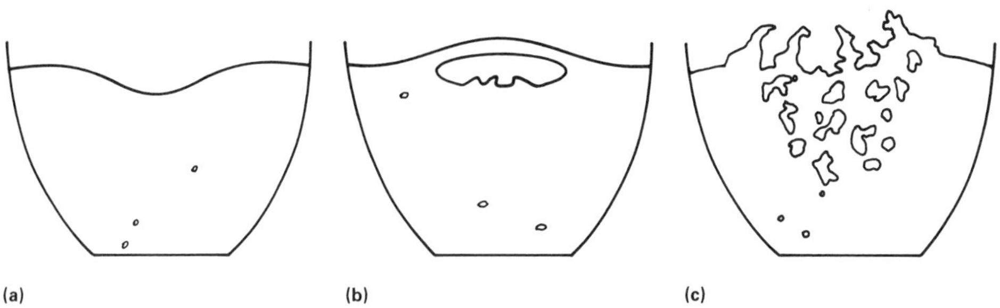

Fig. 38 Effect of pressure on the appearance of copper alloy reduced-pressure test samples containing the same amount of gas. (a) Pressure of 7 kPa (55 torr) results in surface shrinkage. (b) At 6.5 kPa (50 torr), a single bubble forms. (c) Boiling and porosity occur at 6 kPa (45 torr).

which the fluxing tube or lance is plunged, and the size of the collector gas bubble generated. Again, finer bubble sizes have higher surface-area-to-volume ratios and therefore provide better reaction efficiencies. Figure 42 depicts the amount of purge gas necessary to degas a 450 kg (1000 lb) copper melt. Figure 43 shows the response of an aluminum bronze alloy to nitrogen gas purging. The curve for oxygen purging is also shown.

Solid Degassing Fluxes. Other materials can be used to provide inert gas purging. A tabletted granular flux such as calcium carbonate ($CaCO_3$), which liberates CO_2 as the collector gas upon heating, has been successful in degassing a wide variety of copper alloys. This type of degassing is simpler than nitrogen, but must be kept dry and plunged as deep as possible into the melt with a clean, dry plunging rod, preferably made of graphite. This type of degasser may also have the advantage of forming the collector gas by chemical reaction *in situ*. This results in an inherently smaller initial bubble size than injected gas purging for better reaction efficiency. Such an advantage can be inferred from Fig. 44, although after 5 min of degassing the end results are the same.

Other solid degassers in the form of refractory metals or intermetallic compounds, such as calcium-manganese-silicon, nickel-titanium, titanium, and lithium, are effective in eliminating porosity due to nitrogen or hydrogen by their ability to form stable nitrides and hydrides. Again, maintaining dry ingredients, deep plunging, and stirring or mixing will enhance their effectiveness. For best results and optimum reaction efficiency, such degassers should be wrapped or encapsulated in copper materials to provide controlled melting when plunged. Alternatively, copper master alloys containing these elements are available.

For both inert gas fluxing and solid degassing additions, the sparging gas reaction should not be so violent as to splash metal and create an opportunity for gas reabsorption. Furthermore, a melt can be overdegassed; an optimum amount of residual gas remaining in the melt helps to counter localized shrinkage in long freezing range alloys such as leaded tin bronzes.

Vacuum degassing is not generally applied to copper alloys, although it can be very effective. However, the cost of the equipment necessary is relatively high, and there may be a substantial loss of more volatile elements having a high vapor pressure such as zinc and lead. Furthermore, significant superheat (up to 150 °C, or 300 °F) may be required to accommodate the temperature drop during the degassing treatment, further aggravating the vapor losses.

Auxiliary Degassing Methods. There are other ways to degas a copper alloy melt than using a specific treatment. The technique of lowering the melt superheat temperature, if possible, and holding the melt (in a dry, minimal gas environment) provides outgassing simply by lowering the equilibrium liquid state solubility of the gas. During casting, the use of chills to provide directional solidification, particularly for the long freezing range alloys such as the red brasses and tin bronzes, results in less tendency for gas porosity.

In the mold-metal reaction, previously degassed and deoxidized metal containing excess phosphorus or lithium deoxidants can react with green sand containing moisture. Hydrogen is liberated and absorbed as the metal comes into contact with the sand. This can be minimized by correct deoxidant additions (discussed in the following section in this article). In addition, finer facing

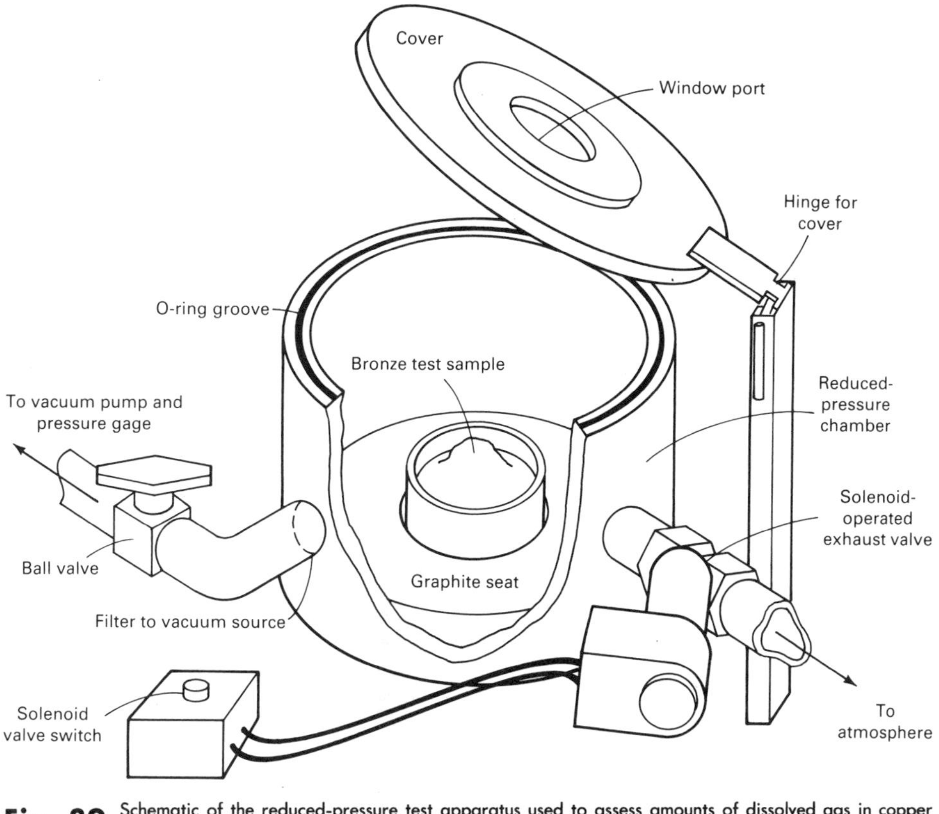

Fig. 39 Schematic of the reduced-pressure test apparatus used to assess amounts of dissolved gas in copper alloys

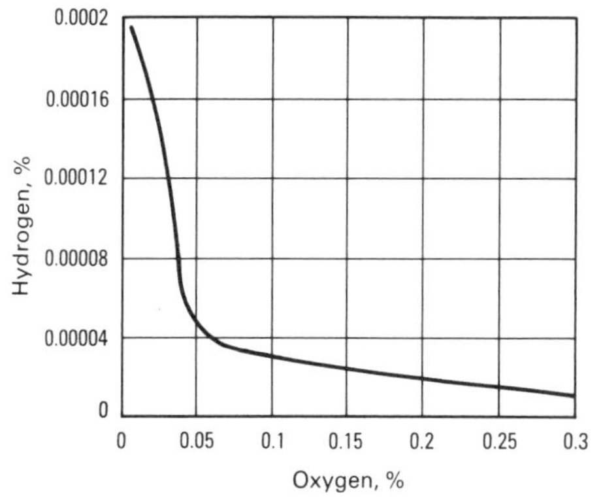

Fig. 40 Hydrogen/oxygen equilibrium in molten copper. Source: Ref 61

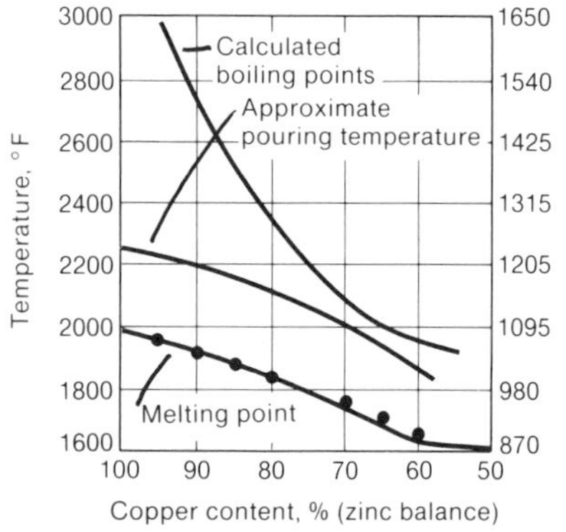

Fig. 41 Influence of zinc content on boiling point or vapor pressure in copper alloys. Source: Ref 62

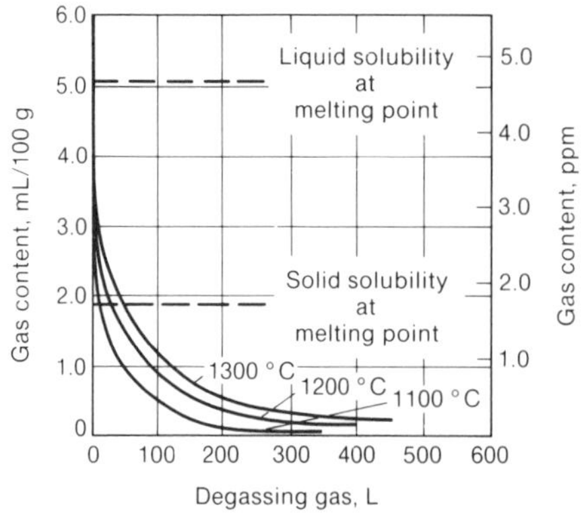

Fig. 42 Amount of purge gas required to degas a 450 kg (1000 lb) copper melt. Source: Ref 70

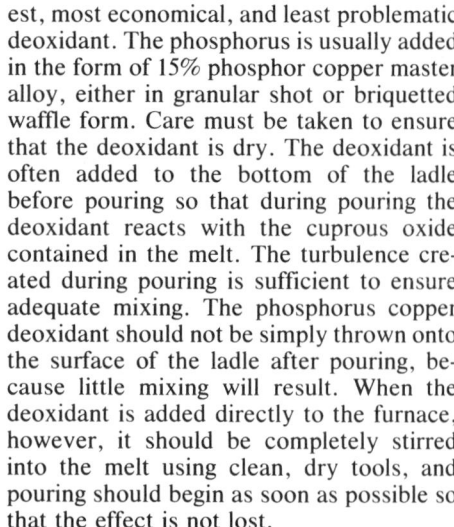

sands or mold coatings (inert or reactive such as sodium silicate or magnesia) can be used; these confine any reaction to the mold/metal interface area and retard hydrogen penetration into the solidifying skin of the cast metal.

Deoxidation of Copper Alloys

All copper alloys are subject to oxidation during most melting operations. Oxygen reacts with copper to form cuprous oxide, which is completely miscible with the molten metal. A eutectic is formed at 1065 °C (1950 °F) and 3.5% Cu_2O, or 0.39% O, as shown in the copper-oxygen phase diagram in Fig. 37.

Therefore, cuprous oxide exists within the melt as a liquid phase and is not generally separated by gravity alone. If not removed, this liquid phase will cause discontinuous solidification during casting, resulting in considerable porosity and low mechanical strength. Thus, some type of deoxidation process is required. In addition, proper deoxidation of all melts enhances fluidity and therefore castability.

Of course, deliberate oxidation treatments (the oxidation-deoxidation process previously described) are still employed. These are designed to preclude hydrogen pickup in copper alloy melts.

Phosphorus Deoxidation. Most copper alloys are deoxidized by a phosphorus reduction of the cuprous oxide. Although several other oxygen scavengers are possible according to the free energy of oxide formation, phosphorus is usually the easiest, most economical, and least problematic deoxidant. The phosphorus is usually added in the form of 15% phosphor copper master alloy, either in granular shot or briquetted waffle form. Care must be taken to ensure that the deoxidant is dry. The deoxidant is often added to the bottom of the ladle before pouring so that during pouring the deoxidant reacts with the cuprous oxide contained in the melt. The turbulence created during pouring is sufficient to ensure adequate mixing. The phosphorus copper deoxidant should not be simply thrown onto the surface of the ladle after pouring, because little mixing will result. When the deoxidant is added directly to the furnace, however, it should be completely stirred into the melt using clean, dry tools, and pouring should begin as soon as possible so that the effect is not lost.

Use of phosphorus deoxidation results in the formation of a liquid slag of cuprous phosphate [$2P + 4Cu + 7O \rightarrow 2Cu_2O{\cdot}P_2O_5$]. This product easily separates from the rest of the melt; therefore, the phosphorus effectively scavenges the oxygen, delivering the product to the surface dross phase, where it can be easily skimmed.

Phosphorus is usually added at a rate of 57 g (2 oz) of 15% master alloy per 45 kg (100 lb) of melt. In cases where oxidation of the melt is deliberately employed to reduce hydrogen absorption, double this amount can be required for full deoxidation. This is usually sufficient to deoxidize completely a melt saturated with 0.39 wt% O. However, the recovery of phosphorus is less than 100% and may be as little as 30 to 60% (Ref 72). It is desirable to maintain a residual level of at least 0.01 to 0.015% P in the melt, especially during pouring, so that reoxidation and potential steam reaction problems are alleviated.

Each foundry must determine the proper addition rate for a given set of conditions.

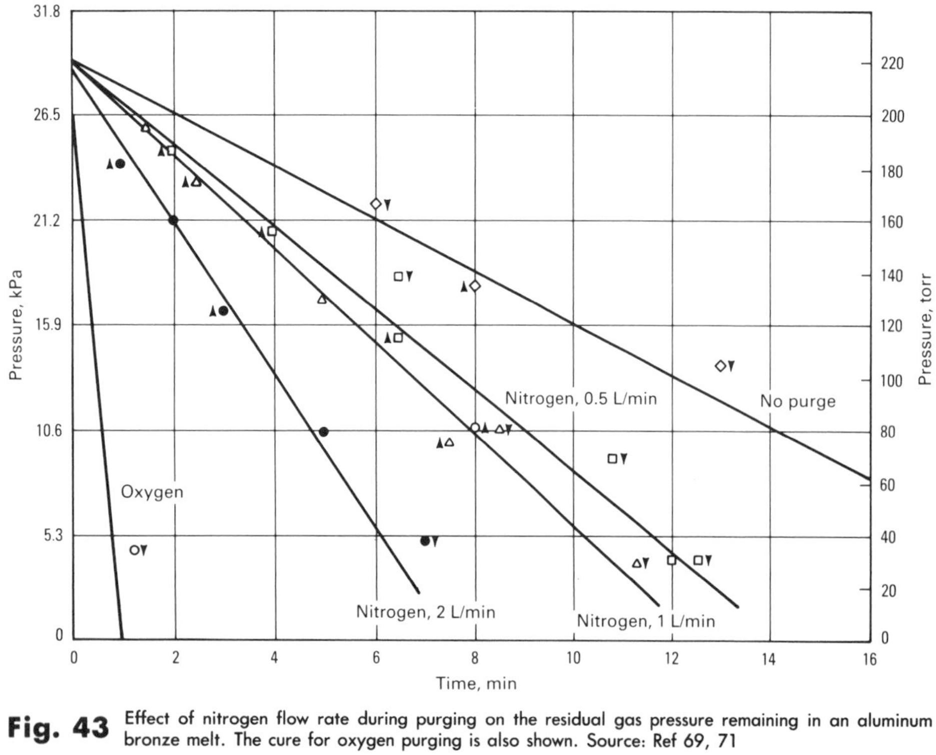

Fig. 43 Effect of nitrogen flow rate during purging on the residual gas pressure remaining in an aluminum bronze melt. The cure for oxygen purging is also shown. Source: Ref 69, 71

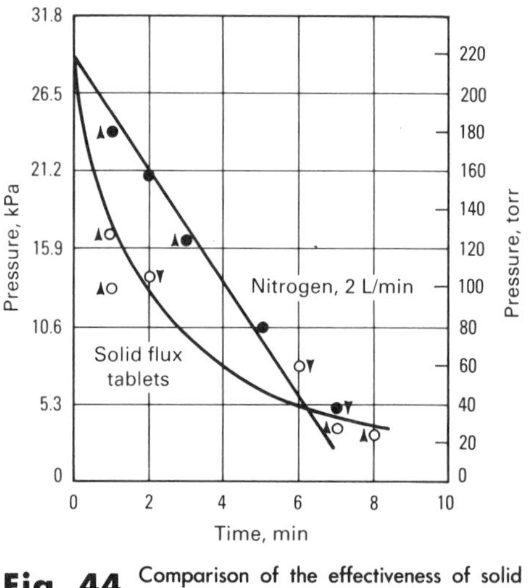

Fig. 44 Comparison of the effectiveness of solid degassing flux versus nitrogen purging. Source: Ref 71

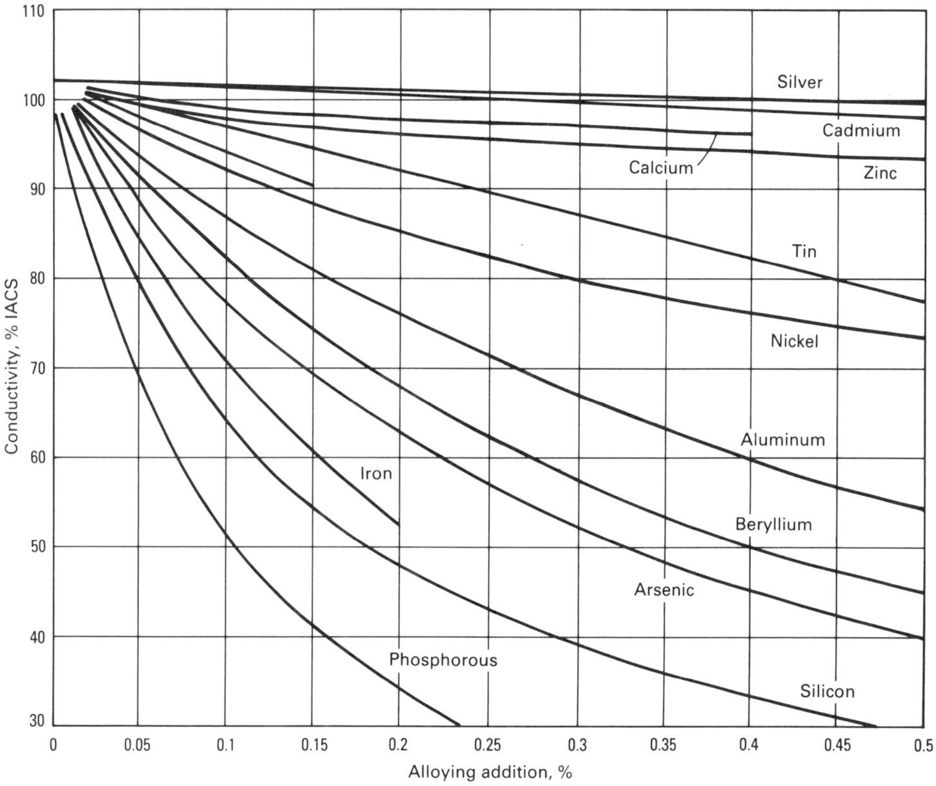

Fig. 45 Effect of alloying elements on the electrical conductivity of copper. Source: Ref 73

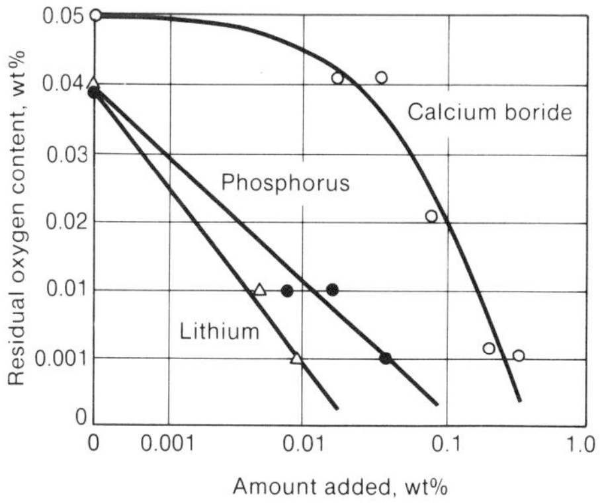

Fig. 46 Deoxidant efficiency in copper alloy melts. Source: Ref 72

Furthermore, with foundry returns, a certain residual amount of phosphorus is usually already present. A routine addition of phosphor copper for deoxidation could then actually result in excessive phosphorus. If the phosphorus content is too high during melting or pouring, the lack of oxygen may invite hydrogen entry and result in the steam reaction during casting. Adding more phosphorus to control the steam reaction can therefore actually aggravate the condition. In addition, when the phosphorus content is 0.03% and beyond, excessive metal fluidity can result in penetration of the molding sand or burn-in during casting.

Phosphorus copper is an effective deoxidant for the red brasses, tin bronzes, and leaded bronzes. However, phosphorus should not be used for deoxidizing high-conductivity copper alloys, because of its deleterious effect on electrical conductivity (as will be discussed shortly), and it should not be used for copper-nickel alloys. In these materials, the presence of phosphorus results in a low-melting constituent that embrittles the grain boundaries. A silicon addition of 0.3% and a magnesium addition of 0.10% serve to deoxidize and desulfurize copper-nickel melts. For nickel-silver alloys (copper-nickel-zinc), the use of 142 g (5 oz) copper-manganese shot per 45 kg (100 lb) of melt, and 57 g (2 oz) manganese coupled with 85 g (3 oz) of 15% P-Cu, is a recommended deoxidation technique.

The yellow brasses, silicon bronzes, manganese bronzes, and aluminum bronzes usually do not require deoxidation *per se*, because of the oxygen-scavenging effects of their respective alloy constituents.

High-Conductivity Copper (Ref 72-75). Where the high-copper alloys (pure copper, silver copper, cadmium copper, tellurium copper, beryllium copper, chromium copper) are employed and electrical conductivity is a desirable property, phosphorus copper cannot be used as a deoxidant. Moreover, the strong oxide formers beryllium and chromium serve as their own deoxidants.

Figure 45 shows the effects of a variety of elements on the electrical conductivity of copper. Clearly, phosphorus even in small amounts significantly decreases conductivity; therefore, alternative deoxidants must be used. Fortunately, both boron and lithium are capable of deoxidizing high-conductivity copper without appreciably affecting electrical conductivity.

Boron Deoxidation. Boron is available either as a copper-boron master alloy or as calcium boride (CaB_6). The boron probably forms a copper-borate slag of the general form $2Cu_2O \cdot B_2O_3$ in much the same fashion as phosphorus produces a cuprous oxide phosphate slag (Ref 72). Theoretically, the boron combines with 60% more oxygen than the stoichiometric amount required to form B_2O_3 and therefore appears to be superefficient. However, practical experience has shown that this theoretical efficiency is not always achieved and that lithium is more effective, as shown in Fig. 46 (Ref 72). Therefore, lithium is often preferred, although there are greater precautions attendant upon its use.

Lithium Deoxidation (Ref 62, 73-76). Lithium has the advantage of serving as both a deoxidant and a degasser because it reacts readily with both oxygen and hydrogen. Lithium is soluble in molten copper but insoluble in the solid state. There is very little residual lithium contained in the casting or in scrap for remelt.

Because lithium metal is very reactive in air, bulk lithium metal must be stored in oil. For foundry applications, lithium is supplied in sealed copper cartridges. These cartridges must be stored in a safe, dry environment and must be preheated (to above 105 °C, or 225 °F) before use to drive off any surface moisture. These preheated cartridges should then be carefully yet firmly and quickly plunged to the bottom of the reacting vessel (furnace or ladle) to achieve full intimate contact and reactivity with the bulk of the melt. Only clean, dry, preheated plunging tools should be used for this task. Graphite rods are usually preferred.

Lithium-copper cartridges are generally available in various sizes ranging from 2.25 g (0.09 oz) for 23 kg (50 lb) melts to 108 g (4 oz). Thus, lithium additions can be made at maximum efficiency.

The specific chemical reactions that can occur with lithium include:

$$2Li + Cu_2O \rightarrow 2Cu + Li_2O \text{ (deoxidation)}$$

$$Li + H \rightarrow LiH \text{ (degassing)}$$

$$LiH + Cu_2O \rightarrow 2Cu + LiOH \text{ (recombination)}$$

$$2Li + H_2O \rightarrow Li_2O + H_2 \text{ (reaction during pouring)}$$

The lithium oxide (Li_2O) and lithium hydroxide (LiOH) products separate cleanly as a low-density fluid slag suitable for skimming (Ref 73). The lithium hydride that forms initially if hydrogen is present is unstable at normal copper melting temperatures and recombines with cuprous oxide (recombination).

When lithium is added in excess of the amount of cuprous oxide present, it will react with moisture in the air during pouring and can therefore generate sufficient hydrogen to regas the melt. This can result in unanticipated additional gas porosity and unsoundness during solidification.

Because lithium is such an effective deoxidant, it can also reduce residual impurity oxides (FeO, P_2O_5, and so on) in high-conductivity copper melts. This allows these elements to redissolve in the molten metal to the extent of their solubility limit and thus reduce electrical conductivity according to Fig. 45. Furthermore, lithium can form intermetallic compounds with silver, lead, tin, and zinc when the residual lithium exceeds that required for deoxidation. These intermetallic compounds, while reducing solid solubility, may improve mechanical properties and electrical conductivity.

Occasionally, it may be desirable to practice a duplex deoxidation treatment using the less expensive phosphorus, followed by a lithium treatment (Ref 73, 75, 76). Care must be taken to do this quickly and not allow phosphorus reversion to occur by letting the copper phosphate deoxidation slag remain on the melt for an extended period of time.

Magnesium Deoxidation. Magnesium behaves similarly to lithium, but may be stored in air rather than oil. It is actually a stronger deoxidant in terms of its free energy of oxide formation (Ref 77), and it is used to deoxidize (and desulfurize) copper-nickel alloys. The deoxidation product (magnesia) is a stable refractory, unlike lithium compounds, but it forms a tenacious oxide skin and can result in inclusions in copper casting alloys.

Testing for Proper Deoxidizer Addition. As stated previously, each foundry should assess its own casting practice for a given alloy and set of melting conditions and should determine the optimal addition of deoxidizer. However, there are two tests that can be used to determine whether a given amount of deoxidizer is adequate.

In the first test, a test plug or shrink bar of metal approximately 75 mm (3 in.) in diameter by 75 mm (3 in.) deep is poured. If a shrinkage cavity results, the metal is deoxidized and ready for pouring, which should then be done immediately. Shrinkage will not occur until about 0.01% residual phosphorus is present (Ref 77). A puffed-out or mushroomed cap on the test plug indicates that deoxidation is incomplete and that more should be added.

The second test involves a carbon or graphite rod immersed in the melt. When the rod surface reaches the molten metal temperature, if there is oxygen present, the rod will vibrate because of the reaction ($2C + O_2 \rightarrow 2CO$) occurring on the bar surface. The intensity of the vibration is a function of the oxygen content, and an experienced foundryman can readily determine the point at which the reaction becomes negligible, that is, when the melt is sufficiently deoxidized. The vibration decreases near the level of 0.01% residual phosphorus, as expected for a deoxidized melt (Ref 77).

Relative Effectiveness of Copper Deoxidizers. Various elements capable of scavenging oxygen from copper alloy melts have been described. The theoretical relative capabilities of several deoxidizers are listed below:

Deoxidizer	Reaction products	Amount of deoxidizer required to remove 0.01% oxygen, g/100 kg	oz/100 lb
Carbon	CO	7.5	0.12
Carbon	CO_2	3.8	0.06
Phosphorus	P_2O_5	7.5	0.12
Phosphorus	$2Cu_2P_2O_5$	5.6	0.09
Cu-15P	P_2O_5	49.3	0.79
Cu-15P	$2Cu_2P_2O_5$	36.8	0.59
Boron	B_2O_3	4.4	0.07
Boron	$2Cu_2OB_2O_3$	2.5	0.04
Cu-2B	B_2O_3	224.5	3.60
Cu-2B	$2Cu_2OB_2O_3$	134.7	2.16
Lithium	Li_2O	8.7	0.14
Magnesium	MgO	15.0	0.24

In practice, selection of the deoxidizer must be based on actual efficiency, economics, ease of use, and the specific metallurgical requirements of the alloy in question.

Melt Refining of Aluminum Alloys

The principal elemental impurities in molten aluminum are alkali metals (lithium, sodium, and calcium) in very small concentrations (<20 ppm) and magnesium in large concentrations (0.20 to 1.5%). In the latter case, the magnesium is commonly removed by demagging processes (see the section "Demagging of Aluminum Alloys" in this article). Phosphorus is sometimes present as an impurity when remelting phosphorus-treated hypereutectic aluminum-silicon alloys, but is generally removed by conventional fluxing processes.

The alkali metals lithium, sodium, and calcium are true impurities arising from primary aluminum production and can have detrimental effects on solidification and casting integrity if not removed. Residual lithium originates from certain electrolytic reduction cell practices in the production of primary aluminum, where lithium salts are used to enhance energy efficiency in the Al_2O_3 reduction process. In the new aluminum-lithium alloys, the element is a deliberate addition (2 to 4%). However, as these alloy products are scrapped, remelt contamination of nonlithium alloys can become a problem, especially for the secondary smelting industry. Sodium may also be present as an impurity, dissolved by the molten aluminum from the cryolite used in the reduction process. Calcium may originate from the Al_2O_3, refractory vessel deterioration, or as an impurity in other additives.

Removal of Alkali Metals

The removal of these impurities is rendered thermodynamically favorable when the halogens chlorine or fluorine are introduced into the melt. Figure 47 shows the free energy of formation of various chloride compounds. It can be seen that the chlorination of aluminum will result in the preferential formation of chlorides of lithium, sodium, potassium, and calcium.

Various methods are available for removing these impurities. Simple gas fluxing or gas or solid flux injection coupled with rotating dispersers can be used. Most of these mechanical processes apply to the primary, secondary, or mill products industries, although flux injection is also applicable to foundry operations.

The TAC process, or the treatment of aluminum in crucible (Fig. 48), employs the injection of AlF_3 powder into a crucible of metal that is stirred with a cast iron paddle mixer. Critical process parameters include mixer paddle design, position within the crucible, amount of flux added, and proper speed (revolutions per minute) to achieve optimum reaction capability and performance (Ref 80). Favorable reactions between dissolved alkali metals and the decomposed fluorides occur to form complex metal fluoride compounds. Figure 49 shows the alkali removal results obtainable with this process.

Other rotating injection systems are also used in-line between the melting or holding furnace and the casting equipment. In general, each of these processes involves chlorine plus an inert carrier gas such as argon or nitrogen, and similar results are obtained. Information on these processes can be found in Ref 47, 52, and 55, which discuss the SNIF, Alcoa, and Alpur processes, respectively.

The metal in-line treatment (MINT) system (Fig. 28) operates with the metal flow counter to the flow of the purge gas being injected through a series of nozzles, thus creating a swirling tank reaction without the use of other moving parts. Table 5 lists the alkali removal capabilities of this system using argon plus 1 to 2% Cl at a rate of 0.4 L/lb/min for metal flow rates near 590 kg/min (1300 lb/min).

General Principles of Gas Injection Refining. A relatively recent investigation has shown how calcium can be removed by chlorination in a stirred tank reactor typical of many in-line metal treatment technologies and has demonstrated several fundamentals that apply to other melt-refining and degassing (hydrogen removal) reactions (Ref 82, 83):

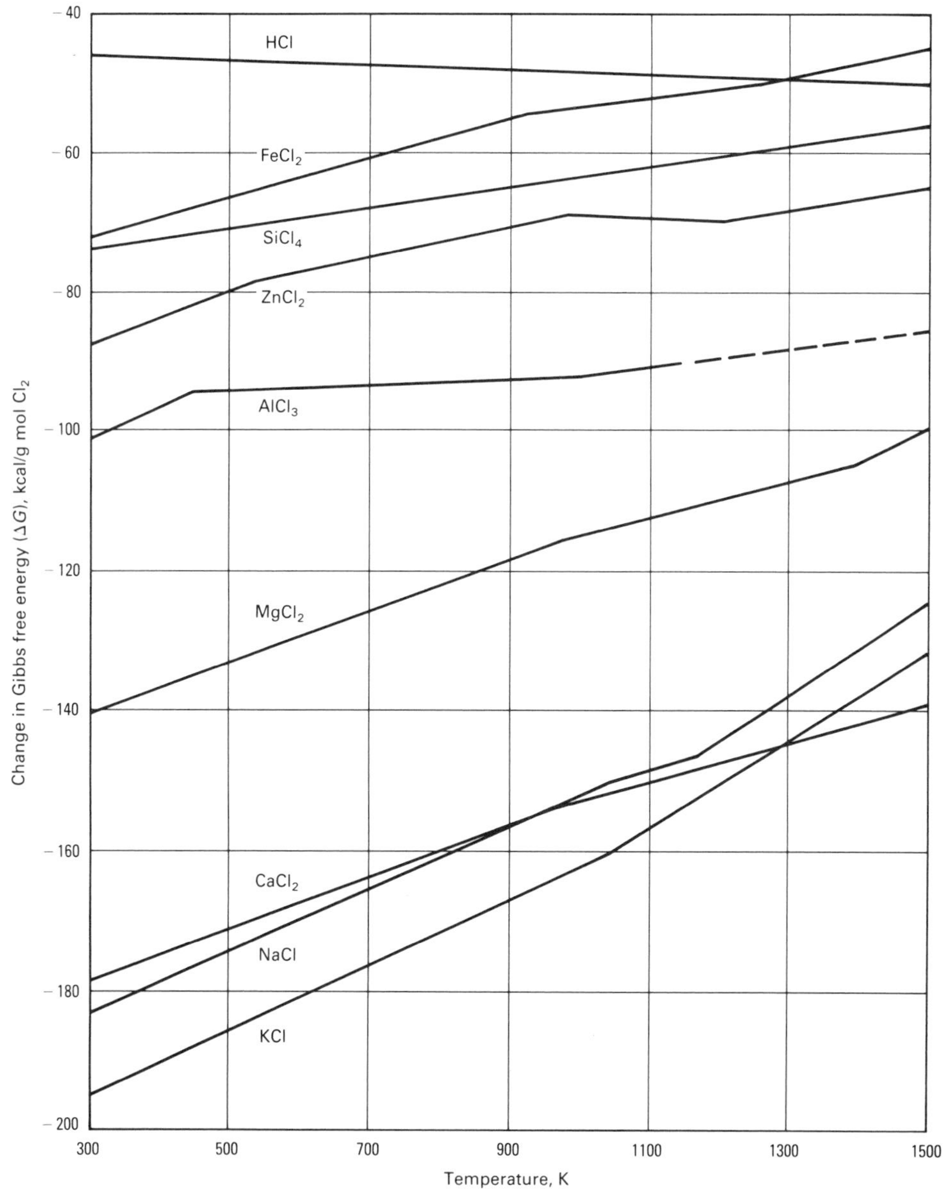

Fig. 47 Thermodynamics of metal chloride formation in aluminum alloys. Source: Ref 78

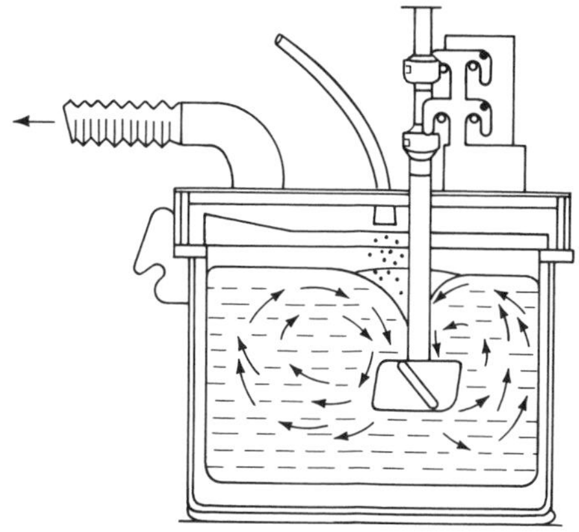

Fig. 48 Schematic of the TAC process for the removal of alkali metal impurities from aluminum. Source: Ref 79

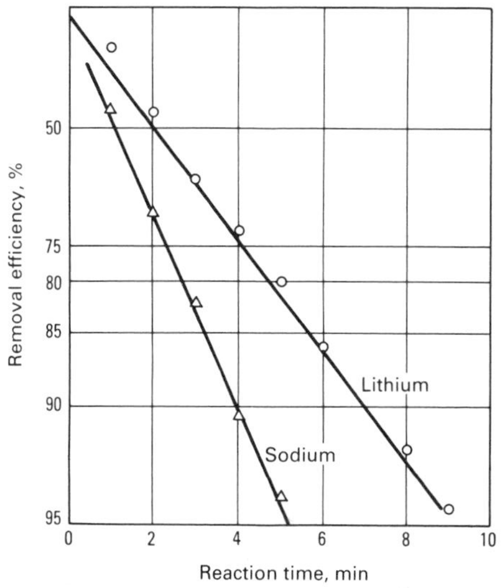

Fig. 49 Efficiency of TAC alkali metal removal versus time. Temperature: 800 °C (1470 °F); AlF_3 addition: 2 kg/Mg (4 lb/ton); rotor speed: 150 rpm. Source: Ref 81

- In each stirred tank reactor, there are limits of fluxing gas, flow rate, and rotor speed, varying with reactor size and geometry, beyond which efficiency is reduced
- Fluxing gas average bubble size in stirred tank reactors varies with fluxing gas flow rate and rotation speed (Fig. 50)
- The fluxing gas interfacial area is the primary reaction rate controlling parameter in these systems
- An increase in the total fluxing gas interfacial area (that is, reduced bubble size and increased bubble numbers) increases the reaction rate more significantly than increasing chlorine concentration alone (Fig. 51)

It has also been found that increased chlorine concentration as a reactive gas may not always produce better trace element removal. This may be related to an already minimized gas bubble size; increased gas volume is beneficial only if mass transfer is not the rate-limiting step.

Filter Bed Refining Process. Another method of removing alkaline impurities is to use a bed filter. One such technique employs petrol coke, which most likely removes alkali metals through the formation of salts or metal carbon compounds such as Li_2C_2, LiC_4, and $NaC_{64}NaC_{12}$ (Ref 84). The process equipment (Fig. 52) consists of a bed of petrol coke and corundum. Argon gas is sparged to flow counter to the molten aluminum. A sodium reduction of up to 75% or down to as little as 1 to 3 ppm can be achieved. Lithium has been removed at rates up to 60%. Removal rate and efficiency increase with metal temperature and decrease with metal flow rate (Ref 84). Efficiency also decreases as the petrol coke becomes saturated.

Demagging of Aluminum Alloys

Many useful aluminum alloys contain magnesium in quantities from 0.10 to 5.5%. Wrought alloys in particular contain higher levels of magnesium, and casting alloys 518 and 520 also contain 8 and 10% Mg, respectively. These materials constitute a significant portion of the scrap market and are therefore largely available for recycling/remelting. Mill recyclers often need to produce new alloy product with lower magnesium content, and secondary smelters producing die cast and foundry ingot also need to produce very low magnesium content alloys (generally <0.2%). Therefore, there may be a need to demag aluminum scrap during these remelting/alloying operations.

Table 5 Removal of alkali metals from aluminum alloys using the MINT system

Alloy	Sodium, ppm In	Sodium, ppm Out	Sodium, ppm Removal efficiency, %	Lithium, ppm In	Lithium, ppm Out	Lithium, ppm Removal efficiency, %	Calcium In	Calcium Out	Calcium Removal efficiency, %
6063	5.9	0.9	85	...	...	...	4.0	1.0	75
	10.2	3.1	70	...	...	...	4.0	2.0	50
	9.5	2.0	79	...	...	...	3.0	1.0	66
	8.9	1.5	83	...	...	...	...	...	...
	12.9	2.1	84	...	...	...	...	...	...
	13.4	1.8	87	...	...	...	...	...	...
	13.6	2.3	83	...	...	...	...	...	...
1050	...	...	...	1.1	0.10	91	...	...	...
	...	...	...	2.5	0.50	80	...	...	...
	...	...	...	5.0	0.40	92	...	...	...
	...	...	...	3.1	0.90	90	...	...	...
	...	...	...	3.3	1.50	55	...	...	...
	...	...	...	3.2	0.30	91	...	...	...
	...	...	...	2.9	0.80	72	...	...	...
	...	...	...	1.1	0.20	82	...	...	...
	...	...	...	2.7	0.60	78	...	...	...
	...	...	...	4.2	1.30	69	...	...	...
	...	...	...	3.8	1.00	74	...	...	...
	...	...	...	3.0	0.80	73	...	...	...

Source: Ref 57

Fortunately, the demagging of aluminum alloy scrap is possible during remelting. There are three general types of processes that can be used to remove magnesium from aluminum: chlorination, use of solid chlorine-containing fluxes, and electrolytic demagging (Ref 85). Although electrolytic demagging does not yet exist commercially, it is worthy of mention because it eliminates the potential environmental problems associated with chlorination. Demagging by chlorination or fluxing is discussed further in the section "Chlorination and Fluxing Processes" in this article.

Principles of Demagging

Magnesium can be removed from molten aluminum alloy by adding halogen compounds containing chlorine and/or fluorine. The reaction between magnesium and chlorine or fluorine occurs because it is thermodynamically favorable; that is, there is a preferred chemical affinity between magnesium and the halogen at normal molten aluminum operating temperatures. Although both halogens can be used, commercial practice, economics, and availability usually favor the use of chlorine.

Figure 47 shows the thermodynamics of metal chloride formation; it can be seen that indeed $MgCl_2$ formation is favorable. Accordingly, when gaseous chlorine is introduced into molten aluminum, $AlCl_3$ is formed; this is a gaseous phase, and it further decomposes to react with the magnesium present. The resultant product—$MgCl_2$—forms as a liquid phase above 712 °C (1313 °F). It is less dense than aluminum and therefore ultimately rises to the surface or dross phase, where it can be removed by skimming. However, favorable thermodynamics alone do not guarantee that magnesium will be efficiently removed. Kinetic factors (rate of mixing, contact area) also have pronounced effects. Therefore, adding chlorine alone does not guarantee effective magnesium removal, particularly if it is injected simply by lances or fluxing tubes.

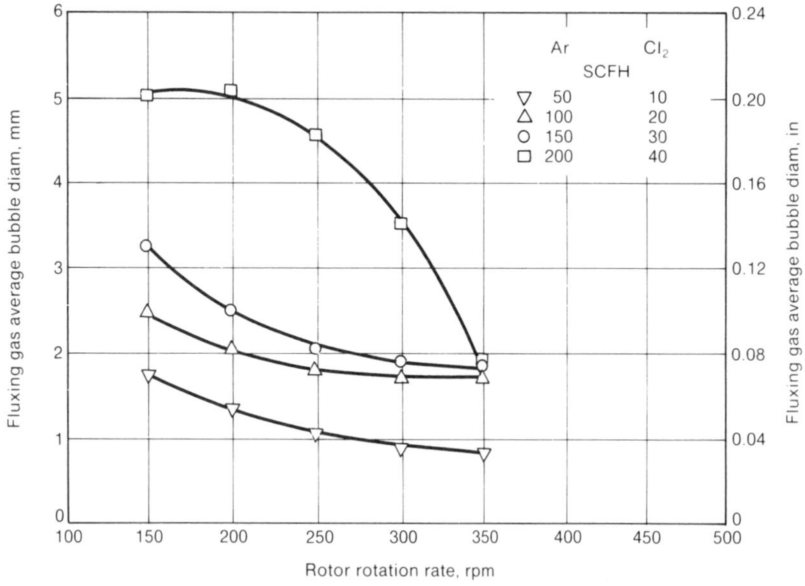

Fig. 50 Effect of gas flow rate (in standard cubic feet per hour, SCFH) and rotor speed on bubble size of a stirred tank reactor. Melt weight: 1500 kg (3300 lb); alloy: aluminum plus calcium; fluxing gas: 16.7% Cl. Source: Ref 82

Figure 53 shows free energy changes for various reactions in the demagging process as a function of magnesium content. The change in free energy ΔF is negative; that is, the demagging reaction is thermodynamically favorable throughout the entire range of magnesium levels to 1% (and beyond, although this is not shown). Kinetic factors, however, determine whether or not the reaction will occur. The most important kinetic factors are the size and residence time of the $AlCl_3$ bubbles. Finer bubble sizes have higher surface-area-to-volume ratios; this results in more contact area and therefore higher reaction rates. Accordingly, processes that provide for finer bubble sizes and better mixing are desired for effective and efficient demagging.

Demagging Efficiency. For a perfect (100% efficient) reaction, 1.34 kg (2.95 lb) of chlorine is required to remove 0.45 kg (1 lb) of magnesium. If process factors are such that the reaction is not completely efficient, substantially more chlorine may be required. Inefficient reactions waste time and chlorine and usually result in substantial emissions and fuming, creating environmental hazards and corrosion problems. It is therefore useful to express actual demagging efficiency as a percentage of the stoichiometry of a complete reaction. Thus:

$$\text{Demagging efficiency (\%)} = \frac{2.95}{\text{Actual chlorine used per pound of magnesium}}$$

When the above-mentioned kinetic factors are not sufficiently satisfied, the reduction of $AlCl_3$ is incomplete. Effluent from the reaction chamber, which contains $AlCl_3$, will react with moisture in the air to form aluminum oxychlorides and chlorohydrates plus aluminum oxide and hydrochloric acid.

Demagging efficiency also varies with magnesium concentration. Figure 54 shows a decreasing efficiency at lower magnesium levels, following the concept of Henry's law for lower free energy available for reaction at low or dilute levels of solute (in this case, magnesium in aluminum). Therefore, reducing magnesium from 1.0 to 0.3% is much easier than reducing it from 0.3 to 0.1%.

Chlorination and Fluxing Processes

Flux Tube Demagging. The simplest method of demagging is the use of flux tubes

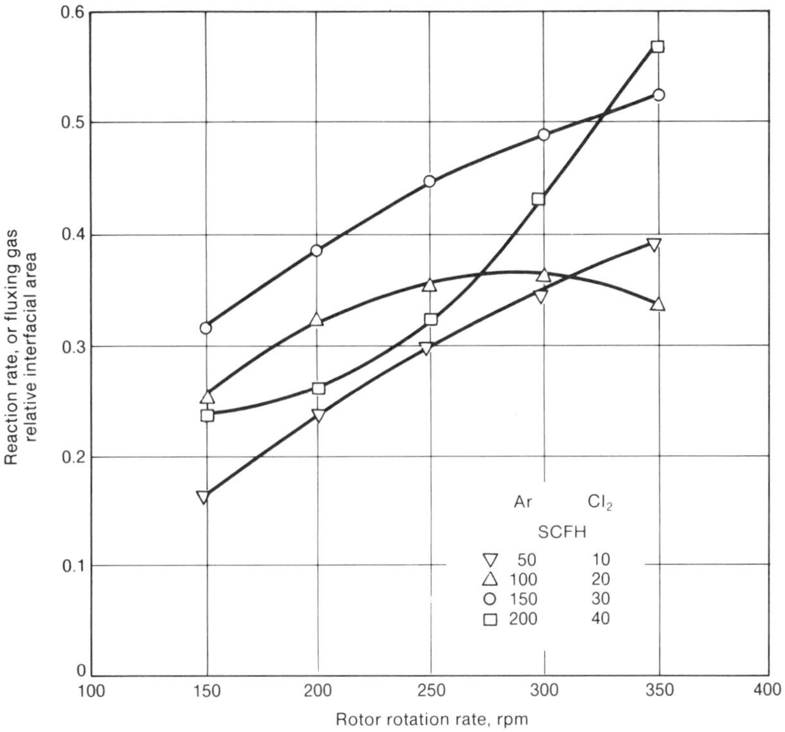

Fig. 51 Effect of gas flow and rotor speed on the reaction rate for a stirred tank reactor. Same alloy and parameters as given for Fig. 50. Source: Ref 82

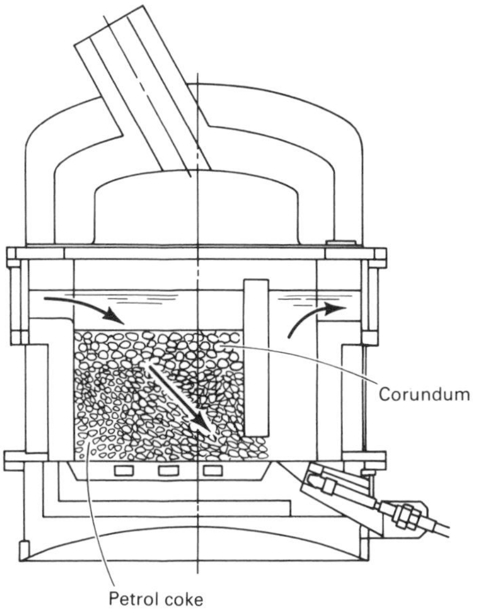

Fig. 52 Schematic of a one-filter bed process for removing alkali metals from aluminum. Source: Ref 84

for chlorination. However, flux tube demagging is not particularly efficient, because the bubble size generated is not sufficiently fine and the effluent may be extensive. Figure 55 illustrates magnesium loss during the simple flux tube chlorination of aluminum melts. As shown, the efficiency for pure chlorine is temperature dependent, optimizing near the melting point of $MgCl_2$ (710 °C, or 1310 °F).

The Bell process is based on the injection of chlorine gas into molten aluminum under a cast iron bell submerged into the melt. The favored chlorine partial pressure under the bell maximizes the reaction time, and the chloride fumes generated and the unused chlorine gas are mostly trapped within the bell. However, a scrubber system and a settling pond are often required to handle the effluent in most applications. Consequently, the Bell process has largely given way to other demagging methods with better environmental acceptance (Ref 87).

The Derham process uses a flux layer, usually chloride salts and/or cryolite, that traps the $AlCl_3$ fumes generated by chlorine injection. This flux layer creates a further reaction between the $AlCl_3$ and the magnesium-bearing molten aluminum. However, a large amount of dross can be created in the process, resulting in disposal problems as well as the emission of salt fumes.

The Derham and Bell processes have been commonly used in the secondary industry and have achieved better than 97% efficiency at magnesium levels of 0.10%. However, the need for close control of chlorine injection and the flux disposal problems of the Derham process, as well as the mechanical system complexities of the Bell process, have limited the usefulness of these systems in modern secondary smelting operations (Ref 87).

Channel Injection Process. The channel induction-injection furnace uses chlorine injection into the loop or throat of the induction furnace. Small gas bubbles are gener-

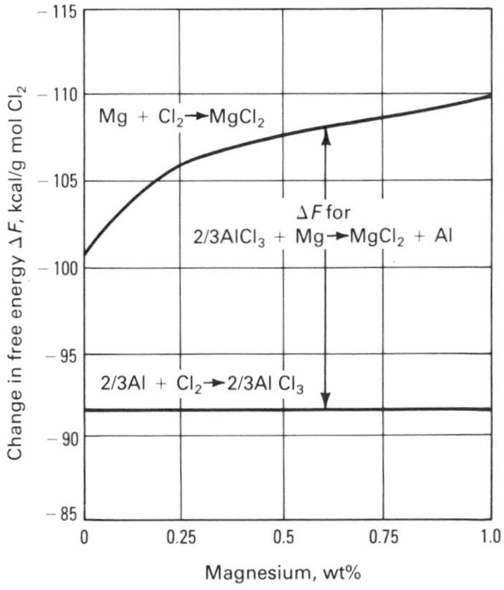

Fig. 53 Free energy changes for reactions occurring during the demagging of aluminum at 700 °C (1290 °F). Source: Ref 86

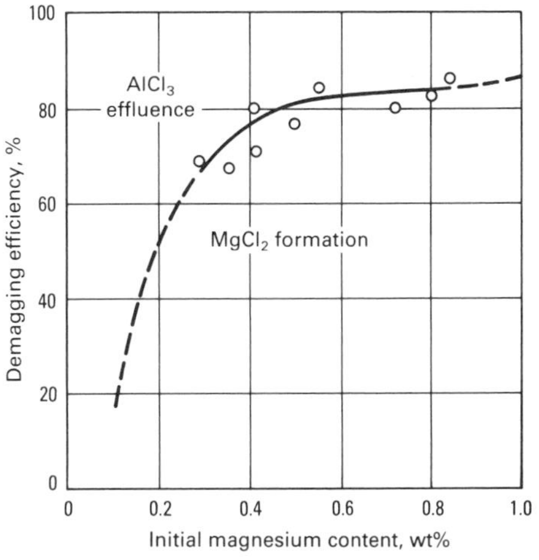

Fig. 54 Efficiency of demagging versus initial magnesium content. Source: Ref 86, 87

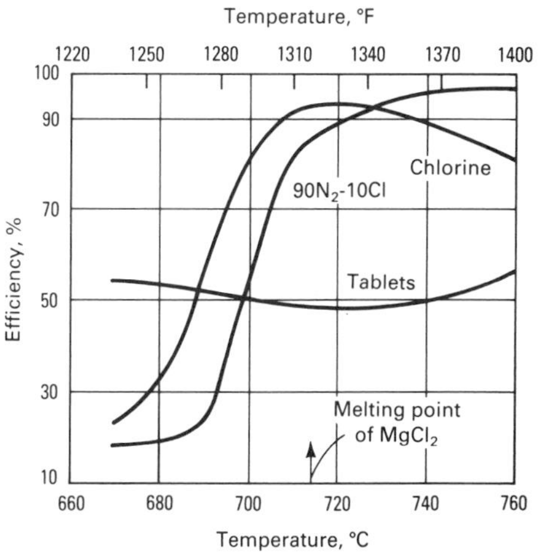

Fig. 55 Effect of temperature on the efficiency of flux tube demagging process associated with the formation of liquid $MgCl_2$. Source: Ref 88

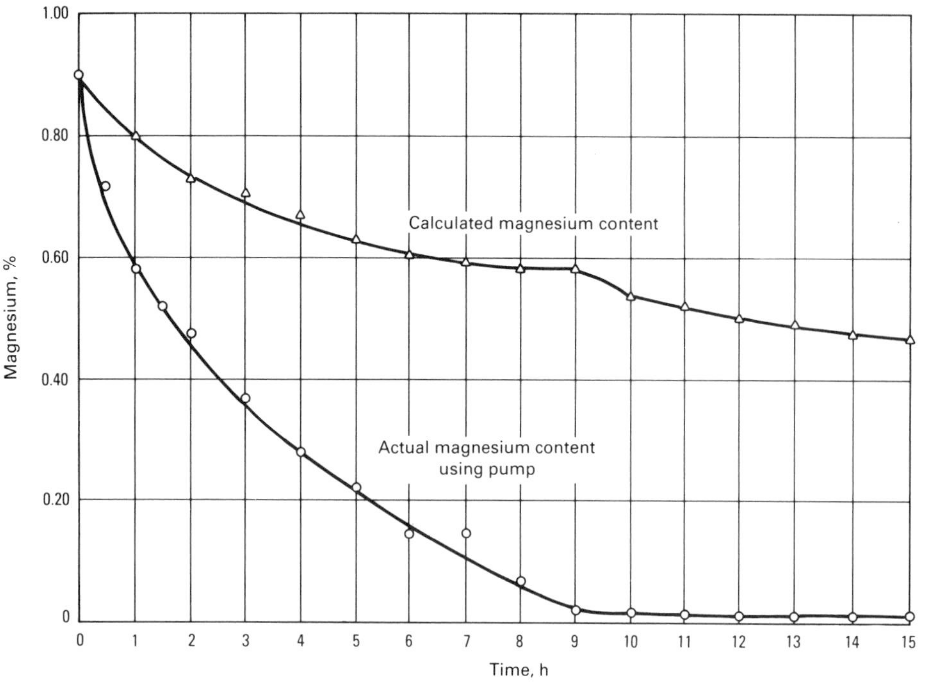

Fig. 56 Demagging of aluminum scrap using a gas injection pump during melting in a 68 Mg (150 000 lb) furnace. Source: Ref 87

ated by the electromagnetic induction stirring action of the furnace (Ref 89). Excellent magnesium removal results have been reported, but the system has not achieved significant commercialization, possibly because of high energy costs and lack of flexibility. Furthermore, the channel induction melting of scrap aluminum and the accompanying problems with dross formation may pose significant operating and maintenance difficulties.

The ALCOA process is often used as an in-line demagging treatment in the production of mill products production (Ref 52). This process is similar to rotary degassing, and it consists of a reactor vessel with one or more rotating dispersers. A flux gas consisting of chlorine plus nitrogen is introduced, and the disperser(s) ensures good mixing. Excellent removal rates and process efficiencies (now 99%) have been achieved with this type of unit, but it is expensive and is primarily suitable for the recycling of mill products.

The gas injection pump has become the most widely used demagging method in secondary smelting operations. Combining forced convection molten metal circulation with chlorine injection through a flux tube discharging just ahead of the rotating impeller, the pump allows the *in situ* demagging of high-magnesium aluminum scrap during the melting process. Demagging efficiencies with this device have been reported to be virtually 100% when the molten metal bath temperature is kept at 760 °C (1400 °F) or slightly above (Ref 87).

The principal advantage of the gas injection pump is its ability to provide demagging directly during the melting process. Figure 56 illustrates the effect of the gas injection pump in the demagging of aluminum scrap in a 68 Mg (150 000 lb) furnace. The magnesium content was continuously reduced during the time the scrap was being charged into the furnace, resulting in a low final magnesium content without a separate treatment.

A tremendous quantity of aluminum used beverage container (UBC) scrap is being generated. Recycling of UBC scrap has therefore become a significant commercial enterprise. The gas injection pump is capable of demagging UBC scrap to levels of approximately 0.05% (Ref 87).

Melting of Lead Alloys

Lead and its alloys constitute a small but commercially important segment of the nonferrous metal casting industry. Principal lead alloys and their areas of importance include Pb-60Sn (solders), antimony-lead and calcium-lead (battery grids), and Pb-7Sn (terne plating).

The melting and casting of pure lead and foundry alloys pose fewer problems than with other nonferrous systems. The major lead alloys have relatively low melting points (315 to 425 °C, or 600 to 800 °F) and are not susceptible to hydrogen absorption and the attendant gas porosity to any great extent. Dross formation can be a problem, especially with alloy constituents, so simple cover fluxes such as graphite or vermiculite are sometimes used. However, oxidation is relatively minor unless temperatures are excessive.

In the terne plating industry, where lead-tin coatings are applied by hot dipping steel articles, principally for use in fuel systems, a cover flux is often used. This is similar to the flux used in galvanizing—a $ZnCl_2$-NH_4Cl mixture, often with other chloride salts such as KCl. The flux used in terne plating serves to preclean the steel, to absorb pretreatment pickling residue, to prevent spitting as the articles enter the plating bath, to alleviate bath dross formation, and to prewet the article being coated.

In the die casting of lead alloy acid battery grids, no treatment is usually necessary at the die cast machine melting pot itself, although a melt cover can be used for heat insulation purposes and to prevent dross buildup. In remelt pots, where trim scrap and other scrap is melted, chloride-base drossing fluxes can be used. Pitch, sawdust, or wood chips are used to agglomerate the residues that accumulate on the surface.

The lead is usually first softened (refined) in the production of the newer calcium-lead-tin alloys used for maintenance-free battery grids from either primary metal or recycled scrap. Calcium is then added as an alloy, often under a protective cover such as sawdust, wood chips, or resin to prevent oxidation at the usual higher temperature at which alloying is facilitated.

Refining of Lead

A number of fluxing procedures and refining processes are used in the smelting and refining of primary and secondary lead production. Several of these involve chemical flux-refining techniques somewhat analogous to those used with other nonferrous metals. Figure 57 is a flow chart showing a number of important sequential process steps in the refining of lead. These steps will be briefly discussed here and are covered in greater detail in Ref 90.

Copper Drossing. In copper drossing, the first step, the elements copper, iron, nickel, cobalt, and zinc can be removed, if present, by cooling the lead almost to its freezing point. The liquid solubility of these elements at that temperature is very low; thus, they can be removed as an intermetallic dross as they separate from the heavier lead melt.

Copper will precipitate with any sulfur, arsenic, antimony, or tin, in that order. Lead sulfide may also form.

These drosses are usually intimately intermingled with the lead metal phase. A molten phase (copper-lead sulfide) may form, and when the newer calcium lead battery scrap is remelted, calcium may be present in this phase, lowering the lead content. If oxides are present, there is also a slag phase, and silica can be added to increase its fluidity, resulting in the creation of iron, zinc, and lead silicates.

It is also possible to remove copper from lead down to a level of 0.05% by stirring

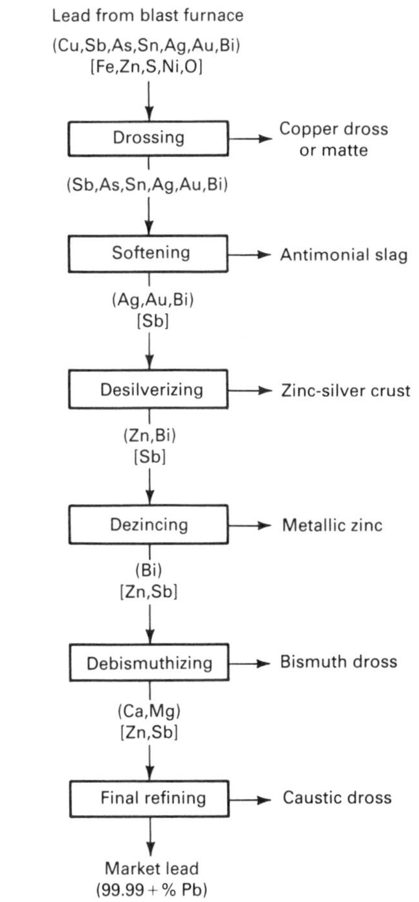

Fig. 57 Flow chart showing steps in the lead-refining process. Impurities remaining are shown in brackets after each step. Source: Ref 90

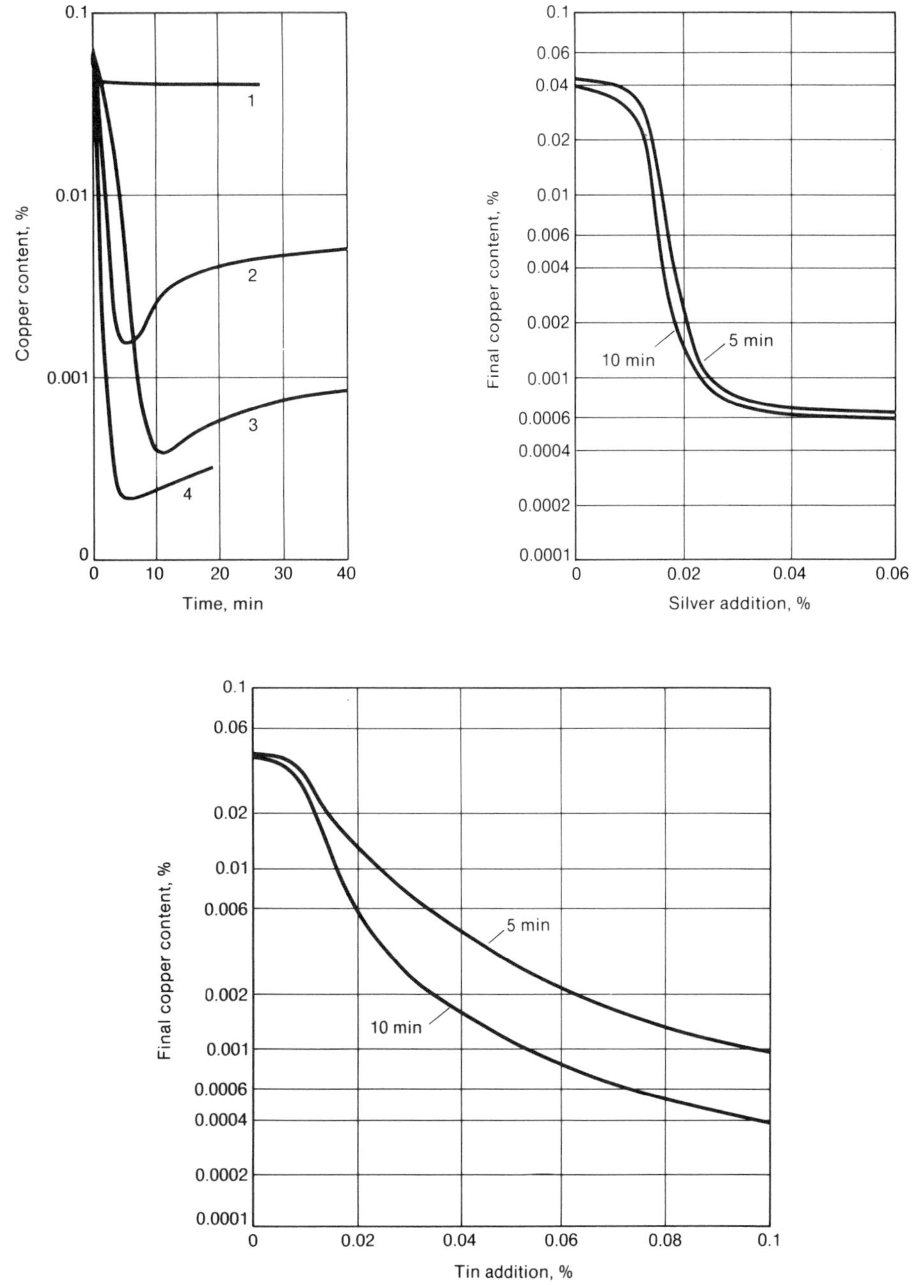

Fig. 58 Decoppering of lead. (a) Effect of sulfur. Curve 1 represents the copper content of silver-free lead as a function of time while stirring in 0.1% S at 330 °C (625 °F), and curve 4 represents similar conditions for a lead containing 0.04% Ag. Curve 2 resulted when only 0.05% S was added, and curve 3 when a total of 0.1% S was added as ten equal additions of 0.01% S each, at 1 min intervals, approximating plant practice in which the sulfur is added gradually over ten minutes. (b) Effect of silver. (c) Effect of tin. Source: Ref 90

sulfur into the lead melt at temperatures near the freezing point. When silver or tin is present, copper can be reduced to even lower levels, but a reversion (copper reentering the metallic phase) can occur if stirring is excessive (over 10 min) or if temperature is increased (Fig. 58).

Softening. In softening, the second process step, elements more readily oxidized than lead are removed by oxidation. These include tin, arsenic, antimony, and indium, whose free energies of oxide formation are greater than that of lead. The term softening relates to the effect these elements have on lead. When present, they contribute to solid-solution strengthening; thus, their removal results in softening.

The Harris process of softening consists of oxidizing the impurities, converting them to sodium salts, and collecting them in a caustic soda melt slag phase that rejects molten lead, providing for relatively easy physical separation. Temperatures of best reactivity are fairly high (700 to 800 °C, or 1290 to 1470 °F), and thorough mixing of the metal with the salt flux is required.

Tellurium, arsenic, and tin can be reacted directly with caustic soda to form lower sodium salts. A further oxidation reaction, employing sodium nitrate ($NaNO_3$) as the oxidizing agent, forms the higher sodium salts (arsenates, stannates, and antimonates). The intermixed metallic oxide products produced by these reactions are then removed and separated by various wet processing and precipitation methods.

A simple softening process involves the refining of lead by oxygen at a somewhat reduced temperature (650 °C, or 1200 °F) directly in a refining kettle. A 22.5 Mg (25 ton) lead melt containing 0.6% Sb was refined to a level of just 0.02% by blowing 10 m^3 (350 ft^3) of oxygen for 1.5 h (Ref 91). Moreover, approximately 100 kg/m^2 (20 lb/ft^2) of antimony was oxidized with oxygen injection, versus a rate of only 10 kg/m^2 (2 lb/ft^2) using air. Because this is in excess of the expected stoichiometric ratio, the oxygen must also act catalytically.

Desilvering is the term applied in lead refining to the removal of elements more noble than lead—silver, gold, and bismuth. These elements can be coprecipitated with lead and zinc by the Parkes process, which is the addition and mixing of zinc with the lead at a temperature just above the freezing point of lead (~320 °C, or 610 °F). Because there is substantial metal value contained by silver- or gold-bearing precipitate residues, technologies (pressing, liquation, and zinc distillation, for example) have evolved for separating the noble metals from the precipitate.

Zinc Removal. Zinc can be removed preferentially from the base lead metal by preferential oxidation or chlorination or by the Harris process without the use of an additional oxidizing agent:

$$Zn + 2NaOH \rightarrow Na_2ZnO_2 + H_2$$

Vacuum dezincing, however, provides much higher zinc recoveries than any of these techniques and is therefore preferred.

Bismuth Removal. Bismuth is more noble than lead and therefore cannot be removed by chemical fluxing techniques. Fractional oxidation or crystallization can yield liquid slag and solid crystals that are purer than the starting metal. Bismuth can also be precipitated from lead by adding alkali or alkaline earth metal reagents, such as calcium and magnesium (the Kroll-Betterton process). Magnesium metal and Pb-5Ca master alloy are stirred into the bismuth-containing molten lead at 420 °C (790 °F). As the lead is cooled back to its freezing point, $CaMg_2Bi_2$ crystals precipitate, leaving the lead refined of bismuth. The precipitate is intermixed with lead, and the latter can be removed by liquation by adding the mixture to the next batch and raising the temperature once again. Further reduction of bismuth is possible by stirring in antimony (or antimonial lead) in the presence of residual calcium and magnesium in the melt after the initial $CaMg_2Bi_2$ precipitate has been removed. The antimony addition forms a similar precipitate, which facilitates easier formation and substitution of bismuth for antimony in the precipitate at more dilute concentrations of bismuth.

Final Refining. Before casting, a final refining step is carried out on the molten lead, using a modified Harris process at 450 to 500 °C (840 to 930 °F). Caustic soda or $NaNO_3$ at the rate of 0.1% is stirred into the lead to decrease the zinc and antimony that may remain after previous refining steps to less than 1 ppm each, as shown in Fig. 59 (Ref 90). Arsenic, tin, calcium, magnesium, and iron will also be removed to very low levels with or before the antimony. The softened and highly purified lead thus produced is required in the production of battery alloys, whose electrochemical performance is highly dependent on the removal of impurities.

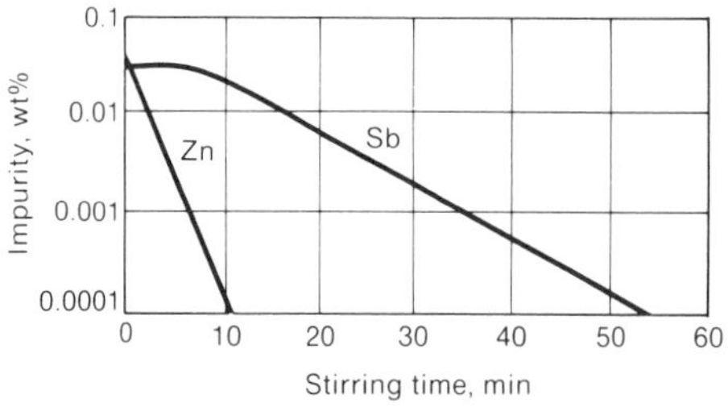

Fig. 59 Final lead refining using the modified Harris process. Source: Ref 90

Grain Refining of Aluminum Alloys

Grain refining is widely practiced in the commercial production of virtually all aluminum alloys, whether wrought or cast. Although many alloys contain alloying constituents that provide for fine-grain cast structures by virtue of their specific compositional effects, other alloys, especially those containing silicon, do not readily develop a fine grain size without assistance.

Benefits of Grain Refining

Grain refining is a necessary processing step in the production of most wrought shapes and foundry castings because:

- Mechanical properties and high-temperature properties are usually enhanced with finer grain structures; heavy-section properties are also improved by deliberate grain-refining practices
- Better castability results from the finer grain structure provided by creating more nuclei as the melt solidifies. Shrinkage porosity and hot tearing of castings is usually minimized. Better feeding characteristics usually mean improved casting soundness
- Better control and uniformity of response to heat treatment occur with finer grain size because there is more alloy homogeneity throughout the cast structure
- Finer grain structures result in enhanced extrudability, sheet formability, and machinability. Surface-finishing operations such as polishing and anodizing are also easier and more successful with finer grain structure

Measuring Grain Size

Grain size can be measured by a number of procedures, which are detailed in ASTM E 112. The grain size of finished castings or cast or wrought products can be determined metallographically. Grain size can be compared with a known and measured set of standards associated with a particular alloy or alloy family.

In the wrought industry, many producers have developed their own specific grain size tests correlating with their specific practices. These tests are designed to simulate the specific solidification conditions found in typical direct chill casting for large ingots and have been found to correlate well one with another.

Another, simpler test has recently evolved with great application for aluminum foundry casting purposes. In the ring test, samples of molten aluminum are poured into one or more iron rings 75 mm (3 in.) in diameter by 25 mm (1 in.) high set on a block of fused silica and allowed to solidify. Three surfaces of each test sample exhibit the grain structures to be found in most aluminum casting methods. The ring itself provides an immediate chill; therefore, the side of the test casting exhibits a grain structure typical of permanent mold casting. The top of the test casting, open to the atmosphere, cools slowly; its grain structure is considered to be indicative of a sand casting. The bottom surface of the test casting is in contact with the insulating and smooth fused silica and will exhibit a grain structure resembling a direct chill cast shape casting (Ref 6, 92).

The test can be used to compare the capabilities of various grain refiners or differing amounts. It can also be used as a process control check with each furnace heat or during prolonged holding times when delays in casting are experienced. Because the effects of grain refinement can be diminished with prolonged holding time, additional grain refiner may need to be added, and this simple test will monitor that need.

The use of thermal analysis has recently found application as a method of determining both sufficient grain size and silicon phase modification treatment immediately before casting. This will be discussed in the section "Modification of Aluminum-Silicon Alloys" in this article.

Grain-Refining Methods

Several methods are used to achieve grain refinement—for example, rapid cooling, mechanical agitation, growth-hindering additions, and addition of nucleating agents (Ref 93).

Rapid cooling produces finer grain size in most aluminum alloys. The mechanisms may involve reduced grain growth, reduced critical size of the nucleus, activation of additional nucleants, increase in nucleation sites, homogeneous nucleation, and recrystallization (Ref 93). Rapid cooling has a significant effect on the other aspects of a cast structure that influence properties, such as decreased dendritic arm spacing, increased solid solubility, decreased segregation, and the appearance of metastable phases. However, additional, deliberate grain refinement may still nevertheless be sought.

Mechanical agitation can be achieved by manual or electromagnetic stirring, gas purging, or ultrasonic vibration. Such motion in the molten metal induced by these techniques most likely breaks up or bends growing dendrite arms. The amount of refinement obtained usually increases with the mechanical energy put into the melt up to a point. Although other benefits such as reduced segregation can result, the high cost, labor, or complexity of the equipment required has precluded the large-scale use

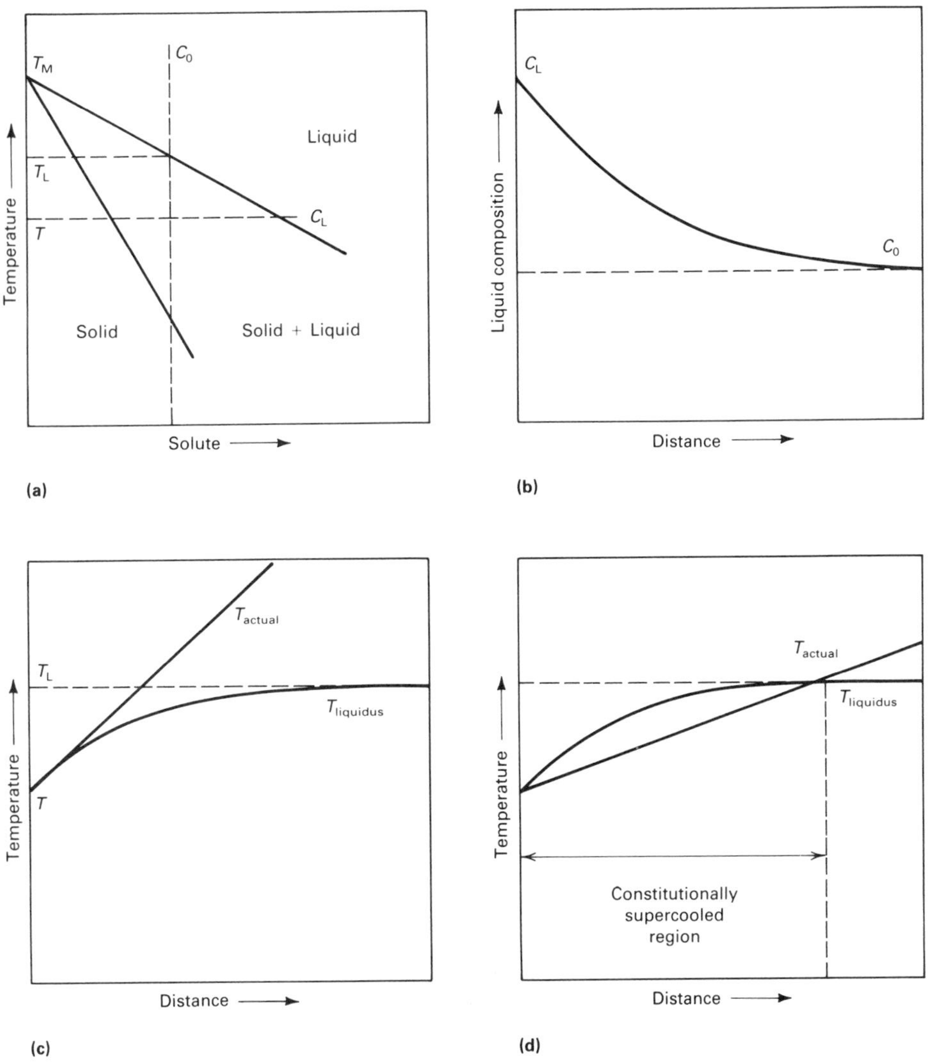

Fig. 60 Constitutional supercooling during solidification of alloy with composition C_0. (a) Phase diagram. (b) Solute-enriched layer ahead of solid/liquid interface. (c) Stable interface. (d) Unstable interface. Source: Ref 94

of such grain-refining methods in actual practice.

Growth-hindering additions are usually alloying elements or solutes in an alloy system in which constitutional supercooling can take place during solidification. Figure 60 illustrates this phenomenon in a binary phase diagram; in this diagram, as an alloy of composition C_0 solidifies, the solute concentrates in the liquid ahead of the growing solid. As a consequence, there is a zone in which the actual temperature is lower than the equilibrium freezing temperature, yet the liquid metal persists. This is the undercooled or supercooled zone, in which the solute enrichment or supersaturation facilitates nucleation and therefore smaller grain size. The lower the eutectic temperature and the smaller the amount of solute element needed, the more effective this means of grain refinement. Copper alloys, with relatively small amounts of the solute that causes this effect, are therefore usually quite adequately grain refined by alloying alone. Although aluminum alloys exhibit similar behavior when solute is added (Fig. 61), additional methods of producing grain refinement are always employed commercially.

Types of Nucleating Additions. By far the most common method of achieving grain refinement in aluminum is the use of specific nucleating additions—salts or master alloys. Grain-refining agents such as titanium, boron, and zirconium are added singly or in combination to aluminum melts in very small quantities that do not alter the base composition of the melt to any great degree. These additions have historically been made as salts (often tabletted) or as master alloys (hardeners) with aluminum in the form of shot, waffles, or briquettes. Carbon (graphite) can be an effective grain refiner in the presence of carbide-forming elements such as titanium (Ref 96).

Salt Flux Grain Refining. Grain-refining elements can be introduced by fluoride and double fluoride salts that react with the melt (Ref 1, 95).

$$3K_2TiF_6 + 4Al \rightarrow 3Ti + 4AlF_3 + 6KF$$

$$2KBF_4 + 3Al \rightarrow AlB_2 + 2AlF_3 + 2KF$$

$$3K_2TiF_6 + 6KBF_4 + 10Al \rightarrow 3TiB_2 + 10AlF_3 + 12KF$$

Various other fluxing salts can be included to improve fluidity and reactivity, and both boron and titanium salts can be included in the same product. Some fluxes may also contain carbon to enhance carbide formation, because carbides will also serve as effective grain nuclei (Ref 96). In addition, it is now known that intermetallic compounds formed by titanium and aluminum (for example, $TiAl_3$) play a major role as grain-nucleating agents.

Master alloys, or hardeners, are alloys of aluminum with certain amounts of titanium and/or boron. Master alloys are usually produced in induction furnaces and are cast as ingot, waffle, sheared rod, or shotted product.

Binary aluminum-titanium hardeners contain only 3 to 10% Ti because of their relatively high melting points even at this low level. Even at such low concentrations, gravity segregation is difficult to control. When boron is added to these alloys, the boron content is kept equally low (0.5 to 3%) because titanium diborides have an even greater tendency to segregate due to their higher density relative to liquid aluminum.

Use of Grain Refiners

The first issue to settle when considering a grain-refining treatment is to determine the degree of refinement needed for each application and then to choose from the variety of grain-refining compositions available. Factors to be considered include economics; the effect of titanium and boron residuals on composition, specifications, and properties; and end use of the products.

In foundry applications, grain refiner is usually added by salt flux tablet or master alloy waffle, sheared rod, or shot to the furnace. The grain-refining capabilities are diminished by degassing or fluxing treatments and by prolonged holding. This loss of refining capability is due to agglomeration of the nucleating particles, dissolution at high temperatures, and the formation of complex compounds with other alloying elements and fluxing agents. It is therefore essential that grain refiner be added immediately before pouring and casting. In general, the amounts to be added are 0.05 to 0.15% Ti, 0.04% B, or 0.01 to 0.08% Ti plus 0.003% B, even though foundry ingot as received from the smelter is usually prerefined.

Grain refinement is diminished by the subsequent remelting of scrap castings, gates and risers, and so on (Ref 95, 97). In general, increased pouring temperature and holding time result in coarser grain size for

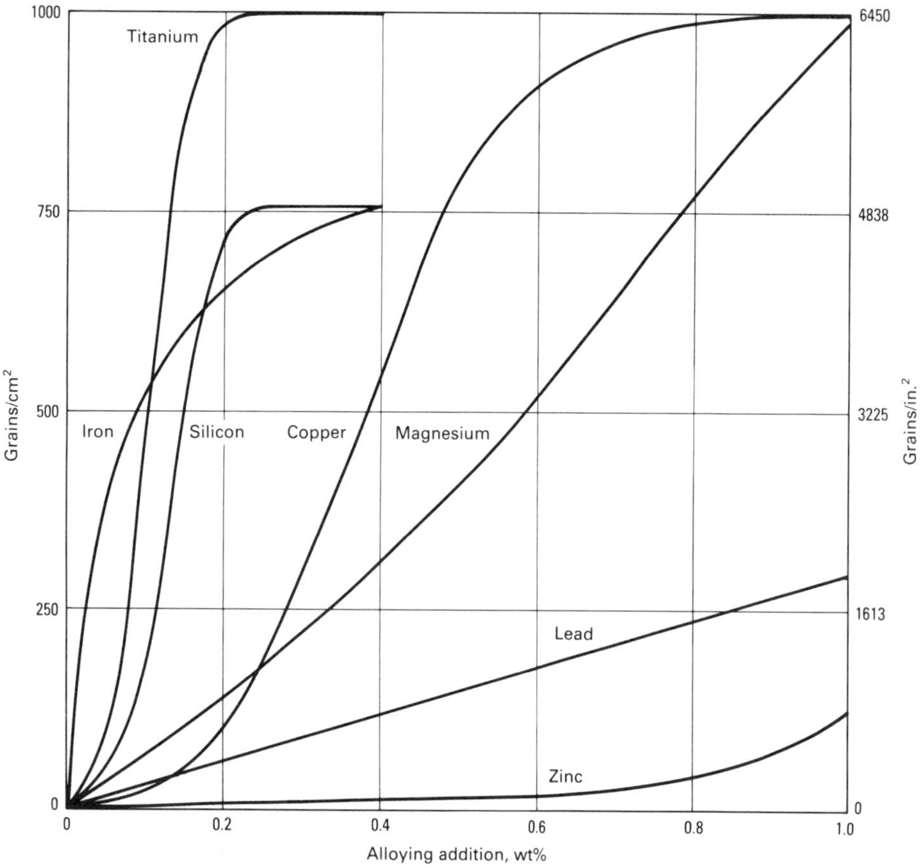

Fig. 61 Effect of alloying elements on the grain size of aluminum. Source: Ref 95

most types of grain-refiner additions (Fig. 62).

Grain refining as a specific treatment does not have great value for die casting and certain permanent mold processes. This is because the high rate of solidification and turbulent flow into the mold in these processes serve to increase the number of grain nuclei.

Shape Casting. In the production of foundry ingot or aluminum cast shapes, grain-refiner additions are made with master alloys either by furnace addition of waffle or briquette or by a rod (9 to 10 mm, or 0.35 to 0.4 in., in diameter) into the casting launder. The specific choice of which type to add depends on requirements, economics, and alloy composition.

In the traditional furnace method, the master alloy is added about 30 min before casting to ensure ample time for contact and dissolution. Potential problems with the furnace addition method include inadequate mixing in a full furnace charge and the settling and agglomeration of the heavier titanium diboride particles to form a sludge.

The second method—adding master alloy in rod form with a wire rod feeder directly into the casting launder—permits a lower addition rate, but relies on the use of a master alloy with a short dissolution or contact time. Depending on the casting route and the type of filter used, the time available between addition and eventual metal solidification varies from 30 s to 10 min (Ref 98). Consequently, the contact time available (determined by the flow distance and metal flow rate) and the efficiency achieved will often dictate which type of grain refiner to add. Figure 63 illustrates the efficiency of three types of grain-refining additions. Clearly, the Al-5Ti-1B alloy addition would be preferred in rod-feeding applications in which minimal contact time is available.

Proper positioning of a casting launder grain-refining addition (before or after a filter) is currently a subject of debate. Filtering can remove or agglomerate grain-refining nuclei. Filtration can also remove impurity constituents from the grain-refiner rod itself and thus have a positive effect. Available contact time, type of product being cast, and properties and experience with a particular grain refiner should determine the flexibility of where the addition point can be positioned in the casting sequence. In many cases, grain refiner is added directly into rotating disperser metal treatment systems in which the increased agitation facilitates thorough mixing and dissolution.

Fade is the decline in grain-refining properties that takes place in both cast foundry alloys or wrought cast aluminum alloys. It occurs with prolonged holding times before casting and may also be aggravated by intervening fluxing or degassing. Fade is usually thought to result from the agglomeration and settling of TiB_2 particles, but it may also be due to a submicron precipitation poisoning effect on particle surfaces, rendering them unsuitable or sluggish in serving as effective nuclei for aluminum grains during solidification. In addition, the gradual dissolution of nucleating particles with prolonged holding at higher temperature may reduce effective grain-refining nuclei.

Figure 64 shows the effects of various boron addition levels on fading in a 99.99% Al melt. Higher addition levels (in this case, 0.008% B to 0.15% Ti) prolong grain-refining effects (that is, retard the fading behavior). For this reason, most secondary smelter foundry ingot production, which in turn is remelted by the foundry, contains boron and titanium as grain-refining agents.

Grain Refining of Foundry Alloys

In foundry alloys with 7% Si, salts or aluminum-titanium master alloy have usually been used for grain refining. An addition level of 0.10 to 0.25% Ti as a specification usually ensures adequate grain refinement. The specific nuclei, in the absence of boron, are $TiAl_3$ particles.

Despite the successful use of titanium as a grain refiner for foundry alloys, it has recently been found that boron alone may be a more effective refiner than titanium in aluminum-silicon alloys (Ref 99). Figure 65 compares the grain size achieved with additions of boron, titanium, and a titanium-boron mixture on Alloy 356. Similar results are reported in Ref 100.

These results are also in contrast to wrought alloy experience, in which aluminum-titanium-boron master alloys are powerful refiners but aluminum-boron alloys alone do not grain refine effectively. In the latter wrought alloys, however, the silicon content is usually less than 0.5%; foundry alloys have 7% Si or greater.

Therefore, it may be that with higher silicon contents the effectiveness of the $TiAl_3$ nucleating particles is poisoned by silicon-titanium compound precipitation (Ref 99). As a consequence, relatively large titanium additions are required to achieve adequate refinement in alloys containing higher silicon.

In melts using aluminum-boron grain refiner alone, the principal refining constituent is an AlB_2 precipitate, which seems to be enhanced with higher dissolved silicon contents. However, the AlB_2 will dissolve and react with residual titanium to form TiB_2, which may settle to the bottom of the furnace as sludge. If this occurs, the grain-refining capabilities are lost. In addition,

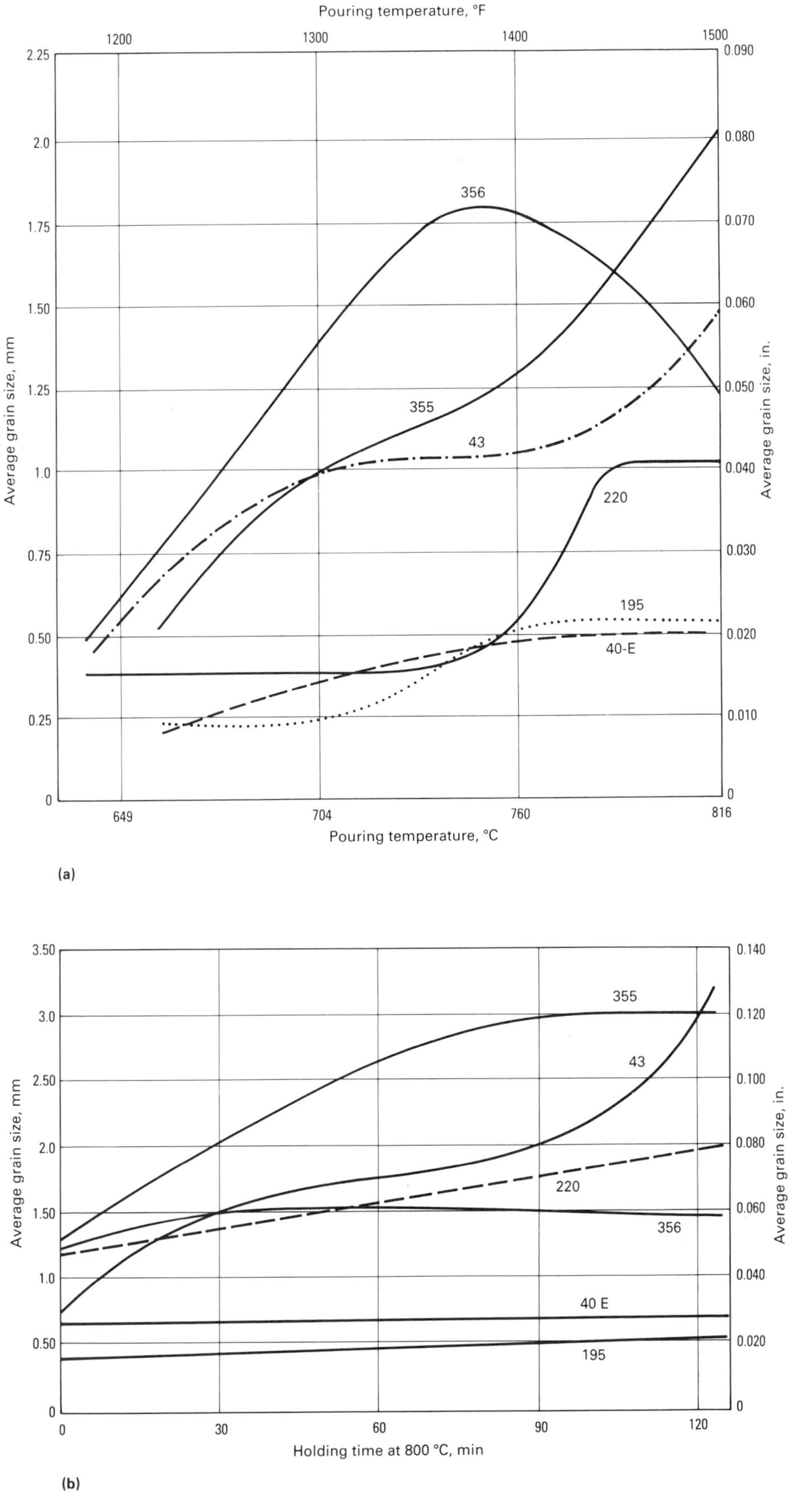

Fig. 62 Effect of pouring temperature (a) and holding time (b) on the grain size of several aluminum casting alloys. Source: Ref 95

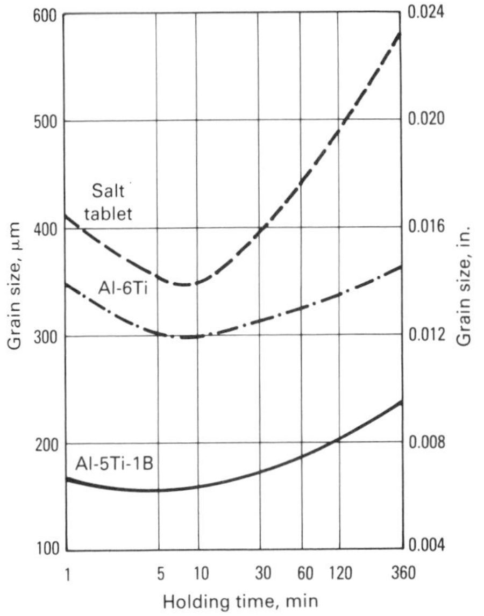

Fig. 63 Comparison of efficiency of three grain-refining additions to 99.7% Al. Source: Ref 98

boron is more expensive than titanium. Consequently, a compromise is sought, employing both titanium and boron, in a master alloy. This yields both AlB_2 and TiB_2 nucleating crystals, comprising a mixed boride $(Al,Ti)B_2$.

Figure 66 demonstrates the effectiveness of Al-2.5Ti-2.5B master alloys to 356 and 319 foundry alloys compared to the Al-5Ti-1B alloys previously used. Clearly, the use of higher levels of boron and lower levels of titanium than those previously used can have substantial benefits for foundry alloys. Figure 67 illustrates the comparison of an adequately grain-refined Al-7Si alloy before and after the grain-refining addition.

Grain Refining of Wrought Alloys

In the grain refining of aluminum alloys that are direct chill or continuously cast as semifinished products (billet, slab, foil, strip, or forging stock), the grain-refining alloy of choice will depend on several factors, as previously discussed. These factors are residence time, available effectiveness, cost, and compositional effects (Ref 101).

Generally, aluminum-titanium-boron refiners are used, except when electrical conductivity is important. In this case, titanium is undesirable, because its presence decreases conductivity. For these alloys, an aluminum-boron master alloy is used.

In wrought alloys, titanium and boron provide more effective grain refinement than either alone. The use of a straight titanium master alloy would require significant concentration (~0.15% Ti). This concentration is unacceptable in wrought aluminum alloys, because it results in the formation of coarse, brittle intermetallic titanides, which interfere with desirable

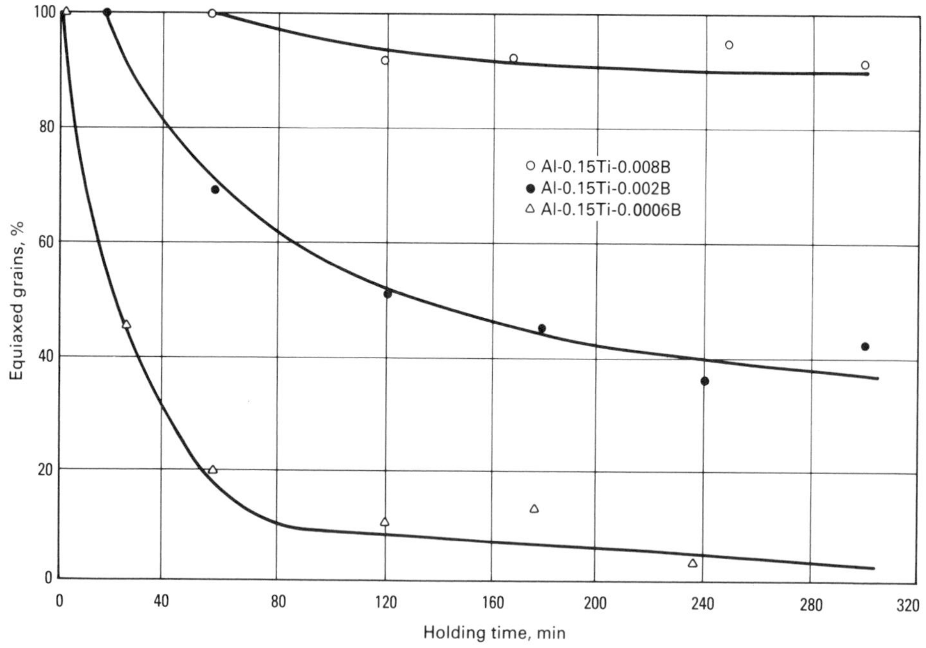

Fig. 64 Effect of various residual boron levels on the fading behavior of a 0.15% Ti grain-refining addition. Source: Ref 97

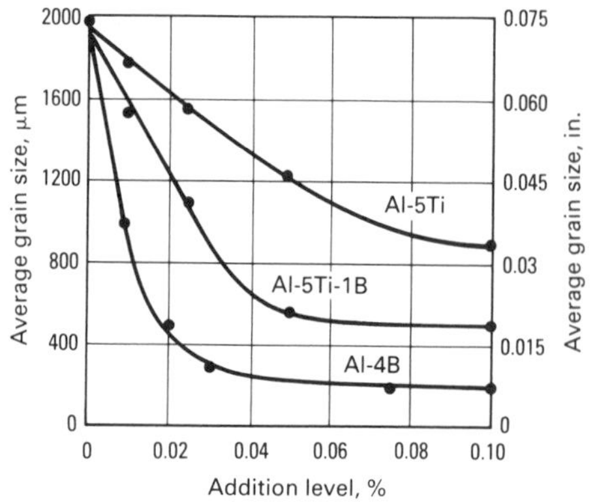

Fig. 65 Comparison of the effectiveness of various master alloy grain refiners in aluminum alloy 356. Source: Ref 99

wrought alloy forming properties. Furthermore, an aluminum-titanium addition is subject to considerable fade because Al_3Ti dissolves rapidly above 700 °C (1290 °F). Boron additions retard or delay the fade phenomenon.

A variety of aluminum-titanium-boron grain-refining alloys are currently in use, including Al-10Ti-0.4B, Al-5Ti-1B, Al-5Ti-0.5B, Al-5Ti-0.2B, Al-3Ti-1B, and Al-3Ti-0.2B. These varieties exist because each alloy and application may require a specific grain refiner or addition rate to achieve the desired result. The typical desired grain size is generally 200 μm (0.008 in.) or finer in certain cases. There appears to be an optimum titanium/boron ratio for achieving the finest grain size in individual alloys (Fig. 68 and 69).

Even within a given alloy, more or less grain refiner may be required, depending on alloy composition. Certain levels of alloying elements may influence the amount of grain refiner needed to achieve a specific result. Figure 70 shows the effect of iron level on the grain size of 99.99% Al when a specific Al-5Ti-1B addition is made. As can be seen, additional refiner would be required to achieve the 200 μm (0.008 in.) grain size desired at low iron levels. Similar results have been observed in commercial alloys 3004 and 6063.

Residual titanium will also influence grain refining when a specific amount of refiner is added. Higher levels of residual titanium result in finer grain size for a given addition. More information on the selection of aluminum alloy grain refiners for specific applications is available in Ref 102 and 103.

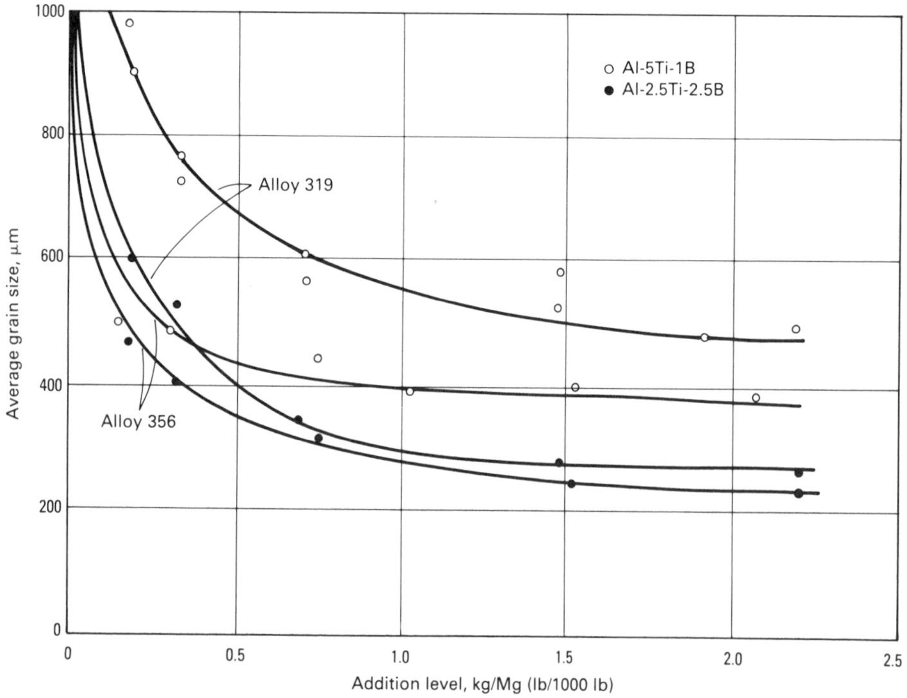

Fig. 66 Effectiveness of titanium-boron grain refiners in aluminum alloys 356 and 319. Source: Ref 99

Grain Refining of Magnesium Alloys (Ref 104)

Pure magnesium solidifies with a relatively coarse grain structure. In most cases, the common alloying elements—aluminum, zinc, rare earths, and thorium—serve to reduce the as-cast grain size. This is most likely accomplished through typical constitutional supercooling effects.

Zirconium additions (0.1 to 0.2%) have a marked effect on grain size. Magnesium alloys directly alloyed with zirconium exhibit a naturally very fine grain size with the attendant increase in mechanical properties.

Superheating of magnesium-aluminum-zinc alloys produces considerable grain refinement through supersaturation and precipitation upon cooling of preferred grain nuclei. To provide the superheat, the temperature of the melt is raised to between 850 and 925 °C (1560 and 1695 °F). Time required at superheat decreases with increasing aluminum content, proportion of scrap in the charge, and maximum melt temperature, and it increases with crucible size and heating rate. Certain elements, such as zirconium, titanium, and the rare earths, retard the grain-refining effect produced by superheating.

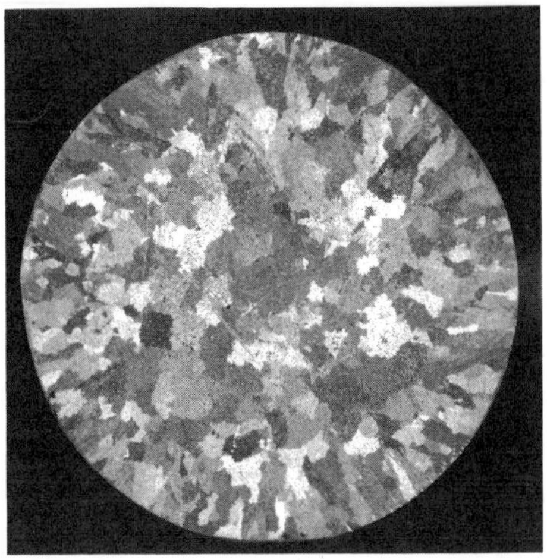

(a)

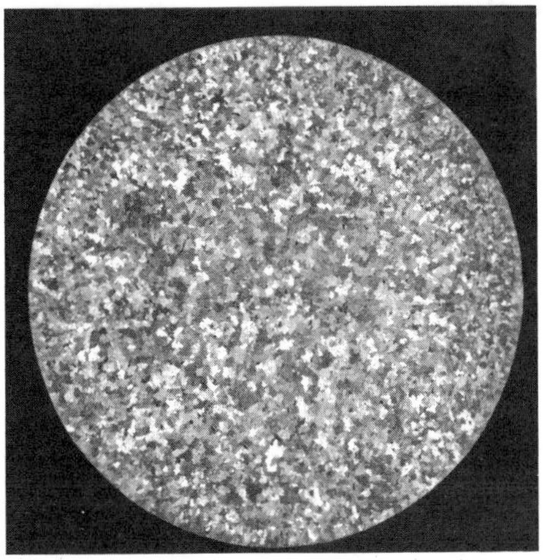

(b)

Fig. 67 Effect of addition of 5Ti-1B grain refiner to Al-7Si alloy. Before (a) and after (b) addition. Both at 2×. Courtesy of KB Alloys, Inc.

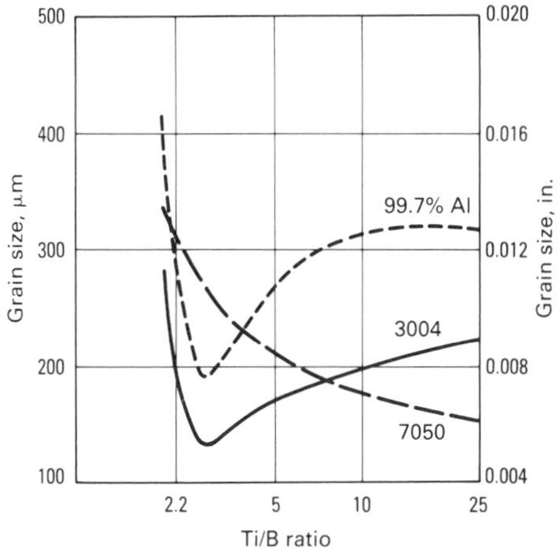

Fig. 68 Effect of titanium/boron ratio on the grain size of three aluminum materials. Source: Ref 97

Hexachloroethane. Grain refinement is possible with hexachloroethane, which enables both degassing and grain refining to be accomplished in one step without superheating. Therefore, melting times, crucible wear, and excessive hydrogen absorption through superheating are avoided. Hexachloroethane tablets are plunged carefully into the melt (above any sludge layer) and held for complete reaction. Approximately 0.1% by weight addition is usually sufficient to achieve maximum grain refinement.

The major grain-refining constituent with hexachloroethane appears to be carbon, which forms an Al_4C_3 intermetallic submicroscopic precipitate. Carbon inoculation is probably also active in superheating, where high temperatures corrode steel melting crucibles, thus generating carbon. The use of graphite fluxing tubes may also provide for carbon grain inoculation, where tube wear yields carbon.

Rare-earth alloying elements, which are strong carbide formers, suppress grain refining by limiting the amount of carbon available for Al_4C_3 nuclei formation. Zinc-iron master alloys can be used for grain refining rare earth containing alloys. Methods for grain refining magnesium alloys are summarized in Table 6.

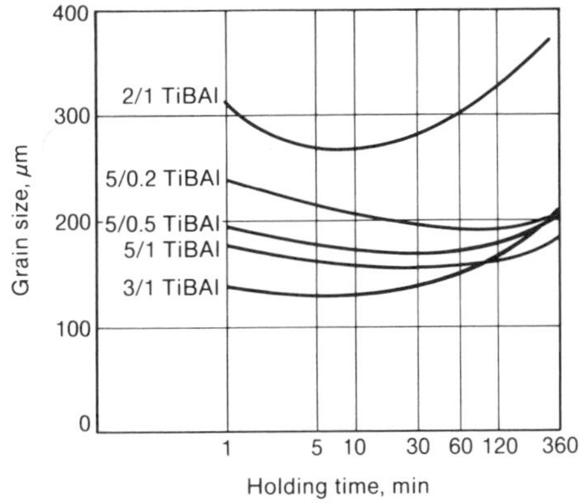

Fig. 69 Effect of addition level and holding time on grain size for 3004 alloy. Source: Ref 98

Grain Refining of Copper Alloys

In general, the grain refinement of copper alloys is not practiced as a specific molten metal processing step *per se*, because a certain degree of refinement can be achieved through normal casting processes. As with aluminum alloys, grain refinement in copper alloys can be achieved by rapid cooling, mechanical vibration, or the addition of nucleating or grain growth restricting agents. Further, many commercial copper alloys have sufficient solute (zinc, aluminum, iron, tin) to achieve constitutional supercooling during solidification. In this case, grain nucleation and growth are naturally retarded. Commercially pure copper can be grain refined by small additions (as little as 0.10%) of lithium, bismuth, lead, or iron, which provide constitutional supercooling effects (Ref 105, 106).

Copper-zinc single-phase alloys can be grain refined by additions of iron or by zirconium and boron (Ref 107). In the latter case, the probable mechanism is the formation of zirconium boride particle nuclei for grain formation. In one case, the vibration of a Cu-32Zn-2Pb-1Sn alloy improved yield and tensile strengths by about 15%, with a 10% reduction in grain size from the unvibrated state. In general, the α copper-zinc alloys (<35% Zn) exhibit grain size reduction and greater improvement in properties, while the α-β alloys do not.

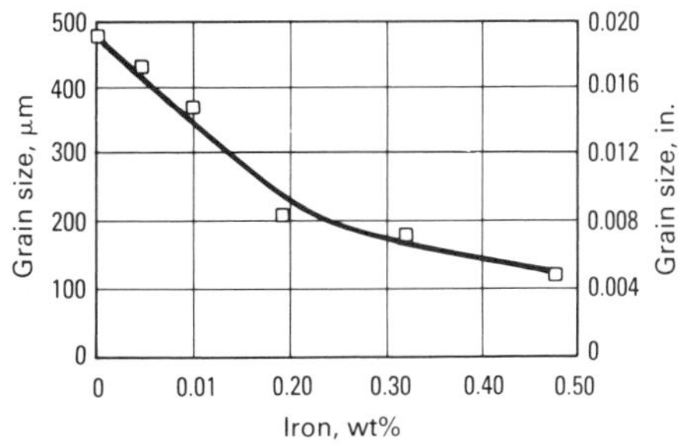

Fig. 70 Effect of iron level in pure aluminum on grain refining using Al-5Ti-1B alloy. Source: Ref 101

Copper-aluminum alloys have been effectively grain refined with additions of 0.02 to 0.05% B; the effective nucleating agent is boron carbide (B_4C). Figure 71 illustrates the improvement in mechanical properties achieved by grain refining a Cu-10Al alloy.

Tin-bronze alloys have been successfully grain refined by the addition of zirconium (0.02%) and boron master alloys (Ref 108). However, pressure tightness is reduced because in these long freezing range alloys, finer grain size concentrates porosity because of gas entrapment.

Modification of Aluminum-Silicon Alloys

In the aluminum foundry alloy family, the aluminum-silicon alloys form the bulk of the commercially important applications. Silicon is the principal alloying element in these materials. For the purposes of this discussion, these alloys will be grouped as either hypoeutectic (~7% Si) or hypereutectic (≥12% Si).

A two-phase structure exists for both classes. The hypoeutectic alloys consist of a primary α-aluminum matrix plus a second phase based on the silicon eutectic. In the hypereutectic alloys, silicon is the primary phase (Fig. 72). In either case, it is the morphology of the silicon phase that gives rise to the practice of modification. The silicon phase in hypoeutectic alloys, which is the eutectic constituent, solidifies as an acicular or platelike structure. In the hyper-

Table 6 Summary of grain-refining methods for magnesium alloys

Alloy	Grain-refining element or treatment	Degree of refinement	Probable mechanism	Interfering elements(a)	Remarks
Magnesium	Al, Zn, rare earths, Th, Si, Ca, Zr, others	Mild	Concentration gradients	...	...
Zirconium		Extreme	Nucleation by zirconium or Zr-enriched Mg	Al, Si, Fe, H, Sn, Sb, Co, Ni, (Mn), (Be), (Li), others	...
Mg-Al, Mg-Al-Zn-Mn	Carbon inoculation	Marked	Carbide nucleation	...	...
	Superheating	Marked	Nucleation by Al-Mn(Fe) compounds or carbon	Be, Zr, Ti, (excess Mn)	...
Mg-Al-Mn-(Zn)	Ferric chloride	Marked	Nucleation by Fe-Al-Mn compounds and carbon	Zr, Be	Requires manganese
Mg-Zn(-Re-Mn)	Ferric chloride, iron-zinc alloy	Very marked	Nucleation by iron compounds	Al, Si, Th	Iron-zinc requires manganese content of ≤1%
	Ammonia	Very marked	Nucleation by hydrogen	Al, Si, Th	Gassy metal; Manganese can exceed 1%.
Mg-Mn	Calcium + nitrogen	Mild	...	...	...
	Zirconium	Increases with decreasing manganese	Nucleation by zirconium or Zr-enriched Mg	Al, Si, Fe, H, Sn, Sb, Co, Ni, Mn, others	...

(a) Elements in parentheses interfere to a lesser extent. Source: Ref 104

eutectic alloys, the primary silicon phase is very blocky. These morphologies are detrimental to most aluminum mechanical properties, causing embrittlement, and also result in poor machinability. In addition, unmodified silicon structures are associated with increased porosity in the raw casting.

Therefore, modifiers are deliberately added to the molten metal to change the shape of the silicon phase (eutectic in hypoeutectic alloys, primary silicon in hypereutectic) to a more well-rounded shape that is also more coherent with the matrix. This change also disperses shrinkage during solidification and greatly improves mechanical properties and machinability. Element additives such as sodium, calcium, strontium, and antimony are currently used to *modify* the silicon phase in hypoeutectic alloys, and phosphorus is used to *refine* the silicon phase in hypereutectic alloys.

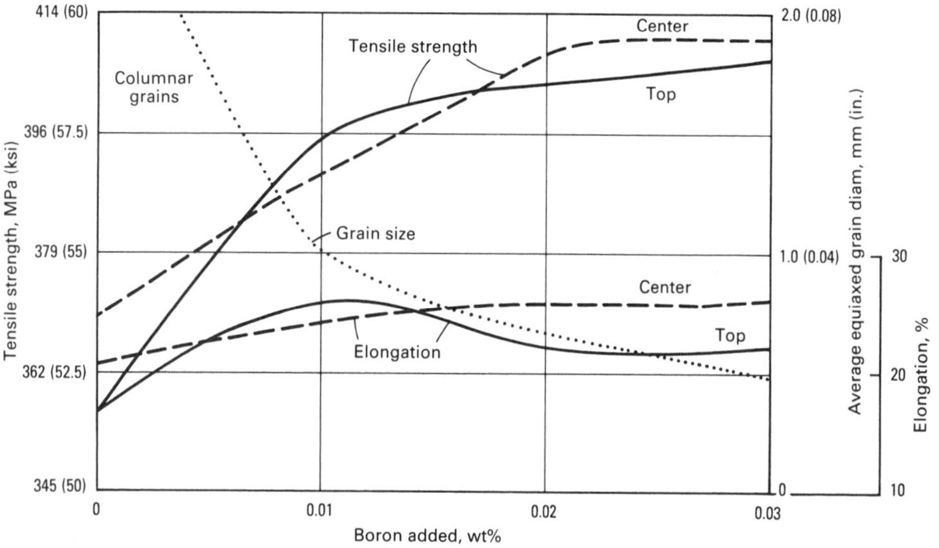

Fig. 71 Effect of boron-refined grain size on the mechanical properties of Cu-10Al alloy. Test specimens were removed from the center or the top of the ingot as indicated. Source: Ref 107

Theory of Modification

The work of Crossley and Mondolfo (Ref 109) has been instrumental in describing how different impurities nucleate different solid phases. When sodium is present, the nucleation of the eutectic silicon takes place at approximately 3 to 12 °C (5 to 20 °F) undercooling (a factor that thermal analysis takes into account; see the section "Thermal Analysis" in this article). The specific degree of supercooling indicative of complete modification is different for each alloy. For 356 alloy, a typical modified eutectic temperature is about 568 °C (1054 °F). In hypoeutectic alloys, an AlSiNa precipitate appears to serve as the nucleation substrate for the silicon eutectic phase (Ref 110, 111). An AlP precipitate seems to be present even in hypoeutectic alloys (due to residual phosphorus). Sodium neutralizes AlP so that easy nucleation of silicon is prevented.

In the hypoeutectic alloys, another contributing mechanism is thought to be the influence of the modifier element (for example, sodium) on restricting the growth of nuclei by reducing liquid surface tension within the eutectic phase which consists of a lamellar structure of both aluminum and silicon (Ref 6). The reduction in surface tension increases the contact angle between the aluminum and the silicon, permitting the aluminum to wet the silicon phase overall more easily and therefore restrict favored growth along one particular axis. Sodium apparently reduces the rate of diffusion of silicon in the liquid, thus restricting nuclei and grain growth.

Hypereutectic Alloys. Silicon phase modification in these alloys by phosphorus, which is referred to as refinement, is theoretically a nucleation phenomenon that influences the size and shape of the primary silicon only. Phosphorus reportedly combines with aluminum to form submicroscopic AlP precipitate whose crystallographic characteristics are similar to those for silicon and that permit the primary silicon crystal to nucleate more easily and grow evenly during solidification. It is also apparent that metal casting temperature, mold temperature, overall metal composition, and turbulence of the flowing metal all play a role in determining the extent of modification (Ref 6, 112, 113).

Thermal Analysis

Thermal analysis is used to predict or analyze the anticipated degree of modification to be expected from a melt (Ref 114). The principles are based on the thermal arrests and degree of undercooling that take place when a melt sample is allowed to solidify under carefully controlled conditions. A thermocouple is placed in a precise position within a mold containing a melt sample, and it monitors the temperature as a function of time as the sample solidifies. A characteristic curve can be obtained, as illustrated in Fig. 73 for an unmodified alloy. The break in the curve occurs at the

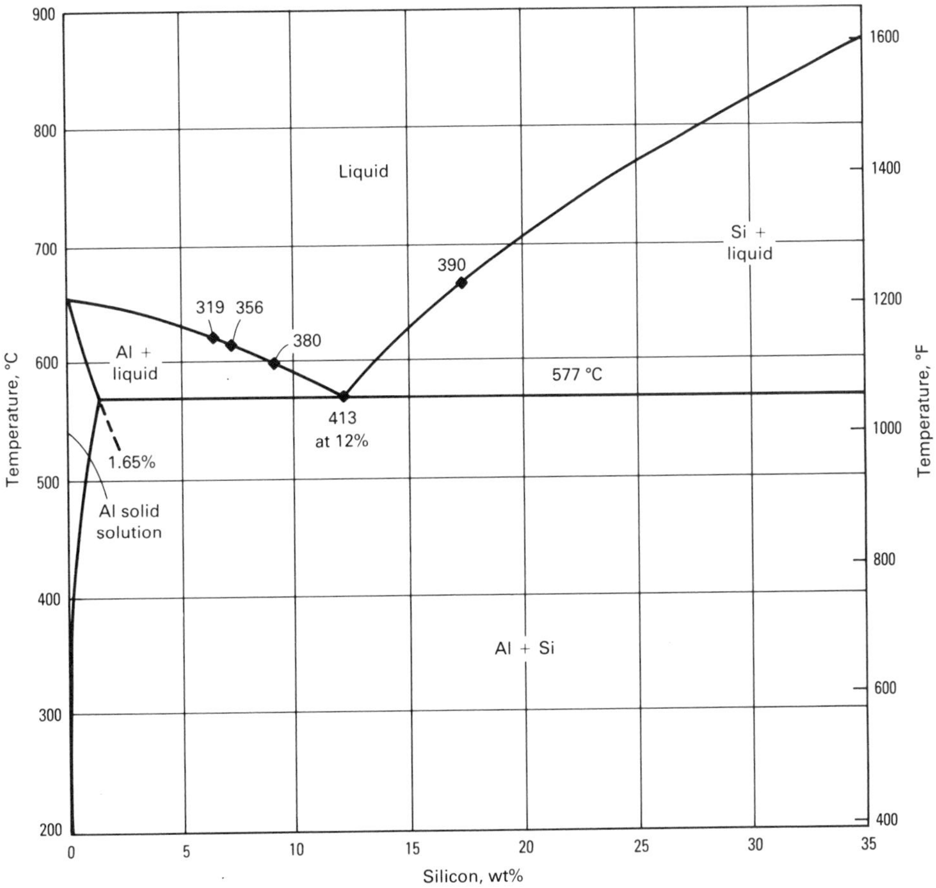

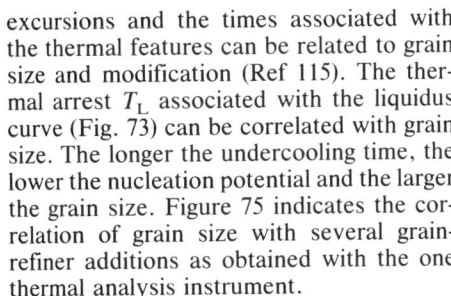

Fig. 72 Aluminum-silicon binary phase diagram showing common foundry and die casting alloy compositions 319, 356, 380, 390, and the eutectic alloy 413. Source: Aluminum Smelters and Refiners Inc.

eutectic temperature T_e when a two-phase liquid plus solid mixture X or B phase solidifies to α plus β. The beta is the eutectic silicon. In the presence of a modifier, the eutectic temperature is depressed, as shown in Fig. 74, and this depression can be correlated with the amount of modification that can be expected to take place in the full casting if the same solidification conditions are met.

Other features of thermal analysis that can be monitored are the times associated with each thermal change. The eutectic temperature depression is influenced by alloy content, with T_e = 1.1 (Mg) + 0.25 (Cu) + 0.18 (Fe) (Ref 114). The temperature excursions and the times associated with the thermal features can be related to grain size and modification (Ref 115). The thermal arrest T_L associated with the liquidus curve (Fig. 73) can be correlated with grain size. The longer the undercooling time, the lower the nucleation potential and the larger the grain size. Figure 75 indicates the correlation of grain size with several grain-refiner additions as obtained with the one thermal analysis instrument.

Two principal instruments are commercially available for thermal analysis. Both consist of a sample cup, thermocouple, and microprocessor unit. Regardless of the instrument used, it is important that the test be conducted consistently each time for accurate results (Ref 114-117).

Alternate Techniques to Determine Modification

There are other nondestructive techniques capable of assessing the degree of modification in a solidified casting or in regions of a casting. Such an analysis can serve as a useful foundry check. Electrical resistivity or conductivity has been found to correlate well with degree of modification (Ref 118 and 119). Conductivity has been found to increase linearly with degree of modification in Alloys A444 and A356, but detection capability decreases as the magnesium level increases. Surface finish is also a major factor. Polished surfaces display greater sensitivity and response to this technique than as-cast or even machined surfaces. Ultrasonic measurement has also been investigated, but thus far the technique has not been found to provide adequate sensitivity.

Modifier Additions

Most aluminum foundry ingot is modified before shipment to the foundry. However, modification is subject to fade, as with grain refinement, and additional modification is therefore usually necessary.

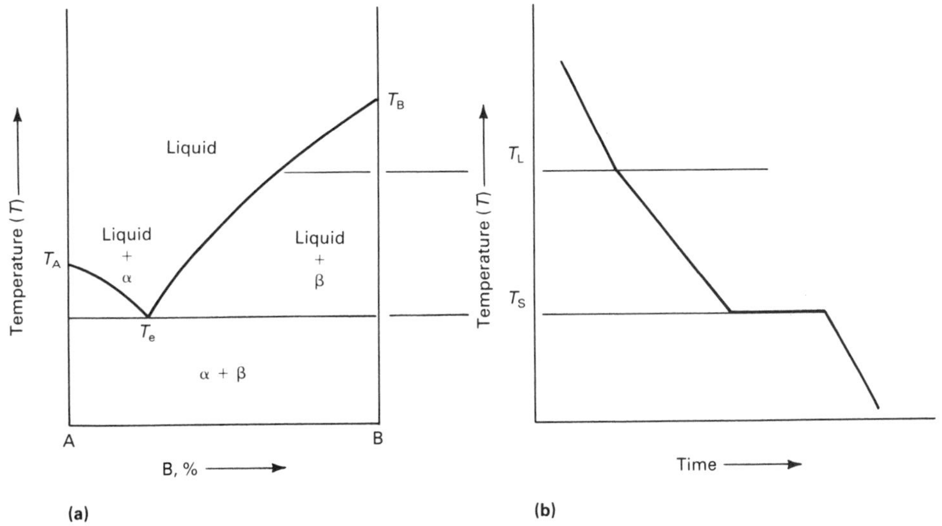

Fig. 73 Binary phase diagram (a) and thermal analysis curve (b) for an alloy having complete solubility in the liquid state and incomplete solubility in the solid state. Source: Ref 112

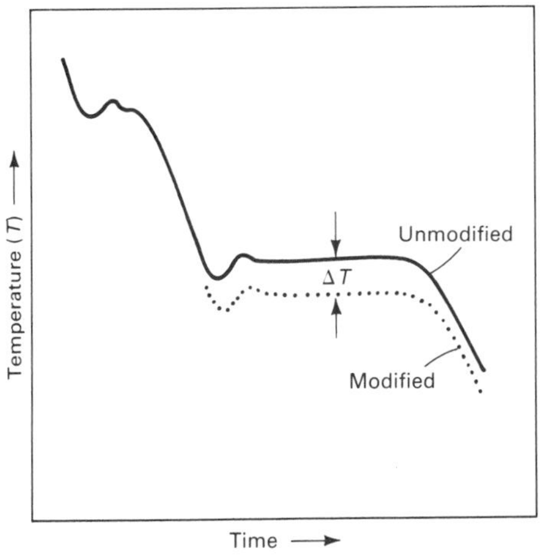

Fig. 74 Characteristic thermal analysis curve showing temperature depression that occurs after modification. Source: Ref 114

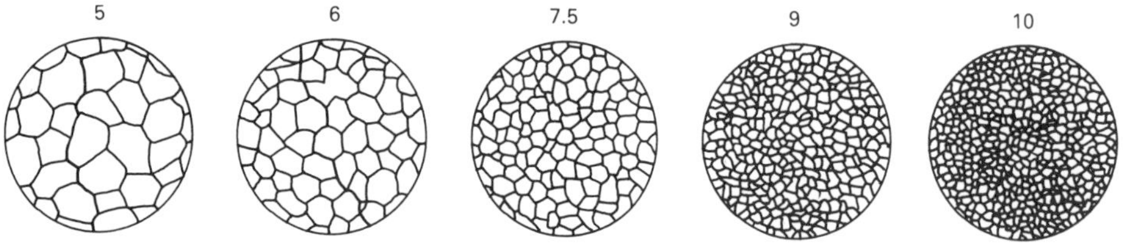

Fig. 75 Schematic showing aluminum grain size (ASTM E 112) as evaluated and displayed by one thermal analysis instrument. Source: Ref 116

Hypoeutectic alloys are modified by additions of sodium, strontium, calcium, or antimony, as described below.

Sodium is usually added in metallic form, encapsulated in an aluminum can to retard its natural burning in air. The sodium should be plunged into the melt and stirred gently for a short time to provide good dissolution and dispersion. Alternatively, several commercial salt flux tablets are available whose decomposition yields sodium for modification. The usual precautions apply regarding proper storage and the use of clean, dry tools. Fade, or loss of modification, increases with excessive agitation, degassing, prolonged holding periods, or excessively high temperatures (Ref 6).

In general, a sodium addition of 0.015 to 0.020% is required for adequate modification, which results in a residual level of sodium in the casting of about 0.002%. Sodium should not be used with alloys containing more than 1% Mg, because of possible embrittlement.

Strontium presents no special storage or handling problems, as does sodium, and is usually added in the form of Al-10Sr-14Si or aluminum-strontium master alloy ingots or waffles. An Al-10Sr rod is also available, which dissolves more easily. Additions of 0.01 to 0.02% in foundry castings are sufficient, particularly when melting returns where typical residual 0.008% Sr is available. Overall, strontium modification seems more resistant to fade than sodium.

Calcium is also added as a modifier in the form of a master alloy that also contains 5% Si. Nominal addition levels for modification are about 0.01%. Calcium also seems more resistant to fade than sodium, although prolonged holding will result in calcium loss (Ref 120).

Antimony is a modifier that is used predominantly in Europe and Japan. It is added at about the 0.12% level and serves as a permanent alloy constituent of the aluminum alloy. It is therefore added by the supplier of foundry ingot and does not require any make up addition at the foundry. Antimony is reported to be unaffected by holding time, remelting, or degassing. Alloys modified with antimony are distinguished by lower gas susceptibilities than those containing sodium or strontium. However, cross contamination problems are possible, and antimony is more toxic.

Hypereutectic Alloys. Although various elements have been found to be capable of modifying (refining) hypereutectic alloys, only phosphorus is used commercially. Other elements, such as magnesium, tungsten, sulfur, and lanthanum, have been reported to be effective modifiers (Ref 113) but thus far none has achieved commercial importance.

Phosphorus is available for the refinement of hypereutectic alloys in the form of master alloys (phoscopper), aluminum phosphide, or silicon phosphide; as phosphorus pentachloride; or as various proprietary salt mixtures. Additions usually range from 0.020 to 0.025% P. Addition should be made carefully, with the customary precautions to ensure completely dry materials and tools.

Modifier Comparisons and Interactions

Dissolution. With the advent of strontium modification for hypoeutectic alloys and the increased use of foreign castings modified with antimony, many questions have arisen concerning the merits, problems, and interactions between sodium, strontium, and antimony. Strontium modification has been examined in several recent papers, and the nature of the strontium addition has been found to be quite important (Ref 121-123). An oxide-free strontium surface increases the rate of dissolution and minimizes the so-called hydrogen pickup that has sometimes been associated with strontium additions (Ref 122).

In general, the strontium master alloys perform better than strontium-bearing salts, and sheared rod produces faster dissolution than waffle ingot. Strontium dissolution is characterized by an exothermic reaction that results in the formation of intermetallic compounds (Ref 121, 122). The rate of this reaction, which is temperature dependent, determines the effectiveness of strontium recovery and therefore the degree of modification achieved.

It has been found that when elemental strontium is present, as in a 90Sr-10Al master alloy, the exothermic reaction proceeds at about 725 °C (1340 °F). However, at lower temperatures, the reaction is slowed, and an intermediate intermetallic precipitate, $SrAl_2Si_2$, forms with good dissolution characteristics. Therefore, lower temperature provides better recovery than higher temperature (Fig. 76a).

On the other hand, lower strontium master alloys such as 10Sr-90Al dissolve by simple dissolution; therefore, recovery increases at higher temperatures (Fig. 76b) because the dissolution rate is greater (Ref 123). A certain amount of time (10 to 20 min) may be necessary for optimal recovery, which is dependent on temperature and the specific strontium alloy addition (Ref 124).

Sodium dissolution occurs more simply and rapidly, providing almost immediate optimal modification. Antimony appears to behave similarly. All modifiers increase the eutectic suppression. At equivalent levels of suppression, strontium yields better modification than sodium.

Fade. It is well known that sodium modification fades; that is, sodium is lost through oxidation and reaction within the melt, perhaps forming some sodium-aluminum-silicon complex that retards the ability for sodium to modify. Degassing or melt agitation easily destroys sodium modification, while such stirring does not seem to affect strontium modification (Ref 125).

Contamination. The recycling of scrap castings poses the problem of intermingling of different modifiers and possible detrimental consequences. Sodium and strontium have been found to be compatible with each other in the range of 0.007 to 0.012% Na and 0.007 to 0.019% Sr (Ref 124, 126). The strontium presence overcomes the effect of sodium fade for up to nearly 2 h. Therefore, strontium can be used to modify melts that may contain some sodium from recycled scrap.

Antimony, however, poisons both sodium and strontium modifications. In the case of sodium, at normal sodium addition levels near 0.01%, the presence of 0.02 to 0.05% Sb eliminates modification. A level of 0.01% Sb might be tolerated, but requires up to 0.04% Na to avoid fade and to retain modification after 1 h (Ref 126). It is thought that this negative interaction is a result of a sodium-antimony (possibly Na_3Sb) intermetallic compound that forms, depleting the alloy of sodium available for modification.

Antimony has a similar negative effect on strontium modification. Figure 77 shows a change in antimony content when it is present as a contaminant in strontium-modified melts. There is reasonably clear evidence that a Mg_2Sb_2Sr intermetallic forms concurrently as the strontium dissolves. This precipitate, having a higher density than molten aluminum, forms a sludge at the bottom of the melting vessel. Figure 78 indicates that much higher amounts of strontium are needed to achieve modification when antimony is present.

When all three modifiers are present, the ternary interactions behave as expected on the basis of the binary interactions. The antimony content decreases as either strontium or sodium is added, reflecting the

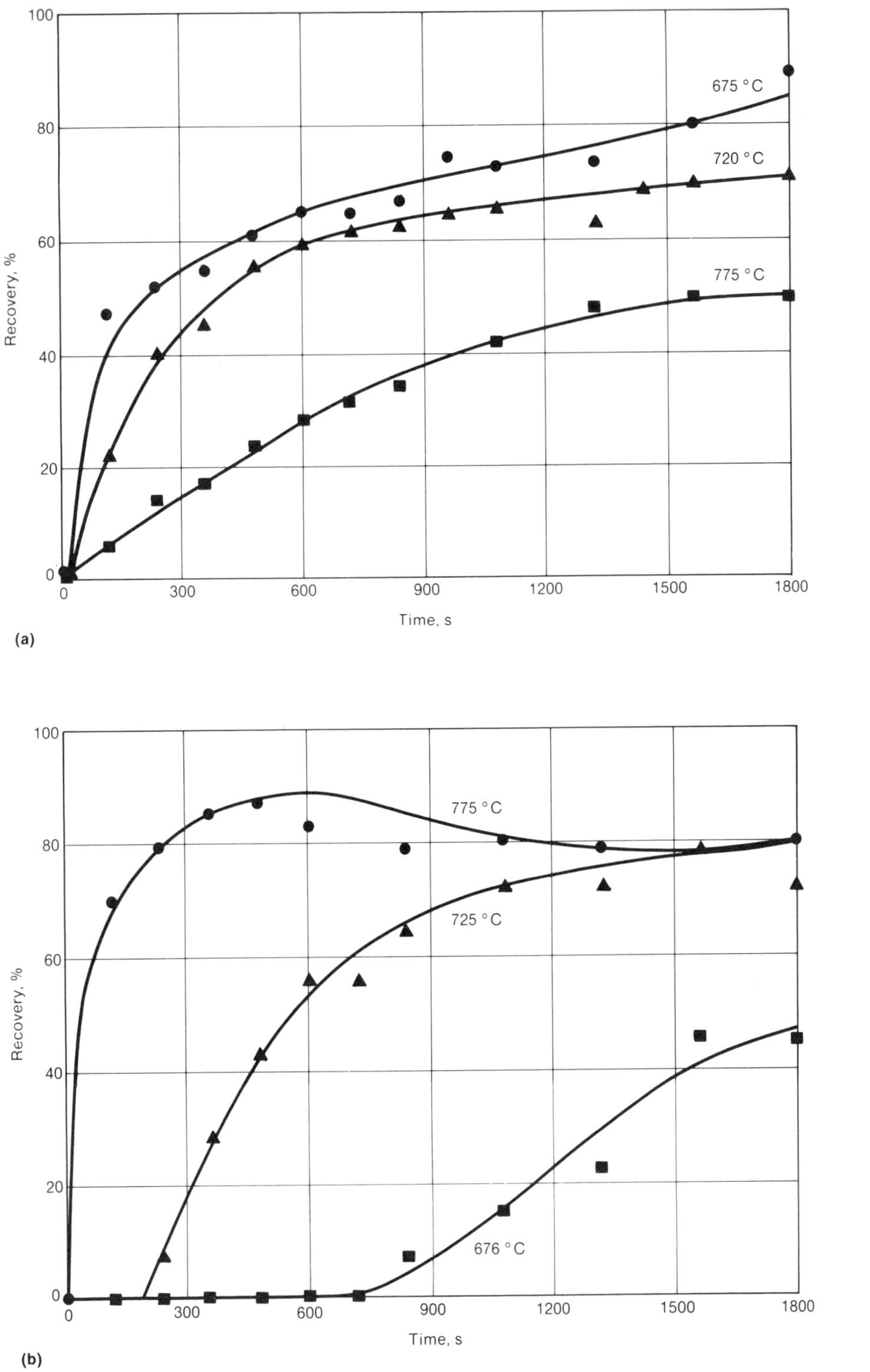

Fig. 76 Comparison of dissolution rates for strontium additions made with 90Sr-10Al (a) and 10Sr-90Al (b) master alloys. Source: Ref 123

intermetallic compound formation. Sodium content decreases rapidly at first, especially when elemental sodium is plunged. If the total modifier concentration is high enough, particularly strontium at 0.010 to 0.020%, the effects of antimony contamination can be overcome and modification maintained.

Phosphorus is chemically similar to antimony and has a similar deleterious effect on either sodium or strontium modification. For a given level of phosphorus (which could be present as an impurity arising from hypereutectic alloy scrap or phosphate-bonded refractory materials), a higher amount of either sodium or strontium is necessary to achieve full modification.

Hydrogen and Degassing. The effect of modification practice on hydrogen gas absorption and the subsequent response in degassing have been commonly observed and studied in some detail. Researchers (Ref 124) have noticed a tendency toward higher gas levels using either sodium or strontium (Fig. 79). The increase in gas content does not necessarily occur immediately, but increases with time. An increase in gas absorption when modifiers are present is especially noticeable at temperatures above 745 °C (1375 °F). Deliberately gassed strontium-modified melts show greater hydrogen uptake and less natural outgassing than unmodified melts. The influence of antimony on gas levels has not been discussed in the literature as frequently, but antimony modification appears less gas prone than strontium modification (Ref 127), with much higher densities (less porosity) reported for a given gas content (Fig. 80).

On the other hand, there is a recent body of work indicating that strontium modification does not necessarily increase the propensity for an A356 alloy melt to absorb hydrogen (Ref 121, 125, 128). Binary master alloy additions may be better than ternary alloy modifier additions; this suggests that the silicon in the latter may be harmful (Ref 121). In another study (Ref 126), there was no evidence to suggest that hydrogen is carried into the melt from any such occlusion in the strontium itself (Fig. 81), nor was there any noticeable difference in regassing behavior with either a clean strontium surface or a corroded strontium hydroxide surface layer. A subsequent study confirmed that there was no significant difference in the rates of regassing for a strontium modification versus an unmodified melt (Ref 128). Stirring, however, increases the rate of regassing for both strontium-modified and unmodified melts.

The degassing of modified melts does illustrate significant differences between sodium and strontium modifiers. In the case of sodium, degassing with either inert gases (argon, nitrogen) or reactive gases (chlorine, freon) removes sodium and depletes modification. Consequently, melts to be sodium modified should always be degassed before modification. On the other hand, strontium-modified melts exhibit markedly different behavior (Ref 125, 128). Argon is effective in degassing a strontium-modified melt, and the rate of degassing is not affected by the presence or absence of strontium. Argon shows no detrimental effect on strontium content. Therefore, in contrast to sodium-modified melts, strontium-modified melts may be, and should be, degassed after the modification treatment. The fade rate during degassing is almost negligible.

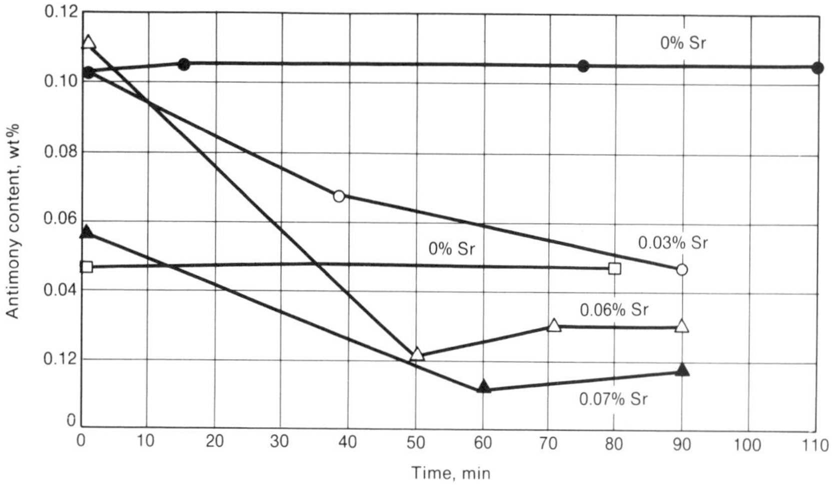

Fig. 77 Interaction between antimony and strontium present in A356 alloy as a function of time. Source: Ref 126

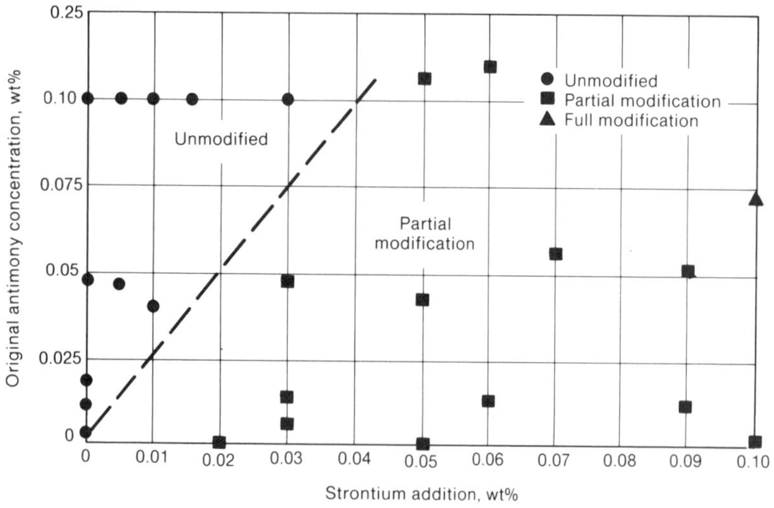

Fig. 78 Effect of antimony on strontium modification in A356 alloy. Source: Ref 126

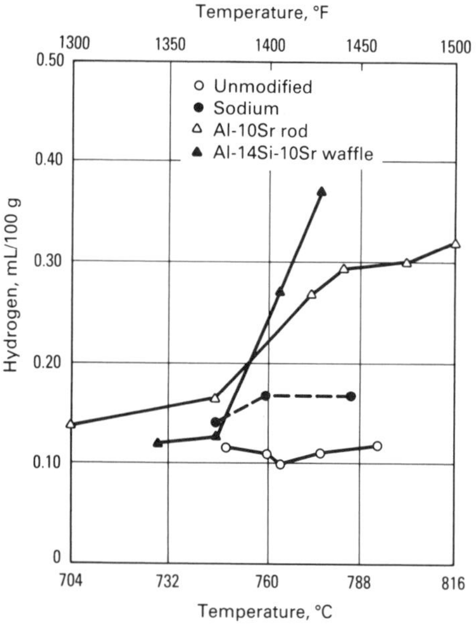

Fig. 79 Hydrogen content as a function of holding temperature for aluminum alloy 356 melts with various modifiers. Source: Ref 124

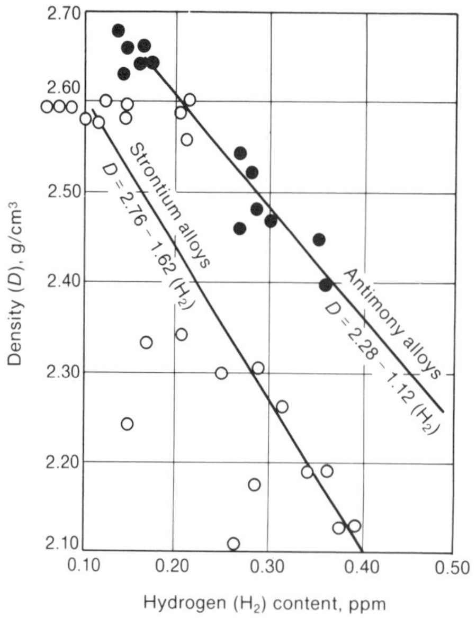

Fig. 80 Effect of modifier used on the density of an aluminum-silicon alloy for a given hydrogen content. Source: Ref 127

Nitrogen-freon mixtures have been found to be ineffective degassers when strontium is present and in fact may actually remove strontium through the possible formation of strontium chloride. It has been postulated that such a compound may retard degassing by blocking hydrogen transport into the purging gas bubble (Ref 125).

Molten Metal Pumping Systems

Molten nonferrous metals have traditionally been transferred from the melting vessel to a mold or casting unit by tapout or lip pouring (essentially gravity flow systems) or by using an intermediate ladle. Mechanical transfer pumps are now commonly used with aluminum, magnesium, zinc, and lead alloys (Ref 25, 26, 129, 130). A mechanical pumping system has several advantages. Pressurized flows permit larger volumes of metal to be transferred per unit time, with resultant productivity gains. The ability to move greater volumes of metal in shorter times also means that less superheat must be put into the metal to accommodate temperature losses during transfer. Thus, shorter overall furnace melting cycles are possible. Because most metal transfer pumping systems have submerged intakes, dross entrainment from the surface is avoided, providing better metal quality. Transfer pumping *per se* does not produce additional oxidation or dross losses, because molten nonferrous metals are always subjected to turbulent flows (high Reynolds numbers) even with gravity flow tapout systems.

Transfer pumps can often be used with closed piping systems to provide transfers surmounting physical obstacles (over other equipment, for example) or overcoming head and height limitations when gravity flow is limited or not possible. Melting and casting facilities can therefore be mixed and matched and need not be directly in line with one another.

Transfer pumps are now commonly used in many segments of the aluminum casting industry to move molten metal from melting furnace to holding furnace, to feed or empty auxiliary refining vessels (used for alloying, degassing, and filtration), to empty and fill transfer ladles, to directly pour castings and fill molds, or to feed casting machines in the wrought industry. In the magnesium, zinc, and lead metal casting industries, mechanical pumps transfer metal between furnaces, fill ingot and pig molds, feed die casting machines, cast electrogalvanizing anodes, and empty galvanizing kettles when repairs or desludging is required. In an unusual application, one foundry has successfully poured a 9000 kg (20 000 lb) casting using hot metal delivery, an intermediate vessel, and a transfer pump (Ref 131). In this way, a much larger casting was produced than the foundry had actual melting capacity for.

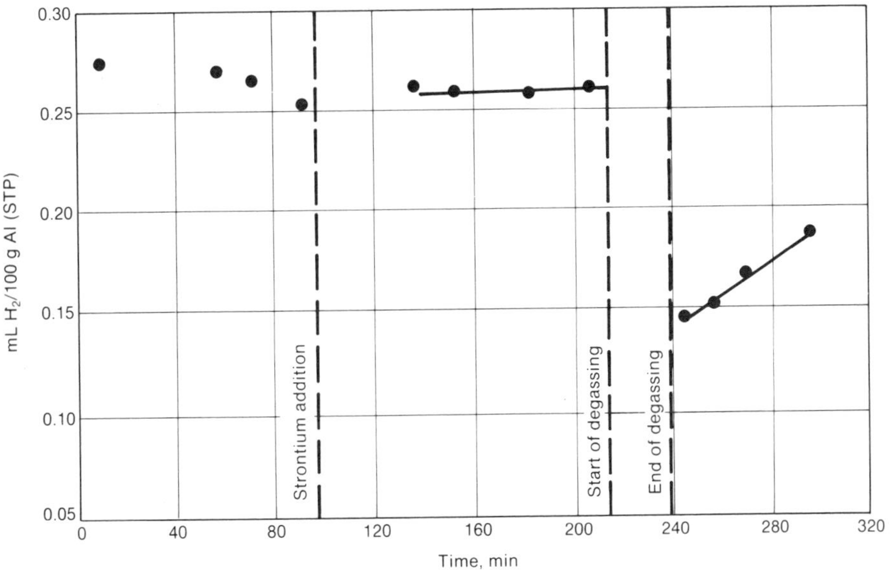

Fig. 81 Effect of 0.03% Sr addition (90% Sr master alloy) on hydrogen level in A356 alloy. Source: Ref 126

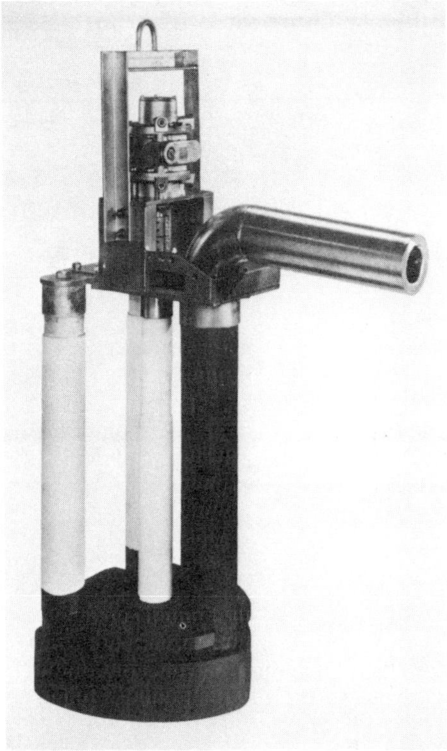

Fig. 82 Centrifugal transfer pump for molten aluminum and zinc. Courtesy of Metaullics Systems

Molten copper alloys have previously been transferred only by small pressure pump systems, but a centrifugal pump system has recently been shown to be successful in pumping 72 600 kg (160 000 lb) of molten copper continuously for 55 min (Ref 130).

Several kinds of mechanical pumping systems exist for nonferrous metals. The first is simply a positive-displacement pressure pump. This is often quite useful for small furnaces or small flows (<45 kg, or 100 lb). When actuated, external air pressure displaces the volume of air in the cylinder, which forces an equivalent volume of liquid metal through the delivery tube.

Another type of pump is the siphon pump. Both of these pumps are quite capable of filling small molds where the accuracy of fill demanded is moderate (several percent variation) and where pressure head requirements are minimal (0.3 to 0.6 m, or 1 to 2 ft, maximum).

When larger volumes of metal, more sustained flow, or higher head is required, a centrifugal pump can be used (Fig. 82). Flow is directed by an impeller up through a riser tube. Operated either by air or electric drive, these pumps are typically constructed of graphite and are easily operated in molten aluminum, magnesium, and zinc alloys. It is important to keep the riser tube open, using insulation and careful external heating if necessary, to prevent freezeups, especially if use is intermittent.

Another type of centrifugal mechanical pump often used with magnesium alloys is shown in Fig. 83. Pump construction is usually cast iron, low-carbon steel, or 400-series stainless steel. Nickel impurities in metal pump components should be specified to be below 0.20% to avoid the dissolution of nickel in the molten magnesium and the subsequent reduced corrosion resistance of magnesium castings.

Electromagnetic pumps have recently been developed that offer promise for very controlled filling capabilities in such applications as permanent mold casting, large die casting, low-pressure die casting, and perhaps small ingot casting. This type of pump is constructed from ceramic materials and delivers metal through the left-hand rule of electromagnetic field theory, utilizing the molten metal as the secondary field in which the aluminum is caused to move upward. One specific advantage of this type of pump is that there are no moving parts. However, the molten metal channel opening must be kept very clean and free of oxide buildup or dross; otherwise, the field strength and resulting pumping capacity can be greatly reduced.

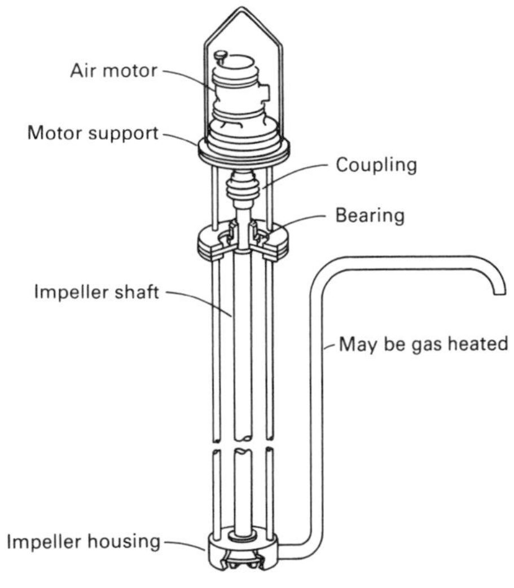

Fig. 83 Schematic of metallic centrifugal pump used to transfer magnesium. Source: Ref 124

Centrifugal pumping systems for zinc are usually constructed of graphite, but metallic centrifugal pumps are often used as well. Pumping systems for lead are usually constructed with cast iron because little corrosion is experienced. Cast iron can also be used to pump zinc as long as the temperature is kept below 480 °C (900 °F), above which molten zinc begins to dissolve iron. Austenitic stainless steels can be used as pump materials with aluminum-containing zinc alloys 0.2 to 3.5% Al. Stainless steel must not be used, however, with pure zinc, because liquid-metal embrittlement and intergranular attack occur quite rapidly. With zinc-aluminum galvanizing alloys (40 to 60% Al), the high operating temperature (595 to 650 °C, or 1100 to 1200 °F) is in the region where austenitic stainless steels become sensitized to greater intergranular carbide precipitation and subsequent accelerated corrosion. Stainless steel pumps should therefore not be used for this application. A useful but expensive alternative for the latter two applications is an Mo-30W or Mo-40W pump shaft, which provides excellent corrosion resistance but is fairly brittle at high temperatures.

Molten Metal Filtration

The most significant modern development in nonferrous molten metal processing

Table 7 Characteristics of inclusions commonly found in aluminum alloys

Inclusion type	Chemical formula	Form	Density g/cm^3	Density lb/in.^{3}	Size range μm	Size range μin.
Oxides						
Alumina	Al_2O_3	Particles, skins	3.97	0.143	0.2–30 (particles) 10–5000 (skins)	8–1200 400–2 × 10^5
Magnesia	MgO	Particles, skins	3.58	0.129	0.1–5 (particles) 10–5000 (skins)	4–200 400–2 × 10^5
Spinel	$MgAl_2O_4$	Particles, skins	3.60	0.130	0.1–5 (particles) 10–5000 (skins)	4–200 400–2 × 10^5
Silica	SiO_2	Particles	2.66	0.096	0.5–5	20–200
Salts						
Chlorides	Varies	Particles	1.98–2.16	0.072–0.078	0.1–5	4–200
Fluorides	Varies	Particles	1.98–2.16	0.072–0.078	0.1–5	4–200
Carbides						
Aluminum carbide	Al_4C_3	Particles	2.36	0.085	0.5–25	20–1000
Silicon carbide	SiC	Particles	3.22	0.116	0.5–25	20–1000
Nitrides						
Aluminum nitride	AlN	Particles, skins	3.26	0.117	10–50	400–2000
Borides						
Titanium boride	TiB_2	Agglomerated particles	4.50	0.163	1–30	40–1200
Aluminum boride	AlB_2	Particles	3.19	0.115	0.1–3	4–120

Source: Ref 132, 133

involves the application of filtration. Aluminum alloys, especially with their tendency toward oxidation, often contain impurities (nonmetallic inclusions) that are deleterious to physical, mechanical, electrical, and aesthetic properties. Filtration has become a primary means of removing such solid impurities. Zinc and copper alloy melts are filtered to a much lesser extent than aluminum, but use of filtration is increasing for these alloys also.

The filtration process consists of passing the molten metal through a porous device (a filter) in which the inclusions contained in the flowing metal are trapped or captured by one or more filtration mechanisms (see the section "Filtration Fundamentals" in this article). The filter material itself must have sufficient integrity (strength, refractoriness, thermal shock resistance, and corrosion resistance) so that it is not destroyed by the molten metal before its task is accomplished. Consequently, most filter media are ceramic materials in a variety of configurations.

Sources of Inclusions

Inclusions arise from many sources. Charge materials can contain nonmetallic inclusions from unfiltered primary or secondary ingot. Scrap materials may contain molding sand debris from back-charged gates and runners, metalworking lubricants, paint and coatings, plastics, or unsorted and unwanted heavy metal contaminant oxides (iron, manganese, chromium, and so on). Furnace tools, such as skimmers, puddlers, and rakes can also add solidified oxide skins back into the melt if they are not cleaned before use.

Excessive temperatures during melting accelerate the oxidation of aluminum and alloying elements. Where fluxes are used, incomplete skimming or separation of fluxing salts often results in contamination and salt-wetted oxide inclusions.

Grain-refiner additions can be a source of inclusions (see the earlier sections of this article on grain refining of aluminum alloys, magnesium alloys, and copper alloys).

Intermetallic compound precipitation may also contribute to inclusions. Iron, chromium, and manganese form intermetallics with aluminum and the formation of intermetallic sludge is a common problem in die casting alloys.

Refractory Inclusions. Finally, the erosion and wear of melting and holding vessels may add alumina, spinel, silica, and silicon carbide inclusions to nonferrous alloy melts.

Inclusions in Aluminum Alloys

There are several types of solid-phase inclusions in molten aluminum alloys:

- Oxides (Al_2O_3, MgO)
- Spinels (Mg_2AlO_4)
- Borides (TiB_2, VB_2, ZrB_2)
- Carbides (Al_3C_4, TiC)
- Intermetallics ($MnAl_3$, $FeAl_3$)
- Nitrides (AlN)
- Exogenous refractory inclusions (oxides and/or carbides of iron, silicon, aluminum, etc.)

These inclusions possess a series of complexes, true formulations, and densities ranging from lighter to heavier than the molten aluminum itself, as well as a range of sizes and shapes (Table 7). Gravity separation alone plus skimming is not always successful for adequate removal. In addition, there are different degrees of wettability of these inclusions with molten aluminum, thus allowing the use of fluxing salts to wet the inclusion properly and render separation possible by detaching the inclusion from the melt.

The size and shape of the inclusions vary from globular (oxides and borides) to stringerlike (oxides and spinels). The latter are particularly insidious in their effect on castability and the properties of the casting.

Liquid-phase inclusions can occur in molten aluminum alloys. The most prevalent is $MgCl_2$ arising from the chlorination of magnesium-containing melts. These inclusions are not generally removed to a great extent by conventional filtration techniques and therefore must be avoided. Preventing excessive and inefficient chlorination of the melts during degassing and thorough skimming at least help to avoid such inclusions. The nature of these and other inclusions in aluminum alloys is discussed in more detail in Ref 133-137.

Inclusions in Magnesium Alloys

The inclusions found in magnesium melts are similar in many cases to those found in aluminum—oxides, spinels, furnace refractory particles, and so on. In addition, alloying elements (zirconium and the rare earths) lead to both oxides of these elements as well as intermetallic compound precipitates. If the fluxing process is used to clean magnesium, fluxing salt inclusions may also be present.

Inclusions in Copper Alloys

In addition to the usual inclusions arising from oxides, fluxing salts, and intermetallics, copper oxide inclusions and phosphorus pentoxide (from deoxidation) may be present in copper alloys if the melt is not allowed to settle or if it is inadequately skimmed before pouring and casting.

Inclusions in Zinc Alloys

In the zinc foundry and die casting alloys containing aluminum, nonmetallic oxide inclusions are usually less important than intermetallic inclusions (Ref 18, 138, 139). Iron-zinc and iron-aluminum intermetallic compounds may also be present, especially above 480 °C (900 °F) in melting in iron-base vessels. These particles may be small crystals less than 5 μm (200 μin.) in size or may agglomerate into millimeter-size aggregates.

Other heavy metals present, such as lead, cadmium, nickel, manganese, and chromium, can also lead to intermetallic compound precipitation (for example, Mn_2Al_5, Ni_2Al_3, and $CrAl_4$). These compounds (often called aluminides) can form when the concentration of impurity exceeds the liquid solubility level of the element, which is often quite low (<0.02% for manganese, silicon, chromium, and nickel in die casting alloys). The intermetallic particles can be lighter than the base alloy and therefore rise to the top dross. Fluxing salt inclusions and exogenous refractory particles can also be present, as with the other nonferrous alloy systems.

Oxide inclusions in zinc are somewhat limited because of the high positive vapor pressure of zinc. However, alloys containing aluminum can be subject to greater oxidation of the aluminum constituent. Oxide inclusions can also result during melting or casting when excessive turbulence is encountered; thus, mechanical stirring or molten metal pumping should be controlled. Other nonmetallic inclusions arise from refractory vessel erosion and include silica, alumina, and silicon carbide. There is an appreciable density difference between these oxides and the base melt, unlike in aluminum; for this reason, gravity separation and thorough removal are substantially easier.

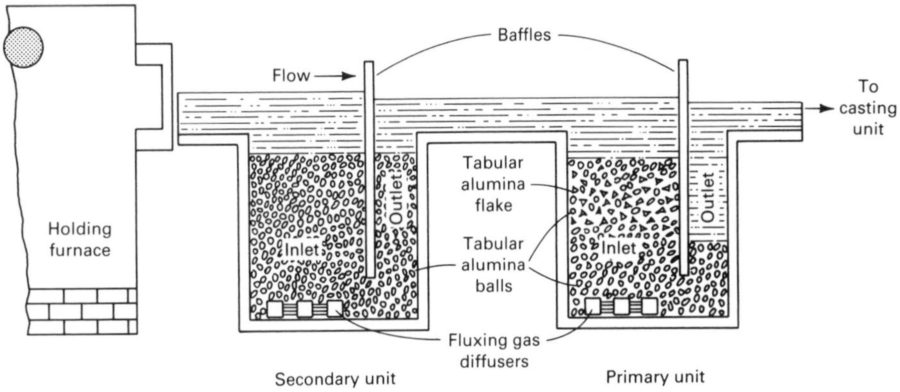

Fig. 84 Example of one type of bed filter used for in-line processing. Source: Ref 140

Filtration Fundamentals

Two types of filtration modes occur when filtering particles from a flowing liquid stream: cake mode and depth mode.

In cake filtration, inclusion removal occurs by the mechanical entrapment of inclusions on the surface of the filter. Particles larger than 30 μm (1200 μin.) especially are captured by this method. These particles agglomerate to form a filter cake on the filter surface, which further aids filtering of inclusions during subsequent molten metal flow. Ceramic foam filters, one of the most prevalent kinds of filters, function largely in this manner.

Depth filtration is the filtering mode most prevalent in bed filtration. Depth filtration is capable of removing particles much smaller than 30 μm (1200 μin.) as they are carried in a molten metal flow stream through a very tortuous path in a deep bed filter (Fig. 84) that is commonly used in mill product casthouses. The tortuosity provides increasing probability for a two-step particle capture mechanism involving transport from the liquid metal flow and attachment to the filter surface itself. Particularly in filters possessing irregular surface characteristics, the fluid transport phenomena of inertia, sedimentation, particle interception, and diffusion come into play (Fig. 85). Particles brought into contact with the walls of the filtering medium then may attach themselves or be retained by one or more mechanisms:

- Gravity, friction
- Physical entrapment
- Chemical bonds
- Van der Waals forces
- Electrostatic forces

Researchers familiar with filtration theory are now concentrating more heavily on studying the nature of the chemical bonding mechanism. It appears that there are peculiar interactions between different types of inclusions and specific ceramic surfaces that give rise to differing filtration capabilities and efficiencies.

Filtration efficiency is influenced by several factors. First, the nature of the molten metal itself must be considered—the amount of inclusions, their size, shape, and distribution. Second, the dynamic conditions encountered by the filter will have a pronounced effect:

- Lower molten metal flow rates generally result in greater filtration efficiency because of better capture probability in/by the filter
- Greater filter surface area increases filtration efficiency
- Filters that provide a stronger balance between cake and depth filtration provide higher filtration efficiency
- Greater length or depth of filter dimension permits greater probability of inclusion capture and therefore greater filtration efficiency

The efficiency of various filters is subject to considerable debate (Ref 132, 141-144). Taking the above-mentioned factors into consideration, however, it can be said that those filters providing more depth plus cake filtration function more efficiently than those that rely more or only on cake mode filtration. Figure 86 compares the efficiencies of ceramic foam and bed filters over a range of metal flow rates.

Types of Filters

The simplest filters are the strainer cores used in downsprues in conventional foundry sand casting. While serving to choke or control metal flow, they also trap some

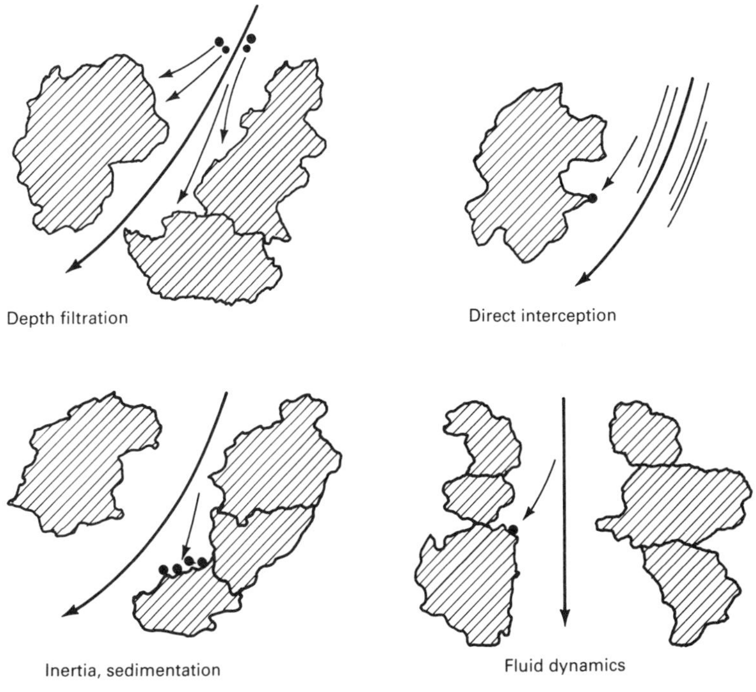

Fig. 85 Transport and capture mechanisms operating during depth filtration

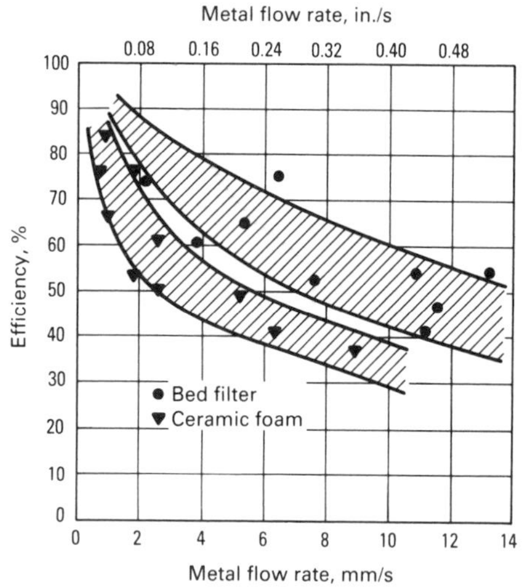

Fig. 86 Comparison of filtration efficiencies of a bed filter and a ceramic foam filter as a function of melt velocity. Source: Ref 144

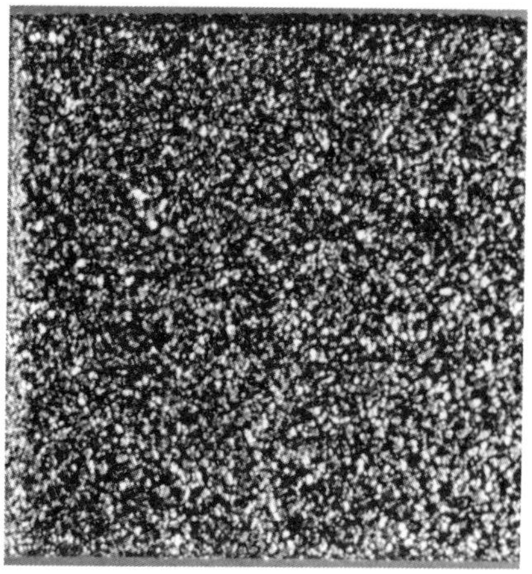

Fig. 87 Typical bonded particle filter

particulate by surface attachment and a limited degree of cake filtration.

Metal or Fiberglass Screens. Steel mesh screens and fiberglass cloth are also commonly used for aluminum, and molybdenum mesh is used for copper alloys. These materials act as planar filters. They provide moderate filtration, particularly of larger particles and inclusions exceeding 100 μm (0.004 in.) in size. The mesh screen, fiberglass cloth, and strainer cores are of course single-use filters.

Rotary degassing can also remove inclusions by flotation. It is usually effective in removing inclusions larger than 40 μm (1600 μin.) in size.

Bed filters composed of tabular Al_2O_3 are quite common in the casting of mill products, where up to 1800 Mg (2000 tons) of a single alloy might be filtered in-line between holding furnace and casting station before replacement is necessary. Bed filters provide the best source of fine particle filtration, but their size, expense, and single-alloy use preclude their economic use in multiple-alloy wrought casting (ingot) or foundry casting operations. Bed filters are capable of providing nearly the highest filtration efficiency attainable, but because the bed is not rigid, movement and flow channeling can occur, which can dump inclusions.

Bonded particle filters are finding greater use in foundry and die casting operations (Ref 131, 145, 146). These filters consist of a refractory grain (either Al_2O_3 or SiC) bonded together to form a rigid structure (Fig. 87). Bonded particle filters provide a relatively tortuous path for metal flow through the filter (good depth filtration). These filters are often used vertically to separate melting or holding furnace hearth areas from the dip-out well (Fig. 88) and thus provide cleaner metal on the dip-out side. Silicon carbide vertical gate filters are quite durable and offer the potential of low thermal gradients between the hot face and cold face because of the relatively high thermal conductivity of silicon carbide compared to other materials, such as oxide ceramics. Temperature gradients of 22 to 42 °C (40 to 75 °F) have been reported for these filters, versus 33 to 72 °C (60 to 130 °F) for oxide ceramic foams, for example (Ref 145).

One proprietary bonded particle filter system employs the bonded particle filter in a three-dimensional configuration (Fig. 89). This system has been successfully used in mill product, foundry, and die casting operations. The filter is housed in its own holding furnace, and it stays primed between casts or pours. The filter is rejuvenated between uses by backflushing to remove much of the accumulated filter cake, thus permitting continued metal flow capability. Up to 455 Mg (500 tons) of throughput has consistently been achieved before filter replacement is necessary, depending on initial metal cleanliness. In this sense, the filter is reusable (unlike a single-use ceramic foam) and offers the potential for substantial cost savings for filtration.

Cartridge Filters. One interesting application of the rigid filter concept is the cartridge filter (Fig. 90). This is simply a shell and tube type construction similar to a heat exchanger. This type of filter provides a great amount of surface area and usually has very fine pore size. Therefore, although substantial priming head (up to 0.3 m, or 1 ft) of metal may be required before flow through begins and although the flow rate is not very high, filtration efficiency is excellent. Cartridge filters are now receiving attention for such premium-quality applications as aircraft plate, computer disks, and foil. Efficiencies of 95% or better in removing particles less than 5 μm (200 μin.) in size have been reported.

Ceramic foam filters are the most common types of filters currently used in nonferrous casting. They are produced by slurry coating a reticulated polyurethane

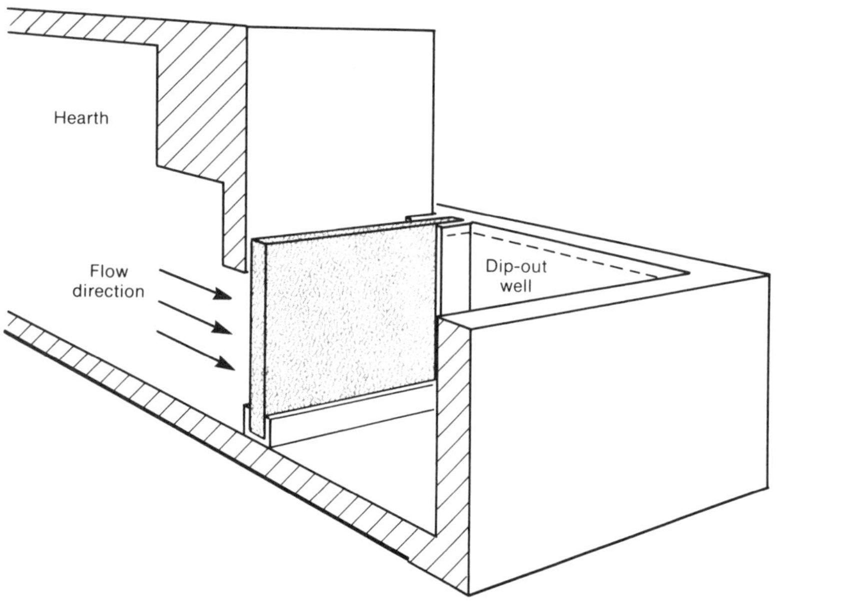

Fig. 88 Schematic of a vertical filter (bonded particle) used for holding furnace filtration for die casting operations

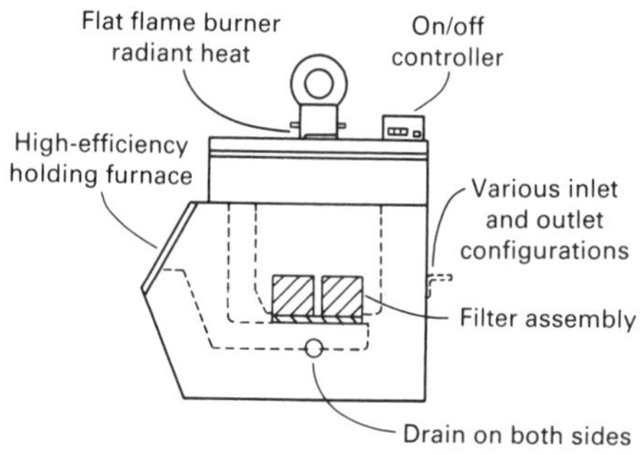

Fig. 89 Schematic of a proprietary bonded particle filter system. Source: Ref 146

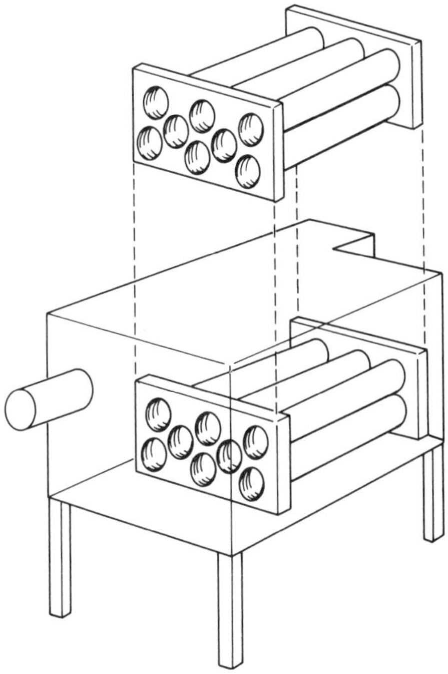

Fig. 90 Schematic of a cartridge bonded particle filter

cellular foam, followed by drying and firing to burn out the precursor foam. This results in a ceramic replica of the original organic foam structure (Fig. 91). Suitable ceramics for these filters are alumina, zirconia, and mullite. Some ceramic filters also employ chromic oxide. These filters are approximately 75% porous and are produced in a variety of pore sizes as measured by pores per lineal inch (ppi). Ceramic foam filters are used in flat, platelike form. In the primary aluminum industry, plate sizes vary from 305 × 305 mm (12 × 12 in.) to 585 × 585 mm (23 × 23 in.), depending on flow rate (up to 680 kg/min, or 1500 lb/min) and casting size (up to 34 000 kg, or 75 000 lb). Smaller filters are used in the die casting and foundry industries.

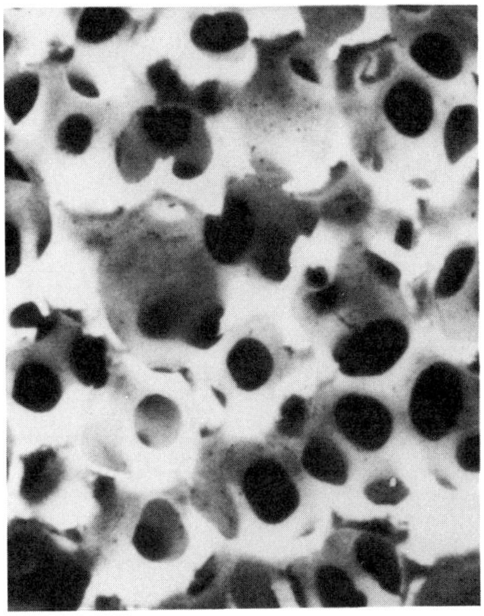

Fig. 91 Structure of a ceramic foam filter. Courtesy of Foseco Inc.

Dual-structure or multilayered ceramic foam filters are also employed. These use a finer pore size for greater inclusion capture capability, coupled with a coarser pore size to permit greater throughput before the filter becomes saturated or clogged.

It is generally believed that ceramic foam filters function primarily through cake filtration, although undoubtedly some depth filtration also occurs. Bonded particle filters possess less porosity (38% versus 75% for a ceramic foam) and greater tortuosity; therefore, they function more as depth filters.

Choosing a Filter. Filter selection often depends on ease of use. Ceramic foams are lighter, easier to preheat, and provide easier flow capability than bonded particle filters, although the filtration efficiency of the latter is reported to be greater. For most foundry applications, however, a ceramic foam filter is most suitable. In the primary and mill products casting industry, filter selection will depend on economics, space constraints, auxiliary metal treatment capabilities, desired filtration efficiency, and end product application.

Ceramic foam filters and bonded particle filters have both been employed in filter baskets, baffled melting crucibles, and pouring ladles with some degree of success in many foundry operations (Fig. 92). Care must be taken to properly install, preheat, and maintain the filters used in these applications. In this way, such filters can be used through a number of repetitions.

Foundry Filters

The typical ceramic foam filter sections used in foundry gating systems are square, rectangular, or round sections up to 150 mm (6 in.) long and 19 to 25 mm (0.75 to 1.0 in.) thick. The filters are placed either tangentially or vertically to the metal flow path in a filter-sized print in the molding medium (sand, plaster, permanent mold). Typical placement schemes for gating or runner system filters are shown in Fig. 93. It is desirable to place filters as close to the actual mold cavity as possible to maximize effectiveness and to minimize turbulence. Bottom filling, when practical, is preferred.

Filter Size. Rules of sizing relate filter size to the flow rate and throughput capability desired. Sizing must also be tailored to the proper mold filling time necessary to achieve the desired solidified cast structure, because any filter produces a restriction to metal flow. Experience has shown that the following general rules apply when sizing gating system filters for nonferrous alloys, principally aluminum (Ref 149, 150):

- For a given required fill rate, enlargement of the gating system is necessary to accommodate the restriction created by the filter
- Total ratio of filter area to runner area must be at least 2:1. Often 5:1 to 10:1 is

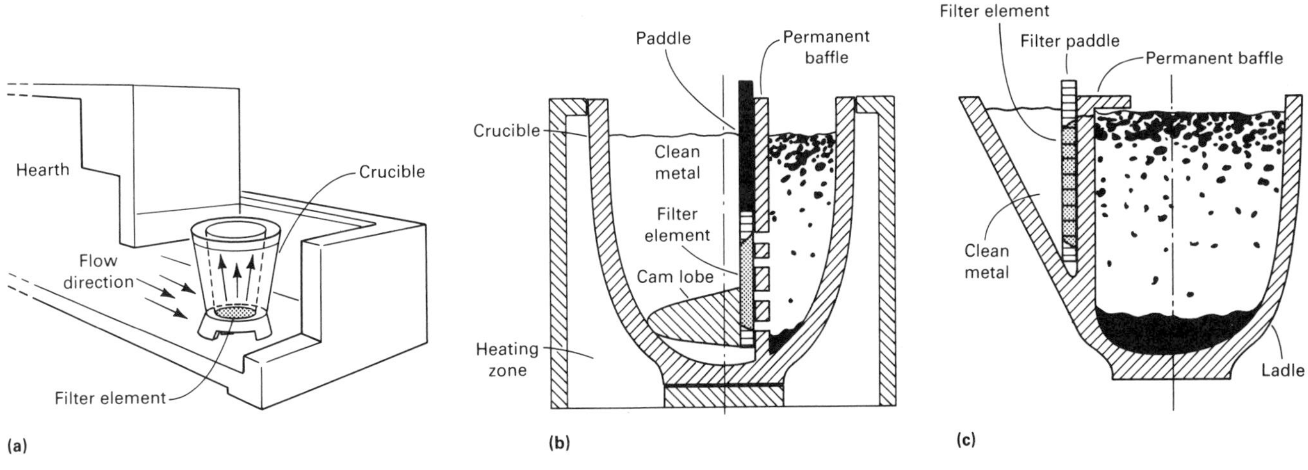

Fig. 92 Three foundry applications for ceramic foam filters. (a) Crucible filter. (b) Crucible with filter baffle. (c) Pouring ladle filter. Source: Ref 147, 148

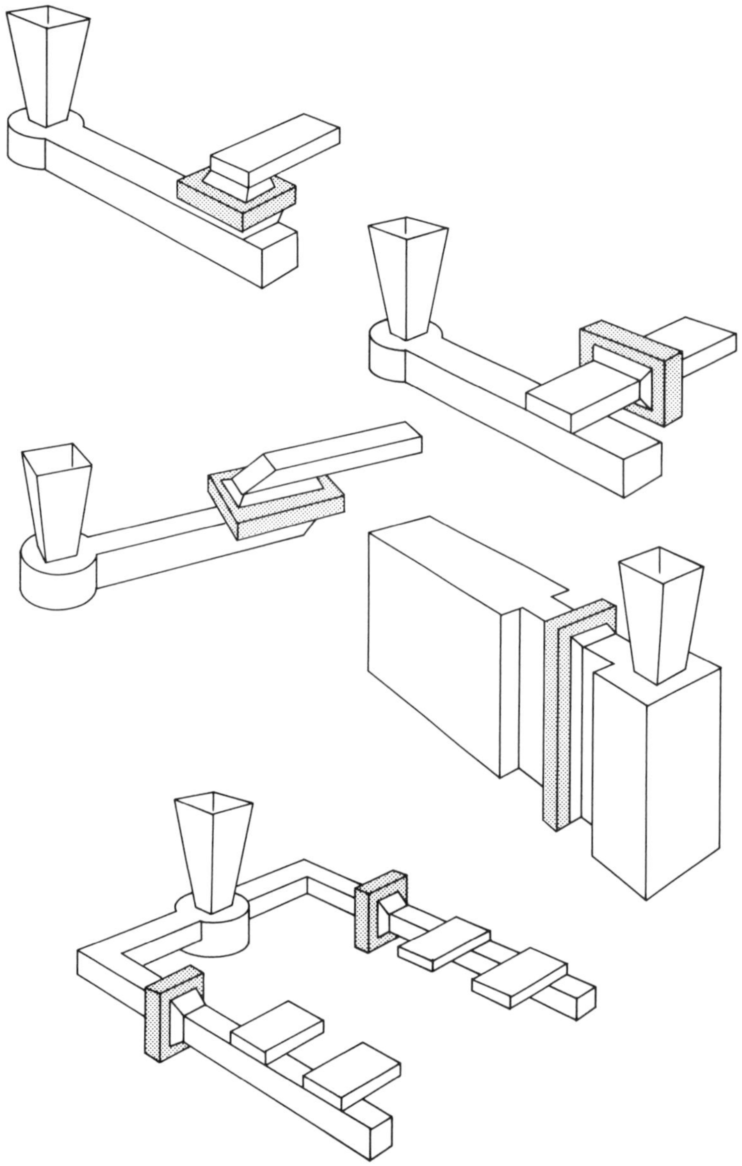

Fig. 93 Typical foundry filter placements (shaded). Courtesy of Selee Corporation

most desirable, particularly to maintain a desired fill rate for soundest cast structure

- To take full advantage of a vertical filter placement, the runner system should be flared (Fig. 92)
- Each square inch of filter area in a (typical) 20 ppi ceramic foam filter is capable of passing (roughly) 2 kg (5 lb) of total metal throughput. Consequently, a 45 kg (100 lb) casting fed with a single gate would require at least a 100 × 100 mm (4 × 4 in.) filter or possibly larger. Of course, multiple gates are often employed in larger castings. Each gate would have its own filter in the runner system ahead of each gate
- Some increase in metal temperature (between 6 and 28 °C, or 10 and 50 °F) is usually necessary to accommodate the temperature drop in the system caused by the filter absorbing heat energy and/or the drop associated with restricted flow
- Flow rate, and thus fill time, is also a function of the metallostatic head (sprue height). Sprue height can often be increased to offset the flow rate restriction caused by a filter

Benefits of Filtration

The presence of any dross, inclusions, or nonmetallic impurities, as well as insoluble intermetallic precipitates, detracts from the natural fluidity of any nonferrous alloy melt. As shown in Fig. 94, filtering of these impurities results in substantial improvement of fluidity (as much as 25%). Such fluidity is required to provide mold filling, especially in the casting of thin, intricate sections.

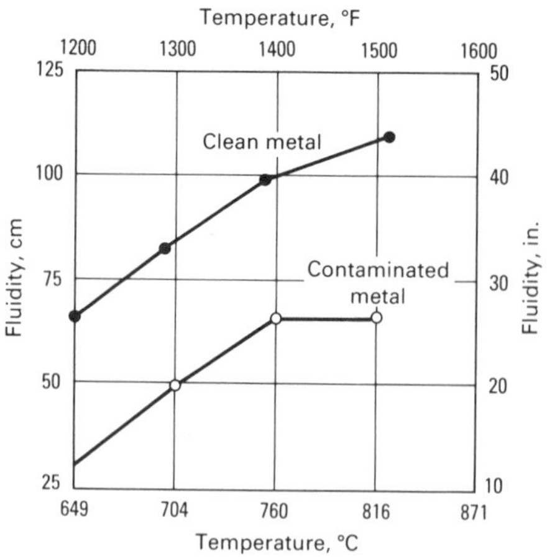

Fig. 94 Effect of metal cleanliness on fluidity of an aluminum die casting alloy. Source: Ref 151

Removal of inclusions by filtration results in lower die wear, decreased tendency toward die soldering, and less scoring in permanent mold or die casting operations. Inclusions also cause hard spots in machining; therefore, filtration provides longer tool life and often permits faster metalworking.

For articles requiring smooth finish, superior corrosion resistance, and better appearance, filtration usually removes the vast majority of inclusions responsible for defects that detract from these characteristics. Inclusions can also nucleate hydrogen porosity, so their removal can result in reduced shrinkage, voids, and microporosity. As might be expected, the mechanical properties of filtered metal castings are usually superior to those of unfiltered metal castings, as detailed in Ref 151.

The greatest benefit of filtration to the operating foundry is the increase in production generated. Filtration reduces the number of defective castings that must be scrapped, and it reduces cleaning, rework, and repair. Reject rates on problem castings have been reduced substantially, virtually to zero in many cases (Ref 145, 151).

In the wrought aluminum casthouse and in subsequent shaping operations, filtration increases yields, reduces hot tears, reduces surface defects (inclusion lines, pinhole, and so on), and improves mechanical properties (Ref 152). Quality improvements and increased productivity are vital benefits achieved in the production of mill products.

Filtration of copper casting alloys is on the increase, using ceramic foam sections in the gating system. Oxide inclusions have been successfully removed from aluminum bronze alloys (Ref 153). Investment castings, both aluminum and copper base, can be successfully filtered using a filter section in the pouring cup or, in the case of larger castings, molded directly into the wax runner bars. Ceramic filters for copper alloys are usually alumina, mullite, or zirconia. Filtration of zinc alloys (foundry ingot, die casting alloys, and sand casting alloys) is not yet as prevalent as filtering of aluminum, but the use of filters has resulted in reduced concentration of intermetallic particles, improved impact resistance, and slightly improved mechanical properties (Ref 139).

Determining Melt Cleanliness

The determination of general melt cleanliness (that is, characterization of inclusion content) is attempted using various techniques. All of these methods have disadvantages, and no single technique has yet been established as the standard throughout the industry.

By far the most common, and certainly one of the most laborious, techniques is conventional metallography and optical microscopy. A polished section of a solidified sample from the melt is examined for inclusions in this manner. Inclusion particulate may be further characterized and identified, if not fully quantified, by x-ray or electron diffraction, microprobe, or other spectrographic analyses.

Analytical techniques such as dissolution and the Coulter Counter analysis of individual particles have been attempted in several instances and may be reasonably accurate with respect to particle size and particle size distribution. However, this method is very tedious, and the nonmetallic particles themselves often become dissolved along with the aluminum matrix, thus rendering the quantified particle and particle size distribution count inaccurate.

Ultrasonic methods have been used by several mill products producers to determine inclusion counts *in situ* in the melt during a casting run (Ref 154). The technique also permits the evaluation and comparison of various metal cleaning parameters, such as holding time, fluxing times, and filtration. The largely proprietary instrumentation for this technique is relatively sophisticated and expensive; therefore, this method is beyond the reach of all but the largest melting and casting installations.

Another technique used is the liquid inclusion sampler described in Ref 155. In this method, inclusions from the melt are concentrated by passing the liquid stream through a precisely controlled porous carbon or ceramic disk. The aluminum-impregnated disk is then solidified and prepared metallographically for examination using typical optical microscopy. The concentration of inclusions from the melt into this relatively small and fine porous disk provides easier analysis and permits quantifiable image analysis techniques to be used. Recently, back-scatter x-ray reflection has been successfully applied (Ref 156). Various samples can be compared and indexed from melt to melt, or differences in cleanliness can be identified as a function of degree of filtration or other metal cleaning processes employed.

REFERENCES

1. I.A. Abreu, Products Used in Treating Metallic Aluminum Baths, *AFS Int. Cast Met. J.*, 1977, p 57
2. R.J. Kissling and J.F. Wallace, Fluxing to Remove Oxides From Aluminum Alloys, *Foundry*, 1963
3. L.P. Semersky and W.F. Joseph, Aluminum Handling Practices Influence Profits, *Die Cast. Eng.*, March-April, 1978, p 22
4. R.W. Bruner, Basics on the Fluxing of Aluminum, *Die Cast Eng.*, Nov-Dec 1986, p 42
5. S.C. Jain and S.C. Tiwari, Fluxing Processes in Production of Aluminum and Its Alloys, *Indian Found. J.*, Vol 25, 1979, p 12-16
6. *Aluminum Casting Technology*, American Foundrymen's Society, 1986
7. E.F. Emley, *Principles of Magnesium Technology*, Pergamon Press, 1966
8. V.B. Kurfman, Melt Loss Prevention for Magnesium Alloys, *Trans. AFS*, 1965, p 317
9. J.N. Reding, A Study of Factors Influencing the Separation of Flux From Magnesium Alloys, *Trans. AFS*, 1968, p 92
10. H.I. Kaplan, Basic Metallurgy of the Magnesium Die Casting Alloys, *Die Cast. Eng.*, Nov-Dec, 1986
11. *Magnesium Die Casting Manual*, Dow Chemical Company, 1980
12. J.F. Wallace and R.J. Kissling, Fluxing of Copper Alloy Castings, *Foundry*, 1963
13. J.F. Wallace and R.J. Kissling, Gases in Copper Base Alloys, *Foundry*, Dec 1962, p 36-39; Jan 1963, p 64-68
14. L.V. Whiting and D.A. Brown, "Air/Oxygen Injection Refining of Secondary Copper Alloys," Report MRP/PMRL 79-SO(J), Physical Metallurgy Research Laboratories, CANMET, 1979
15. J.E. Stalarcyzk, A. Cibula, P. Gregory, and R.W. Ruddle, The Removal of Lead From Copper in Fire Refining, *J. Inst. Met.*, Vol 85, 1957, p 49
16. J.G. Peacey, G.P. Kubanek, and P. Tarasoff, "Arsenic and Antimony Removal From Copper by Blowing and Fluxing," Paper presented at the Annual Meeting of AIME, Las Vegas, NV, Nov 1980
17. A.V. Larsson, Influence of Aluminum on Properties of Cast Gun Metal and Removal of Aluminum by Slag, *Trans. AFS*, 1952, p 75
18. D.V. Neff, Practical Metallurgical Control of Zinc Die Casting Alloys, *Die Cast. Eng.*, Nov-Dec, 1978
19. *Zaclon Galvanizing Handbook*, Dupont Corporation, 1971
20. R.J. Harris, "Aluminum Treatment Technology of the Future," Paper presented at British Nonferrous Sixth International Conference, Birmingham, England, Sept 1986
21. J.M. Fuqua, "An Advanced Injection Treatment System for Aluminum Silicon Alloys," Paper presented at the AFS Congress, St. Louis, MO, American Foundrymen's Society, April 1987
22. "Treatment of Molten Aluminum Alloys by the Flux Injection Process," Product Bulletin, Aikoh America Corporation, 1984
23. T. Pedersen and E. Myerbostäd, Refining and Alloying of Aluminum by Injection, in *Light Metals*, The Metallurgical Society, 1986, p 759
24. R.J. Omalley, C.E. Dremann, and D. Apelian, Alloying of Molten Aluminum by Manganese Powder Injection, *J. Met.*, Feb 1979, p 14
25. D.V. Neff, Molten Metal Pumping Systems—Current Applications and Benefits, in *Light Metals*, The Metallurgical Society, 1987, p 805
26. D.V. Neff, Molten Metal Pumping in Foundry and Foundry Ingot Applications, *Trans. AFS*, 1987, p 273
27. P.J. Bamji and F.W. Pierson, Electromagnetic Circulation of Molten Aluminum, *J. Met.*, 1983, p 44
28. D.V. Neff, Molten Metal Pumping and Pipeline Transport in Aluminum Casthouses, in *Proceedings of Energy Workshop X*, Aluminum Association, 1987
29. T.L. Cunard, Application of Graphite Pump to Direct Charged Aluminum Melting Furnaces, in *Proceedings of Energy Workshop X*, Aluminum Association, 1987
30. P.D. Hess, Measuring Hydrogen in Aluminum Alloys, *J. Met.*, Oct 1973, p 46
31. W.R. Opie and N.J. Grant, Hydrogen Solubility in Aluminum and Some Aluminum Alloys, *Trans. AIME*, Vol 1988, 1950, p 1237
32. R.Y. Lin and M. Hoch, "Solubility of Hydrogen in Molten Aluminum Alloys," Paper presented at the Annual Meeting of AIME, New Orleans, LA, 1986
33. G.K. Sigworth, A Scientific Basis for Degassing Aluminum, *Trans. AFS*, 1987, p 73
34. T.A. Engh and G.K. Sigworth, Molten Aluminum Purification, in *Light Metals*, The Metallurgical Society, 1982, p 983
35. G.K. Sigworth and T.A. Engh, Chemical and Kinetic Factors Related to Hydrogen Removal From Aiuminum, *Metall. Trans.*, Vol 138, 1982, p 447
36. T.A. Engh and T. Pedersen, Removal of Hydrogen From Molten Aluminum

by Gas Purging, in *Light Metals*, The Metallurgical Society, 1984, p 1329
37. J.L. Jorstad, An Overview of the Need for Melt Cleanliness Control, in *Proceedings of the International Molten Aluminum Conference*, American Foundrymen's Society, 1986, p 285
38. R.K. Owens, H.W. Antes, and R.E. Edelman, Effects of Nitrogen and Vacuum Degassing on Properties of a Cast Aluminum-Silicon Magnesium Alloy (Type 356), *Trans. AFS*, 1957, p 424
39. H. Rosenthal and S. Lipson, Measurement of Gas in Molten Aluminum, *Trans. AFS*, 1955, p 301
40. D.J. Neil and A.C. Burr, Initial Bubble Test for Determination of Hydrogen Content in Molten Aluminum, *Trans. AFS*, 1961, p 272
41. D.A. Hilton, Measurement of Hydrogen in Molten Aluminum Alloys Using a Quantitative Reduced Pressure Technique, in *Proceedings of the International Molten Aluminum Processing Conference*, American Foundrymen's Society, 1986, p 381
42. D.A. Granger, Telegas for Determining Hydrogen in the Foundry Industry, in *Proceedings of the International Molten Aluminum Processing Conference*, American Foundrymen's Society, 1986, p 417
43. J.C. Miller, Vacuum Solidification Testing and the Vibrated Vacuum Gas Test, in *Proceedings of the International Molten Aluminum Conference*, American Foundrymen's Society, Feb 1986, p 433
44. D.L. Lebahn, "Effects Observed When Fluxing Aluminum With Mixed Gases," Paper presented at the Pacific Northwest Conference on Advances in Aluminum Casting, Spokane, WA, 1982
45. R.F. Molland, J.E. Dome, and N. Davidson, "A Low Emission Process for the Melt Treatment of Aluminum Alloys," Paper presented at the U.S. EPA Conference on Control of Particulate Emissions in Primary Nonferrous Metals Industries, Monterey, CA, Environmental Protection Agency, 1979
46. "The NOPO Degassing Process," Product Bulletin, Selee Corporation
47. D.E. Groteke, Influence of SNIF Treatment on Characteristics of Aluminum Foundry Alloys, *Trans. AFS*, 1985, p 953
48. A. Aarflot, "Development of Melt Treatment," Paper presented at the International Seminar on Refining and Alloying of Liquid Aluminum and Ferro Alloys, Trondheim, Norway, Aug 1985
49. A.J. Clegg, Aluminum Degassing Practice, in *Proceedings of the International Molten Aluminum Processing Conference*, American Foundrymen's Society, 1986, p 369
50. R.J. Kissling and J.F. Wallace, *Gas Porosity in Aluminum Castings*, Penton Publishing, 1963
51. J.V. Griffin, "SNIF Develops the New R-10 System," Paper presented at the Fifth Yugoslav International Symposium of Aluminum, the Yugoslav Aluminum Industry and the University of Ljubljana, Mostar Yugaslavia, 1986
52. R.F. Miller, L.C. Blayden, M.T. Bruno, and C.E. Brooks, In Line Fumeless Metal Treatment, in *Light Metals*, The Metallurgical Society, 1978, p 491
53. G. Snow, D. Patthe, and G. Walker, R.D.U.—An Efficient Degassing System for the Aluminum Casthouse, in *Light Metals*, The Metallurgical Society, 1987, p 717
54. J.M. Hichter, Alpur Refining Process, in *Light Metals*, The Metallurgical Society, 1983, p 1005
55. J. Bildstein and J.M. Hichter, The Development of the Alpur Systems for the In-Line Treatment of Molten Aluminum and Aluminum Alloys, in *Light Metals*, The Metallurgical Society, 1985, p 1209
56. P.G. Bibby and J.C. Bildstein, Comalco Rolled Products Operating Experience With an Alpur D5000 Degassing Unit, in *Light Metals*, The Metallurgical Society, 1988, p 359
57. A.R. Anderson, Practical Observation Rotary Impeller Degassing, *Trans. AFS*, 1987
58. M.L. Heaman and D.R. Grimm, Utilization of the MINT III System and the Electromagnetic Casting Process in the Production of 3004 Alloy Can Body and 1145 Alloy Fine Gauge Foil Stock, in *Light Metals*, The Metallurgical Society, 1988
59. J.T. Bopp, D.V. Neff, and E.P. Stankiewicz, Degassing Multicast Filtration System (DMC)—New Technology for Producing High Quality Molten Metal, in *Light Metals*, The Metallurgical Society, 1987, p 729
60. V.B. Kurfman, Hydrogen Escape From Magnesium Alloys, *Trans. AFS*, 1964, p 9
61. P.K. Trojan, T.R. Ostrom, and R.A. Flinn, Melt Control Variables in Copper Base Alloys, *Trans. AFS*, 1982, p 729; "High Conductivity Copper Alloys," Bulletin 43, Foseco, Inc., 1972
62. R.J. Kissling and J.F. Wallace, Gases in Copper Base Alloys, *Foundry*, 1962, 1963
63. M.P. Renatv, C.M. Andres, G.P. Douglas, and R.A. Flinn, Solubility Relations in Liquid Copper-Nickel-Carbon-Oxygen Alloys, *Trans. AFS*, 1976, p 641
64. B.N. Ames and N.A. Kahn, Gas Absorption and Degasification of Cast Monel, *Trans. AFS*, 1947, p 558
65. T.R. Ostrom, P.K. Trojan, and R.A. Flinn, Gas Evolution From Copper Alloys—A Summary, *Trans. AFS*, 1981, p 731
66. T.R. Ostrom, P.K. Trojan, and R.A. Flinn, The Effects of Tin, Aluminum, Nickel, and Iron on Dissolved Gases in Molten Copper Alloys, *Trans. AFS*, 1980, p 437
67. *Casting Copper Base Alloys*, American Foundrymen's Society, 1984
68. T.R. Ostrom, R.A. Flinn, and P.K. Trojan, Gas Content of Copper Alloy Melts—Test Equipment and Field Test Results, *Trans. AFS*, 1977, p 357
69. T.R. Ostrom, P.K. Trojan, and R.A. Flinn, Dissolved Gases in Commercial Copper Base Alloys, *Trans. AFS*, 1975, p 485
70. R.J. Cooksey and R.W. Ruddle, New Techniques in Degassing Copper Alloys, *Trans. AFS*, Vol 68, 1960
71. P.K. Trojan, S. Suga, and R.A. Flinn, Influence of Gas Porosity on Mechanical Properties of Aluminum Bronze, *Trans. AFS*, 1973, p 552
72. R.J. Kissling and J.F. Wallace, Fluxing of Copper Alloy Castings, *Foundry*, 1964
73. M.G. Neu and J.E. Gotheridge, Fluxing and Deoxidation Treatments for Copper, *Trans. AFS*, 1956, p 616
74. Y.T. Hsu and B.O. Reilly, Impurity Effects in High Conductivity Copper, *J. Met.*, Dec 1977, p 21
75. "Lithium Cartridges for Treatment of Copper and Copper Alloys," Product Bulletin 304, Lithium Corporation of America, 1986
76. R.C. Harris, Deoxidation Practice for Copper, Shell-Molded Castings, *Trans. AFS*, 1958, p 69
77. J.L. Dion, A. Couture, and J.O. Edwards, "Deoxidation of Copper for High Conductivity Castings," Report MRP/PMRL-78-7(J), Physical Metallurgy Research Laboratories, CANMET, April 1978
78. B.L. Tiwari, Demagging Processes for Aluminum Alloy Scrap, in *Light Metals*, The Metallurgical Society, 1982, p 889
79. G. Dube, TAC—A Novel Process for the Removal of Lithium and Other Alkalis in Primary Aluminum, in *Light Metals*, The Metallurgical Society, 1983, p 991
80. B. Gariepy and G. Dube, TAC, A New Process for Molten Aluminum Refining, in *Proceedings of Aluminum Technology '86*, Institute of Metals, 1986
81. B. Gariepy, G. Dube, C. Simoneau, and G. LeBlanc, The TAC Process: A Proven Technology, in *Light Metals*, The Metallurgical Society, 1984, p 1267
82. H. Yu, A Study of the Reaction Kinet-

ic in the Removal of Calcium From Aluminum Melt in a Batch Stirred Tank Reactor, in *Advances in Aluminum Casting*, Proceedings of the Pacific Northwest Metals & Mineral Conference, 1982
83. J.G. Stevens and H. Yu, Mechanism of Sodium, Calcium, and Hydrogen Removal From an Aluminum Melt in a Stirred Tank Reactor—The Alcoa 622 Process, in *Light Metals*, The Metallurgical Society, 1988, p 437
84. J.D. Bormand and K. Buxman, Dufi: A Concept of Metal Filtration, in *Light Metals*, The Metallurgical Society, 1986, p 1249
85. B.L. Tiwari, B.J. Howie, and R.M. Johnson, Electrolytic Demagging of Secondary Aluminum in a Prototype, *Trans. AFS*, 1986, p 385
86. M.C. Mangalick, Demagging in the Secondary Aluminum Industry, *J. Met.*, June 1975, p 6
87. D.V. Neff, The Use of Gas Injection Pumps in Secondary Aluminum Metal Refining, in *Recycle and Secondary Recovery of Metals*, Conference Proceedings, AIME, 1985, p 73
88. B. Lagowski, Magnesium Loss During Chlorination of Aluminum Melts, *Trans. AFS*, 1969, p 205
89. L. Smith, The Development of the Channel Injection Furnace for Demagging in the Secondary Aluminum Industry, *J. Met.*, Sept 1978, p 15
90. T.R.A. Davey, The Physical Chemistry of Lead Refining, in *Proceedings of the Symposium on Lead-Zinc-Tin*, AIME, 1980, p 477
91. J. Blanderer, The Refining of Lead by Oxygen, *J. Met.*, Dec 1984, p 53
92. G.K. Sigworth, Fundamentals of Grain Refining in Aluminum Alloy Castings, in *International Molten Aluminum Processing*, Conference Proceedings, American Foundrymen's Society, 1986, p 75
93. L.F. Mondolfo, Grain Refinement in the Casting of Non Ferrous Alloys, in *Grain Refining in Castings and Welds*, AIME, 1983
94. D.A. Granger, "Practical Aspects of Grain Refining Aluminum Alloys Melts," Paper presented at the International Seminar on Refining and Alloying of Liquid Aluminum and Ferro Alloys, Trondheim, Norway, Aug 1985
95. R.J. Kissling and J.F. Wallace, Grain Refinement of Aluminum Castings, *Foundry*, 1963
96. A. Banerji and W. Reif, Development of Al-Ti-C Grain Refiner Containing TiC, *Metall. Trans. A*, Dec 1986, p 2127
97. R.M. Kotschi and C.R. Loper, Jr., Grain Refinement in Cast Aluminum Alloys, *Trans. AFS*, 1977, p 425
98. J. Pearson, M.E.J. Birch, and D. Hadlet, Recent Advances in Aluminum Grain Refinement, in *Solidification Technologies in the Foundry and Cast House*, Proceedings of the Conference on Applied Metallurgy and Metals Technology, The Metals Society, Sept 1980
99. G.K. Sigworth and M.M. Guzowski, Grain Refining of Hypoeutectic Al-Si Alloys, *Trans. AFS*, 1985, p 907
100. D. Apelian and J.A. Cheng, Effect of Processing Variables on the Grain Refinement and Eutectic Modification of Al-Si Foundry Alloys, in *Proceedings of the International Molten Aluminum Processing Conference* (City of Industry, CA), American Foundrymen's Society, 1986, p 179
101. F.R. Mollard, W.G. Lidman, and J.C. Bailey, Systematic Selection of the Optimum Grain Refiner in the Aluminum Cast Shop, in *Light Metals*, The Metallurgical Society, 1987, p 749
102. J. Pearson and M.E.J. Birch, Improved Grain Refining With TiBAl Alloys Containing 3% Titanium, in *Light Metals*, The Metallurgical Society, 1984, p 1217
103. J. Pearson and M.E. Birch, Effect of the Titanium:Boron Ratio on the Efficiency of Aluminum Grain-Refining Alloys, *J. Met.*, Nov 1977, p 27
104. E.F. Emley, *Principles of Magnesium Technology*, Pergamon Press, 1966
105. A. Cibula, Grain Refining Additions for Cast Copper Alloys, *J. Inst. Met.*, Vol 82, 1953, p 513
106. G.C. Gould, G.W. Form, and J.F. Wallace, Grain Refinement of Copper, *Trans. AFS*, 1960, p 258
107. R.J. Kissling and J.F. Wallace, Grain Refinement of Copper Alloy Castings, *Foundry*, June-July, 1963
108. A. Couture and J.O. Edwards, Grain Refinement of Sand Cast Bronzes and Its Influence on Their Properties, *Trans. AFS*, 1973, p 453
109. P.B. Crossley and F.L. Mondolfo, The Modification of Aluminum-Silicon Alloys, *Mod. Cast.*, 1966, p 89
110. G.K. Sigworth, Theoretical and Practical Aspects of the Modification of Al-Si Alloys, *Trans. AFS*, 1983, p 7
111. B. Gallois and G.K. Sigworth, An Analysis of Silicon Eutectic Modification, in *Proceedings of the Conference on Thermal Analysis of Molten Aluminum* (Rosemont, IL), American Foundrymen's Society, 1984, p 101
112. A.J. Clegg, Hypereutectic Alloys Primary Silicon Refinement, in *Proceedings of the International Molten Aluminum Processing Conference* (City of Industry, CA), American Foundrymen's Society, 1988, p 269
113. G.K. Sigworth, Observations on the Refinement of Hypereutectic Al-Si Alloys, *Trans. AFS*, 1987
114. B.L. Tuttle, Principles of Thermal Analysis for Molten Metal Process Control, in *Proceedings of the Conference on Thermal Analysis of Molten Aluminum* (Rosemont, IL), American Foundrymen's Society, 1984, p 1
115. D. Apelian and J. Cheng, "Thermal Analysis of Al-Si Foundry Alloys as a Means of Quality Assurance in the Cast Shop," Drexel University
116. N.G. Walker, Thermal Analysis in the Aluminum Casting Industry, in *Proceedings of the Conference on Thermal Analysis of Molten Aluminum* (Rosemont, IL), American Foundrymen's Society, 1984, p 155
117. B.N. Closset, Thermal Analysis Techniques to Monitor the Melt Quality of Aluminum Castings, in *Proceedings of the International Molten Aluminum Processing Conference* (City of Industry, CA), American Foundrymen's Society, 1986, p 219
118. T.J. Hurley, Using Electrical Conductivity and Ultrasonics to Determine Modification in Al-Si Alloys, *Trans. AFS*, 1986, p 159
119. B. Closset, Modification and Quality of Low Pressure Aluminum Castings, Paper 76, *Trans. AFS*, 1988
120. A. March, High Volume Permanent Mold Casting With Calcium Modification—Thirty Years of Use, in *Proceedings of the International Molten Aluminum Processing Conference* (City of Industry, CA), American Foundrymen's Society, 1986, p 245
121. B.M. Closset and S. Kitaoka, Evaluation of Strontium Modifiers for Al-Si Casting Alloys, *Trans. AFS*, 1987
122. M.O. Pekguleryuz, B. Closset, and J.E. Gruzleski, The Dissolution of Metallic Strontium in Liquid Aluminum and Liquid A356 Alloy, *Trans. AFS*, 1984, p 109-118
123. M.O. Pekguleryuz and J.E. Gruzleski, Conditions for Strontium Master Alloy Addition to A356 Melts, Paper 15, *Trans. AFS*, 1988
124. T.J. Hurley and R.G. Atkinson, Effects of Modification Practice on Aluminum A-356 Alloys, *Trans. AFS*, 1985, p 291
125. J.E. Gruzleski, N. Handiak, H. Campbell, and B. Closset, Hydrogen Measurement by Telegas in Strontium Treated A356 Melts, *Trans. AFS*, 1986, p 147
126. N. Handiak, J.E. Gruzleski, and D. Argo, Sodium, Strontium, and Antimony Interactions During the Modification of (A356) Alloys, *Trans. AFS*, 1987
127. J. Bildstein, J. LaCroix, P. Netter, P. Prevost, and G. Zahorke, Degassing Casting Alloys Ingots Using the Alpur System, in *Light Metals*, The Metallurgical Society, 1988, p 431
128. F. Dimayuga, N. Handiak, and J.E.

Gruzleski, The Degassing and Regassing Behavior of Strontium Modified A356 Melts, Paper 17, *Trans. AFS*, 1988
129. D.V. Neff, Modern Systems for Transport of Molten Metal, *Trans. Soc. Die Cast. Eng.*, 1985
130. D.V. Neff, Application of Molten Metal Pumping Technology in Copper Alloy Melting and Casting, *Trans. AFS*, 1986
131. D.V. Neff, Molten Metal Pumping, Degassing, and Filtration Techniques to Produce High Quality Aluminum Castings, in *Advanced Casting Technology*, ASM INTERNATIONAL, 1987
132. L.J. Gauckler and M.M. Wacher, Industrial Application of Open Pore Ceramic Foam for Molten Metal Filtration, in *Light Metals*, The Metallurgical Society, 1985, p 126
133. C.J. Simensen and G. Berg, A Survey of Inclusions in Aluminum, *Aluminum*, Vol 56, 1980, p 335
134. J. Langerweger, Nonmetallic Particles as the Cause of Structural Porosity, Heterogeneous Cell Structure and Surface Cracks in DC Cast Al Products, in *Light Metals*, The Metallurgical Society, 1981, p 685
135. C.J. Simensen, The Effect of Melt Refining Upon Inclusions in Aluminum, *Metall. Trans.*, 1980
136. C.J. Simensen and U. Hartvedt, "Analysis of Oxides in Aluminum By Means of Melt Filtration," Center for Industrial Research, Trondheim, Norway, 1984
137. C.J. Simensen, Sedimentation Analysis of Inclusions in Aluminum and Magnesium, *Metall. Trans. B*, 1981, p 733
138. S.K. Miller, Inclusion Control of Die Casting Alloys, *Die Cast. Eng.*, Sept 1984, p 54
139. J.D. Rutherford, G.P. Walker, and C.A. Loong, Ceramic Foam Filtration of Zinc Alloys, *Trans. Soc. Die Cast. Eng.*, 1985
140. L.C. Blayden and K.J. Brondyke, Alcoa 469 Process, *J. Met.*, 1974, p 25
141. D. Apelian and R. Mutharason, Filtration: A Melt Refining Method, *J. Met.*, Sept 1980, p 14
142. C.E. Eckert, R.E. Miller, D. Apelian, and R. Mutharasan, Molten Aluminum Filtration: Fundamentals and Models, in *Light Metals*, The Metallurgical Society, 1984, p 1281
143. R. Mutharasan and D. Apelian, Flow Behavior of Liquid Deformable Inclusions in Packed Beds, in *Light Metals*, The Metallurgical Society, 1983, p 963
144. R. Mutharasan, D. Apelian, and C. Romanowski, A Laboratory Investigation of Aluminum Filtration Through Deep-Bed and Ceramic Open Pore Filters, *J. Met.*, 1981, p 12
145. D.V. Neff, The Filtering and Degassing of Aluminum Die Casting Alloys, *Die Cast. Eng.*, Sept 1986
146. D.V. Neff and E.P. Stankiewicz, The Multicast Filtration System, in *Light Metals*, The Metallurgical Society, 1986, p 821
147. D.E. Groteke, Eliminating Hard Spots From Aluminum Die Castings, *Die Cast. Eng.*, Sept 1985, p 16
148. D.E. Groteke, Point of Pour Filtration Systems for Aluminum Alloys, in *Proceedings of the International Molten Aluminum Processing Conference*, American Foundrymen's Society, 1986, p 321
149. Filter Product Bulletin, Selee Corporation
150. J.M. Stamper, "Application of Ceramic Foam Molten Metal Filters," Technical Paper, Metals/Materials Technology Series, American Society for Metals, 1982
151. D.E. Groteke, A Production Filtration Process for Aluminum Melts, *Trans. Soc. Die Cast. Eng.*, 1983
152. "Molten Metal Filtration," Product Bulletin, Metaullics Systems, 1983
153. M. Sahoo, J.R. Barry, and K. Kleinschmidt, Use of Ceramic Foam Filters in the Brass and Bronze Foundry, *Trans. AFS*, 1981, p 611
154. T.L. Mansfield and C.L. Bradshaw, Ultrasonic Inspection of Molten Aluminum, *Trans. AFS,* 1985, p 317
155. R.N. Dokken, T.G. Kinisky, and K.B. Reuter, Aluminum Quality Evaluation Using High Resolution Image Analysis, in *Interactions, ET '88*, Vol 2, Proceedings of Fourth International Aluminum Extrusion Technology Seminar, Chicago, April 1988
156. R. Davis and R.N. Dokken, Product Quality Improvements Through In-Line Refining with SNIF, in *Light Metals*, The Metallurgical Society, 1987, p 711

Automatic Pouring Systems

Jack Thielke, ASEA Brown Boveri, Inc.

AUTOMATIC POURING EQUIPMENT includes a wide range of mechanized pouring methods and specialized equipment designed for automatic pouring. Traditional metal transfer from the molten metal supply to the mold is done with open-top ladles ranging from hand held to multi-ton, crane-supported, mechanized tilt ladles. Underpouring from a ladle, either through the incorporation of a teapot spout or bottom stopper rod and nozzle, reduces slag inclusions in castings (Fig. 1). The first semiautomatic pouring units used controlled stopper rods in bottom-pour ladles mounted above indexing molding lines to pour into molds indexed into position beneath the ladles (Fig. 2).

Because the logical application of automatic pouring is in conjunction with automatic molding equipment in foundry applications, the synchronization of controls, interlocking, and safety devices must be considered in the control system. The *x*- and *y*-axis positioning of the pour stream with respect to the pouring basin must be adjustable to compensate for variations in mold production.

Benefits of Automatic Pouring Systems

The benefits of automatic pouring soon became evident and directed the development of the various equipment types currently available. The benefits fall into six categories:

- Increased productivity
- Increased yield
- Reduced scrap costs
- Improved quality control
- Improved process control
- Improved working conditions

Increased Productivity. Automated pouring systems do not necessarily reduce the manpower requirements in all applications, but they do increase the productivity of the personnel involved in the pouring operation. The productivity improvement is related to the consistency of pours from one mold to the next. Manual pouring is subject to inconsistent results due to variations in the workers' performance on repetitive tasks and the learning curves associated with changes in patterns. Modern automatic pouring systems pour faster and more efficiently than any manual system. The consistent control of pour rates and metal levels within the mold during automatic pouring allows operational and design savings that result in increased operational efficiency.

Increased Yield. The most significant benefit of automatic pouring is the increase in yield due to the elimination of nonproductive metal pouring. Nonproductive pouring naturally includes overpours and spills, but also includes the typically larger size of the pour basin and the resultant increase in sprue weight required for manual pouring. The pigging of manual pouring ladle heels is also eliminated. Consistent, controllable pour rates may allow for the design of less complex gating and riser systems as well as for simultaneous multiple-pour streams in the same mold.

Reduced Scrap Costs. The reduction of scrap and the associated costs directly related to the incorporation of automatic pouring include the reduction of slag inclusions in castings and the reduction of pouring failures. Pouring failures include underpours, interrupted pours, and casting defects due to irregular pour rates.

Improved Quality Control. Casting quality is improved in automated systems because of the previously mentioned benefits as well as the benefits of better temperature control within the pouring vessel and more consistent metallurgy. Both of these benefits are due in part to the buffer effect of the volume of the pouring ladle and in part to a more continuous pouring operation. Alloy and inoculant additions can be automatically fed into the pouring ladle or metal pour stream.

Improved Process Control. The improvements in process control include the benefits that occur before, during, and after the pouring process. Melt system pouring temperatures can sometimes be reduced when automatic pouring equipment is used because of the reduction in temperature losses before pouring. Even slight reductions in

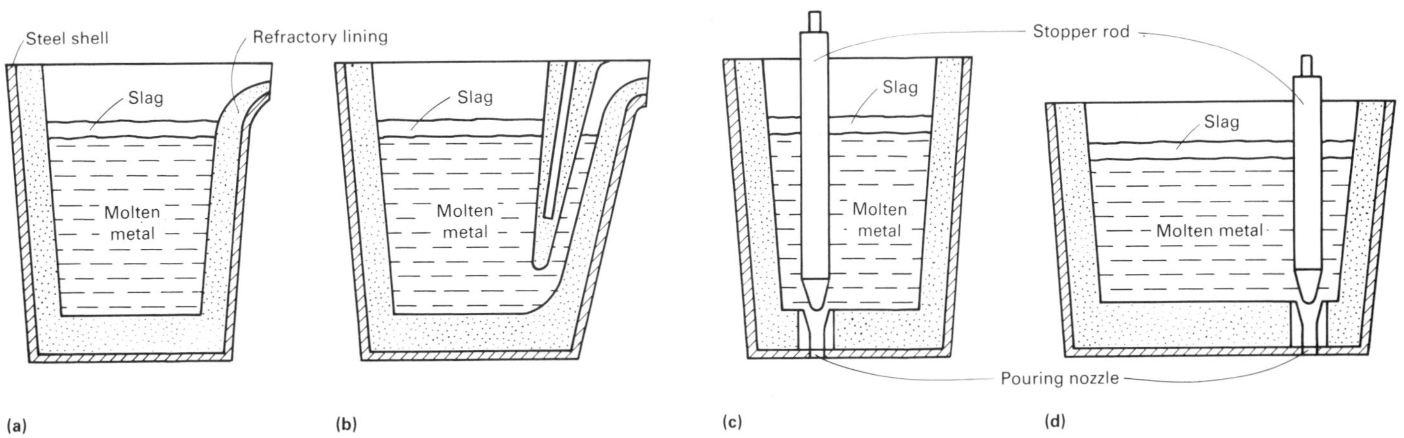

Fig. 1 Four types of pouring ladles. (a) Open lip-pour ladle. (b) Teapot ladle. (c) and (d) Bottom-pour ladles

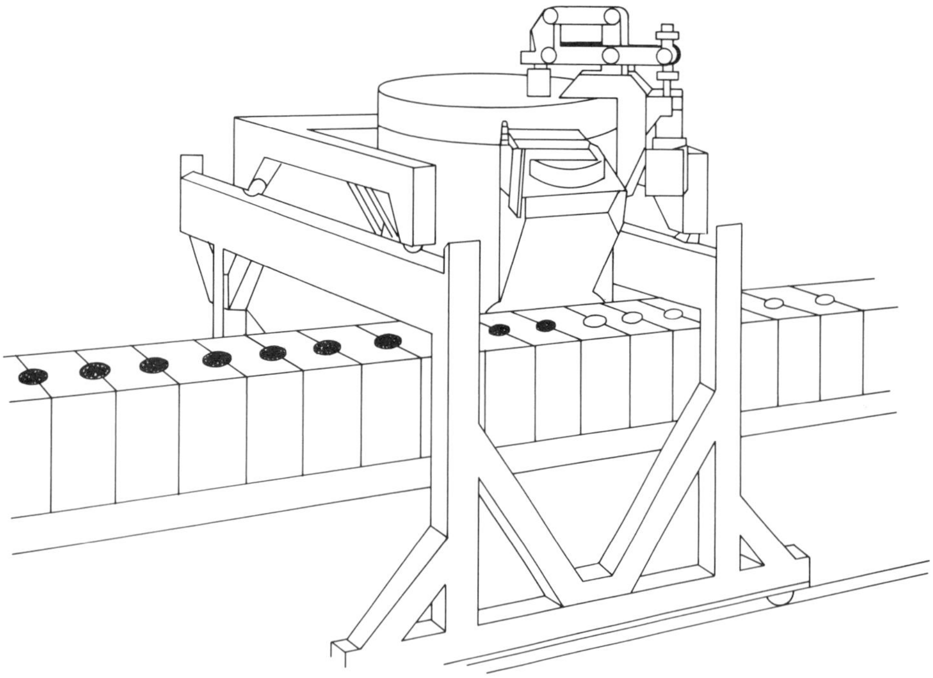

Fig. 2 Stopper rod pouring into flaskless molds on an indexing pouring line

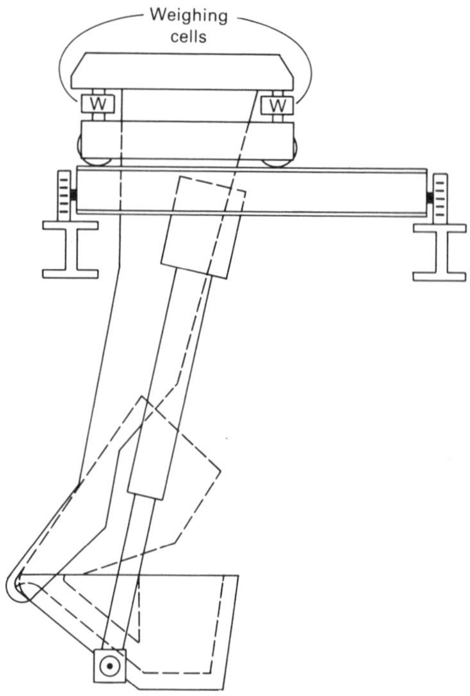

Fig. 3 Schematic showing the operation of a mechanized pouring ladle

melt temperatures can result in significant energy and refractory cost savings. Frequent pattern changes are made more effective with automatic pouring systems. Microprocessor-based process control systems can provide both production and inventory information for defect identification and tracking, as well as statistical data on molten metal and casting production.

Improved Working Conditions. Benefits relating to improved working conditions fall into three areas: health and safety, environmental, and operational. The health and safety benefits include the isolation of the operator and the metal pouring. This protects the operator from the effects of metal spatter, gases, and smoke. Such protection is particularly important when pouring lead-bearing alloys. Environmental benefits include those mentioned above as well as the natural concentration of areas requiring fume collection. Operational benefits include more variability in worker tasks as well as isolation from the pouring area.

Methods of Automatic Pouring

There are two basic methods of automatic pouring: mechanized ladle pouring and direct pouring. Automated mechanized ladles are used to transfer measured amounts of metal from a storage vessel or a controlled pouring unit to the mold. Automated dip and pour mechanized ladles and robotic ladles are used for aluminum and other nonferrous pouring. Mechanized ladles and pouring robots are programmed to duplicate pouring sequences. The use of a mechanized ladle allows for in-ladle alloying and inoculation before pouring. Dripping of molten metal onto the mold is eliminated when a mechanized ladle is used in combination with a stopper rod and nozzle pouring unit. This is very important when pouring lost foam molds. Ladle tilting about an axis in line with the pouring lip can reduce molten metal stream travel during pouring. Mechanized ladles generally have capacities matched to single mold requirements and include load cell weighing systems and controlled tilt rates (Fig. 3).

In the case of direct pouring, metal is metered directly into the mold from the pouring vessel with the appropriate controls to ensure the desired results. The capacities of mechanized pouring ladles are based on anticipated metal requirements. The dispensing of a metal stream from a ladle at a specific point requires that the nozzle or pouring lip not move, relative to the mold, during pouring (Fig. 4).

One method of synchronized pouring on a continuously moving mold line may require the use of a pouring vessel to measure predetermined weights of metal and either one or two (depending on line speed) mechanized pouring ladles. Each of the mechanized pouring ladles must be capable of x- and y-axis travel and synchronization with the pouring line (Fig. 5).

Direct bottom-pour ladles use either slide gates or stopper rods and nozzles to control and stop the flow of metal to the mold. Slide gates and nozzles are most commonly used to control flow rates on bottom-pour ladles for continuous and semicontinuous strand casting operations (see the article "Continuous Casting" in this Volume). Slide gates offer more reliable shutoff than stopper rod valves (Fig. 6); however, service life of the slide gate can be shorter than that of the stopper rod. Stopper rods and nozzles offer more control of pour rates than slide gate valves and are less expensive. For these reasons, bottom-mounted stopper rods and nozzles are more common in foundry pouring operations.

Electric Heating of Pouring Vessels

The metal temperature can be controlled, the heat losses of a pouring vessel replaced

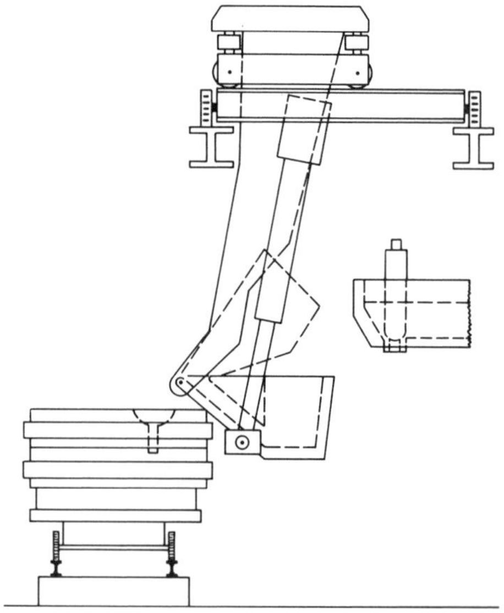

Fig. 4 Schematic of automatic mechanized pouring system

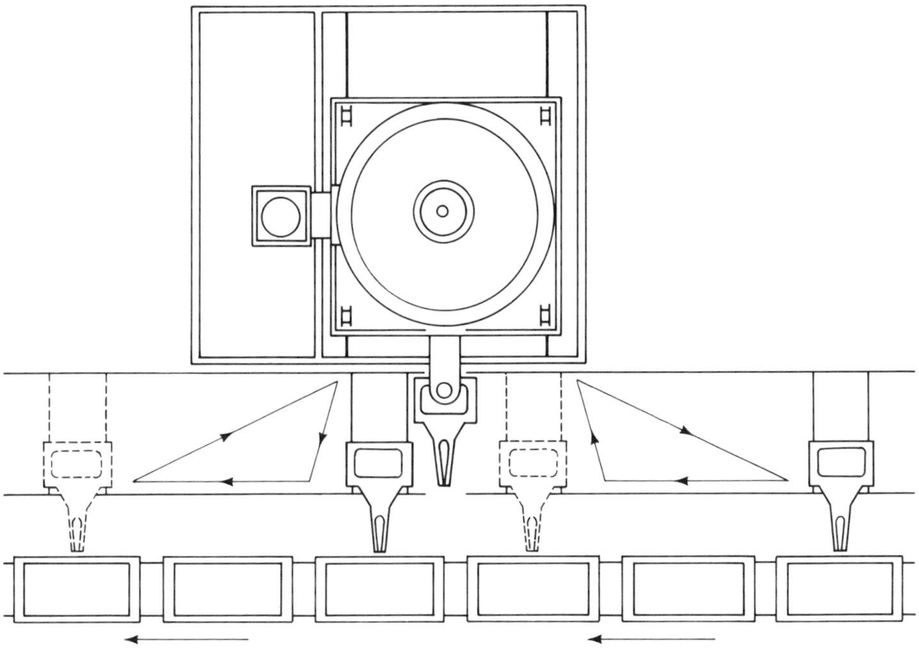

Fig. 5 Synchronized pouring on a continuously moving line

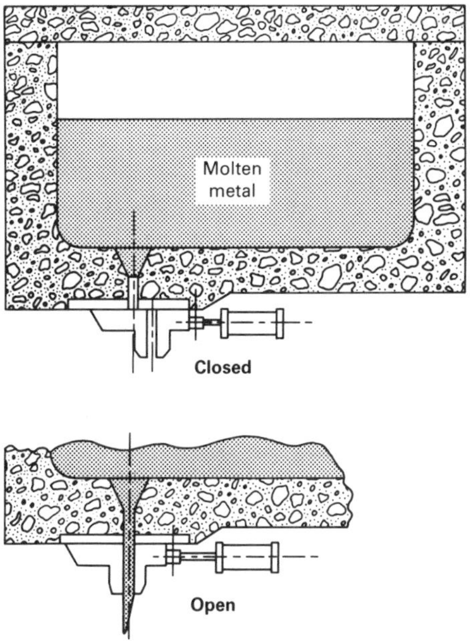

Fig. 6 Schematic of a slide gate pouring system

and the capacity of the vessel increased with the addition of electric heating devices. The holding of molten metal during extended pouring interruptions is possible because of the replacement of heat losses that would reduce the metal temperature. Metal chemistry is more consistent in large capacity pouring vessels, and other improvements in melting and pouring process control are obtained. Electric heating methods will be discussed below.

Electric Resistance Heating. Heat is produced by passing an electric current through a resistance heating element. The heating element may be in the form of a coil surrounding a crucible or in the form of a radiating element or glow bar positioned over the molten metal. In either case, the heat produced replaces heat loss within the vessel, thus maintaining molten metal temperature.

Electric Induction Heating. Heat is produced through the generation of electric currents within the molten metal. Magnetic induction generates electric currents directly into the molten metal, and these induced currents produce heat by passing through the resistance of the molten metal. There are two methods of applying this principle: channel induction and coreless induction.

Channel Induction. A channel or loop of molten metal is passed through a magnetic core that is very similar to a transformer core. The induction coil acts as the transformer primary, and the molten metal acts as the transformer secondary. Magnetic forces acting on the metal in the channel cause molten metal circulation through the loop.

Coreless Induction. An induction coil surrounds the crucible containing the molten metal and induces circulatory currents into the molten bath. Magnetic forces acting on the metal in the crucible cause metal circulation within the crucible. The electric induction heating of a pouring vessel provides metal stirring to reduce alloy segregation and the capability of superheating the metal to a temperature higher than the initial charging temperature before pouring.

Electrically Heated Pouring Furnaces

Radiant element, electric resistance heated pouring furnaces are used in both nonferrous and ferrous pouring applications (see the articles in the Sections "Ferrous Casting Alloys" and "Nonferrous Casting Alloys" in this Volume). For nonferrous, low-temperature applications, metal or silicon carbide heating elements are used. For ferrous high-temperature applications, graphite rod heating elements are used (Fig.

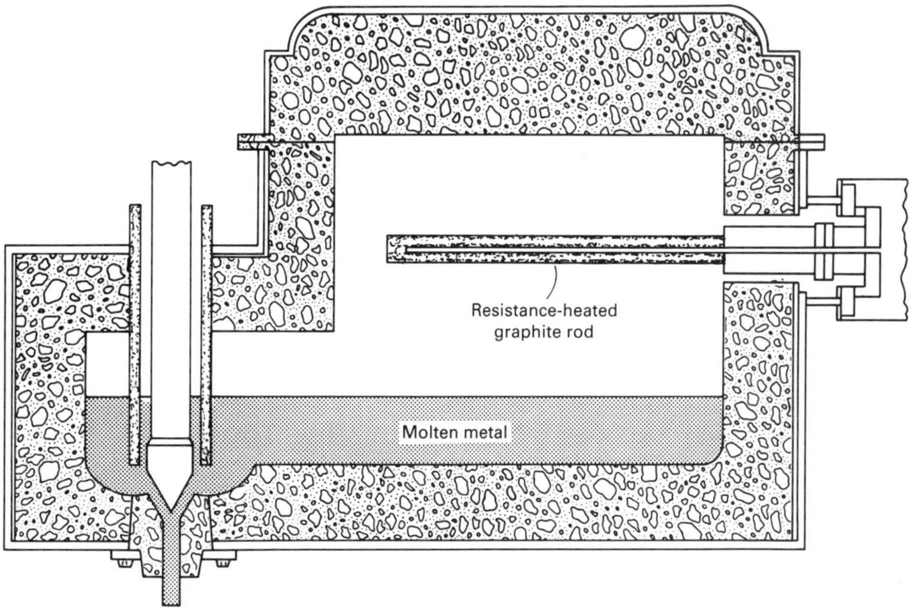

Fig. 7 Pouring furnace heated by resistance-heated graphite rods

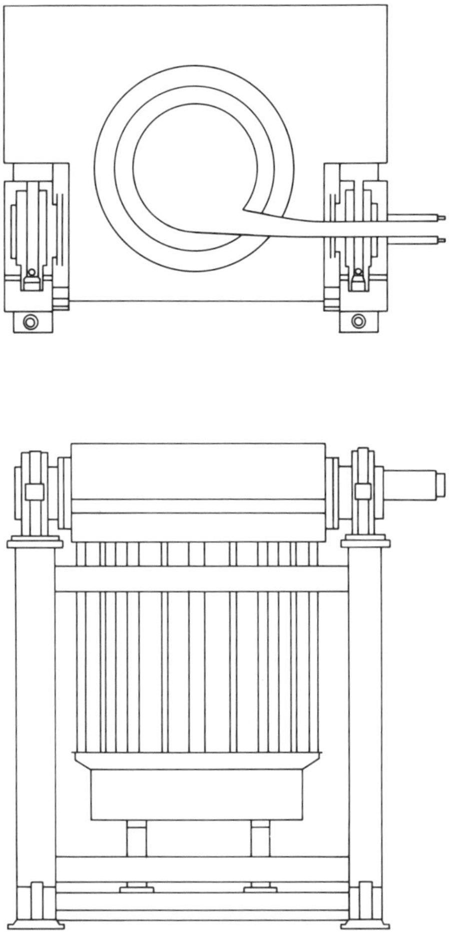

Fig. 8 Schematic of a trunion pouring furnace

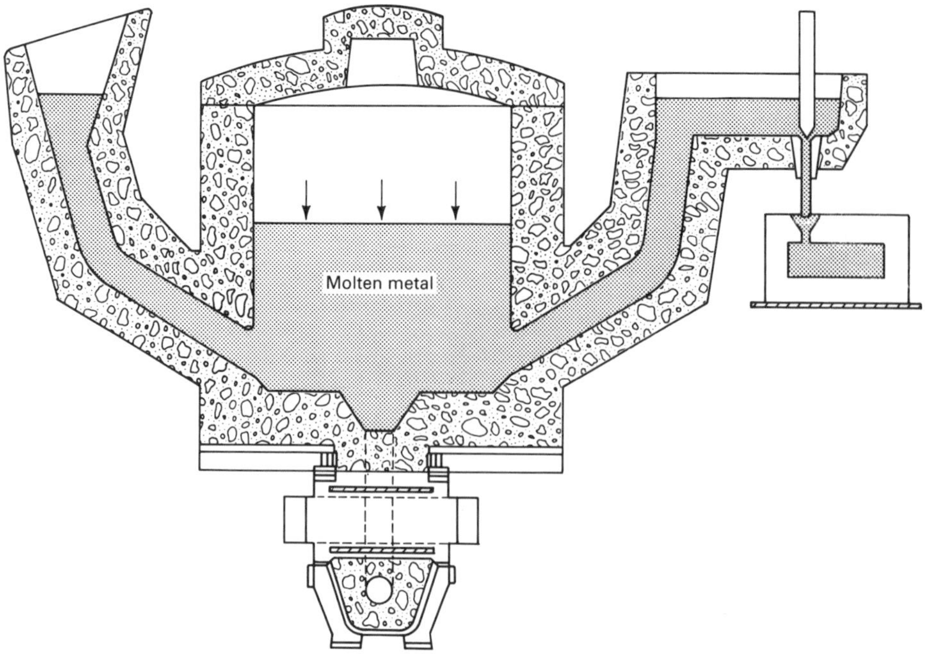

Fig. 9 Schematic of pressurized pouring furnace

7). Radiant-heated furnaces are typically bottom-pour units with stopper rods and nozzles mounted in the furnace bottom or in a detachable pour box. The furnace can be tilted to drain the pour box and to contain the molten metal beneath the radiant heat source.

Crucible-type electric resistance heated and coreless induction-heated furnaces are most commonly used in nonferrous pouring applications. These applications vary from the simple dip and pour mechanized ladle systems to the underpour stream discharge, axial-pour, and trunion-pour furnaces (Fig. 8). Underpour stream discharge in aluminum pouring for continuous and semicontinuous casting processes uses the oxide skin on the metal surface to protect the metal stream from further oxidation while transferring metal from the melting furnace to the automatic pouring ladle.

Channel-type electric induction heated pouring furnaces are sealed and designed to use gas pressure to discharge metal from the furnace vessel. Pouring can be accomplished over the lip of a pour spout or through a nozzle, with or without a stopper rod, in a pour box mounted on the furnace vessel. Vessel gas pressure can be varied to control the rate of pouring or to control the level of metal over the pouring nozzle (Fig. 9). Control of the level of metal in the pour box establishes a constant static head pressure above the nozzle and ensures consistent pour rates. Venting of the pressure in the vessel causes the metal in the pour box to drain back into the furnace. Inert gas can be used as the pressure medium in applications where the metal is particularly susceptible to oxidation.

Pouring Control Parameters

In the final analysis, there are only two variables in pouring: rate and duration. The optimum pour rate is the maximum rate at which a mold will accept metal, and the optimum duration is the time from the start of the pour until the mold will no longer accept metal. Other important considerations in automatic pouring systems include the relative positioning of the mold and pour spout or nozzle and the sequencing of the pour.

In manual pouring, the worker compensates for variables in the mold, ladle, and metal while pouring to produce the best results for each pour. Duplication of this effort is the challenge of automatic pouring.

Control Schemes

Satisfactory pouring results can be obtained in some automatic pouring applications without complex electronic control systems. Some requirements can be met by establishing a flow rate and simply timing the pour. A slightly more complex system includes dual pour rates and pour times. Simple infrared photosensors designed to detect the presence of molten metal can be used to end the pour cycle.

If the exact weight of metal required for a mold is known and a mechanized ladle is weighed while filling, at least one variable is under control. If the movements of the operator controlling the pouring of a mechanized ladle are measured and then stored in a programmable controller, these move-

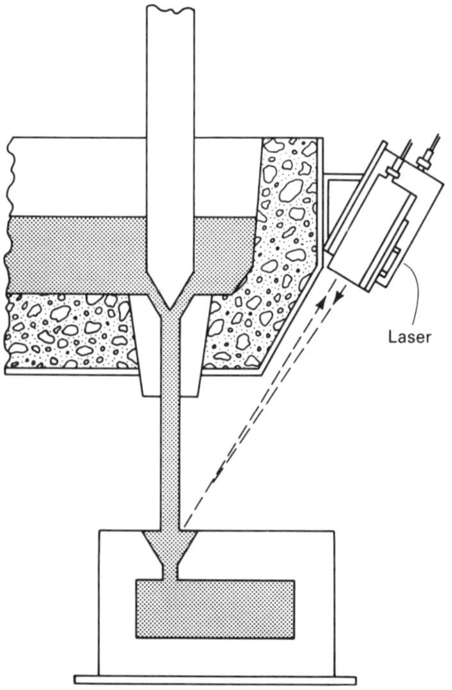

Fig. 10 Schematic showing laser pouring control. See text for details.

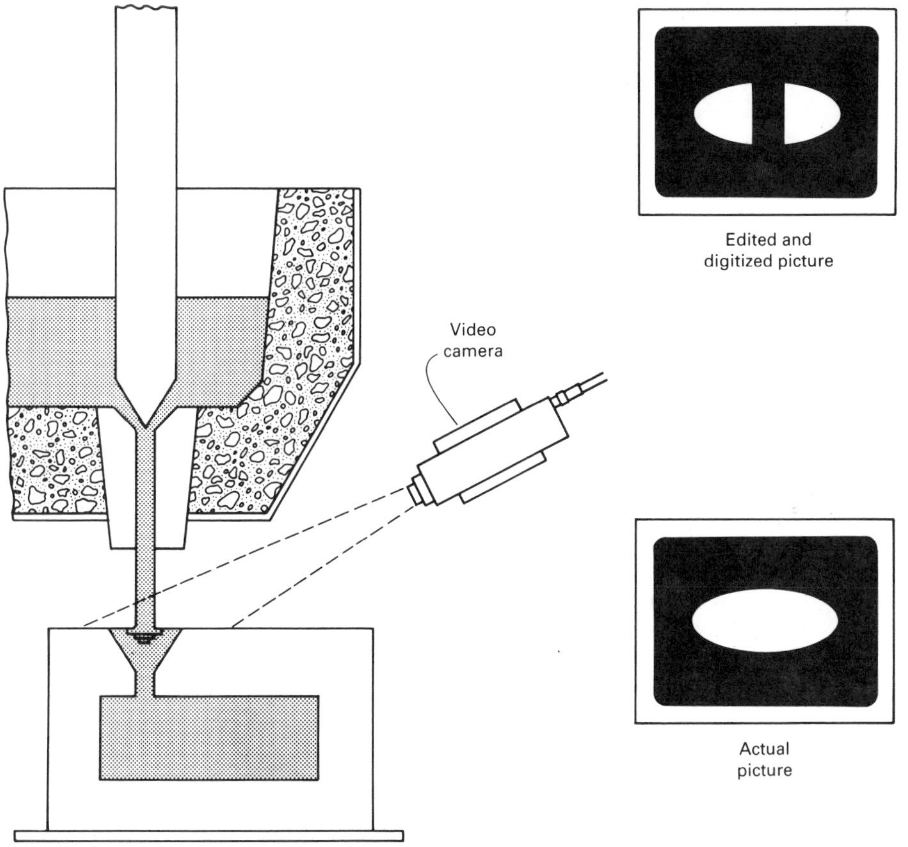

Fig. 11 Schematic of vision-controlled pouring. See text for details.

ments and the results can be duplicated under automatic control. The programming of a controller by duplication of remote control is referred to as a teach-in system. The program is referred to as the pouring profile.

Weighing of a bottom-pour ladle and pouring by weight can also be effective in some cases, although the relatively small capacity of the ladle results in substantial variations in metal head pressure above the stopper and subsequent variations in pour rate. Increasing the capacity of the pouring vessel, facilitated by electric heating, reduces the variation in head pressure. The incorporation of a process control system to compensate for reduced head pressure through stopper positioning by feedback of weight, anticipated weight, or actual bath height is effective in some cases.

If the rate of metal flow through a nozzle is constant for a constant head pressure, the repeated variations in the position of a stopper rod above the nozzle will yield repeatable and controlled pouring rates. A teach-in system is used to program the pouring profile.

If slag buildup on the nozzle and variations in metal head pressure cause flow variations, the system pouring profile program must be corrected by a process control feedback loop. Feedback can be provided by a timing and metal presence sensing system that checks the rate of mold fill or by a more sophisticated control system for mold level measurement.

Mold level measurement and control systems fall into two categories: laser systems and vision systems. Both of these electronic process control systems monitor the level of metal in the pour cup and control the position of the stopper rod to maintain a nearly constant level throughout the pour. This type of feedback system most closely duplicates the efforts of a skilled operator.

The laser reflection angular detection system (Fig. 10) determines the position of a low-power laser beam reflected from the surface of the molten metal in the pouring basin. The laser beam is directed at the pouring basin from an angle. The distance to the molten metal level will affect the incident and reflected angle and therefore the position of the beam on a sensing unit. The comparative value of the beam position on the sensor is used to control the position of the stopper rod; thus, the level of the metal in the pouring basin is controlled.

A vision system uses a video camera to observe the pouring process (Fig. 11). The video signal is displayed on a monitor for the operator and processed electronically in a manner that separates the brightness levels of the molten metal from the other portions of the picture. The molten metal stream is electronically blanked from the composite picture of the metal in the pouring basin and the area of metal fill compared to the actual area of the pouring basin. The comparative value is used to control the position of the stopper rod, and the level of the metal in the pouring basin is controlled.

The most sophisticated pouring systems may incorporate more than one set of controls. The use of integrated control schemes allows the system backup, improved safety, and increased flexibility that are required for pouring a variety of castings. For example, the benefits of mold level measurement and control are enhanced when used in combination with a teach-in duplicated pouring profile in some pouring applications.

Processing of Castings

Shakeout and Core Knockout

Gene J. Maurer, Jr., United States Industries

THE SHAKEOUT OPERATION, that is, the separation of the casting from the mold and core sand, is the first step to finishing a casting after it has solidified and cooled in the mold. Figure 1 shows the position the shakeout operation holds in a typical foundry flow chart. A considerable amount of energy is required to remove the adhering layer of sand and oxide (that formed under the effect of heat from the metal which was poured into the mold) and also to separate the casting from lumps of sand still strongly bonded together.

Regardless of the metal poured, castings must solidify before shakeout can begin. Some castings will crack if shaken out too soon. Others require a controlled cooling rate to attain a specific microstructure. In establishing the time for shakeout, consideration must be given to the metal, pouring temperature, casting size and design, coring, and sand mixture used.

The energy needed to separate the casting from the mold may be supplied by a shaker pan conveyor, vibrating deck or grate, or any other type of vibratory, oscillating, or rotational mechanical action (Fig. 2). For small castings, the vibratory shakeout action may be combined with a means of conveying the casting to the location at which the next operation is performed.

Vibrating conveyors all use some periodic motion of a trough to produce conveying action. Their use has grown for shakeouts and feeders, and even in scrap preheating systems.

In general, the conveyor consists of a pan or trough driven by a vibrating force. The

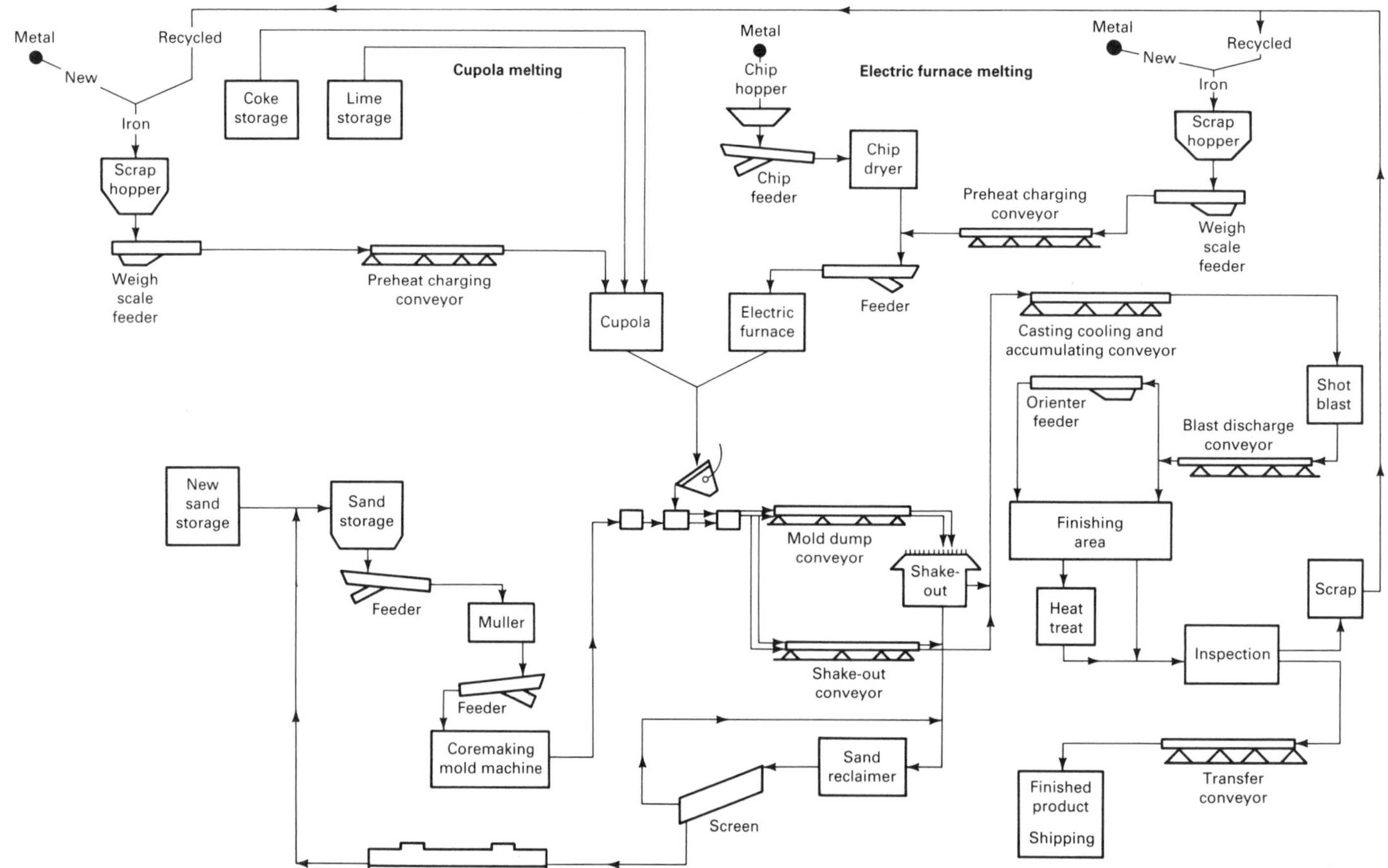

Fig. 1 Typical foundry flow chart

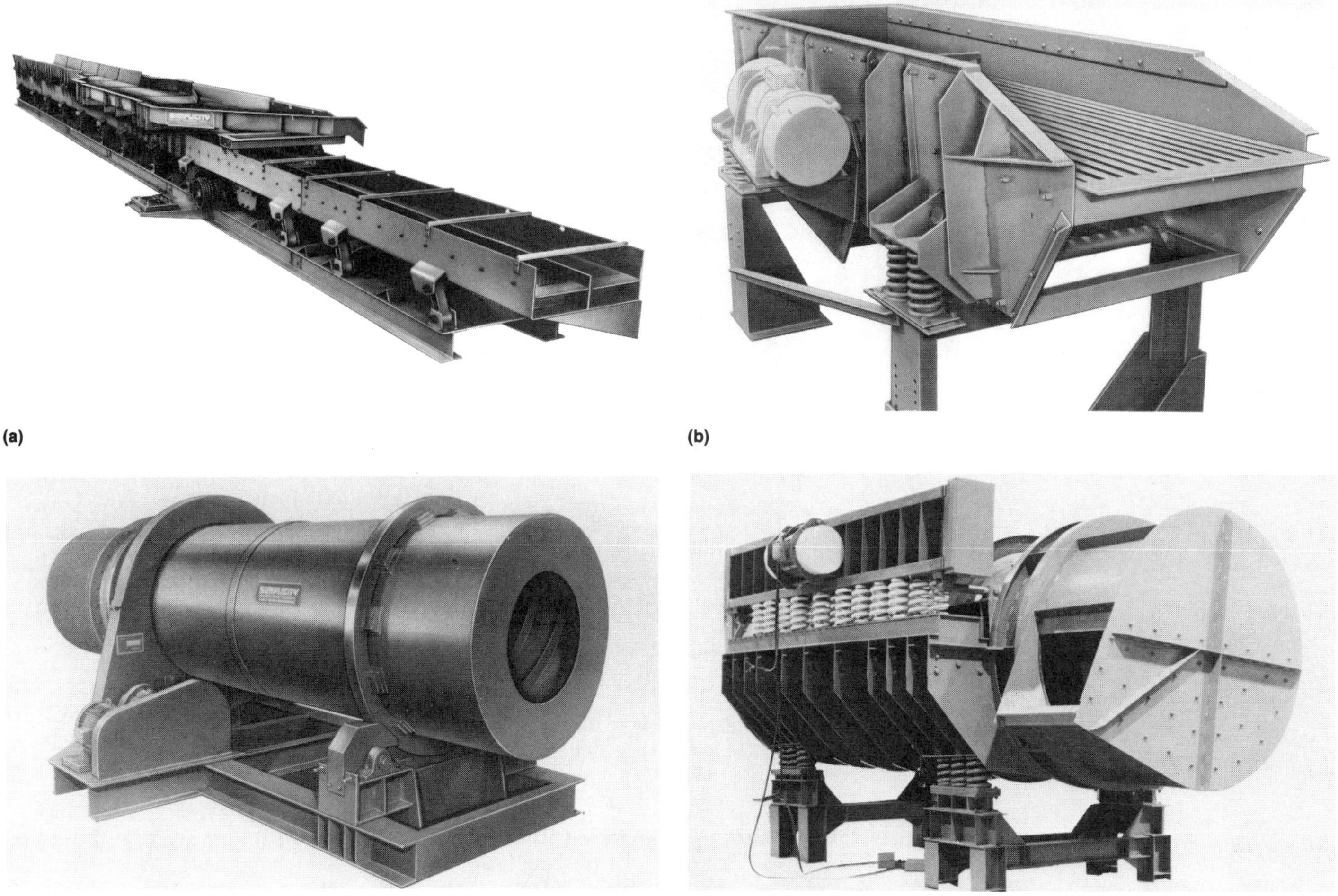

(a) (b) (c) (d)

Fig. 2 Examples of various types of shakeout equipment. (a) Vibrating conveyor. (b) High-frequency shakeout table. (c) Rotary shakeout device. (d) Vibrating drum. Photographs (a), (b), and (c) courtesy of Simplicity Engineering, Inc. Photograph (d) courtesy of General Kinematics Corporation

motion of the pan can be linear, circular, elliptical, or helical. The stroke, frequency, and motion pattern are selected to meet specific requirements, from the high-intensity near-vertical action of a shakeout conveyor to the shuffle action of a transfer-type feeder conveyor.

Direction of flow can be horizontal, up an incline, down a decline, around right-angle turns, or in a complete circle. Vibrating equipment is either a single vibrating mass or multiple vibrating masses, with either brute-force or natural-frequency drives. In a brute force driven system, vibrations are generated continuously and directly by the drive, which is usually electric-motor powered. Outboard rotating weights attached to a concentric shaft produce out-of-balance centrifugal forces required to vibrate the shakeout or conveyor.

Natural- or resonant-frequency systems use springs to amplify a relatively small drive force to create vibrations of large, heavy masses. The force-amplifying spring system can distribute vibrations along great conveyor lengths. The electric-motor power sources are relatively small compared to those of brute-force systems.

For feeder applications, vibrating conveyors that move with a straight-line quick-return motion are available. Castings can be transported without bouncing, and sand can be moved without dusting. More detailed information on the types of shakeout equipment available and their applicability can be found in the section "Green Sand Molding Equipment and Processing" in the article "Sand Processing" in this Volume (see the discussion on shakeout). Information on shakeout practice for specific metals and alloys can be found in the Sections "Ferrous Casting Alloys" and "Nonferrous Casting Alloys" in this Volume.

With the exception of those castings poured from low melting point metals (aluminum or magnesium), the heat of the metal burns out the resin bond in the sand sufficiently for the mechanical action of the shakeout device to cause almost all of the sand to fall away from the casting. Sand remaining on the surfaces of the castings can be removed by abrasive blast cleaning, which is described in the following section, "Blast Cleaning of Castings" in this article.

Core Knockout

Core knockout, or decoring, is the process of removing internal resin-bonded sand cores from castings. Ferrous metal castings are poured at much higher temperatures (~1650 °C, or 3000 °F) than nonferrous castings. Consequently, more energy is required to remove core sand from aluminum castings than from gray iron castings. In gray iron, relatively low energy shaking, vibrating, tumbling, and/or shot blasting is used to force loose core sand out through the casting openings. Core sand in ferrous castings has much less strength because the ambient heat of the metal, as it solidifies in the mold, burns the resin-bonding agents out of the sand. When the resin is burned out, the grains of sand cannot attach to one another, making high-production cleaning of ferrous castings less complex.

The major focus of this section will be the cleaning of aluminum castings, which are molded in a temperature range of 705 to 815 °C (1300 to 1500 °F), approximately one-half the temperature required to cast ferrous metals. This comparatively low molding

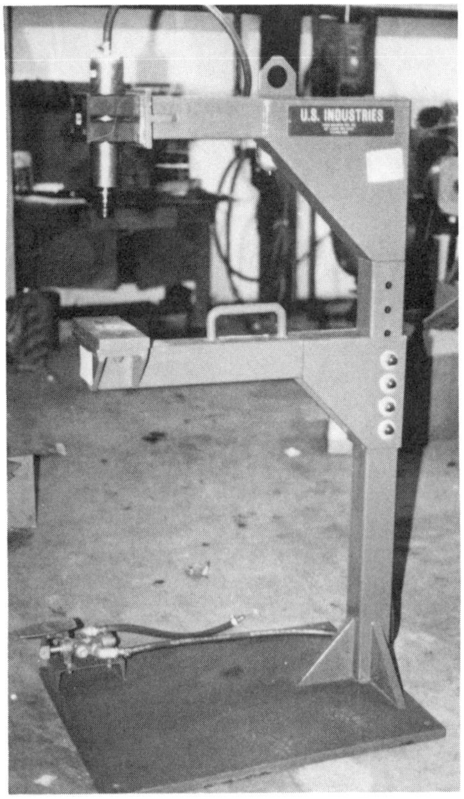

Fig. 3 Vertical core knockout machine. Courtesy of United States Industries

temperature does not allow sufficient heat to transfer from the casting walls to the resin-bonded core sand. Only the cored passageway closest to the metal will show heat penetration and breakdown.

Types of Core Knockout Equipment

Casting complexity, type of resin, type of sand, wall thickness, type of core (solid or hollow), metal to sand ratio, and gating/risering all affect the amount of energy required to remove the sand core from the casting. The most common method used to clean core sand from aluminum castings is mechanical vibration, which is best described as an air tool driving a piston that impacts a chisel. The chisel transmits energy into the risering or gating of the casting. The shock waves move through the metal, creating a momentum that causes the resin-bonded sand to break down. In most instances, the core sand has less inherent strength than does the metal; therefore it breaks down prior to metal fatigue or fracture.

Many foundries use hand-held, vertical, A-frame, or fixtured air tools that operate in the 1000-to-3000-blows-per-minute range. Each type of casting and each core sand is unique. For optimal cleaning, the proper type of vibration and/or blows per minute must be suited for each application. This is not easily determined, however, and because foundrymen normally operate with only one or two settings for all their castings, cracked castings or excessive air tool component failure may result.

The hand-held air tool method of decoring castings involves an operator placing a casting on the floor or on a table and locating the chisel of the air tool on the risering. When the operator pulls the trigger on the tool to allow air to pass through and create vibration, the casting shakes, and core sand breaks down.

The advantages of hand-held air tools include:

- Low cost
- Flexibility (can be used on many different castings)
- Effectiveness for low-volume cleaning applications

The disadvantages are:

- Noise (100 to 120 dB); hearing protection required for personnel
- Vibration, which affects operator's hand, arm, and body
- Required maintenance of air tool
- At times, insufficient energy to clean casting
- Operator required to hold air tool
- Creation of silica dust, which affects operator
- Insufficient speed for high production

The vertical core knockout machine (Fig. 3) involves the use of a specially made air tool that strokes down to clamp the casting for holding and simultaneously vibrates the casting to break down the core sand. When the operator sees that the internal passages are cleared, the foot switch is depressed to stop the vibration and the air tool is retracted from the casting. These tools are commonly referred to as C-frame models because the supporting structure resembles the letter C. This cleaning method uses the vibration-under-compression theory, which is generally applicable for all castings, yet is somewhat slow.

The advantages of vertical core knockout tools include:

- Low cost
- Flexibility (can be effectively used on castings from 2 to 15 kg, or 5 to 35 lb)
- No adverse effect on operator's hands, arms, or body
- Several machines can be run by one operator
- Applicability to low-volume production
- Relatively small amount of floor space required
- Portability
- Easily adjustable frequency and amplitude

The disadvantages are:

- Noise (100 to 120 dB)
- Required maintenance of air tool parts
- Lower effectiveness on larger castings

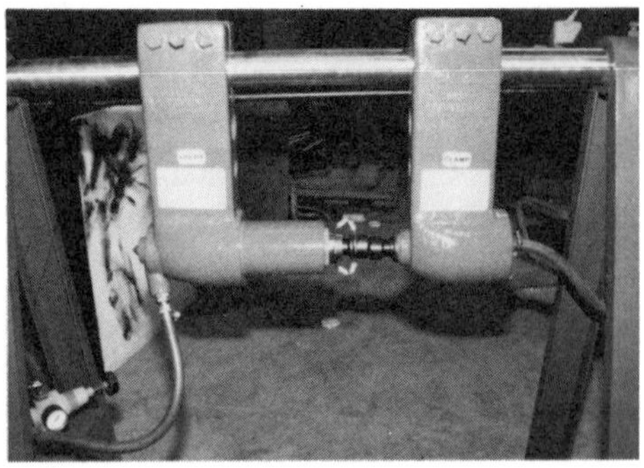

Fig. 4 A-frame core knockout machine. See also Fig. 6(b). Courtesy of United States Industries

- Insufficient speed for high production
- At times, inappropriate for a cleaning method
- Creation of silica dust

The A-frame core knockout machine involves the use of a cast iron clamp/vibrating head and cast iron clamping head mounted on a common shaft and A-frame leg stands and controls (Fig. 4). The cast iron heads are supported by the common mounting shaft and face one another. The left side contains a clamping and vibrating tool, and the right side contains a bumper or an air cushion device to soften the vibration and energy transmission.

To operate the A-frame core knockout machine, an operator positions the casting between the clamp heads and depresses a foot switch to simultaneously hold and vibrate the casting to break down the core sand. This method allows the operator to rotate and probe the casting to reduce cleaning cycle time. When the core sand has been removed, the operator depresses the foot switch to stop vibration and unclamp the casting. The A-frame machines use vibration under compression, which is generally suitable for all types of castings. On this machine, both the clamping pressure and the vibrating pressure are independently adjustable so the vibration can be adapted to suit the casting.

The advantages of A-frame machines include:

- Moderate price
- Flexibility (can be used on both large and small castings, that is, from 2 to 45 kg, or 5 to 100 lb)
- No harmful effect on operator's body
- Several machines can be run by one operator
- Applicability to low- and high-volume production
- Rugged construction to withstand abusive foundry practices
- About 2.5 m^2 (25 ft^2) of floor space required

The disadvantages are:

- Noise (100 to 120 dB)

(a) (b) (c)

Fig. 5 Multiple-section air tool fixtures. (a) Two-station. (b) Three-station. (c) Four-station. Courtesy of United States Industries

- Multiple-unit facilities generate sound waves that can have harmful effects on internal organs
- In medium- to high-production plants, excessive maintenance is required
- High replacement part inventory is necessary
- Castings must be transported to and from special acoustical rooms
- Insufficient speed for high production
- Ineffectiveness on thin-walled heavily cored castings
- Creation of silica dust

The multiple-station air tool fixture involves the use of air tools mounted on a supporting structure with holding and locating blocks to position castings. These fixtures are dedicated to the core knockout of a specific casting. These devices either operate by means of vibration under compression, whereby the casting is clamped and vibrated, or they have a 6 to 10 mm (1/4 to 3/8 in.) gap surrounding the casting, which undergoes vibration. As the casting is impacted, it moves violently within this tight space. This allows more energy to penetrate the casting from many different directions. The major problem with this method is that loading and unloading of the casting can be difficult because of the varying sizes of risers.

The retractable tools employing vibration under compression allow easy loading/unloading and maintenance, yet can reduce cleaning effectiveness. With dedicated fixtures, multiple air tools can be used on a single casting, thus inducing greater amounts of energy to those areas that require it most. Multiple-station fixtures are normally built to clean two (Fig. 5a), three (Fig. 5b), or four castings (Fig. 5c) simultaneously. These core knockout fixtures must be isolated from their support structures by vibration-proof mounts.

Dedicated fixtures are most commonly used on high volume production applications, although certain low to medium production run castings may require dedicated tooling because of the complexity of the casting. These units generate extremely high noise levels; therefore, they are often housed in acoustical-control cabinets that are designed to be portable with interchangeable fixtures (Fig. 6).

The advantages of acoustically enclosed, dedicated fixture units include:

- Low noise levels
- High production and lower per-casting cleaning cost
- Easier cleaning
- Timed cycle
- Interchangeable fixtures
- Sand containment
- Improved safety
- Portability
- Location possible near production line

The disadvantages are as follows:

- High initial capital expenditure
- Acoustical enclosures are a nonincome-producing expense
- Maintenance required for air tool components
- Few available suppliers
- Fixturing dedicated to specific castings (not universal)

New Developments in Core Knockout

Methods that are in the development stage for cleaning both ferrous and nonferrous castings are described below.

(a) (b)

Fig. 6 Acoustical-control cabinets. (a) With three-station dedicated fixture. (b) With A-frame tooling. Courtesy of United States Industries

A high-frequency drive machine involves the use of two motors spinning weights in opposite directions to develop a rapid push-pull movement of a load table. The casting is hydraulically retained on the mounting table. The rapid back and forth movement causes a floating piston to move and contact a strike plate that transmits energy to the casting. This high-frequency drive device is said to develop between 5 and 32 g (~50 and 315 $m \cdot s^{-2}$, or 160 and 1030 $ft \cdot s^{-2}$) of force. The motors operate at between 2900 and 3600 rpm. The optimal amplitude or stroke is between 5 and 10 mm (3⁄16 and 3⁄8 in.). These units have proved to be effective in cleaning some of the most difficult castings. However, they are very loud, and the hydraulic clamping devices have failed in production settings.

The advantages of high-frequency drive machines include:

- Speed
- Adjustability
- Interchangeable fixtures
- Timed cycle

The disadvantages are:

- High cost
- Self-destruction
- Dedicated construction
- Noise (110 to 130 dB)

The high-pressure water blast device incorporates the use of a gimbal-mounted spray gun positioned in a stainless steel enclosure with a high pressure pumping system. The successful operating pressures lie within a 17 to 105 MPa (2.5 to 15 ksi) range. A casting is located on a rotating, tiltable table, and the water is directed at the core material through the spray gun. The operator moves the casting to blast the water at the cored passageways. This system is effective in cleaning ceramic material from investment castings and complex sand-cored castings, and is also effective for removing burnt-in core sand from internal passages.

The advantages of a high-pressure water blast system include:

- Adjustable water pressure
- Water can be focused directly at core
- Safety
- Quiet operation
- Dedicated fixturing can be used

The disadvantages are:

- High cost
- Contaminated water disposal
- Low operating speed
- Maintenance required for nozzle and pump
- Unreliable water recirculation
- Expensive spare parts
- Possible inability to direct water at blind or oblique passageways

In the shot blast cleaning method, castings are hung on a tree and placed in an enclosure. Shot is spun off a wheel at the casting. This method is not very effective for cleaning aluminum castings because the energy is not sufficient to break down the cores and the cycle time is longer than with other methods. Also, mixing sand with spent shot can cause excessive wear to the reclamation system. Nevertheless, some foundrymen do use this process because of the availability of a shot blast machine and/or because the casting requires shot blasting.

With the core bake-out method, castings are loaded into baskets and placed into the ovens. The castings are heated to the temperature that will completely burn out the resin binders without affecting the metallurgical structure of the casting. Eventually the sand becomes loose enough to flow out of the castings when moved. This method is normally used when all other processes fail, but it is expensive and time consuming, and requires heat energy. The castings still hold the loose sand. Either the loose sand that does fall out has to be cleaned from the ovens or a sand reclamation unit must be added to the system.

A vibrating-media-type system involves submerging the castings in a media that is vibrated at a high frequency. These machines are expensive and slow and leave media in the casting. Such systems must also be fine tuned for each casting.

Blast Cleaning of Castings

James H. Carpenter, Pangborn Corporation

Blast cleaning of castings is a process in which abrasive particles, usually steel shot or grit, are propelled at high velocity to impact the casting surface and thereby forcefully remove surface contaminants. The contaminants are usually adhering mold sand, burned-in sand, heat treat scale, and the like. For aluminum castings, the process is often used to provide a uniform cosmetic finish in addition to merely cleaning the workpiece. This is especially true of engine components such as heads and manifolds that are highly visible. In the case of cast aluminum wheels, die cast transmission cases, and so on, the process prevents leakage by healing surface porosity. Additionally, in some cases deburring and a pleasing cosmetic surface are obtained.

The usual methods of imparting high velocity to abrasive particles are by the use of either centrifugal wheels or compressed air nozzles. The performance characteristics of each of these components will be discussed later.

Centrifugal wheels are the most widely used method because of their ability to propel large volumes of abrasive efficiently. For example, a 56 kW (75 hp) centrifugal wheel can accelerate steel shot to 73 m/s (240 ft/s) at 55.8 Mg/h (123 000 lb/h) flow. To do the same with 13 mm (½ in.) direct pressure venturi nozzles at 45 kg (100 lb)/min per nozzle would require approximately 20 nozzles and an air flow of 0.120 m^3/s/nozzle (260 ft^3/min/nozzle) × 20, or a total of 2.45 m^3/s (5200 ft^3/min) at 550 kPa (80 psi). Approximately 700 kW (940 hp) at the air compressor would be required to supply this amount of air, which gives a 700 kW/56 kW = 12.5 to 1 (940 hp/75 hp = 12.5 to 1) efficiency advantage for the centrifugal wheel.

Even though the nozzle blast is not as efficient overall as the wheel blast, in some applications it may be more efficient because the blast stream can be more efficiently applied, when, for example, blasting into small holes to clean the interior areas of a casting. Other reasons for using nozzle blast are requirements for:

- Low production
- Portability
- Suitability for very hard abrasives, such as aluminum oxide

Choosing Blast Cleaning Equipment

In general, the choice of the equipment is based on the following criteria:

- Production requirements, such as high production of a single or similar parts
- Complexity of the parts
- Ability of the parts to be tumbled
- Size, shape, and mix of the parts

All of these factors must be evaluated in order to select the most cost-effective blasting equipment.

Batch-Tumbling Barrel Type. When the production requirement is low to moderate and the parts are of a size and shape that can be tumbled, the machine type most often used is a batch tumbling barrel (Fig. 1a and 1b). This type of machine is probably the most widely used machine for cleaning castings and is available from many vendors in sizes ranging from 0.028 to 2.2 m^3 (1 to 76 ft^3) capacity. Standard-size barrels from 0.028 to 0.34 m^3 (1 to 12 ft^3) can handle parts weighing from approximately 11 to 45 kg/piece (25 to 100 lb/piece). Heavy-duty barrels will handle parts weighing from 230 to

1100 kg/piece (500 to 2500 lb/piece). These machines consist of a continuous rubber belt or steel slat conveyor that continuously tumbles the parts while they are being blasted. By reversing the conveyor, the parts can be unloaded into a tote box or a takeaway conveyor. Sometimes, to increase productivity, several of these machines are positioned side by side, and automatic loading and unloading systems are used to move castings to and from the machine. One advantage of this type of machine is that the blast time can be set to allow sufficient time for every casting to be cleaned.

Continuous Tumbling Barrel Type. When production requirements are high and the parts are of a size and shape that can be tumbled and an in-line continuous flow of parts is desired, the continuous tumbling barrel is used. There are three types of these barrels (Fig. 2). Figure 2(a) shows the type that has been in use since the 1950s and consists of an in-feed drum with screw flights to feed the parts to the tumbling blast section, which is constructed like the previously described batch tumbling barrel. The parts then enter a discharge drum for unloading. These machines have served satisfactorily for years; however, a serious shortcoming is that fragile parts are often broken because of the bed depth required to cause the parts to flow. A further disadvantage of the bed depth is that some parts slip through without being completely blasted. Perhaps the most serious drawback of these machines is that they have many individual parts that are exposed to the blast; therefore, maintenance and downtime can be significant.

A more recent type of equipment consists of a simple, perforated drum with internal screw flights to move the parts and kickers, causing the parts to tumble (Fig. 2b). The centrifugal wheels are mounted at each end of the drum. This type allows for a low bed depth for parts and has fewer parts to wear out. The main disadvantage is that the blast lacks efficiency because the wheels are outside of the drum rather than directly over the work. These drums are 2.1 to 2.7 m (7.0 to 9.0 ft) in diameter. Breakage of fragile parts is still a problem because some parts ride up the side of the drum above the drum centerline before they tumble.

The most recent continuous barrel design is the rocker-type barrel (Fig. 2c). In this piece of equipment, the barrel is a trough-shape perforated barrel that rocks back and forth about its longitudinal axis. The wheels are directly over the work, and the tumbling action is gentle. The parts are moved axially either by sloping the barrel or by a recently introduced system that oscillates the barrel longitudinally to convey the parts while the barrel rocks back and forth simultaneously. The advantage of the oscillating system is that the throughput can be accurately controlled so that the conveying action can be slowed for hard-to-clean parts while the

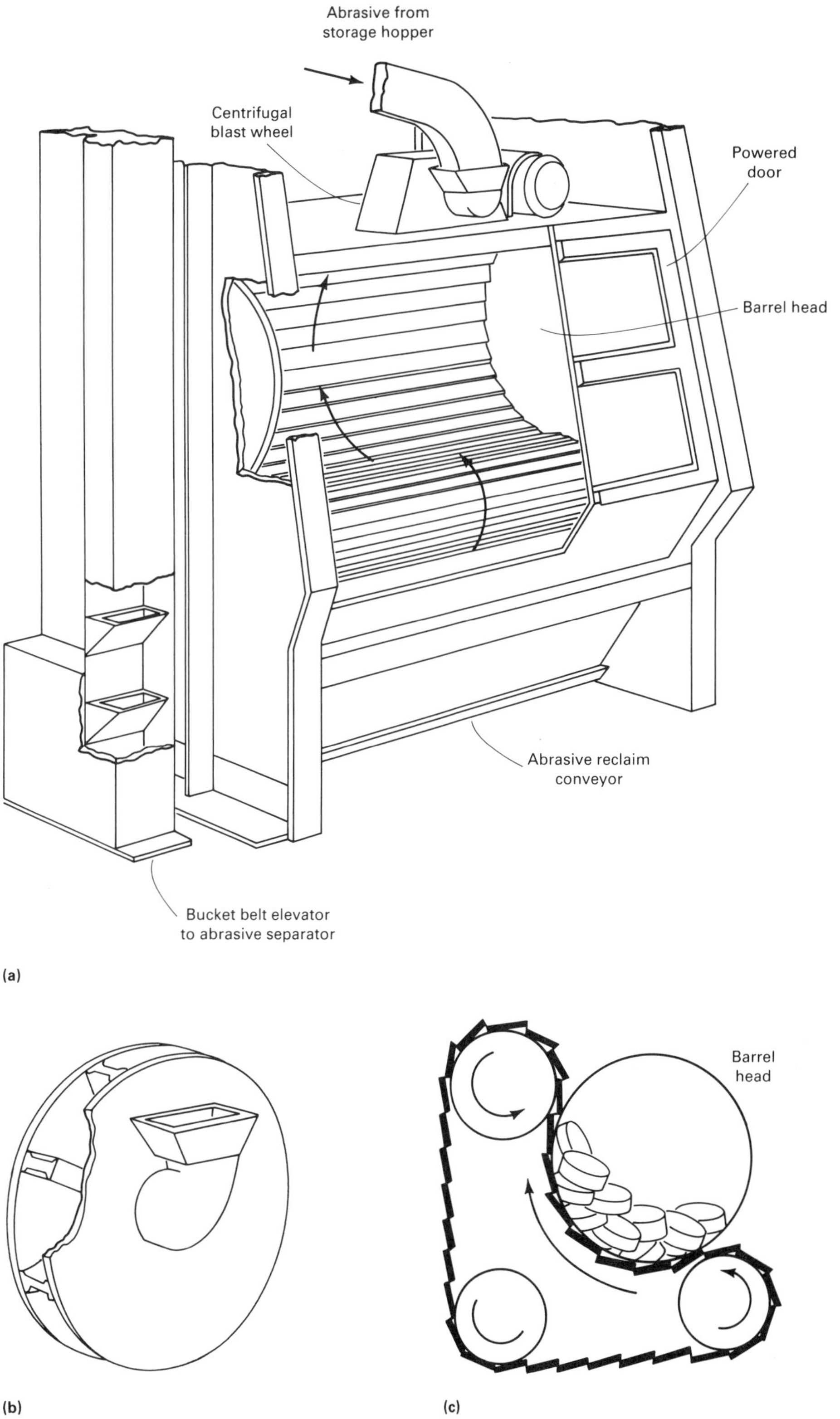

Fig. 1(a) A batch tumbling barrel type abrasive blasting machine showing its principal components. (a) A typical machine. (b) Detail of the centrifugal blast wheel. (c) The continuous rubber belt or endless steel slat conveyor system

tumbling rate remains unchanged. For easy-to-clean parts, the conveying speed can be increased. This barrel concept also allows batch processing of parts because it can empty itself of parts.

For the most part, the three types of continuous machines described are for high production (9 to 36 Mg/h, or 10 to 40 tons/h). Machines of Fig. 2(b) and Fig. 2(c) configurations are currently being devel-

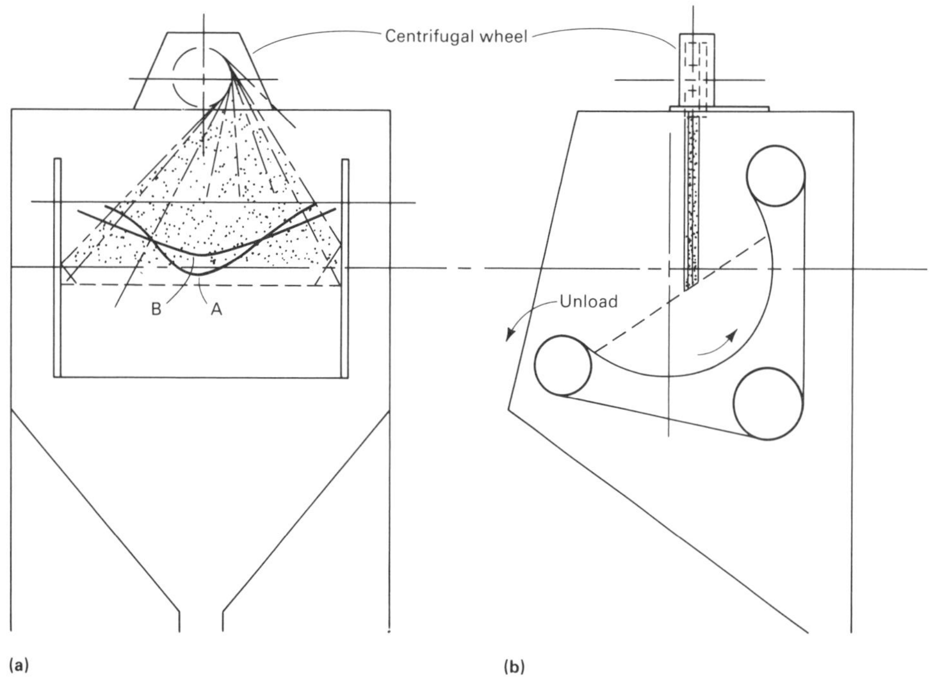

Fig. 1(b) Blast pattern produced by a batch tumbling barrel type abrasive blasting machine. The extent of the pattern is illustrated in (a) the front and (b) side views of the blast pattern. Also illustrated in (a) are two distinct abrasive distribution patterns of the centrifugal wheel; A, short pattern; B, long pattern.

oped for low production (1.8 to 5.4 Mg/h, or 2 to 6 tons/h).

Rotating Hanger Type. For parts that are of moderate size but are too fragile or are of a size and shape that cannot be tumbled, the machine type most often used is the rotating-hanger-type machine. Basically, all of this type have one or many rotating hangers and a hook to hold one or more castings (Fig. 3). The hanger and hook are moved horizontally to a position opposite the centrifugal wheels, where the castings are blasted while rotating about a vertical axis.

These machines come in a number of configurations (Fig. 4). For low-production requirements (20 to 30 hooks/h) and a maximum hook capacity of approximately 2700 kg (6000 lb), the hanger machine may be a simple in-and-out type, as shown in Fig. 4(a). In this machine, one of two hooks is loaded or unloaded at points A or B. After loading, the hanger and hook are moved either manually or with power-operated carriers into the machine to the blast area. Usually, while the part is being blasted, the hanger is indexed a short distance to give three horizontal blast positions for good cleaning coverage. While the part or parts are being blasted, the second hook is being loaded to be ready to enter the machine.

Figure 4(b) shows another low production inexpensive hanger machine. In this machine, hanger and rotating mechanisms are located on two separate doors. When one door is closed, the part attached to the hanger and hook is being blasted. Simultaneously, the hook containing a finished blast-cleaned workpiece on the open door is being unloaded, and another workpiece is being loaded and readied to enter the machine. Because it is not easy to provide three blast positions with this arrangement, the wheels are oscillated to give better blast coverage.

Turnstile Hanger Type. When higher production rates are required, a turnstile-type hanger machine is sometimes used (Fig. 4c). With this machine, the load-and-unload operator station is never exposed to the blast because each turnstile index hooks into the blast station. The only time blast is not impacting the work is during the short index time. Production rates are governed by the blast time needed and the time required to load and unload the hook. Production rates are usually about 40 to 60 hooks/h with a hook capacity of 910 kg (2000 lb).

Powered Carrier Hanger Type. Another hanger machine that is popular and very versatile is the powered carrier or power-and-free carrier concept (Fig. 5). Production rates can be up to 120 hooks/h or more with a hook capacity to 18 Mg (20 tons). These machines may be a simple cabinet with doors at each end and two or three self-powered carriers (Fig. 6a) or a multi-blast station machine with many carriers (Fig. 6b). These carriers move the work to other functions such as through a cooling tunnel and then to a preselected grind station and onto a load station where the carrier rotates to position the hook for loading by a robot. At this point, the operator or robot can program the carrier, depending on the part being loaded, to select the blast hp to be applied, the number of wheels to blast the part, the blast positions, the grind station stop, and so on.

Continuous-Monorail Hanger Type. When production requirements are very high, a continuous monorail (Fig. 7) is sometimes used. These machines can run up to 750 or more hooks per hour at a maximum hook capacity of 1.8 Mg (4000 lb). The conveyor may run at a continuous speed or at a fast speed between blast stations and then at a slow speed for a short distance at the blast station. These machines are hard to automate because the conveyor is always moving and there is no easy way to position the rotator for automatic loading and unloading. These machines have, for the most part, been used by high-production foundries such as automobile foundries. Today, these foundries favor automated, in-line machines.

When parts are too large to either tumble or hang on a hook, or when production requirements are low, the machine types most often used are either a table-type machine or a room-type machine.

Table-type machines (Fig. 8) are furnished in sizes from 0.91 to 3.7 m (3.0 to 12.0 ft) in diameter. The smaller table machines are used to handle smaller castings that cannot be tumbled. They are also used when size, shape, or low production requirements rule out a hanger machine. A large table machine (2.4 to 3.7 m, or 8.0 to 12.0 ft in diameter) can handle castings weighing from 4.5 to 18 Mg (5 to 20 tons). The part or parts are placed on the table by a power lift device, and, after the doors are closed, the parts are blasted by one to three centrifugal wheels. A disadvantage of this type of machine is that usually the parts must be turned over or repositioned and the blast cycle must be repeated to effect complete cleaning.

Room-type machines are used when the requirement is low production on a variety of medium-size to very large parts. These machines usually consist of a room with power-operated entrance doors. A self-powered car with a rotating table carries the work to be blasted into the room to the blast position. Mounted on the room are two or more centrifugal wheels. While the rotating part is being blasted, either the car moves to three blast positions, or, in some cases, the wheels are oscillated for better blast coverage. As with a table machine, the part or parts may need to be turned over or repositioned for complete coverage. Car and tables have been built with a capacity of 230 Mg (250 tons). For very low production of large castings, this type of machine may be furnished with a manual air nozzle blast system. If this is the case, the rotating table is not absolutely necessary, because the operator can walk around the part, which is supported on a work car.

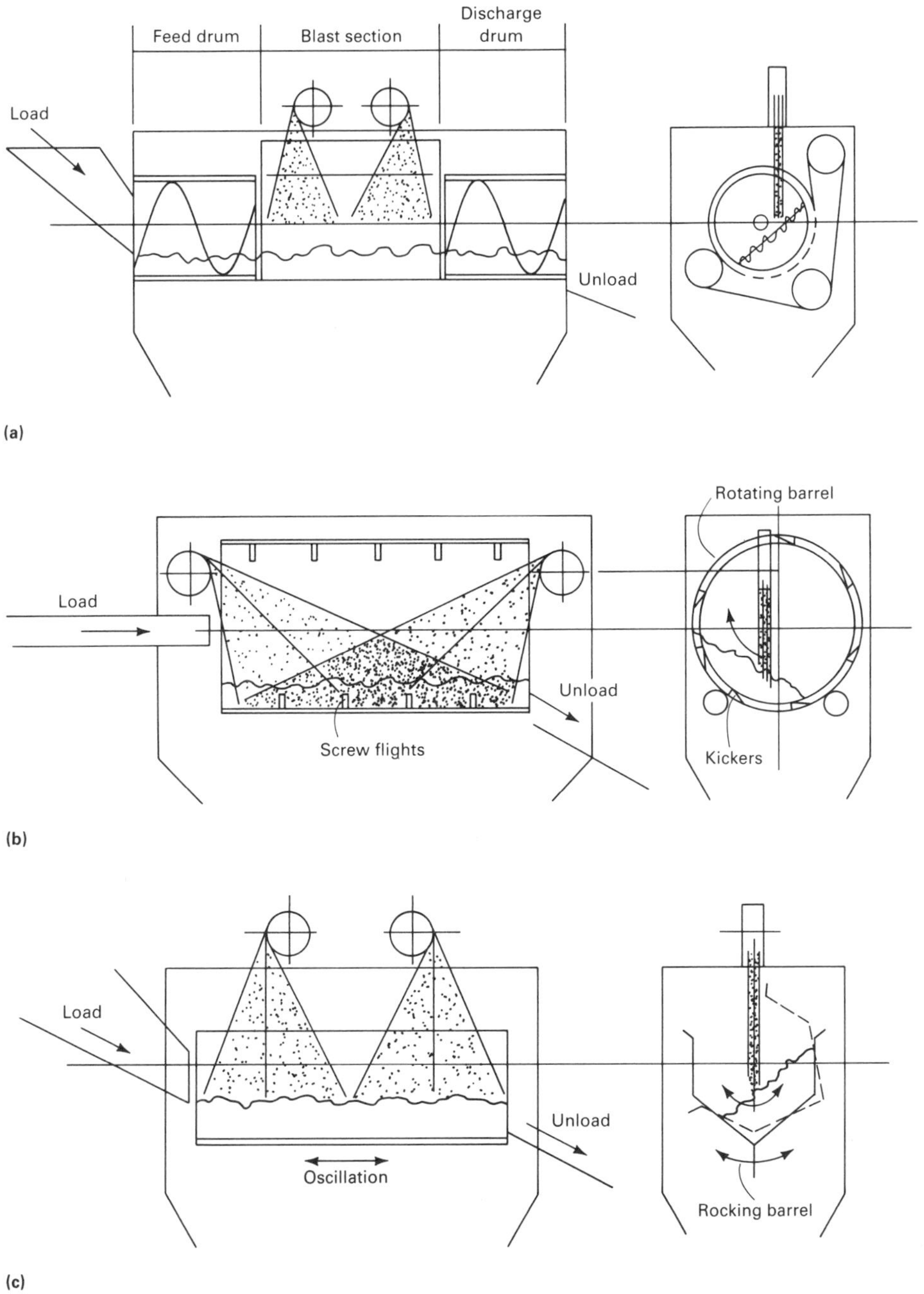

Fig. 2 Three variations of a continuous tumbling barrel type blast cleaning machine, each shown in front and side views. (a) Infeed drum type that uses screw flights to feed each workpiece. (b) Simple perforated-drum type incorporating internal screw flights to move each workpiece and kickers to cause parts to tumble. (c) Rocker type incorporating trough-shape perforated barrel that is rocked back and forth around its transverse axis and moves the parts along the longitudinal axis either by sloping the barrel or by oscillating the barrel longitudinally

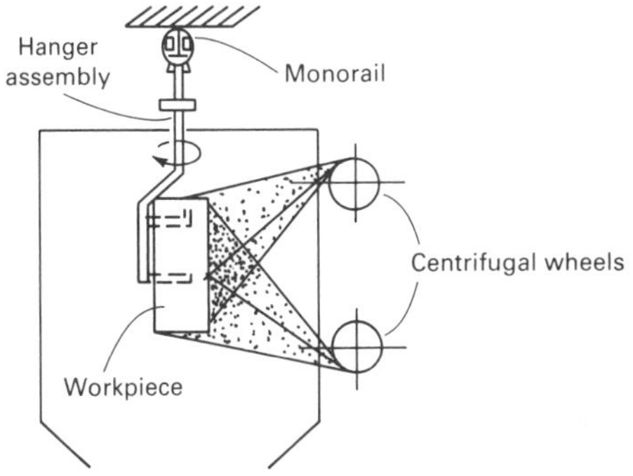

Fig. 3 Front view of a typical rotating hanger assembly. The workpiece is rotated 360° as it is held in position on the monorail track to provide full coverage by the blast pattern produced by the centrifugal wheels.

In order to facilitate the blast cleaning of extremely large castings, some room-type machines incorporate slots in the roof of the room to allow the lowering of the casting into the work area using either an overhead crane or a gantry crane.

Axial-Flow Machine. There are times when the production requirements are so high that special machines must be developed for parts or a family of parts that cannot be successfully tumbled. This is especially true for some parts cast in foundries serving the transportation industry. For example, motor blocks are processed at up to 650 blocks/h, heads at 1000/h, disk brake rotors and brake drums at 1600/h, and so on.

One example of this type of machine is an axial-flow machine for cleaning motor heads (Fig. 9). This machine consists of a skeletal barrel through which parts are rammed one against the other while the barrel is simultaneously rotating. This machine is very efficient because the entire blast stream impacts the parts. Additionally, the interior water passages are effectively cleaned because of the continual draining of the internal passages during the rotation of the part.

These machines are completely automated and give an in-line flow of parts. The main disadvantage of this concept is that the ends of large square-end parts such as motor blocks cannot be effectively cleaned. A recent innovation on the axial flow machine concept (Fig. 10) allows cleaning at the ends by traversing the parts through the skeletal barrel by oscillating the barrel through use of a flat-stroke drive mechanism instead of ramming the parts through an axially stationary barrel.

Special Machines. Because of the demand for quality castings, sometimes called world class castings, there is a trend to preblast castings before grinding and then finish blast them to complete the cleaning. With this technique, motor blocks, for example, are conveyed by an oscillating conveyor system without rotation through a preblast machine. The blast wheels are mounted on the side of the machine and partially clean the castings, with most of the cleaning being done on the sides that require grinding. The castings then move to a shakeout and turnover station to remove loose sand and shot in the casting. Next, they move to a grind station where flash is removed and finally into an oscillating conveyor axial-flow machine where up to ten centrifugal wheels polish clean the castings. This example gives some idea of the special machines that are available. Other dedicated machines clean transmission cases, cam shafts, and so on.

Robotic Holding and Transferring Devices. Robots and similar manipulative devices have recently been incorporated into blast-cleaning operations. These devices travel on tracks and are programmed to stop at a designated centrifugal wheel blasting station as the casting is automatically transported from one workstation to another. When the workpiece is positioned within the abrasive pattern, the holder rotates the casting through a programmed cycle of

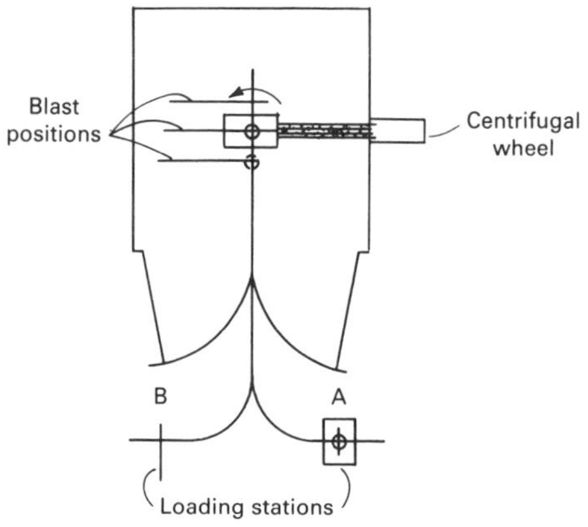

(a)

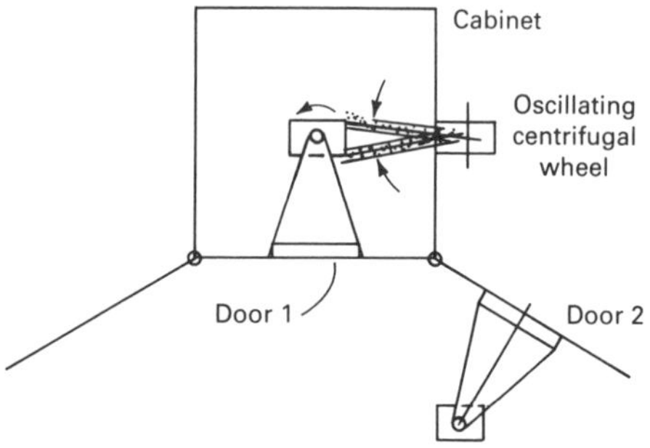

(b)

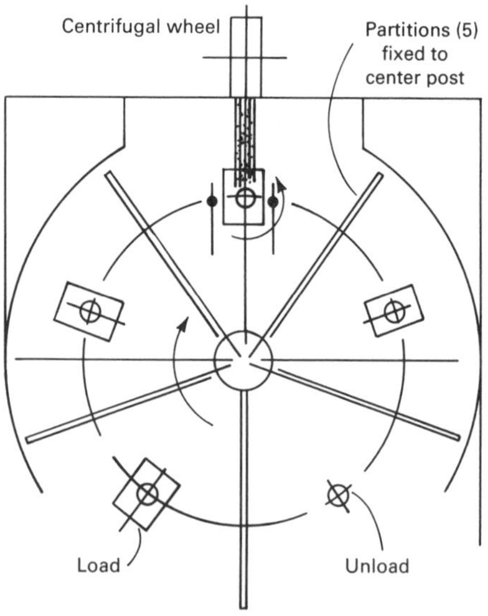

(c)

Fig. 4 Top views of three machine configurations that use hanger and hook assemblies for blast cleaning operations. (a) In-and-out type used in low-production operations. (b) Low-production setup with hanger assemblies mounted on two separate doors to allow loading and unloading of one workpiece while another workpiece is being cleaned. (c) Turnstile type, which indexes each hanger assembly into the blast station, allowing higher production rates than is possible with either of the two previously mentioned configurations

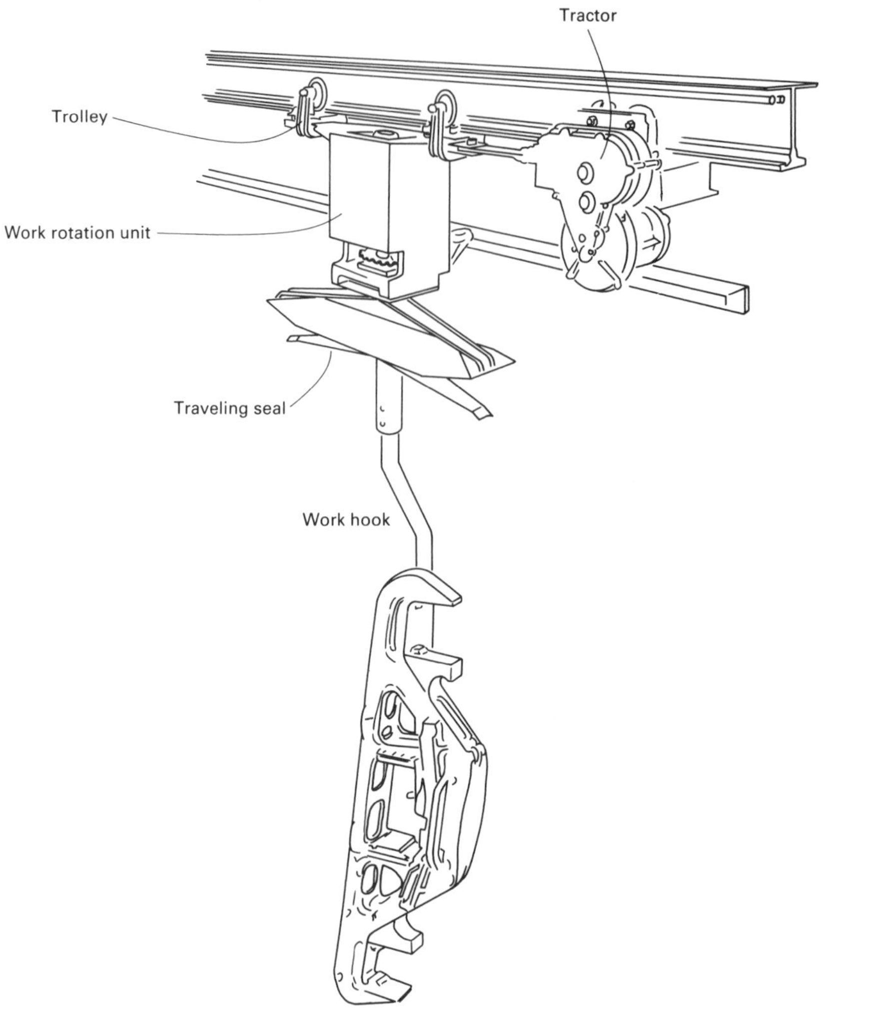

Fig. 5 Powered carrier type hanger capable of producing 120 pieces/h with a capacity to hold workpieces weighing up to 18 Mg (20 tons). Hanger is moved on rail using a 1.1 kW (1.5 hp) friction-drive motor. Work rotation unit is free turning when drive sprocket is disengaged, and it can incorporate optional ac motor drive.

movements. This motion includes oscillating the casting at varying speeds while it is within the abrasive pattern in order to concentrate the pattern on difficult-to-clean surfaces of the workpiece. For detailed information on the application of automation in foundry operations, see the article "Foundry Automation" in this Volume.

Abrasives

Usually abrasive manufacturers suggest asking the following questions in order to choose the correct abrasive size and type:

- What will clean the work rapidly and effectively?
- Will it subject machine parts to relatively gentle wear?
- Will it last a reasonably long time before breaking up?

In the past, chilled iron grit and malleable abrasives were used. Today, however, practically all the shot and grit used is high-carbon cast steel that is heat treated and drawn to give a desired tempered martensite microstructure and hardness. The hardness range of the most commonly used shot and grit is HRC 40 to 50. Harder shot and grit are also produced, with a range of HRC 55 to 65. For faster cleaning or special surface finish requirements, hard abrasive is not often used; but when it is, the wear on machine parts is high, and the abrasive breakdown rate is more rapid. This is especially true of hard grit. Table 1 lists commercially available shot and grit sizes.

Making a choice between shot and grit depends on the surface contaminants or the surface texture required. Grit is used when a chiselling action is required, for example, when removing rust, or perhaps to help remove burned-in sand or provide a good bonding surface for painting, plating, or enameling. Hard grit is used to clean the surface of bathtubs prior to enameling, where a definite tooth is required on the casting surface to provide for better adhesion of the enamel. In cleaning castings, steel shot is typically used; sometimes a shot and grit mixture is used.

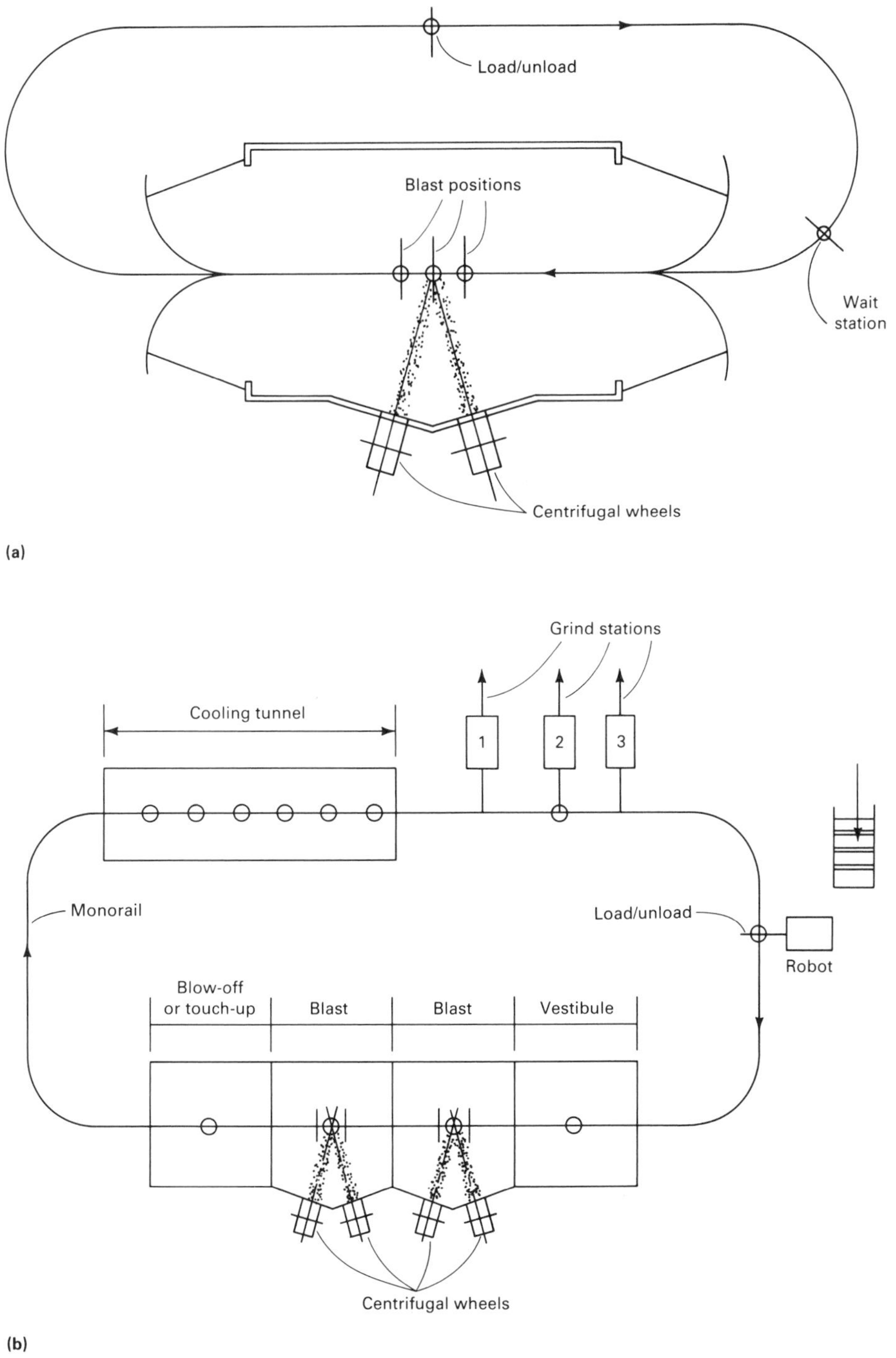

Fig. 6 Top views of two power carrier type configurations. (a) Single blast station with two to three self-powered carriers. (b) Multiblast station with numerous self-powered carriers

The usual rule to follow when choosing the size of shot or grit is to use the smallest-size abrasive that will effectively clean the part. The reason for this is that the smaller sizes give better coverage and longer abrasive life at the same time. This can be illustrated by considering that if blasting were done with S-170 instead of S-460, an increase of $(0.046/0.017)^3$ or approximately 19.8 times the number of particles striking the work would be obtained. However, because the S-460 makes bigger and fewer indentations, the coverage increase with S-170 would be approximately $19.8 \times (0.017/0.046)^2$ or 2.7 times greater than the area coverage with S-460. The photos in Fig. 11 show steel scale blasted at approximately the same abrasive velocity and at the same pounds per square inch coverage. The surface blasted with the finer abrasive, S-170, has much more contamination removed when compared to the surface blasted with S-460.

Another factor that needs to be considered in casting cleaning is that because of the sand in the abrasive mix that results from blasting sand cast parts, the abrasive used can be too small for the abrasive separator to separate the sand from the abrasive. Abrasive sizes smaller than approximately S-330 make it difficult to remove large percentages of sand efficiently without also removing an excessive amount of good abrasive. Usually, parts having large amounts of adhering sand are blasted with abrasives ranging in size from S-330 to S-550.

Another factor that needs to be understood in abrasive size selection is that it is inherent in the process that after a specific abrasive size is added to a machine, the abrasive soon breaks down to a size distribution called an operating mix. For example, an S-460 abrasive as purchased is usually a mix, approximately 85% of which is contained on two screen sizes. The size distribution would be approximately as follows:

Size	New S-460, %	Particles/lb
0.07	2	397
0.06	46	14 526
0.05	45	24 555
0.04	4	4 263
0.03	3	7 578
	Total	**51 319**

After this mix is in use for a period of time, it will break down to an operating mix. This mix can vary considerably depending on the blast application, separator setting, replenishing frequency, and so on. Typically, the mix would be:

Size	S-460 operating mix, %	Particles/lb
0.07	. . .	. . .
0.06	24	7 578
0.05	25	13 641
0.04	24	25 578
0.03	15	37 893
0.02	12	102 313
	Total	**187 003**

It can be seen that the operating mix has approximately 3.6 times more particles per pound so that there are a sufficient number of large particles to break up the surface contaminants and many small particles to remove the residue. Trial and error plus experience usually determine the best operating mix. Once determined, it can be maintained by carefully monitoring the separator performance and making small regular abrasive additions to compensate for losses.

Types of Centrifugal Wheels

In recent years, there has been a proliferation of centrifugal wheel types and sizes. These include the old, standard wheels (usually 495 mm, or 19½ in. in diameter), belt driven, run at 2250 rpm, that give an abrasive velocity of 73 m/s (240 ft/s) and approximately 1035 kg/h/kW (1700 lb/h/hp) flow at hp_{net}, where $hp_{net} = hp_{motor} - hp_{idle}$. A standard centrifugal wheel having

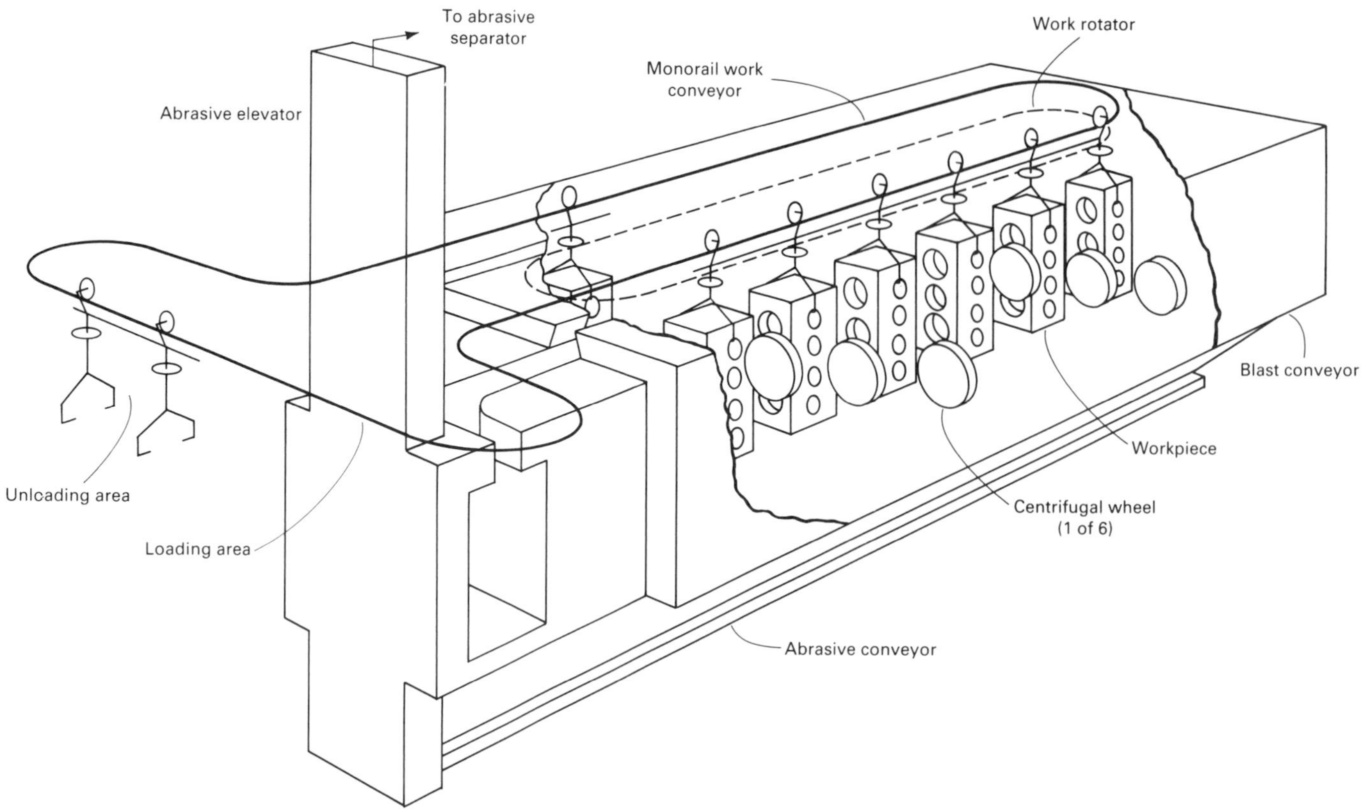

Fig. 7 Continuous monorail hanger type machine used in high-production foundry operations. Hook capacity is 1.8 Mg (4000 lb) with rail capable of transporting 750 hooks/h.

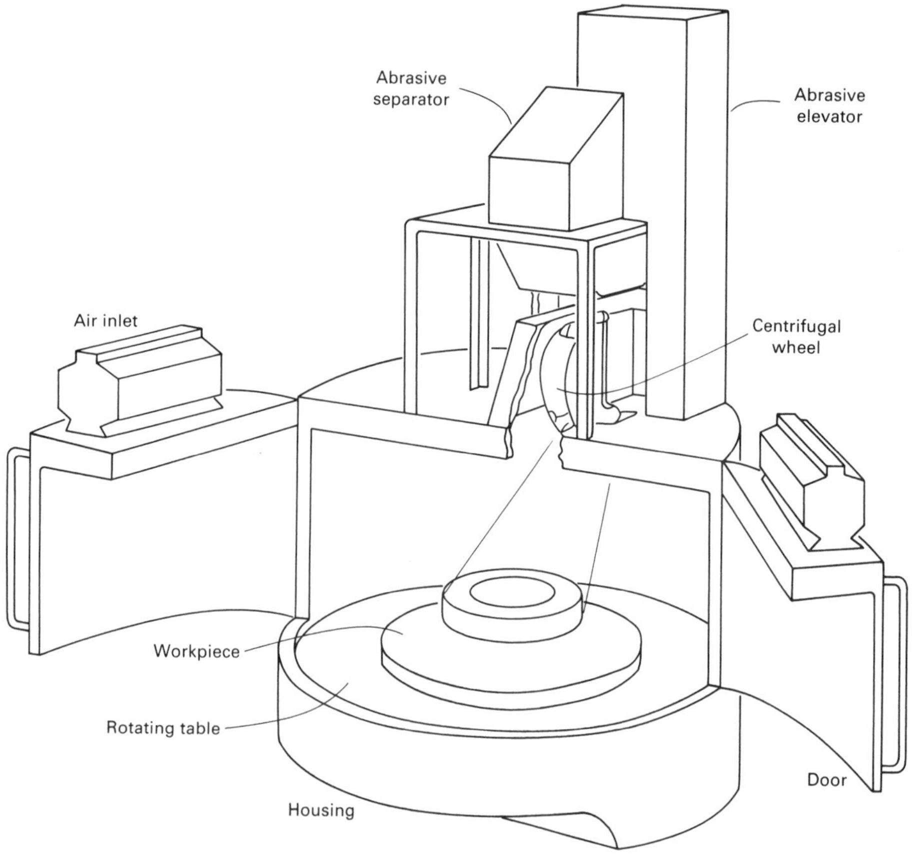

Fig. 8 Components of a typical table-type blast cleaning machine

this dimension and these performance characteristics is shown in Fig. 12. The abrasive velocity was obtained as follows:

The tangential or vane tip velocity (v_t) of the wheel is calculated using the equation:

$$v_t = 2\pi RN$$

$$= 2\pi\left[\frac{9.75}{12}\right]\left[\frac{2250}{60}\right]$$

$$= 191.4 \qquad \text{(Eq 1)}$$

where R is the wheel radius and N is the angular velocity of the wheel. The radial velocity of the wheel, v_R, is given by:

$$v_R = v_t\,(0.76)$$

$$= 145.5 \text{ ft/s} \qquad \text{(Eq 2)}$$

where 0.76 is an empirical factor derived from velocity tests. The resultant abrasive velocity, v, is obtained from v_t and v_R:

$$v = \sqrt{v_t^2 + v_R^2}$$

$$= \sqrt{(191.4)^2 + (145.5)^2}$$

$$= 240 \text{ ft/s} \qquad \text{(Eq 3)}$$

The abrasive flow rate (in pounds per hour) at hp_{net} is given by:

$$\text{Flow rate}_{abrasive} = 1700\,(hp_{motor} - hp_{idle}) \qquad \text{(Eq 4)}$$

In addition to standard wheels, other wheel configurations are increasingly being used: curved-vane, canted-vane, and direct-drive wheels.

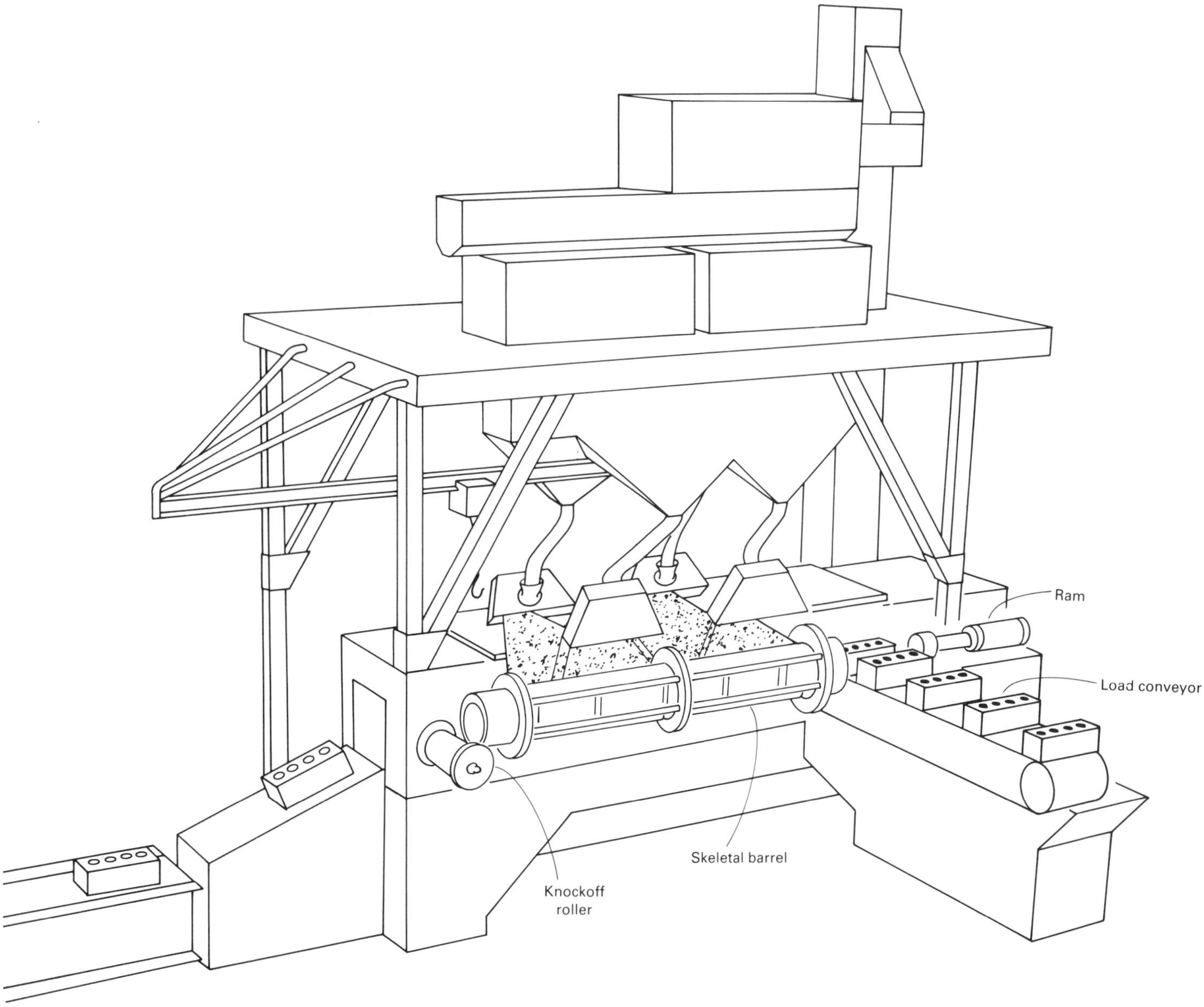

Fig. 9 Axial-flow machine capable of processing 7000 engine heads per hour. Sequence proceeds from load conveyor via ram and skeletal barrel to exit initiated by knockoff roller.

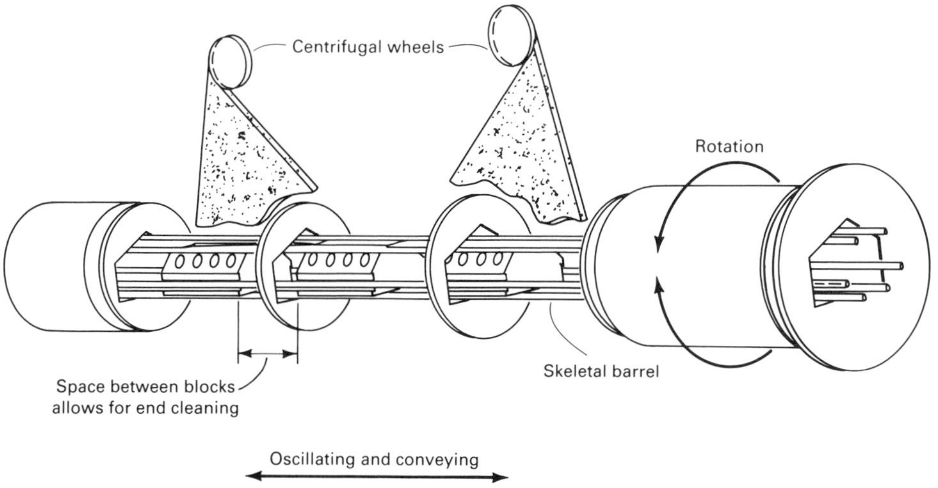

Fig. 10 Oscillating axial-flow machine for blast cleaning of large square-end surfaces of engine block castings, which are difficult if not impossible to clean using an axial flow machine as shown in Fig. 9

Curved-vane wheels were developed in order to reduce the noise source in the centrifugal wheel. In Fig. 12 and 13, it can be seen that a standard 495 mm (19½ in.) diam wheel has a vane tip velocity of 57.9 m/s (191 ft/s) to give an abrasive velocity of 73 m/s (240 ft/s), whereas a 457 mm (18 in.) diam curved vane has a tip velocity of 50.2 m/s (165 ft/s) to obtain the same 73 m/s (240 ft/s) abrasive velocity. The abrasive velocity and flow rate for the curved vane wheel in Fig. 13 are obtained as follows:

Using Eq 1, the vane tip velocity is calculated as:

$$v_t = 2\pi RN$$

$$= 2\pi\left[\frac{9}{12}\right]\left[\frac{2100}{60}\right]$$

$$= 165$$

Table 1 Society of Automotive Engineers shot and grit size specifications

	High-limit screen			Nominal screen			Low-limit screen	
	Maximum % retained	Screen number and aperture (in.)	Maximum % retained	Screen number and aperture (in.)	Minimum % retained	Screen number and aperture (in.)	Minimum % retained	Screen number and aperture (in.)
Shot number								
S-780	1	7 (0.111)	...	...	85	10 (0.0787)	97	12 (0.0661)
S-660	1	8 (0.0937)	...	...	85	12 (0.0661)	97	14 (0.0555)
S-550	1	10 (0.0787)	...	...	85	14 (0.0555)	97	16 (0.0469)
S-460	1	10 (0.0787)	5	12 (0.0661)	85	16 (0.0469)	96	18 (0.0394)
S-390	1	12 (0.0661)	5	14 (0.0555)	85	18 (0.0394)	96	20 (0.0331)
S-330	1	14 (0.0555)	5	16 (0.0469)	85	20 (0.0331)	96	25 (0.0280)
S-280	1	16 (0.0469)	5	18 (0.0394)	85	25 (0.0280)	96	30 (0.0232)
S-230	1	18 (0.0394)	10	20 (0.0331)	85	30 (0.0232)	97	35 (0.0197)
S-170	1	20 (0.0331)	10	25 (0.0280)	85	40 (0.0165)	97	45 (0.0138)
S-110	All pass	30 (0.0232)	10	35 (0.0197)	80	50 (0.0117)	90	80 (0.0070)
S-70	All pass	40 (0.0165)	10	45 (0.0138)	80	80 (0.0070)	90	120 (0.0049)
Grit number								
G-10	1	7 (0.111)	...	...	80	10 (0.0787)	90	12 (0.0661)
G-12	1	8 (0.0937)	...	...	80	12 (0.0661)	90	14 (0.0555)
G-14	1	10 (0.0787)	...	...	80	14 (0.0555)	90	16 (0.0469)
G-16	1	12 (0.0661)	...	...	75	16 (0.0469)	85	18 (0.0394)
G-18	1	14 (0.0555)	...	...	75	18 (0.0394)	85	25 (0.0280)
G-25	1	16 (0.0469)	...	...	70	25 (0.0280)	80	40 (0.0165)
G-40	1	18 (0.0394)	...	...	70	40 (0.0165)	80	50 (0.0117)
G-50	1	25 (0.0280)	...	...	65	50 (0.0117)	75	80 (0.0070)
G-80	All pass	40 (0.0165)	...	...	65	80 (0.0070)	75	120 (0.0049)
G-120	All pass	50 (0.0117)	...	...	60	120 (0.0049)	70	200 (0.0029)
G-200	All pass	80 (0.0070)	...	...	55	200 (0.0029)	65	325 (0.0017)

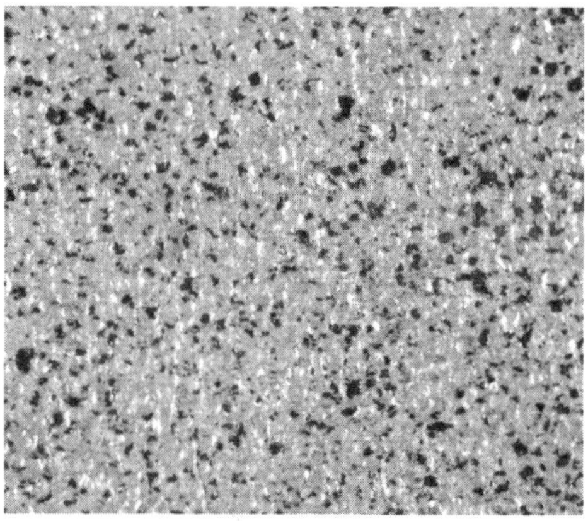

(a)

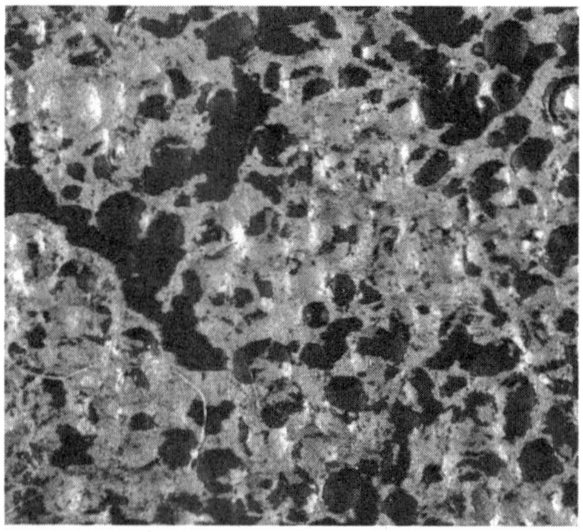

(b)

Fig. 11 Comparison of surface finish obtained using (a) finer S-170 abrasive versus (b) coarser S-460 abrasive. In both cases, abrasive velocity was 73 m/s (240 ft/s), and coverage was 97 kg/m^2 (20 lb/ft^2).

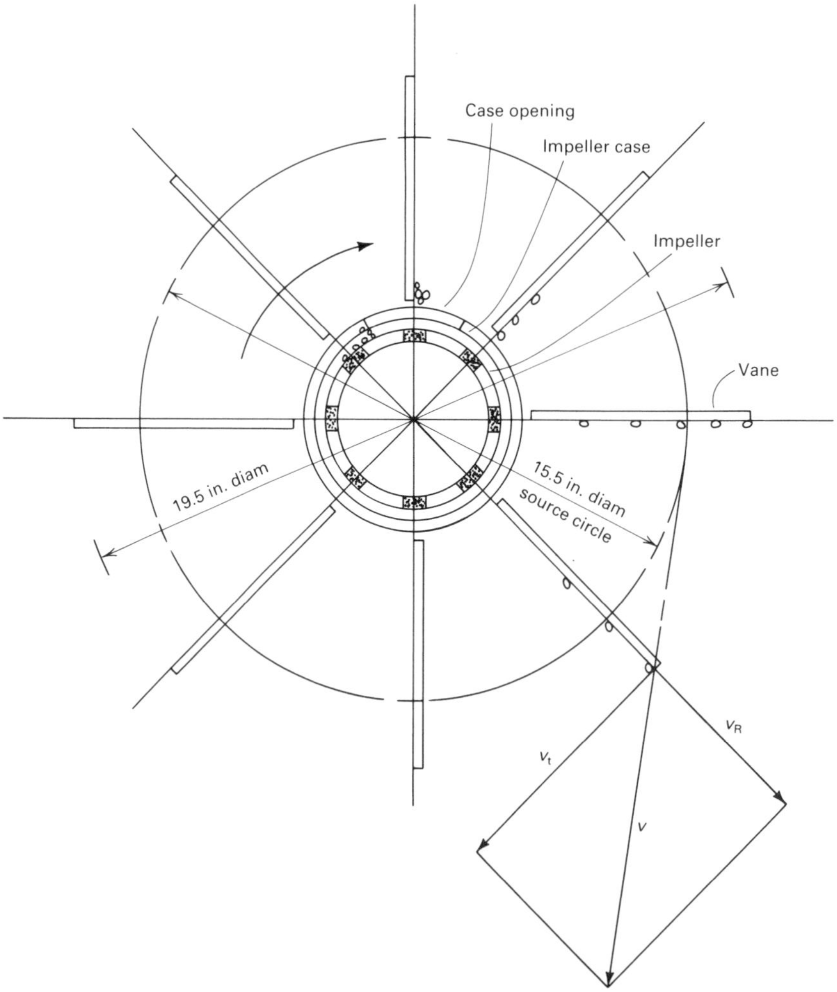

Fig. 12 Cross section of a standard (495 mm, or 19½ in. diam) centrifugal wheel with straight vanes showing radial and tangential components of abrasive velocity. See text for discussion.

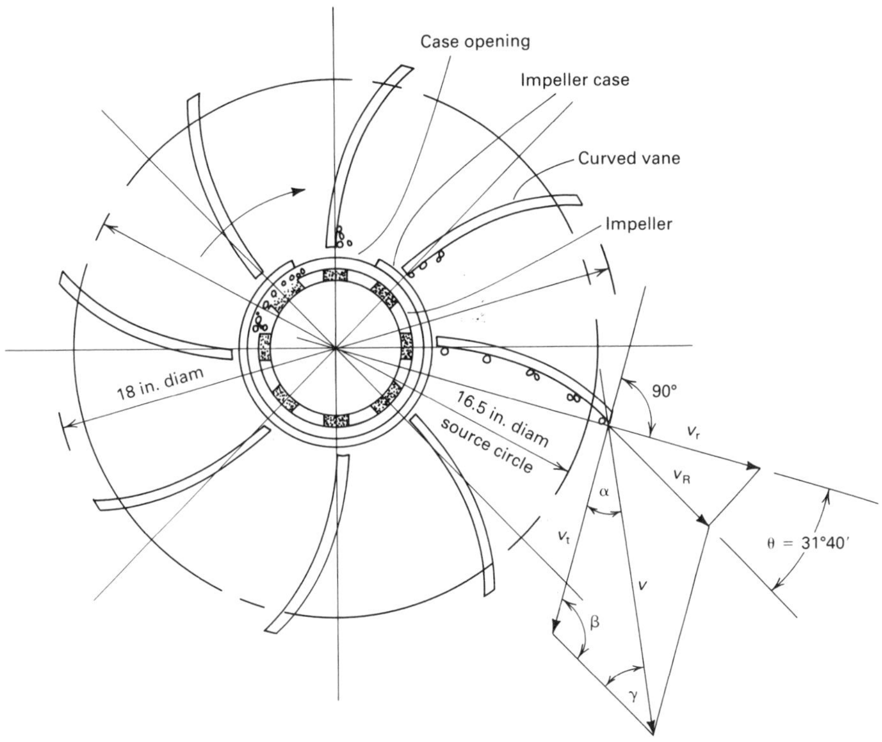

Fig. 13 Cross section of a 457 mm (18 in.) diam curved-vane centrifugal wheel showing radial and tangential components of abrasive velocity. See text for discussion.

The radial component of velocity is obtained using:

$$v_r = v_t\,(0.765)$$
$$= (165)(0.765)$$
$$= 126.23$$

where v_r is perpendicular to v_t and 0.765 is an empirical factor derived from velocity tests. Because v_t and v_R are not at a right angle:

$$v_R = v_r \cos\theta$$
$$= 126.23(\cos 31°40')$$
$$= 107.43 \quad \text{(Eq 5)}$$

The resultant abrasive velocity is calculated as:

$$v = \sqrt{v_t^2 + v_R^2 - 2v_t v_R \cos\beta}$$

$$v = \sqrt{(165)^2+(107.43)^2-(2)(165)(107.43)(\cos 121°40')}$$
$$= 239.54 \text{ ft/s} \quad \text{(Eq 6)}$$

where $\beta = 90° + \theta$, or, with $\theta = 31°\,40'$, $\beta = 121°\,40'$, and the flow rate (in pounds per hour) is:

$$\text{Flow rate}_{\text{abrasive}} = 1700\,[hp_{\text{motor}} - hp_{\text{idle}}]$$

This reduction in rpm and tip speed gives approximately a 7 dB sound power reduction.

The most common curved-vane wheels are:

- 330 mm (13 in.) diam, 3450 rpm, direct drive or 2900 rpm, V-belt driven
- 457 mm (18 in.) diam, 2100 rpm, V-belt driven
- 559 mm (22 in.) diam, 1750 rpm, direct drive

Another advantage of the curved-vane wheel is that, for a given abrasive velocity, the hp_{idle} is less due to the fact that the wheel rpm is less (2100 versus 2250) and the wheel diameter is smaller (457 versus 495 mm, or 18 versus 19.5 in.). For a 102 mm (4 in.) wide vane, 457 mm (18 in.) diam curved vane, hp_{idle} is 1.9 versus 3.6 kW (2.6 versus 4.5 hp) for a 102 mm (4 in.) wide vane 495 mm (19½ in.) diam straight-vane wheel. Laboratory testing and field experience have shown that the wear rate of the curved vane is as good and in some cases better than, an equivalent-width 495 mm (19½ in.) straight vane, provided that the separation system is properly adjusted to remove most of the sand. The curved vane should not be used when using hard grit abrasive (55 to 65 HRC).

Canted-vane wheels consist of a centrifugal wheel in which the eight vanes are alternately canted forward and backward approximately 3 to 6°. This is done to project the abrasive at a slight angle from the wheel in an alternating fashion so that, from a single wheel, two side-by-side patterns are produced at the impact surface (Ref 1). These wheels were originally developed for high-horsepower steel descaling to reduce abrasive interference. However, recent use for casting cleaning has resulted in significant improvement in cleaning efficiency. It can be seen in Fig. 14 that the wider pattern improves the ability of the wheel to clean the sides of cavities, and so on, in a hanger-type machine by enabling the abrasive to cover a wider surface area. In some cases, batch and continuous barrels also can realize better cleaning because of the wider pattern. These wheels are available in straight-vane and curved-vane versions.

Direct-drive wheels have been available for 15 to 20 years. Because of economic pressures to develop less expensive machines, direct-drive wheels are being furnished on new equipment more often. A direct-drive wheel is one in which the centrifugal wheel is mounted directly onto a motor shaft. This design eliminates the standard bearing spindle, V-belts, guards, and belt take-up device.

For the most part, industry experts agree that for high-production rigorous-use machines, the V-belt driven wheel is the most rugged and reliable centrifugal throwing device. There are, however, many applications for which the direct-drive wheel may be a satisfactory choice. This is especially true of wheels mounted on an 1800 rpm motor. These wheels are quieter because of less vibration and lower rpm. Because they are 558 to 610 mm (22 to 24 in.) in diameter in order to give sufficient abrasive velocity, the vane life is increased due to the lower abrasive accelerating forces on the increased area of long vanes running at lower rpm.

Some direct-drive wheels are mounted on 3600 rpm motors. These wheels are usually 330 to 381 mm (13 to 15 in.) in diameter. Because of their higher speed, vibration can be a problem, as can be seen by comparing the vane centrifugal force for the two wheels. The centrifugal force for a 381 mm (15 in.) wheel vane at 1.10 kg/vane (2.42 lb/vane) is 2000 kg (4420 lb) versus only 1420 kg (3130 lb) for a 2.2 kg/vane (4.8 lb/vane) on a 610 mm (24 in.) diam, 1750 rpm wheel. From this it can be seen that if the unbalanced weight is the same on a 381 mm (15 in.) wheel as it is on a 610 mm (24 in.) wheel, the unbalanced load on the motor bearings will be higher on the first wheel. The noise level will also be higher due to more vibration and higher vane tip speed. For a given horsepower, the vane wear will also be higher due primarily to the friction energy acting over a smaller surface area and to higher abrasive accelerating forces on the vane. However, for low-cost, low-horsepower machines, the 3600 rpm direct drive wheel may be the best choice due to space and economic considerations.

The main disadvantages of the direct-drive wheels are:

- The abrasive impact velocity is fixed and cannot be changed without changing the wheel diameter, changing to curved

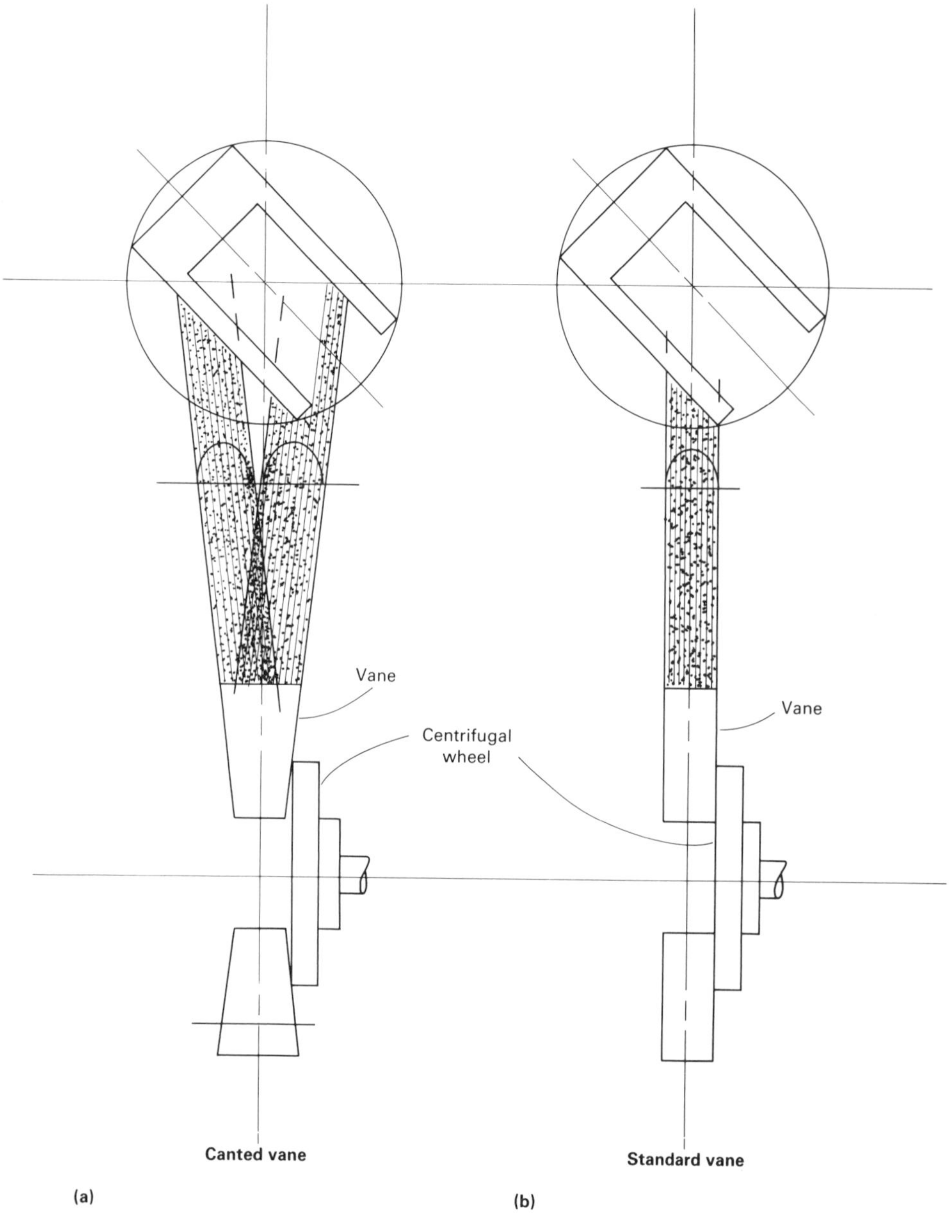

Fig. 14 Comparison of the blast pattern produced by (a) a canted vane centrifugal wheel versus that produced by (b) a standard centrifugal wheel

vanes, or running the motor from an expensive variable-frequency drive

- Motor bearings are not nearly as rugged as standard spindle bearings, and consequently bearing failure can be a problem

Technical Aspects of Blast Cleaning

Centrifugal Wheel Performance. With the proliferation of centrifugal wheels and the recent trend to run wheels at higher and higher speeds or, in other cases, at lower than traditional speed, there is some confusion about the performance characteristics of the different diameter wheels and the effect of both high and low speeds on flow rates and cleaning efficiency.

There are five factors that affect cleaning rates by centrifugal wheels:

- Abrasive velocity
- Abrasive flow rate
- Abrasive distribution in the wheel pattern
- Direction of the angle of impact of the abrasive
- Size, shape, and hardness of the abrasive

Figure 15 shows a cross section of a typical modern V-belt driven centrifugal wheel. All centrifugal wheels used today (except for some air-fed and batter wheels) have basically the same components shown in this photo. The components that affect the performance of a wheel, whether V-belt driven or direct drive are: the feed spout that feeds abrasive into the center of the impeller; and the rotating impeller blades that carry the abrasive around to the fixed impeller case opening, where the abrasive is discharged through the opening to the heel of the rotating vanes, which accelerate the abrasive to a useful velocity.

The rotating vanes discharge the abrasive at high velocity according to the velocity diagrams shown in Fig. 12 and 13. Figure 12 shows the velocity diagram of a 495 mm (19½ in.) diam straight-vane wheel rotating at 2250 rpm. Figure 13 shows the velocity diagram of a 457 mm (18 in.) diam curved-vane wheel operating at 2100 rpm. Velocity calculations for both vane types are given in the discussion "Types of Centrifugal Wheels" in this section of this article. The 31° 40′ discharge angle shown in Fig. 13 has been shown mathematically to be an optimum abrasive discharge angle (Ref 2). Figure 16 shows the approximate abrasive velocity versus rpm for centrifugal wheels most commonly used in the United States.

Abrasive flow rates are dependent on the power available to accelerate the abrasive from zero to its discharge velocity after overcoming the idling hp due to windage, bearing friction, and so on, and the power used to overcome friction of the abrasive on the wheel parts as it passes through the wheel. The following results are given in English units (Ref 3) (metric units can be obtained by consulting the conversion tables in the back of this Volume):

$$P = \frac{W \cdot v^2}{2g \cdot 3600} \qquad \text{(Eq 7)}$$

where W is pounds of abrasive thrown per hour, g is 32.17, and v is abrasive velocity (ft/s). Then

$$\text{hp} = \frac{P}{550} \qquad \text{(Eq 8)}$$

Consider, for example, a typical 19½ in. diam wheel, shown in Fig. 15, driven by a 60 hp motor at 2250 rpm. The flow, W, measured at 60 hp—with the hp being determined accurately with a motor analyzer—is 94 350 lb/h; v, the abrasive velocity, is measured at 240 ft/s; hp_{idle} is measured at 4.5 hp. The hp required to accelerate 94 350 lb/h to 240 ft/s is:

$$\text{hp} = \frac{(94\,350)(240)^2}{(2)(32.17)(3600)550} = 42.66$$

Because it is known that $hp_{\text{idle}} = 4.5$, the remaining hp = 60 − (42.66 + 4.5). Therefore, the energy expended due to friction is 12.84. The total power consumption is 42.66 + 4.5 + 12.84, which totals 60 hp and represents 71.1, 7.5, and 21.4%, respectively, of the motor's 60 hp output.

Laboratory tests run on 13 in. curved-vane, 18 in. curved-vane, 19½ in. straight-vane, and 24 in. straight-vane wheels have

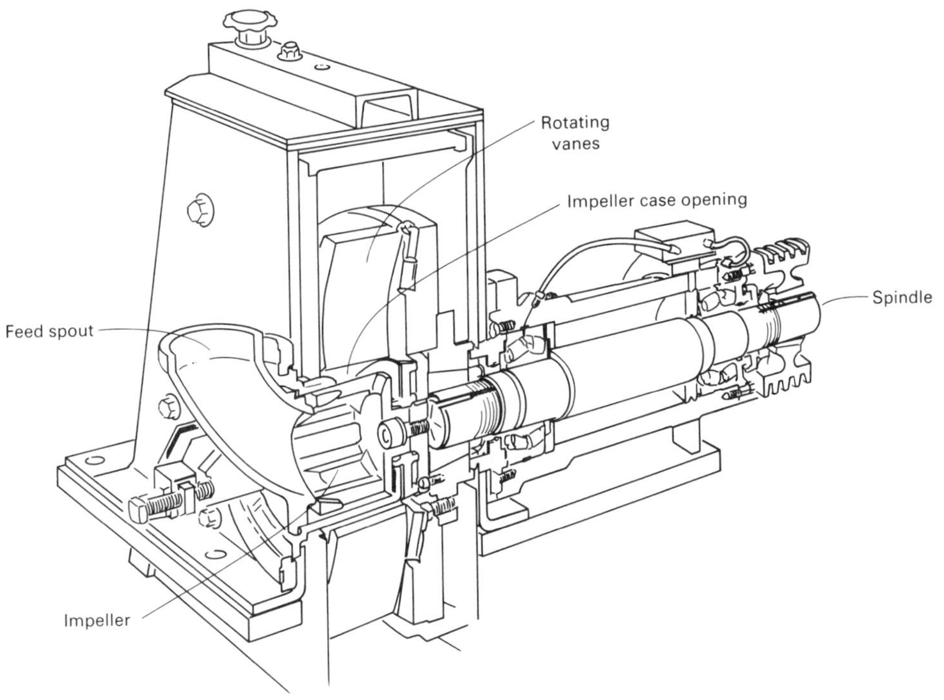

Fig. 15 Cutaway view of curved-vane centrifugal wheel showing primary components involved in feeding and accelerating the abrasive blast medium

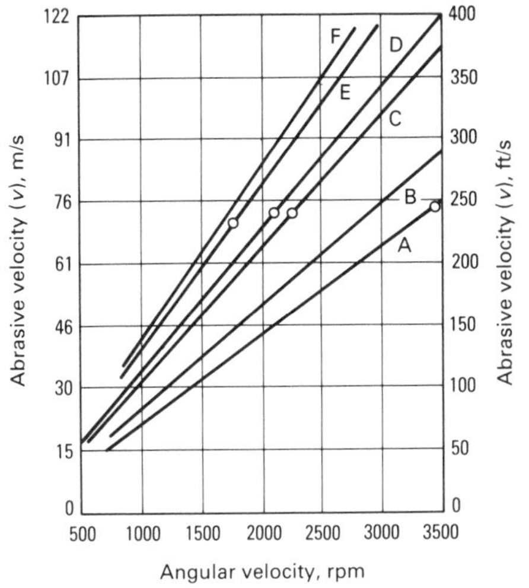

Fig. 16 Plot of abrasive velocity versus angular velocity for both straight-vane and curved-vane centrifugal wheels. Straight vane: A, 330 mm (13 in.); B, 381 mm (15 in.); C, 495 mm (19½ in.); and E, 610 mm (24 in.). Curved vane: B, 330 mm (13 in.); D, 457 mm (18 in.); and F, 559 mm (22 in.). Note that B applies to both a 381 mm (15 in.) straight-vane wheel and a 330 mm (13 in.) curved-vane wheel. Also shown are the standard angular velocities (indicated by a circle) at which some are normally operated.

shown that the average flow in lb/h/hp_{net} to be approximately 1700 lb/h/hp at an abrasive velocity of 240 ft/s, where $hp_{net} = hp_{motor} - hp_{idle}$. Thus for any other abrasive velocity, the flow is:

$$1700 \left(\frac{240}{v}\right)^2 \qquad \text{(Eq 9)}$$

For example, given a 15 in. wheel at an abrasive velocity of 283 ft/s, the flow is:

$$1700 \left(\frac{240}{283}\right)^2 = 1223 \text{ lb/h/}hp_{net}$$

The total flow with a 60 hp motor is 1223 (60 − 3.1 hp_{idle}) = 69 589 lb/h.

It can be seen in Fig. 17 that as the abrasive velocity is reduced, the flow per hp increases; conversely, as the velocity increases, the flow per hp decreases.

Abrasive Distribution in the Wheel Pattern. As a general rule, most vendors and users of blast equipment think of a blast pattern as the pattern produced on a steel sheet that has been blasted for several seconds. The pattern on the sheet is visually examined, and it is determined that the pattern is either a long pattern or a short pattern. A blue area produced in the plate because of the heat generated is called the hot spot. While this system is suitable for targeting or directing the hot spot to a specific area on the part or parts to be cleaned, it does not indicate the distribution of the abrasive in the wheel pattern.

Figure 18 shows some typical wheel stream patterns that show the percentage of abrasive distribution in every 7½° segment of the wheel stream. For example, the short pattern, curve C, has much more of its abrasive in the most intense, or hot spot, 7½° segment than does the long pattern, curve A.

Knowing this distribution enables the manufacturer to apply the centrifugal wheel blast stream more optimally onto the work to be cleaned. Users of existing machines in some cases can profit by having their machines evaluated by blast-cleaning manufacturers to determine whether a different wheel stream distribution would increase their cleaning rates.

The pattern distribution can be changed several ways. Pattern B in Fig. 18 can be considered a standard pattern. This pattern is produced using an impeller case or control cage with a rectangular slot (Fig. 19). Pattern A is generally a pattern from a 610 mm (24 in.) diam wheel with a rectangular slot. An equivalent pattern with a smaller curved-vane or straight-vane wheel can be obtained by using a triangular or modified triangular slot in the case (Fig. 19). Shorter patterns such as C can be obtained by using a patented impeller-to-vane configuration (Ref 4). Some manufacturers also shorten the rectangular slot to shorten the pattern. This is only marginally effective, and then only if the case opening is overly large to begin with.

Figure 1(b) shows a typical barrel application in which a long pattern (curve B) may improve cleaning. Figure 20 shows a typical hanger-type machine application in which a short pattern may be more effective. Figure 14(a) shows a hanger application in which a wider canted-vane transverse pattern is more effective than a standard vane (Fig. 14b).

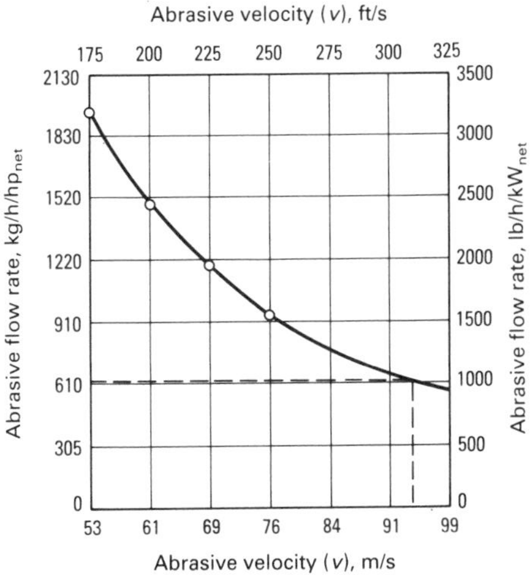

Fig. 17 Plot of abrasive flow rate versus abrasive velocity at hp_{net}. Inverse relationship of these variables is given in English units by flow rate (in lb/h/hp_{net}) being equal to 1700 $[240/v]^2$.

Angle of Impact of the Abrasive. Laboratory testing has shown that in removing steel scale, if the impacts per square inch are the same, the cleaning rate varies directly with the sine of the impact angle (measured from the horizontal). It is not known whether this relationship holds true when blasting a relatively rough casting surface covered with sand; however, it does suggest that the angle of impact of most of the blast stream in any blast application should be kept as high as possible, preferably 45° or more (cleaning ability of the abrasives de-

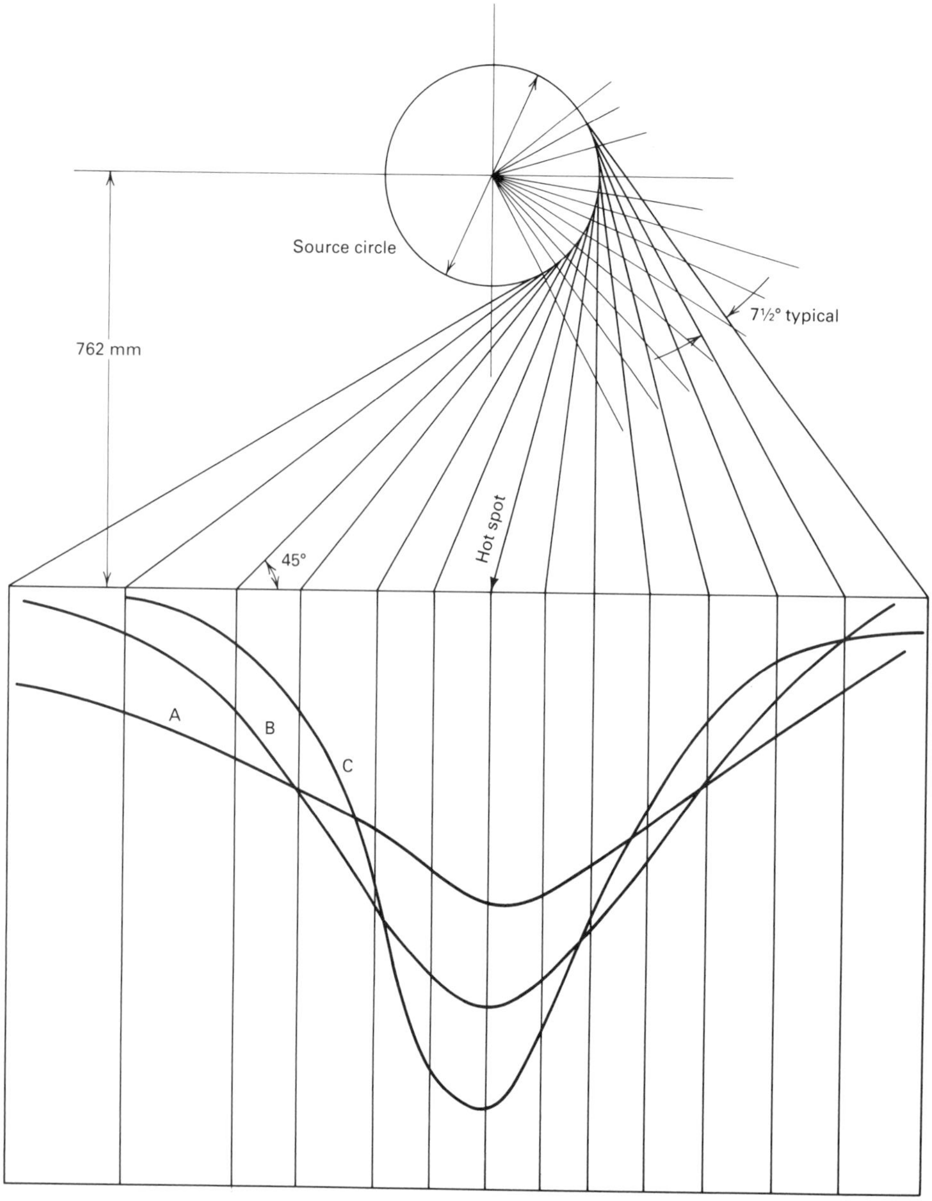

Fig. 18 Plots of abrasive distribution produced by typical centrifugal wheel patterns at 7½° increments of the wheel. The axis of the wheel is located 762 mm (30 in.) from the sample sheet. Typical patterns obtained by varying the size and vanes of the wheels and using either rectangular or modified triangular slot configuration on the impeller case or control cage: A, a 610 mm (24 in.) straight-vane wheel with rectangular case slots, a 457 mm (18 in.) curved-vane wheel and a 495 mm (19½ in.) straight-vane wheel with modified triangular case slots; B, a 457 mm (18 in.) curved-vane wheel or a 495 mm (19½ in.) straight-vane wheel with rectangular case slots; and C, a special impeller (made to specifications in U.S. Patent 4,164,104)

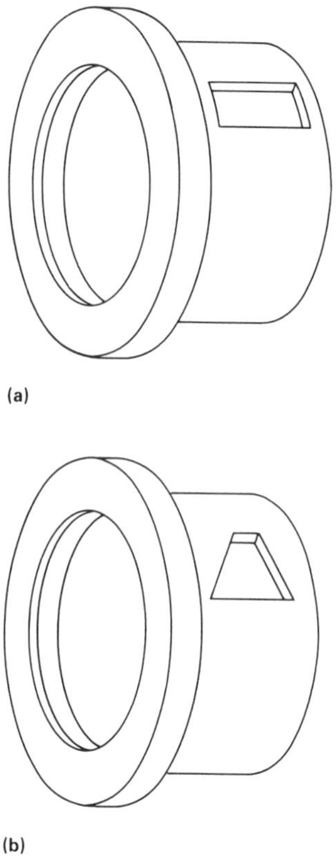

Fig. 19 Abrasive pattern distribution is altered by (a) rectangular slot or (b) modified triangular slot, located on an impeller case or control cage.

clines dramatically at impact angles below 30°). This is especially true when large, flat surfaces are being cleaned such as those encountered with large castings.

Abrasive Parameters. As a general rule, the choice of abrasive size and shape for a given cleaning job is based on past experience with similar castings or by a trial and error specimen blast by the vendor at his demonstration facility. Both methods meet, for the most part, industry needs; however, as with all modern technologies, a better understanding of the dynamics of abrasive blasting can result in a more efficient application of a blast medium to a part.

Cleaning efficiency is determined by supplying a sufficient amount of energy per particle to crush or dig out a surface contaminant and a sufficient number of impacts per area unit to completely remove all the surface contaminant. Figure 21 shows the particles per pound for different size abrasives. In practice, a given shot size consists of a mix of several sizes. However, these charts can be useful in determining the approximate shot size needed in certain cases.

The following equation can be used to obtain the abrasive particle count per pound:

$$\text{particles/lb} = \frac{1}{\left[\frac{\pi\,(\text{size})^3}{6}\right] 0.28} \qquad \text{(Eq 10)}$$

Thus, for an S-280 shot size (the nominal size is approximated as 0.0280 in.), Eq 10 becomes:

$$\text{particles/lb} = \frac{1}{\{[(3.14)(0.0280)^3]/6\}(0.28)}$$

$$= \frac{6.824}{(0.0280)^3}$$

$$= \frac{6.824}{2.19 \times 10^{-5}}$$

$$= 3.12 \times 10^5$$

$$= 312\ 000$$

For example, given a tumbling barrel machine with a 457 mm (18 in.) curved-vane centrifugal wheel, driven by a 37 kW (50 hp) motor, using S-460 shot, at an abrasive velocity of 73 m/s (240 ft/s), the capacity of the abrasive system is 45 Mg/h (100 000 lb/h). Assuming that the parameters, such as the stream pattern, impact intensity, and coverage are optimum, the only way to increase production of this machine is to increase the wheel hp.

The flow in kg/h (lb/h) for a 457 mm (18 in.), 37 kW (50 hp), curved-vane wheel at 73 m/s (240 ft/s) is 36 600 (80 600), whereas, if we increase the motor output to 56 kW (75

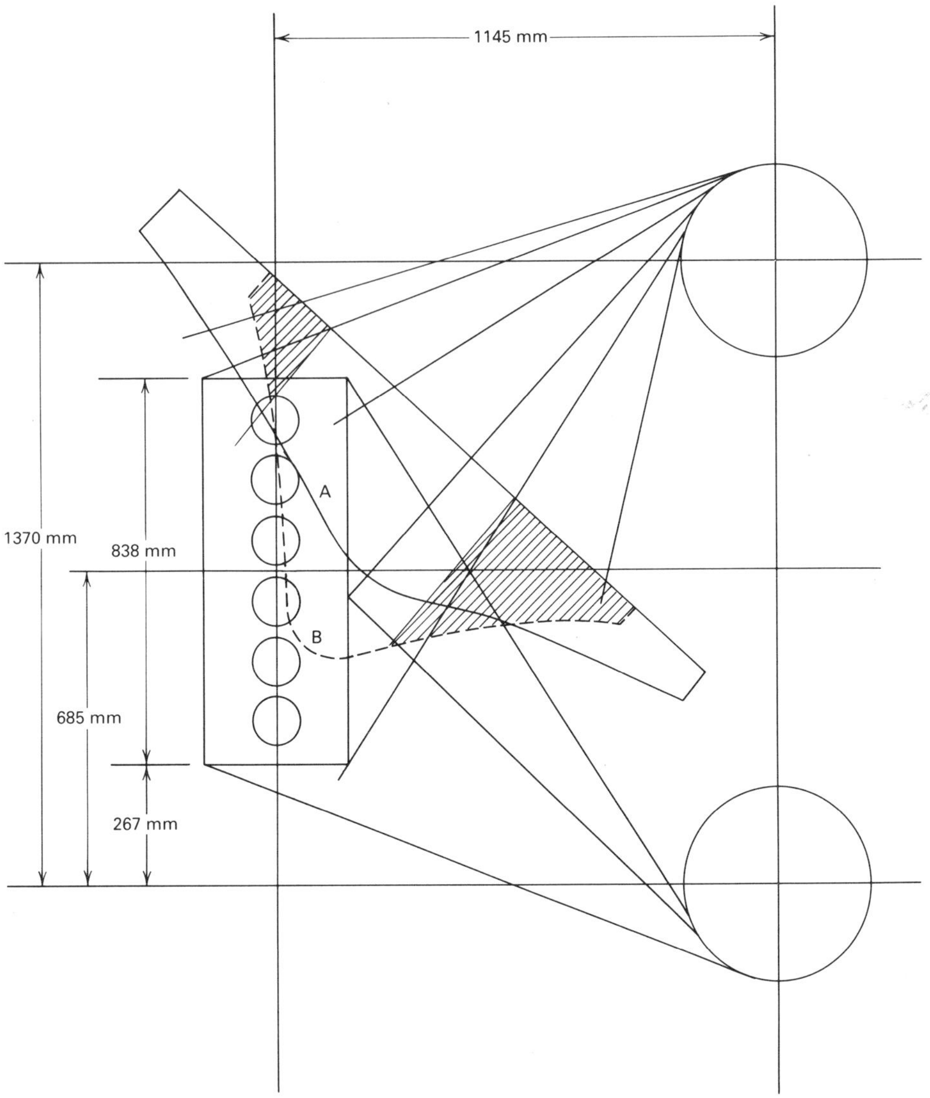

Fig. 20 Wheel pattern coverage provided by two centrifugal wheels working in tandem on a motor casting: A, long pattern, in which 47% of the abrasive impacts the workpiece while 53% misses the part as indicated by the shaded area; B, short pattern, in which 71% of the abrasive impacts the workpiece while only 29% misses the part as indicated by the hatched area

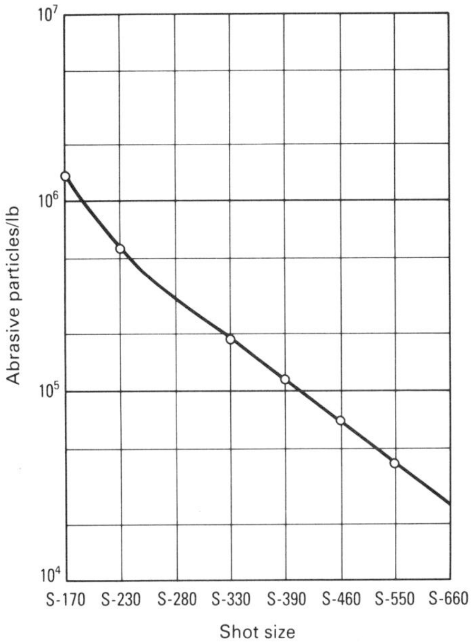

Fig. 21 Abrasive particles per pound versus shot size

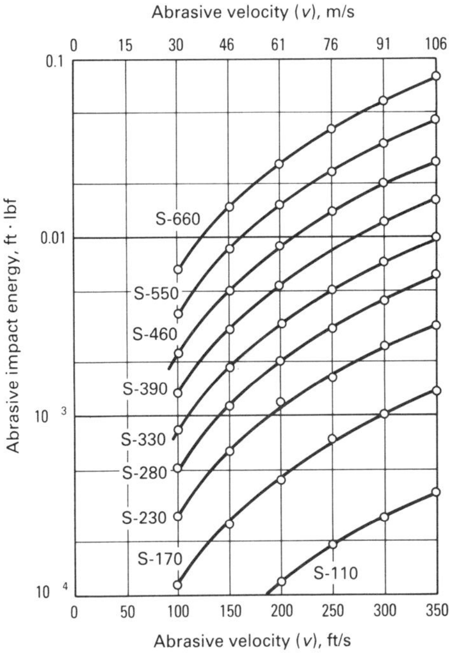

Fig. 22 Abrasive impact energy versus abrasive velocity for various shot sizes. Impact intensity at 73 m/s (240 ft/s) is considered standard.

hp), the flow equals 55.8 Mg/h (123 000 lb/h), which is a 10.4 Mg/h (23 000 lb/h) higher flow rate than the abrasive system can handle. A possible solution to this follows.

Because the wheel is V-belt driven, angular velocity of the wheel rpm and the velocity of the abrasive can be increased. Because the impact intensity of S-460 at 73 m/s (240 ft/s) is assumed to be optimum, Fig. 22 shows that if the velocity of S-390 is increased to approximately 93.6 m/s (307 ft/s), the energy per particle 0.0173 J (0.0128 ft · lbf) is equal to the energy of an S-460 particle at 73 m/s (240 ft/s). Figure 17 shows that for 93.6 m/s (307 ft/s) velocity, the flow is 631 kg/h/kW (1039 lb/h/hp). If the speed of the wheel rpm is increased to 2700 rpm, the total flow will be 33.4 Mg/h (73 665 lb/h), well within the 45 Mg/h (100 000 lb/h) abrasive-system capacity. At 37 kW (50 hp), total impact per second for S-460 is (Fig. 21):

$$\left(\frac{70\ 000 \text{ particles}}{\text{lb}}\right)\left(\frac{80\ 600 \text{ lb/h}}{3600}\right)$$
$$= 1\ 567\ 222 \text{ impacts/s}$$

At 56 kW (75 hp) impacts per second for S-390 is 2 352 921, or 2 352 921/1 567 222 = 1.50. This is 50% more impacts each at the same S-460 impact energy per particle. Thus, we have increased the cleaning energy and coverage by 50% by changing only the speed, horsepower, and abrasive size, without having to change the abrasive handling system.

V-belt driven wheels in the 457 to 533 mm (18 to 21 in.) diam range usually cannot be safely operated above the 2500 to 2700 rpm range (this can be checked with the equipment manufacturer) or above a maximum abrasive velocity of approximately 85 to 95 m/s (280 to 310 ft/s). It can be seen from Fig. 22 that only with sizes S-660 to S-330 can the abrasive size be reduced one size and retain the same kinetic energy per particle as at 73 m/s (240 ft/s). Of course, it may not be necessary or desirable to have the energy levels the same, and in reality it is impossible to regulate energy levels precisely. However, these graphs give an approximate idea of the energy changes made when the abrasive velocity and abrasive size are changed.

All of the above assumes that each abrasive size contains all the same size particles. As already pointed out in the tables in the discussion in "Abrasives" in this section of this article, a designated abrasive size, for example, S-460, contains a mix of sizes within a specified range. Therefore, the above illustration is not precise; however, it is useful as an approximate way to determine size changes that may be required when abrasive velocity is changed.

Other factors to consider when changing abrasive velocity to a higher speed are abrasive breakdown and wear on machine components. For example, if S-460 is used at 73 m/s (240 ft/s), the flow is 1035 kg/h/kW_{net} (1700 lb/h/hp_{net}) (Fig. 17). The breakdown at 73 m/s (240 ft/s), as obtained from test data, is 0.000285 for a breakdown per hp_{net} of (1700) (0.000285) = 0.485 lb/h (0.220 kg/h). If the abrasive velocity is changed to 94 m/s (310 ft/s), the flow will be 617 kg/h/kW_{net} (1015 lb/h/hp_{net}). If the abrasive is changed to S-390 at 94 m/s (310 ft/s), the breakdown per hp_{net}, as obtained from test data, is (1015) (0.00051) = 0.517, or approximately ½ lb/hp_{net}/h, which is approximately the same as the original S-460 breakdown. Laboratory testing has shown that the wear rate on cabinet wear liners in direct-blast operations is approximately proportional to the breakdown rate of shot.

The wear rate of chrome-molybdenum iron or Hadfield manganese steel is 2.7 times greater with S-550 shot than it is with S-110 shot. Again, these data show that if the abrasive velocity is increased, in most cases it is wise to decrease the abrasive size. This usually decreases the wear on the cabinet interior parts and also increases the abrasive coverage.

Cleaning Internal Surfaces

Many times when cleaning castings it is important to clean internal surfaces such as the water passages in engine heads and blocks or other surfaces that cannot be impacted directly because of the casting configuration. In these cases, ricocheting abrasive must do the cleaning. To clean internal surfaces, it is usually important to drain the internal passages while blasting so that pocketed abrasive does not cushion the impact. It is also important to have sufficient ricochet velocity to remove the surface contaminants effectively.

In the majority of cases, the best way to drain internal passages is to rotate the part on a horizontal axis, as is done on axial-flow machines or other machines that incorporate horizontal rotating devices. Recent laboratory tests and field experience have shown that increasing the wheel speeds improves internal cleaning significantly.

At times, the blast from a centrifugal wheel does not effectively reach into all of the internal cavities of a casting. Often, the solution to this situation is to resort to a "spot blast," that is, use of a compressed-air nozzle to direct the blast into selected casting openings. Here again high abrasive velocity can be effective. Changing from straight nozzles to long venturi nozzles can result in abrasive velocity increases of 30% or more, or, if abrasive velocity is more important than coverage, the nozzle flow can be decreased and the velocity increased. Blast equipment manufacturers can advise nozzle users on velocity increases that are possible with different types of nozzle configurations and flow rate changes.

REFERENCES

1. J. Bowling, Abrasive Blasting Wheels and Vanes, U.S. Patent 3,348,339, Oct 1967
2. K.S. Ramaswamy, Curved Vanes for Throwing Wheels, U.S. Patent 3,872,624, March 1975
3. W.E. Rosenberger, *Impact Cleaning*, Penton, p 278
4. J.H. Carpenter and D.G. Corderman, Apparatus and Method for Obtaining a Shortened Blast Pattern With a Centrifugal Throwing Wheel, U.S. Patent 4,164,104, Aug 1979

Welding of Cast Irons and Steels*

Chairman: Ravi Menon, Teledyne McKay

Arc Welding of Cast Irons

Cast irons include a large family of ferrous alloys covering a wide range of chemical compositions and metallurgical microstructures. Some of these materials are readily welded, while others require great care to produce a sound weldment. Some cast irons are considered nonweldable. Consequently, welding procedures must be suited to the type of cast iron to be welded. Various processes for joining iron castings are described in this section, as well as the advantages and disadvantages of each. Iron castings are welded to:

- Repair defects to salvage or upgrade a casting before service
- Repair damaged or worn castings
- Fabricate castings into welded assemblies

Repair of defects in new iron castings represents the largest single application of welding cast irons. Defects such as porosity, sand inclusions, cold shuts, washouts, and shifts are commonly repaired. Fabrication errors, such as mismachining and misaligned holes, are also weld repaired. Restoration of corrosion resistance, structural integrity, or wear resistance can also be achieved by weld overlaying, repair, and/or hardfacing.

Classification of cast iron weldments by design requirements aids in the selection of welding processes and consumables. Three categories of design requirements are:

- Highly stressed welds
- Non-stress-bearing welds
- Weld overlay cladding for corrosion, abrasion, and wear resistance

Highly stressed welds should have mechanical properties that are compatible with the cast iron base metal. Non-stress-bearing welds are common where restoration of proper casting size and shape is the primary concern. Both of these categories of welds may also require good machinability and color match between the weld deposit and the iron casting. Weld overlay cladding for specific surface requirements necessitates careful selection of welding consumables and the deposition process.

The welding of simple castings to form assemblies is often more economical than casting a complex shape. Iron castings can be welded to other iron castings and to steel, nickel alloys, and many wrought shapes.

Welding Metallurgy of Cast Irons

Cast irons have carbon contents in excess of 2% and silicon in excess of ½%. Carbon can be present in the form of eutectic graphite flakes, graphite nodules caused by modifications of eutectic graphite during solidification, pearlitic iron carbide, eutectic iron carbide, or carbon retained in a solid phase such as martensite. Carbon in the form of iron carbide is in a metastable state and can transform to graphite if the kinetics are suitable. Depending on the alloy content, melting practice, and thermal treatment, cast irons represent a wide range of alloys, including the following categories:

- White
- Malleable
- Gray

*The sections "Arc Welding of Cast Irons" and "Oxyacetylene Welding of Cast Irons" were condensed from Volume 6 of the 9th Edition of *Metals Handbook*. The section "Welding of Cast Steels" was prepared by Dr. Menon.

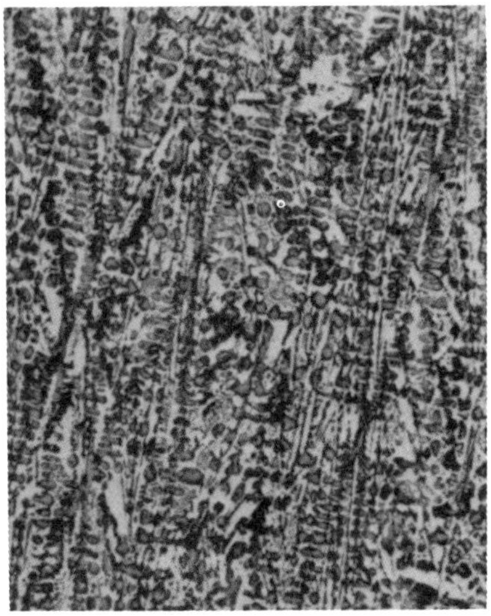

Fig. 1 Unalloyed chilled cast white iron consisting of coarse lamellar pearlite and ferrite in a matrix of M_3C carbides. 100×

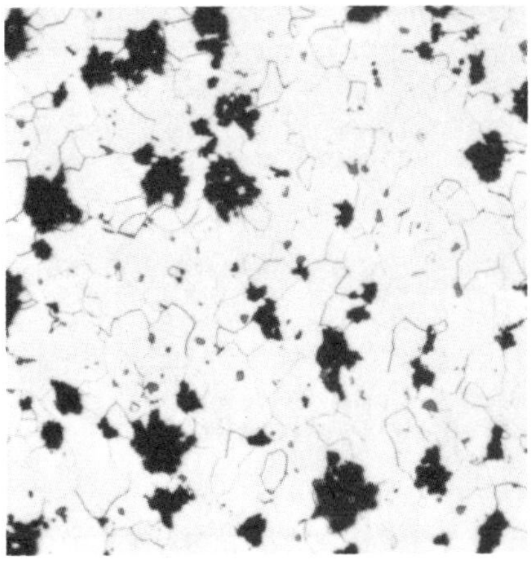

Fig. 2 ASTM A602 grade M3210, ferritic malleable iron. Two-stage annealed by holding 4 h at 955 °C (1750 °F), cooling to 705 °C (1300 °F) in 6 h, and air cooling. Type III graphite (temper carbon) nodules in a matrix of granular ferrite; small gray particles are manganese sulfide. 100×

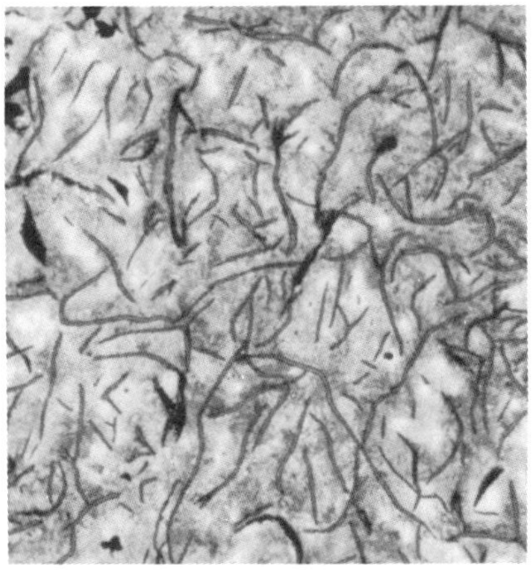

Fig. 3 Class 30 as-cast gray iron in a sand mold. Structure: type A graphite flakes in a matrix of 20% free ferrite (light constituent) and 80% pearlite (dark constituent). 100×

- Ductile
- Compacted graphite irons

Mechanical properties and weldability are dependent on microstructure, which is directly related to the partitioning of carbon during solidification and subsequent cooling. White iron is generally considered to be unweldable because of its extreme hardness and brittleness. Ductile iron is easier to weld than gray iron, partially because of the lower levels of sulfur and phosphorus in ductile iron. Ductile iron offers superior base metal properties; however, weaknesses that are not critical in a gray iron weldment are unacceptable in a ductile iron weldment.

White iron is cast iron in which the majority of carbon occurs as the intermetallic compound cementite (Fe_3C), a very hard constituent. White iron usually consists of a pearlitic matrix with a network of eutectic carbide. Hypereutectic irons have the greatest hardness and are more brittle. Most white irons are of hypoeutectic compositions. The white iron microstructure (Fig. 1) is a function of both alloy composition and cooling rate during solidification. Unalloyed white iron castings have a hardness of about 400 to 450 HB and must be ground to achieve required size and shape.

Malleable iron is heat-treated white iron in which the iron carbides are dissociated when annealed at a temperature above 870 °C (1600 °F) for periods greater than 6 h. Irregularly shaped graphite nodules are precipitated and grow in the solid iron. Figure 2 shows a typical malleable iron microstructure.

Gray cast irons are alloys of iron, carbon, and silicon in which the separation of excess carbon from the liquid during solidification takes the form of graphite flakes, as shown in Fig. 3. Silicon and slow cooling rates promote graphitization in the iron. The low toughness and ductility of graphitic cast iron is neither a direct result of the brittleness of the graphite flakes nor a result of the large proportion of graphite. The continuity of the ductile matrix is interrupted by the shape and distribution of these graphite flakes, resulting in low toughness and ductility. Some castings have insufficient silicon content or cool too quickly to achieve a completely graphitic structure. This results in a white iron microstructure at the surface and a gray iron microstructure at the center; such castings are called mottled iron.

Ductile cast iron, also known as nodular iron or spheroidal graphite iron, has graphite present as small spheroids instead of flakes (Fig. 4). The spheroidal graphite structure is produced by ladle or mold additions (that is, magnesium and/or rare earths such as cerium) to certain molten gray iron compositions. To achieve favorable mechanical properties, graphite in spheroidal form can be dispersed, causing minimal interruption of the continuity of the metallic matrix. Ductile cast irons combine the principal advantages of gray iron, such as excellent machinability, with the engineering advantages of steel, such as high strength, toughness, ductility, hot workability, and hardenability.

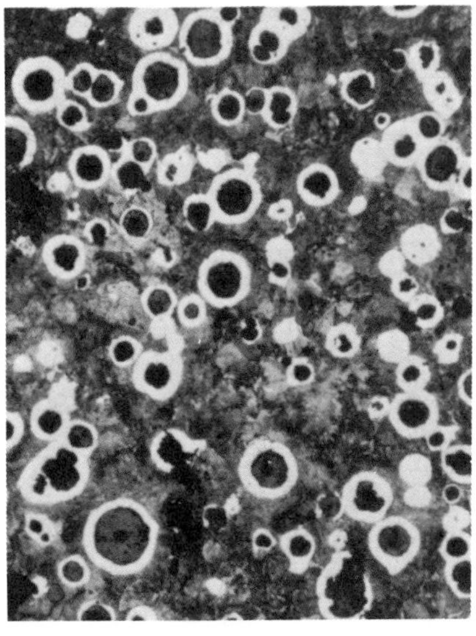

Fig. 4 Grade 80-55-06 as-cast ductile iron. Graphite nodules in envelopes of free ferrite in a matrix of pearlite. 100×

Compacted graphite or vermicular graphite irons have structure and properties that are between those of gray and ductile irons, as shown in Fig. 5. The graphite forms interconnected flakes as in gray iron; but the flakes are thicker, and their edges are blunted. This graphite flake shape provides for higher strength levels than gray irons and greater machinability than ductile irons. Graphitic irons are classified according to their graphite shape (ASTM A 247). The matrix and graphite shape must be

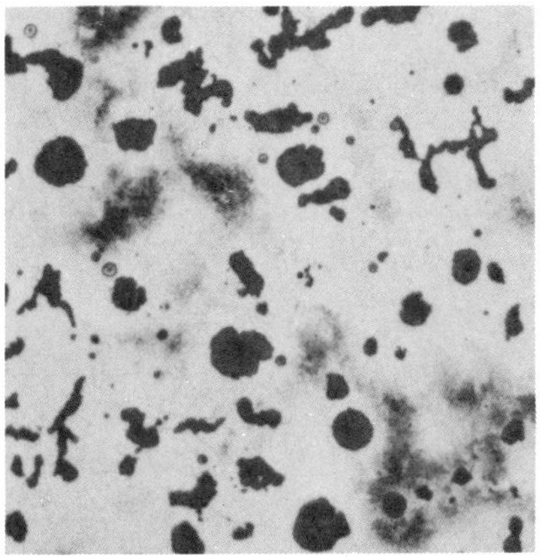

Fig. 5 Compacted graphite cast iron microstructure. Type I graphite (spherulites) and type III graphite (vermicular) in matrix of ferrite and pearlite. 100×

Table 1 Chemical composition limits for cast irons

Type	ASTM specification	Composition limits, %: C	Si	Mn	S	P	Other
White	...	1.8–3.6	0.5–1.9	0.25–0.8	0.06–0.20	0.06–0.18	...
Gray	A 48, A 159, A 278, A 319	2.5–4.0	1.0–3.0	0.25–1.00	0.02–0.25	0.05–1.0	...
Malleable	A 47, A 197, A 220, A 338, A 602	2.0–2.6	1.1–1.6	0.20–1.0	0.04–0.18	0.18 max	...
Ductile	A 536, A 395, A 476	3.2–4.0	1.8–2.8	0.10–0.8	0.03 max	0.10 max	0.030 Mg
Compacted graphite	...	3.5	2.0–3.0	0.20–0.40	...	0.05 max	0.015 Mg, 0–1.5 Cu
High chromium white	A 532	2.3–3.6	0.8–1.0	0.5–1.5	0.06	0.10	1.4–28.0 Cr, 1.5–7.0 Ni, 0.5–3.5 Mo
Austenitic gray	A 436	3.0	1.00–2.8	0.5–1.5	0.12 max	...	28–32 Ni, 1.5–6.0 Cr, 0.50–7.5 Cu
High silicon	A 518	0.7–1.1	14.2–14.8	1.5	...	...	0.5 Cr, 0.5 Mo, 0.5 Cu
Austenitic ductile	A 439, A 571	2.4–3.0	1.0–6.00	0.70–2.5	...	0.08	18.0–36.0 Ni, 0.20–5.5 Cr

defined to identify the specific type of cast iron. For example, flake graphite in a pearlitic matrix is termed pearlitic gray iron.

Chemical composition limits for some cast irons are given in Table 1. The influence of alloy additions and contaminants on cast iron microstructure and properties is important in the selection of the proper welding process, consumables, variables, and procedures. Additional information on cast iron materials and their properties is available in the Section "Ferrous Casting Alloys" in this Volume.

Microstructure

Cast irons have varied microstructures and physical properties, resulting in marked differences in weldability. Variations in thermal gradients across the weldment result in varying microstructures and properties. The various microstructures are classified into different zones and regions, as shown in Fig. 6. The nature and size of these zones in cast iron weldments are determined by the thermal weld cycle, the composition of the base metal, and the welding consumables. To develop welding procedures that minimize the deleterious effects of these zones, the influence of welding variables on mechanical properties must be considered.

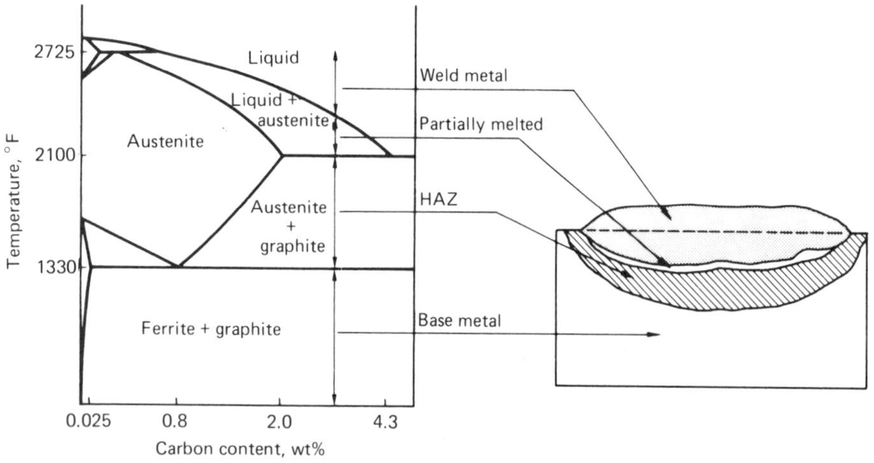

Fig. 6 Schematic representation of the zones in a typical cast iron welding

Heat-Affected Zone (HAZ)

Welding of cast irons is characterized by rapid cooling as compared to cooling rates during casting. Consequently, properties of the weld and the sections of the casting exposed to elevated temperatures (the HAZ) differ from the remainder of the casting. Portions of the cast iron HAZ reach temperatures during welding that cause the carbon to diffuse into the austenite. Upon cooling, this austenite transforms into hard eutectoid decomposition products such as martensite. The amount of martensite formed depends on the cast iron composition and thermal treatment. Ferritic cast irons contain most of their carbon in the form of graphite, which dissolves slowly, thus producing less martensite. The highest percentage of carbon in pearlitic cast irons is finely divided into the pearlitic structure. This carbon dissolves readily, producing a large amount of martensite. The brittle martensite can be tempered to a lower strength, more ductile structure through preheating and interpass temperature control, multiple-pass welding, or postweld heat treatments such as stress-relief annealing. Hardness in the HAZ ranges from 250 to 650 HB, with most of the area below 450 HB. Some welding procedures are designed to reduce the size of the HAZ and thus minimize cracking. Methods of accomplishing this are reduction of heat input, use of small-diameter electrodes, use of low-melting welding rods and wires, and use of lower preheat temperatures. Typical HAZ widths are in the range of 0.8 to 2.5 mm (0.03 to 0.10 in.).

Partially Melted Region

The partially melted region of a weldment is an extension of the HAZ that occurs where a high peak temperature has caused partial melting of the base metal near the fusion line. The effects of this region on the mechanical properties of ductile iron welds must be taken into account to produce successful welds. The liquid in the partially melted region is similar to liquid eutectic cast iron, which freezes as white iron because of the high cooling rates during the weld heat cycle. The microstructure of the partially melted region is complex and consists of a mixture of martensite, austenite, primary carbide, and ledeburite that surrounds partially dissolved nodules or flakes of graphite. If the amount of dissolved graphite is great enough to form an almost continuous molten matrix, the carbide network is also continuous; less extensive melting produces a discontinuous carbide structure. The partially melted region, which contains a great proportion of hard products, is the hardest zone in the weld.

The high hardness and consequent low toughness of the partially melted region adjacent to the fusion line create mechanical problems in the welding of cast iron. The formation of hard structures results from the transport of carbon from graphite particles into the adjacent matrix. Reduction of peak temperatures and time at high temperature is the most effective method of reducing cracking. This can be done by controlling the heat input, preheat, and interpass temperatures.

Low-melting filler materials can also alleviate fusion line cracking by reducing peak temperatures. If a weldment is sufficiently small, a large heat input may raise base metal temperatures enough to cause severe fusion line problems even though no preheat is used. Conversely, a high preheat temperature used to prevent martensite in the HAZ and to reduce the thermal expansion stresses may result in fusion line cracking even with low heat input. Therefore, trial-and-error procedure development may be necessary. A welding procedure that successfully incorporates the partially melted region and HAZ can produce joints with good mechanical properties.

Fusion Zone

The fusion zone microstructure and properties are primarily influenced by the selection of the welding consumables. The fusion zone composition consists of melted welding electrode or wire with some dilution from the iron casting. The weld pool is mixed during the welding process to produce a relatively uniform composition within each weld bead. Dilution should be kept to a minimum. The fusion zone should be machinable, below 300 HB in hardness. If color match between the weld and the base metal is important, filler metal compositions that match the cast iron must be used.

Welding Procedures and Processes

More than 90% of all industrial welding of cast iron is done by arc welding processes. Arc welding has a lower heat input than oxyfuel gas welding because of a faster welding speed and higher deposition rate. The welding operation can be automated to varying degrees, and distortion due to welding heat is more readily controlled. Arc welding achieves temperatures in excess of those required for fusion of the base metal (~3000 °C, or 6000 °F). The intensity of the heat source allows the necessary fusion while heating only a small portion of the weldment. This may cause high cooling rates and may result in large thermal expansion and contraction stresses. However, arc welding processes can produce welds of good quality with proper selection of the welding process, consumables, and procedures. Detailed descriptions of the various processes described below are given in Volume 6 of the 9th Edition of *Metals Handbook*.

Shielded metal arc welding (SMAW), the most widely used and best known process, has the advantages of a large selection of consumables, good availability, low-cost power supplies, and all-position welding. The disadvantages include a higher cost per pound of weld metal deposited and an associated low rate of deposition. These disadvantages are not important in a low-volume or maintenance application.

Gas metal arc welding (GMAW) has limited use, and its advantages include a high deposition rate and efficiency. Several disadvantages are limited selection of consumables, shielding gas, higher power source cost, and need for wire-feeding equipment. Gas metal arc welding can be operated in various metal transfer modes, including globular, pulsing, spray, and short circuiting. The spray transfer mode is characterized by a relatively high heat input, high deposition rate, and use in high-volume operations, but can be used only in the flat or horizontal position. The short circuiting transfer mode is characterized by lower heat input, moderate deposition rate, and the capability of operating in all positions.

Flux cored arc welding (FCAW), which is growing in usage, has advantages and disadvantages similar to those of gas metal arc welding in the spray transfer mode. The slag coverage provides the advantages of better protection during cooling, improved wetting of the bead at corners, an out-of-position weld capability, and chemical refinement of the weld deposit. In many applications, flux cored arc welding has higher deposition rates than gas metal arc welding, particularly for out-of-position welding. Disadvantages include the added step of removing the slag and a lower deposition efficiency.

Gas tungsten arc welding (GTAW) is generally used to repair smaller parts. Advantages include the capacity to operate at low power levels, all-position capability, and superior control of heat input. Gas tungsten arc welding is the only process that can be used to repair a broken part without the addition of filler metal. Some disadvantages are the very low deposition rate, expensive and complicated power supply, the need for greater welder skill level, and the need for a shielding gas.

Submerged arc welding (SAW) has several advantages: very low fume, no arc flash, very high deposition rate and efficiency, and low cost per pound of deposit. Disadvantages include expensive equipment, the need for flux, and operation in the flat position only. This process generates the highest heat input for welding processes used for cast irons, which leads to the greatest dilution and widest HAZ. However, when comparing the same size wire, it has a lower heat input than gas metal arc welding in the spray transfer mode. Submerged arc welding has limited application to iron castings and is usually used for weld overlay cladding of cast iron.

Consumables

Filler Metals for Shielded Metal Arc Welding of Cast Irons. A variety of covered electrodes are used for the shielded metal arc welding of iron castings. Economic considerations and weld requirements determine the appropriate product for each application. Electrodes designed specifically for welding iron castings are described in American Welding Society (AWS) Specification A5.15.

Covered electrodes utilizing a cast iron rod as the core are classified by AWS as ECI. They are of comparative low cost and produce a weldment with chemical and mechanical properties similar to those of the base metal. Color match is excellent; however, ease of operation of these products is marginal, and appropriate procedures must be followed closely to ensure weld quality. If a cast iron electrode is used, graphite flakes or spheroids form in the weld during cooling. The ductility of a repaired malleable or ductile iron casting may be less than that of an unrepaired part.

Electrodes using steel as a core wire but depositing a cast iron type of deposit are similar in usage to the ECI electrodes. These products, classified as ESt, are generally of higher quality because they are produced by extrusion techniques. Welding procedures must be followed closely to avoid brittle or cracked weldments. Because of their ease of operation, widespread availability, and low cost, low-hydrogen types of covered electrodes, for example, E7015, E7106, E7018, and E7028, are used for welding iron castings. Color match is excellent. When these steel products are deposited on cast iron, the resulting first-pass weld deposit has a high carbon content of approximately 0.8 to 1.5%. The weldment is a high-carbon steel and can be quite brittle and crack sensitive. As a result, the use of these products is limited to cosmetic repairs in nonstructural areas and applications where machining is not required.

Austenitic stainless steel electrodes such as AWS A5.4 classes E308, E309, E310, and E312 have seen some application for welding iron castings. However, because of chromium carbide formation in the weld and tearing due to differences in strength and coefficient of expansion, use of these products is marginal, and extreme care should be exercised with their use.

Nickel-base electrodes have been widely accepted for the welding of iron castings. Unlike iron, nickel has a low solubility for carbon in the solid state. Thus, as the weld pool solidifies and cools, carbon is rejected from the solution and precipitates as graphite. This reaction increases the volume of the weld deposit, thus offsetting shrinkage stresses and lessening the likelihood of fusion zone cracking.

Electrodes classified as ENi-CI utilize a commercially pure nickel-core wire and thus produce a deposit of high nickel content. Even when highly diluted by the base metal, the deposit remains ductile and machinable. ENiFe-CI electrodes produce a nickel-iron deposit and, as a result, have four distinct advantages over the ENi-CI electrodes:

- The deposits are stronger and more ductile; this increased strength level makes the product suitable for welding the higher strength gray and ductile irons, as well as for many dissimilar-metal joint applications
- Nickel-iron deposits are more tolerant of phosphorus than nickel deposits; thus,

Table 2 Welding current ranges for the shielded metal arc welding of cast iron with nickel-iron and nickel electrodes

For welding in the flat position, current for overhead welding should be 5–15 A less than shown; for vertical welding, 10–20 A.

Electrode diameter		ENiFe-CI electrodes(a)		ENi-CI electrodes(b)	
mm	in.	Current (dc)(c), A	Current (ac), A	Current (dc)(c), A	Current (ac), A
2.5	3/32	40–70	40–70	40–80	40–80
3.2	1/8	70–100	70–100	80–110	70–110
4.0	5/32	100–140	110–140	110–140	110–150
4.5	3/16	120–180	130–180	120–160	120–170

(a) Percentage composition, based on analysis of deposit on standard test pad: 2.00 C, 4.00 Si, 1.00 Mn, 0.03 S, 45–60 Ni, 2.50 Cu, 1.00 other, rem Fe. (b) Percentage composition, based on analysis of deposit on standard test pad: 2.00 C, 4.00 Si, 1.00 Mn, 0.03 S, 8.00 Fe, 85 min Ni, 2.50 Cu. (c) Either polarity

ENiFe-CI electrodes are preferred for welding gray iron castings with higher phosphorus contents

- The coefficient of expansion of the nickel-iron deposit is somewhat less than that of the nickel deposit; therefore, nickel-iron products can be used to weld heavier sections while still avoiding fusion line cracking due to expansion differences
- ENiFe-CI electrodes are generally less expensive than ENi-CI electrodes

Welding current ranges for the shielded metal arc welding of cast iron with nickel-iron and nickel electrodes are given in Table 2. For nickel-iron electrodes, the welding current should be reduced 5 to 15 A for overhead welding and 10 to 20 A for vertical welding from the values listed in Table 2.

Electrodes with a nickel-copper deposit classified as ENiCu-A and ENiCu-B have limited use for welding iron castings. Because of the sensitivity of their deposit to iron dilution, they have been replaced for the most part by the ENi-CI and ENiFe-CI products.

Copper-Base Electrodes. Bronze covered electrodes, which are of the copper-tin or copper-aluminum type, are used for welding iron castings. A low melting point allows their use at low current level, thus minimizing dilution and the effect of heat on the base metal. Although deposits are soft and machinable, color match is poor. Care must be exercised when using these electrodes because excessive dilution and contamination readily cause cracking.

Filler Metals for Gas Metal Arc, Gas Tungsten Arc, and Submerged Arc Welding of Cast Irons. In general, bare wires of chemical composition similar to that of the covered electrodes are used for cast iron welding. The same cautions hold for the solid wire as those listed for the covered electrodes.

Carbon steel filler metals are used for welding iron castings because they are economical and readily available. Because the diluted weld deposit is brittle, use of these products is limited to cosmetic, non-stress-bearing, and nonmachined welds. American Welding Society Specification A5.18 class ER70S-6 materials are the most frequently used products. The high silicon content of the wire (0.80 to 1.15% Si) helps deoxidize the weld pool and improves weld metal flow and wetting of the base metal.

Austenitic stainless steel filler metals are sometimes applied to the welding of iron castings. However, their use should be limited because of the same problems that occur with stainless steel electrodes. These products are classified in AWS A5.9.

American Welding Society Specification A5.14 class ENi-1 nickel wire is commonly used for welding iron castings. As discussed previously, the deposits are relatively strong, machinable, and free of fissuring. Nickel-iron and nickel-iron-manganese filler metals are sold under proprietary labels. Because these products are not classified by specification, they should be judged on their own merits. Welding condition ranges for the gas metal arc welding of cast irons with nickel-iron and nickel wire electrodes are given in Table 3.

Filler Metals for Flux Cored Arc Welding of Cast Irons. Flux cored filler metal products that produce deposits with compositions paralleling covered electrodes are available. Because they combine the best features of coated electrodes and solid wire processes, they are increasing in popularity and variety. Welding condition ranges for the welding of cast iron with nickel-iron flux cored electrodes are given in Table 3.

A proprietary nickel-iron-manganese flux cored filler metal is gaining widespread use for welding iron castings. This filler metal produces a deposit that is similar in composition to ENiFe-CI except for a higher level of manganese. It is extremely crack resistant and capable of high deposition rates. Applications are similar to those of ENiFe-CI.

Steel flux cored electrodes, specified in AWS A5.20 and A5.22 are also used to weld iron castings. Use of these electrodes is subject to the same precautions as described for covered electrodes with similar composition.

Preparation for Welding

Proper preparation of the casting prior to welding is extremely important. When a defect is being repair welded, the defect must first be removed, usually by grinding, gouging, or machining. Attempting to weld over a defect, instead of removing it completely, usually results in poor weld quality. When the defect is a crack contained within the material, a 3.2 mm (1/8 in.) diam hole should be drilled at each end of the crack prior to preparation to prevent crack propagation. Air carbon arc gouging is successfully used on iron castings for removing defects; however, at least 1.6 mm (1/16 in.) of additional material should be removed from the arc-gouged surface by grinding to eliminate the resultant high-hardness layer.

The second step consists of inspecting the area to ensure that the defect has been completely removed prior to welding. The liquid penetrant inspection process is an excellent method of detecting small defects (cracks, pores, shrinkage) that are not readily visible to the naked eye. Low-phosphorus-containing penetrants should be used with nickel consumables.

When the defect is completely removed, the preparation of the joint should be completed. A minimum of 60° included angle is generally required for virtually all of the filler metals and processes listed in this section, and 70° is recommended for most of them. Ends of narrow grooves completely contained within a surface should be beveled to facilitate good fusion in those areas. The size of the weld joint must be kept to a minimum to reduce stress levels resulting from mismatch of thermal contraction rates between weld metal and base metal. The

Table 3 Welding conditions for the GMAW and FCAW of cast iron

Electrode	Wire diameter, mm	Wire diameter, in.	Process	Voltage	Current
Spray transfer					
Nickel-iron wire	1.1	0.045	GMAW(a)	30–32	170–260 DCEP
	2.0	0.078	FCAW(b)	28–30	275–325
	2.4	0.093	FCAW(b)	28–32	300–375
Nickel wire	0.9	0.035	GMAW(a)	30–32	260–280
	1.1	0.045	GMAW(a)	30–32	260–280
	1.6	0.065	GMAW(a)	30–34	260–300
Short arc transfer					
Nickel-iron wire	0.9	0.035	GMAW(a)	15–19	100–120
Nickel-wire	1.6	0.062	GMAW(a)	18–19	290–300 DCEP

(a) Pure argon shielding gas. (b) Self-shielding, but CO_2 shielding is optional

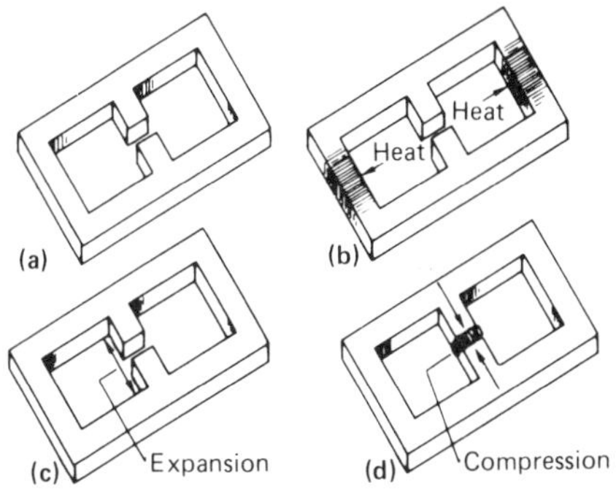

Fig. 7 Preheating of castings. The specific part should be heated such that the weld is under compressive stress upon cooling.

Table 4 Effect of preheat on hardness of shielded metal arc welds in class 20 gray iron welded with 4.8 mm (3/16 in.) nickel-iron (ENiFe-CI) electrodes

Preheat temperature °C	°F	Hardness, HB Weld	HAZ	Base metal
No preheat		342–362	426–480	165–169
105	225	297–362	404–426	165–169
230	450	305–340	362–404	169
315	600	185–228	255–322	169–176

Table 5 Influence of preheat on HAZ of cast irons

Preheat temperature °C	°F	Resulting microstructure
20	70	Martensite
100	210	Pearlite transformation occurs.
200	390	A greater proportion of martensite and carbides are replaced by pearlite.
300	570	Almost all martensite is replaced by pearlite.
400	750	All martensite is prevented.

load-carrying ability of a cast iron weldment can be improved through proper joint design. Joints with incomplete penetration reduce load-carrying cross-sectional thicknesses and create potential stress concentrations beneath the weld bead. If possible, the weld joint should be located away from a highly stressed area, especially where there is a change in section thickness. If possible, welded joints should be loaded in compression, torsion, or shear rather than in tension or bending.

Prior to welding, the casting should be thoroughly cleaned in the area of the weld. If the casting was exposed to dirt, grease, or paint in service, these contaminants must be completely removed from the weld joint, including the back or root side of any full-penetration welds. Repeated degreasing with a good solvent is necessary for castings exposed to grease and oil. Heating the casting to 370 to 480 °C (700 to 900 °F) for 15 to 30 min aids removal of oil and grease impregnated in the casting, followed by mechanical cleaning and additional degreasing with solvent. The surface (skin) of the casting should also be removed in the immediate area of the weld, because of contaminants (burned-in sand, dirt, paint) that cause poor-quality welds. All liquid penetrant and developer applied during the inspection operations should be completely removed before welding.

Prior to repair welding, all closed chambers within the casting should be vented before heating to prevent pressure buildup, and distortion and stresses caused by heating and cooling should be minimized. Drilling small holes into cored areas and forming internal passages for venting relieves excessive pressure buildup. Welding a groove from both sides minimizes distortion. If this is not possible, the part must be restrained during welding to minimize distortion. Peening or preheating can be used to prevent cracking of the weld during cooling.

Preheat and Interpass Temperature

In many cases, preheating is not needed to produce an acceptable weld, but can be used to reduce the thermal gradient and conductivity and to decrease the rate of cooling. Preheating is also useful in reducing differential mass effects, such as those encountered when welds are made between light and heavy sections. A thorough analysis and judicious use of preheating for each shape enable the welder to obtain a uniform cooling rate and thus reduce cooling stresses (Fig. 7). A fundamental preheating rule is illustrated in Fig. 7; the preheat is applied in a manner such that upon cooling, the weld is under compressive stress. Preheating in the areas indicated in Fig. 7(b) expands the metal. Thus, the crack is expanded. After welding, the weld is in compression as the part cools and the metal shrinks (Fig. 7d).

The major concerns with use of excessive preheat temperature are the resulting amount of unnecessary base metal melted, a larger HAZ, and the formation of massive continuous carbides along the fusion line. Cast iron welding requires careful selection of a preheat temperature if the weldment is expected to have the same mechanical properties as the unaffected base metal. Advantages of preheating are most evident in the welding of gray iron. Continuous carbides are not a major concern, because gray iron is already brittle due to its discontinuous flake graphite microstructure. The influence of preheating on the hardness of gray iron weldments made by shielded metal arc welding using ENiFe-CI (nickel-iron) type electrodes is listed in Table 4. Preheating of gray iron is effective in reducing excessive thermal contraction mismatch, thus avoiding fracture during cooling.

The actual preheating temperature range that is best suited to a specific application depends on the hardenability of the base metal, the size and complexity of the weldment, and the type of electrodes to be used. Table 5 lists typical preheat temperatures necessary to achieve a specific type of microstructure in the HAZ of cast iron.

Preheat temperatures for both gray and ductile iron are usually above the martensite start, M_s, temperature; however, to prevent martensite formation in the HAZ, temperatures for gray iron are typically higher than those for ductile iron. If preheat is required, a temperature of 370 °C (700 °F) is typically used for welding gray iron. Temperatures in the 595 to 650 °C (1100 to 1200 °F) range may be necessary when welding very heavy sections because of the heat sinks caused by the large mass of metal. Temperatures as high as 870 °C (1600 °F) can be used in some cases, as in the welding of high-alloy cast irons. The amount of area preheated and the time of preheating become more significant with increasing size of the part to be welded.

When welding ductile iron, the selection of an optimum preheating temperature is more complex. Preheating is not always beneficial or necessary. Preheat temperatures as low as 150 to 175 °C (300 to 350 °F) are preferred when high thermal stresses are expected. At temperatures above 315 °C (600 °F), continuous carbides often form along the fusion line. Preheat temperatures above 315 °C (600 °F) may be necessary, however, for production welding where higher deposition rates are required and for welding of heavy sections. Also, temperatures as high as 480 to 705 °C (900 to 1300 °F) can be used to offset the chill effect caused by decreasing carbon content in the liquid metal when producing a spheroidal-graphite cast iron fusion zone. In such cases, heat input must be reduced to minimize the formation of continuous carbides. The choice of a preheat temperature depends on several factors, and no single temperature is satisfactory for all base metal shapes and compositions under all welding conditions.

The martensite formed in the weld deposit and the HAZ behaves in the same way as the martensite in a quenched steel. Martensite can be transformed into softer products by tempering. In multiple-pass weldments, some tempering of the previous layers occurs as each subsequent layer is deposited, causing a fluctuation in hardness measured over the height of the joint. Heat of welding in single-pass arc welding does not temper the martensite enough to provide reasonable machinability or reduce the danger of postweld cracking. Welds should be made with at least two passes to produce good machinability. The interpass temperature should exceed the preheat temperature by more than 40 °C (100 °F) to prevent additional martensite formation and to capitalize on the tempering effect of later deposits. The alternative to tempering is to decrease

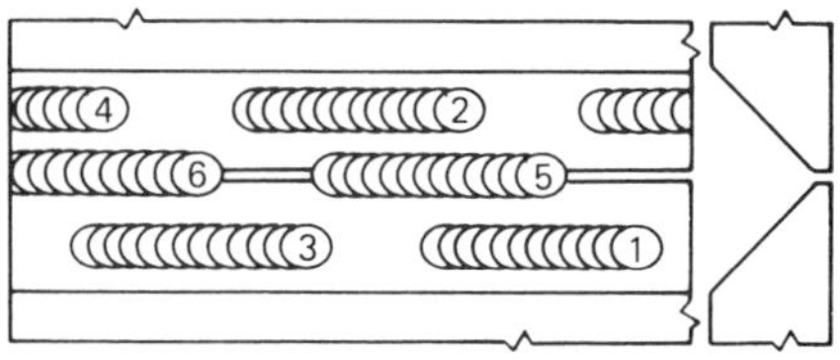

Fig. 8 Use of short weld beads to avoid cumulative stresses

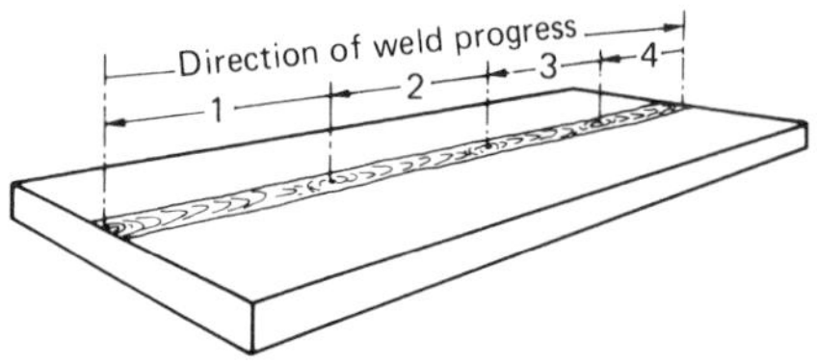

Fig. 9 Backstep sequence

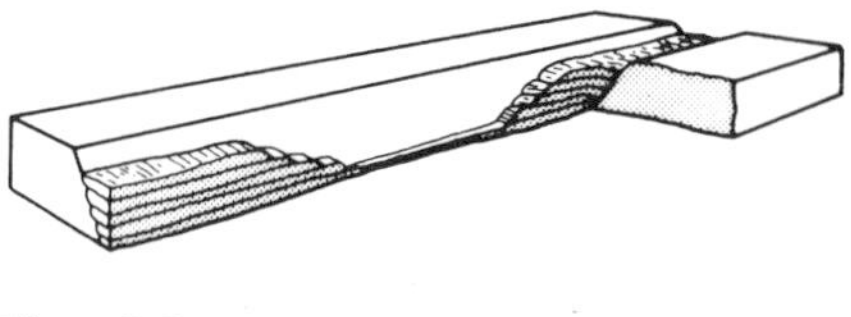

Fig. 10 Block sequence to minimize weld stress

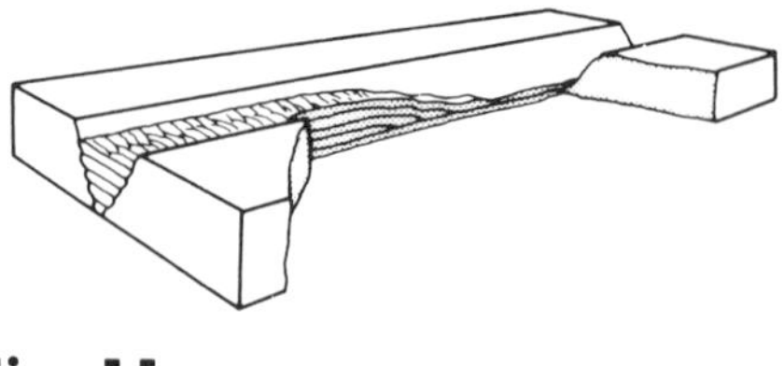

Fig. 11 Cascade sequence of reducing weld stress

the amount of martensite by reducing heat input to produce a narrower HAZ.

Welding Techniques

Heat Input. One method of controlling or minimizing heat input is by the selection of the type of weld bead deposited. Of the two predominant types of weld beads, stringer and weave, the stringer bead is preferred because heat input is minimized. If weave beads are required, the weave should not exceed two or three times the electrode diameter.

Short Weld Sequence. Another method for controlling weld heat buildup is to deposit small weld beads in various portions of the weld joint and to allow each weld bead to cool to approximately 40 °C (100 °F), as shown in Fig. 8. A sequence of short welds helps to avoid the cumulative stresses that occur with long weld beads. Peening these short welds before they cool deforms or works the metal and reduces stress. This technique allows the welder to deposit a weld bead in another location while a previous weld bead cools.

The backstep technique reduces transverse and longitudinal weld shrinkage stress. This technique utilizes short weld beads, which are deposited as shown in Fig. 9. The first weld bead is deposited with a short length of about 50 to 75 mm (2 to 3 in.) and 50 to 75 mm (2 to 3 in.) inward from the edge of the weld joint. Subsequent weld beads are deposited inward from the previously deposited weld bead. All weld beads should overlap and should be deposited in the same direction.

The block and cascade techniques can also be used to minimize shrinkage stresses in large weld grooves. The block sequence utilizes a longitudinal buildup of intermittent weld blocks joined by subsequent weld blocks (Fig. 10). It reduces longitudinal weld stresses, but does not minimize transverse weld stresses. Both longitudinal and transverse shrinkage stresses are reduced with the cascade sequence (Fig. 11). The cascade technique uses a combined longitudinal and overlapping weld bead buildup sequence, as each succeeding bead tempers the one beneath or adjacent to it. The cumulative heating effect of subsequent passes provides additional tempering.

Peening. The principal effects of peening are relief of shrinkage stress, minimizing of distortion, and refinement of microstructure. Peening is best performed at temperatures that turn the metal a dull red. Peening performed below a dull red heat cold works the metal and increases susceptibility to cracking. Peening should be performed with a 13 or 19 mm (½ or ¾ in.) diam round-nosed hammer. Repeated moderate blows should be used instead of a few heavy blows. After peening, all slag should be removed by wire brushing before starting subsequent passes.

Welding Practices

The following general practices have been found to be useful in the welding process:

- When no preheat is used, the interpass temperature should not exceed 95 °C (200 °F)
- When preheat is used, the interpass temperature should not exceed the preheat temperature by more than 40 °C (100 °F)
- To minimize welding stresses, a backstep sequence should be used with stringer beads no more than 50 to 75 mm (2 to 3 in.) in length; each deposit should be allowed to cool to approximately 40 °C (100 °F) before making subsequent deposits
- Avoid melting more of the casting than is necessary
- Whenever possible, deposit two or more layers for best machinability
- Always strike the arc in the weld groove, never on the casting
- The arc length should be kept as short as possible, typically 3 to 5 mm (⅛ to 3/16 in.)

Postweld Heat Treatment

Postweld heat treatment may be done for several reasons and can take many forms. The most common treatments are intended to eliminate the hard structures and thermal stresses formed during welding. Practical considerations may require the finished weldment to cool to room temperature before the postweld heat treatment can be started. When heat treatment is not applied immediately after welding, the casting must be cooled slowly from the welding temperature to room temperature by covering it with insulating materials. Heat treatment should be started as soon as possible, especially when brittle structures and high thermal stresses are present. In such cases, the likelihood of cracking increases as temperature decreases, and welds may fail catastrophically.

The simplest postweld heat treatments for gray, compacted graphite and ductile irons are stress relief and/or tempering, for which 1 or 2 h at 595 to 650 °C (1100 to 1200 °F) is usually sufficient. A heating and cooling rate of 55 °C/h (100 °F/h) above 315 °C (600 °F) is recommended. Holding times and postweld heat treatment temperatures are specified in Table 6. Furnace cooling rates specified in Table 6 to at least 315 °C (600 °F) are recommended before air cooling. This treatment also tempers the martensite and bainite transformation products and reduces hardness of the products by a considerable degree. When carbides are present, a full anneal may be necessary. Longer times may be needed for dissolving massive continuous carbides formed at high peak temperatures.

The weldment can be fully softened by a ferritize annealing process, as specified in Table 6 for gray and ductile iron. This treatment results in a ferritic matrix and tends to dissociate iron carbide and transports the carbon to the graphite. This structure has maximum machinability and ductility. This practice is common for the 60-43-18 and 65-45-12 grades of ductile iron.

Other recommended postweld heat treatment annealing practices are listed in Table 6 for the various types of graphitic cast iron. In critical applications that require radiographic inspection after heat treatment, castings are inspected before heat treatment to save unnecessary costs if an internal defect is present. Additional heat treatment information on specific cast irons can be found in Volume 4 of the 9th Edition of *Metals Handbook*.

Gray Irons

Because gray iron has graphite in flake form, carbon can be easily introduced into the weld pool, thus causing weld metal embrittlement. Consequently, techniques that minimize base metal dilution are recommended. Care must be taken to compensate for shrinkage stresses, and the use of low-strength filler metals helps reduce

Table 6 Recommended postweld heat treatment practice for various types of graphitic cast irons

Treatment	Temperature °C	Temperature °F	Holding time, min/mm (h/in.) of thickness	Cooling rate
Gray iron				
Stress relief	595–650	1100–1200	3.5 (1.5)	Furnace cool to 315 °C (600 °F) at 55 °C/h (100 °F/h), air cool to room temperature.
Ferritize anneal	705–760	1300–1400	2.4 (1)	Furnace cool to 315 °C (600 °F) at 55 °C/h (100 °F/h), air cool to room temperature.
Full anneal	790–900	1450–1650	2.4 (1)	Furnace cool to 315 °C (600 °F) at 55 °C/h (100 °F/h), air cool to room temperature.
Graphitizing anneal	900–955	1650–1750	1–3 h, plus 2.4 min/mm (1 h/in.)	Furnace cool to 315 °C (600 °F) at 55 °C/h (100 °F/h), air cool to room temperature.
Normalizing anneal	870–955	1600–1750	1–3 h, plus 2.4 min/mm (1 h/in.)	Air cool from annealing temperature to below 480 °C (900 °F); may require stress relief
Ductile iron				
Stress relief				
Unalloyed	510–565	950–1050	3.5 (1.5)	Furnace cool to 315 °C (600 °F), at 55 °C/h (100 °F/h) in air from 315 °C (600 °F) to room temperature.
Low alloy	565–595	1050–1100		
High alloy	540–650	1000–1200		
Austenitic	620–675	1150–1250		
Ferritize anneal	900–955	1650–1750	1 h plus 2.4 min/mm (1 h/in.)	Furnace cool to 690 °C (1275 °F), hold at 690 °C (1275 °F) for 5 h plus 2.4 min/mm (1 h/in.) of thickness, furnace cool to 345 °C (650 °F) at 55 °C/h (100 °F/h), air cool to room temperature.
Full anneal	870–900	1600–1650	2.4 (1)	Furnace cool to 345 °C (650 °F) at 55 °C/h (100 °F/h), air cool to room temperature.
Normalizing and tempering anneal	900–940	1650–1725	4.8 (2)(a)	Fast cool with air to 540 to 650 °C (1000 to 1200 °F), furnace cool to 345 °C (650 °F) at 55 °C/h (100 °F/h), air cool to room temperature.

(a) Minimum holding time

cracking without sacrificing overall joint strength.

Gray iron weldments are susceptible to the formation of porosity. This can be controlled by lowering the amount of dilution of the weld metal or by slowing the cooling rate so the gas has time to escape. Preheating helps reduce porosity and reduces the cracking tendency. A minimum preheat of 205 °C (400 °F) is recommended, but 315 °C (600 °F) is generally used.

The most common arc welding electrodes for gray iron are nickel and nickel-iron types (AWS A5.15 class ENi-CI and ENiFe-CI). These electrodes have been used with or without preheating and/or postweld heat treatment. The cast iron (ECI) and steel (ESt) types must be used with high preheats (540 °C, or 1000 °F) to avoid cracking and hard deposits. The copper-aluminum (ECuAl-Al) and copper-tin (ECuSn) types are used, but color match is poor. Copper-base welds have quality comparable to that of a brazed joint, and a preheat temperature of at least 150 °C (300 °F) must be employed. The strength of these joints is often poor, partially because of the difference in thermal expansion between the copper alloy and the gray iron. Typical tensile properties for weldments of various grades of gray iron are given in Table 7 as a function of welding process and consumable.

Table 7 Approximate tensile strength and color match obtainable in welded joints in gray iron using different processes

Based on results reported from various tests and on production experience

Process	Filler metal	Base metal class	Tensile strength of joint(a) MPa	Tensile strength of joint(a) ksi	Color match
SMAW	ENi-CI	30	170–205	25–30	Fair
	ENiFe-CI	30	170–205	25–30	Fair
	ENiFe-CI	40	250	36	Fair
	ESt	30	205	30	Poor to fair
	ESt	40	275	40	Poor to fair
GMAW	Ni or Ni-Fe wire(b)	30	275	40	Fair
	Ni or Ni-Fe wire(b)	40	275	40	Fair
FCAW	Ni-Fe wire	30	170–205	25–30	Fair
		40	205	30	Fair
GTAW	Ni or Ni-Fe wire(b)	30	275	40	Fair
	Ni or Ni-Fe wire(b)	40	275	40	Fair

(a) Approximate strength expected if good welding procedures and skilled operators are used. Wide variations in strength may occur as a result of variation in welding practice or in base metal condition or size. (b) These filler metals are not classified, but wire coils and rods of compositions equivalent to ENi-CI and ENiFe-CI are obtainable.

Ductile Irons

Ductile cast irons have greater weldability than gray irons, but require specialized welding procedures and filler materials. Pearlitic ductile iron produces a larger amount of martensite in the HAZ than ferritic ductile iron and is generally more susceptible to cracking.

Shielded metal arc welding using an ENiFe-CI (nickel-iron) electrode is the most common technique for welding ductile cast irons. Shielded metal arc welding with an ENiFe-CI electrode sometimes employs a 150 to 205 °C (300 to 400 °F) preheat; preheats up to 315 °C (600 °F) are used on large castings. Most ductile iron castings, however, do not require preheating. Electrodes should be baked according to the manufacturer's recommendation to minimize hydrogen damage and porosity. Direct current electrode positive (DCEP) is usually used at a current sufficient to produce stringer beads with a moderate traveling speed. If machinability or optimum joint properties are desired, castings should be annealed immediately after welding. Transverse joint properties of typical ductile iron weldments are listed in Table 8.

Gas metal arc welding utilizing short arc transfer with nickel (AWS ERNi-1) filler metal and pure argon shielding has been successfully used for welding ductile cast irons. Recommended welding current is 130 to 160 A at 18 to 24 V. Based on the wire-feed rate, these controls produce a short arc transfer mode, which reduces heat input and reduces the amount of HAZ carbides and martensite, but requires a preheat of 200 °C (400 °F) on heavier (>13 mm, or ½ in.) wall castings. Ferritic ductile iron, however, can usually be welded without preheat. Typical transverse tensile strength properties obtained in 25 mm (1 in.) thick pearlitic and ferritic ductile iron test plates using this process are listed in Table 9.

Gas tungsten arc welding produces reasonably strong weld joints utilizing either nickel-iron or ductile iron filler metals. The transverse joint properties for ductile iron weldments made with gas tungsten arc welding are given in Table 10.

Table 8 Transverse joint properties of typical SMAW ductile iron weldments using ENiFe electrodes

Plate condition	Tensile strength		Yield strength 0.2% offset,		Elongation, %	Reduction in area, %	Weld hardness, HB	Maximum hardness in HAZ, HV	Bend angle, degrees	Unnotched Charpy impact strength	
	MPa	ksi	MPa	ksi						J	ft · lbf
Ferritic (60-45-10)											
Unwelded	470–520	68.5–72.5	325–395	47.5–57	10–18	. . .	. . .	. . .	78–80	137	101
As-welded	350–470	51–68.5	275–395	40–57	1–9	11–18	205–250	665	14–28	19–27	14–20
Annealed(a)	370–495	54–72	255–395	37–57	5–14	5–28	185–235	420	18–38	38–49	28–36
Ferritize annealed	420–470	61–68	270–360	39–52	6–12	3–5	175–180	175	40–45	45–49	33–36
Pearlitic (80-60-03)											
Unwelded	680	98.5	590	86	3	nil	. . .	. . .	5–10	14–15	10–11
As-welded	455–565	66–82	490–525	71–76	1–2	3–4	190–225	535	14–15	18	13
Annealed(a)	435–490	63–71	470	68	1–2	2–6	175–205	555	. . .	15	11
Ferritize annealed	395–450	57–65	340–350	49–51	6–12	7–15	175–180	185	33–36	49–54	36–40

(a) 595–650 °C (1100–1200 °F) for various times. Source: Ref 1

Table 9 Transverse tensile properties of gas metal arc butt welds on ductile iron plates

Short arc process used on 25 mm (1 in.) thick plates. Test pieces were 14.3 mm (0.564 in.) in diameter over 80 mm (3.15 in.) parallel position.

Filler metal	0.2% proof stress		Maximum stress		Elongation over weld in 50 mm (2 in.), %	Location of fracture
	MPa	ksi	MPa	ksi		
Pearlitic plate						
Unwelded	385–465	56–67.2	615–770	89.6–112	1–3	. . .
Nickel 61	360	52	520	75.7	3	Weld fusion zone
	360	52	575	83.8	4	Heat-affected zone
Ferritic plate						
Unwelded	250–310	36.6–44.8	385–540	56–78.4	15–25	. . .
Nickel 61	300	43.9	425	61.8	11.5	Away from weld
	305	44.4	415	60.5	11.5	Away from weld

Source: Ref 2

Table 10 Transverse joint properties of GTAW ductile iron weldments

Filler metal	Postweld heat treatment	Tensile strength		Yield strength 0.2% offset		Elongation, %
		MPa	ksi	MPa	ksi	
Ductile iron	None	380	55	. . .	. . .	8
	595 °C/2 h (1100 °F/2 h); furnace cool	370	54	. . .	. . .	5
	900 °C/2 h (1650 °F/4 h); furnace cool	400–420	58–61	310–385	(45–56)	4–6
	Full ferritize anneal	310–340	45–49	. . .	. . .	. . .
Nickel-iron (60–40)	None	425–460	62–67	. . .	. . .	1–7
	Full ferritize anneal	420	61	. . .	. . .	8

Malleable Irons

During welding, the ductility of the HAZ of malleable iron is severely reduced because graphite dissolves and precipitates as carbide. Although postweld annealing softens the hardened zone, minimal ductility is regained. Despite these limitations, for certain applications, malleable iron castings can be welded satisfactorily and economically if precautions are taken. Malleable iron castings should not be repaired by welding to correct a failure caused by overstressing of the part.

Because most malleable iron castings are small, preheating is seldom used. If desired, small welded parts can be stress relieved at temperatures up to 540 °C (1000 °F). For heavy sections and highly restrained joints, preheating at temperatures up to 150 to 205 °C (300 to 400 °F) and postweld malleablizing annealing are recommended. However, this costly practice is not always followed, especially when the design of the assembly is based on reduced strength properties of the welded joint. Because no welding procedure can satisfy all types of service conditions, each application for welding malleable iron assemblies should be carefully reviewed and tested before production is begun.

Ferritic malleable grades 32510 and 35018 have the highest weldability of the malleable irons, even though impact strength is reduced. Pearlitic malleable irons, because of their higher combined carbon content, have lower impact strength and higher cracking susceptibility when welded. Small leaks, gas holes, and other small casting defects can be repaired in grades 32510 and 35018 by arc welding. If a repaired area must be machined, arc welding should be done with a nickel-base electrode. Shielded metal arc welding using low-carbon steel and low-hydrogen electrodes at low amperage produces satisfactory welds in malleable iron. If low-carbon steel electrodes are used, the part should be annealed to reduce any increased hardness in the weld (due to carbon pickup) and in the HAZ.

White and Alloy Cast Irons

Chilled and white cast irons are abrasion-resistant cast irons having structures free of graphitic carbon. Because of their extreme hardness and brittleness, they are generally considered unweldable. Alloy irons are used in applications requiring good abrasion, corrosion, or heat resistance properties. The most important consideration of welding alloy cast iron is to achieve equivalent service properties.

Because abrasion-resistant cast irons (such as ASTM A532) have limited resistance to thermal shock, welding is generally not recommended. Welding is sometimes employed in repair operations or to attach parts to other machine components. Gas welding is preferred over arc welding because of the lessened tendency toward cracking.

Arc welding can be done with a type 310 stainless steel electrode (25Cr-20Ni) if the welded area is not subject to abrasion. This electrode minimizes cracking in the parent metal. If the welded area is subject to abrasion, an electrode that produces weld metal of the same abrasion resistance as the base metal should be used. In either case, the casting should be preheated from 315 to 480 °C (600 to 900 °F) and slowly cooled after welding by covering with insulating material. Stress relieving at 205 °C (400 °F) should follow. Unless welding is done very carefully, the weld metal may exhibit very fine cracks that may cause failure if the casting is subjected to heavy loads in service. Under most operating conditions, however, the repaired casting should perform satisfactorily.

Corrosion- and Heat-Resistant Cast Irons

Corrosion-resistant cast irons are high-silicon, high-chromium, or high-nickel irons. Specifications for many of these irons permit welding for the repair of minor casting defects. For more extensive welding, the effect of welding on the service properties of the casting should first be determined; applications of this group of cast irons are highly specialized. Because weld deposits usually need to duplicate base metal compositions, filler metals may not be generally available.

Heat-resistant cast irons provide high strength at elevated temperatures, as well as resistance to scaling. They are normally produced as flake-graphite irons, but can also be produced with the free carbon in the form of nodules or spheroids. Heat-resistant cast irons can be welded for repair of minor casting defects, but much like corrosion-resistant irons, the effects of welding on service properties must be determined before extensive welding is undertaken.

Oxyacetylene Welding of Cast Irons

Oxyacetylene welding (OAW) is widely used on gray iron, to a smaller extent on ductile iron, and only to a minor extent on malleable iron. Cast iron filler metal is melted together with the base metal to form the joint. An oxyacetylene flame has a maximum temperature of about 3300 °C (6000 °F), which is several thousand degrees less than that of a welding arc. Oxyacetylene welding is therefore slower than arc welding and results in greater total heat input and wider HAZs. This heat results in expansion and localized stress, particularly in large castings. For this reason, high preheats of 595 to 650 °C (1100 to 1200 °F) are generally used for the oxyacetylene welding of cast irons. However, for local repairs and small unrestrained castings, lower preheat temperatures, often as low as 425 °C (800 °F), are used. Depending on the mass, composition, and structure of the casting, postheating requirements vary from slow cooling for complete stress relief (620 °C, or 1150 °F) to full annealing (900 °C, or 1650 °F). Because of higher preheat, and thus lower cooling rates, oxyacetylene welding produces less hardening of the HAZ than arc welding. Cast irons that are preheated, welded, and slow cooled are readily machinable, which can be important in repair work.

Oxyacetylene welding is not recommended for the joining of malleable iron (see the discussion "Welding of Malleable and White Irons" in this section). Good results have been obtained in oxyacetylene welding ferritic ductile iron with ductile iron rods. In production welding, however, the speed of oxyacetylene welding cannot compare with that of arc welding. In addition, welds deposited by oxyacetylene welding cannot compare with that of arc welding. In addition, welds deposited by oxyacetylene welding with cast iron rods usually are less machinable than nickel or nickel-iron welds deposited by arc welding. Porosity, which is a common problem in oxyacetylene welding, can be minimized by using a slightly reducing flame.

As a general practice, oxyacetylene welding has been used on cast irons for:

- Repair of minor casting defects in gray iron and ductile iron. Minor surface blemishes in malleable irons are sometimes repaired; generally, however, oxyacetylene welding of malleable iron is avoided whenever possible
- Repair of service-incurred wear and damage mostly on gray and ductile irons
- Production of gray and ductile iron weldments involving either two parts made of cast iron, or one of cast iron and one of another metal, usually steel

Repair of Casting Defects. One of the most common applications of welding cast iron is the repair of rough gray iron castings. Although the majority of this repair work is done by oxyacetylene welding, some is done by arc welding (see the discussion "Arc Welding of Cast Irons" in this section). If repair welding is confined to the correction of small defects that affect only the appearance of the casting, inferior mechanical properties and machinability of the weld are of no consequence. The defect must be in an unstressed area that requires no machining and, as a result, should not extend through the section. Typical defects include sand holes, porosity, washouts, cold shuts, and shift. Reclaiming defective castings by repair welding is common practice.

Castings that have defects resulting from machining errors can also be repaired by oxyacetylene welding, provided that the heat of welding does not cause distortion. Arc welding is usually preferred to oxyacetylene welding for correction of machining errors because arc welding is faster, has a lower total heat input (and therefore causes less distortion), and produces welds with adequate properties. Good color match of weld and base metal generally is an additional requirement.

Repair of Damaged Castings. Iron castings that have become cracked, broken, or worn in service are regularly repaired by oxyacetylene welding. Braze welding is used in many applications because of its simplicity and low preheat requirements and because color match is seldom important in such repair (see the article "Oxyacetylene Braze Welding of Steel and Cast Irons" in Volume 6 of the 9th Edition of *Metals Handbook*). If welding must be done under adverse conditions, extra care and attention to procedural detail are required. Because repair of damaged castings is often a major welding operation, in that a considerable mass of base metal is subjected to high temperatures, preheating is required. A temporary oven around the part or a means of providing localized heating may be needed, depending on size and shape of the part, the required temperature, and the duration of heating.

Repair of Worn Castings by Hardfacing. Oxyacetylene welding is often used to repair (build up by hardfacing) specific areas of worn gray or ductile iron castings. Malleable iron castings are not well suited to repair by hardfacing.

The choice of a welding process depends largely on service requirements and the equipment available. If the properties obtained are acceptable, braze welding is the logical choice because the casting is far less likely to crack than when arc or oxyacetylene welded. The choice between arc and oxyacetylene welding for repairing worn castings by hardfacing usually depends on the equipment available. Similar results can be obtained with both processes.

The cost of repairing worn castings must be weighed against the cost of replacing them. Consider the cost of building up cast iron, a relatively cheap material, with costly hardfacing alloys. In practice, however, there are considerations other than just the price of the casting, such as delays in getting replacement castings and downtime for replacing a casting in a machine. Repair by hardfacing eliminates procurement delays. A casting surface that has been hardfaced by welding often lasts two to five times as long as the original surface. Mill-roll journals, rolling-mill guides, wire-spinning rolls, and cast components of mills that process abrasive materials such as cement and clay products are typical applications of repair by hardfacing.

The overlay (hardfacing) material applied to castings by oxyacetylene welding usually contains at least 3.0%, and more often 4.0 to 5.0%, C. This usually equals or exceeds the carbon content of the base metal. In addition to having high carbon contents, most hardfacing materials also have high alloy contents. Most hardfacing materials are proprietary alloys; three typical ones have nominal (iron-base) compositions:

- 3.9C-32.0Cr-6.0Mo
- 4.1C-16.0Cr-2.0Ni-8.0Mo-1.0V
- 4.3C-16.0Cr-6.0Ni-8.0Mo

Hardfacing alloys such as these are available in rod form for oxyacetylene welding.

Preheating of the casting prior to welding the overlay minimizes cracking. For castings having reasonably uniform sections, a preheat of 345 to 370 °C (650 to 700 °F) is sufficient; however, for castings having a wide variation in section thickness, preheating in the range of 595 to 650 °C (1100 to 1200 °F) is generally recommended. Time at preheating temperature should be sufficient to ensure that the casting has been uniformly heated.

Postheating is not necessary, but the welded casting must be cooled slowly. In preferred practice, the welded casting is immediately placed in a furnace maintained at or near the preheating temperature used and then cooled in the furnace. If a furnace is not available, the welded casting should be buried immediately in an insulating material such as lime or spent carburizing compound and should be allowed to remain buried until it has cooled to near room temperature.

Preparation of Castings

If a casting to be repaired has been in service, preparation of the casting for welding requires, in addition to edge preparation, the removal of surface contaminants. Oil, grease, and paint should be removed with solvents, commercial cleaners, or paint removers. Impregnated oil or other volatile matter can be eliminated by heating the casting or weld groove to approximately 480 °C (900 °F)—a dull red heat—for about 15 min and then wire brushing, grinding, or rotary filling to remove the residue. Casting skin on surfaces adjacent to the joint area can be removed by grinding, chipping, shotblasting, or rotary filing. Defects such as porosity, inclusions, and cold shuts should be gouged out, and the bottom of the cavity should be well rounded rather than V-shaped.

Completely broken sections should be dressed to form a single-V joint, with a 2 to 3 mm (1/16 to 1/8 in.) root face to align the parts. Gas welding requires a V-groove with an included angle of 60 to 90° to permit proper manipulation of the torch and welding rod. For heavy sections, a double-V joint should be prepared whenever feasible, with a root face at or near the center of the workpiece.

Preheating

Adequate preheating decreases the rate of cooling of the weld metal and adjacent base metal, thus minimizing the formation of brittle microstructures in the zone around the weld that might cause cracking immediately after welding or in service. Softer, less brittle microstructures are obtained with high rather than low preheat temperatures. Recommended preheats are discussed in the discussions "Gray Irons" and "Ductile Irons" in this section.

To ensure that the preheat temperature is maintained throughout the welding operations, it may be necessary to insulate the heated casting. Heat input from welding must not be permitted to increase the interpass temperature above the maximum preheat temperature, or welding must be stopped until the temperature drops to the preheat range. The temperature should be measured by contact pyrometers or temperature-indicating crayons at or near the weld zone and at one or more other places as required.

Table 11 Chemical compositions of cast iron filler metals used for OAW of cast iron (AWS A5.15)

Classification	Composition, %							
	C	Si	Mn	Ni	Mo	P	S	Ce
RCI	3.25–3.50	2.75–3.00	0.60–0.75	Trace	Trace	0.50–0.75	0.10 max	. . .
RCI-A	3.25–3.50	2.00–2.50	0.50–0.70	1.20–1.60	0.25–0.45	0.20–0.40	0.10 max	. . .
RCI-B	3.25–4.00	3.25–3.75	0.10–0.40	0.50 max	. . .	0.50 max	0.03 max	0.20 max

Postweld Heat Treatment

Postweld heat treatment can be either stress relieving or full annealing. Stress relieving at 620 °C (1150 °F) and then furnace cooling to 370 °C (700 °F) or lower is recommended whenever feasible. Full annealing at 900 °C (1650 °F) produces greater softening of the weld zone and more nearly complete stress relief, but it also lowers the tensile strength of all but the softest irons.

Welding Rods

Cast iron rods used in the welding of cast iron should contain enough carbon and silicon to allow for losses of these elements during welding. The silicon content of the filler metal must be high enough to permit carbon to precipitate as free carbon during solidification and to promote a soft, machinable matrix as the weld cools to room temperature. Chemical compositions of cast iron filler metals used for the oxyacetylene welding of cast iron are given in Table 11.

Rods made of cast iron are usually 610 mm (24 in.) long and 6 mm (1/4 in.) square, although rods 3 to 12 mm (1/8 to 1/2 in.) square are available. The RCI filler metals and a number of proprietary rod compositions are used for welding gray irons of classes 20 to 35 (140 to 240 MPa, or 20 to 35 ksi, tensile strength). Gray irons that have tensile strengths of 240 to 275 MPa (35 to 40 ksi) can be welded with RCI-A filler metal, which is similar to RCI but also contains 1.20 to 1.60% Ni and 0.25 to 0.45% Mo. Many proprietary rods contain chromium, nickel, molybdenum, copper, or vanadium, either singly or in combination, to produce high-strength welds; carefully controlled procedure is required with these rods to avoid obtaining a hard weld deposit. Two basic types of welding rods have been successfully used for welding ductile iron: RCI-B, which is generally higher in carbon and silicon content and lower in manganese than the gray iron rods and contains cerium as a nodularizing agent, and proprietary rods in which magnesium is used as the nodularizing agent.

Fluxes

A flux is required in the oxyacetylene welding of cast iron to increase the fluidity of the fusible iron-silicate slag and to aid in the removal of the slag. Fluxes for gray iron rods are usually composed of borates or boric acid, soda ash, and small amounts of other compounds such as sodium chloride, ammonium sulfate, and iron oxide.

Fluxes suitable for welding ductile iron rod are similar to those used for welding gray iron, but are formulated to produce a slag with a lower melting point. Some proprietary fluxes contain inoculants.

Welding of Gray Iron

The oxyacetylene welding of gray iron is generally done for repair, but can also be done for production of simple assemblies. The cavity is first prepared for welding by the usual methods (see the discussion "Preparation of Castings" in this section) and then tested by liquid penetrant or magnetic-particle inspection to ensure freedom from defects.

The casting is preferably preheated to 620 °C (1150 °F) in a furnace and then covered with heat-retaining material, exposing only the cavity to be welded. If a furnace is not available, the casting can be covered with heat-retaining material and locally heated by gas flame.

A high-velocity torch that produces a concentrated flame pattern similar to that used for welding low-carbon steel should be used. The cavity surface is dusted with a thin layer of flux, the heated rod is dipped in flux and positioned in the cavity, and both are heated with an oxyacetylene torch adjusted for a neutral or slightly reducing flame. The rod and a small area of the cavity soften under the flame, and as the rod melts off and combines with the base metal, the torch and rod are slowly moved along the cavity. The flame should be directed at the bottom of the cavity with the tip held 6 to 3 mm (1/4 to 1/8 in.) from the metal until a molten pool up to 25 mm (1 in.) long begins to form. The torch is moved from side to side until the walls of the cavity start to melt into the pool. This process is continued until the entire cavity is filled with weld metal 2 to 3 mm (1/16 to 1/8 in.) thick. Playing the welding flame back over the previously deposited metal to retard its cooling rate will reduce residual stress and permit escape of entrapped gas.

Postweld stress relieving is recommended, particularly for complex castings or where accurate machining will follow. Immediately after completion of welding, the casting should be placed in a furnace heated to the same temperature as the casting and gradually heated to 595 to 650 °C (1100 to 1200 °F), where it is held for 1 h per inch of thickness. The casting can then be cooled to 260 °C (500 °F) or below at a rate no faster than 25 °C/h (50 °F/h). When the welded casting is not stress relieved, it should be cooled slowly either in a furnace or by covering it with heat-retaining material, sand, or some other insulated material.

When class 30 irons are preheated to 600 to 620 °C (1100 to 1150 °F) before oxyacetylene welding, some fine pearlite and ferrite are present in the final weld metal and in the HAZ; weld tensile strength approaching 205 MPa (30 ksi) and hardness less than 200 HB can be obtained. When class 40 irons are oxyacetylene welded under similar preheat conditions, higher proportions of fine pearlite, less ferrite, and some cementite are present, and weld strength may reach 240 MPa (35 ksi), with hardness as high as 200 HB. Although higher strength irons containing chromium and molybdenum have been welded, welding rods compatible with these irons are not readily available. Welding is generally not attempted in gray iron stronger than class 50 because of the high probability of cracking.

Difficulty in machining is seldom encountered if high preheat and interpass temperatures are used. Class 40 irons can usually be oxyacetylene welded to a hardness of 200 HB or less in weld metal and the HAZ. Cast iron welding rods provide good color match.

Welding of Ductile Iron

Oxyacetylene welding is often used for the repair of defects in ductile iron castings. It has also been successfully used for the hardfacing of specific areas on castings to increase abrasion resistance. The oxyacetylene welding of ductile iron has been most successful using a ductile iron rod.

The repair of ductile iron castings is complicated by the fact that the only way to obtain graphite in nodular form in the weld deposit is to cause it to precipitate from the liquid. Processes have been developed to cause nodularization by introducing magnesium or cerium, or both, into the weld zone by means of a special cast iron filler rod or special flux. Filler metal RCI-B should be used for welding ductile iron.

Magnesium and cerium are carbide formers and consequently must not be present in amounts beyond those required for nodularization. If these elements are present in excessive amounts, a postweld ferritizing anneal is necessary to restore ductility to the weld area; annealing reduces the strength of pearlitic irons and causes distortion of machined surfaces.

Joint preparation and joint cleanliness require the same careful attention and the same procedures described earlier for welding gray iron. Preheating practice and interpass temperatures are essentially the same for welding ductile iron as those described for gray iron.

In welding ductile iron, a reducing flame is used to minimize oxidation of the volatile nodularizing elements contained in both the base metal and the welding rod. The welding tip should produce a concentrated flame pattern. The weld area or the complete casting is first preheated to a dull red, and flux is applied to the bottom of the weld groove. Heat is directed at the bottom of the weld groove until a pool begins to form. Walls adjacent to the weld groove are then softened to blend into the pool. The recommended length of the molten pool should be limited to 25 mm (1 in.). A major problem in the welding of ductile iron is the complete loss of ductility in the HAZ.

Welding of Malleable and White Irons

The effect of oxyacetylene welding on malleable iron is to create a wide HAZ of white iron, the material from which the malleable iron was originally produced by applying a malleablizing heat treatment. Because of the hardness and brittleness of the white iron, base metal properties are lost, and the joint is prone to cracking. The hardened zones in either ferritic or pearlitic malleable iron can be reduced by annealing, but only by processing the casting through a special heat-treating procedure for which the original foundry is best equipped.

Oxyacetylene welding is used in the foundry repair of small defects on rough castings while the casting is in the white iron condition, before malleablizing. White iron welding rods are used, which are cast by the foundry to match the base metal composition, after allowing for constituent losses in deposition. Repair procedures are the same as for gray iron except that, after welding, the casting is given its normal malleablizing heat treatment.

Braze welding, using a copper alloy welding rod, is a more satisfactory method of obtaining relatively strong, machinable joints in malleable iron. Because the base metal is not melted in braze welding, peak temperatures are relatively low, and there is very little hardening in the HAZ. Color match between the weld and the base metal is poor, however, and the service temperature of the casting is limited to about 260 °C (500 °F). Details of the process are given in the article "Oxyacetylene Braze Welding of Steel and Cast Irons" in Volume 6 of the 9th Edition of *Metals Handbook*.

Gas metal arc welding with bare wire steel electrodes has been used in joining malleable iron where end uses do not require full base metal strength properties. This process produces high peak temperatures, which are of short duration because of high welding speed. Heat-affected zones are therefore narrow. This process is discussed in the discussion "Arc Welding of Cast Irons" in this section.

Welding of Cast Steels

The welding of cast steels is done for many reasons. These include, but are not limited to, repair welding (the upgrading of castings to meet strict quality control standards), cast-weld construction, and weld overlay and facing applications.

Repair of steel castings is necessitated when discontinuities are present that exceed the allowable limits set by the standards to which the casting is being manufactured. Repair of the casting may involve cutting or gouging out the defect, followed by welding with a filler material that is compatible with the base casting composition and strength. The need to produce castings to high quality control standards has involved the use of welding to upgrade castings, especially in regions where narrow and complicated cross sections make it difficult to achieve perfect feeding. Such regions in castings must be removed and replaced with weld metal. To improve the integrity of a fabricated component, it may be necessary to join a casting to another casting or a wrought component. Finally, weld overlay and facing considerations are similar to those in wrought steels, in which the surface properties are to be significantly different from the base material properties. The use of hardfacing materials or corrosion-resistant overlays is the situation typically encountered.

In all of the above-mentioned instances where cast steels are involved with welding, the procedures involved are similar to those for wrought steels; therefore, few standards have been established that deal exclusively with the welding of cast steels. A list of typical specifications covering the welding of steel and alloy castings is summarized in Table 12.

Welding Processes

Many of the processes commonly used in the welding of wrought steels are utilized in welding operations involving cast steels. These include shielded metal arc welding, gas metal arc welding, gas tungsten arc welding, submerged arc welding (all of which were described earlier in the section "Arc Welding of Cast Irons" in this article), and electroslag welding (ESW). The shielded metal arc process (also referred to as stick welding) is the most versatile and popular of the processes for both repair and joining, while the submerged arc and electroslag processes are more suitable for joining operations. Detailed descriptions of these processes can be found in Volume 6 of the 9th Edition of *Metals Handbook*.

Table 12 Typical specifications covering the welding of steel and alloy castings

American Society for Testing and Materials

A 488 Standard Recommended Practice for Qualification of Procedures and Personnel for the Welding of Steel Castings
E 390 Reference Radiographs of Steel Fusion Welds

American Society of Mechanical Engineers

ASME Boiler & Pressure Vessel Code Section II Part C
Welding Rods, Electrodes, and Filler Metals
SFA 5.1 Mild Steel Covered Arc Welding Electrodes
SFA 5.2 Iron and Steel Gas Welding Rods
SFA 5.4 Corrosion Resistant Chrome and Chrome Nickel Steel Covered Welding Electrodes
SFA 5.5 Low Alloy Steel Covered Arc Welding Electrode
SFA 5.9 Corrosion Resisting Chrome and Chrome Nickel Steel Welding Rods and Bare Electrodes
SFA 5.11 Specification for Nickel and Nickel Alloy Covered Welding Electrodes
SFA 5.13 Surfacing Welding Rods and Electrodes
SFA 5.14 Nickel and Nickel Alloy Bare Wire and Electrodes
SFA 5.17 Bare Carbon Steel Electrodes and Fluxes for Submerged Arc Welding
SFA 5.18 Mild Steel Electrodes for Gas Metal-Arc Welding
SFA 5.20 Mild Steel Electrodes for Flux Cored Arc Welding
SFA 5.22 Flux Cored Corrosion Resisting Chrome and Chrome Nickel Steel Electrodes
SFA 5.23 Bare Low Alloy Steel Electrodes and Fluxes for Submerged Arc Welding
SFA 5.25 Consumables Used for Electro-Slag Welding of Carbon and High Strength Low Alloy Steels
SFA 5.26 Consumables Used for Electro-Gas Welding of Carbon and High Strength Low Alloy Steels
ASME Boiler and Pressure Vessel Code Section IX
Welding and Brazing Qualifications

American Welding Society

AWS A2.2 NDT Testing Symbols
AWS A3.0 Welding Terms and Definitions
AWS A5.0 Filler Metal Comparison Charts
AWS A5.1, A5.2, A5.4, A5.4 Ad. 1, 1975, A5.5, A5.9, A5.11, A5.13, A5.14, A5.17, A5.18, A5.20, A5.22, A5.23, A5.25, and A5.26 are identical to ASME Section II Part C Grades SFA 5.2, SFA 5.4, etc.

Military

MIL-STD-22 Welding Joint Design
MIL-STD-00248 (Ships) Welding and Brazing Procedure and Performance Qualification
MIL-STD-278 Fabrication Welding and Inspection and Casting Inspection in Repair for Machinery, Piping, and Pressure Vessels in Ships of the United States Navy
MIL-E-19933 Electrodes and Rods—Welding Bare, Chromium and Chromium Nickel Steels
MIL-E-19922 Fluxes
MIL-E-2200/1 Electrodes Welding—Covered
MIL-E-22200/7 Electrodes Welding—Covered
QQE 450 Electrodes Welding—Covered: Mild Steel
NAVSHIPS 0090-003-9000 Radiographic Standards for Production and Repair Welds

Source: Ref 3

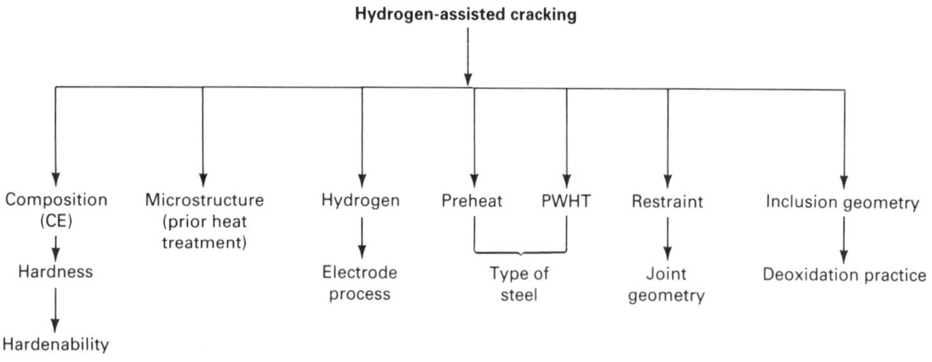

Fig. 12 Factors affecting hydrogen-assisted cracking

Weldability

Weldability, as defined by the American Welding Society, is the capacity of a material to be welded under the fabrication conditions imposed, into a specific, suitably designed structure, and to perform satisfactorily in the intended service. When welding cast steels, the procedures generally used are those that are commonly recommended and used for equivalent-composition wrought steels. However, the unique features of castings versus wrought materials should be taken into account—for example, segregation, porosity, and shrinkage in castings and banded structures in wrought steels. In this section, procedures will be outlined for the welding of cast steels with emphasis on comparison with the weldability of equivalent composition wrought steels.

Welding of Cast Carbon and Low-Alloy Steels

In general, the procedures used for welding wrought carbon and low-alloy steels can be used for cast steels of similar composition. During the welding of as-cast as well as wrought components, a problem commonly encountered in the HAZ is hydrogen-assisted cracking. The three most important factors affecting the susceptibility of a steel to hydrogen-assisted cracking are:

- The composition and therefore the hardenability
- The amount of diffusible hydrogen introduced into the weld during the welding process
- Restraint (stresses associated with welding)

Hydrogen-assisted cracking in the HAZ of a steel is generally found to occur after the weld has cooled to room temperature. Therefore, it is also referred to as cold cracking. Cold cracks need not occur immediately after welding; their formation may be delayed for even up to 1 week after welding has been completed. Therefore, these types of cracks are sometimes referred to as delayed cracks. To avoid confusion, the term hydrogen-assisted cracking will be adopted throughout this section to refer to cold, as well as delayed, cracking. The factors affecting the tendency of a steel toward hydrogen-assisted cracking are summarized in Fig. 12.

Hydrogen-assisted cracks can be observed either in the weld metal or the HAZ. These cracks can be longitudinal or transverse in relation to the weld (Fig. 13). Longitudinal cracks in the HAZ of the base metal that occur a short distance from the weld line are called underbead cracks. Toe cracks originate from the toe of the weld. The toe is a discontinuity in the cross section and therefore acts as a stress concentrator for crack initiation.

It is recognized that the HAZ microstructure governs, to a considerable extent, the susceptibility to hydrogen-assisted cracking. In general, the harder the microstructure, the greater the susceptibility, with twinned martensitic structures being the most susceptible. Other factors that affect the susceptibility to hydrogen-assisted cracking of the HAZ are the hydrogen content of the weldment and the stresses (restraint or applied) acting on the weldment. Considering the situation in which all of the above factors are identical for equivalent composition cast and wrought steels, the reasons for possible differences in hydrogen-assisted cracking between cast and wrought materials may arise from the intrinsic differences between the two types of materials. Cast steels are, in general, coarse

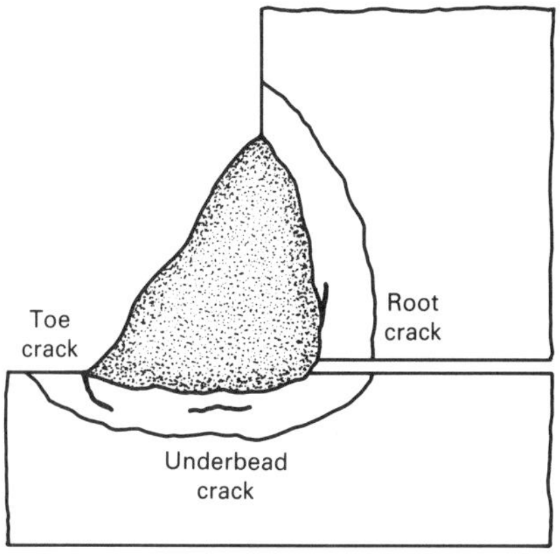

(a)

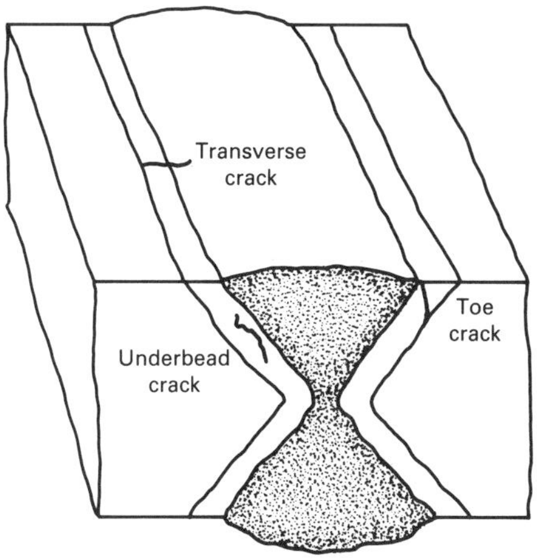

(b)

Fig. 13 Hydrogen-assisted cracks in HAZs of fillet (a) and butt welds (b). Source: Ref 6

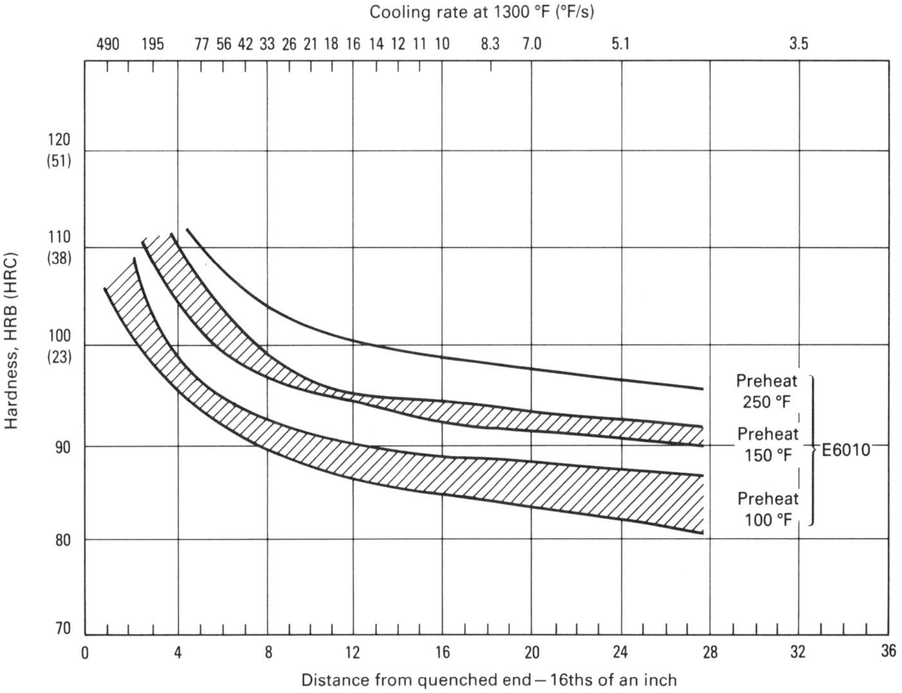

Fig. 14 Preheat temperature required to prevent hydrogen-assisted cracking in the underbead cracking test related to Jominy hardenability for cast-and-normalized carbon-manganese-silicon steels

grained and exhibit relatively isotropic properties compared to their wrought counterparts. Inclusion morphologies in cast steels are generally globular compared to the elongated, stringered inclusions found in wrought steels. Further, as-cast steels may have regions of segregation or may have regions of porosity or interdendritic shrinkage.

The effects of composition on hydrogen-assisted cracking are usually summarized by using a carbon equivalent (CE) to normalize the chemical composition. However, the CE formulas commonly used do not take into account the effects of such factors as prior microstructure or sulfur content, which have been reported to affect the susceptibility of steels to hydrogen-assisted cracking. Attempts have been made to include the hydrogen content into the formulas; however, this entails the precise knowledge of diffusible hydrogen contents for each type of welding process employed. Other elements, such as aluminum (acid soluble and total) and sulfur, do not fall in the category of those that can be included to calculate the carbon equivalent, because they do not have a significant, direct influence on hardenability. The aluminum content, however, has been reported to influence hydrogen-assisted cracking as well as HAZ toughness because of its influence on grain-coarsening characteristics and nucleation/precipitation reactions in the HAZ. Further, the aluminum content, together with the sulfur content, influences the inclusion morphology and distribution. The sulfur content can be taken to be a rough guide to the volume fraction of inclusions present; however, the morphology is determined by the deoxidation practice used prior to casting of the ingot and by the subsequent thermomechanical treatment.

A commonly proposed formula for carbon equivalent is:

$$CE = \%C + \frac{\%Mn}{6} + \frac{\%Cr + \%Mo + \%V}{5} + \frac{\%Cu + \%Ni}{15} \quad \text{(Eq 1)}$$

Recently, an extensive investigation has determined that for normalized cast steels with carbon equivalents up to 0.47 as calculated by:

$$CE = \%C + \frac{\%Mn}{6} + \frac{\%Si}{24} + \frac{\%Cr}{12} + \frac{\%Mo}{10} + \frac{\%Ni}{40} \quad \text{(Eq 2)}$$

a preheat of 40 °C (100 °F) is sufficient to prevent hydrogen-assisted cracking in the HAZ when welded under high-hydrogen conditions with E6010 type electrodes (Ref 4). These results are based on test results using the Battelle Underbead Cracking Test with steels in the range of carbon contents up to 0.35%, manganese contents up to 1.25%, silicon contents up to 0.7% and (%Cr + %Mo + %Ni + %V + %Cu) up to 0.59%. Increasing carbon equivalents, and therefore increasing hardenabilities, require higher preheat temperatures to avoid hydrogen-assisted cracking. A relationship between hardenability and the preheat temperature required to prevent hydrogen-assisted cracking for the above group of cast-and-normalized steels under high-hydrogen conditions is shown in Fig. 14. The change in microstructure with an increase in carbon equivalent from 0.37 to 0.58 is shown in Fig. 15. The microstructure changes from mixed ferrite, pearlite, bainite, and martensite (CE = 0.37) to almost completely martensite (CE = 0.58). The existence of underbead cracking (Fig. 15c) confirms the greater susceptibility of the high-CE steel. In general, a hardness level below 35 HRC (325 HB) in the HAZ will lessen the probability of hydrogen-assisted cracking. The above research also determined that as-cast steels require preheat temperatures 28 °C (50 °F) higher than cast-and-normalized steels to avoid hydrogen-assisted cracking under high-hydrogen conditions. A change from acid to basic melting practice did not decipherably alter susceptibility to hydrogen-assisted cracking.

Preheating has several functions. Preheating:

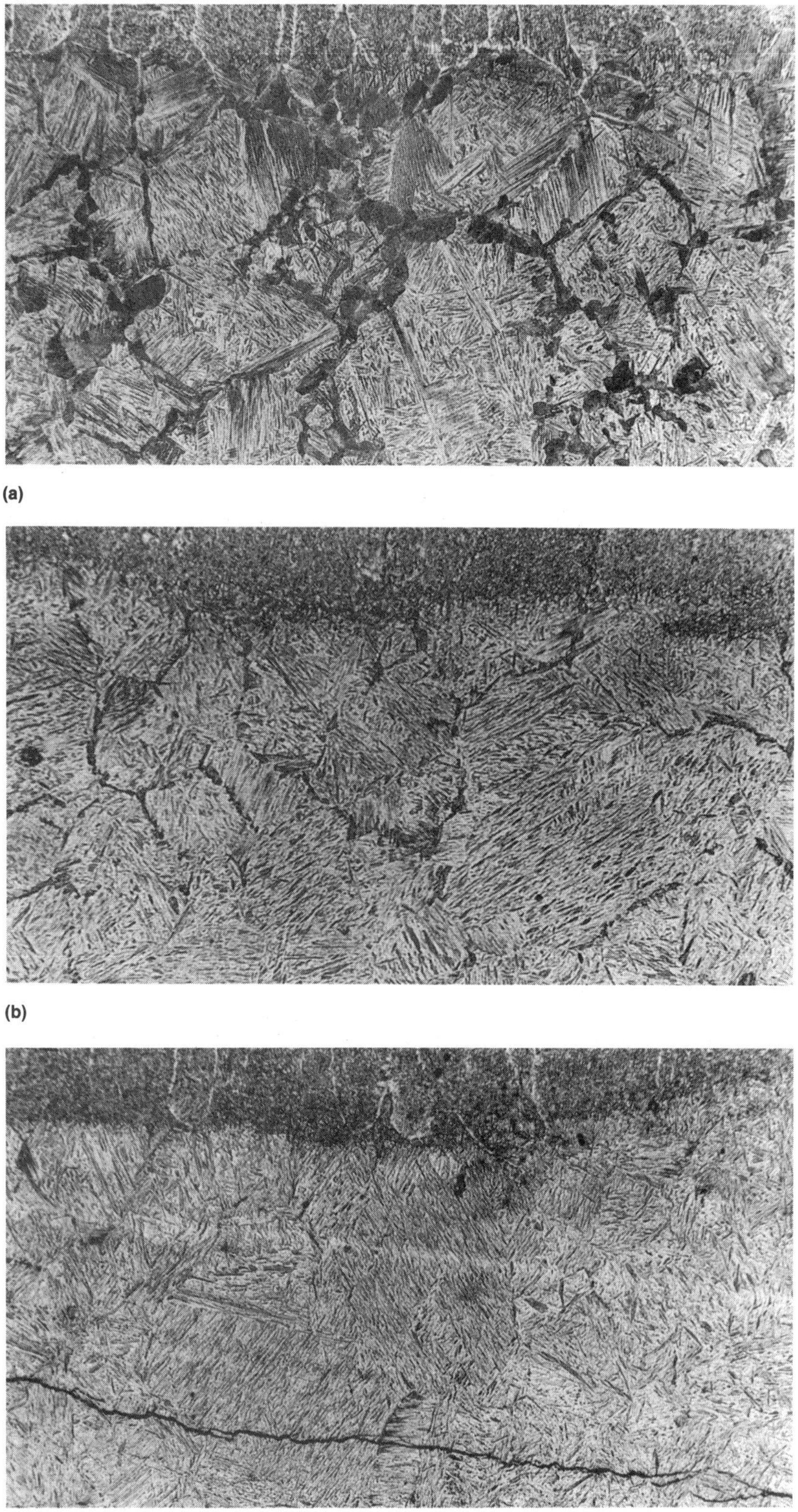

(a) (b) (c)

Fig. 15 Microstructures of grain-coarsened HAZs in cast-and-normalized carbon-manganese-silicon steels. (a) CE = 0.37. (b) CE = 0.47. (c) CE = 0.58. Picral etch. 200×. See also Fig. 16 and 17.

- Reduces the cooling rate and thus promotes higher-temperature transformation products, which result in a relatively lower HAZ hardness
- Promotes the diffusion of hydrogen away from the HAZ
- Aids in the reduction of residual stresses

A question frequently asked is, What preheat temperature should be used to weld a particular steel casting? It is clear that the first approach should be based on composition. Welding Research Council (WRC) Bulletin 191 provides guidelines in the selection of preheat temperatures for cast as well as wrought steels. An updated version of this bulletin is included in Ref 5. The approach in WRC Bulletin 191 is conservative and differentiates between low and non-low hydrogen processes. Material thickness is also taken into account as a factor; thickness will govern the cooling rate and state of stress in the weldment. Table 13 compares the preheat temperatures suggested by American Society for Testing and Materials (ASTM) specifications with those recommended by WRC Bulletin 191 for a selection of cast steels. The ASTM specifications appear to assume low hydrogen conditions. Several other methodologies exist for determining preheat temperatures, most notably, Ref 6 and AWS Structural Welding Code D1.1 (1987). A detailed discussion of these procedures is beyond the scope of this section.

Stress-Relief and Postweld Heat Treatment

Stress-relief treatments are those treatments that are conducted at subcritical temperatures to reduce residual stresses. Postweld heat treatments also reduce HAZ hardness and thus improve ductility and toughness. The need for such treatments depends on the particular welding task under consideration. In fabrications involving large and extensive welds, intermediate subcritical heat treatments can reduce the buildup of stresses. The removal of residual stresses is also important in maintaining the dimensional stability of a casting. Stress relieving can also improve brittle fracture characteristics for service at low temperatures.

Postweld heat treatments combine stress relief and reduction of HAZ hardness. However, weld metals are generally lower in carbon content than the casting and therefore tend to soften more than the casting during heat treatment. Therefore, a filler metal of higher alloy is often desirable. A welding technique that increases dilution of the weld metal from the base metal can also increase the amount of carbon and alloying elements in the weld metal. Gas metal arc, submerged arc, and flux cored arc welding will usually result in enough dilution of the weld metal so that postweld heat treatment will result in relatively uniform properties across the weldment. The toughness of the HAZ and its response to heat treatment strongly depend on carbon equivalent of the parent cast steel.

Typical microstructures of simulated grain-coarsened HAZs in the cast-and-nor-

Table 13 Comparison of ASTM recommended preheat temperatures with those in WRC 191 for selected cast carbon and low-alloy steels

Steel—ASTM specification: Specification	Steel—ASTM specification: CE(a)	Minimum ASTM recommendation, low H_2, 25 mm (1 in.) thickness °C	°F	WRC 191 minimum recommendation, 25 mm (1 in.) thickness: Low H_2 °C	°F	Non-low H_2 °C	°F
A27 Carbon Steel Castings for General Application	0.50 max without residuals	NS(b)		40	100	150	300
A148 High Strength Steel Castings	NS S <0.06 P <0.05	NS		Depending upon composition			
A216-WCB, WCC Castings for Fusion Welding for High Temperature Service	0.55 max	NS		40	100	95	200
A356-Grade 1 Carbon and Low Alloy Castings for Steam Turbines	0.47 max	10	50	40	100	95	200
A486 C1.70 Steel Castings for Highway Bridges	0.50 max	10	50	95	200	175	350
A487 Grade 1 Class A Castings for Pressure Service	0.56 max	95	200	40	100	150	300
A643 Grade A Carbon and Low Alloy Steel Castings for Pressure Vessels	0.53 max	10	50	95	200	150	300

(a) Carbon equivalent (IIW formula); see Eq 1. (b) NS, not specified

malized steels described in Fig. 15 are shown in Fig. 16 and 17. Figure 16 shows untempered grain-coarsened HAZs, while Fig. 17 shows the grain-coarsened HAZs after postweld heat treatment at 650 °C (1200 °F) for 1 h. For the steel with CE = 0.37, the toughness before and after postweld heat treatment remained relatively unchanged at about 11 J (8 ft · lbf). The presence of extensive amounts of proeutectoid ferrite on the grain boundaries both before (Fig. 16a) and after (Fig. 17a) postweld heat treatment is the reason for the relatively low toughness. For the steel with CE = 0.47, a significant improvement in toughness is obtained after postweld heat treatment—from 15 to 79 J (11 to 58 ft · lbf). The cast steel with CE = 0.58 improved in toughness from 4 J (3 ft · lbf) in the as-welded condition to 80 J (59 ft · lbf) in the postweld heat treated condition. For the steels with CE = 0.47 and 0.58, postweld heat treatment resulted in the tempering of the bainitic/martensitic microstructures, which resulted in an improvement of the toughness. The results of these tests demonstrate that microstructures that are not susceptible to hydrogen-assisted cracking may not necessarily possess good as-welded or postweld heat treatment toughnesses. The presence of upper temperature transformation products on prior-austenite grain boundaries is especially detrimental to the impact properties of the HAZ. It should also be noted that the susceptibility to hydrogen-assisted cracking is not governed by the location of the microconstituents in the HAZ, but rather by the type and volume fraction of constituents in the HAZ.

The ASME Boiler and Pressure Vessel Code Section IX and ASTM Standard A 488 provide good guidelines for choosing stress-relieving and postweld heat treatment schedules. Additional information can also be found in the article "Residual Stresses and Distortion" in Volume 6 of the 9th Edition of *Metals Handbook*.

Comparative Weldability of Cast and Wrought Steels

Although general recommendations indicate that the weldabilities of cast and wrought steels are similar, considerable differences of opinion exist in the welding literature. Extensive work conducted at Battelle during the late 1940s and early 1950s indicated that cast steels are more resistant to the occurrence of underbead hydrogen-assisted cracking when compared to wrought steels of equivalent composition. Although no reasons were given for this difference in weldability, the following can be hypothesized:

- The porosity and microshrinkage associated with cast steels may provide sinks for hydrogen
- The transformation behavior in the HAZ may be affected by localized variations in elemental concentration resulting from segregation
- The globular inclusion morphology in cast steels differs from that in wrought steels, where the inclusions are usually elongated

Recent research has shown that the elongated inclusion morphology in wrought steels renders them more susceptible to hydrogen-assisted cracking when compared to equivalent-composition cast steels (Ref 4). This research also determined that cast steels that have been critically deoxidized to have type II inclusions (eutectic sulfides) present at the substructure and grain boundaries are more susceptible to hydrogen-assisted cracking than those deoxidized to have type I (globular sulfides) or type III (angular sulfides) present in them. Typical examples of crack association with type II inclusions in the grain-coarsened and partially refined HAZs are shown in Fig. 18. Thus, it appears that inclusion morphology resulting from a given composition and deoxidation practice can influence the susceptibility to HAZ cracking of cast carbon and low-alloy steels.

Wear-Resistant Manganese Steels

The use of austenitic manganese steels as castings or as a weld deposit is a common method of providing a tough wear-resistant surface on component faces that are subject to abrasion and wear. The role of welding is in casting repair, overlaying, and joining between manganese castings or manganese castings to carbon or alloy steels. During the welding of such castings, exposures at temperatures above 315 °C (600 °F) for extended periods of time result in embrittlement accompanied by a drastic decrease in ductility. At a temperature of 425 °C (800 °F), an exposure period of only 2 h may be enough to cause embrittlement. Preheating of the casting is therefore not recommended and in any case should be restricted to 40 °C (100 °F) maximum. Interpass temperatures should be held to 315 °C (600 °F) maximum. Welding processes, such as gas welding, that involve slow heating and cooling periods are to be avoided.

Cast Stainless Steels

As for carbon and low-alloy steels, the general principles and procedures applicable for wrought stainless steels are also applicable for cast stainless steels. The CF-8 (cast equivalent of AISI 304) and CF-8M (cast equivalent of AISI 316) steels are easily weldable. However, precautions must be taken to avoid sensitization. These include controlling the interpass temperature to 150 °C (300 °F) and the use of stringer bead welding techniques. When higher temperatures and severe corrosive environments are encountered, it may be necessary to solution heat treat the casting at 1095 °C (2000 °F) to put the carbides that precipitated during welding into solution. The use of stabilized grades such as CF-8C (cast equivalent of AISI 347) can help in

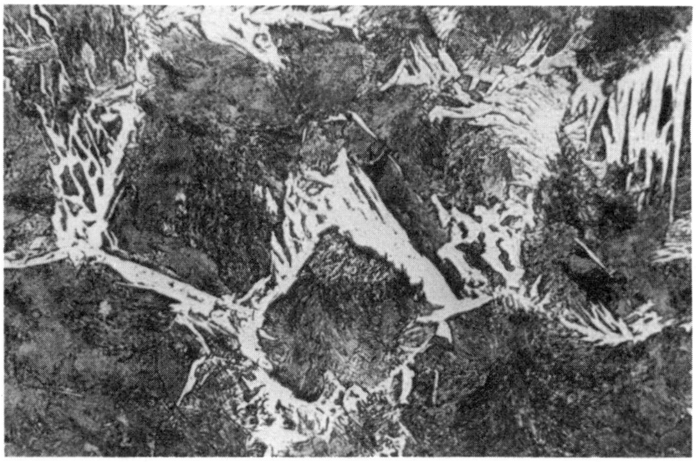

(a)

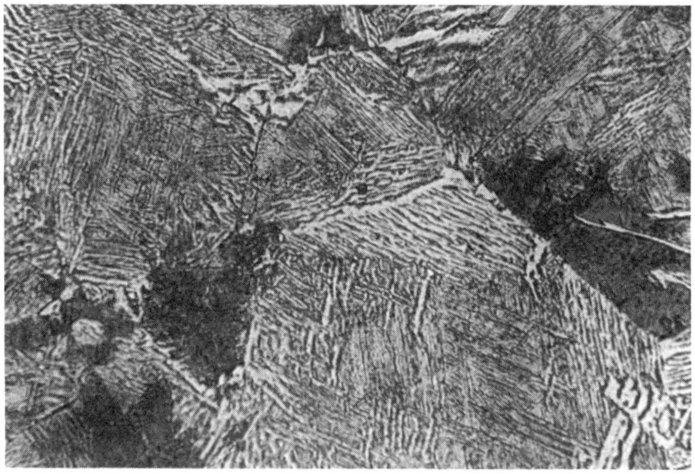

(b)

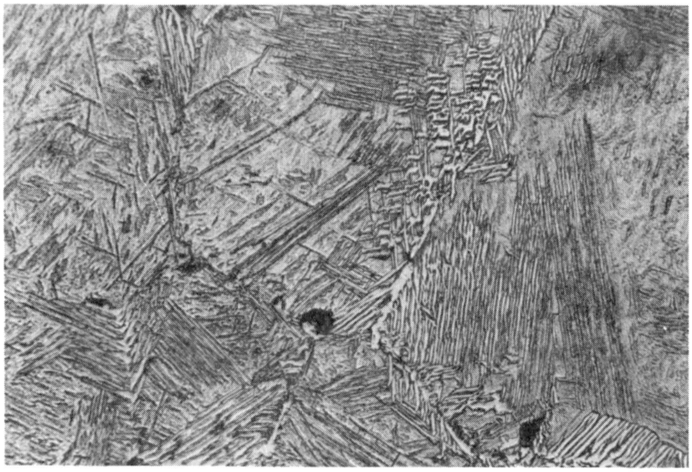

(c)

Fig. 16 Microstructures of simulated grain-coarsened HAZs in cast-and-normalized carbon-manganese-silicon steels in the thermal cycled condition. 2 kJ/mm (50 kJ/in.), 1315 °C (2400 °F) peak temperature. (a) CE = 0.37. (b) CE = 0.47. (c) CE = 0.58. Picral etch. 500×

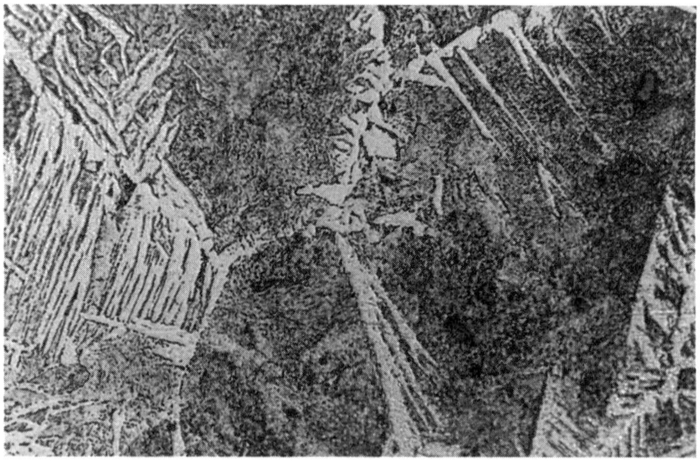

(a)

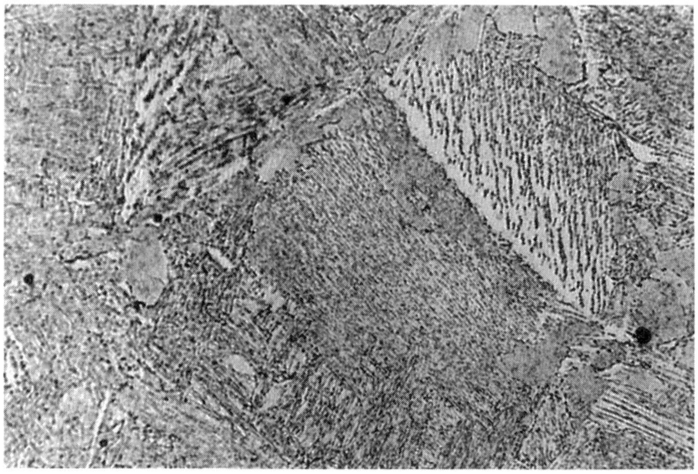

(b)

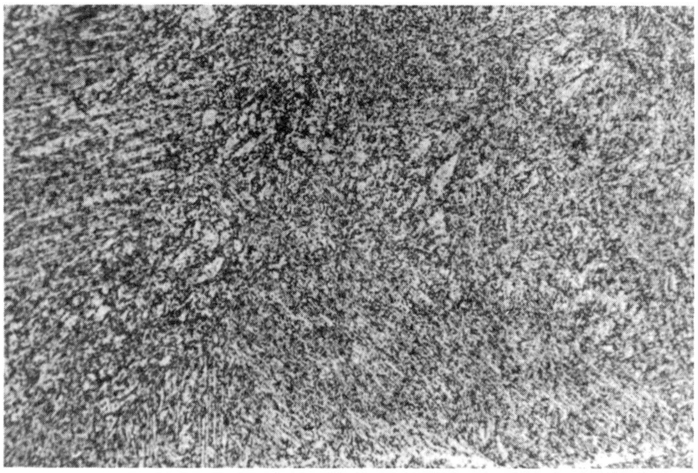

(c)

Fig. 17 Microstructures of simulated grain-coarsened HAZs (same steels as in Fig. 16) in the thermal cycled and postweld heat treatment conditions. 2 kJ/mm (50 kJ/in.), 1315 °C (2400 °F) peak temperature. Postweld heat treatment: 650 °C (1200 °F) for 1 h. (a) CE = 0.37. (b) CE = 0.47. (c) CE = 0.58. Picral etch. 500×

avoiding a solution heat treatment. Another approach is the use of CF-3 and CF-3M grades, which restrict the carbon content to 0.03% maximum and reduce the probability of sensitization. As alloy grades become richer, the structure becomes fully austenitic, and hot cracking is a potential problem. The problem can be alleviated by using low heat inputs and interpass temperatures as well as by peening the weld.

In general, cast stainless steels can be welded with filler metals recommended for

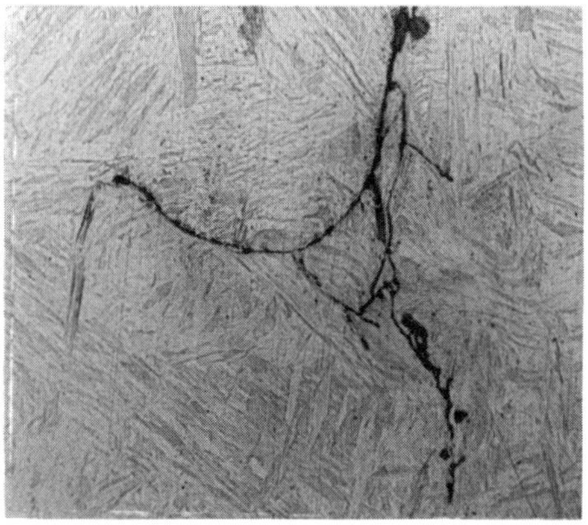

(a)

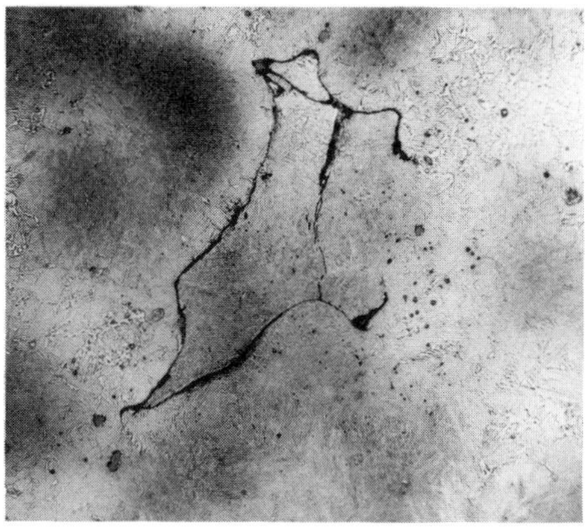

(b)

Fig. 18 Hydrogen-assisted cracking associated with type II inclusions in A216-WCB steel. (a) Grain-coarsened HAZ. 2000×. (b) Partially refined HAZ. 200×. Picral etch

equivalent-composition wrought stainless steels. Table 14 compares cast and equivalent wrought stainless steels and suggests filler metal compositions.

In the dissimilar-metal welding between cast austenitic and cast or wrought ferritic steels, the choice of filler metal is dictated by the temperature and stresses involved in service. Moderate temperatures (<315 °C, or 600 °F) and stresses can allow the use of high-alloy austenitics such as type 309 or 310. Higher temperatures and stresses will entail the use of high-nickel weld metal. The welding process and technique must be adapted to reduce dilution of the weld metal by the base metal.

Table 14 Filler metals suggested for cast stainless alloys

Base metal—Alloy Castings Institute casting alloy designation	Similar to wrought alloy type	Suggested filler metal
CA-6NM	410NiMo	410NiMo
CA-15	12Cr	410, 410NiMo
CA-40	420	410, 410NiMo
CB-7Cu	Precipitation hardening	630
CB-30	431	312
CF-3	304L	308L
CF-3M	316L	316L
CF-8	304	308
CF-8M	316	316, 317
CF-8C	347	347
CF-12M	316	316, 317
CF-20	302	308
CH-20	309	309
CG-8M	317	317
CK-20	310	310
CN-7M	20Cr-30Ni-Mo-Cu	320
HC	446	446, 312
HD	327 (28Cr,5Ni) (Duplex)	Similar composition with enhanced nickel to improve ductility
HE	312	312
HF	320B, 308HiC	308
HH	309	309
HI	28Cr-15Ni	309
HK	310H	310H
HT	330	330
HU	19Cr-39Ni	330

REFERENCES

1. G.R. Pease, The Welding of Ductile Iron, *Weld. J.*, Vol 39 (No. 1), 1960, p 1s
2. *Iron Castings Handbook*, Iron Casting Society, 1981
3. *Steel Castings Handbook*, 5th ed., P.F. Weiser, Ed., Steel Founders' Society of America, 1980
4. "Weldability of Cast Carbon and Low Alloy Steels—Effect of Microstructure and Inclusion Morphology on the Hydrogen Induced HAZ Cracking Susceptibility," Research Report 96, Steel Founders' Society of America, Jan 1985
5. R.D. Stout, *Weldability of Steels*, 4th ed., Welding Research Council, 1987
6. F.R. Coe, *Welding Steels Without Hydrogen Cracking*, The Welding Institute, 1973

SELECTED REFERENCES

- F.A. Ball and D.R. Thorneycroft, Metallic-Arc Welding of Spherical-graphite Cast Iron, *Foundry Trade J.*, Vol 97, Oct 1954, p 499
- R.C. Bates and F.J. Morley, Welding Nodular Iron Without Postweld Annealing, *Weld. J.*, Vol 40 (No. 9), 1961, p 417s
- R.A. Bishel, Flux-cored Electrode for Cast Iron Welding, *Weld. J.*, Vol 52 (No. 6), 1973, p 372
- R.A. Bishel and H.R. Conway, Fluxed Cored Arc Welding for High-Quality Joints in Ductile Iron, *Mod. Cast*, Vol 67 (No. 1), 1977, p 59-60
- C. Cookson, The Metal Arc Welding of Cast Iron for Maintenance and Repair Welding, *Met. Constr. Br. Weld. J.*, Vol 3 (No. 5), 1971, p 179
- C. Cookson, Metal-Arc Welding of White Cast Iron, *Met. Constr. Br. Weld. J.*, Vol 5 (No. 10), 1973, p 370
- A.M. Davila, D.L. Olson, and T.A. Freese, Submerged-arc Welding of Ductile Iron, *Trans. AFS*, Vol 85 (No. 77), 1977
- A.M. Davila and D.L. Olson, Welding Consumable Research for Ductile Iron, *Mod. Cast.*, Vol 70 (No. 11), 1980, p 70
- J.H. Devletian, Weldability of Gray Iron Using Fluxless Gray Iron Electrodes for SMAW, *Weld. J.*, Vol 57, 1978, p 183s
- R.H.T. Dixon and D.R. Thorneycroft, Filler Rod for the Gas Welding of S.g. Iron, *Foundry Trade J.*, Vol 108, May 1960, p 583
- S.A. Forberg, Short-Arc Welding of S.g. Iron in the SKF Katrineholm Works, *Sweden Foundry Trade J.*, Vol 124 (No. 2685), 1968, p 833
- A.G. Hogaboom, Welding of Gray Iron, *Weld. J.*, Vol 56 (No. 2), 1977, p 17
- E.E. Hucke and H. Udin, Welding Metallurgy on Nodular Cast Iron, *Weld. J.*, Vol 32 (No. 8), 1953, p 378s
- S.D. Kiser, Production Welding of Cast Irons, *Trans. AFS*, Vol 85 (No. 37), 1977
- J. Klimek and A.V. Morrison, Gray Cast Iron Welding, *Weld. J.*, Vol 56 (No. 3), 1977, p 29
- D.J. Kotecki, N.R. Braton, and C.R. Loper, Jr., Preheat Effects on Gas Metal-Arc Welded Ductile Cast Iron, *Weld. J.*, Vol 48 (No. 4), 1969, p 161s
- R.L. Kumar, Welding Gray Iron With Mild Steel Electrodes, *Foundry*, Vol 96 (No. 1), 1968, p 64
- R. Mohler, Repairing Cast Iron by Welding, *Plant Eng.*, Vol 31 (No. 23), 1977, p 171-174
- E.F. Nippes, W.F. Savage, and W.A. Owczarski, The Heat-affected Zone of Arc Welded Ductile Iron, *Weld. J.*, Vol 39 (No. 11), 1960, p 465s
- G.R. Pease, The Welding of Ductile Iron, *Weld. J.*, Vol 39 (No. 1), 1960, p 1s
- R.V. Riley and J. Dodd, Ferrous Rod for Welding Nodular Graphite Cast Iron, *Foundry Trade J.*, Vol 93 (No. 1887), 1952, p 555
- C.F. Walton and T.J. Ojar, Ed., *Iron Castings Handbook*, Iron Casting Society, 1981, p 599-665

Hot Isostatic Pressing of Castings

John M. Eridon, Howmet Corporation

Hot isostatic pressing (HIP) is a process that subjects a component to both elevated temperature and isostatic gas pressure in an autoclave. The most widely used pressurizing gas is argon. For the processing of castings, argon is applied at pressures between 103 and 206 MPa (15 and 30 ksi), with 103 MPa (15 ksi) being the most common. Process temperatures vary from 480 °C (900 °F) for aluminum castings to 1315 °C (2400 °F) for single-crystal nickel-base superalloys.

When castings are hot isostatically pressed, the simultaneous application of heat and pressure virtually eliminates internal voids and microporosity through a combination of plastic deformation, creep, and diffusion. The elimination of internal defects leads to improved nondestructive testing ratings, increased mechanical properties, and reduced data scatter.

The HIP process was invented in the mid-1950s at Battelle Columbus Laboratories. The process was specifically developed as a gas pressure diffusion bonding technique for cladding nuclear fuel elements (Ref 1). In 1965, the use of hot isostatic pressing for improving the fatigue life of cast aluminum diesel engine pistons was investigated (Ref 2). Although it was shown that a significant increase in fatigue life could be achieved, the limited availability of commercial production HIP vessels prevented its implementation. In 1967, hot isostatic pressing was evaluated as a means of improving the quality of both titanium and superalloy investment castings for use in the aerospace industry. Further investigations into the effect of hot isostatic pressing on titanium and superalloy castings were conducted in the early 1970s; these investigations demonstrated that significant improvements in mechanical properties could be obtained (Ref 3, 4).

Initially, most of the castings being hot isostatically pressed were critical aerospace components, such as turbine blades and structural hardware for aircraft gas turbine engines. These relatively expensive components could accept the high added cost of the HIP process. Recently, however, the cost of hot isostatic pressing has been reduced to the point at which it is affordable for castings in less critical applications. This reduction in cost has resulted from the increased number and size of vessels dedicated to the hot isostatic pressing of cast-

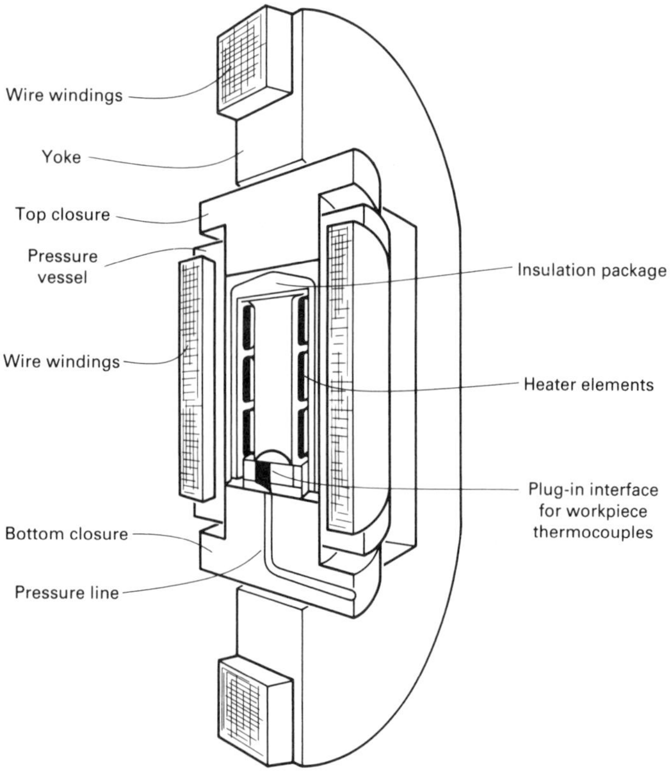

Fig. 1 Schematic of one type of large, production HIP vessel. Source: ASEA

Table 1 Cooling rates for rapid-cool and rapid uniform cool HIP furnaces

Vessel is 500 mm (19 in.) in diameter and 1500 mm (59 in.) long with a 500 kg (1100 lb) load.

	Cooling rate					
	Top		Middle		Bottom	
Type of furnace	°C/min	°F/min	°C/min	°F/min	°C/min	°F/min
Rapid cool	8	14	18	32	50	90
Rapid uniform cool	50	90	50	90	50	90

Source: National Forge Company

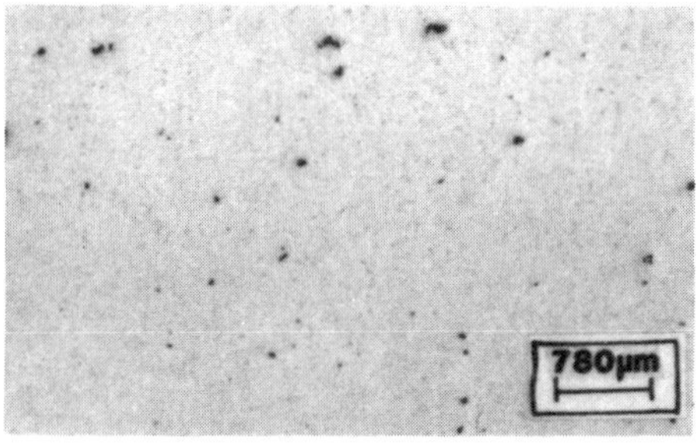

(a)

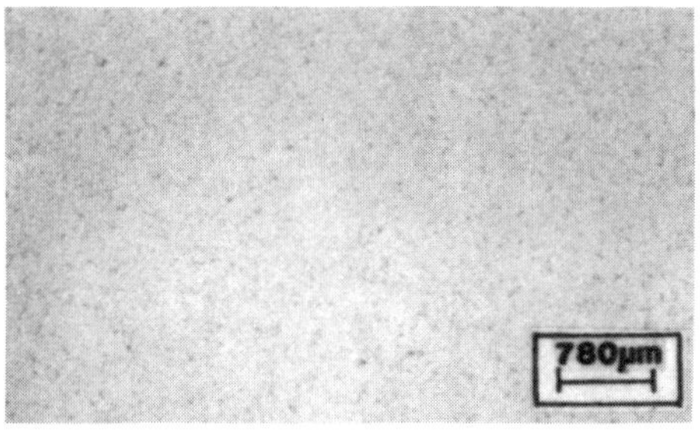

(b)

Mechanical properties	Before HIP	After HIP
At 760 °C (1400 °F)/603.3 MPa (87.5 ksi)		
Rupture life, h	57.4	137.9
Elongation, %	11.4	13.8
At 982 °C (1800 °F)/189.6 MPa (27.5 ksi)		
Rupture life, h	27.1	42.1
Elongation, %	15.4	14.4

Fig. 2 Effect of hot isostatic pressing on porosity levels in investment cast René 80. (a) Before hot isostatic pressing. (b) After hot isostatic pressing. Courtesy of Howmet Corporation

ings and from improvements in equipment (shortened process cycle times).

Equipment

Hot isostatic presses are available in many sizes with varying process capabilities. Sizes vary from small 102 mm (4 in.) diam laboratory units to 1524 mm (60 in.) diam production-size vessels. Units with temperature capability to 2200 °C (3990 °F) and pressure capability to 310 MPa (45 ksi) are available from commercial autoclave manufacturers. A schematic illustrating the construction of a typical hot isostatic press is shown in Fig. 1. A detailed description of HIP equipment can be found in the article "Hot Isostatic Pressing of Metal Powders" in Volume 7 of the 9th Edition of *Metals Handbook*.

Reduced Cycle Time. One of the more significant advancements in HIP equipment has been the introduction of furnace designs that allow uniform rapid cooling from the HIP temperature. Original furnace designs were capable of slow cooling only. The advent of rapid-cool furnace designs allowed reductions in floor-to-floor cycle times of approximately 4 to 6 h, depending on unit size and load mass. However, the early rapid-cool designs were not able to achieve uniform cooling rates over the entire load of parts. Recently, furnaces capable of rapid uniform cooling over the entire load have been introduced (Table 1). Cooling rates are sufficiently rapid to allow the solution heat treatment of many alloys at the end of the HIP cycle, thus eliminating the need for additional post-HIP solution heat treatments.

Effect of HIP on the Mechanical Properties of Castings

Nickel-Base Superalloys. In the past 15 years, hot isostatic pressing has become an integral part of the manufacturing process for high-integrity aerospace castings. The growth of hot isostatic pressing has paralleled the introduction of advanced nickel-base superalloys and increasingly complex casting designs, both of which tend to increase levels of microporosity. In addition, to optimize mechanical properties, turbine engine manufacturers have become more stringent in allowances for microporosity. The requirement for reduced porosity levels and increased mechanical properties has been achieved in many cases through the use of hot isostatic pressing.

The first production use of hot isostatic pressed castings for critical rotating parts in an aircraft gas turbine occurred in the mid-1970s. Figure 2 shows the effect of hot isostatic pressing on porosity levels and rupture properties for investment cast René 80. As can be seen, the elimination of porosity and subsequent increase in both the 760 °C (1400 °F)/603 MPa (87.5 ksi) and the 980 °C (1800 °F)/190 MPa (27.5 ksi) rupture lives are evident. Figure 3 shows the effects of hot isostatic pressing on the rupture properties of Alloy IN738. The data were obtained on test bars machined from land-based industrial gas turbine blades. Of significant interest is the improvement in mechanical properties of hot isostatic pressed material at the -2σ limits (σ = standard deviation).

Titanium. One of the earliest successes of hot isostatic pressing as applied to aerospace castings was with investment cast

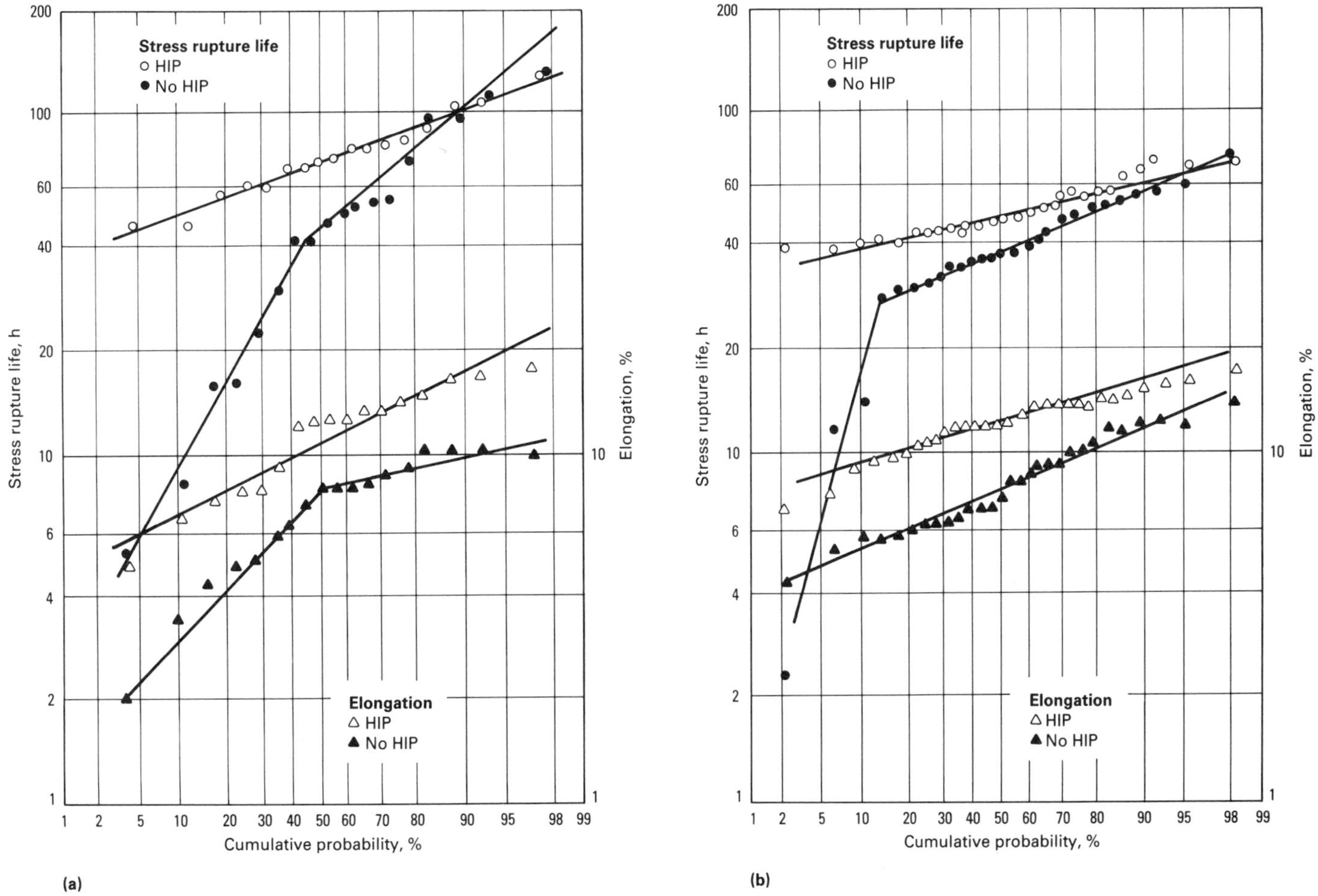

Fig. 3 Stress rupture properties of IN738 hot isostatically pressed at 1205 °C (2200 °F) and 103 MPa (15 ksi) for 4 h. (a) Test at 760 °C (1400 °F) and 586 MPa (85 ksi). (b) Test at 980 °C (1800 °F) and 152 MPa (22 ksi). Source: Howmet Corporation

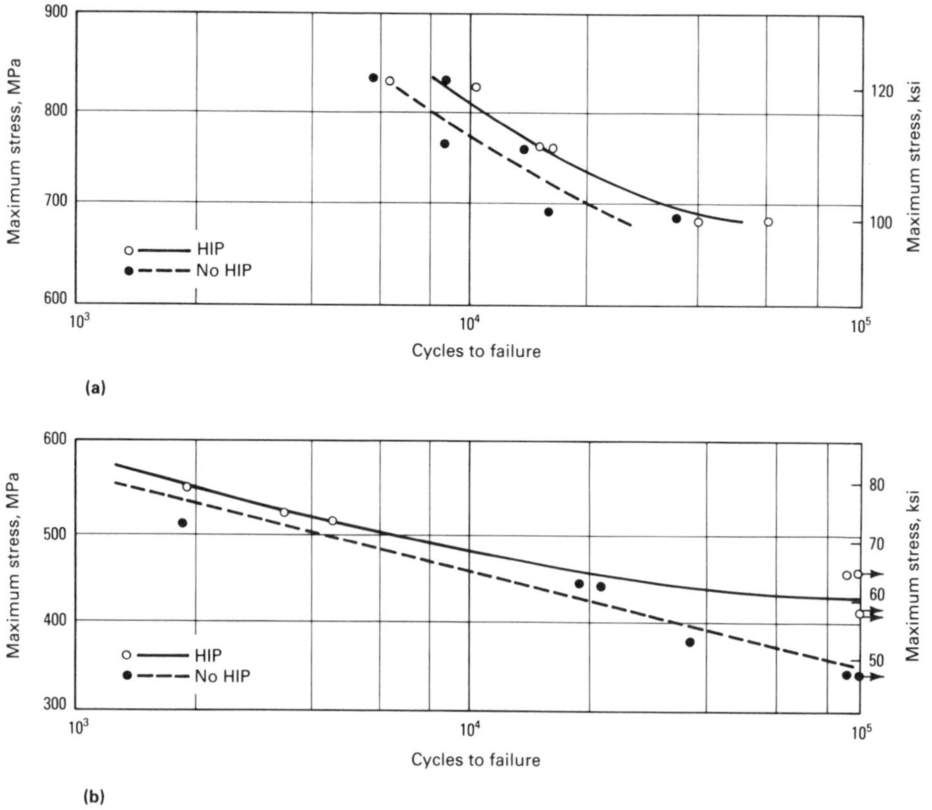

Fig. 4 Effect of hot isostatic pressing on room-temperature fatigue properties of investment cast and annealed Ti-6Al-4V. HIP conditions: 900 °C (1650 °F) and 103 MPa (15 ksi) for 2 h. Test conditions: maximum stress ratio $R = 0.1$, frequency = 0.3 Hz for low-cycle fatigue and 60 Hz for high-cycle fatigue, and theoretical stress concentration factor $K_t = 1.0$. (a) Low-cycle fatigue. (b) High-cycle fatigue. Source: Ref 5

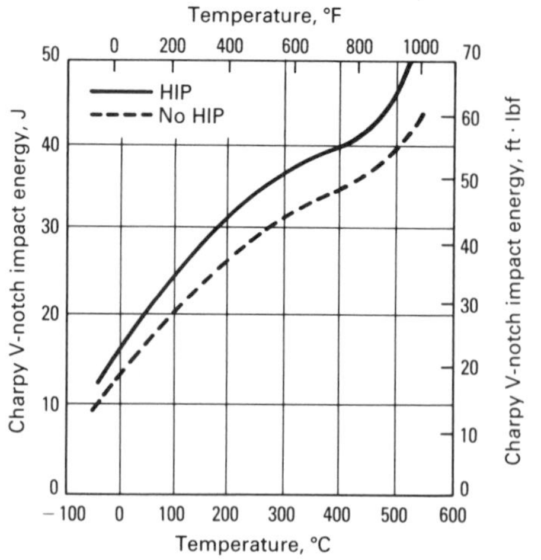

Fig. 5 Charpy V-notch impact energy for hot isostatically pressed and non-hot isostatically pressed Ti-6Al-4V investment castings. Source: Ref 5

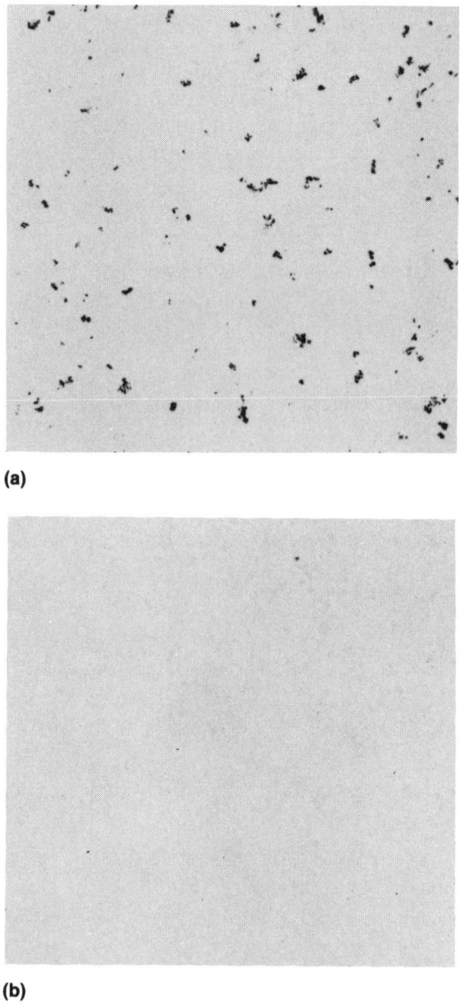

Fig. 6 Effect of hot isostatic pressing on porosity levels of aluminum alloy A356. (a) Before hot isostatic pressing. (b) After hot isostatic pressing. Courtesy of Howmet Corporation. Both 2.5×

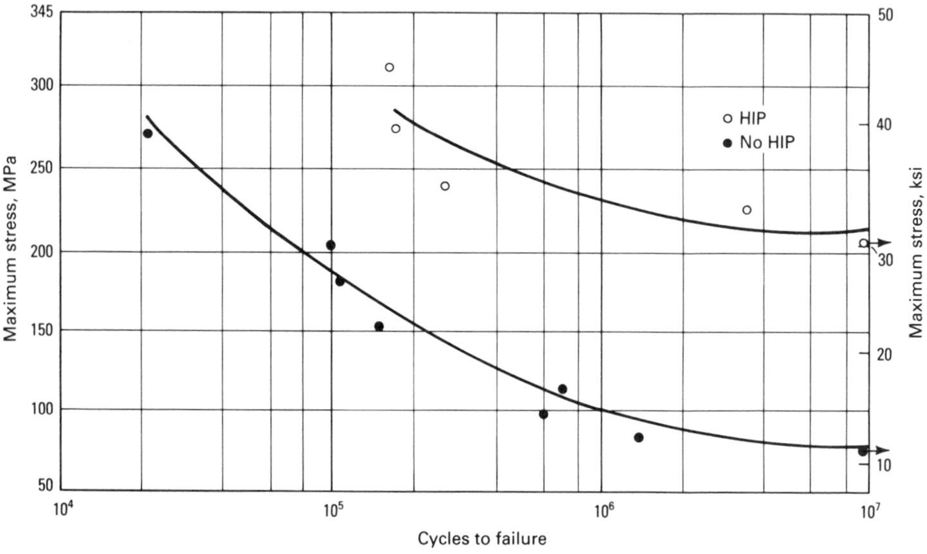

Fig. 7 Effect of hot isostatic pressing on the room-temperature high-cycle fatigue behavior of aluminum alloy A201-T7. HIP conditions: 520 °C (970 °F) and 103 MPa (15 ksi) for 2 h. Stress-controlled test on axially loaded specimen; $R = 0.1$; frequency = 60 Hz; $K_t = 1.0$. Source: Howmet Corporation

titanium. Because of the high reactivity of molten titanium, most investment cast titanium aerospace components are produced in a vacuum furnace utilizing a consumable arc skull remelting process. This process does not allow superheating of the melt; therefore, the ability to produce sound castings with consistent mechanical properties is limited. As shown in Fig. 4 and 5, the application of hot isostatic pressing to investment cast Ti-6Al-4V results in improved room-temperature low-cycle and high-cycle fatigue properties and improved impact strength. The improved properties obtainable with hot isostatic pressing have allowed investment cast titanium components to meet stringent requirements for aerospace and other critical applications.

Table 2 Effect of HIP on the room-temperature tensile properties of three aluminum alloys

Alloy and temper	Ultimate tensile strength: Mean, MPa	Mean, ksi	-2σ(a), MPa	-2σ(a), ksi	0.2% offset yield strength: Mean, MPa	Mean, ksi	-2σ(a), MPa	-2σ(a), ksi	Elongation, %: Mean	-2σ(a)
A356-T6										
HIP	248	36	234	34	172	25	165	24	6.7	3.8
No HIP	193	28	152	22	138	20	124	18	5.4	3.5
A357-T6										
HIP	255	37	248	36	207	30	200	29	5.8	5.1
No HIP	200	29	165	24	179	26	172	25	2.6	0.0
A201-T7										
HIP	441	64	421	61	372	54	358	52	11.2	9.6
No HIP	386	56	365	53	372	54	358	52	1.4	0.8

(a) σ = standard deviation. Source: Howmet Corporation

Table 3 Effect of HIP on the room-temperature tensile properties of Custom 450 stainless steel in the H1050 condition

Condition	Ultimate tensile strength: Mean, MPa	Mean, ksi	-2σ, MPa	-2σ, ksi	0.2% offset yield strength: Mean, MPa	Mean, ksi	-2σ, MPa	-2σ, ksi	Elongation, %: Mean	-2σ	Reduction in area, %: Mean	-2σ
HIP	1069	155	993	144	1000	145	910	132	20.7	16.3	62.0	53.9
No HIP	993	144	972	141	903	131	883	128	19.4	14.5	50.9	30.7

Source: Ref 6

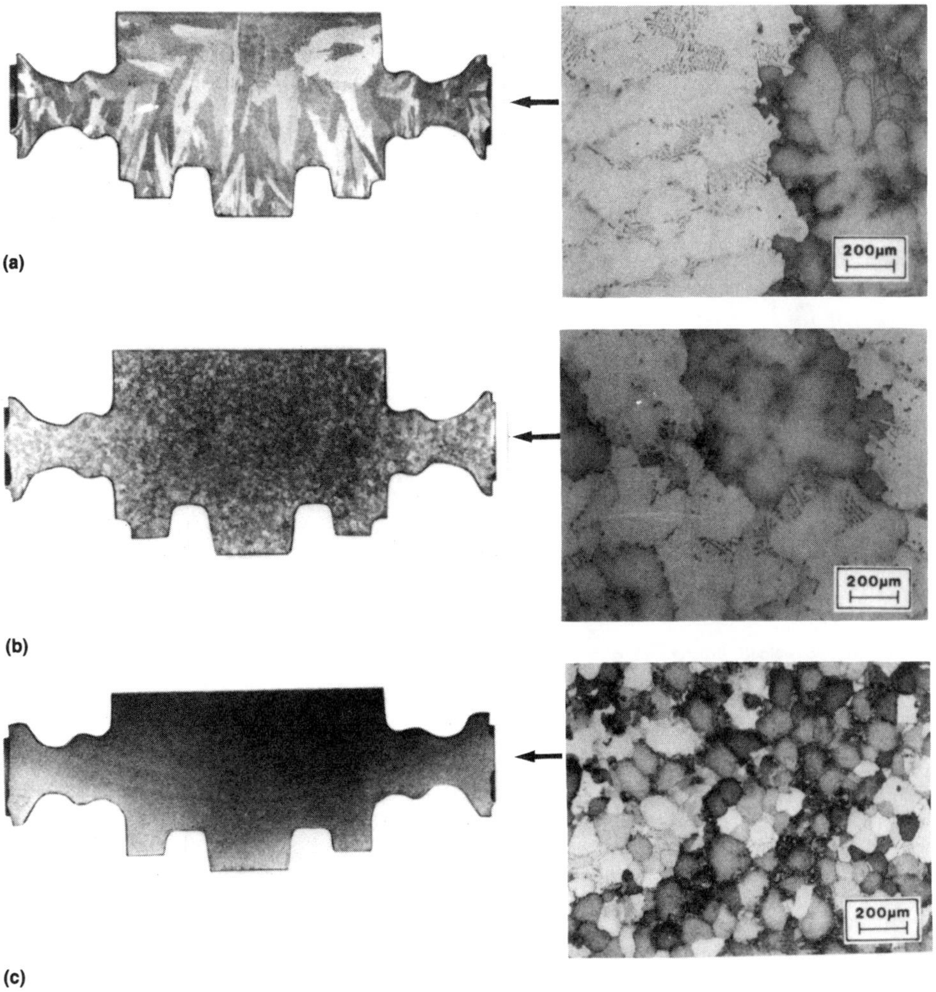

Fig. 8 Structure of conventionally cast turbine wheel (a) compared to wheels cast using the Grainex (b) and Microcast-X (c) processes. Courtesy of Howmet Corporation

Aluminum Alloys. As previously mentioned, the improvement in the fatigue life of cast aluminum pistons as a result of hot isostatic pressing was investigated in 1965. Pistons were processed at or near the solution temperature of the alloy for 2 to 3 h under a pressure of 103 MPa (15 ksi). Figure 6 shows the effect of hot isostatic pressing on eliminating internal porosity in sand cast aluminum alloy A356. Tensile data comparing the effects of hot isostatic pressing on aluminum alloys A356, A357, and A201 are presented in Table 2. The statistical treatment of these data reveals the reduction in data scatter obtainable with hot isostatic pressing. The effect of hot isostatic pressing on the fatigue life of A201 is shown in Fig. 7.

Stainless Steels. The effect of hot isostatic pressing on the room-temperature tensile properties of investment cast Custom 450 precipitation-hardening stainless steel is demonstrated in Table 3. The hot isostatically pressed material displays enhanced properties at both average and -2σ limits compared to the non-hot isostatically pressed material. Data for this comparison were obtained using test specimens machined from airframe support brackets. The hot isostatically pressed brackets were processed at 1120 °C (2050 °F) and 103 MPa (15 ksi) for 4 h to obtain closure of internal porosity. The hot isostatically pressed and non-hot isostatically pressed brackets were both heat treated to the H1050 condition.

Process Considerations

Parameter Selection. In selecting HIP process parameters for a particular alloy,

Table 4 General HIP processing parameters for various cast alloys

Alloy type	Temperature range °C	Temperature range °F	Pressure MPa	Pressure ksi	Time, h
Magnesium alloys	370–400	700–750	103	15	2–4
Aluminum alloys	480–530	900–985	103	15	2–6
Copper alloys	705–980	1300–1800	103	15	2–6
Low-alloy steels	1010–1175	1850–2150	103	15	2–4
Stainless steels	1065–1220	1950–2225	103	15	2–4
Titanium alloys	845–970	1550–1775	103	15	2–4
Nickel-base alloys	1120–1315	2050–2400	103–172	15–25	2–4
Cobalt-base alloys	1190–1230	2175–2250	103	15	2–4

Source: Ref 7

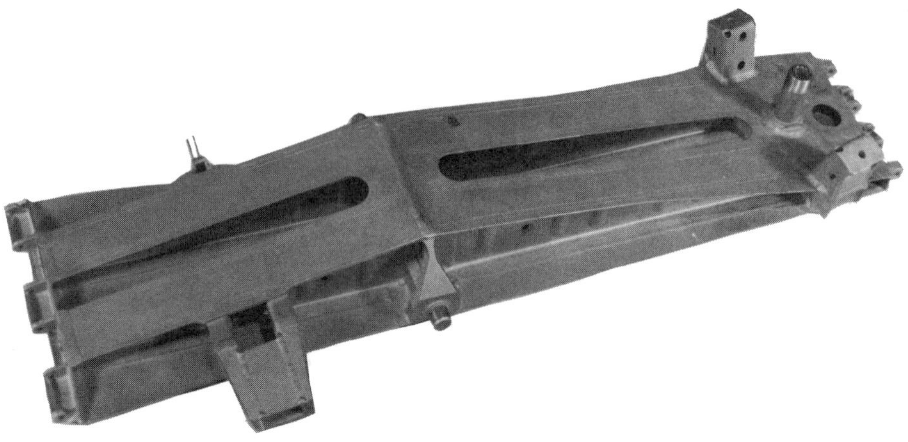

(a)

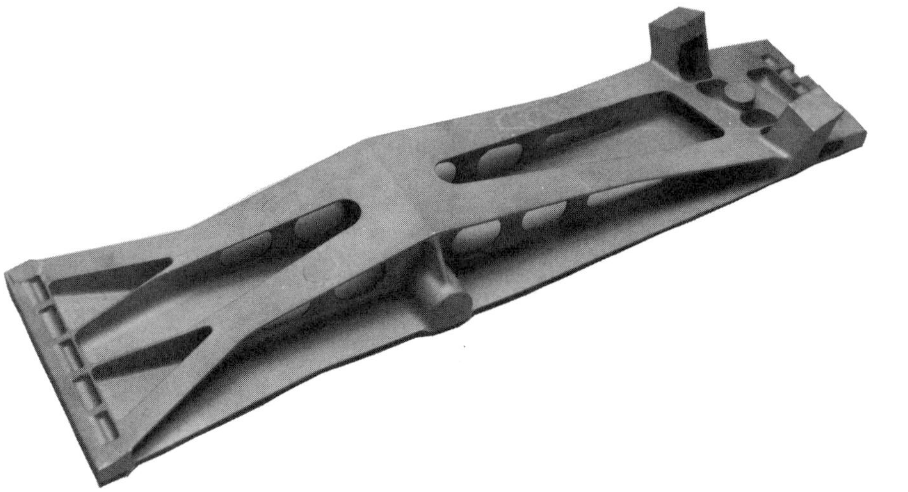

(b)

Fig. 9 Ti-6Al-2Sn-4Zr-2Mo aircraft flap that was converted from a welded structure to a one-piece hot isostatically pressed casting. (a) Original fabrication. (b) Casting. Courtesy of Pratt & Whitney Aircraft

the primary objective is to use a combination of time, temperature, and pressure that is sufficient to achieve closure of internal voids and microporosity in the casting. There are also material considerations for avoiding such deleterious effects as incipient melting, grain growth, or degradation of constituent phases such as carbides.

If encountered, incipient melting can be avoided by pre-HIP homogenization heat treatments or by lowering HIP temperatures. If the temperature is lowered, an increase in processing pressure may be required for obtaining closure in certain alloys. For example, hafnium-bearing nickel-base superalloys such as C-101 and MAR-M247, when cast with heavy sections (for example, integral wheels), have been found to undergo incipient melting when hot isostatically pressed at 1205 °C (2200 °F) and 103 MPa (15 ksi) for 4 h. To prevent incipient melting and still obtain closure, the HIP parameters were changed to 1185 °C (2165 °F) and 172 MPa (25 ksi) for 4 h. This trade-off between temperature and pressure can also be used to prevent grain growth and to prevent or limit carbide degradation while obtaining closure of microporosity.

Time at temperature and pressure will obviously affect processing cost. For most alloys, 2 to 4 h is sufficient. Exceptions are massive section sizes, which require additional thermal soaking time. General processing guidelines for several alloy types are given in Table 4.

Gas Purity. Surface contamination of castings can occur unless extremely pure argon pressurizing gas is used. Impurities such as hydrogen, nitrogen, oxygen, carbon monoxide, water vapor, and hydrocarbons in parts per million concentrations have been found to cause deleterious surface reactions in certain alloys. For nickel-base superalloys, nitrogen forms brittle carbonitrides, oxygen forms surface and intergranular oxides, and carbon compounds (carbon monoxide, methane, and so on) can cause surface carburization. In titanium alloys, oxygen causes the formation of an α case. To prevent contamination, the purity of the process gas should be verified by gas chromatography before each HIP cycle. To assess the entire gas supply system, sampling should be conducted on the gas as it exits the hot isostatic pressing vessel.

Distortion of castings during the hot isostatic pressing process typically results from two sources: the thermal cycle and the deformation associated with void closure. Because hot isostatic pressing is a thermal treatment, the same type of distortion as would occur in a normal heat treat cycle will also occur during hot isostatic pressing. In addition, if large voids exist in the casting, surface depressions may develop as a result of material displacement during void closure. If small, uniformly distributed voids are present in a casting, a dimpled surface may be evident after hot isostatic pressing. Such a condition is typical in many sand cast aluminum parts. In most cases, however, no void-induced distortion can be detected after hot isostatic pressing.

Surface defects in a casting cannot be closed during hot isostatic pressing, because no pressure differential can be established at the defect. To prevent the opening of any subsurface defects prior to hot isostatic pressing, castings should be hot isostatically pressed before the use of any aggressive metal removal techniques such as abrasive blasting, chemical cleaning, or machining. For certain castings, it may be desirable to leave substantial gate witnesses to prevent the opening of pipe shrinkage that leads into the casting. In weldable alloy systems, surface defects can be repair welded before hot isostatic pressing if nondestructive testing techniques can identify the defect. Alloys that are not readily weldable, such as superalloys, can be repaired by vacuum brazing or other techniques to seal surface defects prior to hot isostatic pressing.

(a)

(b)

Fig. 10 IN718 turbine casing for a helicopter engine that was originally produced from four forgings and spinnings (a). The assembly was replaced by a hot isostatically pressed casting, which is shown in cross section in (b). Courtesy of General Electric Company

Casting Salvage. When hot isostatic pressing is used to salvage scrap castings, the presence of surface-connected porosity must be considered. The surface integrity of the scrap castings must be assessed in order to determine if the components are suitable for hot isostatic pressing. If extensive surface-connected porosity is present, an economic analysis should be conducted to determine if the combined costs of sealing surface defects and hot isostatic pressing are economically viable.

Post-HIP Heat Treatment. Most of the HIP furnaces in production are of the slow-cool type, with cooling rates of approximately 4 to 11 °C (7 to 20 °F) per minute. Alloys that need fast cooling for the development of properties will require post-HIP heat treatments. Stainless steels that are susceptible to sensitization must be solution annealed and fast cooled in order to maintain corrosion resistance. Alloy steels can be normalized to refine the grain size of castings that experience grain growth during hot isostatic pressing. The thermal soak time employed during hot isostatic pressing may allow homogenization treatments for certain alloys to be shortened or eliminated.

New Foundry Process Applications

Fine-Grain Castings. One of the most significant foundry advancements made possible through the use of hot isostatic pressing has been the development of fine-grain superalloy castings. In the late 1960s, experiments were conducted on a grain refinement technique for integral turbine wheels. The technique used the mechanical motion of a mold to shear dendrites from the solidifying metal. These dendrites then acted as nucleation sites for additional grains. However, the process was not commercially introduced, because it produced castings with unacceptable levels of porosity.

In the mid-1970s, developmental work on this process resumed when it was realized that hot isostatic pressing could be used to eliminate residual casting porosity. The process that developed from this work, known as Grainex, results in ASTM grain sizes as fine as No. 2. A further development, the Microcast-X process, has led to a greater refinement in grain size (ASTM No. 3 to 5). Figure 8 compares the microstructures of grain-refined rotors with those of a conventionally cast part.

Casting Replacement of Forgings and Fabrications

The demonstrated ability of hot isostatic pressing to increase mechanical properties and to reduce data scatter has made it possible for castings to replace more expen-

Table 5 Stress rupture life of new and HIP-rejuvenated René 80 first-stage turbine blade castings

	Stress rupture life, h			
	760 °C (1400 °F) and 621 MPa (90 ksi)		980 °C (1800 °F) and 183 MPa (26.5 ksi)	
Blade condition	Mean	-2σ	Mean	-2σ
Engine operated	21.8	12.7	11.3	6.0
New	33.7	20.8	18.9	13.4
HIP rejuvenated	65.5	40.8	17.1	9.8

Source: Ref 9

Table 6 Stress rupture life of new and HIP-rejuvenated directionally solidified MAR-M200 + Hf first-stage turbine blade castings

Test conditions: 980 °C (1800 °F) and 172 MPa (25 ksi)

	Stress rupture life, h	
Blade condition	Mean	-2σ
Engine operated	76	0
New	150	91
HIP rejuvenated	151	93

Source: Ref 10

sive wrought and fabricated components. For example, bearing supports were originally produced from AISI 4340 or AM355 bar stock (Ref 8). A substantial amount of machining was required to produce the support. By converting to a near-net shape hot isostatically pressed investment casting, it was possible to produce a support that met the performance requirements and reduced machining costs.

In another case, a Ti-6Al-4V flap (Fig. 9) originally fabricated from welded sheet stock was converted into a less expensive cast and hot isostatically pressed component. Similarly, the four separate IN718 forgings and spinnings shown in Fig. 10(a) are welded together to produce a turbine casing for a helicopter engine (Ref 8). By changing to a net shape investment cast and hot isostatically pressed design (Fig. 10b), it was possible to eliminate many manufacturing and inspection operations. The cast part is less expensive than the assembled forgings and meets performance requirements.

Casting Refurbishment

The hot sections of industrial and aircraft gas turbine engines contain turbine blades that are produced primarily as investment castings in nickel-base superalloys. During engine operation, turbine blades develop both internal and external damage. Internal damage may consist of creep, grain-boundary cavitation, and microcracking, all of which can be closed with hot isostatic pressing to rejuvenate the blade. After traditional refurbishment processes such as coating and heat treatment, blades can be returned to service. Investigations conducted under Air Force contract have demonstrated the feasibility of this process (Ref 9, 10). Tables 5 and 6 present stress rupture properties obtained on rejuvenated René 80 and directionally solidified MAR-M200 + Hf turbine blades. In addition, because turbine blades are manufactured from alloys that contain strategic materials, extending the blade life with hot isostatic pressing can reduce the consumption of these important materials.

REFERENCES

1. H.A. Saller, S.J. Paprocki, R.W. Dayton, and E.S. Hodge, "A Method of Bonding," U.S. Patent 687,842; Canadian patent 680,160; 18 Feb 1964
2. "Casting Densification Process," TMD Report 5, Technology Marketing Division, Aluminum Company of North America
3. G.E. Wasielewski and N.R. Lindblad, Elimination of Casting Defects Using HIP, in *Superalloy Processing*, MCIC 72-10, Proceedings of the 2nd International Conference, Metals and Ceramics Information Center, Sept 1972
4. "Process for High Integrity Castings," AFML TR-74-152, General Electric Company, July 1974
5. "Investment Cast Ti-6Al-4V," Technical Bulletin TB1660, Howmet Turbine Components Corporation
6. R. Smickley, L.E. Dardi, and W.R. Freeman, "Development of High Performance Custom 450 Investment Castings—A Progress Report," Paper presented at the 27th Annual Meeting, Chicago, IL, Investment Casting Institute, Oct 1979
7. "Hot Isostatic Pressing of Castings," Draft B86BC, Proposed Aerospace Material Specification
8. T.M. Regan and J.N. Fleck, Case Studies of Castings Replacing Forgings and Fabrications in a Helicopter Engine, in *Advanced Casting Technology*, J. Easwaran, Ed., Proceedings of an Advanced Casting Technology Conference, Kalamazoo, MI, Nov 1986, ASM INTERNATIONAL, 1987, p 103-110
9. D.C. Stewart and G.T. Bennett, "HIP Rejuvenation of Damaged Blades," AFWAL-TR-80-4043, Air Force Wright Aeronautical Laboratories, May 1980
10. D.J. Kenton, N.M. Madhava, and H. Koven, "Hot Isostatic Pressing Rejuvenation of Life Limited Turbine Hardware," PRAM Project 13278-01, Wright-Patterson Air Force Base, Jan 1981

Testing and Inspection of Casting Defects*

General inspection procedures for castings are established at the foundry to ensure conformance with customer drawings and documents, which are frequently based on various government, technical society, or commercial specifications. For a foundry to ensure casting quality, inspection procedures must be efficiently directed toward the prevention of imperfections, the detection of unsatisfactory trends, and the conservation of material—all of which ultimately lead to reduction in costs. Inspectors should be able to assess on sight the probable strong and weak points of a casting and know where weaknesses and faults would most likely be found.

Inspection of castings normally involves checking for shape and dimensions, coupled with aided and unaided visual inspection for external discontinuities and surface quality. Chemical analyses and tests for mechanical properties are supplemented by various forms of nondestructive inspection, including leak testing and proof loading, all of which are used to evaluate the soundness of the casting. These inspections add to the cost of the product; therefore, the initial consideration must be to determine the amount of inspection needed to maintain adequate control over quality. In some cases, this may require full inspection of each individual casting, but in other cases sampling procedures may be sufficient.

Methods for Determining Surface Quality. Cracks and other imperfections at the surface of a casting can be detected by a number of inspection techniques, including visual inspection, chemical etching, liquid penetrant inspection, eddy current inspection, and magnetic particle inspection (which can also reveal discontinuities situated immediately below the surface). All inspection methods require clean and relatively smooth surfaces for effective results.

Methods for Detecting Internal Discontinuities. The principal nondestructive methods used for detecting internal discontinuities in castings are radiographic, ultrasonic, and

*By the ASM Committee on Nondestructive Inspection of Castings. Frederick A. Morrow, TFI Corporation; Mark J. Alcini and Franklin L. Kiiskila, Williams International; Colin Lewis, William Gavin, and Francis Brozo, Hitchcock Industries, Inc.; Michael Wrysch, Detroit Diesel Allison Division of General Motors Corporation; Alvin F. Maloit, Consulting Metallurgist; Kenneth Whaler, Stahl Specialties Company; Lawrence E. Smiley, Reliable Castings Corporation

eddy current inspection. Of these methods, radiography is the most highly developed technique for detailed inspection; it can provide a pictorial representation of the form and extent of many types of internal discontinuities. Ultrasonic inspection, which is less universally applicable, can give qualitative indications of many discontinuities. It is especially useful in the inspection of castings of fairly simple design, where the signal pattern can be most reliably interpreted. Ultrasonic inspection can also be used to determine the shape of graphite particles in cast iron. Eddy current and other closely related electromagnetic methods are used to sort castings for variations in composition, surface hardness, and structure.

Methods for Dimensional Inspection. There are a number of techniques used to determine the dimensional accuracy of castings. These include manual checks with micrometers, manual and automatic gages, coordinate-measuring machines, and three-dimensional automatic inspection stations (machine vision systems). This section will discuss the use of coordinate-measuring machines. Additional information on methods for dimensional inspection will be provided in Volume 17 of the 9th Edition of *Metals Handbook*.

Casting Defects

Although foundrymen favor referring to the deviations in less-than-perfect castings as discontinuities, these imperfections are more commonly referred to as casting defects. Some casting defects may have no effect on the function or the service life of cast components, but will give an unsatisfactory appearance or will make further processing, such as machining, more costly. Many such defects can be easily corrected by shot blast cleaning or grinding. Other defects that may be more difficult to remove can be acceptable in some locations. It is most critical that the casting designer understand the differences and that he write specifications that meet the true design needs.

Classification of Casting Defects. Foundrymen have traditionally used rather unique names, such as rattail, scab, buckle, snotter, and shut, to describe various casting imperfections (such terms are defined in the "Glossary of Terms" in this Volume). Unfortunately, foundrymen may use different nomenclature to describe the same defect. The International Committee of Foundry Technical Associations has standardized the nomenclature, starting with the identification of seven basic categories of casting defects:

- Metallic projections
- Cavities
- Discontinuities
- Defects
- Incomplete casting
- Incorrect dimension
- Inclusions or structural anomalies

In this scheme, the term discontinuity has the specific meaning of a planar separation of the metal, that is, a crack.

Table 1 presents some of the common defects in each category. In general, defects that can serve as stress raisers or crack promoters are the most serious. These include preexisting cracks, internal voids, and nonmetallic inclusions. The causes of these defects and their correction and prevention are discussed in the Section "Design Considerations" in this Volume.

Common Inspection Procedures

Inspection of castings is most often limited to visual and dimensional inspections, weight testing, and hardness testing. However, for castings that are to be used in critical applications, such as in aerospace components, additional methods of nondestructive inspection are used to determine and to control casting quality.

Visual inspection of each casting ensures that none of its features has been omitted or malformed by molding errors, short running, or mistakes in cleaning. Most surface defects and roughness can be observed at this stage.

Initial sample castings from new pattern equipment should be carefully inspected for obvious defects. Liquid penetrant inspection can be used to detect surface defects. Such casting imperfections as shrinks, cracks, blows, or dross usually indicate the need for adjustment in the gating or foundry techniques. If the casting appears to be satisfactory upon visual inspection, internal quality can be checked by radiographic and ultrasonic inspection.

The first visual inspection operation on the production casting is usually performed immediately after shakeout or knockout of the casting. This ensures that major visible imperfections are detected as quickly as possible. This information, promptly relayed to the foundry, permits early corrective action to be taken with a minimum of scrap loss. The size and complexity of some sand castings require that the gates and risers be removed to permit proper inspection of the casting. Many castings that contain numerous internal cores or have close dimensional tolerances require a rapid but fairly accurate check of critical wall dimensions. In some cases, an indicating-type caliper gage is suitable for this work, and special types are available for casting shapes that do not lend themselves to the standard types. Ultrasonic inspection is also used to determine wall thickness in such components as cored turbine blades made by investment casting (see the article "Investment Casting" in this Volume).

Dimensional Inspection. Dimensional deviations on machined surfaces are relatively simple to evaluate and can be accurately specified. However, it is not so simple to determine the acceptability of dimensions that involve one or more unmachined surfaces. Dimensional inspection can be carried out with the aid of gages, jigs, and templates.

Most initial machining operations on castings use a cast surface as a datum; the exceptions are those large castings that are laid out, before machining, to give the required datum. Therefore, it is important that the cast surface used as a datum be reasonably true and that it be in the correct position relative to other critical machined or unmachined surfaces on the same casting, within clearly defined limits.

The cast surface used as a datum can be a mold surface, and variations can occur because of mold movement. The cast surface can be produced by a core; movement of cores is a frequent cause of casting inaccuracy. Errors involving these surfaces can produce consequential errors or inadequate machining stock elsewhere on the casting.

Where dimensional errors are detected in relation to general drawing tolerances, their true significance must be determined. A particular dimension may be of vital importance, but may have been included in blanket tolerances. This situation stresses the desirability of stating functional dimensions on drawings so that tolerances are not restricted unnecessarily.

Weight Testing. Many intricately cored castings are extremely difficult to measure accurately, particularly the internal sections. It is important to ensure that these sections are correct in thickness for three main reasons:

- There should be no additional weight that would make the finished product heavier than permissible
- Sections must not be thinner than designed to prevent detracting from the strength of the casting
- If hollow cavities have been reduced in area by increasing the metal thickness of the sections, any flow of liquid or gases is reduced

A ready means of testing for these discrepancies is by accurately weighing each casting or by measuring the displacement caused by immersing the casting in a liquid-filled measuring jar or vessel. In certain cases in which extreme accuracy is demanded, a tolerance of only ±1% of a given weight may be allowed.

Hardness testing is often used to verify the effectiveness of heat treatment applied to actual castings. Its general correlation with the tensile strength of many ferrous alloys enables a rough prediction of tensile strength to be made.

Table 1 International classification of common casting defects

Metallic Projections

A 100: Metallic projections in the form of fins or flash

A 110: Metallic projections in the form of fins (or flash) without change in principal casting dimensions

No.	Description	Common name	Sketch
A 111	Thin fins (or flash) at the parting line or at core prints	Joint flash or fins	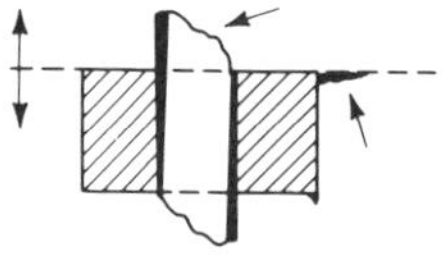
A 112	Projections in the form of veins on the casting surface	Veining or finning	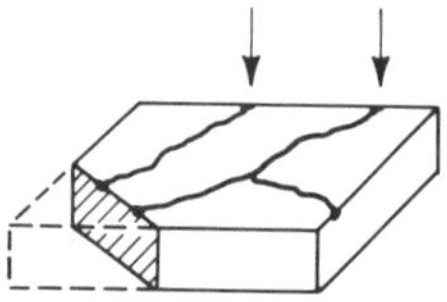
A 113	Network of projections on the surface of die castings	Heat-checked die	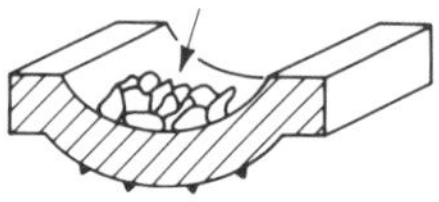
A 114(a)	Thin projection parallel to a casting surface, in re-entrant angles	Fillet scab	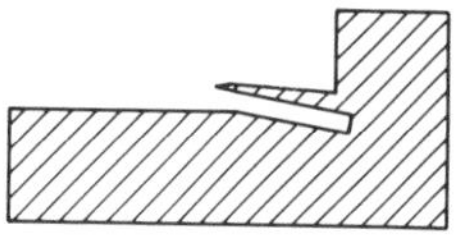
A 115	Thin metallic projection located at a re-entrant angle and dividing the angle in two parts	Fillet vein	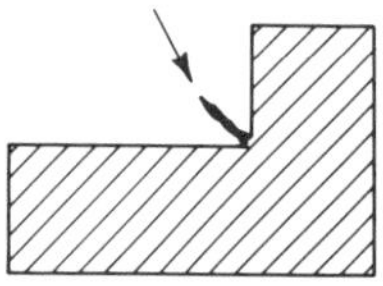

A 120: Metallic projections in the form of fins with changes in principal casting dimensions

No.	Description	Common name	Sketch
A 123(a)	Formation of fins in planes related to direction of mold assembly (precision casting with waste pattern); principal casting dimensions change	Cracked or broken mold	

A 200: Massive projections

A 210: Swells

No.	Description	Common name	Sketch
A 212(a)	Excess metal in the vicinity of the gate or beneath the sprue	Erosion, cut, or wash	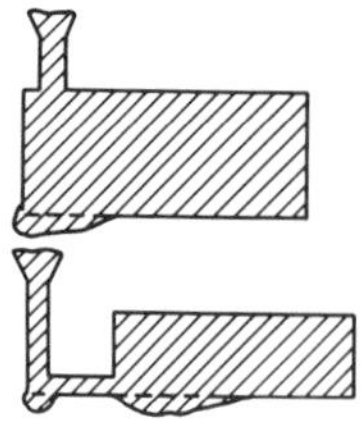
A 213(a)	Metal projections in the form of elongated areas in the direction of mold assembly	Crush	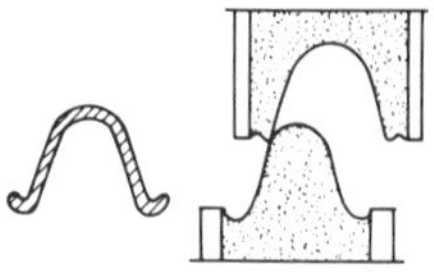

A 220: Projections with rough surfaces

No.	Description	Common name	Sketch
A 221(a)	Projections with rough surfaces on the cope surface of the casting	Mold drop or sticker	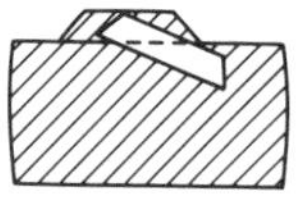
A 222(a)	Projections with rough surfaces on the drag surface of the casting (massive projections)	Raised core or mold element cutoff	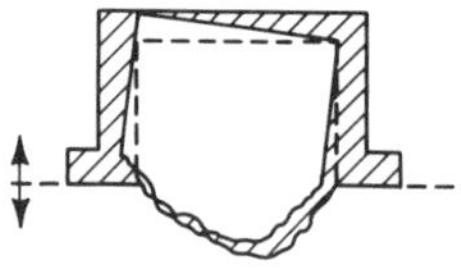
A 223(a)	Projections with rough surfaces on the drag surface of the casting (in dispersed areas)	Raised sand	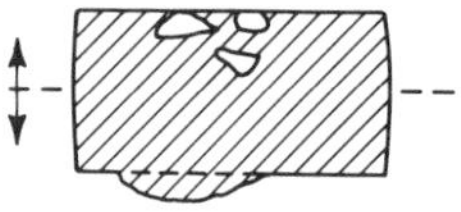
A 224(a)	Projections with rough surfaces on other parts of the casting	Mold drop	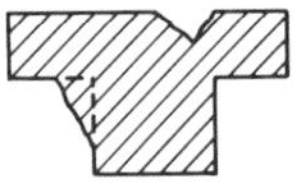
A 225(a)	Projections with rough surfaces over extensive areas of the casting	Corner scab	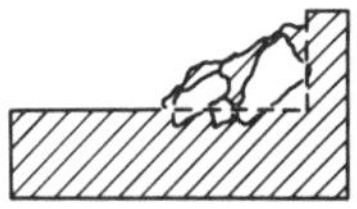
A 226(a)	Projections with rough surfaces in an area formed by a core	Broken or crushed core	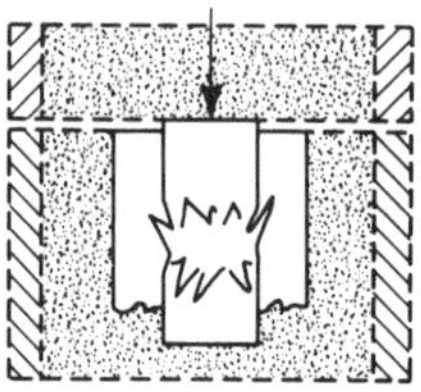

(continued)

(a) Defects that under some circumstances could contribute, either directly or indirectly, to casting failures. Adapted from *International Atlas of Casting Defects*, American Foundrymen's Society, Des Plaines, IL

Table 1 (continued)

Cavities

B 100: Cavities with generally rounded, smooth walls perceptible to the naked eye (blowholes, pinholes)

B 110: Class B 100 cavities internal to the casting, not extending to the surface, discernible only by special methods, machining, or fracture of the casting

No.	Description	Common name	Sketch
B 111(a)	Internal, rounded cavities, usually smooth-walled, of varied size, isolated or grouped irregularly in all areas of the casting	Blowholes, pinholes	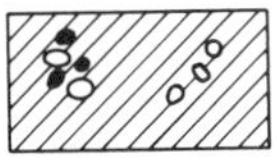 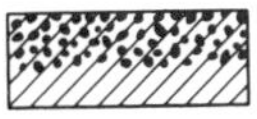
B 112(a)	As above, but limited to the vicinity of metallic pieces placed in the mold (chills, inserts, chaplets, etc.)	Blowholes, adjacent to inserts, chills, chaplets, etc.	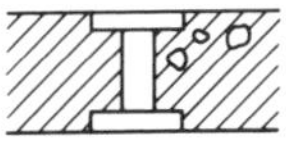 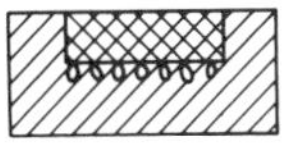 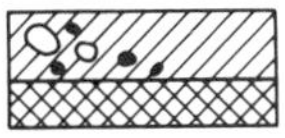
B 113(a)	Like B 111, but accompanied by slag inclusions (G 122)	Slag blowholes	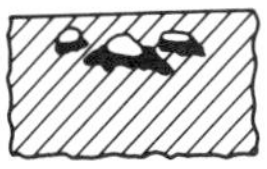

B 120: Class B 100 cavities located at or near the casting surface, largely exposed or at least connected with the exterior

No.	Description	Common name	Sketch
B 121(a)	Exposed cavities of various sizes, isolated or grouped, usually at or near the surface, with shiny walls	Surface or subsurface blowholes	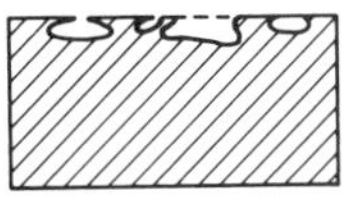
B 122(a)	Exposed cavities, in re-entrant angles of the casting, often extending deeply within	Corner blowholes, draws	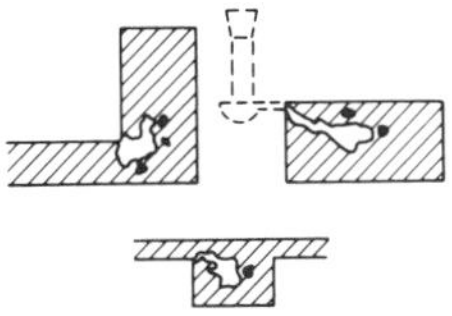
B 123	Fine porosity (cavities) at the casting surface, appearing over more or less extended areas	Surface pinholes	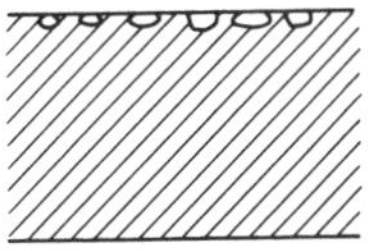
B 124(a)	Small, narrow cavities in the form of cracks, appearing on the faces or along edges, generally only after machining	Dispersed shrinkage	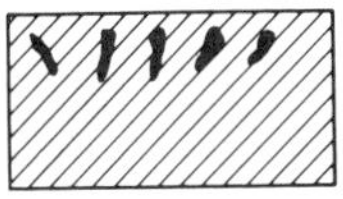

B 200: Cavities with generally rough walls, shrinkage

B 210: Open cavity of Class B 200, sometimes penetrating deeply into the casting

No.	Description	Common name	Sketch
B 211(a)	Open, funnel-shaped cavity; wall usually covered with dendrites	Open or external shrinkage	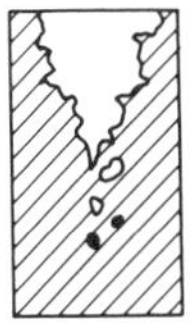
B 212(a)	Open, sharp-edged cavity in fillets of thick castings or at gate locations	Corner or fillet shrinkage	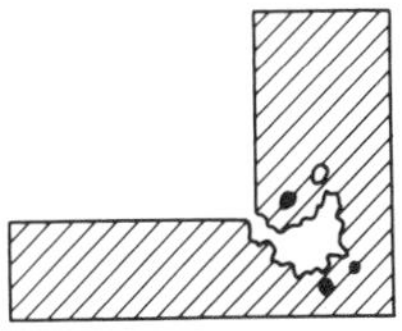
B 213(a)	Open cavity extending from a core	Core shrinkage	

B 220: Class B 200 cavity located completely internal to the casting

No.	Description	Common name	Sketch
B 221(a)	Internal, irregularly shaped cavity; wall often dendritic	Internal or blind shrinkage	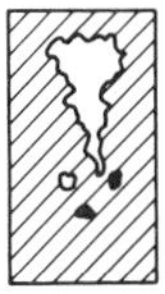
B 222(a)	Internal cavity or porous area along central axis	Centerline or axial shrinkage	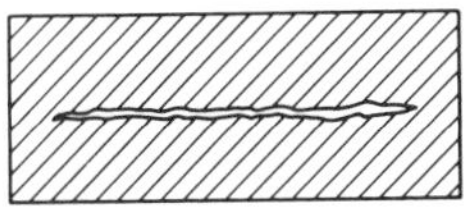

B 300: Porous structures caused by numerous small cavities

B 310: Cavities according to B 300, scarcely perceptible to the naked eye

No.	Description	Common name	Sketch
B 311(a)	Dispersed, spongy dendritic shrinkage within walls of casting; barely perceptible to the naked eye	Macro- or micro-shrinkage, shrinkage porosity, leakers	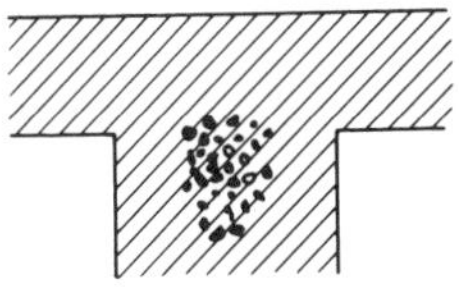

(continued)

(a) Defects that under some circumstances could contribute, either directly or indirectly, to casting failures. Adapted from *International Atlas of Casting Defects*, American Foundrymen's Society, Des Plaines, IL

Table 1 (continued)

Discontinuities

C 100: Discontinuities, generally at intersections, caused by mechanical effects (rupture)

C 110: Normal cracking

No.	Description	Common name	Sketch
C 111(a)	Normal fracture appearance, sometimes with adjacent indentation marks	Breakage (cold)	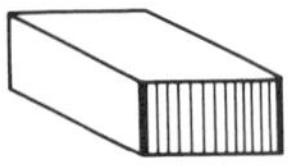

C 120: Cracking with oxidation

No.	Description	Common name	Sketch
C 121(a)	Fracture surface oxidized completely around edges	Hot cracking	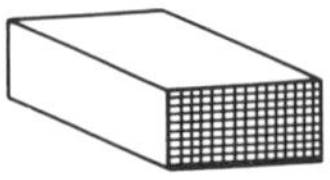

C 200: Discontinuities caused by internal tension and restraints to contraction (cracks and tears)

C 210: Cold cracking or tearing

No.	Description	Common name	Sketch
C 211(a)	Discontinuities with squared edges in areas susceptible to tensile stresses during cooling; surface not oxidized	Cold tearing	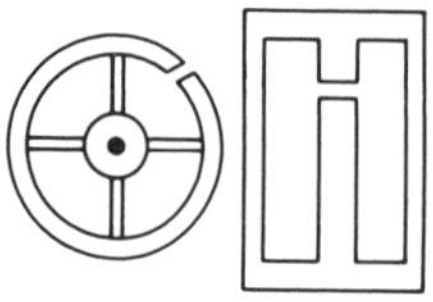

C 220: Hot cracking and tearing

No.	Description	Common name	Sketch
C 221(a)	Irregularly shaped discontinuities in areas susceptible to tension; oxidized fracture surface showing dendritic pattern	Hot tearing	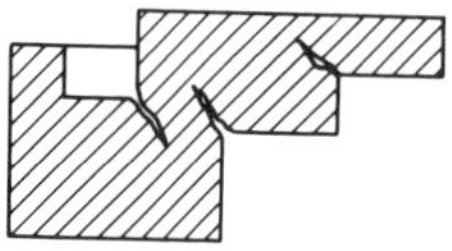
C 222(a)	Rupture after complete solidification, either during cooling or heat treatment	Quench cracking	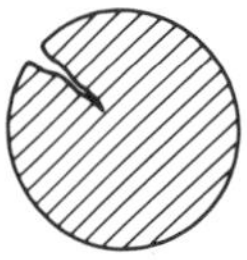

C 300: Discontinuities caused by lack of fusion (cold shuts); edges generally rounded, indicating poor contact between various metal streams during filling of the mold

C 310: Lack of complete fusion in the last portion of the casting to fill

No.	Description	Common name	Sketch
C 311(a)	Complete or partial separation of casting wall, often in a vertical plane	Cold shut or cold lap	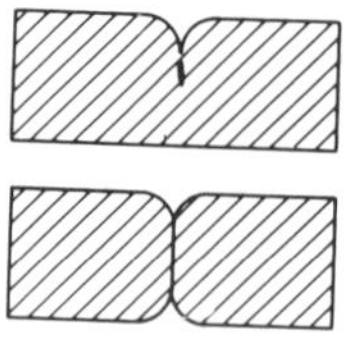

C 320: Lack of fusion between two parts of casting

No.	Description	Common name	Sketch
C 321(a)	Separation of the casting in a horizontal plane	Interrupted pour	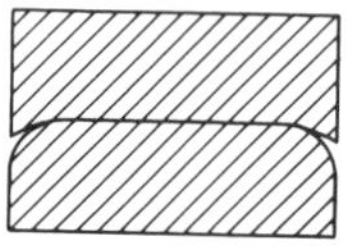

C 330: Lack of fusion around chaplets, internal chills, and inserts

No.	Description	Common name	Sketch
C 331(a)	Local discontinuity in vicinity of metallic insert	Chaplet or insert cold shut, unfused chaplet	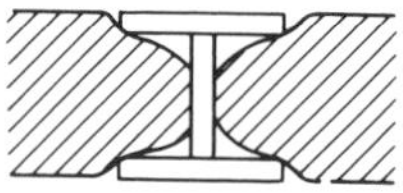

C 400: Discontinuities caused by metallurgical defects

C 410: Separation along grain boundaries

No.	Description	Common name	Sketch
C 411(a)	Separation along grain boundaries of primary crystallization	Conchoidal or "rock candy" fracture	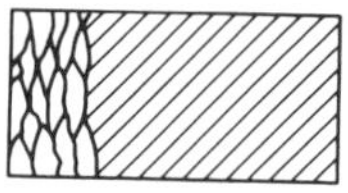
C 412(a)	Network of cracks over entire cross section	Intergranular corrosion	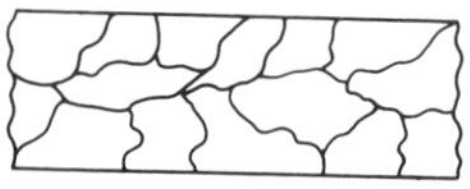

Defective Surface

D 100: Casting surface irregularities

D 110: Fold markings on the skin of the casting

No.	Description	Common name	Sketch
D 111	Fold markings over rather large areas of the casting	Surface folds, gas runs	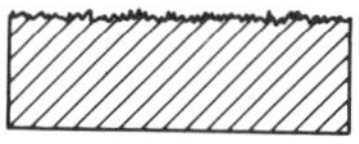

(continued)

(a) Defects that under some circumstances could contribute, either directly or indirectly, to casting failures. Adapted from *International Atlas of Casting Defects*, American Foundrymen's Society, Des Plaines, IL

Table 1 (continued)

No.	Description	Common name	Sketch
D 112	Surface shows a network of jagged folds or wrinkles (ductile iron)	Cope defect, elephant skin, laps	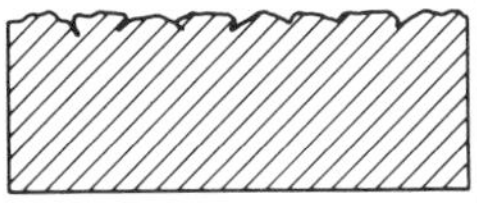
D 113	Wavy fold markings without discontinuities; edges of folds at same level, casting surface is smooth	Seams or scars	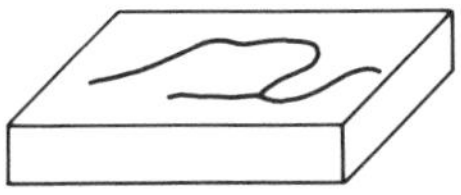
D 114	Casting surface markings showing direction of liquid metal flow (light alloys)	Flow marks	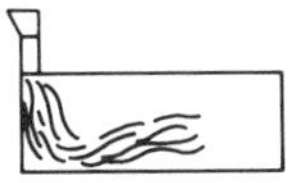
D 120: Surface roughness			
D 121	Depth of surface roughness is approximately that of the dimensions of the sand grains	Rough casting surface	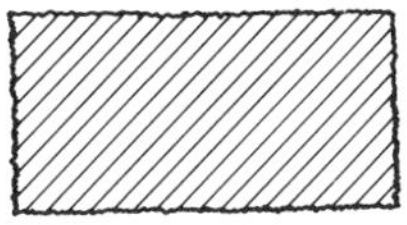
D 122	Depth of surface roughness is greater than that of the sand grain dimensions	Severe roughness, high pressure molding defect	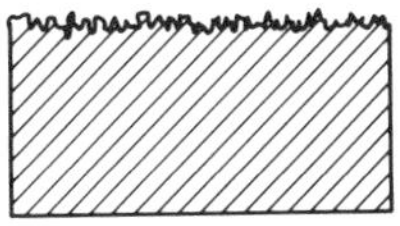
D 130: Grooves on the casting surface			
D 131	Grooves of various lengths, often branched, with smooth bottoms and edges	Buckle	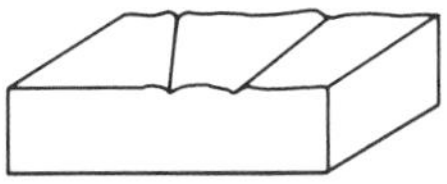
D 132	Grooves up to 5.1 mm (0.2 in.) in depth, one edge forming a fold which more or less completely covers the groove	Rat tail	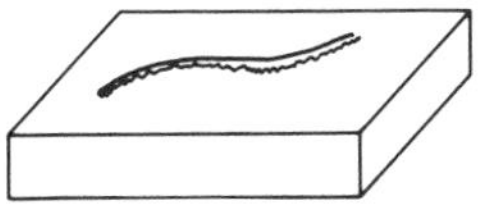
D 133	Irregularly distributed depressions of various dimensions extending over the casting surface, usually along the path of metal flow (cast steel)	Flow marks, crow's feet	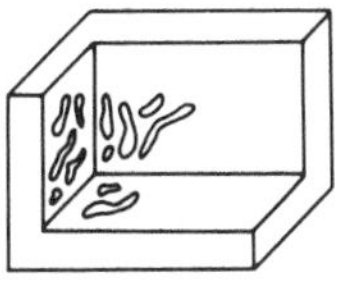
D 134	Casting surface entirely pitted or pock-marked	Orange peel, metal mold reaction, alligator skin	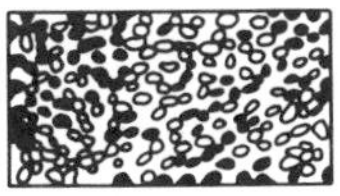
D 135	Grooves and roughness in the vicinity of re-entrant angles on die castings	Soldering, die erosion	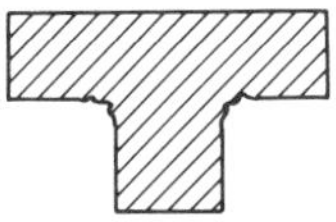
D 140: Depressions in the casting surface			
D 141	Casting surface depressions in the vicinity of a hot spot	Sink marks, draw or suck-in	
D 142	Small, superficial cavities in the form of droplets of shallow spots, generally gray-green in color	Slag inclusions	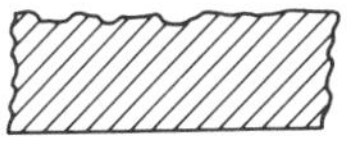
D 200: Serious surface defects			
D 210: Deep indentation of the casting surface			
D 211	Deep indentation, often over large area of drag half of casting	Push-up, clamp-off	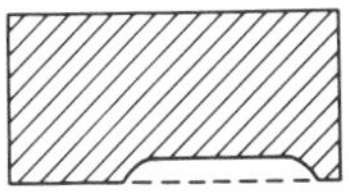
D 220: Adherence of sand, more or less vitrified			
D 221	Sand layer strongly adhering to the casting surface	Burn on	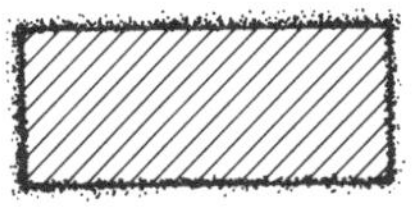
D 222	Very adherent layer of partially fused sand	Burn in	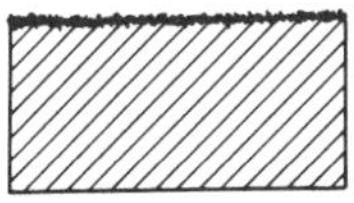

(continued)

(a) Defects that under some circumstances could contribute, either directly or indirectly, to casting failures. Adapted from *International Atlas of Casting Defects*, American Foundrymen's Society, Des Plaines, IL

Table 1 (continued)

No.	Description	Common name	Sketch
D 223	Conglomeration of strongly adhering sand and metal at the hottest points of the casting (re-entrant angles and cores)	Metal penetration	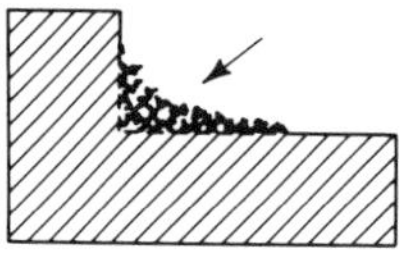
D 224	Fragment of mold material embedded in casting surface	Dip coat spall, scab	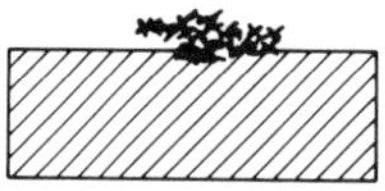

D 230: Plate-like metallic projections with rough surfaces, usually parallel to casting surface

No.	Description	Common name	Sketch
D 231(a)	Plate-like metallic projections with rough surfaces parallel to casting surface; removable by burr or chisel	Scabs, expansion scabs	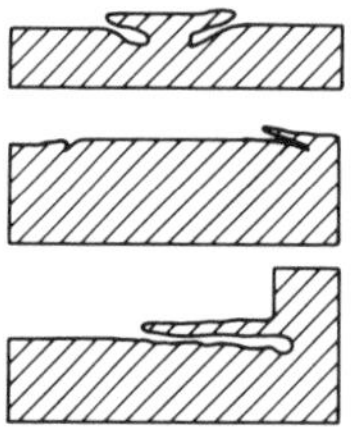
D 232(a)	As above, but impossible to eliminate except by machining or grinding	Cope spall, boil scab, erosion scab	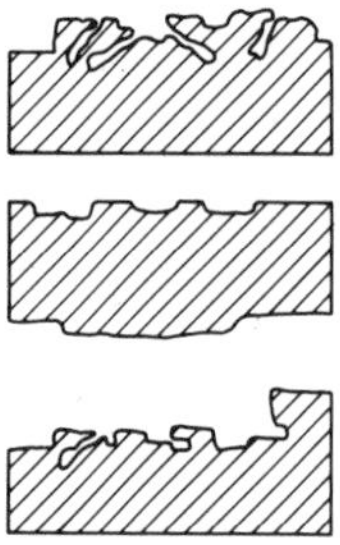
D 233(a)	Flat, metallic projections on the casting where mold or core washes or dressings are used	Blacking scab, wash scab	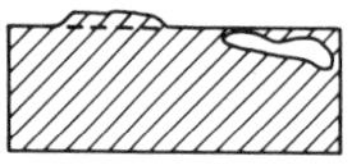

D 240: Oxides adhering after heat treatment (annealing, tempering, malleablizing) by decarburization

No.	Description	Common name	Sketch
D 241	Adherence of oxide after annealing	Oxide scale	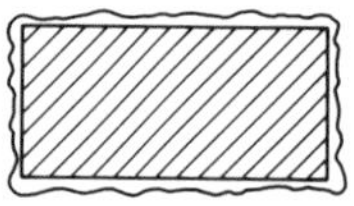
D 242	Adherence of ore after malleablizing (white heart malleable)	Adherent packing material	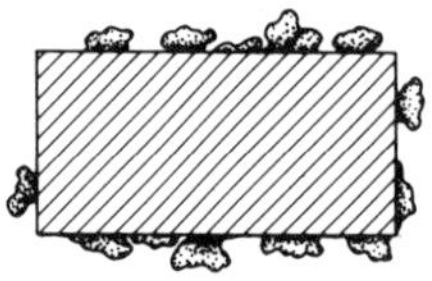
D 243	Scaling after anneal	Scaling	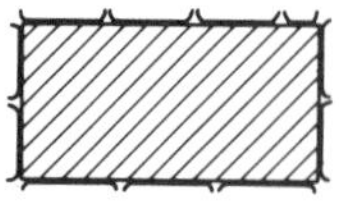

Incomplete Casting

E 100: Missing portion of casting (no fracture)

E 110: Superficial variations from pattern shape

No.	Description	Common name	Sketch
E 111	Casting is essentially complete except for more or less rounded edges and corners	Misrun	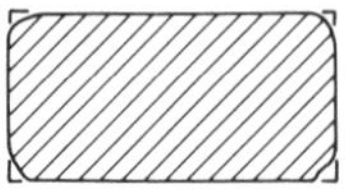
E 112	Deformed edges or contours due to poor mold repair or careless application of wash coatings	Defective coating (tear-dropping) or poor mold repair	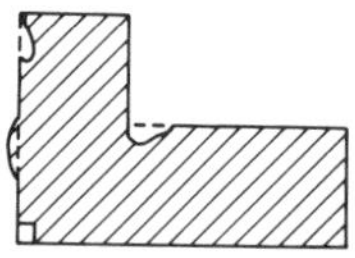

E 120: Serious variations from pattern shape

No.	Description	Common name	Sketch
E 121	Casting incomplete due to premature solidification	Misrun	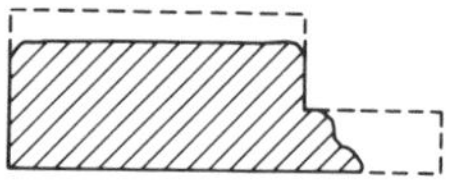
E 122	Casting incomplete due to insufficient metal poured	Poured short	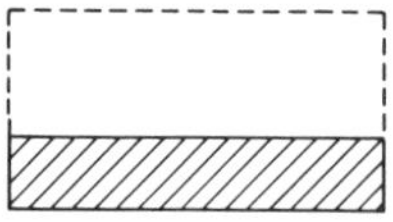
E 123	Casting incomplete due to loss of metal from mold after pouring	Runout	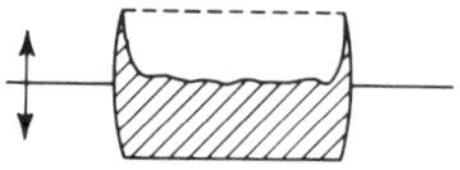
E 124	Significant lack of material due to excessive shot-blasting	Excessive cleaning	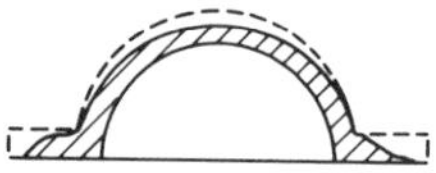

(continued)

(a) Defects that under some circumstances could contribute, either directly or indirectly, to casting failures. Adapted from *International Atlas of Casting Defects*, American Foundrymen's Society, Des Plaines, IL

Table 1 (continued)

No.	Description	Common name	Sketch
E 125	Casting partially melted or seriously deformed during annealing	Fusion or melting during heat treatment	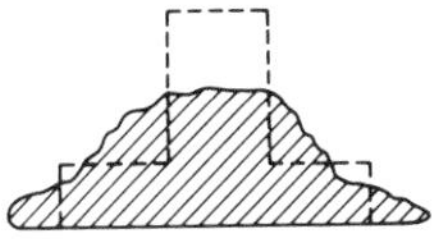

E 200: Missing portion of casting (with fracture)

E 210: Fractured casting

No.	Description	Common name	Sketch
E 211	Casting broken, large piece missing; fractured surface not oxidized	Fractured casting	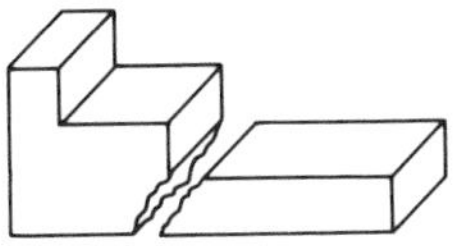

E 220: Piece broken from casting

No.	Description	Common name	Sketch
E 221	Fracture dimensions correspond to those of gates, vents, etc.	Broken casting (at gate, riser, or vent)	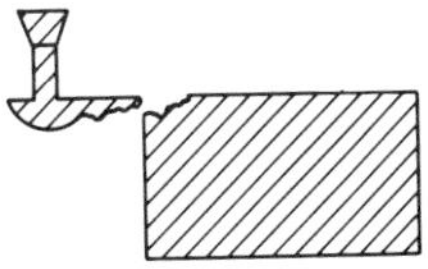

E 230: Fractured casting with oxidized fracture

No.	Description	Common name	Sketch
E 231	Fracture appearance indicates exposure to oxidation while hot	Early shakeout	

Incorrect Dimensions or Shape

F 100: Incorrect dimensions; correct shape

F 110: All casting dimensions incorrect

No.	Description	Common name	Sketch
F 111	All casting dimensions incorrect in the same proportions	Improper shrinkage allowance	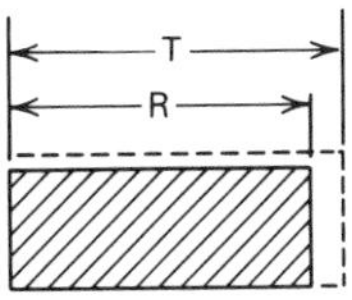

F 120: Certain casting dimensions incorrect

No.	Description	Common name	Sketch
F 121	Distance too great between extended projections	Hindered contraction	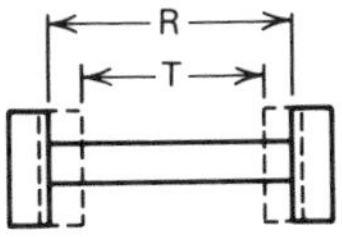
F 122	Certain dimensions inexact	Irregular contraction	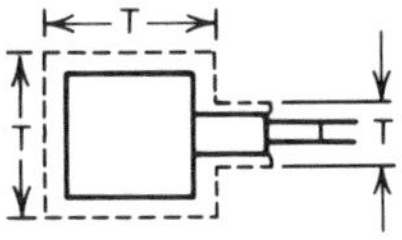
F 123	Dimensions too great in the direction of rapping of pattern	Excess rapping of pattern	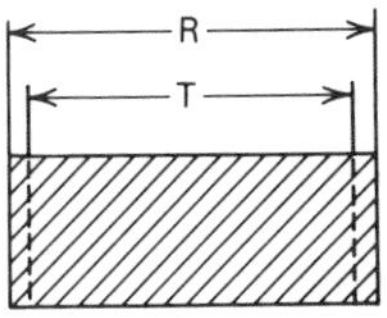
F 124	Dimensions too great in direction perpendicular to parting line	Mold expansion during baking	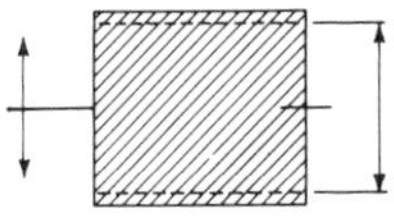
F 125	Excessive metal thickness at irregular locations on casting exterior	Soft or insufficient ramming, mold-wall movement	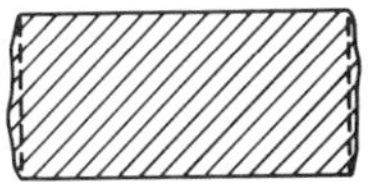
F 126	Thin casting walls over general area, especially on horizontal surfaces	Distorted casting	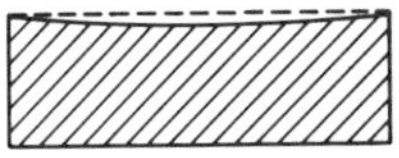

F 200: Casting shape incorrect overall or in certain locations

F 210: Pattern incorrect

No.	Description	Common name	Sketch
F 211	Casting does not conform to the drawing shape in some or many respects; same is true of pattern	Pattern error	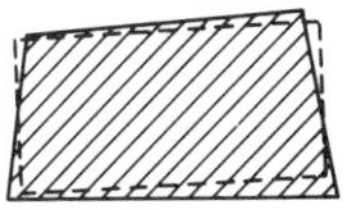
F 212	Casting shape is different from drawing in a particular area; pattern is correct	Pattern mounting error	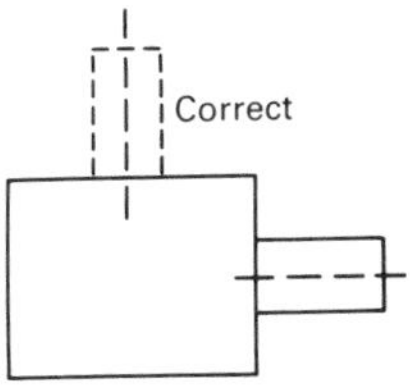

(continued)

(a) Defects that under some circumstances could contribute, either directly or indirectly, to casting failures. Adapted from *International Atlas of Casting Defects*, American Foundrymen's Society, Des Plaines, IL

Table 1 (continued)

F 220: Shift or Mismatch

No.	Description	Common name	Sketch
F 221	Casting appears to have been subjected to a shearling action in the plane of the parting line	Shift	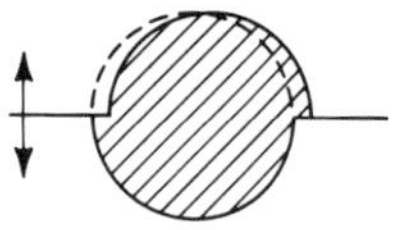
F 222	Variation in shape of an internal casting cavity along the parting line of the core	Shifted core	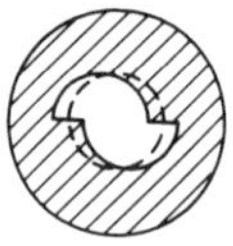
F 223	Irregular projections on vertical surfaces, generally on one side only in the vicinity of the parting line	Ramoff, ramaway	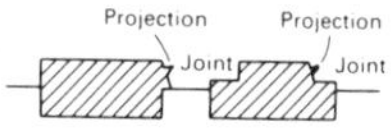

F 230: Deformations from correct shape

No.	Description	Common name	Sketch
F 231	Deformation with respect to drawing proportional for casting, mold, and pattern	Deformed pattern	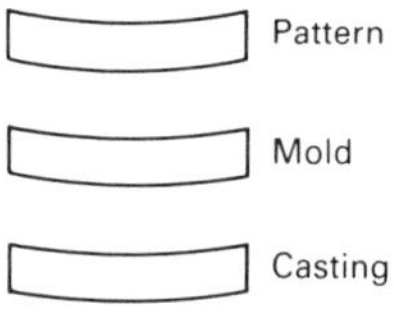
F 232	Deformation with respect to drawing proportional for casting and mold; pattern conforms to drawing	Deformed mold, mold creep, springback	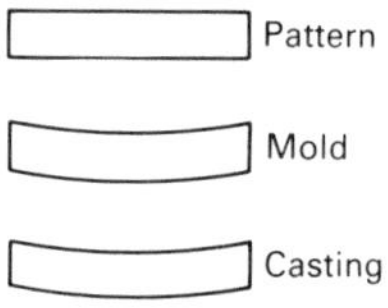
F 233	Casting deformed with respect to drawing; pattern and mold conform to drawing	Casting distortion	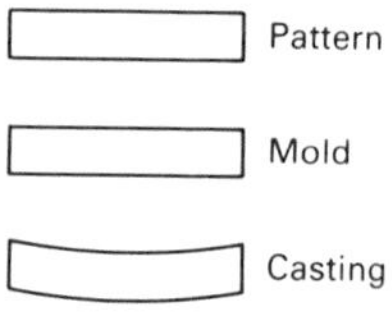
F 234	Casting deformed with respect to drawing after storage, annealing, machining	Warped casting	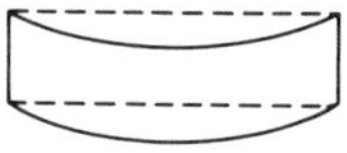

Inclusions or Structural Anomalies

G 100: Inclusions

G 110: Metallic inclusions

No.	Description	Common name	Sketch
G 111(a)	Metallic inclusions whose appearance, chemical analysis or structural examination show to be caused by an element foreign to the alloy	Metallic inclusions	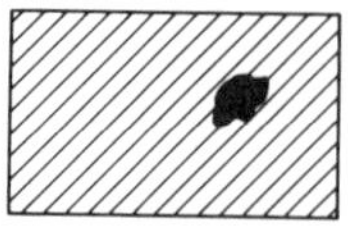
G 112(a)	Metallic inclusions of the same chemical composition as the base metal; generally spherical and often coated with oxide	Cold shot	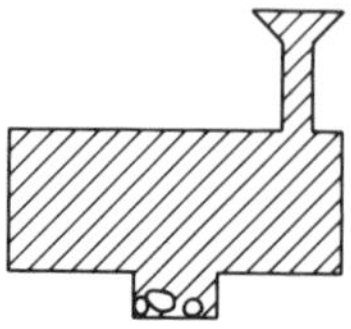
G 113	Spherical metallic inclusions inside blowholes or other cavities or in surface depressions (see A 311). Composition approximates that of the alloy cast but nearer to that of a eutectic	Internal sweating, phosphide sweat	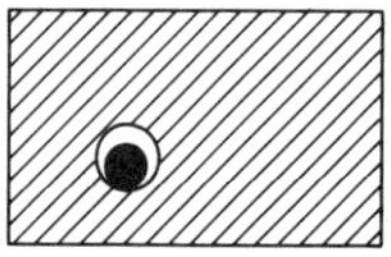

G 120: Nonmetallic inclusions; slag, dross, flux

No.	Description	Common name	Sketch
G 121(a)	Nonmetallic inclusions whose appearance or analysis shows they arise from melting slags, products of metal treatment or fluxes	Slag, dross or flux inclusions, ceroxides	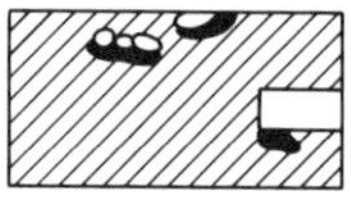
G 122(a)	Nonmetallic inclusions generally impregnated with gas and accompanied by blowholes (B 113)	Slag blowhole defect	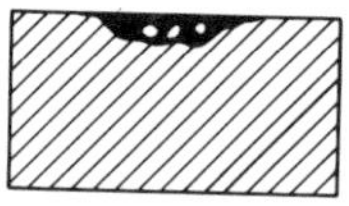

G 130: Nonmetallic inclusions; mold or core materials

No.	Description	Common name	Sketch
G 131(a)	Sand inclusions, generally very close to the surface of the casting	Sand inclusions	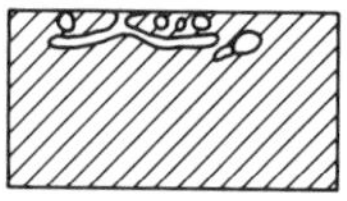
G 132(a)	Inclusions of mold blacking or dressing, generally very close to the casting surface	Blacking or refractory coating inclusions	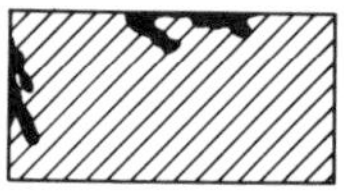

(continued)

(a) Defects that under some circumstances could contribute, either directly or indirectly, to casting failures. Adapted from *International Atlas of Casting Defects*, American Foundrymen's Society, Des Plaines, IL

Table 1 (continued)

No.	Description	Common name	Sketch
G 140:	**Nonmetallic inclusions; oxides and reaction products**		
G 141	Clearly defined, irregular black spots on the fractured surface of ductile cast iron	Black spots	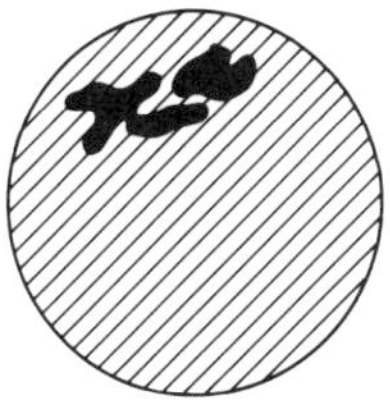
G 142(a)	Inclusions in the form of oxide skins, most often causing a localized seam	Oxide inclusion or skins, seams	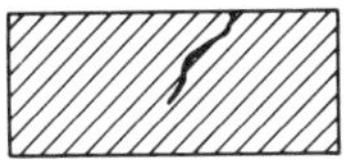
G 143(a)	Folded films of graphitic luster in the wall of the casting	Lustrous carbon films, or kish tracks	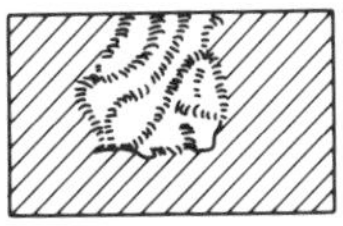
G 144	Hard inclusions in permanent molded and die cast aluminum alloys	Hard spots	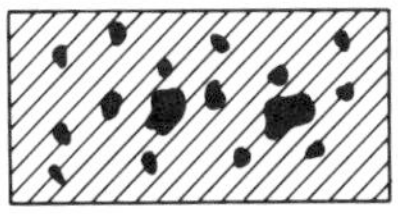

(a) Defects that under some circumstances could contribute, either directly or indirectly, to casting failures. Adapted from *International Atlas of Casting Defects*, American Foundrymen's Society, Des Plaines, IL

The Brinell hardness test is most frequently used for casting alloys. A combination of large-diameter ball (5 or 10 mm) and heavy load (500 to 3000 kgf) is preferred for the most effective representation because a deep impression minimizes the influence of the immediate surface layer and of the relatively coarse microstructure. The Brinell hardness test is unsuitable for use at high hardness levels (above ~600 HB), because distortion of the ball indenter can affect the shape of the indentation.

Either the Rockwell or the Vickers (136° diamond pyramid) hardness test is used for alloys of extreme hardness or for high-quality and precision castings in which the large Brinell indentation cannot be tolerated. Because of the very small indentations produced in Rockwell and Vickers tests, which use loads of 150 kg or less, results must be based on the average of a number of determinations. Portable hardness testers or ultrasonic microhardness testers can be used on large castings that cannot be placed on the platform of a bench-type machine. More detailed information on hardness testing is available in Volume 8 of the 9th Edition of *Metals Handbook*.

The hardness of ferrous castings can be related to the sonic velocity of the metal and determined from it if all other test conditions remain constant. This has been demonstrated on chilled rolls in determining the average hardness of the core.

Liquid Penetrant Inspection

Liquid penetrant inspection essentially involves a liquid wetting the surface of a workpiece, flowing over that surface to form a continuous and uniform coating, and migrating into cracks or cavities that are open to the surface. After a few minutes, the liquid coating is washed off the surface of the casting and a developer is placed on the surface. The developer is stained by the liquid penetrant as it is drawn out of the cracks and cavities. Liquid penetrants will highlight surface defects so that detection is more certain.

Liquid penetrant inspection should not be confined to as-cast surfaces. For example, it is not unusual for castings of various alloys to exhibit cracks, frequently intergranular, on machined surfaces. A pattern of cracks of this type may be the result of intergranular cracking throughout the material because of an error in composition or heat treatment, or the cracks may be on the surface only as a result of machining or grinding. Surface cracking may result from insufficient machining allowance, which does not allow for complete removal of imperfections produced on the as-cast surface, or it may result from faulty machining techniques. If imperfections of this type are detected by visual inspection, liquid penetrant inspection will show the full extent of such imperfections, will give some indication of the depth and size of the defect below the surface by the amount of penetrant absorbed, and will indicate whether cracking is present throughout the section.

Magnetic Particle Inspection

Magnetic particle inspection is a highly effective and sensitive technique for revealing cracks and similar defects at or just beneath the surface of castings made of ferromagnetic metals. The capability of detecting discontinuities just beneath the surface is important because such cleaning methods as shot or abrasive blasting tend to close a surface break that might go undetected in visual or liquid penetrant inspection.

When a magnetic field is generated in and around a casting made of a ferromagnetic metal and the lines of magnetic flux are intersected by a defect such as a crack, magnetic poles are induced on either side of the defect. The resulting local flux disturbance can be detected by its effect on the particles of a ferromagnetic material, which become attracted to the region of the defect as they are dusted on the casting. Maximum sensitivity of indication is obtained when a defect is oriented in a direction perpendicular to the applied magnetic field and when the strength of this field is just enough to saturate the casting being inspected.

Equipment for magnetic particle inspection uses direct or alternating current to generate the necessary magnetic fields. The current can be applied in a variety of ways to control the direction and magnitude of the magnetic field.

In one method of magnetization, a heavy current is passed directly through the casting placed between two solid contacts. The induced magnetic field then runs in the transverse or circumferential direction, producing conditions favorable to the detection of longitudinally oriented defects. A coil encircling the casting will induce a magnetic field that runs in the longitudinal direction, producing conditions favorable to the detection of circumferentially (or transversely) oriented defects. Alternatively, a longitudinal magnetic field can be conveniently generated by passing current through a flexible cable conductor, which can be coiled around any metal section. This method is particularly adaptable to castings of irregular shape. Circumferential magnetic fields can be induced in hollow cylindrical castings by using an axially disposed central conductor threaded through the casting.

Small castings can be magnetic particle inspected directly on bench-type equipment that incorporates both coils and solid contacts. Critical regions of larger castings can be inspected by the use of yokes, coils, or contact probes carried on flexible cables connected to the source of current; this setup enables most regions of castings to be inspected.

Eddy Current Inspection

Eddy current inspection consists of observing the interaction between electromagnetic fields and metals. In a basic system, currents are induced to flow in the testpiece by a coil of wire that carries an alternating current. As the part enters the coil, or as the coil in the form of a probe or yoke is placed on the testpiece, electromagnetic energy produced by the coils is partly absorbed and converted into heat by the effects of resistivity and hysteresis. Part of the remaining energy is reflected back to the test coil, its electrical characteristics having been changed in a manner determined by the properties of the testpiece. Consequently, the currents flowing through the probe coil are the source of information describing the characteristics of the testpiece. These currents can be analyzed and compared with currents flowing through a reference specimen.

Eddy current methods of inspection are effective with both ferromagnetic and nonferromagnetic metals. Eddy current methods are not as sensitive to small, open defects as liquid penetrant or magnetic particle methods are. Because of the skin effect, eddy current inspection is generally restricted to depths less than 6 mm (¼ in.). The results of inspecting ferromagnetic materials can be obscured by changes in the magnetic permeability of the testpiece. Changes in temperature must be avoided to prevent erroneous results if electrical conductivity or other properties, including metallurgical properties, are being determined.

Applications of eddy current and electromagnetic methods of inspection to castings can be divided into the following three categories:

- Detecting near-surface flaws such as cracks, voids, inclusions, blowholes, and pinholes (eddy current inspection)
- Sorting according to alloy, temper, electrical conductivity, hardness, and other metallurgical factors (primarily electromagnetic inspection)
- Gaging according to size, shape, plating thickness, or insulation thickness (eddy current or electromagnetic inspection)

Radiographic Inspection*

Radiographic inspection is a process of testing materials using penetrating radiation from an x-ray generator or a radioactive

*This section was prepared by Frederick A. Morrow, TFI Corporation.

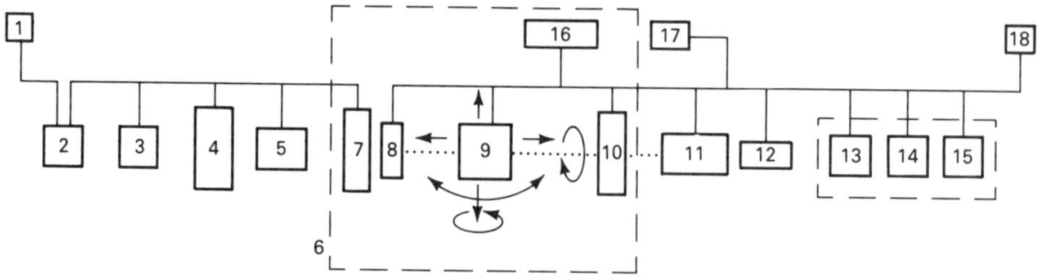

1 Electrical power supply
2 Voltage stabilizer
3 Closed circuit cooling system
4 High-tension x-ray generator
5 X-ray control
6 Radiation protective enclosure
7 X-ray tubehead
8 X-ray shutter
9 Remotely controlled manipulator
10 Imaging system shutters
11 Imaging system
12 Closed circuit video camera
13 System control center
14 Video monitor: imagery display
15 Digital image processor, hard copy device(s), video recorder/player
16 Radiation safety system
17 Warning sign
18 Electrical power supply

Fig. 1 Schematic of real time x-ray inspection system. Source: TFI Corporation

source and an imaging medium, such as x-ray film or an electronic device. In passing through the material, some of the radiation is attenuated, depending on the thickness and the radiographic density of the material, while the radiation that passes through the material forms an image. The radiographic image is generated by variations in the intensity of the emerging beam.

Internal flaws, such as gas entrapment or nonmetallic inclusions, have a direct effect on the attenuation. These flaws create variations in material thickness, resulting in localized dark or light spots on the image.

The term radiography usually implies a radiographic process that produces a permanent image on film (conventional radiography) or paper (paper radiography or xeroradiography), although in a broad sense it refers to all forms of radiographic inspection. When inspection involves viewing an image on a fluorescent screen or image intensifier, the radiographic process is termed filmless or real time inspection (Fig. 1). When electronic nonimaging instruments are used to measure the intensity of radiation, the process is termed radiation gaging. Tomography, a radiation inspection method adapted from the medical computerized axial tomography scanner, provides a cross-sectional view of a testpiece. All of the above terms are primarily used in connection with inspection that involves penetrating electromagnetic radiation in the form of x-rays or γ-rays. Neutron radiography refers to radiographic inspection using neutrons rather than electromagnetic radiation.

The sensitivity, or the ability to detect flaws, of radiographic inspection depends on close control of the inspection technique, including the geometric relationships among the point of x-ray emission, the casting, and the x-ray imaging plane. The smallest detectable variation in metal thickness lies between 0.5 and 2.0% of the total section thickness. Narrow flaws, such as cracks, must lie in a plane approximately parallel to the emergent x-ray beam to be imaged; this requires multiple exposures for x-ray film techniques and a remote control parts manipulator for a real time system.

Real time systems have eliminated the need for multiple exposures of the same casting by dynamically inspecting parts on a manipulator, with the capability of changing the x-ray energy for changes in total material thickness. These capabilities have significantly improved productivity and have reduced costs, thus enabling higher percentages of castings to be inspected and providing instant feedback after repair procedures.

Advances. Several advances have been made to assist the industrial radiographer. These include the computerization of the radiographic standard shooting sketch, which graphically shows areas to be x-rayed and the viewing direction or angle at which the shot is to be taken, and the development of microprocessor-controlled x-ray systems capable of storing different x-ray exposure parameters for rapid retrieval and automatic warm-up of the system prior to use. The advent of digital image processing systems and microfocus x-ray sources (near point source), producing energies capable of penetrating thick material sections, have made real time inspection capable of producing images equal to, and in some cases superior to, x-ray film images by employing geometric relations previously unattainable with macrofocus x-ray systems. The near point source of the microfocus x-ray system virtually eliminates the edge unsharpness associated with larger focus devices.

Digital image processing can be used to enhance imagery by multiple video frame integration and averaging techniques that improve the signal-to-noise ratio of the image. This enables the radiographer to digitally adjust the contrast of the image and to perform various edge enhancements to increase the conspicuity of many linear indications.

Interpretation of the radiographic image requires a skilled specialist who can establish the correct method of exposing the castings with regard to x-ray energies, geometric relationships, and casting orientation and can take all of these factors into account to achieve an acceptable, interpretable image. Interpretation of the image must be performed to establish standards in the form of written or photographic instructions. The inspector must also be capable of determining if the localized indication is a spurious indication, a film artifact, a video aberration, or a surface irregularity.

Ultrasonic Inspection

Ultrasonic inspection is a nondestructive method in which beams of high-frequency acoustic energy are introduced into the material under evaluation to detect surface and subsurface flaws and to measure the thickness of the material or the distance to a flaw. An ultrasonic beam will travel through a material until it strikes an interface or defect. Interfaces and defects interrupt the beam and reflect a portion of the incident acoustic energy. The amount of energy reflected is a function of the nature and orientation of the interface or flaw as well as the acoustic impedance of such a reflector. Energy reflected from various interfaces or defects can be used to define the presence and locations of defects, the thickness of the material, or the depth of a defect beneath a surface.

The advantages of ultrasonic tests are as follows:

- High sensitivity, which permits the detection of minute cracks
- Great penetrating power, which allows the examination of extremely thick sections
- Accuracy in measurement of flaw position and estimation of defect size

Ultrasonic tests have the following limitations:

- Size-contour complexity and unfavorable discontinuity orientation can pose problems in interpretation of the echo pattern
- Undesirable internal structure—for example, grain size, structure, porosity, inclusion content, or fine dispersed precipitates—can similarly hinder interpretation
- Reference standards are required

Ultrasonic inspection is more commonly used for wrought and welded products than for castings. It should be noted, however, that slag, porosity, cold shuts, tears, shrinkage cracks, and inclusions can be detected, particularly in castings that are not complex in shape. Wall thickness examination of cored castings is also conducted by ultrasonic inspection.

Leak Testing

Castings that are intended to withstand pressures can be leak tested in the foundry. Various methods are used, according to the type of metal being tested. One method consists of pumping air at a specified pressure into the inside of the casting and then submerging the casting in water at a given temperature. Any leaks through the casting become apparent by the release of bubbles of air through the faulty portions. An alternative method is to fill the cavities of a casting with paraffin at a specified pressure. Paraffin, which will penetrate the smallest of crevices, will rapidly find any defect, such as porosity, and will show quickly as an oily or moist patch at the position of the fault. Liquid penetrants can be poured into areas of apparent porosity and time allowed for the liquid to seep through the casting wall.

Pressure testing of rough (unmachined) castings at the foundry may not reveal any leaks, but it must be recognized that subsequent machining operations on the casting may cut into porous areas and cause the casting to leak after machining. Minor seepage leaks can be sealed by impregnation of the casting with liquid or filled sodium silicate, a synthetic resin, or other suitable substance. As-cast parts can be impregnated at the foundry to seal leaks if there is to be little machining or if experience has shown that machining does not affect the pressure tightness. However, it is usually preferable to impregnate after final machining of the casting.

Inspection of Ferrous Castings

Ferrous castings can be inspected by most of the nondestructive inspection methods. Magnetic particle inspection can be applied to ferrous metals with excellent sensitivity, although a crack in a ferrous casting can often be seen by visual inspection. Magnetic particle inspection provides good crack delineation, but the method should not be used to detect other defects. Irrelevant magnetic particle indications occasionally occur on ferrous castings, especially with a strong magnetic field. For example, a properly fused-in steel chaplet can be indicated as a defect because of the difference in magnetic response between low-carbon steel and cast iron. Even the graphite in cast iron, which is nonmagnetic, can cause an irrelevant indication.

Standard x-ray and radioactive-source techniques can be used to make radiographs of ferrous castings, but the typical complexity of shape and varying section thicknesses of the castings may require complex procedures. Radiography is sometimes used to inspect critical production castings that will be subjected to high service stresses, but it is more often used to evaluate design or casting procedures.

Ultrasonic inspection for both thickness and defects is practical with most ferrous castings except for the high-carbon gray iron castings, which have a high damping capacity and absorb much of the input energy. The measurement of resonant frequency is a good method for inspecting some ductile iron castings for soundness and graphite shape. Electromagnetic testing can be used to distinguish metallurgical differences between castings. The criteria for separating acceptable from unacceptable castings must be established empirically for each casting lot.

Gray Iron Castings

Gray iron castings are susceptible to most of the imperfections generally associated with castings, with additional problems resulting from the relatively high pouring temperatures. These additional problems result in a higher incidence of gas entrapment, inclusions, poor metal structure, interrupted metal walls, and mold wall deficiencies.

Gas entrapment is a direct result of gas being trapped in the casting wall during solidification. This gas may be in the metal prior to pouring, may be generated from aspiration during pouring, or may be generated from core and mold materials. Internal defects of this type are best detected by radiography, but ultrasonic and eddy current methods of inspection are useful when the defect is large enough to be detected by these methods.

Inclusions are casting defects in which solid foreign materials are trapped in the casting wall. The inclusion material can be slag generated in the melting process, or it can be fragments of refractory, mold sand, core aggregate, or other materials used in the casting process. Inclusions appear most often on the casting surface and are usually detected by visual inspection; however, in many cases, the internal walls of castings contain inclusions that cannot be visually detected. The internal inclusions can be detected by eddy current, radiographic, or ultrasonic inspection; radiography is usually the most reliable method.

Poor Metal Structure. Many casting defects resulting from metal structure are related to shrinkage, which is either a cavity or a spongy area lined with dendrites or is a depression in the casting surface. This type of defect arises from varying rates of contraction while the metal is changing from a liquid to a solid. Other casting defects resulting from varying rates of contraction during solidification include carbide formation, hardness variations, and microporosity.

Internal shrinkage defects are best detected by radiography, although eddy current or ultrasonic inspection can be used. Soft or hard gray iron castings are usually detected by Brinell hardness testing; electromagnetic methods have proved useful on some castings.

Interrupted Metal Walls. Included in this category are such flaws as hot tears, cold shuts, and casting cracks. Cracking of

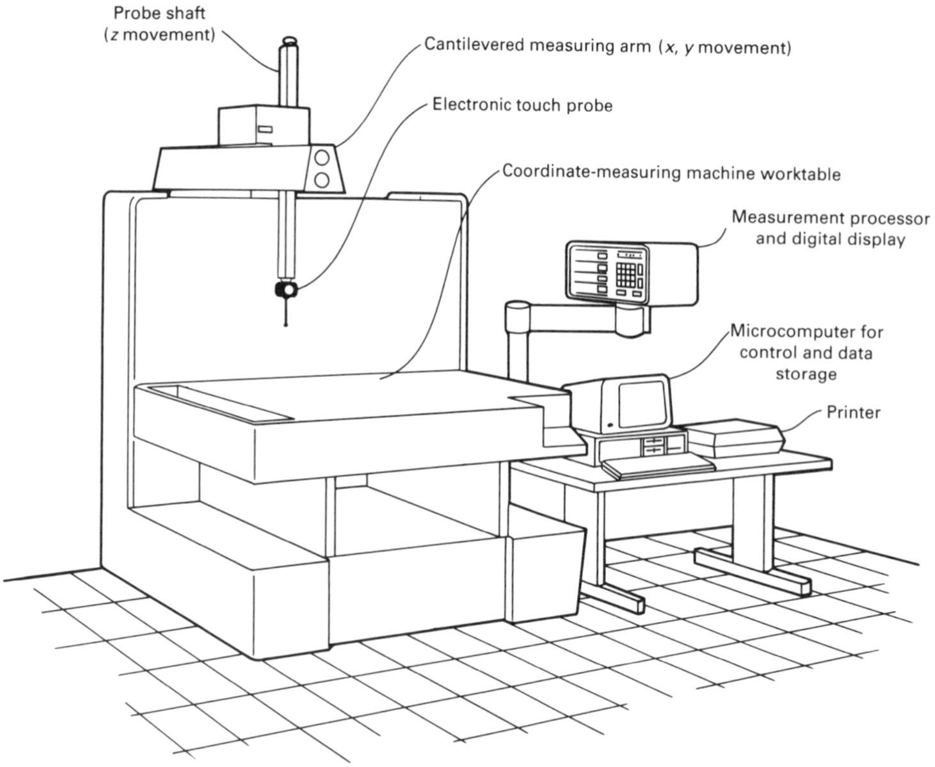

Fig. 2 Equipment used in a typical installation for the computer-aided dimensional inspection of castings showing a coordinate-measuring machine and microcomputer

castings is often a major problem in gray iron foundries as a result of the combination of casting designs and high production rates. Visual inspection, or an aided visual method such as liquid penetrant or magnetic particle inspection, is used to detect cracks and cracklike flaws in castings.

Mold wall deficiencies are common problems in gray iron castings. They result in surface flaws such as scabs, rattails, cuts, washes, buckles, drops, and excessive metal penetration into space between sand grains. These flaws are generally detected by visual inspection.

Malleable Iron Castings

Blowholes and spikes are defects that are often found in malleable iron castings. Spikes are a form of surface shrinkage not normally visible to the naked eye but appear as a multitude of short discontinuous surface cracks when subjected to fluorescent magnetic particle inspection. Unlike true fractures, spikes do not propagate, but they are not acceptable where cyclic loading could result in fatigue failure. Spikes are usually seen as short indications about 1.6 mm (1/16 in.) long or less and never more than 75 μm (0.003 in.) deep. These defects do not have a preferred orientation but a random pattern that may or may not follow the direction of solidification. Shrinkage or open structure in the gated area is a defect often found in malleable iron castings that may be overlooked by visual inspection, although it is readily detected by either liquid penetrant or magnetic particle inspection.

Ductile Iron Castings

Ductile iron is cast iron in which the graphite is present in tiny balls or spherulites instead of flakes (as in gray iron) or compacted aggregates (as in malleable iron). The spheroidal graphite structure is produced by the addition of one or more elements to the molten metal.

Casting defects associated with foundry practice—that is, shrinkage, voids from entrapped gas, nonmetallic inclusions, and failure to fill the mold shape—are essentially the same for ductile iron as for gray iron. Carbon-nodule segregation occurs when the carbon equivalent (CE) of ductile iron [CE = % total carbon + 0.3 (%Si + %P)] is incorrect for the section thickness of the casting.

Subsurface inclusions arise from the formation of nonmetallic compounds (mainly sulfides) following inoculation of the molten iron. Slag inclusions form in ductile iron in appreciable amounts upon inoculation with magnesium because some of the magnesium ignites in the molten metal. Desulfurization also promotes slag formation.

Inspection of Aluminum Alloy Castings

Effective quality control is needed at every step in the production of an aluminum alloy casting, from selection of the casting method, casting design, and alloy to mold production, foundry technique, machining, finishing, and inspection. Visual methods, such as visual inspection, pressure testing, liquid penetrant inspection, ultrasonic inspection, radiographic inspection, and metallographic examination, can be used to inspect for casting quality. The inspection procedure used should be geared toward the specified level of quality.

Stages of Inspection. Inspections can be divided into three stages: preliminary, intermediate, and final. After tests are conducted on the melt for hydrogen content, for adequacy of silicon modification, and for degree of grain refinement, preliminary inspection may consist of the inspection and testing of test bars cast with the molten alloy at the same time the production castings are poured. These test bars are used to check the quality of the alloy and effectiveness of the heat treatment. Preliminary inspection also includes chemical or spectrographic analysis of the casting, thus ensuring that the melting and pouring operations have resulted in an alloy of the desired composition.

Intermediate inspection, or hot inspection, is performed on the casting as it is taken from the mold. This step is essential so that castings that are obviously defective can be discarded at this stage of production. Castings that are judged unacceptable at this stage can then be considered for salvage by impregnation, welding, or other methods, depending on the type of flaw present and the end use of the casting. More complex castings usually undergo visual and dimensional inspection after the removal of gates and risers.

Final inspection establishes the quality of the finished casting, using any of the methods previously mentioned. Visual inspection also includes the final measurement and comparison of specified and actual dimensions. Dimensions of castings from a large production run can be checked using gages, jigs, fixtures, or coordinate-measuring systems (described later).

Liquid penetrant inspection is extensively used as a visual aid for detecting surface flaws in aluminum alloy castings. Liquid penetrant inspection is applicable to castings made from all the aluminum casting alloys as well as to castings produced by all methods. One of its most useful applications, however, is for inspecting small castings produced in permanent molds from alloys such as 296.0, which are characteristically susceptible to hot cracking. For example, in cast connecting rods, hot shortness may result in fine cracks in the shank sections. Such cracks are virtually undetectable by unaided visual inspection, but are readily detectable by liquid penetrant inspection.

All of the well-known liquid penetrant systems (that is, water-washable, postemulsifiable, and solvent-removable) are applicable to inspection of aluminum alloy castings. In some cases, especially for certain high-integrity castings, more than one system can be used. Selection of the system is primarily based on the size and shape of the castings, surface roughness, production quantities, sensitivity level desired, and available inspection facilities.

Pressure testing is used for castings that must be leaktight. Cored-out passages and internal cavities are first sealed off with special fixtures having air inlets. These inlets are used to build up the air pressure on the inside of the casting. The entire casting is then immersed in a tank of water, or it is covered by a soap solution. Bubbles will mark any point of air leakage.

Radiographic inspection is a very effective means of detecting such conditions as cold shuts, internal shrinkage, porosity, core shifts, and inclusions in aluminum alloy castings. Radiography can also be used to measure the thickness of specific sections. Aluminum alloy castings are ideally suited to examination by radiography because of their relatively low density; a given thickness of aluminum alloy can be penetrated with about one-third the power required for penetrating the same thickness of steel.

Aluminum alloy castings are most often radiographed by an x-ray machine, using film to record the results. Real time radiography is also widely used, particularly for examining large numbers of relatively small castings, and is best suited to detecting shrinkage, porosity, and core shift. Gamma-ray radiography is also satisfactory for detecting specific conditions in aluminum castings. Although the γ-ray method is used to a lesser extent than the x-ray method, it is about equally as effective for detecting flaws or measuring specific conditions. Aluminum alloy castings are most often radiographed to detect about the same types of flaws that may exist in other types of castings, that is, conditions such as porosity or shrinkage, which register as low-density spots or areas and appear blacker on the film or fluoroscopic screen than the areas of sound metal.

Aluminum ingots may contain hidden internal cracks of varying dimensions. Depending on size and location, these cracks may cause an ingot to split during mechanical working and thermal treatment, or they may show up as a discontinuity in the final wrought product. Once the size and location of such cracks are determined, an ingot can be scrapped, or sections free from cracks can be sawed out and processed further. Because the major dimensions of the cracks are along the casting direction, they present good reflecting surfaces for sound waves traveling perpendicular to the casting direction. Thus, ultrasonic methods using a wave frequency that gives adequate penetration into the ingot provide excellent sensitivity for 100% inspection of that part of the ingot that contains critical cracks.

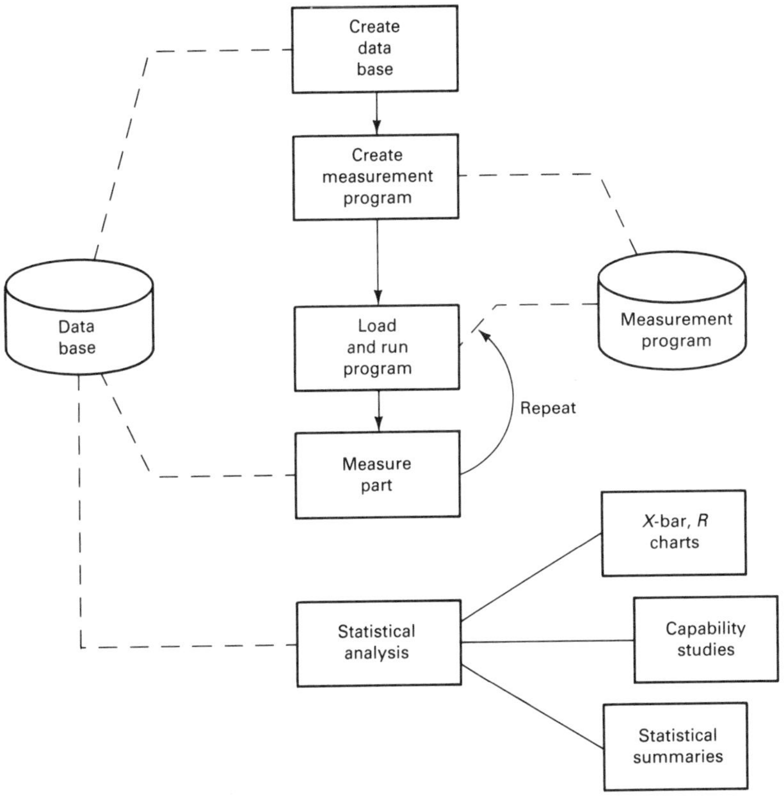

Fig. 3 Flowchart showing typical sequence of operations for computer-aided dimensional inspection

Because of ingot thickness (up to 406 mm, or 16 in.) and the small metal separation across the crack, radiographic methods are impractical for inspection.

Ultrasonic Inspection. Aluminum alloy castings are sometimes inspected by ultrasonic methods to evaluate internal soundness or wall thickness. The principal uses of ultrasonic inspection for aluminum alloy castings include detection of porosity in castings and internal cracks in ingots.

Inspection of Copper and Copper Alloy Castings

Inspection of copper and copper alloy castings is generally limited to visual and liquid penetrant inspection of the surface, along with radiographic inspection for internal discontinuities. In specific cases, electrical conductivity tests and ultrasonic inspection can be applied, although the usual relatively large cast grain size could prevent a successful ultrasonic inspection.

Visual inspection is simple yet informative. A visual inspection would include significant dimensional measurements as well as general appearance. Surface discontinuities often indicate that internal discontinuities are also present.

For small castings produced in reasonable volume, a destructive metallographic inspection on randomly selected samples is practical and economical. This is especially true on a new casting for which foundry practice has not been optimized and a satisfactory repeatability level has not been achieved.

For castings of some of the harder and stronger alloys, a hardness test is a good means of estimating the level of mechanical properties. Hardness tests are of less value for the softer tin bronze alloys because hardness tests do not reflect casting soundness and integrity.

Because copper alloys are nonmagnetic, magnetic particle inspection cannot be used to detect surface cracks. Instead, liquid penetrant inspection is recommended. Ordinarily, liquid penetrant inspection requires some prior cleaning of the casting to highlight the full detail.

For the detection of internal defects, radiographic inspection is recommended. Radiographic methods and standards are well established for some copper alloy castings (for example, ASTM specifications E 272 and E 310).

As a general rule, the methods of inspection applied to some of the first castings made from a new pattern should include all those that provide a basis for judgment of the acceptability of the casting for the application intended. Any deficiencies or defects should be reviewed and the degree of perfection defined. This procedure can be repeated on successive production runs until repeatability has been ensured.

Gas Porosity. Copper and many copper alloys have a high affinity for hydrogen,

LAYOUT REPORT

FILE: FP5.DAT DESCRIPTION: FUEL PUMP MOLD 5 REC. NO: 40
MONTH/DAY: 02/20 YEAR: 87 SHIFT : 1 MOLD : 5

	DESCRIPTION	ACT	MEAN	− DEV 0 + DEV
1	DIM A	2.813	2.790	
2	DIM B	2.501	2.500	
3	DIM C	0.547	0.560	
4	DIM D	4.453	4.440	
5	DIM E	1.622	1.620	
6	DIM F	1.873	1.880	
7	DIM G	1.658	1.670	
8	DIM H	1.576	1.560	
9	DIM I	1.358	1.350	
10	DIM J	1.657	1.670	
11	DIM K	3.456	3.435	
12	DIM L	0.750	0.750	
13	DIM MX	2.791	2.790	
14	DIM MY	0.325	0.370	
15	SNP RG A	2.156	2.190	
16	SNP RG B	3.237	3.235	
17	MLD HF R	0.658	0.655	
18	MLD HF L	3.986	3.970	
19	WALL A	0.206	0.214	
20	WALL B	0.159	0.153	
21	WALL C	0.237	0.228	
22	OAH MAX	3.838	3.770	*******
23	OAH MIN	3.802	3.770	

Fig. 4 Example layout report showing all dimensions measured on a single casting, with visual indication of deviations from print mean. Note out-of-tolerance condition indicated by asterisks.

with an increasing solubility as the temperature of the molten bath is increased. Conversely, as the metal cools in the mold, most of this hydrogen is rejected from the metal. Because all the gas does not necessarily escape to the atmosphere and may become trapped by the solidifying process, gas porosity may be found in the casting.

In most alloys, gas porosity is identified by the presence of voids that are relatively spherical and are bright and shiny inside. Visible upon sectioning or by radiography, they may either be small, numerous, and rather widely dispersed or fewer in number and relatively large. Regardless of size, they are seldom interconnected except in some of the tin bronze alloys, which solidify in a very dendritic mode. In these alloys, the gas porosity tends to be distributed in the interstices between the dendrites.

Shrinkage voids caused by the change in volume from liquid to solid in copper alloys are different only in degree and possibly shape from those found in other metals and alloys. All nonferrous metals exhibit this volume shrinkage when solidifying from the molten condition.

Shrinkage voids may be open to the air when near or exposed to the surface, or they may be deep inside the thicker sections of the casting. They are usually irregular in shape, compared to gas-generated defects, in that their shape frequently reflects the internal temperature gradients induced by the external shape of the casting.

Hot Tearing. The tin bronzes as a class, as well as a few of the leaded yellow brasses, are susceptible to hot shortness; that is, they lack ductility and strength at elevated temperature. This is significant in that tearing and cracking can take place during cooling in the mold because of mold or core restraint. In aggravated instances, the resulting hot tears in the part appear as readily visible cracks. Sometimes, however, the cracks are not visible externally and are not detectable until after machining. In extreme cases, the cracks become evident only through field failure because the tearing was deep inside the casting.

Nonmetallic inclusions in copper alloys, as with all molten alloys, are normally the result of improper melting and/or pouring conditions. In the melting operation, the use of dirty remelt or dirty crucibles, poor furnace linings, or dirty stirring rods can introduce nonmetallic inclusions into the melt. Similarly, poor gating design and pouring practice can produce turbulence and can generate nonmetallic inclusions. Sand inclusions may also be evident as the result of improper sand and core practice. All commercial metals, by the nature of available commercial melting and molding processes, usually contain very minor amounts of small nonmetallic inclusions. These have little or no effect on the casting. Inclusions of significant size or number are considered detrimental.

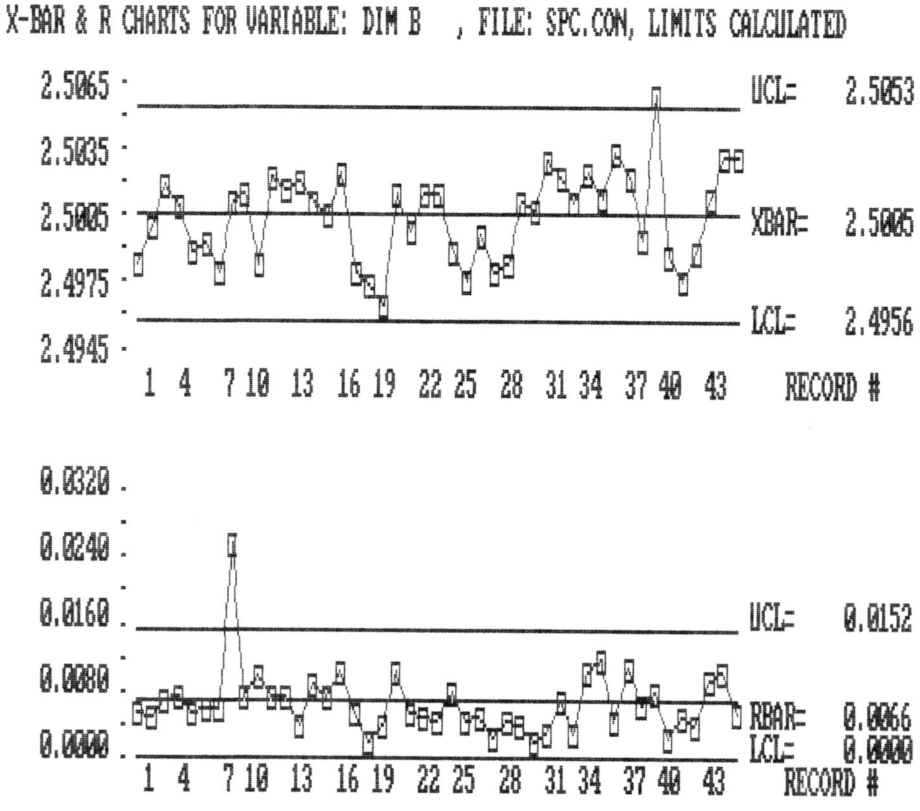

Fig. 5 An example control chart with average of groups of measurements (*X* values) plotted above and ranges within the groups (*R*) plotted below. Control limits have been calculated and placed on the chart by the computer.

Computer-Aided Dimensional Inspection*

The use of computer equipment in foundry inspection operations is finding more

*This section was prepared by Lawrence E. Smiley, Reliable Castings Corporation.

acceptance as the power and usefulness of available hardware and software increase. The computerization of operations can reduce the man-hours required for inspection tasks, can increase accuracy, and can allow the analysis of data in ways that are not possible or practical with manual operations. Perhaps the best example of this, given the currently available equipment, is provided by the application of computer technology to the dimensional inspection of castings.

Importance of Dimensional Inspection

One of the most critical determinants of casting quality in the eyes of the casting buyer is dimensional accuracy (see the article "Dimensional Tolerances and Allowances" in this Volume). Parts that are within dimensional tolerances, given the absence of other casting defects, can be machined, assembled, and used for their intended functions with testing and inspection costs minimized. Major casting buyers are therefore demanding statistical evidence that dimensional tolerances are being maintained. In addition, the statistical analysis of in-house processes has been demonstrated to be effective in keeping those processes under control, thus reducing scrap and rework costs.

The application of computer equipment to the collection and analysis of dimensional inspection data can increase the amount of inspection that can be performed and decrease the time required to record and analyze the results. This furnishes control information for making adjustments to tooling on the foundry floor and statistical information for reporting to customers on the dimensional accuracy of parts.

Statistical Summary Report   Page 1
FILE: FP5.DAT   DESCRIPTION: FUEL PUMP MOLD 5
For SHIFT : ALL   MOLD : ALL   Start Rec: 1   End Rec: 68

	Actual Mean	Spec.	Spec.− Mean	Tol.	Sigma	6 Sigma	Process Capab
1 DIM A	2.8025	2.7900	−0.0125	0.0460	0.0070	0.0419	1.0976
2 DIM B	2.5017	2.5000	−0.0017	0.0460	0.0059	0.0352	1.3072
3 DIM C	0.5537	0.5600	0.0063	0.0460	0.0051	0.0304	1.5152
4 DIM D	4.4517	4.4400	−0.0117	0.0460	0.0083	0.0496	0.9281
5 DIM E	1.6182	1.6200	0.0018	0.0460	0.0074	0.0447	1.0300
6 DIM F	1.8665	1.8800	0.0135	0.0460	0.0218	0.1305	0.3524
7 DIM G	1.6607	1.6700	0.0093	0.0460	0.0088	0.0526	0.8740
8 DIM H	1.5671	1.5600	−0.0071	0.0460	0.0097	0.0580	0.7931
9 DIM I	1.3621	1.3500	−0.0121	0.0460	0.0044	0.0265	1.7354
10 DIM J	1.6597	1.6700	0.0103	0.0460	0.0106	0.0639	0.7201
11 DIM K	3.4469	3.4350	−0.0119	0.0460	0.0093	0.0560	0.8212
12 DIM L	0.7479	0.7500	0.0021	0.0460	0.0047	0.0282	1.6326
13 DIM MX	2.7821	2.7900	0.0079	0.0900	0.0102	0.0609	1.4771
14 DIM MY	0.3338	0.3700	0.0362	0.1000	0.0108	0.0648	1.5433
15 SNP RG A	2.1586	2.1900	0.0314	0.0900	0.0108	0.0648	1.3899
16 SNP RG B	3.2267	3.2350	0.0083	0.0900	0.0114	0.0683	1.3170
17 MLD HF R	0.6615	0.6550	−0.0065	0.0600	0.0036	0.0216	2.7834
18 MLD HF L	3.9807	3.9700	−0.0107	0.0600	0.0083	0.0498	1.2055
19 WALL A	0.2133	0.2140	0.0007	0.0280	0.0084	0.0502	0.5573
20 WALL B	0.1541	0.1530	−0.0011	0.0300	0.0097	0.0581	0.5161
21 WALL C	0.2286	0.2280	−0.0006	0.0400	0.0073	0.0440	0.9097
22 OAH MAX	3.7892	3.7700	−0.0192	0.0800	0.0352	0.2110	0.3791
23 OAH MIN	3.7589	3.7700	0.0111	0.0800	0.0244	0.1464	0.5465

68 Parts included in analysis
Actual Sigma calculated

$$\text{Process Capability} = \frac{\text{Blueprint Tolerance}}{6 \times \text{Sigma}}$$

Fig. 6 An example statistical summary report showing the mean of measured observations, the blueprint specification for the mean, difference between specified and measured means, the tolerance, the standard deviation of the measured dimensions, and the capability of the process. These calculations are performed for all measured dimensions on the part.

Typical Equipment

A typical equipment installation for the dimensional inspection of castings consists of an electronic coordinate-measuring machine, a microcomputer interfaced to the coordinate-measuring machine controller with a data transfer cable, and a software system for the microcomputer. This equipment is illustrated in Fig. 2. The software system should be capable of controlling the functions and storing of the coordinate-measuring machine as well as recalling and analyzing the data it collects. The software serves as the main control element for the dimensional inspection and statistical reporting of results. Such software can be purchased or, if the expertise is available, developed in-house where requirements are highly specialized.

Coordinate-measuring machines typically record dimensions along three axes from datum points specified by the user. Depending on the sophistication of the controller, such functions as center and diameter finds for circular features and electronic rotation of measurement planes can be performed. Complex geometric constructions, such as intersection points of lines and planes and out-of-roundness measurements, are typically off-loaded for calculation into the microcomputer. The contact probe of the coordinate-measuring machine can be manipulated manually, or in the case of direct computer controlled machines, the probe can be driven by servomotors to perform the part measurement with little operator intervention.

The Measurement Process

Figure 3 illustrates the general procedure that is followed in applying semiautomatic dimensional inspection to a given part. The first step is to identify the critical part dimensions that are to be measured and tracked. Nominal dimensions and tolerances are normally taken from the customer's specifications and blueprints. Dimensions that are useful in controlling the foundry process can also be selected. A data base file, including a description and tolerance limits for each dimension to be checked, is then created using the microcomputer software system.

The next step in the setup process is to develop a set of instructions for measuring the part with the coordinate-measuring machine. The instructions consist of commands that the coordinate-measuring machine uses to establish reference planes and to measure such features as center points of circular holes.

This measurement program can be entered in either of two ways. Using the first method, the operator simply types in a list of commands that he wants the coordinate-measuring machine to execute and that give the required dimensions as defined in the part data base. The second method uses a teach mode; the operator actually places a part on the worktable of the coordinate-measuring machine and checks it in the proper sequence, while the computer monitors the process and stores the sequence of commands used. In either case, the result is a measurement program stored on the microcomputer that defines in precise detail how the part is to be measured. Special commands can also be included in the measurement program to display operator instructions on the computer screen while the part is being measured.

In developing the measurement program, consideration must be given to the particular requirements of the part being measured. Customer prints will normally show datum planes from which measurements are to be made. When using a cast surface to establish a datum plane, it is good practice to probe a number of points on the surface and allow the computer to establish a best-fit plane through the points. Similarly, center points of cast holes can best be found by probing multiple points around the circumference of the hole. Machined features can generally be measured with fewer probe contacts. When measuring complex castings, maximum use should be made of the ability of the coordinate-measuring machine to electronically rotate measurement planes without physically moving the part; unclamping and turning a part will lower the accuracy of the overall layout.

Once the setup process is complete, dimensional inspection of parts from the foundry begins. Based on statistical considerations, a sampling procedure and frequency must be developed. Parts are then selected at random from the process according to the agreed-upon frequency. The parts are brought to the coordinate-measuring machine, and the operator calls up the measurement program for that part and executes it. As the part is measured, the dimensions are sent from the coordinate-measuring machine to the data base on the microcomputer. Once the measurement process is complete, information such as mold number, shift, date, or serial number should be entered by the operator so that this particular set of dimensions can be identified later. A layout report can then be generated to show how well the measured part checked out relative to specified dimensions and tolerances. Figure 4 shows a sample report in the form of a bar graph, in which any deviation from print tolerance appears as a line of dashes to the left or right of center. A deviation outside of tolerance limits displays asterisks to flag its condition. Such a report is useful in that it gives a quick visual indication of the measurement of one casting.

```
                          FREQUENCY PLOT/HISTOGRAM
  FILE: FP2.DAT        DESCRIPTION: FUEL PUMP  MOLD 2
                     DATA ITEM NUMBER 2 : DIM B
  For  SHIFT   : ALL       MOLD     : ALL       Start Rec: 229      End Rec: 448

2.468   TO    2.470  :
2.470   TO    2.472  :
2.472   TO    2.474  :
2.474   TO    2.476  :
2.476   TO    2.478  :------------------------------------------------------------
2.478   TO    2.480  :
2.480   TO    2.482  :
2.482   TO    2.484  :
2.484   TO    2.486  :
2.486   TO    2.488  :
2.488   TO    2.490  :
2.490   TO    2.492  :
2.492   TO    2.494  :**
2.494   TO    2.496  :*******
2.496   TO    2.498  :**********************************************
2.498   TO    2.500  :*****************************
2.500   TO    2.502  :*****************************************...................
2.502   TO    2.504  :*****************************
2.504   TO    2.506  :***********************
2.506   TO    2.508  :********************
2.508   TO    2.510  :******
2.510   TO    2.512  :******
2.512   TO    2.514  :**
2.514   TO    2.516  :****
2.516   TO    2.518  :
2.518   TO    2.520  :*
2.520   TO    2.522  :***
2.522   TO    2.524  :*-----------------------------------------------------------
2.524   TO    2.526  :
2.526   TO    2.528  :
2.528   TO    2.530  :
2.530   TO    2.532  :


--------    TOLERANCE LIMITS
........    PRINT MEAN
```

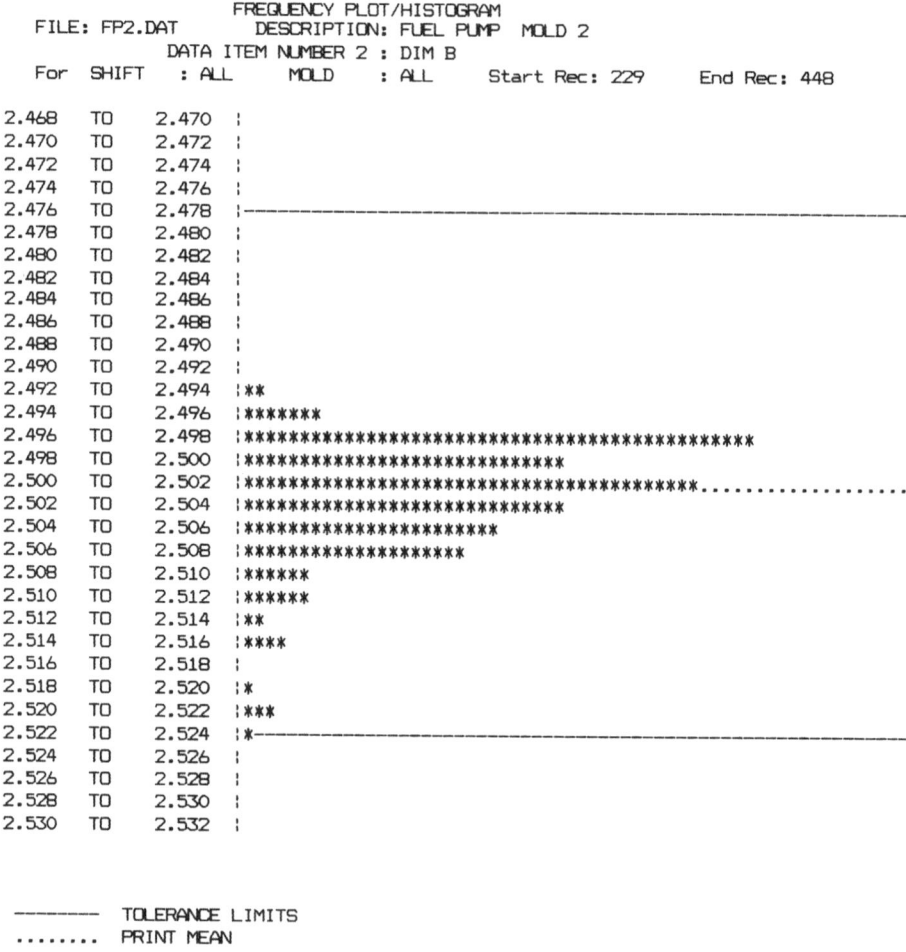

Fig. 7 A frequency plot for one measured dimension showing the distribution of the measurements

Statistical Analysis

The use of statistical analysis permits the mathematical prediction of the characteristics of all the parts produced by measuring only a sample of those parts. All processes are subject to some amount of natural variation; in most processes, this variation follows a normal distribution, the familiar bell-shaped curve, when the probability of occurrence is plotted against the range of possible values. Standard deviation, a measure of the distance from center on the probability curve, is the principal means of expressing the range of measured values. For example, a spread of six standard deviations (plus or minus three standard deviations on either side of the measured mean) represents the range within which one would expect to find 99.73% of observed measurements for a normal process. This allows the natural variation inherent in the process to be quantified.

Control Charts. With statistical software incorporated into the microcomputer system, the results of numerous measurements of the same part can be analyzed to determine, first, how well the process is staying in control, that is, whether the natural variations occurring in a given measurement are within control limits and whether any identifiable trends are occurring. This is done by using a control chart (Fig. 5), which displays the average values and ranges of groups of measurements plotted against time. Single-value charts with a moving range can also be helpful. The control limits can also be calculated and displayed. With the computer, this type of graph can be generated within seconds. Analysis of the graph may show a developing trend that can be corrected by adjusting the tooling before out-of-tolerance parts are made.

Statistical Summary Report. The second type of analysis shows the capability of the process, that is, how the range of natural variation (as measured by a specified multiple of the standard deviation) compares with the tolerance range specified for a given dimension. An example of a useful report of this type is shown in Fig. 6. This information is of great interest both to the customer and the process engineer because it indicates whether or not the process being used to produce the part can hold the dimensions within the required tolerance limits. The user must be aware that different methods of capability analysis are used by different casting buyers, so the software should be flexible enough to accommodate the various methods of calculation that might be required.

Histograms. An alternative method of assessing capability involves the use of a histogram, or frequency plot. This is a graph that plots the number of occurrences within successive, equally spaced ranges of a given measured dimension. Figure 7 shows an example output report of this type. A graph such as this, which has superimposed upon it the tolerance limits for the dimension being analyzed, allows a quick, qualitative evaluation of the variation and capability of the process. It also allows the normality of the process to be judged through comparison with the expected bell-shaped curve of a normal process.

Other Applications for Computer-Aided Inspection

The general sequence described for semiautomatic dimensional inspection can be applied to a number of other inspection criteria. Examples would include pressure testing or defect detection by electronic vision systems. The statistical analysis of scrap by defect types is also very helpful in identifying problem areas. In some cases, direct data input to a computer may not be feasible, but the benefits of entering data manually into a statistical analysis program should not be overlooked. The computer allows rapid analysis of large amounts of data so that statistically significant trends can be detected and proper attention paid to appropriate areas for improvement. The benefits and costs of each anticipated application of automation to a particular situation, as well as the feasibility of applying state-of-the-art equipment, need to be studied as thoroughly as possible prior to implementation.

Coating of Castings

Revised by Charles F. Walton, Consultant

Cast components sometimes require special surface characteristics, such as resistance to deterioration or an appealing appearance. Coatings can extend casting life when resistance to corrosion, wear, and erosion is required. This article will discuss cleaning and coating practices for cast iron components. Similar information on nonferrous casting alloys is available in Volume 5 of the 9th Edition of *Metals Handbook*.

Cleaning

Cleaning of the surface is the most important prerequisite of any coating process. Suitable levels of cleanliness and surface roughness are established by various mechanical and chemical methods. Foundries deliver castings that have been shot or grit blasted (see the section "Blast Cleaning of Castings" in this article). Supplementary nonmechanical cleaning may be needed to reach interior passages or to remove heat-treating scale or machining oil.

The choice of cleaning process depends not only on the types of soils to be removed but also on the characteristics of the coating to be applied. The cleaning process must leave the surface in a condition that is compatible with the coating process. For example, if a casting is to be treated with phosphate and then painted, the cleaning process must remove all oils and oxide scale because these inhibit good phosphating.

If castings are heat treated before they are coated, the choice of heat treatment conditions can influence the properties of the coating, particularly a metallic or conversion coating. In most cases, heat treatment should be done in an atmosphere that is not oxidizing. Oxides and silicates formed during heat treating must be removed before most coating processes.

Molten salt baths are excellent for cleaning complex interior passages in castings. In one electrolytic, molten salt cleaning process, the electrode potential is changed so that the salt bath is alternately oxidizing and reducing. Scale and graphite are easily removed with reducing and oxidizing baths, respectively. Molten salt baths are fast compared to other nonmechanical methods, but castings may crack if they are still hot when salt residues are rinsed off with water.

Pickling in an acid bath is usually done prior to hot dip coating or electroplating. Overpickling should be avoided because a graphite smudge can be formed on the surface. Cast iron contains silicon; therefore, a film of silica can form on the surface as a result of heavy pickling. This can be avoided by adding hydrofluoric acid to the pickling bath. Special safety and environmental protection regulations must be met when using pickling.

Chemical cleaning is different from pickling because, in chemical cleaning, the cleaners attack only the surface contaminants, not the iron substrate. Many chemical cleaners are proprietary formulations, but in general they are alkaline solutions, organic solvents, or emulsifiers. Alkaline cleaners must penetrate contaminants and wet the surface in order to be effective.

Organic solvents that were commonly used in the past include naphtha, benzene, methanol, toluene, and carbon tetrachloride. These have been largely replaced by chlorinated solvents, such as those used for vapor degreasing. Solvents effectively remove lubricants, cutting oils, and coolants, but are ineffective against such inorganic compounds as oxides or salts. Emulsion cleaners are solvents combined with surfactants; they disperse contaminants and solids by emulsification. Emulsion cleaners are most effective against heavy oils, greases, slushes, and solids entrained in hydrocarbon films. They are relatively ineffective against adherent solids such as oxide scale.

After wet cleaning, short-term rust prevention is accomplished by the use of an alkaline rinse. This can be followed by mineral oils, solvents combined with inhibitors and film formers, emulsions of petroleum-base coatings and water, and waxes.

Electroplating

Table 1 lists the common electroplated metals. Other metals—for example, gold, silver, and alloys such as brass and bronze—are used for special decorative effects.

Iron castings are electroplated to impart corrosion resistance or to provide a pleasing appearance. Typical applications include many different types of interior hardware, machine parts such as printing cylinders, decorative trim, and casings. Iron castings are also electroplated to enhance their wear resistance; for example, hard chromium plating is applied to the wear surfaces of piston rings. Areas not requiring a plated surface can be masked or stopped off to prevent coverage. Plating can be applied as a very thin layer for applications requiring only a pleasing appearance or mild corrosion resistance; thicker plating can be applied for more wear resistance, longer corrosion resistance, or to replace lost metal (Table 1).

Electrodeposition is done by making clean iron castings cathodic in an aqueous solution containing a salt of the coating metal and then passing a direct electrical current through the solution. The effective metal content in the plating bath can be replenished by using anodes made of the coating metal to complete the electrical circuit. Variations in the properties of the deposited coating are influenced by the composition, temperature, pH, and agitation of the bath and by current density. In addition, further variations in the coating can result from the design of the casting, the distance of the casting from the anode, and the preparation of the surface before plating.

Nickel is unique among plated coatings because it can be applied without using an impressed electrical current. Table 2 lists nonelectrolytic methods. The high-temperature treatments produce partly diffused coatings containing both nickel and phosphorus. Because no impressed electrical current is involved, the thickness of an electroless nickel plate is exceptionally uniform regardless of the shape of the casting. Preheating large castings prior to immersion prevents a delay in the start of nickel deposition.

Electroless nickel is used not only for its uniformity in coating thickness but also for its ability to be heat treated to increase its hardness. Typically, an as-plated hardness of 49 to 50 HRC can be increased to 67 to 68

Table 1 Properties and characteristics of conventional electroplated metal coatings

Metal	Coating hardness	Appearance	Thickness µm	Thickness mil	Characteristics and uses
Cadmium	30–50 HV	Bright white	3–10	0.15–0.5	Pleasing appearance for indoor applications; less likely to darken than zinc; anodic to ferrous substrate
Chromium	900–1100 HV	White—can be varied	0.2–1(a) 1–300(b)	0.01–0.06(a) 0.05–12.0(b)	Excellent resistance to wear, abrasion, and corrosion; low friction and high reflectance
Cobalt	250–300 HK	Gray	2–25	0.1–1.0	High hardness and reflectance
Copper	41–220 HV	Bright pink	4–50	0.2–2.0	High electrical and thermal conductivities; used as undercoat for other electroplates
Lead	5 HB	Gray	12–200(c) 1300(d)	0.5–8.0(c) 50(d)	Resistant to many acids and hot corrosive gases
Nickel	140–500 HV	White	2–40(a) 130–500(c)	0.1–1.5(a) 5–20(c)	Resistant to many chemicals and corrosive atmospheres; often used in conjunction with copper and chromium; can be applied by electroless plating
Rhodium	400–800 HB	Bright white	0.03–25	0.001–1.0	High electrical conductance; brilliant white appearance is tarnish and corrosion resistant.
Tin	5 HB	Bright white	4–25	0.015–1.0	Corrosion resistant; hygienic applications for food and dairy equipment; good solderability
Zinc	40–50 HB	Matte gray	2–13(a) 12–50(d)	0.1–0.5(a) 0.5–2.0(d)	Easily applied; high corrosion resistance; anodic to ferrous substrate

(a) Decorative. (b) Hard. (c) Wear applications. (d) Corrosion applications

HRC by proper heat treatment. More detailed information on electroplating is available in the Section "Plating and Electropolishing" in Volume 5 of the 9th Edition of *Metals Handbook* and in the articles "Electroplated Coatings" and "Electroplated Hard Chromium" in Volume 13 of the 9th Edition of *Metals Handbook*.

Hot Dip Coatings

Hot dip coating consists of immersing the casting in a bath of molten metal. A flux-coated and/or chemically cleaned surface is necessary to achieve satisfactory results. Aluminum, tin, zinc, and their alloys can be applied from a molten bath. Hot dip coatings are preferred because they are thicker than electroplates and because an alloy layer is formed between the coating metal and the iron. This provides additional durability and adhesion. Castings of complex shape are easily coated by these processes, although air may become trapped in blind holes unless the castings are rotated.

Hot dip zinc coating (galvanizing) is widely used on iron castings, particularly pipe, valves, and fittings. The uniform and adherent coating provides a barrier against corrosive attack and will further protect an iron casting by acting as a sacrificial anode or by undergoing preferential corrosion. Successful galvanizing depends on surface preparation. Pickling followed by dipping in a bath of zinc ammonium chloride or other flux is done prior to dipping in molten zinc. Excess zinc may be drained or centrifuged from the castings before quenching. Quenching improves the brightness of the coating. Iron castings of any type and any composition can be hot dip galvanized.

Hot dip tin coating (hot tinning) provides a protective, decorative, and nontoxic coating for food equipment, a bonding layer for babbitted bearings, or a precoated surface of soldering. Surface preparation is particularly important, and when maximum adherence is desired, such as when tinning is used to prepare a casting for the application of babbitt, electrolytic cleaning in a molten salt is preferred.

For the hot dip lead coating of iron castings, lead-base alloys are preferred over pure lead; with pure lead, bonding is mechanical rather than metallurgical. Tin is the element most widely used to enhance bonding. Lead coatings are noted for their resistance to fumes from sulfuric and sulfurous acids.

The aluminum coating (aluminizing) of iron castings imparts resistance to corrosion and heat. The coating oxidizes rapidly, thus passivating the surface. The resultant aluminum oxide is refractory in nature; it seals the surface and resists degradation at high temperatures. An aluminized surface has limited resistance to sulfur fumes, organic acids, salts, and compounds of nitrate-phosphate chemicals. More detailed information on hot dip coating (continuous and batch processes) is available in the articles "Hot Dip Coatings" and "Corrosion of Zinc" in Volume 13 of the 9th Edition of *Metals Handbook*.

Hardfacing

Hardfacing can be used when a casting requires an unusually hard and wear-resistant surface and when it is impractical to produce a hard surface in the casting process or by selective heat treatment. Frequently, hardfacing is used to repair worn castings by building up an overlay of new material.

Hardfacing is basically a welding operation in which an alloy is fused to the base metal. Both gas and arc methods are used. Ferrous alloys can be either austenitic or hardenable types. Metallic carbides are the chief intermetallic materials used for hardfacing. Carbides are not melted or fused into the surface layer, but are bonded in place by an enveloping metal such as cobalt. The articles "Hardfacing" and "Metal Powders Used for Hardfacing" in Volumes 6 and 7, respectively, of the 9th Edition of *Metals Handbook* contain detailed information on hardfacing

Table 2 Nonelectrolytic nickel plating methods and solution constituents

Method	Solution constituents	Temperature °C	Temperature °F	Plating rate µm/h	Plating rate mil/h	Special considerations
Immersion	Nickel chloride, sodium hypophosphite	70	160	1.2	0.05	Porous coating with moderate adhesion can be improved with postheating at 650 °C (1200 °F)
High-temperature chemical reduction	Slurry of nickel oxide and ammonium phosphite	870–955	1600–1750	. . .	. . .	Selective coverage by reduction of mixture at high temperature
Electroless acid	Nickel chloride, sodium hypophosphite, sodium citrate	95	200	12	0.5	Postplating treatment not necessary for iron castings
Alkaline	Nickel chloride, sodium hypophosphite, ammonium chloride	95	200	8	0.3	Solutions are more difficult to maintain than acid baths

Table 3 Types of diffusion coatings and their characteristics

Type	Coating structure	Properties	Uses
Calorized	Metallic aluminum introduced into surface layer, forming aluminum-iron alloy	High-temperature oxidation resistance	Chemical processes, steam superheaters, and heat transfer
Chromized	Chromium carbide case formed on surface	High hardness and wear resistance	Combustion and mechanical equipment
Cyanided or carbonitrided	Carbon-nitrogen compound formed by diffusion into surface	Wear resistance for surfaces combined with core toughness	Gears, cams, pawls, and shafts
Nickel-phosphorus	Ammonium phosphate and nickel oxide products reduced and diffused	Corrosion resistance comparable to austenitic irons; poor wear resistance	Chemical process pipe and fittings
Nitrided	Nitrogen introduced into surface by contact of ammonia or other nitrogenous material	Wear and corrosion resistance at elevated temperatures	Same as for carbonitrided
Sheradized	Zinc introduced into surface	Corrosion resistance	Atmospheric-corrosion resistance

materials, hardfacing alloy selection, and hardfacing process selection.

Thermal Sprayed Metals and Ceramics

Thermal spraying originally consisted of propelling small molten globules of metal through a flame onto the workpiece. Technological improvements in the process have permitted the development of methods for applying both metallic and nonmetallic materials. Nonmetallics include ceramics, refractories, and carbides.

The thermal spraying of low-carbon iron or steel can be used to build up worn or mismachined surfaces. Nickel combined with other elements can be thermal sprayed to form a hardfaced surface. Zinc, aluminum, and lead can impart general corrosion resistance and can be deposited at very high rates. Babbitt alloys can be thermal sprayed onto bearings. Copper and bronzes are used for electrical and decorative purposes. Thermal barriers in the form of oxides or silicides are generally applied by plasma methods to provide either heat shielding or resistance to attack by molten metals. Wear resistance is enhanced by the application of carbides or borides.

Thermal spraying uses oxyacetylene, oxyhydrogen, detonation plasma, and electric arc techniques to impact molten or semimolten particles onto a substrate casting and to build up the desired thickness. The sprayed coating resembles the source material in composition, but its physical and mechanical properties are different from those of the source material—more closely resembling the properties of sintered metals or refractories. Bonding of the coating to the casting surface occurs principally by mechanical adhesion and only partly by metallurgical bonding. Rough machining, arc spattering, or coarse abrasive blasting prior to thermal spraying improves the mechanical bonding. Selection of the appropriate method of application depends on the form of the source material (wire, rod, or powder), the stability of the coating material at the application temperature, and the temperature required to achieve satisfactory adhesion. Temperatures within the nozzle range from 2760 °C (5000 °F) in the oxyacetylene process to 16 500 °C (30 000 °F) in plasma spraying. More detailed information is available in the article "Thermal Spray Coatings" in Volume 5 of the 9th Edition of *Metals Handbook*.

Diffusion Coatings

Diffusion coatings can be formed on iron castings by heating the castings in contact with a coating medium. The forms of coating materials are powdered metals, volatilized metal or metal salt, fused baths of metal salts, or a gas. Various types of diffusion coatings and their uses are summarized in Table 3. In some of the processes—nitriding and carbonitriding (case hardening), for example—the coating material becomes an integral constituent of the surface layer and is not visible on the surface of the casting. Diffusion coating and case-hardening processes are discussed in Volumes 5 and 4, respectively, of the 9th Edition of *Metals Handbook*.

Conversion Coatings

Chemical reactions at casting surfaces can produce iron-containing compounds that provide wear resistance or an attractive appearance or that serve as excellent bonding agents for subsequent organic coatings. Table 4 details common conversion coatings and their useful properties (chromate and phosphate conversion coatings are discussed in greater detail in Volumes 5 and 13 of the 9th Edition of *Metals Handbook*). Most of the successful processes are proprietary, and reproducibility of consistently good finishes is one of the important features.

Gray, ductile, and malleable iron castings all lend themselves readily to phosphating. The ability of a cast iron to accept a phosphate coating is not affected by alloy content, but hinges primarily on two requirements: a clean surface, and a metal temperature approximately equal to that of the phosphating bath. Dry machined surfaces need no further cleaning; cast surfaces can be prepared by blasting or other cleaning methods to remove scale and sand.

Porcelain Enameling

Porcelain enamels are inorganic vitreous coatings that are matured by heat. The inherently good heat transfer, thermal stability, and rigidity of iron at firing temperatures, coupled with the excellent adherence of vitreous ceramic frits as they fuse onto the cast surface, make the combination of porcelain on iron an excellent product. General corrosion resistance or resistance to specific chemicals can be obtained by selecting the proper porcelain enamel. The scratch resistance and hardness of the enamel coating, which allow the surface to resist abrasion, are almost equal in importance to the corrosion resistance.

Four processes are used to apply enamels: dry, wet, thermal spray (or plasma spray), and electrostatic precipitation. The latter two methods are seldom used on iron castings. Preparations for enameling start by blasting with sand, steel shot, or iron grit.

Dry coating methods use formulations that are mainly silica; these formulations generate a surface with the hardness and abrasion resistance of glass. Fluxes and opacifiers are mixed into the silica, and the

Table 4 Chemical conversion coatings, structures, and characteristics

Type	Coating structure	Properties	Uses
Chromate	Nonporous film acts as moisture barrier.	High corrosion resistance; inhibits corrosion if surface is broken; can be colored	Marine applications; can be decorative; nonporous bond layer for paint
Oxide	Ferric oxide formed from iron	Inhibits formation of ferrous oxide; highly absorbent; some wear resistance	Decorative blue-black coating; readily absorbs overlays of wax or oil
Phosphate	Iron zinc or manganese phosphates are crystalline structures formed on the surface by deposition from chemical solution.	Chemically neutral and high adherence to iron surfaces; highly absorbent	Excellent for bonding paint to iron; prevents abnormal wear or seizing during break-in

Table 5 Properties of organic coatings on iron castings

Resin	Resistance to chemicals and environment(a): Hydrocarbons	Solvents	Acids	Alkalies	Salts	Water	Weathering	Heat	Cold	Abrasion	Application method: Need for primer	Spray	Dip	Fluidized bed	Curing method: Air dry	Bake	Typical applications
Low cost																	
Phenolic	E	E	E	F	E	E	E	G	E	E	no	yes	yes	no	yes	yes	Appliances
Urea	E	G	E	E	E	G	G	G	G	E	no	yes	yes	no	no	yes	Appliances
Polyester	G	G	G	F	E	G	G	G	G	G	no	yes	yes	no	yes	yes	Thick coatings
Alkyd	G	P	F	F	E	G	E	F	G	G	yes	yes	yes	no	yes	yes	General purpose
Epoxy	E	E	E	E	E	G	G	E	E	E	no	yes	yes	yes	yes	yes	Scratch-resistant finish
Polyethylene	G	E	E	E	E	E	G	F	E	F	no	no	no	yes	no	yes	Coatings
Styrene-butadiene	E	G	E	E	E	E	G	G	E	G	no	yes	yes	no	yes	yes	General purpose
Urethane	G	E	G	G	E	E	E	E	G	G	yes	yes	yes	no	yes	yes	Scuff-resistant coatings: chemical and marine finishes
Moderate cost																	
Vinyl chloride	G	F	E	E	E	E	E	G	G	E	yes	yes	yes	yes	yes	yes	Chemical equipment
Melamine	E	G	E	E	E	G	E	G	G	E	no	yes	yes	no	no	yes	Appliances
Polyamide	F	G	F	E	E	F	P	G	G	E	no	yes	yes	yes	yes	yes	Abrasion-resistant coatings
Vinyl butyral	G	F	F	G	G	G	E	G	E	E	no	yes	yes	no	yes	yes	General purpose, primers
Cellulose nitrate	F	F	G	F	E	E	E	P	G	F	yes	yes	yes	no	yes	yes	High-gloss lacquer
Acrylic	F	P	F	G	E	E	E	P	F	F	yes	yes	yes	no	yes	yes	Water-resistant finishes
Vinyl acetate	F	P	F	F	F	G	E	F	F	E	no	yes	yes	no	yes	yes	Decorative
Cellulose acetate butyrate	F	F	F	F	G	F	E	P	G	F	yes	yes	yes	no	yes	yes	Decorative
High cost																	
Chlorinated polyether	G	E	E	E	E	E	E	G	E	F	no	no	no	yes	no	yes	Chemical equipment
Fluorocarbon	E	E	E	E	E	E	E	E	G	P	yes	yes	no	yes	no	yes	Chemical equipment, nonstick surfaces
Silicone	G	F	G	G	G	E	E	E	E	G	yes	yes	yes	no	yes	yes	Heat-resistant finishes

(a) E, excellent; G, good; F, fair; P, poor

mixture is then melted, quenched, and ground to make frit. (Frit is the term applied to the basic coating materials.) After application of a bonding ground coat, relatively heavy, smooth coatings (such as those on sinks and bathtubs) can be obtained by multiple firing. Each firing is followed by hot dusting with additional powdered frit, until the desired finish has been achieved.

Wet methods produce thinner coatings. The powdered frit is suspended in a solution of electrolytes and water or in an organic solvent and is applied by spraying or dipping over the ground coat. The sprayed or dipped coating must be dried prior to firing.

Ground coats that wet iron readily, adhere well, and are compatible with the cover coat are essential to the enameling process. Good ground coats promote adhesion between the enamel layer and the substrate, seal and smooth the irregularities of the surface, and prevent oxidation of the iron casting at firing temperatures. Top coats (cover coats) must provide the desired appearance and must be compatible with the ground coat. Formulation of frits requires the judgment and experience of a frit manufacturer to ensure that the coating provides successful results.

Organic Coatings

Organic coatings have a wide variety of properties, but their primary uses require corrosion resistance combined with a pleasing colored appearance. An organic-base film is often resistant to certain environmental substances but not to others, and so must be chosen for a specific set of well-defined service conditions. For example, a vinyl paint might be used on a pump casing that must operate in contact with acidic industrial waters. However, if the same casing is expected to contact hydrocarbons such as gasoline or solvents, a styrene, epoxy, or phenolic coating would most likely provide superior protection.

The term paint was once commonly used to designate all liquid organic coatings, but it is considered inadequate to describe modern liquid organic coatings, which in general are subdivided into enamels, lacquers, aqueous mixtures, suspensions, bituminous substances, and rubber-base products. Resins dispersed in a vehicle—for example, enamels or lacquers—cure to relatively hard gels by polymerization, oxidation, or solvent evaporation. A comparison of the chemical and environmental resistance of common resins is given in Table 5.

Enamels consist of milled pigments and other additives dispersed in resins and solvents and are converted from liquids to hard films by oxidation or polymerization. Lacquers are thermoplastic resins dissolved in organic solvents that dry rapidly by evaporation. In aqueous coatings, water is the principal vehicle or reducer. The advantages of water-base paints are nominal cost, nonflammability, true odorlessness, and nontoxicity. The disadvantages are difficulties in wettability, flow, and drying. Rubber-base coatings are noted for their mechanical properties and corrosion resistance rather than their decorative effects. Bituminous paints are black materials in which coal tar is dissolved in a solvent that evaporates. The major uses of bituminous paints are those that require extremely low permeability and high resistance to water. Unusual protection against chemical solutions, or special decorative effects, can be obtained by the use of asphaltic coatings or those produced by japanning, both of which are also considered bituminous coatings.

Fluorocarbon coatings produce an unusual combination of properties. They are tough, stain resistant, and nonsticking, and have a very low coefficient of friction. Fluorocarbon coatings resist all common industrial acids and temperatures to 300 °C (570 °F). Domestic cookware and chemical-processing equipment are two major applications of fluorocarbon coatings on iron castings. Fluorocarbon coatings are sprayed as emulsions of proprietary products onto a primed surface and then fused at temperatures of 385 to 425 °C (725 to 800 °F).

Organic coatings are applied by spraying, dipping, flow coating, fluidized bed coating, electrostatic deposition, and electrophoresis (electrocoating).

Spraying is adaptable to both low-volume and high-volume workloads. It is done by propelling the coating material toward the workpiece by compressed air, hot spraying, hydraulic-airless, and airless-electrostatic methods. Overspraying is most troublesome with compressed air methods and least troublesome with electrostatic methods.

Dipping has been used for centuries; modern refinements include flow coating and electrophoresis. Not all shapes can be painted by dipping. Pockets can exclude paint from some surfaces. The shape of the casting should allow easy draining after dipping. The coating should be selected to inhibit sagging or the formation of droplets on edges. The dipping process is easily automated and can be very efficient in use of materials. A thorough review of organic coatings and their applications, advantages, and limitations is available in the article "Organic Coatings and Linings" in Volume 13 of the 9th Edition of *Metals Handbook*.

Fused Dry-Resin Coatings

Dry-resin polymers can be applied (by fusion bonding) to iron castings for many of the same applications for which liquid organic coatings are used. Generally, the fused coatings are thick and can be applied very rapidly—often in minutes; in contrast, several hours is required for the drying and curing of a liquid organic coating.

The fusion bonding of polymers on iron castings can be readily accomplished by the application of dry solvent-free powder by fluidized bed coating or electrostatic deposition. The advantages of the process include the use of resins that are insoluble in ordinary solvents, the elimination of carrier solvents, and the ability to combine various plastics in the coating for maximum effectiveness. Sandblasted castings are excellent bases for this process. The plastic films can be easily machined, which contributes to flexibility in manufacturing. Two disadvantages are that thin films are not easily applied and that finding a suitable holding point on a part to be coated in a fluidized bed may be difficult. The six basic types of plastic fusion-bonded finishes, as well as a comparison of their characteristics, are given in Table 6.

Table 6 Relative effectiveness of fusion-bonding resin coatings

	Effectiveness(a)					
	Vinyl	Cellulose	Epoxy	Nylon	Chlorinated polyether	Polyethylene
Exterior durability	E	VG	F	F	F	F
Salt spray resistance	E	G	VG	VG	F	F
Water resistance	E	VG	G	G	E	F
Acid resistance	VG	G	VG	F	E	E
Alkali resistance	VG	G	VG	G	E	E
Solvent resistance	G	F	F	E	VG	G
Abrasion resistance	E	VG	E	E	VG	P
Impact resistance	E	G	(b)	VG	G	P
Heat resistance	VG	G	VG	VG	VG	P
Flexibility	E	G	(b)	G	F	E
Electrical characteristics	VG	VG	E	G	VG	VG
Color range	E	E	P	F	P	F
Gloss	E	E	VG	VG	G	G

(a) E, excellent; VG, very good; G, good; F, fair; P, poor. (b) Ranges from excellent to poor depending on composition

In a fluidized bed, castings heated to 175 to 310 °C (350 to 590 °F) are placed in a chamber containing resin powder suspended by upward-moving air. The dry resin floats around the casting, adhering to all the surfaces regardless of the complexity of the shape. Heating fuses the coating into a continuous film. This method produces a uniform coating that covers sharp corners, edges, and projections and that can be applied in a wide range of thicknesses (up to 1.5 mm, or 0.06 in.) in a single application.

In electrostatic deposition, the powdered resin is conveyed to a gun, in which it is given an electrostatic charge. The casting has the opposite charge. The charged powder is attracted to the surface, where it is deposited evenly. The electrostatic process is especially useful for applying thinner coatings because the residual charge on the workpiece is discharged by the powder, or leaks off, thus limiting the amount of powder that can be deposited. Preheating of the casting permits thicker coatings because the initial powder layer fuses as it is applied. Curing is done by reheating the casting to fuse the resin coating. Primers may be needed for some types of polymers (such as butyrates and vinyls) in order to achieve adequate bonding.

Foundry Automation

Ronald L. Lewis, and Yeou-Li Chu, The Ohio State University

AUTOMATION in the foundry is an integrated technology concerned with the application of both complex hardware (robots, computers, and electronic controllers) and software (management information software, statistical process control, and material process planning techniques) to control various foundry production activities. Automation applications in the foundry can be divided into the following categories:

- Foundry robotic applications
- Cell applications
- Automatic pouring systems
- Automatic sorting and inspection systems
- Automatic storage and retrieval systems
- Computer-aided design and manufacture

Each of these will be briefly reviewed in the following sections.

Foundry Robotic Applications

A robot is a programmable, multifunctional manipulator designed to move material, parts, tools, or specialized devices through variable programmed motions for the performance of a variety of tasks (Ref 1, 2). Depending on the sensory requirement of the task to be performed, robots can be divided into different classes. One classification scheme is given below (Ref 3):

- *Level 0*: This robot registers no information from the surroundings in which it operates. An example of this level is the pick-and-place device. It is merely capable of being fed movements that are repeated indefinitely. These devices are limited to assembly tasks involving smooth parts. Foundry applications include spraying mold wash and extracting castings from a die casting machine
- *Level 1*: Capable of monitoring contact between two surfaces through force and tactile sensors, these robots provide feedback to allow the mating of pieces without jamming. However, they are restricted to what they feel. They must be taught the location of the parts to be assembled, and the design of the parts cannot change during an assembly run. An example of such a foundry application would be robots attached to coremaking machines
- *Level 2*: Still in the research stage, these robots incorporate visual and nonvisual imaging sensors. Thus, they can select pieces from a moving conveyor line and then inspect and orient each workpiece for assembly. Examples would include adaptively controlled grinding and torch cutting operations

Different foundry applications will dictate different levels of sophistication and cost. Robots may range in price from $5000 for a simple pick-and-place unit to over $200 000 for a large, highly programmable and flexible multiaxis robot (1988 dollars). In addition to this capital investment, the foundry must be prepared to incur other costs, such as maintenance and education. Therefore, a number of benefits must be achieved to offset these costs and to provide a beneficial cost position. These benefits include (Ref 3-10):

- *Increased productivity*: Metalcasters report productivity increases of 10 to 300% (Ref 10) because robots do not require coffee breaks, lunch periods, and so on
- *Reduced costs*: Direct labor costs and labor overhead costs such as parking spaces, pensions, health care, wash rooms, and so on, are saved
- *Reduced injuries and illnesses*: Robots can replace a foundry employee working in dangerous and/or hazardous conditions
- *Improved casting quality*: Properly installed and correctly programmed, a robot can perform the same production function repeatedly, rapidly, and consistently without the variations inherent in human handcraft
- *High-speed precision*: Robots can move and position grippers and attach tools at higher speeds and accuracies than any human operator. Repeatability accuracies for some robot applications approach 0.013 mm (0.0005 in.), but because of the nature of most foundry applications, these higher accuracies are not achieved. Movement speeds of 1270 mm/s (50 in./s) are also possible
- *Performance flexibility*: Unlike hard automation, programmable robots can be retaught for varied, dissimilar tasks
- *High uptime*: Maximizing the use of capital equipment, robots can achieve uptime performance levels as high as 98% in long-run high-production situations
- *Inflation resistance*: The hourly rate for a robot installed today will be the same in 5 years, with no slowdowns or strikes
- *Round-the-clock output*: Robots can work two or three shifts a day, seven days a week, given sufficient parts, materials, and maintenance support
- *Improved worker morale*: Robots have been shown to improve morale, especially when they are installed to perform unpopular foundry tasks

Robotic applications in the foundry industry include cleaning, riser cutting, pick and place, mold venting, and mold spraying.

Cleaning Operations

Cleaning (fettling) operations are traditionally a very labor intensive component of the cost of a casting, often exceeding 60% of the total casting cost (Ref 11). A very high percentage, some estimate over 90%, of all cleaning operations are still performed manually (Ref 12). A number of robotic installations have been completed both in Europe and in the United States. Foundry technical associations and universities such as BCIRA (Ref 13), the Steel Castings Research and Trade Association (Ref 14), Svenska Gjuteriforeningen (Ref 15), the Fraunhofer-Institute in Stuttgart (Ref 16), the University of Aachen, the University of Leuven, and the University of Rhode Island have been conducting research in robotic foundry applications.

Cleaning operations can be divided into two categories: fixed and variable (Ref 17). Fixed grinding is the removal of the material that is present on every casting in a fixed position (that is, feeder pads, flash, gates, and so on). Variable cleaning can occur anywhere on the casting and usually results from defects in the mold or cores. To date, there are no applications of automated variable cleaning. In addition, if an automatic cleaning operation is implemented, some form of inspection is necessary to detect variable cleaning requirements and to ensure that they do not interfere with the automatic cleaning operation.

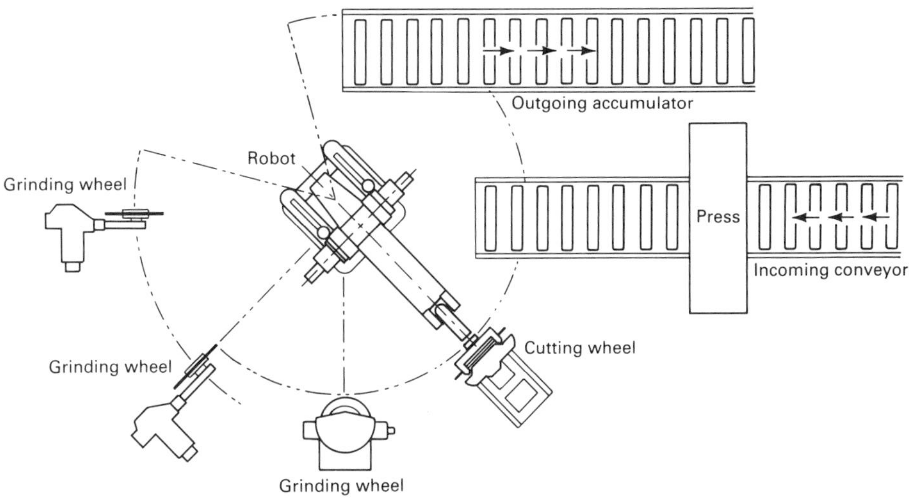

Fig. 1 Schematic of a first-generation robotic system designed to perform fettling operations at a rate of one completely machined casting every 3 min

For high-volume production, automatic fixed-stop machines have been specifically designed to grind castings. Many were designed to allow for the simultaneous grinding of many casting faces. Other machines are rotary in design with multiple heads. Through the use of template-driven machines, a complex casting periphery can be ground. Unfortunately, all of these systems, although automated, are not usually programmable; therefore, they cannot be easily adjusted for widely varying casting shapes or geometries. In addition, they cannot adapt to significant differences in casting irregularities.

Truly programmable cleaning operations are normally conducted by robots. Figure 1 shows a successful robotic installation. The cell consists of a robot and four specially designed grinding wheels. The wheels are hydraulically driven from a central unit. Wheel wear compensation is automatic such that the working point of the wheel is at the exact same location from cycle to cycle. Before entering the cell, castings are introduced to a dedicated press, where the main riser is removed and specific contour locations are stamped that will serve as reference areas for positioning. The casting is picked up by the robot and moved to each grinding machine. The system is now operating three shifts a day with an 89% uptime. The casting finish weight is 18 kg (40 lb). The complete cycle time is 3 min per casting. The average production rate, sampled over a 4-month period, is over 400 castings per day.

Depending on casting size, the robot can either hold the casting and move it to the grinding machine or it can hold the grinding tools and apply them to the casting. The advantages of using a robot to manipulate a casting include the following (Ref 17):

- The robot can pick up the casting from a fixed point that can be magazine fed or from several predetermined positions if the castings are palletized
- The robot can pick up castings within the limits of its capacity that would in many cases be too heavy for a manual operation
- The robot can move the casting to a succession of different tools to perform the cleaning operation to the required standard in the minimum amount of time by using the most efficient tool available at each operation. These sequential operations can be carried out without having to stop the cycle to change tools
- The robot can put the casting down on a fixture and regrip it to allow access to other surfaces of the casting. For example, the robot can turn the casting around or completely over
- Upon completion of cleaning, the robot can place the casting in a fixture, drop it in a bin, or place it on any predetermined position on a pallet

When the casting weight is high or when a variety of grinding wheels are necessary, it is more feasible to allow the robot to handle the tool rather than the casting. Industrial manipulators, although not programmable, are useful for this application (Ref 18). Tool change downtime is obviously very important with such a system.

A number of problems are inherent in any system, regardless of the type used. These problems include the following (Ref 17):

- Grinding develops varying load conditions, which cause inaccuracies in robot positioning. This can be overcome by either an adaptive control system or by a template-driven system such as that shown in Fig. 2
- Imprecision in casting geometries creates inaccuracies in location
- Corrections must be made for wheel wear; simple light beam sensors are often used to sense the location of the edge of the wheel

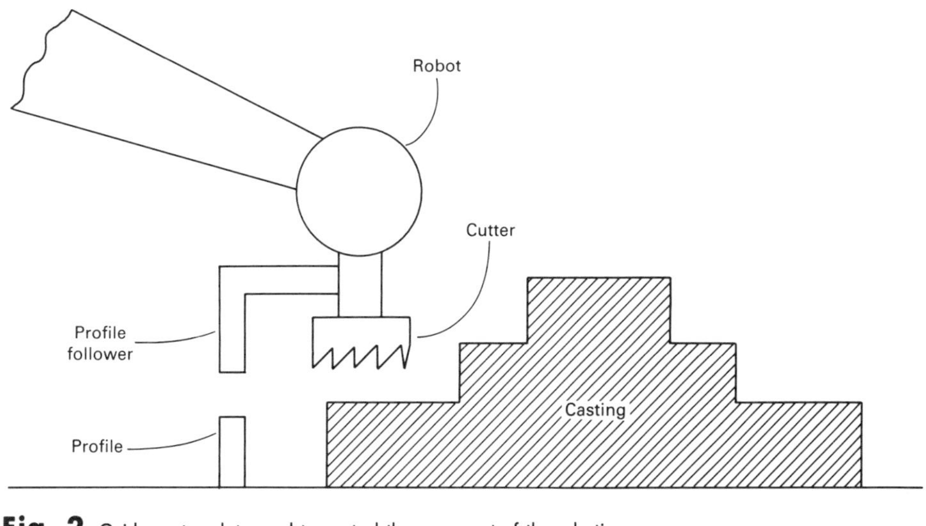

Fig. 2 Guidance template used to control the movement of the robotic arm

Riser Cutting

For the reasons previously outlined, robotic flame cutting for riser and gate removal is now possible. This is particularly important for heavy section steel castings, where risers cannot be removed economically by grinding, impact, and so on. Unlike cleaning, however, the torch must be held by the robot. As in cleaning, such systems range from semiautomatic to fully programmable robotized systems. Installations can use both conventional and gantry-style robots (Ref 19, 20). Steel thicknesses in excess of 610 mm (24 in.) have been successfully cut. To perform the flame cutting operation, the operator, using the control panel, moves the torch into the general position of the riser. The unit is preprogrammed for six riser contact sizes. The operator then selects the proper size, and the cut is executed.

A more automatic system uses a six-axis robot for complete automation of the cut-

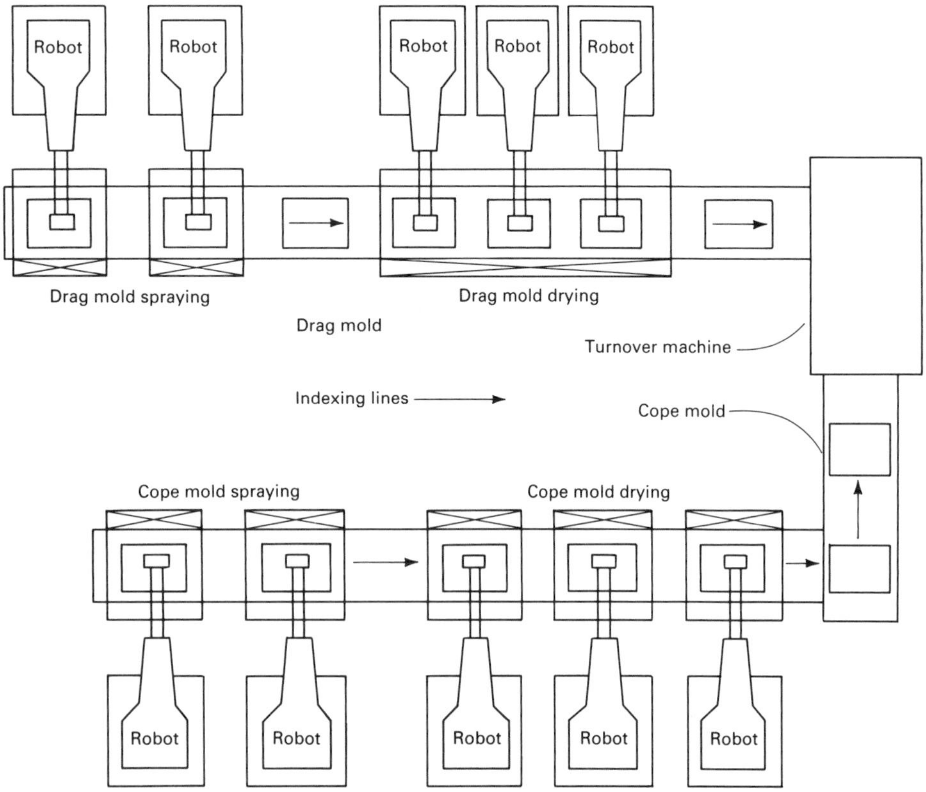

Fig. 3 Robots stationed along two indexing lines to spray a coating on each cope and drag, followed by drying of the foundry mold components prior to assembly in the turnover machine

ting operation (Ref 19). Specially designed locating blocks or pads are cast into each casting. Contact location principles are utilized to provide alignment between the robot and the casting. The robot is programmed using a teach pendant device. In addition to position and path, the operator programming the robot must also define the travel speed between each point, preheat delay, and cutting oxygen pressure for each cut. Results indicate a savings of 120 to 180%, depending on the casting to be cut (Ref 19-21).

The following factors must be taken into account when considering an installation or application (Ref 21):

- *Fixturing*: Distortion of gate position due to rough handling, mold inaccuracy, and so on, all lead to errors in position. Some means of sensing the positions of the gates and risers is necessary
- *Feedback system*: Because of variations in thickness and other defects, such as flash and drops on the casting, an adaptive control system that accounts for this would be very beneficial (Ref 22)
- *Automatic torch ignition*: This provides an added measure of safety and convenience
- *Preheat sensing*: One system is equipped with a fiber optic lens that sights through the oxygen orifice (Ref 19). An infrared sensing measurement is made to the computer controlling the robot. If the signal indicates kindling temperature, oxygen is employed and the cut is continued

Pick and Place Operations

The most common application of robots in any manufacturing environment consists of pick-and-place tasks. This is also true in the foundry industry. Some applications are given below:

- The first application of robotics in the foundry industry was in the die casting process (Ref 21, 23-25). Robots were used for such operations as extracting the casting from the die and loading and unloading the casting to firm presses
- Most attribute the second application of robotics in the foundry to investment casting (Ref 21, 26). In this application, robots are used to dip the pattern assembly into the ceramic slurry to obtain a precise and uniform distribution of the coating. This process has been quite successful because it has greatly increased the consistency as compared to manual dipping
- More recently, robots have been used in the evaporative foam process. Here robots have been used to remove the expandable polystyrene patterns from the mold and then assemble them into a complete pattern, to position the patterns into flasks, and to remove them after pouring. As in investment casting, robots have also been used to dip the castings into mold washes
- Core handling applications can also be found, especially for shell core handling

Mold Venting

To produce sound, high-quality castings, one of the most important manufacturing steps is to vent the mold properly. In green sand molding, this is accomplished either by pressing a metal pin into the mold or by drilling through the mold in precise locations.

A robot can be programmed by a teach pendant device to perform this operation. The program is then stored and retrieved when the pattern is produced again. The venting function requires a robot capable of only an X-Y positioning and an up-and-down motion (Ref 27). Off-line programming functions for the robot have even been incorporated into this operation.

Mold Spraying

One of the most common applications of robotics in manufacturing is paint spraying. Similarly, in sand molding, robots have been used to spray molds and cores with a wash before drying with gas torch flames. One application is shown in Fig. 3.

One advantage of such a system is the consistency of spray and the reduction of overspraying. Robotic-controlled drying has been found to reduce fuel consumption by 67%. In addition, after each spray of the mold wash, the spray gun can be cleaned for the next mold, thus eliminating the downtime that would have been spent to fix clogged spray heads.

Another application of robotic spraying systems is found in die casting (Ref 24, 28-30). Robots are used to spray both water and die lubricants on the die during each cycle.

Cell Applications

Many traditional manufacturing processes use cell concepts to improve productivity, control, and quality (Ref 31). Figure 4 shows a die casting cell. Such cells incorporate many individual automation concepts and techniques into an integral production system (Ref 32, 33). One important feature of cells is that they are usually under the supervision of a computer control. This concept of cell automation is being used in the foundry industry. Examples can be found in both the die casting and the evaporative foam processing areas (Ref 7, 21, 29, 30). The die casting cell consists of a number of components, including the following:

- *Die casting machine*: This may consist of either a vertical or a horizontal system
- *Furnace*: This unit provides a continuous supply of liquid metal that must be maintained at the proper temperature
- *Trim press*: The die casting cell often has a trim press that automatically removes most of the flash and the gating system

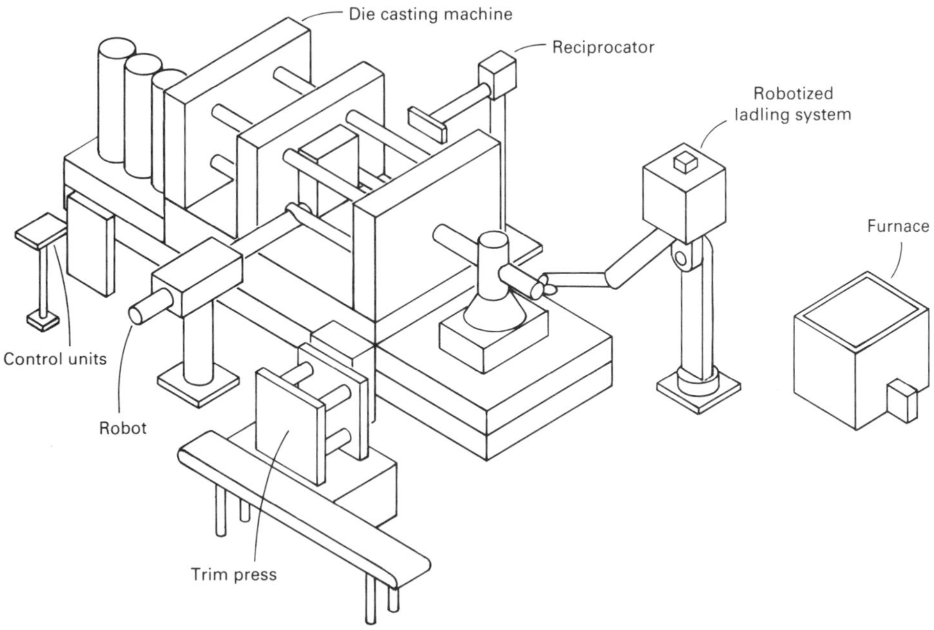

Fig. 4 A typical highly automated cell used in a die casting operation. It includes a die casting machine, a furnace, a trim press, two robots, and a computer control system to coordinate the movement of each cell component.

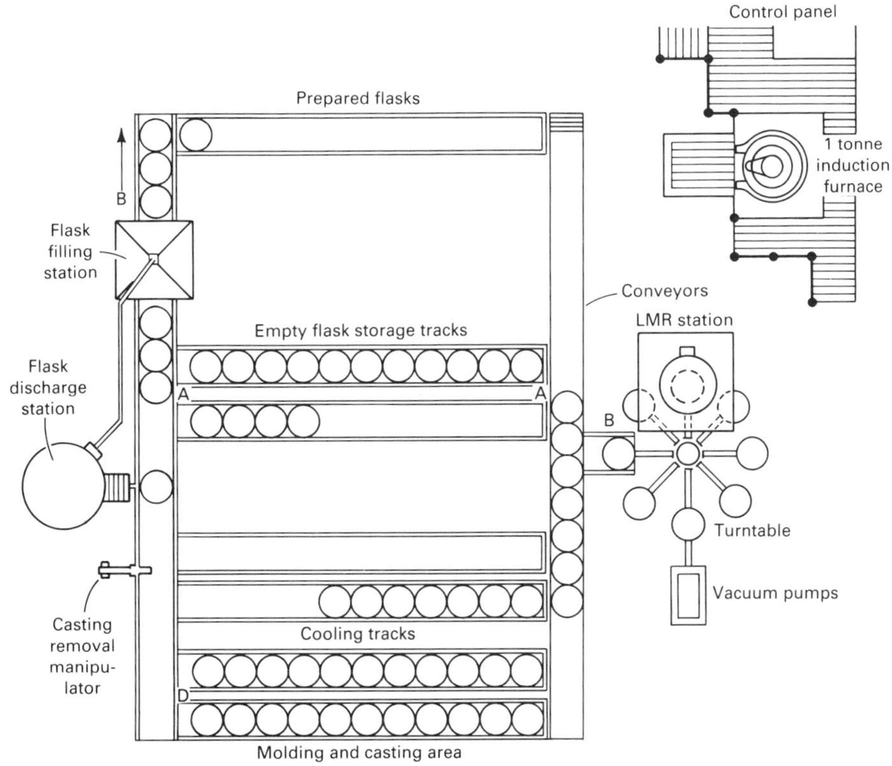

Fig. 5 Schematic of a layout for a CIM system that uses the Replicast ceramic shell process to produce steel valve castings. The loop conveyor system consists of four separate tracks (B–B in a U-shaped configuration along the perimeter of the system and A–A, C–C, and D–D enclosed in a ladder array within B–B) that contain 3 sets of 24 flasks each for a total of 72 flasks. Each track performs the following function in the production routing. A–A is the storage area for empty flasks. B–B transfers empty flasks to the flask filling station, where the thin ceramic shell is placed in the flask with its pouring cup at the center and the flasks vibrated to compact the sand that supports the shell; conveys flasks to liquid-metal refining (LMR) station turntable, where the molten metal is then poured into the mold (either under atmospheric pressure or a vacuum). C–C and D–D are cooling stations where the molten metal solidifies. B–B conveys flasks to a device that separates the casting from the mold and then transports the flasks to a discharge station, where the sand is dumped into hopper to be screened, cooled, and classified. Empty flasks in A–A await reuse to repeat the cycle of operations described.

- *Robots*: These are often used for three separate tasks within the die casting cell. First, they are used to remove the casting from the machine and to load the casting parts into a trim press, cooling tower, quench tank, or pallet. Second, robotic ladling systems handle the liquid metal. Finally, robots are used to spray lubricants on the die surface itself
- *Computer control system*: All of the above systems must be integrated with each other to form a complete die casting cell. Many die casting process variables are controlled and/or monitored by computer controls. This computer control allows the die casting machine to regulate the die casting process more precisely, thus producing higher-quality and more consistent castings

The results of implementing such cells are significant. A savings of $120 000 of direct labor (1986 dollars) cost per year per cell for a three-shift operation has been realized over manual operations (Ref 29). The hourly production of parts has been found to increase by 48.7% (Ref 23).

A totally automatic computer-integrated manufacturing (CIM) system for the Replicast process is currently under development (Ref 33). The proposed layout is shown in Fig. 5. As can be seen, the system is totally automated. The proposed system will use existing technology in each subportion of the cell. The integration of these techniques is the problem. The products to be produced are steel valve castings. The expected production level is 18 Mg (20 tons) of finished steel castings per week. The loop conveyor system consists of 72 flasks; the proposed cycle time is 165 to 190 min per flask.

A second proposed prototype casting cell is shown in Fig. 6. The proposed system uses well-established techniques for the automatic transport of expanded polystyrene type patterns, ceramic shells, and flasks. The system requires two to three robots, depending on the level of automation required. The first would be used to produce the molding material, while the second would position the pattern into the flask and then remove the casting. Robotic-controlled flame cutting stations and automatic cleaning facilities could also be installed.

Automatic Pouring Systems

The pouring control system for any automatic foundry cell is critical for achieving maximum yield per run or shift while still maintaining high quality standards in the castings produced (Ref 35). A number of automatic pouring systems are available; some are discussed below. Additional information can be found in the article "Automatic Pouring Systems" in this Volume.

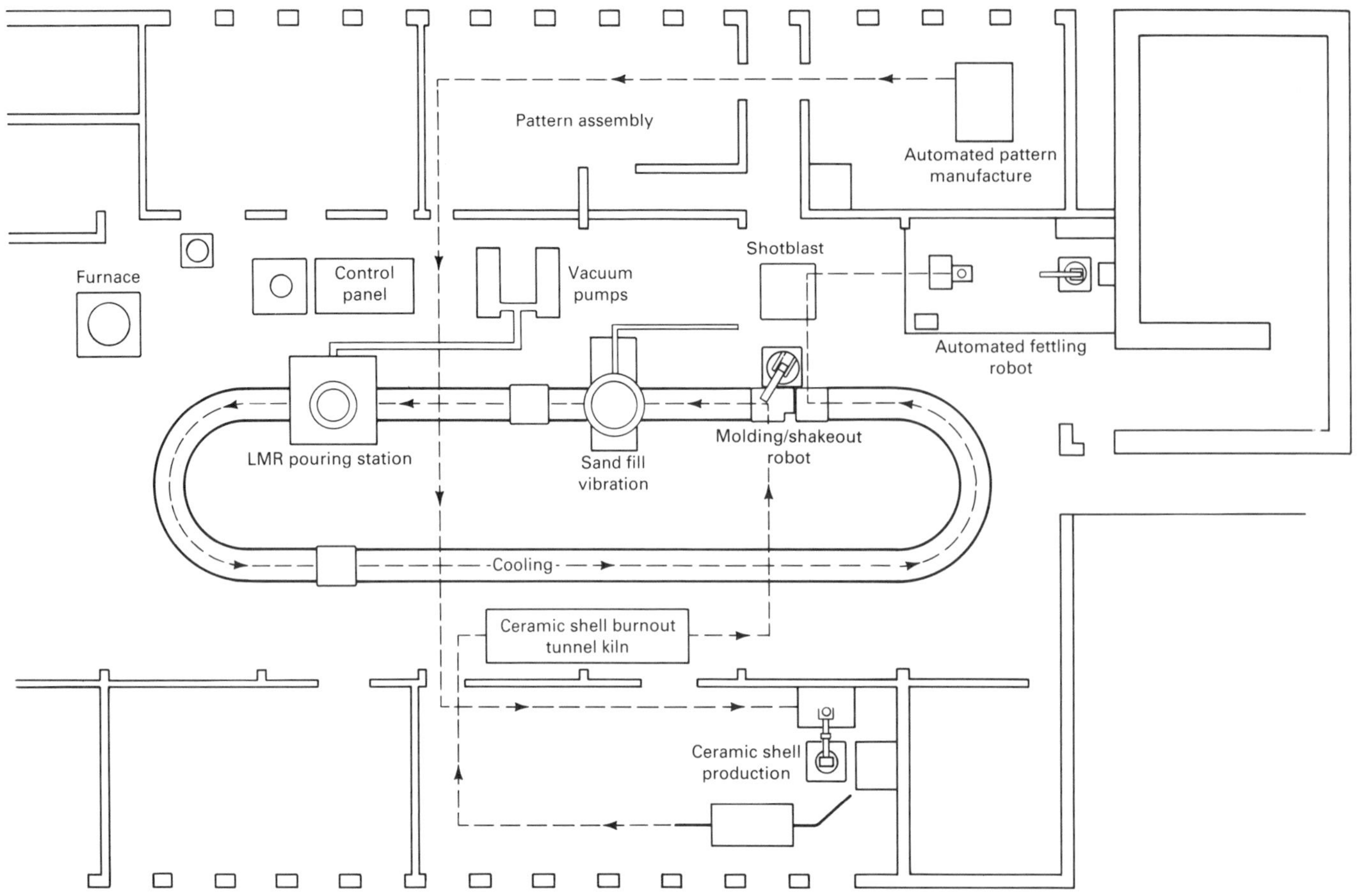

Fig. 6 Schematic of an integrated casting system that incorporates existing installations and technology to demonstrate the operation of a totally automated system. Up to three robots could be employed. The first is used to produce the ceramic shells widely used in investment casting operations; the second, to position shells in molding flasks and to remove castings from the mold after the molten metal has been poured; and the third, to flame cut or abrasive cut the casting in an automated fettling cell.

Robotic System. The die casting industry uses an automatic pouring ladle with robot-like movements that is controlled by a programmable controller. The cycle begins at the holding furnace, the ladle dips precise amounts of molten aluminum from a dip-well, carries it to a position directly over the mold without spillage within 3 to 6 s, and pours it quickly into the mold.

A simple adjustment on the controller determines the amount of liquid metal dipped by the ladle. A sensor detects the level of the liquid-metal surface to ensure accurate shot repeatability throughout the metal-level height range.

Bottom-Pour System. The automatic system combines microprocessor control, fiber optics, and infrared sensing to control bottom-pour ladles. A pulsed infrared beam is bounced off the mold surface to direct and control the hydraulic alignment of the mold with the ladle stopper. Two fiber optic cables sense the leading and trailing edges of the pouring cup. This reduces metal loss due to spillage and results in higher-quality castings. The microprocessor controls the time the stopper is open by computing the diminishing height of metal during the pouring process (Ref 36).

Laser Level Measurement. It is feasible to combine a laser light source with a solid-state electronic camera and a process control computer to control the amount of molten iron dispersed into special pouring ladles. This control has been installed on holding furnaces that supply predetermined quantities of molten iron to rotary mechanical pouring machines. Figure 7 shows a basic schematic of the system.

The laser beam controls the rate at which each ladle is filled based on the heat of metal in the holding furnace. The laser beam is focused on the surface of the molten iron. A video camera picks up the laser beam reflection and transmits the information to a process control computer. Decisions are then made as to the amount of metal poured. It is estimated that 5 to 14 kg (10 to 30 lb) per mold of iron is saved by this system. Payback is estimated to be 1 to 3 months.

Automatic Storage and Retrieval Systems

The storage of cores and patterns is another aspect of foundry operations that will lend itself to automated material handling technology. Automatic storage and retrieval in high-rise narrow-aisle rack space has had wide acceptance in almost every warehousing environment. In recent years, its use as a temporary storage method for workpieces in process and process equipment has brought this technology from the warehouse into the manufacturing plant itself—sometimes right to machine-side. An example of such a system is shown in Fig. 8.

This type of automatic storage and retrieval (AR/RS) system can be used to solve the typical problems involved in securing temporary storage for, or simply locating, cores and/or patterns. Computerized AS/RS systems make maximum use of the plant (square footage set aside for storage by storing at heights impossible or dangerous to reach by fork truck). The computer provides a real time, accurate inventory of all stored materials and the exact location of each storage container.

Automatic guided vehicles (AGV) represent still another recent material handling development that has a place in foundry applications, especially in die casting shops. The battery-operated, driverless vehicle follows a wire guidepath imbedded just below the surface of the plant or warehouse floor. The wire carries a low-voltage signal that is

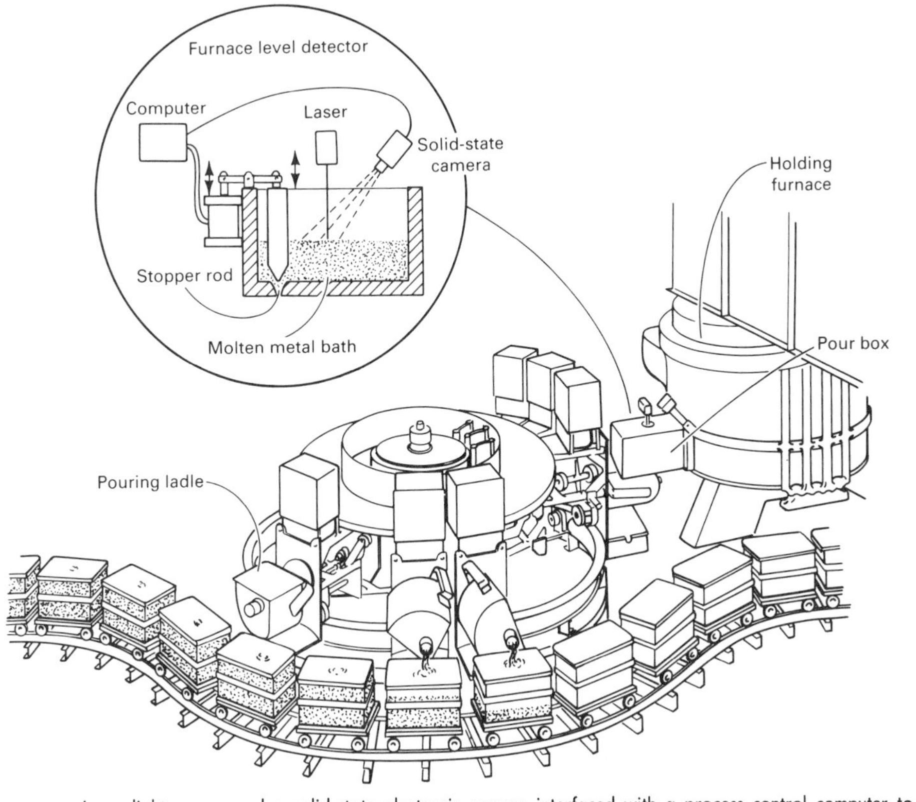

Fig. 7 Laser light source and a solid-state electronic camera interfaced with a process control computer to monitor the furnace level in a state-of-the-art rotary mechanical pourer

picked up by sensors mounted beneath the vehicle. These sensors, in turn, energize small servo motors that steer the vehicle and keep it on track. Guidepath design ranges from simple loops to very complex multipath, multivehicle systems. An AGV that transports palletized workpieces within a flexible manufacturing cell is illustrated in Fig. 9. Such a system could be used for the transportation of dies, molds, and/or patterns in a foundry (Ref 31).

These vehicles interface easily with an AS/RS system. If a central computer is being used to direct the flow of materials and to store and record inventory, the system can become highly automated.

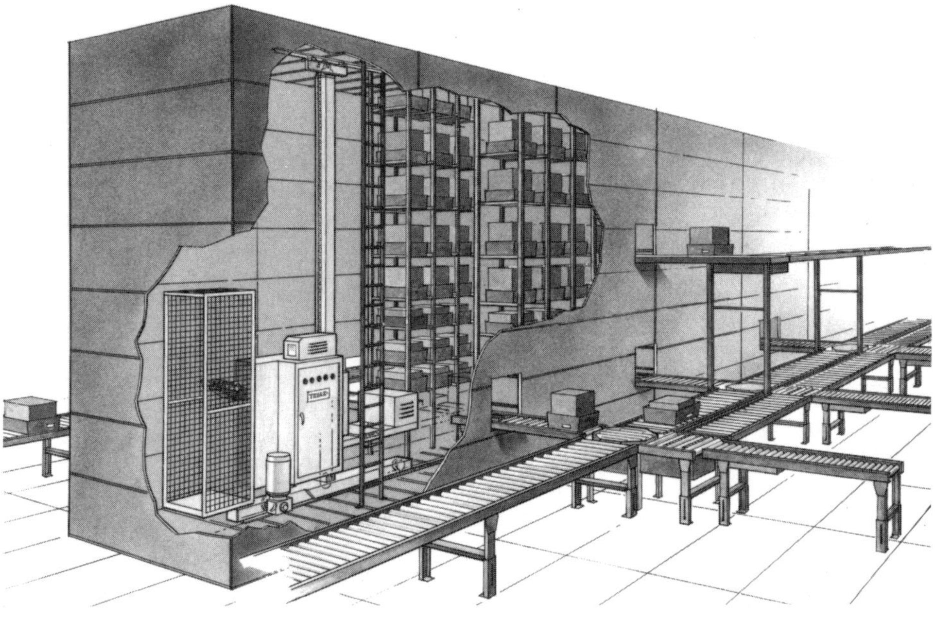

Fig. 8 Components of a typical in-plant automatic storage/retrieval system

The AGV is best suited to intermediate use where transport must be made on a regular basis but not continuously. This would describe the situation in most die casting shops, where dies must be transported from the point of manufacture to storage or point of use.

Automatic Sorting and Inspection

For the complete automation of foundry operations, methods must be developed that can automatically sense and sort castings. One such computer-aided vision system is described in Ref 37. This system is illustrated in Fig. 10.

When the linear photo sensor array is scanned, bright segments correspond to the unobstructed belt, while dark segments correspond to the part. The vision camera compares the acquired image to its memory of various parts. The part location is maintained by a position encoder. After the casting is identified, it is placed into the appropriate container by pneumatic kikers. The sorting unit has the capability of sorting 200 different objects at a rate of 10 000 pieces per hour.

Computer-assisted coordinate-measuring machines are being installed in foundry pattern shops to measure both pattern and casting dimensions. These machines can perform a variety of dimensional checks ranging from basic geometric measurement to parallel and plane projection. The operator simply identifies critical part locations so that the machine can establish a working plane. The coordinate-measuring machine can perform in a few minutes the tedious checks that take 2 to 4 h to be done manually.

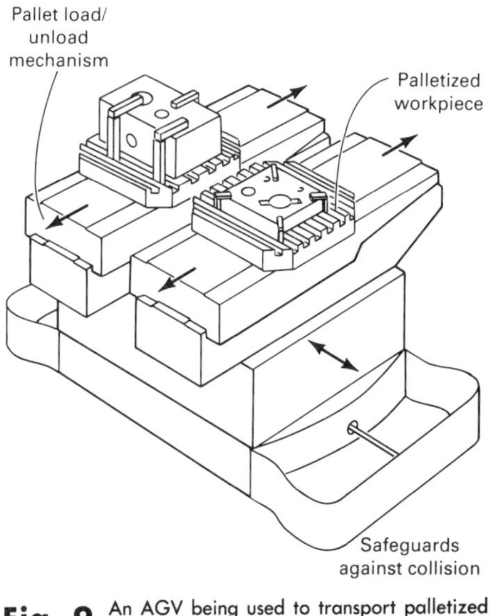

Fig. 9 An AGV being used to transport palletized workpieces within a flexible manufacturing cell

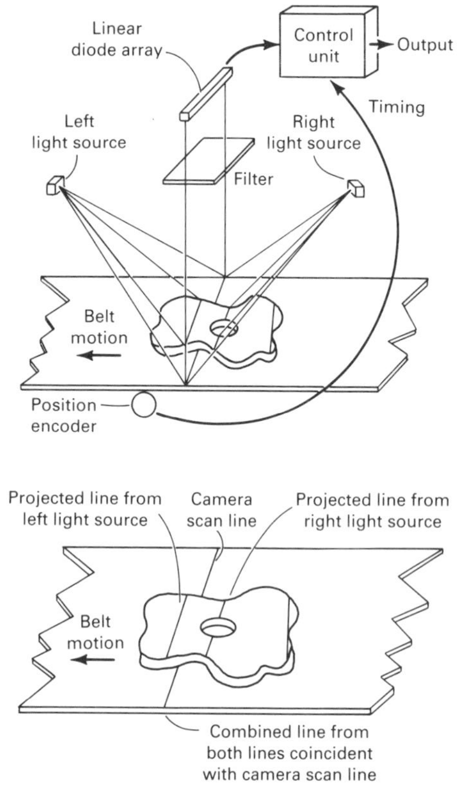

Fig. 10 Schematic of a computer-aided vision system capable of sorting 200 different objects at a rate of 10 000 pieces per hour

Computer-aided design and manufacture (CAD/CAM) offers a way to link the interrelated activities at the various manufacturing facilities, each using information as required from the data base and contributing new information as it is generated. The process begins with a part description, usually an illustration. From this, the designs for the casting, pattern, and mold are developed. At each step, the design information from previous steps is available for the geometric modeling of additional features. The CAD system automates some design tasks. Commonly used components are stored in the data base and recalled as necessary. The pattern design can be replicated and positioned to form the basis for the mold design. Special design calculations can be programmed. As an example, shrinkage allowances and draft angles are automatically calculated (Ref 38-42).

One advantage of developing a computer-aided design of the casting is that relevant data are available for creating foundry tooling. Drawings and material lists can be generated from the CAD system. Where computer numerical control (CNC) machining is to be used, cutter paths will be generated and downloaded directly to machine tools. Future foundry automation would use information on mold and casting geometry for automatic pouring, robotic handling, and verification of manufactured items as required (whether patterns, castings, or machined parts). Again, from an integrated data base, dimensional information is available and can be transferred to the shop floor by the CAM system to generate templates, drawings, checklists, or coordinate-measuring machine programs.

A true marriage of computer-aided design and manufacture will reduce casting implementation time by 40 to 50% (Ref 43, 44). Many pattern shops currently use computer numerical control to produce patterns. This method is proving to be a more accurate and faster approach to building patterns than the cast-to-size or trace machining methods (Ref 45). With the CNC approach, the mathematically defined surface model designed on a three-dimensional CAD system is expanded to include shrinkage requirements. The numerical control (NC) programmer develops tool paths from the surface model geometry and then test machines the part graphically to verify tool path selection. In addition, accurate predictions of the actual machine cutting times can be obtained. The final step is to feed the electronic data to a CNC machine that cuts the final shape. No craftsman-formed master pattern is required. The electronic part data can be compactly stored without distortion. Part data can be retrieved and revised very quickly compared to manual part programming techniques.

REFERENCES

1. R. Asfahl, *Robots and Manufacturing Automation*, John Wiley & Sons, 1985
2. W.R. Tanner, *Industrial Robots*, Society of Manufacturing Engineers, 1979
3. C.W. Meyers and J.T. Berry, The Impact of Robotics on the Foundry Industry, Paper 30, *Trans. AFS*, 1979, p 107-112
4. G.N. Booth, High Tech in a Smokestack Industry, *Foundry Mgmt. Technol.*, April 1985
5. C.F. James and J.G. Sylvia, The Robot's Role in Foundry Mechanization, *Mod. Cast.*, May 1982
6. H.J. Heine, New Ideas for the Cleaning Room—Part I, *Foundry Mgmt. Technol.*, Aug 1983
7. J.C. Miske, Improving Cleaning Room Productivity, *Foundry Mgmt. Technol.*, Oct 1984
8. E. Ford, Automating Britain's Foundries, *Mod. Cast.*, June 1982
9. D. Williamson, Automating Castings: What GM Gets for $214 Million, *Mfg. Eng.*, Aug 1985
10. *Foundry Management and Technology Data Book*, 1988
11. *Costs and Methods for Fettling Castings*, IVF Result 76639, Svenska Gjuteriforeningen, 1976
12. R.G. Godding, Fettling in the Light of Recent Developments, *Br. Foundryman*, Vol 76, Dec 1983
13. R.A. Wragg, *Practical Experiences of Using a Robot to Fettle Casting*, Seminar Proceedings, BCIRA
14. B.J. Sims and R. Wallis, The Role of Automation in Foundry Operation, in *Proceedings of the SCRATA Research Annual Conference*, Steel Castings Research and Trade Association, 1984
15. *Experiments With Robots in Foundries*, Mekanresultat 76009, Svenska Gjuteriforeningen, Nov 1976
16. W. Sturz and D. Boley, *Development in Using Industrial Robots for Deburring and Fettling of Castings*, Fraunhofer Institute
17. R.G. Godding, Fettling in the Light of Recent Developments, *Br. Foundryman*, Vol 76, Dec 1983
18. W.T. Hickman, Cleaning Casting With an Industrial Manipulator, *Mod. Cast.*, Sept 1984
19. M.D. Schneider and R.R. Petersen, Production Applications of Manipulators and Robots for Riser Cutting, Paper 161, *Trans. AFS*, 1986
20. M.D. Schneider, Automatic Riser Removal at Rockwell International, in *Proceedings of the 39th Annual Technical and Operating Conference*, 12 Nov 1984, Steel Founders' Society of America
21. G.E. Munson, *Foundries, Robots and Productivity*, Unimation Inc., 1978
22. R.T. Hughes, S. Lepak, and R. Scholz, Demonstration of Control Technology for Torch Cutting, *Trans. AFS*, Vol 59, 1985
23. V.G. Parodi, Robots in the Automated Production of Aluminium-Alloy Pressure Diecasting, *Foundry Trade J. Int.*, Dec 1978
24. N.W. Rhea, Robots Improve a Die Casting Shop, *Tool. Prod.*, Vol 43, March 1978
25. Die Casting With Robot, in *Die Casting and Metal Molding*, Cutlands Press, p 16-20
26. Robots at Work: Unimation Designs Automated Investment Casting System, *Robotics Today*, Fall 1981
27. A.L. Carr and W.P. O'Neil, Computerized Off-Line Programming for Robotic Mold Venting, Paper 106, *Trans. AFS*, 1984
28. R.C. Rodgers, Robots 9 Show and Conference Highlights Vision Systems, *Foundry Mgmt. Technol.*, Sept 1985
29. W.A. Wiesmueller, Robots in the Real World of Die Cast Foundries, *Robotics Eng.*, March 1986
30. J. Canner, Automated Die Casting—A Concept Comes Full Circle, *Die Cast. Eng.*, March/April 1986
31. P. Ranky, *The Design and Operation of FMS*, IFS Publications, 1983
32. M.P. Groover, *Automation, Production Systems and Computer Aided Manufacturing*, Prentice-Hall, 1983

33. C. Lewis, *State of the Technology in Die Casting*, Ohio State University, 1987
34. M.C. Ashton, B.J. Sims, and S.G. Sharman, The Conceptual Integration of New Technology, in *Proceedings of the SCRATA Research Annual Conference*, Steel Castings Research and Trade Association, 1984
35. R.C. Rodgers, Automatic Pouring for Short Runs—Pro and Con, *Foundry Mgmt. Technol.*, May 1986
36. R.C. Rodgers, Innovations in Automatic Control for Metal Pouring, *Foundry Mgmt. Technol.*, June 1985
37. W.E. Willis, *Consight/Magsight*, *Mod. Cast.*, Sept 1983
38. M. Hoggan, High Tech Innovations Keep St. Catharines Foundry Out Front, *Mod. Cast.*, Aug 1986
39. D.R. Westlund and G.R. Anderson, Applying CAD/CAM to Foundry Tooling, *Mod. Cast.*, Jan 1984
40. G.S. Cole, Using Computers in Metalcasting—Part I, *Foundry Mgmt. Technol.*, April 1986
41. C.T. Smith and K. Lee, Computer-Aided Pattern Design for Casting Processes, Paper 8, *Trans. AFS*, 1986
42. J.A. Blevins, Computer Aided Design of Castings, *Mod. Cast.*, June 1982
43. D.P. Kanicki, New Technologies Shaping Foundries of the Future, *Mod. Cast.*, Oct 1985
44. D.P. Kanicki and R. Bailey, Micros and Minis Lead Metalcasting Into Computer Age, *Mod. Cast.* Feb 1984
45. G. Olson, CAD/CAM Interfaces With Foundry Patternmaking, *Mod. Cast.*, Aug 1985

Design Considerations

Riser Design

Lee A. Plutshack and Anthony L. Suschil, Foseco, Inc.

RISER DESIGN, or risering, deals with the development of suitable reservoirs of feed metal in addition to the desired casting shape so that undesirable shrinkage cavities in the casting are eliminated or moved to locations where they are acceptable for the intended application of the casting. When metals solidify and cool to form a casting, they generally undergo three distinct stages of volume contraction, or shrinkage. (Exceptions to this shrinkage behavior of some graphitic cast irons will be noted later in this article.) These stages, shown schematically in Fig. 1, are:

- Liquid shrinkage: The liquid metal loses volume as it gives up superheat and cools to its solidification temperature
- Solidification shrinkage: The metal freezes, changing from a liquid to a higher-density solid. For pure metals, this contraction will occur at a single temperature, but for alloys it will take place over some temperature range or freezing interval
- Solid shrinkage: The solid casting cools from its solidification temperature to room temperature

The last of these, solid shrinkage (also called patternmaker's shrinkage), is accommodated by making the pattern (and therefore the mold cavity) somewhat larger than the desired dimensions of the final casting. Liquid shrinkage and solidification shrinkage are the concern of risering practice. In the absence of risers, a casting would otherwise solidify as shown in Fig. 2.

Visible signs of shrinkage-induced casting defects include internal shrinkage voids, surface deformation or dishing, and surface puncture. These defects will vary with different alloys; for example, internal shrinkage may be more dispersed, or alloys with strong skin-forming behavior may not exhibit surface deformation.

To eliminate these undesirable defects in the casting, a riser will be added to accommodate the liquid shrinkage and to supply liquid feed metal to compensate for the solidification shrinkage within the casting (Fig. 3). Therefore, the shrinkage in the riser/casting system is concentrated in the riser, which will then be removed from the finished casting.

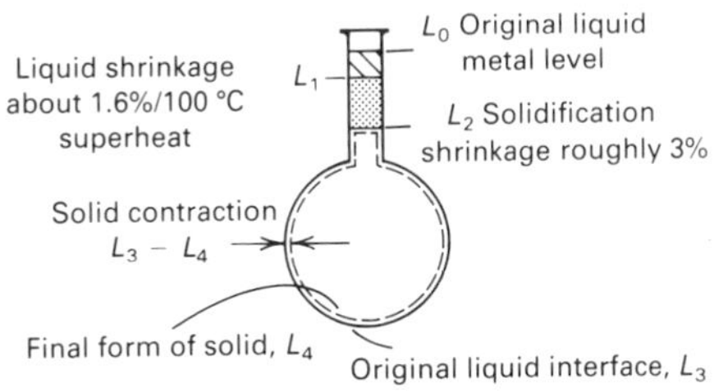

Fig. 1 Schematic of the shrinkage of low-carbon steel. The contribution of each one of the three distinct stages of volume contraction is shown: liquid shrinkage, solidification shrinkage, and solid contraction. Source: Ref 1

As illustrated in Fig. 3, the riser must often be larger than the casting it feeds, because it must supply feed metal for as long as the casting is solidifying. Various methods are used to reduce the size of the required riser, including chilling the casting (that is, reducing its solidification time) or insulating the riser (that is, extending its solidification time).

Optimum Riser Design

The role of the methods engineer in designing risers can be stated simply as making sure that risers will provide the feed metal:

- In the right amount
- At the right place
- At the right time

To this list can be added several other considerations:

- The riser/casting junction should be designed to minimize riser removal costs
- The number and size of risers should be minimized to increase mold yield and to reduce production costs
- Riser placement must be chosen so as not to exaggerate potential problems in a particular casting design (for example, tendencies toward hot tearing or distortion)

In practice, these considerations are often in conflict, and the final riser design and pattern layout represent a compromise.

Feed Metal Volume

The riser must be adequate to satisfy the liquid and solidification shrinkage requirements of the casting. In addition, the riser itself will be solidifying, so the total shrinkage requirement to be met will be for the riser/casting combination. The total feeding requirement will depend on the specific alloy, the amount of superheat, the casting geometry, and the molding medium.

Liquid shrinkage will depend on the alloy and the amount of superheat. As indicated in Fig. 1, liquid shrinkage for carbon steels is generally considered to be in the range of 1.6 to 1.8%/100 °C (0.9 to 1.0%/100 °F) superheat. For graphitic cast irons, liquid shrinkage has been variously reported in the range of 0.68 to 1.8%/100 °C (0.38 to 1.0%/100 °F) (Ref 2-5).

Solidification Shrinkage. Table 1 indicates that solidification shrinkage will vary considerably according to the alloy melted and that, within the graphitic cast irons, expansion may occur. This phenomenon is often ascribed to the precipitation of the less dense graphite phase overcoming the contraction associated with the solidification of austenite. Theoretical calculations indicate that such density differences cannot account for the higher reported expansion percentages (Ref 2, 5). Practice shows that, with proper control of metallurgical and mold conditions, expansion phenomena can be used to reduce greatly or eliminate risers (with the liquid shrinkage accommodated by the gating system instead of the riser system) (Ref 5).

Mold Dilation. Mold wall movement after a mold cavity has been filled with liquid metal can enlarge the casting and thus increase the feed metal requirements. Such mold dilation is a function of the molding medium, the mold filling temperature, and the alloy.

With gray and ductile irons, mold dilation may result partially from expansion pressures within the solidifying casting generated by the precipitation of graphite. In soft green sand molds, such mold dilation may produce an additional 15% feed metal requirement above that needed to satisfy the calculated liquid and solidification shrinkages (Ref 7). In copper-base alloys, it has been suggested that an additional 1% volumetric shrinkage should be expected as a result of mold cavity expansion in green sand molds (Ref 8).

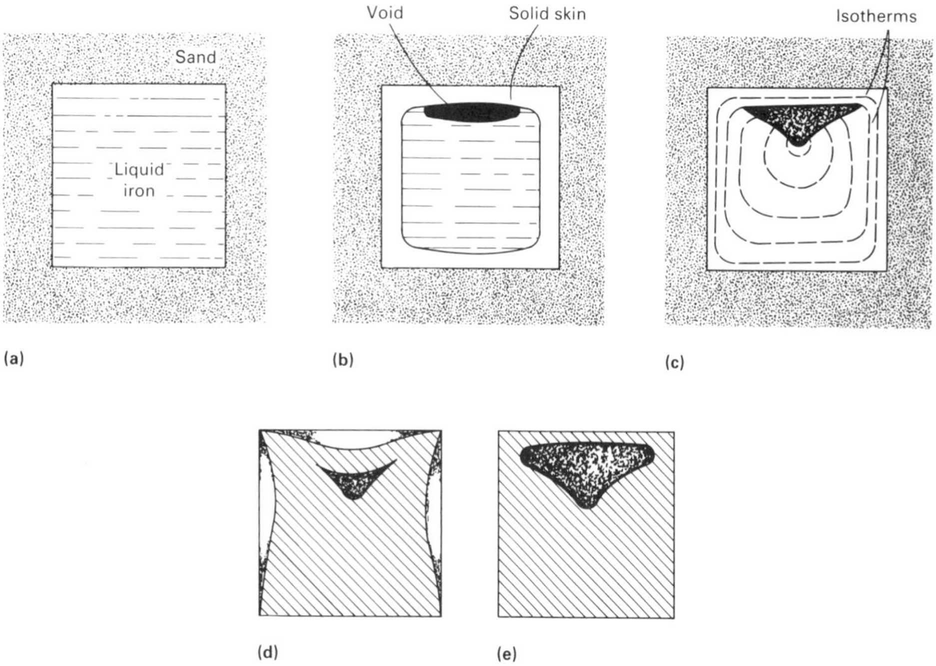

Fig. 2 Schematic of sequence of solidification shrinkage in an iron cube. (a) Initial liquid metal. (b) Solid skin and formation of shrinkage void. (c) Internal shrinkage. (d) Internal shrinkage plus dishing. (e) Surface puncture

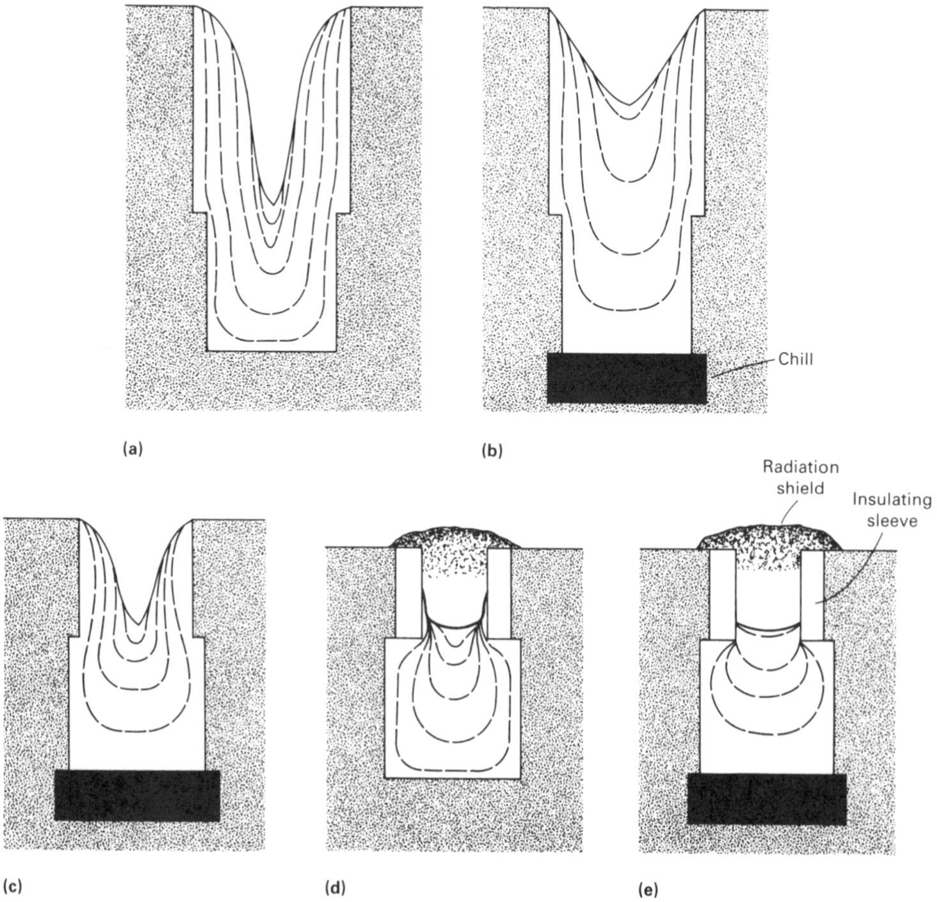

Fig. 3 Methods of controlling shrinkage in an iron cube to reduce riser size. (a) Open-top riser. (b) Open-top riser plus chill. (c) Small open-top riser plus chill. (d) Insulated riser. (e) Insulated riser plus chill

Table 1 Solidification contraction for various cast metals

Metal	Percentage of volumetric solidification contraction
Carbon steel	2.5–3
1% carbon steel	4
White iron	4–5.5
Gray iron	Varies from 1.6 contraction to 2.5 expansion
Ductile iron	Varies from 2.7 contraction to 4.5 expansion
Copper	4.9
Cu-30Zn	4.5
Cu-10Al	4
Aluminum	6.6
Al-4.5Cu	6.3
Al-12Si	3.8
Magnesium	4.2
Zinc	6.5

Source: Ref 6

Casting Geometry. The shape of a casting will affect the size of the riser needed to meet its feed requirements for the obvious reason that the longer the casting takes to solidify, the longer the riser must maintain a reservoir of liquid metal. For rangy, thin-section castings (where solidification will be rapid), feed metal requirements may be smaller than what would ordinarily be calculated. This is because a portion of the liquid and solidification shrinkages will be fed by liquid metal entering the mold from the gating system. Table 2 indicates the effect of differences in casting geometry on minimum riser volume requirements for steel castings.

Riser Location

To determine the correct riser location, the methods engineer must make use of the concept of directional solidification. If shrinkage cavities in the casting are to be avoided, solidification should proceed directionally from those parts of the casting farthest from the riser, through the intermediate portions of the casting, and finally into the riser itself, where the final solidification will occur. Shrinkage at each step of solidification is thus fed by liquid feed metal being drawn out of the riser.

The ability to achieve such directional solidification will depend on:

- The alloy and its mode of solidification
- The mold medium
- The casting design

Two distinct types of castings must be considered: castings with uniform wall thickness and castings with wall sections of varying thickness.

Progressive and Directional Solidification. Figure 4 illustrates the interplay of progressive and directional solidification in a casting. With the mold cavity filled, solidification will generally proceed from the mold wall, where a skin of solid metal will form. As heat is lost to the mold, that skin will grow progressively inward. Two condi-

Table 2 Minimum volume requirements for risers on steel castings

Type of casting	Minimum riser volume/casting volume (V_r/V_c), %: Insulated risers, H/D = 1:1	Insulated risers, H/D = 2:1	Sand risers, H/D = 1:1	Sand risers, H/D = 2:1
Very chunky (cubes, etc.): dimensions in ratio 1:1.33:2	32	40	140	198
Chunky: dimensions in ratio 1:2:4	26	32	106	140
Average: dimensions in ratio 1:3:9	19	22	58	75
Fairly rangy: dimensions in ratio 1:10:10	14	16	30	38
Rangy: dimensions in ratio 1:15:30	9	10	13	15
Very rangy: dimensions in ratio 1:>15:>30	8	8	11	13

Source: Ref 9

tions serve to change the rate of this growth. At the casting edge, where the greater surface area allows more rapid transfer of heat to the mold, the solidification rate will be faster. At the riser, where the mass of the riser provides more heat, and where heat transfer to the mold is reduced at the internal angle of the riser/casting junction, the rate of skin formation will be reduced. This combination of edge effect, or end effect, and riser effect provides directional solidification.

If the wedge-shaped pattern of the solidification front begun at the casting edge can be maintained, a channel of liquid feed metal should be available throughout its progress toward the riser. If, however, the parallel advancing walls progressively solidifying in the intermediate zone begin to meet, movement of liquid feed metal will be restricted, and centerline shrinkage will result.

Solidification Mode. The ability to promote and sustain directional solidification will depend greatly on the manner in which an alloy solidifies. Alloys can be classified into three types based on their freezing ranges:

- *Short*: liquidus-to-solidus interval <50 °C (<90 °F)
- *Intermediate*: interval of 50 to 110 °C (90 to 200 °F)
- *Long*: interval >110 °C (>200 °F)

This classification is not precise, but the general solidification mode of each type is illustrated in Fig. 5 through 8.

For pure metals (Fig. 5), in which the freezing range approaches zero, the solidifying casting walls progress inward as a plane front. Short freezing range alloys (Fig. 6) will show a strong tendency toward skin formation, and the fronts of the crystals solidifying inward (start of freeze) will not advance much faster than their bases (end of freeze). Such relative, short crystalline growth helps keep liquid feed metal in contact with all the solidifying surfaces. Such strong progressive solidification in these short freezing range alloys promotes the development of directional solidification along any temperature gradients in the solidifying casting. For example, in carbon steel, gradients of only 0.022 to 0.045 °C/mm (1 to 2 °F/in.) in plates and 0.135 to 0.269 °C/mm (6 to 12 °F/in.) in bars are sufficient to produce a shrinkage-free casting section through directional solidification.

For long freezing range alloys (Fig. 7), the development of directional solidification is difficult. Although a thin skin may initially form on the mold walls, solidification does not proceed progressively inward. Rather, it develops throughout the solidifying casting at scattered locations. This mushy or pasty mode of solidification results in the development of numerous small channels of liquid metal late in solidification. Feeding through these channels is restricted, and dispersed shrinkage porosity occurs throughout the casting.

Such solidification is typical of many commercial copper-base alloys, in which the difficulty in feeding caused by shrinkage porosity is aggravated, especially in thick sections, by the high thermal conductivity of the alloys, which helps maintain a nearly uniform temperature throughout the solidifying casting. To promote directional solidification in such alloys may require temperature gradients as high as 1.46 °C/mm (65 °F/in.), which can usually be achieved only by severely chilling one portion of the solidifying casting. Generally, the goal in risering such alloys is not to eliminate shrinkage but to ensure that it is finely dispersed (microporosity).

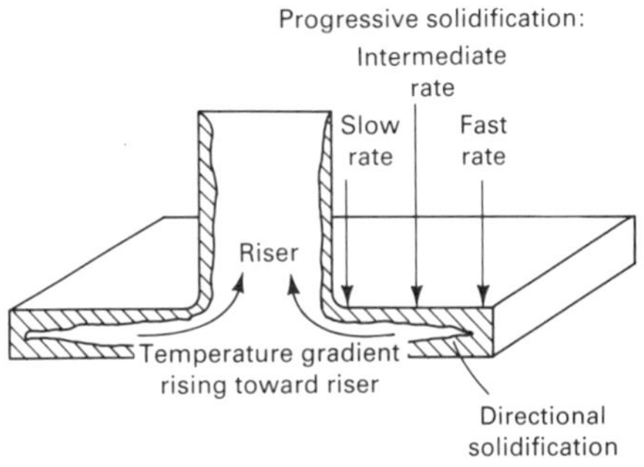

Fig. 4 Directional and progressive solidification in a casting equipped with a riser. Source: Ref 10

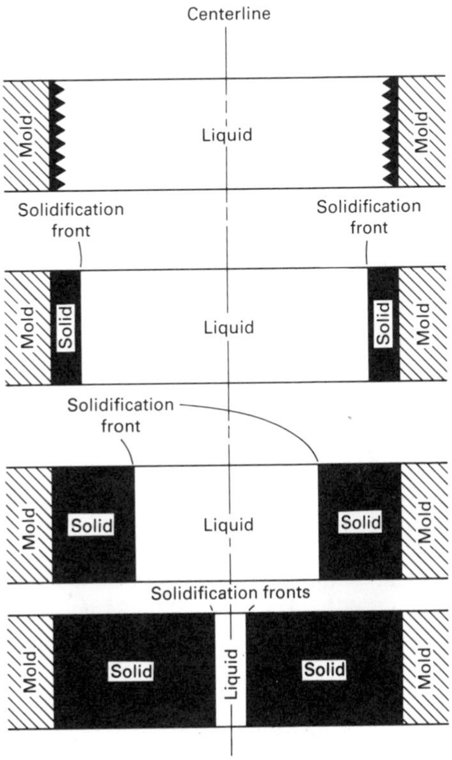

Fig. 5 Schematic of mode of freezing in pure metals. Crystallization begins at the mold wall and advances into the casting interior on a plane solidification front. Source: Ref 11

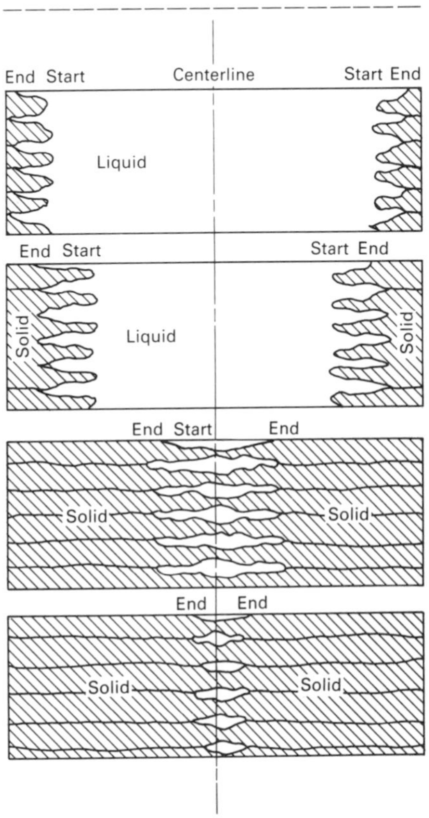

Fig. 6 Diagram of mode of freezing in alloys having a short freezing range

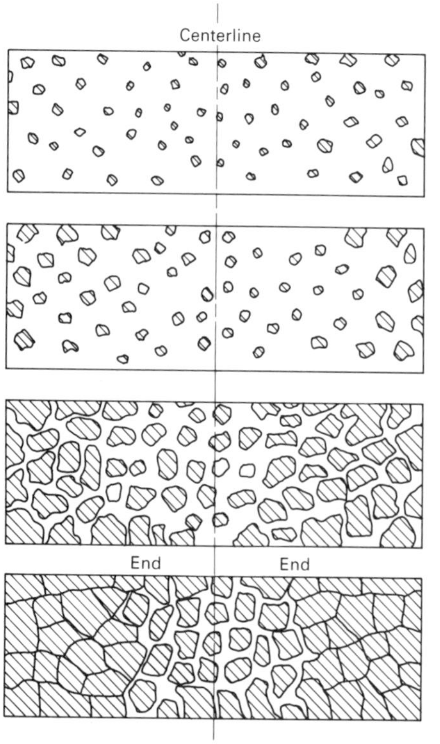

Fig. 7 Schematic of mode of freezing in alloys having a long freezing range

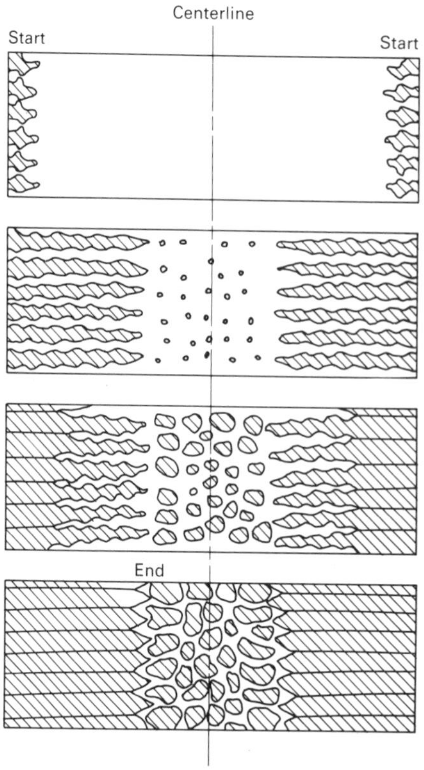

Fig. 8 Schematic of intermediate mode of freezing in alloys having a moderate freezing range

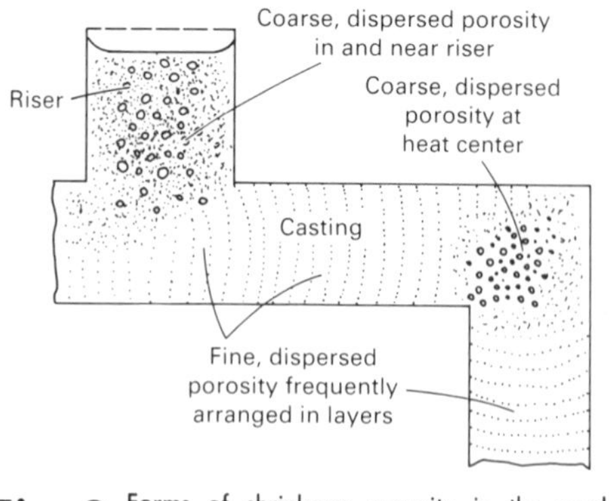

Fig. 9 Forms of shrinkage porosity in the sand castings of alloys that freeze in a pasty manner

For alloys with an intermediate freezing range (Fig. 8), the mode of solidification will combine elements of both the skin-forming and mushy solidification modes. Short freezing range alloys may shift to a more intermediate mode of solidification in heavy casting sections, in which heat loss from the casting surface will be slowed as the molding medium heats up. As temperature gradients from the center of the solidifying section to the casting edge are reduced, crystal growth will change from the columnar pattern growing in from the mold walls to an equiaxial pattern dispersed throughout the still-liquid center.

The various solidification modes result in very different typical shrinkage configurations in the casting and riser (Fig. 9 and 10) and present the methods engineer with distinctly different problems to overcome in riser and casting design. Selection of appropriate methods will depend largely on the possibility of promoting directional solidification. Figure 11 illustrates the effects of several mold and metal variables on the development of progressive (and therefore directional) solidification.

Castings of Uniform Wall Thickness. The specific alloy and the section configuration will combine to impose a limiting feeding distance over which a casting can solidify free of centerline shrinkage. As shown in Fig. 12, total feeding distance in a section with a cooling edge is the sum of the riser effect and edge effect discussed earlier. Figure 12 illustrates several key points:

- The contribution from the edge effect is generally greater than that from riser effect
- In the absence of cooling edges, feeding distance between risers is dramatically reduced
- If the maximum feeding distance in a section is exceeded, the edge effect will give a sound edge to its usual length, but centerline shrinkage may extend for some variable distance into the area that would ordinarily be expected to be sound because of riser effect

Figure 13 illustrates the same relationships in steel bars. When compared with Fig. 12, Fig. 13 also highlights the fact that bar-shaped sections will have shorter feeding distances than platelike sections of the same thickness.

Figure 14 shows the use of chills to extend feeding distance. When applied to the edge of a casting section, the chill will withdraw heat rapidly, enhancing the development of directional solidification away from the edge. This will add to the length of the zone that will be sound due to end effect.

In addition, if a chill is placed between risers in a casting section where there is no natural cooling edge, it can be used to establish an artificial end effect. In this way, the distance between risers can be dramatically increased, thus reducing the number of risers needed to ensure a sound casting.

Such a use of chills is illustrated in Fig. 15. The first attempt at subdividing this casting into feedable sections with riser placement based on the absence of end effect (except at the periphery of the flange) results in the use of eight risers (two on the hub and six on the flange). The overlapping feeding zones of the riser (based on riser effect) cover the feeding requirements of most of the flange, but there still remain unfed regions (in which centerline shrinkage would occur) that would probably require the addition of at least one more riser on the flange.

The second subdivision of the casting, using chills to establish artificial end effects, reduces the number of risers needed to only five (one on the hub and four on the flange). Such an application of chills, in addition to ensuring a sound casting, can provide economic advantages by simplifying molding and patternmaking procedures, increasing casting yield, and reducing riser removal costs.

It should be reemphasized that feeding distances in sections of uniform thickness depend on alloy characteristics and section configuration (that is, whether barlike or platelike) to determine how far directional solidification can be sustained. If the walls of the progressively solidifying intermediate sections begin to come together, disrupting and constricting the feeding channels

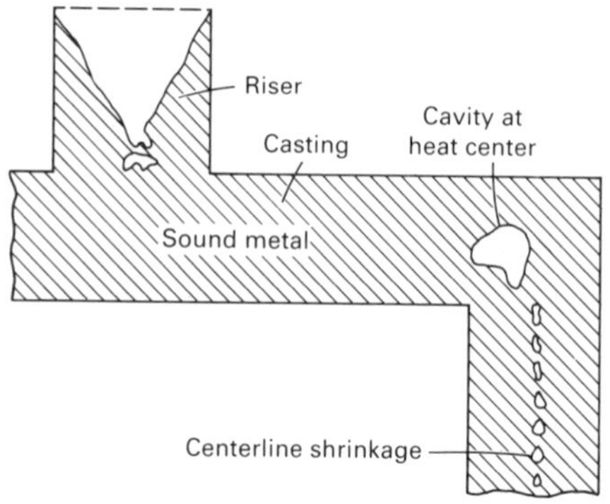

Fig. 10 Shrinkage cavities produced by skin formation in alloys

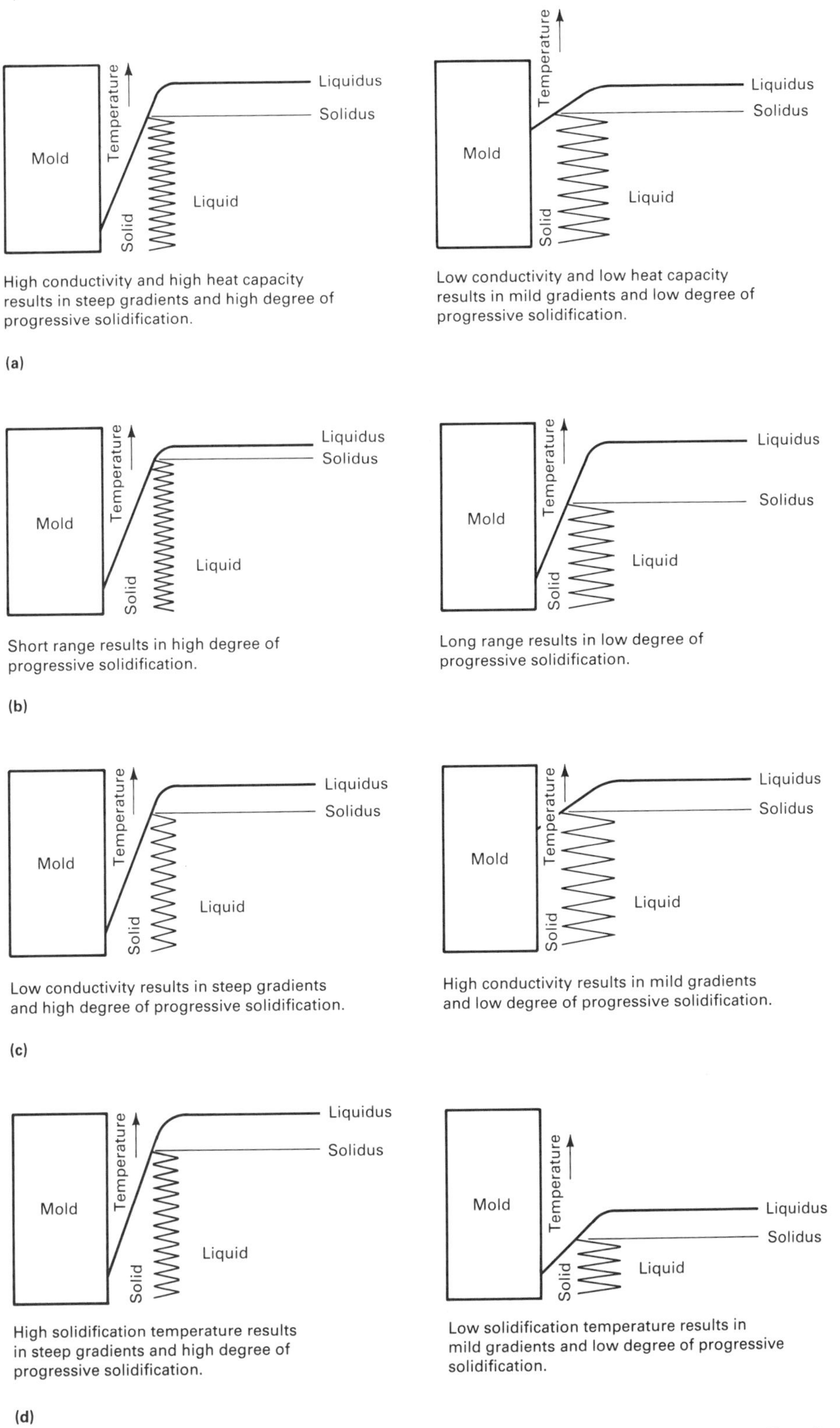
Mold
Temperature
Solid
Liquidus
Solidus
Liquid
High conductivity and high heat capacity results in steep gradients and high degree of progressive solidification.
Low conductivity and low heat capacity results in mild gradients and low degree of progressive solidification.
(a)
Short range results in high degree of progressive solidification.
Long range results in low degree of progressive solidification.
(b)
Low conductivity results in steep gradients and high degree of progressive solidification.
High conductivity results in mild gradients and low degree of progressive solidification.
(c)
High solidification temperature results in steep gradients and high degree of progressive solidification.
Low solidification temperature results in mild gradients and low degree of progressive solidification.
(d)

Fig. 11 Schematic of the effect of mold and metal variables on progressive solidification. (a) Effect of mold conductivity on solidifying metal. (b) Effect of liquidus-to-solidus range of solidifying metal. (c) Effect of conductivity on solidifying metal. (d) Effect of temperature level on solidification. Source: Ref 12

through which feed metal can move, centerline shrinkage will occur. Once the feeding distance of a section has been exceeded, the development of centerline shrinkage will not be overcome by increasing the size of the riser.

It should also be noted that the greatest amount of data on feeding distance is available for carbon steel. A variety of nomograms and tables have been widely used for decades (Ref 13-16). For most other alloys, precise data are not available, so their feeding distances are often characterized by their similarity (or lack of similarity) to carbon steel. One method of making this comparison is by the calculation of centerline feeding resistance (Ref 6). This measurement indicates that some alloys (for example, Monel) will have feeding distances very much like those of carbon steel, and the methods engineer can use feeding distance tables and nomograms established for the latter.

Some alloys, such as 18-8 steel, 12% Cr steel, 99.8% Cu, and 60-40 brass, will have greater feeding distances in similar casting sections, so a multiplier can be applied to steel-base nomograms. In other alloys, such as 88-10-2 bronze, 85-5-5-5 bronze, Al-8Mg, and Al-4.5Cu, the centerline feeding resistance is so great that feeding distance is virtually nonexistent unless severe chilling is used.

Finally, there are the graphitic cast irons. In these materials, the crystallization of the low-density graphite as flakes or nodules should promote self-feeding behavior in the solidifying casting and allow infinite feeding distances as long as the early liquid feed demand of the casting is satisfied by the gating system or by a riser.

Very large gray and ductile iron castings are often produced to meet radiographic or ultrasonic soundness standards with minimal risering. The key to such practice is a rigid molding medium to minimize mold wall movement.

Castings With Sections of Varying Thicknesses. Most commercially produced castings consist of sections of varying thickness and configuration. Thicker, more slowly solidifying sections are separated from each other by thinner, rapidly solidifying sections. The heavier sections will then act as risers, supplying the feed metal demands of the lighter sections.

The selection of a risering method changes from a problem of riser spacing to one of riser placement so that each of the late-solidifying sections has its feed requirements met. Therefore, the methods engineer must divide a casting into sections requiring risers by determining the feeding paths by which solidification will move directionally from early- to late-solidifying sections (Ref 17). These feed paths can often be manipulated by proper application of chilling or insulating materials to minimize risering needs.

Several methods of risering isolated heavy sections are shown in Fig. 16, using a casting with two heavy sections joined by a thinner connection. In Fig. 16(a), with no risers, shrinkage develops in the two heavy sections. When an adequate riser is applied to one side (Fig. 16b), shrinkage remains

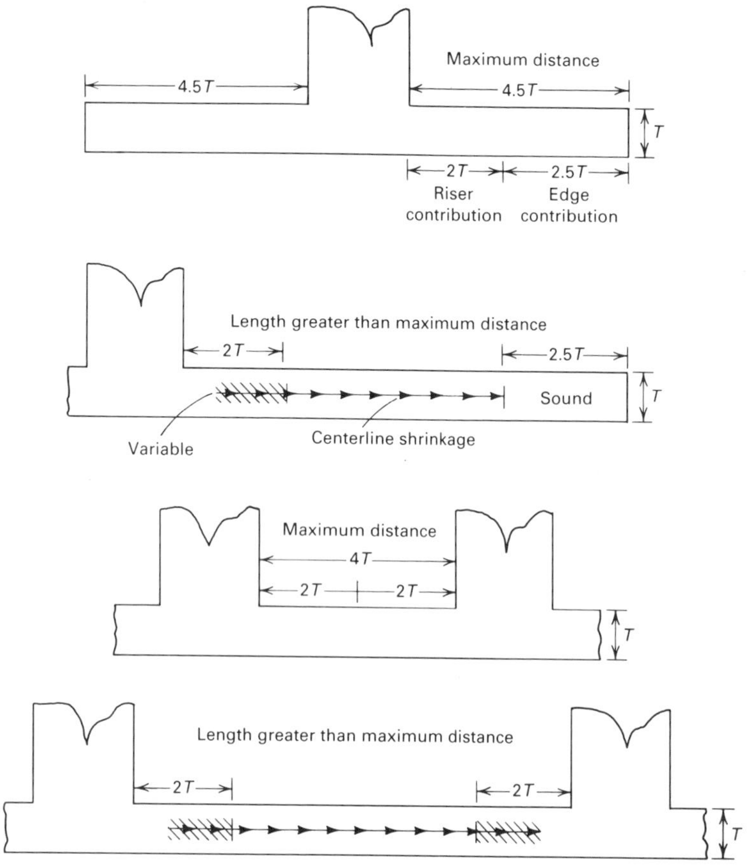

Fig. 12 Feeding distance relationships in steel plates (section width greater than 3T, where T = thickness). Source: Ref 13

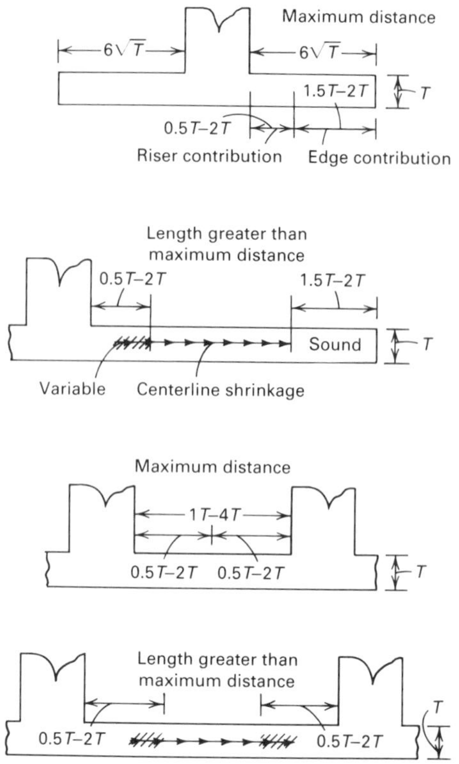

Fig. 13 Feeding distance relationships in steel bars (section width equal to thickness, T). Source: Ref 13

unfed in the other heavy section, or hot spot, because the connecting section freezes first.

A simple solution is to use risers on both sides (Fig. 16c). Two feed paths are thus established, running from the center section outward to the two risers.

Two alternative methods are shown by which a single feed path can be generated. In Fig. 16(d), a chill is applied to the isolated section to reduce its solidification time below that of the connection. In Fig. 16(e), the solidification time of the connection is extended by applying an insulating or exothermic pad to the casting walls.

Duration of Liquid Feed Metal Availability

A variety of methods have been devised to calculate the riser size needed to ensure that liquid feed metal will be available for as long as the solidifying casting requires. Several of the most commonly used methods will be discussed briefly.

Shape Factor Method. Drawing on the theoretical work of Caine (Ref 18), researchers at the U.S. Naval Research Laboratory (NRL) devised a method to determine riser size by calculating a shape factor by adding the length and width of a casting section and dividing this sum by the section thickness (Ref 19). The NRL method is illustrated in Fig. 17 with the example of a 508 mm (20 in.) square plate 50 mm (2 in.) thick. According to the earlier discussion of feeding distance, this casting (at least in steel) should be able to be made free of centerline shrinkage with a single top riser in the center of the casting. Figure 12 shows that the feeding distance of this 50 mm (2 in.) plate, with end effect, will be 229 mm (9 in.) from the edge of the riser. If the proposed central riser is at least 50 mm (2 in.) in diameter, the feeding distance requirements should be met.

Figure 17 shows the calculation of the shape factor and volume of the casting. In Fig. 17(a), one reads up from the shape factor (20) on the x-axis to intersect the NRL-developed curve. From this point, one reads across to the y-axis to find a minimum riser volume, V_r, needed to feed the casting volume, V_c. Figure 17(a) shows that, for this example, the required V_r/V_c ratio is 0.25; that is, the riser must be designed to contain a minimum of 25% as much metal as is required by the casting itself. To find the appropriate riser size, simple nomograms like the one shown in Fig. 17(b) are used; risers of various height-to-diameter ratios can be found to satisfy the calculated volume requirement. As shown in Fig. 17(b), most such nomograms assume a riser H/D ratio of around 1:1.

Geometric Method. Originally devised for determining side risers for malleable iron castings, the geometric method takes the typical conical shape of the shrinkage cavity formed in a riser as it simultaneously feeds metal into the casting and solidifies inward from its walls (Fig. 18) and simplifies it to the configuration of a cylindrical pipe (Ref 20). The size of this pipe depends on the weight of the casting section and the shrinkage percentage (both liquid and solidification shrinkages) of the casting alloy. Simple nomograms and tables are available to determine pipe sizes for a given alloy and casting weight and for a variety of ratios for the height of the pipe, H_p, divided by the diameter of the pipe, D_p.

As shown in Fig. 19, the actual riser size is arrived at by surrounding the pipe by a riser wall, W, that is expected to solidify at the same rate as the walls of the casting. The calculation of the required wall thickness will thus vary with casting configuration:

- For platelike castings, W = casting thickness
- For cube-shaped castings, W = 35% of the edge length of the cube
- For round or square bars, W = twice the bar diameter or thickness

The final riser diameter, D_r, thus equals $D_p + 2W$. The overall riser height, H_r, is determined by adding together:

- The pipe height, H_p

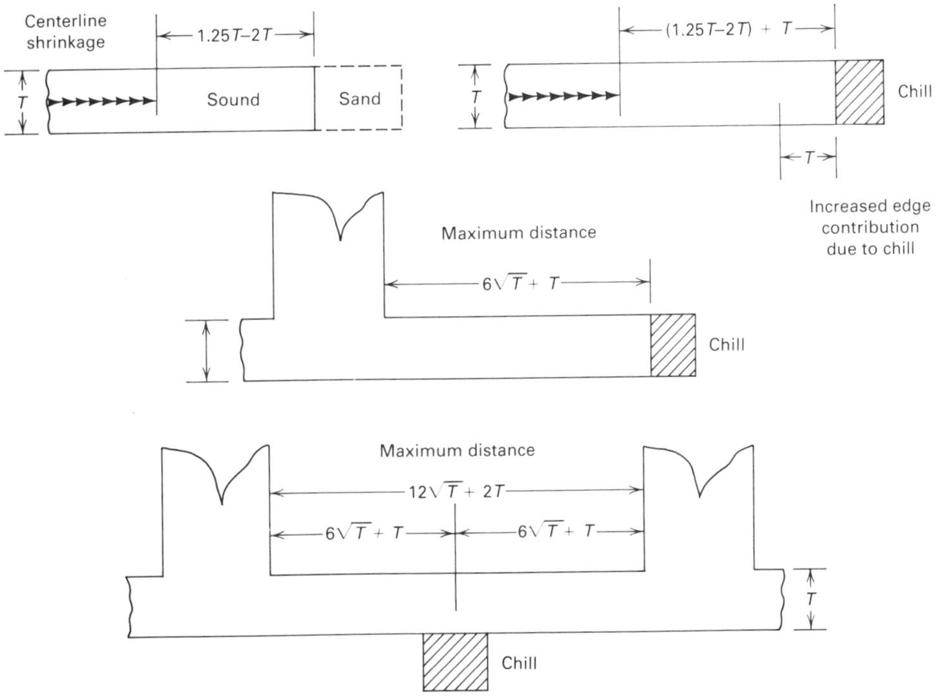

Fig. 14 Effect of chills on feeding distance relationships in steel bars. Source: Ref 13

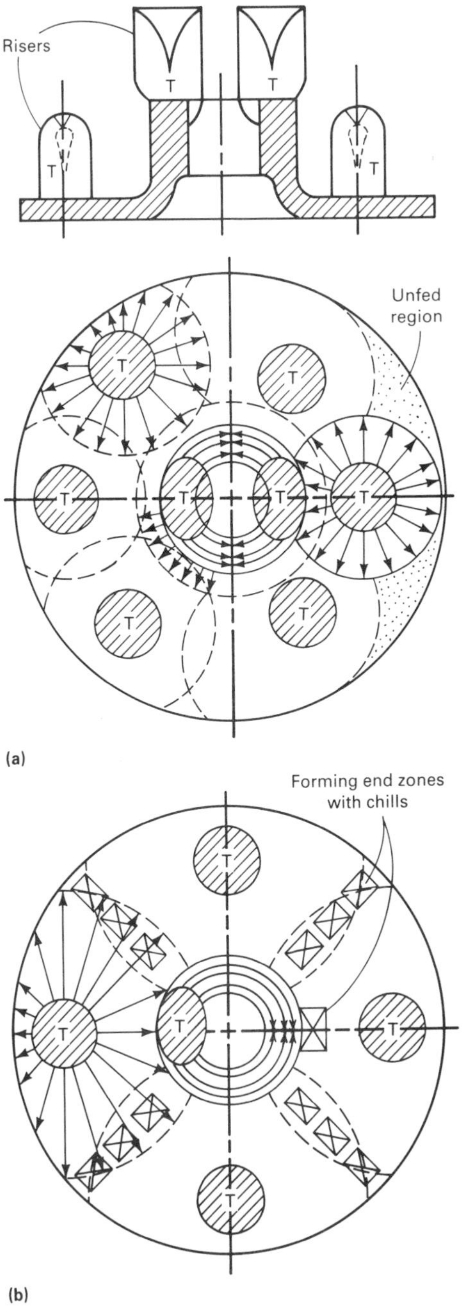

Fig. 15 Use of chills to reduce the number of risers (T) on a steel flange casting. (a) Side and top view of the casting illustrating locations of the eight risers used when the workpiece is divided into feeding areas without considering end effects. (b) Top view of identical casting showing locations of five risers used when the workpiece is divided into feeding areas in which riser effect and end effect considerations are accounted for through the use of chills. Source: Ref 14

- A middle section, H_m, above the riser connection with the casting (neck) to provide a safety margin
- A base section, H_b, including the depth of the riser neck

Casting yield is usually maximized by choosing a ratio for H_p/D_p of 2.5:1, generally resulting in a riser with an H/D ratio (above the neck) of about 1:1.

Design rules for the geometric method have been further developed to design tapered risers, as shown in Fig. 20. This design not only improves casting yield but also promotes the development of definite piping behavior in the riser, ensuring that metal will feed out from riser to casting.

The modulus method is based on the concept that the freezing time of a casting or a casting section can be approximated by using Chvorinov's rule:

$$t = k^2 \left(\frac{V_c}{A_c}\right)^2 = k^2 M_c^2 \quad \text{(Eq 1)}$$

where t is the freezing time of the casting, V_c is the volume of the casting, A_c is the surface area of the casting, and k is the constant governed by metal and mold properties.

This concept was developed by Wlodawer for practical riser calculations by eliminating the need to calculate actual solidification times in favor of simply determining the relative solidification times of casting sections and risers (Ref 14). Chvorinov's rule is thus simplified to:

$$t \sim \frac{V_c}{A_c} \quad \text{(Eq 2)}$$

The volume-to-area ratio of the casting is termed the casting modulus, M_c:

$$M_c = \frac{V_c}{A_c} \quad \text{(Eq 3)}$$

The freezing times of risers and castings are proportional to their respective moduli, and if the modulus of the riser, M_r, is sufficiently greater than the modulus of the casting, M_c, good feeding will be obtained. In steel, if $M_r = 1.2 \cdot M_c$, feeding will be satisfactory.

For other skin-forming alloys, including many aluminum- and copper-base alloys, the M_r/M_c ratio of 1.2:1 is appropriate. With gray and ductile irons, depending on the carbon equivalent, the required M_r/M_c ratio can range from 0.8:1 to 1.2:1 because the riser may not be required to supply feed metal throughout the entire solidification time of the casting.

Wlodawer simplified the modulus method by showing that many casting sections can be reduced to simple geometric shapes for which M_c can be readily found without elaborate calculations of actual surface areas and volumes (for example, for a plate section, M_c = half the plate thickness). Wlodawer further simplified the method by providing simple conversion charts, such as that shown in Fig. 21. This chart allows easy calculation of risers of various shapes if the necessary modulus is known.

One note of caution must be kept in mind when using the modulus approach to riser design. This method can recommend risers that are too small on very rangy, thin-section castings. In such cases, the thermal requirements of the casting would indicate very small risers, but the demand for feed metal volume can still be substantial. For such volume-controlled castings, the riser size arrived at through modulus calculations must be verified against riser volume charts such as that given in Table 2.

Computerized Methods. Over the past decade, a variety of computer programs have become available to assist in riser design.

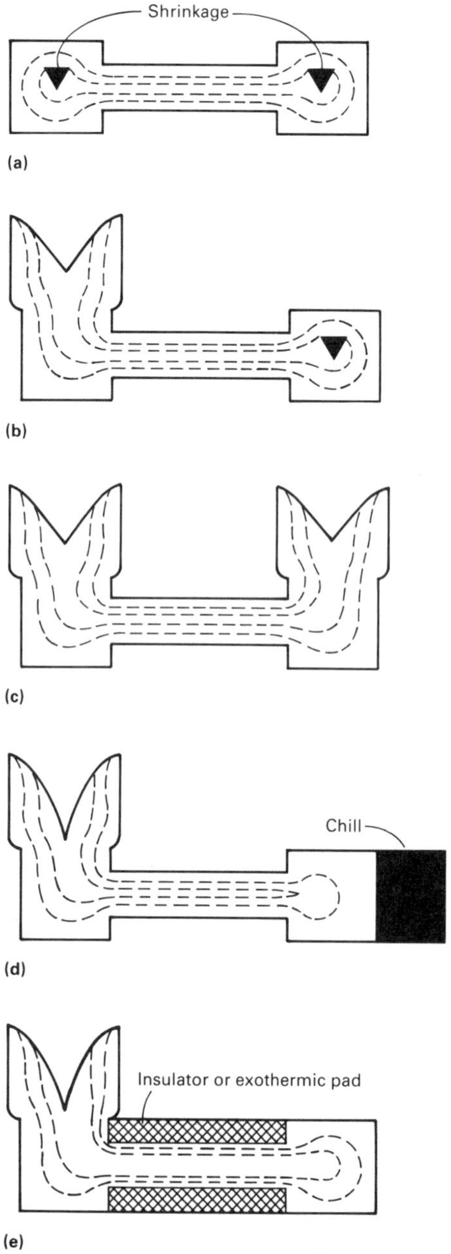

Fig. 16 Risering of isolated heavy sections joined by a thinner section to minimize shrinkage and number of risers. (a) Workpiece with no risers. (b) Riser added to one side. (c) Risers located on both ends. (d) Chill applied to one end and riser to other end. (e) Riser used on one end and insulator or exothermic pad on opposite end

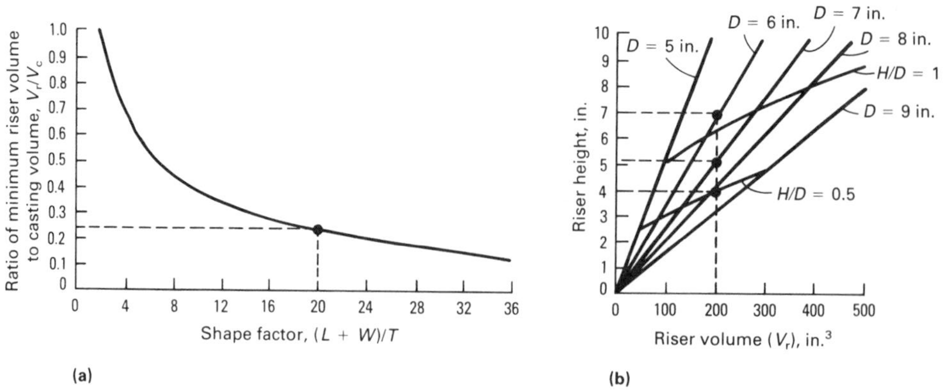

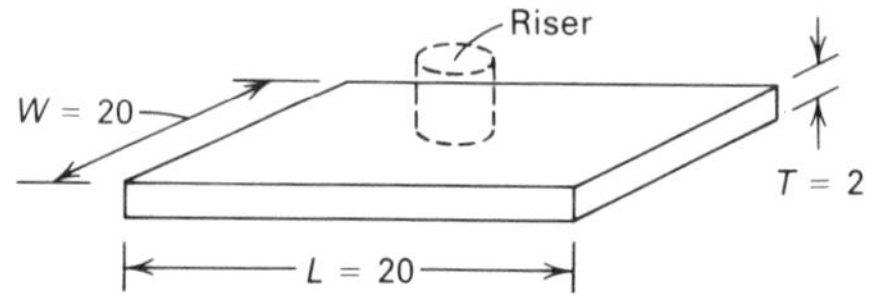

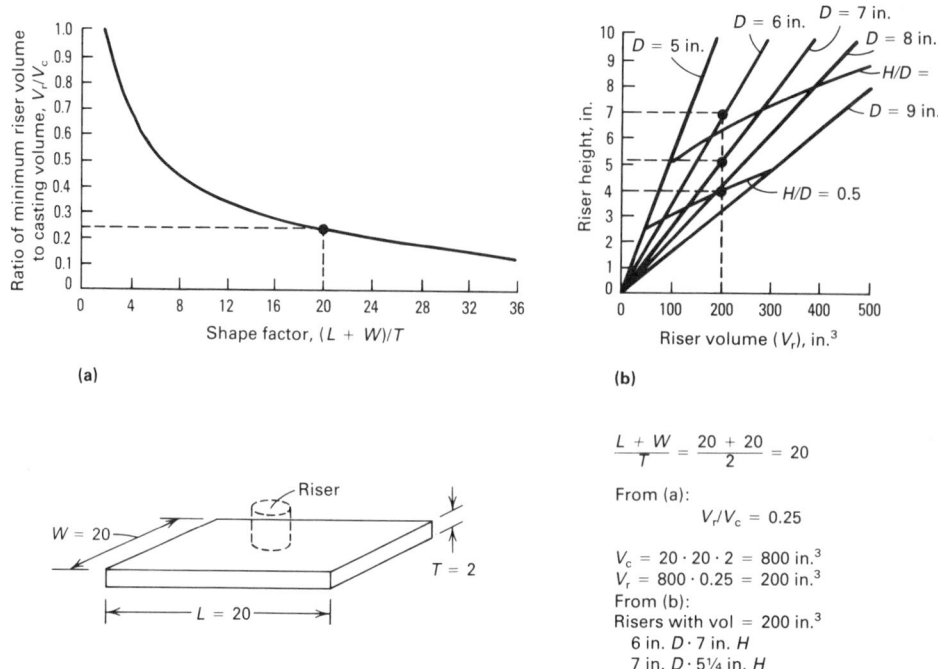

$$\frac{L + W}{T} = \frac{20 + 20}{2} = 20$$

From (a):

$V_r/V_c = 0.25$

$V_c = 20 \cdot 20 \cdot 2 = 800$ in.^{3}
$V_r = 800 \cdot 0.25 = 200$ in.^{3}
From (b):
Risers with vol = 200 in.^{3}
6 in. $D \cdot 7$ in. H
7 in. $D \cdot 5\frac{1}{4}$ in. H
8 in. $D \cdot 4$ in. H

Fig. 17 Procedure for determining a minimum riser size using the shape factor method. (a) The shape factor, derived from the dimensions of the casting, is used to determine the minimum ratio of riser volume to casting volume. (b) The same dimensions used to obtain the shape factor provide the data for casting volume, V_c, from which riser height and riser diameter can be determined. Dimensions given in inches. Source: Ref 19

These can be grouped into two general categories.

The first category includes programs that generate recommendations for riser sizes (Ref 22-26). These programs are generally run on microcomputers, and the basis of the calculation is often one or more of the manual calculation methods discussed above. These programs usually contain subroutines to assist in calculating casting weights, section modulus, and feeding distance. Riser calculations generally require simple inputs of such factors as section weight and shape, shrinkage percentage, section modulus, mold medium, ingate location, and desired riser shape for the program to provide a variety of riser alternatives.

The second category comprises programs that simulate solidification to predict riser performance (Ref 27-30). These generally use heat transfer calculations to simulate the progress of solidification in a casting/riser combination. Some are limited to a two-dimensional analysis, while others provide three-dimensional analysis. The latter can require considerable computing power. These programs seldom provide recommendations for riser sizes, so riser sizes must be determined by other methods. As shown in Fig. 22, these programs are often used to

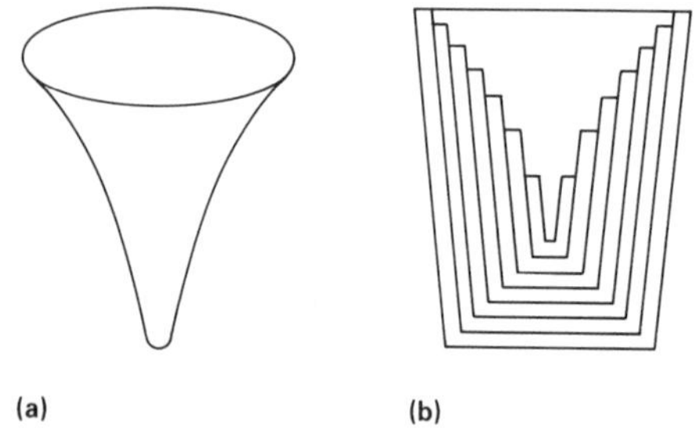

Fig. 18 Formation of the conical type of shrinkage cavity (a) due to the accumulation of solidified layers on the outer walls of the riser (b)

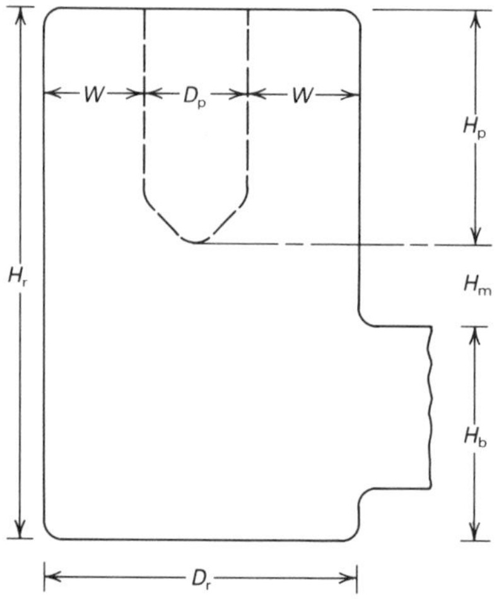

Fig. 19 Cross section of a piping side riser designed by the geometric method

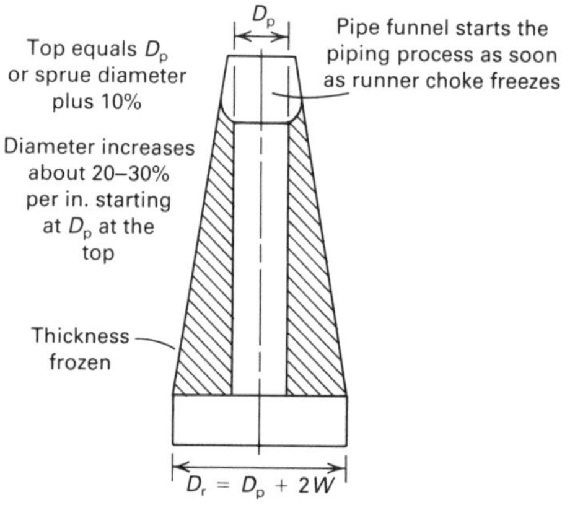

Fig. 20 Example of a design system for tapered tall risers. Source: Ref 21

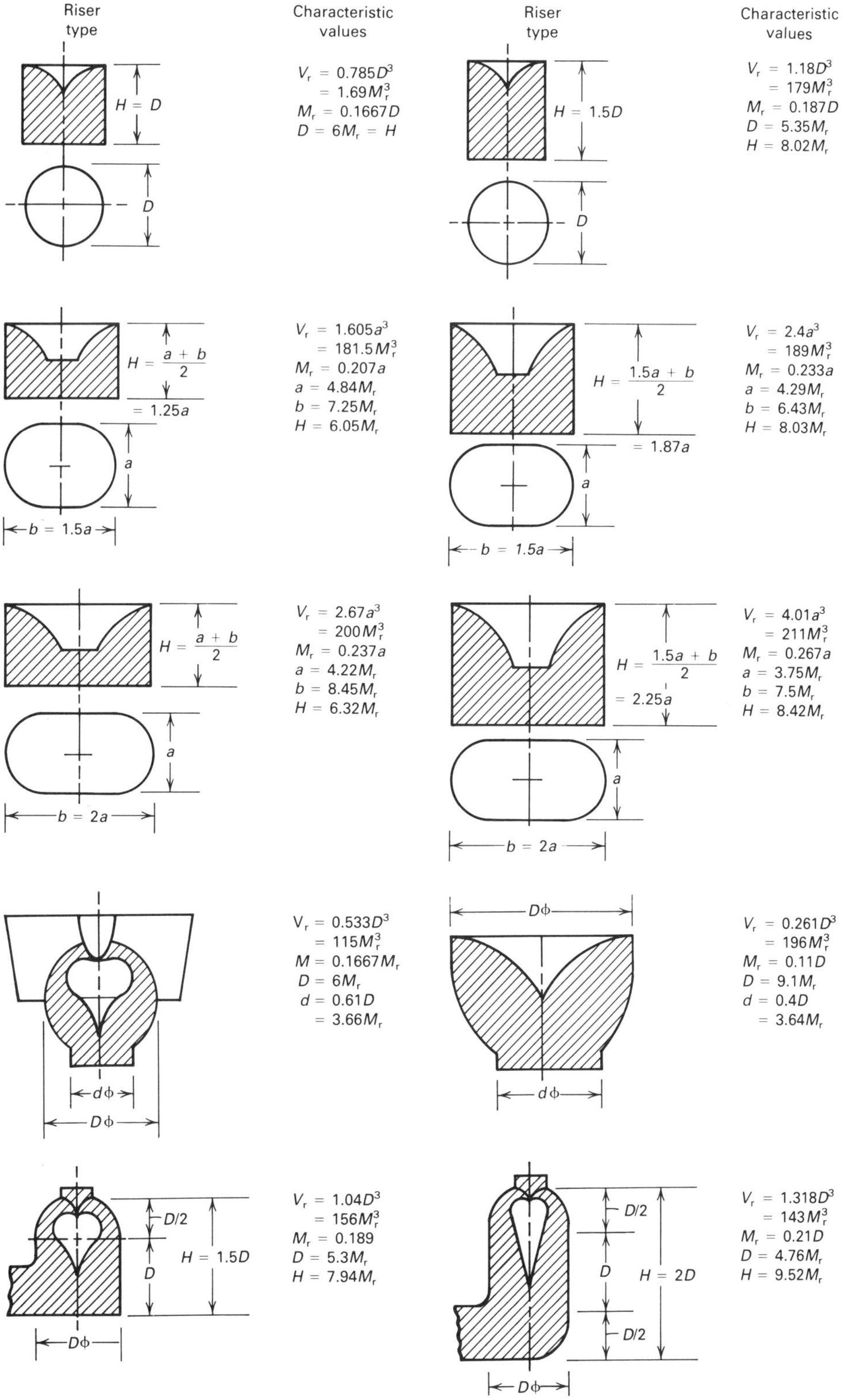

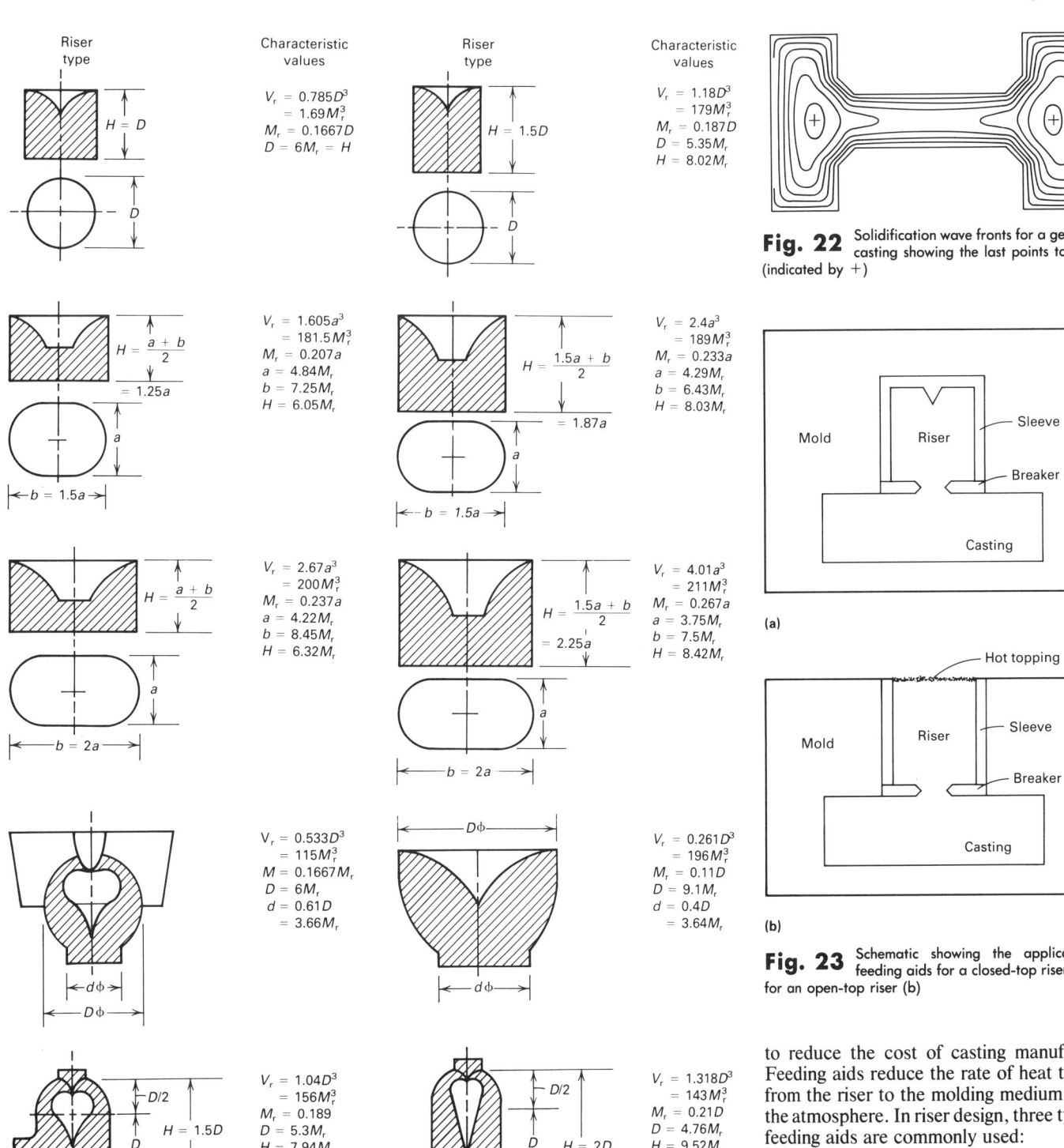

Fig. 21 Riser configurations and their characteristic values (M_r, V_r, D, H, and so on). Source: Ref 14

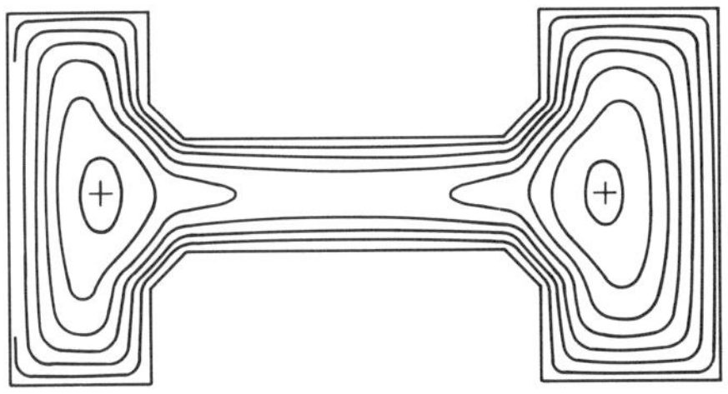

Fig. 22 Solidification wave fronts for a gear blank casting showing the last points to solidify (indicated by +)

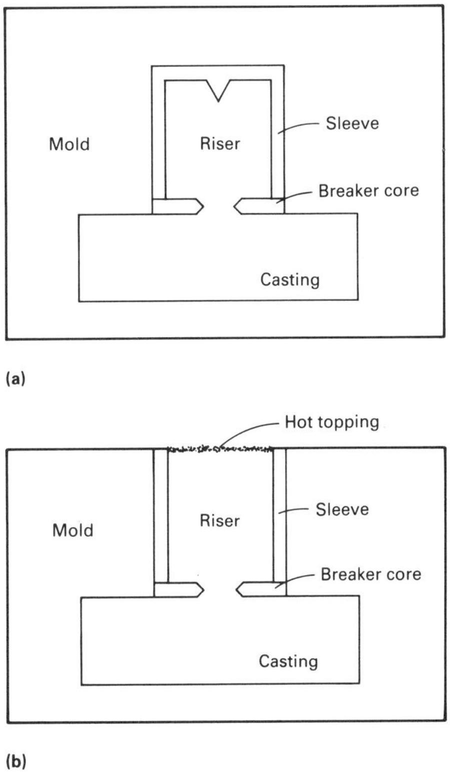

Fig. 23 Schematic showing the application of feeding aids for a closed-top riser (a) and for an open-top riser (b)

analyze the solidification of a casting to predict the location of the last points to solidify, so that risers (or chills, and so on) can be placed to eliminate the expected shrinkage in these locations.

Feeding Aids Used in Riser Design

Feeding aids are widely used by the foundry industry to increase casting soundness and to reduce the cost of casting manufacture. Feeding aids reduce the rate of heat transfer from the riser to the molding medium and to the atmosphere. In riser design, three types of feeding aids are commonly used:

- Riser sleeves or panels are used to insulate the riser sidewall or riser top from the mold
- Topping compounds are used to insulate the top of open risers from the atmosphere
- Breaker cores are used between the riser and the casting to facilitate the removal of the riser from the casting; the use of these materials is illustrated in Fig. 23

Metal is transferred from the riser by three mechanisms: gravity, atmospheric pressure (or pressure applied by other means, as in die-casting), and capillary ac-

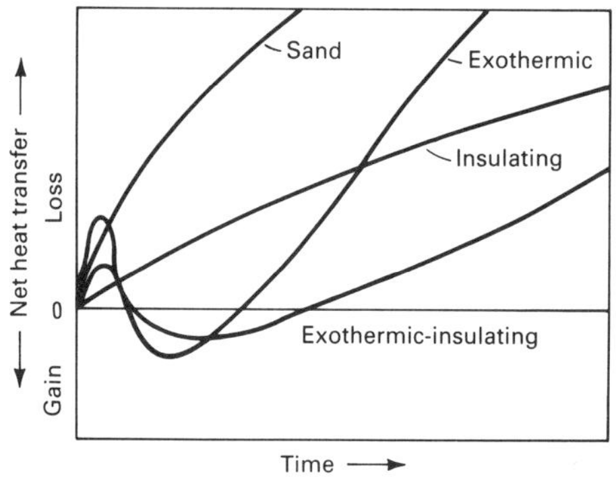

Fig. 24 Heat loss from a riser as a function of time for the major classifications of feeding aids, with sand included as the basis for comparison

Table 3 Effect of risering aids on the solidification times of various alloys

Alloy cast	Radiation loss through top, %	Solidification time, min: Sand riser/ open end	Sleeved riser/ open top	Sand riser/ insulated top	Sleeved riser/ insulated top
		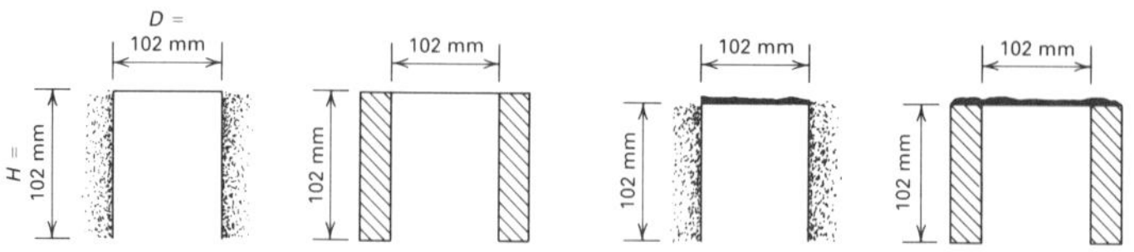			
Steel	42	5	7.5	13.4	43.0
Copper	26	8.2	15.1	14.0	45.0
Aluminum	8	12.3	31.1	14.3	45.6

Source: Ref 32

tion. For most casting processes, atmospheric pressure is the most important.

Feeding Aid Advantages

Feeding aids delay solidification and the formation of a skin of solid metal in the riser. This promotes atmospheric puncture of the riser, assisting in the transfer of metal from the riser to the casting. Alloys, specifically those that have a wide freezing range, benefit the most from feeding aids.

Feeding aids increase the temperature gradient between riser and casting. As a result, directional solidification from casting to riser is promoted; that is, a stronger feeding path is established.

Feeding aids, by reducing the rate of heat transfer between the riser and the mold, increase the effective modulus of a riser when compared with the geometric modulus of that riser. For a given riser modulus, a riser incorporating feeding aids will be smaller than an uninsulated riser. Less liquid metal is needed to produce a given casting, and the cost of producing that casting is reduced (Ref 31). Reducing riser size may allow a given casting to be produced in a smaller mold or may allow the production of more castings in a given mold, both resulting in lower cost.

The value of feeding aids in delaying riser solidification is indicated in Table 3, in which the solidification time of a 102 mm (4 in.) diam, 102 mm (4 in.) high riser is calculated in three alloys with various combinations of insulation on the riser sidewall and top. Also given is the percentage of total heat lost from the solidifying riser through radiation from the riser top. This highlights the fact that hot topping to reduce radiative heat loss is significantly more important in steel casting than in aluminum casting.

Thermal Properties of Materials. Riser sleeves and topping compounds are classified by thermal properties as exothermic, insulating, and exothermic-insulating. The generalized thermal properties of each of these classifications are shown in Fig. 24.

Exothermic feeding aids are based on the oxidation of aluminum to produce heat. These feeding aids tend to be of relatively high density, and the matrix has thermal properties similar to those of a sand mold after the exothermic reaction has subsided. They display an initial chilling effect on the molten metal and rely on a strong exothermic reaction to remelt any metal that may have solidified. Exothermic materials tend to be used on small and intermediate-size risers and are not generally recommended for large risers having a long solidification time. Exothermic materials also have the greatest likelihood of metal contamination.

Metal-producing topping compounds are a special class of exothermic material, and they have a composition based on the thermite reaction. These materials are sometimes used on large risers. Contamination must be carefully monitored; the mctal produced by the thermite reaction should not be allowed to feed into the casting.

Insulating feeding aids are formulated from refractory materials to have a low density. They exhibit a very low initial chill of the molten metal and rely on their low density to minimize heat loss from the riser. Insulating feeding aids tend to be used on small to intermediate-size risers and alloys having lower pouring temperatures. They are not generally recommended for large risers, because these low-density materials thermally degrade when exposed to high pressure and high temperature for extended periods of time. Insulating feeding aids are least likely to cause problems with metal contamination.

Exothermic-insulating feeding aids consist of exothermic materials surrounded by a highly refractory, insulating matrix. These materials are the most versatile, having a low initial chill, extended exothermic reaction, and good insulation after the exothermic reaction has subsided. Exothermic-insulating materials are used for the entire range of riser sizes. Their propensity for metal contamination lies between insulating and exothermic materials.

Metal Contamination. Feeding aids are formulated to minimize the possibility of metal contamination. Depending on the formulation of the feeding aid and its method of application, certain elements may be absorbed by the metal in the riser. Carbon, silicon, aluminum, oxygen, nitrogen, and sulfur may be absorbed. The widespread use of feeding aids in riser design is an indication that contamination is not normally a problem, but the possibility should not be overlooked.

Factors in Riser Size

The thermal properties of feeding aids can be readily incorporated into the modulus method of riser calculation discussed above (Ref 33). Cylindrical risers are the most common because this shape has a high modulus for a given volume of metal and because this shape is easy to mold. The modulus of a cylindrical riser, M_r, is given by:

$$M_r = \frac{V_r}{A_r} = \frac{\pi R^2 H}{2\pi RH + \pi R^2} = \frac{RH}{2H + R} = \frac{DH}{4H + D} \qquad \text{(Eq 4)}$$

where R is the riser radius, D is the riser diameter, and H is the riser height.

Riser insulation in the form of a sleeve and hot topping may be regarded as effectively decreasing the surface area of the riser. The effect of the sidewall (sleeve) insulation may be represented by a factor x and that of a hot topping by a factor y, both relative to sand (where $x = 1$ for sand). The factors x and y have been termed apparent surface alteration factors (ASAF). The effective modulus M_r of a cylindrical riser incorporating feeding aids on both the sidewall and top is given by:

$$M_r = \frac{DH}{4Hx + Dy} \qquad \text{(Eq 5)}$$

The ASAF values of insulating and exothermic feeding aid materials vary and generally range from 0.50 to 0.90. The smaller the ASAF value, the more efficient the insulation. For example, if $x = 0.65$ and $y = 0.7$, then for a cylindrical riser with a height-to-diameter ratio of 1:1 (where $H = D$):

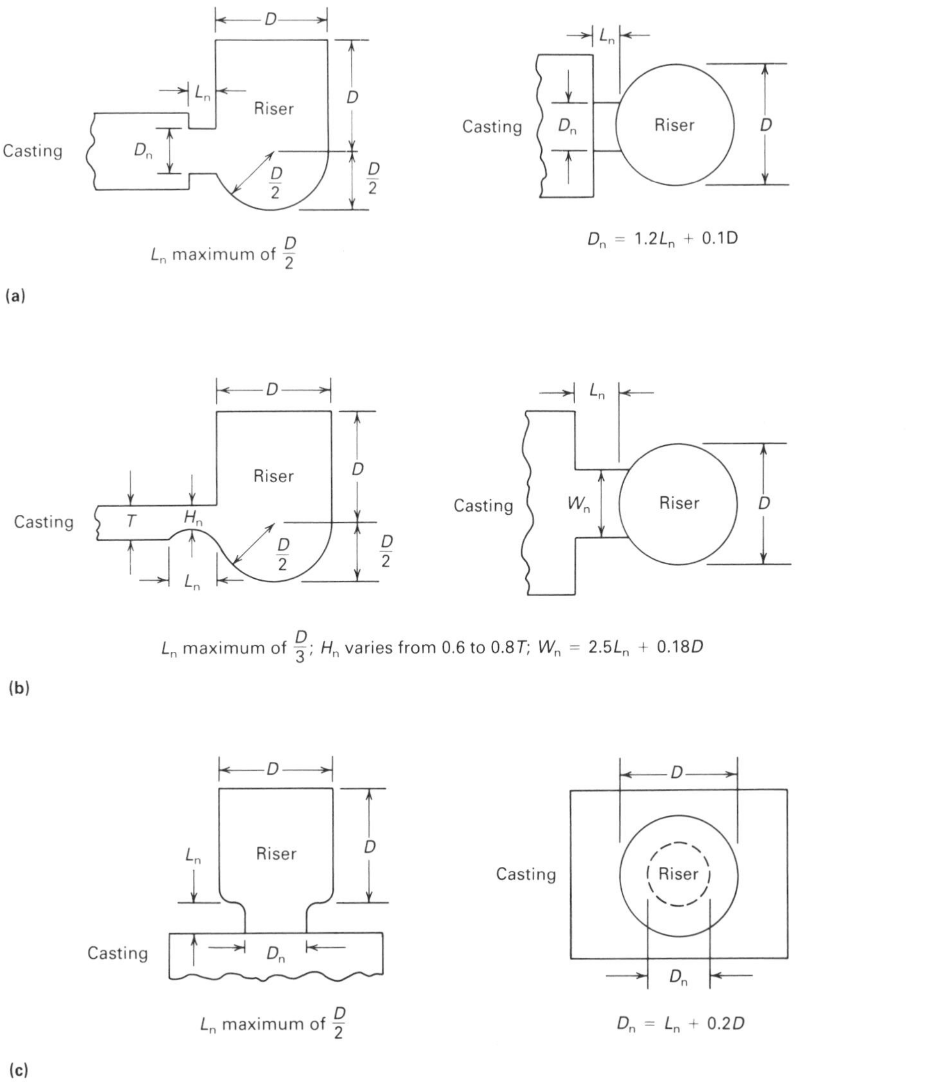

Fig. 25 General design rules for riser necks used in iron casting applications (side view and top view, respectively). (a) General type of side riser. (b) Side riser for plate casting. (c) Top round riser. Source: Ref 34

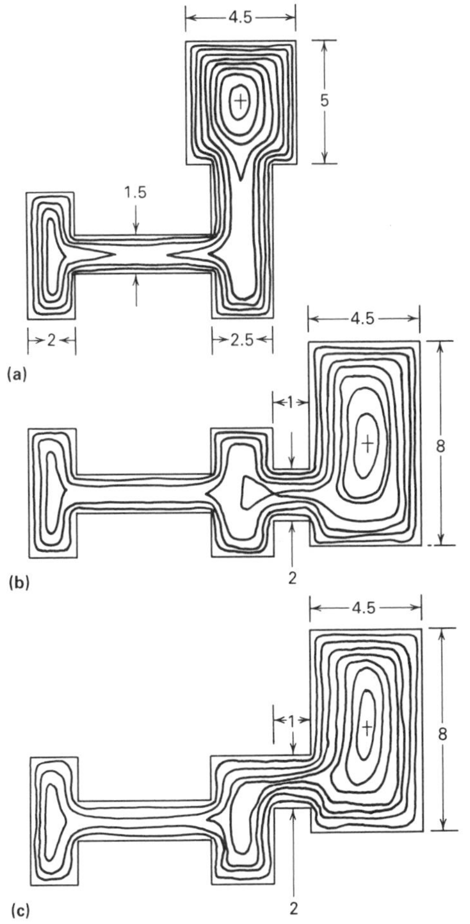

Fig. 26 Effect of risers and riser contacts on solidification wave fronts in the gear blank casting shown in Fig. 22. In (a) the side riser attached directly at rim center generates a hot spot inside the casting itself. The addition of a thin section between the workpiece and the riser in both (b) and (c) overcomes the problem of hot spots inside the casting. Dimensions given in inches

$$M_r = \frac{D^2}{4D\,(0.65) + D\,(0.7)} = \frac{D^2}{3.3D} = 0.303D \qquad \text{(Eq 6)}$$

This should be compared with the modulus of a similar 1:1 sand-lined riser, which is 0.2*D*. The ASAF values for feeding aids are generally available from manufacturers of these products. Although the size of a riser incorporating feeding aids can be readily calculated using the approach described above, computer programs are now widely available to speed the process of designing risers—both conventional sand-lined risers and risers incorporating feeding aids.

Riser Necks and Breaker Cores

To promote directional solidification from the casting into the riser, the modulus of the riser neck, M_n, should be intermediate to the moduli of the casting and riser, M_c and M_r (Ref 14). The general rule given above for the required riser modulus was $M_r = 1.2\ M_c$.

The general rule for riser neck design—at least for skin-forming alloys—is $M_n = 1.1\ M_c$. Once again, the graphitic cast irons are an exception because the graphite expansion phase makes it unnecessary for the riser neck to stay open for feed metal transfer throughout the entire solidification time of the casting. Depending on the chemistry of the casting, riser necks for gray and ductile irons can have moduli in the range of 0.67 to 1.1 times the casting modulus (Ref 5). General design rules for riser necks for iron castings are widely available (Fig. 25).

For the economical removal of risers, breaker cores can be used between the riser and the casting, as shown in Fig. 23. Breaker cores are typically made of bonded sand or fused ceramic materials. For risers lined with sleeves, the breaker core thickness is generally 10% of the riser diameter, and the breaker core opening is generally in the range of 40 to 50% of the riser diameter. By keeping the mass of the breaker core small, the breaker core quickly reaches the temperature of the surrounding metal and does not appreciably affect the solidification of the riser.

Optimum Riser and Neck Configurations

Applying the required riser to a specific casting may generate problems. For example, the necessary riser or neck size may not easily fit the casting configuration. An important problem may be created simply by attaching the riser, because the new riser/casting configuration will have its own solidification pattern, sometimes with unintended results.

This is illustrated in Fig. 26, in which the attachment of a side riser at the center of the rim of a gear blank casting generates a hot spot inside the casting, where shrinkage would be expected to occur after the neck freezes. As seen in Fig. 26, this problem can be avoided either by moving the attachment of the side riser or by using a top riser on the rim. The top riser has the added benefit of improving casting yield.

REFERENCES

1. H.F. Taylor, M.C. Flemings, and J. Wulff, *Foundry Engineering*, John Wiley & Sons, 1959
2. C.E. Bates and B. Patterson, Volumetric Changes Occurring During Freezing of Hypereutectic Ductile Irons, *Trans. AFS*, Vol 87, 1979, p 323-334
3. B.P. Winter, T.R. Ostrom, D.H. Hartman, P.K. Trojan, and R.D. Pehlke, Mold Dilation and Volumetric Shrinkage of White, Gray, and Ductile Cast Irons, *Trans. AFS*, Vol 92, 1984, p 551-560
4. R.W. Heine, The Fe-C-Si Solidification Diagram for Cast Irons, *Trans. AFS*, Vol 94, 1986
5. S.I. Karsay, *Ductile Iron III: Gating and Risering*, QIT—Fer Et Titane Inc., 1981
6. R.A. Flinn, *Fundamentals of Metal Casting*, Addison-Wesley, 1963
7. J. Briggs, Risering Gray and Ductile Iron, in *90% Yield Seminar: Gating and Risering of Iron Castings*, Foseco, Inc., 1983
8. B.P. Winter, R.D. Pehlke, and P.K. Trojan, Volumetric Shrinkage and Gap Formation During Solidification of Copper-Base Alloys, *Trans. AFS*, Vol 91, 1983, p 81-88
9. R.W. Ruddle, Influence of Feeding Practice on Energy Conservation and Production Economics in Steel Foundries, *Br. Foundryman*, Sept 1978, p 197-222
10. J.L. Francis and P.G.A. Pardoe, The Feeding of Iron Castings, in *Applied Science in the Casting of Metals*, K. Strauss, Ed., Pergamon Press, 1970
11. R.W. Ruddle, Solidification of Copper Alloys, *Trans. AFS*, Vol 68, 1960, p 685-690
12. R.W. Heine, C.R. Loper, Jr., and P.C. Rosenthal, *Principles of Metal Casting*, McGraw-Hill, 1967
13. W.S. Pellini, Factors Which Determine Riser Adequacy and Feeding Range, *Trans. AFS*, Vol 61, 1953, p 61-80
14. R. Wlodawer, *Directional Solidification of Steel Castings*, Pergamon Press, 1966
15. H.F. Bishop and W.S. Pellini, The Contribution of Riser and Chill-Edge Effects to Soundness of Cast Steel Plates, *Trans. AFS*, Vol 58, 1950, p 185-197
16. E.T. Myskowski, H.F. Bishop, and W.S. Pellini, Feeding Range of Joined Sections, *Trans. AFS*, Vol 61, 1953, p 302-308
17. R.W. Heine, Feeding Paths for Risering Castings, *Trans. AFS*, Vol 76, 1968, p 463-469
18. J.B. Caine, A Theoretical Approach to the Problem of Dimensioning Risers, *Trans. AFS*, Vol 56, 1948, p 492-501
19. E.T. Myskowski, H.F. Bishop, and W.S. Pellini, A Simplified Method of Determining Riser Dimensions, *Trans. AFS*, Vol 63, 1955, p 271-281
20. R.W. Heine, Riser Design for Mold Dilation, *Mod. Cast.*, Feb 1965
21. R.W. Heine, Design Method for Tapered Riser Feeding of Ductile Iron Castings in Green Sand, *Trans. AFS*, Vol 90, 1982, p 147-158
22. R.W. Ruddle and A.L. Suschil, Riser Sizing by Microcomputer, in *Modeling of Casting and Welding Processes II, 1983 Engineering Foundation Conference*, Conference Proceedings, The Metallurgical Society, 1983, p 403-419
23. C.F. Corbett, Methoding of Castings Using a Microcomputer, *Br. Foundryman*, Aug 1983, p 67-78
24. G.L. Moffat, FEEDERCALC—The Development of a Micro-Computer Program to Determine Optimum Feeder Sizes for Ductile Iron Castings, *Foundry Pract.*, No. 210, Jan 1985, p 3-8
25. J.E. Pickin, The Development of a Computer Assisted Feeding System, in *Proceedings of 1981 Annual Conference*, Steel Castings Research and Trade Association, 1981
26. W.T. Adams and K.W. Murphy, Optimum Full Contact Top Risers to Avoid Severs Under Riser Chemical Segregation in Steel Casting, *Trans. AFS*, Vol 88, 1980, p 389-404
27. R.W. Heine and J.J. Uicker, Risering by Computer Assisted Geometric Modeling, *Trans. AFS*, Vol 91, 1983, p 127-136
28. E. Niyama, T. Uchida, M. Morikawa, and S. Saito, A Method of Shrinkage Prediction and Its Application to Steel Casting Practice, in *Proceedings of the 49th International Foundry Conference*, International Committee of Foundry Technical Associations, 1982
29. C.F. Corbett, Computer Aided Thermal Analysis and Solidification, *Br. Foundryman*, Oct 1987, p 380-389
30. M.K. Walther, Experimental Verification of C.A.S.T., *Trans. AFS*, Vol 95, 1987, p 15-24
31. R.W. Ruddle, Profit Through Steel Risering Technology, *Trans. AFS*, Vol 87, 1979, p 423-432
32. C.M. Adams, Jr., and H.F. Taylor, Fundamentals of Riser Behavior, *Trans. AFS*, Vol 61, 1953, p 686
33. R.W. Ruddle, "Risering of Steel Castings," Foseco, Inc., 1979
34. J.F. Wallace and E.B. Evans, Risering of Gray Iron Castings, *Trans. AFS*, Vol 66, 1958, p 49

Gating Design

Anthony L. Suschil and Lee A. Plutshack, Foseco, Inc.

A GATING SYSTEM is the conduit network through which liquid metal enters a mold and flows to fill the mold cavity, where the metal can then solidify to form the desired casting shape. The basic components of a simple gating system for a horizontally parted mold are shown in Fig. 1. A pouring cup or a pouring basin provides an opening for the introduction of metal from a pouring device. A sprue carries the liquid metal down to join one or more runners, which distribute the metal throughout the mold until it can enter the casting cavity through ingates.

Design Variables

Methods used to promote any of the desirable design considerations discussed below often conflict with another desired effect. For example, attempts to fill a mold rapidly can result in metal velocities that promote mold erosion. As a result, any gating system will generally be a compromise among conflicting design considerations, with the relative importance of the consideration being determined by the specific casting and its molding and pouring conditions.

Rapid mold filling can be important for several reasons. Especially with thin-section castings, heat loss from the liquid metal during mold filling may result in premature freezing, producing surface defects (for example, cold laps) or incompletely filled sections (misruns). Superheating of the molten metal will increase fluidity and retard freezing, but excessive superheat can increase problems of gas pickup by the molten metal and exaggerate the thermal degradation of the mold medium. In addition, the mold filling time should be kept shorter than the mold producing time of the molding equipment to maximize productivity.

Minimizing Turbulence. Turbulent filling and flow in the gating system and mold cavity can increase mechanical and thermal attack on the mold. More important, turbulence may produce casting defects by promoting the entrainment of gases into the flowing metal. These gases may by themselves become defects (for example bubbles), or they may produce dross or inclusions by reacting with the liquid metal.

Turbulent flow increases the surface area of the liquid metal exposed to air within the gating system. The susceptibility of different casting alloys to oxidation varies considerably. For those alloys that are highly sensitive to oxidation, such as aluminum alloys; magnesium alloys; and silicon, aluminum, and manganese bronzes, turbulence can generate extensive oxide films that will be churned into the flowing metal, often causing unacceptable defects.

Avoiding Mold and Core Erosion. High flow velocity or improperly directed flow against a mold or core surface may produce defective castings by eroding the mold surface (thus enlarging the mold cavity) and by entraining the dislodged particles of the mold to produce inclusions in the casting.

Removing Slag, Dross, and Inclusions. This factor includes materials that may be introduced from outside the mold (for example, furnace slags and ladle refractories) and those that may be generated inside the system. Methods can be incorporated into the gating system to trap such particles (for example, filters) or to allow them time to float out of the metal stream before entering the mold cavity.

Promoting Favorable Thermal Gradients. Because the last metal to enter the mold cavity will generally be the hottest, it is usually desirable to introduce metal in those parts of the casting that would already be expected to be the last to solidify. One obvious method of accomplishing this is to direct the metal flow from the gating system into a riser, from which it then enters the mold cavity. Because the riser is generally designed to be the last part of the riser/casting system to solidify, such a gating arrangement will help promote directional solidification from the casting to the riser.

If the gating system cannot be designed to promote some desirable thermal gradient, it should at least be designed so that it will not produce unfavorable gradients. This will often involve introducing metal into the mold cavity through multiple ingates so that no one location becomes a hot spot.

Maximizing Yield. A variety of unrecoverable costs must go into the metal that will fill the gating system and risers. These components must then be removed from the casting and generally returned for remelt,

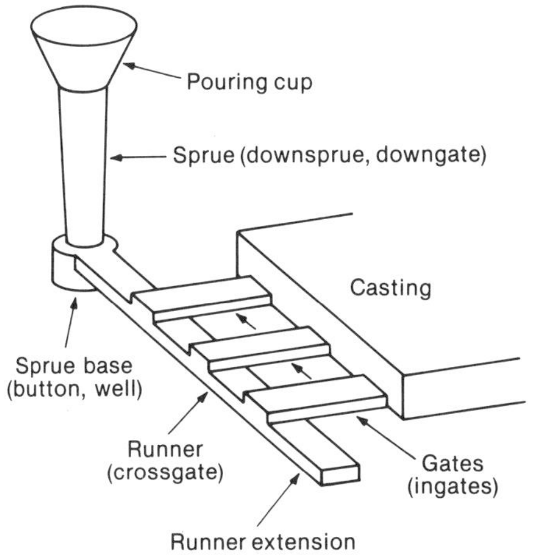

Fig. 1 Basic components of a simple gating system for a horizontally parted mold

where their value is downgraded to that of scrap. Production costs can be significantly reduced by minimizing the amount of metal contained in the gating system. The production capacity of a foundry can also be enhanced by increasing the percentage of salable castings that can be produced from a given volume of melted metal.

Economical Gating Removal. Costs associated with the cleaning and finishing of castings can be reduced if the number and size of ingate connections with the casting can be minimized. Again, it may be advantageous to introduce metal into the mold cavity through a riser, because the riser neck can also serve as an ingate.

Avoiding casting distortion is especially important with rangy, thin-wall castings, in which uneven distribution of heat as the mold cavity is filled may produce undesirable solidification patterns that cause the casting to warp. In addition, the contraction of the gating system as it solidifies can pull on sections of the solidifying casting, resulting in hot tearing or distortion.

Compatibility With Existing Molding/Pouring Methods. Modern high-production molding machines and automated pouring systems often severely limit the flexibility allowed in locating and shaping the pouring cup and sprue for introducing metal into the mold. They also generally place definite limits on the rate at which metal can be poured.

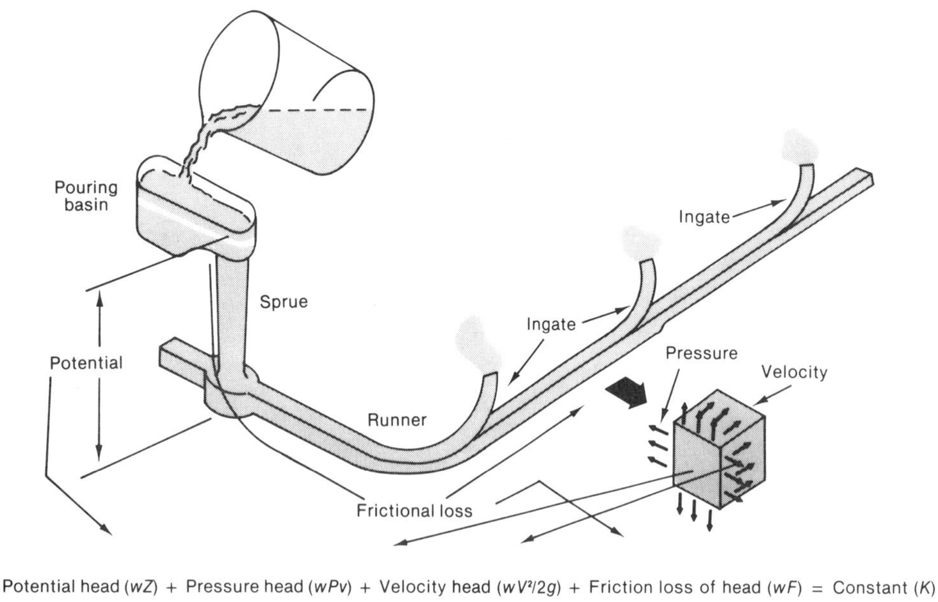

Fig. 2 Schematic illustrating the application of Bernoulli's theorem to a gating system. Source: Ref 1

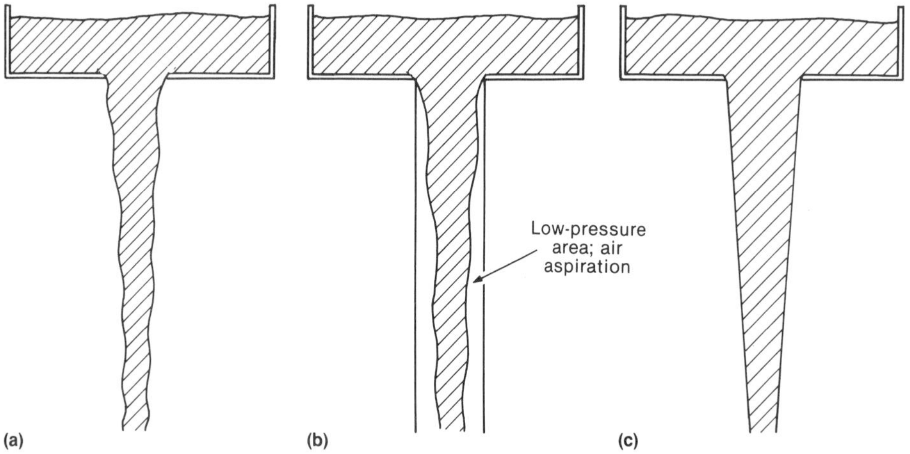

Fig. 3 Schematic showing the advantages of a tapered sprue over a straight-sided sprue. (a) Natural flow of a free-falling liquid. (b) Air aspiration induced by liquid flow in a straight-sided sprue. (c) Liquid flow in a tapered sprue

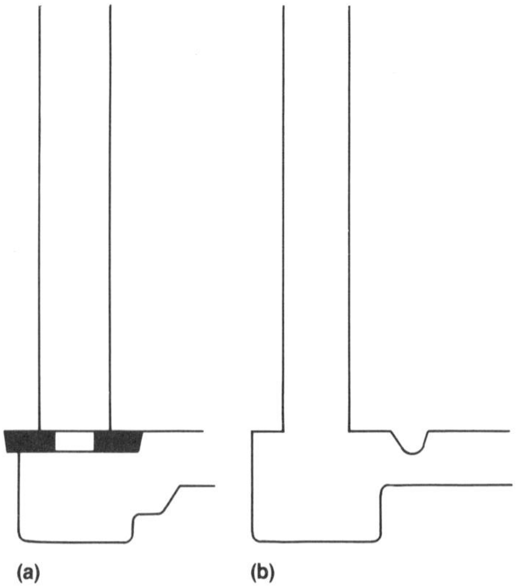

Fig. 4 Choke mechanisms incorporated into straight-sided sprues to approximate liquid flow in tapered sprues. (a) Choke core. (b) Runner choke

Controlled Flow Conditions. A steady flow rate of metal in the gating system should be established as soon as possible during mold filling, and the conditions of flow should be predictably consistent from one mold to the next.

Principles of Fluid Flow

Proper design of an optimized gating system will be made easier by the application of several fundamental principles of fluid flow. Chief among these principles are Bernoulli's theorem, the law of continuity, and the effect of momentum.

Bernoulli's Theorem

This basic law of hydraulics relates the pressure, velocity, and elevation along a line of flow in a way that can be applied to gating systems. The theorem states that, at any point in a full system, the sum of the potential energy, kinetic energy, pressure energy, and frictional energy of a flowing liquid is equal to a constant. The theorem can be expressed as:

$$wZ + wPv + \frac{wV^2}{2g} + wF = K \qquad \text{(Eq 1)}$$

where w is the total weight of the flowing liquid (in pounds), Z is the height of the liquid (in inches), P is the static pressure in liquid (in pounds per square inch), v is the specific volume of liquid (in cubic inches per pound), g is the acceleration due to gravity (386.4 in./s^2), V is the velocity (in inches per second), F is the friction loss per unit weight, and K is a constant.

If Eq 1 is divided by w, all the terms reduce to dimensions of length and will represent:

- Potential head Z
- Pressure head, Pv
- Velocity head $V^2/2g$
- Frictional loss of head F

Equation 1 allows prediction of the effect of the several variables at different points in the gating system, although several conditions inherent in foundry gating systems complicate and modify its strict application. For example:

- Equation 1 is for full systems, and at least at the start of pouring, a gating system is empty. This indicates that a gating system should be designed to establish as quickly as possible the flow conditions of a full system
- Equation 1 assumes an impermeable wall around the flowing metal. In sand foundry practice, the permeability of the mold medium can introduce problems, for example, air aspiration in the flowing liquid
- Additional energy losses due to turbulence or to friction (for example, because of changes in the direction of flow) must be accounted for
- Heat loss from the liquid metal is not considered, which will set a limit on the time over which flow can be maintained. Also, solidifying metal on the walls of the gating system components will alter their design while flow continues

Bernoulli's theorem (Eq 1) is illustrated schematically in Fig. 2, and several practical interpretations can be derived. The potential energy is obviously at a maximum at the highest point in the system, that is, the top of the pouring basin. As metal flows from the basin down the sprue, potential energy changes to kinetic energy as the stream increases in velocity because of gravity. As the sprue fills, a pressure head is developed.

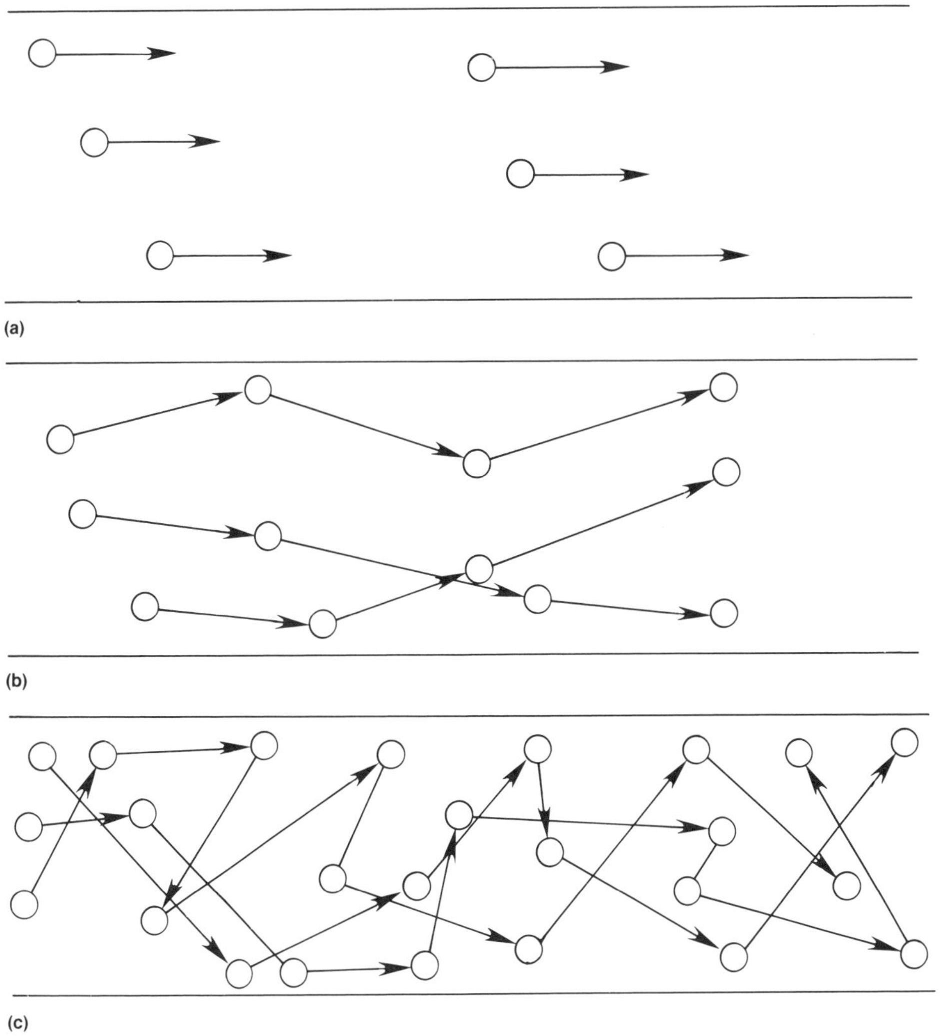

Fig. 5 Reynold's number, N_R, and its relationship to flow characterization. (a) $N_R < 2000$, laminar flow. (b) $2000 \leq N_R < 20\,000$, turbulent flow. (c) $N_R \geq 20\,000$, severe turbulent flow

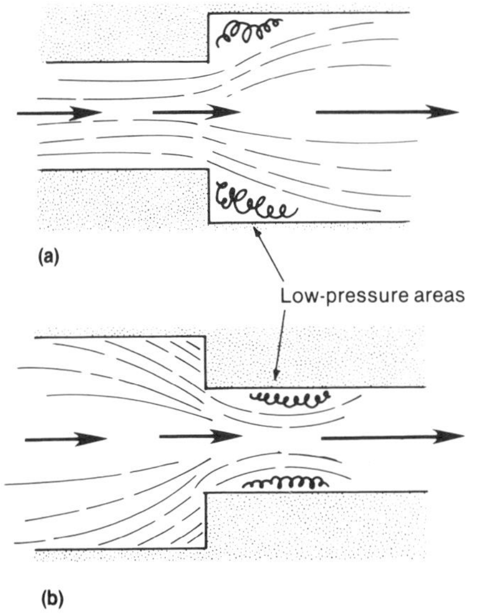

Fig. 6 Schematic showing the formation of low-pressure areas due to abrupt changes in the cross section of a flow channel. (a) Sudden enlargement of the channel. (b) Sudden reduction of the channel

Once flow in a filled system is established, the potential and frictional heads become virtually constant, so conditions within the gating system are determined by the interplay of the remaining factors. The velocity is high where the pressure is low, and vice versa.

The Law of Continuity

This law states that, for a system with impermeable walls and filled with an incompressible fluid, the rate of flow will be the same at all points in the system. This can be expressed as:

$$Q = A_1 v_1 = A_2 v_2 \qquad \text{(Eq 2)}$$

where Q is the rate of flow (in cubic inches per second), A is the cross-sectional area of the stream (in square inches), v is the velocity of the stream (in inches per second), and the subscripts 1 and 2 designate two different locations in the system. Again, the permeability of sand molds can complicate the strict application of this law, introducing potential problems into the casting process.

One practical implication of the law of continuity is illustrated in Fig. 3, which illustrates the flow of metal from a pouring basin. As indicated in Eq 1, potential energy is high but velocity is low as the stream leaves the basin. Velocity increases as the stream falls, so the cross-sectional area of the stream must decrease proportionately to maintain the balance of the flow rate. The result is the tapered shape typical of a free-falling stream shown in Fig. 3(a).

If the same flow is directed down a straight-sided sprue (Fig. 3b), the falling stream will create a low-pressure area as it pulls away from the sprue walls and will probably aspirate air. In addition, the flow will tend to be uneven and turbulent, especially when the stream reaches the base of the sprue.

The tapered sprue shown in Fig. 3(c) is designed to conform to the natural form of the flowing stream and therefore reduces turbulence and the possibility of air aspiration. It also tends to fill quickly, establishing the pressure head characteristic of the full-flow conditions required by Eq 1.

Many types of high-production molding units do not readily accommodate tapered sprues, so the gating system designer will try to approximate the effect of a tapered sprue by placing a restriction, or choke, at or near the base of the sprue to force the falling stream to back up into the sprue (Fig. 4).

Momentum Effects

Newton's first law states that a body in motion will continue to move in a given direction until some force is exerted on it to change its direction.

Reynold's Numbers and Types of Flow. The flow of liquids can be characterized by a special measurement called the Reynold's number, which can be calculated as follows:

$$N_R = \frac{v d \rho}{\mu} \qquad \text{(Eq 3)}$$

where N_R is the Reynold's number, v is the velocity of the liquid, d is the diameter of the liquid channel, ρ is the density of the liquid, and μ is the viscosity of the liquid. As shown in Fig. 5, if the Reynold's number is less than 2000, the flow is characterized as laminar, with the molecules of the liquid tending to move in straight lines without turbulence (Fig. 5a).

If a fluid flow system has a Reynold's number between 2000 and 20 000, some mixing and turbulence will occur (Fig. 5b), but a relatively undisturbed boundary layer will be maintained on the surface of the stream. This type of turbulent flow, common in most foundry gating systems, can be considered relatively harmless so long as the surface is not ruptured, thus avoiding air entrainment in the flowing stream.

With a Reynold's number of about 20 000, flow will be severely turbulent (Fig.

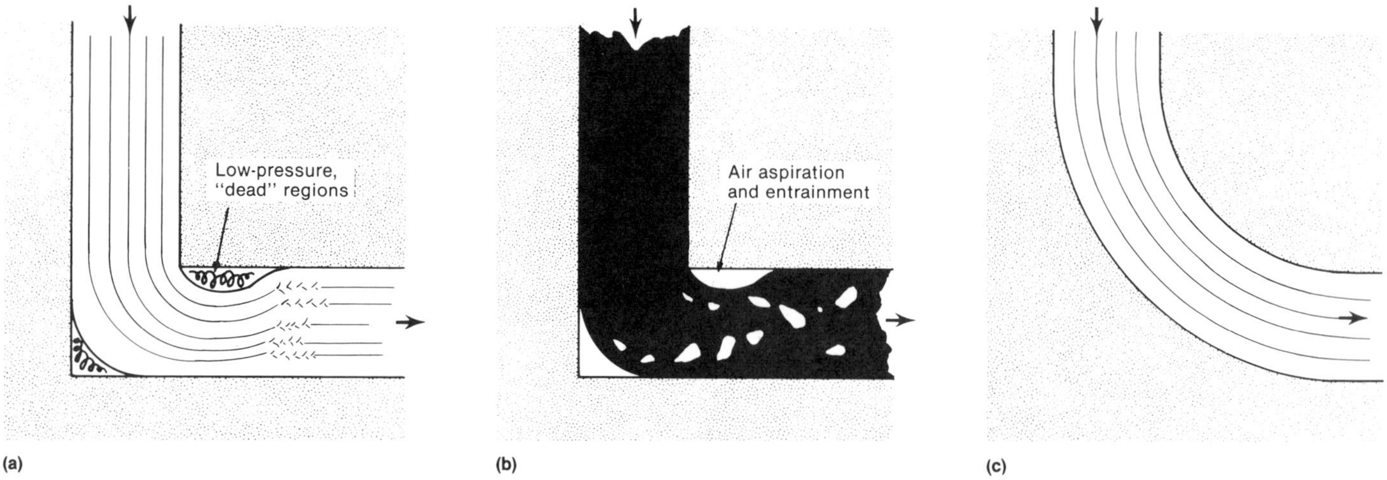

Fig. 7 Schematic illustrating fluid flow around right-angle and curved bends in a gating system. (a) Turbulence resulting from a sharp corner. (b) Metal damage resulting from a sharp corner. (c) Streamlined corner that minimizes turbulence and metal damage

5c). This will lead to rupturing of the stream surface with the strong likelihood of air entrainment and dross formation as the flowing metal reacts with gases.

Abrupt Changes in Flow Channel Cross Section. As shown in Fig. 6, low-pressure zones—with a resulting tendency toward air entrainment—can be created as the metal stream pulls away from the mold wall. With a sudden enlargement of the channel (Fig. 6a), momentum effects will carry the stream forward and create low-pressure zones at the enlargement. With a sudden reduction in the channel (Fig. 6b), the law of continuity shows that the stream velocity must increase rapidly. This spurting flow will create a low-pressure zone directly after the constriction. The problems illustrated in Fig. 6 can be minimized by making gradual changes in the flow channel cross section; abrupt changes should be avoided.

Abrupt Changes in Flow Direction. As shown in Fig. 7, sudden changes in the direction of flow can produce low-pressure zones, as described above. Problems of air entrainment can be minimized by making the change in direction more gradual.

Abrupt changes in flow direction, in addition to increasing the chances of metal damage, will increase the frictional losses during flow. As shown in Fig. 8, a system with high frictional losses will require a greater pressure head to maintain a given flow velocity.

Using a Runner Extension. Use of a runner extension beyond the last ingate is illustrated in Fig. 1. The first metal entering the gating system will generally be the most heavily damaged by contact with the mold medium and with air as it flows. To avoid having this metal enter the casting cavity, momentum effects can be used to carry it past the ingates and into the runner extension. The ingates will then fill with the cleaner, less damaged metal that follows the initial molten metal stream.

Equalizing flow through ingates by decreasing the runner cross section after the ingate is illustrated in Fig. 9; this is done for systems with multiple ingates. As noted earlier, in the filled system shown, potential and frictional energies become constants, so they can be dropped from consideration in Eq 1 to show the interaction of pressure and velocity effects.

At the first ingate, velocity is high as momentum effects carry the flowing stream past the gate. At the second ingate, velocity decreases in the runner as it reaches the end, causing higher pressure and resulting in higher flow through the gate.

By stepping down the runner after the first ingate, metal velocities and pressures at the two ingates can be equalized. This effect can be achieved by gradually tapering the runner to a smaller cross section along its length, but patternmaking limitations usually make it simpler to incorporate actual steps in the runner. Additional information on gating systems and their effect on fluid flow is available in the article "Modeling of Fluid Flow" in this Volume.

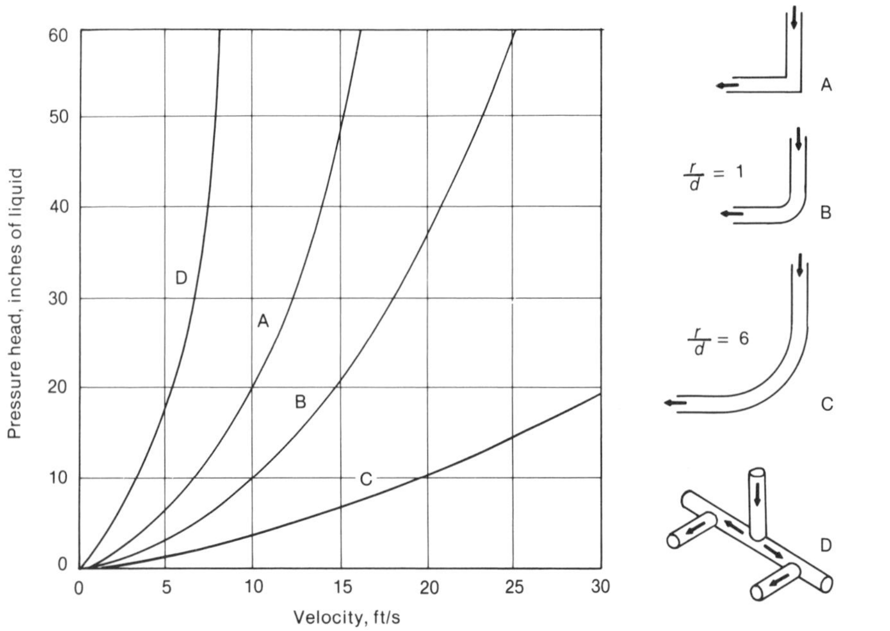

Fig. 8 Effect of pressure head and change in gate design on the velocity of metal flow. A, 90° bend; B, r/d = 1; C, r/d = 6; D, multiple 90° bends. The variables r and d are the radius of curvature and the diameter of the runner, respectively. Source: Ref 2

Design Considerations

In applying fluid flow principles to the design of a specific gating system, several design decisions must be made before the actual sizes of the various components can be calculated.

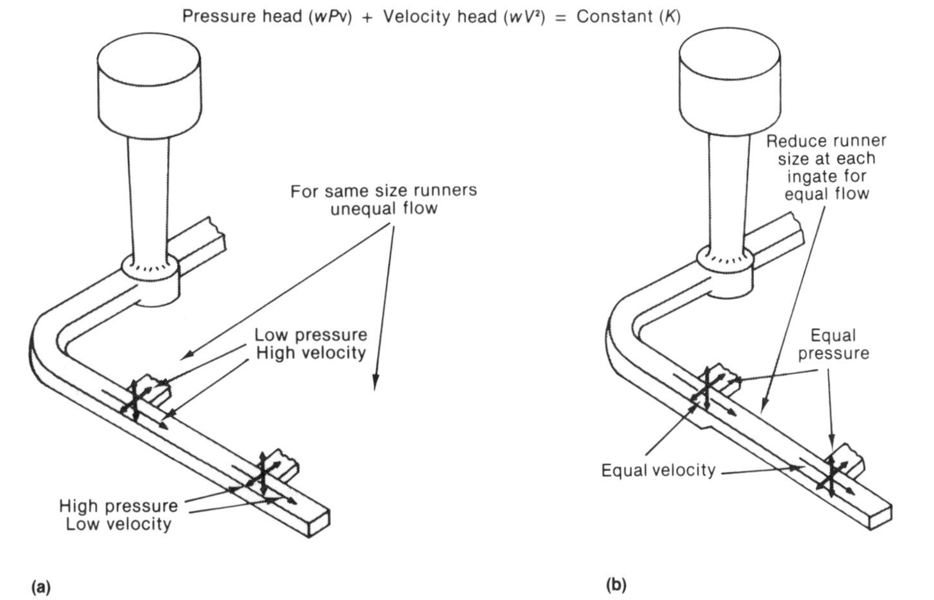

Fig. 9 Applying Bernoulli's theorem to flow from a runner at two ingates for a filled system and comparing velocity and pressure at the ingates for two runner configurations. (a) Same runner cross section at both ingates. (b) Stepped runner providing two different runner cross sections at each ingate. Source: Ref 1

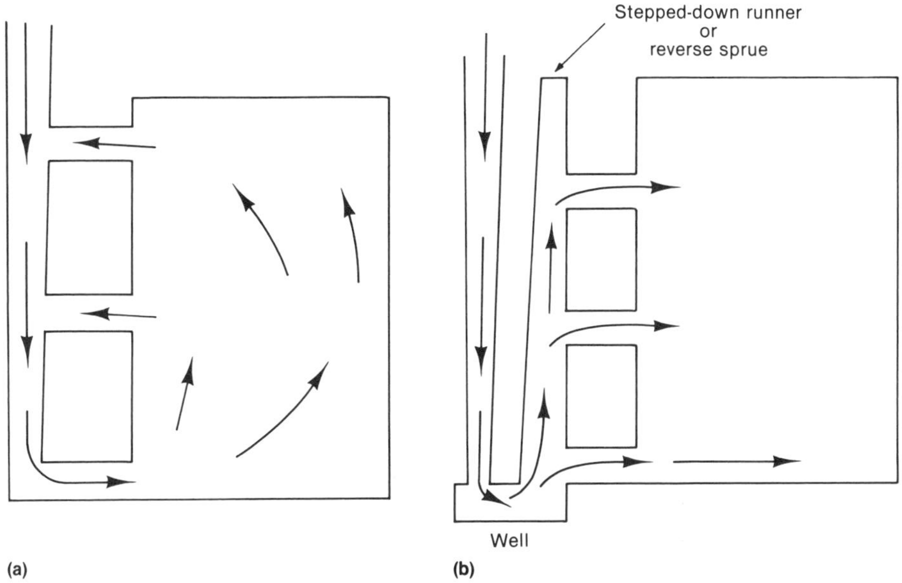

Fig. 10 Comparison of flow patterns in two vertical gating systems. (a) Poorly designed system. (b) Properly designed system utilizing a tapered runner that equalizes flow through the ingates

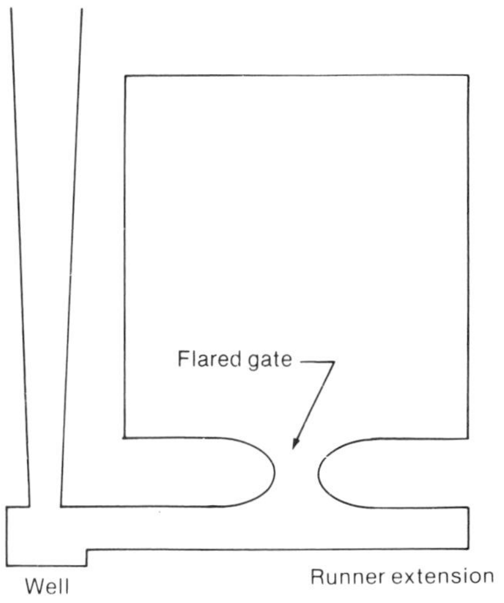

Fig. 11 Properly designed bottom gate that ensures smooth filling of the casting while producing minimal turbulence

Table 1 Densities of metals and metal oxides commonly used in casting alloys

Casting alloy	Density, g/cm³
Aluminum alloys	
Al	2.41
Al_2O_3	3.96
$3Al_2O_3 \cdot 2SiO_2$	3.15
Magnesium alloys	
Mg	1.57
MgO	3.58
Copper alloys	
Cu	8.00
CuO	6.00
ZnO	5.61
SnO	6.45
BeO	3.01
Iron alloys	
Cast iron	6.97
Low-carbon steel	7.81
2% C steel	6.93
FeO	5.70
Fe_2O_3	5.24
Fe_3O_4	5.18
$FeSiO_4$	4.34
MnO	5.45
Cr_2O_3	5.21
SiO_2	2.65

Runner and Ingate. Figures 1 and 2 show gating systems with ingates coming off the top of the runner and then into the casting. This arrangement of cope ingates and drag runners is common and has the advantages that the runner will be full before metal enters the ingates. This establishes the full-flow conditions discussed earlier in this article. A full runner will reduce turbulence and will help to allow any low-density inclusions in the flowing stream to float out and attach themselves to the mold wall.

A system of cope runners and drag ingates (or ingates coming off the base of a cope runner) is also common and has strong proponents (Ref 3). The basis of this design is that momentum effects will carry the first metal past the ingates, and if the runner can be quickly filled (at least above the level of the ingates), clean metal will flow from the bottom of the runner, while inclusions carried along in the metal stream will float above the ingates.

A common element of this system is that the total cross-sectional area of the ingates should be smaller than the cross-sectional area of the runner. Such a pressurized system is intended to force the metal to back up at the ingates and rapidly fill the runner, although complete filling of a cope runner will often depend on at least partial filling of the casting. During this period of incomplete filling, turbulence and the potential for air entrainment and dross generation are increased.

Pressurized Versus Unpressurized Systems. The difference between these two systems is in the choice of the location of the flow-controlling constriction, or choke, that will determine the ultimate flow rate for the gating system. This decision involves the determination of a desired gating ratio, that is, the relative cross-sectional areas of the sprue, runner, and gates. This ratio, numerically expressed in the order sprue: runner:gate, defines whether a gating system is increasing in area (unpressurized) or

Fig. 12 Two crankshafts produced using a ceramic foam filter positioned vertically in the drag just downstream of the sprue. Casting yield is 91%.

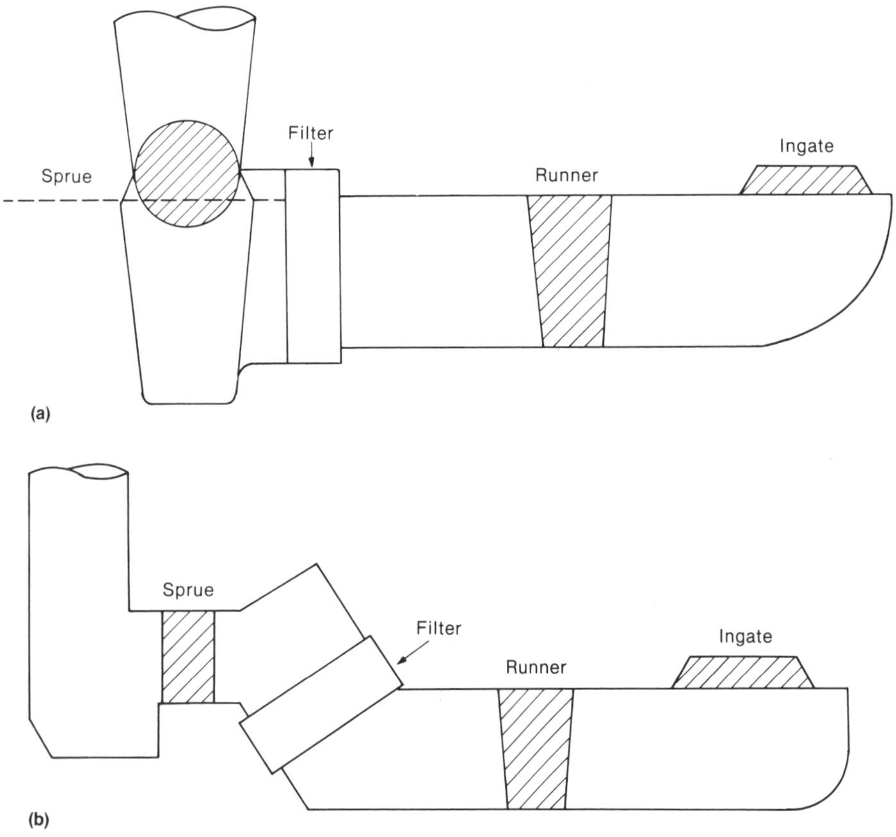

Fig. 13 Gating system designs for optimizing the effectiveness of ceramic filters in horizontally parted molds having sprue:filter:runner:ingate cross-sectional area ratios of 1:3–6:1.1:1.2 (a) and 1.2:3–6:1.0:1.1 (b)

constricting (pressurized). Common unpressurized gating ratios are 1:2:2, 1:2:4, and 1:4:4. A typical pressurized gating ratio is 4:8:3.

Both types of systems are widely used. The unpressurized system has the advantage of reducing metal velocity in the gating system as it approaches and enters the casting. Lower velocities help encourage laminar (or less turbulent) flow, so unpressurized systems are recommended for alloys that are highly sensitive to oxide and dross formation.

Pressurized systems generally have the advantage of reduced size and weight for a given casting, thus increasing mold yield. The single greatest disadvantage of pressurized systems is that, by design, stream velocities are highest at the gates just as the metal enters the casting. This increases the potential for mold or core erosion and places a premium on proper location of ingates to minimize such damage.

Vertical Versus Horizontal Gating Systems. This decision may simply be imposed by the orientation of the mold parting line. Figure 1 shows conventional, horizontally parted molds with the gating system most conveniently arranged along the parting line.

Vertically parted molds impose a vertical placement of gating components, but the design considerations for these components are often the same as those for horizontal systems, for example, the desirability of tapered sprues and runner extensions. In addition, the need for stepped-down sprues to equalize flow through multiple ingates is, if anything, more critical than in a horizontal system. This element of a properly designed vertical gating system is illustrated in Fig. 10.

One form of vertical gating common in both vertically and horizontally parted molds is called bottom gating. Elements of a properly designed bottom gating system are shown in Fig. 11. This method has the particular advantage of introducing metal into the casting cavity at its lowest point, thus helping to ensure smooth filling of the casting with minimal turbulence. Numerous examples of calculations for gating system design are available in the References cited at the end of this article and will not be covered in detail here.

Optimum mold filling times are determined by such factors as metal type, casting weight, and typical casting section thickness. Once the optimum time is established, principles of fluid flow are used to determine the size of the system necessary to deliver the metal with the minimum flow rate required, which in turn determines the cross-sectional area of the choke. Once that is calculated, the rest of the components are easily calculated, moving downstream from the choke in an unpressurized system or upstream from the choke (that is, the gates) in a pressurized system.

Fig. 14 Several common filtration and flow modification devices (from left to right): strainer core, extruded ceramic filter, ceramic foam filter, mica screen, and woven fabric screen. The two types of ceramic filters are by far the most widely used.

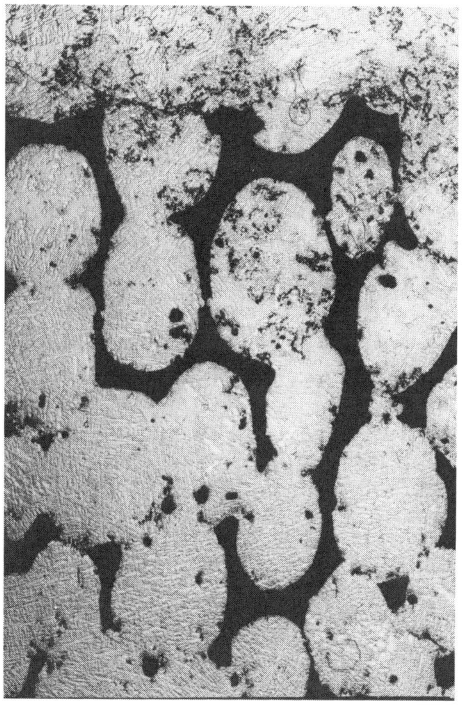

Fig. 15 Cross section through a runner bar containing a ceramic foam filter. Particle capture occurs at the front face (top) and inside the filter. The alloy is aluminum-base LM25.

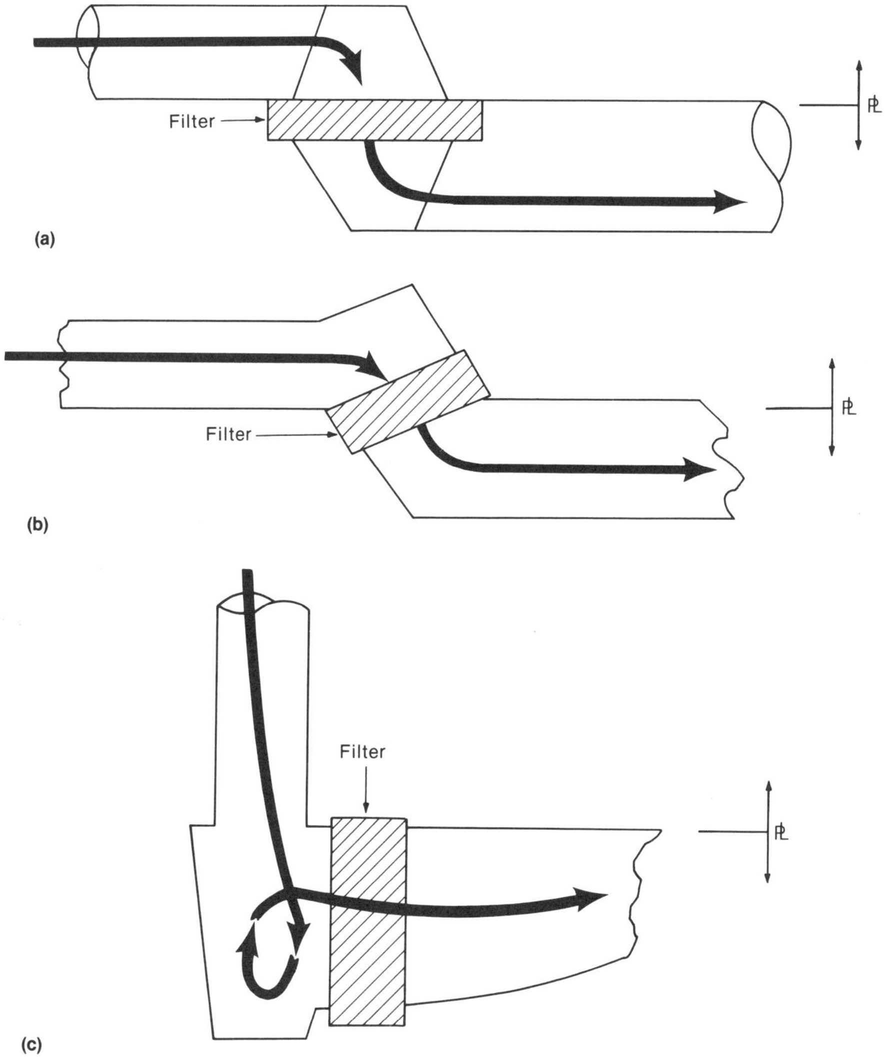

Fig. 16 Common methods of filter placement in horizontally parted molds. (a) Parallel to parting line. (b) Between 0 and 90° to parting line. (c) 90° to parting line. Arrows indicate the direction of metal flow.

Ceramic Filters in Gating Design

Ceramic filters are extensively used in the foundry industry to improve casting cleanliness and to reduce the cost of casting manufacture. Incorporated into the gating system, ceramic filters remove slag, dross, and other nonmetallic particles from the metal stream before the metal enters the mold cavity. Most casting alloys are subject to the presence of particles that can deleteriously affect the physical properties and appearance of the casting. These particles commonly include:

- Oxides formed during melting, metal transfer, and pouring
- Refractory particles from the furnace and ladle
- Refractory particles present in the gating system or dislodged from the mold or cores during pouring
- Reaction products from metallurgical operations
- Undissolved metallic or nonmetallic particles made as additions to the molten metal for metallurgical modifications

These particles, or inclusions, act as discontinuities in the metal matrix of a casting and can have a variety of adverse effects:

- Large inclusions can reduce mechanical properties such as tensile strength and elongation
- Fatigue life can be reduced (Ref 4-6)

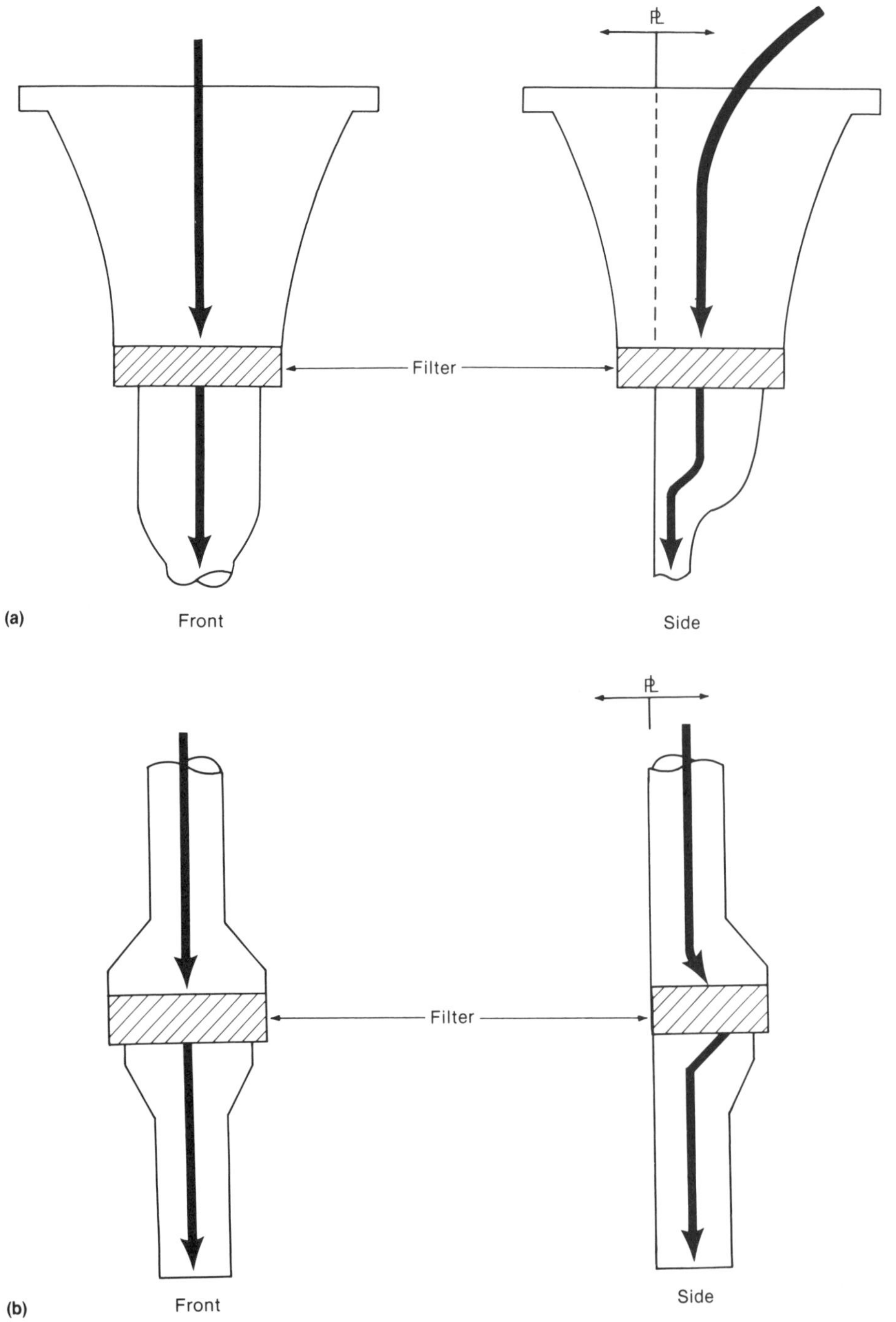

Fig. 17 Common methods of filter placement in vertically parted molds. (a) Filter located in pouring basin. (b) Filter located inside the mold. Arrows indicate the direction of metal flow.

- Machining can become more difficult, and the rate of tool wear can increase
- Surface finish can deteriorate in appearance
- Lack of pressure tightness can occur
- Subsequent surface treatments such as anodizing or ceramic coating can be adversely affected

The conventional approach taken to remove inclusions from the metal stream is through gating system design. Using this approach, gating systems are designed to encourage the separation of particles from the metal due to the density difference between the metal and the inclusion (Table 1).

Both pressurized and unpressurized gating systems are used for this purpose. To be effective, the gating system must have sufficient length to allow the lower-density particles sufficient time to float and adhere to the mold surface before they can enter the casting cavity. In practice, this approach does not always provide castings of adequate quality, and casting yield is often reduced, especially with unpressurized systems.

Filter Advantages. Ceramic filters, when correctly applied, can be relied on to trap particles before they can enter the casting cavity. This feature can result in the following advantages:

- Inclusion-related scrap and inclusion-related rectification can be reduced
- Gating system size can be reduced and casting yield increased (Fig. 12)
- Casting machinability can be improved
- Machine tool life can be increased
- Machine stock allowances can be reduced
- Casting physical properties can be increased
- Casting surface finish can be improved
- Reliability of the casting process can be increased

The use of a ceramic filter in a gating system of conventional design can be effective in reducing inclusion-related defects, but a gating system designed specifically to incorporate ceramic filters will be more effective. Typical gating systems for horizontally parted molds designed to incorporate a ceramic filter are shown in Fig. 13. By relying on a filter to remove particles from the metal stream, the gating system can be designed with the following features and advantages:

- The system can be unpressurized, resulting in reduced metal velocity as metal enters the casting cavity
- Runners can be reduced in size—both length and cross-sectional area—thus increasing casting yield
- Runners are in the drag, resulting in more rapid, complete filling of the runner and reduced opportunity for metal oxidation

Filter Types. Ceramic filters are available in a wide variety of materials and in many different forms. Mullite, alumina, cordierite, silica, zirconia, and silicon carbide are all commonly used. The most common forms are open-weave cloth, reticulated foam, extruded cellular shapes, pressed shapes, and perforated sheets (Fig. 14). Ceramic foams and cellular ceramics have been shown to be the most effective for inclusion removal and are the most widely used. Strainer cores were widely used prior to the advent of ceramic filters, and it is important to note the distinction between these two devices and their applications.

Strainer (choke) cores are designed to function as a restriction or choke in the gating system. The open frontal area—the ratio of the area available for metal passage to total part cross-sectional area—is in the range of 20 to 45%. Strainer cores are designed into a gating system to control the rate of mold filling and can aid rapid filling of the gating system. This action can assist in the flotation and separation of large particles from the molten metal stream, but

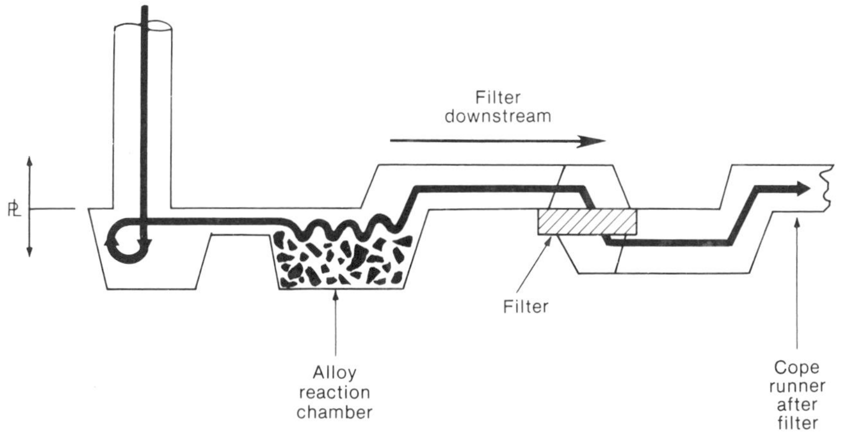

Fig. 18 Filter placement when a metallurgical operation (that is, magnesium treatment or inoculation of iron) occurs in the mold

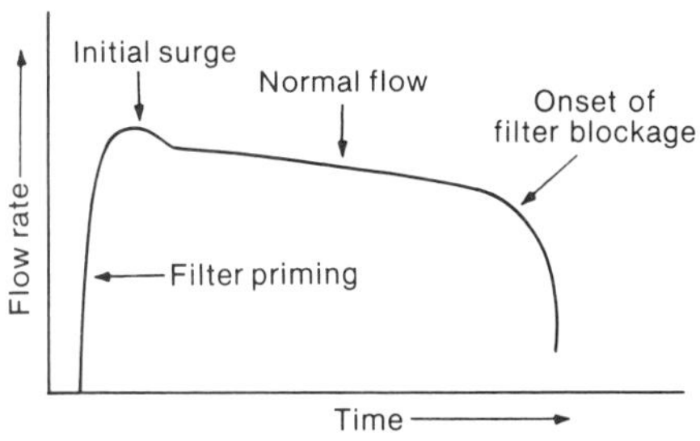

Fig. 19 Plot of flow rate versus time showing the behavior of a ceramic filter during the course of pouring until complete blockage occurs

strainer cores are not intended to remove particles. Their large, widely spaced holes split the metal flow into several separate metal streams, often resulting in aspiration and subsequent oxide formation downstream.

Ceramic filters, on the other hand, are designed to remove inclusions from molten metal. Filtering occurs by two mechanisms: physical screening (or sieving) and chemical attraction (Fig. 15). When properly designed into a gating system, filters do not act as a significant restriction to metal flow. The open frontal area of most ceramic filters is in the range of 60 to 85%, and the flow downstream of the filter is much less turbulent than with strainer cores.

Use of Ceramic Filters. Design of the optimum gating system for ceramic filter use incorporates the following principles:

- Filter placement must be easily accomplished
- Mold filling time must be constant and not affected by the presence of the filter
- Filter type must be correct for the application
- Gating system design must provide minimum metal turbulence downstream from the filter and in the casting cavity
- Gating system size must be kept to a minimum

Filter Placement. The location and position of a ceramic filter are influenced by the molding method, pattern layout, and any metallurgical processes performed inside the mold cavity, such as the nodularizing and inoculating additions made in certain iron castings. The molding method is particularly important in determining filter position. In molding processes that use an expendable pattern, filters are most easily incorporated into the pouring basin. In horizontally parted molds, filters are commonly positioned as shown in Fig. 16. Filters should not be placed at the base of the sprue, because this increases the possibility of filter breakage and reduces filter effectiveness. In vertically parted molds, filters are commonly positioned as shown in Fig. 17. Although the pouring basin location is often used, filters are more effective when located farther down the gating system. When metallurgical additions are made in the mold cavity, filters must be located downstream, as shown in Fig. 18.

Filter Area/Choke Area Ratio. The size and number of ceramic filters required are determined by the rate of mold filling required and the volume of metal to be filtered. As filtration occurs, individual cells in a filter become blocked, and the rate at which the filter can pass metal is reduced. This phenomenon is illustrated in Fig. 19.

Filter size is determined such that the filter operates in the range termed normal flow in Fig. 19. In general, the necessary ratio of filter area to choke area is in the range of 2:1 to 4:1 at the start of pouring. If the filter is greatly undersize, complete filter blockage can occur, and incomplete mold filling will result.

REFERENCES

1. J.F. Wallace and E.B. Evans, Principles of Gating, *Foundry*, Vol 87, Oct 1959
2. J.G. Sylvia, *Cast Metals Technology*, Addison-Wesley, 1972
3. S.I. Karsay, *Ductile Iron III: Gating and Risering*, QIT—Fer et Titane, Inc., 1981
4. P.R. Khan, W.M. Su, H.S. Kim, J.W. Kang, and J.F. Wallace, Flow of Ductile Iron Through Ceramic Filters and the Effects on the Dross and Fatigue Properties, *Trans. AFS*, Vol 95, 1987, p 106-116
5. W. Simmons, The Filtering of Molten Metal to Improve Productivity, Yield, Quality and Properties, in *Proceedings of the 52nd International Foundry Congress*, 1985
6. W. Simmons and H.A. Bowes, Efficient Filtration Improves Casting Quality, *Foundry Pract.*, Vol 209, 1984

SELECTED REFERENCES

- R.A. Flinn, *Fundamentals of Metal Casting*, Addison-Wesley, 1963
- L.F. Porter and P.C. Rosenthal, Fluidity Testing of Gray Cast Irons, *Trans. AFS*, Vol 60, 1952
- J.M. Svoboda, *Basic Principles of Gating and Risering*, American Foundrymen's Society, 1973
- H.F. Taylor, M.C. Flemings, and J. Wulff, *Foundry Engineering*, John Wiley & Sons, 1959

Casting Design

Ronald M. Kotschi, Kotschi's Software & Services, Inc.

A DESIGN ENGINEER committing pencil to paper is not only creating a shape but also strongly influencing the cost required to place that shape into use. In other words, well-designed parts are less costly to obtain than poorly designed parts and are more likely to be delivered to the customer on time. Although this is true of every method of manufacture, it is particularly true of shapes to be made by the casting process. The casting process has gained the reputation as a technique that can be used to create almost any shape the designer can envision, but whether or not a shape can be cast economically is another matter. If the manufacturing process were considered at the design table, an inherently easier to manufacture design would result. Ease of manufacture translates into castings that are purchased at a lower cost, are of high quality, and are delivered on time.

All aspects of this topic cannot be covered in a single article. However, through an understanding of only a few of the most important aspects of this approach to casting design, a great deal of the potential economic savings can be realized.

Solidification

The first aspect that is important to casting designers is an understanding of solidification (the freezing of molten metal inherent in all casting processes). The solidification of a casting can involve as many as three separate contractions as a result of cooling:

- Liquid-liquid contraction
- Liquid-solid contraction
- Solid-solid contraction

Liquid-liquid contraction occurs as a result of the liquid cooling from its pouring temperature (usually 110 to 165 °C, or 200 to 300 °F, above its melting point) down to the melting point or solidification temperature. This particular factor is of little consequence to designers and is fairly easily dealt with by the foundry engineer. However, the design engineer must consider the other two contractions if cost-effective designs and specifications are to be realized.

Solid-solid contraction occurs after a casting has solidified and as it cools from the solidification temperature to room temperature. The design engineer must be concerned with this contraction. To ensure that the dimensions of the casting are correct, the pattern used to produce a given casting usually must be made slightly larger than the casting dimensions at room temperature. The patternmaker compensates for this pattern enlargement for a particular alloy by using a shrink rule specifically for the alloy involved. Further, because the amount of solid contraction is a function of the particular metal to be cast, problems with dimensions can often occur when changing alloys if the same pattern equipment is used. Whenever such a change is contemplated, the foundry engineer should be provided with this information because other factors such as gating and risering could also be involved. Therefore, the designers specifying the alloys should carefully consider any change in alloy to ensure that the cost of new equipment does not cancel out the benefits to be achieved by such changes.

Solid-solid contraction should also be considered in part design, because it is one of the primary causes of warped and cracked castings. A basic concept that governs the way castings cool is the casting modulus (the volume of a portion of a casting divided by the surface area of that portion of the casting). This relationship of geometry to cooling is easily understood by considering the effects of both volume and surface area on cooling rates. As volume increases, more hot metal will be contained within it, and the casting will therefore take longer to cool. Conversely, because all the heat within a casting must pass through a surface at the metal/mold interface, the greater the surface area, the faster the casting will cool. Thus, as the volume-to-surface area ratio (casting modulus) increases, the time required for cooling and solidification is extended.

Using this tool, one can easily see the problems created by the design shown in Fig. 1. Because the design contains a rather wide range of section thicknesses, the various sections will cool and solidify at different rates. The first portion of the casting to solidify would be the sections having a modulus of 1. At some later time, the section with a modulus of 5 would solidify and begin cooling to room temperature. However, the thinner sections will have cooled and contracted by solid-solid contraction before the thicker section, and because of this the thicker section applies a compressive stress to the thinner sections as it cools. Such stresses have been measured to levels as high as 552 MPa (80 ksi), depending on the alloy and section size variations. Therefore, a casting designed in this way will have a strong tendency toward warpage as a result of the imposed stresses. Although a casting may not be warped when taken from the mold, the internal stresses that develop as a result of design can appear at later stages as cracks or warping, often after heat treatment, welding, and machining.

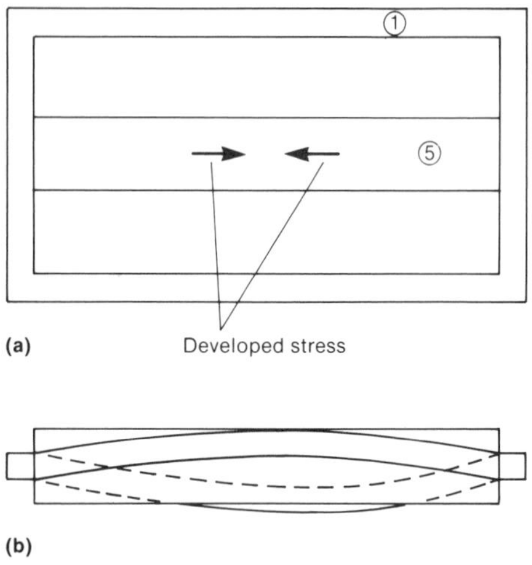

Fig. 1 The effects of design on distortion of castings. (a) Top view of casting; numbers indicate moduli of the two sections. (b) Distortion caused by solidification stresses

If design features result in warpage, foundrymen often compensate for this by using tie bars that are usually small cast-on bars attached to various parts of the casting to brace the casting and therefore prevent warpage. Although such devices may result in castings that are not warped, other problems can result. For example, such devices cannot prevent the stresses that cause the castings to warp and therefore do nothing to prevent the cause of the problem. In some cases, warpage is eliminated at the expense of increased residual internal stresses. Because such stresses cannot be detected with nondestructive testing, they remain undetected, and upon application of heat or other stresses, warpage, cracking, or premature failure can result. In most cases, castings produced with tie bars present no problems in service. However, tie bars do nothing to solve the root cause of the problem, namely, the stresses caused by design features; therefore, it is best to avoid their use by employing uniform section sizes whenever possible.

The importance of solid-solid contraction as applied to design means that an attempt should be made to reduce dramatic variations in the section sizes of castings. Other factors in foundry practice that can lead to the distortion of castings include improper heat-treating practices, difficult-to-collapse molds or cores, shakeout procedures, rigging, and the way in which the casting cools; these factors are under the control of the foundry engineer. Problems with warpage and cracks can often be eliminated through the cooperative effort of casting designers and foundry engineers.

Liquid-solid contraction is by far the greatest difficulty due to solidification that faces the foundry engineer. It is also one of the greatest opportunities for the design engineer to design for low cost, high quality, and timely delivery. Most metals contract as they pass from the liquid to the solid state. Certain compositions of gray and ductile iron are the exceptions to this rule in the major alloys produced by foundries. The entire founding process is possible only because volumetric contraction locates itself in solidifying castings in a systematic way.

Solidification Sequence

The aspect of liquid-solid contraction that allows castings to be produced is that all of the contraction is concentrated in the last portion(s) of the casting to solidify. The foundryman uses this principle to produce sound castings by attaching a volume of metal to the last portion of the casting to solidify. This technique is illustrated in Fig. 2. Such feed metal reservoirs are called risers. Proper placement of risers on castings changes the way in which both casting and riser(s) solidify such that the riser is the last to solidify. When used properly, this produces a casting free of shrinkage because all the shrinkage for the entire mass of both casting and riser will be concentrated in the riser. However, in many cases, the design of the casting restricts the proper placement of risers, making the production of sound castings difficult if not impossible. In this way, casting designers have a significant impact on quality and cost.

The technique of designing castings that are easily risered can be best understood by considering a simple shape such as a wedge (Fig. 3a). The solidification direction, as

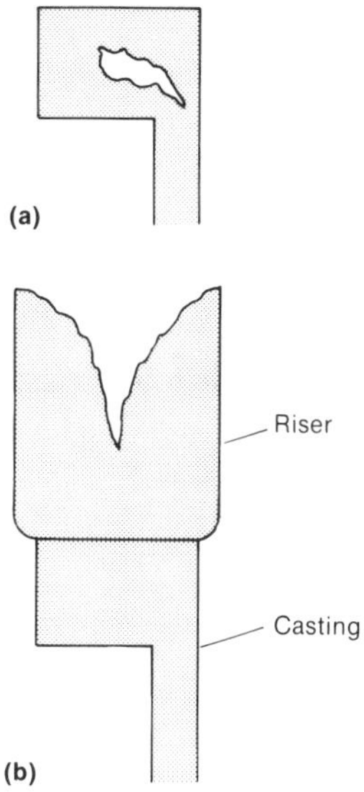

Fig. 2 The function of risers. (a) Shrinkage in casting at last portion to solidify without a riser. (b) Shrinkage in riser allows the production of a sound casting.

seen in Fig. 3(b), is related to the casting modulus, which is the volume-to-surface-area ratio. Because a wedge casting has a natural solidification direction, if the design is such that a riser cannot be placed at the wide end of the wedge, a shrinkage cavity will develop in the casting even though a riser was attached. As shown in Fig. 3(a), this occurs because solidification begins somewhere near the middle of the wedge/riser combination. Thus, from a solidification or thermal design point of view, there are actually two castings in this example, each having a last place to solidify. If the design allows for the placement of the riser at the proper end of the wedge, as in Fig. 3(b), a sound casting will result because of the unidirectional solidification pattern. Shrinkage defects may not always represent a problem, but in cases where they jeopardize mechanical strength, appearance, or pressure tightness, the castings would have to be repaired or scrapped. Under such circumstances, the final costs (including delays in delivery) can be severe.

Very few castings are as simple as the wedge shown in Fig. 3; therefore, a technique capable of determining the solidification sequence of complex shapes is needed if design engineers are to take advantage of the potential cost savings. Regardless of their apparent complexity, castings can usually be resolved into a series of rather simple shapes that make up the complex shape. These simple shapes are:

- T-sections (plate intersections forming a T)
- X-sections (plate intersections forming an X)
- L-sections (plate intersections forming an L)

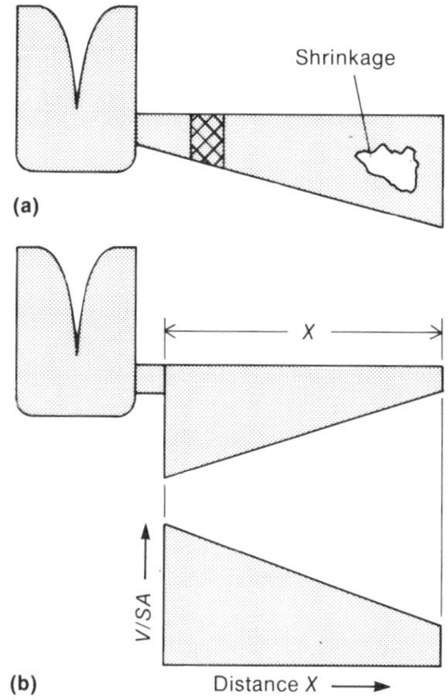

Fig. 3 Casting design and solidification of a simple wedge. (a) Riser placed at narrow end of wedge; shrinkage occurs at wide end. The crosshatched region represents the approximate area of the casting where solidification is first complete, thus cutting off the feeding path of the casting. (b) Correct riser placement. *V/SA* is the volume-to-surface-area ratio (casting modulus).

- Plates
- Cylinders
- Cylinder/plate intersections

Most casting geometries can be covered by using these basic shapes. For example, a box shape can be simulated by a series of L-sections, and a hub casting can be simulated as a T-section revolved about a central axis. By solving the relationship between casting design parameters and the manner in which these simple shapes solidify, designers can obtain a useful guide to cost reduction and improved quality through casting design.

T-Sections

Figure 4(a) shows a hub casting. There are four possible ways to position the risers, depending on the casting design. These four alternatives are shown in Fig. 4(b) and (c). Selection of one of these alternatives by the foundry engineer depends on the casting design. Figure 4(d) shows the dimensions and critical sections involved. In Fig. 4(d), as well as in each figure in this article that illustrates the dimensions of a given section type, a special notation is used to differentiate between the dimensions describing the size of the section components and the section itself. Although the same letter is used, for example, T and *T*, the roman T refers to the section, while the italic *T* refers to the dimension of the section.

Factors Influencing the Solidification Sequence. There are three basic design dimensions for a T-section: *t*, *T*, and *R*. If

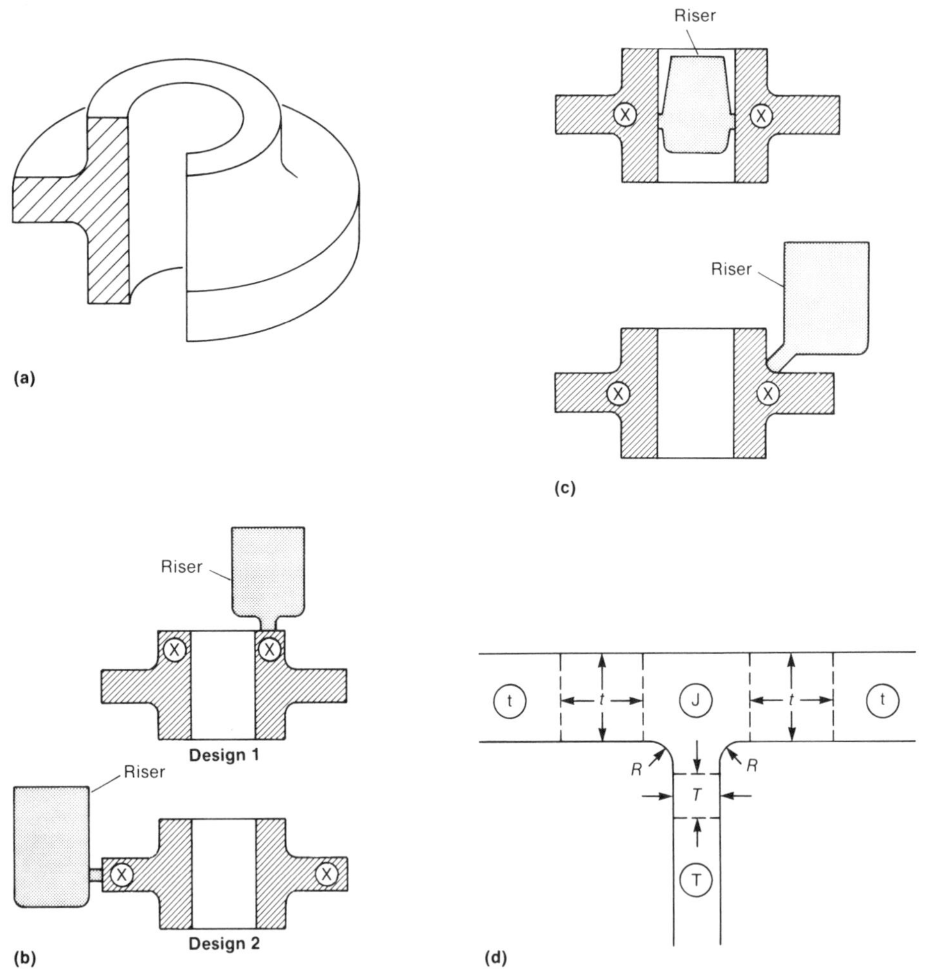

Fig. 4 Cross section (a) of a hub casting showing that it is actually a T-section revolved around an axis. (b) and (c) Possible riser placements; X indicates the last portion of the casting to solidify. See text for discussion. (d) Principal design dimensions of T-section and component sections; see text for discussion.

the cross plate dimension t is large with respect to the intersection plate dimension T, the last portion of the casting to solidify will be area t in Fig. 4(d). This will require the riser placement labeled Design 1 in Fig. 4(b). If the intersecting plate dimension T is large compared to that of the cross plate dimension t, the last portion to solidify will be area T in Fig. 4(d). This type of design will require the riser placement labeled Design 2 in Fig. 4(b). Finally, if the dimensions t and T are nearly the same, either of the two possibilities shown in Fig. 4(c) can be used, because area J in Fig. 4(d) will tend to be the last area of the casting to solidify.

In a solidification sequence in which the last portion of a hub casting to solidify is the J portion of the T-section, as shown in Fig. 4(d), one of two difficulties usually arises. In many designs of this type, the hole cavity formed by the rotation of the T-section about its central axis is too small to allow a riser of sufficient size to feed the section. When this cavity is too small, there is no alternative but to place the riser outside of the casting, as shown in Fig. 4(c), which necessitates the use of a core, thus adding cost.

If the enclosed cavity should be large enough to contain a riser of sufficient size, the problem then usually involves the removal of the riser from within the cavity. In a high-production environment, such as automotive applications, special equipment to extract such internal risers can be economically feasible if the design allows placement in the internal cavity of the hub.

Such a special case is the manner in which cast automotive connecting rods are produced. For this particular type of casting, a small wafer riser is placed within the crankshaft receiver end of the connecting rod, which adequately feeds this portion of the casting. However, the use of this casting design and the wafer riser is predicated on high production requirements, which allow the use of specially designed punch presses to remove the riser and the associated gating system. Without such special equipment, the internal riser placement would be highly undesirable.

This case is an exception to the overall undesirable nature of solidification sequences in which the J portion of a hub casting is the last section to solidify. This points out a very important concept: There are no definitive rules that always apply to casting design. It is this potential for alternative solutions based on specialized conditions that makes the topic of casting design so interesting but very difficult to confine within a rigid set of guidelines. The goal is to design the casting with a particular solidification sequence in mind so that the cast part can be risered easily and subsequent riser removal can be accomplished with ease.

In general, however, either of the two riser placements shown in Fig. 4(c) can usually be considered undesirable. Because of the added costs imposed by a solidification sequence of this type, such designs generally should be avoided. However, if such a design is unavoidable, some modifications can be made to facilitate manufacture. The key is to recognize the problem, namely, that area J of the T-section of the hub (Fig. 4d) is the last to solidify. Once this is known, appropriate measures can be taken to modify the solidification sequence.

One alternative would be to alter the casting modulus by changing the shape of the hub to reduce the volume contained and/or to increase the surface area. This would be an effective solution to the problem if the design already required a core to form the hole in the hub before the change in shape. If such a design change increases the number of cores required to produce the casting, this technique may not be cost effective, because additional cores usually necessitate increased casting costs.

Another possible solution to the same solidification sequence problem might be the use of a feed pad (Fig. 5). Once again, by recognizing the solidification sequence problem imposed by the design, the designer could elect to add metal to the flange of the hub to change the solidification sequence. Foundrymen refer to this technique as padding. This is a good alternative solution if the pad can either remain on the casting in use or does not affect machining costs adversely. There are many other possible design alternatives, but the designer must first be aware of the problem caused by the solidification sequence.

Either of the riser placements shown in Fig. 4(b) is an effective alternative if large production quantities are not required. Because of the need to extract the riser through the top of the cope half of the mold, Design 1 is less compatible with high-production molding techniques than the side riser used in Design 2. Therefore, if high production is expected, the preferred solidification sequence would be that in which area T in Fig. 4(d) is the last portion of the casting to solidify.

Development of Solidification Sequence Graphs. A series of graphs has been developed to assist designers and foundry engineers in determining the solid-

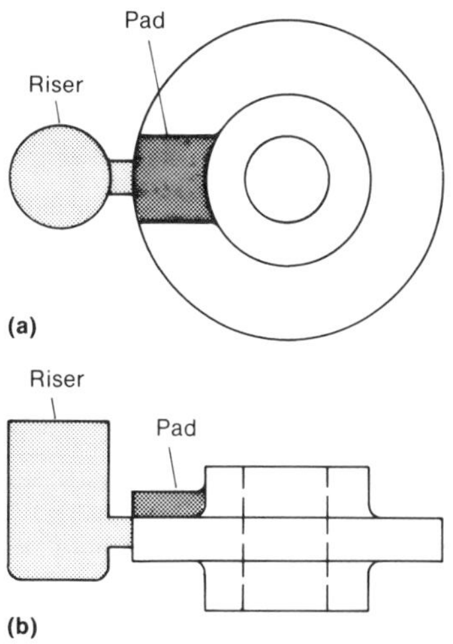

Fig. 5 A hub casting with a pad to provide a feeding channel from the riser. (a) Top view. (b) Side view

ification sequence of the basic shapes listed earlier. These graphs show the solidification sequences as a function of the design parameters constituting the sections. Considering the T-section shown in Fig. 4(d), one can easily see that there are three parameters governing the manner in which the section solidifies, namely, dimensions T, t, and R. As each of these parameters changes, the casting modulus of each component making up the T-section changes. The solidification sequence can be estimated by comparing the modulus of each component (t, T, and J) of the T-section. With the use of a computer, this process was employed to evaluate all the possible designs of the T-section from $t = 0$ to 10 and $T = 0$ to 10 while holding $R = 0$. The results of this analysis are shown in Fig. 6(a).

The technique used in this analysis was the well-known Chvorinov's rule, which is commonly used to compare the solidification times of simple casting shapes. Although originally developed for the solidification of pure metals and alloys solidifying over a very narrow temperature interval, the concept is more broadly applicable and states that the total solidification time of a casting (or casting section) is proportional to the square of the volume-to-area ratio of the casting (or casting section). This rule can be stated mathematically as follows:

$$t_f = K \cdot \left(\frac{V_c}{A_c}\right)^2$$

where t_f is the total solidification time for the casting or casting section, V_c is the volume of the casting or casting section, A_c is the surface area of the casting or casting section, and K is a constant for a given metal and mold combination.

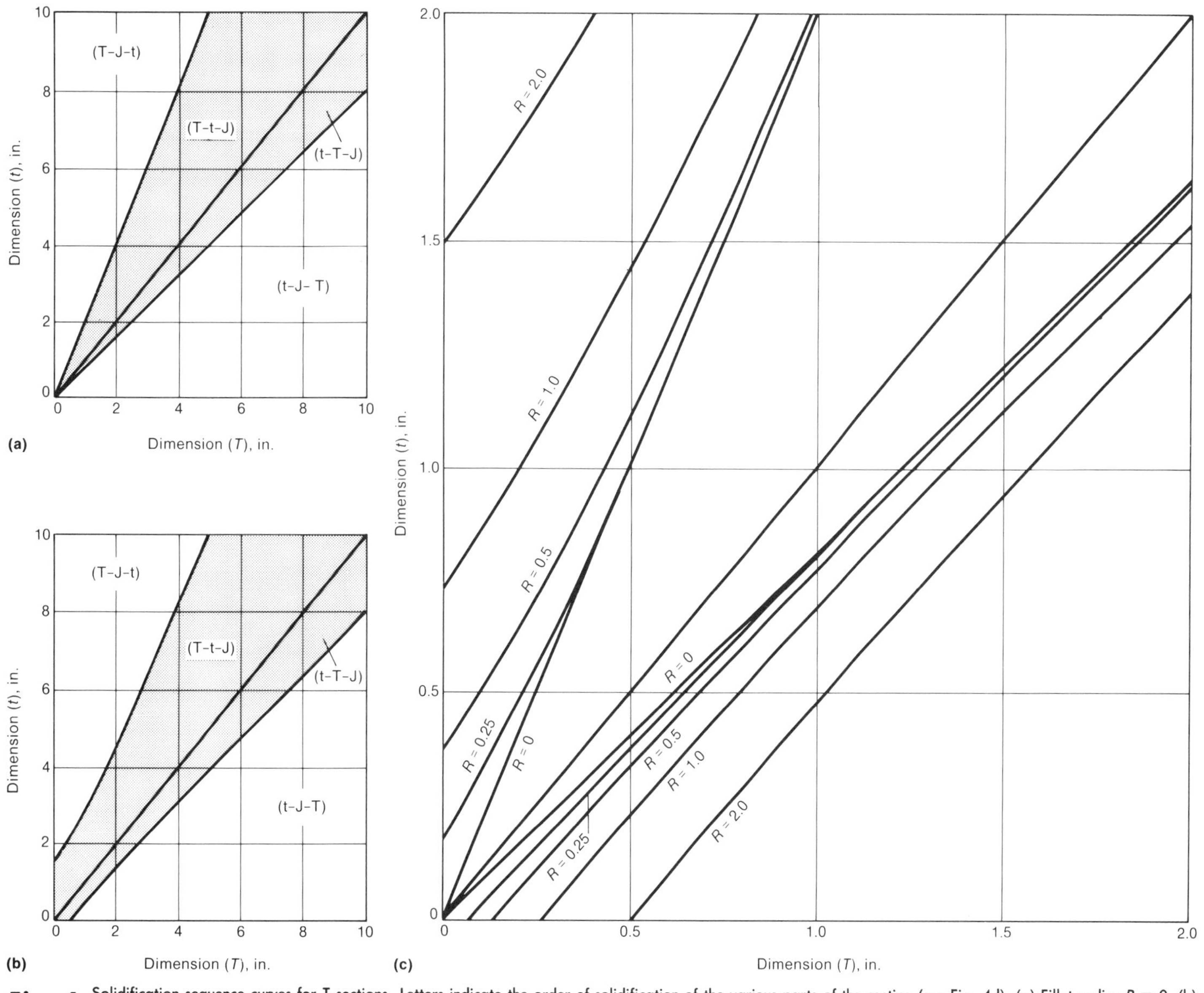

Fig. 6 Solidification sequence curves for T-sections. Letters indicate the order of solidification of the various parts of the casting (see Fig. 4d). (a) Fillet radius $R = 0$. (b) $R = 2$. (c) Composite of design curves for various fillet radii

The factors that relate to the constant K are mold temperature, metal temperature, the density of the solidifying metal, the specific heat of the mold, the density of the mold material(s), the thermal conductivity of the metal, the thermal conductivity of the mold, and the specific heat of the metal. These factors involve the specifics of a given casting manufactured in a given mold. For a mold that consists of a single material, these factors can be considered a constant throughout the mold.

For factors that are constant throughout the part of the mold containing a given portion of the casting, Chvorinov's rule becomes a proportional equation in which the time for solidification is proportional to the square of the volume divided by the surface area. In this form, the results of the equation are independent of the type of mold and metal being considered, provided they are the same over the casting section to which the results are to be applied. Furthermore, for the purposes of providing a proper solidification sequence, the actual time for a given portion of a casting to solidify is not needed. In this type of analysis, it is important only to determine the order in which each component solidifies with respect to the other components being studied. By remembering these facts, the proportionality equation discussed above (that is, solidification time is proportional to the volume/surface area ratio) is all that is required to study the solidification sequence of various section components relative to one another.

Consider the components making up the T-section in Fig. 4(d). There are three basic dimensions (T, t, and R) that describe the section, and the section consists of two plates labeled T and t whose intersection forms another section labeled J. Assuming from a thermal point of view that the formed section is infinite (that is, the ends of plates in the z-direction perpendicular to the drawn section do not influence the solidification of the section or the plates), the volume of the section and plates can be determined from the perimeter and enclosed cross-sectional areas of the sections. Such an assumption is valid if the thickness of the section in the z-direction is five times the thickness of the thickest section.

For this case, the volume of the plate having a thickness T is equal to T^2. The surface area is equal to $2T$; therefore, the volume-to-surface-area ratio (casting modulus) of the plate is $T/2$. Similarly, it can be shown that the casting modulus for the other plate making up the section is $t/2$. To study the relationship of the solidification sequence of all the components making up the T-junction, all that remains is to develop the casting modulus for the junction of the two plates referred to as J in Fig. 4(d). Assuming that the thermal effect of the junction extends one plate thickness beyond the tangent point of the radii (that is, the junction length along the t plate is considered to be $2t + 2R + T$), the volume and surface area of the junction can be written as follows:

$$\text{Volume} = 2t^2 + t(2R + T) + RT + T^2 + \frac{(4R^2 - 3.1416\,R^2)}{2} \qquad \text{(Eq 1)}$$

$$\text{Surface area} = 4t + 2R + T + 3.1416\,R + 2T \qquad \text{(Eq 2)}$$

Therefore, the casting moduli for the T-junction J and the plates making up the junction are as follows. For the t-plate, the casting modulus is $t/2$. For the T-plate, the casting modulus is $T/2$. For the J section (the T-junction itself), the casting modulus is:

$$\frac{2t^2 + 2Rt + tT + RT + T^2 + 0.4292\,R^2}{4t + 3T + 5.1416\,R} \qquad \text{(Eq 3)}$$

Based on Eq 1 to 3 and the design parameters of R, t, and T, it is a simple matter to determine the solidification sequence of a T-section. By entering the dimensions of t, T, and R into the respective equations and calculating the resulting casting moduli, one obtains the solidification sequences based on the design parameters of the T-section and the plates making up the section. The results of such calculations for values of $R = 0$ and $R = 2.0$ for plate thicknesses from 0 to 10 are shown in Fig. 6(a) and (b).

A series of T-section design curves for R values of 0, 0.25, 0.5, 1.0, and 2.0 for values of t and T from 0 to 2 are presented in Fig. 6(c). For design curves not covered, one can easily use Eq 1 to 3 for the junction and the respective plates of thickness t and T.

Equations 1 to 3 are not specific for the use of inches only, because the form of Eq 1 to 3 and the results are purely numeric. Thus, one can use metric units such as centimeters or millimeters as the dimensions and use any of the graphs directly. Equations 1 to 3 also allow the user to convert any of the graphs to any radius. All that is needed is to determine a ratio of the desired radius to a specific radius on the graph and then use this ratio as a scaling factor to adjust the scale of the two dimensional axes t and T. For example, assume a graph for a 1 in. or 1 cm radius was needed but only Fig. 6(b) was available. Because the scaling ratio for this case is ½, the T dimension of the horizontal axis could be labeled 0, 1, 2, 3, 4, and 5, replacing the 0, 2, 4, 6, 8, and 10 currently on the graph. By following this same procedure for the t dimension on the vertical scale, one would produce the graph form necessary for the $R = 1$ problem. Thus, the user of these graphs is not limited to the cases presented, and any combination based on the required R value can be used.

The following example further illustrates the use of such curves. Figure 4(a) shows the cross section of a hub casting. With reference to Fig. 4(d), it can be determined that the T-section dimensions are as follows:

- $T = 4$ in.
- $t = 6$ in.
- $R = 0$

Note that T-sections with $R = 0$ are not recommended, because of the stress concentration and the potential for hot tears or cracks.

By referring to Fig. 6(a) and plotting the T and t dimensions, it can be seen that for this design the $T = 4$, $t = 6$ point on the graph lies in the zone of the graph labeled T-t-J. Thus, based on casting modulus, a solidification sequence of T-t-J would be expected. Because the J portion of the section is the last to solidify, the riser would have to be attached to this portion of the casting. Therefore, because of the solidification sequence imposed, a riser placement scheme similar to that shown in Fig. 4(c) must be used. However, this type of solidification sequence is generally undesirable, and design alternatives should be considered. Some metal can be removed from the junction, or the T dimension should be increased somewhat above 8 in. as either a pad (Fig. 5) or a complete increase of the entire flange of the casting. Yet another approach would be to increase or decrease the hub thickness dimension t.

These are a few of the potential solutions to the problem that are available to the design engineer. The designer knows both the stresses and the environment in which the part will function; therefore, the designer alone must make the final decision on any proposed changes. It is important to inform the design engineer of a potential manufacturing problem early in the designing process while flexibility still exists, not at later stages when the flexibility is lost.

The previous example considered a T-section in which the radius R was 0. This case is somewhat academic because, in terms of well-known design principles, the design discussed leads to undue stress concentrations at the sharp corners. This is to be avoided from the standpoints of pure design and casting manufacture. Because of the contraction of the metal as it solidifies and cools, such stress concentration can lead to cracking of the casting (either hot tearing or cold cracking). *Therefore, the first rule of economical casting design is to use generous fillet radii whenever possible.*

Figure 6(b) was obtained by following the same procedure discussed for the development of Fig. 6(a). In Fig. 6(b), the radius dimension R of Fig. 4(a) was set to 2 in. A comparison of Fig. 6(a) and (b) reveals the effects of adding a 2 in. fillet radius. Very little difference in the solidification sequence is observed at the heavy plate thicknesses (the upper right-hand corner of the graph). However, in those designs in which thin plates are used (the lower left-hand corner of the graph), there is a much stronger tendency for the plate junction J to be the last portion of the T-section to solidify. It must be remembered that Fig. 6(b) represents all the design variations of the T-section from *T* and $t = 0$ to *T* and $t = 10$ to which a constant 2 in. fillet radius is applied. A series of design curves for various fillet radii are shown in Fig. 6(c).

Whenever any fillet radius is applied, an effect on the solidification sequence must be expected because the fillet radius causes an increase in volume and a decrease in surface area in the vicinity of the fillet. The increase in volume and the decrease in surface area caused by the use of a fillet are only functions of the fillet radius; therefore, the effect of adding a given fillet radius to the casting modulus of the T-junction is a constant based on the radius. Furthermore, because the casting modulus of the junction is a function of the plate thickness, the addition of a 2 in. fillet to thin plate designs increases the casting modulus of the junction far more with respect to the modulus of the plate than for thick plates. Thus, the effect of the 2 in. fillet addition is far greater in thin plate designs than in heavier plate T-sections.

Using these graphs, the designer can recognize potential problem areas by remembering *the second rule of casting design for economy, which states that the last place of any casting to solidify must be accessible for riser placement*. Ideal riser placement is at the parting line and at the outer perimeter of the casting. If this rule is not followed, weld repair, scrapped castings or sudden catastrophic failure could result, and these outcomes will cause increased casting costs, delivery delays, and a serious danger of premature failure.

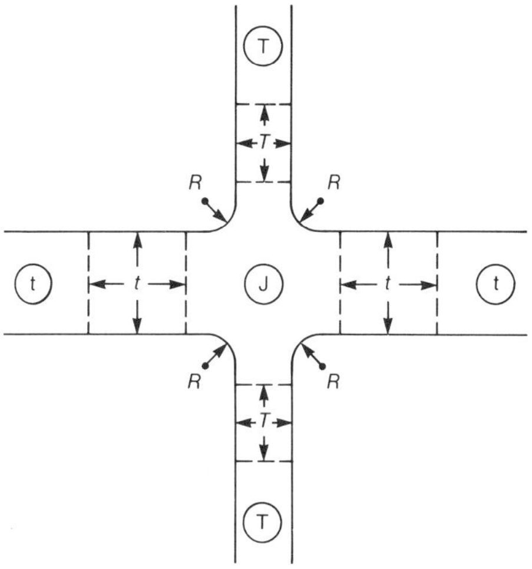

Fig. 7 Model of X-section with opposite legs equal. See Fig. 8 for solidification sequence curves.

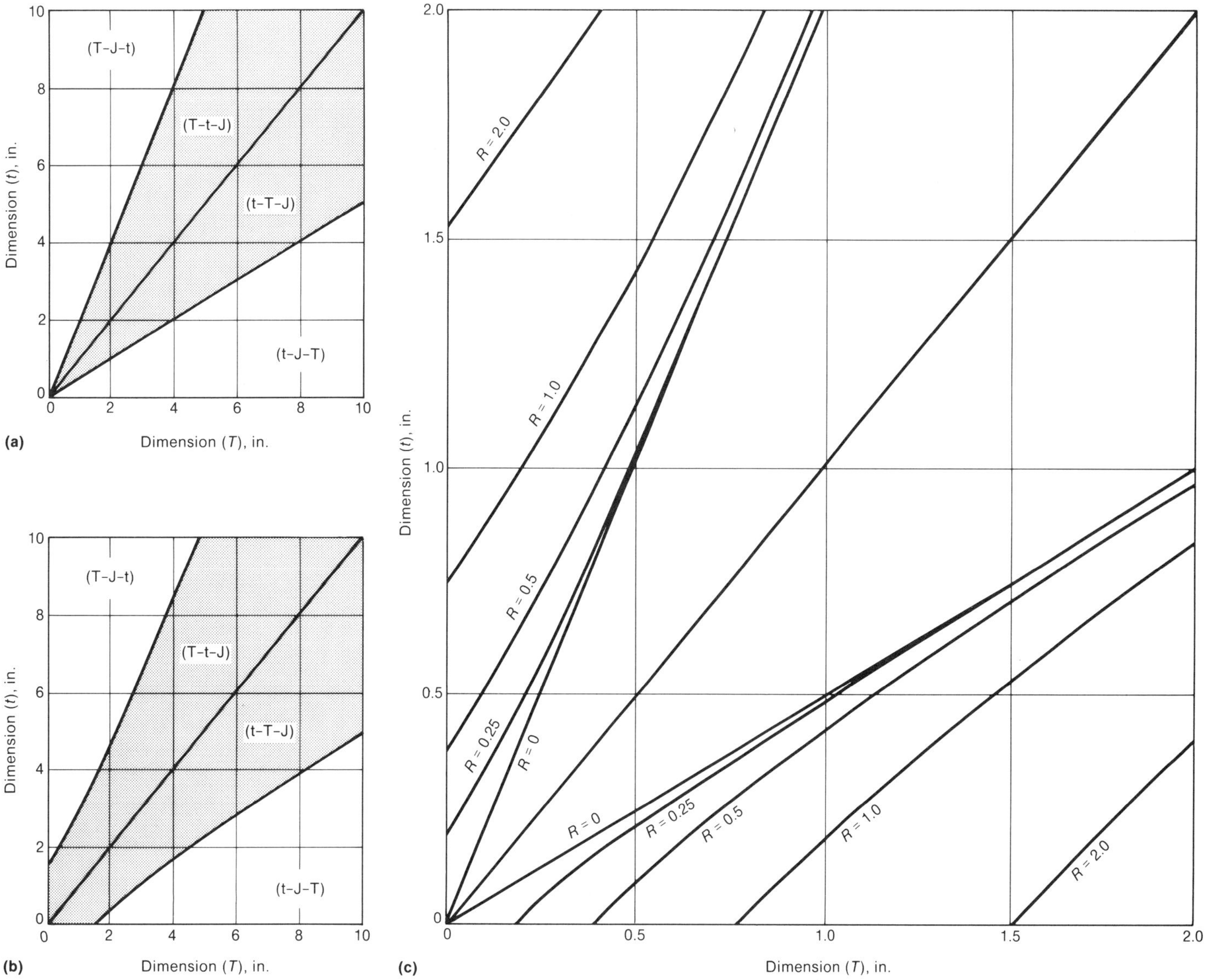

Fig. 8 Solidification sequence curves for X-section with opposite legs equal (see Fig. 7). (a) $R = 0$. (b) $R = 2$. (c) Composite of design curves for various fillet radii

X-Sections

Another simple shape of importance to designers is the X-section. However, for X-sections, the designer must consider two conditions in order to cover all the possible design combinations that might be encountered.

Equal Opposite Legs. The first type of X-section design is that in which the opposite legs of the X are equal. The model for this case is shown in Fig. 7. The solidification sequence curves of this model, which contain fillet radii of $R = 0$ and $R = 2$, are presented in Fig. 8(a) and (b). Design curves for various fillet radii are shown in Fig. 8(c). As before, the casting modulus equations for the plates making up the X-section are $T/2$ and $t/2$. For X-sections in which the opposite legs of the section are equal, the casting modulus for the J- or X-junction is as follows:

$$\frac{\text{Volume}}{\text{Surface area}} = \frac{2t^2 + 2T^2 + 2RT + 2Rt + 0.8584\,R^2}{4t + 4T + 6.2832\,R} \quad \text{(Eq 4)}$$

Because of the greater volume of metal at the junction, there are far more designs in which the J- or X-junction is the last to solidify compared to the T-junction curves discussed earlier. This is why many authors do not recommend the use of X-sections in casting design. However, this type of thinking can often cause excellent opportunities for cost reduction to be overlooked.

Merely because the junction portion of an X, T, L, or boss section solidifies last does not mean that the section is poorly designed. The most economically produced design will often take advantage of this very fact. The junction of a section solidifying last becomes a problem only when it violates the second rule of economical casting design; that is, it leaves that portion of the casting solidifying last in an inaccessible area for riser placement. Although accessibility for riser placement is essential, limiting the number of risers required can further improve economy by design. *Thus, the third rule for economical design is to create designs requiring as few risers as possible.*

Solidification proceeds through the casting in what is called the solidification or feeding path. Each feeding path will require a riser, and each riser adds to the cost of the casting. Less expensive and better-quality castings will result from reducing the number of feeding paths and ensuring that the end of each feeding path is accessible. Therefore, designs in which the junctions solidify last can be advantageously used to combine feeding paths, thus reducing the overall cost while increasing the quality of the part.

Three Legs Equal. The second type of X-section is shown in Fig. 9. In this design, three of the plates constituting the section have the same thickness. The corresponding solidification sequence curves for this case in which $R = 0$ and $R = 2$ are shown in Fig. 10(a) and (b), respectively. There is a difference in the curves across the isothickness line running diagonally from left to right on the curves. The upper left portions of the curves represent those designs in which a thick plate of dimension t is being chilled by the three equal thickness plates whose dimensions are T. This portion of the curve is similar to the curves discussed previously and is fairly straightforward.

The lower right-hand portion of the curve represents the case in which a T-section is to be chilled by the addition of a thin plate. In this case, if the thin plate is too small, it does not provide sufficient surface area to chill the T-junction, thus causing it to solidify before the plates making up the T-junction. This is the t-T-J sequence portion near the horizontal axis of the curve. Only when the t plate provides sufficient surface area to chill the T-junction can the t-J-T sequence be achieved. Finally, as the thinner plate continues to increase, the t-T-J sequence is again encountered.

When one of the legs that make up the X-section has a different dimension than the remaining three legs, two separate casting moduli equations must be used. For the case in which the odd plate labeled t on Fig. 9 is larger than the other three plates labeled T, the following casting modulus equation applies for the resulting J junction:

$$\frac{\text{Volume}}{\text{Surface area}} = \frac{t^2 + 3T^2 + 3RT + Rt + Tt + 0.8584\,R^2}{5T + 3t + 6.2832\,R} \quad \text{(Eq 5)}$$

Equation 5 will apply to the upper left corner of the resulting design curves in which t is greater than T. For the case in which the dimension T is greater than t, the following casting modulus applies for the J junction:

$$\frac{\text{Volume}}{\text{Surface area}} = \frac{6t^2 + 3T^2 + tR + 3RT + 0.8584\,R^2}{11t + 7T + 6.2832\,R} \quad \text{(Eq 6)}$$

The effect of adding a 2 in. radius to this X-section is identical to that seen on the T-section previously discussed. This can be demonstrated by comparing Fig. 10(a) and (b). Design curves for various fillet radii are shown in Fig. 10(c) and (d).

L-Sections

The third basic shape of importance to design engineers is the L-section. The model of an L-section is shown in Fig. 11. Two different radii are used to describe this section: internal radius r and external radius R. Because two different radii are required in describing this type of section, two design cases arise: one in which R is greater than r and one in which R is less than r.

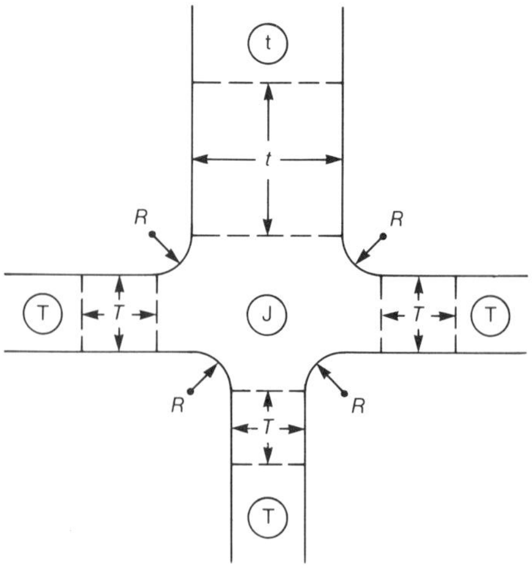

Fig. 9 Model of X-section with three legs equal. See Fig. 10 for solidification sequences.

For the case in which the internal radius r is less than the external radius R, an L-section similar to that shown in Fig. 11 will result. For cases in which r is greater than R, L-sections of the form shown in Fig. 12 will result.

External Radius Greater than Internal Radius. The casting moduli equations for the J- or L-junction in which $R > r$ are also a function of the design parameters being considered, and four equations must be used. For $R > t + r$ and $R < T + r$, the casting modulus of the junction is:

$$\frac{\text{Volume}}{\text{Surface area}} = \frac{T^2 + t^2 + TR + tr + tT + 0.2146\,r^2 - 0.2146\,R^2}{t + 3T + 1.5707\,r + 1.5707\,R} \quad \text{(Eq 7)}$$

For $R > T + r$ and $R > t + r$, the casting modulus of the junction is:

$$\frac{\text{Volume}}{\text{Surface area}} = \frac{T^2 + t^2 + RT - tT + tR + 0.2146\,r^2 + 0.2146\,R^2}{t + T + 2.8584\,R - 0.4292\,r} \quad \text{(Eq 8)}$$

For $R < T + r$ and $R < t + r$, the casting modulus of the junction is:

$$\frac{\text{Volume}}{\text{Surface area}} = \frac{T^2 + t^2 + Tt + rt + rT + 0.2146\,r^2 - 0.2146\,R^2}{3T + 3t + 2.8584\,r - 0.4292\,R} \quad \text{(Eq 9)}$$

For $R > T + r$ and $R < t + r$, the casting modulus of the junction is:

$$\frac{\text{Volume}}{\text{Surface area}} = \frac{T^2 + t^2 + Rt + Tr + tT + 0.2146\,r^2 - 0.2146\,R^2}{3t + T + 1.5707\,R + 1.5707\,r} \quad \text{(Eq 10)}$$

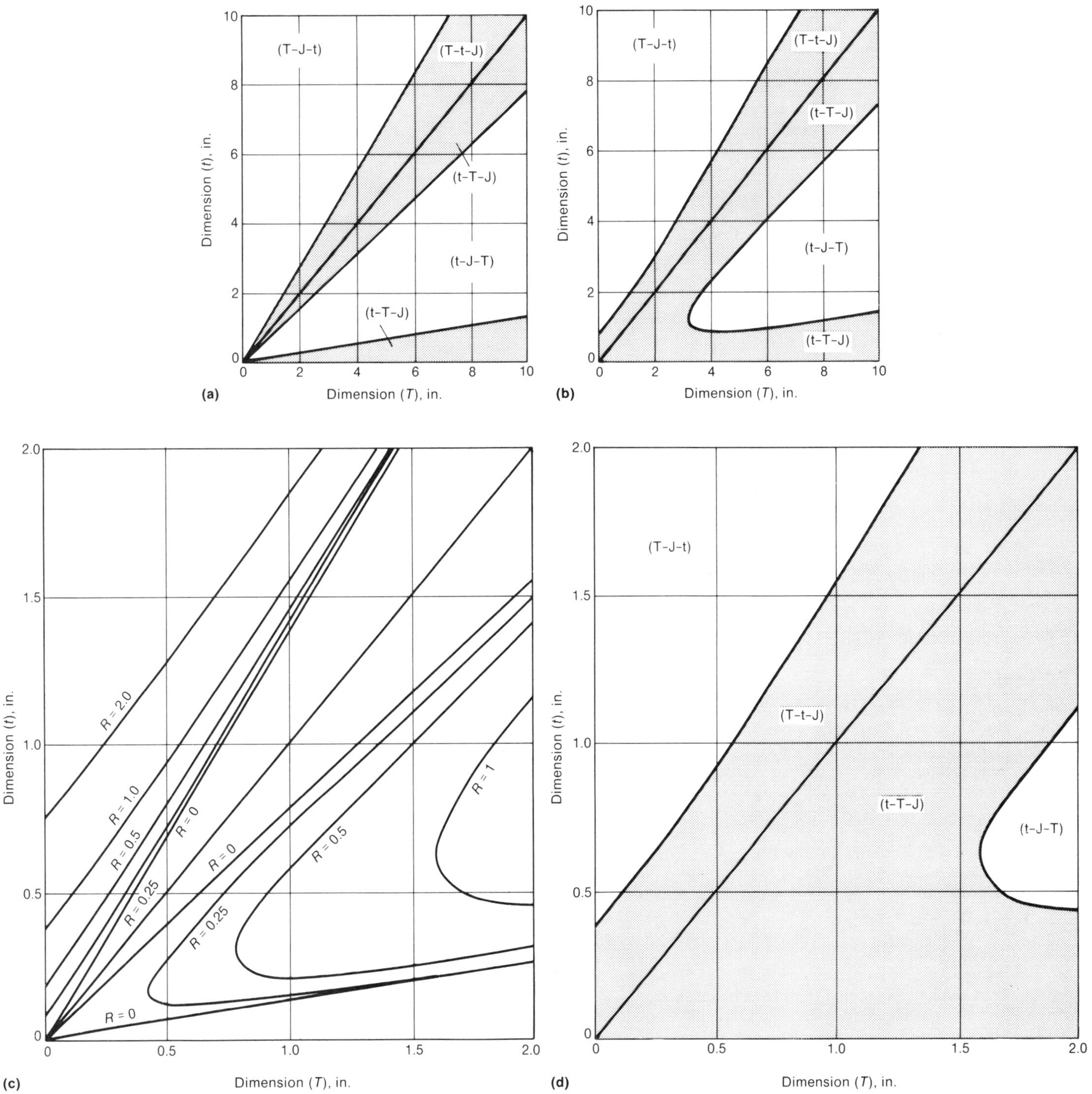

Fig. 10 Solidification sequence curves for X-section with three legs equal. (a) $R = 0$. (b) $R = 2$. (c) Composite of design curves for various fillet radii. (d) One segment ($R = 1$) of the composite design curve shown in (c)

Although the sequencing curves for $R > r$ shown in Fig. 13(a) and (b) seem complex, the complexity is easily understood by remembering that both the internal and external radii (r and R, respectively) are held constant on any given graph. The first unusual feature of graphs of this type is the forbidden zone, that is, the dark area in the lower left corner of the graph. In this area, L-sections using the radii combinations given in the figure caption are impossible, and the internal radius punctures the external radius (Fig. 14a).

Another unusual feature of these curves is the crossover point, which in Fig. 13(a) occurs at $t = T = 0.25$. At this point, the casting moduli of all the components of the L-section (t, T, and J) are equal. This occurs because a uniformly thick plate results from these particular design conditions, as shown in Fig. 14(b). The crossover point always occurs when $t = T = (R - r)$ for L-sections in which $R > r$. The effects of increasing the radii can be seen by comparing Fig. 13(a) with 13(d).

External Radius Less Than Internal Radius. If the external radius R of an L-section is designed to be less than the internal radius r, completely different sequencing curves result. The simplified geometry of L-sections in which $r > R$ also considerably simplifies the casting modulus of the resulting J- or L-junctions. Because

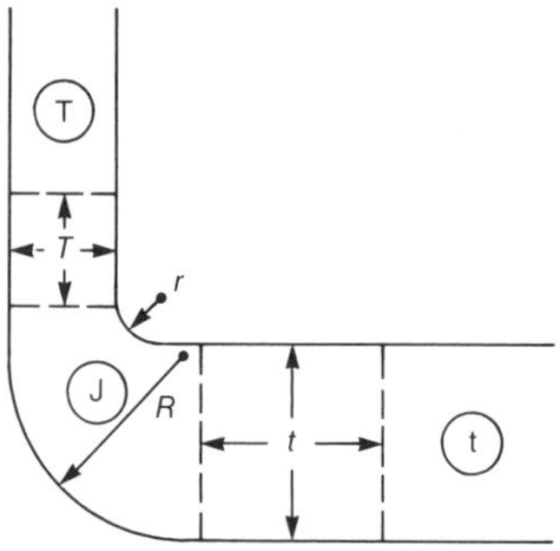

Fig. 11 Model of L-section for $R > r$. Compare with Fig. 12. See also Fig. 13 for solidification sequences.

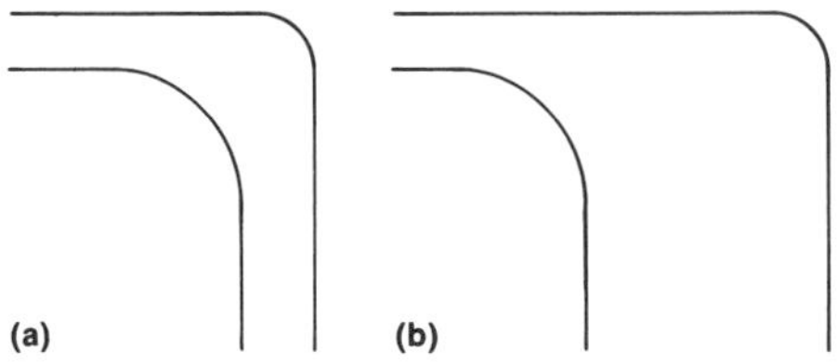

Fig. 12 Two L-section designs that can occur when $R < r$. (a) L-section has a higher modulus than the plates when plate thicknesses are nearly equal. (b) Plate thicknesses are very unequal; the modulus of the L-section is less than that of the thick plate.

of the geometry, only the following casting modulus equation is required:

$$\frac{\text{Volume}}{\text{Surface area}} = \frac{T^2 + t^2 + rT + rt + Tt + 0.2146\,r^2 - 0.2146\,R^2}{3T + 3t + 3.5707\,r - 0.4292\,R} \quad \text{(Eq 11)}$$

Curves for L-sections with two such $R < r$ combinations are shown in Fig. 15. No crossover point occurs, nor does the curve contain a forbidden zone. The reasons for this can be determined by noting the features of such L-section designs (Fig. 12).

Feeding

Solidification of Flat Plates. The L-, X-, and T-sections discussed to this point consist of a combination of plates and the junctions themselves. Any discussion of casting design must also consider the solidification of simple flat plates as well as their more complex junctions.

The solidification of metals does not occur as a flat moving front, but rather as a series of protuberances ranging in shape from cone points to very complex, branched shapes called dendrites. The complexity of these shapes is a function of the speed of growth of the front, the cooling rates involved, and the composition of the alloy. For the purposes of design, the nature of these shapes concerns the ability of metal to develop a feeding path through uniform flat plates or cylinders. Because these protuberances in most alloys of interest at commonly found growth rates and thermal gradients usually manifest themselves as dendrite-type shapes, they can provide a significant barrier to the continued flow of metal. This phenomenon can lead to shrinkage defects in castings because the feed metal supply cannot reach into the solidifying front at which the liquid-solid contraction is causing a feed metal demand. In particular, this effect influences casting design in the area of feeding distances.

In a simple plate having a constant thickness T, a set of solidification fronts will move from each of its surfaces until they meet at its centerline. However, as these fronts move together, their dendritic character begins to cause a blockage of the metal flow from a feeding path, which will result in some degree of shrinkage. This phenomenon has been quantified, and it has been determined that the ability of a riser to feed through a plate is very limited (Ref 1). Often, sound casting production can be achieved only for a few thicknesses along the plate even when chills are used. This is very important because long flat plates will require extensive risering for this reason alone. This limited ability to feed can be overcome by simply reducing the tendency for the front closure to occur as a straight line. This is accomplished by using a plate whose cross section is not uniform in thickness but is continuously changing in thickness, such as the wedge casting discussed earlier.

Another way of accomplishing the same objective without changing the geometry of the plate is to alter the thermal shape of the casting through the use of an insulating or chilling material to change the rate of heat extraction. Still another way of accomplishing feeding through an extensive constant plate thickness is to create a series of feeding paths close to each other.

Consider a circular flat plate with a single riser at the center of the plate, as shown in Fig. 16(a). If the feeding distance creates a problem, multiple feeding paths could be considered as a solution. The design of feeding paths involves the development of shape geometries that permit solidification to be directed to a portion of the casting that can be easily fed by a riser.

For this example, it is necessary to create a feeding path to the center of the plate where it can be fed by the riser. If the addition of radial ribs presents no problem, wedge-shaped ribs could be added, with the taper increasing as the rib nears the center of the plate (Fig. 16b). Because feeding distance is a problem only for continuous-dimensioned flat plates if the distance between these ribs does not exceed the feeding distance for the plate thickness, the ribs will ensure feeding of the plates to soundness. Because the ribs themselves are tapered to create a path to the riser, they should also be sound (free of shrink). If feeding distance again becomes a problem at the outer circumference of the circular plate, the ribs could be branched to ensure that the feeding distance for the flat plate is maintained (Fig. 16c). If the geometry of the rib causes difficulty with the design of the resulting casting, the same idea could be applied through the use of insulating or chilling materials to create the same effect thermally without the addition of the added metal of the ribs. However, because the use of molding materials to create the needed feeding path would no doubt increase the cost and complexity of the solution, it should be avoided if possible.

Solidification of Cylinders. A restriction of feeding distances similar to that in plates also occurs in cylinders. Once again, this can be overcome through a change in the thermal shape by developing a cone shape either geometrically or through the use of chilling or insulating materials. Cylinders, however, do create a new condition.

The new condition of concern is the junctions produced by the intersection of a cylinder and other cylinders or plates. If the cylinders are hollow, it is not necessary to consider them separately, because these conditions can be visualized as L- or T-sections merely revolved about an axis. However, a separate geometric case develops when the cylinder is solid. Figure 17 shows an example of such a case as well as the resulting design curve. On this curve, the solidification sequence is either D-J-T or T-J-D, and no case exists in which the junction J is the last to freeze. For such a junction, either the cylinder acts as a chill for the junction or the plate tends to act as a chill dependent on the cylinder diameter D and the plate thickness T. Thus, such intersections are straightforward and should cause little problem for designers or casting engineers if the feeding distance relationships and other thermal shape factors for promoting directional design are followed.

Changing Thermal Shape

Thus far, this article has considered the designs of plates, cylinders, and T-, X-, and L-sections, in which all of the components of the mold extract heat uniformly. However, there are many instances in casting manufacture in which this is not the case. Such situations occur during the normal course of the casting process and in some cases are done intentionally to change the solidification sequence—for example, the use of a core or a chill.

Use of Cores and Chills. Cores are often used in casting, but their effect on the solidification sequence of a section depends on the relative heat extractive capability or heat diffusivity differences between the mold and core components. If the heat extractive character of the core components is greater than that of the mold components,

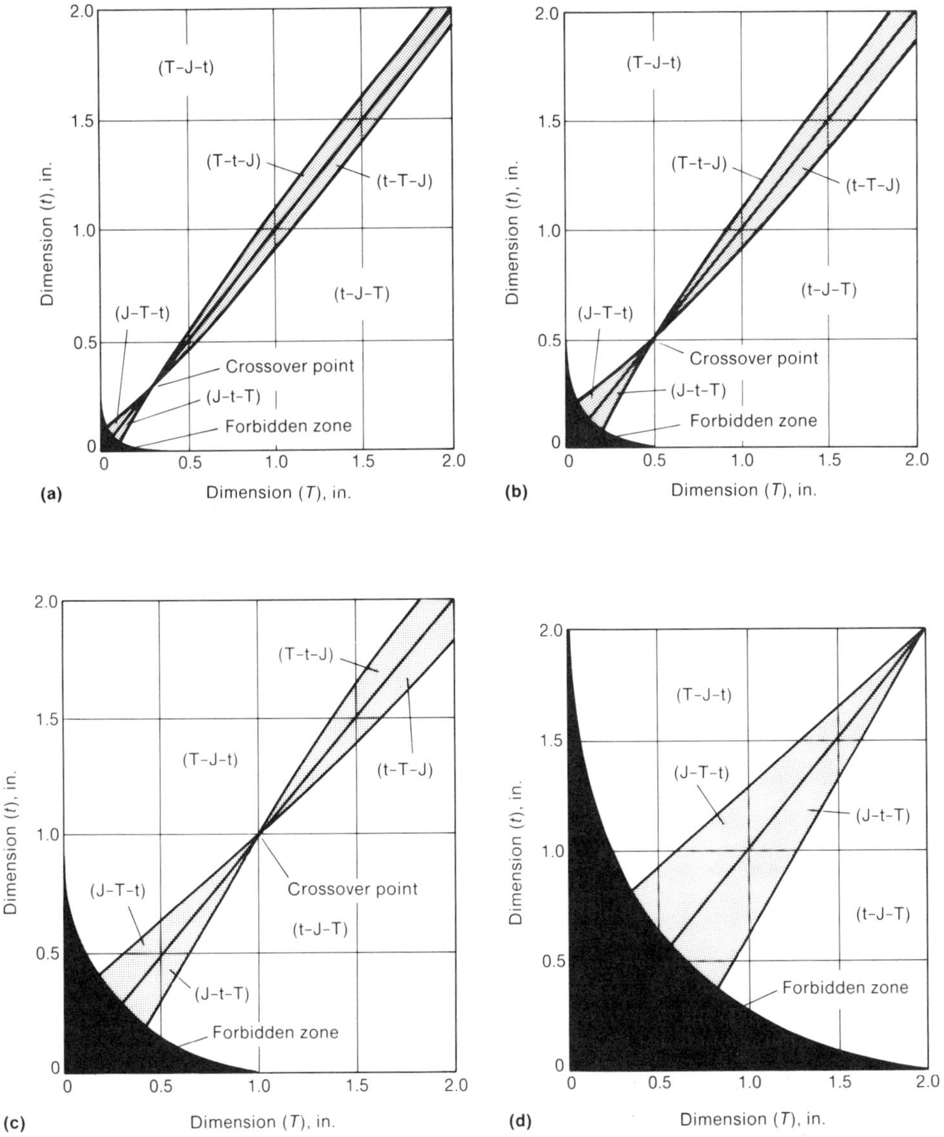

Fig. 13 Solidification sequence curves for the L-section shown in Fig. 11 with $R > r$. (a) $r = 0.25$, $R = 0.5$. (b) $r = 0.5$, $R = 1$. (c) $r = 1.0$, $R = 2.0$. (d) $r = 2.0$, $R = 4.0$

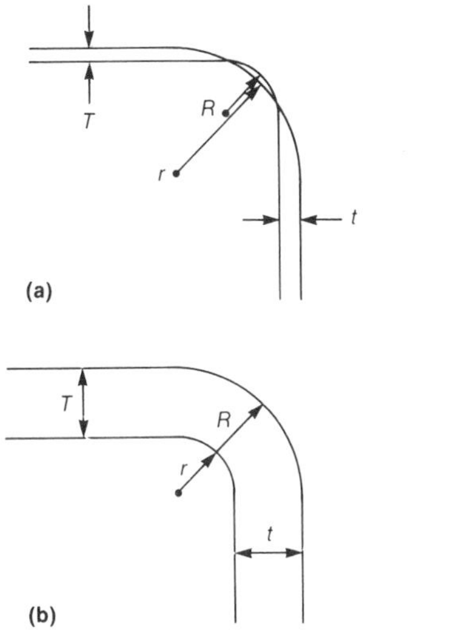

Fig. 14 L-section designs that fall in the forbidden zones of Fig. 13. (a) Internal radius r punctures external radius R. (b) Crossover point occurs when moduli of all the components are equal—in this case at $t = T = 0.25$, as shown in Fig. 13(a).

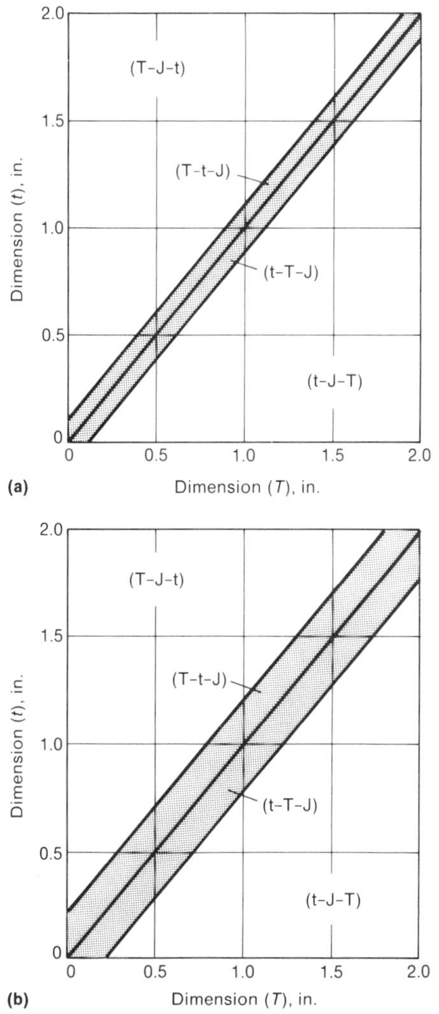

Fig. 15 Solidification sequence curves for L-sections with $r > R$. (a) $r = 0.5$, $R = 0.25$. (b) $r = 1$, $R = 0.5$. Compare with Fig. 13.

the core will tend to act as a chill. This occurs because heat is removed more efficiently from the portions of the casting in contact with the chill than from those in contact with mold components.

More commonly, the diffusivity of the core when made of resin bonded silica sand is lower than that of the green sand mold components, and the core acts as an insulator with respect to the mold components. It must be remembered that the relative heat extractive capacity of the various mold components determines the resulting solidification sequence. The results shown here are independent of the actual mold and cores being considered because the solidification sequence is important in casting design, not the time required for an individual portion of the shape to solidify.

Only in a few rare cases, as in continuous casting and slush casting (the manufacture of hollow castings through decantation of unsolidified liquid), is the actual time for solidification important. In these cases, the actual position of the solidification front at a particular time is important. This is not true for most casting design situations, in which only the sequence of events is important because the order affects the formation of a feeding path. However, the actual time required for an individual component to solidify is usually relatively unimportant if its position in the overall solidification sequence can be determined.

Because of this comparative unimportance of the actual solidification time of a given component, only the relative ratios of the thermal diffusivities of the two materials involved, rather than the actual diffusivity of each, need be considered in the design curves. Thus, to understand the major effects on the design of T-, X-, and L-sections, only two cases must be considered: one in which one mold component is insu-

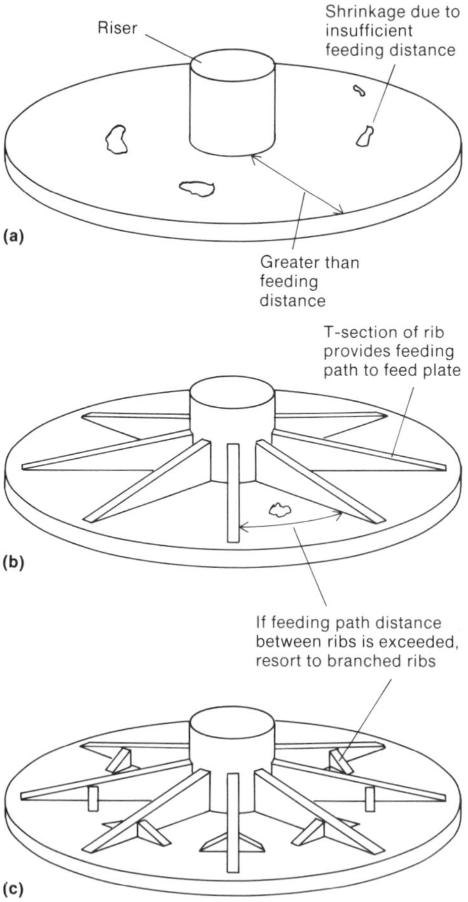

Fig. 16 Feeding path design considerations. (a) Circular flat plate with a single riser. (b) Addition of wedge-shaped ribs to ensure proper solidification. (c) Branched ribs to overcome feeding problems at the circumference of the plate

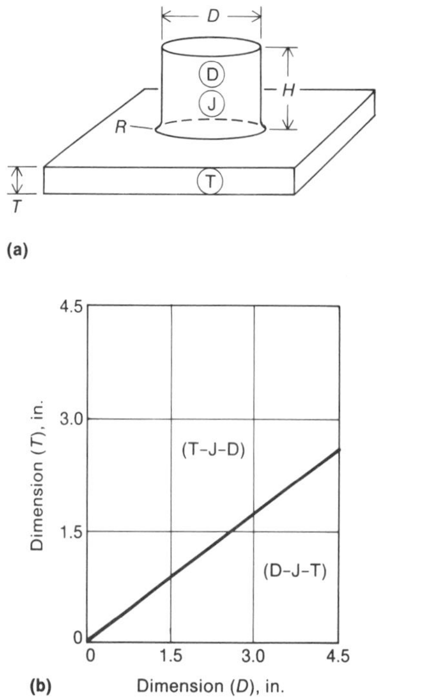

Fig. 17 Example of a design using a solid cylinder (a) and the solidification sequence curves for such a design (b)

lating with respect to the other, and a second in which one component has a chilling effect compared to the other. By convention, the diffusivity ratio is always defined as the ratio of a material divided by the diffusivity of the mold. For example, in a green sand mold, the ratio is the diffusivity of the non-green sand component and the diffusivity of green sand, and for a permanent mold application, the ratio equals the diffusivity of a given component/diffusivity of die material.

If one mold component is insulating with respect to the mold material, the diffusivity ratio would be less than 1. A common condition of this type exists for the case in which a resin bonded solid sand core is placed in a green sand mold. The diffusivity ratio would be approximately 0.85 for this case. However, as stated above, the actual materials involved are relatively unimportant provided the same diffusivity ratio exists; the principal concern is the ratio of diffusivities rather than their actual values.

L-Sections. Before discussing the design curves for L-sections, it is important to consider the geometric consequences of such placements. An L-section is shown in Fig. 18(a). In such a section, there are two possible positions, A and B, that might be occupied by the insulating material. The basic difference between these two possible placements has to do with the amount of casting surface area exposed to each. For placements of insulation material in the inner corner (the A position), less area at the J-junction is exposed to the insulation material per unit volume than is exposed along the plates making up the section. This is important because one might expect the placement of insulation material along a junction to slow the solidification of the section.

However, when the resulting L-section design curve (Fig. 18b) for this situation is observed, the exact opposite seems to be the case; that is, there are far fewer designs in which the junction solidifies last. This occurs because in position A the material extracts heat differently from the junction than the material along its outer area (position B). Because a greater percentage of surface area is contacted by a material in position A along the plates making up the L-section than from the junction itself, the plates are insulated to a greater degree than the junction. This means that insulation material placed in the A position tends to act as a chill from the point of view of the junction with respect to the plates making up the section. This occurs because a greater effect of insulation is experienced by the plates due to the increase in surface area exposed to the insulation material than by the junction. This results in a slowing of

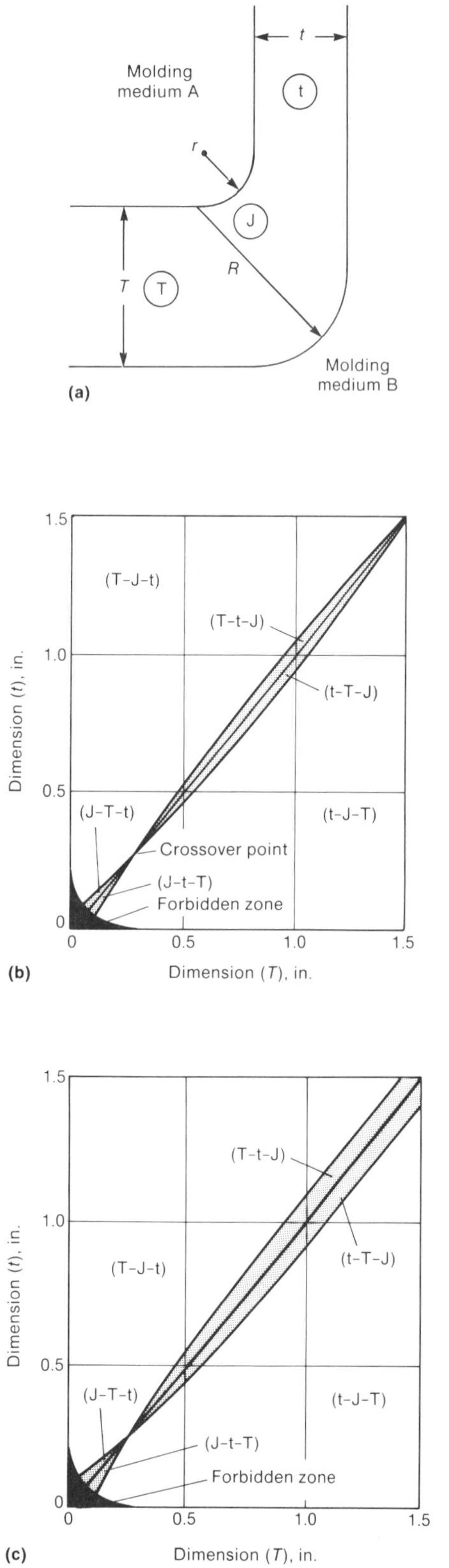

Fig. 18 Model for an L-section (a) cast using two different molding materials A and B. (b) Solidification sequence model for L-section shown in (a) in which the insulating material is in the A position; r = 0.25, R = 0.5, diffusivity of mold material A = 2, diffusivity of mold material B = 1. (c) Baseline curve for comparison with Fig. 18(b), 19, 20, and 21 in which r = 0.25, R = 0.5, A = 1, and B = 1 where insulating materials are the same for positions A and B. See also Fig. 19 to 21.

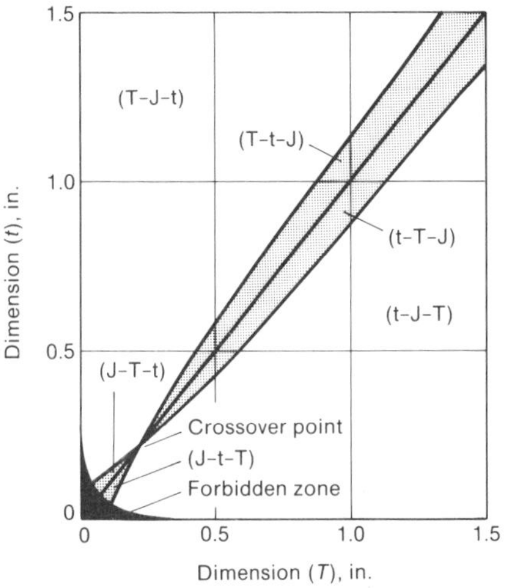

Fig. 19 Another solidification curve for the L-section shown in Fig. 18(a) in which a chill has been placed in the A position. In this case, $r = 0.25$, $R = 0.5$, A = 2, and B = 1. See also Fig. 18(b) and (c), 20, and 21.

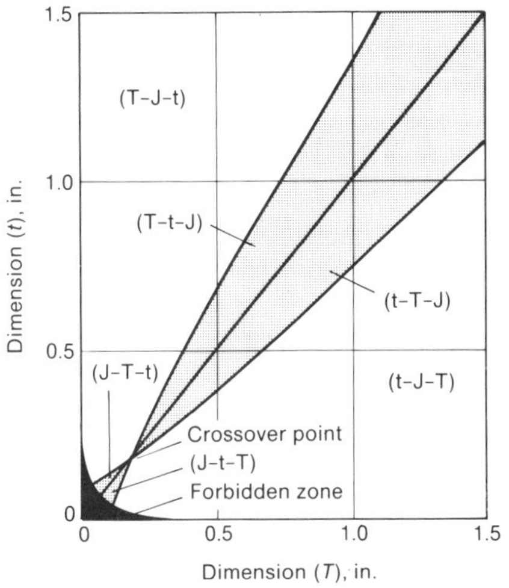

Fig. 20 Another solidification curve for the L-section shown in Fig. 18(a) in which the insulating material has been placed in the B position. $r = 0.25$, $R = 0.5$, A = 1, and B = 0.85. See also Fig. 18(b) and (c), 19, and 21.

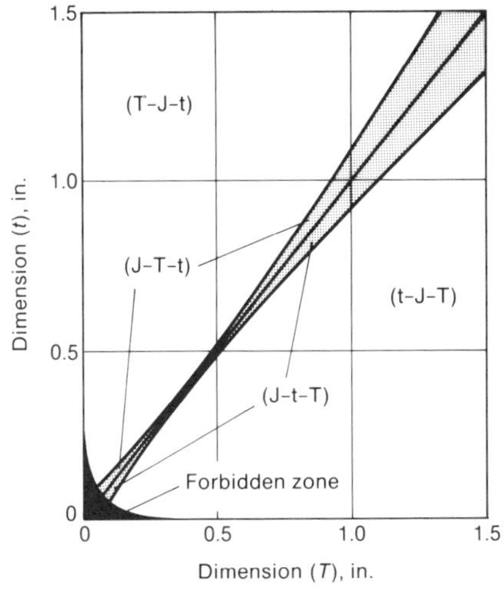

Fig. 21 Another solidification curve for the L-section shown in Fig. 18(a) in which a chill has been placed in the B position. $r = 0.25$, $R = 0.5$, A = 1, and B = 2. Note that there is no crossover point in this figure. See also Fig. 18(b) and (c), 19, and 20.

solidification of the plates more than its effect on the junction. Note that there are fewer designs in which J solidifies last in Fig. 18(b) than in Fig. 18(c), in which the mold (insulating) materials are the same for the A and B positions.

A similar apparent reversal of the expected effect when a chilling material is placed in the A position of an L-section can be seen in Fig. 19 for a relative diffusivity ratio of 2.0. Once again, the chill has a greater effect on the plates making up the L-section than on the junction itself. Such chill materials, when placed in the A position, tend to act as insulation materials from the point of view of the L-section. That is, because the plates are chilled to a greater degree than the junction due to the exposed surface area to the chill, chill placement in the A position creates more designs in which the junction tends to be the last portion to solidify, as can be seen in Fig. 19.

If material having an insulation character is placed in the B position of an L-section, it contacts a greater amount of surface area to volume in the junction than is contacted by the plates making up the section; the result of this is the reverse of that discussed for materials placed in the A position. This placement is illustrated in Fig. 20. Because the insulation effect is greater in this case on the junction due to the greater surface area contact, it creates more designs in which the L-junction solidifies after the plates. The effect of a chill placed in the B position can be seen in Fig. 21. It is clear from Fig. 18 that, because the chill affects the junction to a greater degree than the plates, the resulting design curves contain more designs in which the junction is the first to solidify. In addition, a radius in the A position placements tends to increase the area in contact with materials in the A position. However, external radii tend to lessen the surface area in contact with a material in the B position.

T-Sections. In the case of a T-section, there are three potential positions in which insulating or chilling materials can be placed, namely, positions A, B, and C, as can be seen in Fig. 22. The T-shape causes several interesting effects. The first of these is similar to that observed for the L-sections. Materials placed in either the A or B position have a greater influence on the plates making up the section than on the T-section itself. Similarly, material in the C position has a greater influence on the solidification of the T-junction than on the plates making up the section.

These effects are similar to those discussed for the L-sections, but because of the particular geometry involved in a T-section, several others also occur. For example, when the material in the C position is different from that in the A and B positions, the effects on the plates ta and tb are indistinguishable (Fig. 23). If the materials in the A and B positions are different, there will be a difference in the solidification of the two halves of the cross plates ta and tb. Figure 24 illustrates this effect, in which each of the plates ta and tb are shown to provide an independent section on the design curve. This effect is a result of the imbalance of the heat extraction across the intersecting plate labeled *T* in Fig. 22, and it is also evident in the use of chilling as well as insulating materials (Fig. 25). In addition, if A and C or B and C are similar materials, an imbalance still exists across the intersecting plate of the T-section, resulting in a design curve in which the plates ta and tb separate into individual effects (Fig. 24 and 25).

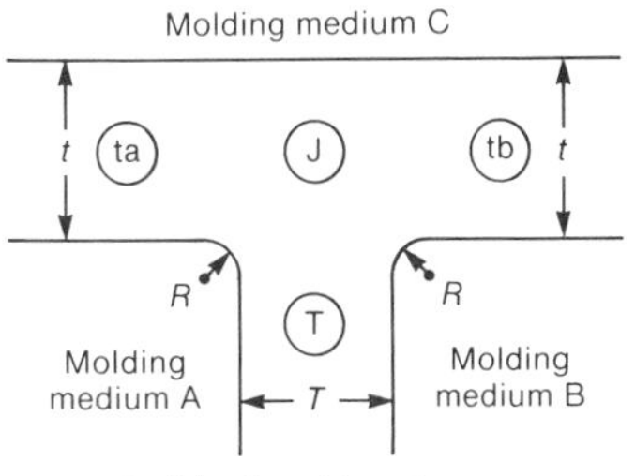

Fig. 22 Model of a T-section cast using three different mold materials. See Fig. 23 to 25 for possible solidification sequences.

X-Sections. In the case of X-sections, a geometry-related effect exists. When an insulating or chilling material is placed in a single position, it influences two of the plates but not the other two plates. It also

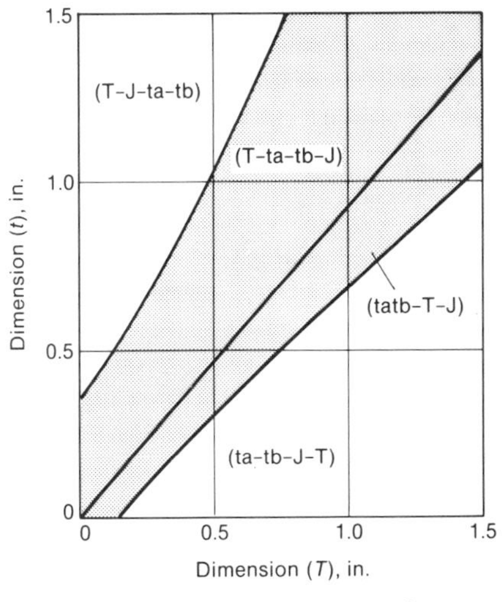

Fig. 23 Solidification sequence curve for T-section shown in Fig. 22 with $R = 0.5$, A = 1, B = 1, and C = 0.85. Note that in one case (area marked tatb-T-J), ta and tb solidify at the same time. See also Fig. 24 and 25.

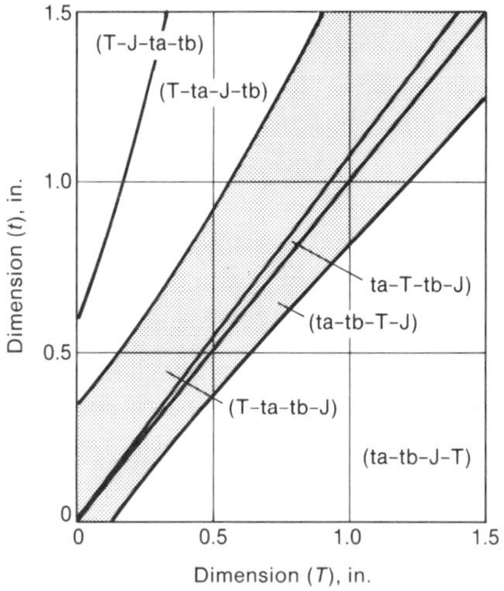

Fig. 24 Solidification sequence curve for T-section shown in Fig. 22 with $R = 0.5$, A = 1, B = 0.85, and C = 1. See also Fig. 23 and 25.

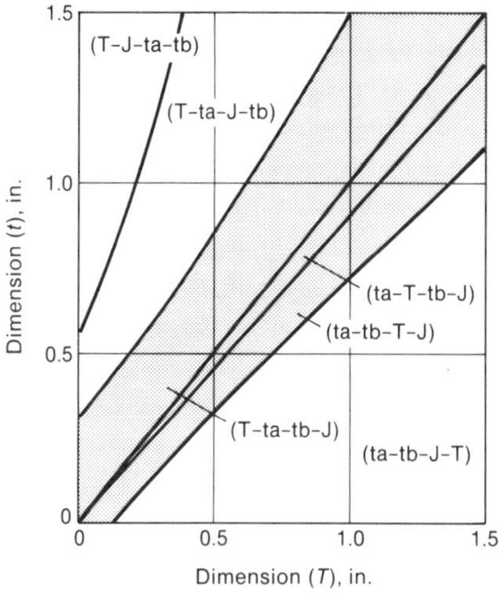

Fig. 25 Solidification sequence curve for T-section shown in Fig. 22 with $R = 0.5$, A = 0.85, B = 1, and C = 0.85. See also Fig. 23 and 24.

has an effect similar to placements on L-sections in their inner corner in that the material has a greater influence on the plates making up the section than on the X-section itself. This means that the effects are similar to those seen in L-sections discussed previously; chills tend to act as insulating materials from the point of view of the X-section. In addition, diagonal placements across the section cause a matched effect on all of the plates making up the section.

Therefore, when mold materials of different diffusivities are employed, the results can be predicted, but geometric effects can often produce results that may not be obvious. Thus, to predict the results of a given set of conditions, one must consider the effects very carefully. However, the amount of contact area of each portion of a section in direct contact with a chill or insulator plays a significant role in the eventual solidification behavior.

Computer-Aided Casting Design

Computer-aided design and manufacture (CAD/CAM) systems are becoming increasingly prevalent in industrial design, and their use offers both opportunities and problems for casting designers. These systems greatly reduce the cost of the engineering and design time involved in preparing engineering drawings. When using CAD systems, most designers still use primarily straight lines and chordal parts of circles. This is unfortunate, because the true power of such systems (as related to casting design) would be the full use of the superior design capability of such shapes as sines and parabolas. Designers should make the effort to consider such shapes for both improved stress distribution and better casting designs.

Another problem with the use of CAD/CAM systems is that many companies are using wireframe systems rather than the newer solid modelers. This situation will probably not change for some time, because solid modeler CAD/CAM systems require the designer to work with three-dimensional shapes instead of the more familiar two-dimensional features. Further, the thought process taught to designers is not one of removing area from an existing shape, as required by solid modelers, but rather of developing a shape through a construction procedure. Thus, even though solid modelers are better algorithms for casting design, it is unlikely that they will be extensively used in the near future.

Because wireframes seem to be the preferred technique for the present, designers must recognize and deal with the problems of this algorithm. Wireframe algorithms were designed to create pictures and graphic representations of the actual three-dimensional shape, not necessarily to produce shapes of true three-dimensional integrity. Until such three-dimensional representations of the desired shape are attempted to be used to create pattern equipment, the problem with these algorithms may not even be recognized. Because the shapes are not truly three dimensional, nonmatches of straight lines and arcs can go unnoticed until the pattern equipment attempts to use the CAD information to cut the pattern. This problem is recognized, and software is available to solve it and similar problems; but such software requires transfer of data from computer-aided design to pattern cutting, which is often difficult since data formats have not been standardized but are computer and software dependent.

If the full cost savings available through CAD-generated pattern equipment is to be realized, improved wireframe algorithms or solid modelers that eliminate these kinds of problems must be developed and used widely. There is also a significant need for better software that truly allows the error-free passage of CAD/CAM data to other computers and other software.

One very effective computer tool is available that uses CAD casting designs to simulate the solidification of a casting through solidification modeling. The developed algorithms of such simulations can be divided into three basic types:

- Those based on classical heat transfer equations
- Those based on finite-element methods or finite-difference methods (FEM or FDM), which are iterative techniques based on classical heat transfer
- Those based on ratios of surface area to volume, as discussed by Kotschi and Plutshack (Ref 2)

Heat Transfer Analysis. The first of these approaches involves attempts to write a single set of heat transfer equations to describe a solidification problem. Because of the complexity of the differential equations involved, this approach has not led to substantial developments. Further, many of the differential equations required for such a solution cannot be solved without significant simplification of the equations; therefore, the potential for this technique is limited until such techniques can be developed. Solutions to these problems have been attempted, but the sheer magnitude and complexity of the equations and iterations required even for supercomputers holds little hope of a viable solution of economic interest to industrial designers. This leaves the two remaining approaches, both of which have provided workable and usable algorithms to solve this problem.

FEM and FDM Analysis. Use of either the FDM or FEM technique involves the development of a three-dimensional grid. The grid blocks are then used as a set of differential elements to calculate the heating and cooling effects. The heat content of each element and its relationship to all the other grid elements in its influence are calculated. In this way, a solution of the entire shape is developed through a series of iterations as a function of time.

Selection of the grid points is critical to the accuracy of FEM or FDM solutions. The greater the accuracy desired from such calculations, the finer the mesh of the required grid. However, the cost of the solution increases dramatically as the number of grid points increases. Thus, to reduce the cost of such solutions, grid points are usually increased in areas of interest or areas having a significant effect on the solution. Similarly, the number of grid points is reduced in areas deemed less significant to the final solution. This has been a major problem in the extensive use of such systems in industry.

Unfortunately, the high-accuracy FEM or FDM systems employed to date have had no grid selection algorithm; therefore, selection is done by personnel skilled in this selection process and in FEM or FDM techniques. Such talent is not readily available, and even when it is, the expenditure of several weeks by several individuals on such a problem is hardly cost effective, especially since this is only 1 of 10 to 15 solutions required to develop an optimum manufacturable casting design.

An example of the cost of finite-element and finite-difference methods is provided by the analysis recently performed on a three-dimensional section of an engine block. A single cylinder (one of six in the complete engine) and its associated valve cavities were studied. Development of the grid and setup of the problem took four individuals a month. The actual analysis required 8 h on a supercomputer. With setup and grid development costs of $10 000 and computer time costs of $25 per minute (a very low estimate), the cost of a single analysis was $22 000. Considering that ten or more analyses may be required to obtain the optimal design, it can be seen that such analyses are not economically feasible for most companies.

The complexity of such problems, and therefore the cost of the solutions, can be reduced by simplifying the problem through, for example, the use of fewer grid points. In this simplified form, particularly in the case of only two-dimensional slices through a critical area, the computer overhead can be greatly reduced, and economically useful solutions can be found.

An example of such software using an FDM approach is the AFSolid Software available from the American Foundrymen's Society. This software analyzes only in two dimensions, incorporating a uniform pattern of grid points for ease of use, and it limits the number of grid points to 5000 as a maximum to hold down the iterative overhead. At the maximum number of grid points, the program can take approximately 15 to 20 min on a high-end desktop computer such as an IBM PS/2 model 80 with a math co-processor without including heat of fusion in the calculation; the program can take 2 h if heat of fusion is included. At this level of sophistication of both hardware and software, economically available answers are practical and cost effective. However, this type of software is of more value to foundrymen than to design engineers because it requires knowledge of the potential solidification problem to such a degree that it can be reduced to a two-dimensional problem.

Unfortunately, to assist design engineers at the drawing board or CAD station, three-dimensional solidification analysis software would be of more value, particularly if it were easily accessed from the CAD software. However, to go from two dimensions to three dimensions with the FDM or FEM approach, the complexity of the problem increases geometrically. For this reason, without substantial increases in hardware and software sophistication and speed, this approach will see only limited use in the future for the three-dimensional problem.

Volume-to-Surface-Area Ratios. Another algorithm was first proposed by Kotschi and Plutshack based on the use of volume-to-surface-area contour lines. These lines were shown to simulate the progression of solidification throughout the cross section of a two-dimensional slice of a solidifying shape. The importance of this approach is in the simplification of the algorithm and the resulting higher speed and lower cost for analysis. It must be remembered that, without low-cost analysis techniques, solidification simulation is of only academic interest. The method can also be applied to three-dimensional shapes with less increase in computer software overhead than with FEM or FDM techniques.

Software based on the use of the casting modulus contours suggested by Kotschi and Plutshack has been developed at the University of Wisconsin through a foundry consortium. This software, called SWIFT, holds much promise as a potential solution to the need for a three-dimensional solidification analysis that is useful to design engineers, but at present the marketing rights to the three-dimensional software are available only to consortium members. However, a two-dimensional system is being marketed, but is not available for MS-DOS or OS/2 machines. Most of the two-dimensional software packages of this type are currently running on VAX and Mini-VAX type machines.

An alternative software program developed for three-dimensional solidification analyses is SOLSTAR*. Although the algorithm is proprietary, the method is based on a type of FDM approach in which the time differential variable is eliminated from the model and each node is in some way compared with other nodes to determine the solidification sequence. The maximum capacity of SOLSTAR is 300 000 node points for a three-dimensional problem. Because of the simplification of the FDM approach, the computer running time for a 300 000 point simulation is reported to be 45 min on a system similar to the IBM PS/2 discussed above.

Only by simplifying the FDM/FEM models or by implementing new approaches can the computer overhead be reduced enough to provide economical solutions that can be used by the majority of the foundry and casting design industry. If the puristic FDM/FEM solutions are all that are available, the benefits of such techniques will affect only those corporations significantly large enough and requiring sufficient production quantities to afford them. This will of course change if supercomputers are available in the future at personal computer prices.

*SOLSTAR is a registered trademark of Foseco Ltd.

Mold Complexity

Casting solidification is the first major factor a designer should consider in reducing costs. However, there is a second major factor that should also be considered, namely, the mold complexity factor. This is usually related to a requirement for an excessive number of cores. *Thus, the fourth rule for economical casting design is to eliminate as many design features requiring cores as possible.*

The use of cores in the casting process provides a unique feature that is not available in most other methods of manufacture, yet each core required adds to the final casting cost. Thus, only those cores that are absolutely necessary for producing the desired shape should be used if design is to be directed toward lower cost castings.

The first principle that must be understood to eliminate cores is the method of mold manufacture. Even investment and expendable pattern (lost foam) molds require that the molds or tooling used to make the patterns be fabricated as separate pieces or mold halves. For example, in sand processes, a boxlike device called a flask is set over the pattern and filled with sand. The sand is hardened either through chemical or mechanical means, and the pattern is then removed. The pattern must be removed by drawing it away from the mold in a direction perpendicular to the parting line; therefore, the design factors related to this common practice must be considered by the design or casting engineers.

The required factor that assists pattern withdrawal is called draft. As can be seen in Fig. 26, the need to withdraw the pattern from the mold requires that some taper or draft be added to the pattern. Although pattern draft is usually not a problem, the requirement of removing the pattern by withdrawing it in a direction 90° from the parting line does restrict total design freedom if casting costs are to be minimized.

Because the pattern is to be withdrawn straight out of the mold, no protrusions that restrict this movement can be allowed in the construction of the pattern. If the geometry of the part requires such protrusions, there are only two alternatives for the casting engineer:

- Use a loose piece that remains in the mold or core after the pattern is drawn and then is removed separately

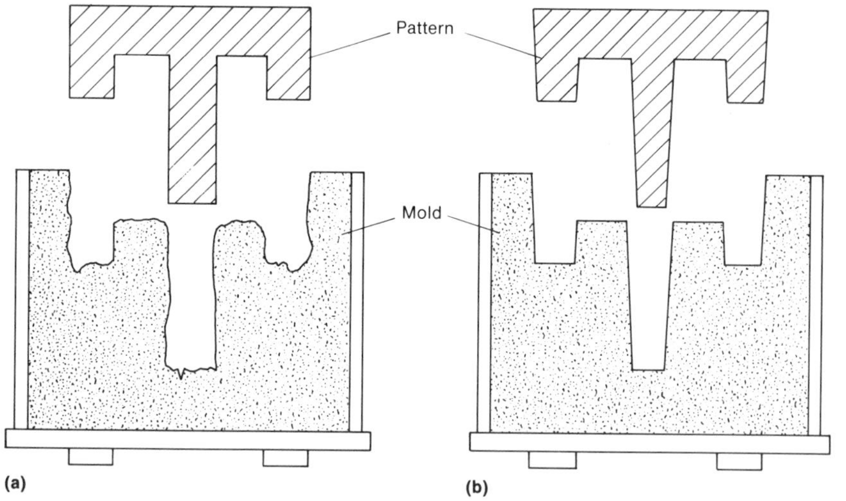

Fig. 26 Use of draft to facilitate withdrawal of the pattern from the mold material. (a) Lack of draft causes damage to the mold when the pattern is removed. (b) Ample draft allows easy pattern removal.

- Provide a portion of the pattern that creates a cavity for the setting of a core to create the geometry

The first alternative has limitations in that sufficient room and access to remove the loose pieces must exist. Loose pieces can be used only in moderately low volume casting production because they limit productivity in high production. Furthermore, because of the added handwork and potential for mold breakage during extraction, the use of loose pieces can be expected to increase casting costs even in low production. The second alternative involves the addition of a core and therefore also increases casting costs. The problem and some proposed solutions to undercuts are shown in Fig. 27.

Finally, because part geometry can require more complex patterns due to irregular parting lines, *the final rule for reducing casting costs is to design straight parting lines whenever possible*. This rule can be seen in practice in Fig. 28. Because much thicker patterns are required for offset parting lines, the cost of the pattern is also greater.

The factors that affect the resulting cost of castings and that have sufficient common traits to be independent of the casting process selected include:

- Requirements for surface finish and dimensional tolerances
- The number of castings to be produced
- The material selected for the casting
- The pattern material selected
- Any special or unusual testing/inspection or property requirements
- The casting design
- The casting process selected

Although casting costs are affected by these seven points, casting cost is only one element in the equation that determines the shipping dock cost of the final part headed to the consumer. If subsequent costs, for example, in machining, can be decreased by purchasing a more expensive casting, then the customer must weigh the potential cost savings against the cost of a less expensive casting.

The way in which consideration of casting and pattern costs alone can lead to false economies can be easily demonstrated by an example of the manufacture of a simple cube. The choice of parting line is usually not even considered by a casting customer, but is left to the quoting foundry. Because of this, many casting customers miss opportunities for innovative engineering solutions to shipping dock cost reduction.

For the cube example, there is a need to have the opposite sides of each face of the cube parallel and at right angles to one another. However, the need for draft angles on the tooling to allow removal of the pattern is in opposition to this. If this requirement is not conveyed to the casting supplier, he will undoubtedly quote the pattern equipment with a parting line as shown in Fig. 29(a). This parting line will be chosen by a casting supplier since it will produce the least costly pattern because it is the simplest pattern to make. Casting quotes are often rejected because of pattern prices; therefore, the casting supplier will generally opt for the lowest cost pattern.

This choice of parting line will produce a cube in which only two of the opposing sides will be parallel to each other. A different parting that results in a slightly more expensive pattern is one in which the diagonal of two sides of the cube is used as the parting line (Fig. 29b). This would produce a part in which four of the opposing sides of the cube are parallel and at right angles to one another. However, the complicated parting line shown in Fig. 29(c) eliminates the need for draft on the tooling. Thus, all

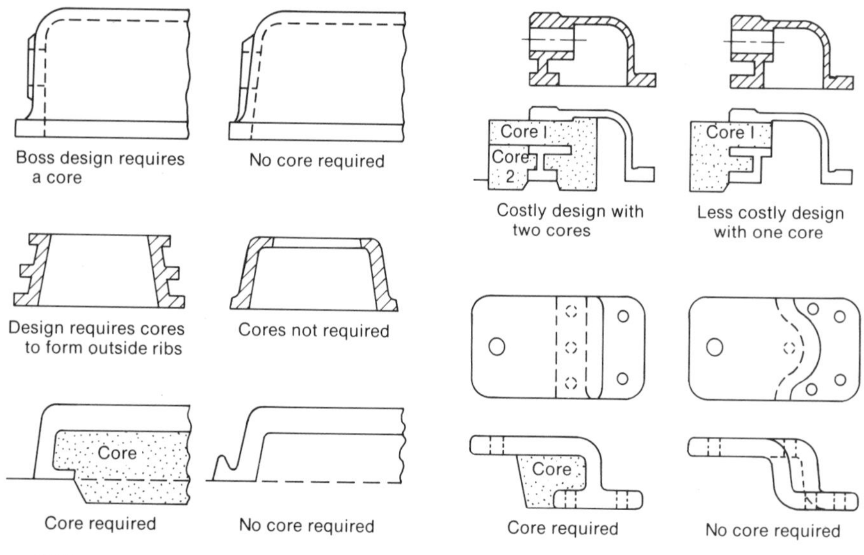

Fig. 27 Restrictions to pattern removal, and some potential solutions

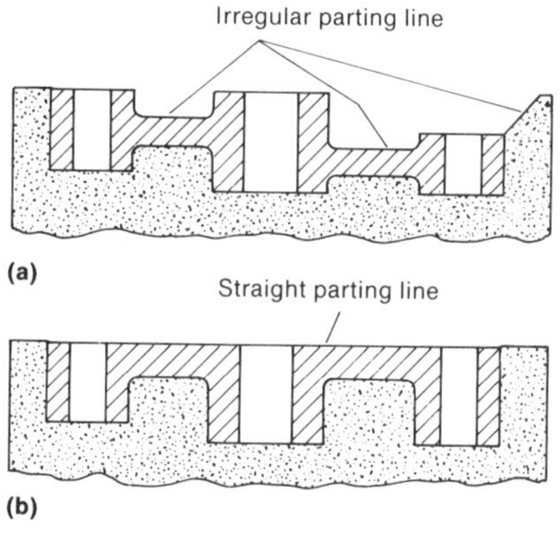

Fig. 28 Designing for a straight parting line to reduce pattern and casting costs. (a) Irregular parting line is a costly design. (b) Straight parting line is less expensive.

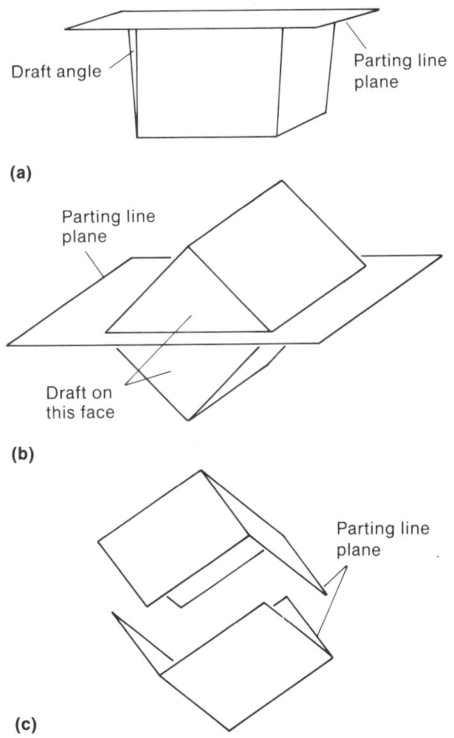

Fig. 29 Parting line options for a cube that must have as many sides as possible parallel and at 90° to each other. (a) Cube parted with four drafted sides is the least expensive option but does not meet design requirements. (b) Parting along the diagonal is a moderate-cost solution that results in four faces in compliance with design requirements. (c) Irregular parting is the most costly option, but is the only one that allows all sides to be parallel and at 90° to each other.

six sides of the cube can be made parallel to their opposing sides, and all of the sides will be at right angles to each other. This will be the most expensive pattern because of the complexity of the geometry involved. However, based on the requirement for parallel sides, it is the least costly method of producing the required shape.

The parting line should be considered in the design phase of a casting because of its importance in the final cost of the part. If it is not specified on the drawing, it should be requested to be returned with any casting quote; only in this way can the customer be assured of comparing equivalent manufacturing methods on the returned quotes. The customer should also request information on cores and other related production tooling in a quote. Only with this information can the best alternative be found for very low shipping dock cost. Without such information, the potential of various tooling designs cannot be determined, let alone used to develop a unique manufacturing approach to reduce the shipping dock cost. However, foundries usually consider such information to be proprietary.

If such manufacturing tooling design is of importance to the shipping dock cost of the final component, then it should be specified by the casting customer. This approach is virtually nonexistent in the industrial design of casting components, but it is essential if truly low cost component design is to be achieved.

REFERENCES

1. R.A. Flinn, *Fundamentals of Metal Casting*, Addison Wesley, 1963
2. R.M. Kotschi and L.A. Plutshack, An Easy and Inexpensive Technique to Study the Solidification of Castings in Three Dimensions, *Trans. AFS*, Vol 89, 1981, p 601-610

SELECTED REFERENCES

- J.B. Caine, *Design of Ferrous Castings*, American Foundrymen's Society, 1979
- *Casting Design Handbook*, American Society for Metals, 1962
- N. Chvorinov, Theory of the Solidification of Castings, *Giesserei*, Vol 27, 1940, p 177-225
- *Design of Aluminum Castings*, American Foundrymen's Society, 1973
- G.H. Geiger and D.R. Poirier, *Transport Phenomena In Metallurgy*, Addison-Wesley, 1973
- R.W. Heine, Feeding Paths for Risering Castings, *Trans. AFS*, Vol 76, 1968, p 463-469
- R.M. Kotschi and C.R. Loper, Jr., Design of T and X Sections for Castings, *Trans. AFS*, Vol 82, 1974, p 535-542
- R.M. Kotschi and C.R. Loper, Jr., Effect of Chills and Cores on the Design of Junctions in Castings, *Trans. AFS*, Vol 84, 1976, p 631-640
- R.M. Kotschi, C.R. Loper, Jr., R.E. Frankenberg, and L. Janowski, Elimination of Shrinkage Defects Through Casting Redesign, *Trans. AFS*, Vol 85, 1977, p 571-576
- C.R. Loper, Jr., and R.M. Kotschi, Design of Bosses and L Sections for Casting, *Trans. AFS*, Vol 83, 1975, p 173-184
- C.T. Marek, *Fundamentals in the Production and Design of Castings*, John Wiley & Sons, 1950
- O.W. Smalley, *Fundamentals of Casting Design as Influenced by Foundry Practice*, Meehanite Metal Corporation, 1950

Dimensional Tolerances and Allowances

Daniel E. Groteke, Metcast Associates, Inc.

NO TWO MANUFACTURED PARTS are ever exactly alike, regardless of the process by which they are made. For example, castings made with supposedly identical processing in the same foundry will differ slightly from one another in terms of dimensions. This article will discuss the nature and causes of as-cast dimensional variations and the consequences of these variations in relation to tolerances.

Even more dramatic differences in dimensional reproducibility are seen when two casting sources manufacture the same casting. This is true even if both sources are using the same tooling to manufacture the shape, and it is a reflection of the total process capability of the source. The major variables contributing to the dimensional spread will be discussed in detail for each process method, and although they have some commonality among casting methods, the impact on the final shape can vary greatly, depending on the level of planning and process control exercised. It is acknowledged that the designer cannot exercise total control over the processing variables that cause dimensional variations, but an awareness of their existence and an understanding of their nature will assist him in specifying economically attainable tolerances.

An engineering decision to select a manufacturing process for the fabrication of a particular component involves many factors, including the quantity requirements, available processes, choice of materials, contact with qualified vendors, secondary operations after the initial fabrication, final component delivery schedules, and the basic manufacturing and economic decisions dealing with anticipated component product life. Because the cost effectiveness of a product is of primary concern, the designer is rarely handed a piece of blank drafting paper with a completely free choice of processes to produce any given component. Instead, the choices are limited to a relatively narrow spectrum by the existing corporate facilities of the employer, relations with vendors, production timetables and deadlines, and targeted manufacturing costs.

Therefore, unless a completely new product line is being considered for design and manufacture, general decisions have been made for the designer before the first line is drawn to define the shape of the component. This should not be thought of as an oppressive limitation on product design, because all processes and materials, including the traditional casting processes, are in a constant state of flux in order to improve their competitive positions and to expand their markets. Many previous assumptions about the process capability of any particular casting method should be challenged, whether on the basis of properties obtained in the final shape or even the dimensional reproducibility of any feature. Factors influencing these considerations include the availability of precision tooling, automation of operations to eliminate cost and human factors in production, improvements in materials, and stabilization of processes through statistical process control. Another significant factor is the attitude of the personnel in the producing organization and their interest in working with the designer to maximize process potential while emphasizing cost reduction for the final product configuration.

It is especially important to recognize that achieving the lowest possible cost in the final shape (while meeting function, service conditions, and product life requirements) is the ultimate goal, and not the lowest cost in any initial manufacturing step. Sacrifices can be made in casting cost if they will be reflected as lower finished component prices. For example, additional cores, inspection, or gaging operations; more careful grinding and finishing; and so on, can all be easily justified if they will eliminate or reduce subsequent machining or finishing operations.

Process Variables

To determine the specific casting method that will be used to produce the part to customer specifications, the engineer must consider such process variables as tolerance requirements, process capability, and casting distortion attributable to both metal solidification and mold restraint.

Tolerance Requirements

The engineer is always advised to allow the most generous tolerances consistent with the proper functioning of the part. Normally, the dimensions and tolerances specified will represent a compromise between the function of the casting and the capability of the process selected. The goal of the designer and foundry staff should be to bring the casting as close as possible to a near-net shape as the limitations of function and process will allow.

If the function of a casting permits the assignment of dimensional tolerances that are normal with respect to the capabilities of the process by which the casting is to be made, cost and delivery will not be adversely affected by the attainment of such tolerances. If the tolerances chosen are not realistically matched to the casting process, higher costs and delivery delays will almost certainly be the result. The producing foundry will experience higher scrap in the manufacture of the shape or will necessarily add gaging and straightening operations to provide castings that meet the requirements. In extreme cases, a casting may not be producible in the initial foundry chosen to the close tolerances assigned. This will require selecting another foundry as vendor of the part, with all of the attendant delays and possible retooling costs.

It should be noted that many of these problems can be avoided by a free exchange of information with the casting source. If the foundry engineers are aware of the performance criteria for the final shape and the manufacturing steps or operations that are required to bring it to that stage, it is possible that the total cost of the final product can be significantly reduced if the full capabilities of the process method are utilized. This can be especially true if the casting source is challenged to produce a shape that meets the criterion of maximum economy for the level of function provided, and not merely the lowest cost casting. With the enhanced process capability of all

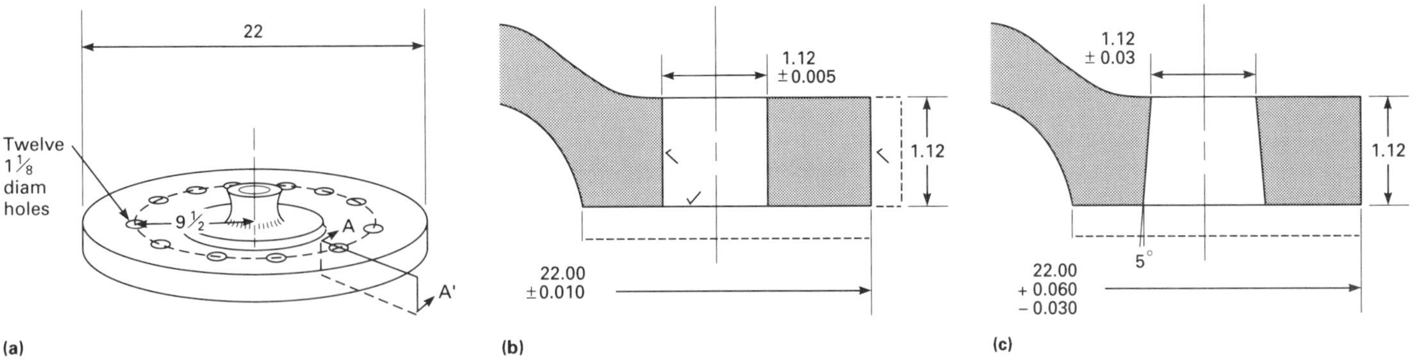

Fig. 1 Advantages of producing a commercial near-net shape casting of a bronze end cover used on a chemical reactor. The finished piece (a) is 559 mm (22 in.) in diameter. By eliminating the turning operation needed to form the center hub, as well as the drilling operation to form the 12 holes in the machined version (b), the as-cast version of the casting (c) was made more cost effective. Both the cross sections in (b) and (c) were taken at A–A′. Dimensions given in inches

casting methods, many users are finding that a small additional investment in the price of the cast shape can minimize secondary processing expense and result in a lower cost final component.

An example of the potential savings is illustrated in Fig. 1, which shows bronze end cover casting used on a chemical reactor. The casting requires a series of flange mounting holes machined/cast around the outer circumference to assemble the cover to the main body of the reactor. When faced with the decision to machine the holes on a numerical control (NC) setup or to attempt to cast them to size using green sand cores formed directly in the molding operation, the engineers had an easy choice. The savings associated with casting the holes and eliminating a turning operation on the outside diameter are listed in Table 1.

Process Capability

One mistake frequently made by purchasing departments charged with the narrow responsibility of securing parts at the lowest unit cost is the assumption that all casting operations are equal in a generic group (for example, sand foundries). Nothing could be further from reality than this assumption, with extreme differences existing between casting sources and even within molding operations/lines at the same source.

An analogy would be to assume that the process capability of a 1940 machine tool is equal to a current piece of NC equipment able to perform the same function. Even presuming that the level of maintenance is such that the machines are performing at their new capability, it is extremely doubtful that the tolerance potential of the two machine tools would be even close, much less equal.

Table 1 Cost differentials in bronze end cover

Item	Machined casting	Cored casting
Metal	$61.09	$57.75
Melting expense	6.80	5.77
Machining	8.75	...
Cleaning room	...	2.08
Scrap/rework	...	0.60
Total	**$76.64**	**$66.20**

Cost savings of cored casting over machined casting = $10.44

On the same basis, foundry equipment built during the same time periods can be expected to have similar ranges of process capability and are certainly not equal. The expected tolerances that can be held economically in a sand casting vary dramatically depending on the molding process used.

Green sand floor molding, traditionally the process chosen for the low-volume manufacture of castings, will normally use an unmounted, loose wood pattern with very limited accuracy. Sand is packed around the pattern by hand or with an air rammer, and the pattern is withdrawn to form the cavity for the molten metal. Much of the tolerance potential of the process is lost by molder operations that fall into the category of art rather than science. Examples include variations in density of the hand-rammed mold (which can yield and swell when molten metal is poured into the cavity) to the need to manually rap the pattern to loosen it in the sand before withdrawing it. All of these factors contribute to a great loss of tolerance control in the final shape.

The more modern foundries will frequently use more automated equipment to minimize the labor/skill elements in the preparation of a mold. Machines perform the many tasks necessary to prepare the sand mass to receive metal, but the sequence and method of performing individual steps in the molding operation can vary widely. All of these steps will influence the ultimate tolerance capability of the molding process, as will the compatibility of the part design to the actual molding equipment selected. Depending on the chosen method of filling the mold preform before compaction and the means available to achieve compaction, the tolerance capability of the equipment can vary widely for any one type of casting. A flat casting can be manufactured to very close dimensional tolerances on one type of molding machine. This same machine might not be able to achieve equal hardness in the mold mass for a part with deep pockets or changes in elevation and will thus sacrifice much of the tolerance capability gained.

The problem is very similar to the necessity for machine capability studies of machining centers. If the maximum advantage is to be developed from the casting process, it is necessary to understand the limitations of the method chosen. If this understanding is developed and if a spirit of cooperation exists between the casting supplier and user, then significant economies may be within reach. The magnitude of these economies become apparent when it is realized that the conversion of parts from a low-pressure squeeze molding method to a high-pressure automatic molding operation will frequently produce a reduction in finished casting weight that might approach 3 or even 5%.

Casting Distortion

Distortion Induced by Metal Solidification. One problem common to all casting processes, and potentially a large factor in dimensional tolerance control, is casting distortion. When metal masses solidify and cool to room temperature, depending on process control and manufacturing methods, very high stresses can be generated. These stresses are the result of thermal differences that can be created during the solidification or cooling of unequal section thicknesses or the resistance to contraction offered by unyielding mold sections.

The stresses can take many forms:

- They may be hidden and remain in the form of residual stresses in the cast shape (with a high potential for premature failure in service)
- They may be partially relieved in subsequent fabrication or machining operations, with a resultant change in dimensions or flatness of the casting as a new balance is reached with the remaining residual stresses
- They can reach a magnitude during solidification or cooling that results in hot tearing or cracking of the casting

- Although the design of the casting is a major factor in the formation of these stresses (and is addressed in the article "Casting Design" in this Volume), processing variables can play a major role in adding to or even reducing the stress level in the final shape. The achievement of low residual stress levels is the best guarantee that dimensional accuracy will not be affected by subsequent processing

The existence of residual stresses has long been recognized by users of castings, and a variety of methods have been adopted to lessen their effects. One of the first techniques was simply to allow the poured casting to remain in the mold for a longer time and to slow cool through the critical temperature ranges to develop a form of stress relief. This was not sufficient for manufacturers of large machine tools, who found that the machining of their large gray iron cast machine beds immediately after casting frequently resulted in distortion and out-of-flat conditions on the frames. They discovered that the problem could be alleviated considerably by seasoning the casting before machining or by allowing it to weather in a storage yard for 1 to 3 years. Daily temperature cycles and seasonal changes produced enough stress relief in the cast shapes to minimize subsequent distortion upon machining.

Current manufacturing practices using automated molding lines (with very short holding times in the mold) and just-in-time inventory practices do not permit the luxury of leisurely processing. More modern techniques rely on thermal treatments at a temperature chosen to allow plastic deformation of the shape, thus relieving the stresses. Of course, other detrimental metallurgical effects can be created if the temperature and processing are not selected and handled with extreme caution. These can include a loss of strength through an annealing effect or the formation of undesirable microstructures—for example, grain-boundary carbides in stainless steel and the accompanying loss of corrosion resistance. Thermal treatments must also be controlled carefully on both the heating and cooling cycles, or the stress level in the casting may be added to rather than reduced. An alternative to thermal treatment relies on a mechanical relief of the stresses through high-frequency vibration. Depending on the techniques used, the nonthermal treatments may be less effective, but are simpler and less costly to perform.

From the point of view of the casting designer or foundry engineer, it is important to understand how the residual stresses are formed and what options might be available to minimize their impact on the final shape. These stresses, which are thermal in origin, are the result of temperature differences in the product that occur during solidification or subsequent cooling. The thermal differences, regardless of their source, can induce plastic deformation of the shape and will be reflected as elevated residual stresses when the casting is at constant temperature. Depending on the area of the casting, the stresses may be either tensile or compressive in nature and will be in balance. Achieving this balance is one of the sources of initial casting distortion, and the dimensional errors may be increased if the stress balance is altered by stock removal during machining.

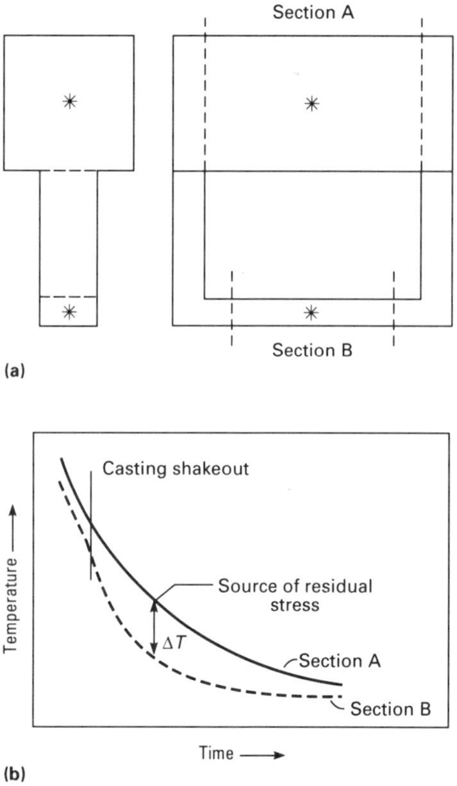

Fig. 2 Effect of casting mass on cooling rate. (a) Side and front views of a nonuniform section thickness having a major difference in section moduli; thermocouple location is indicated by *. (b) Time-temperature plot showing that temperature variation, ΔT, between the two sections is the source of deformation and residual stress buildup

The mode of formation of the stresses can be better understood by consideration of Fig. 2, which shows a simple shape of nonuniform section thickness. When metal is poured into the mold cavity and solidification begins, it will follow the rules described in the article "Casting Design" in this Volume. The heavier section (A) has more mass and heat content than the light section (B) and will cool more slowly (maintain a higher temperature) than the lighter section. This temperature differential, ΔT, is the source of all stresses (Fig. 2b).

During the cooling process, Section B is contracting at a greater rate than Section A and is placing the material in Section A in compression. If the temperature differential is sufficiently high, the stresses exceed the strength of the material and plastic deformation occurs. Section A is then slightly shorter than B, and when the temperature differential no longer exists, Section A will be placed in tension. The stresses are always in balance in a shape, and the tensile stresses in A are offset by an equal compressive stress in Section B. If the compressive stress is greater than the strength of the material, the stresses will be relieved by distortion, and the high level of compressive stresses will be manifested by the familiar oil canning.

The residual tensile stresses in the heavy section can be high enough to crack the casting during cooling, or they may be increased by subsequent machining or handling. This is frequently the cause of seemingly inexplicable cracking that occurs during postcasting operations on brittle materials such as cast iron.

It is important to recognize that the cause of the stress and distortion is the temperature differential, ΔT, that can exist upon cooling, and not the actual section thickness itself. This means that castings with uniform sections but with distributions of concentrated mass that cool differentially may also be susceptible to high levels of stress or distortion. Some typical shapes with this sensitivity are shown in Fig. 3. The solution lies in finding a way to reduce ΔT, thus proportionally reducing both stresses and distortion. Foundry engineers have found a variety of innovative methods of achieving uniform cooling, including the use of chills, rapid shakeout and selective forced cooling, or removing sand from some areas of the casting and retaining it on other surfaces.

Distortion Due to Mold Restraint. A second major cause of casting distortion, particularly in fragile castings, is the restraint imposed by the mold as the casting cools and contracts. In its most severe form, this type of distortion is difficult to eliminate except by casting redesign or by adding tie bars to minimize the free movement of unrestrained sections. The condition can also lead to an even more serious problem if the casting is being manufactured in an alloy that is hot short. With these alloy systems, random cracking can occur that may go undetected until field failures result.

The sand casting shown in Fig. 4 was made without difficulty with regard to soundness or filling of the thin sections. Significant distortion was encountered, however, when the casting cooled in the sand and was restrained by the relatively solid mass of sand between the flanges. This unyielding mass caused the two vertical flanges to be bent outward as the bottom plate cooled and contracted. Two options in design are shown in Fig. 4(b) and (c), with the preferred method incorporating the tie bar. A third choice for a solution would not involve a change in the casting itself but

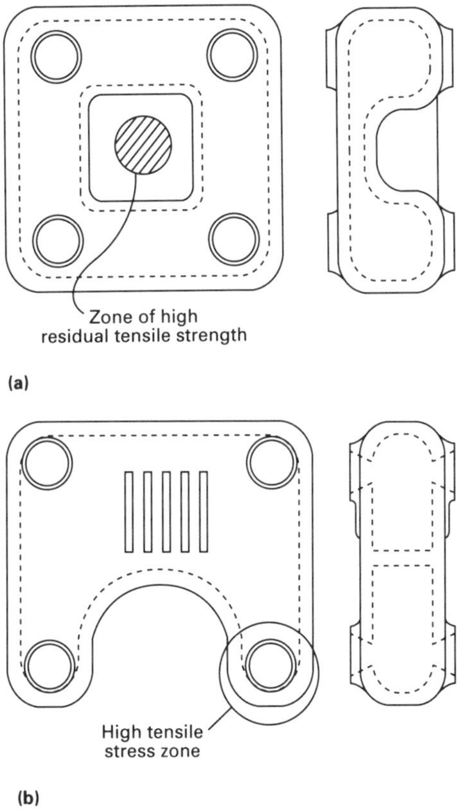

Fig. 3 Uniform section casting configurations yielding zones of high residual tensile stress. (a) Stresses concentrated in central interior of casting. (b) Stresses concentrated on outer projections of casting

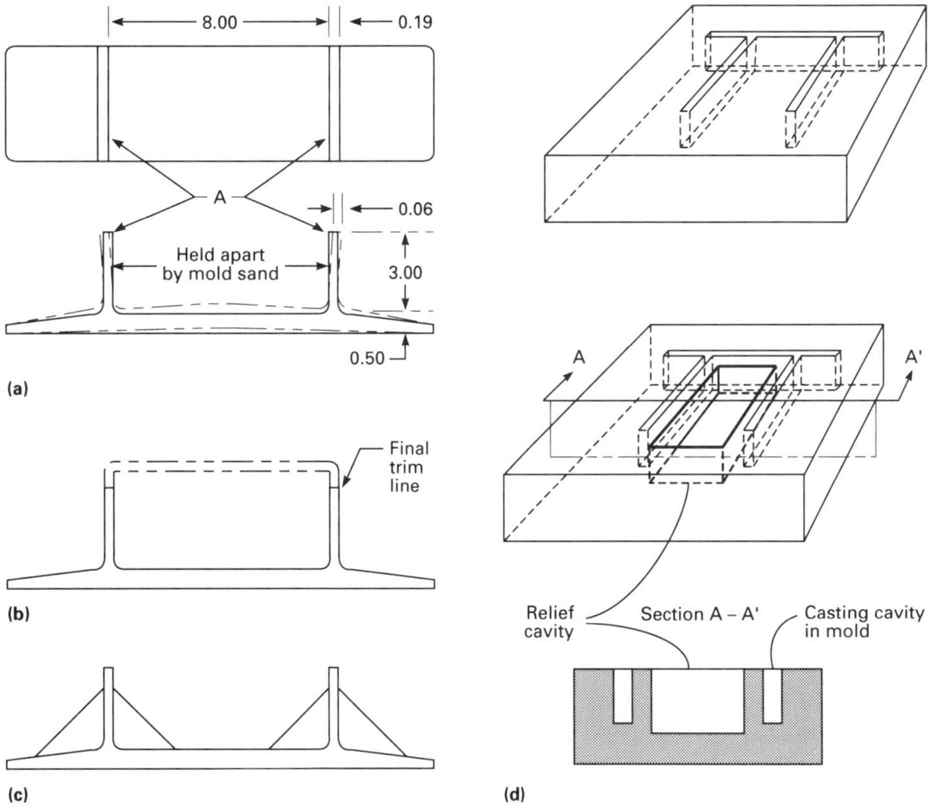

Fig. 4 Preventing distortion in a casting caused by mold restraint. The original design, shown in top and front views in (a), was altered to three possible preventative designs, as follows. (b) Preferred method incorporating a tie bar. (c) Less effective method than that shown in (b). (d) Open cavity created in the molding media to relieve restraint upon flanges during casting solidification by allowing solid sand mass to collapse during shrinkage and minimize restraint. Dimensions given in inches

rather a change in the sand mold. It has been found that the sand mass between the two flanges can be relieved so that it will yield under the shrinkage stresses. This is accomplished by adding a block to the pattern so that a hollow relief cavity is formed between the flanges (Fig. 4d). This practice reduces the effective compressive strength of the molding media between the flanges and allows the sand mass to collapse during shrinkage of the casting, thus minimizing the imposed restraint.

An example of a casting containing the design potential for hot tears is shown in Fig. 5. This shape was being manufactured in 535.0 aluminum alloy as a permanent mold casting. If the casting and molding sequences were not carefully controlled with respect to the time the mold was opened after pouring, the casting would develop hot tears in the locations shown. The problem was resolved by establishing very restrictive controls on mold temperature, pouring temperature, and mold opening time following the end of the pouring operation. Even with these controls, the scrap frequently exceeded 10% of the pieces poured.

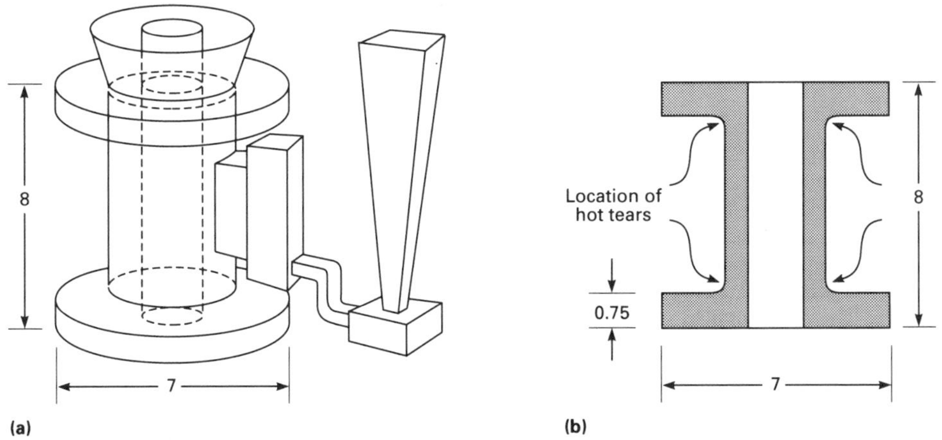

Fig. 5 Permanent mold casting of 535.0 aluminum alloy that is prone to hot tear sensitivity. (a) Casting shown with gate-riser system attached. (b) Casting with gate-riser system removed showing location of hot tears. Dimensions given in inches

Sand Castings

As noted previously, there are many processes used to mold castings in an aggregate media. Typically, sand castings are defined as castings that are molded in an aggregate media, which may be bonded with a variety of different binders. These may range from clay and water mixtures (green sand castings) to chemically bonded sand/resin mixtures that harden around the pattern. Discussion of these processes for making the sand mass rigid is beyond the scope of this article, but may be found in the Section "Molding and Casting Processes" in this Volume. It is sufficient to recognize that the molding method chosen can greatly affect the accuracy of the final product.

Regardless of the final molding method selected to fabricate the sand casting, the accuracy of the shape will not be any better than the pattern equipment used to form the cavity. The better the casting process, the more faithful the dimensional reproduction of the mold cavity created by the pattern equipment and the more accurate the final shape produced. It is at this point in the manufacturing sequence that tolerance decisions made by the designer first come into play. If low production is expected and tooling costs are to be minimized, the assignment of wide tolerances will allow the use of loose (unmounted) wood patterns to fabricate the casting. As the anticipated production run lengthens, tooling with a longer life is required, and more restrictive casting tolerances may be specified without greatly influencing casting prices or lead times.

The types of patterns used may range from unmounted, loose wood patterns through mounted wood patterns, impressions that have been replicated in a plaster molding operation and cast on an aluminum match plate, impressions machined from solid and replicated in plastic materials before mounting on a metal plate, through metal impressions that have all been duplicated by an accurate machining process and then mounted on metal plates, and finally to multiple impressions machined from a solid blank of material with the operations and dimensions controlled by direct transfer of computer-aided design and manufacture (CAD-CAM) data. In general, increased accuracy will follow through each of the above stages of pattern equipment options, and the costs progressively increase as well.

The advent of CAD-CAM processing of pattern equipment and core boxes promises to revolutionize the industry as the methods receive wider application and use. Until recently, increased accuracy in pattern equipment went hand in hand with increased cost. If multiple cavities are needed and can be produced to help amortize the programming costs, the final tooling costs using CAD/CAM may be reduced by approximately 50% over competing replication processes. Further, the accuracy of reproduction is such that the designed casting wall thickness can be taken to the minimum for the process without fear that tooling errors will increase scrap for misruns or incomplete fills. Reports of savings of up to 10% on the weight of thin-wall castings are common in the literature when the tooling processes are properly applied and utilized.

Production Variables Influencing Dimensions

The manufacture of sand castings is influenced by a number of variables that will have a great impact on the final dimensions of a cast part. Some of these variables are common to other casting processes, but many are unique to the molding method. Metal contraction, mold influences, and metal variables are described below.

Metal Contraction. Most alloys undergo a large reduction in volume during the change from the molten state to the form of a solid casting at room temperature. In aluminum alloys, the transition may be as much as 10%, requiring careful planning on the part of the patternmaker and foundry engineer to ensure the production of a dimensionally accurate casting with the required integrity.

Average values for the contraction of each metal are known and can be built into the pattern construction. The skill of the patternmaker enters into the equation when it becomes necessary to estimate where nonstandard shrinkage allowances are required because of mold restraint imposed by external gating systems, rigid cores, chills to influence directional solidification, and so on. The usual practice is to refine the gating and riser system until a sound casting meeting all of the quality requirements is produced. At this point, adjustments are made to the pattern equipment and locating surfaces to bring the final shape into the profile required by the casting designer. These adjustments are, by necessity, a trial-and-error process that can be greatly complicated by multiple pattern impressions and core box cavities. If the true potential of the process is to be realized, the adjustments are an absolute necessity.

Mold Influences. The packed density of the mold aggregate is a major factor in maintaining uniform dimensions in the final casting. If variations in density, as reflected by mold hardness, are allowed to occur, the molten metal mass will force the mold surface to yield, and inconsistent dimensions will result. The magnitude of the problem is a function of the specific gravity and temperature of the alloy being poured, with steel castings showing the greatest variation and magnesium showing the least of the common alloys.

A second problem is to ensure that the mold assembly maintains the relative position of each component during closing, handling between molding and pouring stations, and the entry of molten metal. If the match between mold sections is not maintained, it will be reflected as a shift in the final casting, as shown in Fig. 6. The only effective method of preventing this error is to pay constant attention to careful handling and maintenance practices in the foundry. It is suggested that the designer anticipate the problem, tolerance the part accordingly, and specify a parting line that will minimize the impact upon the finished item.

Cores may be considered a second mold component in the assembly, and they will introduce an additional element of variation. Again, depending on the nature and condition of the core equipment, the impact on the total tolerance range of the casting

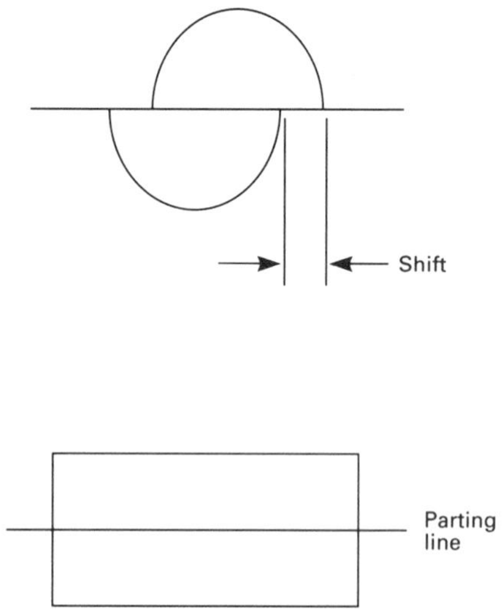

Fig. 6 Patterns in the two mold halves—the cope and the drag—must be aligned to prevent shift (a) and misalignment (b) of parting line.

will be a function of the core manufacturing process as well as the way the core is positioned and located in the mold assembly. If multiple cores are used in the mold and are prefabricated into subassemblies, the potential does exist for careful gaging and even dimensional corrections by selective stock removal.

Metal Variables. As with all other elements of process control, those dealing with the alloy, its composition, its temperature, and the practices used to introduce the metal into the mold can have a large influence on the dimensions of the final casting. The tolerance potential of each major alloy group can vary significantly for any one molding method. This is illustrated in Fig. 7, which shows the normal expected linear deviation from blueprint dimensions for steel and malleable castings made in green sand. Although the reported variation is large, it must again be remembered that it represents a data base from a family of molding operations and should not be construed as an average capability for the process or alloy family.

Table 2 lists a second set of typical noncritical tolerances for steels and castings that has some significant differences. If general tolerances are sought for an alloy family, it is suggested that the industry group association producing that alloy be solicited for their recommendations. An example of the data that might be provided is the list of 18 engineering standards published by the Aluminum Association for sand and permanent mold casting or a similar group published by the American Die Casting Institute (ADCI). The best source of information concerning tolerances is the prospective production foundry under consideration as a final source for the casting. There is no single source that can better define the capability of their processes and

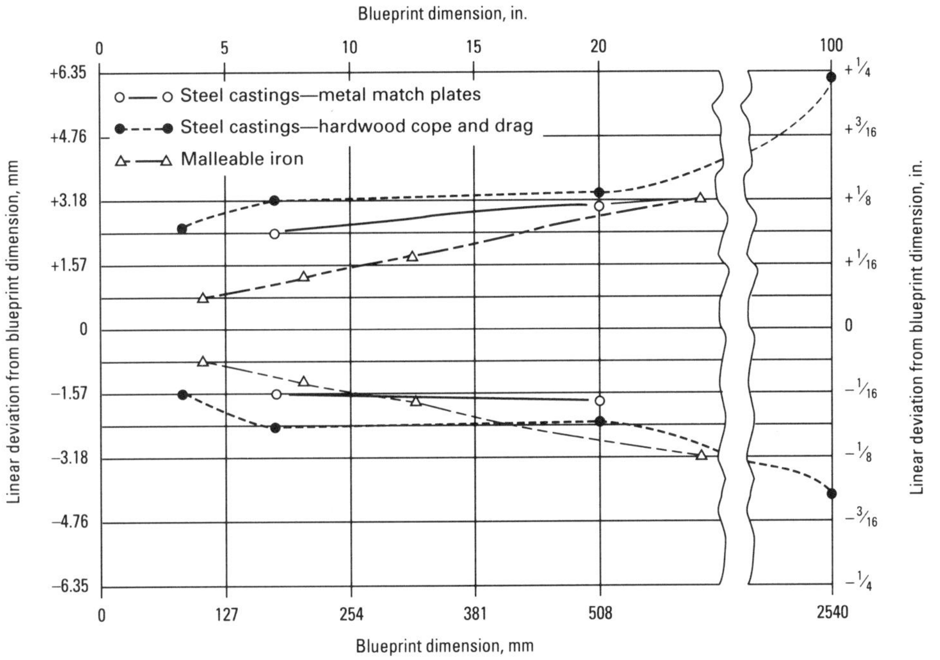

Fig. 7 Normal expected linear deviation from blueprint dimensions of steel and malleable castings made from green molding sand. Source: Ref 1

controls than the operation responsible for manufacturing the casting.

Beyond the overall tolerance limitations for an alloy family, there are many factors in the processing of the molten material that can influence the final dimensions. A partial list would include:

- Subtle changes in alloy chemistry that can produce a major change in the linear shrinkage of the material (for example, trace elements in iron-base materials that can induce the formation of white iron structures in an otherwise gray iron casting)
- Variations in pouring temperature or metal cleanliness that would cause nonstandard solidification, feeding, or casting density changes
- Improper clamping or weighting of the mold mass before pouring. The metallostatic head associated with even nominal mold heights can exert very large hydraulic lifting forces if the projected area of the casting is large or if the mold is not tightly sealed. This is one of the sources for variation in dimensions measured across parting lines

No-Bake Molds

No-bake molding practices are described in detail in the article "Resin Binder Processes" in this Volume. The technology offers a significant improvement in dimensional tolerance control because of the rigidity of the aggregate mass, which is normally hardened in place around the pattern equipment. The strength conferred by the chemical binders helps prevent distortion of the aggregate mass when metal is poured into the mold and has the potential to offer more uniform castings when compared over the duration of a production run. This has frequently been evidenced by a weight reduction in castings converted from green sand practices to no-bake. Depending on the level of control exercised over the green sand, the weight reduction can exceed 5% and can produce financial savings that will more than cover the cost of the resins and the disposal/reclamation of the sand materials.

The increased strength of the mold mass does create a different set of problems in that casting restraint on solidification is frequently higher. In addition to lower collapsibility in the media, castings molded by this method are frequently shaken out of the molds at near room temperature, which means that the shape is under restraint over a much longer temperature interval than even die casting or permanent mold practices. These conditions would suggest that shrinkage allowances and dimensions would require verification if the pattern equipment was not constructed with the anticipation of use in no-bake molding.

A second problem when using green sand pattern equipment for no-bake processes is that dimensional control across the parting line may suffer if the pattern platens are not flat and true. The rigid mold mass will replicate the pattern plane, and if the surfaces are not flat, the mass will not have a tight closure. Again, the dimensional accuracy of the process is largely influenced by the quality of the pattern equipment.

One major advantage of the process is that the molds are not friable in the cured state and degrade very slowly in ambient atmospheres. This feature allows an almost unlimited time for mold assembly, careful gaging of core assemblies, venting, surface preparation, treatment of chills, spot repairs, and so on, before mold closure. All of the above can contribute to greater dimensional control in the final product and can minimize scrap during production.

Die Casting

This family of processes encompasses all metal casting operations done under high pressure, including conventional die casting, the Acurad process, and squeeze casting. (Acurad, an acronym for accurate, rapid, and dense, designates a proprietary die casting method developed to eliminate some of the casting quality problems associated with high-pressure castings.)

All die castings can be made to a very high level of accuracy if the production shop exercises the proper control of processes and tooling. For example, zinc castings are routinely manufactured with wall thicknesses as thin as 0.51 mm (0.020 in.) and cast-in-place hole tolerances as tight as ±0.015 mm (±0.0006 in.).

These levels of accuracy can be achieved only with precision tooling and absolute process control over all aspects of the operation. The controls start with the alloy composition and extend through the metal pouring temperature, die temperatures, injection parameters, die lock-up, cycle time, and all of the other factors that control casting quality. The resultant shapes will have an excellent finish, a high degree of dimensional uniformity from piece to piece, adequate mechanical properties for most applications, and a moderate price.

Internal porosity can be reduced through careful control of casting conditions during injection in order to obtain a sound, fine-

Table 2 Typical noncritical tolerances for steel sand castings

Dimension		Typical tolerances							
		Across parting line				Between points formed by core and mold		Between points in one part of mold	
		Horizontal		Vertical					
mm	in.	± mm	± in.	± mm	± in.	± mm	± in.	± mm	± in.
25–178	1–7	1.5	0.06	1.3	0.05	1.3	0.05	0.8	0.03
203–305	8–12	1.8	0.07	1.5	0.06	1.5	0.06	1.0	0.04
330–406	13–16	2.0	0.08	1.8	0.07	1.8	0.07	1.3	0.05
432–508	17–20	2.3	0.09	2.0	0.08	2.0	0.08	1.5	0.06

grain surface layer. This is vital for castings destined for surface treatments such as plating, enameling, and anodizing.

The processes are not without limitations, which for the most part center around the relatively high tooling cost, long lead times for die construction, and limitations on shapes that are readily castable with straight core pulls. Although widely publicized as a method for forming ferrous materials, most commercial applications are limited to the nonferrous alloys of zinc, aluminum, magnesium, and copper. Future developments are expected to include the wider use of expendable cores, the fabrication of larger shapes with thinner walls, improved die materials with longer lives, and the production of composite shapes reinforced with carbon, silicon carbide, or refractory filaments.

With die castings, as with any precision casting process, the challenge to the designer is to determine where close tolerances are required as well as where they are not required. The potential for net shape production is very high, but the abuse of specifying overly restrictive tolerances will add greatly to lead time and cost in the final casting.

Normal commercial limits for die castings are defined in standards published by ADCI. These standards cover values consistent with speed, uninterrupted production, reasonable die and tool life, and moderate maintenance practice. In most cases, they are capable of being improved upon by an individual die caster, often at little or no increase in cost. Examples of the standards are given below. For as-cast components, linear dimensional tolerances are (ADCI-E1-61):

Tolerance	Critical dimensions, mm (in.)	Noncritical dimensions, mm (in.)
Basic tolerance up to 25 mm (1 in.)	±0.10 (±0.004)	±0.25 (±0.010)
Additional tolerance for each additional mm (in.)		
>25–305 mm (1–12 in.)	±0.038 (±0.0015)	±0.051 (±0.002)
>305 mm (12 in.)	±0.025 (±0.001)	±0.025 (±0.001)

Note: Tolerances must be modified if dimension is affected by parting line or moving die part. Source: Ref 2

Parting line tolerances, based on a single-cavity die, are (ADCI-E2-61):

Projected area of casting, mm² (in.²)	Parting line tolerances, mm (in.)
Up to 32 000 (50)	±0.13 (±0.005)
32 000–64 000 (50–100)	±0.20 (±0.008)
64 000–129 000 (100–200)	±0.30 (±0.012)
129 000–194 000 (200–300)	±0.38 (±0.015)

Note: Tolerances to be added to linear tolerances specified in above Table. Source: Ref 2

Part tolerances of the moving die part are as follows (ADCI-E3-61):

Projected area of die casting portion(a), mm² (in.²)	Part tolerances, mm (in.)
6400 (10)	±0.13 (±0.005)
6400–12 900 (10–20)	±0.20 (±0.008)
12 900–32 300 (20–50)	±0.30 (±0.012)
32 300–64 500 (50–100)	±0.38 (±0.015)

Note: Moving die part tolerances, in addition to linear dimensions and parting line tolerances, must be provided when moving die part affects a linear dimension. (a) Projected area is area (in mm² or in.²) of that portion of casting affected by moving die part. Source: Ref 2

Permanent Mold Castings

Permanent mold processes are widely used for all alloy systems except steel and superalloys. These processes encompass conventional gravity die casting, low-pressure casting operations, slush casting, and some special purpose processes that offer advantages in the production of specific end products. The processes may be readily automated, offer the potential of much better tolerance control than is possible with some sand casting operations, are considerably more flexible than die casting in terms of the size and shape of parts that can be fabricated, and can be tooled at much lower cost than parts made with die casting methods.

The tolerance potential for the processes is somewhat comparable to die casting in that a high percentage of the castings are made in all metal tooling. Although casting detail may not be as fine, because of the lower fill and injection pressures, the tooling need not be as sturdy to contain the injection process. Also, with the absence of high injection stresses, the die materials are chosen for their conductivity, availability, and low cost rather than their ability to withstand the shock loading stresses at the elevated temperatures that are common in die casting. The more tolerant process conditions encourage the use of gray iron molds, which are frequently cast to near-final dimensions to minimize machining and tool construction expense.

Permanent mold casting is used for the volume manufacture of castings with methods that are highly automated. Many mold installations are currently operating with all elements of the molding operation being performed automatically. The operator is an observer and an inspector and does not participate in any actual casting functions. Internal cores of increasing complexity are stripped from the castings automatically and then reassembled to pour the next shape. This permits a high level of accuracy, and many piston castings are being produced with 0.38 mm (0.015 in.) finish stock on some machined surfaces.

As with the other casting processes, the dimensional capability of permanent mold casting is largely limited by the accuracy of the dies used. Tooling wear and erosion are more serious problems than with other processes because of the need to interrupt production periodically to remove drags (cast metal adhering to the die in areas of low draft) and to renew the mold coatings used to protect the molds. The coatings are also a source of dimensional variation in that their thickness and composition are frequently changed to modify solidification patterns in the mold. By design, some areas of the mold can be operated with very thin coatings that are more lubricating than insulating. Other areas can be covered with multiple layers of insulating materials with thicknesses to 0.76 mm (0.030 in.). Because the coatings are refractory, they have an expansion coefficient that is different from that of the metal base, and they will spall and chip if uniform mold temperatures are not maintained. When this occurs, the coating must be removed down to the base material and renewed.

This removal and cleaning process is the source of most dimensional change over the service life of a permanent mold. The coatings are usually formulated with a sodium silicate base to promote adhesion, and they become very hard in service. Removal is accomplished by a light sandblasting or wire brushing of the entire mold cavity. This process also wears away a portion of the base material, enlarging the mold cavity with each cleaning cycle until it eventually becomes unserviceable.

An example of the effect of tool wear on component weight is given in Fig. 8, which shows the change in casting weight of an aluminum engine manifold casting. The die was reconditioned several times during the period, but this repair was limited to the die faces rather than the total cavity. This is an area in which the new tooling and repair processes offered by CAD/CAM systems are expected to develop a large savings. Large automotive casting shops are scheduling a complete reconditioning of the die face and cavity on some dies as frequently as every 15 000 cycles. This expense is justified by the reduction in average casting weight produced, which is reported to save over 7% of the total metal required to make the parts. In addition, because the wall thickness can be accurately defined and controlled, new tools are manufactured to the bottom tolerance limit for wall thickness. An automotive gearcase designed in this manner led to a savings of 10% of the casting weight.

General casting tolerances are again available from the producing foundries and industry associations. An example of those published for aluminum permanent mold castings is shown in Fig. 9. As with all standards of this type, they are intended to establish the individual requirements involved in producing a usable, unmachined casting consistent with normal production practices, reproducibility, reasonable mold or pattern life, main-

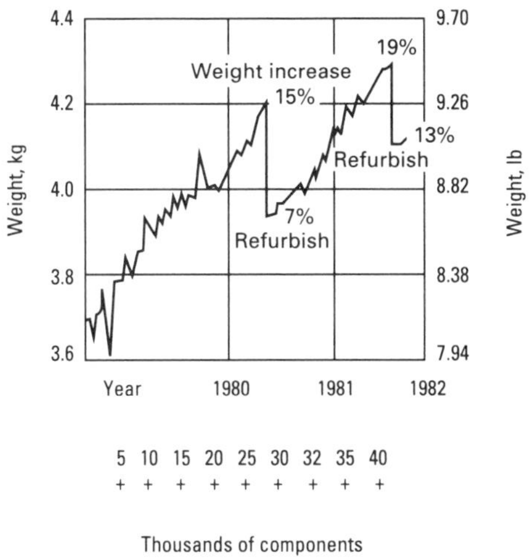

Fig. 8 Effect of mold wear on casting weight. The weight of the casting, an aluminum engine inlet manifold component, increases over the life of the mold. Source: Ref 3

tenance costs, and so on. Consultation with the foundry is strongly recommended.

Investment Castings

One of the casting processes that offer the most hope for near-net shape production is the ceramic mold production of investment cast shapes. These are made in virtually all castable alloys and in a size range from a fraction of a kilogram up to 45 kg (100 lb), and larger. Wall sections as thin as 0.76 mm (0.030 in.) can be cast with the precision process, and intricate coring and detail can be incorporated into production castings that cannot even be approached by other casting methods.

The tolerance capability of the process is limited by many of the same factors that influence other casting methods. Changes in tooling to correct for dimensional variation are sometimes costly, but once made, the reproducibility is excellent because the process lends itself to automation and very restrictive process control.

As with any process, costs are increased when tighter tolerances than required are specified. Typical investment casting tolerances are listed in Table 3, and with the cooperation of the producing foundry, it may be possible to improve these tolerances substantially.

One of the features of the process that is rapidly gaining acceptance and new markets for castings is the ability to assemble individual wax preforms into larger shapes with a high degree of complexity. This minimizes tooling costs and makes the process capable of making almost any shape and configuration. When coupled with precise gaging operations, the assemblies can result in final cast shapes that are without equal in design flexibility and tight tolerances.

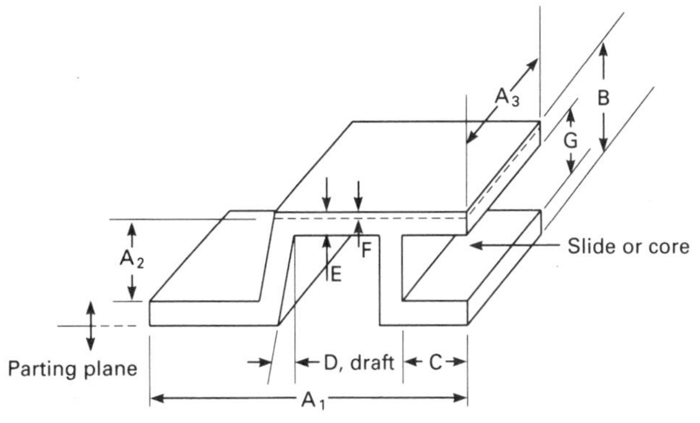

Minimum diameter of cored holes 9.53 mm (0.375 in.)

Between two points in the same part of the mold; not affected by parting plane or core

Specified dimension, mm	Specified dimension, in.	Tolerance, mm	Tolerance, in.
≤25	≤1	±0.38	±0.015
>25	>1	0.38 ± 0.002 mm/mm over 25 mm	0.015 ± 0.002 in./in. over 1 in.

Flatness tolerance, as-cast

Flatness tolerance, the total allowable deviation from a plane, consists of the total distance between two parallel planes embracing the entire tolerated surface.

Greatest dimension(a), mm	Greatest dimension(a), in.	Tolerance, permanent and semipermanent mold, mm	Tolerance, permanent and semipermanent mold, in.
0–152	0–6	Within 0.51	Within 0.020
Each additional mm (in.)(b)		0.003 mm/mm	0.003 in./in.

(a) Section diameter or diagonal. (b) For castings over 610 mm (24 in.), consult foundry.

Across parting plane; A-type dimension plus below

Projected area of casting, $A_1 \cdot A_3 = X$, mm²	in.²	Additional tolerance for parting plane, mm	in.
$X \leq 6450$	$X \leq 10$	±0.25	±0.010
$6450 < X \leq 31\,600$	$10 < X \leq 49$	±0.38	±0.015
$32\,300 < X \leq 63\,900$	$50 < X \leq 99$	±0.51	±0.020
$64\,500 < X \leq 161\,000$	$100 < X \leq 249$	±0.64	±0.025
$161\,000 < X \leq 323\,000$	$250 < X \leq 500$	±0.76	±0.030

Affected by moving parts; A-type dimension plus below

Projected area of casting affected by moving part, $A_3 \cdot G = YA_1 \cdot A_3 = X$, mm²	in.²	Additional tolerance, Metal core or mold, mm	Metal core or mold, in.	Sand core, mm	Sand core, in.
$Y \leq 6450$	$Y \leq 10$	±0.25	±0.010	±0.38	±0.015
$6450 < Y \leq 31\,600$	$10 < Y \leq 49$	±0.38	±0.015	±0.64	±0.025
$32\,300 < Y \leq 63\,900$	$50 < Y \leq 99$	±0.38	±0.015	±0.76	±0.030
$64\,500 < Y \leq 322\,000$	$100 < Y \leq 499$	±0.56	±0.022	±1.02	±0.040
$323\,000 < Y \leq 645\,000$	$500 < Y \leq 1000$	±0.81	±0.032	±1.52	±0.060
$Y > 645\,000$	$Y > 1000$, consult foundry	...	...	...	...

Allowance for finish

Maximum dimension, x, mm	in.	Nominal allowance, Metal core or mold, mm	Metal core or mold, in.	Sand core, mm	Sand core, in.
$x < 152$	$x \leq 6$	0.76	0.030	1.52	0.060
$152 < x \leq 305$	$6 < x \leq 12$	1.14	0.045	2.29	0.090
$305 < x \leq 457$	$12 < x \leq 18$	1.52	0.060	3.05	0.120
$457 < x \leq 610$	$18 < x \leq 24$	2.29	0.090	4.57	0.180
$x > 610$	$x > 24$, consult foundry				

Fig. 9 Suggested dimensional tolerances for permanent and semipermanent mold castings. Normally, an illustration does not show draft. Standard foundry practice is to add draft to the part. Source: Ref 4

Table 3 Typical linear tolerances in investment castings

Dimension, mm	Dimension, in.	Normal tolerance, mm	Normal tolerance, in.
Up to 13	Up to ½	±0.13	±0.005
Up to 25	Up to 1	±0.25	±0.010
Up to 50	Up to 2	±0.33	±0.013
Up to 75	Up to 3	±0.41	±0.016
Up to 102	Up to 4	±0.48	±0.019
Up to 127	Up to 5	±0.56	±0.022
Up to 152	Up to 6	±0.63	±0.025
Up to 178	Up to 7	±0.71	±0.028
Maximum variation	Maximum variation	±1.02	±0.040

Source: Ref 5

Table 4 Typical tolerance relations for shell mold steel castings

Dimension, mm	Dimension, in.	Typical tolerance: Across parting line, ± mm	Across parting line, ± in.	Between points in one part of mold, ± mm	Between points in one part of mold, ± in.
25	1	0.51	0.020	0.25	0.010
50	2	0.51	0.020	0.25	0.010
75	3	0.64	0.025	0.38	0.015
102	4	0.64	0.025	0.38	0.015
127	5	0.76	0.030	0.51	0.020
152	6	0.76	0.030	0.51	0.020
178	7	0.76	0.030	0.51	0.020
203	8	0.89	0.035	0.64	0.025
229	9	0.89	0.035	0.64	0.025
254	10	1.02	0.040	0.76	0.030
279	11	1.02	0.040	0.76	0.030
305	12	1.02	0.040	0.76	0.030
330	13	1.14	0.045	0.89	0.035
356	14	1.14	0.045	0.89	0.035
381	15	1.27	0.050	1.02	0.040
407	16	1.27	0.050	1.02	0.040
432	17	1.27	0.050	1.02	0.040
457	18	1.40	0.055	1.14	0.045
483	19	1.40	0.055	1.14	0.045
508	20	1.52	0.060	1.27	0.050

Newly available waxlike materials offer the capability of direct machining to the final shape for prototypes, followed by assembly of the shape into an investment mold for subsequent casting. The waxlike material is completely heat disposable and offers the same tolerance capability of production tooling for the final investment casting.

Evaporative Foam Castings

This process offers much in the way of potential cost reductions in manufacturing cast shapes with a requirement for internal shapes that would normally be formed with separate cores. Through paste-up practices that are similar to investment casting methods, individual foam shapes can be built up to very complex assemblies that will result in a casting that is an accurate replica of the foam.

The process does require metal tooling for production of the foam preshapes, and again, the final casting will only be as accurate as the initial tool. The manufacture of the intermediate foam pattern does introduce an additional set of variables that can influence casting dimensions. The density of the foam shape must be closely controlled, as well as the paste-up, coating with washes, sand compaction in the mold while avoiding distortion of the foam shape, and pouring.

Users of the processes claim that it is possible to manufacture shapes that cannot be cast by another method and that very tight tolerances can be held upon selected dimensions. The process is very sensitive to manufacturing techniques, and the accuracy has been found to vary directly with the amount of process control that is exercised.

Shell Molded Castings

Castings made by the shell molding process are generally more accurate dimensionally than some of the sand casting processes, although high-pressure sand molding has often been found to approximate the casting accuracy of shell molded casting. Problems encountered in shell casting manufacture are similar to those encountered in sand molding, and variables that influence tolerance capability are common to both processes.

The tolerance capability of shell molded castings is directly related to the size of the casting; small castings are capable of holding much tighter tolerances than the larger shapes. In part, this can be traced to variations induced by the use of heated patterns and the difficulty encountered in anticipating the contraction of the phenolic bonded shell when it cools to room temperature. Pattern adjustments can minimize these discrepancies, but they are still a source of variation.

The second difficulty encountered in the manufacture of large shapes is the deflection and distortion that can occur when high-melting alloys are poured into the resin bonded molds. This can be a significant source of dimensional change and will vary across the cross section of the casting, depending on the support and restraint provided by the backup material. The magnitude of the problem is reduced if low-density nonferrous alloys are used with lower melting temperatures.

Table 4 lists typical tolerance relations for shell molded steel castings in two directions of measurement. The improved accuracy of shell molded castings relative to that of sand castings can be inferred by comparing Table 2 with Table 4. With the cooperation of the producing foundry, it is frequently possible to halve the projected tolerance, if required for a particular dimension.

Plaster Mold Castings

This process is normally judged to have a dimensional accuracy somewhere near the midpoint between shell molded and investment castings, but may be tooled and put into production at a considerably lower cost. The inert mold material has very high insulating qualities and, when coupled with a mild preheat and pressure assist on pouring, can produce thin-wall shapes with exceptional tolerances.

The process is frequently used to replicate patterns from low cost wood masters and will produce equipment that can approach the tolerances of machined metal impressions. Although the plaster cast aluminum impressions may sacrifice some dimensional accuracy, they are capable of being produced very economically. This feature makes the process the most popular method of tooling construction for medium-volume parts.

Another popular use of the process, again making use of the dimensional capabilities, is to prototype die casting designs. Thin-wall shapes can be evaluated and field tested before entering into the expense of constructing die cast tools.

Plaster mold castings are normally found to contain a minimum of warpage and distortion because of the very low cooling rates and high plasticity of the materials cast. Near-net shape castings are frequently produced with machining operations eliminated or minimized because of the accuracy of the cast shapes. Typical examples of these parts include impellers that have vane thicknesses down to 0.76 mm (0.030 in.) or large wave guides where the alignment of the sections is maintained within very close limits as-cast. Detailed information on plaster mold castings and the other molding and casting processes mentioned in this article is available in the Section "Molding and Casting Processes" in this Volume.

Casting Dimensioning and Tolerancing

Regardless of the process chosen to manufacture the casting, the practices used to dimension the part are critical to the ultimate utility and cost of production. Two

Table 5 Symbols used in geometric dimensioning and tolerancing

Type of feature	Type of tolerance	Characteristic	Symbol
Individual (no datum reference)	Form	Flatness	⏥
		Straightness	—
		Roundness	○
		Cylindricity	⌭
Individual or related	Profile	Profile of a line	⌒
		Profile of a surface	⌓
Related (datum reference required)	Orientation	Perpendicularity	⊥
		Angularity	∠
		Parallelism	∥
	Location	Position	⌖
		Concentricity	◎
	Runout	Circular	↗
		Total	⌰

Source: Ref 6

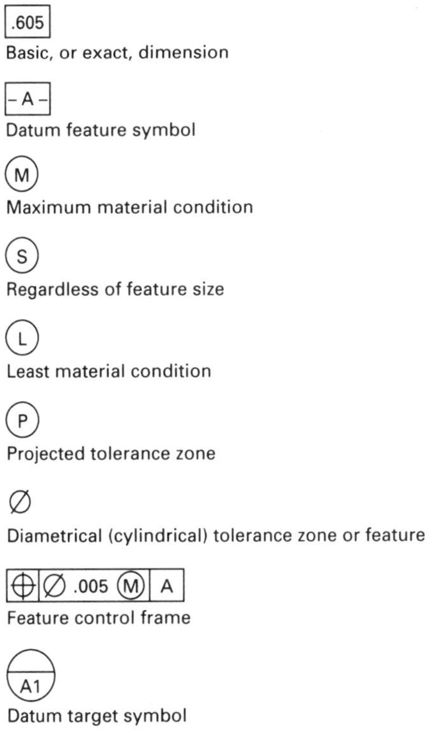

Fig. 10 Related geometric characteristic symbols and terms

conventional methods are normally available:

- *Coordinate tolerancing*, which incorporates conventional plus and minus allowances on all referenced dimensions
- *Geometric dimensioning and tolerancing*, which designates the true position of features controlled from referenced datums

Coordinate Tolerancing. The antiquated coordinate tolerancing system, although still used in some design areas, has been a source of many misunderstandings among design engineers, patternmakers, and foundrymen. When used on complex drawings, the stack-up of tolerances often makes it impossible for the patternmaker to interpret the intentions of the designer without frequent consultations and discussions. These discussions are often held after the manufacture of prototype parts, resulting in delays and additional expense in retooling or correcting the cast shape to meet the intended function.

Geometric dimensioning and tolerancing is an international system of design language. It enables the true functional limits of the production variability of any part to be expressed in a drawing.

The development of international drafting standards for clearly specifying the acceptable limits of production variability has evolved worldwide over the last 40 years. The system these standards define, known as geometric dimensioning and tolerancing, provides a drafting language by which sophisticated design requirements can be clearly stated and uniformly understood. These relationships are expressed through international, standardized symbols that clearly define part geometry, without relying on lengthy and often misinterpreted notes or the subtleties of view interrelationships.

Engineering drawings conventionally have two aspects: views that show the shape of the object and dimensions that show size. Problems occur most frequently when describing the elements of size, resulting in costly errors in which design features do not fit mating assemblies, and so on. Problems can be avoided by using geometric dimensioning and tolerancing, as documented by the American National Standards Institute (ANSI) and the International Standards Organization (ISO). The system breaks down language barriers while adding precision and clarity to the engineering drafting language.

Characteristic Symbols. There are 13 basic geometric symbols that describe the relationships of design features to each other and the reference datum planes. These are presented in Table 5, and their use is described in detail in ANSI Y14.5M. An example of their use, when combined with descriptive terms, is shown in Fig. 10.

REFERENCES

1. *Design of Ferrous Castings*, American Foundrymen's Society, 1963, p 103
2. *Aluminum Casting Technology*, American Foundrymen's Society, 1986, p 90
3. D.B. Welbourn, CAD/CAM Plays Major Role in Foundry Economics, *Mod. Cast.*, Sept 1987, p 41
4. Standards for Aluminum Sand and Permanent Mold Castings, Aluminum Association, Inc., Washington, D.C.
5. T.G. Coghill, International Casting Offers Economy for Many Applications, *Precis. Met.*, March 1983, p 23-31
6. "Dimensioning and Tolerancing," Y14-5M, American National Standards Institute

SELECTED REFERENCES

- *Casting Design Handbook*, American Society for Metals, 1962
- Die Castings Future Seen in High Precision Parts, *Mod. Met.*, Aug 1987
- L.W. Foster, *Geo-Metrics II*, Addison-Wesley, 1986
- E. Swing, "Using Near Net Shape Cast Structure in Aerospace," Paper 8501-002, Metals/Materials Technology Series, American Society for Metals, 1985

Ferrous Casting Alloys

Section Chairmen:
D.M. Stefanescu, The University of Alabama
Norman Lillybeck, Deere and Company

Classification of Ferrous Casting Alloys

Doru M. Stefanescu, University of Alabama

CAST IRONS AND STEELS (ferrous alloys) represent some of the most complex alloy systems. A wide variety of microstructures and resulting properties are possible, depending on composition, solidification conditions, and appropriate heat treatment. The intent of this article is to provide a classification system for ferrous alloys. Figures 1 and 2 classify cast irons and steels according to their commercial name or application and their structure. The articles that constitute this Section discuss key aspects for each alloy system. These include:

- Chemical composition
- Structure and property correlations
- Melting practice and melt treatment
- Specifics of foundry practice (pouring, gating, and risering)
- Heat treatment
- Applications

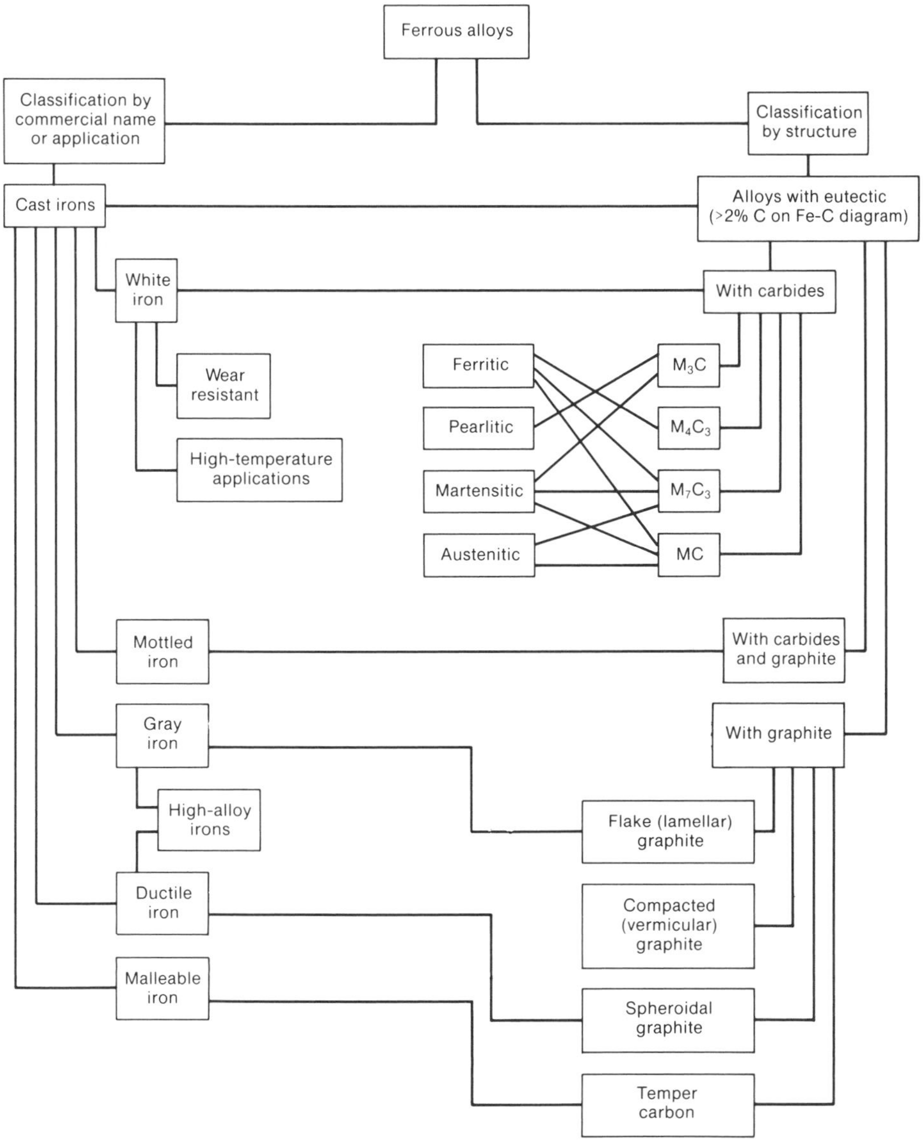

Fig. 1 Classification of cast irons

In addition to the contributions on cast irons and steels, an article on "Cast Alnico Alloys" is provided. These difficult-to-cast, precipitation-hardenable materials, which contain high amounts of aluminum, nickel, and cobalt as well as other alloying elements, are used for permanent magnetic applications.

Cast Irons. The term cast iron, like the term steel, identifies a large family of ferrous alloys. Cast irons are iron-carbon base alloys that solidify with a eutectic. They contain various amounts of Si, Mn, P, S, and trace elements such as Ti, Sb, and Sn. They may also contain various amounts of alloying elements. Wide variations in properties can be achieved by varying the balance between carbon and silicon, by alloying with various metallic or nonmetallic elements, and by varying melting, casting, and heat-treating practices.

The five basic types of cast iron are white iron, gray iron, mottled iron, ductile iron, and malleable iron (Fig. 1). White iron and gray iron derive their names from the appearance of their respective fracture surfaces: white iron exhibits a white, crystalline fracture surface, and gray iron exhibits a gray fracture surface with exceedingly tiny facets. Mottled iron falls between gray and

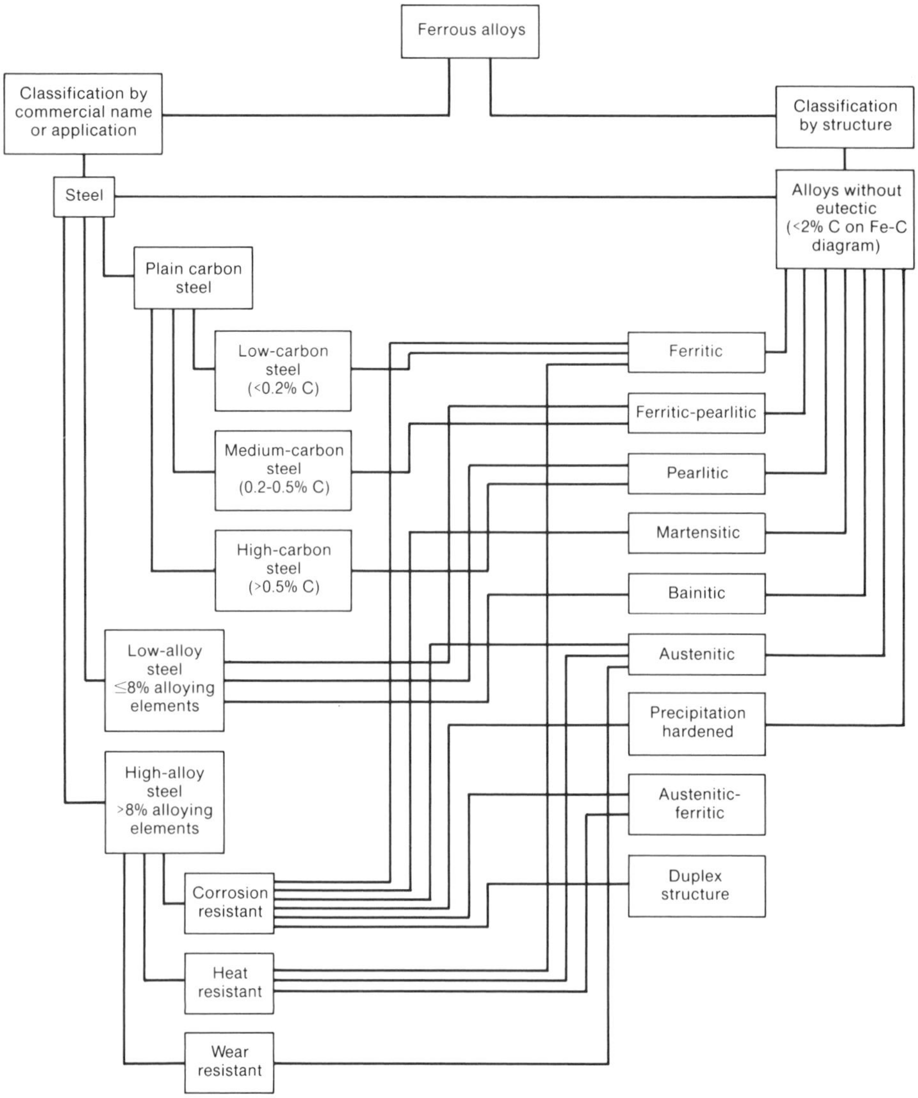

Fig. 2 Classification of steels

white iron, with the fracture showing both gray and white zones. Ductile iron is so named because in the as-cast form it exhibits its measurable ductility. By contrast, neither white nor gray iron exhibit significant ductility in a standard tensile test. Malleable iron is initially cast as white iron, then "malleablized," that is, heat treated to impart ductility to an otherwise exceedingly brittle material.

Two additional subdivisions of these five basic types include high-alloy graphitic irons and compacted graphite irons. The high-alloy graphitic irons, which are primarily used for applications requiring corrosion resistance or a combination of strength and oxidation resistance, are produced in both flake graphite (gray iron) and spheroidal graphite (ductile iron). Compacted graphite (CG) cast iron is characterized by graphite that is interconnected within eutectic cells, as is the flake graphite in gray iron. Compared with the graphite in gray iron, however, the graphite in CG iron is coarser and more rounded.

Steels can be classified on the basis of composition, such as carbon, low-alloy, and high-alloy steel (Fig. 2); microstructure, such as ferritic, austenitic, martensitic, and so forth; or product form, such as bar, plate, sheet, strip, tubing, or structural shape.

Common use has further subdivided these broad classifications. For example, carbon steels are often classified according to carbon content as low-, medium-, or high-carbon steels. Alloy steels are often classified according to the principal alloying element (or elements) present. Thus, there are nickel steels, chromium steels, chromium-vanadium steels, and so on.

Gray Iron

D.B. Craig, M.J. Hornung, and T.K. McCluhan, Elkem Metals Company

THE TERM GRAY IRON refers to a broad class of ferrous casting alloys normally characterized by a microstructure of flake graphite in a ferrous matrix. Gray irons are in essence iron-carbon-silicon alloys containing small quantities of other elements. As a class, they vary widely in physical and mechanical properties.

The metallurgy of gray irons is extremely complex because of a wide variety of factors that influence their solidification and subsequent solid-state transformations. In spite of this complexity, gray irons have found wide acceptance based on a combination of outstanding castability, excellent machinability, economics, and unique properties.

Metallurgy

Crucial to understanding the production, properties, and applications of gray iron is an understanding of its metallurgy. While it is beyond the scope of this article to detail gray iron metallurgy, it is important to understand the metallurgical background of this group of ferrous casting alloys. The importance of composition and processing variables in product performance cannot be overemphasized.

Composition

For purposes of clarity and simplicity, the chemical analyses of gray iron can be broken down into three categories. The first category includes the major elements. In the second group are minor, normally low-level alloying elements that are critically related to gray iron solidification. Finally, there are a number of trace elements that affect the microstructure and/or properties of the material.

Major Elements. The three major elements in gray iron are carbon, silicon, and iron. Carbon and silicon levels found in commercial irons vary widely, as shown below:

Type of iron	Total carbon, %	Silicon, %
Class 20	3.40–3.60	2.30–2.50
Class 30	3.10–3.30	2.10–2.30
Class 40	2.95–3.15	1.70–2.00
Class 50	2.70–3.00	1.70–2.00
Class 60	2.50–2.85	1.90–2.10

Primarily because of the development of ductile iron and some specialized grades of alloyed irons, most gray irons are produced with total carbon levels from 3.0 to 3.5%. Normal silicon levels vary from 1.8 to 2.4%.

Gray irons are normally viewed as iron-carbon-silicon ternary alloys. A section from the equilibrium phase diagram at 2.5% Si is shown in Fig. 1. As can be seen, the material exhibits eutectic solidification and is subject to a solid-state eutectoid transformation. These two factors dominate the metallurgy of gray iron.

Both carbon and silicon influence the nature of iron castings. It is therefore necessary to develop an approximation of their impact on solidification. This has been accomplished through development of the concept of carbon equivalence (CE). Using this approach, carbon equivalence is calculated as:

$$CE = \%C + \frac{\%Si}{3} \quad \text{(Eq 1)}$$

or more precisely, taking phosphorus into consideration:

$$CE = \%C + \frac{(\%Si + \%P)}{3} \quad \text{(Eq 2)}$$

Using Eq 1 and 2, it is possible to relate the effect of carbon, silicon, and phosphorus to the binary iron-carbon system. Irons with a carbon equivalent of 4.3 are considered to be of eutectic composition. Most gray irons

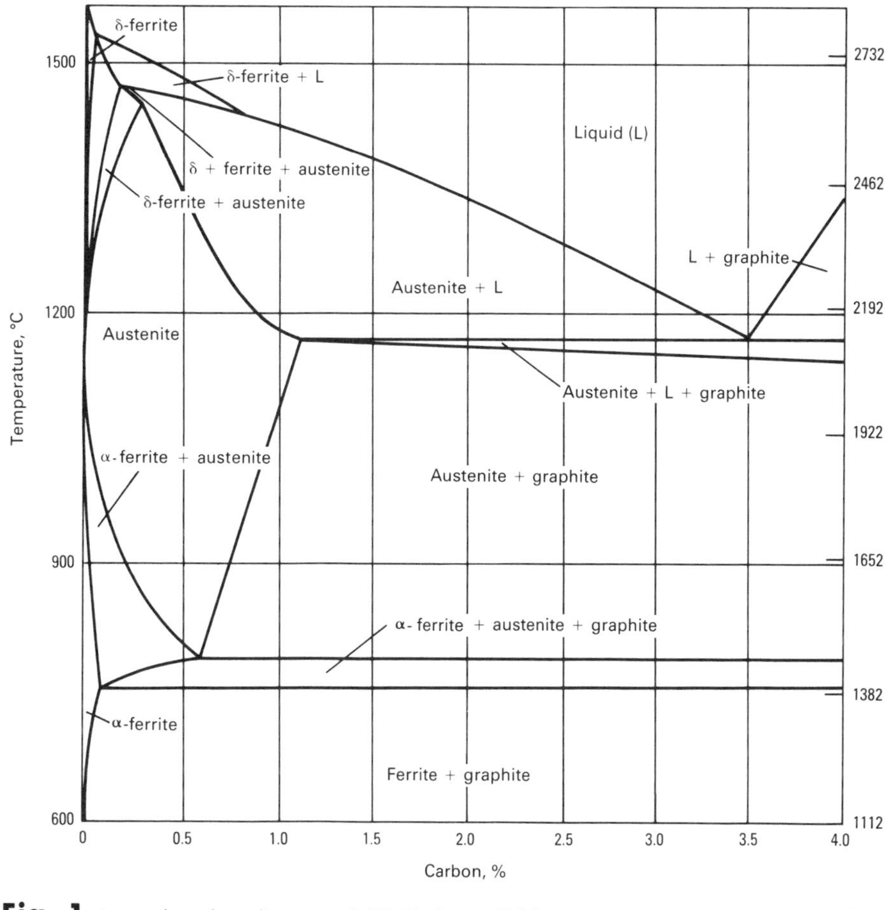

Fig. 1 Iron-carbon phase diagram at 2.5% Si. Source: Ref 1

are hypoeutectic (that is, CE < 4.3). Nearly all of the mechanical and physical properties of gray iron are closely related to CE value.

The minor elements in gray iron are phosphorus and the two interrelated elements manganese and sulfur. These elements, like carbon and silicon, are of significant importance in gray iron metallurgy. Control is required for product consistency. Absolute levels vary somewhat with application and foundry process variables.

Phosphorus is found in all gray irons. It is rarely added intentionally, but tends to come from pig iron or scrap. To some extent, it increases the fluidity of iron. Phosphorus forms a low-melting phosphide phase in gray iron that is commonly referred to as steadite. At high levels, it can promote shrinkage porosity, while very low levels can increase metal penetration into the mold (Ref 2, 3). As a result, most castings are produced with 0.02 to 0.10% P. In critical castings involving pressure tightness, it may be necessary to develop optimum levels for the application.

Sulfur levels in gray iron are very important and to some extent are an area of current technical controversy. Numerous investigators have shown that sulfur plays a significant role in the nucleation of graphite in gray iron. The impact of sulfur on cell counts and chill depth in gray iron can be seen in Fig. 2 for uninoculated and inoculated gray irons. This work indicates that sulfur levels in gray iron should be in the approximate range of 0.05 to 0.12% for optimum benefit.

It is important that the sulfur content of iron be balanced with manganese to promote the formation of manganese sulfides. This is normally accomplished by using Eq 3:

$$\%Mn \geq 1.7\%\ S + 0.3\% \quad \text{(Eq 3)}$$

Recent work has indicated that the 0.3% level may be reduced slightly; some foundries add only 0.2% excess manganese.

Trace Elements. In addition to these primary elements, there are a number of minor elements that affect the nature and properties of gray iron. Table 1, extracted in part from a tabulation by BCIRA, shows the effects of some trace elements on gray iron as well as their possible sources. Depending on property requirements, many of these elements can be intentionally added to gray iron. For example, tin and copper are often added to promote pearlite.

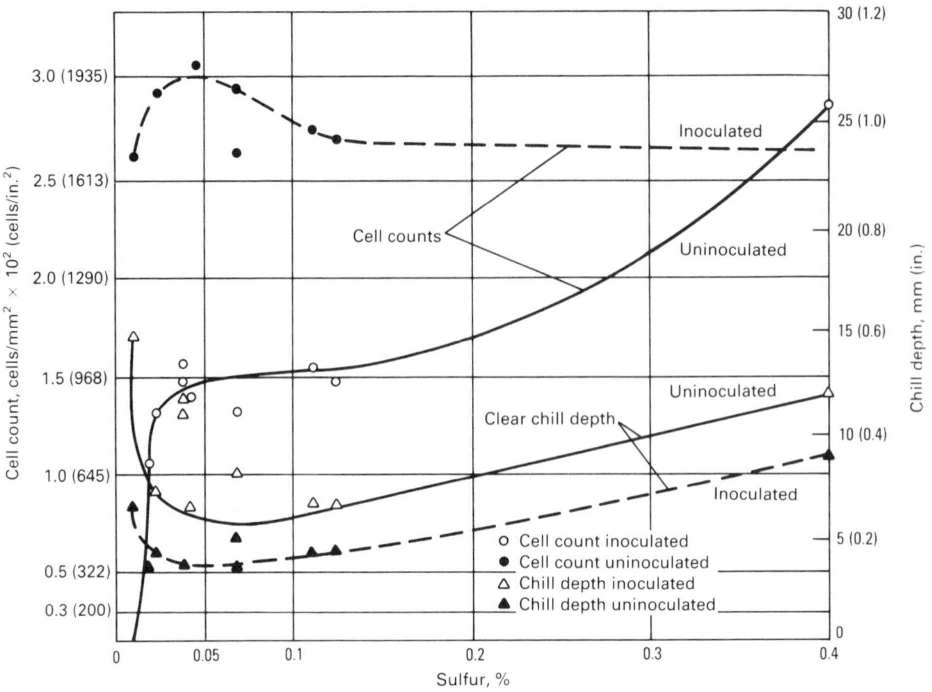

Fig. 2 Effect of sulfur on eutectic cell count and clear chill depth for inoculated and uninoculated gray irons. Source: Ref 4

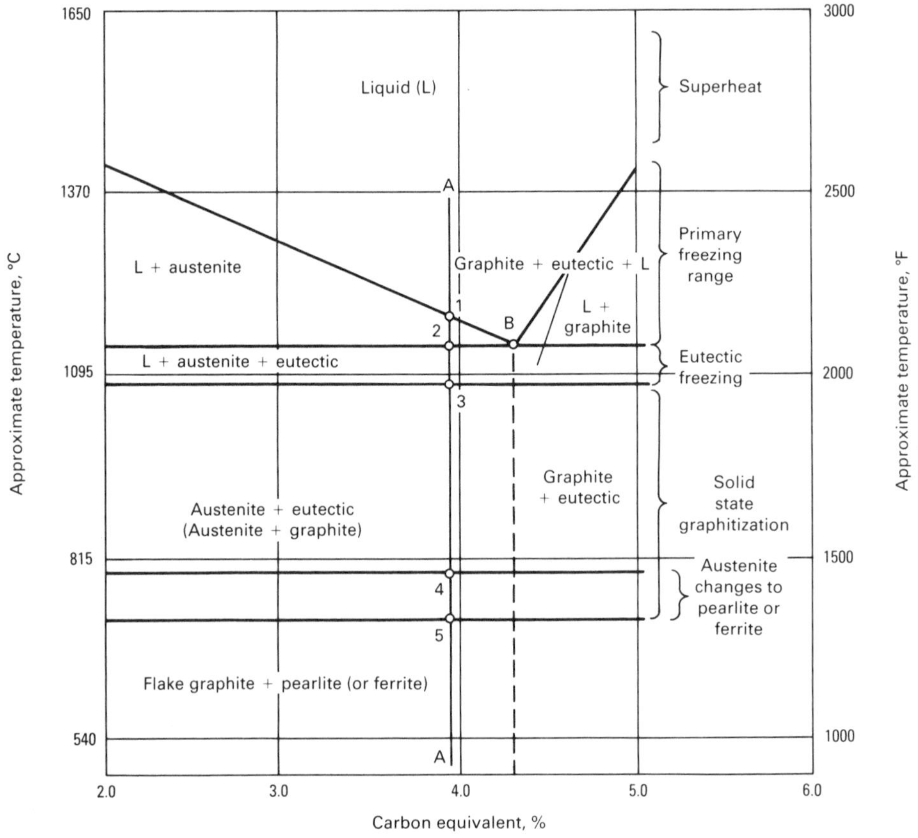

Fig. 3 Simplified iron-carbon-silicon phase diagram at 2% Si. Source: Ref 6

Solidification

Eutectic solidification and the accompanying transformations responsible for the development of the graphite and matrix structure of a cast iron are discussed in the section "Solidification of Eutectics" of the article "Fundamentals of Growth" in this Volume. A review of these principles as they pertain to gray iron will be presented in the following paragraphs.

Most gray irons are hypoeutectic in nature (that is, CE < 4.3). The sequence of events associated with the solidification of hypoeutectic irons can be studied with the simplified version of the iron-carbon-silicon ternary phase diagram taken at 2% Si (Fig. 3).

Table 1 Effects, levels, and sources of some trace elements in gray iron

Element	Trace level, %	Effects	Sources
Aluminum	≤0.03	Promotes hydrogen pinhole defects, especially when using green sand molds and at levels above 0.005%. Neutralizes nitrogen	Deliberate addition, ferrous alloys, inoculants, scrap contaminated with aluminum components
Antimony	≤0.02	Promotes pearlite. Addition of 0.01% reduces the amount of ferrite sometimes found adjacent to cored surfaces.	Vitreous enameled scrap, steel scrap, white metal bearing shells, deliberate addition
Arsenic	≤0.05	Promotes pearlite. Addition of 0.05% reduces the amount of ferrite sometimes found adjacent to cored surfaces.	Pig iron, steel scrap
Bismuth	≤0.02	Promotes carbides and undesirable graphite forms that reduce tensile properties	Deliberate addition, bismuth-containing molds and core coatings
Boron	≤0.01	Promotes carbides, particularly in light-section parts. Effects become significant above about 0.001%.	Deliberate addition, vitreous enameled scrap
Chromium	≤0.2	Promotes chill in thin sections	Alloy steel, chromium plate, some refined pig iron
Copper	≤0.3	Trace amounts have no significant effect and can be ignored.	Copper wire, nonferrous alloys, steel scrap, some refined pig iron
Hydrogen	≤0.0004	Produces subsurface pinholes and (less often) fissures or gross blowing through a section. Mild chill promoter. Promotes inverse chill when insufficient manganese is present. Promotes coarse graphite	Damp refractories, mold materials, and additions
Lead	≤0.005	Results in Widmanstätten and "spiky" graphite, especially in heavy sections with high hydrogen. Can reduce tensile strength 50% at low levels (≥0.0004%). Promotes pearlite	Some vitreous enamels, paints, free-cutting steels, nonferrous alloys, terne plate, white metal, solder, some pig irons
Molybdenum	≤0.05	Promotes pearlite	Some refined pig iron, steel scrap
Nickel	≤0.01	Trace amounts have no major effect and can be ignored.	Refined pig iron, steel scrap
Nitrogen	≤0.02	Compacts graphite and increases strength. Promotes pearlite. Increases chill. Can cause pinhole and fissure defects. Can be neutralized by aluminum or titanium	Coke, carburizers, mold and core binders, some ferroalloys, steel scrap
Tellurium	≤0.003	Not usually found, but a potent carbide former	Free-cutting brasses, mold and core coatings, deliberate addition
Tin	≤0.15	Strong pearlite promoter; sometimes deliberately added to promote pearlitic structures	Solder, steel scrap, nonferrous alloys, refined pig iron, deliberate addition
Titanium	≤0.15	Promotes undercooled graphite. Promotes hydrogen pinholing when aluminum is present. Combines with nitrogen to neutralize its effects	Some pig irons, steel scrap, some vitreous enamels and paints, deliberate addition
Tungsten	≤0.05	Promotes pearlite	Tool steel
Vanadium	≤0.08	Forms carbides; promotes pearlite	Steel scrap; some pig irons

Source: Ref 5

At temperatures above point 1 in Fig. 3, the iron is entirely molten. As the temperature is decreased and the liquidus line is crossed, primary freezing begins with the formation of proeutectic austenite dendrites. These dendrites grow and new dendrites form as the temperature drops through the primary freezing range, which is marked by points 1 and 2. Dendrite size is governed by the carbon equivalent of the iron and the solidification rate. Lower carbon equivalents produce large dendrites because the temperature interval between the liquidus and eutectic lines is greater for these irons than for those with a higher carbon equivalent. As expected, rapid cooling promotes a finer dendrite size.

During the formation of the austenite dendrites, carbon is rejected into the remaining liquid. The carbon content of the liquid increases until it reaches the eutectic composition of 4.3%. Once this composition is attained, the liquid transforms into two solids. This takes place between points 2 and 3. The type of solid formed depends on whether solidification is following the metastable or stable eutectic reaction. Iron carbide plus austenite form during the metastable reaction. Graphite plus austenite form during the stable reaction. When eutectic solidification is complete, no liquid metal remains, and any further reaction takes place in the solid state.

Although not shown in Fig. 3, in the temperature interval between the eutectic and eutectoid transformations, marked by points 3 and 4, the high-carbon austenite rejects carbon, which diffuses to the graphite flakes. This allows the austenite to acquire the composition needed for the eutectoid transformation, which, under equilibrium conditions, takes place between points 4 and 5. This transformation involves the decomposition of austenite into pearlite or pearlite plus ferrite, depending on such factors as the cooling rate and alloy content of the iron. In unalloyed gray irons, no significant changes in microstructure occur below the eutectoid transformation line.

Graphite Morphology

The mechanical and physical properties of gray iron are governed in part by the shape, size, amount, and distribution of the graphite flakes. A method for evaluating graphite flake distribution and size is given in ASTM A 247, and the metallography of cast irons is discussed in Ref 7 to 9.

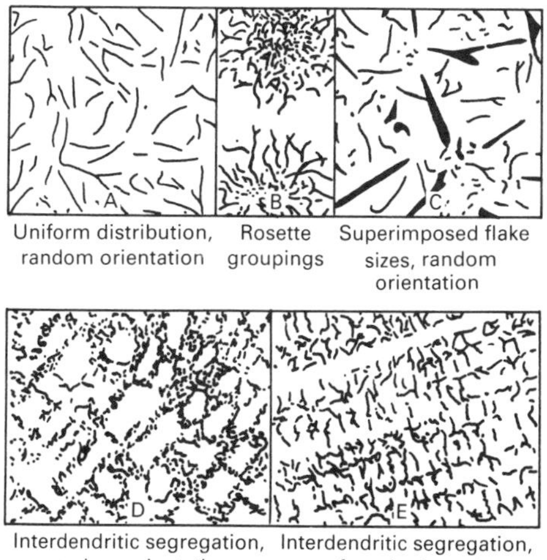

Fig. 4 Graphite distributions specified in ASTM A 247

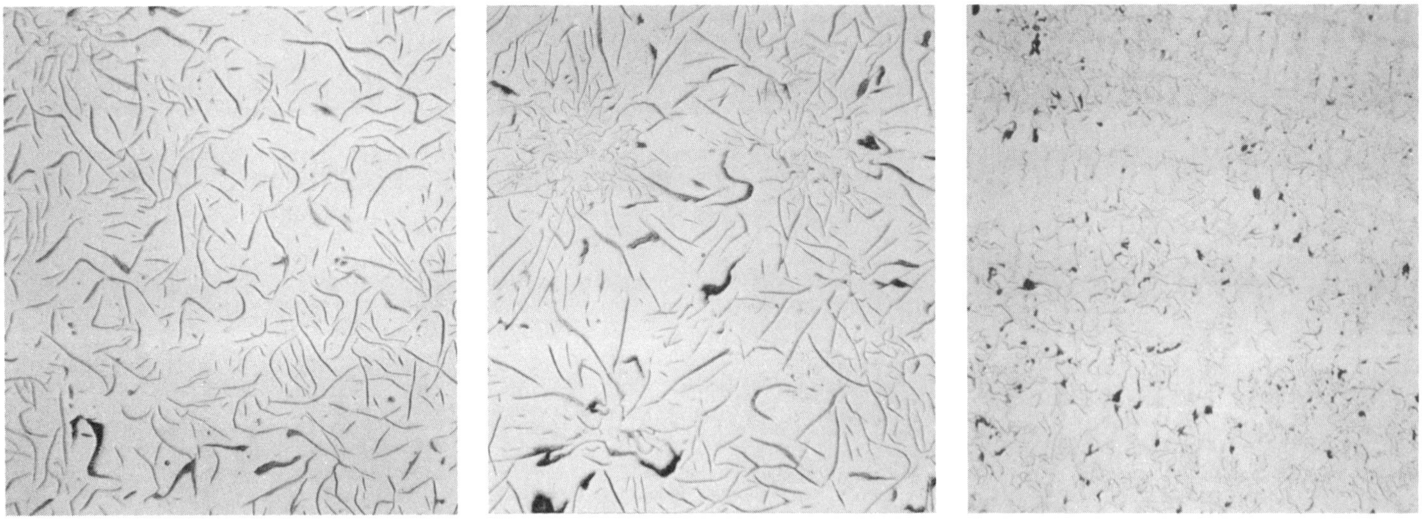

Fig. 5 Examples of type A (a), type B (b), and type D (c) graphite from foundry-produced gray irons. As-polished. 100×

There are five graphite flake distributions: A to E (Fig. 4). Type A graphite flakes are randomly distributed and oriented throughout the iron matrix. This type of graphite is found in irons that solidify with a minimum amount of undercooling, and type A is the structure desired if mechanical properties are to be optimized.

Type B graphite is formed in irons of near-eutectic composition that solidify with a greater amount of undercooling than that associated with type A graphite. Rosettes containing fine graphite, which are characteristic of type B, precipitate at the start of eutectic solidification. The heat of fusion associated with their formation increases the temperature of the surrounding liquid, thus decreasing the undercooling and resulting in the formation of type A graphite.

Types D and E graphite form when the amount of undercooling is high but is not sufficient to cause carbide formation. Both types are found in interdendritic regions. Type D graphite is randomly distributed, while the type E flakes have a preferred orientation. The manner in which the plane of polish intersects the graphite flakes may be responsible for this difference in orientation. Elements such as titanium and aluminum have been found to promote undercooled graphite structures. The iron matrix associated with undercooled graphite is usually ferrite because formation of the fine, highly branched flakes reduces carbon diffusion distances and results in a low-carbon matrix. Because ferrite has a lower tensile strength than pearlite, there is a reduction in the anticipated strength of the iron. Examples of type A, B, and D graphite found in commercial irons are shown in Fig. 5.

Type C graphite occurs in hypereutectic irons, particularly those with a high carbon content. Type C graphite precipitates during the primary freezing of the iron. Kish graphite, as it is often called, appears as straight, coarse plates. It greatly reduces the mechanical properties of the iron and produces a rough surface finish when machined. Type C graphite is, however, desirable in applications requiring a high degree of heat transfer.

Graphite flake sizes as categorized in ASTM A247 are shown in Fig. 6. Large flakes are associated with irons having high carbon equivalents and slow cooling rates. Strongly hypoeutectic irons and irons subjected to rapid solidification generally exhibit small, short flakes. The large flakes are desirable in applications requiring high thermal conductivity and damping capacity. Small flakes, because they disrupt the matrix to a lesser extent, are desired when maximum tensile properties and a fine, smooth surface finish are needed.

Matrix Structure

An etchant such as 2% nital is required to reveal the matrix phases in which the graphite flakes reside. Commonly found phases in cast iron are ferrite, cementite, and pearlite.

Ferrite is the soft, low-carbon α-iron phase that exhibits low tensile strength but high ductility. It is promoted by graphitizers such as silicon as well as slow cooling rates such as those found in heavy sections. As previously mentioned, ferrite is often found in conjunction with undercooled graphite (Fig. 7).

Cementite, or eutectic carbide, is a hard, brittle intermetallic compound of iron and carbon. Its formation is favored in areas of a casting where rapid cooling takes place, such as in thin sections, at corners, and along the cast surface. Irons with low carbon equivalences, particularly those with low silicon contents, are likely to contain cementite. An example of eutectic carbide found in a mottled iron is shown in Fig. 8.

Pearlite is the eutectoid transformation product and in gray iron consists of lamellar plates of ferrite and cementite. It possesses higher hardness and tensile strength than

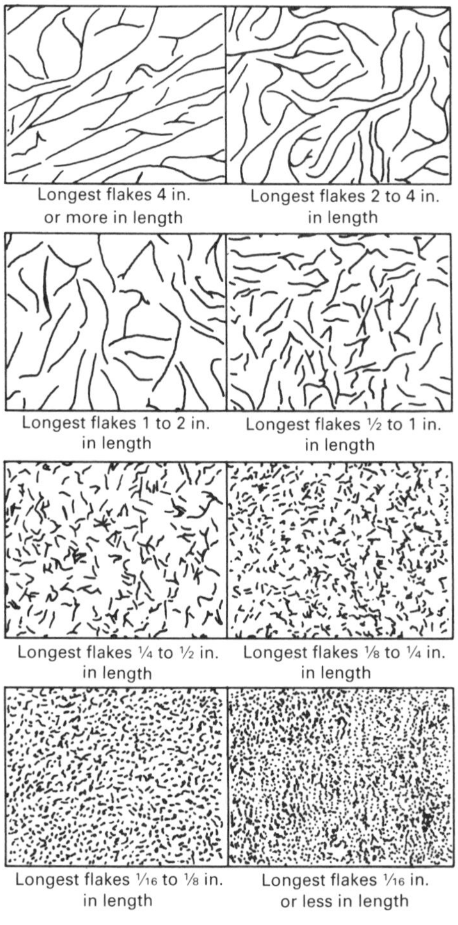

Fig. 6 Graphite flake sizes as specified in ASTM A 247

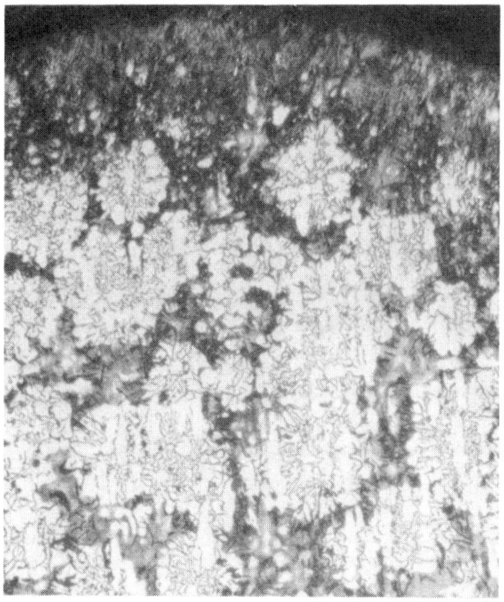

Fig. 7 Graphite and pearlite revealed by etching with 2% nital. The ferrite is found with undercooled graphite. 100×

ferrite but lower ductility. The hardness and tensile strength associated with pearlite depend primarily on the spacing of the plates. Higher values are found in pearlite with fine interlamellar spacing, which is associated with more rapid cooling rates or alloying. A comparison of tensile strength, ductility, and hardness values for these matrix phases is given below:

Phase	Tensile strength, MPa (ksi)	Elongation, %	Hardness, HB
Ferrite	272–290 (39.5–42)	61	75
Pearlite	862 (125)	10	200
Cementite	. . .	. . .	550

Source: Ref 6

Other microconstituents can also be formed in gray iron by changing the solidification rate or by adding alloying elements. Bainite can be produced by subjecting the iron to an isothermal heat treatment. Quenching the iron from the austenite region can induce martensite formation. Alloying elements such as nickel can be used to produce austenitic gray irons. Hardness values for various combinations of graphite and other matrix phases are given in Table 2.

Table 2 Hardness ranges for various combinations of gray iron microstructures

Microstructure	Hardness, HB
Ferrite + graphite	110–140
Pearlite + graphite	200–260
Pearlite + graphite + massive carbides	300–450
Bainite + graphite	260–350
Tempered martensite + graphite	350–550
Austenite + graphite	140–160

Source: Ref 11

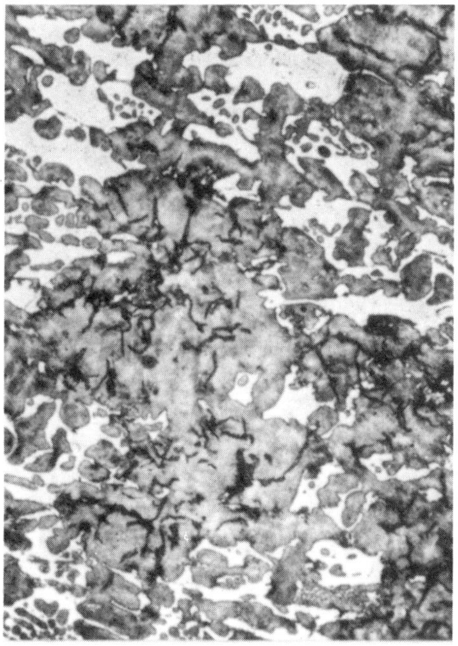

Fig. 8 Mottled cast iron etched using 4% picral. The white phase is eutectic carbide. 50×. Source: Ref 10

Steadite, the iron-phosphide eutectic, is commonly found in gray irons with phosphorus contents in excess of the 0.02% level considered to be soluble in austenite. It has a low melting point (about 930 °C, or 1705 °F) and is typically the last constituent to solidify. This explains its presence at cell boundaries, where it can assume a concave triangle appearance (Fig. 9). Steadite, like iron carbides, can decrease the mechanical properties of the iron. Elements such as chromium and molybdenum can concentrate in the phosphide phase, thus increasing its volume (Ref 13).

Manganese sulfides are commonly found evenly distributed in the matrix of gray iron, as shown in Fig. 10. They are dove gray, geometrically shaped inclusions that are formed before final solidification. The presence of manganese sulfide is a result of deliberate additions of manganese to prevent the formation of brittle iron sulfides that would otherwise form at the grain boundaries. Sufficient manganese must be added to tie up the sulfur to prevent this from occurring. Equation 3 is used to determine the manganese needed to balance the sulfur. Additional manganese is sometimes added, and a general rule is to add three times as much manganese as there is sulfur to ensure neutralization.

Titanium carbides or titanium carbonitrides are often observed in gray iron. This is particularly true for irons to which deliberate titanium additions have been made to prevent the formation of nitrogen fissure defects. These inclusions are angular, often cubic in appearance, and are found throughout the matrix but are concentrated in the

Fig. 9 Steadite in gray cast iron. Etched using 2% nital. 400×. Source: Ref 12

intercellular regions (Fig. 10). They usually possess an orange color when viewed under reflected light, but other colors, including blue-gray, violet, pink, and yellow, have been observed, depending on nitrogen content (Ref 14).

Section Sensitivity

The solidification of a gray iron casting is controlled by the composition and the cooling rate of the iron within the casting. Each variable has a considerable effect on solidification. However, once the iron is poured into the mold, the composition is fixed and, except for localized segregation, remains

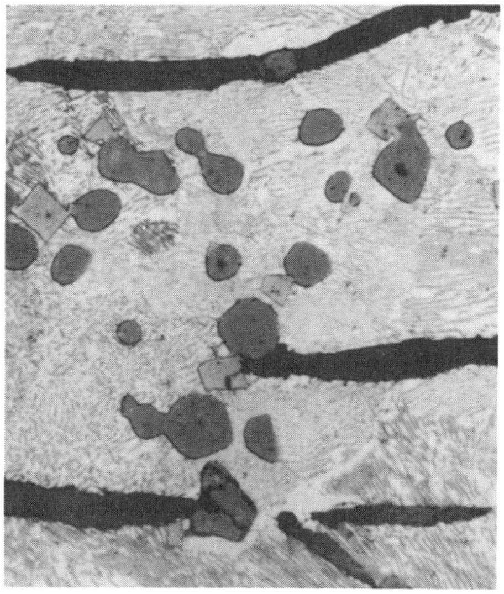

Fig. 10 Manganese sulfide (dark gray, rounded) and titanium carbonitride (light gray, angular) inclusions. Etched using 2% nital. 500×

relatively homogeneous throughout the casting. Therefore, during solidification, the cooling rate becomes the controlling variable. The cooling rate influences the microstructure from the inception of solidification until the iron passes through the eutectoid transformation. It is a controlling factor in the amount of carbon remaining in solution and therefore affects the resulting microstructure.

Cooling rate is influenced by a number of variables that include pouring temperature, pouring rate, volume of iron to be cooled, surface area of the iron, thermal conductivity of mold material, amount of mold material surrounding the casting, the number of castings in a mold, location of cores, and the position of gates and risers (Ref 15). Within a mold, a number of these variables will remain constant. However, as volume-to-surface-area ratio of the casting varies from section to section, so does the corresponding cooling rate. This variation in cooling rate results in a changing solidification pattern for each section, which can create a variation in mechanical properties within the casting. Therefore, the solidification of a gray iron casting is said to be section sensitive.

It is important for both the casting designer and the foundryman to recognize the section sensitivity of gray iron. The casting designer should indicate which mechanical properties are required in each section of a casting as well as which sections are critical. The foundryman can then select the iron composition that will develop the desired mechanical properties during solidification.

The effect of section thickness on the hardness of gray iron is illustrated in Fig. 11. A wedge-shaped bar with a taper of 10° was cast in a sand mold and sectioned near the center of the bar. Rockwell Hardness determinations were made progressively from the tip to the base of the wedge. The effect of increasing section size is shown by the change in hardness associated with the increasing width of the wedge.

The cooling rate is highest at the tip of the wedge. This results in the formation of white iron, a mixture of iron carbide and pearlite that is considerably harder than gray iron (Fig. 11). When the cooling rate has decreased sufficiently to allow the formation of some graphite, a mottled zone appears. This mottled zone, which is a mixture of gray and white iron, has a lower hardness than the white iron tip. As the width continues to increase, the white iron gradually disappears, and there is a corresponding drop in hardness. As the white iron disappears, the microstructure becomes a mixture of ferrite and type D graphite, resulting in the minimum hardness shown in Fig. 11. A further reduction in the cooling rate results in an increase in hardness as the microstructure shifts from type D to type A graphite and the matrix converts from ferrite to pearlite. As the cooling

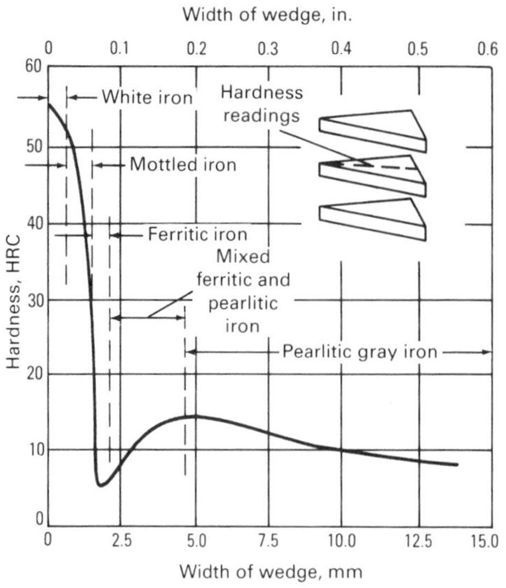

Fig. 11 Effect of section thickness on the hardness and microstructure of gray iron. Hardness readings were taken at increasing distance from the tip of a cast wedge (see inset). Iron composition was Fe-3.52C-2.55Si-1.01Mn-0.215P-0.086S.

rate continues to decrease, the hardness is reduced because of a gradual conversion of the pearlite to ferrite and the formation of a coarser graphite structure.

Figure 11 shows the hardness profile for one composition of gray iron. A change in composition or foundry practice can shift this curve to the right or left; therefore, the wedge can be a useful indicator of the tendency of the iron to chill. By measuring the depth of chill in the wedge, the foundryman can monitor gray iron for variations in the foundry process.

The cooling rate influences the amount of time allowed for the diffusion of carbon from the austenite to the graphite and therefore determines the level of combined carbon retained in the iron. Slow cooling rates allow more time for carbon diffusion. As the amount of combined carbon decreases, the amount of ferrite increases, resulting in an overall reduction of the mechanical strength of the iron. Therefore, the cooling rate must be controlled until the critical sections of the casting pass through the eutectoid temperature to ensure that the desired mechanical properties have been achieved.

In ASTM A 48, gray iron is classified based on the tensile strength of the iron. The categories range from class 20 to class 60 (minimum tensile strength of 138 to 414 MPa, or 20 to 60 ksi, respectively) in a 30.5 mm (1.2 in.) test bar. Figure 12 shows the effect of varying section size on tensile strength for various classes of iron. Tensile strength decreases with increasing section size for all classes of gray iron. For example, an iron with a tensile strength of 310 MPa (45 ksi) in a 25 mm (1 in.) section will develop only 207 MPa (30 ksi) in a 76 mm (3 in.) section because of the decreased cooling rate associated with the larger section.

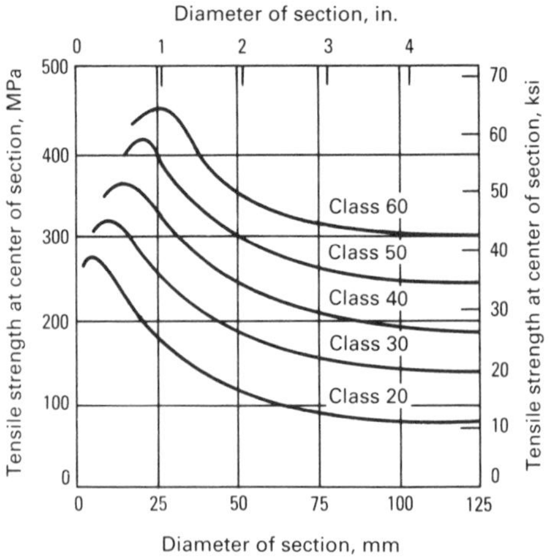

Fig. 12 Effect of section size on tensile strength of specimens cast from five classes of gray iron

This reduction in tensile strength is caused by the presence of larger graphite flakes and by a reduction in combined carbon. Decreasing combined carbon results in an increase in the amount of ferrite found as either a coarser pearlite spacing, or in the appearance of a ferrite phase.

It should be noted that the maximum strength of class 20 iron occurs at a smaller section size than class 60 iron. Each class has a minimum section size that can be cast without the formation of iron carbide; recommended minimum section sizes are listed in Table 3 for each class of unalloyed gray iron. It is important for both the foundryman and the designer to recognize that each class of iron has a minimum section size in which it can be cast without the presence of chill and that this minimum increases with higher classes of gray iron.

Cooling rates are difficult to predict for complex shapes. However, the cooling rate for the casting can be approximated by comparing the various sections of the casting to the simplified shapes listed in Table 4. The volume-to-surface-area ratios have been calculated for these shapes and can be compared to the minimum volume-to-surface-area ratios for each class of iron listed in Table 3.

Table 3 Minimum recommended section sizes for unalloyed gray irons

Class	Minimum section thickness		Volume-to-surface-area ratio(a)	
	mm	in.	mm	in.
20	3.2	⅛	1.5	0.06
25	6.4	¼	3.0	0.12
30	9.5	⅜	4.3	0.17
35	9.5	⅜	4.3	0.17
40	15.9	⅝	7.1	0.28
50	19.0	¾	8.4	0.33
60	25.4	1	10.7	0.42

(a) Volume-to-surface-area ratios are for square plates.

Table 4 Volume-to-surface-area ratios for bars and plates

Cast form and size	Volume-to-surface-area ratio mm	in.
Bar 12.7 mm (½ in.) diam × 533 mm (21 in.) long	3.0	0.12
Bar 12.7 mm (½ in.) square × 533 mm (21 in.) long	3.0	0.12
Plate 6.4 mm (¼ in.) × 305 × 305 mm (12 × 12 in.)	3.0	0.12
Bar 30.5 mm (1.2 in.) diam × 533 mm (21 in.)	7.4	0.29
Bar 30.5 mm (1.2 in.) square × 533 mm (21 in.)	7.4	0.29
Plate 15.9 mm (⅝ in.) × 305 × 305 mm (12 × 12 in.)	7.1	0.28
Bar 51 mm (2 in.) diam × 559 mm (22 in.)	12.2	0.48
Bar 51 mm (2 in.) square × 559 mm (22 in.)	12.2	0.48
Plate 28.6 mm (1⅛ in.) × 305 × 305 mm (12 × 12 in.)	11.9	0.47
Bar 102 mm (4 in.) diam × 457 mm (18 in.)	22.9	0.90
Bar 102 mm (4 in.) square × 457 mm (18 in.)	22.9	0.90
Plate 65 mm (2$\frac{9}{16}$ in.) × 305 × 305 mm (12 × 12 in.)	22.8	0.90
Bar 152 mm (6 in.) diam × 457 mm (18 in.)	32.7	1.29
Bar 152 mm (6 in.) square × 457 mm (18 in.)	32.7	1.29
Plate 114 mm (4½ in.) × 305 × 305 mm (12 × 12 in.)	32.7	1.29

The minimum section sizes were developed for plates and can vary depending on foundry practice and mass effects. Mass effects occur when a very large section of a casting cools very slowly and influences the cooling rate of nearby smaller sections.

Commercial foundries often use separately cast test bars to monitor the mechanical properties of the iron. Because gray iron is section sensitive, it is important for the foundry to select a test bar with a cooling rate similar to that of the controlling section of the casting. Specification ASTM A 48 covers five categories of test bars that represent various section sizes within a casting. Foundries typically select a test bar 30.5 mm (1.2 in.) in diameter to monitor the mechanical properties of gray iron. It is interesting to note that this 30.5 mm (1.2 in.) diam bar is used in determining the minimum tensile strength required for each class of iron. It is important that the designer not only designate which class of iron is required but also in which sections this strength is required, so that the foundryman can accurately control the mechanical properties of the casting.

Foundry Practice

Specific aspects of foundry practice for gray iron that will be discussed include melting, inoculation, alloying, pouring, and molding. Foundry practices for other cast irons are detailed in the articles "Ductile Iron," "Compacted Graphite Irons," "High-Alloy White Irons," "Malleable Iron," and "High-Alloy Graphitic Irons" in this Volume.

Melting

The essential purpose of melting is to produce molten iron of the desired composition and temperature. For gray iron, this can be accomplished with various types of melting equipment. Cupolas and induction furnaces tend to be the types most commonly found in the gray iron foundry. The cupola was traditionally the major source of molten iron. However, gradual acceptance of electric melting has reduced the dominance of the cupola. Gray iron can be melted with a single furnace or a combination of furnaces. Detailed information on the operating parameters of each furnace type can be found in the Section "Foundry Equipment and Processing" in this Volume.

Regardless of the type of furnace used, the basic melting process is a physical transformation from solid to liquid rather than a complex reduction or oxidation reaction. Therefore, the composition of all the materials charged into the furnace determines the composition of the slag/iron mixture. As mentioned previously, the composition of the iron will affect the resulting microstruc-

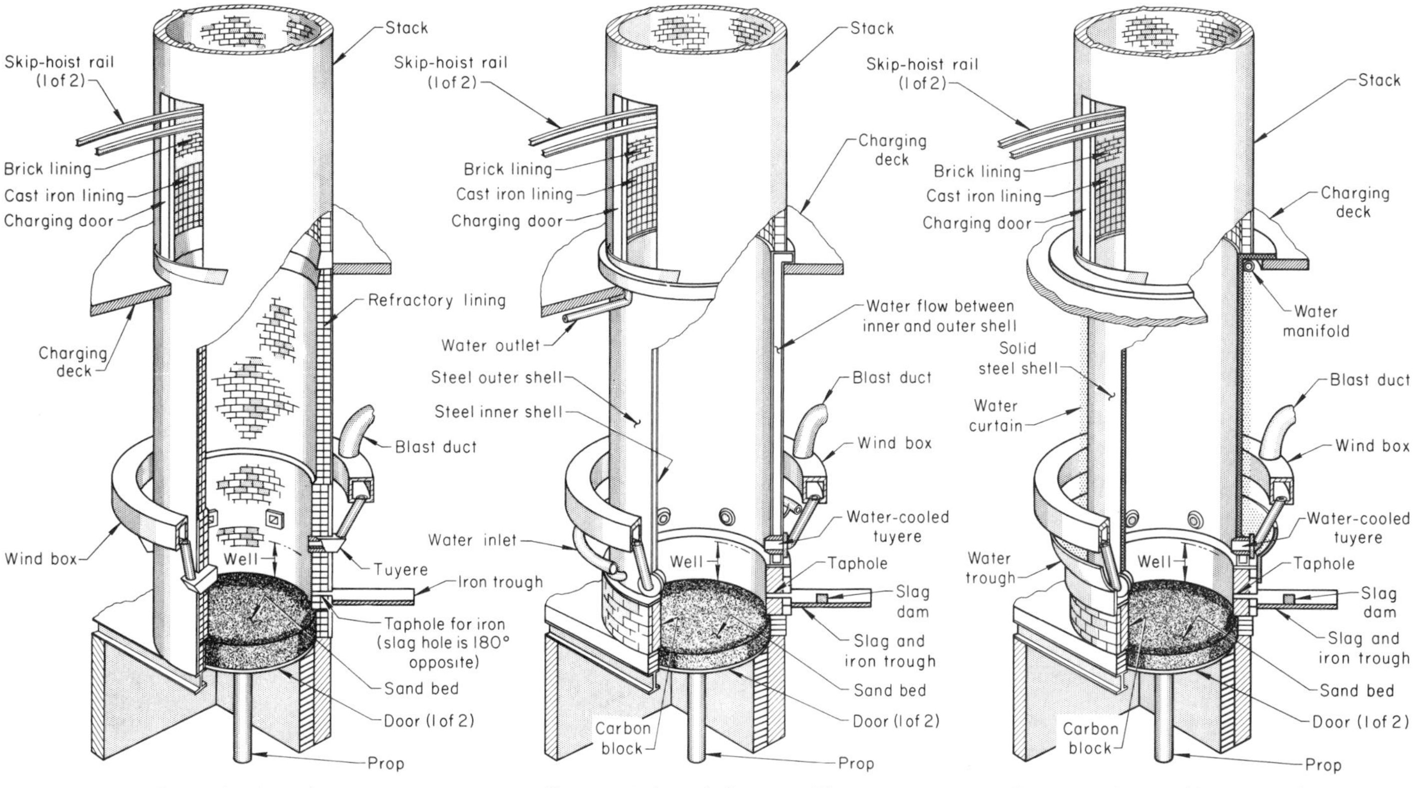

Fig. 13 Sectional views of conventional and water-cooled cupolas

ture and mechanical properties. Therefore, control of the major, minor, and trace elements in the charge materials will ultimately influence the properties of the iron.

The cupola is a vertical steel shaft that can be refractory lined or water cooled, as shown in Fig. 13. It consists of a furnace hearth (or well), a melting zone, and an upper stack. The furnace hearth includes the bottom doors, the sand bed, and the iron trough. The tuyeres are located near the bottom of the melting zone, and the charging doors are located in the upper stack.

The cupola is prepared for operation by closing the bottom doors and supporting them with a prop. A bed of rammed sand 150 to 200 mm (6 to 8 in.) thick is placed on the doors. The sand bed is sloped such that the molten iron and slag that collect on the bed flow toward the taphole, which is located at the edge of the sand bed. The hearth and the melting zone are filled with a coke bed to a level of approximately 1 to 1.5 m (40 to 60 in.) above the tuyere level. The coke bed is lit, and the shaft is filled with alternating layers of iron-bearing materials, coke, fluxes, and alloys. The shaft is filled to the bottom of the charging doors. The air supply or blast is turned on, causing a flow of air to impinge on the coke bed through the tuyeres. The coke burns, liberating heat that superheats a mixture of air and combustion products.

The hot gases from the combustion reaction rise through the stack, heating the charge materials. As the charge materials descend through the stack, the metallics and the fluxes begin to melt. Small iron droplets form and percolate down through the coke bed. The coke bed remains solid up to temperatures of approximately 2000 °C (3630 °F) and thus supports the charge. The iron droplets are superheated as they filter through the coke bed and collect in the hearth. The coke, which is charged in alternate layers with the metallic materials, replenishes that which is consumed during combustion. Therefore, melting is continuous as long as the air blast is on.

As the metal droplets cascade through the coke bed, their surfaces become carburized through contact with the coke. The iron also absorbs sulfur from the coke and oxidizes slightly in the region above the tuyeres. The oxidized iron is then reduced in the region below the tuyeres. Portions of the silicon and manganese alloys are also oxidized in the cupola. However, carbon reduces some of these oxides to their metallic form (Ref 16). Any oxides not reduced in the lower region of the furnace are entrained in the slag. The molten droplets collect within the hearth, and the iron and slag separate due to their different densities. Iron can be tapped from the cupola continuously or intermittently, depending on design.

Cupola slag is formed from coke ash and impurities arising from the charge, together with oxides of silicon, manganese, and other alloying elements (Ref 16). Cupola slag is usually acidic. Acidity is determined by the ratio of CaO to SiO_2 in the slag. Limestone is added to the cupola charge to control the acid/base ratio of the slag and to flux away the coke ash. This fluxing action intensifies the combustion and carburization action of the coke. The acid/base ratio of the slag affects the final composition of the gray iron. An increase in basicity reduces the final sulfur and increases the carbon in the iron. A decrease in silicon can accompany this increase in basicity because of increased oxidation losses. Therefore, control of the fluxing materials charged can influence the resulting composition of cupola-melted gray iron.

The composition of gray iron melted in a cupola depends on the composition of all materials charged and the degree of oxidation each element experiences during melting. The approximate gain or loss various charge constituents experience during the melting process is shown below:

Charge constituent	Loss or gain, % of weight charged
Silicon(a)	−7 to 12
Manganese(a)	−10 to 20
Ferrosilicon (lump)	−10 to 15
Ferromanganese (lump)	−15 to 25
Spiegeleisen	−15 to 25
Phosphorus	+Trace
Ferrochromium (lump)	−10 to 20
Nickel (shot or ingot)	−2 to 5
Copper(b)	−2 to 5
Alloys in briquets	−5 to 10
Sulfur	+40 to 60

(a) In pig iron or scrap. (b) As shot or scrap with minimum thickness of 4.8 mm (3/16 in.)

The metallic charge composition is based on the estimated losses and the compositional requirements of the gray iron. Compositions for gray iron melted in a cupola will vary among foundries, and some experimentation is usually required to develop a charge composition.

Induction Melting. Commercial induction furnaces are classified as either coreless or channel induction based on the design feature used to induce energy into the metal. Melting is accomplished in an induction furnace through the conversion of an electrically induced magnetic field into heat within the metallic charge materials. The passing of a magnetic field through an electrically insulated metal generates electrical currents that are converted to heat due to resistance within the metal. Melting occurs when sufficient energy has been induced into the metallic charge. A further influx of energy results in the superheating of the liquid metal. Controlled energy input gives excellent temperature control of the molten metal.

A coreless induction furnace (Fig. 14a) consists of a refractory-lined crucible encircled by a hollow copper coil. The copper coil is water cooled and is connected to a high-energy electrical source. The coil is commonly held in place by vertical iron bars that also act as magnetic yokes. The crucible is usually deeper than it is wide for good electrical coupling. The induction field creates a stirring action in the furnace crucible that is directly proportional to the energy input. A coreless induction furnace can melt as either a batch or semibatch unit. The semibatch unit requires about 50 to 85% of the molten iron remain in the crucible, while the batch method allows for complete removal of the molten iron prior to charging of the cold material. The excellent mixing due to the induced energy makes the coreless induction furnace an excellent primary melter that delivers iron with uniform chemical composition at the desired temperature.

A channel induction furnace (Fig. 14b) consists of a refractory-lined metal reservoir connected to a refractory-lined transformer core. A small channel filled with molten iron acts as one leg of the transformer, and the electrical energy is induced into this channel of iron. The induced energy superheats and circulates the iron within the channel. The superheated iron is returned to the reservoir, where it mixes into the bath by convection, creating a gentle stirring action. A channel induction furnace is highly efficient electrically and generally operates on a low-frequency power supply. Due to the lower stirring action, a channel induction furnace is inefficient as a primary melting unit. Because of its design, this furnace requires a molten metal charge. The channel furnace is generally used in conjunction with another type of melting furnace as a duplexing unit. It is used to smooth out fluctuations in chemistry and temperature from the primary melter and to superheat the iron.

Compositional control of gray iron is greater in induction furnaces than cupolas because of the reduced oxidation losses during melting and the elimination of coke as fuel. The composition of the metallic charge materials is directly related to the final composition of the iron. Therefore, accurate weights and analysis of charge materials are required for good compositional control. Typical charge materials include steel scrap, cast iron scrap, foundry returns, ferrosilicon, and carbon. Pig iron is rarely used because of cost, although it does offer advantages in terms of purity. The type of scrap employed will depend on economics. The elimination of coke from the charge also reduces the sulfur absorbed by the iron; therefore, resulfurization is often desirable in electrically melted gray iron.

Inoculation

Inoculation is defined as the late addition of certain silicon alloys to molten iron to produce changes in graphite distribution, improvements in mechanical properties,

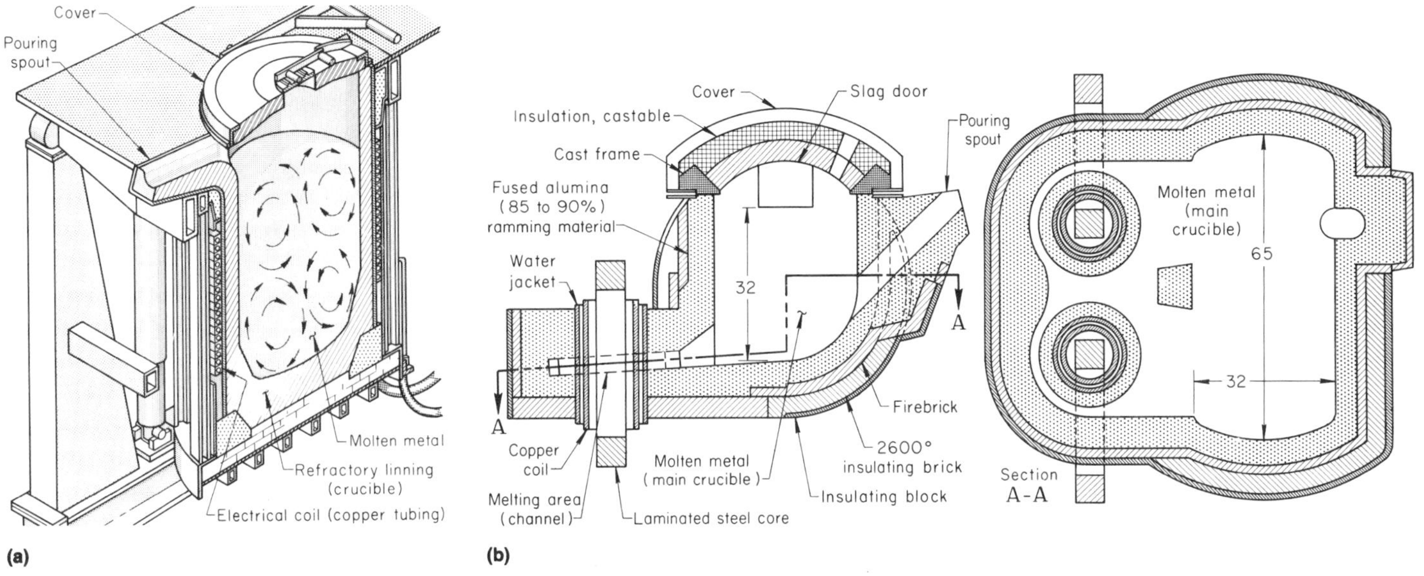

Fig. 14 Sectional views of coreless (a) and cored (b) induction furnaces. Dimensions given in inches

and a reduction of the chilling tendency that are not explainable on the basis of composition change with respect to silicon (Ref 17). Graphite, added alone or in combination with ferrosilicon, will also produce these changes without significantly altering the chemistry of the iron. It is recognized that two irons with the same apparent composition can have dramatically different microstructures and properties if one is inoculated and the other is not. A great deal of research has been dedicated to determining the mechanism behind inoculation (Ref 18, 19). Although many theories exist, no conclusions have been reached regarding possible mechanisms.

The purpose of an inoculant is to increase the number of nuclei in molten iron so that eutectic solidification, specifically graphite precipitation, can begin with a minimum amount of undercooling. When undercooling is minimized, there is a corresponding reduction in the tendency to form eutectic carbide or white iron, which is referred to as chill. Instead, a more uniform microstructure consisting of small type A graphite flakes is produced. These microstructural changes can result in improved machinability and mechanical properties.

Irons are inoculated for various reasons. The primary reason is to control chill in areas of castings that experience rapid solidification, such as in thin sections, at corners, and along edges. Tensile strength can be improved through inoculation. This is particularly true for the lower carbon equivalent irons, which are selected for applications requiring a higher tensile strength (tensile strength decreases as carbon equivalence increases) (Ref 20). Low carbon equivalent irons, however, are also the grades that are most susceptible to carbide formation. Inoculation helps overcome this problem by minimizing the chill-forming tendency of the iron, thus allowing low carbon equivalent irons to be poured in thin sections. Irons that are stored in holding furnaces or in pouring systems for an extended period of time are also more susceptible to chill formation. This susceptibility can be attributed to the reduction in nuclei in the melt that takes place during extended holding. This effect is accelerated if holding occurs at high temperatures. Melting method influences white iron formation; electric-melted irons are generally more prone to carbide formation than cupola-melted irons.

Types of Inoculants. Graphite- or ferrosilicon-base alloys can be used to inoculate gray iron. The graphite used must be highly crystalline to give the best effect. Examples of highly crystalline graphite include some naturally occurring graphite and graphite electrode scrap. Amorphous forms of carbon, such as metallurgical coke, petroleum coke, and carbon electrode scrap, are not suitable for inoculation. Graphite is rarely used by itself and is usually added in combination with a crushed ferrosilicon. Because inconsistent results have been obtained with graphite, careful addition methods and relatively higher temperatures are needed to ensure its complete solution. Graphite has been found to promote extremely high eutectic cell counts.

Ferrosilicon alloys are also used to treat gray iron. They are typically based on 50 or 75% ferrosilicon and act as carriers for the inoculating (reactive) elements, which include aluminum, barium, calcium, cerium or other rare earths, magnesium, strontium, titanium, and zirconium. The silicon in the alloys does not cause significant chill reduction unless added to a level that produces a marked increase in carbon equivalence. Because ferrosilicon dissolves readily, it helps to distribute the reactive elements uniformly throughout the melt. It is important to note that the reactive elements, in addition to reacting with iron, react readily with sulfur and oxygen. Their addition, therefore, can result in dross formation. The

Table 5 Compositions of ferrosilicon inoculants for gray cast iron

Performance category of inoculant	Composition, %(a)									
	Si	Al	Ca	Ba	Ce	TRE(b)	Ti	Mn	Sr	Others
Standard	46–50	0.5–1.25	0.60–0.90	...	...	...	...	...	...	...
	74–79	1.25 max	0.50–1.0	...	...	...	...	...	...	...
	74–79	0.75–1.5	1.0–1.5	...	...	...	...	...	...	...
Intermediate	46–50	1.25 max	0.75–1.25	0.75–1.25	...	...	...	1.25 max	...	...
	60–65	0.8–1.5	1.5–3.0	4–6	...	...	...	7–12	...	...
	70–74	0.8–1.5	0.8–1.5	0.7–1.3	...	...	...	...	...	...
				0.75–1.25						
	42–44	...	0.75–1.25		...	...	9–11	...	...	...
	50–55	...	5–7	...	...	...	9–11	...	...	...
	50–55	...	0.5–1.5	...	...	...	9–11	...	...	...
High	36–40	...	...	...	9–11	10.5–15	...	...	...	...
	46–50	0.50 max	0.10 max	...	...	...	...	...	0.60–1.0	...
	73–78	0.50 max	0.10 max	...	...	...	...	...	0.60–1.0	...
Stabilizing	6–11	0.50 max	0.50 max	...	...	...	...	...	...	48–52 Cr

(a) All compositions contain balance of iron. (b) TRE, total rare earths

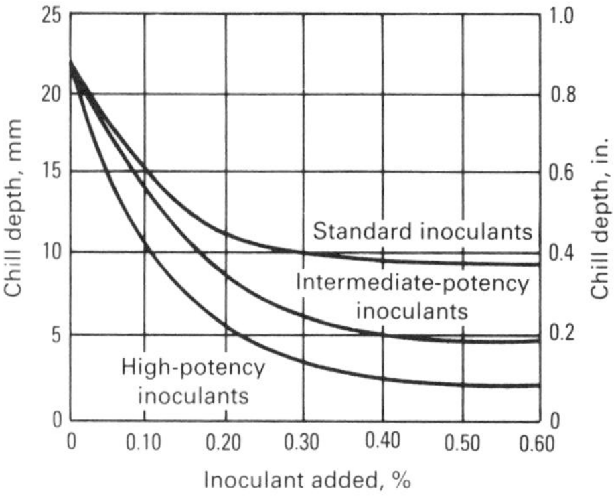

Fig. 15 General classification of inoculants showing chill reduction in iron with carbon equivalence of 4.0

quantity of dross produced is directly proportional to the amount of reactive elements in the alloy. Compositions of some of the commercially available ferrosilicon inoculants are given in Table 5.

As indicated in Table 5, it is convenient to group inoculants into four performance categories: standard, intermediate, high potency, and stabilizing. The calcium-bearing alloys fall into the standard category. Improvement in chill reduction is obtained by pairing calcium with barium (Ref 21). This type of inoculant falls into the intermediate group. The strontium or calcium plus cerium alloys are the strongest chill reducers. Figure 15 shows the performance obtainable with these three inoculant groups. It also shows that as the amount of inoculant added increases, a reduction in chill is realized until a point of diminishing returns is reached. It is not correct to assume that if a little is good, a lot is better. Addition levels above those needed to control chill and produce the desired mechanical and microstructural changes result in higher costs and can lead to a variety of problems, including inclusion defects caused by inoculant dross, hydrogen pinholing, and shrinkage.

Aluminum is found in all ferrosilicon inoculants in amounts that can vary considerably. This is because the effectiveness of aluminum in controlling chill in gray iron is still under debate. Because aluminum has been linked to hydrogen pinhole formation in castings produced in green sand molds, it is best to keep its level rather low.

Stabilizing inoculants are also available. They are designed to promote pearlite and at the same time provide graphitization during solidification. They are useful in producing high-strength castings with a minimum of chill, and they help to eliminate ferrite in thick sections (Ref 22). Stabilizing inoculants normally employ chromium as the pearlite stabilizer. Because these alloys can be difficult to dissolve, they are not suggested for mold addition.

Inoculant Evaluation. A variety of methods can be employed to evaluate the effectiveness of an inoculant. Because chill reduction is often of primary concern, one way to test an alloy is by pouring chill bars or wedges. The depth of chill produced in the treated iron can be compared to that of the base iron. Details of chill testing can be found in ASTM A 367.

Inoculation is known to increase the eutectic cell count of gray iron. Therefore, eutectic cell count can be used as an indication of the nucleation state of the melt. Samples for this type of evaluation are typically polished through 400-grit paper and then etched with Stead's Reagent (Ref 23). Results should be interpreted with caution because no standard relationship exists between cell count and chill depth. For example, strontium-bearing ferrosilicon inoculants reduce chill to very low levels without dramatically increasing the eutectic cell count.

Evaluation of the microstructure and mechanical properties provides information regarding the performance of an inoculant. The presence of type A graphite rather than carbides and undercooled graphite, along with attainment of the desired tensile properties, indicates that inoculation has been successful. Cooling curves can also be used to evaluate inoculant performance; the relative amount of undercooling can be used as an indicator of effectiveness.

Each technique of alloy evaluation has advantages and disadvantages. The use of a suitable combination of these techniques is suggested for obtaining an adequate assessment of inoculant performance in a particular base iron under a given set of foundry conditions.

Addition Methods. Ladle inoculation is a common method of treating gray iron. In this method, the alloy is added to the metal stream as it flows from the transfer ladle into the pouring ladle. A small heel of metal should be allowed to accumulate in the bottom of the ladle prior to the addition. This allows the inoculant to be mixed and evenly distributed in the iron. Addition of the alloy to the bottom of an empty ladle is not recommended, because this may cause sintering and a reduction in inoculant effectiveness. Problems can also arise if the alloy is added to a full ladle because the material can become entrapped in the slag layer that forms on the surface.

The amount of inoculant needed in this treatment normally varies between 0.15 and 0.4%, depending on the potency of the inoculant. If graphite alone is used, the addition level is about 0.1 to 0.2%. Excessive additions should be avoided for the reasons cited earlier. Inoculants for this method typically have a 6 or 12 mm (¼ or ½ in.) maximum size. Minimum size has not been found to be as critical, although excessive fines should be avoided because they can float on the surface and lose their effectiveness through oxidation. All additions should be weighed or measured accurately, and the use of proper metal temperatures ensures good dissolution. Minimizing the time between treatment and pouring helps avoid loss of the inoculating effect, which is known as fade.

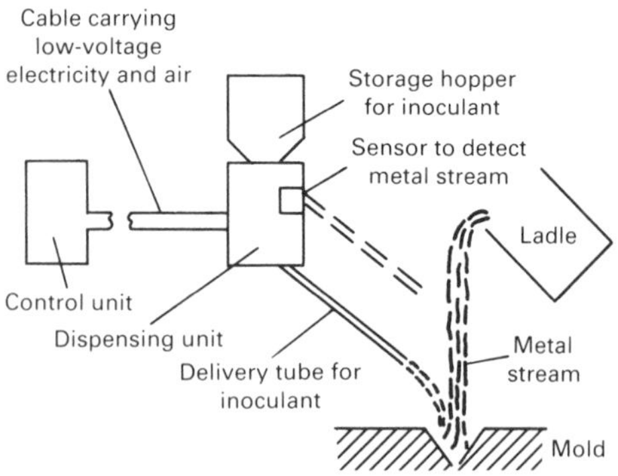

Fig. 16 Schematic showing the principle of stream inoculation. Source: Ref 25

The maximum effect of an inoculant is realized immediately after the alloy is dissolved in the metal. More than half of the effect of inoculation can be lost, because of fade, in the first 5 min after the addition. Complete loss can occur if the iron is held for 15 to 30 min (Ref 24). Although the composition of the iron does not change dramatically, an increase in chilling tendency occurs, along with an accompanying decrease in mechanical properties. All inoculants fade to some degree.

Methods that can help avoid or minimize the loss of the inoculation effect have been gaining acceptance. By adding inoculant late in the production process, the effect of time can be greatly reduced. Stream and mold inoculation are two late methods of iron treatment that are believed to promote more uniform quality from casting to casting.

Stream inoculation requires that the alloy be added to the stream of metal flowing from the pouring ladle into the mold. One of the electropneumatic devices used to sense when the metal flow starts and stops is shown schematically in Fig. 16. This device ensures that the alloy is dispensed in such a manner that the last metal entering the mold is treated similarly to the first metal. The same inoculants used to treat iron in the ladle can be used for stream inoculation, but less of a performance distinction has been observed among them. Graphite is not recommended, because of its relatively poor solution characteristics. A uniform and consistent size seems to be a very important factor in stream inoculation. Too large a size can cause plugging of the equipment and incomplete dissolution. A maximum particle size of 8 to 30 mesh and a minimum size of about 100 mesh are recommended. Addition levels range from 0.10 to 0.15%.

Mold inoculation involves placement of the alloy in the mold, such as in pouring basins, at the base of the sprue or in suitable chambers in the runner system (Ref 26). Inoculants for this method can be crushed material, powder bonded into a pellet, or precast slugs or blocks. As in stream inoculation, alloy dissolution rate is an important factor. Crushed alloy is for this application typically 20 to 70 mesh in size, and the addition rate can be as little as 0.05%. The precast and bonded alloys are designed to dissolve at a controlled rate throughout the entire pouring cycle (Ref 27). Mold inoculation is often used as a supplement to ladle inoculation.

There are several advantages of late inoculation over ladle inoculation. As previously stated, fading is virtually eliminated, and because the castings are inoculated to the same extent, there is greater consistency in structure from casting to casting. It has also been observed that late inoculation is more successful in preventing carbide formation in thin sections, thus eliminating heat treatment.

Alloying

Alloying is used to a fairly large degree in the production of gray iron. Alloyed irons are discussed in the article "High-Alloy White Irons" in this Volume. However, it is important to recognize the minor alloying of gray iron that is conducted at levels below that considered in the category of alloyed irons.

In most cases, minor alloying of gray iron is done to increase strength and promote pearlite. The minor elements normally used in gray iron alloying are chromium, copper, nickel, molybdenum, and tin. Recently, beneficial effects of vanadium as an alloying element have also been demonstrated (Ref 28).

Chromium. Small chromium additions (up to about 0.5 to 0.75%) cause significant increases in the strength of gray iron. Chromium also promotes a pearlitic matrix and an associated increase in hardness. Chromium is a carbide promoter, and in light-sectioned castings or at heavy addition rates, it can cause chill formation. Chromium is normally added as a ferrochromium alloy. Care should be taken to ensure that the alloy is completely dissolved.

Copper also increases the tensile strength of gray iron by promoting a pearlitic matrix. Its effect is most pronounced at lower addition levels of 0.25% to 0.5%. At higher addition rates, its effects are not as dramatic. Copper has a mild graphitizing effect and therefore does not promote carbides in light sections. Copper should be added as high-purity material to avoid the introduction of tramp elements such as lead.

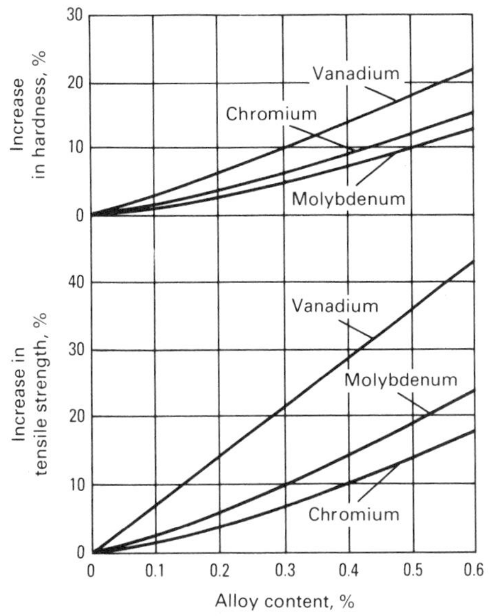

Fig. 17 Effects of alloying elements on the properties of gray cast iron. Source: Ref 28

Nickel additions of up to 2% cause only a minor increase in the tensile strength of gray iron. Nickel does not promote the formation of carbides and in fact has a minor graphitizing effect. Nickel is normally added as elemental material, and no problems with dissolution have been reported.

Molybdenum. Small molybdenum additions in the range of 0.25 to 0.75% have a significant impact on the strength of gray iron. This appears to be the result of both matrix strengthening and graphite flake refinement (Ref 29). Molybdenum does not promote carbides. It is normally added as a ferromolybdenum alloy.

Tin in the range of 0.025 to 0.1% is a strong pearlite stabilizer. It does not appear to increase the strength of a fully pearlitic gray iron. It can, however, give a small strength increase in irons that would otherwise contain free ferrite. Additions above the levels required for pearlite stabilization should be avoided to prevent embrittlement. Tin is normally added as commercially pure tin. Care should be taken to avoid material contaminated with such elements as antimony, bismuth, and lead.

Vanadium has recently been suggested as a minor alloying element for gray iron (Ref 28). As shown in Fig. 17, vanadium has a significant effect on the hardness and strength of gray iron. It was reported that the strength increases were sustained even after annealing; this is a significant advantage. Vanadium at higher levels and in light sections can promote the formation of carbides, so good inoculation practices are suggested. Alloying can be accomplished by using ferrovanadium.

Pouring

Metal temperature, cleanliness, and delivery technique are the essential variables to be controlled when pouring gray iron. Removal of slag and dross from the liquid iron surface reduces the possibility of inclusion-type defects. A variety of metal-pouring techniques are employed in the gray iron foundry. Each technique should enable the operator to maintain a constant flow of metal sufficient to keep the gating system full of metal until the pour is completed.

The molten iron should possess sufficient superheat to enable the cavity to be completely filled with iron without the formation of temperature-related defects. Superheat is the differential between the molten iron temperature and the liquidus temperature. The fluidity of the iron is directly proportional to the amount of superheat contained in the iron. Therefore, thinner sections that require greater fluidity for successful casting also require more superheat than thicker sections. The liquidus of gray iron is determined by the composition of the iron. For hypoeutectic irons, the liquidus is inversely proportional to the carbon equivalent of the iron. Therefore, the higher-strength irons with lower carbon equivalents require higher pouring temperatures to maintain the necessary level of superheat. Table 6 lists the typical pouring temperatures required for the various classes of gray iron.

Iron containing insufficient superheat can lead to partially filled cavities, misruns, blowholes, and chill. An excess of superheat can lead to shrinkage, metal penetration, veining, and scabbing. The amount of superheat required will vary for individual castings, and some experimentation will be required to determine the optimum pouring temperature for each casting.

Table 6 Typical pouring temperatures for some classes of gray iron

Class	Approximate liquidus temperature		Pouring temperature: Small castings, Thin sections		Small castings, Thick sections		Large castings, Thin sections		Large castings, Thick sections	
	°C	°F	°C	°F	°C	°F	°C	°F	°C	°F
30	1150	2100	1400	2550	1370	2500	1345	2450	1315	2400
35	1175	2150	1425	2600	1400	2550	1370	2500	1345	2450
40	1200	2190	1450	2640	1420	2590	1395	2540	1365	2490
45	1220	2230	1470	2680	1445	2630	1415	2580	1390	2530

Molding

Molds. The primary function of the molding material is to produce a dimensionally stable cavity that will withstand the thermal and mechanical stresses exerted by the liquid iron. Gray iron can be produced by using most of the molding processes, which are covered in detail in the Section "Molding and Casting Processes" in this Volume.

Green sand molding is the most widely used technique for the production of gray iron castings (see the article "Sand Molding" in this Volume). Gray iron expands slightly because of the formation of graphite during eutectic solidification. This is most pronounced in higher carbon equivalent irons, in which more graphite is precipitated. This expansion stresses the molding material, causing an enlargement of the mold cavity if the sand is insufficiently compacted. This can lead to shrinkage defects. Therefore, it is important when producing gray iron castings that the mold hardness be sufficient to withstand the eutectic expansion of the gray iron.

If the dimensional accuracy and surface finish requirements of the casting exceed the limits of the green sand molding process, other processes, such as shell molding, chemical bond molding, and permanent molding, can be used. Each process creates a rigid mold surface that withstands the eutectic expansion of gray iron, thus improving dimensional accuracy. In addition, an excellent surface finish can be obtained with each of these processes. Shell molds are produced from a mixture of sand and a thermosetting resin. Chemically bonded molds are produced from a mixture of sand and a resin binder that requires a catalyst for room-temperature curing. Permanent molds are often produced from gray iron and generally require some form of coating to assist in the removal of the casting.

Cores. The purpose of a core is to create a dimensionally stable cavity within the casting during solidification. All commercially available coremaking processes can be used in the production of gray iron castings. These processes include oil-sand, shell, hot box, carbon dioxide, and chemical bonding. The type of core does not have to be related to the type of molding material. Selection of the core material depends on core size, complexity, dimensional accuracy, and cost.

The gating and feeding system serves the same purpose in the casting of gray iron as in the other metal casting systems. These functions include:

- Filling the mold cavity without turbulence
- Preventing slag, dross, or mold material from entering the mold
- Preventing the introduction of air or mold gases into the stream of metal
- Producing heat transfer characteristics that will aid in the progressive solidification of the casting
- Enabling production of the casting with the use of a minimum amount of metal

Gating and feeding practice for gray iron is less complicated than for other materials. The precipitation of graphite during eutectic solidification causes an expansion, which offsets some of the contraction usually experienced during the solidification of a metal. In fact, a class 20 iron precipitates enough graphite to create sufficient expansion so that feeding is not required. As the carbon equivalent of the iron decreases, less graphite is precipitated and feeding requirements increase. A class 50 iron requires approximately 4% volumetric feeding to offset the contraction of the iron.

Dimensional Control. The production of gray iron castings with good dimensional stability requires that the foundryman produce a mold rigid enough to withstand eutectic expansion and that the pattern be constructed in such a way as to compensate for the metal contraction during cooling. The contraction depends on the amount of graphite precipitated during eutectic solidification. Irons with a higher carbon equivalent will precipitate more graphite, resulting in less shrinkage. Typical shrinkage values of the various grades of iron are:

Class	Carbon equivalent, %	Pattern shrinkage allowance, mm/m (in./ft)
30	...	8.3 (1/10)
35	3.7–4.1	10.4 (1/8)
40	3.5–3.9	10.4 (1/8)
45	3.45–3.8	10.4 (1/8)
50	3.3–3.6	13.0 (5/32)
55	...	13.0 (5/32)

Defects

Matrix discontinuities in gray iron castings caused by cavities and inclusions increase the scrap rate and decrease the productivity of a foundry. In many cases, these defects are subsurface and are revealed only after machining in-house or at the customer's facility. Accurate defect identification is required before steps can be taken to eliminate or minimize these problems. The location, shape, and size of a defect provide valuable clues about its origin. These three factors will be considered in the following review of common defects found in gray iron.

Shrinkage cavities can appear as either isolated or interconnected irregularly shaped voids, as shown in Fig. 18. When examined at low magnification, they are often found to contain dendrites that possess a treelike form (Fig. 19). Heavy sections and hot spots, such as areas adjacent to ingates and feeders or regions experiencing changes in section size, are most susceptible to this type of defect. Shrinkage, because it is subsurface, is usually revealed during machining or pressure testing. Factors that promote shrinkage formation include lack of mold rigidity, unsuitable metal composition, incorrect pouring temperature, and a high degree of nucleation. These factors may operate independently or in combination.

Almost all liquids contract during freezing. In gray iron, however, expansion occurs during the formation of the austenite-graphite eutectic. This expansion increases if the iron is highly nucleated, a state that is produced by inoculation. Molds, particularly green sand molds, that are not rammed to sufficient hardness are incapable of containing this expansion. This leads to enlargement of the mold cavity. Shrinkage cavities are formed if the metal supply is not sufficient to accommodate this enlargement. Soft molds can be detected by measuring the dimensions and weight of the casting.

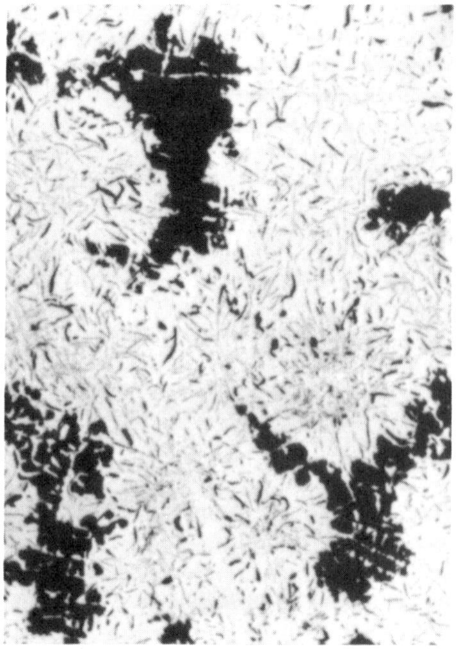

Fig. 18 Typical shrinkage defect in gray cast iron. As-polished. 50×

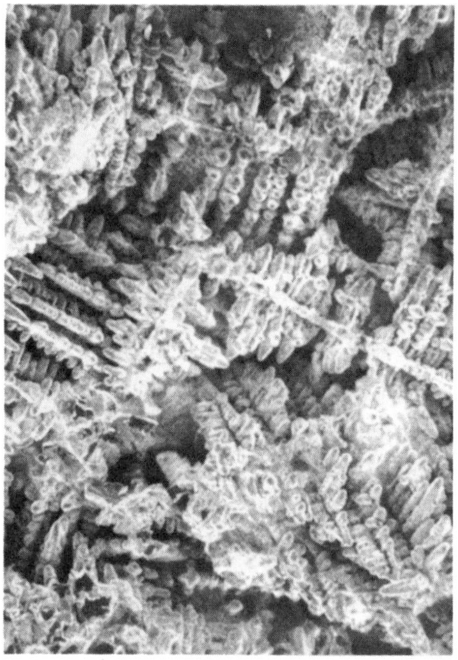

Fig. 19 SEM view of shrinkage cavities in heavy-section gray iron. 20×

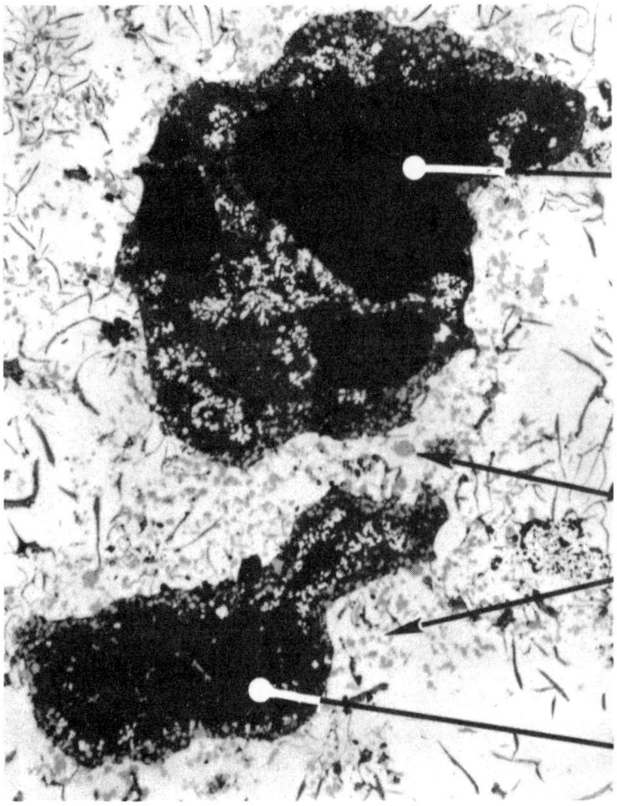

Fig. 20 Blowhole defect associated with manganese sulfide segregation. 100×

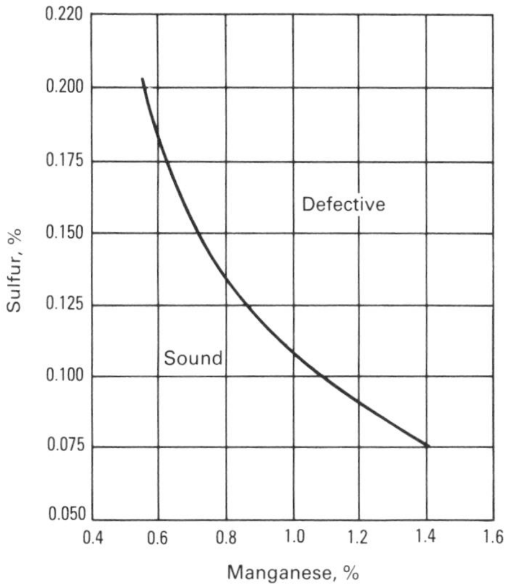

Fig. 21 Occurrence of blowhole defects in gray iron castings as a function of sulfur and manganese contents. Pouring temperature was constant at 1280 °C (2335 °F). Defective castings had blowholes associated with manganese sulfide inclusions. Source: Ref 31

Oversize or overweight castings signal a potential problem. This problem can be minimized by avoiding excessive moisture levels, by maintaining adequate levels of carbonaceous material, and by using only the amount of inoculant needed to control chill and bring about the desired properties.

Iron composition can have a pronounced effect on shrinkage. Very low carbon contents have occasionally been related to unsound castings. Phosphorus levels in the iron as low as 0.02% can cause fine porosity at eutectic cell boundaries and hot spots (Ref 3). Adjustment of the phosphorus level is usually a tradeoff between avoiding shrinkage and preventing finning and penetration defects. A maximum level of 0.3% is suggested for heavy-section castings where slow solidification times result in segregation (Ref 30). Phosphorus levels in castings that must pass pressure tests should not exceed 0.10%

Excessively high pouring temperatures can also increase the contraction of the metal as it cools to the solidification temperature, thus encouraging shrinkage. In addition, because green sand molds are not dimensionally stable under heat, the higher temperatures increase the chances of mold wall movement. A compromise exists between too high and too low a pouring temperature. If too low a pouring temperature is used, blowhole-type defects, cold shuts, and carbides can occur. Therefore, experience is the best way of determining the optimum pouring temperature for a particular casting.

The blowholes referred to in this section are those found below the cope surface of a casting or where a core forms a ceiling in the mold. They are usually revealed by machining or by heavy shotblasting. These blowholes can be spherical or irregular in shape and have been reported to have a gray or blue-gray lining. Many of the holes contain slag, and some cavities contain exuded metal beads. Manganese sulfide inclusions are usually found clustered in the iron matrix near the defect and are sometimes present in the slag itself. An example is shown in Fig. 20.

Cold metal resulting from low pouring temperatures is the primary cause of blowholes. This explains why the last castings poured from a ladle are most likely to be unsound. Excessive sulfur and manganese levels, however, compound the problem. Figure 21 shows the sulfur and manganese levels at which sound and defective castings are produced. The higher the sulfur and manganese levels, the higher the pouring temperature must be to avoid blowholes.

The following sequence of events leads to the formation of the blowhole. As the temperature of the molten metal falls, manganese sulfides form and separate from the melt. They float to the surface, where they mix with the ladle slag (iron and manganese silicate), creating a slag of higher fluidity. This slag enters the mold cavity, reacts with the graphite precipitating during the eutectic reaction, and results in the evolution of carbon monoxide and the formation of blowholes. Proper metal temperature, balanced manganese and sulfur levels, clean ladles, and good skimming practices help minimize blowholes.

Hydrogen pinholes (Fig. 22) are small, spherical, or pear-shaped cavities about 3 mm (1/8 in.) or less in diameter. A continuous graphite film is often found in these holes, although an iron oxide layer may be present if the castings are heat treated. Nonmetallic inclusions are not present in the voids, but some have been found to contain a small bead of metal. Pinholes are typically subsurface in nature and are therefore not visible until after heavy shotblasting or more likely after machining. Thin sections are more susceptible to this defect than thick sections, as are areas remote from the ingates.

A number of mold and metal factors have been associated with hydrogen pinhole formation. It is likely that these factors function in combination rather than individually. Aluminum levels in the iron as low as 0.005% have been found to encourage dissociation of the water vapor arising from the mold, thus increasing the hydrogen content of the metal. Aluminum sources include ferroalloys used in iron treatment and scrap. Excessive moisture in the mold and lack of active carbonaceous mold additions also favor pinholing. Sand grain size and mold permeability have not been found to affect this problem. Damp furnace and ladle refractories, long runners and downsprues that increase the time the metal is in contact

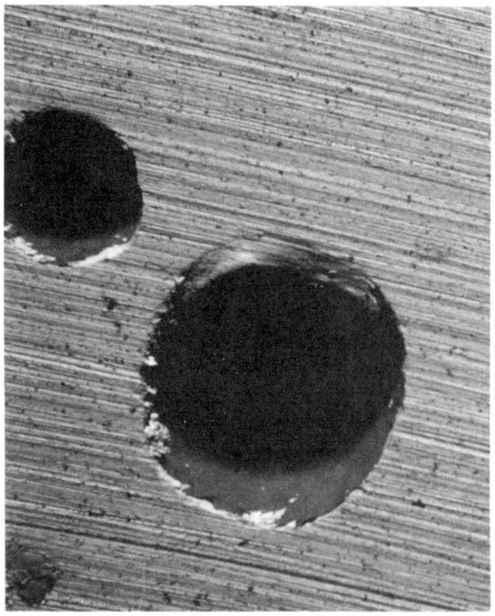

Fig. 22 SEM view of hydrogen pinholes in gray iron. 20×

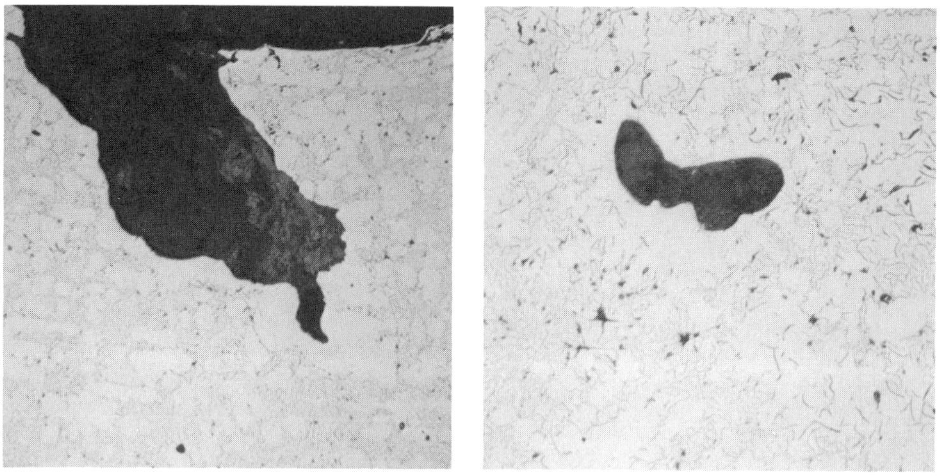

(a) (b)

Fig. 23 Surface (a) and subsurface (b) nitrogen defects in gray iron with nitrogen content of 135 ppm. Both 100×

with the mold, and turbulent pouring are other factors that can contribute to this defect.

Nitrogen Defects. Nitrogen levels of 20 to 80 ppm are normal in gray iron. At higher levels, the iron becomes less able to contain the nitrogen, resulting in its liberation during solidification. This causes the formation of interdendritic voids or blowholes (Fig. 23). These voids tend to be clear and bright and at times contain a graphite layer. An oxide film, however, may be present in castings that have been heat treated. Compacted graphite is sometimes found in the matrix along the defect perimeter. The amount of nitrogen needed to produce unsound castings varies. Light sections may not be affected until nitrogen levels reach 130 ppm, while heavy sections can experience problems at levels of only 80 ppm. Irons made from a charge with a high proportion of steel scrap (50% or greater) are quite susceptible to this problem (Ref 32). Recarburizing materials containing nitrogenous compounds, molds, and cores produced with high nitrogen content resins, as well as mold and core coatings containing carbonaceous and resin components, all act as nitrogen sources (Ref 33). The effect of nitrogen can be neutralized with the addition of 0.02 to 0.03% Ti. Acceptable levels of nitrogen may become dangerous if hydrogen is also present.

Abnormal graphite forms have been found to account for a high proportion of cracked castings. In extreme cases, they have resulted in premature, catastrophic failure. Lead is one of the agents that have been found to be responsible for these decreases in the physical and mechanical properties of cast iron that are associated with a change in graphite morphology. Microstructurally, lead contamination often results in fine, spiky graphite on the normal graphite flakes (Fig. 24). Degenerate graphite structures ranging from a Widmanstätten-like structure in heavy-section castings to a mesh type with long interconnected flakes in thinner sections have also been observed (Ref 35). These structures are often difficult to observe at low magnification and can be detected only when examined at high magnifications.

Heavy-section castings are particularly vulnerable to the influence of lead because their slow cooling rates favor lead segregation. A residual lead level as low as 0.004% in a heavy-section casting can result in failure. The presence of other elements, such as hydrogen and aluminum, which has been found to increase the amount of hydrogen in iron, adds to the effect. This is why castings produced in green sand molds are more susceptible to this type of defect.

There is no known neutralizing agent for lead in gray iron. Although some lead loss is experienced during melting and holding, it is best to keep lead out of the iron. This can be done by being aware of the sources of lead contamination, including leaded steel scrap, heavily painted components, vitreous enameled steels, and copper alloys.

Heat Treatment

Although most gray iron castings are used in the as-cast condition, heat treatment can be employed to meet specific casting requirements. Detailed information on the heat treatment of gray iron is available in the article "Heat Treating of Gray Irons" in Volume 4 of the 9th Edition of *Metals Handbook*.

The three most commonly used forms of heat treatment for gray iron are annealing, stress relieving, and normalizing. Other standard heat treatments, such as hardening and tempering, austempering, and martempering, are used on limited occasions. Gray irons can also be flame or induction hardened.

In considering the heat treatment of gray iron castings, it is important to recognize the complexity of the relationship between metallurgical and thermal phenomena. In terms of heat treatment, gray iron can essentially be considered a composite material made up of free graphite (flakes) and eutectoid steel (matrix). The situation can be further complicated by the variety of sections, and therefore thermal responses, found in most castings. For this reason, it is often necessary to develop experimentally the precise process for given castings if optimal results are desired.

Annealing. Three principal annealing processes are used for gray iron. They have

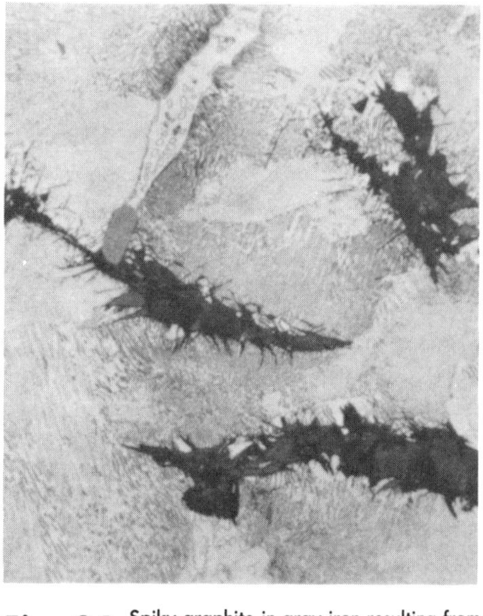

Fig. 24 Spiky graphite in gray iron resulting from lead contamination. Etched using 4% picral. 250×. Source: Ref 34

a common purpose and metallurgy in that they are employed primarily to improve machinability. Therefore, all three treatments involve the production of a ferritic matrix. Figure 1 shows a section of the iron-carbon-silicon diagram at 2.5% Si. As can be seen, the eutectoid reaction consists of the transformation of austenite into ferrite plus graphite. The annealing processes are all designed to take advantage of this reaction.

The first of these processes is normally called a ferritizing or subcritical anneal. It consists of heating the casting to a temperature of 700 to 760 °C (1290 to 1400 °F), or just below the eutectoid transformation temperature. At this temperature, pearlite decomposes into a ferritic matrix. Holding times at this temperature vary to some degree with section size and iron chemistry. As a general rule, time at temperature should be approximately 1 h for each 25 mm (1 in.) of casting thickness. Longer times may be required if alloying elements are present. Cooling rates after ferritization are not critical but should not exceed 100 °C/h (180 °F/h) to avoid inducing stresses. The reduction in hardness and improved machinability associated with this process also result in a reduction in strength.

Medium or full annealing is used in situations where, because of alloying or the presence of minor amounts of chill, the subcritical anneal does not convert the iron to a ferritic matrix. In this process, the iron is heated to a temperature of 800 to 900 °C (1470 to 1650 °F), which is above the eutectoid transformation temperature. After soaking for approximately 1 h for each 25 mm (1 in.) in cross section, the iron is slow cooled through the eutectoid transformation region to promote the formation of ferrite. If the iron contains minor alloying elements such as chromium, manganese, copper, nickel, or tin, longer holding times or higher soaking temperatures may be necessary. Following transformation, the castings can be air cooled from approximately 675 °C (1245 °F).

The third form of annealing is designated as a graphitizing anneal. It is used in gray iron only when the removal of massive carbides or chilled iron is required. It consists of heating the casting to a temperature of approximately 900 to 925 °C (1650 to 1700 °F). Time at temperature should be minimized based on microstructural evaluation to avoid scaling unless a controlled-atmosphere furnace is employed. After the decomposition of carbides, the desired cooling rate depends on the microstructure desired. If a highly machinable ferritic structure is desired, furnace cooling is recommended with very slow cooling through the eutectoid transformation. If a stronger, pearlitic matrix is desired, air cooling can be used. Experimentation may be necessary to compensate for the effects of casting geometry and iron composition.

Table 7 As-cast mechanical properties of standard gray iron test bars

Class	Tensile strength		Torsional shear strength		Compressive strength		Reversed bending fatigue limit		Transverse load on test bar		Hardness, HB
	MPa	ksi	MPa	ksi	MPa	ksi	MPa	ksi	kgf	lbf	
20	152	22	179	26	572	83	69	10	839	1850	156
25	179	26	220	32	669	97	79	11.5	987	2175	174
30	214	31	276	40	752	109	97	14	1145	2525	210
35	252	36.5	334	48.5	855	124	110	16	1293	2850	212
40	293	42.5	393	57	965	140	128	18.5	1440	3175	235
50	362	52.5	503	73	1130	164	148	21.5	1633	3600	262
60	431	62.5	610	88.5	1293	187.5	169	24.5	1678	3700	302

Normalizing. In principle, the normalizing of gray iron is a relatively straightforward process. The castings are heated to a temperature of 875 to 900 °C (1605 to 1650 °F) and held for about 1 h per 25 mm (1 in.) of cross section. This should result in the transformation of the matrix to austenite. The castings are then air cooled to form a pearlitic matrix.

As expected, the presence of alloying elements and variations in casting geometry complicate this practice and may alter the final results. It may be necessary to adjust holding times and/or cooling cycles to obtain the desired results. For example, forced air cooling may be desirable to avoid the formation of ferrite in heavy sections. In the case of alloyed castings, it may be possible to produce small increases in strength and hardness by normalizing.

Stress relieving is also used in gray iron castings. The purpose of this practice is to reduce stresses induced in the casting during solidification. In essence, the process consists of heating the casting to a temperature ranging from 500 to 650 °C (930 to 1200 °F), depending on composition. The casting is held in this temperature range for 2 to 8 h, then air cooled.

Properties and Applications

Graphite morphology and matrix characteristics affect the physical and mechanical properties of gray iron. Large flakes, common in high carbon equivalent irons and heavy-section castings, impart desirable properties, such as good damping capacity, dimensional stability, resistance to thermal shock, and ease of machining. Higher tensile strength and modulus of elasticity values, resistance to crazing, and smooth machined surfaces are obtainable with irons containing small flakes, which are promoted by low carbon equivalents and faster cooling rates. Pearlite refinement and stabilization of acicular structures result in an increase in hardness, tensile strength, and wear resistance. In addition to composition (particularly carbon equivalent) and section size, factors such as alloy additions, heat treatment, thermal properties of the mold, and casting geometry affect the microstructure and therefore the properties of the iron.

Mechanical Properties

The tensile properties of gray cast iron (Table 7) include tensile strength, yield strength, ductility, and modulus of elasticity. Minimum tensile strength is used to classify gray iron in ASTM A 48. Tensile strength is inversely proportional to carbon equivalence.

Yield strength can be determined from the stress-strain diagram by using either the 0.1% or 0.2% offset method. Elongation values for gray iron are very low, typically of the order of about 0.6%. The modulus of elasticity of gray iron is not a single number, because gray iron does not possess an elastic range in which stress and strain exhibit a straight line relationship. Values for modulus of tension are usually estimated by either the tangent modulus or the secant modulus method. A higher modulus of elasticity is desired for applications requiring rigidity and minimum deflection values. A low modulus of elasticity is preferred for vibration damping and severe heat shock applications.

Transverse Properties. Strength in bending and deflection are determined with the transverse bend test. Results from this method of testing are usually used in combination with the tensile test. Samples require little preparation, and the method used in loading can be more characteristic of what the part will experience in service. Modulus of rupture values can be calculated from the transverse breaking load by using the standard beam formula. Transverse breaking loads can be found in Table 7.

Hardness tests are routinely performed because they are simple, rapid methods of determining the approximate strength characteristics and machinability of a gray iron casting. The Brinell hardness test is most often used on gray iron because the larger diameter of the indentor provides an average of the various microconstituents present in the iron. Figure 25 shows an example of the relationship between hardness and tensile strength of gray iron for a particular foundry and process.

Compression. Gray cast iron is about three to four times stronger in compression than in tension (Ref 36). This is because the characteristics of the graphite flakes have less of an influence in compression than in tension. Matrix structure is the determining

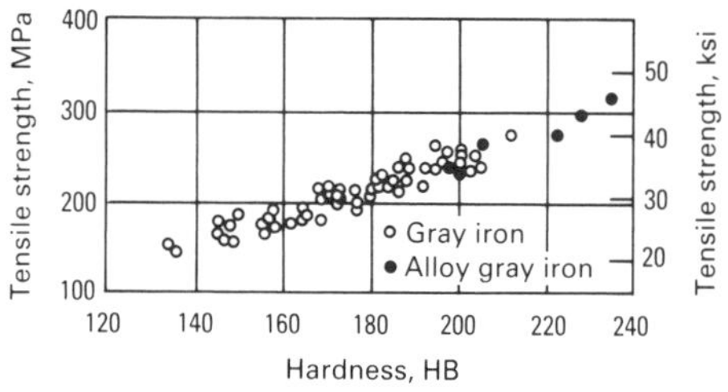

Fig. 25 Relationship between tensile strength and hardness for a series of inoculated irons from a single foundry

factor in compressive strength; a considerable increase in strength is associated with a change from ferrite to pearlite to martensite. Compressive strength values for a number of gray iron classes are given in Table 7.

Torsional Properties. High torsional shear strength, which is required in such applications as crankshafts, camshafts, and axles, is obtainable with gray iron (Table 7). Shear strength has been reported to range from 1.1 to 1.5 times tensile strength (Ref 36). Torsion tests can be carried out on testpieces or full-size parts.

Damping Capacity. The interconnected nature of the graphite flakes in gray iron imparts excellent damping capacity and makes gray iron ideal for applications such as machine bases and supports, cylinder blocks, and brake components. Damping capacity refers to the ability of a material to quell vibrations and to dissipate the energy as heat. Relative damping capacities for a number of ferrous materials and for aluminum are given below, where damping capacity is taken as the natural logarithm of the ratio of successive amplitude:

Material	Relative damping capacity
White iron	2–4
Malleable iron	8–15
Ductile iron	5–20
Gray iron, fine flake	20–100
Gray iron, coarse flake	100–500
Eutectoid steel	4
Armco iron	5
Aluminum	0.4

Source: Ref 36

Damping capacity increases as carbon equivalence and section size increase.

Machinability. The excellent machinability of gray iron compared to that of other ferrous materials is attributed to the presence of graphite flakes, which act as chip breakers and serve to lubricate the cutting tool. Factors such as chill, surface inclusions (sand or slag), swells, and shrinkage adversely affect the machining characteristics of gray iron. More information on this subject and on other properties of gray iron, such as fatigue strength, impact strength, residual stresses, and wear resistance, can be found in the article "Gray Iron" in Volume 1 of the 9th Edition of *Metals Handbook*.

Table 8 Physical properties of gray iron as a function of tensile strength

Tensile strength, MPa	Tensile strength, ksi	Density(a), g/cm³	Density(a), lb/in.³	Thermal conductivity, W/m·K(b), 100 °C (212 °F)	300 °C (572 °F)	500 °C (932 °F)	Electrical resistivity at 20 °C (68 °F), μΩ·m(c)
150	22.0	7.05	0.255	65.7	53.3	40.9	0.80
180	26.0	7.10	0.257	59.5	50.3	40.0	0.78
220	32.0	7.15	0.258	53.6	47.3	38.9	0.76
260	38.0	7.20	0.260	50.2	45.2	38.0	0.73
300	43.5	7.25	0.262	47.7	43.8	37.4	0.70
350	51.0	7.30	0.264	45.3	42.3	36.7	0.67
400	68.0	7.30	0.264	43.5	41.0	36.0	0.64

(a) Source: Ref 37. (b) Source: Ref 38. (c) Source: Ref 39

Table 9 Typical applications for gray iron castings

Specification	Grade or class	Typical applications
ASTM A 48	20, 25	Small or thin-sectioned castings requiring good appearance, good machinability, and close dimensional tolerances
	30, 35	General machinery, municipal and waterworks, light compressors, automotive applications
	40, 45	Machine tools, medium-duty gear blanks, heavy compressors, heavy motor blocks
	50, 55, 60	Dies, crankshafts, high-pressure cylinders, heavy-duty machine tool parts, large gears, press frames
ASTM A 159, SAE J431	G1800	Miscellaneous soft iron castings in which strength is the primary consideration; exhaust manifolds
	G2500	Small cylinder blocks and heads, air-cooled cylinders, pistons, clutch plates, oil pump bodies, transmission cases, gear boxes, light-duty brake drums
	G2500a	Brake drums and clutch plates for moderate service where high carbon is desirable to minimize heat checking
	G3000	Cylinder blocks, heads, liners, flywheels, pistons, clutch plates
	G3500	Truck cylinder blocks and heads, heavy flywheels, differential carriers
	G3500b	Brake drums and clutch plates for heavy-duty service that require heat resistance and high strength
	G4000	Truck and tractor cylinder blocks and heads, heavy flywheels, tractor transmission cases, differential carriers, heavy gear boxes
	G3500c	Extraheavy-duty brake drums
	G4000d	Alloyed automotive engine camshafts
	G4500	Diesel engine castings, liners, cylinders, and pistons; heavy-duty parts for general industry
ASTM A 278	40, 50, 60, 70, 80	Valve bodies, paper mill dryer rollers, chemical process equipment, pressure vessel castings
ASTM A 319	I, II, III	Stoker and firebox parts, grate bars, process furnace parts, ingot molds, glass molds, caustic pots, metal melting pots
ASTM A 823	...	Automobile, truck, appliance, and machinery castings in quantity
ASTM A 436	1	Valve guides, insecticide pumps, flood gates, piston ring bands
	1b	Seawater valve and pump bodies, pump section belts
	2	Fertilizer applicator parts, pump impellers, pump casings, plug valves
	2b	Caustic pump casings, valves, pump impellers
	3	Turbocharger housings, pumps and liners, stove tops, steam piston valve rings, caustic pumps and valves
	4	Range tops
	5	Glass rolls and molds, machine tools, gages, optical parts requiring minimal expansion and good damping qualities, solder rails and pots
	6	Valves

Source: Ref 40

Physical Properties

The density of gray iron is composition and temperature dependent. Increases in free graphite, which has a low density, result in a decrease in the density of the iron. This is why high-strength iron has a greater density than low-strength iron (Table 8). The density of molten cast iron falls in the range of 6.65 to 7.27 g/cm³.

Thermal conductivity is an important consideration in applications requiring heat transfer, such as brake drums, internal combustion engines, and ingot molds. Thermal conductivity is strongly influenced by the amount and form of graphite. An increase in thermal conductivity is obtained as the amount of free graphite increases and as the flakes become coarser and longer. Matrix structure and composition also exert an influence, as does temperature. Thermal conductivity values as a function of tensile strength and temperature are listed in Table 8.

Electrical resistivity is a function of graphite structure, matrix constituents, and temperature. Increases in free graphite, coarse flakes, pearlite, and temperature result in increases in electrical resistivity. Resistivity values at 20 °C (68 °F) for various classes of gray irons are given in Table 8.

Thermal expansion is measured by the coefficient of linear expansion. It is primarily dependent on the matrix structure of the iron. Ferritic and martensitic matrices have a slightly higher coefficient of linear expansion than pearlitic ones. Values for austenitic iron can be even greater, depending on nickel content. Graphite structure has little effect on this property. The coefficient of linear expansion increases with temperature; a value of 10×10^{-6}/°C is commonly used at room temperature.

Applications

The changes in physical and mechanical properties that can be produced in gray iron by controlling the characteristics of its free graphite and matrix structures lead to versatility in its application. Gray iron can be effectively used in highly competitive, low cost applications where its founding properties are of paramount importance. Such applications include implement weights, elevator counterweights, guards and frames, enclosures for electrical equipment, and fire hydrants. A variety of iron grades can be used in these applications. Gray iron is also employed in more critical applications in which mechanical or physical property requirements determine iron selection, such as in pressure-sensitive castings, automotive castings, and process furnace parts.

Standards established by the American Society for Testing and Materials, the Society of Automotive Engineers, the federal government, and the military provide assistance in the selection of the appropriate grade or class of iron to meet specific mechanical or physical requirements. Automotive castings requiring resistance to heat checking are covered by ASTM A 159. Specifications for pressure-sensitive parts can be found in ASTM A 278, while composition information for castings requiring thermal shock resistance is available in ASTM A 319. A summary of typical applications for gray iron, based on specifications and information available in the literature, is provided in Table 9.

REFERENCES

1. E. Piwowarsky, *Hochwertiges Gusseisen*, 2nd ed., Springer-Verlag, 1958
2. J.C. Hamaker, W.P. Wood, and F.B. Rote, Internal Porosity in Gray Iron, *Trans. AFS*, Vol 60, 1952, p 401-427
3. "The Importance of Controlling Low Phosphorus Contents in Gray Iron," Broadsheet 162, BCIRA, 1977
4. K.M. Muzumdar and J.F. Wallace, Effect of Sulfur in Cast Iron, *Trans. AFS*, Vol 81, 1973, p 412-423
5. "Effect of Some Residual or Trace Elements on Cast Iron," Broadsheet 192, BCIRA, 1981
6. R.W. Heine, C.R. Loper, Jr., and P.C. Rosenthal, *Principles of Metal Casting*, 2nd ed., McGraw-Hill, 1967, p 496-497
7. G. Petzow, *Metallographic Etching*, American Society for Metals, 1978, p 61-68
8. J. Nelson and G.M. Goodrich, Metallography ... A Necessity for Even the Small Foundry, *Mod. Cast.*, June 1978, p 60-63
9. W.V. Ahmed and L.J. Gawlick, A Technique for Retaining Graphite in Cast Irons During Polishing, *Mod. Cast.*, Jan 1983, p 20-21
10. G. Lambert, Ed., *Typical Microstructures of Cast Metals*, 2nd ed., The Institute of British Foundrymen, 1966, p 47
11. C.F. Walton, *Gray and Ductile Iron Castings Handbook*, Gray and Ductile Iron Founder's Society, 1971, p 193
12. H.T. Angus, *Physical and Engineering Properties of Cast Iron*, BCIRA, 1960, p 21
13. W.D. Forgeng and T.K. McCluhan, Electron Microprobe Study of Distribution of Elements in Tungsten and Molybdenum Alloyed High Strength Gray Irons, *Cast Met. Res. J.*, Vol 1 (No. 3), Dec 1965, p 1-8
14. G.V. Sun and C.R. Loper, Jr., Titanium Carbonitrides in Cast Irons, *Trans. AFS*, Vol 91, 1983, p 639-646
15. H.C. Winte, Gray Iron Casting Section Sensitivity, *Trans. AFS*, Vol 54, 1946, p 436-443
16. H.G. Rachner, The Cupola and Its State of Development, *Foundry Trade J.*, Vol 161 (No. 3357), 8 Oct 1987
17. N.C. McClure *et al.*, Inoculation of Gray Cast Iron, *Trans. AFS*, Vol 65, 1957, p 340-351
18. H.W. Lownie, Theories of Gray Cast Iron Inoculation, *Trans. AFS*, Vol 54, 1946, p 837-844
19. B. Lux, Nucleation of Eutectic Graphite in Inoculated Gray Iron by Saltlike Carbides, *Mod. Cast.*, May 1964, p 222-232
20. C.F. Walton, *The Gray Iron Castings Handbook*, Gray Iron Founder's Society, 1958, p 119
21. P.J. Bilek, J.M. Dong, and T.K. McCluhan, The Roles of Calcium and Aluminum in the Inoculation of Gray Iron, *Trans. AFS*, Vol 80, 1972, p 183-188
22. J. Briggs, R.W. Newman, and M.D. Bryant, Control of Inoculation—How Much? What Size? When?, in *Proceedings of AFS-CMI Conference*, American Foundrymen's Society, Feb 1979
23. "Counting Eutectic Cells in Flake Graphite Iron Castings," Broadsheet 94-1, BCIRA, 1974
24. A.G. Fuller, Fading of Inoculants, in *Proceedings of AFS-CMI Conference*, American Foundrymen's Society, Feb 1979
25. G.F. Sergeant, Late Metal Stream Inoculation BCIRA Developments, in *Proceedings of AFS-CMI Conference*, American Foundrymen's Society, Feb 1979
26. G.Fr. Hillner and K.H. Kleeman, Mold Inoculation of Gray and Ductile Cast Iron—New Solutions to Old Problems, *Trans. AFS*, Vol 83, 1975, p 167
27. R.E. Eppich, Solid Inserts, in *Proceedings of AFS-CMI Conference*, American Foundrymen's Society, Feb 1979
28. J. Powell, Ferroalloys in the Production of Cast Iron, in *AIME Electric Furnace Conference Proceedings*, Vol 44, Iron and Steel Society, 1986, p 215-231
29. C.E. Bates, Alloy Element Effects on Gray Iron Properties: Part II, *Trans. AFS*, Vol 94, 1986, p 889-912
30. J.M. Greenhill, Diagnosing Defects in Gray Iron, *Foundry Trade J.*, Nov 1971, p 56-60
31. "Subsurface Blowholes Associated With Segregation of Manganese Sulphide Inclusions," Broadsheet 6, BCIRA, 1975
32. J.M. Greenhill and N.M. Reynolds, Nitrogen Defects in Iron Castings, *Foundry Trade J.*, 16 July 1981, p 111-122
33. "Nitrogen in Cast Iron," Broadsheet 41, BCIRA, 1975
34. "Lead Contamination of Cast Iron," Broadsheet 50**, BCIRA, 1986
35. C.E. Bates and J.F. Wallace, Trace Elements in Gray Iron, *Trans. AFS*, Vol 74, 1966, p 513
36. C.F. Walton and T.J. Opar, *Iron Castings Handbook*, Iron Castings Society, 1981, p 203-295
37. "Density of Cast Irons," Broadsheet 203-5, BCIRA, 1984
38. "Thermal Conductivity of Unalloyed Cast Irons," Broadsheet 203, BCIRA, 1987
39. "Electrical Resistivity of Unalloyed Cast Irons," Broadsheet 203-4, BCIRA, 1984
40. *Metal Progress: Materials & Processing Databook*, American Society for Metals, 1985, p 38-39

SELECTED REFERENCES

- C.E. Bates, Alloy Element Effects on Gray Iron Properties: Part II, *Trans. AFS*, Vol 94, 1986, p 889-912

- A. Boyles, *The Structure of Cast Iron*, American Society for Metals, 1947
- J.V. Dawson, J.N. Kilshaw, and A.D. Morgon, The Nature and Origin of Gas Holes in Iron Castings, *Trans. AFS*, Vol 63, 1965, p 224-240
- A. McLean and J.M. Svoboda, *Physical Chemistry of Ferrous Melting*, American Foundrymen's Society Cast Metals Institute, 1975
- H. Morrough, The Status of the Metallurgy of Cast Irons, *J. Iron Steel Inst.*, Jan 1968
- M.T. Rowley *et al.*, *International Atlas of Casting Defects*, American Foundrymen's Society, 1974
- G.F. Ruff and J.F. Wallace, Control of Graphite Structure and Its Effect on Mechanical Properties of Gray Iron, *Trans. AFS*, Vol 84, 1976, p 705-728
- P.F. Wieser, C.E. Bates, and J.F. Wallace, *Mechanism of Graphite Formation in Iron-Silicon-Carbon Alloys*, Malleable Founders' Society, 1967

Ductile Iron

I.C.H. Hughes, BCIRA International Centre for Cast Metals Technology, Great Britain

DUCTILE IRON has been known only since the late 1940s, but it has grown in relative importance and currently represents about 20 to 30% of the cast iron production of most industrial countries. Ductile iron is also known as nodular iron or spheroidal graphite iron. Unlike gray iron, which contains graphite flakes, ductile iron has an as-cast structure containing graphite particles in the form of small, rounded, spheroidal nodules in a ductile metallic matrix. Therefore, ductile iron has much higher strength than gray iron and a considerable degree of ductility; both of these properties of ductile iron, as well as many of its others, can be further enhanced by heat treatment.

Ductile iron also supplements and extends the properties and applications of malleable irons. It has the advantage of not having to be cast as a white iron and then annealed for castings having section thicknesses of about 6 mm (¼ in.) and above, and it can be manufactured in much thicker section sizes. However, it cannot be routinely produced in very thin sections with as-cast ductility, and such sections usually need to be heat treated to develop ductility. It has the advantage, in common with gray iron, of excellent fluidity, but it requires more care to ensure sound castings and to avoid hard edges and carbides in thin sections, and it usually has a lower casting yield than gray iron. Compared to steel and malleable iron, it is easier to make sound castings, and a higher casting yield is obtained; however, more care is often required in molding and casting.

Ductile iron is made by treating low-sulfur liquid cast iron with an additive containing magnesium (or occasionally cerium) and is usually finally inoculated just before or during casting with a silicon-containing alloy (inoculant). There are many variations in commercial treatment practice. In general, the composition range is similar to that of gray iron, but there are a number of important differences.

Raw Materials for Ductile Iron Production

The spheroidal form of graphite that characterizes ductile iron is usually produced by a magnesium content of about 0.04 to 0.06%. Magnesium is a highly reactive element at molten iron temperatures, combining readily with oxygen and sulfur. For magnesium economy and metal cleanliness, the sulfur content of the iron to be treated should be low (preferably <0.02%); this is readily achieved in an electric furnace by melting charges based on steel scrap or special-quality pig iron supplied for ductile iron production, together with ductile iron returned scrap. Low sulfur content can also be achieved by melting in a basic cupola, but acid cupola melted iron has a higher sulfur content and normally needs to be desulfurized before treatment by continuous or batch desulfurization in a ladle or special vessel. Treatment of acid cupola melted iron with magnesium without prior desulfurization is not recommended, because the iron consumes more magnesium and produces excessive magnesium sulfide slag, which is difficult to remove thoroughly and may lead to dross defects in castings.

To produce ductile iron with the best combination of strength, high ductility, and toughness, raw materials must be chosen that are low in many trace elements, particularly those that promote a pearlitic matrix structure. A low manganese content is also needed to achieve as-cast ductility and to facilitate successful heat treatment to produce a ferritic structure. For this purpose, it is necessary to use deep-drawing or other special grades of steel scrap and pig iron of a quality produced specially for ductile iron production.

Draft International Standard ISO/DIS 9147 specifies two grades of pig iron for ductile iron production: Grade 3.1 (Nodular (SG) base) and Grade 3.2 (Nodular (SG) base, higher manganese). The compositions of these grades are:

Grade	Composition C	Si	Mn	P	S
3.1	3.5–4.6	<3.0	<0.1	<0.08	0.03 max
3.2	3.5–4.6	<4.0	<0.1–0.4	<0.08	0.03 max

The contents of elements that affect the formation of nodular graphite and promote the formation of carbide are low according to the application of the grade concerned.

Higher-strength grades of ductile iron can be made with common grades of constructional steel scrap, pig iron, and foundry returns, but certain trace elements, notably, lead, antimony, and titanium, are usually kept as low as possible to achieve good graphite structure. Their undesirable effects can, however, be offset by the addition of a small amount of cerium to give a residual cerium content of 0.003 to 0.01%. An important control of raw materials involves the exclusion of aluminum, which may promote unsoundness and dross defects. Table 1 lists the typical minor element contents of three raw materials used in ductile iron manufacture.

Control of the Composition of Ductile Iron

All elements in the composition of ductile iron need to be controlled, and this section will outline the most common and important requirements.

Carbon. In electric furnace melting practice, carbon is derived from pig iron, carburizers, and cast iron scrap. The carburization of steel scrap charges is achieved by adding low-sulfur graphite or graphitized coke, and the rate of solution and the recovery of carbon increase with the purity of the carbon source used. In cupola melting, carbon is also derived from the coke charged. The optimum range for this element is usually 3.4 to 3.8%, depending on silicon content. Above this range, there is a danger of graphite flotation, especially in heavy sections, and of increased casting expansion during solidification, leading to unsoundness, particularly in soft green sand molds. Below this range, unsoundness may also occur because of lack of feeding, and at very low carbon contents, carbides may appear in the castings, particularly in thin sections.

Silicon enters ductile iron from raw materials, including cast iron scrap, pig iron, and ferroalloys, and to a small extent from silicon-containing alloys added during inoculation. The preferred range is about 2.0 to 2.8%. Lower silicon levels lead to high ductility in heat-treated irons but to a danger of carbides in thin sections as-cast, while high silicon accelerates annealing and

Table 1 Typical minor element contents of raw materials for ductile iron

Raw material	Minor elements, %											
	Mn	S	P	Ni	Cr	Cu	Mo	Al	Sn	As	B	V
High-purity pig iron	0.04	0.015	0.013	0.06	0.01	0.02	0.01	<0.005	<0.01	<0.01	0.0005	<0.01
Deep-drawing steel	0.26	0.015	0.016	0.01	0.01	0.02	0.01	<0.005	<0.01	<0.01	0.0006	0.01
Constructional steel	0.4–0.9	0.023	0.015	0.08	0.2	0.08	0.02	<0.005	0.02–0.04	0.01–0.04	0.0008	0.02

helps to avoid carbides in thin sections. As the silicon content rises, the ductile-to-brittle transition temperature in ferritic iron increases. Hardness, proof stress, and tensile strength also increase.

Carbon Equivalent (CE). The carbon, silicon, and phosphorus contents can be considered together as a CE value, which can be a useful guide to foundry behavior and some properties. There are several CE formulas, and they are useful in assessing the casting properties and solidified structure of the iron. When the carbon equivalent CE = C% + ⅓(%Si + %P) is equal to 4.3, the iron will be of wholly eutectic composition and structure, and the deviation of the value of CE from this value is a measure of the relative amount of eutectic. If CE is lower than 4.3, there will be a proportion of dendrites; if CE is higher than 4.3, there will be primary graphite nodules in the structure. The degree of saturation *Sc* is sometimes used, particularly in the German literature, to express the nearness to the eutectic composition. The value of *Sc* can be determined from Eq 1:

$$Sc = \frac{\text{Actual }\%\text{C}}{4.23 - 0.3(\%\text{Si} + \%\text{P})} \quad \text{(Eq 1)}$$

When *Sc* is less than 1, the iron is hypoeutectic and will contain primary dendrites. If *Sc* is greater than 1, there will be primary graphite in the structure. The carbon equivalent liquidus (CEL) is a measure of the liquidus temperature, which has a minimum value at the eutectic composition; that is, CEL = %C + %Si/4 + %P/2. Maximum fluidity occurs when this value is reached. The CEL is the only carbon equivalent that can be conveniently measured in a shop-floor thermal analysis test for metal composition, used on untreated iron prior to magnesium treatment. It is usual to aim for a CEL value of about 4.4 to 4.5, and values much higher than this are restricted in order to avoid graphite flotation. Other carbon equivalent formulas are sometimes used, and Fig. 1 is an example of a nomogram constructed to provide guidance on an optimum range of composition.

Manganese. The main source of manganese is steel scrap used in the charge. This element should be limited in order to obtain maximum ductility. In as-cast ferritic irons, it should be 0.2% or less. For irons to be heat treated to the ferritic condition, it should be 0.5% or less, but in irons to be used in the as-cast pearlitic condition, it may be present up to 1%. Manganese is subject to undesirable microsegregation. This is especially true in heavy sections, in which manganese encourages grain-boundary carbides, which promote low ductility, low toughness, and persistent pearlite.

Magnesium. The magnesium content required to produce spheroidal graphite usually ranges from 0.04 to 0.06%. If the initial sulfur content is below 0.015%, a lower magnesium content (in the range of 0.035 to 0.04%), may be satisfactory. Compacted graphite structures with inferior properties may be produced if the magnesium content is too low, while too high a magnesium content may promote dross defects.

Sulfur is derived from the charged metallic raw materials. In cupola melting, it is also absorbed from the coke. Before magnesium treatment, the sulfur content should be as low as possible, preferably below 0.02%. The final sulfur content of ductile iron is usually below 0.015%, but if cerium is present, it may be higher because of the presence of cerium sulfides in the iron. Excessive final sulfur contents are usually associated with magnesium sulfide slag and dross. When using cupola-melted iron, it is common to desulfurize the metal—usually with lime or calcium carbide, either continuously or in batches, before magnesium treatment—to levels of 0.02% or less. For treatments in the mold or in the metal stream, the initial sulfur may with advantage first be reduced to 0.01% or less.

Cerium can be added to neutralize undesirable trace elements that interfere with the formation of spheroidal graphite and to aid inoculation. It may then be desirably present to the extent of 0.003 to 0.01%. In irons of very low minor element content, cerium may be undesirable and may promote chunky nonspheroidal graphite, especially in thick sections; the deliberate addition of impurities may be necessary to avoid this effect. Cerium is added as a minor constituent of a number of magnesium addition alloys and inoculants to improve graphite structure. It is removed during the remelting of scrap ductile iron.

Minor Elements Promoting Nonspheroidal Graphite. Lead, antimony, bismuth, and titanium are undesirable elements that may be introduced in trace amounts with raw materials in the charge, but their effects can be neutralized by a cerium addition.

Minor Elements Promoting Pearlite. Nickel, copper, manganese, tin, arsenic, and antimony all promote pearlite and are listed here in order of increasing potency. They can enter the iron as trace constituents of raw materials. Copper up to 0.3% and tin up to 0.1% can be deliberately used as ladle additives when fully pearlitic structures are required. Almost all trace elements promote pearlite formation, and their effects are cumulative; therefore, a charge of high purity is essential for achieving fully

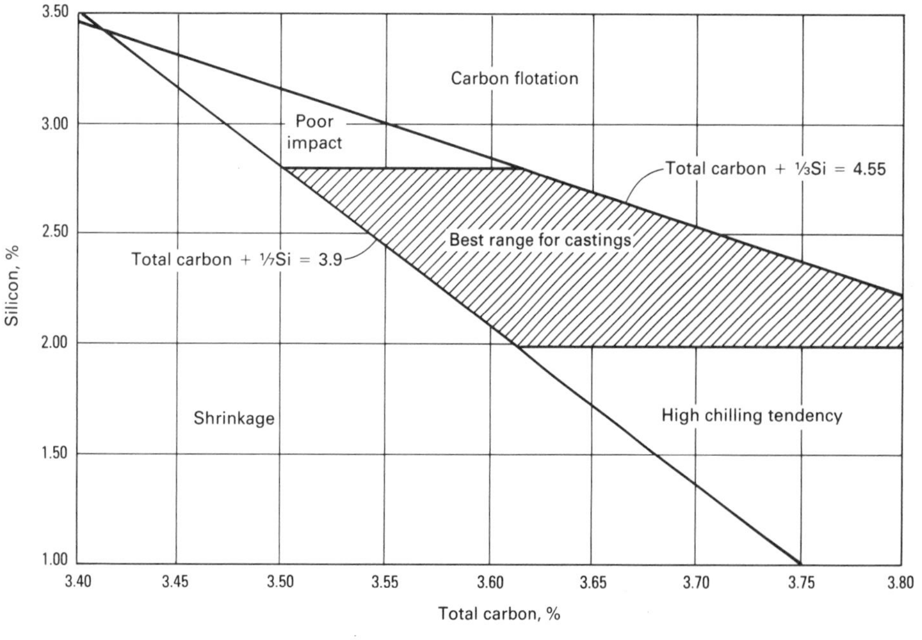

Fig. 1 Typical carbon and silicon ranges for ductile iron castings. Source: Ref 1

ferritic structures as-cast or with minimal annealing.

Aluminum. The presence of even trace amounts of aluminum in ductile iron may promote subsurface pinhole porosity and dross formation and should therefore be avoided. The most common sources of aluminum are contaminants in steel and cast iron scrap, notably, in the form of aluminum pistons from scrap automobile engines. Another source is aluminum-containing inoculants, and the use of inoculants of low aluminum content is advisable whenever possible. Late metal stream inoculation, in which only very small quantities of inoculant are added, is beneficial. Aluminum as low as 0.01% may be sufficient to cause pinholes in magnesium-containing ductile iron.

Phosphorus is normally kept below 0.05% because it promotes unsoundness and lowers ductility.

Minor Elements Promoting Carbides. Chromium, vanadium, and boron are all carbide promoters. Manganese may also accentuate the carbide-stabilizing effects of these elements, especially in heavy sections in which segregation promotes grain-boundary carbides. They are controlled by careful selection of metallic raw materials for melting.

Alloying Elements Promoting Hardenability. Nickel up to about 2% and molybdenum up to about 0.75% are the usual elements added deliberately to promote hardenability when heat treatment is to be used. Small amounts of manganese and copper also promote hardenability, but are normally used only in combination with other elements. Copper has limited solubility and must be kept below 1.5%.

Alloying Elements to Achieve Special Properties. Austenitic matrix structures are achieved by the addition of 20% Ni or more when resistance to heat, corrosion, or oxidation is required, and up to 5% Cr can be added to such irons. Nickel contents to 36% produce irons of controlled low-expansion properties. Up to 10% Mn in austenitic irons leads to low magnetic permeability and allows a lower nickel content to be used to achieve a stable austenite.

Silicon contents to 6% produce ferritic matrix structures with reduced growth, scaling and thermal distortion, and cracking at elevated temperatures. The addition of up to 2% Mo to pearlitic, ferritic, and austenitic irons confers improved creep and elevated-temperature strength. Irons containing deliberately high alloying element contents are discussed in the article "High-Alloy Graphitic Irons" in this Volume.

Molten Metal Treatment

Treatment to produce ductile iron involves the addition of magnesium to change the form of the graphite, followed by or combined with inoculation of a silicon-containing material to ensure a graphitic structure with freedom from carbides. This section will discuss methods of magnesium treatment, control of magnesium content, and inoculation following magnesium treatment.

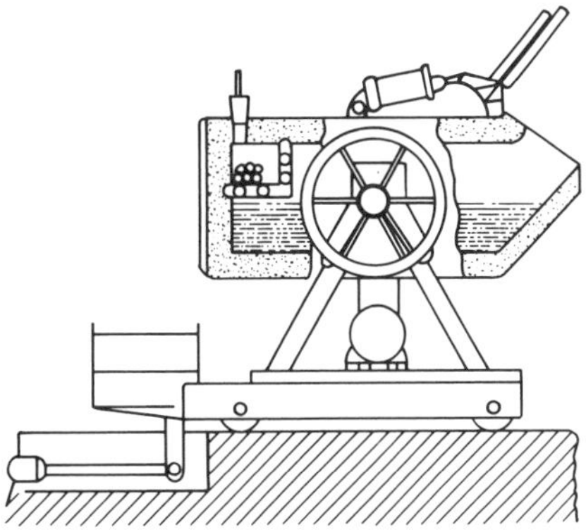

(a)

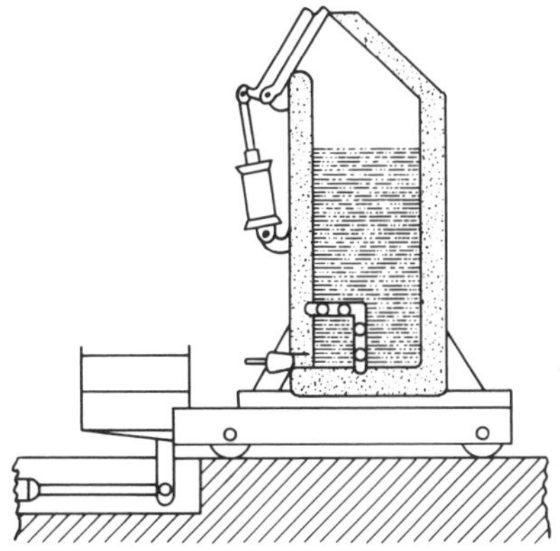

(b)

Fig. 2 Schematic of the Fischer converter. (a) Vessel in filling position. (b) Vessel in treating position

Magnesium Treatment

For magnesium treatment, the iron will normally be at a temperature of 1450 to 1510 °C (2640 to 2750 °F). Many methods of treatment are practiced; these are discussed more fully in Ref 2 and 3. Three of the most important methods will be outlined in this section.

Metallic Magnesium Treatment. The reaction between metallic magnesium and molten iron is violent. The magnesium is vaporized and burns vigorously in air. Special enclosed reaction vessels are needed, and one of the best established processes uses a converter (Fig. 2) in which ingots of magnesium are introduced into the bottom of the liquid metal in a tiltable vessel under atmospheric pressure. Other methods have involved introducing magnesium powder, rods, or wire into the molten metal with atmospheric or pressurized vessels designed to exclude air and to prevent the ejection of molten metal and fumes during the solution of the magnesium.

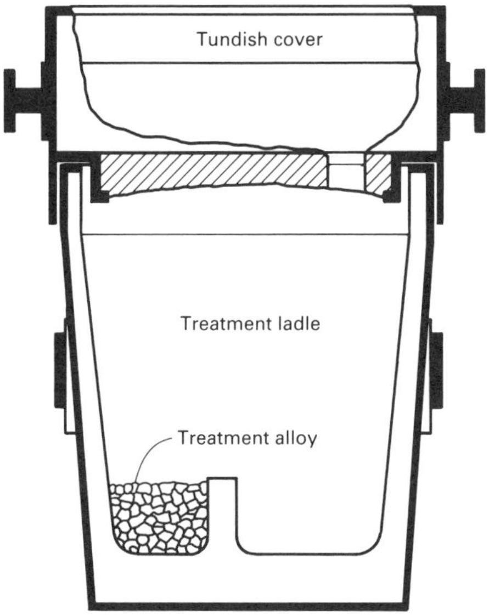

Fig. 3 Treatment ladle with tundish cover used in the magnesium treatment of ductile iron

Magnesium particles have been made into compacts with iron powder or swarf, which reduces somewhat the violence of the reaction. Such materials can be added in special vessels or by plunging into the ladle in a refractory bell. In another technique, coke impregnated with magnesium is plunged into the liquid iron.

Magnesium Alloy Addition. A nickel alloy with 14 to 16% Mg can be added to the ladle during filling or by plunging. The reaction is spectacular but not violent, and a very consistent recovery is obtained. A disadvantage lies in the accompanying increase in nickel and the cost of the alloy. Other nickel alloys containing much lower magnesium contents (down to 4%) have also been used and involve a much quieter reaction. Most alloys used to introduce magnesium into molten iron are based on ferrosilicon containing 3 to 10% Mg. The reaction varies from fairly violent (with 10% Mg) to quiet (with 3% Mg). The alloy can be plunged in a refractory bell or added in the ladle using a number of different techniques, including pouring the molten iron onto the alloy in the bottom of the ladle. When this method is used, the alloy is often placed in a specially designed pocket and covered with a layer of steel turnings (this is known as the sandwich process). Treatment usually takes place in ladles having a height-to-diameter ratio of about 2:1 to contain any splashes of metal, while the use of a cover with a tundish through which the iron can be poured reduces any ejection of metal, fume, and flame and improves the yield of magnesium (Fig. 3).

A method of treating the molten metal as it fills the ladle is the Flotret process, in which the magnesium alloy is placed in a cavity in a closed vessel through which the molten metal is poured between the furnaces and the ladle. This method can be used when pouring the metal from one ladle to another. Another design of flow-through vessel is used in the Imconod process.

In-The-Mold Treatment (Inmold Process). Magnesium-containing alloys can be added in the mold. The alloy, usually of the ferrosilicon magnesium type, is placed in a specially designed chamber or enlargement of the running system before closing the mold, and the alloy dissolves as the metal is cast. This method contains the reaction in the mold and avoids ejection of fume and flare. Low-sulfur iron (≤0.01% S) is used to avoid problems resulting from sulfide inclusions in the castings.

Control of Magnesium Content

In all methods of magnesium treatment, accurate measurement of the weight of metal treated and the amount of additive is essential. The initial sulfur content of the iron and the treatment temperature must also be known because they influence the final magnesium content. The recovery of magnesium can be expressed as:

$$\text{Recovery \%} = \left[\frac{(\text{\% final magnesium content})}{\left(\begin{array}{c}\text{\% added}\\ \text{magnesium}\end{array}\right) - \frac{3}{4}\left(\begin{array}{c}\text{\% initial}\\ \text{sulfur content}\end{array}\right)} \right] \times 100 \quad \text{(Eq 2)}$$

The magnesium sulfide produced by the reaction flows into the slag on the top of the ladle and is removed by skimming or by use of a teapot ladle. Typical values of recovery at a treatment temperature of 1450 °C (2640 °F) are:

- 50% for a 16% Ni-Mg alloy added in the ladle
- 40% for a 9% Mg ferrosilicon added using the sandwich process
- 60% for a 5% Mg ferrosilicon added using the sandwich process
- 50% for pure magnesium added in the converter
- 45% for the inmold or Flotret process

During magnesium treatment, there is usually a decrease in metal temperature of 35 to 50 °C (65 to 90 °F).

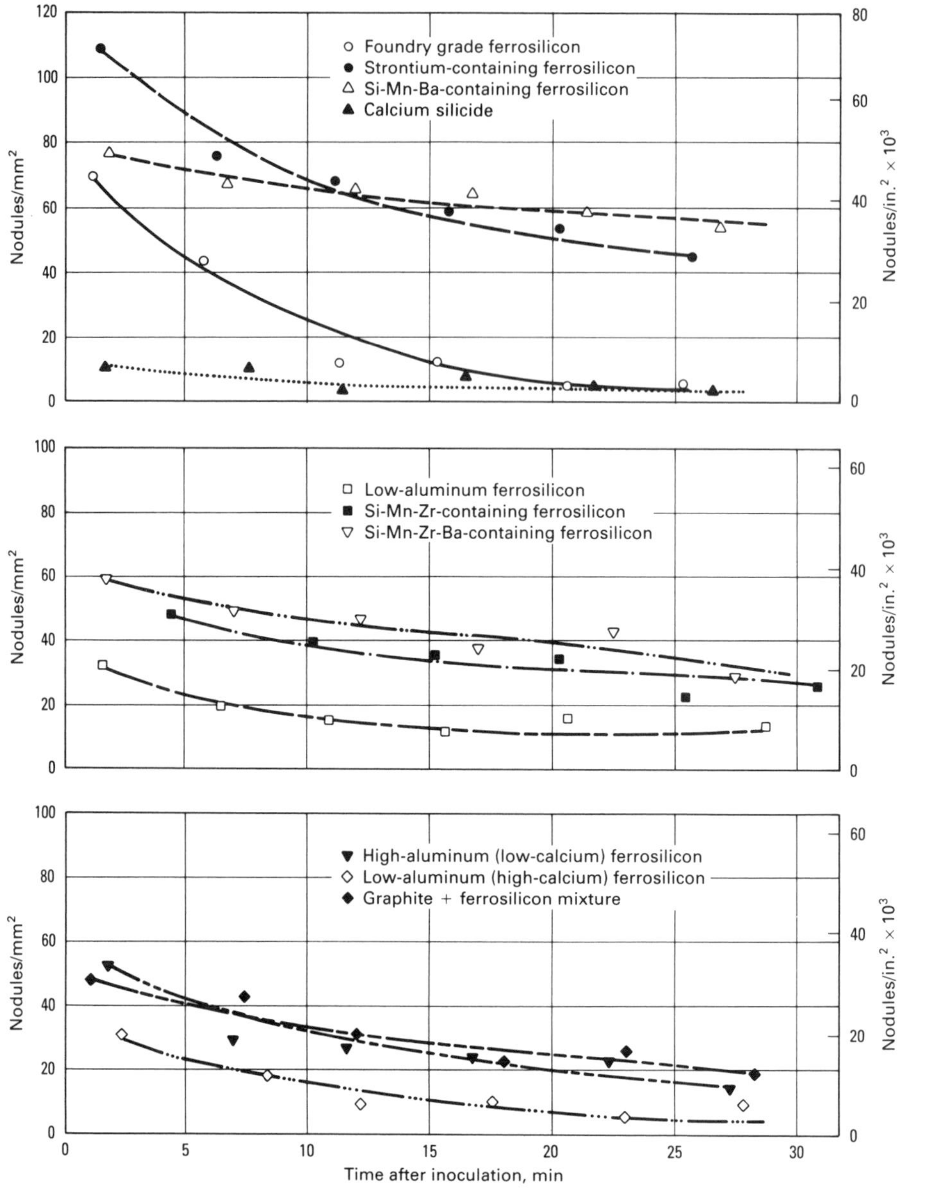

Fig. 4 Variations in nodule number with time for 22 mm (0.875 in.) bars of cerium-free ductile irons after addition of various inoculants. Source: Ref 4

Inoculation (Ref 4)

Following magnesium treatment, the iron is usually subjected to final inoculation, sometimes referred to as postinoculation. Inoculation is commonly carried out in the ladle using a granular inoculant, which may be commercial ferrosilicon containing 75% Si or one of a wide range of proprietary alloys usually containing 60 to 80% Si. The amount of inoculant added will usually range from about 0.25 to 1.0%. A higher percentage of silicon in the magnesium-addition alloy may permit less inoculation. The inoculant can be added during reladling, stirred into the metal, placed on the bottom of the ladle before filling, or plunged in a refractory bell, as late as possible before casting. Effective stirring is necessary, and one way of achieving this is by bubbling air or nitrogen through the melt using a porous plug in the bottom of the ladle.

Inoculation reduces undercooling during solidification and helps to avoid the presence of carbides in the structure, especially in thin sections. It increases the number of graphite nodules, thus improving homogeneity, assisting in the formation of ferrite, and promoting ductility. It assists in reducing annealing time and reduces hardness.

The effect of an inoculant is greatest as soon as it has dissolved, after which it fades over a period of 20 to 30 min. Both the initial potency and the rate of fading are influenced by trace elements, which include calcium, aluminum, cerium, strontium, barium, and bismuth. Figure 4 shows how the number of nodules in test bars decreased with time after ladle inoculation for ten different inoculants.

Late addition of an inoculant as the metal is being cast is much more effective and can be achieved by placing granular inoculant or shaped inoculant particles in the mold in an enlargement of the runner or a special chamber in the running system. Alternatively, fine granular inoculant can be added to the metal pouring stream either by dispensing it to the sprue or by encasing it in hollow wire and passing this into the sprue. These late methods of inoculation require only 1/5 to 1/10 as much inoculant as ladle inoculation and are sometimes also used to supplement ladle inoculation, particularly when fading has occurred through long holding times in the ladle. The superior effect of late inocu-

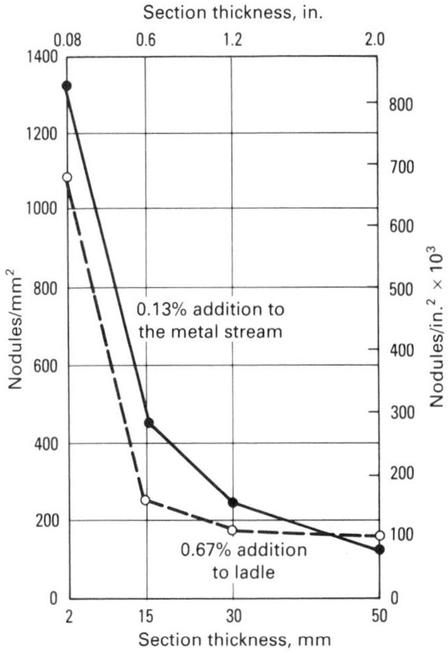

Fig. 5 Nodule numbers in ductile irons inoculated with foundry grade ferrosilicon (75% Si) added either to the ladle or to the metal stream. Source: Ref 4

lation in increasing nodule number is illustrated in Fig. 5, which also shows that the advantage is greatest in thinner section castings.

Late inoculation is sometimes practiced in addition to other methods of inoculation as a means of intensifying the inoculation effect or as a safeguard against fading of inoculation in the ladle, especially when making very thin castings. When magnesium treatment is done in the mold, the magnesium alloy usually contains enough silicon to produce the required inoculation effect, but may also be supplemented by mixing additional inoculant with the magnesium alloy.

Casting and Solidification

Ductile iron shares with gray irons the properties of very good fluidity and castability, but its casting and solidification characteristics have some important differences that require special attention.

Gating Systems

Because ductile iron normally contains magnesium, oxidation of this element, together with silicon, may give rise to small inclusions and dross both in the ladle and in the gating system. This dross must not enter the casting, where it can cause laps, rough surfaces, or pinholes. Other recommendations include the following:

- Metal held in the ladle must be skimmed clean; the use of teapot ladles is advisable
- Magnesium content should be kept to a minimum, preferably less than 0.06%

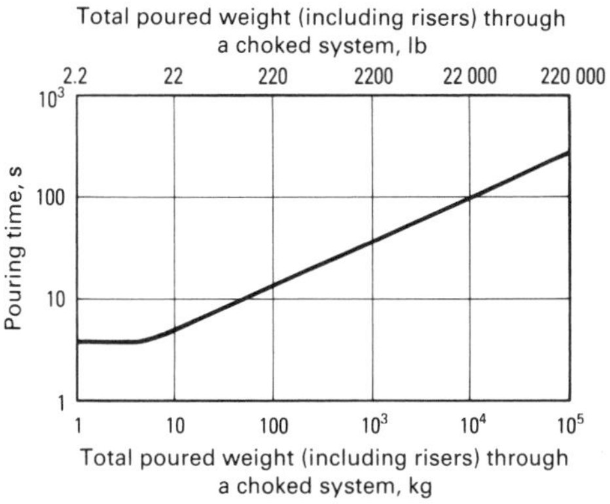

Fig. 6 Example of pouring rates found to be successful for ductile iron castings. Source: Ref 5

- A gating system of adequate size should be used, designed to achieve rapid mold filling with minimum turbulence of metal
- A slightly pressurized gating ratio is desirable; for example, suitable ratios of cross sections of sprue-runner-ingate may be 4:8:3 to ensure a short filling time and a full sprue during casting
- Placing ingates in the drag is recommended, with ingates as near to the bottom of the casting as possible
- The use of a ceramic filter in the gating system removes small slag inclusions and assists smooth metal flow, and it has been shown to reduce dross defects and to improve the mechanical properties of the iron

Pouring Temperature and Speed. Excessive temperature gradients in the casting and pouring of cold metal are undesirable and may promote carbide formation in thin sections and casting extremities. These can be avoided by pouring at not less than 1315 °C (2400 °F) for castings of 25 mm (1 in.) section and above up to 1425 °C (2600 °F) for castings of 6 mm (¼ in.) section and by modifying the gating system to include such features as flow-offs and runner bar extensions. There are no absolute rules for the speed of pouring, but Fig. 6 illustrates typical values found to be successful.

Solidification and Feeding (Risering)

The change from flake to nodular graphite is accompanied by important differences in solidification behavior. In hypereutectic irons, graphite segregation and flotation occur more easily than in gray iron. Such segregation can aggravate dross problems and give rise to local high-carbon regions, which adversely affect mechanical properties and the appearance of machined surfaces. During solidification, the precipitation of graphite in a nodular form causes ductile iron castings to expand to a greater degree and with more force than gray iron. Consequently, they require more rigid molds and more attention to feeding if the castings are to be sound.

In green sand molds in particular, ductile iron castings are likely to be oversize compared with the pattern and compared with gray iron castings. It is necessary to make dense molds, and flasks may need internal bars and will require good weighting and clamping, especially for larger castings. The use of chemically hardened or dry sand molds is a great advantage, provided they are adequately cured (Ref 6).

It is sometimes possible to make sound ductile iron castings of simple shapes without the use of risers by taking advantage of graphite expansion, which occurs during eutectic solidification. The requirements are:

- A strong, well-compacted, chemically bonded sand mold contained in a rigid molding flask
- A carbon equivalent of about 4.3 and a carbon content of about 3.6%
- A low casting temperature, probably below 1350 °C (2460 °F)
- A gating system and running speed to minimize temperature gradients developed in the mold

Nevertheless, for most purposes, ductile iron castings, even in rigid molds, will require the use of risers to ensure fully sound castings. The principles of directional solidification should be employed in casting design with regard to the placement of gates and the locations of risers.

There are three stages of solidification volume change, and rational risering practice takes account of them. They are:

- A volume decrease of the liquid metal as soon as it has been poured, while it cools to the temperature of eutectic solidification
- A volume increase during eutectic solidification, when graphite nodules grow and exert considerable expansion pressure on the mold; depending on casting design and mold strength, this can lead to oversize castings
- A volume decrease during the last stage of solidification (secondary shrinkage)

For the most efficient use of risers to achieve sound castings, advantage is taken of the second-stage volume increase to reduce the amount of feed metal required. This is done with pressure control, in which riser (feeder) necks are designed to freeze during solidification, causing the solidification volume increase to compensate for any casting expansion and late-stage volume decrease or secondary shrinkage.

There are a number of approaches to the feeding of ductile iron castings. A pressure control approach relates the cooling modulus of the casting to those of the riser and the riser neck. A typical nomogram is shown in Fig. 7, in which the modulus of the

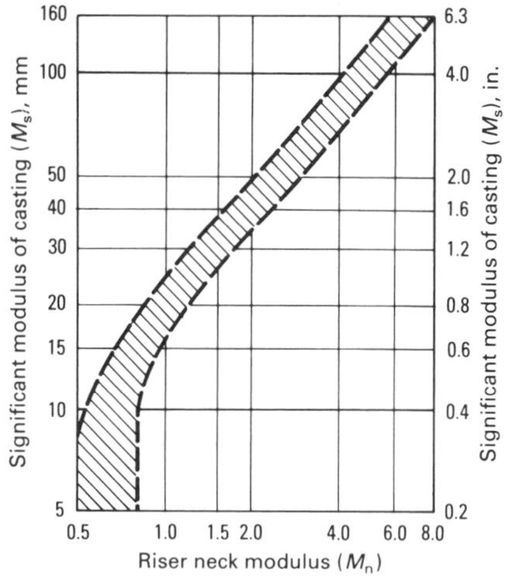

Fig. 7 A system for calculating riser dimensions. M_s = volume/effective cooling surface of largest section of casting; M_n = circumference of riser neck/cross section area of riser neck. Modulus of riser = 1.2 M_n. Source: Ref 7

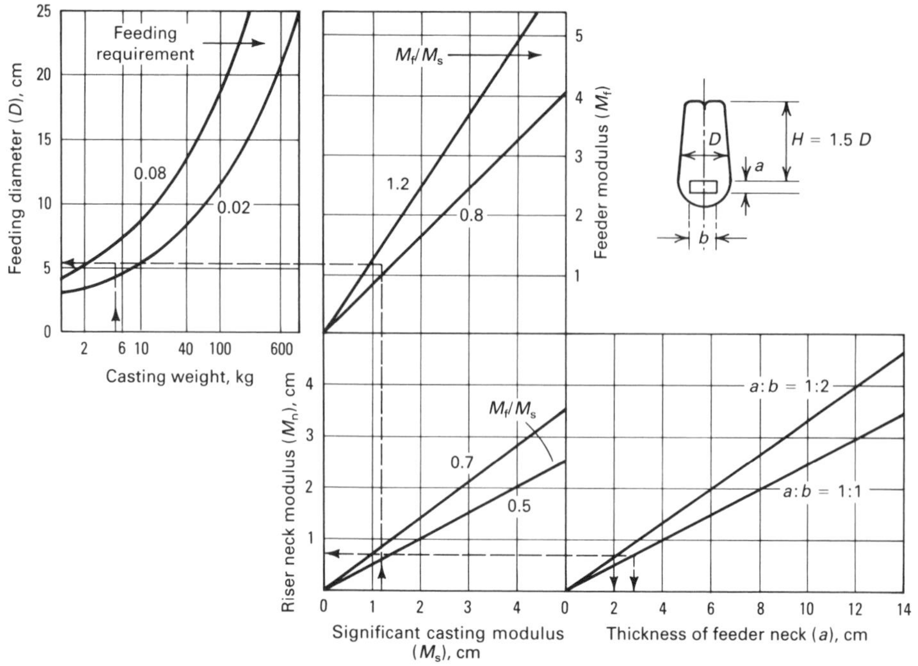

Fig. 8 System of calculating feeder dimensions. First, calculate significant modulus M_s. Second, read off from M_s the values of M_f, M_n, D, and a:b. Third, select feeding requirements depending on mold type and pouring temperature. Finally, select M_f/M_s based on foundry experience. Source: Ref 8

riser neck, M_n, is related to the significant modulus of the casting, M_s.

More advanced calculations of feeding requirements for ductile iron take into account variables affecting mold rigidity, metal composition (including carbon equivalent and content), and pouring temperature. These can be combined to provide a factor that can be included in a nomogram such as that shown in Fig. 8.

Because the risers themselves can promote or accentuate hot spots and excessive mold wall movement, the use of smaller risers by the application of exothermic feeder sleeves can be an advantage and can improve yield, often to more than 80%. Nomograms and computer programs for the use of such sleeves are generally supplied by the manufacturers. Excessive ferrostatic head should also be avoided because this increases the pressure on the mold, leading to casting swelling and further feed metal requirement. Several computer-aided methods are available for determining the best risering system, but for their use to be successful, all of the production variables affecting the shrinkage behavior of ductile iron must be controlled.

Ductile iron risers are more difficult to remove than those of gray iron. Notched riser necks and the use of breaker cores can facilitate breaking off, and mechanical devices are available to assist the breaking or levering off of risers. However, abrasive cutting wheels sometimes must be used to avoid damage to the castings.

The change in shape of graphite from flake to nodular makes ductile iron more prone to carbide formation in thin sections, and it may be difficult to obtain graphite throughout sections less than 6 mm (¼ in.) thick. There is also a tendency for inverse chill to occur in the heat centers of round and symmetrical sections. This is related to the cooling conditions of such sections and is most likely to be cured by adjusting the gating system and by using late inoculation in the metal stream or in the mold.

Metallurgical Controls of Ductile Iron Production

Metal Composition. Before treatment with magnesium, it is necessary to know the sulfur content of the liquid iron. This is best determined rapidly by taking a chilled coin sample for spectroscopic analysis. If required, the sulfur and the carbon content can be determined rapidly by the combustion of crushed solid samples taken from the pins attached to the test coin. Spectroscopic analysis for other elements can be carried out either before or after magnesium treatment, but it is unlikely that a completely white sample suitable for analysis will be obtained after inoculation or if a heavy addition of a silicon-containing magnesium alloy has been made. Final silicon content after inoculation may need to be determined chemically with drillings taken from a sample cast after treatment. A reliable determination of carbon in treated ductile iron cannot be obtained chemically except on a solid sample obtained, for example, from pins attached to a coin sample, because of loss of carbon during drilling and the likely presence of graphite in the coin. Rapid shopfloor control of carbon equivalent, carbon, and silicon can be carried out by thermal analysis of the metal in the same manner for gray iron, but this must be done before magnesium treatment.

Other properties of the iron can be deduced from thermal analysis data, including the undercooling of the eutectic, which is a guide to nodule number and chilling tendency, but in general such developments are still taking place. For this reason, nodularity and nodule number are generally determined by other methods.

Nodularity and Nodule Number. The success of the magnesium treatment, freedom from undesirable trace elements, and adequate postinoculation in producing an iron with a good nodular graphite structure and of adequate nodule number to ensure freedom from carbides and to yield correct properties are generally determined rapidly by casting and examining a test coupon of a design such as that recommended by American Foundrymen's Society (AFS) Committee 12K and shown in Fig. 9. The mold is made in hardened sand and consists of a central block with two ears, which are broken off as soon as the casting has solidified and the ears have cooled to a dull red or black heat. After cooling to room temperature, an ear is broken or cut in two, and the cross section is ground, polished, and examined under the microscope. Alternatively, small samples for metallographic examination can be cut from selected castings or from other types of agreed test bar. The number of nodules in a specified area is counted using a magnification usually of 100×; the specified area is typically a square millimeter or a square inch of the field of view or along a specified length of line. The shape of the nodules is evaluated in accordance with the classification of the

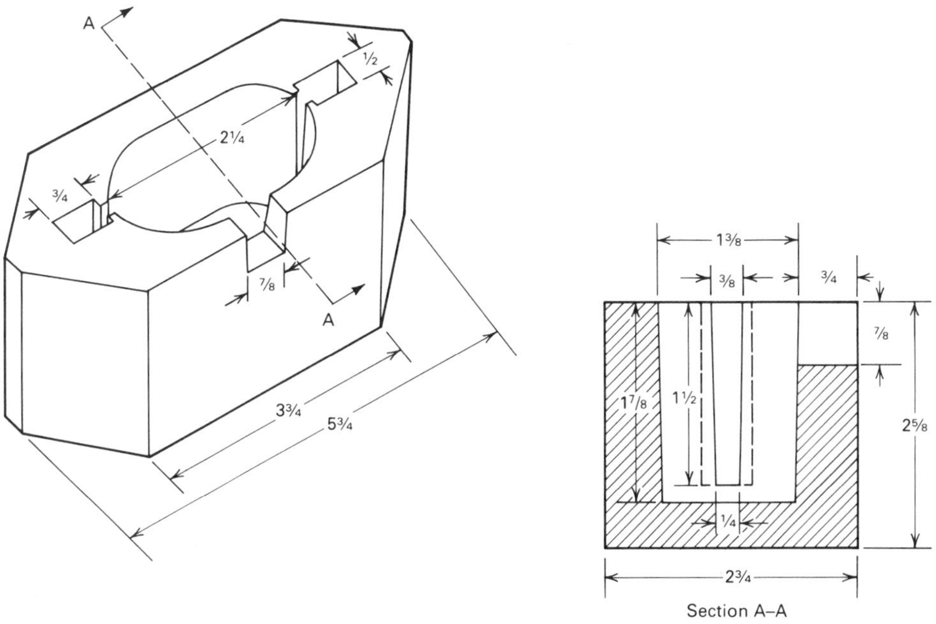

Fig. 9 Test coupon recommended by AFS Committee 12K for rapid examination of ductile iron microstructures. Dimensions given in inches

standards ISO 945 or ASTM A 247. Nodule number and perfection of shape can also be evaluated using quantitative metallographic instruments.

Good practice requires a nodularity sample to be taken from each batch of iron treated and poured or, when using automated holding and pouring of metal, at regular intervals of time. For most purposes, 85 to 100% of the graphite should be in a fully nodular form. Samples from castings or test bars will be periodically examined to verify the graphite and matrix structure of the iron produced, to evaluate ferrite and pearlite contents, and to verify freedom from undesirable carbides, inclusions, and dross as a general control of the production process.

Mechanical Test Coupons. To ensure that the castings meet the specified requirements, mechanical properties are checked periodically on test pieces machined from cast test coupons. Recommended test bars cast separately or attached to castings are in the form of keel blocks, Y blocks, or horizontal bars with knock-off heads and are indicated in most specifications. If the castings are to be used in the heat-treated condition, the test bars will be heat treated with the castings they represent before mechanical test specimens are obtained. The actual properties of castings are sometimes checked by machining test bars from the castings with agreement between the foundry and its customer.

Other Rapid Tests of Metal Quality. If a sand cast test bar is made from the treated iron, its nodularity can be evaluated by measuring either its resonant frequency or the velocity of ultrasonic radiation through it. Suitable tests of this kind have been described and are sometimes used as rapid shopfloor control tests. Similar tests for casting quality control are described in the section "Nondestructive Evaluation of Ductile Iron Castings" in this article.

Specifications for Ductile Iron

Standard specifications for engineering grades of ductile iron castings classify the grades according to the tensile strength of a test bar cut from a prescribed test casting. The International Standards Organization (ISO) specification ISO 1083:1976 and most national specifications also specify the ductility in terms of percentage of elongation and the 0.2% proof strength or offset yield strength. The impact values of those grades with the highest ductility are frequently specified in the ISO, UK, and West German specifications, and a guide to microstructure is included in most specifications. Hardness is usually specified, but is only mandatory in SAE J434C. Table 2 summa-

Table 2 Ductile iron property requirements of various national and international standards

	Tensile strength		0.2% offset yield strength			Impact energy					
						Mean(a)		Individual			
Grade	MPa	ksi	MPa	ksi	Elongation (min), %	J	ft · lbf	J	ft · lbf	Hardness, HB	Structure
ISO Standard 1083 (International)											
800-2	800	116	480	70	2	...	...	...	...	248–352	Pearlite or tempered
700-2	700	102	420	61	2	...	...	...	...	229–302	Pearlite
600-3	600	87	370	54	3	...	...	...	...	192–269	Pearlite + ferrite
500-7	500	73	320	46	7	...	...	...	...	170–241	Ferrite + pearlite
400-12	400	58	250	36	12	...	...	...	...	<201	Ferrite
370-17	370	54	230	33	17	13	9.5	11	8.1	<179	Ferrite
ASTM A 536 (United States)											
60-40-18	414	60	276	40	18	...	...	...	...	...	...
60-42-10	414	60	290	42	10	...	...	...	...	...	...
65-45-12	448	65	310	45	12	...	...	...	...	...	...
70-50-05	485	70	345	50	5	...	...	...	...	...	...
80-55-06	552	80	379	55	6	...	...	...	...	...	...
80-60-03	552	80	414	60	3	...	...	...	...	...	...
100-70-03	690	100	483	70	3	...	...	...	...	...	...
120-90-02	827	120	621	90	2	...	...	...	...	...	...
SAE J434C (United States)(b)											
D4018	414	60	276	40	18	...	...	...	...	170 max	Ferrite
D4512	448	65	310	45	12	...	...	...	...	156–217	Ferrite + pearlite
D5506	552	80	379	55	6	...	...	...	...	187–255	Ferrite + pearlite
D7003	690	100	483	70	3	...	...	...	...	241–302	Pearlite
DQ&T(c)	...	...	...	...	...	...	...	...	...	...	Martensite

(a) Mean value from three tests. (b) These irons are specified primarily based on hardness and structure. Mechanical properties are given for information only. (c) Quenched-and-tempered grade; hardness subject to agreement between supplier and purchaser

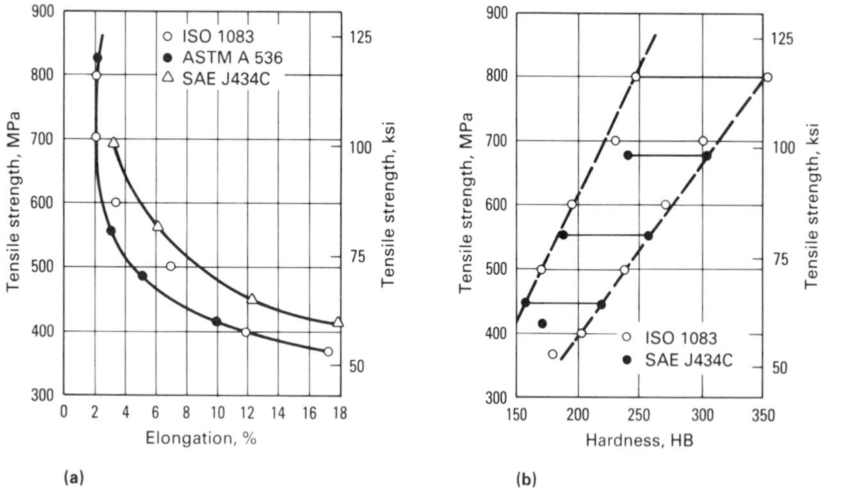

Fig. 10 (a) Tensile strength versus elongation. (b) Tensile strength versus hardness

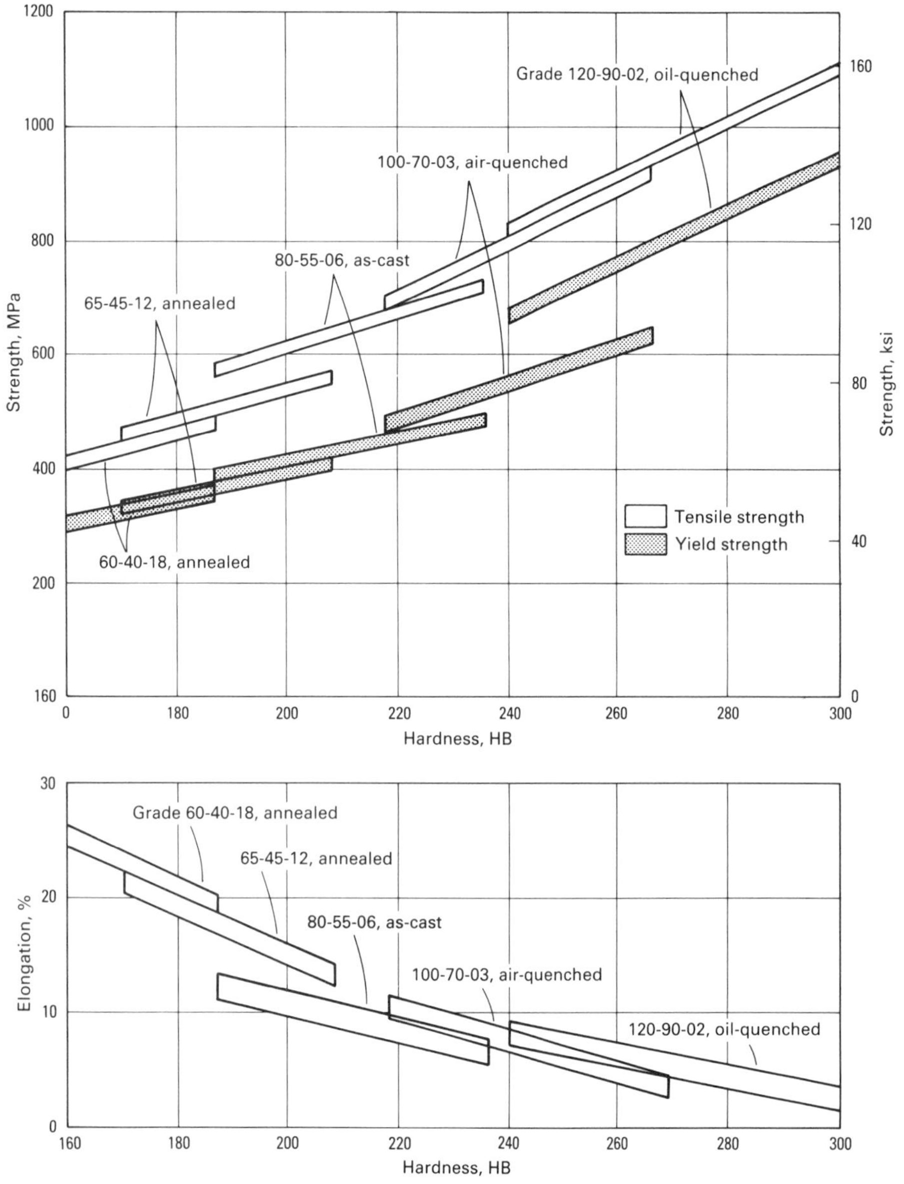

Fig. 11 Strength and ductility versus hardness ranges for ductile iron

rizes the ISO 1083:1976, ASTM A 536, and SAE J434C specifications.

Most national specifications follow approximately the grades of the ISO specification. There are some exceptions, for example, SAE J434C and Japan JIS G 5502-1982. The standard specifications of a number of other countries are listed in BCIRA Broadsheet 156-1. An engineer's guide to many of the properties to be expected from the various grades of the British specification has been published (Ref 9) and provides information that enables an estimate to be made of a variety of other mechanical and physical properties of irons covered by this and most other specifications.

Figure 10(a) shows the relationship between minimum specified values for tensile strength and elongation representing the range of values covered by the ISO, ASTM, and SAE specifications. Figure 10(b) shows the relationship between minimum values for tensile strength and hardness ranges represented by the three specifications.

The actual values of properties to be expected from good-quality ductile irons produced to meet any given specified grade will normally cover a range that more than satisfies the requirements of the specification. Data on strength, hardness, and elongation from actual production are recorded in Fig. 11 and illustrate the typical ranges achievable.

Specifications for the highest-strength grades usually mention the possibility of hardened-and-tempered structures, but for the most recently reported austempered ductile irons, which have the highest combinations of tensile strength and ductility, there are as yet only tentative unofficial specifications. Table 3 summarizes tentative values of grades suggested in the United States, the United Kingdom, and Switzerland. Figure 12 compares the properties of austempered ductile irons with those of other ductile irons.

Factors That Affect Properties

Graphite Structures. The amount and form of the graphite in ductile iron are determined during solidification and cannot be altered by subsequent heat treatment. All of the mechanical and physical properties characteristic of this class of materials are a result of the graphite being substantially or wholly in the spheroidal nodular shape, and any departure from this shape in a proportion of the graphite will cause some deviation from these properties. It is common to attempt to produce greater than 90% of the graphite in this form (>90% nodularity), although structures between 80 and 100% nodularity are sometimes acceptable. Figure 13 illustrates microstructures containing estimated graphite nodularities of 99, 80, and 50%.

Table 3 Some tentative specifications for austempered ductile iron

Grade	Tensile strength (min) MPa	Tensile strength (min) ksi	0.2% offset yield strength (min) MPa	0.2% offset yield strength (min) ksi	Elongation (min), %	Hardness, HB
Ductile Iron Society (United States)						
1	860	125	550	80	10	269–321
2	1035	165	690	100	7	302–363
3	1205	175	827	120	4	363–444
4	1380	200	965	140	2	388–477
Sulzer (Switzerland)						
GGG80 BAF	800	116	500	73	8	250–310
GGG100	1000	160	700	102	5	280–340
GGG120	1200	174	950	138	2	330–390
BCIRA (Great Britain)						
950/6	950	138	670	97	6	300–310
1050/3	1050	152	780	113	3	345–355
1200/1	1200	174	940	136	1	390–400

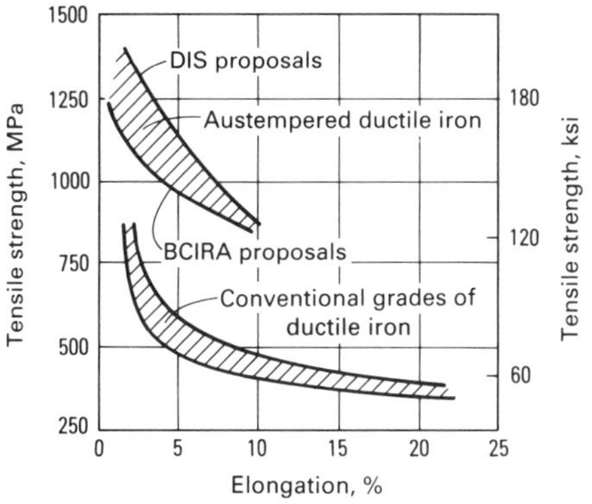

Fig. 12 Proposed strength and elongation ranges of austempered ductile iron compared to established criteria for other grades of ductile iron

All properties relating to strength and ductility decrease as the proportion of non-nodular graphite increases, and those relating to failure, such as tensile strength and fatigue strength, are more affected by small amounts of such graphite than properties not involving failure, such as proof strength. This is illustrated in Fig. 14, which shows the typical effects of graphite form on tensile and offset yield (proof) strength.

The form of non-nodular graphite is important because thin flakes of graphite with sharp edges have a more adverse effect on strength properties than compacted forms of graphite with rounded ends. For this reason, visual estimates of percentage of nodularity are only a rough guide to properties. Graphite form also affects modulus of elasticity, which can be measured by resonant frequency and ultrasonic velocity measurements, and such measurements are therefore often a better guide to nodularity and its effects on other properties (see the section "Nondestructive Evaluation of Ductile Iron Castings" in this article). A low percentage of nodularity also lowers impact energy in the ductile condition, reduces fatigue strength, increases damping capacity, increases thermal conductivity, and reduces electrical resistivity.

Inoculation has the effect of increasing nodule number, which prevents the formation of carbides and increases ferrite, thus avoiding hard, brittle castings. Figure 15 shows how nodule number increased in thin plate castings as the amount of inoculant added was increased.

Graphite Amount. As the amount of graphite increases, there is a relatively small decrease in strength and elongation, a decrease in modulus of elasticity, and a decrease in density. In general, these effects are small compared with the effects of other variables because the carbon equivalent content of spheroidal graphite iron is not a major variable and is generally maintained close to the eutectic value.

Matrix Structure. The principal factor in determining the different grades of ductile iron in the specifications is the matrix structure. In the as-cast condition, the matrix will consist of varying proportions of pearlite and ferrite, and as the amount of pearlite increases, the strength and hardness of the iron also increase.

Ductility and impact properties are principally determined by the proportions of ferrite and pearlite in the matrix, as illustrated in the article "Ductile Iron" in Volume 1 of the 9th Edition of *Metals Handbook*. As the amount of pearlite decreases, the maximum impact energy in the ductile condition increases, and the ductile-to-brittle transition temperature range falls.

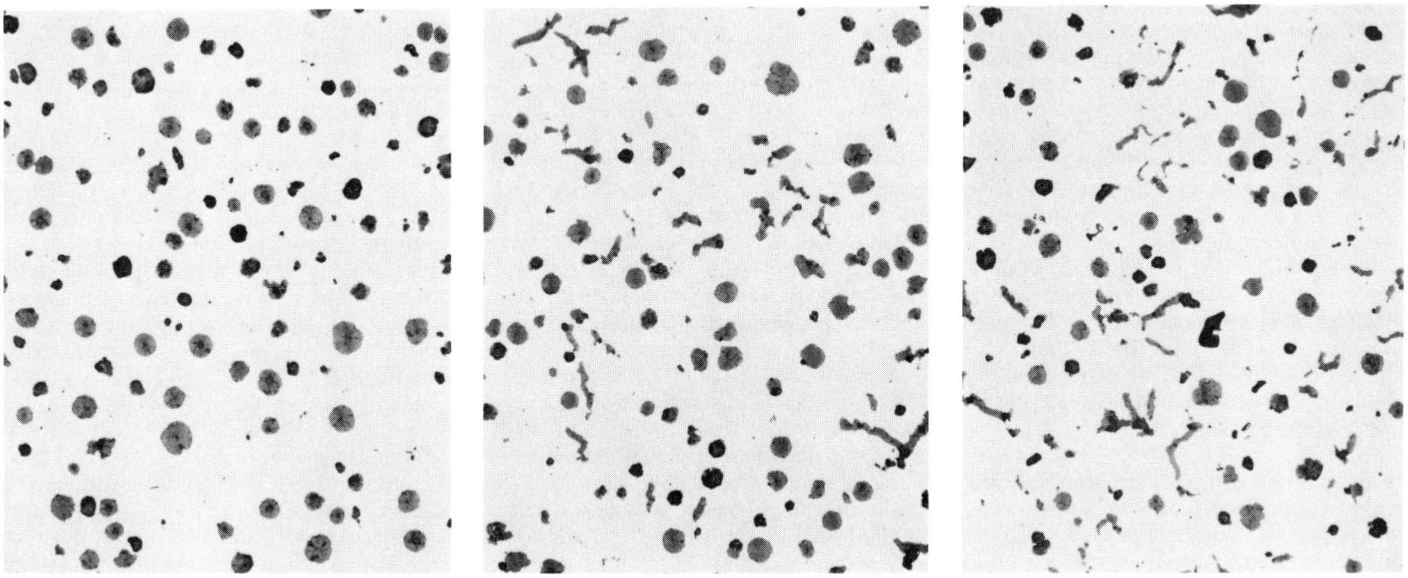

(a) (b) (c)

Fig. 13 Microstructures of ductile irons of varying degrees of nodularity. (a) 99% nodularity. (b) 80% nodularity. (c) 50% nodularity. All unetched. 36×

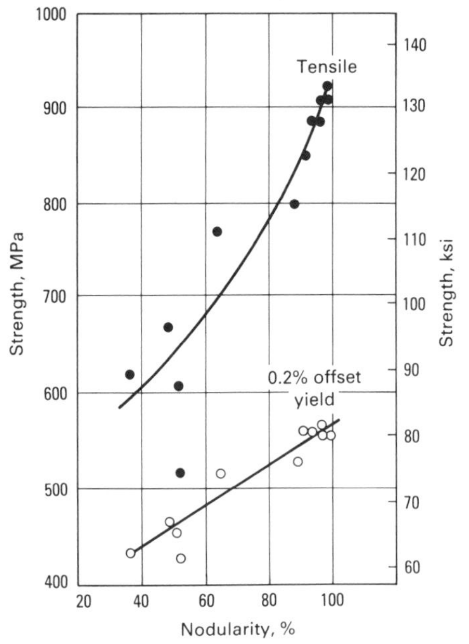

Fig. 14 Tensile and yield strength of ductile iron versus visually assessed nodularity. Source: Ref 10

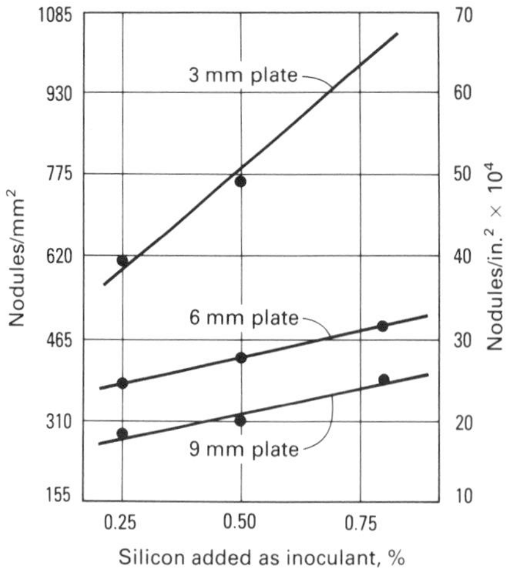

Fig. 15 Nodule number versus amount of silicon inoculant added in the ladle

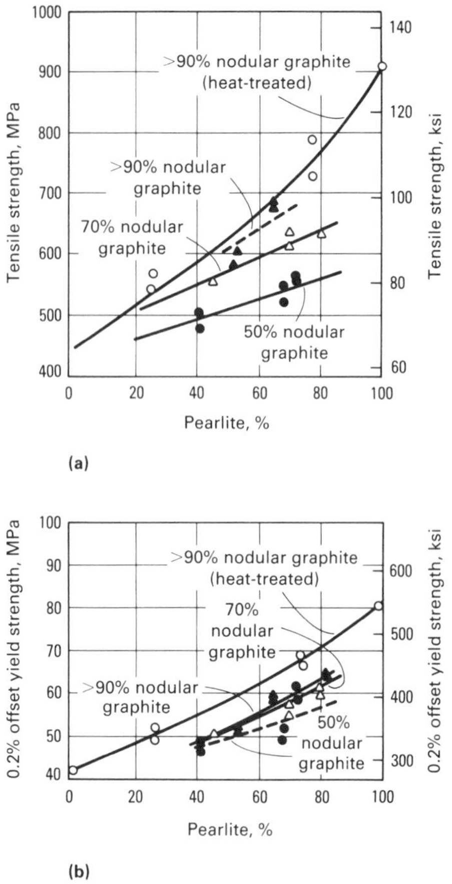

Fig. 16 Relationships between strength and amount of pearlite. (a) Tensile strength versus amount of pearlite in irons having varying properties of graphite in a nodular form. (b) 0.2% offset yield strength versus amount of pearlite in irons having varying proportions of graphite in a nodular form. Source: Ref 11

The matrix structure can be changed by heat treatments, and those most often carried out are annealing to produce a fully ferritic matrix and normalizing to produce a substantially pearlitic matrix (see the section "Heat Treatment of Ductile Iron" in this article). Other heat treatments will be described later. In general, annealing produces a more ductile matrix with a lower impact transition temperature than is obtained in as-cast ferritic irons. Normalizing produces a higher tensile strength with a higher amount of elongation than is obtained in fully pearlitic as-cast irons. In the former case, properties are due to a refined ferrite grain structure; in the latter case, increased strength and ductility result from homogenization and a finer pearlite structure than occurs in the as-cast condition. Figure 16 shows how pearlite content influences tensile strength and yield strength, how heat treatment increases strength, and how reduced nodularity lowers strength at given pearlite content.

The graphite structure can affect the matrix structure. An increased nodule number, achieved by better inoculation, will tend to increase the amount of ferrite in the as-cast condition and will lead to more rapid annealing with less chance of retained pearlite after a given annealing time.

The presence of carbides reduces ductility, increases hardness, and promotes premature failure in tension and in fatigue and impact loading. Special care is necessary to avoid carbides because they are difficult to detect by nondestructive means. It is desirable to ensure adequate silicon content and inoculation for the section size being cast and freedom from carbide-forming trace elements, particularly chromium.

Section Size. As section size decreases, the solidification and cooling rates in the mold increase. This results in a fine-grain structure that can be annealed more rapidly. In thinner sections, however, carbides may be present, which will increase hardness, decrease machinability, and lead to brittleness. To achieve soft ductile structures in thin sections, heavy inoculation, probably at a late stage, is desirable to promote graphite formation through a high nodule number.

As the section size increases, the nodule number decreases, and microsegregation becomes more pronounced. This results in a large nodule size, a reduction in the proportion of as-cast ferrite, and increasing resistance to the formation of a fully ferritic structure upon annealing. In heavier sections, minor elements, especially carbide formers such as chromium, titanium, and vanadium, segregate to produce a segregation pattern that reduces ductility, toughness, and strength (Fig. 17). The effect on proof strength is much less pronounced. It is important for heavy sections to be well inoculated and to be made from a composition low in trace elements.

Composition. In addition to the effects of elements in stabilizing pearlite or retarding transformation (which facilitates heat treatment to change matrix structure and properties), certain aspects of composition have an important influence on some properties. Silicon hardens and strengthens ferrite and raises its impact transition temperature; therefore, silicon content should be kept as low as practical, even below 2%, to achieve maximum ductility and toughness. Figure 18 summarizes information reported more fully earlier, showing how increasing ferrite lowers impact transition temperature and raises the ductile impact value. Figure 18 also shows how, in the ferritic condition, a range of values is obtained with improving properties as the silicon content is lowered. Further improvement in impact properties is obtained with phosphorus contents below about 0.05%; phosphorus also has a powerful embrittling effect in ferritic ductile iron and is therefore normally kept low.

Nickel also strengthens ferrite, but has much less effect than silicon in reducing ductility. When producing as-cast grades of iron requiring fairly high ductility and strength such as ISO Grade 500-7, it is necessary to keep silicon low to obtain high ductility, but it may also be necessary to add some nickel to strengthen the iron sufficiently to obtain the required tensile strength.

Almost all elements present in trace amounts combine to reduce ferrite formation, and high-purity charges must be used for irons to be produced in the ferritic as-cast condition. Similarly, all carbide-forming elements and manganese must be kept low to achieve maximum ductility and low hardness. Silicon is added to avoid carbides and to promote ferrite as-cast in thin sections.

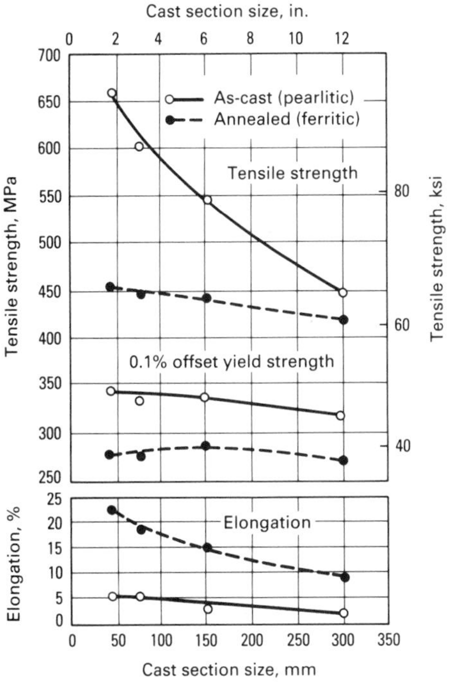

Fig. 17 Effect of cast section size on the properties of ductile iron

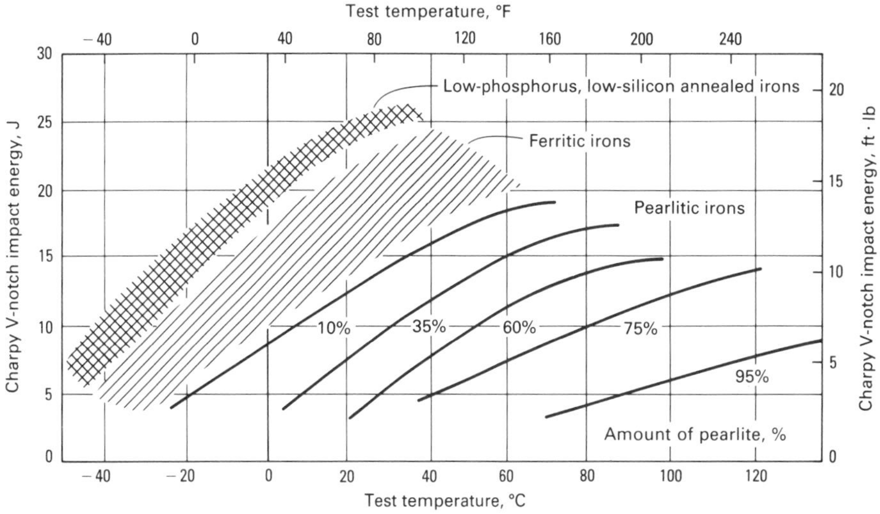

Fig. 18 Charpy V-notch impact energies of ductile irons. Source: Ref 12

The electrical, magnetic, and thermal properties of ductile irons are influenced by the composition of the matrix. In general, as the amount of alloying elements increases, resistivity increases, thermal conductivity decreases, and the magnetic hardness of the material increases.

Heat Treatment of Ductile Iron (Ref 13, 14)

The first stage of most heat treatments designed to change the structure and properties of ductile iron consists of heating to, and holding at, a temperature between 850 and 950 °C (1560 and 1740 °F) for about 1 h plus 1 h for each 25 mm (1 in.) of section thickness to homogenize the iron. When carbides are present in the structure, the temperature should be approximately 900 to 950 °C (1650 to 1740 °F), which decomposes the carbides prior to subsequent stages of heat treatment. The time may have to be extended to 6 or 8 h if carbide-stabilizing elements are present. In castings of complex shape in which stresses could be produced by nonuniform heating, the initial heating to 600 °C (1110 °F) should be slow, preferably 50 to 100 °C (90 to 180 °F) per hour.

To prevent scaling and surface decarburization during this stage of treatment, it is recommended that a nonoxidizing furnace temperature be maintained using a sealed furnace; a controlled atmosphere may be necessary. Care must also be taken to support castings susceptible to distortion and to avoid packing so that castings are not distorted by the weight of other castings placed above them.

The most important heat treatments and their purposes are:

- Stress relieving, a low-temperature treatment, to reduce or relieve internal stresses remaining after casting
- Annealing to improve ductility and toughness, to reduce hardness and to remove carbides
- Normalizing to improve strength with some ductility
- Hardening and tempering to increase hardness or to give improved strength and higher proof stress ratio
- Austempering to yield bainitic structures of high strength, with some ductility and good wear resistance
- Surface hardening by induction, flame, or laser to produce a local wear-resistant hard surface

Stress Relieving. The object of stress-relieving heat treatment is to remove residual stress without causing any change in structure or properties. High stresses can be present as-cast in ductile iron castings of complex shape and can be substantially removed by heat treatment at approximately 500 to 600 °C (930 to 1110 °F). The casting is typically heated at 50 °C (90 °F) per hour from 200 to 600 °C (390 to 1110 °F), held at 600 °C (1110 °F) for 1 h for each 25 mm (1 in.) of section thickness plus 1 h, and then furnace cooled at 50 °C (90 °F) per hour to below 200 °C (390 °F), after which the castings can be air cooled to room temperature. It is of great importance to ensure that the heating and cooling rates are slow enough to avoid thermal shock and to avoid the generation of more stress through the formation of high-temperature gradients in the casting. Stress relief is not likely to be necessary for annealed castings, but may be required for as-cast pearlitic castings and for those that have been air cooled during normalizing.

Annealing. The primary purpose of annealing is to generate a ferritic structure and to remove pearlite and carbides, thus achieving maximum ductility and toughness. Annealing would normally be used to achieve properties in grades specifying 15% or more elongation. The treatment may take one of several forms, but interrupted cooling, controlled slow cooling, and single-stage treatment are typical.

Interrupted Cooling. The first stage is to homogenize the iron as described above. This is followed by cooling to 680 to 700 °C (1255 to 1290 °F) and holding at this temperature for 4 to 12 h to develop ferrite. The higher the purity of the iron, the shorter the time required. Castings of simple shape can be furnace cooled to below 650 °C (1200 °F) and air cooled, but complex castings that could develop residual stresses should be furnace cooled according to the recommendations for stress relieving.

Controlled Slow Cooling. The first stage is homogenization, as above. This is followed by cooling at 30 to 60 °C (55 to 110 °F) per hour through the temperature range of 800 to 650 °C (1470 to 1200 °F). Lower-purity irons require slower cooling rates. Cooling to room temperature is then carried out as in the interrupted method.

Single-Stage Treatment. The casting is heated from room temperature to 680 to 700 °C (1255 to 1290 °F) without a prior first-stage austenitizing treatment, then held at this temperature for 2 to 16 h to graphitize the pearlite. The time increases with decreased metal purity and is generally longer than for other methods because of a lack of prior homogenization. Cooling to room temperature is carried out as in the interrupted cooling method. This treatment is used only to break down pearlite in irons that have no eutectic carbide. If the iron

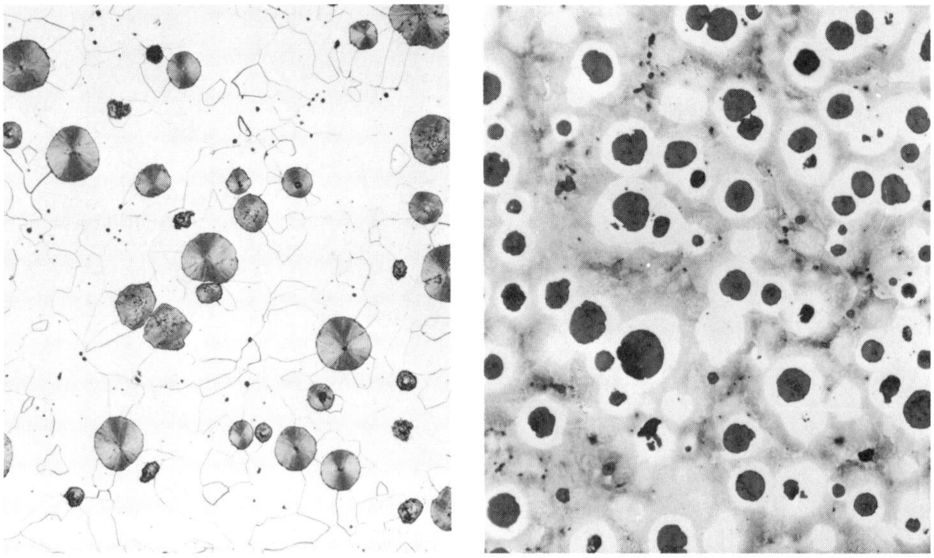

(a) (b)

Fig. 19 Microstructures of annealed irons. (a) Fully annealed ferritic matrix. (b) Partially annealed bull's-eye structure. Both etched in picral. 100×

Fig. 20 Microstructure of normalized iron showing pearlite matrix. Compare with Fig. 22 and 23. Etched in picral. 500×

contains carbides, interrupted cooling or controlled slow cooling treatments should be used.

Selection of Annealing Treatment. Most rapid annealing occurs in irons of higher silicon content, of low manganese, copper, tin, arsenic, and antimony contents, and of generally low trace element content. If the iron contains no carbides, any of the treatments outlined above can be used, but for optimum ductility, interrupted cooling should be chosen. It should be noted that if single-stage treatment is used the ferrite grain structure of the iron will be inferior to that produced by the other structures and ductility and toughness will be lower. The ferrite-forming temperature in the range of 680 to 700 °C (1255 to 1290 °F) can be increased with increased silicon content, and the rate of annealing is then increased.

The annealing cycle can be varied to produce structures containing mixed pearlite and ferrite matrices with higher strength and intermediate ductility. For example, the time in the temperature range of 700 to 720 °C (1290 to 1330 °F) can be reduced to achieve this, but a very good combination of strength with improved ductility can be obtained by austenitizing at 900 to 925 °C (1650 to 1695 °F), air cooling to room temperature, and then reheating and holding at 680 to 700 °C (1255 to 1290 °F) for a sufficient time to form ferrite around the nodules in a bull's-eye structure. Figures 19(a) and (b) show annealed iron microstructures with a fully ferritic and a bull's-eye ferrite structure, respectively. A form of embrittlement may occur in ferritic irons if they are rapidly cooled from the range of 500 to 600 °C (930 to 1110 °F); the quenching of castings from this temperature range should be avoided.

An overall increase in casting dimensions occurs during annealing because of the graphitization of pearlite and carbides. This increase in size can be as much as 0.005 cm/cm (0.005 in./in.) when annealing a fully pearlitic iron.

Normalizing consists of soaking the castings at a high temperature at which they are fully austenitized and any carbides decomposed, followed by cooling in air at a rate that produces a fine pearlitic matrix with traces of ferrite and free from other transformation products. Normalizing can be used to achieve grades with strengths of 700 to 900 MPa (100 to 130 ksi), and it improves the ratio of proof stress to tensile strength. A typical cycle is as follows.

The first-stage treatment is homogenization. Castings are then removed from the furnace and cooled in air to room temperature. The air cooling rate through the range of 780 to 650 °C (1435 to 1200 °F) must be fast enough to produce a fully pearlitic matrix in the casting section being treated. This may require the use of a forced air blast, especially for thicker sections. In any case, castings will probably have to be cooled suspended individually, jigged, or vibrated over a grid, but not merely placed on the floor nor in batches in a basket or in other containers. This completes the cycle.

To achieve a substantially pearlitic structure, the iron matrix must be saturated in carbon at the austenitizing temperature before air cooling; this is most quickly achieved if the iron is substantially pearlitic as-cast. If the iron contains a ferritic matrix as-cast, a longer time at the high temperature or the same time at a higher temperature will be required to ensure adequate solution of carbon from the graphite nodules. As the cooling rate increases, the pearlite will become finer, the strength and hardness will increase, and elongation may decrease. As the austenitizing temperature is increased, the strength increases and elongation decreases because of a higher carbon content of the matrix. Elements that promote pearlite in the as-cast condition, notably, manganese, nickel, copper, and tin, will shorten the time required at the soaking temperature and will enable fully pearlitic structures to be obtained in thicker section sizes. Figure 20 shows a typical normalized microstructure.

Hardening and Tempering. Ductile iron of high strength, generally in excess of 700 MPa (100 ksi), and with low elongation is obtained by heating to 875 to 925 °C (1605 to 1695 °F), holding at this temperature for 2 to 4 h (or longer if required to break down carbides), quenching in an oil bath to produce a matrix structure of martensite, and then tempering at 400 to 600 °C (750 to 1110 °F) to produce a matrix structure of tempered martensite. Care must be taken to avoid cracking intricate castings during quenching, and this can be prevented by quenching into warm oil at, for example, 100 °C (212 °F), followed by final cooling to room temperature. Marquenching can also be used, involving quenching in hot oil at approximately 200 °C (390 °F), but this must be followed by cooling to room temperature, preferably in a cold water bath, to obtain the desired structure and properties.

For successful hardening during quenching, a fully martensitic structure must be obtained, and except in very thin sections, this will require alloying with elements that impart hardenability. Copper, nickel, manganese, and molybdenum increase hardenability in ascending efficiency. Copper can

Table 4 Examples of alloying combinations used to increase the hardenability of ductile iron

Alloying elements used, %					Maximum diameter of bar that could be hardened by oil quenching	
C	Si	Mn	Ni	Mo	mm	in.
3.4	2.0	0.3	. . .	. . .	25	1
3.4	2.5	0.3	. . .	. . .	28	1.1
3.4	2.0	0.3	1.0	. . .	30	1.2
3.4	2.0	1.3	. . .	. . .	38	1.5
3.4	2.0	0.3	. . .	0.5	51	2.0
3.4	2.0	0.9	1.5	0.25	63	2.5

only be used sparingly in ductile iron because of its limited solubility. Although silicon increases hardenability in steels, it has the opposite effect in ductile iron through decreasing carbon solubility, while increasing carbon content also decreases hardenability slightly through increasing the amount of graphite rather than increasing carbon in solution.

In practice, increased hardenability is achieved by combinations of alloying elements. The combinations listed in Table 4 are examples showing the effects of manganese, nickel, and molybdenum in increasing hardenability. Additional data on the hardenability of ductile irons are available in Ref 14.

Tempering should normally be carried out in an air circulation furnace for at least 4 h, during which there is a progressive decrease in strength and hardness and an increase in ductility. The effect of tempering temperature on the properties of hardened-and-tempered ductile iron is illustrated by the examples shown in Fig. 21 for 25 mm (1 in.) bars of an unalloyed iron. Figure 22 illustrates the microstructure of an iron that was hardened and tempered at 550 °C (1020 °F).

Austempering. If ductile iron is austenitized and quenched into a salt bath or a hot oil transformation bath at a temperature of 320 to 550 °C (610 to 1020 °F) and held at this temperature, transformation to a structure containing mainly bainite with a minor proportion of austenite takes place. Irons that are transformed in this manner are referred to as austempered ductile irons. Austempering generates a range of structures, depending on the time of transformation and the temperature of the transformation bath. The properties are characterized by very high strength, some ductility and toughness, and often an ability to work harden, giving appreciably higher wear resistance than that of other ductile irons. The properties depend principally on austempering temperature and time, and typical austempering treatment fall into two categories:

- Heat to 875 to 925 °C (1605 to 1695 °F), hold for 2 to 4 h, quench into a salt bath at 400 to 450 °C (750 to 840 °F), hold for 1 to 6 h, and cool to room temperature
- The same as above but hold for 1 to 6 h at 235 to 350 °C (455 to 660 °F)

The first treatment listed above would yield high ductility and high strength with medium hardness but a very good ability to work harden. The second treatment would yield very high strength with some ductility and a fairly high hardness. Austempering is successful only if the quench avoids the formation of pearlite. This may require the presence of alloying elements in sections greater than about 15 mm (0.6 in.). Typical alloys are nickel, copper, and molybdenum. Manganese is generally not recommended, because it creates segregation, which can lead to failure to achieve the optimum combination of properties. Maximum wear resistance and work hardening occur when there is a high amount of residual austenite. Residual austenite results from the use of short austempering times and incomplete transformation to bainite and is favored by higher alloying element contents and especially by the relatively high silicon content of ductile iron. Figure 23 shows the microstructure of an iron austempered at 375 °C (705 °F) containing about 15% austenite.

Useful information on austempered ductile irons is available in Ref 15 and 16. Figure 24 illustrates the influence of austempering temperature on the properties of an unalloyed ductile iron in section size of about 25 mm (1 in.).

The surface-hardening treatments that will be discussed in this section are flame and induction hardening, nitriding, and laser or plasma surface remelting.

Flame and induction hardening are usually employed to produce a hard surface layer on the surface of a casting. A flame or a specially shaped induction coil is passed over the surface of the casting at a rate that raises the temperature of the surface to 850 to 950 °C (1560 to 1740 °F) to a depth of about 2 to 4 mm (0.08 to 0.16 in.). The flame or induction source is followed by a water-quenching arrangement, and this produces a martensitic layer having a hardness of 600 to 700 HV.

The development of maximum hardness depends on the carbon content of the matrix, which transforms to austenite upon heating and to martensite during quenching. The time allowed does not normally permit

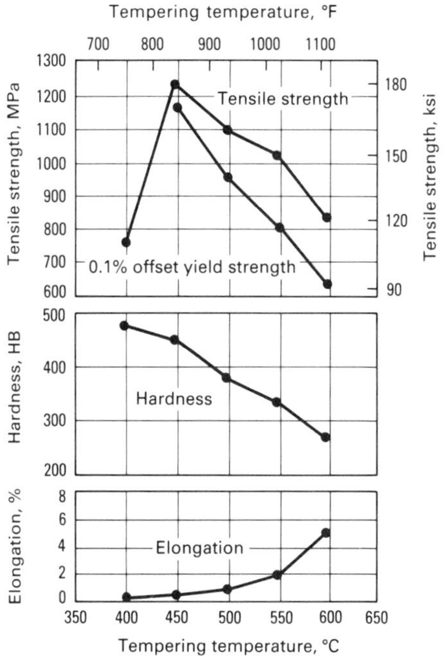

Fig. 21 Properties of hardened-and-tempered iron (Fe-3.2C-1.96Si-0.29Mn) quenched from 900 °C (1650 °F) and tempered 4 h at different temperatures

adequate solution of carbon in initially ferritic matrix structures; therefore, it is important to use fully pearlitic grades of iron for flame or induction hardening. The depth of hardening achieved can be increased by alloying, as indicated below:

Composition	Initial hardness, HRC	Surface hardness after treatment, HRC	Depth of hardened layer mm	in.
Fe-0.4Mn-0.07Ni-0.05Mo-0.1Cu	60	62	1.5	0.06
Fe-0.32Mn-0.75Ni-0.44Mo-0.56Cu	61	62.5	3.5	0.14

Flame and induction hardening are used to harden components requiring wear resistance, such as tappets, rolls, and gears, and may reduce the amount of wear by a factor of five or six.

Nitriding is a case-hardening process that involves the diffusion of nitrogen into the surface at a temperature of about 550 to 600 °C (1020 to 1110 °F). Most commonly, the source of nitrogen is ammonia, and the process produces a surface layer about 0.1 mm (0.004 in.) deep with a surface hardness approaching 1100 HV. The surface layer is usually white and featureless in an etched microstructure, but nitride needles can be found just below it. Alloying elements can be used to increase case hardness, and 0.5 to 1% of aluminum, nickel, and molybdenum have been reported to achieve useful results. Nitrided cases provide, in addition to very high hardness, increased wear resistance and antiscuffing properties, improved fatigue life, and improved corrosion resis-

Fig. 22 Microstructure (tempered martensite) of hardened-and-tempered iron. Compare with Fig. 20 and 23. Etched in picral. 500×

Fig. 23 Microstructure of austempered iron showing matrix of upper bainite and retained austenite. Compare with Fig. 20 and 22. Etched in picral. 500×

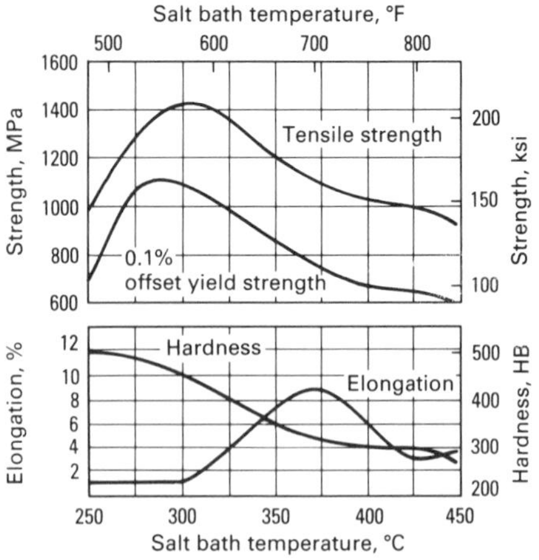

Fig. 24 Properties achieved by austempering an unalloyed iron for 7 h at different temperatures after austenitizing for 1 h at 900 °C (1650 °F)

tance. Typical applications are for cylinder liners, bearing pins, and small shafts.

Nitriding can also be carried out in liquid salt baths based on cyanide salts. Such processes have lower temperatures of treatment, although case depth may be decreased. More recently, processes for nitriding in a plasma have been developed and applied to ductile iron with success, but the process may be more restricted because of the special equipment and cost likely to be involved.

Remelt Hardening. With the very high local heating achievable with plasma torches or lasers, it is possible to produce a very small melted area on the surface of a ductile iron component. This area then rapidly resolidifies because of the self-quenching effect of the casting mass. The remelted and resolidified region has a structure of white iron that is substantially graphite free and therefore has high hardness and wear resistance. The area that is remelted by a 2 kW laser is very small, typically 1.5 mm (0.06 in.) in diameter, 0.5 to 2 mm (0.02 to 0.08 in.) deep, and having a hardness of about 900 HV without cracking. By traversing the casting surface, the area hardened by this method can be of useful size and is likely to find application in tappets, cams, and other small components subjected to sliding wear. Figure 25 shows the microstructure of a pearlitic iron traversed by a 1.5 kW laser at 456 mm/s (18.25 in./s).

Mechanical Properties

Tensile, Compressive, and Torsional Properties. Tensile properties are given in Table 2 and Fig. 10 and 11 in relation to standard specifications, which refer only to the minimum values of strength and ductility and mean hardness ranges to be expected. Higher strengths and greater elongations will generally be expected in commercial castings of good graphite structure. The ranges of values to be expected from these properties in commercial irons and the applicable multiplying factors are discussed in Ref 9. These factors can be applied to any grades in any national specifications.

Tables 5 and 6 summarize several measurements of the major mechanical and physical properties of irons conforming to British specifications, most grades of which are the same as those of the ISO specifications. These are typical values for unalloyed irons. Other typical tensile, compressive, and torsional data are reported in the article "Ductile Iron" in Volume 1 of the 9th Edition of *Metals Handbook*.

Compressive strength is higher than tensile strength, and in Table 5 the 0.2% proof strengths in compression reported are generally approximately 1.05 times the 0.2% proof strength in tension, although values have been reported that range up to 1.2 times greater. Similar ratios have been reported for 0.1% and 0.5% proof strengths.

Shear strengths have been reported at around 0.9 times the values of tensile strength. Few data are available, and it is difficult to make accurate measurements on ductile materials because of bending during testing.

Modulus of elasticity in both tension and compression is generally in the range of 162 to 176 GPa, depending on the graphite content and the perfection of the graphite structure. It is, in general, equal in tension and compression.

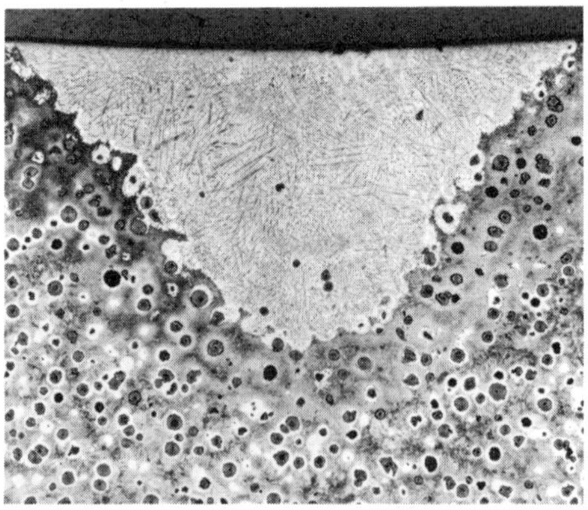

Fig. 25 Remelt-hardened and transition zones in a pearlitic iron after treatment with a 1.6 kW, 1.5 mm (0.06 in.) diam laser beam at 4.56 mm/s (0.18 in./s). Etched in picral. 50×

Poisson's ratio ν does not vary significantly from 0.275 for most ductile irons. Modulus of rigidity G is related to ν and to modulus of elasticity E by the formula $E = 2G(1 + \nu)$. A typical value of G is about $0.39E$, and values of 62 to 68.6 GPa (23.5×10^3 to 25.5×10^3 ksi) have been reported.

All ductile irons deform considerably under torsional shear stresses, and it is difficult to obtain satisfactory test data to failure. It is estimated that torsional strength is about 0.9 times the tensile strength and that values measured for limit of proportionality and proof strength in torsion are generally between 0.7 and 0.775 times the corresponding values in tension.

A damping capacity of ductile iron greater than 90% nodularity exceeds that of steels by a factor of six or seven. It is, however, much less than that of flake graphite irons by a factor of about ten. Values have been reported from about 6.1 to 8.3×10^{-4} (measured by resonant frequency as a decay of vibration) for irons

Table 5 Some mechanical properties expected in ductile iron grades covered by UK standard B52789

Grade	Tensile strength MPa	Tensile strength ksi	Yield strength MPa	Yield strength ksi	Elongation, %	Yield strength in compression MPa	Yield strength in compression ksi	Shear strength MPa	Shear strength ksi	Torsional strength MPa	Torsional strength ksi	Modulus of elasticity (E) GPa	Modulus of elasticity (E) 10^6 psi	Modulus of rigidity (G) GPa	Modulus of rigidity (G) 10^6 psi	Poisson's ratio, ν	Hardness, HB	Fatigue limit(a) Notched MPa	Fatigue limit(a) Notched ksi	Fatigue limit(a) Unnotched MPa	Fatigue limit(a) Unnotched ksi
Ferritic grades																					
350/22; 350/22L40	350	51	215	31	22	229	33	315	46	315	46	169	24.5	65.9	9.6	0.275	107–130	114	17	180	26
400/18; 400/18L20	400	58	259	38	18	273	40	360	52	360	52	169	24.5	65.9	9.6	0.275	120–140	122	18	195	28
420/12	420	61	278	40	12	292	42	378	55	378	55	169	24.5	65.9	9.6	0.275	140–155	124	18	201	29
Intermediate grades																					
450/10	450	65	305	44	10	319	46	405	59	405	59	169	24.5	65.9	9.6	0.275	150–172	128	19	210	30
500/7	500	73	339	49	7	351	57	450	65	450	65	169	24.5	65.9	9.6	0.275	172–216	134	20	224	32
600/3	600	87	372	54	3	382	55	540	78	540	78	174	25.2	67.9	9.8	0.275	216–247	149	22	248	36
Pearlitic as-cast and normalized																					
700/2	700	102	416	60	2	425	62	630	91	630	91	176	25.5	68.6	9.9	0.275	247–265	168	41	280	41
800/2	800	116	471	68	2	480	70	720	104	720	104	176	25.5	68.6	9.9	0.275	>265	182	44	304	44
900/2	900	131	526	76	2	535	78	810	117	810	117	176	25.5	68.6	9.9	0.275	>265	190	46	317	46
Hardened-and-tempered grades																					
700/2	700	102	550	80	2	559	81	630	91	630	91	172	24.9	67.1	9.7	0.275	232–259	168	41	280	41
800/2	800	116	630	91	2	639	93	720	104	720	104	172	24.9	67.1	9.7	0.275	>259	182	44	304	44
900/2	900	131	710	103	2	719	104	810	117	810	117	172	24.9	67.1	9.7	0.275	>259	190	46	317	46

(a) Wöhler specimen 10.6 mm (0.42 in.) in diameter unnotched; 10.6 mm (0.042 in.) in diameter at root of notch in notched tests. Circumferential 45° V-notch with 25 mm (1 in.) root radius and notch depth of 3.6 mm (0.14 in.). Source: Ref 9

with more than 90% nodularity. The value is very sensitive to the graphite structure and falls markedly if a small proportion of flake graphite is present.

Impact Properties. The energy needed to cause fracture in an impact test is generally measured in a Charpy test using a V-notch specimen 10 mm (0.4 in.) square in cross section. Earlier work undertaken to compare variables was carried out on unnotched specimens that had impact energies in the ductile condition some six times those of unnotched specimens.

The article "Ductile Iron" in Volume 1 of the 9th Edition of *Metals Handbook* reports the effects of a number of variables on impact energy, and much of the data are summarized in Fig. 18, which shows how changing from a fully pearlitic to a fully ferritic matrix structure lowers the ductile-to-brittle transition temperature and raises the maximum impact energy in the ductile condition. Decreasing the silicon content also lowers transition temperature and raises maximum impact energy in the fully ferritic annealed condition by an important amount, and for optimum impact properties, silicon contents below 2.5% and preferably below 2%, are desirable. Phosphorus also has a powerful embrittling effect in the impact test, and for optimum properties, a value well below 0.1%, and preferably below 0.05%, is desirable. In the ferritic annealed condition, optimum impact properties are obtained by a two-stage anneal involving austenitization and then subcritical graphitization. By combining all of these variables and maintaining all other alloying and trace elements at low levels, it is possible to obtain impact transition temperature below −25 °C (−13 °F), which is superior to that of most steels and is critical in determining if a component will fail in a ductile manner at low temperature. The maximum impact energy in the ductile range in ferritic irons is generally 12 to 20 J (8.9 to 14.8 ft · lbf).

The effects of various heat treatments have been compared with regard to the as-cast condition on a number of ductile irons (Ref 17). More recent data on austempered ductile iron confirm that a very good combination of low transition temperature and high impact energy can be obtained in

Table 6 Some physical properties expected of ductile iron grades covered by UK specification BS2789

Grade	Thermal conductivity, 100 °C (212 °F)	W/m · K (Btu/ft · °F) at 500 °C (930 °F)	Specific heat capacity, 20–700 °C (68–1290 °F) J/kg · K	Specific heat capacity, 20–700 °C (68–1290 °F) Btu/lb · °F	Coefficient of thermal expansion at 20–400 °C (68–750 °F), 10^{-6}/K	Electrical resistivity, μΩ/m
Ferritic grades						
350/22; 350/22L40	36.5 (21.1)	35.8 (20.7)	603	0.144	12.5	0.500
400/18; 400/18L20	36.5 (21.1)	35.8 (20.7)	603	0.144	12.5	0.500
420/12	36.5 (21.1)	35.8 (20.7)	603	0.144	12.5	0.500
Intermediate grades						
450/10	36.5 (21.1)	35.8 (20.7)	603	0.144	12.5	0.500
500/7	36.5 (21.1)	34.9 (20.2)	603	0.144	12.5	0.510
600/3	32.8 (18.9)	32.2 (18.6)	603	0.144	12.5	0.530
Pearlitic as-cast and normalized grades						
700/2	31.4 (18.2)	30.8 (17.8)	603	0.144	12.5	0.540
800/2	31.4 (18.2)	30.8 (17.8)	603	0.144	12.5	0.540
900/2	31.4 (18.2)	30.8 (17.8)	603	0.144	12.5	0.540
Hardened-and-tempered grades						
700/2	33.5 (19.4)	32.9 (19.0)	603	0.144	12.5	>0.540
800/2	33.5 (19.4)	32.9 (19.0)	603	0.144	12.5	>0.540
900/2	33.5 (19.4)	32.9 (19.0)	603	0.144	12.5	>0.540

Source: Ref 9

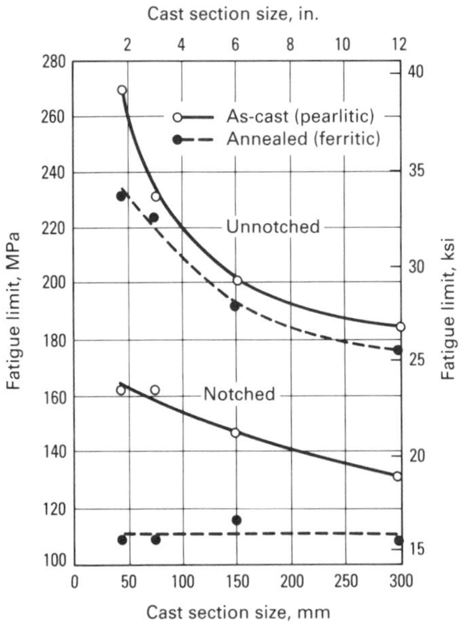

Fig. 26 Effect of cast section size on the fatigue properties of ductile irons

Table 7 Tensile design stresses for three nodular irons at elevated temperatures

Grade	Structure	Temperature °C	Temperature °F	Maximum tensile design stress MPa	Maximum tensile design stress ksi	Basis for maximum design stress
420/12	Ferrite	20	68	136	20	0.52 × (0.1% offset yield strength)
		100	212	130	19	
		200	390	124	18	
		300	570	124	18	
600/3	Pearlite	20	68	148	21	0.45 × (0.1% offset yield strength)
		100	212	131	19	
		200	390	130	19	
		300	570	130	19	
700/2	Pearlite	20	68	173	25	0.45 × (0.1% offset yield strength)
		100	212	154	22	
		200	390	150	22	
		300	570	147	21	

Source: Ref 9

irons containing retained austenite (Ref 14, 15).

Fracture Toughness. Many figures have been reported for the stress intensity factor K_{Ic} obtained under plane-strain linear-elastic conditions. However, because of the ductility of ductile irons, most investigations of ferritic and pearlitic/ferritic irons have been conducted at subzero temperatures to obtain the necessary criteria of elastic behavior. Values ranging from 25 to 54 MPa$\sqrt{m}$ (23 to 49 ksi$\sqrt{in.}$) have been reported with the higher values for ferritic irons. In general, the value of K_{Ic} has been observed to increase with testing temperature. Values of crack-opening displacement obtained from cast irons have also shown wide-ranging values increasing from 0.015 to about 0.2 mm (6×10^{-4} to 0.008 in.) upon passing from pearlitic to ferritic structure. More information on the fracture toughness testing of ductile iron is available in Ref 18.

Fatigue. In the unnotched condition, the fatigue limit for a ferritic iron is about 0.5 times the tensile strength obtained in a Wöhler reversed-bending test. The fatigue limit decreases with increasing hardness to about 0.4 times the tensile strength in hardened irons up to 740 MPa (107 ksi) and may be even lower in stronger irons. Ductile irons are notch sensitive, and in a Wöhler specimen 10.6 mm (0.42 in.) in diameter with a 45° notch of root radius measuring 0.25 mm (0.01 in.), the fatigue limit decreases to about 0.63 times the unnotched value for ferritic irons and 0.6 times the unnotched value for hardened-and-tempered irons.

The fatigue strength decreases as the cast section size is increased, as shown in Fig. 26 for both ferritic and pearlitic irons notched and unnotched. As the size of the component subjected to reversed-bending fatigue increases, there is a decrease in fatigue strength up to about 50 mm (2 in.) in diameter. This effect is not thought to occur in axial fatigue in tension/compression.

Fatigue test results generally refer to machined and polished surfaces. Surfaces of poorer quality and as-cast surfaces show lower values of fatigue strength, and the presence of minor surface defects has an even more detrimental influence on fatigue strength. Some improvement in the fatigue life of the cast surface is obtained by shot-peening. Because of the notch sensitivity of ductile iron, fatigue strength is reduced if nodularity decreases.

Corrosion Fatigue (Ref 19). The fatigue strengths of pearlitic and ferritic ductile irons are reduced in a water environment by some 15 to 19% as a result of corrosion. In a saltwater environment, this reduction may increase to 68 to 83%. Corrosion inhibitors generally do not prevent corrosion fatigue; but alkaline sodium nitrite solutions can reduce it, and a solution of 0.05% of sodium chromate protects against corrosion fatigue in a water spray environment.

Mechanical Properties at Elevated Temperatures. The short-term tensile strength of unalloyed pearlitic ductile irons decreases more or less continuously with increasing temperature and at 400 °C (750 °F) is about two-thirds the room-temperature strength. For ferritic irons, the decrease is less pronounced and at 400 °C (750 °F) the strength is about three-fourths the room-temperature value. Proof stress, however, for both ferritic and pearlitic irons is more or less maintained up to 350 to 400 °C (660 to 750 °F), above which it falls rapidly. The hot hardness of ductile iron is also maintained up to about 400 °C (750 °F), above which it falls rapidly.

For temperatures to 300 °C (570 °F), static design stress can, as at room temperature, be based on proof stress values obtained at room temperature. Design stresses are given in Table 7. At temperatures above 350 °C (660 °F), design stresses should be based on creep data.

The growth and oxidation for scaling of ductile iron at all temperatures are much less than for gray irons, and data have been obtained at 350 and 400 °C (660 and 750 °F) for times well in excess of 100 000 h. These data are given in Table 8.

Creep and stress rupture data have been obtained to support the use of pearlitic and ferritic ductile irons to 350, 375, and 400 °C (660, 705, and 750 °F). Table 9 lists the stresses needed to produce 0.1, 0.2, or 1% strain or rupture in 1000, 10 000, 30 000, and 100 000 (extrapolation) at 350 and 400 °C (660 and 750 °F).

A small addition of molybdenum considerably improves the short-term hot strength and creep properties of both ferritic and pearlitic irons. The improvement brought about by molybdenum enables useful creep and rupture properties to be extended to temperatures of 450 °C (840 °F).

Low-Temperature Tensile Properties (Ref 20). As for impact properties, there is also a transition temperature range below which the tensile elongation decreases. Proof stress increases continuously with decreasing temperature, but tensile strength also exhibits a transition. Above the transition temperature range, tensile strength tends to remain constant or increases with decreasing temperature, but once the transition temperature is passed, tensile strength decreases with further lowering of temperature. In ferritic irons in the ductile range, because there is some reduction in area, the true fracture stress is higher than that normally reported, and if this were to be taken into consideration, the transition in tensile strength and in elongation would be seen to occur over the same temperature range. As with impact properties, silicon and phosphorus increase the tensile transition temperature and reduce the maximum tensile strength in the ductile range.

Table 8 Growth and scaling of ductile irons in air

Grade	Exposure temperature °C	Exposure temperature °F	Exposure time, years	Growth mm/mm	Growth in./in.	Scaling weight gain, g/m²
700/2 (pearlite)	350	660	0	0.000	0.000	0.00
			4.9	0.005	0.005	8.77
			10.4	−0.003	−0.003	12.61
			21.3	−0.003	−0.003	13.71
	400	750	0	0.000	0.000	0.00
			4.9	0.020	0.020	21.93
			10.4	0.018	0.018	29.60
			21.3	0.015	0.015	33.44
500/7 (pearlite + ferrite)	350	660	0	0.000	0.000	0.00
			4.9	0.003	0.003	5.48
			10.4	0.000	0.000	9.87
			21.3	−0.005	−0.005	9.87
	400	750	0	0.000	0.000	0.00
			4.9	0.005	0.005	16.45
			10.4	0.075	0.075	26.31
			21.3	0.047	0.047	32.34
400/12 (ferrite)	350	660	0	0.000	0.000	0.00
			4.9	0.005	0.005	6.03
			10.4	0.000	0.000	10.42
			21.3	−0.017	−0.017	9.87
	400	750	0	0.000	0.000	0.00
			4.9	0.008	0.008	17.54
			10.4	0.007	0.007	24.12
			21.3	0.000	0.000	27.96

Physical Properties

Data on the physical properties of ductile iron are less well established than mechanical properties and fewer figures are available than for mechanical properties. The following sections summarize the best available values.

Density decreases slightly with increasing graphite content and decreases slightly with increasing ferrite content. Typical values for ISO and British grades at 20 °C (68 °F) are:

Grade BS 2789:1973	370/17	420/12	500/7	600/3	700/2
Density, g/cm³	7.10	7.10	7.10–7.17	7.17–7.20	7.20

Coefficient of Thermal Expansion. Typical values are given in Table 6. The expansion characteristics of cast irons are complex because of the transformations that take place involving the solution and precipitation of graphite, the graphitization of pearlite, and austenite formation above 700 °C (1290 °F). Upon repeated heating and cooling above about 700 °C (1290 °F), irreversible expansion occurs because of growth, and in air, additional expansion may occur because of oxidation. Any carbides present in the as-cast structure may be graphitized, giving rise to further nonreversible expansion.

Other Physical Properties. Typical values of electrical, magnetic, and thermal properties for ISO or BS grades of ductile iron are given in Table 6.

Corrosion Resistance. In many applications, the corrosion resistance of ductile iron is similar to that of gray iron and is often superior to that of steels. Ductile iron pipes normally perform well in soils and can be further protected by the use of sacrificial anodes, zinc coating, plastic sleeving, and in some cases polyurethane coating. A useful summary of published data on corrosion rates in various environments is given in Ref 21, and information on the corrosion of cast irons is also available in the article "Corrosion of Cast Irons" in Volume 13 of the 9th Edition of *Metals Handbook*.

Nondestructive Evaluation of Ductile Iron Castings

Routine nondestructive tests can be applied to ductile iron castings to confirm their soundness and integrity, to ensure freedom from physical defects, and to confirm graphite structure and properties dependent on graphite structure.

Soundness and Integrity. Cracks and fine tears that break the surface of the casting but are difficult to detect visually can be revealed by the use of dye penetrants or by magnetic-particle inspection. Modern techniques of magnetizing the casting, followed by application of fluorescent magnetic inks, are very effective and widely used.

Methods of sonic testing that involve vibrating the casting and noting electronically the rate of decay of resonant frequency or damping behavior are also used to detect cracked or flawed castings. Internal unsoundness, when not immediately subsurface, can be detected by ultrasonic inspection. This can be detected by failure to observe a back-wall echo when using reflected radiation or by a weakening of the signal in the transmission through the casting. Coupling of the probes and interpretation of the results involve operator skill, but methods are available that involve partial or total immersion of the casting in a liquid, automatic or semiautomatic handling of the probes, and computer signal processing to ensure more reliable and consistent interpretation of results. Problems arise in detecting very-near-surface defects and when examining thin castings, but the use of angled probes and shear wave techniques has yielded good results.

The soundness of the ductile iron can also be assessed by x-ray or γ-ray examination. The presence of graphite, especially in heavy sections, makes the method more difficult to evaluate than for steels, but the use of image intensification by electronic means offers considerable promise, especially for sections up to 50 mm (2 in.) thick.

Confirmation of Graphite Structure (Ref 11). Both the velocity of ultrasonic transmission and the resonant frequency of a casting can be related to the modulus of

Table 9 Stress required to produce specific creep strain or rupture in ductile iron

Type of ductile iron	Creep strain or rupture, %	Stress required at indicated time and temperature, MPa (ksi): 350 °C (660 °F) 1000 h	350 °C (660 °F) 10 000 h	350 °C (660 °F) 30 000 h	350 °C (660 °F) 100 000 h(a)	400 °C (750 °F) 1000 h	400 °C (750 °F) 10 000 h	400 °C (750 °F) 30 000 h	400 °C (750 °F) 100 000 h(a)
Pearlitic grade 700/2	0.1	239 (34.5)	178 (26)	145 (21)	124 (18)	120 (17.5)	70 (10)	50 (7.5)	28 (4)
	0.2	276 (40)	219 (32)	199 (29)	151 (22)	147 (21.5)	93 (13.5)	77 (11)	40 (6)
	0.5	312 (45)	270 (39)	246 (35.5)	222 (32)	199 (29)	140 (20.5)	114 (16.5)	80 (11.5)
	1.0	355 (51.5)	297 (43)	278 (40.5)	256 (37)	239 (34.5)	184 (26.5)	150 (22)	128 (18.5)
	Rupture	430 (62.5)	370 (53.5)	352 (51)	317 (46)	309 (45)	255 (37)	195 (28.5)	160 (23)
Ferritic grade 400/12	0.1	185 (27)	159 (23)	142 (20.5)	120 (17.5)	96 (14)	60 (8.5)	43 (6.5)	26 (4)
	0.2	204 (29.5)	171 (25)	158 (23)	137 (20)	111 (16)	75 (11)	59 (8.5)	35 (5)
	0.5	222 (32)	195 (28.5)	176 (25.5)	167 (24)	130 (19)	94 (13.5)	77 (11)	59 (8.5)
	1.0	241 (39.5)	210 (30.5)	192 (28)	175 (25.5)	142 (20.5)	106 (15.5)	88 (13)	71 (10.5)
	Rupture	298 (43)	264 (35.5)	246 (35.5)	225 (32.5)	195 (28)	154 (22.5)	136 (20)	114 (16.5)

(a) Extrapolated from stress/log time curve

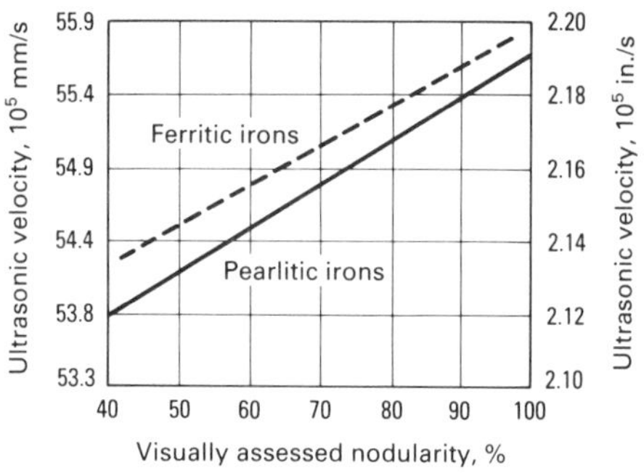

Fig. 27 Ultrasonic velocity versus visually assessed nodularity in ductile iron castings. Source: Ref 10

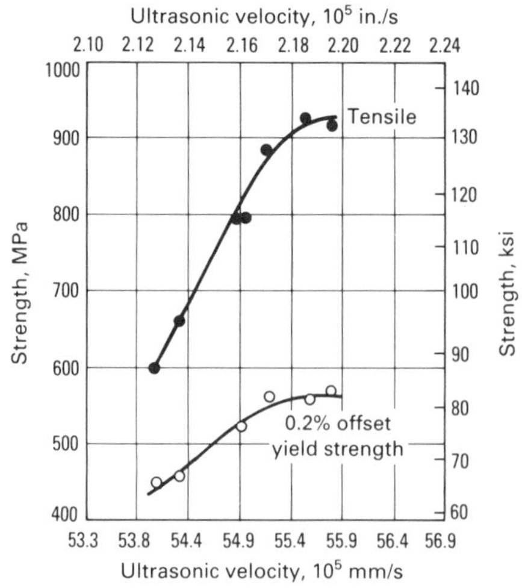

Fig. 28 Ultrasonic velocity versus strength in ductile iron castings. Source: Ref 10

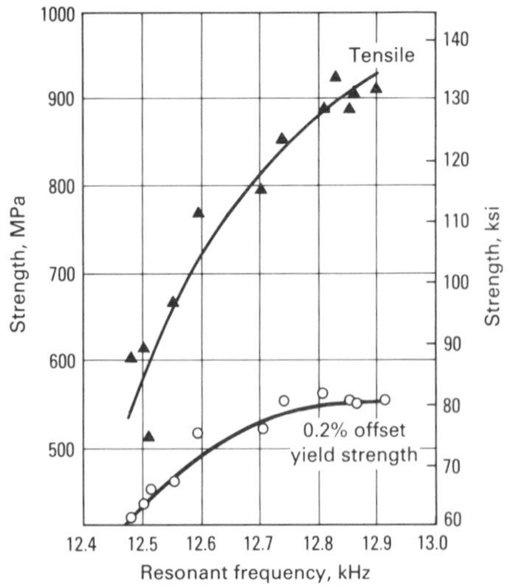

Fig. 29 Strength versus resonant frequency in nondestructive evaluation of ductile iron test bars. Source: Ref 11

elasticity. In cast iron, the change from flake graphite to nodular graphite is related to an increase in both modulus of elasticity and strength; therefore, ultrasonic velocity or resonant frequency measurement can be employed as a guide to nodularity, strength, and other related properties. Because microscopic estimation of nodularity is a subjective measurement, these other nondestructive methods of examination may provide a better guide to some properties provided the matrix remains constant (Ref 10). Figure 27 illustrates how ultrasonic velocity may vary with graphite nodularity.

Ultrasonic transmission measurement can be done with two probes on either side of the casting and provides a guide to the local properties. It must be coupled with a thickness measurement, and automatic equipment is commonly used, often involving immersion of the casting in a tank of fluid. The calculation does not require calibration of the castings. Simple caliper devices have also been used for examining castings and simultaneously measuring their thickness to provide a calculated value of ultrasonic velocity and a guarantee of nodularity (Ref 22).

Sonic testing involves measurement of the resonant frequency of the casting or rate of decay of resonance of a casting that has first been excited by a mechanical or electrical means. This method evaluates the graphite structure of the entire casting and requires calibration of the castings against standard castings of known structure. It is also necessary to maintain casting dimensions within a well-controlled narrow range. Some foundries use sonic testing as a routine method of final inspection and structure guarantee (Ref 23).

The relationship between nodularity, resonant frequency, or ultrasonic transmission velocity and properties has been documented for tensile strength, proof strength, fatigue, and impact strength. Examples are shown in Fig. 28 and 29.

The presence of carbides can also be detected with sonic or ultrasonic measurements provided enough carbides are present to reduce the graphite sufficiently to affect modulus of elasticity. Discrimination between the effects of graphite variation and carbide amount would require an additional test, such as hardness, to be carried out.

Properties Partially Dependent on Graphite Structure (Ref 23). When the matrix structure of ductile iron varies, this variation cannot be detected as easily as variations in graphite structure, and sonic and ultrasonic readings may not be able to reflect variations in mechanical properties. A second measurement, such as a hardness measurement, is then needed to detect matrix variations in the same way as would be necessary to confirm the presence of carbides.

Eddy current or coercive force measurements can be used to detect many changes in casting structure and properties; but the indications from such measurements are difficult to interpret, and the test is difficult to apply to many castings unless they are quite small and may be passed through a coil 100 to 200 mm (4 to 8 in.) in diameter. Eddy current indications are, however, useful for evaluating pearlite and carbide in the iron matrix. Multifrequency eddy current testing uses probes that do not require the casting to pass through a coil, it is less sensitive to casting size, and it allows automatic measurements and calculations to be made; but the results remain difficult to interpret with reliability in all cases. It may, however, be a very good way to detect chill and hard edges on castings of reproducible dimensions.

Trends in Nondestructive Examination (Ref 22). There is a growing trend toward combining nondestructive testing with quality assurance and statistical process control (SPC) procedures. For this reason, progress is rapid in the automatic and semiautomatic applications of tests to reduce operator fatigue and the computer interpretation of results to eliminate human error. Results are increasingly being read directly into computers provided with programs to report and record not only the results but also their statistical interpretation and the information needed to correct any departures from specification. Such developments are of particular importance when ductile iron castings are treated individually with magnesium in the mold because, even with very well-controlled reproducibility of treatment, a higher proportion of castings may require examination to provide confidence in the uniformity of the product.

Joining of Ductile Iron Castings

Welding. Ductile iron castings can be joined to each other and to steel components by welding, and many examples have been published. Successful welding depends on using a technique such as dip-transfer MIG welding with electrodes usually containing a high proportion (40 to 60%) of nickel together with iron and sometimes manganese. By restricting the heat input, the heat-affected zone, which contains carbides and martensite, is kept to a minimum, and the strength and proof stress of the joint will approach that of the parent iron (Table 10). However, it is unlikely that the ductility of the joint will match that of the ductile iron, and good practice will involve joint designs in which bending stresses and stress concentrations are minimized. Automated welding processes are likely to achieve more consistent quality of joints than hand welding. In some cases, it may be feasible to anneal the assembly after welding, and pre- and post-heating are recommended for obtaining improved joint properties (Ref 25).

Brazing. Strong joints between ductile iron components can be obtained by brazing with copper-, silver-, and nickel-base brazing alloys. By using capillary joints and designs to promote shear stressing rather

Table 10 Mechanical properties of weld joints in ASTM A 536, grade 60/45/10, ductile iron

Welds were made using high-nickel (55% Ni) flux-cored wire.

Specimen	Type of shielding	Tensile strength		0.2% offset yield strength		Elongation, %	Reduction in area, %	Hardness, HRB
		MPa	ksi	MPa	ksi			
All weld metal	None	476	69	310	45	15.5	14.5	81
All weld metal	CO_2	496	72	314	45.5	21.0	18.8	80
All weld metal	Sub-arc flux	510	74	338	49	18.5	20.6	86
Transverse	None	455	70	300	43.5	...	...	...
Transverse	CO_2	455	70	303	44	...	...	...
Transverse	Sub-arc flux	441	64	310	45	...	...	...
All weld metal(a)	CO_2	468	68	303	44	15.0	16.2	80
Transverse(a)	CO_2	467	68	300	43.5	...	...	...

(a) Pulsing arc power source. Source: Ref 24

than tension or bending, it is possible to obtain joint properties of strengths similar to those of the parent iron at costs comparable with those of welding, but skilled operation is desirable (Ref 24).

Adhesive Bonding. Ductile iron castings can be adhesively bonded if the joints are carefully designed to employ shear loading and freedom from bubbles in the adhesive.

Machinability. Unalloyed ductile iron is a readily machinable material, and the cost of machining components will often be less than that for cast low-carbon steels. Ductile iron is somewhat less machinable than gray iron, although at a similar hardness relatively little difference is sometimes reported. Many data have been published for machining under a wide range of conditions (see the article "Ductile Iron" in Volume 1 of the 9th Edition of *Metals Handbook* and Ref 26-28). In general, machinability improves upon moving from fully pearlitic irons of higher hardness to fully ferritic irons of lower hardness.

Apart from variations in machinability, tool life, machining time, and cutting speeds with cutting and tool parameters, the surface condition and depth of initial cut can have a major influence on machinability. The as-cast surface of ductile iron may be hard and abrasive because of a thin graphite-free layer and because of oxide particles. It is important that the initial cut be deep enough to remove this layer completely to avoid rapid blunting of the tool, and an initial depth of at least 2 mm (0.08 in.) is recommended when turning.

Applications of Ductile Iron Castings

Ductile iron is finding increasing application for a very wide range of components in which it can replace gray iron because of its superior properties. Examples of automotive applications are crankshafts, exhaust manifolds, piston rings, and cylinder liners. The use of ductile iron in these applications provides increased strength and permits weight savings. Gray iron spun pipes have been largely superseded by ductile iron pipes having a high degree of ductility and thinner walls, and many fittings are also made of this material. In agricultural and earth-moving applications, brackets, couplings, rollers, hydraulic valves, sprocket wheels, and track components of improved strength and toughness are made of ductile iron. General engineering applications include hydraulic cylinders, mandrels, machine frames, switch gear, rolling mill rolls, tunnel segments, low-cost rolls, bar stock, rubber molds, street furniture such as covers and frames, and railway railclip supports. For these applications, ductile iron has provided increased performance or weight savings.

Ductile iron can be used to replace many more expensive components previously made in wrought or cast steels or other metals because of its higher strength-to-weight ratio, lower damping capacity, better machinability, and better castability. Examples include brake calipers and cylinders, steering gear and other safety-critical components, turbochargers, connecting rods, gear boxes and gears, valve bodies, pump components, bulldozer parts, nuclear fuel containers and transporters, bridge rollers and railing supports, coal and mineral crushing components, crane wheels, oil well equipment, mining roof supports, overhead switchgear, shafts and spindles, railway axle boxes and fittings for rolling stock, and low-pressure turbine casings.

Ductile iron gears have performed well in noncritical engineering and agricultural applications, but austempered ductile iron offers a combination of strength, fatigue properties, and wear resistance that makes it of great interest for heavy engineering and automotive gears—applications in which the use of austempered ductile iron is likely to increase. A number of well-established applications of ductile iron are listed in Ref 21. However, many new engineering components are likely to be amenable to design with ductile iron. Illustrations of these are described, and their advantages illustrated, in Ref 29.

REFERENCES

1. H.E. Henderson, Compliance With Specifications for Ductile Iron Castings Assures Quality, *Met. Prog.*, Vol 89, May 1966, p 82-86; Ductile Iron—Our Most Versatile Ferrous Material, in *Gray, Ductile and Malleable Iron Castings—Current Capabilities*, STP 455, American Society for Testing and Materials, 1969, p 29-53
2. H.J. Heine, "An Overview of Magnesium Treatment Processes Which Have Stood the Test of Time in America," Paper 9, presented at the BCIRA conference, SG Iron—The Next 40 Years, University of Warwick, BCIRA, April 1987
3. *Ductile Iron Molten Metal Processing*, 2nd ed., American Foundrymen's Society, 1986
4. *Modern Inoculating Practices for Gray and Ductile Iron*, Conference Proceedings, Rosemont, IL, Feb 1979, American Foundrymen's Society/Cast Metals Institute, 1979
5. J.V. Anderson and S.I. Karsay, Pouring Rate, Pouring Time and Choke Design for SG Iron Castings, *Br. Foundryman*, Vol 78 (No. 10), Dec 1985, p 492-498
6. P.J. Rickards, Factors Affecting the Soundness and Dimensions of Iron Castings Made in Cold-Curing Chemically Bonded Sand Moulds, *Br. Foundryman*, Vol 75 (No. 11), Nov 1982, p 213-223
7. H. Roedter, An Alternative Method of Pressure-Control Feeding for Ductile Iron Castings, *Foundry Trade J. Int.*, Vol 9 (No. 31), Sept 1986, p 174-179
8. R. Hummer, Feeding Requirements and Dilation During Solidification of Spheroidal Graphite Cast Iron—Conclusions for Feeder Dimensioning, *Giesserei-Prax.*, No. 17/18, 16 Sept 1985, p 241-254
9. G.N.J. Gilbert, *Engineering Data on Nodular Cast Irons-SI Units*, BCIRA, 1986
10. A.G. Fuller, Evaluation of the Graphite Form in Pearlitic Ductile Iron by Ultrasonic and Sonic Testing and the Effect of Graphite Form on Mechanical Properties, *Trans. AFS*, Vol 85, 1977, p 509-526
11. A.G. Fuller, P.J. Emerson, and G.F. Sergeant, A Report on the Effect Upon Mechanical Properties of Variation in

Graphite Form in Irons Having Varying Amounts of Ferrite and Pearlite in the Matrix Structure and the Use of Nondestructive Tests in the Assessments of Mechanical Properties of Such Irons, *Trans. AFS*, Vol 88, 1980, p 21-50
12. W.S. Pellini, G. Sandoz, and H.F. Bishop, Notch Ductility of Nodular Irons, *Trans. ASM*, Vol 46, 1954, p 418-445
13. *The Heat Treatment of SG Cast Iron*, British S.G. Iron Producers' Association, 1969
14. Heat Treating of Ductile Irons, in *Heat Treating*, Vol 4, 9th ed., *Metals Handbook*, American Society for Metals, 1981, p 545-551
15. *Austempered Ductile Iron: Your Means to Improved Performance, Productivity and Cost*, Proceedings of the 1st International Conference, Rosemont, IL, April 1984, American Society for Metals, 1984
16. *Austempered Ductile Iron: Your Means to Improved Performance, Productivity and Cost*, Proceedings of the 2nd International Conference, Ann Arbor, MI, March 1986, Gear Research Institute, American Society of Mechanical Engineers, 1986
17. C. Vishnevsky and J.F. Wallace, The Effect of Heat Treatment on the Impact Properties of Ductile Iron, *Gray Iron News*, July 1962, p 4-10
18. S. Wolfensberger, P. Uggowitzer, and M.O. Speidel, The Fracture Toughness of Cast Iron. Part II: Cast Iron With Spheroidal Graphite, *Giessereiforschung*, Vol 39 (No. 2), 1987, p 71-80 (in German)
19. K.B. Palmer, Effect of Cast Section Size on Fatigue Properties and the Prevention of Corrosion Fatigue in Nodular Irons, *Proceedings of the 79th Annual Conference of the Institute of British Foundrymen (Brighton)*, Institute of British Foundrymen, June 1982, p 9-20
20. P.J. Rickards, Low Temperature Properties of Cast Irons, in *Engineering Properties and Performance of Modern Iron Castings*, BCIRA, 1970, p 251-282
21. S.I. Karsay, *Ductile Iron II. Engineering Design, Properties, Applications*, Sorel, Canada, Quebec Iron and Titanium Corporation, 1971
22. P.J. Rickards, "Progress in Guaranteeing Quality Through Non-Destructive Methods of Evaluation," Paper 21, presented at the 54th International Foundry Congress, New Delhi, The International Committee of Foundry Technical Associations (CIATF), Nov 1987
23. A.G. Fuller, Nondestructive Assessment of the Properties of Ductile Iron Castings, *Trans. AFS*, Vol 88, 1980, p 751-768
24. R.A. Harding, Progress in Joining Iron Castings, *Foundryman*, Vol 80, Nov 1987
25. R.A. Bishel, Flux-Cored Electrode for Cast Iron Welding, *Weld. J.*, Vol 52, June 1973, p 372-381
26. *Machining Data Handbook*, Vol 1 and 2, 3rd ed., Machinability Data Center, 1980
27. H.P. Staudinger, Machining Nodular Cast Iron Using a Lathe, *VDI Z.*, Vol 126 (No. 4), Feb 1984, p 45-50 (in German)
28. H.P. Staudinger, Machining Nodular Cast Iron Using Drilling, *VDI Z.*, Vol 126 (No. 11), June 1984, p 398-402 (in German)
29. *Konstruieren und Giessen*, Zentrale für Gussverwendung, published quarterly

Compacted Graphite Irons

D.M. Stefanescu, University of Alabama
R. Hummer and E. Nechtelberger, Austrian Foundry Research Institute

COMPACTED (VERMICULAR) GRAPHITE (CG) IRONS have inadvertently been produced in the past as a result of insufficient magnesium or cerium levels in melts intended to produce spheroidal graphite iron; however, it has only been since 1965 that CG iron has occupied its place in the cast iron family as a material with distinct properties requiring distinct manufacturing technologies. The first patent was obtained by R.D. Schelleng (Ref 1). Since that time, an impressive number of publications have been written on this subject, as summarized by the review work in Ref 2 and 3.

As discussed in the section "Cast Iron" of the article "Solidification of Eutectic Alloys" in this Volume, the shape of compacted graphite is rather complex. An acceptable CG iron is one in which there is no flake graphite (FG) in the structure and for which the amount of spheroidal graphite (SG) is less than 20%; that is, 80% of all graphite is compacted (vermicular) (ASTM A 247, type IV). Typical CG iron microstructures are shown in Fig. 1. It can be seen that although the two-dimensional appearance of compacted graphite is that of flakes with a length:thickness ratio of 2:10 (Fig. 1a), the three-dimensional SEM structures (Fig. 1b and c) show that graphite does not appear in flakes but rather in clusters interconnected within the eutectic cell.

Production Techniques

Chemical Composition. The characteristic properties of CG irons have been demonstrated over a rather wide range of carbon equivalent (CE) values, extending from hypoeutectic (CE = 3.7) to hypereutectic (CE = 4.7), with carbon contents of 3.1 to 4.0% and silicon in amounts of 1.7 to 3.0% (Ref 4-6). At constant silicon levels, a lower CE slightly increases the chilling tendency and results in lower nodularity. At constant CE, higher silicon increases nodularity (Ref 7). The optimum carbon and silicon contents can be selected from Fig. 2.

The optimum CE must be selected as a function of section size. For a given section size, too high a CE will result in graphite flotation, as in the case of spheroidal graphite cast iron, while too low a CE may result in increased chilling tendency. For wall thickness ranging from 10 to 40 mm (0.4 to 1.6 in.), eutectic composition (CE = 4.3%) is recommended in order to obtain optimum casting properties. Manganese content can vary between 0.1 and 0.6%, while phosphorus content should be less than 0.06% in order to take advantage of the ductility of this material.

Although CG iron has been produced from base irons having sulfur contents as

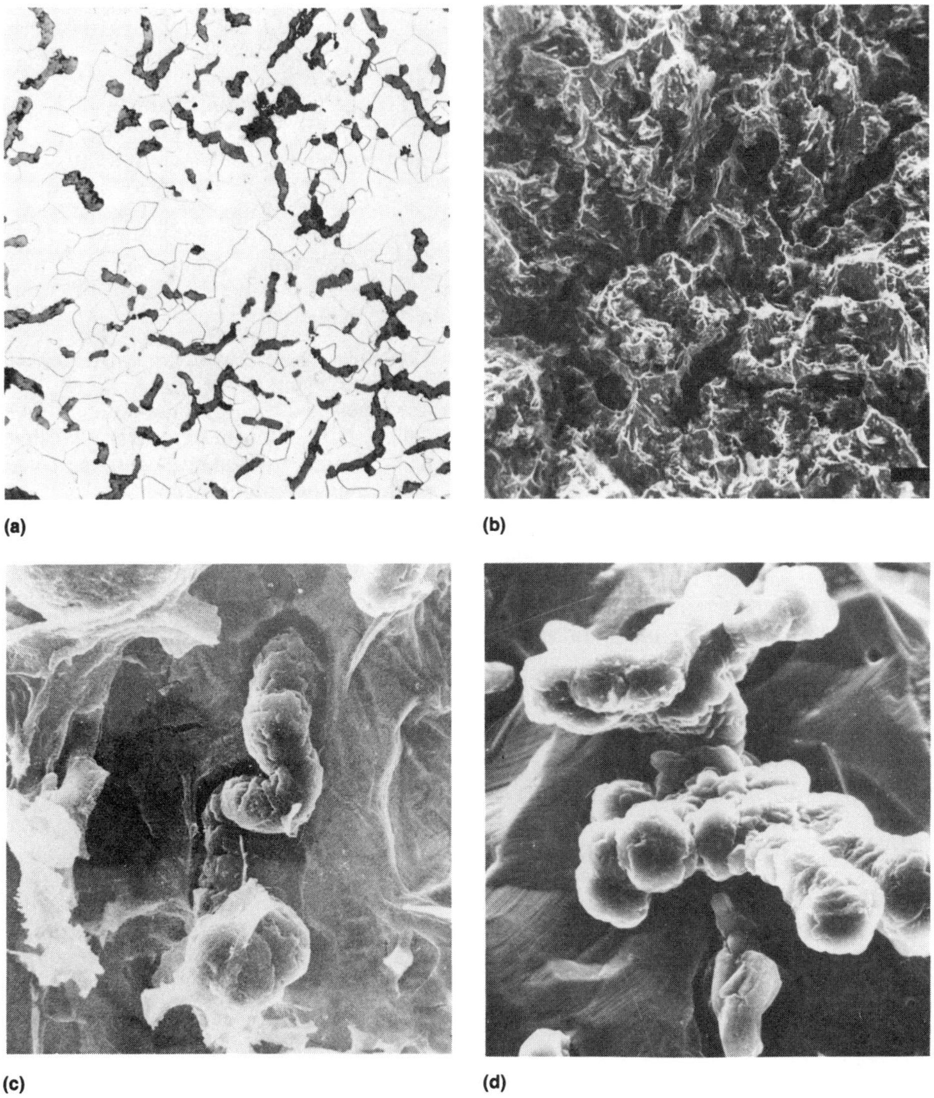

Fig. 1 Typical microstructures of CG irons. (a) Optical micrograph. Etched with nital. (b) Tensile load fracture surface. Overall view. Ion bombardment etched. SEM, 65×. (c) and (d) Examples of true shape of graphite in CG irons. Full deep etch. SEM, 395×. Courtesy of Austrian Foundry Research Institute

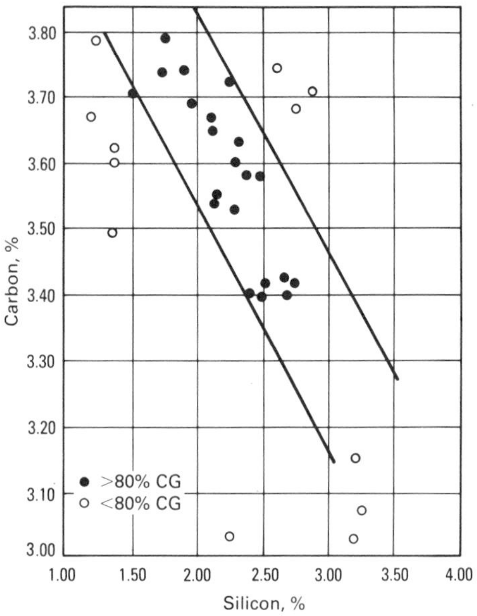

Fig. 2 Range of optimum carbon and silicon contents for CG iron. Source: Ref 7

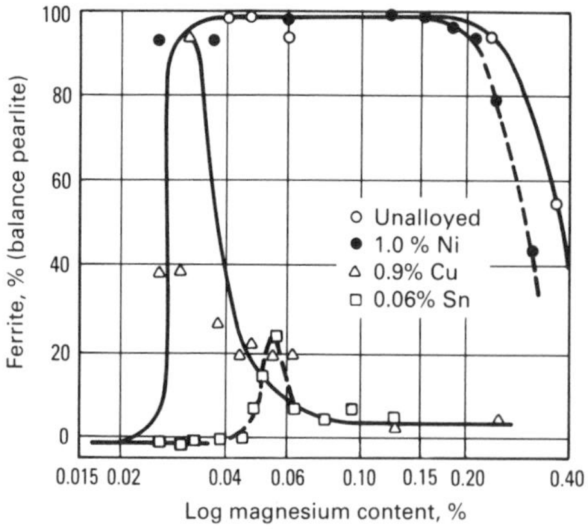

Fig. 3 The effect of copper, nickel, and tin on the type of matrix in the composition range between CG and FG iron of 25 mm (1 in.) wall thickness. Source: Ref 43

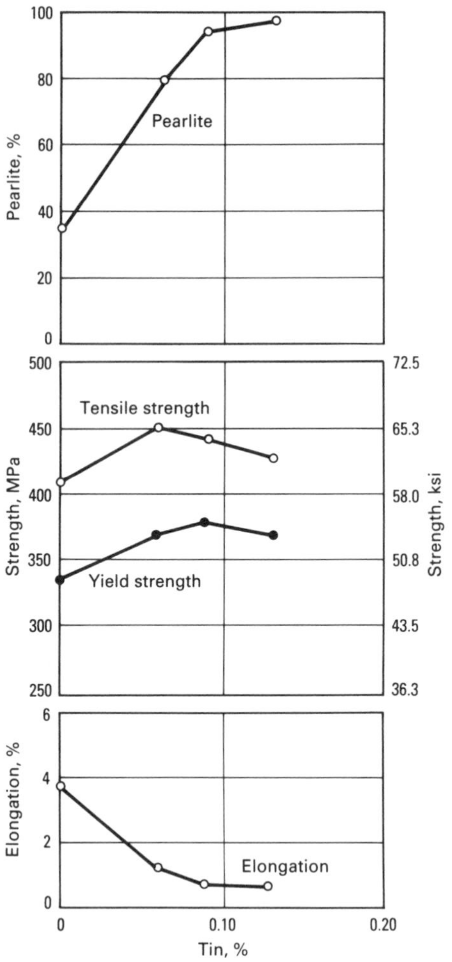

Fig. 4 The effect of tin on pearlite content and tensile properties of as-cast CG iron 25 mm (1 in.) thick. Source: Ref 43

high as 0.07 to 0.12% (Ref 8), it is probably more economical to desulfurize the iron to a level of 0.01 to 0.025% before liquid treatment. The higher the sulfur content, the more alloy is required for melt treatment. Also, the risk of missing the composition window for CG iron is increased, because the residual treatment elements must be balanced with the residual sulfur. Typical residual sulfur levels after treatment are 0.01 to 0.02% (Ref 6).

In order to ensure the compacted (vermicular) graphite shape, it is necessary to use some treatment elements, just as in the case of SG iron. These elements include magnesium, rare earths (cerium, lanthanum, and so on), calcium, titanium, and aluminum. Their influence and use will be discussed in the section "Melt Treatment" in this article.

Alloying elements, such as copper, tin, molybdenum, and even aluminum can be used to change the as-cast matrix from ferrite to pearlite. Typical ranges are 0.48% Cu or 0.033% Sn (Ref 9), 0.5 to 1% Mo (Ref 10), and up to 4.55% Al (Ref 11 and 12). Regardless of the alloying elements used, the high ferritizing tendency of CG iron should be taken into account. Figure 3 shows that even strong pearlite promoters such as copper or tin show reduced effectiveness in CG iron (Ref 12). The same is true when treating with cerium-mischmetal. A fully pearlitic structure could not be obtained even at 1.7% Cu when using high-purity charge materials. Tin, however, is effective. As shown in Fig. 4, 95% pearlite was obtained with 0.13% Sn. To produce a pearlitic matrix, therefore, it may be necessary to add higher-than-usual levels of alloying elements or to make multiple additions of two or three elements.

Melting. The melting aggregates to produce CG iron are in principle the same as those used for SG iron. Induction furnaces, cupolas, and arc furnaces have been reported as being adequate for melting. Similar requirements of raw materials, superheating, and desulfurization before melt treatment apply. If a ferritic structure in the as-cast condition is desired, a pure pig iron with low manganese, phosphorus, and sulfur contents is recommended. If some pearlite is acceptable, steel scrap can be used.

Melt Treatment. The most important methods used for manufacturing CG iron can be classified as:

- Controlled undertreatment with magnesium-containing alloys
- Treatment with alloys containing both compactizing (magnesium, rare earths, and calcium) and anticompactizing (titanium and aluminum) elements
- Treatment with rare earth base alloy or magnesium-rare earth alloys
- Treatment of a base iron containing rather high amounts of anticompactizing elements (sulfur and aluminum) with alloys containing compactizing elements (magnesium and cerium)

Compositions of typical treatment alloys used in the production of iron are listed in Table 1.

As shown in Fig. 5, CG can be obtained when treating an SG-type base iron with magnesium-iron-silicon alloy (for example, alloy 1 in Table 1), when residual magnesium is controlled in the range of 0.013 to 0.022% (Ref 13). This type of control is difficult to achieve when ladle treatment is used; overtreatment (resulting in too much SG) or undertreatment (resulting in flake graphite) can occur. Nevertheless, instances in which this process is applied are reported in Ref 3.

Better control of residual magnesium can be achieved if the in-mold process is used. Indeed, good quality CG iron has been produced using alloys similar to alloy 1 (Table 1) from iron with 0.008 to 0.017% base sulfur. Residual Mg was 0.015 to 0.025% in section sizes of 8 to 63 mm (0.3 to 2.5 in.) (Ref 9). As the base sulfur content increases, higher residual magnesium is required.

The treatment of iron with alloys containing compactizing elements balanced by anticompactizing elements has much wider industrial application. A treatment alloy containing about 5% Mg and 0.3% Ce balanced by about 9% Ti (for example, alloy 2 in Table 1) is used in order to obtain 0.015 to 0.035% residual Mg, 0.06 to 0.13% residual Ti, and low levels of cerium in the final cast iron (Ref 4). As shown in Fig. 6, using a magnesium-titanium combination rather than magnesium alone dramatically increases the range of residual magnesium over which CG can be obtained.

This treatment alloy was later improved by the inclusion of 4 to 5.5% Ca (alloy 3 in Table 1) to extend the working range of residual magnesium and the tolerable range of base sulfur (Ref 14). A wide range of carbon and silicon contents (3.15 to 4% and

Table 1 Nominal compositions of typical treatment alloys for CG iron

Alloy number	Composition, %: Compactizing elements: Mg	Ce	La	TRE(a)	Ca	Anticompactizing elements: Ti	Al	Neutral elements: Si	Fe	Reference
1	5	...	...	...	1	...	<1.2	45	Balance	9, 13
2	5	0.3	...	0.3	<1	9	<1.5	52	Balance	4
3	5	0.3	...	0.3	4.5	9	1.2	50	Balance	14
4	...	24	14.4	48	7.5	...	4.3	33.2	Balance	15
5	...	30	50	80	...	...	...	...	Balance	16, 17
6	...	16	80	96	...	...	...	...	Balance	18
7	...	2.9	26.5	29.4	0.63	...	0.13	30.5	Balance	18
8	3.7	0.8	0.5	1.7	1.05	...	0.88	45.3	Balance	20
9	4.3	0.7	1.8	2.9	0.66	...	0.9	45.3	Balance	20

(a) TRE, total rare earth elements

1.7 to 3%, respectively) can be used to produce CG iron in section thicknesses of 12 to 65 mm (½ to 2½ in.). Base iron sulfur contents can vary from 0.02 to 0.04%, resulting in final sulfur levels of 0.01 to 0.02% (Ref 6). Excessive sulfur has been shown to promote a gray rim (flake graphite), while too low a sulfur content promotes higher nodularity. The amount of magnesium-titanium alloy used for CG treatment is not critical, but depends on the base sulfur content. As for SG iron, a ladle postinoculation is indicated. It appears important to keep a treatment of about 1400 °C (2550 °F) because lower temperatures result in higher nodularity (Ref 5). The main disadvantage of this method seems to be titanium contamination (0.1 to 0.15% Ti) of returns and castings.

Alloy 4 in Table 1 can also be included in the category of alloys relying on a balance of compactizing and anticompactizing elements. Additions of 0.2% of this alloy resulted in CG structures in bar castings ranging in thickness from 25 to 127 mm (1 to 5 in.) (Ref 15). A rather high chilling tendency was observed.

The treatment of liquid iron with rare earth alloys was actually one of the first processes to gain industrial application. Cerium-mischmetal (for example, alloy 5 in Table 1) (Ref 16, 17) is used for production of medium- and heavy-section CG iron castings. Residual cerium contents of 0.013 to 0.075% are required, depending on base sulfur. The dependency of alloy addition on base sulfur level is shown in Fig. 7.

Although rare earth treatment offers a number of advantages over magnesium treatment, including lower fading time and no smoke during the reaction time, the high chilling tendency in thin sections seems to be its biggest drawback. Enhanced postinoculation with ferrosilicon does not seem to solve the problem entirely, because it results in increased nodularity.

Rare earth alloys containing a higher lanthanum/rare earth ratio than typical mischmetal have been suggested to alleviate the chilling tendency (alloys 6 and 7 in Table 1) (Ref 18). As a general observation, it should be noted that postinoculation is required when silicon-free alloys are used.

Good results in foundry practice have also been claimed when using a cerium-calcium-silicon alloy of undisclosed composition (Ref 19). With low base sulfur, good CG structure could be achieved when at least 0.016% Ce and 0.24% Ca were added to the melt. When base sulfur was raised to 0.059%, the cerium and calcium contents had to be increased to 0.091% and 0.53%, respectively. Fading times of the order of 22 min for a 500 kg (1100 lb) ladle, as well as good response to the recycling of returns were also claimed as advantages of this process.

Another way to avoid the chilling problems associated with mischmetal treatment is to use a magnesium-rare earth-ferrosilicon-type alloy such as alloy 8 or 9 in Table 1. When these alloys were used in conjunction with the in-mold process (Ref 20), a residual magnesium, cerium, and lanthanum content of 0.018 to 0.028% was required for CG structure when the base sulfur was 0.016%, which correlates with data given in Ref 9.

Finally, another group of treatment methods addresses the problem of high levels of anticompactizing elements such as sulfur or aluminum in the base iron. A special alloy with undisclosed composition, containing high levels of magnesium and cerium and conventional quantities of aluminum and calcium is claimed to allow CG iron production at virtually all sulfur levels.

Regardless of the treatment alloy used, for a specific alloy addition and base sulfur content there is an optimum range of treatment temperatures. For example, for an iron with base sulfur of 0.12 to 0.13% and a treatment addition of 1.75% alloy, the treatment temperature must be between 1468

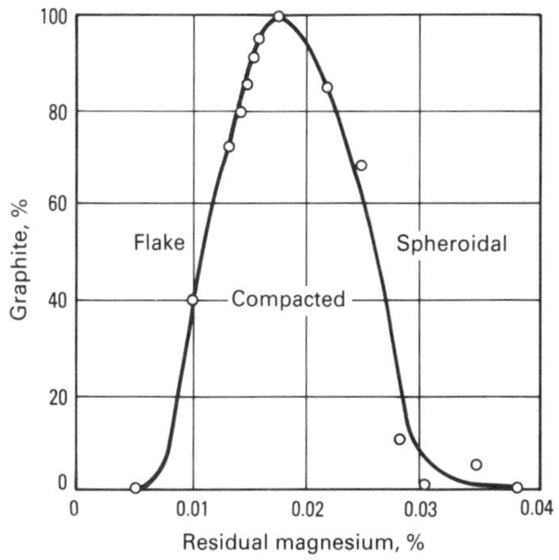

Fig. 5 The influence of residual magnesium on graphite shape. Source: Ref 3, 13

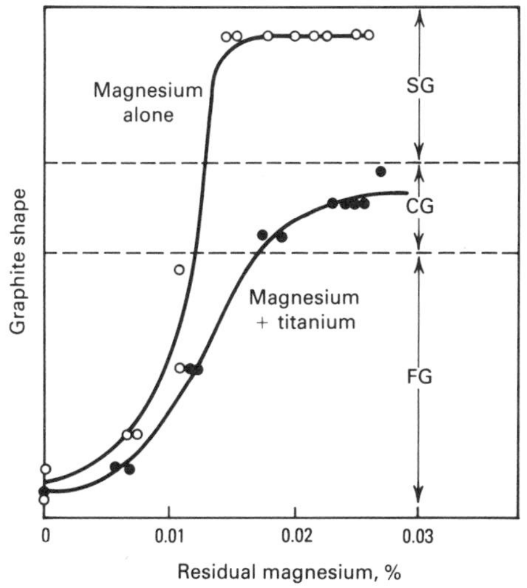

Fig. 6 Range of residual magnesium that produces compacted graphite. Source: Ref 5

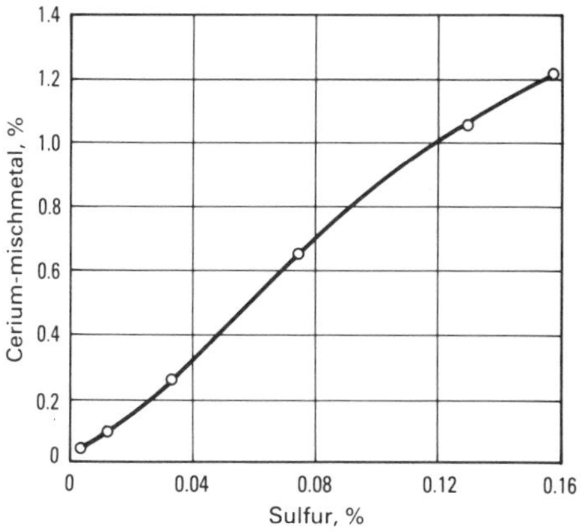

Fig. 7 The correlation between base sulfur content and amount of cerium-mischmetal treatment alloy required to produce compacted graphite. Source: Ref 17

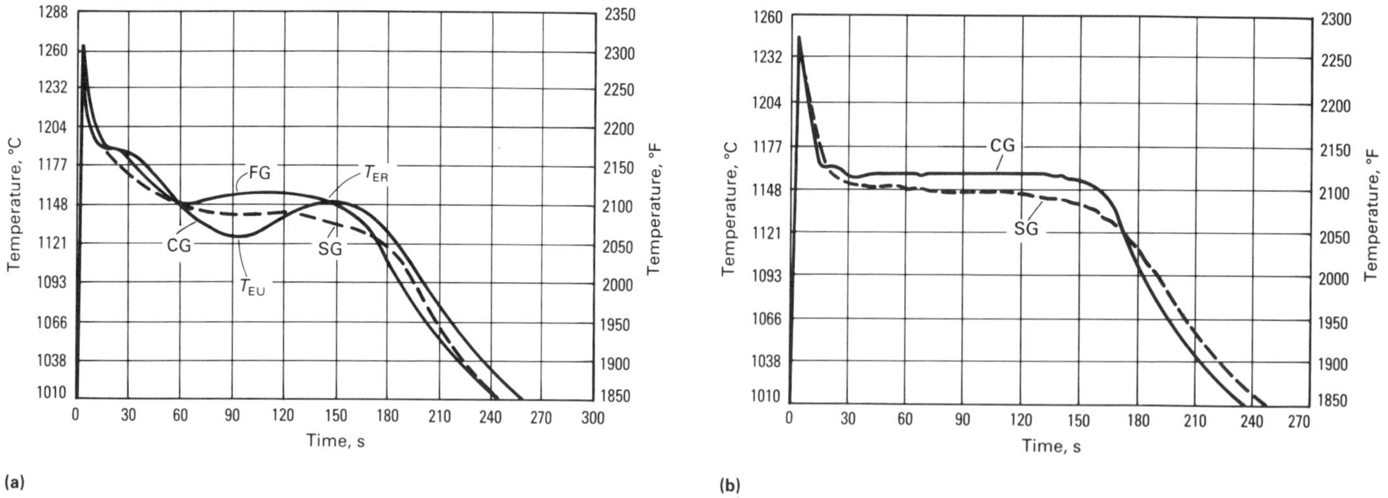

Fig. 8 Cooling curves. (a) For hypoeutectic irons. (b) For hypereutectic irons. T_{EU}, temperature of eutectic undercooling; T_{ER}, temperature of eutectic recalescence. Source: Ref 23

and 1510 °C (2675 and 2750 °F). Too low a temperature results in increased nodularity; an excessive temperature causes premature alloy dissolution, and the iron remains gray with FG.

Another approach to producing CG iron is to substitute aluminum for most of the silicon in the iron. This has the advantage of allowing the use of a magnesium-ferrosilicon alloy, because aluminum has an anticompactizing effect and can counterbalance the influence of magnesium. Thus, a correct balance between aluminum and magnesium should result in CG. This idea has been applied using both ladle and in-mold treatment (Ref 12). A 5% magnesium-ferrosilicon alloy with 0.3% Ce was used. Good CG structures with no chill on wedge tests were achieved for aluminum levels of 2.68 to 4.48% by using the trigger treatment method. A 1.4% addition of magnesium-ferrosilicon alloy resulted in residual magnesium of 0.016 to 0.028%, and less than 20% nodularity was achieved over this range. No postinoculation was used. It is interesting to note that holding times in excess of 30 min have proved to be acceptable even when the treated metal was held in the induction furnace. Excellent results were obtained when the in-mold process was used with a chamber area to result in 0.013 to 0.023% residual Mg and 0.022 to 0.032% residual Ce, for irons containing 1.62 to 4.43% Al.

Regarding postinoculation, it is generally accepted that CG iron exhibits a high chilling tendency, especially in sections thinner than 6 mm (¼ in.). To counteract this negative effect, it is necessary to use either silicon in the treatment alloy or some type of postinoculation. Most of the time, 0.2 to 0.3% ferrosilicon containing 75% Si is used for postinoculation. Nevertheless, in thin sections of CG iron castings, it is difficult to counteract chilling effects properly by postinoculation, because increasing the amount of inoculant normally results in SG rather than in CG. It has been suggested that 0.1 to 0.2% Al (Ref 21) or a ferrosilicon-titanium-aluminum alloy of undisclosed composition (Ref 2, 22) can be used as postinoculants to decrease chill without increasing nodularity.

Process Control. Production of good quality CG iron requires careful control at every stage. The chemical composition must be kept within rather rigid limits, particularly in regard to sulfur content. Sulfur content must be determined with ±0.001% accuracy before melt treatment (Ref 3), because excess sulfur can result in undertreatment, while too low a sulfur content can produce iron with excessive nodularity. Infrared determination methods are recommended for sulfur. Carbide stabilizing elements must also be checked accurately, because they increase the already high chilling tendency of CG iron. For example, chromium should be kept under 0.01%.

A variety of methods can be used for the control of melt treatment. When cerium-containing treatment alloys are used, a classic wedge test approach is recommended (Ref 3). Wedge tests are poured before treatment, after treatment with mischmetal, and after postinoculation, with required fractures being dark gray, completely white, and light gray, respectively.

Cooling curve analysis is a good way to predict microstructure before pouring and thus allow corrections if necessary. Typical cooling curves for flake, compacted, and nodular graphite irons and for white iron are shown in Fig. 8. It can be seen from Fig. 8(a) that hypoeutectic CG iron has a temperature of eutectic undercooling (T_{EU}) lower than that of FG or SG iron, although its temperature of eutectic recalescence (T_{ER}) is between the other two. Nevertheless, depending on the level of postinoculation, it is sometimes possible for CG to have a higher T_{EU} or a lower T_{ER} than SG, although this is the exception rather than the rule. It is apparent that the easiest way to identify a good CG structure is based on the value of $\Delta T = T_{ER} - T_{EU}$. The value of ΔT should be in excess of 25 to 30 °C (77 to 86 °F) for rare earth-treated iron, or in excess of 10 °C (50 °F) for magnesium-treated irons (Ref 24). For hypereutectic composition (Fig. 8b), the undercooling can be higher for either SG or CG iron, so that simple cooling curve analysis cannot be used to differentiate between these two irons.

Improved microstructure predictions can be made by computer-aided differential thermal analysis (CADTA), which consists of numerical processing of cooling curve data. Based on first derivative curves obtained using CADTA, hypoeutectic CG differs from other irons by the fact that it has only one slope on the second part of the curve (after recalescence) while the other irons have two. For hypereutectic irons, the first derivative curve of CG iron exhibits two maxima compared with only one for SG iron.

When large quantities of compacted graphite iron are poured, it may be possible to run a metallographic sample, as is done in some SG iron foundries, immediately after melt treatment and before pouring. To date, a universally accepted method for estimating the amount of CG in cast iron is not available. Certainly the simplest method is one in which the nodularity is determined by the ratio between the number of type I and II (nodular) graphite particles (ASTM A247) and the total number of graphite particles (Ref 25):

$$\text{Nodularity (\%)} = \frac{\text{type I} + \text{type II}}{\text{type I} + \text{type II} + \text{type III} + \text{type IV}}$$

The amount of CG is considered to be the difference between 100% and the percent of nodularity.

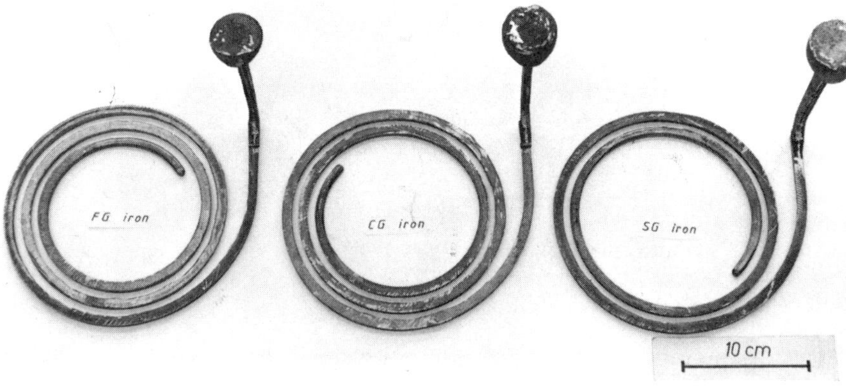

Fig. 9 Spiral fluidity samples poured from FG, CG, and SG iron. Source: Ref 27

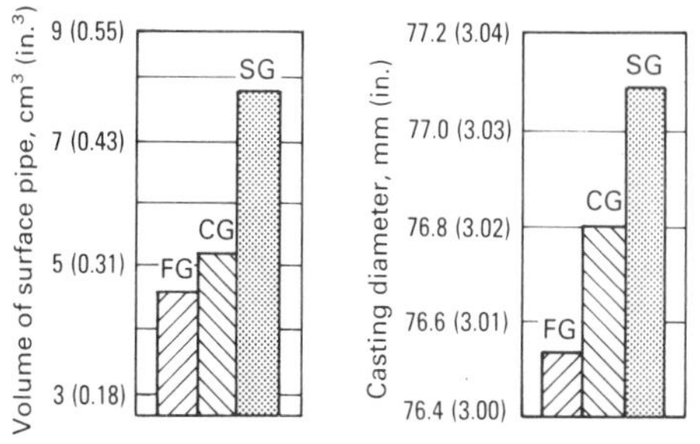

Fig. 10 Comparison of shrinkage and dimensions of spherical FG, CG, and SG iron castings poured in green sand molds. Source: Ref 5

This quantitative method is likely to be satisfactory for graphite shape estimation in production conditions, but has the disadvantage that it neglects the fact that several CG particles can belong to the same graphite aggregate, or that graphite considered to be spheroidal is in fact part of a CG aggregate. Another method suggests the use of different standard microphotographs (Ref 4, 15).

Molding Materials. As for SG iron, all molding materials that can produce rigid molds can be used. They include bentonite-bonded sands, cement-bonded sands, and resin-bonded sands.

Compacted graphite iron melts are more sensitive to sulfur pickup from the mold than are SG iron melts: unlike SG iron, CG iron must not be overtreated (Ref 3). Particular attention should be payed when using reclaimed resin-bonded sand with paratoluosulfonic acid (PTS) as a hardening catalyst. While such a new sand contains 0.086% S, at an amount of 80% reclaimed sand, the sulfur content increases to 0.31%, which can cause sulfur pickup in the casting of up to 0.05% S in the surface area, resulting in a FG rim up to several millimeters in thickness. To avoid graphite deterioration in the surface layer, especially for fatigue-stressed components, it may be useful to substitute phosphoric acid in part for PTS as a hardener. Also, the application of protective mold coatings with lime, magnesia, or talc base, is recommended (Ref 3).

Quality Control. Typical quality control procedures may include verification of structure, mechanical properties, physical properties, casting soundness, and dimensional accuracy. Structure can be assessed by classic metallographic techniques on coupons but also, to a certain extent, by ultrasonic velocity measurements on castings. While FG is easily separated from CG using the ultrasonic technique, it is virtually impossible to differentiate between CG and low-nodularity SG (Ref 9). Nevertheless, good correlations between mechanical properties and ultrasonic velocity are reported and will be discussed in the section "Mechanical Properties" in this article. When ultrasonic tests are applied to castings, the ultrasonic velocity for CG structures is independent of the shape of the casting, but should be calibrated for the section thickness. For example, for 30 mm (1.2 in.) diam bars, the ultrasonic velocity range associated with CG is 5.2 to 5.45 km/s (3.2 to 3.4 mi/s), but for very large castings such as ingot molds, the ultrasonic velocity for good CG structures lies between 4.85 and 5.1 km/s (3.0 and 3.15 mi/s). Resonant frequency (sonic testing) on the other hand, must be calibrated for a particular casting design, for which examples of satisfactory and unsatisfactory structures must be previously checked to provide a calibration range.

Separately cast samples are recommended for determination of mechanical properties. They can be 30 mm (1.2 in.) diam cylindrical bars poured in a vertical position or keel (U) blocks with 25 mm (1 in.) wide legs. For larger castings, however, sample bars cast on the appropriate location on the casting are preferred (Ref 3, 26).

Castability

The fluidity of molten cast iron is broadly governed by carbon and silicon contents and temperature. In addition, solidification morphology plays a role (Ref 27). Figure 9 shows three fluidity spiral samples. They were poured from the same base metal at the same temperature and have comparable chemical compositions. They differ only in their melt treatment. Flake graphite iron shows the best fluidity, and SG iron shows the worst. As expected, CG iron occupies an intermediate position. However, because CG irons are stronger than gray iron with the same CE, a higher CE can be used to obtain the same strength, allowing greater fluidity and easier pouring of thin sections.

Shrinkage Characteristics. Obtaining sound castings in CG irons, free from external and internal shrinkage porosity, is easier than with SG irons and slightly more difficult than with FG irons. This is because the tendency for mold wall movement to occur also lies between that of SG and FG irons. Figure 10 shows the results of experiments that demonstrate these effects using a 76 mm (3 in.) diam spherical test (Ref 5).

For cast iron of eutectic composition cast in sand molds, it is reported that the shrinkage volume is 4.1% for FG, 4.8% for CG, and 7.0% for SG irons (Ref 28). A higher-grade FG iron will have a shrinkage volume very close to that of CG iron. In relative numbers, solidification expansion has been found to be 4.4 for SG iron and 1 to 1.8 for CG iron if FG iron is 1 (Ref 22).

Because of the rather low shrinkage of CG iron, it can sometimes be cast riserless. Expensive pattern changes are therefore not necessary when converting from gray iron to CG iron, because the same gating and risering techniques can be applied.

Yields for large castings vary between 65 and 75%, depending on casting complexity and weight (Ref 3).

Dross Formation. Magnesium, cerium, and calcium are dross-forming elements, and the problem of dross exists in CG irons just as it does in SG irons. Lower magnesium contents should be an advantage, but the best conditions for producing dross-free CG irons are the same as for SG irons: low CE composition, low sulfur content before treatment, treatment alloy low in calcium, high pouring temperature, adequate slag traps in the gating system, and smooth, nonturbulent pouring.

Chilling Tendency. Contrary to the belief of some, the chilling tendency of CG iron does not fall between that of FG and SG irons. This correlates with cooling curve data and is explained by the combination of low nucleation rate and low growth rate occurring during solidification of CG iron.

Mechanical Properties

Mechanical properties of CG irons have been summarized in a number of publications (Ref 2, 3, 29, 30). More than 80% of all failures of parts can be attributed to fatigue, which is not caused by insufficient mechanical strength, but by the initiation of cracks due to lack of plasticity, particularly at

Table 2 Typical mechanical properties of CG iron

Tests were performed on 30 mm (1.2 in.) diam test bars.

Iron matrix	Tensile strength MPa	Tensile strength ksi	0.2% offset yield strength MPa	0.2% offset yield strength ksi	Elongation, %	Modulus of elasticity GPa	Modulus of elasticity ksi × 10^3	Hardness, HB
Ferritic	250–380	36–55	175–300	25–43	3–8	120–126	17.5–18.3	130–179
Pearlitic	405–620	59–90	315–425	45–63	1–2	127–165	18.5–24	207–269

elevated temperatures. For high-duty gray iron, elongation is normally less than 1%, which in many cases is insufficient. Replacement of gray iron with a material of higher strength, ductility, and toughness, such as ductile iron, is not always possible because of poorer casting properties, lower thermal conductivity, higher modulus of elasticity, and so forth. In such cases, CG iron can close the gap between the two materials, because its mechanical properties are closer to those of SG iron, while its physical (expansion, conductivity) and elevated-temperature properties (thermal fatigue, thermal shock) are closer to those of FG iron.

Tensile Properties and Hardness. The typical range of room-temperature properties for CG irons with ferritic or pearlitic matrices is given in Table 2. Compacted graphite irons exhibit linear elasticity for both pearlitic and ferritic matrices but to a lower limit of proportionality than SG iron (Fig. 11). The yield point ratio (ratio of yield strength to tensile strength) ranges from 0.72 to 0.82 (higher than for SG iron of the same composition). This makes possible a higher loading capacity.

The graph in Fig. 12 compares hardness with tensile strength for SG, CG, and FG irons. The average tensile strength to hardness ratio of gray iron is 1.3, compared with 2.08 for compacted graphite, and 2.75 for ductile iron. The intermediate behavior of CG iron and the apparent effect of sulfur content are worth noting (Ref 31).

The main factors affecting mechanical properties are:

- Composition
- Structure (nodularity and matrix)
- Section size

All of these factors are interdependent.

It has been demonstrated that the tensile properties of CG irons are much less sensitive to variations in carbon equivalent than are those of FG irons. Even at CE near the eutectic value of 4.3, both pearlitic and ferritic CG irons have higher strengths than low CE, high-duty, unalloyed FG cast iron (Ref 5).

An increase in the silicon content up to 2.6% benefits strength and hardness in both the as-cast and annealed conditions (Fig. 13). This is true even though the matrix becomes more ferritic, because silicon strengthens the ferrite. The same is true in the case of as-cast elongation because there is a decrease in the amount of pearlite with an increase in Si, while elongation in the annealed condition decreases (Ref 31). Although increasing the phosphorus content slightly improves strength, a maximum of 0.04% P is desirable to avoid lower ductility and impact strength.

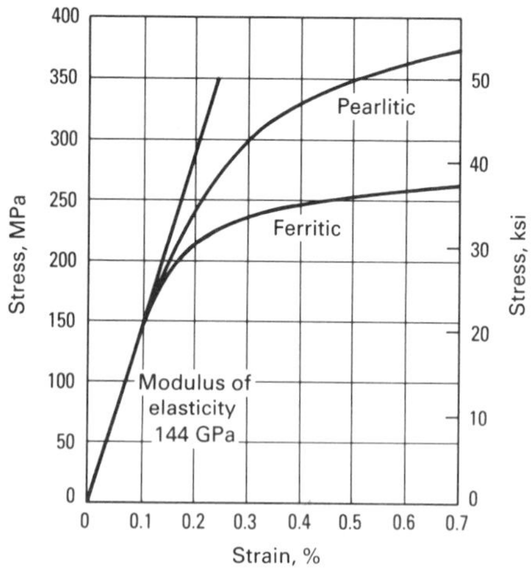

Fig. 11 Typical stress-strain curves for CG irons. The pearlitic iron had a tensile strength of 410 MPa (59.5 ksi) and elongation of 1%. The ferritic iron had a tensile strength of 320 MPa (46.5 ksi) and elongation of 3.5%. Source: Ref 3, 17

A number of alloying elements, such as copper, molybdenum, tin, manganese, and aluminum (Ref 9, 10, 12, 25, 32), or heat treatment can be used to increase the pearlite/ferrite ratio. Some microstructure-properties correlations for iron-carbon-aluminum CG irons produced by the in-mold process are shown in Fig. 14.

Both the pearlite/ferrite ratio and nodularity influence mechanical properties. Tensile strength and elongation increase as the nodularity of CG irons increases (Table 3), but nodularity must be maintained at less than

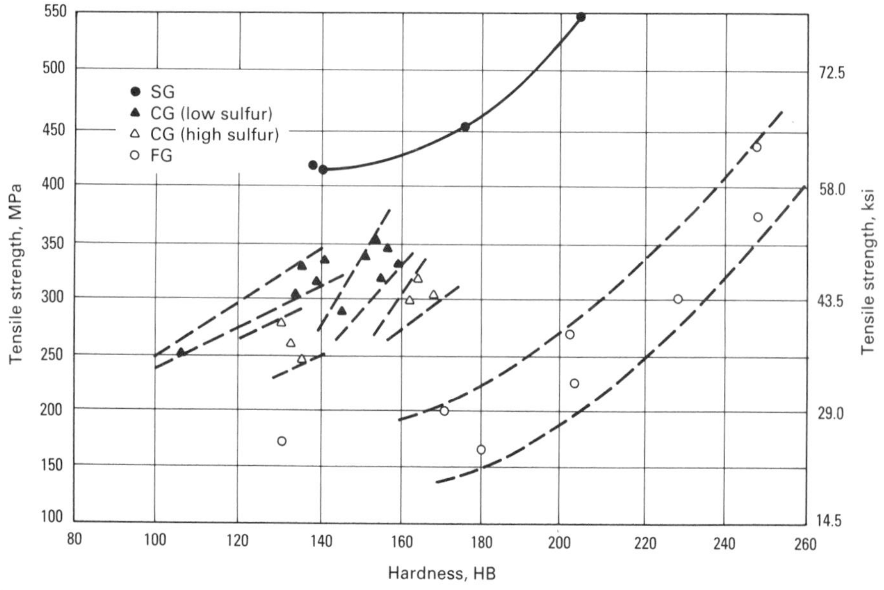

Fig. 12 Tensile strength versus hardness for SG, CG, and FG irons. Source: Ref 31

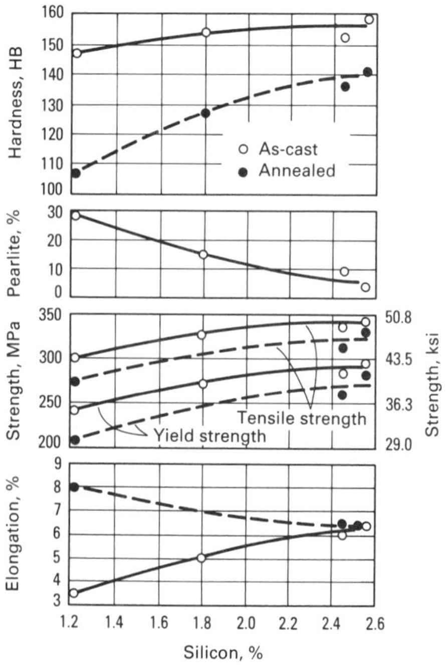

Fig. 13 The effect of silicon on mechanical properties of CG irons produced by the in-mold process. Carbon equivalents ranged from 4.33 to 4.45. Source: Ref 12

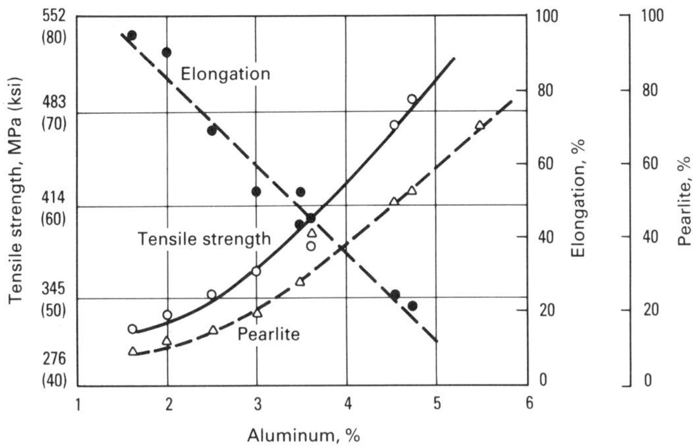

Fig. 14 The effect of aluminum on the structure and mechanical properties of CG irons produced by the in-mold process. Source: Ref 12

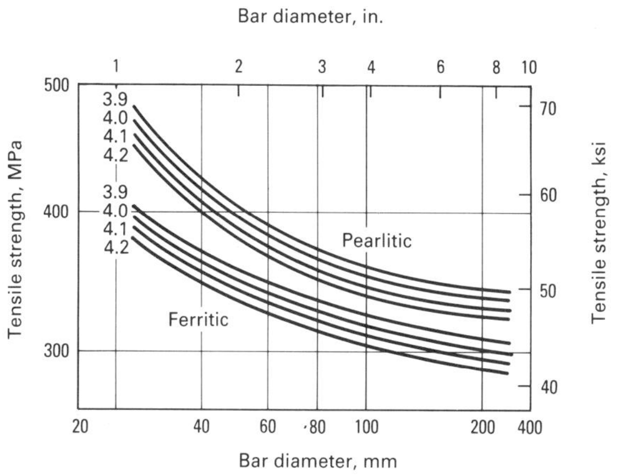

Fig. 15 The effect of section size on tensile strength of CG irons with various structures and carbon equivalents. Numbers beside curves indicate carbon equivalents. Source: Ref 5

20% for the iron to qualify as CG iron. However, SG contents of up to 30% and even more must be expected in the thin sections of castings with considerable variation in wall thickness.

As for gray and ductile iron, tensile strength, yield strength, and hardness of CG iron increase with an increasing pearlite/ferrite ratio, whereas elongation decreases. The influence of heat treatment on mechanical properties of CG irons with 20% nodularity is illustrated in Table 4.

Mechanical properties of CG irons are less sensitive to wall thickness than are those of FG irons (Ref 2), but the influence of wall thickness may still be significant (Fig. 15). Other factors influencing the cooling of the casting, such as shakeout temperature, can also influence properties.

Tensile strength correlates well with resonant frequency (sonic testing) and ultrasonic velocity measurements. Nevertheless, it must be noted that it is difficult to distinguish between high-nodularity CG and low-nodularity SG (Ref 3).

Compressive Properties. The ultimate compressive strength of predominantly pearlitic CG iron is approximately three times the tensile strength. More details are available in Ref 3, 29, and 30.

Impact Properties. While ductile iron exhibits substantially greater toughness at low pearlite contents, pearlitic CG irons have impact strengths equivalent to those of SG irons (Fig. 16). Charpy impact energy measurements at 21 °C (70 °F) and −40 °C (−40 °F) showed that CG irons produced from a ductile iron base absorbed greater energy than those made from gray iron base iron (Ref 31). This is attributed to tramp elements in the gray iron having solute hardening effects in the matrix.

Studies on crack initiation and growth under impact loading conditions showed that, in general, initiation of matrix cracking was preceded by graphite fracture, either at the graphite-matrix interface or through the graphite, or both. The most dominant form of graphite fracture appeared to be that occurring along the boundaries between graphite crystallites (Ref 34). Matrix cracks were usually initiated in the ferrite by transgranular cleavage (graphite was nearly always surrounded by ferrite), although in some instances intergranular ferrite fracture appeared to be the initiating mechanism. Matrix crack propagation generally occurred by a brittle cleavage mechanism, transgranular in ferrite and interlamellar in pearlite. In general, the impact resistance of CG irons increases with carbon equivalent, and decreases with phosphorus or increasing pearlite.

Fatigue Properties. Typical fatigue properties for CG irons are shown in Fig. 17. It is evident that pearlitic structures, higher nodularity, and unnotched samples resulted in better fatigue properties. The fatigue endurance ratio was 0.46 for ferritic matrix, 0.45 for pearlitic matrix, and 0.44 for pearlitic higher-nodularity CG iron (Ref 35).

Elevated-Temperature Properties

Tensile Properties. The variation with temperature of tensile strength, yield strength, and elongation of CG iron produced with cerium-mischmetal treatment alloys is similar to that typical for SG iron (Fig. 18), but the values are somewhat lower (Ref 36). Similar results are reported for CG irons produced with magnesium-titanium-ferrosilicon alloys (Ref 2). As expected, a slight increase in nodularity led to higher tensile strength values at all temperatures.

Growth and Scaling. Tests conducted for 32 weeks in air have shown that at 500 °C (930 °F) growth and scaling of CG iron was not significantly different from that exhibited by FG irons of similar composition. However, at 600 °C (1110 °F), the growth of CG irons was less than that of FG iron, and their scaling resistance was superior (Ref 5).

Other oxidation studies of cast irons conducted at 600 °C (1110 °F) concluded that weight gains due to oxidation are 10 to 15% higher for CG irons than for SG irons, but 30 to 60% lower than for FG irons (Ref 37).

Table 3 Properties of CG as a function of nodularity

Nodularity, %	Tensile strength MPa	Tensile strength ksi	Elongation, %	Thermal conductivity, W/m · K	Shrinkage, %
10–20	320–380	46–55	2–5	50–52	1.8–2.2
20–30	380–450	55–65	2–6	48–50	2.0–2.6
40–50	450–500	65–73	3–6	38–42	3.2–4.6

Source: Ref 33

Table 4 Effect of heat treatment on tensile properties of CG iron

Heat treatment	Iron matrix	Tensile strength MPa	Tensile strength ksi	Yield strength MPa	Yield strength ksi	Elongation, %	Hardness, HB
As-cast	60% ferrite	425	62	330	48	3.5	170
Annealed	100% ferrite	393	57	317	46	5.0	156
Normalized	90% pearlite	490	71	420	61	2.5	229

Source: Ref 33

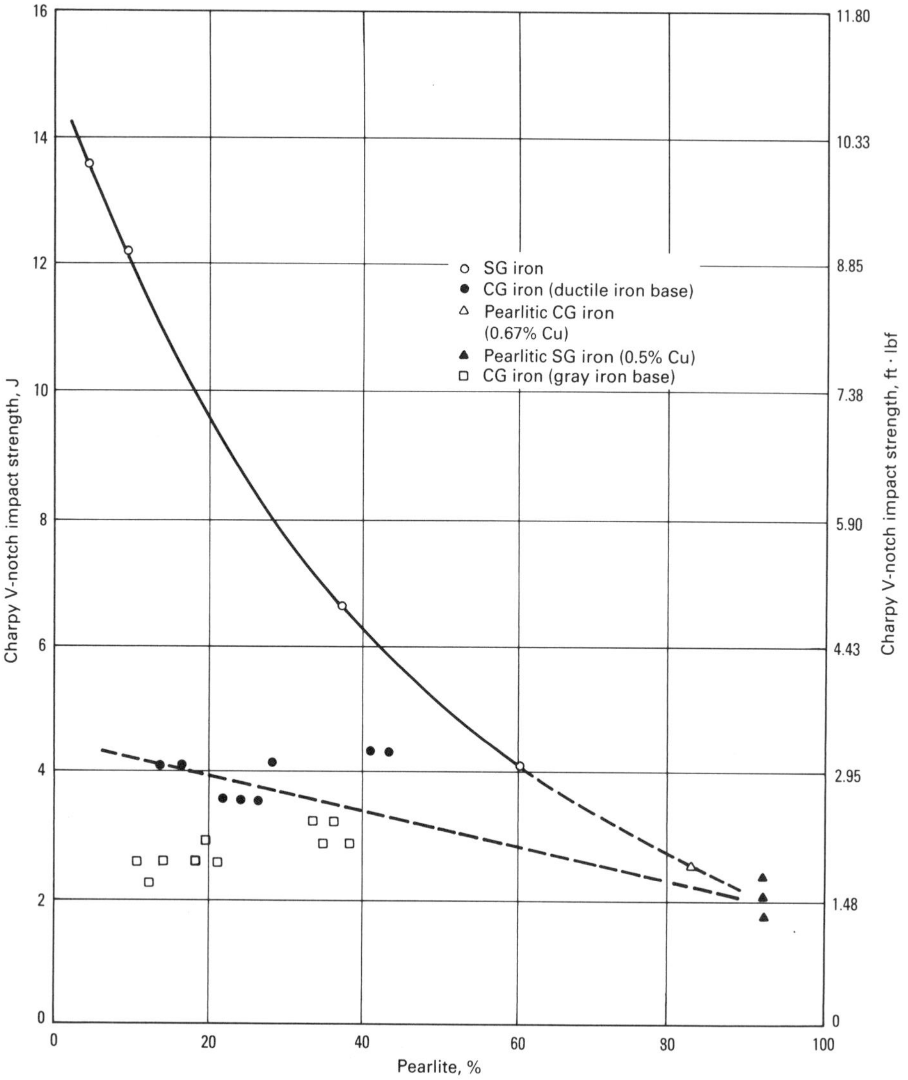

Fig. 16 The effect of pearlite content on the room-temperature Charpy V-notch impact strength of as-cast CG irons compared to that of SG iron. Source: Ref 31

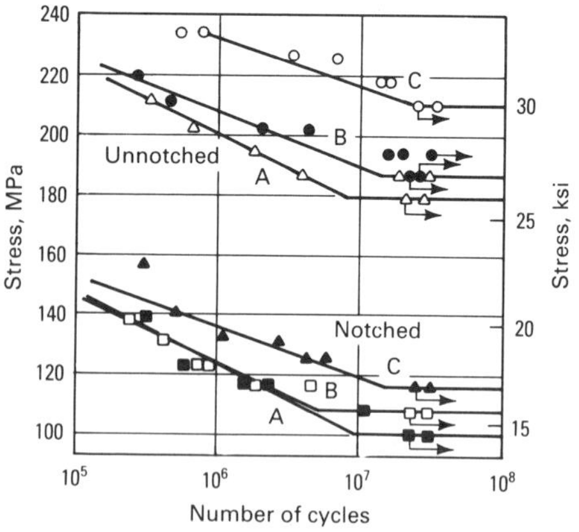

Fig. 17 Fatigue curves for CG irons. (a) With ferritic matrix. (b) With pearlitic matrix. (c) With higher nodularity. Source: Ref 29, 35

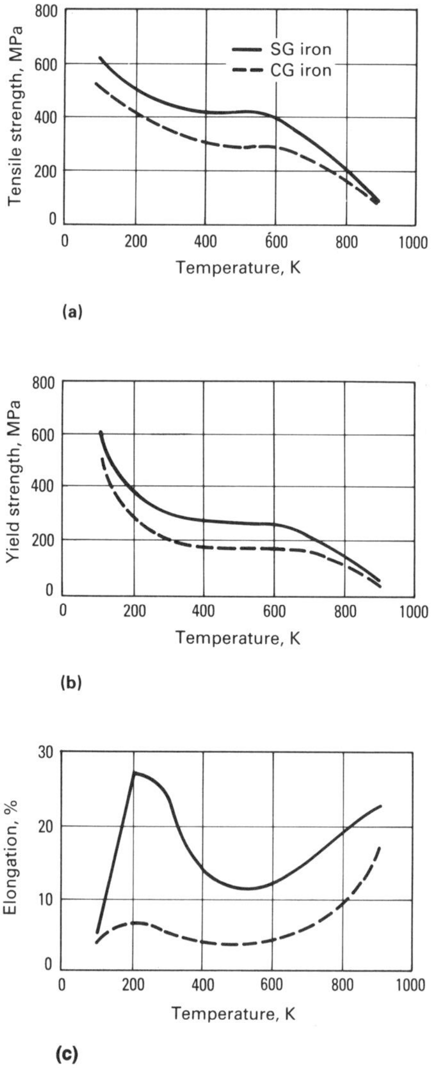

Fig. 18 Variation with temperature of tensile properties of CG and SG irons. Source: Ref 35

Thermal Fatigue. Interpretation of thermal fatigue tests is complicated by the many different methods employed by various investigators. The two widely accepted methods are constrained thermal fatigue (12.7 mm, or 0.5 in., diam bars) and finned disk thermal shock tests (Ref 38, 39).

During thermal cycling, in the constrained thermal fatigue test for which the specimen is mounted between two stationary plates, compressive stresses develop upon heating, and tensile stresses develop upon cooling. As thermal cycling continues, the specimen is accumulating fatigue damage similar to that in mechanical fatigue testing; ultimately, the specimen fails by fatigue. Experimental results (Fig. 19) point to higher thermal fatigue for CG iron than for FG iron, and to the beneficial effect of molybdenum. In fact, regression analysis of experimental results indicates that the main factors influencing thermal fatigue are tensile strength (TS) and molybdenum content:

$$\log N = 0.934 + 0.026\ \mathrm{TS} + 0.861\ \mathrm{Mo}$$

where N is the number of thermal cycles to failure, tensile strength is in kips per square inch (ksi), and molybdenum is in percent.

In the finned disk thermal shock test, the specimen is cycled between a moderate-temperature environment and a high-temperature environment, which causes thermal expansion and contraction. For good resistance to thermal fatigue, cast irons must have high thermal conductivity, low modulus of elasticity, high strength at room and elevated temperatures and, for use above 500 to 550 °C (930 to 1020 °F), resistance to oxidation and structural change. Relative ranking of the irons varies with test conditions. When high cooling rates are encountered, experimental data and commercial experience show that thermal conductivity and low modulus of elasticity are most important. Consequently, gray irons of high carbon content (3.6 to 4%) are superior (Ref 38, 39). When intermediate cooling rates exist, ferritic SG and CG irons

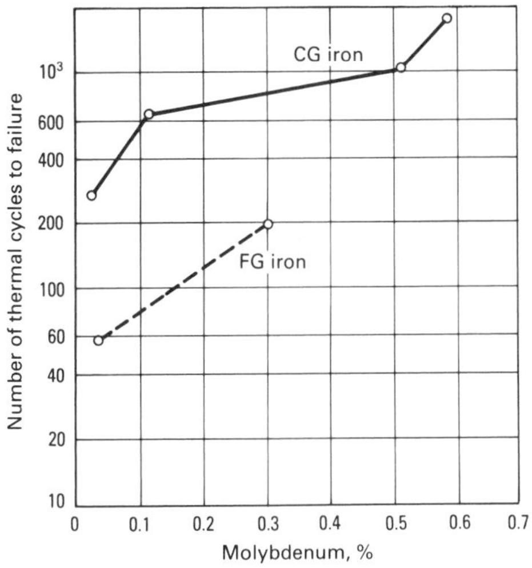

Fig. 19 Results of constrained thermal fatigue tests conducted between 100 and 540 °C (212 and 1000 °F). Source: Ref 39

have the highest resistance to cracking, but are subject to distortion. When low cooling rates exist, high-strength pearlitic SG irons or SG irons alloyed with silicon and molybdenum are best with regard to cracking and distortion (Ref 38).

Physical Properties

Thermal Conductivity. As can be seen from Table 5, the thermal conductivity of CG iron is very close to that of gray cast iron, and considerably higher than that of SG iron (Ref 3, 5). This behavior is explained by the fact that much like FG, CG is interconnected. As for FG irons, increasing the carbon equivalent results in higher thermal conductivity for CG iron. As the temperature is increased, the thermal conductivity reaches a maximum at about 200 °C (390 °F), an effect also shown by SG irons but differing from FG iron. Increasing nodularity results in lower thermal conductivity (Ref 3, 40).

Linear Expansion. For irons of similar chemical composition, there seems to be no difference in total expansion regardless of graphite shape (Ref 40). However, when different compositions are used in order for an iron to fall in the typical range for a given type of cast iron, the linear expansion of CG iron is between that of SG and FG irons (Ref 3).

Other Properties

Corrosion Resistance. At room temperature, the corrosion rate of CG iron in 5% sulfuric acid is nearly half that of FG iron. With increasing temperature, the difference becomes smaller. The pearlitic matrix has higher corrosion resistance than the ferritic one. As expected, corrosion accelerates when stress is applied (Ref 41). Detailed information on the corrosion resistance of cast irons is available in the article "Corrosion of Cast Irons" in Volume 13 of the 9th Edition of *Metals Handbook*.

Machinability. In general, from both experimental data and practical experience in machine shops, it can be concluded that for a given matrix the machinability of CG iron is between that of gray and ductile iron (Ref 3, 5, 37, 42, 43). The CG morphology makes the iron sufficiently brittle for the machine swarf to break into small chips, yet strong enough to prevent it from forming powdery chips. Neither large swarf nor fine powdery swarf is ideal for high machinability (Ref 31).

Damping Capacity. The relative damping capacity of various irons, obtained by measuring the relative rates at which the amplitude of an imposed vibration decreased with time (Ref 35), is:

FG iron : CG iron : SG iron = 1.0 : 0.6 : 0.34

Typical Applications

Primary in the selection of CG iron for practical application is its position between FG and SG iron. CG iron can be substituted for FG iron in all cases in which the strength of FG iron has become insufficient but in which a change to SG iron is undesirable because of the less favorable casting properties of the latter. Examples are: bed plates for large diesel engines, crank cases, gearbox and turbocharger housings, connecting forks, bearing brackets, pulleys for truck servo drives, sprocket wheels, and eccentric gears. Figure 20 shows three of these applications.

Because the thermal conductivity of CG iron is higher than that of SG iron, CG iron is preferred for castings operating at elevated temperature and/or under thermal fatigue conditions. Applications include ingot molds, crank cases, cylinder heads, exhaust manifolds, and brake disks (Fig. 21). Other engineering applications are summarized in Ref 3.

Table 5 Thermal conductivities of FG, CG, and SG irons at various temperatures

Graphite shape	Carbon equivalent	Thermal conductivity, W/m · αK (Btu/ft · h · °F) 100 °C (212 °F)	200 °C (390 °F)	300 °C (570 °F)	400 °C (750 °F)	500 °C (930 °F)
Flake	3.8	50.24 (29.02)	48.99 (28.30)	45.22 (26.12)	41.87 (24.19)	38.52 (22.25)
	4.8	53.39 (30.84)	50.66 (29.27)	47.31 (27.33)	43.12 (24.91)	38.94 (22.49)
Compacted	3.9	38.10 (22.01)	41.0 (23.69)	39.40 (22.76)	37.30 (21.55)	35.20 (20.34)
	4.1	43.54 (25.15)	43.12 (24.91)	40.19 (23.22)	37.68 (21.77)	35.17 (20.32)
Spheroidal	4.2	32.34 (18.68)	34.75 (20.08)	33.08 (19.11)	31.40 (18.14)	29.31 (16.93)

Source: Ref 5

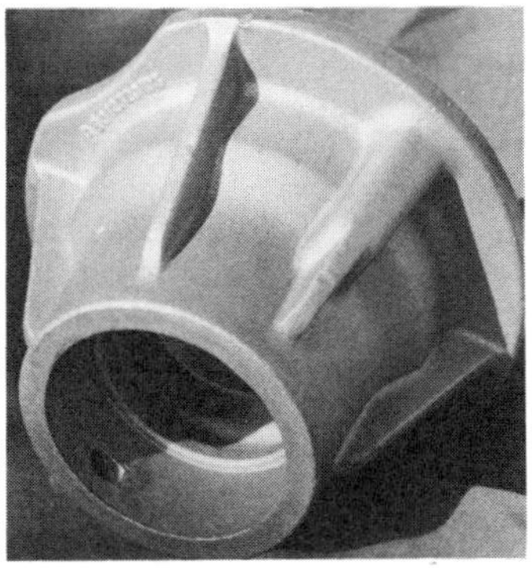

(a)

(b)

(c)

Fig. 20 Three typical applications for CG iron castings. (a) Intermediate gear box; weight, 17.9 kg (39.5 lb). (b) Connecting forks; weight, 2 and 2.6 kg (4.4 and 5.7 lb). (c) Sprocket wheels; weight, 1.1 and 1.6 kg (2.4 and 3.5 lb). Source: Ref 3

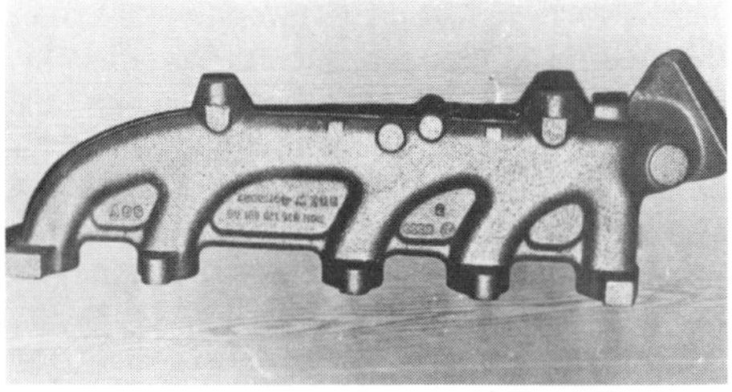

(a)

(b)

(c)

Fig. 21 Elevated-temperature applications for CG iron castings. (a) Exhaust manifold. Courtesy of Buderus AG. (b) Railroad brake disk consisting of four half disks (two on each side of the wheel). (c) MaK-M 551 cylinder cover with cast-on prismatic sample bar. Source: Ref 3, 46

REFERENCES

1. R.D. Schelleng, Cast Iron (With Vermicular Graphite), U.S. Patent 3,421,886, May 1965
2. D.M. Stefanescu and C.R. Loper, Recent Progress in the Compacted/Vermicular Graphite Cast Iron Field, *Giesserei-Prax.*, (No. 5), 1981, p 73
3. E. Nechtelberger, H. Puhr, J.B. von Nesselrode, and A. Nakayasu, "Cast Iron with Vermicular/Compacted Graphite—State of the Art Development, Production, Properties, Applications," Paper presented at the International Foundry Congress, Chicago, April 1982
4. E.R. Evans, J.V. Dawson, and M.J. Lalich, Compacted Graphite Cast Irons and Their Production by a Single Alloy Addition, *Trans. AFS*, Vol 84, 1976, p 215
5. G.F. Sergeant and E.R. Evans, The Production and Properties of Compacted Graphite Irons, *Br. Foundryman*, Vol 71, 1978, p 115
6. K.P. Cooper and C.R. Loper, Jr., A Critical Evaluation of the Production of Compacted Graphite Cast Iron, *Trans. AFS*, Vol 86, 1978, p 267
7. H.H. Cornell and C.R. Loper, Jr., Variables Involved in the Production of Compacted Graphite Cast Iron Using Rare Earth-Containing Alloys, *Trans. AFS*, Vol 93, 1985, p 435
8. G.F. Ruff and T.C. Vert, Investigation of Compacted Graphite Iron Using a High Sulfur Gray Iron Base, *Trans. AFS*, Vol 87, 1979, p 459
9. J. Fowler, D.M. Stefanescu, and T. Prucha, Production of Ferritic and Pearlitic Grades of Compacted Graphite Cast Iron by the In-Mold Process, *Trans. AFS*, Vol 92, 1984, p 361
10. K.R. Ziegler and J.F. Wallace, The Effect of Matrix Structure and Alloying on the Properties of Compacted Graphite Iron, *Trans. AFS*, Vol 92, 1984, p 735
11. E. Nechtelberger, R. Hummer, and W. Thury, Aluminum-Alloyed Cast Iron With Vermicular Graphite, *Giesserei-Prax.*, Vol 24, 1970, p 387
12. F. Martinez, D.M. Stefanescu, Properties of Compacted/Vermicular Graphite Cast Irons in the Fe-C-Al System Produced by Ladle and In-Mold Treatment, *Trans. AFS*, Vol 91, 1983, p 593
13. W. Dünki, Gusseisen mit Vermiculargraphit—Herstellung und Eigenschaften, Escher Wyss Mitteilungen, Vol 53 (No. 1/2), 1980, p 215
14. M.J. Lalich, Compacted Graphite Cast Iron—Its Properties and Production With a New Alloy, *Mod. Cast*, July 1976, p 50
15. T. Kimura, C.R. Loper, Jr., and H. Cornell, Rare Earth Silicide Additions to Cast Iron, *Trans. AFS*, Vol 88, 1980, p 67
16. W. Thury, R. Hummer, and E. Nechtelberger, Austrian Patent 290,592, 1968
17. J. Sissener, W. Thury, R. Hummer, and E. Nechtelberger, Cast Iron With Vermicular Graphite, *AFS Cast Met. Res. J.*, 1972, p 178
18. D.M. Stefanescu, R.C. Voigt, and C.R. Loper, Jr., The Importance of the Lanthanum/Rare Earth Ratio in the Production of Compacted Graphite Cast Irons, *Trans. AFS*, Vol 89, 1981, p 119
19. W. Simmons and J. Briggs, Compacted Graphite Irons Produced With a Cerium-Calcium Treatment, *Trans. AFS*, Vol 90, 1982, p 367
20. C.G. Chao, W.H. Lu, J.L. Mercer, and J.F. Wallace, Effect of Treatment Alloys and Section Size on the Compacted Graphite Structures Produced by the In-the-Mold Process, *Trans. AFS*, Vol 93, 1985, p 651
21. D.M. Stefanescu, F. Martinez, I.G. Chen, Solidification Behavior of Hypoeutectic and Eutectic Compacted Graphite Cast Irons—Chilling Tendency and Eutectic Cells, *Trans. AFS*, Vol 91, 1983, p 205
22. D.M. Stefanescu, I. Dinescu, S. Craciun, and M. Popescu, "Production of Vermicular Graphite Cast Irons by Operative Control and Correction of Graphite Shape," Paper 37 presented at the 46th International Foundry Congress, Madrid, 1979
23. D.M. Stefanescu, Solidification of Flake, Compacted/Vermicular and Spheroidal Graphite Cast Irons as Revealed by Thermal Analysis and Directional Solidification Experiments, in *The Physical Metallurgy of Cast Iron*, H. Fredriksson and M. Hillert, Ed., Elsevier, 1985, p 151
24. D.M. Stefanescu, C.R. Loper, Jr., R.C. Voigt, and I.G. Chen, Cooling Curve Structure Analysis of Compacted Vermicular Graphite Cast Irons Produced by Different Melt Treatments, *Trans. AFS*, Vol 90, 1982, p 333
25. C.R. Loper, M.J. Lalich, H.K. Park, and A.M. Gyarmaty, "Microstructure-Mechanical Property Relationship in Compacted/Vermicular Graphite Cast Iron," Paper 35 presented at the 46th International Foundry Congress, Madrid, 1979
26. J.B. von Nesselrode, Gusseisen mit Vermiculargraphit, ein Werkstoff fur Zylinderköpfe, *Giesserei-Prax.*, (No. 23/24), 1979, p 445
27. R. Hummer, Schmelzekontrollverfahren, unpublished research at the Austrian Foundry Research Institute, A.-Nr. 31.253, 1987
28. G. Nandori and J. Dul, Untersuchungen Uber den Abklingeffekt bei Gusseisen mit Kugelgraphit durch Messung der Längen—und Temperaturänderung wahrend der Erstarrung, *Giesserei-Prax.*, (No. 18), 1978, p 284
29. C.F. Walton, T.J. Opar, Ed., *Iron Castings Handbook*, Iron Casting Society Inc., 1981
30. E. Nechtelberger, *The Properties of Cast Iron up to 500 °C*, Austrian Foundry Research Institute, Technicopy Ltd., Stonehouse, England, 1980
31. K.P. Cooper and C.R. Loper, Jr., Some Properties of Compacted Graphite Cast Iron, *Trans. AFS*, Vol 86, 1978, p 241
32. N.N. Aleksandrov, B.S. Milman, L.V. Ilicheva, M.G. Osada, and V.V. Andreev, Production and Properties of High-Duty Iron With Compacted Graphite, *Russ. Cast. Prod.*, Aug 1976, p 319
33. *Spravotchnik po Tshugunomu Litiu (Cast Iron Handbook)*, 3rd ed., Mashinostrojenie, 1978
34. A.F. Heiber, Fracture in Compacted Graphite Iron, *Trans. AFS*, Vol 87, 1979, p 569

35. K.B. Palmer, Mechanical Properties of Compacted Graphite Irons, *BCIRA J.*, Jan 1976, p 31
36. K. Hütterbräucker, O. Vöhringer, and E. Macherauch, Verformunungsverhalten von Ferritischem Gusseisen mit Vermiculargraphit, *Giessereiforschung*, (No. 2), 1978, p 39
37. I. Riposan, M. Chisamera, L. Sofroni, Contribution to the Study of Some Technical and Applicational Properties of Compacted Graphite Cast Iron, *Trans. AFS*, Vol 93, 1985, p 35
38. K. Roehrig, Thermal Fatigue of Gray and Ductile Irons, *Trans. AFS*, Vol 86, 1978, p 75
39. Y.J. Park, R.B. Gundlach, R.G. Thomas, and J.F. Janowak, Thermal Fatigue Resistance of Gray and Compacted Graphite Irons, *Trans. AFS*, Vol 93, 1985, p 415
40. R.W. Monroe and C.E. Bates, Some Thermal and Mechanical Properties of Compacted Graphite Iron, *Trans. AFS*, Vol 90, 1982, p 615
41. A.E. Krivosheev, B.V. Marinchenkov, and N.M. Fettisow, Corrosion Resistance of Cast Iron Without Additive Treatment, *Russ. Cast. Prod.*, 1973, p 86
42. C.W. Phillips, Machinability of Compacted Graphite Iron, *Trans. AFS*, Vol 90, 1982, p 47
43. W. Thury and R. Hummer, Herstellung von Perlitischem Gusseisen mit Vermiculargraphit, unpublished research at the Austrian Foundry Research Institute, A.-Nr. 19.175 from 8.29.1972 and A.-Nr. 18.692 from 12.12.1972
44. *Giessereikalender 1981*, Giesserei-Verlag, Dusseldorf, Federal Republic of Germany, 1981, p 83
45. J.B. von Nesselrode, Gusseisen mit Vermiculargraphit-Seine Anwendung für gegossene Bauteile im Grossmotorenbau, *Giesserei*, Vol 72 (No. 13), 1985, p 390
46. Foote CGJ Report 4, Ferroalloys Division, Foote Mineral Company

High-Alloy White Irons

Richard B. Gundlach, Climax Research Services

HIGH-ALLOY WHITE CAST IRONS are an important group of materials whose production must be considered separately from that of ordinary types of cast irons. In these cast iron alloys, the alloy content is well above 4%, and consequently they cannot be produced by ladle additions to irons of otherwise standard compositions. They are usually produced in foundries specially equipped to produce highly alloyed irons. These iron alloys are most often melted in electric furnaces, specifically electric arc furnaces and induction furnaces, in which the precise control of composition and temperature can be achieved. The foundries usually have the equipment needed to handle the heat treatment and other thermal processing unique to the production of these alloys.

The high-alloy white irons are primarily used for abrasion-resistant applications and are readily cast into the parts needed in machinery for crushing, grinding, and handling of abrasive materials. The chromium content of high-alloy white irons also enhances their corrosion-resistant properties. The large volume fraction of primary and/or eutectic carbides in their microstructures provides the high hardness needed for crushing and grinding other materials. The metallic matrix supporting the carbide phase in these irons can be adjusted by alloy content and heat treatment to develop the proper balance between the resistance to abrasion and the toughness needed to withstand repeated impact.

All high-alloy white irons contain chromium to prevent the formation of graphite upon solidification and to ensure the stability of the carbide phase. Most also contain nickel, molybdenum, copper, or combinations of these alloying elements to prevent the formation of pearlite in the microstructure. While low-alloy white iron castings, which have an alloy content below 4%, develop hardnesses in the range of 350 to 550 HB, the high-alloy irons range in hardness from 450 to 800 HB. In addition, several grades contain alloy eutectic carbides (M_7C_3 chromium carbides), which are substantially harder than the M_3C iron carbides in low-alloy irons. For many applications, the increased abrasion resistance of the more expensive high-alloy white irons adds significantly to wear life, enabling them to provide the most cost-effective performance.

Alloy Grades. Specification ASTM A 532 covers the composition and hardness of the abrasion-resistant white iron grades (see Table 1). Many castings are ordered according to these specifications. However, a large number of castings are produced with composition modifications for specific applications. It is most desirable that the designer, metallurgist, and foundryman work together to specify the composition, heat treatment, and foundry practice to develop the most suitable alloy and casting design for a specific application.

The high-alloy white cast irons fall into two major groups:

- Nickel-chromium white irons, which are low-chromium alloys containing 3 to 5% Ni and 1 to 4% Cr, with one alloy modification that contains 7 to 11% Cr
- The chromium-molybdenum irons containing 11 to 23% Cr, up to 3% Mo and often additionally alloyed with nickel or copper.

A third group comprises the 25% or 28% Cr white irons, which may contain other alloying additions of molybdenum and/or nickel up to 1.5%. The nickel-chromium irons are also commonly identified as Ni-Hard types 1 to 4.

Nickel-Chromium White Irons

The oldest group of high-alloy irons of industrial importance, the nickel-chromium white irons, or Ni-Hard irons, have been produced for more than 50 years and are very cost-effective materials for crushing and grinding. In these martensitic white irons, nickel is the primary alloying element because at levels of 3 to 5% it is effective in suppressing the transformation of the austenite matrix to pearlite, thus ensuring that a hard, martensitic structure (usually containing significant amounts of retained austenite) will develop upon cooling in the mold. Chromium is included in these alloys, at levels from 1.4 to 4%, to ensure that the irons solidify carbidic, that is, to counteract the graphitizing effect of nickel. A typical microstructure is shown in Fig. 1.

The optimum composition of a nickel-chromium white iron alloy depends on the properties required for the service conditions and the dimensions and weight of the casting. Abrasion resistance is generally a function of the bulk hardness and the volume of carbide in the microstructure. When abrasion resistance is the principal requirement and resistance to impact loading is secondary, alloys having high carbon contents, ASTM A 532 class I type A (Ni-Hard 1), are recommended. When conditions of repeated impact are anticipated, the lower-carbon alloys, class I type B (Ni-Hard 2), are recommended because they have less carbide and, therefore, greater toughness. A special grade, class I type C, has been developed for producing grinding balls and slugs. Here, the nickel-chromium alloy composition has been adapted for chill casting and specialized sand casting processes.

The class I type D (Ni-Hard 4) alloy is a modified nickel-chromium iron that contains higher levels of chromium, ranging from 7 to 11%, and increased levels of nickel, ranging from 5 to 7%. Whereas the eutectic carbide phase in the lower-alloyed nickel-chromium irons is M_3C (iron carbide), which forms as a continuous network in these irons, the higher chromium in the type D alloy promotes M_7C_3 chromium carbides, which form a relatively discontinuous eutectic carbide distribution (Fig. 2). This modification in the eutectic carbide pattern provides an appreciable improvement in resistance to fracture by impact. The higher alloy content of this iron grade also results in improved corrosion resistance, which has proved to be useful in the handling of corrosive slurries.

Melting Practice

A principal advantage of the nickel-chromium irons is that they can be melted in a cupola; however, better control of composition and temperature are achieved with electric furnace melting. Consequently, electric arc furnace melting and induction furnace melting are most common. The nickel-chromium irons are readily made in either acid-, neutral-, or basic-lined electric furnaces. They are normally dead melted, and there is usually no reason to use oxygen lancing except as a means of slightly reduc-

Table 1 Composition and mechanical requirements of abrasion-resistant cast irons per ASTM A 532

Class	Type	Designation	Composition: C	Mn	Si	Ni	Cr	Mo
I	A	Ni-Cr-HC	3.0–3.6	1.3 max	0.8 max	3.3–5.0	1.4–4.0	1.0 max(a)
I	B	Ni-Cr-LC	2.5–3.0	1.3 max	0.8 max	3.3–5.0	1.4–4.0	1.0 max(a)
I	C	Ni-Cr-GB	2.9–3.7	1.3 max	0.8 max	2.7–4.0	1.1–1.5	1.0 max(a)
I	D	Ni-Hi Cr	2.5–3.6	1.3 max	1.0–2.2	5.0–7.0	7.0–11.0	1.0 max(b)
II	A	12% Cr	2.4–2.8	0.5–1.5	1.0 max	0.5 max	11.0–14.0	0.5–1.0(c)
II	B	15% Cr-Mo-LC	2.4–2.8	0.5–1.5	1.0 max	0.5 max	14.0–18.0	1.0–3.0(c)
II	C	15% Cr-Mo-HC	2.8–3.6	0.5–1.5	1.0 max	0.5 max	14.0–18.0	2.3–3.5(c)
II	D	20% Cr-Mo-LC	2.0–2.6	0.5–1.5	1.0 max	1.5 max	18.0–23.0	1.5 max(c)
II	E	20% Cr-Mo-HC	2.6–3.2	0.5–1.5	1.0 max	1.5 max	18.0–23.0	1.0–2.0(c)
III	A	25% Cr	2.3–3.0	0.5–1.5	1.0 max	1.5 max	23.0–28.0	1.5 max(c)

Class	Type	Designation	Mechanical requirements: Hardness, HB: Sand cast, min	Chill cast, min	Hardened, min	Softened, max	Typical section thickness, max: in.	mm
I	A	Ni-Cr-HC	550	600	. . .	. . .	8	200
I	B	Ni-Cr-LC	550	600	. . .	. . .	8	200
I	C	Ni-Cr-GB	550	600	. . .	. . .	3 diam ball	75 diam ball
I	D	Ni-Hi Cr	550	500	600	400	12	300
II	A	12% Cr	550	. . .	600	400	1 diam ball	25 diam ball
II	B	15% Cr-Mo-LC	450	. . .	600	400	4	100
II	C	15% Cr-Mo-HC	550	. . .	600	400	3	75
II	D	20% Cr-Mo-LC	450	. . .	600	400	8	200
II	E	20% Cr-Mo-HC	450	. . .	600	400	12	300
III	A	25% Cr	450	. . .	600	400	8	200

(a) Maximum: 0.30% P, 0.15% S. (b) Maximum: 0.10% P, 0.15% S. (c) Maximum: 0.10% P, 0.06% S, 1.2% Cu

ing carbon content. Acid linings are generally more economical than basic linings. The type of lining affects silicon and chromium losses. Very little adjustment to slag composition is necessary in acid melting, which is used for the majority of current production.

Normal charge materials are various kinds of steel scrap, foundry returns, or returns of similar alloy from service. For foundries that lack the facilities for rapid melt analysis, it may be necessary to select steel scrap rather carefully to ensure that the residual levels of alloying elements, such as manganese and copper, which have a potent effect on austenite retention, are consistent and under control.

Chromium is obtained in the form of high-carbon ferrochrome and is generally added near the end of the heat to avoid excessive oxidation losses. Carbon is obtained from electrode graphite, petroleum coke, and other sources. Pig iron is also used to carburize the melt and can be an additional source of silicon. Silicon and manganese are added as ferroalloys. Molybdenum is usually added as ferromolybdenum; however, with arc furnaces, molybdenum oxide briquettes may be used. Sulfur is limited to 0.06%, and phosphorus is kept to 0.10%, as specified by ASTM A 532.

High superheating temperatures have not been necessary when melting in induction furnaces, which have good stirring action, and a final temperature of 1480 °C (2700 °F) is usually adequate for thick-section castings. Final temperatures of up to 1565 °C (2850 °F) are commonly used in arc furnace practice to ensure homogeneity of the bath composition and to accelerate the solution of carbon, added after meltdown, and late alloy additions. No particular problems have been associated with high super-heat temperatures, but it is more difficult to predict melting losses.

As in steel melting, deoxidation of the bath with aluminum was common, but it is no longer practiced by most producers, with no adverse effects on the soundness or mechanical properties of the casting. A late addition/inoculation of foundry-grade ferrosilicon is often made to improve toughness.

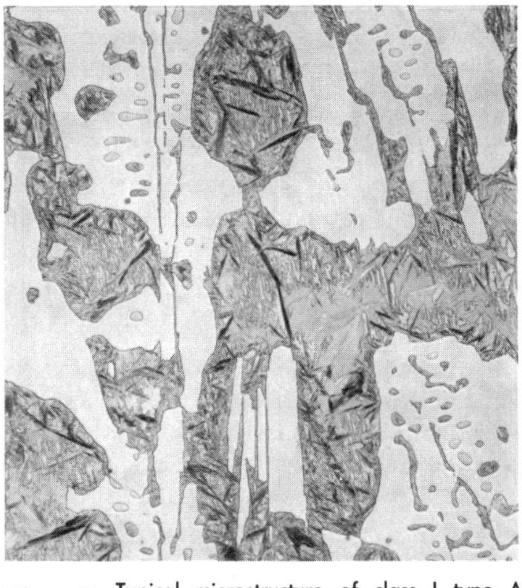

Fig. 1 Typical microstructure of class I type A nickel-chromium white cast iron. 340×

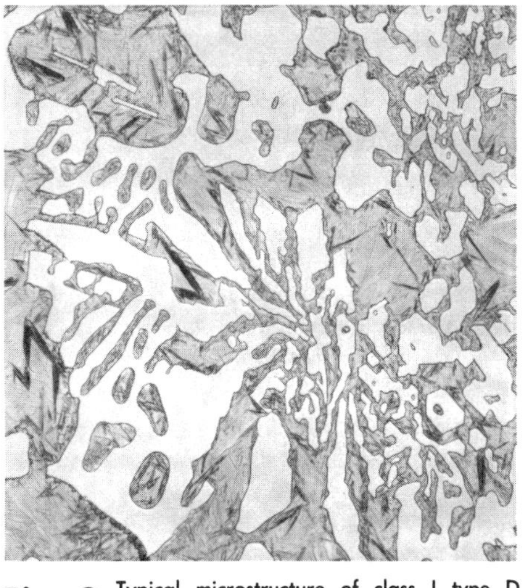

Fig. 2 Typical microstructure of class I type D nickel-chromium white cast iron. 340×

Composition Control

Carbon is varied according to the properties needed for the intended service. Carbon contents in the range of 3.2 to 3.6% are prescribed when maximum abrasion resistance is desired. When impact loading is expected, carbon content should be held in the range of 2.7 to 3.2%.

Nickel content increases with section size or cooling time of the casting to inhibit pearlitic transformation. For castings of 38 to 50 mm (1.5 to 2 in.) thick, 3.4 to 4.2% Ni is sufficient to suppress pearlite formation upon mold cooling. Heavier sections may require nickel levels up to 5.5% to avoid the formation of pearlite. It is important to limit nickel content to the level needed for control of pearlite; excess nickel increases the amount of retained austenite and lowers hardness.

Silicon is needed for two reasons. A minimum amount of silicon is necessary to improve fluidity of the melt and to produce a fluid slag, but of equal importance is its effect on as-cast hardness. Increased levels of silicon, in the range of 1 to 1.5%, have been found to increase the amount of martensite and the resulting hardness. Late additions of ferrosilicon (0.2% as 75% Si grade ferrosilicon) have been reported to increase toughness. It is important to note that higher silicon contents can promote pearlite and may increase the nickel requirement.

Chromium, primarily added to offset the graphitizing effects of nickel and silicon in types A, B, and C alloys, ranges from 1.4 to 3.5%. Chromium content must increase with increasing section size. In type D alloy, chromium levels range from 7 to 11% (typically 9%) for the purpose of producing eutectic carbides of the M_7C_3 chromium carbide type, which are harder and less deleterious to toughness.

Manganese is typically held to a maximum of 0.8% even though 1.3% maximum is allowed according to ASTM A 532 specification. While it provides increased hardenability to avoid pearlite formation, it is a more potent austenite stabilizer than nickel, and promotes increased amounts of retained austenite and lower as-cast hardness. For this reason, higher manganese levels are undesirable. When considering the nickel content required to avoid pearlite in a given casting, the level of manganese present should be a factor.

Copper increases both hardenability and the retention of austenite and therefore must be controlled for the same reason that manganese must be limited. Copper should be treated as a nickel substitute and, when properly included in the calculation of the amount of nickel required to inhibit pearlite, it reduces the nickel requirement.

Molybdenum is a potent hardenability agent in these alloys and is used in heavy-section castings to augment hardenability and inhibit pearlite.

Pouring Practices

High pouring temperatures aggravate shrinkage under feeder heads and other hot spots and can lead to microshrinkage and coarse dendritic structures. Careful control of pouring temperature is particularly important in the consistent production of thick-section castings. Low pouring temperatures are necessary not only to avoid shrinkage defects but also to avoid problems such as metal penetration and burned-on sand. Low pouring temperatures are also effective in controlling dendrite size and the coarseness of the eutectic carbide structure.

The eutectic temperature for the various nickel-chromium iron grades is approximately 1200 °C (2190 °F), and solidification generally begins at temperatures of 1200 to 1280 °C (2190 to 2340 °F), depending on composition. When selecting a pouring temperature, common rules apply. Pouring temperatures are seldom lower than 100 °C (180 °F) above the liquidus temperature. Higher pouring temperatures may be necessary when pouring smaller castings. Of course, casting configurations must be considered when selecting optimum pouring temperatures.

Molds, Patterns, and Casting Design

These alloys may be sand cast or chill cast in permanent molds. The chill cast microstructures develop higher hardness, strength, and impact toughness over the sand cast structures, because of the finer carbides. It is recommended that, whenever practical, the wearing faces of the casting be cast against a chill to improve abrasion resistance and toughness.

Sand molds may be made either of green sand or of dry sand, oil sand, or steel casting sand. Air-setting and thermal-setting resin sands are also commonly used. Molds must be rigid to minimize shrinkage defects.

White irons are subject to hot tearing, unless suitable precautions are taken. Occasionally when castings are extracted from the mold, they are found to be cracked. Stresses large enough to cause fracture can arise when either the mold or the cores are excessively strong and there is inadequate mold or core breakdown to allow normal thermal contraction with cooling. Large cores, such as those used in pump volutes, and even small cores used to form bolt holes, can cause cracking of this type.

Shrinkage allowance is 13 to 21 mm/m (5⁄32 to 1⁄4 in./ft); a commonly used figure is 16 mm/m (3⁄16 in./ft). The actual pattern allowance depends on the choice of alloy and the geometry of the casting. Patterns should employ generous radii and avoid sharp section changes to avoid initiating cracking upon solidification and subsequent cooling.

These alloys exhibit relatively high liquid-to-solid shrinkage (5%); therefore, large gates and risers are required for feeding. Special consideration should be given to the design of end gates and the positioning of risers so that they can be readily removed. Risers cannot be removed by burning and are often notched or necked down to facilitate removal by impact. They are most easily removed after the castings have cooled to room temperature, that is, when martensite is present. Martensitic transformation begins at temperatures below 230 °C (450 °F).

Shakeout Practices

The shakeout practice is a critical step in the successful production of nickel-chromium iron castings. High residual stresses and cracking can result from extracting castings from a mold at too high a temperature. Cooling all the way to room temperature in the mold is desirable because, as stated above, transformation to martensite occurs below 230 °C (450 °F) during the last stages of cooling. This precaution is mandatory in heavy-section castings; the volume increase that occurs with martensite formation results in the development of high transformation stresses in castings having steep temperature gradients due to rapid cooling. Thin-section castings and castings having simple shapes are less susceptible to stresses and are often removed from the mold once the castings have reached black heat. With these alloys, slow cooling favors higher as-cast hardness.

Heat Treatment

All nickel-chromium white iron castings are given a stress-relief heat treatment because, properly made, they have a martensitic matrix structure, as-cast. Tempering is performed between 205 to 260 °C (400 to 500 °F) for at least 4 h. This tempers the martensite, relieves some of the transformation stresses, and increases the strength and impact toughness by 50 to 80%. Some additional martensite may form upon cooling from the tempering temperature. This heat treatment does not reduce hardness or abrasion resistance.

High-Temperature Heat Treatment. In the past, hardening of the class I type D, Ni-Hard 4, was performed by supercritical heat treatment when as-cast hardness was insufficient. An austenitizing heat treatment usually comprised heating the casting at temperatures between 750 and 790 °C (1380 and 1450 °F) with a soak time of 8 h. Air or furnace cooling, not over 30 °C/h (50 °F/h), was conducted, followed by a tempering/stress-relief heat treatment.

Refrigeration treatment is the more commonly practiced remedy for low hardness today. To achieve a hardness of 550 HB, it is necessary for the as-cast austenite-martensite microstructure to have at least 60% martensite present. When martensite content is increased to 80 to 90%, however, hardness values exceed 650 HB. To reduce the amount of retained austenite, that is, form more martensite, deep freeze treatments are commonly applied. Refrigeration to temperatures between −70 and −185 °C (−90 and −300 °F) for 1⁄2 to 1 h will usually raise the hardness level 100 HB units. A tempering/stress-relief heat treatment usually follows. The typical refrigerated nickel-chromium iron microstructure is shown in Fig. 3.

In the heat treatment of any white cast iron, care must be taken to avoid cracking by thermal shock; the castings must never be placed in a hot furnace or otherwise subjected to rapid heating or cooling. The risk of cracking increases with the complexity of the casting shape and section thickness.

Machining

Alloyed white irons are generally ground to finish size; care must be taken to avoid cracking by overheating. Single-point turn-

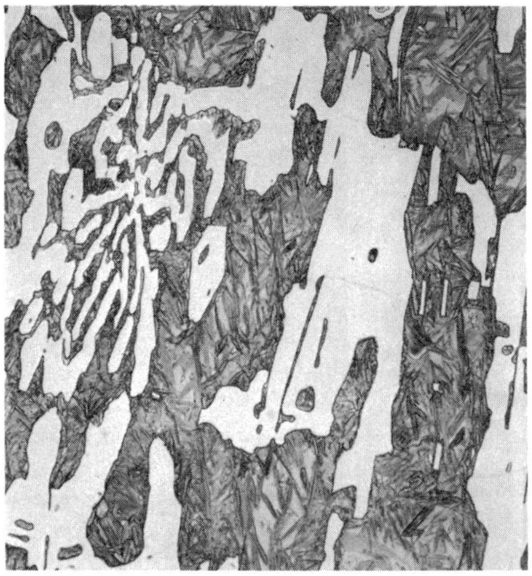

Fig. 3 Microstructure of class I type D nickel-chromium white cast iron after refrigeration. 340×

ing and boring can be done using heavy tooling, rigid setups, sufficient power, and carbide- or boride-tipped cutters or special ceramic inserts. Electrical discharge machining and electrochemical machining are also feasible.

Applications

Because of their low cost, the martensitic nickel-chromium white irons are consumed in large tonnages in mining operations as ball mill liners and grinding balls. Class I type A castings are used in applications requiring maximum abrasion resistance, such as ash pipes, slurry pumps, roll heads, muller tires, augers, coke-crusher segments, classifier shoes, brick molds, pipe elbows carrying abrasive slurries, and grizzly disks. Type B is recommended for applications requiring more strength and exerting moderate impact, such as crusher plates, crusher concaves, and pulverizer pegs.

Class I type D, Ni-Hard type 4, has a higher level of strength and toughness and is therefore used for the more severe applications that justify its added alloy costs. It is commonly used for pump volutes handling abrasive slurries and coal pulverizer table segments and tires.

The class I type C alloy (Ni-Hard 3) is specifically designed for the production of grinding balls. This grade is both sand cast and chill cast. Chill casting has the advantage of lower alloy cost, and, more important, provides a 15 to 30% improvement in life. All grinding balls require tempering for 8 h at 260 to 315 °C (500 to 600 °F) to develop adequate impact toughness.

Special Nickel-Chromium White Iron Alloys

Certain proprietary grades of type A alloy have been developed by the rolling mill industry. The compositions of these alloys have been modified to produce mottled structures, containing some graphite. The graphite inclusions are reported to improve resistance to thermal cracking. These indefinite chill rolls are cast in thick-wall gray iron chiller molds in roll diameters of up to 1015 mm (40 in.) or more. The silicon to chromium ratios and inoculation with ferrosilicon are carefully controlled to control the amount and distribution of the graphite particles. The rolls can be double poured with a gray iron core. With molybdenum modification, the matrix of the chill cast shell becomes martensitic. Some roll alloys are designed to be heat treated, that is, by a modified normalizing heat treatment, to obtain a bainitic microstructure.

High-Chromium White Irons

The high-chromium white irons have excellent abrasion resistance and are used effectively in slurry pumps, brick molds, coal-grinding mills, shot-blasting equipment, and components for quarrying, hard-rock mining, and milling. In some applications they must also be able to withstand heavy impact loading. These alloyed white irons are recognized as providing the best combination of toughness and abrasion resistance attainable among the white cast irons.

In the high-chromium irons, as with most abrasion-resistant materials, there is a trade-off between wear resistance and toughness. By varying composition and heat treatment, these properties can be adjusted to meet the needs of most abrasive applications.

As a class of alloyed irons, the high-chromium irons are distinguished by the hard, relatively discontinuous M_7C_3 eutectic carbides present in the microstructure, as opposed to the softer, more continuous M_3C eutectic carbides present in the alloyed irons containing less chromium. These alloys are usually produced as hypoeutectic compositions.

Classes of High-Chromium Irons

Specification ASTM A 532 covers the compositions and hardnesses of two general classes of the high-chromium irons (see Table 1). The chromium-molybdenum irons (class II of ASTM A532) contain 11 to 23% Cr and up to 3.5% Mo and can be supplied either as-cast with an austenitic or austenitic-martensitic matrix, or heat treated with a martensitic matrix microstructure for maximum abrasion resistance and toughness. They are usually considered the hardest of all grades of white cast irons. Compared to the lower-alloy nickel-chromium white irons, the eutectic carbides are harder and can be heat treated to achieve castings of higher hardness. Molybdenum, as well as nickel and copper when needed, is added to prevent pearlite and to ensure maximum hardness.

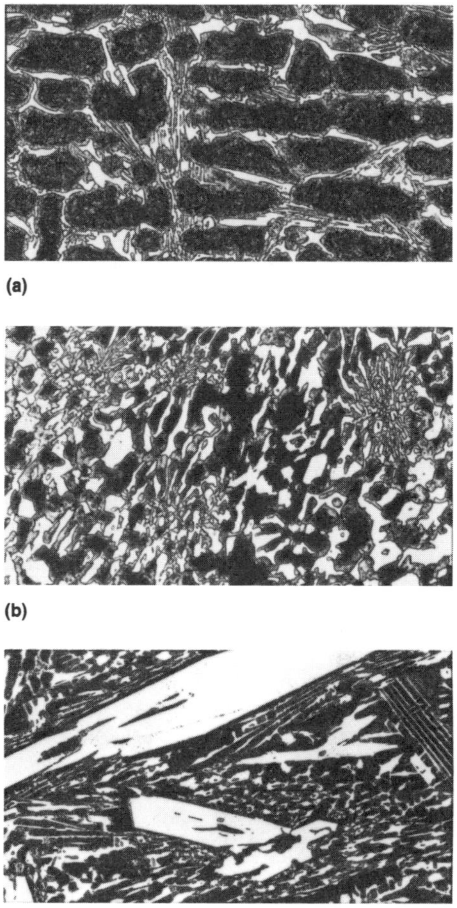

(a) (b) (c)

Fig. 4 Microstructures of high-chromium white iron compositions. (a) Low carbon (hypoeutectic). (b) Eutectic. (c) High-carbon (hypereutectic). All 75×. Courtesy of Climax Molybdenum Company

The high-chromium irons (class III of ASTM A 532) represent the oldest grade of high-chromium irons, with the earliest patents dating back to 1917. These general-purpose irons, also called 25% Cr and 28% Cr irons, contain 23 to 28% Cr with up to 1.5% Mo. To prevent pearlite and attain maximum hardness, molybdenum is added in all but the lightest-cast sections. Alloying with nickel and copper up to 1% is also practiced. Although the maximum attainable hardness is not as high as in the class II chromium-molybdenum white irons, these alloys are selected when resistance to corrosion is also desired.

Microstructures

Carbide. The carbides in high-chromium irons are very hard and wear resistant but are also brittle. In general, wear resistance is improved by increasing the amount of carbide (increasing the carbon content), while toughness is improved by increasing the proportion of metallic matrix (reducing the carbon content). The influence of carbon content on the shape and distribution of the carbide phase in these alloys is illustrated in the photomicrographs of Fig. 4.

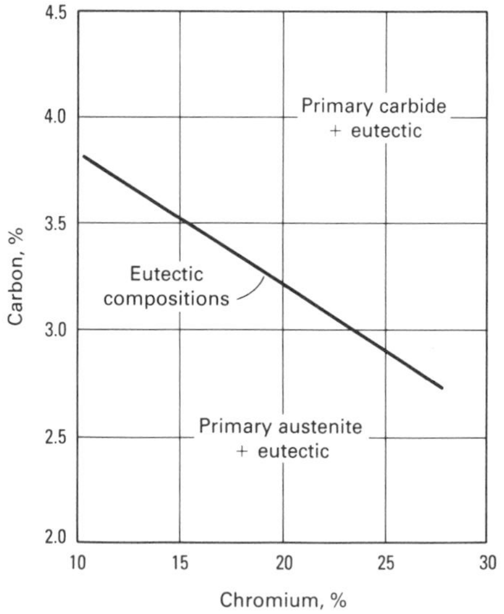

Fig. 5 Relationship between the chromium and carbon contents and the eutectic composition in high-chromium irons

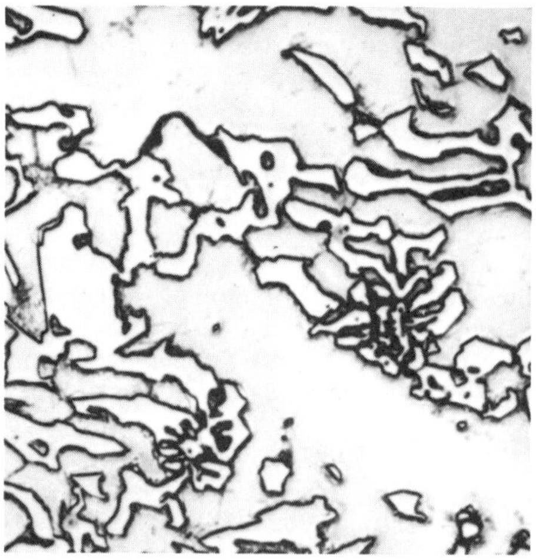

Fig. 6 High-chromium iron with an as-cast austenitic matrix microstructure. 500×. Courtesy of Climax Molybdenum Company

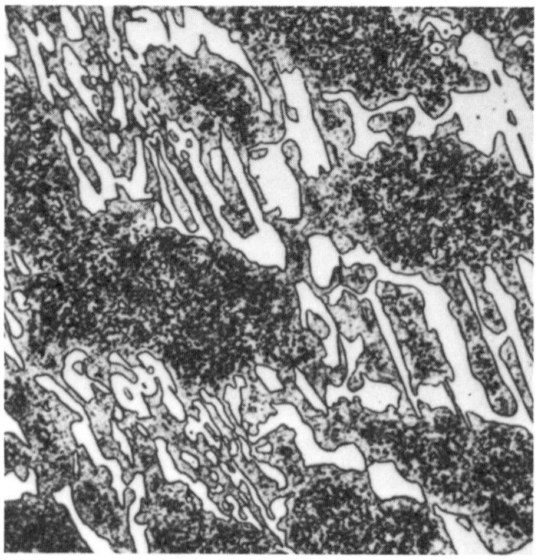

Fig. 7 High-chromium iron with an as-cast austenitic-martensitic matrix microstructure. 500×. Courtesy of Climax Molybdenum Company

Large hexagonal carbide rods occur when carbon contents exceed the eutectic carbon content (Fig. 4c). These primary chromium carbides, which precipitate from the melt before eutectic solidification, are quite deleterious to impact toughness and should be avoided in castings subjected to any impact in service. The eutectic carbon content changes inversely with the chromium content in these alloys. The relationship between eutectic carbon content and chromium content is shown in Fig. 5.

As-Cast Austenitic. Solidification in the hypoeutectic alloys occurs by the formation of austenite dendrites followed by the eutectic formation of austenite and M_7C_3 chromium carbides. Under equilibrium conditions, additional chromium carbide precipitates from the austenite matrix upon cooling from the eutectic to the critical temperature, that is, about 760 °C (1400 °F), and transformation to ferrite and carbide occurs with subsequent cooling. However, when cooling under nonequilibrium conditions such as those encountered in most commercial castings, the austenite becomes supersaturated in carbon and chromium. Because of elevated carbon and chromium contents, a metastable austenitic cast iron normally develops, provided pearlitic and bainitic transformations have been inhibited (Fig. 6). With sufficient alloying with molybdenum, manganese, nickel, and copper, pearlitic transformation can be avoided in virtually any cast section.

As-Cast Martensitic. Martensitic structures can be obtained as-cast in heavy-section castings that cool slowly in the mold. With slow cooling rates, austenite stabilization is incomplete, and partial transformation to martensite occurs. But in these castings, martensite is mixed with large amounts of retained austenite (Fig. 7), and therefore hardness levels are lower than can be achieved in heat-treated martensitic castings. These castings must contain sufficient alloy to suppress pearlite upon cooling. Some compositions (higher silicon) have been developed to assist martensite formation in refrigeration treatments. Subcritical heat treatment has been used to reduce austenite content and, at the same time, increase hardness and toughness.

Heat-Treated Martensitic. To obtain maximum hardness and abrasion resistance, martensitic matrix structures must be produced by full heat treatment. The casting must contain sufficient alloy to avoid pearlite formation upon cooling from the heat-treatment temperature.

Selecting Compositions to Obtain Desired Structures

Many complex sections, such as slurry pump components, are often used in the as-cast austenitic/martensitic condition to avoid the possibility of cracking and distortion when heat treated. To prevent pearlite in mold cooling, alloying additions are usually required. As the carbon content is increased, more chromium is consumed, forming additional carbide, and therefore, larger alloying additions are required. Table 2 presents a guide for appropriate alloying to prevent pearlite in the various classes of as-cast irons.

Optimum performance is usually achieved with heat-treated martensitic structures. Again, alloying must be sufficient to ensure a pearlitefree microstructure, with heat treatment. Of necessity, the heat treatment, discussed later in detail (see the section "Quenching" in this article), requires an air quench from the austenitizing temperature. Faster cooling rates should not be used, because the casting can develop cracks due to high thermal and/or transformation stresses. Thus, the alloy must have sufficient hardenability to allow air hardening. However, overalloying with manganese, nickel, and copper promotes retained austenite, which detracts from resistance to abrasion and spalling. Therefore, it is best to obtain adequate hardenability primarily with molybdenum. Table 3 is offered as a guide to alloying for air quenching heat-treated castings of various sections.

Melting Practice

The high level of carbon pickup with the high chromium contents in these alloys precludes the use of cupola melting. Consequently, electric arc and induction furnaces are normally used. The high-chromium

Table 2 Minimum alloy content to avoid pearlite in mold-cooled castings for indicated effective section size (plate thickness or radius of rounds)

ASTM A532 class	Cr(a), %	C(a), %	Plate thickness or radius of rounds: 25 mm (1 in.)	50 mm (2 in.)	100 mm (4 in.)
IIB, C	14–18	2.0	1.0 Mo	1.5 Mo	1.5 Mo + 1.0 (Ni + Cu)
		3.5	2.0 Mo	2.5 Mo	2.5 Mo + 1.0 (Ni + Cu)
IID, E	18–23	2.0	0.5 Mo	1.0 Mo	1.0 Mo + 1.0 (Ni + Cu)
		3.2	1.5 Mo	2.0 Mo	2.0 Mo + 1.0 (Ni + Cu)
IIIA	23–28	2.0	...	0.5 Mo	1.0 Mo
		3.0	1.0 Mo	1.5 Mo	1.5 Mo + 1.0 (Ni + Cu)

(a) In base irons containing 0.6% Si and 0.8% Mn

Table 3 Minimum alloy content to avoid pearlite in heat treatment for indicated effective section size (plate thickness or radius of rounds)

ASTM A532 class	Cr(a), %	C(a), %	Plate thickness or radius of rounds: 50 mm (2 in.)	125 mm (5 in.)	150–255 mm (6–10 in.)
IIB, C	14–18	2.0	1.5 Mo	1.5 Mo + 0.5 (Ni + Cu)	2.0 Mo + 1.0 (Ni + Cu)
		3.5	3.0 Mo	2.0 Mo + 1.0 (Ni + Cu)	2.5 Mo + 1.2 (Ni + Cu)(b)
IID, E	18–23	2.0	1.0 Mo	2.0 Mo	2.0 Mo + 0.5 (Ni + Cu)
		3.2	1.5 Mo	2.0 Mo + 0.7 (Ni + Cu)	2.0 Mo + 1.2 (Ni + Cu)(b)
IIIA	23–28	2.0	0.5 Mo	1.5 Mo	1.5 Mo + 0.5 (Ni + Cu)
		3.0	1.5 Mo	1.5 Mo + 0.6 (Ni + Cu)	1.5 Mo + 1.2 (Ni + Cu)(b)

(a) In base irons containing 0.6% Si and 0.8% Mn. (b) Nickel and copper promote retained austenite and should be restricted to combined levels of 1.2% maximum; manganese behaves similarly and should be restricted to 1.0% maximum.

irons are readily made in acid-, neutral-, or basic-lined electric furnaces. They are normally dead melted, and there is usually no reason to use oxygen lancing except as a means of slightly reducing carbon content. Despite rapid refractory wear due to chromium-bearing slag, acid linings are generally more economical than basic linings. Very little adjustment to slag composition is necessary in acid melting, which is used for the majority of current production. Viscous or nonfluid slags should be avoided.

Normal charge materials are various kinds of steel scrap, foundry returns, or returns of similar alloy from service. For foundries that lack the facilities for rapid melt analysis, it may be necessary to select steel scrap rather carefully to ensure that the total content of alloying elements that have a potent effect on hardenability and austenite retention is consistent and under control.

While some furnace operators prefer to add a portion of high-carbon ferrochrome to the charge, it is generally added near the end of the heat to avoid excessive oxidation losses. Carbon is obtained from electrode graphite, petroleum coke, and other sources. If pig iron is used to carburize the melt, it should have a low silicon content. The ideal silicon content is 0.6%. Less than 0.4% Si in the bath can cause difficulties with viscous slag, while higher silicon contents can promote pearlite. Selection of an incorrect grade of ferrochromium with high silicon is a common cause for an excessively high silicon content. Manganese can range from 0.5 to 1.5%, according to ASTM A 532 (see Table 1). Manganese on the high side of this range can erode acid furnace linings. Therefore, manganese content should be limited to 1% in meltdown; the remainder can be added to the ladle as crushed ferroalloy. Molybdenum is usually added as ferromolybdenum; however, with arc furnaces, molybdenum oxide briquettes may be used. Sulfur is limited to 0.06%, and phosphorus is kept to 0.10%, as specified by ASTM A 532.

High superheating temperatures have not been necessary when melting in induction furnaces, which have good stirring action, and a final temperature of 1480 °C (2700 °F) is usually adequate for thick-section castings. Final temperatures of up to 1565 °C (2850 °F) are commonly used in arc furnace practice to ensure homogeneity of the bath composition and to accelerate the solution of carbon, added after meltdown, and late alloy additions. No particular problems have been associated with high superheat temperatures, but it is more difficult to predict melting losses.

As in steel melting, deoxidation of the bath with aluminum was common in the past, but this practice has been discontinued by most producers with no adverse effects on the soundness or mechanical properties of the casting. Titanium is sometimes added to limit dendrite size. There have been reports that bath additions of aluminum and titanium aggravate feeding problems. For heats in which hardenability is marginal, there is evidence that these elements tend to promote pearlite with mold cooling or in heat-treated castings.

Pouring Practices

High pouring temperatures aggravate shrinkage under feeder heads and other hot spots and can lead to microshrinkage and coarse dendritic structures. Careful control of pouring temperature is particularly important in the consistent production of thick-section castings. Low pouring temperatures are necessary to avoid shrinkage defects and prevent problems such as metal penetration and burned-on sand. Low pouring temperatures are also effective in controlling dendrite size and the coarseness of the eutectic carbide structure.

The eutectic temperature for the various high-chromium iron grades ranges from 1230 to 1270 °C (2245 to 2315 °F), and solidification generally begins at temperatures up to 1350 °C (2460 °F), depending on composition. Pouring temperatures are seldom lower than 100 °C (180 °F) above the liquidus temperature. Castings thicker than 102 mm (4 in.) are generally poured between 1345 and 1400 °C (2450 and 2550 °F). Higher pouring temperatures may be necessary for smaller castings. Casting configurations also must be considered when selecting optimum pouring temperatures.

In the ladle or during pouring, the metal appears cold and sluggish because an oxide film readily forms on the surface; but the metal is actually quite fluid and may be poured into intricate shapes. This rather viscous surface oxide is less liable to cause surface defects on castings that are poured quickly. Even when the metal is poured on the cold side, flasks have to be clamped, or weighted, to prevent the metal from running out at the parting line.

Molds, Patterns, and Casting Design

Molds must be rigid to minimize shrinkage defects. Molds may be made of either green sand, or of dry sand, oil sand, or steel casting sand. Air-setting and thermal-setting resin sands are also commonly used. Because white irons are subject to hot tearing, cores should break down readily.

Unless suitable precautions are taken, the high-chromium irons are somewhat more prone to crack in the foundry than are white irons of lower alloy content. Castings are occasionally found to be cracked when extracted from the mold. Stresses large enough in magnitude to cause fracture can arise when either the mold or cores are excessively strong and there is inadequate mold or core breakdown to allow normal thermal contractions upon cooling. Large cores such as those used in pump volutes, and even small cores used to form bolt holes, can cause cracking of this type.

Shrinkage allowance is 18 to 21 mm/m (7/32 to 1/4 in./ft). The actual pattern allowance depends on the choice of subsequent heat treatment procedures, if any, and the geometry of the casting. Patterns should employ generous radii and avoid sharp section changes to avoid initiating cracking on solidification and subsequent cooling.

These alloys exhibit relatively high liquid-to-solid shrinkage; therefore, large gates and risers are required for feeding. Special consideration should be given to the design of end gates and to the positioning of risers so that they can be readily removed. Risers are often notched or necked down to facilitate removal by impact.

Shakeout Practices

The shakeout practice is probably the most critical step in the successful production of high-chromium iron castings. A frequent cause of high residual stresses and of cracking is the common practice of extracting castings from the mold at too high a temperature. Cooling all the way to room temperature in the mold is desirable and can be a requirement to avoid cracking, especially if martensite forms during the last stages of cooling. This precaution is mandatory in heavy-section castings to be used in the as-cast condition where the desired, mold-cooled structure is a mixture of austenite and martensite.

For irons that are heat treated, the desired mold-cooled structure is often pearlite. This softer structure facilitates removal

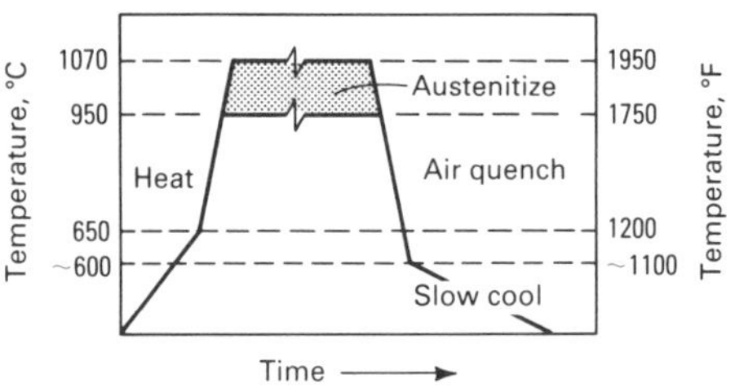

Fig. 8 Heat treatment schedule for hardening high-chromium irons

of gates and risers, minimizes the transformational and thermal stresses that cause cracking, and shortens the response to heat treatment. Careful design of alloy composition ensures the development of a substantially pearlitic structure in the casting after mold cooling, yet provides enough hardenability to prevent pearlite formation during subsequent heat treatment. Heavy-section castings made pearlitic by such an alloy content can often be removed from the mold once the castings have reached black heat.

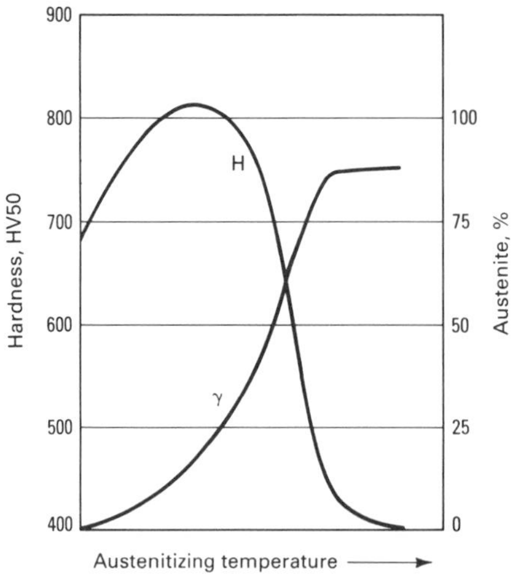

Fig. 9 Influence of austenitizing temperature on hardness (H) and retained austenite (γ) in high-chromium irons

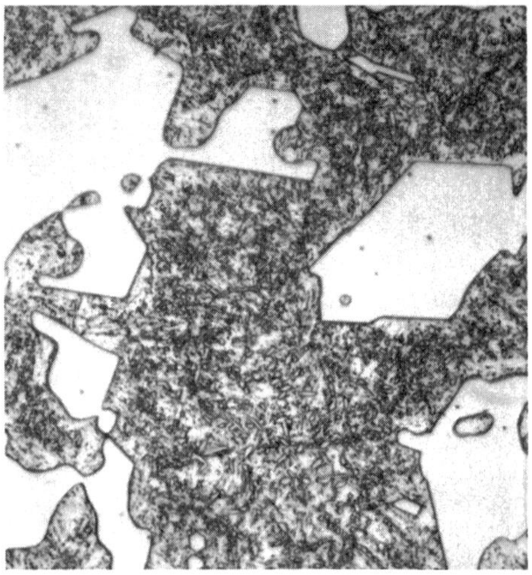

Fig. 10 Microstructure of heat-treated martensitic high-chromium iron illustrating fine secondary M_7C_3 carbides. 680×

Heat Treatment

Toughness and abrasion resistance are improved by heat treatment to a martensitic microstructure. Figure 8, which illustrates the heat treatment process, shows the importance of slow heating to 650 °C (1200 °F) to avoid cracking. For complex shapes, a maximum rate of 30 °C/h (50 °F/h) is recommended. Simple shapes and fully pearlitic castings can be heated at faster rates. The heating rate can be accelerated above red heat.

Austenitization. There is an optimum austenitizing temperature for achieving maximum hardness (Fig. 9), which varies for each composition. The austenitizing temperature determines the amount of carbon that remains in solution in the austenite matrix. Too high a temperature increases the stability of the austenite, while the higher retained austenite content reduces hardness. Low temperatures result in low-carbon martensite, reducing both hardness and abrasion resistance. Because of this sensitivity to temperature, furnaces that can produce accurate and uniform temperatures are most desirable. The successful heat treatment produces austenite destabilization by precipitation of fine, secondary M_7C_3 carbides within the austenite matrix (Fig. 10).

Class II irons containing 12 to 20% Cr are austenitized in the temperature range of 955 to 1010 °C (1750 to 1850 °F). Class III irons containing 23 to 28% Cr are austenitized in the temperature range of 1010 to 1095 °C (1850 to 2000 °F). Heavy sections usually require temperatures on the upper end of the range.

Castings should be held at temperature long enough to accomplish equilibrium dissolution of chromium carbides to ensure a proper hardening response. A minimum of 4 h at temperature is necessary. For heavy sections, the rule of 1 h/in. of section thickness is usually adequate. For castings that are fully pearlitic before heat treatment, the holding time at temperature can be reduced.

Quenching. Air quenching (vigorous fan cooling) the castings from the austenitizing temperature to below the pearlite temperature range, that is, to between 550 and 600 °C (1020 and 1110 °F), is highly recommended. The subsequent cooling rate should be substantially reduced to minimize stresses; still-air or even furnace cooling to ambient temperature is common. Complex and heavy-section castings are often re-placed in the furnace, which is at 550 to 600 °C (1020 to 1110 °F), and allowed sufficient time to reach a uniform temperature within the casting. After the temperature is equalized, the castings are either furnace or still-air cooled to ambient temperature.

Tempering. Castings can be put into service in the hardened (as-cooled) condition without further tempering or subcritical heat treatments; however, tempering in the range of 205 to 230 °C (400 to 450 °F) for 2 to 4 h is recommended to restore some toughness in the martensitic matrix and to further relieve residual stresses. The microstructure after hardening always contains retained austenite in the range of 10 to 30%. Some retained austenite is transformed following tempering at low temperatures, but if spalling is a problem, higher-temperature tempering can be used to further reduce austenite content.

Subcritical Heat Treatment. Subcritical heat treatment (tempering) is sometimes performed, particularly in large heat-treated martensitic castings, to reduce retained austenite content and increase resistance to spalling. The tempering parameters necessary to eliminate retained austenite are very sensitive to time and temperature and depend on the casting composition and prior thermal history. Typical tempering temperatures range from 480 to 540 °C (900 to 1000 °F), and times range from 8 to 12 h. Excess time or temperature results in softening and a drastic reduction in abrasion resistance. Insufficient tempering results in incomplete elimination of austenite. The amount of retained austenite cannot be determined metallographically; those experienced with this heat treatment practice have developed techniques using specialized magnetic instruments to determine the level of retained austenite after tempering.

Annealing. Castings can be annealed to make them more machinable, either by subcritical annealing or by a full anneal. Subcritical annealing is accomplished by pearlitizing by means of soaking in the narrow range between 695 and 705 °C (1280 and 1300 °F) for 4 to 12 h, which produces hardnesses ranging from 400 to 450 HB. Lower hardnesses often can be achieved with full annealing, in which castings are heated in the range of 955 to 1010 °C (1750 to 1850 °F), followed by slow cooling to 760 °C (1400 °F), and holding at this temperature for 10 to 50 h, depending on composition. Annealing affects neither the primary carbides nor the potential for subsequent hardening; guidelines for hardening as-cast castings also apply to annealed castings.

Machining

These alloyed white irons are generally ground to finish size; care must be exercised to avoid cracking by overheating. Single-point turning and boring can be done using heavy tooling, rigid setups, sufficient power, and carbide- or boride-tipped cutters or special ceramic inserts. Electrical discharge machining and electrochemical machining are also feasible.

Applications

The high-chromium white irons are superior in abrasion resistance and are used effectively in impellers and volutes in slurry pumps, classifier wear shoes, brick molds, impeller blades and liners for shot blasting equipment, and refiner disks in pulp refiners. In many applications they withstand heavy impact loading, such as from impact hammers, roller segments and ring segments in coal-grinding mills, feed-end lifter bars and mill liners in ball mills for hard-rock mining, pulverizer rolls, and rolling mill rolls.

Special High-Chromium Iron Alloys

Irons for Corrosion Resistance. Alloys with improved resistance to corrosion, for applications such as pumps for handling fly ash, are produced with high-chromium content (26 to 28% Cr) and low-carbon content (1.6 to 2.0% C). These high-chromium, low-carbon irons provide the maximum chromium content in the matrix. The addition of 2% Mo is recommended for improving resistance to chloride-containing environments. For this application, fully austenitic matrix structures provide the best resistance to corrosion, but some reduction in abrasion resistance must be expected. Castings are normally supplied in the as-cast condition.

Irons for High-Temperature Service. Because of castability and cost, high-chromium white iron castings can often be used for complex and intricate parts in high-temperature applications at considerable savings compared to stainless steel. These cast iron grades are alloyed with 12 to 39% Cr at temperatures up to 1040 °C (1900 °F) for scaling resistance. Chromium causes the formation of an adherent, complex, chromium-rich oxide film at high temperatures. The high-chromium irons designated for use at elevated temperatures fall into one of three categories, depending on the matrix structure:

- Martensitic irons alloyed with 12 to 28% Cr
- Ferritic irons alloyed with 30 to 34% Cr
- Austenitic irons which, in addition to containing 15 to 30% Cr, also contain 10 to 15% Ni to stabilize the austenite phase

The carbon contents of these alloys range from 1 to 2%. The choice of an exact composition is critical to the prevention of σ phase formation at intermediate temperatures and at the same time avoids the ferrite-to-austenite transformation during thermal cycling, which leads to distortion and cracking. Typical applications include recuperator tubes, breaker bars and trays in sinter furnaces, grates, burner nozzles and other furnace parts, glass bottle molds, and valve seats for combustion engines.

ACKNOWLEDGMENTS

The author wishes to thank Warren Spear of the Nickel Development Institute and Ralph Nelson of Thomas Foundries Inc. for helpful discussions.

SELECTED REFERENCES

- *Abrasion Resistant Castings for Handling Coal*, The International Nickel Company, Inc.
- "Chrome-Moly White Cast Irons," Publication M-630, AMAX Inc.
- J. Dodd and J.L. Parks, "Factors Affecting the Production and Performance of Thick Section High Chromium-Molybdenum Alloy Iron Castings," Publication M-383, AMAX Inc.
- *Engineering Properties and Applications of Ni-Hard*, The International Nickel Company, Inc.
- W. Fairhurst and K. Rohrig, Abrasion-Resistant High-Chromium White Cast Irons, *Foundry Trade J.*, May 1974
- F. Maratray, Choice of Appropriate Compositions for Chromium-Molybdenum White Irons, *Trans. AFS*, Vol 79, 1971, p 121-124

Malleable Iron

MALLEABLE IRON is a cast ferrous metal that is initially produced as white cast iron and is then heat treated to convert the carbon-containing phase from iron carbide to a nodular form of graphite called temper carbon. There are two types of ferritic malleable iron: blackheart and whiteheart. Only the blackheart type is produced in the United States. This material has a matrix of ferrite with interspersed nodules of temper carbon. Cupola malleable iron is a blackheart malleable iron that is produced by cupola melting and is used for pipe fittings and similar thin-section castings. Because of its low strength and ductility, cupola malleable iron is usually not specified for structural applications. Pearlitic malleable iron is designed to have combined carbon in the matrix, resulting in higher strength and hardness than ferritic malleable iron. Martensitic malleable iron is produced by quenching and tempering pearlitic malleable iron.

Malleable iron, like ductile iron, possesses considerable ductility and toughness because of its combination of nodular graphite and low-carbon metallic matrix. Because of the way in which graphite is formed in malleable iron, however, the nodules are not truly spherical as they are in ductile iron but are irregularly shaped aggregates.

Malleable iron and ductile iron are used for some of the same applications in which ductility and toughness are important. In many cases, the choice between malleable and ductile iron is based on economy or availability rather than on properties. In certain applications, however, malleable iron has a distinct advantage. It is preferred for thin-section castings; for parts that are to be pierced, coined, or cold formed; for parts requiring maximum machinability; for parts that must retain good impact resistance at low temperatures; and for some parts requiring wear resistance (martensitic malleable iron only).

Ductile iron has a clear advantage where low solidification shrinkage is needed to avoid hot tears or where the section is too thick to permit solidification as white iron. (Solidification as white iron throughout a section is essential to the production of malleable iron.) Malleable iron castings are produced in section thicknesses ranging from about 1.5 to 100 mm (1/16 to 4 in.) and in weights from less than 0.03 to 180 kg (1/16 to 400 lb) or more.

Composition. The chemical composition of malleable iron generally conforms to the ranges given in Table 1. Small amounts of chromium (0.01 to 0.03%), boron (0.0020%), copper (~1.0%), nickel (0.5 to 0.8%), and molybdenum (0.35 to 0.5%) are also sometimes present.

Five-digit designations are assigned in ASTM A 47 (Ref 1), which covers two grades of ferritic malleable iron, and in A 220 (Ref 2), which deals with pearlitic malleable iron. The first three digits indicate the minimum yield strength (in kips per square inch), and the last two indicate the minimum percentage of elongation in 50 mm (2 in.). Another standard, ASTM A 602 (Ref 3), covers both types of malleable iron; it assigns an "M" and four digits. The first two digits are typical yield strength in kips per square inch, and the last two are typical percentage of elongation in 50 mm (2 in.) test specimens cut from actual cast parts.

Melting Practices

Melting can be accomplished by batch cold melting or by duplexing. Cold melting is done in coreless or channel-type induction furnaces, electric arc furnaces, or cupola furnaces. In duplexing, the iron is melted in a cupola or electric arc furnace, and the molten metal is transferred to a coreless or channel-type induction furnace for holding and pouring. Charge materials (foundry returns, steel scrap, ferroalloys, and, except in cupola melting, carbon) are carefully selected, and the melting operation is well controlled to produce metal having the desired composition and properties. Minor corrections in composition and pouring temperature are made in the second stage of duplex melting, but most of the process control is done in the primary melting furnace. Detailed information on induction furnaces, electric arc furnaces, and cupolas is available in the article "Melting Furnaces" in this Volume.

Table 1 Chemical composition of malleable iron

Element	Composition, %
Carbon	2.16–2.90
Silicon	0.90–1.90
Manganese	0.15–1.25
Sulfur	0.02–0.20
Phosphorus	0.02–0.15

Molds are produced in green sand, silicate CO_2 bonded sand, or resin bonded sand (shell molds) on equipment ranging from highly mechanized or automated machines to that required for floor or hand molding methods, depending on the size and number of castings to be produced (see the article "Sand Molding" in this Volume). In general, the technology of molding and pouring malleable iron is similar to that used to produce gray iron (see the article "Gray Iron" in this Volume). Heat treating is done in high-production controlled-atmosphere continuous furnaces or batch-type furnaces, again depending on production requirements.

After it solidifies and cools, the metal is in a white iron state, and gates, sprues, and feeders can be easily removed from the castings by impact. This operation, called spruing, is generally performed manually with a hammer because the diversity of castings produced in the foundry makes the mechanization or automation of spruing very difficult. After spruing, the castings proceed to heat treatment, while gates and risers are returned to the melting department for reprocessing.

Malleable iron castings are produced from the white iron by a two-stage annealing process. First- and second-stage annealing processes are described in detail in the section "Control of Annealing" in this article.

After heat treatment, ferritic or pearlitic malleable castings are cleaned by shotblasting, gates are removed by shearing or grinding, and, where necessary, the castings are coined or punched. Close dimensional tolerances can be maintained in ferritic malleable iron and in the lower-hardness types of pearlitic malleable iron, both of which can be easily straightened in dies. The harder pearlitic malleable irons are more difficult to press because of higher yield strength and a greater tendency toward springback after die pressing. However, even the highest-strength pearlitic malleable can be straightened to achieve good dimensional tolerances.

Control of Melting. Metallurgical control of the melting operation is designed to ensure that the molten iron will have a certain composition and will:

- Solidify white in the castings to be produced

- Anneal on an established time-temperature cycle set to minimum values in the interest of economy
- Produce the desired graphite distribution (nodule count) upon annealing

Changes in melting practice or composition that would satisfy the first requirement listed above are generally opposed to satisfaction of the second and third, while attempts to improve annealability beyond a certain point may result in an unacceptable tendency for the as-cast iron to be mottled instead of white.

The common elements in malleable iron are generally controlled within about ±0.05 to ±0.15%. A limiting minimum carbon content is required in the interest of mechanical quality and annealability because decreasing carbon content reduces the fluidity of the molten iron, increases shrinkage during solidification, and reduces annealability. A limiting maximum carbon content is imposed by the requirement that the casting be white as-cast. The range in silicon content is limited to ensure proper annealing during a short-cycle high-production annealing process and to avoid the formation of primary graphite during solidification. Manganese and sulfur contents are balanced to ensure that all sulfur is combined with manganese and that only a safe, minimum quantity of excess manganese is present in the iron. An excess of either sulfur or manganese will retard annealing in the second stage and therefore increase annealing costs. The chromium content is kept low because of the carbide-stabilizing effect of this element and because it retards both the first-stage and second-stage annealing reactions.

A mixture of gray iron and white iron in variable proportions that produces a mottled (speckled) appearance is particularly damaging to the mechanical properties of the annealed casting, whether ferritic or pearlitic malleable iron. Primary control of mottle is achieved by maintaining a balance of carbon and silicon contents.

Because economy and castability are enhanced when the carbon and silicon contents of the base iron are in the higher portions of their respective ranges, some malleable iron foundries produce iron with carbon and silicon contents at levels that might produce mottle and then add a balanced, mild carbide stabilizer to prevent mottle during casting. Bismuth and boron in balanced amounts accomplish this control. A typical addition is 0.01% Bi (as metal) and 0.001% B (as ferroboron). Bismuth retards graphitization during solidification; small amounts of boron have little effect on graphitizing tendency during solidification, but accelerate carbide decomposition during annealing. The balanced addition of bismuth and boron permits the production of heavier sections for a given base iron or the utilization of a higher-carbon higher-silicon base iron for a given section thickness.

Tellurium can be added in amounts of 0.0005 to 0.001% to suppress mottle. Tellurium is a much stronger carbide stabilizer than bismuth during solidification, but also strongly retards annealing if the residual exceeds 0.003%. Less than 0.003% residual tellurium has little effect on annealing, but has a significant influence on mottle control. Tellurium is more effective if added together with copper or bismuth.

Residual boron should not exceed 0.0035% in order to avoid nodule alignment and carbide formation. Also, the addition of 0.005% Al to the pouring ladle significantly improves annealability without promoting mottle.

Microstructure

Malleable iron is characterized by microstructures consisting of uniformly dispersed fine particles of free carbon in a matrix of ferrite or tempered martensite. These microstructures can be produced in base metal of essentially the same composition. Structural differences between ferritic malleable iron and the various grades of pearlitic or martensitic malleable iron are achieved through variations in heat treatment. The microstructure of a casting of any type of malleable iron is derived by controlled annealing of white iron of suitable composition. During the annealing cycle, carbon that exists in combined form, either as massive carbides or as a microconstituent in pearlite, is converted to a form of free graphite known as temper carbon.

Ferritic malleable iron requires a two-stage annealing cycle. The first stage converts primary carbides to temper carbon, and the second stage converts the carbon dissolved in austenite at the first-stage annealing temperature to temper carbon and ferrite.

The microstructure of ferritic malleable iron is shown in Fig. 1. A satisfactory structure consists of temper carbon in a matrix of ferrite. There should be no flake graphite and essentially no combined carbon in ferritic malleable iron. Pearlitic and martensitic malleable irons contain a controlled quantity of combined carbon, which, depending on heat treatment, may appear in

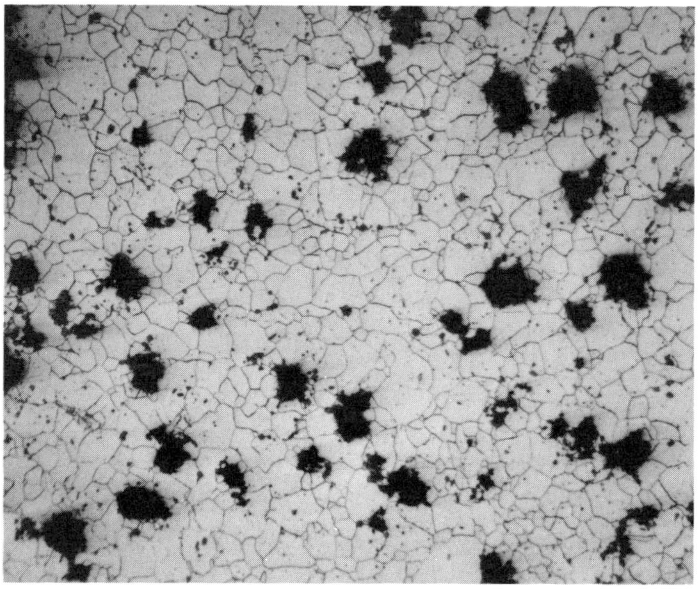

Fig. 1 Structure of annealed ferritic malleable iron showing temper carbon in ferrite. 100×

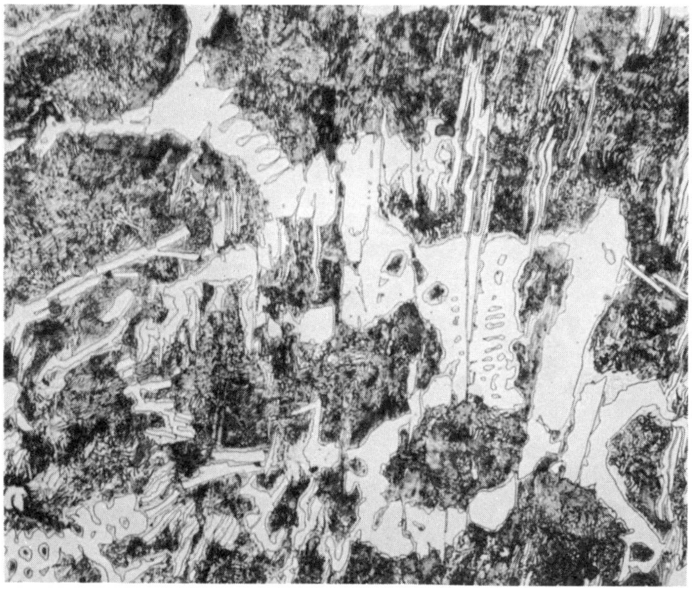

Fig. 2 Structure of as-cast malleable white iron showing a mixture of pearlite and eutectic carbides. 400×

the metallic matrix as lamellar pearlite, tempered martensite, or spheroidite.

Molten iron produced under properly controlled melting conditions solidifies with all carbon in the combined form, producing the white iron structure fundamental to the manufacture of either ferritic or pearlitic malleable iron (Fig. 2). The base iron must contain balanced quantities of carbon and silicon to simultaneously provide castability, white iron in even the thickest sections of the castings, and annealability; therefore, precise metallurgical control is necessary for quality production. Thick metal sections cool slowly during solidification and tend to graphitize, producing mottled or gray iron. This is undesirable, because the graphite formed in mottled iron or rapidly cooled gray iron is generally of the type D configuration, a flake form in a dense, lacy structure, which is particularly damaging to the strength, ductility, and stiffness characteristics of both ferritic and pearlitic malleable iron.

Control of Nodule Count. Proper annealing in short-term cycles and the attainment of high levels of casting quality require that controlled distribution of graphite particles be obtained during first-stage heat treatment. With low nodule count (few graphite particles per unit area or volume), mechanical properties are reduced from optimum, and second-stage annealing time is unnecessarily long because of long diffusion distances. Excessive nodule count is also undesirable, because graphite particles may become aligned in a configuration corresponding to the boundaries of the original primary cementite. In martensitic malleable iron, very high nodule counts are sometimes associated with low hardenability and nonuniform tempering. Generally, a nodule count of 80 to 150 discrete graphite particles per square millimeter (80 to 150 in 15.5 in.^{2} of a photomicrograph at 100×) appears to be optimum. This produces random particle distribution, with short distances between particles.

Temper carbon is formed predominantly at the interface between primary carbide and saturated austenite at the first-stage annealing temperature, with growth around the nuclei taking place by a reaction involving diffusion and carbide decomposition. Although new nuclei undoubtedly form at the interfaces during holding at the first-stage annealing temperature, nucleation and graphitization are accelerated by the presence of nuclei that are created by appropriate melting practice. High silicon and carbon contents promote nucleation and graphitization, but these elements must be restricted to certain maximum levels because of the necessity that the iron solidify white.

Control of Annealing. The rate of annealing of a hard iron casting depends on chemical composition, nucleation tendency (as discussed above), and annealing temperature. With the proper balance of boron content and graphitic materials in the charge, optimum number and distribution of graphite nuclei are developed in the early part of first-stage annealing, and growth of the temper carbon particles proceeds rapidly at any annealing temperature. An optimum iron will anneal completely through the first-stage reaction in approximately 3½ h at 940 °C (1720 °F). Irons with lower silicon contents or less-than-optimum nodule counts may require as much as 20 h for completion of first-stage annealing.

The temperature of first-stage annealing exercises considerable influence on the rate of annealing and the number of graphite particles produced. Increasing the annealing temperature accelerates the rate of decomposition of primary carbide and produces more graphite particles per unit volume. However, high first-stage annealing temperatures can result in excessive distortion of castings during annealing, which leads to straightening of the casting

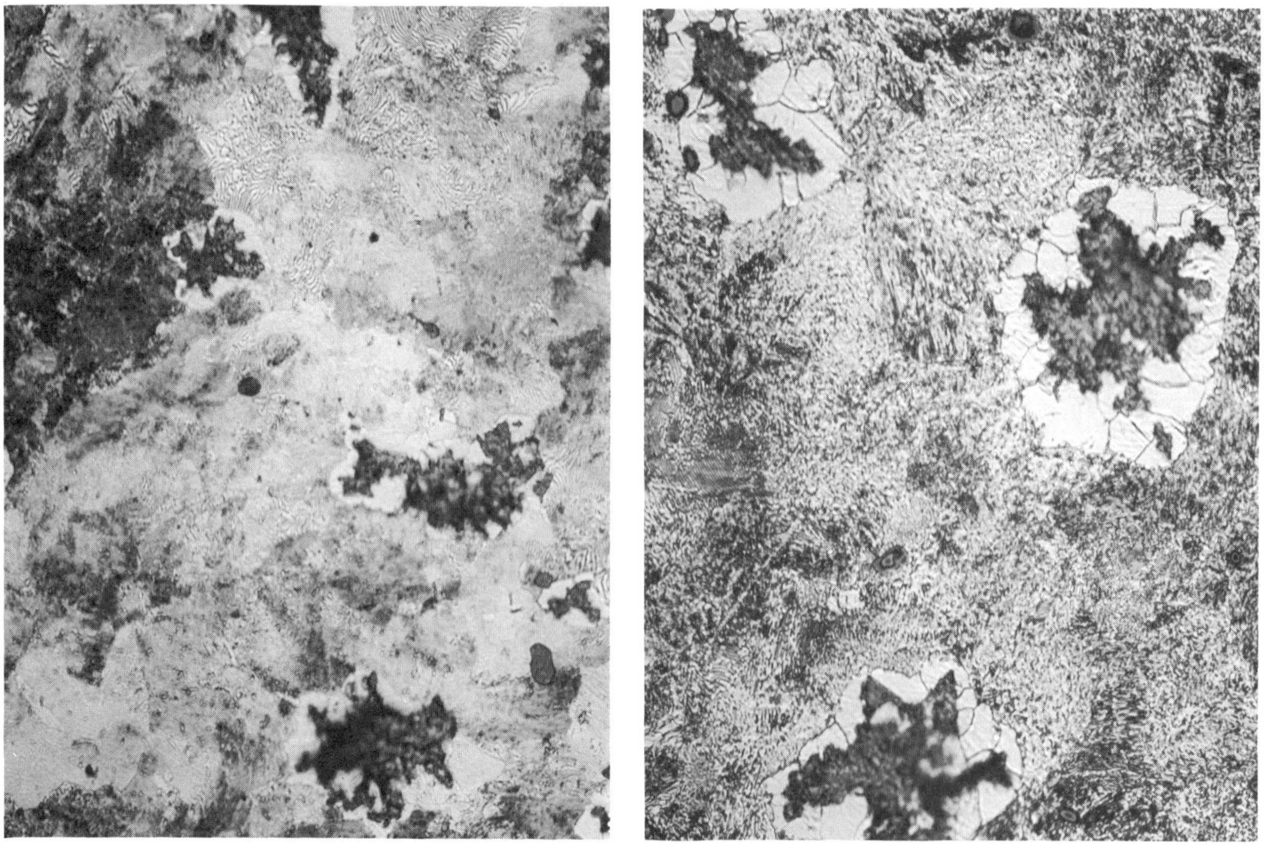

(a) (b)

Fig. 3 Structure of air-cooled pearlitic malleable iron. (a) Slowly air cooled. 400×. (b) Cooled in an air blast. 400×

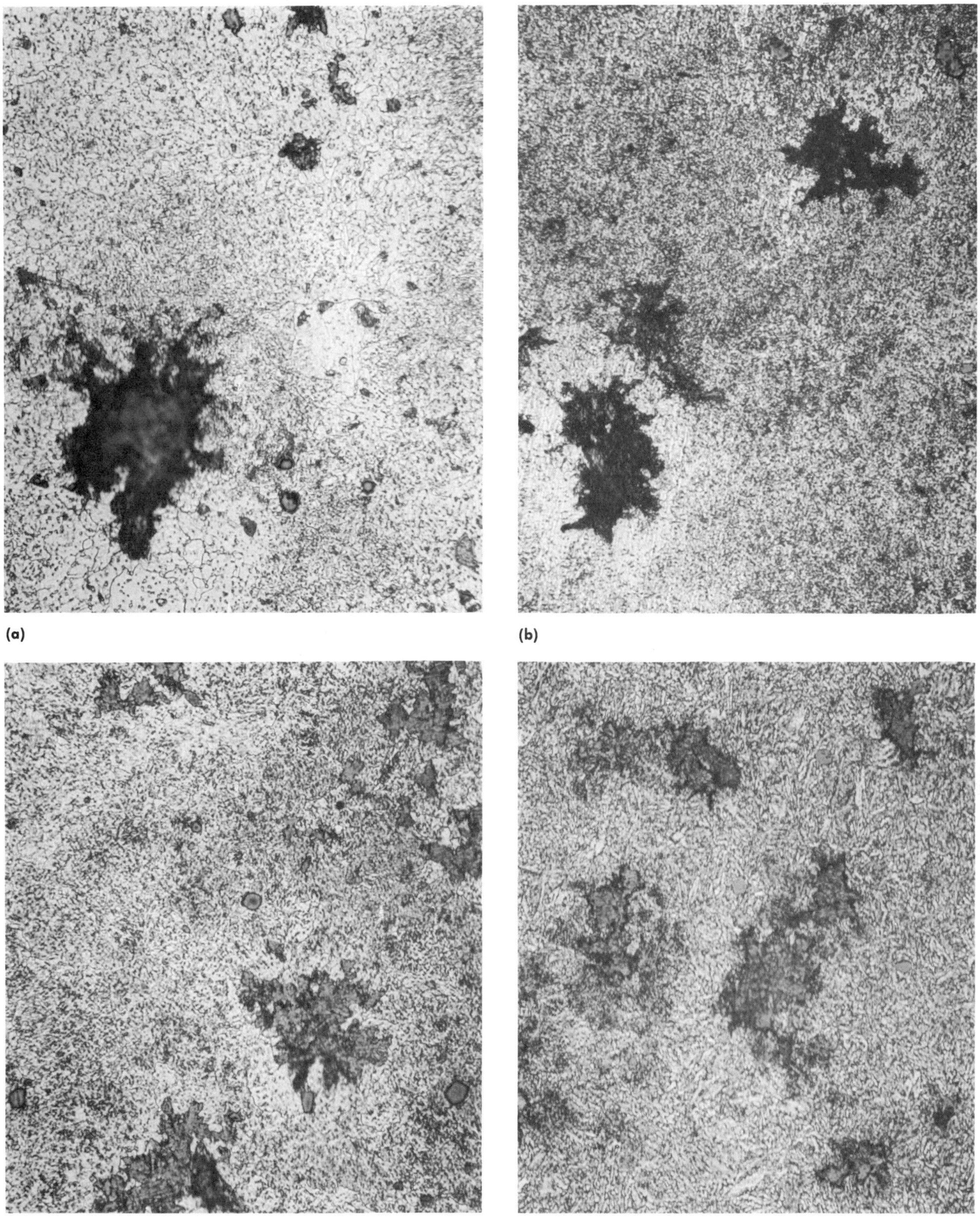

(a) (b) (c) (d)

Fig. 4 Structure of all-quenched and tempered martensitic malleable iron. (a) 163 HB. 500×. (b) 179 HB. 500×. (c) 207 HB. 500×. (d) 229 HB. 500×

after heat treatment. Annealing temperatures are adjusted to provide maximum practical annealing rates and minimum distortion and are therefore controlled within the range of 900 to 970 °C (1650 to 1780 °F). Lower temperatures result in excessively long annealing times, while higher temperatures produce excessive distortion.

After first-stage annealing, the castings are cooled as rapidly as practical to 740 to 760 °C (1360 to 1400 °F) in preparation for second-stage annealing. The fast cooling

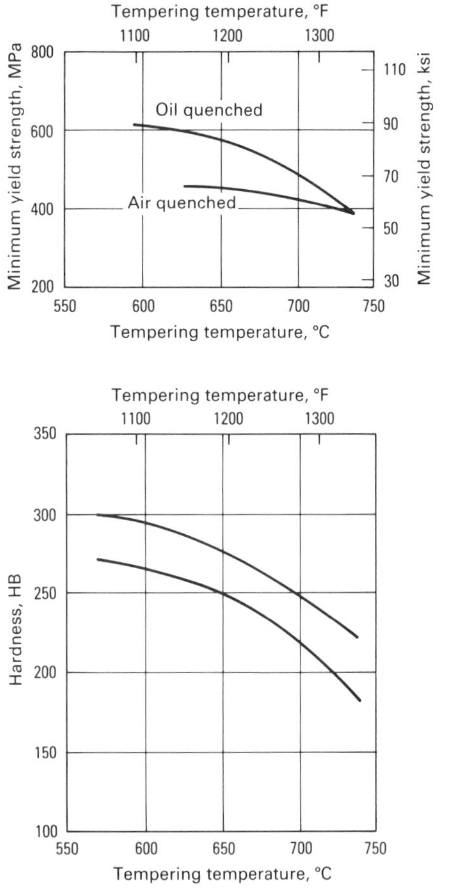

Fig. 5 Hardness and minimum yield strength of pearlitic malleable iron. Relationships of tempering time and temperature to hardness and minimum yield strength are given.

step requires 1 to 6 h, depending on the equipment being employed. Castings are then cooled slowly at a rate of about 3 to 11 °C (5 to 20 °F) per hour. During cooling, the carbon dissolved in the austenite is converted to graphite and deposited on the existing particles of temper carbon. This results in a fully ferritic matrix.

In the production of pearlitic malleable iron, the first-stage heat treatment is identical to that used for ferritic malleable iron. However, some foundries then slowly cool the castings to about 870 °C (1600 °F). During cooling, the combined carbon content of the austenite is reduced to about 0.75%, and the castings are then air cooled. Air cooling is accelerated by an air blast to avoid the formation of ferrite envelopes around the temper carbon particles (bull's-eye structure) and to produce a fine pearlitic matrix (Fig. 3). The castings are then tempered to specification, or they are reheated to reaustenitize at about 870 °C (1600 °F), oil quenched, and tempered to specification. Large foundries usually eliminate the reaustenitizing step and quench the castings in oil directly from the first-stage annealing furnace after stabilizing the temperature at 845 to 870 °C (1550 to 1600 °F).

The furnace atmosphere for producing malleable iron in continuous furnaces is controlled so that the ratio of CO to CO_2 is between 1:1 and 20:1. In addition, any sources of water vapor or hydrogen are eliminated; the presence of hydrogen is thought to retard annealing, and it produces excessive decarburization of casting surfaces. Proper control of the gas atmosphere is important for avoiding an undesirable surface structure. A high ratio of CO to CO_2 retains a high level of combined carbon on the surface of the casting and produces a pearlitic rim, or picture frame, on a ferritic malleable iron part. A low ratio of CO to CO_2 permits excessive decarburization, which forms a ferritic skin on the casting with an underlying rim of pearlite. The latter condition is produced when a significant portion of the subsurface metal is decarburized to the degree that no temper carbon nodules can be developed during first-stage annealing. When this occurs, the dissolved carbon cannot precipitate from the austenite, except as the cementite plates in pearlite.

The rate of cooling after first-stage annealing is important in the formation of a uniform pearlitic matrix in the air-cooled casting, because slow rates permit partial decomposition of carbon in the immediate vicinity of the temper carbon nodules, which results in the formation of films of ferrite around the temper carbon (bull's-eye structure). When the extent of these films becomes excessive, a carbon gradient is developed in the matrix. Air cooling is usually done at a rate not less than about 80 °C (150 °F) per minute.

Air-quenched malleable iron castings have hardnesses ranging from 269 to 321 HB, depending on casting size and cooling rate. Such castings can be tempered immediately after air cooling to obtain pearlitic malleable iron with a hardness of 241 HB or less.

High-strength malleable iron castings of uniformly high quality are usually produced by liquid quenching and tempering, using any of the three procedures. The most economical procedure is direct quenching after first-stage annealing. In this procedure, castings are cooled in the furnace to the quenching temperature of 845 to 870 °C (1550 to 1600 °F) and held for 15 to 30 min to homogenize the matrix. The castings are then quenched in agitated oil to develop a matrix microstructure of martensite having a hardness of 415 to 601 HB. Finally, the castings are tempered at an appropriate temperature between 590 and 725 °C (1100 and 1340 °F) to develop the specified mechanical properties. The final microstructure consists of tempered martensite plus temper carbon, as shown in Fig. 4. In heavy sections, higher-temperature transformation products such as fine pearlite are usually present.

Some foundries produce high-strength malleable iron by an alternative procedure in which the castings are forced-air cooled after first-stage annealing, retaining about 0.75% C as pearlite, and then reheated to 840 to 870 °C (1545 to 1600 °F) for 15 to 30 min, followed by quenching and tempering as described above for the direct-quench process.

Rehardened-and-tempered malleable iron can also be produced from fully annealed ferritic malleable iron with a slight variation from the heat treatment used for arrested-annealed (air-quenched) malleable. The matrix of fully annealed ferritic malleable iron is essentially carbon free, but can be recarburized by heating at 840 to 870 °C (1545 to 1600 °F) for 1 h. In general, the combined carbon content of the matrix produced by this procedure is slightly lower than that of arrested-annealed pearlitic malleable iron, and the final tempering temperatures required for the development of specific hardnesses are lower. Rehardened malleable iron made from ferritic malleable may not be capable of meeting certain specifications.

Tempering times of 2 h or more are needed for uniformity. In general, control of final hardness of the castings is precise, with process limitations approximately the same as those encountered in the heat treatment of medium- or high-carbon steels. This is particularly true when specifications require hardnesses of 241 to 321 HB where control limits of ±0.2 mm Brinell diameter can be maintained with ease. At lower hardnesses, a wider process control limit is required because of certain unique characteristics of the pearlitic malleable iron microstructure. The relationships between tempering conditions and properties (yield strength and hardness) are illustrated in Fig. 5.

Applications

The requirement that any iron produced for conversion to malleable iron must solidify white places definite section thickness limitations on the malleable iron industry. Thick metal sections can be produced by melting a base iron of low carbon and silicon contents or by alloying the molten iron with a carbide stabilizer. However, when carbon and silicon are maintained at low levels, difficulty is invariably encountered in annealing, and the time required to convert primary and pearlitic carbides to temper carbon becomes excessively long. High-production foundries are usually reluctant to produce castings more than about 40 mm (1½ in.) thick. Some foundries, however, routinely produce castings as thick as 100 mm (4 in.).

Automotive and associated applications of ferritic and pearlitic malleable irons include many essential parts in vehicle power trains, frames, suspensions, and wheels. A partial list includes differential carriers, differential cases, bearing caps, steering-gear housings, spring hangers, universal-joint yokes, automatic-transmission parts, rocker

Fig. 6 Examples of malleable iron automotive applications. (a) Driveline yokes. (b) Connecting rods. (c) Diesel pistons. (d) Steering gear housing. Courtesy of Central Foundry Division, General Motors Corporation

arms, disc brake calipers, wheel hubs, and many other miscellaneous castings. Examples are shown in Fig. 6. Ferritic and pearlitic malleable irons are also used in the railroad industry and in agricultural equipment, chain links, ordnance material, electrical pole line hardware, hand tools, and other parts requiring section thicknesses and properties obtainable in these materials.

Malleable iron castings are often selected because the material has excellent machinability in addition to significant ductility. In other applications, malleable iron is chosen because it combines castability with good toughness and machinability. Malleable iron is often chosen because of shock resistance alone. Table 2 lists some of the typical applications of malleable iron castings.

Ferritic Malleable Iron

Because ferritic malleable iron consists of only ferrite and temper carbon, the properties of ferritic malleable castings depend on the quantity, size, shape, and distribution of the temper carbon particles and on the composition of the ferrite. Fully annealed ferritic malleable iron castings contain 2.00 to 2.70% graphitic carbon by weight, which is equivalent to about 6 to 8% by volume. Because the graphitic carbon contributes nothing to the strength of the castings, those with the lesser amount of graphite are somewhat stronger and more ductile than those containing the greater amount (assuming equal size and distribution of graphite particles). Elements such as silicon and manganese in solid solution in the ferritic matrix contribute to the strength and reduce the elongation of the ferrite. Therefore, by varying base metal composition, slightly different strength levels can be obtained in

Table 2 Applications of malleable iron castings

Mechanical properties are given in Table 3.

Specification No.	Class or grade	Microstructure	Typical applications
Ferritic			
ASTM A47, ANSI G48.1, FED QQ-1-666c	32510 35018	Temper carbon and ferrite	General engineering service at normal and elevated temperatures for good machinability and excellent shock resistance
ASTM A338	32510 35018	Temper carbon and ferrite	Flanges, pipe fittings, and valve parts for railroad, marine, and other heavy-duty service to 345 °C (650 °F)
ASTM A197, ANSI G49.1	· · ·	Free of primary graphite	Pipe fittings and valve parts for pressure service
Pearlitic and martensitic			
ASTM A220, ANSI G48.2, MIL-I-11444B	40010 45008 45006 50005 60004 70003 80002 90001	Temper carbon in necessary matrix without primary cementite or graphite	General engineering service at normal and elevated temperatures. Dimensional tolerance range for castings is stipulated
Automotive			
ASTM A602, SAE J158	M3210	Ferritic	For low-stress parts requiring good machinability: steering-gear housings, carriers, and mounting brackets
	M4504	Ferrite and tempered pearlite(a)	Compressor crankshafts and hubs
	M5003	Ferrite and tempered pearlite(a)	For selective hardening: planet carriers, transmission gears, and differential cases
	M5503	Tempered martensite	For machinability and improved response to induction hardening
	M7002	Tempered martensite	For high-strength parts: connecting rods and universal-joint yokes
	M8501	Tempered martensite	For high strength plus good wear resistance: certain gears

(a) May be all tempered martensite for some applications

the fully annealed ferritic product. Table 3 lists specifications and properties applicable to malleable iron castings.

The mechanical properties that are most important for design purposes are tensile strength, yield strength, modulus of elasticity, fatigue strength, and impact strength. Hardness can be considered no more than an approximate indicator that the ferritizing anneal was complete, and it is seldom used for any other purpose. The hardness of ferritic malleable iron almost always ranges from 110 to 156 HB.

The tensile properties of ferritic malleable iron (Table 3) are usually measured on unmachined test bars. Machined test bars are normally used for pearlitic malleable iron.

The fatigue limit of unnotched ferritic malleable iron is about 50% of the tensile strength, or from 170 to 205 MPa (25 to 30 ksi). Figure 7 summarizes the effects of notches on fatigue strength. Notch radius generally has little effect on fatigue strength, but fatigue strength decreases with increasing notch depth.

Modulus of elasticity in tension is about 170 GPa (25 × 10^6 psi). The modulus in compression ranges from 150 to 170 GPa (22 × 10^6 to 25 × 10^6 psi); in torsion, from 65 to 75 GPa (9.5 × 10^6 to 11 × 10^6 psi).

Because brittle fractures are most likely to occur at high strain rates, at low temperatures, and with a high restraint on metal deformation, notch tests such as the Charpy V-notch test are conducted over a range of test temperatures to establish the toughness behavior and the temperature range of transition from a ductile to a brittle fracture. Figure 8 illustrates the behavior of ferritic malleable iron and several types of pearlitic malleable iron in the Charpy V-notch test. This shows that ferritic malleable iron has a higher upper shelf energy and a lower transition temperature to a brittle fracture than pearlitic malleable iron does. Additional information on fracture toughness of malleable irons can be found later in the discussion of "Pearlitic and Martensitic Malleable Iron" in this article.

Short-term tensile properties show no significant change to 370 °C (700 °F). Sustained-load stress rupture data from 425 to 650 °C (800 to 1200 °F) are given in Fig. 9.

The corrosion resistance of ferritic malleable iron is increased by the addition of

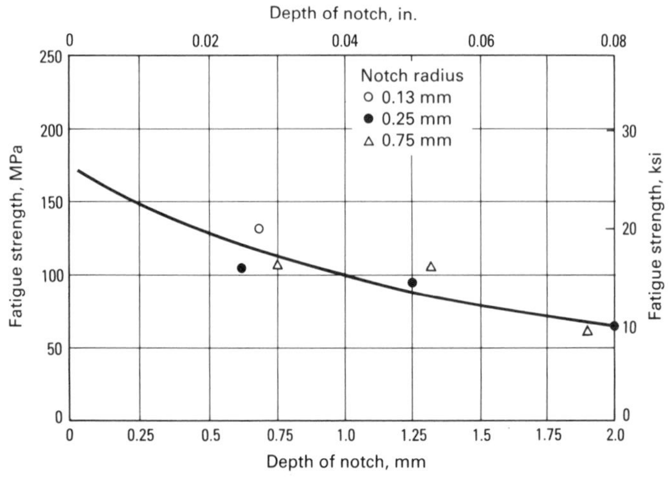

Fig. 7 Effects of notch radius and notch depth on the fatigue strength of ferritic malleable iron

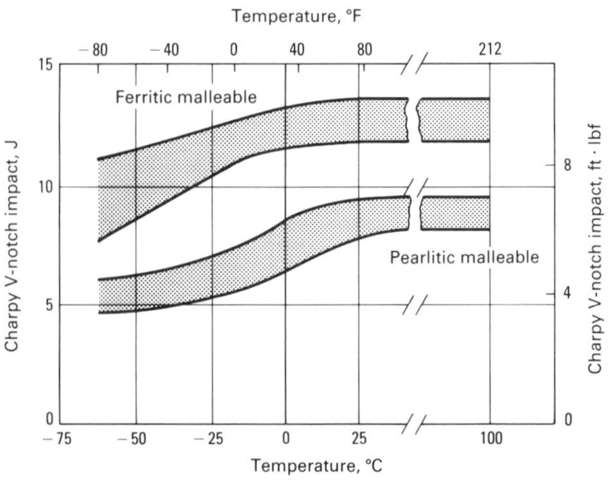

Fig. 8 Charpy V-notch transition curves for ferritic and pearlitic malleable irons. Source: Ref 4

Table 3 Properties of malleable iron castings

Microstructures and typical applications are given in Table 2.

Specification No.	Class or grade	Tensile strength MPa	Tensile strength ksi	Yield strength MPa	Yield strength ksi	Hardness, HB	Elongation(a), %
Ferritic							
ASTM A47 and A338, ANSI G48.1, FED QQ-I-666c	32510	345	50	224	32	156 max	10
	35018	365	53	241	35	156 max	18
ASTM A197	...	276	40	207	30	156 max	5
Pearlitic and martensitic							
ASTM A220, ANSI G48.2, MIL-I-11444B	40010	414	60	276	40	149–197	10
	45008	448	65	310	45	156–197	8
	45006	448	65	310	45	156–207	6
	50005	483	70	345	50	179–229	5
	60004	552	80	414	60	197–241	4
	70003	586	85	483	70	217–269	3
	80002	655	95	552	80	241–285	2
	90001	724	105	621	90	269–321	1
Automotive							
ASTM A602, SAE J158	M3210(b)	345	50	224	32	156 max	10
	M4504(c)	448	65	310	45	163–217	4
	M5003(c)	517	75	345	50	187–241	3
	M5503(d)	517	75	379	55	187–241	3
	M7002(d)	621	90	483	70	229–269	2
	M8501(d)	724	105	586	85	269–302	1

(a) Minimum in 50 mm (2 in.). (b) Annealed. (c) Air quenched and tempered. (d) Liquid quenched and tempered

copper, usually about 1%, in certain applications, for example, conveyor buckets, bridge castings, pipe fittings, railroad switch stands, and freight-car hardware. One important use for copper-bearing ferritic malleable iron is chain links. Ferritic malleable iron can be galvanized to provide added protection. The effects of copper on the corrosion resistance of ferrous alloys are well documented in Volume 13 of the 9th Edition of *Metals Handbook*.

Welding and Brazing. Welding of ferritic malleable iron almost always produces brittle white iron in the weld zone and the portion of the heat-affected zone immediately adjacent to the weld zone. During welding, temper carbon is dissolved, and upon cooling it is reprecipitated as carbide rather than graphite. In some cases, welding with a cast iron electrode may produce a brittle gray iron weld zone. The loss of ductility due to welding may not be serious in some applications. However, welding is usually not recommended unless the castings are subsequently annealed to convert the carbide to temper carbon and ferrite. Ferritic malleable iron can be fusion welded to steel without subsequent annealing if a completely decarburized zone as deep as the normal heat-affected zone is produced at the faying surface of the malleable iron part before welding. Silver brazing and tin-lead soldering can be satisfactorily used. Additional information on welding can be found in the article "Processing of Castings" in this Volume (see the section "Welding of Cast Irons and Steels").

Pearlitic and Martensitic Malleable Iron

Pearlitic malleable iron is produced either by controlled heat treatment of the same base white iron used to produce ferritic malleable iron or by alloying to prevent the decomposition of carbides dissolved in austenite during cooling from the first-stage annealing temperature. It can be produced by air cooling after first-stage annealing and subsequent tempering to develop specified properties. Martensitic malleable iron is produced by liquid quenching and tempering to develop specified properties (see the section "Control of Annealing" in this article). Variations in heat treatment, coupled with variations in base composition and melting practice, make it possible to obtain a wide range of properties in pearlitic or martensitic malleable iron.

The mechanical properties of pearlitic and martensitic malleable iron vary in a substantially linear relationship with Brinell hardness (Fig. 10 and 11). In the low hardness ranges, below about 207 HB, the properties of air-quenched and tempered pearlitic malleable are essentially the same as those of oil-quenched and tempered martensitic malleable. This results from the fact that attaining the low hardnesses requires considerable coarsening of the matrix carbides and partial second-stage graphitization. Either an air-quenched pearlitic structure or an oil-quenched martensitic structure can be coarsened and decarburized to meet this hardness requirement.

At higher hardnesses, oil-quenched and tempered malleable iron has higher yield strength and elongation than air-quenched and tempered malleable because of greater uniformity of matrix structure and finer distribution of carbide particles. Oil-quenched and tempered pearlitic malleable

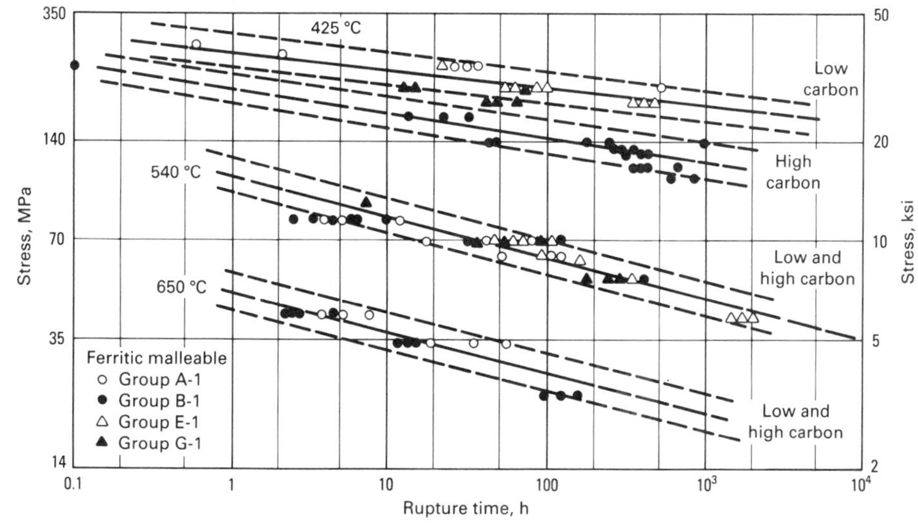

Group	Grade	C	Si	Mn	P	S	Cr
		Composition, %					
A-1	35018	2.21	1.14	0.35	0.161	0.081	...
B-1	32510	2.50	1.32	0.43	0.024	0.159	0.029
E-1	35018	2.16	1.17	0.38	0.137	0.095	0.017
G-1	35018	2.29	1.01	0.38	0.11	0.086	...

Fig. 9 Stress rupture plot for various grades of ferritic malleable iron. The solid lines are curves determined by the method of least squares from the existing data and are least squares fit to data. The dashed lines define the 90% symmetrical tolerance interval. The lower dashed curve defines time and load for 95% survivors, and the upper dashed curve is the boundary for 5% survivors. Normal distribution is assumed. Source: Ref 1

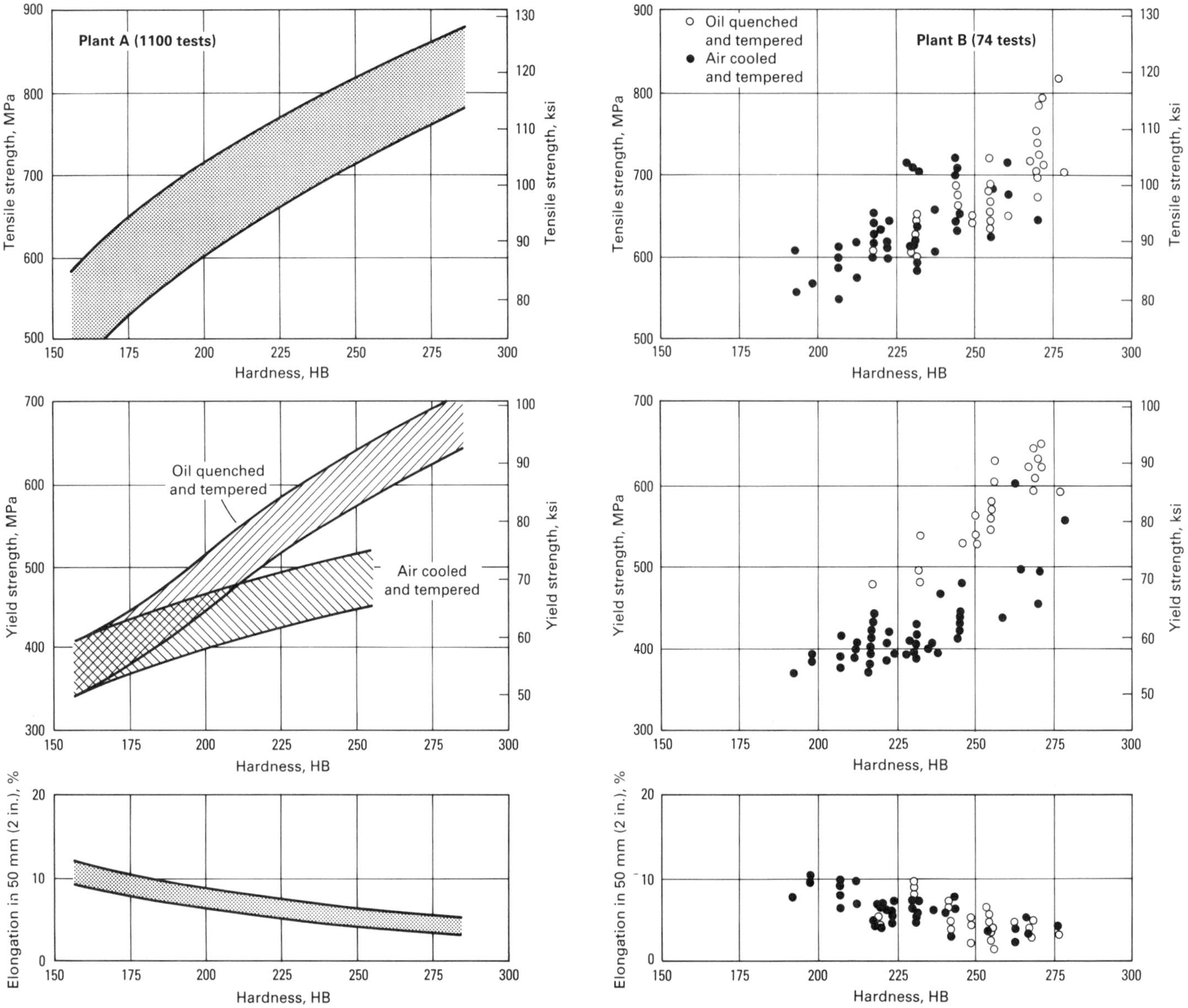

Fig. 10 Relationships of tensile properties to Brinell hardness for pearlitic malleable irons from two foundries. The mechanical properties of these irons vary in substantially a linear relationship with Brinell hardness, and in the low hardness ranges (below about 207 HB), the properties of air-quenched and tempered material are essentially the same as those produced by oil quenching and tempering.

iron is produced commercially to hardnesses as high as 321 HB, while the maximum hardness for high-production air-quenched and tempered pearlitic malleable iron is about 255 HB. The lower maximum hardness is applied to the air-quenched material for several reasons:

- Because hardness on air quenching normally does not exceed 321 HB and may be as low as 269 HB, attempts to temper to a hardness range above 255 HB produce nonuniform hardness and make the process control limits excessive
- Very little structural alteration occurs during the tempering heat treatment to a higher hardness, and the resulting structure is more difficult to machine than an oil-quenched and tempered structure at the same hardness
- There is only a slight improvement in other mechanical properties with increased hardness above 255 HB

Because of these considerations, applications for air-quenched and tempered pearlitic malleable iron are usually those requiring moderate strength levels, while the higher-strength applications need the oil-quenched and tempered material.

The modulus of elasticity in tension of pearlitic malleable iron is 175 to 195 GPa (25.5 × 10^6 to 28.0 × 10^6 psi). For automobile crankshafts, the modulus is important and must be determined with greater precision.

The results of Charpy V-notch tests on pearlitic malleable iron are presented in Fig. 8. The fracture toughness of ferritic and pearlitic malleable irons has not been widely studied, but one researcher has estimated K_{Ic} values for these materials by using a *J*-integral approach (Ref 5). Table 4 summarizes the fracture toughness values obtained for the various grades of malleable iron at various temperatures. All of the materials exhibited stable crack extension prior to fracture for 25 mm (1 in.) wide compact-tension specimens.

As for ductile irons, fracture toughness testing indicates that malleable irons possess considerably more toughness than is indicated by Charpy impact toughness results. Although the fracture toughness values for pearlitic grades are similar to those obtained for ferritic grades, the higher yield strengths of the pearlitic grades indicate that their critical flaw sizes, which are proportional to $(K_{Ic}/\sigma_y)^2$, are less than those of the ferritic grades of malleable iron. Detailed information on the principles of frac-

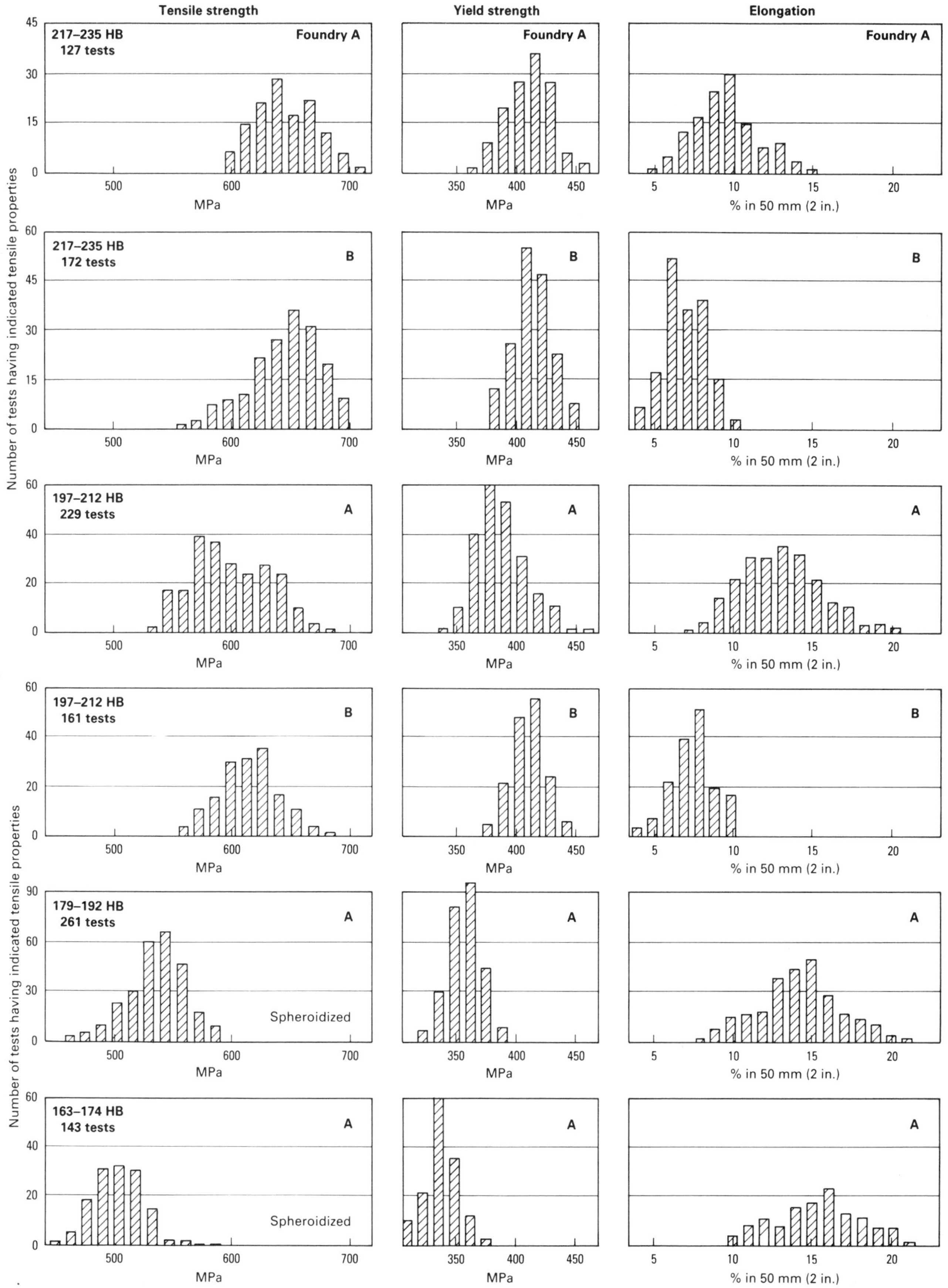

Fig. 11 Tensile properties of pearlitic malleable iron at various hardness levels. At foundry A, the iron was made by alloying with manganese, with completion of first-stage graphitization, air cooling under air blast from 938 °C (1720 °F), and subcritical tempering for spheroidizing.

Table 4 Fracture toughness of malleable irons

Malleable iron grade	Test temperature °C	Test temperature °F	Yield strength MPa	Yield strength ksi	K_{Ic} MPa $\sqrt{m}$	K_{Ic} ksi $\sqrt{in.}$
Ferritic						
M3210	24	75	230	33	44	40
	−19	−3	240	35	41	38
	−59	−74	250	36	44	40
Pearlitic						
M4504 (normalized)	24	75	360	52	55	50
	−19	−2	380	55	48	44
	−57	−70	390	57	30	27
M5503 (quenched and tempered)	24	75	410	60	45	41
	−19	−3	430	64	51	47
	−58	−73	440	66	30	27
M7002 (quenched and tempered)	24	75	520	75	54	49
	−19	−3	550	80	39	35
	−58	−72	570	83	39	36

Source: Ref 5

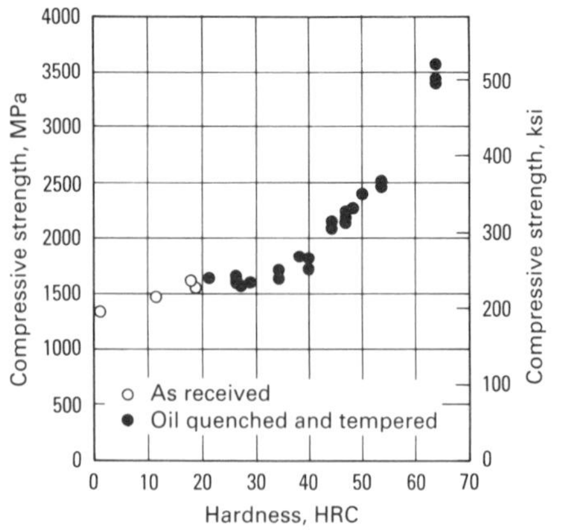

Fig. 12 Compressive strength of pearlitic and martensitic malleable irons

ture toughness and the nomenclature associated with fracture mechanics studies is available in the Section "Fracture Mechanics" and the article "Dynamic Fracture Testing" in Volume 8 of the 9th Edition of *Metals Handbook*.

The compressive strength of pearlitic and martensitic malleable irons is plotted in Fig. 12. Torsional strength values are presented in Table 5.

Sustained-load stress rupture data for eight different grades of pearlitic malleable iron are shown in Fig. 13. Results of high temperature Charpy V-notch tests showing the effect of hardness on impact energy are given in Fig. 14.

The fatigue limits of pearlitic and martensitic malleable irons are about 40 to 50% of tensile strength. Oil-quenched and tempered martensitic iron usually has a higher fatigue ratio than pearlitic iron made by the arrested anneal method.

Wear Resistance. Because of its structure and hardness, pearlitic and martensitic malleable irons have excellent wear resistance. In some moving parts where bushings are normally inserted at pivot points, heat-treated malleable iron has proved to be so wear resistant that the bushings have been eliminated. One example of this is the rocker arm for an overhead-valve automotive engine.

Welding and Brazing. Welding of pearlitic or martensitic malleable iron is difficult because the high temperatures used can cause the formation of a brittle layer of graphite-free white iron. Pearlitic and mar-

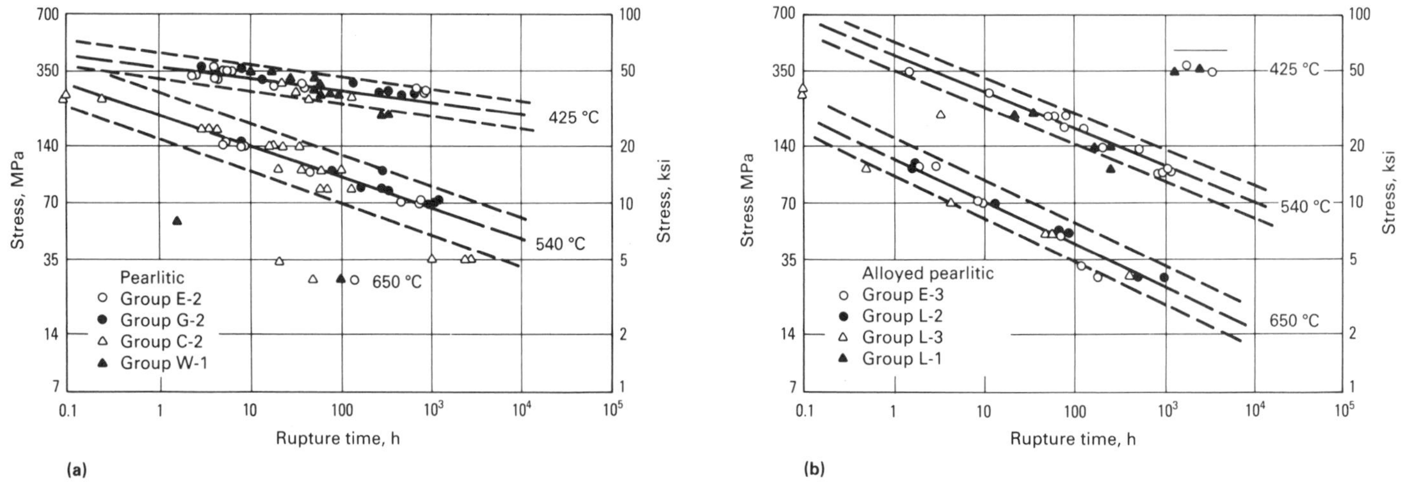

Material	C	Si	Mn	S	P	Cr	Others
Pearlitic (low carbon-high phosphorus)							
Group E-2	2.27	1.15	0.89	0.098	0.135	0.019	. . .
Group G-2	2.29	1.01	0.75	0.086	0.11	. . .	. . .
Pearlitic (high carbon-low phosphorus)							
Group C-2	2.65	1.35	0.41	0.15	. . .	0.018	0.0020 B
Group W-1	2.45	1.38	0.41	0.12	0.04	0.032	. . .
Alloyed pearlitic (low carbon-high phosphorus)							
Group E-3	2.21	1.13	0.88	0.110	0.122	0.021	0.47Mo,1.03Cu
Group L-1	2.16	1.18	0.72	0.120	0.128	. . .	0.34Mo,0.83 Ni
Group L-2	2.16	1.18	0.80	0.123	0.128	. . .	0.40Mo,0.62 Ni
Group L-3	2.32	1.14	0.82	0.117	0.128	. . .	0.38Mo,0.65 Ni

(Composition, %: C, Si, Mn, S, P, Cr, Others)

Fig. 13 Stress rupture plot for pearlitic malleable iron (a) and alloyed pearlitic malleable iron (b). The solid lines are curves determined by the method of least squares from the existing data. The dashed lines define the 90% symmetrical tolerance interval. The lower dashed curve defines time and load for 95% survivors, and the upper dashed curve is the boundary for 5% survivors. Normal distribution is assumed. Source: Ref 1

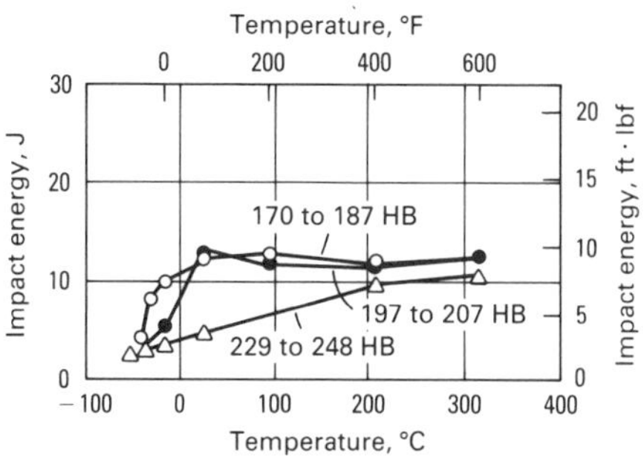

Fig. 14 Charpy V-notch impact energy of one heat of air-quenched and tempered pearlitic malleable iron

Table 5 Torsion test values of malleable irons

Material(a)	Modulus of rupture MPa	ksi	Yield strength MPa	ksi	Ultimate twist, degree	Modulus of elasticity GPa	10^6 psi
Ferritic	375	54	130	19	540	65.15	9.45
45010(b)	570	83	215	31	350	66.75	9.68
60003(b)	665	97	320	46	280	70.3	10.20
80002(b)	715	104	410	59	140	69.22	10.04
High strength(c)	850	123	490	71	70	67.5	9.79

(a) Specimens 20 mm (0.750 in.) in diameter. (b) Pearlitic malleable iron. (c) High-strength malleable iron with yield strength of about 620 MPa (90 ksi), tensile strength of 830 MPa (120 ksi), and 2% elongation in 50 mm (2 in.)

tensitic malleable iron can be successfully welded if the surface to be welded has been heavily decarburized.

Pearlitic or martensitic malleable iron can be brazed by various commercial processes. One application is the induction silver brazing of a pearlitic malleable casting and a steel shaft to form a planetary output shaft for an automotive transmission. In another automotive application, two steel shafts are induction copper brazed to a pearlitic malleable iron shifter shaft plate.

Selective Surface Hardening. Pearlitic malleable iron can be surface hardened by either induction heating and quenching or flame heating and quenching to develop high hardness at the heat-affected surface. Considerable research has been done to determine the surface-hardening characteristics of pearlitic malleable and its capability of developing high hardness over relatively narrow surface bands. Generally, little difficulty is encountered in obtaining hardnesses in the range of 55 to 60 HRC, with the depth of penetration being controlled by the rate of heating and by the temperature at the surface of the part being hardened (Fig. 15).

The maximum hardness obtainable in the matrix of a properly hardened pearlitic malleable part is 67 HRC. However, conventional hardness measurements made on castings show less than 67 HRC because of the presence of the graphite particles, which are averaged into the hardness. Generally, a casting with a matrix microhardness of 67 HRC will have about 62 HRC average hardness, as measured with the standard Rockwell tester. Similarly, a Rockwell or Brinell hardness test on softer structures will show less than matrix microhardness because of the presence of graphite.

Two examples of automobile production parts hardened by induction heating are rocker arms and clutch hubs. An example of a flame-hardened pearlitic malleable iron part is a pinion spacer used to support the cup of a roller bearing. To preclude service failures, the ends of the pinion spacer are flame hardened to file hardness to a depth of about 2.3 mm (3/32 in.).

Malleable iron can be carburized, carbonitrided, or nitrided to produce a surface with improved wear resistance. In addition, heat treatments such as austempering have been used in specialized applications.

REFERENCES

1. "Standard Specification for Malleable Iron Castings," A 47, *Annual Book of ASTM Standards*, American Society for Testing and Materials
2. "Standard Specification for Pearlitic Malleable Iron Castings," A 220, *Annual Book of ASTM Standards*, American Society for Testing and Materials
3. "Standard Specification for Automotive Malleable Iron Castings," A 602, *Annual Book of ASTM Standards*, American Society for Testing and Materials
4. C.F. Walton and T.J. Opar, Ed., *Iron Castings Handbook*, Iron Castings Society, 1981, p 297-321
5. W.L. Bradley, Fracture Toughness Studies of Gray, Malleable and Ductile Cast Iron, *Trans. AFS*, Vol 89, 1981, p 837-848

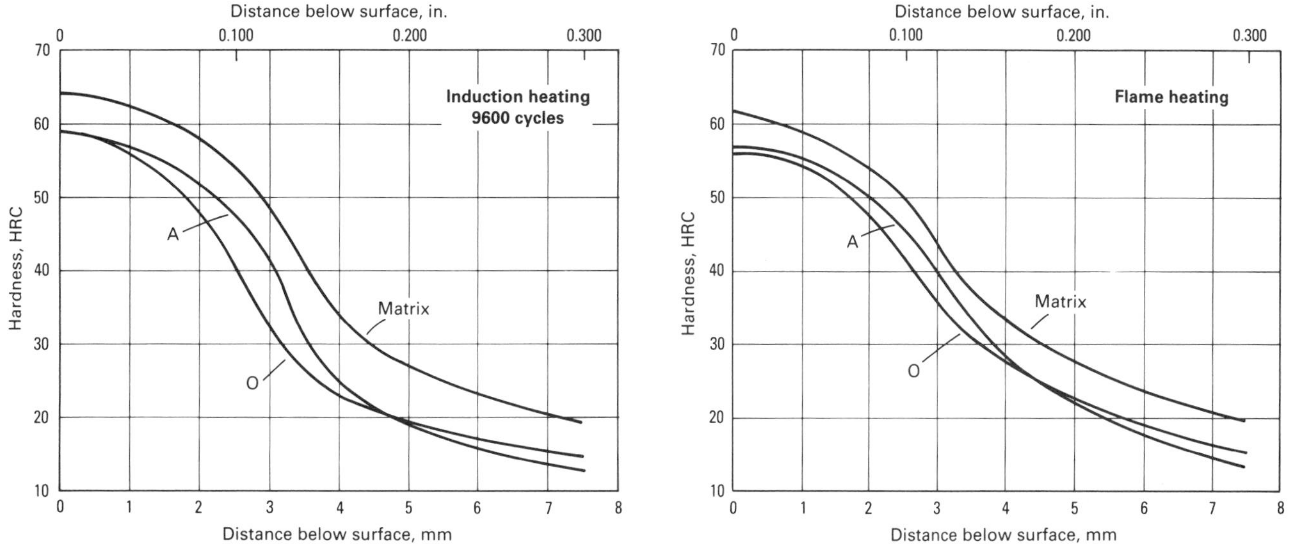

Fig. 15 Hardness versus depth for surface-hardened pearlitic malleable irons. Curves labeled "Matrix" show hardness of the matrix, converted from microhardness tests. O, oil quenched and tempered to 207 HB before surface hardening; A, air cooled and tempered to 207 HB before surface hardening

High-Alloy Graphitic Irons

Richard B. Gundlach, Climax Research Services

HIGH-ALLOY GRAPHITIC IRONS have found special use primarily in applications requiring corrosion resistance or strength and oxidation resistance in high-temperature service. They are commonly produced in both flake graphite and nodular graphite versions. Those alloys used in applications requiring corrosion resistance are the nickel-alloyed (13 to 36% Ni) gray and ductile irons (also called Ni-Resist irons) and the high-silicon (14.5% Si) gray irons. The alloyed irons produced for high-temperature service are the nickel-alloyed gray and ductile irons, the high-silicon (4 to 6% Si) gray and ductile irons, and the aluminum-alloyed gray and ductile irons. Two groups of aluminum-alloyed irons are recognized: the 1 to 7% Al irons and the 18 to 25% Al irons. Neither the 4 to 6% Si irons nor the aluminum-alloyed irons are covered by American Society for Testing and Materials (ASTM) standards. Although the oxidation resistance of the aluminum-alloyed irons is exceptional, problems in melting and casting the alloys are considerable, and commercial production of the aluminum-alloyed irons is uncommon.

In general, the molten metal processing of high-alloy graphitic irons follows that of conventional gray and ductile iron production. The same attention to chill control and inoculation practice is necessary, and the conditions for successful magnesium treatment for producing nodular graphite are similar or identical to those for conventional gray and ductile irons. The high alloy contents influence the eutectic carbon content; therefore, carbon levels in these alloys are lower. In many cases, hypereutectic compositions (high carbon contents) in the ductile irons are avoided because of a greater tendency toward the formation of graphite flotation and degenerate graphite forms in sections larger than 25 mm (1 in.).

The higher alloy contents affect the constitution of the irons, creating conditions that favor the formation of third phases and/or secondary eutectics during solidification. Therefore, many of the alloys commonly contain interdendritic carbides or silicocarbides in the as-cast structure. These constituents remain after heat treatment and are an accepted part of the microstructure.

High-Silicon Irons for High-Temperature Service

Graphitic irons alloyed with 4 to 6% Si provide good service and low cost in many elevated-temperature applications. These irons, whether gray or ductile, provide good oxidation resistance and stable ferritic matrix structures that will not go through a phase change at temperatures up to 900 °C (1650 °F). The elevated silicon contents of these otherwise normal cast iron alloys reduce the rate of oxidation at elevated temperatures because it promotes the formation of a dense, adherent oxide at the surface, which consists of iron silicate rather than iron oxide. This oxide layer is much more resistant to oxygen penetration, and its effectiveness improves with increasing silicon content. Detailed information on the corrosion-resistant properties of high-silicon cast irons is provided in the article "Corrosion of Cast Irons" in Volume 13 of the 9th Edition of *Metals Handbook*.

High-Silicon Gray Irons

The high-silicon gray irons were developed in the 1930s by BCIRA and are commonly called Silal. In Silal, the advantages of a high critical (A1) temperature, a stable ferritic matrix, and a fine, undercooled type D graphite structure are combined to provide good growth and oxidation resistance. Oxidation resistance is further improved with additions of chromium, which in these grades can approach levels of 2% Cr. An austenitic grade called Nicrosilal was also developed, but the Ni-Resist irons have replaced this alloy.

Applications. Although quite brittle at room temperature, the high-silicon gray irons are reasonably tough at temperatures above 260 °C (500 °F). They have been successfully used for furnace and stoker parts, burner nozzles, and heat treatment trays.

High-Silicon Ductile Irons

The advent of ductile iron led to the development of high-silicon ductile irons, which currently constitute the greatest tonnage of these types of iron being produced. Converting the eutectic flake graphite network into isolated graphite nodules further improves oxidation resistance and growth. The higher strength and ductility of the ductile iron version of these alloys qualify it for more rigorous service.

The high-silicon ductile iron alloys are designed to extend the upper end of the range of service temperatures for ferritic ductile irons. These irons are used to temperatures of 900 °C (1650 °F). Raising the silicon content to 4% raises the A1 temperature to 815 °C (1500 °F), and at 5% Si the A1 is above 870 °C (1600 °F). The mechanical properties of these alloyed irons at the lower end of the range (4 to 4.5% Si) are similar to those of standard ferritic ductile irons. At 5 to 6% Si, oxidation resistance is improved and critical temperature increased, but the iron can be very brittle at room temperature. At higher silicon levels, impact transition temperature rises well above room temperature, and upper shelf energy is dramatically reduced. Ductility is restored when temperatures exceed 425 °C (800 °F).

For most applications, alloying with 0.5 to 1% Mo provides adequate elevated-temperature strength and creep resistance (see Fig. 1). Higher molybdenum additions are used when maximum elevated-temperature strength is needed. High molybdenum additions (>1%) tend to generate interdendritic carbides of the Mo_2C type, which persist even through annealing, and tend to reduce toughness and ductility at room temperature.

Silicon lowers the eutectic carbon content, which must be controlled to avoid graphite flotation. For 4% Si irons, carbon content should range from 3.2 to 3.5% C, depending on section size, and at 5% Si it should be around 2.9% C. The unique properties of each element and its contribution to the total alloy system of a ductile graphitic cast iron are described in the article "Ductile Iron" in this Volume.

Melting Practice. For the high-silicon ductile irons, standard ductile iron melting practices apply. Cupola melting is acceptable, but these irons are commonly electric melted. Acid, neutral, or basic linings are used. Conventional ductile iron charge materials are also used; however, care should be taken to minimize chromium, manganese, and phosphorus.

Manganese content should not exceed 0.5%, and it is preferable to keep it below 0.3%. Chromium should be no more than 0.1% and preferably below 0.6%. These

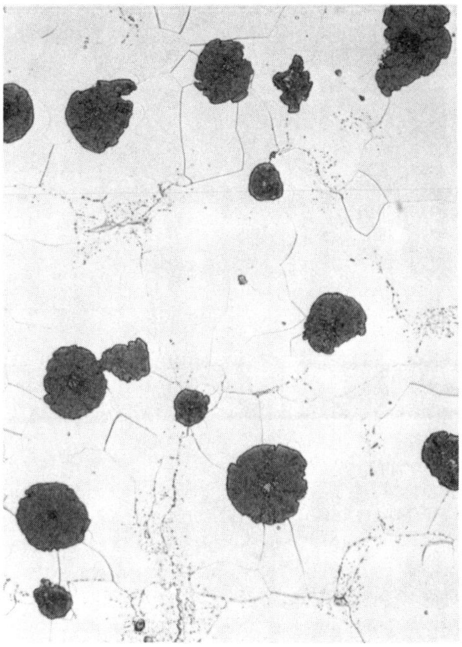

Fig. 1 Photomicrograph of a 4Si-Mo ductile iron showing nodular graphite structure. 400×

elements promote as-cast embrittlement due to carbides and pearlite, which form networks in the interdendritic regions. These microstructural constituents are difficult to break down through subsequent heat treatment, and they degrade toughness and machinability. They are a particular problem in turbocharger applications, in which minimum ductility requirements must be met. Consequently, a larger proportion of pig iron in the charge is common.

Silicon additions can be made in the ladle, but it is highly recommended that all silicon be added to the furnace, except for that required for magnesium treatment and postinoculation. The same is true for molybdenum additions; either ferromolybdenum or molyoxide briquettes can be used.

Conventional magnesium treatment techniques are utilized, and the same rules for balancing sulfur and setting residual magnesium levels apply to the high-silicon irons. These irons are more susceptible to the formation of intercellular carbides when magnesium levels exceed those needed to balance sulfur and to nodularize the graphite. Although these irons respond like standard ductile irons to most inoculants, foundry grade ferrosilicon, containing 75 to 85% Si, is most often used for postinoculation.

Pouring Practice. Due to increased silicon contents, higher pouring temperatures (>1425 °C, or 2600 °F) are required to develop a clean melt surface in the ladle. Therefore, pouring temperatures somewhat higher than those for ductile irons should be used to minimize dross and slag defects.

Molds, Patterns, and Casting Design. These alloys are normally sand cast; sand molds can be made of green sand, shell mold, and air-set chemically bonded sands suitable for gray and ductile irons. As with any graphitic iron, high mold rigidity is necessary to minimize mold wall movement and shrinkage porosity. Softer molds will produce bigger castings, but in general little or no allowance for shrinkage is required in the pattern.

As with conventional ductile irons, the highly reactive magnesium dissolved in these irons, along with the high levels of silicon, gives rise to the formation of dross and inclusions due to the reaction with oxygen in the air during pouring and filling of the mold. Consequently, clean ladles with teapot designs or skimmers are recommended. Furthermore, measures must be taken in designing the runners and gating system to minimize turbulence and to trap dross before it enters the mold cavity (see the articles "Gating Design" and "Riser Design" in this Volume).

Shakeout Practice. As mentioned previously, room-temperature impact resistance is low; therefore, riser and gate removal is somewhat easier with these alloys than with standard ductile iron grades. These irons are quite ductile at elevated temperatures, and they should be allowed to cool before cleaning. They do exhibit a ductility trough in the temperature range of 315 to 425 °C (600 to 800 °F), and riser and gate removal is aided when performed upon cooling through this temperature range.

Heat Treatment. The high-silicon ductile irons are predominantly ferritic as-cast, but the presence of carbide-stabilizing elements will result in a certain amount of pearlite and often intercellular carbides. These alloys are inherently more brittle than standard grades of ductile iron and usually have higher levels of internal stress due to lower thermal conductivity and higher elevated-temperature strength. These factors should be taken into account when selecting heat treatment requirements.

High-temperature heat treatment is advised in all cases to anneal any pearlite and to stabilize the casting against growth in service. A normal graphitizing (full) anneal in the austenitic temperature range is recommended when undesirable amounts of carbide are present. For the 4 to 5% Si irons, this will require heating to at least 900 °C (1650 °F) for several hours, followed by slow cooling to below 705 °C (1300 °F). At higher silicon contents (>5%), in which carbides readily break down, and in castings that are relatively carbide free, subcritical annealing in the temperature range of 720 to 790 °C (1325 to 1450 °F) for 4 h is effective in ferritizing the matrix. Compared to full annealing, the subcritically annealed material will have somewhat higher strength, but ductility and toughness will be reduced.

Applications. The high-silicon and silicon-molybdenum ductile irons are currently produced as manifolds and turbocharger housings for trucks and some automobiles. They are also used in heat-treating racks.

Austenitic Nickel-Alloyed Gray and Ductile Irons

The nickel-alloyed austenitic irons are produced in both gray and ductile cast iron versions for high-temperature service. Austenitic gray irons date back to the 1930s, when they were specialized materials of minor importance. After the invention of ductile iron, austenitic grades of ductile iron were also developed. These nickel-alloyed austenitic irons have been used in applications requiring corrosion resistance, wear resistance, and high-temperature stablility and strength (see the article "Corrosion of Cast Irons" in Volume 13 of the 9th Edition of *Metals Handbook*). Additional advantages include low thermal expansion coefficients, nonmagnetic properties, and cast iron materials having good toughness at low temperatures. When compared with corrosion and heat-resistant steels, nickel-alloyed irons have excellent castability and machinability.

Applications of Nickel-Alloyed Irons. The nickel-alloyed (Ni-Resist) irons have found wide application in chemical process related equipment such as compressors and blowers; condenser parts; phosphate furnace parts; pipe, valves, and fittings; pots and retorts; and pump casings and impellers. Similarly, in food-handling equipment, the various alloy components include bottling and brewing equipment, canning machinery, distillery equipment, feed screws, metal grinders, and salt filters. In high-temperature applications, nickel-alloyed irons are used as cylinder liners, exhaust manifolds, valve guides, gas turbine housings, turbocharger housings, nozzle rings, water pump bodies, and piston rings in aluminum pistons. Figure 2 shows the typical microstructure obtained in Ni-Resist irons.

Austenitic Gray Irons

Specification ASTM A 436 defines eight types of austenitic gray iron alloys, four of which are designed to be used in elevated-temperature applications and four in applications requiring corrosion resistance (Table 1). The nickel produces a stable austenitic microstructure with good corrosion resistance and strength at elevated temperatures. The nickel-alloyed irons are also alloyed with chromium and silicon for wear resistance and oxidation resistance at elevated temperatures. Types 1 and 1b, which are designed for corrosion-resistant applications, are alloyed with 13.5 to 17.5% Ni and 6.5% Cu. Type 1 alloys are used to produce ring carriers utilized in conjunction with aluminum pistons in diesel engines. Types 2b, 3, and 5, which are principally used for high-temperature service, contain 18 to 36% Ni, 1 to 2.8% Si, and 0 to 6% Cr. With the development of ductile iron, most high-temperature applications shifted to similar Ni-Resist ductile iron grades. Type 4

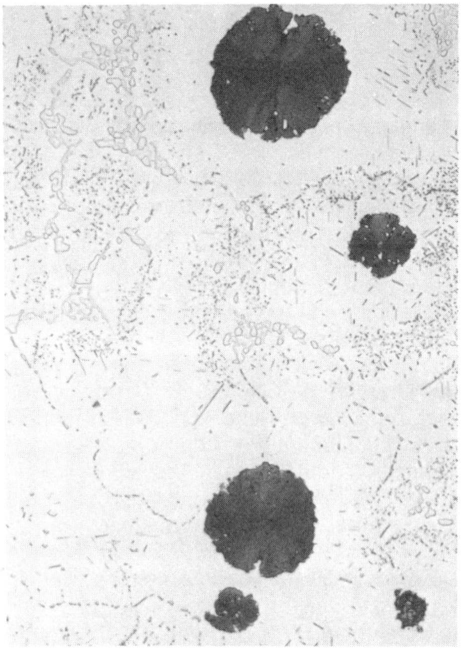

Fig. 2 Photomicrograph of a D5S Ni-Resist ductile iron showing nodular graphite structure. 400×

Table 1 Compositions of austenitic gray iron alloys

Alloy	Composition, % C	Si	Mn	Ni	Cu	Cr	S	Mo
Type 1	3.00 max	1.00–2.80	0.5–1.5	13.50–17.50	5.50–7.50	1.5–2.5	0.12 max	...
Type 1b	3.00 max	1.00–2.80	0.5–1.5	13.50–17.50	5.50–7.50	2.50–3.50	0.12 max	...
Type 2	3.00 max	1.00–2.80	0.5–1.5	18.00–22.00	0.50 max	1.5–2.5	0.12 max	...
Type 2b	3.00 max	1.00–2.80	0.5–1.5	18.00–22.00	0.50 max	3.00–6.00(a)	0.12 max	...
Type 3	2.60 max	1.00–2.00	0.5–1.5	28.00–32.00	0.50 max	2.50–3.50	0.12 max	...
Type 4	2.60 max	5.00–6.00	0.5–1.5	29.00–32.00	0.50 max	4.50–5.50	0.12 max	...
Type 5	2.40 max	1.00–2.00	0.5–1.5	34.00–36.00	0.50 max	0.10 max	0.12 max	...
Type 6	3.00 max	1.50–2.50	0.5–1.5	18.00–22.00	3.50–5.50	1.00–2.00	0.12 max	1.00 max

(a) Where some machining is required, the 3.00–4.00% Cr range is recommended. Source: Ref 1

alloys are alloyed with 29 to 32% Ni, 5 to 6% Si, and 4.5 to 5.5% Cr and are recommended for their stain-resistant properties.

Austenitic Ductile Irons

Specification ASTM A 439 defines the group of austenitic ductile irons (Table 2). The austenitic ductile iron alloys have similar compositions to the austenitic gray iron alloys but have been treated with magnesium to produce nodular graphite irons. Ductile alloys are available in every type but type 1; this is because of its high copper content, which is not compatible with the production of nodular graphite. The ductile iron alloys have high strength and ductility, combined with the same desirable properties of the gray iron alloys. They provide frictional wear resistance, corrosion resistance, strength and oxidation resistance at high temperatures, nonmagnetic characteristics, and, in some alloys, low thermal expansivity at ambient temperatures.

Melting Practice. In the past, these iron alloys were generally cupola melted, but today melting is almost exclusively done in electric furnaces. Choice of furnace linings is usually based on other alloys being melted in the shop; acid, neutral, or basic linings are used. Selection of charge materials is more critical in melting the higher-nickel alloys of the ductile iron versions because they are more sensitive to the tramp elements that affect graphite structure. The high nickel content causes the materials to be more prone to hydrogen gas defects; therefore, charge materials should be thoroughly dry, and melting times should be quick. The molten iron should only be superheated to the temperature necessary to treat and pour, and for as short a time as possible. Pouring is generally done above 1400 °C (2550 °F) to keep the molten iron surface clean and free of oxide.

Magnesium treatment of the ductile iron alloys is normally accomplished with nickel-magnesium alloys. The nickel-magnesium treatment alloy is often added in the furnace; there is no concern for pyrotechnics. The same rules for balancing sulfur and setting residual magnesium levels apply to these nickel-alloyed irons. These irons are more susceptible to the formation of intercellular carbides when magnesium levels exceed those needed to balance sulfur and to nodularize the graphite. Although these irons respond like standard ductile irons to most inoculants, foundry grade ferrosilicon, containing 75 to 85% Si, is most often used for postinoculation. Postinoculation is performed when tapping the furnace. Stream inoculation, where feasible, is also recommended for improved machinability.

Carbon content for the type D-5S alloy must be monitored carefully, because section sensitivity is high. Although the ASTM specification allows up to 2.4% C, sections over 25 mm (1 in.) are susceptible to exploded and chunky graphite formation. Carbon levels of 1.6 to 1.8% are recommended for heavy-section castings.

Molding and Casting Design. Conventional ferrous molding sands are used, including green sand, shell mold, and chemically bonded sands. The same precautions taken for high-strength irons apply to these alloys. Solidification should progress from thin to thick sections without interruption. Abrupt changes in section thickness should be avoided. Riser location should allow convenient access for riser removal. The shrinkage allowance is generally 21 mm/m (¼ in./ft), or 2.1%.

Shakeout. These alloys can develop large thermal stresses upon cooling because of a relatively low thermal conductivity, combined with high elevated-temperature strength and high thermal expansion rates. Consequently, care should be taken in shaking out too hot, and mold cooling is recommended for intricately shaped castings and castings of widely varying section thickness.

Heat treatment of nickel-alloyed ductile irons serves to strengthen the casting and to stabilize the microstructure of the casting for increased durability. Additional information on the heat treatment of ductile iron is available in the article "Ductile Iron" in this Volume.

Stress relief heat treatments are typically conducted at temperatures between 620 and 675 °C (1150 and 1250 °F) to remove residual casting stresses. Mold cooling to 315 °C (600 °F) is a satisfactory alternative to furnace stress relief.

Annealing of some castings may be necessary to reduce hardness. Annealing is performed at 955 to 1040 °C (1750 to 1900 °F) for 30 min to 5 h, and this treatment will normally break down some of the carbides and spheroidize the rest.

Heat treatment for stability of the microstructure for service at temperatures of 480 °C (900 °F) and above is performed by heating at 760 °C (1400 °F) for a minimum of 4 h and furnace cooling to 540 °C (1000 °F), followed by air cooling. An alternative treatment is to heat at 900 °C (1650 °F) for 2 h and furnace cool to 540 °C (1000 °F). These treatments stabilize the microstruc-

Table 2 Compositions of austenitic nodular irons

Alloy	Composition, % C	Si	Mn	P	Ni	Cr
Type D-2(a)	3.00 max	1.50–3.00	0.70–1.25	0.08 max	18.00–22.00	1.75–2.75
Type D-2B	3.00 max	1.50–3.00	0.70–1.25	0.08 max	18.00–22.00	2.75–4.00
Type D-2C	2.90 max	1.00–3.00	1.80–2.40	0.08 max	21.00–24.00	0.50 max
Type D-3(a)	2.60 max	1.00–2.80	1.00 max(b)	0.08 max	28.00–32.00	2.50–3.50
Type D3-A	2.60 max	1.00–2.80	1.00 max(b)	0.08 max	28.00–32.00	1.00–1.50
Type D-4	2.60 max	5.00–6.00	1.00 max(b)	0.08 max	28.00–32.00	4.50–5.50
Type D-5	2.40 max	1.00–2.80	1.00 max(b)	0.08 max	34.00–36.00	0.10 max
Type D-5B	2.40 max	1.00–2.80	1.00 max(b)	0.08 max	34.00–36.00	2.00–3.00
Type D-5S	2.30 max	4.90–5.50	1.00 max(b)	0.08 max	34.00–37.00	1.75–2.25

(a) Additions of 0.7–1.0% Mo will increase the mechanical properties above 425 °C (800 °F). (b) Not intentionally added. Source: Ref 2

Table 3 Compositions of high-silicon iron alloys

Alloy	Composition, % C	Mn	Si	Cr	Mo	Cu
Grade 1	0.70–1.10	1.50 max	14.20–14.75	0.50 max	0.50 max	0.50 max
Grade 2	0.75–1.15	1.50 max	14.20–14.75	3.25–5.00	0.40–0.60	0.50 max
Grade 3	0.70–1.10	1.50 max	14.20–14.75	3.25–5.00	0.20 max	0.50 max

Source: Ref 3

ture and minimize growth and warpage in service. The treatments are designed to reduce carbon levels in the matrix and some growth and distortion often accompany heat treatment. Type 1 alloys are not generally amenable to this stabilizing treatment.

Dimensional stability, when truly critical, can be ensured by heat treating at 870 °C (1600 °F) or higher for 2 h plus an additional hour per 25 mm (1 in.) of cross section, furnace cooling not faster than 55 °C/h (100 °F/h) to 540 °C (1000 °F), and then holding for 1 h per 25 mm (1 in.) of cross section and cooling uniformly. After rough machining, the casting should be reheated to 455 to 480 °C (850 to 900 °F) and held 1 h per 25 mm (1 in.) of cross section.

Refrigeration and reaustenitization heat treatments are applied to type D-2 alloys to increase yield strength. Solution heat treating at 925 °C (1700 °F), refrigerating at −195 °C (−320 °F), and then reheating between 650 and 760 °C (1200 and 1400 °F) will increase yield strength considerably without materially affecting magnetic properties or corrosion resistance in seawater or dilute sulfuric acid. Detailed information on the heat treatment of ductile iron is available in the article "Ductile Iron" in this Volume.

Aluminum-Alloyed Irons

The aluminum-alloyed irons consist of two groups of gray and ductile irons. The low-alloyed group contains 1 to 7% Al, and the aluminum essentially replaces silicon as the graphitizing element in these alloys. The high-alloyed group contains 18 to 22% Al. Irons alloyed with aluminum in between these two ranges will be white irons as-cast and will have no commercial importance.

The aluminum greatly enhances oxidation resistance at elevated temperatures and also strongly stabilizes the ferrite phase to very high temperatures—up to and beyond 980 °C (1800 °F). Like the silicon-alloyed irons, the aluminum irons form a tight, adherent oxide on the surface of the casting that is very resistant to further oxygen penetration.

Unfortunately, the aluminum-alloyed irons are very difficult to cast without dross inclusions and laps (cold shuts). The aluminum in the iron is very reactive at the temperatures of the molten iron, and contact with air and moisture must be negligible. Care must be taken not to draw the oxide skin, which forms during pouring, into the mold in order to avoid dross inclusions. Methods for overcoming these problems in commercial practice are under development.

At present, there is no ASTM standard covering the chemistry and expected properties of these alloys, and commercial production is very limited. In the past, the 1.5 to 2.0% irons have been used in the production of truck exhaust manifolds.

High-Silicon Irons for Corrosion Resistance

Irons with high silicon contents (14.5% Si) constitute a unique corrosion-resistant ferritic cast iron group. These alloys are widely used in the chemical industry for processing and for transporting highly corrosive liquids. They are particularly suited to handling sulfuric and nitric acids (see the article "Corrosion of Cast Irons" in Volume 13 of the 9th Edition of *Metals Handbook*). The three most common high-silicon iron alloys are covered in ASTM A 518M (Table 3). These alloys contain 14.2 to 14.75% Si and 0.7 to 1.15% C. Grades 2 and 3 are also alloyed with 3.25 to 5% Cr, and grade 2 contains 0.4 to 0.6% Mo. Other compositions are also commercially produced with up to 17% Si.

Melting Practice. Induction melting is the preferred method for these alloys. Induction melting permits the very tight control of chemistry needed in melting these materials in order to minimize scrap losses due to cracking. The alloys have a very low tolerance for hydrogen and nitrogen; thus, charge materials must be carefully controlled to minimize the levels of these gases. Vacuum treatment can be used to increase strength and density.

The melting point of the eutectic 14.3% Si cast iron is approximately 1180 °C (2160 °F), and the alloy is generally poured at approximately 1345 °C (2450 °F). Because of the very brittle nature of high-silicon cast iron, castings are usually shaken out after cooling to ambient temperature. However, some casting geometries demand hot shakeout while the castings are still red hot, so that the castings can be immediately stress relieved and furnace cooled to prevent cracking.

Molding and Casting Design. These alloys are routinely cast in sand molds, investment molds, and permanent steel molds. Cores must have good collapsibility to prevent fracture during solidification. Flash must be minimized because it will be chilled iron and can readily become an ideal nucleation site for cracks that propagate into the casting.

Casting design for these high-silicon irons requires special considerations. To avoid cracking, sharp corners and abrupt changes in section size must be avoided. Casting designs should have tapered ingates to permit easy removal of gates and risers by impact and to minimize the amount of grinding required. Unless vacuum degassing is employed, conventional risers are not generally used or desired.

High-silicon cast irons are generally cast in sections ranging from 4.8 to 38 mm (3/16 to 1.5 in.). Thin sections fill well because of the excellent fluidity of these cast irons, but tend to chill and become extremely brittle white iron. Sections over 38 mm (1.5 in.) are prone to segregation and porosity, which reduce strength and corrosion resistance. Castings are generally designed with a patternmaker's rule of 18 mm shrink to the meter (7/32 in. shrink to the foot), or 1.8%.

Heat Treatment. High-silicon irons can be stress relieved by heating in the range of 870 to 900 °C (1600 to 1650 °F), followed by slow cooling to ambient temperatures to minimize the likelihood of cracking. Heat treatments have no significant effect on corrosion resistance.

Machinability. High-silicon cast irons have hardnesses of approximately 500 HB and are normally considered machinable only by grinding. Machinability can be improved with higher additions of carbon and/or phosphorus, but only at the expense of other properties, such as strength and corrosion resistance.

Applications. High-silicon irons are extensively used in equipment for the production of sulfuric and nitric acids, for sewage disposal and water treatment, for handling mineral acids in petroleum refining, and in the manufacture of fertilizer, textiles, and explosives. Specific components include pump rotors, agitators, crucibles, and pipe fittings in chemical laboratories.

ACKNOWLEDGMENTS

The author wishes to thank Warren Spear of The Nickel Development Institute and Don Stickle of the Duriron Company for their valuable assistance.

REFERENCES

1. "Standard Specification for Austenitic Gray Iron Castings," A 436, *Annual Book of ASTM Standards*, American Society for Testing and Materials
2. "Standard Specification for Austenitic Ductile Iron Castings," A 439, *Annual Book of ASTM Standards*, American Society for Testing and Materials
3. "Standard Specification for Corrosion-Resistant High-Silicon Iron Castings," A 518M, *Annual Book of ASTM Standards*, American Society for Testing and Materials

Plain Carbon Steels

John M. Svoboda, Steel Founders' Society of America

CARBON STEELS contain only carbon as the principal alloying element. Other elements are present in small quantities, including those added for deoxidation. Silicon and manganese in cast carbon steels typically range from 0.25 to about 0.80% Si, and 0.50 to about 1.00% Mn. Carbon steels can be classified according to their carbon content into three broad groups:

- *Low-carbon steels*: ≤0.20% C
- *Medium-carbon steels*: 0.20 to 0.50% C
- *High-carbon steels*: ≥0.50% C

Low-alloy steels contain alloying elements, in addition to carbon, up to a total alloy content of 8%. Cast steels containing more than the following amounts of a single alloying element are considered low-alloy cast steels:

Element	Amount, %
Manganese	1.00
Silicon	0.80
Nickel	0.50
Copper	0.50
Chromium	0.25
Molybdenum	0.10
Vanadium	0.05
Tungsten	0.05

Detailed information on cast low-alloy steels is available in the article "Low-Alloy Steels" in this Volume.

For the deoxidation of carbon and low-alloy steels (that is, for control of their oxygen content), aluminum, titanium, and zirconium are used. Of these, aluminum is used more frequently because of its effectiveness and low cost. Unless otherwise specified, the normal sulfur limit for carbon and low-alloy steels is 0.06%, and the normal phosphorus limit is 0.05%.

Structure and Property Correlations

Carbon steel castings are produced to a great variety of properties because composition and heat treatment can be selected to achieve specific combinations of properties, including hardness, strength, ductility, fatigue resistance, and toughness. Although selections can be made from a wide range of properties, it is important to recognize the interrelationships among these properties. For example, higher hardness, lower toughness, and lower ductility values are associated with higher strength values. The relationships among these properties and mechanical properties are discussed further in this section.

Property trends among carbon steels are illustrated as a function of the carbon content in Fig. 1. Unless otherwise noted, the properties discussed refer to those obtained from specimens that have been removed from standard ASTM keel blocks, which are made with a 32 mm (1.25 in.) section size. The effect of larger section sizes on these properties is discussed in the section "Section Size and Mass Effects" in this article.

Strength and Hardness. Depending on alloy choice and heat treatment, ultimate tensile strength levels from 414 to 1724 MPa (60 to 250 ksi) can be achieved with cast carbon and low-alloy steels. Figure 2 illustrates the tensile strength and tensile ductility values that can be expected from normalized and quenched-and-tempered cast carbon steels having a range of Brinell hardness values. For carbon steels, the hardness and strength values are largely determined by carbon content and the heat treatment (Fig. 1c). The effect of tempering normalized carbon steel is shown in Fig. 3.

Strength and Ductility. Ductility depends greatly on the strength, or hardness, of the cast steel (Fig. 2). Actual ductility requirements vary with the strength level and the specification to which a steel is ordered. Because yield strength is a primary design criterion for structural applications, the relationships shown in Fig. 1(a) and (b) are replotted in Fig. 4 to reveal the major trends for cast carbon steels. Quenched-and-tempered steels exhibit higher ductility values for a given yield strength level than normalized, normalized-and-tempered, or annealed steels.

Strength and Toughness. Several test methods are available for evaluating the toughness of steels or the resistance to sudden or brittle fracture. These include the Charpy V-notch impact test, the drop-weight test, the dynamic tear test, and specialized procedures to determine plane-strain fracture toughness. The results of all of these tests are in use and will be reviewed in this section because each of these tests offers specific advantages that are unique to the test method.

Charpy V-notch impact energy trends at room temperature (Fig. 4) reveal the distinct effect of strength and heat treatment on toughness. Higher toughness is obtained when a steel is quenched and tempered, rather than normalized and tempered. The effect of heat treatment and testing temperature on Charpy V-notch toughness is further illustrated in Fig. 5 for a carbon steel. Quenching, followed by tempering, produces superior toughness as indicated by the shift of the impact energy transition curve to lower temperatures.

Nil ductility transition temperatures (NDTT) ranging from 38 °C (100 °F) to as low as −90 °C (−130 °F) have been recorded in tests on normalized-and-tempered cast carbon and low-alloy steels in the yield strength range of 207 to 655 MPa (30 to 95 ksi) (Fig. 6). Comparison of the data in Fig. 6 with those of Fig. 7 shows the superior toughness values at equal strength levels that low-alloy steels offer compared to carbon steels. When cast steels are quenched and tempered, the range of strength and of toughness is broadened. Depending on alloy selection, NDTT values of as high as 10 °C (50 °F) to as low as −107 °C (−160 °F) can be obtained in the yield strength range of 345 to 1345 MPa (50 to 195 ksi) (Fig. 7).

An approximate relationship exists between the Charpy V-notch impact energy-temperature behavior and the NDTT value. The NDTT value frequently coincides with the energy transition temperature determined in Charpy V-notch tests.

Plane-strain fracture toughness (K_{Ic}) data for a variety of steels reflect the important strength-toughness relationship. Fracture mechanics tests have the advantage over conventional toughness tests of being able to yield material property values that can be used in design equations. Additional information is available in the Selected References at the end of this article.

Strength and Fatigue. The most basic method of presenting engineering fatigue data is by means of the *S-N* curve, which relates the dependence of the life of the fatigue specimen in terms of the number of cycles to failure *N* to the maximum applied

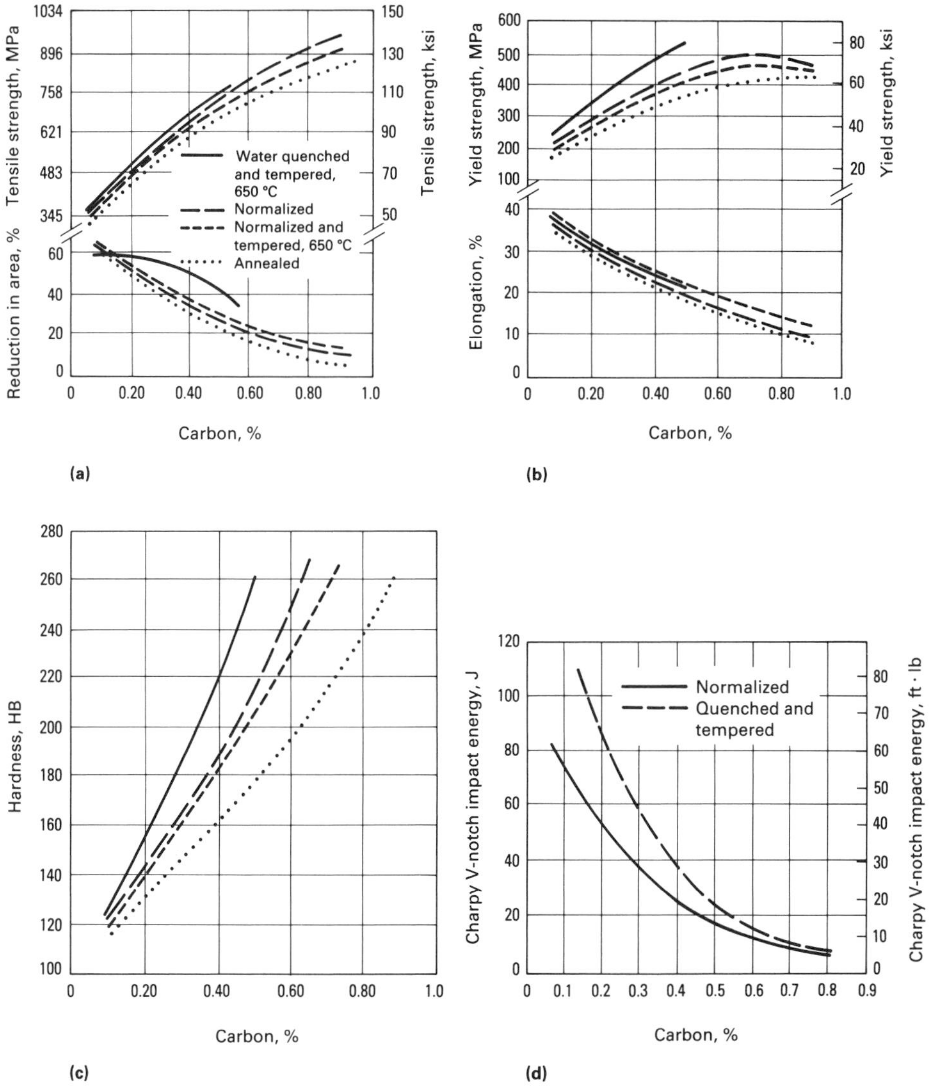

Fig. 1 Properties of cast carbon steels as a function of carbon content and heat treatment. (a) Tensile strength and reduction of area. (b) Yield strength and elongation. (c) Brinell hardness. (d) Charpy V-notch impact energy

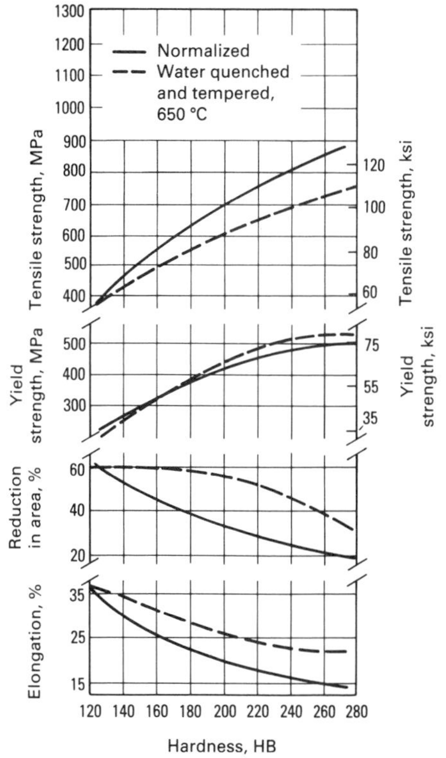

Fig. 2 Tensile properties of cast carbon steels as a function of Brinell hardness

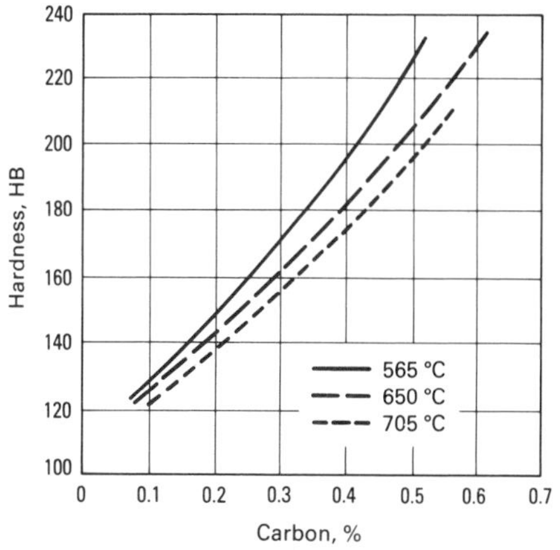

Fig. 3 Hardness versus carbon content of normalized cast carbon steels tempered at various temperatures for 2 h. Tempering temperatures are indicated on the graph.

stress *S*. Other tests have been used, and the principal findings for cast carbon steels are highlighted in the following sections.

Constant Amplitude Tests. The endurance ratio (endurance limit divided by the tensile strength) of cast carbon and low-alloy steels as determined by rotating-beam bending fatigue tests (mean stress = 0) is generally taken to be approximately 0.40 to 0.50 for smooth bars. The data given in Table 1 indicate that this endurance ratio is largely independent of strength, alloying additions, and heat treatment.

The fatigue notch sensitivity factor, q, determined in rotating-beam bending fatigue tests is related to the microstructure of the steel (composition and heat treatment) and the strength. Table 2 shows that q generally increases with strength—from 0.23 for annealed carbon steel at a tensile strength of 577 MPa (83.5 ksi) to 0.68 for the higher-strength normalized-and-tempered low-alloy steels. The quenched-and-tempered steels with a martensitic structure are less notch sensitive than the normalized-and-tempered steels with a ferrite-pearlite microstructure. Similar results and trends in notch sensitivity have been reported for tests with sharper notches.

Cast steels suffer less degradation of fatigue properties due to notches than equivalent wrought steels. When the ideal laboratory test conditions are replaced with more realistic service conditions, the cast steels exhibit much less notch sensitivity to variations in the values of the test parameters than wrought steels. Table 3 shows that the q values for wrought steels are 1.4 to 2.3 times higher than those for cast steels. Under laboratory test conditions (uniform specimen section size, polished and honed surfaces, and so on), the endurance limit of wrought steel is higher than that of cast steel. The same fatigue characteristics as those of cast steel, however, are obtained when a notch is introduced or when standard lathe-turned surfaces are employed in the rotating-beam bending fatigue test. These effects are illustrated in Fig. 8 and 9.

The cyclic stress-strain characteristics shown in Fig. 10 and Table 4 indicate a reduction of the strain-hardening exponent n of the normalized-and-tempered cast carbon steel (SAE 1030) from 0.3 in monotonic tension to 0.13 under cyclic-strain-controlled tests. Figure 11 shows that the strain life characteristics of normalized-and-tempered cast carbon steel (SAE 1030) and

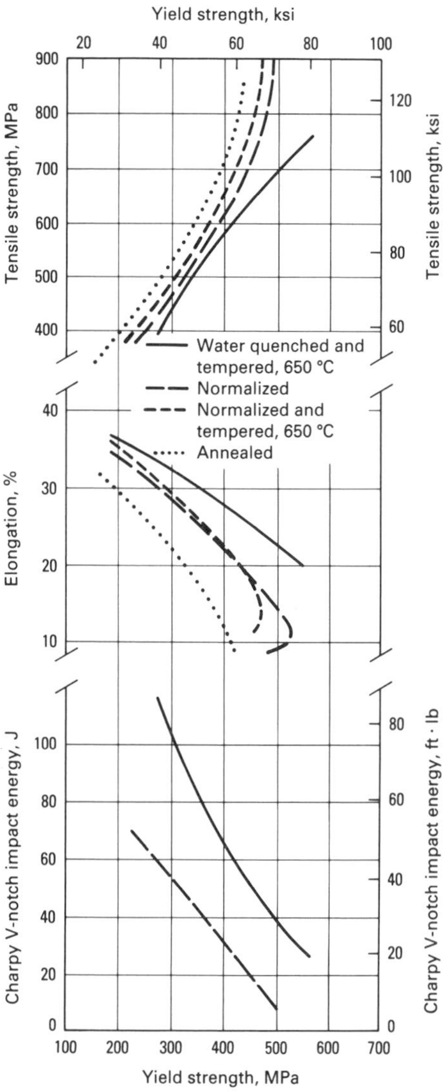

Fig. 4 Room-temperature properties of cast carbon steels after different heat treatments

wrought steel are similar. These data were obtained from strain-controlled constant-amplitude low-cycle fatigue tests (0.001 to 0.02 strain range amplitudes, with a constant strain rate triangle wave form of 2.5 × 10^{-4} s^{-1} at 0.5 to 3.3 Hz).

Constant load amplitude fatigue crack growth properties for load ratios of $R = 0$ (Fig. 12a) indicate comparable properties for cast and wrought steel and slightly better properties for normalized-and-tempered cast carbon steel (SAE 1030) under load ratios of $R = -1$ (Fig. 12b). These tests were conducted in air at 10 to 30 Hz, depending on load ratio, initial stress intensity, and crack length.

Variable load amplitude fatigue tests indicate equal total life for cast and wrought carbon steels (cast 1030 and wrought 1020, respectively) (Fig. 13). The slower crack growth rate in the cast material compensated for the longer crack initiation life of the wrought carbon steel.

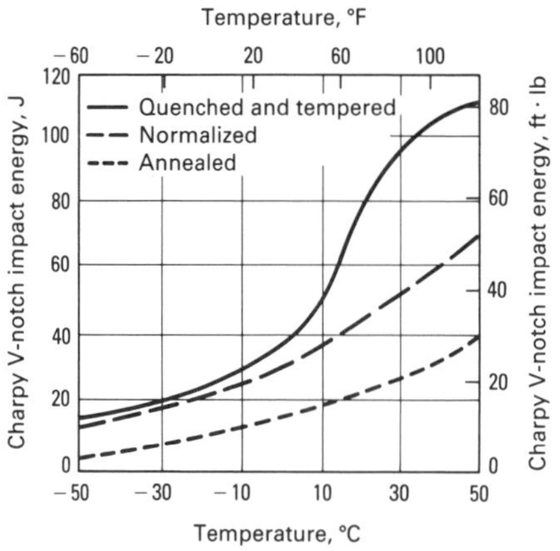

Fig. 5 Effect of various heat treatments on the Charpy V-notch impact energy of a 0.30% C steel

Section Size and Mass Effects. Mass effects are common to steels, whether rolled, forged, or cast, because the cooling rate during heat treating varies with section size and because the microstructure constituents, grain size, and nonmetallic inclusions increase in size from surface to center. Mass effects are metallurgical in nature and are distinct from the effect of discontinuities, which are discussed in the following section in this article.

The section size or mass effect is of particular importance in steel castings because mechanical properties are typically assessed from test bars machined from standardized coupons having fixed dimensions and are cast separately from or attached to the castings. The removal of test bars from the casting is impractical because removal of material for testing would destroy the usefulness of the component.

Test specimens removed from a casting will not routinely exhibit the same properties as test specimens machined from the standard test coupon designs for which minimum properties are established in specifications. The mass effect discussed above, that is, the difference in cooling rate between the test coupons and the part being produced, is the fundamental reason for this situation. Several specifications provide for the mass effect by permitting the testing of coupons that are larger than normal and that have cooling rates more representative of those experienced by the part being produced. Among these specifications are ASTM E 208, A 356, and A 757.

Discontinuity Effects. The treatment of discontinuities in design is undergoing major changes because of the wider use of fracture mechanics in the industry. If the plane-strain fracture toughness K_{Ic} of a material is known at the temperature of interest, designers can determine the critical combination of flaw size and stress required to cause failure in one load application. In addition, designers can calculate the remaining life of a component having a discontinuity, or they can compute the largest acceptable flow from knowledge of the crack growth rate da/dN of a material and other fracture mechanics parameters. In the absence of suitable plane-strain fracture mechanics data, approximations can be made on the basis of results obtained from various tensile, impact, and fatigue tests.

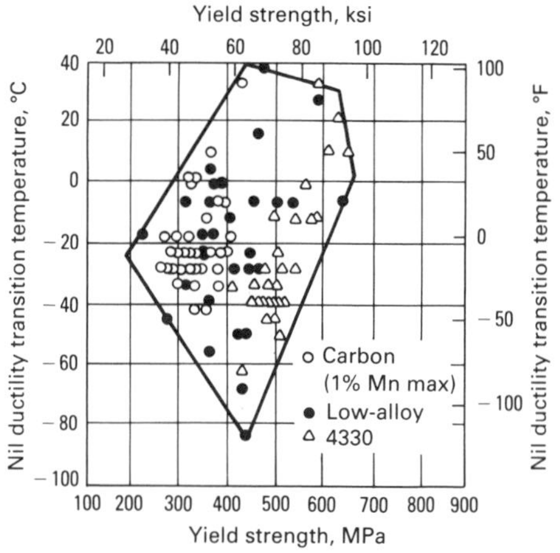

Fig. 6 Nil ductility transition temperatures and yield strengths of normalized-and-tempered commercial cast steels

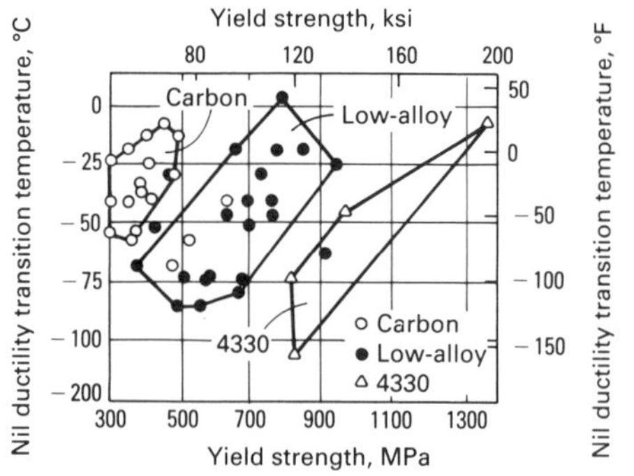

Fig. 7 Nil ductility transition temperatures and yield strengths of quenched-and-tempered commercial cast steels

Test Coupon Versus Casting Properties. Coupon properties refer to the properties of specimens cut and machined from a separately cast coupon or a coupon that is attached to and cast integrally with the casting. Typically, the legs of the ASTM standard double-leg keel block (A 370) serve as the coupons; the legs are 32 mm (1.25 in.) thick.

Test Coupons. The ASTM double-leg keel block is the most prominent design for test coupons. Table 5 provides information on the reliability of tensile test results obtained from the double-leg keel block. The data indicate that for two tests there is 95% assurance that the actual strength is within ±6.9 MPa (±1 ksi) of the actual ultimate tensile strength and within ±11 MPa (±1.6 ksi) of the actual yield strength. For tensile ductility, the data show that two tests pro-

Table 1 Properties of various classes of cast carbon and low-alloy steels

Class(a)	Heat treatment(b)	Tensile strength MPa	Tensile strength ksi	Yield strength MPa	Yield strength ksi	Reduction in area, %	Elongation, %	Hardness, HB	Fatigue endurance limit MPa	Fatigue endurance limit ksi	Ratio of endurance limit to tensile strength
Carbon steels											
60	A	434	63	241	35	54	30	131	207	30	0.48
65	N	469	68	262	38	48	28	131	207	30	0.44
70	N	517	75	290	42	45	27	143	241	35	0.47
80	NT	565	82	331	48	40	23	163	255	37	0.45
85	NT	621	90	379	55	38	20	179	269	39	0.43
100	QT	724	205	517	75	41	19	212	310	45	0.47
Low-alloy steels(c)											
65	NT	469	68	262	38	55	32	137	221	32	0.47
70	NT	510	74	303	44	50	28	143	241	35	0.47
80	NT	593	86	372	54	46	24	170	269	39	0.45
90	NT	655	95	441	64	44	20	192	290	42	0.44
105	NT	758	110	627	91	48	21	217	365	53	0.48
120	QT	883	128	772	112	38	16	262	427	62	0.48
150	QT	1089	158	979	142	30	13	311	510	74	0.47
175	QT	1234	179	1103	160	25	11	352	579	84	0.47
200	QT	1413	205	1172	170	21	8	401	607	88	0.43

(a) Class of steel based on tensile strength (ksi). (b) A, annealed; N, normalized; NT, normalized and tempered; QT, quenched and tempered. (c) Below 8% total alloy content

Table 2 Fatigue notch sensitivity of various cast steels

Steel	Tensile strength MPa	Tensile strength ksi	Fatigue endurance limit, Unnotched MPa	Unnotched ksi	Notched(a) MPa	Notched(a) ksi	Ratio of fatigue endurance limit to tensile strength, Unnotched	Notched
Normalized and tempered								
1040	648	94.2	260	37.7	193	28	0.40	0.30
1330	685	99.3	334	48.4	219	31.7	0.49	0.32
1330	669	97	288	41.7	215	31.2	0.43	0.32
4135	777	112.7	353	51.2	230	33.3	0.45	0.30
4335	872	126.5	434	63	241	34.9	0.50	0.28
8630	762	110.5	372	54	228	33.1	0.49	0.30
Quenched and tempered								
1330	843	122.2	403	58.5	257	37.3	0.48	0.31
4135	1009	146.4	423	61.3	280	40.6	0.42	0.28
4335	1160	168.2	535	77.6	332	48.2	0.46	0.29
8630	948	137.5	447	64.9	266	38.6	0.47	0.27
Annealed								
1040	576	83.5	229	33.2	179	26	0.40	0.31

(a) Notched tests run with theoretical stress concentration factor of 2.2

Table 3 Fatigue notch sensitivity factors for cast and wrought steels at various strength levels

Steel	Tensile strength MPa	Tensile strength ksi	Fatigue notch sensitivity factor q(a)
Annealed			
1040 cast	576	83.5	0.23
1040 wrought	561	81.4	0.43
Normalized and tempered			
1040 cast	649	94.2	0.29
1040 wrought	620	90.0	0.50
1330 cast	669	97.0	0.28
1340 wrought	702	101.8	0.65
4135 cast	777	112.7	0.45
4140 wrought	766	111.1	0.81
4335 cast	872	126.5	0.68
4340 wrought	859	124.6	0.97
8630 cast	762	110.5	0.53
8640 wrought	748	108.5	0.85
Quenched and tempered			
1330 cast	843	122.2	0.48
1340 wrought	836	121.2	0.73
4135 cast	1009	146.4	0.43
4140 wrought	1012	146.8	0.93
4335 cast	1160	168.2	0.51
4340 wrought	1161	168.4	0.92
8630 cast	948	137.5	0.57
8640 wrought	953	138.2	0.90

(a) $q = (K_f - 1)/(K_t - 1)$, where K_f is the notch fatigue factor (endurance limit unnotched/endurance limit notched) and K_t is the theoretical stress concentration factor.

duce, with 95% assurance, the elongation results within ±3% and the reduction in area value within ±5%. When 32 mm (1.25 in.) thick test coupons are suitably attached to the casting and cast integrally with the production casting, the tensile properties determined for the coupon will be comparable to those from a separately cast keel block.

The properties determined from keel block legs with dimensions exceeding those of the ASTM double-leg keel block, that is, thicker than 32 mm (1.25 in.), may differ, especially if the steel involved is of insufficient hardenability for the heat treatment employed to produce a microstructure similar to that in 32 mm (1.25 in.) section keel block legs. The data given in Table 6 show slightly decreasing strength and ductility with increasing keel block section size of annealed 0.26% C steel.

The weldability of carbon steels is primarily a function of composition and heat treatment. Carbon steels having low manganese and silicon contents and carbon contents below 0.30% can be welded without any special precautions. When the carbon content exceeds 0.30%, preheating of the casting prior to welding may be advisable. The low-temperature preheat (120 to 205 °C, or 250 to 400 °F) reduces the rate at which heat is extracted from the heat-affected zone (HAZ) adjacent to the weld. Preheating also helps to relieve mechanical stresses and to prevent underbead cracking, because hydrogen is still relatively mobile and can diffuse away from the last areas to undergo a metallurgical transformation. Preheating minimizes the chances of a hardening transformation occurring in the HAZ and therefore reduces the hardness adjacent to the weld. Stress raisers that could result from the hardened area can be eliminated in this manner. Generally, if the hardness of the HAZ after welding does not exceed 35 HRC or 327 HB, preheating of the casting for welding is not required.

Melting Practice

One major difference between melting practice in a steel foundry and in a steel mill is the higher tapping temperature used for foundry melting to attain better fluidity of the molten steel. The producer of ingots for rolling is less concerned with fluidity because mold filling is simpler in ingot molds than in the molds used for producing relatively complex shapes.

The melting furnaces used in steel foundries are essentially the same as those used

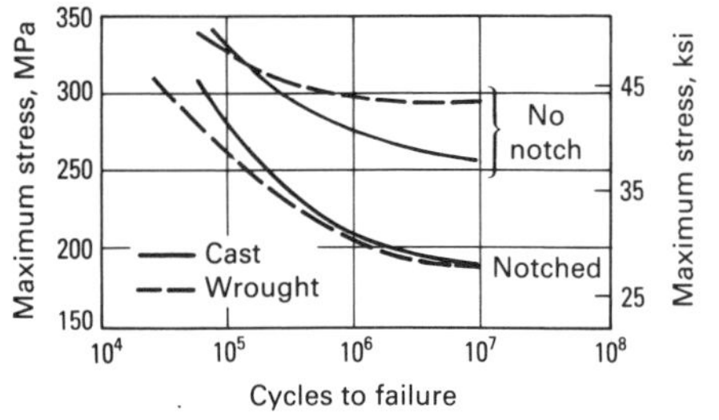

	Tensile strength, MPa (ksi)	Yield strength, MPa (ksi)	Elongation, %	Hardness, HB
Cast	648 (94)	386 (56)	25	187
Wrought	620 (90)	386 (56)	27	170

Fig. 8 Fatigue characteristics of normalized and tempered cast and wrought 1040 steels. Notched and unnotched specimens were tested in R.R. Moore rotating beam tests

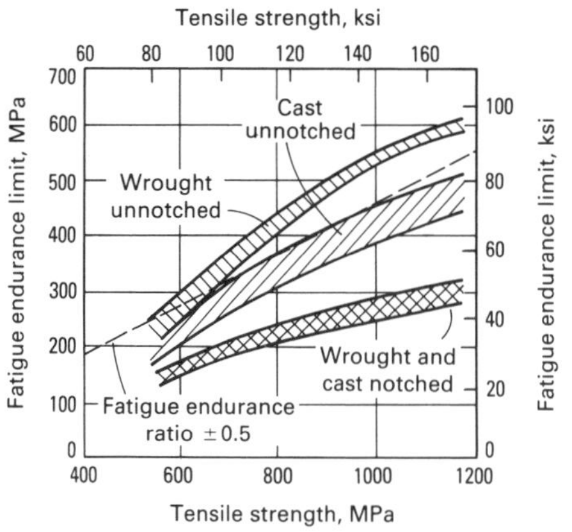

Fig. 9 Fatigue endurance limit versus tensile strength for notched and unnotched cast and wrought steels with various heat treatments. Data obtained in R.R. Moore rotating beam fatigue tests; theoretical stress concentration factor = 2.2

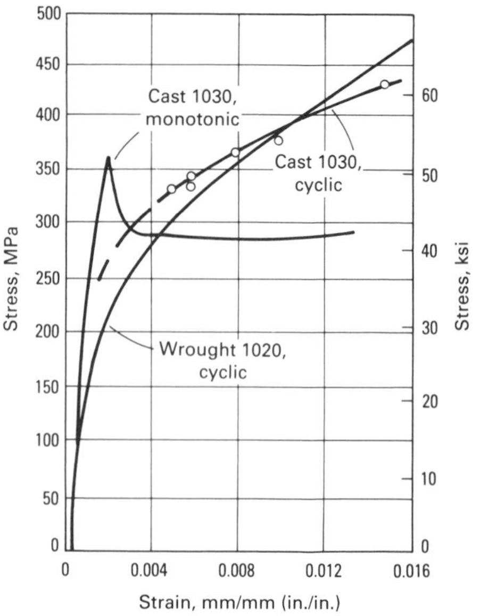

Fig. 10 Monotonic tensile and cyclic stress-strain behavior of comparable cast and wrought normalized-and-tempered carbon steels

for the production of steel ingots except that most foundry melting units are smaller. Although the equipment is the same, the processes are often different. Steel ingots can be made as rimmed, semikilled, or killed steel. Only thoroughly killed steel is used for steel foundry products. The method of production of the killed steel used for castings may differ from that used for wrought products because of the fluidity requirement. However, the salient features of making steel in a foundry are the same as those for producing fully killed steel ingots.

Foundries that have access to a good grade of scrap, and therefore do not need to reduce the phosphorus and sulfur contents of the steel to meet specifications, usually prefer to use a furnace lined with silica refractories (acid lined). Foundries that do not have a good source of low-phosphorus, low-sulfur scrap use refractories such as magnesite and dolomite for furnace lining (basic lining). The compositions of certain steels, such as austenitic manganese steels, require that they be made in a furnace with a basic lining.

Direct Arc Melting

The direct arc furnace consists essentially of metal shell lined with refractories. This lining forms a melting chamber, the hearth of which is bowl shaped. Three carbon or graphite electrodes carry the current into the furnace. Steel, either solid or molten, is the common conductor for the current flowing between the electrodes. The metal is melted by arcs from the electrodes to the metal charge—both by direct impingement of the arcs and by radiation from the roof and walls. The electrodes are controlled automatically so that an arc of proper height can be maintained.

In acid electric practice, the furnace hearth is composed of silica sand or ganister rammed into place. The furnace is charged with selected scrap low in phosphorus and sulfur because the acid process cannot eliminate these elements. About 40% of the charge usually consists of foundry scrap (gates and risers). The general practice is to charge small pieces first in order to form a compact mass in the furnace, thus aiding electrical conductivity. The heavy, lumpy portion of the charge is placed over the smaller pieces, followed by the lightest portion.

The charge is melted down as quickly as possible. Small amounts of sand and limestone are occasionally added to the bath during the melting period to form a protective slag of proper consistency. Once melting is complete, oxygen is used as an oxidizing agent to produce excess oxygen. Iron oxide in slag reacts with silicon and manganese in the metal bath and produces oxides and silicates in the slag. After most of the silicon and manganese have been oxidized, the bath begins to boil. The boil is evidence of carbon elimination; it results from the interaction of dissolved carbon and free oxygen, giving rise to carbon monoxide bubbles.

In the oxidizing stage of melting, the slag is rich in iron oxide. As carbon is eliminated

Table 4 Monotonic tensile and cyclic stress-strain properties of cast and wrought steels

Property	Cast 1030	Wrought 1020
Monotonic tension		
0.2% yield strength, MPa (ksi)	303 (44)	262 (38)
Ultimate strength, MPa (ksi)	496 (72)	414 (60.0)
True fracture strength, MPa (ksi)	750 (109)	1000 (145)
Reduction in area, %	46	58
True fracture ductility	0.62	0.87
Modulus of elasticity E, GPa (ksi)	207 (3 × 10^4)	203 (2.9 × 10^4)
Strain-hardening exponent n	0.3	0.19
Strength coefficient K, MPa (ksi)	1090 (158)	738 (107)
Cyclic stress-strain		
0.2% yield strength, MPa (ksi)	317 (46)	241 (35)
Strength coefficient K', MPa (ksi)	710 (103)	772 (112)
Strain-hardening coefficient n'	0.13	0.18

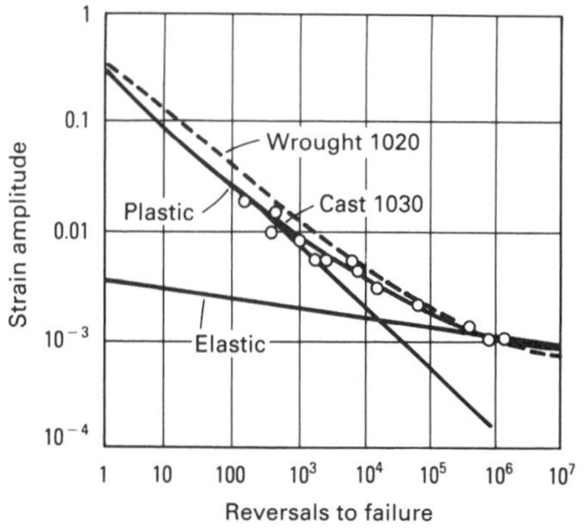

Fig. 11 Low-cycle strain-control fatigue behavior of cast and wrought carbon steels in the normalized-and-tempered condition

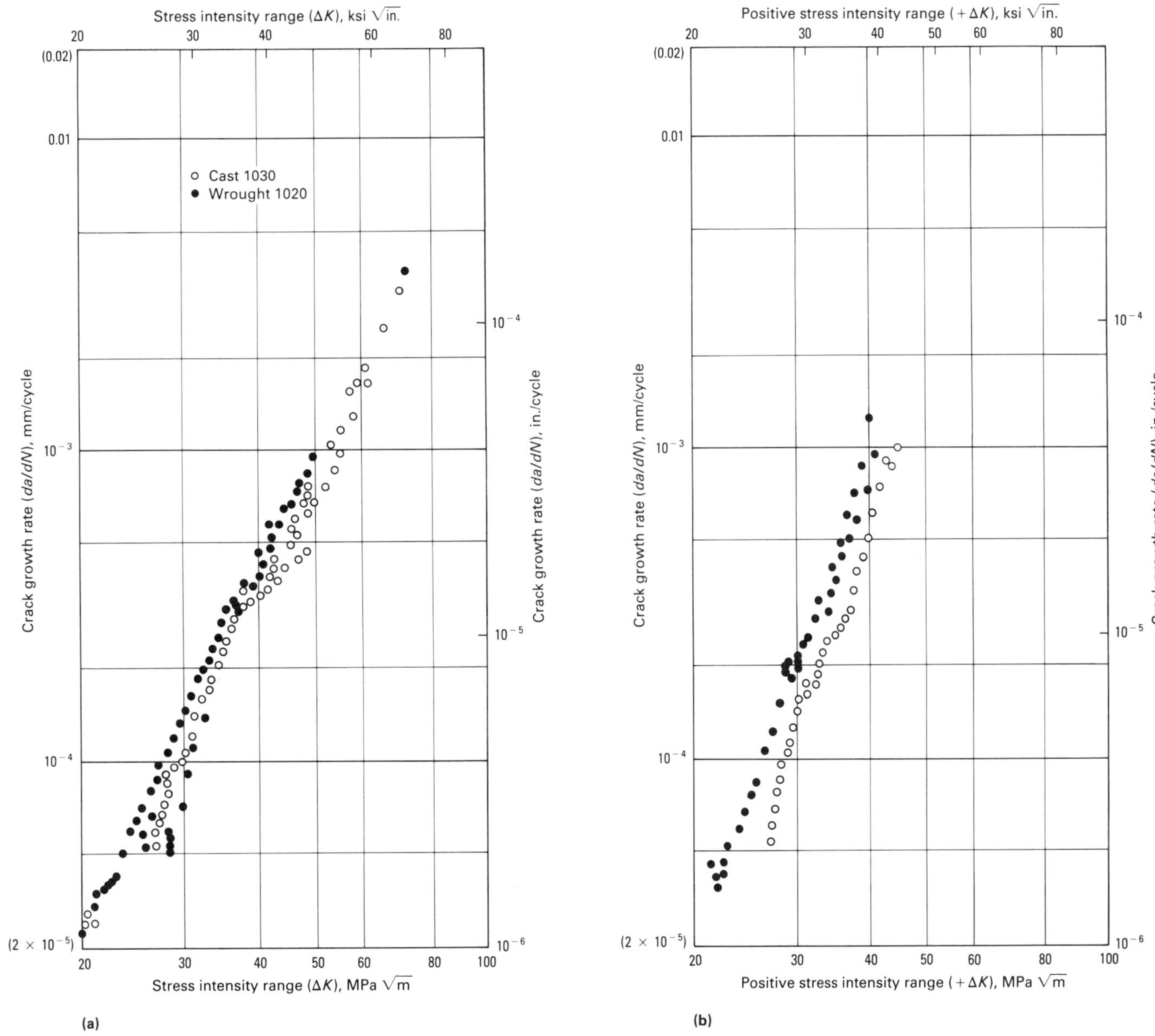

Fig. 12 Constant-amplitude fatigue behavior of normalized-and-tempered cast and wrought carbon steels. (a) Load ratio $R = 0$. (b) $R = -1$

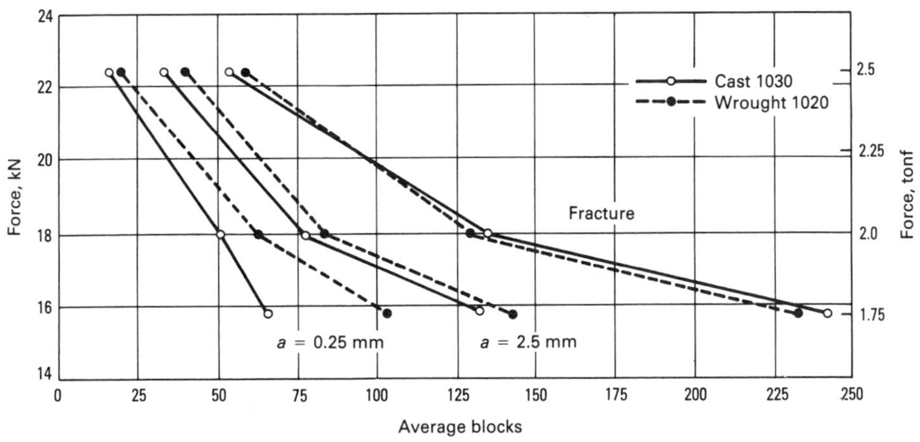

Fig. 13 Average blocks to specific crack lengths and fracture for comparable cast and wrought carbon steels in the normalized-and-tempered condition

from the molten iron, the iron oxide content of the slag decreases. The finishing slag contains approximately 55 to 60% SiO_2, 12 to 16% MnO, 4 to 6% Al_2O_3, 7 to 10% CaO, and 12 to 20% FeO.

The carbon content is reduced to approximately 0.20 to 0.25% during the vigorous carbon boil. The boil is stopped by the addition of carbon in the form of a carburizing iron of low sulfur and phosphorus content. Further deoxidizers (ferromanganese and ferrosilicon) are then added to the bath. Next, the metal is tapped into the ladle. A small amount of aluminum is added to the ladle as a final deoxidizer.

For basic electric furnace melting, the furnace lining is a basic refractory such as magnesite or dolomite. The charge is usually composed of purchased scrap steel and foundry returns. During the melting period,

Table 5 Number of tests required for various degrees of accuracy using double-leg keel block test specimens

Property tested	Acceptable variation MPa	Acceptable variation ksi	Tests required for indicated probability 99%	95%	90%	80%	70%
Tensile strength	0.69	0.1	266	166	177	72	47
	1.38	0.2	72	42	30	18	12
	2.07	0.3	32	19	13	8	6
	2.76	0.4	18	11	8	5	3
	3.45	0.5	12	7	5	3	2
	4.14	0.6	8	5	4	2	...
	4.83	0.7	6	4	2	...	...
	5.52	0.8	5	3	...	...	...
	6.89	1	3	2	...	...	...
Yield strength	3.45	0.5	34	20	14	9	6
	4.14	0.6	24	14	10	6	4
	4.83	0.7	18	10	8	5	3
	5.52	0.8	14	8	6	4	3
	6.21	0.9	11	6	5	3	2
	6.89	1	9	5	4	3	...
	8.27	1.2	6	4	3	2	...
	9.65	1.4	5	3	2	...	...
	11.03	1.6	4	2	...	...	...
Elongation in 50 mm (2 in.)	±1%		17	10	7	5	3
	±2%		5	3	3	2	2
	±3%		2	2	2	...	...
Reduction in area	1%		60	35	24	15	10
	2%		15	9	6	4	3
	3%		7	4	2	2	2
	4%		4	3	...	...	...
	5%		3	2	...	...	...
	6%		2	...	...	...	...

Table 6 Mechanical property variations with specimen size and location in annealed carbon steel bars

Cross section of bar mm	in.	Location of specimen	Tensile strength MPa	ksi	Yield strength MPa	ksi	Elongation, %	Reduction in area, %
75 × 75	3 × 3	Center	496	72	310	45	29	39
		Top	496	72	310	45	28	35
		Bottom	510	74	310	45	28	40
		Corner	510	74	310	45	29	42
100 × 100	4 × 4	Center	490	71	310	45	29	39
		Top	490	71	317	46	29	43
		Bottom	496	72	310	45	30	46
		Corner	503	73	317	46	30	46
200 × 200	8 × 8	Center	476	69	290	42	27	36
		Top	483	70	290	42	26	40
		Top center	469	68	296	43	26	40
		Lower center	476	69	290	42	28	41
		Bottom	490	71	303	44	29	44
		Corner	496	72	303	44	29	44

small quantities of lime are occasionally added to form a protective slag over the molten metal. Iron ore is added to the bath just as melting is complete. The slag is then highly oxidizing and in the correct condition to take up phosphorus from the metal. Shortly after all of the steel has melted, this first slag is taken off (if a two-slag process is to be used), and a new slag composed of lime, fluorspar, and sometimes a little sand is added.

As soon as the second slag is melted, the current is reduced, and pulverized coke, carbon or ferrosilicon, or a combination of these is spread at intervals over the surface of the bath. This period of furnace operation is known as the refining period, and its purposes are to reduce the oxides of iron and manganese in the slag and to form a calcium carbide slag, which is essential to the removal of sulfur from the metal. The refining slag has approximately the following composition:

Element	Composition, %
CaO	45–55
SiO_2	15–20
FeO	0.50–1.5
CaF	5–15

Adjustments are made in the carbon content of the bath by the addition of a low-phosphorus pig iron. After the proper bath temperature is obtained, ferromanganese and ferrosilicon are added and the furnace is tapped. Aluminum is generally added in the ladle as a final deoxidizer.

Induction Melting

The high-frequency induction furnace is essentially an air transformer in which the primary is a coil of water-cooled copper tubing and the secondary is the metal charge. Inside the shell is placed the circular winding of copper tubing. Firebrick is placed on the bottom of the shell, and the space between that and the coil is rammed with grain refractory. The furnace chamber may be a refractory crucible or a rammed and sintered lining. General practice is to use ganister rammed around a steel shell that melts down with the first heat, leaving a sintered lining. Basic linings are often preferred; a rammed lining of magnesia grain or a clay-bonded magnesia crucible can be used.

The process consists of charging the furnace with steel scrap and then passing a high-frequency current through the primary coil, thus inducing a much heavier secondary current in the charge, which heats the furnace to the desired temperature. As soon as a pool of liquid metal is formed, a pronounced stirring action takes place in the molten metal, which helps to accelerate melting. In this process, melting is rapid, and there is only a slight loss of the easily oxidized elements. If a capacity melt is required, steel scrap is continually added during the melting-down period. Once melting is complete, the desired superheat temperature is obtained, and the metal is deoxidized and tapped.

Because only 10 to 15 min elapses from the time the charge is melted down until the heat is tapped, there is not sufficient time for chemical analysis. Therefore, the charge is usually carefully selected from scrap and alloys of known composition in order to produce the desired composition in the finished steel. Composition can be closely controlled in this manner.

In most induction furnaces, no attempt is made to melt under a slag cover, because the stirring action of the bath makes it difficult to maintain a slag blanket on the metal. However, slag is not required, because oxidation is slight. The induction furnace is especially valuable because of its flexibility in operation, particularly in the production of small lots of alloy steel castings. Induction melting is also well adapted to the melting of low-carbon steels because no carbon is picked up from electrodes, as may occur in an electric arc furnace.

Induction furnaces used in the steel foundry range in capacity from 14 to 22 680 kg (30 to 50 000 lb), but most have capacities of 45 to 4500 kg (100 to 10 000 lb).

Deoxidation Practice

Proper melting practice and, to a lesser degree, proper heat treatment can limit the gas content of conventionally melted steel to acceptable levels, regardless of the type of furnace equipment employed. Failure to control oxygen, hydrogen, and nitrogen contents may result in porosity or a serious decrease in ductility or both.

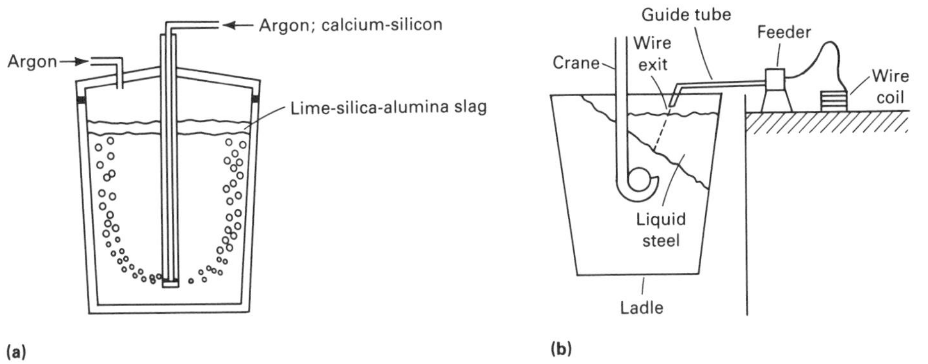

Fig. 14 Schematics showing powder injection and wire injection methods used for desulfurization. (a) Powder injection. (b) Wire injection

Gas content is largely adjusted during the oxygen boil. After the cold charge is melted and the bath is in the temperature range of 1510 to 1540 °C (2750 to 2800 °F), oxygen is introduced into the molten metal, usually by means of a piping arrangement. The oxygen combines with the dissolved carbon in the steel to form bubbles of carbon monoxide. As the bubbles form, dissolved hydrogen and nitrogen are caught up in the bubbles in much the same way that dissolved oxygen finds its way into bubbles of boiling water. Thus, the bubbles of gas contaminants are boiled out.

The normal hydrogen content in acid-melted steel is in the range of 2 to 4 ppm. Hydrogen content resulting from basic practice is slightly higher. Austenitizing and tempering treatments serve to lower the hydrogen content further in sections of average thickness. However, these treatments may not suffice when very heavy sections are involved. The alternative is to degas the molten steel under vacuum, particularly when pouring heavy castings that will be subjected to severe dynamic loading.

The lowest nitrogen contents are obtained with either acid or basic open-hearth practice. Nitrogen contents in electric arc melted steels range from 3 to 10 ppm.

The primary role of deoxidation is to prevent pinholes due to carbon monoxide formation during solidification. Oxygen content should be less than 100 ppm to prevent porosity. Silicon and manganese are mild deoxidizers; they are added to stop the carbon boil and to adjust the chemistry. Manganese (>0.6%) and silicon (0.3 to 0.8%) additions are limited by other alloy effects and are normally inadequate alone to prevent pinholes.

Aluminum is the most common supplemental deoxidizer used to prevent pinholes. As little as 0.01% Al will prevent pinholes. Aluminum is normally added at the tap—about 1 kg/Mg (2 lb/ton) (0.10% added) with a recovery of 30 to 50% for a final content of about 0.03 to 0.05%. This is normally supplemented at the pour ladle with additional deoxidation, which could be more aluminum or calcium, barium, silicon, manganese, rare earths, titanium, or zirconium. More than the minimum amount of deoxidizer required for preventing porosity is needed to maintain sulfide inclusion shape control, but excessive amounts can cause intergranular failures or dirty metal. The second addition at the pour ladle is normally 1 to 3 kg/Mg (2 to 7 lb/ton). The commonly used elements for deoxidation are (in order of decreasing power) zirconium, aluminum, titanium, silicon, carbon, and manganese.

Inclusion Shape Control

Nonmetallic inclusions that form during solidification depend on the oxygen and sulfur contents of the casting. Deoxidation decreases the amount of nonmetallic inclusions by eliminating the oxides, and it affects the shape of the sulfide inclusions. High levels of oxygen (>0.012% in the metal) form FeO, which decreases the solubility of manganese sulfides and causes the manganese sulfides to freeze early in solidification as globules. The use of silicon deoxidation alone normally causes the formation of globular (Type I) sulfides.

Decreasing the oxygen to between 0.008 and 0.012% in the metal, through the use of aluminum, titanium, or zirconium increases the solubility of the manganese sulfides so that they solidify last as grain-boundary films. These grain-boundary sulfide films (Type II inclusions) are detrimental to the ductility and toughness of the steel and cause increased susceptibility to hot tearing. If the oxygen content is very low (<0.008% in the metal), then the deoxidizer forms complex sulfides that are crystalline and form early in solidification. These Type III inclusions are less harmful than Type II, but are more harmful than Type I.

The actual sulfide shape depends on the type and amount of deoxidizer. There are complex interactions, so care must be taken to avoid porosity and Type II sulfides. If a high level of rare-earth metals (RE/S > 1.5) is added, then galaxies of sulfides (Type IV inclusions) form; these are only for rare-earth additions. The most common inclusion modifier is calcium. Methods of addition will be covered later in this article.

Argon-Oxygen Decarburization (AOD)

Some foundries have recently installed AOD units to achieve some of the results that vacuum melting can produce. These units look very much like Bessemer converters with tuyeres in the lower sidewalls for the injection of argon or nitrogen and oxygen. They are processing units that must be charged with molten metal from an arc or induction furnace. Up to about 20% cold charge can be added to an AOD unit; however, the cold charge is usually less than 20% and consists of solid ferroalloys. The continuous injection of gases causes a violent stirring action and intimate mixing of slag and metal, which can lower sulfur values to below 0.005%. The gas contents approach or may be even lower than those of vacuum induction melted steel. The dilution of oxygen with inert gas, argon, or nitrogen causes the carbon-oxygen reaction to go to completion in favor of the oxidation reaction of iron and the oxidizable elements, notably, chromium in stainless steel. Therefore, superior chromium recoveries from less expensive high-carbon ferrochromium are obtained compared to those of electric arc melting practices.

Argon-oxygen decarburization units are used in the production of high-alloy castings, particularly of grades that are prone to defects due to high gas contents. Carbon and low-alloy steels for castings with heavy wall sections may be subject to hydrogen embrittlement and are also processed in these units with good results.

Powder Injection/Wire Injection

The powder injection or wire injection of reactive metals (calcium, magnesium, or rare earths) into liquid steel for desulfurization has come to the forefront of technology in the past decade. Ultralow levels of sulfur (<0.005%) can be obtained with this technique. Powder injection and wire injection operations are shown schematically in Fig. 14.

The basis of reactive metal injection desulfurization is the chemical combination of the reactive metal with sulfur dissolved in the liquid steel. Because the reactive metals are strong deoxidizers, it is necessary to have a low oxygen level in the liquid steel before the start of the injection process. Otherwise, the reactive metal would combine with the dissolved oxygen and not the dissolved sulfur. Aluminum is used to obtain the necessary low oxygen levels.

It is also necessary to have a slag present to hold the sulfur removed by the reactive metal. If such a slag is not present, the metal sulfide will be oxidized, and the sulfur will return to the liquid metal.

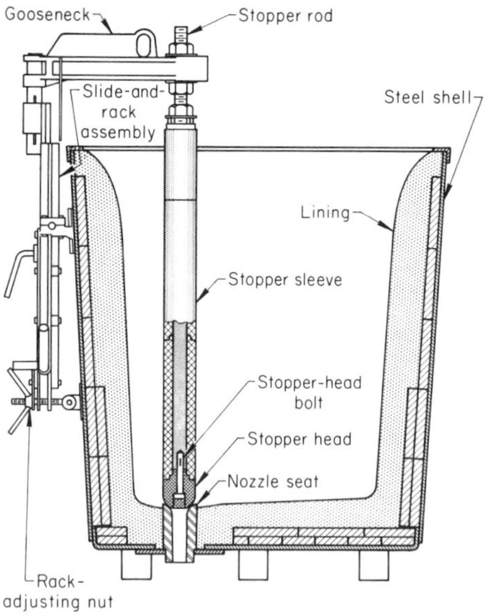

Fig. 15 Schematic of a bottom-pour ladle. This type of ladle is most often used to pour large castings.

Plasma Refining

Steadily increasing requirements for steel cleanliness have led producers to adopt several novel refining technologies and process routes, many of which involve increased use of the ladle as a refining vessel. Such procedures require longer holding times in the ladle, which necessitate either increased superheats in the furnace or external heating in the ladle in order to avoid early solidification. Higher superheat, in addition to requiring excessive energy expenditure, can contribute to the problem of dirty steel. The preferred solution is to supply heat to the ladle, maintaining the steel at minimal superheat during refining. This can be accomplished by the transferred arc plasma torch, with the added benefit of enhanced refining reactions that aid in the production of clean steel with low levels of residual elements. In this work, experiments have been carried out in an induction furnace equipped with a gas-stabilized graphite electrode to investigate the control of oxygen and inclusion levels and the enhancement of desulfurization afforded by the transferred arc plasma.

Specifics of Foundry Practice

The foundry practice used for carbon and low-alloy steels is also used for cast stainless steels and cast nickel-base alloys (see the articles "High-Alloy Steels" and "Nickel and Nickel Alloys" in this Volume). Specific aspects of foundry practice discussed in this article include pouring, gating, and risering. The cleaning and heat treatment of carbon steel castings will also be discussed.

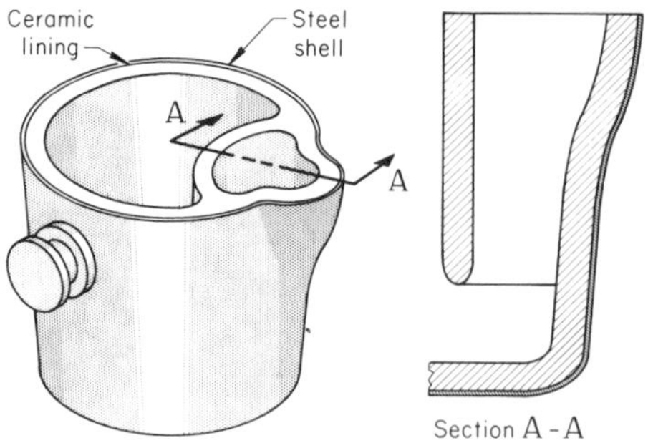

Fig. 16 Typical teapot ladle used for pouring small-to-medium size steel castings

Pouring Practice

Three types of ladles are used for pouring steel castings: bottom-pour, teapot, and lip-pour. Ladle capacities normally range from 45 kg to 36.3 Mg (100 lb to 40 tons), although ladles having much larger capacities are available.

The bottom-pour ladle (Fig. 15) has an opening in the bottom that is fitted with a refractory nozzle. A stopper rod, suspended inside the ladle, pulls the stopper head up from its seat in the nozzle, allowing the molten steel to flow from the ladle. When the stopper head is returned to the position shown in Fig. 15, the flow is cut off. The position of the stopper head is controlled manually by the slide-and-rack mechanism shown at left in Fig. 15.

Bottom pouring is preferred for pouring large castings from large ladles because it is difficult to tip a large ladle and still control the stream of molten steel. In addition, the bottom-pour ladle delivers cleaner metal to the mold. Inclusions, pieces of ladle lining, and slag float to the top of the ladle; therefore, pouring from the bottom greatly reduces the risk of passing nonmetallic particles into the mold cavity.

On the other hand, it is impractical to pour from a large bottom-pour ladle into small molds. The pressure head created by the metal remaining in the ladle delivers the molten steel too fast. The sand mold is not strong enough to withstand this heavy onrush of molten steel. In addition, because little time is needed to fill a small mold, a large bottom-pour ladle must be opened and closed many times in order to empty it. This may cause the ladle to leak; however, special nozzles have been developed to minimize leakage. Although bottom-pour ladles could be scaled down in size for pouring smaller castings, this is unnecessary because of the almost equal ability of the teapot ladle to deliver clean metal.

The teapot ladle (Fig. 16) incorporates a ceramic wall, or baffle, that separates the bowl of the ladle from the spout. The baffle extends almost four-fifths of the distance to the bottom of the ladle. As the ladle is tipped, hot metal flows from the bottom of the ladle up the spout and over the lip. Because the metal is taken from near the bottom of the ladle, it is free of slag and pieces of eroded refractory, although it may pick up foreign materials in the spout section or at the lip. The teapot design is feasible in various sizes, generally covering the entire range of casting sizes that are below the minimum size for which the bottom-pour ladle is used.

Lip-pour ladles are similar in external form to the teapot type. Because lip-pour ladles have no baffles to hold back the slag and because the hot metal is not taken from the bottom of the ladle, this type of ladle pours a dirtier steel and is seldom used to pour steel castings. Nevertheless, it is widely used as a tapping ladle (at the melting furnace) and as a transfer ladle to feed smaller ladles of the teapot type.

Ladle Linings. Most ladle problems in the steel foundry are caused by deficiencies in preparing and maintaining the linings. Ladles with capacities above about 360 kg (800 lb) are lined with layers of firebrick covered with a thick refractory coating. When used to pour steel made by the acid process and when properly maintained, these linings will last for 60 to 75 heats. However, when pouring steel made by the basic process, the life of the lining is only 10 to 25 heats because of the erosion of refractory by the slag.

Smaller ladles are protected by monolithic linings prepared by ramming or by pouring and vibrating refractory mixtures around forms. The optimum refractory for this purpose is high-purity alumina, but because of the high cost of alumina, alternative materials such as olivine and zircon are more widely used. The service lives of these monolithic linings, which are most commonly used for small ladles, vary considerably, depending on the refractory mixture and the type of service. Thermal expansion and contraction can cause hairline cracks, into which hot metal enters. This results in erosion of the lining. Erosion not only shortens the life of the lining but also contributes to excessive dirt in the steel. Disposable fiber linings that do not require preheating are gaining acceptance.

Preheating the ladle helps to prolong the life of the ladle lining. Heating the lining to a temperature close to that of molten steel serves to lessen the thermal shock to the lining when the molten steel contacts it. Common practice is to preheat ladle linings to 815 to 1095 °C (1500 to 2000 °F) with oil-fired or gas-fired torches. Ladles lined with high-purity alumina can be preheated to temperatures as high as 1370 °C (2500 °F).

Pouring Time. Ideally, the optimum pouring time for a given steel casting would be established on the basis of the weight and shape of the casting, the temperature of the molten metal and its solidification characteristics, and the heat transfer characteris-

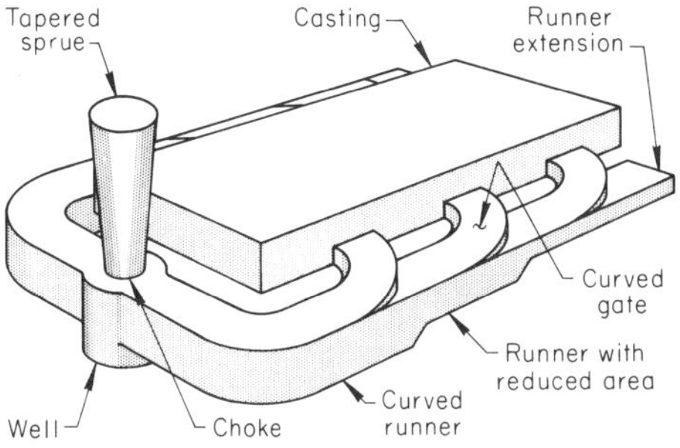

Fig. 17 Gating system that allows good metal flow. The desired metal flow is obtained in this system by proportioning the cross-sectional area of the choke of the spur to all the runners emanating from the sprue and to all of the gates in accordance with a 1:4:4 ratio.

tics and thermal stability of the mold. Most foundries, however, are required to pour many different castings from one heat, or even from one ladle. Therefore, these foundries do not attempt to control pouring time, but they do control the speed at which molten steel enters the mold cavity. This control is achieved through the design of the gating system.

Gating Design

An effective gating system for pouring steel and other metals into sand molds is one that fills the mold as rapidly as possible without developing pronounced turbulence. It is essential that the mold be filled rapidly, primarily because heat is radiated to the walls and top surface of the mold as the molten steel rises in the mold cavity; this heat can destroy the binder in the molding sand, causing the mold to collapse unless the metal rises rapidly to provide the necessary support.

Turbulent metal flow is harmful because it breaks up the metal stream, exposing more surface area to air and forming metallic oxides that rise to the top of the mold cavity; this produces a rough surface or inclusions in the casting. In addition, turbulent flow erodes the mold material. These eroded particles also float to the top of the mold cavity.

Preferred Metal Flow. According to preferred practice, the pourer directs the metal stream toward the pouring cup at the top of the mold, controlling the pouring rate to keep the cup full of molten steel throughout the pouring cycle. The opening in the bottom of the cup is directly over the sprue, or downgate, which is tapered with the large end up, thus reducing the diameter of the stream of descending metal. The taper prevents the stream from pulling away from the walls and drawing air into the gating system. The descending metal impinges on the sprue well at the bottom of the sprue, and the direction of flow changes from vertical to horizontal, with the metal flowing along runners to gates (ingates) and then to the main body of the casting. A gating system that incorporates these features is shown in Fig. 17.

The design of the gating system will largely determine the manner in which molten steel is fed into the mold, as well as the feed rate. Therefore, a system having several gates will influence the distribution of the flow between gates. A good design will have even distribution between gates, both initially and while the mold is filling. The distribution of flow during filling of the mold will affect the type of flow that occurs in the main body of the casting. A large difference in the flow between gates will create a swirl of metal in the mold about a vertical axis, in addition to that occurring about a horizontal axis.

The gating system shown in Fig. 17 is an example of a finger-type parting line system, in which the fingers feed metal to the casting just above the parting line. Other major types of gating systems used in steel foundries include the bottom gate, which feeds metal to the bottom of the casting, and the step gate, which feeds metal through a number of stepped gates, one above another. In the system shown in Fig. 17, the ratio of the cross-sectional area of the choke of the sprue to all of the runners emanating from the sprue basin and to all of the gates is 1:4:4. The runner area decreases progressively by an amount equal to the area of each gate it passes. This practice ensures that once the system is filled with metal it will remain full during the pouring cycle and will feed equally to each gate.

Further, the gates are located in the cope, while the runner, which extends beyond the last gate, is located in the drag. The extension of the runner serves as a trap for the first, and usually the dirtiest, metal to enter the system. The entire runner must fill before the metal will rise to the level of the gates. Thus, each gate begins feeding at the same time. The runners and gates are curved wherever a change in direction occurs. This streamlining reduces turbulence in the metal stream and minimizes mold erosion.

In contrast to the ratio of the system shown in Fig. 17 (1:4:4), if the total cross-sectional area of the gates is less than that of the runners (1:2:0.8, for example), the result is a pressurized system. With a pressurized system, the metal squirts into the mold cavity and flows turbulently over the mold bottom. This may cause roughening of the bottom surfaces.

Conversely, if the total cross-sectional area of the gates is significantly greater than that of the runners (1:2:3, for example), the gating system will be incompletely filled, and flow from the gates will be uneven. This condition increases the likelihood of mold erosion. When required, other more complicated additions to gating systems are used, including whirl gates, horn gates, strainer cores, tangential gates, and slit gates. However, any addition to the gating system usually increases the cost of the casting because all gating must be removed.

Mold Erosion. In addition to the contribution of gating design to a reduction in mold erosion, further reduction can be achieved by making the gating system out of tile, which is superior to green sand in resistance to erosion. However, the use of tile is generally limited to gating systems for large castings; in such cases, the quantity and speed of molten metal passing through the gating system would seriously erode green sand. The automated production of smaller molds is hampered by the addition of tile gates and runners. Thus, gating systems for these smaller castings are rammed in sand, usually with a semicircular or rectangular cross section for the gates and runners.

Riser Design

Molten steel contracts 0.9% per 55 °C (100 °F) as it cools from the pouring temperature to the solidification temperature. It then undergoes solidification contraction of 3% during freezing, and finally the solidified metal contracts 7.2% during cooling to room temperature. Therefore, when casting steel, an ample supply of molten metal must be available from risers (reservoirs) to compensate for the volume decrease, or shrinkage cavities will develop in the locations that solidify last.

Because feeding from the riser depends on gravity, risers are usually located at the top of the casting. Riser forms are placed on the pattern and molded into the cope half of the mold. The riser cavity is usually open to the top of the mold, although blind risers are sometimes used.

Riser Size and Shape. Formulas based on surface area, volume, and freezing time of the casting are used to determine riser size. Most risers are cylindrical, with their height approximately equal to their diameter. This configuration provides a low ratio of surface area to volume, which prolongs the time the steel remains liquid. A spherical riser would constitute an optimum design, but spherical shapes are difficult to mold.

The placement of a riser, in conjunction with its size, determines its effectiveness. The thicker sections of a casting act as reservoirs for feeding the thinner sections, which solidify first. Therefore, risers are placed over thick sections that cannot be fed by other areas of the casting.

Feeding Distance. Castings of uniform thickness present a different problem. Studies have established the feeding distances of a riser for various rectangular shapes in both the horizontal and vertical planes, with and without an end effect. An end effect is the additional cooling provided by the sand cover of an end surface.

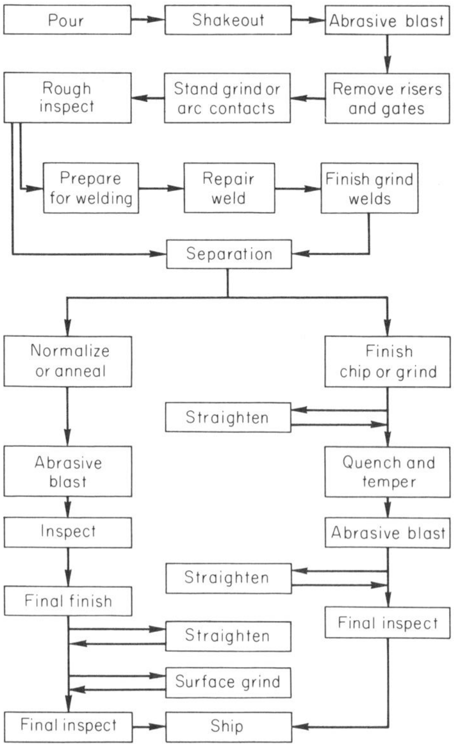

Fig. 18 Flow chart showing steps taken in the cleaning of small (up to 18 kg, or 40 lb) castings

The maximum feeding distance can be extended for a uniform section by adding a taper. The progressively thicker section solidifies in a progressively longer time, so that a favorable temperature gradient is established from the end of the section to the riser. A tapered pad of exothermic material placed in the mold along the length of the casting will also produce a favorable temperature gradient.

Cleaning Operations

The choice of equipment and processing method for operations in the cleaning room of a steel foundry depends largely on the level of surface finish and dimensional accuracy required and on the cost that can be justified for the particular casting.

Cleaning of Small Castings. Figure 18 shows a flow chart for the cleaning (and other postshakeout operations) of small castings. This sequence of operations is suitable when the casting is small enough (up to about 18 kg, or 40 lb) to be processed on a floor-standing grinder rather than requiring a portable grinder.

Small castings are usually shaken out (removed from their molds) by a semiautomatic operation in which the molds are placed on a conveyor and taken to a vibrating grate. At the vibrating grate, sand from the broken molds and some of that adhering to the castings falls through the grate while the casting vibrates to the far end of a table and is removed. Certain very fragile castings may require special processing to prevent mechanical damage at this and subsequent stages in the cleaning operation. Castings are then allowed to cool to below about 205 °C (400 °F). At this point, the castings and the attached gates and risers still contain some adhering sand and a thin layer of scale.

The pressure abrasive blasting of small steel castings is generally accomplished in machines that repeatedly tumble the castings on a continuous belt in order to expose the surfaces to a stream of steel (or iron) shot or grit abrasive. Pressure abrasive blasting is a batch operation that usually requires 8 to 20 min per batch.

After the abrasive cleaning operation, gates and risers are removed by gas cutting, sawing, shearing, or breaking off. The equipment used for this operation includes gas-oxygen torches, abrasive cutoff wheels, friction or band saws, sledge hammers, or small power shears. The fuel gases used for gas cutting include acetylene, natural gas, propane, and special gases. The shape of the casting and the locations at which the gates and risers are attached exert a large influence in determining the method of removal.

Castings that are small enough to be readily handled by one man and manipulated manually (up to about 14 to 18 kg, or 30 to 40 lb) are sent to floor-standing grinders to finish contact areas, provided the shape of the casting at the contact areas is such that extraneous metal can be removed by this type of equipment. Flat surfaces and convex curves that are accessible to the wheel can be shaped with the stand grinder. At the same time, parting line or core fins and other bumps or raised blemishes can also be removed from the castings. If the contours at the contact areas are complex or not readily accessible or if the casting is too heavy or awkward to handle, the contact areas can be smoothed down by the air carbon-arc cutting process. In this process, a carbon electrode approximately 9.5 to 25 mm (3/8 to 1 in.) in diameter strikes an arc with the surface of the contact area, and compressed air, issuing from jets in the electrode holder, continuously blows the steel droplets away as they are formed.

Castings can then be inspected, and those requiring weld repair can be removed from the flow and sent to weld preparation areas. Discontinuities are usually best removed prior to welding by using the air carbon-arc cutting process; other methods such as chipping, grinding, or torch scarfing are also employed. Welded areas are finished by grinding or by the air carbon-arc process.

Most carbon steel castings and some low-alloy steel castings require only normalizing or annealing heat treatment, rather than hardening and tempering. After normalizing or annealing, castings are again blast cleaned to remove the scale formed during heat treatment. Further finishing operations, such as grinding, may then be required. Castings that do not need additional finishing are ready for shipment or for machining.

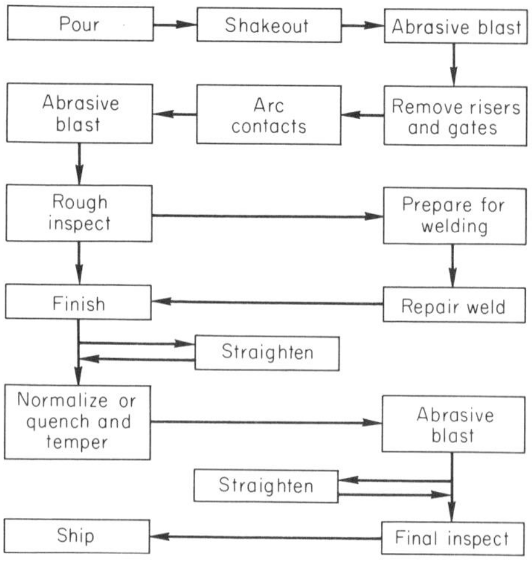

Fig. 19 Flow chart showing steps taken in the cleaning of medium and large castings

Some castings require straightening, which is done by hydraulic press or hand hammer and die operations. Surface grinding may also be necessary on certain castings that require very flat or smooth surfaces in particular areas. If any of the above operations is performed after the inspection that follows the abrasive cleaning operation, another inspection is required in order to qualify castings for shipment.

Some carbon steel castings and most low-alloy steel castings must be quenched and tempered to achieve the specified strength. Castings so treated can be finished by chipping and grinding before heat treating (see alternate flow procedure in Fig. 18). Castings that are heat treated to only moderate strength levels and are simpler in design can be straightened after heat treatment, if straightening is required.

Cleaning of Large Castings. Figure 19 illustrates typical sequences of operation for cleaning large castings, for which portable equipment must be used while the casting rests on a worktable. These castings can be moved to the shakeout deck on a conveyor, or the individual flasks can be transported to the shakeout site with an overhead crane. When large castings must be shaken out while still at a red heat, the flask can be emptied by setting it on the edge of the shakeout deck so that the casting falls free of the deck and is allowed to cool before being placed on the shakeout deck for more complete removal of the adhering sand. This practice helps to prevent mechanical damage to the castings while they are red hot and soft.

Abrasive blasting follows rough cleaning on the shakeout deck. The blasting equipment can be of the same type as that used for small castings but of larger capacity. Castings heavier than about 450 kg (1000 lb) each are usually blasted with a different type of equipment—a large room-type cabinet with a turntable. The casting is placed on the turntable in the desired position, and blasting begins with the turntable revolving to expose different surfaces of the casting to the blast. It may be necessary to reposition the casting once or twice in order to clean all of the surfaces. Blast cabinets equipped with hand-held hoses that propel shot in a high-velocity air stream are also common, especially for cleaning the cored internal cavities of castings. After blast cleaning, large castings are usually transported directly to the gate-and-riser removal area by overhead crane.

Risers and gates are usually removed by gas cutting. The stub of the contact remaining after cutting is removed by grinding with a portable grinder or by the air carbon-arc cutting process, depending on the contour of the surface that is contacted and the overall shape of the casting. The air carbon-arc method is preferred whenever possible because it is faster than grinding. Carbon electrodes 19 or 25 mm (3/4 or 1 in.) in diameter are usually used on castings of medium and large size. Following this operation, a short (about 3 min) blast-cleaning cycle can be used to remove slag, arc spatter, or oxide films from the casting surface, and discontinuities and blemishes can be marked for removal and weld repair. Weld preparation can be done by the air carbon-arc process or by chipping or grinding; the former is usually preferred.

After welding, castings are sent to finishing stations where blemishes, bumps, parting line fins, or core fins are removed and welded areas are dressed down. Portable grinding tools powered by compressed air or self-contained electric motors are generally used. Pneumatic chipping hammers and a variety of special chisels may also be required because of the shape of the casting and the location of the fins. Remaining spatter from gas cutting or carbon-arc cutting is also removed at this time.

Unless the castings require straightening, heat treatment is usually the next operation. After heat treatment, castings are blast cleaned to remove scale and then sent to the press area for straightening, if necessary, and on to final inspection.

Heat Treatment

Heat treatment is an important step in the production of steel castings because it develops the mechanical properties of a hardenable steel. Several types of heat treatment are available. The essential elements of any heat treatment are the heating cycle and the cooling cycle. Figure 20 shows schematically a heating cycle and three different cooling cycles. The length of time that a casting is held at temperature and the cooling rate are important factors. The holding time should be long enough to complete the desired microstructural transformation.

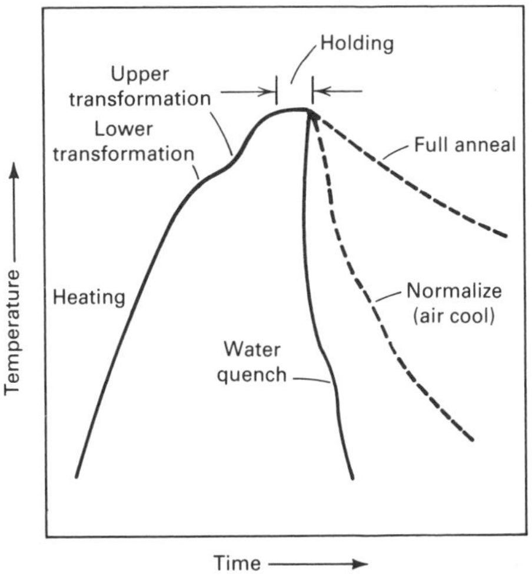

Fig. 20 Schematic showing heating/cooling cycles used for various heat treatments

Annealing is practiced on low-carbon steels to provide a soft, readily machinable structure. Annealed castings have relatively low strength but good ductility.

In a full anneal, the castings are heated above the upper critical temperature, held there long enough to complete the transformation to austenite, and then furnace cooled at a controlled rate to obtain a stress-relieved casting with a pearlite-ferrite structure that is ductile and readily machinable. There are variations of the annealing heat treatment for specialized purposes, for example, to achieve a spheroidized pearlite structure. The annealing temperatures of such specialized treatments may differ substantially from those of a full anneal. Full annealing and spheroidizing heat treatments are costly and should not be specified unless maximum ductility is actually required.

Cooling is accomplished by reducing or by simply turning off the heat input to the furnace. When the castings have cooled to below the lower critical temperature (about 425 °C, or 800 °F), the transformation of austenite is usually complete. The castings can then be removed from the furnace and air cooled or even quenched. Furnace cooling to lower temperatures merely wastes furnace time and requires additional heat to bring the furnace up to temperature for the next load.

Normalizing consists of heating the steel to a suitable temperature above the upper critical transformation temperature, holding long enough to complete the transformation to austenite, removing the work from the furnace, and cooling it in still air. Some castings are tempered after normalizing.

The castings must be placed so that the air can circulate freely around every casting. If air flow is restricted, the operation will be more like annealing. On the other hand, accelerated cooling by fans or forced-air flow may produce a result more like quenching.

The microstructure that results from normalizing is a mixture of ferrite and pearlite, associated with only low residual stresses and almost no distortion. Tensile strengths up to 655 MPa (95 ksi) can be obtained in this way. The normalizing and tempering operation is used to meet a number of standard specifications in this strength range. Because of the uniform structure obtained upon normalizing, machinability is good.

The cost of normalizing makes this heat treatment attractive. It requires less furnace time than annealing, and its cooling cycle is less expensive than quenching.

In hardening by quenching, the work is heated above the transition temperature (austenitized) as in annealing or normalizing (Fig. 20). The work is cooled more rapidly than in the other heat treatments—fast enough so that pearlite and ferrite do not have time to form.

Water and oil are the media most commonly used for quenching steel castings. Water is used whenever possible. Higher-carbon steels require oil quenching. Some complicated shapes also demand oil quenching to minimize quench cracking.

Oil quenches a little more slowly than water at all temperatures. At lower temperatures, the cooling curves for oil taper off; this means even slower rates than for water. Therefore, martensite forms more slowly in oil than in water.

Certain organic chemicals can be added to water to produce a quenching solution that resembles oil in its heat-removal characteristics. The main advantage of these solutions is that they have the behavior of oil without the fire hazard. Their greatest disadvantage is that they coat the work and so change the composition of the bath. The quench severity of these baths varies widely with small changes in composition. Tight control is necessary and sometimes difficult.

The quenching of carbon steels is always followed by tempering in order to adjust the mechanical properties of the quenched steel. The higher tensile strength levels of carbon steels can be obtained only by quenching and tempering. The quenching and tempering operation produces the optimum combination of strength and toughness.

Tempering is a heat treatment that follows quenching and sometimes normalizing. One purpose of tempering is to reduce the residual stresses that develop during cooling and transformation. Another objective of tempering is to modify the metallurgical structure of martensite and thus adjust strength and other mechanical properties to specified levels.

Tempering consists of heating the work to a temperature below the transformation range, holding for a specified time, and finally cooling. Carbon steels are tempered in the range of 175 to 705 °C (350 to 1300 °F). The holding time at temperature may vary from 30 min to several hours. A longer time at a given tempering temperature or a higher tempering temperature for a given time produces greater tempering.

Martensite softens more than pearlite at a given tempering temperature. Composition also affects the tempering response; carbide-forming elements cause the steel to exhibit greater resistance to tempering.

Tempering below 595 °C (1100 °F) may cause temper embrittlement in certain steels. Tempering is usually not done in this temper embrittlement range, and when higher tempering temperatures are used, the work can be quenched from the tempering temperature to minimize time in the embrittling zone during cooling.

Stress Relief. Because tempering is a stress-relief treatment, there is no need for a special stress relief after tempering. Stress relief is obtained by heating to temperatures above 260 °C (500 °F). A stress relief treatment is sometimes required when operations are performed after heat treatment that leave residual stresses in the casting. Welding, induction hardening, and grinding are examples of such operations. The maximum temperature for stress relief is generally limited to 30 °C (50 °F) below the tempering temperature that had been used in heat treating the casting to prevent it from softening.

Hydrogen Removal. Hydrogen has been found to cause low elongation and reduction in area in steel. If the steel contains 4 or 5 ppm of hydrogen, the ductility will be about 20% of that of a hydrogen-free steel. Hydrogen in steel is a mobile element. Above room temperature, the hydrogen will diffuse from the steel, and ductility will be restored.

Hydrogen removal can be accelerated by heating to 205 to 315 °C (400 to 600 °F). This heat treatment is commonly referred to as aging. The aging time is proportional to section thickness. Generally, 25 mm (1 in.) equals 20 h. For heavy sections (≥250 mm, or 10 in.), hydrogen removal by aging becomes impractical because of time requirements.

Applications

Two methods of identifying grades of cast steels are extensively used in the United States. The cast steel designations presented at the beginning of this article and the AISI designations for wrought steels are examples of the first method of identifying steel grades by alloy type. In this system, the first two digits indicate the alloy type, and the second two digits represent the carbon content. For example, a 1010 steel represents a carbon steel with 0.10% C, while a 1320 steel represents a manganese steel with 0.20% C. This system does not include mechanical properties or heat treatment. Accordingly, a cast 1040 steel (0.40% C) can exhibit a yield strength of 330 MPa (48 ksi) or of 496 MPa (72 ksi), depending on the choice of heat treatment.

In the second method, letters and numbers are arbitrarily assigned to steels with well-defined compositions, which index the heat treatment as well as the mechanical properties. Specification-writing bodies utilize this second system. Most specifications are written for specific end uses and identify each grade of steel by a unique set of letters and numbers. Because many steel grades are suitable for more than one end use, there are usually many steel grade designations that represent a single type of steel. For example, there are four ASTM specifications that together include 16 grades of chromium-molybdenum steels. These 16 grades, however, are made up of only three different steels. Although such a system may appear confusing because of the redundancy of designations, the system does offer the advantage of characterizing the cast steel end product as thoroughly as is needed for its end use.

Low-carbon cast steels are those that have a carbon content less than 0.20%. The elements present in low-carbon cast steel, other than carbon, are manganese (0.50 to 1.00%), silicon (0.25 to 0.80%), and sulfur and phosphorus (up to 0.05% maximum each). Residual elements such as nickel, chromium, copper, and molybdenum will be present in small amounts. Cast carbon steels containing less than 0.10% C are mainly produced for electrical and magnetic equipment and are normally given a full anneal heat treatment.

Some castings for the railroad industry are produced from low-carbon cast steel. Castings for the automotive industry are also produced from this class of steel, as are annealing boxes and hot metal ladles. Steel castings in this class are also produced for case carburizing, by which process the castings are given a hard wear-resistant exterior and a tough, ductile core. The magnetic properties of cast low-carbon steels make them useful in the manufacture of electrical equipment.

The medium-carbon cast steels have carbon contents of 0.20 to 0.50%. This class of steels represents the bulk of steel casting production. A very large proportion of cast steel in the medium carbon class is heat treated by normalizing, which consists of cooling the castings in air from approximately 55 °C (100 °F) above the upper critical temperature. A stress-relief treatment can be used to relieve stresses set up in the casting by cooling conditions or welding operations and to soften the HAZ resulting from welding.

More than half of all the steel castings produced are of the medium-carbon type. These parts are used in a wide variety of ways, including applications in the railroad and other transportation industries, machinery and tools, equipment for rolling mills, mining and construction equipment, and many other miscellaneous applications.

High-Carbon Cast Steels. Cast steels containing more than 0.50% C are considered to be high-carbon cast steels. Other elements present in these grades are manganese (0.50 to 1.0%), silicon (0.30 to 0.80%), and phosphorus and sulfur (0.05% maximum each). Because of their high carbon contents, these grades are the most hardenable of the plain carbon cast steels. They are therefore used in applications that require relatively high strength levels.

SELECTED REFERENCES

- W.J. Jackson, Ed., *Steel Castings Design Properties and Applications*, Steel Castings Research and Trade Association, 1983
- R.I. Stephens, "Fatigue and Fracture Toughness of Five Carbon or Low Alloy Cast Steels at Room or Low Climatic Temperatures," *Research Report 94 (A & B)*, Steel Founders' Society of America, 1982
- P.F. Wieser, Ed., *Steel Castings Handbook*, 5th ed., Steel Founders' Society of America, 1980

Low-Alloy Steels

John M. Svoboda, Steel Founders' Society of America

LOW-ALLOY STEELS contain alloying elements, in addition to carbon, up to a total alloy content of 8%. Cast steels containing more than the following amounts of a single alloying element are considered low-alloy cast steel:

Element	Amount, %
Manganese	1.00
Silicon	0.80
Nickel	0.50
Copper	0.50
Chromium	0.25
Molybdenum	0.10
Vanadium	0.05
Tungsten	0.05

Aluminum, titanium, and zirconium are used for the deoxidation of low-alloy steels. Of these elements, aluminum is used most frequently because of its effectiveness and low cost.

Numerous types of cast low-alloy steel grades exist to meet the specific requirements of the end use, such as structural strength and resistance to wear, heat, and corrosion. The designations of the American Iron and Steel Institute (AISI) and the Society of Automotive Engineers (SAE) have historically been used to identify the various types of steel by their principal alloy content. Cast steels, however, do not follow precisely the composition ranges specified by AISI and SAE designations for wrought steels. In most cases, the cast steel grades will contain 0.30 to 0.65% Si and 0.50 to 1.00% Mn unless otherwise specified. The principal low-alloy cast steel designations, their AISI and SAE equivalents, and their alloy type are:

Cast steel designation	Nearest wrought equivalent	Alloying elements
1300	1300	Manganese
8000, 8400	8000, 8400	Manganese, molybdenum
80B00	80B00	Manganese, molybdenum, boron
2300	2300	Nickel
8600, 4300	8600, 4300	Nickel, chromium, molybdenum
9500	9500	Manganese, nickel, chromium, molybdenum
4100	4100	Chromium, molybdenum

The 8000, 8400, 2300, and 9500 designations are no longer used by AISI. However, because these alloy types are extensively used as cast steels, their cast steel designation numbers are continued in the steel casting industry. There are additional alloy types that are infrequently specified as cast steels; namely, 3100 (nickel-chromium), 3300 (nickel-chromium), 4000 (molybdenum), 5100 (chromium), 6100 (chromium-vanadium), 4600 (nickel-molybdenum) and 9200 (silicon).

Hardenability

Hardenability is the property of steel that governs the depth to which hardening occurs in a section during quenching. It should not be confused with hardness, which is the resistance to penetration as measured by Rockwell, Brinell, or other hardness tests. Hardenability is of considerable importance because it relates directly to the strength of steel and to many other mechanical properties, notably, toughness and fatigue properties.

The principal method of hardening carbon and low-alloy steels consists of quenching the steel from the austenitizing temperature. Steels vary in their response to this quenching operation because the depth below the surface of a part to which the part hardens will depend on the composition of the steel and the severity of the quench. Because cooling during a quench is fastest at the surface of a part and slowest at its center and because the hardening reaction is time dependent, hardenability is a vital consideration in alloy selection. Figure 1 shows schematically the cooling curves for the center and surface of a hypothetical steel section superimposed on the continuous cooling transformation diagram. The curve labeled "50% martensite" intersects the pearlite area briefly; thus, half of the mass is transformed to pearlite and half transforms later to martensite. At some point closer to the surface, the cooling rate is given by the curve labeled "Critical cooling rate." This curve represents the cooling rate that is just fast enough to avoid transformation to any type of metallurgical structure except martensite.

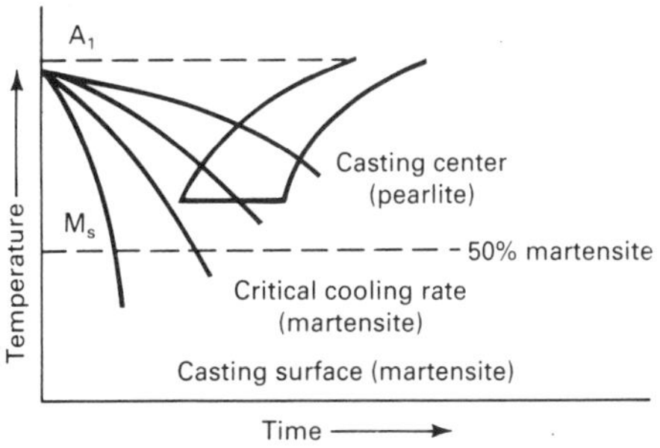

Fig. 1 Difference in cooling rates at the surface and center of a steel casting and the resulting microstructures obtained

Alloying Elements

The compositions of low-alloy cast steels are characterized by carbon contents primarily under 0.45% and by small amounts of alloying elements, which are added to produce certain definite properties. Low-alloy steels are applied when strength requirements are higher than those obtainable with carbon steels. Low-alloy steels also have better toughness and hardenability than carbon steels.

Carbon-Manganese Cast Steels. Cast steels containing 1.00 to 1.75% Mn and 0.20 to 0.50% C have received considerable attention from engineers in the past because of the excellent properties that can be developed with a single, relatively inexpensive alloying element and by a single normalizing or normalizing and tempering heat treatment.

Carbon-manganese steels are also referred to as medium-manganese steels and are represented by the cast 1300 series of steels (1.60 to 1.90% Mn).

Manganese-molybdenum cast steels are very similar to the medium-manganese steels with the added characteristics of high yield strength at elevated temperatures, higher ratio of yield strength to tensile strength at room temperature, greater freedom from temper embrittlement, and greater hardenability. Therefore, these steels have replaced medium-manganese steel for certain applications.

There are two general grades of manganese-molybdenum cast steels:

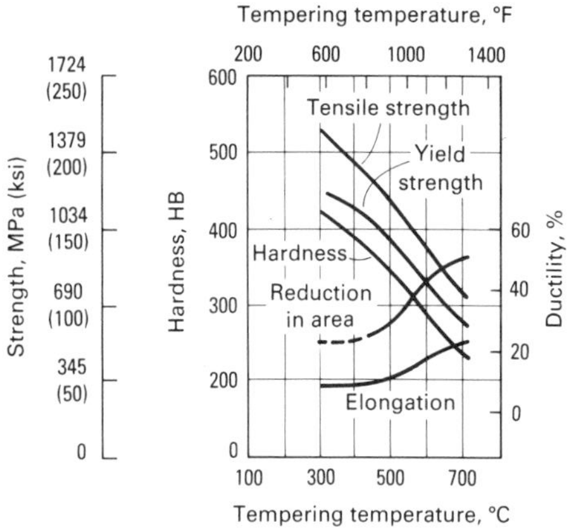

Fig. 2 Mechanical properties of water-quenched cast 8630 steel

- 8000 series (1.0 to 1.35% Mn, 0.10 to 0.30% Mo)
- 8400 series (1.35 to 1.75% Mn, 0.25 to 0.55% Mo)

For both of these alloy types, the carbon content is frequently selected between 0.20 and 0.35%, depending on the heat treatment employed and the strength characteristics desired.

Manganese-Nickel-Chromium-Molybdenum Cast Steels. The cast 9500 series low-alloy steels are primarily produced for their high hardenability. Sections exceeding 127 mm (5 in.) in thickness can be quenched and tempered to obtain a fully tempered martensitic structure. The composition range employed for the 9500 series is:

Element	Composition, %
Manganese	1.30–1.60
Nickel	0.40–0.70
Chromium	0.55–0.75
Molybdenum	0.30–0.40

Nickel Cast Steels. Among the oldest alloy cast steels are those containing nickel. These steels are characterized by high tensile strength and elastic limit, good ductility, and excellent resistance to impact. The cast steels of the 2300 series contain 2.0 to 4.0% Ni, depending on the grade required.

Nickel-Chromium-Molybdenum Cast Steels. The addition of molybdenum to nickel-chromium steel significantly improves hardenability and makes the steel relatively immune to temper embrittlement. Nickel-chromium-molybdenum cast steel is particularly well suited to the production of large castings because of its deep-hardening properties. In addition, the ability of these steels to retain strength at elevated temperatures extends their usefulness in many industrial applications.

Chromium-Molybdenum Cast Steels. Chromium contents of about 1.00% or more provide a nominal improvement in elevated-temperature properties. The chromium cast steels (5100 series; 0.70 to 1.10% Cr) are not in common use in the steel casting industry. However, the chromium-molybdenum low-alloy cast steels are widely used.

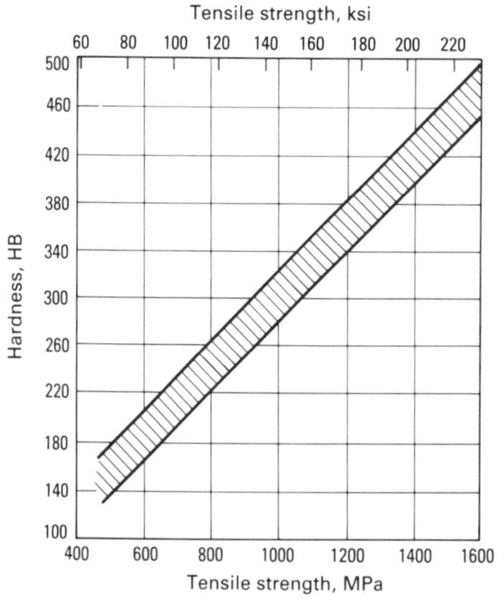

Fig. 3 Hardness versus tensile strength of low-alloy cast steels

Copper Bearing Cast Steels. There are several types of copper-containing steels. Selection among these various types is primarily based on either their atmospheric corrosion resistance (weathering steels) or the age-hardening characteristics that copper adds to steel.

High-strength cast steels cover the tensile strength range of 1207 to 2068 MPa (175 to 300 ksi). Cast steels with these strength levels and with considerable toughness and weldability were originally developed for ordnance applications. These cast steels can be produced from any of the above medium-alloy compositions by heat treating with liquid-quenching techniques and low tempering temperatures. Cast 4300 series steels or modifications thereof are usually employed.

Structure and Property Correlations

Carbon and low-alloy steel castings are produced to a great variety of properties because composition and heat treatment can be selected to achieve specific combinations of properties, including hardness, strength, ductility, fatigue, and toughness. Although selections can be made from a wide range of properties, it is important to recognize the interrelationships among these properties. For example, higher hardness, lower toughness, and lower ductility values are associated with higher strength values.

Figure 2 illustrates the range of properties that can be achieved with a single steel (in this case, nickel-chromium-molybdenum cast 8630) and the interrelationship of its mechanical properties. These relationships and mechanical property ranges will be further discussed in the following sections.

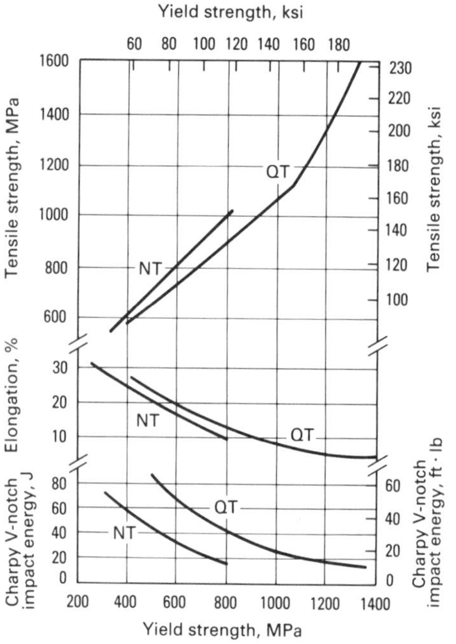

Fig. 4 Room-temperature properties of cast low-alloy steels. QT, quenched and tempered; NT, normalized and tempered

Strength and Hardness. Depending on alloy selection and heat treatment, ultimate tensile strength levels ranging from 414 to 1724 MPa (60 to 250 ksi) can be achieved with cast low-alloy steels. The normally expected Brinell hardness-ultimate tensile strength combinations of cast low-alloy steels are shown in Fig. 3. As can be seen, these two properties are directly proportional.

Strength and Ductility. Ductility is inversely proportional to the strength, or hardness, of the cast steel. Actual ductility requirements vary with the strength level and the specification to which a steel is ordered. Yield strength is a primary design criterion for structural applications. Therefore, Fig. 4 plots tensile strength, ductility (as measured by elongation), and toughness (based on Charpy V-notch impact energy) in terms of yield strength. Quenched-and-tempered steels exhibit higher ductility values for a given yield strength level than normalized-and-tempered steels.

Strength and Toughness. Several test methods are available for evaluating the toughness of steel, or the resistance to sudden or brittle fracture. These include the Charpy V-notch impact test, the drop-weight tear test, the dynamic tear test, and specialized procedures to determine plane-strain fracture toughness.

Room-temperature Charpy V-notch impact energy trends (Fig. 4) reveal the dis-

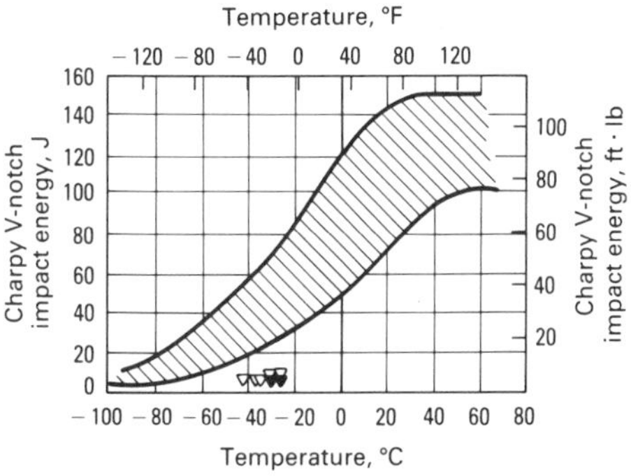

Fig. 5 NDTT values (triangles) and the scatter band of Charpy V-notch energy transition curves of several quenched-and-tempered carbon-manganese cast steels. Nominal ultimate tensile strength: 552 MPa (80 ksi)

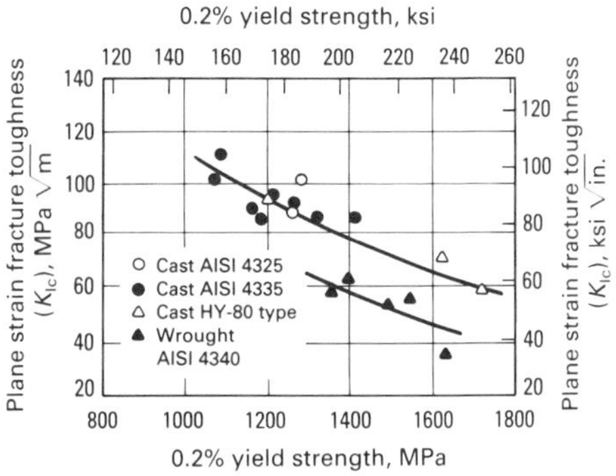

Fig. 6 Plane-strain fracture toughness K_{Ic} and strength relationships at room temperature for quenched-and-tempered nickel-chromium-molybdenum steels

tinct effect of strength and heat treatment on toughness. Higher toughness is obtained when a steel is quenched and tempered, rather than normalized and tempered; quenching, followed by tempering, produces superior toughness.

Nil ductility transition temperatures (NDTT) from 38 °C (100 °F) to as low as −90 °C (−130 °F) have been recorded in tests on normalized-and-tempered cast carbon and low-alloy steels in the yield strength range of 207 to 655 MPa (30 to 95 ksi). Nil ductility transition temperature values of as high as 10 °C (50 °F) to as low as −107 °C (−160 °F) can be obtained in the yield strength range of 345 to 1345 MPa (50 to 195 ksi), depending on alloy selection.

An approximate relationship exists between Charpy V-notch impact energy-temperature behavior and the nil ductility transition temperature. The nil ductility transition temperature frequently coincides with the energy transition temperature determined in Charpy V-notch tests. Applicable data for cast carbon-manganese steels are shown in Fig. 5.

Fracture Mechanics Tests. Plane-strain fracture toughness (K_{Ic}) data for the various steels listed in Table 1 reflect the important strength-toughness relationship. For quenched-and-tempered nickel-chromium-molybdenum steels, Fig. 6 indicates high K_{Ic} values of about 110 MPa$\sqrt{m}$ (100 ksi$\sqrt{in.}$). at a 0.2% offset yield strength level of 1034 MPa (150 ksi). At a yield strength level of 1655 MPa (240 ksi), K_{Ic} values level off to about 66 MPa$\sqrt{m}$ (60 ksi$\sqrt{in.}$). Data are also plotted in Fig. 6 for wrought plates made of comparable steel of somewhat higher carbon content.

Fracture mechanics tests have the advantage over conventional toughness tests of being able to yield material property values that can be used in design equations. Additional information is available in the Selected References at the end of this article.

Table 1 Plane-strain fracture toughness of cast low-alloy steels at room temperature

Alloy type	Heat treatment(a)	Yield strength, 0.2% offset, MPa	ksi	Plane strain fracture toughness, K_{Ic}, MPa$\sqrt{m}$	ksi$\sqrt{in.}$
Fe-1.25Cr-0.5Mo	SRANTSR	275	40	88	80
Cast 1030	NT	303	44	127	116
Fe-0.5Cr-0.5Mo-0.25V	NT	367	53	55	50
Fe-0.5C-1.5Mn	NT	412	59	107	98
Fe-0.5C-1Cr	NT	413	60	58	53
Fe-0.5C	NT	425	61	65	59
Cast 9535	NT	614	89	67	61
Fe-0.35C-0.6Ni-0.7Cr-0.4Mo	NT	683	99	64	58
Cast 4335	SLQT	747	108	69	63
Cast 9536	NT	752	109	59	54
Fe-0.3C-1Ni-1Cr-0.3Mo	NT	787	114	66	60
Cast 4335	SLQT	814	118	97	87
Cast 4335	SLQT	903	131	105	96
Cast 4335	QT	1090	158	115	105
Cast 4335	QT	1166	169	92	84
Ni-Cr-Mo	QT	1207	175	98	89
Cast 4340	QT	1207	175	115	105
Cast 4325	QT	1263	183	75	82
Cast 4325	QT	1280	186	104	95
Cr-Mo	QT	1379	200	84	76
Cast 4340	QT	1450	210	67	61

(a) SR, stress relieved; A, annealed; N, normalized; T, tempered; Q, quenched; SLQ, slack quenched

Strength and Fatigue. The most basic method of presenting engineering fatigue data is by means of the *S-N* curve. The *S-N* curve shows the life of the fatigue specimen in terms of the number of cycles to failure *N* and the maximum applied stress *S*. Additional tests have been used, and the principal findings for cast steels are highlighted in the following sections.

Constant Amplitude Tests. The endurance ratio (fatigue endurance limit divided by the tensile strength) of cast carbon and low-alloy steels, determined in Moore rotating-beam bending fatigue tests (mean stress = 0), is generally taken to be about 0.40 to 0.50 for smooth bars. The data in Table 2 show that the endurance ratio is largely independent of strength, alloying elements, and heat treatment.

Fatigue notch sensitivity *q*, determined in rotating-beam bending fatigue tests, is related to the microstructure of the steel (composition and heat treatment) and to strength. Table 3 indicates that *q* generally increases with strength—from 0.23 for annealed carbon steel at a tensile strength of 577 MPa (83.5 ksi) to 0.68 for the higher-strength normalized-and-tempered low-alloy steels. The quenched-and-tempered steels with a martensitic structure are less notch sensitive than the normalized-and-tempered steels with a ferrite-pearlite microstructure. Similar results and trends on notch sensitivity have been reported for tests with sharper notches.

Cast steels suffer less degradation of fatigue properties due to notches than equivalent wrought steels. When laboratory test conditions are replaced with more realistic service conditions, the cast steels show much less notch sensitivity than wrought steels. Table 4 indicates that wrought steels are 1.5 to 2.3 times as notch sensitive as cast steels. Under laboratory test conditions and with specimen preparation, the unnotched endurance limit of wrought steels is higher. However, the same fatigue characteristics as those of cast steel are obtained when a notch is introduced (Fig. 7).

Section Size and Mass Effects

Mass effects are common to steel, whether rolled, forged, or cast, because the cooling rate during the heat-treating operation varies with section size and because microstructural components, grain size, and nonmetallic inclusions vary from surface to center. Mass effects are metallurgical in nature and are distinct from the effect of discontinuities, which are discussed in the next section of this article. Figure 8 shows how the mass of a component lowers the strength properties of wrought AISI 8630 steel plate. The properties are plotted for the ¼*T* location, halfway between the surface and the center of the plate.

Table 2 Fatigue properties of cast carbon and low-alloy steels

Class(a) and heat treatment(b)	Tensile strength MPa	Tensile strength ksi	Yield strength MPa	Yield strength ksi	Reduction in area, %	Elongation, %	Hardness, HB	Fatigue endurance limit MPa	Fatigue endurance limit ksi	Endurance ratio
Carbon steels										
60 A	434	63	241	35	54	30	131	207	30	0.48
65 N	469	68	262	38	48	28	131	207	30	0.44
70 N	517	75	290	42	45	27	143	241	35	0.47
80 NT	565	82	331	48	40	23	163	255	37	0.45
85 NT	621	90	379	55	38	20	179	269	39	0.43
100 QT	724	105	517	75	41	19	212	310	45	0.47
Low-alloy steels(c)										
65 NT	469	68	262	38	55	32	137	221	32	0.47
70 NT	510	74	303	44	50	28	143	241	35	0.47
80 NT	593	86	372	54	46	24	170	269	39	0.45
90 NT	655	95	441	64	44	20	192	290	42	0.44
105 NT	758	110	627	91	48	21	217	365	53	0.48
120 QT	883	128	772	112	38	16	262	427	62	0.48
150 QT	1089	158	979	142	30	13	311	510	74	0.47
175 QT	1234	179	1103	160	25	11	352	579	84	0.47
200 QT	1413	205	1172	170	21	8	401	607	88	0.43

(a) Class of steel based on tensile strength. (b) A, annealed; N, normalized; NT, normalized and tempered; QT, quenched and tempered. (c) Below 8% total alloy content

Table 3 Fatigue notch sensitivity of several cast carbon and low-alloy steels

Steel	Tensile strength MPa	Tensile strength ksi	Endurance limit: Unnotched MPa	Endurance limit: Unnotched ksi	Endurance limit: Notched MPa	Endurance limit: Notched ksi	Fatigue endurance ratio Unnotched	Fatigue endurance ratio Notched	Fatigue notch sensitivity factor, q
Normalized and tempered									
1040	648	94.2	260	37.7	193	28	0.40	0.30	0.29
1330	685	99.3	334	48.4	219	31.7	0.49	0.32	0.44
1330	669	97	288	41.7	215	31.2	0.43	0.32	0.28
4135	777	112.7	353	51.2	230	33.3	0.45	0.30	0.45
4335	872	126.5	434	63	241	34.9	0.50	0.28	0.68
8630	762	110.5	372	54	228	33.1	0.49	0.30	0.53
Quenched and tempered									
1330	843	122.2	403	58.5	257	37.3	0.48	0.31	0.48
4135	1009	146.4	423	61.3	280	40.6	0.42	0.28	0.43
4335	1160	168.2	535	77.6	332	48.2	0.46	0.29	0.51
8630	948	137.5	447	64.9	266	38.6	0.47	0.27	0.57
Annealed									
1040	576	83.5	229	33.2	179	26	0.40	0.31	0.23

(a) Notched tests run with theoretical stress concentration factor of 2.2. (b) $q = (K_f - 1)/(K_t - 1)$, where K_f is the endurance limit notched/endurance limit unnotched and K_t is the theoretical stress concentration factor

The section size or mass effect is of particular importance in steel castings because mechanical properties are typically assessed from test bars machined from standardized coupons that have fixed dimensions and are cast separately from or attached to the castings (Fig. 9). Removal of the test bars from the casting is impractical, because removal of material for testing would destroy the usefulness of the component or would necessitate costly weld repairs for replacing the material removed for testing purposes.

Specimens removed from a casting cannot be routinely expected to exhibit the same properties as test specimens machined from the standard test coupon designs for which minimum properties are established in specifications. The mass effect discussed above, that is, the differences in cooling rate between test coupons and the part being produced, is the fundamental reason for this situation. Several specifications, such as ASTM E 208, A 356, and A 757, provide for the mass effect by permitting test coupons that are larger than the basic keel block shown in Fig. 9 and whose cooling rate is therefore more representative of that experienced by the part being produced.

Discontinuity Effects

The treatment of discontinuities in design is undergoing major changes as a result of the wider use of fracture mechanics in the industry. If the plane-strain fracture toughness K_{Ic} of a material is known at the temperature of interest, designers can determine the critical combination of flaw size and stress required to cause failure in one load application. In addition, designers can calculate the remaining life of a component having a discontinuity, or they can compute the largest acceptable flaw from a knowledge of the crack growth properties (da/dN) and fracture mechanics parameters (K, n, and C) of a material. In the absence of suitable plane-strain fracture mechanics data, approximations can be made on the basis of test results obtained from a variety of tensile, impact, and fatigue tests.

Table 4 Fatigue notch sensitivity factors for cast and wrought carbon and low-alloy steels at a number of strength levels

Steel	Tensile strength MPa	Tensile strength ksi	Fatigue notch sensitivity factor, q
Annealed			
1040 cast	576	83.5	0.23
1040 wrought	561	81.4	0.43
Normalized and tempered			
1040 cast	649	94.2	0.29
1040 wrought	620	90.0	0.50
1330 cast	669	97.0	0.28
1340 wrought	702	101.8	0.65
4135 cast	777	112.7	0.45
4140 wrought	766	111.1	0.81
4335 cast	872	126.5	0.68
4340 wrought	859	124.6	0.97
8630 cast	762	110.5	0.53
8640 wrought	748	108.5	0.85
Quenched and tempered			
1330 cast	843	122.2	0.48
1340 wrought	836	121.2	0.73
4135 cast	1009	146.4	0.43
4140 wrought	1012	146.8	0.93
4335 cast	1160	168.2	0.51
4340 wrought	1161	168.4	0.92
8630 cast	948	137.5	0.57
8640 wrought	953	138.2	0.90

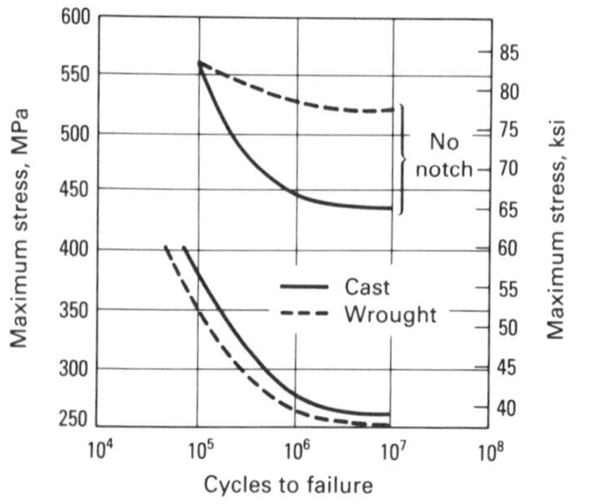

	Tensile strength, MPa (ksi)	Yield strength, MPa (ksi)	Elongation, %	Hardness, HB
Cast 8630	952 (138)	869 (126)	15	286
Wrought 8640	952 (138)	855 (124)	22	286

Fig. 7 Fatigue characteristics (S-N curves) for cast and wrought 8600 series steels, quenched and tempered to the same hardness, both notched and unnotched. Moore rotating-beam tests; stress concentration factor K: 2.2

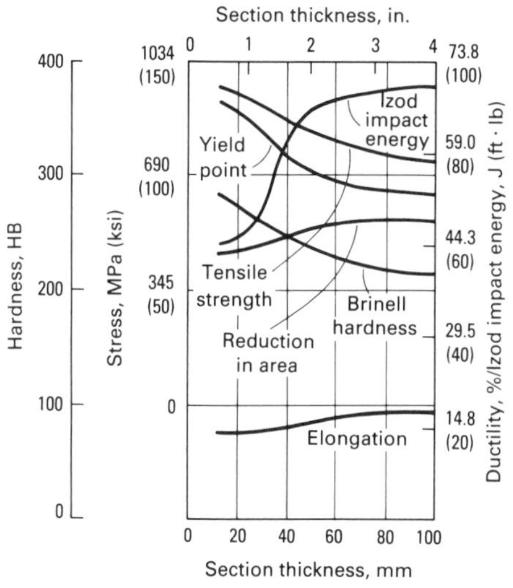

Fig. 8 Section size effects on water-quenched and tempered wrought AISI 8630 steel in sizes over 25 mm (1 in.). The properties reported are those at the 1/4*T* location (midway between surface and center).

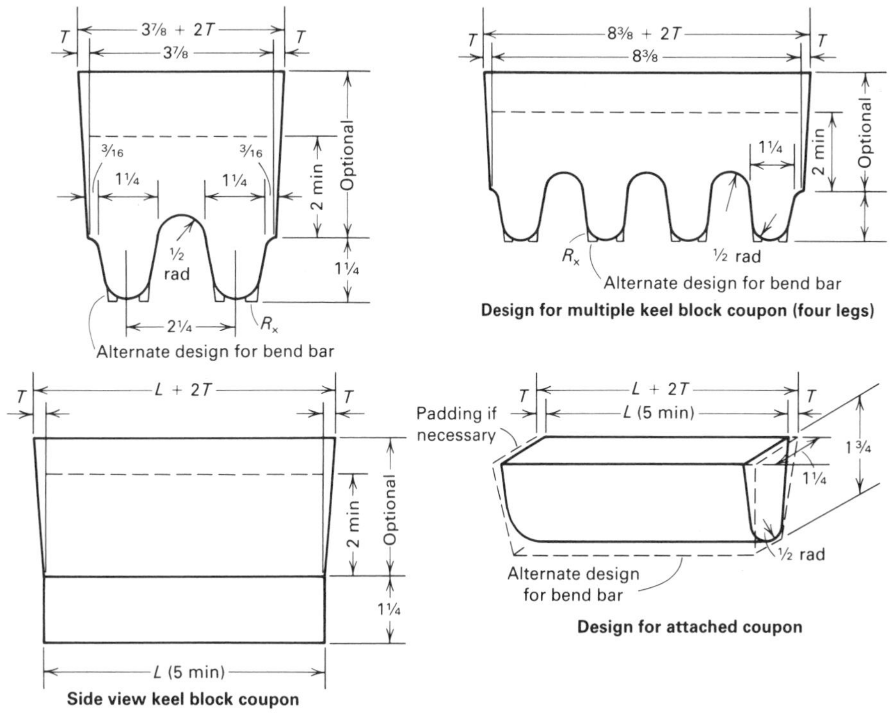

Fig. 9 Keel block coupon according to ASTM A 370. Dimensions given in inches

Test Coupon Versus Casting Properties

Coupon properties refer to the properties of specimens cut and machined from either a separately cast coupon or a coupon that is attached to and cast integrally with the casting. Typically, the legs of the ASTM standard keel block (A 370) serve as the coupons. The legs of this keel block are 32 mm (1.25 in.) thick.

Casting properties refer to the properties of specimens cut and machined from the production casting itself. A casting from which properties are determined in this manner is either destroyed in the process or requires repair welding to replace the metal removed for testing.

Test Coupons. The ASTM double-leg keel block (Fig. 9) is the most prominent design for test coupons among those in use and among those recognized in ASTM A 370. Table 5 in the article "Plain Carbon Steels" in this Volume offers information on the reliability of tensile test results obtained from the double-leg keel block. The data indicate that for two tests there is 95% assurance that the strength is within 6.9 MPa (±1 ksi) of the actual ultimate tensile strength and within 11 MPa (±1.6 ksi) of the actual yield strength. For tensile ductility, the data show that two tests produce, with 95% assurance, the elongation results within ±3% and the reduction in area value within ±5%.

When 32 mm (1.25 in.) thick test coupons are suitably attached to the casting and cast integrally with the production casting, the tensile properties determined for the coupon will be comparable to those from a separately cast keel block. Table 5 contains data on this conclusion for several grades of cast steel.

The properties determined from keel block legs whose dimensions exceed those of the ASTM double-leg keel block (that is, thicker than 32 mm, or 1.25 in.) may differ, especially if the steel involved is of insufficient hardenability for the heat treatment employed to produce a microstructure similar to that in 32 mm (1.25 in.) section keel block legs. Large mass effects are evident in Table 6 for several of the quenched-and-tempered materials and for the normalized-and-tempered materials. These data apply to low-alloy steels of similar carbon content.

Weldability

The weldability of low-alloy steels is primarily a function of composition and heat treatment. Carbon steels having low manganese and silicon contents and a carbon content below 0.30% can be welded without any special precautions. When the carbon content exceeds 0.30%, preheating of the casting prior to welding may be advisable. The low-temperature preheat (120 to 205 °C, or 250 to 400 °F) reduces the rate at which heat is extracted from the heat-affected zone (HAZ) adjacent to the weld. Preheating also helps to relieve mechanical stresses and to prevent underbead cracking because hydrogen is still relatively mobile and can diffuse away from the last areas to undergo a metallurgical transformation. Preheating minimizes the chances of a hardening transformation occurring in the HAZ and thus reduces the hardness adjacent to the weld. The metallurgical notch that could result from the hardened area can be eliminated in this manner. Generally speaking, if the hardness of the HAZ after welding does not exceed 35 HRC or 327 HB, preheating of the casting for welding is not required.

As additional alloying elements are added to the steel, the need for preheating increases. Most of the low-alloy steels, such

Table 5 Comparison of properties of keel block and separately cast test specimens

Steel type	Coupon type	Tensile strength MPa	Tensile strength ksi	Yield strength MPa	Yield strength ksi	Elongation, %	Reduction in area, %	Hardness, HB
No. 1 carbon steel	Keel block	524	76	317	46	34.5	54.4	153
	Attached	510	74	303	44	31.0	50.3	148
No. 2 Mn-Mo steel	Keel block	807	117	648	94	22.5	51.7	241
	Attached	800	116	641	93	21.0	47.2	241
No. 3 carbon steel	Keel block	503	73	338	49	34.0	57.8	...
	Attached	503	73	324	47	35.5	60.8	...
Grade B steel	Keel block	538	78	324	47	27.5	47.1	...
	Attached	552	80	324	47	27.5	42.7	...

Table 6 Effect of mass on the tensile and impact properties of cast low-alloy steels with similar carbon contents

Specimens were taken from the center of bars.

Steel	Heat treatment(a)	Bar size mm	Bar size in.	Tensile strength MPa	Tensile strength ksi	Yield strength MPa	Yield strength ksi	Elongation, %	Reduction in area, %	Charpy V-notch impact energy(b) J	Charpy V-notch impact energy(b) ft · lb
Cast 1330 (0.31% C, 1.50% Mn)	WQ	25	1	710	103	490	71	26	56	54	40
	NT	25	1	655	95	379	55	28	55	41	30
	WQT	51	2	676	98	469	68	27	58	52	38
	NT	51	2	641	93	379	55	27	55	34	25
	WQT	102	4	641	93	400	58	23	53	49	36
	NT	102	4	607	88	365	58	23	52	30	22
Cast 8030 (0.32% C, 1.20% Mn, 0.16% Mo)	WQT	25	1	793	115	655	95	21	52	41	30
	NT	25	1	669	97	448	65	24	49	27	20
	WQT	51	2	738	107	621	90	20	50	34	25
	NT	51	2	648	94	427	62	23	46	30	22
	WQT	102	4	641	93	414	60	20	45	30	22
	NT	102	4	641	93	421	61	19	40	23	17
Cast 8430 (0.32% C, 1.43% Mn, 0.34% Mo)	WQT	25	1	841	122	717	104	20	50	45	33
	NT	25	1	717	104	517	75	22	50	28	21
	WQT	51	2	807	117	676	98	21	51	46	34
	NT	51	2	738	107	524	76	22	47	32	24
	WQT	102	4	717	104	558	81	20	48	43	32
	NT	102	4	710	103	531	77	18	42	28	21
Cast 9530 (0.29% C, 1.41% Mn, 0.60% Ni, 0.60% Cr, 0.37% Mo)	WQT	25	1	896	130	772	112	18	42	34	25
	NT	25	1	786	114	614	89	18	44	27	20
	WQT	51	2	88	128	758	110	20	45	38	28
	NT	51	2	8	116	641	93	18	42	33	24
	WQT	102	4	862	125	758	110	17	44	33	24
	NT	102	4	779	113	627	91	16	40	24	18

(a) WQT, water quenched and tempered; NT, normalized and tempered. (b) Keyhole notch

as the cast 8630, 8730, or 4130 steels, require some preheating. When properly preheated, such steels are readily welded. A comparison can be made between the weldability of two steels by comparing their carbon equivalents, CE. Of the various formulas available, one that appears to be quite useful is:

$$CE = \%C + \frac{\%Mn}{6} + \frac{\%Cr + \%Mo + \%V}{5} + \frac{\%Cu + \%Ni}{15}$$

Steels having the same carbon equivalent will have approximately the same weldability and will require similar preheating and other precautions.

The alloy steels of higher hardenability, for example, the cast 4300 series, require more care and must be preheated in the range of 205 to 315 °C (400 to 600 °F). In fact, a few of the ASTM standards, such as A 217 and A 487, specify preheat temperatures that must be employed. The preheat temperatures specified are conservative and are based on the preheating temperatures normally used in the fabrication of piping assemblies, in which the stresses resulting from the shrinkage of the weld and restraint of the assembly must also be considered.

In the case of some low-alloy steels, especially those subject to toughness requirements, the interpass temperature (maximum or minimum temperature of the weld bead before additional beads are laid) must be maintained at a prescribed low level to prevent embrittlement of the parent metal.

Corrosion Properties

Low-alloy steels are generally not considered to be corrosion resistant, and casting compositions are not normally selected on the basis of corrosion resistance. In some environments, however, significant differences are observed in corrosion behavior such that the corrosion rate of one steel may be half that of another grade. In general, steels alloyed with small amounts of copper tend to have somewhat lower corrosion rates than copper-free alloys. As little as 0.05% Cu has been shown to exert a significant effect. In some environments, nominal levels of nickel, chromium, phosphorus, and silicon may also bring about modest improvements, but when these four elements are present, the presence of copper holds little if any additional advantage. Detailed information on the corrosion resistance of steels is available in the articles "Corrosion of Carbon Steels," "Corrosion of Alloy Steels," and "Corrosion of Cast Steels" in Volume 13 of the 9th Edition of *Metals Handbook*.

Elevated-Temperature Properties

Steels operating at temperatures above ambient are subject to failure by a number of mechanisms other than mechanical stress or impact. These include oxidation, hydrogen damage, sulfide scaling, and carbide instability, which manifests itself as graphitization.

The environmental factors involved in elevated-temperature service (370 to 650 °C, or 700 to 1200 °F) require that steels used in this temperature range be carefully characterized. As a consequence, four ASTM specifications have been developed for cast carbon and low-alloy steels for elevated-temperature service. One of these specifications, ASTM A 216, describes carbon steels; the other three, A 217, A 356, A 389, cover low-alloy steels.

The two alloying elements common to nearly all the steel compositions used at elevated temperatures are molybdenum and chromium. Molybdenum contributes strongly to creep resistance. Depending on microstructure, it has been shown that 0.5% Mo reduces the creep rate of steels by a factor of at least 10^3 at 600 °C (1110 °F).

Chromium also reduces the creep rate modestly at levels to approximately 2.25%. At higher chromium levels, creep resistance is somewhat reduced. Vanadium improves creep strength and is indicated in some specifications. Other elements that improve creep resistance include tungsten, titanium, and niobium. The effect of tungsten is similar to that of molybdenum, but on a weight percent basis more tungsten is required in order to be equally beneficial. Titanium and niobium have been shown to improve the creep properties of carbon-free alloys, but because they remove carbon from solid solution, their effect tends to be variable.

None of the latter three elements appears in U.S. specifications for cast steels for elevated-temperature service.

Temper embrittlement is normally characterized by an increase in the Charpy V-notch impact transition temperature. It is generally associated with quenched-and-tempered steels, but it may also occur in normalized-and-tempered steels and may develop during either tempering or elevated-temperature service. The tempering or service temperature range in which embrittlement occurs is considered to be 425 to 595 °C (800 to 1100 °F). During heat treatment, embrittlement can be avoided by rapid cooling through this range.

The elements generally considered to cause embrittlement are phosphorus, tin, antimony, and arsenic; manganese and silicon enhance the effect of phosphorus. An addition of 0.05% Mo to alloy steels is very effective in reducing temper embrittlement. All of the alloy steels normally used for elevated-temperature service contain at least this quantity of molybdenum; therefore, temper embrittlement during heat treatment is seldom a problem. However, in those cases in which castings must be cooled from the tempering temperature at a rate below 20 °C (35 °F) per hour, it may be necessary to control the amount of temper-embrittling elements present in the steel. When embrittlement occurs during service, no reduction in impact properties at elevated temperature is apparent. Instead, diminished impact toughness may be observed during periods of shutdown when the equipment is at ambient temperatures.

Melting, Deoxidation, and Foundry Practices

Melting and deoxidation practices for low-alloy steels are the same as those employed for plain carbon steels and are discussed in detail in the article "Plain Carbon Steels" in this Volume. Pouring practice and gating and risering are also discussed in that article. Foundry practices, including the cleaning and heat treatment of steel castings, also are discussed in the article "Plain Carbon Steels" in this Volume.

SELECTED REFERENCES

- W.J. Jackson, Ed., *Steel Castings: Design Properties and Applications*, Steel Castings Research and Trade Association, 1983
- R.I. Stephens, "Fatigue and Fracture Toughness of Five Carbon or Low Alloy Cast Steels at Room or Low Climatic Temperatures," Research Report 94 (A & B), Steel Founders' Society of America, 1982
- P.F. Wieser, Ed., *Steel Castings Handbook*, 5th ed., Steel Founders' Society of America, 1980

High-Alloy Steels

John M. Svoboda, Steel Founders' Society of America

CAST HIGH-ALLOY STEELS are widely used for their corrosion resistance in aqueous media at or near room temperature and for service in hot gases and liquids at elevated temperatures (>650 °C, or 1200 °F). High-alloy cast steels are most often specified on the basis of composition using the designation system of the High Alloy Product Group of the Steel Founders' Society of America. (The High Alloy Product Group has replaced the Alloy Casting Institute (ACI), which formerly administered these designations.). These alloy designations (for example, CF-8M) have been adopted by the American Society for Testing and Materials (ASTM) and are preferred for cast high-alloy steels.

The first letter of the designation indicates whether the alloy is intended primarily for liquid corrosion service (C) or high-temperature service (H). The second letter denotes the nominal chromium-nickel type of the alloy (Fig. 1). As nickel content increases, the second letter of the designation is changed from A to Z. The numeral or numerals following the first two letters indicate maximum carbon content (percentage × 100) of the alloy. Finally, if further alloying elements are present, these are indicated by the addition of one or more letters as a suffix. Thus, the designation of CF-8M refers to an alloy for corrosion-resistant service (C) of the 19Cr-9Ni type (Fig. 1), with a maximum carbon content of 0.08% and containing molybdenum (M).

Some of the high-alloy cast steels exhibit many of the same properties of cast carbon and low-alloy steels (see the articles "Plain Carbon Steels" and "Low-Alloy Steels" in this Volume). Some of the mechanical properties of these grades (for example, hardness and tensile strength) can be altered by suitable heat treatment. The cast high-alloy grades that contain more than 20 to 30% Cr + Ni, however, do not show the phase changes observed in plain carbon and low-alloy steels during heating or cooling between room temperature and the melting point. These materials are therefore nonhardenable, and their properties depend on composition rather than heat treatment. Therefore, special consideration must be given to each grade of high-alloy cast steel with regard to casting design, foundry practice, and subsequent thermal processing (if any).

Corrosion-Resistant High-Alloy Steels

Corrosion-resistant high-alloy cast steels, more commonly referred to as cast stainless steels, have grown steadily in technological and commercial importance during the past 40 years. The principal applications for these steels are as materials of construction for chemical-processing and power-generating equipment involving corrosion service in aqueous or liquid-vapor environments at temperatures normally below 315 °C (600 °F). These alloys are also used for special services at temperatures to 650 °C (1200 °F).

An appropriate definition for cast stainless steels is the familiar one based on the discovery in 1910 that a minimum of 12% Cr will impart resistance to corrosion and oxidation to steel. Cast stainless steels are defined as ferrous alloys that contain a minimum of 12% Cr for corrosion resistance. Most cast stainless steels are of course considerably more complex compositionally than this simple definition implies. Stainless steels typically contain one or more alloying elements in addition to chromium (for example, nickel, molybdenum, copper, niobium, and nitrogen) to produce a specific microstructure, corrosion resistance, or mechanical properties for particular service requirements.

Corrosion-resistant high-alloy cast steels are usually classified on the basis of composition or microstructure. It should be recognized that these bases for classification are not completely independent in most cases; that is, classification by composition also often involves microstructural distinctions.

Table 1 lists the compositions of the commercial cast corrosion-resistant alloys. Alloys are grouped as chromium steels, chromium-nickel steels in which chromium is the predominant alloying element, and nickel-chromium steels in which nickel is the predominant alloying element. The serviceability of cast corrosion-resistant steels depends greatly on the absence of carbon, and especially precipitated carbides, in the alloy microstructure. Therefore, cast corrosion-resistant alloys are generally low in carbon (typically <0.08% C).

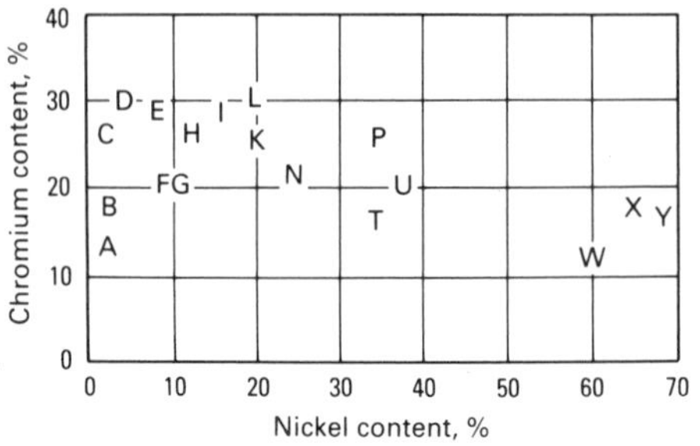

Fig. 1 Chromium and nickel contents in ACI standard grades of heat- and corrosion-resistant steel castings. See text for details.

As shown in Table 1, the high-alloy cast steels can also be classified on the basis of microstructure. Structures may be austenitic, ferritic, martensitic, or duplex; the structure of a particular grade is primarily determined by composition. Chromium, nickel, and carbon contents are particularly important in this regard (see the section "Ferrite in Cast Stainless Steel" in this article). In general, straight chromium grades of high-alloy cast steel are either martensitic or ferritic, the chromium-nickel grades are either duplex or austenitic, and the nickel-chromium steels are fully austenitic.

Martensitic grades include Alloys CA-15, CA-40, CA-15M, and CA-6NM. The CA-15 alloy contains the minimum amount of chromium necessary to make it essentially rustproof. It has good resistance to atmospheric corrosion as well as to many organic media in relatively mild service. A higher-carbon modification of CA-15, CA-40 can be heat treated to higher strength and hardness levels. Alloy CA-15M is a molybdenum-containing modification of CA-15 that provides improved elevated-temperature strength. Alloy CA-6NM is an iron-chromium-nickel-molybdenum alloy of low carbon content.

Austenitic grades include CH-20, CK-20, and CN-7M. The CH-20 and CK-20 alloys are high-chromium, high-carbon, wholly austenitic compositions in which the chromium exceeds the nickel content. The more highly alloyed CN-7M has excellent

Table 1 Compositions and microstructures of corrosion-resistant high-alloy cast steels

Alloy	Wrought alloy type(a)	Most common end-use microstructure	Composition, %(b)								
			Cr	Ni	Mo	Si	Mn	P	S	C	Others
Chromium steels											
CA-15	410	Martensite	11.5–14.0	1.00	0.50	1.50	1.00	0.04	0.04	0.15	...
CA-15M	...	Martensite	11.5–14.0	1.00	0.15–1.0	0.65	1.00	0.04	0.04	0.15	...
CA-40	420	Martensite	11.5–14.0	1.00	0.5	1.50	1.00	0.04	0.04	0.20–0.40	...
CB-30	431,442	Ferrite + carbides	18.0–21.0	2.00	...	1.50	1.00	0.04	0.04	0.30	...
CC-50	446	Ferrite + carbides	26.0–30.0	4.00	...	1.50	1.00	0.04	0.04	0.50	...
Chromium-nickel steels											
CA-6NM	...	Martensite	11.5–14.0	3.5–4.5	0.40–1.0	1.00	1.00	0.04	0.03	0.06	...
CB-7Cu	17-4PH	Martensite-age hardenable	15.5–17.0	3.6–4.6	...	1.50	1.00	0.04	0.04	0.07	2.3–3.3 Cu
CD-4MCu	...	Austenite in ferrite-age hardenable	25.0–26.5	4.75–6.0	1.75–2.25	1.00	1.00	0.04	0.04	0.04	2.75–3.25 Cu
CE-30	...	Ferrite in austenite	26.0–30.0	8.0–11.0	...	2.00	1.50	0.04	0.04	0.30	...
CF-3	304L	Ferrite in austenite	17.0–21.0	8.0–12.0	...	2.00	1.50	0.04	0.04	0.03	...
CF-8	304	Ferrite in austenite	18.0–21.0	8.0–11.0	...	2.00	1.50	0.04	0.04	0.08	...
CF-20	302	Austenite	18.0–21.0	8.0–11.0	...	2.00	1.50	0.04	0.04	0.20	...
CF-3M	316L	Ferrite in austenite	17.0–21.0	9.0–13.0	2.0–3.0	1.50	1.50	0.04	0.04	0.03	...
CF-8M	316	Ferrite in austenite	18.0–21.0	9.0–12.0	2.0–3.0	1.50	1.50	0.04	0.04	0.08	...
CF-12M	...	Ferrite in austenite or austenite	18.0–21.0	9.0–12.0	2.0–3.0	2.00	1.50	0.04	0.04	0.12	...
CF-8C	347	Ferrite in austenite	18.0–21.0	9.0–12.0	...	2.00	1.50	0.04	0.04	0.08	Nb = 8 × C, 1.0 max
CF-16F	303	Austenite	18.0–21.0	9.0–12.0	1.50	2.00	1.50	0.17	0.04	0.16	0.20–0.35 Se
CG-8M	317	Ferrite in austenite	18.0–21.0	9.0–13.0	3.0–4.0	1.50	1.50	0.04	0.04	0.08	...
CH-20	309	Austenite	22.0–26.0	12.0–15.0	...	2.00	1.50	0.04	0.04	0.20	...
CK-20	310	Austenite	23.0–27.0	19.0–22.0	...	1.75	1.50	0.04	0.04	0.20	...
Nickel-chromium steel											
CN-7M	320	Austenite	19.0–22.0	27.5–30.5	2.0–3.0	1.50	1.50	0.04	0.04	0.07	3.0–4.0 Cu

(a) Wrought alloy type numbers are AISI designations for grades most closely corresponding to casting alloys. Wrought alloy type numbers are given only as a guide for determining corresponding cast and wrought grades. Buyers should use cast alloy designations when specifying castings. (b) Maximum unless range is given. All compositions contain balance of iron.

corrosion resistance in many environments and is often used in sulfuric acid service.

Ferritic grades are designated CB-30 and CC-50. Alloy CB-30 is practically nonhardenable by heat treatment. As this alloy is normally made, the balance among the elements in the composition results in a wholly ferritic structure similar to wrought AISI type 442 stainless steel. Alloy CC-50 has substantially more chromium than CB-30 and has relatively high resistance to localized corrosion in many environments.

Austenitic-ferritic alloys include CE-30, CF-3, CF-3A, CF-8, CF-8A, CF-20, CF-3M, CF-3MA, CF-8M, CF-8C, CF-16F, and CG-8M. The microstructures of these alloys usually contain 5 to 40% ferrite, depending on the particular grade and the balance among the ferrite-promoting and austenite-promoting elements in the chemical composition (see the section "Ferrite in Cast Stainless Steels" in this article).

Duplex Alloys. Two duplex alloys—CD-4MCu and Ferralium—are currently of interest. Alloy CD-4MCu is the most highly alloyed duplex alloy. Ferralium was developed by Langley Alloys and is essentially CD-4MCu with about 0.15% N added. With high levels of ferrite (about 40 to 50%) and low nickel, the duplex alloys have better resistance to stress-corrosion cracking (SCC) than CF-3M. Alloy CD-4MCu, which contains no nitrogen and has a relatively low molybdenum content, has only slightly better resistance to localized corrosion than CF-3M. Ferralium, which has nitrogen and slightly higher molybdenum than CD-4MCu, exhibits better localized corrosion resistance than either CF-3M or CD-4MCu.

Improvements in stainless steel production practices (for example, electron beam refining, vacuum and argon-oxygen decarburization, and vacuum induction melting) have created a second generation of duplex stainless steels. These steels offer excellent resistance to pitting and crevice corrosion, significantly better resistance to chloride SCC than the austenitic stainless steels, good toughness, and yield strengths two to three times higher than those of type 304 or 316 stainless steels.

First generation duplex stainless steels, for example, AISI type 329 and CD-4MCu, have been in use for many years. The need for improvement in the weldability and corrosion resistance of these alloys resulted in the second generation alloys, which are characterized by the addition of nitrogen as an alloying element.

Second generation duplex stainless steels are usually about a 50-50 blend of ferrite and austenite. The new duplex alloys combine the near immunity to chloride SCC of the ferritic grades with the toughness and ease of fabrication of the austenitics. Among the second generation duplexes, Alloy 2205 seems to have become the general-purpose stainless. Table 2 lists the nominal compositions of first and second generation duplex alloys.

Precipitation-Hardening Grades. The alloys in this group are CB-7Cu and CD-4MCu. Alloy CB-7Cu is a low-carbon martensitic alloy that may contain minor amounts of retained austenite or ferrite. The copper precipitates in the martensite when the alloy is heat treated to the hardened (aged) condition.

Heat-Resistant High-Alloy Steels

Heat-resistant high-alloy steel castings are extensively used for applications involving service temperatures in excess of 650 °C (1200 °F). Strength at these elevated temperatures is only one of the criteria by which these materials are selected, because applications often involve aggressive environments to which the steel must be resistant. The atmospheres most commonly encountered are air, flue gases, or process gases; such atmospheres may be either oxidizing or reducing and may be sulfidizing or carburizing if sulfur or carbon are present.

Carbon and low-alloy steels seldom have adequate strength and corrosion resistance at elevated temperatures in the environments for which heat-resistant cast steels are normally selected. Only heat-resistant steels exhibit the required mechanical properties and corrosion resistance over long periods of time without excessive or unpredictable degradation. In addition to long-term strength and corrosion resistance, some cast heat-resistant steels exhibit special resistance to the effects of cyclic tem-

Table 2 Nominal compositions of first and second generation duplex stainless steels

UNS designation	Common name	Composition, %(a) Cr	Ni	Mo	Cu	N	Others
First generation steels							
S31500	3RE60	18.5	4.7	2.7	...	...	1.7Si
S32404	Uranus 50	21	7.0	2.5	1.5	...	...
S32900	Type 329	26	4.5	1.5	...	...	...
J93370	CD-4MCu	25	5	2	3	...	...
Second generation steels							
S31200	44LN	25	6	1.7	...	0.15	...
S31260	DP-3	25	7	3	0.5	0.15	0.3W
S31803	Alloy 2205	22	5	3	...	0.15	...
S32550	Ferralium 255	25	6	3	2	0.20	...
S32950	7-Mo PLUS	26.5	4.8	1.5	...	0.20	...
J93404	Atlas 958, COR 25	25	7	4.5	...	0.25	...

(a) All compositions contain balance of iron.

peratures and changes in the nature of the operating environment.

A number of cast high-alloy grades have been developed and successfully used for a variety of service requirements. There are three principal categories, based on composition:

- Iron-chromium alloys
- Iron-chromium-nickel alloys
- Iron-nickel-chromium alloys

These alloy types resemble high-alloy corrosion-resistant steels except for their higher carbon contents, which impart greater strength at elevated temperature. The higher carbon content and, to a lesser extent, alloy composition ranges distinguish cast heat-resistant steel grades from their wrought counterparts. Table 3 summarizes the compositions of standard cast heat-resistant grades.

Iron-chromium alloys contain 8 to 30% Cr and little or no nickel. They are ferritic in structure and exhibit low ductility at ambient temperatures. Iron-chromium alloys are primarily used where resistance to gaseous corrosion is the predominant consideration because they possess relatively low strength at elevated temperatures. Examples of such alloys are the cast HA, HC, and HD grades listed in Table 3.

Iron-chromium-nickel alloys contain more than 18% Cr and more than 8% Ni, with the chromium content always exceeding that of nickel. They exhibit an austenitic matrix, although several grades also contain some ferrite. These alloys exhibit greater strength and ductility at elevated temperatures than those in the iron-chromium group and withstand moderate thermal cycling. Examples of these alloys are the HE, HF, HH, HI, HK, and HL grades listed in Table 3.

Iron-nickel-chromium alloys contain more than 10% Cr and more than 23% Ni, with the nickel content always exceeding that of chromium. These alloys are wholly austenitic and exhibit high strength at elevated temperatures. They can withstand considerable temperature cycling and severe thermal gradients and are well suited to many reducing, as well as oxidizing, environments. Examples of iron-nickel-chromium alloys are the HN, HP, HT, HU, HW, and HX grades listed in Table 3. Even though nickel is the major element in the HW and HX grades, these grades are ordinarily referred to as high-alloy steels rather than nickel-base alloys (see the article "Nickel and Nickel Alloys" in this Volume).

Ferrite in Cast Stainless Steels

The CF alloys constitute the most technologically important and highest-tonnage segment of corrosion-resistant casting production. These 19Cr-9Ni alloys are the cast counterparts of the AISI 300-series wrought stainless steels (Table 1). In general, the cast and wrought alloys possess equivalent resistance to corrosive media, and they are frequently used in conjunction with each other.

Important differences do exist, however, between the cast CF alloys and their wrought AISI counterparts. Most significant among these is the difference in alloy microstructure in the end-use condition. The CF grade cast alloys have duplex structures (Table 1) and usually contain 5 to 40% ferrite, depending on the particular alloy. Their wrought counterparts are fully austenitic. The ferrite in cast stainless with duplex structures is magnetic, a point that is often confusing when cast stainless steels are compared to their wrought counterparts by checking their attraction to a magnet. This difference in microstructures is attributable to the fact that the chemical compositions of the cast and wrought alloys are not identical by intent. The differences in composition will be discussed further in this section.

Significance of Ferrite. Ferrite is intentionally present in cast CF grade stainless steels for three principal reasons: to provide strength, to improve weldability, and to

Table 3 Compositions of cast heat-resistant high-alloy steels

Cast alloy designation	Wrought alloy type(a)	Composition, %(b) C	Mn	Si	P	S	Cr	Ni	Mo
HA	...	0.20	0.35–0.65	1.00	0.04	0.04	8–10	...	0.90–1.20
HC	446	0.50	1.00	2.00	0.04	0.04	26–30	4	0.5(c)
HD	327	0.50	1.50	2.00	0.04	0.04	26–30	4–7	0.5(c)
HE	...	0.20–0.50	2.00	2.00	0.04	0.04	26–30	8–11	0.5(c)
HF	302B	0.20–0.40	2.00	2.00	0.04	0.04	18–23	8–12	0.5(c)
HH	309	0.20–0.50	2.00	2.00	0.04	0.04	24–28	11–14	0.5(c)
HI	...	0.20–0.50	2.00	2.00	0.04	0.04	26–30	14–18	0.5(c)
HK	310	0.20–0.60	2.00	2.00	0.04	0.04	24–28	18–22	0.5(c)
HL	...	0.20–0.60	2.00	2.00	0.04	0.04	28–32	18–22	0.5(c)
HN	...	0.20–0.50	2.00	2.00	0.04	0.04	19–23	23–27	0.5(c)
HP	...	0.35–0.75	2.00	2.50	0.04	0.04	24–28	33–37	0.5(c)
HT	330	0.35–0.75	2.00	2.50	0.04	0.04	15–19	33–37	0.5(c)
HU	...	0.35–0.75	2.00	2.50	0.04	0.04	17–21	37–41	0.5(c)
HW	...	0.35–0.75	2.00	2.50	0.04	0.04	10–14	58–62	0.5(c)
HX	...	0.35–0.75	2.00	2.50	0.04	0.04	15–19	64–68	0.5(c)

(a) Wrought alloy type numbers are listed only as a guide for determining equivalent cast and wrought grades. Buyers should use cast alloy designations when specifying castings. (b) Maximum unless range is given. All compositions contain balance of iron. (c) Molybdenum not intentionally added

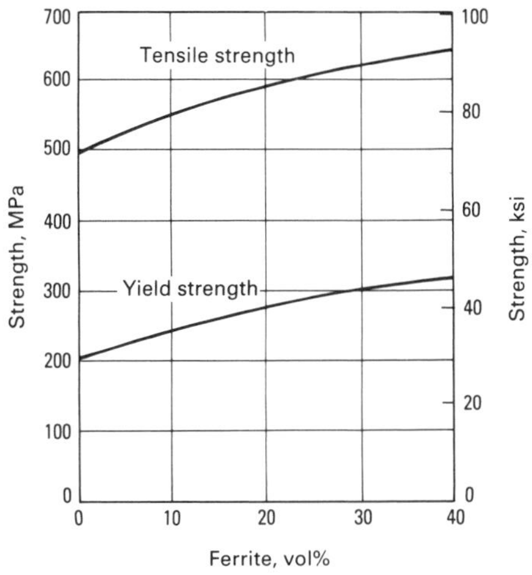

Fig. 2 Yield strength and tensile strength versus percentage of ferrite for CF-8 and CF-8M alloys. Curves are mean values for 277 heats of CF-8 and 62 heats of CF-8M.

maximize resistance to corrosion in specific environments. Strengthening in the cast CF grade alloys is limited essentially to that which can be gained by incorporating ferrite into the austenite matrix phase. These alloys cannot be strengthened by thermal treatment, as with the cast ferritic or martensitic alloys, nor by hot or cold working, as with the wrought austenitic alloys. Strengthening by carbide precipitation is also out of the question because of the detrimental effect of carbides on corrosion resistance in most aqueous environments. Thus, the alloys are effectively strengthened by balancing alloy composition to produce a duplex microstructure consisting of ferrite (up to 40% by volume) distributed in an austenite matrix. It has been shown that the incorporation of ferrite into 19Cr-9Ni cast steels improves yield and tensile strengths substantially without loss of ductility or impact toughness at temperatures below 425 °C (800 °F). The magnitude of this strengthening effect for CF-8 and CF-8M alloys at room temperature is illustrated in Fig. 2.

Fully austenitic stainless steels are susceptible to a weldability problem known as hot cracking or microfissuring. The intergranular cracking occurs in the weld deposit and/or in the weld heat-affected zone and can be avoided if the composition of the filler metal is controlled to produce about 4% ferrite in the austenitic weld deposit. Duplex CF grade alloy castings are immune to this problem.

The presence of ferrite in duplex CF alloys improves resistance to SCC and generally to intergranular attack. Although failures of high-alloy castings due to these two types of corrosion are not common, SCC and intergranular attack are concerns because they can occur unexpectedly, particularly in castings that have been sensitized by welding in the field, where postweld heat treatment to restore corrosion resistance is impractical or impossible. In the case of SCC, the presence of ferrite pools in the austenite matrix is thought to block or make more difficult the propagation of cracks. In the case of intergranular corrosion, ferrite is helpful in sensitized castings because it promotes preferential precipitation of carbides in the ferrite phase rather than at the austenite grain boundaries, where they would increase susceptibility to intergranular attack.

The presence of ferrite also places additional grain boundaries in the austenite matrix, and there is evidence that intergranular attack is arrested at austenite-ferrite boundaries. The most comprehensive study of the effect of ferrite on the corrosion resistance of cast stainless steels indicates that ferrite:

- Improves the resistance of CF alloys to chloride SCC
- Improves resistance of these alloys to intergranular attack
- Affords greater operating safety for the CF alloys with respect to both types of attack at ferrite contents exceeding 10%

It is important to note, however, that not all studies have shown ferrite to be unconditionally beneficial to the general corrosion resistance of cast stainless steels. Whether or not corrosion resistance is improved by ferrite and to what degree depends on the specific alloy composition and heat treatment and the service conditions (environment and stress state).

Ferrite Control. From the preceding discussion, it is apparent that controlled ferrite contents in predominantly austenitic cast chromium-nickel steels, notably, the CF alloys, offer certain property advantages and that the amount of ferrite present will depend primarily on the compositional balance of the alloy. The underlying causes for the dependence of ferrite content on composition are found in the phase equilibria for the iron-chromium-nickel system. These phase equilibria have been exhaustively documented and related to commercial stainless steels.

The major elemental components of cast stainless steels are in competition to promote austenite or ferrite phases in the alloy microstructure. Chromium, silicon, molybdenum, and niobium promote the presence of ferrite in the alloy microstructure; nickel, carbon, nitrogen, and manganese promote the presence of austenite. By balancing the contents of ferrite- and austenite-forming elements within the specified ranges for the elements in a given alloy, it is possible to control the amount of ferrite present in the austenite matrix. The alloy can usually be made fully austenitic or with ferrite contents up to 30% or more in the austenite matrix.

The relationship between composition and microstructure in cast stainless steels permits the foundryman to predict and control the ferrite content of an alloy, as well as its resultant properties, by adjusting the composition of the alloy. This is accomplished with the Schoefer constitution diagram for cast chromium-nickel alloys (Fig. 3). This diagram was derived from an earlier diagram developed by Schaeffler for stainless steel weld metal. Use of Fig. 3 requires that all ferrite-stabilizing elements in the composition be converted into chromium equivalents and that all austenite-stabilizing elements be converted into nickel equivalents by means of empirically derived coefficients representing the ferritizing or austenitizing power of each element. A composition ratio is then obtained from the total chromium equivalent Cr_e and nickel equivalent Ni_e, calculated for the alloy composition according to the following:

$$Cr_e = \%Cr + 1.5(\%Si) + 1.4(\%Mo) + \%Nb - 4.99 \quad \text{(Eq 1)}$$

$$Ni_e = \%Ni + 30(\%C) + 0.5(\%Mn) + 26(\%N - 0.02) + 2.77 \quad \text{(Eq 2)}$$

where the elemental concentrations are given in weight percent. Although similar expressions have been derived that take into account additional alloying elements and different compositional ranges in the iron-chromium-nickel alloy system, use of the Schoefer diagram has become standard for estimating and controlling ferrite content in stainless steel castings.

The Schoefer diagram possesses obvious utility for casting users and the foundryman. It is useful for estimating or predicting

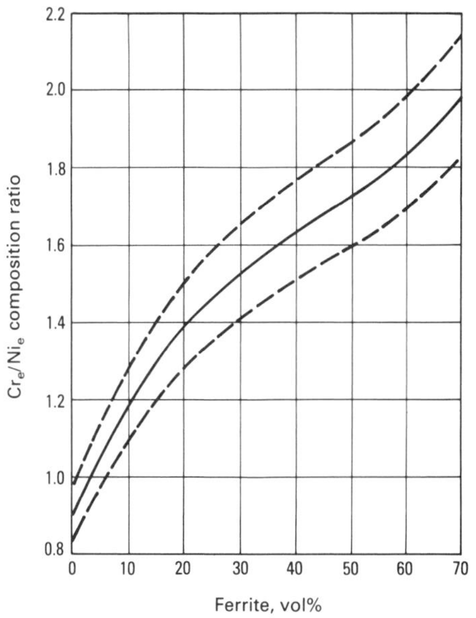

Fig. 3 Schoefer diagram for estimating the ferrite content of steel castings in the composition range 16 to 26% Cr, 6 to 14% Ni, 4% Mo (max), 1% Nb (max), 0.2% C (max), 0.19% N (max), 2% Mn (max), and 2% Si (max). Dashed lines denote scatter bands caused by the uncertainty of the chemical analysis of individual elements. See text for equations used to calculate Cr_e and Ni_e.

Table 4 Room-temperature mechanical properties of cast corrosion-resistant alloys

Alloy	Heat treatment(a)	Tensile strength MPa	Tensile strength ksi	Yield strength (0.2% offset) MPa	Yield strength (0.2% offset) ksi	Elongation in 50 mm (2 in.), %	Reduction in area, %	Hardness, HB	Charpy impact energy J	Charpy impact energy ft · lb	Charpy impact energy Specimen
CA-6NM	>955 °C (1750 °F), AC, T	827	120	689	100	24	60	269	94.9	70	V-notch
CA-15	980 °C (1800 °F), AC, T	793	115	689	100	22	55	225	27.1	20	Keyhole notch
CA-40	980 °C (1800 °F), AC, T	1034	150	862	125	10	30	310	2.7	2	Keyhole notch
CB-7Cu	1040 °C (1900 °F), OQ, A	1310	190	1172	170	14	54	400	33.9	25	V-notch
CB-30	790 °C (1450 °F), AC	655	95	414	60	15	...	195	2.7	2	Keyhole notch
CC-50	1040 °C (1900 °F), AC	669	97	448	65	18	...	210			
CD-4MCu	1120 °C (2050 °F), FC to 1040 (1900), WQ	745	108	558	81	25	...	253	74.6	55	V-notch
	1120 °C (2050 °F), FC to 1040 °C (1900 °F), A	896	130	634	92	20	...	305	35.3	26	V-notch
CE-30	1095 °C (2000 °F), WQ	669	97	434	63	18	...	190	9.5	7	Keyhole notch
CF-3	>1040 °C (1900 °F), WQ	531	77	248	36	60	...	140	149.2	110	V-notch
CF-3A	>1040 °C (1900 °F), WQ	600	87	290	42	50	...	160	135.6	100	V-notch
CF-8	>1040 °C (1900 °F), WQ	531	77	255	37	55	...	140	100.3	74	Keyhole notch
CF-8A	>1040 °C (1900 °F), WQ	586	85	310	45	50	...	156	94.9	70	Keyhole notch
CF-20	>1095 °C (2000 °F), WQ	531	77	248	36	50	...	163	81.4	60	Keyhole notch
CF-3M	>1040 °C (1900 °F), WQ	552	80	262	38	55	...	150	162.7	120	V-notch
CF-3MA	>1040 °C (1900 °F), WQ	621	90	310	45	45	...	170	135.6	100	V-notch
CF-8M	>1065 °C (1950 °F), WQ	552	80	290	42	50	...	170	94.9	70	Keyhole notch
CF-8C	>1065 °C (1950 °F), WQ	531	77	262	38	39	...	149	40.7	30	Keyhole notch
CF-16F	>1095 °C (2000 °F), WQ	531	77	276	40	52	...	150	101.7	75	Keyhole notch
CG-8M	>1040 °C (1900 °F), WQ	565	82	303	44	45	...	176	108.5	80	V-notch
CH-20	>1095 °C (2000 °F), WQ	607	88	345	50	38	...	190	40.7	30	Keyhole notch
CK-20	1150 °C (2100 °F), WQ	524	76	262	38	37	...	144	67.8	50	Izod V-notch
CN-7M	1120 °C (2050 °F), WQ	476	69	214	31	48	...	130	94.9	70	Keyhole notch

(a) AC, air cool; FC, furnace cool; OQ, oil quench; WQ, water quench; T, temper; A, age

ferrite content if the alloy composition is known, and it is useful for setting nominal values for individual elements in calculating the furnace charge for an alloy in which a specified ferrite range is desired.

Limits of Ferrite Control. Although ferrite content can be estimated and controlled on the basis of alloy composition only, there are limits to the accuracy with which this can be done. The reasons for this are many. First, there is an unavoidable degree of uncertainty in the chemical analysis of an alloy (note the scatter band in Fig. 3). Second, the ferrite content depends on thermal history in addition to composition, although to a lesser extent. Third, ferrite contents at different locations in individual castings can vary considerably, depending on section size, ferrite orientation, presence of alloying element segregation, and other factors.

Measurements of ferrite content in stainless steel castings are also subject to significant limitations. Magnetic measurements of ferrite content are limited to small material volumes and require simple casting geometries. In addition, careful calibration with primary and secondary standards is required for accuracy of measurement. Quantitative metallographic determinations of ferrite content on polished surfaces are essentially impossible to conduct in a nondestructive fashion with respect to the casting. The metallographic approach is also quite time consuming, is limited by alloy etching characteristics and microscope resolution, and is complicated by the fact that it is a two-dimensional technique while ferrite pools and colonies in the alloy structure are three-dimensional.

Both the foundryman and user of stainless steel castings should recognize that the factors mentioned above place significant limits on the degree to which ferrite content (either as ferrite number or ferrite percentage) can be specified and controlled in stainless steel castings. In general, the accuracy of ferrite measurement and the precision of ferrite control diminish as the ferrite number increases. As a working rule, it is suggested that ±6 about the mean or desired ferrite number be viewed as a limit of ferrite control under ordinary circumstances, with ±3 possible under ideal circumstances.

Mechanical Properties: Corrosion-Resistant Alloys

The importance of mechanical properties in the selection of corrosion-resistant cast steels is established by the casting application. The paramount basis for alloy selection is normally the resistance of the alloy to the specific corrosive media or environment of interest. The mechanical properties of the alloy are usually, but not always, secondary considerations in these applications. The corrosion resistance of these materials is discussed in detail in the article "Corrosion of Cast Steels" in Volume 13 of the 9th Edition of *Metals Handbook*.

Strength and Hardness. Representative room-temperature tensile properties, hardness, and Charpy impact values for corrosion-resistant alloys are given in Table 4 and Fig. 4. These properties are representative of the alloys rather than the specification requirements. Minimum specified mechanical properties for these alloys are given in ASTM standards A 351, A 743, A 744, and A 747. A wide range of mechanical properties are attainable in high-alloy steel grades, depending on the selection of alloy composition and heat treatment. Tensile strengths ranging from 476 to 1310 MPa (69 to 190 ksi) and hardnesses from 130 to 400 HB are available among the cast corrosion-resistant alloys. Similarly, wide ranges exist in yield strength, elongation, and impact toughness.

The straight chromium steels (CA-15, CA-40, CB-30, and CC-50) possess either martensitic or ferritic microstructures in the end-use condition (Table 1). The CA-15 and CA-40 alloys, which contain nominally 12% Cr, are hardenable through heat treatment by means of the martensite transformation and are often selected as much or more for their high strength as for their comparatively modest corrosion resistance. Castings of these alloys are heated to a temperature at which the structure is fully austenitic and then cooled at a rate (usually in air) adapted to the casting composition so that the austenite transforms to martensite. Strengths in this condition are quite high (for example, 1034 to 1379 MPa, or 150 to 200 ksi), but tensile ductility and impact toughness are limited. Consequently, martensitic castings are usually tempered at 315 to 650 °C (600 to 1200 °F) to restore ductility and toughness at some sacrifice in strength. It follows, then, that significant ranges of tensile properties, hardness, and impact toughness are possible with the martensitic CA-15 and CA-40 grades, depending on the choice of tempering temperature.

The higher-chromium CB-30 and CC-50 alloys, on the other hand, are fully ferritic alloys that are not hardenable by heat treatment. These alloys are generally used in the

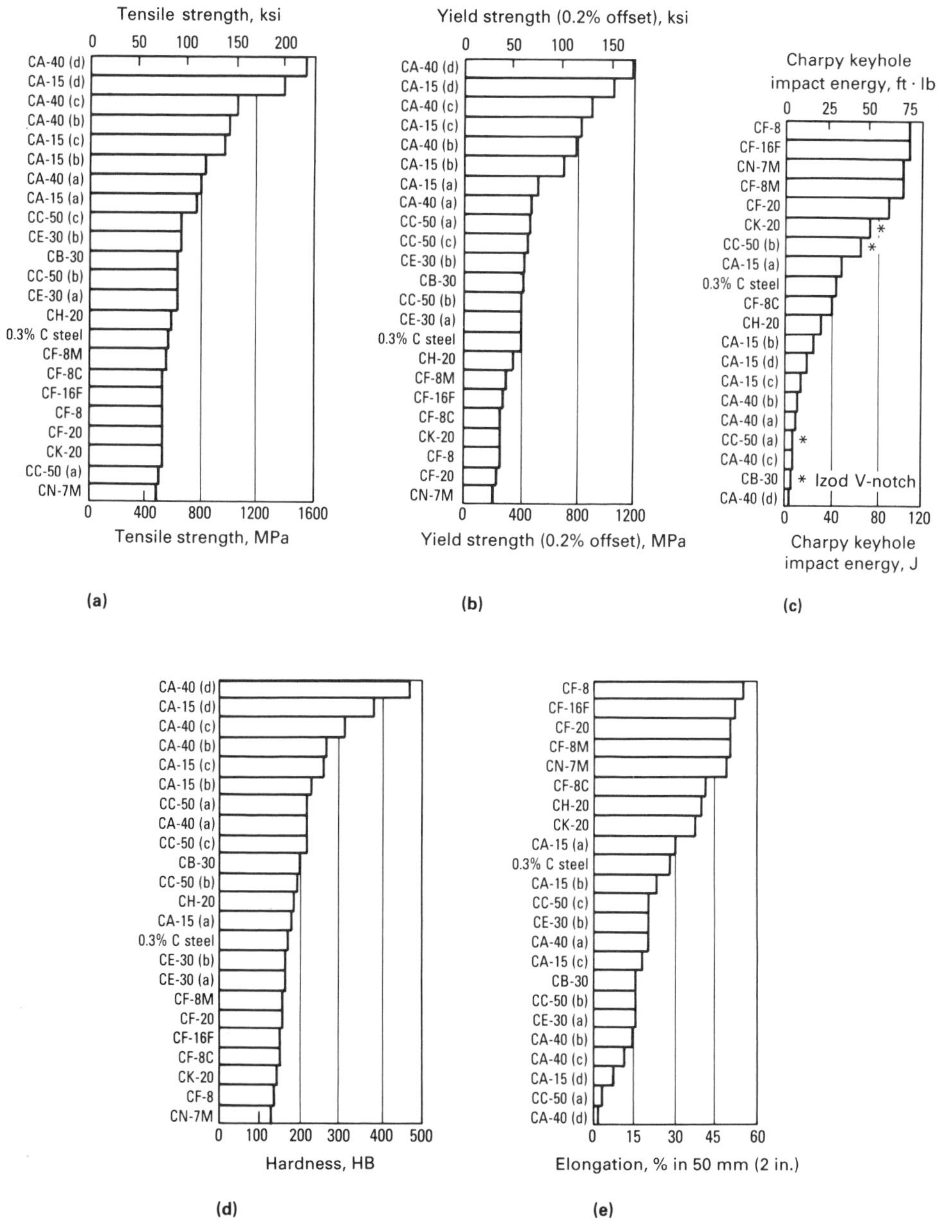

Alloy	Heat treatment	Alloy	Heat treatment
CA-15 (a)	AC from 980 °C (1800 °F), T at 790 °C (1450 °F)	CC-30 (a)	As-cast (<1% Ni)
(b)	AC from 980 °C (1800 °F), T at 650 °C (1200 °F)	(b)	As-cast (>2% Ni; >0.15% N)
(c)	AC from 980 °C (1800 °F), T at 595 °C (1100 °F)	(c)	AC from 1040 °C (1900 °F) (>2% Ni; >0.15% N)
(d)	AC from 980 °C (1800 °F), T at 315 °C (600 °F)	CE-30 (a)	As-cast
CA-40 (a)	AC from 980 °C (1800 °F), T at 760 °C (1400 °F)	(b)	WQ from 1065–1120 °C (1950–2050 °F)
(b)	AC from 980 °C (1800 °F), T at 650 °C (1200 °F)	CF-8	WQ from 1065–1120 °C (1950–2050 °F)
(c)	AC from 980 °C (1800 °F), T at 595 °C (1100 °F)	CF-20	WQ from above 1095 °C (2000 °F)
(d)	AC from 980 °C (1800 °F), T at 315 °C (600 °F)	CF-8M, CF-12M	WQ from 1065–1150 °C (1950–2100 °F)
CB-30	A at 790 °C (1450 °F), FC to 540 °C (1000 °F), AC	CF-8C	WQ from 1065–1120 °C (1950–2050 °F)
		CF-16F	WQ from above 1095 °C (2000 °F)
		CH-20	WQ from above 1095 °C (2000 °F)
		CK-30	WQ from above 1150 °C (2100 °F)
		CN-7M	WQ from above 1065–1120 °C (1950–2050 °F)

Fig. 4 Mechanical properties of cast corrosion-resistant steels at room temperature. (a) Tensile strength. (b) 0.2% offset yield strength. (c) Charpy keyhole impact energy. (d) Brinell hardness. (e) Elongation. Also given are the heat treatments used for test materials: AC, air cool; FC, furnace cool; WQ, water quench; A, anneal; T, temper.

annealed condition and exhibit moderate tensile properties and hardness (Table 4). Like most ferritic alloys, CB-30 and CC-50 possess limited impact toughness, especially at low temperatures.

Three chromium-nickel alloys—CA-6NM, CB-7Cu, and CD-4MCu—are exceptional in their response to heat treatment and the resultant mechanical properties. Alloy CA-6NM is balanced compositionally for martensitic hardening response. This alloy was developed as an alternative to CA-15 and has improved impact toughness and weldability. The CB-7Cu and CD-4MCu alloys both contain copper and can be strengthened by age hardening. These alloys are initially solution heat treated and then cooled rapidly (usually by quenching in oil or water); thus, the phases that would normally precipitate at slow cooling rates cannot form. The casting is then heated to an intermediate aging temperature at which the precipitation reaction can occur under controlled conditions until the desired combination of strength and other properties is achieved. The CB-7Cu alloy possesses a martensitic matrix, while the CD-4MCu alloy possesses a duplex microstructure, consisting of approximately 40% austenite in a ferritic matrix. Alloy CB-7Cu is applied in the aged condition to obtain the benefit of its excellent combination of strength and corrosion resistance, but alloy CD-4MCu is seldom applied in the aged condition, because of its relatively low resistance to SCC in this condition compared to its superior corrosion resistance in the solution-annealed condition.

The CE, CF, CG, CH, CN, and CK alloys are essentially not hardenable by heat treatment. To ensure maximum corrosion resistance, however, it is necessary that castings of these grades receive a high-temperature solution anneal. This treatment consists of holding the casting at a temperature that is high enough to dissolve all chromium carbides, which are damaging to intergranular corrosion resistance, and then cooling them rapidly enough to avoid reprecipitation of the carbides by quenching in water, oil, or air. Although this can be accomplished throughout in the lower-carbon grades (<0.08% C), heavy sections of alloys with higher carbon contents may have carbides present at some distance below the surface where the cooling rate is slow. Subsurface carbides could be exposed by subsequent machining of the casting, resulting in lowered corrosion resistance in service.

By virtue of their microstructures, which are fully austenitic or duplex without significant carbide precipitation, the alloys exhibit generally excellent impact toughness at low temperatures. The tensile strength range represented by these alloys typically extends from 476 to 669 MPa (69 to 97 ksi). As indicated earlier in this article, the alloys

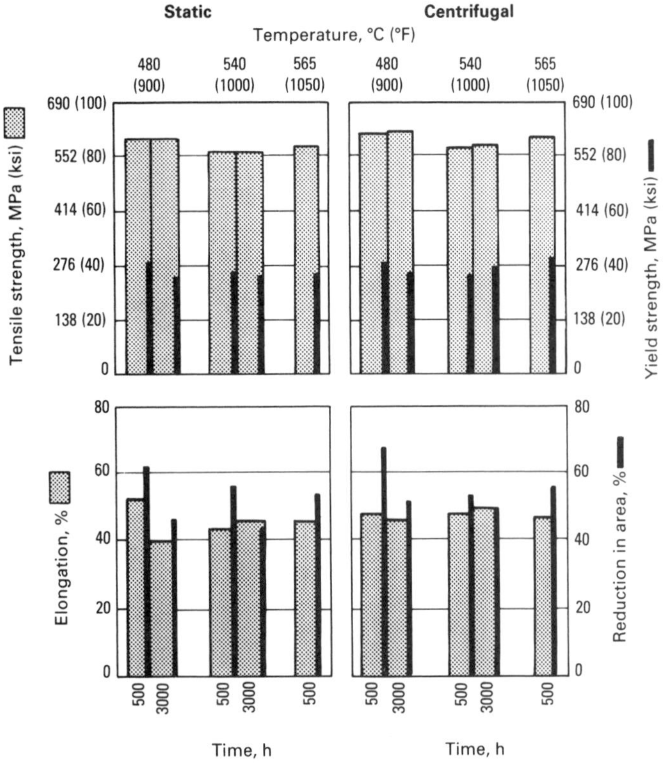

Fig. 5 Effects of time at elevated temperature on the tensile properties of static and centrifugal CF-8 alloy castings. Parts had a ferrite number of 9 to 11 and contained 0.081% N.

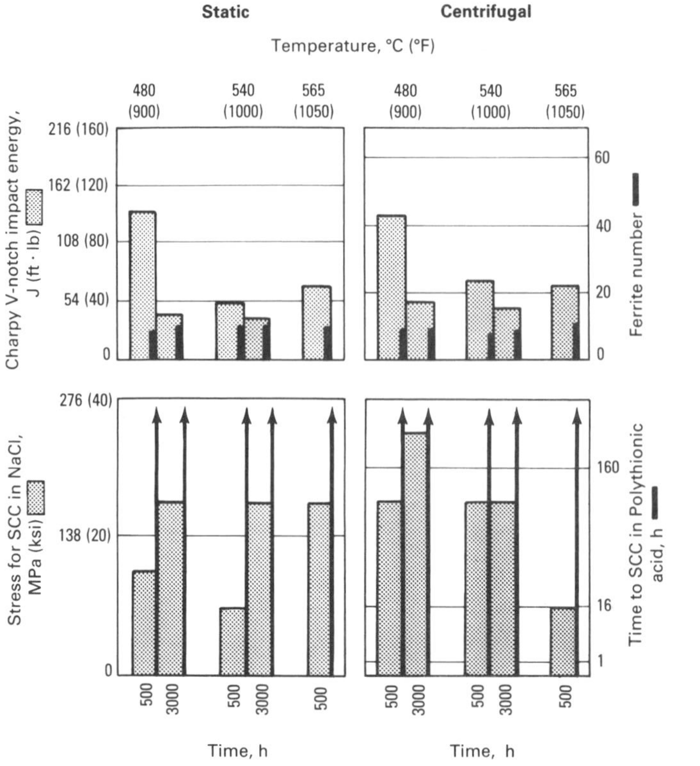

Fig. 6 Effect of time at elevated temperature on the room-temperature impact strength, ferrite number and SCC resistance of CF-8 castings with a ferrite number of 9 to 11 and nitrogen content of 0.081%. Arrows indicate no failure in SCC testing after 336 h.

with duplex structures can be strengthened by balancing the composition for higher ferrite levels. The tensile and yield strength of CF alloys with a ferrite number of 35 are typically 150 MPa (22 ksi) higher than those of fully austenitic alloys.

Fatigue properties can become a design factor in applications in which the casting experiences cyclic loading conditions in service. The resistance of cast stainless steels to fatigue depends on a sizable number of material, design, and environmental factors. For example, design factors of importance include the stress distribution within the casting (residual and applied stresses), the location and severity of stress concentrators (surface integrity), and the environment and service temperature. Material factors of importance include strength and microstructure. It is generally found that fatigue strength increases with the tensile strength of a material. Both fatigue strength and tensile strength generally increase with decreasing temperature. Under equivalent conditions of stress, stress concentration, and strength, evidence suggests that austenitic materials are less notch sensitive than martensitic or ferritic materials.

Toughness. The fully austenitic corrosion-resistant alloys and those with duplex microstructures exhibit excellent toughness. Figure 4 indicates the magnitude of toughness attainable in various grades in Charpy keyhole impact testing.

Effects of Aging. Cast corrosion-resistant high-alloy steels are extensively used at moderately elevated temperatures (up to 650 °C, or 1200 °F). Elevated-temperature properties are important selection criteria for these applications. In addition, room-temperature properties after service at elevated temperatures are increasingly considered because of the aging effect that these exposures may have. For example, cast alloys CF-8C, CF-8M, CE-30A, and CA-15 are currently used in high-pressure service at temperatures to 540 °C (1000 °F) in sulfurous acid environments in the petrochemical industry. Other uses are in the power-generating industry at temperatures to 565 °C (1050 °F).

Room-temperature properties in the aged condition, that is, after exposure to elevated service temperatures, may differ from those in the as-heat-treated condition because of the microstructural changes that may take place at the service temperature. Microstructural changes in iron-nickel-chromium-(molybdenum) alloys may involve the formation of carbides and such phases as σ, χ, and η (Laves). The extent to which these phases form depends on the composition as well as the time at elevated temperature.

The martensitic alloys CA-15 and CA-6NM are subject to minor changes in mechanical properties and SCC resistance in NaCl and polythionic acid environments upon exposure for 3000 h at up to 565 °C (1050 °F). In CF-type chromium-nickel-(molybdenum) steels, only negligible changes in ferrite content occur during 10 000 h exposure at 400 °C (750 °F) and during 3000 h exposure at 425 °C (800 °F). Carbide precipitation, however, does occur at these temperatures, and noticeable Charpy V-notch energy losses have been reported. These effects are illustrated in Fig. 5 and 6 for CF-8 cast corrosion-resistant steel.

Above 425 °C (800 °F), microstructural changes in chromium-nickel-(molybdenum) alloys take place at an increased rate. Carbides and σ phase form rapidly at 650 °C (1200 °F) at the expense of ferrite (Fig. 7). Tensile ductility and Charpy V-notch impact energy are prone to significant losses under these conditions. Density changes, resulting in contraction, have been reported as a result of these high-temperature exposures.

Properties of Heat-Resistant Alloys

Elevated-Temperature Tensile Properties. The short-term elevated-temperature test, in which a standard tension test bar is heated to a designated uniform temperature and then strained to fracture at a standardized rate, identifies the stress due to a short-term overload that will cause fracture in uniaxial loading. The manner in which

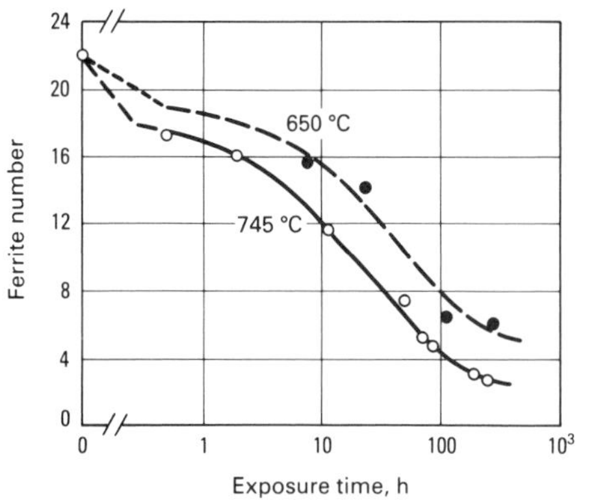

Fig. 7 Transformation of δ-ferrite to austenite and σ phase upon exposure of a solution-treated CF-8 casting to elevated temperature

the values of tensile strength and ductility change with increasing temperature is shown in Fig. 8 for alloy HP-50WZ. Representative tensile properties at temperatures between 650 and 1095 °C (1200 and 2000 °F) are shown in Table 5 for several heat-resistant alloy steel grades.

Creep and Stress Rupture Properties. Creep is defined as the time-dependent strain that occurs under load at elevated temperature. Creep is operative in most applications of heat-resistant high-alloy castings at the normal service temperatures. In time, creep may lead to excessive deformation and even fracture at stresses considerably below those determined in room-temperature and elevated-temperature short-term tension tests. The designer must usually determine whether the serviceability of the component in question is limited by the rate or the degree of deformation. When the rate or degree of deformation is the limiting factor, the design stress is based on the minimum creep rate and design life after allowing for initial transient creep. The stress that produces a specified minimum creep rate of an alloy or a specified amount of creep deformation in a given time (for example, 1% total creep in 100 000 h) is referred to as the limiting creep strength or

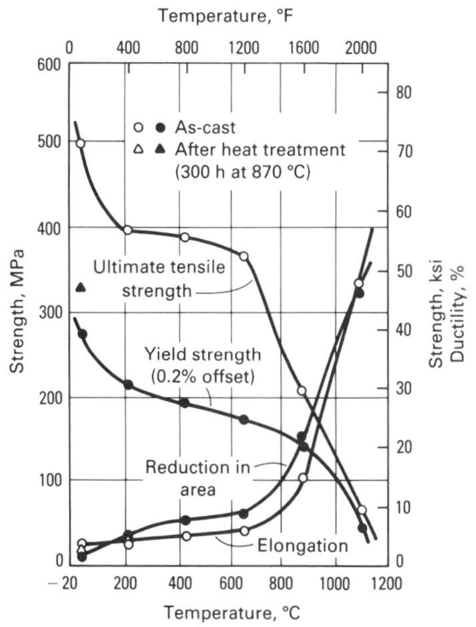

Fig. 8 Tensile properties versus temperature for heat-resistant alloy HP-50WZ

limiting stress. The manner in which the minimum creep rate depends on the applied stress is illustrated in Fig. 9 by data for ACI alloy HP-50WZ.

When fracture is the limiting factor, stress-to-rupture values can be used in design (Fig. 10). The stress-to-rupture values can be combined with those for minimum creep rate, as shown in Fig. 11.

It should be recognized that long-term creep and stress rupture values (for example, 100 000 h) are often extrapolated from shorter-term tests. Whether these property values are extrapolated or determined directly often has little bearing on the operating life of high-temperature parts. The actual material behavior is often difficult to predict accurately because of the complexity of the service stresses relative to the idealized, uniaxial loading conditions in the standardized tests and because of the attenuating factors such as cyclic loading, temperature fluctuations, or metal loss from

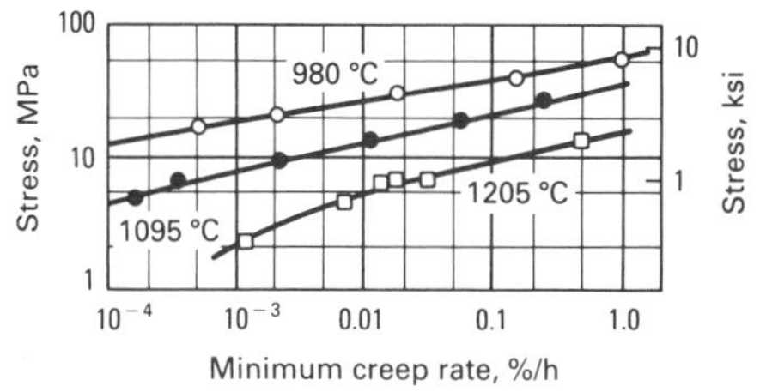

Fig. 9 Minimum creep rate versus stressed temperature for alloy HP-50WZ

corrosion. The designer should anticipate the synergistic effects of these variables.

Thermal Fatigue Resistance. The design of components that are subject to considerable temperature cycling must also include consideration of thermal fatigue. This is particularly true if the temperature changes are frequent or rapid and nonuniform within or between casting sections. Fatigue is a condition in which failure results from alternating load applications in shorter times or at lower stresses than would be expected from constant-load properties. Thermal fatigue denotes the conditions in which the stresses are primarily due to hindered thermal expansion or contraction. Good design helps minimize external restraints to expansion and contraction. Rapid heating and cooling may, however, impose temperature gradients within the part, causing the relatively cool elements of the component to restrain the hotter elements. Finite-element computer analysis has shown that, for some industrial applications, these thermally induced stresses may exceed those resulting from mechanical loads. Nevertheless, such test results have been useful in considering alloy selection questions, identifying the superior thermal fatigue resistance of nickel-containing grades, and documenting the good performance of some HH type compositions.

Thermal Shock Resistance. Thermal shock failure may occur as a result of a single, rapid temperature change or as a result of rapid cyclic temperature changes, which induce stresses that are high enough

Table 5 Representative short-term tensile properties of cast heat-resistant alloys at elevated temperatures

	Property at indicated temperature														
	650 °C (1200 °F)					870 °C (1600 °F)					1095 °C (2000 °F)				
	Ultimate tensile strength		Yield strength		Elongation,	Ultimate tensile strength		Yield strength		Elongation,	Ultimate tensile strength		Yield strength		Elongation,
Alloy	MPa	ksi	MPa	ksi	%	MPa	ksi	MPa	ksi	%	MPa	ksi	MPa	ksi	%
HD	...	...	...	...	...	159	23	...	...	18	...	...	...	...	...
HF	414	60	217	31.5	10	145	21	107	15.5	16	...	...	...	...	...
HH (type I)	...	...	...	...	...	127	18.5	93	13.5	30	...	...	...	...	...
HH (type II)	417	60.5	222	32	14	148	21.5	110	16	18	38	5.5	...	...	...
HI	...	...	...	...	...	179	26	...	...	12	...	...	...	...	...
HK	...	...	...	...	...	161	23	101	15	16	39	5.5	34	5	55
HL	...	...	...	...	...	210	30.5	...	...	...	...	...	...	...	...
HN	...	...	...	...	...	140	20	100	14.5	37	43	6	34	5	55
HP	...	...	...	...	...	179	26	121	17.5	27	52	7.5	43	6	69
HT	292	42.5	193	28	5	130	19	103	15	24	41	6	...	...	...
HU	...	...	...	...	...	135	19.5	...	...	20	...	...	...	...	...
HW	...	...	...	...	...	131	19	103	15	...	...	...	...	...	...
HX	303	45	138	20	8	141	20.5	121	17.5	48	...	...	...	...	...

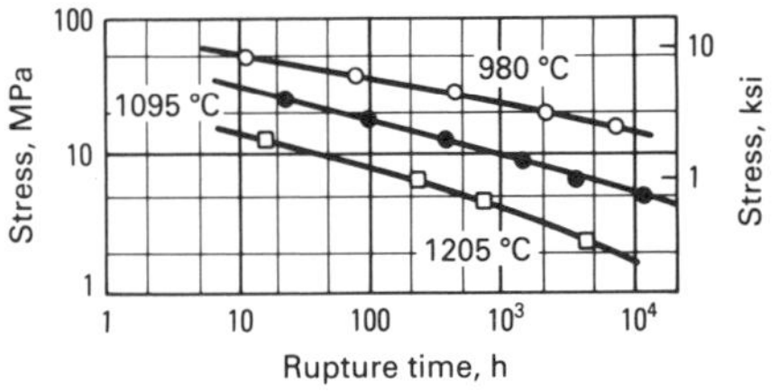

Fig. 10 Rupture time versus stress and temperature for alloy HP-50WZ

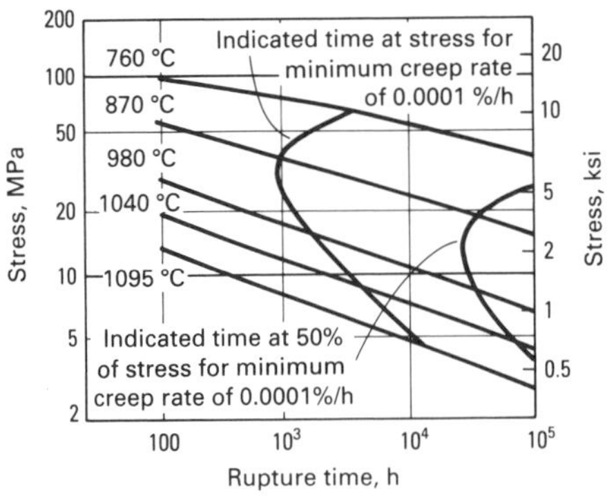

Fig. 11 Creep rupture properties of Alloy HK-40. Scatter bands are ±20% of the central tendency line. Although such a range usually encompasses data for similar alloy compositions, scatter of values may be much higher, especially at longer times and high temperatures.

to cause failure. Brittle ceramic materials are subject to such failure in a single temperature cycle, but only very infrequently are conditions encountered that would cause failure in a single thermal cycle of the high-alloy cast heat-resistant grades under discussion. The nickel-predominating grades are generally the most resistant to thermal shock.

Resistance to Hot Gas Corrosion. The corrosion of heat-resistant alloys, that is, their attack by the environment at elevated temperatures, varies significantly with alloy type, temperature, velocity, and the nature of the specific environment to which the part is exposed. Table 6 presents a general ranking of the standard cast heat-resistant grades in various environments.

Foundry Practice

Foundry practices for cast high-alloy steels are essentially the same as those used for cast plain carbon steels. Details on melting practice, metal treatment, and foundry practices, including gating, risering, and cleaning of castings, are available in the article "Plain Carbon Steels" in this Volume.

Weldability

Corrosion-Resistant High-Alloy Steels. Most of the corrosion-resistant cast stainless steels, such as the CF-8 (the cast equivalent of wrought AISI type 304) or CF-8M (the cast equivalent of wrought AISI type 316), are readily weldable, especially if their microstructures contain small percentages of δ-ferrite. These grades of stainless can become sensitized and lose their corrosion resistance if subjected to temperatures above 425 °C (800 °F). Great care must therefore be used in welding to be certain that the casting or fabricated component is not heated excessively. For this reason, these stainless steels are almost never preheated. In many cases, the weld is cooled with a water spray between passes to reduce the interpass temperature to 150 °C (300 °F) or below.

Any welding performed on the corrosion-resistant grades will affect the corrosion resistance of the casting, but for many services the castings will perform satisfactorily in the as-welded condition. Where extremely corrosive conditions exist or where SCC may be a problem, complete reheat treatment may be required after welding. Heating the casting above 1065 °C (1950 °F) and then cooling it rapidly redissolves the carbides precipitated during the welding operation and restores corrosion resistance.

Where maximum corrosion resistance is desired and postweld heat treatment (solution annealing) cannot be performed, alloying elements can be added to form stable carbides. Although niobium and titanium both form stable carbides, titanium is readily oxidized during the casting operation and therefore is seldom used. The niobium-stabilized grade CF-8C (the cast equivalent of wrought AISI type 347) is the most commonly used cast grade. The stability of the niobium carbides prevents the formation of chromium carbides and the consequent chromium depletion of the base metal. This grade may therefore be welded without postweld heat treatment. Another approach where postweld heat treatment is undesirable or impossible is to keep the carbon content below 0.03%, as in the CF-3 and CF-3M grades. At this low carbon level, the depletion of the chromium due to carbide precipitation is so slight that the corrosion resistance of the grade is unaffected by the welding operation.

As the alloy content of the corrosion-resistant grades is increased to produce a fully austenitic structure, welding without cracking becomes more difficult. The fully austenitic low-carbon grades tend to form microfissures adjacent to the weld. This tendency toward microfissuring increases as nickel and silicon contents increase and carbon content decreases. Microfissuring is most evident in coarse-grain alloys with carbon contents of approximately 0.10 to 0.20% and nickel contents exceeding 13%. The microfissuring is reduced by extremely low sulfur contents. In welding these grades, low interpass temperatures, low heat inputs, and peening of the weld to relieve mechanical stresses are all effective. Where strength is not a great factor, an initial weld deposit or "buttering of the weld" with AISI type 304 material is also occasionally used.

The heat-resistant grades are also subject to microfissuring in heavy sections, but to a lesser extent than the high-nickel lower-carbon corrosion grades. The same precautions used for the high-nickel corrosion grades (low heat input, low interpass temperature, and peening when required) should be exercised.

Table 6 Corrosion resistance of heat-resistant cast steels at 980 °C (1800 °F) in 100 h tests in various atmospheres

Alloy	Corrosion rating(a) in indicated atmosphere: Air	Oxidizing flue gas(b)	Reducing flue gas(b)	Reducing flue gas(c)	Reducing flue gas (constant temperature)(d)	Reducing flue gas cooled to 150 °C (300 °F) every 12 h(d)
HA	U	U	U	U	U	U
HC	G	G	G	S	G	G
HD	G	G	G	S	G	G
HE	G	G	G	. . .	G	. . .
HF	S	G	S	U	S	S
HH	G	G	G	S	G	G
HI	G	G	G	S	G	G
HK	G	G	G	U	G	G
HL	G	G	G	S	G	G
HN	G	G	G	U	S	S
HP	G	G	G	G	G	. . .
HT	G	G	G	U	S	U
HU	G	G	G	U	S	U
HW	G	G	G	U	U	U
HX	G	G	G	S	G	U

(a) G, good (corrosion rate $r < 1.27$ mm/yr, or 50 mils/yr); S, satisfactory ($r < 2.54$ mm/yr, or 100 mils/yr); U, unsatisfactory ($r > 2.54$ mm/yr, or 100 mils/yr). (b) Contained 2 g of sulfur/m³ (5 grains S/100 ft³). (c) Contained 120 g S/m³ (300 grains S/100 ft³). (d) Contained 40 g S/m³ (100 grains S/100 ft³)

Heat Treatment

The heat treatment of stainless steel castings is very similar in purpose and procedure to the thermal processing of comparable wrought materials (see the article "Heat Treating of Stainless Steels" in Volume 4 of the 9th Edition of *Metals Handbook*). However, the differences in detail warrant separate consideration here.

Martensitic alloys CA-15 and CA-40 do not require subcritical annealing to remove the effects of cold working. However, in work-hardenable ferritic alloys, machining and grinding stresses are relieved at temperatures from about 260 to 540 °C (500 to 1000 °F). Casting stresses in the martensitic alloys noted above should be relieved by subcritical annealing prior to further heat treatment. When these hardened martensitic castings are stress relieved, the stress-relieving temperature must be kept below the final tempering or aging temperature.

Alloy CA-6NM (UNS J91540) possesses better casting behavior and improved weldability, it equals or exceeds all of the mechanical, corrosion, and cavitation resistance properties of CA-15, and it has largely replaced the older alloy. Both CA-6NM and CA-15 castings are normally supplied in the normalized condition at 955 °C (1750 °F) minimum and tempered at 595 °C (1100 °F) minimum. However, when it is necessary or desirable to anneal CA-6NM castings, a temperature of 790 to 815 °C (1450 to 1500 °F) should be used. The alloy should be furnace cooled or otherwise slow cooled to 595 °C (1100 °F), after which it can be air cooled. When stress relieving is required, CA-6NM can be heated to 620 °C (1150 °F) maximum, followed by slow cooling to prevent martensite formation.

Homogenization. Alloy segregation and dendritic structures may occur in castings and may be particularly pronounced in heavy sections. Because castings are not subjected to the high-temperature mechanical reduction and soaking treatments involved in the mill processing of wrought alloys, it is frequently necessary to homogenize some alloys at temperatures above 1095 °C (2000 °F) to promote uniformity of chemical composition and microstructure. The full annealing of martensitic castings results in recrystallization and maximum softness, but it is less effective than homogenization in eliminating segregation. Homogenization is a common procedure in the heat treatment of precipitation-hardening castings.

Ferritic and Austenitic Alloys. The ferritic, austenitic, and mixed ferritic-austenitic alloys are not hardenable by heat treatment. They can be heat treated to improve their corrosion resistance and machining characteristics. The ferritic alloys CB-30 and CC-50 are annealed to relieve stresses and to reduce hardness by heating above 790 °C (1450 °F).

The austenitic alloys achieve maximum resistance to intergranular corrosion by solution annealing. As-cast structures, or castings exposed to temperatures from 425 to 870 °C (800 to 1600 °F), may contain complex chromium carbides precipitated preferentially along grain boundaries in wholly austenitic alloys. This microstructure is susceptible to intergranular corrosion, especially in oxidizing solutions. (In partially ferritic alloys, carbides tend to precipitate in the discontinuous ferrite pools; thus, these alloys are less susceptible to intergranular attack.) The purpose of solution annealing is to ensure the complete solution of carbides in the matrix and to retain these carbides in solid solution.

Solution-annealing procedures for all austenitic alloys are similar and consist of heating to a temperature of about 1095 °C (2000 °F), holding for a time sufficient to accomplish complete solution of carbides, and quenching at a rate fast enough to prevent reprecipitation of the carbides—particularly while cooling through the range from 870 to 540 °C (1600 to 1000 °F). The temperatures to which castings should be heated prior to quenching vary somewhat, depending on the alloy.

A two-step heat-treating procedure can be applied to the niobium-containing CF-8C alloy. The first treatment consists of solution annealing. This is followed by a stabilizing treatment at 870 to 925 °C (1600 to 1700 °F), which precipitates niobium carbides, prevents formation of the damaging chromium carbides, and provides maximum resistance to intergranular attack.

Because of their low carbon contents, CF-3 and CF-3M as-cast do not contain enough chromium carbides to cause selective intergranular attack; therefore, these alloys can be used in some environments in this condition. However, for maximum corrosion resistance, these grades require solution annealing.

Martensitic Alloys. Alloy CA-6NM should be hardened by air cooling or oil quenching from a temperature of 1010 to 1065 °C (1850 to 1950 °F). Even though the carbon content of this alloy is lower than that of CA-15, this fact in itself and the addition of molybdenum and nickel enable the alloy to harden completely without significant austenite retention when cooled as suggested.

The choice of cooling medium is primarily determined by the maximum section size. Section sizes exceeding 125 mm (5 in.) will harden completely when cooled in air. Alloy CA-6NM is not susceptible to cracking during cooling from elevated temperatures. For this reason, no problem should arise in the air cooling or oil quenching of configurations that include thick as well as thin sections.

A wide selection of mechanical properties is available through the choice of tempering temperatures. Alloy CA-6NM is normally supplied, normalized, and tempered at 595 to 620 °C (1100 to 1150 °F). Reaustenitizing occurs upon tempering above 620 °C (1150 °F); the amount of reaustenitization increases with temperature. Depending on the amount of this transformation, cooling from such tempering temperatures may adversely affect both ductility and toughness through the transformation to untempered martensite.

Even though the alloy is characterized by a decrease in impact strength when tempered in the range of 370 to 595 °C (700 to 1100 °F), the minimum reached is significantly higher than that of CA-15. This improvement in impact toughness results from the presence of molybdenum and nickel in the composition and from the lower carbon content. The best combination of strength with toughness is obtained when the alloy is tempered above 510 °C (950 °F).

The minor loss of toughness and ductility that does occur is associated with the lesser degree of tempering that takes place at the lower temperature and not with embrittlement, as might be the situation with other 12% Cr steels that contain no molybdenum. The addition of molybdenum to 12% Cr steels makes them unusually stable thermally and normally not susceptible to embrittlement in the annealed or annealed and cold-worked conditions, even when exposed for long periods of time at 370 to 480 °C (700 to 900 °F). No data are currently available on such steels in the quenched-and-tempered or normalized-and-tempered conditions.

The hardening procedures for CA-15 castings are similar to those used for the comparable wrought alloy (type 410). Austenitizing consists of heating to 955 to 1010 °C (1750 to 1850 °F) and soaking for at least 30 min; the high side of this temperature range is normally employed. Parts are then cooled in air or quenched in oil. To reduce the probability of cracking in the brittle, untempered martensitic condition, tempering should take place immediately after quenching.

Tempering is performed in two temperature ranges: up to 370 °C (700 °F) for maximum strength and corrosion resistance, and from 595 to 760 °C (1100 to 1400 °F) for improved ductility at lower strength levels. Tempering in the range of 370 to 595 °C (700 to 1100 °F) is normally avoided because of the resultant low impact strength.

In the hardened-and-tempered condition, CA-40 provides higher tensile strength and lower ductility than CA-15 tempered at the same temperature. Both alloys can be annealed by cooling slowly from the range 845 to 900 °C (1550 to 1650 °F).

Precipitation-Hardening Alloys. It is desirable to subject precipitation-hardenable castings to a high-temperature homogenization treatment to reduce alloy segregation and to obtain more uniform response to subsequent heat treatment. Even investment castings that are slowly cooled from the pouring temperature exhibit more nearly uniform properties when they have been homogenized.

Applications of C-Type Alloys

Martensitic grades include CA-15, CA-40, CA-15M, and CA-6NM. Alloy CA-15 contains the minimum amount of chromium necessary to make it essentially rustproof. It has good resistance to atmospheric corrosion as well as to many organic media in relatively mild service. Alloy CA-40 is a higher-carbon modification of CA-15 that can be heat treated to higher strength and hardness levels. A molybdenum-containing modification of CA-15, alloy CA-15M provides improved elevated-temperature strength properties. Alloy CA-6NM is an iron-chromium-nickel-molybdenum alloy of low carbon content. The presence of nickel offsets the ferritizing effect of the low carbon content so that strength and hardness properties are comparable to those of CA-15 and impact strength is substantially improved. The molybdenum addition improves the resistance of the alloy in seawater.

A wide range of mechanical properties can be obtained in the martensitic alloy group. Tensile strengths from 620 to 1520 MPa (90 to 220 ksi) and hardnesses as high as 500 HB can be obtained through heat treatment. The alloys have fair to good weldability and machinability if proper techniques are employed; CA-40 is considered the poorest and CA-6NM is the best in this regard. The martensitic alloys are used in pumps, compressors, valves, hydraulic turbines, propellers, and machinery components.

Austenitic grades include Alloys CH-20, CK-20, and CN-7M. The CH-20 and CK-20 alloys are high-chromium, high-carbon, wholly austenitic compositions in which the chromium exceeds the nickel content. They have better resistance to dilute sulfuric acid than CF-8 and have improved strength at elevated temperatures. These alloys are used for specialized applications in the chemical-processing and pulp and paper industries for handling pulp solutions and nitric acid. The high-nickel CN-7M grade containing molybdenum and copper is widely used for handling hot sulfuric acid. This alloy also offers resistance to dilute hydrochloric acid and hot chloride solutions. It is used in steel mills as containers for nitric-hydrofluoric pickling solutions and in many industries for severe-service applications for which the high-chromium CF-type alloys are inadequate.

Ferritic grades are designated CB-30 and CC-50. Alloy CB-30 is practically nonhardenable by heat treatment. As this alloy is normally made, the balance among the elements in the composition results in a wholly ferritic structure similar to that of wrought type 442 stainless steel. By balancing the composition toward the low end of the chromium and the high ends of the nickel and carbon ranges, however, some martensite can be formed through heat treatment, and the properties of the alloy approach those of the hardenable wrought type 431. Alloy CB-30 castings have greater resistance to most corrosives than the CA grades and are used for valve bodies and trim in general chemical production and food processing. Because of its low impact strength, however, CB-30 has been supplanted in many applications by the higher-nickel-containing austenitic grades of the CF type. The high-chromium CC-50 alloy has good resistance to oxidizing corrosives, mixed nitric and sulfuric acids, and alkaline liquors. It is used for castings in contact with acid mine waters and in nitrocellulose production. For best impact strength, the alloy is made with more than 2% Ni and more than 0.15% N.

Austenitic-ferritic grades include CE-30, CF-3, CF-3A, CF-8, CF-8A, CF-20, CF-3M, CF-3MA, CF-8M, CF-8C, CF-16F, and CG-8M. These alloys usually contain 5 to 40% ferrite, depending on the particular grade and the balance among the ferrite-promoting and austenite-promoting elements in the chemical composition. This ferrite content improves the weldability of the alloys and increases their mechanical strength and resistance to SCC. The amount of ferrite in a corrosion-resistant casting can be estimated from its composition by using the Schoefer diagram (Fig. 3) or from its response to magnetic measuring instruments.

Alloy CE-30 is a high-carbon high-chromium alloy that has good resistance to sulfurous acid and can be used in the as-cast condition. It has been extensively used in the pulp and paper industry for castings and welded assemblies that cannot be effectively heat treated. A controlled ferrite grade designated CE-30A, is used in the petroleum industry for its high strength and resistance to SCC in polythionic acid.

The CF alloys as a group constitute the major segment of corrosion-resistant casting production. When properly heat treated, the alloys are resistant to a great variety of corrosives and are usually considered the best general-purpose types. They have good castability, machinability, and weldability and are tough and strong at temperatures down to −255 °C (−425 °F). Alloy CF-8, the cast equivalent of AISI type 304 stainless steel, can be viewed as the base grade, and all the others as variants of this basic type. The CF-8 alloy has excellent resistance to nitric acid and all strongly oxidizing conditions. The higher-carbon CF-20 grade is satisfactorily used for less corrosive service than that requiring CF-8, and the low-carbon type CF-3 is specifically designed for use where castings are to be welded without subsequent heat treatment.

The molybdenum-containing grades CF-8M and CF-3M have improved resistance to reducing chemicals and are used to handle dilute sulfuric and acetic acids, paper mill liquors, and a wide variety of industrial corrosives. Alloy CF-8M has become the most frequently used grade for corrosion-resistant pumps and valves because of its versatility in meeting many corrosive service demands. Because CF-3M has a low carbon content, it can be used without heat treatment after welding. The niobium-stabilized CF-8C alloy is the cast equivalent of AISI type 347. Castings of this alloy, therefore, are used to resist the same corrosives as CF-8 but where field welding or service temperatures of 650 °C (1200 °F) are involved.

Higher mechanical properties are specified for grades CF-3A, CF-8A, and CF-3MA than for the CF-3, CF-8, and CF-3M alloys because the compositions are balanced to provide a controlled amount of ferrite that will ensure the required strength. These alloys are being used in nuclear power plant equipment.

The CF-16F grade has an addition of selenium to improve the machinability of castings that require extensive drilling, threading, and the like. It is used in service similar to that for which CF-20 is used. Type CG-8M has a higher molybdenum content than CF-8M and is preferred to the latter in service where improved resistance to sulfuric and sulfurous acid solutions and to the pitting action of halogen compounds is needed. Unlike CF-8M, however, it is not suitable for use in nitric acid or other strongly oxidizing environments.

The duplex alloys have higher yield strength than the austenitics. This difference gives the duplex alloys an economic edge in, for example, the chemical process industry; higher process flow rates and operating pressures are possible without a major equipment modification. Cost savings can also be realized when the higher strength allows the downgaging of the wall thickness of piping, heat exchanger tubing, tanks, columns, and pressure vessels. For rotating equipment, such as centrifuges, the mass of equipment can be reduced by using a duplex stainless steel. Further savings in motors and gearing results because of smaller loads. For some time, cast duplex valves and pumps have utilized the strength of these materials either to allow higher pressures or to lower costs by using thinner walls.

Precipitation-hardening alloys include CB-7Cu and CD-4MCu. Alloy CB-7Cu is a low-carbon martensitic alloy that may contain minor amounts of retained austenite or ferrite. The corrosion resistance of CV-7Cu lies between that of the CA types and the nonhardenable CF alloys, so it is used when both high strength and improved corrosion resistance are required. Castings of CB-7Cu are machined in the solution-treated condition and then through hardened by a low-temperature aging treatment (480 to 595 °C, or 900 to 1000 °F). Because of this capability, the CB-7Cu grade has found wide application in highly stressed, machined castings in the aircraft and food-processing industries.

Type CD-4MCu is a two-phase alloy with an austenite-ferrite structure that, because

Table 7 Compositions of austenitic manganese steels

Grade	Composition, % C	Mn	Cr	Mo	Other
Standard	1.0–1.4	12.0–14.0	. . .	. . .	. . .
Chromium	1.0–1.4	12.0–14.0	1.5–2.5	. . .	. . .
1% Mo	0.8–1.3	12.0–15.0	. . .	0.8–1.2	. . .
1% Mo (lean)	1.1–1.4	5.0–7.0	. . .	0.8–1.2	. . .
2% Mo	1.0–1.5	12.0–15.0	. . .	1.8–2.2	. . .
High yield strength	0.4–0.7	12.0–15.0	. . .	1.8–2.2	2.0–4.0Ni, 0.5–1.0V
Machinable	0.3–0.6	18.0–20.0	. . .	. . .	2.0–4.0Ni, 0.2–0.4Bi

of its high chromium and low carbon contents, does not develop martensite when heat treated. Like the CB-7Cu grade, CD-4MCu can be hardened by a low-temperature aging treatment, but it is normally used in the solution-annealed condition. In this condition, its strength is double that of the CF grades, and its corrosion resistance is optimized. This alloy has corrosion resistance equal to, or better than, the CF types and has excellent resistance to SCC in chloride-containing media. It is highly resistant to sulfuric and nitric acids and is used for pumps, valves, and stressed components in the marine, chemical, textile, and pulp and paper industries, for which a combination of superior corrosion resistance and high strength is essential.

Applications of H-Type Alloys

The iron-chromium alloys include grades HA, HC, and HD. Alloy HA has limited application because of its low strength and limited resistance to gaseous corrosion at high temperature. It has been used in valves, flanges, and fittings where light stresses are encountered.

Alloys HC and HD can be used for load-bearing applications up to 650 °C (1200 °F) and where only light loads are involved up to 1040 °C (1900 °F). These grades, however, become embrittled by σ phase in the 650 to 870 °C (1200 to 1600 °F) temperature range. They are similar to each other in corrosion resistance, with HD exhibiting higher strength. Both are used in ore-roasting furnaces for such parts as rabble arms and blades, for salt pots and grate bars, and in high-sulfur applications in which high strength is not required.

The iron-chromium-nickel alloys include HE, HF, HH, HI, HK, and HL. They are predominantly or completely austenitic and exhibit greater strength and ductility than the iron-chromium alloys.

Alloy HE has excellent corrosion resistance at high temperatures coupled with moderate strength. This combination of corrosion resistance and strength makes HE suitable for service to 1095 °C (2000 °F). Grade HE can be used in high-sulfur applications and is often found in ore-roasting and steel mill furnaces. The alloy is prone to σ-phase formation, however, at temperatures of 650 to 870 °C (1200 to 1600 °F).

Alloy HF is essentially immune to σ-phase formation and can be used at temperatures to 870 °C (1600 °F). It is used for tube supports and beams in oil refinery heaters and in cement kilns, ore-roasting ovens, and heat-treating furnaces.

Grade HH exhibits high strength and excellent resistance to oxidation at temperatures to 1095 °C (2000 °F). Its composition can be balanced to yield a partially ferritic or a completely austenitic structure and a wide range of properties. Because of this, the composition of the HH alloy should be tailored to the application. The partially ferritic alloy, type I, has a somewhat lower creep strength and a higher ductility at elevated temperature than the wholly austenitic alloy type II. Type I is also more prone to σ-phase formation between 650 and 870 °C (1200 and 1600 °F). Type II is preferred for application in this temperature range. Both types are widely used for furnace parts of many kinds, but are not recommended for severe-temperature cycling service, such as that experienced by quenching fixtures.

Grade HI is similar to the fully austenitic alloy HH, but its higher chromium content confers sufficient scaling resistance for use up to 1180 °C (2150 °F). Its major application has been in cast retorts for calcium and magnesium production.

Alloy HK has high creep and rupture strengths and can be used in structural applications to 1150 °C (2100 °F). Its resistance to hot gas corrosion is excellent. It is often used for furnace rolls and parts as well as for steam reformer and ethylene pyrolysis tubing.

Grade HL has properties similar to those of HK and exhibits the best resistance to corrosion in high-sulfur environments to 980 °C (1800 °F) of the alloys in this group. It is typically used in gas dissociation equipment.

The iron-nickel-chromium alloys include grades HN, HP, HT, HU, HW, and HX. These materials employ nickel as a predominant alloying (or base) element and remain austenitic throughout their temperature range of application. They are generally suitable for use to 1150 °C (2100 °F) and resist thermal fatigue and shock induced by severe temperature cycling. However, the nickel-base grades are not considered suitable for high-sulfur environments.

Alloy HN has properties similar to those of HK. It is employed in brazing fixtures, furnace rolls, and parts.

Grade HP is extremely resistant to oxidizing and carburizing atmospheres. It has good strength in the temperature range of 900 to 1095 °C (1650 to 2000 °F) and is often specified for heat treat fixtures, radiant tubes, and coils for ethylene pyrolysis heaters.

Alloy HT can withstand oxidizing conditions to 1150 °C (2100 °F) and reducing conditions to 1095 °C (2000 °F). It is widely used to heat treat furnace parts subject to cyclic heating, such as rails, rolls, disks, chains, boxes, pots, and fixtures. It has also found application for glass rolls, enameling racks, and radiant tubes.

Alloy HU has excellent resistance to hot gas corrosion and thermal fatigue, and it has good high-temperature strength. It is often used for severe applications, such as burner tubes, lead and cyanide pots, retorts, and furnace rolls.

Grades HW and HX are extremely resistant to oxidation, thermal shock, and fatigue. Their high electrical resistivities make them suitable for the production of cast electrical heating elements. Both are highly resistant to carburization when in contact with tempering and cyaniding salts. The higher alloy content of HX confers better gas corrosion resistance, particularly in reducing gases containing sulfur, where HW is not recommended. Both grades are typically used for hearths, mufflers, retorts, trays, burner parts, enameling fixtures, quenching fixtures, and containers for molten lead.

Austenitic Manganese Steels

The compositions of the austenitic manganese steels (Table 7) can be varied to achieve differing combinations of strength, ductility, wear resistance, and machinability (see ASTM A128). Manganese steel castings are used in all sizes for a wide variety of applications.

The chromium-alloyed manganese steels are the most extensively used, following the standard A128 grade. Medium-thickness (50 to 125 mm, or 2 to 5 in.) crusher castings generally exhibit increased wear life, which is attributable to the chromium alloying. In some installations, however, it is difficult to substantiate the degree of wear improvement, and lowered ductility can lead to premature breakage in severe service. Figure 12 shows that the tensile elongation in 150 mm (6 in.) thick chromium-manganese steels is 30 to 40% lower than that of the standard grades. Impact property comparisons (Fig. 13) also show a decided impairment.

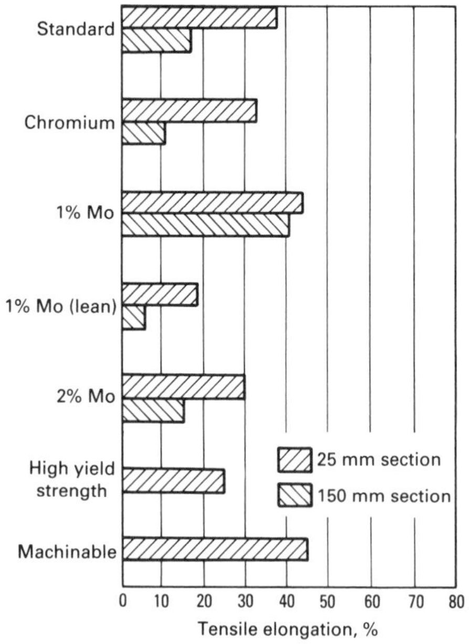

Fig. 12 Typical tensile elongations of austenitic manganese steels

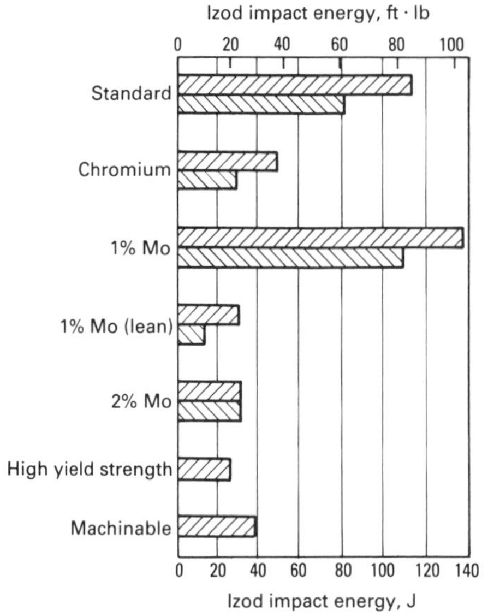

Fig. 13 Typical Izod impact energies of austenitic manganese steels. See Fig. 12 for key.

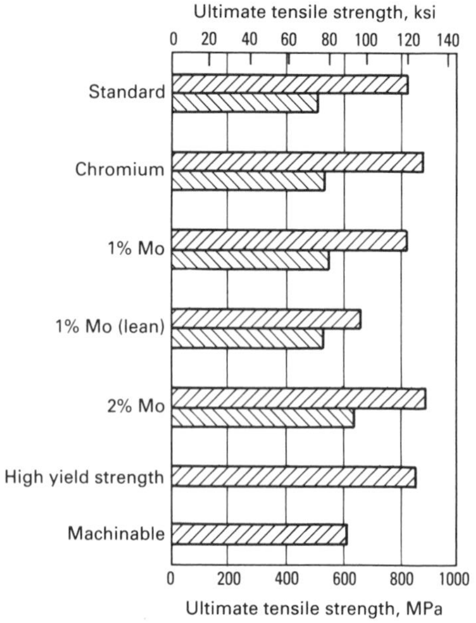

Fig. 14 Typical values of ultimate tensile strength for austenitic manganese steels. See Fig. 12 for key.

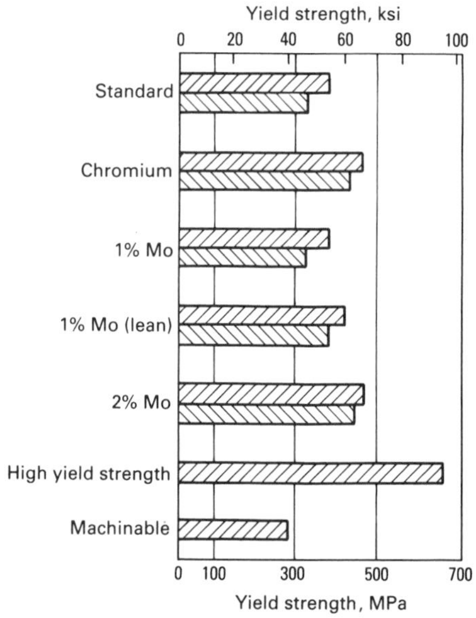

Fig. 15 Typical yield strength values for austenitic manganese steels. See Fig. 12 for key.

The molybdenum-alloyed manganese steels can be divided into two groups: those with 1% Mo and those with 2% Mo. The 1% Mo category is further divided into a normal alloy segment and a so-called lean alloy category. The lean alloy austenitic grade has manganese contents from 5 to 7%; this contrasts with the 12 to 14% level usually associated with austenitic manganese steels.

A normal 1% Mo alloyed grade can involve carbon levels as low as 0.80% and as high as 1.30%. Each of the extremes of the carbon range provides a distinct benefit. Heavy-section mechanical property improvement and improved resistance to mechanical property degradation as a result of elevated temperatures can lead to the use of lower-carbon materials. The higher-carbon grades can be used where improved abrasive wear resistance is needed.

The lean grade of the 1% Mo steels finds limited use in crushing applications, in which lower ductility and lower impact levels are permissible. The more rapid work-hardening tendency of the material has been claimed to account for improved wear life in certain applications, such as ball mills and rod mills.

The 2% Mo grades follow the more classical lines in that manganese contents tend to fall within the 12 to 15% range. Carbon content can vary from 1 to 1.5% in alloys that may either undergo a standard austenitizing heat treatment of a special two-step dispersion-hardening treatment. Before the availability of the multiple-alloyed, high yield strength manganese steels, the 2% Mo grades were used in the dispersion-hardened condition, in which they develop high yield strengths. The desirability of high yield strength is evident in such casting applications as large jaw plates and concave segments; this service results in a high degree of kneading of the working surface, which in turn develops extensive surface expansion. The cumulative effect of this growth can be sufficient to raise the casting from its seat or backing plate. In extreme cases, the wear castings can be physically displaced or can crack restraining structures. The 2% Mo grade can often provide a discernible improvement in overall wear life if premature failure is not encountered. However, economic considerations become involved to an extent that judgment must be exercised regarding the added cost in relation to additional wear life.

The machinable grade of manganese steel listed in Table 7 plays a somewhat parallel role, but in a different property area. Austenitic manganese steels exhibit poor machinability, certainly with regard to the drilling and tapping of small holes. The machinable grade, however, exhibits drilling and tapping advantages at a level somewhat higher than that of a wrought type 308 stainless steel.

Mechanical properties for the compositions shown in Table 7 are illustrated in Fig. 12 to 15. Property impairment as a result of increasing metal section and the reduced response to heat treatment is indicated. All grades, with the exception of the lean austenitic and the machinable grades, approach or exceed 827 MPa (120 ksi) in ultimate tensile strength in 25 mm (1 in.) sections. The chromium-alloyed grades and the 2% Mo grades show yield strengths of 414 MPa (60 ksi) or greater, while the high yield strength alloy develops 655 MPa (95 ksi) in 25 mm (1 in.) metal sections. Elongation and impact strength data favor the 1% Mo grades in heavy metal sections, while the 1% Mo lean alloys exhibit the poorest properties in this regard.

Low magnetic permeabilities can be attained quite readily and economically. The combination of gouging wear resistance and low magnetic permeability is used to good advantage in manganese steel magnet cover wear plates.

Machinability. Austenitic manganese steels are difficult to machine largely because they harden under and in front of the tool. Even the smallest amount of tool chatter on the casting results in high hardness at the point of contact and in the dulling and rapid wear of the tool. Machining is performed with cemented carbide tools using heavy, rigid equipment; slow, steady feed; and deep cuts.

A well-equipped machine shop, through the use of grinding techniques, can perform all basic machine tool functions. Boring, planing, keyway cutting, and similar operations are efficiently done by grinding.

Weldability. Weld repair, rebuilding, and assembly operations are routinely performed. Chromium-nickel- and molybdenum-alloyed manganese steel welding wires and electrodes are readily available. With reasonable caution, all normal welding operations can be performed. Work-hardened metal should be removed by grinding before welding. Use of the 1% Mo grades in casting affords added protection against embrittlement.

Applications. Austenitic manganese steels are widely used in crushing and grinding applications. Primary and secondary crusher mantles can vary in weight from 3.6 to 22.7 Mg (4 to 25 tons). Upper and lower sections are used to reduce overall gross weights. Smaller cone crusher castings range from 180 to 3400 kg (400 to 7500 lb). Crusher rolls, jaw crusher wear plates, hammers, impacter bars, and cage rings frequently use austenitic manganese steels as well.

Power shovel dipper buckets of 19 m^3 (25 yd^3) capacity are being manufactured using cast-weld assembly techniques. Gross weights approach 45 Mg (50 tons), while large shovel shoes approach 2720 kg (6000 lb) each. Idlers, sprockets, sheaves, and racking are also widely used.

Large pumps with 2270 to 9070 kg (5000 to 20 000 lb) cast casings are extensively used. The pumping of head-sized rock is not uncommon. Other involved wear components are impellers as well as engine and section side plates. Weld rebuilding can be accomplished if desired. Dredge buckets, tumblers, and bushings also commonly use austenitic manganese steels.

Railroad frog and crossing castings are extensively used in trackwork assemblies. Repetitive wheel impact on points can lead to point batter and flow into flangeways, which require removal by grinding. Prehardening retards these occurrences, but does not always eliminate them. Nevertheless, austenitic manganese steels are almost exclusively used in this type of service. Other applications include steel plant crane wheels and chains, heavy-duty apron feeder pans, electric magnet cover wear plates, and coal conveyor chains.

SELECTED REFERENCES

- "High Alloy Data Sheets—Corrosion Series, Supplement 8, *Steel Castings Handbook*, Steel Founders' Society of America, 1981
- "High Alloy Data Sheets—Heat Series, *Steel Castings Handbook,* Steel Founders' Society of America, 1981
- W.J. Jackson, Ed., *Steel Castings Design Properties and Applications*, Steel Castings Research and Trade Association, 1983
- J.D. Redmond, Selecting Second-Generation Duplex Stainless Steels, *Chem. Eng.*, 27 Oct and 24 Nov 1986
- P.R. Wieser, Ed., *Steel Castings Handbook*, 5th ed., Steel Founders' Society of America, 1980

Cast Alnico Alloys

Robert A. Schmucker, Jr., Thomas & Skinner, Inc.

ALNICO ALLOYS (aluminum-nickel-copper-cobalt-iron) constitute a group of industrial permanent magnet materials developed in the 1940s and 1950s. The high-coercive high-titanium-bearing versions were added and perfected in the 1960s and 1970s. Although the dominant position of these materials has been challenged in the last decade by the magnetically stronger rare-earth powder metallurgy magnets, and earlier by the much less expensive ferrites, the Alnicos remain an important group.

The idea of using aluminum-nickel-iron alloys for a permanent magnet originated in Japan in the 1930s. Toward the end of that decade, greatly improved modifications were developed in England. Additional modifications were developed in Holland in 1940. The idea of increasing the magnetic strength in the most important alloy (Alnico 5) by casting to produce directional grains, that is, columnar crystals, was patented in the United States in 1951. Improvements in the heat treatment of the high-coercive high-titanium Alnico (Alnico 8) were patented in Holland in 1959 and 1960. Composition modifications to this type of alloy to permit columnar solidification (Alnico 9) were developed in Britain in 1962. Although the development of the Alnico system has been an international effort, the greatest production of the cast material has been in the United States. Alnico permanent magnets are currently used in a wide variety of applications, such as motors, tachometers, magnetos, generators, electron focusing or deflecting devices such as traveling wave and magnetron tubes, and the more familiar holding and lifting devices.

Finally, it should be emphasized that cast Alnico is a hard, brittle material that lacks ductility and can be surface finished only by grinding. It is produced primarily for its magnetic properties and should not be used as a structural component. Minor surface and internal imperfections rarely impair its magnetic strength and should not be considered detrimental.

Foundry Practice

There have been marked advances in the foundry technology of Alnico castings since the early manufacturing of the 1940s. A number of improved molding systems developed for other materials have been adapted for Alnico. Alnico magnets are cast essentially to shape in sizes that range in weight from a fraction of a pound to over 23 kg (50 lb), although most are under 4.5 kg (10 lb).

Patterns

A variety of patterns can be used to produce molds for Alnico castings (see the article "Patterns and Patternmaking" in this Volume). The molding system will determine the type of pattern. For the older baked sand cores and the current shell cores, a composite mold is used that consists of several cores and contains a common sprue and gating system (see the section "Bonded Sand Molds" in the article "Sand Molding" in this Volume). A single core may contain impressions for several dozen magnets. The production pattern is usually made of brass or aluminum. In most cases, the full impressions of the magnet contour are in one sand core. The bottom of the upper core, upon stacking, serves to form one flat side of the magnet. When holes are required in Alnico magnets, they are produced by casting around sand or graphite inserts.

Molds for producing directional-grained Alnico, as will be explained later in this article, are entirely different from shell molds and do not consist of stacks of cores. The patterns are mounted separately on a plate, have a greater height, and are individually gated. They can be made of metal or wood.

For precision investment casting, patterns are actually wax replicas of the magnet itself; these replicas are formed by injecting wax into a metal die. The patterns are made slightly larger to compensate for volumetric shrinkage in the pattern production stage and during solidification of the metal (see the section "Basic Concepts in Crystal Growth and Solidification" in the article "Fundamentals of Growth" in this Volume).

Molding

A wide variety of molding techniques are available for the casting of Alnico alloys, including baked sand, shell, cold-set phenolic, carbon dioxide, precision investment, and exothermic processes. No-bake processes have recently been added to both the shell sand and oil sand molding methods.

Baked sand molding was the predominant technique in the manufacture of Alnico castings until the 1970s (see the articles "Resin Binder Processes" and "Coremaking" in this Volume). This technique has been largely replaced by shell molding, but is still used for certain repetitive parts, such as pouring cups. Silica sand is mixed in a muller with a suitable cereal binder and core oil, molded in molding machines (jolt/rollover or core blower), and baked in an oven at about 205 °C (400 °F).

Shell molding (the Croning process) was developed in the United States after World War II on the basis of earlier German technology. This process was not extensively used for Alnico castings until somewhat later. It permits the manufacture of castings with a better surface finish, slightly closer tolerances, and at a lower overall cost than the baked sand method. The process involves the use of a mixture of round-grain silica sand and a thermosetting resin, usually a phenol formaldehyde. Some Alnico manufacturers mix their own sand, and others purchase commercially precoated sand. A brass or aluminum pattern with ejector pins is used for making the shell core. The sand mixture is dropped onto the preheated pattern and held for 10 to 30 s. The core is further polymerized by heating to 290 to 425 °C (550 to 800 °F) for a few minutes with the pattern attached. The core can, at this stage, be pushed away from the pattern assembly by actuation of the ejector system. A typical individual core would be 305 × 381 × 38 mm (12 × 15 × 1½ in.). A stack of these cores about 508 mm (20 in.) high clamped together with a pouring cup would constitute the composite mold. Additional information on shell molding is available in the article "Resin Binder Processes" in this Volume.

Cold-set Phenolic Urethane Molds. A recent alternative to the shell molding system for small Alnico castings is the use of cold-set phenolic urethane molds. The cores are similar in size and shape to shell cores, and similar patterns can be used; but the process and molding equipment are entirely different. Phenolic urethane resins cure at room temperature. The process consists of combining a dry round-grain sand (AFS 60) with two liquid resin components and a liquid catalyst in a high-speed mixer. The mixture is blown onto the pattern and permeated with an amine gas to cause a chemical reaction, forming a phenol bonded core. The core can be stripped from the pattern in less than 1 min. The individual cores are stacked and clamped to form a composite mold similar to that of shell cores (see the article "Sand Molding" in this Volume).

Carbon dioxide silicate molding has a restricted usage in some Alnico foundries. It has been used to produce very large Alnico magnets. Such molds have heavy walls, and they are harder and less collapsible than shell. The process consists of combining round-grain silica sand, liquid sodium silicate, and certain additives; mixing in a muller; and then ramming the mix into molds with conventional molding machines. The molds are placed into a chamber and gassed with carbon dioxide; during this process, a silica hydrogel bond is formed (see the article "Sand Molding" in this Volume).

Precision Investment Molding. The precision investment process is used to produce only a small percentage of Alnico castings. Such castings are usually less than 0.45 kg (1 lb) in weight. The process is limited to small intricate shapes or to cases where smooth cast surfaces or restricted cast tolerances are required. The usual tolerance is ±0.005 mm/mm (±0.005 in./in.) of length. There are also certain military applications that require freedom from internal inclusions for which the investment casting process is advantageous.

In the investment molding process, the wax patterns (actually replicas of the magnet) are removed from the die, gated, and mounted on wax sprues attached to a pouring cup. The entire cluster is then dipped into a ceramic slurry, drained, and coated with a fine ceramic sand. After drying, this process is repeated again and again, using progressively coarser grades of ceramic material until a self-supporting shell is formed. The coated cluster is then placed in a steam autoclave, where the patterns melt and run out through the gates, leaving a ceramic shell. This ceramic shell mold is preheated in a furnace at 870 to 1040 °C (1600 to 1900 °F) and then removed and placed upright in a supporting container when it is ready for the casting process. Additional information is available in the article "Investment Casting" in this Volume).

Exothermic Molds. The molding processes described above are used for a wide variety of metals and alloys and are of course not restricted to Alnico castings. The resulting structure is so-called random-grain Alnico, that is, essentially equiaxed crystals. Techniques for producing fully directional-grained or columnar Alnico are unique for this material and enhance its magnetic properties. Columnar crystals are obtained by casting into hot molds placed on top of a cold chill, with a reservoir of molten metal feeding the gate (this technique has been recently applied to the production of turbine blades for jet engines). This prevents radial heat flow and ensures the growth of columnar crystals directed from the chill end to the top gated end. The method is restricted to simple shapes such as cylinders or cuboids. Mold materials must be more refractory to withstand the high preheating temperatures.

The titanium-free Alnico 5 alloy is more readily cast in columnar form, designated Alnico 5-7. Several types of proprietary mold materials can be used. One consists of an aluminum silicate refractory with a sodium silicate binder, the mold being reacted with carbon dioxide gas. Such molds are preheated in a furnace at 1260 to 1370 °C (2300 to 2500 °F), removed, and placed on a chill; the molten metal is then poured directly in.

For the high-titanium Alnico 9 alloy, columnar crystals are more difficult to obtain because of the nucleating effect of the titanium oxide particles in the melt. The usual method involves the use of exothermic molds, which produce a higher mold cavity temperature of 1425 to 1480 °C (2600 to 2700 °F). One such mixture combines an alumina-silica sand, aluminum powder, sodium nitrate, and a small amount of sodium silicofluoride. This is bonded with a sodium silicate binder, and the resulting molds are reacted with carbon dioxide gas. The molds are placed on a chill and ignited with a torch; the molten metal is poured in after the exothermic reaction is complete.

Melting and Casting Practice

Alnico alloys are melted in high-frequency induction furnaces with capacities ranging from 120 to 450 kg (250 to 1000 lb). An ideal frequency is about 960 Hz because of the resultant stirring action of the molten metal. The furnace charge usually consists of Armco iron, cobalt in the form of broken cathodes or granules, electrolytic nickel, secondary electrolytic copper, virgin aluminum, commercially pure titanium (ferrotitanium is unsatisfactory), ferrosilicon, and certain additives such as zirconium or niobium. Sulfur is sometimes added to improve fluidity of the molten metal. In the case of Alnico 9, a sulfur addition is essential to counteract the nucleating effect of the titanium oxide particles in the melt. Clean Alnico revert scrap and refined cobalt-nickel-copper-iron granules are also frequently used in the charge.

Certain impurities must be avoided in the charge, especially carbon; when carbon is present above 0.03%, it increases susceptibility to the formation of γ phase in the alloy, which degrades the magnetic properties of Alnico 5 and 6. Chromium and molybdenum above 0.25% should also be avoided.

During the melting cycle, the highly oxidizable aluminum, titanium, silicon, zirconium and niobium are added during the final stages to prevent excessive losses of these elements. Air melting is usually used, but atmospheres of nitrogen or argon are occasionally substituted.

The casting temperature must be closely controlled to ensure sound well-formed castings. Immersion thermocouples provide the best temperature measurement. The temperature is usually in the range of 1540 to 1760 °C (2800 to 3200 °F). Casting can be performed by pouring the heat directly from the furnace into the mold stacks. The alternative method consists of tapping the heat into one or several preheated ladles and pouring from these ladles into the molds. As previously mentioned, when castings with columnar crystals are desired, the heat is poured into preheated or exothermic fired molds placed on metal chills.

Chemical Analysis and Magnetic Property Control. Control of chemical composition is essential in order to obtain optimum magnetic properties. It is important to conduct a chemical analysis of an individual heat a short time after pouring. X-ray fluorescence spectrography is the preferred method of analysis. An alternative control method consists of accelerated heat treatment and magnetic testing of a sample taken from each heat at the time of casting. All Alnico producers have used some combination of these two methods for many years. A more recent and precise method involves the extraction of a prepour sample directly from the melt, rapid analysis of this sample, and correction of the composition by additions of individual elements to the bath before the heat is poured. Regardless of the type of control, all castings from individual heats are identified in the foundry until their magnetic quality and/or chemical analysis is verified.

Processing of Castings and Preliminary Inspection. Shortly after casting, most grades of Alnico magnets are stripped from their molds and air cooled to room temperature. Because Alnico is brittle, the castings are readily broken away from the sprues manually or in a shotblast machine (see the section "Blast Cleaning of Castings" in the article "Processing of Castings" in this Volume). Castings are then subjected to 100% preliminary inspection for visible physical defects, such as shrinkage, pipe, porosity, and hot tears (see the section "Testing and Inspection of Casting Defects" in the article "Processing of Castings" in this Volume).

Gate and flash grinding operations are usually carried out before heat treatment. Gate and flash are removed by manual grinding on pedestal-type grinders. In the case of large volumes of small cylinders, gates are often removed in centerless grinding machines. At this stage, those magnets with graphite inserts are subjected to drilling operations to remove this material.

Heat Treatment

All Alnico magnets must be heat treated in order to develop their magnetic properties. This consists of three stages: a high-temperature solution treatment, a controlled cooling at a specified rate with or without a magnetic field, and a coercive aging or drawing operation that develops the necessary coercive force and energy product values. The high-temperature solution treatment can be omitted in the lower grades of Alnico if adequate cooling is applied at the initial casting stage.

Isotropic Alloys. The original Alnicos 1 through 4 have limited commercial importance today. Alnico 2 is occasionally made, and its treatment consists of a 980 °C (1800 °F) solution, cooling in about 5 min to a black color (~590 °C, or 1100 °F) without a magnetic field, and an aging draw of about 635 °C (1175 °F) for at least 2 h.

Anisotropic Alloys (Alnico 5 and 6). In these important commercial alloys, directional magnetic properties result from cooling through the Curie temperature region in a magnetic field of at least 9.5×10^4 A · m^{-1} (1200 Oe). If the alloy has been protected against γ phase formation by the addition of silicon and zirconium, it can be solution treated at about 915 °C (1680 °F), followed by the field cooling. Linear magnetic fields are provided by solenoids or electromagnets, but certain magnet configurations require the use of curved fields. The final coercive aging can be carried out at 580 °C (1075 °F) for 24 h, but variations in the form of two-step or continuous cycles can be implemented to shorten the time required.

Anisotropic Alloys (Alnico 8 and 9). For these two high-titanium alloys (Alnico 9 being essentially a directional-grained Alnico 8), a high-temperature full solution treatment in the range of 1230 to 1275 °C (2250 to 2330 °F) must be used. Also, in the field treatment stage, an isothermal holding about 45 °C (80 °F) below the Curie temperature is required, and the field strength should be at least 2.2×10^5 A · m^{-1} (2800 Oe). The isothermal furnace itself must be within this linear field. Curved field configurations have not been developed. A longer draw cycle than that used for Alnico 5 is required, typically 650 °C (1200 °F) for 5 h plus 550 °C (1020 °F) for 24 h.

Final Grinding, Inspection, and Testing

Because Alnico is a hard (45 to 58 HRC), brittle, and coarse-grained material, it can be finished to close tolerances only by grinding. These operations, which employ a variety of grinding equipment, must be carried out with great care using proper grinding wheels, coolants, and speeds to prevent cracking, heat checking, and chipping.

Final inspection and magnetic testing involve many procedures. Dimensional inspection is carried out using a variety of precision measuring devices. Physical inspection involves visual examination of parts for excessive cracks, chips, and porosity. However, minor imperfections do not impair the magnetic performance of a part and are not considered cause for rejection unless a standard is mutually agreed upon by producer and user. Magnetic testing involves dozens of procedures and techniques that are, in general, designed to duplicate the performance of a given magnet in its final operating circuit. The general

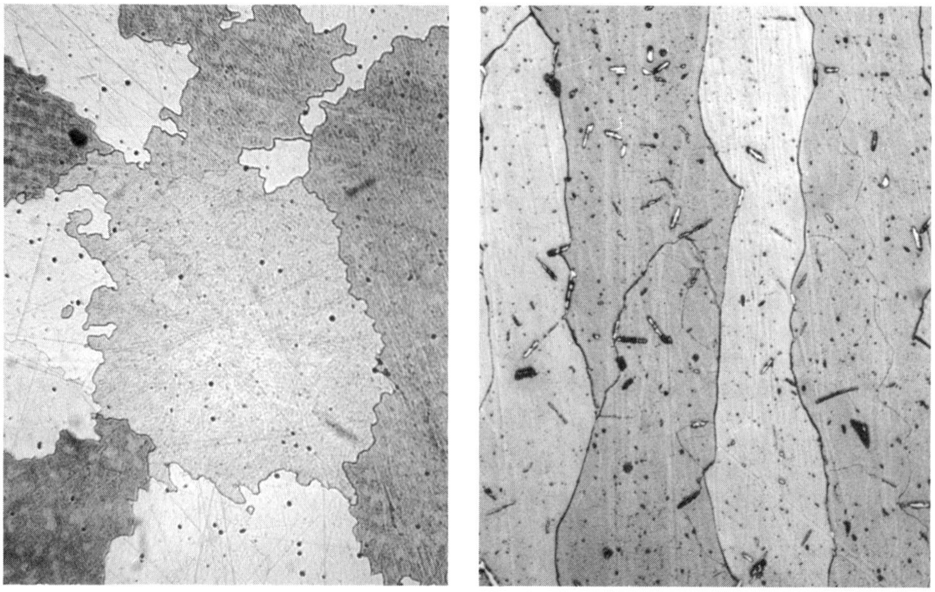

(a) (b)

Fig. 1 Photomicrographs of cast and heat-treated anisotropic Alnico alloys showing bcc structure. (a) Equiaxed-grain Alnico 5 with random grain structure. 30×. (b) Columnar-grained Alnico 9. 100×. Both etched with Marble's reagent

Table 1 Magnetic, mechanical and physical properties of major cast Alnico alloys

The value $(BH)_{max}$ is the most important because it represents the maximum magnetic energy that a unit volume of the material can produce in an air gap.

Cast alloy(a)	Nominal composition, wt%(b)					Nominal magnetic properties(c)					
						B_r		H_c		$(BH)_{max}$	
	Al	Ni	Co	Cu	Ti	kG	T	Oe	A · m^{-1} × 10^4	MG · Oe	J · m^{-3} × 10^4
Alnico 2	10	19	13	3	...	7.5	0.75	560	4.46	1.7	1.4
Alnico 5	8	14	24	3	...	12.8	1.28	640	5.09	5.5	4.4
Alnico 5–7	8	14	24	3	...	13.5	1.35	740	5.88	7.5	6.0
Alnico 6	8	16	24	3	1	10.5	1.05	780	6.21	3.9	3.1
Alnico 8	7	15	35	4	5	8.2	0.82	1650	13.1	5.3	4.2
Alnico 9	7	15	35	4	5	10.6	1.06	1500	11.9	9.0	7.2

Cast alloy(a)	Mechanical and physical properties(d)							
	Density, g/cm^3	Tensile strength		Transverse modulus of rupture		Hardness, HRC	Curie temperature	
		MPa	ksi	MPa	ksi		°C	°F
Alnico 2	7.1	21	3.0	48	7.0	45	810	1490
Alnico 5	7.3	37	5.4	72	10.5	50	860	1580
Alnico 5–7	7.3	34	5.0	55	8.0	50	860	1580
Alnico 6	7.3	160	23	310	45	50	860	1580
Alnico 8	7.3	70	10	210	30	55	860	1580
Alnico 9	7.3	48	7.0	55	8.0	55	860	1580

(a) Alnico 2 is isotropic. Alnico 5, 5–7, 6, 8, and 9 are anisotropic. Alnico 5–7 and Alnico 9 are also directional grain (columnar crystals). (b) The composition balance for all alloys is iron. Small percentages of silicon, zirconium, niobium, and sulfur may also be present in certain alloys. (c) B_r, remanent magnetization; H_c, normal coercive force; $(BH)_{max}$, maximum magnetic energy = magnetic induction × magnetic field strength. (d) Measurement of properties such as hardness plus tensile and rupture strength can be determined only under laboratory conditions and only for comparison purposes. Source: MMPA Standard 0100-87 guide to all inspection criteria is provided in MMPA Standard 0100-87.

Structure and Properties

The magnetic properties of anisotropic Alnico alloys are associated with directional, submicroscopic two-phase structures. These are visible only in the electron microscope, typically at magnifications of 25 000 to 100 000×. Conventional light microscopy at magnifications below 1000× is occasionally used to diagnose such problems as the presence of spoiling γ phase or to reveal the details of grain structures. Two such structures, a normal equiaxed-grain Alnico 5 and a columnar-grained Alnico 9, are shown in Fig. 1. The lattice structure is body-centered cubic (bcc). Table 1 lists various properties of the six most important Alnico alloys.

SELECTED REFERENCES

- K.J. DeVos, "The Relationship Between Microstructure and Magnetic Properties of Alnico Alloys," Thesis, Eindhoven University, 1966
- J.E. Gould, Magnets With Columnar Crystallization, *Cobalt*, Vol 23, 1964
- *The Investment Casting Process*, The Investment Casting Institute, 1978
- E.L. Kotzin, Ed., *Metalcasting and Molding Processes*, American Foundrymen's Society, 1981
- *Materials for Permanent Magnets*, in *Properties and Selection: Stainless Steels, Tool Materials, and Special Purpose Metals*, Vol 3, 9th ed., *Metals Handbook*, American Society for Metals, 1980, p 615-639
- M. McCaig and A.G. Clogg, *Permanent Magnets in Theory and Practice*, 2nd ed., Pentech Press, 1987
- K.E.L. Nicholas, *The CO_2-Silicate Process in Foundries*, BCIRA, 1972
- "Standard Specifications for Permanent Magnet Materials," MMPA 0100-85, Magnetic Materials Producers Association
- E.M. Underhill, Ed., *The Permanent Magnet Handbook*, Crucible Steel Company of America, 1957

Nonferrous Casting Alloys

Section Chairman:
Elwin L. Rooy, Aluminum Company of America

Aluminum and Aluminum Alloys

Elwin L. Rooy, Aluminum Company of America

ALUMINUM CASTINGS have played an integral role in the growth of the aluminum industry since its inception in the late 19th century. The first commercial aluminum products were castings, such as cooking utensils and decorative parts, which exploited the novelty and utility of the new metal. Those early applications rapidly expanded to address the requirements of a wide range of engineering specifications. Alloy development and characterization of physical and mechanical characteristics provided the basis for new product development through the decades that followed. Casting processes were developed to extend the capabilities of foundries in new commercial and technical applications. The technology of molten metal processing, solidification, and property development has been advanced to assist the foundryman with the means of economical and reliable production of parts that consistently meet specified requirements.

Today, aluminum alloy castings are produced in hundreds of compositions by all commercial casting processes, including green sand, dry sand, composite mold, plaster mold, investment casting, permanent mold, counter-gravity low-pressure casting, and pressure die casting.

Casting processes are normally divided into two categories: expendable mold processes and those processes in which castings are produced repetitively in tooling of extended life. Examples are green and dry sand molding for the former and die and permanent mold casting for the latter.

Alloys can also be divided into two groups: those most suitable for gravity casting by any process and those used in pressure die casting. A finer distinction is made between alloys suitable for permanent mold application and those for other gravity processes. In general, the most alloy-versatile processes (that is, those in which the largest number of alloys can be used) are those in which mold collapse accompanies pouring and solidification. The least forgiving processes, which require special consideration in alloy selection, are more rigid or permanent mold processes. The least alloy-versatile casting process based on process requirements is pressure die casting, in which more than the usual measures of castability apply. The process demands a high level of fluidity, hot strength, hot tear resistance, and die soldering resistance.

Material constraints that formerly limited the design engineer's alloy choice once a casting process had been selected are increasingly being blurred by advances in foundry technique. In the same way, process selection is also less restricted today. For example, many alloys thought to be unusable in permanent molds because of casting characteristics are in production by that process.

Chemical Compositions

Systems used to designate casting compositions are not internationally standardized. In the United States, comprehensive listings are maintained by general procurement specifications issued through government agencies (federal, military, and so on) and by technical societies such as the American Society for Testing and Materials and the Society of Automotive Engineers. Alloy registrations by the Aluminum Association are in broadest use; its nomenclature is decimalized to define foundry alloy composition variations.

Designations in the form *xxx*.1 and *xxx*.2 include the composition of specific alloys in remelt ingot form suitable for foundry use. Designations in the form *xxx*.0 in all cases define composition limits applicable to castings. Further variations in specified compositions are denoted by prefix letters used primarily to define differences in impurity limits. Accordingly, one of the most common gravity cast alloys, 356, is shown in variations A356, B356, and C356; each of these alloys has identical major alloy contents but has decreasing specification limits applicable to impurities, especially iron content. Aluminum Association composition limits for registered aluminum foundry alloys used to cast shapes are given in Table 1. Table 1 does not include alloys that are cast into ingots intended for subsequent working.

Although the nomenclature and designations for various casting alloys are standardized in North America, many important alloys have been developed for engineered casting production worldwide. For the most part, each nation (and in many cases the individual firm) has developed its own alloy nomenclature. Excellent references are available that correlate, cross reference, or otherwise define significant compositions in international use (Ref 2, 3).

Because differences in process capabilities exist, not all alloys can be cast by all methods. Alloys are usually separated into families in which alloy characteristics are considered as a function or process requirements. Table 1 lists common compositions and conventional use according to process type.

The Aluminum Association designation system attempts alloy family recognition by the following scheme:

- 1*xx*.*x*: Controlled unalloyed compositions
- 2*xx*.*x*: Aluminum alloys containing copper as the major alloying element
- 3*xx*.*x*: Aluminum-silicon alloys also containing magnesium and/or copper
- 4*xx*.*x*: Binary aluminum-silicon alloys
- 5*xx*.*x*: Aluminum alloys containing magnesium as the major alloying element
- 6*xx*.*x*: Currently unused
- 7*xx*.*x*: Aluminum alloys containing zinc as the major alloying element, usually also containing additions of either copper, magnesium, chromium, manganese, or combinations of these elements
- 8*xx*.*x*: Aluminum alloys containing tin as the major alloying element
- 9*xx*.*x*: Currently unused

Effects of Alloying Elements

Antimony. At concentration levels equal to or greater than 0.05%, antimony refines eutectic aluminum-silicon phase to lamellar form in hypoeutectic compositions. The effectiveness of antimony in altering the eutectic structure depends on an absence of phosphorus and on an adequately rapid rate of solidification. Antimony also reacts with either sodium or strontium to form coarse intermetallics with adverse effects on castability and eutectic structure.

Antimony is classified as a heavy metal with potential toxicity and hygiene implications, especially as associated with the possibility of stibine gas formation and the effects of human exposure to other antimony compounds.

Table 1 Compositions of registered aluminum casting alloys used to cast shapes

Compositions of alloys used to cast primary ingots are not shown.

		Composition, %(b)											
												Others	
Alloy	Products(a)	Si	Fe	Cu	Mn	Mg	Cr	Ni	Zn	Sn	Ti	Each	Total
201.1	S	0.10	0.15	4.0–5.2	0.20–0.50	0.15–0.55	. . .	. . .	. . .	. . .	0.15–0.35	0.05(c)	0.10
A201.0	S	0.05	0.10	4.0–5.0	0.20–0.40	0.15–0.35	. . .	. . .	. . .	. . .	0.15–0.35	0.03(c)	0.10
B201.0	S	0.05	0.05	4.5–5.0	0.20–0.50	0.25–0.35	. . .	. . .	. . .	. . .	0.15–0.35	0.05(d)	0.15
202.0	S	0.10	0.15	4.0–5.2	0.20–0.8	0.15–0.55	0.20–0.6	. . .	. . .	. . .	0.15–0.35	0.05(c)	0.10
203.0	S	0.30	0.50	4.5–5.5	0.20–0.30	0.10	. . .	1.3–1.7	0.10	. . .	0.15–0.25(e)	0.05(f)	0.20
204.0	S, P	0.20	0.35	4.2–5.0	0.10	0.15–0.35	. . .	0.05	0.10	0.05	0.15–0.30	0.05	0.15
206.0	S, P	0.10	0.15	4.2–5.0	0.20–0.50	0.15–0.35	. . .	0.05	0.10	0.05	0.15–0.30	0.05	0.15
A206.0	S, P	0.05	0.10	4.2–5.0	0.20–0.50	0.15–0.35	. . .	0.05	0.10	0.05	0.15–0.30	0.05	0.15
208.0	S, P	2.5–3.5	1.2	3.5–4.5	0.50	0.10	. . .	0.35	1.0	. . .	0.25	. . .	0.50
213.0	S, P	1.0–3.0	1.2	6.0–8.0	0.6	0.10	. . .	0.35	2.5	. . .	0.25	. . .	0.50
222.0	S, P	2.0	1.5	9.2–10.7	0.50	0.15–0.35	. . .	0.50	0.8	. . .	0.25	. . .	0.35
224.0	S, P	0.06	0.10	4.5–5.5	0.20–0.50	. . .	. . .	. . .	. . .	. . .	0.35	0.03(g)	0.10
238.0	P	3.5–4.5	1.5	9.0–11.0	0.6	0.15–0.35	. . .	1.0	1.5	. . .	0.25	. . .	0.50
240.0	S	0.50	0.50	7.0–9.0	0.30–0.7	5.5–6.5	. . .	0.30–0.7	0.10	. . .	0.20	0.05	0.15
242.0	S, P	0.7	1.0	3.5–4.5	0.35	1.2–1.8	0.25	1.7–2.3	0.35	. . .	0.25	0.05	0.15
A242.0	S	0.6	0.8	3.7–4.5	0.10	1.2–1.7	0.15–0.25	1.8–2.3	0.10	. . .	0.07–0.20	0.05	0.15
243.0	S	0.35	0.40	3.5–4.5	0.15–0.45	1.8–2.3	0.20–0.40	1.9–2.3	0.05	. . .	0.06–0.20	0.05(h)	0.15
249.0	P	0.05	0.10	3.8–4.6	0.25–0.50	0.25–0.50	. . .	. . .	2.5–3.5	. . .	0.02–0.35	0.03	0.10
295.0	S	0.7–1.5	1.0	4.0–5.0	0.35	0.03	. . .	. . .	0.35	. . .	0.25	0.05	0.15
296.0	P	2.0–3.0	1.2	4.0–5.0	0.35	0.05	. . .	0.35	0.50	. . .	0.25	. . .	0.35
305.0	S, P	4.5–5.5	0.6	1.0–1.5	0.50	0.10	0.25	. . .	0.35	. . .	0.25	0.05	0.15
A305.0	S, P	4.5–5.5	0.20	1.0–1.5	0.10	0.10	. . .	. . .	0.10	. . .	0.20	0.05	0.15
308.0	S, P	5.0–6.0	1.0	4.0–5.0	0.50	0.10	. . .	. . .	1.0	. . .	0.25	. . .	0.50
319.0	S, P	5.5–6.5	1.0	3.0–4.0	0.50	0.10	. . .	0.35	1.0	. . .	0.25	. . .	0.50
A319.0	S, P	5.5–6.5	1.0	3.0–4.0	0.50	0.10	. . .	0.35	3.0	. . .	0.25	. . .	0.50
B319.0	S, P	5.5–6.5	1.2	3.0–4.0	0.8	0.10–0.50	. . .	0.50	1.0	. . .	0.25	. . .	0.50
320.0	S, P	5.0–8.0	1.2	2.0–4.0	0.8	0.05–0.6	. . .	0.35	3.0	. . .	0.25	. . .	0.50
324.0	P	7.0–8.0	1.2	0.40–0.6	0.50	0.40–0.7	. . .	0.30	1.0	. . .	0.20	0.15	0.20
328.0	S	7.5–8.5	1.0	1.0–2.0	0.20–0.6	0.20–0.6	0.35	0.25	1.5	. . .	0.25	. . .	0.50
332.0	P	8.5–10.5	1.2	2.0–4.0	0.50	0.50–1.5	. . .	0.50	1.0	. . .	0.25	. . .	0.50
333.0	P	8.0–10.0	1.0	3.0–4.0	0.50	0.05–0.50	. . .	0.50	1.0	. . .	0.25	. . .	0.50
A333.0	P	8.0–10.0	1.0	3.0–4.0	0.50	0.05–0.50	. . .	0.50	3.0	. . .	0.25	. . .	0.50
336.0	P	11.0–13.0	1.2	0.50–1.5	0.35	0.7–1.3	. . .	2.0–3.0	0.35	. . .	0.25	0.05	. . .
339.0	P	11.0–13.0	1.2	1.5–3.0	0.50	0.50–1.5	. . .	0.50–1.5	1.0	. . .	0.25	. . .	0.50
343.0	D	6.7–7.7	1.2	0.50–0.9	0.50	0.10	0.10	. . .	1.2–2.0	0.50	. . .	0.10	0.35
354.0	P	8.6–9.4	0.20	1.6–2.0	0.10	0.40–0.6	. . .	. . .	0.10	. . .	0.20	0.05	0.15
355.0	S, P	4.5–5.5	0.6(i)	1.0–1.5	0.50(i)	0.40–0.6	0.25	. . .	0.35	. . .	0.25	0.05	0.15
A355.0	S, P	4.5–5.5	0.09	1.0–1.5	0.05	0.45–0.6	. . .	. . .	0.05	. . .	0.04–0.20	0.05	0.15
C355.0	S, P	4.5–5.5	0.20	1.0–1.5	0.10	0.40–0.6	. . .	. . .	0.10	. . .	0.20	0.05	0.15
356.0	S, P	6.5–7.5	0.6(i)	0.25	0.35(i)	0.20–0.45	. . .	. . .	0.35	. . .	0.25	0.05	0.15
A356.0	S, P	6.5–7.5	0.20	0.20	0.10	0.25–0.45	. . .	. . .	0.10	. . .	0.20	0.05	0.15
B356.0	S, P	6.5–7.5	0.09	0.05	0.05	0.25–0.45	. . .	. . .	0.05	. . .	0.04–0.20	0.05	0.15
C356.0	S, P	6.5–7.5	0.07	0.05	0.05	0.25–0.45	. . .	. . .	0.05	. . .	0.04–0.20	0.05	0.15
F356.0	S, P	6.5–7.5	0.20	0.20	0.10	0.17–0.25	. . .	. . .	0.10	. . .	0.04–0.20	0.05	0.15
357.0	S, P	6.5–7.5	0.15	0.05	0.03	0.45–0.6	. . .	. . .	0.05	. . .	0.20	0.05	0.15
A357.0	S, P	6.5–7.5	0.20	0.20	0.10	0.40–0.7	. . .	. . .	0.10	. . .	0.04–0.20	0.05(j)	0.15
B357.0	S, P	6.5–7.5	0.09	0.05	0.05	0.40–0.6	. . .	. . .	0.05	. . .	0.04–0.20	0.05	0.15
C357.0	S, P	6.5–7.5	0.09	0.05	0.05	0.45–0.7	. . .	. . .	0.05	. . .	0.04–0.20	0.05(j)	0.15
D357.0	S	6.5–7.5	0.20	. . .	0.10	0.55–0.6	. . .	. . .	. . .	. . .	0.10–0.20	0.05(j)	0.15
358.0	S, P	7.6–8.6	0.30	0.20	0.20	0.40–0.6	0.20	. . .	0.20	. . .	0.10–0.20	0.05(k)	0.15
359.0	S, P	8.5–9.5	0.20	0.20	0.10	0.50–0.7	. . .	. . .	0.10	. . .	0.20	0.05	0.15
360.0	D	9.0–10.0	2.0	0.6	0.35	0.40–0.6	. . .	0.50	0.50	0.15	. . .	. . .	0.25

(continued)

(a) D, die casting; P, permanent mold; S, sand. Other products may pertain to the composition but are not listed. (b) Weight percent; maximum unless range is given or otherwise indicated. All compositions contain balance of aluminum. (c) 0.40–1.0 Ag. (d) 0.50–1.0 Ag. (e) 0.50 max Ti + Zr. (f) 0.20–0.30 Sb, 0.20–0.30 Co, 0.10–0.30 Zr. (g) 0.05–0.15 V, 0.10–0.25 Zr. (h) 0.06–0.20 V. (i) If iron exceeds 0.45%, manganese content shall not be less than one-half of iron content. (j) 0.04–0.07 Be. (k) 0.10–0.30 Be. (l) 0.8 max Mn + Cr. (m) 0.25 max Pb. (n) 0.02–0.04 Be. (o) 0.08–0.15 V. (p) 0.10 max Pb. (q) 0.003–0.007 Be, 0.005 max B. Source: Ref 1

Beryllium additions of as low as a few parts per million may be effective in reducing oxidation losses and associated inclusions in magnesium-containing compositions. Studies have shown that proportionally increased beryllium concentrations are required for oxidation suppression as magnesium content increases.

At higher concentrations (>0.04%), beryllium affects the form and composition of iron-containing intermetallics, markedly improving strength and ductility. In addition to changing beneficially the morphology of the insoluble phase, beryllium changes its composition, rejecting magnesium from the Al-Fe-Si complex and thus permitting its full use for hardening purposes.

Beryllium-containing compounds are, however, numbered among the known carcinogens that require specific precautions in the melting, molten metal handling, dross handling and disposition, and welding of alloys. Standards define the maximum beryllium in welding rod and weld base metal as 0.008 and 0.010%, respectively.

Bismuth improves the machinability of cast aluminum alloys at concentrations greater than 0.1%.

Boron combines with other metals to form borides, such as Al_2 and TiB_2. Titanium boride forms stable nucleation sites for interaction with active grain-refining phases such as $TiAl_3$ in molten aluminum.

Metallic borides reduce tool life in machining operations, and in coarse particle form they consist of objectionable inclusions with detrimental effects on mechanical properties and ductility. At high boron concentrations, borides contribute to furnace sludging, particle agglomeration, and increased risk of casting inclusions. However, boron treatment of aluminum-contain-

Table 1 (continued)

Alloy	Products(a)	Composition, %(b) Si	Fe	Cu	Mn	Mg	Cr	Ni	Zn	Sn	Ti	Others Each	Others Total
A360.0	D	9.0–10.0	1.3	0.6	0.35	0.40–0.6	...	0.50	0.50	0.15	...	...	0.25
361.0	D	9.5–10.5	1.1	0.50	0.25	0.40–0.6	0.20–0.30	0.20–0.30	0.50	0.10	0.20	0.05	0.15
363.0	S, P	4.5–6.0	1.1	2.5–3.5	(l)	0.15–0.40	(l)	0.25	3.0–4.5	0.25	0.20	(m)	0.30
364.0	D	7.5–9.5	1.5	0.20	0.10	0.20–0.40	0.25–0.50	0.15	0.15	0.15	...	0.05(n)	0.15
369.0	D	11.0–12.0	1.3	0.50	0.35	0.25–0.45	0.30–0.40	0.05	1.0	0.10	...	0.05	0.15
380.0	D	7.5–9.5	2.0	3.0–4.0	0.50	0.10	...	0.50	3.0	0.35	...	...	0.50
A380.0	D	7.5–9.5	1.3	3.0–4.0	0.50	0.10	...	0.50	3.0	0.35	...	...	0.50
B380.0	D	7.5–9.5	1.3	3.0–4.0	0.50	0.10	...	0.50	1.0	0.35	...	...	0.50
383.0	D	9.5–11.5	1.3	2.0–3.0	0.50	0.10	...	0.30	3.0	0.15	...	...	0.50
384.0	D	10.5–12.0	1.3	3.0–4.5	0.50	0.10	...	0.50	3.0	0.35	...	...	0.50
A384.0	D	10.5–12.0	1.3	3.0–4.5	0.50	0.10	...	0.50	1.0	0.35	...	...	0.50
385.0	D	11.0–13.0	2.0	2.0–4.0	0.50	0.30	...	0.50	3.0	0.30	...	...	0.50
390.0	D	16.0–18.0	1.3	4.0–5.0	0.10	0.45–0.65	...	...	0.10	...	0.20	0.10	0.20
A390.0	S, P	16.0–18.0	0.50	4.0–5.0	0.10	0.45–0.65	...	...	0.10	...	0.20	0.10	0.20
B390.0	D	16.0–18.0	1.3	4.0–5.0	0.50	0.45–0.65	...	0.10	1.5	...	0.20	0.10	0.20
392.0	D	18.0–20.0	1.5	0.40–0.8	0.20–0.6	0.8–1.2	...	0.50	0.50	0.30	0.20	0.15	0.50
393.0	S, P, D	21.0–23.0	1.3	0.7–1.1	0.10	0.7–1.3	...	2.0–2.5	0.10	...	0.10–0.20	0.05(o)	0.15
413.0	D	11.0–13.0	2.0	1.0	0.35	0.10	...	0.50	0.50	0.15	...	...	0.25
A413.0	D	11.0–13.0	1.3	1.0	0.35	0.10	...	0.50	0.50	0.15	...	...	0.25
B413.0	S, P	11.0–13.0	0.50	0.10	0.35	0.05	...	0.05	0.10	...	0.25	0.05	0.20
443.0	S, P	4.5–6.0	0.8	0.6	0.50	0.05	0.25	...	0.50	...	0.25	...	0.35
A443.0	S	4.5–6.0	0.8	0.30	0.50	0.05	0.25	...	0.50	...	0.25	...	0.35
B443.0	S, P	4.5–6.0	0.8	0.15	0.35	0.05	...	...	0.35	...	0.25	0.05	0.15
C443.0	D	4.5–6.0	2.0	0.6	0.35	0.10	...	0.50	0.50	0.15	...	...	0.25
444.0	S, P	6.5–7.5	0.6	0.25	0.35	0.10	...	...	0.35	...	0.25	0.05	0.15
A444.0	P	6.5–7.5	0.20	0.10	0.10	0.05	...	...	0.10	...	0.20	0.05	0.15
511.0	S	0.30–0.7	0.50	0.15	0.35	3.5–4.5	...	...	0.15	...	0.25	0.05	0.15
512.0	S	1.4–2.2	0.6	0.35	0.8	3.5–4.5	0.25	...	0.35	...	0.25	0.05	0.15
513.0	P	0.30	0.40	0.10	0.30	3.5–4.5	...	...	1.4–2.2	...	0.20	0.05	0.15
514.0	S	0.35	0.50	0.15	0.35	3.5–4.5	...	...	0.15	...	0.25	0.05	0.15
515.0	D	0.50–1.0	1.3	0.20	0.40–0.6	2.5–4.0	...	...	0.10	...	...	0.05	0.15
516.0	D	0.30–1.5	0.35–1.0	0.30	0.15–0.40	2.5–4.5	...	0.25–0.04	0.20	0.10	0.10–0.20	0.05(p)	...
518.0	D	0.35	1.8	0.25	0.35	7.5–8.5	...	0.15	0.15	0.15	...	...	0.25
520.0	S	0.25	0.30	0.25	0.15	9.5–10.6	...	...	0.15	...	0.25	0.05	0.15
535.0	S	0.15	0.15	0.05	0.10–0.25	6.2–7.5	...	...	...	...	0.10–0.25	0.05(q)	0.15
A535.0	S	0.20	0.20	0.10	0.10–0.25	6.5–7.5	...	...	...	...	0.25	0.05	0.15
B535.0	S	0.15	0.15	0.10	0.05	6.5–7.5	...	...	...	...	0.10–0.25	0.05	0.15
705.0	S, P	0.20	0.8	0.20	0.40–0.6	1.4–1.8	0.20–0.40	...	2.7–3.3	...	0.25	0.05	0.15
707.0	S, P	0.20	0.8	0.20	0.40–0.6	1.8–2.4	0.20–0.40	...	4.0–4.5	...	0.25	0.05	0.15
710.0	S	0.15	0.50	0.35–0.65	0.05	0.6–0.8	...	...	6.0–7.0	...	0.25	0.05	0.15
711.0	P	0.30	0.7–1.4	0.35–0.65	0.05	0.25–0.45	...	...	6.0–7.0	...	0.20	0.05	0.15
712.0	S	0.30	0.50	0.25	0.10	0.50–0.65	0.40–0.6	...	5.0–6.5	...	0.15–0.25	0.05	0.20
713.0	S, P	0.25	1.1	0.40–1.0	0.6	0.20–0.50	0.35	0.15	7.0–8.0	...	0.25	0.10	0.25
771.0	S	0.15	0.15	0.10	0.10	0.8–1.0	0.06–0.20	...	6.5–7.5	...	0.10–0.20	0.05	0.15
772.0	S	0.15	0.15	0.10	0.10	0.6–0.8	0.06–0.20	...	6.0–7.0	...	0.10–0.20	0.05	0.15
850.0	S, P	0.7	0.7	0.7–1.3	0.10	0.10	...	0.7–1.3	...	5.5–7.0	0.20	...	0.30
851.0	S, P	2.0–3.0	0.7	0.7–1.3	0.10	0.10	...	0.30–0.7	...	5.5–7.0	0.20	...	0.30
852.0	S, P	0.40	0.7	1.7–2.3	0.10	0.6–0.9	...	0.9–1.5	...	5.5–7.0	0.20	...	0.30
853.0	S, P	5.5–6.5	0.7	3.0–4.0	0.50	...	...	...	...	5.5–7.0	0.20	...	0.30

(a) D, die casting; P, permanent mold; S, sand. Other products may pertain to the composition but are not listed. (b) Weight percent; maximum unless range is given or otherwise indicated. All compositions contain balance of aluminum. (c) 0.40–1.0 Ag. (d) 0.50–1.0 Ag. (e) 0.50 max Ti + Zr. (f) 0.20–0.30 Sb, 0.20–0.30 Co, 0.10–0.30 Zr. (g) 0.05–0.15 V, 0.10–0.25 Zr. (h) 0.06–0.20 V. (i) If iron exceeds 0.45%, manganese content shall not be less than one-half of iron content. (j) 0.04–0.07 Be. (k) 0.10–0.30 Be. (l) 0.8 max Mn + Cr. (m) 0.25 max Pb. (n) 0.02–0.04 Be. (o) 0.08–0.15 V. (p) 0.10 max Pb. (q) 0.003–0.007 Be, 0.005 max B. Source: Ref 1

ing peritectic elements is practiced to improve purity and electrical conductivity in rotor casting. Higher rotor alloy grades may specify boron to exceed titanium and vanadium contents to ensure either the complexing or precipitation of these elements for improved electrical performance (see the section "Rotor Castings" in this article).

Cadmium in concentrations exceeding 0.1% improves machinability. Precautions that acknowledge volatilization at 767 °C (1413 °F) are essential.

Calcium is a weak aluminum-silicon eutectic modifier. It increases hydrogen solubility and is often responsible for casting porosity at trace concentration levels. Calcium concentrations greater than approximately 0.005% also adversely affect ductility in aluminum-magnesium alloys.

Chromium additions are commonly made in low concentrations to room temperature aging and thermally unstable compositions in which germination and grain growth are known to occur. Chromium typically forms the compound $CrAl_7$, which displays extremely limited solid-state solubility and is therefore useful in suppressing grain growth tendencies. Sludge that contains iron, manganese, and chromium is sometimes encountered in die casting compositions, but it is rarely encountered in gravity casting alloys. Chromium improves corrosion resistance in certain alloys and increases quench sensitivity at higher concentrations.

Copper. The first and most widely used aluminum alloys were those containing 4 to 10% Cu. Copper substantially improves strength and hardness in the as-cast and heat-treated conditions. Alloys containing 4 to 6% Cu respond most strongly to thermal treatment. Copper generally reduces resistance to general corrosion and, in specific compositions and material conditions, stress corrosion susceptibility. Additions of copper also reduce hot tear resistance and decrease castability.

Iron improves hot tear resistance and decreases the tendency for die sticking or soldering in die casting. Increases in iron content are, however, accompanied by substantially decreased ductility. Iron reacts to form a myriad of insoluble phases in aluminum alloy melts, the most common of which are $FeAl_3$, $FeMnAl_6$, and $\alpha AlFeSi$. These essentially insoluble phases are responsible for improvements in strength, especially at elevated temperature. As the fraction of insoluble phase increases with increased iron content, casting considerations such as flowability and feeding characteristics are adversely affected. Iron participates in the formation of sludging phases with manganese, chromium, and other elements.

Lead is commonly used in aluminum casting alloys at greater than 0.1% for improved machinability.

Magnesium is the basis for strength and hardness development in heat-treated Al-Si alloys and is commonly used in more complex Al-Si alloys containing copper, nickel, and other elements for the same purpose. The hardening phase Mg_2Si displays a useful solubility limit corresponding to approximately 0.70% Mg, beyond which either no further strengthening occurs or matrix softening takes place. Common premium-strength compositions in the Al-Si family employ magnesium in the range of 0.40 to 0.070% (see the section "Premium Castings" in this article).

Binary Al-Mg alloys are widely used in applications requiring a bright surface finish and corrosion resistance, as well as attractive combinations of strength and ductility. Common compositions range from 4 to 10% Mg, and compositions containing more than 7% Mg are heat treatable. Instability and room-temperature aging characteristics at higher magnesium concentrations encourage heat treatment.

Manganese is normally considered an impurity in casting compositions and is controlled to low levels in most gravity cast compositions. Manganese is an important alloying element in wrought compositions, through which secondary foundry compositions may contain higher manganese levels. In the absence of work hardening, manganese offers no significant benefits in cast aluminum alloys. Some evidence exists, however, that a high volume fraction of $MnAl_6$ in alloys containing more than 0.5% Mn may beneficially influence internal casting soundness. Manganese can also be employed to alter response in chemical finishing and anodizing.

Mercury. Compositions containing mercury were developed as sacrificial anode materials for cathodic protection systems, especially in marine environments. The use of these optimally electronegative alloys, which did not passivate in seawater, was severely restricted for environmental reasons.

Nickel is usually employed with copper to enhance elevated-temperature properties. It also reduces the coefficient of thermal expansion.

Phosphorus. In AlP_3 form, phosphorus nucleates and refines primary silicon-phase formation in hypereutectic Al-Si alloys. At parts per million concentrations, phosphorus coarsens the eutectic structure in hypoeutectic Al-Si alloys. Phosphorus diminishes the effectiveness of the common eutectic modifiers sodium and strontium.

Silicon. The outstanding effect of silicon in aluminum alloys is the improvement of casting characteristics. Additions of silicon to pure aluminum dramatically improve fluidity, hot tear resistance, and feeding characteristics. The most prominently used compositions in all casting processes are those of the aluminum-silicon family. Commercial alloys span the hypoeutectic and hypereutectic ranges up to about 25% Si.

In general, an optimum range of silicon content can be assigned to casting processes. For slow cooling rate processes (such as plaster, investment, and sand), the range is 5 to 7%, for permanent mold 7 to 9%, and for die casting 8 to 12%. The bases for these recommendations are the relationship between cooling rate and fluidity and the effect of percentage of eutectic on feeding. Silicon additions are also accompanied by a reduction in specific gravity and coefficient of thermal expansion.

Silver is used in only a limited range of aluminum-copper premium-strength alloys at concentrations of 0.5 to 1.0%. Silver contributes to precipitation hardening and stress corrosion resistance.

Sodium modifies the aluminum-silicon eutectic. Its presence is embrittling in aluminum-magnesium alloys. Sodium interacts with phosphorus to reduce its effectiveness in modifying the eutectic and that of phosphorus in the refinement of the primary silicon phase.

Strontium is used to modify the aluminum-silicon eutectic. Effective modification can be achieved at very low addition levels, but a range of recovered strontium of 0.008 to 0.04% is commonly used. Higher addition levels are associated with casting porosity, especially in processes or in thick-section parts in which solidification occurs more slowly. Degassing efficiency may also be adversely affected at higher strontium levels.

Tin is effective in improving antifriction characteristics, and is therefore useful in bearing applications. Casting alloys may contain up to 25% Sn. Additions can also be made to improve machinability. Tin may influence precipitation-hardening response in some alloy systems.

Titanium is extensively used to refine the grain structure of aluminum casting alloys, often in combination with smaller amounts of boron. Titanium in excess of the stoichiometry of TiB_2 is necessary for effective grain refinement. Titanium is often employed at concentrations greater than those required for grain refinement to reduce cracking tendencies in hot short compositions.

Zinc. No significant benefits are obtained by the addition of zinc to aluminum. Accompanied by the addition of copper and/or magnesium, however, zinc results in attractive heat treatable or naturally aging compositions. A number of such compositions are in common use. Zinc is also commonly found in secondary gravity and die casting compositions.

Casting Processes

The wide applicability of casting processes and process variations in the production of aluminum-base compositions necessitates a comprehensive understanding of process characteristics and capabilities. The selection of casting method is based on the capabilities of each process relative to the design and the specified requirements for each part. In most cases, castings can be readily produced by more than one technique. In these cases, economics largely based on volume of production dictate the process choice. For other examples, specific quality or engineering requirements restrict the process choice.

Although the rate of new alloy development has declined in recent years, activity in new casting process development has dramatically increased. These developments are primarily variations of three broad categories: expendable mold processes, including green and dry sand casting, as well as plaster and investment molding; gravity permanent mold, including centrifugal and low-pressure casting; and pressure die casting.

Progress in pattern and mold materials and in pouring techniques in each of these processes is documented in the articles "Sand Molding," "Plaster Molding," "Ceramic Molding," "Investment Casting," "Permanent Mold Casting," "Centrifugal Casting," "Die Casting," "Replicast Process," and "New and Emerging Processes" in this Volume.

Melting and Metal Treatment

Aluminum and aluminum alloys can be melted in a variety of ways (see the article "Nonferrous Molten Metal Processes" in this Volume). Coreless and channel induction furnaces, crucible and open-hearth reverberatory furnaces fired by natural gas or fuel oil, and electric resistance and electric radiation furnaces are all in routine use. The nature of the furnace charge is as varied and important as the choice of furnace type for metal casting operations. The furnace charge may range from prealloyed ingot of high quality to charges made up exclusively from low-grade scrap. Even under optimum melting and melt-holding conditions, molten aluminum is susceptible to three types of degradation:

- With time at temperature, adsorption of hydrogen results in increased dissolved hydrogen content up to an equilibrium

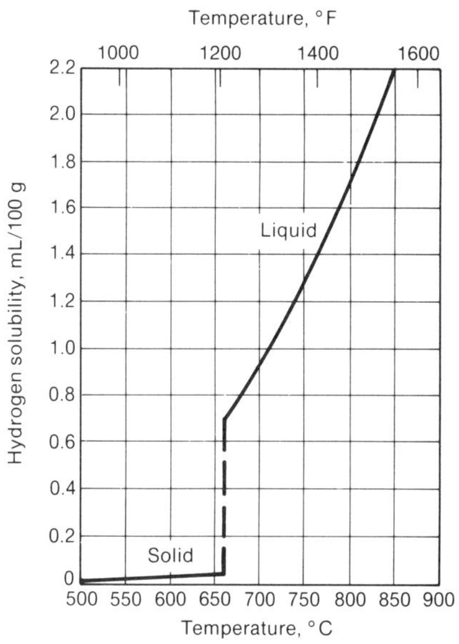

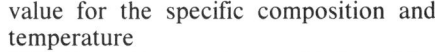

Fig. 1 Solubility of hydrogen in aluminum at 1 atm hydrogen pressure

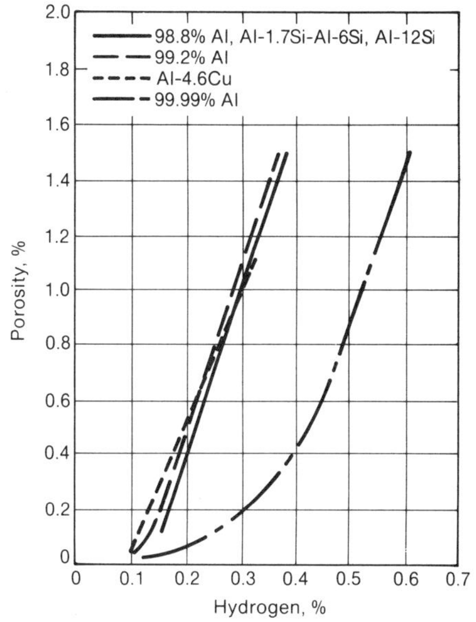

Fig. 2 Porosity as a function of hydrogen content in sand-cast aluminum and aluminum alloy bars

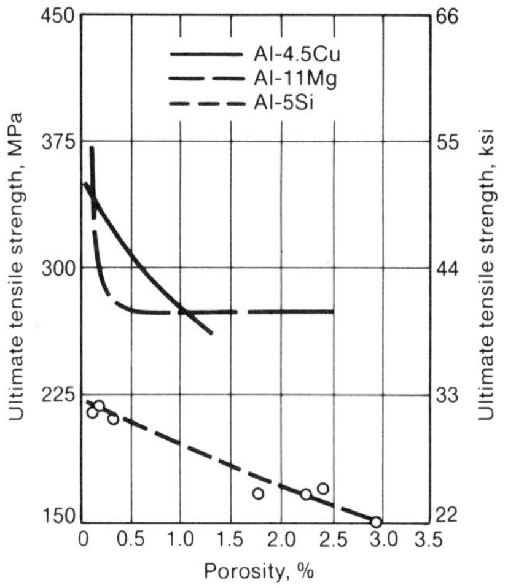

Fig. 3 Ultimate tensile strength versus hydrogen porosity for sand cast bars of three aluminum alloys

value for the specific composition and temperature

- With time at temperature, oxidation of the melt occurs; in alloys containing magnesium, oxidation losses and the formation of complex oxides may not be self-limiting
- Transient elements characterized by low vapor pressure and high reactivity are reduced as a function of time at temperature; magnesium, sodium, calcium, and strontium, upon which mechanical properties directly or indirectly rely, are examples of elements that display transient characteristics

Turbulence or agitation of the melt and increased holding temperature significantly increase the rate of hydrogen solution, oxidation, and transient element loss. The mechanical properties of aluminum alloys depend on casting soundness, which is strongly influenced by hydrogen porosity and entrained nonmetallic inclusions. Reductions in dissolved hydrogen content and in suspended included matter are normally accomplished through treatment of the melt before pouring.

Hydrogen

Hydrogen is the only gas that is appreciably soluble in aluminum and its alloys. Its solubility varies directly with temperature and the square root of pressure. As shown in Fig. 1, hydrogen solubility is considerably greater in the liquid than in the solid state. Actual liquid and solid solubilities in pure aluminum just above and below the solidus are 0.65 and 0.034 mL/100 g, respectively. These values vary slightly with alloy content. During the cooling and solidification of molten aluminum, dissolved hydrogen in excess of the extremely low solid solubility may precipitate in molecular form, resulting in the formation of primary and/or secondary voids.

Hydrogen bubble formation is strongly resisted by surface tension forces, by liquid cooling and solidification rates, and by an absence of nucleation sites for hydrogen precipitation such as entrained oxides. Dissolved hydrogen concentrations significantly in excess of solid solubility are therefore required for porosity formation. In the absence of nucleating oxides, relatively high concentrations (of the order of 0.30 mL/100 g) are required for hydrogen precipitation. No porosity was found to occur in a range of common alloys at hydrogen concentrations as high as 0.15 mL/100 g.

Hydrogen Sources. There are numerous sources of hydrogen in aluminum. Moisture in the atmosphere dissociates at the molten metal surface, offering a concentration of atomic hydrogen capable of diffusing into the melt. The barrier oxide of aluminum resists hydrogen solution by this mechanism, but disturbances of the melt surface that break the oxide barrier result in rapid hydrogen dissolution. Alloying elements, especially magnesium, may also affect hydrogen absorption by forming oxidation reaction products that offer reduced resistance to the diffusion of hydrogen into the melt and by altering liquid solubility. The introduction of moisture-contaminated tools to the melt dramatically increases dissolved hydrogen levels. Salt fluxes, which may have hydroscopically adsorbed moisture, and unprotected fluxing tubes coated by reaction product salts both increase dissolved hydrogen content. The melt charge may contain both dissolved hydrogen and, when not preheated before charging, moisture-contaminated surfaces.

An additional source of hydrogen, especially in green sand casting, is the reaction involving molten metal and water in the mold itself. In addition, the turbulence that inevitably occurs during drawing, pouring, and to some extent within the gating system increases the potential for hydrogen solution and subsequent precipitation.

Hydrogen Porosity. Two types or forms of hydrogen porosity may occur in cast aluminum. Of greater importance is interdendritic porosity, which is encountered when hydrogen contents are sufficiently high that hydrogen rejected at the solidification front results in solution pressures above atmospheric. Secondary (micron-size) porosity occurs when dissolved hydrogen contents are low, and void formation is characteristically subcritical.

Finely distributed hydrogen porosity may not always be undesirable. Hydrogen precipitation may alter the form and distribution of shrinkage porosity in poorly fed parts or part sections. Shrinkage is generally more harmful to casting properties. In isolated cases, hydrogen may actually be intentionally introduced and controlled in specific concentrations compatible with the application requirements of the casting in order to promote superficial soundness.

Nevertheless, hydrogen porosity adversely affects mechanical properties in a manner that varies with the alloy. Figure 2 shows the relationship between actual hydrogen content and observed porosity. Figure 3 defines the effect of porosity on the ultimate tensile strength of selected compositions.

It is often assumed that hydrogen may be desirable or tolerable in pressure-tight applications. The assumption is that hydrogen porosity is always present in the cast structure as integrally enclosed rounded voids. In fact, hydrogen porosity may occur as rounded or elongated voids and in the presence of shrinkage may decrease rather than increase resistance to pressure leakage.

Hydrogen in Solid Solution. The disposition of hydrogen in a solidified structure depends on the dissolved hydrogen level and the conditions under which solidification occurs. Because the presence of hydrogen porosity is a result of diffusion-controlled nucleation and growth, decreasing the hydrogen concentration and increasing the rate of solidification act to suppress void formation and growth. For this reason, castings made in expendable mold processes are more susceptible to hydrogen-related defects than parts produced by permanent mold or pressure die casting.

Hydrogen Removal. Dissolved hydrogen levels can be reduced by a number of methods, the most important of which is fluxing with dry, chemically pure nitrogen, argon, chlorine, and freon. Compounds such as hexachloroethane are in common use; these compounds dissociate at molten metal temperatures to provide the generation of fluxing gas.

Gas fluxing reduces the dissolved hydrogen content of molten aluminum by partial pressure diffusion. The use of reactive gases such as chlorine improves the rate of degassing by altering the gas/metal interface to improve diffusion kinetics. Holding the melt undisturbed for long periods of time at or near the liquidus also reduces hydrogen content to a level no greater than that defined for the alloy as the temperature-dependent liquid solubility.

Measurement of Hydrogen. It is possible to define quantitatively the concentration of hydrogen in an aluminum melt by liquid and solid-state extraction techniques. One instrument offers real time, accurate measurement of hydrogen in molten aluminum. A variety of solid-state extraction techniques exist for measuring hydrogen in aluminum after solidification. A number of devices have also been developed to quantify hydrogen content in molten aluminum. Commercially available systems of this type rely on the correlation of parameters governing the evolution of hydrogen from samples solidifying under vacuum (time, temperature, and pressure) or may employ other relationships to approximate or define hydrogen content. Semiquantitative tests, also based on the application of vacuum, are sometimes used.

Samples placed in a vacuum chamber should be visually observed during solidification. The rapid evolution of gas bubbles with the application of vacuum indicates a high level of oxide contamination and an unknown hydrogen level. The emergence of bubbles from the sample surface only in the last stages of solidification indicates an absence of oxides and moderately high hydrogen levels. In many foundries, vacuum solidification samples are not observed during solidification but are subsequently sectioned to obtain comparative ratings of void content. The latter technique is inferior to the analysis obtained by the visual interpretation of events that occur during sample solidification.

The selection of vacuum pressure for performing vacuum solidification tests is important. For very high quality castings, pressures of 2 to 5 mm (0.08 to 0.2 in.) of mercury are employed as the most rigorous standard. In these cases, any evolution from the visually observed sample indicates unacceptable melt quality. Different absolute test pressures are selected for lower quality acceptance limits.

For foundries relying on specific gravity determinations performed on the solidified sample, 102 mm (4 in.) of mercury is a desirable test pressure. Although not accurate for the assessment of hydrogen *per se*, the density control system has proved remarkably effective in predicting acceptance limits based on the combined influence of dissolved hydrogen and oxides on porosity formation and offers, along with more accurate measures of hydrogen content, the opportunity of statistical treatment.

Because hydrogen solubility is inversely dependent on absolute pressure during solidification, any reduction in vacuum pressure results in reduced sensitivity in whatever approach is employed in vacuum testing for hydrogen detection. It is essential that reproducible conditions be ensured regardless of the test and test pressure employed. Systems that rely on the differential from atmospheric pressure are markedly inferior to those that measure absolute and controllable test chamber pressure, preferably in millimeters of mercury.

Hydrogen porosity is normally distinguishable in casting structures by radiography, by machining tests, and by low-power or microscopic examination. Discrimination between porosity caused by shrinkage and by hydrogen is difficult and is often subject to misinterpretation. Hydrogen porosity is generally found to affect the casting cross section uniformly, with only small variations in void size and shape occurring within the cast structure.

Oxidation

Oxide Formation. Aluminum and its alloys oxidize readily in both the solid and molten states to provide a continuous self-limiting film. The rate of oxidation increases with temperature and is substantially greater in molten than in solid aluminum. The reactive elements contained in alloys such as magnesium, strontium, sodium, calcium, beryllium, and titanium are also factors in oxide formation. In both the molten and solid states, oxide formed at the surface offers benefits in self-limitation and as a barrier to hydrogen diffusion and solution. Induced turbulence, however, results in the entrainment of oxide particles, which resist gravity separation because their density is similar to that of molten aluminum.

Oxides are formed by direct oxidation in air, by reaction with water vapor, or by aluminothermic reaction with oxides of other metals, such as iron or silicon, contained in tools and refractories. Aluminum oxide is polymorphic, but at molten metal temperature the common forms of oxide encountered are crystalline and of a variety of types depending on exposure, temperature, and time. Some crystallographic oxide forms affect the appearance and coloration of castings, without other significant effects.

Special consideration must be given to alloys containing magnesium, which oxidize more readily and more continuously as a function of environment, time, and temperature. The result can be increased melt loss and oxide generation, which increases with magnesium concentration. Magnesium oxide occurs most frequently in the form of micron-size particles. At higher holding temperatures (>745 °C, or 1375 °F), complex aluminum-magnesium oxide (spinel) is formed with a potential for rapid growth. In some forms, spinel assumes a hard, black crystalline form that contaminates furnaces, holding crucibles, and ladles. Although spinels occur only at temperatures exceeding conventional holding temperatures, it is important to consider localized rather than bulk temperatures in metal melting, holding, handling, and treatment. The impingement of gas burners, excessive temperatures at the melt surface regardless of heat source, and exothermic reactions that occur during fluxing treatments may result in the thermal conditions required for spinel formation.

Oxide Separation and Removal. It is usually necessary to treat melts of aluminum and its alloys to remove suspended nonmetallics. This is normally accomplished by using either solid or chemically active gaseous fluxes containing chlorine, fluorine, chlorides, and/or fluorides. In each case, the objective is the dewetting of the oxide/melt interface to provide effective separation of oxides and other included matter and the flotation of these nonmetallics by attachment to either solid or gaseous elements or compounds introduced or formed during flux treatment.

Fluxes can also be used to minimize oxide formation. For this reason, melts containing magnesium are often protected by the use of salts that form liquid layers, most often of magnesium chloride, on the melt surface. These fluxes, termed covering fluxes, must be periodically removed and replaced. Carbon, graphite, and boron powder also effectively retard oxidation when applied to the melt surface.

It is increasingly common to employ filtration in the treatment of molten metal to remove suspended nonmetallic particles. Such processes can be used in the transfer of metal from furnace to ladle or crucible, within crucibles and furnaces, in holding or drawing chambers, and within the mold gating system. Porous foam ceramics, bonded rigid media elements, fused permeable refractories, and depth loose medium filter systems are all in use in aluminum foundry operations. Not to be confused with ceramic strainers, metal screens, and metallic wools useful only for gross oxide separation and as gating system chokes, true molten metal filtration is capable of substantially reducing the nonmetallic content of metal introduced to the casting, measured at times in microns.

Effects of Inclusions. In addition to oxides, a number of additional compounds can be considered inclusions in cast structures. All aluminum contains aluminum carbide (Al_4C_3) formed during reduction. Borides may also be present; by agglomeration, borides can assume sufficient size to represent a significant factor in the metal structure, with especially adverse effects in machining.

Under all conditions, inclusions—whether in film or particle form—are damaging to mechanical properties. The gross effect of inclusions is to reduce the effective cross section of metal under load. The more devastating effect on properties is that of stress concentration when inclusions appear at or near the surface of parts or specimens. Fatigue performance is reduced under the latter condition by the notch effect. Ultimate and yield strengths are typically lower, and ductility may be substantially reduced when inclusions are present.

Hard particle inclusions are frequently found in association with film-type oxides. Borides, carbides, oxides, and nonmetallic particles in the melt are scavenged and then concentrated in localized regions within the cast structure.

Inclusion content directly affects fluidity and feeding capability. The effect on shrinkage can be both general (based on feeding effectiveness) and localized (the barrier effect of some oxides on molten metal flow during solidification). The tendency for nonfills also increases with oxide content. Misruns can be minimized by ensuring a low level of oxide contamination.

Tests for Oxides. The presence of inclusions can be determined by vacuum solidification tests, radiography, metallography, and machining tests. The presence of oxide is frequently detected in mechanical test specimens because failure in standard tensile testing will occur through the weakest (inclusion-affected) plane. More sophisticated quantitative tests are available, such as specimen leaching, atomic and neutron absorption, and neutron activation. Quantitative determinations of oxide content are rarely performed because practical samples are of limited size, for which test results may not be meaningfully correlated with casting quality. Ultrasonic and electrical conductivity tests for molten aluminum are under development, but at this time are unproven in production casting operations.

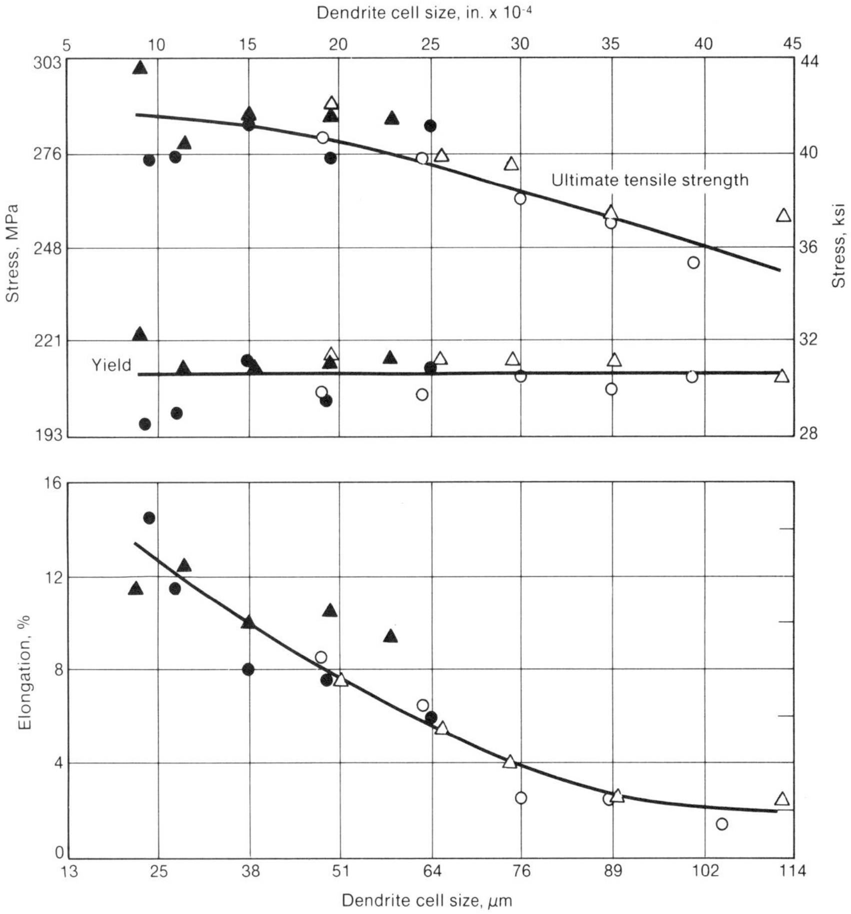

Fig. 4 Tensile properties versus dendrite cell size for four heats of aluminum alloy A356-T62 plaster cast plates

Structure Control

A number of factors define the metallurgical structure in aluminum castings. Of primary importance are dendrite cell size or dendrite arm spacing, the form and distribution of microstructural phases, and grain size. The foundryman can control the fineness of dendrite structure by controlling the rate of solidification.

Microstructural features such as the size and distribution of primary and intermetallic phases are considerably more complex. However, chemistry control (particularly control of impurity element concentrations), control of element ratios based on the stoichiometry of intermetallic phases, and control of solidification conditions to ensure uniform size and distribution of intermetallics are all means to this end. Effective grain refinement shares and strongly influences the same objectives. The use of modifiers and refiners to influence eutectic and hypereutectic structures in aluminum-silicon alloys is also an example of the manner in which microstructures and macrostructures can be optimized in foundry operations.

Dendrite Arm Spacing

In all commercial processes, solidification takes place through the formation of dendrites in the liquid solution. The cells contained within the dendrite structure correspond to the dimensions separating the arms of primary dendrites and are controlled for a given composition exclusively by solidification rate.

Through microstructural examination, it is possible to define the rate at which a given region of a casting has solidified by reference to data obtained from unidirectionally solidified samples spanning solidification rates represented by the full range of various casting processes. Figure 4 illustrates the improvement in mechanical properties achievable by the change in dendrite formation controlled by solidification rate.

In premium engineered castings and in many other casting applications, careful attention is given to obtaining solidification rates corresponding to optimum mechanical property development. Solidification rate affects more than dendrite cell size, but

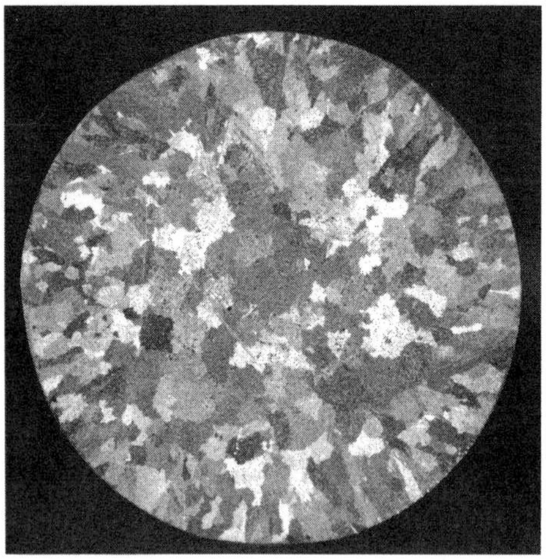

(a)

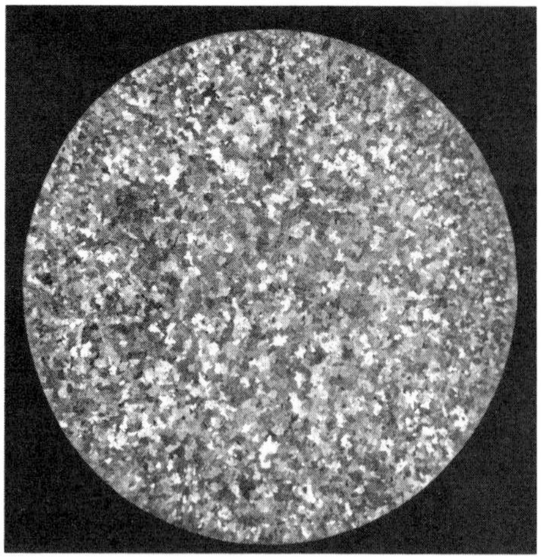

(b)

Fig. 5 As-cast Al-7Si ingots showing the effects of grain refinement. (a) No grain refiner. (b) Grain refined. Both etched using Poulton's etch; both 2×. Courtesy of W.G. Lidman, KB Alloys, Inc.

dendrite cell size measurements are becoming increasingly important.

Grain Structure

A fine, equiaxed grain structure is normally desired in aluminum castings. The type and size of grains formed are determined by alloy composition, solidification rate, and the addition of master alloys (grain refiners) containing intermetallic phase particles, which provide sites for heterogeneous grain nucleation.

Grain Refinement Effects. A finer grain size promotes improved casting soundness by minimizing shrinkage, hot cracking, and hydrogen porosity. The advantages of effective grain refinement are:

- Improved feeding characteristics
- Increased tear resistance
- Improved mechanical properties
- Increased pressure tightness
- Improved response to thermal treatment
- Improved appearance following chemical, electrochemical, and mechanical finishing

Under normal solidification conditions spanning the full range of commercial casting processes, aluminum alloys without grain refiners exhibit coarse columnar and/or coarse equiaxed structures. The coarse columnar grain structure (Fig. 5a) is less resistant to cracking during solidification and postsolidification cooling than the well-refined grain structure of the same alloy shown in Fig. 5(b). This is because reduced resistance to tension forces at elevated temperature may be expected as a result of increased sensitivity to grain-boundary formations in coarse-grain structures.

A fine grain structure also minimizes the effects on castability and properties associated with the size and distribution of normally occurring intermetallics. Large, insoluble intermetallic particles that are present or form in the temperature range between liquidus and solidus reduce feeding. A finer grain size promotes the formation of finer, more evenly distributed intermetallic particles with corresponding improvements in feeding characteristics. Because most of these more brittle phases precipitate late in the solidification process, their preferential formation at grain boundaries also profoundly affects tear resistance and mechanical properties in coarse-grain structures. By reducing the magnitude of grain-boundary effects through grain refinement, the hot cracking tendencies of some predominantly solid-solution alloys, such as those of the 2*xx* and 5*xx* families, can be substantially reduced.

Porosity, if present, is of smaller discrete void size in fine-grain parts. The size of interdendritic shrinkage voids is directly influenced by grain size. The previously mentioned effects of structural refinement on feeding characteristics minimize the potential formation of larger shrinkage cavities, and when hydrogen porosity is present, larger pore size with more damaging impact on properties will be experienced in coarse-grain rather than fine-grain castings.

The finer distribution of soluble intermetallics throughout grain-refined castings results in faster and more complete response to thermal treatment. More consistent mechanical properties can be expected following thermal treatment.

Large grains frequently emphasize different reflectances based on random crystal orientation, resulting in an orange peel or spangled appearance following chemical finishing, anodizing, or machining. Tearing also becomes more pronounced in the machining of soft coarse-grain compositions.

Grain Refinement. All aluminum alloys can be made to solidify with a fully equiaxed, fine grain structure through the use of suitable grain-refining additions. The most widely used grain refiners are master alloys of titanium, or of titanium and boron, in aluminum. Aluminum-titanium refiners generally contain from 3 to 10% Ti. The same range of titanium concentrations is used in Al-Ti-B refiners with boron contents from 0.2 to 1% and titanium-to-boron ratios ranging from about 5 to 50. Although grain refiners of these types can be considered conventional hardeners or master alloys, they differ from master alloys added to the melt for alloying purposes alone. To be effective, grain refiners must introduce controlled, predictable, and operative quantities of aluminides (and borides) in the correct form, size, and distribution for grain nucleation. Wrought refiner in rod form, developed for the continuous treatment of aluminum in primary operations, is available in sheared lengths for foundry use. The same grain-refining compositions are furnished in waffle form. In addition to grain-refining master alloys, salts (usually in compacted form) that react with molten aluminum to form combinations of $TiAl_3$ and TiB_2 are also available.

Grain Refinement Mechanisms. Despite much progress in understanding the fundamentals of grain refinement, no universally accepted theory or mechanism exists to satisfy laboratory and industrial experience. It is known that $TiAl_3$ is an active phase in the nucleation of aluminum crystals, ostensibly because of similarities in crystallographic lattice spacing. Nucleation may occur on $TiAl_3$ substrates that are undissolved or precipitate at sufficiently high titanium concentrations by peritectic reaction. Grain refinement can be achieved at much lower titanium concentrations than those predicted by the binary Al-Ti peritectic point of 0.15%. For this reason, other theories, such as conucleation of the aluminide by TiB_2 or carbides and constitutional effects on the peritectic reaction, are presumed to be influential. Recent findings also suggest the active role of more complex borides of the Ti-Al-B type in grain nucleation.

Additions of titanium in the form of master alloys to aluminum casting compositions normally result in significantly finer and equiaxed grain structure. The period of effectiveness following grain-refiner addition and the potency of grain-refining action are enhanced by the presence of TiB_2. In the testing of some compositions, notably those of the aluminum-silicon family, aluminum borides and titanium boride in the absence of excess titanium have been found to provide effective grain refinement. However, the requirement of an excess of titanium compared to stoichiometric balance with boron in TiB_2 is commonly accepted for optimum grain-refining results, and titanium or higher-ratio titanium-boron master alloys are used almost exclusively for grain size control.

The role of boride in enhancing grain refinement effectiveness and extending its useful duration is observed in both casting and wrought alloys, forming the basis for its use. However, when the boride is present in the form of large, agglomerated particles, it assumes the character of a highly objectionable inclusion with especially damaging effects in machining. Particle agglomeration is found in master alloys of poor quality, or it may occur as a result of long, quiescent holding periods. For the latter reason, it is essential that holding furnaces be routinely and thoroughly cleaned when boron-containing master alloys are used.

The objective in every case in which master alloys or other grain refiners are added to the melt is the release of constituent particles capable of nucleating grain formation to ensure uniform, fine, equiaxed grain structure. The selection of an appropriate grain refiner, practices for grain-refiner addition, and practices covering holding and pouring of castings following grain-refiner addition are usually developed by the foundry after considering casting and product requirements and after referral to the performance characteristics of commercial grain refiners furnished by the supplier. However, grain refiners of the 5Ti-1B and 5Ti-0.6B types, which are characterized by cleanliness and fine, uniform distribution of aluminide and boride phases when added to the melt at 0.01 to 0.03% Ti, should be expected to provide acceptable grain refinement under most conditions.

Grain Size Tests. Various tests have been devised to sample molten aluminum for the purpose of determining the effectiveness of grain refinement. These tests employ principles of controlled solidification to ensure accurate, meaningful, and reproducible grain size determination following polishing and etching. For many foundries, a useful standardized test can use a dry sand mold of fixed design or routine samples, such as those poured in vacuum solidification or for mechanical property tests. The grain size of cast structures obtained by sectioning the casting itself, polishing, and etching is also determined under magnification. Determinations are usually comparative and judgmental, but quantitative grain size measurement by intercept methods is also practiced. Thermoanalytical and electrical conductivity methods are under development for the nondestructive and predictive assessment of grain structure.

Modification and Refinement of Aluminum-Silicon Alloys

Hypoeutectic aluminum-silicon alloys can be improved by inducing structural modification of the normally occurring eutectic. In general, the greatest benefits are achieved in alloys containing from 5% Si to the eutectic concentration; this range includes most common gravity cast compositions.

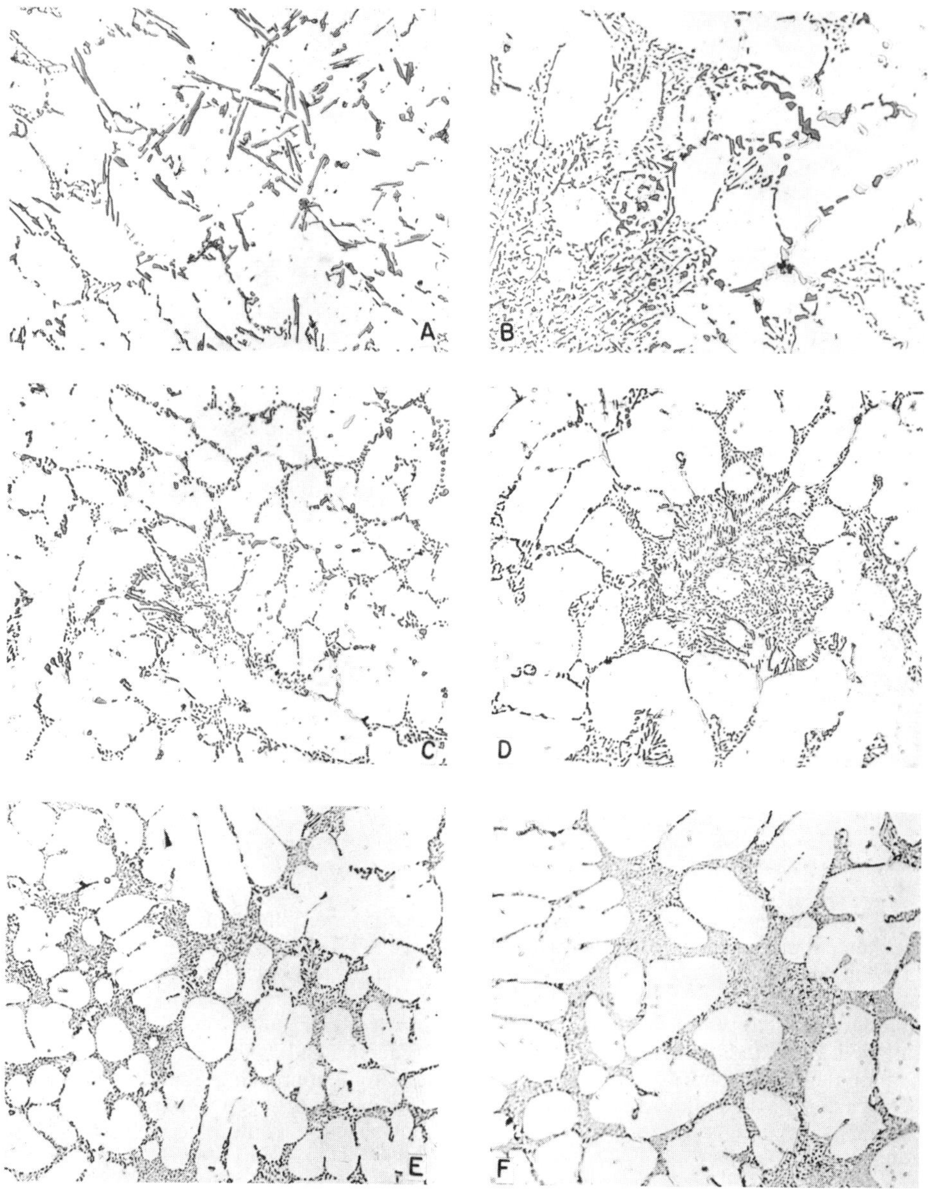

Fig. 6 Varying degrees of aluminum-silicon eutectic modification ranging from unmodified (A) to well modified (F). See Fig. 7 for the effectiveness of various modifiers.

Chemical Modifiers

The addition of certain elements, such as calcium, sodium, strontium, and antimony, to hypoeutectic aluminum-silicon alloys results in a finer lamellar or fibrous eutectic network. It is also understood that increased solidification rates are useful in providing similar structures. There is, however, no agreement on the mechanisms involved. The most popular explanations suggest that modifying additions suppress the growth of silicon crystals within the eutectic, providing a finer distribution of lamellae relative to the growth of the eutectic. Various degrees of eutectic modification are shown in Fig. 6.

The results of modification by strontium, sodium, and calcium are similar. Sodium has been shown to be the superior modifier, followed by strontium and calcium, respectively. Each of these elements is mutually compatible so that combinations of modification additions can be made without adverse effects. Eutectic modification is, however, transient when artificially promoted by additions of these elements. Figure 7 illustrates the relative effectiveness of various modifiers as a function of time at temperature.

Antimony has been advocated as a permanent means of achieving structural modification. In this case, the modified structure differs; a more acicular refined eutectic is obtained compared to the uniform lace-like dispersed structures of sodium-, calcium-, or strontium-modified metal. As a re-

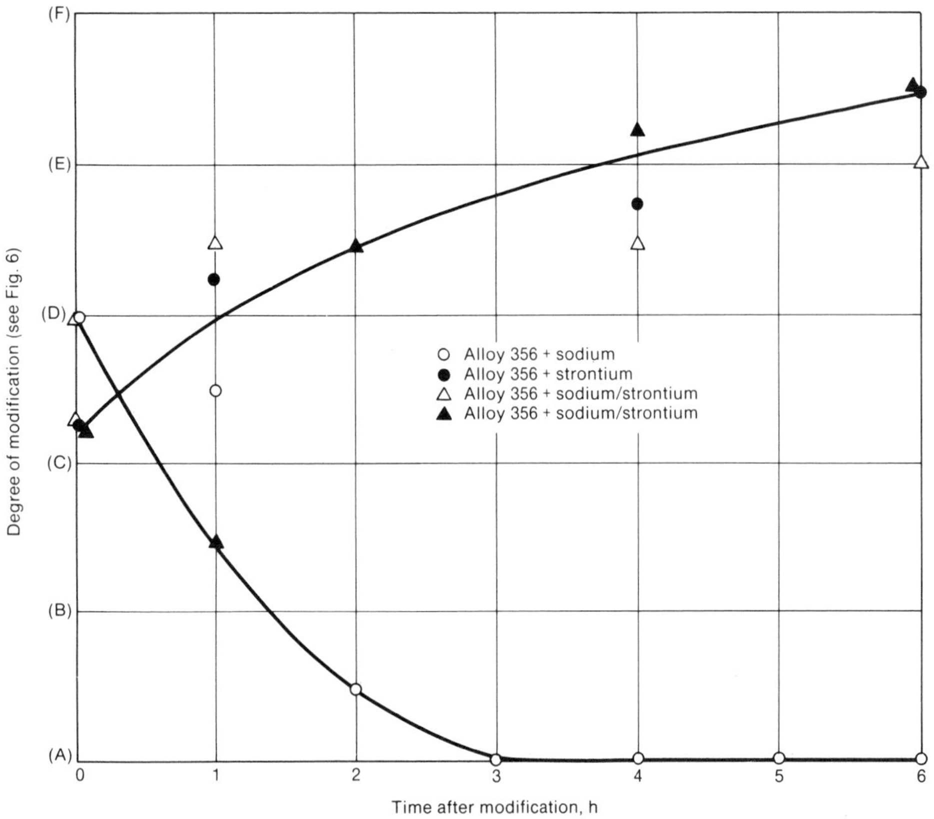

Fig. 7 Effectiveness of sodium and strontium modifiers as a function of time. See Fig. 6 for degrees of modification.

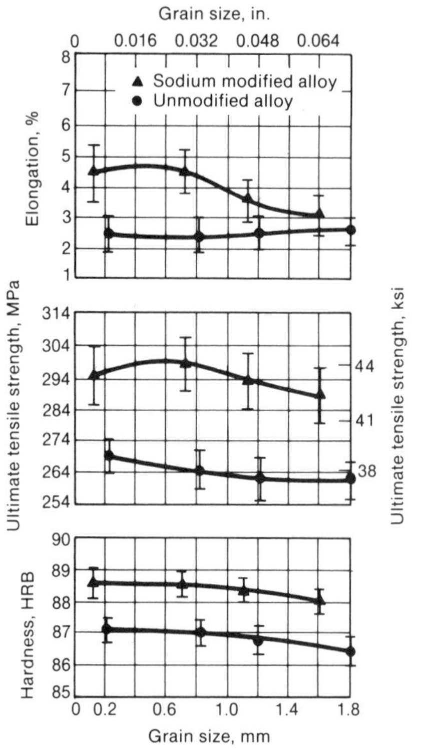

Fig. 8 Mechanical properties of as-cast A356 alloy tensile specimens as a function of modification and grain size

sult, the improvements in castability and mechanical properties offered by this group of elements are not completely achieved. Structural refinement is obtained that is time independent when two conditions are satisfied. First, the metal to be treated must be essentially phosphorus free, and second, the velocity of the solidification front must exceed a minimum value approximately equal to that obtained in conventional permanent mold casting.

Antimony is not compatible with other modifying elements. In cases in which antimony and other modifiers are present, coarse antimony-containing intermetallics are formed that preclude the attainment of an effectively modified structure and adversely affect casting results.

Modifier additions are usually accompanied by an increase in hydrogen content. In the case of sodium and calcium, the reactions involved in element solution are invariably turbulent or are accompanied by compound reactions that by their nature increase dissolved hydrogen levels. In the case of strontium, master alloys may be highly contaminated with hydrogen, and there are numerous indications that hydrogen solubility is increased after alloying.

For sodium, calcium, and strontium modifiers, the removal of hydrogen by reactive gases also results in the removal of the modifying element. Recommended practices are to obtain modification through additions of modifying elements added to well-processed melts, followed by inert gas fluxing to acceptable hydrogen levels. No such disadvantages accompany antimony use.

Calcium and sodium can be added to molten aluminum in metallic or salt form. Vacuum-prepackaged sodium metal is commonly used. Strontium is currently available in many forms, including aluminum-strontium master alloys ranging from approximately 10 to 90% Sr and Al-Si-Sr master alloys of varying strontium content.

Very low sodium concentrations (~0.001%) are required for effective modification. More typically, additions are made to obtain a sodium content in the melt of 0.005 to 0.015%. Remodification is performed as required to maintain the desired modification level.

A much wider range of strontium concentrations is in use. In general, addition rates far exceed those required for effective sodium modification. A range of 0.015 to 0.050% is standard industry practice. Normally, good modification is achievable in the range of 0.008 to 0.015% Sr. Remodification through strontium additions may be required, although retreatment is less frequent than for sodium.

To be effective in modification, antimony must be alloyed to approximately 0.06%. In practice, antimony is employed in the much higher range of 0.10 to 0.50%.

It is possible to achieve a state of overmodification, in which eutectic coarsening occurs, when sodium and/or strontium are used in excessive amounts. The corollary effects of reduced fluidity and susceptibility to hydrogen-related problems are usually encountered well before overmodification may be experienced.

The Importance of Phosphorus. It has been well established that phosphorus interferes with the modification mechanism. Phosphorus reacts with sodium and probably with strontium and calcium to form phosphides that nullify the intended modification additions. It is therefore desirable to use low-phosphorus metal when modification is a process objective and to make larger modifier additions to compensate for phosphorus-related losses.

Primary producers may control phosphorus contents in smelting and processing to provide less than 5 ppm of phosphorus in alloyed ingot. At these levels, normal additions of modification agents are effective in achieving modified structures. However, phosphorus contamination may occur in the foundry through contamination by phosphate-bonded refractories and mortars and by phosphorus contained in other melt additions, such as master alloys and alloying elements including silicon.

Effects of Modification. Typically, modified structures display somewhat higher tensile properties and appreciably improved ductility when compared to similar but unmodified structures. Figure 8 illustrates the desirable effects on mechanical properties that can be achieved by modification. Improved performance in casting is characterized by improved flow and feeding as well as by superior resistance to elevated-temperature cracking.

Refinement of Hypereutectic Aluminum-Silicon Alloys

The elimination of large, coarse primary silicon crystals that are harmful in the casting and machining of hypereutectic silicon alloy compositions is a function of primary silicon refinement. Phosphorus added to molten alloys containing more than the eutectic concentration of silicon, made in the form of metallic phosphorus or phosphorus-containing compounds such as phosphor-copper and phosphorus pentachloride, has a marked effect on the distribution and form of the primary silicon phase. Investigations have shown that retained trace concentrations as low as 0.0015 through 0.03% P are effective in achieving the refined structure. Disagreements on recommended phosphorus ranges and addition rates have been caused by the extreme difficulty of accurately sampling and analyzing for phosphorus. More recent developments employing vacuum stage spectrographic or quantometric analysis now provide rapid and accurate phosphorus measurements.

Following melt treatment by phosphorus-containing compounds, refinement can be expected to be less transient than the effects of conventional modifiers on hypoeutectic modification. Furthermore, the solidification of phosphorus-treated melts, cooling to room temperature, reheating, remelting, and resampling in repetitive tests have shown that refinement is not lost; however, primary silicon particle size increases gradually, responding to a loss in phosphorus concentration. Common degassing methods accelerate phosphorus loss, especially when chlorine or freon is used. In fact, brief inert gas fluxing is frequently employed to reactivate aluminum phosphide nuclei, presumably by resuspension.

Practices that are recommended for melt refinement are as follows:

- Melting and holding temperature should be held to a minimum
- The alloy should be thoroughly chlorine or freon fluxed before refining to remove phosphorus-scavenging impurities such as calcium and sodium
- Brief fluxing after the addition of phosphorus is recommended to remove the hydrogen introduced during the addition and to distribute the aluminum phosphide nuclei uniformly in the melt

Figure 9 illustrates the microstructural differences between refined and unrefined structures.

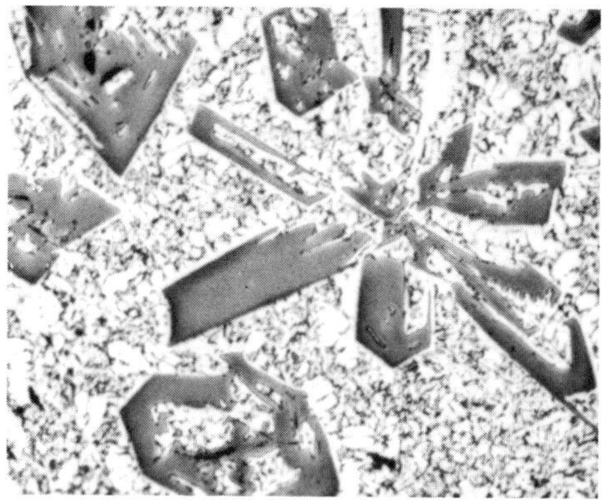

(a)

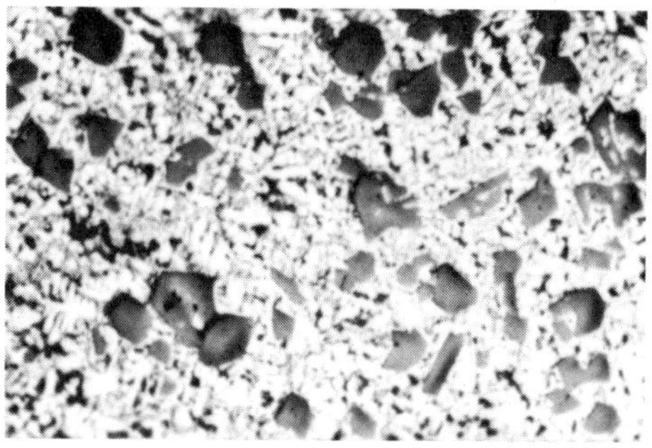

(b)

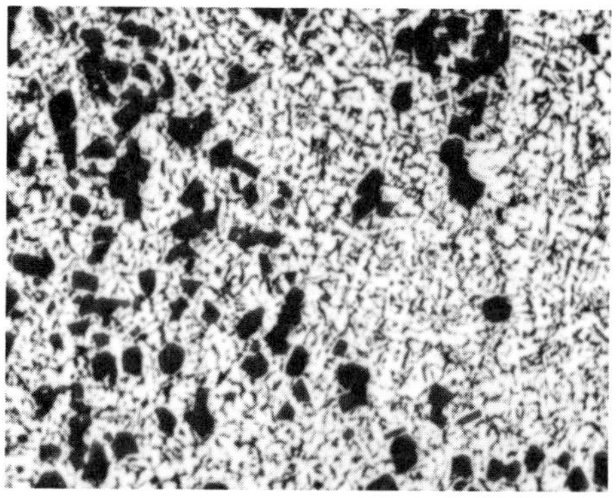

(c)

Fig. 9 Effect of phosphorus refinement on the microstructure of Al-22Si-1Ni-1Cu alloy. (a) Unrefined. (b) Phosphorus-refined. (c) Refined and fluxed. All 100×

Effects of Refinement. Refinement substantially improves mechanical properties and castability. In some cases, especially at higher silicon concentrations, refinement forms the basis for acceptable foundry results.

Modification and Refinement. No elements are known that beneficially affect both eutectic and hypereutectic phases. The potential negative consequences of employing modifying and refining additions in the melt are characterized by the interaction of phosphorus with calcium, sodium, and strontium. Strontium has been claimed to benefit hypoeutectic and hypereutectic structures, but this claim has not been substantiated.

Metal Preparation

Gravity Casting. Regardless of the types of melting and holding furnaces and the particular gravity casting process used, there is great concern for reducing or eliminating dissolved hydrogen and entrained oxides. These procedures are less frequently employed for pressure die casting, in which concerns are focused on the dominant process-related causes of casting unsoundness, namely, entrapped gas and pouring injection-associated inclusions.

Sensitivity to melt quality varies with the casting process and part design and necessitates special consideration of relevant criteria for each application. In general, the melt is processed to achieve hydrogen reductions and the removal of oxides to meet specific casting requirements. Modification and grain-refiner additions are made as appropriate to the given alloy and end product.

Die Casting. Different melt preparation practices are employed in die casting operations because process-related conditions are more dominant in the control of product quality than those controlled by melt treatment. For this reason, degassing for the removal of hydrogen, grain refinement, and modification or silicon refinement in the case of hypereutectic silicon alloys are often intentionally neglected. The movement toward higher-integrity die castings has brought into focus the importance of the same melt quality parameters established and used in the gravity casting of aluminum alloys.

In high-production die casting operations, the consumption of internal and external scrap is of primary importance in reducing base metal costs for the predominantly secondary alloy compositions that are consumed. Scrap crushing, shredding, and pretreatment of various types precede melting, often in efficient induction systems. Oxides entrained in the melt as a result of this sequence of operations are dealt with through the use of salts and/or reactive gas fluxing. Melt treatment is typically confined to this and to rudimentary fluxing in holding furnaces to remove gross oxide and to facilitate the maintenance of minimum furnace cleanliness.

A concern in die casting is the formation of complex intermetallics that are insoluble at melt-holding temperatures and/or precipitate under holding conditions or during transfer to and injection from the hot or cold chamber. These intermetallics (sludge) affect furnaces, transfer systems, and, by inclusion, the quality of the castings produced. Die casters are familiar with composition limits that prevent sludge formation. A common rule is that iron content plus two times manganese content plus three times chromium content should not exceed the sum of 1.7%. This limit is arbitrary and inexact, it is often assigned values from 1.5 through 1.9%, and it is subject to the specific composition and actual minimum process temperature.

Pouring

It is critically important that the metal be drawn and poured according to the best manual or automatic procedures. These procedures avoid excessive turbulence, minimize oxide generation and entrainment, and limit regassing of hydrogen. Frequent skimming of the melt surface from which metal is drawn may be necessary to minimize oxide contamination in the ladle. Siphon ladles that fill from below the melt surface are used for these purposes, but most often, coated and preheated ladles of simple design are employed. The process of repetitive drawing and skimming inevitably degrades melt quality, and this necessitates reprocessing if required melt quality limits are exceeded.

Pouring should take place at the lowest position possible relative to the pouring basin or sprue opening. Once pouring is initiated, the sprue must be continuously filled to minimize aspiration and to maintain the integrity of flow in gates and runners.

Counter-gravity mold filling methods inherently overcome most of the disadvantages of manual pouring. Proprietary casting processes based on low pressure, displacement, or pumping mechanisms might be considered optimum for preserving the processed quality of the melt through mold filling, but some important and relevant considerations apply. Melt processing by fluxing is more difficult in some cases because the crucible or metal source may be confined. If, as in the case of low-pressure casting, the passage for introducing metal into the mold is used repetitively, its inner surface becomes oxide contaminated and a source of casting inclusions. In other counter-gravity casting, mold intrusion into the melt and devices employed to displace or pump metal to the cavity may be the source of turbulence, moisture reactions, and the possibility of hydrogen regassing.

Automated pouring systems are common in the die casting industry. Robotized ladle transfer, as well as metered pumping, may nevertheless incorporate features and reflect provisions to protect molten metal quality through sound drawing, transfer, and pouring techniques. The same techniques have application in die casting operations in which these operations are performed manually.

Hot chamber operation offers apparent metal transfer advantages over cold chamber operation. Recent developments in the use of siphons or vacuum legs to the cold chamber in pressure die casting offer new and interesting opportunities for upgrading the quality of the metal deliverable to the die cavity.

Gating and Risering

It is beyond the scope of this article to discuss comprehensively the subject of gating and risering for the processes employed in the casting of aluminum. In fact, it is in the practices, methods, and designs of gating and risering that individual foundries most differentiate their capabilities. For the most part, the evolution of these systems has been based on experience, and effective and imaginative solutions incorporating refined fluid and solidification dynamics have been developed.

More recently, technical societies and associations have fostered and developed sophisticated techniques for the design of gates, runners, and risers in gravity casting and for the design of gates, runners, injection models, and analytical control schemes for pressure die casting.

There is general agreement on the principles applicable to this highly individualistic and vital phase of metal casting. Ultimately, the challenge to the foundryman is the transfer of metal at the desired quality level to the casting cavity while retaining an acceptable level of metal quality through transfer and ensuring that solidification occurs in such a way that an acceptable level of surface, dimensional, and internal quality is attained. Transfer initiates with drawing and/or pouring, and it concludes with the final compensation of volumetric shrinkage by the riser system.

Gating and Risering Principles. The methods for introducing metal into the casting cavity, for minimizing degradation in metal quality, and for minimizing the occurrence of shrinkage porosity in the solidifying casting differ among the various casting processes, primarily as a function of process limitations. However, the objectives and principles of gating and risering are universally applicable:

- Establish nonturbulent metal flow
- Systematically fill the mold cavity with metal of minimally degraded quality
- In conjunction with the selection of an appropriate pouring temperature, provide conditions for mold filling consistent with misrun avoidance
- Establish thermal gradients within the cavity to promote directional solidification and to enhance riser effectiveness
- Design riser size and geometry, and locate risers and riser inlets to minimize the ratio of gross weight to net weight
- Minimize to the extent possible the vertical distance the metal must travel from the lowest position of metal entry to the base of the sprue
- Taper the sprue or use a sprue geometry other than cylindrical to minimize vortexing and aspiration
- Keep the sprue continuously filled during pouring
- Avoid abrupt changes in the direction of metal flow; gate and runner passages should be streamlined for minimum induced turbulence at angles or points of divergence in the system
- Provide contoured transitions in gate, runner, and infeed cross sections at points of cross-sectional area changes
- Employ multiple gates to improve thermal distribution and to reduce metal velocity at entry points
- Avoid molten metal impingement on mold surfaces or cores by appropriate gate location
- Design runners and gates, if two or more sprues are used, to prevent the turbulence associated with the collision of flow patterns
- Design risers to be of sufficient size and effectiveness to compensate for volumetric shrinkage. Riser position, shape, and filling from the gating system relative to the casting cavity are interrelated and critical considerations. In general, risers should be placed to achieve the maximum pressure differential and, when possible, should be open to the mold surface. Blind or enclosed risers must be adequately vented
- Observe the principles of directional solidification. The use of chills, riser insulation, and casting design changes may be required. The effects of inadequate gating and riser design can in some cases be corrected only by complete redesign
- Provide runner overruns, dross traps, or in-system filtration to avoid the impact of degraded metal on casting quality
- Locate the runners in the drag and locate the gates in the cope for horizontal mold orientation. This rule is subject to intelligent variation by the uniqueness of each part
- Place the riser cavities in the gate path for maximum effectiveness whenever possible
- Never place filters (if used) between riser and cavity
- Design the gates so that metal entry occurs near the lowest surface of the casting cavity
- Geometrically contour the runners to maintain uniform fluid pressure throughout. Formulas applicable to all gravity casting methods have been developed for this purpose
- Consider the ease and economics of trimming operations in gate and riser design

Somewhat different techniques in gating and risering are used for different alloys. In general, riser size and the need for stronger thermal gradients increase with more difficult-to-cast alloys.

Crack sensitivity or hot shortness forces compromises in the steps normally taken to achieve directional solidification. Extensive localized chilling may aggravate crack formation. In these cases, more uniform casting section thickness, larger fillets, more gradual section thickness changes, larger risers and in some cases riser insulation, and more graduated chilling offer the best prospects for success.

In alloys that are more difficult to feed but are relatively insensitive to cracking at elevated temperature, establishing thermal gradients by selective chilling (and heating as in permanent mold casting) usually provides good casting results. Examples are alloys high in eutectic content, as well as purer compositions, in which the solidification range is limited. In these cases, localized areas of shrinkage are inevitable in the absence of adequate thermal gradients accompanied by effective risering.

Gating of Die Castings. Conventional pressure die casting uses gating principles that are different from those for gravity casting. Although the fundamental principles employed in gravity casting remain desirable, injection under high pressure at significant metal velocity precludes application of many of the rules that govern gravity casting processes.

In pressure die casting, gates and runners are the means by which molten metal is transferred from the shot chamber or injection system to the die impression. The gating system must be designed to permit the attainment of cavity pressurization without reducing cycle time by its own mass and solidification time. It must also be of minimum size to maximize the gross-to-net-weight ratio and to minimize trimming and finishing costs.

The objective of gating in die casting is filling of the die cavity by establishing uninterrupted frontal flow. This requires the prevention of excessive turbulence and mixing within the die cavity, and it minimizes the entrainment of air and volatiles derived from the injection system and from lubricants contained in the die cavity. Runners are usually semiellipsoidal and decrease in cross-sectional area in the direction of metal flow. Their respective cross-sectional areas should exceed corresponding gate dimensions. At no time should gate thickness exceed that of the casting. Gates are normally severely tapered at the entry point(s) to facilitate removal during trimming with minimum risk of damage to the casting.

The greatest progress has been made in research dedicated to die design and the design of metal entry conditions, including sprue, runner, gating, and parameters of injection. Based on carefully analyzed applied research as well as modeling of the die filling sequence, mathematical and instrumented programs are now available to die casters for the development, modification, and control of gating design and operation. Much less sophisticated mathematical relationships and nomographs have been employed for many years to determine the relative dimensions of plunger diameters, runners, gates, and process parameters, including fill rate and velocity, plunger velocity, gate velocity, and system pressures.

Die Casting Compositions

The most important compositions in use for pressure die casting are the highly castable and forgiving alloys of the aluminum-silicon family. Of these, Alloy 380 and its variations predominate.

Magnesium content is usually controlled at low levels to minimize oxidation and the generation of oxides in the casting process. Nevertheless, alloys containing appreciable magnesium concentrations are routinely produced.

Iron content of 0.7% or greater is preferred in most die casting operations to maximize elevated-temperature strength, to facilitate ejection, and to minimize soldering to the die face. Improved ductility through reduced iron content has been an incentive resulting in widespread efforts to develop a tolerance for iron as low as approximately 0.25%. These efforts focus on process refinements and improved die lubrication. Hypereutectic aluminum-silicon alloys are growing in importance as their valuable characteristics and excellent die casting properties are exploited in automotive and other applications.

Foundry Practices for Specialty Castings

Rotor Castings. Most squirrel cage induction-type electric motors employ an integrally cast aluminum rotor. There are many economic and manufacturing process advantages that constitute an improvement of this type of rotor construction over wire-wound assemblies. The rotors produced by the casting process range in diameter from 25 to approximately 760 mm (1 to 30 in.).

Several casting processes are used to produce cast aluminum rotors. The principal processes used are vertical and horizontal cold chamber die casting and, to a lesser extent, centrifugal and permanent mold casting.

Aluminum alloys display a wide range of electrical characteristics, and the choice of a specific alloy for a cast motor rotor depends on the specified operating characteristics of the motor. Standardized alloys of high, intermediate, and low conductivity have been developed for this industry.

Conductivity. Most cast aluminum motor rotors produced are in carefully controlled, more pure compositions 100.0, 150.0, and 170.0 (99.0, 99.5, and 99.7% Al, respectively). Impurities in these alloys are controlled to minimize variations in electrical performance based on conductivity and to minimize the occurrence of microshrinkage and cracks during casting.

Rotor alloy 100.0 contains a significantly larger amount of iron and other impurities, and this generally improves castability. With higher iron content, crack resistance is improved, and a lower tendency toward shrinkage formation will be observed. This alloy is recommended when the maximum dimension of the part is greater than 127 mm (5 in.). For the same reasons, Alloy 150.0 is preferred over 170.0 in casting performance.

Minimum and typical conductivities for each rotor alloy grade are:

Alloy	Minimum conductivity, % IACS	Typical conductivity, % IACS
100.1	54	56
150.1	57	59
170.1	59	60

IAS, International Copper Annealed Standard

For motor rotors requiring high resistivity—for example, motors with high starting torque—more highly alloyed die casting compositions are commonly used. The most popular are Alloys 443.2 and A380.2. By choosing alloys such as these, conductivities from 25 to 35% IACS can be obtained; in fact, highly experimental alloys with even higher resistivities have been developed for motor rotor applications.

Although gross casting defects may adversely influence electrical performance, the conductivity of alloys employed in rotor manufacture is more exclusively controlled by composition. Table 2 lists the effects of various elements in and out of solution on the resistivity of aluminum. Simple calculation using these values accurately predicts total resistivity and its reciprocal conductivity for any composition. A more general and easy-to-use formula for conductivity calculation that offers sufficient accuracy for most purposes is:

$$\text{Conductivity} = 63.50 - 6.9x - 83y$$

where electrical conductivity is in percent IACS, x = iron + silicon (in weight percent), and y = titanium + vanadium + manganese + chromium (in weight percent).

Reference to specified composition limits for rotor alloys shows the use of composition controls that reflect electrical considerations. The peritectic elements are limited because their presence is harmful to electrical conductivity. Prealloyed ingot produced to these specifications is produced by boron addition, which complexes and precipitates these elements before casting. In addition, iron and silicon contents are subject to control by ratio with the objective of promoting the α Al-Fe-Si phase intermetallics

Table 2 Effect of elements in and out of solution on the electrical resistivity of aluminum

Element	Maximum solubility in aluminum, %	Average increase in resistivity per weight percent, μΩ · cm(a) In solution	Out of solution(b)
Chromium	0.77	4.00	0.18
Copper	5.65	0.344	0.030
Iron	0.052	2.56	0.058
Lithium	4.0	3.31	0.68
Magnesium	14.9	0.54(c)	0.22(c)
Manganese	1.82	2.94	0.34
Nickel	0.05	0.81	0.061
Silicon	1.65	0.65	0.059
Titanium	1.0	2.88	0.12
Vanadium	0.5	3.58	0.28
Zinc	82.8	0.094(d)	0.023(d)
Zirconium	0.28	1.74	0.044

(a) Add indicated increase to the base resistivity for high-purity aluminum (2.65 μΩ · cm at 20 °C, or 68 °F, or 2.71 μΩ · cm at 25 °C, or 77 °F). (b) Limited to about twice the concentration given for the maximum solid solubility, except as noted. (c) Limited to approximately 10%. (d) Limited to approximately 20%. Source: Ref 4

least harmful to castability. Ignoring these important relationships results in variable electrical performance and, of equal importance, variable casting results.

As in all casting operations, but especially in die casting, establishment of reproducible casting conditions depends heavily on the rhythm of the process itself. Variable chemical composition—for example, when unfurnaced and untreated primary aluminum grades are used for cost advantage over rotor alloys—will always introduce a degree of unpredictable casting results that will adversely effect both process and product performance. In contrast, the use of rotor alloys ensures optimum casting and controlled product results.

Casting Processes. The horizontal cold chamber die casting method is recommended and is the most widely used for the high-volume production of fractional-horsepower motor rotors. Multiple-cavity dies are usually employed to cast several rotors of the same or different design simultaneously.

The vertical pressure die casting method has been successfully used for many years to cast both fractional and integral horsepower units. In this process, the lower press platen contains a sump or depression into which molten metal is automatically or manually poured. The sump is typically lined with a refractory material, such as fiberfrax paper, mica, or other insulating paper. Sprayed refractory coatings can also be used to insulate and to protect the basin from attack. The mold and indexed mounting for the laminate stack are located in the upper platen. Molten metal is forced into the die through a network of small gates separating the lower and upper platens. These gates are normally tapered in the direction of metal flow through the base plate and into the collector ring of the part.

For either horizontal or vertical casting processes, preheating of the laminate pack is recommended, but in practice preheating is rarely performed. The preheat temperature for the laminate assembly is from 205 to 540 °C (400 to 1000 °F). When preheating is not performed, individual ferrous laminations must be free of surface contamination, especially oil and moisture. Preheating also oxidizes the laminate and its sheared outer surface to inhibit metallurgical bonding during the casting process. A post-casting thermal treatment of heating to a temperature in excess of 260 °C (500 °F) will effectively shear the brittle intermetallics formed when bonding does occur. Advantages are attributed to more severe postcasting thermal treatment to ensure bond separation and to oxidize the interface surfaces. Annealing at 425 to 510 °C (800 to 950 °F) followed by still air cooling is conventional, with cooling practices dictated by the electrical advantages of rejecting solute phases from solid solution.

The foundryman should be certain that the laminations are clean and free of excessive burrs that might result in current leakage. The preheating of poorly sheared laminate disks at higher temperature is recommended.

Good die venting is essential in rotor casting. Failure to provide and to maintain vents will always result in excessive end ring porosity and unsoundness in the conductor bars. Another defect encountered in rotor casting that is related to venting is backfilled conductor bars. This condition is induced by preferential flow in one region of the cavity and simultaneous molten metal entry to a single bar from both end rings. Accurate laminate assembly and placement in the mold cavity, clean laminates, minimum die lubrication, effective gate design, and adequate die venting are essential in the production of quality rotors.

Gating scrap is often directly recharged to the furnace in rotor casting. This results in the potentially heavy oxide contamination in the metal source that is a major cause of casting defects.

The most common defects, apart from misruns or nonfills, are broken bars resulting from different rates of thermal contraction between the core and the conductor bars and from massive oxide inclusions, which significantly influence flowability and casting recovery and obstruct current flow in the conductor bars. The direct recharging of scrap may include the inadvertent or intentional addition of iron, particularly in the form of steel laminates readily available to operators for the purposes of improving casting results at the expense of electrical characteristics.

The direct recharging of scrap also influences temperature control with obvious harmful effects on process variability. For best results, a temperature control range of ±10 °C (±18 °F) is recommended.

Premium engineered castings provide higher levels of quality and reliability than are found in conventionally produced parts. These castings may display optimum performance in one or more of the following characteristics: mechanical properties (determined by test coupons machined from representative parts), soundness (determined radiographically), dimensional accuracy, and finish. However, castings of this classification are notable primarily for mechanical property attainment that reflects extreme soundness, fine dendrite arm spacing, and well-refined grain structure. These technical objectives require the use of chemical compositions competent to display premium engineering properties. Alloys considered to be premium engineered compositions appear in separately negotiated specifications or in specifications such as military specification MIL-A-21180, which is extensively used in the United States for premium engineered casting procurement.

Alloys commonly considered premium by definition and specification are 201.0, C355.0, A206.0, A356.0, 224.0, A357.0, 249.0, 358.0, and 354.0. All alloys employed in premium engineered casting work are characterized by optimum concentrations of hardening elements, and restrictively controlled impurities. Although any alloy can be produced in cast form with properties and soundness conforming to a general description of premium values relative to corresponding commercial limits, only those alloys demonstrating yield strength, tensile strength, and especially elongation in a premium range belong in this discussion. They fall into two categories: high-strength aluminum-silicon compositions and those alloys of the 2*xx* series, which, by restricting impurity element concentrations, provide outstanding ductility, toughness, and tensile properties with notably poorer castability.

In all premium casting alloys, impurities are strictly limited for the purposes of improving ductility. In aluminum-silicon alloys, this translates to control of iron at or below ~0.010% with measurable advantag-

Table 3 Mechanical property specifications for premium engineered castings

Alloy	Class	Ultimate tensile strength (min), MPa	ksi	0.2% offset yield strength (min) MPa	ksi	Elongation in 50 mm (2 in.), %
Specimens cut from designated casting areas						
249.0	10	379	55	310	45	3
	11	345	50	276	40	2
354.0	10(a)	324	47	248	36	3
	11(a)	296	43	228	33	2
355.0	10(a)	283	41	214	31	3
	11(a)	255	37	207	30	1
	12(a)	241	35	193	28	1
A356.0	10(a)	262	38	193	28	5
	11(a)	228	33	186	27	3
	12(a)	221	32	152	22	2
A357.0	10(a)	262	38	193	28	5
	11(a)	283	41	214	31	3
224.0	10	310	45	241	35	2
	11	345	50	255	37	3
XA201.0	10	386	56	331	48	3
	11	386	56	331	48	1.5
Specimens cut from any casting area						
249.0	1	345	50	276	40	2
	2	379	55	310	45	3
	3	414	60	345	50	5
354.0	1(a)(b)	324	47	248	36	3
	2(a)(b)	345	50	290	42	2
C355.0	1(b)	283	41	214	31	3
	2(a)(b)	303	44	228	33	3
	3(a)(b)	345	50	276	40	2
A356.0	1(b)	262	38	193	28	5
	2(a)(b)	276	40	207	30	3
	3(a)(b)	310	45	234	34	3
A357.0	1(a)(b)	310	45	241	35	3
	2(a)(b)	345	50	276	40	5
224.0	1	345	50	255	37	3
	2	379	55	255	37	5
XA201.0	1	414	60	345	50	5
	2	414	60	345	50	3

(a) Values from specification MIL-A-21180. (b) This class is obtainable in favorable casting configurations and must be negotiated with the foundry for the particular configuration desired.

es in further reductions to the range of 0.03 to 0.05%, the practical limit of commercial smelting capability.

Beryllium is present in A357.0 and 358.0 alloys not to inhibit oxidation (although that is a corollary benefit) but to alter the form of the insoluble phase to a more nodular form least detrimental to ductility. Beryllium also alters the chemistry of the insoluble phase to exclude magnesium, which then becomes fully available for hardening purposes. Magnesium is normally controlled in the upper specification range to maximize Mg_2Si formation for strength development at some loss in ductility.

The presence of iron and silicon in alloys such as 295.0 contributes to reasonably good castability. For the development of premium properties in the 2*xx* family of alloys, these impurities are severely restricted. As a result of these composition restrictions, all premium engineered casting alloys of the 2*xx* type can be characterized as extremely sensitive to crack formation. They are also highly susceptible to shrinkage, requiring unusual foundry expertise in gating and risering. Design engineers and the producing foundry usually must collaborate to develop configurations offering maximum compatibility with casting requirements. The development of hot isostatic pressing for aluminum alloys is pertinent to the broad range of premium castings, but is especially relevant for the more difficult-to-cast aluminum-copper series (see the section "Hot Isostatic Pressing of Castings" in the article "Processing of Castings" in this Volume).

Table 3 defines mechanical property limits normally applied to premium engineered castings. It must be emphasized that the negotiation of limits for specific parts is usual practice, with higher as well as lower specific limits based on part design criteria and foundry capabilities.

Melt Preparation. Metal treatment is performed for this class of casting to achieve the highest possible quality. Reactive gas fluxing to hydrogen levels lower than 0.10 mL/100 g and effective removal of oxides and other nonmetallics by active gas fluxing and/or filtration are essential in meeting specified soundness levels and mechanical property limits.

Extreme care is also exercised in all phases of metal handling and pouring to preserve metal quality. Proprietary pouring methods have been developed for this purpose, including mechanical pumping, displacement, and nonturbulent transfer from metal source to the casting cavity.

Casting Processes. Molds are usually dry sand and other materials comprising composite mold construction, such as plaster and metallic sections. Elaborate planning and control of solidification through mold design and controlled chilling are normal.

Many premium engineered castings are produced by other gravity casting processes, notably, permanent mold and plaster molding. In every case, however, the same exacting approach is necessary for control of solidification to achieve the structures and soundness needed to furnish castings within specification.

Dimensional Control. Dimensional variations depend on part size, complexity, and alloy, but the use of highly accurate mold and core patterns and the dimensional control of molding media provide for extreme dimensional accuracy. Tolerances applicable to typical casting dimensions can be specified at ±0.254 mm (±0.010 in.), and in special cases, tolerances as small as ±0.127 mm (±0.005 in.) may apply. Even the inspection methods used in these cases represent an unusual capability on the part of premium engineered casting foundries. Datum plane measuring concepts are required that are agreed upon by both customer and vendor. The datum plane system is a cost-effective means of establishing dimensions, and it is highly adaptable to automated coordinate measurement.

Another advantage of premium engineered castings is the attainment of structural integrity in thin walls. This ranges from 1.52 to 2.03 mm (0.060 to 0.080 in.) for large plane areas and is as low as 0.51 to 0.76 mm (0.020 to 0.030 in.) for more limited casting sections.

Casting surface finish is always a function of mold surface and characteristics. The selection of mold materials and the accuracy of mold finish maintained in premium casting operations ensure that specified requirements are met.

Heat Treatment

The metallurgy of aluminum and its alloys fortunately offers a wide range of opportunities for employing thermal treatment practices to obtain desirable combinations of mechanical and physical properties. Through alloying and temper selection, it is possible to achieve an impressive array of features that are largely responsible for the current use of aluminum alloy castings in virtually every field of application. Although the term heat treatment is often used to describe the procedures required to achieve maximum strength in any suitable composition through the sequence of solution heat treatment, quenching, and precip-

Table 4 Typical heat treatments for aluminum alloy sand and permanent mold castings

Alloy	Temper	Type of casting(a)	Solution heat treatment(b) Temperature(c) °C	°F	Time, h	Aging treatment Temperature(c) °C	°F	Time, h
201.0	T6	S	510–515;	950–960;	2			
			525–530	980–990	14–20	155	310	20
	T7	S	510–515;	950–960;	2			
			525–530	980–990	14–20	190	370	5
204.0	T4	S or P	520	970	10	...	...	...
208.0	T55	S	...	...	...	155	310	16
222.0	O(d)	S	...	...	...	315	600	3
	T61	S	510	950	12	155	310	11
	T551	P	...	...	...	170	340	16–22
	T65	...	510	950	4–12	170	340	7–9
242.0	O(e)	S	...	...	...	345	650	3
	T571	S	...	...	...	205	400	8
		P	...	...	...	165–170	330–340	22–26
	T77	S	515	960	5(f)	330–355	625–675	2 (min)
	T61	S or P	515	960	4–12(f)	205–230	400–450	3–5
295.0	T4	S	515	960	12	...	...	...
	T6	S	515	960	12	155	310	3–6
	T62	S	515	960	12	155	310	12–24
	T7	S	515	960	12	260	500	4–6
296.0	T4	P	510	950	8	...	...	...
	T6	P	510	950	8	155	310	1–8
	T7	P	510	950	8	260	500	4–6
319.0	T5	S	...	...	...	205	400	8
	T6	S	505	940	12	155	310	2–5
		P	505	940	4–12	155	310	2–5
328.0	T6	S	515	960	12	155	310	2–5
332.0	T5	P	...	...	...	205	400	7–9
333.0	T5	P	...	...	...	205	400	7–9
	T6	P	505	950	6–12	155	310	2–5
	T7	P	505	940	6–12	260	500	4–6
336.0	T551	P	...	...	...	205	400	7–9
	T65	P	515	960	8	205	400	7–9
354.0	...	(g)	525–535	980–995	10–12	(h)	(h)	(h)
355.0	T51	S or P	...	...	...	225	440	7–9
	T6	S	525	980	12	155	310	3–5
		P	525	980	4–12	155	310	2–5
	T62	P	525	980	4–12	170	340	14–18
	T7	S	525	980	12	225	440	3–5
		P	525	980	4–12	225	440	3–9
	T71	S	525	980	12	245	475	4–6
		P	525	980	4–12	245	475	3–6
C355.0	T6	S	525	980	12	155	310	3–5
	T61	P	525	980	6–12	Room temperature		8 (min)
						155	310	10–12

(continued)

(a) S, sand; P, permanent mold. (b) Unless otherwise indicated, solution treating is followed by quenching in water at 65–100 °C (150–212 °F). (c) Except where ranges are given, listed temperatures are ±6 °C or ±10 °F. (d) Stress relieve for dimensional stability as follows: hold 5 h at 413 ± 14 °C (775 ± 25 °F); furnace cool to 345 °C (650 °F) over a period of 2 h or more; furnace cool to 230 °C (450 °F) over a period of not more than ½ h; furnace cool to 120 °C (250 °F) over a period of approximately 2 h; cool to room temperature in still air outside the furnace. (e) No quench required; cool in still air outside furnace. (f) Air-blast quench from solution-treating temperature. (g) Casting process varies (sand, permanent mold, or composite) depending on desired mechanical properties. (h) Solution heat treat as indicated, then artificially age by heating uniformly at the temperature and for the time necessary to develop the desired mechanical properties. (j) Quench in water at 65–100 °C (150–212 °F) for 10–20 s only. (k) Cool to room temperature in still air outside furnace.

itation hardening, in its broadest meaning, heat treatment comprises all thermal practices intended to modify the metallurgical structure of products in such a way that physical and mechanical characteristics are controllably altered to meet specific engineering criteria. In all cases, one or more of the following objectives form the basis for temper selection:

- Increase hardness for improved machinability
- Increase strength and/or produce the mechanical properties associated with a particular material condition
- Stabilize mechanical and physical properties
- Ensure dimensional stability as a function of time under service conditions
- Relieve residual stresses induced by casting, quenching, machining, welding, or other operations

To achieve any of these objectives, parts can be annealed, solution heat treated, quenched, precipitation hardened, overaged, or treated with combinations of these practices. In some simple shapes (for example, bearings), thermal treatment can also include plastic deformation in the form of cold work.

Temper Designations and Practices

The Aluminum Association has standardized the definitions and nomenclature applicable to thermal practice and maintains a registry of standard heat treatment practices and designations for industry use. Standardized temper designations applicable to castings are:

- O (formerly T2, T2*x*): annealed (thermally stress relieved)
- T4: solution heat treated and quenched
- T5: artificially aged
- T6: solution heat treated, quenched, and artificially aged
- T7: solution heat treated, quenched, and overaged
- T8: cold reduced before aging to improve compressive yield strength (bearings only)

Variations in thermal treatment practice are shown as the second and third digits in the standard designations; for example,

Table 4 (continued)

Alloy	Temper	Type of casting(a)	Solution heat treatment(b): Temperature(c) °C	°F	Time, h	Aging treatment: Temperature(c) °C	°F	Time, h
356.0	T51	S or P	...	...	...	225	440	7–9
	T6	S	540	1000	12	155	310	3–5
		P	540	1000	4–12	155	310	2–5
	T7	S	540	1000	12	205	400	3–5
		P	540	1000	4–12	225	440	7–9
	T71	S	540	1000	10–12	245	475	3
		P	540	1000	4–12	245	475	3–6
A356.0	T6	S	540	1000	12	155	310	3–5
	T61	P	540	1000	6–12	Room temperature		8 (min)
						155	310	6–12
357.0	T6	P	540	1000	8	175	350	6
	T61	S	540	1000	10–12	155	310	10–12
A357.0	...	(g)	540	1000	8–12	(h)	(h)	(h)
359.0	...	(g)	540	1000	10–14	(h)	(h)	(h)
A444.0	T4	P	540	1000	8–12	...	...	...
520.0	T4	S	430	810	18(j)	...	...	...
535.0	T5(d)	S	400	750	5	...	...	...
705.0	T5	S	...	...	...	Room temperature		21 days
						100	210	8
		P	...	...	...	Room temperature		21 days
						100	210	10
707.0	T5	S	...	...	...	155	310	3–5
		P	...	...	...	Room temperature or		21 days
						100	210	8
	T7	S	530	990	8–16	175	350	4–10
		P	530	990	4–8	175	350	4–10
710.0	T5	S	...	...	...	Room temperature		21 days
711.0	T1	P	...	...	...	Room temperature		21 days
712.0	T5	S	...	...	...	Room temperature or		21 days
						155	315	6–8
713.0	T5	S or P	...	...	...	Room temperature or		21 days
						120	250	16
771.0	T53(d)	S	415(k)	775(k)	5(k)	180(k)	360(k)	4(k)
	T5	S	...	...	...	180(k)	355(k)	3–5(k)
	T51	S	...	...	...	205	405	6
	T52	S	...	...	...	(d)	(d)	(d)
	T6	S	590(k)	1090(k)	6(k)	130	265	3
	T71	S	590(e)	1090(e)	6(e)	140	285	15
850.0	T5	S or P	...	...	...	220	430	7–9
851.0	T5	S or P	...	...	...	220	430	7–9
	T6	P	480	900	6	220	430	4
852.0	T5	S or P	...	...	...	220	430	7–9

(a) S, sand; P, permanent mold. (b) Unless otherwise indicated, solution treating is followed by quenching in water at 65–100 °C (150–212 °F). (c) Except where ranges are given, listed temperatures are ±6 °C or ±10 °F. (d) Stress relieve for dimensional stability as follows: hold 5 h at 413 ± 14 °C (775 ± 25 °F); furnace cool to 345 °C (650 °F) over a period of 2 h or more; furnace cool to 230 °C (450 °F) over a period of not more than ½ h; furnace cool to 120 °C (250 °F) over a period of approximately 2 h; cool to room temperature in still air outside the furnace. (e) No quench required; cool in still air outside furnace. (f) Air-blast quench from solution-treating temperature. (g) Casting process varies (sand, permanent mold, or composite) depending on desired mechanical properties. (h) Solution heat treat as indicated, then artificially age by heating uniformly at the temperature and for the time necessary to develop the desired mechanical properties. (j) Quench in water at 65–100 °C (150–212 °F) for 10–20 s only. (k) Cool to room temperature in still air outside furnace.

T61, T62, T572, and so on. There is no consistent convention for the assignment of temper designation variations except that for solution heat treated alloys general practice has been to assign tempers T6, T61, and T62 to an ascending order in strength development up to full hardness. Recommended thermal treatment practices for aluminum casting alloys are listed in Table 4.

Principles of Heat Treatment

The heat treatment of aluminum alloys is based on the varying solubilities of metallurgical phases in a crystallographically isotropic system. Because the solubility of the eutectic increases with increasing temperature to the solidus, as in the binary eutectic system shown in Fig. 10, the variations in degree of solution and the formation and distribution of precipitated phases can be used to influence material properties.

In addition to the phase and morphology changes associated with soluble elements and compounds, other (sometimes desirable) effects accompany elevated-temperature treatment. The microsegregation characteristic of all cast structures is minimized or eliminated. The residual stresses caused by solidification or by prior quenching are reduced, insoluble phases may be physically altered, and susceptibility to corrosion, especially in certain compositions, may be affected.

Solution Heat Treatment. Exposure to temperatures corresponding to maximum safe limits relative to the lowest melting temperature for a specific composition results in the dissolution of soluble phases formed during and after casting solidification. The rate of heating to solution temperature is technically unimportant in casting compositions except when more than one soluble phase is present, as in the Al-Si-Cu-Mg and Al-Zn-Cu-Mg systems. In these cases, stepped heat treatment is sometimes essential to avoid the melting of lower-melting phases. Very rapid heating can result in nonequilibrium melting in highly segregated structures, but the conditions required for such occurrences are most unlikely in engineered castings.

In all cases, the most complete degree of solution is desirable for subsequent hardening, and a number of factors are involved. Different casting processes and foundry practices result in microstructural differences with relevance to heat treatment practice. The coarser microstructures associated with slow solidification rate processes require a longer solution heat treatment exposure. The time required at temperature to achieve solution is progressively shorter for investment, sand, and permanent mold castings, but thin-wall sand castings produced with extensive use of chills can often display finer microstructures than heavy-section permanent mold parts produced in such a way that process advantages are not exploited. For these reasons, solution heat treatment practices can be optimized for any specific part to

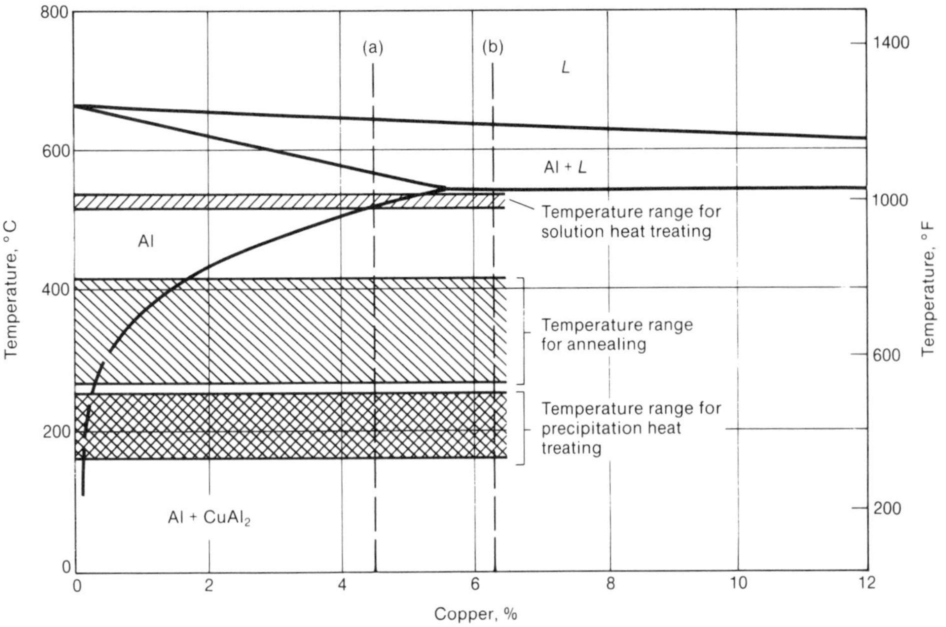

Fig. 10 Portion of aluminum-copper binary phase diagram. Temperature ranges for annealing, precipitation heat treating, and solution heat treating are indicated.

achieve solution with the shortest reasonable cycle once production practice is finalized, even though most foundries and heat treaters will standardize a practice with a large margin of safety.

Because the slope of the characteristic solvus illustrated in Fig. 10 changes as temperature approaches the eutectic melting point, it is apparent that temperature is critical in determining the degree of solution that can be attained. In addition, temperature affects diffusion rates, and this directly influences the degree of solution as a function of time at temperature. Within the temperature ranges defined for solution heat treatment by applicable specifications lies a significant corresponding range of solution values.

The knowledgeable heat treater/foundry seeking to obtain superior properties based on solution heat treatment will bias temperature selection within specification limits to obtain the highest degree of solution. Superior furnaces, thermocouples, and furnace controls, representing the state-of-the-art, along with recognition of the normal resistance of cast structures to melting based on diffusion considerations, offer the potential of superior property development compared to historically more conservative temperature selection. However, although temperatures just below the eutectic melting point are desirable for optimum property development, it is critical that eutectic melting resulting in brittle intergranular eutectic networks be avoided.

Insoluble phases, including those containing impurity elements, are normally thought to be unaffected by solution heat treatment, but limited changes do occur. The surfaces of primary and eutectic silicon particles are characteristically rounded during solution heat treatment. The solution heat treatment of Alloy A444, which contains no soluble phase, is justified solely by this phenomenon and its effect on ductility. Limited solubility also results in similar boundary physical changes in other insoluble intermetallics.

Quenching. This discussion separates quenching as a distinct step in thermal practice leading to the metastable solution heat treated (T4) condition. Specific parameters can be associated with the heating of parts to achieve solution, and separate parameters apply to the steps required to achieve the highest postquench degree of retained solution. Rapid cooling from solution temperature to room temperature is essential, and it is possible to describe quenching as the most difficult and often least controlled step in thermal treatment.

For optimum results, quench delay must be minimized. Although specifications often define quench delay limits, in practice, the shortest possible delay is observed. This may require specialized equipment, such as bottom-drop or continuous furnaces. Excessive delays result in temperature drop and the rapid formation of coarse precipitate in a temperature range in which the effects of precipitation are lost for hardening purposes. Even though castings are characteristically more tolerant of quench delay than wrought products based on structure and diffusion, excessive quench delays result in less than optimum strengthening potential.

Water is the quench medium of choice for aluminum alloys, and its temperature has a major effect on results. Most commercial quenching is accomplished in water near the boiling point, but room temperature, 65 °C (150 °F), and 80 °C (180 °F) have also become common standardized alternatives. Because higher potential strength is generally thought to be associated with the most rapid quenching and because corrosion and stress corrosion performance are usually enhanced by rapid quenching, it would appear that room-temperature water should routinely be employed. However, the selection of quench temperature becomes less obvious when the objective of quenching is to retain the highest possible degree of solution with the least warpage or distortion and the lowest level of induced residual stresses consistent with commercial or specified requirements.

The key to the compromise between goals involving property development and the physical consequences of quenching is uniformity of heat extraction, which is in turn a complex function of the operable heat extraction mechanism. Nucleate, vapor film, and convective boiling occur with dramatically different heat extraction rates at different intervals. Differences in section thickness, load density, positioning, and casting geometry also influence the results. As section thicknesses increase, the metallurgical advantages of quench rates obtained with water temperatures less than 65 °C (150 °F) diminish, but the cooling rate advantage of the 65 versus the 100 °C (150 versus 212 °F) quench temperature is retained independently of section thickness. In addition to developing racking and loading methods that space and orient parts for the most uniform quenching, quenchant additions can be made for the following purposes:

- To promote stable vapor film boiling by the deposition of compounds on the surface of parts as they are submerged in the quench solution
- To suppress variations in heat flux by increasing vapor film boiling stability through chemically decreased quench solution surface tension
- To moderate quench rate for a given water temperature

Quenching rates are also affected by the surface condition of the parts. More rapid quenching occurs with oxidized, stained, and rough surfaces, while bright, freshly machined, and etched surfaces quench more slowly.

The common use of water as a quenching medium is largely based on its superiority in terms of quenching characteristics relative to other materials. Nevertheless, quenching has been accomplished in oil, salt baths, and organic solutions. For many compositions, fan or mist quenching is feasible as a means of obtaining dramatic reductions in residual stress levels with considerable sacrifices in hardening potential.

Precipitation Heat Treating/Aging. In general, parts that have been solution heat treated and quenched display tensile properties and elongation superior to those of the as-cast (F) condition. However, the T4 condition is rarely employed. Instead, the advantages of aging or precipitation hardening are obtained by thermal treatment following quenching. Among these advantages are increased strength and hardness with a corresponding loss in ductility, improved machinability, the development of more stable mechanical properties, and reduced residual stresses. It is through precipitation treatment that cast aluminum products may be most powerfully differentiated.

Most aluminum alloys age harden naturally to some extent after quenching; that is, properties change as a function of time at room temperature solely as a result of zone formation within the solid solution. The extent of change is highly alloy dependent. For example, room-temperature aging in alloys such as A356 and C355 occurs within 48 h, with insignificant changes taking place thereafter. Alloy 520, normally used in the T4 condition, age hardens over a period of years, and a number of aluminum-zinc-magnesium alloys that are used without heat treatment exhibit rapid changes in properties over 3 or 4 weeks and harden at progressively reduced rates thereafter.

The process of hardening is accelerated by artificial aging at temperatures ranging from approximately 95 to 260 °C (200 to 500 °F), depending on the alloy and the properties desired. In artificial aging, supersaturation, which characterized the room-temperature solution condition, is relieved by the precipitation of solute, which proceeds in stages with specific structural effects.

At low aging temperatures or during transition at higher temperatures, the principal change is the diffusion of solute atoms to high-energy sites within the lattice, producing distortion of the lattice planes and forming concentrations of subcritical crystal nuclei. With continued exposure to aging temperature, these sites reach or fail to reach critical nucleation size, a stage leading to the formation of discrete particles displaying the identifiable crystallographic character of the precipitated phase. These transitional phase particles grow with an increase in coherency strains until, with sufficient time and temperature, interfacial bond strength is exceeded.

Coherence is lost and with it the strengthening effects associated with precipitate formation and growth. Continued growth of the now equilibrium phase occurs without strength benefit, corresponding to the overaged condition.

The practice to be employed in artificial aging depends entirely on the desired level of property development. Aging curves have been developed to facilitate process selection. The results of a large number of aging curve studies are reflected in specification recommendations and industry references. The heat treater can reasonably predict the results of aging by reference to these curves. It should be noted that longer times at lower aging temperatures generally result in higher peak strength values. The behavior of aging response or rate of property change as a function of time at peak strength points is also of interest. The flatter curves associated with lower aging temperatures allow greater tolerance in the effects of time-temperature variations. Unlike solution heat treatment, the time required to reach the aging temperature may be significant, but is seldom included in age cycle control. The energy imparted during heating to the precipitation-hardening temperature can be integrated into the control sequence to control results more accurately with minimum cycle time.

The overaged T7 condition is less common than the T6 temper, but there are good reasons for its use. Precipitation hardening as practiced for the T6 condition results in reductions in residual stress levels imposed by quenching of 10 to 35%. By definition, overaging consists of carrying the aging cycle to a point beyond peak hardness, but it is most often conducted at higher temperatures than those used for the fully hardened condition. A substantial further decrease in residual stresses is associated with the higher-temperature aging treatment. Furthermore, parts become more dimensionally stable as a result of more complete degrowthing, and increased stability in properties and performance is ensured when service involves exposure at elevated temperatures.

Annealing was originally assigned the designation T2, but is now known as the O temper. It is rarely employed, but is useful in providing parts with extreme dimensional and physical stability and the lowest practical level of residual stresses. The annealed condition is also characterized by extremely low strength and a correspondingly poor level of machinability. Typical annealing practices are for relatively short (2 to 4 h) exposures at a minimum temperature of 345 °C (650 °F). Higher-temperature practices are employed for more complete relaxation of residual stresses. The cooling rate from annealing temperatures must be controlled such that residual stresses are not reinduced and resolution effects are avoided.

Heat Treating Problems. Under optimum conditions, the measurable results of thermal practices may be other than those anticipated. When specification limits are not met, analytical procedures and judgments are applied to establish a corrective course of action based on evidence or assumptions concerning the cause of failure.

Mechanical property limits statistically define normalcy for a given composition heat treated by specific practices. Chemical composition is a major variable in mechanical property development, and when mechanical properties have failed specification limits, chemistry is the logical starting point for investigation.

The role of trace elements in mechanical property development is important because these elements are often not separately specified in alloy specifications except as others each and others total. Sodium and calcium are embrittling in 5*xx* alloys. Low-melting elements such as lead, tin, and bismuth may under some conditions form embrittling intergranular networks with similar effects. Insoluble impurity elements are generally responsible for decreases in elongation.

Low concentrations of soluble elements in heat-treatable compositions naturally result in the more frequent distribution of mechanical property values in the lower specification range. Element relationships such as copper-magnesium, silicon-magnesium, iron-silicon, iron-manganese, and zinc-copper-magnesium are also important considerations in defining the causes of abnormal mechanical property response to thermal treatment.

A second important consideration is that of casting soundness. Unsound castings will not consistently meet specified mechanical property or other limits sensitive to structural integrity. All defects adversely affect strength and elongation. Shrinkage, hydrogen porosity, cracks, inclusions, and other casting-related defects influence mechanical properties adversely, and their effects should be considered before addressing the possibility of heat treatment problems.

The quality of solution heat treatment can be assessed in several ways. There is, of course, the rounding effect on insoluble phases, which serves as evidence of elevated-temperature exposure. The elimination of coring in many alloys is another indication of elevated-temperature treatment. The effective solution of soluble phases can be determined microscopically. Undissolved solute can be distinguished from the appearance of precipitate that forms at high temperatures and results from quench delay or an inadequate or incomplete quench. There is also a tendency for the precipitates formed as a result of quench delay or inadequate quench to concentrate at grain boundaries as opposed to more normal distribution through the microstructure for properly solution heat treated and aged material.

Although the overaged condition is microscopically apparent, underaging is difficult to assess because of the submicroscopic nature of transitional precipitation. Evidence of acceptable aging practice is best obtained from aging furnace records, which might indicate errors in the age cycle. Underaging can be corrected by additional aging, but for all other heat treatment aberrations except those associated with objectionable conditions such as high-temperature oxidation or eutectic melting, re-solution heat treatment is an acceptable corrective action despite many myths to the contrary. Eutectic melting occurs when the eutectic melting temperature is exceeded, resulting in the characteristic rosettes of resolidified eutectic and/or intergranular eutectic.

High-temperature oxidation is a misnamed condition of hydrogen diffusion that affects surface layers during elevated-temperature treatment. This condition can result from moisture contamination in the furnace atmosphere and is sometimes aggravated by sulfur (as in heat treatment furnaces also used for magnesium alloy castings) or other furnace refractory contamination.

There are no technical reasons for discouraging even repeated heat treatments to obtain acceptable mechanical properties. In the case of aluminum-copper alloys, it is essential that re-solution heat treatment be conducted at a temperature equivalent to or higher than the original practice to ensure effective re-solution. However, when the results of repeated heat treatment prove unsatisfactory, it should be apparent that other conditions are responsible for mechanical property failure.

Metallurgical structure also plays a role in property development. Modification of the eutectic is effective in improving elongation in hypoeutectic aluminum-silicon alloys and has little effect on tensile and yield strengths. Refinement of the primary phase in hypoeutectic silicon alloys is even more important in reducing brittle behavior.

Fine grain size promotes improved mechanical properties, while coarse grain size, by emphasizing grain-boundary effects, results in lower mechanical property performance. Cell size has been recognized as highly significant in improving ductility for given strength levels.

Quality Control

The most effective method of determining the combined consequences of chemistry, material condition, and heat treatment is the determination of tensile strength, yield strength, and elongation. This is often done by testing separate cast tensile specimens. These values most conclusively indicate the acceptability of a product relative to specification requirements. Hardness can be established as an acceptance criterion through negotiation, but it is less adequate in aluminum alloys than in other metal systems for control purposes. Hardness in aluminum corresponds only approximately to yield strength, and although hardness can be viewed as an easily measurable indicator of material condition, it is not normally accurate enough to serve as a guaranteed limit. Electrical conductivity only approximates material condition in cast or cast and heat treated structures, and it remains excessively variable for most control purposes.

Stability

Because thermal treatment affects the stability of mechanical properties and directly influences residual stress levels, additional discussion of these two considerations is warranted.

Stability is defined as the condition of unchanging structural and physical characteristics as a function of time under specific conditions of application. As previously mentioned, the metastable T4 condition is subject to hardening (extensive in some alloys and limited in others) at room and higher temperatures. Under service conditions involving temperatures greater than room temperature, significant physical and mechanical changes are to be expected.

Although these changes may be acceptable for a given application, changes in susceptibility to corrosion and stress corrosion, which can be associated with transitional states in some alloy systems, must be carefully considered.

The most stable conditions obtainable are associated with the (in order of decreasing stability) annealed, overaged, and as-cast conditions. Underaged and solution heat treated parts are least stable.

Residual stresses are caused by differing rates of cooling, especially postsolidification cooling, quenching from solution heat treatment temperature, and drastic changes in temperature at any intermediate step. Residual stress development depends on the differential rate of cooling, section thickness, and material strength. Stresses imposed by quenching from the solution heat treatment temperature are much more important than casting stresses or stresses normally obtained in conventional processing. Decreasing the severity of the quench from solution heat treatment results in a lower level of residual stresses but with correspondingly decreased material strength. Air quenching may provide a useful compromise in applications requiring unusual dimensional stability.

Residual stresses can also be relaxed by exposure to elevated temperature followed by slow cooling, or by plastic deformation. Plastic deformation that is routinely practiced for stress relief in wrought products has little application in the complex designs of engineered products such as castings; therefore, stress relief becomes more exclusively a function of postquench thermal treatment. Overaging results in significant reductions in residual stresses, and annealing provides a practical minimum in residual stress levels.

Natural Versus Artificial Property Development

When foundrymen work closely with design engineers, alloys and tempers can be selected that are capable of meeting both the manufacturing objectives of the foundry and the engineering criteria at the lowest possible cost. A frequent point of discussion is whether a superior selection might be a naturally hardening alloy (a composition that hardens naturally at room temperature after casting) or an alloy requiring heat treatment for mechanical property development. Unfortunately, room temperature hardening alloys, which offer a wide range of attractive properties, are also characterized by relatively inferior casting characteristics so that final material selection may become an issue of foundry versus application considerations. The selection of a heat-treatable composition with good casting characteristics in combination with effective heat treatment often offers significant advantages in cost, performance, uniformity, and reliability.

Cleaning

Alkaline and acid solutions are used to clean aluminum. Alkaline solutions are formulated in various degrees of etching activity ranging from nonetching silicates and borates to trisodium phosphate and highly aggressive sodium hydroxide. Acid cleaners are frequently based on phosphoric and sulfuric acids, often with appropriate detergents to emulsify contaminants.

Oxides are removed prior to other operations by using so-called deoxidizers. These are usually acid solutions containing phosphates, sulfates, and fluorides, often in conjunction with an oxidizing agent.

Chemical Finishing. Chemical treatments are applied to aluminum to remove surface contamination, staining, and oxides; to develop specular and matte appearances; to remove metal selectively as in chemical milling; and to develop various types of decorative or protective conversion coatings. Although sometimes used as the final finish, chemical treatments are widely used to prepare aluminum for other finishing operations, such as anodizing, electroplating, painting, and adhesive bonding.

Specular and Matte Appearances. Concentrated phosphoric-nitric acid solutions at 90 °C (195 °F) or, occasionally, nitric-fluoride solutions at lower temperatures are used on relatively pure metal to develop specularity. A matte surface is obtained by etching the aluminum with sodium hydroxide and then removing the etching smut with an oxidizing acid solution. High-silicon alloys are treated most effectively in a concentrated nitric-hydrofluoric acid solution.

Chemical milling and selective etching are conducted with both alkaline and acid solutions. Hot 15% sodium hydroxide solutions are generally used for chemical milling, with a removal rate of about 0.025 mm (0.001 in.) per minute. Pattern etching and engraving are often performed in ferric chloride or in acid fluoride solutions.

Chemical conversion coatings are formed when a portion of the aluminum surface is dissolved and a reaction film is formed. Examples of such films are aluminum oxide, chromic phosphate, and chromium aluminum chromate. Although there are various applications, the most important use for conversion coatings has been as a preparation for organic finishes. The clean, uniform, and inert coating provides good adhesion for the organic film and protects against undermining corrosion.

Welding

Foundry welding is employed to salvage scrap castings primarily by correcting surface and dimensional defects or irregularities. The inert gas shielded arc welding processes are normally utilized, rather than the processes using flux mixtures or flux-coated filler rods. The welding techniques used for castings are similar to those for wrought products. However, special consideration must be given to the thicker-surface aluminum oxide film and the void content of castings. The quality of weldments is closely related to the soundness of the casting in the weld area. The quality of conventional welds is generally very poor in castings displaying hydrogen porosity and in most die castings.

Welding Method. Casting flaws or areas requiring correction are usually small; therefore, the gas tungsten arc welding (GTAW) process is normally used. The gas metal arc welding process can be used if economics and operating conditions indicate that satisfactory results can be achieved.

Equipment for the GTAW Process. Alternating current welding transformers that deliver a balanced wave and incorporate high-frequency stabilization minimize the production of tungsten inclusions and produce the highest-quality welds. In general, the ac-dc and unbalanced wave types of power supply tend to spit tungsten. Tungsten inclusions in a weld are considered as detrimental as other casting defects of similar dimension.

Pure tungsten is preferred for facilitating a stable arc and minimizing tungsten spitting. The second choice is zirconiated tungsten.

Shielding Gas. Argon and mixtures of argon and helium gases are used to gas tungsten arc weld aluminum castings. When a gas mixture is used, helium usually constitutes 25 to 50% of the total mixture. The helium helps produce a hotter arc at a particular current setting, clean up the weld puddle if casting quality is a problem, and reduce the size of gas porosity voids in the weld. A y-tube connector can be used to blend the appropriate volumes of gases directly from the individual gas bottles and flowraters rather than maintaining a supply of a variety of premixed gas ratios.

Preparation for Welding. Flaws should be removed by pneumatic chipping or hand tool milling. This operation should be performed without the use of a lubricant, if possible. The leading edge of the chipping tool should be u-shaped rather than v-shaped to facilitate welding. The milling or deburring tool should have a coarse cutting bit to prevent it from becoming loaded with aluminum.

Cavity Preparation. The cavity that is to be repaired should be smooth, that is, without exaggerated ridges, and the sides of the cavity should be chamfered rather than being a steep-sided hole. Chamfering promotes good fusion in the casting repair. It also provides a means for any gas in the casting that is released during welding to become uniformly distributed throughout the weld rather than becoming concentrated as linear porosity around the edge of the weld.

Cleaning. Oil and grease, if present, should be removed with a locally applied solvent or by a vapor degreasing operation. The type of solvent used—toluene, for example—should not leave a residue that may gasify the weld or produce harmful fumes during the welding operation. Etchants should not be used to clean the casting before welding unless absolutely necessary. Etchants are usually corrosive and tend to be retained in the surface pores of the casting. They can also contribute to gas porosity in the weld if the casting surface containing these trapped liquids becomes involved in the welding operation.

Impregnated Castings. Castings should not be impregnated for pressure tightness before welding. If an impregnated casting must be welded, the impregnant can be partially removed by prolonged heating at a temperature of 150 to 205 °C (300 to 400 °F).

Surface Preparation. After cleaning with a solvent and just before welding, the area to be welded and the immediately adjacent area (a 13 mm, or ½ in., band around the weld area) should be hand brushed with a stainless steel bristle brush to reduce the thickness of the aluminum oxide on the casting surface and to remove any sand or foreign material. In cases in which the as-cast surface is to be welded, the surface should be scraped or scarfed to a depth of 1.5 mm (0.06 in.) or more to overcome the effects of the oxidized, rough casting surface.

The filler metal is often the same alloy as the casting being welded. Filler material can be cast in rod-shaped molds when production castings are being poured. If the castings are to be heat treated after welding, the filler material selected may be different from that used if ease of welding is the only consideration.

The size of the filler rod used for welding is usually the largest one that will not freeze the weld puddle. To minimize hot short weld cracking, weld composition should be maintained as closely as possible to a ratio of 70% filler alloy to 30% base alloy.

Preheating. Although it can be beneficial in reducing the size of gas porosity in the weld, preheating of the casting is not necessary before making a weld repair. The ability to maintain a puddle is the primary consideration. In some cases, however, preheating may be necessary to overcome cracking, distortion, or welding speed problems. Each case is different, and the need for preheating depends on the alloy, section thickness, casting size and geometry, residual stress, and potential thermal gradients between hot and cool areas. When it is required, preheating can be accomplished by placing the entire casting in a furnace, or if the size of the casting does not permit this method, the casting can be preheated with a gas torch. The casting is usually heated to 150 to 205 °C (300 to 400 °F).

Welding Technique. Weld quality is directly related to casting quality. The general problems encountered in making good welds with castings are the same as those encountered with wrought alloys except that the aluminum oxide on casting surfaces is thicker than on wrought alloys and must be reduced to diminish welding problems.

A good general rule of welding technique is to "get in and then get out quickly." To accomplish fast melting and weld puddle cleanup, helium shielding gas in a mixture with argon gas should be used. Cracks cannot be melted out, however, because the melting point of the aluminum oxide that lines the cracks is prohibitively high. The cracks must be chipped out and the cavity filled with weld metal.

Postweld Heat Treatment. Castings that require heat treatment should be heat treated after welding. Castings that have been heat treated and then welded will suffer localized loss of tensile properties. However, postweld heat treatment will restore their properties if the proper filler alloy has been used in welding.

Properties of Aluminum Casting Alloys

The physical and mechanical properties of aluminum casting alloys are well documented and are listed in Tables 5 to 7 in this article (see also the Selected References). The discussion in this article is not intended as a comprehensive discussion of all the commonly tested properties of aluminum

Table 5 Typical physical properties of aluminum casting alloys

Alloy	Temper and product form(a)	Specific gravity(b)	Density(b) kg/m³	Density(b) lb/in.³	Approximate melting range °C	Approximate melting range °F	Electrical conductivity, % IACS	Thermal conductivity at 25 °C (77 °F), W/m · K	Coefficient of thermal expansion, per °C × 10⁻⁶ (per °F × 10⁻⁶) 20–100 °C (68–212 °F)	Coefficient of thermal expansion, per °C × 10⁻⁶ (per °F × 10⁻⁶) 20–300 °C (68–570 °F)
201.0	T6 (S)	2.80	2796	0.101	570–650	1060–1200	27–32	0.29	34.7 (19.3)	44.5 (24.7)
	T7 (P)	2.80	2796	0.101	570–650	1060–1200	32–34	0.29	34.7 (19.3)	44.5 (24.7)
206.0	...	2.8	2796	0.101	570–650	1060–1200	...	0.29	... ...	... ...
A206.0	...	2.8	2796	0.101	570–650	1060–1200	...	0.29	... ...	... ...
208.0	F (S)	2.79	2796	0.101	520–630	970–1170	31	0.29	22.0 (12.2)	23.9 (13.3)
	O (S)	2.79	2796	0.101	520–630	970–1170	38	0.35	... ...	... ...
222.0	F (P)	2.95	2962	0.107	520–625	970–1160	34	0.32	22.1 (12.3)	23.6 (13.1)
	O (S)	2.95	2962	0.107	520–625	970–1160	41	0.38	... ...	... ...
	T61(S)	2.95	2962	0.107	520–625	970–1160	33	0.31	22.1 (12.3)	23.6 (13.1)
224.0	T62 (S)	2.81	2824	0.102	550–645	1020–1190	30	0.28	... ...	... ...
238.0	F (P)	2.95	1938	0.107	510–600	950–1110	25	0.25	21.4 (11.9)	22.9 (12.7)
240.0	F (S)	2.78	2768	0.100	515–605	960–1120	23	0.23	22.1 (12.3)	24.3 (13.5)
242.0	O (S)	2.81	2823	0.102	530–635	990–1180	44	0.40	... ...	... ...
	T77 (S)	2.81	2823	0.102	525–635	980–1180	38	0.36	22.1 (12.3)	23.6 (13.1)
	T571 (P)	2.81	2823	0.102	525–635	980–1180	34	0.32	22.5 (12.5)	24.5 (13.6)
	T61 (P)	2.81	2823	0.102	525–635	980–1180	33	0.32	22.5 (12.5)	24.5 (13.6)
295.0	T4 (S)	2.81	2823	0.102	520–645	970–1190	35	0.33	22.9 (12.7)	24.8 (13.8)
	T62 (S)	2.81	2823	0.102	520–645	970–1190	35	0.34	22.9 (12.7)	24.8 (13.8)
296.0	T4 (P)	2.80	2796	0.101	520–630	970–1170	33	0.32	22.0 (12.2)	23.9 (13.3)
	T6 (P)	2.80	2796	0.101	520–630	970–1170	33	0.32	22.0 (12.2)	23.9 (13.3)
	T62 (S)	2.80	2796	0.101	520–630	970–1170	33	0.32	... ...	... ...
308.0	F (P)	2.79	2796	0.101	520–615	970–1140	37	0.34	21.4 (11.9)	22.9 (12.7)
319.0	F (S)	2.79	2796	0.101	520–605	970–1120	27	0.27	21.6 (12.0)	24.1 (13.4)
	F (P)	2.79	2796	0.101	520–605	970–1120	28	0.28	21.6 (12.0)	24.1 (13.4)
324.0	F (P)	2.67	2658	0.096	545–605	1010–1120	34	0.37	21.4 (11.9)	23.2 (12.9)
332.0	T5 (P)	2.76	2768	0.100	520–580	970–1080	26	0.25	20.7 (11.5)	22.3 (12.4)
333.0	F (P)	2.77	2768	0.100	520–585	970–1090	26	0.25	20.7 (11.5)	22.7 (12.6)
	T5 (P)	2.77	2768	0.100	520–585	970–1090	29	0.29	20.7 (11.5)	22.7 (12.6)
	T6 (P)	2.77	2768	0.100	520–585	970–1090	29	0.28	20.7 (11.5)	22.7 (12.6)
	T7 (P)	2.77	2768	0.100	520–585	970–1090	35	0.34	20.7 (11.5)	22.7 (12.6)
336.0	T551 (P)	2.72	2713	0.098	540–570	1000–1060	29	0.28	18.9 (10.5)	20.9 (11.6)
354.0	F (P)	2.71	2713	0.098	540–600	1000–1110	32	0.30	20.9 (11.6)	22.9 (12.7)
355.0	T51 (S)	2.71	2713	0.098	550–620	1020–1150	43	0.40	22.3 (12.4)	24.7 (13.7)
	T6 (S)	2.71	2713	0.098	550–620	1020–1150	36	0.34	22.3 (12.4)	24.7 (13.7)
	T61 (S)	2.71	2713	0.098	550–620	1020–1150	37	0.35	22.3 (12.4)	24.7 (13.7)
	T7 (S)	2.71	2713	0.098	550–620	1020–1150	42	0.39	22.3 (12.4)	24.7 (13.7)
	T6 (P)	2.71	2713	0.098	550–620	1020–1150	39	0.36	22.3 (12.4)	24.7 (13.7)
C355.0	T61 (S)	2.71	2713	0.098	550–620	1020–1150	39	0.35	22.3 (12.4)	24.7 (13.7)
356.0	T51 (S)	2.68	2685	0.097	560–615	1040–1140	43	0.40	21.4 (11.9)	23.4 (13.0)
	T6 (S)	2.68	2685	0.097	560–615	1040–1140	39	0.36	21.4 (11.9)	23.4 (13.0)
	T7 (S)	2.68	2685	0.097	560–615	1040–1140	40	0.37	21.4 (11.9)	23.4 (13.0)
	T6 (P)	2.68	2685	0.097	560–615	1040–1140	41	0.37	21.4 (11.9)	23.4 (13.0)

(continued)

(a) S, sand cast; P, permanent mold; D, die cast. (b) The specific gravity and weight data in this table assume solid (void-free) metal. Because some porosity cannot be avoided in commercial castings, their specific gravity or weight is slightly less than the theoretical value.

casting alloys (for example, tensile strength). Rather, certain properties, such as fatigue resistance and corrosion resistance, will be discussed in terms of testing, problems in testing, and how the properties of cast aluminum may differ from those of wrought alloys. Melt fluidity, shrinkage during cooling, and hot cracking will also be discussed.

It must be noted here that the data given in Tables 5 to 7 are useful for comparing alloys, but they should not be used for design purposes. Properties for design must be obtained from pertinent specifications or design standards or by negotiation with the producer.

Fatigue

When a part is subjected to repeated loads, cracks may develop in highly stressed areas and grow as cyclic or repetitive loading occurs. Such cracks may occur at stress levels below the characteristic ultimate strength of the material. Failure by such a mechanism is termed fatigue failure. Fatigue cracks frequently initiate at stress raisers such as abrupt changes in section, holes, notches, machining gouges, stamped emblems, the edges of a weld bead, porosity, surface oxide concentrations, dross inclusions, or other surface discontinuities.

Fatigue Testing. Separately cast test bars are often employed to avoid the destruction of production parts. Unfortunately, the fatigue resistance of such specimens is usually higher than that of most cast products due to factors such as relatively low porosity and the presence of beneficial residual surface compressive stresses. It may therefore be appropriate to machine specimens from critical casting areas when fatigue properties are important.

The fatigue strength of castings is normally lower than that of wrought materials. However, discontinuities that considerably reduce the fatigue strength of a wrought material (for example, holes) have much less effect on the fatigue performance of a cast part (that is, castings are less notch-sensitive than wrought materials).

Premium engineered castings combine a number of conventional casting procedures in a selective manner to provide premium strength, soundness, dimensional tolerances, surface finish, or a combination of these characteristics. As a result, premium engineered castings typically have somewhat higher fatigue resistance than conventional castings. Improved surface finish, better soundness at the surface, finer grain size, and adjustments in heat treated condition may all contribute to this enhanced resistance.

Avoiding Fatigue Cracking. Because castings are less notch-sensitive than wrought materials, it is not so critical to avoid discontinuities with castings as it is with wrought products. The following general considerations, however, should be kept in mind for castings that are to be subject to repeated loading:

Table 5 (continued)

Alloy	Temper and product form(a)	Specific gravity(b)	Density(b) kg/m³	Density(b) lb/in.³	Approximate melting range °C	Approximate melting range °F	Electrical conductivity, % IACS	Thermal conductivity at 25 °C (77 °F), W/m · K	Coefficient of thermal expansion, per °C × 10^{-6} (per °F × 10^{-6}) 20–100 °C (68–212 °F)	Coefficient of thermal expansion, per °C × 10^{-6} (per °F × 10^{-6}) 20–300 °C (68–570 °F)
A356.0	T6 (S)	2.69	2713	0.098	560–610	1040–1130	40	0.36	21.4 (11.9)	23.4 (13.0)
357.0	T6 (S)	2.68	2713	0.098	560–615	1040–1140	39	0.36	21.4 (11.9)	23.4 (13.0)
A357.0	T6 (S)	2.69	2713	0.098	555–610	1030–1130	40	0.38	21.4 (11.9)	23.6 (13.1)
358.0	T6 (S)	2.68	2658	0.096	560–600	1040–1110	39	0.36	21.4 (11.9)	23.4 (13.0)
359.0	T6 (S)	2.67	2685	0.097	565–600	1050–1110	35	0.33	20.9 (11.6)	22.9 (12.7)
360.0	F (D)	2.68	2685	0.097	570–590	1060–1090	37	0.35	20.9 (11.6)	22.9 (12.7)
A360.0	F (D)	2.68	2685	0.097	570–590	1060–1090	37	0.35	21.1 (11.7)	22.9 (12.7)
364.0	F (D)	2.63	2630	0.095	560–600	1040–1110	30	0.29	20.9 (11.6)	22.9 (12.7)
380.0	F (D)	2.76	2740	0.099	520–590	970–1090	27	0.26	21.2 (11.8)	22.5 (12.5)
A380.0	F (D)	2.76	2740	0.099	520–590	970–1090	27	0.26	21.1 (11.7)	22.7 (12.6)
384.0	F (D)	2.70	2713	0.098	480–580	900–1080	23	0.23	20.3 (11.3)	22.1 (12.3)
390.0	F (D)	2.73	2740	0.099	510–650	950–1200	25	0.32	18.5 (10.3)	... ...
	T5 (D)	2.73	2740	0.099	510–650	950–1200	24	0.32	18.0 (10.0)	... ...
392.0	F (P)	2.64	2630	0.095	550–670	1020–1240	22	0.22	18.5 (10.3)	20.2 (11.2)
413.0	F (D)	2.66	2657	0.096	575–585	1070–1090	39	0.37	20.5 (11.4)	22.5 (12.5)
A413.0	F (D)	2.66	2657	0.096	575–585	1070–1090	39	0.37	... ...	... ...
443.0	F (S)	2.69	2685	0.097	575–630	1070–1170	37	0.35	22.1 (12.3)	24.1 (13.4)
	O (S)	2.69	2685	0.097	575–630	1070–1170	42	0.39	... ...	... ...
	F (D)	2.69	2685	0.097	575–630	1070–1170	37	0.34	... ...	... ...
A444.0	F (P)	2.68	2685	0.097	575–630	1070–1170	41	0.38	21.8 (12.1)	23.8 (13.2)
511.0	F (S)	2.66	2657	0.096	590–640	1090–1180	36	0.34	23.6 (13.1)	25.7 (14.3)
512.0	F (S)	2.65	2657	0.096	590–630	1090–1170	38	0.35	22.9 (12.7)	24.8 (13.8)
513.0	F (P)	2.68	2685	0.097	580–640	1080–1180	34	0.32	23.9 (13.3)	25.9 (14.4)
514.0	F (S)	2.65	2657	0.096	600–640	1110–1180	35	0.33	23.9 (13.3)	25.9 (14.4)
518.0	F (D)	2.53	2519	0.091	540–620	1000–1150	24	0.24	24.1 (13.4)	26.1 (14.5)
520.0	T4 (S)	2.57	2574	0.093	450–600	840–1110	21	0.21	25.2 (14.0)	27.0 (15.0)
535.0	F (S)	2.62	2519	0.091	550–630	1020–1170	23	0.24	23.6 (13.1)	26.5 (14.7)
A535.0	F (D)	2.54	2547	0.092	550–620	1020–1150	23	0.24	24.1 (13.4)	26.1 (14.5)
B535.0	F (S)	2.62	2630	0.095	550–630	1020–1170	24	0.23	24.5 (13.6)	26.5 (14.7)
705.0	F (S)	2.76	2768	0.100	600–640	1110–1180	25	0.25	23.6 (13.1)	25.7 (14.3)
707.0	F (S)	2.77	2768	0.100	585–630	1090–1170	25	0.25	23.8 (13.2)	25.9 (14.4)
710.0	F (S)	2.81	2823	0.102	600–650	1110–1200	35	0.33	24.1 (13.4)	26.3 (14.6)
711.0	F (P)	2.84	2851	0.103	600–645	1110–1190	40	0.38	23.6 (13.1)	25.6 (14.2)
712.0	F (S)	2.82	2823	0.102	600–640	1110–1180	40	0.38	23.6 (13.1)	25.6 (14.2)
713.0	F (S)	2.84	2879	0.104	595–630	1100–1170	37	0.37	23.9 (13.3)	25.9 (14.4)
850.0	T5 (S)	2.87	2851	0.103	225–650	440–1200	47	0.44	... ...	... ...
851.0	T5 (S)	2.83	2823	0.102	230–630	450–1170	43	0.40	22.7 (12.6)	... ...
852.0	T5 (S)	2.88	2879	0.104	210–635	410–1180	45	0.42	23.2 (12.9)	... ...

(a) S, sand cast; P, permanent mold; D, die cast. (b) The specific gravity and weight data in this table assume solid (void-free) metal. Because some porosity cannot be avoided in commercial castings, their specific gravity or weight is slightly less than the theoretical value.

- Use gradual changes in section and symmetry of design where practical
- Minimize scratches and the presence and magnitude of surface defects or discontinuities of any kind in highly stressed areas
- Alleviate possible fretting between contacting surfaces
- Provide proper maintenance and suitable protection against corrosion

Corrosion

The corrosion resistance of aluminum casting alloys in service depends on such factors as alloy composition and the environment to which the alloy is exposed. The mechanical and physical conditions under which the alloy is exposed are also important. Detailed information on the corrosion resistance of wrought aluminum alloys is available in the article "Corrosion of Aluminum and Aluminum Alloys" in Volume 13 of the 9th Edition of *Metals Handbook*.

Minimizing Corrosion Problems. Corrosion can be minimized by designing the application to include protective measures and by selecting the alloy that will economically meet all design requirements, including those pertaining to corrosion. In many applications, the corrosion resistance of various casting alloys is similar enough that alloy selection can be based on other factors. Only in severe environments, such as marine atmospheres or heavily polluted urban industrial atmospheres, do differences in the general corrosion resistances of various aluminum casting alloys become apparent.

Mechanical Properties

Mechanical Properties and Casting Design. Any casting design is based on two expectations:

- The casting will meet certain minimum requirements for mechanical properties that are characteristic of the alloy and temper selected
- The combination of design and properties will provide reliable service performance

The mechanical properties of castings produced by a particular process in a specific alloy and temper are critical considerations in alloy and temper selection.

Mechanical Properties and Quality Assurance. Some type of quality assurance program is required to ascertain that metal composition, molten metal quality, foundry practices, and heat treatment are consistently under control, and that parts with the expected properties are being consistently produced. Specimens for mechanical property testing may be cast separately, integrally cast, or cut from representative castings.

Minimum Values for Mechanical Tests. Minimum mechanical property limits are usually defined by the terms of general procurement specifications, such as those developed by government agencies and technical societies. These documents often specify testing frequency, tensile bar type and design, lot definitions, testing procedures, and the limits applicable to test results. By references to general process specifications, these documents also invoke standards and limits for many additional supplier obligations, such as specific practices and controls in melt preparation, heat treatment, radiographic and liquid penetrant inspection, and test procedures and interpretation.

Table 6 Ratings of castability, corrosion resistance, machinability, and weldability for aluminum casting alloys

1, best; 5, worst. Individual alloys may have different ratings for other casting processes.

Alloy	Resistance to hot cracking(a)	Pressure tightness	Fluidity(b)	Shrinkage tendency(c)	Corrosion resistance(d)	Machin-ability(e)	Weld-ability(f)
Sand casting alloys							
201.0	4	3	3	4	4	1	2
208.0	2	2	2	2	4	3	3
213.0	3	3	2	3	4	2	2
222.0	4	4	3	4	4	1	3
240.0	4	4	3	4	4	3	4
242.0	4	3	4	4	4	2	3
A242.0	4	4	3	4	4	2	3
295.0	4	4	4	3	3	2	2
319.0	2	2	2	2	3	3	2
354.0	1	1	1	1	3	3	2
355.0	1	1	1	1	3	3	2
A356.0	1	1	1	1	2	3	2
357.0	1	1	1	1	2	3	2
359.0	1	1	1	1	2	3	1
A390.0	3	3	3	3	2	4	2
A443.0	1	1	1	1	2	4	4
444.0	1	1	1	1	2	4	1
511.0	4	5	4	5	1	1	4
512.0	3	4	4	4	1	2	4
514.0	4	5	4	5	1	1	4
520.0	2	5	4	5	1	1	5
535.0	4	5	4	5	1	1	3
A535.0	4	5	4	4	1	1	4
B535.0	4	5	4	4	1	1	4
705.0	5	4	4	4	2	1	4
707.0	5	4	4	4	2	1	4
710.0	5	3	4	4	2	1	4
711.0	5	4	5	4	3	1	3
712.0	4	4	3	3	3	1	4
713.0	4	4	3	4	2	1	3
771.0	4	4	3	3	2	1	...
772.0	4	4	3	3	2	1	...
850.0	4	4	4	4	3	1	4
851.0	4	4	4	4	3	1	4
852.0	4	4	4	4	3	1	4
Permanent mold casting alloys							
201.0	4	3	3	4	4	1	2
213.0	3	3	2	3	4	2	2
222.0	4	4	3	4	4	1	3
238.0	2	3	2	2	4	2	3
240.0	4	4	3	4	4	3	4
296.0	4	3	4	3	4	3	4
308.0	2	2	2	2	4	3	3
319.0	2	2	2	2	3	3	2
332.0	1	2	1	2	3	4	2
333.0	1	1	2	2	3	3	3
336.0	1	2	2	3	3	4	2
354.0	1	1	1	1	3	3	2
355.0	1	1	1	2	3	3	2
C355.0	1	1	1	2	3	3	2
356.0	1	1	1	1	2	3	2
A356.0	1	1	1	1	2	3	2
357.0	1	1	1	1	2	3	2
A357.0	1	1	1	1	2	3	2
359.0	1	1	1	1	2	3	1
A390.0	2	2	2	3	2	4	2
443.0	1	1	2	1	2	5	1
A444.0	1	1	1	1	2	3	1
512.0	3	4	4	4	1	2	4
513.0	4	5	4	4	1	1	5
711.0	5	4	5	4	3	1	3
771.0	4	4	3	3	2	1	...
772.0	4	4	3	3	2	1	...
850.0	4	4	4	4	3	1	4
851.0	4	4	4	4	3	1	4
852.0	4	4	4	4	3	1	4
Die casting alloys							
360.0	1	1	2	2	3	4	
A360.0	1	1	2	2	3	4	
364.0	2	2	1	3	4	3	
380.0	2	1	2	5	3	4	
A380.0	2	2	2	4	3	4	
384.0	2	2	1	3	3	4	
390.0	2	2	2	2	4	2	
413.0	1	2	1	2	4	4	
C443.0	2	3	3	2	5	4	
515.0	4	5	5	1	2	4	
518.0	5	5	5	1	1	4	

(a) Ability of alloy to withstand stresses from contraction while cooling through hot short or brittle temperature range. (b) Ability of liquid alloy to flow readily in mold and to fill thin sections. (c) Based on resistance of alloy in standard salt spray test. (d) Composite rating based on ease of cutting, chip characteristics, quality of finish, and tool life. (e) Based on ability of material to be fusion welded with filler rod of same alloy

Some specifications for sand, permanent mold, plaster, and investment castings have defined the correlation between test results from specimens cut from the casting and separately cast specimens. A frequent error is the assumption that test values determined from these sources should agree. Rather, the properties of separately cast specimens should be expected to be superior to those of specimens machined from the casting. In the absence of more specific guidelines, one rule of thumb defines the average tensile and yield strengths of machined specimens as not less than 75% of the minimum requirements for separately cast specimens, and elongation as not less than 25% of the minimum requirement. These relations may be useful in establishing the commercial acceptability of parts in dispute.

Premium Engineered Castings. Quality assurance for premium engineered castings frequently involves tensile testing of specimens machined from castings or cast integrally with the casting. Property requirements are established by negotiation and agreement between customer and supplier, and may include guarantees only for critical areas of the casting or for the entire part.

Molten Metal Fluidity

The Nature of Fluidity. Fluidity is the characteristic of a molten alloy that enables mold filling. Different alloys have different fluidities, and therefore different abilities to provide definition of detail and integrity in thin sections.

Factors Affecting Fluidity. Fluidity depends on two major factors: the intrinsic fluid properties of the molten metal, and casting conditions. The properties usually thought to influence fluidity are viscosity, surface tension, the character of the surface oxide film, inclusion content, and manner in which the particular alloy solidifies.

Casting conditions that influence fluidity include part configuration; physical measures of the fluid dynamics of the system, such as liquidstatic pressure drops, casting head, and velocities; mold material; mold surface characteristics; heat flux; rate of pouring; and degree of superheat.

Viscosity. The measured viscosities of molten aluminum alloys are quite low and fall within a relatively narrow range. Kinematic viscosity (viscosity/specific gravity) is less than that of water. It is evident on this basis that viscosity is not strongly influential in determining casting behavior and therefore is an unlikely source of variability in casting results.

Surface Tension and Oxide Film. A high surface tension has the effect of increasing the pressure required for liquid metal flow. A number of elements influence surface tension, primarily through their effects on the surface tension of the oxide. Figure 11 illustrates the effect of selected elements on surface tension. In aluminum alloys, the true effect of surface tension is overpowered by the influence of surface oxide film characteristics. The oxide film on pure aluminum, for example, triples apparent surface tension.

Table 7 Typical mechanical properties of aluminum casting alloys

Alloy	Temper	Ultimate tensile strength, MPa	Ultimate tensile strength, ksi	0.2% offset yield strength, MPa	0.2% offset yield strength, ksi	Elongation in 50 mm (2 in.), %	Hardness, HB(a)
Sand casting alloys							
201.0	T43	414	60	255	37	17.0	. . .
	T6	448	65	379	55	8.0	130
	T7	467	68	414	60	5.5	. . .
A206.0	T4	354	51	250	36	7.0	. . .
208.0	F	145	21	97	14	2.5	55
213.0	F	165	24	103	15	1.5	70
222.0	O	186	27	138	20	1.0	80
	T61	283	41	276	40	<0.5	115
	T62	421	61	331	48	4.0	. . .
224.0	T72	380	55	276	40	10.0	123
240.0	F	235	34	200	28	1.0	90
242.0	F	214	31	217	30	0.5	. . .
	O	186	27	124	18	1.0	70
	T571	221	32	207	30	0.5	85
	T77	207	30	159	23	2.0	75
A242.0	T75	214	31	. . .	. . .	2.0	. . .
295.0	T4	221	32	110	16	8.5	60
	T6	250	36	165	24	5.0	75
	T62	283	41	221	32	2.0	90
319.0	F	186	27	124	18	2.0	70
	T5	207	30	179	26	1.5	80
	T6	250	36	164	24	2.0	80
355.0	F	159	23	83	12	3.0	. . .
	T51	193	28	159	23	1.5	65
	T6	241	35	172	25	3.0	80
	T61	269	39	241	35	1.0	90
	T7	264	38	250	26	0.5	85
	T71	241	35	200	29	1.5	75
C355.0	T6	269	39	200	29	5.0	85
356.0	F	164	24	124	18	6.0	. . .
	T51	172	25	138	20	2.0	60
	T6	228	33	164	24	3.5	70
	T7	234	34	207	30	2.0	75
	T71	193	28	145	21	3.5	60
A356.0	F	159	23	83	12	6.0	. . .
	T51	179	26	124	18	3.0	. . .
	T6	278	40	207	30	6.0	75
	T71	207	30	138	20	3.0	. . .
357.0	F	172	25	90	13	5.0	. . .
	T51	179	26	117	17	3.0	. . .
	T6	345	50	296	43	2.0	90
	T7	278	40	234	34	3.0	60
A357.0	T6	317	46	248	36	3.0	85
A390.0	F	179	26	179	26	<1.0	100
	T5	179	26	179	26	<1.0	100
	T6	278	40	278	40	<1.0	140
	T7	250	36	250	36	<1.0	115
443.0	F	131	19	55	8	8.0	40
A444.0	F	145	21	62	9	9.0	. . .
	T4	159	23	62	9	12.0	. . .
511.0	F	145	21	83	12	3.0	50
512.0	F	138	20	90	13	2.0	50
514.0	F	172	25	83	12	9.0	50
520.0	T4	331	48	179	26	16.0	75
A535.0	F	250	36	124	18	9.0	65
710.0	F	241	35	172	25	5.0	75
712.0(h)	F	241	35	172	25	5.0	75
713.0(h)	F	241	35	172	25	5.0	74
850.0	T5	138	20	76	11	8.0	45
851.0	T5	138	20	76	11	5.0	45
852.0	T5	186	27	152	22	2.0	65
Permanent mold casting alloys							
201.0	T43	414	60	255	37	17.0	. . .
	T6	448	65	379	55	8.0	130
	T7	469	68	414	60	5.0	. . .
A206.0	T4	431	62	264	38	17.0	. . .
	T7	436	63	347	50	11.7	. . .
213.0	F	207	30	165	24	1.5	85
222.0	T52	241	35	214	31	1.0	100
	T551	255	37	241	35	<0.5	115
	T65	331	48	248	36	<0.5	140
238.0	F	207	30	165	24	1.5	100
242.0	T571	276	40	234	34	1.0	105
	T61	324	47	290	42	0.5	110
249.0	T63	476	69	414	60	6.0	. . .
	T7	278	62	359	52	9.0	. . .
296.0	T4	255	37	131	19	9.0	75
	T6	276	40	179	26	5.0	90
	T7	270	39	138	20	4.5	80
308.0	F	193	28	110	16	2.0	70
319.0	F	234	34	131	19	2.5	85
	T6	276	40	186	27	3.0	95
324.0	F	207	30	110	16	4.0	70
	T5	248	36	179	26	3.0	90
	T62	310	45	269	39	3.0	105
332.0	T5	248	36	193	28	1.0	105
333.0	F	234	34	131	19	2.0	90
	T5	234	34	172	25	1.0	100
	T6	290	42	207	30	1.5	105
	T7	255	37	193	28	2.0	90
336.0	T551	248	36	193	28	0.5	105
	T65	324	47	296	43	0.5	125
356.0	F	179	26	124	18	5.0	. . .
	T51	186	27	138	20	2.0	. . .
	T6	262	38	186	27	5.0	80
	T7	221	32	165	24	6.0	70
A356.0	T61	283	41	207	30	10.0	90
357.0	F	193	28	103	15	6.0	. . .
	T51	200	29	145	21	4.0	. . .
	T6	359	52	296	43	5.0	100
A357.0	T6	359	52	290	42	5.0	100
359.0	T62	345	50	290	42	5.5	. . .
A390.0	F	200	29	200	29	<1.0	110
	T5	200	29	200	29	<1.0	110
	T6	310	45	310	45	<1.0	145
	T7	262	38	262	38	<1.0	120
443.0	F	159	23	62	9	10.0	45
A444.0	F	165	24	76	11	13.0	44
	T4	159	23	69	10	21.0	45
513.0	F	186	27	110	16	7.0	60
711.0	F	241	35	124	18	8.0	70
850.0	T5	159	23	76	11	12.0	45
851.0	T5	138	20	76	11	5.0	45
852.0	T5	221	32	159	23	5.0	70
Die casting alloys							
360.0	F	324	47	172	25	3.0	75
A360.0	F	317	46	165	24	5.0	75
364.0	F	296	43	159	23	7.5	. . .
380.0	F	331	48	165	24	3.0	80
A380.0	F	324	47	159	23	4.0	75
384.0	F	324	47	172	25	1.0	. . .
390.0	F	279	40.5	241	35	1.0	120
	T5	296	43	265	38.5	1.0	. . .
392.0	F	290	42	262	38	<0.5	. . .
413.0	F	296	43	145	21	2.5	80
A413.0	F	241	35	110	16	3.5	80
443.0	F	228	33	110	16	9.0	50
513.0	F	276	40	152	22	10.0	. . .
515.0	F	283	41	. . .	. . .	10.0	. . .
518.0	F	310	45	186	27	8.0	80

(a) 500 kg (1100 lb) load on 10 mm (0.4 in.) ball

Inclusions in the form of suspended insoluble nonmetallic particles dramatically reduce the fluidity of molten aluminum.

Solidification. It has been shown that fluidity is inversely proportional to freezing range (that is, fluidity is highest for pure metals and eutectics, and lowest for solid solution alloys). The manner in which solidification occurs may also influence fluidity.

Fluidity Testing. Correlation between fluidity test results (usually of the spiral or variable-dimension type) and foundry experience has been poor. No accepted method for the determination or prediction of fluid-

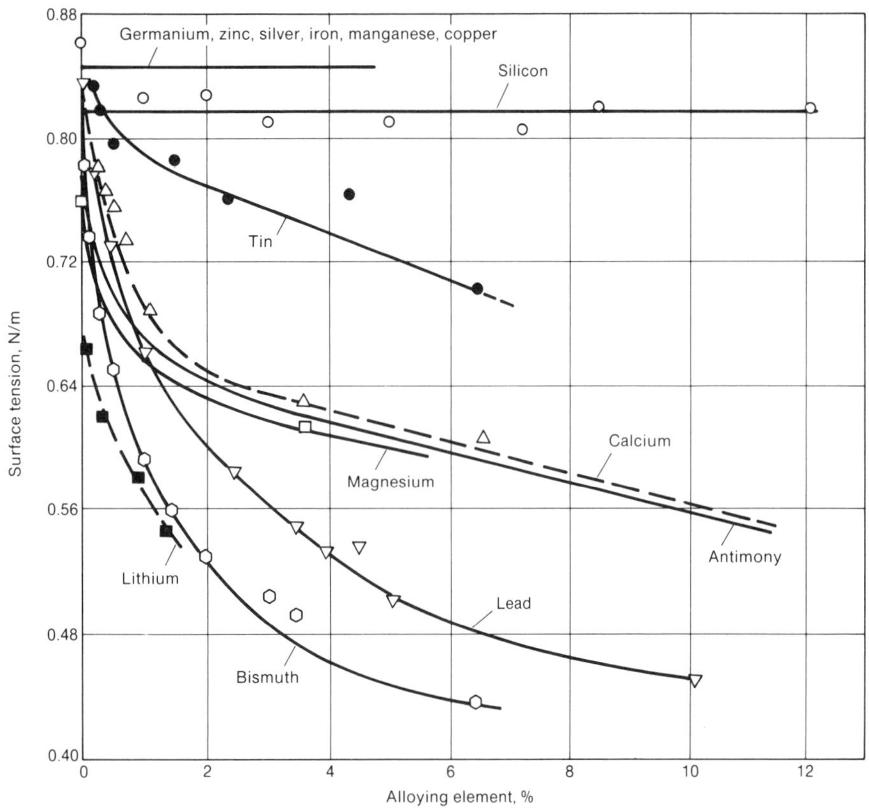

Fig. 11 Effect of various elements on surface tension of 99.99% Al in argon at 700 to 740 °C (1290 to 1365 °F)

Table 8 Representative applications for aluminum casting alloys

Alloy	Representative applications
100.0	Electrical rotors larger than 152 mm (6 in.) in diameter
201.0	Structural members; cylinder heads and pistons; gear, pump, and aerospace housings
208.0	General-purpose castings; valve bodies, manifolds, and other pressure-tight parts
222.0	Bushings; meter parts; bearings; bearing caps; automotive pistons; cylinder heads
238.0	Sole plates for electric hand irons
242.0	Heavy-duty pistons; air-cooled cylinder heads; aircraft generator housings
A242.0	Diesel and aircraft pistons; air-cooled cylinder heads; aircraft generator housings
B295.0	Gear housings; aircraft fittings; compressor connecting rods; railway car seat frames
308.0	General-purpose permanent mold castings; ornamental grilles and reflectors
319.0	Engine crankcases; gasoline and oil tanks; oil pans; typewriter frames; engine parts
332.0	Automotive and heavy-duty pistons; pulleys, sheaves
333.0	Gas meter and regulator parts; gear blocks; pistons; general automotive castings
354.0	Premium-strength castings for the aerospace industry
355.0	Sand: air compressor pistons; printing press bedplates; water jackets; crankcases. Permanent: impellers; aircraft fittings; timing gears; jet engine compressor cases
356.0	Sand: flywheel castings; automotive transmission cases; oil pans; pump bodies. Permanent: machine tool parts; aircraft wheels; airframe castings; bridge railings
A356.0	Structural parts requiring high strength; machine parts; truck chassis parts
357.0	Corrosion-resistant and pressure-tight applications
359.0	High-strength castings for the aerospace industry
360.0	Outboard motor parts; instrument cases; cover plates; marine and aircraft castings
A360.0	Cover plates; instrument cases; irrigation system parts; outboard motor parts; hinges
380.0	Housings for lawn mowers and radio transmitters; air brake castings; gear cases
A380.0	Applications requiring strength at elevated temperature
384.0	Pistons and other severe service applications; automatic transmissions
390.0	Internal combustion engine pistons, blocks, manifolds, and cylinder heads
413.0	Architectural, ornamental, marine, and food and dairy equipment applications
A413.0	Outboard motor pistons; dental equipment; typewriter frames; street lamp housings
443.0	Cookware; pipe fittings; marine fittings; tire molds; carburetor bodies
514.0	Fittings for chemical and sewage use; dairy and food handling equipment; tire molds
A514.0	Permanent mold casting of architectural fittings and ornamental hardware
518.0	Architectural and ornamental castings; conveyor parts; aircraft and marine castings
520.0	Aircraft fittings; railway passenger car frames; truck and bus frame sections
535.0	Instrument parts and other applications where dimensional stability is important
A712.0	General-purpose castings that require subsequent brazing
713.0	Automotive parts; pumps; trailer parts; mining equipment
850.0	Bushings and journal bearings for railroads
A850.0	Rolling mill bearings and similar applications

Source: Compiled from *Aluminum Casting Technology*, American Foundrymen's Society, 1986

ity, with relevance to the casting process, has been developed.

Shrinkage

For most metals, the transformation from the liquid to the solid state is accompanied by a decrease in volume. In aluminum alloys, volumetric solidification shrinkage can range from 3.5 to 8.5%. The tendency for formation of shrinkage porosity is related to both the liquid/solid volume fraction and the solidification temperature range of the alloy. Riser requirements relative to the casting weight can be expected to increase with increasing solidification temperature range. Requirements for the establishment of more severe thermal gradients, such as by the use of chills or antichills, also increase.

Hot Cracking

The Source of Hot Cracking. As previously mentioned, aluminum alloys shrink from 3.5 to 8.5% during solidification. A further, much smaller contraction occurs in post-solidification cooling to room temperature. Tearing or hot cracking occurs during solidification, when the greatest amount of shrinkage occurs and when the casting is least resistant to stresses imposed by the geometrical constraints of the mold. This type of cracking, sometimes termed hot shortness, is always intergranular and is a feature to some extent of all casting alloys. Hot strength (resistance to cracking at solidification temperature), however, is alloy-dependent.

Minimizing Hot Cracking. Hot cracking problems can be minimized by proper selection of casting process, alloy selection, part and mold design, solidification control, and grain refinement. Aluminum casting alloys demonstrate varying degrees of resistance to cracking; therefore, the crack sensitivity of alloys capable of meeting part requirements should be an important criterion in final alloy selection.

Mold Design. Improper mold design is a frequent cause of hot cracking. It is important to avoid abrupt changes in cross sectional area and inadequately radiused or filleted angles or corners. These are design features that can sharply increase and concentrate stresses resulting from restrained contraction during and following solidification. The intelligent use of chills and/or antichills can also be beneficial in preventing hot cracks.

Effective grain refinement to ensure a fine, equiaxed grain structure is especially important in hot-short alloys or in casting designs with features that aggravate cracking tendencies. Fine grain size reduces local stress concentrations by minimizing grain boundary effects.

Applications

Aluminum alloy castings are economical in many applications. They are used in the automotive industry, in construction of machines, appliances, and structures, as cooking utensils, as covers and housings for electronic equipment, and in innumerable other areas. Table 8 lists typical applications of some aluminum casting alloys, and Fig. 12 illustrates some common applications.

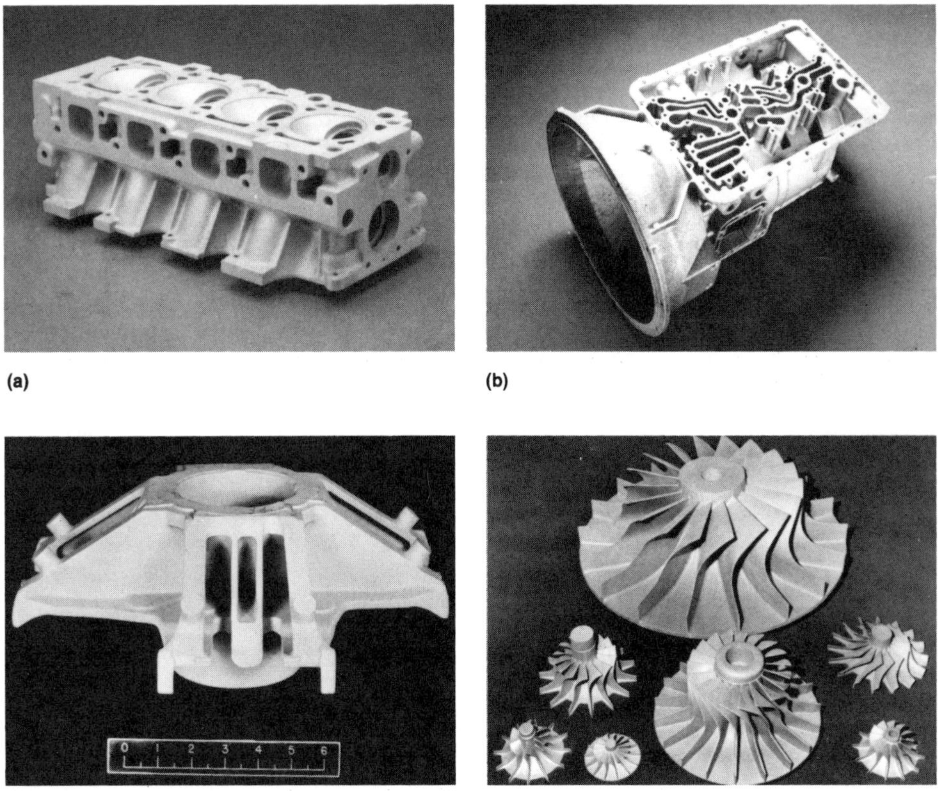

(a) (b) (c) (d)

Fig. 12 Some applications for aluminum alloy castings. (a) Alloy 319 automotive cylinder head. (b) Alloy 380 automotive transmission case. (c) Alloy A357 helicopter rotor hub. (d) Close-tolerance alloy 224 pump impellers

REFERENCES

1. *Registration Record of Aluminum Association Alloy Designations and Chemical Composition Limits for Aluminum Alloys in the Form of Castings and Ingot,* The Aluminum Association, July 1985
2. R.C. Gibbons, Ed., *Woldman's Engineering Alloys*, 6th ed., American Society for Metals, 1979
3. *Handbook of International Alloy Compositions and Designations*, Metals and Ceramics Information Center, Battelle Memorial Institute, 1976
4. K.R. Van Horn, Ed., *Properties, Physical Metallurgy, and Phase Diagrams*, Vol 1, *Aluminum*, American Society for Metals, 1967, p 174

SELECTED REFERENCES

General References

- *Aluminum Casting Technology*, American Foundrymen's Society, 1986
- J.E. Hatch, Ed., *Aluminum: Properties and Physical Metallurgy*, American Society for Metals, 1984
- K.R. Van Horn, Ed., *Aluminum*, Vol 1-3, American Society for Metals, 1967

Alloys

- *Aluminum Standards and Data*, The Aluminum Association, 1984
- *Registration Record of Aluminum Association Alloy Designations and Composition Limits for Aluminum Alloys in the Form of Castings and Ingot*, The Aluminum Association, 1987

Casting Processes

- R. Bruner, *The Metallurgy of Die Casting Alloys*, Society of Die Casting Engineers, 1986
- *Core and Mold Process Control*, American Foundrymen's Society, 1977
- *Foundry Core Practice*, American Foundrymen's Society, 1966
- *Fundamental Molding Sand Technology*, American Foundrymen's Society, 1973
- E.A. Herman, *Die Casting Handbook*, Society of Die Casting Engineers, 1982
- *Plaster Mold Handbook*, American Foundrymen's Society, 1984
- Precision Ceramic Castings, Paper 107, *Trans. AFS*, 1976
- A.C. Street, *The Die Casting Book*, Portcullis Press, 1977
- B. Upton, *Pressure Die Casting*, Society of Die Casting Engineers, 1982

Solidification

- J. Burke, M. Flemings, and A. Gorum, *Solidification Technology*, Brook Hill Publishing, 1974
- W. Kurz and E. Fisher, *Fundamentals of Solidification*, TransTech Publications, 1986
- A. Montgomery, Metallography of Aluminum Casting Alloys, *Am. Foundryman*, April 1949
- *Solidification*, American Society for Metals, 1971

Hydrogen in Aluminum

- Correlation of Tensile Properties to the Amounts of Porosity in Permanent Mold Test Bars, Paper 44, *Trans. AFS*, 1981
- E. Rooy, Control of Aluminum Casting Quality by Vacuum Solidification Tests, *Mod. Cast.*, July 1968
- D. Talbot, Effects of Hydrogen in Aluminum, Magnesium, Copper, and Their Alloys, *Int. Metall. Rev.*, Vol 20, 1975
- D. Talbot and D. Granger, Secondary Hydrogen Porosity in Aluminum, *J. Inst. Met.*, Vol 92, 1963-1964
- P. Thomas and J. Gruzleski, Threshold Hydrogen for Pore Formation During the Solidification of Aluminum Alloys, *Metall. Trans. B*, March 1978

Oxidation

- K. Brondyke and P. Hess, *Filtering and Fluxing Processes for Aluminum Alloys*, American Institute of Mining, Metallurgical, and Petroleum Engineers, 1964
- E. Rooy, The Use of Molten Metal Filters to Eliminate Air Pollution and Improve Melt Quality, *Trans. AFS*, 1971
- W. Thiele, The Oxidation of Melts of Aluminum and Aluminum Alloys, *Aluminum*, Vol 38, 1962, p 707-715

Dendrite Arm Spacing

- K. Oswalt and M. Misra, Dendrite Arm Spacing, Paper 51, *Trans. AFS*, 1980
- K. Radhakrishna, S. Seshan, and M. Seshadri, *Dendrite Arm Spacing and Mechanical Properties of Aluminum Alloy Castings*, Metallurgical Society, Indian Institute of Science, 1979
- R. Spear and G. Gardner, Dendrite Cell Size, Paper 65, *Trans. AFS*, 1963

Grain Refinement

- M. Guzowski and G. Sigworth, Grain Refining of Hypoeutectic Al-Si Alloys, Paper 172, *Trans. AFS*, 1985, p 85-172

- M. Guzowski, G. Sigworth, and D. Sentner, The Role of Boron in the Grain Refinement of Aluminum With Titanium, *Metall. Trans. A*, Vol 10A, April 1987
- R. Kotschi and C. Loper, Grain Refinement in Cast Aluminum Alloys, Paper 113, *Trans. AFS*, 1977
- *Solidification Characteristics of Aluminum Alloys*, Skan Aluminum, 1986

Modification and Refinement

- D. Apelian, G. Sigworth, and K. Whaler, Assessment of Modification and Grain Refining of Aluminum Silicon Casting Alloys by Thermal Analysis, Paper 161, *Trans. AFS*, 1984
- O. Atasoy, F. Yilmaz, and R. Elliot, Growth Structures in Aluminum-Silicon Alloys, *J. Cryst. Growth*, Jan-Feb 1984
- S. Bercovici, Solidification, Structure, and Properties of Aluminum Silicon Alloys, *Giesserei*, Vol 67 (No. 17), 1980
- J. Charbonnier and A. Kearney, French Experience and Developments in Thermal Analysis of Aluminum Casting Alloys, Paper 133, *Trans. AFS*, 1984
- J. Charbonnier, J. Perrier, and R. Portalier, Recent Developments in Aluminum-Silicon Alloys Having Guaranteed Structures or Properties, *Cast Met. J.*, Dec 1978
- N. Handiak, J. Gruzleski, and D. Argo, Sodium, Strontium, and Antimony Interactions During the Modification of AS7G03 (A356) Alloys, Paper 20, *Trans. AFS*, 1987
- E. Rooy, Summary of Technical Information on Hypereutectic Al-Si Alloys, Paper 44, *Trans. AFS*, 1972
- M. Shamsuzzoha and L. Hogan, The Crystal Morphology of Fibrous Silicon in Strontium Modified Al-Si Eutectic, *Philos. Mag.*, Vol 54 (No. 4), 1986
- G. Sigworth, Observations on the Refinement of Hypereutectic Silicon Alloys, Paper 82, *Trans. AFS*, 1982
- O. Vorrent, J. Evensen, and T. Pedersen, Microstructure and Mechanical Properties of Al-Si (Mg) Casting Alloys, Paper 162, *Trans. AFS*, 1984
- J. Weiss and C. Loper, Primary Silicon in Hypereutectic Aluminum-Silicon Casting Alloys, Paper 32, *Trans. AFS*, 1987
- C. Zheng, L. Yao, and Q. Zhang, Effects of Cooling Rate and Modifier Concentration on Modification of Al-Si Eutectic Alloys, *Acta Metall.*, Vol 18 (No. 6), Dec 1982

Gating and Risering

- *Basic Principles of Gating*, American Foundrymen's Society, 1967
- *Basic Principles of Risering*, American Foundrymen's Society, 1968
- Computer Gating Program, Society of Die Casting Engineers
- J. Davies and V. Kondic, Mechanics of Formation of Shrinkage Cavities in Castings, *Br. Foundryman*, Feb 1976
- E. Herman, *Heat Flow in the Die Casting Process*, Society of Die Casting Engineers, 1985
- H. Pokorny and P. Thukkaram, *Gating Die Casting Dies*, Society of Die Casting Engineers, 1984

Heat Treatment

- E. Rooy, "Practical Aspects of Heat Treatment," Paper presented at the Fall Meeting, Toronto, Canada, American Institute of Mining, Metallurgical, and Petroleum Engineers, Oct 1985

Properties

- G. Bouse and M. Behrendt, Metallurgical and Mechanical Property Characterization of Premium Quality Vacuum Investment Cast 200 and 300 Series Aluminum Alloys, *Adv. Cast. Technol.*, Nov 1986
- Directionally Solidified Aluminum Foundry Alloys, Paper 69, *Trans. AFS*, 1987
- M. Holt and K. Bogardus, The "Hot" Aluminum Alloys, *Prod. Eng.*, Aug 1965
- F. Mollard, Understanding Fluidity, Paper 33, *Trans. AFS*, 1987
- E. Rooy, Improved Casting Properties and Integrity With Hot Isostatic Processing, *Mod. Cast.*, Dec 1983
- G. Scott, D. Granger, and B. Cheney, Fracture Toughness and Tensile Properties of Directionally Solidified Aluminum Foundry Alloys, *Trans. AFS*, 1987, p 69
- J. Tirpak, "Elevated Temperature Properties of Cast Aluminum Alloys A201-T7 and A357-T6," AFWAL-TR-85-4114, Air Force Wright Aeronautical Laboratories, 1985

Copper and Copper Alloys

Robert F. Schmidt, Colonial Metals Company
Donald G. Schmidt, R. Lavin & Sons, Inc.
Mahi Sahoo, Canadian Centre for Minerals and Energy Technology, Canada

COPPER IS ALLOYED with other elements because pure copper is extremely difficult to cast as well as being prone to surface cracking, porosity problems, and to the formation of internal cavities. The casting characteristics of copper can be improved by the addition of small amounts of elements including beryllium, silicon, nickel, tin, zinc, chromium, and silver. Alloy coppers, for example, constituted to have improved strength properties over those of high-purity copper, while maintaining a minimum of 85% conductivity, are widely used for cast electrical conducting members.

When casting copper and its alloys, the lowest possible pouring temperature needed to suit the size and form of the solid metal should be adopted to encourage as small a grain size as possible as well as to create a minimum of turbulence of the metal during pouring.

Copper alloys in cast form (designated in UNS numbering system as C80000 to C99999) are specified when factors such as tensile and compressive strength, wear qualities when subjected to metal-to-metal contact, machinability, thermal and electrical conductivity, appearance, and corrosion resistance are considerations for maximizing product performance. Such is the case when using cast copper alloys in applications such as bearings, bushings, gears, fittings, valve bodies, and miscellaneous components for the chemical processing industry.

Types of Copper Alloys

Copper alloys are poured into many types of castings such as sand, shell, investment, permanent mold, chemical sand, centrifugal, and die. Additional information on these casting processes can be found in the articles "Sand Molding," "Permanent Mold Casting," "Centrifugal Casting," and "Die Casting" in this Volume.

The copper-base casting alloy family can be subdivided into three groups according to solidification (freezing range). Unlike pure metals, alloys solidify over a range of temperatures. Solidification begins when the temperature drops below the liquidus; it is completed when the temperature reaches the solidus. The liquidus is the temperature at which the metal begins to freeze, and the solidus is the temperature at which the metal is completely frozen. The three groups are as follows:

Group I alloys are alloys that have a narrow freezing range, that is, a range of 50 °C (90 °F) between the liquidus and solidus.

Group II alloys are those that have an intermediate freezing range, that is, a freezing range of 50 to 110 °C (90 to 200 °F) between the liquidus and the solidus.

Group III alloys have a wide freezing range. These alloys have a freezing range of well over 110 °C (200 °F), even up to 170 °C (300 °F).

Concern is felt about the freezing ranges given. Factually, the freezing range of cupronickel is probably less than 50 °C (90 °F); this

Table 1 Nominal chemical composition and typical mechanical properties for group I alloys

Alloy type	UNS No.	Composition, %										Yield strength(a), 0.5%		Tensile strength(a)		Elongation(a), %
		Cu	Sn	Pb	Zn	Ni	Fe	Al	Mn	Si	Other	MPa	ksi	MPa	ksi	
Copper	C81100	100	...	...	...	...	...	...	...	...	...	28	4	124	18	40
Chrome copper	C81500	99	...	...	...	...	...	...	...	...	1.0 Cr	276	40 (HT)	34	5 (HT)	17 (HT)
Yellow brass	C85200	72	1	3	24	...	...	...	...	...	...	90	13	262	38	35
	C85400	67	1	3	29	...	...	...	...	...	...	83	12	234	34	35
	C85700	61	1	1	37	...	...	...	...	...	...	124	18	345	50	40
	C85800	62	1	1	36	...	...	...	...	...	...	207	30	379	55	15
	C87900	65	...	...	34	...	...	...	...	1	...	241	35	483	70	25
Manganese bronze	C86200	63	...	...	27		3	4	3	...	...	331	48	654	95	20
	C86300	61	...	...	27	...	3	6	3	...	...	476	69	793	115	15
	C86400	58	1	1	38	...	1	5	5	...	...	172	25	448	65	20
	C86500	58	...	...	39	...	1	1	1	...	...	207	30	489	71	30
	C86700	58	1	1	34	...	2	2	2	...	...	290	42	586	85	20
	C86800	55	...	...	36	3	2	1	3	...	...	262	38	565	82	22
Aluminum bronze	C95200	88	...	...	...	...	3	9	...	...	...	186	27	552	80	35
	C95300	89	...	...	...	...	1	10	...	...	...	186–290	27–42 (HT)	517–586	75–85 (HT)	25–18 (HT)
	C95400	86	...	...	...	...	4	10	...	...	...	241–317	35–46 (HT)	586–758	85–110 (HT)	20–12 (HT)
	C95410	84	...	...	...	2	4	10	...	...	...	248–400	36–58 (HT)	662–800	96–116 (HT)	15–10 (HT)
	C95500	81	...	...	...	4	4	11	...	...	...	303–496	44–72 (HT)	717–827	104–120 (HT)	12–6 (HT)
	C95600	91	...	...	...	...	...	7	...	2	...	234	34	517	75	18
	C95700	75	...	...	...	2	3	8	12	...	...	310	45	655	95	26
	C95800	81	...	...	...	4.5	4	9	1.5	...	...	262	38	655	95	25
Nickel bronze	C97300	57	2	9	20	12	...	...	...	...	...	117	17	241	35	30
	C97600	64	4	4	8	20	...	...	...	...	...	165	24	310	45	20
	C97800	66	5	2	2	25	...	...	...	...	...	207	30	379	55	15
White brass	C99700	58	...	2	22	5	...	1	12	...	...	172	25	379	55	25
	C99750	58	...	1	20	...	...	1	20	...	...	221	32	448	65	30

(a) HT, heat treated

Table 2 Nominal chemical composition and typical mechanical properties for group II alloys

Alloy type	UNS No.	Composition, % Cu	Zn	Ni	Fe	Mn	Si	Nb	Other	Yield strength(b)(c), 0.5% MPa	ksi	Tensile strength(b) MPa	ksi	Elongation(b), %
Beryllium copper	C81400	99.1	...	...	...	...	...	...	0.6 Be	248	36 (HT)	365	53 (HT)	11 (HT)
									0.8 Cr	...		...		...
	C82000	97	...	...	...	...	...	...	0.5 Be	121	17.6	243	35.2	20
									2.5 Co	517	75 (HT)	689	100 (HT)	3 (HT)
	C82200	98	...	1.5	...	...	...	...	0.5 Be	145	21.1	276	40.1	20
									...	517	75 (HT)	654	95 (HT)	8 (HT)
	C82400	97.8	...	...	...	...	...	...	1.7 Be	179	26.0	349	50.6	20
									0.5 Co	965	140 (HT)	1035	150 (HT)	1 (HT)
	C82500	97.2	...	...	...	...	0.3	...	2.0 Be	218	31.6	387	56.2	20
									0.5 Co	...		1105	160 (HT)	1 (HT)
	C82600	96.8	...	...	...	...	0.3	...	2.4 Be	228	33.0	397	57.6	20
									0.5 Co	1070	155 (HT)	1140	165 (HT)	1 (HT)
	C82800	96.6	...	...	...	...	0.3	...	2.6 Be	267	38.7	470	68.2	20
									0.5 Co	1000	145 (HT)	1140	165 (HT)	1 (HT)
Silicon brass	C87500	82	14	...	...	...	4	...	30	462	67	145	21	...
Silicon bronze	C87300	9.5	...	...	...	1	4	...	25	400	58	241	35	...
	C87600	91	5	...	...	...	4	...	32	455	66	138	20	...
	C87610	92	4	...	...	...	4	...	25	400	58	207	30	...
	C87800(a)	82	14	...	...	...	4	...	50	586	85	172	25	...
Copper-nickel	C96200	87	...	10	1.5	1	...	1	27	345	50	152	22	...
	C96400	66	...	30.5	0.5	1	...	1	37	469	68	193	28	...

(a) Die cast properties. (b) HT, heat treated. (c) 0.2% offset

would place this alloy in group I rather than in group II. It might be better to simply have three groups—short, medium, and long—without specifying exact freezing ranges.

The Major Copper-Base Alloy Ranges

The alloys in group I (with narrow solidification temperature range of 50 °C, or 90 °F, or less) have the nominal chemical composition and typical mechanical properties shown in Table 1. These are the yellow brasses, manganese and aluminum bronzes, nickel bronze, manganese bronze alloys, chromium copper, and copper.

The alloys in group II (intermediate solidification temperature range of 50 to 110 °C, or 90 to 200 °F) have the nominal chemical composition and typical mechanical properties shown in Table 2. These are the beryllium coppers, silicon bronzes, silicon brass, and cupro-nickel alloys.

The alloys in group III (wide solidification temperature range of over 110 °C, or 200 °F) have the nominal chemical composition and typical mechanical properties shown in Table 3. These are the leaded red and semired brasses, tin and leaded tin bronzes, and high leaded tin bronze alloys.

Melting Practice

Fuel-Fired Furnaces. Copper-base alloys are melted in oil- and gas-fired crucible and open-flame furnaces. Crucible furnaces, either tilting or stationary, incorporate a removable cover or lid for removal of the crucible, which is transported to the pouring area where the molds are poured. The contents of the tilting furnace are poured into a ladle, which is then used to pour the molds (Fig. 1 and 2).

These furnaces melt the raw materials by burning oil or gas with sufficient air to achieve complete combustion. The heat from the burner heats the crucible by conduction and convection; the charge melts and then is superheated to a particular temperature at which either the crucible is removed or the furnace is tilted to pour into a ladle. While the molten metal is in the crucible or ladle, it is skimmed, fluxed, and transferred to the pouring area, where the molds are poured.

The other type of fuel-fired furnace is the open-flame furnace, which is usually a large rotary-type furnace with a refractory lined steel shell containing a burner at one end and a flue at the other. The furnace is

Table 3 Nominal chemical composition and typical mechanical properties for group III alloys

Alloy type	UNS No.	Composition, % Cu	Sn	Pb	Zn	Ni	Yield strength, 0.5% MPa	ksi	Tensile strength, 0.5% MPa	ksi	Elongation, %
Leaded red brass	C83450	88	2.5	2	6.5	1	103	15	255	37	34
	C83600	85	5	5	5	...	110	16	248	36	32
	C83800	83	4	6	7	...	110	16	241	35	28
Leaded semired brass	C84400	81	3	7	9	...	96	14	234	34	28
	C84800	76	2.5	6.5	15	...	103	15	255	37	29
Tin bronze	C90300	88	8	...	4	...	138	20	310	45	30
	C90500	88	10	...	2	...	152	22	317	46	30
	C90700	89	11	...	...	...	152	22	303	44	20
	C91100	84	16	...	...	...	172	25	241	35	2
	C91300	81	19	...	...	...	241	35	207	30	0.5
Leaded tin bronze	C92200	86	6	1.5	4.5	...	110	16	283	41	45
	C92300	87	8	1	4	...	138	20	290	42	32
	C92600	87	10	1	2	...	138	20	303	44	30
	C92700	88	10	2	...	...	142	21	300	42	20
High-leaded tin bronze	C92900	84	10	2.5	...	3.5	179	26	324	47	20
	C93200	83	7	7	3	...	117	17	262	38	30
	C93400	84	8	8	...	...	110	16	248	36	25
	C93500	85	5	9	1	...	110	16	221	32	20
	C93700	80	10	10	...	...	124	18	276	40	30
	C93800	78	7	15	...	...	110	16	221	32	20
	C94300	70	5	25	...	...	110	16	207	30	18

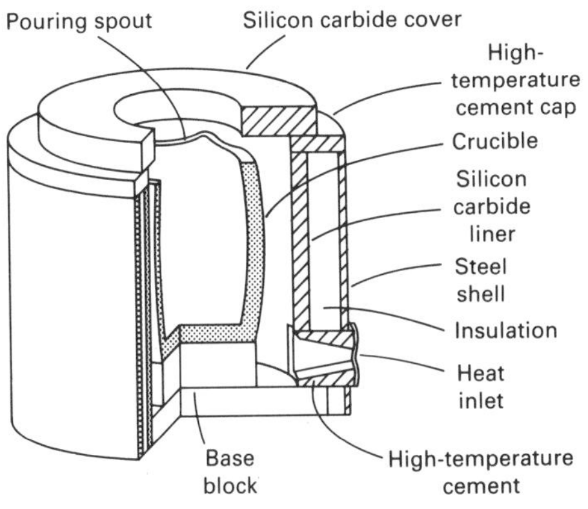

Fig. 1 Typical lift-out type of fuel-fired crucible furnace, especially well adapted to foundry melting of smaller quantities of copper alloys (usually less than 140 kg, or 300 lb)

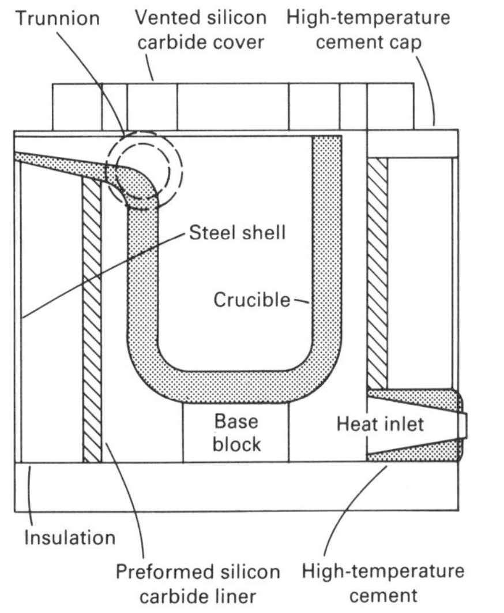

Fig. 2 Typical lip-axis tilting crucible furnace used for fuel-fired furnace melting of copper alloys. Similar furnaces are available that tilt on a central axis.

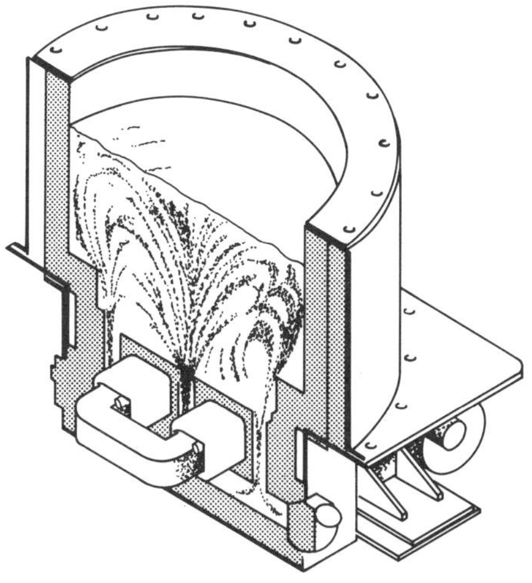

Fig. 3 Cutaway drawing of a twin-channel induction melting furnace

rotated slowly around the horizontal axis, and the rotary movement helps to heat and melt the furnace charge. Melting is accomplished both by the action of the flame directly on the metal and by heat transfer from the hot lining as this shell rotates. These furnaces usually tilt so that they can be charged and poured from the flue opening. At the present time, these furnaces are not used often because of the requirement that a baghouse be installed to capture all the flue dust emitted during melting and superheating. While these furnaces are able to melt large amounts of metal quickly, there is a need for operator skill to control the melting atmosphere within the furnace. Also, the refractory walls become impregnated with the melting metal, causing a contamination problem when switching from one alloy family to another.

Electric Induction Furnaces. In the past 20 years, there has been a marked changeover from fuel-fired melting to electric induction melting in the copper-base foundry industry. While this type of melting equipment has been available for more than 50 years, very few were actually used due to the large investment required for the capital equipment. Because of higher prices and the question of availability of fossil fuels and because of new regulations on health and safety imposed by the Occupational Safety and Health Administration (OSHA), many foundries have made the changeover to electric induction furnaces.

When melting alloys in group III, fumes of lead and zinc are given off during melting and superheating. The emission of these harmful oxides is much lower when the charge is melted in an induction furnace because the duration of the melting cycle is only about 25% as long when melting the same amount of metal in a fuel-fired furnace. By the use of electric induction melting, compliance with OSHA regulations can be met in many foundries without the need for expensive air pollution control equipment. Additional information on electric induction furnaces is available in the article "Induction Furnaces" in this Volume.

The two types of electric induction furnaces are the core type, better known as the channel furnace, and the coreless type.

Core Type. This furnace (shown in Fig. 3) is a large furnace used in foundries for pouring large quantities of one alloy when a constant source of molten metal is required. This furnace has a primary coil, interfaced with a laminated iron core, surrounded by a secondary channel, which is imbedded in a V- or U-shape refractory lining located at the bottom of a cylindrical hearth. Here the channel forms the secondary of a transformer circuit. This furnace stirs and circulates molten metal through the channel at all times, except when the furnace is emptied

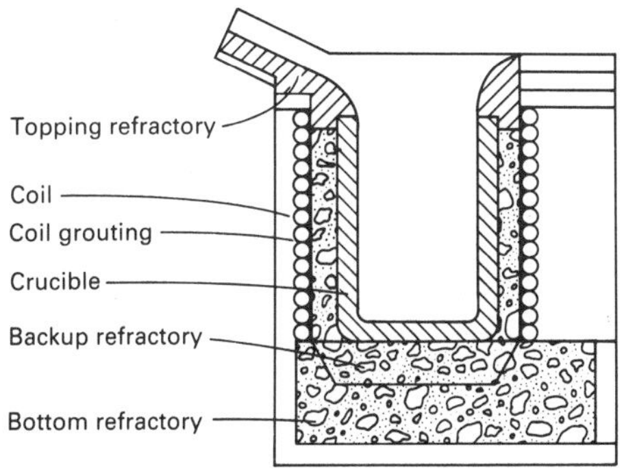

Fig. 4 Cross section of a tilt furnace for high-frequency induction melting of brass and bronze alloys. Crucible is of clay graphite composition.

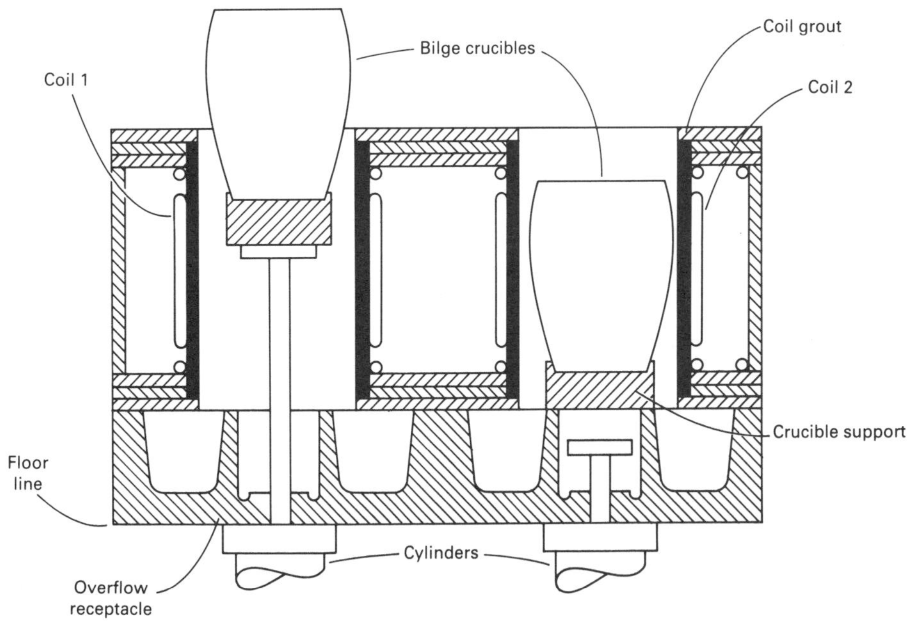

Fig. 5 Cross section of a double push-out furnace. Bilge crucibles are placed on refractory pedestals and raised and lowered into position within the coils by hydraulic cylinders.

Fig. 6 Foundry installation of high-frequency induction lift swing furnaces

and shut down. When starting up, molten metal must be poured into the furnace to fill up the "heel" on the bottom of the bath. Because these furnaces are very efficient and simple to operate with lining life in the millions of pounds poured, they are best suited for continuous production runs in foundries making plumbing alloys of group III. They are not recommended for the dross-forming alloys of group I. The channel furnace is at its best when an inert, floating, cover flux is used and charges of ingot, clean remelt, and clean and dry turnings are added periodically.

Coreless Type. This furnace has become the most popular melting unit in the copper alloy foundry industry. In earlier years, the coreless furnace was powered by a motor generator unit, usually at 980 Hz. The present coreless induction furnaces draw 440 volt, 60 cycle power and, by means of solid state electronic devices, convert the power to 440 volts and 1000 or 3000 Hz. These furnaces are either tilting furnaces (Fig. 4) or crucible lift-out units (Fig. 5 and Fig. 6).

A coreless induction furnace is comprised of a water-cooled copper coil in a furnace box made of steel or Transite. The metal is contained in a crucible or in a refractory lining rammed up to the coil. Crucibles used in these furnaces are made of clay graphite; silicon carbide crucibles cannot be used because they become overheated when inserted in a magnetic field. Clay graphite crucibles do a good job of conducting the electromagnetic currents from the coil into the metal being melted.

Induction furnaces are characterized by electromagnetic stirring of the metal bath. Because the amount of stirring is affected by both power input and power frequency, the power unit size and frequency should be coordinated with the furnace size in order to obtain the optimum-size equipment for the specific operation. In general, the smaller the unit, the higher the frequency and the lower the power input.

Large tilting units are used in foundries requiring large amounts of metal at one time. These furnaces, if over 4.5 Mg (10 000 lb) capacity, operate at line frequency (60 Hz). They are very efficient and will melt large quantities of metal in a very short time if powered with the proper-size power unit.

Stationary lift-out furnaces are often designed as shown in Fig. 5. Here the crucible sits on a refractory pedestal, which can be raised or lowered by a hydraulic cylinder. This unit, also called a push-out furnace, operates by lowering the crucible into the coil for melting and then raising the crucible out of the coil for pickup and pouring. While one crucible is melting, the other crucible can be charged and ready to melt when the knife switch is pulled as the completed heat is being pushed up for skimming and pouring.

The other common type of coreless induction melting is the lift swing furnace (Fig. 6). Here the coil (and box) is cantilevered from a center post to move up or down vertically and swing horizontally about the post in a 90° arc. Because there are two crucible positions, one crucible can be poured, recharged, and placed into position to melt, while the other is melting. When the metal is ready to pour, the furnace box is lifted (by hydraulic or air cylinder), pivoted to the side, and lowered over the second crucible. The ready crucible is then standing free and can be picked up and poured, while melting is taking place in the second furnace.

Melt Treatment

For group I to III alloys, the melting procedure and fluxing vary considerably from one alloy family to another. Pouring temperatures for these alloys can be found in Table 4. These groupings can be categorized as follows:

Group I Alloys

Pure Copper and Chromium Copper. Commercially pure copper and high copper alloys are very difficult to melt and are very susceptible to gassing. In the case of chromium copper, oxidation loss of chromium during melting is a problem. Copper and chromium copper should be melted under a floating flux cover to prevent both oxidation and the pickup of hydrogen from moisture in the atmosphere. In the case of copper, crushed graphite should cover the melt. With chromium copper, the cover should be a proprietary flux made for this alloy. When the molten metal reaches 1260 °C (2300 °F), either calcium boride or lithium should be plunged into the molten bath to deoxidize the melt. The metal should then be poured without removing the floating cover.

Yellow Brasses. These alloys flare, or lose zinc, due to vaporization at temperatures relatively close to the melting point. For this reason, aluminum is added to increase fluidity and keep zinc vaporization to a minimum. The proper amount of aluminum to be retained in the brass is 0.15 to 0.35%. Above this amount, shrinkage takes place during freezing, and the use of risers becomes necessary. Other than the addition of aluminum, the melting of yellow brass is very simple, and no fluxing is necessary. Zinc should be added before pouring to compensate for the zinc lost in melting.

Manganese Bronzes. These alloys are carefully compounded yellow brasses with measured quantities of iron, manganese, and aluminum. The metal should be melted and heated to the flare temperature or to the point at which zinc oxide vapor can be detected. At this point, the metal should be removed from the furnace and poured. No fluxing is required with these alloys. The only addition required with these alloys is zinc. The amount required is that which is needed to bring the zinc content back to the original analysis. This varies from very little, if any, when an all-ingot heat is being poured, to several percent if the heat contains a high percentage of remelt.

Aluminum Bronzes. These alloys must be melted carefully under an oxidizing atmosphere and heated to the proper furnace temperature. If needed, degasifiers can be stirred into the melt as the furnace is being tapped. By pouring a blind sprue before tapping and examining the metal after freezing, it is possible to tell whether it shrank or exuded gas. If the sample purged or overflowed the blind sprue during solidification,

Table 4 Pouring temperatures of copper alloys

Alloy type	UNS No.	Light castings °C	Light castings °F	Heavy castings °C	Heavy castings °F
Group I alloys					
Copper	C81100	1230–1290	2250–2350	1150–1230	2100–2250
Chromium copper	C81500	1230–1260	2250–2300	1205–1230	2200–2250
Yellow brass	C85200	1095–1150	2000–2100	1010–1095	1850–2000
	C85400	1065–1150	1950–2100	1010–1065	1850–1950
	C85800	1150–1175	1950–2150	1010–1095	1850–2000
	C87900	1150–1175	1950–2150	1010–1095	1850–2000
Manganese bronze	C86200	1150–1175	1950–2150	980–1065	1800–1950
	C86300	1150–1175	1950–2150	980–1065	1800–1950
	C86400	1040–1120	1900–2050	950–1040	1750–1900
	C86500	1040–1120	1900–2050	950–1040	1750–1900
	C86700	1040–1095	1900–2000	950–1040	1750–1900
	C86800	1150–1175	1950–2150	980–1065	1800–1950
Aluminum bronze	C95200	1120–1205	2050–2200	1095–1150	2000–2100
	C95300	1120–1205	2050–2200	1095–1150	2000–2100
	C95400	1150–1230	2100–2250	1095–1175	2000–2150
	C95410	1150–1230	2100–2250	1095–1175	2000–2150
	C95500	1230–1290	2250–2350	1175–1230	2150–2250
	C95600	1120–1205	2050–2200	1095–1205	2000–2200
	C95700	1065–1150	1950–2100	1010–1205	1850–2200
	C95800	1230–1290	2250–2350	1175–1230	2150–2250
Nickel bronze	C97300	1205–1225	2200–2240	1095–1205	2000–2200
	C97600	1260–1425	2300–2600	1205–1315	2250–2400
	C97800	1315–1425	2400–2600	1260–1315	2300–2400
White brass	C99700	1040–1095	1900–2000	980–1040	1800–1900
	C99750	1040–1095	1900–2000	980–1040	1800–1900
Group II alloys					
Beryllium copper	C81400	1175–1220	2150–2225	1220–1260	2225–2300
	C82000	1175–1230	2150–2250	1120–1175	2050–2150
	C82400	1080–1120	1975–2050	1040–1080	1900–1975
	C82500	1065–1120	1950–2050	1010–1065	1850–1950
	C82600	1050–1095	1925–2000	1010–1050	1850–1925
	C82800	995–1025	1825–1875	1025–1050	1875–1925
Silicon brass	C87500	1040–1095	1900–2000	980–1040	1800–1900
	C87800	1040–1095	1900–2000	980–1040	1800–1900
Silicon bronze	C87300	1095–1175	2000–2150	1010–1095	1850–2000
	C87600	1095–1175	2000–2150	1010–1095	1850–2000
	C87610	1095–1175	2000–2150	1010–1095	1850–2000
Copper nickel	C96200	1315–1370	2400–2500	1230–1315	2250–2400
	C96400	1370–1480	2500–2700	1290–1370	2350–2500
Group III alloys					
Leaded red brass	C83450	1175–1290	2150–2350	1095–1175	2000–2150
	C83600	1150–1290	2100–2350	1065–1175	1950–2150
	C83800	1150–1260	2100–2300	1065–1175	1950–2150
Leaded semired brass	C84400	1150–1260	2100–2300	1065–1175	1950–2150
	C84800	1150–1260	2100–2300	1065–1175	1950–2150
Tin bronze	C90300	1150–1260	2100–2300	1040–1150	1900–2100
	C90500	1150–1260	2100–2300	1040–1150	1900–2100
	C90700	1040–1095	1900–2000	980–1040	1800–1900
	C91100	1040–1095	1900–2000	980–1040	1800–1900
	C91300	1040–1095	1900–2000	980–1040	1800–1900
Leaded tin bronze	C92200	1150–1260	2100–2300	1040–1175	1900–2150
	C92300	1150–1260	2100–2300	1040–1150	1900–2100
	C92600	1150–1260	2100–2300	1050–1150	1920–2100
	C92700	1175–1260	2150–2300	1065–1175	1950–2150
High-leaded tin bronze	C92900	1095–1205	2000–2200	1040–1095	1900–2000
	C93200	1095–1230	2000–2250	1040–1121	1900–2050
	C93400	1095–1230	2000–2250	1010–1150	1850–2100
	C93500	1095–1205	2000–2200	1040–1150	1900–2100
	C93700	1095–1230	2000–2250	1010–1150	1850–2100
	C93800	1095–1230	2000–2250	1040–1150	1900–2100
	C94300	1095–1205	2000–2200	1010–1095	1850–2000

degassing is necessary. Degasifiers remove hydrogen and oxygen. Also available are fluxes that convert the molten bath. These are in powder form and are usually fluorides. They aid in the elimination of oxides, which normally form on top of the melt during melting and superheating.

Nickel Bronzes. These alloys, also known as nickel silver, are difficult alloys to melt. They gas readily if not melted properly because the presence of nickel increases the hydrogen solubility. Then, too, the higher pouring temperatures shown in Table 4 aggravate hydrogen pickup. These alloys must be melted under an oxidizing atmosphere and quickly superheated to the proper furnace temperature to allow for temperature losses during fluxing and handling. Proprietary fluxes are available and should be stirred into the melt after tapping the furnace. These fluxes contain manganese, calcium, silicon, magnesium, and phosphorus and do an excellent job in removing hydrogen and oxygen.

White Manganese Bronze. There are two alloys in this family, both of which are copper-zinc alloys containing a large amount of manganese and, in one case, nickel. They are manganese bronze type alloys, are simple to melt, and can be poured at low temperatures because they are very fluid (Table 4). They should not be overheated, as this serves no purpose. If the alloys are unduly superheated, zinc is vaporized and the chemistry of the alloy is changed. Normally, no fluxes are used with these alloys.

Group II Alloys

Information on melting and fluxing of group II alloys is found below. Again, pouring temperatures for these alloys are found in Table 4.

Beryllium Coppers. These alloys are very toxic and dangerous if beryllium fumes are not captured and exhausted by proper ventilating equipment. They should be melted quickly under a slightly oxidizing atmosphere to minimize beryllium losses. They can be melted and poured successfully at relatively low temperatures (Table 4). They are very fluid and pour well.

Silicon Bronzes and Brasses. The alloys known as silicon bronzes, UNS alloys C87300, C87600, and 87610, are relatively easy to melt and should be poured at the proper pouring temperatures (Table 4). If overheated, they can pick up hydrogen. While degassing is seldom required, if necessary, one of the proprietary degasifiers used with aluminum bronze can be successfully used. Normally no cover fluxes are used here. The silicon brasses (UNS alloys C87500 and C87800) have excellent fluidity and can be poured slightly above their freezing range. Nothing is gained by excessive heating, and in some cases, heats can be gassed if this occurs. Here again, no cover fluxes are required.

Copper-Nickel Alloys. These alloys (90Cu-10Ni, UNS C96200 and 70Cu-30Ni, UNS C96400) must be melted carefully because the presence of nickel in high percentages raises not only the melting point but also the susceptibility to hydrogen pickup. In virtually all foundries, these alloys are melted in coreless electric induction furnaces, because the melting rate is much faster than it is with a fuel-fired furnace. When ingot is melted in this manner, the metal should be quickly heated to a temperature slightly above the pouring temperature (Table 4) and deoxidized either by the use of one of the proprietary degasifiers used with nickel bronzes or, better yet, by plunging 0.1% Mg stick to the bottom of the ladle. The purpose of this is to remove all the oxygen to prevent any possibility of steam-reaction porosity from occurring. Normally there is little need to use cover fluxes if the gates and risers are cleaned by shotblasting prior to melting.

Group III Alloys

The following discussion covers melting and fluxing of group III alloys. Again, pouring temperatures are found in Table 4.

These alloys, namely leaded red and semired brasses, tin and leaded tin bronzes, and high-leaded tin bronzes, are treated the same in regard to melting and fluxing and thus can be discussed together. Because of the long freezing ranges involved, it has been found that chilling, or the creation of a steep thermal gradient, is far better than using only feeders or risers. Chills and risers should be used in conjunction with each other for these alloys. For this reason, the best pouring temperature is the lowest one that will pour the molds without having misruns or cold shuts. In a well-operated foundry, each pattern should have a pouring temperature which is maintained by use of an immersion pyrometer.

Fluxing. In regard to fluxing, these alloys should be melted from charges comprised of ingot and clean, sandfree gates and risers. The melting should be done quickly in a slightly oxidizing atmosphere. When at the proper furnace temperature to allow for handling and cooling to the proper pouring temperature, the crucible is removed or the metal is tapped into a ladle. At this point, a deoxidizer (15% phosphor copper) is added. The phosphorus is a reducing agent (deoxidizer). This product must be carefully measured so that enough oxygen is removed, yet a small amount remains to improve fluidity. This residual level of phosphorus must be closely controlled by chemical analysis to a range between 0.010 and 0.020% P. If more is present, internal porosity may occur and cause leakage if castings are machined and pressure tested.

In addition to phosphor copper, pure zinc should be added at the point at which skimming and temperature testing take place prior to pouring. This replaces the zinc lost by vaporization during melting and superheating. With these alloys, cover fluxes are seldom used. In some foundries in which combustion cannot be properly controlled, oxidizing fluxes are added during melting, followed by final deoxidation by phosphor copper.

Pouring and Gating (Ref 1)

The major function of a gating system is to deliver clean metal from the ladle into the mold cavity without adversely affecting the quality of the metal. Secondary considerations are the ease of molding, removal of gates, and high casting yield. However, these factors should not dictate a design that contributes to the production of castings of unacceptable quality.

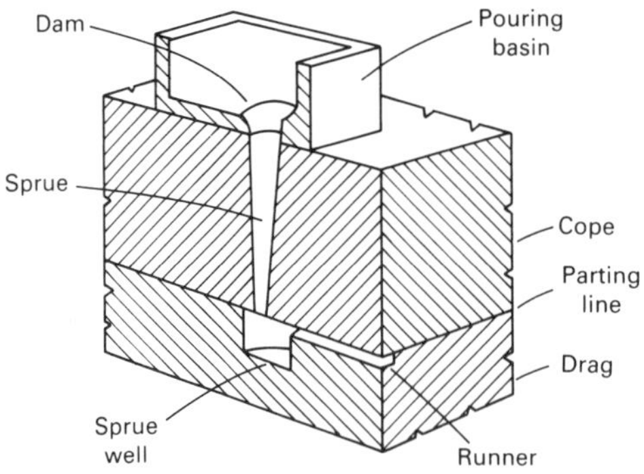

Fig. 7 Section of a typical sand mold with pouring basin

The Pouring Basin. The production of high-quality castings requires not only proper melting and molding operations and properly designed pattern equipment but also an understanding of the principles of gating so that clean metal can be delivered to the mold cavity with a minimum amount of turbulence. A pouring basin allows a sprue to be filled quickly and maintains a constant head throughout the pour (Fig. 7).

When the weight of poured metal in a mold exceeds 14 kg (30 lb), use of a pouring basin offers many advantages. The pourer can better direct the flow of metal from the ladle into the basin, with less chance of spillages; also, the sprue need not be located near the edge of the mold. The pouring ladle can be brought within 25 to 50 mm (1 to 2 in.) of the basin, and a continued flow rate may be more easily maintained through a larger pouring head. If there are any brief interruptions in pouring the metal into the basin, the surplus metal will take up the slack until pouring has resumed. The major disadvantage of the pouring basin is that the yield is lowered, thereby requiring more metal to be recycled.

Sprue. The correct sprue size is the single most important part of the gating system. If a wrong size is selected, or an improper taper is used, the damage done to the metal in the mold cavity is extensive and cannot be corrected regardless of the quality of the runner and gating systems.

Because most molds under about 14 kg (30 lb) of poured weight are made on a high production scale in flasks of 102 to 152 mm (4 to 6 in.) in cope height, a fairly standard sprue size may be used for all copper-base alloys. The top third of the sprue should be the pouring part, with about a 50 mm (2 in.) diam opening. The remaining portion of the sprue should taper down to 13 to 22 mm (½ to ⅞ in.) in final diameter depending on the pouring rate to be used.

Figure 8 shows a sketch of a sprue that will do an excellent job of conveying brass or bronze into the gating system. There are many charts and formulas available to determine the entry diameter of a tapered sprue, but for the most part this diameter should be just sufficient to provide about a 10 to 20° slope on the side of the sprue. When the sprue height is over 305 mm (12 in.), the top diameter of the sprue is much more important and should be about 50% larger than the diameter at the base of the sprue. When designing a pouring system for sprues of 102 to 152 mm (4 to 6 in.) in height, it is best to select the desired pouring rate first in order to determine the proper sprue base to be used.

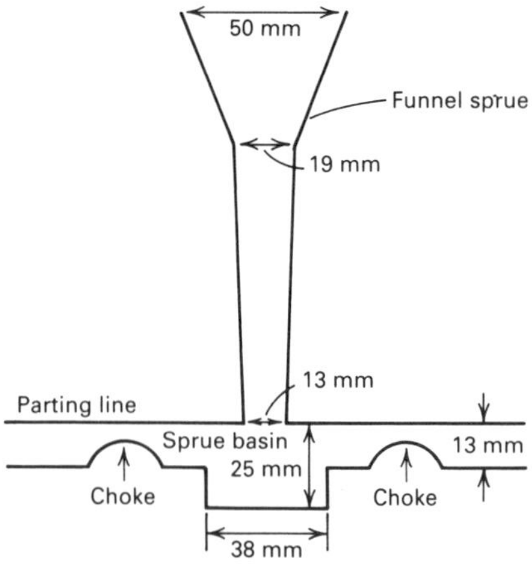

Fig. 8 Funnel sprue, sprue basin, and chokes for reducing turbulence

The Sprue Base. Because the velocity of the stream is at its maximum at the bottom of the sprue and is proportional to the square root of the height of the fall of the metal, it is mandatory that a sprue base or well be used as a cushion for the stream flowing down the sprue. The base also helps change the vertical flow of metal into a horizontal flow with the least amount of turbulence. Recommended sprue basin sizes are about twice as deep as the drag runner and two to three times as wide as the base of the sprue. In most cases a well 25 to 38 mm (1 to 1½ in.) deep with a width of 38 to 50 mm (1½ to 2 in.) on each side is usually adequate for the majority of sprues being used for most normal pouring rates. Little damage is done if the sprue base is larger than necessary except that the overall casting yield will be lowered slightly.

Chokes should be used only when the proper pouring rate cannot be controlled by the correct sprue size. If clean metal is delivered into the sprue, a strainer core serves the sole function of retarding the metal flow rate. Conventional strainer cores, whether of tinned steel, mica, glass fiber, or ceramic, usually reduce the flow of metal by about 70%, depending on the size and number of holes that are open to the sprue area. The best strainer is one that has only one hole with a diameter of the correct sprue size. This avoids the turbulence caused by the metal being divided into many streams as it enters the runner. In no

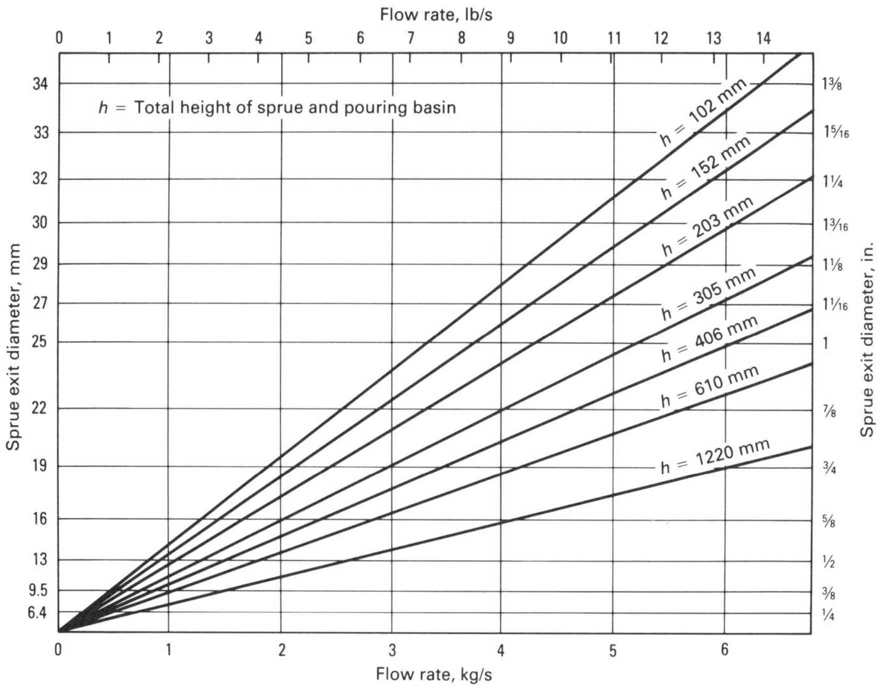

Fig. 9 Flow rates of copper-base alloys through tapered sprues of varying diameter and height

instance should a strainer be placed into the top of the sprue; if one must be used, the only suitable place is just above the sprue base at the parting line. Tinned steel strainers are the least acceptable because the remelted runners can introduce iron and hard spots to the copper-base alloys if they are not properly skimmed during melting. The mica and glass fiber strainers are popular because they can be laid on the parting line just above the sprue base before the mold is closed, thereby requiring no "prints" or recesses, as do the thicker ceramic or sand strainer cores, which are usually about 3.2 mm (⅛ in.) thick.

A choke in the runner pattern is often the only consistent way to achieve a proper pouring rate. In no instance should the choke be put in the gate area. When necessary, it should be placed in the drag runner as close to the sprue as possible (Fig. 8). The chokes should have a smooth radiused contour and be located at the bottom of the drag runner. Choke depth may vary from ¼ to ¾ of the total runner depth with a cross-sectional area not exceeding ¾ of the area of the sprue base. The chokes should be located within an inch of the sprue base to ensure rapid filling of the sprue and maintenance of full capacity throughout the pour. This also permits dissipation of the turbulence before the stream reaches the gates.

Pouring rate depends on many factors, such as weight of the casting, section size, height of the sprue, and alloy system. Most alloys in group III for small work weighing 14 kg (30 lb) or less are poured with a hand ladle at about 1.8 kg/s (4 lb/s). Light memorial plaque castings are being successfully poured at 4.5 kg/s (10 lb/s), while many automatic pouring units operate at mold pouring rates of 3.6 to 4.5 kg/s (8 to 10 lb/s). Alloys in group I, if the poured weight is under 14 kg (30 lb), should be poured at 0.9 to 1.8 kg/s (2 to 4 lb/s) in order to obtain a clean, nonturbulent metal flow in the mold. Sprue exit diameters required for specific flow rates and various sprue heights are shown in Fig. 9. Table 5 shows flow rates from the bottom of the sprue for a number of commonly used sprue heights and diameters. As an example, for a gross casting weight of 14 kg (30 lb) or less and a sprue height of 102 to 152 mm (4 to 6 in.), a sprue diameter of 13 to 19 mm (½ to ¾ in.) is adequate to obtain a flow rate of 0.9 to 1.8 kg/s (2 to 4 lb/s). It should not be necessary to use a 22 to 29 mm (⅞ to 1⅛ in.) diam sprue base unless pouring plaque work or using automatic pouring. A quite popular sprue for most production work is the 16 to 19 mm (⅝ or ¾ in.) diam size, which will deliver enough hot metal to fill most molds up to 14 kg (30 lb) weight in 8 to 10 s. The total pouring time in seconds may be calculated by dividing the total weight of the mold poured (castings plus gates and risers) by the flow rate at the base of the sprue, or:

$$\frac{\text{Total weight of castings including gates and risers (lb)}}{\text{Flow rate at base of sprue (lb/s)}}$$

$$= \text{Calculated pouring time in seconds}$$

Runners and Gates. For alloys in groups I and II, it is mandatory that all runners be placed in the drag and as much of the casting as possible be placed in the cope. In this way, all runners will be completely filled before any metal enters the gates, and the metal will drop the least amount or will rise to enter the mold cavity from the gates. Although this practice is also excellent for alloys in group III, experience has shown that quality castings may be obtained by using more traditional casting techniques because the alloys in group III are less sensitive to drossing and have a tendency to self heal when dross is formed in the gating system. Runners should be as rectangular in shape as possible, and their total maximum cross-sectional area should be two to four times that of the tapered sprue or the choke, if chokes are used in the runner system. Care must be taken to ensure that the cross section of the runners is adequate in order to prevent premature chilling. Experience has shown that a rectangular runner with the wide side laying horizontal works best. The next best is a square runner, and the least desirable is a rectangular runner with the wide side being vertical, although sometimes space limitations necessitate use of this type of runner in order to obtain the proper ratios. The rectangular runner should be about twice as wide as it is deep.

The cross-sectional area of the runner must be reduced by that of each gate as it is passed, so that metal enters the mold cavity simultaneously from each gate (Fig. 10).

Table 5 Flow rates of copper-base alloys through tapered sprues of varying diameter and height

Sprue				Flow rate for sprue height, mm (in.)									
Area		Diameter		102 (4)		152 (6)		305 (12)		610 (24)		1220 (48)	
mm^2	$in.^2$	mm	in.	kg/s	lb/s	kg/s	lb/s	kg/s	lb/s	kg/s	lb/s	kg/s	lb/s
129	0.2	13	½	0.82	1.8	0.91	2.0	1.36	3.0	1.81	4.0	2.72	6.0
194	0.3	16	⅝	1.27	2.8	1.50	3.3	2.04	4.5	2.72	6.0	4.08	9.0
284	0.44	19	¾	1.81	4.0	2.04	4.5	2.95	6.5	4.08	9.0	6.12	13.5
387	0.60	22	⅞	2.49	5.5	2.95	6.5	4.08	9.0	5.67	12.5	8.16	18.0
506	0.785	25	1	3.40	7.5	3.86	8.5	5.22	11.5	7.48	16.5	11.11	24.5
645	1.0	29	1⅛	4.30	9.5	4.76	10.5	7.71	17.0	9.30	20.5	13.61	30.0

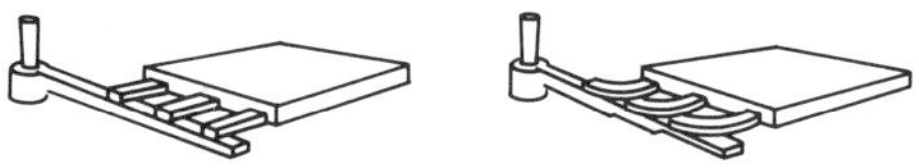

Fig. 10 Typical single-cavity gating systems. (a) Tapered runner. (b) Stepped runner

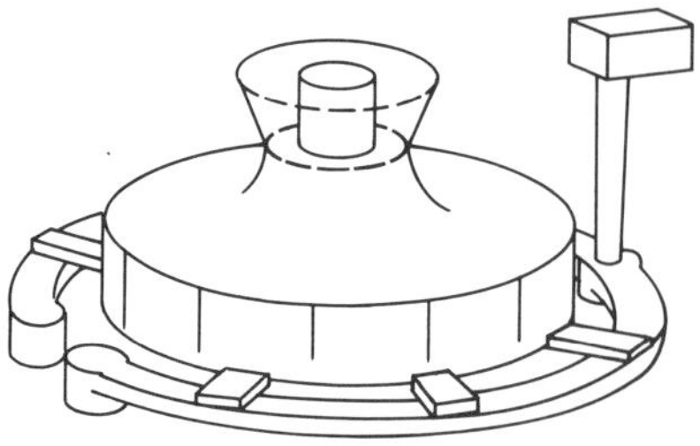

Fig. 11 Method of running a pump impeller with a well at the end of the runner

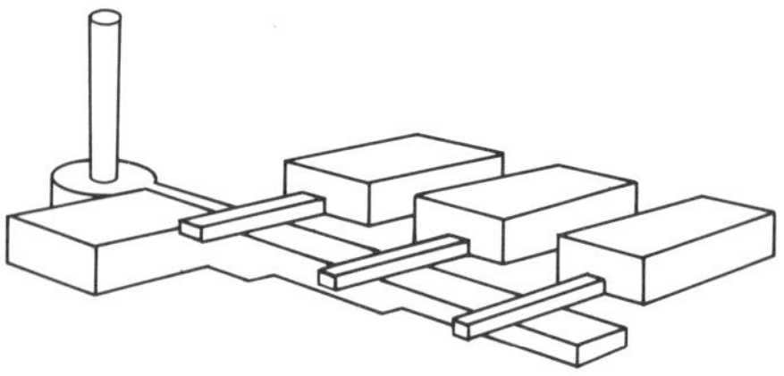

Fig. 12 Recommended multiple-cavity gating system with stepped runner

Because back filling is seldom desired from the runner system, a well at the end of the runner can be used (Fig. 11), particularly if the runner does not have any taper. A good example of multiple-cavity gating may be seen in Fig. 11. X-ray movies of metal flow in sand molds show that relatively uniform gate discharge rates are achieved only when stepped or tapered runners are used.

Multiple gates are shown in Fig. 10 to 12. The preferred location is in the cope just above the runner at the parting line. A rectangular flat gate is more desirable than a square gate, and a gate that has its wide dimension in the vertical plane is the least desirable, just as is the case for runners. In order to avoid a pressurized gating system, it is important that the total gate area be at least as large as the total runner area. If a pattern has an excessive amount of small castings, it might be necessary to have the total gate area many times the runner area in order to obtain a sufficient gate to each casting. This deviation is acceptable because the gating system remains unpressurized. Figure 10(a) and (b) shows good gating systems with streamlined gates and right-angle gates, with Fig. 10(a) producing the minimum amount of turbulence. Gates should enter the casting cavity at the lowest possible level in order to avoid the erosion and turbulence associated with a falling stream of molten metal. To ensure nonturbulent filling of the casting nearest to the sprue, its gate should be at least 50 mm (2 in.) away from the base of the sprue.

Regardless of the excellence of a gating system design, castings of acceptable quality will not be produced if the ladle is not positioned as close as is practical to the pouring basin or sprue, and if the sprue is not filled quickly and kept at full capacity throughout the pour.

Knife and Kiss Gating Systems. Special applications of gating systems work in many cases for specific castings. Knife and kiss gating are popular when group III alloys are used but not recommended for groups I and II because these alloys form too much dross with this system and cannot be fed adequately to eliminate surface shrinkage. Advantages are a high casting yield, easy removal of runner systems, and minimum grinding of gates. The major disadvantage is that many small castings become detached during shakeout, necessitating their manual retrieval from mechanized systems. Figure 13 shows a graphic representation of the arrangement of knife and kiss gating. In kiss gating, the casting must be completely in the cope or the drag with the runner overlapping the casting by 0.8 to 2.4 mm (1/32 to 3/32 in.). Actually there is no gate in this system because the metal goes directly from the runner into the casting.

Knife gating is used when the casting is in both the cope and the drag and there is a contact at the parting line of 0.8 to 2.4 mm (1/32 to 3/32 in.) thickness just at the casting surface. Knife gating systems work well when the runner is in just the cope or just the drag or in both the cope and the drag.

Maximizing Casting Quality. Excellent, clean, high quality castings may be obtained for the copper-base alloy groups of narrow, intermediate, and wide freezing range alloys if the basic principles discussed for the pouring basin, sprue, sprue base, chokes, pouring rates, runners, and gates are applied. By following these recommendations, the maximum ease in molding, casting yield, and ease of removal of gates and runners may be obtained.

Feeding

The objectives of feeding or risering are to eliminate surface sinks or draws and to reduce internal shrinkage porosity to acceptable levels (less than 1%).

To minimize porosity, the feeding system must establish:

- Directional solidification, as shown in Fig. 14, in which the solidification front is substantially V-shaped in a longitudinal cross section with the large end of the V directed toward the riser
- Steep temperature gradients along the casting toward the riser

The feeding techniques for group I (short freezing range) alloys and group II (medium freezing range) alloys can be discussed together. The basic principles of risering of group III (long freezing range) alloys will be described separately.

Group I and II Alloys

The feeding technique for these alloys is similar to that used in the manufacture of steel castings. Gates and risers are positioned such that directional solidification is ensured, with due consideration being given to the feeding range of the alloy in question.

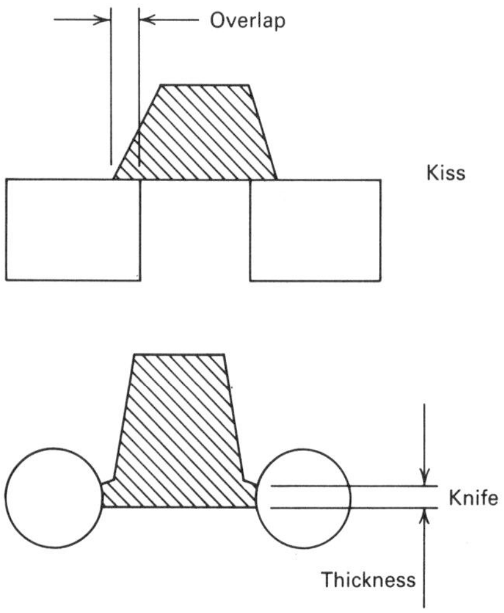

Fig. 13 Basic kiss and knife gates

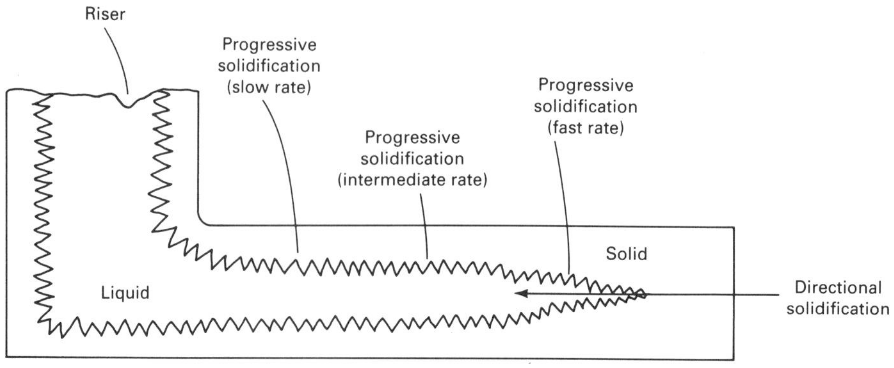

Fig. 14 Features of progressive and directional solidification. Source: Ref 2

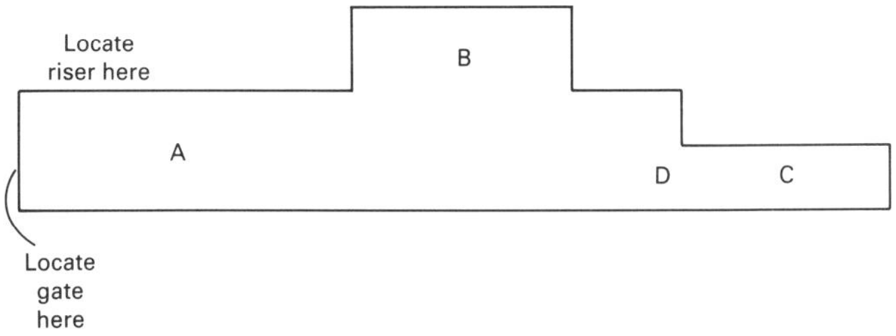

Fig. 15 Hypothetical casting used to illustrate the principles of feeding technique. Source: Ref 3

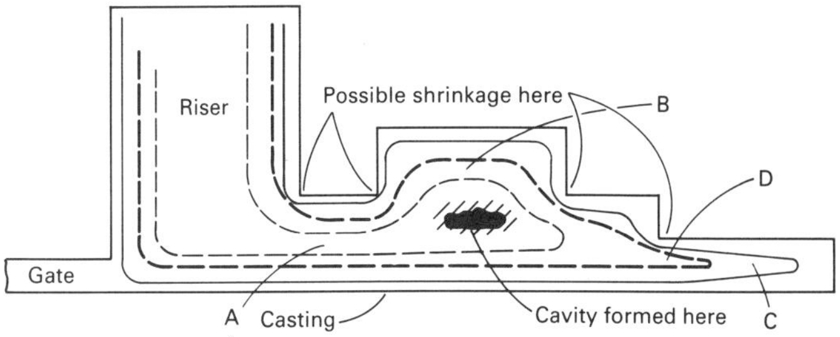

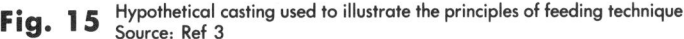

Fig. 16 Mode of freezing the casting in Fig. 15 without special precaution to avoid shrinkage. Source: Ref 3

To avoid hot spots, local chills may be applied to bosses, ribs, and to other sections having sudden changes in thickness.

Solidification Contours. The first step in determining riser placement is to draw the solidification contours. This is illustrated by the hypothetical casting shown in Fig. 15, which consists essentially of a plate to which is attached a thinner section, C, and a boss, B. The thin end of the casting, C, would normally undergo rapid cooling after pouring as the result of edge-cooling effects. Thus, it is possible to place the riser at the heavy section, A, and gate through the riser to provide favorable temperature gradients. The dotted lines in Fig. 16 show successive positions of the solidification front. As shown, porosity will develop in the boss unless a chill is placed on the boss or the riser is relocated there. A chill is a block of metal or other material with a higher heat conductivity and heat capacity than sand.

Feeding Ranges. The number and location of the feeders to be used must be consistent with the feeding range of the alloy. The feeding range is the distance that can be fed by a feeder on a bar or plate. It is generally desirable to divide the casting into a number of sections to determine the number of risers to be used. Because all parts of a casting must be within the feeding range of at least one of the risers, it is important to have quantitative information regarding feeding ranges. The feeding range values for group I and II copper-base alloys have not been well documented. In the absence of specific data for particular alloys, satisfactory results can often be attained by applying values that have been developed for carbon steels. The following approximate values for feeding ranges have been quoted in the literature, but should be used with caution:

Alloy	Shape	Feeding distance, T
Manganese bronze ...	Square bars	$4\ T$ to $10\ \sqrt{T}$, depending on thickness
	Plates	$5.5\ T$ to $8\ T$, depending on thickness
Aluminum bronze ...	Square bars	$8\ \sqrt{T}$
Nickel-aluminum bronze	Square bars	$<8\ \sqrt{T}$
Copper-nickel	Square bars	$5.5\ \sqrt{T}$

Use of chills can further increase feeding range. Consequently, the spacing between risers may be increased to about ten times the section thickness if chills are located midway between each pair of risers (Fig. 17).

Riser Size. From time to time, various methods have been proposed for the calculation of the optimum riser size to be used to feed a particular casting or casting section. One of the earlier methods was developed for steel castings at the Naval Research Laboratories (NRL). In this technique, an empirical "shape factor" defined as the length (L) plus the width (W) of the casting divided by the thickness (T), that is, $(L + W)/T$, is first determined. The correct riser size is obtained from a plot of V_R/V_C versus $(L + W)/T$, where V_R and V_C are the riser and casting volume, respectively.

Work sponsored by the American Foundrymen's Society has led to the development of a series of curves for aluminum bronzes, copper-nickel, and manganese bronzes (Fig. 18 to 21).

Figures 18 to 21 indicate that the riser volumes necessary for sound castings can be reduced and the effectiveness of feeding improved by the use of exothermic sleeves. Insulating sleeves also can be used to increase the effectiveness of the risers. Exothermic sleeves, which were once popular, have now largely been discontinued in favor of insulating sleeves, which are more economical to use and cause fewer problems.

In recent years, greater attention has been given to what has become known as the modulus method for calculating riser size, which includes the development of pertinent data and computer programs to check that riser volume is adequate. This method is now, and no doubt will continue to be, the most widely used technique in the industry.

Chvorinov's rule states that the freezing time, t, of a cast shape is given by the relationship:

$$t = k \cdot (V/A)^2 \quad \text{(Eq 1)}$$

where V and A are the volume and surface area, respectively, of the cast shape, and k is a constant proportionality whose value is dependent on the thermal properties of the metal and the mold. For additional information on the thermal properties of copper, see the article "Aluminum-Base and Copper-Base Alloys" in this Volume.

For convenience, the term (V/A) in Chvorinov's equation is generally replaced by the symbol M, a value referred to as the

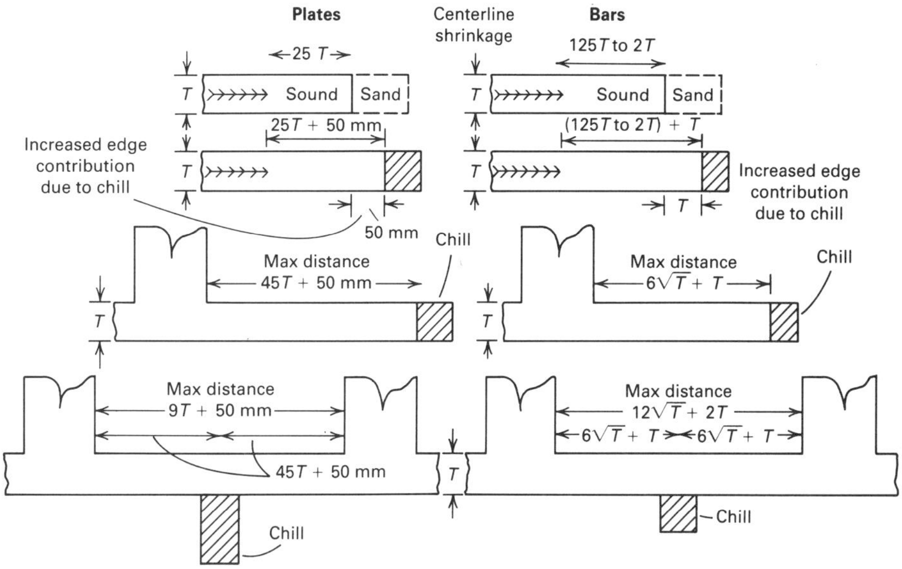

Fig. 17 Effect of chills in increasing feeding range of risers. Source: Ref 2

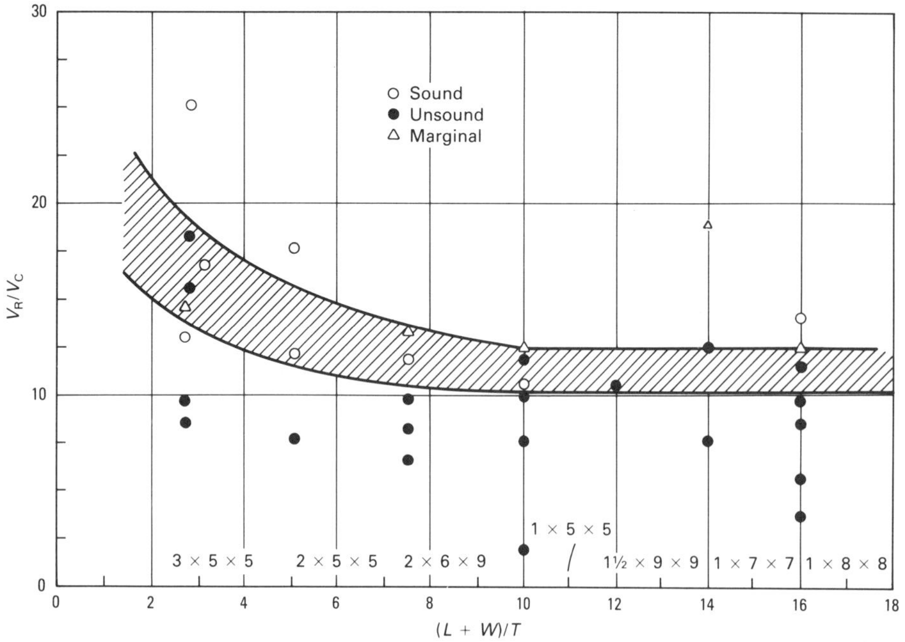

Fig. 18 Naval Research Laboratories (NRL)-type riser size curve for manganese bronze (alloy C86500). Source: Ref 4

modulus of the shape. Equation 1, above, can be rewritten more simply to read:

$$t = k \cdot M^2 \quad \text{(Eq 2)}$$

Because Chvorinov's equation can be applied to any cast shape, it applies equally to that which is intended to be the casting itself and to the attached riser. With connected shapes, such as a riser and a casting, the surface area of each shape to be considered includes only those portions that contribute to the loss of heat during freezing.

For the riser to be effective in feeding, its solidification time, t_R, must be greater than the solidification time, t_C, for the casting. This can be written:

$$\frac{t_R}{t_C} = \frac{k \cdot M_R^2}{k \cdot M_C^2} = F^2, \text{ or } M_R^2 = F^2 \cdot M_C^2 \quad \text{(Eq 3)}$$

Further simplified, this becomes:

$$M_R = F \cdot M_C \quad \text{(Eq 4)}$$

This means that the modulus for the riser, M_R, must be greater than the modulus for the casting, M_C, by some factor, F. Experience has shown that the proper value of F depends on the metal being cast. A value of about 1.3 is preferred for the short freezing range copper-base alloys.

As a practical working equation, therefore, we may say that with these alloys the modulus of the riser should be about 1.3 times that of casting (or casting section) to be fed, or:

$$M_R = 1.3\ M_C \quad \text{(Eq 5)}$$

Equation 5 merely gives us an empirical way of proportioning a riser so that it freezes more slowly than the casting. The other basic requirement of any riser is that it must have sufficient volume to provide the necessary amount of feed metal to the casting or casting section to which it is attached. These values can be calculated if necessary; however, it is much easier (though less precise) to use data of the type shown in Table 6. The numbers listed in the table indicate the minimum values for the ratio between riser volume and casting volume (as percentages) to ensure that the riser can, indeed, supply the necessary amount of feed metal to the casting. Five general classes of castings are shown, ranging from "very chunky" to "rangy." Notice that risers having a height-to-diameter ratio (H/D) of 1:1 are more efficient than when the H/D is 2:1. More important, it can be seen that insulated risers are far more efficient than those formed directly in the sand mold.

Feeder Shape. One of the requirements of the riser is to remain liquid longer than the casting; that is, from Chvorinov's rule:

$$(V/A)_R > (V/A)_C \quad \text{(Eq 6)}$$

The shape with the highest possible V/A ratio is the sphere. However, spherical risers are rarely used in industry because of molding considerations.

The next best shape for a riser is the cylinder. The H/D for cylindrical risers is in the range of 0.5 to 1.0.

Riser Neck Dimensions. The ideal riser neck should be dimensioned such that it solidifies after the casting but slightly before the riser. With this arrangement, the shrinkage cavity is entirely within the riser, this being the last part of the casting-riser combination to solidify.

Specific recommendations for the dimensions of riser necks are contained in the literature for ferrous alloys. These should apply to short freezing range copper alloys and are given in Table 7.

Hot Topping. About 25 to 50% of the total heat from a copper-base alloy riser is lost from the exposed surface by radiation. In order to minimize this radiation loss and thereby increase the efficiency of the riser, some sort of cover should be used on the top surface. Any cover, even dry sand, is better than nothing at all. A reliable exo-

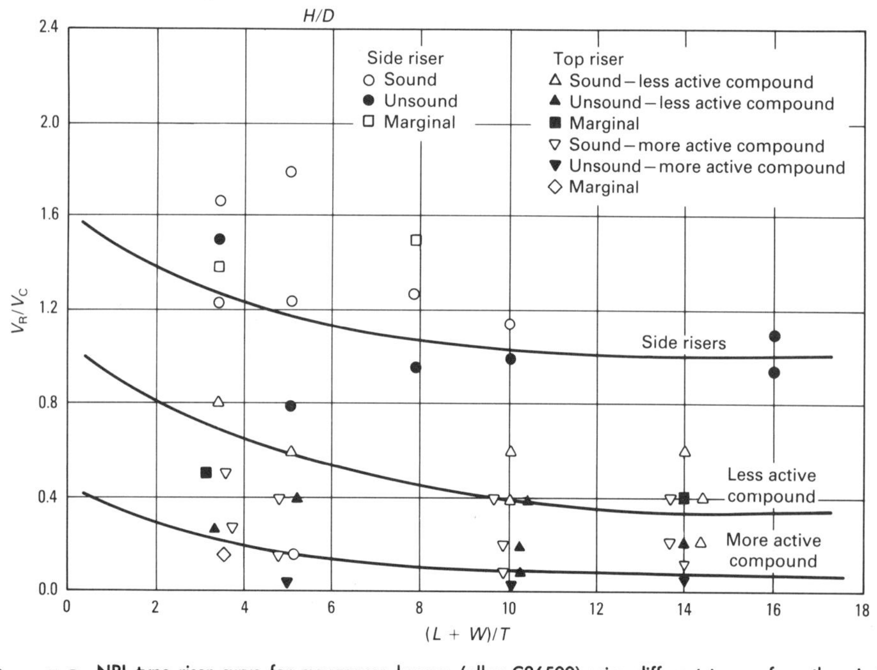

Fig. 19 NRL-type riser curve for manganese bronze (alloy C86500) using different types of exothermic hot topping and top risers. Source: Ref 5

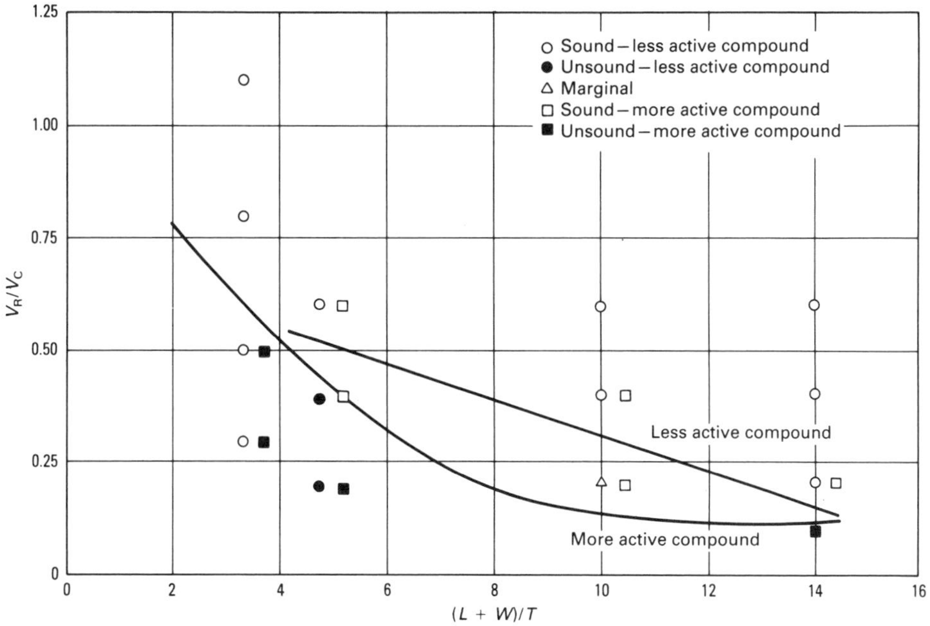

Fig. 20 NRL-type riser curve for aluminum bronze (alloy C95300) using different types of exothermic hot topping and top risers. Source: Ref 5

thermic hot topping is one form of usable cover.

Chills. The heat abstraction of the mold walls can be increased locally by the use of chills. Though expensive, metal chills are particularly effective because they reduce the solidification time by a factor of more than 55. As mentioned earlier, chills can be used to increase feeding distances and thereby reduce the number of feeders required. When it is impractical to attach feeders at certain locations, chills are particularly useful for initiating directional solidification, for example, at junctions, and so on, which would otherwise be porous.

Padding. The process of solidification can also be controlled by means of padding. Padding is the added section thickness (usually tapered) to promote directional solidification, and the bulk of it should be as close to the riser as possible.

Interaction of Gates and Risers. The effectiveness of side risers can be increased considerably by using a gating system that enters the mold cavity through the riser. The advantages of this arrangement are:

- Cleaner molten metal enters the mold cavity
- Because the metal in the riser remains liquid for a longer time, steep thermal gradients are established to improve the soundness of the casting

Detailed information on computer programs for pouring and gating can be found in the Section "Computer Applications in Metal Casting" of this Volume, particularly in the articles "Modeling of Solidification Heat Transfer" and "Modeling of Fluid Flow."

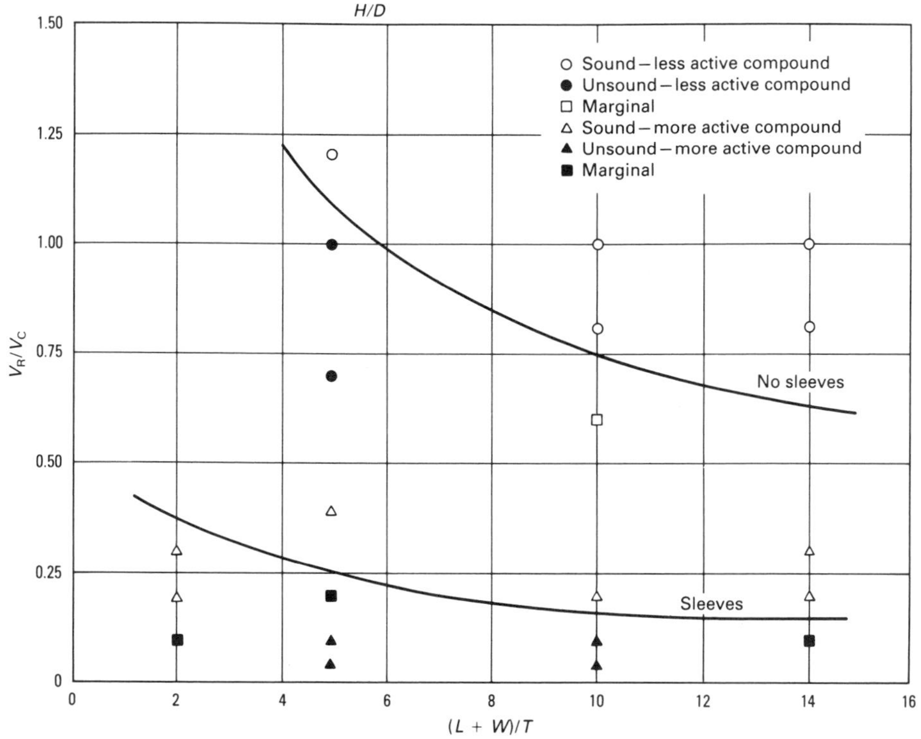

Fig. 21 NRL-type riser curve for Cu-30Ni (alloy C96400) using exothermic sleeves and hot topping versus hot topping alone. Source: Ref 5

Group III—Wide Freezing Range Alloys

The "workhorse" alloys of the copper-base group are the leaded red brasses and tin bronzes, virtually all of which have wide freezing ranges. These alloys have practically no feeding range, and it is extremely difficult to get fully sound castings. The average run of castings in these alloys contains 1 to 2% porosity. Only small castings may exhibit porosity below 1%. Attempts to reduce it more by increasing the size of the risers are often disastrous and actually decrease the soundness of the casting rather than increase it.

Experience has shown that success in achieving internal soundness depends on avoiding slow cooling rates. The foundryman has three possible means for doing this, within the limitations of casting design and available molding processes:

- Minimize casting section thickness
- Reduce and/or evenly distribute the heat of the metal entering the mold cavity
- Use chills and mold materials of high chilling power

In order to produce relatively sound castings, the following points should be considered.

Directional solidification, best used for relatively large, thick castings, can be promoted in various ways:

- Gate into hot spot
- Riser into hot spot
- Ensure that riser freezes last (consider riser size, insulation, and chills)
- Promote high thermal gradients by the use of chills, preferably tapered chills unless casting section is light (less than 12.5 mm, or ½ in. thick)
- Make sure risers are not so large that they unduly extend the solidification time of

Table 6 Minimum volume requirements of risers

Type of casting	Minimum V_R/V_C, %: Insulated risers H/D = 1:1	H/D = 2:1	Sand risers H/D = 1.1	H/D = 2.1
Very chunky; cubes, and so on; dimensions in ratio 1:1.33:2(a)	32	40	140	198
Chunky; dimensions in ratio 1:2:4(a)	26	32	106	140
Average; dimensions in ratio 1:3:9(a)	19	22	58	75
Fairly rangy; dimensions in ratio 1:10:10(a)	13	15	30	38
Rangy; dimensions in ratio 1:15:30 or larger(a)	8	9	12	14

(a) Ratio of thickness:width:length. Source: Ref 2

Table 7 Riser neck dimensions

Type of riser	Length, L_N	Cross section
General side	Short as feasible, not over $D/2$	Round, D_N = 1.2 L_N + 0.1D
Plate side	Short as feasible, not over $D/3$	Rectangular, H_N = 0.6 to 0.8T as neck length increases. W_N = 2.5 L_N + 0.18D
Top	Short as feasible, not over $D/2$	Round, D_N = L_N + 0.2D

L_N, D_N, H_N, W_N: length, diameter, height, and width of riser neck, respectively. D, diameter of riser. T, thickness of plate casting. Source: Ref 6

the casting, which would generate porosity beneath or behind the riser

Uniform solidification, best used for smaller, thin wall castings, can be promoted in various ways:

- Gate into cold spots, using several gates for uniform temperature distribution
- Use no risers, except perhaps on gate areas
- Use chills on hot spots to ensure that they cool at the same rate as the rest of the casting
- Use chills on areas that must be machined, thereby moving porosity to areas where the cast skin will be left unmachined; that is, maintain pressure tightness
- Gate into areas away from machined sections to maintain pressure tightness
- Use low pouring temperature (care should be taken to avoid misruns)
- See whether increased gas content (no degassing, reduced deoxidation) or induced metal mold reaction increases pressure tightness
- Make castings as thin as possible to increase cooling rate and reduce machining

Heat Treatment

The only copper-base alloys susceptible to heat treatment are beryllium copper alloys, chromium copper alloys, and aluminum bronze alloys containing more than 10% aluminum.

Beryllium copper alloys can be heat treated by solution treating and aging. Solution-treating temperature limits must be observed if optimum properties are to be obtained from the precipitation hardening treatment. Exceeding the upper limit causes grain growth, which results in a brittle metal that does not fully respond to precipitation hardening. Solution heating below the specified minimum temperature results in insufficient solution of the beryllium rich phase, a cause of lower hardness after precipitation hardening.

After castings are solution treated, they are quenched in water. All castings, except those of alloy C82000, may be solution heated in air and water quenched immediately after removal from the furnace. Alloy C82000 must be solution treated in a protective atmosphere such as cracked ammonia or natural gas. The duration of solution treating depends on section thickness. For castings greater than 25 mm (1 in.) in thickness, the time depends on section thickness.

Following solution heating, the castings are precipitation hardened. Table 8 shows the heat-treating cycles for beryllium copper alloys.

Chromium Copper. This alloy of Cu-1Cr can be heat treated in the same manner as beryllium copper. Here the solution treatment is 1 h at 980 to 1015 °C (1800 to 1860 °F), followed by a water quench. Next, the castings are precipitation hardened at 500 °C (930 °F) for 2 h. Because chromium is sensitive to oxidation, a protective atmosphere should be used to avoid an oxidized zone of approximately 3.05 mm (0.12 in.) on the casting surface. If heat treating is done in an air furnace, the castings must be machined after treatment to remove this oxide in order to obtain accurate conductivity and hardness measurements.

Aluminum bronze casting alloys containing more than 10% aluminum are heat treatable. These are alloys whose normal microstructures contain more than one phase to the extent that beneficial quench and temper treatments are possible. The copper aluminum alloys normally containing iron are heat treated by procedures somewhat similar to those used for heat treatment of steel, and have isothermal transformation diagrams that resemble those of carbon steels. For these alloys, the quench-hardening treatment is essentially a high-temperature soak intended to dissolve all of the α phase into the β phase. Quenching results in a hard room-temperature β martensite, and subsequent tempering reprecipitates fine α needles in the structure, forming a tempered β martensite. Table 9 shows typical heat treatments for three major aluminum bronze alloys.

Table 8 Heat treatment of beryllium copper alloys

Alloy	Solution heat treatment	Aging treatment
C81400	1 h at 980–1010 °C (1800–1850 °F)	2 h at 480 °C (900 °F)
C82000	1 h at 915–925 °C (1675–1700 °F)	3 h at 480 °C (900 °F)
C82200	1 h at 925 °C (1700 °F)	3 h at 455 °C (850 °F)
C82400	1 h at 800–815 °C (1475–1500 °F)	3 h at 345 °C (650 °F)
C82500	1 h at 800–815 °C (1475–1500 °F)	3 h at 345 °C (650 °F)
C82600	1 h at 800–815 °C (1475–1500 °F)	3 h at 345 °C (650 °F)
C82800	1 h at 800–815 °C (1475–1500 °F)	3 h at 345 °C (650 °F)

Specific Applications

Copper alloy castings are used in applications that require superior corrosion resistance, good bearing-surface qualities, high thermal or electrical conductivity, and other special properties. These applications may be divided into six principal groups:

- Plumbing hardware, pump parts, and valves and fittings
- Bearings and bushings
- Gears
- Marine castings
- Electrical components
- Architectural and ornamental parts

Figure 22 illustrates the wide variety of intricate shapes and sizes into which copper and its alloys are typically cast.

Plumbing Hardware, Pump Parts, Valves, and Fittings. The requirements for such fluid-handling components are pressure tightness to avoid leakage; reasonable mechanical strength at low, room, and high temperatures to avoid bursting; good corrosion resistance; and ease of machining. In addition, for a pleasing appearance, as in water fixtures, the parts must be easily platable.

Table 9 Heat treatment of aluminum bronze alloys

Alloy	Solution treatment	Tempering treatment
C95300	2 h at 900 °C (1650 °F)	1 h at 540–595 °C (1000–1100 °F)
C95400	2 h at 900 °C (1650 °F)	1 h at 565–620 °C (1050–1150 °F)
C95500	2 h at 900 °C (1650 °F)	1 h at 565–620 °C (1050–1150 °F)

Fig. 22 Variety of intricate shapes and sizes obtained by using continuous casting methods to produce brass and bronze alloy parts. Courtesy of ASARCO, Inc.

Plumbing fixtures and pump parts for the waterworks industry are usually produced in red brasses and semired brasses (alloys C83300 to C84800). Yellow brass (C85200) is sometimes used to cast plumbing fixtures. Similarly, pump parts are cast in silicon bronze (C87200).

A variety of alloys, however, are used to produce valves and fittings. These alloys are specified in ASTM B 763, and the list includes leaded red brasses, leaded semired brasses, silicon bronzes, silicon brasses, tin bronzes, leaded tin bronzes, high-leaded tin bronzes, nickel-tin bronzes, leaded nickel-tin bronzes, aluminum bronzes, leaded nickel bronzes, and so on.

Parts that do not require high strength are usually produced in red brasses, semired brasses, tin bronzes, and so forth, but when higher strength is required, the nickel-tin bronzes, high-strength yellow brasses, and so on, are preferred. For example, the valve stem in a control valve is cast in nickel-tin bronze (alloy C94700), whereas the facing is cast in alloy C83600 (Fig. 23).

Fig. 23 Cutaway views of an as-cast and finish machined/threaded body of a 50 mm (2 in.) gate valve-union bonnet assembly rated at 1.0 MPa (150 psi). The body section was sand cast of C83600 alloy (Cu-5Sn-5Pb-5Zn composition) and weighs 2.4 kg (5.2 lb). Courtesy of Crane Company

Equipment for handling more corrosive fluids, such as crude oil and salt water encountered in the oil field industry, is different from that of the waterworks industry. The requirements are corrosion resistance, pressure tightness, and better mechanical properties. The aluminum bronzes are widely used in the oil field industry to meet these requirements. Similar specifications apply to valves used in hydroelectric generating plants (Fig. 24). One such example is the reciprocating pump, in which all areas exposed to the corrosive fluids being pumped are made of aluminum bronzes (C95300 or C95800). Check valves and diaphragm backs for use in oil wells and chemical-processing equipment are cast in nickel-tin bronze (C94700).

The requirements for pressure-tight valves and fittings for different gases are higher than those for liquids. Such components can still be produced in the semired brass (C83600). However, care must be exercised in the casting process to ensure that shrinkage porosity is avoided.

Pump parts, valves, and fittings are also produced for marine application. Alloys used for such applications are discussed in the section "Marine Castings" in this article.

Bearings and Bushings. Copper alloys have long been used for bearings because of their combination of moderate-to-high strength, corrosion resistance, and either wear resistance or self-lubrication properties. The choice of an alloy depends on required corrosion resistance and fatigue strength, rigidity of backing material, lubrication, thicknesses of bearing material, load, speed of rotation, atmospheric conditions, and other factors. Copper alloys may be cast into plain bearings, cast on steel backs, cast on rolled strip, made into sin-

Fig. 24 A 1.37 m (54 in.) diam aluminum bronze stop valve for power generation cooling loop applications using a centrifugally cast body and a sand cast disc assembly. Seating surface is cold-rolled plate welded to the valve body with aluminum bronze spooled wire. Courtesy of Ampco Metal

Fig. 25 Centrifugally cast aluminum bronze worm gear blanks being inspected. Courtesy of Wisconsin Centrifugal, Inc.

Fig. 26 Propeller for a 114 000 ton tanker measures 7.5 m (24.7 ft) in diameter and weighs 37.52 Mg (82 725 lb). Part was machined and polished from a single 53.75 Mg (118 500 lb) nickel-aluminum bronze casting. Courtesy of Baldwin-Lima-Hamilton Corporation

Fig. 27 Vertical centrifugally cast ship propeller hub for controllable-pitch propeller blades is made of nickel-aluminum bronze, weighs 8.44 Mg (18 600 lb), and measures 1575 mm (62 in.) in diameter and 1270 mm (50 in.) in length. Courtesy of Wisconsin Centrifugal, Inc.

tered powder metallurgy shapes, or pressed and sintered onto a backing material.

Three groups of alloys are used for bearings and wear-resistant applications:

- Phosphor bronzes (copper-tin)
- Copper-tin-lead (low-zinc) alloys
- Manganese, aluminum, and silicon bronzes

Some of these applications are described below.

Phosphor bronzes (copper-tin-phosphorus or copper-tin-lead-phosphorus alloys) have residual phosphorus, ranging from a few hundredths of 1% (for deoxidation and slight hardening) to a maximum of 1%, which imparts great hardness. Often nickel is added to refine grain size and disperse the lead. Phosphor bronzes of higher tin content, such as C91100 and C91300, are used in bridge turntables, where loads are high and rotational movement is slow.

High-leaded tin bronzes are used when a softer metal is required at slow-to-moderate speeds and at loads not exceeding 5.5 MPa (800 psi). Alloys of this type include C93200, C93500, C93700, and C94100. C93700 is widely used in machine tools, electrical and railroad equipment, steel mill machinery, and automotive applications. Alloys C93200 and C93500 are less costly than C93700 and are used chiefly for replacement bearings in machinery. Alloy C93800 (15% Pb) and C94300 (24% Pb) are used when high loads are encountered under conditions of poor or nonexistent lubrication; under corrosion conditions, such as in mining equipment (pumps and car bearings); or in dusty atmospheres, as in stone-crushing operations and cement plants. These alloys replace the tin bronzes or low-leaded tin bronzes when operating conditions are unsuitable for alloys containing little or no lead.

Aluminum bronzes with 8 to 9% Al are used widely for bushings and bearings in light-duty or high-speed machinery. Aluminum bronzes containing 11% Al, either as cast or heat treated, are suitable for heavy-duty service (such as valve guides, rolling mill bearings, nuts, and slippers) and precision machinery applications.

Gears. When gears are highly loaded and well lubricated, the tin bronzes and nickel-tin bronzes are used. Specification ASTM B 427 gives the chemical compositions and mechanical properties of the five commonly used alloys; namely, C90700, C90800, C91600, C91700, and C92900. These are particularly advantageous when operating against hardened steel. It appears that the dispersion hardening of the δ phase in a solution-hardened matrix (by tin) provides the required strength. There is enough ductility to permit corrosion of minor misalignment with the hard steel mating part. Also, because of the dissimilarity of materials, no galling or scuffing is encountered.

When lubrication is irregular or omitted as in chemical applications, leaded materials are used. One such alloy is the leaded nickel-tin bronze containing Cu-20Pb-5Ni-5Sn. For gears exposed to harsh atmospheric conditions, manganese bronze (alloy 86500) has been successful.

Some typical applications are worm gears for rolling mills (alloy C91700), worm gears for oil well equipment (alloy C90700), and gearing of the stripper crane that removes the ingot from the ingot mold in the steel making industry (alloy C91700). Aluminum bronze is also used in worm gear applications (Fig. 25).

Marine Castings. The selection of materials for marine applications such as ship construction, desalination plants, and so forth, is governed by surrounding corrosive environments, which may include salt water, fresh water, or various corrosive cargoes such as oils, chemicals, and so on. Copper alloys generally give the greatest service life per dollar because of their excellent corrosion resistance in fresh water, salt water, alkaline solutions (except those containing ammonia), and many organic chemicals. The most commonly used alloys are the high-strength copper-nickels (both Cu-10Ni and Cu-30Ni, that is, alloys C96200 and C96400), aluminum bronzes (especially the nickel-aluminum bronze, alloy C95800, and manganese-nickel-aluminum bronzes, alloy C95700), and manganese bronzes (alloys C86100, C86200, C86400, C86500, and C86800). These are used in pump bodies, valves, tees, elbows, propellers (Fig. 26), propeller shafts, propeller hubs (Fig. 27), hull gear, impellers, turbines, and the like. A new addition to the group of copper-nickels is alloy IN768, which contains chromium instead of niobium, as in alloys C96200 and C96400.

The most important alloys to cast propellers are the nickel-aluminum bronzes (C95800), manganese-nickel-aluminum bronzes (C95700), and manganese bronzes or high-strength yellow brasses (C86500). Manganese bronze propellers dezincify in salt water, and the aluminum bronzes should be preferred for such applications. Bearings for propellers and ship rudders, however, are produced in tin and leaded-tin bronzes such as alloys C90500 and C92200.

Fig. 28 Twelve foot high bronze sand casting of the Great Seal of the United States located in lobby of the Federal Deposit Insurance Corporation building in Washington, D.C. Weight, 1.8 Mg (4000 lb)

Electrical Components. Copper and copper alloys are used extensively in the electrical industry because of their current-carrying capacity. They are used for substation, transformer, and pole hardware components for power transmission, plating and welding of electrical-equipment parts, and turbine runners for hydroelectric-power generation.

Cast copper is soft and low in strength. Increased strength and hardness and good conductivity can be obtained with heat-treated alloys containing beryllium, nickel, chromium, and so on, in various combinations.

Pure copper and beryllium copper are used to cast complex shapes for current conductors, often with water-cooled passages. Conductivity of the fittings is not important, the principal requirements being corrosion resistance and strength. Such fittings can be produced in leaded red brasses (alloys C83300 or C83600), heat-treated nickel-tin bronze (alloy C94700), or manganese bronze (alloy C86500).

Beryllium copper (alloy C82500) is also used to cast carriers for plating work and a variety of shapes and sizes for the welding industry.

Aluminum bronzes are the most important alloys for producing components for the hydroelectric-power generation plants because of their good corrosion resistance. Although parts have been produced from alloys C95200 and C95400, heat-treated nickel-aluminum bronze (alloy C95800) and manganese-nickel-aluminum bronzes (alloy C95700) are the most useful because of their resistance to dealuminification.

Architectural and Ornamental Applications. The aesthetic applications of copper-base alloys in artistic, musical, and ornamental work are due to their excellent corrosion resistance, remarkable castability, and variety of colors. Bronze statues are cast in silicon bronze alloy (UNS C87200) because it has good fluidity and is free from pitting and corrosion, and the development of an adherent patina reduces the corrosion rate. Figure 28, a bronze casting which dominates the lobby of a federal building in Washington, D.C., shows the fine detail which can be produced using copper in ornamental applications. For this reason, yellow and leaded yellow brasses (alloys C85200, C85300, C85400, C85500, and C85700) are also used for a variety of internal and external hardware. Church bells are usually cast in copper-tin alloys containing about 19% Sn. These alloys contain a network of the brittle δ phase in the matrix, which reduces the damping capacity and produces a better tone.

A complete range of colors, from red to bronze and gold to silvery yellow and silver can be obtained by adjusting the composition. The artist can take advantage of these color combinations to produce ornamental castings such as door handles in red and semired brasses (C83600 and C84400), yellow and leaded yellow brasses, and nickel silvers (alloys C97300, C97400, C97600, and C97800).

REFERENCES

1. D.G. Schmidt, Gating of Copper Base Alloys, *Trans. AFS*, Vol 88, 1980, p 805-816
2. *Casting Copper-Base Alloys*, American Foundrymen's Society, 1964
3. R.W. Ruddle, Risering Copper Alloy Castings, *Foundry*, Vol 88, Jan 1960, p 78-83
4. R.A. Flinn, Copper, Brass and Bronze Castings—Their Structures, Properties and Applications, Non-Ferrous Founders' Society, 1963
5. R.A. Flinn, R.E. Rote, and P.J. Guichelaar, Risering Design for Copper Alloys of Narrow and Extended Freezing Range, *Trans. AFS*, Vol 74, 1966, p 380-388
6. J.W. Wallace, Risering of Castings, *Foundry*, Vol 87, Nov 1959, p 74-81

SELECTED REFERENCES

- *The Aluminum Bronzes*, Copper Development Association, 1966
- *Casting Copper-Base Alloys*, American Foundrymen's Society, 1984
- Cast Products Alloy Data, in *Standards Handbook*, Part 7, Copper Development Association, Inc., 1978
- *Foundry Handbook*, Colonial Metals Company, 1984
- *Metals Handbook*, Vol 4, 9th ed., *Heat Treating*, American Society for Metals, 1981
- *Metals Handbook*, Vol 5, 8th ed., *Forging and Casting*, American Society for Metals, 1970

Zinc and Zinc Alloys

Dale C.H. Nevison, Zinc Information Center, Ltd.

DIE CASTING is the process most often used for shaping zinc alloys. The most commonly used zinc die casting alloys are UNS Z33521 (Alloy 3) and a modification of this alloy (UNS Z33522) distinguished by the commercial designation 7 instead of 3. The compositions of these alloys are given in Table 1, while the mechanical properties of zinc casting alloys are compared to those of other materials in Table 2. Although Alloy 3 is more frequently specified, the properties of the two alloys are generally similar. The second alloy listed in Table 1 (UNS Z35531) is used when higher tensile strength and/or hardness is required.

To address the increasing demand for high-performance high-quality die castings, a new family of zinc-base engineering casting alloys has been developed. For the last few years, market development emphasis for these alloys has focused on gravity casting, for which they have found increasing acceptance as engineering materials that provide superior properties, outstanding field performance, and excellent cost savings. Alloys such as aluminum, brass, bronze, and cast and malleable iron have been substituted in uses ranging from plumbing fixtures, pumps, and impellers to automotive vehicle parts and, recently, bronze bearing substitutes.

Three members of this family of alloys are generically identified industry-wide as ZA-8, ZA-12, and ZA-27. The composition is zinc plus aluminum, with small amounts of copper and magnesium.

The numerical components of the alloy designation indicate the approximate aluminum content. Small amounts of copper and magnesium are also added to produce the best combination of properties, stability, and castability. Commercial acceptance of ZA alloys has resulted in the issuance of national and international standards under ASTM B 669.

Alloy ZA-8 is the preferred choice for permanent mold casting, and it can be hot chamber die cast, providing further cost benefits. It offers excellent machinability, is antisparking, and has the best finishing characteristics for decorative parts.

Alloy ZA-12 is the general-purpose alloy, and it is often the first choice when converting from iron, brass, or aluminum. Usually cast in sand molds, it also performs well in graphite permanent molds and can be cold chamber die cast. The alloy has excellent pressure tightness, is antisparking, and is easily machined. There is growing evidence that ZA-12 has excellent bearing and wear characteristics.

Alloy ZA-27 is the ultrahigh-performance material, offering the highest strength and elongation. It is generally cast in sand molds, and like ZA-12 can be cold chamber die cast. It has excellent machinability and shows considerable promise for bearing and wear applications.

Die cast ZA-8, ZA-12, and ZA-27 alloys provide significant improvements in mechanical properties over current zinc materials. The ZA-8 alloy has been successfully used in the hot chamber die cast process; however, ZA-12 and ZA-27 alloys are usually cast using the cold chamber process.

The ZA alloys deliver the highest strength among the most widely used nonferrous alloys and match or exceed the strength of some cast irons. Ultimate tensile strengths range up to 441 MPa (64 ksi), depending on alloy and condition. Tensile yield strengths are as high as 379 MPa (55 ksi), which is roughly twice that of most commonly used casting alloys.

All three ZA alloys show hardness superior or equivalent to that of aluminum, brass, and bronze. This high hardness usually results in improved wear resistance and resistance to galling.

The allowable design stress or resistance to creep of the ZA alloys is significantly better than that of conventional zinc die casting alloys. The room-temperature design stress of ZA-27, for example, is approximately 90 MPa (13 ksi). Resistance to sustained loads allows for their use in many stressed applications at temperatures to 120 to 150 °C (250 to 300 °F).

The ZA-12 and ZA-27 alloys can provide equivalent, and in many cases superior, performance compared to the traditional cast bearing bronzes at a significantly lower cost. High load-carrying capacity, low wear rate, and good emergency running capability are well documented for high-load, low-speed, lubricated journal bearing conditions.

The 3, 5, 7, and ZA-8 and ZA-12 alloys are considered nonincendiary and sparkproof. This characteristic means that these zinc alloys will not ignite hazardous fuel-air mixtures, vapors, or particulate matter when struck by rusted ferrous materials. The nonmagnetic properties of zinc make it suitable for use in electronics and for delicate moving parts that would otherwise be adversely affected by magnetic disturbances.

Zinc-base alloys have good corrosion resistance in normal atmospheric conditions, in aqueous solutions, and when used with petroleum products. The corrosion resistance is enhanced by painting, plating, or

Table 1 Compositions of zinc casting alloys

Alloy	Applicable standards	Composition, %(a)								
		Al	Cu	Mg	Fe	Pb	Cd	Sn	Ni	Zn
No. 3 (UNS Z33521)	ASTM B 86	3.5–4.3	0.25	0.02–0.05	0.100 0.100	0.005	0.004	0.003	···	rem
No. 5 (UNS Z35531)	ASTM B 86	3.5–4.3	0.75–1.25	0.03–0.08	0.075	0.005	0.004	0.003	···	rem
No. 7 (UNS Z33522)	ASTM B 86	3.5–4.3	0.25	0.005–0.02	0.10	0.003	0.002	0.001	0.005–0.02	rem
ZA-8 (UNS Z25630)	ASTM B 669	8.0–8.8	0.8–1.3	0.015–0.03	0.075	0.004	0.003	0.002	···	rem
ZA-12 (UNS Z35630)	ASTM B 669	10.5–11.5	0.5–1.25	0.015–0.03	0.10	0.004	0.003	0.002	···	rem
ZA-27 (UNS Z35840)	ASTM B 669	25.0–28.0	2.0–2.5	0.01–0.02	···	0.004	0.003	0.002	···	rem

(a) Maximum unless range is given or otherwise indicated.

Table 2 Comparison of typical mechanical properties of casting alloys

Alloy and product form(a)	Ultimate tensile strength, MPa	Ultimate tensile strength, ksi	0.2% offset yield strength, MPa	0.2% offset yield strength, ksi	Elongation, % in 50 mm (2 in.)	Hardness, HB	Impact strength, J	Impact strength, ft · lbf	Fatigue strength, MPa	Fatigue strength, ksi	Young's modulus, GPa	Young's modulus, ksi × 10^3
Zinc alloys												
No. 3 (D)	283	41	...	...	10	82	58(c)	43	47.6	6.9	...	...
No. 5 (D)	331	48	...	...	7	91	65(c)	48	56.5	8.2	...	...
No. 7 (D)	283	41	...	...	13	80	58(c)	43	...	...	...	...
ZA-8 (S)	248–276	36–40	200	29	1–2	80–90	20(c)	15	...	...	85.5	12.4
ZA-8 (P)	221–255	32–37	207	30	1–2	85–90	...	...	51.8	7.5	85.5	12.4
ZA-8 (D)	372	54	290	42	6–10	95–110	42(c)	31	...	...	...	...
ZA-12 (S)	276–317	40–46	207	30	1–3	90–105	25(c)	19	103.5	15	83.0	12.0
ZA-12 (P)	310–345	45–50	207	30	1–3	90–105	...	...	...	...	83.0	12.0
ZA-12 (D)	400	58	317	46	4–7	95–115	28(c)	21	...	...	...	...
ZA-27 (S)(b)	400–440	58–64	365	53	3–6	110–120	47(c)	35	172.5	25	75.2	10.9
ZA-27 (P)	421–427	61–62	365	53	1	110–120	...	...	...	...	75.2	10.9
ZA-27 (D)	421	61	365	53	1–3	105–125	...	...	...	...	...	...
Aluminum alloys												
319 (S)	185	27	124	18	2	70	5(c)	4	69	10	74.0	10.7
356-T6 (P)	262	38	185	27	5	80	11(c)	8	90	13	72.4	10.5
380 (D)	325	47	159	23	3.5	80–85	4(c)	3	138	20	71.0	10.3
Copper alloys												
C83600 brass (S)	255	37	117	17	30	60	15(d)	11	76	11	83.0	12.0
C86500 bronze (S)	490	71	193	28	30	98	42(c)	31	145	21	103.5	15.0
C93200 bronze (S)	240	35	124	18	20	65	8(e)	6	110	16	100	14.5
C93700 bronze (S)	240	35	124	18	20	60	15(d)	11	90	13	80	11.5
Cast irons												
Class 30 gray iron (S)	214	31	124	18	...	210	...		97	14	90–113	13.0–16.4
Malleable iron (S)	345	50	221	32	10	110–156	54–88(c)	40–65	172–207	25–30	172	25

(a) D, die cast; S, sand cast; P, permanent mold cast. (b) As-cast. (c) Unnotched Charpy. (d) Notched Charpy. (e) Izod

chromate or phosphate treatment and is substantially improved by anodizing.

The ZA alloys readily accept a wide variety of decorative and corrosion-resistant surface finishes. Low cost components are painted to match adjacent parts, chromium plated to offer a durable luster (except ZA-27), and brush finished to take on the rich appearance of brass, bronze, or stainless steel at a fraction of the cost. Anodizing produces a thin, ceramiclike, abrasion-resistant film exhibiting excellent resistance to most natural and industrial corrosive materials.

Zinc die castings ranging in weight from a fraction of an ounce to 23 kg (50 lb) have been successfully produced. However, most zinc die castings comprise a wide variety of hardware items having weights near the low end of the above range.

Control of Alloy Composition

Zinc alloys for die casting are sensitive to variations in composition and impurity levels—generally more so than aluminum alloys. However, limitations on the permissible amounts of added elements or impurities in zinc are liberal enough that a reasonable program of control and generally sound shop practice are sufficient to maintain adequate consistency in alloy composition.

Agitation. Although agitation during melting will not affect alloy composition, it results in melt oxidation and should therefore be minimized. Overheating often results in loss of aluminum and magnesium through oxidation and in an increase in iron due to a decrease in the scavenging action provided by aluminum.

Use of Foundry Scrap. Clean scrap of acceptable composition returned from the foundry can be charged into the furnace in unlimited proportions, although use of 50% maximum of scrap per charge is recommended. Not surprisingly, the usual practice is to remelt zinc scrap runners, overflow wells, sprues, and castings. However, there are safeguards that should be employed to ensure that remelt does not disrupt the fine balance of additives in the melt.

The ability to remelt zinc process scrap many times without losing properties is a significant advantage to the die caster. However, caution should be exercised to keep this material clean and free of unwanted substances. It should be stored separately away from other metals if it is accumulated in batches. If conveyed back to a central melter, conveyors and the furnaces should be covered when maintenance work is being done nearby or overhead. Floors and tables should be kept clean.

If there is doubt as to the purity of the scrap, it should not be used in any proportion until it has been analyzed. Scrap returned for recycling must be free of oil and moisture. A safety hazard is created when oil or moisture is present on the metal being charged into the furnace.

Some zinc scrap can be electroplated. This material should be remelted separately and added back in small quantities or, better still, sold outright. If this is not practical, the scrap should be fed back moderately. The electrodeposits will separate and float to the top of the bath, where they can be skimmed off. Agitation will increase copper, nickel, and chromium levels and should therefore be avoided.

Castings that have had chemical surface treatments can generally be remelted as clean scrap. Under no circumstances should cadmium-plated, tin-plated, or soldered die castings be remelted. Caution should also be exercised when remelting returns from customer plants. These castings may contain lead plugs or other undesirable materials. It is usually recommended that the scrap not exceed 30 to 40% of the amount of newly prepared alloy.

Melt Temperature and Fluxing. High temperatures and flux can change the percentage of the alloying elements. No flux is needed when the melting stock is clean ingot, but 1.4 to 2.3 kg (3 to 5 lb) of a chloride or fluoride flux is added for each 450 kg (1000 lb) of metal when the melting stock is partly or totally comprised of trimmings, gates, and rejected castings. A few pounds of flux per ton of alloy will reduce the magnesium content, and greater flux additions can make the magnesium disappear completely. Consequently, it is necessary to control temperatures continuously, to flux properly, and to check the analysis.

Alloying Elements. The purposes served by the alloying elements and the effects of using these elements in amounts exceeding specified limits are summarized in the following paragraphs. Strict control of chemical composition is absolutely essential for avoiding any chance of intergranular (intercrystalline) corrosion, dimensional changes, or loss of mechanical properties. Specified compositions for zinc alloys are given in Table 1.

Aluminum is added to zinc for die casting to strengthen the alloy, to reduce grain size, and to minimize the attack of the molten metal on the iron and steel in the casting and handling equipment. Aluminum adds to the fluidity of the molten metal and improves its castability. As indicated in Table 1, aluminum contents range from 3.5 to 4.3% for Alloys 3 and 5. An aluminum content lower than 3.5% requires higher-than-normal metal temperatures for satisfactory castability. The higher temperatures result in undue attack on the dies. Other disadvantages of low aluminum are lower strength and less dimensional stability than alloys containing aluminum within the specified range.

When aluminum content is higher than 4.3%, it lowers the impact strength of the castings. The zinc-aluminum eutectic forms at about 5% Al. This eutectic alloy is extremely brittle and must be avoided.

Magnesium content must be carefully maintained within the ranges shown in Table 1. Magnesium is added primarily to minimize susceptibility to intergranular corrosion caused by the presence of impurities. Excessive amounts of magnesium lower the fluidity of the melt, promote hot cracking, increase hardness, and decrease elongation. Cracking is generally confined to castings of complex form that are free to shrink in the die.

Copper, like magnesium, minimizes the undesirable effects of impurities and, to a small extent, increases the hardness and strength of the castings. Castings containing more than about 1.25% Cu are less stable dimensionally than those with less copper. The copper range for Alloy 5 is 0.6 to 1.25%. The lower limit places the alloy into the high-tensile and high-hardness range, while the upper limit is safely under the copper content that produces aging changes in castings at room temperature.

Iron in amounts up to 0.10% has little detrimental effect, but may contribute to problems in buffing or machining. Iron under 0.02% is in solid solution. Greater amounts form hard iron-aluminum compounds, which can produce comet tails during buffing and can dull tools during machining.

Nickel, chromium, silicon, and manganese are not harmful in amounts up to the solubility limit of each (0.02% Ni, 0.02% Cr, 0.035% Si, and 0.5% Mn). When these metals exceed their solubility limits, they form light intermetallic compounds with aluminum and can be skimmed off the surface of the melt.

Lead, cadmium, and tin at levels exceeding the limits shown in Table 1 can cause die cast parts to swell, crack, or distort. These defects can occur within 1 year. The maximum limit for lead, which can promote the occurrence of subsurface network corrosion, is 0.006%. Cadmium is detrimental in its effect at some concentrations and is neutral at others. As such, the maximum limit for cadmium is set at 0.005%. Tin, like lead, can promote subsurface network corrosion, and therefore is also restricted to the maximum safe limit of 0.005%.

Furnaces

In the past, the standard furnace at the hot chamber zinc die casting machine was a gas-fired unit that held a cast iron pot. Quite often, the furnace at the machine was also used for melting. Large installations generally had a central melting facility, usually gas fired and accommodating a cast iron pot. Hot metal was pumped or siphoned from the furnace into the transport crucible or ladle. The transport ladle was suspended from an overhead conveyor and filled and emptied by mechanical tipping mechanisms. Only one worker was required. However, the working conditions were far from satisfactory. Manual handling of the metal was heavy, hot, smoky, and dangerous.

Furnaces are frequently an integral part of the die casting machine. Most furnaces for melting and alloying, as well as for holding, are fuel-fired open-pot, immersion tube heated, or induction heated. Most pots for melting and containing molten zinc alloys are cast from gray or ductile cast iron. Ladles, if used, are of cast iron or pressed steel. Regardless of its type, the furnace should be equipped with controls so that the temperature of the molten zinc can be maintained within 6 °C (10 °F).

The furnace capacity required depends on the size of the casting machine, workpiece size, and production rate. Generally, a holding furnace should be able to hold at least four times the amount of metal required for 1 h of operation. The capacity range of melting furnaces is usually 450 to 9000 kg (1000 to 20 000 lb), although immersion tube furnaces can range up to 18 Mg (40 000 lb). Total furnace capacity is usually five to seven times the amount of metal required per hour.

A major innovation was the introduction of gas-fired immersion tube furnaces. Eliminating the need for pots, these furnaces use metal tubes immersed in the metal bath. The results are excellent; furnace efficiencies are increased considerably by this almost-direct method of heating. Immersion tubes are used for melting as well as holding furnaces and for heating molten metal launder systems. Many installations began using complete systems with immersion tubes to heat the melter, holders, and interconnecting launders.

The Launder System

The launder system consists of three main components: a central furnace, a number of metal feed furnaces (one for each die casting machine), and a launder system connecting the furnaces. The central melting furnace is arranged to feed the main launder continuously with molten zinc. Each casting station is equipped with a holding furnace that has a branch launder connecting it to the main launder.

The furnaces are fully enclosed, with tight-sealing lids and extremely thick walls. The castable main refractory does not contaminate zinc alloys and is backed up with heavy insulating material. The objective is to achieve such efficient sealing and insulation that very little heat is felt when the furnace is touched. The heat source is either immersed in the melt or located underneath the furnace lid (or a combination of both). The openings for charging and discharging are narrow and have heat locks to keep heat loss to a minimum. As long as the melting furnace metal is maintained at a recommended level, molten zinc flows by gravity from the ladling chamber, through an exit cast into the side of the furnace, and into the main launder. Because the channels are located well below the surface of the bath, surface dross from the charge end does not enter the ladling end.

This type of system—immersion zinc alloying, remelting, laundering, and holding—can be used to gravity feed a number of die casting machines. In this type of system, cold-charged metal has little, if any, effect on the temperature of the metal in the holding furnace of the die cast machine.

The laundering and holding system requires skimming only once a month, which reduces metal loss due to dross formation/removal. The minimal effort required to maintain the immersion system increases efficiency, while also improving the working conditions.

In a launder, the metal is not directly exposed to the atmosphere. The trough is lined with heavy insulation having extremely low thermal conductivity. This insulation is also nonwettable by molten zinc. The launders are also well insulated and sealed, and the cross-sectional area of the metal stream in the launder is small (26 cm^2, or 4 $in.^2$). The covers are heavily insulated and are hinged to permit inspection of any portion of the launder.

Metal conveyance throughout the launder system into the machine-feed furnace is free from turbulence. The metal is moved smoothly and continuously under a protective layer of stationary metal oxides. Most important, metal temperature fluctuations are virtually eliminated. In addition, almost all manual work is eliminated.

Scrap Return

Directly below the trim press is the scrap conveyor, which goes to the melting furnace. Also leaving the trim press is the

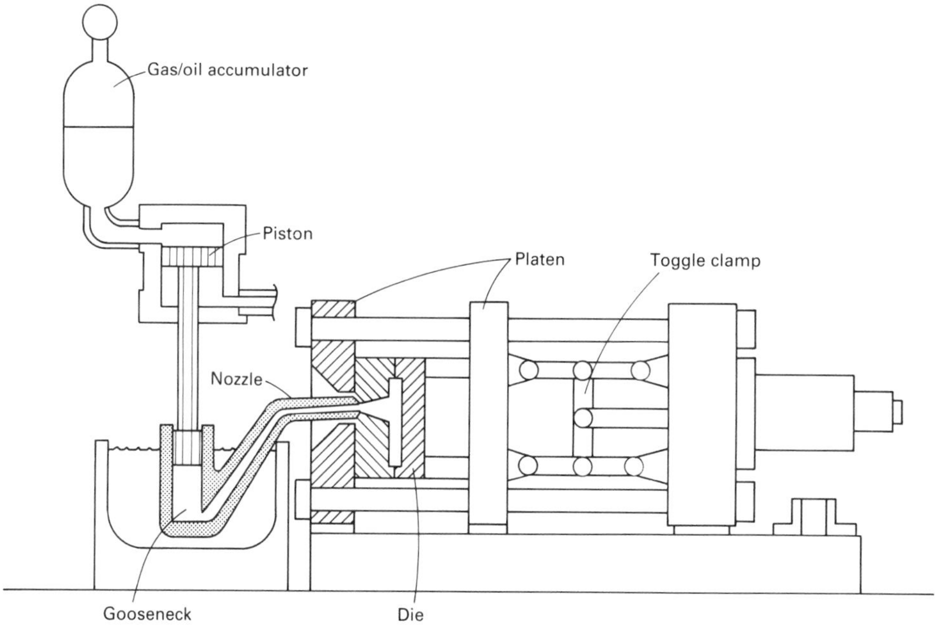

Fig. 1 Hot chamber machine used to make zinc die castings

finished parts conveyor, which directs the parts for further processing. The transfer conveyors are typically steel belts or oscillating conveyors.

An innovation is the overlapping steel belt. Each section of this belt is hinged on only one side, allowing the belt surface to swing free at the discharge point. This motion sheds the material being conveyed, minimizes carryover, and prevents the accumulation of material in the hinge points, which can be a primary cause of wear and jamming in the conventional hinged steel belt. The overlapping belt is designed to use standard belt components and can be used in conventional hinged steel belt frames.

Die Casting Machines

Die casting machines used for zinc alloys are usually of the hot chamber type (Fig. 1), in which the pressure chamber, or gooseneck, is submerged in the molten metal. With this type of machine, the metal is injected into the die in the shortest time and with the least decrease in temperature.

Selection of materials for the component parts of a die casting machine is less of a problem in casting zinc alloys 3, 5, and ZA-8 than in casting aluminum or copper alloys. Material requirements are less rigorous because zinc alloys are cast at relatively low temperatures and because molten zinc alloys do not rapidly attack ferrous metals. Pots, goosenecks, and other components of the machine that come in contact with the molten metal are usually made of gray or ductile cast iron, although they can be made of cast steel. Availability and cost usually determine selection. Sleeves and nozzle seats, because they receive high wear, have been made from nitrided alloy steel, hot-work tool steel (such as H13), high-speed tool steel, and stainless steel. Frequently, the sleeve and nozzle seat of the gooseneck are replaceable to permit inexpensive repair.

The injection cylinder can be either hydraulic or pneumatic. With the same stroke length, the amount of metal injected can be changed by increasing or decreasing the size of the cylinder bore. Injection pressures used for the die casting of zinc alloys usually range from 10.3 to 20.6 MPa (1500 to 3000 psi). The lower pressures are used for simpler castings; the higher pressures, for more complex ones. Good practice is to use the lowest pressure that will produce acceptable castings; however, a minimum pressure of 10.3 MPa (1500 psi) is essential for obtaining an acceptable combination of soundness, surface finish, and mechanical properties.

Ideally die casting installations that are to be automated would include the following features:

- Die casting machine designed for automation and equipped with closed-loop control of machine variables, reliable automatic central lubrication and hydraulic system, repeatable and recordable machine functions, interlocked cycle stop, and shutdown circuitry triggered by out-of-tolerance machine variables
- Constant-temperature molten metal supply with interlocked shutdown circuitry monitoring temperature tolerances
- Optimum designed and precision-built dies with closed-loop cooling and heating control
- Flexible, programmable multifunction die lubricator system to provide die release, surface cooling, and efficient lubrication of the moving die parts
- Integrated auxiliary equipment that matches the die casting machine in performance and physical properties and is designed to endure the die casting environment

Early machines were designed and built for operator-controlled die casting and are not suitable for automation.

Dies

Zinc alloys can be cast in single-cavity, multiple-cavity, combination, or unit dies. Dies that are designed to produce zinc alloy castings can seldom be used to produce castings of aluminum alloys or other metals that are cast at higher temperatures, because zinc alloys can be cast to thinner sections, smaller radii, and closer tolerances than aluminum, magnesium, or copper alloys. However, a die designed for casting the higher-melting alloys can be used for casting zinc alloys.

The cover half of a die must be equipped with a nozzle seat that will provide a good seal with the gooseneck of the machine. A sprue hole, or bushing, and spreader must be incorporated into the die to ensure feeding of the molten metal to the runners and to permit removal of the solidified sprue with the casting.

Die Materials. In the die casting of zinc alloys, die temperature is relatively low, usually ranging from 160 to 245 °C (325 to 475 °F). Therefore hot-work tool steels are not generally required for dies. However, for extremely long runs and for high dimensional accuracy, hot-worked tool steels such as H13 will provide optimum die life.

Die hardness in the casting of zinc alloys is less critical than for alloys of higher casting temperature. Steels prehardened by the manufacturer to any maximum hardness consistent with reasonable machinability can generally be used. The typical hardness is 29 to 34 HRC (280 to 320 HB).

Slides and Cores. Hardenable stainless steels such as type 440B are often used for cores. Hot-work tool steels such as H13 can be used for both cores and slides. Because these tool steels respond readily to nitriding, they can be selectively hardened; a component made from one of them can be treated for different properties in different areas. For example, the main portion of cores and slides can be nitrided for wear resistance, and the end portions can be masked to resist nitriding for better resistance to heat checking and spalling. Lubricating the slides and cores with molybdenum disulfide or colloidal graphite in oil helps to ensure smooth action and to minimize wear.

Clearance between the slides and the guides should be kept to a minimum to prevent the molten zinc from wedging between them. Ejector pins of nitrided H11 tool steel or 7140 alloy steel are available as stock items for insertion in the dies.

Die Life. The service life of a die casting die is directly related to the temperature of the metal being cast, thermal gradients within the die, and frequency of exposure to high temperature. Because of the relatively low temperatures of the dies (see the section "Die Temperature" in this article) and of the molten metal, die life for casting zinc alloys is generally much longer than for casting aluminum, magnesium, or copper alloys; it is not unusual for dies to last for 1 million shots. Maximum die life depends largely on having well-designed dies, trained operators, and a rigidly enforced program of machine and die maintenance. A deficiency in any of these areas will result in decreased die life.

Die Temperature

The temperature at which a die will run (level out) during continuous operation depends on the weight of the shot, the surface area of the shot, the cycle speed, and the shape of the die. When dies are too cold, cold shuts, laminations, internal porosity, incomplete filling, and poor finish with excessive flow marks are likely to result. Dies that are too hot cause shrinkage, heat sinks, excessive flash, spitting, poor ejection, soldering, and die erosion. For a new application, some experimentation is usually required to establish a satisfactory optimum die temperature.

The optimum die temperature for zinc is usually between 160 and 245 °C (325 and 475 °F). The lower end of this range is used for thick-section castings, and the higher end is for thin-section castings. When hardware finish is required, higher die temperatures (near 245 °C, or 475 °F) are generally required, regardless of the casting thickness. Once established, die temperature should be maintained within 6 °C (10 °F).

Some casting shapes require localized heating or cooling of the die above or below the established temperature. Metal overflows are often used to heat die areas surrounding the perimeters of castings that have thin sections far from the main runner. This method of local heating helps to fill thin sections and to improve casting finish. Conversely, water channels are frequently placed behind the runner area immediately adjacent to the sprue to provide localized cooling and to prevent soldering of the molten metal to the die.

Control of Casting Temperature

Research has shown that the variation in die temperature is one of the most important parameters affecting production rates and casting quality. To eliminate defects caused by excessively low die temperature, the die caster should design the die with excess cooling capacity and then use a temperature control system to modulate the flow of coolant in the die to achieve the desired die temperature. This incorporates the use of thermocouples, solenoid valves, and controllers. If necessary, heat can be added to the die through the use of heating elements inserted in the die. A complete system can be purchased that will control both heating and cooling using oil as the heat transfer medium. Some die casters have found the oil controller to be superior in overall performance and flexibility.

Metal temperatures for casting the zinc alloys range from 400 to 440 °C (755 to 825 °F). Generally, the lower end of this range is used for castings with thick sections, and the higher end is for castings in which section thickness is near minimum. In practice, a temperature of 415 °C (780 °F) is used for a wide variety of casting sizes and shapes.

For best results, including best casting finish, some experimentation is usually required to arrive at the optimum metal temperature for a given application. When the optimum temperature is established, it should be controlled within 6 °C (10 °F).

Die Lubricants

Selection of the optimum lubricant (die release agent) for a specific application often requires some experimentation because of the various operating temperatures and die shapes. An optimum lubricant is one that carbonizes at the operating temperature. A lubricant that carbonizes above the operating temperature will stain the casting, and one that carbonizes below the operating temperature will be used up on the first shot. The black oil and graphite lubricants have been replaced by water-base lubricants; this has reduced the fire hazard and smoky environments commonly found in die casting plants. Water-base lubricants are formulated to be an effective release agent and aid in cooling the dies.

Die Lubrication System

Because of the long periods spent spraying the die and the variety of spray patterns the operators will use in a day's production, a considerable amount of time can be saved by automating this operation. Installation of the spray unit also reduces the work load of the operator.

A number of die spray reciprocators have appeared on the market in recent years. Automatic die spray has been the most cost-effective measure in the die casting process. It resulted in a production increase of up to 25% and has a significant reduction in rejects (primarily because of improved surface finish). The die spray serves to release the casting, to lubricate moving die parts, and to cool the die surface. Selection of a spray system should involve consideration of the following requirements:

- Moving die parts should be lubricated by an appropriate central lubricating system built into the die and coupled to an external lubricator
- Die cooling should be accurately calculated and achieved by internal water channels. In marginal cases, additional cooling can be accomplished by the die spray. In such cases, the dilution is extended to avoid excessive use of the release agent and resultant buildup
- The die sprayer must be easily programmable to perform its function with reasonable repetitive accuracy
- Essential features of the sprayer include the ability to spray two different media and air blast
- Lateral movement of the spray nozzles may be essential for reaching deep die cavities

Considerations of only the basic requirements when selecting a spray system will result in dissatisfaction. Knowledge of die size, special spray patterns, and so on, is also important. Once the system is operational, adjustments should require a minimum of time.

The spray system should be selected to allow for ease in movement when installing dies and/or making machine repairs. Each system has advantages and limitations. Most use air for spraying and blow-off. Some systems use air for movement, while others use hydraulic power either from a separate hydraulic power supply or from the hydraulic system of the die casting machine. Because of the air pressure fluctuation in most plants (anywhere from 550 to 825 kPa, or 80 to 120 psi), a hydraulic system gives more constant movement when spraying, and this in turn provides more control. Other areas that must be investigated before purchasing a spray system include the adequacy of the plant air supply, the need for a tank for die release, and the size of the area the spray pattern should cover (which depends on the die size to be run in the system).

Casting Removal

At this point in the installation, everything is operating semiautomatically. The only item needed to complete the system is a unit that physically removes the casting from the die cast machine, known as an unloader, extractor, robot, grabber, or drop system.

Unloaders vary widely in function. Some units will unload the machine, spray the die, and trim the castings. Others will only unload. Some systems are programmable, and others require the adjustment of limit switches and timers.

A drop-through system is the simplest means of unloading a die casting machine. However, there are several problems with the system. When a drop-through system is decided upon, a method of removing the dropped casting must be considered. Normally, a pit is dug under the machine, and a conveyor is installed to remove the castings. The pit is filled with water to be used as a quench. Quenching ensures the solidification of the casting when it reaches the conveyor, thus preventing bending. If the water is too shallow, the casting will hit the conveyor and will be bent.

Sensing a falling part in one way is quite simple, using a limit switch, photo detector, infrared detector, or a radio wave device. The problem is to determine whether or not the complete shot has left the die.

There are several methods of determining if the entire die is clear for the next cycle. One is to weigh each shot as it leaves the die. This method appears simple, but it must be accomplished under water or after the casting leaves the water. Another method is to use a radio wave device to scan both halves of the die in conjunction with a heat detector for the sprue bushing to determine whether the sprue has been removed.

The equipment mentioned is not easily set up, and it requires frequent servicing. The only positive way to determine whether or not the die is clear for another cycle is to have an operator standing by, watching each cycle and clearing anything that sticks to the die.

After considering the problems and economics involved, the drop system is not as simple as it may seem. It does, however, perform well for many companies.

Grabbers. A slightly more complex method of casting removal is the grabber. This equipment, although not inexpensive, is relatively low in cost compared to other good casting removal units. Its ability to handle a wide variety of tasks, however, is limited by its lack of mobility. This type of equipment performs well, but the purchaser must take the limitations into consideration. One advantage of some grabbers is their ability to spray and sense, which reduces the total capital investment. The grabbing mechanism on most units can be modified to handle a variety of dies; this is sufficient for the average die caster. The grabber design is compatible with other die casting industry equipment; controls, both electrical and hydraulic, are standard, using basic design concepts to minimize complexity. These units can be purchased with their own hydraulic power sources, or they can be connected to the hydraulic system of the die casting machine. Whatever system is used, the unit has been designed with ease of maintenance in mind.

The unloader is not attached to the die casting machine, but supports itself. This type of equipment is sometimes more sophisticated than the above types and is generally self-sufficient, relying only on input and/or output signals to control its operation. Some unloaders can be connected to the hydraulic system of the die casting machine, reducing the overall cost. Various accessories are available with unloaders. Some unloaders can be purchased with a trim press, or spray system and sensing unit. Others are strictly unloaders that can be reprogrammed by the changing of limit switches and rotating cams. The unloader differs from the grabber in its versatility and its programmability.

Each system has been developed with the die casting machine in mind. Except for one or two specific items, these machines are designed using the standard relay logic and hydraulic systems of the die casting machine industry. For the most part, the unloader is a more sophisticated and versatile grabber.

The robot, or programmable unloader, is the most versatile of all the types of unloading equipment. The robot can be moved through a series of operations and can store each operation in memory. It is the most expensive of all unloaders and the most sophisticated. Robots can usually be serviced by trained maintenance personnel because most of the work needed to correct a malfunction consists of removal and replacement. However, with all of its versatility, the robot can handle only 70 to 80% of the average dies. With additional equipment, the robot can handle approximately 90% of the average dies.

The most important aspects of the robot are the reduced time it takes to program and the speed at which it operates. Some robots can be programmed to operate several pieces of equipment at the same time. The robot can also be programmed to spray the die with a very precise pattern.

The choice between a robot and an extractor is subject to individual consideration. Generally, a single-purpose extractor is more reliable, more cost effective, and simpler to program and operate than a robot. Although the repetitive accuracy of robots and extractors is well known, placing any casting accurately in a trim press is not an easy task.

As with any type of casting removal, there must be a method of detecting the casting once it has been removed from the die. This detecting can be accomplished with limit switches, photocells, infrared probes, or tactile sensors. Infrared sensors can sometimes be triggered by the surrounding environment instead of the casting and therefore may not be completely reliable.

Tactile sensors consist of multiple probes or antennas connected to a low-voltage detection system. When the probes make contact with the casting, the detection circuit is complete, and the next step of the casting cycle is initiated. Detection systems having as many as 12 probes are common; therefore, multiple-cavity gates and different portions of the same casting can be sensed. For example, deep bosses or similar features that tend to stick in the die can be individually monitored.

The above sensing systems (infrared or tactile) are possible only when a robot is used because it is necessary to bring the complete shot accurately to the probes. Newer sensing systems have memory capability in that all the multiple probes do not need to contact the casting simultaneously. This feature allows a "fly through" of the shot to permit quicker sensing with no loss in reliability. In the fast-running zinc machines, such a capability is economically significant. Any of these detection devices must be mounted in some predetermined position near the casting removal location.

Trimming. In finish trimming (not mere breaking of the casting from the gate), the part must almost always be quenched in water-soluble coolant/lubricant similar to that used in machining zinc castings. Failure to include soluble oil usually causes the trim die plates to solder, with consequent tearing of the trimmed part. Quenching several hundred pounds of shots per hour in a small tank requires either a heat exchanger for the quenching fluid or connection to a larger remote reservoir so that the heat can be dissipated naturally. Another method would be to use a spray and to recycle from the tank. If the robot or extractor dips the shot in a quench tank and then places it in a trim press, a drag-out conveyor can be fitted into the tank to remove the flash that settles to the bottom.

The primary reason for mating a robot to the die casting machine is that the die casting process is a single-position oriented process; that is, the process begins with molten metal and ends with a part always in the same location and identically oriented with respect to fixed points in space. This sequence can be compared to the operation of a trimming press, which is a two-position process:

- Random-oriented positions in containers ahead of the press
- Fixed-oriented position on the location of the trimming die

Present-day robots, on the other hand, can only be effective in transferring parts that have fixed or identical orientation relative to some reference point. Therefore, although a robot can operate or service a die casting machine, it cannot operate a trim press in the same manner as a trim operator.

However, if advantage is taken of the fact that the robot preserves position utility (part orientation) after it has unloaded a die casting machine, then a robot can be used as the connecting link between the tandem processes of casting and trimming. To remove the trimmed part, gates, runners and overflows, conveyors, or chutes are required.

An alternative to the cast-trim operation using a robot is the removal of the part from the casting machine to pipe- or screw-type conveyors that feature a buffer storage capacity because the robot has the capability of placing a shot on such a conveyor, which is usually at least 2 to 2.5 m (6 to 8 ft) above the floor level. With a buffer storage conveyor, the entire system is not shut down for trim press problems.

Automation of the process requires some method of trimming the casting. Three basic choices can be considered:

- Die casting and in-die degating
- Die casting machine and automatically loaded trim press
- Die casting automatically with separate trimming department

Separate trimming was a preferred method in the past; this method suited the manual casting shop. The current range of sophisticated machinery makes this choice less attractive. The first and second choices listed above result in decidedly higher productivity. Because of the limited production use of in-die degating, the second choice is preferred at this time. The production line concept also favors the integrated or interconnected trimming, where the orientation of the cast component is retained; thus, the casting can be easily transferred to a following machine for further machining operation.

The frequently heard objection regarding the breakdown of one unit stopping the entire line is less valid with the currently offered machinery than it was in the past. Solid-state technology, improved limit switches, and state-of-the-art hydraulics possess proven track records. The statistics of the early 1970s are no longer valid. One can state that the single largest cause of stoppages is die failures.

Hydraulically operated presses are used for cast-trim operations, and various designs (vertical, inclined, and horizontal) have been used to help overcome the problem of part, flash, overflow, and gate removal from the trimming location. The robot can accurately place the shot of castings on a location and can remove the part, gate, and flash off the location, but such a procedure almost always causes a delay in the casting cycle, forcing some economic tradeoffs. The use of a shuttle press can overcome this problem. Another advantage of the robot in a cast-trim operation is in meeting the requirements of the Occupational Safety and Health Administration Power Press Standard concerning no hands at the point of operation by elimination of the press operator.

Conveyors

The last item required for any installation is some method of removing castings from the trim press. A conveyor may not seem very important, but it can be the weak link in a trouble-free system. The conveyor must give workers time to perform their jobs and must supply adequate storage capacity. The most widely used conveyor is the belt type, which can withstand the most adverse conditions. However, pipe- or screw-type conveyors, which feature a buffer storage capacity, should not be ruled out. With a buffer storage conveyor, it is not necessary to have personnel always stationed at the end of the conveyor. This feature not only allows ease of administration but also permits secondary operations linked to the conveyor to be done at an incentive pace. Other conveyors would probably be flat-top or roller conveyors, depending on the specifics of the particular job.

Design Advantages

Some of the design advantages of die castings in general can be realized to a greater extent in zinc alloy die castings than in die castings of aluminum or copper alloys.

Section Thickness. The minimum section thickness for zinc alloy die castings depends on the surface area of the casting, the metal flow in the die, and the location of the gate. Even in large castings, a relatively thin section can be cast if it does not extend over the entire cross section of the casting and throttle the flow of the metal. Sections should be as thin as possible (consistent with castability and adequate strength and stiffness). Thin sections reduce metal costs and improve productivity as the castings solidify faster in the die, thus shortening the production cycle.

The sections of a die casting should always be as uniform as possible, with gradual transitions where the function demands differences in thickness. Cores can often be incorporated into a die to maintain uniformity of thickness. Metal saver cores are used to avoid a thick mass of metal, which would be difficult to cast and unnecessarily heavy.

For some applications, the mass of a zinc die casting is an advantage. Phonograph turntables have been die cast with weights up to 5 kg (11 lb). The heavy rim section contributes to the steadiness of rotation. On a smaller scale, zinc alloy flywheels have been used in computer tape decks and radio tuning mechanisms.

The die parting must usually be at the maximum diameter or section of the casting. The designer of a casting should visualize the casting in the die, shaping the part to facilitate its removal from the die and arranging for resulting flash to be in a convenient position for efficient removal. Die costs and flash removal costs are minimized when the parting is in one plane at right angles to die motion. By parting the casting on a face that must be machined, the flash can be removed simultaneously.

The ejector pins will leave small marks on casting surfaces unless special lugs are incorporated or the ejectors can act on the feed metal and overflow. The die should be designed so that these marks will not leave disfiguring blemishes on visible faces of the finished casting.

Ejector pin marks on most die castings can be raised or depressed by not more than 0.4 mm (0.015 in.). Ejector pin marks are surrounded by a flash of metal. If end use permits, ejector pin mark flash will not be removed but can be crushed or flattened. Complete removal of ejector pin marks and flash by machining or hand scraping operations should be specified only when requirements justify the expense involved in the additional operations.

Wall taper is normally between 1 and 2° per side. Shallow ribs, however, require more taper (5 to 10°), although small tapers are more acceptable for ribs in line with shrinkage, as for the spokes of a wheel.

Ribs. A thin section that requires reinforcement with ribs (rib thickness should not exceed the section thickness of the area it adjoins) may still provide lower overall weight than unribbed sections of greater thickness. The judicious placement of ribs often aids metal flow into thin sections, and ribs or beads that are discreetly placed at thin sections where trimming is required and where the casting is to be gated diminish the chances of warping and reduce trimming costs. Many die castings can be made as thin as metal stampings in shapes that cannot be duplicated in one-piece stampings and at lower tool cost.

Bosses or similar metal concentrations that are heavier than adjacent thin walls can result in unequal shrinkage. This sometimes gives rise to so-called shrink marks or shadow marks, which are actually shallow depressions on the face of the casting opposite the thickened section. Such marks may be unsightly, especially if the surface is to receive a lustrous finish. The effect can be minimized by making the variation in thickness as small and as gradual as conditions permit. Shadow marks can be masked by ribs or low-relief designs and seldom occur in sections over 2.5 mm (0.1 in.) thick. Ribs are often faired to bosses where load concentrations occur in service and help to distribute the load over a larger area of the casting.

Tapped bosses are stronger than threaded studs because external threads cause a notch effect under shock loads. Therefore, tapped bosses are always preferable and are sometimes as economical as threaded cast studs. However, tap and chip clearance must be allowed beyond the last thread of the hole, or a through hole must be provided. Holes to be tapped should usually be countersunk 2.5 mm (0.1 in.) larger than the thread for ease of tapping and assembly, especially when the hole is cored.

Undercuts. Because die costs are often greatly increased and casting rates decreased when undercuts are cast, the general rule is to design to avoid undercuts. If undercuts exist on the exterior of the die casting, slides or movable cores that substantially increase die costs are needed to eject the casting. The interior of a die casting or undercuts require the use of a loose piece that is withdrawn from the die with the casting and must be replaced in the die for subsequent castings.

When a loose piece is judged worthwhile, several are made to avoid delays. Zinc die castings can be used as loose pieces to form undercuts.

When the quantity of castings required is large, a costly and complex die may be fully justified by even a small net savings per casting. A comparatively uniform section can be obtained, despite complex shapes, by the judicious use of cores and slides that form undercuts. The metal saved can justify the additional die cost.

Fillets and Blends. Sharp corners are always a source of weakness and should be avoided by the use of fillets. Even the smallest fillets have an appreciable strengthening effect. A minimum radius of 0.4 mm (0.015 in.) is suggested in place of sharp corners, and larger radii are desirable when conditions permit their use. Fillets of 0.4 mm (0.015 in.) radius are barely noticeable even on outside edges, and a 0.8 mm (0.03 in.) radius is seldom evident except on close inspection.

It is common die casting practice to use a fillet having a minimum radius of 1.5 mm (0.06 in.) on inside edges. A slight radius on outside corners of castings reduces die cost and promotes the durability of any subsequent finish. Buffing or polishing is likely to cut through the finish at sharp outside edges, while organic finishes tend to thin out and give inadequate protection along sharp edges.

Plain Flat Surfaces. Large areas of plain flat surfaces should be avoided if very smooth finishes are required. Such surfaces lead to many rejections and increased costs. Broad surfaces should be slightly curved, crowned, or broken by beads, steps, or low relief so that they can be cast without imperfections, which will be magnified by glossy finishes. Such simple expedients mask these slight imperfections. Textured finishes can be applied to such surfaces in the die by photoengraving or other means.

Lettering. When die cast lettering, numerals, trademarks, diagrams, or instructions are required, they should be designed and placed to facilitate die construction and removal of the casting from the die. Normally, the designer should specify raised lettering because it is easier to cut a design into a die surface than to make a raised design on the surface. Depressed lettering on the casting is much more expensive and deteriorates with time because of erosion by the molten alloy.

When the engraving may not project above the surrounding surface of the casting, raised engraving on a panel sunk into the surface of the casting can generally be used. Engraving is preferably done on surfaces parallel or nearly parallel to the die parting. It should never constitute an undercut, which could interfere with the ejection of the casting from the die. In many designs, engraving is effectively used for scale or graduation markings. When the engraving is depressed in the casting, the recesses are often filled with paint or are wiped in to provide contrast with surrounding areas.

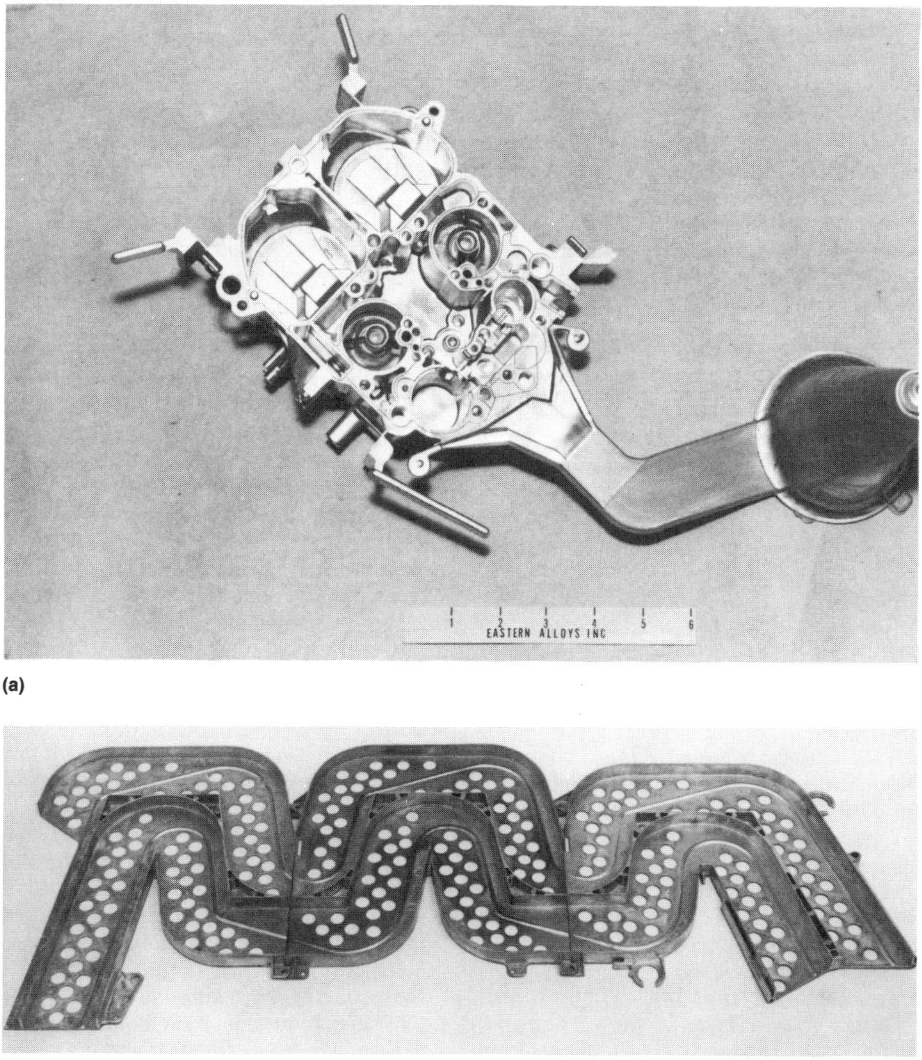

(b)

Fig. 2 Applications of zinc die castings. (a) Automotive carburetor bodies; gating is still attached. Courtesy of Eastern Alloys, Inc. (b) Vending machine feed track for beverage cans fabricated from three die castings. See also Fig. 3.

Bending and Forming. The ductility of zinc makes it possible to incorporate integral rivets, to shape integral flanges to curving contours, to bend hollow arms, to spin out undercuts, to form projections, or to twist parts of the casting through 90° or more. It is possible to use a flat parting that provides parallel bosses, cored holes, or studs at right angles to the parting and then to form the casting so that the axes of these elements are no longer parallel. Thin plates with cast bosses or holes at right angles to the surface require much less expensive dies than if cast to a curved shape.

Inserts are generally used for one or both of the following reasons:

- To provide greater strength, hardness, wear resistance, ductility, flexibility, or some other property not possessed, at least in the same degree, by the die casting alloy

- To provide shapes of parts or passages that cannot be cored or cast or are less expensive or better as inserts

Inserts can usually be cast in place, but there are many cases in which they are applied after casting in holes cored for the purpose. The object of casting the insert in place is either to anchor it securely or to locate it in a position where it could not be placed after casting.

When inserts are designed for casting in place, they should be provided with knurling, holes, or grooves to ensure firm anchorage. Provision should be made for a sufficient thickness of the casting alloy around the insert to give the required support.

In the case of inserted studs, the thread should end at least 2.5 mm (0.1 in.) from the casting; otherwise, it may be filled by molten metal. Because of possible variations in the diameter of inserts, a shoulder or other sealing surface should be provided between the end of the thread and the casting to prevent any flash around the insert from entering the threads. In service, a washer resting against such a shoulder will avoid any tendency for tightening of the nut to pull the stud from the casting.

Machining. Zinc die castings are manufactured to very fine tolerances as-cast, and if any machining is necessary, the cuts required are usually light. A minimum machining allowance of 0.25 mm (0.01 in.) is recommended, along with a maximum of 0.5 to 1 mm (0.02 to 0.4 in.). Machining operations are made easier by the good machining qualities of zinc die casting alloys.

Design drawings should show where machining is to be done and should indicate how much metal is to be removed in machining, unless this is left to the judgment of the die caster. Surfaces to be machined should be of minimum area, consistent with other requirements, and when possible should be positioned to simplify machining. If possible, location should be from the fixed die half. For example, flats can often be trued by simple grinding if the surfaces to be ground are accessible. Placing flats (such as boss faces) all in one plane expedites grinding. Such surfaces should be slightly above surrounding areas that do not require machining.

When the number of castings to be produced from a die is small, the die cost must be kept low. In such cases, it is sometimes preferable to avoid expensive machining on the die and to perform additional machining on the castings.

Small holes (≤3mm, or 0.12 in.) in thin sections are often drilled or punched in preference to coring because the flash from cored holes must normally be removed by such operations. It is almost as quick to drill or punch the full depth of the hole as to remove flash. The drilling operation can be simplified by casting a start for the drill.

Minimizing Trimming Costs. Castings should be designed to minimize trimming costs. Flash occurs at die partings, and its removal usually constitutes a considerable factor in the cost of the casting. This is one phase of machining that is practically unavoidable, but the cost can be minimized by positioning the parting so that flash removal is easily and quickly accomplished. In the production of very long runs of small, thin castings, carefully made dies can be economically employed to produce flash-free castings.

Flash is commonly removed from larger castings by a trim die through which the casting is forced by a press. If the parting is in a single plane, preferably, at right angles to the motion of the die, the flash is easily sheared, but if the parting is not in a single plane, greater cost is incurred in flash removal.

When the flash occurs at a flange or bead, rather than in a recess or on a flat surface, flash removal is facilitated. Flash can often be designed to occur on a surface or edge where machining is required, eliminating a separate flash removal operation.

Coring of the internal form of a die casting can vastly increase the complexity and cost of tooling and should therefore be avoided if possible. This is not to say that coring is to be completely eliminated, because without it many of the advantages of die casting cannot be realized. Coring takes two different forms. On the one hand, there is the coring of internal shapes, while on the other there is the coring of holes.

Holes cast in die castings produced by the use of cores require taper. Very small cored holes should be avoided where possible. The cores tend to overheat and are easily broken or bent. It is normally more economical to drill small holes. Drilling out small cored holes can be troublesome because any misalignment of drill and hole can break the drill.

Gears and Components With Irregular Outlines. When a die casting is removed from the die, it often has a surrounding flash, which is usually removed by a trimming pool. If the component has an irregular outline, the production of this tool can be costly, and the problem can be alleviated if it is possible to incorporate a shroud following a regular form. This has the additional advantage of strengthening the casting.

Means of Attachment. Studs formed as an integral part of the casting usually cost less than inserted studs and in general constitute a highly economical means of fastening a casting to a mating part. Production rates are seriously impaired when separate inserts must be placed in hot dies before each shot.

Integral studs should not be so small in diameter as to be fragile or easily damaged in handling; if such studs are made at least 6 mm (0.24 in.) in diameter on large or medium-size castings, little trouble is experienced. With small, light castings, proportionately smaller studs can be safely used.

All studs should have a liberal fillet where they join the body of the casting. When the radius at the base of a stud interferes with a square edge in the hole of the mating part, the radius can be formed in a recess at the base of the stud.

Threads. The ability to cast threads is a major advantage of the die casting process. Cast threads should be specified wherever

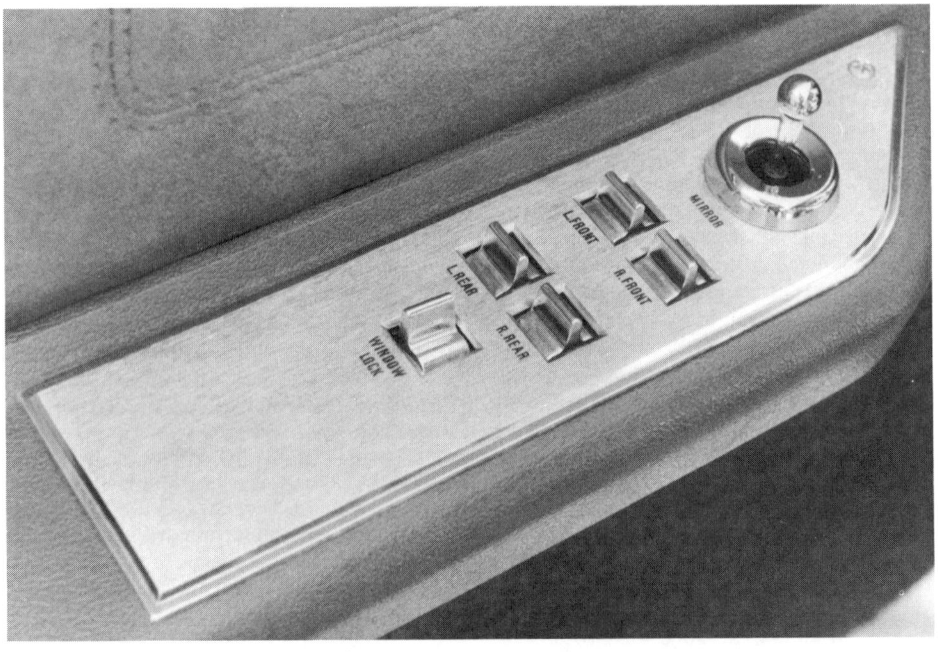

Fig. 3 Zinc die cast automotive arm rest control panel with a chromium plated surface for decorative and corrosion-resistant purposes

their use reduces cost over that for cut threads.

Most external threads can be cast. It is common practice to cast coarse external threads over 20 mm (0.8 in.) in diameter if they are located at a die parting. Threads at a parting usually have flash removed by the trim die, but chasing is sometimes done to produce a truer thread. The weight of a component with a cast screw thread is important. If it is heavy, mechanical bruising may damage fine threads. Pitch errors are likely to be greater than with cut or rolled threads. It is advisable to limit the length of engagment to one-half a diameter. Tooling costs will be higher than for unthreaded components because of the accuracy necessary in die sinking and the increased die maintenance.

Most internal threads are more expensive cast than cut. Cast internal threads are occasionally useful for very steep pitches, and whatever the pitch, the thread can be carried down to a shoulder or to the bottom of a blind hole. All holes requiring fine threads are tapped, and cast interior threads under 20 mm (0.8 in.) in diameter are rarely economical.

Soldering, Welding, and Use of Adhesives. Zinc alloy die castings are not easy to solder, because of the aluminum content of the alloy. Furthermore, the soldered joint may be subject to intergranular corrosion arising from the lead and tin of the solder diffusing into the alloy. These disadvantages can be overcome by plating the casting with a metal such as copper and soldering onto the plated surface. Care must be taken to ensure that no soldered die castings (plated or nonplated) are remelted with other scrap die castings. A very small amount of solder can spoil a large batch of alloy when the two are melted together.

The repair welding of zinc alloy die castings is recommended only when the damaged casting cannot be replaced. In such cases, and for emergency repairs, a procedure has been developed for building up the damaged part by welding, using an alloy rod of the same composition as the casting with the gentle heat of a slightly reducing oxyacetylene flame. Stud welding has been used, but it is not satisfactory, because the alloy adjacent to the stud is weakened and may break.

When appropriate, zinc die castings can be bonded with a range of modern high-strength adhesives, particularly those based on epoxy and phenolic resins. The final choice of adhesive for any application should be discussed with the adhesive supplier.

Finishes for Die Castings. For many applications, surface finish is not important, but where necessary, a wide range of applied finishes can be used. Finishes for die castings can be functional, decorative, or both. Almost any desired texture, such as simulated cloth, leather, or woodgrain, can be cast simply and economically.

Chromium plating is the most widely used finish for zinc alloy die castings. Normally, to obtain a high-quality bright chromium finish on castings, they must be buffed and polished before plating. However, a combination of modern zinc die casting and plating techniques substantially reduces or, in some cases, eliminates the need for mechanical polishing.

If a finish is decorative rather than protective, this does not imply that it requires only casual consideration; electroplating must be carried out to an exacting specification if it is to be satisfactory. For example, an unplated die cast door handle on an automobile would suffer little corrosion

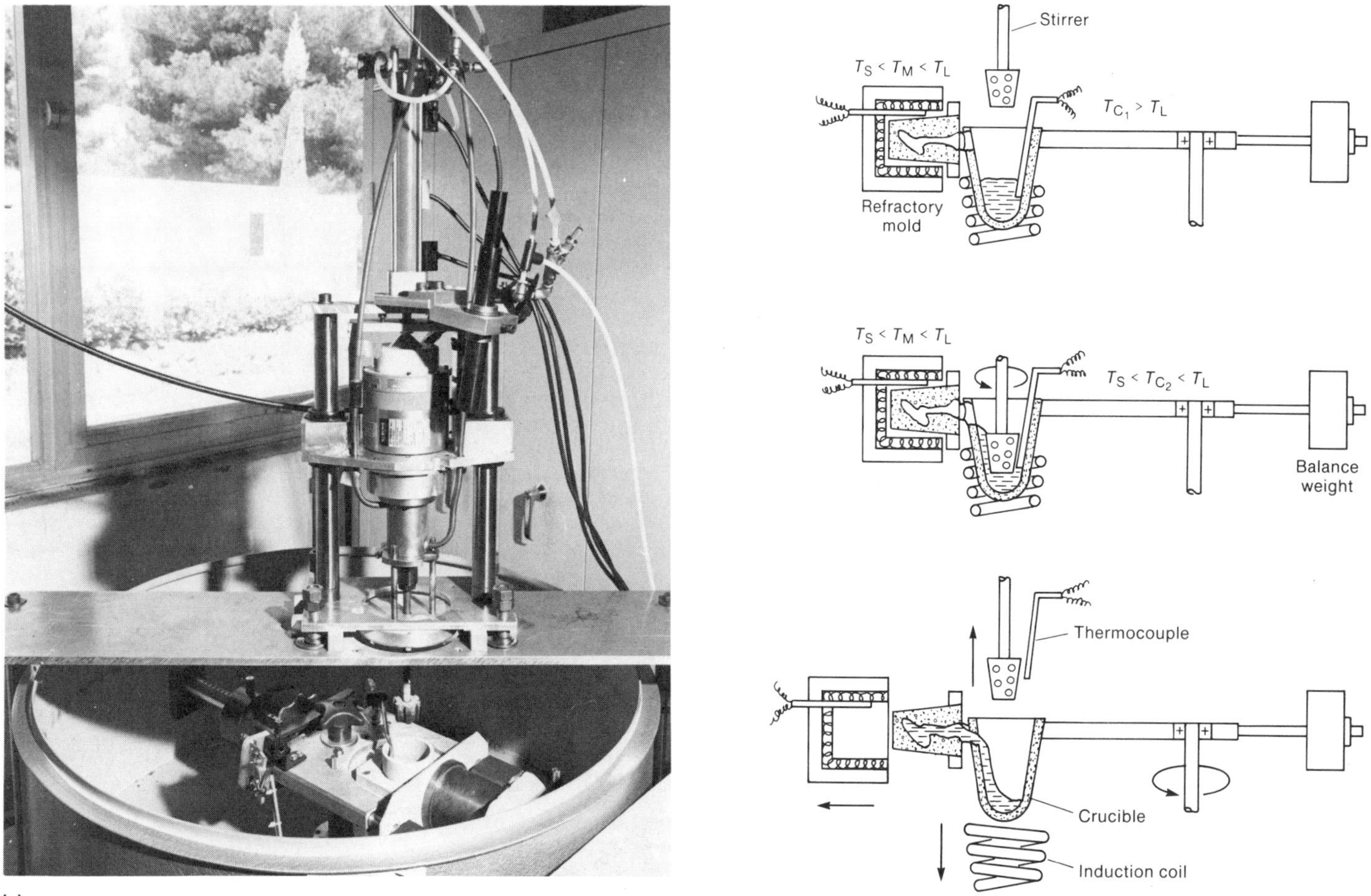

(a) (b)

Fig. 4 The Gircast process apparatus (a) and schematic of process (b). Source: J. Collot, Ecole Nationale Supérieare des Mines de Paris (apart from discoloration) in service, while a poorly plated one would soon exhibit blistering and pitting. Therefore, plating is normally required only for appearance, but when it is used, it must be of sufficient quality to provide long-term durability.

Polishing. Where polishing is necessary, the following points must be kept in mind. Sharp corners and edges are difficult to polish without damaging their outlines; a radius of 0.8 mm (0.03 in.) will help polishing and plating. Large flat surfaces are difficult to polish evenly, but undulations will be less noticeable if the surface has a slight crown (1.5 mm/100 mm, or 0.015 in./in.). Small recesses and acute angles are impossible to reach with a polishing wheel.

Chromate Finishes. Chemical treatments prevent the growth of white corrosion products on the surface when the castings are exposed to stagnant moisture, such as condensation. This treatment results in a dull green-yellow finish.

Painting. Most types of paint can be successfully applied, but surface preparation and pretreatment are very important.

Clear Lacquers. Many attempts have been made to formulate clear lacquers to preserve the bright finish of polished zinc. The recently developed nonyellowing polyurethane lacquers containing rubeanic acid show promise for this application. Acrylic lacquers are also being used, but they are less resistant to mechanical damage.

Electropainting (Electrophoretic Painting). The electropainting of die castings is well established. The process requires specially formulated water-base soluble resin paints, and this imposes some limitations on the colored pigments that can be used. The advantage of electropainting, in which the parts to be painted are made cathodic relative to the steel tank containing the paint, is that a very dense and uniform paint film can be applied to a complex surface, so that one coat of electropaint can replace two coats of conventional paint. It is possible to achieve a high gloss with electrodeposited paints, but color matching with other paints still presents problems, especially over long production runs. Only one coat can be electrodeposited because the dry paint is an insulator. Some paint manufacturers recommend giving zinc die castings a chromate treatment before electropainting.

Plastic Coatings. Epoxy powder coatings are being increasingly used for zinc die castings as an alternative to heat-curing paints. The process is economical because the only labor required is loading and unloading the conveyor that carries the castings past the spray gun and through the curing oven. Matte or glossy coatings are formed about 0.04 mm (1.5 mils) thick.

Plated Finishes. Where zinc alloy die castings are correctly designed, properly produced and carefully prepared plated finishes are very satisfactory. They have excellent decorative value and, given a coating of sufficient thickness and good adherence, a very long life. Copper and nickel can both be plated directly onto zinc alloy, but it is nickel applied over a preliminary deposit of copper that is most commonly used to enhance the corrosion resistance of the casting, coupled with a chromium finish.

For special effects, die castings plated with copper can be treated to simulate an-

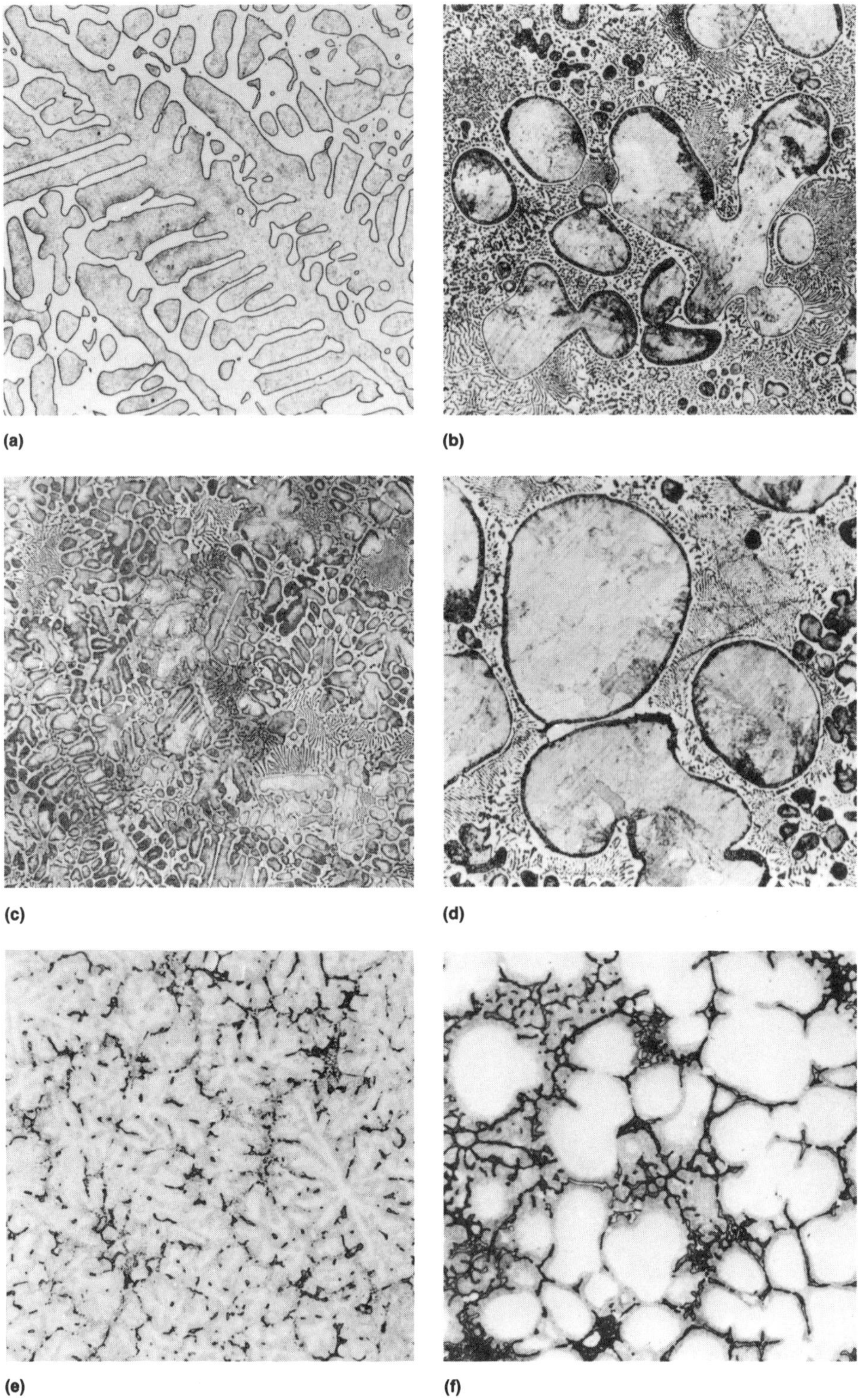

Fig. 5 Comparison of microstructures of conventionally cast (gravity cast into a permanent mold and shown on the left-hand side) and semisolid cast (into a permanent mold and shown on the right-hand side) ZA alloys. (a) and (b) Alloy ZA-8. (c) and (d) Alloy ZA-12. (e) and (f) Alloy ZA-27. All 56×. Courtesy of S. Murphy, Aston University

tique bronze. Nickel can also be wire brushed or polished with a coarse abrasive to give various attractive satin finishes before chromium plating. Black chromium is another finish with decorative possibilities.

Vacuum Metallizing. Die castings that need a bright finish but are unlikely to be damaged by rubbing or knocks can be vacuum metallized. Color effects, such as those of brass, can be created by immersing parts in a dye solution that tints the protective lacquer very evenly.

Anodizing. A process for anodizing zinc and zinc alloy die castings has been developed that provides excellent resistance to salt water and detergent solutions and good resistance to abrasion. The finish is available only in matte, light green, gray, and brownish tints.

Applications for Zinc Die Castings

The automotive industry is the largest user of zinc die castings. Some of the important mechanical components made as zinc alloy die castings are carburetor bodies (Fig. 2a), bodies for fuel pumps, windshield wiper parts, control panels (Fig. 3), grilles, horns, and parts for hydraulic brakes. Structural and decorative zinc alloy castings include grilles for radios and radiators, lamp and instrument bezels, steering wheel hubs, interior and exterior hardware, instrument panels, and body moldings.

Other applications include the electrical, electronic, and appliance industries, business machines and other light machines of all types (including beverage vending machines, Fig. 2b), and tools. Building hardware, padlocks, and toys and novelties are major areas of application for zinc die castings.

Other Casting Processes for Zinc Alloys

Although the vast majority of zinc castings are produced by hot chamber die casting, several other casting processes are also employed. These include sand casting, permanent mold casting, and plaster casting. Also beginning to be applied are such emerging processes as squeeze casting and semisolid casting. Zinc-matrix metal-matrix composites (MMCs) have also been produced by various foundry methods.

Sand Casting. The ZA alloys, especially ZA-12 and ZA-27, are being increasingly used in gravity sand casting operations. The wide freezing range of the ZA-27 alloy (~109 °C, or 200 °F) means that control of solidification is especially important for this alloy. The use of chills or patterns that promote directional solidification is recommended.

Permanent mold casting is done using both metallic and machined graphite molds. Cast iron or steel is most commonly used for metallic permanent molds. The use of graphite molds permits as-cast tolerances similar to those obtained in die casting. Machining time is reduced or eliminated, making the graphite process attractive for intermediate production volumes (500 to 20 000 parts per year). More information on permanent mold casting is available in the article "Permanent Mold Casting" in this Volume.

Squeeze casting is a process in which the liquid metal solidifies under pressure in closed dies held together by a hydraulic press. Essentially, the metal is forged to near-net or net shape while it solidifies. Metal-matrix composites are manufactured by squeeze casting by infiltrating a porous ceramic preform with the liquid metal under pressure. This process has been employed to cast MMCs with ZA alloy matrices and silicon carbide or alumina chopped-fiber reinforcements. More information on this process is available in the section "Squeeze Casting" in the article "New and Emerging Processes" in this Volume.

Semisolid Casting. Zinc casting alloys are also being processed by semisolid casting. In this process, alloys are poured with negative superheat (that is, the pouring temperature is between the liquidus and the solidus). Vigorous mechanical agitation of the cooling metal melt prevents the formation of normal dendrites and maintains the solid fraction of the melt in the form of rounded, primary particles (see the section "Semisolid Metal Casting and Forging" of the article "New and Emerging Processes" in this Volume). One semisolid metal processing method that has been applied to ZA alloys is the Gircast process. As shown in Fig. 4, this stir casting process involves three major steps:

- The alloy temperature is elevated to T_{C_1}, which is higher than the liquidus temperature $T_L (T_{C_1} > T_L)$
- The agitator is lowered, and the paddles are rotated to stir the metal at a temperature T_{C_1}. The crucible is then cooled to a temperature T_{C_2} intermediate between the solidus and the liquidus temperature $(T_S < T_{C_2} < T_L)$
- The following operations are then performed simultaneously: (1) the agitator is stopped, (2) the paddles are raised, and (3) the induction heating means are retracted downward to release the crucible. The thermocouple is retracted upward. As soon as operations 1 to 3 have been executed, the centrifuged casting motor is started

Figure 5 compares the microstructures of conventionally cast and semisolid cast ZA alloys.

Magnesium and Magnesium Alloys

Henry Proffitt, Haley Industries Ltd., Canada

MAGNESIUM ALLOY CASTINGS can be produced by nearly all of the conventional casting methods, namely, sand, permanent, and semipermanent mold and shell, investment, and die casting. The choice of a casting method for a particular part depends upon factors such as the configuration of the proposed design, the application, the properties required, the total number of castings required, and the properties of the alloy. The discussion here will focus on the variety of alloys, furnaces, and associated melting equipment, and on the casting methods available for manufacturing magnesium castings.

Magnesium Alloys

A large range of magnesium-base alloys is available for the production of castings. Sand castings (and investment castings) can be made in all of the available alloys (see the article "Sand Molding" in this Volume). However, not all alloys are suitable for production by all casting methods. For example, alloys normally cast by the permold process are somewhat limited in number, and those used in the die casting process are even more restricted.

The method of codification used in North America to designate magnesium alloy castings is taken from ASTM Standard Practice B 275 (Table 1). It gives an immediate, approximate idea of the chemical composition of an alloy, with letters representing the main constituents and figures representing the percentages of these constituents.

As an example, consider the three alloys AZ91A, AZ91B, and AZ91C. In these designations:

- A represents aluminum, the alloying element specified in the greatest amount
- Z represents zinc, the alloying element specified in the second greatest amount
- 9 indicates that the rounded mean aluminum percentage lies between 8.6 and 9.4
- 1 signifies that the rounded mean of the zinc lies between 0.6 and 1.4
- A as the final letter in the first example indicates that this is the first alloy whose composition qualified assignment of the designation AZ91
- The final serial letters B and C in the second and third examples signify alloys subsequently developed whose specified compositions differ slightly from the first and from one another but do not differ sufficiently to effect a change in the basic designation.

The nominal compositions of the alloys used for sand, investment, and permold castings are shown in Table 2, and those for die castings are shown in Table 3.

Although alloys used for the die casting process are somewhat limited in number, more of the aluminum-zinc-manganese alloys (for example, the AZ91 type, particularly the high-purity grade) are now being used. A large, growing application for die castings is the automotive market.

Magnesium castings of all types have found use in many commercial applications, especially where their lightness and rigidity are a major advantage, such as for chain saw bodies, computer components, camera bodies, and certain portable tools and equipment. Magnesium alloy sand castings are used extensively in aerospace components.

Sand Casting. Magnesium alloy sand castings are used in aerospace applications because they offer a clear weight advantage over aluminum and other materials. A considerable amount of research and development on these alloys has resulted in some spectacular improvements in general properties compared with the earlier AZ types (Ref 1).

Although there has been, and still is, a large volume of castings for aerospace applications being produced in the older, conventional AZ-type alloys, the trend is toward the production of a greater proportion of aerospace castings in the newer zirconium types.

Although the magnesium-aluminum and magnesium-aluminum-zinc alloys are generally easy to cast, they are limited in certain respects. They exhibit microshrinkage when sand cast, and they are not suitable for applications in which temperatures of over 95 °C (200 °F) are experienced. The magnesium-rare earth-zirconium alloys were developed to overcome these limitations. Sand castings in the EZ33A alloy do in fact show excellent pressure tightness. The greater tendency of

Table 1 Standard three-part ASTM system of alloy designations for magnesium alloys

First part	Second part	Third part
Indicates the two principal alloying elements	Indicates the amounts of the two principal elements	Distinguishes between different alloys with the same percentages of the two principal alloying elements
Consists of two code letters representing the two main alloying elements arranged in order of decreasing percentage (or alphabetically if percentages are equal)	Consists of two numbers corresponding to rounded-off percentages of the two main alloying elements and arranged in same order as alloy designations in first part	Consists of a letter of the alphabet assigned in order as compositions become standard
A–Aluminum E–Rare Earth H–Thorium K–Zirconium M–Manganese Q–Silver S–Silicon T–Tin Z–Zinc	Whole numbers	A–First compositions, registered ASTM B–Second compositions, registered ASTM C–Third compositions, registered ASTM D–High-purity, registered ASTM E–High corrosion resistant, registered ASTM X1–Not registered with ASTM

Table 2 Nominal compositions of magnesium casting alloys for sand, investment, and permanent mold castings

Alloy	Composition, % Al	Zn	Mn	Rare earths	Th	Y	Zr
AM100A	10.0	...	0.1 min	...	...	...	...
AZ63A	6.0	3.0	0.15	...	...	...	...
AZ81A	8.0	0.7	0.13	...	...	...	...
AZ91C	9.0	0.7	0.13	...	...	...	...
AZ91E	9.0	2.0	0.10	...	...	...	...
AZ92A	9.0	2.0	0.10	...	...	...	...
EZ33A	...	2.7	...	3.3	...	...	0.60
HK31A	...	...	...	...	3.3	...	0.70
HZ32A	...	2.1	...	...	3.3	...	0.70
QE22A(a)	...	...	...	2.0	...	...	0.60
EQ21A(a,b)	...	...	...	2.0	...	...	0.60
ZE41A	...	4.2	...	1.2	...	...	0.70
ZE63A	...	5.7	...	2.5	...	...	0.70
ZH62A	...	5.7	...	...	1.8	...	0.70
ZK51A	...	4.6	...	...	...	...	0.70
ZK61A	...	6.0	...	...	...	...	0.70
WE54A	...	...	...	3.50(c)	...	5.25	0.50

(a) These alloys also contain silver, that is, 2.5% in QE22A and 1.5% in EQ21A. (b) EQ21A also contains 0.10% Cu. (c) Comprising 1.75% other heavy rare earths in addition to the 1.75% Nd present

the zirconium-containing alloys to oxidize is overcome by the use of specially developed melting processes.

The two magnesium-zinc-zirconium alloys originally developed, ZK51A and ZK61A, exhibit high mechanical properties, but suffer from hot-shortness cracking and are nonweldable.

For normal, fairly moderate temperature applications (up to 160 °C, or 320 °F), the two alloys ZE41A and EZ33A are finding the greatest use. They are very castable and can be used to make very satisfactory castings of considerable complexity. In addition, they have the advantage of requiring only a T5 heat treatment (that is, precipitation treatment).

When a demand arose in some aerospace engine applications for the retention of high mechanical properties at higher elevated temperatures (up to 205 °C, or 400 °F), thorium was substituted for the rare earth metal content in alloys of the ZE and EZ type, giving rise to the alloys of the type ZH62A and HZ32. Not only were there substantial improvements in mechanical properties at elevated temperatures in these alloys, but good castability and welding characteristics also were retained.

The thorium-containing alloys, however, exhibited a greater tendency for oxidation, requiring greater care in meltdown and pouring. The mildly radioactive nature of the dross and sludges from processing these alloys and the disposal of these byproducts are associated problems with this alloy group.

A further development aimed at improving both room-temperature and elevated-temperature mechanical properties produced an alloy designated QE22A. In this alloy, silver replaced some of the zinc, and the high mechanical properties were obtained by grain-refinement with zirconium and by a heat treatment to the full T6 condition (that is, solution heat treated, quenched H_2O, and precipitation aged). However, problems were experienced with both of these alloys. The use of thorium has become increasingly unpopular environmentally, and the price of silver has become very unstable in recent years. Hence, there has been a considerable amount of research and development work on alternative alloy types.

The most recent alloy emerging from this research was an alloy containing about 5.0% Y in combination with other rare earth metals (that is, WE54A), replacing both thorium and silver (Ref 2). This alloy has better elevated-temperature properties and a corrosion resistance almost as good as the high-purity magnesium-aluminum-zinc types (AZ91C). The alloys used for investment casting are very similar to those used for the sand casting process.

Permanent Mold Casting. In general, the alloys that are normally sand cast are also suitable for permanent mold casting (see the article "Permanent Mold Casting" in this Volume). The exception to this are the alloys of the magnesium-zinc-zirconium type (for example, AZ51 and ZK61A), which exhibit strong hot-shortness tendencies and are consequently unsuitable for processing by this method.

Die Casting. The alloys from which die castings are normally made are mainly of the magnesium-aluminum-zinc type, for example, AZ91 (see the article "Die Casting" in this Volume). Two versions of this alloy from which die castings have been made for many years are AZ91A and AZ91B. The only difference between these two versions is the higher allowable copper impurity in AZ91B.

More recent development work on this alloy type has produced the high-purity version of the alloy in which the nickel, iron, and copper impurity levels are very low and the iron-to-manganese ratio in the alloy is strictly controlled. This high-purity alloy shows a much higher corrosion resistance than the earlier grades.

General Applications

The most important feature of magnesium castings, which gives rise to their preferred use compared with other metals and materials, is their light weight. Because of this, magnesium castings have found considerable use since World War II in aircraft and aerospace applications, both military and commercial. More recently, as a result of a general requirement for lighter weight automobiles to conserve energy, there has been a growing use of magnesium in the automotive field, principally as die castings.

Magnesium, however, has other important casting advantages over other metals:

- It is an abundantly available metal
- It is easier to machine than aluminum
- It can be machined much faster than aluminum, preferably dry

In the die casting process, it can be cast up to four times faster than aluminum. Die lives are considerably longer than with the aluminum alloys, because much less welding onto the die surfaces takes place. When protected correctly, particularly against galvanic effects, it behaves in a very satisfactory manner. Modern casting methods and the application of protective coatings currently available ensure long life for well-designed components.

Today's state-of-the-art technology makes it possible to produce parts of considerable complexity having thin-wall sections. The end product has a high degree of stability as well as being light in weight.

Table 3 Nominal compositions of magnesium casting alloys for die castings

Alloy	Alloying element Mg	Al	Mn	Si	Zn
AM60A	rem	6.0	0.13 min	...	...
AS41A	rem	4.25	0.35	1.0	...
AZ91A	rem	9.0	0.13 min	...	0.7
AZ91B	rem	9.0	0.13 min	...	0.7
AZ91D (HP)(b)	rem	9.0	(a)	...	0.7

(a) Manganese content to be dependent upon iron content. (b) The proposed alloy to have very low limits for iron, nickel, and copper. HP, high purity

Melting Furnaces and Auxiliary Pouring Equipment

Furnaces for melting and holding molten magnesium casting alloys are generally the indirectly heated crucible type, of a design similar to those employed for the aluminum casting alloys. The different chemical and physical properties of the magnesium alloys in comparison to aluminum alloys, however, necessitate the use of different crucible materials and refractory linings and the modification of process equipment design.

When magnesium becomes molten, it tends to oxidize and burn, unless care is taken to protect the molten metal surface against oxidation. Molten magnesium alloys behave differently from aluminum alloys, which tend to form a continuous, impervious oxide skin on the molten bath, limiting further oxidation. Magnesium alloys, on the other hand, form a loose, permeable oxide coating on the molten metal surface. This allows oxygen to pass through and support burning below the oxide at the surface. Protection of the molten alloy using either a flux or a protective gas cover to exclude oxygen is therefore necessary.

Molten magnesium does not attack iron in the same way as molten aluminum, and the metal can therefore be melted and held at temperature in crucibles fabricated from ferrous materials. It is common practice, therefore, especially with larger castings, to melt and process the molten magnesium alloy and to pour the casting from the same steel crucible.

Figure 1 shows the cross-sectional design of a typical fuel-fired stationary crucible furnace of the bale-out type, from which metal for small castings can be hand poured using ladles (see the section "Reverberatory Furnaces and Crucible Furnaces" in the article "Melting Furnaces" in this Volume). This use of metallic crucibles allows the crucible to be supported from the top by means of a flange, leaving a space below the crucible. Not only is this a distinct advantage in the transfer of heat to the crucible charge, it also ensures that there is room for easy removal of any detached scale that might form on the outer surface of the crucible during the melting operation. The furnace chamber has a base that slopes toward a cleanout door.

A progressive thinning of the crucible walls can occur, which may tend to be localized in fuel-fired furnaces because of flame impingement. There is the possibility of molten-metal leakage if the wall thickness is not checked regularly.

With scale, there is also the possibility of reaction between the iron oxide and the molten magnesium, which can be explosive in character. Furnace bottoms must there-

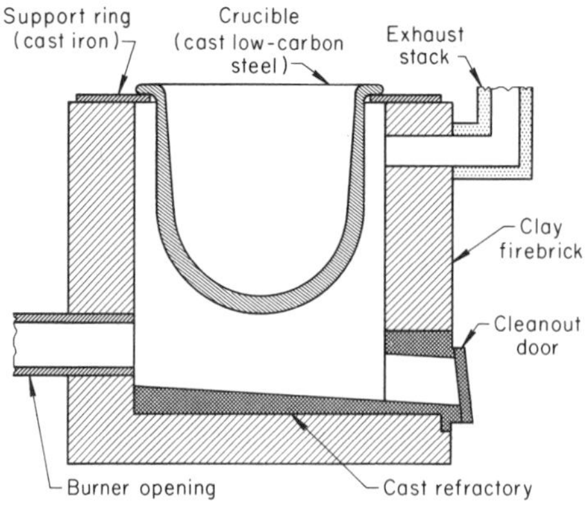

Fig. 1 Cross section of a stationary fuel-fired furnace for open-crucible melting of magnesium alloys

fore be kept free from scale buildup. It is important also to have a run-out pan capable of taking the full crucible content in the event of leakage.

More recently, and especially in cases where it is difficult to check on scale formation, it is possible to use steel crucibles that are clad with a nickel-chrome alloy on the outside heating surface in order to eliminate scaling and still not detract from the heating efficiency of the furnace.

The furnace lining refractories are important, because molten magnesium reacts violently with some refractories. High-alumina refractories and high-density "super-duty" firebrick of a typical 57% Si and 43% Al composition have been found to give satisfactory results.

The cleanout door on fuel-fired furnaces can be easily opened for its intended purpose. With electric-resistance crucible furnaces it is common practice to seal the cleanout door with a sheet of low melting point material, such as zinc, which will not act as a barrier to molten magnesium alloy in the event of a run-out, but will prevent the "chimney" effect, which can accelerate oxidation of the crucible.

Burning, which can occur at or above the melting point of the alloy, is prevented by the use of either a flux sprinkled on the molten metal or a suitable fluxless technique using a gas mix containing 1% SF_6. Both procedures will be described in detail later in this article. Increasingly stringent environmental controls in the foundry have rendered the older sulfur-dioxide domed bale-out furnace unacceptable.

The type and size of furnace used is largely dependent on the type of casting operation. A small jobbing foundry, carrying out a batch-type operation in a wide range of different alloys, normally uses the lift-out crucible technique. Larger-scale operations, typically operating on a more restricted range of casting alloys, may employ a larger bulk melting unit from which the molten alloy for the casting operation is distributed to holding crucible furnaces. In the crucible furnaces, metal treatments are carried out, and the metal is either poured directly from the crucible or hand ladled to the casting molds.

In cold-chamber die casting operations, the supply of molten alloy to the machine is maintained by hand ladling or by automated means. The hot-chamber process is becoming more popular. A cold-chamber machine differs from a hot-chamber machine in that the cylinder and injection plunger are not submerged in the molten metal. Because the injection mechanism in a hot-chamber machine is submerged in the furnace bath, the operation is faster and the plunger pressure can be adjusted to force the molten metal into the finest die detail. Electromagnetic pumps can also be used for this purpose. The overriding consideration in metal transfer is that the metal must be transferred in as nonturbulent a manner as possible in order to avoid oxidation, which can give rise to oxide skins and inclusions in the final casting. Excessive oxidation has ruled against the use of direct-fired reverberatory furnaces similar to those used quite satisfactorily for aluminum alloys.

Crucibles. The indirect heating crucible method of melting is of comparatively low thermal efficiency. Electric coreless induction furnaces, although much higher in initial capital cost, have lower running costs and occupy less floor space than fuel-fired units (see the section "Induction Furnaces" in the article "Melting Furnaces" in this Volume).

Crucibles range in size from about 30 to 910 kg (60 to 2000 lb). In the lower range, they may be constructed as steel weldments. The steels normally used are of low carbon content, that is, less than 0.12% C. Because of the extremely adverse effect of nickel and copper on the corrosion resistance of magnesium alloys, these two elements in the steel must be restricted to less than 0.10% each.

Magnesium melting operations, particularly if a flux melting procedure is used, normally result in the formation of a sludge with a comparatively low thermal conductivity at the bottom of the crucible. If this material is not periodically removed, overheating of the crucible can result in this area, accompanied by excessive scaling of the crucible. Excessive oxide buildup on the crucible walls can have the same effect. Records of the number of charges being melted should be kept for each crucible as a routine safety measure.

Less buildup is normally experienced with the fluxless method of melting. It is a very desirable procedure, however, to periodically withdraw the crucibles from use and allow them to soak, filled with water, to remove all buildup.

Pouring Ladles. For smaller castings, hand ladling can be conventionally used, with pouring ladles taking molten alloy from

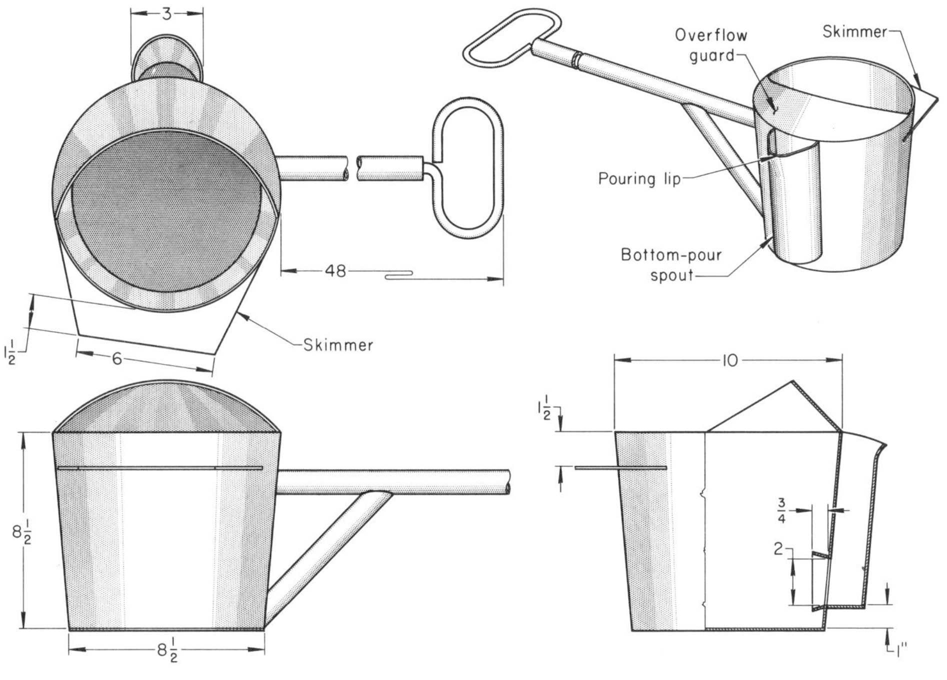

Fig. 2 Construction details of a typical ladle used for pouring magnesium alloys. Dimensions given in inches

a bale-out-type furnace. Pouring ladles can be bucket shaped for the slightly larger range of castings and hemispheric shaped for the smaller ones. Both types of ladles however, should be constructed from low-carbon, low-nickel steel and be about 12 gage (2.67 mm, or 0.105 in., thick). A typical design of a bucket type ladle is shown in Fig. 2. Essential design features include an overflow guard and a bottom-pour spout to avoid the possibility of the pour being flux contaminated.

Other essential items of metal handling equipment include sludge removal ladles, sludge pans to contain this material, stirrers, puddling tools, and skimmers. All of these pieces of equipment should have the same steel composition as the crucible.

Thermocouples. Accurate temperature control is critical to the processing of magnesium alloys. Iron-constantan- or chromel-alumel-type thermocouples are recommended. There should be a permanent type of installation such that temperature determinations can be made at appropriate stages in the melting and metal treatment process. Light-gage thermocouples in thermocouple protection tubes of either mild steel or nickel-free stainless steel should be used.

Melting Procedures and Process Parameters

There are basically two main systems, flux and fluxless, for the melting and pouring of magnesium alloys. Each of them, if done correctly, is perfectly capable of producing good metal for the casting process. Many of the precautions taken apply equally to both methods.

The Flux Process. The basic requirement in melting magnesium is to exclude oxygen from the molten magnesium as it melts. Because early attempts to develop a system using gaseous protection were not completely satisfactory, successful melting procedures became possible only when flux methods were developed. In the course of time, suitable fluxes for handling both the magnesium-aluminum-zinc-manganese and the magnesium-zirconium-containing alloys emerged, and corresponding techniques for producing clean, flux-free castings were developed. These procedures have been successfully used for several decades (Ref 3). A typical flux-melting procedure would be for the crucible with a small quantity of flux (about 1½% of charge weight) placed in the bottom, to be preheated to dull red heat.

The metal charge to be loaded must be clean, dry, and free from oil, oxide, sand, and corrosion. There should be no foreign metals present.

Contamination by even tiny pieces of sand or debris from other alloys must be prevented by close control of melt parameters. No oxide-contaminated material should be allowed to enter the melt charge. All materials that contain dirt or oxide should be separately refined and ingotted before being recycled into production melts.

"Bridging" of the metal charge in the crucible should be avoided, the object being to feed the remainder of the charge progressively into the crucible and maintain an advancing level of liquid alloy. During this procedure, additional flux is lightly sprinkled onto the melt surface.

There are separate proprietary fluxes for each type of magnesium alloy (that is, AZ types and magnesium-zirconium types). Supplier instructions for these fluxes must be precisely followed, and their use must be restricted to the type of alloy for which they were developed. During the meltdown process, localized overheating of the charge must be prevented. The process of chlorination of the melt for refining purposes is no longer considered to be an acceptable practice unless effective steps are taken to collect chlorine fumes.

The Fluxless Process. With the flux technique, particularly in the field of die castings, the presence of flux gave some operational difficulties even with hot-chamber die casting processes. Flux inclusions in the castings were not uncommon, creating a major hindrance to the greater use of magnesium.

A significant breakthrough in this area resulted in the development of a fluxless process for use in the melting, holding, and pouring of magnesium alloys (Ref 4-9). This involved the use of air/sulfur-hexafluoride gas or air/carbon-dioxide/sulfur-hexafluoride gas mixes of low (≤2%) SF_6 concentration.

The protection afforded the magnesium being melted was very effective, with the added advantage that the mix was nontoxic and odorless. This process became immediately accepted by both the ingot producers and the die casting sections of the foundry industry, because it answered this objection to the flux melting method.

The new melting process was next extended to the sand casting process (Ref 10). There were two aspects of the sand process that had to be allowed for. The temperatures for pouring the magnesium alloys, particularly the zirconium alloy (see Fig. 3), were appreciably higher than for the die casting alloys. Also, the sand process is generally a more open method, in that the molten alloy cannot be enclosed to the same degree as with die castings. For these reasons the gas mixture used by the sand caster is generally richer in sulfur-hexafluoride content, and sometimes, particularly with the magnesium-zirconium alloys, the parent gas is carbon dioxide.

The melting process using fluxless methods proved to be much more acceptable to the die casting process, and in practice melting losses were considerably reduced (by about half in some cases). The reasons for the much higher melting losses with flux melting are, first, the entrapment of small globules of metal in the sludge at the bottom of the crucible and the difficulty of recovering such metal. With the absence of flux in the fluxless method, the amount of sludge at the bottom of the crucibles is greatly reduced.

Grain Refinement. The earlier procedure of superheating the melt to 870 to 925 °C (1600 to 1700 °F) followed by a rapid cooling to process temperature is no longer

Fig. 3 Pouring of a sand mold for a magnesium-zirconium alloy casting using the fluxless technique. Courtesy of Haley Industries Ltd.

popular or acceptable because it considerably shortens crucible life and can increase the iron content of the alloy melted.

With these AZ-type alloys, the present-day system of grain refinement is by the use of tablets of hexachlorethane or hexachlorbenzine for grain refinement and degassing. With the magnesium-zirconium alloys, very effective grain refinement is achieved by the zirconium addition. To achieve the optimum fineness of grain, it is necessary for the melt to be supersaturated with soluble zirconium. Insoluble zirconium complexes can also be present in the melt, resulting from contaminations of different types. Aluminum and silicon are very undesirable for this reason. Thus, it is necessary for an excess of zirconium to be added to the melt, over and above that which is theoretically required, to provide the required soluble zirconium level.

It is also necessary, for the same reason, to maintain the heel of zirconium-bearing material at the bottom of the crucible into which the insoluble zirconium complexes settle. To prevent any of this liquid heel from being poured off into the castings, an adequate amount of molten alloy (that is, about 15% of the charge weight) is left behind after the casting molds have been poured.

Undue disturbance of the melt during pouring must be avoided. Great care must be taken not to overpour, and the melting procedure must allow adequate settling time. The normal control check carried out on the AZ alloy is to fracture a small test sample cast in sand for visual examination. For the magnesium-zirconium alloys, a small chill cast bar is fractured and visually examined. Comparison is made with norms on which the grain size has been checked metallographically. The grain size value regarded as satisfactory is 0.03 mm (0.0012 in.). A very important factor, which normally requires continuous surveillance, is the possibility of cross-contamination by alloy mixing.

Alloying. Most foundries purchase prealloyed ingot, which is subsequently charged into the melting furnace with a proportion of the process scrap. In the case of some die casting operations, the amount of process scrap generated is low, and it becomes economically feasible to have this scrap remelted and ingotted before it is reused. With the AZ alloys used in the sand casting and die casting operations, little correction to the composition is necessary. However, the magnesium-zirconium alloys contain alloying constituents that tend to be lost during each remelt operation and need to be added each time the material is remelted. Such corrections can be made by adding the pure metals themselves (such as zinc, misch metal, and so forth), or hardener alloys with a fairly high content of the alloying element. Zirconium added as a master alloy of about 30% in magnesium and cerium rare earths added as a master alloy with a content of 20% rare earths in magnesium are examples.

Composition control must allow for the fact that the addition of a master alloy to correct one element may lead to a dilution of the melt, causing the content of other elements to be reduced. Normally, these master alloys are added into the melt by placing them into a preheated steel basket, from which they readily dissolve into the melt.

A "puddling" technique involving either manual or mechanical stirring, followed by a settling procedure to allow insoluble zirconium complexes to settle, produces the required degree of supersaturation of the melt with zirconium. Care must be taken not to hold the melt too long or to allow the melt temperature to fall, however, because losses of zirconium will result.

Melt Treatments. In the melt-down operation, oxidation of the metal must be prevented by the sprinkling of flux on the melting metal or, in the case of fluxless techniques, by the efficient use of sulfur hexafluoride/carbon dioxide atmosphere. Depending largely on the type of alloy and the melting and casting process used, the gas may consist of sulfur hexafluoride/air, or sulfur hexafluoride/carbon dioxide/air mixtures. For example, with the die casting process, in which comparatively low casting temperatures are used and the molten metals can be effectively enclosed, sulfur hexafluoride/air mixture with a low sulfur hexafluoride content (typically <0.25%) provides adequate protection.

With sand castings, particularly with the magnesium-zirconium alloys, higher melt temperatures are used. To provide adequate protection, it is normal to employ a carbon dioxide/sulfur hexafluoride or carbon dioxide/air/sulfur hexafluoride mixture. It is also normal for the sulfur hexafluoride content of the mixture to be increased up to a maximum of 2%.

Gas Content and Grain Size. Currently, the process of chlorination with the AZ-type alloys to clean and degas the melt is rarely acceptable on environmental and safety grounds. Also, the earlier process of grain refinement by superheating these alloys to a high temperature (that is, 850 °C, or 1560 °F), followed by a rapid cool to casting temperature is unacceptable on both technical grounds, because of the increased iron pickup from the crucible, and economic grounds, because of the shortening of crucible life due to the high furnace chamber temperatures. These earlier techniques have been replaced by the use of carbon inoculation, using hexachlorethane or hexachlorbenzine. These materials, added in the form of compressed tablets, simultaneously provide degassing and grain-refining effects.

With the zirconium alloys, the situation is different in that a separate degassing or grain refinement treatment is no longer necessary, both of these features being taken care of by the addition of zirconium.

With flux melting of the aluminum-zinc-manganese-type alloys, fluxes high in magnesium chloride are employed, but there are inherent dangers if these fluxes are used with the zirconium alloys. However, special proprietary fluxes have been developed for the zirconium alloy to avoid these problems.

Foundry Control Procedures. When sand casting an AZ-type alloy, apart from conducting standard spectrographic control of composition, it is normal to pour a sand cast round bar approximately 19 mm (3/4 in.) in diameter and then fracture it for visual examination. Examination of this fractured surface and comparison with sand cast standards serve as a useful check of the degree of grain refinement achieved.

With the zirconium alloys, a grain size check is usually carried out on a standard chill cast bar that is fractured and examined in a similar way. This test satisfactorily determines whether the zirconium content is at an acceptable level for a prescribed degree of grain refinement. Comparison standards are set up to ensure that the grain size meets a 0.03 mm (0.0012 in.) size limit. Chemical composition control is normally carried out spectrometrically using cast slugs poured into a steel mold.

A large proportion of the sand castings made in magnesium alloys are for aircraft or military requirements, for which very stringent checks and inspections are required. Normally, samples for tensile testing must be provided from each melt. These are either separately cast test bars made in a sand mold or coupons attached to the parent casting. Frequently, complete destructive testing of sample castings is called for to further demonstrate the quality level of the workpiece. For the die castings, the standard control procedures involve not only visual and dimensional checks but also inspection of the castings by fluoroscopic screening.

Pouring Methods. The method of pouring depends on the casting process. Thus, pouring can vary from dip ladling from a bale-out crucible, to more automated systems using metered shots of molten alloy, to a die casting machine. The automated hot chamber machine is another possibility, and large sand castings are currently being poured directly from the crucible in single or multiple crucible pours.

In the flux-melting method, the dried flux on the top of the molten metal must be completely removed by careful skimming. Just as important is the complete removal with a steel wire brush of any loose flux on the rim of the crucible or the pouring lip.

During this operation, and until pouring is complete, oxidation is controlled by dusting the surface of the molten metal with a mixture of equal parts of coarse sulfur and fine boric acid. Proprietary material is available for this purpose. The surface of the metal should be left as undisturbed as possible until the melt is ready to pour. At this stage the skin formed by the protective agent is pushed back to ensure that none of it enters the molten metal stream. During the actual pour, the metal stream can be adequately protected by dusting with sulfur.

When pouring, it is undesirable for the molten alloy to enter the sprue direct from the pouring ladle, because this tends to cause any oxide on the metal stream to be carried directly into the mold. If a properly designed pouring box that allows for a degree of separation is used, this will not occur.

Because the flux melting process results in a quantity of residual flux-bearing material at the bottom of the crucible, the pouring crucible should never be completely emptied to avoid pouring some of this material off into the molds. With the zirconium alloys, this is even more important, and about 15% of the charged weight is retained as a heel. With these alloys, the heel will contain residual zirconium and zirconium complexes in addition to the flux and some entrapped magnesium particles.

In the fluxless melting method, there is still some surface oxide, which should be carefully skimmed from the melt surface, while maintaining the flow of the protective gas. The gas flow can be temporarily discontinued while the crucible is removed from the furnace, but the gas flow must be continued until the pouring is completed.

The same protective gas mix used for melting is used on the metal stream during pouring and is used to flush the molds to supplement the action of the inhibitors. The heel left behind can be smaller in volume than the sludge from the flux melting procedure because there is no residual flux. The clean, sludge-free portion of the heel can be recycled into subsequent melts.

Pouring. Small sand, permold, or die castings can be poured from hand ladles. For this purpose, clean, preheated bottom-pour ladles can be used to dip the molten alloy from open crucibles. The same precautions against oxidation must be observed as with direct pouring from the crucible. Oxidation and burning of the metal in the hand ladle can be minimized by sprinkling powdered sulfur on the metal surface.

For hand ladling a special flux pot can be used into which the ladle is immersed between pours to dissolve nonmetallics and also keep the ladle hot. The ladle must be drained until it is free of flux. "Double-filling" the ladle, avoiding undue agitation, and taking more metal than is actually needed for the pour are measures that help prevent the carry-over of flux into the castings.

The possibility of flux contamination and associated corrosion problems and the tedium of the precautions listed above have spawned the development of automated pouring and fluxless methods in both the low- and high-pressure die casting fields.

Safety Precautions. The usual precautions for handling other molten metals are even more exacting with magnesium. These include the use of face shields and fireproof clothing by plant personnel. With magnesium there is a greater hazard in that moisture, from whatever source, increases the danger of explosion and fire: when moisture comes in contact with molten magnesium, it generates hydrogen, a potential source of explosion. Certain minimum precautions must therefore be observed:

- All scrap must be clean and dry; corroded material should be precleaned
- Any fluxes, which tend to be hygroscopic, must be kept dry and stored in airtight containers
- Ladles, tools, and anything that will come into contact with molten magnesium must be completely dry and preheated
- The danger of molten magnesium coming into contact with iron oxide scale must be avoided

Sand Casting

Sand castings are produced in a wide range of alloys, weighing from a few ounces to as much as 1400 kg (3000 lb). Successful production of sand castings became possible only when the means for preventing metal/mold reactions from occurring were developed. This was achieved by the addition of suitable inhibitors to the sand mix used in the manufacture of the mold and the cores. These inhibitors include the following, used singly or in combination: sulfur, boric acid, potassium fluoborate, and ammonium fluosilicate.

The amount of inhibitor that must be added to the sand in order to prevent metal/mold reaction depends on the moisture content of the sand. The original process was called the green sand method (see the article "Sand Processing" in this Volume), in which the bond is produced by the natural clay present in the sand being activated by the addition of water or by the use of the clay products southern or western bentonite. Diethylene glycol is frequently added to the latter type of mix in order to minimize the amount of water necessary and to prevent the sand from drying out. In these green sand mixes, the water content can be in the range 2.0 to 4.0%. A correspondingly high inhibitor content is therefore required.

The amount of inhibitor is also dependent upon the pouring temperature, the type of alloy being cast, and the casting section thickness. The higher the pouring temperature, the higher the reactivity and the need for larger inhibitor additions. The heavier the casting section, the slower the cooling rate. The inhibitors, particularly the more volatile ones, are more easily lost from the mold surface and need to be replenished from the mold areas at a distance from the heavy section.

The green sand molding process is still used extensively for the manufacture of a wide range of sand castings. It has certain limitations, but is economical because the sand can be reused repeatedly. To reuse the sand, it is necessary only to restore the moisture, inhibitor, and glycol contents and

Table 4 Typical compositions and properties of magnesium molding sands

Ingredient or property	British		American						
	Synthetic	Natural	Synthetic					Semisynthetic	
Clay	...	8	...	...	...	...	...	1	1
Bentonite	4	...	3	3–7	4	3	3	3	5
Moisture	2¾	5–6	2½–3½	1½–2½	2¼–3¼	2½	1¼–2¼	3	2½–3
Sulfur	4	6	4½	1¼–2¼	2	3	2	1½	3
Boric acid	½	0.3	0.3	¼	2	...	1½	1½	...
Potassium fluoborate	...	...	...	...	...	1½	...	½	1½
Diethylene glycol	0.1	...	¼	½–1¼	¾	2½	⅓	1½	¾
Compressive strength (green, psi)	8–10	10–13	11–15	7–11	8–12	8–10	9–12	10–12	8–10
Permeability (green)	100	30–40	80–100	120–180	50	90–150	80–90	80–100	100–140

then to suitably mull the sand mixture. There is no loss of expensive binder materials used in the modern chemically bonded sands. Automated sand-mixing equipment, with delivery of the sand mix to the point of use, is commonly employed.

Molding Sands. It is essential to carry out various sand control tests, as described elsewhere, in order to maintain a consistent sand mix. In most magnesium foundries making a wide range of sizes of castings, the molding sand and inhibitor type and amount are normally adjusted to suit the heaviest section being cast.

Molding sands must have high permeability in order to allow free passage of the generated mold gases away from the metal/mold interface, that is, 60 to 90 AFS (see the article "Sand Molding" in this Volume). For this reason, the basic sand used should be comparatively coarse grained. This is especially important because most additives tend to reduce permeability.

Unfortunately, these relatively coarse sands tend to produce rough casting surfaces. The choice is therefore between a good casting surface on the one hand and the venting of generated mold gases on the other. For this reason, it is of the utmost importance to assist the venting of mold or cores and any other reduction in gas generation.

The natural venting through the body of the core must be augmented by additional vent channels drilled into the core. These channels must be capable of rapidly taking gases outside of the mold by means of core prints. Assisted evacuation of these gases is sometimes used.

Natural-bonded sand can and has been used quite satisfactorily for small molds, but very careful control is needed to maintain satisfactory results because of its relatively high clay content and poor uniformity. Better and more consistent results are obtained by the use of a synthetic sand mix. A washed and graded silica sand is used, with the bond being provided by carefully controlled additions of western or southern bentonite. The essential properties of these two types of bentonite are quite different, and by joint use and controlled blending an optimum balance of desired properties can be obtained. Western bentonite imparts toughness (a combination of high green strength and low deformation). Because these green sand mixes are repeatedly reused, losses of moisture, inhibitors, and so forth can occur, and these materials must be restored at each remulling (see the article "Sand Processing" in this Volume). Frequent and regular testing of the properties and composition of the sand is imperative. Also, mulling must be carried out on a controlled basis. The composition and typical properties of some molding sands are shown in Table 4 (Ref 11). A variant of these molding sand mixes, which has also been used successfully, employs a modified type of bentonite using an oil as a mixing agent for bonding.

Control over the properties and additives is as important as it is with the conventional green sand process. More efficient air extraction is necessary to remove burnt oil fumes and render the process environmentally acceptable for present-day standards. Essentially, the type of sand mix chosen is largely dependent upon the molding process used and the sand cast product being made.

Advantages and Disadvantages of Green Sand Molding. Green sand molding is the least expensive of the sand casting processes. This is because only small amounts of additives are needed each time the sand is mulled, and it is not necessary to use expensive reclamation systems in order to recycle the sand. However, the process does not lend itself to the production of cast components of high complexity. Also, it is not capable of producing castings that satisfy the level of dimensional accuracy currently being demanded for many applications.

Fortunately, technology in the mold- and core-making fields has advanced sufficiently to keep pace with these challenges. The first breakthrough, which occurred during the World War II era, was the development of the shell molding process (also known as the Croning process) for the manufacture of shell cores and molds. Almost concurrently the carbon dioxide/silicate process came into general use. Both of these processes improved the quality and dimensional accuracy of castings. Also, the core breakdown characteristics were much improved from those of the earlier oil sand cores.

The carbon dioxide/silicate process, originally developed for cores, was adapted to the manufacture of molds of much improved dimensional accuracy and good breakdown. The reclamation of these, to enable sand to be recycled, proved to be difficult. With the carbon dioxide/silicate process, there was appreciably less moisture present after the molds were gassed than there was with green sand. It was therefore possible to considerably reduce the amount of inhibitor addition.

In the carbon dioxide/silicate process, the sand is fully hardened by passing carbon dioxide through the core *in situ* in the core box. This eliminates most of the distortions that occurred with the oil sand core and affords a significant energy saving. It is also possible to use the cores in a mold very soon after they are produced.

Following this development there were a number of further advances involving various types of phenolic, furan, and epoxy resins of the chemical self-set or gas-hardening type. The latter group involves hardening by the use of air, carbon dioxide, sulfur dioxide, methyl formate, or organic amines.

Preference for the different proprietary systems vary from foundry to foundry, but in general they all produce molds and cores much superior to those obtained from the green sand molds or oil-sand/baked cores. It is largely these improvements in materials and processes that have enabled the foundry industry to produce complex castings such as those illustrated in Fig. 4 to 6 (Ref 12).

The aircraft gearbox housing shown in Fig. 6 is a sand casting in ZE41A-T5 alloy made by the furan chemical-set process. It contains 56 cores within its configuration. A number of these are small-diameter cores forming complex oil passageway configurations. The accurate positioning of these cores and the thin-wall sections in this casting are made possible only by the use of this type of process.

Casting solidification rates can be increased locally by the use of metallic chills or by the use of zircon sand to help in producing an optimum solidification pattern in the casting. Proprietary sprays, which are compatible with molten magnesium alloy,

Fig. 4 Typical main transmission housing for a large military/commercial helicopter that was sand cast in ZE41A magnesium alloy having T5 temper. Courtesy of Haley Industries Ltd.

Fig. 5 Typical engine gearbox for a private or commuter turboprop aircraft cast from QE22A magnesium alloy with T6 temper. Courtesy of Haley Industries Ltd.

Fig. 6 Gearbox housing for a military fighter aircraft composed of ZE41A magnesium alloy of T5 temper. Courtesy of Haley Industries Ltd.

are often used to improve the surface hardness of the mold and thereby reduce the effects of molten metal impingement.

The gap between the cope and drag may be sealed by the use of a proprietary core paste material, to prevent metal leakage or flashing in the casting. Light torching of the mold, using a nonoxidizing acetylene flame, which deposits carbon on the metal contact surface, improves fluidity.

Coremaking Procedures. Major cores, other than small pipe cores, are normally produced in a silica sand mix and should have approximately the same grain fineness as the molds themselves. Frequently, smaller cores, particularly those of small diameter and long length, may be of a different type of sand mix than that used for the molds. This factor becomes particularly important if the process allows a substantial amount of core sand to enter the mold sand system during shakeout. Subsequent recycling of the main sand burden may therefore change in grain size distribution. In fact, the introduction of new sand used for the cores may have a desirable ''sweetening'' effect on the main sand system. It would be normal for the molds to be made from a sand mix with a high content of recycled/reclaimed sand.

The sand mixes used for core production must also contain an inhibitor addition, usually of a type similar to that used for the mold. The binders used for cores are similar to the types used in the mold. However, because magnesium alloys have a comparatively low heat content, not as much heat is available for the breakdown of the core as for other metals. Careful choice of binder type and of the percentage of binder is important.

Other desirable features of cores are:

- They must be capable of being handled and drilled for venting without deterioration
- They must have good shelf life and not be affected by atmospheric conditions of storage
- They must be entirely stable and not distort in storage
- The larger cores must be capable of incorporating chills, when necessary, in the core to produce directional solidification within the casting

For these reasons it is desirable to choose a core binder and core system that produce a degree of strength and dimensional stability to prevent sagging without the necessity for core dryers or bottom boards.

The carbon dioxide/silicate process has been used extensively for cores for magnesium alloys, and is a good example of the earlier gas-hardening process. Details of the carbon dioxide/silicate and shell molding methods can be found elsewhere in this Volume (see the article ''Sand Molding'').

Modern resin binder systems in which the core sand mix is hardened by an acidic activating agent or by the passage of acid gases such as sulfur dioxide *in situ* in the core box have gained considerable popularity. Cores can be made by automated equipment that can be very closely controlled to perform all of the necessary functions of coremaking.

Hot-box methods of core production, also used extensively for magnesium castings, are essentially the same as for other metals and are described in the articles ''Sand Molding'' and ''Coremaking'' in this Volume.

Magnesium is widely used for aerospace and race car engine applications, where the particular attributes of the metal and its weight-saving potential are required. The ability to incorporate complex small-diameter oil passageways by coring is an important requirement. Using some of the improved core binders and coremaking processes described, those requirements can be met. If much higher production costs can be tolerated, it is possible to produce even smaller-diameter cores by the use of special processes described in Ref 13. These include the production of ceramic cores, salt cores, wire cores with wash coatings, quartz glass cores, fiberglass cores with ceramic coatings, and wire helix cores with wash coatings. The latter process can be used to manually produce cores with a diameter as small as 2 mm (0.079 in.). For many casting designs, it would be quite impossible to produce these passageways by machining operations.

Shrinkage Allowances. Nominal shrinkage allowance data for the various alloys is available and can serve as a general guide. It is generally much safer, however, to establish factors for shrinkage under a particular set of foundry conditions.

Gating practices for magnesium have been largely developed to accommodate some of the specific chemical and physical characteristics of the metal and its alloys. Probably the most important of these is the pronounced tendency of the alloys to oxidize. This is particularly the case with those alloying constituents such as rare earth metals and thorium. The second factor is the very low density of these alloys. Magnesium has only two-thirds the weight of aluminum at a comparable volume.

Because of these two factors, minimizing turbulent effects in a mold must be a major objective in any casting method. Turbulence causes the formation of oxide skins and films (dross) being folded into the flowing metal and appearing later in the casting as inclusions or surface pits. The technique for introducing the molten alloy into the mold cavity must therefore be one designed to minimize turbulence. For example, the metal must be introduced at the level of the lowest part of the casting.

Gravity pour systems are normally used, with the metal passing from a pouring cup down a sprue and into a runner system situated at the bottom of the mold. One or more sprues may be used, these normally being tapered so that the metal is choked at their bases. This prevents the aspiration of air. Particularly long sprues should be broken by an intermediate step.

During the pour, the pouring cup or box should be kept full. From the bottom of the sprue, the design is such that the metal is allowed to take only a horizontal or upward flow. The normal procedure is for the metal to pass through a runner channel and then through gates into the bottom-most region of the casting.

It is standard practice to position screens or filters to interrupt the metal flow before it enters the mold cavity to remove any oxide resulting from pour turbulence. The local cross section of runner bar is enlarged locally so that the flow of the molten alloy is not impeded. The screens may consist of a piece of perforated steel sheet, often with coarse steel wool as a backup. There are also proprietary woven screens of stainless steel wire. These are preformed into a wedge shape to fit easily into a matching cavity formed in the running system. The use of these filters ensures a degree of consistency and is normally a preferred technique. Apart from their essential function of removing oxide, filters also regulate the metal flow into the mold cavity.

The design of the runner bar itself can also help to remove dross (slag). Thus, if the cross-sectional area in the runner bar is increased (that is, compared to the sprue), the flow rate decreases, and oxide can be skimmed from the metal stream before it enters the mold cavity. To enhance this effect, the cross section of the runner channel is made fairly large and long, and is taken beyond the last gate into the casting.

The design of the runner system should be such that the runner bar is completely filled before any metal can flow through the gates into the casting. It is desirable, therefore, for as much of the casting as is possible to be formed in the cope, or to use a three-part mold with the runner system in the lowest section. At all times, metal spurting must be prevented. This can be achieved by making the runner bar cross section larger than the sprue cross section, and the total cross section of the gates larger than the runner channel cross section. It is customary to use a ratio of dimensions such as 1:2:4 or 1:4:8, according to the experience of the particular foundry.

Molten metal should never be allowed to fall within the mold from one level to a lower level, because the entrapment of gas/air and oxide inclusions will result. In order to avoid misruns, particularly in thin-cast sections, it is accepted practice to have fast pouring rates, high pouring temperatures, and more numerous gates that are closely clustered around the casting.

A particular problem to avoid if possible is a design mold in which a thick section and a thin wall are at the same level as the metal being poured. The metal flow will be preferential to the thick section, which may result in misruns in the thin section. Proper design of the gating system can overcome this problem. For additional information, see the article "Gating Design" in this Volume.

Risers. The function of risers in magnesium castings is similar to that in other metals (see the article "Sand Molding" in this Volume). Magnesium alloy sand castings have a high propensity for shrinkage. For this reason, a larger amount of risering is required than with most other cast metals. Because of this and the relatively high cost of some alloys, it is necessary to increase the feeding efficiency of risers, while still reducing their total volume. This can be achieved by the use of insulating sleeves that encase the risers. Another important advantage of this technique, resulting from the improved casting yield, is the reduction in the amount of recycled process scrap. For additional information, see the article "Riser Design" in this Volume.

Chills, which may consist of cast-to-form iron castings, zircon sand, or cores containing steel shot, are used extensively to promote progressive or directional solidification of the casting from the bottom to the top in the gravity pouring system. Heavy sections or bosses in the casting that are remote from direct riser feed should be chilled so that they conform to the general pattern of directional solidification within the casting. The very fine grain of the magnesium-zirconium alloys is mainly derived from the very marked grain refining effect of the zirconium content. However, with the nonzirconium alloys, chilling, in combination with a good grain refinement technique, produces a fine grain and superior mechanical properties.

Shake-Out. Small castings, up to about 90 kg (200 lb) cast weight, should be allowed to stand after casting until they have cooled to below 260 °C (500 °F). This is because the magnesium alloys tend to exhibit hot-shortness cracking tendencies; if the castings are removed from the molds too soon, they may show cracks.

The larger the casting, the longer they should be allowed to cool before shake-out. Traditional methods for removal of the casting from the mold use vibratory screens equipped to separate chills by the use of electromagnets, a noisy and dusty process, however.

Modern methods include placing the mold on a revolving table in an airtight chamber and bombarding it with steel shot to break down the mold. The process must be controlled to avoid abrasion of casting surfaces. It is very important, subsequently in the process, to acid etch the castings to remove any possible surface contamination, thereby precluding reduced corrosion resistance.

In order to preserve optimum corrosion resistance, the use of a nonmetallic abrasive medium, such as alumina, instead of steel shot for cleaning the casting is desirable.

With the resin-bonded sand, thermal reclamation is becoming popular because the product obtained is very clean and is equivalent to new sand. It can therefore be used at a high percentage for mold production. For additional information, see the section "Shakeout and Core Knockout" in the article "Processing of Castings" in this Volume.

Current Applications. The production of sand castings to the extremely high level of dimensional quality currently demanded by the aircraft designer is dependent upon the quality of the pattern and core box equipment. These normally consist of precision equipment manufactured in metal and plastic. Wooden pattern equipment, although appreciably cheaper, is rarely used for this type of application. It is, however, quite suitable for castings where the production numbers are low and the dimensional quality requirement is not as stringent.

One particular feature of modern mold and core production that has encouraged its widespread use for the manufacture of complex gearbox-type castings for aerospace application is the capability to include complex coring systems within these designs. It is possible, for example, to incorporate very complex systems of small-diameter oil passageway cores and, often surrounding these, some very thin wall sections. In some cases, complex designs that previously could only be made by the investment casting method can now be made by these refined sand casting methods. The examples shown in Fig. 4 to 6 show applications that use these techniques.

Investment Casting

Throughout the history of magnesium alloy castings, designs involving cast components that were very complex, embraced very thin wall sections, had fine surface finish, and attained tight dimensional tolerances became the market for investment casting (see the article "Investment Casting" in this Volume).

As already mentioned, the gap between modern sand casting and investment casting has been considerably narrowed. However, the investment casting process is still capable of producing castings of tighter tolerances and thinner wall sections than those of sand castings.

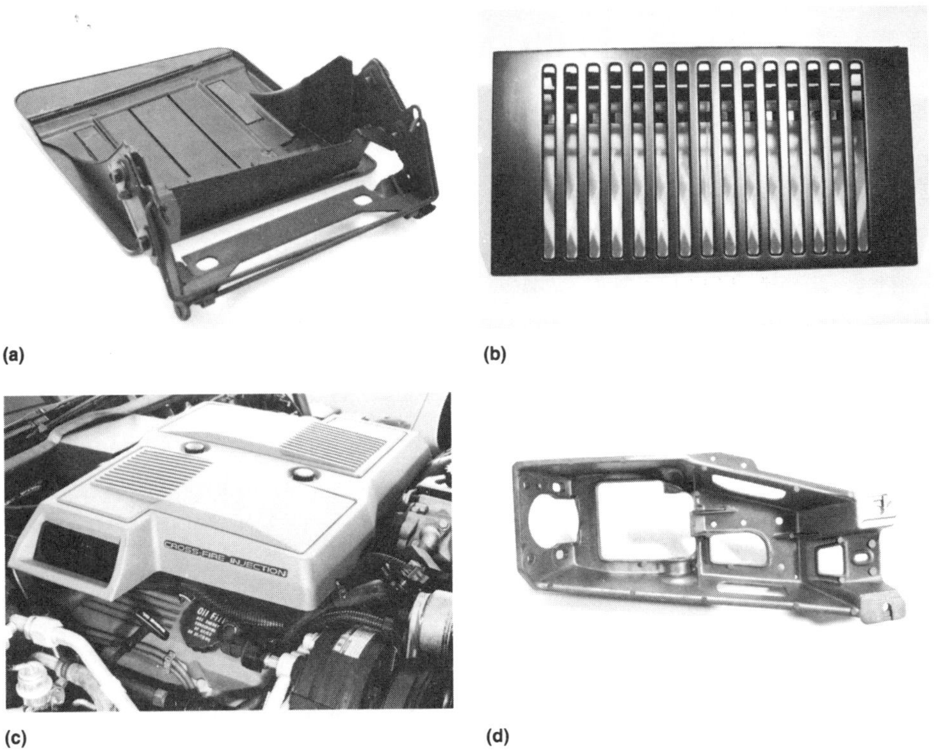

Fig. 7 AZ91 die cast magnesium alloy used in automotive applications. (a) Door frame for hidden headlight assembly weighing 0.370 kg (0.816 lb). (b) Air intake grille weighing 3.240 kg (7.143 lb). (c) Air cleaner cover (shown mounted on a vehicle engine) weighing 2.307 kg (5.086 lb). (d) Brake and clutch pedal bracket weighing 0.637 kg (1.40 lb). Courtesy of International Magnesium Association

The disadvantages of the investment process are the cost of the casting in terms of both capital equipment and cost per casting. There is also a much greater restriction on the size of the casting that can be produced.

The same alloys can be cast by this process as with sand casting. The processes used are mainly "jointless" in character (that is, lost wax or frozen mercury). This gives, among other things, a general improvement in dimensional accuracy by the elimination of the parting line.

Permanent Mold Casting

The salient features of the permanent mold casting method are:

- This process should be considered when the design of the part is within the capability of the process. There are two main types of permanent mold casting, depending on whether metal cores (permold) or destructible sand cores (semipermanent mold) are used. It is not currently possible to equal the size or complexity obtainable by the sand casting process
- The fact that the mold or die can be repeatedly reused is a very big advantage over sand casting; however, the total number of castings required from a given die must be capable of amortising the high initial capital cost of the die, and so forth
- In general, the surface finish and dimensional tolerances are superior to those obtainable from a comparable sand casting
- A distinct disadvantage to the process is the fact that there is very little scope for modifying the casting design or the running system, once the die is constructed
- Nearly all of the alloys that can be sand cast are suitable for use in this process. One exception is the high-zinc alloy with no rare earths present (that is, ZK61A), which is extremely hot short and is deemed uncastable

If steps are taken to accommodate the general tendency for hot shortness in the other alloys, a whole range of castings can be economically produced. The precautions that should be taken to minimize the effects of hot-shortness cracking include provisions for adequate draft. Extreme care is needed when retracting the metal cores not to put undue stress on the hot casting; when two or more cores are being used, they should be extracted simultaneously.

There are also practical limitations on the intricacy of the shapes that can be cast. In particular, deep ribs and complex coring are not practical.

In general, because of faster solidification, mechanical properties tend to be superior to those of sand castings.

Low-Pressure Die Casting Process. In this process the metal dies that are used are basically similar to those employed for permanent mold casting. The main difference between the two processes is that in the low-pressure process the molten metal enters the mold from below under a low pressure. The solidification pattern is inverted. The feed to the solidifying casting comes from below. In recent years, this method has become popular for the manufacture of high-quality automotive wheels (Ref 14).

Die Casting

The die casting process is ideally suited to high-volume production of the type of cast components shown in Fig. 7 and 8. The high cost of the die can be amortised by the large production volume.

The process is a fast production method capable of a high degree of automation, one for which certain magnesium alloys are ideally suited. The number of alloys that can successfully be cast by this process is much more restricted than is the case with sand castings. The alloys typically cast include AZ91A, B, and E; AM60; AS41; and AZ61. Four hundred shots per hour from a hot-chamber machine have been reported.

Two important advantages shown by magnesium die castings are that molten magnesium does not weld onto the die in the same way as does aluminum; therefore, the expensive dies last much longer. Dies last two to three times longer than with aluminum. Also, magnesium can be machined four times faster than aluminum. The relative machinability ratings of magnesium compared to other metals are:

Metal	Relative power required
Magnesium alloys	1.0
Aluminum alloys	1.3
Brass	2.3
Cast iron	3.5
Mild steel	6.3
Nickel alloys	10.0

These factors, combined with the fact that magnesium is only two-thirds the weight of aluminum (Fig. 9), account for the growing popularity of magnesium die casting, particularly in the automotive field.

The excellent machining characteristics of magnesium offer the following advantages:

- Reduction in machining time
- Tool life as much as ten times greater than for other commercial metals
- Excellent surface finishes with one cut
- Minimum tool buildup
- Well-broken chips, minimizing handling
- Lower machining costs

The same methods of supplying metal to the die, such as the hand-ladling and autopour methods, can be applied.

(a) (b)

Fig. 8 Components illustrating wide diversity of applications using AZ91 magnesium die castings in both commercial and consumer products. (a) Carriage frame for computer daisy wheel. (b) Wheel frame assembly for a wheelchair. Courtesy of International Magnesium Association

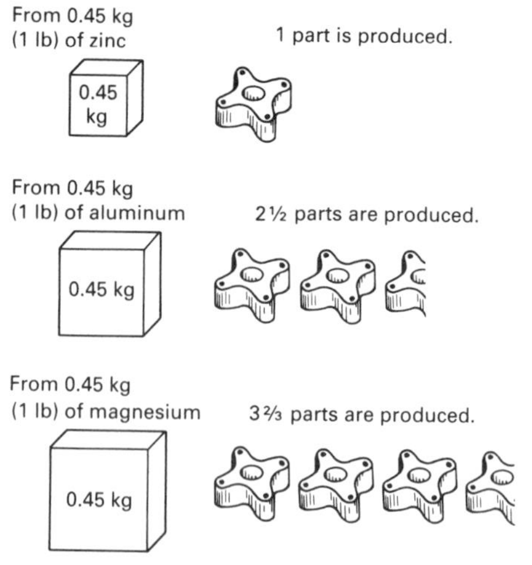
From 0.45 kg
(1 lb) of zinc
1 part is produced.
0.45
kg
From 0.45 kg
(1 lb) of aluminum
2½ parts are produced.
0.45 kg
From 0.45 kg
(1 lb) of magnesium
3⅔ parts are produced.
0.45 kg

Fig. 9 Comparison of the density of magnesium with the densities of zinc and aluminum to illustrate the light weight properties of magnesium alloy die castings

One other process that can also be advantageously applied to magnesium is the hot chamber process, in which the supplying container is immersed in the bath of molten alloy, thereby keeping it replenished and at the correct pouring temperature. This latter process is made possible only by the fact that iron is not attacked by the molten magnesium alloy at the bath and pouring temperatures involved in die casting magnesium.

In this high-production die casting process, economics demand that wherever possible the delivery of the metal into the die be automated. This can be achieved very readily in the hot chamber process (Ref 15), but in the cold chamber process the molten metal must first be delivered into the shot chamber before being injected into the die.

Machines. A large number of machines of both the hot and cold chamber type are currently in use. The cold chamber method is often preferred because of the higher injection pressure that can be employed. The machines must be equipped with a means for ensuring that no water comes into contact with the magnesium.

Dies. The materials used for the manufacture of the dies are similar to those used for aluminum, with dies and cores typically of hot-work tool steels (for example, H11 and H13) heat treated to 44 to 48 HRC. Nitriding the die surfaces is often done. The die life can be appreciably lengthened if the die surface is finished to a 0.38 μm (15 μin.) finish. Stress relieving as well as repolishing the surface of the die after approximately 25 000 shots is sometimes carried out to retain this high finish. Under modern conditions, any process variables (such as too rapid an injection speed, giving rise to larger dimensions across the parting line) can be fairly easily controlled.

Lubrication of the surface of the die by the use of a fine mist of one of the water-free proprietary lubricants is the standard method and is best done using automated lubricating systems.

Die Temperature. Modern processes have facilities to control the die temperature, which should be in the optimum die temperature range of 260 °C ± 14 °C (500 °F ± 25 °F). After the dies are preheated into this temperature range, the heat supplied by repeated casting is normally sufficient to maintain the required temperature. In fact, there is usually a need to cool the die to stay within this range. Thus, dies are provided with a means for water cooling. Such control over die temperature is necessary to maintain casting quality. If die temperature is too low, misruns are likely to occur; if it is too high, die soldering, hot tears, and shrinkage can result.

Control of Metal Composition. The production of sound magnesium alloy die castings that are free from any contamination from oxide or flux and show superior corrosion resistance depends largely on the degree of control exercised over metal cleanliness and on the method of melting and handling the metal.

Very rarely is it possible or desirable to recycle the typical die casting process scrap directly into the casting furnace. This is because scrap is normally highly contaminated with lubricant, and the slug or biscuit from the runner system normally requires a refining process, casting off into ingot, before it can be reused. Some die casters carry out this procedure themselves, while others have the material treated on a contract basis.

Whatever method of recycling is used, the basic requirement is that the process scrap be absorbed into the melt charges as it arises. Each charge should consist of the same ratio of rerun material (normally ingot) and new-metal ingot as that which is being currently created in production.

As with other methods of casting, both methods of melting (that is, flux and fluxless) are in current use. Most die casters, however, favor the fluxless procedure for the reasons already stated.

Much closer attention must presently be paid to the whole aspect of metal handling because of the growing importance of the newer high-purity AZ91D alloy. The very good corrosion resistance, which is a feature of this alloy, is dependent upon retaining a very low nickel, iron, and copper content.

Flux Process Parameters. Refining the magnesium die casting alloys consists of heating the melt charge to approximately 705 °C (1300 °F), adding flux, and stirring the melt thoroughly to allow the flux to absorb the oxide present. The melt is then allowed to stand for 10 to 15 min to allow the oxide-laden flux to settle as sludge to the bottom of the crucible.

The flux generally used is a mixture of magnesium chloride, potassium and other chlorides, plus a smaller amount of calcium fluoride. The amount of flux varies from 1 to 3% of the melt, depending on the cleanliness of the charge.

Care must be taken to make sure that any recycled turnings are dry before attempts are made to melt them. Also, in the casting process, contamination of the melt with flux may result in corrosion of the final casting. Charred lubricant products or oxides impair the castability and mechanical properties of the casting. During the metal-handling procedure, any excessive tendency to oxidize and any noticeable difference in viscosity are indications that the metal has not been refined properly.

Fluxless Process Parameters. This method of melting, which has already been described, has become very popular with the die caster. It avoids many of the problems associated with flux, including the risk of corrosion due to flux inclusions.

There is also an economic advantage to the use of the fluxless process in that melting losses, primarily due to the entrapment of metal globules in the sludge formed with flux melting, are greatly reduced. Problems with sludge removal from the flux-melting process are either eliminated entirely or very much reduced.

Periodic analysis of the melt is necessary because with repeated melting, losses of beryllium, manganese, and aluminum can occur. Aluminum is present in the alloy to provide strength, hardness, and castability, but too much aluminum results in brittleness. Zinc improves the castability of the alloy and, to a degree, its corrosion resistance. Zinc, like aluminum, improves the mechanical properties. Hot shortness, however, increases with increasing zinc content. Manganese, up to its limit of solubility, improves the corrosion resistance.

With the high-purity AZ91D alloy the limits for the contaminating elements are (Ref 16):

- 0.005% Fe (Fe = 0.032 Mn maximum)
- 0.0010% Ni (10 ppm)
- 0.0400% Cu (400 ppm)

The iron to manganese ratio must be controlled. Beryllium is sometimes added in quantities up to 0.001% to reduce the burning tendency of the molten alloy.

Design Parameters. Considerable effort has been applied to the formulation of design criteria (Ref 17). As a general rule of thumb for the designer, wall thicknesses can be 1/100 the distance to the gate:

$$t = \frac{D}{100} \qquad \text{(Eq 1)}$$

where t is the uniform wall thickness and D is the distance from the gate. These thicknesses can be as small as 0.9 mm (0.035 in.). Typical relationships between minimum wall thickness and surface area are:

Minimum wall thickness, mm	in.	Surface area of die cast magnesium part m²	in.²
1.78–2.51	0.070–0.099	≤0.052	≤80
≥2.51	≥0.100	>0.052	>80

The various relationships between draft and wall depth for interior walls and between draft and cored hole depths are all calculable.

Most surfaces vertical to the parting plane of the die should be tapered so that the casting can be readily ejected. Shrinkage forces are considerable, and it is important to have sufficient ejectors to ensure that the casting is not damaged. Draftless magnesium castings can also be made to the competitive advantage of magnesium.

Also, since many of the die casting alloys show a tendency toward stress corrosion, it is important for designers to be able to calculate the induced stress levels to ensure that these fall below limiting values for the alloy in question.

Some of these important design features are shown in Fig. 10. Today's design data replace much of the earlier rule-of-thumb practices. Some basic requirements still apply, such as:

- Ample fillets
- Rounded corners
- Blended sections
- Avoidance of remote heavy sections
- Strengthening thin sections by the use of ribs
- Radii that are as large as possible (that is, larger than for zinc alloy die castings)

Casting Finish. The best surface finish on magnesium alloy die castings is obtained by close control over the process variables. These include die temperature, cavity filling, die lubrication, metal temperature, holding time in the die, and the smoothness of the die surfaces.

Tolerances. Typical tolerances are shown in the following tables. Typical tolerances for noncritical dimensions are:

Dimensions	Tolerance mm	in.
Length up to 25 mm (1 in.)	±0.25	±0.010
Additional tolerance for each 25 mm (1 in.) in length		
25–305 mm (1–12 in.)	±0.038	±0.0015
>305 mm (12 in.)	±0.025	±0.001

Typical tolerances for critical dimensions are:

Dimensions	Tolerance mm	in.
Length up to 25 mm (1 in.)	±0.10	±0.004
Additional tolerance for each 25 mm (1 in.) in length		
25–305 mm (1–12 in.)	±0.038	±0.0015
>305 mm (12 in.)	±0.025	±0.001

These tolerances represent ordinary production practices and are based on dimensions that are not affected by die parting or moving parts. If those factors cannot be ignored, these tolerances must be increased to compensate for their effect on the die casting dimensions.

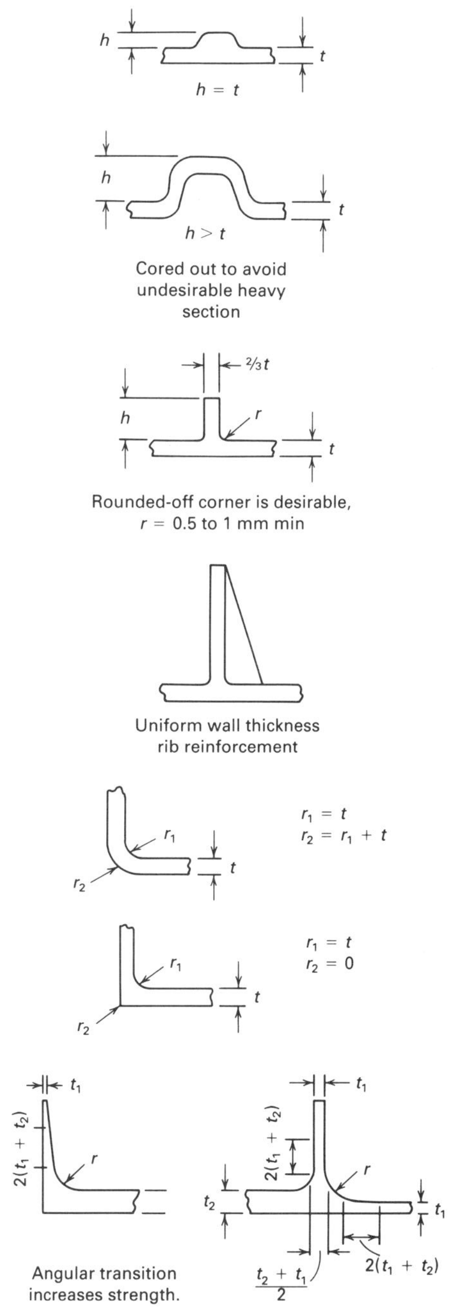

Fig. 10 Design parameters to aid stress reduction in magnesium alloy castings

Casting Defects. In die castings, misruns and cold shuts probably represent the most frequent form of defect. There can be several causes of these two types of defects, mostly process related. They may, for example, be caused by extremely slow filling of the die, excessive application of lubrication, incorrect metal or die temperature, dirty metal (that is, excessively oxide laden), slow casting speed, or incorrect shot weight in the cold chamber process.

There is also a greater tendency for porosity in die castings, most often resulting from gas trapped within the casting. If this effect is serious, it can be minimized by conducting an evaluation of the running, gating, venting, and lubricating system. An increase in total vent area to at least half the gate area often lessens the number of cold shuts. An increased injection pressure may eliminate cold shuts altogether.

Because of the possibility of gas-related porosity, die castings are not normally thermally treated other than with the occasional use of a low-temperature stabilizing treatment.

For additional information on casting defects, see the section "Testing and Inspection of Casting Defects" in the article "Processing of Castings" in this Volume.

REFERENCES

1. W. Unsworth, "Application Guide Lines for Various Magnesium Alloys," Paper presented at Magnesium Symposium, Westinghouse Electric Corporation, Lima, Sept 1984
2. W. Unsworth, "Meeting the High Temperature Aerospace Challenge," in *Proceedings of the 43rd Annual World Magnesium Conference*, International Magnesium Association, June 1986
3. E.F. Emley and P.A. Fisher, "The Control of Quality of Magnesium-Base Alloy Castings," *J. Inst. Met.*, Vol 85, Part 6, 1956-1957
4. J.W. Frueling, "Protective Atmospheres for Molten Magnesium," Ph.D. thesis, University of Michigan, 1970
5. J.W. Frueling and J.D. Hannawalt, Protective Atmosphere for Melting Magnesium Alloys, *Trans. AFS*, 1969, p 159
6. J.D. Hannawalt, Practical Protective Atmospheres for Molten Magnesium, *Met. Eng.*, Nov 1972, p 6
7. G. Schemm, Sulphur Hexafluoride as a Protection Against Oxidation, *Giesserei*, Vol 58, 1971, p 558
8. J.D. Hannawalt, "SF6—Protective Atmospheres for Molten Magnesium," Paper G-775-111, Society of Die Casting Engineers, 1975
9. R.S. Busk and R.B. Jackson, Use of SF6 in the Magnesium Industry, in *Proceedings of the 37th Annual World Magnesium Conference*, International Magnesium Association, June 1980
10. H.J. Proffitt, Magnesium Sand Casting Technology, in *Proceedings of a Special Conference on Recent Advances in Magnesium Technology*, AFS/CMI, June 1985
11. E.F. Emley, *Principles of Magnesium Technology*, Pergamon Press, 1966
12. L. Petro, "Premium Quality Magnesium Sand Castings—Production and Application," Paper presented at Magnesium Symposium, Westinghouse Electric Corporation, Lima, Sept 1984
13. G. Betz and N. Zeumer, Production of Light Metal Castings With Precast Lubrication Passageways, in *Proceedings of the 43rd Annual World Magnesium Conference*, International Magnesium Association, June 1986
14. J. Bolstad, What Improved Metal Handling Practices Can Do to Automotive Castings, in *Proceedings of the 44th Annual World Magnesium Conference*, International Magnesium Association, May 1987
15. W.G. Treiber Jr., High Technology With Hot Chamber Magnesium Die Castings, in *Proceedings of the 44th Annual World Magnesium Conference*, International Magnesium Association, May 1987
16. K.N. Riechek, K.J. Clark, and J.E. Hillis, Controlling the Salt Water Corrosion Performance of Magnesium ZA91 Alloy in High and Low Pressure Cast Forms, in *Proceedings of a Special Conference on Recent Advances in Magnesium Technology*, AFS/CMI, June 1985
17. R.S. Busk, *Magnesium Products Design*, Marcel Dekker, 1987

Cobalt-Base Alloys

Timothy J. Pruitt, Zimmer, Inc.
Michael J. Hanslits, Precision Castparts Corporation

COBALT-BASE ALLOYS were developed in the late 1930s for applications in aircraft turbochargers. The common alloys of today were born from Vitallium, a high carbon dental alloy of the composition Co-27Cr-5Mo-0.5C. These alloys offer better wear- and heat-resistant properties than those associated with the iron-base alloys. Many of the hot section turbine blades and vanes originally developed from cobalt-base alloys have since been replaced by high-temperature nickel-base alloys. Yet the cobalt alloys are still widely used for their excellent high-temperature strength properties up to 815 °C (1500 °F). The development of new cobalt-base alloys has not kept pace with other alloy systems, but approximately 20 cobalt-base alloys are commonly used today. These are typically segregated into either heat- or wear-resistant grades. Nominal compositions and applications of the most common alloys are presented in Table 1, and Fig. 1 shows some applications of cobalt-base alloys.

These alloys still find wide application in the aircraft industry, but have seen extensive development as both medical implant devices and chemical-resistant hardware, such as pump housings, valves, impellers, wear plates, and cutting tools. The investment casting process is well suited to making complicated configurations for these applications, as is emphasized in this article. In general, the cobalt-base alloys are easy to work with in the foundry and exhibit good casting properties, including:

- Good fluidity
- Low melting points
- Freedom from dissolved-gas defects
- Low alloy losses due to oxidation

Areas that may be of concern to a caster include:

- Cost
- Contamination of subsequent heats (furnace lining)
- Crack prone characteristic of certain wear-resistant grades
- Generally poor machining qualities

Manufacture and Remelting of Master Ingots

The manufacture of master ingots for cobalt-base alloys has progressed well, and now includes technologies such as vacuum induction melting, vacuum arc remelting, argon-oxygen decarburization, and electroslag remelting. These processes produce clean, chemically homogeneous ingots. As with any alloy, it is important to begin the investment casting process with the highest quality alloy economically feasible for the end product. Remelting cobalt-base alloys is typically conducted in air, although a few alloys are melted and poured in vacuum. Alloys that contain reactive elements such as titanium, aluminum, tantalum, and zirconium should be melted and poured in vacuum. These include MAR-M 302, MAR-M

Table 1 Nominal compositions and some applications for cobalt-base alloys

Alloy	Composition, %(a) C	Mn	Si	Cr	Ni	W	Fe	Others	Applications
Wear-resistant alloys									
Alloy No. 3	2.0–2.7	1.00	1.00	29.0–33.0	3.00	11.0–14.0	3.00	. . .	Cutting tools, wear strips, valve seats
Alloy No. 6	0.9–1.4	1.00	1.50	27.0–31.0	3.00	3.5–5.5	3.00	1.5 Mo	Valve seats, punches, wear plates
Alloy No. 12	1.1–1.7	1.00	1.00	28.0–32.0	3.00	7.0–9.5	3.00	. . .	Bushings, saw teeth
Alloy No. 19	1.5–2.1	1.00	1.00	29.5–32.5	3.00	9.5–11.5	3.00	. . .	Cutting tools, bearings, rollers
Star-J	2.20	1.00	1.00	31.0–34.0	2.50	16.0–19.0	3.00	. . .	Cutting tools, wear parts
Alloy 98M2	1.7–2.2	1.00	1.00	28.0–32.0	2.0–5.0	17.0–20.0	2.50	0.8 Mo, 1.1 B, 4.2 V	Cutting tools, wear parts
Heat-resistant alloys									
Alloy No. 21	0.2–0.3	1.00	1.00	25.0–29.0	1.75–3.75	. . .	3.00	5.5 Mo, 0.007 B	Turbine blades, combustion chambers to 815 °C (1500 °F)
Alloy No. 25	0.05–0.15	1.0–2.0	1.00	19.0–21.0	9.0–11.0	14.0–16.0	3.00	. . .	Gas turbine rotors and buckets
Alloy No. 31	0.45–0.55	1.00	1.00	24.5–26.5	9.5–11.5	7.0–8.0	2.00	0.5 Mo	Turbine blades
Alloy X-40	0.45–0.55	1.00	1.00	24.5–26.5	9.5–11.5	7.0–8.0	2.00	0.01 B	Gas turbine parts, nozzle vanes
Alloy X-45	0.20–0.30	0.4–1.0	0.75–1.0	24.5–26.5	9.5–11.5	7.0–8.0	2.00	0.01 B	Nozzle vanes
Alloy FSX-414	0.20–0.30	0.4–1.0	0.5–1.0	28.5–30.5	9.5–11.5	6.5–7.5	2.00	0.01 B	Gas turbine vanes
Alloy WI-52	0.40–0.50	0.50	0.50	20.0–22.0	1.00	10.0–12.0	2.00	2.0 Nb	Gas turbine parts, nozzle valves
MAR M 302	0.78–0.93	0.20	0.40	20.0–23.0	. . .	9.0–11.0	1.50	0.01 B, 0.20 Zr, 9.0 Ta	Turbine vanes (815–1095 °C, or 1500–2000 °F)
MAR M 509	0.55–0.65	0.10	0.40	21.0–24.0	9.0–11.0	6.5–7.5	1.50	0.20 Ti, 0.10 B, 0.50 Zr, 3.5 Ta	Gas turbine parts
Biomedical alloy									
ASTM F75	0.35	1.00	0.40	27.0–30.0	1.00	. . .	1.50	6.0 Mo	Orthopedic implants

(a) In all compositions, cobalt makes up the balance.

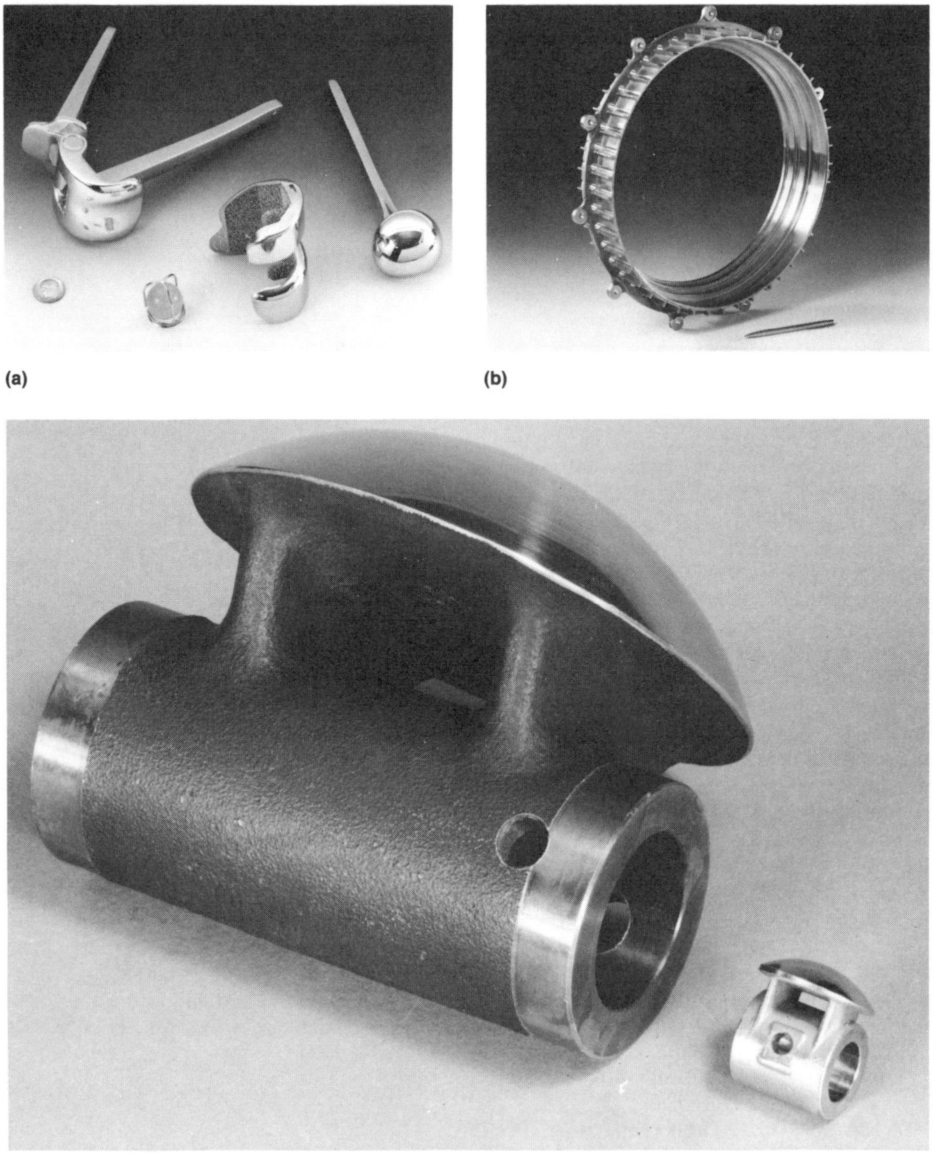

(a) (b) (c)

Fig. 1 Applications for cast-cobalt-base alloys. (a) Surgical implants. (b) Turbine case. Courtesy of Cannon-Muskegon Corporation. (c) Valve plugs for use in erosive applications where abrasive solids are entrained in the flowing media. Courtesy of Fisher Controls International, Inc.

Table 2 Melting ranges for cobalt-base alloys

Alloy	Melting range °C	Melting range °F
Wear-resistant alloys		
Alloy No. 3	1215–1285	2215–2345
Alloy No. 6	1260–1355	2300–2475
Alloy No. 12	1255–1340	2290–2445
Alloy No. 19	1240–1300	2260–2370
Star-J	1215–1300	2220–2370
Alloy 98M2	1225–1275	2235–2330
Heat-resistant alloys		
Alloy No. 21	1340–1365	2440–2490
Alloy No. 25	1330–1410	2425–2570
Alloy No. 31	1340–1395	2445–2545
Alloy X-40	1340–1395	2445–2545
Alloy X-45	1340–1395	2445–2545
Alloy FSX-414	1340–1395	2445–2545
WI-52	1315–1345	2400–2450
MAR M 302	1315–1370	2400–2500
MAR M 509	1290–1400	2350–2550
Biomedical alloy		
ASTM F75	1315–1345	2400–2450

Table 3 Phases in cobalt-base superalloys

Alloy	Phases
S-816	$M_{23}C_6$, Nb(C,N), M_6C, laves
Alloy No. 25	M_6C, $M_{23}C_6$, laves
Alloy No. 31	M_7C_3, M_6C, $M_{23}C_6$
Alloy No. 21	M_7C_3, $M_{23}C_6$, M_6C, Cr_2C_3
MAR M 302	MC, M_6C, $M_{23}C_6$
MAR M 509	MC, $M_{23}C_6$
WI-52	MC, M_6C, $M_{23}C_6$
Alloy No. 188	M_6C, $M_{23}C_6$, laves

509, and AiResist 215. Vacuum casting processes are sometimes used even when they are not needed to meet chemical composition requirements. Examples of alloys that do not contain reactive elements and are adaptable for air casting are HS-21, HS-25, and WI-52.

Air Melting Practices

The Crucible. When air melting cobalt alloys, consideration must be given to the

Table 4 Properties of cobalt-base superalloys

Alloy	Ultimate tensile strength, MPa	Ultimate tensile strength, ksi	0.2% offset yield strength, MPa	0.2% offset yield strength, ksi	Elongation, %	Stress to rupture at 815 °C (1500 °F), 100 h, MPa	100 h, ksi	1000 h, MPa	1000 h, ksi
Alloy No. 3	441	64	Near tensile strength		Nil	...	...	...	...
Alloy No. 6	793	115	662	96	3	...	...	...	...
Alloy No. 12	738	107	Near tensile strength		Nil	...	...	...	...
Alloy No. 19	724	105	Near tensile strength		1	...	...	...	...
Star-J	414	60	Near tensile strength		Nil	...	...	...	...
98M2	552	80	Near tensile strength		Nil	...	...	...	...
Alloy No. 21	710	103	565	82	8	152	22	98	14.2
Alloy No. 25	621	90	448	65	15	152	22	121	17.5
Alloy No. 31	779	113	552	80	8	193	28	162	23.5
X-40	745	108	524	76	9	179	26	138	20
X-45	745	108	524	76	9	131	19	103	15
WI-52	752	109	586	85	7	200	29	172	25
FSX-414	738	107	441	64	2	152	22	117	17
MAR M 302	965	140	690	100	2	276	40	207	30
MAR M 509	779	113	586	85	3	269	39	228	33
ASTM F75	758	110	579	84	9	...	...	...	...

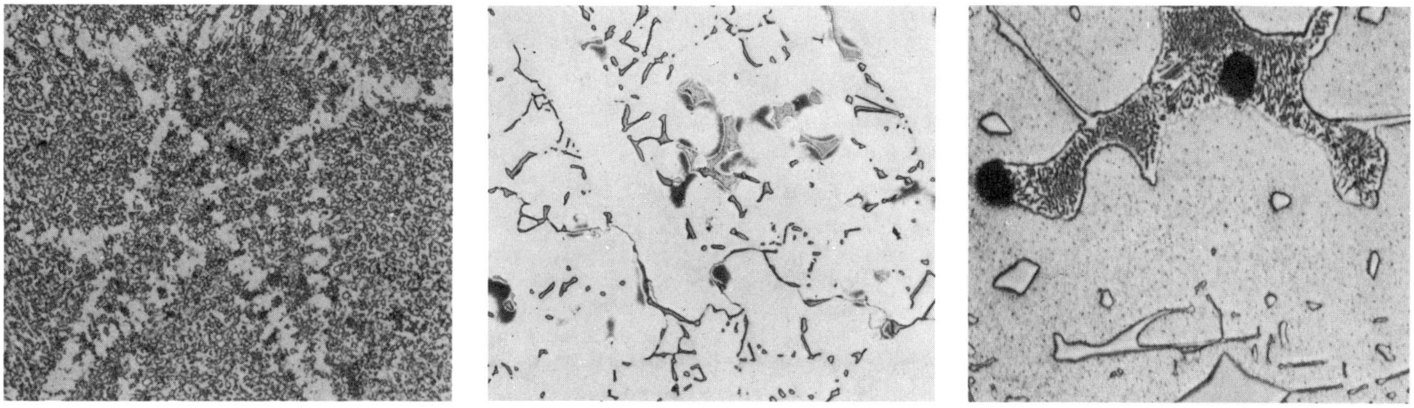

(a) (b) (c)

Fig. 2 As-cast microstructures of three cobalt-base alloys. (a) Alloy 98M2 investment cast ring with large primary carbides in a matrix of secondary carbides and Co-Cr-W solid solution. (b) MAR M 509 alloy showing MC and $M_{23}C_6$ carbides. (c) MAR M 302 alloy showing gray islands of primary eutectic carbide, light MC particles, and small, dark $M_{23}C_6$ precipitates in the matrix. All three 500×

type of crucible or crucible lining to be used. Typically, melting is conducted in a rammed furnace lining or a prefired ceramic crucible. Common ceramic compositions include alumina, silica, zirconia, and magnesia. The use of silica materials may result in a pickup of silicon from the ceramic lining and is not strongly recommended.

When using rammed linings, care must be taken to ensure that the lining has been properly cured and fired before the melting of cobalt alloys. Residual moisture in the lining can result in hydrogen pickup in the first few charges of a new lining. This gas can diffuse to the molten metal and result in internal or external defects similar in appearance to those caused by improper feeding practice. It is therefore recommended that an addition of 0.5 wt% ferrosilicon or ferrotitanium be added to the first two or three melts of the lining.

The use of prefired crucibles is the more desirable alternative. Many of these considerations are of no concern in a vacuum application in which oxidation and other reactions are minimized.

Gas Covers. Air melting should be conducted with an inert-gas cover, when possible. Argon gas is well suited for this and can be dispensed over the melt surface by a variety of methods. A single port to blow argon over the melt is often preferred, while elaborate sealed chimneys also have been successfully used. Whatever the method, the purpose is to keep oxygen from reacting with the molten metal.

Deslagging

The surface reactions that take place in air melting produce slag, a common condition in all air melting operations. Deslagging procedures vary considerably throughout the industry, but the use of slag rods is common for this purpose. It is suggested that rods be of the same or similar alloy as that being melted in order to reduce melt contamination. It is also recommended that a cold rod be used for each attempt at removing slag. A cold rod tends to pick up the slag more readily than one in continuous use. Some casters prefer to use slag coagulants or surface conditioners to aid in slag removal. Slag formation can be further reduced by precise process controls to limit the amount of time the metal is in the molten state before pouring. Slag should also be removed just before the pour to avoid the formation of a new layer requiring a repetition of the deslagging process. Teapot-type ladles and pouring spouts that incorporate dams are also quite successful at minimizing slag defects in castings.

The temperature of the melt can be monitored using a dip tip thermocouple. Although the use of radiation pyrometry is desirable for a hands-off operation, the slag content of the air-melted surface can greatly affect temperature readings. Therefore this approach is not recommended.

Preheat and Pouring

Ceramic mold preheat and the metal pouring temperature can affect the castability of the alloy and the resulting quality of the product. Mold temperature ranges common to the casting of cobalt-base alloys are 760 to 1150 °C (1400 to 2100 °F); the metal pouring temperature varies from 1425 to 1595 °C (2600 to 2900 °F). Melting ranges for cobalt-base alloys are listed in Table 2. Naturally, these are process ranges, and the exact process parameters depend on the casting configuration, quality requirements, and specific alloy. The parameters selected do have a pronounced effect on the resulting casting: Consideration must be given to the effect on the ability of the metal to fill the mold cavity without creating a hot tear or undesirable solidification shrinkage. Hot tears are common to the higher carbon (>0.40 wt%) cobalt alloys and can be avoided with alloy chemistry adjustments, less rigid mold materials, and slower cooling rates from the casting temperature. Wrapping molds in selected areas with an insulating material often reduces hot tears and does not greatly alter the solidification of the total system.

Fluidity of cobalt alloys can also be adjusted chemically. A standard spiral casting test can aid in determining the castability of an alloy before chemical adjustments are made. In air casting, silicon and manga-

Table 5 Recommended heat treatments for cobalt-base alloys

Alloy	Heat treatment
Alloy No. 3	900 °C (1650 °F) for 4 h, furnace cool
Alloy No. 6	900 °C (1650 °F) for 4 h, furnace cool
Alloy No. 12	900 °C (1650 °F) for 4 h, furnace cool
Alloy No. 19	900 °C (1650 °F) for 4 h, furnace cool
Star-J	As-cast
98M2	As-cast
Alloy No. 21	815 °C (1500 °F) for 5 to 50 h, air cool
Alloy No. 25	1205 °C (2200 °F) for 1 h, air cool
Alloy No. 31	As-cast
X-40	1175 °C (2150 °F) for 1 h and water quench + 815 °C (1500 °F) for 4 h and air cool + 980 °C (1800 °F) for 2 to 4 h and oil quench + 730 °C (1350 °F) for 16 h and air cool
X-45	1175 °C (2150 °F) for 1 h and water quench + 815 °C (1500 °F) for 4 h and air cool + 980 °C (1800 °F) for 2 to 4 h and oil quench + 730 °C (1350 °F) for 16 h and air cool
WI-52	1010 °C (1850 °F) for 2 h and oil quench + 1290 °C (2350 °F) for 20 h and air cool + 650 °C (1200 °F) for 20 h and air cool
FSX-414	As-cast
MAR M 302	As-cast
MAR M 509	1095 °C (2000 °F) for 2 h and water quench + 790 °C (1450 °F) for 2 h and air cool + 720 °C (1325 °F) for 24 h and air cool
ASTM F75	1230 °C (2250 °F) for 3 h and air cool

nese levels should be adjusted to the high end of the specification range to maximize their effects. In vacuum casting, these types of adjustments are unnecessary and can be more easily controlled with casting parameters. The option of adding these constituents to the master ingot or directly to the melt at the time of casting is usually determined based on the experience of individual foundries.

Gating and mold configuration also can contribute greatly to the success of producing an acceptable casting. Proper gating of cobalt alloy castings involves the same principles as those used for other superalloys and will not be expanded upon in this article.

Cobalt-base superalloys tend to be complex combinations of elements; each alloy is designed for a specific purpose. The matrix of these alloys is face-centered cubic. Modern alloys typically gain their strength through a number of complex carbides (depending on chemistry) and, in some alloys, through the use of solid-solution strengthening. Other alloys, such as J-1570, gain strength through intermetallic compounds (in the case of J-1570, Ni_3Ti).

A list of alloys and their possible phases is provided in Table 3. These phases represent a wide variety of alloying concepts that are used to produce a desired carbide morphology and matrix combination to yield the best properties for specific applications. Many different types of carbides have been identified in these alloys, including MC, M_6C, M_7C_3, $M_{23}C_6$, and Cr_2C_3, where M represents the metal atom. Mechanical property data are provided in Table 4. Figure 2 shows the as-cast microstructures of three alloys.

These carbides can be greatly affected by the casting parameters and resulting solidification rates. For example, there is a common lamellar carbide phase comprised of both an $M_{23}C_6$ and probably M_6C in as-cast HS-21. This phase is often considered to reduce ductility due to its brittle nature. It is possible to control the frequency and size of the phase through chemistry modification to reduce total carbon or to revise casting parameters to increase solidification rates and thus retard precipitation. Several of the cobalt alloys can be altered by similar techniques to obtain microstructural control of the precipitates.

Heat Treatment

Several options are available in the thermal treatment of cast cobalt-base alloys. In general, the choices are:

- Solutioning at temperatures above 1095 °C (2000 °F) for several hours to put the secondary phases into solution
- Solutioning plus aging at a lower temperature to precipitate a desired phase effectively
- Homogenization, that is, prolonged treatment at a temperature near the incipient melting point, thus reducing composition gradients to a minimum
- Homogenization plus aging
- Aging from the as-cast condition

True solutioning of the alloys is rare. It is possible to see marked effects in the microstructures at temperatures in excess of 1175 °C (2150 °F), but processing temperatures are typically in the 1205 to 1260 °C (2200 to 2300 °F) range. Aging can be done at temperatures as low as 760 °C (1400 °F), but it is necessary to ensure that the effects of aging are not lost because of an operating temperature in excess of the aging temperature. The ultimate thermal processing alternatives employed depend on the application, alloy, and desired properties. Suggested thermal treatments are listed in Table 5.

Nickel and Nickel Alloys

John M. Svoboda, Steel Founders' Society of America

NICKEL-BASE ALLOY CASTINGS are widely used in corrosive-media and high-temperature applications. The principal alloys are identified by designations of the Alloy Casting Institute (ACI) (now called the High Alloy Product Group) of the Steel Founders' Society of America and are included in ASTM specifications A 494 and A 297 and Federal specifications (QQ). There are also many proprietary grades for severe-corrosion applications, as well as heat-resistant alloys. In addition to these conventionally processed alloys, directionally solidified (DS) and single-crystal (SC) alloys are also being processed (see the section "Directional and Monocrystal Solidification" in the article "New and Emerging Processes" in this Volume). The various types of cast alloys can be classified as:

- Nickel
- Nickel-copper
- Nickel-chromium-iron
- Nickel-chromium-molybdenum
- Nickel-molybdenum
- Nickel-base proprietary
- Directional solidification/single crystal

The cast nickel-base alloys, with the exception of some high-silicon and proprietary grades, have equivalent wrought grades and are frequently specified as the cast components of a wrought-cast system. Compositions of cast and equivalent wrought grades differ in minor elements because workability is the dominant factor in wrought alloys, while castability and soundness are the dominant factors in cast alloys. The differences in minor elements do not result in significant differences in serviceability.

Tables 1 and 2 provide designations and compositions for corrosion-resistant, heat-resistant, and DS/SC alloys. It should also be noted that extensive data on mechanical properties, microstructural characteristics, and corrosion properties of nickel-base castings can be found in Volumes 3, 9, and 13, respectively, of the 9th Edition of *Metals Handbook*.

Compositions

Cast Nickel. The ASTM A 494 CZ-100 grade covers the requirement for a cast nickel alloy. A higher-carbon or a higher-silicon grade is occasionally specified for greater resistance to wear and galling. The minor elements within specified limits (Table 1) provide for excellent castability and the production of sound, pressure-tight castings. The usual practice is to produce the alloy with 0.75% C and 1.0% Si. When properly produced, the carbon is present in the matrix as a finely distributed spheroidal graphite. A reduction in mechanical properties results if flake graphite is present. A maximum carbon content of 0.10% or less is occasionally specified when castings are welded into wrought nickel systems. Low-carbon CZ-100, however, is a difficult material to cast, with no significant advantage over the higher-carbon option under any known service conditions.

Cast nickel-copper alloys are designated in ASTM A 494 as M-35-1, M-35-2, M-30H,

Table 1 Compositions of cast corrosion-resistant nickel-base alloys

Alloy	Composition, %								
	C	Si	Mn	Cu	Fe	Cr	P	S	Mo	Others
Cast nickel										
CZ-100	1.0 max	2.0 max	1.5	1.25	3.0	. . .	0.03	0.03	. . .	. . .
Nickel-copper										
M-35-1	0.35	1.25	1.5	26.0–33.0	3.50 max	. . .	0.03	0.03	. . .	. . .
M-35-2	0.35	2.0	1.5	26.0–33.0	3.50 max	. . .	0.03	0.03	. . .	. . .
M-30H	0.30	2.7–3.7	1.50	27.0–33.0	3.50 max	. . .	0.03	0.03	. . .	. . .
M-25S	0.25	3.5–4.5	1.50	27.0–33.0	3.50 max	. . .	0.03	0.03	. . .	. . .
M-30C	0.30	1.0–2.0	1.50	26.0–33.0	3.50 max	. . .	0.03	0.03	. . .	1.0–3.0Nb
QQ-N-288-A	0.35	2.0	1.5	26.0–33.0	2.5	. . .	. . .	. . .	. . .	. . .
QQ-N-288-B	0.30	2.7–3.7	1.5	27.0–33.0	2.5	. . .	. . .	. . .	. . .	. . .
QQ-N-288-C	0.20	3.3–4.3	1.5	27.0–31.0	2.5	. . .	. . .	. . .	. . .	. . .
QQ-N-288-D	0.25	3.5–4.5	1.5	27.0–31.0	2.5	. . .	. . .	. . .	. . .	. . .
QQ-N-288-E	0.30	1.0–2.0	1.5	26.0–33.0	3.5	. . .	. . .	. . .	. . .	1.0–3.0 Nb + Ta
QQ-N-288-F	0.40–0.70	2.3–3.0	1.5	29.0–34.0	2.5	. . .	. . .	. . .	. . .	. . .
Nickel-chromium-iron										
CY-40	0.40	3.0	1.5	. . .	11.0 max	14.0–17.0	0.03	0.03	. . .	. . .
Nickel-chromium-molybdenum										
CW-12MW	0.12	1.0	1.0	. . .	4.5–7.5	15.5–17.5	0.04	0.03	16.0–18.0	0.20–0.40V, 3.75–5.25W
CW-7M	0.07	1.0	1.0	. . .	3.0 max	17.0–20.0	0.04	0.03	17.0–20.0	. . .
CW-2M	0.02	0.8	1.0	. . .	2.0 max	15.0–17.5	0.03	0.03	15.0–17.5	0.20–0.60V
CW-6MC	0.06	1.0	1.0	. . .	5.0	20–23.0	0.015	0.015	8.0–10.0	3.15–4.50Nb
Nickel-molybdenum										
N-12MV	0.12	1.0	1.0	. . .	4.0–6.0	1.0	0.04	0.03	26.0–30.0	0.20–0.60V
N-7M	0.07	1.0	1.0	. . .	3.0 max	1.0	0.04	0.03	30.0–33.0	. . .

Table 2 Compositions of heat-resistant nickel-base casting alloys

Alloy designation	Nominal composition, %											
	C	Ni	Cr	Co	Mo	Fe	Al	B	Ti	W	Zr	Others
B-1900	0.1	64	8	10	6	...	6	0.015	1	...	0.10	4Ta(a)
CMSX-2 (SC)	<30 ppm	66	8	4.6	0.6	...	5.6	...	1.0	8	...	6.0Ta
CMSX-3 (SC)	<30 ppm	66	8	4.6	0.6	...	5.6	...	1.0	8	...	6.0Ta, 0.10Hf
CM-247-LC (SC)	0.07	62	8.1	9.2	0.5	...	5.6	0.015	0.7	9.5	0.015	3.2Ta
HW(b)	0.55	60	12	...	0.5	23	...	...	...	...	...	2.0Mn, 2.5Si
HX(b)	0.55	66	17	...	0.5	12	...	...	...	...	...	2.0Mn, 2.5Si
Alloy X	0.1	50	21	1	9	18	...	...	...	1	...	...
Alloy 100	0.18	60.5	10	15	3	...	5.5	0.01	5	...	0.06	1V
Alloy 738X	0.17	61.5	16	8.5	1.75	...	3.4	0.01	3.4	2.6	0.1	1.75Ta, 0.9Nb
Alloy 792	0.2	60	13	9	2.0	...	3.2	0.02	4.2	4	0.1	4Ta
Alloy 713C	0.12	74	12.5	...	4.2	...	6	0.012	0.8	...	0.1	2Nb
Alloy 713LC	0.05	75	12	...	4.5	...	6	0.01	0.6	...	0.1	2Nb
Alloy 718	0.04	53	19	...	3	18	0.5	...	0.9	...	...	0.1Cu, 5Nb
Alloy X-750	0.04	73	15	...	...	7	0.7	...	2.5	...	...	0.25Cu, 0.9Nb
M-252	0.15	56	20	10	10	...	1	0.005	2.6	...	...	...
MAR-M 200 (DS)	0.15	59	9	10	...	1	5	0.015	2	12.5	0.05	1Nb(c)
MAR-M 246	0.15	60	9	10	2.5	...	5.5	0.015	1.5	10	0.05	1.5Ta
MAR-M 247 (DS)	0.15	...	9.4	10	0.7	...	5.5	0.015	1	10	0.05	1.5Hf, 3Ta
NX 188 (DS)	0.04	74	...	...	18	...	8	...	...	...	...	...
René 77	0.07	58	15	15	4.2	...	4.3	0.015	3.3	...	0.04	...
René 80	0.17	60	14	9.5	4	...	3	0.015	5	4	0.03	...
René 100	0.18	61	9.5	15	3	...	5.5	0.015	4.2	...	0.06	1V
TRW-NASA VIA	0.13	61	6	7.5	2	...	5.5	0.02	1	6	0.13	0.4Hf, 0.5Nb, 0.5Re, 9Ta
Udimet 500	0.1	53	18	17	4	2	3	...	3	...	...	...
Udimet 700	0.1	53.5	15	18.5	5.25	...	4.25	0.03	3.5	...	...	...
Udimet 710	0.13	55	18	15	3	...	2.5	...	5	1.5	0.08	...
Waspaloy	0.07	57.5	19.5	13.5	4.2	1	1.2	0.005	3	...	0.09	...
WAZ-20 (DS)	0.20	72	...	...	...	...	6.5	...	...	20	1.5	...

(a) B-1900 + Hf also contains 1.5% Hf. (b) ACI designation. (c) MAR-M 200 + Hf also contains 1.5% Hf.

M-25S, and M-30C and in Federal specification QQ-N-288 as A, B, C, D, E, and F (see Table 1). The lower-silicon grades (≤2.0% Si) are commonly used in corrosion applications in conjunction with wrought nickel-copper and copper-nickel alloys as pump, valve, and fitting castings. The higher-silicon grades (>2.0% Si) combine corrosion resistance with high strength and wear resistance. Grades containing the highest silicon contents (>4% Si) are used when exceptional resistance to galling is desired. The high-carbon composition (QQ-N-288-F) is used where improved machinability is more important than ductility and/or weldability.

Nickel-Chromium-Iron Alloys. The cast nickel-chromium-iron CY-40 is included in ASTM A 494. Its composition is given in Table 1. The cast alloy differs from the parallel wrought grade in carbon, manganese, and silicon contents for improved castability and pressure tightness.

Heat-resistant nickel-chromium-iron grades HW and HX (Table 2) are covered in ASTM A 297. HW alloy (60Ni-12Cr-23Fe) is especially well suited to applications in which wide and/or rapid fluctuations in temperature are encountered. In addition, HW exhibits excellent resistance to carburization and high-temperature oxidation. This alloy performs satisfactorily at temperatures up to about 1120 °C (2050 °F) in strongly oxidizing atmospheres and up to 1040 °C (1900 °F) in oxidizing or reducing products of combustion, provided that sulfur is not present in the gas.

HW alloy is widely used for intricate heat-treating fixtures that are quenched with the load and for many other applications (such as furnace retorts and muffles) that involve thermal shock, steep temperature gradients, and high stresses. Its structure is austenitic and contains carbides in amounts that vary with carbon content and thermal history. In the as-cast condition, the microstructure consists of a continuous interdendritic network of elongated eutectic carbides. Upon prolonged exposure at service temperatures, the austenitic matrix becomes uniformly peppered with small carbide particles except in the immediate vicinity of eutectic carbides. This change in structure is accompanied by an increase in room-temperature strength, but no change in ductility.

HX alloy (66Ni-17Cr-12Fe) is similar to HW but contains more nickel and chromium. Its higher chromium content gives it substantially better resistance to corrosion by hot gases (even sulfur-bearing gases), which permits it to be used in severe service applications at temperatures up to 1150 °C (2100 °F).

The cast nickel-chromium-molybdenum alloys, designated CW-12MW, CW-7M, CW-2M, and CW-6MC in ASTM A 494 (Table 1), are used in severe service conditions that usually involve combinations of acids at elevated temperatures. The molybdenum in these compositions improves both resistance to nonoxidizing acids and high-temperature strength.

The cast nickel-molybdenum alloys, designated N-12MV and N-7M in ASTM A 494 (Table 1), are most frequently applied in handling hydrochloric acid at all concentrations and temperatures, including the boiling point. Nickel-molybdenum alloys are also produced under proprietary names.

Nickel-Base Proprietary Alloys. In addition to the more common nickel-base alloys, there are a number of trademarked, proprietary alloys and other nickel-base alloys that are widely used in corrosive service. Many of these alloys have excellent general corrosion resistance and are most commonly used in applications for which stainless steels are inadequate. Others are used in specialized applications and should not be considered substitutes for stainless steel. Producers should be consulted when applying these alloys, particularly in applications for which there is no history of use.

Heat-Resistant Alloys. Casting are classified as heat resistant if they are capable of sustained operation while exposed, either continuously or intermittently, to operating temperatures that result in metal temperatures in excess of 650 °C (1200 °F). Many alloys of the same general types are also used for their resistance to corrosive media at temperatures below 650 °C (1200 °F), and castings intended for such service are classified as corrosion resistant. Although there is usually a distinction between heat-resistant alloys and corrosion-resistant alloys, based on carbon content, the line of demarcation is vague—particularly for alloys used in the range from 480 to 650 °C (900 to 1200 °F).

Table 2 lists a number of nickel-base casting alloys used for high-temperature applications. In addition to the HW and HX grades mentioned above in the discussion on nickel-chromium-iron alloys, a number of proprietary alloys are listed. These ma-

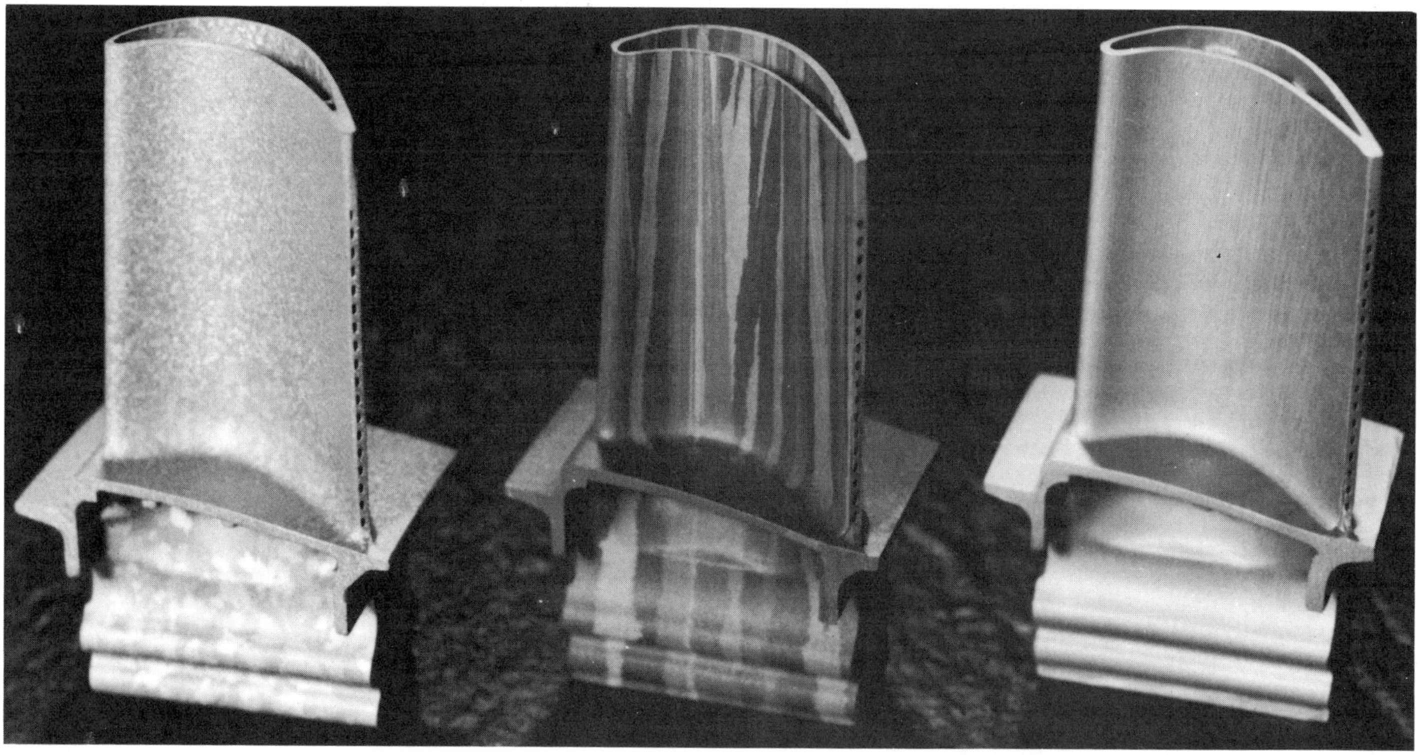

Fig. 1 Comparison of equiaxed (left), directionally solidified (center), and single-crystal (right) nickel-base alloy turbine blades for an aircraft engine. Courtesy of Howmet Corporation, Whitehall Casting Division

terials, often referred to as superalloys, contain appreciable amounts of chromium and cobalt, with aluminum and titanium added for strengthening. The effects of aluminum and titanium on the structure and the resulting properties of nickel-base alloys are discussed in the section "Structure and Property Correlations" in this article.

Directionally solidified and single-crystal alloys have the highest elevated-temperature strengths of any of the nickel-base alloys. Directional solidification is accomplished by removing all of the heat from one end of the casting. This is done by establishing a strong thermal gradient and passing it from one end of the casting to the other. In this way, large, columnar grains are produced that are oriented such that they provide maximum strength in service.

Alloys developed for single-crystal casting are characterized by the absence of grain-boundary strengthening elements such as carbon, boron, zirconium, and hafnium. The removal of these alloying elements results in materials with very high incipient melting temperatures. Figure 1 compares the macro grain structures of equiaxed (conventional), directionally solidified, and single-crystal nickel-base alloy turbine blades. Table 2 lists several compositions of DS/SC alloys.

Table 3 Tensile requirements for cast nickel-base alloys

Alloy	Tensile strength MPa	Tensile strength ksi	Yield strength MPa	Yield strength ksi	Elongation in 50 mm (2 in.), %	Hardness, HB
Corrosion-resistant						
CZ-100	345	50	125	18	10	. . .
M-35-1	450	65	170	25	25	. . .
M-35-2	450	65	205	30	25	. . .
M-30H	690	100	415	60	10	243–294
M-25S	. . .	. . .	. . .	. . .	. . .	300 (min)
M-30C	450	65	225	32.5	25	125–150
N-12MV	525	76	275	40	6	. . .
N-7M	525	76	275	40	20	. . .
CY-40	485	70	195	28	30	. . .
CW-12MW	495	72	275	40	4	. . .
CW-7M	495	72	275	40	25	. . .
Heat-resistant						
HW	415	60	415	60	. . .	. . .
HX	415	60	415	60	. . .	. . .

Structure and Property Correlations

Cast Nickel. The mechanical property requirements for CZ-100 are listed in Table 3. Cast nickel has excellent toughness, thermal resistance, and heat transfer characteristics.

Nickel-Copper Alloys. Tensile properties of the nickel-copper castings are controlled by the solution-hardening effect of silicon or silicon plus niobium. The tensile properties of M-35-1, M-35-2, and composition A are controlled by a carbon-plus-silicon relationship; composition E and M-30C tensile properties are determined by a silicon-plus-niobium relationship.

At approximately 3.5% Si, an age-hardening effect appears; at approximately 3.8% Si, the solubility of silicon in the nickel-copper matrix is exceeded, and hard, brittle silicides begin to appear. The combination of an aging effect plus silicides in composition D results in an alloy with exceptional resistance to galling. As the silicon content is increased above 3.8%, the amounts of hard, brittle silicides in the tough nickel-copper matrix increase; ductility decreases sharply; and tensile and yield strengths increase. As a result, hardness is the only mechanical property specified for composition D. The toughness of nickel-copper alloys decreases with increasing silicon content, but all grades retain their room-temperature toughness down to at least −195 °C (−320 °F). Tensile require-

Table 4 Elevated-temperature mechanical properties of alloy CY-40

Temperature °C	Temperature °F	Tensile strength MPa	Tensile strength ksi	Yield strength MPa	Yield strength ksi	Elongation in 25 mm (1 in.), %	Stress to rupture in 100 h MPa	Stress to rupture in 100 h ksi
Room		486	70.5	293	42	16	...	...
480	900	427	62	...	...	20	...	...
650	1200	372	54.5	...	...	21	165	24
730	1350	314	45.5	...	...	25	103	15
815	1500	187	27.1	...	...	34	62	9
925	1700	...	...	...	...	...	38	5.5

ments for nickel-copper alloys are listed in Table 3.

Nickel-Chromium-Iron Alloys. Minimum mechanical properties for both corrosion-resistant and heat-resistant alloys are given in Table 3. Alloy CY-40 is frequently used for elevated-temperature fittings in conjunction with the wrought alloy of similar base composition. Typical elevated-temperature properties are listed in Table 4. Applications for HW and HX alloys were discussed above.

Nickel-Chromium-Molybdenum Alloys. The CW-12MW and CW-7M grades have a relatively high yield strength (Table 3) due to the solution-hardening effects of chromium, molybdenum, and silicon in CW-7M and of tungsten and vanadium in CW-12MW. Ductility is excellent up to the limit of solid solubility. Inadequate heat treatment or improper composition balance, however, may result in the formation of a hard, brittle phase and in a significant loss of ductility. Careful control within the specified composition range is therefore necessary to meet the specified ductility. Carbon and sulfur contents should be kept as low as practicable.

Nickel-Molybdenum Alloys. The N-12MV and N-7M grades have good yield strengths because of the solution-hardening effect of molybdenum (Table 3). Ductility is controlled by the carbon and molybdenum contents. For optimum ductility, carbon content should be as low as practicable, and molybdenum content should be adjusted to avoid the formation of intermetallic phases.

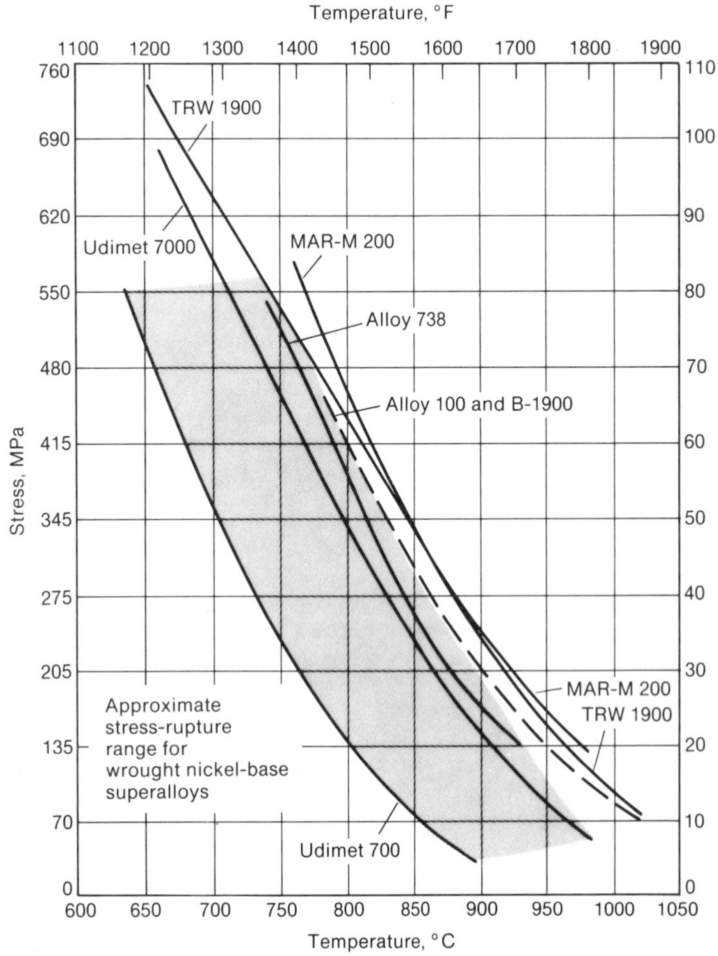

Fig. 2 Stress-rupture curves for 1000-h life of selected cast nickel-base heat-resistant alloys

Heat-Resistant Alloys. Nickel-base heat-resistant casting alloys, often referred to as superalloys, generally contain substantial levels of aluminum and titanium (Table 2). These elements strengthen the austenitic matrix through precipitation of $Ni_3(Al,Ti)$, an ordered face-centered cubic (fcc) compound referred to as gamma prime (γ'). Various ratios of aluminum and titanium are used in the different nickel-base heat-resistant alloys; generally, titanium atoms can replace aluminum atoms up to a ratio of 3 Ti to 1 Al without altering the ordered fcc crystallographic structure of γ'. When excess titanium is present, Ni_3Ti, an ordered close-packed hexagonal compound known as eta phase (η), precipitates. Because γ' is coherent with the matrix, precipitation of this phase has a greater strengthening effect than precipitation of η.

In addition to the strengthening imparted by γ' precipitation, solid-solution strengthening is conferred by the addition of refractory elements, and grain-boundary strengthening by additions of boron, zirconium, carbon, and hafnium. Hafnium also enhances grain-boundary ductility. Stress-rupture curves for various nickel-base alloys are shown in Fig. 2.

The strength of these alloys is complemented by superior corrosion resistance, which is conferred by chromium and aluminum (titanium may be more favorable than aluminum under hot-corrosion conditions). Coatings are used on most nickel alloys for temperatures exceeding about 815 °C (1500 °F) to provide adequate protection from oxidation and corrosion at these temperatures.

Nickel-base heat-resistant alloy castings are produced by investment casting under vacuum, and improvements in properties have been made not only through control of composition but also through more precise control of microstructure. A significant advance in microstructure control was the development of a columnar grain structure produced by directional solidification and single-crystal technology (see discussion below). Extensive use of nickel alloy castings essentially began with Alloy 713, and alloys are available that can be used at temperatures up to about 1040 °C (1900 °F).

In addition to creep strength and corrosion resistance, two other properties—stability, and resistance to thermal fatigue—are important considerations in the selection of nickel-base heat-resistant casting alloys. Thermal-fatigue resistance is partially controlled by composition, but it is also significantly affected by grain-boundary area and alignment relative to applied stresses. The crystallographic orientation of grains also influences thermal stresses because the modulus of elasticity, which directly influences thermal stresses, varies with grain orientation. The stability of property values is directly influenced by metallurgical stability; any microstructural

Fig. 3 Various radial and axial turbine wheels made from Mar-M-247 alloy. Courtesy of Howmet Corporation, Whitehall Casting Division

Fig. 4 Directionally solidified turbine vane made from CM-247-LC alloy. Courtesy of Thyssen Guss AG

changes that take place during long-term exposure at high temperatures under stress cause attendant changes in properties. For example, if the γ' phase coarsens, strength decreases. Also, potentially deleterious topologically close-packed (tcp) secondary phases, such as σ, Laves, and μ, may form. Coarsening of γ' can be controlled to some degree by adjusting alloy additions. Formation of tcp phases is controlled by adjusting the composition of the matrix to minimize the electron vacancy number, N_v. A high N_v indicates a tendency toward the formation of tcp phases. In general, an N_v value below 2.4 indicates minimal formation of deleterious phases; however, this relationship varies with base-alloy composition. The metallurgical structures of both cast and wrought heat-resistant alloys are discussed in greater detail in Volume 9 of the 9th Edition of *Metals Handbook*.

Alloys 713C and 713LC are closely related investment casting alloys used principally for low-pressure turbine airfoils in gas turbines. Intended for operation at intermediate temperatures from 790 to 870 °C (1450 to 1600 °F), these alloys are generally used in uncooled airfoil designs.

Alloy 738X is an investment casting alloy similar in strength to Alloy 713C and Udimet 700 but with outstanding sulfidation resistance. It is used principally for latter-stage turbine airfoils and for hot-corrosion-prone applications such as industrial and marine engines.

Udimet 700, although primarily a wrought alloy, is also used in investment cast high-pressure turbine blades. In cast form, it is similar in strength to Alloy 713C but offers better hot-corrosion resistance. It is designed for operation at intermediate temperatures from 730 to 900 °C (1350 to 1650 °F). A stability-controlled version of U-700 is known as René 77.

Alloy 100 is designed for use at metal temperatures up to about 980 °C (1800 °F) in cooled and uncooled airfoils. A stability-controlled version of Alloy 100 is known as René 100.

B-1900, to which 1% Hf is usually added to improve ductility and thermal-fatigue resistance, is designed for use at metal temperatures up to about 980 °C (1800 °F) in cooled and uncooled airfoils.

René 80 offers excellent corrosion resistance in sulfur-bearing environments. It is designed for use at metal temperatures up to about 950 °C (1750 °F).

Alloy 792 is designed for use in applications similar to those of René 80. It is one of the most sulfidation-resistant nickel alloys available.

MAR-M 246 and MAR-M 247 are designed for use at metal temperatures of about 980 to 1010 °C (1800 to 1850 °F) in cooled and uncooled airfoils and radial and axial wheels (Fig. 3).

DS MAR-M 200 + Hf is produced by directional solidification (see discussion below) and is designed for metal temperatures of about 1010 to 1040 °C (1850 to 1900 °F). It is used in cooled airfoils.

Other alloys (such as Udimet 500) are occasionally used in turbine airfoil applications, and Alloy 718 has been cast into large static structures for gas turbines. Additional information on the applications and processing of investment cast nickel-base heat-resistant alloys can be found in the articles "Classification of Processes and Flow Chart of Foundry Operations" and "Investment Casting" in this Volume.

Alloys for directional and single-crystal solidification possess high elevated-temperature strengths. Directionally solidified turbine blades have high strength in the direction of principal stress (the longitudinal direction) because grain boundaries are aligned parallel to this direction. Thus, the effect of grain boundaries on properties is minimized.

Single-crystal alloys have no grain boundaries and therefore require no grain-boundary strengthening elements. For this reason, they can be solution heat treated at higher temperatures than conventional alloys, precipitating a greater amount of the γ' strengthening phase. The lack of grain boundaries also enhances the corrosion resistance of these materials. Table 2 lists several DS/SC alloy compositions. A turbine vane made from CM-247-LC DS alloy is shown in Fig. 4. Properties and performance of DS/SC alloys are detailed in Ref 1 to 4.

Melting Practice

Electric induction furnaces have become the mainstay of the foundry industry for small heat sizes, especially when a number of different alloys are produced. They are also the least expensive of the major furnace types to install. The foundry industry uses these furnaces in sizes ranging from 9 kg to 18 Mg (20 lb to 20 tons); however, most electric induction furnaces are in the 25 to 1350 kg (50 to 3000 lb) range.

Figure 5 shows a cross section of an induction furnace. The furnace shell rests on trunnions, which tilt the furnace during tapping. A copper coil surrounds the furnace lining and the charge materials inside. The metal charge is melted by its resistance to the current induced by a magnetic field when current flows through the coil. More detailed information on induction furnaces can be found in the article "Melting Furnaces" in this Volume.

Vacuum Melting. Nickel-base alloys containing more than about 0.2% of the reactive elements aluminum, titanium, and zirconium are not suitable for melting and casting in oxidizing environments such as air. At the higher alloying levels, these elements readily oxidize, resulting in gross inclusions, oxide laps, and poor composition control. Consequently, such alloys generally require inert gas injection or vacuum melting and casting methods. Extralow gas contents, which can be obtained by vacuum melting, are also required for certain nickel-base alloys. Vacuum melting processes, which are described in the article "Vacuum Melting and Remelting Processes" are always used for directional solidification and single-crystal casting alloys.

Metal Treatment

Argon Oxygen Decarburization (AOD). Some foundries have recently installed AOD units to achieve some of the results that vacuum melting can produce. The AOD unit looks very much like a Bessemer converter with tuyeres in the lower sidewalls for the injection of argon or nitrogen and oxygen. These units must be charged with molten metal from an arc or induction furnace. About 20%, but usually less, cold charge consisting of solid virgin material can be added to an AOD unit. The continuous injection of gases causes a violent stirring action and intimate mixing of slag and metal, which can lower sulfur values to below 0.005%. The gas contents (hydrogen, nitrogen, and oxygen) may be even lower than those of vacuum induction melted alloys. More information on AOD processing is available in the section "Argon Oxygen Decarburization" of the article "Degassing Processes (Converter Metallurgy)" in this Volume.

Electroslag remelting furnaces represent another type of equipment that may see

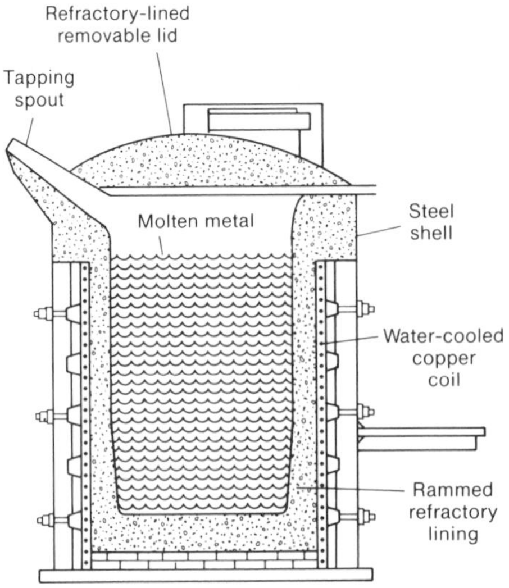

Fig. 5 Cross section of a coreless electric induction furnace

some use in the high-alloy foundry in the next decade. Electroslag remelting machines have been used for many years by the wrought steel companies to produce refined ingots. In the Soviet Union, electroslag remelting is being used to cast shapes, and the technology is being evaluated in the United States as well. The process works by taking an ingot (which becomes the electrode), remelting it in stages under molten slag to refine it, and then resolidifying the metal in a water-cooled mold [see the section "Electroslag Remelting (ESR)" in the article "Vacuum Melting and Remelting Processes" in this Volume].

Plasma Refining. Steadily increasing requirements for alloy cleanliness have led producers to adopt several novel refining technologies and process routes, many involving increased use of the ladle as a refining vessel. Such procedures require longer holding times in the ladle, which necessitate either increased superheats in the furnace or external heating in the ladle to avoid early solidification. Higher superheat, in addition to requiring excessive energy expenditure, can contribute to the problem of melt contamination. The preferred solution is to supply heat to the ladle, maintaining the alloy at minimal superheat during refining. This can be accomplished by the transferred arc plasma torch, with the added benefit of enhanced refining reactions that aid in the production of clean metal with low levels of residual elements. In this work, experiments have been carried out in an induction furnace equipped with a gas-stabilized graphite electrode to investigate the control of oxygen and induction levels and the enhancement of desulfurization afforded by the transferred arc plasma [see the section "Plasma Heating and Degassing" in the article "Degassing Processes (Ladle Metallurgy)" in this Volume].

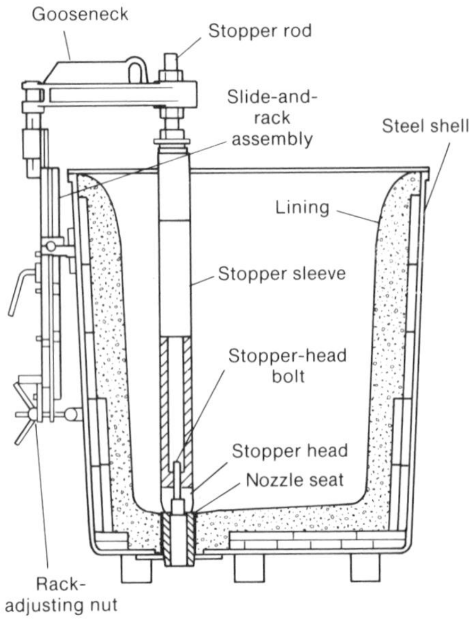

Fig. 6 Typical bottom-pour ladle used to pour large castings

Foundry Practice

Foundry practice for nickel-base alloys is for the most part similar to that used for cast stainless steels (see the article "High-Alloy Steels" in this Volume). Specific aspects of foundry practice discussed here include pouring, gating and risering, cleaning, welding, and heat treatment of conventional corrosion-resistant nickel-base alloy castings. The processing of investment cast and DS/SC alloys is reviewed in the articles "Investment Casting" and "New and Emerging Processes" (see the section "Directional and Monocrystal Solidification"), respectively, in this Volume.

Pouring Practice

Three types of ladles are used for pouring nickel-base castings: bottom pour, teapot, and lip pour. Ladle capacity normally ranges from 45 kg to 36 Mg (100 lb to 40 tons), although ladles having much larger capacities are available.

The bottom-pour ladle has an opening in the bottom that is fitted with a refractory nozzle (Fig. 6). A stopper rod, suspended inside the ladle, pulls the stopper head up from its seat in the nozzle, allowing the molten alloy to flow from the ladle. When the stopper head is returned to the position shown in Fig. 6, the flow is cut off. The position of the stopper head is controlled manually by the slide-and-rack mechanism shown at the left in Fig. 6.

Bottom pouring is preferred for pouring large castings from large ladles, because it is difficult to tip a large ladle and still control the stream of molten steel. Also, the bottom-pour ladle delivers cleaner metal to the mold. Inclusions, pieces of ladle lining, and slag float to the top of the ladle; thus, bottom pouring greatly reduces the risk of

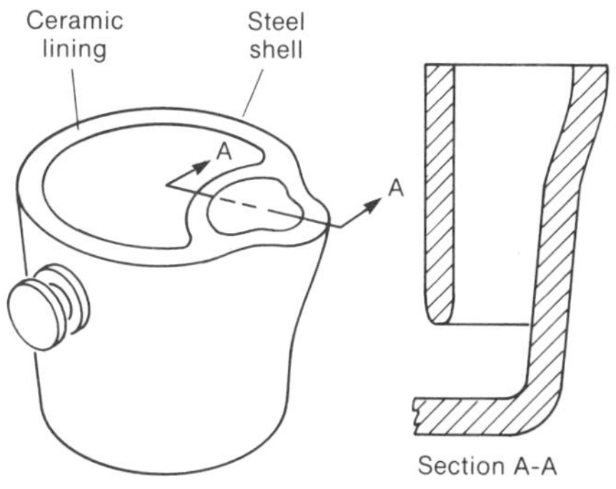

Fig. 7 Typical teapot ladle used to pour small- to medium-size castings

passing nonmetallic particles into the mold cavity.

On the other hand, it is impractical to pour molten metal into small molds from a large bottom-pour ladle. The pressure head created by the metal remaining in the ladle delivers the molten metal too fast. Also, the time required to fill a small mold is short, thus requiring that a large bottom-pour ladle be opened and closed many times in order to empty it. This may cause the ladle to leak, although special nozzles have been developed to minimize leakage. Despite the fact that the size of bottom-pour ladles could be scaled down for pouring smaller castings, this is unnecessary because of the almost equal ability of the teapot ladle to deliver clean metal.

The teapot ladle incorporates a ceramic wall, or baffle, that separates the bowl of the ladle from the spout. The baffle extends almost four-fifths of the distance to the bottom of the ladle (Fig. 7). As the ladle is tipped, hot metal flows from the bottom of the ladle up the spout and over the lip. Because the metal is taken from near the bottom of the ladle, it is free of slag and pieces of eroded refractory. The teapot design is feasible in various sizes, generally covering the entire range of casting sizes that are below the minimum size for which the bottom-pour ladle is used.

Lip-pour ladles resemble the teapot type but have no baffles to hold back the slag. Because the hot metal is not taken from the bottom of the ladle, this type of ladle pours a more contaminated metal and is seldom used to pour high-alloy castings. Nevertheless, it is widely used as a tapping ladle (at the melting furnace) and as a transfer ladle to feed smaller ladles of the teapot type.

Pouring Time. Ideally, the optimum pouring time for a given casting would be determined by the weight and shape of the casting, the temperature and solidification characteristics of the molten metal, and the heat transfer and thermal stability characteristics of the mold. However, most foundries are required to pour may different castings from one heat or even from one ladle. Therefore, rather than attempting to control

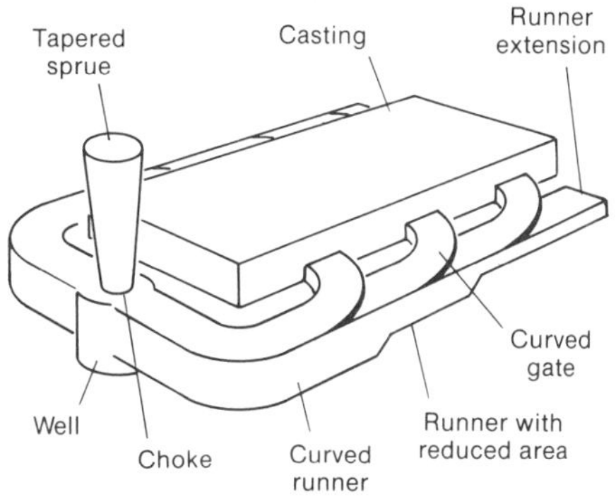

Fig. 8 Gating system for good metal flow

pouring time directly, foundries control the speed with which molten steel enters the mold cavity. This control is achieved through the design of the gating system.

Gating Systems

An effective gating system for pouring nickel-base alloys, as well as other metals, into green sand molds is one that fills the mold as rapidly as possible without developing pronounced turbulence. It is essential that the mold be filled rapidly to minimize temperature variations within the metal; this results in optimized control of solidification.

Turbulent metal flow is harmful because it breaks up the metal stream, exposing more surface area to air and forming metallic oxides. The oxides can rise to the top of the mold cavity, resulting in a rough surface of inclusions in the casting. In addition, turbulent flow erodes the mold material. These eroded particles also float to the top of the mold cavity.

Preferred Metal Flow. According to preferred practice, the pourer directs the metal stream toward the pouring cup at the top of the mold, controlling the pouring rate to keep the cup full of molten metal throughout the pouring cycle. The opening in the bottom of the cup is directly over the sprue, or downgate, which is tapered at the bottom, thus reducing the diameter of the stream of descending metal. The taper prevents the stream from pulling away from the walls and drawing air into the gating system. The descending metal impinges on the sprue well at the bottom of the sprue, and the direction of flow changes from vertical to horizontal, with the metal flowing along runners to gates (ingates), and then to the main body of the casting. A gating system that incorporates these features is shown in Fig. 8.

Gating system design largely determines the manner in which molten metal is fed into the mold, as well as the rate of feeding. The number of gates influences the distribution of the flow between gates. A good design has even distribution between gates both initially and while the mold is filling. The distribution of flow in the gating system affects the type of flow that occurs in the main body of the casting. A large difference in the flow between gates creates a swirl of metal in the mold about a vertical axis, in addition to that occurring about a horizontal axis.

The gating system shown in Fig. 8 is an example of a so-called finger-type parting line system, in which the fingers feed metal to the casting just above the parting line. Other major types of gating systems used in alloy foundries include the bottom gate, which feeds metal to the bottom of the casting, and the step gate, which feeds metal through a number of stepped gates, one above another. In the system shown in Fig. 8, the ratio of the cross-sectional area of the choke of the sprue to that of all of the runners emanating from the sprue basin and to all of the gates is 1:4:4. As shown in Fig. 8, the runner area decreases progressively by an amount equal to the area of each gate it passes. This practice ensures that, once the system is filled with metal, it remains full during the pouring cycle and feeds equally to each gate.

Furthermore, the gates are located in the cope, while the runner, which extends beyond the last gate, is located in the drag. Extension of the runner serves as a trap for the first, and usually the most contaminated, metal to enter the system. The entire runner must fill before the metal will rise to the level of the gates. Thus, each gate begins feeding at the same time. The runners and gates are curved wherever a change in direction occurs. This streamlining reduces turbulence in the metal stream and minimizes mold erosion.

In contrast to the ratio of the system shown in Fig. 8 (1:4:4), if the total cross-sectional area of the gates is less than that of the runners (1:2:1, for example), the result is a pressurized system. The metal squirts into the mold cavity and flows turbulently over the mold bottom, which can cause roughening of the bottom surfaces.

Conversely, if the total cross-sectional area of the gates is significantly greater than that of the runners (1:2:3, for example), the gating system will be incompletely filled, and flow from the gates will be uneven. This condition increases the likelihood of mold erosion. When this type of system is required, complicated additions to gating systems are used, including whirlgates, horn gates, strainer cores, tangential gates, and slit gates. However, any addition to the gating system usually increases the cost of the casting because all gating must be removed. More detailed information on gating practice can be found in the article "Gating Design" in this Volume.

Mold Erosion. In addition to the contribution of gating design to a reduction in mold erosion, further reduction can be achieved by making the gating system out of tile, which is superior to green sand in erosion resistance. However, the use of tile is generally limited to gating systems for large castings, where the quantity and speed of molten metal passing through the gating system would seriously erode green sand and where precise control of the flow rate is less critical. Thus, gating systems for smaller castings are rammed in sand, usually with a semicircular or rectangular cross section for the gates and runners.

Risers

Molten nickel-base alloys contract approximately 0.9% per 55 °C (100 °F) as they cool from the pouring temperature to the solidification temperature. They then undergo solidification contraction of 3% during freezing, and finally the solidified metal contracts 2.2% during cooling to room temperature. Therefore, when casting nickel alloys, an ample supply of molten metal must be available from risers (reservoirs) to compensate for the volume decrease, or shrinkage cavities will develop in the locations that solidify last.

Because feed from the riser occurs by gravity, risers are usually located at the top of the casting. Riser forms are placed on the pattern and molded into the cope half of the mold. The riser cavity is usually open to the top of the mold, although blind risers are sometimes used.

Riser Size and Shape. As described in the article "Riser Design" in this Volume, formulas based on surface area, volume, and freezing time of the casting are used to determine riser size. Most risers are cylindrical in shape, with their heights approximately equal to their diameters. This configuration provides a low ratio of surface area to volume, which prolongs the time the metal in the riser remains liquid.

Placement of a riser, in conjunction with its size, determines its effectiveness. The thicker sections of a casting act as reservoirs for feeding the thinner sections, which solidify first. Thus, risers are placed over thick sections that cannot be fed by other areas of the casting. Demonstrating this principle, the gear blank casting shown in Fig. 9 is provided with a large riser over the central hub and six smaller risers spaced equally around the rim of the gear to ensure adequate feeding. Metal enters the mold at the two gates, which are 180° apart.

Feeding Distance. Castings of uniform thickness present a different problem. Studies have established the feeding distances of a riser for various rectangular shapes in both the horizontal and vertical planes, with and without an end effect (that is, the extra cooling provided by the sand cover of an end surface).

For a uniform section, the maximum feeding distance can be extended by adding a taper. The progressively thicker section

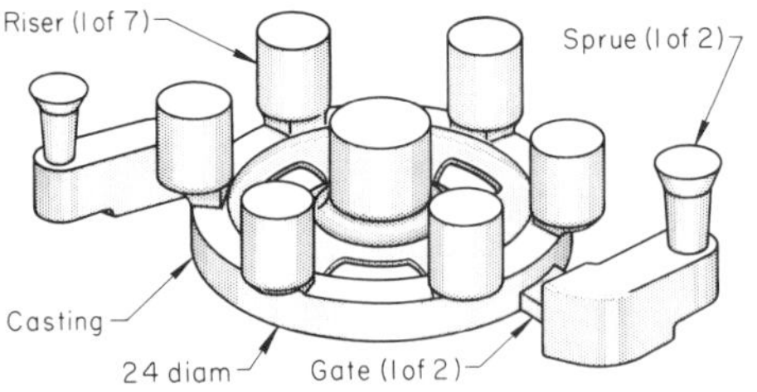

Fig. 9 Gating and feeding system used to cast gear blanks

solidifies in a progressively longer time, so that a favorable temperature gradient is established from the end of the section to the riser. A tapered pad of exothermic material placed in the mold along the length of the casting will also produce a favorable temperature gradient.

Welding

Cast Nickel. Alloy CZ-100 can be readily repair welded or joined to other castings or to wrought forms by using any of the usual welding processes with suitable nickel rod and wire. Joints or cavities must be carefully prepared for welding because small amounts of sulfur or lead cause weld embrittlement.

Nickel-Copper Alloys. The weldability of the nickel-copper alloys decreases with increasing silicon content, but is adequate up to at least 1.5% Si. Niobium can enhance weldability, particularly when small amounts of low-melting residuals are present. Careful raw material selection and proper foundry practice, however, have largely eliminated any difference in weldability between niobium-containing and niobium-free grades.

The higher-silicon compositions (≥3.5% Si) are not considered weldable. They can be brazed or soldered, as can the lower-silicon grades.

Nickel-Chromium-Iron Alloys. The CY-40 castings can be repair welded or fabrication welded to matching wrought alloys by any of the usual welding processes. Rod and wire of matching nickel-chromium contents are available. Postweld heat treatment is not required after repair welding or fabrication, because the heat-affected zone is not sensitized by the weld heat.

Nickel-Chromium-Molybdenum Alloys. Alloys CW-12MW and CW-7M can be welded by any of the usual welding processes, using wire or rod of matching composition. For optimum weldability, carbon content should be as low as practicable. The usual practice is to solution treat and quench after repair welding. Heat treatment after welding is generally necessary because these alloys are subject to sensitization in the heat-affected zone and because intermetallic precipitates may form in the heat-affected zone.

Nickel-Molybdenum Alloys. Alloys N-12MV and N-7M can be welded by using any of the usual welding processes with wire or rod of matching composition. Postweld heat treatment is usually performed because these alloys are subject to the precipitation of intermetallic compounds in the heat-affected zone.

Heat Treatment

Cast nickel (alloy CZ-100) is used in the as-cast condition. Some other alloys are also used as-cast, but most require some type of thermal treatment to develop optimum properties.

Nickel-copper alloys are used in the as-cast condition. Homogenization at 815 to 925 °C (1500 to 1700 °F) may, under some conditions, improve corrosion resistance slightly, but in most corrosive conditions, alloy performance is not affected by the minor segregation present in the as-cast alloy.

At about 3.5% Si, silicon begins to have an age-hardening effect. The resultant combination of aging and the formation of hard silicides when the silicon content exceeds about 3.8% can cause considerable difficulty in machining. Softening is accomplished by a solution heat treatment, which consists of heating to 900 °C (1650 °F), holding at temperature for 1 h per 25 mm (1 in.) of section thickness, and oil quenching. Maximum softening is obtained by oil quenching from 900 °C (1650 °F), but such treatment is likely to result in quench cracks in castings with complex shapes or varying section thickness.

In the solution heat treatment of complicated or varying-section castings, it is advisable to charge them into a furnace below 315 °C (600 °F) and heat to 900 °C (1650 °F) at a rate that limits the maximum temperature difference within the casting to about 56 °C (100 °F). After being soaked, castings should be transferred to a furnace held at 730 °C (1350 °F), allowed to equalize in temperature, and then oil quenched. Alternatively, the furnace can be rapidly cooled to 730 °C (1350 °F), the casting temperature can be equalized, and the castings can be quenched in oil. Solution heat treated castings are age hardened by placing them in a furnace held at 315 °C (600 °F), heating uniformly to 595 °C (1100 °F), holding at 595 °C (1100 °F) for 4 to 6 h, and air or furnace cooling.

Nickel-Chromium-Iron Alloys. Alloy CY-40 is used in the as-cast condition because it is insensitive to the intergranular attack encountered in as-cast or sensitized stainless steels. A modified cast nickel-chromium-iron alloy for nuclear applications with 0.12% C (max) is usually solution treated as an additional precaution.

Sensitization in the heat-affected zone is not a problem with CY-40. Unless residual stresses pose a problem, postweld heat treatment is therefore not required.

Nickel-Chromium-Molybdenum Alloys. The high chromium and molybdenum contents of CW-12MW and CW-7M result in the precipitation of carbides and intermetallic compounds in the as-cast condition, which can be detrimental to corrosion resistance, ductility, and weldability. These alloys should therefore be solution treated at a temperature of 1175 to 1230 °C (2150 to 2250 °F) and water or spray quenched.

Nickel-Molybdenum Alloys. Slow cooling in the mold is detrimental to the corrosion resistance, ductility, and weldability of N-12MV and N-7M. These alloys should therefore be solution treated at a minimum temperature of 1175 °C (2150 °F) and water quenched.

Specific Applications

Corrosion-resistant nickel-base castings are primarily used in fluid-handling systems with matching wrought alloys; they are also commonly used for pump and valve components or for applications with crevices and velocity effects requiring a superior material in a wrought stainless system. Because of their relatively high cost, nickel-base alloys are usually selected only for severe service conditions in which maintenance of product purity is of great importance and for which less costly stainless steels or other alternative materials are inadequate. Detailed information on the corrosion resistance of nickel-base alloys in aqueous media is available in the article "Corrosion of Nickel-Base Alloys" in Volume 13 of the 9th Edition of *Metals Handbook*.

In the application of heat-resistant alloys, considerations include:

- Resistance to corrosion (oxidation) at elevated temperatures
- Stability (resistance to warping, cracking, or thermal fatigue)
- Creep strength (resistance to plastic flow)

Numerous applications of cast heat-resistant nickel-base alloys were discussed earlier in this article. Information on the high-temperature corrosion resistance of these alloys is available in the articles "Fundamentals of Corrosion in Gases," "General Corrosion" (see the section "High-Temperature Corrosion"), and "Corrosion of Metal Processing Equipment" (see the section "Corrosion of Heat-Treating Furnace Accessories") in Volume 13 of the 9th Edition of *Metals Handbook*.

Cast Nickel. Nickel castings are most commonly used in the manufacture of caustic soda and in processing with caustic (see the section "Corrosion by Alkalies and Hypochlorite" in the article "Corrosion in the Chemical Processing Industry" in Volume 13 of the 9th Edition of *Metals Handbook*). As the temperature and caustic soda concentration increase, austenitic stainless steels are useful only up to a point. The nickel-copper and nickel-chromium-iron alloys take over as useful alloys under these conditions, while cast nickel is selected for the higher caustic concentrations, including fused anhydrous soda. Minor amounts of such elements as oxygen and sulfur can have profound effects on the corrosion rate of nickel in caustic. Detailed corrosion data should therefore be consulted before making a final alloy selection.

Nickel-Copper Alloys. The principal advantages of the Ni-30Cu alloys are high strength and toughness, coupled with excellent resistance to mineral acids, organic acids, salt solution, food acids, strong alkalies, and marine environments. The most common applications for nickel-copper castings are in the manufacture of, and processing with, hydrofluoric acid and the handling of salt water, neutral and alkaline salt solutions, and reducing acids.

Nickel-chromium-iron alloys are commonly used under oxidizing conditions to handle high-temperature corrosives or corrosive vapors where stainless steels might be subject to intergranular attack or stress-corrosion cracking. In recent years, the CY-40-type alloy has found large-scale application in handling hot boiler feedwater in nuclear plants because of a greater margin of safety over stainless steels. More information on this application is available in the article "Corrosion in the Nuclear Power Industry" in Volume 13 of the 9th Edition of *Metals Handbook*.

Nickel-chromium-molybdenum alloys are probably the most common materials for upgrading a system in which service conditions are too demanding for either standard or special stainless steels because of severe combinations of acids and elevated temperatures. These cast alloys can be used in conjunction with similar wrought materials, or they can serve to upgrade pump and valve components in a wrought stainless steel system.

Nickel-molybdenum alloys have specialized application areas, primarily in the handling of hydrochloric acid at all temperatures and concentrations. Applications should not be based on upgrading in areas where stainless steels are inadequate, because the nickel-molybdenum alloys are unsuitable for handling most oxidizing solutions for which stainless steels are used.

Alloys for directional and single-crystal solidification are used as blades for aircraft and some land-based turbines (Fig. 1 and 4). Under elevated temperatures, they have very high strength in the direction of primary stress.

REFERENCES

1. K. Harris, G.L. Erickson, and R.E. Schwer, "Development of the Single-Crystal Alloys CM SX-2 and CM SX-3 for Advanced Technology Turbine Engines," Technical Paper 83-GT-244, American Society of Mechanical Engineers
2. K. Harris, G.L. Erickson, and R.E. Schwer, "Directionally Solidified DS CM 247 LC—Optimized Mechanical Properties Resulting From Extensive γ' Solutioning," Paper presented at the Gas Turbine Conference and Exhibit, Houston, TX, March 1985
3. K. Harris, G.L. Erickson, R.E. Schwer, J. Wortmann, and D. Froschhammer, "Development of Low-Density Single-Crystal Superalloy CMSX-6," Technical Paper, Cannon-Muskegon Corporation
4. K. Harris, G.L. Erickson, and R.E. Schwer, "CMSX Single Crystal, CM DS & Integral Wheel Alloys Properties and Performance," Paper presented at the Cost 50/501 Conference, High Temperature Alloys for Gas Turbines and Other Applications, Liège, Oct 1986

SELECTED REFERENCES

- W.J. Jackson, Ed., *Steel Castings Design Properties and Applications*, Steel Castings Research and Trade Association, 1983
- J.D. Redmond, Selecting Second-Generation Duplex Stainless Steels, *Chem. Eng.*, Oct 1986 and Nov 1986
- *Steel Castings Handbook*, Supplement 8, High Alloy Data Sheets, Corrosion Series, Steel Founders' Society of America, 1981
- *Steel Castings Handbook*, Supplement 9, High Alloy Data Sheets, Heat Series, Steel Founders' Society of America, 1981
- P.F. Wieser, Ed., *Steel Castings Handbook*, 5th ed., Steel Founders' Society of America, 1980

Titanium and Titanium Alloys

Jeremy R. Newman, Titech International Inc.
Daniel Eylon, University of Dayton
John K. Thorne, Precision Castparts Corporation

SINCE THE INTRODUCTION OF TITANIUM and titanium alloys in the early 1950s, these materials have in a relatively short time become one of the backbone materials for the aerospace, energy, and chemical industries (Ref 1). The combination of high strength-to-weight ratio, excellent mechanical properties, and corrosion resistance makes titanium the best material for many critical applications. Today, titanium alloys are used for static and rotating gas turbine engine components. Some of the most critical and highly stressed civilian and military airframe parts are made of these alloys.

Net Shape Technology Development. The use of titanium has expanded in recent years from applications in nuclear power plants to food processing plants, from oil refinery heat exchangers to marine components and medical prostheses (Ref 2). However, the high cost of titanium alloy components may limit their use to applications for which lower-cost alloys, such as aluminum and stainless steels, cannot be used. The relatively high cost is often the result of the intrinsic raw material cost of the metal, fabricating costs, and, usually most important, the metal removal costs incurred in obtaining the desired end-shape. As a result, in recent years a substantial effort has been focused on the development of net shape or near-net shape technologies to make titanium alloy components more competitive (Ref 3). These titanium net shape technologies include powder metallurgy (PM), superplastic forming (SPF), precision forging, and precision casting. Precision casting is by far the most fully developed and the most widely used net shape technology.

Casting Industry Growth. The annual shipment of titanium castings increased by 240% between 1978 and 1986 (Fig. 1) and titanium casting is the fastest growing segment of titanium technology.

Even at current levels (approaching 450 Mg, or 1×10^6 lb, annually), castings still represent less than 2% of total titanium mill product shipments. This is in sharp contrast to the ferrous and aluminum industries, where foundry output is 9% (Ref 5) and 14% (Ref 6) of total output, respectively. This suggests that the growth trend of titanium castings will continue as users become more aware of industry capability, suitability of cast components in a wide variety of applications, and the net shape cost advantages.

Properties Comparable to Wrought. The term castings often connotes products with properties generally inferior to wrought products. This is not true with titanium cast parts. They are generally comparable to wrought products in all respects and quite often superior. Properties associated with crack propagation and creep resistance can be superior to those of wrought products. As a result, titanium castings can be reliably substituted for forged and machined parts in many demanding applications (Ref 7, 8). This is due to several unique properties of titanium alloys. One is the $\alpha + \beta$-to-β allotropic phase transformation at a temperature range of 705 to 1040 °C (1300 to 1900 °F), which is well below the solidification temperature of the alloys. As a result, the cast dendritic β structure is wiped out during the solid state cooling stage, leading to an $\alpha + \beta$ platelet structure (Fig. 2a), which is also typical of β processed wrought alloy. Further, the convenient allotropic transformation temperature range of most titanium alloys enables the as-cast microstructure to be improved by means of postcast cooling rate changes and subsequent heat treatment.

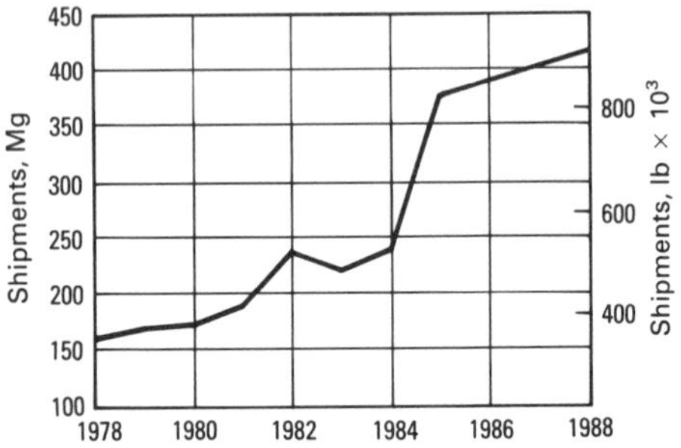

Fig. 1 Growth of 240% in United States titanium casting shipments from 1978 to 1986. Source: Ref 4

Reactivity. Another unique property is the high reactivity of titanium at elevated temperatures, leading to an ease of diffusion bonding. As a result, hot isostatic pressing (HIP) of titanium castings yields components with no subsurface porosity. At the HIP temperature range (820 to 980 °C, or

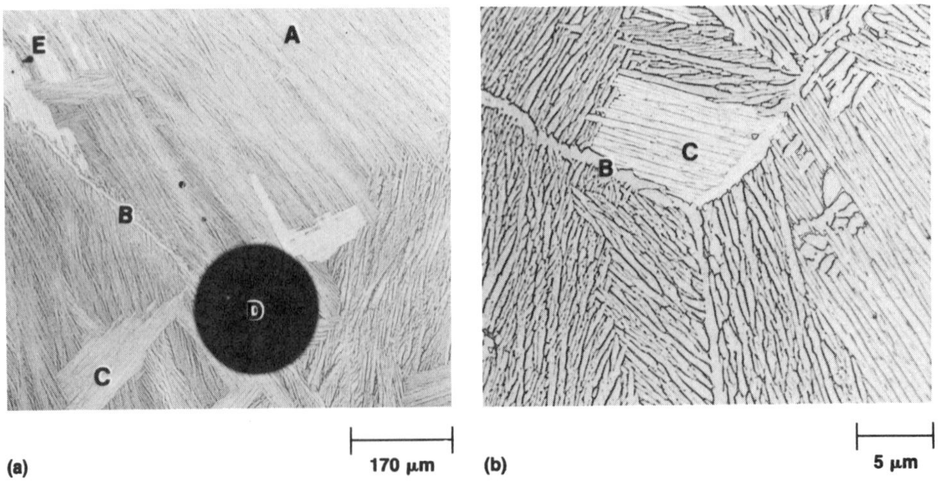

Fig. 2 Comparison of the microstructures of (a) as-cast versus (b) cast + HIP Ti-6Al-4V alloys illustrating lack of porosity in (b). Grain boundary α (B) and α plate colonies (C) are common to both alloys; β grains (A), gas (D), and shrinkage voids (E) are present only in the as-cast alloy.

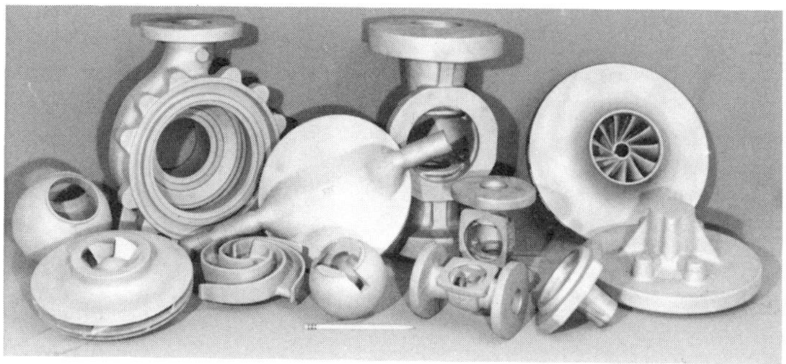

(a)

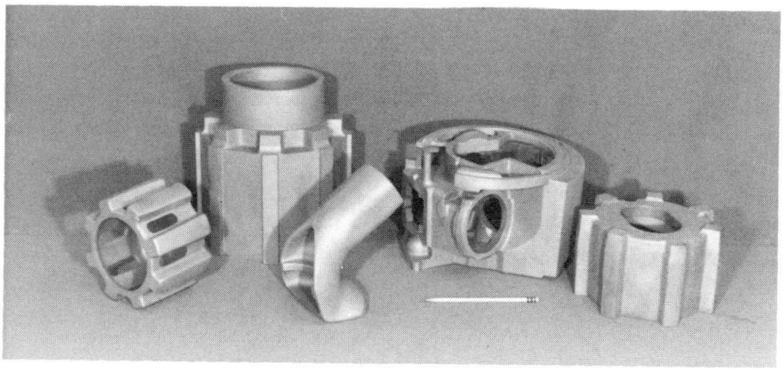

(b)

Fig. 3 Typical titanium parts produced by the rammed graphite process. (a) Pump and valve components for marine and chemical-processing applications. (b) Brake torque tubes, landing arrestor hook, and optic housing components used in aerospace applications

1500 to 1800 °F) titanium dissolves any microconstituents deposited upon internal pore surfaces, leading to complete healing of casting porosity as the pores are collapsed during the pressure and heat cycle. Both the elimination of casting porosity and the promotion of a favorable microstructure improve mechanical properties. However, the high reactivity of titanium, especially in the molten state, presents a special challenge to the foundry. Special, and sometimes relatively expensive, methods of melting (Ref 9), moldmaking, and surface cleaning (Ref 7, 8) may be required to maintain metal integrity. Additional information on HIP of castings may be found in the section "Hot Isostatic Pressing of Castings" in the article "Processing of Castings" in this Volume.

Historical Perspective of Casting Technology

Although titanium is the fourth most abundant structural metal in the earth's crust (0.4 to 0.6 wt%) (Ref 9), it has emerged only recently as a technical metal. This is the result of the high reactivity of titanium, which requires complex methods and high energy input to win the metal from the oxide ores. The required energy per ton is 1.7 times that of aluminum and 16 times that of steel (Ref 10). From 1930 to 1947, metallic titanium extracted from the ore as a powder or sponge form was processed into useful shapes by P/M methods to circumvent the high reactivity in the molten form (Ref 11).

Melting Methods. The melting of small quantities of titanium was first experimented with in 1948 using methods such as resistance heating, induction heating, and tungsten arc melting (Ref 12, 13). However, these methods never developed into industrial processes. The development during the early 1950s of the cold crucible, consumable-electrode vacuum arc melting process, "skull melting," by the U.S. Bureau of Mines (Ref 13, 14) made it possible to melt large quantities of contamination-free titanium into ingots or net shapes. Additional information on numerous melting methods is available in the articles "Melting Furnaces" and "Vacuum Melting and Remelting Processes" in this Volume.

First Castings. Shape casting of titanium was first demonstrated in the United States in 1954 at the U.S. Bureau of Mines using machined high-density graphite molds (Ref 13, 15). The rammed graphite process developed later, also by the U.S. Bureau of Mines (Ref 16), lead to the production of complex shapes. This process, and its derivations, are used today to produce parts for marine and chemical-plant components such as the pump and valve components shown in Fig. 3(a). Some aerospace components such as the aircraft brake torque tubes, landing arrestor hook, and optic housing shown in Fig. 3(b) have also been produced by this method.

Molding Methods

Rammed Graphite Molding. The traditional rammed graphite molding process uses powdered graphite mixed with organic binders (see the article "Rammed Graphite Molds" in this Volume). Patterns typically are made of wood. The mold material is pneumatically rammed around the pattern and cured at high temperature in a reducing atmosphere to convert the organic binders to pure carbon. The molding process and the tooling are essentially the same as for cope and drag sand molding in ferrous and nonferrous foundries. In the 1970s, derivations of rammed graphite mold materials were developed using components of more traditional sand foundries along with inorganic binders. This resulted in more dimensionally stable and less costly molds that were capable of containing molten titanium without undue metal/mold reaction.

Lost Wax Investment Molding. The principal technology that allowed the proliferation of titanium alloy castings in the aerospace industry was the investment casting method, introduced in the mid-1960s (see the article "Investment Casting" in this Volume). This method, used at the dawn of the metallurgical age for casting copper and bronze tools and ornaments, was later adapted to enable production of high-quality steel and nickel base cast parts. The adaptation of this method to titanium casting technology required the development of ceramic slurry materials with minimum reaction with the extremely reactive molten titanium.

Proprietary lost wax ceramic shell systems have been developed by the several foundries engaged in titanium casting manufacture. Of necessity, these shell systems must be relatively inert to molten titanium and cannot be made with the conventional foundry ceramics used in the ferrous and nonferrous industries. Usually, the face coats are made with special refractory oxides and appropriate binders. After the initial face coat ceramic is applied to the wax pattern, more traditional refractory systems are used to add shell strength from repeated backup ceramic coatings. Regardless of face coat composition, some metal/mold reaction inevitably occurs from titanium reduction of the ceramic oxides. The oxygen-rich surface of the casting stabilizes the α phase, usually forming a metallographically distinct α case layer on the cast surface, which may be removed later by means of chemical milling.

Foundry practice focuses on methods to control both the extent of the metal/mold reaction and the subsequent diffusion of reaction products inward from the cast surface. Diffusion of reaction products into the cast surface is time-at-temperature dependent. Depth of surface contamination can vary from nil on very thin sections to more than 1.5 mm (0.06 in.) on heavy sections. On critical aerospace structures, the brittle α case is removed by chemical milling. The depth of surface contamination must be taken into consideration in the initial wax pattern tool design. Hence, the wax pattern and casting are made slightly oversize, and final dimensions are achieved through careful chemical milling. Metal superheat, mold

Table 1 Status and capacity of titanium foundries in the United States, Japan, and Western Europe in 1987

Foundry	Maximum pour weight kg	Maximum pour weight lb	Approximate maximum envelope size: Rammed graphite, mm	Rammed graphite, in.	Investment casting, mm	Investment casting, in.	Melt stock	Use of postcast HIP
Howmet Corp. (MI and VA)	730	1600	. . .		1525 diam × 1525	60 diam × 60	Billet	Always
Oremet Corp. (OR)	750	1650	1525 diam × 1830	60 diam × 72	. . .		Billet and revert	Seldom
PCC (OR)	770	1700	. . .		1525 diam × 1220	60 diam × 48	Billet and revert	Always
Rem Products (OR)	180	400	. . .		815 diam × 508	32 diam × 20	Billet	Often
Tiline, Inc. (OR)	750	1650	. . .		1370 diam × 610	54 diam × 24	Billet and revert	Always
Titech International, Inc. (CA)	400	875	915 diam × 610	36 diam × 24	915 diam × 610	36 diam × 24	Billet and revert	Often
PCC France (France)	270	600	990 diam × 990	39 diam × 39	1220 diam × 1220	48 diam × 48	Billet and revert	Always
Tital (West Germany)	180	400	1145 diam × 760	45 diam × 30	1015 diam × 635	40 diam × 25	Billet	Always
Settas (Belgium)	820	1800	1525 diam × 1220	60 diam × 48	610 diam × 610	24 diam × 24	Billet and revert	Often
VMC (Japan)	180	400	1270 diam × 635	50 diam × 25	Research and development		Billet and revert	Seldom

temperature and thermal conductivity, *g* force (if centrifugally cast), and rapid postcast heat removal are other key factors in producing a satisfactory product. These parameters are interrelated, that is, a high *g* force centrifugal pour into cold molds may achieve the same relative fluidity as a static pour into heated molds.

Other Molding Systems. The combination of graphite powder, stucco, and organic binders has also been used as a shell system for the investment casting of titanium. After dewax, the shell is fired in a reducing atmosphere to remove or pyrolyze the binders before casting. This technology has not been promoted as much as the use of refractory oxide shell systems and is presently primarily of historic interest.

In addition to the rammed graphite and investment molding methods, a poured ceramic mold has also been used to produce large parts that require good dimensional accuracy. This method, developed in the late 1970s, was used to a limited extent for several years.

Semipermanent, reusable molds, frequently made from machined graphite, have been used successfully since the earliest U.S. Bureau of Mines work, but only on relatively simple-shaped parts that allow metal volumetric shrinkage to occur without restriction. The method is economical only when reasonably high volumes are required, that is, thousands of parts, because of the high cost of the solid mold material.

Currently, a titanium sand casting technique based on conventional foundry mold-making practices is under development at the U.S. Bureau of Mines (Ref 17). Because the mold materials are less costly and the cast part is easier to remove from the sand mold than from other methods of titanium casting, this development could lower production costs. However, surface quality problems are restricting the use of this method thus far.

Foundries and Capacities. Table 1 summarizes the use and capacities of the various titanium casting practices by a number of foundries in several countries.

Alloys

All production titanium castings to date are based on traditional wrought product compositions. As such, the Ti-6Al-4V alloy dominates structural casting applications. This alloy similarly has dominated wrought industry production since its introduction in the early 1950s, becoming the benchmark alloy against which others are compared. However, other wrought alloys have been developed, for special applications, with better room-temperature or elevated-temperature strength, creep, or fracture toughness characteristics than those of Ti-6Al-4V. These same alloys are also being cast when net shape casting technology is the most economical method of manufacture. As with Ti-6Al-4V, other cast titanium alloys have properties generally comparable to those of their wrought counterparts.

Chemistry and Demand. Table 2 lists the most prevalent casting alloy chemistries and the most unique attribute of each in comparison with Ti-6Al-4V, plus current approximate market share.

Typical Properties. Table 3 is a summary of room-temperature tensile properties for various alloys. These properties, which are typical, vary depending on microstructure as influenced by foundry parameters such as solidification rate and any postcast HIP and heat treatments.

Specifications. Industry-wide specifications, listed in Table 4 for reference, give more detail on mechanical property guarantees and process control features. In addition, most major aerospace companies have comparable specifications. *MIL Handbook V, Aerospace Design Specifications* does not presently include titanium alloy castings, but it is expected that such information will be incorporated in the near future. As with wrought products, commercially pure titanium castings are used almost entirely in corrosion service. Commercially pure titanium pumps and valves are the principal components made using titanium casting technology for the corrosion resistance field, and are used extensively in chemical and petrochemical plants.

Newer Alloys. As aircraft engine manufacturers seek to use cast titanium at higher operating temperatures, Ti-6Al-2Sn-4Zr-2Mo and Ti-6Al-2Sn-4Zr-6Mo are being specified more frequently. Other titanium alloys for service to 595 °C (1100 °F) are being developed as castings. Extra low in-

Table 2 Comparison of cast titanium alloys

Alloy	Estimated relative usage of castings	Nominal composition, wt %: O	N	H	Al	Fe	V	Cr	Sn	Mo	Zr	Special properties(a)
Ti-6Al-4V	90%	0.18	0.015	0.006	6	0.13	4	. . .	. . .	. . .	. . .	General purpose
Ti-6Al-4V ELI	2%	0.11	0.010	0.006	6	0.10	4	. . .	. . .	. . .	. . .	Cryogenic toughness
Commercially pure titanium Gr2	5%	0.25	0.015	0.006	. . .	0.15	. . .	. . .	. . .	. . .	. . .	Corrosion resistance
Ti-6Al-2Sn-4Zr-2Mo	2%	0.10	0.010	0.006	6	0.15	. . .	. . .	2	2	4	Elevated-temperature creep
Ti-6Al-2Sn-4Zr-6Mo	<1%	0.10	0.010	0.006	6	0.15	. . .	. . .	2	6	4	Elevated-temperature strength
Ti-5Al-2.5Sn	<1%	0.16	0.015	0.006	5	0.2	. . .	. . .	2.5	. . .	. . .	Cryogenic toughness
Ti-3Al-8V-6Cr-4Zr-4Mo	<1%	0.10	0.015	0.006	3.5	0.2	8.5	6	. . .	4	4	Strength
Ti-15V-3Al-3Cr-3Sn	<1%	0.11	0.015	0.006	3	0.2	15	3	3	. . .	. . .	Strength
Total	**100%**											

(a) Superior, relative to Ti-6Al-4V

Table 3 Typical room-temperature tensile properties of titanium alloy castings (bars machined from castings)

Specification minimums are less than these typical properties.

Alloy(a)	Yield strength MPa	Yield strength ksi	Ultimate strength MPa	Ultimate strength ksi	Elongation, %	Reduction of area, %
Commercially pure (Grade 2)	448	65	552	80	18	32
Ti-6Al-4V, annealed	855	124	930	135	12	20
Ti-6Al-4V-ELI	758	110	827	120	13	22
Ti-6Al-2Sn-4Zr-2Mo, annealed	910	132	1006	146	10	21
Ti-6Al-2Sn-4Zr-6Mo, STA	1269	184	1345	195	1	1
Ti-3Al-8V-6Cr-4Zr-4Mo, STA	1241	180	1330	193	7	12
Ti-15V-3Al-3Cr-3Sn, STA	1200	174	1275	185	6	12

(a) Solution-treated and aged (STA) heat treatments may be varied to produce alternate properties.

terstitial grade Ti-6Al-4V has been used for critical cryogenic space shuttle service where fracture toughness is an important design criteria. The most recent alloys to receive attention in the casting industry are the metastable β alloys Ti-3Al-8V-6Cr-4Zr-4Mo (Beta C) and Ti-15V-3Al-3Cr-3Sn (Ti 15-3). Originally developed as a highly cold-formable and subsequently age-hardened sheet material, these alloys are highly castable and are readily heat treated to an 1170 MPa (170 ksi) strength level, making them serious candidates for the replacement of high-strength precipitation-hardened stainless steels such as 17-4PH. The full density advantage of titanium of about 40% is preserved because strength levels are comparable in both materials. Titanium-aluminide castings are being developed for application in the compressor sections of aircraft gas turbine engines subjected to the highest temperatures. Compositions based upon both the α_2 (Ti_3Al) and γ(TiAl) ordered phases have been cast experimentally, with the former being closer to limited-production status. The low ductility of these alloys at room temperature has been the major producibility challenge. It is anticipated that the service potential for titanium aluminides in the 595 to 925 °C (1100 to 1700 °F) temperature range will eventually be realized. The difficulty in machining shapes in these brittle alloys may increase the advantage of net shape methods such as castings.

Because Ti-6Al-4V dominates the industry, much more metallurgical work has been accomplished with this alloy and is discussed below.

Microstructure of Ti-6Al-4V

Cast Microstructure. To understand the relatively high mechanical property levels of titanium alloy castings and the many improvements made in recent years, it is necessary to understand the microstructures of castings and their influence on the mechanical behavior of titanium. The phase transformation from β to α + β leads to the elimination of the dendritic cast structure. The existence of such dendrites is evident in the surface morphology of shrinkage pores (Fig. 4). The phase transformation, which in the alloy Ti-6Al-4V is typically initiated at 995 °C (1825 °F), results in the microstructural features shown in Fig. 2(a). This microstructure, which will be discussed in detail, is very similar to a β-processed wrought microstructure and has similar properties. Thus, in the study and development of titanium alloy castings, it is possible to draw much information from conventional ingot metallurgy.

Hot isostatic pressing is now becoming almost a standard practice for all titanium cast parts produced for the aerospace industry (Table 1). As a result, cast + HIP microstructure also needs to be considered (Fig. 2b). Because the HIP temperature is typically well below the β_t temperature, the as-cast (Fig. 2a) and the cast + HIP microstructures look very much alike, except for the lack of porosity in the latter.

As-Cast and Cast + HIP Microstructures. Because most castings for demanding applications are produced with Ti-6Al-4V alloy (Table 2), only microstructures of this α + β alloy will be reviewed here.

Beta Grain Size. Beta grains (A, in Fig. 2a) develop during the solid state cooling stage between the solidus/liquidus temperature and the β_t temperature. As a result, large sections, which cool slower, show larger β grains. The size range of the β grains is 0.5 to 5 mm (0.02 to 0.2 in.). As will be further discussed, large β grains may lead to large α plate colonies. This is beneficial for fracture toughness, creep resistance, and fatigue crack propagation resistance (Ref 18, 19) and detrimental for low- and high-cycle fatigue strength (Ref 20, 21).

Grain Boundary α. This α phase (B, in Fig. 2a) is formed along the β grain boundaries when cast material is cooled through the α + β phase field (in Ti-6Al-4V this is typically from 995 °C, or 1825 °F, down to room temperature). This phase is typically plate shaped and represents the largest α plates in the cast structure. The length of these plates can equal the β grain radius. This has been found to be very detrimental to fatigue crack initiation at room temperature (Ref 22, 23) and at elevated temperatures (Ref 23, 24). Many postcast thermal treatments eliminate this phase to improve fatigue life.

Alpha Plate Colonies. Alpha platelets (C, in Fig. 2a) are the transformation products of the β phase when cooled below the β_t temperature. The hexagonal close-packed (hcp) orientation of these plates is related to the parent body-centered cubic (bcc) β phase orientation through one of the 12 possible variants of the Burgers relationship (Ref 25, 26):

$$\{110\}\beta \parallel (0001)\alpha$$
$$\langle 111\rangle\beta \parallel \langle 1120\rangle\alpha$$

Table 4 Standard industry specifications applicable to titanium castings

MIL-T-81915	Titanium and titanium alloy castings, investment
AMS-4985A	Titanium alloy castings, investment or rammed graphite
AMS-4991	Titanium alloy castings, investment
ASTM B 367	Titanium and titanium alloy castings
MIL-STD-2175	Castings, classification and inspection of
MIL-STD-271	Nondestructive testing requirements for metals
MIL-STD-453	Inspection, radiographic
MIL-Q-9858	Quality program requirement
MIL-I-6866B	Inspection, penetrant method of
MIL-H-81200	Heat treatment of titanium and titanium alloys
ASTM E 155	Reference radiographs for inspection of aluminum and magnesium castings
ASTM E 192	Reference radiographs, investment steel castings
ASTM E 186	Reference radiographs, steel castings 50 to 102 mm (2 to 4 in.)
ASTM E 446	Reference radiographs, steel castings up to 50 mm (2 in.)
ASTM E 120	Standard methods for chemical analysis of titanium and titanium alloys
ASTM E 8	Methods of tension testing of metallic materials
AMS-2249B	Chemical-check analysis limits for titanium and titanium alloys
AMS-4954	Titanium alloy welding wire Ti-6Al-4V
AMS-4956	Titanium alloy welding wire Ti-6Al-4V, extra low interstitial

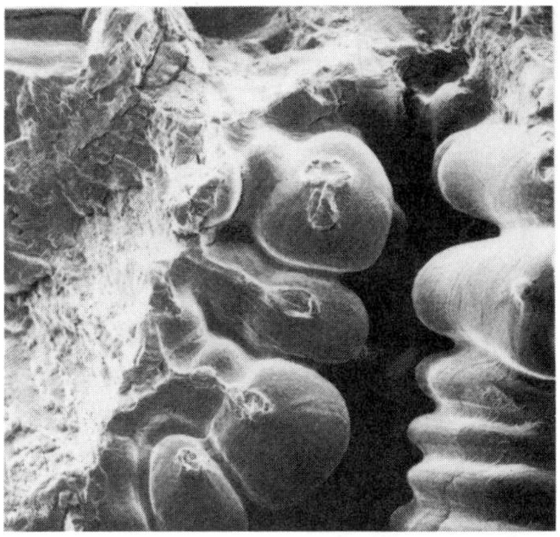

20 μm

Fig. 4 Dendritic structures present in the surface morphology of an as-cast titanium component

Table 5 Methods for modifying the microstructure of α + β titanium alloy net shape products

Method(a)	Typical solution treatment	Hydrogenation temperature		Intermediate treatment		Dehydrogenation temperature		Typical annealing or aging treatment	Applied to product forms	Ref
		°C	°F	°C	°F	°C	°F			
BUS	1040 °C (1900 °F) for ½ h	...	...	...	...	...	...	845 °C (1550 °F) for 24 h	Cast, P/M, I/M	34–37
GTEC	1050 °C (1925 °F) for ½ h	...	...	...	...	...	...	845 °C (1550 °F) for ½ h and 705 °C (1300 °F) for 2 h	Cast	38
BST	1040 °C (1900 °F) for ½ h and GFC(b)	...	...	...	...	...	...	540 °C (1000 °F) for 8 h	Cast, I/M	39
ABST	955 °C (1750 °F) for 1 h and GFC	...	...	...	...	...	...	540 °C (1000 °F) for 8 h	Cast, I/M	39
HVC (Hydrovac process)	...	650	1200	870 (glass encapsulated)	1600	760	1400	...	P/M, I/M	40, 41
TCT	1040 °C (1900 °F) for ½ h	595	1100	Cool to RT(c)		760	1400	...	Cast, P/M, I/M	41–43
CST	...	870	1600	No intermediate step (continuous process)		815	1500	...	Cast	44
HTH	...	900	1650	Cool to RT		705	1300	...	Cast, P/M, I/M	45

(a) Most data apply to Ti-6Al-4V. β_t temperature approximately 995 °C (1825 °F). (b) GFC, gas fan cooled. (c) RT, room temperature

When cooling rates are relatively slow, such as in thick-section castings, many adjacent α platelets transform into the same Burgers variant and form a colony of similarly aligned and crystallographically oriented platelets. The large colonies (like those marked C in Fig. 2a) may be associated with early fatigue crack initiation (Ref 21), the result of heterogeneous basal slip across the plates (Ref 27). At the same time, the large colony structure is beneficial for fatigue crack propagation resistance (Ref 28, 29). Because α platelet colonies cannot grow larger than the β grains, titanium castings with large β grain typically have large colonies. The α platelets are typically 1 to 3 μm (40 to 120 μin.) in thickness and 20 to 100 μm (800 to 4000 μin.) in length (Ref 30, 31), and the typical colony size in castings is 50 to 500 μm (10.002 to 0.02 in.) (Ref 22, 30, 31).

As a general rule, slower cooling rates, such as in thick cast sections, result in microstructures with larger β grains, a longer and thicker grain boundary α phase, thicker α platelets, and larger α platelet colonies.

Porosity. Gas (D, in Fig. 2a) and shrinkage (E) voids are typical phenomena in as-cast titanium products. Hot isostatic pressing, however, closes and heals these pores. This is demonstrated by comparing the as-cast microstructure in Fig. 2(a) with the cast + HIP structure in Fig. 2(b). Hot isostatic pressing also causes a degree of α plate coarsening.

Modification of Microstructure. Virtually all titanium castings produced commercially today are supplied in the annealed condition. However, much microstructural modification development work has recently been done, and it can be expected that solution-treated and aged or other postcast thermal processing will eventually become specified on cast parts requiring certain property enhancement. The following section reviews several of the developmental procedures and their results.

Modification of microstructure is one of the most versatile tools available in metallurgy for improving mechanical properties of alloys. This is commonly achieved through a combination of cold or hot working followed by heat treatment. Net shapes such as castings or P/M products cannot be worked, which limits the options for controlling microstructures. A substantial amount of work has been done in recent years to improve the microstructures of titanium alloy net shape products, with an emphasis on Ti-6Al-4V material. Most treatment schemes can be successfully applied to both cast parts (Ref 8) and P/M compacts (Ref 32, 33). In the case of titanium alloy castings, the main goal has been to eliminate the grain boundary α phase, the large α plate colonies, and the individual α plates. This is accomplished either by solution treatments or by a temporary alloying with hydrogen. In some cases, the hydrogen and solution treatments are combined. The details of these methods, including the appropriate references, are listed in Table 5. The typical resulting microstructures of the α-β solution treatment (ABST), β solution treatment (BST), broken-up structure (BUS), and high-temperature hydrogenation (HTH) methods are shown in Fig. 5(a), (b), (c), and (d), respectively. As can be seen from the photomicrographs, these treatments are successful in eliminating the large α plate colonies and the grain boundary α phase. As discussed below, a substantial improvement of both tensile and fatigue properties is achieved with these processes.

Mechanical Properties of Ti-6Al-4V

Oxygen Influence. Figure 6 is a frequency distribution of tensile properties from separately cast test bars representing hot isostatic pressed and annealed Ti-6Al-4V castings. Note that oxygen, a carefully controlled alloy addition, is in the 0.16 to 0.20% range, which is common for many aerospace specifications.

Some specifications allow a 0.25% maximum oxygen content. The resultant properties with oxygen in the 0.20 to 0.25% range are typically about 69 to 83 MPa (10 to 12 ksi) higher than those shown in Fig. 6 with slightly lower ductility levels. In this case, it is possible to guarantee 827 MPa (120 ksi) yield strength and 896 MPa (130 ksi) ultimate tensile strength levels with 6% minimum elongation. This strength level is the same minimum guarantee for wrought-annealed Ti-6Al-4V.

Microstructure Influence. Because the microstructure of titanium alloy cast parts is very similar to β processed wrought or ingot metallurgy (I/M) material, many properties of hot isostatic pressed castings, such as tensile strength, fracture toughness, fatigue crack propagation, and creep, are at the same levels as forged and machined parts. Tensile strength and fracture toughness properties of cast, cast + HIP, and cast + HIP + heat-treated material (Table 5) are

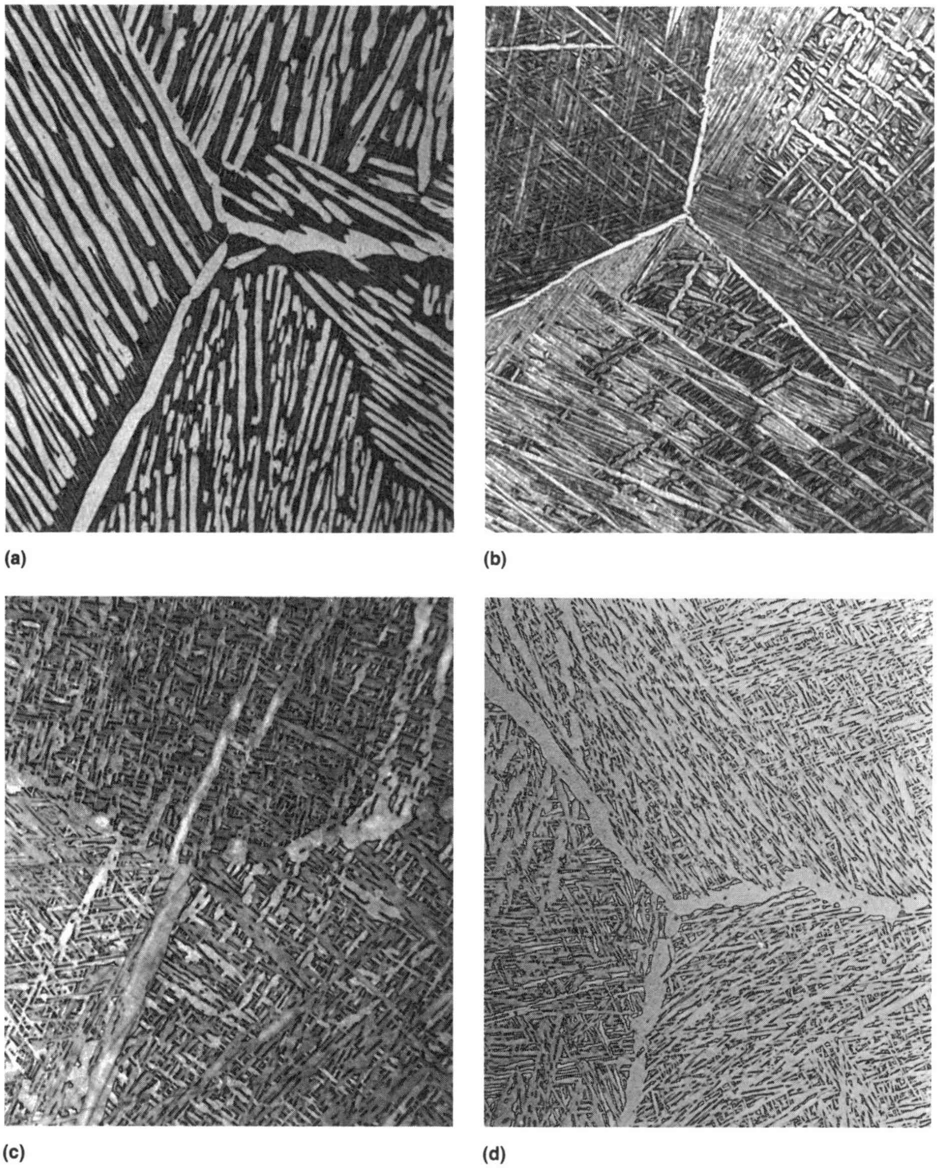

Fig. 5 Photomicrographs of microstructures resulting from a variety of hydrogen and solution heat treatments used to eliminate large α plate colonies and grain boundary α phase in titanium alloys. (a) ABST. (b) BST. (c) BUS. (d) HTH

compared in Table 6 to wrought β-annealed data. To provide a complete review, properties of castings treated by many of the methods listed in Table 5 are also included. At the present time, fracture toughness data are available for only a few of the conditions. As can be seen, some of the treated conditions present properties in excess of I/M β-annealed material. However, it should be noted that tests were done on relatively small cast coupons. Properties of actual cast parts, especially large components, could be somewhat lower, the result of coarser grain structure or slower quench rates. Of special interest are the hydrogen-treated conditions (such as thermo-chemical treatment, or TCT; constitutional solution treatment, or CST; and HTH, in Table 5) that result in very high tensile strength (as high as 1124 MPa, or 163 ksi) with tensile elongation as high as 8%.

Fatigue and Fatigue Crack Growth Rate. The fatigue crack growth rate (FCGR) behavior of cast Ti-6Al-4V is also, as expected, very similar to that of β-processed wrought Ti-6Al-4V (Ref 50-52). This is demonstrated in Fig. 7 in which the scatterband of FCGR of cast and cast-HIP alloy is compared to β-processed I/M (Ref 18, 53).

The scatterbands of smooth axial fatigue results of cast, cast-HIP (Ref 15, 47, 48, 54-57), and wrought Ti-6Al-4V are shown in Fig. 8. This figure clearly indicates that the HIP process results in a substantially improved fatigue life well into the wrought-annealed region. The fatigue properties of aerospace quality castings have always been an important issue, because in most other alloy systems this is the property that is most degraded when compared to wrought products. However, because of the complete closure and healing of gas (D, in Fig. 2a) and shrinkage (E, in Fig. 2a) pores by HIP and the β-annealed microstructure, it is possible to get fatigue life comparable to wrought material in premium investment cast and hot isostatic pressed parts. As indicated previously (Table 5), substantial work has been done in recent years to modify the microstructure of cast parts to produce fatigue properties either equivalent or superior to the best wrought-annealed products. Figure 9 compares the smooth fatigue life of Ti-6Al-4V treated by ABST, BST, BUS, CST, Garrett treatment (GTEC), and HTH (Table 5) to wrought material scatterband. As can be seen, all of these treatments were successful in improving fatigue life above wrought levels. The hydrogen treatments (CST and HTH) resulted in the highest improvement in fatigue strength. However, it should be noted that wrought products subjected to the same treatments result in comparable improvements in fatigue strength.

Casting Design

The best casting design is usually achieved by means of a thorough review by the manufacturer and user when the component is still in the preliminary design stage (see the article "Casting Design" in this Volume). Additional features may be incorporated to reduce machining cost, and components may be integrated to eliminate later fabrication. Specifications and tolerances may be reviewed vis-à-vis foundry capabilities, producibility, and pattern tool concepts to achieve the most practical and cost-effective design (see the articles "Dimensional Tolerances and Allowances" and "Patterns and Patternmaking" in this Volume). When minimum cast part weight is critical, such as in aerospace components, the capability of the foundry to produce varying wall thicknesses, for example, may be beneficial. Often, cast features that cannot be economically duplicated by any other method may be readily produced.

Titanium castings present the designer with few differences in design criteria, compared with other metals. Ideal designs do not contain isolated heavy sections or uniform heavy walls of large area so that centerline shrinkage cavities and regions with a coarse microstructure may be avoided. From a practical sense, however, ideal tapered walls to promote directional solidification are not usually a reality. The advent of hot isostatic pressing to heal internal as-cast shrinkage cavities has offered the designer much more freedom; however, there still is a practical limit to the size of

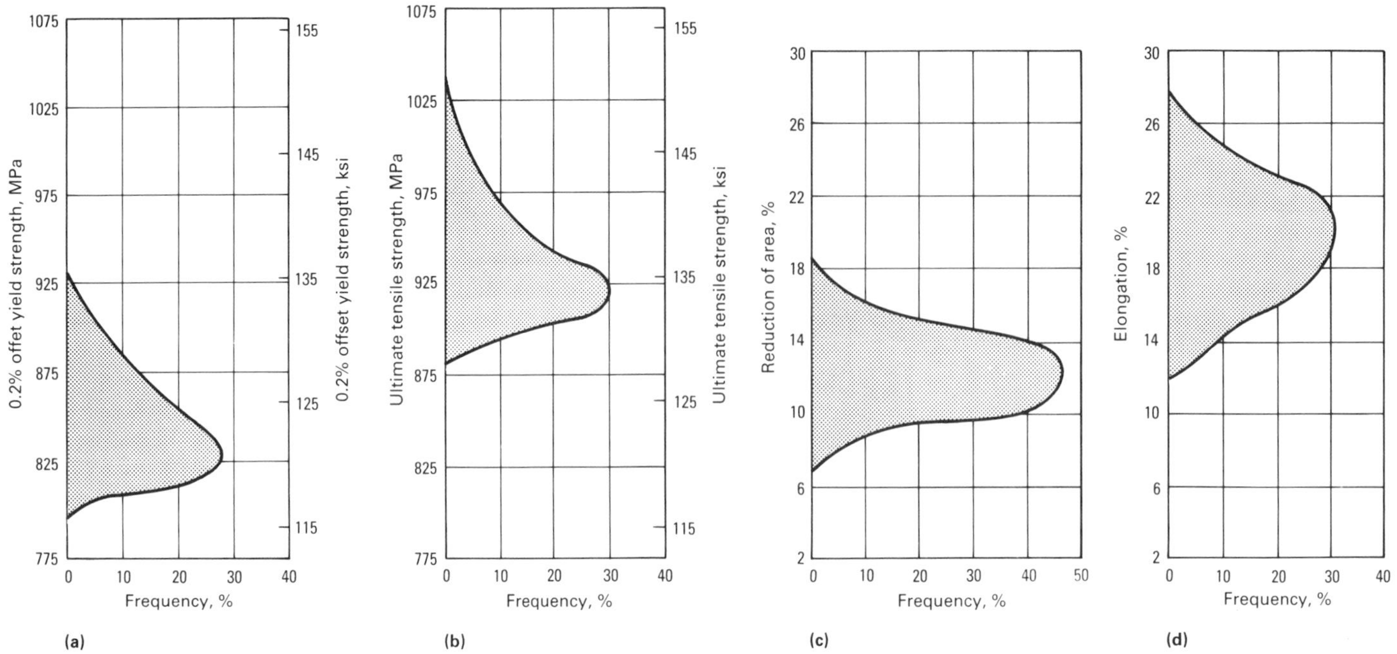

Fig. 6 Frequency distribution of tensile properties of HIP'ed and annealed Ti-6Al-4V casting test bars. Percent frequency is plotted versus (a) 0.2% offset yield strength, (b) ultimate tensile strength, (c) percent elongation, (d), and percent reduction of area. Alloy composition is 0.16 to 0.20% O_2; sample size is 500 heats. Source: Ref 46

internal cavity that can be healed through hot isostatic pressing without contributing significant surface or structural deformation due to the collapse of internal pores.

The lost wax investment process provides more design freedom for the foundry to properly feed a casting than does the traditional sand or rammed graphite approach. It is normal practice to add a gate and riser to hot isostatic pressed investment castings to achieve reasonably good internal x-ray quality so that hot isostatic pressing will not cause extensive surface or structural deformation.

The usual required minimum practical wall thickness for investment castings is 2.0 mm (0.080 in.); however, sections as small as 1.1 mm (0.045 in.) are routinely made. Even thinner walls may be achieved by chemical milling beyond that required for α case removal; however, as-cast wall variation is not improved and becomes a larger percentage of the resultant wall thickness. Sand or rammed graphite molded castings have a usual minimum wall thickness of 4.75 mm (0.187 in.), although 3.0 mm (0.12 in.) is not unreasonable for short sections.

Fillet radii should be as generous as possible to minimize the occurrence of hot tears. While 0.76 mm (0.030 in.) radii are produced, the preferred minimum is 3.0 mm (0.12 in.). A rule of thumb is that a fillet radius should be 0.5 times the sum of the thicknesses of the two adjoining walls.

With proper tool design, zero draft walls are possible. To promote directional solidification, a 3° included draft angle may be preferred. Hot isostatic pressing will close any centerline shrinkage cavities in zero draft walls, making it unnecessary to provide draft. Draft requirements are also dependent upon foundry practice, with rammed graphite tooling usually requiring draft, and investment casting typically not requiring draft.

Tolerances. Typically, the major area of concern is true position of a thin-section surface with respect to a datum. Surface areas of approximately 129 cm^2 (20 $in.^2$) or greater in sections of less than approximately 5.08 mm (0.200 in.) thickness are susceptible to distortion, depending on adjoining sections. The high strength of titanium com-

Table 6 Tensile properties and fracture toughness of Ti-6Al-4V cast coupons compared to typical wrought β-annealed material

Material condition(a)	Yield strength MPa	Yield strength ksi	Ultimate tensile strength MPa	Ultimate tensile strength ksi	Elongation, %	Reduction of area, %	K_{Ic} ksi$\sqrt{in.}$	K_{Ic} MPa$\sqrt{m}$	Ref
As-cast	896	130	1000	145	8	16	97	107	37, 47
Cast HIP	869	126	958	139	10	18	99	109	37, 39, 48
BUS(b)	938	136	1041	151	8	12	...	...	37, 39
GTEC(b)	938	136	1027	149	8	11	...	...	38
BST(b)	931	135	1055	153	9	15	...	...	39
ABST(b)	931	135	1020	148	8	12	...	...	39
TCT(b)	1055	153	1124	163	6	9	...	...	41, 49
CST(b)	986	143	1055	153	8	15	...	...	44
HTH(b)	1055	153	1103	160	8	15	...	...	45
Typical wrought β-annealed	860	125	955	139	9	21	83	91	18, 19

(a) All conditions (except as-cast) are cast plus HIP. (b) See Table 5 for process details.

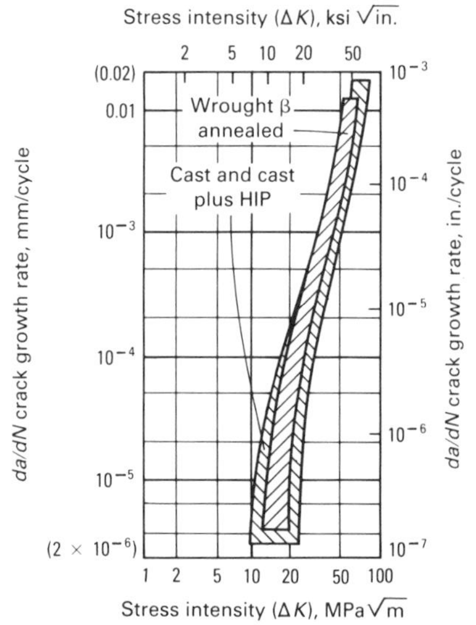

Fig. 7 Scatterband comparison of FCGR behavior of wrought β annealed Ti-6Al-4V to cast and cast HIP Ti-6Al-4V

Table 7 General linear and diametrical tolerance guideline for titanium castings

Size, mm	Size, in.	Total tolerance band(a): Investment cast	Total tolerance band(a): Rammed graphite process
25 to <102	1 to <4	0.76 mm (0.030 in.) or 1.0%, whichever is greater	1.52 mm (0.060 in.)
102 to <305	4 to <12	1.02 mm (0.040 in.) or 0.7%, whichever is greater	1.78 mm (0.070 in.) or 1.0%, whichever is greater
305 to <610	12 to <24	1.52 mm (0.060 in.) or 0.6%, whichever is greater	1.0%
≥610	≥24	0.5%	1.0%
Examples			
254 mm (10 in.)		1.78 mm (0.070 in.) total tolerance band or ±0.89 mm (±0.035 in.)	2.54 mm (0.100 in.) total tolerance band or ±1.27 mm (±0.050 in.)
508 mm (20 in.)		3.05 mm (0.120 in.) total tolerance band or ±1.52 mm (±0.060 in.)	5.08 mm (0.200 in.) total tolerance band or ±2.54 mm (±0.100 in.)

(a) Improved tolerances may be possible depending on the specific foundry capabilities and overall part-specific requirements.

Table 8 Surface finish of titanium castings

Process	NAS 823 surface comparator	RMS equivalent µm	RMS equivalent µin.
Investment			
As-cast	C-12	3.2	125
Occasional areas of	C-25	6.3	250
Rammed graphite			
As-cast	C-30-40	7.5–10	300–400
Occasional areas of	C-50	12.5	500
Hand finished	C-12-25	3.2–6.3	125–250

RMS, root mean square

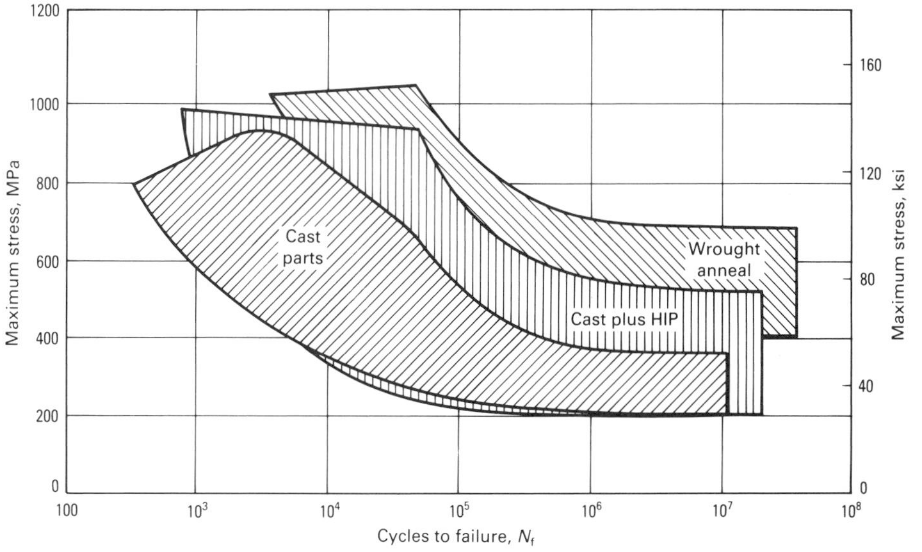

Fig. 8 Comparison of smooth axial fatigue rate in cast and wrought Ti-6Al-4V at room temperature with R = +0.1

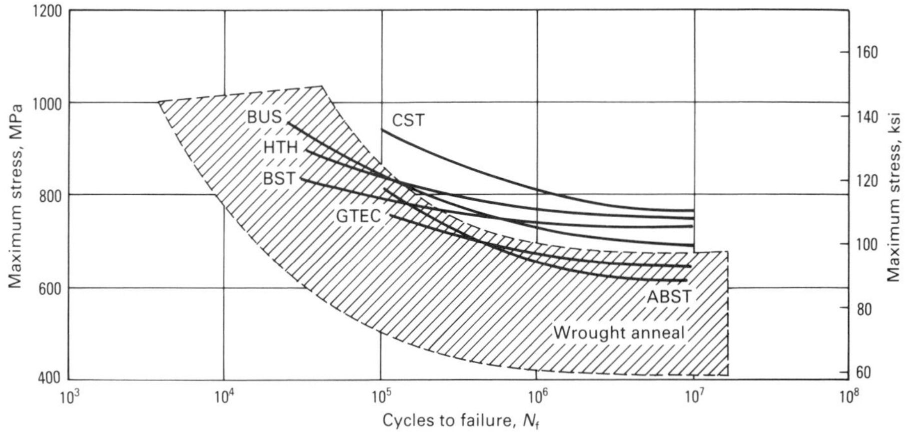

Fig. 9 Comparison of Ti-6Al-4V investment castings subjected to various thermal and hydrogen treatments. Smooth axial fatigue measured at room temperature. R = +0.1; frequency = 5 Hz using triangular wave form

pared with aluminum and low elastic modulus compared with steel present challenges in straightening and in maintaining extremely tight, true positions. General tolerance band capabilities for linear dimensions are shown in Table 7.

Hot sizing fixtures have been increasingly used to help control critical casting dimensions. This technique typically involves the use of steel fixtures to "creep" the casting into final tolerances in an anneal or stress relief heat treatment by the weight of the steel or the use of differential thermal expansion of the steel relative to the titanium.

Standard casting industry thickness tolerances of ±0.76 mm (±0.030 in.) for rammed graphite and ±0.25 mm (±0.010 in.) for investment cast walls are more difficult to maintain with titanium primarily because of the influence of chemical milling. As mentioned earlier, for critical applications it is necessary to mill all surfaces chemically to remove the residue α case. This operation is subject to variation because of part geometry and bath variables, and because it is usually manually controlled. Standard industry surface finishes are shown in Table 8.

Melting and Pouring Practice

Vacuum Consumable Electrode. The dominant, almost universal, method of melting titanium is with a consumable titanium electrode lowered into water-cooled copper crucibles while confined in a vacuum chamber. This skull melting (see the section "Vacuum Arc Skull Melting and Casting" in the article "Vacuum Melting and Remelting Processes" in this Volume) technique prevents the highly reactive liquid titanium from dissolving the crucible because it is contained in a solid "skull" frozen against the water-cooled crucible wall. When an adequate melt quantity has been obtained, the residual electrode is quickly retracted, and the crucible is tilted for pouring into the molds. A "skull" of solid titanium remains in the crucible for reuse in a subsequent pour or for later removal.

Superheating. The consumable electrode practice affords little opportunity for superheating the molten pool because of the cooling effect of the water-cooled crucible. Because of limited superheating, it is common either to pour castings centrifugally, forcing the metal into the mold cavity, or to pour statically into preheated molds to obtain adequate fluidity. Postcast cooling

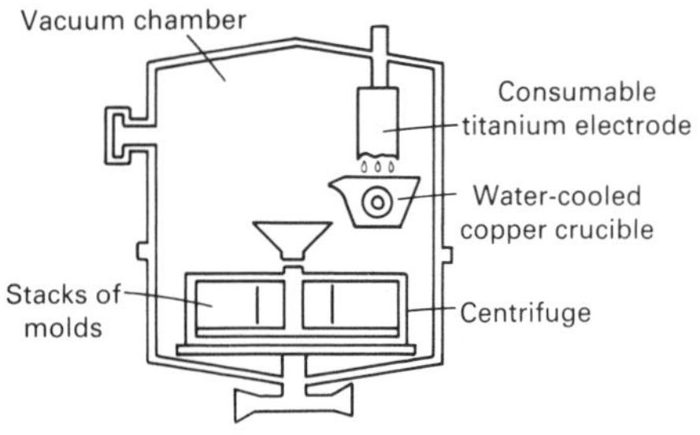

Fig. 10 A centrifugal vacuum casting furnace

takes place in a vacuum or in an inert gas atmosphere until the molds can be safely removed to air without oxidation of the titanium.

Electrode Composition. Consumable titanium electrodes are either ingot metallurgy forged billet, consolidated revert wrought material, selected foundry returns, or a combination of all of these. Casting specifications or user requirements can dictate the composition of revert materials used in electrode construction. Figure 10 shows a typical centrifugal casting furnace arrangement.

Chemical Milling

Residual surface contamination, or α case, is typically removed from as-cast parts before further processing. This is to eliminate the possibility of diffusion of these contaminants into the part during subsequent HIP or heat treatment. Chemical milling is normally conducted in solutions based on hydrofluoric and nitric acid mixtures plus additives designed to enhance surface finish and control hydrogen pickup. Hydrogen pickup is more likely the greater the β phase content of the alloy and is also influenced by etch rate and bath temperature. Subsequent vacuum anneals may be used to remove hydrogen picked up in chemical milling. The general objective is to remove the entire as-cast surface uniformly to the extent of maximum α case depth, and to retain the dimensional integrity of the part.

Hot Isostatic Pressing

Hot isostatic pressing may be used to ensure complete elimination of internal gas (D, in Fig. 2a) and shrinkage (E, in Fig. 2a) porosity. The cast part is chemically cleaned and typically subjected to argon pressure of 103 MPa (15 ksi) at 900 to 955 °C (1650 to 1750 °F) for a 2 h hold time (Ti-6Al-4V alloy) for void closure and diffusion bonding. This practice has been shown to reduce the scatterband of fatigue property test results and significantly improve fatigue life (Fig. 8). HIP temperature may coarsen the microstructure, causing a slight debit in tensile strength, but the benefits of HIP normally exceed this slight decrease in

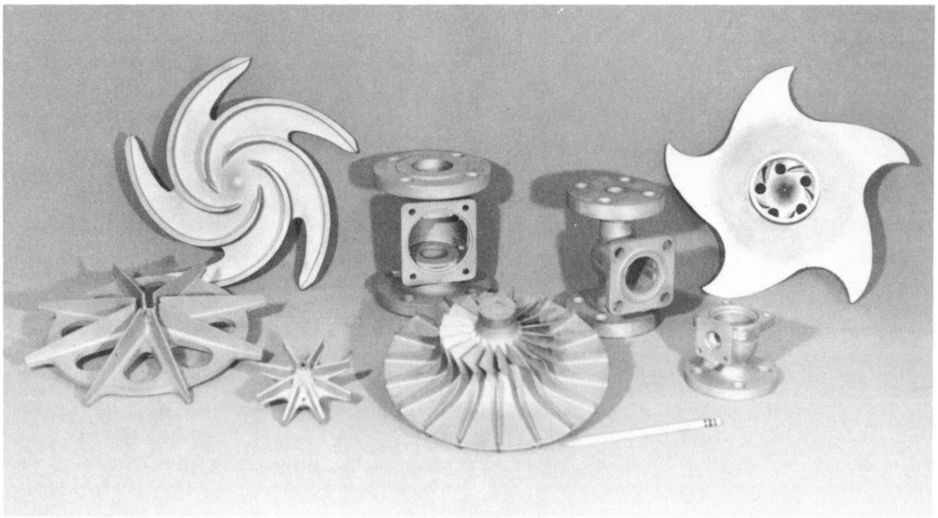

Fig. 11 Investment cast titanium components for use in corrosive environments

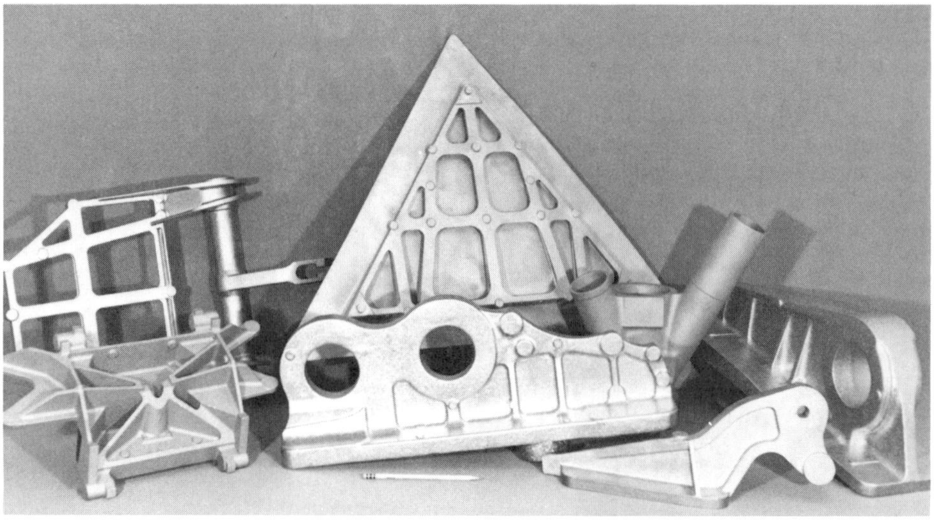

Fig. 12 Investment cast titanium airframe parts

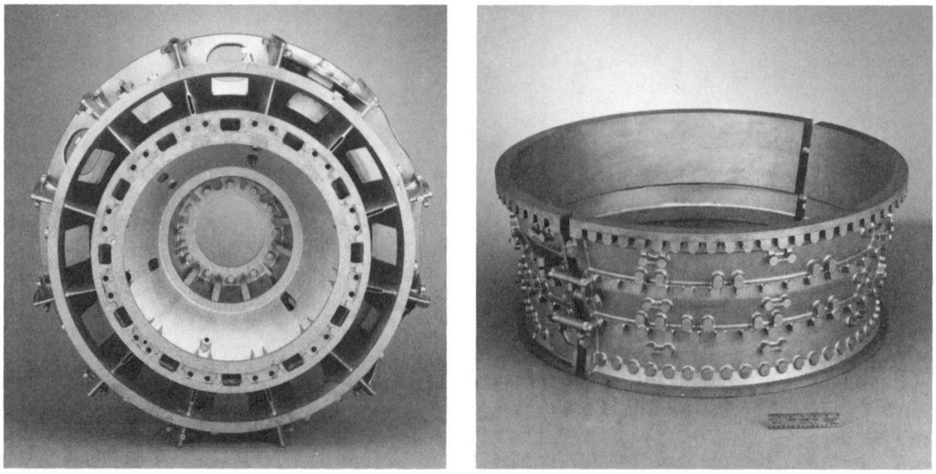

Fig. 13 Typical investment cast titanium components used for gas turbine applications

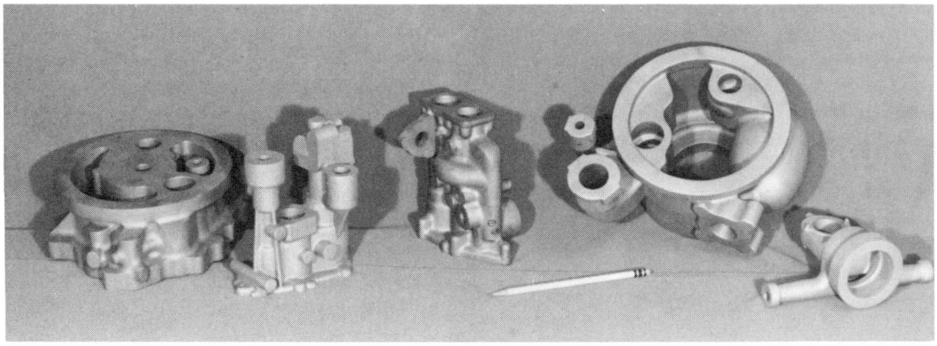

Fig. 14 Titanium hydraulic housings produced by the investment casting process

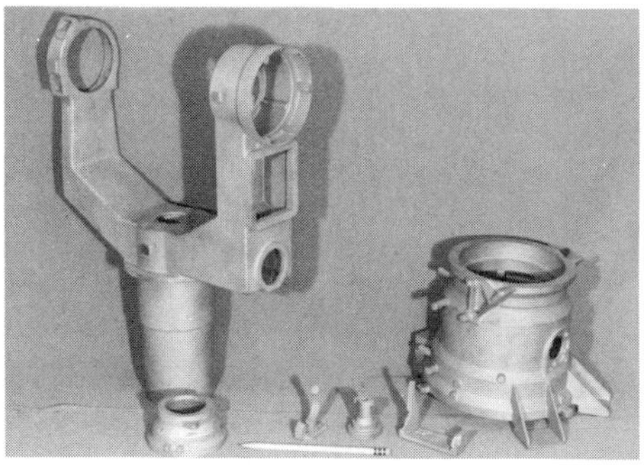

Fig. 15 Titanium housings for military optical applications produced by the investment casting process

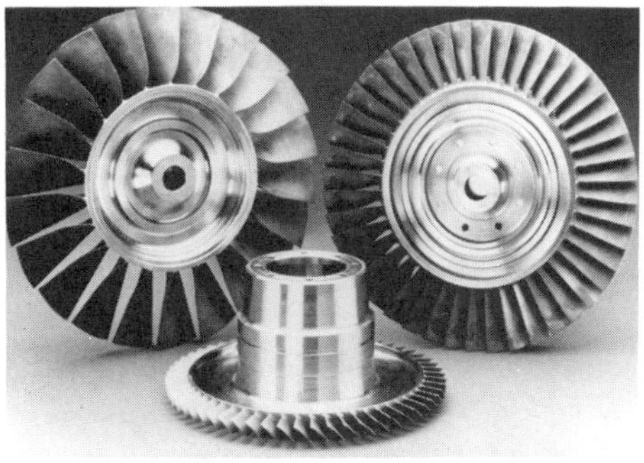

Fig. 16 Investment cast titanium engine airfoil components

strength, and the practice is widely used for aerospace titanium alloy cast parts.

Weld Repair

Weld repair of titanium castings is a normal facet of the manufacturing process and is used to eliminate surface-related defects, such as HIP-induced surface depressions. Tungsten inert-gas (TIG) welding practice in argon-filled glove boxes is used with weld filler wire of the same composition as the parent metal. Generally, all weld-repaired castings are stress relief annealed. Excellent quality weld deposits are routinely obtained in proper practice. Weld deposits may have higher strength but lower ductility than the parent metal because of microstructural differences due to the fast cooling rate of the welding process. Those differences may be eliminated by a postweld solution heat treatment, but standard practice is for stress relief or anneal only.

Heat Treatment

Conventional heat treatment of titanium castings is for stress relief anneal after any weld repair. The Ti-6Al-4V alloy is typically heat treated at 730 to 845 °C (1350 to 1550 °F). This is done in a vacuum to ensure removal of any hydrogen pickup from chemical milling and to protect the titanium from oxidation. As with HIP, castings must be chemically clean prior to heat treatment if diffusion of surface contaminants is to be avoided. Alternate heat treatments for property improvement, such as solution-treating and aging of Ti-6Al-4V alloy castings, are available. Numerous other heat treatments are in various stages of development, as discussed in an earlier section (Table 5).

Final Evaluation and Certification

Titanium castings are produced to numerous quality specifications. Typically, these require some type of x-ray and dye penetrant inspection in addition to dimensional checks using layout equipment, dimensional inspection fixtures, and coordinate measuring machines. Metallurgical certifications may include HIP and heat treatment run certifications, and chemistry, tensile properties, and microstructural examination of representative coupons for absence of surface contamination.

In the absence of universally accepted x-ray standards, it is common practice to use steel or aluminum reference radiographs (Table 4). Because internal discontinuities in titanium do not necessarily appear the same as they do in other metals, it is necessary to have an expert evaluation of radiographs for proper interpretation. Currently, an industry task force is working on the development of radiographic standards for titanium castings through the American Society for Testing and Materials (ASTM).

Product Applications

The titanium castings industry is relatively young by most foundry standards. The earliest commercial applications, in the 1960s, were for use in corrosion-resistant service in pump and valve components. These applications continue to dominate the rammed graphite production method; however, in more recent years, some users have justified the expense of lost wax investment tooling for some commercial corrosion-resistant casting applications (see Fig. 11).

Aerospace use of rammed-graphite-type castings became a production reality in the early 1970s for aircraft brake torque tubes, missile wings, and hot gas nozzles. As the more precise investment casting technology developed and the commercial use of HIP became a reality in the mid-1970s, titanium casting applications quickly expanded into critical airframe (Fig. 12) and gas turbine engine (Fig. 13) components. The first applications were primarily in Ti-6Al-4V, the workhorse alloy for wrought aerospace products, and castings were often substituted for forgings, with the addition of some net shape features; this trend continues. With continuing experience in manufacturing and specifying titanium castings, applications expanded from relatively simple less-critical components for military engines and airframes to large, complex structural shapes for both military and commercial engine and airframe programs. Today, titanium cast parts are routinely produced for critical structural applications such as space shuttle attachment fittings, complex airframe structures, engine mounts, compressor cases and frames of many types, missile bodies and wings, and hydraulic housings (Fig. 14). Quality and dimensional capabilities continue to be advanced. Titanium castings are used for framework for very sensitive optical equipment due to their stiffness and the compatibility of the coefficient of thermal expansion of titanium with that of glass (Fig. 15). Applications evolving for engine airfoil shapes (Fig. 16) include individual vanes and integral vane rings for stators, as well as a few rotating parts that would otherwise be made from wrought product. Growth will continue as users seek to take advantage of the flexibility of design inherent in the investment casting process and the improvement in economics of net and near-net shapes.

In spite of the wide acceptance of titanium castings for airframe applications, growth has been somewhat hindered because of the lack of an industry-wide data base to establish casting factors accurately, relative to wrought material. Such standards are now being established with prob-

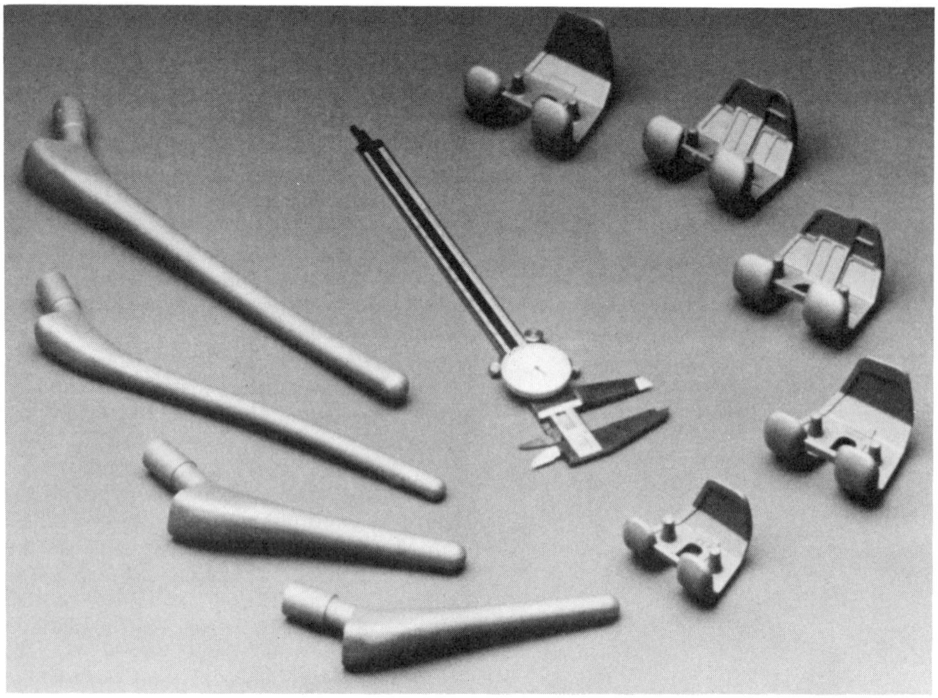

Fig. 17 Titanium knee and hip implant prostheses manufactured by the investment casting process

able elimination of design casting factors (Ref 58). Foundry size capabilities are expanding to allow the manufacture of larger airframe and static gas turbine engine structures. Widespread routine use of aerospace titanium castings is anticipated as the titanium foundry industry matches with well-established quality and product standards, and user understanding and confidence continue to be gained from satisfactory product performance.

Concurrent with the above trend, investment cast titanium is increasingly being specified for medical prostheses because of its inertness to body fluids, elastic modulus approaching that of bone, and the net shape design flexibility of the casting process. Custom-designed knee and hip implant components (Fig. 17) are routinely produced in volume. Some of these are subsequently coated with a diffusion bonded porous surface to facilitate bone ingrowth or an eventual fixation of the metal implant with the patient's bone structure.

REFERENCES

1. H.B. Bomberger, F.H. Froes, and P.H. Morton, Titanium—A Historical Perspective, in *Titanium Technology: Present Status and Future Trends*, F.H. Froes, D. Eylon, and H.B. Bomberger, Ed., Titanium Development Association, 1985, p 3-17
2. *Titanium for Energy and Industrial Applications*, D. Eylon, Ed., The Metallurgical Society, 1981, p 1-403
3. *Titanium Net Shape Technologies*, F.H. Froes and D. Eylon, Ed., The Metallurgical Society, 1984, p 1-299
4. "Titanium 1986, Statistical Review 1978-1986," Annual Report of the Titanium Development Association, 1987
5. American Foundrymen's Society, private conversation, 1987
6. Aluminum Association, private conversation, 1987
7. D. Eylon, F.H. Froes, and R.W. Gardiner, Developments in Titanium Alloy Casting Technology, *J. Met.*, Vol 35 (No. 2), Feb 1983, p 35-47; also, in *Titanium Technology: Present Status and Future Trends*, F.H. Froes, D. Eylon, and H.B. Bomberger, Ed., Titanium Development Association, 1985, p 35-47
8. D. Eylon and F.H. Froes, "Titanium Casting—A Review," in *Titanium Net Shape Technologies*, F.H. Froes and D. Eylon, Ed., The Metallurgical Society, 1984, p 155-178
9. H.B. Bomberger and F.H. Froes, The Melting of Titanium, *J. Met.*, Vol 36 (No. 12), Dec 1984, p 39-47; also, in *Titanium Technology: Present Status and Future Trends*, F.H. Froes, D. Eylon, and H.B. Bomberger, Ed., Titanium Development Association, 1985, p 25-33
10. E.W. Collings, *Physical Metallurgy of Titanium Alloys*, American Society for Metals, 1984
11. "Titanium: Past, Present and Future," NMAR-392, National Materials Advisory Board, National Academy Press, 1983; also, PB83-171132, National Technical Information Service
12. W.J. Kroll, C.T. Anderson, and H.L. Gilbert, A New Graphite Resistor Vacuum Furnace and Its Application in Melting Zirconium, *Trans. AIME*, Vol 175, 1948, p 766-773
13. R.A. Beahl, F.W. Wood, J.O. Borg, and H.L. Gilbert, "Production of Titanium Castings," Report 5265, U.S. Bureau of Mines, Aug 1956, p 42
14. A.R. Beall, J.O. Borg, and F.W. Wood, "A Study of Consumable Electrode Arc Melting," Report 5144, U.S. Bureau of Mines, 1955
15. R.A. Beahl, F.W. Wood, and A.H. Robertson, Large Titanium Castings Produced Successfully, *J. Met.*, Vol 7 (No. 7), July 1955, p 801-804
16. S.L. Ausmus and R.A. Beahl, "Expendable Casting Molds for Reactive Metals," Report 6509, U.S. Bureau of Mines, 1964, p 44
17. R.K. Koch and J.M. Burrus, "Bezonite-Bonded Rammed Olivine and Zircon Molds for Titanium Casting," Report 8587, U.S. Bureau of Mines, 1981
18. G.R. Yoder, L.A. Cooley, and T.W. Crooker, "Fatigue Crack Propagation Resistance of Beta-Annealed Ti6Al-4V Alloys of Differing Interstitial Oxygen Content," *Metall. Trans. A*, Vol 9A, 1978, p 1413-1420
19. R.R. Boyer and R. Bajoraitis, "Standardization of Ti6Al-4V Processing Conditions," AFML-TR-78-131, Air Force Materials Laboratory, Boeing Commercial Airplane Company, Sept 1978
20. D. Eylon, T.L. Bartel, and M.E. Rosenblum, High Temperature Low Cycle Fatigue of Beta-Annealed Titanium Alloy, *Metall. Trans. A*, Vol 11A, 1980, p 1361-1367
21. D. Eylon and J.A. Hall, Fatigue Behavior of Beta-Processed Titanium Alloy IMI-685, *Metall. Trans. A*, Vol 8A, 1977, p 981-990
22. D. Eylon, Fatigue Crack Initiation in Hot Isostatically Pressed Ti6Al-4V Castings, *J. Mat. Sci.*, Vol 14, 1979, p 1914-1920
23. D. Eylon and W.R. Kerr, The Fractographic and Metallographic Morphology of Fatigue Initiation Sites, in *Fractography in Failure Analysis*, STP 645, American Society for Testing and Materials, 1978, p 235-248
24. D. Eylon and M.E. Rosenblum, Effects of Dwell on High Temperature Low Cycle Fatigue of a Titanium Alloy, *Metall. Trans. A*, Vol 13A, 1982, p 322-324
25. W.G. Burgers, *Physics*, Vol 1, 1934, p 561-586
26. J.C. Williams, Kinetics and Phase Transformation, in *Titanium Science and Technology*, Vol 3, R.I. Jaffee and H.M. Burte, Ed., Plenum Press, 1973, p 1433-1494

27. D. Schechtman and D. Eylon, On the Unstable Shear in Fatigued Beta-Annealed Ti-11 and IMI-685 Alloys, *Metall. Trans. A*, Vol 9A, 1978, p 1273-1279
28. G.R. Yoder and D. Eylon, On the Effect of Colony Size on Fatigue Crack Growth in Widmanstätten Structure Alpha+Beta Alloys, *Metall. Trans. A*, Vol 10A, 1979, p 1808-1810
29. D. Eylon and P.J. Bania, Fatigue Cracking Characteristics of Beta-Annealed Large Colony Ti-11 Alloy, *Metall. Trans. A*, Vol 9A, 1978, p 1273-1279
30. R.J. Smickley and L.P. Bednarz, Processing and Mechanical Properties of Investment Cast Ti6Al-4V ELI Alloy for Surgical Implants: A Progress Report, in *Titanium Alloys in Surgical Implants*, STP 796, H.A. Luckey and F. Kubli, Ed., American Society for Testing and Materials, 1983, p 16-32
31. R.J. Smickley, Heat Treatment Response of HIP'd Cast Ti6Al-4V, in the *Proceedings of the WesTech Conference*, ASM INTERNATIONAL and Society of Manufacturing Engineers, 1981
32. F.H. Froes, D. Eylon, G.E. Eichelman, and H.M. Burte, Developments in Titanium Powder Metallurgy, *J. Met.*, Vol 32 (No. 2), 1980, p 47-54
33. F.H. Froes and D. Eylon, Powder Metallurgy of Titanium Alloys—A Review, in *Titanium, Science and Technology*, Vol 1, G. Lutjering, U. Zwicker, and W. Bunk, Ed., DGM, 1985, p 267-286; also, in *Powder Metall. Int.*, Vol 17 (No. 4), 1985, p 163-167 and continued in Vol 17 (No. 5), 1985, p 235-238; also, in *Titanium Technology: Present Status and Future Trends*, F.H. Froes, D. Eylon, and H.B. Bomberger, Ed., Titanium Development Association, 1985, p 49-59
34. D. Eylon and F.H. Froes, Method for Refining Microstructures of Cast Titanium Articles, U.S. Patent 4,482,398, Nov 1984
35. D. Eylon and F.H. Froes, Method for Refining Microstructures of Prealloyed Powder Metallurgy Titanium Articles, U.S. Patent 4,534,808, Aug 1985
36. D. Eylon and F.H. Froes, Method for Refining Microstructure of Blended Elemental Powder Metallurgy Titanium Articles, U.S. Patent 4,536,234, Aug 1985
37. D. Eylon, F.H. Froes, and L. Levin, Effect of Hot Isostatic Pressing and Heat Treatment on Fatigue Properties of Ti6Al-4V Castings, in *Titanium, Science and Technology*, Vol 1, G. Lutjering, U. Zwicker, and W. Bunk, Ed., 1985, p 179-186
38. D.L. Ruckle and P.P. Millan, Method for Heat Treating Cast Titanium Articles to Improve Their Mechanical Properties, U.S. Patent 4,631,092, Dec 1986
39. D. Eylon, W.J. Barice, and F.H. Froes, Microstructure Modification of Ti6Al-4V Castings, in *Overcoming Material Boundaries*, Vol 17, Society for the Advancement of Material and Process Engineering, 1985, p 585-595
40. W.R. Kerr, P.R. Smith, M.E. Rosenblum, F.J. Gurney, Y.R. Mahajan, and L.R. Bidwell, Hydrogen as an Alloying Element in Titanium (Hydrofac), in *Titanium '80, Science and Technology*, H. Kimura and O. Izumi, Ed., The Metallurgical Society, 1980, p 2477-2486
41. R.G. Vogt, F.H. Froes, D. Eylon, and L. Levin, Thermo-Chemical Treatment (TCT) of Titanium Alloy Net Shapes, in *Titanium Net Shape Technologies*, F.H. Froes and D. Eylon, Ed., The Metallurgical Society, 1984, p 145-154
42. L. Levin, R.G. Vogt, D. Eylon, and F.H. Froes, Method for Refining Microstructures of Titanium Alloy Castings, U.S. Patent 4,612,066, Sept 1986
43. L. Levin, R.G. Vogt, D. Eylon, and F.H. Froes, Method for Refining Microstructures of Prealloyed Powder Compacted Articles, U.S. Patent 4,655,855, April 1987
44. R.J. Smickley and L.E. Dardi, Microstructure Refinement of Cast Titanium, U.S. Patent 4,505,764, March 1985
45. C.F. Yolton, D. Eylon, and F.H. Froes, High Temperature Thermo-Chemical Treatment (TCT) of Titanium With Hydrogen, in the *Proceedings of the Fall Meeting*, The Metallurgical Society, 1986, p 42
46. Titech International Inc., unpublished research
47. F.C. Teifke, N.H. Marshall, D. Eylon, and F.H. Froes, Effect of Processing on Fatigue Life of Ti6Al-4V Castings, in *Advanced Processing Methods for Titanium*, D. Hasson, Ed., The Metallurgical Society, 1982, p 147-159
48. R.R. Wright, J.K. Thorne, and R.J. Smickley, Howmet Turbine Components Corporation, Ti-Cast Division, private communication, 1982; also, Technical Bulletin TB 1660, Howmet Corporation
49. L. Levin, R.G. Vogt, D. Eylon, and F.H. Froes, Fatigue Resistance Improvement of Ti6Al-4V by Thermo-Chemical Treatment, in *Titanium, Science and Technology*, Vol 4, G. Lutjering, U. Zwicker, and W. Bunk, Ed., 1985, p 2107-2114
50. L.J. Maidment and H. Paweltz, An Evaluation of Vacuum Centrifuged Titanium Castings for Helicopter Components, in *Titanium '80, Science and Technology*, H. Kimura and O. Izumi, Ed., The Metallurgical Society, 1980, p 467-475
51. J.-P. Herteman, "Properties d'Emploi de l'Alliage de Titane T.A6V Moule Densifie ou non par Compaction Isostatic à Chaud," CEAT, Technical Report 30/M/79, July 1979
52. W.H. Ficht, "Centrifugal Cast Titanium Compressor Case," Paper presented at the Manufacturing Technology Advisory Group Meeting, General Electric Company, Aircraft Engine Group, Lynn, MA, 1979
53. D. Eylon, P.R. Smith, S.W. Schwenker, and F.H. Froes, Status of Titanium Powder Metallurgy, in *Industrial Applications of Titanium and Zirconium: Third Conference*, STP 830, R.T. Webster and C.S. Young, Ed., American Society for Testing and Materials, 1984, p 48-65
54. J.K. Kura, "Titanium Casting Today," MCIC-73-16, Metals and Ceramics Information Center, Dec 1973
55. J.R. Humphrey, Report IR-162, REM Metals Corporation, Nov 1973
56. M.J. Wynne, Report TN-4301, British Aircraft Corporation, Nov 1972
57. H.D. Hanes, D.A. Seifert, and C.R. Watts, *Hot Isostatic Processing*, Battelle Press, 1979, p 55
58. R.J. Tisler, Fatigue and Fracture Characteristics of Ti6Al-4V HIP'ed Investment Castings, in the *Proceedings of the International Conference on Titanium*, Titanium Development Association, Oct 1986, p 23-41

Zirconium and Zirconium Alloys

John P. Laughlin, Oregon Metallurgical Corporation

ZIRCONIUM casting technology was developed at the U.S. Bureau of Mines Albany Research Center in the mid-1950s (Ref 1-3). The program was sponsored by the Atomic Energy Commission to develop casting technology capable of producing simple shapes that could be substituted for more expensive machined parts in nuclear applications. These researchers were also engaged in the development of titanium casting technology. Because of the similarities between these reactive metals, the molding, melting and casting, and processing technologies produced were identical. These similarities continue today.

No commercial zirconium casting alloys have yet been developed. Existing wrought alloys have been used with allowances for higher interstitial (oxygen, nitrogen, and carbon) and trace element contaminants.

Although the initial usage of zirconium castings was in nuclear applications, zirconium castings are currently used exclusively in the chemical processing industry, where their corrosion properties are utilized. The typical shapes produced are shown in Fig. 1. This article will focus on foundry practices, such as fabricating patterns and molds, as well as the melting, welding, and hot isostatic pressing techniques utilized to produce zirconium and zirconium alloy castings.

Patterns

A variety of patterns can be used to produce molds for the casting of zirconium. The molding system will determine the usable pattern type. With rammed graphite systems, match plate patterns, cope and drag patterns, core boxes, and loose patterns can be utilized, with the same pattern being used to produce zirconium, titanium, and other alloys. Wax patterns are used with refractory/ceramic shell molds. Permanent molds are compatible with zirconium casting, but they are not in use at this time. Shell molds are produced with gates and risers as integral components of the mold. Graphite molds are rammed without gates and risers. The gates and risers are cut in when the molds are assembled. Compared to other reactive metals, zirconium has very poor fluidity. Many large gates must be provided when gating a mold.

Molds

As with other reactive metals, zirconium must be handled very carefully when in the molten state. It will react with all conventional molding media. The extent of the mold-metal reaction depends on several factors:

- Size and configuration of the casting
- Mass of the casting
- Mold media
- Surface roughness of the mold

The first commercial zirconium casting utilized a machined graphite mold with a rammed graphite core.

Sand Systems. Although machined graphite is still used for base plates and spacers, most castings utilize a granular graphite sand with a hydrocarbon binder consisting of corn syrup, starch, and powdered pitch. The molds are rammed up by hand and allowed to set for several hours to develop an acceptable green strength. The mold is dried at 200 °C (390 °F) to remove moisture and then fired at 900 °C (1650 °F), under a reducing atmosphere, to carburize the binder constituents. Typical mold shrinkage is 16 to 26 mm/m ($^{3}/_{16}$ to $^{5}/_{16}$ in./ft). Care must be taken to provide a smooth surface for the mold/molten metal interface. Rough-surfaced molds will allow greater reaction between the mold and the metal. Graphite, however, has a good chilling effect on molten zirconium, and in general only internal core areas are susceptible to metal-mold reactions.

The typical reaction surface is 0.000 to 0.13 mm (0.000 to 0.005 in.) thick, but large masses of metal will be prone to greater metal-mold reaction. After casting, the graphite mold is physically broken off from the casting and recycled, with only minimal

(a)

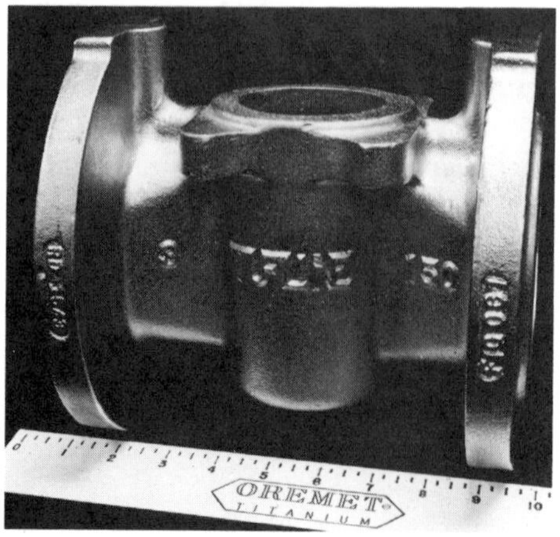

(b)

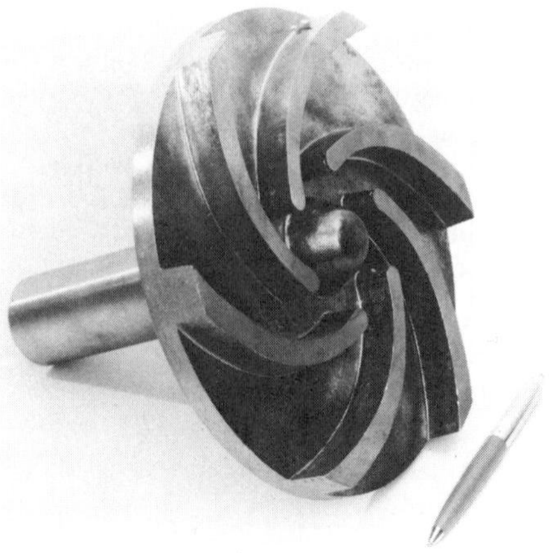

(c)

Fig. 1 Zirconium cast components used in chemical processing equipment. (a) Nuclear fuel tube top. (b) Valve body. (c) Impeller

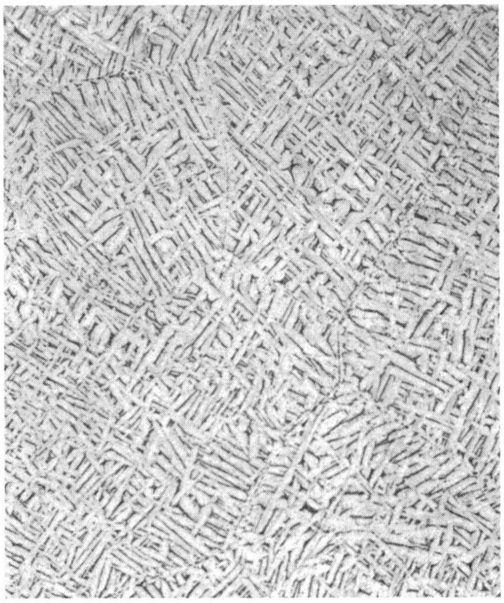

Fig. 2 Photomicrograph of as-cast zirconium Alloy 702. The etchant used was 45% HNO_3, 10% HF, and 45% H_2O. 100×

losses from fines generated in the regrinding. Cast zirconium is not a forging material, and care must be used when removing mold and core. One misplaced hammer blow can ruin a casting. Castings with thicknesses as great as 127 mm (5 in.) have been produced commercially with the rammed graphite mold system.

Two other sand systems have been identified as having potential use as mold media for zirconium (Ref 4, 5). The first is zircon sand/waterglass utilizing zirconium silicate sand and sodium silicate as the binder. The advantage of this system is reduced shrinkage of the mold to provide a more precise casting and a no-fire binder (200 °C, or 390 °F, cure), which would eliminate the multiple-stage firing process needed for graphite. This mold system requires a zirconia mold wash to protect the molten metal from reaction with the sand. Castings of a commercial size (23 kg, or 50 lb) were poured in 1979 with moderate success. The casting surface was rougher than with graphite molds, and the cores showed higher metal reaction. The second system, an olivine sand/bentonite binder, also requires the use of a zirconia mold wash to minimize the mold-metal reaction. No commercial-size castings have been produced with this system.

Shell Molds. For investment castings, shell molds are produced from wax patterns dipped in slurried ceramic, and a suitably strong shell is manufactured by repeated buildup of refractory layers. Upon curing of the molds, the wax is melted out and the shell is ready for pouring. Although shell mold systems are proprietary and may vary from foundry to foundry, zirconium is being successfully cast into shells produced from zirconia inner layers and aluminum silicate outer layers. Casting thicknesses as great as 25 mm (1 in.) have been successfully poured. Exact mold reaction layers have not been determined, but because zirconium is more reactive in the molten state than titanium, it is anticipated that the reaction layer of investment cast zirconium will be at least equal to that experienced in titanium.

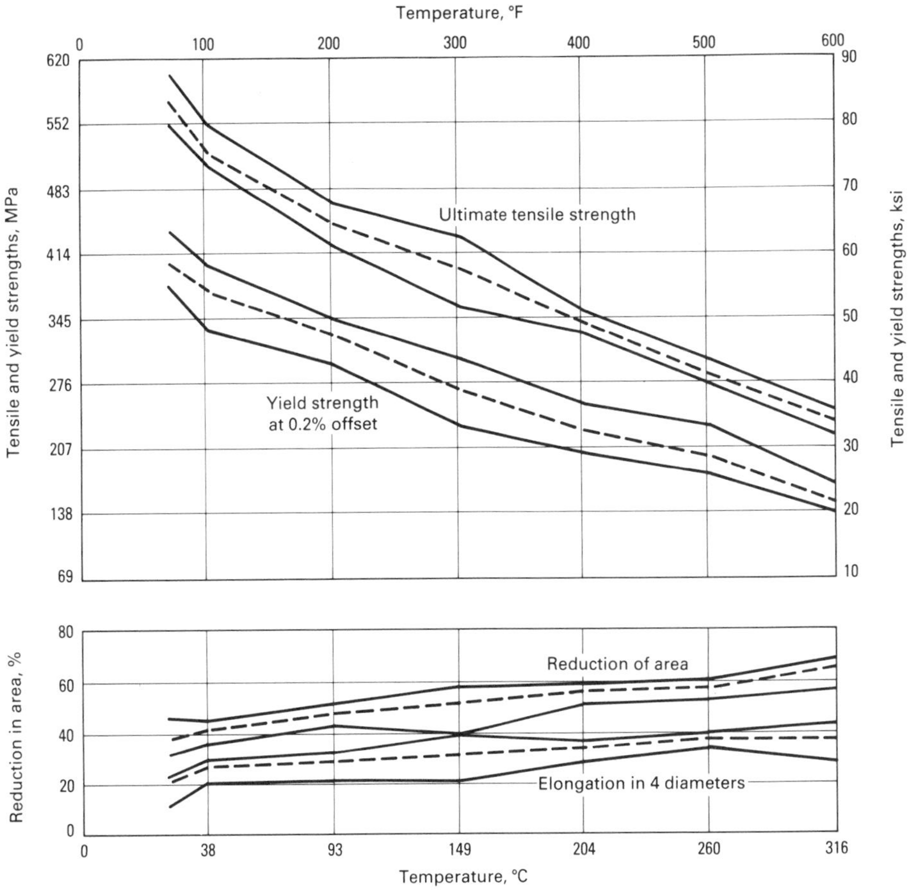

Fig. 3 Typical tensile properties of cast zirconium Alloy 702 at temperatures of 25 to 316 °C (70 to 600 °F)

Melting

Zirconium casting utilizes two melting methods: vacuum arc skull melting and vacuum induction melting. Both furnace systems are capable of melting all reactive alloys.

Vacuum arc skull melting furnaces use consumable electrodes melting into water-cooled copper crucibles. When sufficient metal has been melted, the crucible is tipped and the metal is poured into the mold positioned below. Skull melting furnaces are capable of using all types of mold systems and are limited only by the weight of the melted metal and the size constraint of the furnace. Zirconium melts of 955 kg (2100 lb) have been poured, yielding finished castings as large as 635 kg (1400 lb).

Castings can be produced with the receiving molds in a static mode as well as by centrifugal casting. Centrifugal casting is accomplished by mounting the molds on a turntable. This setup utilizes a center sprue with a runner system to feed from the outside of the mold in. The mold is filled against the centrifugal forces, allowing a slower fill rate and reducing the potential for entrapped gases in the casting. There are limitations for centrifugal casting in a vacuum arc skull melt furnace. The pour must have a balanced setup, that is, molds equally spaced around the center sprue. The mold media and mold design must be strong enough to support the forces exerted on the setup by the molten metal.

Vacuum induction melting (VIM) is used to produce small casting pours. The current furnace design utilizes a split copper crucible with a melting capacity of approximately 27 kg (60 lb). The advantage of a VIM system is the ability to pour small single casting pours with minimal skull losses. As with skull melting, VIM furnaces have the capacity to use all types of molding systems (Ref 6).

Processing

After knockout, gates and risers are removed by abrasive cutoff or by oxyacetylene torch. If torch cutting is used, the slag area and subslag-contaminated material must be removed by grinding.

Table 1 Minimum room-temperature properties of cast zirconium alloys

Grade	Minimum tensile strength		Minimum yield strength, 0.2% offset		Minimum elongation in 25 mm (1 in.), %	Maximum hardness, HB	Maximum hardness, Rockwell scale
	MPa	ksi	MPa	ksi			
702C	380	55	276	40	12	210	96 HRB
704C	413	60	276	40	10	235	36 HRC
705C	483	70	345	50	12	235	36 HRC

Source: Ref 8

All metal surfaces in contact with the mold will have a contaminated layer. For graphite systems, it is a carbon-stabilized layer; for refractory systems, oxygen stabilized. Depending on the application, this hard, brittle layer should be removed from the surface because it may be a source of contaminants during any subsequent weld repair and it may adversely affect the mechanical properties of the casting. In some applications, corrosion resistance may be affected.

Abrasive grit or shotblasting can be used to remove most of this contaminated layer.

Chemical Milling. For internal surfaces or blind areas, chemical milling is more efficient than abrasive grit or shotblasting and is the preferred method. Chemical milling solutions are of the following general chemical composition: 25 to 50% nitric acid, 3 to 10% hydrofluoric acid, and the remainder water. Temperature for the solution should be kept below 55 °C (130 °F) to minimize hydrogen pickup.

Weld Repair of Casting Defects

Casting Defects. As-cast zirconium typically has three types of casting defects: gas hole porosity, cold shuts (or surface laps), and shrinkage cavities.

Gas hole porosity is generated from the metal-mold reaction and the evolution of gases from the mold. This type of defect can be minimized by proper firing of the mold, but certain portions of a casting may still contain these small spherical voids.

Cold Shunts. Because of the high melting point of zirconium (1850 °C, or 3360 °F) and the chilling effect the mold media has on the metal, zirconium will exhibit more cold shunts than titanium or other castings. However, these defects are surface phenomena only and, depending on application, may be left in with no detrimental effect.

Shrinkage Cavities. Zirconium has less fluidity than titanium, which is the only material to compare zirconium to for melting parameters and mold material. This low fluidity will result in shrinkage cavities forming in areas of a casting where flow conditions are upset. Therefore, zirconium requires larger or greater numbers of metal distribution gates to provide castings with minimum shrinkage.

Weld Repair. It is probable that all castings will have some shrinkage, and unless a given casting is designed for use with shrinkage, it will be necessary to remove the voids. Suitable excavation and weld repair then become necessary steps in the production of zirconium castings. Weld repair is performed using gas tungsten arc welding. Welding can be done on a bench setup with sufficient inert-gas protection, but for production it is best to perform the weld repair in a weld chamber. Zirconium is extremely reactive to atmospheric gases, and if contaminated, the weld will be embrittled, will possess adverse mechanical properties, and will perhaps exhibit reduced corrosion resistance.

Preparation of the weld area is necessary for sound welds. The area must be cleaned of any residual grease or oil. The filler rod should also be degreased before use. For repair of internal defects, the defect is exposed, usually by drilling or machining. This should be done without lubricants or coolants in order to minimize contamination of the weld area; the surrounding area must be cleaned after the machining operation. The cavity itself need not be cleaned. After placing the casting in the weld chamber, either a vacuum pumpdown followed by backfilling with inert gas or a flow-through purge of inert gas is required to protect the metal during welding. It is also advisable to getter the weld chamber by striking an arc on a sacrificial plate or piece of scrap before working on the part. The getter pool should be held until the resultant weld bead is bright with no sign of discoloration. Filler metal should be of the same nominal composition as the casting.

Postweld stress relieving, if necessary, will be carried out in the 535 to 595 °C (1000 to 1100 °F) temperature range. Some specific applications, such as high sulfuric acid concentrations, may require a 745 °C (1375 °F) stress-relief temperature for optimum corrosion resistance properties.

Hot Isostatic Pressing

Hot isostatic pressing (HIP) has been used as an alternative to weld repair for improving internal integrity. A HIP schedule has been developed for zirconium; zirconium has been included in the same HIP load with titanium, and acceptable results have been obtained. Castings should be x-rayed before hot isostatic pressing to evaluate the location of shrink voids for possible distortion or warpage.

Table 2 Typical chemical analysis and mechanical properties of cast zirconium Alloy 702

Chemical analysis	
Carbon, %	0.011
Hydrogen, %	0.0008
Oxygen, %	0.139
Nitrogen, %	0.006
Hafnium, %	0.071
Aluminum, %	0.010
Copper, %	0.004
Iron, %	0.107
Nickel, %	0.001
Titanium, %	0.028
Mechanical properties	
Ultimate tensile strength, MPa (ksi)	469 (68)
0.2% offset yield strength, MPa (ksi)	330 (47.8)
Elongation in 4 diameters, %	20.1
Reduction in area, %	32.4

Machining

Cast zirconium has machining characteristics similar to those of the wrought product. As with any zirconium machining, care should be taken in the handling and storage of turnings and dust because of their pyrophoric nature.

Structure and Properties

All zirconium initially solidifies into a body-centered cubic structure and then, upon further cooling, transforms into a hexagonal close-packed Widmanstätten α phase (Fig. 2). Cast zirconium alloys are equal in strength to wrought alloys of the same composition. Ductilities, however, are below typical values for comparable wrought alloys. Table 1 lists the minimum room-temperature tensile properties of cast zirconium alloys (Ref 7). Table 2 lists typical room-temperature tensile properties for commercially pure zirconium (grade 702). There are no data for the typical properties for Alloys 704 and 705. Figure 3 shows typical elevated-temperature tensile properties for cast Alloy 702. Impact strength is not a property that is normally tested for. However, cast zirconium has been tested infrequently for Charpy V-notch impact strengths with average values of 9.5 J (7 ft ·

lbf) with a range of 5.4 to 15 J (4 to 11 ft · lbf). Wrought Alloy 702 tested in the longitudinal direction produces a typical Charpy value of 35 J (26 ft · lbf) (Ref 8).

REFERENCES

1. R.A. Beall, F.W. Wood, J.O. Borg, and H.L. Gilbert, "Production of Titanium Casting," Report of Investigation 5265, Bureau of Mines, 1956
2. S.L. Ausmus, F.W. Wood, and R.A. Beall, "Casting Technology for Titanium, Zirconium and Hafnium," Report of Investigation 5686, Bureau of Mines, 1960
3. S.L. Ausmus and R.A. Beall, "Expendable Casting Molds for Reactive Metals," Report of Investigation 6509, Bureau of Mines, 1964
4. R.K. Koch, J.L. Hoffman, M.L. Transue, and R.A. Beall, "Casting Titanium and Zirconium in Zircon Sand Molds," Report of Investigation 8208, Bureau of Mines, 1977
5. R.K. Koch and J.M. Burrus, "Shape Casting Titanium in Olivine, Garnet, Chromite and Zircon Rammed and Shell Molds," Report of Investigation 8443, Bureau of Mines, 1980
6. D. Chronister, The Duriron Company, private communication, 1987
7. "Standard Specification for Casting, Zirconium Base, Corrosion Resistant, for General Application," B 752-85, *Annual Book of ASTM Standards*, American Society for Testing and Materials
8. R. Sutherlin, Teledyne Wah Chang Albany, private communication, 1987

Cast Metal-Matrix Composites

Pradeep Rohatgi, The University of Wisconsin—Milwaukee

METAL-MATRIX COMPOSITES (MMCs) are engineered combinations of two or more materials (one of which is a metal) in which tailored properties are achieved by systematic combinations of different constituents. Conventional monolithic materials have limitations in terms of achievable combinations of strength, stiffness, coefficient of expansion, and density. Engineered MMCs consisting of continuous or discontinuous fibers (designated by an f), whiskers (w), or particles in a metal result in combinations of very high specific strength and specific modulus (Table 1). Furthermore, systematic design and synthesis procedures can be developed to achieve unique combinations of engineering properties, such as high elevated-temperature strengths, fatigue strength, damping properties, electrical conductivity, thermal conductivity, and coefficient of thermal expansion. In a broader sense, several cast materials with two-phase microstructures in which the volume and shape of the phases are governed by phase diagrams (that is, cast iron and aluminum-silicon alloys) have long been produced in foundries. Modern cast MMCs differ from these older materials in that any selected volume, shape, and size of reinforcement can be introduced into the matrix (Table 2); these tailored materials represent a new product opportunity for foundrymen.

A variety of methods for producing MMCs, including foundry techniques, have recently become available. The potential advantage of preparing these composite materials by foundry techniques is near-net shape fabrication in a simple and cost-effective manner. In addition, foundry processes lend themselves to the manufacture of large numbers of complexly shaped components at high production rates, which is required by automotive and other consumer-oriented industries.

Structurally, cast MMCs consist of continuous or discontinuous fibers, whiskers, or particles in an alloy matrix that solidifies in the restricted spaces between the reinforcing phase to form the bulk of the matrix. By carefully controlling the relative amounts and distributions of the ingredients constituting a composite and by controlling the solidification conditions, MMCs can be imparted a tailored set of useful engineering properties that cannot be realized with conventional monolithic materials. In addition, the solidification microstructure of the matrix is refined and modified because of the fibers and particles, indicating the possibility of controlling microsegregation, macrosegregation, and grain size in the matrix. This represents an opportunity to develop new matrix alloys.

From the standpoint of property-performance relationships and from the viewpoint of ease of processing and manufacturing, the interface between the matrix and the reinforcing phase (fiber or particle) is of central importance. Improvements in wetting between the matrix and the reinforcing phase in MMCs are an important goal of surface engineering. Other interrelated goals of surface engineering are control of fracture path (toughness), protection of filaments, and their role as nucleation catalysts to control the matrix microstructure. The solidification processing of MMCs in a foundry allows tailoring of the interface between the matrix and the fiber in order to suit specific property-performance requirements. The cost of producing cast MMCs has decreased rapidly, especially with the use of low cost particulate reinforcements such as graphite and silicon carbide. Low cost composites such as metal-silicon carbide (SiC) particle and metal-graphite particle are commercially available.

Table 1 Specific strength and specific modulus of some metal matrices, reinforcements, and MMCs

Material	Amount of fiber reinforcement, vol%	Specific strength, N · m/kg	Specific modulus, N · m/kg
Al-Li/Al_2O_3			
0°	60	20 000	7.59×10^7
90°	60	4986–6000	4.406×10^7
Ti-6Al-4V/SiC			
0°	35	45 337	7.77×10^7
90°	35	10 622	. . .
Mg/carbon (Thornel)	38	28 333	. . .
Al/carbon	30	28 163	6.53×10^7
6061 Al	. . .	11 481	2.53×10^7
2014 Al	. . .	17 143	2.59×10^7
SiC(f)	100	78 431	1.567×10^8
SiC(w)	100	6.67×10^5	2.19×10^8 to 3.29×10^8
Al_2O_3(f)	100	50 000	1.175×10^8
B(f)	100	1.538×10^5	1.62×10^8
C(f)	100	1.618×10^5	1.35×10^8
Be(f)	100	59 459	1.68×10^8
W(f)	100	14 974	1.79×10^7
Al/boron			
0°	50	56 604	7.92×10^7
90°	50	5 283	5.66×10^7
Al/SiC			
0°	50	8 803	1.092×10^8
90°	50	3 697	. . .

Fiber-Metal Wettability and Its Effect on Casting Quality

Cast MMCs are made by introducing fibers or particles into molten or partially solidified metals, followed by casting of these slurries in molds. Alternatively, a preform of fibers or a prepacked bed of particles is made and infiltrated by molten alloys, which then freeze in the interfiber spaces to form the composite. In both these processes, mixing and wetting between molten alloys and dispersoids are desirable in terms of ease of fabrication and ultimate distribution of particulates and fibers (Ref 1, 2).

Graphite/magnesium, graphite/aluminum, and several other fiber-reinforced metals are valuable structural materials because they combine high specific strength and

Table 2 Matrix-dispersoid combinations used to make cast MMCs

Matrix	Dispersoid	Dispersoid size, μm	Amount of dispersoid, %
Aluminum	Graphite flakes	20–60	0.9–0.815
Aluminum	Graphite granules	15–100	1–8
Aluminum	Carbon microballoons	40, thickness 1–2	...
Aluminum	Shell char	125	15
Aluminum	Alumina particles	3–200	3–30
Aluminum	Discontinuous alumina	3–6 mm long; 15 μm diam	0–23 vol%
Aluminum	Silicon carbide particles	16–120	3–20
Aluminum	Silicon carbide whiskers	5–10	10%; 0–0.5 vol%
Aluminum	Mica	40–180	3–10
Aluminum	Silica	5–53	5
Aluminum	Zircon	40	0–30
Aluminum	Glass particles	100–150	8
Aluminum	Glass beads (spherical)	100	30
Aluminum	Magnesia	40	10
Aluminum	Sand	75–120	36 vol%
Aluminum	Titanium carbide particles	46	15
Aluminum	Boron nitride particles	46	8
Aluminum	Silicon nitride particles	40	10
Aluminum	Chilled iron	75–120	36 vol%
Aluminum	Zirconia	5–80	4
Aluminum	Titania	5–80	4
Aluminum	Lead	...	10
Copper	Graphite	...	...
Copper	Alumina	11	Volume fraction 0.74
Copper	Zirconia	5	2.12 vol%
Steel	Titania	8	...
Steel	Cerium dioxide	10	...
Steel	Illite clay	753	3
Steel	Graphite microballoons	...	...

Source: Ref 1

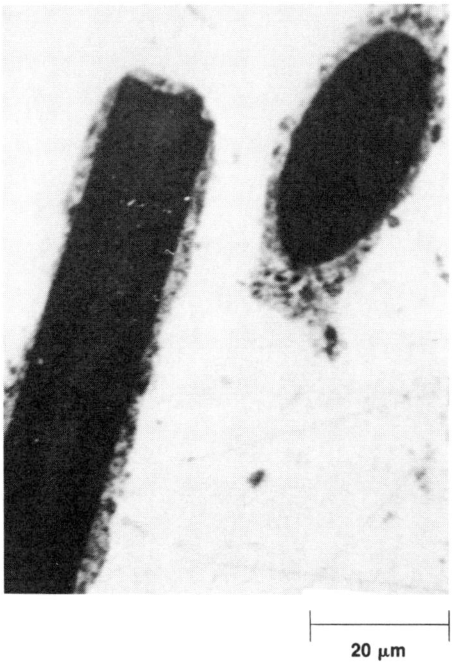

Fig. 1 Micrograph showing apparent interaction zones in Al-4Mg cast alloy containing discontinuous Al_2O_3 fibers. Courtesy of R. Mehrabian

stiffness with a near-zero coefficient of thermal expansion and high electrical and thermal conductivities. The primary difficulty in fabricating these fiber-reinforced metals is poor wetting and bonding between fibers and metals (Ref 3). However, compatibility and bonding between the fiber and the metal in these systems are induced by the chemical vapor deposition of a thin layer of titanium and boron onto the fibers to achieve wetting. These coated fibers are air-unstable because the titanium-boron coating is rapidly oxidized when exposed to air and because molten metal does not wet the fiber. To circumvent this difficulty, air-stable coatings of silicon dioxide that are wetted by magnesium have been used for graphite fibers (Ref 4).

These coatings are deposited on the fiber surfaces using silicon-base organometallic compounds. The fibers are simply passed through the organometallic solution, which is then chemically converted by either hydrolysis or pyrolysis to form the silicon dioxide coating. The flexible coated fibers can then be wound or laid-up and held in place with a removable binder for selective reinforcement. They are then incorporated into magnesium near-net shape structures by the pressure infiltration of magnesium. Complex structural components with high volume fractions of carbon fibers can be fabricated in this manner in a foundry. High-strength high stiffness 100% polycrystalline α-alumina (Al_2O_3)/magnesium composites containing up to 70 vol% Al_2O_3 have been prepared by a pressure infiltration process (described later in this article).

For nonwetting metals, the α-Al_2O_3 is coated with the metal by vapor deposition or by electroless plating before infiltration. Titanium-boron coatings have also been used for graphite/aluminum, Al_2O_3/aluminum, and Al_2O_3/lead MMCs (Ref 5). However, in terms of ease of fabrication and cost, modification of the matrix alloy by the addition of small amounts of reactive elements such as magnesium, calcium, lithium, or sodium is preferred. Alumina-reinforced aluminum, copper, lead, and zinc composites, as well as several particle-filled MMCs, have been synthesized by using reactive agents (Ref 1, 6). Careful control of reactions at the fiber/metal interface is required for improving the wettability and bonding between the two, without causing excessive degradation of the fiber surface. The formation of brittle intermetallic compounds at the interface (for example, carbides in graphite/aluminum and silicon carbide/aluminum composites) during processing can cause degradation of fibers and loss of strength in the longitudinal direction (Ref 7, 8).

Improved wettability and bonding at the interface can be achieved in some systems by a chemical reaction that forms spinels or oxides isostructural with spinels ($MgAl_2O_4$). Spinel promotes bonding at the interface because it forms strong bonds with metals and ceramics. The thickness and uniformity of the interfacial reaction layer (Fig. 1) can be controlled by optimizing melt temperature, residence time of particles or fibers in the melt (Fig. 2), and the degree of melt agitation (Ref 9). Continuous adherent metallic coatings (for example, copper and nickel) on several nonwetting particles such as graphite, shell char, and mica improve the melt-particle wettability and allow high percentages of these particles to be recovered in the solidified castings (Ref 10, 11).

The wetting properties of ceramics by liquid metals are governed by a number of variables, including heat of formation, stoichiometry, valence electron concentration in the ceramic phase, interfacial chemical reactions, temperature, and contact time (Ref 12). The work of adhesion between a ceramic and a melt decreases with increasing heat of formation of carbides. The high energy of formation of a stable carbide implies strong interatomic bonds and correspondingly weak interaction with melts. This leads to a high interfacial energy and a small work of immersion, resulting in poor wetting. The wetting angles of copper and the carbides of group IVA metals (titanium and zirconium) decrease with increasing carbon content, indicating the influence of stoichiometry on the surface energies that govern wetting behavior.

High valence electron concentration generally implies lower stability of carbides and improved wettability between ceramics and metals. High temperature and long contact times promote melt-ceramic wettability due to reactions at the melt/ceramic interface, resulting in reduced contact angle. Figure 3 shows the dependence of contact angle on temperature in the Al_2O_3/aluminum and carbon/aluminum systems.

Therefore, although MMCs are not restricted by phase diagram considerations (fixed proportions, chemistry, and morphol-

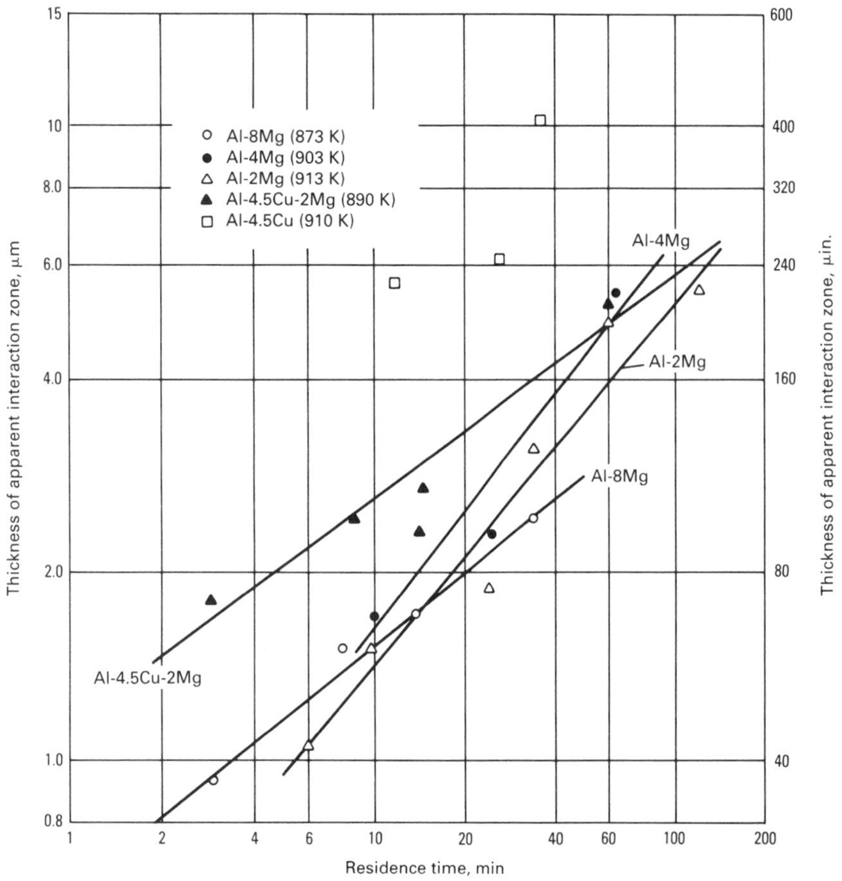

Fig. 2 Average thickness of apparent interaction zone as a function of residence time of Al_2O_3 fibers in various aluminum alloy slurries

ogy of solidifying phases), thermodynamic free energy and kinetic barriers still exist in their processing in the form of poor wettability and rates of mixing, and they must be addressed in the production of these composites (Ref 1, 2).

Casting Techniques

A basic requirement of the foundry processing of MMCs is the initial intimate contact and the intimate bonding between the ceramic phase and the molten alloy. This is achieved either by mixing the ceramic dispersoids into molten alloys or by pressure infiltration of molten alloys into preforms of the ceramic phase. As mentioned earlier, because of the poor wettability of most ceramics by molten metals, intimate contact between fiber and alloy can be promoted only by artificially inducing wettability or by using external forces to overcome the thermodynamic surface energy barrier and viscous drag. Mixing techniques generally used for introducing and homogeneously dispersing a discontinuous phase in a melt are:

- Addition of particles to a vigorously agitated fully or partially molten alloy (Ref 13-16). Figure 4 shows a schematic of an agitation vessel
- Injection of discontinuous phase into the melt with an injection gun (Ref 17)
- Dispersion of pellets or briquettes, formed by compressing powders of base

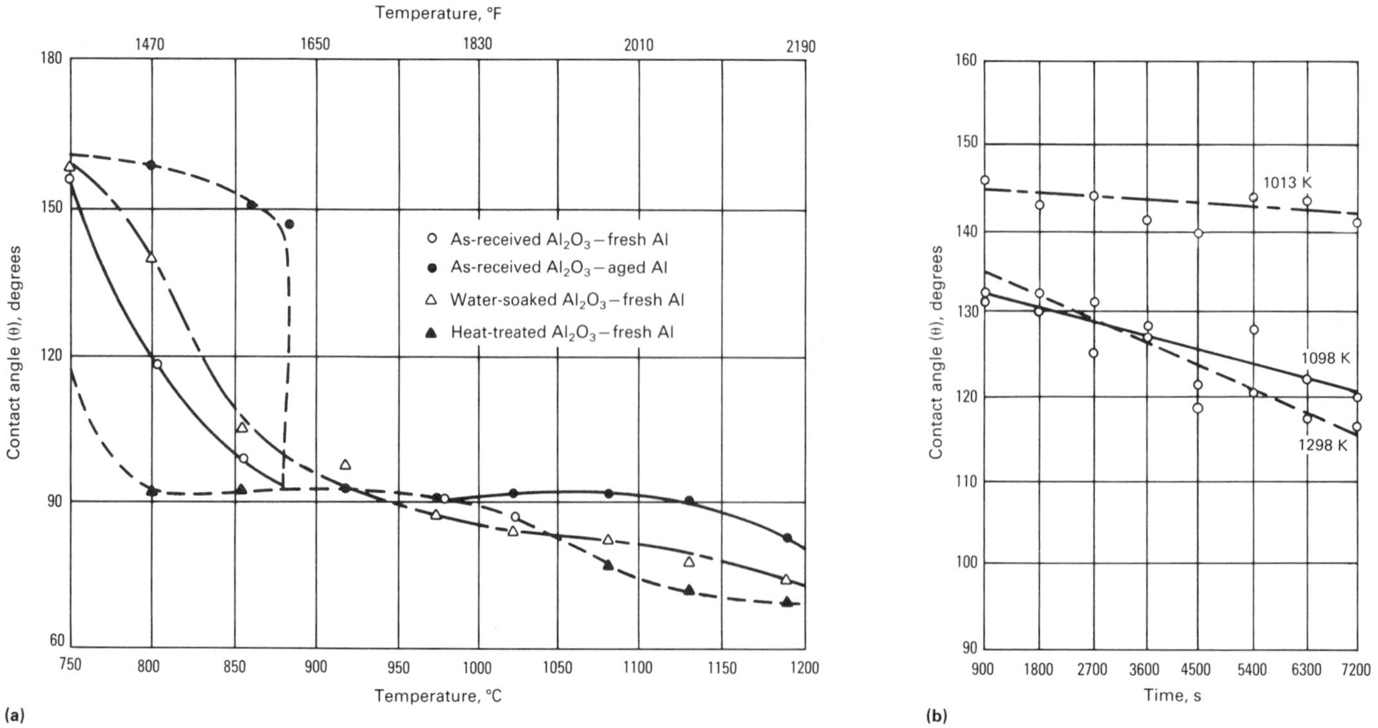

Fig. 3 Contact angle (a) in Al_2O_3/Al composite as a function of temperature. (b) Contact angle in aluminum/graphite composite as a function of time

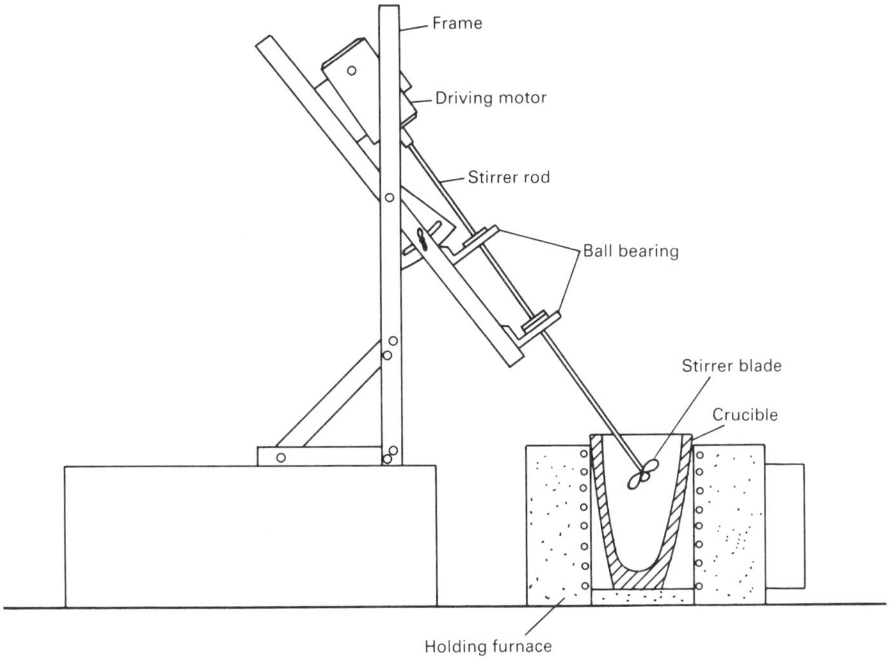

Fig. 4 Schematic showing the mechanical stirring device used in the vortex technique of dispersing particles in melts

alloys and the ceramic phase, into a mildly agitated melt (Ref 18)

- Addition of powders to an ultrasonically irradiated melt. The pressure gradients caused by cavitation phenomena promote homogeneous mixing of ceramics in metallic melts
- Addition of powders to an electromagnetically stirred melt. The turbulent flow caused by electromagnetic stirring is used to obtain a uniform suspension
- Centrifugal dispersion of particles in a melt. This technique has been used for carbon microballoons

In all the above techniques, external force is used to transfer a nonwettable ceramic phase into a melt and to create a homogeneous suspension of the ceramic in the melt. The uniformity of particle dispersion in a melt before solidification is controlled by the dynamics of particle movement in agitated vessels. Because of the inherent complexity of dispersion, it is difficult to optimize the initial processing conditions to achieve a specific particle distribution in the cast matrix. Visual observations on transparent models of agitated melt-particle systems have been found to provide useful guidelines for systematic design of processing schedules.

The melt-particle slurry can be cast either by conventional foundry techniques, such as gravity, pressure die, or centrifugal casting, or by novel techniques such as squeeze casting (liquid-forging), spray codeposition, melt spinning, or laser melt-particle injection. The choice of casting technique and mold configuration is of central importance to the quality (soundness, particle distribution) of a composite casting because the suspended particles experience buoyancy-driven movement in the solidifying melt until they are encapsulated in the solidifying structure by crystallizing phases. Particles such as graphite, mica, talc, porous alumina, and hollow microballoons are lighter than most aluminum alloys and tend to segregate near the top portion of gravity castings, leaving behind a particle-impoverished region near the bottom of the casting. Similarly, heavier particles such as zircon, glass, SiC, SiO_2, TiO_2, and ZrO_2 tend to settle and segregate near the bottom portion of the gravity castings (Ref 1).

The spatial arrangement of the discontinuous ceramic phase in the cast structure principally determines the properties of the cast composite. The distribution of phases depends on the quality of melt-particle slurry prior to casting and the following additional variables:

- Cooling rate

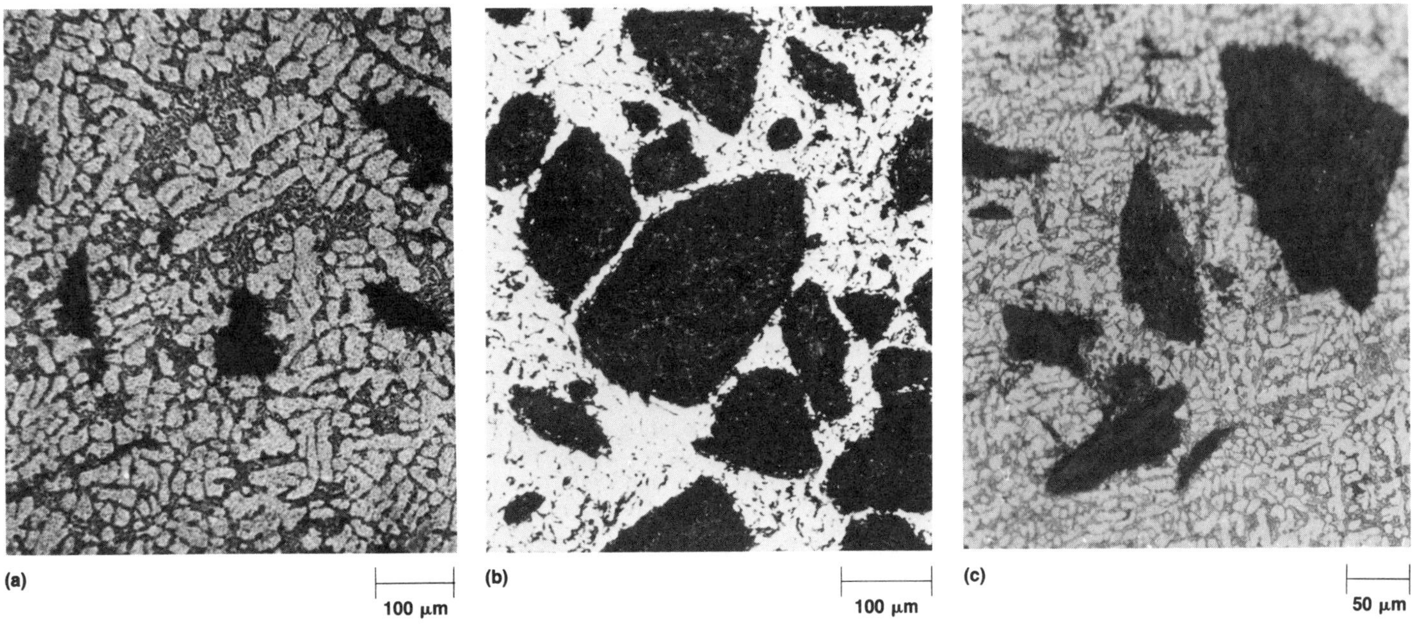

Fig. 5 Micrographs showing the uniform particle distribution obtained in the permanent mold casting of MMCs. (a) Graphite particles in aluminum-silicon alloy matrix. (b) Zircon particles in aluminum-silicon alloy matrix. (c) Talc in aluminum-silicon alloy matrix

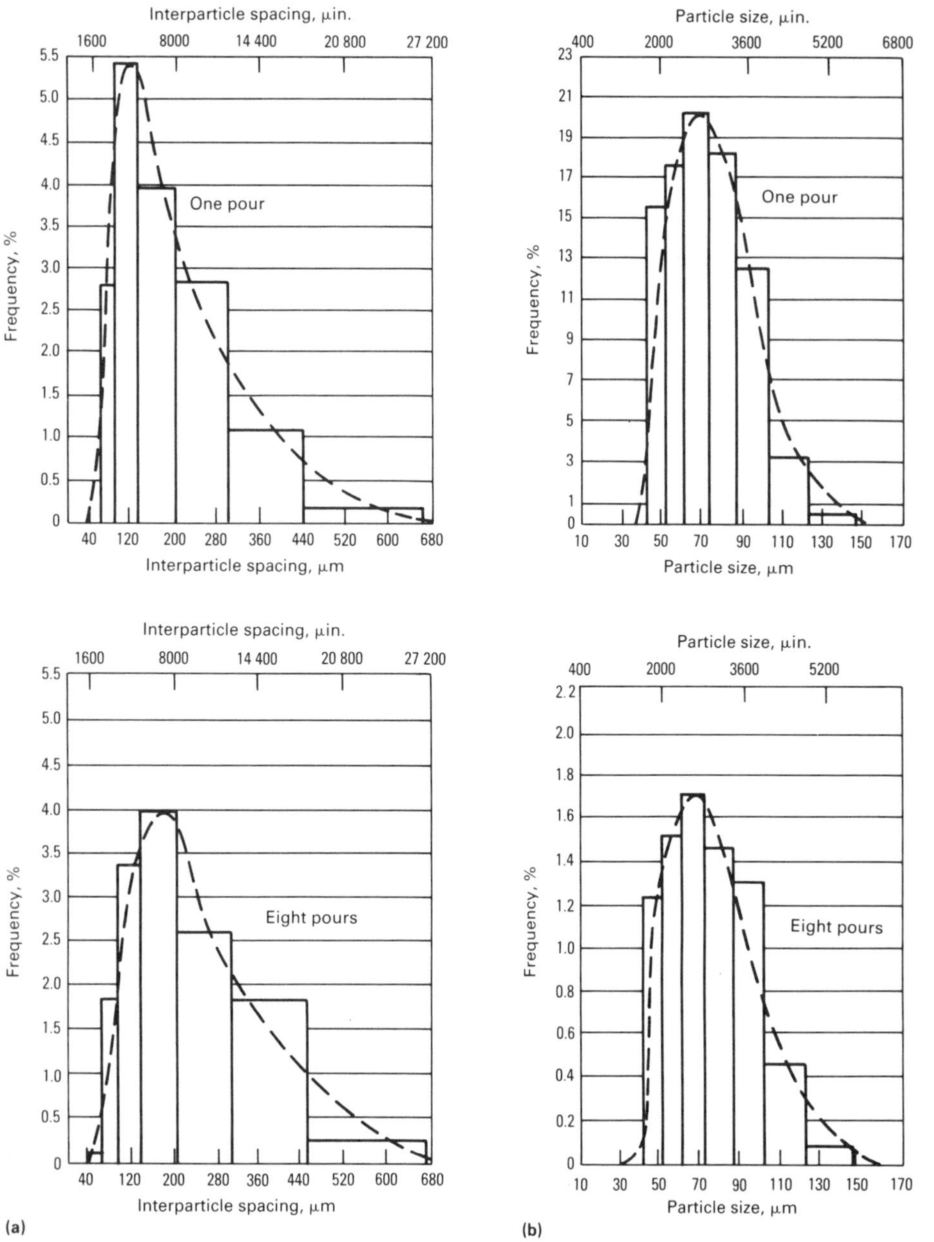

Fig. 6 Histograms showing graphite particle spacing and size distribution in aluminum alloy/graphite composites made in permanent molds after one and eight pours in the same mold. (a) Particle spacing. (b) Particle size distribution

- Viscosity of the solidifying melt
- Shape, size, and volume fraction of particles
- Particle and melt specific gravities
- Thermal properties of particles and matrix alloy
- Chemistry and morphology of crystallizing phases and their interactions with particles, nucleation of primary phases on ceramics, entrapment or pushing of particles by solidifying interfaces
- The flocculation (clustering) of particles
- The presence of any external forces during solidification

The various techniques used to solidify melt-particle slurries are discussed below.

Sand Casting. The slow solidification rates obtained in insulating sand molds permit considerable buoyancy-driven segregation of particles. Depending on the intrinsic hardness of the dispersed particles, high particle volume fraction surfaces can serve as selectively reinforced surfaces—for example, tailor-made lubricating or abrasion-resistant contacting surfaces for various tribological applications. Thin sections of sand cast composites or the codispersion of ceramic particles and a suitable micro-chill powder (normally the same as base alloy) reduce particle segregation in specific areas by decreasing solidification time.

Die Casting. The relatively rapid solidification rates in metallic molds generally give rise to a more homogeneous distribution of particles in the cast matrix (Ref 1, 19). Figure 5 shows microstructures of the permanent mold gravity die casting of aluminum alloys containing dispersions of graphite, zircon, and talc particles. Figure 6 shows frequency distributions of interparticle spacing and particle size in a gravity die cast graphite/aluminum MMC. These measurements were made across three different sections of the composite ingot (top, middle, and bottom) for two pourings during a batch process. The computed mean particle size and spacing are very close to the corresponding value for maximum frequency, indicating in statistical terms an absence of appreciable clustering or segregation of particles in the permanent mold casting of particle composites (Ref 20). Further improvements in particle distribution can be achieved by using water-cooled molds, copper chills, and other agitation techniques.

Centrifugal casting of composite melts containing particle dispersions results in the formation of two distinct zones in the solidified material: a particle-rich zone and a particle-impoverished zone. If the particles are lighter than the melt (for example, graphite, mica, or carbon microballoons in aluminum), the particle-rich zone forms at the inner circumference. The outer zone is particle rich if the particles are denser than the melt (for example, zircon or silicon carbide in aluminum).

Centrifugal acceleration in the rotating mold causes the lighter graphite, mica, or carbon to segregate near the axis of rotation, producing high particle volume fraction surfaces for bearing applications. The thickness of these particle-rich zones decreases with increasing pouring temperature and speed of rotation, but remains adequate for machining. Up to 8% by weight mica and graphite and up to 30% by weight zircon particles can be incorporated into selected zones of an aluminum alloy by this technique. Figure 7 shows sections of typical centrifugal castings of aluminum/zircon (Fig. 7a) and aluminum/graphite (Fig. 7b) composites. In Fig. 7(a), the heavier zircon segregates near the outer circumference, producing hard, abrasion-resistant surfaces; in Fig. 7(b), the lighter graphite concentrates near the inner periphery, producing wear-resistant, solid-lubricated surfaces for such antifriction applications as bearings or cylinder liners.

Compocasting. Particles and discontinuous fibers of SiC, Al_2O_3, TiC, silicon nitride, graphite, mica, glass, slag, MgO, and boron carbide have been incorporated into vigorously agitated, partially solidified aluminum alloy slurries by the compocasting technique (Ref 1, 9). The discontinuous ceramic phase is mechanically entrapped between the proeutectic phase present in the alloy slurry, which is held between its liquidus and solidus temperatures. Under

(a)

(b)

Fig. 7 Centrifugally cast MMCs showing segregation of particles caused by centrifugal force. (a) Aluminum/zircon composite with zircon particles segregated at outer periphery. (b) Aluminum/graphite particle composite with segregation of graphite particles at inner rim

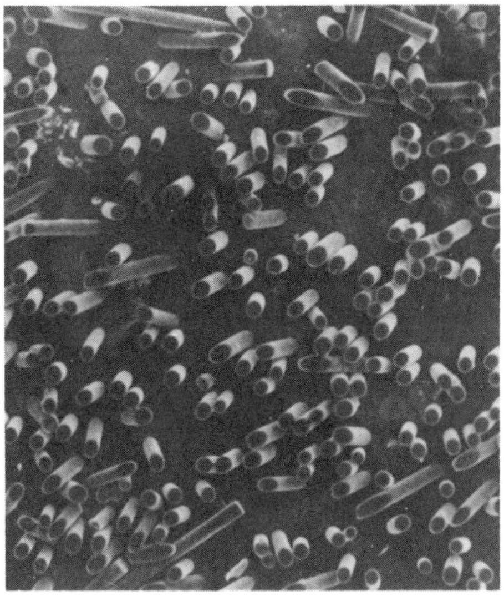

Fig. 8 SEM micrograph of electrochemically etched vertical section of cast Al-4Mg/23 vol% alumina fiber composite showing random planar orientation of fibers. Courtesy of R. Mehrabian

mechanical agitation, such an alloy slurry exhibits thixotropy in that the viscosity decreases with increasing shear rate. This effect appears to be time dependent and reversible. Increasing the particle residence time in the slurry promotes wetting and bond formation due to chemical reactions at the interface, as shown in Fig. 2 and discussed in the section "Fiber-Metal Wettability and Casting Quality" in this article. This semifusion process allows near-net shape fabrication by extrusion or forging because deformation resistance is considerably reduced due to the semifused state of the composite slurry.

Aluminum-silicon alloys (4 to 16% Si) containing 2.8 wt% graphite particles prepared by compocasting followed by squeeze casting show better particle distribution in hypoeutectic alloys (<12.6 wt% Si) than in near-eutectic and hypereutectic alloys. The last two alloys show some evidence of graphite clustering. Apparently, the significantly higher proportion of the proeutectic solid present in the hypoeutectic alloy slurries aids the separation and dispersion of particles.

Compocasting generally results in significant fiber breakage because of the grinding effect of the stirrer and the partially solid alloy slurry. This can be avoided by first mixing the discontinuous fibers in fully liquid alloy, followed by solidification, remelting, and squeeze filtering of the composite melt in a porous ceramic filter using a set of forging dies. This process can remove up to 70% excess melt, depending on squeeze conditions, and it allows a high volume fraction of fibers having random-planar alignment (Fig. 8) to be obtained. However, the chances of fiber breakage and degradation of properties of the composites because of the squeezing operation could offset its advantages at high volume fractions of fibers.

Pressure die casting of composites allows larger and more intricate component shapes to be produced rapidly at relatively low pressures (≤15 MPa, or 2.2 ksi) and equivalent capital expenditure. Pressurized gas and hydraulic ram in a die casting machine have been employed to synthesize porosity-free fiber and particle composites. It has been reported that high pressures, short infiltration paths, and columnar solidification toward the gate produce void-free composite castings. The pressure die cast particle composites exhibit lower bulk and interfacial porosities, more uniform particle distribution, less agglomeration of particles, and occasional exfoliation and/or fragmentation of soft particles (for example, graphite in aluminum alloy), with the melt penetrating into fine exfoliation cleavages (Ref 21). High concentrations (60 wt% or more) of zircon ($ZrSiO_4$) particles can be uniformly dispersed in pressure die cast aluminum-silicon-magnesium alloys. Pressure die castings of Al-12Si-1Cu-1Mg-1.5Ni-0.8Fe-0.5Mn/7 wt% graphite and Al-8Si-0.75Mg/Al_2O_3 particle composites showed considerable improvement in particle distribution, good particle-matrix bonding, and elimination of porosity, as confirmed by scanning electron microscopy and ultrasonic velocity measurements.

Squeeze casting or liquid forging of MMCs is a recent development that involves unidirectional pressure infiltration of fiber preforms or powder beds in order to produce void-free, near-net shape castings of composites. Aluminosilicate fiber reinforced aluminum alloy pistons made by a Japanese automaker have been in use in heavy diesel engines for some years, and a considerable amount of work has been done on the pressure casting of ceramic particle and fiber-reinforced MMCs for industrial applications (Ref 22-32). The fibers or particles exert a considerable influence on grain size, coarsening kinetics, and microsegregation in the matrix alloy. The fundamental phenomena involved are surface thermodynamics, surface chemistry, fluid dynamics, and heat and solute transport. The processing variables governing the evolution of microstructures in squeeze cast MMCs, in decreasing order of importance, are (Ref 31, 33):

- Fiber preheat temperature
- Interfiber spacing
- Infiltration pressure
- Infiltration speed
- Metal-superheat temperature

If the metal or fiber temperature is too low, poorly infiltrated or very porous castings will be produced. If the temperatures are too high, excessive fiber/metal reaction will degrade casting properties. Finally, if the plunger speed is too fast, the preform can be deformed on infiltration. A threshold pressure is required to initiate liquid-metal flow through a fibrous preform or powder bed. This pressure begins in the thermodynamic surface energy barrier because of nonwetting, as discussed earlier. Because wetting is a function of time, the threshold pressure increases with increasing infiltration rates (Ref 33). Table 3 lists some experimental results on the pressure infiltration of B_4C particles and SiC fibers by aluminum alloys.

The squeeze casting of composite melts into finished shapes promotes a fine, equiaxed grain structure because of large

Table 3 Wetting characteristics and threshold infiltration pressures for aluminum/SiC and aluminum/B_4C systems

Alloy	Temperature °C	Temperature °F	Surface energy γ, $J/m^2 \times 10^{-3}$	Al/SiC: Contact angle θ, °	Al/SiC: Threshold infiltration pressure P_{th}, kPa	Al/SiC: Threshold infiltration pressure P_{th}, psi	Al/B_4C: Contact angle θ, °	Al/B_4C: Threshold infiltration pressure P_{th}, kPa	Al/B_4C: Threshold infiltration pressure P_{th}, psi
Pure aluminum	700	1290	851	106.3	917	133	120.0	800	116
	800	1470	840	102.3	686.1	99.5	115.6	751.7	109
	900	1650	830	101.2	620.6	90	109.4	572.4	83
Al-2Cu	700	1290	843	106.7	934.3	135.5	116.7	786.2	114
	800	1470	832	103.7	758.5	110	114.3	710.3	103
	900	1650	822	100.2	558.5	81	110.1	586.2	85
Al-4.5Cu	800	1470	831	103.0	717.1	104	112.9	668.9	97
Al-2Mg	700	1290	767	104.6	744.7	108	115.1	675.9	98
	800	1470	757	101.2	565.4	82	98.9	241.4	35
	900	1650	747	99.3	462.0	67	95.1	137.9	20
Al-4.5Mg	800	1470	652	102.1	524.0	76	92.9	68.9	10
Al-2Si	700	1290	847	105.7	882.6	128	...	...	...
	800	1470	836	103.6	737.8	107	...	...	...
	900	1650	826	99.1	503.3	73	...	...	...
Al-4.5Si	800	1470	831	103.2	730.0	106	...	...	...

Source: Ref 34, 35

undercoolings and rapid heat extraction. Preheated particles or fiber preforms are inserted into a metal die and infiltrated with molten metal under high pressure (70 to 200 MPa, or 10 to 29 ksi), followed by solidification under pressure. Alternatively, whiskers or particles can be mixed with molten metal before squeeze casting. Aluminum alloy composites containing SiC and Al_2O_3 powders, aluminosilicate fibers, and silicon nitride whiskers have been fabricated by squeeze casting. These ceramic particles and fibers are poorly wetted by metallic alloys (contact angle > 90°), and their infiltration requires considerable hydrostatic pressure to overcome the capillarity pressure.

Further, frictional forces arising from the viscosity of the melt tend to oppose fluid flow through interfiber channels or interparticle corridors. Additional pressure is therefore required to overcome the viscous friction. Several theoretical analyses for modeling and analyzing the frictional forces have been proposed (Ref 24, 28, 31). These relate the infiltration velocity to applied pressure, capillarity, viscosity, and interfiber spacing as well as to fiber preform permeability and the change of permeability due to concurrent solidification in preforms preheated to below the melting point of the metal (Ref 33).

Freeze choking is an important aspect to be considered during the pressure infiltration of fiber preforms or powder beds. Freeze choking depends on the melt superheat, die temperature, and fiber temperature. Initially, a layer of metal solidifies on the prepacked fiber surface, which is of sufficient thickness to bring the fiber temperature to the melting point of the metal as a result of latent heat evolution. This results in an increase in the effective fiber diameter and volume fraction and a concomitant reduction in the permeability of the preform. The infiltration process ceases when the liquid front advances through the interfiber channels slowly enough to enable the fibers to extract latent heat with sufficient rapidity, leading to near-complete freezing of the front. This results in metal-starved regions in the composite and the formation of porosity.

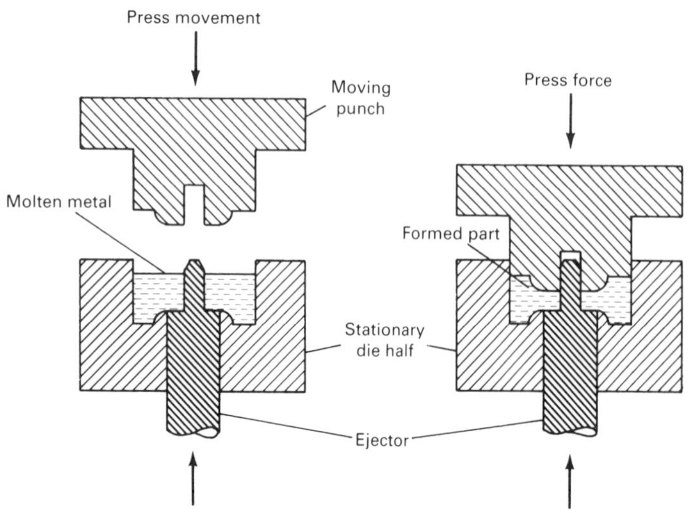

Fig. 9 Schematic of the squeeze casting process for synthesizing MMCs

In earlier studies on the squeeze casting of premixed ceramic-melt slurries, the wettability problem was overcome by codispersing SiC whiskers and aluminum alloy powder in an aqueous solution of isopropyl alcohol, followed by infiltration, compaction into small briquettes, and vacuum degassing. These briquettes were then disintegrated into a mechanically stirred base alloy melt, followed by squeeze casting under a pressure of 207 MPa (30 ksi). The resultant strengthening effects of composites are attributable to several factors, including fine grain size, elimination of bulk and interfacial porosities, increased solid solubility due to hydrostatic pressure, and the presence of high-strength SiC whiskers. Figure 9 shows a schematic illustrating the principle of squeeze casting for particle composites. The threshold pressures, P_{th} obtained experimentally for SiC fiber reinforced and B_4C particle filled aluminum alloys are given in Table 3. Combinations of compocasting and squeeze casting have been used to produce aluminum/SiC particle composites containing 20 wt% SiC (Ref 36) and aluminum/graphite particle composites containing 8 wt% particles (Ref 37).

Packed beds or aggregates of single or hybrid particles kept inside a mold can be pressure infiltrated to produce cast metal-hybrid particle composites. The particles must be preheated to a temperature equal to or greater than critical preheating temperature (cpt) in order to avoid freeze choking. For hybrid particle systems (for example, glass, carbon, aluminum, or copper particles in tin), cpt is given by (Ref 26):

$$\text{cpt} = T_E - \frac{(1 - V_v^{mix})\, D^M N^M [1 - (L_m)^{mix}]}{V_v^{mix} \Sigma X^i D^i C^i} \quad \text{(Eq 1)}$$

where T_E, D^M, and N^M are the melting point, density, and latent heat of the melt infiltrated; L_m is the critical liquid fraction; V_v^{mix} is the packing density of particles; and X^i, D^i, and C^i are the volume fraction, density, and specific heat of particle i in the hybrid particle assembly.

For single-particle systems, cpt is expressed as (Ref 24):

$$\text{cpt} = T_E - \left(\frac{0.233\, D^M N^M}{V_v^P D^P C^P}\right) \quad \text{(Eq 2)}$$

where V_v^P, D^P, and C^P are the packing density, the density, and the specific heat of the particles, respectively. For composite systems using particles smaller than a critical value, cpt does exist; however, for systems using particles larger than that size, the identification of cpt becomes difficult

Table 4 Typical casting conditions for SiC fiber-reinforced aluminum and aluminum alloy (Al-4.5Cu, Al-11.8Si, and Al-4.8Mg) composites

Type of composite	Metal mold temperature, K	Time between pouring and pressurizing, s	Applied pressure MPa	Applied pressure ksi	Pressure time, s	Casting temperature, K
Plate type	573	10	49	7.1	90	1073
Tubular type	573	13	40	5.8	90	1073

Source: Ref 25

Table 5 Elastic moduli of SiC fiber-reinforced plate-shaped MMCs produced by squeeze casting

Type of MMC	Fiber packing density V_f	Elastic modulus GPa	Elastic modulus psi × 10^3
Aluminum matrix	0	3.79	0.55
Al-4.5Cu/continuous SiC fiber	0.35	10.85	1.57
Al-11.8Si/continuous SiC fiber	0.35	10.50	1.52
Al-4.8Mg/continuous SiC fiber	0.35	10.25	1.49
Aluminum/discontinuous SiC fiber	0.44	11.6	1.68

Source: Ref 25

with increasing sizes of particles. Systems using angular particles have cpt lower than that of spherical particle systems because the packing density of particles in the former system is lower than that of the latter. The pressure applied to the melt, as well as the melt superheat, does not influence the cpt of the particles. Similar observations on the effect of metal superheat on the sound composite lengths in pressure-infiltrated fiber-reinforced metals are discussed in the section "Fluidity of Composites" in this article (Ref 33).

Plate and tubular composites of aluminum alloys containing continuous or discontinuous SiC fibers can be synthesized by squeeze casting (Ref 25). The SiC yarn, consisting of about 500 monofilaments (average diameter: 13 μm, or 510 μin.), is mechanically wound around a steel frame or aligned unidirectionally in an aluminum vessel. In the case of discontinuous SiC fibers, fiber can be chopped and packed in the vessel. The vessel with fiber is preheated in air for good penetration of the molten metal matrix into the interfiber space. The vessel is then placed into the mold, which is preheated to 500 to 700 K. The fiber volume fraction of composites is controlled by selecting the winding conditions (for continuous fiber) or packing conditions (for discontinuous fibers) before casting. Typical casting conditions for plate-type SiC fiber-reinforced aluminum alloy composites are given in Table 4.

With the above casting conditions and fiber-matrix combinations, high moduli and strengths (both room temperature and high temperature) of SiC fiber-reinforced MMCs can be obtained. Elastic moduli of plate-shaped MMCs produced in this manner are given in Table 5.

During the fabrication of fiber-reinforced MMCs, the fibers are exposed to several environments that may cause thermal, chemical, or mechanical degradation of the fibers. This could alter the tensile characteristics and fracture mode of MMCs. Isothermal heating at elevated temperatures (up to 700 °C, or 1290 °F, for 110 h), hydrostatic pressurizing (in oil at 49 or 196 MPa, or 7 or 28 ksi, for 10 min), and chemical vapor deposited aluminum coatings on fibers, followed by isothermal heating (700 °C, or 1290 °F, for 10 min), do not cause fiber degradation in the first two treatments. However, the fibers decrease in average strength and coefficient of variation with exposure time after exposure to liquid aluminum. Aluminum-treated fibers fracture because of the formation of reaction layers, leading to a decrease in tensile strength. As-received fibers fracture because of inner defects or surface flaws; fibers failing from inner defects have the highest strength, followed by those failing from surface defects and those failing from the formation of reaction layers, in that order.

Vacuum Infiltration. Several fiber-reinforced metals are prepared by the vacuum infiltration process. In the first step, the fiber yarn is made into an easy-to-handle tape with a fugitive binder in a manner similar to that used to produce a resin-matrix composite prepreg. Fiber tapes are then laid out in the desired orientation, fiber volume fraction, and shape, and they are inserted into a suitable casting mold. The fugitive organic binder is burned away, and the mold is infiltrated with molten matrix metal. The best castings in terms of soundness (voids, porosity, distribution of fibers) and mechanical properties are obtained if the mold assembly is heated and evacuated before introducing the liquid metal. In the case of α-Al_2O_3 reinforced aluminum-lithium alloys prepared using steel molds, it has been found that thin (0.13 mm, or 0.005 in.) rupturable steel membranes attached ahead of the mold gating system provide considerable flexibility for infiltration through direct immersion of the mold in the melt or through piping the melt to the mold assembly.

The liquid infiltration process used for making aluminum/graphite composites differs from the above process for preparing α-Al_2O_3/aluminum composites. Graphite fibers are first surface treated and then infiltrated with molten metal in the form of wires. These coated graphite wires are then diffusion bonded together to form larger sections.

In the case in which a wetting agent such as lithium or magnesium is used to promote wetting during liquid-metal infiltration, careful control of the concentration of additive/time/temperature and other process parameters is required to prevent excessive fiber reaction and the resulting strength degradation. The best mechanical performance of MMCs is achieved by optimizing the concentration of wetting agents to obtain fully wetted (low-porosity) composites while maintaining a minimum amount of reaction with the fibers. In the case of α-Al_2O_3/aluminum-lithium composites, lithium concentrations in excess of approximately 3.5 to 4.0 wt% significantly reduce tensile strength. Composites with 3.3 wt% Li, however, show significant retention of strength after 10 min of contact time with molten metal at an infiltration temperature of 700 °C (1290 °F).

The α-Al_2O_3 fibers extracted from vacuum-infiltrated aluminum-lithium composites are gray rather than the original white color, and the intensity of the color increases with increasing reaction time. The fiber-matrix reaction products contain aluminum as an oxygen-deficient suboxide (lithium can reduce Al_2O_3). The extracted gray fibers change back to their original color (white) when heated in air, which is consistent with the reoxidation of the reduced fiber surface to Al_2O_3. The reaction product in these MMCs is the compound $LiAlO_2$ (Ref 5).

The secondary ion mass spectroscopy of α-Al_2O_3 extracted from an Al-1.7Li matrix shows the presence of lithium to depths greater than 8 nm (80 Å) in the fiber. Electron diffraction analysis has shown the presence of $Li_2O \cdot 5Al_2O_3$ spinel as well as a second unidentified phase. Further, the thin (~5 nm, or 50 Å) silica coating applied to the fiber surface before fabrication is mostly removed during infiltration with aluminum-lithium alloy.

Investment casting of MMCs uses the filament-winding of prepreg-handling procedures developed for fiber-reinforced plastics to position and orient the proper volume fraction of continuous fibers within the casting. The layers of reinforcing fibers are glued together with an appropriate plastic adhesive (fugitive binder), which burns away without contaminating either the matrix or the fiber/matrix interface. These layers are stacked in the proper sequence and orientation, and the fiber preform thus produced is either infiltrated under pressure or by creating vacuum in the permeable preform.

The advantage of investment casting of composites is that a machine mold capable

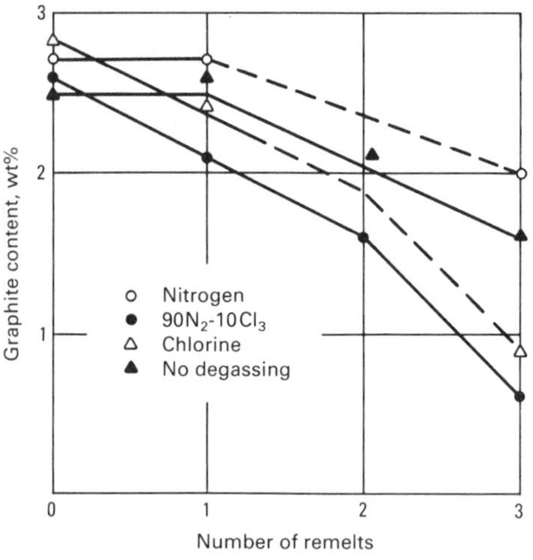

Fig. 10 Graphite loss in an Al-12Si-2.7Ni/graphite composite as a function of remelting and degassing

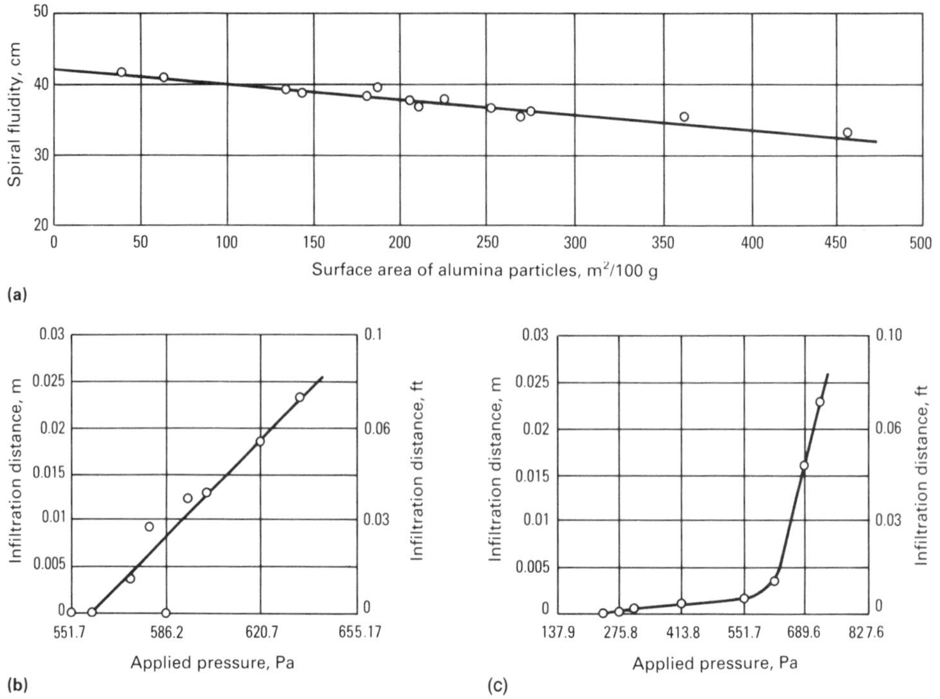

Fig. 11 Spiral fluidity (a) of aluminum-silicon alloy as a function of surface area per unit weight of Al_2O_3 particles. (b) Infiltration distance as a function of applied pressure for an Al-2Cu/SiC fiber composite. (c) Infiltration distance as a function of applied pressure for an Al-2Mg/boron carbide particle composite. Data in (b) and (c) supplied by J.A. Cornie, Massachusetts Institute of Technology

of withstanding high temperatures can be dispensed with because the fiber preform becomes the basic pattern for the component. Continuous graphite fiber-reinforced magnesium has been produced by this method (Ref 38-41).

Effects of Remelting and Degassing

Many of the studies on the remelting and degassing of MMCs have been reported for aluminum/graphite particle composites (Ref 42). No noticeable loss of graphite occurs as a result of remelting and degassing aluminum-graphite particle composites with nitrogen (Fig. 10), and graphite distribution remains reasonably uniform even after three remelts. This occurs for both uncoated and nickel- or copper-coated graphite particles in aluminum. However, degassing with chlorine or chlorine-bearing compounds such as hexachloroethane adversely affects the recovery and distribution of particles in the casting, because chlorine presumably impairs the wetting between graphite and aluminum.

The addition of modifiers such as sodium or sodium-bearing compounds to aluminum-silicon alloy melts containing graphite does not adversely affect the process of graphite dispersion in the melt when these modifiers are added before the addition of graphite. However, the addition of compound modifiers that release chlorine after graphite dispersion causes rejection of all graphite particles. The presence of graphite leads to partial refinement and modification of silicon in aluminum-silicon alloys, reducing the amount of sodium or phosphorus required for modification (Ref 43).

The hot tearing tendency of aluminum/graphite composite alloys is not increased by the presence of graphite particles (Ref 42). The Al-11.8Si alloy is inherently resis-

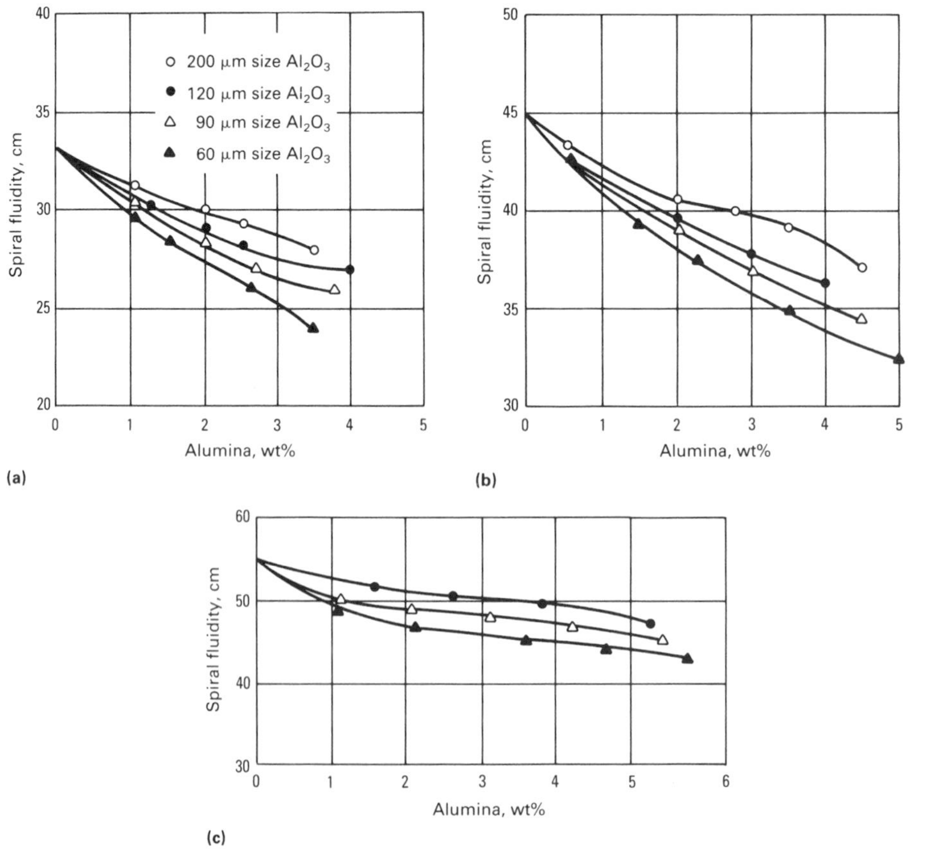

Fig. 12 Effect of particle size and pouring temperature on the spiral fluidity of Al-11.8Si/Al_2O_3 MMCs. (a) Pouring temperature: 680 °C (1255 °F). (b) Pouring temperature: 700 °C (1290 °F). (c) Pouring temperature: 740 °C (1365 °F)

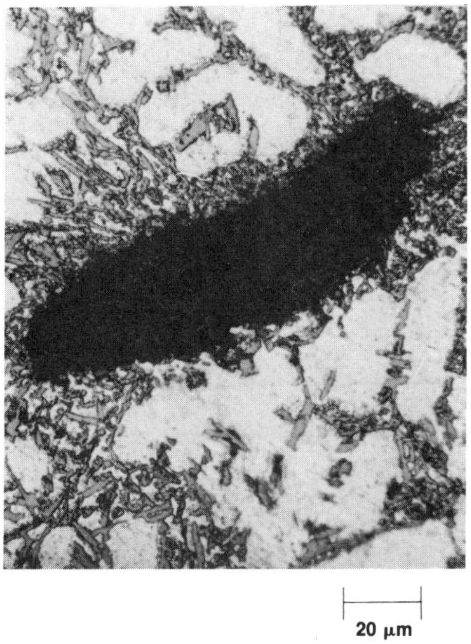

Fig. 13 Micrograph showing a graphite particle in the last freezing interdendritic liquid in an aluminum-silicon alloy

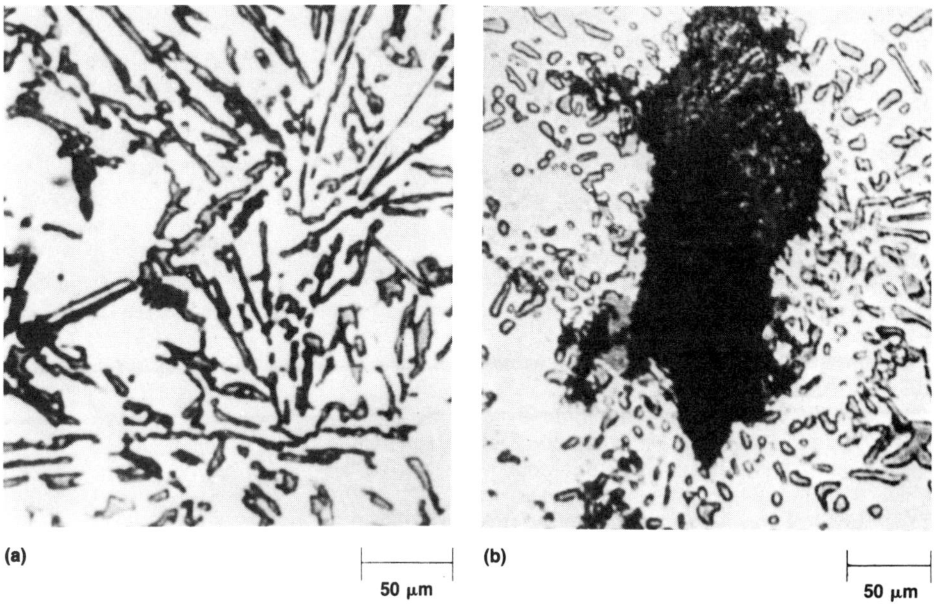

Fig. 14 Effect of graphite on the eutectic phase in aluminum-silicon alloys. (a) Needlelike eutectic silicon in as-cast alloy. (b) As-cast alloy with graphite particles; note shape modification of the eutectic

tant to hot tearing, and the addition of up to 4 wt% Cu coated graphite particles (equivalent alloy composition: Al-11.8Si-2.75Cu) does not increase susceptibility to hot tearing in testpieces of up to 200 mm (8 in.) in length.

Fluidity of Composites

Discontinuous Fibers and Particles. The dispersion of discontinuous fibers and particles into metallic melts imparts viscous, slurrylike characteristics to the latter and therefore affects their flowability or fluidity (Ref 1). Additions of discontinuous particles such as alumina, graphite, mica, and other ceramic particles to aluminum alloys cause a reduction in spiral fluidity, which is measured as the distance from the center of a standard spiral plate mold that a liquid front from a given quantity of melt flows before freezing. The spiral fluidity of these alloys decreases linearly with increasing particle surface area per unit weight, as shown in Fig. 11(a). This occurs presumably because of the progressive increase in effective viscosity of the suspension with increasing particle volume fraction.

The casting fluidity of aluminum alloys containing suspended particles decreases with a decrease in melt temperature in a manner similar to the fluidity of alloys without particles. The fluidity of composite slurries is extremely sensitive to the variations in volume fraction of dispersed particles, and it decreases sharply with progressive additions of particles. The fluidity of composite melts also depends on the shape, size, flocculation, and segregation of particles in a melt. For example, the spiral fluidity of an Al-11.8Si alloy containing dispersions of Al_2O_3 particles decreases with decreasing size of Al_2O_3 at constant weight percent of particles (Fig. 12). Figure 12 also shows that the spiral fluidity of composite slurry decreases with decreasing pouring temperature for constant particle size and weight percentage. Reciprocal relationships between fluidity and the apparent viscosity of melt-particle suspensions have been noted in aluminum-silicon alloys, iron-carbon-sulfur alloys, and Al-4.5Cu-1.5Mg alloy/2.5 wt% flaky mica composites. The fluidity values of particle-filled composite melts are adequate for making gravity cast composites at low volume fraction of particles.

Continuous Fibers. In MMCs reinforced with continuous fibers, infiltration length and preform permeability are indicators of fluidity. Experiments show a strong dependence of permeability and infiltration length on fiber volume fraction and fiber temperature, and a weaker dependence on metal temperature. Figure 11(b) and (c) show the influence of applied pressure on infiltration lengths in SiC fiber-reinforced Al-2Cu alloy and in B_4C particle filled Al-2Mg alloy, respectively; a threshold pressure (Table 3) is required to initiate infiltration in both cases. The infiltration distance increases with increasing applied pressure, with B_4C particle reinforced alloys showing a slow initial increase. This effect is apparently due to greater tortuosity and greater fluid drag in the packed bed of particles compared to the continuous-fiber preforms.

The relative magnitudes of fiber preheat temperature and metal preheat temperature can significantly influence the infiltration behavior of fiber preforms by a pure metal (Ref 33). In the presence of metal superheat, the solid metal sheath initially formed around cold fibers (fiber temperature below the melting point of the metal) remelts, and

Table 6 Selected potential applications of cast metal-matrix composites

Composite	Applications	Special features
Aluminum/graphite	Bearings	Cheaper, lighter, self-lubricating, conserve Cu, Pb, Sn, Zn, etc.
Aluminum/graphite, aluminum/α-Al_2O_3, aluminum/SiC-Al_2O_3	Automobile pistons, cylinder liners, piston rings, connecting rods	Reduced wear, antiseizing, cold start, lighter, conserves fuel, improved efficiency
Copper/graphite	Sliding electrical contacts	Excellent conductivity and antiseizing properties
Aluminum/SiC	Turbocharger impellers	High-temperature use
Aluminum/glass or carbon microballoons	...	Ultralight materials
Magnesium/carbon fiber	Tubular composites for space structures	Zero thermal expansion, High-temperature strength, good specific strength and specific stiffness
Aluminum/zircon, aluminum/SiC, aluminum/silica	Cutting tools, machine shrouds, impellers	Hard, abrasion-resistant materials
Aluminum/char, aluminum/clay	Low-cost, low-energy materials	...

the length of the remelted region always remains a fixed fraction of the total infiltration length for the case of constant applied pressure, no external heat extraction, and instantaneous heat transfer between metal and fiber. For the case of preheated fibers (fiber temperature greater than the melting point of the metal) under identical conditions, metal flow continues indefinitely unless external heat extraction causes cessation of flow to the point where the flow channel closes due to solidification from an external heat sink.

Impurity content in the metal significantly affects infiltration length. For example, a decrease in the purity of aluminum from 99.999% to 99.9% results in a decrease of 50% in infiltration length through aluminosilicate fibers. This is presumably due to a breakdown of the planar solid/liquid interfaces in the presence of impurities, which create greater drag on fluid flow.

Microstructures

Aluminum-silicon and aluminum-copper alloys have been primarily used as matrix materials in a wide variety of cast MMCs containing graphite and ceramic particles, carbon and glass microballoons, and discontinuous or continuous ceramic fibers (Ref 44). The microstructures of these MMCs show that primary aluminum in hypoeutectic alloys tends to avoid the discontinuous ceramic phase (SiC, alumina, graphite, mica, and so on) and nucleates in the interstices between particles or fibers unless special surface modification techniques are used to promote heterogeneous nucleation on the fiber surface. Primary aluminum crystallizing from the melt does not nucleate on the particle or fiber surface, because of surface thermodynamics and solute and thermal fields in the finite spaces between solidifying interfaces and the constraining second phase. The primary solid (α-aluminum) grows by rejecting solute in the melt, while the discontinuous ceramic phase tends to restrict diffusion and fluid flow; therefore, α-aluminum tends to avoid it, as shown in Fig. 13. Primary silicon and the eutectic in aluminum-silicon alloys tend to concentrate on the particle or fiber surface (Ref 1).

A closely related feature of these microstructures is the phenomenon of second-phase avoidance in MMCs, which is exemplified by unidirectional solidification experiments in which solidifying interfaces tend to push a freely suspended ceramic particle in a melt. These experiments show the existence of a minimum critical growth velocity of the front that must be exceeded to attain engulfment of the ceramic particle by the growth front.

The discontinuous ceramic phase also tends to modify or refine the structure; for example, eutectic silicon in aluminum-sili-

(a) (b) (c) (d)

Fig. 15 Automotive applications for cast aluminum/graphite MMCs. (a) Piston. (b) Bearing surface of cylinder after wear testing. (c) Cylinder block. (d) Journal bearings con alloys is modified, while primary silicon is refined when solidification occurs in the presence of a high volume fraction of ceramic phase. Figure 14 shows modification of eutectic silicon due to the presence of graphite particles in aluminum-silicon allcys (Ref 19, 42).

Microstructures in fiber-reinforced MMCs can be modulated in a predetermined manner by controlling interfiber spacing and cooling rate. If the cooling rate is sufficiently high or if fiber volume fraction is sufficiently low, the matrix alloy solidifies uninfluenced by the fibers. At sufficiently slow cooling rates, when the secondary dendrite arm spacing in the unreinforced alloy is comparable to interfiber spacing, the grain size becomes large compared to interfiber spacing. In this case, fibers do not enhance the nucleation of the solid phase. With further decrease in the cooling rates, the extent of microsegregation is reduced, and at sufficiently slow cooling rates, the matrix can be rendered free of microsegregation. The underlying mechanisms range from restricted diffusion in the solid state to dendrite coalescence during solidification (Ref 29-32).

Most of the above results are based on studies of two common alloy matrices: aluminum-copper and aluminum-silicon. The Al-4.5Cu alloy is well characterized and undergoes dendritic solidification in normal unreinforced castings, with coring and some second-phase $CuAl_2$ between dendrite arms. Aluminum-silicon alloys do not show microsegregation patterns in normal castings.

Properties and Applications

Modern fiber-reinforced or particle-filled MMCs produced by foundry techniques find a wide variety of applications (Table 6) because of the low cost of their fabrication and the specificity of achievable engineering properties (Ref 39-41, 45-57). Some of these properties are high longitudinal and transverse strengths at normal and elevated temperatures; near-zero coefficients of thermal expansion; good electrical and thermal conductivities; and excellent antifriction, antiabrasion, damping, and machinability properties.

Several generalized theoretical models predicting the properties of MMCs are available, although simple equations and rules of thumb often give excellent results (Ref 48, 49). The tensile strengths, elastic moduli, and densities of MMCs can generally be estimated from a rule of mixture. The rule of mixture for elastic modulus, for example, is a simple superposition of the effective contributions of the constituent phases to the overall property:

$$E_c = E_m \cdot V_m + E_f \cdot V_f \qquad \text{(Eq 3)}$$

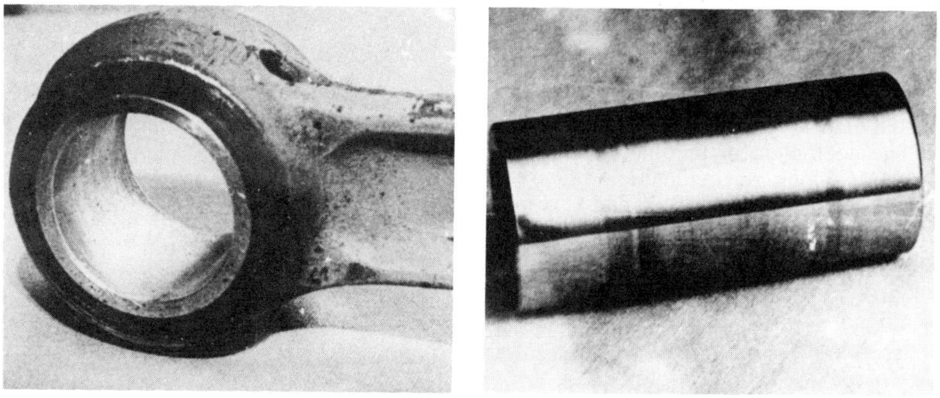

(a) (b)

Fig. 16 Al-4.5Cu/graphite bearing (a) in connecting rod and corresponding pin (b) showing surface scoring after 100 h of testing

where V_m and V_f are the volume fractions of the matrix and the fibers or particles, respectively, and E_c, E_m, and E_f represent the elastic moduli of the composite, matrix, and fiber, respectively. Similar relationships hold for the tensile strength of α-Al_2O_3 reinforced aluminum-lithium alloys, zircon particle dispersed aluminum-silicon alloys, and many other systems. Equation 3 holds along the longitudinal (fiber) direction. The composite modulus along the transverse direction in fiber composites follows (Ref 49):

$$\frac{1}{E_T} = \frac{V_f}{E_f} + \frac{(1 - V_f)}{E_m} \qquad \text{(Eq 4)}$$

The composite strength σ_L along the fiber direction can be calculated using Eq 5:

$$\sigma_L = \sigma_{fu} \left\{ \left[\frac{(K_0 - 1)}{K_0} - \frac{S_c}{2S} \right] V_f + \frac{1}{K_0} \right\} \qquad \text{(Eq 5)}$$

where σ_{fu} is the ultimate tensile strength of the fiber, S is half of the fiber length-to-diameter ratio $(L/2d)$, $S_c = \sigma_{fu}/2\ \tau_i$ (τ_i is interfacial shear stress between fiber and matrix), and $K_0 = E_f/E_m$.

Equation 5 is applicable to MMCs with relatively long fibers ($S \geq S_c$). For short fiber ($S < S_c$) reinforced MMCs, the yield strength σ_{Ly} of composite along longitudinal direction is given by:

$$\sigma_{Ly} = V_f\, S\, \tau_i + (1 - V_f)\, \sigma_{my} \qquad \text{(Eq 6)}$$

where σ_{my} is the yield strength of the matrix material. For imperfectly bonded MMCs, transverse yield strength σ_{Ty} is expressed as:

(a) (b)

Fig. 17 Comparison of wear of standard piston after a 30 h run (a) and aluminum alloy/graphite piston after a 60 h run (b)

$$\sigma_{Ty} = V_f \sigma_i + (1 - V_f)\sigma_m \quad \text{(Eq 7)}$$

where σ_i is the interfacial tensile strength of the composite and σ_m is the tensile strength of the matrix.

The coefficient of thermal expansion of an MMC along longitudinal and transverse directions are given by Eq 8 and 9, respectively (Ref 49):

$$\alpha_L = \frac{\alpha_f V_f K_0 + \alpha_m (1 - V_f)}{(K_0 - 1) V_f + 1} \quad \text{(Eq 8)}$$

$$\alpha_T = \frac{\alpha_{Tf} V_f + (1 - V_f)\alpha_{Tm} + V_f(1 - V_f)(\nu_f E_m - \nu_m E_f)}{V_f E_f + (1 - V_f) E_m (\alpha_{Lf} - \alpha_{Lm})} \quad \text{(Eq 9)}$$

where α_{Tf} and α_{Tm} are thermal expansion coefficients of fiber and matrix along transverse directions and ν_f and ν_m are Poissons ratios for fiber and matrix, respectively. The coefficient of thermal expansion of composites is an important property for various structural applications, such as aerospace, in which dimensional integrity over a wide temperature range is important. Modern fiber-reinforced metals such as magnesium/graphite can achieve zero thermal expansion coefficient until very high temperatures and are therefore ideally suited to various structural applications in space.

The mechanical strength of many MMCs reinforced with continuous fibers is anisotropic as well as extremely sensitive to changes in volume fraction of ceramic phase. The high-temperature strength of MMCs is enhanced by reinforcements, such as SiC whiskers or continuous Borsic (boron fibers coated with SiC) fibers. Aluminum/silicon carbide composites have excellent high-temperature strength up to 500 °C (930 °F); above this temperature, however, debonding and decohesion between fibers and matrix cause fiber pullout and failure of material. Aluminum/carbon MMCs combine very high stiffness with very low thermal expansion due to the almost zero expansion coefficient of carbon fibers in the longitudinal direction. Magnesium/carbon composites also have nearly zero expansion coefficients (Ref 39, 41).

In the case of particle-filled MMCs, the mechanical properties are not as significantly altered (as in continuous-fiber composites), but tribological properties (wear, friction, galling) show marked improvements. Soft, solid lubricant particles such as graphite and mica improve the antiseizing properties of aluminum alloys, while hard particles such as SiC, Al_2O_3, WC, TiC, zircon, silica, and boron carbide greatly improve the resistance to abrasion of aluminum alloys (Ref 51). Particle additions can also give rise to better damping and conductivity of the matrix alloy. For example, the damping capacity of aluminum and copper alloys is considerably enhanced when graphite powder is dispersed in them. A Japanese company has produced a high-damping MMC of aluminum or copper and 20% graphite. The damping capacity of the composite is considerably more stable at high temperature than conventional vibration-insulating alloys, including cast irons (Ref 58). Sliding electrical contacts made from the Cu/20 graphite alloy perform better than the sintered materials generally used, because the alloy combines excellent resistance to seizure with high electrical conductivity.

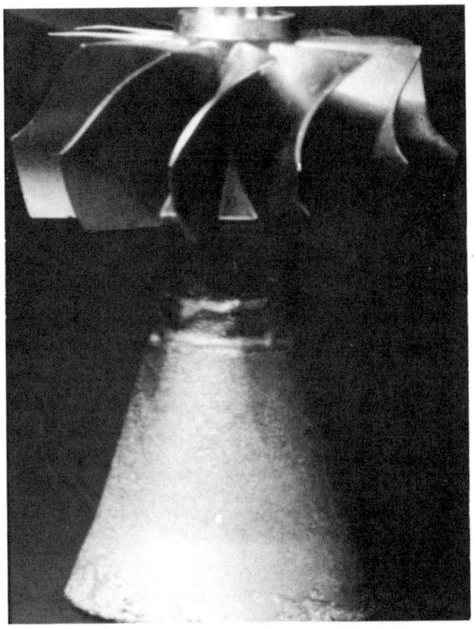

Fig. 18 SiC particle reinforced turbocharger impeller. Courtesy of D.M. Schuster, DACC

Fig. 19 Magnesium/graphite fiber tubes produced by filament winding and vacuum casting. Courtesy of Martin-Marietta Company

Selective reinforcement of metals with ceramic fibers (for example, SiC or Al_2O_3 in aluminum) is used in automotive parts such as diesel engines (Ref 53). Enhanced wear resistance and higher use temperature at equivalent cost have made such selectively reinforced MMCs a potentially very useful class of modern composite materials. Figure 15(a) shows an automotive piston selectively reinforced with Al_2O_3 fibers and produced by squeeze infiltrating a fiber assembly with molten aluminum alloy (Ref 46, 47). Other automotive applications are also shown in Fig. 15. Figures 16 and 17 show bearing-type and wear-resistant applications for graphite-containing composites. Connecting rods have been made from α-Al_2O_3/aluminum-silicon carbide composites by casting techniques (Ref 49). In all these cases, graphite particles were stirred in molten aluminum alloys, followed by the solidification of suspensions using conventional foundry techniques. The use of graphite in automobile engine parts considerably reduces the wear of the cylinder liner and improves fuel efficiency and engine horsepower at equivalent cost (Ref 45, 46, 50, 52, 53, 59).

The most promising application for cast aluminum/graphite particle composite alloys is for bearings, which would be less expensive, lighter, and self-lubricating, compared to the bearings currently being made out of copper-, lead-, tin-, and cadmium-containing alloys. Cast aluminum-graphite fan bushings experience considerably lower wear as well as reduced temperature rise during trials at 1400 rpm for 1500 h.

Cast aluminum-graphite alloy pistons used in single-cylinder diesel engines with a cast iron bore reduce fuel consumption and frictional horsepower losses. Because of their lower density, aluminum/graphite composites reduce the overall weight of internal combustion engines. Such an engine will not seize during cold start or failure of lubricant, because of the excellent antiseizing properties of aluminum/graphite MMCs (Ref 46-50). In the Brazilian Gran Prix 42-lap race in 1975, a racing car with a cast aluminum/graphite particle composite engine liner won the race even though the radiator broke and cooling fluid was completely lost after 27 laps (Ref 56).

Alloys with a dispersed ceramic phase are finding application in impellers and other tribological systems that operate at high temperatures where there is a possibility of liquid lubricant failure. A proprietary process has been developed for producing cast aluminum/SiC MMCs that shows acceptably uniform particle distribution, high strength, and high stiffness (Ref 60, 61). Cast aluminum alloys reinforced with ceramic phase are being proposed for use as turbocharger impellers, which run at high temperatures. Figure 18 shows an investment cast aluminum alloy A357/SiC composite cast to the net shape of a prototype turbocharger impeller (Ref 54). Figure 19 shows cast magnesium/graphite composite tubes produced for space structure applications.

ACKNOWLEDGMENTS

The author wishes to thank James Cornie, D.M. Schuster, K.K. Chawla, and R.

Asthana for their valuable assistance. The author would also like to thank several researchers at MIT, the University of Wisconsin—Milwaukee, the Regional Research Laboratories at Bhopal and Trivandrum, I.I.T.s at Kanpur and Delhi, the Indian Institute of Sciences at Bangalore, the University of Banaras (Jaipur and Roorkee), the National Metallurgical Laboratory at Jamshedpur, the Defence Metallurgical Research Laboratory at Hyderabad, India Pistons at Madras, Bharat Heavy Electricals Ltd., Escorts at Patiala, Hindustan Aluminum at Renukoot, and La Prenca at Bombay.

REFERENCES

1. P.K. Rohatgi, R. Asthana, and S. Das, Solidification, Structures and Properties of Cast Metal-Ceramic Particle Composites, *Int. Met. Rev.*, Vol 31, 1986, p 115
2. P.K. Rohatgi, Interfacial Phenomenon in Cast Metal-Ceramic Particle Composites, in *Interfaces in Metal-Matrix Composites*, A.K. Dhingra and S.G. Fishman, Ed., Conference Proceedings, The Metallurgical Society, 1986, p 186
3. F. Delannay, L. Frozen, and A. Deruyttere, The Wetting of Solids by Molten Metals and Its Relation to the Preparation of Metal-Matrix Composites, *J. Mater. Sci.*, Vol 22, 1987, p 1
4. H.A. Katzman, Carbon-Reinforced Metal-Matrix Composites, U.S. Patent 4,376,808, 1983
5. A.R. Champion, W.H. Krueger, H.S. Hartman, and A.K. Dhingra, "Fiber FP Reinforced Metal-Matrix Composites," Paper presented at the Second International Conference on Composite Materials, Toronto, Canada, April 1978
6. G.R. Cappleman, J.F. Watts, and T.W. Clyne, The Interface Region in Squeeze-Infiltrated Composites Containing δ-Alumina Fiber in an Aluminum Matrix, *J. Mater. Sci.*, Vol 20, 1985, p 2159
7. V. Laurent, D. Chatain, and N. Eustathopoulas, Wettability of SiC by Al and Al-Si Alloys, *J. Mater. Sci.*, Vol 22, 1987, p 244
8. J.A. Cornie, Y.M. Chiang, D.R. Uhlmann, A. Mortensen, and J.M. Collins, Processing of Metal and Ceramic Matrix Composites, *Ceram. Bull.*, Vol 65, 1986, p 293
9. A. Sato and R. Mehrabian, Aluminum Matrix Composites—Fabrication and Properties, *Metall. Trans. B*, Vol 7B, 1976, p 443
10. A. Banerjee, P.K. Rohatgi, and W. Reif, Role of Wettability in the Preparation of Metal-Matrix Composites, *Metallography*, Vol 38, 1984, p 356
11. M.K. Surappa and P.K. Rohatgi, Production of Aluminum-Graphite Particle Composites Using Copper Coated Graphite Particles, *Met. Technol.*, Vol 5, 1978, p 358
12. K.C. Russell, J.A. Cornie, and S.Y. Oh, Particulate Wetting and Particle-Solid Interface Phenomenon in Casting Metal-Matrix Composites, in *Interfaces in Metal-Matrix Composites*, A.K. Dhingra and S.G. Fishman, Ed., Conference Proceedings, The Metallurgical Society, 1986, p 61
13. A. Banerjee, M.K. Surappa, and P.K. Rohatgi, Preparation and Properties of Cast Aluminum Alloy-Zircon Particle Composites, *Metall. Trans. B*, Vol 14B, 1983, p 273
14. P.K. Rohatgi, B.C. Pai, and S.C. Panda, Preparation of Cast Aluminum-Silica Particulate Composites, *J. Mater. Sci.*, Vol 14, 1979, p 2277
15. M.K. Surappa and P.K. Rohatgi, Preparation and Properties of Cast Aluminum-Ceramic Particle Composites, *J. Mater. Sci.*, Vol 16, 1981, p 983
16. D. Nath, R.T. Bhat, and P.K. Rohatgi, Preparation of Cast Aluminum Alloy-Mica Particle Composites, *J. Mater. Sci.*, Vol 15, 1980, p 1241
17. F.A. Badia and P.K. Rohatgi, Dispersion of Graphite Particles in Aluminum Castings Through Injection of the Melt, *Trans. AFS*, Vol 77, 1969, p 402
18. B.C. Pai and P.K. Rohatgi, Production of Cast Aluminum-Graphite Particle Composites Using a Pellet Method, *J. Mater. Sci.*, Vol 13, 1978, p 329
19. P.K. Rohatgi, S. Das, and R. Asthana, "Synthesis, Structure and Properties of Cast-Metal Ceramic Particle Composites," Paper 8408-032, Metals/Materials Technology Series, American Society for Metals, 1985
20. R. Asthana, S. Das, T.K. Dan, and P.K. Rohatgi, Solidification of Al-Si Alloys in the Presence of Graphite Particles, *J. Mater. Sci. Lett.*, Vol 5, 1986, p 1083
21. T.K. Dan, S.C. Arya, B.K. Prasad, and P.K. Rohatgi, Study on the Suitability of LM 13-Graphite Composite as Pressure Die Cast Alloy, *Br. Foundryman*, 1987
22. T.W. Clyne and M.G. Bader, Analysis of a Squeeze Infiltration Process for Fabrication of Metal-Matrix Composites, *Bull. Jpn. Soc. Mech. Eng.*, p 755
23. M. Fukunaga and K. Goda, Composite Structure of Silicon Carbide Reinforced Metals by Squeeze Casting, *Bull. Jpn. Soc. Mech. Eng.*, Vol 28, 1985, p 1
24. S. Nagata and K. Matsuda, Effects of Some Factors on the Critical Preheating Temperature of Particles in Producing Metal-Particle Composites by Pressure Casting, *J. Jpn. Foundrymen's Soc.*, Vol 53, 1981, p 46
25. E. Nakata, Y. Kagawa, and M. Terao, "Fabrication Method of SiC Fiber Reinforced Aluminum and Aluminum Alloy Composites by Squeeze Casting Method," Report 34, Castings Research Laboratory, Waseda University, 1983, p 27
26. S. Nagata and K. Matsuda, Pressure Casting Conditions in Producing Metal-Hybrid Particle Composites, *J. Jpn. Foundrymen's Soc.*, Vol 54 (No. 10), 1982
27. H. Fukunaga, K. Goda, and N. Tabata, Tensile Characteristics of Silicon Carbide Fiber in Fabrication Process of FRM by Liquid Infiltration Method, *Bull. Jpn. Soc. Mech. Eng.*, Vol 28, 1985, p 2224
28. H. Fukunaga and K. Goda, Fabrication of Fiber Reinforced Metal by Squeeze Casting, *Bull. Jpn. Soc. Mech. Eng.*, Vol 27, 1984, p 1245
29. A. Mortensen, M.N. Gungor, J.A. Cornie, and M.C. Flemings, Alloy Microstructure in Cast Metal-Matrix Composites, *J. Met.*, March 1986, p 30
30. J.A. Cornie, A. Mortensen, M. Gungor, and M.C. Flemings, The Solidification Process During Pressure Casting SiC and Al_2O_3 Reinforced Al-4.5% Cu Metal-Matrix Composites, in *Proceedings of the Fifth International Conference on Composite Materials (ICCM-V)*, W.C. Harrigan, J. Strife, and A.K. Dhingra, Ed., American Institute of Mining, Metallurgical, and Petroleum Engineers, 1985, p 809
31. L.J. Masur, A. Mortensen, J.A. Cornie, and M.C. Flemings, Pressure Casting of Fiber Reinforced Metals, in *Proceedings of the Sixth International Conference on Composite Materials (ICCM-VI)*, American Institute of Mining, Metallurgical, and Petroleum Engineers, 1987
32. A. Mortensen, M.C. Flemings, and J.A. Cornie, "Columnar Dendritic Solidification in a Metal-Matrix Composite," Annual Report of the Centre for the Processing and Evaluation of Metal and Ceramic Matrix Composites, Massachusetts Institute of Technology, Feb 1987
33. L.J. Masur, "Infiltration of Fibrous Preforms by a Pure Metal," Ph.D. thesis, Massachusetts Institute of Technology, 1988
34. S.Y. Oh, J.A. Cornie, and K.C. Russell, Particulate Wetting and Metal-Ceramic Interface Phenomena, in *Proceedings of the 11th Annual Conference on Composites and Advanced Ceramic Materials*, Massachusetts Institute of Technology, Jan 1987
35. S.Y. Oh, "Wetting of Ceramic Particles With Liquid Al Alloys," Ph.D. thesis, Massachusetts Institute of Technology, 1987
36. M.A. Bayoumi and M. Surey, Structure

and Mechanical Properties of SiC Particle Reinforced Al Alloy Composites, in *Proceedings of the Sixth International Conference on Composite Materials (ICCM-VI)*, American Institute of Mining, Metallurgical, and Petroleum Engineers, 1987
37. P.R. Gibson, A.J. Clegg, and A.A. Das, Production and Evaluation of Squeeze Cast Graphite Al-Si Alloys, *Mater. Sci. Technol.*, Vol 1, July 1985, p 559
38. K.M. Prewo, Report R77-912245-3, United Technologies Research Center, May 1977
39. D.M. Goddard, Report on Graphite/Magnesium Castings, *Met. Prog.*, April 1984, p 49
40. M. Misra, S.P. Rawal, D.M. Goddard, and J. Jackson, "Novel Processing Techniques in Fabricating Graphite-Magnesium Composites for Space Applications," Phase I Technical Report MCR-85-711, for Naval Sea Systems Command Contract No. 24-84-C-5306, Nov 1985
41. D.M. Goddard, W.R. Whitman, and R.L. Humphrey, "Graphite/Magnesium Composites for Advanced Lightweight Rotary Engines," Technical Paper 860564, Society of Automotive Engineers, Feb 1986
42. M.K. Surappa and P.K. Rohatgi, Melting, Degassing and Casting Characteristics of Al-11.8Si Alloys Containing Dispersions of Copper Coated Graphite Particles, *Met. Technol.*, Vol 7, 1980, p 378
43. B.P. Krishnan and P.K. Rohatgi, Modification of Al-Si Alloy Melts Containing Graphite Particle Dispersions, *Met. Technol.*, Vol 11, 1984, p 41
44. C. Vaidyanathan and P.K. Rohatgi, "Ultralight Metal-Matrix Composites Using Ceramic Microballoons," Report, University of Wisconsin—Milwaukee, 1988
45. L. Ackermann, J. Charbonuier, G. Desplancher, and H. Koslowski, Properties of Reinforced Aluminum Foundry Alloys, in *Proceedings of the Fifth International Conference on Composite Materials (ICCM-V)*, W.C. Harrigan, Jr., J. Strife, and A.K. Dhingra, Ed., American Institute of Mining, Metallurgical, and Petroleum Engineers, 1985, p 687
46. J.W. Holt, Silgraf—A New Silicon/Graphite/Aluminum Alloy for Cylinder Liners, *Copia Ricevuta*, 16.05 G3, Sept 1987, p 20.7
47. T. Donomoto, K. Funatani, N. Miura, and N. Miyake, "Ceramic Fiber-Reinforced Piston for High Performance Diesel Engines," SAE Technical Paper 830252, presented at the International Congress and Exposition, Detroit, MI, Feb-March 1983
48. T.W. Chou, A. Kelly, and A. Okura, Fiber-Reinforced Metal-Matrix Composites, *Composites*, Vol 16, 1985, p 187
49. K.K. Chawla, *Composite Materials—Science and Engineering*, Springer-Verlag, 1987
50. P.K. Rohatgi *et al.*, Performance of an Al-Si-Graphite Particle Composite Piston in a Diesel Engine, *Wear*, Vol 60, 1980, p 205
51. S.V. Prasad and P.K. Rohatgi, Tribological Properties of Aluminum Alloy Particulate Composites, *J. Met.*, Nov 1987, p 22
52. L. Bruni and P. Iguera, *Automot. Eng.*, Vol 3, 1978, p 29
53. M.W. Toaz and M.D. Smale, *Diesel Prog.*, June 1985
54. D.M. Schuster, *Low Cost, High Performance Silicon Carbide Reinforced Aluminum Castings, Forgings, Extrusions and Rolled Sheet*, Dural Aluminum Composite Corporation, 1987
55. W.G. Spengler, *Material Properties of Metal-Matrix Composites*, Engine Products Division, DANA Corporation, 1987
56. R. Casellato, A.E. Borgo (Italy), private communication, 1988
57. "Magnesium Matrix Composites at DOW," Paper presented at the TMS/AIME Conference, Phoenix, AZ, The Metallurgical Society, Jan 1988
58. *Hitachi Graphite Dispersed Cast Alloy-Gradia*, Hitachi Chemical Company, 1981
59. L. Bruni and P. Iguera, *L' Impiego Dell' Alluminio Nelle Canne Cilindro Tre Modj Di Applicazione*, Associated Engineering, S.P.A. (Alpgrano), 1979
60. "DURAL MMC—Tomorrow's Metal Matrix Composite Available and Affordable Today," Publication 5/87, Dural Aluminum Composites Corporation, 1987
61. D.M. Schuster, M. Skibo, and F. Yep, SiC Particle Reinforced Aluminum by Casting, *J. Met.*, Nov 1987, p 60

Computer Applications in Metal Casting

Section Chairmen:
John T. Berry, The University of Alabama
Robert D. Pehlke, The University of Michigan

Introduction

John T. Berry, University of Alabama

THE ARTICLES in this Section describe some of the developments taking place in the field of computer development in casting technology. Computers are being increasingly applied for design of patterns and castings and for other tasks in the casting industry, but these applications are discussed in another part of this Volume (see the Sections "Patterns" and "Design Considerations"). The focus of this Section will be the computer modeling of phenomena associated with the solidification of molten metals.

The first article in this Section, "Modeling of Solidification Heat Transfer," describes the increase in computer applications to what is now termed the macroscopic scale of modeling casting solidification, that is, the movement of freezing fronts. The last article, "Modeling of Microstructural Evolution," looks at a different, more recently developed aspect—macro-microscopic modeling—from which the evolution of microstructure can be deduced.

As the art and science of solidification modeling has progressed, however, it has become clear that neither macroscopic nor microscopic aspects of solidification can be studied in all potential applications without consideration of the free and forced convection phenomena in zones of liquid or partially liquid metal. For this reason, investigators have been looking more carefully into the question of modeling the flow of molten metal in the casting/rigging system and into the heat exchange taking place during both forced and free convection. Thus, the second article in this Section, "Modeling of Fluid Flow," describes how the computer can be applied to study the flow of metal through the gating system into the casting mold cavity.

The third article, "Modeling of Combined Fluid Flow and Heat/Mass Transfer," explains how the combined action of the fluid flow and heat/mass transfer that occurs during the pouring and filling of the castings can affect the initial temperature distribution of the molten metal.

The methods currently used in all of these types of analyses generally require considerable time to set up and run, but this is largely a function of the computing power of currently available machines. Because this field is one that continues to develop, the authors have not been unduly harnessed in their thinking with regard to what is currently economically feasible. Rather, they have felt encouraged to look toward a near-term future of increasingly efficient methods and means of computation and information processing. Each author has emphasized the basic principles and results of the technique in question rather than presenting a rigorous treatment of the details of the analysis and associated theory; the latter can be found in the references provided with each article.

Advantages of Computer Modeling. Computer-aided design (CAD), computer-aided manufacturing (CAM), and computer-aided engineering (CAE) offer a number of advantages for castings. These include:

- Increased casting yield per pound of metal poured
- Improved casting quality (absence of unsoundness)
- Enhanced productivity of casting system
- Geometric models provide casting volume, weight, and surface area data, allowing rapid cost-estimating and permitting efficient rigging design
- Automated enmeshment for general purpose heat transfer simulators permits shorter design time
- Automated machining of patterns, which in turn reduces costs
- Fewer prototypes to be experimentally evaluated; shorter lead times from design concept to product
- Easier implementation and evaluation of engineering changes
- Enhanced ability to deal with batch production of castings of different design

Issues still to be resolved involve:

- Combination of fluid flow/solidification modeling in three dimensions for limited computer capacities
- Enhancement of user friendliness of modeling and simulating systems
- Closure of feedback loop in control of the casting process

Modeling of Solidification Heat Transfer

John T. Berry, University of Alabama
Robert D. Pehlke, University of Michigan

THE DIGITAL COMPUTER has had a singularly important impact on the engineering design of complex, multicomponent, high value-added products (for example, aerospace structures). Since the early 1960s, applications of this device to stress analysis and heat transfer problems have spread to many other industries, ranging from those involved in naval architecture, offshore structures, and pressure vessel fabrication through to metalworking and machine tools and on to automobile component design and performance.

The application of this device to process engineering, in particular to casting production, has not been as pervasive, being confined to a limited number of large metal casting organizations around the world. In many ways, this is surprising because the potential power of computing machinery was much in evidence as early as the 1940s, when under the sponsorship of the American Foundrymen's Society the classic papers of Paschkis describing studies employing the analog computer to predict freezing patterns in castings appeared (Ref 1). It is difficult to determine why this work was not held in higher esteem by practicing foundrymen, although it clearly would have been quite impractical to simulate the freezing of castings of any complexity with the large networks of resistors and capacitors involved in the analog technique.

The first published reference (1962) to the use of the digital computer in a foundry-related application is undoubtedly that of Fursund (Ref 2) in Denmark, who studied the diffusion of heat into foundry sands as it affected surface finish problems in steel castings. Application to the prediction of freezing patterns appeared slightly later (1965). Using the Transient Heat Transfer program developed previously for the U.S. space program, Henzel and Keverian (Ref 3, 4) discussed the successful application of a finite-difference method (FDM) to heavy steel casting production.

The Heat Transfer Committee of the American Foundrymen's Society sought to encourage further research in the area of computer applications during the late 1960s. Sponsorship of research at the University of Michigan resulted in several important publications (Ref 5). Castings simulated by Pehlke, Wilkes *et al.* were poured in a variety of materials (carbon steel, aluminum, and silicon and leaded brasses). All the configurations were sand castings. An important conclusion drawn at the time was that the predictions of the FDM program concerned were strictly limited by the thermophysical properties of the molding and casting media.

Contemporary with this research, the Norwegian group of Davies *et al.* extended their 1973 FDM work on aluminum sand cast bars (Ref 6) to cover permanent mold (gravity die) and low-pressure die casting (Ref 7). A comprehensive review of the developments during this period until 1983 is provided in Ref 8, which, together with Ref 9 and 10, describes most of the important work in the area of solidification modeling using the computer.

Since the early 1980s, there has been a significant growth in the number of publications in this area. Many of these publications are produced by research groups formed specifically for the purpose of studying the techniques of computer simulation and implementing its use in metal casting production around the world. Notable among the sustained research efforts have been several research collaborations, such as those of Sahm, Hansen, and coworkers in Europe; Ohnaka, Aizawa, and colleagues in Japan; and Desai and the present authors in the United States. The proceedings of a recent international conference include brief reports of progress from most of the groups currently active in solidification modeling (Ref 11). More extensive accounts of progress in this area and in the parallel field of weldment simulation can be found in the published proceedings of successive conferences on the modeling of castings and weldments held under the sponsorship of the Engineering Foundation (Ref 12-15).

Geometric Description and Discretization

The form of the differential equations describing the diffusion of heat into the mold, together with the progress of freezing within the casting, is given in the article "Modeling of Microstructural Evolution" in this Volume. To predict freezing history in complex industrial castings, the three-dimensional form must be provided and the correct boundary conditions specified. Even prior to this, there are several valid reasons for defining accurately and unambiguously the geometry and topography of the casting and its rigging:

- Performing the analysis to generate preliminary rigging design
- Estimating the overall costs for casting production for the purpose of quotation
- Defining the melting capacity and molding equipment needs within the casting plant
- Generating the tapes or other information storage forms required to drive machine tools needed for the fabrication of patterns and core boxes and the sinking of wax pattern, permanent mold, and die casting dies

The need to provide this geometric representation and to link the chosen form of representation with the process of discretization, that is, the breaking up of the whole into discrete elements, has been recognized as being of prime importance by only a limited number of research groups (Ref 16, 17). Turning first to the problem of overall geometry, much can be gained by examining the capabilities of the many geometric modelers now available. The three most established types of geometric representation are (Fig. 1):

- Constructive solid geometry (csg)
- Boundary re presentation (b-rep)
- Wireframe (wf)

The wireframe is not a true solid modeler, but because it normally forms an important part of many preprocessing packages for commercial finite-element method (FEM) computational codes, it is sometimes listed as a form of geometric modeling. The csg and b-rep approaches are both often incorporated into commercially developed computer-aided design (CAD) packages. Figure

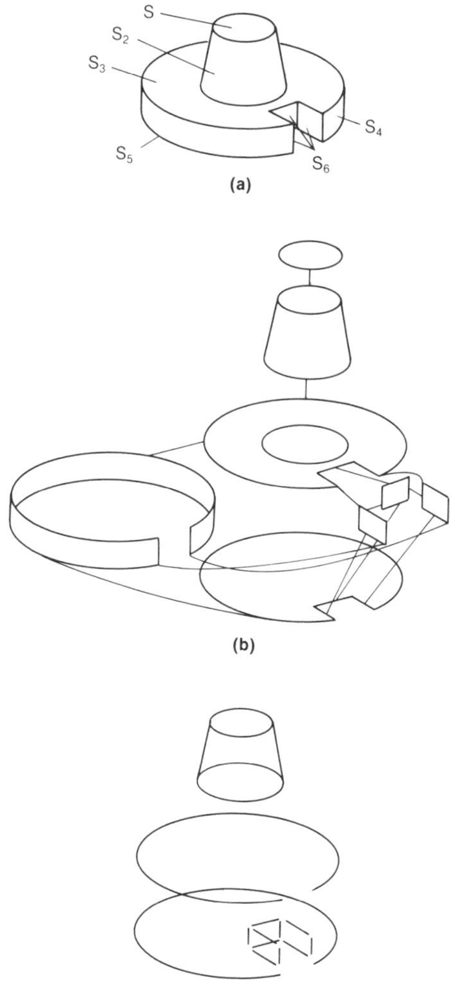

Fig. 1 Three types of three-dimensional geometric representation of a simple casting. (a) Constructive solid geometry. (b) Boundary representation. (c) Wireframe. Source: Ref 18

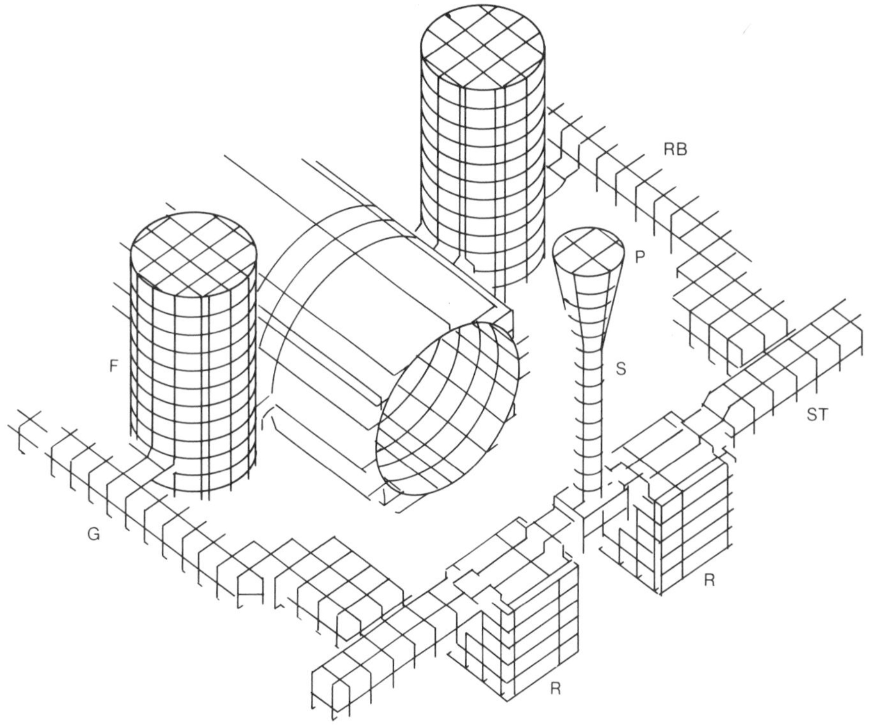

Fig. 2 Portion of a geometric model of a casting with rigging, constructed using a csg geometric modeler. F, feeder or riser; G, ingate location; P, pouring basin; R, reaction chamber; RB, runner bar; S, sprue; ST, slag trap. Source: Ref 19

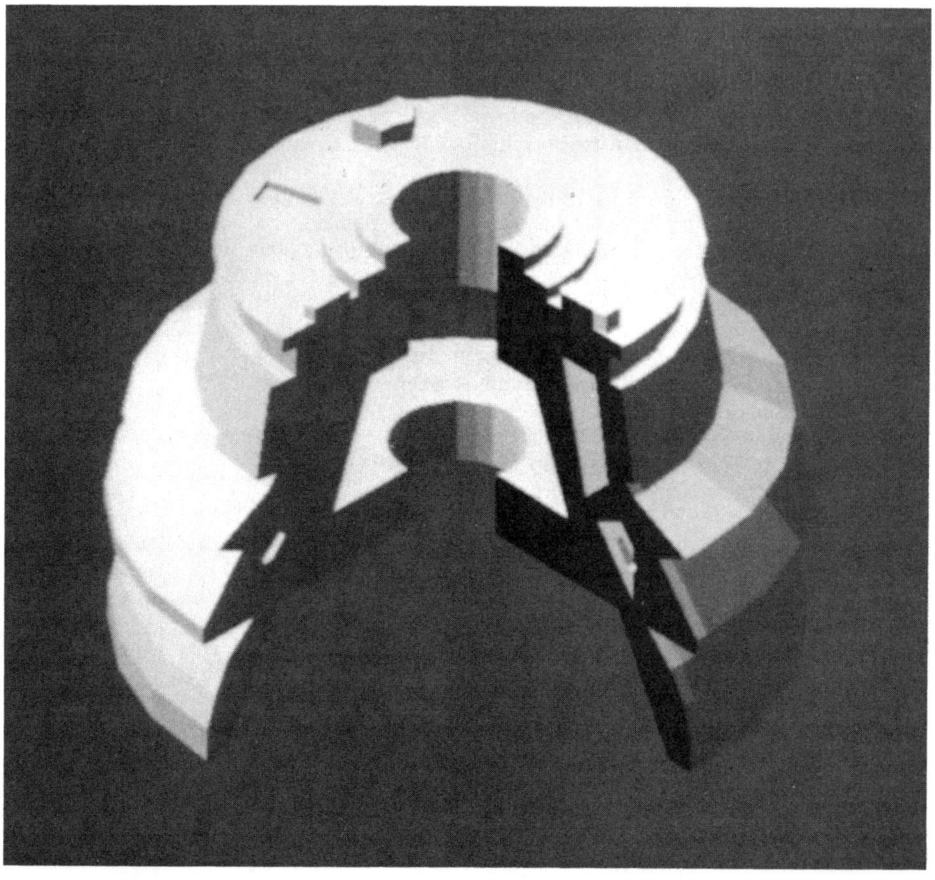

Fig. 3 A partially sectioned three-dimensional geometric model of a small aerospace casting. Courtesy of J.L. Hill and J.T. Berry, University of Alabama

2 shows a csg-based model for a partial section of a casting in the form of a thick-wall cylinder together with its rigging. Many contemporary modeling routines permit not only hidden-line removal but also the use of color and shading. Figure 3 shows such a representation of a small aerospace casting. Perhaps the true test of the utility of a geometric modeling package lies in its ability to represent the full range of complex shapes involved in commercial casting. A particular feature, which is likely to appear in most common casting configurations, is that of fillets, which are a necessary patternmaking expedient and something seen when one solid intersects another (Fig. 4).

Although the surface-modeling technique does not provide a complete geometric description of an object, it has been the basis of one of the most successful attempts thus far to link together computer-aided design and manufacture in the metal casting industry. The DUCT program, which originated at Cambridge University (Ref 20) is able to produce a surface model from a given spine

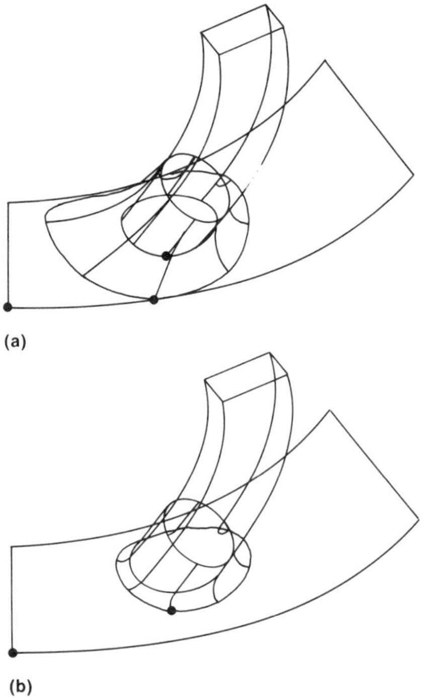

Fig. 4 Blending together of surfaces using fillets in preparation of models for patternmaking or for simulation studies. (a) Normal blend with edge limits. (b) Constant-radius blend. Source: Ref 20

and an orthogonal section set (Fig. 5). Such information can then be conveyed to patternmaking and core box fabricating machinery on a routine basis.

Before a model of any description can be built and a freezing simulation entered into, the foundry methods engineer must have a reasonably accurate picture of the location and dimensioning of the various components of the rigging system (the gating and risering subsystems) that are to be employed during casting. Two approaches are being followed as alternatives to the age-old art of cut and try. These techniques are both computer based, and one in particular uses (most effectively) the readily available personal computer rather than the mainframe machine demanded by both geometric representational (csg, b-rep, etc.) and numerical computational (FDM, FEM, etc.) programs. The techniques can be referred to as:

- Special-purpose foundry rigging engineering programs
- Knowledge-based expert systems programs

Special-Purpose Programs. Many programs fall under the category of special-purpose programs (Ref 21-23). One of the most versatile, the Novacast program (Ref 23), determines riser sizes by using the well-known Chvorinov rule (Ref 24) and locates them by using various other empirically derived feeding range rules. The program will also approximate the geometric features of the gating system using the Bernoulli approximation, together with other empirical axioms. This type of program has met with relatively wide acceptance in foundries around the world. In the United States, the AFSoftware programs and other routines prepared at the University of Wisconsin are in widespread use (Ref 21).

Knowledge-Based Expert Systems. Microcomputer programs such as those discussed above are often sufficient for designing the rigging of castings of a less critical nature. However, for those cast components that form parts of aerospace systems or for the safety features of pressure vessels, and so on, further engineering measures are often necessary.

In such cases, the need arises for a knowledge-based expert system and its associated data base. In any particular organization, a vast amount of the expertise involved in the art of foundry practice often disappears with the retirement of certain employees. Although there is no unique solution to the problem of rigging a particular casting, there are many valuable, although sometimes conflicting, opinions available for consideration. Discussions are beginning to appear in the literature on this approach to preparing a rigging system design that will later be tested by simulation (Ref 25). It is expected that this technique will grow extensively over the next 10 years.

FEM Software. Turning to the question of discretization of space within and around the casting (the mold cavity and the mold itself), a number of commercial FEM packages are associated with preprocessing routines that perform the subdivision involved, together with the parallel preparation of files describing that discretization (numbering of elements and nodes, and so on). The principles involved in this important phase of modeling are discussed in Ref 26. More sophisticated enmeshment schemes, now commercially available, will allow the application of higher mesh density to regions where the temperature gradients are expected to be steep. A simple example of the enmeshment of both mold cavity and the mold itself is shown in Fig. 6. No special attempt was made to vary mesh density in particular locations. It should also be mentioned that the configuration shown in Fig. 6

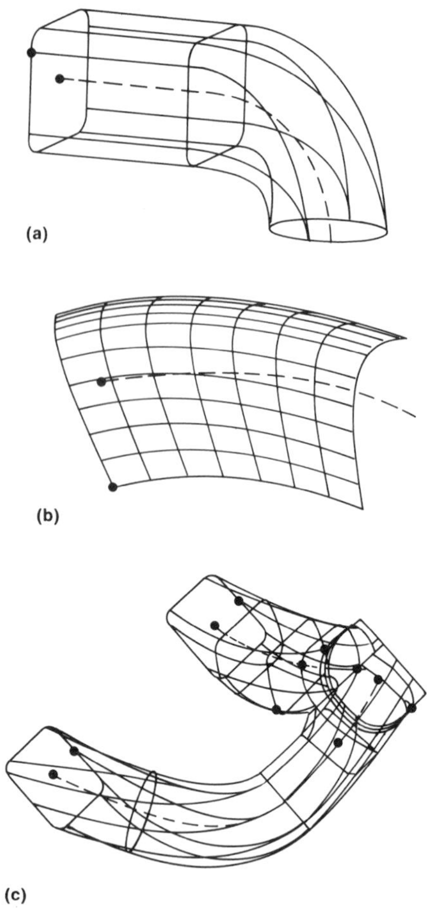

Fig. 5 Method by which a surface-modeling program (DUCT, Ref 20) is used to build a model from a spine (dotted line) and a series of orthogonal sections. The sections are joined by Bezier curves. This technique has been used in the production of automobile manifolds. (a) Single closed DUCT. (b) Single open DUCT. (c) Combination of several closed DUCTs

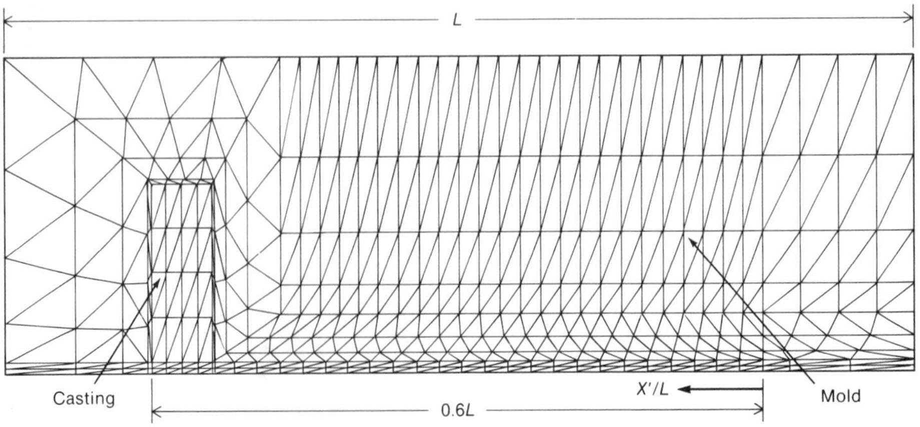

Fig. 6 Discretization (enmeshment) of a plate casting and riser, together with the mold into which they are to be cast

Table 1 Comparison of ANSYS, MARC, and MITAS-II with the ideal goal for metalcasting simulation software

Feature	Ideal	ANSYS	MARC	MITAS-II
Ease of setup and use	Easy	Difficult	Moderately difficult	Moderately difficult
Running cost	Low	High	High	Moderate
Ability to account directly for latent heat	Yes	No	Yes	Yes
Dedicated heat transfer code	Yes	No	No	Yes
Accuracy	Good	Very good	Very good	Very good
Pre- and postprocessing capabilities	Good	Good	Good	Poor

Source: Ref 27

represents a two-dimensional object. Nonetheless, Fig. 6 serves to indicate how rapidly the file-keeping capacity of the computer system will be occupied as the complexity of component design increases.

Figure 6 also shows that a preponderance of elements (or nodes) exists within the mold, per force of its enveloping nature with respect to the casting. This will be discussed in the section "The Data Base."

The Computational System

Many successful casting simulations have employed large-scale commercially available, general-purpose programs. The pitfalls of this approach are discussed in Ref 27. Table 1 lists the attributes of certain large-scale programs in the context of their use in solving heat transfer problems. In view of these limitations, some research groups have chosen to build their own simulation programs or solvers. The development of such a program, a dedicated special-purpose FEM code, is discussed in Ref 28. This particular system—The Michigan Solidification Simulator (MSS)—was written exclusively for modeling the general problems of steady-state or transient energy transfer with nonlinear material properties, phase change, and imposed bulk-flow velocity fields. Two-dimensional Cartesian and cylindrical, three-dimensional Cartesian element types (including interfacial contact resistance elements), totaling three coordinate systems and ten element types, can be supported. In solving the basic governing equations, the simulator permits the use of the following boundary conditions:

1. Temperature specified
2. Heat flux specified
3. Radiative heat transfer specified
4. Convection heat transfer specified

Because heat transfer expressions for the third and fourth conditions have the same mathematical form, only three expressions are required to represent all four boundary conditions. Currently, a uniform initial temperature field must also be specified. The computations involve the use of the Galerkin method of weighted residuals. The discretization of the geometry involved in the simulation uses triangles, distorted bricks, and distorted rectangles according to whether two- or three-dimensional castings are to be modeled. At present, the division of a particular domain is undertaken manually. The validation of the program has been performed, first, using cases where analytical solutions exist, principally for one-, two-, and three-dimensional transient heat transfer with various boundary conditions but without solidification; second, for a case involving a latent heat source term but zero superheat, for which an analytical solution exists; and finally, for a variety of axisymmetric and three-dimensional examples involving the actual solidification of metals, where validation was undertaken experimentally from thermocouple measurements or sectioning (Ref 28).

The actual castings simulated were:

- A cylindrical gray iron casting poured in dry sand
- A cylindrical Al-13Si casting poured in dry sand
- A carbon steel railwheel casting
- An investment cast Alloy IN-100 plate

In the first three cases, very good agreement was obtained between simulation and experiment.

The railwheel casting was chosen as an example of an industrial casting (Ref 29). There are no experimental thermocouple data available, but radiographs and casting sections allow inference of cooling patterns and analysis of defects.

Seven different casting configurations were investigated (Ref 29). For the present article, two (casting shapes C and F; see Fig. 7) were selected that are the best documented. The railwheel casting is shown in Fig. 7. Casting F has a web portion padded with 13 mm (½ in.) of extra metal on both sides.

The finite-element results for casting C are plotted as isotherms in Fig. 8. The isotherms are the liquidus (1515 °C, or 2760 °F) and the solidus (1485 °C, or 2075 °F). The flange region has a pool of liquid metal that is cut off from the riser by the solidified web. One would expect to see defects in the sectioned casting near this point, and in fact there are defects in this region, as predicted by the numerical simulation (Ref 28, 29).

The same simulation was run for casting geometry F, which has 13 mm (½ in.) of additional metal padding on each side of the web. Isotherms from that simulation are shown in Fig. 9 and indicate that the casting should be sound due to the directional solidification. Actual casting sections of cast-

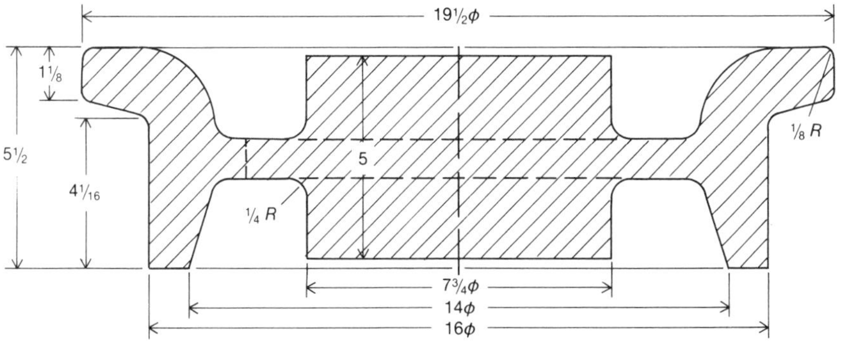

Details of test wheel castings

Casting	Riser diameter mm	Riser diameter in.	Riser height mm	Riser height in.	Pouring temperature °C	Pouring temperature °F	Special features
A	190	7.5	165	6.5	1675	3050	...
B	215	8.5	180	7.0	1675	3050	...
C	190	7.5	230	9.0	1610	2930	...
D	150	6.0	230	9.0	1610	2930	25 mm (1 in.) thick riser sleeve
E	190	7.5	230	9.0	1610	2930	75 mm (3 in.) thick chromite sand facing around rim and flange
F	190	7.5	230	9.0	1605	2920	Metal padding web thickness of 50 mm (2 in.)
G	190	7.5	190	7.5	1610	2930	25 mm (1 in.) thick exothermic padding over the web

Fig. 7 Cross sectional view of a railwheel casting, with dimensions given in inches. The casting was poured in plain carbon steel. The table details the various riser configurations used in pouring. See also Fig. 8 and 9. Source: Ref 28, 29

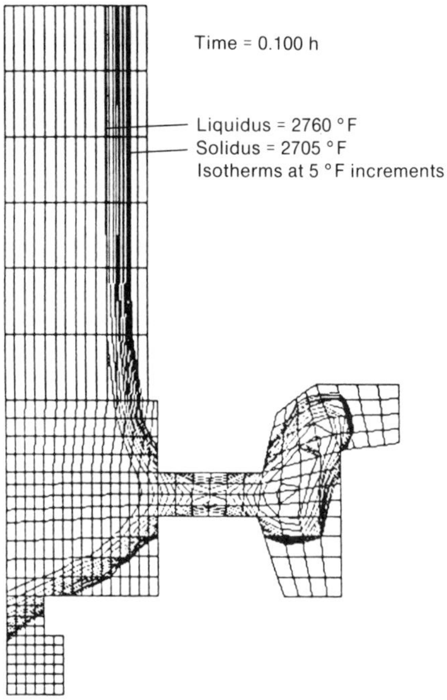

Fig. 8 Results of FEM simulation of railwheel casting using configuration C for riser (see Fig. 7). Pattern of liquidus and solidus isotherms suggests that shrinkage will occur in the rim. Source: Ref 28, 29

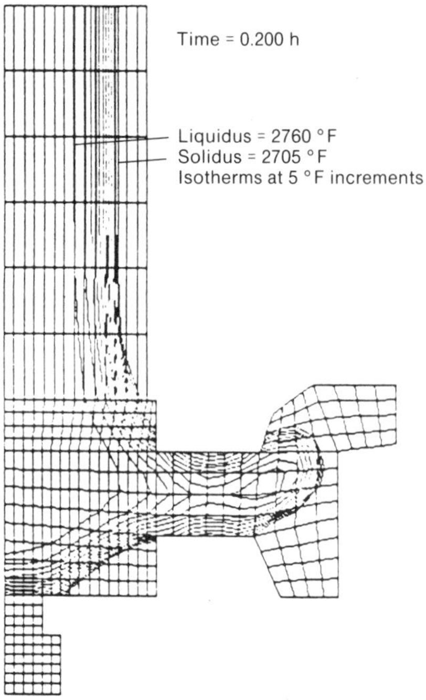

Fig. 9 Results of FEM simulation of railwheel casting using configuration F for riser (see Fig. 7). Web between hub and rim has also been thickened. Isotherms suggest that the casting should be sound. Source: Ref 28, 29

ing F showed no shrinkage defects, thus validating the simulation.

The investment cast plate proved to be more complex to simulate because the surroundings of the investment play an important role. In this case, the agreement was found to depend on the condition assumed for radiation exchange between the investment shell/kaowool wrap and the surrounding environment (an evacuated furnace). In particular, the radiation view factors were seen to be important (Ref 30).

The finite-element method is not the only system in use for performing simulations of shaped casting solidification. Many previous models have been successfully run using the finite-difference method, while more recently publications have appeared describing the use of the boundary-element method, the control volume method, and new developments of the finite-difference method (Ref 31-35). At this time, the FEM-based techniques still possess the greatest all-around flexibility.

Nonetheless, the reader should examine carefully publications describing these new and alternative computational approaches. Particular attention should be paid to the:

- Ability to handle complexities of external shape
- Ability to handle totally enclosed portions of the mold, such as coring
- Speed of computation and type of computer on which the simulation is run
- Linkages provided with pre- and postprocessing packages, in particular the existing commercially available geometric modeler based CAD systems

The Data Base

All mathematical models of the solidification process should possess (Ref 36):

- An accurate representation of geometry
- An adequate treatment for evolution of latent heat
- A sensitivity to the thermophysical properties of the materials involved in this process

The first item listed above has been discussed in this article. The second item, an important part of the data base, is discussed in the article "Modeling of Microstructural Evolution" in this Volume. The thermophysical property data base for solidification modeling is a vast but sometimes sparsely populated region. By considering the most common molding material in shaped casting (bentonite-bonded silica sand), particularly the relationship that its apparent (or effective) thermal conductivity k has with temperature, one can appreciate part of the problem associated with data base development or expansion. As shown in Fig. 10, k is a complex function of temperature. In addition, one must specify the moisture content (even when considering dried sand), ramming density, average grain fineness, and the sand source (Ref 37). Recognizing the burgeoning number of mold materials, which now include zircon, olivine, and chromite sands (Fig. 11), the potential for data base depth seems almost limitless. A collection of such data has been made that includes curve-fitted relationships linking temperature with thermal conductivity, spe-

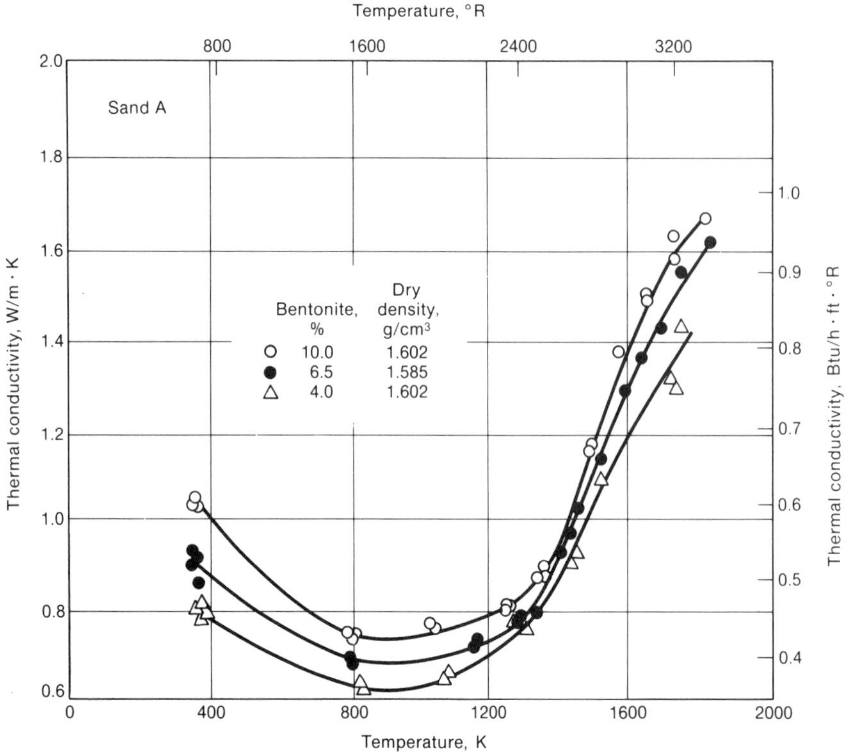

Fig. 10 Variation of apparent thermal conductivity with temperature for compacts of silica sand containing various binder contents. Source: Ref 37, 38

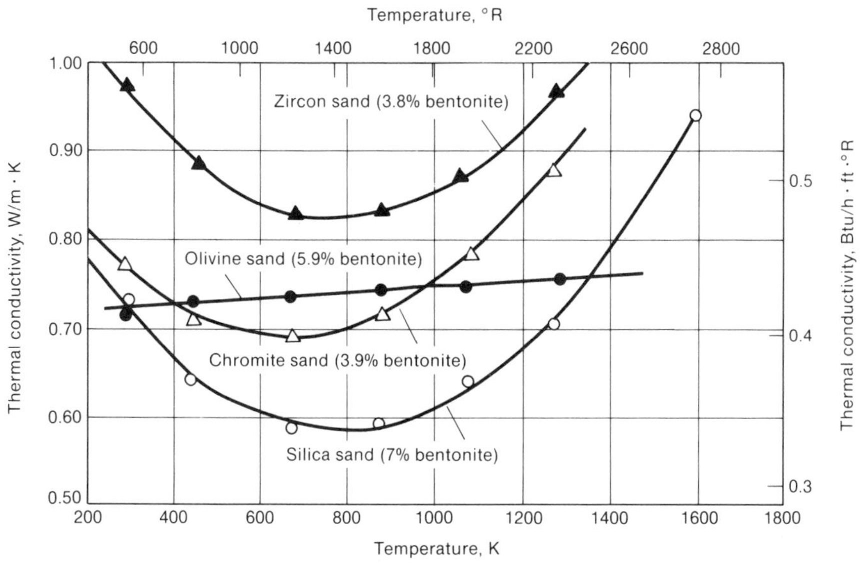

Fig. 11 Apparent thermal conductivities of four types of clay-bonded molding sands. Source: Ref 38

cific heat, and other properties of molding and casting media (Ref 38). If such data were not found to be readily available, estimates were made. This was especially true for the thermophysical properties of metals and alloys above their melting points or ranges.

It was mentioned earlier that in most modeling problems in this area, where numerical analysis is used, the number of elements (or nodes) located within the mold itself vastly outnumbers those within the casting. In recent years, there has been a determined effort on the part of several workers to replace the mold in the simulation with either a heat flux data map or a tabulated series of equivalent heat transfer coefficient values, thus essentially changing the boundary condition (Ref 39-44). Initially proposed independently by Niyama (Ref 39) and Berry (Ref 40), this approach has since been developed extensively, in particular by Wei and Berry (Ref 41, 42) and most recently and with considerable practical success by Dantzig and coworkers (Ref 43). Figures 12 and 13 compare conventional (mold enmeshment constrained) and these alternative (boundary heat flux or boundary curvature governed) calculations.

Interpreting the Output of the Computer Simulation

A common failing of many early computer applications to design engineering lay in the inability of the software to summarize the meaning of the many lines of data emerging as output. This led to the appearance of efficient postprocessing software capable of displaying, for example, two- or three-dimensional stress plots or isotherms, depending on the nature of the numerical analysis. Commercially available postprocessing routines display such information superimposed on the geometric outline of the component (Fig. 14).

With suitable interaction, certain routines will permit the generation of maps displaying the change of specific criteria functions affecting casting soundness. Such functions are normally selected on the basis of theoretical considerations and may, for example, involve some combination of local temperature gradient, G; freezing front speed, V; or cooling rate, R or $\dot{T}$. The effect of combinations of such parameters on governing feeding characteristics of steels is illustrated in Fig. 15. Similar criterion functions for other cast metals, although data are currently unavailable, are the focus of extensive research and development.

Outlook for the Future

As has been indicated, the most successful phase of bringing the computer to the foundry industry has involved the use of first-level personal-computer-supported planning-type software—for example, weight estimation, freezing order, and simple gating and risering calculations rather than full-scale simulation of freezing. Some such software items are now very comprehensive groups of programs and also permit the planning of pattern plate arrangement and part nesting.

Currently, a second, separate growth period is occurring that involves the application of minicomputer-supported, first-generation or subsequent CAD systems. Most of the software used is not tailored to a foundry, die sinking, or patternmaking environment. Therefore, to implement such systems, metal casting plants are undergoing serious changes. These changes involve the recruitment of a new type of computer-literate/technical person into the foundry team. Turning to the software itself, the majority of installations have been concerned with the improvement of tooling design and have only rarely involved solidification simulation. Some installations, however, have incorporated results of the personal-computer-supported software referred to above in connection with rigging design.

Although many organizations possess successfully applied CAD systems that have resulted in substantial savings in pattern and die tooling, other organizations procrastinate while asking:

- Will computer acquisition and maintenance costs decrease?
- Will software acquisition costs decrease?

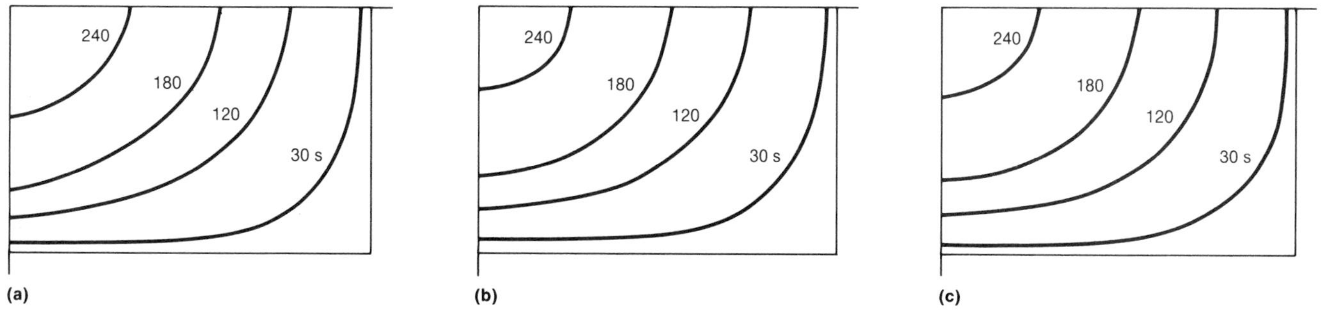

Fig. 12 Computed solidification front movement for pure aluminum in a sand mold. One quarter of a rectangular bar section is shown. (a) Results of FDM simulation using conventional approach, in which the mold is enmeshed and mold properties depend on temperature. (b) Same as (a), except mold properties are assumed to be constant. (c) Mold is not enmeshed, and heat flux q is specified at the mold/metal interface.

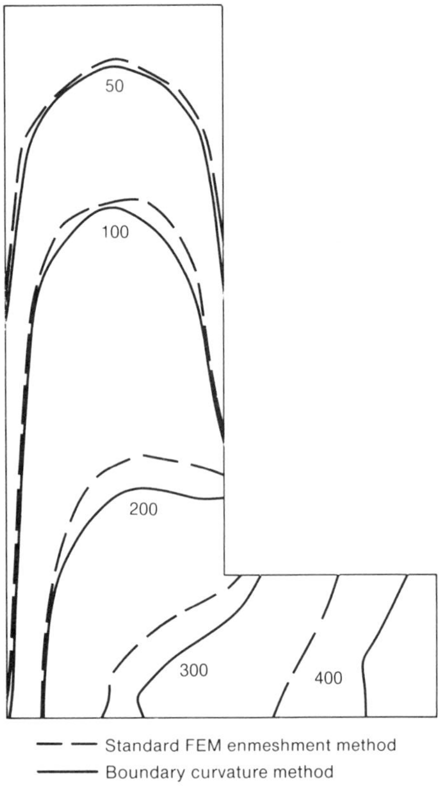

Fig. 13 Successive positions of the solidus isotherms in a one-quarter section of an H-shaped casting (low-carbon steel in sand). Time is in seconds, H half-height is 50 mm (2 in.), and half-width is 30 mm (1.2 in.). Results using standard FEM method are compared with those obtained using boundary curvature method. Source: Ref 43

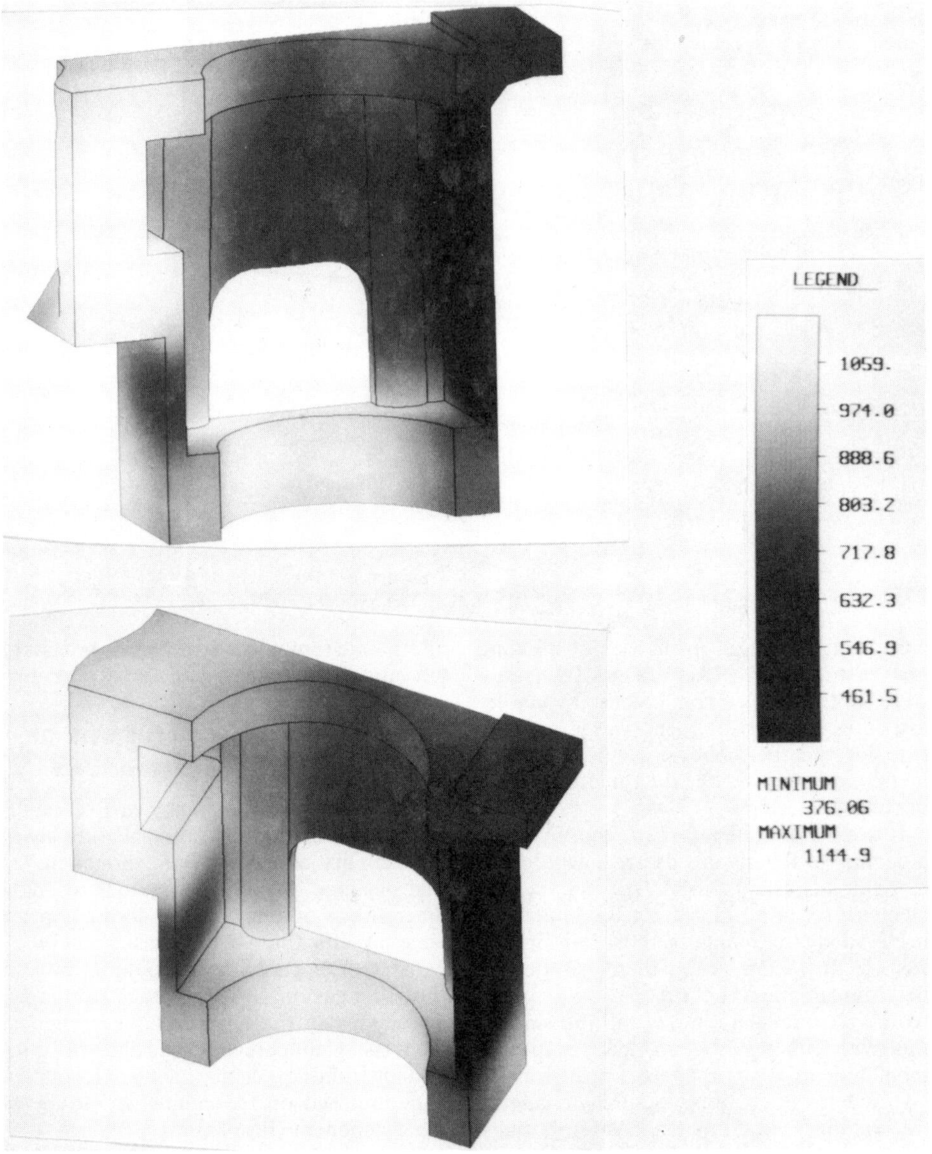

Fig. 14 Two views from a simulation of the solidification of a John Deere engine block 253 s after pouring. The material is eutectic gray cast iron with a freezing point of 1140 °C (2085 °F). The back surface is a plane of symmetry. Courtesy of J.A. Dantzig

- Will software capabilities increase?
- Will total user-friendliness increase?
- Will model-building times decrease?

Although these are not trivial questions, much is to be gained by early exposure to this new technology. At the same time, it must be observed that some of the CAD systems implemented have been examples of the first generation of such systems. They are mostly of a two-dimensional or two and a half dimensional type and often have associated volume-calculation-related limitations. Several of these systems have recently acquired solid-modeling capability, while other new systems are entering (or are about to enter) the market that are completely based on the solid-modeling approach discussed earlier in this article.

A third phase which is now beginning and is especially pertinent to this article, represents the implementation and growth of the low-cost high-flexibility mini/micro-supported computer-aided engineering system. This is most often based on the solid-modeler approach and is fully capable of being networked or linked with separate or parallel processing systems. With such systems, one can also undertake enmeshment for FEM-based stress analysis (performance checks) or with solidification simulation packages (castability checks).

One such system that is available and operating on personal computers permits the user to build a solid model of the part under development, to cut sections, to view from various angles, and to obtain drawings as hard copy if desired. More important, the two-dimensional or two and a half dimensional features of the part can then be enmeshed by accessing a new file that works on the information already implanted in the system describing external boundary geometry. All of the enmeshment and node-numbering information is also held in the small host computer memory. This system can be locally interfaced with a larger machine (for example, a mini-type computer) that can support an FEM program, where stress analysis or solidification simulation can be done. Upon completion of the chosen task, the computed information can be postprocessed on the large machine for viewing.

However, small (non-mainframe) machines generally involve difficulty with part complexity and long execution times. Part complexity is merely a function of contemporary computer capability, and long execution times may not be of primary importance for a foundry in which not all parts to be cast are of a supercritical nature. Consequently, an analysis of either stress distribution or heat transfer that runs overnight on a small machine may not be a great detriment. Similarly, the representation of the casting and the rigging using a personal-computer-based solid modeler may present no problems; the solid modeler can exchange files with a parallel system capable

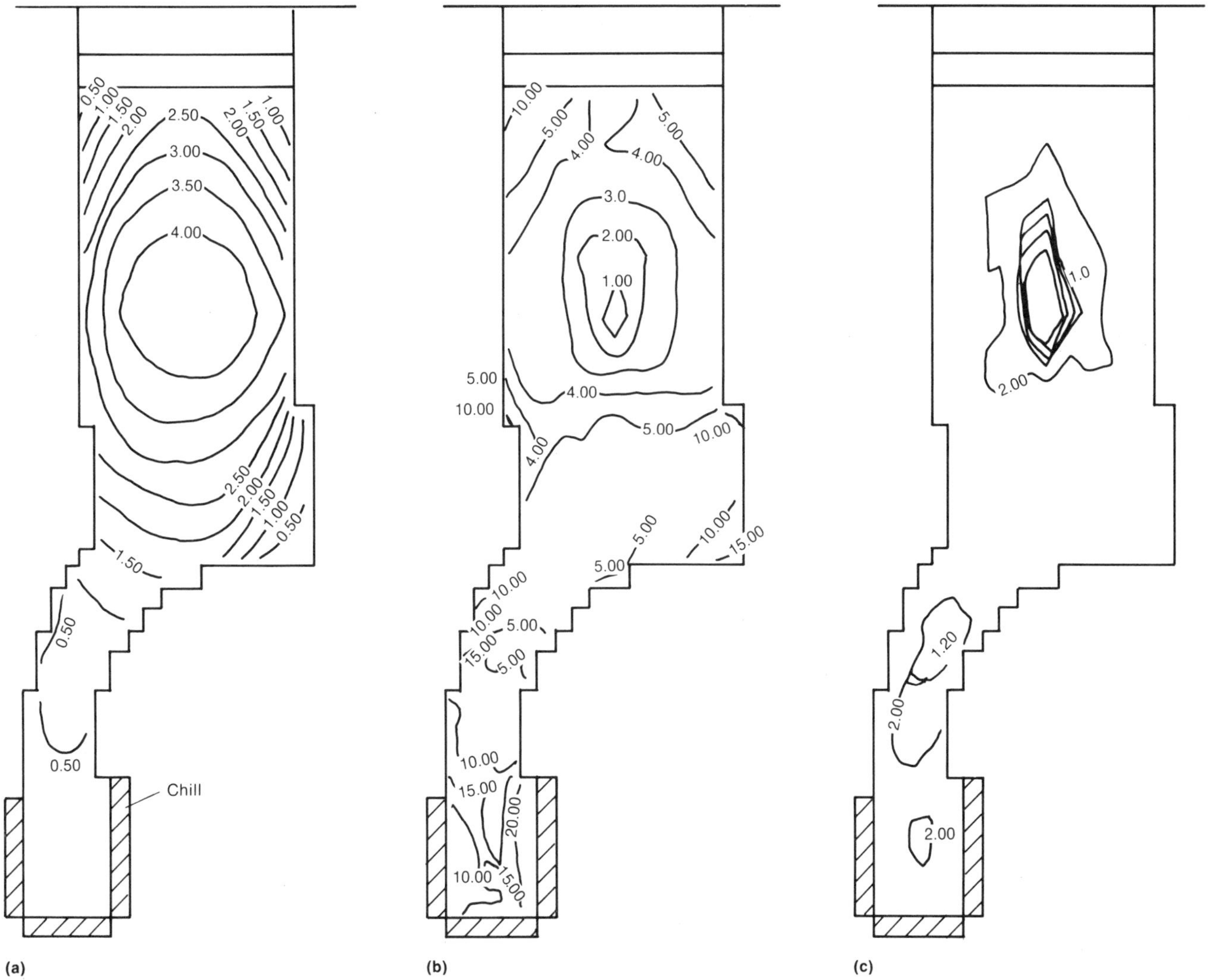

Fig. 15 Results of an FDM simulation of portions of a steel casting. Shown are local solidification time (hours), local temperature gradient during freezing (°C/cm), and a criterion function $G/\sqrt{R}$, where R is rate of cooling during freezing. The freezing isotherm suggests that shrinkage could occur in the upper half of the casting. The criterion function does not suggest shrinkage, and the casting was sound. Source: Ref 39

of generating the numerical control code needed to drive patternmaking or diemaking machinery. Although large-scale, mainframe-supported integrated systems are available with many or all of these capabilities, a significant expansion is expected in the area of micro/mini-based systems employing user-oriented software.

Looking ahead to the turn of the century, the dramatic advances currently taking place in this area should be fully realized, particularly the developments expected in the area of computing power. As chip technology moves into its next phase (10^9 devices per chip), the possibility of the personal computer handling 10 million instructions per second (mips) will become a reality. Because these specifications represent the limits of the capabilities of super-minis, computer mainframes will still be required for determining parameters of most complex components. In combination with developments in computer architecture, including parallel processing and beyond, one can anticipate that multiple simultaneous operations will become commonplace. Software specifically designed to exploit these developments is already evolving. Castings will be computer designed, planned, and rehearsed. On-line monitoring and control of melting, liquid-metal pouring, solidification, and cooling of castings will be in place. Castings will be produced directly to near-net shapes and sections. Surface finishes of the cast product will be optimized. Finishing properties will also be predictable, leaving little to cut and try.

ACKNOWLEDGMENTS

Some of the work described in this article, undertaken at The University of Alabama, The Georgia Institute of Technology, or The University of Michigan, has been supported by the National Science Foundation.

REFERENCES

1. V. Paschkis, *Trans. AFS* (various papers on the analog simulation of solidification), 1944-1961; R.D. Pehlke, R.E. Marrone, and J.O. Wilkes, *Computer Simulation of Solidification*, American Foundrymen's Society, 1976
2. K. Fursund, Das Eindringen von Stahl in Formsand. Einfluss der Oberflachenreaktion und der Temperatur, *Giesserei Tech. Wiss. Beih.*, Vol 14, 1962, p 51-61
3. J.G. Henzel and J. Keverian, Computer Programs for Heat Flow Calculations, *Trans. AFS*, Vol 74, 1966, p 661-679
4. J.G. Henzel and J. Keverian, Comparison of Calculated and Measured Solidification Patterns in a Variety of Castings, *AFS Cast Met. Res. J.*, Vol 1, 1965, p 19-30
5. R.D. Pehlke, J.T. Berry, W. Erickson, and C.H. Jacobs, Simulation of Shaped Casting Solidification, in *Solidification*

and Casting of Metals, The Metals Society, 1979, p 371-379

6. V. de L. Davies, S. Stokke, and O. Westby, Numerical Computation of Heat and Temperature Distribution in Castings, *Br. Foundryman*, Vol 66, 1973, p 305-313
7. V. de L. Davies, Heat Transfer in Gravity Die Castings, *Br. Foundryman*, Vol 73, 1980, p 331-334
8. V. de L. Davies, Modelling Solidification Processes and Computer Aided Design of Castings, in *Proceedings of the Centennial Celebration*, University of Sheffield, 1984, p 101-113
9. D.R. Durham and J.T. Berry, Role of the Mold-Metal Interface During Solidification of a Pure Metal Against a Chill, *Trans. AFS*, Vol 82, 1974, p 101-110
10. W.C. Erickson, Computer Simulation of Solidification, in *Proceedings of the 25th Sagamore Army Materials Research Conference*, July 1978
11. H. Jones, Ed., *Proceedings of the Conference on Solidification Processing*, 1987, The Metals Society, to be published
12. H.D. Brody and D. Apelian, Ed., *Modeling of Casting and Welding Processes*, Vol I, The Metallurgical Society, 1984
13. J.A. Dantzig and J.T. Berry, Ed., *Modeling of Casting and Welding Processes*, Vol II, The Metallurgical Society, 1987
14. S. Kou and R. Mehrabian, Ed., *Modeling of Casting and Welding Processes*, Vol III, The Metallurgical Society, 1987
15. *Conference on Modeling of Casting and Welding Processes IV*, April 1988, The Metallurgical Society, to be published
16. Annual Progress Reports of NSF CADCAST Project, University of Michigan, Georgia Institute of Technology, and University of Alabama
17. Annual Reports of Scandinavian HUBERT Project
18. T.C. Chang and R.A. Wysk, *An Introduction to Automated Process Planning Systems*, Prentice-Hall, 1985, p 72
19. J.T. Berry and J.A.M. Boulet, The Application of Geometric Modeling to Metal Casting Technology, in *Solid Modeling by Computer—From Theory to Application*, M.S. Pickett and J.W. Boyse, Ed., Plenum, 1984, p 105-160
20. D.B. Welbourn, Computer-Aided Engineering (CADCAM/DUCT) in the Foundry, in *Proceedings of the BCIRA Conference on Development for Future Foundry Prosperity*, University of Warwick, April 1984
21. D.C. Schmidt, *User's Manuals for AF-Software*, American Foundrymen's Society
22. J.E. Pickin and A. Beattie, *Crusader Method Design System User's Manual*, Steel Castings Research and Trade Association, 1984
23. R. Sillen, *User's Manual for NOVA-CAST*, NOVACAST Ltd., Ronneby, Sweden
24. N. Chvorinov, Theory of Casting Solidification, *Giesserei*, Vol 27, 1940, p 177-180, 201-208, 222-225
25. J.L. Hill, J.T. Berry, and C. Jordan, Use of Expert Systems in Cast Metals Technology, in *Proceedings of the Conference on Artificial Intelligence in Minerals and Materials Technology*, U.S. Department of the Interior, Bureau of Mines, 1987
26. B. Dalton-Motter, J.A.M. Boulet, and J.T. Berry, Interfacing a Geometric Modeling Package With a Heat Transfer Simulator, *Trans. AFS*, Vol 95, 1987, p 841-846
27. M.J. Beffel, J.O. Wilkes, R.D. Pehlke, and J.T. Berry, Software for Transient Heat Flow Simulation, in *Modeling of Casting and Welding Processes*, Vol II, The Metallurgical Society, 1987
28. M.J. Beffel, J.O. Wilkes, and R.D. Pehlke, Finite-Element Simulation of Casting Processes, *Trans. AFS*, Vol 94, 1986, p 757-764
29. A. Jeyarajan and R.D. Pehlke, Application of Computer-Aided Design to a Steel Wheel Casting, *Mod. Cast.*, Vol 69, 1979, p 72-73
30. M.J. Beffel, Ph.D. dissertation, The University of Michigan, 1986
31. R.D. Pehlke, R.E. Marrone, and J.O. Wilkes, *Computer Simulation of Solidification*, American Foundrymen's Society, 1976
32. C.P. Hong and T. Umeda, Numerical Simulation of Solidification Processes by Boundary Element Method, in *Proceedings of the Conference on Solidification Processing*, The Metals Society, 1987, p 278-280
33. C.S. Kanetkar, D.M. Stefanescu, N. El-Kaddah, and I.G. Chen, Macro-Microscopic Simulation of Equiaxed Solidification of Eutectic and Off-Eutectic Alloys, in *Proceedings of the Conference on Solidification Processing*, The Metals Society, 1987, p 404-407
34. I. Ohnaka, T. Aizawa, K. Namekawa, M. Komiya, and M. Kaiso, Computer Simulation of Solidification of Castings, in *Proceedings of the Conference on Solidification Processing*, The Metals Society, 1987, p 268-270
35. K. Ho and R.D. Pehlke, A Comparison of the Finite Element and the General Finite Difference Method for Transient Heat Flow, *Mater. Sci. Technol.*, Vol 3, 1987, p 466-476
36. J.A. Dantzig, Mathematical Modeling of Solidification Processes, in *Interdisciplinary Issues in Materials Processing and Manufacturing*, Vol II, American Society of Mechanical Engineers, 1987
37. D.V. Atterton, The Apparent Thermal Conductivities of Moulding Materials at High Temperatures, *J. Iron Steel Inst.*, Vol 174, 1953, p 201-211
38. R.D. Pehlke, A. Jeyarajan, and H. Wada, *Summary of Thermal Properties for Casting Alloys and Mold Materials*, The University of Michigan, 1982
39. E. Niyama, Calculation of Solidification Rate of Shape Castings by the Flux-Boundary Method, *J. Jpn. Foundrymen's Soc.*, Vol 49, 1977, p 16-31
40. J.T. Berry, NSF CADCAST Project, private communication, 1981
41. C.S. Wei, P.N. Hansen, and J.T. Berry, The Q-Method—A Compact Technique for Describing the Heat Flux Present at the Mold-Metal Interface in Solidification Problems, in *Numerical Methods in Heat Transfer*, Vol 2, R.W. Lewis, K. Morgan, and B.A. Schrefler, Ed., John Wiley & Sons, 1983, p 461-471
42. C.S. Wei and J.T. Berry, An Analysis of The Transient Edge Effect on Heat Conduction in Wedges, *Int. J. Heat Mass Transfer*, Vol 25, 1982, p 590-592
43. J.A. Dantzig, J. Wiese, and S.C. Lu, Modeling of Heat Flow in Sand Castings, Parts I and II, *Metall. Trans. B*, Vol 16B, 1985
44. M.N. Ozisik, *Heat Conduction*, John Wiley & Sons, 1980, p 13

Modeling of Fluid Flow

W.-S. Hwang, National Cheng Kung University, Taiwan
R.A. Stoehr, University of Pittsburgh

FLUID FLOW MODELING is a technique that uses computers to investigate flow phenomena. These flow phenomena, particularly during the initial filling stage, have major effects on the quality of castings. Designers have commonly relied on experience, rule of thumb, and handbook information to achieve their objectives of smooth flow, proper filling time, minimum gas entrapment, elimination of inclusions and dross, and the desired distribution of metal during mold filling. The goal of improving the quality and cost-effectiveness of castings by means of computer-aided design (CAD) and modern process control requires that data be expressed more scientifically, so that they are amenable to computation.

Knowledge about fluid flow during the filling of castings is important not only in itself, but because it affects heat transfer both during and after filling. This must be taken into account in models of heat transfer and stress analysis in castings if they are to give the most accurate results.

It is difficult to make direct observations of fluid flow inside molds, because the molds and the molten metal are opaque, the temperatures are high, and the conditions are highly transient. Even when observations are made, as by the techniques described later in this article, the location of the metal as a function of time is usually the only information that can be obtained. Not only is fluid flow modeling using computers usually the most economical and practical way to get information about what is going on inside a mold during filling, it is often the only feasible way. Furthermore, it can give information about the velocity and pressure distributions within the molten metal which cannot be obtained by direct observation.

Computational techniques for modeling flow during mold filling can be divided into two categories (Ref 1):

- Energy balance techniques based on the Bernoulli equation and the Saint-Venant equations
- Momentum balance techniques based on the Navier-Stokes equations as embodied in the Marker-and-Cell group of programs which include the Marker-and-Cell (MAC), Simplified Marker-and-Cell (SMAC), and Solution Algorithm (SOLA) techniques

The energy balance techniques are most useful for modeling flow through sprues, runners, and gates when the direction of flow is dictated by the configuration of the system. The momentum balance techniques are needed for calculating flow inside mold cavities where the direction of flow and the location of the fluid must be calculated.

Energy Balance Methods

Because energy is a scalar rather than a vector quantity, these methods are primarily useful in determining flow rates in cases in which the direction of flow is established by the configuration of the system. The Bernoulli equation is used for calculating flow in completely filled channels such as sprues, pressurized runners, and gates, while the Saint-Venant equations are used for partially filled channels such as nonpressurized runners and troughs. Information provided by these calculations is extremely important to the design and manufacture of castings, and industry has used them routinely for many years. Now, CAD programs can apply these techniques to complicated systems with speed and accuracy.

A schematic of a gravity-filled casting system is shown in Fig. 1. The Bernoulli equation applicable to this system may be written (Ref 2):

$$\frac{P_j - P_i}{\rho} + \frac{V_j^2}{2b_j} - \frac{V_i^2}{2b_i} + g(z_j - z_i) + E_{f_{i,j}} = 0 \quad \text{(Eq 1)}$$

where P_i and P_j are pressures at positions i and j, V_i and V_j are mean velocities at positions i and j, z_i and z_j are elevations at

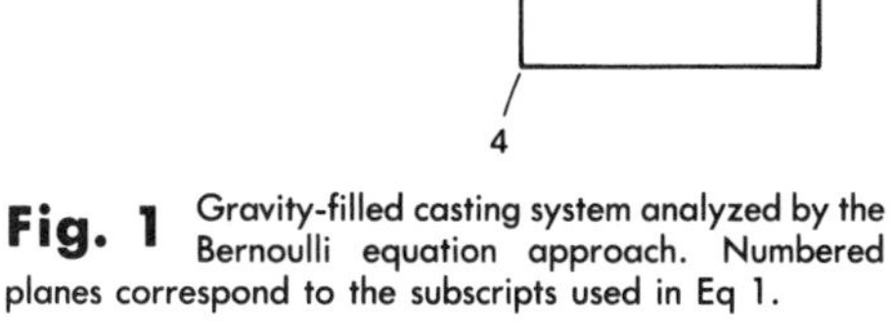

Fig. 1 Gravity-filled casting system analyzed by the Bernoulli equation approach. Numbered planes correspond to the subscripts used in Eq 1.

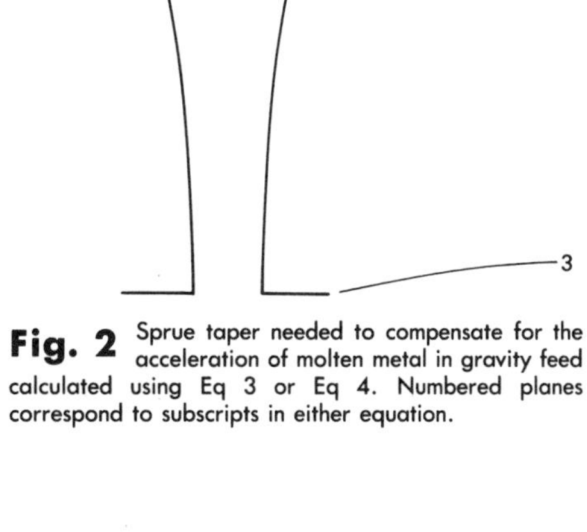

Fig. 2 Sprue taper needed to compensate for the acceleration of molten metal in gravity feed calculated using Eq 3 or Eq 4. Numbered planes correspond to subscripts in either equation.

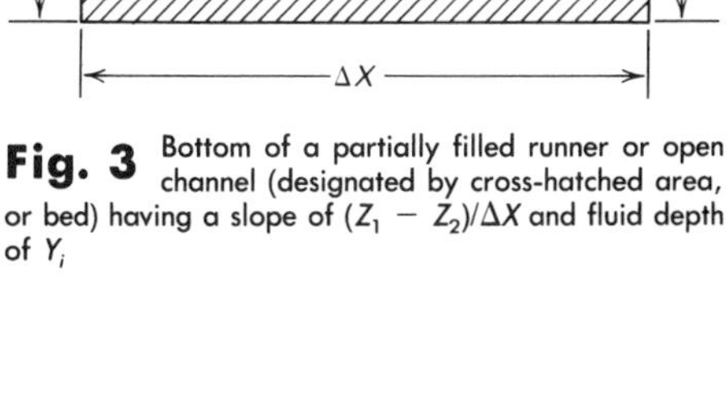

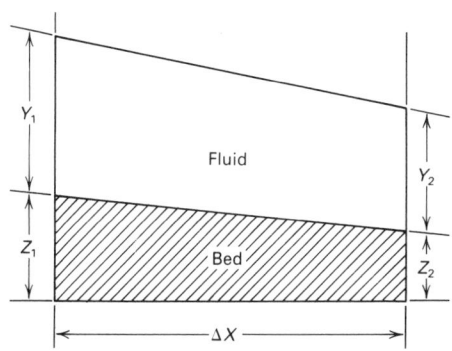

Fig. 3 Bottom of a partially filled runner or open channel (designated by cross-hatched area, or bed) having a slope of $(Z_1 - Z_2)/\Delta X$ and fluid depth of Y_i

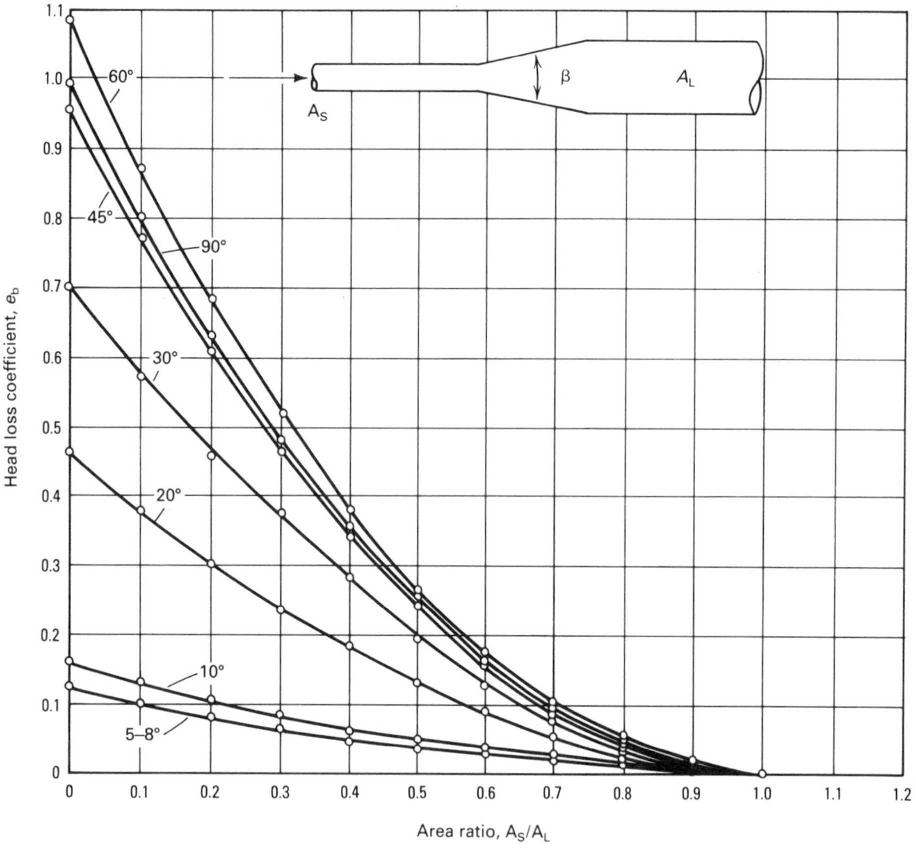

Fig. 4 Head loss coefficient for a gradual enlargement plotted as a function of the area ratio ($\sigma = A_S/A_L$) and the angle β. Open circles indicate selected e_b values obtained using Eq 10. Source: Ref 7

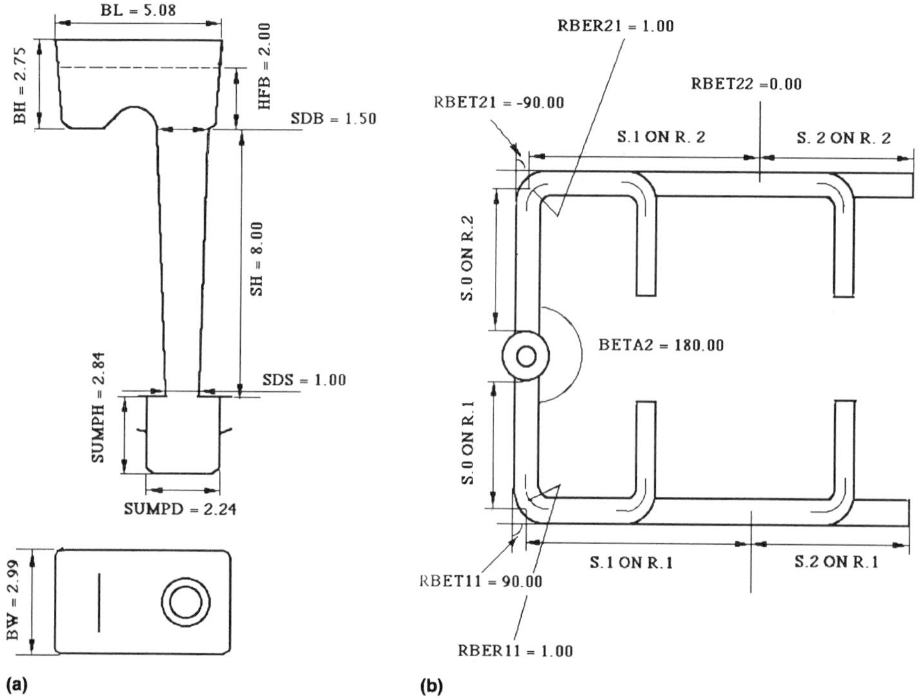

Fig. 5 Typical layouts of a mold-filling system obtained from a CAD program based on sound foundry industry practice and refinements by Bernoulli and Saint-Venant equations. (a) Pouring basin, sprue, and sprue base layout. (b) Runner-gate layout. Dimensions given in inches

positions i and j, b_i and b_j are velocity distribution factors (0.5 for laminar flow and 1.0 for highly turbulent flow), g is the acceleration due to gravity, ρ is the fluid density, and $E_{f_{i,j}}$ is the friction energy loss between positions i and j.

The friction energy loss term $E_{f_{i,j}}$ is a function of the velocity, the Reynolds number, and the configuration of the system between positions i and j (Ref 2-4). The Reynolds (N_{Re}) number is defined by:

$$N_{Re} = \frac{VD_e}{\nu} \qquad \text{(Eq 2)}$$

where ν is the kinematic viscosity and D_e is the equivalent diameter.

The configuration of the system may change from point to point resulting in variations of the velocity and the N_{Re}. Thus, $E_{f_{i,j}}$ must be calculated for each segment of the system and then summed for the whole system.

For some uses, it is not necessary to consider friction energy loss. For example, the calculation of sprue taper (Fig. 2) is often done without it (Ref 5). The objective is to calculate the change in the cross section of a vertical sprue that will compensate for the acceleration of the molten metal as it drops through the sprue under the influence of gravity. This yields the following equations for the ratio of the cross-sectional areas at the top (A_2) and bottom (A_3) of the sprue, or for the diameters of a round sprue at the same points:

$$\frac{A_3}{A_2} = \sqrt{\frac{z_1 - z_2}{z_1 - z_3}} \quad \text{or} \quad \frac{D_3}{D_2} = \sqrt[4]{\frac{z_1 - z_2}{z_1 - z_3}} \qquad \text{(Eq 3)}$$

If friction is considered, this equation is modified to:

$$\frac{A_3}{A_2} = \{(z_1 - z_2)/[z_1 - z_3 - (z_1 - z_2)e_{f_{2,3}} + (z_2 - z_3)e_{f_{1,2}}]\}^{1/2} \qquad \text{(Eq 4)}$$

where the $e_{f_{i,j}}$ are friction energy loss coefficients such that: $E_{f_{i,j}} = \frac{1}{2}e_{f_{i,j}} V_2^2$. Introduction of a friction energy loss $e_{f_{2,3}}$ within the sprue reduces the taper required, while a friction energy loss $e_{f_{1,2}}$ above the sprue increases it.

In nonferrous casting, it is common practice to use nonpressurized filling systems in which the runners are only partially filled with liquid metal. These can be modeled with the Saint-Venant equations (Ref 6), which are useful when the bottom of the channel and the free top surface of the stream slope at different angles (Fig. 3). They are also used when the flow rate and the top surface of the stream vary with time. The equation of continuity is written:

$$W_T \frac{\partial Y}{\partial t} + \frac{\partial (VA)}{\partial X} = 0 \qquad \text{(Eq 5)}$$

where W_T is the width of the channel at flow depth Y, t is time, and VA is the product of velocity and fluid cross-sectional area (that is, the volumetric flow rate).

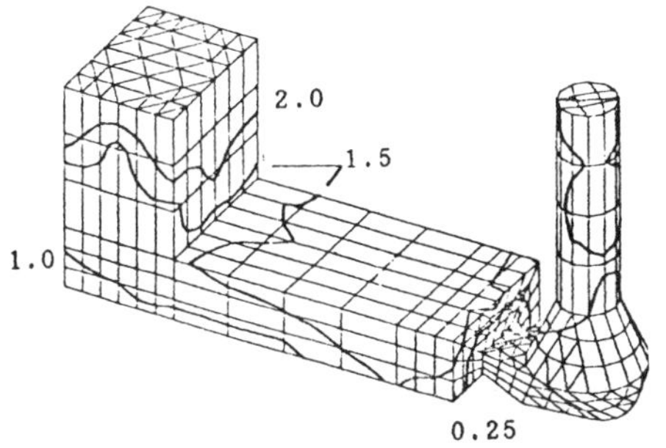

Fig. 6 Three-dimensional visualization of molten metal locations determined by the contact wire method and displayed by a finite-element postprocessor. Numbers in the figure are filling times in seconds. Source: Ref 14

The equation of motion for the steady state case may be written:

$$\frac{V}{g}\frac{\partial V}{\partial X} + \frac{\partial Y}{\partial X} - S_o + S_f = 0 \quad \text{(Eq 6)}$$

where the actual slope of the bottom of the runner S_o is:

$$S_o = \frac{Z_1 - Z_2}{\Delta X} \quad \text{(Eq 7)}$$

and the so-called friction slope S_f is defined by:

$$S_f = \frac{n^2 V^2}{R_h^{4/3}} \quad \text{(Eq 8)}$$

where R_h is the hydraulic radius and n is Manning's roughness coefficient, the values of which are determined empirically using similarity principles.

For the time-dependent case, the equation of motion may be written:

$$d\left(Z + Y + \frac{V^2}{2g}\right) = -S_f\,dX - \frac{1}{g}\cdot\frac{\partial V}{\partial t}\,dX \quad \text{(Eq 9)}$$

in which the last term represents the acceleration.

Note that the roughness factor n takes into account the channel configuration as well as the surface roughness and may change frequently along the course of the channel. It is larger at the tip of the entering stream than in areas already containing a layer of molten metal.

For CAD and analysis of mold-filling systems, the Bernoulli and/or Saint-Venant representations of each of the individual sections of the system can be combined into a network of series and parallel paths. The friction energy loss factors and the friction slopes must be obtained from empirically derived correlations with the N_{Re} and other dimensionless numbers. Their use commonly calls for iterative solutions, because they are functions of the velocity and free surface height, which are dependent variables one seeks to determine. The literature most often presents the necessary empirical correlations in graphs (Ref 7). The relationships represented by the graphs can be reduced to a set of parametric equations for use by the computer. For example, the information shown in Fig. 4 may be reduced to:

$$\begin{aligned} e_b = \{&[(-0.194\times10^{-6}\beta - 0.246\times10^{-3})\beta + 0.0368]\beta \\ &- 0.184 + 0.03125/[(\beta - 6.1)^4 + 1.25] \\ &+ (4.529\beta^2 - 91.75\beta + 464.9)/(\beta^4 - 20\beta^3 \\ &+ 126.7\beta^2 - 500\beta + 2542)\}(1 - \sigma^2) \\ &+ 1.5/[(51\sigma - 17.87)^2 + (\beta - 29)^2 + 33.3] \\ &- 2/[(17.32\sigma - 5.196)^2 + (\beta - 60)^2 + 100)] \end{aligned} \quad \text{(Eq 10)}$$

where e_b is the head loss coefficient, β is the angle shown in Fig. 4, and σ is A_S/A_L. In spite of its complicated appearance, the solution may be found very swiftly on any computer.

The Bernoulli and Saint-Venant equations can be used in a design program that includes the rules of good practice frequently used in the foundry industry. From experience, for example, rules have been established that specify the acceptable range of filling times for castings of a given type as a function of the casting weight. These rules also suggest a certain size for the sprue, one for the runner, and one for the gate, which can be used to produce an initial design. The accuracy can then be verified by performing the energy balance calculation described above (Eq 10). The design can be modified and the energy balance calculation performed again until the calculated result agrees with the desired filling rate. This technique allows special plant conditions to be considered, such as limitations on the amount of molten metal available at one time, and limitations on the flasks and patterns available for making the mold.

This type of modeling of metal flow during mold filling has been integrated into CAD systems that produce finished drawings for the pattern and mold shops. In some cases they even produce numerical-control (NC) tapes for cutting the pattern. An example of a drawing for the pouring basin, sprue, runner, and gates for a particular casting is shown in Fig. 5.

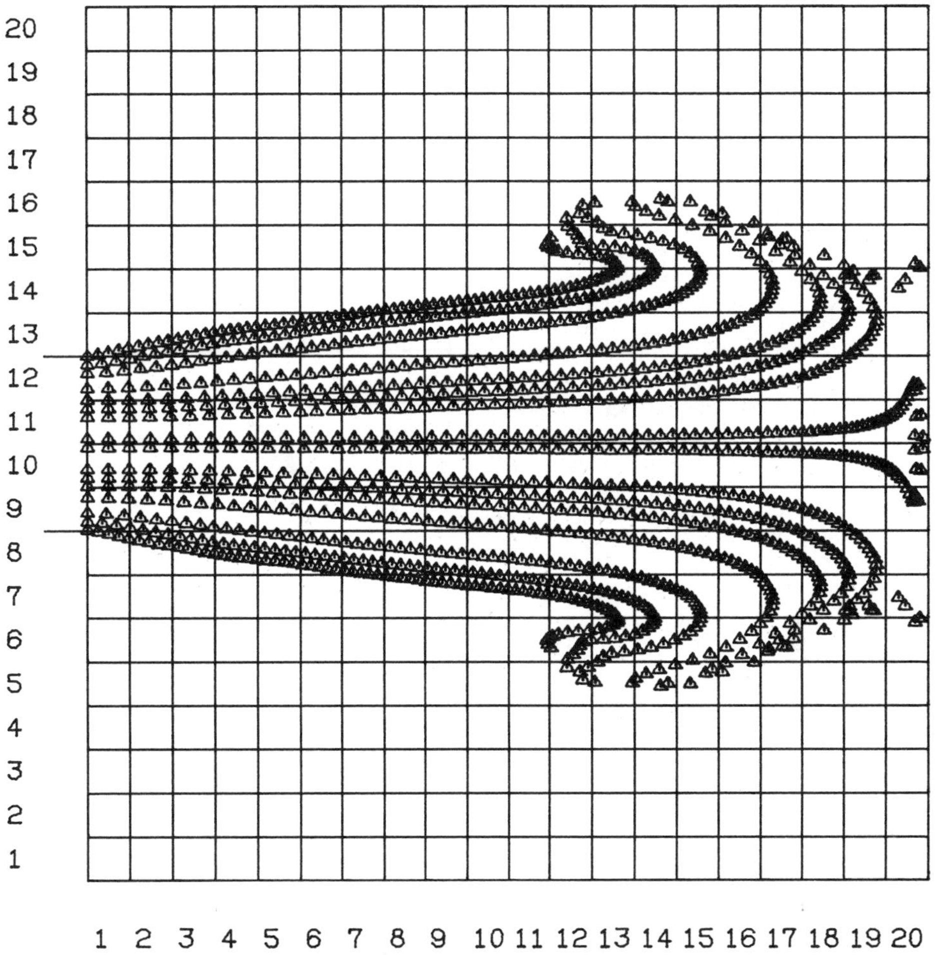

Fig. 7 Flow pattern obtained 3.95 s after pouring begins showing MAC cell divisions and fluid domain locations (indicated by triangular markers)

Physical Modeling of Mold Filling

It is important to verify computational models of mold filling with physical experiments. Physical modeling of metal flow is most often done with water in transparent molds. Water is a suitable fluid model because its kinematic viscosity is nearly the same as that of common metals at their normal pouring temperatures, especially if the temperature of the water is properly controlled. For example, the kinematic viscosity of water at 35 °C (95 °F) is the same as that of 0.5%C steel at 870 °C (1600 °F) (Ref 8). To model the flow of liquids influenced by inertial, viscous, and gravitational forces requires N_{Re} and Froude number, N_{Fr}, similarities, in which the Froude number is defined as $N_{Fr} = V^2/gL$, where L is the characteristic dimension of the system. This is obtained when water at the temperature to achieve proper kinematic viscosity is used in models of the same size as the real system. The application of water models to mold filling is illustrated in several high-speed motion picture studies of gravity-filling (Ref 9, 10) and die casting systems (Ref 11).

It is possible to observe the flow of actual metals into molds from x-ray cinematography studies (Ref 12). Greater detail of the flow of metal into sand molds was observed by filming with a high speed motion picture camera focused through a plate glass window on the side of the mold (Ref 13). Molten cast iron entering a silica sand mold was filmed with this technique. The molten metal locations traced from these films will be shown in the section "Correlations With Actual Metal Flow."

Recently, computerized data acquisition systems have been used to observe the flow of molten metal into nontransparent three-dimensional molds (Ref 14, 15). Two types of sensors have been used: simple contact wires and thermocouples. The simple contact wires are connected to the digital inputs of the data acquisition unit. Molten metal contacting a bare copper wire completes a circuit, and this is then detected by the data acquisition unit. The digital inputs can be read very rapidly by the computer; it is possible to read hundreds of such contact points many times during the filling of a single mold. An accurate picture of the location of the metal at any time can be developed from these data. A clever system for using a finite element postprocessor for presenting the results of such experiments has been reported (Ref 14). Figure 6 shows such a result.

Thermocouples can be used alone or in conjunction with the contact wire sensors. Although they give more information than do simple contact wires, thermocouples produce an analog signal that takes longer to read. The combination of contact wires and thermocouples is particularly useful for observing the flow and solidification patterns in thin-wall castings having the problem of premature freezing (namely, cold shuts).

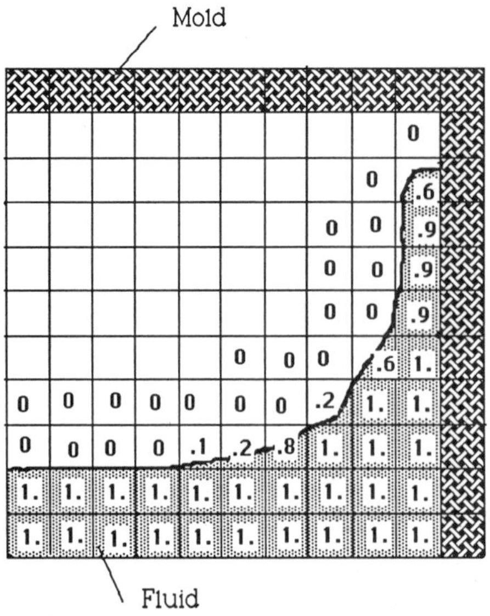

Fig. 8 Fluid domain plot obtained with the fluid function F using the SOLA-VOF technique. $F = 1.0$, $0.0 < F < 1.0$, and $F = 0$ indicate full, surface, and empty cells, respectively.

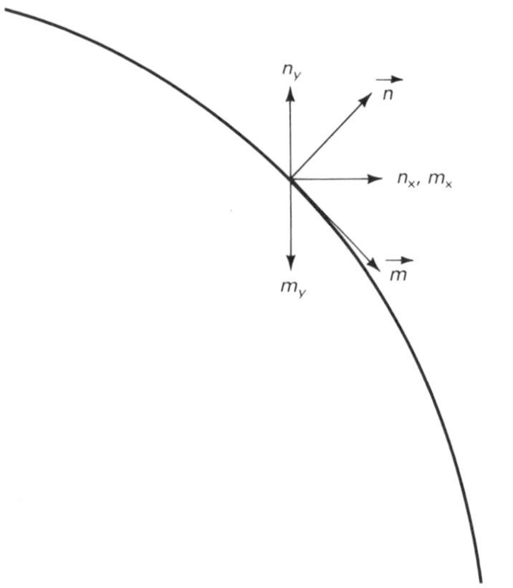

Fig. 9 Orientation of the free surface represented in Eq 14 and Eq 15

Momentum Balance Techniques

Fluid flow within the mold cavity during filling is transient; the amount and location of the liquid changes rapidly. Calculation of the location of the liquid and the orientation of its free surface must be an integral part of the computational techniques used to model it. The family of computational techniques called MAC (Ref 16), SMAC (Ref 17), and solution algorithm-volume of fluid (SOLA-VOF) (Ref 18) are well suited for handling these problems. Although they differ from each other in the way they keep track of the location of the free surface and the way in which they perform some of the internal iterations, they are based on the same principles. To simplify the discussion, the acronym MAC is used to represent this whole family of computational fluid dynamics techniques.

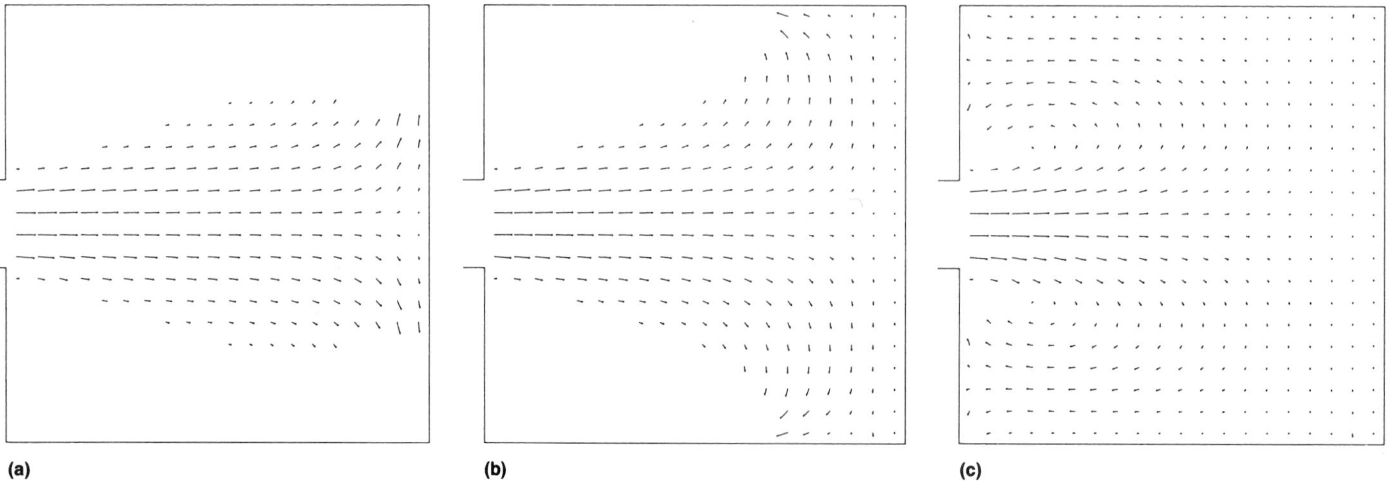

Fig. 10 Vector plot showing flow patterns and velocity profiles obtained when filling a horizontal 610 × 610 mm (2 × 2 ft) square plate casting. Entrance velocity was 305 mm/s (1 ft/s), and vector plots shown are at elapsed times. (a) 3.95 s. (b) 5.45 s. (c) 9.95 s

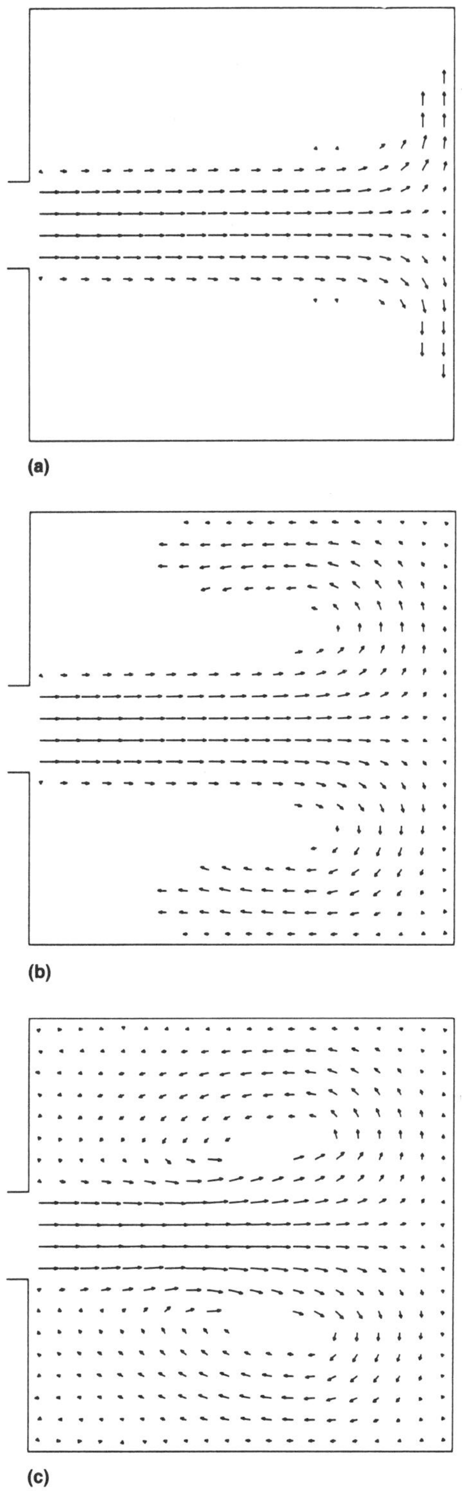

Fig. 11 Vector plot showing flow patterns and velocity profiles obtained when filling a horizontal 61.0 × 61.0 mm (0.2 × 0.2 ft) square plate casting. Entrance velocity was 305 mm/s (1 ft/s), and vector plots are shown at elapsed times. (a) 0.299 s. (b) 0.599 s. (c) 0.879 s

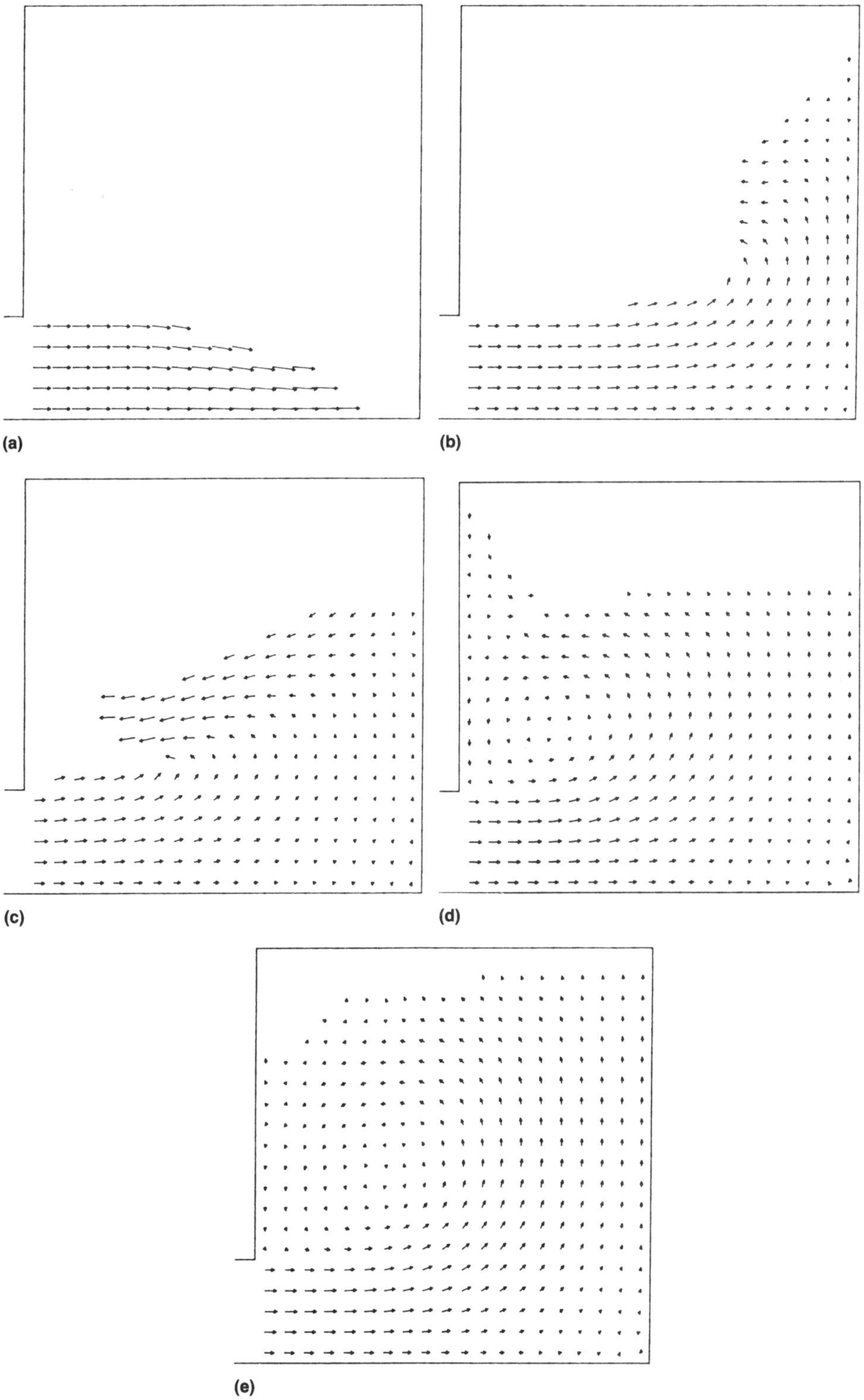

Fig. 12 Vector plot showing flow patterns and velocity profiles obtained when filling a vertical 152 × 152 mm (0.5 × 0.5 ft) square plate casting. Entrance velocity is determined by Bernoulli equation calculation and varies from 975 mm/s (3.2 ft/s) to 518 mm/s (1.7 ft/s). Plots are shown at elapsed time. (a) 0.10 s. (b) 0.32 s. (c) 0.42 s. (d) 0.62 s. (e) 0.76 s

MAC Technique Highlights

MAC uses a finite-difference scheme for the mathematical analysis of fluid flow problems (Ref 19). Like most of these techniques, MAC first divides the system (that is, the configuration of the casting cavity under discussion) into a number of subdivisions, called cells, which are usually rectangular. Then a set of imaginary markers (in MAC and SMAC) or fluid function values called F (in SOLA-VOF) is introduced into the system to represent the location of the fluid at any instant. Figure 7 shows the division of the system into cells and the representation of the fluid domain with markers, while Fig. 8 shows a portion of the

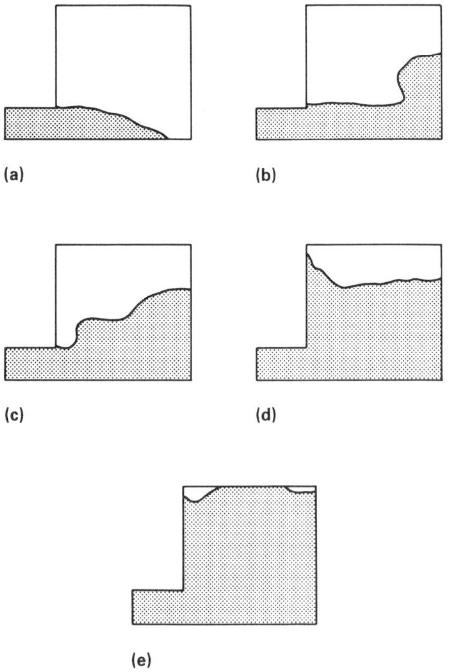

Fig. 13 Tracings of still photographs from a high-speed motion picture of a water model of the vertical 152 × 152 mm (0.5 × 0.5 ft) square plate casting calculated in Fig. 12. (a) 0.10 s. (b) 0.30 s. (c) 0.45 s. (d) 0.65 s. (e) 0.90 s

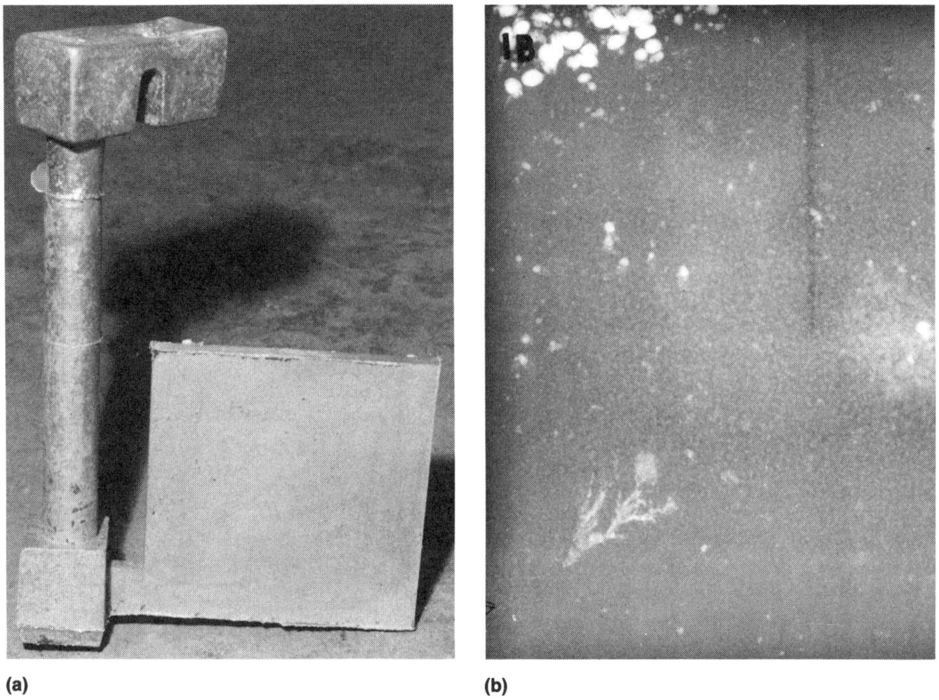

Fig. 14 Photograph (a) and radiograph (b) of an Al-7.5Si alloy casting produced in a vertical square plate mold identical to that used to obtain Fig. 12 and Fig. 13 data. The radiograph shows a large accumulation of gas bubbles in the last area to be filled by the alloy.

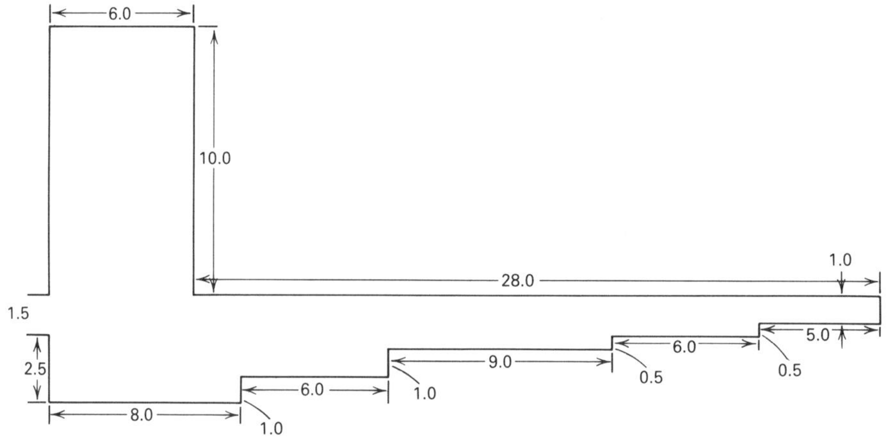

Fig. 15 Configuration and dimensions of a vertical stepped plate mold. Dimensions given in inches

fluid domain with the fluid function F (see the section "Fluid Domain Identification" below). The velocity field of the moving fluid domain can be calculated by the application of fluid dynamics principles. Next, the markers are moved, or the fluid function is updated, according to the calculated velocity field in order to represent the new location of the fluid domain. This procedure can be repeated from the beginning when the cavity is empty until it is completely filled.

Fluid Domain Identification

In MAC, the cells are designated as full, surface, or empty, based on the location of markers or the distribution of the fluid function. With the marker approach, a full cell is one that contains at least one marker, if all of its neighboring cells contain markers as well. A surface cell contains at least one marker, but has at least one neighbor without any markers. An empty cell is any cell with no markers.

With the fluid function technique, F represents the fraction of the volume of a cell that is filled with fluid. F can have values from 0 to 1. F is 1 for a full cell, 0 for an empty cell, and some fractional value for surface cells. The fluid function F can also be used to calculate the approximate location and orientation of the free surface of the fluid. Collectively, the full cells constitute the interior region, and the surface cells constitute the surface regions.

Calculating Velocity Field in Moving Molten Metal

After the flow domain and the corresponding interior and surface regions of the domain have been identified, the velocity and pressure fields within the flow domain are calculated. The physical conditions that govern the flow behavior in the interior regions are somewhat different from those in the surface regions.

Interior-Region Flow Behavior. In the interior regions, the following principles should be obeyed in a cell volume:

Mass Flow In Equals Mass Flow Out. This is due to the incompressible nature of molten metal. This gives the following form of the continuity equation in two dimensions:

$$\frac{\partial u}{\partial x} + \frac{\partial v}{\partial y} = 0 \qquad \text{(Eq 11)}$$

where u and v are the velocity components in the x- and y-directions.

Momentum Change Equals Momentum In Minus Momentum Out. In mathematical form this is:

In the x-direction:

$$\frac{\partial u}{\partial t} + u\frac{\partial u}{\partial x} + v\frac{\partial u}{\partial y} = -\frac{\partial p}{\partial x} + \nu\left(\frac{\partial^2 u}{\partial x^2} + \frac{\partial^2 u}{\partial y^2}\right) + g_x \qquad \text{(Eq 12)}$$

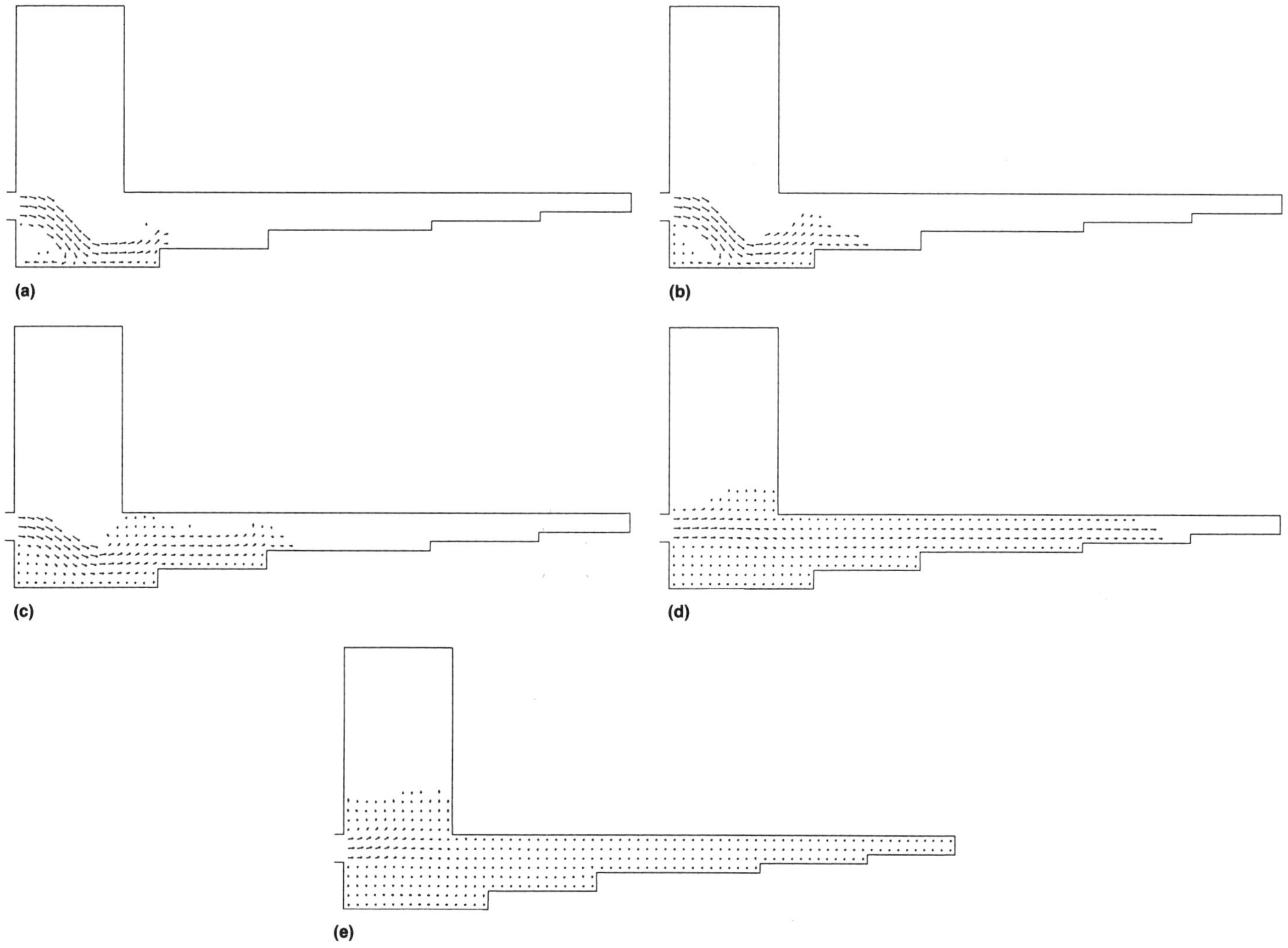

Fig. 16 Vector plot showing flow patterns and velocity profiles of a vertical stepped plate casting produced by using the mold illustrated in Fig. 15. The initial entrance velocity was 975 mm/s (3.2 ft/s), and the plot is shown at elapsed times. (a) 0.27 s. (b) 0.39 s. (c) 0.72 s. (d) 1.47 s. (e) 1.80 s

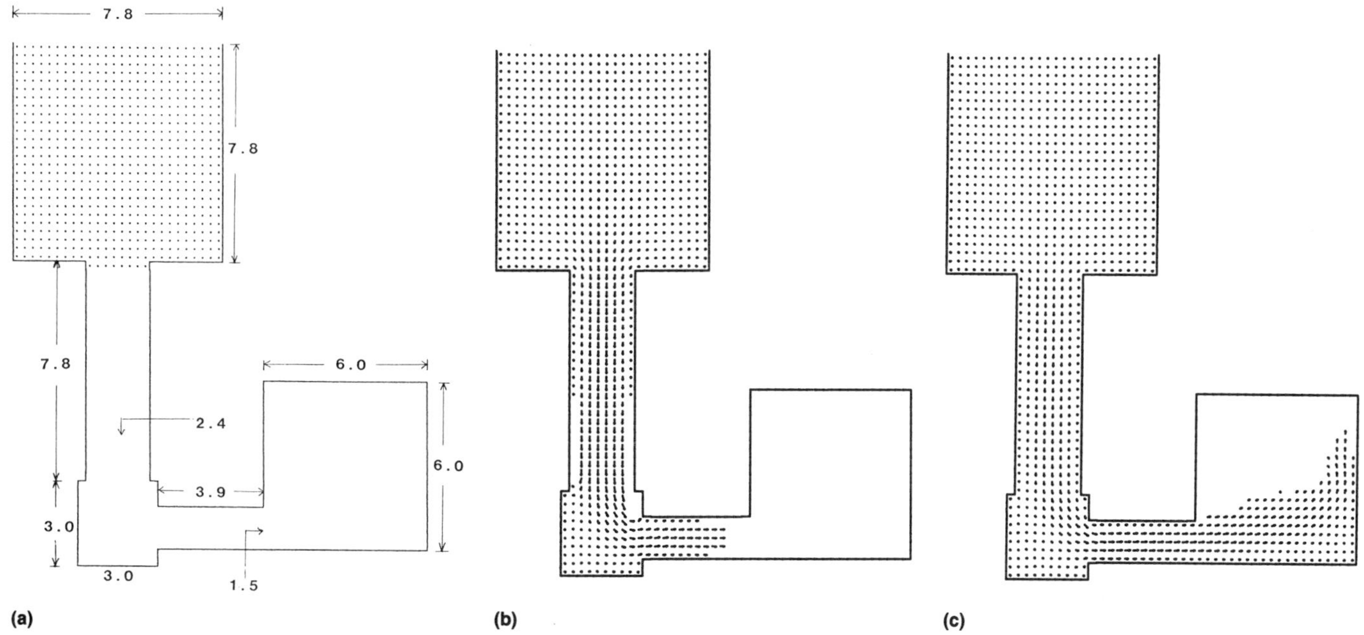

Fig. 17 Geometry, dimensions, and vector plots (showing flow patterns and velocity profiles) of the filling system attached to the vertical 152 × 152 mm (6 × 6 in.) square plate mold discussed in Fig. 12 to 14. The vector plot is shown at elapsed times. (a) 0.0 s. (b) 0.290 s. (c) 0.570 s. Dimensions given in inches

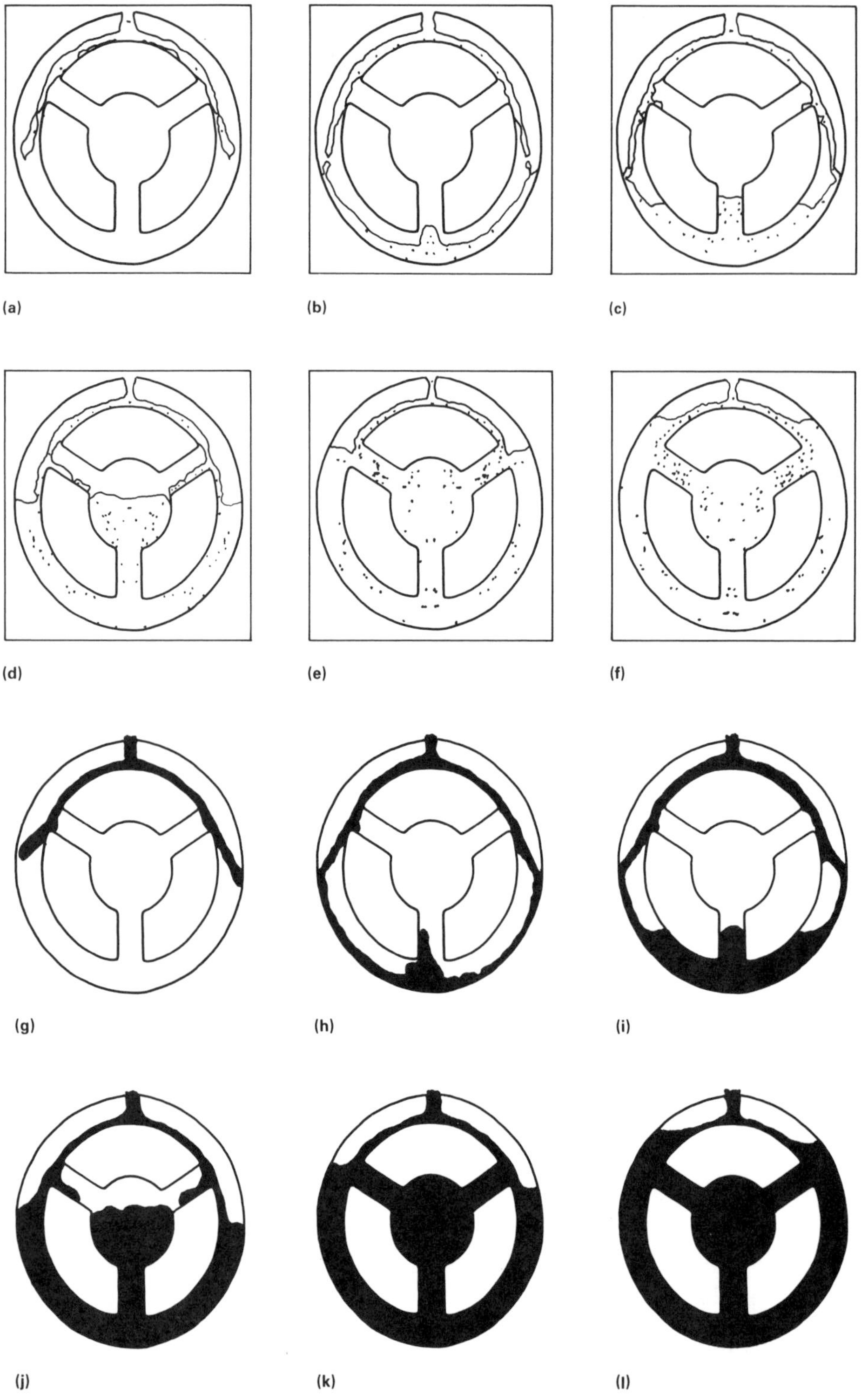

Fig. 18 Calculated flow patterns, (a) to (f), and comparable high-speed motion picture frame tracings, (g) to (l), of molten cast iron entering a top-gated sand mold to produce a three-spoke wheel. Elapsed times for the calculated patterns. (a) 0.250 s. (b) 0.600 s. (c) 0.900 s. (d) 1.600 s. (e) 2.650 s. (f) 3.000 s

In the *y*-direction:

$$\frac{\partial v}{\partial t} + u\frac{\partial v}{\partial x} + v\frac{\partial v}{\partial y} = -\frac{\partial p}{\partial y} + \nu\left(\frac{\partial^2 v}{\partial x^2} + \frac{\partial^2 v}{\partial y^2}\right) + g_y \qquad \text{(Eq 13)}$$

where t is time, p is pressure/density, ν is viscosity/density, and g is the acceleration of gravity. Equations 12 and 13 are commonly called the Navier-Stokes equations. In case of turbulent flow, ν may be replaced by the effective viscosity. The velocity and pressure fields for the interior region are calculated using these three governing equations and the finite-difference technique.

Surface Region Flow Behavior. In the surface region, the momentum balance principle, and thus Eq 12 and 13, still applies. However, Eq 11 is not valid because the mass within the cells of the surface region is changing. Instead, the surface region contains the interface between the molten metal and the atmosphere surrounding it, and free surface boundary conditions should be obeyed. This means that:

Tangential Stress on the Free Surface Should Vanish. This may be expressed by:

$$\nu\left[2n_x m_x \frac{\partial u}{\partial x} + (n_x m_y + n_y m_x)\left(\frac{\partial u}{\partial y} + \frac{\partial v}{\partial x}\right) + 2n_y m_y \frac{\partial v}{\partial y}\right] = 0 \qquad \text{(Eq 14)}$$

where n_x, n_y, m_x, m_y are the x and y components of the unit vectors normal to and tangent to the free surface, as shown in Fig. 9.

Normal Stress Should Balance the Applied Pressure Plus the Surface Tension. This may be expressed by:

$$2\nu\left[n_x n_x \frac{\partial u}{\partial x} + n_x n_y\left(\frac{\partial u}{\partial y} + \frac{\partial v}{\partial x}\right) + n_y n_y \frac{\partial v}{\partial y}\right] = p + p_s \qquad \text{(Eq 15)}$$

where p_a is the applied gas pressure in the empty region and p_s is the surface tension pressure. The flow field in the surface region is calculated using these principles.

Fluid Flow Phenomena in the Filling of Metal Castings

Fluid flow calculations can help gain an understanding of flow phenomena occurring during mold filling (Ref 10). To illustrate this, the results of MAC calculations applied to some simple mold designs are presented below. All of the designs have thin cross sections so that the metal flow is restricted to two dimensions. (Although the MAC technique can be applied to three dimensions, it requires a lot of computer time, so most of the computations of metal flow have been done in two dimensions.)

Example 1: Filling of a Large, Horizontal Square Plate Casting. The first case is a horizontal square plate 610 × 610 mm (2 × 2 ft), with a 122 mm (0.4 ft) wide ingate at the center of the left wall, through which metal enters at 305 mm/s (1 ft/s). The mold fills in 10 s. For numerical analysis, the casting was divided into 400 square cells (20 in each direction). Results of the computation (Fig. 10) show that as the metal enters the mold, the stream expands slightly before reaching the far wall. Upon reaching the wall, the stream splits in two, building up along the far side of the cavity and then reflecting back toward the ingate. The two vortices that form, one on either side of the gate, are the last regions to fill, and any gases in the mold are squeezed to these areas near the end of the filling process.

Example 2: Filling of a Small, Horizontal Square Plate Casting. The second casting is similar but smaller, and it demonstrates the effect of size (and/or relative entrance velocity) on the flow pattern (Fig. 11). The cavity is 61.0 × 61.0 mm (0.2 × 0.2 ft) with an ingate 12.2 mm (0.04 ft) wide. The entrance velocity, again, is 305 mm/s (1 ft/s), so it fills in 1.0 s. For computation, the system is divided into 400 cells. The flow pattern is quite different. Metal enters the mold like a jet, hitting the opposite wall before spreading. When the jet hits the wall, it flows rapidly outward in two symmetrical streams that cling much closer to the wall than in the previous case. These streams race back against the side walls, then along the ingate wall, where they encounter the incoming jet. In contrast to Example 1, the last areas to fill are much farther from the ingate, and the vortices, one on either side of the centerline, are somewhat larger. Also, the calculations show that the initial momentum is directed at a small area of the opposite wall, resulting in high pressures and the possibility of mold erosion.

Example 3: Filling of a Vertical Square Plate Casting to Illustrate the Effect of Gravity on Flow Pattern. The third design demonstrates the capability of the model to include the influence of gravity, or other body forces. It is a vertical square plate 152 × 152 mm (0.5 × 0.5 ft) with a 38.1 mm (0.125 ft) wide ingate located at the lower left corner. The velocity varies from 975 mm/s (3.2 ft/s) to 518 mm/s (1.7 ft/s), depending on the metal level and gas pressure in the mold. The entrance velocity is determined by coupling a Bernoulli equation calculation for flow through the filling system with the MAC calculation for flow within the mold. This casting also was divided into 400 equal-size square cells for computation. The results of the simulations (Fig. 12) show the melt enters as a bore with a sloping top surface. When it hits the opposite wall, it jumps up the wall about 0.8 of the distance to the top, rolling back over on the incoming stream and forming a strong vortex. Then it jumps up the wall above the ingate, and a wave works its way from side to side. The last area to fill is the top left corner. This flow behavior has been verified by high-speed movies of a water model, some frames of which are shown in Fig. 13. An actual casting of Al-7.5Si alloy made in a sand mold of this design showed low density in the area where the vortex was last located and entrapped gas porosity in the last area to fill. A photograph and a radiograph of this casting are shown in Fig. 14.

Example 4: Filling of a Vertical Stepped Plate Casting. The design and dimensions of a somewhat more complicated mold are shown in Fig. 15. It is a plate casting with steps on the bottom. The entrance velocity starts at 975 mm/s (3.2 ft/s) and gradually decreases as the level of molten metal and the gas pressure build up in the mold. The calculated results shown in Fig. 16 indicate that the metal forms a vortex beneath the ingate (which is located above the bottom in this case) and standing waves above each of the steps. Before the flow reaches the far end of the mold, metal begins to rise in the vertical section (which represents a riser), and another vortex forms in this region. This flow pattern has been verified by high-speed movies of a water model.

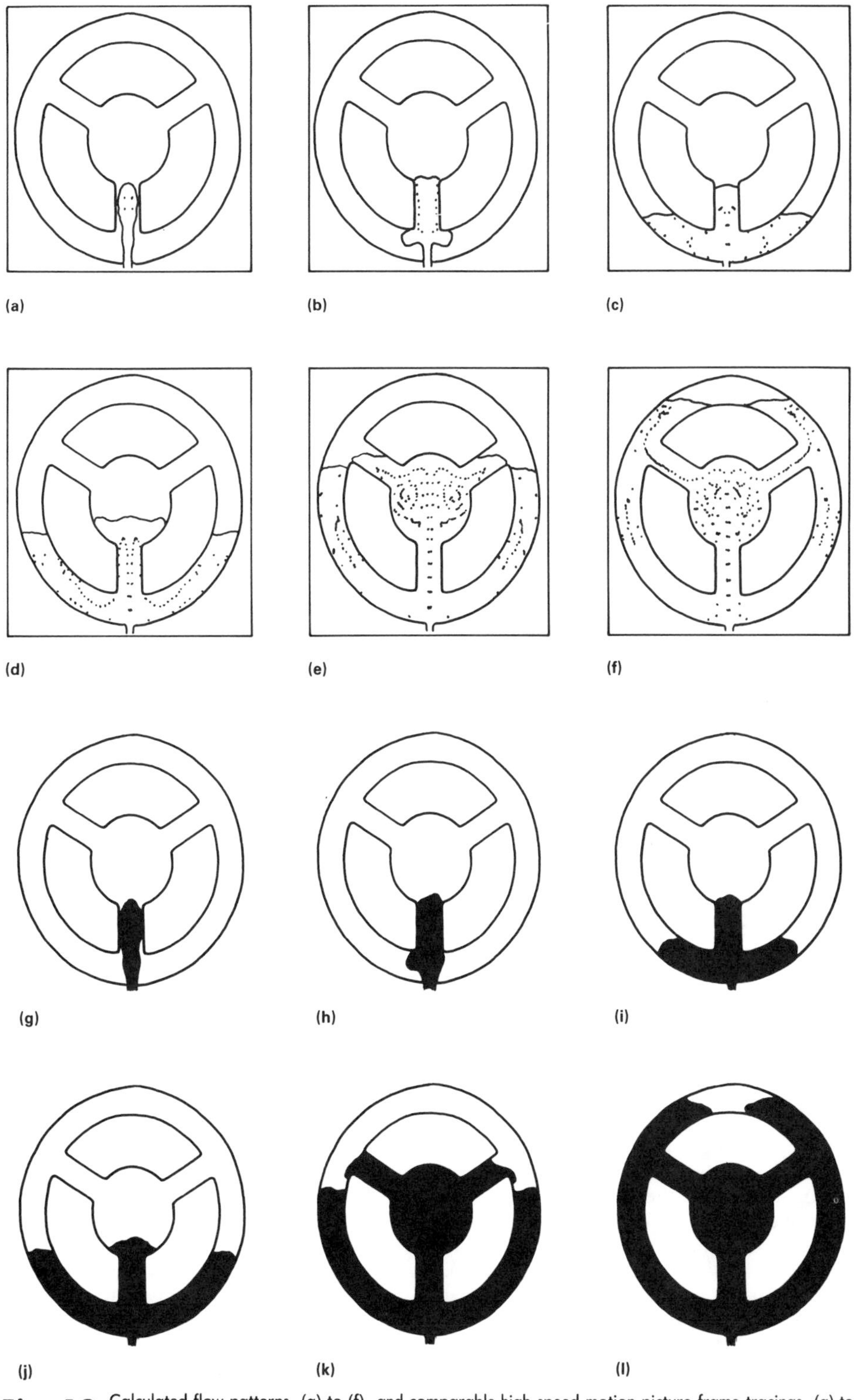

Fig. 19 Calculated flow patterns, (a) to (f), and comparable high-speed motion picture frame tracings, (g) to (l), of molten cast iron entering a bottom-gated sand mold to produce a three-spoke wheel. Elapsed times for the calculated patterns. (a) 0.100 s. (b) 0.200 s. (c) 0.500 s. (d) 1.000 s. (e) 2.200 s. (f) 2.900 s

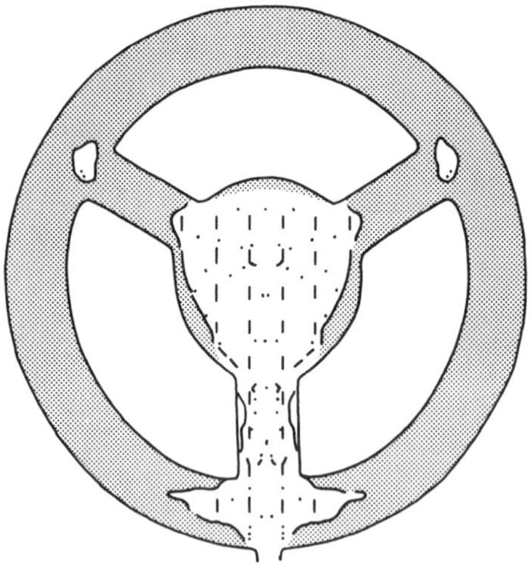

Fig. 20 Calculated solidification pattern in the bottom-gated three-spoke wheel casting in Fig. 19 after an elapsed time of 36.192 s. Fraction solid is greater than 0.5 in shaded area and less than 0.5 in unshaded area to show the residual effect of fresh hot metal flow through the bottom spoke throughout the filling period.

The MAC technique can be applied to flow through the filling system as well as in the mold cavity. Figure 17 shows flow in such a system and demonstrates the undesirable condition in which the descending metal pulls away from the walls of an untapered sprue. Some detailed studies of flow in the sprue and runner have been made using the SOLA-VOF version of these techniques (Ref 20).

Correlations With Actual Metal Flow

Correlations of calculated flow with observations of actual metal flow are shown in Fig. 18 and 19 (the flow of cast iron into silicate-bonded sand molds was recorded by a high-speed motion picture camera through a glass window on the side of the mold) (Ref 21). The casting is a three-spoke wheel. In one of the configurations, the metal enters through a gate at the top, and the gate is positioned midway between two spokes (Fig. 18). In the other, the gate is at the bottom directly opposite a spoke (Fig. 19). The results, calculated using a modified SOLA-VOF technique, are shown. The agreement between the observed patterns and the calculated ones is very good, particularly in the way in which the metal misses the diagonal spokes in the top-gated castings and pours around the rim until filling is about two-thirds complete. It can clearly be seen from these experiments that the bottom-gated casting fills more smoothly and uniformly and probably results in a better-quality casting.

Figure 20 shows the solidification pattern that occurs in the bottom-gated casting 36 s after the start of the pour. The residual influence of the entering stream of hot metal is shown by the molten state of the bottom spoke after the two upper spokes have solidified. These calculations were done by a program that uses SOLA-VOF to calculate fluid flow and couples it with a finite-difference heat transfer program to calculate temperature distributions. This is particularly useful in modeling thin castings, in which solidification may begin before the mold is completely filled.

REFERENCES

1. W.S. Hwang and R.A. Stoehr, "Fluid Flow Modeling for Computer-Aided Design of Castings," *J. Met.*, Vol 35, Oct 1983, p 22-30
2. G.H. Geiger and D.R. Poirier, chapters 3 and 4 in *Transport Phenomena in Metallurgy*, Addison-Wesley, 1973
3. L.F. Moody, Friction Factors in Pipe Flow, *Trans. ASME*, Vol 66, 1944, p 671-684
4. L.F. Moody, An Approximate Formula for Pipe Friction Factors, *Mech. Eng.*, Vol 66, 1947, p 1005-1006
5. S.I. Karsey, *Ductile Iron III, Gating and Risering*, Fer et Titane, 1986
6. D.H. St. John, K.G. Davis, and J.G. Magny, "Computer Modelling and Testing of Fluid Flow in Gating Systems," Internal Report MRP/PMRL 80-12(J), Energy, Mines, and Resources, Canmet, 1980
7. *Basic Principles of Gating and Risering*, Cast Metals Institute, American Foundrymen's Society, 1973
8. G.H. Geiger and D.R. Poirier, chapter 1 in *Transport Phenomena in Metallurgy*, Addison-Wesley, 1973
9. K. Grube, J.G. Kur, and J.H. Jackson, "The Effect of Gating and Risering on Casting Quality," Film produced by Battelle Memorial Institute, for the American Foundrymen's Society
10. R.A. Stoehr and W.S. Hwang, Modeling the Flow of Molten Metal Having a Free Surface During the Entry Into Molds, in *Proceedings of the Engineering Foundation, Modeling and Control of Casting and Welding Processes, II*, The Metallurgical Society, 1983
11. "Water Analogy Studies—Flow and Gating of Castings," Film produced by Case Institute of Technology, for the Training and Research Institute, American Foundrymen's Society, and the Die Casting Foundation, Inc.
12. M.C. Ashton and R.K. Buhr, "Direct Observation of the Flow of Molten Steel in Sand Molds," Internal Report PM-M-73-5, Energy, Mines, and Resources, Canmet, 1973
13. S.T. Andersen and P. Ingerslev, "A Study of Pouring a Symmetrical Casting by Means of Film Shots and Pressure Measurements," Paper presented at the 50th World Foundry Congress, Cairo, 1983
14. C. Galaup, U. Dieterle, and H. Luehr, "3-D Visualization of Foundry Molds Filling," Paper presented at the 53rd World Foundry Congress, Prague, 1986
15. R. Hamar, "Optimal Gating of Thin-Wall Parts," Paper presented at the 53rd World Foundry Congress, Prague, 1986
16. J.E. Welch, F.H. Harlow, P.J. Shannon, and B.T. Dally, "The MAC Method—A Computing Technique for Solving Viscous, Incompressible, Transient Fluid Flow Problems Involving Free Surfaces," Technical Report LA-3425, Los Alamos Scientific Laboratory, 1965
17. A.A. Amsden and F.H. Harlow, "The SMAC Method, A Numerical Technique for Calculating Incompressible Flows," Technical Report LA-4370, Los Alamos Scientific Laboratory, 1970
18. B.D. Nichols, C.W. Hirt, and R.S. Hotchkiss, "SOLA-VOF, A Solution Algorithm for Transient Fluid Flow With Multiple Free Boundaries," Technical Report LA-8355, Los Alamos Scientific Laboratory, 1980
19. R.J. Roache, *Computational Fluid Dynamics*, Hermosa, 1976
20. H. Walther and P.R. Sahm, A Model for the Computer Simulation of Flow of Molten Metal Into Foundry Molds, *Giessereiforschung*, Vol 38, 1986, p 119-124 (in German)
21. R.A. Stoehr and P. Ingerslev, "Flow Analysis of Mold Filling Using Marker-and-Cell," Publication TM 86.09, Laboratory for Thermal Processing, Process Technical Institute, Technical University of Denmark, 1986

Modeling of Combined Fluid Flow and Heat/Mass Transfer

Prateen V. Desai, Georgia Institute of Technology
K.V. Pagalthivarthi, GIW Industries, Inc.

FLUID FLOW AND HEAT/MASS TRANSFER principles are increasingly gaining acceptance as a means of improving the quality and yield of castings. The benefits to be derived from adopting such an approach range from slag- and dross-free gating system design to a desired microstructure of the finished product. In general, the transport of heat, mass, and momentum during solidification processing controls such varied phenomena as solute macrosegregation, distribution of voids and porosity, shrinkage effects, and overall solidification time. These parameters, in turn, result in a variation of the mechanical, thermophysical, and electrical properties of the solidified product.

The complex nature of the coupling between heat and mass transport with fluid flow during solidification necessitates a fundamental understanding of the processes and the mechanisms of interaction in relation to empirical formulas and charts. Heat transfer by forced convection predominates during the filling stages. Once the mold cavity is filled, buoyancy-generated natural convective heat and mass transfer occur before the phase change (Ref 1).

The principles of heat transfer by forced convection are shown schematically in Fig. 1, which is a representation of flowing metal at a superheated pouring temperature T_{01} and a velocity u advancing into a mold channel of width $2d$ and length L, initially at an ambient temperature T_{02}. Stage 1 shows the channel just before the liquid metal enters. Stages 2 and 3 show the liquid region R_I occupying half and almost full lengths of the channel, respectively. The temperature at the liquid metal/mold wall interface keeps evolving as the flowing metal front advances into the channel. The portion of the mold wall not yet covered by the flowing metal remains at a considerably lower temperature. Stage 4 shows the completely filled channel with conventional steady flow and heat transfer processes.

Subsequent stages during the solidification of a binary alloy involve both phase change heat and mass transfer as well as buoyant thermosolutal convection (Ref 3).

Figure 2 shows a schematic representation of the stages in the solidification of a binary alloy. Solidification begins with cooling across boundary B_{II} between the liquid metal region R_{IV} and the mold region R_I, together with cooling at boundary B_I between the mold and the ambient (Fig. 2a). Next, the solid/liquid mushy region R_{III} evolves between boundary B_{II} with the mold and B_{IV} with the liquid metal (Fig. 2b). Further cooling leads to a typical steady-state picture showing a solidified crust region R_{II} with boundaries B_{II} and B_{III} with the mold and the mush, respectively, and the mushy region R_{III} with boundaries B_{III} and B_{IV} with the crust and the liquid, respectively (Fig. 2c). The final transient stage shows the depletion of the all-liquid region R_{IV} (Fig. 2d).

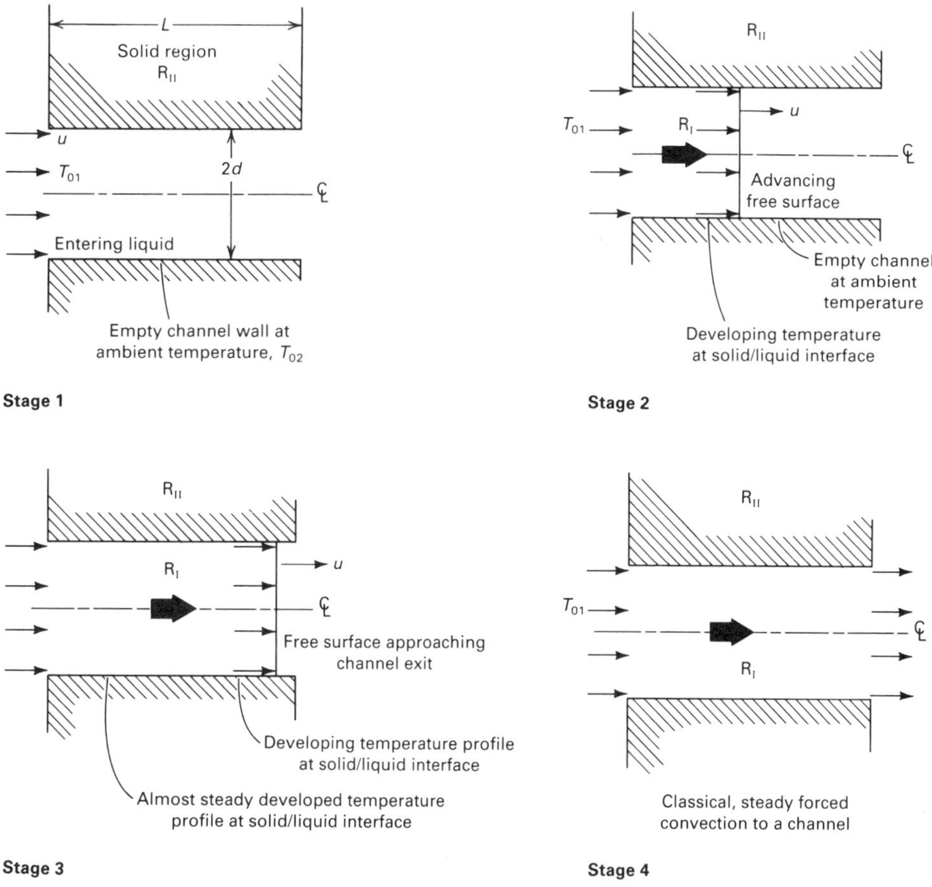

Fig. 1 Filling stages in an empty channel. See the corresponding text for a description of Stages 1 through 4. Source: Ref 2

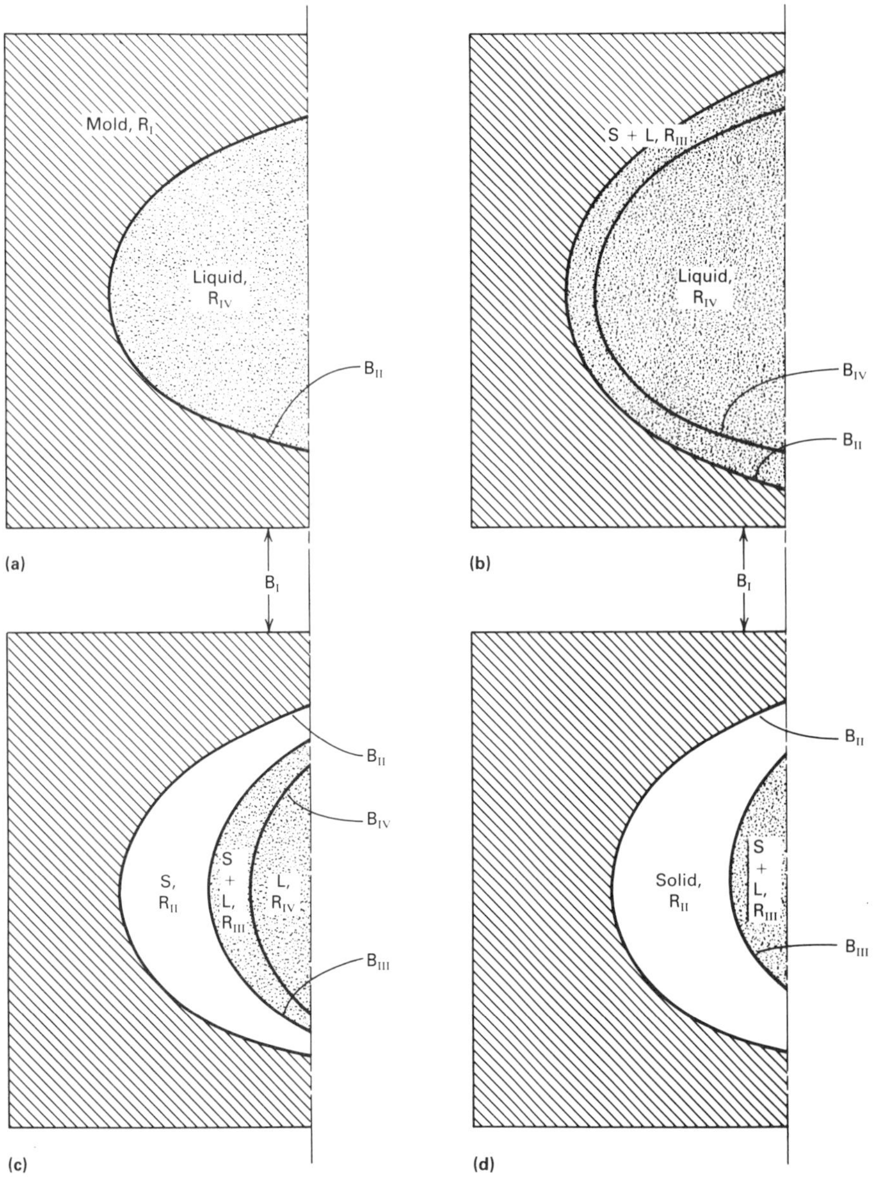

Fig. 2 Stages in binary alloy solidification. (a) All liquid. (b) Initial transient. (c) Steady state. (d) Final transient. See the corresponding text for details. Source: Ref 3

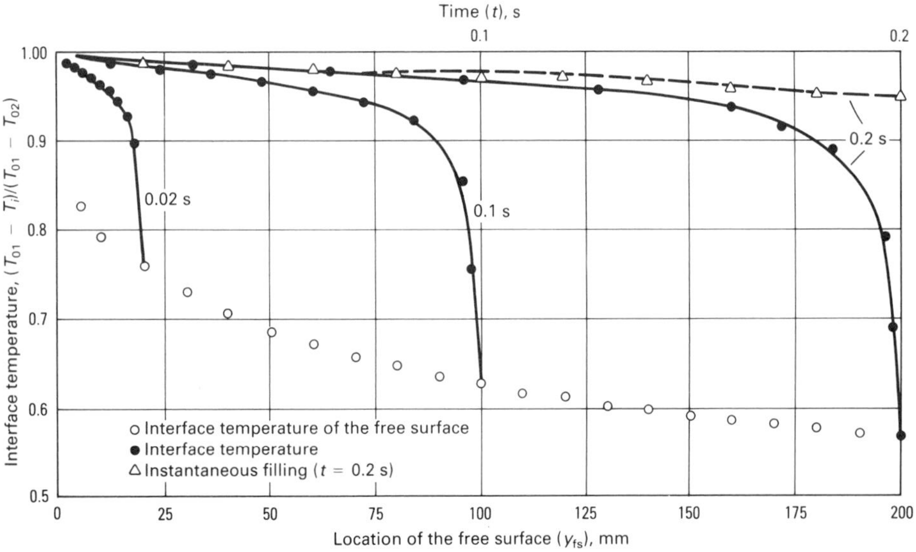

Fig. 3 Interface temperature distribution in a filling channel

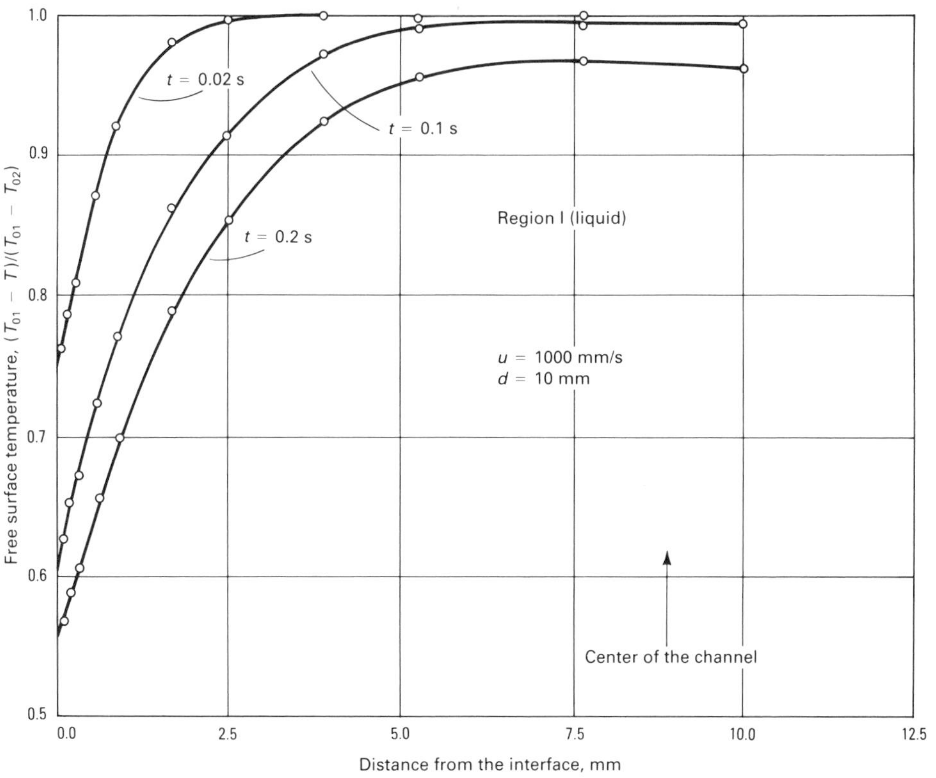

Fig. 4 Transient temperature profiles across filling channel

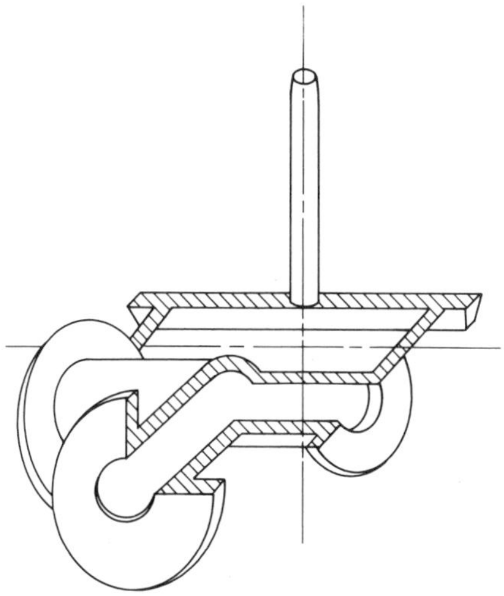

(a)

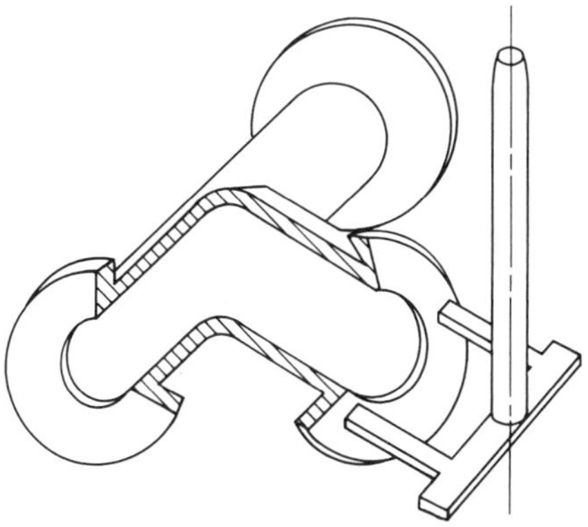

(b)

Fig. 5 Gating systems used during the experiments on mold/metal interface temperature variations. (a) Gating system I; metal enters from side flange. (b) Gating system II; metal enters from middle flange. See also Table 1 and Fig. 6. Source: Ref 8

Table 1 Summary of experimental conditions

See also Fig. 5 and 6, which show the gating systems used and the mold-metal temperature variations, respectively.

Gating system	Test condition	Height/length mm	Height/length in.	Section size mm	Section size in.	Area mm²	Area in.²	Ratio
Sprue								
I	1	300	12	14 × 14	0.55 × 0.55	196	0.30	1
I	2	300	12	31 × 22(a)	1.22 × 0.87	380	0.59	2
I	3	300	12	37.5 × 30(a)	1.48 × 1.18	706	1.09	2
II	4	300	12	14 × 14	0.55 × 0.55	196	0.30	1
II	5	300	12	31 × 22(a)	1.22 × 0.87	380	0.59	2
II	6	300	12	37.5 × 30(a)	1.48 × 1.18	706	1.09	2
Runner								
I	1	400	16	16/8 × 16	0.63/0.32 × 0.63	192 × 2	0.29 × 2	2
I	2	400	16	16/8 × 16	0.63/0.32 × 0.63	192 × 2	0.29 × 2	2
I	3	400	16	20/10 × 20	0.79/0.40 × 0.79	300 × 2	0.47 × 2	2
II	4	180	7	16/8 × 16	0.63/0.32 × 0.63	192 × 2	0.29 × 2	2
II	5	180	7	16/8 × 16	0.63/0.32 × 0.63	192 × 2	0.29 × 2	2
II	6	180	7	20/10 × 20	0.79/0.40 × 0.79	300 × 2	0.47 × 2	2
Ingate								
I	1	35	1.4	11/9 × 10	0.43/0.35 × 0.39	100 × 2	0.16 × 2	1
I	2	35	1.4	11/9 × 10	0.43/0.35 × 0.39	100 × 2	0.16 × 2	1
I	3	35	1.4	11/9 × 15	0.43/0.35 × 0.59	150 × 2	0.24 × 2	1
II	4	110	4.4	11/9 × 10	0.43/0.35 × 0.39	100 × 2	0.16 × 2	1
II	5	110	4.4	11/9 × 10	0.43/0.35 × 0.39	100 × 2	0.16 × 2	1
II	6	106	4.2	11/9 × 15	0.43/0.35 × 0.59	150 × 2	0.24 × 2	1

(a) Diameters of round section. Source: Ref 8

Heat Loss During Filling

Pioneering work on heat loss from the flowing metal to the sand mold runners consisted of obtaining plots of temperature loss versus time, with the ratio of surface area to flow rate used as a parameter (Ref 4). Assuming instantaneous filling and negligible contact resistance between the sand mold and the flowing metal, this method gives good results. A modified version of the method, intended for a constant or linearly decreasing average velocity of the molten metal, indicates that the temperature loss in the runner decreases with time and is proportional to the residence time of the fluid element in the runner (Ref 5). These results can be used to calculate temperature loss for a variety of runner lengths, flow rates, runner diameters, pouring temperatures, and sand mold thermal properties. However, the results are not accurate for metallic molds, nor for very short times after pouring.

Based on more detailed mathematical models of the transient filling process, finite-element solutions for the mold/metal interface temperature for very short times after pouring have been developed (Ref 6). Typical results of such calculations for a horizontal runner channel are shown in Fig. 3. In this case, the maximum temperature

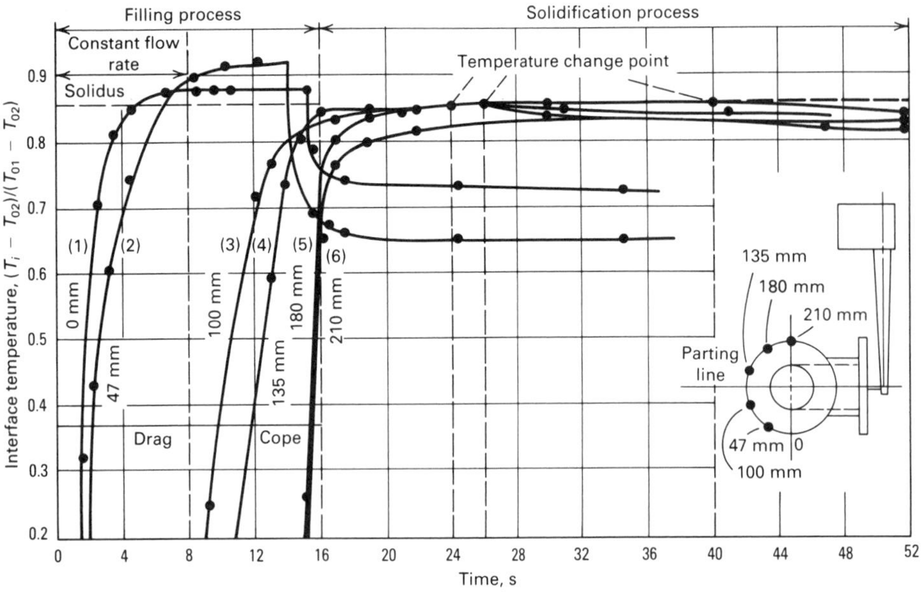

Fig. 6 Mold/casting interface temperature variation at different locations during filling and solidification. Parenthetical values represent the six test conditions described in Table 1. See also Fig. 5. Source: Ref 8

difference in the system, that is, the difference between the pouring temperature and the initial mold temperature, is represented by $T_{01} - T_{02}$. Figure 3 shows that the instantaneous filling assumption (open triangles) yields a midchannel value of wall temperature at a time of 0.1 s after pouring as $[T_{01} - 0.96(T_{01} - T_{02})]$, while the corrected value after accounting for the fluid flow (convection) effects during filling (closed circles) is $[T_{01} - 0.64(T_{01} - T_{02})]$. Although this may indicate an instantaneous freezing, the subsequent flow of superheated metal causes remelting. The temperature profiles across the channel at various times are shown in Fig. 4. Similar calculations have been performed for the vertical gating sprue for a variety of geometries (lengths, tapers, and diameters), flow rates, pouring temperatures, and properties of the melt and the sand mold (Ref 7). The competing effects of conduction heat transfer to the mold and convection due to the flow determine the interface temperature. In other words, both the thermal conductivity of the sand and the thermal capacity of the flow are important.

Experimental verification of the calculated results during the transient filling of a gating system for eutectic iron sand castings is provided in Ref 8. Temperature variations at the mold/metal interface and along the centerline of the longitudinal flow section have been recorded for sprues, runners, and ingates. Table 1 lists the experimental conditions. The furnace charge and chemical analysis of the resulting alloy are shown below:

Constituent	Charge 1	Charge 2
Brazil pig iron, kg (lb)	56 (123)	60 (132)
Steel billet, kg (lb)	24 (53)	20 (44)
Carbon powder, g	980	800
Ferrosilicon, g	32	1000

Charge number	Composition C	Si	Mn	S	P
8590(a)	3.64	2.24	0.177	0.020	0.01
8576(a)	3.60	2.13	0.45	0.009	0.06
8570(a)	3.30	2.19	0.495	0.010	0.08

(a) Balance iron

Figure 5 shows the two gating systems used in the experiments. Typical results for the interface temperature variation at different locations are shown in Fig. 6 for the filling and solidification stages. The number beside each curve indicates the height of the measurement point from the bottom to the top of the melt. The earlier the interface is in contact with the hot melt, the earlier the temperature changes. Most interface temperatures are near the solidus value when the filling is complete. Therefore, any subsequent solidification calculations must include the initial temperature distribution.

Postfilling Buoyant Convection

The loss of liquid metal superheat in the casting cavity of the mold after the filling transients have died out occurs by buoyancy-generated convection currents. These currents tend to redistribute the melt temperature and composition until solidification begins. Subsequent solidification sequences (Fig. 2) also involves heat loss by thermosolutal buoyant convection during the phase change. Figure 7(a) shows patterns of calculated convection currents in a pure melt for a vertical rectangular cavity. The indicated Rayleigh number in Fig. 7(a) characterizes the strength of the buoyant transport in relation to that by pure diffusion. Although miniscule in comparison with the patterns of flow during filling, these currents significantly shift the hot metal to the top and then redistribute the thermal state of the melt before solidification begins. A typical upward shift of the hot isotherm is shown in Fig. 7(b). Any subsequent simulation of the solidification sequence must account for this initial temperature distribution within the cavity.

In the case of alloy melts, the difference in atomic weight of the constituent metals causes an additional convection pattern. The temperature and solute redistributions due to buoyant thermal convection occur in a coupled fashion, each driving and being driven by the other. Both the scale of the convective motion and the time during the solidification sequence when buoyant convection becomes significant must be determined to identify the process parameters governing subsequent microstructural behavior. Mathematical models of the buoyant transport processes during binary alloy solidification can be examined to obtain the applicable time, length, and motion scales for the liquid, the solid, and the mushy zones depicted in Fig. 2. Of these, the most crucial issue from the metallurgical viewpoint is the evolution of the mushy zone and the factors that govern it.

The mushy zone in a tall vertical sample casting (Fig. 8a) is viewed as a forest of dendrites submerged in a melt pool and is characterized by a field distribution of the solidified mass fraction ϕ. The solid-liquid mixture under local thermodynamic phase equilibrium is in turn characterized by the pressure, the temperature, and the average mixture composition. The lever rule then yields the local liquid and solid concentrations. Balance statements for mass, momentum, species, thermal energy, and vorticity for a variable-mass-fraction partially solidified mixture of this type, together with the liquidus equation, are detailed in Ref 3. The buoyant flow can be driven by heat or mass transfer, depending on the magnitude of the buoyancy parameter N:

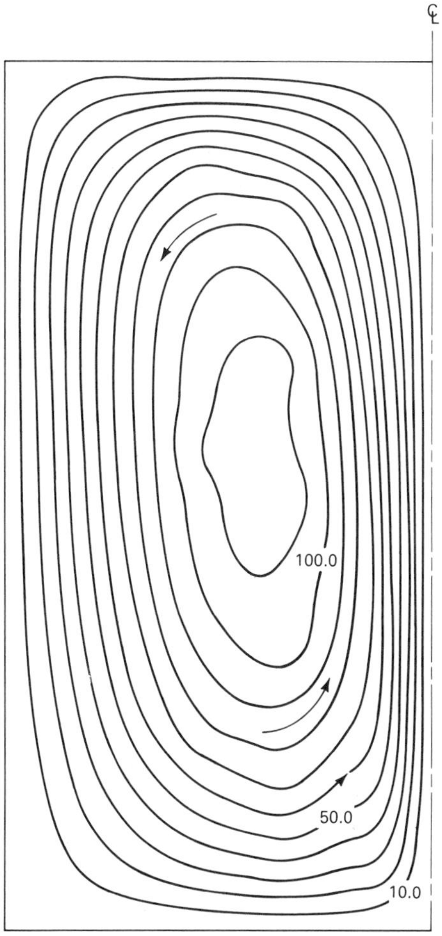

Fig. 7(a) Final steam function distribution in cavity filled with liquid copper. Cavity aspect ratio is 2, and the Rayleigh number of the process is 7000. Only the left half of the two symmetric halves of the cavity is shown. Moving horizontally to the right from center of the illustration to the centerline of the cavity, the distance between the adjacent stream lines decreases, indicating faster upward flow between adjacent stream lines. The centerline forms an insulating boundary between the two symmetric halves of the cavity.

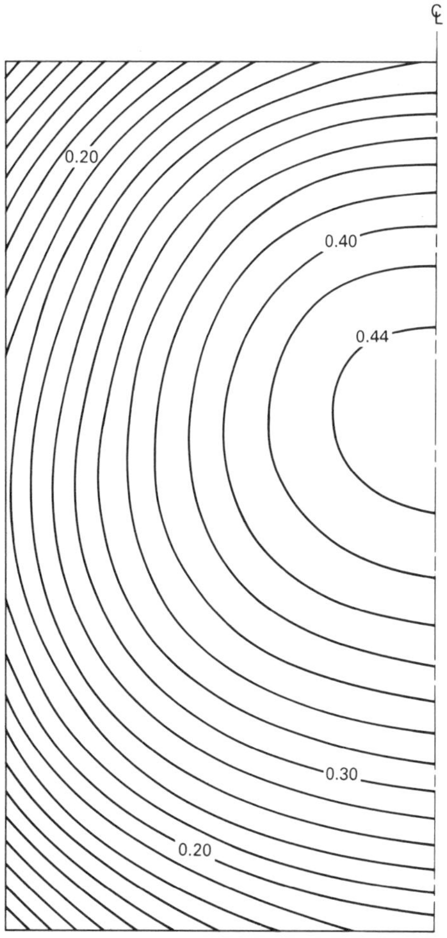

Fig. 7(b) Final temperature distribution in cavity filled with liquid copper. Cavity aspect ratio is 2, and the Rayleigh number of the process is 7000. Only the left half of the two symmetric halves of the cavity is shown. The curve labeled 0.44 is the hottest of the isotherms shown. Its temperature excess over the solidification temperature equals 44% of the maximum temperature difference that existed between the superheated liquid and the solidification temperature at the start of the natural convection process in the cavity.

$$N = \frac{\beta_C \, \Delta_C}{\beta_T \, \Delta_T}$$

where β_C is the fractional change in density due to concentration changes and β_T is the fractional change in density due to temperature changes. However, for most cases of binary alloy solidification of practical interest, the buoyant current is mass transfer driven.

The boundary layer provides concentration, temperature, and velocity in a mass transfer driven flow for the tall, vertical casting, as shown in Fig. 8(b) and (c). In the boundary layer regime, the orders of magnitude x, y, ΔT, and ΔC can be represented as:

$$x \sim B,\ y \sim \delta_C,\ \Delta T \sim (T_0 - T_c),\ \Delta C \sim C_0 - C_e$$

where the subscripts 0, c, and e are the initial, liquidus, and eutectic states, respectively. For short times after solidification begins, all transport processes are diffusive, until a time t_f, when convection overtakes diffusion. A scale analysis of the mixture equations for the balance of mass, species, and vorticity yields order of magnitude of this time (the end of initial diffusion transport period) as:

$$t_f \sim \frac{B^2}{D(R_{SB})^{2/5}}$$

where B is the semi-width of the casting, D is the mass diffusivity of the alloy mixture, and R_{SB} is the solutal Rayleigh number, which is defined as:

$$R_{SB} = \frac{g\beta_C \Delta C B^3}{\nu D}$$

where ν is the kinematic viscosity and g is the acceleration due to gravity. The order of magnitude of the concentration boundary layer thickness at this time is given as:

$$\delta_{Cf} \sim B(R_{SB})^{-1/5}$$

and the steady-state mass transfer rate in the mushy zone is estimated from the Sherwood number, Sh, as:

$$\mathrm{Sh} \sim (R_{SB})^{1/5}$$

The corresponding values for the thermal boundary layer thickness and the Nusselt number for heat transfer rate calculations in the mushy zone can be obtained by a similar scale analysis of the thermal energy equation and the vorticity transport equation. Detailed computations for the actual values for alloy solidification are being performed and are expected to appear in the open literature in the near future.

REFERENCES

1. P.V. Desai and F. Rastegar, Convection in Mold Cavities, in *Modeling of Casting and Welding Processes*, H.D. Brody and D. Apelian, Ed., The Metallurgical Society, 1981, p 351-359
2. P.V. Desai *et al.*, Computer Simulation of Forced and Natural Convection During Filling of a Casting, Paper 97, *Trans. AFS*, 1984, p 519-528
3. K. Pagalthivarthi and P.V. Desai, Modeling Thermosolutal Convection in Binary Alloy Solidification, in *Modeling and Control of Casting and Welding Processes*, S. Kou and R. Mehrabian, Ed., The Metallurgical Society, 1986, p 121-132
4. J.W. Hlinka *et al.*, How Much Super-Heat is Lost in the Runner, Paper 69, *Trans. AFS*, 1961, p 527-534
5. E.W. Jones *et al.*, Heat Transfer From Molten Metals to Sand Mold Runners, Paper 71, *Trans. AFS*, 1963, p 817-825
6. C.W. Kim *et al.*, Moving Free Surface Heat Transfer Analysis by Continuously Deforming Finite Elements, *Numer. Heat Transfer*, Vol 10, 1986, p 147-163
7. P.V. Desai *et al.*, The Thermal Performance of Gating Sprues in Sand Casting Systems, Paper 113, *Trans. AFS*, 1985, p 751-756
8. P.V. Desai *et al.*, Heat Transfer and Flow Experiments During Filling of Gating Systems, Paper 163, *Trans. AFS*, 1987, p 435-442

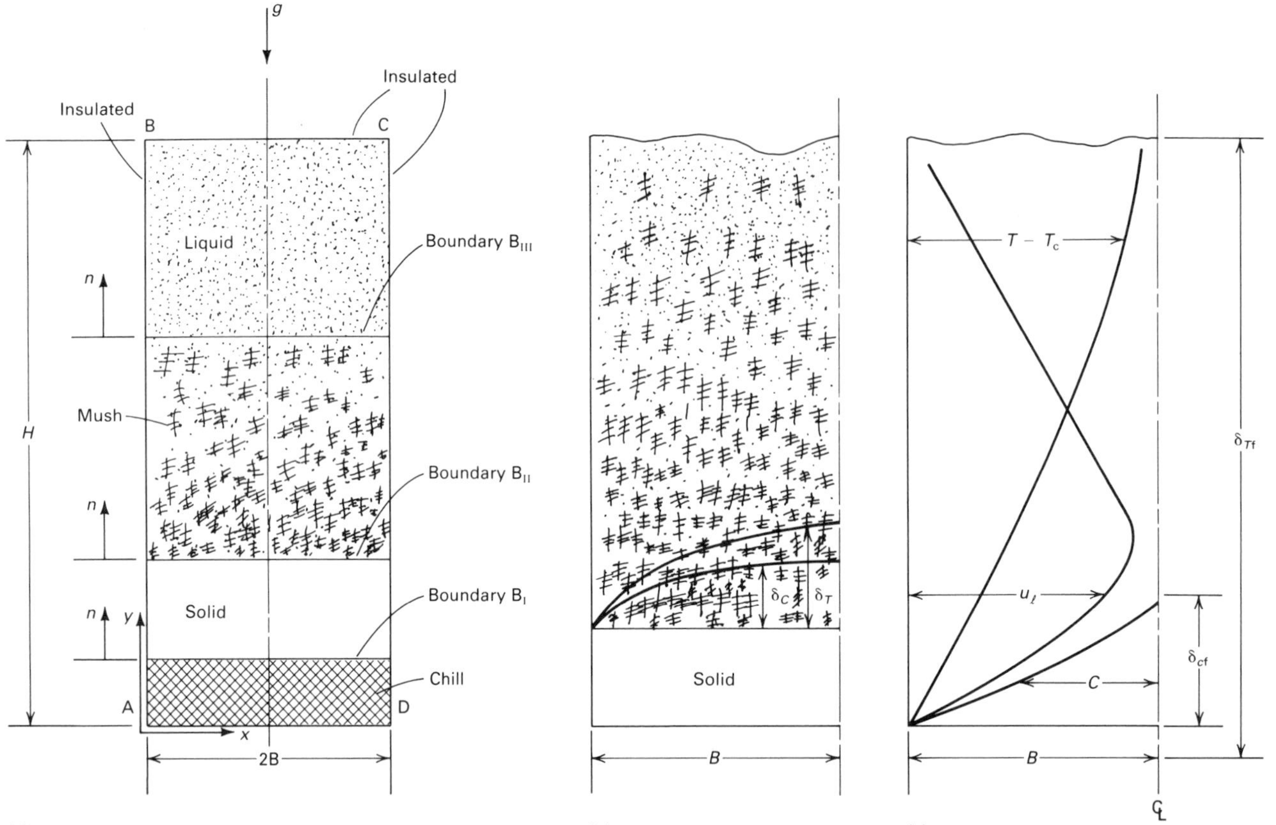

Fig. 8 Solidification of a tall, vertical casting. (a) Stages of a solidifying alloy. (b) Evolution of boundary layer in the mushy zone. (c) Boundary layer concentration, velocity, and temperature profiles

Modeling of Microstructural Evolution

M. Rappaz, Ecole Polytechnique Fédérale de Lausanne
D.M. Stefanescu, University of Alabama

THE MODELING OF SOLIDIFICATION of castings has received increased attention as the computer revolution has matured. The main application of this technique has traditionally been calculation of the path of the isotherms (lines of constant temperature) through shaped castings. In turn, this was used to predict the locations of hot spots in castings and thus to check, using the computer, a proposed gating and risering system, rather than following the classical trial-and-error technique used in foundries.

A pioneering paper by Henzel and Keverian (Ref 1) describing the application of the Transient Heat Transfer program was published in 1965. The Transient Heat Transfer program, a finite-difference method (FDM) program developed in 1959, was used for castings. Soon after this paper appeared, a variety of publications dealt with the use of different numerical techniques in the area of solidification of castings (Ref 2-5). The next logical step forward was to include nucleation and growth kinetics in simulations to generate information about the microstructure of the alloys and to treat the heat generation problem; although this step was envisioned by Oldfield as early as 1966 (Ref 6), progress in this direction was slow. A follow-up paper by Stefanescu and Trufinescu (Ref 7) in 1974 seems to be the only one until 1984, although nucleation and growth kinetics were used in modeling cooling curves to interpret inoculation in cast iron (Ref 8).

Beginning in 1984, a renewed interest in the modeling of microstructural evolution as part of the simulation of casting solidification is manifested in the literature. An analytical approach was applied by Fredriksson and Svensson (Ref 9) to the solidification of eutectic gray, ductile, and white iron and then extended to hypoeutectic irons and to the eutectoid transformation by Stefanescu and Kanetkar (Ref 10). Su *et al.* (Ref 11) used an inner nodal direct FDM scheme combined with a model assuming growth of graphite spheroids controlled by diffusion of carbon through the austenite shell. More recently, similar developments have been made for the dendrite growth of equiaxed grains (Ref 12-14). Hunt has also addressed the question of columnar-to-equiaxed transition (Ref 15).

This article will discuss techniques used for the simulation of solidification of castings. These techniques combine the modeling of heat transfer (macromodeling) with the modeling of microstructural evolution (micromodeling).

Macroscopic Modeling

Solidification of alloys is primarily controlled by heat diffusion and to some extent by convection within the liquid region. In most approaches to the solidification modeling of complex-shaped castings, the continuity equation of motion is not solved explicitly. Instead, correction is taken into account by increasing the heat conductivity above the melting point or the liquidus temperature. Under this assumption, the basic continuity equation governing solidification at the macroscopic scale is that of conservation of energy:

$$\text{div}\,[k(T)\cdot\overrightarrow{\text{grad}}\;T(\vec{x},t)] + \dot{Q} = \rho C_p\,(T)\,\frac{\partial T(\vec{x},t)}{\partial t} \qquad \text{(Eq 1)}$$

where $T(\vec{x},t)$ is the temperature field, $k(T)$ is the thermal conductivity, $\rho C_p(T)$ is the volumetric specific heat, and $\dot{Q}$ is the source term associated with the phase change. In solidification modeling, $\dot{Q}$ can be written as:

$$\dot{Q} = L\,\frac{\partial f_s\,(\vec{x},t)}{\partial t} \qquad \text{(Eq 2)}$$

where $f_s(\vec{x},t)$ is the solid fraction and L is the volumetric latent heat.

To solve Eq 1, a relationship between the fields $T(\vec{x},t)$ and $f_s(\vec{x},t)$ must be found. A simple and widely used approach is to assume that the fraction of solid f_s depends only on the temperature T and not upon cooling rate or growth rate. For pure metals or eutectic alloys, one can assume that f_s = 0 above the melting point or the eutectic temperature and that $f_s = 1$ below the equilibrium temperature. For dendritic alloys, various models of solute diffusion have been developed (Ref 16, 17). They all assume complete mixing of solute within the liquid, thus resulting in a unique $f_s(T)$ curve.

Assuming that f_s depends only on T, Eq 1 and 2 can be combined to give:

$$\text{div}\,[k(T)\cdot\overrightarrow{\text{grad}}\;T(\vec{x},t)] = \left[\rho C_p\,(T) - L\,\frac{df_s}{dT}\right]\frac{\partial T(\vec{x},t)}{\partial t} \qquad \text{(Eq 3)}$$

defining enthalpy H as:

$$H(T) = \int_0^T \rho C_p\,(\theta)\cdot d\theta + L\,[1 - f_s\,(T)] \qquad \text{(Eq 4)}$$

Equation 3 can also be written as:

$$\text{div}\,[k(T)\cdot\overrightarrow{\text{grad}}\;T(\vec{x},t)] = \frac{\partial H(\vec{x},t)}{\partial t} \qquad \text{(Eq 5)}$$

This is known as the enthalpy method.

An effective specific heat, ρC_p^*, can be derived from Eq 4:

$$\rho C_p^*\,(T) = \frac{dH}{dT} = \rho C_p\,(T) - L\,\frac{df_s}{dT} \qquad \text{(Eq 6)}$$

which, when introduced in Eq 3, gives:

$$\text{div}\,[k(T)\cdot\overrightarrow{\text{grad}}\;T(\vec{x},t)] = \rho C_p^*(T)\cdot\frac{\partial T(\vec{x},t)}{\partial t} \qquad \text{(Eq 7)}$$

This is known as the specific heat method.

The curves $H(T)$ and $\rho C_p^*(T)$ can be calculated by various methods, for example, using a Brody-Flemings model of solute diffusion (Ref 16, 17). A detailed comparison of the enthalpy method (Eq 5) and the effective specific heat method (Eq 7) is beyond the scope of this article.

Both methods can be used in macro-micro modeling. Other techniques, such as the latent heat method (Ref 18) or the micro-enthalpy method (Ref 19), have been specifically developed for macro-micro modeling. These methods, which are reviewed in Ref 20, will be discussed in the section ''Macro-Microscopic Modeling of Equiaxed Solidification'' in this article.

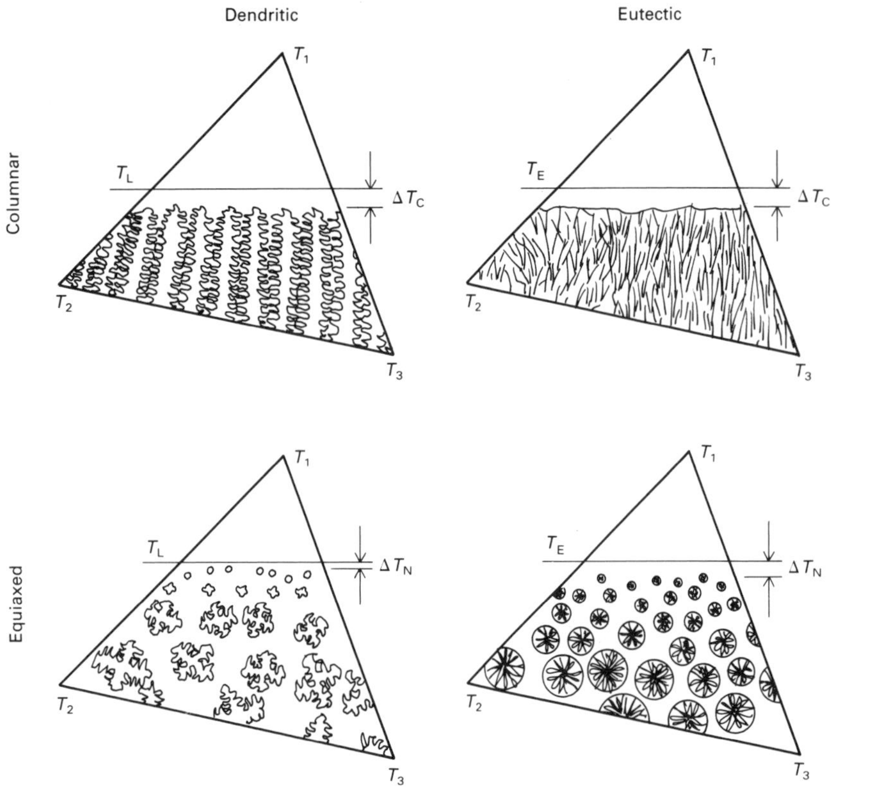

Fig. 1 Schematic of columnar and equiaxed growth of dendritic and eutectic alloys in a thermal gradient G

Modeling of Columnar Structures

The macroscopic approach described briefly in the previous section can be reasonably applied to columnar solidification because the growth rate of the microstructure (eutectic front or dendrite tips) is more or less equal to the speed at which the corresponding isotherms move (eutectic or liquidus isolines). Therefore, microstructural parameters and undercooling can be directly calculated from the temperature field in this case.

Columnar growth morphologies are encountered in both dendritic and eutectic alloys (Ref 16, 17). Solidification occurs in a columnar fashion when the growth speed of the dendrite tip or the eutectic front v_s is directly related to the speed v_m of the isotherms calculated from a macroscopic approach (Fig. 1). It is therefore necessary to have a positive thermal gradient G at the solid/liquid interface. However, this condition is not sufficient to ensure the formation of columnar structures (Ref 13, 15).

The competition between columnar and equiaxed morphologies, in particular the columnar-to-equiaxed transition, is analyzed in Ref 15. In a given macroscopic thermal environment, one can calculate:

- The undercooling ΔT_C associated with the formation of a columnar structure
- The nucleation undercooling ΔT_N at which nuclei are formed within the melt
- The undercooling ΔT_E required to drive equiaxed solidification

Under steady-state growth conditions and considering these three undercoolings, a simple criterion to obtain a fully columnar structure can be defined by:

$$G > A \cdot N_0^{1/3} \left[1 - \left(\frac{\Delta T_N}{\Delta T_C}\right)^3\right] \cdot \Delta T_C \qquad \text{(Eq 8)}$$

where N_0 is the density of grains nucleated at the undercooling ΔT_N, and A is a constant.

Assuming that the thermal gradient G is large enough to ensure that a columnar structure is produced, microstructure formation theories can be easily implemented into macroscopic heat flow calculations if one makes the following hypotheses:

- The kinetics of the eutectic front or the dendrite tip are given by the steady-state growth analysis
- The velocity of the microstructure v_s is related to the velocity v_m of the corresponding equilibrium isotherm, as shown in Fig. 2

In Fig. 2, four different microstructures frequently encountered in solidification are shown: regular and irregular eutectics and cellular and dendritic morphologies. In the first three cases, one has simply:

$$v_s = v_m \qquad \text{(Eq 9a)}$$

For dendritic alloys, the velocity of the dendrite tip is essentially dictated by the trunk orientation, which is imposed more or less by the crystallographic orientation of the solid (for example, ⟨100⟩ for cubic metals). If α is the angle between the trunk orientation and the heat flow direction, then:

$$v_s = \frac{v_m}{\cos \alpha} \qquad \text{(Eq 9b)}$$

In castings, grain selection will occur such that those grains whose angle α is close to zero will grow preferentially. However, dendritic single-crystal growth (Ref 21) or epitaxial dendritic growth from single-crystal substrates (Ref 22) can be characterized by an α value that can deviate substantially from zero.

Based on the two hypotheses mentioned previously, the kinetics of microstructure formation can be implemented into macroscopic heat flow calculations according to the following simple scheme. One first calculates the temperature field evolution without taking into account any undercooling (see the section "Macroscopic Modeling" in this article). Once the temperature field is known, the velocity of the corresponding isotherms (liquidus or eutectic temperature) can be deduced as well as the thermal gradient at the interface. From these values, the undercooling of the columnar microstructure and the associated parameters of the microstructure (eutectic or dendrite trunk spacings) can be calculated using recent theories of microstructure formation.

Calculation of the undercooling of columnar microstructures for one-dimensional heat flow is described in Ref 23. Nucleation and columnar growth have also been considered in the modeling of rapid solidification (Ref 24).

Two researchers have achieved a real coupling between dendritic microstructure formation theory and one-dimensional nonstationary heat flow calculations in the case of spot laser remelting of material surfaces (Ref 25). In particular, they have developed a model of solid fraction that takes into account the large undercooling experienced by the dendrite tips under rapid solidification conditions. They have shown that even under such circumstances the results predicted by this detailed approach do not differ significantly from the simplest model that neglects the undercooling at the macroscopic scale.

This last approach has been applied to the laser treatment of materials surfaces to predict the lamellar spacings of a eutectic aluminum-copper alloy from the calculated stationary shape of the liquid pool (Fig. 3). More recently, this approach has been used to analyze dendritic microstructures produced in electron beam welding of stainless steel single crystals (Fig. 4). It was shown,

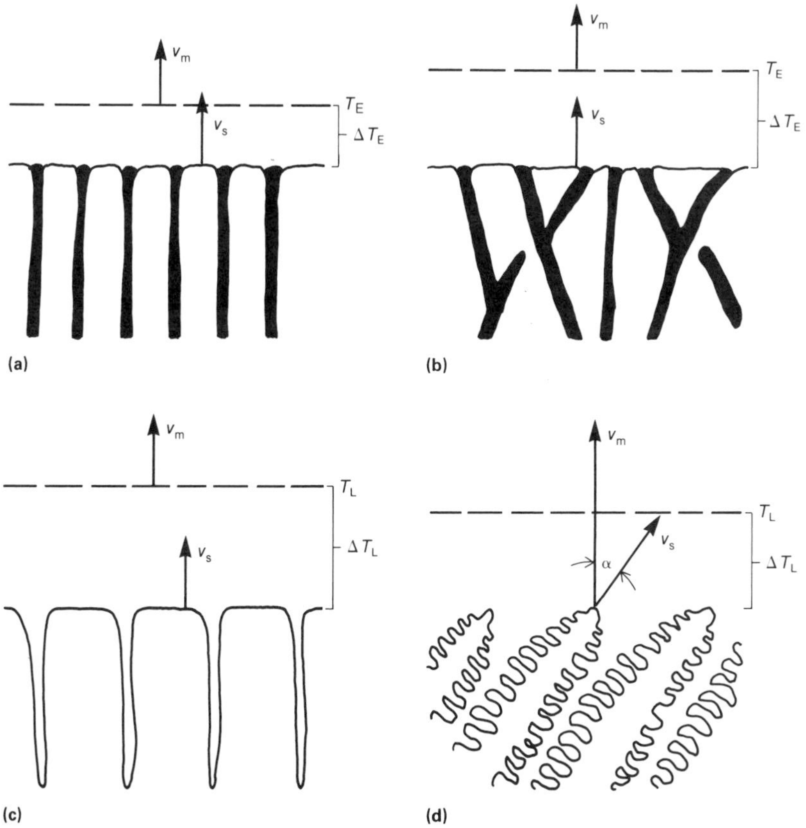

Fig. 2 Relationship between growth velocity of the macroscopic isotherms v_m and growth velocity v_s of four different columnar microstructures. (a) and (b) Regular and irregular eutectics, respectively. (c) Cells. (d) Dendrites

from the macroscopic shape of the liquid pool, that a simple criterion of minimum undercooling, that is, of minimum speed, can be applied to determine which dendrite trunk orientation is selected.

Modeling of Equiaxed Structures

When dealing with equiaxed microstructures, the growth speed of the grains is no longer related to the speed of the isotherms, but rather to local undercooling (Fig. 1). Furthermore, the solidification path is also dependent on the number of grains that have been nucleated within the undercooled melt. In such a case, the approach used must relate the fraction that has solidified to the local undercooling.

Microscopic Modeling of Equiaxed Structures

Consider a small volume element V of uniform temperature T, within which equiaxed solidification (Fig. 5) is proceeding. At a given time t, the fraction of solid $f_s(t)$ is given by (Ref 27):

$$f_s(t) = n(t) \cdot 4/3\ \pi\ R^3(t) \cdot f_i(t) \qquad \text{(Eq 10)}$$

where $n(t)$ is the density of the grains, $R(t)$ is the average equiaxed grain radius characterizing the position of the dendrite tips or that of the eutectic front, and $f_i(t)$ is the internal fraction of solid. For eutectics, the grains are fully solid, and accordingly $f_i(t)$ = 1 at any time. For dendritic alloys, $f_i(t)$ represents the fraction of the grains that is really solid.

To predict the evolution of the solid fraction $f_s(t)$, one must relate the three variables $n(t)$, $R(t)$, and $f_i(t)$ to the undercooling ΔT. This can be done by considering nucleation kinetics, growth kinetics, and, for dendrites, solute diffusion.

Nucleation Kinetics. The rate $\dot{n}(t)$ at which new grains are heterogeneously nucleated within the liquid can be given at low undercooling by (Ref 28):

$$\dot{n}(t) = K_1\ [n_0 - n(t)]\ \exp\left\{\frac{-K_2}{\Delta T(t)^2}\right\} \qquad \text{(Eq 11)}$$

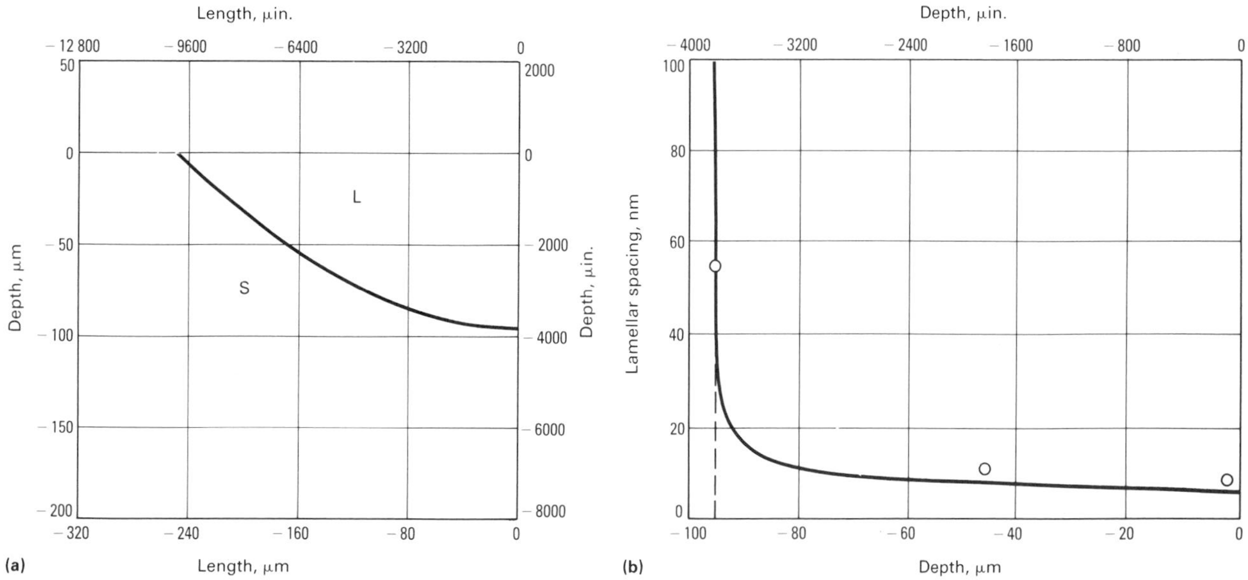

Fig. 3 Calculated stationary shape (a) of the liquid pool that forms during laser treatment of an aluminum-copper eutectic alloy surface. The laser, with 1500 W of total power focused onto a spot 0.2 mm (0.008 in.) in diameter, is moving to the right with a velocity v_b of 1 m/s (3.2 ft/s). Absorption coefficient is 0.15. Although the calculation was made in three dimensions, only the resolidifying back part of the pool within a longitudinal section is shown. (b) Lamellar spacing of the aluminum-copper eutectic alloy versus depth of the laser-treated surface as calculated from the shape of the liquid pool (a) and using the recent theory of eutectic formation. Source: Ref 26

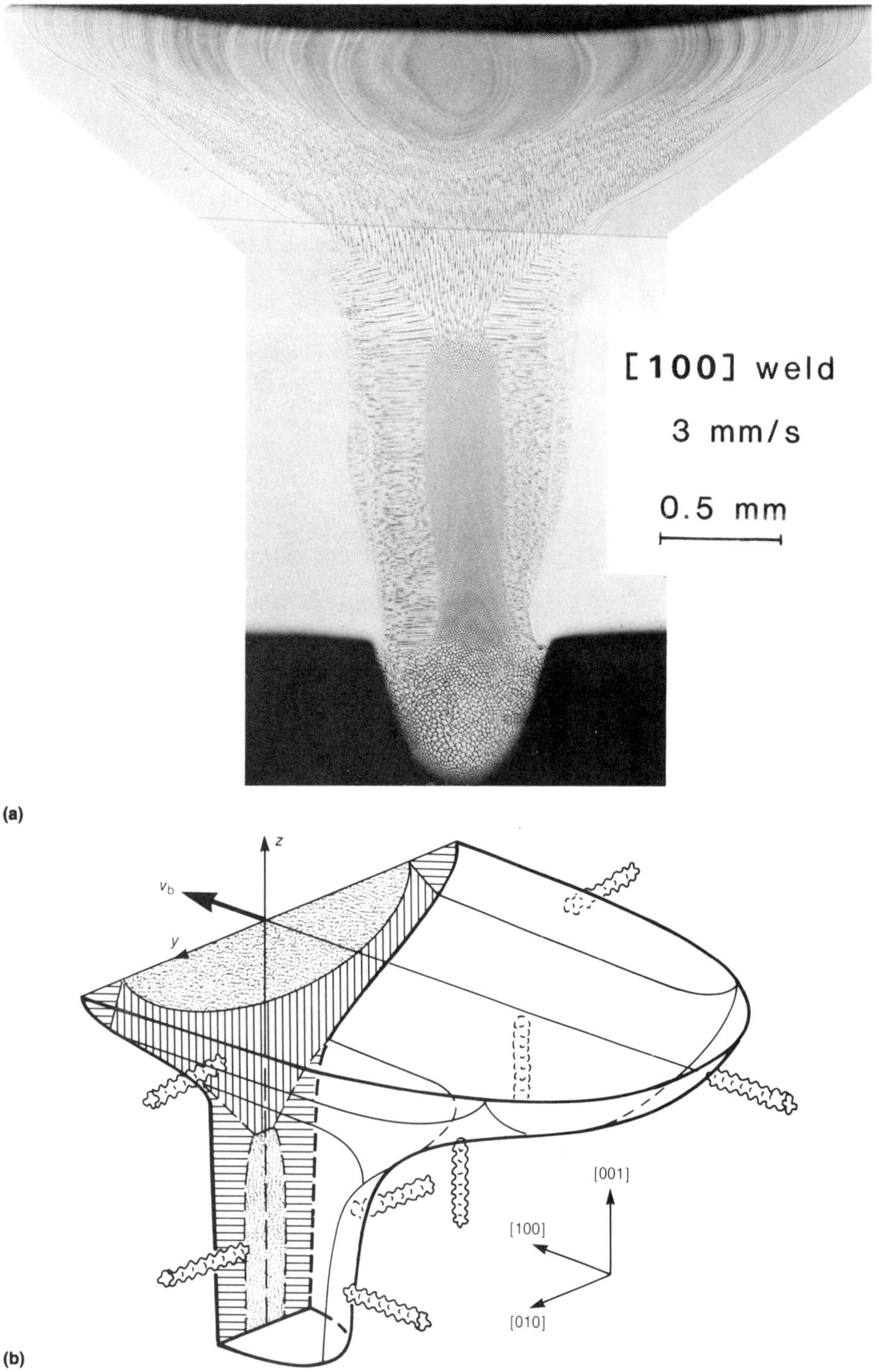

Fig. 4 Transverse section micrograph (a) of electron beam weld of an Fe-15Ni-15Cr single crystal. The electron beam was moved at a velocity of 3 mm/s (0.12 in./s) over the (001) surface along a [100] crystallographic orientation. The dendrites epitaxially grown at the monocrystalline surface of the weld can have their trunks aligned along one of the three ⟨100⟩ orientations. Microstructure selection is made according to a criterion of minimum undercooling (or of minimum speed); therefore, the information in this micrograph can be used to reconstruct the three-dimensional shape of the weld pool, as shown in (b). Source: Ref 22

where K_1 is proportional to a collision frequency with nucleation sites, n_0 is the total number of sites present in the melt before solidification, and K_2 is a constant related to the interfacial energy between substrate and nucleated grain. The constants K_1, n_0, and K_2 must be deduced from experiment. Once they are known, the grain density $n(t)$ can be predicted at each time by integrating Eq 11 over time or temperature:

$$n(t) = \int_{t_0}^{t} \dot{n}(\tau)d\tau = \int_{0}^{\Delta T(t)} \dot{n}(T) \cdot \frac{dT}{dT/dt} \quad \text{(Eq 12)}$$

However, this approach fails to predict the correct grain density, in part because the temperature interval within which nucleation proceeds is very narrow. For an undercooling ΔT smaller than a critical value, $\Delta T_N = \sqrt{K_2}$, there is no significant nucleation. When ΔT_N is reached, $n(t)$ increases very rapidly to its saturation limit n_0 (Fig. 6 and 7). Therefore, it is suggested to replace the complex nucleation law of Eq 11 by a Dirac function in solidification modeling:

$$\frac{dn}{dT} = n_0 \cdot \delta(T - T_N) = n_0 \cdot \delta(\Delta T - \sqrt{K_2}) \quad \text{(Eq 13)}$$

If more than one type of nucleation site is present, one can introduce a set of Dirac functions (Fig. 6):

$$\frac{dn}{dT} = \sum_i n_{0,i}\, \delta(T - T_{N,i}) \quad \text{(Eq 14)}$$

This discrete distribution of nucleation site types can also be replaced by a continuous distribution (Fig. 8). Although this last approach may not reflect the complex phenomena of heterogeneous nucleation, it has some advantages in microscopic modeling of solidification (Ref 20, 27).

In fact, a continuous distribution of nucleation site types can be replaced by a very narrow distribution if one only wants to simulate heterogeneous nucleation occurring at a given undercooling ΔT_N with a given density of sites n_0 (Eq 13). This last approach can be used for eutectic solidification based on the fact that, as previously discussed, the nucleation interval is very narrow. For example, for cast iron, the nucleation interval was calculated to be about 0.1 °C (0.2 °F) (Ref 29). Thus, a nucleation temperature ΔT_N at which all eutectic grains nucleate at the same time can be chosen. For the case of alloys with nonuniform grains, it must be assumed that different types of substrates become active at different nucleation temperatures. Accordingly, several nucleation temperatures must be selected, at which fractions of the final number of nuclei are generated.

Growth. Evolution of the grain radius $R(t)$ can also be related to the undercooling ΔT of the volume element. The speed v of a eutectic front is related to the undercooling through the relationship (Ref 30):

$$v = \frac{dR}{dt} = \mu \cdot (\Delta T)^2 \quad \text{(Eq 15)}$$

where μ is a constant depending on the characteristics of the alloy.

For dendritic alloys, a similar relationship has been deduced in the approximation of a hemispherical dendrite tip, which relates the square of the undercooling ΔT to the velocity v of the dendrite tips (Ref 31). Therefore, Eq 15, with a different μ value, can be used to predict the evolution of grain size. However, in the case of dendrites, one must still calculate the evolution of the interval vol-

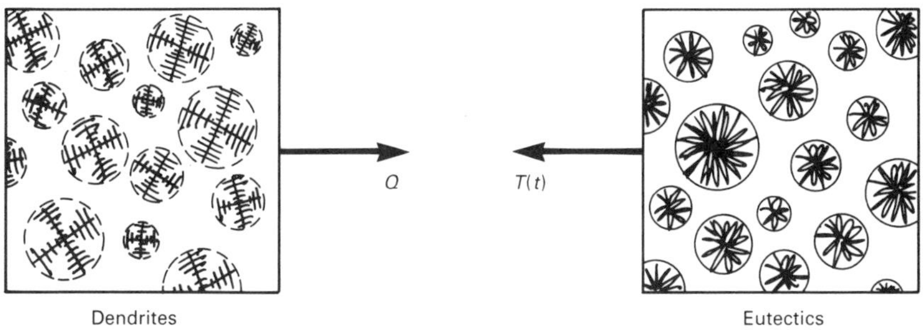

Fig. 5 Schematic showing equiaxed dendritic and eutectic solidification

ume fraction of solid $f_i(t)$ (Eq 10). For that purpose, a solute diffusion model has recently been developed (Fig. 9). Assuming that there is complete mixing of solute within the interdendritic liquid of the spherical grain envelope outlined by the dendrite tip position, the researchers considered the solute balance at the scale of the equiaxed grain and the solute flow leaving out the grain envelope. They found that:

$$f_i(t) = \Omega(t) \cdot g(\delta,R) \quad \text{(Eq 16)}$$

where $\Omega = (C^* - C_0)/[C^*(1 - k)]$ is the supersaturation, $g(\delta,R)$ is a correction function that takes into account the solute layer δ around the grain envelope, C^* is the concentration within the interdendritic liquid (Fig. 9), C_0 is the initial concentration, and k is the partition coefficient. Because the undercooling ΔT is equal to $m(C^* - C_0)$, where m is the slope of the liquidus, $f_i(t)$ is again directly related to ΔT through Eq 16.

From the solute flux balance, it has been shown that the solute layer δ is simply given by the ratio $2D/v$, where D is the diffusion coefficient. The effect of δ on solidification is most noticed when the solute layers of neighboring dendritic grains overlap, thus changing the concentration C_0 in the supersaturation expression.

Grain Impingement. Equation 10 assumes that the grains are spherical during the entire solidification process. It is valid as long as the grains do not impinge on each other. For dendritic alloys, the diffusion layer δ outside of the grain envelope R somehow already takes grain impingement into account. For eutectic grains, grain impingement must be introduced.

The Johnson-Mehl correction for grain impingement predicts that (Ref 33):

$$f_s = 1 - \exp(-n \cdot \tfrac{4}{3}\pi R^3) \quad \text{(Eq 17)}$$

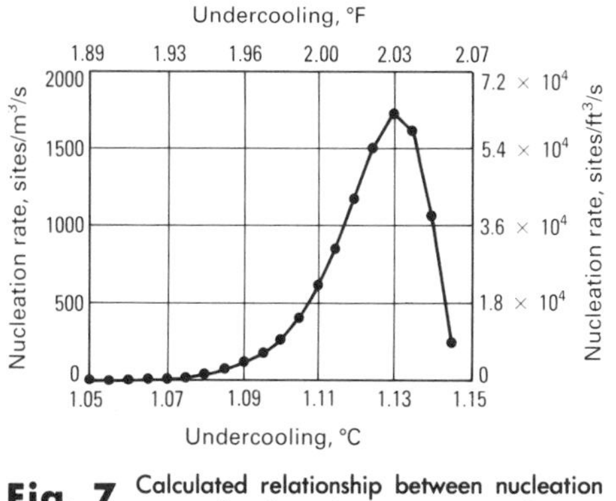

Fig. 7 Calculated relationship between nucleation rate and undercooling in cast iron. Source: Ref 29

Macro-Microscopic Modeling of Equiaxed Solidification

The coupling between the macroscopic heat flow equation and the microscopic models of equiaxed solidification can be achieved according to various schemes. A detailed description of a possible procedure is given in Ref 20. Two basic coupling schemes for equiaxed solidification are shown in Fig. 10.

The latent heat method shown in Fig. 10(a) is the most straightforward one (Ref 18, 29). Formulating Eq 3 with finite-difference method or finite-element method (FEM), the variations $\Delta\{f_s\}$ between t and $t + \Delta t$ at all nodes are calculated according to the microscopic model of solidification (Eq 10). In both dendritic and eutectic alloys, the variation $\Delta\{T\}$ can be derived explicitly or implicitly, while the variation $\Delta\{f_s\}$ is given explicitly by the undercoolings at each node at time t.

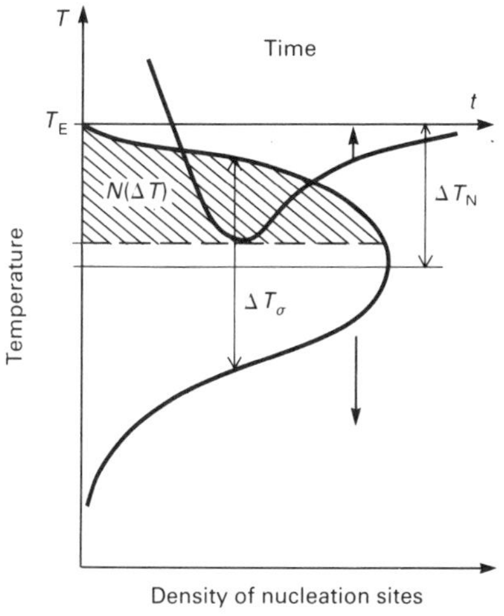

Fig. 8 Continuous distribution of nucleation site types. Source: Ref 27

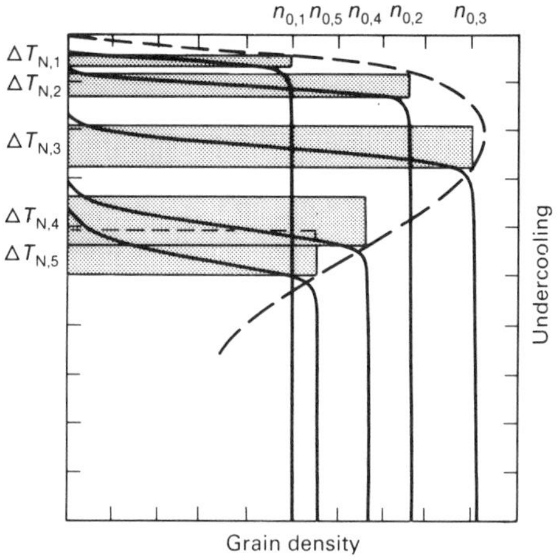

Fig. 6 Schematic of heterogeneous nucleation occurring on a family of inoculant sites, characterized by a density of sites $n_{0,i}$ and by a critical temperature $T_{N,i}$ at which nucleation occurs. Source: Ref 27

The source term $\dot{Q}$ (Eq 2) in the heat conduction equation (Eq 3) is directly coupling the macroscopic heat flow and the microscopic growth kinetics. The latent heat evolved is calculated so as to remain finite in the solidification region. Therefore, no special treatment is required for the latent heat term in solving Eq 3, such as those employed in specific heat and enthalpy formulation of heat conduction equations.

The departure from equilibrium solidification for a cast iron sample can be readily seen from the enthalpy-temperature diagram shown in Fig. 11. The predicted and experimental cooling curves at the middle

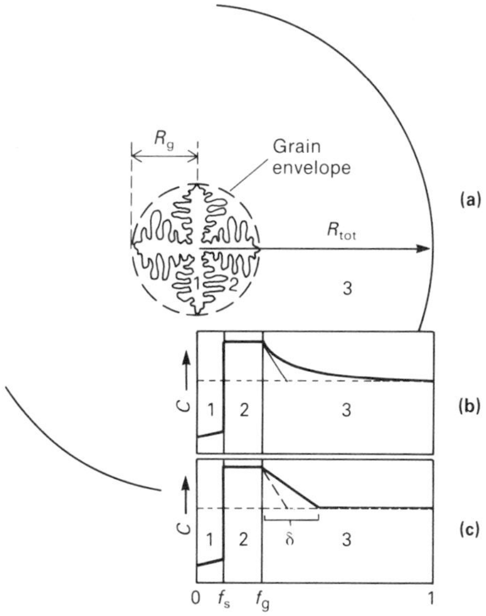

Fig. 9 Schematic showing the solute diffusion model developed for equiaxed dendritic growth. (a) The three regions that can be distinguished are (1) solid dendrite, (2) interdendritic liquid, where complete mixing of solute is assumed, and (3) liquid outside the grain envelope where diffusion occurs. (b) and (c) Concentration profiles corresponding to (a). Source: Ref 32

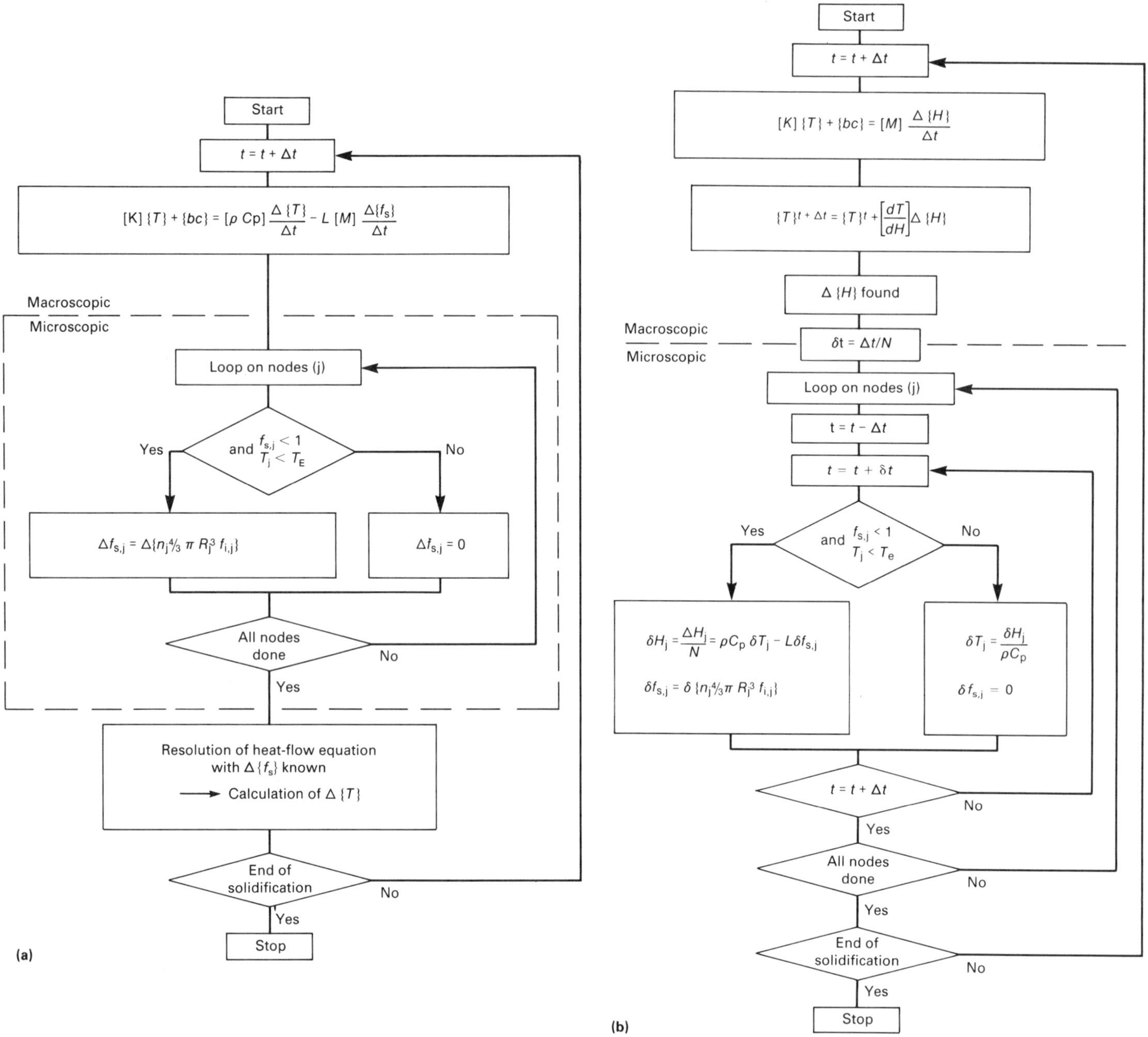

Fig. 10 Flow charts of the macroscopic-microscopic modeling of solidification based on two different schemes. (a) Latent heat method. (b) Microenthalpy method

of a cylindrical mold for the same eutectic cast iron are given in Fig. 12. Two computer programs, EUCAST and BAMACAST, have been used for calculation (Ref 34, 35). It is obvious from Fig. 12 that the macro-micro eutectic model not only accurately predicts the degree of undercooling and the arrest temperature but also the solidification time.

Figure 13 gives theoretical predictions of the width of the mushy zone for the cast iron sample shown in Fig. 12. The data are in good agreement with the experimental values for the beginning and end of solidification for the thermocouple in the center of the sample.

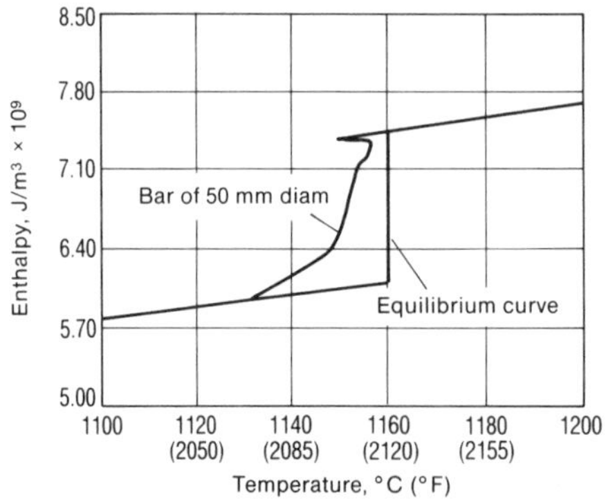

Fig. 11 Calculated and theoretical enthalpy versus temperature curves for cast iron of eutectic composition. Source: Ref 18

A macro-micro modeling approach can have many structure-related applications. For example, macro-micro modeling has been used to attempt to predict the gray/white structural transition in cast irons (Ref 35, 36).

As previously discussed, applications of this method can also be extended to the primary phase. Typical calculated and experimental cooling curves for a hypoeutectic Al-8.5Si alloy are given in Fig. 14.

The microenthalpy scheme (Fig. 10b) has been incorporated into the 3-MOS program, an FEM code developed in Switzerland from the library Modulef (Ref 19, 38, 39). It is essentially based on an enthalpy method.

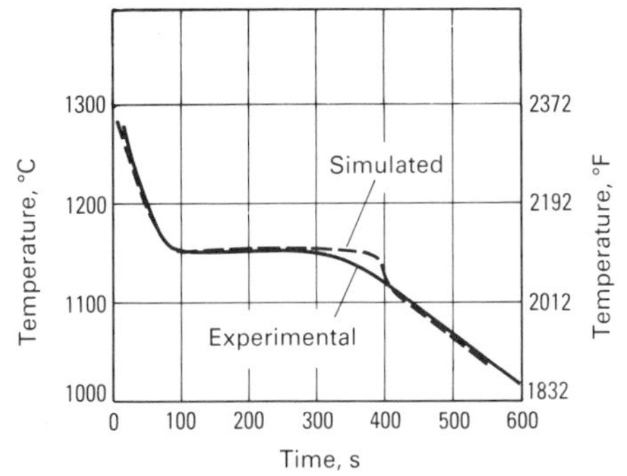

Fig. 12 Calculated and experimental cooling curves for eutectic gray iron poured in a 50 mm (2 in.) diam bar molded in resin bonded sand. Thermocouples were inserted in the middle of the casting. Source: Ref 34

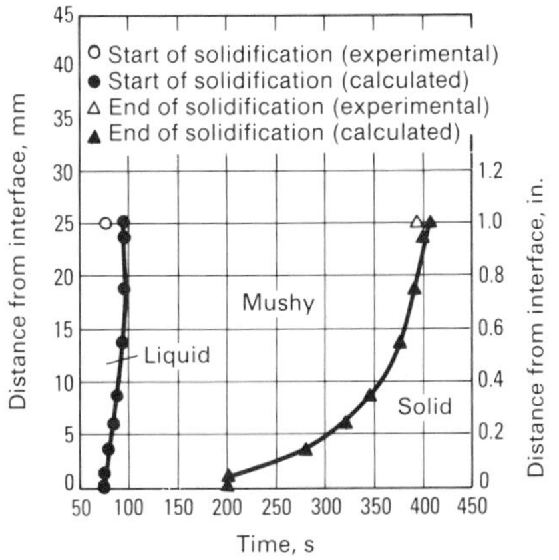

Fig. 13 Calculated beginning and end of solidification wave fronts for a 50 mm (2 in.) diam bar, and experimental points for a thermocouple placed at the center of the bar. Source: Ref 34

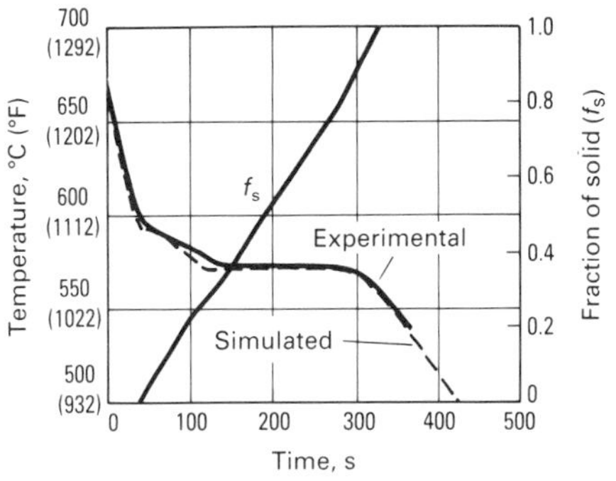

Fig. 14 Experimental and simulated cooling curves and calculated fraction of solid for an Al-8.5Si alloy. Source: Ref 37

Because the variation of enthalpy is independent of the solidification path once the heat flow is known, the macro- and microscopic calculations can be somehow decoupled. At the macro level, one can still solve the heat flow equation, as mentioned in the section "Macroscopic Modeling" in this article. Once the variations of enthalpy $\Delta\{H\}$ at all nodes are known, the solidification path can be computed. As shown in Fig. 10(b), the macroscopic time-step Δt can be subdivided into many smaller time-steps δt to perform the microscopic calculations, assuming that heat removal is made at a constant rate during Δt. The micro-macroscopic coupling scheme seems to give good convergence of the calculated values (undercooling or grain size) (Ref 19). The results (discussed below) illustrate the possibilities of integrating microscopic modeling of solidification into macroscopic heat flow calculations by using an enthalpy method.

Figure 15 shows the recalescences of two Al-7Si specimens. The dotted curves have been measured at the center of two small volumes containing the alloy. The solid curves shown in Fig. 15 have been computed with the analytical model of solute diffusion and are based on the measured grain sizes.

The six cooling curves shown in Fig. 16 have been measured for a one-dimensional gray cast iron (3% C, 2.5% Si) casting poured in a ceramic mold over a copper chill plate (Ref 39). The effect of silicon on the mechanism of eutectic growth was taken into account by modifying the equilibrium eutectic temperature according to a Scheil model of silicon segregation. Although the agreement between modeling and experiment is poor in the liquid region (above 1160 °C, or 2120 °F), solidification is very well predicted with the macro-micro model. In particular, calculated recalescence undercooling and end of solidification are in good agreement with the experimental curves. However, the solidification of the primary phase close to 1190 °C (2175 °F) was not included in the modeling.

One of the primary applications of the macro-microscopic modeling of solidification is the prediction of microstructural features. Figure 17 compares the grain radii measured and calculated at the six locations of the thermocouples where the cooling curves shown in Fig. 16 are recorded. These radii are plotted as a function of the distance from the copper chill plate. The distribution of nucleation sites was a Gaussian line shape whose parameters were deduced from microcastings of the same alloy. Although the discrepancy between experiment and modeling may be substantial (especially for thermocouple No. 5), the trend of increasing the grain size with increasing distances from the chill (or decreasing cooling rates) is correctly predicted. Figure 18

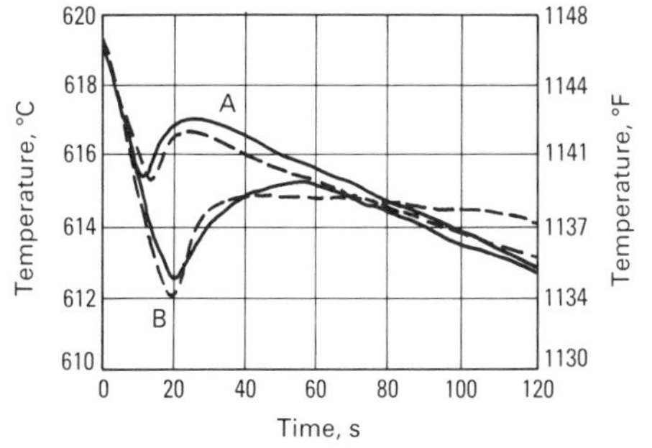

Fig. 15 Measured (dashed lines) and calculated (solid lines) recalescences for two Al-7Si alloys. With 50 ppm Ti inoculant (curve A), the final grain radius was 0.5 mm (0.02 in.). Without inoculant (curve B), the final grain radius was 2 mm (0.08 in.). Source: Ref 32

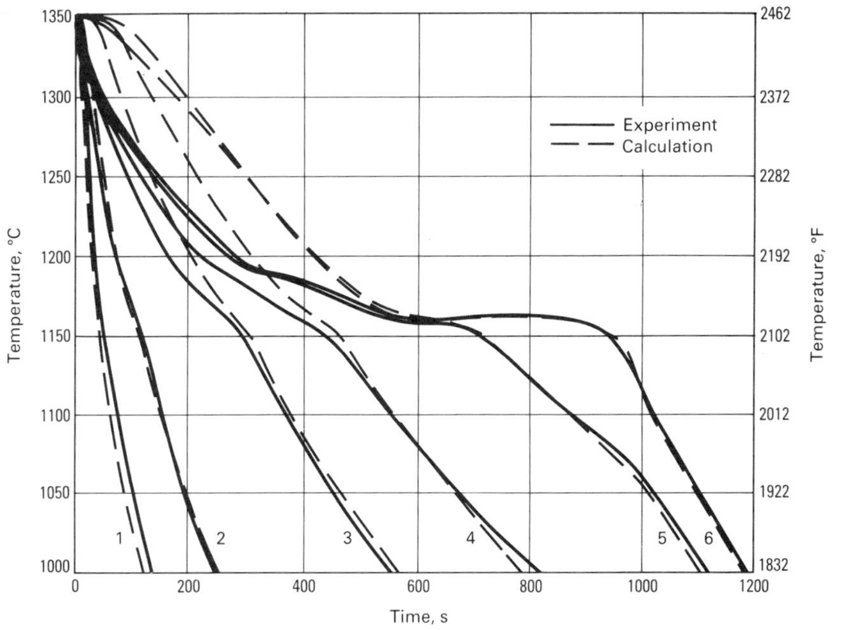

Fig. 16 Measured (dashed lines) and calculated (solid lines) cooling curves for cast iron. Numbers on curves indicate locations of thermocouples in the casting. Height of castings: 120 mm (4.7 in.); number of meshes: 120. The parameters of nucleation deduced from separate microcasting experiments are the following: Gaussian distribution, center at 20 K undercooling, standard deviation: 4.75 K, and total density of sites: 1.2×10^{11}/m^3. Source: Ref 39

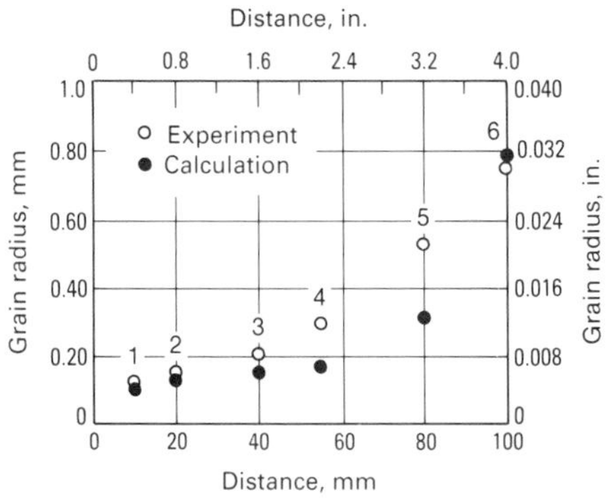

Fig. 17 Experimental and calculated grain radii at the locations of the thermocouples that recorded the cast iron cooling curves shown in Fig. 16. Source: Ref 39

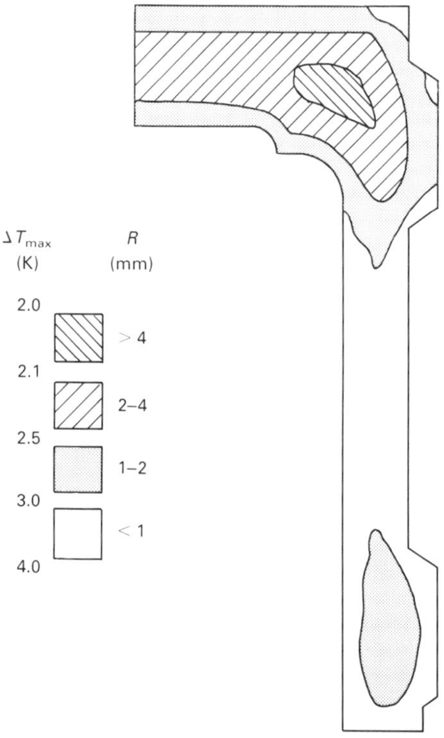

Fig. 18 Map of calculated maximum undercooling ΔT_{max} within a longitudinal section of an axisymmetric casting. Because undercooling can be directly related to the average grain size using the nucleation law, this figure also maps the average grain radius *R* within the casting.

shows a map of grain sizes, calculated with the same micro-macroscopic approach for a two-dimensional Al-7Si casting (Ref 19). As can be seen, the trend of larger grain size at the center of the casting is correctly predicted from the model.

ACKNOWLEDGMENT

The Swiss project is supported by the Office Fédéral de l'Education et de la Science, Switzerland, and the American project is supported by NSF—EPSCoR in Alabama.

REFERENCES

1. J.G. Henzel, Jr. and J. Keverian, Comparison of Calculated and Measured Solidification Patterns for a Variety of Steel Castings, *Trans. AFS*, Vol 73, 1965, p 661-672
2. R.D. Pehlke, R.E. Marrone, and J.O. Wilkes, *Computer Simulation of Solidification*, American Foundrymen's Society, 1976
3. H.D. Brody and D. Apelian, Ed., *Modeling of Casting and Welding Processes*, The Metallurgical Society, 1981
4. J.A. Dantzig and J.T. Berry, Ed., *Modeling of Casting and Welding Processes*, Vol II, The Metallurgical Society, 1984
5. H. Fredriksson, Ed., *State of the Art of Computer Simulation of Casting and Solidification Processes*, Les Editions de Physique, 1986
6. W. Oldfield, A Quantitative Approach to Casting Solidification: Freezing of Cast Iron, *Trans. ASM*, Vol 59, 1966, p 945-959
7. D.M. Stefanescu and S. Trufinescu, Zur Kristallisationskinetik von Grauguss, *Z. Metallkd.*, Vol 65 (No. 9), 1974, p 610-666
8. O. Yanagisawa and M. Maruyama, "Silicon Inoculation Mechanism in Cast Iron," Paper 21, presented at the 46th International Foundry Congress, 1979
9. H. Fredriksson and I.L. Svensson, Computer Simulation of the Structure Formed During Solidification of Cast Iron, in *The Physical Metallurgy of Cast Iron*, H. Fredriksson and M. Hillert, Ed., North Holland, 1984, p 273-284
10. D.M. Stefanescu and C. Kanetkar, Computer Modeling of the Solidification of Eutectic Alloys: The Case of Cast Iron, in *Computer Simulation of Microstructural Evolution*, D.J. Srolovitz, Ed., The Metallurgical Society, 1985, p 171-188
11. K.C. Su, I. Ohnaka, I. Yaunauchi, and T. Fukusako, Computer Simulation of Solidification of Nodular Cast Iron, in *The Physical Metallurgy of Cast Iron*, H. Fredriksson and M. Hillert, Ed., North Holland, 1984, p 181-189
12. I. Dustin and W. Kurz, Modeling of Cooling Curves and Microstructures During Equiaxed Dendritic Solidification, *Z. Metallkunde.*, Vol 77, 1986, p 265
13. S.C. Flood and J.D. Hunt, Columnar and Equiaxed Growth I and II, *J. Cryst. Growth*, Vol 82, 1987, p 543, 552
14. M. Rappaz and P. Thévoz, Solute Diffusion Model for Equiaxed Dendritic Growth, *Acta Metall.*, Vol 353, 1987, p 1487
15. J.D. Hunt, Steady State Columnar and Equiaxed Growth of Dendrites and Eutectic, *Mater. Sci. Eng.*, Vol 65 (No. 1), 1984, p 75
16. W. Kurz and D.J. Fisher, *Fundamentals of Solidification,* Trans Tech, 1986
17. M.C. Flemings, *Solidification Processing*, McGraw-Hill, 1974
18. C.S. Kanetkar, I.G. Chen, D.M. Stefanescu, and N. El-Kaddah, A Latent Heat Method for Macro-Micro Modeling of Eutectic Solidification, submitted to *Trans. Iron Steel Inst. Jpn.*, 1987
19. Ph. Thévoz, J.L. Desbiolles, and M. Rappaz, Modeling of Equiaxed Microstructure Formation in Casting, submitted to *Metall. Trans.*, 1988
20. M. Rappaz and D.M. Stefanescu, Modeling of Equiaxed Primary and Eutectic Solidification, in *Solidification Processing of Eutectic Alloys*, The Metallurgical Society, 1988
21. M. Rappaz and E. Blank, Simulation of Oriented Dendritic Microstructures Using the Concept of Dendritic Lattice, *J. Cryst. Growth*, Vol 74, 1986, p 67
22. M. Rappaz, S.A. David, L.A. Boatner, and J.M. Vitek, Development of Microstructures in Fe-15Ni-15Cr Single-Crystal E-Beam Welds, *Metall. Trans.*, to be published
23. T.W. Clyne, The Use of Heat Flow Modeling to Explore Solidification Phenomena, *Metall. Trans. B*, Vol 13B, 1982, p 471
24. T.W. Clyne, Numerical Treatment of Rapid Solidification, *Metall. Trans. B*, Vol 15B, 1984, p 369
25. B. Giovanola and W. Kurz, Modeling Dendritic Growth Under Rapid Solidification Conditions, in *State of the Art of Computer Simulation of Solidification*, H. Fredriksson, Ed., Proceedings of the E-MRS Conference, Strasbourg, Les Editions de Physique, 1986, p 129-135
26. M. Rappaz, B. Carrupt, M. Zimmermann, and W. Kurz, Numerical Simulation of Eutectic Solidification in the Laser Treatment of Materials, *Helvet. Phys. Acta*, Vol 60, 1987, p 924
27. M. Rappaz, Ph. Thévoz, Zou Jie, J.P. Gabathuler, and H. Lindscheid, Micro-Macroscopic Modeling of Equiaxed Solidification, in *State of the Art of Computer Simulation of Casting and Solidification Processes*, Les Editions de Physique, 1986, p 277-284
28. D. Turnbull, Kinetics of Heterogeneous Nucleation, *J. Chem. Phys.*, Vol 18, 1950, p 198
29. D.M. Stefanescu and C. Kanetkar, Computer Modeling of the Solidification of Eutectic Alloys: Comparison of Various Models for Eutectic Growth of Cast Iron, in *State of the Art of Computer Simulation of Casting and Solidification Processes*, Les Editions de Physique, 1986, p 255-266
30. K.A. Jackson and J.D. Hunt, Lamellar and Rod Eutectic Growth, *Trans. Me-*

tall. Soc. AIME, Vol 236, 1966, p 1129-1142

31. H. Esaka and W. Kurz, Columnar Dendrite Growth: A Comparison of Theory, *J. Cryst. Growth*, Vol 69, 1984, p 362
32. M. Rappaz and Ph. Thévoz, Analytical Model of Equiaxed Dendritic Solidification, in *Solidification Processing*, H. Jones, Ed., Institute of Metals, 1987
33. W.A. Johnson and R.F. Mehl, "Reaction Kinetics in Processes of Nucleation and Growth," AIME Technical Publication 1089, American Institute of Mining, Metallurgical, and Petroleum Engineers, 1939, p 5
34. C.S. Kanetkar, D.M. Stefanescu, N. El-Kaddah, and I.G. Chen, Macro-Microscopic Simulation of Equiaxed Solidification of Eutectic and Off-Eutectic Alloys, in *Solidification Processing*, H. Jones, Ed., Institute of Metals, 1987
35. D.M. Stefanescu and C.S. Kanetkar, "Modeling of Microstructural Evolution of Cast Iron and Aluminum-Silicon Alloys," Paper 19, presented at the 54th International Foundry Congress, New Delhi, India, 1987
36. D.M. Stefanescu and C.S. Kanetkar, Modeling of Microstructural Evolution of Eutectic Cast Iron and of the Gray/White Transition, Paper 68, *Trans. AFS*, Vol 95, 1987
37. C.S. Kanetkar, Ph.D. dissertation, The University of Alabama, 1988
38. J.L. Desbiolles, M. Rappaz, J.J. Droux, and J. Rappaz, Simulation of Solidification of Alloys Using the FEM Code Modulef, in *State of the Art of Computer Simulation of Casting and Solidification Processes*, Les Editions de Physique, 1986, p 49-55
39. Ph. Thévoz, Zou Jie, and M. Rappaz, Modeling of Equiaxed Dendritic and Eutectic Solidification in Castings, in *Solidification Processing*, H. Jones, Ed., Institute of Metals, 1987

Metric Conversion Guide

This Section is intended as a guide for expressing weights and measures in the Système International d'Unités (SI). The purpose of SI units, developed and maintained by the General Conference of Weights and Measures, is to provide a basis for world-wide standardization of units and measure. For more information on metric conversions, the reader should consult the following references:

- "Standard for Metric Practice," E 380, *Annual Book of ASTM Standards*, American Society for Testing and Materials, 1916 Race Street, Philadelphia, PA 19103
- "Metric Practice," ANSI/IEEE 268–1982, American National Standards Institute, 1430 Broadway, New York, NY 10018
- *Metric Practice Guide—Units and Conversion Factors for the Steel Industry,* 1978, American Iron and Steel Institute, 1133 15th Street NW, Suite 300, Washington, DC 20005
- *The International System of Units*, SP 330, 1986, National Bureau of Standards. Order from Superintendent of Documents, U.S. Government Printing Office, Washington, DC 20402-9325
- *Metric Editorial Guide*, 4th ed. (revised), 1985, American National Metric Council, 1010 Vermont Avenue NW, Suite 320, Washington, DC 20005–4960
- *ASME Orientation and Guide for Use of SI (Metric) Units*, ASME Guide SI 1, 9th ed., 1982, The American Society of Mechanical Engineers, 345 East 47th Street, New York, NY 10017

Base, supplementary, and derived SI units

Measure	Unit	Symbol
Base units		
Amount of substance	mole	mol
Electric current	ampere	A
Length	meter	m
Luminous intensity	candela	cd
Mass	kilogram	kg
Thermodynamic temperature	kelvin	K
Time	second	s
Supplementary units		
Plane angle	radian	rad
Solid angle	steradian	sr
Derived units		
Absorbed dose	gray	Gy
Acceleration	meter per second squared	m/s^2
Activity (of radionuclides)	becquerel	Bq
Angular acceleration	radian per second squared	rad/s^2
Angular velocity	radian per second	rad/s
Area	square meter	m^2
Capacitance	farad	F
Concentration (of amount of substance)	mole per cubic meter	mol/m^3
Conductance	siemens	S
Current density	ampere per square meter	A/m^2
Density, mass	kilogram per cubic meter	kg/m^3
Electric charge density	coulomb per cubic meter	C/m^3
Electric field strength	volt per meter	V/m
Electric flux density	coulomb per square meter	C/m^2
Electric potential, potential difference, electromotive force	volt	V
Electric resistance	ohm	Ω
Energy, work, quantity of heat	joule	J
Energy density	joule per cubic meter	J/m^3
Entropy	joule per kelvin	J/K
Force	newton	N
Frequency	hertz	Hz
Heat capacity	joule per kelvin	J/K
Heat flux density	watt per square meter	W/m^2
Illuminance	lux	lx
Inductance	henry	H
Irradiance	watt per square meter	W/m^2
Luminance	candela per square meter	cd/m^2
Luminous flux	lumen	lm
Magnetic field strength	ampere per meter	A/m
Magnetic flux	weber	Wb
Magnetic flux density	tesla	T
Molar energy	joule per mole	J/mol
Molar entropy	joule per mole kelvin	J/mol · K
Molar heat capacity	joule per mole kelvin	J/mol · K
Moment of force	newton meter	N · m
Permeability	henry per meter	H/m
Permittivity	farad per meter	F/m
Power, radiant flux	watt	W
Pressure, stress	pascal	Pa
Quantity of electricity, electric charge	coulomb	C
Radiance	watt per square meter steradian	$W/m^2 \cdot sr$
Radiant intensity	watt per steradian	W/sr
Specific heat capacity	joule per kilogram kelvin	J/kg · K
Specific energy	joule per kilogram	J/kg
Specific entropy	joule per kilogram kelvin	J/kg · K
Specific volume	cubic meter per kilogram	m^3/kg
Surface tension	newton per meter	N/m
Thermal conductivity	watt per meter kelvin	W/m · K
Velocity	meter per second	m/s
Viscosity, dynamic	pascal second	Pa · s
Viscosity, kinematic	square meter per second	m^2/s
Volume	cubic meter	m^3
Wavenumber	1 per meter	1/m

Conversion factors

To convert from	to	multiply by
Angle		
degree	rad	1.745 329 E − 02
Area		
in.^{2}	mm^2	6.451 600 E + 02
in.^{2}	cm^2	6.451 600 E + 00
in.^{2}	m^2	6.451 600 E − 04
ft^2	m^2	9.290 304 E − 02
Bending moment or torque		
lbf · in.	N · m	1.129 848 E − 01
lbf · ft	N · m	1.355 818 E + 00
kgf · m	N · m	9.806 650 E + 00
ozf · in.	N · m	7.061 552 E − 03
Bending moment or torque per unit length		
lbf · in./in.	N · m/m	4.448 222 E + 00
lbf · ft/in.	N · m/m	5.337 866 E + 01
Current density		
A/in.^{2}	A/cm^2	1.550 003 E − 01
A/in.^{2}	A/mm^2	1.550 003 E − 03
A/ft^2	A/m^2	1.076 400 E + 01
Electricity and magnetism		
gauss	T	1.000 000 E − 04
maxwell	μWb	1.000 000 E − 02
mho	S	1.000 000 E + 00
Oersted	A/m	7.957 700 E + 01
Ω · cm	Ω · m	1.000 000 E − 02
Ω circular-mil/ft	μΩ · m	1.662 426 E − 03
Energy (impact, other)		
ft · lbf	J	1.355 818 E + 00
Btu (thermochemical)	J	1.054 350 E + 03
cal (thermochemical)	J	4.184 000 E + 00
kW · h	J	3.600 000 E + 06
W · h	J	3.600 000 E + 03
Flow rate		
ft^3/h	L/min	4.719 475 E − 01
ft^3/min	L/min	2.831 000 E + 01
gal/h	L/min	6.309 020 E − 02
gal/min	L/min	3.785 412 E + 00
Force		
lbf	N	4.448 222 E + 00
kip (1000 lbf)	N	4.448 222 E + 03
tonf	kN	8.896 443 E + 00
kgf	N	9.806 650 E + 00
Force per unit length		
lbf/ft	N/m	1.459 390 E + 01
lbf/in.	N/m	1.751 268 E + 02
Fracture toughness		
ksi $\sqrt{\text{in.}}$	MPA $\sqrt{\text{m}}$	1.098 800 E + 00
Heat content		
Btu/lb	kJ/kg	2.326 000 E + 00
cal/g	kJ/kg	4.186 800 E + 00
Heat input		
J/in.	J/m	3.937 008 E + 01
kJ/in.	kJ/m	3.937 008 E + 01
Length		
Å	nm	1.000 000 E − 01
μin.	μm	2.540 000 E − 02
mil	μm	2.540 000 E + 01
in.	mm	2.540 000 E + 01
in.	cm	2.540 000 E + 00
ft	m	3.048 000 E − 01
yd	m	9.144 000 E − 01
mile	km	1.609 300 E + 00
Mass		
oz	kg	2.834 952 E − 02
lb	kg	4.535 924 E − 01
ton (short, 2000 lb)	kg	9.071 847 E + 02
ton (short, 2000 lb)	kg × 10^3(a)	9.071 847 E − 01
ton (long, 2240 lb)	kg	1.016 047 E + 03
Mass per unit area		
oz/in.^{2}	kg/m^2	4.395 000 E + 01
oz/ft^2	kg/m^2	3.051 517 E − 01
oz/yd^2	kg/m^2	3.390 575 E − 02
lb/ft^2	kg/m^2	4.882 428 E + 00
Mass per unit length		
lb/ft	kg/m	1.488 164 E + 00
lb/in.	kg/m	1.785 797 E + 01
Mass per unit time		
lb/h	kg/s	1.259 979 E − 04
lb/min	kg/s	7.559 873 E − 03
lb/s	kg/s	4.535 924 E − 01
Mass per unit volume (includes density)		
g/cm^3	kg/m^3	1.000 000 E + 03
lb/ft^3	g/cm^3	1.601 846 E − 02
lb/ft^3	kg/m^3	1.601 846 E + 01
lb/in.^{3}	g/cm^3	2.767 990 E + 01
lb/in.^{3}	kg/m^3	2.767 990 E + 04
Power		
Btu/s	kW	1.055 056 E + 00
Btu/min	kW	1.758 426 E − 02
Btu/h	W	2.928 751 E − 01
erg/s	W	1.000 000 E − 07
ft · lbf/s	W	1.355 818 E + 00
ft · lbf/min	W	2.259 697 E − 02
ft · lbf/h	W	3.766 161 E − 04
hp (550 ft · lbf/s)	kW	7.456 999 E − 01
hp (electric)	kW	7.460 000 E − 01
Power density		
W/in.^{2}	W/m^2	1.550 003 E + 03
Press capacity		
See Force		
Pressure (fluid)		
atm (standard)	Pa	1.013 250 E + 05
bar	Pa	1.000 000 E + 05
in. Hg (32 °F)	Pa	3.386 380 E + 03
in. Hg (60 °F)	Pa	3.376 850 E + 03
lbf/in.^{2} (psi)	Pa	6.894 757 E + 03
torr (mm Hg, 0 °C)	Pa	1.333 220 E + 02
Specific heat		
Btu/lb · °F	J/kg · K	4.186 800 E + 03
cal/g · °C	J/kg · K	4.186 800 E + 03
Stress (force per unit area)		
tonf/in.^{2}(tsi)	MPa	1.378 951 E + 01
kgf/mm^2	MPa	9.806 650 E + 00
ksi	MPa	6.894 757 E + 00
lbf/in.^{2} (psi)	MPa	6.894 757 E − 03
MN/m^2	MPa	1.000 000 E + 00
Temperature		
°F	°C	5/9 · (°F − 32)
°R	°K	5/9
Temperature interval		
°F	°C	5/9
Thermal conductivity		
Btu · in./s · ft^2 · °F	W/m · K	5.192 204 E + 02
Btu/ft · h · °F	W/m · K	1.730 735 E + 00
Btu · in./h · ft^2 · °F	W/m · K	1.442 279 E − 01
cal/cm · s · °C	W/m · K	4.184 000 E + 02
Thermal expansion		
in./in. · °C	m/m · K	1.000 000 E + 00
in./in. · °F	m/m · K	1.800 000 E + 00
Velocity		
ft/h	m/s	8.466 667 E − 05
ft/min	m/s	5.080 000 E − 03
ft/s	m/s	3.048 000 E − 01
in./s	m/s	2.540 000 E − 02
km/h	m/s	2.777 778 E − 01
mph	km/h	1.609 344 E + 00
Velocity of rotation		
rev/min (rpm)	rad/s	1.047 164 E − 01
rev/s	rad/s	6.283 185 E + 00
Viscosity		
poise	Pa · s	1.000 000 E − 01
stokes	m^2/s	1.000 000 E − 04
ft^2/s	m^2/s	9.290 304 E − 02
in.^{2}/s	mm^2/s	6.451 600 E + 02
Volume		
in.^{3}	m^3	1.638 706 E − 05
ft^3	m^3	2.831 685 E − 02
fluid oz	m^3	2.957 353 E − 05
gal (U.S. liquid)	m^3	3.785 412 E − 03
Volume per unit time		
ft^3/min	m^3/s	4.719 474 E − 04
ft^3/s	m^3/s	2.831 685 E − 02
in.^{3}/min	m^3/s	2.731 177 E − 07
Wavelength		
Å	nm	1.000 000 E − 01

(a) kg × 10^3 = 1 metric ton or 1 megagram (Mg)

SI prefixes—names and symbols

Exponential expression	Multiplication factor	Prefix	Symbol
10^{18}	1 000 000 000 000 000 000	exa	E
10^{15}	1 000 000 000 000 000	peta	P
10^{12}	1 000 000 000 000	tera	T
10^{9}	1 000 000 000	giga	G
10^{6}	1 000 000	mega	M
10^{3}	1 000	kilo	k
10^{2}	100	hecto(a)	h
10^{1}	10	deka(a)	da
10^{0}	1	BASE UNIT	
10^{-1}	0.1	deci(a)	d
10^{-2}	0.01	centi(a)	c
10^{-3}	0.001	milli	m
10^{-6}	0.000 001	micro	μ
10^{-9}	0.000 000 001	nano	n
10^{-12}	0.000 000 000 001	pico	p
10^{-15}	0.000 000 000 000 001	femto	f
10^{-18}	0.000 000 000 000 000 001	atto	a

(a) Nonpreferred. Prefixes should be selected in steps of 10^3 so that the resultant number before the prefix is between 0.1 and 1000. These prefixes should not be used for units of linear measurement, but may be used for higher order units. For example, the linear measurement, decimeter, is nonpreferred, but square decimeter is acceptable.

Abbreviations and Symbols

a crystal lattice length along the *a* axis; activity
A ampere
A area
Å angstrom
ABST alpha-beta solution treatment
ac alternating current
AC air cool
ACI Alloy Casting Institute
ADCI American Die Casting Institute
ADI austempered ductile iron
AFS American Foundrymen's Society
AGV automatic guided vehicle
AISI American Iron and Steel Institute
ANSI American National Standards Institute
AOD argon oxygen decarburization
ASTM American Society for Testing and Materials
AS/RS automatic storage and retrieval systems
at.% atomic percent
atm atmosphere (pressure)
AWS American Welding Society
b crystal lattice length along the *b* axis
bcc body-centered cubic
BCIRA British Cast Iron Research Association
BEM boundary element method
BOP basic oxygen process
BST beta solution treatment
BUS broken-up structure
c crystal lattice length along the *c* axis
C concentration
C_0 initial concentration
CAB calcium argon blowing
CAD computer-aided design
CADTA computer-aided differential thermal analysis
CAE computer-aided engineering
CAM computer-aided manufacturing
CE carbon equivalent
CET columnar-equiaxed transition
CG compacted graphite
CIM computer-integrated manufacturing
CLA counter-gravity low-pressure casting of air-melted alloys
CLAS counter-gravity low-pressure air-melted sand casting
CLV counter-gravity low-pressure casting of vacuum-melted alloys
CMM coordinate measuring machine(s)
CNC computerized numerical control
cpm cycles per minute
cps cycles per second
cpt critical preheating temperature
CRE carbon removal efficiency
CRR carbon removal rate
CS ceramic shell; constitutional supercooling
csg constructive solid geometry
CSP constitutional supercooling parameter
CST constitutional solution treatment
CV check valve casting
CVD chemical vapor deposition
CVM control volume method
CVN Charpy V-notch (impact test or specimen)
d used in mathematical expressions involving a derivative (denotes rate of change); depth; diameter
D diameter; distance; diffusivity; density
da/dN fatigue crack growth rate
DAS dendrite arm spacing
dc direct current
DCEP direct current electrode positive
DCRF Die Casting Research Foundation
diam diameter
DIS Draft International Standard
DOC Department of Commerce
DS directional solidification
DT drop tower
e natural log base, 2.71828; electron
E modulus of elasticity
E_c elastic modulus of composite
E_f elastic modulus of fiber
E_m elastic modulus of matrix
EAF electric arc furnace
EB electron beam
EDM electrical discharge machining
EPC evaporative pattern casting
Eq equation
EPS expanded polystyrene pattern
ESR electroslag remelting
ESW electroslag welding
et al. and others
EVA ethylene-vinyl acetate co-polymer
f fraction
FC furnace cool
fcc face-centered cubic
FCAW flux cored arc welding
FDM finite difference method
FEM finite element method
FG flake graphic
Fig. figure
FM full mold
FM (process) *fonte mince* (thin iron) process
FRC free radical cure
FRM fiber-reinforced metals
ft foot
g gram; gas
g acceleration due to gravity
G modulus of rigidity; thermal gradient
gal. gallon
GFN grain fineness number
GMAW gas metal arc welding
GPa gigapascal
Gr graphite
GTAW gas tungsten arc welding
h hour
h height
H enthalpy; height; magnetic field strength
HAZ heat-affected zone
HB Brinell hardness
hcp hexagonal close-packed
HIP hot isostatic pressing
HK Knoop hardness
hp horsepower
HR Rockwell hardness (requires scale designation, such as HRC for Rockwell C hardness)
HSLA high-strength low-alloy (steel)
HTH high-temperature hydrogenation
HV Vickers hardness
HVC hydrovac process
Hz hertz
I inductor current
IACS International Annealed Copper Standard
ICFTA International Committee of Foundry Technical Associations
ID inner diameter

in. inch
ISO International Organization for Standardization
JIS Japanese Industrial Standard
K Kelvin
K stress intensity factor
K_{Ic} plane-strain fracture toughness
K_t theoretical stress concentration factor
kg kilogram
km kilometer
kPa kilopascal
ksi kips (1000 lb) per square inch
kV kilovolt
L liter
L length
lb pound
LBE lance bubble equilibrium
LF ladle furnace
LF/VD-VAD ladle furnace and vacuum arc degassing
LF/VD ladle furnace vacuum degassing
LMR liquid metal refining
ln natural logarithm (base *e*)
m meter
M metal
M_s martensite start temperature
MDI methyl di-isocyanate
mg milligram
Mg megagram (metric tonne)
MHD magnetohydrodynamic (casting)
min minimum; minute
MINT metal in-line treatment
mips million instructions per second
mL milliliter
mm millimeter
MMC metal-matrix composite
MPa megapascal
mph miles per hour
n strain-hardening exponent
n' strain-hardening coefficient
N newton
N number of cycles to failure
NASA National Aeronautics and Space Administration
NC numerical control
NDTT nil ductility transition temperatures
nm nanometer
No. number
NRL Naval Research Laboratories
NT normalized and tempered
OAW oxyacetylene welding
OD outside diameter
Oe oersted
OQ oil quench
OSHA Occupational Safety and Health Administration
OTB oxygen top blown
oz ounce
p page
P pressure
Pa pascal
PECB phenolic ester cold box
pH negative logarithm of hydrogen-ion activity
PH precipitation hardenable
P/M powder metallurgy
ppi pores per lineal inch
ppm parts per million
psi pounds per square inch
PTS para-toluosulfonic acid
PUCB phenolic urethane cold box
PUN phenolic urethane no-bake
PWHT postweld heat treatment
q fatigue notch sensitivity factor
QLR quick lining remover
QT quenched and tempered
r radius
R stress (load) ratio; radius; gas constant
RE rare earth
Ref reference
rem remainder
RF radio frequency
s second
S applied stress
SAE Society of Automotive Engineers
SAW submerged arc welding
SC single-crystal
SCC stress corrosion cracking
scfm standard cubic feet per minute
SCFH standard cubic feet per hour
SCRATA Steel Castings Research and Trade Association
SEM scanning electron microscopy
SG spheroidal graphite
Sh Sherwood number
SIC standard industry codes
SIMA strain induced melt activated
SIMS secondary ion mass spectrometry
SLQ slack quenched
SMAW shielded metal arc welding
SNIF spinning nozzle inert flotation
SOLA solution algorithm
SPAR Space Processing Applications Rocket
SR stress relieved
SSVOD strong stirred vacuum oxygen decarburization
STP standard temperature and pressure
t time; thickness
T temperature
T_L liquidus temperature
T_m melting temperature
T_S solidus temperature
TAC treatment of aluminum in crucible
TCT thermochemical treatment
THT transient heat transfer
TNT trinitrotoluene
TPRE twin-plane reentrant edge mechanism
UBC used beverage container
UNS Unified Numbering System (ASTM-SAE)
UTS ultimate tensile strength
v volume; velocity
V voltage
V volume; velocity
VAC-ESR electroslag remelting under reduced pressure
VAD vacuum arc degassing
VADER vacuum arc double electric remelting
VAR vacuum arc remelting
V-D vacuum degassing
VID vacuum induction degassing
VIDP vacuum induction degassing and pouring
VIM vacuum induction melting
VIM/VID vacuum induction melting and degassing
VOD vacuum oxygen decarburization (ladle metallurgy)
VODC vacuum oxygen decarburization (converter metallurgy)
VOID vacuum oxygen induction decarburization
vol volume
vol% volume percent
W watt
W width; weight
WQ water quench
WQT water quenched and tempered
WRC Welding Research Council
wt% weight percent
yr year
° angular measure; degree
°C degree Celsius (centigrade)
°F degree Fahrenheit
$\leftrightarrows$ direction of reaction
÷ divided by
= equals
≈ approximately equals
≠ not equal to
≡ identical with
> greater than
≫ much greater than
≥ greater than or equal to
∞ infinity
∝ is proportional to; varies as
∫ integral of
< less than
≪ much less than

- ≤ less than or equal to
- ± maximum deviation
- − minus; negative ion charge
- × diameters (magnification); multiplied by
- · multiplied by
- Ω ohm
- / per
- % percent
- + plus; positive ion charge
- √ square root of
- ~ approximately; similar to
- α angle
- Δ change in quantity; an increment; a range
- ϵ strain
- $\dot{\epsilon}$ strain rate
- μin. microinch
- μm micron (micrometer)
- ν Poisson's ratio
- π pi (3.141592)
- ρ density
- σ tensile stress
- τ shear stress

Greek Alphabet

Α, α	alpha	**Ι, ι**	iota	**Ρ, ρ**	rho
Β, β	beta	**Κ, κ**	kappa	**Σ, σ**	sigma
Γ, γ	gamma	**Λ, λ**	lambda	**Τ, τ**	tau
Δ, δ	delta	**Μ, μ**	mu	**Υ, υ**	upsilon
Ε, ϵ	epsilon	**Ν, ν**	nu	**Φ, φ**	phi
Ζ, ζ	zeta	**Ξ, ξ**	xi	**Χ, χ**	chi
Η, η	eta	**Ο, ο**	omicron	**Ψ, ψ**	psi
Θ, θ	theta	**Π, π**	pi	**Ω, ω**	omega

Index

C

D

E

F

G

H

J

K

L

N

O

S

T

U

V

W

X

Y

Z